NEWTON's TELECOM DICTIONARY

Published in the United States by
Miller Freeman, Inc.
Tenth floor
12 West 21 Street
New York, NY 10010
212-691-8215 Fax 212-691-1191
1-800-999-0345 and 1-800-LIBRARY

ISBN Number 1-57820-031-8

February, 1999

Manufactured in the United States of America

Fifteenth Expanded and Updated Edition
Cover Design by Saul Roldan and Regula Hoffman
Matt Kelsey, Publisher
Christine Kern, Manager

Printed at Command Web, Secaucus, New Jersey
www.commandweb.com

NEWTON's TELECOM DICTIONARY

The Official Dictionary of Telecommunications & the Internet

15th Updated, Expanded and Much Improved Edition

HELP MAKE THIS DICTIONARY
EVEN BETTER

We're adding 100 new words a week. We're updating, expanding and fixing another 100. Still, we can't keep up.

If we're missing a definition, or if a definition is unclear or (God forbid) wrong, please email me.

I'll reward you with a big "Thank you" and a free copy of the next edition where you'll find yourself immortalized.

Thank you.

Harry Newton

Harry_Newton@HarryNewton.com
Harry Newton
205 West 19 Street
New York, NY 10011
212-206-7140
Fax 212-691-1011

OUR INCREDIBLE JOURNEY

By Harry Newton
Harry_Newton@HarryNewton.com

Enhancement: We added an Appendix. It covers Standards Organizations, Special Telecom Interest Groups, Publications and Country Dialing Codes.

Growth: I'm adding over 100 new definitions a week to this dictionary and fixing and updating another 100. I am obsessed with keeping it up to date. If I'm missing words or got some wrong, please let me know. Harry_Newton@HarryNewton.com

Apology: I'm still apologizing for cutting the size of the type in the dictionary. For 10 editions, I've believed in "readability," i.e. BIG TYPE. But my dictionary blossomed to 1,400 pages, 5 1/2 lbs and was costing $5.50 to ship domestically and $35+ internationally. We spent days trying different type faces, different type sizes, different leading (space between lines) and different spacing between letters. This is the most readable we could create and keep the weight down. It ain't perfect. I'm sorry.

Technology: I wrote this book for those of us trying to keep up. Not easy. No other industry is exploding as telecommunications, networking and the Internet are. The excitement is coming from:

1. The Hardware. The excitement centers on the dramatically improving price performance of fiber optics (switching and transmission), digital signal processors and the high-speed routers of the Internet, now switching terabits of information each second. Telecom's three building blocks — fiber, DSPs and routers — are improving far more rapidly in cost performance than computing's microprocessors and memory.

2. New Standards. Ten years ago, the telecom industry was entirely closed. Telecom standards existed only at the very lowest levels — basic analog phone lines, basic digital lines. Since then a bunch of pioneers, coming chiefly from the computer industry, has pushed open the industry, promulgating real, open standards that we can all use to build, Lego-style, creative new telecom products. We see these pioneers working in a hodge-podge of volunteer and semi-government bodies — from the ATM Forum to the ECTF, from the MVIP to the ITU-T, from ANSI to the Internet Engineering Task Force and to private companies, from Microsoft to Intel, from Dialogic to Natural MicroSystems. There's been an explosion of new telecom standards, defining everything from telecom operating systems, to buses that carry voice inside PCs, to high-speed lines, to new telecom "building block" software (called applications generators) and, of course, to IP Telephony (Internet Protocol Telephony). Open standards lead to low prices, low barriers to entry, fast creation of new products and explosive growth.

3. New Government Awareness That Competition Is Better Than Monopoly. The telecom industry has historically been closed — closed in architecture, closed to new suppliers, closed to new entrants, with its users mauled by high monopoly prices. No longer. Europe de-regulated in January, 1998. Europe, Japan, Australia, Israel, Russia, China — governments everywhere are waking to one realization: Telecom is economic infrastructure. Companies and business go where infrastructure is strongest. You can't make modern infrastructure when you have one bloated, glacial, over-priced, government-run phone company. And you can't limit your entire country's telecom purchases to two or three hand-chosen, gigantic, cumbersome suppliers (as each country in Europe, Japan and Australia has historically done). By the end of this century, European telecom should be as open as North America is today. That's another 300 million people hungry for decent, cheap, flexible communications. Then comes India and China, also opening up.

4. New Startups. Telecommunications is the hottest place for venture capital. No industry in the world — neither software nor petroleum — can match the incredible profitability of the zero-marginal-cost telephone industry. If I call from New York to California, my call costs my supplier — AT&T, MCI, Sprint or whoever — nothing. Nobody to this day knows how many conversations or how much information we can put down a single strand of fiber — see the incredible developments in Dense Wave Division Multiplexing. Add more photonics and electronics at the end. Put more calls down the fiber. More electronics. More photonics switching. More calls. As the world is fibered up, costs will plummet, communications will be postalized (one low price wherever you call) and people will live and work where they choose. I see the day coming when we'll all pay $25 a month and call anywhere in the world for as long as we want, for as often as we want. Flat rate calling everywhere. I see the day when everyone will be connected to the Internet all the time. On The New All The Time. Instant email. I see the day when everyone will have high-speed channels to their houses and their offices. If over the air TV brought us 10 channels, and CATV (cable TV) brought us 100, and satellite TV brought us 250, Internet TV will bring us 50,000 — maybe more. It's logical technology progression. And the technology is here.

Smart engineers are leaving big, sluggish telecom manufacturers in droves and starting new companies. There are hundreds of "new telecom" millionaires already. There will be thousands. The new telecom environment is driving radically new paradigms into a erstwhile staid industry. Among them:

● **Because of fiber, long distance and international calling are becoming incredibly cheap**. The industrial world will export its service jobs in the next 20 years, just as it exported its manufacturing jobs in the past 20. More people will telecommute. Over 50% of Americans will soon not work out of conventional offices. U.S. airlines will answer their U.S. reservation calls from U.S. customers in India. Already, Microsoft, HP and Swissair have call centers in India and answer calls there which come from North America.

● **Telecom switching hardware will go through the same mainframe to client/server PC revolution that the computer industry went through in the past 20 years** — with one big difference: The telecom industry will skip the mini-computer generation. In telecom, we are very close to the "central office inside the PC." As competition and new technology arrive, the time to market for new telecom and Internet features will drop from three years to three weeks.

● **Voice, fax, imaging and video will migrate to and join data in one common IP network — the Internet and the corporate Intranet (the corporate equivalent of the Internet)**. This will drive telecommunications pricing further down, shrinking space, shrinking time and vastly expanding (and improving) every aspect of human endeavor — from remote education to business telecommuting, from remote medical diagnosis to entertainment (all those TV channels in high definition TV).

● **Ecommerce will explode.** Who would have imagined the 1992-1995 phenomenon which produced the democratization and commercialization of the Internet? Who could have imagined prime time TV prominently showing Web site addresses for advertisers? The Internet happening is as important to the dissemination of knowledge as the invention of the Gutenberg Press was in 1453. The Internet and the Web as a buying mechanism is immature — maybe five years old. Physical shops are 5,000 years old. As ecommerce matures in the next few years, it will become a wondrous experience, eclipsing in emotion and excitement anything physical shops have ever delivered. But it will take time (and bandwidth). And the journey will be unbelievably rewarding, exciting and overflowing with opportunities.

I wish I were 30 years younger. The next 20 in this — the most fantastic of all industries — are going to be totally incredible. Adjectives can't describe the excitement I feel.

Writing telecom dictionary updates every nine months is exhausting, exhilarating and very time-consuming. Claire and Michael, my children, and Susan, my wife, nag me, "Get a life!" I apologize to them for seeing so much of Daddy's back as he worked on this dictionary.

HOW TO USE THIS DICTIONARY

This is a dictionary to work every day with. Companies give it to their new employees to bring them up the telecom and Internet learning curve. Salespeople include the definitions in proposals to customers. Novices love it because it cuts through the clutter. Users explain telecom things to their boss. Management uses it to understand telecom technicalities. Lawyers even use it in court. Sometimes they rely on it. God help the justice system.

Give my dictionary to your new employees, your users, your customers, your prospective customers, to your boss. You can even give it to your kids to let them understand what you do. They'll understand why you, too, have no life.

Most technical dictionaries define terms tersely, often in other technical terms. As a result they leave you more confused. This dictionary is different, deliberately so. My definitions tell you what the term is, how it works, how you use it, what its benefits are, what its negatives are. I tell you how it fits into the greater scheme of things, and I occasionally sound warnings or give you buying checklists.

My dictionary is not comprehensive. But each edition gets bigger and better. I add new terms. I re-work and I update old definitions. I'm always looking for new ones. That's an invitation. Send me your in-house glossaries, your corrections, your new words. The best way to get me is Harry_Newton@HarryNewton.com

HOW DICTIONARY IS ORGANIZED

My definitions are in a combination ASCII / alphabetical order. Definitions containing non-letters -- like @, #, / and numbers -- are in ASCII order. Definitions with real letters are in alphabetical order, i.e. capital letters and lower case letters are mixed. I use ASCII because it gives order to hyphens, periods, forward slashes, numbers, etc. Here is the order you'll find in this dictionary:

Blank Space	= ASCII 32	5	= ASCII 53
!	= ASCII 33	6	= ASCII 54
#	= ASCII 35	7	= ASCII 55
& (Ampersand)	= ASCII 38	8	= ASCII 56
-(Hyphen or dash)	= ASCII 45	9	= ASCII 57
. (Period)	= ASCII 46	: (colon)	= ASCII 58
/ (Forward slash)	= ASCII 47	; (semi colon)	= ASCII 59
0 (zero)	= ASCII 48	Letters	= ASCII 65
1	= ASCII 49		
2	= ASCII 50		
3	= ASCII 51		
4	= ASCII 52		

HOW I SPELL

My dictionary conforms to American spelling. To convert American spelling to British and Canadian spelling typically requires adding a second "L" in words like signaling and dialing (they're American) and changing "Z" in words like analyze to analyse. Center in American is Center. In Britain, Europe, Australia and Canada, it's Centre. This dictionary contains more British, Australian and European words than my previous editions — a result of several overseas lecture tours. Fiber is fiber in North America. Elsewhere it's fibre, except for Fibre Channel which is always correct everywhere.

THIS DICTIONARY'S STYLE

All high-tech industries make up new words by joining words together. They typically start by putting two words next to each other. Later, they join them with a hyphen. Then, with age and familiarity, the hyphen tends to disappear. An example: Data base, data-base and now database. Sometimes it's a matter of personal choice. Some people spell database as one word. Some as two, i.e. data base. I prefer it as one, since it has acquired its own logic by now. Sometimes it's a matter of how it looks. I prefer T-1 (T-one), not T1, simply because T-1 is easier to recognize on paper. I define co-location as co-location. Websters spells it collocation, with two Ls, one more than mine. I think mine is more logical. And Mr. Webster is dead. He can't argue with me.

Sometimes the experts don't even get it right. Take something as common as 10Base-T. Or is it 10BaseT? The thing is an IEEE standard. So you'd think they'd know. Forget it. Go to their web site, www.ieee.org and you'll find as many hits for 10Base-T as for 10BaseT. Anyway, Ray and I checked every known and unknown expert in the Western world (i.e. those living within a block or two of Ray). We now believe the correct spelling is 10Base-T. Also we checked www.excite.com (my favorite Web search engine).

Sometimes, I don't simply know. So I may list the definition twice — once as two separate words and once as one complete word. As words and terms evolve, I change them in each edition. I try to conform each new edition to "telecomese" and "Internetese" as it's spoken and written at that time.

BITS, BYTES and BITS and BYTES PER SECOND?

The telecom and computer literature is loaded with references to Bps and bps. You'll see them as Kbps or KBps. You'll see them as Mbps or MBps. You'll see them as Gbps or GBps. Not much consistency. Let's explain this. First, K means Kilo or 1,000. M means mega or one million. And G means giga, which is a thousand million, or 1,000,000,000. Now, if we're really being accurate (and pedantic), Kbps means thousand bits per second. And that's a telecom transmission term meaning that you're transmitting (and/or receiving) one thousand bits in one second. KBps means thousand bytes per second. That's a computer term. And it usually refers to speeds inside the computer, e.g. from your hard disk to your CPU (central processing unit — your main microprocessor). That's the way it's meant to be. But, of course, there's a lot of sloppy writing out there. You'll see MBps or MB/s also meaning one million bits per second as a transmission speed. You really have to figure out if the writer means telecom transmission — i.e. anything outside the computer. Or if it's internal to the computer. You can usually tell from the context. Some other points:

Virtually all telecom transmission is full duplex and symmetrical. This means that if you read that T-1 is 1,544,000 bits per second, it's full duplex (both ways simultaneously) and symmetrical (both directions the same speed). That means it's 1,544,000 bits per second in both directions simultaneously. If the circuit is not full duplex or not symmetrical, this dictionary points that out. For now, the major asymmetrical (but still full duplex) circuit is the xDSL family, starting with ADSL, which stands for asymmetric, which means unbalanced. The DSL "family" no longer starts with "A," but most of it is still asymmetrical. Our definitions point out which is which.

There's one more complication. Inside computers, they measure storage in bytes. Your hard disk contains this many bytes, let's say eight gigabytes. That's fine. But they're not bytes the way we think of them in internal computer transmission terms. They're different and they have to do with a way computer stores material — on hard disks or in RAM. They're what I call "storage bytes." When we talk 1 Kb of storage bytes, we really mean 1,024 bytes. Which comes from the way storage is actually handled inside a computer, and calculated thus: two raised to the power of ten, thus $2 \times 2 \times 2 \times 2 \times 2 \times 2 \times 2 \times 2 \times 2 \times 2 = 1,024$. Ditto for one million, two raised to the power of twenty, thus 1,048,576 bytes. See also BPs.

WHICH WORDS GET DEFINED?

Which words get defined? All good dictionary writers (all ten of them) start with rules: All the important terms in the field. No proprietary products. No proprietary terms. Then, because the rules are no vague, we violate them all over the place. Writing a dictionary is a very personal and very lonely business. You read — I get over 100 magazines a month. You study. You cogitate. You don't have much human interaction. Your wife calls, "Enough with the words, already. It's 2:00 AM. Time to sleep."

You write definitions first to explain things to yourself. That's intellectually exciting. Then you write definitions to amuse yourself. That, you think, shows how clever you are. I suspect it's the same "clever" I got with marijuana 25 years ago. It was nonsense clever. A fake.

As it's grown, my dictionary has become intensely satisfying and (of all things) relaxing. Despite it not being complete (it will never be), it's pleasing to me that this edition — the 15th — is a whole lot better than the ones that came before it. I bet the 20th will be better.

Once again, this dictionary is meant to be a cooperative effort. I can't sit here, all by my own lonesome. Please send me your comments, suggestions, omissions, ideas and complaints. I'm here on email, writing definitions. Harry_Newton@email.msn.com, Harry_Newton@IBM.net or HarryNewton@HarryNewton.com. I have three email addresses, because I'm obsessed with reliability in communications. You should be, too.

PLURALS
Plurals give trouble. The plural of PBX is PBXs, not PBX's. The plural of PC is PCs, not PC's, despite what the New York Times says. The Wall Street Journal and all the major computer magazines agree with me. The plural possessive is PBXs' and PCs', which looks a little strange, but is correct. In this dictionary, I spell the numbers one through nine. Above nine, I write the numbers as arabic numerals, i.e. 10, 11, 12, etc. That conforms to most magazines' style.

There are no rights or wrongs in the spelling business, except that my dictionary is now the correct way of spelling telecom words. My dictionary is correct, because it's the biggest seller (by far). Lawyers use it in court. They get judgments based on what's in my dictionary. God help the justice system.

A or AN? HERE'S THE LOGIC
OK. I admit my fallibility. This edition of this book is riddled with "a" when it should be "an" and "an" when it should be "a." I've never been confused. I always believe "an" is used before vowels, and "a" before consonants. Not so, says my friend, Jay Delmar, who edits technical documentation for Alcatel USA, Inc. in Plano, TX. Here's his explanation.

Concerning the problem of what article ("a" or "an") should be used with a word or an acronym, it all depends on how the acronym is pronounced, that is, whether it's pronounced as a string of letters or as a word. In some cases, the article would be the same. In others, the form would have to switch. Usually "an" is used before vowels, but some consonants require it as well, and some vowels require an "a." It all depends on the sound. Whether a letter is intrinsically a vowel or a consonant doesn't really matter; what matters is if it's pronounced as a vowel or a consonant in the particular context.

If an acronym is pronounced as a string of letters, the following shows the appropriate article to use with the first letter of the acronym:

An A	A J	An S
A B	A K	A T
A C	An L	A U
A D	An M	A V
An E	An N	A W
An F	An O	An X
A G	A P	A Y
An H	A Q	A Z
An I	An R	

v

If an acronym is pronounced as a word, the article might need to change:

An RS-232, but a RAM (pronounced "ram")
An STP, but a SRDM (pronounced "sardem") and a SLC (pronounced "slick")
An FTP, but a FAIC (pronounced "fackey")
An HIC, but a HICUP (pronounced "hiccup")

According to The New York Public Library Writer's Guide to Style and Usage, "The article a is used before all consonant sounds, including a sounded h, a long u, and an o with sound of w (as in one). The article an is used before all vowel sounds except a long u and before words beginning with a silent h." This definition has never helped me because I've never really understood why in "an STP" the "s" sound is a vowel sound and in "a SRDM" the s sound is a consonant sound. Basically, I rely on my ear.

The real trouble, of course, is that unless one is really, really familiar with the acronym, one doesn't know how it's actually used: pronounced as a string of letters or as a word. I thought SIPL would be "an SIPL" (an ess-eye-pea-ell) until fairly recently. I didn't know it was pronounced "a sipple"-or "a sighpull" (I've heard it both ways).

I think Jay makes sense. I'm going to try to be more in line with his concepts in upcoming editions. But this one may have a few inconsistencies.

THANK YOUs

A big "Thank You" to the dozens of people and dozens of companies who helped. If I left you out, I apologize.

Among the manufacturers, special thanks to Amdahl, Anixter, Aspect Telecommunications, AT&T, Bellcore (now owned by SAIC), Dialogic, Ecos Electronics, General Cable, Intel, Lucent, Micom, Microsoft, MCI, NEC, Newbridge Networks, New York Telephone (now Bell Atlantic), Northern Telecom (now Nortel), Racal Data, Ricoh, Sigma Designs, Sharp and Teknekron. Among the magazines I borrowed (or stole), the best were PC Magazine and TELECONNECT, Call Center, Computer Telephony and IMAGING Magazines. Special thank yous also to internetworking expert, Tad Witkowicz at CrossComm, Marlboro MA.; to Stephan Beckert of The Strategis Group in Washington, D.C.; to Ken Guy erstwhile of Micom, Simi Valley (near LA); to Robert M. Slade, who does a wonderful job reviewing books (including this one); to Michael Marcus, president of Able Communications, an excellent interconnect company based in Scarsdale NY, Frank Derfler of PC Magazine; Chris Gahan of 3Com; to Bob Rich of Boeing's System Engineering Group, who's studying for his MCSE; to Glenda Drizos, Enhanced 911 Project Leader, at Sprint PCS, Overland Park, Kansas; to Jeff Deneen of the Norstar Division of Northern Telecom in Nashville; Stephen Doster of Telco Research in Nashville; bugging expert Jim Ross of Ross Engineering, Adamstown, MD; wiring experts John and Carl Siemon of The Siemon Company, Watertown CT; to Jim Gordon and Parker Ladd at TCS Communications, Nashville, TN, the people who do workforce management software for automatic call distributors; to Jun Sun of Cisco, the people who make the Internet routers; to Judy Marterie and the electricity wiring, grounding and test experts at Ecos Electronics Corporation in Oak Park, Il; to Brian Newman of MCI, who understands wireless; to John Perri of SoftCom, NYC; to John Taylor of GammaLink, a Sunnyvale, CA company which produces beautiful fax products (but which is now owned by Dialogic); to Charles Fitzgerald at Microsoft and Herman D'Hooge at Intel who jointly helped created Windows Telephony; to everyone else at Microsoft (including Mark Lee, Toby Nixon, Bill Anderson, Lloyd Spencer and Mitch Goldberg) and Waggener Edstrom (Microsoft's PR agency) who produce such great White Papers and keep pushing the state of telecommunications standards further; to Bill Flanagan who's written fine books on T-1 and voice and data networking; to Jane Laino of Corporate Communications Consultants, NYC; to Jon L. Forsyth, Manager at Cambridge Strategic Management Group, Cambridge, MA.; to Henry Baird of Seattle consultants Baird & Associates; to Sharon O'Brien formerly of Hayes Microcomputer Products in Norcross (Atlanta); to Howard Bubb, John Landau, Jim Shinn, Nick Zwick and Sam Liss at leading voice processing component manufacturer, Dialogic Corporation of Parsippany, NJ;

to David Perez and Nick Nance of Com2001 Technologies, Carlsbad, CA; to Al Wokas of MediaGate, San Jose, CA; to Nayel S. Shafei of Qwest Communications; to Alison Golan of networking company, Interphase Corporation in Dallas, which allowed me to steal some of the definitions from their excellent booklet, "A Hitchhiker's Guide to Internetworking Terms and Acronyms;" to Ian Angus at the Angus TeleManagement Group in Ajax, Ontario, who embarrassed me into expanding my Canadian coverage; to Glenn Estridge, one of the world's leading experts on dense wave division multiplexing and the whole wonderful world of fiber optics; to John Arias, a seriously good technician Bell Atlantic who came to fix a busted line and left educating me on the intricacies of cable naming at his company; John Moring for his wireless terms; to Ed Margulies at Miller Freeman, New York, NY who's written so many fine books on computer telephony; to Charlie Peresta, P.E., PMP, Telecommunications Manager, Intellisource who helped me with some of the telecom energy terms.

I'm very grateful to The ATM Forum of Mountain View, CA (www.atmforum.com) for allowing me to use many of their definitions from their really well-done ATM Forum Glossary.

I'm particularly grateful to Black Box Corporation, Pittsburgh, PA for allowing me to steal some of their excellent RS-XXX definitions and their diagrams on connector pinouts, which adorn my Appendix.

At Information Access Corporation, which publishes this dictionary on CD-ROM discs (for your copy, call 800-321-6388) I'm very grateful to Dave Rubin and Pamela Argandona. At my own office, I'm very grateful to Muriel Fullam. At Miller Freeman, I'm grateful to Matt Kelsey, Christine Kern, Saul Roldan and Jennifer Cooper-Farrow. Without all these wonderful people, this dictionary wouldn't be as good as it's actually turning out. If I sound surprised, you're right. It's now the largest-selling telecom and networking dictionary in the world.

I wrote this dictionary on a series of ever-newer, ever-faster Toshiba laptops (very reliable machines) using The Semware Editor, a very beautiful text editor, which Sammy Mitchell of Marietta, GA wrote, and which I wholeheartedly recommend — www.semware.com. The Toshiba laptop for this dictionary was the Toshiba Tecra 780DVD. It's a 266 Mhz Pentium II machine with 192 Meg of RAM and a five gigabyte removable hard disk. Jennifer Cooper-Farrow and Christine Kern typeset it on a Power PC Macintosh using QuarkXpress and Adobe Illustrator. Saul Roldan and Regula Hoffman did the cover. It's gorgeous. Those people are wonderfully talented.

I welcome comments, corrections, suggestions, improvements.

Harry Newton
205 West 19 Street
New York, NY 10011
Tel 212-206-7140 Fax 212-691-1011
Web site: www.HarryNewton.com
Email: Harry_Newton@HarryNewton.com

RAY HORAK, Senior Contributing Editor

Ray Horak is internationally recognized for developing and delivering world class seminars on telecommunications technologies, services and management systems. He speaks annually before thousands of telecom, networking and IT professionals in public and private seminars, workshops and conferences.

He has been acclaimed for his ability to unravel the intricacies of voice, data, video systems and networks. His seminars provide technical depth, while being delivered in a common sense, plain-English, and thoroughly understandable fashion. I rediscovered Ray through a seminar he taught for TCA (TeleCommunications Association) some years ago. I asked Ray to work with me on this dictionary because he believes in explaining complex technologies in a way normal, intelligent business people can understand.

Ray also is a regular contributor to numerous leading industry trade publications and is a member of the Advisory Boards of Datapro/Gartner Group and Computer Telephony Expos. He is also a member of a number of several prestigious Editorial Boards, including "The Connectivity Management Handbook," "The Journal of Telecommunications in Higher Education," "Telecom Business," and Teleconnect Magazine. Ray also has written his own best-selling book, "Communications Systems and Networks," published by M&T Books in 1996 and now in its fourth printing; he is working on the Second Edition. It's a great book, and a perfect companion to this dictionary. Get it from www.amazon.com, www.barnesandnoble.com, or idahoans.com.

Ray's public seminars are offered in the U.S. through Network World Technical Seminars. Overseas, they are offered through AIC and IDG. He also teaches seminars at a number of major industry conferences, including ComNet, CT Expo, Software Developers Conference, and Networld+Interop. Ray also teaches in-house courses for private corporations; I can't tell you who his clients are, but they are among the most prestigious in the business.

Ray's experience in communications dates way back to 1970, when he joined Southwestern Bell. His experience also includes AT&T, Bell Labs, and CONTEL, where he served as region Vice President. He founded several companies for CONTEL, before serving as General Manager for the company's Houston Executone operation. Ray has been on his own for the last 10 years or so, and wouldn't have it any other way.

Ray is President of The Context Corporation, an independent consultancy and training organization in Mt. Vernon, Washington state (near Microsoft).

According to Ray, his greatest literary accomplishment was an article he wrote for Teleconnect in June 1990. A long time friend from his CONTEL days read it and called him, after having lost touch with him for years. They got together during one of his seminar tours, and discovered that they were still very much in love. They kept in touch. He and the friend, Margaret Blanford, were married on Friday, December 13, 1996, in defiance of traditional superstitions. They are the happiest two people on the face of the Earth.

Ray Horak, President
The Context Corporation
1500A East College Way, Suite 443
Mt. Vernon, WA 98273
Tel 360-336-3448 Fax 360-336-3759
Email: ray@contextcrp.com

NUMBERS

" Double quotation marks. Typically used to signify something your computer should print (to screen, disk or paper), as in PRINT "Thanks for being a good guy."

Some programs allow you to use single quotation marks interchangeably with double ones. Some programs don't. Try one or the other if in doubt.

The character on the bottom right of your touchtone keypad, which is also typically above the 3 on your computer keyboard. The # sign is correctly called an "octothorpe," but sometimes it is also spelled without the "e" on the end. There is even an International Society of Octothorpians who maintains a web page at http://www.nynews.com/octothorpe/home.htm. The octothorpe is commonly called the pound sign, but it's also called the number sign, the crosshatch sign, the tic-tack-toe sign, the enter key, the octothorpe (also spelled octathorp) and the hash. Musicians call the # sign a "sharp."

On some phones the # key represents an "Enter" key like the Enter key on a computer. On some phones it represents "NO." And on others it represents "YES." MCI, AT&T and some other long distance companies use it as the key for making another long distance credit card call without having to redial. Hold down the # key for at least two seconds before the person at the other end has hung up, you'll get a dial tone, punch in your phone number and you can make another long distance call — without having to punch in your credit card number again. (This service is often called Call Reorigination or just plain Next Call.) The # key is used in the paging industry — national and local. When you dial a phone number which represents someone's beeper, you will typically hear a double beep. At that point you punch in your phone number, ending it with a #. At that point, the machine hangs up on you and sends out the numbers you punched in to the pager you just dialed. Many digital phone pagers allow people to send actual text messages to pagers. Many use the # sign to signal the use of certain digits (c, f, i, l, o, s, v, y) as well as to signal the end of the transmittal. The # character is also the comment character used in UNIX files (configuration, code, etc.).

& The "and" sign. Its real name is an "ampersand."

* The star sign. On IVR systems it typically means "No." In computer languages, it often means a multiplication sign. It's also used to represent a wild card or a joker. For example, the command

    ERASE JOHN.*

will erase all the files on your disk beginning with JOHN, e.g. JOHN.TXT, JOHN.NEW, JOHN.OLD, JOHN.BAK, etc.

IEX in Richardson uses * key as a cancel character for its Call Valet application.

*57 The North American universal dialing code which you touchtone in immediately after receiving a harassing, obscene or annoying phone call. By touchtoning that number in, you have alerted your central office to "tag" that phone call. Should a law enforcement agency get involved in investigating your annoying calls, they would be able to go into your records and find the phone number from which the annoying call was made. See TRAP and TRACE.

*67 Dial *67 before you make a call in North America and the person you call won't see your CallerID, i.e. the number you're calling from.

++ In C programming, the expression i++ means use the current variable i and add 1 to it. In more contemporary usage, ++ has come to mean an expansion, an improvement, an upgrading, etc. So, C++ is an improved version of the C programming language. MAE-East++ is an expansion of MAE-West located nearby. See MAE-East.

- The dash. The minus sign. Often we take two words and join them with a dash into a new word. As the word becomes more and more common, we remove the dash and the double word now becomes a single word.

/ The forward slash. Lotus made it famous. UNIX uses it as a directory separator. You see the use of forward slashes in Web addresses, e.g. www.ctexpo/index.html

// The double forward slash. Filenames or other resource names that begin with the string // mean that they exist on a remote computer. They are called UNC (as in Universal Naming Convention) Names. The convention // is commonly seen on the Internet. See \\.

**** The backslash. Used for designating directories on your MS-DOS / Windows machine. This dictionary is located in C:\Work\Dictiona

That means it's in the "dictiona" subdirectory of the "work" directory. If it were under Windows 95, 98 or Windows NT, the file name could be as long as 256 characters and could say something more descriptive, like c:\work\harry's great dictionary.

**** The double backslash. Filenames or other resource names that begin with the string \\ mean that they exist on a remote computer. They are called UNC (as in Universal Naming Convention) Names. The double backslash is more common in the UNIX world, while the // (double forward slash) is more common in the Windows NT world. See //.

^ The character typically above the 6 on your keyboard. It was originally a circumflex. In computer language it became the symbol that was written to is often referred to as the "caret" in computer programming.In typography, represent the Control (Ctrl) key. It's also called the "hat." The ^ symbol the ^symbol is used to show where something is to be inserted.

~ This character is called a tilde. It is used as the UNIX shortcut for "home directory for this account. In Spanish, it tells you how to pronounce the n in senor. According to William Safire, it's a Spanish word from the Latin term for a tiny diacritical mark used to change the phonetic value of a letter.

0+ Calls Called "OH PLUS." 0+ plus calls are calls made by dialing zero plus the desired telephone number. Calls made this way may be interrupted by a live operator requesting billing information, or a recorded announcement requesting the caller to enter the billing information.

0- Calls Called "OH MINUS." 0- calls are operator-assisted calls. The caller dials zero and waits for the operator to pick up the line and talk to the caller.

0B+D An ISDN BRI circuit missing the voice B channels, and only provisioned with the signalling channel. This is common for automated teller machines, travel agency terminals and authorization services. This uses the D channel for a 9600 baud x.25 connection to a provider instead of a leased line. Monthly cost starts around $20. Some providers offer low cost LATA wide packet services for $1-$4 per month, with worldwide x.25 costs starting about $20+ per month usage fees. Anywhere a low speed data connection is needed, this is usually a low cost alternative. All it takes is an NT-1 that trips off the D channel and supports a serial port for your hardware.

011 The prefix you use in the United States to dial a number to another country, except Canada and most countries in the Caribbean. Must be followed by a country code and the area code and the local phone number.

0345 Numbers A British Telecom LinkLine service in England where the caller is charged at the local rate irrespective of the distance of the call. The subscriber pays installation and rental charges in addition to a charge for each call.

0800 Numbers 1. A British Telecom LinkLine service in England where the caller is not charged for the call. Similar to the North American 800 IN-WATS service.
2. Ericsson has 0800-type service on cellular systems it's put in. Ericsson describes it as a "network-oriented service — based on time of day, day of week, or special day — which allows calls to be redirected to other numbers.

0891 and 0898 Numbers A British Telecom Premium rate service in England where the caller is charged at a premium rate for the call. The calls are normally made to receive information or a service. The service provides revenue for the information provider who receives part of the call charge.

0839 and 0881 Numbers Mercury's premium rate service numbers in England.

1+ Pronounced "One plus." In North America, dialing 1 as the first digit has come to signal to your local phone company that the phone number you are dialing is long distance, i.e. is designed to reach a long distance number in the United States, Canada, or several of the Caribbean islands, (including Bermuda, Puerto Rico, the Virgin Islands, Barbados and the Dominican Republic). The number 1 will typically be followed by an area code and then seven digits. For example, to dial me from outside New York City, you would dial 1-212-691-8215. To reach other international countries, from the United States, you dial the international access code "011." This "1+" will work with your local phone company, signaling it that you want to reach another local area code. (In New York City, calling from Manhattan to Brooklyn means dialing 1+718+the seven digit Brooklyn number.) It also will work with the long distance company you have pre subscribed to, through the process known as equal access. To reach another long distance carrier and route calls over their network, you will need to dial 1-0XXX and then the area code and number.

1.544 Mbps The speed of a North American T-1 circuit. 1,544,000 bits per second. See T Carrier and T-1.

1/4/80 The strange starting date embedded in the original IBM PC.

10-baggers Venture capital jargon for companies returning 1000% on their investment.

10-Net An original local area network invented by a company called Fox Research, Dayton, OH. 10-Net is a baseband, Ethernet CSMA/CD peer-to-peer LAN running on one twisted pair at one megabit per second. It is easy to install and has many advantages. It's also slow. Very slow. And it's no longer being made. See ETHERNET.

10 Base X See 10Base X below, e.g. 10Base-2, 10Base-T.

10 Base T See 10Base-T.

100 Base T See 100Base-T

100 Test Line A Northern Telecom switching term. The 100 test line, also known as a quiet or balanced termination, is used for noise and loss measurements. The S100 provides a quiet termination for noise measurements only. In this 100 family, there is the T100, S100 and N100 tests. The N100, a more recent version of the test, also includes a milliwatt test (i.e., a 102 test line) and therefore can be used for far-to-near loss measurements. The T100 is used when the equipment at the terminating office is unknown. When the T100 test line is performed, a two-second time-out is introduced to detect the presence or absence of a milliwatt tone. If the T100 test detects the milliwatt tone, it executes the N100 version of the test; otherwise, the S100 version is initiated. If the version of the distant office test line is known, then that version of the test line can be performed directly, and thus the two-second delay per trunk of the T100 test line test is eliminated.

The 101 test line is used to establish two-way communications between the test position and any trunk incoming to the system. The connection to the 101 test line is established through the switching network.

The 102 test line, also known as the milliwatt line, applies a 1004 Hz test tone towards the originating office to facilitate simple one-way or automatic transmission loss measurements. The test tone is applied for a timed duration of nine-seconds during which answer (off-hook) signal is provided. Then an on-hook signal followed by a quiet termination is transmitted to the originating end until the connection is released by the originating end.

The T103 is used for the overall testing of supervisory and signaling features on intertoll trunks. The test is performed to the far end to check overall supervisory and signaling features of the trunk. If the test fails or if a false tone signal is detected, the test is abandoned and the condition indicated.

The T104 test is used for two-way transmission loss measurements, far-to-near noise measurements and near-to-far noise checks. Normally used in testing toll trunks.

When a 105 test line at a far-end office is called and seized, timing functions are initiated and an off-hook supervisory signal and test progress tone are returned to the originating office. If the responder is idle, the test line is connected to the responder and test progress tone is removed. Transmission tests are then initiated.

The T108 test provides far-end loop-around terminations to which a near-end echo suppression measuring set is connected for the purpose of testing echo suppressors.

1000Base-CX A developing standard for Gigabit Ethernet (GE) connectivity. 1000 means 1,000 Mbps (Megabits per second); Baseband means single-channel transmission; C means Copper; and X is the generic "whatever." The current specification calls for a very specialized copper cable in the form of an electrically balanced, shielded 150-ohm twinaxial cable which is limited to 25 meters in distance; the distance can be extended to 50 meters with a single repeater. Both conductors share a common ground in order to minimize concerns about safety and interference which could be caused by voltage differences. This cable is intended for use as a short jumper to interconnect clustered GE switches in wiring closets or computer rooms. Over time, the IEEE intends to develop a standard for 1000BaseCX connectivity over distances as long as 200 meters, but likely involving a much better form of copper; hence the X "whatever." See also 1000BaseLX, 1000BaseSX, Gigabit Ethernet, Twinax and UTP.

1000Base-LX A developing standard for Gigabit Ethernet

2

(GE) connectivity. 1000 means 1,000 Mbps (Megabits per second); Baseband means single-channel transmission; L means Long-wave laser; and X is the generic "whatever." Long-wave lasers are more expensive than short-wave lasers, but can transverse longer distances as they use low-energy lasers over single-mode fiber at a wavelength of approximately 1,300nm (nanometers). For example, a long-wave laser system can transmit reliably over distances of approximately 3 kilometers through fiber with an inner core diameter of 5 microns; long-wave laser transmission also is supported over multi-mode fiber, and generally over longer distances than short-wave laser. See also 1000Base-CX, 1000Base-SX, Gigabit Ethernet and Fiber Optics.

1000BaseSX A developing standard for Gigabit Ethernet (GE) connectivity. 1000 means 1,000 Mbps (Megabits per second); Baseband means single-channel transmission; S means Short-wave laser; and X is the generic "whatever." Short-wave lasers are less expensive than short-wave lasers, but cannot transverse the same long distances as they use high-energy lasers over multi-mode fiber at a wavelength of approximately 850nm (nanometers). For example, a short-wave laser system can transmit reliably over distances no farther than approximately 550 meters through multi-mode fiber with an inner core diameter of 50 microns; short-wave laser transmission is not supported over single-mode fiber. See also 1000-BaseCX, 1000Base-LX, Gigabit Ethernet and Fiber Optics.

100Base-FX IEEE standard (802.3u) for 100 megabit per second Ethernet implementation over fiber. The MAC layer is compatible with the 802.3 MAC layer. 100Base-FX runs on fiber-optic cable, making the specification best suited for a backbone or long cable segment. 100Base-TX and 100Base-T4 as well as existing 10Base-T hubs can all be connected to a fiber backbone using appropriate hardware such as a bridge or router. 100Base-FX also supports full duplex operation. See 100Base-T for fuller explanation of 100 megabit per second LANs.

100Base-T 100 stands for 100 million bits per second. B stands for baseband. And T stands for trunk. In sort, 100Base-T is a 100 megabit-per-second local area network known by the generic name of Fast Ethernet. There are three basic implementations of Fast Ethernet — 100Base-TX, 100Base-T4 and 100Base-FX. In PC Week, 9/30/96, Eric Peterson wrote, companies moving to Fast Ethernet to accommodate ever-increasing network traffic loads have three from which to choose: 100Base-TX, 100Base-T4 and 100Base-FX. Each specification is identical except for its interface circuitry, which determines what type of cabling it runs on. As a result, the technologies currently aren't interchangeable; each must be connected to its own type of hub. For example, a 100Base-T4 NIC must be connected to a 100Base-T4 interface on either a hub or switch; likewise, a TX NIC must be connected to a TX interface at the other end. Mr. Peterson wrote, 100Base-T is emerging as one of the most popular and cost-effective high-speed network technologies because it is designed to integrate with existing networks with minimal disruption. Essentially an extension of 10Base-T, 100Base-T achieves increased throughput by decreasing the latency period between bits, which effectively increases the packet speed by a factor of 10. The standard adheres to the 802.3u media access controls specification, which builds on the 802.3 Ethernet standard to assure compatibility with existing 10Base-T installations. As a result, an upgrade from 10Base-T to 100base-T is invisible to users, the NOS and current network management applications.

My own experience with upgrading to 100Base-T has not been 100% overwhelming positive. First, we tried upgrading our Macintosh network to 100 Mbps (i.e. Fast Ethernet), but found zero improvement in the speed of transferring a file from the server to a Mac. Problem? Allegedly, the Macintosh network operating system simply won't work that fast. Then we started to upgrade some of our PC workstations to 100 Mbps, but found it made more sense to break our big PC LAN apart into smaller LANs, each run by an Ethernet switch and then have each Ethernet switch join to each other by 100 Mbps over fiber running 100Base-FX. That has worked wonderfully. The point of all this? Designing networks is not trivial. The speed of the individual links — 10 megabits per second or 100 megabits per second — may not be the gating factor. Check before you spend the money. See 100Base-TX, 100Base-T4 and 100Base-FX.

100Base-T2 A LAN that transmits data at 100 megabits per second over copper cabling. It is a half-duplex version of 100Base-T that uses two pairs of category 3, 4 or 5 UTP cable. Officially called 802.3y. See 100Base-T.

100Base-T4 100-megabit per second (one hundred million bits per second) Ethernet implementation using four-pair category 3,4, or 5 cabling. The MAC layer is compatible with the 802.3 MAC layer. Now read 100Base-T. Then come back and read this: In PC Week, 9/30/96, Eric Peterson wrote 100Base-T4 runs over four pairs of wire Category 3, 4 or 5 data grade cable. Like 100Base-TX, 100Base-T4 networks require the use of a matched set of NICs and hubs. This specification is ideal for desktop installations because it is very robust in noisy environments, and it works with the widest variety of unshielded twisted-pair wiring, The technology allows sites that are wired with older types of cabling to implement Fast Ethernet without incurring the expense and disruption of running Category 5 to the desktop. Because 100Base-T4 is limited to half duplex operation, the specification is not as well suited as 100Base-TX or 100Base-FX for linking servers.

100Base-TX 100-Mbps Ethernet implementation over Category 5 cabling. The MAC layer is compatible with the 802.3 MAC layer. In PC Week, 9/30/96, Eric Peterson wrote that 100Base-TX is the most widely implemented Fast Ethernet specification, 100Base-TX requires Category 5 cabling, the type used in almost all new network installations. 100Base-TX uses two pairs of wire — one pair for transmitting and another for receiving. Upgrading a network from 10 megabits per second Ethernet to 100Base-TX requires 100Base-TX NICs and hubs as well as the Category 5 wiring to connect them. PC Week Labs strongly advises people to test end-to-end to verify that their Category 5 cabling has been properly installed and connected to Category 5-compliant patch panels and wall jacks. 100Base-TX is the best choice for connections among servers, hubs, switches and routers because it supports full duplex operation, meaning that it can simultaneously send and receive data. In addition, the cost and effort of upgrading these connections to Category 5 is minimal since servers are often located very near these devices. See 100Base-T for a general discussion of this technology.

100VG-AnyLAN A 100 megabit-per-second local area network standardized by the IEEE 802.12 committee in 1996, and originally known simply as AnyLAN. A joint development of AT&T Microelectronics, Hewlett-Packard and IBM, VG=Voice Grade, meaning that voice grade UTP (Unshielded Twisted Pair) generally is used as the transmission medium. AnyLAN means that LAN networking and internetworking can be accomplished, accommodating any LAN standard or combination of LAN standards, including Ethernet and Token Ring. 100VG-AnyLAN provides medium flexibility, including Cat 3 (Category 3) UTP (4 pairs) and Cat 5 (2 pairs), as well as Level 1 STP (Shielded Twisted Pair) and fiber optic cable. All pairs are used for transmission in half-duplex mode, with the signal being split across the pairs at 25 MHz each. DPMA (Demand Priority Media Access) provides access priority and collisionless transmission, which is an improvement over Ethernet. 100VG-AnyLAN, however, requires equipment upgrade, which does not position it well relative to 100Base-T, also known as Fast Ethernet. See 100Base-T and FDDI for explanation of the competing 100-Mbps LAN standards.

1010XXX See 101XXX.

101B Closure Housing used to protect service wire splices.

101XXXX To connect with a long distance carrier in the United States other than your preselected carrier, as of July, 1998, you must now dial 101-XXXX, where X is any number between 0 and 9. Each long distance carrier in the United States now has a unique four digit code represented by XXXX. That code is called a CIC code, which stands for Carrier Identification Code. AT&T's code is 0288. Thus, to reach AT&T you dial 101-0288, or as they advertise "Ten-Ten ATT." MCI's is 101-0222. The main reason for wanting to use 101-XXXX to dial a different long distance to the one you're subscribed is simple — to save money. You can dial some carriers (e.g. 101-0457) and receive the bill for your phone calls on your monthly bill from your 101-XXXX LEC (local exchange carrier). Others you have to contact in advance and set up an account. Dialing via 101-XXXX is a little complicated. Let's say you want to call my sister, Barbara, in Sydney Australia from the U.S. You would dial 101-0457-011-612-9-663-0411. As of July 1, 1998, this 101-XXXX code replaced the original 10-XXX. The reason 101-XXXX replaced 10-XXX is clearly that 10-XXX lets you dial only 1000 long distance carriers, while 101-XXXX allows you to dial 10,000 long distance carriers. Deregulation and competition simply resulted in more carriers than could be accommodated by the old dialing scheme. In a very forward thinking move, the FCC has already planned for the next expansion, whenever needed, to 10-XXXXX. This allows for up to 100,000 long distance carriers, while preserving the current CICs exactly as they are now. For more information: www.fcc.gov/Bureaus/Common_Carrier/FAQ/cic_faq.html. Kevin Ross, KevinR@seed-berry.com helped significantly on this definition. Thank you.

10Base-2 10Base-2 is the implementation of the IEEE 802.3 Ethernet standard on thin coaxial cable. It's common called thin Ethernet or thinnet because the cable is half the diameter of 10Base-5 Ethernet cable. 10Base-2 LANs, which run at 10 ten million bits per second, have their PCs daisy-chained along a terminated bus topology. The maximum segment length is 185 meters. Connectors are typically BNC. Also called thinwire Ethernet.

10Base-5 A transmission medium specified by IEEE 802.3 that carries information at 10 Mbps in baseband form using bus topology, using 50-ohm coaxial cable and using AUI connectors. 10Base-5 was specified by the original Ethernet standards and is sometimes called ThickWire Ethernet. The maximum segment length (i.e. without a repeater) is 500 meters.

10Base-F Standard for Fiber optic Active and Passive Star based Ethernet segments. Described in IEEE 802.1j-1993 (not in an 802.3 supplement, as you might expect). 10Base-F includes the 10Base-FL standards.

10Base-FB Part of the new IEEE 802.3 10Base-F specification, "Synchronous Ethernet" which is a special-purpose link that links repeaters and allows the limit on segments and repeaters to be enlarged. It is not used to connect user stations. See 10Base-F.

10Base-FL A part of the IEEE Base-F specification that covers Ethernet over fiber. It is interoperable with FOIRL. See 10Base-F.

10Base-T An Ethernet local area network which works on twisted pair wiring that looks and feels remarkably like telephone cabling. In fact, 10Base-T was invented to run on telephone cable. 10Base-T Ethernet local area networks work on home runs in which the wire from each workstation goes directly to the 10Base-T hub (like the wiring of a phone system), just like a phone system. 10Base-T cards which fit inside PCs typically cost the same as those for Ethernet running on coaxial cable. The advantages of 10Base-T (which has become the most commonly installed local area network in the world) are twofold — namely if one machine crashes, it doesn't bring down the whole network (coax Ethernet LANs are typically in one long line, looping from one machine to another. One crash. Every machine goes down.); and secondly, a 10Base-T Ethernet network is easier to manage because the 10Base-T hubs often come with sophisticated management software. Though 10Base-T is designed to work on "normal" telephone lines, no one in their right mind would install "normal" phone wiring. The preferred method of installing 10Base-T networks is to use new Category 5 wiring. If you're forced (because you don't want to open up a pretty wall), then connect it old phone cabling. It will probably work. Remember 10Base-T uses two pairs. Most phones need only one pair. 10Base-T's maximum segment length is 100 meters running on unshielded twisted pairs. See 802.3 10Base-T.

10Broad36 An IEEE 802.3 network specification employing a broadband transmission scheme using thick coax cable and running at 10 million bits per second.

10XXX Calling The original access code that you dialed in North America to reach carrier that you had not equal accessed to. On July 1, 1998, the 10XXX access code was changed to 101XXXX. You must now dial the full seven digits. See 101XXXX for a full explanation.

110-type Connecting Block The part of a 110-type cross connect, developed by AT&T (now Lucent), that terminates twisted-pair wiring and can be used with either jumper wires or patch cords to establish circuit connections.

110-type Cross Connect A compact cross connect, developed by AT&T (now Lucent), that can be arranged for use with either jumper wires or patch cords. Jumper wires, used for more permanent circuits, must be cut down to make circuit connections. Patch cords allow ease of circuit administration for frequently rearranged circuits. The 110-type cross connect also provides straightforward labeling methods to identify circuits.

119 Japan's equivalent of the United States' emergency 911 number.

12-Pack Coax Cable A bundle of 12 50-ohm coaxial cables that often run from a SONET carrier to a Digital Cross-Connect System (DCS), They carry a STS-1 (synchronous transport signal-1).

1394 An IEEE data transport bus that supports up to 63 nodes per bus, and up to 1023 buses. The bus can be tree, daisy chained or any combination. It supports both asynchronous and isochronous data. 1394 is a complimentary technology with higher bandwidth (and associated cost) than Universal Serial Bus. Intel told me in summer of 1996 that it was supporting USB for most devices that attach to PC up through audio and video conferencing. Intel told that they are "supporting IEEE 1394 as the preferred interface for higher bandwidth applications such as high quality digital video editing, and connection to new digital consumer electronics equipment. We expect that 1394 will show up in low volumes in 97, and ramp into high volumes in 99." See USB.

144-line Weighting In telephone systems, a noise weighting used in a noise measuring set to measure noise on a line that would be terminated by an instrument with a No 144-receiver, or a similar instrument.

1453 Johann Gutenberg, a goldsmith from Mainz, Germany, prints his Mazarin Bible, which is believed to be the first book printed with movable type, i.e. printed on his famous Gutenberg Press. It took Gutenberg two years to compose the type for his first bible. But once he had done that he could print multiple copies. Before Gutenberg, all books were copied by hand. Monks usually did the copying. They seldom managed to make more than one book a year. The Gutenberg press was a major advance. Before Gutenberg, there were only 30,000 books on the continent of Europe. By the year 1500, there were nine million. Some people (including me) have likened the invention of the Internet to the Gutenberg Press. Johann Gutenberg lived from 1397 to 1468.

16-bit An adjective that describes systems and software that handle information in words that are 2 bytes (16 bits) wide.

16-bit Computer A computer that uses a central processing unit (CPU) with a 16-bit data bus and processes two bytes (16 bits) of information at a time. The IBM Personal Computer AT, introduced in 1984, was the first true 16-bit PC.

16-CAP An ATM term. Carrierless Amplitude/Phase Modulation with 16 constellation points: The modulation technique used in the 51.84 Mb mid-range Physical Layer Specification for Category 3 Unshielded Twisted-Pair (UTP-3).

16450/8250A Found in most current PCs, these older UART chips use a 1-byte buffer that must be serviced immediately by the CPU. If not, interrupt overruns will result. See 16550 and UART.

16550 An enhanced version of the original National Semiconductor 16xxx series UART, which sits in and controls the flow of information into and out of virtually every PC serial port in the world. The older version contains only a one-byte buffer. This can slow down the transmission of high-speed data especially when you're using a multitasking program, like Windows. The "solution" is to get a serial card or port containing the 16550. This chip contains two 16-byte FIFO buffers, one each for incoming and outgoing data. Also new is the 16550's level-sensitive interrupt-triggering mechanism, which controls the amount of incoming data the buffer can store before generating an interrupt request. Together, these features help reduce your CPU's interrupt overhead and thus speed up your communications. See 16450/8250A and UART.

1753 Benjamin Franklin invents the lightning rod in 1753. It was the first practical victory of science over a natural phenomenon. Two years later, when Lisbon, Portugal, was destroyed by an earthquake and a tidal wave, some ministers in Boston proclaimed it was a punishment for the sacrilege of using lightning rods to avert the wrath of God.

1787 According to a bill for a celebration party thrown September 15, 1787, the 55 framers of the U.S. Constitution drank 54 bottles of Madeira, 60 bottles of claret, 8 bottles of whiskey, 22 bottles of port, 8 bottles of cider, 12 bottles of beer, and 7 large bowls of spiked punch big enough "that ducks could swim in them." Sixteen players provided the background music for the bash. This would appear to explain why the Constitution was signed on the 17th, and not the 16th, of September.

1791 April 27, Samuel Finley Breese Morse born.

1793 Semaphore invented.

1800 First battery invented by Alessandro Volta, an Italian physicist.

1822 Historic term which refers to the original ARPANET host-to-IMP interface. The specifications for this were given in BBN report 1822.

1833 Analytical engine by Charles Babbage.

1836 Elisha Gray born. He invented the telephone at around the same time Alexander Graham Bell did. But Bell got his phone patented before Gray.

1837 Telegraphy by Samuel F. B. Morse. Morse invents American Morse Code.

1840 Samuel Morse patents the telegraph.

1843 1. First commercial test of Morse's telegraph. The US Government paid for a telegraph line between Baltimore and Washington, D.C. It worked.
2. First successful fax machine patented by Scottish inventor, Alexander Bain. His "Recording Telegraph" worked over a telegraph line, using electromagnetically controlled pendulums for both a driving mechanism and timing. At the sending end, a style swept across a block of metal type, providing a voltage to be applied to a similar stylus at the receiving end, reproducing an arc of the image on a block holding a paper saturated with electrolytic solution which discolored when an electric current was applied through it. The blocks at both ends were lowered a fraction of an inch after each pendulum sweep until the image was completed. Bain's device transmitted strictly black and white images.

1844 Samuel Morse sends first public telegraph message.

1845 First rotary printing press by Richard M. Hoe.

1851 The Continental (more commonly called the International) Morse Code is adopted for European telegraphs, but American telegraphers reject it. See Morse Code.

1860 Pony Express formed to carry mail to the Wild West. The Pony Express lasted around a year before the telegraph took over.

1861 Pony Express disbanded. The telegraph took over.

1865 First commercial fax service started by Giovanni Casselli, using his "Pantelegraph" machine, with a circuit between Paris and Lyon, which was later extended to other cities.

1866 1. First experimental wireless by Mahlon Loomis.
2. Two successful transatlantic submarine telegraph cables (one eastbound, one westbound) are laid by Cyrus Field between Valencia, Ireland and White Stand Bay, Newfoundland, Canada.

1867 Typewriter invented by Christopher L. Sholes.

1869 Elisha Gray and Enos Barton form small manufacturing firm in Cleveland, OH.

1871 First British submarine telegraph cable laid in Hong Kong.

1872 Gray & Barton's firm re-named Western Electric Manufacturing Company. See 1869.

1874 April 25, 1874, Guglielmo Marconi born in Bologna, Italy.

1875 February, 1875. Alexander Graham Bell signs a agreement with two partners (one his father-in-law) to start a company to oversee his patents. The deal covered the young man's telepgrahic inventions, but also included "further improvements," one of them later turned out to the transmission of human voice.

1876 1. March 7, Telephone patent issued to 29-year old Boston University professor, Alexander Graham Bell. The patent was number 174,465. Three days later he sent the landmark message, "Mr. Watson, come here. I want you." The telephone has become the most profitable invention in the history of mankind. Bell successfully defended himself against all 600 lawsuits claiming rights to his invention. See 1877.

2. Western Union issues its famous internal memo which contains the incredibly wonderful words," "This 'telephone' has too many shortcomings to be seriously considered as a means of communication. The device is inherently of no value to us."

3. Braving a hostile ocean, the men of the Faraday, a steam-driven ship with three masts, laid the first transatlantic cable between Ireland and America. The cable was made by Siemens. It could carry 22 telegraph messages at one time. It carried the world into a new era of communications.

1877 1. First telephone in a private home. First telephone in New York City.

2. Phonograph invented by Thomas Edison.

3. Western Union turns down a chance to buy the patent rights to the invention of the telephone for $100,000. Western Union believed the telegraph superior technology. They were flat out wrong. It was clearly one of the dumbest decisions made in American business history.

1878 1. Theodore N. Vail begins his career with the Bell System as general manager of the Bell Telephone Company. He later became the first president of the American Telephone & Telegraph Company in 1885. He left AT&T two years later. After pursuing other interests for 20 years, he returned as president of AT&T in 1907, retiring in 1919 as chairman of the board. Vail believed in "One policy, one system, universal service." He regarded telephony as a natural monopoly. He saw the necessity for regulation and welcomed it.

2. The New Haven Telephone Company publishes the first telephone directory. It has 75 pages and consists of one page with 50 listings. In 1996, some 6,200 telephone directories will be published in the United States and will generate about $10 billion in advertising revenues.

1880 Alexander Graham Bell develops the photophone which uses sunlight to carry messages. It was never commercially produced.

1881 1. First long distance line, Boston to Providence.

2. American Bell purchases controlling interest in Western Electric and makes it the manufacturer of equipment for the Bell Telephone companies.

1885 Theodore N. Vail becomes the first president of the American Telephone & Telegraph Company. He left AT&T two years later. After pursuing other interests for 20 years, he returned as president of AT&T in 1907, retiring in 1919 as chairman of the board. Vail believed in "One policy, one system, universal service." He regarded telephony as a natural monopoly. He saw the necessity for regulation and welcomed it.

1887 AT&T (American Telephone & Telegraph Co.) starts business.

1889 1. A. B. Strowger invents the telephone switch, dial telephone.

2. Punch card tabulating machine invented by Herman Hollerith.

1890 Congress passes Sherman Act.

1891 First underseas telephone cable, England to France.

1895 1. Guglielmo Marconi of Italy invents wireless telegraph.

2. When the X-ray was discovered by Wilhelm Roentgen in 1895, some journalists were convinced that the primary user of the revealing shortwave radiation would be the "peeping Tom." The titillating publicity led to: A law introduced in New Jersey which forbade the use of "X-ray opera glasses," and merchants in London sold X-ray- proof underwear for modest ladies.

1899 1. Magnetic voice recorder by Vlademar Poulsen.

2. AT&T, created in 1885, takes over American Bell Telephone and becomes parent to Western Electric and the Bell System companies.

1906 Motion picture sound by Eugene Augustin Lauste.

1909 Paris' best-known monument, the Eiffel Tower, was saved from demolition in 1909 because there was an antenna, of great importance to French radio telegraphy, mounted at the top of the nearly 1000-foot-high structure.

1911 Multiplying and dividing calculating machine by Jay R. Monroe.

1914 Congress passes Clayton Act.

1915 First transcontinental phone call in USA.

1917 VHF Transatlantic radio by Guglielmo Marconi.

1920 First commercial AM radio broadcast in the U.S. KDKA, Pittsburgh.

1921 Facsimile technology (Wirephoto) from Western Union.

1922 First dial exchange in New York City — PE-6 from PEnnsylvania 6.

1924 1. Thomas J. Watson renames Computing-Tabulating-Recording (CTR) the International Business Machines Corporation.

2. The work of Herbert Ives at Bell Labs on the photoelectric effect leads to the first demonstration of the transmission of pictures over telephone lines.

1925 1. IBM begins selling punch-card machinery in Japan.

2. Bell Laboratories is created from the AT&T and Western Electric engineering department, which had been combined in 1907. Frank B. Jewett becomes the first president of Bell Labs.

1926 1. AT&T Bell Labs invents sound motion pictures.

2. First public test of trans-atlantic radiotelephone service — between New York and London.

1927 On January 17, 1927, transatlantic telephone service between London and New York opened, charging $25 a minute, or 15 English pounds for three minutes. Time later reported the event as such "Walter Sherman Gifford, president of the American Telephone and Telegraph Co. picked up a telephone receiver in Manhattan. Said Gifford into the transmitter, 'Good morning, Sir. This is Mr. gifford in New York.' Sir George Evelyn Pemberton Murray, Secretary of the General Post Office of Great Britain in London replied, 'Good morning, Mr. Gifford. Yes. I can hear you perfectly. Can you hear me?' The distinction of talking to London on the first day of transatlantic service was also taken by Adolph S. Ochs, publisher of the New York Times, who let it be known that he was the first private speaker with editor Geoffrey Dawson of the London Times.

1928 January 9. Take a message machine. As Time reported it, "A clever inventor came to the U.S. with the news that he had found a means of clipping the telephone's claws or at least removing

1929 1. April 7, 1929. First public demonstration of long distance TV transmission. Moving black and white pictures were sent over telephone wires between Secretary of Commerce Herbert Hoover in Washington DC and AT&T executives in New York. They went at 18 frames per second.

Further development of this technology led to the creation of TV.

2. Harold S. Black's negative feedback amplifier cuts distortion in long distance telephony. Black is at Bell Labs. 3. Coaxial cable invented in Bell Telephone Laboratories; Herbert Hoover first president to have phone installed on his desk. 4. AT&T Bell Laboratories and Western Electric introduced the Sound Newsreel Camera. It used an AT&T "Light Valve" to record sound directly on the film as it passed through the camera. It was the first single system sound camera.

1933 1. Karl G. Jansky at Bell Labs discovers radio waves from the Milky Way. His discovery leads to the science of radio astronomy.

2. Bell Labs transmits first stereo sound, a symphony concert, over phone lines from Philadelphia to Washington.

3. FM radio invented by Edwin H. Armstrong.

1936 First TV broadcast by the BBC in Great Britain.

193rd Bit The frame b t for a T-1 frame. See Robbed Bit Signaling.

1938 Xerography invented by Chester Carlson.

1940 February 19. DuPont introduces "nylon," an artificial silk billed as a formidable rival to natural Japanese silk. Nylon is technically described as "synthetic fiber-forming polymeric amides." Its basic materials are coal, air and water. DuPont announces that nylon will be put to many other uses besides stockings: non-cracking patent leather, weatherproof clothing and flexible window panes.

1942 Harry Newton born Sydney, Australia on June 10.

1944 Electronic Numerical Integrator and Computor (spelled with an O). Early computer, built in 1944.

1946 1. Electronic Numerical Integrator and Computor (spelled with an O) built at the University of Pennsylvania. ENIAC, the first general-purpose electronic computer, was 30ft x 50ft. and weighed 30 tons. It included 18,000 vacuum tubes, 6,000 switches and 3,000 blinking lights. Although it did not store programs, it could multiply two 6-digit numbers in half a second and could hold an astounding 200 bytes of memory. The ENIAC was turned off for the last time on October 2, 1955.

2. Dr. Robert N. Metcalfe, co-inventor of Ethernet, born Brooklyn, New York on April 7.

3. Ray Horak born Niagara Falls, New York on November 10.

1947 Transistor invented at AT&T's Bell Labs in New Jersey. Inventors were John Bardeen, Walter Brattain and William Shockley. In 1956 they shared the Nobel Prize in Physics for creating the transistor.

1948 1. May 11: Birth of the International Communications Association, among the larger groups of telecommunications users in North America.

2. Claude E. Shannon announces the discovery of information theory, the cornerstone of current understanding of the communication process. Shannon was a Bell Labs employee. See Shannon's Law.

1951 1. First direct distance calling. Phone users can dial long distance without an operator's assistance.

2. Sony unveils the first transistor radio.

1953 May 9. Edmund Hillary and Tenzing reach the summit of Mt. Everest, the first climbers to do so.

1954 1. The solar cell invented by Gerald L. Pearson, Daryl M. Chapin and Calvin S. Fuller at Bell Labs.

2. The year William Shockley left Bell Labs to pursue the commercial opportunities offered by his invention of the transistor.

3. IBM introduces the first business computer, a mainframe. IBM management states that there will never be a need for more than 6 mainframes in the entire world.

1955 Bill Gates, founder and chairman of Microsoft, born October 28, 1955.

1956 1. First transatlantic repeatered telephone cable.

2. First modem was invented by AT&T Bell Laboratories, according to AT&T.

3. Videotape recorder invented by Ampex.

4. AT&T signs consent decree limiting Western Electric to manufacturing equipment for the Bell system and the U.S. government.

1957 1. The U.S.S.R. launched Sputnik on October 4, 1957. It embarrassed the U.S. Government into a frenzy of space investments, culminating in the U.S. being the first country to have people walk on the moon.

2. Dr. Gordon E. Moore and Dr. Robert N. Noyce leave Fairchild Semiconductor and form Intel Corporation. Intel was founded with fewer than $5 million in startup monies using a two-page business plan written by legendary venture capitalist, Arthur Rock.

1958 Integrated circuit invented by Jack S. Kilby at Texas Instruments.

1959 September 28. A desk-sized machine that reproduces documents on ordinary instead of specially treated paper was introduced by Haloid Xerox, Inc. fixing dry ink permanently onto paper, the Xerox 914 turns out reproductions at the rate of six a minute.

1960 1. First test of an electronic switch.

2. MITI creates the Japan Electronic Computer Corporation to promote its domestic computer industry.

3. Laser invented by Theodore Maiman of the U.S. Laser stands for Light Amplification by the Stimulated Emission of Radiation.

1961 1. Leonard Kleinrock first details packet switching, the critical technology of the Internet.

2. T-1 created.

1962 1. LEDs — Light Emitting Diodes — invented.

2. First commercial communications satellite (Telstar). Owned by AT&T.

3. August 31, 1962 President Kennedy signed Communications Satellite Act.

4. Semiconductor laser invented.

5. Ross Perot forms EDS with a reputed $1,000. In 1984, he sold it to General Motors for $2.6 billion.

1963 1. Touch Tone service introduced.

2. Audio cassette tape introduced by Philips.

3. C. Kumar N. Patel at Bell Labs develops the carbon dioxide laser now used around the world as a cutting tool in surgery and industry.

1964 1. Prototype of the first video phone made by the Bell System shown at The World's Fair in Queens, New York City. Pictures were black and white and the technology was very expensive. It was called the Picturephone.

2. IBM showed the first word processor.

1965 1. PDP-8 minicomputer introduced by Digital Equipment Corporation.

2. First trial offers for reversing telephone charges. Telephone bills start to go awry.

1966 October, 1966 the Electronic Industries Association issues its first fax standard: the EIA Standard RS-328, Message Facsimile Equipment for Operation on Switched Voice Facilities Using Data Communications Equipment. The Group 1 standard, as it later became known, made possible the more generalized business use of fax. Transmission was analog and it took four to six minutes to send a page.

1967 1. First 800 call made in the United States.
2. Electronic handheld calculator introduced by Texas Instruments.
1968 1. In a landmark decision, the FCC for the first time allows non-AT&T equipment to attach to the Bell System. The FCC rules that equipment which is privately beneficial, but not publicly harmful is OK for connection. The Carterfone connected two-way radios to the phone network.
2. Fiber optics for communications invented by Robert Maurer.
1969 1. Ken Thompson and Dennis Ritchie, computer scientists at AT&T Bell Laboratories, create the Unix software operating system.
2. ARPANET introduced by the Advanced Research Projects Agency of the U.S. Defense Department, comprising a 50 kilobit per second backbone and four computer hosts.
3. Traffic Service Position System replaces traditional cord switchboards. The system automates many operator functions for the first time.
4. In a landmark decision, Federal Communications Commission authorizes MCI Communications Corporation to be the first long distance company allowed to compete with AT&T in the U.S. long distance market. The route chosen for the competition is Chicago to St Louis. The route was originally Chicago to Springfield, Illinois. But that meant it was an intrastate route and the Illinois Commerce Commission, which had jurisdiction, hinted it would not rule in MCI's favor, and suggested to MCI it would fare better if it moved the venue to Washington, D.C. by extending the route to St. Louis in Missouri.
1970 1. Optical fiber for long-range communications developed.
2. Relational database invented by Dr. E. F. "Ted" Codd at IBM.
3. Floppy disk invented by IBM. It was designed originally to carry the latest IBM mainframe software to mainframe computers, each of which had a floppy drive for the sole purpose of uploading new software. Once the floppies arrived and their software uploaded, they were physically thrown away. It was only later than someone figured you could use floppy disks for permanent storage.
4. Gilbert Chin creates a new type of magnetic alloy now used in most telephone handset speakers.
1971 Ted Hoff at Intel invents the microprocessor — a single chip that contained most of the logic elements used to make a computer. Intel's twin innovations with the device were to put most of the transistors that make up a computer's logic circuits on to a single chip and to make that chip programmable. Here, for the first time, according to Robert X. Cringely's book "Accidental Empires," was a programmable device to which a clever engineer could add a few memory chips and a support chip or two and turn it into a real computer you could hold in your hands. Intel's first microprocessor, the 4004, was released in November, 1971. See 4004. See also MICROPROCESSOR.
1972 1. Microwave Communications, Inc, later called MCI, wins an FCC license to transmit calls between Chicago and St Louis.
2. First commercial video game (Pong) introduced by Nolan Bushnell at Atari.
3. Email introduced on Arpanet, precursor to the Internet.
4. IBM announces SNA.
1973 1. Computerized Axial Tomography (CAT Scan) invented by Allan Cormack and Godfrey N. Hounsfield.
2. Ethernet invented by Dr. Robert N. Metcalfe on May 22, 1973 at the Xerox Palo Alto Research Center (PARC). Dave

Boggs (Dr. David R. Boggs) was the co-inventor. Metcalfe and Boggs (in that order) were the authors of THE Ethernet paper, published July 76 in CACM. CACM is the Communications of the ACM. ACM is the Association for Computing Machinery.
3. Vinton Cerf, computer scientist, invents the basic design of the Internet - the intermediate level gateways (now called routers), the global address space and the concept of end/end acknowledgement.
4. Gerhard Sessler and James E. West of Bell Labs receive a patent for their unidirectional microphone that improves hands-free telephone conversations.
1974 1. AT&T introduces Picturephone, a two-way color videoconferencing service at 12 locations around the country. Businesses rented meeting rooms equipped with the technology.
2. Hewlett-Packard introduces the first programmable pocket calculator, the HP-65.
3. Structured Query Language (SQL) invented by Don Chamberlain and colleagues at IBM Research.
1975 1. Bill Gates co-founds Microsoft.
2. MITS Altair 8800 personal computer kit from MITS. The Altair was the first commercially available personal computer kit. It was on the cover of the Popular Electronics January 1975 issue.
3. Live TV satellite feed. Ali-Frazier fight from HBO (Home Box Office).
4. Sony introduces Betamax, which doesn't do as well as the Video Home System (VHS) introduced later by Matsushita/JVC.
5. Telephones go mobile. The FCC reallocates a swath of the radio spectrum for mobile communications. Cellular is born.
1976 1. First digital electronic central office switch installed.
2. Apple Computer founded in a Cupertino, CA garage by Steve Jobs and Steve Wozniak.
1977 1. First lightwave system installed and begins operation. It's under the streets of Chicago.
2. Interactive cable system (Qube) installed by Warner Cable.
3. Commodore PET was among the hot PCs of 1977.
4. Hayes introduces 300 bit per second modem for $280.
5. Datapoint introduces Arcnet, a 2.5 megabit per second local area network that, at one stage, was the world's largest selling LAN.
1978 Bell Labs invents cellular technology. ITU comes out with Group 2 recommendation on fax. Intel introduces the 8086 chip, with 29,000 transistors and processing 16 bits of data at one time. A variation of this chip, the 8088, introduced in 1980, caught IBM's eye and IBM used it in its first PC.
1979 1. Chapter 11 Federal bankruptcy provision introduced. Chapter 11 is reorganization. Chapter 7 is liquidation.
2. CompuServe Information Service starts and goes on-line.
3. Gordon Matthews invents corporate voice mail. See VMX.
4. A Federal Communications Commission inquiry restricts AT&T from selling enhanced services except through an AT&T subsidiary, American Bell, which begins operations in 1983 and closes down shortly thereafter.
1980 1. ITU comes out with Group 3 recommendation on fax. Group 3 machines are much faster than Group 2 or 1. With Group 3 machines, after an initial 15-second handshake that is not repeated, they can send an average page of text in 30 seconds or less.
2. Supreme Court of the United States rules that patents for software can be issued.
3. May, 1980 CompuServe becomes a wholly-owned subsidiary of H&R Block, Inc.

1981 1. IBM introduces its first personal computer August 12 and soon has 75 percent of the market. Its PC uses a Microsoft disk-operating system called PC-DOS (for PC-Disk Operating System). Microsoft is intelligent and keeps the right to a virtually identical operating system called MS-DOS and competitors quickly develop lower-priced PC "clones" running on MS-DOS.

2. First portable computer, by Osborne.

3. National electronic phone directory (minitel) starts in France.

4. 3Com introduces the first 10 megabit per second Ethernet adapter — $950.

5. France Telecom starts to deploy Minitel.

6. The Telecommunications Act of the U.K. is passed. It is the first step towards liberalizing the teleconnunications market in the U.K. and has four main consequences:

• The General Post Office (the erstwhile monopoly provider of telecommunications services in the U.K.) was divided into two separate entities: The Post Office and British Telecommunications (BT), which retained the monopoly over existing telecommunicaticns networks.

• It determined that a duopoly would be created as a first step towards the introduction of competition in telecommunications.

• The Secretary of State for Trade and Industry was empowered to license other organizations to be known as Public Telecommunications Operators (PTOs), to operate public telecommunications networks (including cellular networks) in the U.K.

• It paved the way for the gradual deregulation of equipment supply, installation and maintenance which had previously been the monopoly of the GPO.

Following the Act, Mercury Communications, majority-owned by Cable & Wireless was created to compete with British Telecommunications.

1982 January 8, 1982 the consent decree to break up AT&T into seven regional holding companies and what was left (long distance and manufacturing) is announced. The divestiture takes place two years later on January 1, 1984.

March 3, 1982, the FCC formally approved the startup of cellular phone services. The FCC indicated that it would accept applications for licenses in the top 30 markets 90 days after procedures were published and for smaller markets, 180 days after publication. The FCC subsequently gave one license in each market to the local phone company (the "wireline") and one for a competitor (the "non-wireline") carrier.

October 21, 1982, the FCC awards the first construction permit for a cellular radio license to AT&T's Ameritech subsidiary (this was prior to the AT&T breakup).

1983 1. Novell introduces its first local area network software called NetWare. It was originally introduced to allow a handful of personal computers to share a single hard disk, which at that stage was a costly and scarce resource. As hard disks became more available the product evolved to allow the sharing of printers and file servers.

2. Nintendo introduces Famicom, a computer turned video game.

3. Cellular radio in the United States gets its first subscriber.

4. Sony introduces the Camcorder.

5. October 13, 1983, Ameritech turns on its new cellular radio system in Chicago, the first in the nation.

6. IBM introduces the PC XT, the first IBM PC to contain a hard disk.

7. Bill Gates of Microsoft announces Windows at November's Comdex.

8. IEEE approves 802.3 — Ethernet local area network.

1984 January 2, 1984. The breakup of the Bell System. AT&T gave up its local operating phone companies, which got formed into seven, roughly equal holding companies. In turn, AT&T got the Justice Department off its back for an antitrust suit and got the right to get into industries other than telecommunications. Its chosen industry was the computer industry for which it felt it had unique skills.

January 24, 1984 Apple Computer Inc.'s Steve Jobs introduces the first Macintosh computer. It was the machine that changed the world of PC computing. Mr. Jobs often described the little machine as "insanely great."

March, 1984, Motorola introduced the DynaTAC 8000X, the first portable cellular phone. It listed for $3,995 and it weighed two pounds.

Ken Oshman sells Rolm to IBM for $1.26 billion. It was not one of IBM's better investments. Rolm is now part of Siemens, which understands telecommunications.

Prodigy Information Service, a service of IBM and Sears, starts.

1984 Telecommunications Act passed in the U.K.

1985 1. CD-ROM introduced by Philips and Sony.

2. Steve Jobs driven from Apple Computer by John Sculley.

3. John Sculley, head of Apple Computer, licenses the copyrights protecting the "look and feel" of the Apple Macintosh operating system to Microsoft in order to get Microsoft to write more applications for the Mac. The license allowed Microsoft to launch its hugely successful Windows operating system in November of 1985 and to defend itself against a lawsuit brought by Apple alleging Windows was so similar to the Mac that it violated Apple's copyrights.

4. IBM introduces four megabit per second token ring local area network.

1986 1. Novell's SFT NetWare, first fault tolerant local area network operating system.

2. McDonalds becomes first commercial customers to trial ISDN. It's provided by Illinois Bell.

1987 1. October, 1987, the one-millionth cellular subscriber signs up for service in America.

2. Cable TV reaches the halfway mark, penetrating 50.5% of U.S. homes.

3. George Forrester Colony of Forrester Research is believed to have coined the term "client-server" computing. See CLIENT-SERVER.

4. First InterOp trade show, Monterey, California.

5. Synoptics ships the first Ethernet hub.

1988 1. First transatlantic optical fiber cable.

2. Robert E. Allen takes over as CEO of AT&T.

3. The European Union chooses GSM as the general standard for mobile communications, ensuring that even through the 15-nation EU's electrical plus and TV sets are compatible, at least the cellular phone system would be.

4. IBM speeds up its token ring local area network to 16 megabits per seconds.

1989 1. Fiber to the home field trial, Cerritos, CA.

2. Novell releases NetWare 3.0, the first 32-bit network operating system for Intel 80386/486-based servers.

3. 1989: Panasonic's household-size video phone with moving color images debuts in Tokyo.

4. IETF established.

1990 1. Demonstration of 2,000 kilometer link using optical amplifiers without repeaters.

2. MVIP formed and first product shipped.

3. Arpanet officially called the Internet.

1991 1. AT&T, under chairman Bob Allen, buys NCR for a

gigantic $7.4 billion and soon renames it AT&T Global Information Solutions. Later AT&T took hundreds of millions of dollars in restructuring and other charges related to the fact that NCR lost pots of money after AT&T bought it. Part of the problem, according to analysts, was that AT&T bought NCR right at the time NCR was making the transition from traditional mainframe computers to so-called massively parallel computers powered by collections of small, cheaper processors run in tandem. NCR also got hit by a decline in its traditional cash register business as low-margin PCs came in. The skinny around the industry at the time AT&T bought NCR was that AT&T bought NCR to disguise the fact that its own computer operations at that time were losing so much money. And the senior management of AT&T at that time wanted to retire with the glories of booming long distance revenues and not lousy computer results.

2. Motorola introduces the lightest cellular phone yet, the MicroTAC Lite for about $1,000 retail.

3. The Electronic Industries Association approves and publishes on July 9, 1991, the Commercial Building Telecommunications Wiring Standard, the most important wiring standard ever published in the history of telecommunications.

4. Scott Hinton at Bell Labs heads a team that builds the first photonic switching fabric, bringing light-based switching technology in telecommunications networks closer to reality.

5. Wiltel introduces frame relay service.

6. Linus Torvalds, a student at the University of Helsinki, invents Linus, the computer operating system. Linus rhymes with cynics.

1992 1. AT&T introduces VideoPhone 2500 marketed as the first home-model color video phone which works on normal dial up analog phone lines. It meets cool reception because of poor image quality and its high price, namely $1,500.

2. Microsoft Windows 3.1 and IBM's OS/2 2.0 operating systems introduced. Windows NT (32-bit operating system) debuts in beta form.

3. Wang files for Chapter 11.

4. MCI introduces VideoPhone for normal dial-up analog phone lines. It retails for $750. It is not compatible with the AT&T video phone. It and the MCI phone promptly bomb.

5. The cellular industry signs its ten millionth subscriber on November 23, 1992. At least that's what the press releases said. Some carriers claimed that at year end, cellular subscribers in the United States had actually hit 11 million, way ahead of all predictions.

6. Apple, EO and others introduce the PDA, the Personal Digital Assistant. Later they called it the Newton.

7. RMON ratified by IETF.

1993 FCC announces its intention to auction off a chunk of spectrum larger than that used in 1993 for cellular radio. The new airwaves will be used for new types of wireless communications, including portable digital communications devices from phones to laptops, palmtops and PDAs equipped to receive and transmit data of all types, including faxes and video. Sprint introduces first ATM service.

Microsoft releases Windows NT.

Jan, 1993. Marc Andreessen and Eric Bina introduce first graphical Web browser called Mosaic. It's from the National Center for Supercomputing Applications.

February 25, McCaw Cellular announces North America's first all-digital cellular service, in Orlando, Florida.

March 17. The Clinton administration urges Congress to eschew comparative hearings and institute a lottery for awarding new radio spectrum. See PCS.

August. AT&T agrees to buy McCaw Cellular for $12.6 billion. The idea is to help get AT&T back into local phone service.

October. Bell Atlantic sets $21.7 billion merger with TCI, cable TV giant. The assumption is that cable tv and telephone networks are "converging" into an information highway for transporting video, voice and data. The deal later fell apart after the FCC cut cable TV rates and TCI's profitability fell part.

December 31. Thomas J. Watson dies, age 79.

AT&T introduces the AT&T EO Personal Communicator 440, based on the Bell Labs-developed Hobbit microprocessor. This hand-held device combines the features of pen-based personal computers, telephones and fax machines. The device is later withdrawn from the market because of poor demand.

1994 GO-MVIP, Inc. formed. Trade association for developers and manufacturers of MVIP computer telephony products.

AT&T pays $12.6 billion for McCaw Cellular Communications Inc. Robert E. Allen is CEO of AT&T. This is his second expensive purchase, the previous one being NCR.

Hughes Satellite starts DirecTV, a direct-broadcast satellite service that beams 175 channels to a home satellite antenna dish 18 inches in diameter. It snags 1.3 million subscribers in less than a year.

April 4. Netscape Communications Corp which will go on to create the Navigator version of a browser, is founded.

July 17. Microsoft signs a consent decree with the Justice Department agreeing to give computer manufacturers more freedom to install programs from other companies. As a result, Microsoft slightly alters its licensing contracts.

September 12. Netscape ships its first Internet / Web browser.

October 8. A team of six programmers and a veteran Microsoft software developer begin writing the code that will become Internet Explorer version 1.0.

Microsoft licenses technology from Spyglass to help it quickly develop a Web browser.

1995 IBM buys Lotus for $3.5 billion, the main attraction being Lotus Notes. One of the key attractions of Lotus Notes is that it saves on phone bills by substituting electronic messaging for calling.

May 26. William H. Gates sends "The Internet Tidal Wave" memo to Microsoft's top executives, making the Internet the company's top priority.

Sun introduces Java.

August 24, Windows 95 finally ships. It contains heavy computer telephony features, including TAPI, VoiceView, fax on demand and binary file transfers using

August. Disney agrees to buy Capital Cities/ABC for $19 billion. The idea of the merger was, according to the New York Times, that "biggest is best." the idea was to combine the most profitable TV network with a name-brand family entertainment empire.

August. CBS accepts Westinghouse's $5.4 billion takeover offer. According to the New York Times, the idea of the takeover is that "even an ailing Big 3 TV network is worth owning, if it comes with the collection of radio and TV stations reaching one-third of U.S. households."

September. Time Warner agrees to buy Turner Broadcasting for $7.5 billion.

September 20, 1995, AT&T announces it will split itself into three companies — long distance, equipment manufacturing and computers. Wall Street applauds the decision and in one day lifts the price of AT&T's stock by 10%, or about $6 1/2 billion. Meantime, AT&T announces that it will substantially reduce the size of its failed computer activities, which were called AT&T Global Information Solutions. See 1991.

November. Microsoft ships Internet Explorer 2.0.

December 7. Microsoft publicly unveils its Internet strategy.

December 8. Digital Versatile Disk (DVD) is announced. DVD is a specification announced by nine companies for a new type of digital videodisk, similar to CD-ROMs but able to store far more music, video or data in a common format. DVDs will be 5 inches in diameter and will be able to store 4.7 gigabytes on each side, equivalent to 133 minutes of motion picture and sound, or enough to hold most feature-length movies. The companies announcing DVD were Philips, Toshiba, Matsushita Electric Industrial, Sony, Time Warner, Pioneer Electronic, the JVC unit of Matsushita, Hitachi and Mitsubishi Electric.

AT&T announces its plan for restructuring into three separate, publicly traded companies: a services company that will retain the name AT&T; a systems and technology company (Lucent Technologies) composed of Bell Labs, Network Systems, Business Communications Systems, Consumer Products and Microelectronics; and a computer company, which recently returned to the NCR name.

1996 January 3. After 10 years of trying, Congress finally passes a bill deregulating most segments of the communications industry. Telephone companies, broadcasters and cable operators are all free to enter each others markets. It's called The Telecommunications Reform Act of 1996.

February. US West signs $5.3 billion deal for Continental Cablevision. The assumption, according to the New York Times, is that it will now have a wire into many homes, by dominating local phone service in 14 states and reaching 16.3 million cable subscribers.

April. SBC Communications (the name for the holding company owning Southwestern Bell Telephone) buys Pacific Telesis (the holding company for Pacific Bell) for $16.7 billion. The assumption of this merger, according to the New York Times, is that regional telephone companies can become national players by combining 30 million phone lines in five states with potential coast-to-coast cellular market of 80 million people.

April. Bell Atlantic buys NYNEX (the holding company for New York Telephone and New England Telephone) for $22.1 billion. The new company will be called Bell Atlantic.

IP Telephony introduced.

December 30. AT&T spins off NCR, the firm it had bought in 1991 for $7.4 billion.

1997 IP Telephony starts to become a reality. Microsoft announces TAPI 3.0, whose cornerstone is IP telephony. See TAPI 3.0.

Worldcom Inc. buys MCI for $30 billion in its own stock for MCI. The offer for $41.50 an MCI share was at a 41% premium over MCI's stockmarket price before the bid was announced.

October 20. Justice Department filed suit against Microsoft in late October, alleging that the company violated its 1995 consent decree — one section of which banned Microsoft from tying the licensing of one product to the acceptance of another.

December 11. Judge Thomas Penfield Jackson issues a temporary restraining order against Microsoft. Microsoft must at least temporarily halt its practice of requiring PC vendors to bundle Internet Explorer with Windows 95 while the case is being decided. Judge Thomas Penfield Jackson also appointed a special master to gather additional evidence. However, Jackson declined to hold Microsoft in contempt — a move which could have cost Microsoft as much as $1 million a day. He also turned down a Justice Department request to strike

down non-disclosure agreements between Microsoft and PC vendors that the department claims has hampered its attempts to solicit testimony against the software vendor.

1998 1. January 1, 1998. The market for fixed telecommunications services in the EU (European Union) is opened to all competitors. European countries see a proliferation of long distance phone companies.

2. SBC Communications, Inc. the new name for the regional Bell holding company, Southwestern Bell, agrees to buy Southern New England Telephone Company — commonly called SNET. See Southern New England Telephone Corporation.

3. MCI gets bought by Worldcom. New company called MCI Worldcom. Worldcom paid for the acquisition by issuing paper script it printed itself. And you thought the U.S. government had a monopoly on printing money.

4. Microsoft sued by the U.S. Justice Department for antitrust violations.

1999 January 1, 1999. The new European currency called the euro is officially introduced. Eleven European countries have pegged their currency to it. Those countries are Austria, Belgium, Finland, France, Germany, Ireland, Italy, Luxembourg, Netherlands, Portugal and Spain. The euro started at around $US1.17.

2002 January 1, 2002. Euro bills and coins go into circulation.

July 1, 2002. National bills and coins of those countries with the euro as currency will no longer be legal tender. Only the euro will be legal tender. See 1999.

1A AT&T's first generation of standardized key telephony system equipment based on a variety of interconnected phone-line-powered relays. Prior to 1A1, key systems were often patched together from a variety of non-standard parts, with varying wiring schemes, making repairs and upgrades very difficult.

1A-ESS An analog central office, made by AT&T and widely deployed by the Bell Operating Companies prior to divestiture.

1A1 AT&T's second generation of standardized KEY TELEPHONE SYSTEM equipment. Unlike the phone-line powered 1A1, it used commercial AC power for added features such as illuminated buttons to indicate line status.

1A2 AT&T's third generation of standardized KEY TELEPHONE SYSTEMS. It was distinctive for its use of plug-in circuit cards, making it much easier to add features or diagnose and cure problems.

1A3 A cute term for an historic TIE electronic key system that provided advanced features, but was priced competitively with 1A2 electromechanical key systems.

1Base-5 Defined in IEEE 802.3, 1Base-5 was the first LAN standard to make use of UTP (Unshielded Twisted Pair). Running at one megabits per second with Manchester encoding, this Ethernet standard operates on a baseband basis, providing for a single transmission at a time. Connection to the centralized hub is accomplished over UTP of 22, 24 or 26 gauge at distances of up to 500 meters in a star topology. Two pair are used, with one providing upstream connectivity to the hub and the other providing downstream connectivity. AT&T's StarLAN adhered to the 1Base-5 standard, which has been eclipsed by 10/100Base-T. See also 10Base-T and Manchester encoding.

1FB One Flat rate analog Business phone line. A phone line you pay a single monthly charge for and you may make as many local phone calls as you wish during that month. A 1FB is an increasing rarity in the United States. See also 1MB.

1FR One Flat rate residential phone line. A phone line you pay a single monthly charge for and you may make as many local phone calls as you wish during that month. See also 1MB.

1MB 1. One Message rate Business phone line. A phone line you pay a single monthly charge for. That charge typically allows you to make a small number of local calls for free. But that each additional local call will cost you. That cost may be by the minute and/or by the distance, or just by the call. See 1FB. 2. Slang for a T-1 line. This derives from the fact that a T-1 data line will deliver 1.544 million bits per second.

1PSS Packet Switching System. The AT&T Western Electric 1PSS is a high-capacity, X.25 packet switch.

2-Line Network Interface. Old type interchangeable lightning protectors. The top is painted white to indicate gas type instead of carbon type.

2-way Trunk A trunk that can be seized at either end.

2-wire Facility A 2-wire facility is characterized by supporting transmission in two directions simultaneously, where the only method of separating the two signals is by the propagation directions. Impedance mismatches cause signal energy passing in each direction to mix with the signal passing in the opposite direction. See 4-WIRE FACILITY.

2000 The year of potential computer apocalypse. Older computer systems and application software were not developed to deal with years ending in "00." Rather, the manufacturers assumed that the systems would be retired before the year 2000, and developed them with internal calendars based on "19xx." At the stroke of midnight on January 1, 2000, the systems will assume that it is the year 1900. Computer systems supporting financial and insurance applications will be devastated unless steps are taken to modify the systems and software. Additionally, 2000 is a leap year. The computers also will fail to add the extra day of February 29. Mainframe systems and software, in particular, are affected, as many of them are of older vintage. Literally billions of lines of computer code must be examined and adjusted in order to correct the problem. The Securities Industry Association (SIA) has formed a Year 2000 committee to address the problem; the 140+ members of this committee will meet monthly until the Spring of 1999. The committee has identified specific problem areas of platform concern, including hardware, operating systems and applications. See also Millennium Bug.

2038 On January 20, 2038 the 32-bit signed time_t integer which clocks seconds in UNIX systems will expire. UNIX-based applications and embedded chips, and some other systems based on a 32-bit architecture, count seconds from midnight January 1, 1970, which is the "UNIX Epoch Start Date." On roughly January 20, 2038, the integer will roll over from a zero followed by 31 ones to a one followed by 31 zeros; the system will interpret the date as January 1, 1970. As most application software queries the OS for date information, rather than calculating it internally, the impact of this oversight could be very significant, indeed. Actually, the effect could be felt much earlier. For instance, 30-year mortgages may not be calculated correctly beginning in the year 2008. On the other hand, a UNIX-based system in a restaurant may calculate your check based on the cost of a meal in 1970 dollars-a pleasant thought, although an unlikely result.

214 Licence Licence from the FCC (Federal Communications Commission) which lets you offer international communications services to customers in the United States.

23B+D An easy way of saying the ISDN Primary Rate Interface circuit. 23B+D has 23 64 Kbps (kilobits per second) paths for carrying voice, data, video or other information and one 64 Kbps channel for carrying out-of-band signaling information. ISDN PRI bears a remarkable similarity to today's T-1 line, except that T-1 can carry 24 voice channels. In ISDN 23B+D, the one D channel is out-of-band signaling. In T-1, signaling is handled in-band using robbed bit signaling. Increasingly, 23B+D is the preferred way of getting T-1 service since the out of band signaling is richer (delivers more information — like ANI and DNIS) and is more reliable than the in-band signaling on the older T-1. One good thing about PRI: You can now organize with your phone company to deliver the signaling for a bunch of ISDN PRI cards on one D channel. Thus your first line would have 23 voice channels, for example. Your second would have 24 voice channels, etc. Several of the more modern voice cards will accept the signaling for up to eight ISDN PRI channels on the D channel of the first one. See ISDN PRI, ROBBED BIT SIGNALING and T-1.

24-bit Mode The standard addressing mode of Apple Macintosh's System 6 operating system, where only 24 bits are used to designate addresses. Limits address space to 16MB (2 to the 24th power), of which only 8MB is normally available for application memory. This mode is also used under System 7 (the Mac's more modern operating system) if 32-bit addressing is turned off. See 32-BIT.

24-bit Video Adapter A color video adapter that can display more than 16 million colors simultaneously. With a 24-bit video card and monitor, a PC can display photographic-quality images.

24-Hour Format Sometimes known as military time. Using twenty-four hours to designate the time of day, rather than two, twelve-hour segments.

2500 Set The "normal" single-line touchtone desk telephone. It has replaced the rotary dial 500 set in most — but definitely not all — areas of the United States and Canada. No one seems to know why the addition of a "2" in front of a model number came to denote touchtone in the old Bell System.

258A Adapter A device about 12 inches long and six inches wide and two inches deep that is used to connect a 25-pair Amphenol cable to RJ-45 patch cords.

2600 Tone Until the late 1960s, America's telephone network was run 100% by AT&T and used 100% in-band signaling, whereby the circuit you talked over was the circuit used for signalling.For in-band signaling to work there needs to be a way to figure when a channel is NOT being used. You can't have nothing on the line, because that "nothing" might be a pause in the conversation. So, in the old days, AT&T put a tone on its vacant long distance lines, those between its switching offices. That tone was 2600 Hertz. If its switching offices heard a 2600 Hz, it knew that that line was not being used. At one point in the 1960s, a breakfast cereal included a small promotion in its cereal boxes. It was a toy whistle. When you blew the whistle, it let out a precise 2,600 Hz tone. If you blew that whistle into the mouthpiece of a telephone after dialing any long distance number, it terminated the call as far as the AT&T long distance phone system knew, while still allowing the connection to remain open. If you dialed an 800 number, blew the whistle and then touchtoned in a series of tones (called MF — multi-frequency — tones) you could make long distance and international calls for free. The man who discovered the whistle was called John Draper and he picked up the handle of Cap'n Crunch in the nether world of the late 1960s phone phreaks. Since then, in-band signaling has been replaced by out-of-band signaling, the newest incarnation being called Signaling System 7. See 2600, CAPTAIN CRUNCH, MULTI-FREQUENCY SIGNALING and SIGNALING SYSTEM 7.

2780 A batch standard used to communicate with IBM mainframes or compatible systems.

2B+D A shortened way of saying ISDN's Basic Rate Interface interface, namely two bearer channels and one data channel. A single ISDN circuit divided into two 64 Kbps digital channels for voice or data and one 16Kbps channel for low speed data (up to 9,600 baud) and signaling. Either or both of the 64 Kbps channels may be used for voice or data. In ISDN 2B+D is known as the Basic Rate Interface. In ISDN, 2B+D is carried on one or two pairs of wires (depending on the interface) — the same wire pairs that today bring a single voice circuit into your home or office. See ISDN.

2B1Q Two Binary, One Quaternary. An line encoding technique used in ISDN BRI; it also is used extensively in the U.S. in first-generation HDSL systems. 2B1Q is a four-level PAM (Pulse Amplitude Modulation) technique which maps two bits of data into one quarternary symbol, with each symbol comprising one of four variations in amplitude and polarity over a circuit. See also HDSL, ISDN and PAM.

2FR A flat-rate party line with two subscribers.

2W Two-Wire. See 2-WIRE FACILITY.

3:2 Pull-down A method for overcoming the incompatibility of film and video frame rates when converting or transferring film (shot at 24 frames per second) to video (shot at 30 frames per second).

311 The Services Code now available for non-emergency access to police, fire and other governmental departments. The FCC instructed the North American Numbering Plan Administration to make 311 available in order to relieve the load on the 911 emergency number. See also 911 and 711.

3172 IBM's network controller. It connects to the mainframe channel on one end and the LAN media (Ethernet, Token Ring, FDDI) on the other.

3174 IBM's cluster controller. It connects to terminals and other I/O devices on one end, and a mainframe channel on the other.

32-bit An adjective that describes hardware or software that manages data, program code, and program address information in 32-bit-wide words. What is the significance of 32-bit? With 32-bit memory, each program can address up to 4 gigabytes (2 to the 32nd power) of memory. This is in contrast to Windows 3.x where programs are limited to 16 MB of memory. Possibly more significant than the amount of memory that is available to a 32-bit application is how that memory is accessed. Under Windows 3.x, memory is accessed by using two 16-bit values that are combined to form a 24-bit memory address. (24-bits is the size of the memory addressing path of the Intel 80286. The 80286 is the architecture that Windows 3.x was designed for.) The first 16-bit value (selector) is used to determine a base address. The second 16-bit value (offset) indicates the offset from the base address. One of the side effects of this architecture is that the maximum size of a single chunk of memory is 64 KB. Windows 95 and Windows NT are 32-bit operating systems. Windows 95 and Windows NT developers can address memory with a single 32-bit value. Such an addressing scheme allows developers to view memory as one flat, linear space with no artificial limits on the size of a single segment. No longer are programmers concerned about selectors and offsets and the 64 KB segment limit. Also, Windows 95 and Windows NT take full advantage of the protection features of the Intel 80386 microprocessor. 32-bit applications are given their own protected address space which tends to prevent applications from inadvertently overwriting each other.

32-bit Addressing See 32-BIT.

32-bit Computer A computer that uses a central processing unit (CPU) with a 32-bit data bus and central processing unit (CPU) which processes four bytes (32 bits) of information at a time. Personal computers advertised as 32-bit machines — such as Macintosh SE, and PCs based on the 80386X microprocessor — aren't true 32-bit computers. These computers use microprocessors (such as the Motorola 68000 and Intel 80386SX) that can process four bytes at a time internally, but the external data bus is only 16 bits wide. 32-bit microprocessors, such as the Intel 80386DX, the Pentium and the Motorola 68030, use a true 32-bit external data bus and can use 32-bit peripherals.

3270 IBM class of terminals (or printers) used in SNA networks.

3270 Gateway An electronic link which uses 3270 terminals to handle data communications between PCs and IBM mainframes.

3270SNA A specific variation of IBM's System Network Architecture for controlling communications between a 3270 terminal connected to an IBM mainframe.

3274 IBM series of Control Units or Cluster Controllers provide a control interface between host computers and clusters of 3270 compatible terminals.

327X Belonging to IBM's 3270 collection of data communications terminals.

3299 A communications device for an IBM mainframe computer.

347x The type of fixed function computer terminals used with IBM mainframe computers.

3745 IBM's communications controllers, often called front-end processors. 3745 devices channel-attach to the mainframe and support connections to LANs and other FEPS.

3780 A batch protocol used to communicate with an IBM mainframe or compatible system.

3B2-400 A UNIX-based minicomputer, manufactured by AT&T and widely deployed by the Bell Operating Companies prior to divestiture.

3D API 3D Application Programming Interface. This generic term refers to any API that supports the creation of standard 3D objects, lights, cameras, perspectives, etc. APIs include Argonaut's BRender and Microsoft's Reality Lab.

3DGF 3-D Geometry File. A platform independent format for exchanging 3-D geometry data among applications. Developed by Macromind.

3FR A flat-rate party line with three subscribers.

3G Third generation, the next generation of wireless technology beyond personal communications service (PCS). 3G networks will transmit data at 144 kilobits per second, or up to 2 megabits per second from fixed locations.

4-wire Facility A 4-wire facility supports transmission in two directions, but isolates the signals by frequency division, time division, space division, or other techniques that enable reflections to occur without causing the signals to mix together. A facility is also called 4-wire if its interfaces to other equipment meet this 4-wire criteria (even if 2-wire facilities are used internally), as long as crosstalk between the two transmission directions, as measured at the interface, is negligible. See 2-WIRE FACILITY.

4.9% No regional Bell Operating Company is presently allowed to own more than 4.9% of the stock of a telecommunications manufacturing company. See DIVESTITURE.

4004 The world's first general-purpose microprocessor (computer on a chip). The 4004 was made by Intel, was 4-bit, was released on November 15, 1971 and contained 2,300 transistors. It executed 60,000 instructions per second. The tiny 4004 had as much computing power as the first electronic computer, ENIAC, which filled 3,000 cubic feet with 18,000 vacuum tubes when it was built in 1946. The 4004 found a home in desktop calculators, traffic lights and electronic scales. Despite its power, its 4-bit structure was too small to process all the bits of data at one time to handle all the letters of the alphabet. It was followed by the 8-bit 8008. See also 1971 and 1978.

411 The number dialed for information in many (not all) cities in North America.

41449 AT&T's specifications for its ISDN PRI (Primary Rate Interface). It is different to the ANSI standard T1.607.

419A A famous old Bell System tool which many installers found very convenient to hold a diminishing marijuana cigarette (called a 'roach') in the 1960s.

42A An early terminal block. The Model 42A is a plastic mounting base about two inches square with four screws and a cover. Before modular connections became widespread, the 42A was used to connect a phone's line cord to the wire inside a wall or running around the baseboard. Adapters, such as the No. 725A made by AT&T and Suttle Apparatus, can be used to convert a 42A into a 4-conductor modular jack. See also TERMINAL BLOCK.

46-49 In North America, most cordless phones operate within the band 46-49 MHz. That band contains only 10 channels and is horribly overcrowded. Recently, the FCC authorized a new frequency range — 905-928 MHz — for use by, amongst other things, cordless phones. The 900 Mhz contains 50 channels.

4A The last generation of "telco-quality" add-on speakerphones, with separately-housed microphone and speaker; made by both Western Electric (AT&T) and Precision Components, Inc.

4FR A flat-rate party line with four subscribers.

4GL Fourth Generation Language.

4WL-WDM Four Wavelength Wave Division Multiplexing, also called Quad-WDM. MCI announced this technology in the Spring of 1996 as a method of allowing a single fiber to accommodate four light signals instead of one, by routing them at different wavelengths through the use of narrow-band wave division multiplexing equipment. The technology allowed MCI to transmit four times the amount of traffic along existing fiber. At that time MCI's backbone network operated at 2.5 gigabits per second (2.5 billion bits) over a single strand of fiber optic glass. Using Quad-WDM the same fiber's capacity will rise to 10 gigabits — enough capacity to carry 64,500 simultaneous transmissions over one single strand of fiber.

500 Service A non-geographic area code specifically assigned for Personal Communications Services (PCS), as originally defined — in other words, not necessarily the cellular-like PCS we hear so much about. 500 numbers provide for follow-me services, which allow the subscriber to define a priority sequence of telephone numbers which the network will use to search for him. For instance, the search might begin at your business phone, progressing to your cellular/PCS phone, then to your home phone, and then to your voice mailbox, assuming that you can't be found or don't want to be found. Options might include distinctive ringing for pre-defined callers of significance such as your spouse, significant other, or boss. Further options might include billing, such as caller pays any long distance charges (hopefully, with pre-notification), call blocking, and selective call blocking.

500 Service promises to offer a single telephone number which can find you anywhere, for life. All available 500 numbers were assigned in 1995; plans exist to expand 500 Service area codes, to include area codes such as 520 and 533. 500 Services, clearly, are network-based. CPE solutions recently have emerged, as well. See also WILDFIRE.

500 Set The old rotary dial telephone deskset. The touchtone version was called a 2500 set.

5250 IBM class of terminals for midrange (System 3x and AS/400) environments.

5250 Gateway An electronic link which uses 5250 terminals to handle communications between PCs and IBM minicomputers.

56 Kbps A 64,000 bit per second digital circuit with 8,000 bits per second used for signaling. Sometimes called Switched 56, DDS (Digital Data Service) or ADN (Advanced Digital Network). Each carrier has its own name for this service. The phone companies are obsoleting this service in favor of the more modern ISDN BRI, which has two 64 Kbps circuits (called Bearer circuits) and one 16 Kbps packet circuit. See 56 Kbps Modem, K56flex and ISDN.

56 Kbps Modem A 56 Kbps is the fastest speed modem that will work on normal dial-up phone lines. The basics of the technology can be attributed to Brent Townshend, an independent inventor. V.pcm was the working name which was used while the ITU-T came up with a new standard for 56 kilobit per second (Kbps) modems. That standard is now called V.90. Until the ITU-T standard, there were two competing, non-compatible 56 Kbps modem pseudo-standards — x2 and K56flex. All 56 Kbps modems (including V.90 modems) are asymmetric. They provide a maximum of 56 downstream and 33.6 Kbps (V.34 ITU-T standard) upstream. While that may seem odd, it's about the best you can do with a modem running on an analog phone line. The asymmetric nature of 56 Kbps modems is justified thus: When you're on the Internet or your corporate Intranet, most of the bandwidth you need is for downstream transmission (i.e. information flowing at you). You need this speed to download graphics-intensive (read bandwidth-intensive) files or large software-intensive files (e.g. updates to your software). Upstream you transmit only a few keystrokes or mouse-clicks, i.e. instructions to retrieve the information. 56 Kbps modems are able to achieve this minor miracle by virtue of the fact that today's PSTN is largely digital; in fact, they depend on the PSTN's digital nature.

56 Kbps modems work like this. The user installs a 56 Kbps modem on his PC or laptop, which connects to the PSTN via an analog local loop. The serving central office must be digital, as must be the entire carrier network(s), as must be the terminating central office, and as must be the local loop connection (T-1 or ISDN) to the terminating location (e.g., corporate Intranet site or ISP, Internet Service Provider). Matching 56-Kbps technology must be in place at the terminating location, typically in the form of an access server or switch. As the transmissions must suffer only one A-to-D (analog-to-digital) conversion process, the potential amount of quantizing noise is limited and the higher speeds can be supported with a satisfactory level of error performance. As the PSTN uses PCM (Pulse Code Modulation) for voice-grade A-to-D conversions, the theoretical transmission speed limit is 64 Kbps. Bit robbing consumed 8 Kbps for signaling and control, thereby limiting effective speed to 56 Kbps. The U.S.'s Federal Communications Commission effectively set a limit on the speed of 56 Kbps modem by regulating the maximum power of the signal transmitted, which in effects puts a limit on

speed. The FCC's mandated power limit of -12dBm — about 63 microwatts — for phones was set to prevent interference or noise bleeding into adjacent phone lines. Imagine trying to talk while loud rock music is playing nearby. At that upper power limit, the maximum attainable speed with a 56 Kbps modem is thus about 53 kilobits per second — though there are now rumbles to beg to the FCC to remove this restriction. The actual transmission speed depends on the quality of your analog local loop. The performance of the local loop is sensitive to anomalies such as bad splices, bridge taps, and poorly insulated splice casings. Additionally, transmission speed is affected by EMI (electromagnetic interference) caused by electrical storms, radio transmissions and other sources of electromagnetic energy. Finally, other transmissions taking place on twisted pairs in close proximity to the same cable may affect transmission speed. In other words, 53 Kbps is the maximum, rather than the norm. Published surveys of 56 Kbps modems show typical achieved transmission speeds of between 40 Kbps and 46 Kbps. I achieve around 48 Kbps on my dial-up phone line in New York City and around 44 Kbps at my phone line in midstate New York (the Albany area).

56 Kbps modems are also V.34 modems. Assuming that the terminating modem is V.34 (and it always is), the 56 Kbps modem will "fall back" (if things get really awful) to that standard, which supports symmetric transmission at 33.6 Kbps. Until the ITU-T came out with V.90, its standard for asymmetric 56 Kbps modems, there were two competing standards — x2 and K56flex. x2 was developed by US Robotics. K56flex was developed by Rockwell Semiconductor and Lucent Technologies. x2 was not compatible with K56flex. Neither x2 nor K56flex conform to V.90. Though some of these modems x2 and K56flex modems are upgradable to V.90, most are not. Sorry. See also K56flex, 56 Kbps Modem, V.90 and V.PCM.

56Flex See K56flex.

5ESS A digital central office switching system made by AT&T. It is typically used as an "end-office," serving local subscribers.

5XB 5 X-Bar central office equipment

611 Phone number used by many cellular and PCS providers in North America for reporting problems to or asking service-related questions of wireless carriers.

64 Bit Architecture The wide data path over which instructions (words composed of bits) are moved to and from Intel's i860's internal registers and memory. Most conventional mainframes use a word length of 32 bits. Intel's i860 RISC-based microprocessor line, introduced in 1989, incorporates many firsts, including more than 1 million transistors on a single chip. The line currently comprises the i860 XR, with 1.2 million transistors, and the i860 XP, with 2.5 million transistors. See also 8-BIT, 16-BIT and 32-BIT.

66 Block The most common type of connecting block used to terminate and cross-connect twisted-pair cables. It was invented by Western Electric eons ago and has stood the test of time. It's still being installed. Its main claims to fame: Simplicity, speed, economy of space. You don't need to strip your cable of its plastic insulation covering. You simply lay each single conductor down inside the 66 block's two metal teeth and punch the conductor down with a special tool, called a punch-down tool. As you punch it down, the cable descends between the two metal teeth, which remove its plastic insulation (it's called insulation displacement) and the cable is cut. The installation is then neat and secure. 66 Blocks are typically rated Category 3 and as such are used mostly for voice applications, although Category 5 66 blocks are available. 66 blocks are open plastic troughs with four

pins across, and the conductors tend to be more susceptible to being snagged or pulled than the conductors terminated on 110, Krone or BIX.

64 Kbps 64,000 bits per second. The standard speed for V.35 interface, DDS service, and also the effective top speed of a robbed-bit 64 Kbps channel. A 64 kbps circuit (DS0). "Clear Channel" is 64 kbps where entire bandwidth is used. See also ISDN.

64-cap An ATM term. Carrierless Amplitude/Phase Modulation with 64 constellation points.

64QAM 64-state quadrature amplitude modulation. This digital frequency modulation technique is primarily used for sending data downstream over a coaxial cable network. 64QAM is very efficient, supporting up to 28-mbps peak transfer rates over a single 6-MHz channel. But 64QAM's susceptibility to interfering signals makes it poorly-suited to noisy upstream transmissions (from the cable subscriber to the Internet). See also QPSK, DQPSK, CDMA, S-CDMA, BPSK and VSB.

66-type Connecting Block A type of connecting block used to terminate twisted-pair cables. All wires are manually cut down with a special tool to terminate or connect them. See 66 BLOCK.

66-type Cross Connect A cross connect made up of the 66-type connecting blocks and jumper wires for administering circuits. All wires, including jumper wires, must be cut down (or punched down) and seated with a special tool. See 66 BLOCK.

6611 IBM's multi protocol router, which supports APPN in addition to TCP/IP, DECnet, AppleTalk, IPX, NetBIOS, and other protocols.

6bone An informal collaborative project intended to develop an Internet-wide IPv6 backbone infrastructure, 6bone is focused on providing early policy and procedures necessary to provide IPv6 transport in order that testing and experience can be carried out. 6bone is an independent outgrowth of the IETF IPng project that resulted in the creation of the IPv6 protocols intended to replace the current IPv4. 6bone will cover North America, Europe and Japan.

7-bit ASCII The standard code for text in which a byte (eight bits) holds the seven ASCII digits that define the character plus one bit for parity.

700 Service A non-geographic area code reserved for the provisioning of special IXC services. AT&T originally marketed 700 Service in the form of Easyreach, which allows your calls to follow you in the same fashion as would 500 Service. Some carriers (who will remain unnamed) once used 700 numbers for user access to the network for purposes of intraLATA long distance calling, such as those from Manhattan to Westchester County — the cost presumably was less than the cost of the same call through the serving LEC. This practice was illegal and no longer is necessary, as local competition now is in place in most states. 700 Service is still evolving, with each carrier having the right to create whatever services it wants with its 700 numbers. Currently, 700 Service commonly is used in both voice and data VPNs (Virtual Private Networks). See also 500 Service and VPN.

711 The Services Code now available for Telecommunications Relay Access (TRS) to aid those with speech and/or hearing disabilities to access police, fire and other governmental departments for both emergency and non-emergency purposes. See also 911 and 311.

8.3 1. Under the MS-DOS naming structure, a file's name can be eight letters in front of the period and three after it, e.g. LAZARUS8.TXT.

2. 8.3 minutes. The time it takes for light to travel from the Sun to the Earth.

8-bit Computer A computer that uses a central processing unit (CPU) with an 8-bit data bus and that processes one byte (8 bits) of information at a time. The first microprocessors used in personal computers, such as the MOS Technology 6502, Intel 8080, and Zilog Z-80, were installed in 8-bit computers such as the Apple II, the MSAI 8080, and the Commodore 64.

800 Portability 800 Portability refers to the fact that you can take your 800 number to any long distance carrier. A case example, we had 1-800-LIBRARY. For many years, that number was serviced by AT&T. When portability came along, we were able to change it from AT&T to MCI and still keep 1-800-LIBRARY. 800 Portability is provided by a series of complex databases the local phone companies, under FCC mandate, have built. 800 Portability started on May 1, 1993.

800 Service Eight-hundred service. A generic and common (and not trademarked) term for AT&T's, MCI's, Sprint's and the Bell operating companies' IN-WATS service. All these IN-WATS services have 800 and 888 as their "area code," with 877 numbers scheduled for 1998 release. Dialing an 800-number is free to the person making the call. The call is billed to the person or company being called. Suppliers of 800 services use various ways to configure and bill their 800-services. One way: you can buy an 800 line which will ring on your normal phone line. You'll only pay per call, but you won't receive any incoming call if you're making an outgoing one. You can terminate an 800 number on your cellular phone or your home number. You might pay a flat monthly rate plus "so-much" (i.e. timed usage) per call. That timed usage may include some calculation for the distance the incoming call traveled. 800-Service into the United States is now available for calls from Canada and many countries overseas including Europe, though people in many of those countries pay a normal toll call to reach an American "toll-free" 800-number.

800 Service works like this: You're somewhere in North America. You dial 1-800 or 1-888 and seven digits. The LEC (Local Exchange Carrier, i.e. the local phone company) central office sees the "1" and recognizes the call as long distance. It ships that call to a bigger central office (or perhaps processes the call itself). The processing central office recognizes the 800 or 888 "area code" and examines the next seven digits. At this point, the LEC switch holds the call, while it queries a centralized database, usually over a SS7 (Signaling System 7) link. That centralized database identifies the LEC or IXC (InterExchange Carrier) providing the 800 number, as well as translating the 800 number into a "real" telephone number. Based on that information, the LEC switch (the local phone company) will route the call to the proper IXC (long distance company). Once the long distance company has the 800 / 888 call, there are many ways it can send the call to the phone it's intended for. The customer who buys the 800 / 888 service typically specifies that way. The simplest way is that the long distance company might simply dial the customer's normal long distance number and connect the call. Or it might connect it via the customer's dedicated T-1 circuit. Or it might connect via the T-1 circuit, but if the T-1 is busy, it might overflow the call to the customer's normal long distance number.

As a real-life example, the publisher of this book, has an 800 number, 800-LIBRARY (or 800-542-7279). When you call that number, MCI routes that number to the first available channel on our dedicated T-1 circuit which we have rented from MCI's New York City POP (Point Of Presence) to our New York City office.

Because 800 long distance service is essentially a database lookup and translation service for incoming phone calls, there are endless "800 services" you can create. You can put permanent instructions into the company to change the routing patterns based on time of day, day of week, number called, number calling. Some long distance companies allow you to change your routing instructions from one minute to another. For example, you might have two call centers into which 800 phone calls are pouring. When one gets busy, you may tell your long distance company to route all the 800 inbound phone calls to the call center which isn't busy. See EIGHT HUNDRED SERVICE and ONE NUMBER CALLING for more, especially all the features you can now get on 800 service.

In May of 1993 the FCC mandated that all 800 (and by extension all 888) numbers became "portable." That means that customers can take their 800 / 888 telephone number from one long distance company to another, and still keep the same 800 / 888 number. See 800 PORTABILITY.

800 Services are known internationally as "Freefone Services." In other countries, the dialing scheme may vary, with examples being 0-800 and 0-500. Such services also go under the name "Greenfone." In June 1996, the ITU-T approved the E.169 standard Universal International Freefone Number (UIFN) numbers, also known as "Global 800." UIFN will work across national boundaries, based on a standard numbering scheme of 800, 888 or (soon) 877 plus an 8-digit telephone number. See also UIFN and VANITY NUMBERS.

802 IEEE 802. The main IEEE standard for local area networking (LAN) and metropolitan area networking (MAN), including an overview of networking architecture. It was approved in 1990. According to the IEEE, the numbering for IEEE's 902-series LAN standards follows a unique pattern. If the number is followed by a capital letter, the designation refers to a stand-alone standard. If it is followed by a lowercase letter, it is a supplement to a standard or part of a multiple-number standard. See IEEE 802 standards.

802 Standards The 802 Standards are a set of standards for LAN (Local Area Network) and MAN (Metropolitan Area Network) data communications developed through the IEEE's Project 802. The standards also include an overview of recommended networking architectures, approved in 1990. The 802 standards follow a unique numbering convention. A number followed by a an upper case letter denotes a stand-alone standard; a number followed by a lower case letter denotes either a supplement to a standard, or a part of a multiple-number standard (e.g., 802.1 & 802.3). The 802 standards segment the data link layer into two sublayers:

1. A Medium Access Control (MAC) layer that includes specific methods for gaining access to the LAN (Local Area Network). These methods — such as Ethernet's random access method and Token Ring's token procedure — are in the 802.3, 802.5 and 802.6 standards.

2. A Logical Link Control (LLC) Layer, described in 802.2 standard, that provides for connection establishment, data transfer, and connection termination services. LLC specifies three types of communications links:

• An Unacknowledged connectionless Link, where the sending and receiving devices do not set up a connection before transmitting. Instead, messages are sent on a "best try" basis and there is no provision for error detection, error recovery, and message sequencing. This type of link is best suited for applications where the higher layer protocols can provide the error correction and functions, or where the loss of broadcast messages is not critical.

• A Connection-mode Link, where a connection between message source and destination is established prior to transmission. This type of link works best in applications, such as file transfer, where large amounts of data are being transmitted at one time.

• An Acknowledged-connectionless Link that, as its name indicates, provides for acknowledgement of messages without burdening the receiving devices with maintaining a connection. For this reason, it is most often used for applications where a central processor communicates with a large number of devices with limited processing capabilities.

802.1 IEEE standard for overall architecture of LANs and internetworking.

802.11 IEEE standard for wireless LANs; intended to provide for interoperability of wireless LAN products from different manufacturers. It was ratified on June 26, 1997. The standard defines the over-the-air interface — the protocol that wireless stations will use to talk to each other or to access points. 802.11 does not define the protocol for backbone or wired networks. 802.11 defines both the Physical (PHY) and Medium Access Control (MAC) protocols for wireless LANs. Specifically, the PHY spec includes three transmission options — one infrared (IR), two radio frequency — RF — which include direct sequence spread spectrum and frequency hopping spread spectrum. Direct sequence data rates clock in at 2Mbps and frequency hopping clocks in at 1Mbps. The IR runs at 1Mbps with an optional implementation to 2Mbps. The MAC protocol works well with standard Ethernet, making wired and wireless nodes on an enterprise LAN logically indistinguishable.

802.12 Standard for 100VG-AnyLAN. Addresses 100 Mbps demand-priority access method physical-layer and repeater specifications. Approved in 1995.

802.1B Standard for LAN/WAN management, approved in 1992 and, along with 802.1k, became the basis of ISO/IEC 15802-2.

802.1D IEEE standard for interconnecting LANs through MAC bridges (specifically between 802.3, 802.4, and 802.5 networks). The standard was approved in 1990, and was incorporated into ISO/IEC 10038. Works at the MAC level.

802.1E IEEE standard for LAN and MAN load protocols. Approved in 1990, and formed the basis for ISO/IEC 15802-4.

802.1F Standard for defining network management information specified in 802 umbrella standards. Approved in 1993.

802.1G A developing standard for remote bridging at the MAC layer.

802.1H IEE practices recommended for bridging Ethernet LANs at the MAC layer. Approved in 1995.

802.1I IEEE standard for using FDDI (Fiber Distributed Data Interface) as a MAC-layer bridge. Approved in 1992, the standard was incorporated into ISO/IEC 10038.

802.1J IEEE standard for LAN connectivity using MAC-layer bridges. A supplement to 802 1D, it was approved in 1996.

802.1K IEEE standard for the discovery and dynamic control of network management information. Approved in 1993. In conjunction with 802.1B, was the basis for ISO/IEC 15802-2.

802.1M A conformance statement for 802.1E, it addresses definitions and protocols for system load management. Approved in 1993, it was incorporated into ISO/IEC 15802-4.

802.1P IEEE extension of 802.1D. Proposed standard for the use of MAC-layer bridges in filtering and expediting multicast traffic. Prioritization of traffic is accomplished through the addition of a 3-bit, priority value in the frame header. Eight topology-independent priority values (0-7) are specified, with all eight values mapping directly into 802.4 and 802.6.

802.1Q IEEE draft standard for implementation of VLANs in Layer 2 LAN switches, with emphasis on Ethernet. Similar to 802.1P, prioritization of traffic is accomplished through an additional 4 bytes of data in the frame header. Most data fields in this addition to the header are specific to VLAN operation. Also included is a field which provides the same 3-bit priority flag specified in 802.1P's priority-mapping scheme. In addition to conventional data traffic, 802.1Q supports voice and video transmission through Ethernet switches.

802.2 The IEEE standard for Logical Link Control, primarily using MAC-layer bridges, in LAN and MAN domains. Originally approved in 1989, and updated in 1994. A format used for frames of data sent on Ethernet, token ring and several other types of local area networks. Now the format favored by Novell for NetWare 4.x LANs over the 802.3.

802.2 SNAP (Cub-Network Access Protocol). A variation on the 802.2/802.3 scheme which expands the 802.2 LLAMA header to provide sufficient space in the header to identify almost any network protocol.

802.3 IEEE standard for carrier sense multiple access with collision detection (CSMA/CD). A physical layer standard specifying a LAN with a CSMA/CD access method on a bus topology. Ethernet and Starlan both follow the 802.3 standard. Typically they transmit at 10 megabits per second (Mbps). The theoretical limit of Ethernet, measured in 64 byte packets, is 14,800 packets per second (PPS). By comparison, Token Ring is 30,000 and FDDI is 170,000. 802.3 forms the basis for ISO/IEC 8802-3.

802.3 1Base5 IEEE standard for baseband Ethernet at 1 Mbps over twisted pair wire to a maximum distance of 500 meters. Also called Starlan.

802.3 10Base-5 IEEE standard for baseband Ethernet at 10 Mbps over coaxial cable to a maximum distance of 500 meters.

802.3 10Base-T Also called 802.3i. 10Base-T is an IEEE standard for operating Ethernet local area networks (LANs) on twisted-pair cabling using the home run method of wiring (exactly the same as a phone system uses) and a wiring hub that contains electronics performing similar functions to a central telephone switch. The full name for the standard is IEEE 802.3 10Base-T. The 10Base-T standard, issued in the fall of 1990, defined the requirements for sending information at 10 million bits per second on ordinary unshielded twisted-pair cabling. The 10Base-T standard defines various aspects of running Ethernet on twisted-pair cabling such as:

• Connector types (typically eight-pin RJ-45),
• Pin connections (1 and 2 for transmit, 3 and 6 for receive),
• Voltage levels (2.2 volts to 2.8 volts peak), and
• Noise immunity requirements to filter outside interference from telephone lines or other electronic equipment.

Ethernet is the most popular LAN in the world. Ethernet running on loop coaxial cable — typically called thin Ethernet or thinnet — is the most popular way of running Ethernet local area networks. Loop networks suffer from the major problem that one cut in the cable can destroy the complete network. 10Base-T is a much more reliable — though more expensive — way of connecting LANs, since it requires electronics at the center of the home run. As I write this, the most common form of 10Base-T electronics is a small box joining about 12 workstations together. To get more on the LAN, you simply daisy chain the boxes together. The boxes are unbelievably reliable. They're easy to install and they often come with LAN management software, which gives you statistics on who's using the network, for how long, what the performance is, and what potential problems might crop up, etc. The cable 10Base-T networks use to connect between their central electronics and their attached workstations is typically standard twisted pair phone wiring, which is a lot easier to install than coaxial cable. 10Base-T networks are now becoming most popular and are being installed at faster rate than old-style loop coaxial wired LANs. For a fuller explanation see ETHERNET.

802.3 10Broad36 IEEE standard for broadband Ethernet at 10 Mbps over broadband cable to a maximum distance of 3600 meters.

802.3b IEEE standard for 10Broad36. Approved in 1985, it is the standard for broadband Ethernet at 10 Mbps over coaxial cable to a maximum distance of 3,600 meters. It was incorporated into ISO/IEC 8802-3.

802.3c Standard for 10 Mbps baseband repeaters. The standard was approved in 1985, and is incorporated into ISO/IEC 8802-3.

802.3d IEEE standard for media attachment devices and baseband media over fiber optic repeater links. Approved in 1987, it has been incorporated into ISO/IEC 8802-3.

802.3e IEEE standard for 1Base-5, baseband Ethernet at 1 Mbps over twisted pair wire to a maximum distance of 500 meters. Also called Starlan. The standard addresses physical media, physical signaling and media attachment. Approved in 1987, it is incorporated into ISO/IEC 8802-3.

802.3h Standard for layer management ins CSMA/CD networks. Approved in 1990, it has been incorporated into ISO/IEC 8802-3.

802.3i The IEEE standard addressing multisegment 10 Mbps networks, and twisted-pair media for 10Base-T networks. 10Base-T is an IEEE standard for operating Ethernet local area networks (LANs) on twisted-pair cabling using the home run method of wiring (exactly the same as a phone system uses) and a wiring hub that contains electronics performing similar functions to a central telephone switch. The full name for the standard is IEEE 802.3 10Base-T. The 10Base-T standard, issued in the fall of 1990, defined the requirements for sending information at 10 million bits per second on ordinary unshielded twisted-pair cabling; the standard has been incorporated into ISO/IEC 8802-3. The 10Base-T standard defines various aspects of running Ethernet on twisted-pair cabling such as:

- Connector types (typically eight-pin RJ-45),
- Pin connections (1 and 2 for transmit, 3 and 6 for receive),
- Voltage levels (2.2 volts to 2.8 volts peak), and
- Noise immunity requirements to filter outside interference from telephone lines or other electronic equipment.

Ethernet is the most popular LAN in the world. Ethernet running on loop coaxial cable — typically called thin Ethernet or thinnet — is the most popular way of running Ethernet local area networks. Loop networks suffer from the major problem that one cut in the cable can destroy the complete network. 10Base-T is a much more reliable — though more expensive — way of connecting LANs, since it requires electronics at the center of the home run. As I write this, the most common form of 10Base-T electronics is a small box joining about 12 workstations together. To get more on the LAN, you simply daisy chain the boxes together. The boxes are unbelievably reliably. They're easy to install and they often come with LAN management software, which gives you statistics on who's using the network, for how long, what the performance is, and what potential problems might crop up, etc. The cable 10Base-T networks use to connect between their central electronics and their attached workstations is typically standard twisted pair phone wiring, which is a lot easier to install than coaxial cable.

10Base-T networks are now becoming most popular and are being installed at faster rate than old-style loop coaxial wired LANs. For a fuller explanation see ETHERNET.

802.3j IEEE standard for 10Base-F, which provides for fiber optics links connecting 10 Mbps active and passive starbased baseband networks. The standard was approved in 1993 and is incorporated into ISO/IEC 8802-3.

802.3k IEEE standard for layer management for repeaters in 10 Mbps baseband networks. It was approved in 1992 and is incorporated into ISO/IEC 8802-3.

802.3l A conformance statement for the media attachment unit protocol for 10Base-T networks. The statement was approved in 1992 and is incorporated into ISO/IEC 8802-3.

802.3p IEEE standard for media attachment unit layer management for 10 Mbps baseband networks. The standard was approved in 1992 and is incorporated into ISO/IEC 8802-3.

802.3q Provides guidelines for the development of managed objects. Approved in 1993, it was incorporated into ISO/IEC 8802-3.

802.3r IEEE standard for CSMA/CD and physical media specifications for 10Base-5, which is baseband Ethernet at 10 Mbps over fat coaxial cable to a maximum distance of 500 meters. This version of the original Ethernet standard was updated in 1996.

802.3t Standard for 120-ohm cables in 10Base-T simplex links. Approved in 1995, it was incorporated into ISO/IEC 8802-3.

802.3u A supplement to 802.3 that governs Carrier Sense Multiple Access/Collision Detection (CSNA/CD) for 100 Mbps networks, i.e., 100Base-T, commonly known as Fast Ethernet. Approved in 1995, this supplement covers the specifications MAC parameters, the physical layer, and repeaters for 100Base-T4, TX, and FX.

802.3v IEEE standard for 150-ohm cables in 10Base-T link segments. Approved in 1995, it was incorporated into ISO/IEC 8802-3.

802.3z Gigabit Ehternet over fiber standard, ratified on June 29, 1998.

802.4 IEEE physical layer standard specifying a LAN with a token-passing access method on a bus topology. It is typically used with Manufacturing Automation Protocol (MAP) LANs. MAP was developed by General Motors. Typical transmission speed is 10 megabits per second.

802.5 IEEE physical layer standard specifying a LAN with a token-passing access method on a ring topology using unshielded twisted pair. Used by IBM's Token Ring hardware. Typical transmission speed is 4 or 16 megabits per second. IEEE physical layer standard specifying a LAN with a token-passing access method on a ring topology using unshielded twisted pair. Used by IBM's Token Ring hardware. Typical transmission speed is 4 or 16 megabits per second. The standard became the basis for ISO/IEC 8802-5. The current version of 802.5 was approved in 1995.

802.6 IEEE standard for MANs (Metropolitan Area Networks). Formerly known as QPSX (Queued Packet and Synchronous Exchange), now known as DQDB (Distributed Queue Dual Bus). It was approved in 1990.

802.7 IEEE technical advisory group on broadband LANs.

802.8 IEEE technical advisory group for fiber-optic LANs.

802.9 IEEE technical advisory on ISLAN, which stands for Integrated Services LAN. ISLAN is Isoethernet with switched or packetized voice on an Ethernet LAN, which is 10 megabits per seconds of Ethernet (used for data) plus six megabits per second of ISDN B channels, which gives you 96 B ISDN channels plus a D channel. You can used the B channels for voice.

802.X The Institute of Electrical and Electronics Engineers (IEEE) committee that developed a set of standards describing the cabling, electrical topology, physical topology, and access scheme of network products; in other words, the 802.X standards define the physical and data-link layers of LAN architectures. Or, in simple language, the set of IEEE standards for the definition of LAN protocols. IEEE 802.3 is the work of an 802 subcommittee that describes the cabling and signaling for a system nearly identical to classic Ethernet. IEEE 802.5 comes from another subcommittee and similarly describes IBM's Token-Ring architecture.

822 Short form of RFC 822. Refers to the format of Internet style e-mail as defined in RFC 822.

82596 The 82596 is an intelligent, 16-/32-bit local area network coprocessor from Intel. The 82596 implements the CSMA/CD access method and can be configured to support all existing 802.3 standards. Coupled with the 82503 Dual Serial Transceiver, the 82596 provides the optimal Ethernet connection to Intel1386 and Intel486 client PCs and servers. The board space required for an 82596/82503 motherboard implementation is less than six square inches. provides full Ethernet bandwidth performance while allowing the CPU to work independently. An on-board four-channel DMA controller along with an intelligent micro machine automatically manages memory structures and provide command chaining and autonomous block transfers while two large independent FIFOs accommodate long bus latencies and provide programmable thresholds.

877 Service When North America ran out of 800 numbers for toll-free service 800 Service, the NANP (North American Numbering Plan) was expanded in April 1996 to include 888 numbers. Now we've run out of 888 numbers. So, the NANC (North American Numbering Council) has responded by assigning 877 numbers to relieve the pressure, with an implementation date of April 4, 1998. See 800 SERVICE and 800 SERVICE.

888 Service When North America ran out of 800 numbers, it adopted a new prefix — 888. The first 888 number came in around April, 1996. 877 numbers will begin in April of 1998. The 877 and 888 prefixes have all the characteristics of today's 800 Service. See 800 Service and 877 SERVICE.

8th-Floor Decision Refers to the 8th floor at the Washington offices of the FCC, where the commissioner's offices and meeting rooms are located. Decisions made on the 8th floor sometimes have a profound effect on new communication services.

9-track A standard for 1/2" magnetic tape designed for data storage. Its nine tracks hold a byte (eight bits) plus one parity bit in a row across the tape width-wise.

900 Number Rule A rule passed by the FTC (Federal Trade Commission) and which became effective November 1, 1993. The Rule requires that advertisements for 900 numbers contain certain disclosures, including information about the cost of the call. This information also must be included in an introductory message, or preamble, at the beginning of any 900-number program where the cost of the call could exceed $2.00. Any caller must be afforded the opportunity to hang up at the conclusion of the preamble, without incurring any cost for the call. The Rule also requires that all preambles state that individuals under the age of 18 must have the permission of a parent or guardian to complete the call. The 900-Number Rule has been very instrumental in reducing the level of 900-number abuse.

900 Service A generic and common (and not trademarked)

term for AT&T's, MCI's, Sprint's and other long distance companies' 900 services. All these services have "900" as their "area code." Dialing a 900-number is free to the company or person receiving the call, but costs money to the person making the call. Here's the story: 900 service was introduced as the industry's "information service" area code. You'd dial 1-900-WEATHER, for example, and punch in some touchtones in response to prompts and you could hear the weather in Sydney, Australia or Paris, France, wherever you might be planning your next vacation. For this service, you'd be charged perhaps 75 to 95 cents a minute. And you'd get the bill as part of your normal monthly phone bill. That was the original idea. Then some people got the idea that 900 would make a wonderful porn number and they started advertising "Call 900-666-3333 and speak with Diana. She really wants you." And they started charging $5 a minute. When huge 900-call bills started appearing on people's bills, there was an outcry from many subscribers who wouldn't pay the bills. Some children called on their parents' phones. Employees made calls from work and the company's accountants went nuts. So the industry retreated from 900 porn. Then someone thought — "Why not sell things through an 900 number?" We could sell a set of ginzu knives for just calling this 900 number. No messing with credit cards or checks. The bill goes straight on your phone bill. At about the same time someone thought that 900 numbers would be great for running sweepstakes. "Call up, register your name for a free trip with a racing car team to the Australian Indianapolis 500. Three lucky people will be chosen. The call will cost you only $2.75." So 900 services became a new type of gambling.

The long distance companies providing 900 services reacted predictably to some of the newer services. They clamped down on who they would sign up, which service and/or product you could, or could not sell. And, rather than charging "a piece of the action" as they did in the beginning, the long distance companies began to charge for them as if they were normal long distance calls: charge a set-up fee, a fee for carrying the call, a fee for collecting the money and a fee for the possibility of bad debts. There are variations on these themes.

The 900-service business is rife with stories of people who are alleged to have made millions overnight with innovative 900 numbers. Clearly enormous monies have been made — especially in the beginning when there was novelty to 900 calls. The prognosis is that the 900 number business will grow, that it will mature and that the North American public will wake up to its various scams and discover real value in many of its services. For example, one of the author's "genuine value" and favorite 900 services is fax-back. Dial a 900 number, punch in some touchtone digits, hang up and within seconds your fax machine begins to churn out useful information.

In the summer of 1991, AT&T issued guidelines for its EXPRESS900 service. Those guidelines included:
• The predominant purpose of the calls does not include Entertainment, Children's Programming, Credit/Loan Information, Fulfillment, Political Fundraising, Games of Chance, Postcard Sweepstakes, Job Lines and Personal Lines;
• Every program must have a Preamble and Caller Grace Period, with notification to callers of the opportunity to hang up before charging begins.
• Sponsors may not route calls to any telecommunications equipment or arrangements which allow charging to begin before the caller realizes any value on the call, e.g., Automatic Call Distribution (ACD) with call queuing, or Caller Hold.

9001 ISO 9001 is a rigorous international quality standard covering a company's design, development, production, installation and service procedures. Compliance with the standard is of increasing significance for vendors trading in international markets, in particular in Europe where ISO 9001 registration is widely recognized as an indication of the integrity of a supplier's quality processes. ISO is the International Standards Organization in Paris.

902-928 MHz A frequency range for use by, amongst other things, cordless phones. such frequency offers much greater range than traditional cordless phones. Some new 900MHz phones also use spread-spectrum technology to further increase range and call security. Previous cordless phones operated within the 46-49 MHz band. The 900 contains 50 channels for cordless telephone transmission. The 46-49 MHz band contains only 10.

911 Service 911 is an emergency reporting system whereby a caller call dial a common number — 911 — for all emergency services. The caller will be answered at a common answering location which will figure the nature of the emergency and dispatch the proper response teams. The first 911 service came on line in 1968. Here are the reasons why 911 benefits a community: Only one number for all emergency services. It's an easy number to remember. It's an easy number to dial. It's great for travelers and new residents. Calls are received by trained personnel. See also E-911, which stands for Enhanced 911 service and typically includes ANI (Automatic Number Identification) and ALI (Automatic Location Information). 911 service is sometimes called B-911 which stands for basic 911 service. See also B-911 and E-911.

9145 A common term in the southern part of America for a customer service representative (i.e. a salesman) of the local telephone company.

950 Local exchange used by some North American long-distance carriers to let their customers access their calling card and other services.

958 Dial 958 in New York City and a computer run by Nynex (now called Bell Atlantic) will tell you the phone number you're calling from. This is very useful. Imagine having a jack on your wall. You've lost track of which phone line it's connected to. Dial 958 and bingo! You know. Other phone companies have similar services but they often have different numbers. In some parts of Pennsylvania, the phone number is 958-4100.

976 A local information, pay-per-call phone exchange. An information service that lets callers listen to recorded messages such as sports scores or adult conversations, at higher rates than normal calls. The sponsor of the 976 service splits revenue with the phone service provider.

999 Great Britain's equivalent of the United States emergency number 911.

@ The character typically above the 2 on your keyboard. It's called the "at sign." In English, its biggest use is in "two apples @ 50 cents each equals $1 total." But in computerese, its big use is in electronic mail addressing. It is used to separate the domain name and the user name in an Internet address and is pronounced "at." For example, Bill Gates' e-mail address is billg@Microsoft.com. It is pronounced "Bill G at Microsoft.com."

@HOME Network Pronounced "at HOME." A new broadband system designed and developed for delivering high-speed information and Internet services using cable television lines.

A 1. Abbreviation for AMP or AMPERE, a unit of electric current. For a longer explanation, see Ampere.

2. The non-local wireline cellular carrier. In one of its less intelligent decisions, the FCC decided to issue two cellular franchises in each city in the United States. They gave one to the local phone company (the B carrier) and one to a competitor (the A carrier). This duopoly has naturally meant little real competition. But with the issuing of PCS licences in recent years, competition has begun to heat up, i.e. prices have dropped. Meantime, the "A" carrier on your cellular phone is the non-local wireline carrier, i.e. the competitor to the local phone company and "B" is the other one, namely the local phone company's cellular company. Recently, many of the "A" carriers have been purchased by other large telecom carriers, notably AT&T.

3. Abbreviation for atto, which is ten to the minus 18th. Which means it's very very small.

A & A1 Control leads that come from 1A2 key telephone sets to operate features like flashing of lights to indicate on hold, line ringing, etc.

A & B Bits Bits used in digital environments to convey signaling information. A bit value of one generally corresponds to loop current flowing in an analog environment. A bit zero corresponds to no loop current. Other signals are made by changing bit values; for example a flash-hook is set by briefly setting the A bit to zero.

A & B Leads Additional leads used typically with a channel bank two-wire E&M interface to certain types of PBXs (also used to return talk battery to the PBX).

A & B Signaling Procedure used in most T-1 transmission links where one bit, robbed from each of the 24 subchannels in every sixth frame, is used for carrying dialing and controlling information. A type of in-band signaling used in T-1 transmission. A and B signaling reduces the available user bandwidth from 1.544 million bits per second to 1.536 Mbps.

A Battery Another term for Talk Battery.

A Block Cellular licenses received from the FCC with no initial association to a telephone company. Also referred to as non-wireline.

A Law The PCM coding and companding standard used in Europe and in areas outside of North American influence. A Law Encoding is the method of encoding sampled audio waveforms used in the 2.048 Mbps, 30-channel PCM primary system known as E-carrier. See MU LAW and PCM.

A Links A SS7 term. A-Links connect an end office or signal point (SP) to a mated pair of Signal Transfer Points (STPs). Two-way path diversity is recommended so that one common disaster does not isolate a signal point from the rest of the network. The other location for A-Links is between mated pairs of STPs and the SCPs. Typically, this would occur at the regional level with the A-Links assigned in a quad arrangement. See B, C and D Links.

A Number A cellular term for the number of the calling party. The originating switch analyzes the number in order to route the call to the B Number, the number of the called party.

A Port Refers to the port in an FDDI topology which connects the incoming primary ring and the outgoing secondary ring of the FDDI dual ring. This port is part of the dual attachment station or a dual attachment concentrator.

A-B Rolls A technique by which audio/video information is played back from two videotape machines rolled sequentially, often for the purpose of dubbing the sequential information onto a third tape, usually a composite master.

A-B Test Direct comparison of the sound/picture quality of two pieces of audio/TV equipment by playing one, then the other.

A-Condition In a start-stop teletypewriter system, the significant condition of the signal element that immediately precedes a character signal or block signal and prepares the receiving equipment for the reception of the code elements. Contrast with start signal.

A-Interface The network (air) interface between a Mobile End System (M-ES) and the Cellular Digital Packet Data (CDPD)-based wireless packet data service provider network.

A-Key Authentication Key. An authentication mechanism of the ANSI-41 (formerly TIA IS-41C) standard for cellular inter-system inter-operability. The A-key is sent between the cellular phone and the Authentication Center (AC), and is known only to those two entities. Subsequently, the cellular phone and the AC generate a second secret key, known as a SSD (Shared Secret Data). Both the A-key and the SSD are transmitted around the ANSI-41 network to be used by switches to perform the challenge-response process of authentication. Both keys are encrypted through the CAVE (Cellular Authentication and Voice Encryption) algorithm. See also CAVE.

A-law The ITU-T companding standard used in the conversion between analog and digital signals in PCM systems. A-law is used primarily in European telephone networks and is similar to the North American mu-law standard. See also Companding and Mu-law.

A/B Switch 1. A switch that allows manual or remote switching between one input and two outputs. See A/B Switch Box.

2. A feature found on all new cellular telephones permitting the user to select either the "A" (non-wireline) carrier or the "B" (wireline) carrier when roaming away from home.

A/B Switch Box A device used to switch one input between two devices, such as printers, modems, plotters, mice, phone lines, etc. An example of how you use such a box: You plug one phone line into the "C" (for Common) jack. You plug a fax machine into the "A" jack. You plug a modem into the "B" jack. By turning the switch, you can use one phone line for either a modem or a fax. A/B switch boxes come in many flavors, including also serial and parallel port versions. There are also A/B/C switches that switch among three devices, e.g. a fax, a modem and a phone. There is also a Crossover Switch that connects two inputs and two outputs. In one position the switch might connect input A with output D and input B with output C. In the other position, it might connect input A with output C and input B with output D. See also A/B Switch.

A/B Switches Input Selector Switch. A switch used by cable customers to alternate between cable and over-the-air television reception through a cable box.

A/D Analog to Digital conversion.

A/D Converter Analog to Digital converter, or digitizer. It is a device which converts analog signals (such as sound or voice from microphone), to digital data so that the signal can be processed by digital circuit such as a digital signal processor. See CODEC.

A/UX An alternate operating system for the Macintosh based on UNIX. A/UX has its own, unique 32-bit addressing mode.

A20 Line A control line on the Intel 80386 microprocessor that allows MS-DOS and an extended memory manager to create the High Memory Area, or HMA. Only one program can claim control over the A20 at a time.

A4 The Basic Group 3 standard that defines the scanning and printing of a page 215 mm (8.5 in) wide. An A5 page is 151 mm (5.9 in) wide, and the A6 is 107mm (4.2 in) wide.

A5 See A4

A6 See A4

AA 1. Automated Attendant. A device which answers callers with a digital recording, and allows callers to route themselves to an extension.
2. Auto Answer. A modem indicator light that is meant to tell you the modem is ready to pick up the phone, so long as there's a communication program running and prepared to handle the call. See also Modem.

AABS Automated Attendant Billing System. A feature which allows collect and third-number billed toll calls to be placed on an automated basis. A synthesized voice prompt guides the caller through the process, the system then seeks approval of the prospective billed party, and either completes or denies the call based on that authorization or lack thereof. AABS is automated in much the same way as calling card services have been automated, through the use of an Intelligent Peripheral (IP) device.

AAL ATM Adaptation Layer of the ATM Protocol Reference Model, which is divided into the Convergence Sublayer (CS) and the Segmentation and Reassembly (SAR) sublayer. The AAL accomplishes conversion from the higher layer, native data format and service specifications of the user data into the ATM layer. On the originating side, the process includes segmentation of the original and larger set of data into the size and format of an ATM cell, which comprises 48 octets of data payload and 5 octets of overhead. On the termination side of the connect, the AAL accomplishes reassembly of the data. Taken together, these processes are known as Segmentation and Reassembly. AAL is defined in terms of Types supported by the Convergence Sublayer. Each type supports certain specific types of traffic, and each offers an appropriate Quality of Service (QoS), based on traditional network references. See the next five definitions for AAL specifics.

	Class A	Class B	Class C	Class D
Timing relation between source and destination	Required		Not Required	
Bit Rate	Constant		Variable	
Connection Mode	Connection-Oriented			Connectionless
Applications	Voice, Video, Circuit Emulation	Compressed Voice or Video	Frame Relay, X.25 Traffic	SMDS, LAN Traffic

AAL Service Classes, with specific attributes and example applications. Source: ITU-T Recommendation I.362 (March, 1993)

AAL-1 ATM Adaptation Layer Type 1: AAL functions in support of Class A traffic, which is connection-oriented, Constant Bit Rate (CBR), time-dependent traffic such as uncompressed, digitized voice and video. Such traffic is isochronous, i.e., stream-oriented and highly intolerant of delay.

AAL-2 ATM Adaptation Layer Type 2: This AAL supports Class B traffic, which is connection-oriented, Variable Bit Rate (VBR), isochronous traffic requiring precise timing between source and sink. Examples include compressed voice and video.

AAL-3/4 ATM Adaptation Layer Type 3/4: AAL support of Class C and D traffic, which is Variable Bit Rate (VBR), delay-tolerant data traffic requiring some sequencing and/or error detection support, but no precise timing between source and sink. Originally two AAL types, AAL types 3 and 4 were combined in support of both connection-oriented and connectionless traffic. Examples include X.25 packet and Frame Relay traffic.

AAL-5 AAL-5 ATM Adaptation Layer Type 5. AAL functions in support of Class C traffic, which is of Variable Bit Rate (VBR), and which is delay-tolerant connection-oriented data traffic requiring minimal sequencing or error detection support. Such traffic involves only a single datagram in Message Mode. Examples of AAL-5 data include signaling and control data, and network management data. AAL-5 also is known as SEAL (Simple and Efficient AAL Layer). AAL-5 traffic originates in the form of a native data payload unit which is known as an IDU (Interface Data Unit). The IDU is of variable length, up to 65,536 octets. At the Convergence Layer the IDU is appended with a trailer including the UU, CPI, Length and CRC fields. The UU (User-to-User) field of one octet contains data to be transferred transparently between users. The CPI (Common Part Indicator) field of one octet aligns the trailer in the total bit stream. The Length field of two octets indicates the length of the total IDU payload. The CRC (Cyclic Redundancy Check) of four octets is used for purposes of error detection and correction in the trailer, only. When the payload user data plus the trailer data hit the ATM Adaptation

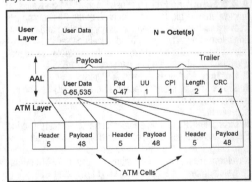

AAL Type 5 operation. User data is appended with trailer at ATM Adaptation Layer and segmented into 48-octet cells at ATM Layer.

Layer, the entire set of data is segmented into 48-octet payloads, with each being prepended with a 5-octet header to form a 53-octet ATM cell. See ATM.

AAL Connection An ATM term. Association established by the AAL between two or more next higher layer entities.

AALn See AAL-1, AAL-2, AAL-3/4, AAL-5.

AAR Automatic Alternate Routing.

AARP Probe Packets Packets transmitted by AARP that

determine if a randomly selected node ID is being used by another node in a nonextended AppleTalk network. If the node ID is not being used, the sending node uses that node ID. If the node ID is being used, the sending node chooses a different ID and sends more AARP probe packets. See also AARP.

AAV Alternative Access Provider. Another name for a CAP. See CAP.

ABAM A designation of Lucent Technologies (nee Western Electric) for 22 gauge, 110 ohm, insulated, twisted pair cable normally used in central offices.

Abandon On Ugly A video call center term coined by Andrew Waite. The term refers to a caller seeing the not-attractive answering agent, and hanging up instantly. Often the caller redials in the hope of reaching someone more attractive.

Abandoned Call The non-technical explanation is: A call that is answered, but disconnected before any conversation happens. The technical explanation is: A call which has been offered unto a communications network or telephone system, but which is terminated by the person originating the call before it is answered by the person being called. Follow this sequence for an explanation: You call an airline. You hear ringing. Their phone rings. A machine, called an Automatic Call Distributor (ACD), answers the call, plays you some dumb message like, "Please don't hang up. A real human will answer eventually. Dial 1 to order a pizza, Dial 2 for anchovies,...." You, the caller, are put on Eternity Hold. You get bored waiting and hang up before a live operator answers. You have just abandoned your phone call. Hence, the term Abandoned Call. Information about abandoned calls is highly useful for planning the number of people (also called operators, agents or telephone attendants) an owner of an automatic call distributor or other phone system should employ on what days, during what times of the day, and at what specific locations. Thus the company can organize its resources (i.e., schedule its people) to ensure that the "right" percentage of incoming calls are answered within the "right" amount of time to provide the caller (i.e., you, the customer) the service you deserve (or the service they think you deserve, or, the service they think you should deserve, or the level of service they can afford to provide to "optimize" the cost/benefit relationship).

Abandoned Call Cost The amount of revenue lost because of abandoned calls. This is calculated based on the number of calls, your estimate of the percentage abandoning, and your estimate of the revenue per call. It's an impossible number to calculate since many callers do, in fact, call back and place their orders on another later call.

Abbreviated Address Calling A calling method that allows the user to employ a logical address (e.g., telephone number) involving fewer characters. The destination's assigned device addresses these characters when initiating a connection. May also be called Abbreviated Dialing when specifically used in connection with telephone systems.

Abbreviated Dialing A feature that permits the calling party to dial the destination telephone number with fewer than the normal digits. Abbreviated Dialing numbers must be set up before using them. Speed Dialing is a typical example of Abbreviated Dialing. See Speed Dialing.

ABC 1. Automatic Bill Calling — a method of billing for payphone calls. Changed in 1982 to Calling Card service.
2. Automated Business Connection.

ABCD Signaling Bits These are bits robbed from bytes in each DS-0 or T-1 channel in particular subframes and used to carry in band all status information such as E&M signaling states.

ABD See Average Business Day.

ABDN Attendant Blocking of Directory Number.

ABEND ABnormal END, or ABortive END, or system "crash," and almost always very bad news. When an operating system detects a serious problem, such as a hardware or software failure, the system issues an abend (abnormal end) message. An abend recognized by Novell's NetWare, for instance, would stop the file server...and you're dead in the water. Usually caused by input or data presented to a computer which is beyond its ability to cope. If an abend happens in a single-task program (like MS-DOS) the machine will cease to take input ("lock up") and must be restarted ("re-booted"). Abends can be caused by a variety of factors, such as poorly functioning NetWare Loadable Modules (NLMs), power problems and heavy network traffic. They are the bane of a NetWare operating manager's existence. Multitasking operating systems (like UNIX) allow other programs to continue running while only stopping the one causing trouble. See VREPAIR which explains one way to repair your ABEND problem.

Ablation Optical memory data writing technique in which a laser burns holes (or pits) into thin metal film.

ABM Asynchronous Balanced Mode. A service of the data link level (Logical Link Control) in IBM's token-passing ring. ABM operates at the Systems Network Architecture data link control level and allows devices to send data link commands at any time.

ABN ABNormal alarm status.

Abort To stop doing something. Often to get out of a software program. Also, to discontinue sending or receiving a message.

Abort Delimiter A local area network term. A signal sent by a Token Ring station indicating that the message it was sending was terminated part way through the transmission, and is known to be incomplete. The station will then increment the soft error counter for Abort Delimiter Transmitted, and send a soft error report within two seconds.

Abort Sequence A series of 12 to 18 1-bits appearing at the end of an AppleTalk LLAP frame. The sequence delineates the end of the frame.

Above 890 Decision The 1959 FCC decision which allowed companies to build their own private microwave communications systems. The decision resulted from AT&T Long Lines' reluctance to provide companies with long distance service to remote places, such as oil wells, gas pipelines, power stations and paper plants. The decision got its name because the FCC allowed privately-owned microwave systems using radio frequencies "above 890" megahertz — which are naturally called "microwave." See also MICROWAVE and ENTELEC.

Above The Line Expenses incurred by telephone company that are charged to the ratepayer by being allowed in the company's rate-base.

ABR 1. Available Bit Rate. As defined by the ATM Forum, ABR is an ATM layer service category for which the limiting ATM layer transfer characteristics provided by the network may change subsequent to connection establishment. A flow control mechanism is specified which supports several types of feedback to control the source rate in response to changing ATM layer transfer characteristics. It is expected that an end-system that adapts its traffic in accordance with the feedback will experience a low cell loss ratio and obtain a fair share of the available bandwidth according to a network specific allocation policy. Cell delay variation is not controlled in this service, although admitted cells are not delayed unnecessarily. In short, ABR provides for transport of traffic at the bit rate available at the time, and on a dynamic basis.

2. Area Border Router. Router located on the border of one or more OSPF areas that connects those areas to the backbone network. ABRs are considered members of both the OSPF backbone and the attached areas. They therefore maintain routing tables describing both the backbone topology and the topology of the other areas.

3. AutoBaud Rate Detect. A process by which a receiving data device determines the speed, code level, and stop bits of incoming data by examining the first character — usually a preselected sign-on character (often a carriage return). ABR allows the receiving device to accept data from a variety of transmitting devices operating at different speeds without needing to establish data rates in advance.

Abrasion Resistance Ability of material or cable to resist surface wear.

Abrupt Close Close of a connection on a network without any attempt to prevent any loss of data.

ABS 1. See Average Busy Season.

2. Alternate Billing Services. These are IN (Intelligent Network) services that allow subscribers to charge a call to a number or telephone other than the one they are using. For example, by using a charge card, credit card or personal identification number.

Absent Subscriber Service A service offered by local telephone companies to subscribers who will be away. A live operator or a machine intercepts the calls and delivers a message. When you come back, you get your old number. But in the meantime, while you're away, you pay less money per month than you would for normal phone service. Also known as Vacation Service.

Abscissa A horizontal coordinate.

Absolute Delay The time interval or phase difference between transmission and reception of a signal.

Absolute Gain In simple language, absolute gain measures how much a device improves the power of a signal. The absolute gain of an antenna, for a given direction and polarization, is the ratio of (a) the power that would be required at the input of an ideal isotropic radiator to (b) the power actually supplied to the given antenna, to produce the same radiation intensity in the far-field region. If no direction is given, the absolute gain of an antenna corresponds to the direction of maximum effective radiated power. Absolute gain is usually expressed in dB. See Isotropic Gain.

The absolute gain of a device is the ratio of (a) the signal level at the output of the device to (b) that of its input under a specified set of operating conditions. Examples of absolute gain are no-load gain, full-load gain, and small-signal gain. Absolute gain is usually expressed in dB.

Absolute URL A URL that contains a scheme (for example, http) and a server address (for example www.harrynewton.com).

Absolute Zero A temperature about 460 degrees below zero in Fahrenheit.

Absorption Attenuation (reduction in strength of a signal) caused by dissipation of energy. In the transmission of electrical, electromagnetic, or acoustic signals, absorption is the conversion of the transmitted energy into another form, usually thermal. Absorption is one cause of signal attenuation. The conversion takes place as a result of interaction between the incident energy and the material medium, at the molecular or atomic level.

Absorption Band A spectral region in which the absorp-

tion coefficient reaches a relative maximum, by virtue of the physical properties of the matter in which the absorption process takes place.

Absorption Coefficient A measure of the attenuation caused by absorption of energy that results from its passage through a medium. Absorption coefficients are usually expressed in units of reciprocal distance. The sum of the absorption coefficient and the scattering coefficient is the attenuation coefficient.

Absorption Index A measure of the attenuation caused by absorption of energy per unit of distance that occurs in an electromagnetic wave of given wavelength propagating in a material medium of given refractive index.

Absorption Loss That part of the transmission loss caused by the dissipation or conversion of electrical, electromagnetic, or acoustic energy into other forms of energy as a result of its interaction with a material medium.

Absorption Modulation Amplitude modulation of the output of a radio transmitter by means of a variable-impedance circuit that is caused to absorb carrier power in accordance with the modulating wave.

Abstract Syntax In open systems architecture, the specification of application-layer data or application-protocol control information by using notation rules that are independent of the encoding technique used to represent the information.

Abstract Syntax Notation One (ASN.1). LAN "grammar," with rules and symbols, that is used to describe and define protocols and programming languages. ASN.1 is the OSI standard language to describe data types. More formally, ASN.1 is a standard, flexible method that (a) describes data structures for representing, encoding, transmitting, and decoding data, (b) provides a set of formal rules for describing the structure of objects independent of machine-specific encoding techniques, (c) is a formal network-management Transmission Control Protocol/Internet Protocol (TCP/IP) language that uses human-readable notation and a compact, encoded representation of the same information used in communications protocols, and (d) is a precise, formal notation that removes ambiguities.

ABT Advanced Broadcast Television.

AC 1. Access Customer.

2. See AuC, Authentication Center.

3. Alternating Current. Typically refers to the 120 volt electricity delivered by your local power utility to the three-pin power outlet in your wall. Called "alternating current" because the polarity of the current alternates between plus and minus, 60 times a second. In an AC power system, current (AMPS) is delivered to a load through a wire called the "hot" wire and returns through a wire called the "neutral" wire. The other form of electricity is DC, or direct current, in which the polarity of the current stays constant. Direct current, for example, is what comes from batteries. Outside North America, electricity typically alternates at 50 times a second — which is neither better nor worse, just different. In North America, standard 120 volt AC may be also be referred to as 110 volts, 115 volts, 117 volts or 125 volts. Con Edison, the electricity supplier to New York City, told me that they are only obliged to deliver voltage to 120 volts plus or minus 10%. This means your outlet may deliver anywhere from 118 volts to 132 volts before your power company will get concerned. But you probably should. Telephone and computer are sensitive to voltage fluctuations — some more to high voltages; some more to low voltages. My suggestion: If your stuff is valuable, protect it with a voltage regulator. Or better, power it with stable DC

current, which you've converted from fluctuating AC power. Your AC electrical circuit consists of two supply conductors — hot and neutral. There is also a "load." That is the term for the device you're running. The hot, energized or live conductor is ungrounded and delivers energy to the load. The hot conductor is connected to the fuse or circuit breaker at the main service entrance. The neutral or common conductor is grounded and completes the circuit from the load back to the utility transformer. The load is any electric or electronic appliance or gadget plugged into the AC electrical outlet. It completes the circuit from the transformer through the hot conductor, to the load, through the neutral conductor and back to the utility transformer. Standard 120 volt circuits also include an equipment ground conductor. This equipment grounding conductor provides an intended path for fault current and is never intended to be a part of the load circuit. The equipment ground serves three very important purposes:

1. It maintains metal appliance cases at zero volts, thus protecting people who touch the cases from receiving an electrical shock.

2. It provides an intentional fault path of low impedance path for current flow when the hot conductor contacts equipment cases (ground fault). This current causes the fuse or circuit breaker to open the circuit to protect people from electric shock.

3. Any electronic equipment (not electrical) uses the equipment ground as a zero volt reference for logic circuits to provide proper equipment performance.

This is a true story: Thomas Edison helped develop the electric chair in order to "prove" the "deadly dangers" of alternating current (AC) electrical systems. Edison was in direct competition with Westinghouse's brilliant Nikola Tesla, whom he detested, and whose efficient AC system was rapidly becoming the preferred method for transmission of electricity over long distances. This threatened Edison's direct current (DC) system. Realizing he was losing the war, Edison began holding demonstrations in which he electrocuted large numbers of dogs and cats by luring them onto a metal plate wired to a 1,000-volt AC generator. This display, although it attracted people, did not work to sway the public to his side and the AC system became the electricity standard. The legacy of all of Edison's efforts directly led to the development of the electric chair, which uses AC electricity.
See AC POWER, GROUNDING, BATTERY and SURGE ARRESTOR.

AC Field The Access Control field of a token. See Token.

AC Power Phone systems typically run on AC, Alternating Current. Except for very small systems, phone systems typically need their own dedicated (i.e., shared with no other device) AC power line. This line should be "cleaned" with a power conditioner, or voltage regulator. It also should be protected with a surge arrestor. If possible, the phone system should also be protected by a battery-based UPS (Uninterruptible Power Supply). The most reliable battery backup is lead acid, the same technology as used in your car. Phone systems consume more power as they process more phone calls. For example, a PBX brochure says that at minimum capacity, it needs less power than eight 100-watt light bulbs. But that at its maximum duplex capacity, it needs the same power as 26 100-watt light bulbs — or 2600 watts. Telephone sets typically draw their power from the central telephone switch, be it a PBX or a central office. Few phones require their own AC power. ISDN devices and phones typically require may require a local AC power source. See AC, GROUND and GROUNDING.

AC To DC Converter An electronic device which converts alternating current (AC) to direct current (DC). Most phone systems, computers and consumer electronic devices (from answering machines to TVs) run on DC. Most phone systems have an AC to DC converter in them. Hint: it's probably buried in the power supply.

AC-DC Ringing A type of telephone signaling that uses both AC and DC components — alternating current to operate a ringer and direct current to aid the relay action that stops the ringing when the called telephone is answered.

ACA Automatic Circuit Assurance.

Academic Computing Research Facility Network ACRFNET. A network connecting various research units such as colleges and research and development laboratories in the U.S.

ACADEMMET A network within Russia which connects universities.

ACAT Additional Cooperative Acceptance Testing. A method of testing switched access service that provides a telephone company technician at the central office and a carrier's technician at its location with suitable test equipment to perform the required tests. ACAT may, for example, consist of the following tests:
• Impulse Noise
• Phase Jitter
• Signal-to-C-Notched Noise Ratio
• Intermodulation (Nonlinear) Distortion
• Frequency Shift (Offset)
• Envelope Delay Distortion
• Dial Pulse Percent Break

ACB 1. Annoyance Call Bureau.
2. Automatic Call Back.

ACC Analog Control Channel. A wireless term. The ACC is the analog signaling and control channel used in some cellular systems in support of call setup and certain features. Defined in IS-54B, the ACC is a radio frequency channel distinct from those used to support conversations in some systems using TDMA (Time Division Multiple Access). A DCCH (Digital Control CHannel) is preferred, for obvious reasons of error performance; the DCCH is specified in IS-136, the successor to the IS-54 series. See IS-54B and IS-136.

Accelerated Aging A series of tests performed on material or cable meant to duplicate long term environmental conditions in a short period of time. Such tests might include exposure to extreme cold, heat, variations in humidity (including "wet"), and mechanical stress. Also refers to the impact of the stress associated with revising and maintaining Newton's Telecom Dictionary, according to Ray Horak, Consulting Editor.

Accelerated Depreciation A method which allows greater depreciation charges in the early years of an asset's life and progressively smaller ones later on. The total amount of depreciation charged is still equal to 100% of the asset's value. By taking the charges early on in the asset's life, you get the time value of money, i.e. depreciation charged today (and tax saved today) is worth more than the same amount of depreciation charged (or tax saved) tomorrow.

Accelerated Graphics Port An Intel product which bypasses a common traffic jam between the Pentium processor and graphics chips.

Accelerator 1. A chemical additive which hastens a chemical reaction under specific conditions. A term used in the telecommunications cable manufacturing industry.
2. In a Windows program, an accelerator is a keystroke that

dispatches a message to a program, invoking one of its functions. For example, Alt-F4 tells the current Windows application of Windows itself to quit.

Accelerator Board A board added onto a personal computer's main board and designed to increase the PC's performance in writing to screen or disk, etc. See also Accelerator Card.

Accelerator Card An Apple term for an add-on product that upgrades the CPU of a Macintosh to a higher speed or more powerful generation of processor. An accelerator card is usually a "daughter board" that clips onto the original CPU or is inserted into the socket that held the original CPU. You need an accelerator card to use virtual memory or enhanced 24-bit addressing on LC, Mac Plus, SE and Classic. See Accelerator Board.

Acceptable Angle The maximum angle that a fiber optic cable accepts light for further transmission.

Acceptable Use Policy AUP. Many transit networks have policies which restrict the nature of their use or the basis on which access privileges are granted. A well known example is that of the now defunct NSFNET's (National Science Foundation NETwork) AUP which traditionally did not allow commercial use. Subsequently, of course, NSFNET did become commercialized as the Internet. AUP enforcement varies with the specific network.

Acceptance Acceptance refers to the amount of time within which a buyer has to decide whether a software or hardware element is acceptable. Different from a warranty on performance, acceptance applies to the appearance and performance of the element as initially configured and installed, and as compared to the specifications to which the seller and purchaser have agreed. Acceptance periods usually range from two to four weeks, but are determined on a case-by-case basis. It's commonly used in the secondary telecom equipment marketplace, as well. See Acceptance Test.

Acceptance Angle The angle over which the core of an optical fiber accepts incoming light. It's usually measured from the fiber axis.

Acceptance Cone In fiber optics, the cone within which optical power may be coupled into the bound modes of an optical fiber. The acceptance cone is derived by rotating the acceptance angle about the fiber axis.

Acceptance Pattern Of an antenna, for a given plane, a distribution plot of the off-axis power relative to the on-axis power as a function of angle or position. The acceptance pattern is the equivalent of a horizontal or vertical antenna pattern. The acceptance pattern of an optical fiber or fiber bundle is the curve of total transmitted power plotted against the launch angle.

Acceptance Test The final test of a new telephone system. If the system passes the test — i.e. it meets all specifications laid down in the sales contract — and is working well, then, and only then, will the customer finish paying for it. See also Acceptance and Acceptance Testing.

Acceptance Testing Operating and testing of a communication system, subsystem, or component, to ensure that the specified performance characteristics have been met. See Acceptance and Acceptance Test.

Acceptance Trial A military term. A trial carried out by nominated representatives of the eventual military users of the weapon or equipment to determine if the specified performance and characteristics have been met.

Access A series of digits or characters which must be dialed, typed or entered in some way to get use of something. That "something" might be a PBX or KTS telephone system, a long distance carrier, an electronic mail service, a private corporate network, a mainframe computer, or a local area network. Once the user dials the main number for the service he must then enter his assigned Access Code to get permission use the system. An Access Code becomes an Authorization Code when it is used for identifying the caller; it becomes an Account Code when used for purposes of identifying the client for billback of associated charges. Access Code may also mean the digit, or digits, a user must dial to be connected to an outgoing trunk. For example, the user picks up his phone and dials "9" for a local line, dials "8" for long distance, dials "76" for the tie line to Chicago, etc. In programming a phone system there are unique Access Codes for Startup, Configuration programming, Administration programming and other functions; technicians who have been certified at various levels are provided with the appropriate Access Codes which allow them to invoke the appropriate level of system privilege.

Access Attempt The process by which one or more users interact with a telecommunications system to start to transfer information. An access attempt begins with an issuance of an access request by an access originator. An access attempt ends either in successful access or in access failure.

Access Bus Access Bus is correctly spelled Access.bus. It is a 100 Kbps bus currently being implemented as part of the Video Electronics Standards Association's Display Data Channel for controlling PC monitors. Access.bus allows bidirectional communication between compatible systems and displays, allowing on-the-go installation. Also supports daisychaining to reduce cable snarl. Access Bus has four pins per connector. See also USB and Firewire.

Access Channel Capacity Cable television channel capacity dedicated to cablecasting by entities not affiliated with the cable system operator, and over which the cable system operator does not exercise editorial control. Categories of access channel capacity are:
- Public Access, dedicated for use by the general public.
- Educational Access, dedicated to local educational authorities.
- Government Access, dedicated to local government.
- Leased Access, dedicated to commercial users.

The significant distinction here is "editorial control": subject to certain narrowly-defined exceptions (obscenity, "unlawful conduct"), the cable operator is prohibited from exercising editorial control over programming carried on access channels. This distinction affects the records which must be kept in the system's PIF:
- The operator is exempt from the requirements of Section 76.225c of the FCC rules regarding limits on commercial matter in children's programming carried on access channels.
- The operator is required by Section 76.701(h) to maintain records to verify compliance with rules governing leased-access channels carrying indecent programming.

Access Charge As part and parcel of the Modified Final Judgement (MFJ) which mandated the breakup of the Bell System though AT&T's divestiture of the Bell Operating Companies, the FCC declared that all end users have easy access to the long distance carrier of their choice. Further, such access to the InterExchange Carrier (IEC, or IXC) was to be provided with equal ease, i.e. on an equal 1+ basis, as traditionally enjoyed by AT&T alone. Additionally, the breakup of the Bell System invalidated the complex settlements structure which served to subsidize the Local Exchange Carriers (LECs) for providing local access to the long distance network(s). In order to replace the traditional settlements process and to compensate the LECs for the use of the vital, expensive and relatively unprofitable local access network, Access Charges were mandated, falling into two general categories.

1. The end user Access Charge, also known as Customer Access Line Charge (CALC) applies to every local loop, and is sensitive to the nature of the circuit. Subject to review and adjustment by the FCC on an annual basis, such Access Charges differ for residential and business users, single lines versus trunks, leased lines versus local access circuits, and so on. A mandatory charge appearing as a separate line item on the user's bill, it applies whether or not the user ever, in fact, places a long distance call over either a wired or wireless network. This surcharge is levied per the Code of Federal Regulations, Title 47, Part 69.

2. The Carrier Access Charge (CAC) applies to all IXCs which connect to the LEC network. Paid by the IXC to the LEC, such charges are determined by special tariffs subject to regulatory approval, and are sensitive to factors including the distance between the IXC Point of Presence (POP) and the point of termination into the LEC network. Additionally, the IXC pays to the LEC a usage charge sensitive to the traffic passed to the IXC as measured by Minutes of Use. The Telecommunications Act of 1996 greatly modifies the basis on which such charges are determined and levied. Most especially, The Act will eliminate such charges to the extent that the IXCs provide local exchange service directly, effectively bypassing the incumbent LECs.

Access Code A series of digits or characters which must be dialed, typed or entered in some way to get use of something. That "something" might be the programming of a telephone system, a long distance company, an electronic mail service, a private corporate network, a mainframe computer, a local area network. Once the user dials the main number for the service he must then enter his assigned Access Code to get permission use the system. An Access Code becomes an Account Code when it is used for identifying the caller and doing the billing. Access Code may also mean the digit, or digits, a user must dial to be connected to an outgoing trunk. For example, the user picks up his phone and dials "9" for a local line, dials "8" for long distance, dials "76" for the tie line to Chicago, etc. In programming a phone system such as Northern Telecom's Norstar, there are Access Codes to begin Startup, Configuration programming, and Administration programming.

Access Control A technique used to define or restrict the rights of individuals or application programs to obtain data from, or place data into, a storage device. Similarly, access to system logic is controlled on the basis of appropriate Access Codes. See Access Code.

Access Control Field 1. A term specific to Synchronous Multimegabit Data Service (SMDS), the Access Control Field controls access to the shared DQDB (Distributed Queue Dual Bus) bus which, in turn, provides access to the SMDS network. It consists of a single octet which is a portion of the 5-octet header of an SMDS cell. See also SMDS.

2. A Token Ring term. A field comprising a single octet (eight bits) in the header of a Token Ring LAN frame. Three Priority (P) bits set the priority of the token, a single Token (T) bit denotes either token or a frame, a Monitor (M) bit prevents frames or high-priority tokens from continuously circling the ring, and three Priority Reservation r bits allow a device to reserve the token for network access the next time the token circles the ring. See also Token Ring.

Access Control List ACL. Most network security systems operate by allowing selective use of services. An Access Control List is the usual means by which access to, and denial of, services is controlled. It is simply a list of the services available, each with a list of the computers and users permitted to use the service.

Access Control Method Sets of rules which determine the basis on which devices are afforded access to a shared physical element, such as a circuit or device. In a Local Area Network environment, it regulates each workstation's physical access to the transmission medium (normally cable), directs traffic around the network and determines the order in which nodes gain access so that each device is afforded an appropriate level of access. By way of example, token passing is the technique used by Token Ring, ARCnet, and FDDI. Ethernet makes use of CSMA/CD or CSMA/CA; DDS makes use of a polling technique. See Media Access Control. (MAC).

Access Control System A system designed to provide secure access to services, resources, or data; for computers, telephone switches or LANs.

Access Controls An electronic messaging term. Controls that enable a system to restrict access to a directory entry or mailbox either inclusively or exclusively.

Access Coordination An MCI definition. The process of ordering, installing, and maintaining the local access channel for MCI customers.

Access Coupler A device placed between two fiber optic ends to allow signals to be withdrawn from or entered into one of the fibers.

Access Customer Name Abbreviation ACNA. A three-character abbreviation assigned to each IntereXchange Carrier (IXC) and listed in the Local Exchange Routing Guide (LERG).

Access Event Bellcore definition for information with a logical content that the functional user and the Network Access FE (Functional Entity) exchange.

Access Floor A system consisting of completely removable and interchangeable floor panels that are supported on adjustable pedestals or stringers (or both) to allow access to the area beneath.

Access Function An intelligent network term. A set of processes in a network that provide for interaction between the user and a network.

Access Group All terminals or phones have identical rights to use the computer, the network, the phone system, etc.

Access Level Used interchangeably with Access Code. "Level" in dialing tends to mean a number.

Access Line A telephone line reaching from the telephone company central office to a point usually on your premises. Beyond this point the wire is considered inside wiring. See LOCAL LOOP and ACCESS LINK.

Access Link The local access connection between a customer's premises and a carrier's POP (Point Of Presence), which is the carrier's switching central office or closest point of local termination. That carrier might be a LEC, IXC or CAP/AAV; in a convergence scenario, the carrier might also be a CATV provider.

Access List List kept by routers to control access to or from the router for a number of services (for example, to prevent packets with a certain IP address from leaving a particular interface on the router).

Access Manager An means of authorization security which employs scripting.

Access Method The technique or the program code in a computer operating system that provides input/output services. By concentrating the control instruction sequences in a common sub-routine, the programmer's task of producing a program is simplified. The access method typically carries with it an implied data and/or file structure with logically similar devices sharing access methods. The term was coined, along with Data Set, by IBM in the 1964 introduction of the System/360 family. It provides a logical, rather than physical, set of references. Early communications access methods were primitive; recently they have gained enough sophistication to be very useful to programmers. Communications access methods have always required large amounts of main memory. In a medium size system supporting a few dozen terminals of dissimilar types, 80K to 100K bytes of storage is not an unusual requirement. The IEEE's 802.x standards for LANs and MANs. See Access Methods.

Access Methods Techniques and rules for figuring which of several communications devices — e.g. computers — will be the next to use a shared transmission medium. This term relates especially to Local Area Networks (LANs). Access method is one of the main methods used to distinguish between LAN hardware. How a LAN governs users' physical (electrical or radio) access to the shared medium significantly affects its features and performance. Examples of access methods are token passing (e.g., ARCnet, Token Ring and FDDI) and Carrier Sense Multiple Access with Collision Detection (CSMA/CD) (Ethernet). See Access Method and Media Access Control.

Access Node Access nodes are points on the edge of a network which provide a means for individual subscriber access to a network. At the Access Node, individual subscriber traffic is concentrated onto a smaller number of feeder trunks for delivery to the core of the network. Additionally, the access nodes may perform various forms of protocol conversion or adaptation (e.g. X.25, Frame Relay, and ATM). Access Nodes include ATM Edge Switches, Digital Loop Carrier (DLC) systems concentrating individual voice lines to T-1 trunks, cellular antenna sites, PBXs, and Optical Network Units (ONUs).

Access Number The telephone number you use to dial into your local Internet Service Provider (ISP). To connect to the Internet you must first establish an account with an ISP your area. Usually you will receive a list of telephone numbers you can use to "dial-in" to the service.

Access Organization An entity which originates program material for transmission over the access channel capacity of a cable television system. An access organization may be an individual, a non-profit corporation, an unincorporated non-profit association, or a for-profit corporation. However, under most cable franchises, commercial advertising is prohibited on Public, Educational, and Government Access channels.

Access Phase In an information-transfer transaction, the phase during which an access attempt is made. The access phase is the first phase of an information-transfer transaction.

Access Point 1. A point where connections may be made for testing or using particular communications circuits.

2. A junction point in outside plant consisting of a semipermanent splice at a junction between a branch feeder cable and distribution cables.

3. AP. A cross-box where telephone cables are cross connected.

Access Protection Refers to the process of protecting a local loop from network outages and failures. Access protection can take many forms, such as purchasing two geographically diverse local facilities, adding protection switches to the ends of geographically diverse local loops, or buying service from a local access provider which offers a survivable ring-based architecture to automatically route around network failures.

Access Protocols The set of procedures which enable a user to obtain services from a network.

Access Provider A company, such as a telephone company, that hooks your computer up to the Internet.

Access Rate 1. The maximum data rate of the access channel, typically referring to access to broadband networks and network services.

2. A Frame Relay term which addresses the maximum transmission rate supported by the access link into the network. The Access Rate defines the maximum rate for data transmission or receipt.

Access Request A message issued by an access originator to initiate an access attempt.

Access Response Channel ARCH. Specified in IS-136, ARCH carries wireless system responses from the cell site to the user terminal equipment. ARCH is a logical subchannel of SPACH (SMS (Short Message Service) point-to-point messaging, Paging, and Access response CHannel), which is a logical channel of the DCCH (Digital Control CHannel), a signaling and control channel which is employed in cellular systems based on TDMA (Time Division Multiple Access). The DCCH operates on a set of frequencies separate from those used to support cellular conversations. See also DCCH, IS-136, SPACH and TDMA.

Access Server Communications processor that connects asynchronous devices to a LAN or WAN through network and terminal emulation software. Performs both synchronous and asynchronous routing of supported protocols. Sometimes called a network access server.

Access Service Request ASR. A forma request for service — typically this happens when a frame relay, long distance provider or CLEC asks your local phone company to provide a line from your business or home to their equipment. Their equipment is sometimes in the local phone company's offices. See CLEC.

Access Signaling A term which Northern Telecom's Norstar telephones use to indicate their ability to access a remote system (such as a Centrex or a PBX), or dial a number on an alternate carrier by means of Access Signaling (also referred to as "End-to-End" Signaling).

Access Switch Feeder node to Enterprise Network Switches that perform multiprotocol bridge/routing and support a wide range of serial-link (e.g., SDLC BSC, asynchronous) attached devices. Also know as Gateways, such devices currently are known as Routers and Encapsulating Bridges, although the differences between them are most significant.

Access Tandem A Local Exchange Carrier switching system that provides a concentration and distribution function for originating and/or terminating traffic between a LEC end office network and IXC POPs. In short, a distinct type of local phone company switching system specifically designed to provide access between the local exchange network and the interexchange networks for long-distance carriers in that area. The Access Tandem provides the interexchange carrier with access to multiple end offices within the LATA. More than one Access Tandem may be needed to provide access to all end offices within any given LATA. Currently, the Access Tandem function may be in the form of a physical and logical partition of a LEC Central Office switch, which also serves end users for purposes of satisfying local calling requirements. Additionally, the IXC may extend the reach of the POP through a high-speed channel extension via dedicated circuits; thereby achieving interconnection with the LEC though collocation of termination facilities in the LEC CO.

Access Time There are many definitions of access time:

1. In a telecommunications system, the elapsed time between the start of an access attempt and successful access. Note: Access time values are measured only on access attempts that result in successful access.

2. In a computer, the time interval between the instant at which an instruction control unit initiates a call for data and the instant at which delivery of the data is completed.

3. The time interval between the instant at which storage of data is requested and the instant at which storage is started.

4. In magnetic disk devices, the time for the access arm to reach the desired track and for the rotation of the disk to bring the required sector under the read-write mechanism.

5. The amount of time that lapses between a request for information from memory and the delivery of the information; usually stated in nanoseconds (ns). When accessing data from a disk, access time includes only the time the disk heads take to settle down reaching the correct track (seek time) and the time required for the correct sector to move over the head (latency). Disk access times range between 9ms (fast) and 100 ms (slow).

Access Unit 1. AU. An electronic messaging term, used for implementing value-added services such as fax, Telex, and Physical Delivery via X.400.

2. In the token ring LAN community, an access unit is a wiring concentrator. See Media Access Unit (MAU).

ACCOLC Access Overload Class. A term used in the cellular phone business to allow the cellular system some way of choosing which calls to complete based on some sort of priority. Originally, when the Federal government began designing cellular systems, the government intended to give certain emergency vehicles (such as police, ambulances, and fire departments) codes in their cellular phones that would allow them priority over other subscribers to communicate during emergencies. There is no standard in use within the United States at this time.

Account Code (Voluntary or Enforced) A code assigned to a customer, a project, a department, a division — whatever. Typically, a person dialing a long distance phone call must enter that code so the Call Accounting system can calculate and report on the cost of that call at the end of the month or designated time period. Many service companies, such as law offices, engineering firms and advertising agencies use account codes to track costs and bill their clients accordingly. Some account codes are very complicated. They include the client's number and the number of the particular project. The Account Code then includes Client and Matter number. These long codes can tax many call accounting systems, even some very sophisticated ones.

Account Executive AE. A fancy, schmanzy name for a salesperson. The idea is that the customer is an "account," and the salesman is the executive running the account. Telephone companies call their salespeople account executives — especially on the equipment and non-long distance side.

Accountant Someone who figures your numbers, then numbers your figures and then sends you a bill. See also Economist.

Accounting Management In network management, a set of functions that enables network service use to be measured and the costs for such use to be determined and includes all the resources consumed, the facilities used to collect accounting data, the facilities used to set billing parameters for the services used by customers, maintenance of the data bases used for billing purposes, and the preparation of resource usage and billing reports.

Accounting management is one of five categories of network management defined by ISO for management of OSI networks. Accounting management subsystems are responsible for collecting network data relating to resource usage. See also configuration management, fault management, performance management and security management.

Accounting Rate A price used between long distance companies to "balance up" what they owe each other. For example, if AT&T sends France Telecom one million minutes of calls, but France sends only 500,000 minutes back to AT&T, then AT&T will have to pay France Telecom for the imbalance. If the accounting rate between France and the United States is $1 per minute, then AT&T will pay France Telecom $250,000 for its work in completing the 500,000 extra calls. AT&T pays France Telecom only half the cost because AT&T does half the work itself. The Accounting Rate is evenly applied, regardless of which carrier originates and which carrier terminates the call. See Billing Rate.

Accounting Servers A Local Area Network costs money to set up and run. Thus it may make sense to charge for usage on it. In LANs which rely on the Novell NetWare Network Operating System, the network supervisor sets up accounting through a program called SYSCON. When this happens, the current file server automatically begins to charge for services. The supervisor can authorize other network services (print servers, database servers, or gateways) to charge for services, or can revoke a server's right to charge.

Accounting Traffic Matrix A mobile term. A traffic matrix is a collection of information, gathered over a period of time, containing statistics on Mobile End System (M-ES) registration, de-registration, and Network Protocol Data Unit (NPDU) traffic.

ACCS Automatic Calling Card Service.

Accumulator 1. A register in which one operand can be stored and subsequently replaced by the result of an arithmetic or logic operation. A term used in computing.

2. A storage register.

3. A storage battery.

Accuracy Absence of error. The extent to which a transmission or mathematical computation is error-free. There are obvious ways of measuring accuracy, such as the percentage of accurate information received compared to the total transmitted.

AC/DC Ringing A common way of signaling a telephone. An alternating current (AC) rings the phone bell and a direct current (DC) is used to work a relay to stop ringing when the called person answers.

ACD See the next seven definitions, ACD, Automatic Call Distributor and ACIS.

ACD Agent A telephony end user that is a member of an inbound, outbound, skills based, or programmable Automatic Call Distribution group. ACD Agents are distinguished from other users by their ability to sign on (i.e., login) to phone systems that coordinate and distribute calls to them.

ACD Agent Identifier The identifier of an ACD agent. An agent identifier uniquely identifies an agent within an ACD group. See ACD Agent.

ACD Application Bridge Refers to the link between an ACD and a database of information resident on a user's data system. It allows the ACD to communicate with a data system and gain access to a database of call processing information such as Data Directed Call Routing.

ACD Application-Based Call Routing In addition to the traditional methods of routing and tracking calls by trunk and agent group, the latest ACDs route and track calls by application. An application is a type of call, e.g. sales vs. service. Tracking calls in this manner allows accurately reported calls especially when they are overflowed to different agent groups.

ACD Call Back Messaging This ACD capability allows callers to leave messages for agents rather than wait for a live agent. It helps to balance agent workloads between peak and off-peak hours. In specific applications, it offers callers the option of waiting on hold. A good example is someone who only wishes to receive a catalog. Rather than wait while people place extensive orders, they leave their name and address as a message for later follow-up by an agent. This makes things simpler for them and speeds up service to those wanting to place orders.

ACD Caller Directed Call Routing Sometimes referred to as an auto attendant capability within the industry, this ACD capability allows callers to direct themselves to the appropriate agent group without the intervention of an operator. The caller responds to prompts (Press 1 for sales, Press 2 for service) and is automatically routed to the designated agent group.

ACD Central Office An Automatic Call Distributor (ACD), usually located in a central office and supplied to the customer by the telephone company (telco) with tariffed pricing structures. Some data gathering equipment is often located on customer premises.

ACD Conditional Routing The ability of an ACD to monitor various parameters within the system and call center and to intelligently route calls based on that information. Parameters include volume levels of calls in queue, the number of agents available in designated overflow agent groups, or the length of the longest call. Calls are routed on a conditional basis. "If the number of calls in queue for agent group #1 exceeds 25 and there are at least 4 agents available in agent group #2, then route the call to agent group #2.

ACD Data Directed Call Routing A capability whereby an ACD can automatically process calls based on data provided by a database of information resident in a separate data system. For example, a caller inputs an account number via touch tone phone. The number is sent to a data system holding a database of information on customers. The number is identified, validated and the call is distributed automatically based on the specific account type (VIP vs. regular business subscriber, as an example).

ACD DN A Nortel term for an Automatic Call Distribution Directory Number (ACD DN), which refers to the queue where incoming calls wait until they are answered. Calls are answered in order in which they entered the queue.

ACD Group Multiple agents assigned to process incoming calls that are directed to the same dialed number. The ACD feature of the telephone switch routes the incoming to one of the agents in the ACD group based upon such properties as availability of the agent and length of time since the agent last completed an incoming call.

ACD Intelligent Call Processing The ability of the latest ACDs to intelligently route calls based on information provided by the caller, a database on callers and system parameters within the ACD such as call volumes within agent groups and number of agents available.

ACD Number The telephone number that calling devices dial to access any of the multiple agents in an ACD group. Once the incoming call arrives at the ACD number, the ACD service can then route the call to one of multiple agents in the ACD group.

ACD Skills-Based Routing See Skills-Based Routing

ACF Advanced Communication Function. A family of software products used by IBM allowing its computers to communicate.

ACF/NCP Advanced Communication Function/ Network Control Program. In host-based IBM SNA networks, ACF/NCP is the control software running on a communications controller that supports the operation of the SNA backbone network.

ACF/VTAM Advanced Communication Function/Virtual Terminal Access Method. In host-based IBM SNA networks the ACF/VTAM is the control software running on a host computer that allows the host to communicate with terminals on an SNA network.

ACFG Short for AutoConFiGuration. The Plug and Play BIOS extensions, now turning up on PCs, are also known as the ACFG BIOS extensions.

ACH Attempts per Circuit per Hour. This is a term you often see in call centers. It refers to the number of times someone tried to reach a circuit in one hour. In a normal phone system, ACH refers to the number of times someone tried to make a call on a circuit in one hour. Measuring ACH is useful for figuring how many inbound or outbound trunks you may need. See also CCH, which is connections per hour.

ACIS Automatic Customer/Caller Identification. This is a feature of many sophisticated ACD systems. ACIS allows the capture of incoming network identification digits such as DID or DNIS and interprets them to identify the call type or caller. With greater information, such as provided by ANI, this data can identify a calling subscriber number. You can also capture caller identity by using a voice response device to request an inbound caller to identify themselves with a unique code. This could be a phone number, a subscriber number or some other identifying factor. This data can be used to route the call, inform the agent of the call type and even pre-stage the first data screen associated with this call type automatically. See also ANI, Caller ID and Skills-Based Routing.

ACK In data communications, ACK is a character transmitted by the receiver of data to ACKnowledge a signal, information or packet received from the sender. In the de facto standard IBM Binary Synchronous Communications (also known as BSC or Bisync) protocol, an ACK is transmitted to indicate the receipt of a block of data without any detected transmission errors. This positive acknowledgement reassures the transmitting device of that fact in order that the next block or data may be transmitted. Binary code for an ACK is 00110000. Hex is 60. See also Acknowledgment.

ACK Ahead A variation of the XMODEM protocol that speeds up file transmission across error-free links. See XMODEM.

ACK1 Bisync acknowledgment for odd-numbered message.

Acknowledgment In data communications, the transmission of acknowledgment (ACK) characters from the receiving device to the sending device indicates the data sent has been received correctly.

ACL 1. Access Control List. A roster of users and groups of users, along with their rights. See Access Control List.
2. Applications Connectivity Link. Siemens' protocol for linking its PBX to an external computer and having that computer control the movement of calls within a Siemens PBX. See also Open Application Interface.

ACM 1. An ATM term. Address Complete Message. One of the ISUP call set-up messages. A message sent in the backward direction indicating that all the address signals required for routing the call to the called party have been received. See
2. Association for Computing Machinery. www.acm.org

3. **Automatic Call Manager.** The integration of both inbound call distribution and automated outbound call placement from a list of phone contacts to be made from a database. Telemarketing and collections applications are targets for this type of system.

ACNA Access Customer Name Abbreviation.

ACO 1. Additional Call Offering

2. Alarm Cut Off

ACONET A research network in Austria.

Acoustic Coupler An acoustic modem. A modem designed to transfer data to the telephone network acoustically (i.e by sound), rather than electronically. An acoustic coupler looks like the reverse of a telephone handset and is typically made of rubber. The data communications link is achieved through acoustic (sound) signals rather than through direct electrical connection. It is attached to the computer or data terminal through an RS-232-C connector. To work the acoustic coupler, start the computer's communications program, dial the distant computer on a single line telephone with a normal (e.g. old-fashioned) handset. When the distant computer answers with a higher pitched "carrier tone," you place the telephone handset in the acoustic coupler and transmit data. Since the data is transmitted by sound between the handset and the acoustic coupler (and vice versa), the quality isn't always reliable. You can usually transmit up to 300 baud. People use acoustic couplers when they're short of time or cannot physically connect their modem electrically, e.g. they're using a payphone without an RJ-11 jack. (There are precious few.)

Acoustic Model In automatic speech recognition, an acoustic model models acoustic behavior of words by gluing together models of smaller units, such as phonemes. (Sorry for the definition of the word model with the word model. But it's actually the best way of defining this term. HN)

Acoustic Noise An undesired audible disturbance in the audio frequency range.

Acoustic (or Air) Suspension A loudspeaker system that uses an air-tight sealed enclosure.

Acoustics That branch of science pertaining to the transmission of sound. The qualities of an enclosed space describing how sound is transmitted, e.g. its clarity.

Acousto-optic The interactions between acoustic waves and light in a solid medium. Acoustic waves can be made to modulate, deflect, and focus light waves by causing a variation in the refractive index of the medium. See also FIBER OPTICS.

Acquisition 1. In satellite communications, the process of locking tracking equipment on a signal from a communications satellite.

2. The process of achieving synchronization.

3. In servo systems, the process of entering the boundary conditions that will allow the loop to capture the signal and achieve lock-on. See also phase-locked loop.

4. In mobile, the process by which a Mobile End System (M-ES) locates a radio Frequency (RF) channel carrying a channel stream, synchronizes to the data transmissions on that channel stream, and determines whether the channel stream is acceptable to the M-ES for network access.

Acquisition Time 1. In a communication system, the amount of time required to attain synchronization.

2. In satellite control communications, the time required for locking tracking equipment on a signal from a communications satellite. See also satellite.

ACP Activity Concentration Point.

ACPI Advanced Configuration and Power Interface, a replacement for APM (Advanced Power Management). This power management standard, proposed by Microsoft, Intel, and Toshiba, lets the PC control power to peripherals like CD-ROMs and printers, as well as consumer devices hooked up to the PC. Peripherals can also use ACPI to turn on the PC. For example: you could insert a CD-ROM into a drive, and the computer would automatically boot up. See APM and WFM for a fuller explanation.

ACR 1. Attenuation to Crosstalk Ratio: One of the factors that limits the distance a signal may be sent through a given media. ACR is the ratio of the power of the received signal, attenuated by the media, over the power of the NEXT crosstalk from the local transmitter, usually expressed in decibels (db). To achieve a desired bit error rate, the received signal power must usually be several times larger than the NEXT power or plus several db. Increasing a marginal ACR may decrease the bit error rate.

2. An ATM term. Allowed Cell Rate: An ABR service parameter, ACR is the current rate in cells/sec at which a source is allowed to send. ACR is a parameter defined by the ATM Forum for ATM traffic management. ACR varies between the MCR and the PCR, and is dynamically controlled using congestion control mechanisms.

ACRFNET Academic Computing Research Facility Network. A network connecting various research units such as colleges and research and development laboratories in the U.S.

Acrobat A standardized way of viewing a file without needing the associated software. For example, you run Word 6.0 or QuarkXPress, make a pretty desktop published document, replete with diagrams, photos and diagrams. Now you want to send the file to someone to view it in all its glory. Simple. Convert the file to an Acrobat file (which has a .PDF extension) and modem it or send it on disk. The receiving person will run an Acrobat viewer program and see your beautiful work. They won't be able to change your work. But they will be able to see it. Acrobat is from Adobe, the Los Altos, CA company which produces PostScript. Acrobat has three benefits: The Acrobat viewing program is very cheap. You can use Acrobat to view virtually any Windows software created file. Third, an Acrobat file can be up to 75% smaller than the original file in its native form, i.e. the original Word or QuarkXpress file.

Acronym A pronounceable artificial word formed from the first (or first few) letters of each word or group of words. For example, BASIC, the Beginner's All-purpose Symbolic Instruction Code, or COBOL, the computer language, which comes from COmmon Business Oriented Language.

ACS 1. Automatic Call Sequencer. A rudimentary automatic call distributor. See Automatic Call Sequencer.

2. Advanced Communication System. An old name for AT&T's data communications/data processing service, originally known as BDN (Bell Data Network) and later called Net 1000. ACS supported multiprotocol communications, offering protocol translation (much like X.25 packet switching), as well. In late 1986, after 10 years in birth, AT&T quietly buried ACS, which offered too little in the face of what — by then — had become cheap, powerful desktop microcomputers and 1200 bps modems priced at less than $200. ACS was depicted in AT&T presentations as a cloud — user data entered the cloud of the network on the originating end and exited the cloud on the terminating end. What went on in the cloud of the network was obscured from the user. The thinking was that the user needn't be concerned with what went on inside the cloud; rather, that was AT&T's responsibility and concern. This clever

conceptual sell was never successful — it was way too obscure and offered way too little. See also Cloud.

3. ATM Circuit Steering. A means of routing ATM traffic to test facilities built in the ATM device (e.g., switch, router, or concentrator), as ATM networks have no point of entry for a test device.

ACSE Association Control Service Element. An OSI application-layer protocol. The method used in OSI for establishing a connection between two applications.

ACT Applied Computer Telephony is Hewlett Packard's program that is a strategy and set of open architecture commands and interfaces for integrating voice and database technologies. The idea is that with ACT a call will arrive at the telephone simultaneously with the database record of the caller. And such call and database record can be transferred simultaneously to an expert, a supervisor, etc. ACT works on both HP 3000 and HP 9000 computers. ACT essentially controls the telephone call movement within PBXs it connects to. See also Open Application Interface.

ACTA America's Carriers Telecommunications Association, a Casselberry, FL organization founded in 1985 by 15 small long distance companies wishing to create an association in which the members controlled the direction of the organization. (That's their words.) "The focus established was to provide national representation before legislative and regulatory bodies, while continuing to improve industry business relations." There are now more than 165 members, which include MCI and Sprint, but not AT&T. ACTA's latest and greatest claim to fame is its 1996 petition to the FCC to outlaw Voice Over the Net (VON). Sprint is among the members which broke ranks over that issue, upon which the FCC so far has decline to act.

ACTAS Alliance of Computer-Based Telephony Application Suppliers — a part of the North American Telecommunications Association (NATA), which has now changed its name to MMTA. ACTAS's mission, according to ACTAS, is to deliver the benefits of computer-based telephone applications to the broadest possible range of customers. ACTAS works to lower the threshold for delivering the benefits of these applications, which include integrated voice and database processing systems automation and customer service, to the general business market. See MMTA.

ACTGA Attendant Control of Trunk Group Access. A complicated term for a simple concept, namely that your operator completes long distance calls. A primitive form of toll control.

ACTIUS Association of Computer Telephone Integration Users and Suppliers. A British organization, ACTIUS provides an open industry forum for interchange and discussion on Computer Telephone Integration (CTI) between its members. Subscriptions are UK stlg275.00 for full membership and UK stlg150 for corresponding membership. A key aim of ACTIUS is to explain the benefits of CTI applications to the broadest possible range of users. ACTIUS also represents its members' interests on relevant regulatory and standards issues, both in the UK and the EU (European Union) Address : 11 Nicholas Road, Henley-on-Thames, OXON, United Kingdom RG9 1RB. Contact: Brian Robson, Secretary Tel: 011-44-1491-575295 Fax; 011-44-1491-410201 email brian.robson@BTInternet.com

Activated Return Capacity A cable TV term. The capability of transmitting signals from a subscriber or user premises to the cable headend. The typical information that can be sent back inlcudes the ID number of the cable TV set-top box and what station you are watching. See also Cable Modem.

Activation A one time initial connection fee to get cellular phone service. As competition has intensified, so more and more carriers are dropping or severely reducing their activation fees. They do this in order to attract more new subscribers.

Activation Fee Fee for the initial connection to the cellular system.

Active Call 1. A definition used in Call Centers. An active call is when the connection is in any state except Hold, Null, or Queued. In other words, any state during the establishment of the connection and/or call. This also includes the actual establishment of the connection and/or call itself.

2. A term which Hayes defines in its Hayes ISDN System adapter manual. An active call is a voice call to which you are connected that is not on hold.

Active Campaign A call center/marketing term. An outbound calling project that is currently running.

Active Channel An Active Channel is what Microsoft calls a Web site that has been enabled for push delivery to Internet Explorer 4.0 browsers. To create a channel, developers write and upload a CDF (channel definition format) file to their Web site. New content is delivered to users automatically when the site is updated. Developers and subscribers can control the update frequency; which channels, subchannels, and items (sections) are subscribed to; and other channel characteristics. Most Active Channels use dynamic HTML (DHTML) and other effects to spice up content and make it more interactive. See also: channel definition format, DHTM and push.

Active Circuits An MCI definition. MCI circuits for which are there is a completed "install order" and a "completed date."

Active Contract One you must sign. See Contract.

Active Coupler A fiber optic coupler that includes a receiver and one or more transmitters. It regenerates (thus "active") input signals and sends them through output fibers, instead of passively dividing input light.

Active Device Electronic components that require external power to manipulate or react to electronic output. These include transistors, op amps, diodes, cathode ray tubes and ICs. Passive devices include capacitors, resisters and coils (inductors).

Active Directory A feature of Windows NT server and first introduced in NT 5.0, which will be called Windows 2000. Think of it as a real-time, super-fast Directory Information service — like they have when you call 411 or 555-1212 in North America. But instead of being answered by a person, it answers requests for peoples' phone numbers and addresses by sending instant messages back to the PC workstation which is asking the question. Here's a simple example, imagine you want to call someone in your organization using your IP telephone. Instead of dialing a number as we do today, we dial a person. Our PC talks to the Windows NT server which we've logged onto and says, in essence, "I want to call Helen. Where is she?" It comes back and says this is "Helen's address. Call there." My PC says "thank you" and then dials Helen at that number. It's called "Active" Directory because its address entries change from moment to moment, as Helen moves around. See TAPI 3.0.

Active Hub A device used to amplify transmission signals in certain local area network topologies. You can use an active hub to add workstations to a network or to lengthen the cable distance between workstations and the file saver.

Active Line A voice or data communications channel currently in use.

Active Link A logical communications circuit that is established only for the duration of communications. An active line needs a call-setup and call-clearing procedure for every connection.

Active Matrix Liquid Crystal Display A technique of making liquid crystal displays for computers in which each of the screen's pixels — the tiny elements that make up a picture — is controlled by its own transistor. Active matrix LCD display technique uses a transistor for each monochrome or each red, green and blue pixel. It provides sharp contrast, speedier screen refresh and doesn't lose your cursor when you move it fast (also knowing as submarining). Some active matrix CD screens are as fast as normal glass CRTs.

Active Medium The material in fiber optic transmission, such as crystal, gas, glass, liquid or semiconductor, which actually "lases." It's also called laser medium, lasing medium, or active material.

Active Monitor Device responsible for managing a Token Ring. A network node is selected to be the active monitor if it has the highest MAC address on the ring. The active monitor is responsible for such management tasks as ensuring that tokens are not lost, or that frames do not circulate indefinitely. See also ring monitor and standby monitor.

Active Open Used in TCP to request connection with another node.

Active Participation A feature in an automatic call distributor, a piece of equipment used in call centers. This feature is typically used to allow intrusion with the ability to speak and listen by a supervisor into an Agent Call. The resultant call is a conference. A Versit definition.

Active Pixel Region On a computer display, the area of the screen used for actual display of pixel information.

Active Push The server on the Web interacts with the client by sending all the content to the client upon the client's request (polling), essentially the way that a client/server application might. PointCast s an example of this. See Push and Directed Push.

Active Splicing Aligning the ends of two optical fibers with the aim of minimizing the splice loss.

Active Terminator A terminator that can compensate for variations in the terminator power supplied by the host adapter through means of a built-in voltage regulator.

Active Video Lines All video lines not occurring in the horizontal and vertical blanking intervals. In other words, the lines conveying the video and audio signals.

Active Vocabulary A phrase used in voice recognition to mean a group of words which a recognizer has been trained to understand and recognize. When a voice says ... PAIR, it knows the word means two, not the fruit, pear. That's one example of why voice recognition is so difficult. English, particularly, is a very complex and confusing language.

Active Window A Windows term. The active windows is the window in which the user is currently working. An active window is typically at the top of the window order and is distinguished by the color of its title bar, typically dark blue.

Active/Passive Device On a local area network, a device that supplies current for the loop is considered active. Such a device is s Token Ring MAU (Multistation Access Unit). A device which does not supply current is considered passive.

Actives A call center/marketing term. Refers to customers who have purchased within a time period defined by the company instigating the marketing. Customers who have purchased outside the specified time period are considered inactive.

ActiveX In its simplest terms, said InfoWorld, May 19, 1997, ActiveX is an architecture that lets a program (the ActiveX control) interact with other programs over a network (such as the Internet). It's quite a different animal than Java, which is an entire new programming language plus a specification for a virtual machine. The ActiveX architecture, according to InfoWorld, uses Microsoft's Component Object Model (COM) and Distributed COM (DCOM) standards. COM allows different applications to talk to each other locally. DCOM allows them to talk over a network.

ActiveX, formerly known as Object Linking and Embedding, or OLE, is an umbrella of mechanisms designed to bring sound bytes, animation and interactivity to Web documents, similar to plug-in technology for Netscape Communications Corp.'s Navigator and Sun Microsystems Inc.'s Java applets....They all provide ways to send small programs to a browser, without the involvement of any other special software on the desktop. In short, ActiveX is software code from Microsoft which allows a developer to add move things to an otherwise static Web page. ActiveX is positioned by Microsoft as a competitive move against Java.

Microsoft's ActiveX enables software components to interact with one another in a networked environment, regardless of the language in which they were created. Kind of like the Olympics. ActiveX is built on the Component Object Model (COM). Elements of ActiveX are:

• ActiveX Controls—the interactive objects in a Web page that provide interactive and user-controllable functions.

• ActiveX Documents—enable users to view non-HTML documents, such as Microsoft Excel or Word files, through a Web browser.

• Active Scripting—controls the integrated behavior of several ActiveX controls and/or Java Applets from the browser or server.

• Java Virtual Machine—the code that enables any ActiveX-supported browser such as Internet Explorer 3.0 to run Java applets and to integrate Java applets with ActiveX controls.

• ActiveX Server Framework—provides a number of Web server-based functions such as security, database access, and others.

ActiveX Controls A component that can be inserted in a page to provide functionality not directly available in HTML, such as animation sequences, credit-card transactions, real-time video sequences or spreadsheet calculations. ActiveX controls can be implemented in a variety of programming languages using C++, Visual Basic or Java. Over 1,000 ActiveX controls are available today. These include the Macromedia Shockwave for Director control and the Adobe Acrobat control.

Activity Concentration Point ACP. A location on a telecommunications network where there is high communications traffic, including voice, data, document distribution and teleconferencing. Generally, there will be some switching equipment present at the ACP.

Activity Costing A call center term. The costing methodology used to determine the specific cost of a given activity. (See Cost Per Phone Hour.)

Activity Factor A decimal fraction which represents the percentage of speech on a voice channel versus those periods of (non-talking) silence on that channel. Most voice channels carry actual speech 30% to 40% of the total available time. This represents an activity factor of 0.3 to 0.4.

Activity Report A report printed by a facsimile machine which lists all transmissions and receptions — their time, date, and number of documents; the remote unit type, diagnostic codes; and machine identification.

ACTS 1. Association of Competitive Telecommunications Suppliers. Trade association of telephone equipment dealers in Canada.

2. Automatic Coin Telephone Service includes a telephone

company central office that can complete all types of payphone calls automatically without an operator. Recorded announcements are used to convey instructions to the customer.

ACU Automatic Calling Unit. Also an 801 ACU. A telephone company-provided device instructed by a computer to place a call on behalf of the computer. The call is then connected to a telephone company-provided Data Set. Anyone other than an IBM shop would simply buy a Hayes or Hayes-compatible modem, and not bother with the trouble and expense of an ACU.

ACUTA The Association for Telecommunications Professionals in Higher Education. Prior to 1998, it was the Association of College & University Telecommunications Administrators. The new name is more meaningful, but the acronym stuck. ACUTA is an international, not-for-profit educational association serving approximately 800 institutions of higher learning. All members are director level or higher, and are responsible for data, video, communications, and all variety of networks, in addition to traditional telephony. Corporate affiliate members are welcome, as well. www.acuta.org

ADA 1. Average Delay to Abandon. Average time a caller is held in queue before they get frustrated and decide to hang up. 2. A high level computer language which the Department of Defense has been trying to foist on its suppliers and thus, make a standard. Ada is named for British mathematician Ada Lovelace, known at the time as Lady Lovelace. She was the girlfriend of Charles Babbage, the inventor of the computer.

ADACC Automatic Directory Assistance Call Completion.

ADAD Automatic Dialing and Announcing Device. Device which automatically places calls and connects them to a recording or agent. A Canadian term for an automatic dialer.

Adaptable Digital Filtering A way of fixing twisted-pair telephone lines so they carry data more efficiently up to 12,000 feet before the need to regenerate the signal. The filter can be customized to meet the needs of a twisted pair.

Adapter 1. A device used to connect a terminal to some circuit or channel so it wile compatible with the system to which it is attached. An adapter converts one type of jack or plug to another, for example, from old 4-prong telephone jacks to new modular. An adapter may also combine various items, such as putting three plugs in one jack.
2. Another name for a NIC — Network Interface Card — a card which fits into a computer and joins the computer to a local area network.

Adapter Card A printed circuit card installed inside of a computer. It takes data from memory and transmits it over cable to connected devices such as a modem, or printer.

Adapter Segment A name sometimes used for the upper memory area of a PC, at hexadecimal addresses A000 through EFFF (640K to 1024K).

Adaptive Antenna Array An antenna array in which the received signal is continually monitored in respect of interference (usually adjacent or co-channel). Its directional characteristics are then automatically adjusted to null out the interference. Such a concept often employs computer control of a planar type antenna.

Adaptive Channel Allocation A method of multiplexing wherein the information-handling capacities of channels are not predetermined but are assigned on demand.

Adaptive Communication Any communication system, or portion thereof, that automatically uses feedback information obtained from the system itself or from the signals carried by the system to modify dynamically one or more of the system operational parameters to improve system performance or to resist degradation.

Adaptive Compression Data compression software that continuously analyzes and compensates its algorithm (technique), depending on the type and content of the data and the storage medium.

Adaptive Differential Pulse Code Modulation See ADPCM.

Adaptive Equalization An electronic technique that allows a modem to continuously analyze and compensate for variations in the quality of a telephone line.

Adaptive Interframe Transform Coding A class of compression algorithms commonly used in video codecs to reduce the data transmission rate.

Adaptive Pulse Code Modulation A way of encoding analog voice signals into digital signals by adaptively predicting future encodings by looking at the immediate past. The adaptive part reduces the number of bits per second that another rival and more common method called PCM (Pulse Code Modulation) requires to encode voice. Adaptive PCM is not common because, even though it reduces the number of bits required to encode voice, the electronics to do it are expensive. See Pulse Code Modulation.

Adaptive Retransmission Algorithms Used by self-adjusting timers to determine and dynamically set timers to effectively adjust data traffic in the event the link is slower than usual due to congestion or their network conditions.

Adaptive Routing A method of routing packets of data or data messages in which the system's intelligence selects the best path. This path might change with altered traffic patterns or link failures.

ADAS Automated Directory Assistance Service. A service from Northern Telecom which automates the greeting and inquiry portion of the directory assistance call. With ADAS, directory assistance callers are greeted by the automated system and asked to state the name of the city and the listing they are seeking. They are then connected with an operator. The ADAS service knocks a few seconds off each directory assistance call.

ADB Apple Desktop Bus. A low-speed serial bus used on Apple Macintosh computers to connect input devices to the Macintosh CPU (central processing unit). Normally the ADP connects via an 8-pin round or DIN connector.

ADC Analog-to-Digital Converter.

ADCCP Advanced Data Communications Control Procedures, A bit-oriented ANSI-standard communications protocol. It is a link-layer protocol. ADCCP is ANSI's version of SDLC/HDLC.

ADCU Association of Data Communications Users.

Add-in Card An expansion board that fits into the computer's slots and is used to expand the system's memory or extend the operation of another device.

Add-on 1. A telephone system feature which allows connecting a third telephone to an existing conversation. This "add-on" feature is initiated by the originator of the call. The feature is also known as "Three-Way Calling."
2. Hardware, often referred to as peripheral equipment, that is added to a system to improve its performance, add memory or increase its capabilities. Voice mail, Automated Attendant and Call Detail Recording Equipment are examples of PBX add-on devices. Lucent, Nortel and some other manufacturers call them applications processors.
3. A call center/marketing term. A technique to increase the revenue of an order, for example, two dozen instead of one dozen or, two green shirts bought and sold with matching green tie.

Add-on Conference A PBX feature. Almost always used in conjunction with another feature called consultation hold, this feature allows an extension user to add a third person to an existing two-person conversation. The user places an existing central office call or internal call on Hold, and obtains system dial tone. The user can then call another internal extension or an outside party. After speaking with the "consulted" party, the originating phone reactivates the initiating command (typically a button push) and creates a three-party conference with the call previously placed on Hold.

Add-on Conference — Intercom Only Allows a telephone user to add someone else to an existing intercom (within-the-same office) conversation.

Add-on Data Module Plug-in circuit cards which allow a PBX to send and receive analog (voice) and digital (data) signals.

Added Bit A bit delivered to the intended destination user in addition to intended user information bits and delivered overhead bit. An added bit might be used to round out the number of bits to some error checking scheme, for example.

Additional Cooperative Acceptance Testing ACAT. A method of testing switched access service that provides a telephone company technician at the central office and a carrier's technician at its location, with suitable test equipment to perform the required tests. ACAT may, for example, consist of the following tests:

- Impulse Noise
- Phase Jitter
- Signal-to-C-Notched Noise Ratio
- Intermodulation (Nonlinear) Distortion
- Frequency Shift (Offset)
- Envelope Delay Distortion
- Dial Pulse Percent Break

Additional Period Billing periods charged after initial, first or minimum period on a call. Usually, long distance toll/DDD has a one-minute initial period at premium rate; subsequent "additional" minutes (period) are billed at a lower rate. Additional period billing increments vary by long distance company.

Additive Primaries By definition, three primary colors result when light is viewed directly as opposed to being reflected: red, green and blue (RGB). According to the tri-stimulus theory of color perception, all other colors can be adequately approximated by blending some mixture of these three lights together. This theory is harnessed in color television and video communications. It doesn't work so well in color printing where special colors are often printed separately.

ADDMD Administrative Directory Management Domain. A X.500 directory management domain run by a PTT (Posts, Telegraph, and Telephone administration) or other public network provider.

Address Characters identifying the recipient or originator of transmitted data. An address is the destination of a message sent through a communications system. A telephone number is considered the address of the called person. In computer terms, an address is a set of numbers that uniquely identifies the physical or logical location of something — a workstation on a LAN, a location in computer memory, a packet of data traveling through a network. This location can be specifically referred to in a software program. IEEE 802.3 and 802.5 recommend having a unique address for each device worldwide. An address may also denote the position of data in computer memory or the data packet itself while in transit through a network.

Address Complete Message ACM. A CCS/SS7 signaling message that contains call-status information. This message is sent prior to the called customer going off-hook.

Address Field In data transmission, the sequence of bits immediately following the opening flag of a frame identifying the secondary station sending, or designated to receive, the frame.

Address Field Extension EA. A Frame Relay term defining a 2-bit field in the Address Field, identifying the fact that the address structure is extended beyond the 2-octet default. Frame Relay standards provide for extension of the address field up to 60 bits, which extension will be implemented as the popularity of Frame Relay grows, placing pressure on the standard addressing convention.

Address Filtering A way of deciding which data packets are allowed through a device. The decision is based on the source and destination MAC (Media Access Control, the lower part of ISO layer two) addresses of the data packet.

Address Mapping Technique that allows different protocols to interoperate by translating addresses from one format to another. For example, when routing IP over X.25, the IP addresses must be mapped to the X.25 addresses so that the IP packets can be transmitted by the X.25 network. See also address resolution.

Address Mask An electronic messaging term. A bit mask used to select bits from a network address (e.g. Internet) for sub-net addressing. The mask is 32 bits long and selects the network portion of the address and one or more bits of the local portion. Sometimes called sub-net mask.

Address Prefix An ATM term. A string of 0 or more bits up to a maximum of 152 bits that is the lead portion of one or more ATM addresses.

Address Resolution 1. An internetworking term. A discovery process used when, as in LAN protocols such as TCP/IP and IBM NetBIOS, only the Network Layer address is known and the MAC address is needed to enable delivery to the correct device. The originating end station sends broadcast packets with the device's NLA to all nodes on the LAN; the end station with the specified NLA address responds with a unicast packet, addressed to the originating end station, and containing the MAC address. See Address Resolution Protocol.

2. An ATM term. Address Resolution is the procedure by which a client associates a LAN destination with the ATM address of another client or the BUS.

Address Resolution Protocol The Internet protocol used to map dynamic Internet addresses to physical (hardware) addresses on local area networks. Limited to networks that support hardware broadcasts.

Address Signaling Signals either the end user's telephone or the central office switching equipment that a call is coming in.

Address Signals Address signals provide information concerning the desired destination of the call. This is usually the dialed digits of the called telephone number or access codes. Typical types of address signals are DP (Dial Pulse), DTMF, and MF.

Address Space The amount of memory a PC can use directly is called its address space. MS-DOS can directly access 1024K of memory (one megabyte). A protected mode control program like Microsoft Windows 3.x or OS/2 can directly address up to 16 megabytes of memory. Here is a definition of address space, as supplied by the Personal Computer Memory Card International Association (PCMCIA) as address space applies to PCMCIA cards: "An address space is a collection of registers and storage locations contained on a PC Card which are distinguished from each other by the value of the Address Lines applied to the Card. There are three, separate, address spaces possible for a card. These are the Common Memory space, the Attribute Memory space and the I/O space."

Addressable Programming A cable TV (CATV) industry term. A subscriber orders a movie or sports event. He does that calling a phone number (generally an 800 number). A computer answers, grabs the calling number, confirms the request, then hangs up. The computer passes the request onto the cable company's computer, which checks the calling phone number against its accounting records. If the subscriber has good credit, the cable company sends a coded message down its cable network to the caller's set-top cable box/converter. The message temporarily enables that particular converter to descramble the channel offering the desired program.

Addressability 1. In computer graphics, the number of addressable points on a display surface or in storage.
2. In micrographics, the number of addressable points, within a specified film frame, written as follows: the number of addressable horizontal points by the number of addressable vertical points, for example, 3000 by 4000.
3. A cable TV term. The capability of controlling the operation of cable subscriber set-top converters by sending commands from a central computer. Such addressability is absolutely require for a cable system to offer pay-per-view services.

Addressable Point In computer graphics, any point of a device that can be addressed. See Addressability.

Addressee The intended recipient of a message.

Addressing Refers to the way that the operating system knows where to find a specific piece of information or software in the application memory. Every memory location has an address.

ADF Automatic Document Feeder

ADH Average Delay to Handle. Average time a caller to an automatic call distributor waits before being connected to an agent.

Adherence A term used in telephone call centers to connote whether the people working in the center are doing what they're meant to be doing. Are they at work? Are they on break? Are they answering the phone? Are they at lunch? All these activities are scheduled by workforce management software. If they're in line, the workers are "in adherence." If not, they're "out of adherence." See Adherence Monitoring.

Adherence Monitoring Adherence monitoring means comparing real-time data coming out an ACD with forecast call volumes, forecast service levels and forecast workforce employment levels. The idea is to see if the people, the calls and the system are working as forecast. This a measure of how well your forecasting work. You need to know how well it works since it's your forecasting on which you base your employment. See Adherence.

Adjacency Relationship formed between selected neighboring routers and end nodes for the purpose of exchanging routing information. Adjacency is based upon the use of a common media segment.

Adjacent Cell A cellular radio term. Two cells are adjacent if it is possible for a Mobile End System (M-ES) to maintain continuous service while switching from one cell to the other.

Adjacent Channel Interference When two or more carrier channels are placed too close together, they interfere with each other and mess up each other's conversations.

Adjacent MD-IS A cellular radio term. Two Mobile Data Intermediate Systems (MD-ISs) are adjacent if each MD-IS controls one of a pair of adjacent cells.

Adjacent MTA An MTA (Message Transfer Agent) that directly connects to another MTA. A Message Transfer Agent operated by a public service provider or PTT (Post, Telegraph, and Telephone administration), or a client MTA.

Adjacent Nodes 1. In SNA, nodes that are connected to a given node with no intervening nodes.
2. In DECnet and OSI, nodes that share a common network segment (in Ethernet, FDDI, or Token Ring networks).

Adjacent Signaling Points Two CCS/SS7 signaling points that are directly interconnected by signaling links.

Adjunct 1. Network system in the Advanced Intelligent Network Release 1 architecture that contains SLEE (Service Logic Execution Environment) functionality, and that communicates with an Advanced Intelligent Network Release 1 Switching System in processing AIN Release 1 calls. Definition from Bellcore. See also Adjunct Processor.
2. An auxiliary device connected to the ISDN set, such as a speakerphone, headset adapter, or an analog interface.

Adjunct Key System A system installed behind a PBX or a Centrex. Such key system provides the users with several more features than the PBX or Centrex. Not a common term today.

Adjunct Processor 1. A computer outside a telephone switching system that "talks" to the switch and gives it switching commands. An adjunct processor might include a database of customers and their recent buying activities. If the database shows that a customer lives in Indiana, the call from the customer might be switched to the group of agents handling Indiana customers. Adjunct processors might also be concerned with energy management, building security etc.
2. An AIN (Advanced Intelligent Network) term for a decentralized SCP (Signal Control Point). An Adjunct Processor supports AIN services which are limited to one or more SSPs (Service Switching Points), which are SS7-equipped Central Office PSTN switches. Where multiple SSPs are supported, they typically comprise a regional network grouping. Adjunct Processors can include routing logic or call authorization security specific to a particular geographic area, providing switches with switching commands.

Adjunct System Application Interface See ASAI.

Adjusted Ring Length When a segment of Token Ring (in practice a dual ring) trunk cable fails, a function known as the Wrap connects the main path to the backup path. In the worse case — the longest path — would occur if the shortest trunk cable segment failed, so ARL is calculated during network design to ensure the network will always work.

ADK Application Definable Keys

ADM Add/Drop Multiplexer. A SONET/SDH term for a device which can either insert or drop DS1, DS2, and DS3 channels or SONET signals into/from a SONET bit stream. The ADM literally can reach up into the SONET pipe and extract a DS1-level signal, without going through the rigorous process of demultiplexing and remultiplexing which is required in the traditional T/E-carrier world. While the devices are much more complex than are TDMs, the process is much faster, induces no signal delay, creates no signal errors. ADM also provide for dynamic bandwidth allocation, optical hubbing, and ring protection.

ADMD Administration Management Domain. An X.400 Message Handling System public carrier. Examples include MCImail and ATTmail in the U.S., British Telecom's Gold400mail in the U.K. The ADMDs in all countries worldwide together provide the X.400 backbone.

Admin Administration.

Adminisphere The rarefied organizational layers beginning just above the rank and file. Decisions that fall from the adminisphere are often profoundly inappropriate or irrelevant to the problems they were designed to solve. This definition from Wired Magazine.

Administrable Service Provider An SCSA definition. A service provider which supports administrable services (for example, SCSA Call Router).

Administration 1. The method of labeling, identifying and documenting an organization's voice/data communications cabling infrastructure.

2. A term used by the telephone industry to program features into a phone system. On a Northern Telecom Norstar system, administration includes making settings on 1. System speed dial; 2. Names or phones; 3. Time and date; 4. Restrictions; 5. Overrides; 6. Permissions; 7. Night Service and 8. Passwords.

Administration By Telephone The capability for the system administrator to perform most routine system administrative functions remotely from any Touch Tone pad. Such functions include mailbox maintenance (e.g. create, delete, set password, set class of service, etc.) and disk maintenance.

Administration Directory Management Domain Administrative Directory Management Domain. A X.500 directory management domain run by a PTT (Posts, Telegraph, and Telephone administration) or other public network provider.

Administration Sub-system Part of Lucent's premises distribution system that distributes hardware components for the addition or rearrangement of circuits.

Administrative Alerts A Window NT term. Administrative alerts relate to server and resource use; they warn about problems in areas such as security and access, user sessions, server shutdown because of power loss (when UPS is available), directory replication, and printing. When a computer generates an administrative alert, a message is sent to a predefined list of users and computers.

Administrative Distance Cisco defines administrative distance as a rating of the trustworthiness of a routing information source. Administrative distance is often expressed as a numerical value between 0 and 255. The higher the value, the lower the trustworthiness rating.

Administrative Domain 1. AD. A group of hosts and networks operated and managed by a single organization. An Internet term.

2. An ATM term. A collection of managed entities grouped for administrative reasons.

Administrative Management Domain An X.400 electronic mail term: A network domain maintained by a telecommunications carrier.

Administrative Point A location at which communication circuits are administered, i.e. rearranged or rerouted, by means of cross connections, interconnections, or information outlets.

Administrative Service Logic Program ASLP. The SLP responsible for managing the feature interactions between Advanced Intelligent Network AIN Release 1 features resident on a single SLEE (Service Logic Execution Environment).

Administrative Subsystem That part of a premises distribution system where circuits can be rearranged or rerouted. It includes cross connect hardware, and jacks used as information outlets.

Administrative Weight A value set by the network administrator to indicate the desirability of a network link. One of four link metrics exchanged by PTSPs to determine the available resources of an ATM network. See PTSP.

Administrator The individual responsible for managing the local area network (LAN). This person configures the network, maintains the network's shared resources and security, assigns passwords and privileges, and helps users.

Administrivia A silly term for administrative tasks, most often related to the maintenance of mailing lists, digests, news gateways, etc. An Internet term.

ADML Asymmetric Digital Microcell Link. A Bellcore standard for Wireless Local Loop (WLL). Using low-power, omnidirectional radio systems, ADML can be deployed to cover an area as large as 1 mile in radius. ADML supports as much as 1 Gbps aggregate bandwidth, providing individual users with bandwidth in radio channels as great as T-1 (1.544 Mbps). See also Wireless Local Loop and LMDS.

ADN Advanced Digital Network. ADN is Pacific Bell of California's low-cost leased 56 Kbps digital service. ADN is available for intraLATA calls.

ADP 1. Apple Desktop Bus. A synchronous serial bus allowing connection of the Mac keyboard, mouse and other items to the CPU. A Mac keyboard or mouse is called an ADB device. Contrast with peripherals, which attach through the SCSI interface.

2. Automatic Data Processing. The same as DP, data processing.

3. The name of a company which processes my pay check.

ADPCM Adaptive Differential Pulse Code Modulation. A speech coding method which uses fewer bits than the traditional PCM (Pulse Code Modulation). ADPCM calculates the difference between two consecutive speech samples in standard PCM coded telecom voice signals. This calculation is encoded using an adaptive filter and therefore, is transmitted at a lower rate than the standard 64 Kbps technique. Typically, ADPCM allows an analog voice conversation to be carried within a 32Kbit digital channel; 3 or 4 bits are used to describe each sample, which represents the difference between two adjacent samples. Sampling is done 8,000 times a second. ADPCM, which many voice processing makers use, allows encoding of voice signals in half the space PCM allows. In short, ADPCM is a reduced bit rate variant of PCM audio encoding (see also PCM).

ADQ Average Delay in Queue. An important measure of the customer responsiveness of a call center. See also ASA, Average Speed of Answer.

ADRMP (pronounced add-rump) AutoDialing Recorded Message Player. A device that calls a bunch of telephone numbers and upon connection will play a message to the answering person. ADRMPs are used for lead solicitation and message delivery. They are often unpopular due to their indiscriminate dialing pattern and random message playing.

ADS AudioGram Delivery Services.

ADSI Analog Display Services Interface. ADSI is a Bellcore standard defining a protocol on the flow of information between something (a switch, a server, a voice mail system, a service bureau) and a subscriber's telephone, PC, data terminal or other communicating device with a screen. The simple idea of ADSI is to add words to, and therefore a modicum of simplicity of use to a system that usually uses only touchtones. Imagine a normal voice mail system. You call it. It answers with a voice menu. Push 1 to listen to your messages, 2 to erase them, 3 to store them, 4 to forward them, etc. It's confusing. You have to remember which is which. ADSI is designed to solve that. It's designed to send to your phone's screen the choices in words that you're hearing. You then have the choice of responding to what you hear or what you see. Your response is the same — a touchtone button. ADSI's signaling is DTMF and standard Bell 202 modem signals from the service to your 202-modem equipped phone. From the phone to the service it's only touchtone. With ADSI, you don't hear the modem signaling because every time the service gets ready to send you information, it first sends a "mute" tone. ADSI works on every phone line in the world. For ADSI to work visually, you'll need a special ADSI-equipped phone (Nortel has one showing 8 lines by 20 characters) or a piece of ADSI software in your PC. The nice feature of ADSI is that the standard is so flexible, it can work on cheap phones with a small display and more expensive phones with a bigger display and on a PC with a real big display.

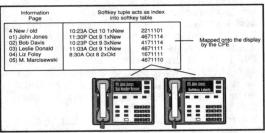

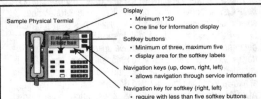

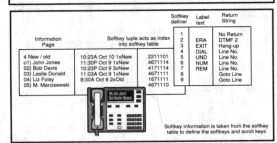

ADSL Asymmetric Digital Subscriber Line. An evolving high-speed transmission technology originally developed by Bellcore and now standardized by ANSI as T1.413; ETSI (European Technical Standards Committee) contributed an Annex to the standards to reflect European requirements. ADSL (a) uses existing UTP copper wires from the telephone company central office to the subscriber's premises, (b) involves electronic equipment in the form of ADSL modems at both the central office and the subscriber's premises, (c) sends high speed digital signals up and down those copper wires, and (d) sends more information one way than the other — hence the term "asymmetric." The original speed specs for ADSL was T-1 (1.536 Mbps) downstream from the carrier to the subscriber's premises and 16 Kbps upstream. While still in field trial status, ADSL is available in a variety of configurations.

The downstream bearer (information-bearing) channel provides bandwidth in T-1/E1 increments (NxT-1/E1), which is designed to maintain consistency with the existing digital hierarchies and which, therefore, offers the advantage of

backward compatibility with the legacy networks. In the U.S., the minimum configuration is T-1 at 1.536 Mbps of usable bandwidth; higher-speed configurations are 3.072 Mbps, 4.608 Mbps, and 6.144 Mbps, which is the maximum rate specified. Outside the U.S., the downstream channel is specified at either 2.048 Mbps or 4.096 Mbps. However, products are available with downstream rates of as much as 8 Mbps. The upstream capabilities of ADSL are through full duplex (bidirectional) channels. These channels include the "C Channel" for POTS, and an optional channel for purposes which include data communications and videoconferencing. The C Channel is either 16 Kbps or 64 Kbps, the latter being appropriate for digital voice over the PSTN in the U.S. based on the PCM encoding standard. The optional channel can be configured at 160 Kbps, 384 Kbps, 544 Kbps, or 576 Kbps. In total, therefore, the total capacity of the bidirectional channels can be as much as 640 Kbps. The POTS C Channel is split off from the digital ADSL modem by filters, thereby ensuring the survival of POTS service in the event that the ADSL devices fail.

Performance at these incredible transmission rates, of course, is subject to the condition of the twisted-pair cable plant. Factors which affect performance include length of the loop, wire gauge (diameter), presence of bridge taps, and cross-coupled interference (NEXT and FEXT). Assuming no bridge taps and assuming 24-gauge copper, ADSL will deliver downstream 1.5/2.0 Mbps over a distance of 18,000 ft. (5.5 km.); at 6.1 Mbps, 12,000 ft. (3.7 km.) is the maximum length of the loop. Where the length of the loop exceeds those maximums, the achievable transmission rate drops precipitously due to signal attenuation and associated error performance. Error performance is addressed through FEC (Forward Error Correction), thereby maximizing throughput.

RADSL (Rate-Adaptive DSL), a non-standard variation on the ADSL theme, is in development. RADSL employs intelligent ADSL modems which can sense the performance of the copper loop and adjust transmission speed accordingly. In the future, it is suggested that these devices will be able adjust dynamically as the performance of the loop varies during a session, much as does a V.34 modem.

Higher bit rate DSL versions, known as VDSL (Very High Data Rate Subscriber Line) also exist. VDSL standards provide downstream configurations of 13-52 Mbps, and upstream configurations of 1.5-2.3 Mbps. Non-standard versions, offering still higher-speed speeds, can provide as much as 155 Mbps downstream, with a 15 Mbps bidirectional channel, although the maximum distance is reduced to 500 meters.

Special electronics at both ends of the connection are required in order to accomplish the minor miracle of ADSL. At the carrier end of the connection is placed an ATU-C (ADSL Termination Unit-Central Office), while an ATU-R (ADSL Termination Unit—Remote) is placed at the customer premise. In order to achieve such a high data rate over UTP, relatively sophisticated compression techniques must be employed. While the standard calls for use of DMT (Discrete Multi-Tone), DMT implementations have experienced some difficulty. Lucent and others manufacture ADSL gear employing CAP (Carrierless Amplitude/Phase modulation). In fact, most RBOC field trials are based on CAP ADSL. CAP is being proposed to ANSI as an alternative ADSL standards.

The application is in the world of the incumbent LECs, which feel the pressure to extend the life of the embedded cable plant, of which there reportedly are several billion miles installed in the domestic local loop. Rather than being forced to upgrade that cable plant to coax or fiber, the LECs would like to use it in a convergence scenario for support not only of voice, but also high(er) speed data, Internet access, remote LAN access, video-on-demand, and entertainment TV (cable TV). The very substantial downstream channel provided by ADSL certainly will support a digital video channel, the primary reason for its development. However, recent competition from DBS (Direct Broadcast Satellite) systems have eroded the CATV business, making the widespread implementation of ADSL somewhat more questionable than it was only a few years ago.

CLECs (Competitive Local Exchange Carriers) and ISPs (Internet Service Providers) also have expressed interest in ADSL. IN fact, a number of them have leased dry copper loops from the LECs, which they have terminated on their own ATU-Cs collocated in the LEC COs under the terms of interconnection and collocation agreements. As the LECs see this as direct competition, several of them lately have refused to lease dry copper.

According to Nortel, initial ADSL field trails and business cases have focused on ADSL's potential for Video on Demand service, in competition with cable pay-per-view and neighborhood video rental stores. But ADSL offers a wide range of other applications, including education and healthcare. Once telephone companies are able to deliver megabits to the home, Nortel expects an explosion in potential applications including work-at-home access to corporate LANs, interactive services such as home shopping and home banking and even multi-party video gaming, interactive travelogues, and remote medical diagnosis. Multimedia retrieval will also become possible, enabling the home user to browse through libraries of text, audio, and image data — or simply subscribe to CD-quality music services. In the field of education, ADSL could make it possible to provide a low-cost "scholar's workstation" — little more than a keyboard, mouse, and screen — to every student, providing access to unlimited computer processing resources from their home. See also ADSL Forum, ADSL Lite, ATU-C, ATU-R, CDSL, HDSL, IDSL, RADSL, SDSL, Splitter, VDSL and Asymmetric Digital Subscriber Line Transceiver and G.990.

ADSL Forum The ADSL Forum is an industry association formed to promote the ADSL concept and to facilitate the development of ADSL system architectures and protocols for major ADSL applications. As the forum comprises competing companies, it does not publish material that discusses line codes or basic modulation systems, or any other material that addresses individual company or product attributes. www.adsl.com See also ADSL and VDSL.

ADSL Lite Also known as G.lite, Universal ADSL and Splitterless ADSL, a developing proposal of the UAWG (Universal ADSL Working Group) for a simplified version of ADSL. The standard is anticipated to be in the form of an interoperable extension of ANSI T1.413 ADSL. Notably, ADSL Lite will be application-specific, specifically for Internet access, which is very much unlike the original ADSL concept. ADSL Lite will allow access to the Internet through a modem (either internal or external) operating at speeds of as much as 1.5 Mbps over existing twisted pair local loops of relatively short length and of good quality. See also ADSL and UAWG.

ADSP Apple Datastream Protocol. A transport mechanism for interprocess communications between Apple Macintosh and Dec Vax minicomputers.

ADSTAR Automated Document STorage And Retrieval.

ADSU ATM DSU. Terminal adapter used to access an ATM network via an HSSI-compatible device. See also DSU.

ADT Abstract Data Type.

ADTF ACR Decrease Time Factor: This is the time permitted between sending RM-cells before the rate is decreased to ICR (Initial Cell Rate). The ADTF range is .01 to 10.23 sec. with granularity of 10 ms.

ADU Asynchronous Data Unit.

Advance Replacement See Advance Replacement.

Advance Replacement Warranty A warranty service whereby the dealer sends the customer a replacement component before the customer returns the defective product. This not only accelerates the replacement time, but also helps the buyer if the component is vital. When you buy vital telecom gear, it's good to check that your equipment has an Advance Replacement Warranty or Guarantee.

Advanced Branch Exchange ABX. An uncommon term meaning a private branch exchange (PBX) with advanced features normally including the ability to handle both voice and data in some sort of integrated way.

Advanced Intelligent Network AIN. The local Bell telephone companies' architecture for the 1990s and beyond. See AIN for a much fuller explanation.

Advanced Interactive Executive AIX. An IBM version of UNIX. AIX runs on PS/2 computers, IBM workstations, minicomputers, and mainframes.

Advanced Interactive Video AIV. Interactive videodisc format and system using LV-ROM, a method of storing analog video, digital audio and digital data on a single videodisc. The system was developed by Philips UK, the British Broadcasting Corporation, Acorn Computer, and Logica Ltd. Most prominent application was the BBC's Domesday Project.

Advanced Power Management An industry standard for taking advantage of a computer's power saving features. Used particularly in battery-powered laptops.

Advanced Private Line Termination An AT&T/Lucent term which means the PBX user gets access to all the services of an Enhanced Private Switched Communications Services (EPCS) network. It also works when it is associated with AT&T's Common Control Switching Arrangement (CCSA) network.

Advanced Television System ATV. Any television technology that provides audio and video quality that is better than can be provided by the current television broadcast system, or that otherwise enhances the current system. This definition, courtesy the FCC. Your taxpayer monies paid for it.

Advancenet An Ethernet-based local area network from Hewlett Packard, Palo Alto, CA. See Ethernet.

Advertising A packet switched networking term. Advertising is a process in which routing or service updates are sent at specified intervals so that other routers on the network can maintain lists of usable routes.

Advisory Tones Signals such as dial tone, busy, ringing, fast-busy, call-waiting, camp-on and all the other tones your telephone system uses to tell you that something is happening or about to happen.

AE Account Executive. A fancy, schmanzy name for a salesperson. The idea is that the customer is an "account," and the salesman is the executive running the account. Telephone companies call their salespeople account executives — especially on the equipment and non-long distance side.

AEB Analog Expansion Bus. The analog voice processing bus designed by Dialogic which allows multiple cards to route audio signals within a PC. It is used to interface DTI/124 and D/4x voice response component boards which fit in an AT-expansion slot of a PC. See also PEB and SCSA which are more modern digital expansion buses.

AEC Acoustic Echo Cancellation

AECS Plan Aeronautical Emergency Communications System Plan. The AECS Plan provides for the operation of aeronautical communications stations on a voluntary, organized basis to provide the President and the Federal Government, as well as heads of state and local governments, or their designated representatives, and the aeronautical industry, with a means of communicating during an emergency.

AEMIS Automatic Electronic Management Information System. This was the first computerized UCD/ACD reporting system introduced by AT&T for CO UCD (Uniform Call Distribution). This package was updated to become the PRO 150/500 system for UCD management on the Dimension PBX/UCD. AEMIS was the successor to the FADS or Force Administration Data System. It was an electro-mechanical system of peg counters and different colored busy lamp fields used to note trunk and position status.

AEP AppleTalk Echo Protocol. Used to test connectivity between two AppleTalk nodes. One node sends a packet to another node and receives a duplicate, or echo, of that packet.

Aerial Cable Cables strung outside and overhead. They're called aerial even though they only hang from poles or buildings. Some aerial cable hangs by its own strength. Some is supported by steel wire above it. Stringing aerial cable is cheaper than burying it, though buried cable lasts longer.

Aerial Cross Box Also called a tree stand. A cross box on a pole. Used when there's a narrow easement.

Aerial Distribution Method A method of running cables through the air, typically pole-to-pole. The old fashioned way. Some phone companies say aerial cable is more reliable than underground. Certainly, it's cheaper to fix or add to. It just looks less appetizing.

Aerial Insert In a direct-buried or underground cable run, an aerial insert is a cable rise to a point above ground, followed by an overhead run, e.g., on poles, followed by a drop back into the ground. An aerial insert is used in places where it is not possible or practical to remain underground, such as might be encountered in crossing a deep ditch, canal, river, or subway line.

Aerial Service Wire Splice A device used to splice aerial service wire and attached to the aerial wire. It's also called a football or a potato. Why? Because that's it shape.

Aeronautical Advisory Station An aeronautical station used for advisory and civil defense communications primarily with private aircraft.

Aeronautical Broadcast Station An aeronautical station which makes scheduled broadcasts of meteorological information and notices to airmen. In certain instances, an aeronautical broadcast station may be placed on board a ship.

Aeronautical Earth Station An earth station in the fixed-satellite service, or, in some cases, in the aeronautical mobile-satellite service, located at a specified fixed point on land to provide a feeder link for the aeronautical mobile-satellite service.

Aeronautical Emergency Communications System (AECS) Plan. The AECS Plan provides for the operation of aeronautical communications stations, on a voluntary, organized basis, to provide the President and the Federal Government, as well as heads of state and local governments, or their designated representatives, and the aeronautical industry with communications in an emergency.

Aeronautical Fixed Service A radiocommunication service between specified fixed points provided primarily for the safety of air navigation and for the regular, efficient and

economical operation of air transport.

Aeronautical Fixed Station A station in the aeronautical fixed service.

Aeronautical Mobile OR (off-route) Service An aeronautical mobile service intended for communications, including those relating to flight coordination, primarily outside national or international civil air routes.

Aeronautical Mobile R (route) Service An aeronautical mobile service reserved for communications relating to safety and regularity of flight, primarily along national or international civil air routes.

Aeronautical Mobile Satellite Service A mobile satellite service in which mobile Earth stations are located on board aircraft; survival craft stations and emergency position-indicating radiobeacon stations may also participate in this service.

Aeronautical Mobile Satellite (OR) (off-route) Service An aeronautical mobile-satellite service intended for communications, including those relating to flight coordination, primarily outside national and international civil air routes.

Aeronautical Mobile-Satellite (R) (route) Service An aeronautical mobile-satellite service reserved for communications relating to safety and regularity of flight, primarily along national or international civil air routes.

Aeronautical Mobile Service A mobile service between aeronautical stations and aircraft stations, or between aircraft stations, in which survival craft stations may participate; emergency position-indicating radiobeacon stations may also participate in this service on designated distress and emergency frequencies.

Aeronautical Multicom Service A mobile service not open to public correspondence, used to provide communications essential to conduct activities being performed by or directed from private aircraft.

Aeronautical Radio Inc. ARINC. The organization that coordinates the design and management of telecommunications systems for the airline industry. It's one of the largest buyers of telecommunications services and equipment in the world.

Aeronautical Radionavigation-Satellite Service A radionavigation-satellite service in which Earth stations are located on board aircraft.

Aeronautical Radionavigation Service A radionavigation service intended for the benefit and for the safe operation of aircraft.

Aeronautical Station A land station in the aeronautical mobile service. In certain instances, an aeronautical station may be located on board ship or on a platform at sea.

Aerospace Air force publicists coined the term "aerospace" to convince everyone that space was the business of those who fly in the air. According to the Economist Magazine, the "aerospace industry" was quickly accepted into the language, perhaps because President Eisenhower's alternative, the "military industrial complex," sounded rather more sinister. After the Apollo program, which ended in 1972, the "space" in aerospace often seemed like a syllable tacked on to make building airplanes sound grander. But the growth in satellite use in the 1980s made space a respectable business in its own right. In America as of writing in the fall of 1991, the annual sales of space hardware are now bigger than those of civilian aircraft.

AES Advanced Encryption Standard. A standard for encryption which is intended to replace DES (Data Encryption Standard), a standard developed by IBM in 1977 and thought to be virtually uncrackable until 1997. The AES standard, which is expected to be developed and released in 2000, will be a symmetric, or private key, algorithm. It also will be a block cipher supporting key lengths ranging from 128 to 256 bits, and variable-length blocks of data. See also Block Cipher, DES, Encryption, and Private Key.

AET Application Entity Title. The authoritative name of an OSI application entity, usually a Distinguished Name from the Directory.

AF 1. Audio Frequency. The range of frequencies which theoretically are audible to the human ear; i.e., 30 Hz - 20 KHz. Truly high fidelity audio covers the entire range. Full AF is not practical over the PSTN, as too much bandwidth required. Most of us can't hear the full AF range, anyway. As you get older, your hearing deteriorates. See also Bandwidth.
2. Assigned Frame. Motorola definition.

AFAIK As Far As I Know.

AFCEA Armed Forces Communications and Electronics Association. An organization of military communications personnel and suppliers who fulfill the specialized needs of government and military communications. They run an big convention each year in Washington in May-June. www.afcea.org

AFE See Analog Front End.

AFI An ATM term. Authority and Format Identifier: This identifier is part of the network level address header.

Affiliate 1. This definition from the Telecommunications Act of 1996. The term `affiliate' means a person that (directly or indirectly) owns or controls, is owned or controlled by, or is under common ownership or control with, another person. For purposes of this paragraph, the term `own' means to own an equity interest (or the equivalent thereof) of more than 10 percent. See the Telecommunications Act Of 1996.
2. A broadcast TV station not owned by a network, but one which includes the network's programs and commercials in its programming schedule.

Affiliated Sales Agency ASA. A term for a company which resells the service of a phone company. Typically, the phone company pays the ASA a commission. Sometimes the commission is so large that it blurs the thinking of the ASA into recommending to its customers telecom products and services they would be better without.

AFI Authority and Format Identifier. The portion of an NSAP format ATM address that identifies the type and format of the IDI portion of an ATM address. See also IDI and NSAP.

AFIPS American Federation of Information Processing Societies. A national, highly-respected organization formed by data processing societies to keep abreast of advances in the field. AFIPS organizes one of the biggest trade shows in the data processing industry — the NCC (National Computer Conference).

AFNOR Acronym for Association Francais Normal. France's national standards-setting organization.

AFP AppleTalk File Protocol. Apple's network protocol, used to provide access between file servers and clients in an AppleShare network. AFP is also used by Novell's products for the Macintosh.

AFT Automatic Fine Tuning; SEE AFC.

After-call Wrap-up The time an employee spends completing a transaction after the call has been disconnected. Sometimes it's a few seconds. Sometimes it can be minutes. Depends on what the caller wants.

AGC Automatic Gain Control. There are two electronic ways you can control the recording of something — Manual or Automatic Gain Control (AGC). AGC is an electronic circuit in tape recorders, speakerphones, and other voice devices which is used to maintain volume. AGC is not always a brilliant idea since it will attempt to produce a constant volume level, that is,

it will try to equalize all sounds — the volume of your voice, and, when you stop talking, the circuit static and/or general room noise which you do not want amplified. Never record a seminar or speech using AGC. The recording will be decidedly amateurish. Manual Gain Control means there is record volume control and is thus, preferred in professional applications.

AGCOMNET US Department of Agriculture's voice and data communications network.

Aged Packet A data packet which has exceeded its maximum predefined node visit count or time in the network.

Agent 1.The classic definition of an agent is an entity acting on behalf of another.
2. This term comes from the huge telephone call-in reservation centers which the airlines, hotels and car rental services run. An agent is the person who answers your call, takes your order or answers your question. Agents are also called Telephone Sales Representatives or Communicators. The term "agent" was first used in the airline business. It came from gate or counter ticket agent.
3. An "Agent" is the person or persons you have legally authorized to order your telephone service and equipment from telephone companies.
4. In the computer programming sense of the word, an agent acts on behalf of another person or thing, with delegated authority. The agent's goals are those of the entity that created it. An agent is an active object with a mission, but agents are abstractions that can be implemented in any way, whereas an object has a formal definition.
Business Week in its February 14, 1994 issue wrote, "It's what computer scientists call an 'agent' — a kind of software program that's powerful and autonomous enough to do what all goods robots should: help the harried humans by carrying out tedious, time-consuming, and complex tasks. Software agents just now emerging from the research labs can scan data banks by the dozen, schedule meetings, tidy up electronic in-boxes, and handle a growing list of clerical jobs."
4. Windows 95 Resource Kit defined agent slightly differently. It said that an agent was software that runs on a client computer for use by administrative software running on a server. Agents are typically used to support administrative actions, such as detecting system information or running services.

Agent Logon/Logoff A call center term. The agent begins their day by punching some buttons on their phone. This indicates to the automatic call distributor that they are now ready to take calls. Later in the day, they punch some other buttons and indicate to the ACD that they are now ready to stop working. This is called logoff.

Agent Sign On/Sign Off A feature which allows any ACD agent to occupy any position in the ACD without losing his or her personal identity. Statistics are collected and consolidated about this agent and calls are routed to this agent no matter where he sits or how many positions he may occupy at one time.

Aggregate AMA Record A telephone company AIN term. An AMA record generated to record multiple instances of service usage within a specified aggregation interval. It is created by formatting peg counts of AMA events.

Aggregate Bandwidth The total bandwidth of channel carrying a multiplexed bit stream. It includes the payload and the overhead. For example, a T-1 line has an aggregate full duplex bandwidth of 1.544 million bits per second.

Aggregate Rate The sum of the channel data rates for a given application.

Aggregation 1. An AMA (Automatic Message Accounting) function that accumulates AMA data, resulting in a less than detailed AMA record. Definition from Bellcore.

2. An ATM term. Token A number assigned to an outside link by the border nodes at the ends of the outside link. The same number is associated with all uplinks and induced uplinks associated with the outside link. In the parent and all higher-level peer group, all uplinks with the same aggregation token are aggregated.

Aggregation Device A specialized ISDN terminal adapter that can aggregate, or bond, the two B channels "on the fly" into a single higher-speed connection. some aggregation devices also include an Ethernet bridge, i.e. a connection to a local area network.

Aggregator A new breed of long distance reseller. An aggregator is essentially a sales agent for a long distance company. Here's how it works: The aggregator goes to a long distance company and says "May I sell your long distance service at a discount?" The long distance company says Yes! The aggregator hits the street and sells cut-rate long distance service to any and everyone. The long distance provider installs the service and bills it. The aggregator makes his profit by charging a fixed monthly service fee, a percentage of savings or some other arrangement. The key to it: The end user saves some money because his calls are "aggregated" with those of ALL the customers of the aggregator and the long distance company extends a bulk savings to the aggregator. Here's what TELECONNECT Magazine wrote about aggregators under the headline, "Aggregator Warning."

"Aggregators are companies which buy long distance wholesale and sell it retail. Aggregators exist because AT&T decided it wanted to win back long distance business it had lost. AT&T sliced its rates, liberalized its bulk billing rules and encouraged those consultants who had recommended their clients switch to MCI and Sprint to become aggregators. These consultant-turned aggregators simply solicit anyone's long distance business and add it to their collection. AT&T sends their end-user a bill and the consultant-turned-aggregator a commission check.

"Should you — as an end-user — consider buying your long distance from an aggregator? The simple answer is YES? AT&T's discounts are so deep it's not uncommon for a company using AT&T today directly to switch to billing through an aggregator and save 20% to 25% — with nothing of substance happening. They still get their bills from AT&T and they still place and receive calls on AT&T as they had been doing. No wires are touched. No routing is changed.

"What about the pitfalls? There are some: First, don't buy long distance that isn't billed directly by the long distance carrier providing the service. If the aggregator does the billing, there's too much opportunity for "mischief," says Dick Kuehn, Cleveland consultant. "There's opportunity for doing things like increasing each of your calls by 30 seconds. And because a user has no answer supervision on his call detail records, it's very hard for the user to figure his exact timing." The problem, says Dick, is there's no way for a user to verify his own bill. Dick says "Carriers are honest. Resellers (aggregators that bill) are open to question."

"TELECONNECT also believes you probably shouldn't deal with an aggregator who bills you a percentage of "savings." This is also open to abuse. There are so many rates, so many changes monthly, so many options that it's virtually impossible for the user to figure out what he would have paid had he not gone with the aggregator. The calculation is too open to abuse. TELECONNECT's feeling: pay a flat service fee.

"P.S. Imagine buying an automobile tire. The local garage car-

ries two options: a Bridgestone for $30 and a Pirelli for $100. You opt for the Bridgestone at $30. You've thus saved $70. The garage proprietor splits your "savings" with you and charges you $35.

By late 1992, the panapoly of companies in the long distance business — not only aggregators — had expanded dramatically. And confusion between companies and what they did became rife. All, of course, purport to save you money on your long distance bills. And many do. Here's a simple explanation of the major categories:

• CARRIER. Owns most of its circuits. Has own sales force and possibly independent sales agents. Best examples: AT&T, MCI, Allnet and Sprint.

• TRADITIONAL RESELLER. Rents/leases most circuits or buys bulk time from carrier. Resells under own brand name, has published prices, sends own bills. Appears to be (and for all practical purposes is) same as the carriers.

• AGGREGATOR. "Sponsor" who buys carrier's (typically AT&T) multi-location 800 or outbound service; enrolls other businesses as sites; volume discounts for all based on total calling at all sites. End user is still the carrier's, not the aggregator's. The carrier typically does the billing.

• REBILLER: (Also called "Switchless Reseller"). Buys service as multi-location customer from carrier. Signs up individual sites (just like aggregator). Generates own end-user bills. No switch or network, but does sales, customer service, billing for long distance calls. Sometimes the rebiller's bills are more detailed than the bills you get directly from the carrier.

• SALES AGENTS: Businesses or groups who are not direct employees of carrier, but who receive sales commissions from carrier. Customers belong to carrier and carrier does billing.

• OTHER THIRD-PARTY MARKETERS. Buying co-ops, user groups, long distance brokers, pyramid (legal) marketing systems, shared tenant providers, Centrex aggregators, affinity groups (like college alumni and church congregation groups).

Aging The change in properties of a material with time under specific conditions.

AGP Accelerated Graphics Port is based on PCI, but is designed especially for the throughput demands of 3-D graphics. Rather than using the PCI bus for graphics data, AGP introduces a dedicated point-to-point channel so that the graphics controller can directly access main memory. The AGP channel is 32 bits wide and runs at 66 MHz. This translates into a total bandwidth of 266 MBps, as opposed to the PCI bandwidth of 133 MBps. AGP also supports two optional faster modes, with throughputs of 533 MBps and 1.07 GBps. In addition, AGP allows 3-D textures to be stored in main memory rather than video memory. AGP has a couple important system requirements:

1. The chipset must support AGP.

2. The motherboard must be equipped with an AGP bus slot or must have an integrated AGP graphics system. The operating system must be the OSR 2.1 version of Windows 95, Windows 98 or Windows NT 5.0, soon to be known as Windows 2000..

AGP-enabled computers and graphics accelerators hit the market in August, 1997. However, there are several different levels of AGP compliance. The following features are considered optional:

Texturing: Also called Direct Memory Execute mode, allows textures to be stored in main memory.

Throughput: Various levels of throughput are offered: 1X is 266 MBps, 2X is 533 MBps; and 4X provides 1.07 GBps.

Sideband Addressing: Speeds up data transfers by sending

command instructions in a separate, parallel channel.

Pipelining: Enables the graphics card to send several instructions together instead of sending one at a time.

AGTK Application Generator ToolKit. A set of tools that are used to implement and modify a voice-processing application. It includes software to create the script and packages for the creation and editing of prompts. See Application Generator.

AHD Audio High Density. System of digital audio recording on grooveless discs, employing an electronically-guided capacitance pickup.

AHFG ATM-attached Host Functional Group: The group of functions performed by an ATM-attached host that supports the ATM Forum's specification for MPOA (Multiprotocol over ATM).

AHR Abbreviation for ampere hour, measurement of battery power: how much current may be drawn for an hour. Important specification for portable computers, cellular phones, etc.

AHT 1. Average Handle Time. The amount of time an employee is occupied with an incoming call. This is the sum of talk time and after-call-work time. Contrast with Average Holding Time. 2. See Average Holding Time.

AHT Distribution Average Handle Time Distribution. A set of factors (either 48 or 96) for each day of the week that defines the typical distribution of average handle times throughout the day. Each factor measure how far AHT in the half or quarter hour deviates from the AHT for day as a whole.

Ahoy See Hello.

AI Artificial Intelligence. Perhaps the next phase of computing. The present forms of AI in computer software are called Expert or Knowledge Based systems.

Ai An ATM term. Signaling ID assigned by Exchange A.

AIA 1. An M.100/S.100 definition. Application Interface Adapter: a component providing the client side of a client server connection to an S.100 server. See M.100 and S.100. 2. Automatic Internal Administration.

AICC Automatic Incoming Call Connection. A Rolm term for connecting an incoming call to the person's phone, without requiring them to press any keys.

AIDS 1. Access IDentifier. 2. A Trojan Horse software program (a virus) which caused extensive damage in December 1989.

AIFF Audio Interchange File Format. This audio file format was developed by Apple Computer for storing high-quality sampled audio and musical instrument information. It is also used by Silicon Graphics and in several professional audio packages. Played by a variety of downloadable software on both the PC and the Mac. See also ADPCM, PCM, sound, TrueSpeech, VOC, WAV and waveform.

AIIM Association for Information and Image Management.

AIM 1. Amplitude Intensity Modulation. 2. Association for Interactive Media. Originally called the Interactive Television Association (ITA). The AIM is a Washington association of companies and organizations involved with interactive media. According to the AIM CEO, "ITA has long been the industry's most forceful proponents of the view that high speed Internet and interactive television development are so interrelated that, from the customer's perspective, these services will be seamless." www.interactive hq.org

AIMS An Acronym for Auto Indexing Mass Storage. Indicates the AIMS Specification which is a standard card interface for storing large data such as image and multimedia files.

AIMUX ATM Inverse Multiplexing: A device that allows multiple T-1 or E1 communications facilities to be combined into a single broadband facility for the transmission of ATM cells.

AIN Advanced Intelligent Network. A term promoted by Bellcore (Bell Communications Research Inc.), adopted by Bellcore's original owners, the regional Bell holding companies, and by AT&T and virtually every other phone company to indicate the architecture of their networks for the 1990s and beyond. While every phone company has a different interpretation of what their AIN is, there seems to be two consistent threads. First, the network can affect (i.e. change) the routing of calls within it from moment to moment based on some criteria other than the normal, old-time criteria of simply finding a path through the network for the call, based on the number originally dialed. Second, the originator or the ultimate receiver of the call can somehow inject intelligence into the network and affect the flow of his call (either outbound or inbound). The concept of AIN is simple. Before calls are sent to their final destination, the network queries a database. "What should I do at this very moment with this phone call?" Depending on the response, depends the disposition of the call. That database may belong to the phone company. Or it may belong to the customer. It makes no difference, so long as they're connected. And various carriers (phone companies) have proposed and implemented various ways of joining these databases. Initial AIN services tend to be focused on inbound 800 toll-free calls. Although no two phone companies seem to have the same idea as to what an Advanced Intelligent Network is, (some call it just an Intelligent Network), it generally includes three basic elements:

1. Signal Control Points. SCPs. Computers that hold databases in which customer-specific information used by the network to route calls is stored.

2. Signal Switching Points. SSPs. Digital telephone switches, which can talk to SCPs and ask them for customer-specific instructions as to how the call should be completed.

3. Signal Transfer Points. STPs. Packet switches that shuttle messages between SSPs and SCPs.

All three communicate via out of band signaling, typically using Signaling System 7 (SS7) protocol. The AIN has increased in complexity, as carriers have added voice response equipment that can prompt callers to enter further instructions as to how they'd like their call handled. Despite the differences between AIN networks, all work fundamentally the same, according to Mark Langner at the time with TeleChoice, Verona, NJ. The SS7 identifies that a call requires intelligent network processing. The SSP creates a query to find out how this call should be handled. The query is passed via out-of-band signaling through STPs to an SCP. That interprets the query based on the criteria in its database and information provided by the SSP. Once the SCP has determined how the call is to be handled, it returns a message through STPs to the SSP. This message instructs the SSP how the call should be handled in the network. According to Langner, the number of actions that could take place at the SCP are truly infinite. The call could be translated into a different number for completion. It could be routed to a user's private network for on-net handling. It could be sent to a voice response unit in the carrier network, where a message is played to the caller. Or it could even be blocked, preventing completion of the call. Ericsson has done focus groups on Mobile Intelligent Network Services. Among the IN (Intelligent Network) services, Ericsson identified: • Enhanced number translation services functions • Enhanced screening services, i.e. selective call diversion • Selective forwarding of calls • Location-dependent call forwarding • Improvements to voice announcements

- Services to support fixed and mobile integration, i.e. personal communications services, PCS and universal personal telecommunications, UPT, and • Enhanced billing.

See IN, NCD, SCP, SiteRP, SS7, SSP, STP and the AIN definitions below.

AIN Release 0 Advanced Intelligent Network Release defined by individual Bell Operating Companies for initial deployment in 1991, or so. See AIN.

AIN Release 0.0 Advanced Intelligent Network Release based on Ameritech specifications with input from Bellcore and some vendors. Contains three trigger detection points. Deployed in 1992 (US) and end of 1993 (Canada). First service for this architecture was "Switch Redirect" for Bell Atlantic (for switch or line failure.) See AIN.

AIN Release 0.1 Advanced Intelligent Network Release provides for some additional functionality and more extensions to Rel 0.0. Contains 5 trigger detection points. See AIN.

AIN Rel 1.0 Advanced Intelligent Network Release target architecture for AIN. Contains 32 trigger detection points. (Hence Rel 0.0 & 0.1). See AIN.

AIN Release 1 Logical Resources For Bell Operating Companies, the logical network resources configured and updated to provide Advanced Intelligent Network Release 1 subscriber services (e.g., SLP and trigger data). Definition from Bellcore. See AIN.

AIN Release 1 Switching System An access tandem, local tandem or end office that contains an ASC (Advanced Intelligent Network Release 1 Switch Capabilities) functional group. Definition from Bellcore. See AIN.

AIN Release 2 An Advanced Intelligent Network Release for initial deployment in 1995, involving from AIN Release 1 and supporting an expanded range of information networking services from the Bell operating telephone companies. Definition from Bellcore. See AIN.

AIN Switch Capabilities ASC. A functional group residing in an Advanced Intelligent Network Release 1 Switching System that contains the Network Access, Service Switching, Information Management, Service Assistance and Operations FEs (Functional Entities). Definition from Bellcore. See AIN.

AINTCC Automated INTercept Call Completion. A new feature of Northern Telecom's central offices. The AINTCC feature provides options for connecting a caller automatically to an intercepted number after hearing an announcement, or connecting a caller to an intercepted number without an announcement. Not using an announcement makes the number change transparent to the caller. The called (intercepted) party then has the option of informing the caller of the number change.

AIOD Automatic Identification of Outward Dialing is the ability of the telephone system to know the specific extension placing a call. It's used as part of the process of recording the detail of each telephone call for billback and cost control purposes. See AIOD LEADS and CALL ACCOUNTING SYSTEM.

AIOD Leads Terminal equipment leads used solely to transmit automatic identified outward dialing (AIOD) data from a PBX to the public switched telephone network or to switched service networks (e.g., EPSDS), so that a vendor can provide a detailed monthly bill identifying long-distance usage by individual PBX extensions, tie-trunks, or the attendant.

AIR An ATM term. Additive Increase Rate: An ABR service parameter, AIR controls the rate at which the cell transmission rate increases.

Air Conditioning In the Department of Defense, air conditioning is a synonym for the term "environmental control," which is the process of simultaneously controlling the tem-

perature, relative humidity, air cleanliness, and air motion in a space to meet the requirements of the occupants, a process, or equipment.

Air Pressure Cable Telephone cable equipped with air-pressure equipment so the phone company can determine when there's a problem with the line. When a cable is cut, the pressure drops and the company is notified of the problem. Nitrogen is often used instead of air because nitrogen is non-corrosive. Nitrogen also prevents water entering the cable when there's a break.

Air Space Coaxial Cable One in which air is the essential dielectric material. A spirally wound synthetic filament of spacer may be used to center the conductor.

Air Time Time spent talking on a cellular network to calculate billing. See also AIRTIME.

Airbrush A computer imaging term. A fine-mist paint tool used to create halos, fog, clouds, and similar effects. Most paint programs let you control the size and shape of the application area. Some packages provide a transparency adjustment that determines the density of the applied color.

Aircraft Earth Station A mobile Earth station in the aeronautical mobile-satellite service located on board an aircraft.

AIRF Additive Increase Rate Factor: Refer to AIR.

Airline Mileage The monthly charge for many leased circuits is billed on the basis of "airline mileage" between the two points. Though it sounds as if it's the distance a crow would fly directly between the two points, when in reality, it is the distance in mileage between two Rate Centers whose position is laid down according to industry standards, originally created by AT&T. The entire U.S. is divided by a vertical and horizontal grid. The coordinates — vertical and horizontal — of each rate center are defined and applied to a square root formula which yields the distance between the two points. Think back to school. There's a right-angled triangle. At the top is one Rate Center. At the side is the other Rate Center. The horizontal is the horizontal coordinate. The vertical is the vertical coordinate. The formula is simple: Square the vertical distance. Square the horizontal distance. Add the two together. Then take their square root. That will give you the distance across the hypotenuse — the side opposite the right angle in the triangle. Thus, your "airline" mileage. For sample V and H city coordinates and the formula on how to calculate airline mileage, see V & H under the letter V.

Airline Miles See Airline Mileage.

Airlink A cellular radio term. Airlink is the physical layer radio frequency channel pair used for communication between the Mobile End System (M-ES) and the Mobile Data Base Station (MDBS).

Airlink Interface The Cellular Digital Pack Data (CDPD)-based wireless packet data service provider's interface for providing services over the airlink to mobile subscribers.

Airtime Actual time spent talking on a cellular telephone. Most cellular carriers bill their customers based on how many minutes of airtime they use each month. Whether the calls are incoming or outgoing makes no difference, the customer is still billed. Whether the calls are going to a toll-number or a toll-free 800 number also makes no difference. The customer racks up airtime and he pays. The more minutes of time spent talking on the phone, the higher the bill. Airtime charges during peak periods of the day in North America vary from 25 to 80 cents per minute. Most carriers offer a discount on these rates for off-peak usage. Some carriers offer a discount on these rates if the customer pays a higher minimum usage charge each month.

Airwave Airwave systems are transmission systems that use the "airwaves," rather than conductors, to transmit information. Airwave systems actually send information across "space," rather than through conductors. The term "airwave" comes from the fact that human speech, in its native form, is an acoustic means of communications which makes use of the physical matter in the air to conduct compression waves. From mouth (transmitter) to ear (receiver) the physical matter (e.g., molecules of oxygen, carbon dioxide and such) in the air actually carries, or conducts, the signal. Airwave transmission systems (e.g., microwave, satellite and infrared) support information transfer from transmitter to receiver through space, using electromagnetic energy in the form of radio or light signals. The presence of the physical matter which occupies the space between transmitter and receiver actually causes the signal to attenuate, or weaken. The term "airwave" persists, however. See also Free Space Communications.

AIS 1. Alarm Indication Signal. Formerly called a "Blue Alarm" or "Blue Signal." An AIS is a signal transmitted downstream informing that an upstream failure has been detected. AIS is a signal that replaces the normal traffic signal when a maintenance alarm indication has been activated. In ATM, an alarm indication signal is an all ones signal sent down or up stream by a device when it detects an error condition or receives an error condition or receives an error notification from another unit in the transmission path. See also Squelching.
2. Automatic Intercept System.

AIX Advanced Interactive eXecutive: IBM's implementation of UNIX. The Open Software Foundation (OSF) based its first operating system (OSF-1) on AIX. The next revision of the OSF operation system (OSF-2) will also be based on AIX with a Mach kernel (Mach was developed by Carnegie Mellon University).

AIW Application Implementer's Workshop. A group of vendors working with IBM to develop software and hardware consistent IBM's Advanced Peer-to-Peer Networking protocol.

AKO Bisync acknowledgment for even-numbered message.

Aka Denwa Japanese for a red telephone. Some coin phones in Japan are red and are known as aka denwas.

AL ATM Adaptation Layer. The third layer of the ATM Protocol Reference Model. The AAL layer is comprises the Convergence Sublayer (CS) and the Segmentation and Reassembly sublayer (SAR). In total, it is at this layer that multiple applications are converted to and from the ATM cell format. ATM Adaptation Layer sits above the ATM Layer, supporting higher-layer service requirements. For data communications services, the AAL defines a segmentation/reassembly protocol for mapping large data packets into the 48-octet payload field of an ATM cell. See the next the five definitions. SEE ALSO CS AND SAR.

ALAP AppleTalk Link Access Protocol. In an AppleTalk network, this link access-layer (or data link-layer) protocol governs packet transmission on LocalTalk.

Alarm Display Attendant console indicators show the status (i.e. what's happening) in the telephone system. There are usually two types of alarms — minor and major. Minor displays may be something as "minor" as a "hung" trunk, i.e. one that didn't hang up when the person speaking on it hung up. They can often be remedied by turning the PBX off, counting to ten, and then turning it on. (Before you do, check it will load itself.) Major problems — such as a blown line card in the PBX, one console out or half the trunks out — often require a service call and are often covered under the Emergency Conditions section of telephone service contracts.

Alarm Indication Signal AIS. A signal that replaces the normal traffic signal when a maintenance alarm indication has been activated. See the next definition.

Alarm Indicating Signal AIS. A code sent downstream indicating an upstream failure has occurred.

Albatross Manager He hangs around you constantly. You can't get rid of him. This definition courtesy Tom Henderson. See also SEAGULL MANAGER.

ALBO Automatic Line BuildOut. ALBO is a means of automatic cable equalization used in T-1 span-line interface equipment.

ALC 1. Automatic Level Control.
2. See Automatic Light Control.
3. Airline Line Control. A full-duplex, synchronous communications protocol used in airline reservations systems. ALC is a packet polling protocol which adheres to a strict master/slave relationship between the central host and the remote terminals. ALC relies on IBM's IPARS (International Airline Passenger Reservation System) character set, which comprises 6 data bits and no parity bits. Error detection is accomplished through a Cyclic Check Character (CCC), but only very limited procedures are identified for error correction. ALC was designed for use in X.25 networks, but also works in a Frame Relay environment.

ALE 1. Approvals Liaison Engineer. This engineer acts on your behalf to asses design and component changes to your BABT approved products. BABT is the British Approvals Board for Telecommunications.
2. See Automatic Link Establishment.

Alerter Service A Windows 2000 term. Notifies selected users and computers of administrative alerts that occur on a computer. Used by the Server and other services. Requires the Messenger service.

Alerting A signal sent to a customer, PBX or switching system to indicate an incoming call. A common form is the signal that rings a bell in the telephone set. Others signals can trigger such devices as whistles, gongs and chimes.

Alerting Call A call for which the subject connection is in the Alerting state. This usually implies that the telephone instrument is ringing.

Alerting Pattern An intelligent network term. Alerting pattern is a specific pattern used to alert a subscriber (e.g. distinctive ringing, tones etc.). See Q.931.

Alerting Signal A ringing signal put on subscriber access lines to indicate an incoming call. Bellcore defines an alerting signals more broadly; thus: "Alerting signals (for example, ringing, receiver off-hook) are transmitted over the loop to notify the customer of some activity on the line."

Alerting State A state in which a device is alerting (e.g., ringing) or is being presented (offered) to a device. This indicates an attempt to connect a call to a device. The device may be a device such as a telephone station. The device may also be a routing or distribution type of device. This includes a ACD or Hunt Group device.

ALEX Software which provides Internet users with a transparent read capability of remote files at anonymous FTP sites.

Algorithm A prescribed finite set of well defined rules or processes for the solution of a problem in a finite number of steps. Explained in normal English, it is the mathematical formula for an operation, such as computing the check digits in packets of data that travel via packet switched networks. Algorithm derives from the name of the ninth-century Persian mathematician al-Khomeini, who also had a lot to do with the invention of algebra. The word algebra comes from the Arabic al-jabr, which first appeared in a treatise by al-Khwarizmi.

ALI Automatic Location Information. ALI is a feature of E-911 (Enhanced 911) systems. ALI is provided to agents answering E-911 calls. It may include information such as name, phone number, address, nearest cross street and special pre-existing conditions (i.e. hazardous materials). On some systems it may also provide the appropriate emergency service address for the particular address. ALI is retrieved from a computer database. The database may be held on site or at a remote location and may be maintained by the local phone company (or its parent) or another agency.

Alias 1. A feature of the Apple Macintosh System 7 allowing the user to create a file that points to the original file. When you click on an alias, the original application is launched. Aliases can work across a network; so you can access a program residing on a file server or a Mac that runs System 7 file sharing.
2. An assumed name under which users of an electronic bulletin board may post messages. For example, Jane Smith may post as "Marketing Group." The system usually provides a list of aliases and the names of the users to which they belong. Some BBS packages allow anonymous message posting.
3. Unwanted signals generated during the A-to-D (Analog to Digital) conversion process. This is typically caused by a sampling rate that is too low to faithfully represent the original analog signal in digital form. Typically, a rate that is less than half the highest frequency to be sampled.
4. A nickname for a domain or host computer on the Internet.

Aliasing Distortion in a video signal. It shows up in different ways depending on the type of aliasing in question. When the sampling rate interferes with the frequency of program material the aliasing takes the form of artifact frequencies that are known as sidebands. Spectral aliasing is caused by interference between two frequencies such the luminance and chrominance signals. It appears as herringbone patterns, wavy lines where straight lines should be and lack of color fidelity. Temporal aliasing is caused when information is lost between line or field scans. It appears when a video camera is focused on a CRT and the lack of scanning synchronization produces a very annoying flickering on the screen of the receiving device.

Aliasing Noise A distortion component that is created when frequencies present in a sampled signal are greater that one-half the sample rate. See Anti-aliasing Filter.

Aligned Bundle A bundle of optical fibers in which the relative spatial coordinates of each fiber are the same at the two ends of the bundle. Also called "Coherent Bundle."

Alignment The adjustment of components in a system for optimum performance.

Alignment Error In IEEE 802.3 networks, an error that occurs when the total number of bits of a received frame is not divisible by eight, i.e. not properly framed. Alignment errors are usually caused by frame damage due to collisions.

ALIS Access Lines In Service. See ACCESS LINE.

ALIT Automatic Line Insulation Testing. Equipment located in a Central Office which sequentially tests lines in the office for battery crosses and grounds.

ALJ An Administrative Law Judge appointed by a State Commission to review a Commission docket, such as a rate case or incentive regulation proposal, and to make recommendations to the Commissioners.

All Call Paging With this feature, a user can broadcast an announcement — a page — to someone through the speakers of all the telephones on the system and, possibly, any external loudspeakers. If you want instant fame, ask your secretary to

call all the airports in the country and page you. Mike Todd, the movie mogul, used to have this secretary perform this wonderful task. Mr. Todd gave gigantic egos a whole new meaning.

All Channel Tuning Ability of a television set to receive all assigned channels. VHF and UHF, channels 2 through 83.

All Dielectric Cable Cable made entirely of dielectric (insulating) materials without any metal conductors.

All Inputs Hostile Measurement technique for troubleshooting networks, particularly for crosstalk, using worst case conditions (typically, full chroma signal on all inputs other than the one under test).

All Number Calling Once upon a time, the first two digits of telephone exchanges sort of corresponded to their location. For example, MU-8 meant Murray Hill 8 in Murray Hill, Manhattan, New York City. Then the phone company started running out of letters, so it went to All Number Calling. The All Number Calling provides a theoretical maximum of 792 central office exchange (NNX) codes per area code (NPA). This is derived on the basis of 800 NXX code combinations (8x10x10) leaving out eight special service combinations, including 411, 611, 911.

All Trunks Busy When a user tries to make an outside call through a telephone system and receives a "fast" busy signal (twice as many signals as a normal busy in the same amount of time), he is usually experiencing the joy of All Trunks Busy. No trunks are available to handle that call. The trunks are all being used at that time for other calls or are out of service. These days, many long distance companies are replacing a "fast" busy signal with a recording that might say something like, "I'm sorry. All circuits are busy. Please try your call later."

Allen, Robert Chairman of the board of AT&T from 1988. In 1980, he headed a task force to look into AT&T's future. The recommendation of the task force: keep equipment manufacturing at all costs. His successor broke the company up and sold the manufacturing arm off. It's now called Lucent Technologies.

Allocate To assign space or resources for a specific task. This is often used to refer to memory or disk space.

Allocated Channel A Radio Frequency (RF) channel that is configured to allow use by Cellular Digital Packet Data (CDPD) transmissions.

Allocation Of Resources A reason which CEOs give for not doing what they should be doing to grow their company.

Allowed Cell Rate ACR. An ATM term. An ABR service parameter, ACR is the current rate in cells/sec at which a source is allowed to send. ACR is a parameter defined by the ATM Forum for ATM traffic management. ACR varies between the MCR and the PCR, and is dynamically controlled using congestion control mechanisms.

Alloy A combination of two or more metals that forms a new or different metal with specific or desirable qualities.

ALM 1. AppWare Loadable Module. A visual computer telephony applications generator that works on Novell's NetWare. An ALM works by tying into Novell NetWare's NLMs. See Appware.
2. Automated Loan Machine. Like an ATM (Automated Teller Machine), an ALM sits in the wall of a building or inside a building on a wall. However, instead of giving money you own, an ALM dispenses money in the form of an instant loan. One of the leading ALM manufacturing companies is Affinity Technologies of Columbia, South Carolina. Alan Fishman of Columbia Financial Partners contributed this definition. Mr. Fishman is a leading New York City venture capitalist, who helped Affinity get started. Mr. Fishman's company is

Columbia Financial Partners.

3. Airline Miles. The method used to calculate the distance (for pricing purposes) of the point-to-point long distance lines in long distance telephone networks. See Airline Mileage.

ALO Transaction An ATP transaction (AppleTalk Transaction Protocol) in which the request is repeated until a response is received by the requester or until a maximum retry count is reached. This recovery mechanism ensures that the transaction request is executed at least once. See also ATP.

Aloha A method of data transmission in which the device transmits whenever it wants to. If it gets an acknowledgement from the device it's trying to reach, it continues to transmit. If not (as in the case of a collision with someone else trying to transmit simultaneously), it starts all over again. The ALOHA method get its name from a dying satellite that was donated to university researchers in the Pacific. It was used to transmit data by satellites among South Sea islands, especially Hawaii. The ALOHA "method" — called "transmit at will" — was invented because the users were short of funds to develop more sophisticated data transmission protocols, and they had a free satellite, which typically had more bandwidth than they had stuff to send. See ALOHANET.

Alohanet An experimental form of frequency modulation radio network developed by the University of Hawaii. Alohanet is implemented by creating transmission frames containing data, control information, and source and destination addresses which are broadcast for reception by the destination receiver and ignored by all others. Actually, Alohanet is an early version of Ethernet, the local area network technique. See ALOHA.

Alpeth Aluminum-polyethylene primary covering known as the sheath for aerial cable.

Alpha 1. Only alphabetic characters.

2. The first (A) version of hardware or software. It typically has so many bugs you only let your employees play with it. A beta is the next version. It's a pre-release version and selected customers (and the press, sometimes) become your guinea pigs. After beta, and when the bugs are removed, comes "general availability" or "genera release. That's when the product is finally available for buying by the general public."

Alpha Channel The upper 8 bits of the 32-bit data path in some 24-bit graphics adapters. The alpha channel is used by some software for controlling the color information in the lower 24 bits.

Alpha Geek The most knowledgeable, technically proficient person in an office or work group. "Ask Harry, he's the alpha geek around here."

Alpha Test The first testing phase of a software version. Alpha tests are conducted in-house, before being promoted to beta test, which typically involves a real customer.

Alphabetic Only alphabetic characters. See also ALPHANUMERIC.

Alphanumeric A set of characters that contains both letters and numbers — either individually or in combination. Numeric is 12345. Alphabetic is ABCDEF: Alphanumeric is 1A4F6HH8. American and Australian zip codes are numeric. Canadian and English postal codes are alphanumeric. No one knows why.

Alphanumeric Display A display on a phone or console showing calling phone number, called number, trunk number, type of call, class of service and perhaps, some other characteristics of the call. It may also contain instructions as to how to move the call around, set up a conference call, etc. The display may be liquid crystal or light emitting diode. Typically,

it's liquid crystal.

Alphanumeric Memory A cellular radio feature that allows you to store names with auto-dial phone numbers.

ALS Active Line State, one possible state of an FDDI optical fiber.

Altair Ethernet Motorola's name for its wireless local area network, which transmits at the very high frequency of 18 to 18 megahertz. Altair users need to fill out a small, one-page FCC application in order to use the system.

ALTEL Association of Long distance TELephone companies. A trade association composed of alternative (to AT&T) long distance carriers and resellers of long distance services.

Alternate Access Provider A carrier providing local access and transport other than the primary local exchange carrier.

Alternate Answering Position Usually refers to a second receptionist's desk which has a telephone switchboard or console functioning like the main one. Also refers to when the main receptionist is away from his/her desk, or is very busy taking calls, the telephone system automatically sends the calls to another console or to a phone that will be answered.

Alternate Buffer In a data communications device, the section of memory set aside for the transmission or receipt of data after the primary buffer is full. This helps the device control the flow of data so transmission is not interrupted due to lack of space for the incoming or outgoing data.

Alternate Entrance A supplementary entrance facility into a building using a different routing to provide diversity of service and for assurance of service continuity.

Alternate Lock Code A three-digit lock code to be used with the partial lock feature in some cellular phones.

Alternate Media Connector A buildings network term. An optional module that plugs into a 1016, 2016, 3024 or 3124 repeater to provide an AUI, BNC, or fiber Media Expansion Port (MEP).

Alternate Recipient An electronic messaging term. In X.400 terms, a user or distribution list that a recipient MTA (Message Transfer Agent) delivers a message to (if allowed) when the message cannot be sent to the preferred recipient.

Alternate Route A second or subsequent choice path between two exchanges, usually consisting of two or more circuit groups in tandem. Sometimes called "alternative route" or "second-choice route."

Alternate Routing AR. Redirecting a call over alternate facilities when the first choice route for that call is unavailable. AR is a mechanism that supports the use of a new path after an attempt to set up a connection along a previously selected path fails. It's a feature used in network design and also in PBXs. For example, with PBXs it's a feature used with long distance calls that permits the telephone system (typically a PBX) to send calls over different (alternate) phone lines. It might do this because of congestion of the primary phone lines the calls would normally be sent over. Alternate routing is often confused with Least Cost Routing in which the telephone system chooses the least expensive way (available at that time) to route that call. Least Cost Routing typically works with so-called "look-up" tables in the memory of the PBX. These tables are put into the PBX by the user. The PBX does not automatically know how to route each call. It must be told by the user. That "telling" might be as simple as saying "all 312 area codes will go via the AT&T FX line." Or it might be as complex as actually listing which exchanges in the 312 area code go by which method. Least Cost Routing tells the calls to go over the lines which are perceived by the user to be the least cost way of getting the call from point A to point B. Alternate routing happens when the least cost routes get congested and alternate routes (typically more expensive) are found from the look-up tables in the PBX's memory.

Alternate Use The ability to switch communications facilities from one type of service to another, i.e., voice to data, etc.

Alternate Voice Data AVD. An older service which is a single transmission facility which can be used for either voice or data (up to 9600 bps). Arrangement includes a manually operated switch (on each end) to allow customers to alternately connect the line to their modem or PBX.

Alternating Current See AC.

Alternative Access Provider AAV. Another name for a CAP. See CAP.

Alternative Channel A call center/marketing term. A competitive marketing strategy to expand the means by which a company can reach its customers. Direct marketing and specifically telephone-based marketing are major examples that have emerged in the highly competitive era of the '80s and '90s.

Alternative Non-Traffic Sensitive Cost-Recovery Plans New charges proposed by the regional Bell holding companies to supplement subscriber line charges. In short, another charge on the subscriber with an interesting, though dubious, justification. They have not been fully implemented.

Alternative Regulatory Framework ARF.

Alternator A machine which generates electricity which is alternating current. See AC.

ALTS Alternative Access Providers to the local telephone network i.e., Teleport.

ALU Arithmetic Logic Unit. The part of the CPU (Central Processing Unit) that performs the arithmetic and logical operations. See Microprocessor.

Alumina Aluminum Oxide, Al2O3. Alumina ceramic is used as the substrate material on which is deposited thin conductive and resistive layers for thin film microwave integrated circuits.

Alvyn Aluminum-polyethylene, the sheath used for riser cable where a flame retardant sheath is required.

Always On/Dynamic ISDN See AO/DI.

AM See Amplitude Modulation. Also Access Module.

AM Noise The random and/or systematic variations in output power amplitude. Usually expressed in terms of dBc in a specified video bandwidth at a specified frequency removed from the carrier.

AM-PM Conversion AM-PM conversion represents a shift in the phase delay of a signal when a transistor changes from small-signal to large-signal operating conditions. This parameter is specified for communications amplifiers, since AM-PM conversion results in distortion of a signal waveform.

AM/VSB Amplitude-Modulated Vestigal Sideband.

AMA Automatic Message Accounting. AMA is another name for Call Detail Recording or Station Message Detail Recording (SMDR). See AMA Tape.

AMA Tape A telephone company machine-readable magnetic tape which contains the customer's long distance calling and billing data for a given month.

AMA Teleprocessing System AMATPS. The primary method for delivery of AMA data from the network to billing systems. The current AMATPS architecture consists of an AMA Transmitter (AMAT) and a collector. Definition from Bellcore.

AMADNS AMA Data Networking System. In OSS, the next generation Bellcore system for the collection and transport of AMA data from central office switches to a billing system. See

also AMA.

Amateur Radio Operators Also known as HAM Radio Operators. A class of noncommercial private radio operators who generally use interactive radio as a hobby. There are five classes of licenses that are earned by examination, which the FCC sponsors.

AMATPS AMA Teleprocessing System. In OSS, the Bellcore legacy system for collecting and transporting AMA data from central office switches to a billing system. The AMATPS consists of an AMA Transmitter and a collector. See also AMA.

Ambient Lighting The general level of illumination throughout a room or area.

Ambient Noise The level of noise present all the time. There is always noise, unless you're in an anechoic chamber. When measured with a sound level meter, it is usually measured in decibels above a reference pressure level of 0.00002 pascal in SI units, or 0.00002 dyne per square centimeter in cgs units.

AMD Air Moving Device. IBM-speak for a fan with a feedback sensor to let the system know if the fan has failed.

America Online The largest North American on-line computer service. AOL provides e-mail, forums, software downloads, news, weather, sports, financial information, conferences, on-line gaming, an encyclopedia, and other features, to its subscribers.

American Bell, Inc. 1. The predecessor to AT&T. American Bell was formed in 1880 as a Massachusetts corporation to supersede National Bell Telephone Company, which had consolidated the interests of the original Bell Telephone and New England Telephone Company. In the face of unfriendly Massachusetts corporate law, American Bell on December 30, 1899 was folded into American Telephone and Telegraph Corporation (AT&T), its wholly owned long distance subsidiary incorporated in New York state.

2. The old name for the unregulated telephone equipment supply subsidiary of American Telephone & Telegraph. American Bell (deja vu all over again) was formed on January 1, 1983 by FCC mandate to market, sell and maintain newly manufactured equipment and enhanced services. American Bell Inc. had its name changed to AT&T Information Systems, becoming a division of AT&T. It has been reorganized many times. When it was American Bell, it was only selling telecommunications products and services to end users. When it become AT&T Information Systems, it sold AT&T phone systems and AT&T computer systems. It then merged with AT&T Long Lines, which was then called AT&T Communications and later called simply AT&T. Subsequently, it became known as AT&T Technologies, including the old Western Electric. In 1996, AT&T decided to split into three separate companies, with AT&T Technologies and Bell Telephone Laboratories becoming Lucent Technologies. Sadly, old gadgetry, knickknacks and momentos bearing the name American Bell, Inc. have no marketable value as antiques or examples of American folk art. See also Lucent Technologies.

American National Standards Institute See ANSI.

American National Standards Institute Character Set The set of characters available. The character set includes, letters, numbers, symbols and foreign language characters.

American Registry for Internet Numbers See ARIN.

American Standard Code For Information Interchange ASCII. The standard 7-bit code for transferring information asynchronously on local and long distance telecommunications lines. The ASCII code enables you to represent 128 separate numbers, letters, and control characters. By using an eighth bit — as in extended ASCII or IBM's EBCDIC — you can represent 256 different characters. ASCII often uses an eighth bit as a parity check or a way of encoding word processing symbols, not as a way of broadening the number of characters and symbols which it can represent. See also ASCII.

American Wire Gauge AWG. Standard measuring gauge for non-ferrous conductors (i.e. non-iron and non-steel). AWG covers copper, aluminum, and other conductors. Gauge is a measure of the diameter of the conductor. See AWG for a bigger explanation.

Ameritech Ameritech Corp is one of the Regional Bell operating companies formed as a result of the AT&T Divestiture. Ameritech covers five states and includes the operating telephone companies: Illinois Bell, Indiana Bell, Michigan Bell, Ohio Bell, and Wisconsin Bell. It also includes some other subsidiaries, which fit into two classifications — administrative (centralized buying, real estate, etc.) or entrepreneurial (cellular mobile radio, venture capital, etc.).

AMI Alternate Mark Inversion. The line-coding format in T-1 transmission systems whereby successive ones (marks) are alternately inverted (sent with polarity opposite that of the preceding mark). Here's AT&T definition: A line code that employs a ternary signal to convey binary digits, in which successive binary ones are represented by signal elements that are normally of alternating, positive and negative polarity but equal in amplitude, and in which binary zeros are represented by signal elements that have zero amplitude. This is an AT&T definition.

AMIS See Audio Messaging Interchange Specification. A standard for networking voice mail systems.

AML Analog Microwave Link

Amp hour AH. A rating system telling you how long a battery will last. A battery with an amp-hour rating of 100 will supply 100 amps for one hour, 50 amps for two hours or 25 amps for four hours. It will run the four amp laptop I'm writing this on for at least 25 hours. That would be wonderful. However, the battery would weigh much more than my laptop. I probably could barely lift it. That would not be so wonderful. As usual, life is a bunch of tradeoffs.

Ampacity The maximum current an insulated wire or cable can safely carry without excluding either the insulation or jacket materials limitations.

Amperage Rating The amperage which may be safely applied to a circuit, service or equipment. See also AMPERE.

Ampere The unit of measurement of electric current or the flow of electrons. One volt of potential across a one ohm impedance causes a current flow of one ampere. AMP is the abbreviation for ampere. It is mathematically equal to watts divided by volts. Note that in the electrical context, WATTS is spelled with two "Ts." In telecommunications, WATS, meaning Wide Area Telecommunications Service, is spelled with only one "T."

Ampere-hour Unit Measurement of battery capacity, determined by multiplying the current delivered by the time it is delivered for. See AMPERE.

Amphenol Connector Amphenol is a manufacturer of electrical and electronic connectors. They make many different models, many of which are compatible with products made by other companies. Their most famous connector is the 25-pair connector used on 1A2 key telephones and for connecting cables to many electronic key systems and PBXs.

The telephone companies call the 25-pair Amphenol connector used as a demarcation point the RJ-21X. The RJ-21X connector is made by other companies including 3M, AMP and TRW. People in the phone business often call non-Amphenol-made 25-pair connectors, amphenol connectors.

Amplified Handset An amplified handset is the best phone gadget you can buy. You use it to crank up the volume of incoming calls (and in some cases the volume of outgoing calls) and save yourself enormous amounts of money on call-backs. "We have a bad line. I'll call you back." There are three types of amplified handsets: 1. The handset with a built-in amplifier. These devices suck their power from the phone line and since the phone line doesn't have much power, you won't have much amplification. I'm not overly impressed with amplified handsets. 2. The handset with amplifying circuits built into the phone. Ditto for our comments about power. 3. The handset with the little external box amplifier which is powered by either AC or by several batteries, typically AA alkalines. Such an external amplifier will produce much greater amplification. This is the type I prefer.

Amplifier When telephone conversations travel through a medium, such as a copper wire, they encounter resistance and thus become weaker and more difficult to hear. An amplifier is an electrical device which strengthens the signal. Unfortunately, amplifiers in analog circuits also strengthen noise and other extraneous garbage on the line. Amplifiers are used in all telephone systems, analog and digital. But in digital systems, signals are regenerated and then amplified. As a result, noise is much less prevalent and less likely to be amplified in digital systems.

Amplitude The distance between high or low points of a waveform or signal. Also referred to as the wave "height." See Amplitude Modulation.

Amplitude Distortion The difference between the output wave shape and the input wave shape.

Amplitude Equalizer A corrective network that is designed to modify the amplitude characteristics of a circuit or system over a desired frequency range. Such devices may be fixed, manually adjustable, or automatic.

Amplitude Modulation Also called AM, it's a method of adding information to an electronic signal in which the signal is varied by its height to impose information on it. "Modulation" is the term given to imposing information on an electrical signal. The information being carried causes the amplitude (height of the sine wave) to vary. In the case of LANs, the change in the signal is registered by the receiving device as a 1 or a 0. A combination of these conveys different information, such as letters, numbers, punctuation marks, or control characters. In the world of modems, digital bit streams can be transmitted over an analog network by amplitude modulation, with the carrier frequency being modulated to reflect a 1 bit by a high amplitude sine wave (or series of sine waves) and a 0 bit with a low amplitude sine wave or (series of sine waves). The principal forms of Amplitude Modulation are QDM: Double-band Amplitude Modulation

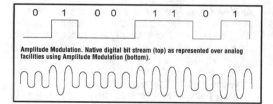

Amplitude Modulation. Native digital bit stream (top) as represented over analog facilities using Amplitude Modulation (bottom).

QAM: Quadrature Amplitude Modulation
SSB: Single-sideband Modulation
VSB: Vestigial Sideband Modulation
Contrast with Frequency Modulation and Phase Shift Keying.

AMPS Advanced Mobile Phone Service. It's another word for the North American analog cellular phone system. The spectrum allocated to AMPS is shared by two cellular phone companies in each area or region (geographic market). This system was deployed during the 1980s in North America. Most other parts of the world deployed cellular later and went straight to digital. See GSM.

AMS 1. Account Management System.
2. Attendant Management System. An NEC term. With the NEAX2400 IMS, the AMS is an on-screen, dynamic Open Applications Interface (OAI) computer application that emulates and enhances attendant console capabilities. A typical AMS workstation combines the NEC HDAC console and headset with a color display and standard keyboard. The computer is equipped with the UNIX System V operating system, the NEC Applications Manager support platform, and a comprehensive package of software components. Communication between the HDAC and the AMS computer software is made possible by the OAI connection between the computer and the NEAX2400 IMS (ICS).

AN Access network.

ANA 1. Assigned Night Answer.
2. Automatic Network Analyzer. A computer controlled test system that measures microwave devices in terms of their small signal S-parameters. The use of this instrument by both engineering and production permits quick and accurate characterization of the input and output impedance, gain, reverse isolation of individual units and the degree of match between units.

Analog Comes from the word "analogous," which means "similar to." In telephone transmission, the signal being transmitted — voice, video, or image — is "analogous" to the original signal. In other words, if you speak into a microphone and see your voice on an oscilloscope and you take the same voice as it is transmitted on the phone line and ran that signal into the oscilloscope, the two signals would look essentially the same. The only difference is that the electrically transmitted signal (the one over the phone line) is at a higher frequency. In correct English usage, "analog" is meaningless as a word by itself. But in telecommunications, analog means telephone transmission and/or switching which is not digital. Outside the telecom industry, analog is often called linear and covers the physical world of time, temperature, pressure, sound, which are represented by time-variant electrical characteristics, such as frequency and voltage. See ANALOG TRANSMISSION.

Analog Bridge A circuit which allows a normal two-person voice conversation to be extended to include a third person without degrading the quality of the call.

Analog Cellular The current standard for cellular communications.

Analog Channel A channel which transmits in analog waveforms. See ANALOG.

Analog Chips Analog chips translate real-world phenomena — like motion, sound, temperature, pressure and light — into electronic or mechanical digital patterns that can be understood by their digital counterparts as for example when a song is recorded for a CD. An analog processor performs the opposite, taking the digital signals, for exmaple off the CD, and converting them into analog sounds which we can enjoy.

Analog Computer A computer that performs its tasks by measuring continuous physical variables — pressure, voltage, flow — and manipulating these variables to produce a solution, which is then converted into a numerical equivalent. Analog computers are largely used as special purpose machines in scientific or technical applications. The earliest analog computers were purely mechanical devices with levers, cogs, cams, etc., representing the data or operator values. Modern analog computers typically employ electrical parameters such as voltage, resistance, or current to represent the quantities being manipulated.

Analog Digital Converter An A/D Converter. Pronounced: "A to D Converter." A device which converts an analog signal to a digital signal.

Analog Driver An accessory circuit for an oscillator of filter which permits its frequency to be changed by a continuously varying signal.

Analog Facsimile Facsimile which can transmit and receive grey shadings — not just black and white. It is called analog because of its ability to transmit what appear to be continuous shades of grey. "Analog" facsimile is usually transmitted digitally.

Analog Fiber Video Network. Implies a network in which fiber optic cable links entities together, with the video transmitted in its original analog format. Many of these systems use an "everybody on" philosophy, in which each participating entity on the network has its own channel and the other entities simply purchase a modulator (like a television tuner, for about $250) for each of the other parties, or channels with which they wish to interact. The practical limit to such a system is 16 channels, though most analog fiber networks link up no more than four entities at a time so that participants see four monitors (TV sets) . . . one for each of the "live" participating sites. The advantage of analog fiber over digital fiber is that an expensive CODEC at each site is not required. The disadvantage is the relatively large band width requirements that result in higher costs of fiber.

Analog Front End The part of the fax machine that converts between the digitally modulated signal and the analog signal used on the telephone line.

Analog Loop-back A method of testing modems and data terminals by disconnecting the device from the telephone line and looping a signal out through the device's transmit side and in through its receive side. The test tells if the trouble is with the telephone line or with the modem.

Analog Microwave A microwave system in which the digital bit stream is modulated by a modem and then frequency shifted up to the appropriate microwave carrier frequency. Contrast with Digital Microwave.

Analog Monitor A computer screen that uses an analog signal, a smoothly varying value of current or voltage that varies continuously. VGA, SVGA and Macintosh models are examples of analog monitors. Most computer screens are analog. Most analog monitors are designed to accept input signals at a precise frequency. Higher frequencies are needed to carry higher-resolution images to the monitor. For this reason, multiscanning monitors have been developed that automatically adjust themselves to the incoming frequency. See also ANALOG and DIGITAL MONITOR.

Analog Private-Line Service A dedicated circuit that transmits information between two or more points. It uses analog transmission signals and is engineered for 300 to 3,000 Hz with a net maximum loss of 16 dB.

Analog Processor See Analog Circuits.

Analog Recording System of recording in which music is

converted into electrical impulses that form "patterns" in the grooves of phonograph record masters or in the oxide particles of master tapes representing (or analogous to) musical waveforms.

Analog Switch Telephone switching equipment that switches signals without changing the analog form. The major form of analog switching is circuit switching.

Analog Synchronization A synchronization control system in which the relationship between the actual phase error between clocks and the error signal device is a continuous function over a given range.

Analog Transmission A way of sending signals — voice, video, data — in which the transmitted signal is analogous to the original signal. In other words, if you spoke into a microphone and saw your voice on an oscilloscope and you took the same voice as it was transmitted on the phone line and threw that signal onto the oscilloscope, the two signals would look essentially the same. The only difference would be that the electrically transmitted signal would be at a higher frequency.

Analog Video Signals represented by an infinite number of smooth transitions between video levels. TV signals are analog. By contrast, a digital video signal assigns a finite set of levels. Because computer signals are digital, analog video must be converted into a digital form before it can be shown on a computer screen.

Analog Wireless The dominant radio transmission standard in the United States; a so called AMPS.

Analogue An English/European way of spelling ANALOG, which is the correct North American spelling. See ANALOG.

Anamorphic Unequally scaled in vertical and horizontal dimensions.

ANC All Number Calling. The dialing plan used in telephone networks. Consisting of all numbers, ANC replaced the old U.S. system which consisted of two letters and five numbers (2L + 5N). In other words, the GR (Greenwood) exchange became 47, PA (Pennsylvania) became 72, and UL (Ulysses) became 85. Remember the Glenn Miller hit, "Pennsylvania 6 5,000?" Those old exchanges were charming, often reflecting the character of the communities to which they were assigned. However, and as the number of telephone numbers grew, we ran out of alpha prefixes that included the first two letters of meaningful words. Eventually, we had to use the 1 on the dial in the second position of the prefix; there are no letters associated with that number. Hence, All Number Calling.

ANCARA Advanced Networked Cities And Regions Association. A formal association of cities and regions exploring advanced uses of information technology. ANCARA was founded in 1996 by the regions/cities of Eindhoven (The Netherlands), Kansai (Japan), Orlando (Florida), Silicon Valley (California), Singapore, and Stockholm (Sweden). The intent is to accelerate the development of the Global Information Infrastructure (GII). www.ancara.nl See also GII.

Ancestor Node An ATM term. A logical group node that has a direct parent relationship to a given node (i.e., it is the parent of that node, or the parent's parent.)

Anchor A hyperlinked word or group of words. An anchor is the same as a hyperlink — the underlined words or phrases you click on in World Wide Web documents to jump to another screen or page. The word anchor is used less often than hyperlink, but it maintains the seafaring theme of navigating and surfing the Net. See also hyperlink.

Anchorage Accord A milestone ATM Forum document (April 12, 1998) so named because of the meeting location, the Anchorage Accord outlines which versions of ATM Forum specifications vendors should implement. ATM Forum specifications comprise approximately 60 baseline specifications for successful market entry of ATM products and services. Included are Broadband InterCarrier Interface (BICI), Interim Local Management Interface (ILMI), LAN Emulation (LANE), network management, Private Network Node Interface (PNNI), signaling, SMDS (Switched Multimegabit Data Service) and IP (Internet Protocol) over ATM, traffic management, and a number of physical interfaces. The accord also limits the conditions under which specifications are revised in order to cut down on future confusion. See also ATM Forum.

Ancillary Charges Charges for supplementary services comprised of optional features, which may consist of both non-recurring and monthly charges.

AND Automatic Network Dialing.

Anechoic Chamber A perfectly quiet room. A room in which sound or radio waves do not reflect off the walls. An anechoic chamber is the only place in which a speakerphone will work perfectly. The more a room resembles an anechoic chamber — i.e. lots of drapes, plush carpet, etc. — the better a speakerphone will work.

Angel Investor in an early stage technology start-up. Typically an angel invests when the company is little more than an idea, a simple business plan and several management people, but rarely a full management team.

Angle Bracket The term for these two brackets < and >. These two brackets have major use in the HTML language. See HTML.

Angle Modulation Modulation in which phase angle or frequency of a sine wave carrier is varied.

Angle of Arrival AOA. A class of Position Determination Technology in which a mobile radio unit's position is calculated based on the direction of its transmitted signal measured from two or more receiving sites; also known as triangulation. Employed in certain wireless E9-1-1 solutions; see also Time Difference of Arrival.

Angle Of Deviation In fiber optics, the net angular deflection experienced by a light ray after one or more refractions or reflections. The term is generally used in reference to prisms, assuming air interfaces. The angle of deviation is then the angle between the incident ray and the emergent ray.

Angle of Incidence The angle between an incident ray and the normal to a reflecting or refracting surface.

Angled End An optical fiber whose end is deliberately polished at an angle to reduce reflections.

Angstrom 1. One ten-billionth of a meter. It is said that the 128-bit addressing scheme of the proposed IPv6 will provide enough unique IP addresses to theoretically provide 1,500 per square angstrom of the earth's surface.
2. A unit of length in optical measurement.

Angular circumference The measurement of the amount of bend in a fiber-optic cable.

Angular Misalignment Loss The optical power loss caused by angular deviation from the optimum alignment of source to optical fiber — fiber-to-fiber, or fiber-to-detector.

ANI Automatic Number Identification. ANI provides for the transmission through the network of the BN (Billing Number), versus the telephone number, of the originating party (i.e., the calling person, also called party in the phone business). ANI originally was intended exclusively for the use of the long distance and local phone carriers for billing purposes. ANI information is sent through the network, from the originating central office, through all intermediate tandem offices, to the terminating central office. The information originally was sent over analog trunks in the form of DTMF (Dual Tone MultiFrequency) signals, although contemporary networks usually pass the information through the digital SS7 (Signaling System 7) network. For some years, ANI has been available to end user organizations, as well. In order to gain access to ANI data, you must have a "trunk side" connection, which carries an additional charge. Much like CLID (Calling Line IDentification), ANI delivers the number of the calling party. Unlike CLID, ANI does not depend on the presence of SS7 throughout the entire network. Also unlike CLID, ANI cannot be blocked by the calling party. So, let's pretend that you are running a large call center. A customer calls you. Before the call is even connected to your ACD (Automatic Call Distributor), ANI presents the BN of the calling party to the ACD. Your ACD captures the BN, dips into a computer database and matches that number with the profile of the caller. As your telephone agent answers the call, he gets a "screen pop" with information about the caller, and he answers the call with "Good morning, Mr. Newton. This is Ray. I read about ANI in your Dictionary. Isn't it wonderful!" Some large users say they save as much as 30 seconds on the average IN-WATS call by knowing the phone number of the person calling them and being able to use that information to access information about them in the company database. They avoid asking regular customers for routine identification information (like their address and phone number) since it is all there in the database. See also Caller ID, CLASS, Common Channel Signaling, DNIS, Flex ANI, ISDN and ISUP.

Anima Someone who communicates with you telepathically.

Animation The process of displaying a sequential series of still images to achieve a motion effect.

Anisochronous Pertaining to transmission in which the time interval separating any two significant instants in sequential signals is not necessarily related to the time interval separating any other two significant instants. Isochronous and anisochronous are characteristics, while synchronous and asynchronous are relationships.

Anisotropic Pertaining to a material whose electrical or optical properties vary with the direction of propagation or with different polarizations of a traveling wave.

ANM ANswer Message. The fourth of the ISUP call set-up messages. A message sent in the backward direction indicating that the call has been answered. See ISUP and Common Channel Signaling.

Anneal The act of using heat to soften a metal such as copper, making it less brittle.

Annex A The first of the frame relay standard extensions, Annex A outlines provisions for a Local Management Interface (LMI) between customer premises equipment and the frame relay network for the purpose of querying network status information

Annex D The second frame relay standard extension dealing with the communication and signaling between customer premises equipment and frame relay network equipment for the purpose of querying network status information.

Annie A Web homepage which seems to have been abandoned for some time. Most of the links are out of date. It's been orphaned.

Annotation Marking such as that done by highlighting, underlining, text, or freehand drawing. see Annotations.

Annotations Notes that you can add to Web documents. These notes are stored on your local hard disk and are available each time that you access a document. This feature is found in NCSA Mosaic, but not Netscape Communicator or Microsoft's Internet Explorer.

Announcement Service Allows a phone user to hear a recording when he dials a certain phone number or extension. These days, announcement services are provided increasingly by totally solid-state digital announcers. These gadgets are more reliable, deliver a clearer message and last much longer than analog tape-based machines (like answering machines), which use recording tape.

Announcement System An arrangement for providing information by means of recorded announcements.

Annoyance Call Bureau The department in your local phone company which you call when you need help with annoying or harassing phone calls you are receiving. The Bureau will recommend you file a report with your local phone company. And then it may apply TRAP and TRACE equipment and techniques to try to locate the source of your annoying phone calls. The Annoyance Call Bureau is the stepchild of the phone industry, which means it is typically underfunded. See TRAP and TRACE.

Annual Percentage Rate APR. A percentage calculation of the finance charge portion of financing contact.

Annular Ring An indicator (or ring) around the circumference of the coaxial cable every so many feet — often 2.5 meters (8.2 feet) — to indicate a point where transceivers are to be connected. Same as transceiver attachment mark.

Annunciator Original name for the indicator on magnetic switchboards which indicates the particular line that is calling the exchange. Now it is simply a light, a bell or a device that tells you something. That something might be the ringing of a phone or it might be a problem that you're having with some piece of remote equipment. A communicating annunciator is a sophisticated device that is connected to a phone line and gets on that line (dial-up or leased) to let you know that something is broken.

Anomaly An impedance discontinuity causing and undesired signal reflection in a transmission cable.

Anonymous Call Rejection ACR. A service some local phone companies are providing their subscribers. It allows subscribers to automatically stop certain calls from ringing their phone. The calls stopped are "restricted," namely they would be displayed as "P" or "Private" on a subscriber's Caller ID device, meaning that the calling person did not send you his calling number. The person who makes such a call would hear, "We're sorry. The party you have reached is not accepting private calls. To make your call, hang up, dial *82 or 1182 on a rotary phone and re-dial." The caller will be able to reach you only by re-dialing without restricting display of his or her number.

Anonymous FTP A way of logging in anonymously to distant hosts on the Internet and often freeware (free software) from the Internet. With an implementation of the FTP protocol, users can get public domain software from Internet sites, using the word "anonymous" for a login ID, and their userid@hostname.domain as the password. A database called Archie contains a list of what is available from anonymous FTP sites, and can be reached at "archie.mcgill.ca" and at "archie.sura.net." See FTP.

Anonymous Telephone Number A telephone number that should not be displayed or voiced back to the called party. Such a designation is stored in switch memory and is included in signaling information sent to the terminating switch for interSPCS calls.

ANOVA ANalysis Of VAriance

ANS Answer. (What else?) (Which is a question, the ANS to which is provided below.)

ANSA Alternate Network Service Agreement. An ISDN term. Under ANSA, customers who reside in areas where the central office switch does not support ISDN can be serviced from a neighboring central office at no additional charge. From the customer's perspective, ISDN is readily available and affordable, but the customer MUST agree to migrate to the local central office if and when service becomes available. In most cases this will involve a change in phone number. This agreement pertains to BellSouth customers only.

ANSI American National Standards Institute. A standards-setting, non-government organization founded in 1918, which develops and publishes standards for transmission codes, protocols and high-level languages for "voluntary" use in the United States. In a press release, ANSI described itself as "a private non-profit membership organization that coordinates the U.S. voluntary standards system, bringing together interests from the private and public sectors to develop voluntary standards for a wide array of U.S. industries. ANSI is the official U.S. member body to the world's leading standards bodies — the International Organization for Standardization (IOS or ISO) and the International Electronic Commission (IEC) via the U.S. National Committee. The Institute's membership includes approximately 1,300 national and international companies, 30 government agencies, 20 institutions and 250 professional, technical, trade, labor and consumer organizations." ANSI is located at 11 West 42 Street, 13th Floor, New York NY 10036 212-642-4900. ANSI puts out a biweekly newsletter called "ANSI Standards in Action. See also ANSI Character Set, CCITT, ECMA, IEEE, and ISO. www.ansi.org

ANSI T1.110-1987 Signaling system number 7 (SS7) - General Information.

ANSI T1.111-1988 Signaling system number 7 (SS7) - Message Transfer Part (MTF)

ANSI T1.112-1988 Signaling System number 7 (SS7) - Signaling Connection Control Part (SCCP)

ANSI T1.113-1988 Signaling System 7 (SS7) - Integrated Services Digital Network (ISDN) user part.

ANSI T1.114-1988 Signaling System 7 (SS7) - Transaction Capability Application Part (TCAP)

ANSI T1.206 Digital Exchanges and PBXs - Digital circuit loopback test lines.

ANSI T1.227-1995 Telecommunications Operations Administration Maintenance and Provisioning

ANSI T1.301 ANSI ADPCM standard.

ANSI T1.401-1988 Interface between carriers and customer installations - Analog voice grade switched access lines using loop-start and ground-start signaling.

ANSI T1.501-1988 Network performance - Tandem encoding limits for 32 Kbit/s Adaptive Differential Pulse-Code Modulation (ADPCM).

ANSI T1.601-1988 Integrated Services Digital Network (ISDN) - Basic access interface for use on metallic loops for application on the network side of the NT (Layer 1 specification).

ANSI T1.605-1989 Integrated Services Digital Network (ISDN) - Basic access interface for S and T reference points (Layer 1 specification).

ANSI T1.Q1 ANSI's standard for telecommunications network performance standards, switched exchange access network transmission performance standard exchange carrier-to-interexchange carrier standards.

ANSI TIX9.4 ANSI's Sonet standard.

ANSI X3T9.5 A committee sponsored by the American National Standards Institute (ANSI) that is responsible for a variety of system interconnection standards. The committee has produced draft standards for high-speed coaxial cable bus and fiber optic ring local networks.

ANSI X3T9.5 TPDDI Twisted-Pair Distributed Data Interface (TPDDI) is a new technology that allows users to run the FDDI standard 100 Mbps transmission speed over twisted-pair wiring. Unshielded twisted-pair has been tested for distances over 50 meters (164 feet). TPDDI is designed to help users make an earlier transition to 100 Mbps at the workstation.

ANSI Character Set The American National Standards Institute 8-bit character set. It contains 256 characters.

Answer Back A signal or tone sent by a receiving equipment or data set to the sending station to indicate that it is ready to accept transmission. Or a signal or tone sent to acknowledge receipt of a transmission. See ANSWER SUPERVISION.

Answer Back Supervision Another word for answer supervision. See Answer Supervision.

Answer Call The name of a Bell Atlantic service. Here are Bell Atlantic's words. Answer Call is an answering machine without the machine. This automated messaging service answers your calls right through your touch-tone phone — 24 hours a day — even when you're on the phone. And since it's on Bell Atlantic network, there's no equipment to buy...nothing to turn to...no wires to connect... and no maintenance. By simply dialing a private passcode, you can listen to your messages, replay them or even change your greeting. What's more, Answer Call gives you the option of providing your employees (who share one line) with up to eight private "mailboxes" to receive and retrieve their own messages.

Answer Detect The use of a digital signal processing technique to determine the presence of voice energy on a telephone line. It is used with call (answer) supervision, to identify an answered line. It's beginning to be used with computerized dialing equipment as it eliminates the need for a telephone representative to constantly monitor call set-up progress on each telephone line in the event a call is answered. See ANSWER SUPERVISION and ANSWER SIGNAL.

Answer Message ANM. A CCS/SS7 signaling message that informs the signaling points involved in a telephone call that the call has been answered and that call charging should start.

Answer Mode When a modem is set by the user to receive data, it is in Answer Mode. In any conversation involving two computers, two terminals or one computer and one terminal, one side of the conversation must always be in Answer Mode. Putting a modem/computer in answer mode is sometimes done through software and sometimes through hardware, i.e. a switch on the side of the machine. You cannot run a data communications "conversation" if both sending and receiving equipment are in "Answer Mode." Computers — mainframe and mini — which receive a lot of phone calls are typically put in "Answer Mode." The terminals or computers calling them are typically in transmit mode.

Answer Signal A supervisory signal, usually in the form of a closed loop, returned from the called telephone to the originating switch when the called party answers. This signal stops the ringback signal from being returned to the caller.

Answer Supervision Follow this scenario: I call you long distance. My central office must know when you answer your phone so my central office can start billing me for the call. It works like this: when you, the called party, answer your phone, your central office sends a signal back to my central office (the originating CO). This tells my central office to start billing me for the call. This signal is called Answer Supervision. Before the Divestiture of the Bell System in early

1984, most of the nation's long distance companies — with the exception of AT&T Communications — did not receive Answer Supervision. They did not know precisely when the called party answered. So they started their billing cycle after some time — 20 or 30 seconds after the caller completed dialing. These long distance companies presumed that after this time, some one will have answered and the call will be in progress and can then be timed and billed. Without Answer Supervision, their billing of calls is inaccurate. They may bill for calls which didn't occur. And you may pay more for calls which did occur.

With the Divestiture of the Bell System, and the introduction of Equal Access, the local phone companies have been told by the FCC that they must provide accurate answer supervision to all long distance calls. And with that answer supervision, the pricing of your long distance calls should be accurate. Not all long distance companies, however, choose to buy answer supervision (it costs a little more). And thus your long distance calls may still be billed inaccurately.

Check. If you are "accessing" your preferred long distance carrier by dialing a seven digit local number, then dialing your number and your account code, your carrier is probably not receiving Answer Supervision and the timing and billing of your long distance calls may be inaccurate. Check this out. Remember: just because your town has equal access doesn't mean your preferred long distance phone company has opted for it because it is expensive or for some other reason.

Virtually no hotels have answer supervision. So they start billing you arbitrarily. Some start billing you after three rings. Some after four. When you check out, carefully check your phone bill. You will, in most instances, find you have been billed for many uncompleted calls. Tell your family to pick up the phone quickly when you're out of town and may be calling them. Don't let the phone ring too many times as you're likely to be billed for the dubious pleasure of listening to ringing signals.

"Answer supervision" is getting better, however, as the electronics of "listening" to sounds on phone lines get better. Electronics are now available to do — to a 95% accuracy — what we as humans do — to a 100% accuracy — namely distinguish between a normal ringing sound, a fast busy sound and a person or fax machine answering the phone and saying "Hello." These electronics are getting better and less expensive, by the month. See Answer Supervision-Line Side.

Answer Supervision-Line Side Answer Supervision-Line Side is a service I first read about in a US West publication, which describes it as "providing an electrical signal that is passed back to the originating end of a switched connection. This signal indicates that the called line has gone off-hook. This service offering has applicability for record start and end, announcement start and end, dialtone reorigination prevention, call progress sequence indications, and other uses. This service offering may be used by terminal equipment (PBX, pay telephone, call diverter, etc). connected to the calling line to determine that the call has been answered."

Answerback In data communications, answerback is a response programmed into a data terminal to identify itself when polled by a remote computer or terminal. This response is usually in reply to a Control-E (ASCII Character 5, Inquiry), which is known on the Telex and TWX networks as a "Who Are You?" character, or "WRU." The Answerback allows a remote computer to verify it has dialed correctly (usually on the Telex or TWX networks) by matching the Answerback received with the Answerback expected.

Answering Machine Detection The ability of outbound dialing systems, inside of the host ACD, the PBX, or as part of the predictive dialer product, to detect and filter out calls answered by answering machines. These systems may place the associated telephone number in a callback (see Callback) queue and may also play a recorded message.

Answering Tone The tone an asynchronous modem will transmit when it answers the phone. The tone indicates that it is willing to accept data.

ANT 1. Access Network Termination.
2. Alternate Number Translation. The ability to reroute 1-800 calls on NCP failure.

Ant Farm Gigantic multiscreen movie theater complex with glass facade, often found near American malls. Also called multiplexes or gigaplexes.

ANTC Advanced Networking Test Center. An FDDI interoperability testing center established in 1990.

Antediluvian Before the flood. Very old fashioned.

Antenna A device for transmitting, receiving or transmitting and receiving signals. Antennas come in all shapes and sizes. Their shape depends on the frequency of the signal they're receiving or transmitting and the use to which their communications is being put to. Antennas can broadcast signals in all directions. They're called omnidirectional. They can also broadcast signals in a fine straight line — like a flashlight. Electrical signals with frequencies higher on the spectrum, for example, are shorter and more directional. As they get higher on the spectrum, they look more like light. These must be focused and thus, require antennas which are shaped like the mirror reflector of a focusing flashlight. This parabolic shape focuses the broad beam (of the bulb or the electrical signal) into a narrow, focused beam. The weaker the received signal, the bigger the antenna must be. Antennas come in many varieties and have cute names, like parabola, caresgrain, helix, lens and horn. See antenna beam.

Antenna Beam The radio frequency energy pattern emitted by an antenna. Imagine a flashlight. Turn the head one way and the light becomes more focused, and thus more intense. Now turn it the other and it becomes less focused. Radio and microwave antennas are designed to be less or more focused. A broadcast TV satellite will have an antenna whose beam covers the entire continental United States. A satellite like Iridium which needs to send individual signals to individual cellular-like phones will have tightly focused radio beams.

Antenna Entrance A pathway facility from the antenna to the associated equipment.

Antenna Gain The ratio, usually expressed in decibels, of the power required at the input of a loss-free reference antenna to the power supplied to the input of the given antenna to produce, in a given direction, the same field strength, or the same irradiance, at the same distance. When not specified otherwise, the gain refers to the direction of maximum radiation. The gain may be considered for a specified polarization.

Antenna Lobe A picture showing an antenna's radiation pattern. A more technical explanation: A three-dimensional radiation pattern of a directional antenna bounded by one or more cones of nulls (regions of diminished intensity).

Antenna Matching The process of adjusting impedance so that the input impedance of an antenna equals or approximates the characteristic impedance of its transmission line over a specified range of frequencies. The impedance of either the transmission line, or the antenna, or both, may be adjusted to effect the match.

Anthropomorphism The process of giving human qualities to inanimate objects. For example, getting a file cabinet to talk about what's inside it, or getting a modem to explain how to do communications.

Anti Aliasing See Antialiasing

Anti Curl A feature marketed by manufacturers of slimy paper fax machines (i.e. thermal paper). As the paper emerges from the fax machine, "anti-curl" simply sends the paper through a path which causes it to bend slightly in the opposite direction to which it was rolled over the roll. This bending purports to make the paper less curly when it emerges. It works to an extent. Virtually all slimy fax machines now have the "feature," though most don't advertise it.

Anti Digit Dialing League A group of people that resisted the move from named exchanges to all number dialing. The Bell System fought against the League because there was no global standardizaton between numbers and digits on rotary dial phones. Thus, IDDD was impossible until the advent of all digit numbers.

Anti Reflection Coating A thin, dielectric or metallic film (or several such films) applied to an optical surface to reduce its reflectance and thereby increase the transmittance of the optical fiber. The ideal value of the refractive index of a single layer film is the square root of the product of the refractive indices on either side of the film, the ideal optical thickness being one quarter of a wavelength.

Anti Static A material, such as packing material, that is treated to prevent the build-up of static electricity. The static charges gradually dissipate instead of building up a sudden discharge.

Anti Stuffing A mechanical flap in a coin phone which prevents the blocking by paper or other material of coin chutes. An anti-stuffing flap is meant to assure that you, the user, get your money back after you've tried to make a call but didn't get through.

Anti Viral Programs Programs which scan disks looking for the tell-tale signatures of computer viruses.

Antialiasing 1. A filter (normally low pass) that band limits an input signal before sampling to prevent aliasing noise. See ALIASING NOSE.

2. A computer imaging term. A blending effect that smooths sharp contrasts between two regions of different colors. Properly done, this eliminates the jagged edges of text or colored objects and images appear smoother. Used in voice processing, anti-aliasing usually refers to the process of removing spurious frequencies from waveforms produced by converting digital signals back to analog.

Anticipointment Raising people's levels of anticipation and then disappointing them. A definition contributed by Gerald Taylor, president, of MCI.

Antistatic A material, such as packing material, that is treated to prevent the build-up of static electricity. The static charges gradually dissipate instead of building up a sudden discharge.

Anycast 1. A term associated with IPv6, the proposed new protocol for the Internet. Anycast refers to the ability of a device to establish a communication with the closest member of a group of devices. By way of example, a host might establish a communication with the closest member of a group of routers for purposes of updating a routing table. That router would then assume responsibility for retransmitting that update to all members of the router group on the basis of a Multicast. Through this approach, the host is relieved of the mundane task of addressing each router in the network, a task clearly best accomplished by a lesser device with lesser responsibilities. See IPv6.

2. In ATM, an address that can be shared by multiple end systems. An anycast address can be used to route a request to a node that provides a particular service.

AnyLAN A high-speed local area network technology for which Hewlett-Packard Co. and International Business Machines Corp. created the original specifications. Announced in 1993, AnyLAN was intended to allow work groups operating Token Ring and Ethernet LANs to swap many more data-intensive applications at much greater speeds than possible at the time, increasing the speed and capacity of LANs sixfold to tenfold. HP and IBM announced they would give the technology of AnyLAN to any competitor free of charge, in order to establish it as a standard and expand the size of the total market. In 1996, the IEEE 802.12 committee officially recognized AnyLAN as fast LAN standard, renaming it 100VG-AnyLAN. See 100VG-ANYLAN for a detailed explanation.

Anynet/MVS IBM product name for the ACF/VTAM feature that implements IBM's "Networking Blueprint" technology on hosts and OS/2 workstations and permits SNA LU 6.2 applications to work over TCP/IP or TCP/IP-oriented sockets applications to run over SNA.

Anywhere Fix The ability of a Global Positioning System (GPS) receiver to start position calculations without being given an approximate location and approximate time. See GPS.

AO/DI Always On/Dynamic ISDN. AO/DI takes advantage of the D channel in an ISDN BRI channel to maintain a constant virtual connection to the central office switch. In this mode, the 16 Kbps D channel is capable of receiving and transmitting data at 9.6 Kbps for low speed data transfer, while still handling its ISDN channel signaling and control chores. Even at 9.6 Kbps, the bandwidth is quite suitable for e-mail, stock quotes or news updates. When higher speed data transfer is required for applications such as downloading Web pages, one or both of the B channels can be activated automatically. Some phone companies are tariffed the AO/DI service for a flat monthly rate of under $10 a month.

AoC Advice of Charge. A wireless telecommunications term. A supplementary service provided to a customer under GSM, Global System for Mobile Communications.

AOCN Administrative Operating Company Number. The term "AOCN" also refers to the company that updates TRA databases under contract to a code holder.

AOHell America OnLine Hell. Hacker programs that allow one to mess with AOL's software. Using these programs you can get access to, inter alia, personal electronic mail accounts.

AOL See America On Line.

AOS Alternate Operator Services. Today there are many Operator Services Providers not owned by the Bell Telephone Companies or AT&T. The AOS industry is dropping the descriptive term "alternate" and communicating that they be known as OSPs. AOS was coined by AT&T. See AOSP and Operator Service Providers.

AOSP Alternate Operator Service Provider. A new breed of long distance phone company. It handles operator-assisted calls, in particular Credit Card, Collect, Third Party Billed and Person to Person. Phone calls provided by OSP companies are typically fare more expensive than phone calls provided by "normal" long distance companies, i.e. those which have their own long distance networks and which you see advertised on TV. You normally encounter an OSP only when you're making a phone call from a hotel or hospital phone, or a privately-owned payphone. It's a good idea to ask the operator the cost of your call before you make it.

AOSSVR Auxiliary Operator Services System Voice Response.

AOW Asia and Oceania Workshop. One of the three regional OSI Implementors' Workshops.

AP 1. See ADD-ON or Applications Processor. AP is an AT&T word for a piece of equipment which hangs off the side of their PBX and makes it do more things, like voice mail.

2. See Adjunct Processor, an AIN term for a decentralized SCP. See Adjunct Processor.

3. Access Providers.

APA All Points Addressable (APA) method of host graphics implementation which uses vertical and horizontal pixel coordinates to create a more graphic image. An SNA definition.

Apache The web server software on about half of the world's existing web sites is Apache. Apache is UNIX freeware. Apache was originally based on code and ideas found in the most popular HTTP server of the time: NCSA httpd 1.3. It has since evolved into a far superior system that can rival — some say surpass — any other UNIX-based HTTP server in terms of functionality, efficiency, and speed. Apache includes several features not found in the free NCSA server, among which are highly configurable error messages, DBM-based authentication databases, and content negotiation. It also offers dramatically improved performance and fixes many bugs in the NCSA 1.3 code.

APAD Asynchronous Packet Assembler/Disassembler.

APC Adaptive Predictive Coding. A narrowband analog-to-digital conversion technique employing a one-level or multi-level sampling system in which the value of the signal at each sample time is adaptively predicted to be a linear function of the past values of the quantized signals. APC is related to linear predictive coding (LPC) in that both use adaptive predictors. However, APC uses fewer prediction coefficients, thus requiring a higher bit rate than LPC.

APCC The American Public Communications Council, which is part of the North American Telecommunications Association (NATA).

APD Avalanche PhotoDiode. A diode that, when hit by light, increases its electrical conductivity by a multiplication effect. APDs are used in lightwave receivers because the APDs have great sensitivity to weakened light signals (i.e. those which have traveled long distances over fiber). APDs are designed to take advantage of avalanche multiplication of photocurrent.

Aperiodic Antenna An antenna designed to have an approximately constant input impedance over a wide range of frequencies; e.g., terminated rhombic antennas and wave antennas.

Aperture For a parabolic reflector or a horn antenna, aperture is the dimension of the open mouth and represents a surface over which it is possible to calculate the radiation pattern. For a series of n stacked transmitting elements such as dipoles or slots, the vertical aperture is usually defined as n times the element spacing in wavelengths.

Aperture Distortion In facsimile, the distortions in resolution, density, and shape of the recorded image caused by the shape and finite size of the scanning and recording apertures or spots.

Aperture Grille A type of monitor screen made up of thin vertical wires. Said to be less susceptible to doming than iron shadow mask.

API An Application Programming Interface is software that an application program uses to request and carry out lower-level services performed by the computer's or a telephone system's operating system. For Windows, the API also helps applica-

tions manage windows, menus, icons, and other GUI elements. In short, an API is a "hook" into software. An API is a set of standard software interrupts, calls, and data formats that application programs use to initiate contact with network services, mainframe communications programs, telephone equipment or program-to-program communications. For example, applications use APIs to call services that transport data across a network. Standardization of APIs at various layers of a communications protocol stack provides a uniform way to write applications. NetBIOS is an early example of a network API. Applications use APIs to call services that transport data across a network.

API_connection An ATM term. Native ATM Application Program Interface Connection: API_connection is a relationship between an API_endpoint and other ATM devices that has the following characteristics:

1. Data communication may occur between the API_endpoint and the other ATM devices comprising the API_connection

2. Each API_connection may occur over a duration of time only once; the same set of communicating ATM devices may form a new connection after a prior connection is released

3. The API_connection may be presently active (able to transfer data), or merely anticipated for the future

APL Automatic Program Load in telecom. In data processing, it's a popular programming language.

APLT Advanced Private Line Termination. Provides the PBX user with access to all the services of an associated enhanced private switched communications services (EPSCS) network. it also functions when associated with a common control switching arrangement (CCSA) network. See Advanced Private Line Termination.

APM 1. Average Positions Manned, the average number of ACD positions manned during the reporting period for a particular group.

2. Advanced Power Management. A specification originally sponsored by Intel and Microsoft to extend the life of batteries in battery-powered computers. The idea of the specification is for the application programs, the system BIOS and the hardware to work together to reduce power consumption. An APM-compliant BIOS provides built-in power management services to the operating system of your PC. The operating system passes calls and information between the BIOS and the application programs. It also arbitrates power management calls in a multi-tasking environment (such as Windows) and identifies power-saving opportunities not apparent to applications. The application software communicates power-saving data via predefined APM interfaces. Windows 95 adopted APM to shut down the computer. It uses a special mode of the latest Intel processors — System Management Mode, or SMM. SMM lets the BIOS take control of the machine at any time and manage power to peripherals. A BIOS' APM support can't be circumvented by other software. This could cause a crash. Microsoft, Intel, Toshiba and others are now working on a new spec, called ACPI — Advanced Configuration and Power Interface. www.intel.com/IAL/powermgm/apmovr.htm and www.ata.or/~acpi/

APNIC Asia Pacific Network Information Center. A group formed to coordinate and promote TCP/IP based networks in the Asia-Pacific region. APNIC is responsible for management and assignment of IP (Internet Protocol) addresses in the Asia-Pacific, just as are ARIN and RIPE in the regions of the Americas and Europe, respectively. See also ARIN, IP, and RIPE.

APO Adaptive Performance Optimization. A technology used on the Texas Instruments ThunderLAN chipset, which was jointly developed by Compaq and Texas Instruments. APO dynamically adjusts critical parameters for minimum latency, minimum host CPU utilization and maximum system performance. This technology ensures that the capabilities of the PCI interface are used for automatically tuning the controller to the specific system in which it is operating.

Apocalypse, Four Horsemen Of The four horsemen of the Apocalypse were War, Plague, Famine and Death.

Apogee The point on a satellite orbit that is most distant from the center of the gravitational field of the Earth. The point in an orbit at which the satellite is closest to the Earth is known as the perigee. In commercial application, the terms have most significance with respect to LEOs (Low Earth Orbiting) and MEOs (Middle Earth Orbiting) satellite constellations, which travel in elliptical orbits. See LEO and MEO.

Apologize To lay the foundation for a future offense.

APON ATM Passive Optical Network. A passive optical network running ATM. See also ATM, Fiber Optics and SONET.

Apparent Power The mathematical product of the RMS current and the RMS voltage. Identical to the VA rating.

APPC Advanced Program-To-Program Communications. In SNA, the architectural component that allows sessions between peer-level application transaction programs. The LUs (Logical Units) that communicate during these sessions are known as LU type 6.2. APFC is an IBM protocol analogous to the OSI model's session layer: it sets up the necessary conditions that enable application programs to send data to each other through the network.

APPC/PC An IBM product that implements APPC on a PC.

Appearance Usually refers to a private branch exchange line or extension which is on (i.e. "appears") on a multi-button key telephone. For example, extension 445 appears on three key systems.

Appearance Test Point The point at which a circuit may be measured by test equipment.

Append To add the contents of a list, or file, to those of another.

APPGEN A shortened form of the words APPlications GENerator.

Apple Computer, Inc. Cupertino, CA. Manufacturer of personal computers. Heavy penetration in the graphics/desktop publishing business and in education. Apple was formed on April Fool's Day, 1976, by Steve Wozniak and Steve Jobs, aided greatly by Mike Markkula.

Apple Desktop Bus The interface on a Mac where non-peripheral devices, such as the keyboard, attaches. A Mac keyboard or mouse is called an ADB device. Contrast with peripherals, which attach through the SCSI interface. See also USB, which is a new bus for use on PCs but fulfilling essentially the same function as the Apple Desktop Bus.

Apple Desktop Interface ADI. A set of user-interface guidelines, developed by Apple Computer and published by Addison-Wesley, intended to ensure that the appearance and operation of all Macintosh applications are similar.

Apple Menu The Apple icon in the upper left hand corner of the Apple Macintosh screen. The Apple menu contains aliases, control panels, the chooser and other desk accessories.

Apple Pie Both an American icon, and the name chosen for Apple Computer's Personal Interactive Electronics (PIE) division, chartered with extending the company into new growth areas such as Personal Digital Assistants (PDAs), e.g. the Apple Newton. The PIE division includes Apple Online Services, Newton and Telecommunications group, publishing activities, and ScriptX-based multimedia PDA development.

Apple Remote Access ARA is Apple Computer's dial-in client software for Macintosh users allowing remote access to Apple and third party servers.

Apple URP Apple Update Routing Protocol. The network routing protocol developed by Apple for use with Appletalk.

AppleShare Apple Computer's local area network. It uses AppleTalk protocols. AppleShare is Apple system software that allows sharing of files and network services via a file server in the Apple Macintosh environment. See APPLETALK.

Applet Mini-programs that can be downloaded quickly and used by any computer equipped with a Java-capable browser. Applets carry their own software players. See JAVA.

AppleTalk Apple Computer's proprietary networking protocol for linking Macintosh computers and peripherals, especially printers. This protocol is independent of what network it is layered on. Current implementations exist for LocalTalk (230.4 Kbps) and EtherTalk (10 Mbps).

AppleTalk Zone and Device Filtering Provides an additional level of security for AppleTalk networks. On AppleTalk networks, network managers can selectively hide or show devices and/or zones to ARA clients. See ARA.

Application A software program that carries out some useful task. Database managers, spreadsheets, communications packages, graphics programs and word processors are all applications.

Application Based Call Routing In addition to the traditional methods of routing and tracking calls by trunk and agent group, the latest Automatic Call Distributors route and track calls by application. An application is a type of call, for example, sales or service. Tracking calls in this manner allows accurately reported calls, especially when they are overflowed to different agent groups. See ACD.

Application Binary Interface ABI. The rules by which software code is written to operate specific computer hardware. Application software, written to conform to an ABI, is able to be run on a wide variety of system platforms that use the computer hardware for which the ABI is designed.

Application Bridge Aspect Telecommunications' ACD to host computer link. Originally it ran only over R2-232 serial connections, but it now runs over Ethernet, using the TCP/IP link protocol. See also OPEN APPLICATION INTERFACE.

Application Class An SCSA term. A group of client applications that perform similar services, such as voice messaging or fax-back services.

Application Entity A cellular radio term. An Application Entity provides the service desired for communication. An Application Entity may exist in an M-ES (Mobile End System) (i.e., mobile application entity) or an F-ES (Fixed End System). An Application Entity is named with an application entity title.

Application Equipment Module AEM. A Northern Telecom term for a device within the Meridian 1 Universal Equipment Module that supports Meridian Link Modules. The Meridian Link Module (MLM) is an Application Module, specially configured to support the Meridian Link interface to host computers.

Application For Service A standard telephone company order form that includes pertinent billing, technical and other descriptive information which enables the company to provide communications network service to the customer and its authorized users.

Application Framework This usually means a class library with a fundamental base class for defining a complete program. The framework provides at least some of the facilities through which a program interfaces with the user, such as menus and windows, in a style that is internally consistent and abstracted from the specific environment for which it has been developed.

This is an explanation I received from Borland. I don't quite understand it, yet. An application framework is an object-oriented class library that integrates user-interface building blocks, fundamental data structures, and support for object-oriented input and output. It defines an application's standard user interface and behavior so that the programmer can concentrate on implementing the specifics of the application. An application framework allows developers to reuse the abstract design of an entire application by modeling each major component of an applications as an abstract class.

Application Generator AG. A program to generate actual programming code. An applications generator will let you produce software quickly, but it will not allow you the flexibility had you programmed it from scratch. Voice processing "applications generators," despite the name, often do not generate programming code. Instead they are self-contained environments which allow a user to define and execute applications. They are more commonly called applications generator, since one generator can define and execute many applications. See Applications Generator for a longer explanation.

Application Module A Northern Telecom term for a computer that can be attached to a Northern Telecom phone system and add intelligence and programmability to the phone system. Often, the AM will be a computer conforming to some standards, such as DOS or Windows, or it may be VME-based.

Application Module Link AML. A Northern Telecom internal and proprietary link that connects the Meridian 1 (via EDSI or MSDL port) to the Meridian Link Module.

Application Program A computer software program designed for a specific job, such as word processing, accounting, spreadsheet, etc.

Application Program Interface API. A set of formalized software calls and routines that can be referenced by an application program to access underlying network services.

Application Programming Interface API. A set of functions and values used by one program to communicate with another program or with an operating system. See API for a better explanation.

Application Profile As SCSA term. A description of the kinds of resources and services required by a client application (or an application class). An application profile is defined once for an instance of an application; then system services such as the SCR will be able to fulfill the needs of the application without the application having to state its needs explicitly.

Application Server As a Sun Microsystems term, it is the foundation of Solaris' client-server Server Suite. It's used to offload legacy systems. Acts as front-end to mainframes by augmenting them with distributed database servers. Adds NetWare's IPX/SPX stack to Solaris 2.5. With this software, NetWare users can run 32-bit multiprocessor, multithreaded databases such as Oracle and Informix without a NetWare Loadable Module (NLM).

Application Service Element ASE. A messaging term. A module or portion of a protocol in the application layer 7 of the OSI (Open Systems Interconnection) protocol stack. Several ASEs are usually combined to form a complete protocol, e.g., the X.400 P1 protocol which consists of the MTSE (Message Transfer Service Element), and the RTSE (Reliable Transfer Service Element).

Application Sharing Feature of many document-conferencing packages that lets two or more users on different (and usually distant) computers simultaneously use an application that resides on only one of the machines. Imagine, there are three of us and we have to jointly present a PowerPoint presentation to the "big bosses" tomorrow. Today, we have to work on the presentation. But, sadly we're in different cities. So, one of us loads application sharing software and dials the other two in a conference call. That "dialing" may be done over normal phone lines or through the Internet. See NetMeeting, the most popular application sharing.

Application Software Interface ASI. The Application Software Interface is a product of the Application Software Interface Expert Working Group of the ISDN Implementor's Workshop. The Interface focuses on the definition of a common application interface for accessing and administering ISDN services provided by hardware commonly referred to in the vendor community as Network Adapters (NAs) and responds to the applications requirements generated by the ISDN Users Workshop (IUW). The characteristics of this Application Interface shall be

• Portable across the broadest range of system architectures;
• Extensible (their words, not mine)
• Abstracted beyond ISDN to facilitate interworking;
• Defined in terms of services and facilities consistent with OSI layer interface standards.

According to Application Software Interface Group, the primary goal of the ASI is to provide a consistent set of application software interface services and application software interface implementation agreement(s) in order that an ISDN application may operate across a broad range of ISDN vendor products and platforms. The application software interface implementation agreements will be referenced by (and tested against) the IUW (ISDN Users Workshop) generated applications. It is anticipated that the vendor companies involved in the development of these implementation agreements will build products for the ISDN user marketplace which conform to them. ASI Implementation Agreements are likely to become a US Government Federal Information Processing Standard (FIPS).

Applications Engineering Applications engineering is the process of analyzing your telephone network to find products and services that will reduce your monthly bill without sacrificing network quality. It can be as simple as calling the telephone company to convert a particular service to a Rate Stabilization Plan (RSP). In many instances, the use of applications engineering concepts will increase the quality of your network. For example, putting DIDs onto a T1 will save you money and provide your network with a digital backbone. Unfortunately, most applications engineering is done by the telephone company or by their sales agents. Their main goal is not to save you money, but rather to sell telephone company products. Therefore, they are unlikely to advise you of all the hidden costs of converting to a particular service. A true application engineer will provide you with a complete cost analysis that includes all the conversion costs, and provides you with the "break-even date." The break-even date is the date that your monthly saving offsets the initial conversion cost of the service. It is often used synonymously with the term break-even point.

Applications Generator An application generator (AG) is a software tool that, in response to your input, writes code a computer can understand. In simple terms, it is software that writes software. Applications generators have three major benefits: 1. They save time. You can write software faster. 2.

They are perfect for quickly demonstrating an application. 3. They can often be used by non-programmers. Applications generators have two disadvantages. 1. The code they produce is often not as efficient as the code produced by a good programmer. 2. They are often limited in what they can produce. Applications generators tend to be either general purpose tools or very specific tools, providing support for specific applications, such as connecting voice response units to mainframe databases, voice messaging system development, audiotex system development, etc. There are simple AGs. There are complex AGs. There are general purpose AGs. There are specialized AGs. There are character-based AGs. There are GUI-based AGS. In researching AGs to write computer telephony and interactive voice response applications, I found three different levels of AG packages. First, there are the sort of non-generator generators. They don't really create new software, but they allow you to tweak existing application blocks. There's no compiling and they're pretty simple to use (though they often lack database and host connectivity). Then there are the pretty GUI forms-based app gens. They usually entail building a call-flow picture, using either pretty icons or easy to understand templates. When you're done filling in all the blanks, you compile it and actually "generate" new software. They're very cute. Finally, there's the script level language of a company like Parity Software, San Francisco. Real programmers dig this. They often feel it gives them a lot more power and flexibility. For very complex apps (with T-1/ISDN, ANI, host connection, speech recognition, multimedia capabilities, etc.) you'll probably need the power and flexibility of a script language. Most of the better GUI application generators let you drop down to a script-level language (and C too).

Applications Layer The seventh and highest layer of the Open Systems Interconnection (OSI) data communications model of the International Standards Organization (ISO). It supplies functions to applications or nodes allowing them to communicate with other applications or nodes. File transfer and electronic mail work at this layer. See OSI Model.

Applications Partner An Applications Partner is AT&T's new name for an outside company which will write software to work on AT&T phone systems, such as the Merlin, Legend and the Definitely. AT&T is setting up an Applications Partner Program to work with companies to help them develop programs and distribute their products. See also Desktop Connection.

Applications Processor A special purpose computer which attaches to a telephone system and allows it (and the people using it) to perform different "applications," such as voice mail, electronic mail or packet switching. We think AT&T invented the term. See also Add-On.

Applique Circuit components added to an existing system to provide additional or alternate functions. Some carrier telephone equipment designed for ringdown manual operation can be modified with applique to allow for use between points having dial equipment.

APPN Advanced Peer-to-Peer Networking (APPN) is, according to its creator IBM, a leading-edge distributed networking feature IBM has added to its Systems Network Architecture (SNA). It provides optimized routing of communications between devices. In addition to simplifying the addition of workstations and systems to a network and enabling users to send data and messages to each other faster, APPN is designed to support efficient and transparent sharing of applications in a distributed computing environment. Because APPN permits direct communication between users anywhere on a network, it facilitates the development of client/server computing, in which workstation users anywhere on a network can share processing power, applications and data without regard to where the information is located. Workstations on an APPN network are dynamically defined so they can be relocated easily on the network without extensive re-programming. APPN also allows remote workstations to communicate with each other, without intervention by a central computer. Also, IBM's Advanced Peer-to-Peer Networking software.

APPN End Node An APPN end node is the final destination of user data and cannot function as an intermediate node in an APPN network and cannot perform routing functions. See APPN.

Approved Ground Grounds that meet the requirements of the NEC (National Electrical Code), such as building steel, concrete-encased electrodes, ground rings, and other devices. See AC and Grounding.

AppServer A SCSA term. AppServer defines the software environment that enables voice processing applications to run on any computing platform. AppServer sits on a PC equipped with call processing hardware and allows a remotely hosted application to control the call processing hardware.

APR Annual Percentage Rate. A percentage calculation of the finance charge portion of financing contact.

APS Automatic Protection Switching. A means of achieving network resiliency through switching devices which automatically switch from a primary circuit to a secondary (usually geographically diverse) circuit. This switching process would take place when the primary circuit fails or when the error rate on the primary line exceeds a set threshold. There are two basic APS architectures in SONET optical fiber networks: 1+1 and 1:N. A 1+1 architecture is characterized by permanent electrical bridging to service and protection equipment, which is placed at both ends of the circuit. At the head end, or transmitting end, the same payload signal is sent over both the primary and the secondary optical circuit. The optical signal is monitored for failures at the tail end independently and identically over both optical circuits. The receiving equipment at the tail end selects either the service channel (primary circuit) or the protection channel (secondary circuit), based on predefined switching performance criteria. A 1:N protection switch architecture is one in which any of "N" (i.e., any Number of) service channels (primary circuits) can be bridged to a single optical protection channel (secondary circuit).

AR Automatic Recall.

ARA AppleTalk Remote Access. Provides an asynchronous AppleTalk connection to another Macintosh and its network services through a modem. A remote user using ARA can log on to a remote server and mount the volume on his desktop as if he were connected locally.

ARAB Attendant Release Loop. A feature of the PBX console. See RELEASE.

ARAM Audio grade DRAM. DRAMS are low cost integrated circuits that are widely use in consumer electronic's products to store digital data.

Aramid Aramid is a synthetic textile material which is lightweight, nonflammable, and highly impact-resistant. Dupont markets it under the trademark Kevlar. In addition to being used in construction of some fiber optic cables to provide tensile strength, aramid fibers are used in bulletproof vests, sailboat sails, and industrial-strength shoelaces. See Tight Buffer Fiber Optic Cables.

Arbitrated Loop Topology A Fibre Channel topology that provides a (FC-AL) low-cost solution to attach multiple communicating ports in a loop. Nodes are linked together in a closed loop. Traffic is managed with a token-acquisition protocol, and only one connection can be maintained within the loop at a time. See Fibre Channel.

Arbitrage The price of gold in London equals the price of gold in New York. If it didn't, traders would step in, buy in one place and sell in another. This process is called arbitrage. There are huge price differences in long distance and international calling from different carriers. In the late 1990s, there was talk of "on-line arbitrage" — whereby I would indicate that I wanted to make a call, and offer that call to on-line traders who would instantly find me a cheap rate.

Arbitration A Fibre Channel term. The process of selecting one respondent from a collection of several candidates that request service concurrently.

ARCH Access Response CHannel. Specified in IS-136, ARCH carries wireless system responses from the cell site to the user terminal equipment. ARCH is a logical subchannel of SPACH (SMS (Short Message Service) point-to-point messaging, Paging, and Access response CHannel), which is a logical channel of the DCCH (Digital Control CHannel), a signaling and control channel which is employed in cellular systems based on TDMA (Time Division Multiple Access). The DCCH operates on a set of frequencies separate from those used to support cellular conversations. See also DCCH, IS-136, SPACH and TDMA.

Archie An Internet term. A corruption of "archive," Archie is a FTP search engine located on several computers around the country. It's sort of a superdirectory to the files on the internet. If you're looking for a file or even a particular topic, Archie provides its specific location. Veronica, Jughead and WAIS (Wide Area Information Servers) are other tools for searching the huge libraries of information on the Internet. Some companies, such as Hayes, make Archie software which give you a menu driven interface that lets you browse through the various Archie servers on the Internet as though browsing through card catalogs of remote libraries.

Architect One who drafts a plan of your office and then plans a draft of your money.

Architectural Assemblies Walls, partitions, or other barriers that are not load bearing. In contrast, Architectural Structures are load bearing.

Architectural Freedom An AT&T term for flexibility in locating functions, such as control, storage or processing of information, at any site in or around a network, such as customer premises, central offices or regional service bureaus. Architectural freedom also means the ability to distribute functions among combinations of locations and have them interrelate through a high-throughput, low-delay, transparent network. See also Architecture.

Architectural Structures Walls, floors, floor/ceilings and roof/ceilings that are load bearing. In contrast, Architectural Assemblies are not load bearing.

Architecture The architecture of a system refers to how it is designed and how the components of the system are connected to, and operate with, each other. It covers voice, video, data and text. Architecture also includes the ability of the system to carry narrow, medium and broadband signals. It also includes the ability of the system to grow "seamlessly" (i.e. without too many large jumps in price).

Architecture Police An individual or group within a company that makes sure software and hardware development follows established corporate guidelines. The architecture police tend to rein in creative development efforts.

Archival Readable (and sometimes writable) media. Archival media have defined minimum life-spans over which the information will remain stable (i.e, accurate without degradation).

Archive A backup of a file. An archived file may contain backup copies of programs and files in use or data and materials no longer in use, but perhaps needed for historical or tax purposes. Archive files are kept on paper, on microfilm, on disk, on floppies, etc. They may be kept in compressed or uncompressed form. See Archiver.

Archive Bit A Window NT (soon to be Windows 2000) term. Backup programs use the archive bit to mark files after backing them up, using the normal or incremental backup types.

Archive Server An email-based file transfer facility offered by some computers on Internet.

Archiver A software program for compressing files. If you compress files, you will save on communications charges, since you will be able to transmit those files faster as they're now smaller. My favorite MS-DOS archiver, also called file compression utility is Phil Katz's PKZIP.EXE and PKUNZIP.EXE. You can cut a database by as much as 90% and a word processed file by maybe 30% by using PKZIP. How much you can cut is determined by how much fluff is in the file. PKZIP is the most widely-used archive and compression utility today. You can recognized "zipped" files because their extension is always ZIP. There are other compression programs out there which you will recognize by these extensions, ARC, AR7, ARJ, LZH, PAK and ZOO.

Archiving Files This is a process where the information contained in an active computer file is made ready for storing in a non-active file, perhaps in off-line or near-line storage. Typically when files are archived, they are compressed to reduce their size. To restore the file to its original size requires a process known as unarchiving. See also Archiver.

ARCNET Attached Resource Computer NETwork. One of the earliest and most popular local area networks. A 2.5M-bits-per-second LAN that uses a modified token-passing protocol. Developed by Datapoint, San Antonio, TX, Arcnet interface cards are now manufactured by many vendors, including Standard Microsystems and Pure Data, Ltd. Arcnet has lost popularity in recent years to Ethernet (IEEE 802.3) and Token Ring (IEEE 802.5).

ARD Automatic Ring Down. A private line connecting a telephone in one location to a distant telephone with automatic two-way signaling. The automatic two-way signaling used on these circuits causes the station instrument on one end of the circuit to ring when the station instrument on the other end goes off-hook. This circuit is sometimes called a "hot-line" because urgent communications are typically associated with this service. ARD circuits are commonly used in the financial industry, but you see them at airports, where they're used to call hotels. May also have one way signaling. Station "A" rings Station "B" when Station "A" goes off hook, but Station "B" cannot ring Station "A".

ARDIS A public data communications wireless network that allows people carrying handheld devices to send and receive short data messages. Such messages might be from sheriff standing in the street searching his department's data base for unpaid parking tickets. ARDIS is jointly owned by Motorola and IBM. It is an outgrowth of a network originally created for IBM service technicians. A competitor to Ardis is RAM Mobile Data.

ARE An ATM term. All Routes Explorer: A specific frame initiated by a source which is sent on all possible routes in Source Route Bridging.

Area Code A three-digit code designating a "toll" center in the United States and Canada. Until January, 1995 the first digit of an area code was any number from 2 through 9. The second digit was always a "1" or "0". In January 1995, North America (i.e. the US and Canada) adopted the North American Numbering Plan (NANP) and second digits could be any number. This dramatically increased the number of possible area codes — from 152 to 792 and the number of phone numbers to more than six billion. For a full explanation, see North American Numbering Plan. For a full listing of area codes, see North American Area Codes.

Area Code Expansion The new North American Numbering Plan (NANP) allowed basically any three numbers to become an area code. This exploded the number of area codes now possible. Some manufacturers of phone equipment, e.g. Rockwell, choose to call this happening "Area Code Expansion." They claimed that their switch would accommodate all future permutations and combinations of area codes.

Area Code Restriction The ability of the telephone equipment (or its ancillary devices) to selectively deny calls to specific (but not all) area codes. Area code restriction is often confused with "0/1" (zero/one) restriction which denies calls to all area codes by sampling the first and second dialed digits (is it a 0 or 1?) and thus, identifying and blocking an attempt at making a toll call. For a full listing of area codes, see North American Area Codes.

Area Color Code A cellular radio term. A color code that is shared by all cells controlled through a single Mobile Data Intermediate System (MD-IS). The value of the Area Color Code must be different between any two adjacent cells controlled by adjacent MD-ISs. Refer to color code.

Area Transfer A rerouting, by splicing, of subscriber cable facilities from one Central office to another, usually within the same exchange area. An area transfer normally requires a change of telephone number for the subscribers involved and is, therefore, scheduled to occur on, or near, the Directory delivery date.

ARF Alternative Regulatory Framework.

Argument Separator In spreadsheet programs and programming languages, a comma or other punctuation mark that sets off one argument from another in a command or statement. The argument separator is essential in commands that require more than one argument. Without the separator, the program can't tell one argument from another.

ARI Automatic Room Identification. In Hotel/Motel telephone system applications, the ability to display the room number on the console.

Ariane The name of a family of rockets used, amongst other things, for sending communications satellites into space. Ariane is a product of the European Space Agency, the equivalent of the U.S.'s NASA.

ARIN American Registry for Internet Numbers. A not-for-profit, voluntary, association charged with the responsibility of management of IP (Internet Protocol) addresses in the geographic areas of North America, South America, the Caribbean, and sub-Saharan Africa. ARIN membership comprises end users, including ISPs (Internet Service Providers), corporate entities, universities, and individuals. ARIN became operational on December 22, 1997 as a result of a broad-based industry agreement to separate management of IP addresses from that of URLs (Uniform Resource Locators). URL administration now is the responsibility of CORE. Both previously were the sole responsibility of InterNIC. ARIN's counterparts are APNIC (Asia Pacific Network Information

Center) and RIPE (Reseaux IP Europeens). See also APNIC, CORE, InterNIC, IP, RIPE, and URL.

ARINC Aeronautical Radio INC. The collective organization that coordinates the design and management of telecommunications systems for the airline industry. It's one of the largest buyers of telecommunications services and equipment in the world. In its own words, "ARINC develops and operates communications and information processing systems for the aviation and travel industries and provides systems engineering and integration solutions to government and industry. Founded in 1929 to provide reliable and efficient radio communications for the airlines, ARINC is a $280 million company headquartered Annapolis, MD with over 2,000 employees worldwide." www.arinc.com

Arithmetic Coding A compression technique which produces code for an entire message, rather than encoding each character in a message. Arithmetic coding improves on Huffman Encoding, although it is slower. See also Compression and Huffman Encoding.

Arithmetic Logic Unit ALU. The part of the CPU (Central Processing Unit) that performs the arithmetic and logical operations. See Microprocessor.

Arithmetic Operation The process that results in a mathematically correct solution during the execution of an arithmetic statement or the evaluation of an arithmetic expression.

Arithmetic Register A register (i.e. short-term storage location) that holds the operands or the results of operations such as arithmetic operations, logic operations, and shifts.

Arithmetic Unit The part of a computing system which contains the circuits that perform the arithmetic operations. See also ALU.

arj The extension .arj shows that a file or program has been "compressed," and must be "exploded" with the arj program before being either read or used. Groups of files may be compressed together, but this is more commonly done with the zip program. See zip.

ARM Asynchronous Response Mode. A communication mode involving one primary station and at least one secondary station, where either the primary or one of the secondaries can initiate transmission.

Armageddon The fabled battlefield where God's heavenly forces are to defeat the demon-led forces of evil. The final battle.

Armor Mechanical protection usually accomplished by a metallic layer of tape, braid or served wires or by a combination of jute, steel tapes or wires applied over a cable sheath for additional protection. It is normally found only over the outer sheath. Armor is used mostly on cables lying on lake or river bottoms or on the shore ends of oceans. See Armored Cable.

Armored Cable 1. A stainless steel handset cord which is meant to resist vandalism. Typically used on a coin phone, most stainless steel handset cords are too short. This is said to be because they were first ordered for use in prisons, where guards wanted to be certain they would not be used by the prisoners as hanging devices. Thus, they requested Western Electric to make them too short for such a use. Whether there is any validity to this story is dubious, however, it is part of telephone industry folk history and therefore, worth preserving. 2. In outside cable an armored cable has its sheath covered with three protective layers: a vinyl jacket, a steel wrap, and another vinyl jacket. Armored cable is intended for use in direct-burial applications; the steel armor protects the sheath from damage during installation. See also Hard Cable.

ARP Address Resolution Protocol. 1. A low-level protocol within the Transmission Control Protocol/Internet Protocol (TCP/IP) suite that "maps" IP addresses to the corresponding Ethernet addresses. In other words, ARP is used to obtain the physical address when only the logical address is known. An ARP request with the IP address is broadcast onto the network. The node on which the IP address resides responds with the hardware address in order that the packets can be transmitted. By way of example, TCP/IP requires ARP for use with Ethernet, in which case the physical address would be defined by the MAC address hard-coded on the NIC (Network Interface Card) of the target workstation. See also RARP.

2. A low-level protocol which serves to map IP addresses, or other non-ATM addresses, to the corresponding address of the target ATM device. Once the ATM address has been identified, the ARP server can stream data to the target device as long as the session is maintained.

ARPA Advanced Research Projects Agency of the U.S. Department of Defense. (The whole DOD annual telecommunications bill exceeds $1 billion.) Much of the country's early work on packet switching was done at ARPA. At one stage it was called DARPA, which stands for Defense Advanced Research Projects Agency. ARPA was the US government agency that funded research and experimentation with the ARPANET and later the Internet. The group within DARPA responsible for the ARPANET is ISTO (Information Systems Techniques Office), formerly IPTO (Information Processing Techniques Office). See also DARPA INTERNET. DARPA has changed its name to ARPA and back again. It's hard to keep up.

ARPANET Advanced Research Projects Agency NETwork. A Department of Defense data network, developed by ARPA, which tied together many users and computers in universities, government and businesses. ARPANET was the forerunner of many developments in commercial data communications, including packet switching, which was first tested on a large scale on this network. The predecessor of the Internet, it was started in 1969 with funds from the Defense Department's Advanced Projects Research Agency (ARPA). ARPANET was split into DARPANET (Defense ARPANET) and MILNET (MILitary NETwork) in 1983. ARPANET was officially retired in 1990.

ARQ Automatic Retransmission reQuest. The standard method of checking transmitted data, used on virtually all high-speed data communications systems. The sender encodes an error-detection field based on the contents of the message. The receiver recalculates the check field and compares it with that received. If they match, an "ACK" (acknowledgment) is transmitted to the sender. If they don't match, a "NAK" (negative acknowledgment) is returned, and the sender retransmits the message. Note: this method of error correction assumes the sender temporarily or permanently stores the data it has sent. Otherwise, it couldn't possibly retransmit the data. No error detection scheme in data transmission is foolproof. This one is no exception.

Array 1. The description of a location of points by coordinates. A 2-D array is described with x,y coordinates. A 3-D array is described with x,y,z coordinates.

2. A named, ordered collection of data elements that have identical attributes; or an ordered collection of identical structures.

3. Two or more hard disks that read and write the same data. In a RAID system, the operating system treats the array as if it were a single hard disk.

Array Antenna Take a bunch of directional antennas. Aim them at the same transmitting source. Join them together. Presto, you now have a very powerful giant antenna. Array antennas are used for picking up weak signals. They are often used in astronomical and defense communications systems.

Array Connector A connector for use with ribbon fiber cable that joins 12 fibers simultaneously. A fan-out array design can be used to connect ribbon fiber cables to non-ribbon cables.

Array Processor A processor capable of executing instructions in which the operands may be arrays rather than data elements.

Arrestor A device used to protect telephone equipment from lightning, electrical storms, etc. An arrestor is typically gas filled so when lightning strikes, the gas ionizes and, bingo, a low resistance to the ground that drains the damaging high voltage elements of the lightning away.

Arrival Rate A call center term. The pattern in which calls arrive. Call Arrival Rates can be smooth, like outgoing telemarketing calls, or random, like incoming toll-free number calls, or peaked, where calls escalate in response to advertising.

ARS Automatic Route Selection, also called Least Cost Routing. A way that your phone system automatically chooses the least expensive way of making the call that it is presented with. That least expensive way may be a tie line or a WATS line, etc. It may even be dial-up. See Least Cost Routing and Alternate Routing.

Article An Internet term. An article is a USENET conversation element. It is a computer file that contains a question or piece of information made available to the USENET community by posting to a newsgroup.

Artifacts Distortions in a video signal. Unintended, unwanted visual aberrations in a video image. In all kinds of computer graphics, including any display on a monitor, artifacts are things you don't want to see. They fall into many categories (such as speckles in scanned pictures), but they all have one thing in common: they are chunks of stray pixels that don't belong in the image.

Artificial Intelligence In 1950, Alan Turing, a British mathematician, challenged scientists to create a machine that could trick people into thinking it was one of them. And this for long was THE classic definition of artificial intelligence. One way to trick people is to have the computer make typing mistakes, like real humans do. The real challenge these days with artificial intelligence, now more commonly called "expert systems," is not to recreate people but to recognize the uniqueness of machine intelligence and learn to work with it in intelligent, useful ways.

Artificial Line Interface In T-1 transmission, refers to the ability of a piece of transmission equipment to attenuate its output level to meet the required loop loss of 15-22.5 dB normally switch selectable between 0,7.5, and dB.

ARU Audio Response Unit. A device which gives audible information to someone calling on the phone. "Press 1 for the train timetable to Boston." The ARU reads the timetable. The caller responds to questions by punching buttons on his telephone keypad. If this sounds like Interactive Voice Response — IVR, you're 100% right because that's exactly what it is. See IVR.

AS Autonomous System. An Internet term. An Autonomous System is just that — a system which is autonomous. Typically, an AS is an ISP, an Internet Service Provider. Within the ISP, routers exchange information freely — all systems are trusted, as they are under a single administration in the same domain. Therefore, such systems can run an IGP (Interior Gateway Protocol) such as IGRP (Interior Gateway Routing Protocol) or OSPF (Open Shortest Path First). As the same level of trust does not exist between ASs, they must run an EGP (Exterior Gateway Protocol) such as BGP (Border

Gateway Protocol) or IDRP (InterDomain Routing Protocol). See also BGP, EGP, IDRP, IGP, IGRP and OSPF.

As Is A term used in the secondary telecom equipment business. "As is" is equipment that is bought or sold with no stated or implied warranties. You should expect any condition from good to bad, from complete to incomplete. Buy As Is equipment at your own risk.

As Is Tested or As Is Working A term used in the secondary telecom equipment business. One step up from "as is" condition. The product has been tested. It works and is complete, unless otherwise specified. Buyer should test upon receipt. There is no warranty beyond receipt. Seller is guaranteeing the product will work upon arrival. After that, the buyer is responsible for any problems.

AS&C Alarm Surveillance and Control

AS/400 IBM's mid-range mini-computer. AS/400 stands for Application System/400. IBM has a product called CallPath/400 which allows AS/400 computers to link to PBXs from the leading manufacturers.

ASA 1. Average Speed of Answer. How long the average caller has to wait before they speak to an agent. The time can vary, even over the course of one day, due to call volumes and staff levels. An important measure of service quality. ASA is used in most call centers.
2. Affiliated Sales Agency. A term for a company which resells the service of a phone company. Typically, the phone company pays the ASA a commission. Sometimes the commission is so large that it blurs the thinking of the ASA into recommending to its customers telecom products and services they would be better without.

ASC AIN Switch Capabilities. See AIN.

ASCI-Assisted Routing A layer 3 switch that has some of its routing functionality built within ASCIS.

ASCII Pronounced: as'-kee. American Standard Code for Information Interchange. It's the most popular coding method used by small computers for converting letters, numbers, punctuation and control codes into digital form. (Computers can only understands zeros or ones.) Once defined, ASCII characters can be recognized and understood by other computers and by communications devices. ASCII defines 128 characters, including alpha characters, numbers, punctuation marks or signals in seven on-off bits and a parity bit (used for data). A capital "C", for example, is 1000011, while a "3" is 0110011. As a seven-bit code, and since each bit can only be a "one" or a "zero,"
ASCII can represent 128 "things," i.e. $2 \times 2 \times 2 \times 2 \times 2 \times 2 \times 2$ which equals 128. ASCII is the code virtually every personal computer in the world encodes "things," including IBM, Apple and Radio Shack/Tandy. This compatible encoding (it was developed by ANSI — the American National Standards Institute) allows virtually all personal computers to talk to each other, if they use a compatible modem, or null modem cable and transmit and receive at the same speed. There are variations of ASCII. (Nothing is totally standard anymore.) The most important variation — one originally from IBM — is called Extended ASCII. It codes characters into eight bits (or one byte) and uses those ASCII characters above 127 to represent foreign language letters, and other useful symbols, such as those to draw boxes. But at 127 and below, extended 8-bit ASCII is identical to standard 7-bit ASCII. The ITU (now called the ITU-T) calls ASCII International Telegraph Alphabet 5.
The other major method of encoding is IBM's EBCDIC (pronounced ebb'-si-dick). It's largely used on IBM and IBM-compatible mainframe computers (but not their PCs, which use ASCII and extended ASCII.) EBCDIC is an eight-bit encoding scheme, thus allowing up to 256 "things" to be encoded, i.e. $2 \times 2 \times 2 \times 2 \times 2 \times 2 \times 2 \times 2 = 256$. EBCDIC codes letters, characters and punctuation marks in a totally different way than ASCII. For ASCII files to be read by an IBM mainframe (one that reads EBCDIC), those ASCII files must be translated into EBCDIC by one of the many translation programs available. See also ASCII Editor, Baudot, EBCDIC, Extended Graphics Character Set, Morse Code and Unicode.

ASCII Editor An ASCII editor (also called a "text," "DOS" or "non-document mode" editor) does NOT use extended ASCII and printer [ESCAPE] codes, which are used by word processor to create advanced features such as bold, italic, underlining, and super/subscript printing effects; and fancy formatting such as automatic paragraph reformat, pagination, hyphenation, footers, headers, and margins. I initially wrote this dictionary using an ASCII editor called ZEdit, which is a customized version of QEdit, undoubtedly the best editor ever written. Then, the author QEdit, produced a new and more powerful editor, called The Semware Editor. And I'm now using it to write this edition. Since an ASCII editor can't do so much, why would anyone use one? Well, its strength is in the lack of those very things a word processor has, which clutter it and slow it down! Here are my benefits:
1. It's lightning fast. No word processor can match an ASCII editor's speed at loading itself, loading files, finding things in files, etc.
2. A file produced by an ASCII editor can be read and edited by any word processor (absolutely any). Thus it's the universal word processing file. A WordPerfect file typically can't be read by WordStar and vice versa. The reason is that every word processor uses different high-level codes for the same features (underlining, bolding, etc.) There is no consistency among word processors as to how they encode their text so they can tell printers to do bolding, etc.
3. An ASCII editor is better to type programming languages, such as EDLIN (for batch files), BASIC, FORTRAN, PASCAL, etc. If QEDIT used extended ASCII and printer codes, it could not be used by these programs...for each program interprets these "high level" codes differently from another program. An ASCII editor types straight, "vanilla" text...nothing fancy about it.

ASCII File An ASCII file consists solely of ASCII 127 and below ASCII characters that are visible. You create an ASCII file using a simple editor, also called an ASCII editor. An ASCII file is also called a text file. See ASCII.

ASCII-To-Fax Conversion Allows the transfer of a word-processed file directly to your fax board so it can be faxed without being scanned from a hard copy print-out. Documents faxed with ASCII-TO-FAX conversion come out much cleaner at the other end, since the scanning process always degrades the image.

ASDS Accunet Spectrum of Digital Services. AT&T's leased line (also called private line) digital service at 56 Kbps. MCI and Sprint have similar services. It is available in N x 56/64 Kbps, for N = 1, 2, 4, 6, 8, 12. The 56/64 Kbps POP-POP service (between long distance carrier central offices) costs the same as an analog line.

ASE A messaging term. Application Service Element. A module or portion of a protocol in the application layer 7 of the OSI (Open Systems Interconnection) protocol stack. Several ASEs are usually combined to form a complete protocol, e.g., the X.400 P1 protocol which consists of the MTSE (Message Transfer Service Element), and the RTSE (Reliable Transfer Service Element).

ASH Ardire-Stratigakis-Hayduk, a synchronous compression algorithm that is said to offer four times throughput on a typical synchronous channel. It can be used in bridges, routers, ISDN and modems. Transcend of Cleveland, OH said at one point that it was the exclusive licensor of ASH.

ASI 1. Advanced Services Implementation.

2. Application Software Interface. An important ISDN term. See Application Software Interface.

3. Adapter Support Interface. The driver specification developed by IBM for networking over IEEE 802.5 Token-Rings.

ASIC Application Specific Integrated Circuit. This is a chip that has been built for a specific application. Manufacturers use it to consolidate many chips into a single package, reducing system board size and power consumption. Many video boards and modems use ASICs. ASICs span programmable array logic (PAL) devices, electrically programmable logic devices (EPLDs), field programmable logic devices (FPGAs), gate arrays, standard cell-based devices, and full custom, designed from scratch ICs. See also ASSP.

ASIC Chip Application Specific Integrated Circuit Chip. A fancy name for microprocessor chips which do specific tasks. For example, an ASIC chip might be responsible for a graphics display.

ASL Adaptive Speed Leveling. A US Robotics term for adjusting the transmission speed of a modem up or down, depending on the conditions on the line. US Robotics says it can adjust speed in 2 or 3 seconds after detecting changed line conditions. It requires like modems on either end of the transmission.

ASN Abstract Syntax Notation.

ASN.1 Abstract Syntax Notation One. LAN "grammar," with rules and symbols, that is used to describe and define protocols and programming languages. ASN.1 is the OSI standard language to describe data types. The Abstract Syntax Notation is a formal language defined by ITU X.208 and ISO 8824. Under both CMIP and SNMP, ASN.1 defines the syntax and format of communication between managed devices and management applications. See CMIP, SNMP. For a fuller definition, see Abstract Syntax Notation One.

ASP 1. A Northern Telecom term for Attached Support Processor.

2. Adjunct Service Point. An intelligent-network feature that resides at the intelligent peripheral equipment and responds to service logic interpreter requests for service processing. See also AIN.. 3. Administrable Service Provider. A SCSA term.

3. Abstract Service Primitive. An ATM term. An implementation-independent description of an interaction between a service-user and a service-provider at a particular service boundary, as defined by Open Systems Interconnection (OSI).

4. Active Server Page. A dynamic HTML scheme designed by Microsoft that is a combination of HTML and VBScript (Visual Basic Script)

Aspect Ratio The ratio of width to height of a computer display or TV screen. The aspect ratio of NTSC and PAL TV is four units of width to every three units of height. This is expressed as 4 x 3 aspect ratio. A 35mm frame measures 12 x 24 mm, which means it has two units of width to one unit of height. It is different in size from a TV screen. This is why the side parts of movies are chopped off on TV. For VGA and Indeo video technology, the aspect ratio is 4:3 yielding today's standard PC screen sizes in pixels of 640 x 480 and 1024 x 768.

ASPI ASPI stands for Advanced SCSI Programming Interface set, which is software primitives and data structures which allow software using the ASPI interface to be SCSI host adapter-independent. SCSI stands for Small Computer System Interface. (Pronounced Scuzzie.) ASPI is software that

acts as a liaison between SCSI device drivers (the software that drives the SCSI devices) and the interface card (also known as the host adapter). Whenever a new device is added to a computer system, a software program called a "driver" must tell the computer how to talk to the new device. Instead of forcing vendors to write drivers for every host adapter, ASPI lets them write a driver to ASPI standards, supposedly guaranteeing that the device the driver controls will work with all ASPI-compatible host adapters.

The idea behind ASPI is to create a "black box" software interface - one which allows programmers to create software without having to know anything about the details of the SCSI interface hardware used in your computer. With ASPI, it's possible to write programs that can be used with any SCSI-based device used on a computer system that supports ASPI. While things are not always 100% perfect in all cases, ASPI greatly reduces potential compatibility problems for you, the user.

How does ASPI work? Essentially, there are two parts to an ASPI implementation. First, there's the ASPI "manager" which is a device driver supplied by the hardware manufacturer, and the ASPI software application. It's important to note that without an ASPI manager, ASPI compatibility is not possible. It's the manager that creates the standard ASPI-compatibility layer between the SCSI host adapter hardware and the ASPI-compatibility application. The manager is very hardware-specific, and is almost always supplied by the manufacturer of your SCSI host adapter.

ASQ Automated Status Query.

ASR 1. Automatic Speech Recognition. See Interactive Voice Response.

2. Automatic Send-Receive teletype or telex machine. Such a machine, if left on and loaded with paper, will receive incoming messages and print them, even when nobody is present. See also Automatic Send Receive.

3. Access Service Request. This is a request that a telehone company gives to another telephone company for any of many kinds of interconnectivity or data sharing needs. These requests can be between Local carriers or long distance carriers and can originate with either an incumbent or an alternative company.

4. Authorized Sales Representative. Many phone companies have programs which allow interconnect companies to resell their services — from simple local lines to T-1 lines.

Assembler A program which translates an assembly programming language into the code of ones and zeros used by computers. See also Assembly code and Assembly language.

Assembler code A programming language that is a close approximation of the binary machine code. Difficult to write and different for each processor.

Assembly Pertaining to the translation of a program from symbolic language into machine code. See Assembly Language.

Assembly Language A computer language for writing software. It is a language which is converted by programs called compilers or interpreters into machine language programs which consist of only 1s and 0s and which a computer can understand. Even though an assembly language consists of recognizable menonics and meaningful words, it's not easy to program in. It is referred to as a "low-level language". Assembly language programs run faster than high-level language programs, such as BASIC, COBOL or FORTRAN, which are much easier to learn and program in. Choosing a programming language is a tradeoff of ease for speed.

Asserted A signal is asserted when it is in the state which is indicated by the name of the signal. Opposite of Negated.

Assignation A secret romantic rendezvous. The invitation to an assignation doesn't work if she doesn't know the meaning of the word. Are you listening Jane Laino?

Assigned Cell An ATM term. Cell that provides a service to an upper layer entity or ATM Layer Management entity (ATM-entity).

Assigned Frequency The center of the assigned frequency band assigned to a station.

Assigned Frequency Band The frequency band within which the emission of a station is authorized; the width of the band equals the necessary bandwidth plus twice the absolute value of the frequency tolerance. Where space stations are concerned, the assigned frequency band includes twice the maximum Doppler shift that may occur in relation to any point of the Earth's surface.

Assigned Night Answer ANA. After business hours or when you place your phone system on "Night Answer," this feature sends calls from specified trunks to designated extensions or departments. You may use this feature to send calls directly to modems, or to emergency numbers, or even to outside home numbers.

Assigned Plant Concept A pair is dedicated from the central office to the subscriber home and maintained at that address, even when idle. See Reassignment.

Assignment A call center term. The process of assigning individual employees to specific schedules in a Master File or Daily Workfile. Master File assignment can be done either manually or automatically (based on employee schedule preference and seniority). See Assignment Lists.

Assignment Lists In a non-mechanized line assignment environment in a telephone company, assignment lists of lines and numbers are prepared by the Network Administrator as a means of providing input to the Service Center for service order preparation. The lines and numbers made available for assignment are determined by the guidelines for overall loading plan and load balance objectives. The age of telephone numbers is also a consideration. Also see Intercept Interval.

Associate A verb used in Windows by File Manager. You associate a three character extension with an application. This tells File Manager that, when you click twice on the file, File Manager will know which application to launch. For example, you may tell Windows that the .QXD extension is associated with QuarkXpress. When you click on a QXD file, File Manager will launch Quark and load that particular file.

Associated Common-Channel Signaling A form of common-channel signaling in which the signaling channel is associated with a specific trunk group and terminates at the same pair of switches as the trunk group. The signal channel is usually transmitted by the same facilities as the trunk group.

Association A relationship between two connection segments that share a common Leg O (i.e., a common subscriber is in control of connection segments). Definition from Bellcore.

Association Control Service Element ACSE. The International Standards Organization's Open Systems Interconnect (OSI) application layer services used, for example, in Manufacturing Automation Protocol V3.0 (MAP).

ASSP Application Specific Standard Product is an integrated circuit that performs functions for a single application (e.g., keyboard controller). ASSP is a more precise term for a device often referred to as an ASIC.

Assurance Level Probability expressed as a percent. Example: There is 90% Assurance (probability) that the mean holding time on the trunk group is between 168.5 and 191.5

seconds.

AST Automatic Scheduled Testing. A method of testing switched access service (Feature Groups B, C, and D) where the customer provides remote office test lines and 105 test lines with associated responders or their functions' equivalent; consists of monthly loss and C-message noise tests and annual balance test.

ASTM American Society for Testing and Materials, a non-profit industry-wide organization which publishes standards, methods of test, recommended practices, definitions and other related material.

ASU Application-Specific Unit.

Asymmetric Not symmetric, i.e., unbalanced. A asymmetric telecom channel has more bandwidth (i.e. speed) in one direction than in the other. Its bandwidth is unbalanced. There are reasons for this. Take the Internet. Grabbing stuff from the Internet to your PC needs more bandwidth than sending stuff back from your PC. At least that's one theory. To accommodate this theory, for example, there's ADSL (Asymmetric Digital Subscriber Line). ADSL provides asymmetric bandwidth, as the downstream (from the network to the user premises) bandwidth of as much as 6.144 Mbps, and a return channel (from the user premises to the central office) of something like 608 Kbps. Asymmetric can also refer to the physical topology of the network. For example, a point-to-multipoint circuit might connect one device on the East Coast directly to three devices on the West Coast through the use of a bridge. For example, Miller Freeman might lease a multi-point circuit which connects its New York office to its office in San Francisco. At the San Francisco office is a bridge which has three drops, 1 for the San Francisco office and one for each of its two offices in Menlo Park. All communications between the sites take place through the multidrop bridge. The circuit is assymetric as it lacks symmetry. There is one site connected on the East Coast and there are three sites connected on the West Coast. Multipoint circuits also are known as multi-drop circuits and fan-tail circuits, as they fan out like the tail of a fish on the distant end. See the next several definitions. See also ADSL, Full Duplex and Symmetric.

Asymmetric Digital Subscriber Line See ADSL and Asymmetric Digital Subscriber Line Transceiver.

Asymmetric Digital Subscriber Line Transceiver A microprocessor chip that is the crux of asymmetric digital subscriber line service. I found the following description of just such a chip in Motorola literature describing their MC145650 144-pin transceiver. "The MC145650 is a single integrated circuit transceiver device for ANSI (American National Standard Institute) T1.413 category 2 ADSL modems, based on the Discrete Multi-Tone (DMT) line code. The category 2 specification requires payload rates of (6.144 Mbps + 640 Kbps) downstream and 640 Kbps upstream, with crosstalk, over carrier serving area (CSA) range loops, and to achieve (1.544 Mbps + 176 Kbps) downstream and 176 Kbps upstream with crosstalk, over selected ANSI integrated services digital network (ISDN) loops. The payload makeup is flexible, thereby allowing multiple data streams to be multiplexed and demultiplexed. The MC145650 is capable of data rates up to 8 Mbps downstream and 1 Mbps bidirectionally; however, actual data rates obtained in any system are dependent on loop length, impairments, and transmitted power. The ADSL and DMT techniques are adaptive, changing system parameters based on loop characteristics in order to optimize the data route."

Asymmetrical Compression Techniques where the

decompression process is not the reverse of the compression process. Asymmetrical compression is more computer-intensive on the compression side so that the decompression of video images can be easily performed at the desktop or in applications where sophisticated codecs are not cost effective. In short, any compression technique that requires a lot of processing on the compression end, but little processing to decompress the image. Used in CD-ROM creation, where time and costs can be incurred on the production end, but playback must be inexpensive and easy. See ASYN.

Asymmetrical Modem A type of modem which uses most of the available bandwidth for transmission and only a small part for reception.

Asymmetrical Modulation A duplex transmission technique which splits the communications channel into one high speed channel and one slower channel. During a call under asymmetrical modulation, the modem with the greatest amount of data to transmit is allocated the high speed channel. The modem with less data is allocated the slow, or back channel. The modems dynamically reverse the channels during a call if the volume of data transfer changes.

Asymmetrical Multiprocessing A relatively simple implementation of multiprocessing in which the operating system kernel runs on one dedicated CPU and assigns tasks as they come in to other "slave processors." It is also known as "master/slave" processing.

Asymmetrical PVC This terms refers to a PVC (Private Virtual Circuit) which supports simplex, or asymmetrical, assignments of committed information rate in each direction of transmission. A PVC transmission path is duplex, meaning that there must be a communications path in each direction between the two points being connected. However with an asymmetrical PVC, the network capacity in each direction does not necessarily have to be equal.

Asyn Greek prefix meaning "not."

Asynchronous See ASYNCHRONOUS TRANSMISSION.

Asynchronous Balanced Mode ABM. Used in the IBM Token Ring's Logical Link Control (LLC), ABM operates at the SNA data link control and allows devices on a Token Ring to send data link commands at any time and to initiate responses independently.

Asynchronous Completion A Versit definition. A domain issues a service request and need not wait for it to complete. If the application waits for this completion, this is known as synchronous, but if it is sent off to another system entity and the domain goes on to other activities before the service request completes (and the system later sends a message to the domain announcing the service's completion), that

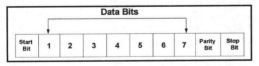

completion is known as Asynchronous.

Asynchronous Gateway A routing device used for dial-up services such as modem communications.

Asynchronous Mapping A SONET term. SONET optical fiber transmission systems run at a very high rate of speed, of course. In fact, SONET runs at a minimum of 51.84 Mbps, which is the foundation transmission level known as OC-1 (Optical Carrier Level 1), the OC-1 frame begins as a T-3 electrical signal at 44.736 Mbps. The native format of the incoming signals always is electrical in nature, and originates at

various speeds. Examples are 64 Kbps (DS-0), 1.544 Mbps (DS-1 — specifically, T-1), 2.048 Mbps (DS-1 — specifically, E-1), or 44.736 (DS-3 — specifically, T-3). As these incoming signals of various speeds are presented to the SONET facility, they are multiplexed to form a T-3 frame and are converted from the T-3 electrical format to the OC-1 optical format. The OC-1 frames then are mapped into (presented to, accepted by, and fit into) the SONET facility in an asynchronous fashion. While the SONET transmission facility, itself, is highly synchronized, it deals with inputs on an asynchronous (start-stop) fashion. These mappings are defined for clear channel transport of digital signals that meet the standard DSX cross connect requirements, typically DS-1 and DS-3 in most practical applications, although DS-2 is also supported. See also SONET.

Asynchronous Request An SCSA term. A request where the client does not wait for completion of the request, but does intend to accept results later. Contrast with synchronous request.

Asynchronous Teleconferencing. An interactive group communication that allows individuals to communicate as a group without being present together in time or place. Participants to join and exit the conference when it is convenient for them, leaving messages for others and receiving messages left for them. Computer conferencing is an example of asynchronous teleconferencing.

Asynchronous Terminal A terminal which uses asynchronous transmissions. See Asynchronous Transmission.

Asynchronous Time Division Multiplexing A multiplexing technique in which a transmission capability is organized in a priori unassigned time slots. The time slots are assigned to cells upon request of each application's instantaneous real need.

Asynchronous Transfer Mode ATM is the technology selected by the Consultative Committee on International Telephone & Telegraph (ITU) International standards organization in 1988 (now called the ITU-T) to realize a Broadband Integrated Services Digital Network (B-ISDN). It is a fast, cell-switched technology based on a fixed-length 53-byte cell. All broadband transmissions (whether audio, data, imaging or video) are divided into a series of cells and routed across an ATM network consisting of links connected by ATM switches. Each ATM link comprises a constant stream of ATM cell slots into which transmissions are placed or left idle, if unused. The most significant benefit of ATM is its uniform handling of services, allowing one network to meet the needs of many broadband services. ATM accomplishes this because its cell-switching technology combines the best advantages of both circuit-switching (for constant bit rate services such as voice and image) and packet-switching (for variable bit rate services such as data and full motion video) technologies. The result is the bandwidth guarantee of circuit switching combined with the high efficiency of packet switching. For a longer explanation, see ATM.

Asynchronous Transmission Literally, not synchronous. A method of data transmission which allows characters to be sent at irregular intervals by preceding each character with a start bit, and following it with a stop bit. It is the method most small computers (especially PCs) use to communicate with each other and with mainframes today. In every form of data transmission, every letter, number or punctuation mark is transmitted digitally as "ons" or "offs." These characters are also represented as "zeros" and "ones" (See ASCII). The problem in data transmission is to define when the letter, the number or the punctuation mark begins. Without knowing when it begins, the receiving computer or terminal won't be able to figure out what the transmission means.

One way to do this is by using some form of clocking signal. At a precise time, the transmission starts, etc. This is called synchronous transmission. In asynchronous transmission there's no clocking signal. The receiving terminal or computer knows what's what because each letter, number or punctuation mark begins with a start bit and ends with a stop bit. Transmission of data is called synchronous if the exact sending or receiving of each bit is determined before it is transmitted or received. It is called asynchronous if the timing of the transmission is not determined by the timing of a previous character.

Asynchronous is used in lower speed transmission and by less expensive computer transmission systems. Large systems and computer networks typically use more sophisticated methods of transmission, such as synchronous or bisynchronous, because of the large overhead penalty of 20% in asynchronous transmission. This is caused by adding one start bit and one stop bit to an eight bit word — thus 2 bits out of ten. The second problem with large transfers is error checking. The user sitting in front of his own screen checks his asynchronous transmission by looking at the screen and re-typing his mistakes. This is impractical for transferring long files at high speed if there is not a person in attendance.

In synchronous transmission start and stop bits are not used. According to the book Understanding Data Communications, characters are sent in groups called blocks with special synchronization characters placed at the beginning of the block and within it to ensure that enough 0 to 1 or 1 to 0 transitions occur for the receiver clock to remain accurate. Error checking is done automatically on the entire block. If any errors occur, then the entire block is retransmitted. This technique also carries an overhead penalty (nothing is free), but the overhead is far less than 20% for blocks or more than a few dozen characters.

AT 1. Access Tandem.

2. Advanced Technology. Refers to a 16 bit Personal Computer architecture using the 80X86 processor family which formed the basis for the ISA Bus as found in the first IBM PC.

3. AudioTex. See AudioTex.

4. See AT COMMAND SET.

AT Bus The electrical channel used by the IBM AT and compatible computers to connect the computer's motherboard and peripheral devices, such as memory boards, video controllers, PC card modems, bus mouse boards, hard and floppy disk controllers and serial/parallel input/output devices. The AT bus supports 16 bits of data in one slug, whereas the original IBM PC supported only 8 bits (and was called the ISA bus for Industry Standard Architecture. These days there are much faster "buses," including the EISA, MCA (MicroChannel Architecture), Local Bus, PCI, VESA, etc.

AT Command Set Also known as the Hayes Standard AT Command Set. A language that enables PC communications software to get an asynchronous and "Hayes-compatible modem" to do what you want it to do. So called "AT" because all the commands begin with "AT," which is short for ATtention. The most common commands include ATDT (touchtone a number), ATA (manually answer the phone), ATZ (reset modem — it will answer OK), ATSO=0 (disable auto-answer), and ATH (hang up the phone).

To avoid having yourself knocked off your data call by the

beep that comes in on the phone company's call waiting, put the following line in your modem setup: ATS10=20. That will increase your S10 register to two seconds. This register sets the time between loss of carrier (caused by the 1.5 second call waiting signal) and internal modem disconnect. Factory default on most modems is 1.4 seconds — just perfect to be cut off by the wall waiting tone! (Dumb.)

If you have to dial through several phone systems, waiting for dial tone on the way and/or going through fax/modem switches, you may consider a dial stream that looks like ATDT 1-800-433-9800 [W]212-989-4675 [W]22, where [W] means (in some software programs) "Wait for any key. When you get it, touchtone out the next digits." In other software programs — pure Hayes command — W means wait for second dialtone.

If W in square brackets doesn't work for you, then change X3 in your setup line to X1; change your computer's dialed number to 9; and dial your distant computer with your phone. When you hear the modem at the other end answer, tell your computer's software to dial 9. It will dial 9, hear the modem tone at the other end and connect as though it had dialed it all by itself. X1 tells your modem to dial (or touchtone) immediately — without waiting for dial tone.

You can use several AT commands on one line. You only need AT before the first one. Some modems require commands typed in capital letters. When your dialing fails and you can't figure why, get out of your communications software program and start again. Or in total desperation, turn your computer and modem completely off and start again. The word "Hayes" comes from the manufacturer of modems called Hayes Microcomputer, Norcross, GA, the creator of the command set. Not all Hayes compatible modems are. See also AT+V and Hayes Command Set.

At Local Mode One of the command modes available on the ISDN set. It is used for compatibility with existing communications packages for analog modems or for data-only application programs. See AT Command Set.

At Work Pronounced "At Work." Microsoft's office equipment architecture announced on June 9, 1993. Microsoft's idea was to put a set of software building blocks into both office machines and PC products, including Desktop and network-connected printers; Digital monochrome and color copiers; Telephones and voice messaging systems; Fax machines and PC fax products; Handheld systems and Hybrid combinations of the above. At Work didn't go very far. But Windows CE came out and became popular.

AT#V See AT+V below.

AT&T Consent Decree The Telecommunications Act of 1996 defined it as follows. The term `AT&T Consent Decree' means the order entered August 24, 1982, in the antitrust action styled United States v. Western Electric, Civil Action No. 82-0192, in the United States District Court for the District of Columbia, and includes any judgment or order with respect to such action entered on or after August 24, 1982. See Telecommunications Act of 1996.

AT+V V standards for voice. AT+V is a new ANSI standard for voice modems. It's a superset of the Hayes AT command set which worked so well in modems. AT+V combines pre-fixed Hayes AT commands with a new set of voice-related +V commands. The specification is detailed in ANSI/TIA/EIA IS-101 "Facsimile Digital Interfaces — Voice Control Interim Standard for Asynchronous DCE." The TIA TR-29.2 subcommittee details the specification in their PN-3131. Rockwell's voice modem chipset does not comply with this standard, but uses another called AT#V, which is similar. In Windows 95, the variance between these command sets is ratified by the Win 95 system registry and vendor-supplied INF files. See also WWindows Telephony.

ATA 1. American Telemarketing Association. The professional industry association for telephone sales and marketing.
2. Analog Terminal Adapter. A device for a Northern Telecom Norstar phone system that lets it use analog devices, for example FAX, answering machines, modems and single line phones, behind the Norstar's central telephone unit (its KSU). Before you buy the analog terminal adapter, check that its speed is fast enough for you. In mid-1995, it was constrained to 9,600 bps, or 14,400 bps if the phone line was clear.
3. AT Attachment. Refers to the interface and protocol used to access a hard disk on AT compatible computers. Disk drives adhering to the ATA protocol are commonly referred to as IDE interfaced drives for PC compatible computers. The ATA specification is fully backward compatible with the ST-506 standard it superseded. IDE drives are sometimes referred to as ATA drives or AT bus drives. The newer ATA-2 specification defines the EIDE interface, which improves upon the IDE standard. See ATA2, IDE and Enhanced IDE.

ATA2 The second generation AT attachment specification for IDE devices that defines faster transfer speeds and LBA (Logical Block Address) sector-locating method. See ATA, IDE and Enhanced IDE.

ATA Document The latest draft of the ANSI X3.T9 subcommittee AT Attachment document.

ATA Registers These registers are accessed by a host to implement the ATA protocol for transferring data, control and status information to and from the PC Card. They are defined in the ATA Document. These registers include the Cylinder High, Cylinder Low, Sector Number, Sector Count, DriveHead, Drive Address, Device Control, Error, Feature, Status and Data registers. The I/O and memory address decoding options for these registers are defined within this specification.

ATAPI Attachment Packet Interface specification does for CD-ROM and tape drives why ATA-2 does for hard drives. It defines device-side characteristics for an IDE-connected peripheral. The benefits of having a single interface for the most common non-disk storage device in the desktop world, the CD-ROM are obvious. For the manufacturer, there is no need to add a separate controller card for the CD-ROM. For the end-user it means no more fussing with interrupts, cards and proprietary driver software. ATAPI essentially adapts the established SCSI command set to the IDE interface.

ATB All Trunks Busy. One measure which your phone company or phone systems might give you of telephone traffic in and out of your office. See All Trunks Busy.

ATD 1. Asynchronous Time Division.
2. ATtention Dial the phone. The first three letters in the most frequently-used command in the Hayes command set for asynchronous modems — typically those used with microcomputers.

ATDNet Advanced Technology Demonstration Network. A joint research effort of Bellcore, Bell Atlantic, and the U.S. Government, this network is aimed at demonstrating the efficacy of advanced technologies in the network of the future.

ATIS Alliance for Telecommunications Industry Solutions, a trade group based in Washington, D.C. and open to membership of North American and World Zone 1 Caribbean telecommunications carriers, resellers, manufacturers, and providers of enhanced services. Originally called the Exchange Carriers Standards Association (ECSA), the ATIS is heavily involved in

standards issues including interconnection and interoperability issues. www.atis.org

ATM 1. Automated Teller Machine. The street corner banking machine which is usually hooked up to a central computer through leased local lines and a multiplexed data network. For the most part, ATM machines traditionally worked over multipoint DDS circuits, although a wide variety of network technologies could be employed. In fact, some ATM networks work over ATM. See 2.
2. Asynchronous Transfer Mode. Very high speed transmission technology. ATM is a high bandwidth, low-delay, connection-oriented, packet-like switching and multiplexing technique. Usable capacity is segmented into 53-byte fixed-size cells, consisting of header and information fields, allocated to services on demand. The term "asynchronous" applies, as each cell is presented to the network on a "start-stop" basis—in other words, asynchronously. The access devices, switches and interlinking transmission facilities, of course, are all highly synchronized.

Here's some history on ATM from the Networking Alliance: The ATM method of moving information is not completely new. Like most things it is an evolution of earlier methods. The key difference between ATM and "X.25 packet switching" and the popular "Frame Relay" technologies is that the packets of the earlier technologies varied in size. Engineers realized that as the speed was dramatically increased to be able to carry "real time" voice and video, the varied length packets would become unmanageable. During the 1980s the ITU, now the ITU-T (International Telecommunications Union-Telecommunications Services Sector), adopted ATM as the transport technology of the future. Ultimately and after a great deal of debate, the ITU-T determined that each cell would be 53 octets long. To meet current and future demands, networking technologies and protocols have evolved to optimize network performance based on traffic characteristics. ATM represents the first world-wide standard to be embraced by the computer, communications and entertainment industries.

Each ATM cell contains a 48-octet payload field, the size of which has an interesting background. Data people prefer to move data in huge blocks or frames, which are more efficient for large file transfers. Voice people, on the other hand prefer tiny blasts of data, which are more effective for moving digitized voice samples (ala PCM in a T-Carrier environment). Since ATM is positioned as the ultimate service offering in support of data , voice data, video data, image data, and multimedia data, the small payload prevailed. With that battle out of the way, the European and U.S. camps clashed, with the

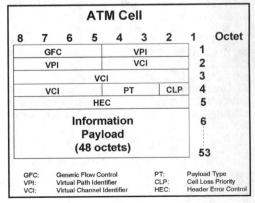

ATM Cell

8	7	6	5	4	3	2	1	Octet
GFC				VPI				1
VPI				VCI				2
VCI								3
VCI				PT			CLP	4
HEC								5
Information Payload (48 octets)								6 ⋮ 53

GFC:	Generic Flow Control	PT:	Payload Type
VPI:	Virtual Path Identifier	CLP:	Cell Loss Priority
VCI:	Virtual Channel Identifier	HEC:	Header Error Control

European Telecommunications Standards Institute (ETSI) proposed a 32-octet cell and the U.S. Exchange Carriers Standards Association (ECSA) proposed a 64-octet cell—the issue was the difference in standard PCM voice encoding techniques. After lengthy wrangling, it was decided that a 48-octet cell would be the perfect mathematical compromise. Although neither camp was perfectly pleased (such tends to be the nature of a compromise, I am told), it was a solution that all could accept.

In any event, each cell also is prepended with a 5-octet Header which identifies the Virtual Path (Virtual Circuit), Virtual Channel, payload type, and cell loss priority; as well as providing for flow control, and header error control.

The small, fixed-length cells require lower processing overhead and allow higher transmission speeds than traditional packet switching methods. ATM allocates bandwidth on demand, making it suitable for high-speed connection of voice, data, and video services. ATM services will be available at access speeds up to 622 Mbps, with the backbone carrier networks operating at speeds currently as high as 2.5 Gbps. The ATM edge and core backbone switches operate at very high speeds, and typically contain multiple busses providing aggregate bandwidth of as much as 200+ Gbps. ATM core switches currently are available with capacities of as much as one terabit per second, although none have been deployed at this level.

Here's a full explanation: Conventional networks carry data in a synchronous manner. Because empty slots are circulating even when the link is not needed, network capacity is wasted. The ATM concept which has been developed for use in broadband networks and optical fiber based systems is supported by both ITU-T (nee ITU) and ANSI standards, can also be interfaced to SONET (Synchronous Optical Network). ATM automatically adjusts the network capacity to meet the system needs and can handle data, voice, video and television signals. These are transferred in a sequence of fixed length data units called cells. Common standards definitions are provided for both private and public networks so that ATM systems can be interfaced to either or both. ATM is therefore a wideband, low delay, packet-like switching and multiplexing concept that allows flexible use of the transmission bandwidth and capable of working at data rates as high as 622.08 Mbps, with even higher rates planned. Each data packet consists of five octets of header field plus 48 octets for user data. The header contains data that identifies the related cell, a logical address that identifies the routing, header error correction bits, plus bits for priority handling and network management functions. Error correction applies only to the header as it is assumed that the network medium will not degrade the error rate below an acceptable level. All the cells of a Virtual Path (VP) follow the same path through the network that was determined during call set-up. (Note that ATM is a connection-oriented network service.) As there are no fixed time slots in the system, any user can access the transmission medium whenever an empty cell is available. ATM is capable of operating at bit rates of 155.52 and 622.03 Mbps; the cell stream is continuous and without gaps. The position of the cells associated with a particular VC is random, and depends upon the activity of the network. Cells produced by different streams to the ATM multiplexer are stored in queues awaiting cell assignment. Since a call is accepted only when the necessary bandwidth is available, there is a probability of queue overflow. Cell loss due to this forms one ATM impairment. However, this can be minimized through the use of statistical multiplexers. Bit errors in the header which are beyond the FEC capability can lead to misrouting.

While ATM was developed as a backbone WAN technology, a 25.6 Mbps version of ATM was reluctantly approved by the ATM Forum for use in a LAN workgroup environment. The Desktop ATM25 Alliance, which promoted the standard, disbanded in 1996 due to lack of interest. ATM has continued to march to the desktop, however slowly and at the higher speeds. ATM also has found its way into the LAN world through the development of cost-effective, high-performance ATM LAN backbone switches. PBX manufacturers also are working diligently to determine how best to incorporate ATM switching fabrics into voice/data/video/multimedia PBX systems, resulting in an ATM-based communications controller for premise application. See also ATM Forum, ATM Access Switch and ATM Forum UNI V3.0.

ATM Access Switch A specialized ATM switch which sits on the end user premise, providing access into a carrier ATM network. The ATM Access Switch is used for such applications as distance learning and telemedicine. It is a high-capacity, cell-based switch designed to support broadband networking. Its fully integrated access, multiplexing and switching functions provide the capability for a variety of combined data, video, imaging and voice services on a single platform. See ATM.

ATM Address Defined in the UNI Specification as three formats, each having 20 bytes in length including country, area and end-system identifiers. See ATM.

ATM Backbone Switch A specialized ATM switch which sits in the carrier backbone network. The ATM Backbone Switch is claimed to be ideal for backbone networks supporting multiple services in corporations, telcos, cellular and internet public service providers. Network operators can aggregate all of their traffic over a single backbone of ATM. It is ideal for service provider backbones supporting multiple services such as cell relay, permanent virtual circuits (PVCs), switched virtual circuits (SVCs) circuit emulation, LAN interconnectivity and frame relay. The Backbone Switch has throughput traffic and traffic management features needed for large-scale ATM deployment and service offerings. ATM backbone switches include internal busses providing bandwidth of as much as 200+ Gbps, and are interconnected by SONET fiber optic transmission facilities currently operating at speeds of as much as 2.5 Gbps. See ATM.

ATM Edge Switch An ATM cell switch which sits at the edge of the carrier network, providing access from the end users' world to the carriers' ATM backbone network. It is analogous to a Central Office providing access to a Tandem network in the traditional, circuit-switched voice and data world. ATM Edge Switches also are known as Access Nodes and Service Nodes.

ATM Ethernet LAN Service Unit An ATM ELSU provides 12 independent virtual Ethernet bridges for running over ATM networks. ELSUs are designed for flexible deployment, either local to an ATM switch or at a remote site. ELSUs are designed for LAN internetworking services over ATM networks.

ATM Forum, The An industry organization with some 800 members, co-founded by N.E.T. and three other leading networking companies, which focuses on speeding the development, standardization and deployment of ATM (Asynchronous Transfer Mode) products. It has been remarkably successful. The ATM Forum is based in Mountain View, CA. Their phone number is 415-949-6700. See ATM. www.atmforum.com

ATM Forum UNI V3.0 The ATM Forum UNI V3.0 implementation agreement is based on a subset of the ITU-TS broadband access signaling protocol standards. Additions to this subset have been made where necessary to support early deployment and interoperability of ATM equipment. The procedures and protocol defined in the agreement apply to both public and private UNIs. Moreover, since the protocol is symmetrical, it also applies in the configuration ATM-end-point to ATM-end-point. See ATM and ATM Forum.

ATM Inverse Multiplexing AIMUX. A device used to combined multiple T-1 or E-1 links into a single broadband facility, over which ATM cells can then be transmitted.

ATM-25 Workgroup ATM running at 25 million bits per second. ATM-25 is mainly used on internal corporate local area networks. For a much fuller explanation, see ATM.

ATMARP ATM Address Resolution Protocol. The means of mapping IETF classical IP addresses to ATM hardware addresses. The process works in much the same way as conventional ARP, which maps network-layer addresses to the MAC (Media Access Control) layer in a LAN.

ATM Adaptation Layer SEE AAL

ATM Layer ATM. The second layer of the ATM Protocol Reference Model. At this layer are included such functions as cell multiplexing, creation of headers, flow control and selection of VPIs (Virtual Path Identifiers) and VCIs (Virtual Channel Identifiers). See ATM Layer Link.

ATM Layer Link A section of an ATM Layer connection between two adjacent active ATM Layer entities (ATM-entities).

ATM Link A virtual path link (VPL) or a virtual channel link (VCL).

ATM Peer-to-Peer Connection A virtual channel connection (VCC) or a virtual path connection (VPC).

ATM Protocol Reference Model A multidimensional protocol model consisting of 4 layers and 3 planes and serving as a point of reference for understanding, developing and implementing ATM technology. Each layer addresses a discrete set of related functions, with all layers closely interrelated. The layers include the Physical Layer, ATM Layer, ATM Adaptation Layer, and Higher Layers related to the specifics of the native user data protocol. The planes include the Control Plane, User Plane and Management Plane.

ATM Traffic Descriptor A generic list of traffic parameters that can be used to capture the intrinsic traffic characteristics of a requested ATM connection.

ATM Token Ring Lan Service Unit The ATM TLSU provides a powerful tool for offering internetworking services over ATM networks. Emulated token rings consist of up to 64 TLSU token ring ports located anywhere in the ATM network, interconnected with PVCs. These emulated token ring networks can be completely isolated form one another to ensure security and fairness among the attached LANs. The TLSUs are designed for flexible deployment, either local to an ATM switch or at a remote site. See ATM Ethernet LAN Service Unit.

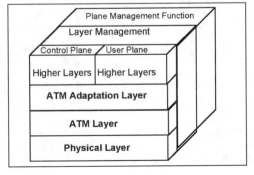

ATM User-User Connection An association established by the ATM Layer to support communication between two or more ATM service users (i.e., between two or more next higher entities or between two or more ATM-entities). The communications over an ATM Layer connection may be either bidirectional or unidirectional. The same Virtual Channel Identifier (VCI) issued for both directions of a connection at an interface.

ATP AppleTalk Transaction Protocol. Transport-level protocol that provides a loss-free transaction service between sockets. This service allows exchanges between two socket clients in which one client requests the other to perform a particular task and to report the results; ATP binds the request and response together to ensure the reliable exchange of request-response pairs.

ATS Abstract Test Suite: A set of abstract test cases for testing a particular protocol. An "executable" test suite may be derived from an abstract test suite.

ATSC Advanced Television Systems Committee. Formed by the Joint Committee on Inter-Society Coordination (JCIC) to establish voluntary standards for Advanced TV (ATV) systems, the ATSC focuses on digital television, interactive systems and broadband multimedia communications standards. Membership is open to American (North and South America, including the Caribbean) entities directly affected by the work of the committee. ATSC comprises 54 members including television networks, motion picture and television program producers, trade associations, television and other electronic equipment manufacturers and segments of the academic community. ATSC's proposal for ATV (Advanced TV) was accepted by the FCC on November 28, 1995 and was formally adopted, in the most part, as the U.S. DTV (Digital TV) standard on December 24, 1996. www.atsc.org. See JCIC, HDTV and ATV.

ATT Automatic Toll Ticketing. A system which telephone companies use to automatically keep call detail records including calling number, number called, time of day and length of call. The phone company uses this information, together with the cost of phone calls, to generate an invoice to its customers.

Attach A command that assigns a connection number to a workstation and attaches the workstation to the LOGIN directory on the default (or specified) file server. As many as 100 workstations can be attached to a file server running NetWare v2.2. When loaded the NetWare shell (workstation file NETx.COM) automatically attaches your workstation to the nearest file server. You can also specify in SHELL.CFG which server you prefer to attach to.

Attach Terminal To assign a terminal for exclusive use by the application program. Contrast with Detach Terminal.

Attachment The process of attaching a file to an electronic message.

Attack Time The time interval between the instant that a signal at the input of a device or circuit exceeds the activation threshold of the device or circuit, and the instant that the device or circuit reacts in a specified manner, or to a specified degree, to the input. The term often implies a protective action such as the provided by a clipper (peak limiter) or compressor, but may be used to describe the action of a device such as a vox (Voice Operated circuit), where the action is not protective.

Attempt Trying to make a telephone call. Also defined as a call offered to a telecommunications system, regardless of whether it is completed. More technically, an attempt is a seizure of a component of equipment. Even a momentary offhook condition of a telephone may cause a seizure (attempt).

Attendant The "operator" of a phone system console. Typically, the first person to answer an incoming call. That person usually directs incoming calls to the proper person or department. That person may also assign outgoing lines or trunks to people requesting them. Few companies spend any time training their attendants. They should. There are two types of things attendants should be trained for: 1. Manners, including the correct way to keep people waiting and to screen incoming calls, and 2. The structure of the company. If a caller asks for some help, the attendant should know which department or person might be responsible for providing that help. Increasingly in North America, company phone systems are answered by devices called "automated attendants." They allow a caller to punch in the extension he wants or the department he wants. See Automated Attendant.

Attendant Busy Lamp Field Lamps, lights or LEDs that show whether a PBX or key system extension is busy or not. These days, many attendant busy lamp fields are being incorporated into CRT displays. We hope more will do this as many lamp-based attendant busy lamp fields are difficult to read.

Attendant Call Waiting Indication An unusual feature on a PBX console. The call waiting button on the attendant console lights to indicate a predetermined number of calls in queue. The light flashes when a second (programmable) threshold is reached.

Attendant Camp-On If the extension is busy, the attendant or operator can place the call in a queue behind the call already in progress. When the call is over, the "camped-on" call will automatically ring the extension.

Attendant Conference PBX feature that allows the attendant (or operator) to establish a conference connection between central office trunks and internal phones.

Attendant Console An attendant console is the larger, specialized telephone set used by the operator or attendant to answer incoming calls and send those calls to the proper extension. Consoles are becoming more sophisticated these days in several ways. Operators need to punch fewer buttons to move calls around, while the information they present to the attendant is more useful for keeping tabs on calls and letting people know what's happening. Many consoles are acquiring TV screens that report the status of each extension, who's speaking, where the call is going, and whether there are problems, such as broken lines or trunks, etc. anywhere on the system. Some of the more modern screens will allow the operator to send messages around the company that can alert someone as to who's calling before he/she picks up the phone. You can also easily program switches through consoles with CRT (also called TV) screens. In the old days you needed to punch in complex codes. Now you can respond to "Yes/No" decisions on a screen with lots of explanatory words and help menus.

Attendant Control of Trunk Group Access The telephone operator or attendant controls the users' access to trunks for making local and/or long distance calls. This may reduce long distance call abuse.

Attendant Direct Station Select This feature gives an operator the ability to reach an extension by simply pushing one button. In direct station select, every extension has its own button. Direct Station Select usually comes with some form of Attendant Busy Lamp Field which shows whether the extensions are busy. Some attendants like direct station select. Others don't, preferring to simply punch in 345, instead of hunting for the button which corresponds to extension 345. The best consoles these days are using some form of easy-to-read screen prompts.

Attendant Exclusion A PBX feature which stops the attendant from listening in on a phone call once she or he has passed the call to the correct extension.

Attendant Forced Release An attendant-activated (pushbutton) facility that will automatically "disconnect" all parties on a given circuit when that circuit is "entered" by the attendant.

Attendant Incoming Call Control A PBX feature which diverts incoming trunk calls automatically to a predetermined phone after a predesignated period of time or number of rings.

Attendant Key Pad Allows the attendant to perform all functions using a standard touch tone key pad on the console or adjacent to it.

Attendant Locked Loop Operation PBX feature which allows the attendant at a console to retain supervision or recall capability of any particular call which has been processed.

Attendant Lockout This feature denies an attendant the ability to re-enter a phone call unless specifically recalled by that PBX extension.

Attendant Loop Transfer Allows the attendant to transfer any call to another attendant.

Attendant Monitor A special attendant circuit which allows "listening in" on all circuits with the console handset/headset transmitter deactivated.

Attendant Override A feature that allows an attendant to enter a busy trunk connection and key the trunk number within the PBX. A warning tone will be heard by the connected parties, after which they connected parties and the attendant will be in a three-way connection.

Attendant Position Where a telephone operator sits to answer calls and send them on to the people in the company. This is usually in front of a telephone system with buttons, toggle switches, etc. that facilitate this process.

Attendant Recall When a phone call has been transferred to a telephone extension and not answered, this telephone system feature sends the call back to the attendant. Sometimes the call will return to a special part of the attendant console which will indicate to the attendant that it is a "returned" call. It's a good idea to pay attention to the speed of recall back to the operator. People hate to be extended into endless ringing. Think of calling a hotel and how aggravating it is to wait until the call comes back to an operator after she/he extended it to the room...and it rings and rings.

Attendant Recall On Trunk Hold The system will recall the attendant if a trunk placed on hold is not re-entered within a predetermined time.

Attendant Transfer of Incoming Calls A PBX and Centrex feature. A telephone extension is talking on a line but that person wants to transfer the call to someone else. The person hits their hookswitch a couple of times. (The hookswitch is the toggle switch the handset depresses when you replace it.) This flashing of the hookswitch signals the attendant to join the call. The person asks the attendant "to please transfer this call." The attendant then transfers the call to the new extension. This feature is totally inefficient as it's hard to reach the attendant, who's always busy, etc. . All newer phones can transfer both incoming and outgoing calls automatically by just flashing the hookswitch, dialing the extension number and hanging up.

Attended A telephone system having an attendant or receptionist whose primary job is to answer all incoming calls. Many smaller systems, such as key systems, are not centrally "attended." The phone is simply answered by whoever is near. A non-attended phone system should be set up so anyone can answer an incoming call. Some systems, such as most key systems, come this way from the factory. Others, such as PBXs, have to be specially set up. Some systems can be set up so an attendant will get first shot at answering the incoming call, but then, after a couple of rings, anyone else can answer the call (perhaps a loud "night" bell will ring).

Attended Mode Imagine a communications situation where your computer is connected over a phone line to another user on another computer and you are uploading and downloading files. Attended mode refers to a situation where both users manually enter the commands required to send or receive a file concurrently, usually while conversing over the phone. Compare this to Unattended Mode.

Attention Key A key or combination of keys on a computer or terminal which signals the main computer to stop its present task and wait for a new command. The ESCape key is often the Attention Key. In Crosstalk, it's Control A.

Attention Management A new term covering the whole area of push technology at the desktop.

Attenuate To decrease electrical current, voltage or power in communicating channel. Refers to audio, radio or carrier frequencies. See ATTENUATION.

Attenuation The decrease in power of a signal, light beam, or lightwave, either absolutely or as a fraction of a reference value. The decrease usually occurs as a result of absorption, reflection, diffusion, scattering, deflection or dispersion from an original level and usually not as a result of geometric spreading, i.e., the inverse square of the distance effect. Optical fibers have been classified as high-loss (over 100 dB/km), medium- loss (20 to 100 dB/km), and low-loss (less than 20 dB/km). In other words, attenuation is the loss of volume during transmission. The received signal is lower in volume than the transmitted signal due to losses in the transmission medium (such as that caused by resistance in the cable). Attenuation is measured in decibels. It is the opposite of Gain. Some electrical components are listed as "with attenuation" which means they will compensate for irregular electrical supply (e.g. surges). See GAIN.

Attenuation Coefficient The rate at which average power decreases with distance.

Attenuation Equalizer Any device inserted in a transmission line or amplifier circuit to improve the shape of its frequency response.

Attenuator A device to reduce signal amplitude by a known amount without introducing distortion.

ATTND Attendant. (What else?)

Atto Atto means one quintillion, which is 10 to the power of minus 18. See also FEMTO, which is 10 to the power of minus 15. See FEMTOSECOND.

Attribute The form of information items provided by the X.500 Directory Service. The directory information base consists of entries, each containing one or more attributes. Each attribute consists of a type identifier together with one or more values. See Attributes.

Attributes Information about an MS-DOS or Windows file that indicates whether the file is read-only, hidden, or system, and whether it has been changed since it was last backed up. You can assign attributes to a file using the ATTRIB command. You can identify a file as read-only (meaning others can't change it, but can read it) and/or as a file you want to archive when using the BACKUP, RESTORE, and XCOPY commands. The command to make a file read only is typically

ATTRIB +R filename
By using the ATTRIB command to make a file "read only," you also make it impossible to erase the file from your disk. If you want to remove the "read only" protection, i.e. make the file "read and write, the command is
ATTRIB -R filename

ATU ADSL Transceiver Unit. The ADSL Forum uses terminology for DSL equipment based on the ADSL model for which the Forum was originally created. Thus, the DSL endpoint is known as the ATU-R and the CO unit is know as the ATU-C. These terms have since come to be used for other types of DSL services, like RADSL and SDSL. ATU generally represents xDSL services.

ATU-C ADSL Transmission Unit-Central Office. Special electronics in support of ADSL (Asymmetrical Digital Subscriber Line) and placed in the carrier's CO. The ATU-C has a matching unit on the subscriber premise in the form of an ATU-R. The two units, in combination, support a high data rate over standard UTP copper cable local loops. See ADSL and ATU-R.

ATU-R ADSL Transmission Unit-Remote. Special electronics in support of ADSL (Asymmetrical Digital Subscriber Line) and placed at the customer's premise. The ATU-R has a matching unit at the carrier's Central Office in the form of an ATU-C. The two units, in combination, support a high data rate over standard UTP copper cable local loops. See ADSL and ATU-C.

ATUG Australian Telecommunications Users Group, based in Milsons Point, Sydney, Australia.

ATV Advanced TV. Refers to any system of distributing television programming that generally results in better video and audio quality than that offered by the NTSC 525-line standard. This group of techniques is based on digital signal processing and transmission. HDTV (High Definition TV) can be considered one type of ATV. Although ATV systems are collectively considered to offer better quality than the NTSC signal, they can carry multiple pictures of lower-quality and can also support the cancellation of artifacts in ordinary NTSC signals.

ATX Audiotex. Interactive voice response systems that deliver information or entertainment to general telephone callers, i.e. anyone with a phone. Audiotex services are typically widely advertised. They include everything from sex to the weather. See also Audiotex.

au A UNIX sound file format, i.e. filename.au. When a Sun Microsystems or other Unix computer makes a noise, it does so in AU file format. And because the Internet is dominated by Unix boxes, you'll find a lot of AU files there. Macintosh and PC browsers are usually able to play AU file. in contrast, a sound file that originated on a PC is likely to be in WAV or MIDI format instead.

AuC AuC is also called AC. It stands for Authentication Center. A mobile term. The AuC is a piece of HLR, which is the Home Location Register — a permanent database used in GSM mobile systems (like those in Europe) to identify a subscriber and to contain subscriber data related to features and services. HLR is used to authenticate the user of mobile station equipment, i.e. a cell phone. The AuC performs secret, mathematical computations to verify the authenticity of the cell phone and thus to allow the user to make a phone call. AuC is no longer strictly coupled with the HLR. IS-41 Rev. C (now ANSI-41) defines the AuC as a stand-alone network element. See also CAVE.

Auction An FCC definition. A procedure for choosing the users of spectrum space.In auction, the federal government treasury receives the profits.

Audible Indication Control Three fancy words for the ability to turn up or down the bell or beeper on your PBX attendant console.

Audible Ring A sound sent from the called party's switch to inform the calling party that the called line is being rung. A long explanation for a bell or buzzer that tells you it's for you.

Audible Ringing Tone The information tone sent back to the calling telephone subscriber as an indication that the called line is being rung.

Audible Sound Audible sound spans a huge range of frequencies from around 20 hertz (vibrations per second) to 20 kilohertz. See Sound.

Audible Tones Audible tones are the sounds provided by the network or an attached switch to inform callers of the status of the line or of an event. Audible tones most frequently encountered in computer telephony applications include: ringing, busy tones (called party busy and network busy), SIT tones (Special Information Tones), and special tones used by computer telephony systems such as the "record at the beep" (which is usually a 1,000 HZ tone).

Audio Sound you hear which may be converted to electrical signals for transmission. A human being who hasn't had his ears blown by listening to a Sony Walkman or a ghetto blaster can hear sounds from about 15 to 20,000 hertz.

Audio Bridge In telecommunications, a device that mixes multiple audio inputs and feeds back composite audio to each stations, minus that station's input. Also known as a mix-minus audio system.

Audio Crosspoint Module Circuit board containing crosspoints for audio signal switching.

Audio Frequencies Those frequencies which the human ear can detect (usually in the range of 20 to 20,000 hertz). Only those from 300 to 3,000 hertz are transmitted through the phone, which is why the phone doesn't sound "Hi-fi."

Audio Frequency The band of frequencies (approximately 30 hertz to 20,000 hertz) that can be heard by the healthy human ear.

Audio Interchange File Format AIFF. A sound file format common to the Macintosh platform used to store digital audio data. This file format also supports position and loop point markers used by audio recording/editing software. Can be compressed into the AIFF-C format.

Audio Menu Options spoken by a voice processing system. The user·can choose what he wants done by simply choosing a menu option by hitting a touchtone on his phone or speaking a word or two. Computer or voice processing software can be organized in two basic ways — menu-driven and non-menu driven. Menu-driven programs are easier for users to use, but they can only present as many options as can be reasonably spoken in a few seconds. Audio menus are typically played to callers in automated attendant/voice messaging, voice response and transaction processing applications. See also MENU and PROMPTS.

Audio Messaging Interchange Specification AMIS. Issued in February 1990, AMIS is a series of standards aimed at addressing the problem of how voice messaging systems produced by different vendors can network or inter-network. Before AMIS, systems from different vendors could not exchange voice messages. AMIS deals only with the interaction between two systems for the purpose of exchange voice messages. It does not describe the user interface to a voice messaging system, specify how to implement AMIS in a particular systems or limit the features a vendor may implement.

AMIS is really two specifications. One, called AMIS-Digital, is based on completely digital interaction between two voice messaging systems. All the control information and the voice message itself, is conveyed between systems in digital form. By contrast, the AMIS-Analog specification calls for the use of DTMF tones to convey control information and transmission of the message itself is in analog form. AMIS was discussed in detail in the October 1990 issue of Business Communications Review, a monthly magazine out of Hinsdale, IL. AMIS specifications are available from Alison Caughman 404-355-7785.

Audio Response Unit A device which translates computer output into spoken voice. Let's say you dial a computer and it said "If you want the weather in Chicago, push 123, then it would give you the weather. But that weather would be "spoken" by an audio response unit. Here's a slightly more technical explanation: An audio response unit is a device that provides synthesized voice responses to dual-tone multi-frequency signaling input. These devices process calls based on the caller's input, information received from a host data base, and information carried with the incoming call (e.g., time of day). ARUs are used to increase the number of information calls handled and to provide consistent quality in information retrieval. See also Audiotex and Interactive Voice Response.

Audio Track The section of a videodisc or tape which contains the sound signal that accompanies the video signal. Systems with two separate audio tracks (most videodiscs) can offer either stereo sound or two independent soundtracks.

Audioconference Another term for Teleconference. Teleconferences make use of conference bridges to allow participants to join a voice conference over the PSTN. See Teleconference.

Audiographic Conferencing Teleconferencing that also allows participants to share and interact through graphics, figures and printed text. Hardware used during audiographic conferences includes facsimile, telewriters and film-based projectors. Transmission is via a narrowband telecommunications channel such as a telephone line or a radio subcarrier.

Audiographics The technology which allows sound and visual images to be transmitted simultaneously. According to AT&T, audiographics generally refers to single frame or slow frame visual images as opposed to continuous frame image transmission (e.g. television). Audiographic transmission is often used to teach or train people in remote locations from an educational institution or business training center, saving travel and housing expense.

Audiotex A generic term for interactive voice response equipment and services. Audiotex is to voice what on-line data processing is to data terminals. The idea is you call a phone number. A machine answers, presenting you with several options, "Push 1 for information on Plays, Push 2 for information on movies, Push 3 for information on Museums." If you push 2, the machine may come back, "Push 1 for movies on the south side of town, Push 2 for movies on the north side of town, etc." See also Information Center Mailbox.

Audiotext A different, and less preferred, spelling of Audiotex. See Audiotex.

Audio/visual Multimedia Services AMS is an ATM term. It specifies service requirements and defines application requirements and application program interfaces (APIs) for broadcast video, videoconferencing, and multimedia traffic. AMS is being developed by the ATM Forum's Service Aspects and Applications (SAA) working group. An important debate in the SAA concerns how MPEG-2 applications will travel over ATM (asynchronous transfer mode). Early developers chose to carry MPEG-2 over ATM adaptation layer 1 (AAL 1); others found AAL 5 a more workable solution. Recently, some have suggested coming up with a new video-only AAL using the still-undefined AAL 2.

Auditory Pattern Recognition Auditory pattern recognition is the ability to recognize spoken words.

Audit To conduct an independent review and examination of system records and activities in order to test the adequacy and effectiveness of data security and data integrity procedures, to ensure compliance with established policy and operational procedures, and to recommend any necessary changes.

Audit File On some systems, each time a billing file is generated, an audit file is created to record the details of the generation process.

Audit Trail A record of all the events that occur when users request and use specific resources. An audit trail gives you the ability to trace who did what and who was responsible for what. An audit trail is a chronological record of system activities that is sufficient to enable the reconstruction, review, and examination of the sequence of environments and activities surrounding or leading to an operation, a procedure or an event in a transaction from its inception to final results. Audit trail may apply to information in an automated information system, to the routing of messages in a communications system, or to material exchange transactions, such as in financial audit trails. Audit trails are great for finding out what activity caused the disaster, the network to crash, etc. Audit trails are great for tracing crooks, i.e. unauthorized break-ins.

Auditing 1. Checking to see if the phone bill you got from your carrier is accurate. You do this by comparing it to your own telephone system's records.
2. Tracking activities of users by record and selected types of events in the security log of a server or a workstation.

Auger A type of drill bit typically used to make large, deep holes for passing wire or cable through wood.

Augmentation Of Bandwidth Bandwidth augmentation is the ability to add another communications channel to an already existing communications channel.

AUI Autonomous Unit Interface or Attachment Unit Interface. Most commonly used in reference to the 15 pin D type connector and cables used to connect single and multiple channel equipment to an Ethernet transceiver.

AUI Cable Attachment Unit Interface Cable is usually a four-twisted pair cable that connects an Ethernet device to an Ethernet external receiver (XCVR).

AUP Acceptable Use Policy. The term used to refer to the restrictions placed on use of a network; usually refers to restrictions on use for commercial purposes, most commonly with respect to the Internet.

Aural Relating to the sense of hearing.

AURP AppleTalk Update-Based Routing Protocol. Method of encapsulating AppleTalk traffic in the header of a foreign protocol, allowing the connection of two or more discontiguous AppleTalk internetworks through a foreign network (such as TCP/IP) to form an AppleTalk WAN. This connection is called an AURP tunnel. In addition to its encapsulation function, AURP maintains routing tables for the entire AppleTalk WAN by exchanging routing information between exterior routers. See also AURP Tunnel.

AURP tunnel Connection created in an AURP WAN that functions as a single, virtual data link between AppleTalk internetworks physically separated by a foreign network (a TCP/IP network, for example). See also AURP.

Authenticate To establish, usually by challenge and response, that a transmission attempt is authorized and valid. To verify the identity of a user, device, or other entity in a computer system, or to verify the integrity of data that have been stored, transmitted, or otherwise exposed to possible unauthorized modification. A challenge given by voice or electrical means to attest to the authenticity of a message or transmission.

Authentication The process whereby a user or information source proves they are who they claim to be. In other words, the process of determining the identify of a user attempting to access a system. Authentication codes are exchanged between two computers that use SMTP. I believed that defining authentication was pretty simple until I came across the following definition in a book on cryptography. Here it is: Authentication is any technique enabling the receiver to automatically identify and reject messages that have been altered deliberately or by channel errors. Also, can be used to provide positive identification of the sender of a message. Although secret (symmetric) key algorithms may be used for some authentication without the prior sharing of any secrets between the messaging parties. See Authentication Token.

Authentication Center See AuC.

Authentication Random Number A random value used in authentication procedures.

Authentication Sequence Number A sequence count that is incriminated for each change of Authentication Random Number (ARN).

Authentication Token A portable device used for authenticating a user. Authentication tokens operate by challenge/response, time-based code sequences, or other techniques. An example is Security Dynamics Technologies Inc.'s SecurID card. See Authentication.

Auto Responder Auto responder is just what it sounds like. You send an email to an email address on the Internet. The person who owns the address is away. He has set up an "auto responder," which is a piece of software. That software automatically sends you a response. The most common use of auto responders is for vacation messages. "I'm away. I'll be back in the office on December 25 and will reply then. If you need help earlier, contact Joe Plumpudding on TastyMorsel@Plumpudding.com. Auto responders are also used for sales@plumpudding.com and your return email is price list of all the plumpuddings we sell.

AutoAvailable An ACD feature whereby the ACD is programmed to automatically put agents into Available after they finish Talk Time and disconnect calls. If they need to go into After-Call Work, they have to manually put themselves there. See Auto Wrap-up.

AutoChanger A jukebox-style optical media system that permits the storage and playback multiple discs. Autochangers are available for both 5-,8, and 12-inch optical discs.

AutoDiscovery 1. Auto discovery is what takes place when a new device is added to a network. That device (e.g., an ethernet board, a modem, a router) communicates to the network management system that it is online and defines its key characteristics.
2. The process by which MPOA (MultiProtocol Over ATM) edge devices automatically find each other through the ATM network. See MPOA.

AutoGreeting Agent's pre-recorded greeting that plays automatically when a call arrives.

AutoNegotiation The algorithm that allows two devices at either end of a link segment to negotiate common features and functions.

AutoPlay The Microsoft AutoPlay feature of the Windows 95 operating system. This lets your CD run an installation program or the game itself immediately upon insertion of the CD-ROM.

AutoResponder The preferred spelling is two words — auto responder. Auto responder is just what it sounds like. You send an email to an email address on the Internet. The person who owns the address is away. He has set up an "auto responder," which is a piece of software. That software automatically sends you a response. The most common use of auto responders is for vacation messages. "I'm away. I'll be back in the office on December 25 and will reply then. If you need help earlier, contact Joe Plumpudding on TastyMorsel@Plumpudding.com. Auto responders are also used for sales@plumpudding.com and your return email is price list of all the plumpuddings we sell.

AutoSpid When you install an ISDN device on an ISDN line you have to give that device the phone line's SPIDs — Service Profile IDentifier numbers. The SPID is actually a label identifier that points to a particular location in your telephone company's central office memory where the relevant details of your ISDN service are stored. Auto-SPID is an industry-wide initiative of ISDN equipment makers that automatically "negotiates" a connection to the telecom provider by downloading the SPID numbers from your digital central office switch. this means that in future, when you get an ISDN line, you won't have to worry about begging SPID numbers out of the telephone company.

AutoWrap-up An ACD feature whereby the ACD is programmed to automatically put agents into After-Call Work after they finish Talk Time and disconnect calls. When they have completed any After-Call Work required, they put themselves back into Available. See AutoAvailable.

Authoring Authoring is the process of using multimedia applications to create multimedia materials (including Web pages) for others to view. Multimedia authoring uses many tools, from the more familiar text editor or desktop publishing application, to tools for capturing and manipulating video images or editing audio files. Authors might include specialized creators of training, sales, or corporate applications such as insurance claims processing. Or, they might be creators of everyday business communications like voice-annotated email. Over time, everyone involved in business communications will probably have some level of multimedia authoring capability.

Authoring System Software which helps developers design interactive courseware easily, without heavy computer programming. A specialized, high-level, plain-English computer language which permits non-programmers to perform the programming function of courseware development. The program logic and program content are combined. See AUTHORING.

Authorization Think of charging things on your MasterCard, Visa, or American Express card. If the store cannot authorize the amount of your purchase, your Visa card will not allow you to make the purchase. Authorization is needed for many long distance calls, especially those made using credit cards, telephone company calling card, etc. Authorization is done by the operator's computer checking with the remote validation database service. See BVA, BVS and Validation.

Authorization Code A code in numbers and/or letters employed by a user to gain access to a system or service. If you are making a call out on a restricted line, the PBX will ask you for an authorization code. If you have one, your call will go through. If not, your call will be denied (i.e. not go through). Authorization codes come in various flavors. Some

can be used for making long distance calls. Some can be used also for international calls, etc. See Authorized User.

Authorized Agent Also called Authorized Sales Agent. A term chosen by some of the Bell operating companies and many of the cellular phone companies to refer to companies which sell their network services on commission. Some of these companies have specific industry knowledge and have written specialized software. The idea is to work with businesses to arm them with the absolute best package of telecommunications hardware, software and services.

Authorized Bandwidth The necessary bandwidth required for transmission and reception of intelligence. This definition does not include allowance for transmitter drift or Doppler shift.

Authorized Dealer See DEALER.

Authorized Frequency A frequency that is allocated and assigned by an authority to a specific user for a specific purpose.

Authorized User A person, firm, corporation or any other legal entity authorized by the provider of the service to use the service being provided.

Auto Answer The capability of a phone, a terminal, a video phone, a modem or a computer to answer an incoming call and to set up a connection without anyone actually doing anything to physically answer the call. I have auto answer on my video phone. When it rings, my phone answers and I see who's calling. It's nice.

Auto Attendant A shortened name for an automated attendant, a device which answers a company's phones, encourages you to touchtone in the extension you want, and rings that extension. If that extension doesn't answer, it may send the call to voice mail or back to the attendant. It may also allow you to punch in digits and hear information, e.g. the company's hours of business, addresses of local branches, etc. See also Automated Attendant.

Auto Baud Automatic speed recognition. The ability of a device to adapt to the data rate of a companion device at the other end of the link.

Auto Baud Detect See Auto Baud.

Auto Busy Redial A feature of a phone or phone system where the phone has the ability to keep trying a busy number until answered. The circuit actually recognizes the busy tone, hangs up, and dials again. One of the greatest time-savers ever invented.

Auto Call Automatic Calling; a machine feature that allows a transmission control unit or a station to automatically initiate access to (i.e. dial) a remote system over a switched line.

Auto Dial A feature of phone systems and modems which allows them to dial a long phone number (usually long distance) by punching fewer buttons than there are numbers to dial. One button auto dial on electronic phones is very common these days. Most communications software programs will allow you to auto dial a string of 35 to 40 digits, which you may need if you're dialing through a complex network.

Auto Dial Auto Answer A modem feature. Auto Dial lets you dial a phone number through your modem, using your personal computer or data terminal keyboard. Auto Answer permits the modem to automatically answer the incoming call without anybody having to be there.

Auto Dialer See Automatic Dialer.

Auto Fax Tone Also called CNG, or Calling Tone. This tone is the sound produced by virtually all Group 3 fax machines when they dial another fax machine. CNG is a medium pitch tone (1100 Hz) that lasts 1/2 second and repeats every 3 1/2 seconds. A FAX machine will produce CNG for about 45 seconds after its dials. See also CNG.

Auto Line Feed An instruction in a communications program which causes the program to perform a Line Fee (LF) when you hit a carriage return or the "Enter" key.

Auto Partition A feature of 10Base-T. When 32 consecutive collisions are sensed by a port in a hub or concentrator from its attached workstation or network segment, or when a packet that far exceeds the maximum allowable length is received, the port stops forwarding packets. The port continues to monitor traffic and will automatically begin normal packet forwarding when the first correct packet is received.

Auto Recognition A term used in file conversion in which your conversion software figures out by itself in what form the original file was — WordPerfect 5.0, Word 6.0, Wordstar 5.5 etc. See also AUTO STYLING.

Auto Selection Tool An imaging term. A tool that selects an entire area within a specified range of color values around a selected pixel.

Auto Sensing See Auto Styling.

Auto Start A standby electrical power system that starts up when the normal supply of commercial power fails.

Auto Stream An AT&T ISDN term. The method of data flow in which both channels between the ISDN set and the application are in use simultaneously.

Auto Styling Auto styling is a term we found in a database conversion software program. What it means is that the program looks at the data in a field and determines from that data if the field is a numeric, character or memo, etc. The problem with auto styling is that it's frequently wrong. For example, it might check one field, find all numbers and decide it's a numeric field. Such a field might be a zip code, which actually is normally a character field. One reason why you might want you zip code to be a character field is that character fields are set left. Numeric fields are set right. (They line up at the decimal point.) Another name for auto styling is auto sensing.

Autoanswer A feature of a telephone which automatically answers incoming calls without the user of the phone lifting a handset or otherwise answering the call. Modems and fax machines also autoanswer. In North America, it's spelled AUTO ANSWER. In Britain, it's spelled Autoanswer.

Autoattendant See Automated Attendant.

Autobauding The process by which the terminal software determines the line speed on a dial-up line.

AUTODIN The worldwide data communications network of the U.S. Department of Defense. Acronym for "AUTOmatic DIgital Network."

Autodial Button An Autodial button on a phone provides one-touch dialing of outside numbers, intercom numbers, or feature codes.

Automated Attendant A device which answers callers with a digital recording, and allows callers to route themselves to an extension through touch tone input, in response to a voice prompt. An automated attendant avoids the intervention of a human being in the form of a console attendant, thereby avoiding related personnel costs. Commonly implemented in Voice Processor systems and software, front-ending PBXs and ACDs.

An automated attendant is typically connected to a PBX or a Centrex service. When a call comes in, this device answers it and says something like, "Thanks for calling the ABC Company. If you know the extension number you'd like, push-button that extension now and you'll be transferred. If you don't know your extension, pushbutton "0" (zero) and the live

operator will come on. Or, wait a few seconds and the operator will come on anyway." Sometimes the automated attendant might give you other options, such as, "dial 3" for a directory of last names and dial 4 for a directory of first names. Automated attendants are also connected also to voice mail systems ("I'm not here. Leave a message for me."). Some people react well to automated attendants. Others don't. A good rule to remember is before you spring an automated attendant on your people/customers/subscribers, etc., let them know. Train them a little. Ease them into it. They'll probably react more favorably than if it comes as a complete surprise. The first impression is rarely forgotten, so try to make it a good experience for the caller. See also DIAL BY NAME.

Automated Coin Toll Service ACTS. In the old days, operators handled routine toll calls by counting the sound of coins hitting the box, checking prices, putting calls through, figuring and collecting overtime charges, etc. ACTS does all this automatically. It figures charges, tells those charges by digitized computerized voice to the customer, counts the coins as they are deposited and then sets up the call.

Automated Intercept Call Completion AINTCC. A new feature of Northern Telecom's central offices. The AINTCC feature provides options for

• connecting a caller automatically to an intercepted number after hearing an announcement, or

• connecting a caller to an intercepted number without an announcement.

Not using an announcement makes the number change transparent to the caller. The called (intercepted) party then has the option of informing the caller of the number change.

Automated Maritime Telecommunications System An automatic, integrated and interconnected maritime communications system serving ship stations on specified inland and coastal waters of the United States.

Automated Radio A radio with the capability for automatically controlled operation by electronic devices that requires little or no operator intervention.

Automated Tactical Command And Control System A command and control system or part thereof which manipulates the movement of information from source to user without intervention. Automated execution of a decision without human intervention is not mandatory.

Automated Voice Response Systems AVRS. Devices which automatically answer calls. They may simply inform the caller that the call is in a queue and will be answered soon; alternatively they can prompt the caller to use voice commands, or touchtones to seek more information.

Automatic Address Discovery This refers to the process by which a network device can poll other network devices to discover the network addresses which each device supports. Automatic address discovery makes the set up and on-going maintenance of complex internetworks much simpler than if all address updates were performed manually.

Automatic Button Restoration When the telephone handset of a multi-line instrument (typically a 1A2 multi-line key set) is placed back in its cradle, the line button being used automatically "pops" back up. Conversely, when a user picks up the handset, he must always push down a line button to make a call. Most phones with this feature can be disabled, so the buttons stay down when the handset sits on the cradle. A twist of a single screw inside the instrument will usually solve the aggravation of the automatic button restoration. Some people like automatic button restoration because it saves a user from accidentally barging into someone else's call. This was a much greater problem with 1A2 key systems. It no longer is a problem with most electronic key systems since they usually extend the user automatic privacy once they get on a call so no one else can barge in, even if they want to.

Automatic Call Distributor ACD. A specialized phone system designed originally for handling many incoming calls, now increasingly used by companies also making outgoing calls. You receive and make lots of phone calls typically to customers. You need an ACD. Once used only by airlines, rent-a-car companies, mail order companies, hotels, etc., it is now used by any company that has many incoming calls (e.g. order taking, dispatching of service technicians, taxis, railroads, help desks answering technical questions, etc.). There are very few large companies today that don't have at least one ACD. Many smaller companies, like the company that publishes this dictionary, also have one.

An ACD performs four functions. 1. It will recognize and answer an incoming call. 2. It will look in its database for instructions on what to do with that call. 3. Based on these instructions, it will send the call to a recording that "somebody will be with you soon, please don't hang up!" or to a voice response unit (VRU). 4. It will send the call to an agent as soon as that operator has completed his/her previous call, and/or the caller has heard the canned message.

The term Automatic Call Distributor comes from distributing the incoming calls in some logical pattern to a group of operators. That pattern might be Uniform (to distribute the work uniformly) or it may be Top-down (the same agents in the same order get the calls and are kept busy. The ones on the top are kept busier than the ones on the bottom). Or it may be Specialty Routing, where the calls are routed to answerers who are most likely to be able to help the caller the most. Distributing calls logically is the function most people associate with an ACD, though it's not the most important.

The management information which the ACD produces is much more valuable. This information is of three sorts: 1. The arrival of incoming calls (when, how many, which lines, from where, etc.) 2. How many callers were put on hold, asked to wait and didn't. This is called information on ABANDONED CALLS. This information is very important for staffing, buying lines from the phone company, figuring what level of service to provide to the customer and what different levels of service (how long for people to answer the phone) might cost. And 3. Information on the origination of the call. That information will typically include ANI (Automatic Number Identification — picking up the calling number and DNIS (Direct Number Identification Service) picking up the called number. Knowing the ANI allows the ACD and its associated computer to look up the caller's record and thus offer the caller much faster service. Knowing the DNIS may allow the ACD to route the caller to particular agent or keep track of the success of various advertising campaigns. Ad agencies will routinely run the same ad in different towns using different 800 phone numbers. Picking up which number was called identifies which TV station the ad ran on.

The seven definitions that follow the definition "ACD" show some of the features which newer ACDs have. See also 800 Service, ACD and Automatic Call Sequencer.

Automatic Call Intercept A feature of a Rolm ACD. This feature automatically forwards calls to an attendant if the dialed number is not installed or out of order. It can also intercept an attempted trunk call that is in violation of a Class of Service restriction. Automatic Call Intercept will also recall the attendant after a predetermined period of offshoot inactivity (e.g. flash or hold).

Automatic Call Rescheduling When a call is unsuccessful, either no reply, or busy, the system will automatically dial the number again after a pre-determined time.

Automatic Call Sequencer A device for handling incoming calls. Typically it performs three functions. 1. It answers an incoming call, gives the caller a message, and puts them on "Hold." 2. It signals the agent (the person who will answer the call) which call on which line to answer. Typically, the call which it signals to be answered is the call which has been on "hold" the longest. 3. It provides management information, such as how many abandoned calls there were, how long the longest person was kept on hold, how long the average "on hold" was, etc.

There are three types of devices which handle incoming calls. The least expensive is the Automatic Call Sequencer which is traditionally used with key systems. It differs from Uniform Call Distributors (UCDs) and Automatic Call Distributors (ACDs) in that it has no internal switching mechanism and does not affect the call in any way. It simply recommends which call should be picked up and keeps statistical information on the progress of calls. A more expensive type of device is the UCD.

The most full-featured and expensive is the ACD. Distinctions between ACDs and UCDs and/or PBXs with features called UCDs and ACDs are blurring as UCDs get more sophisticated. The main difference, as we understand it, is that a UCD offers fewer options for routing an incoming call and answering calls in any particular order. ACDs typically produce the most detailed management information reports. One company also makes something called an Electronic Call Distributor. It is essentially an automatic call distributor.

Automatic Callback When a caller dials another internal extension and finds it busy, the caller dials some digits on his phone or presses a special "automatic callback" button. When the person he's calling hangs up, the phone system rings his number and the number of the original caller and the phone system automatically connects the two together. This feature saves a lot of time by automatically retrying the call until the extension is free. See also CAMP ON. Wouldn't it be nice if they had this feature on long distance calls?

Automatic Calling Unit ACU. A device that places a telephone call on behalf of a computer.

Automatic Circuit Assurance ACA is a PBX feature that helps you find bad trunks. The PBX keeps records of calls of very short and very long duration. If these calls exceed a certain parameter, the attendant is notified. The logic is that a lot of very short calls or one very long call may suggest that a trunk is hung, broken or out of order. The attendant can then physically dial into that trunk and check it.

Automatic Cover Letter In a fax transmission, an automatic cover letter allows the user to automatically attach a cover letter to the document being sent. This is especially convenient when sending material directly from your PC.

Automatic Dialer Or Autodialer A device which allows the user to dial pre-programmed telephone numbers by pushing one or two buttons. Sometimes referred to as a "repertory" dialer. Dialers can be bought as a separate device and added to a phone, however, today most telephone sets are outfitted with autodialers. There are four basic measures of an automatic dialer's efficiency. 1. What's the longest number it will dial automatically? This is important because using some of AT&T's long distance competitors requires dialing lots of numbers, with lots of pauses. 2. How many numbers will it dial? Some people like to have a dialer which dials hundreds of numbers. Others like a small one, just for their most frequently called numbers. 3. Will the dialer recognize dial tone? This is important because using a long distance company or dialing through a PBX requires one to recognize consecutive dial tones. 4. Can you "chain" dial? In other words, can you hit one speed dial button after another and have the machine dial through a complex network and throw in authorization codes, etc.?

Automatic Dialing See Speed Dialing.

Automatic Directory Propagation In electronic mail, automatic directory propagation is the ability to update addresses automatically in one domain after manually entering address changes in another domain, whether on the same LAN or another LAN connected by a gateway. In general, automatic directory propagation can be peer-to-peer, where changes in any post office are sent to all other post offices, or master-to-slave, where changes in the master post office are sent to the slaves, but changes in the slave post office do not go to the master.

Automatic Equalization The process of compensating for distortion of data communications signals over an analog circuit.

Automatic Exchange A term for a central office which automatically and electronically switches calls between subscribers without using an operator. Not a common term.

Automatic Facilities Test System AFACTS is a Rolm CBX feature. It is an automatic testing system for identifying faulty tie and central office trunks. AFACTS can pinpoint faulty trunks and generate exception and summary reports.

Automatic Fallback A modem's ability to negotiate an appropriate data rate with the modem on the other end of the link, depending on line quality. For example, if two 2400 baud modems can not pass data at 2400 baud, they would "fall back" to 1200 baud automatically in order to transmit data without excessive errors.

Automatic Frequency Control A circuit in a radio receiver which automatically brings the tuning units of the set into resonance with a wave which is partially tuned in.

Automatic Gain This is an electronic circuit which automatically increases the volume when someone is speaking quietly and drops it when someone is speaking loudly. The idea is to keep the transmitted signal even. Most tape recorders, for example, have automatic gain circuits. This allows them to pick up voices of people in a room, even though the volume of each person's conversation arriving at the tape recorder is different. The problem with automatic gain circuits is they're always looking for something to amplify. Such that when it's quiet (and meant to be) the automatic gain circuit will also try and amplify the ambient noise in the room — to keep the sound level constant. All professionally recorded tapes are done on tape reorders with manual volume controls.

Automatic Hold — Station or Intercom When a user is having a conversation and receives another call, he may press the button to answer that new call. The call he was on originally is automatically put on hold.

Automatic Identified Outward Dialing AIOD. The toll calls placed by all extensions on the telephone systems are automatically recorded. This information allows bills to be sent, long distance lines to be chosen, etc. Some central offices, for example, can provide an itemized breakdown of charges (including individual charges for toll calls) for calls made by each CPE telephone extension. See Call Accounting, Call Detail Recording, SMDR and AIOD.

Automatic Intercept Center AIC. A Bellcore definition: "The centrally located set of equipment that is part of

an Automatic Intercept System (AIS) that automatically advises the calling customer, by means of their recorded or electronically assembled announcements, of the prevailing situation that prevents completion of connection to the called number."

Automatic Level Control ALC. A control system that adjusts the incoming signal to a predetermined level. Somewhat similar to automatic gain control. See Automatic Gain Control.

Automatic Light Control ALC. Vidicon camera control which automatically adjusts the target voltage to compensate for variations in light levels. See also Automatic Gain.

Automatic Line Hold A PBX feature. As long as a phone does not go "on-hook," activation of various line pushbuttons will automatically place the first line on hold without the use of special "hold" button.

Automatic Line Insulation Testing ALIT. Equipment located in a Central Office which sequentially tests lines in the office for battery crosses and grounds.

Automatic Link Establishment ALE. The capability of an HF radio station to contact, or initiate a circuit, between itself and another specified radio station, without operator assistance and usually under computer control. ALE techniques include automatic signaling, selective calling, and automatic handshaking. Other automatic techniques that are related to ALE are channel scanning and selection, Link Quality Analysis (LQA), polling, sounding, message store and forward, address protection, and anti-spoofing.

Automatic Message Accounting The network functionality that measures, collects, formats, and outputs subscriber network-usage data to upstream billing OSs and other OSs (Operations Systems).

Automatic Message Switching A technique of sending messages to their appropriate destination through information contained in the message itself — typically in its "address."

Automatic Network Restoral Automatic network restoral is a term which reflects the ability of a network to restore service rapidly and automatically following a catastrophic failure, such as that of a network cable cut. The result is higher network availability and reliability.

Automatic Number Identification ANI. Being able to recognize the phone number of the person calling you. You must have equipment at your office. And the network must have the ability to send the calling number to you. For a much longer explanation, see ANI, CALLER ID, CLASS, ISDN and System Signaling 7.

Automatic Overflow To DDD Toll calls jump to expensive direct distance dialed calls, when all lower cost FX, WATS lines, etc. are busy.

Automatic Phone Relocation The Automatic Phone Relocation feature now available on some phone systems allows a telephone to retain its personal and system programming when it is reconnected to another physical location.

Automatic Privacy When someone is speaking on a phone line or on an intercom, this feature ensures no one else can accidentally or deliberately butt into that conversation. If you did, however, want somebody else to come into the conversation (for example, someone to provide some additional information), there's usually another feature called Privacy Release. By pushing this button on the phone, other people who have the same extension button or intercom button on their phones can then push their buttons and join the conversation. Or you can bring them into your conversation by dialing them in. Most modern key systems come with

Automatic Privacy. Many people don't like it, especially those who live in small offices. Some newer phone systems are coming standard without it. And you have to program it in, if you want it.

Automatic Program Load APL is a PBX feature that allows it to load its own software into RAM from a local device such as a hard disk or a floppy disk. All this takes place automatically without human intervention. APL is an important feature since it often determines how fast a PBX can get back into service after some sort of failure — usually a failure in commercial power.

Automatic Protection Switching Switching architecture designed for SONET to perform error protection and network management from any point on the signal path.

Automatic Queuing Queuing is exactly as it sounds. Something you want is being used. So you get placed in line for that device. There are two types of queuing — automatic and manual. Manual is when you're put in queue by a person, for example an operator. Automatic is when you're put in queue by a machine, for example a PBX aided by its software.

Automatic Recall 1. A central office feature which gives telephone subscribers the ability to automatically redial their last incoming call — without actually knowing that number. On some central offices it is now possible for the calling person to block the ability of someone they've called to automatically call them back.
2. A PBX feature which returns a call to the PBX attendant (or alerts the attendant) if a call extended to a telephone is not answered within a pre-set period of time. The most logical time is three rings, or 18 seconds. This feature allows the attendant to give the caller some information, take a message or connect the caller to someone else. Most hotel switches have this feature. And when the call doesn't get answered, the switch sends it back to the operator. The sad thing is that the hotel operator is usually so busy, he/she keeps you waiting another 20 or 30 seconds, irritating you.

Automatic Recovery Your telephone system dies — typically because its power is cut off. Once the power comes on, instructions in the machine direct it to reload its software so that within minutes the system can be back and running normally. Those "instructions" are normally not affected by power drops.

Automatic Replication Also known as data replication. It refers to the process of automatically duplicating and updating data in multiple computers on a network. The word "automatically" in this case means that the process of duplication is handled automatically by the software responsible for the data. The idea is that if one or several of the machines do down, the company using the data will still have reliable delivery of data.

Automatic Rerouting This refers to the process by which an intelligent voice or data network can automatically route a call, or virtual circuit, around a network failure. With frame relay, PVCs represent a fixed path through the network. However, in the event of a network failure along the primary path over which the PVC is routed, the PVC will be automatically routed to a secondary network path until the primary path is physically restored.

Automatic Restart 1. The mechanism whereby, after a power failure and then power resumes, a process automatically restarts. Restart is from the exact point of interruption.
2. System facilities that allow restart from the point of departure after system failure.

Automatic Ring Down ARD. A private line connecting a station instrument in one location to a station instrument in a distant location with automatic two-way signaling. The automatic two-way signaling used on these circuits causes the station instrument on one end of the circuit to ring when the station instrument on the other end goes off-hook. This circuit is sometimes called a "hot-line" because urgent communications are typically associated with this service. ARD circuits are commonly used in the financial industry. May also have one way signaling only.

Automatic Ringdown Tie Trunk A direct path signaling facility to a distant phone. Signaling happens automatically when you lift the receiver on either phone. See also Manual Ringdown TIE Trunk.

Automatic Rollback A feature of the Transaction Tracking System (TTS) that returns a database on a Novell NetWare local area network to its original state. When a network running under TTS fails during a transaction, the database is "rolled back" to its most recent complete state. This prevents the database from being corrupted by the incomplete transaction.

Automatic Route Selection Your phone system automatically chooses the least cost way of sending a long distance call. See Least Cost Routing and Alternate Routing.

Automatic Scheduled Testing AST. A method of testing switched access service (Feature Groups B, C, and D) where the customer provides remote office test lines and 105 test lines with associated responders or their functions' equivalent; consists of monthly loss and C-message noise tests and annual balance test.

Automatic Secure Voice Communications Network AUTOSEVOCOM. A worldwide, switched, secure voice network developed to fulfill DoD long-haul, secure voice needs.

Automatic Send/Receive ASR. A data device in which the transmitting thing is different from the receiving part, thus enabling the device to receive calls and transmit them simultaneously. See ASR for a different definition. We're not sure precisely which one is right. Both could be.

Automatic Sequential Connection A service feature provided by a data service to connect automatically, in a predetermined sequence, the terminals at each of a set of specified addresses to a single terminal at a specified address.

Automatic Set Relocation A phone system feature which allows a telephone to retain its personal and system programming when it is reconnected to another physical location.

Automatic Speed Matching The ability of an asynchronous modem to automatically determine whether it is expected to communicate at 300, 1200 or 2400 bps.

Automatic Time-Out On Uncompleted Call A PBX feature. If a phone stays "off-hook" without dialing for a predetermined time interval, or stays connected to a busy signal longer than the predetermined time interval, the intercom switching equipment will automatically connect this phone to intercept.

Automatic Toll Ticketing A system which makes a record of the calling phone number, the called number, the time of day, the length of the call, etc. and then generates an instant phone bill for that call. Often used in hotel/motels.

Automatic Traffic Overload Protection ATOP. A Rolm feature defined as a dynamic form of line-load control, which automatically denies a dial tone during those periods when the Rolm CBX may become overloaded. One wonders why someone would create this feature.

Automatic Vehicle Location See AVL.

Automatic Voice Network AUTOVON. The principal

long-haul, unsecure (meaning it's not secure) voice communications network within the Defense Communications System.

Automatic Volume Control A circuit in a radio receiver; automatically maintains various received transmissions at approximately the same volume.

Automatic Wakeup The capability for the user to schedule a wake-up call to a predetermined telephone number, either one time or daily.

Automatic Wakeup Service The guest or the operator dials into a machine which records a request for a guest wakeup call the following morning. The auto wakeup machine is a glorified, programmable auto-dial answering machine. The machine is said to save hotels money and make wakeup calls more reliable, and certainly more anonymous.

Autonomous System A collection of routers under a single administrative authority using a common Interior Gateway Protocol for routing packets.

Autonomous System Number ASN. A unique number assigned by the InterNIC that identifies an autonomous system in the Internet. ASNs are used by routing protocols (like BGP — Border Gateway Protocol) to uniquely define an autonomous system.

Autoscaling A drawing feature that automatically adjusts the axis units of a graph to the minimum and maximum numerical values of a set of data.

Autosearch See Recorder.

Autoseed A United Kingdom definition. The process of selecting records from a database using pre-defined criteria and allocating the records to outbound or mailing campaigns.

AUTOSEVOCOM The AUTOmatic SEcure VOice COMmunications system of the U.S. Department of Defense. A worldwide, switched, secure voice network developed to fulfill DoD long-haul, secure voice needs.

Autotimed Recall When a user places a call on hold and forgets about it, Autotimed Recall will ring that user or the receptionist after a predetermined time. That time is usually programmable. It shouldn't be longer than 30 seconds, otherwise your customers, sitting endlessly on your eternity hold, will go nuts and go elsewhere.

Autotimed Transfer This telephone system feature switches unanswered incoming calls to a backup answering position after a predetermined (usually adjustable) interval of time.

Autotype Protocol Hayes Microcomputer definition for a file transfer protocol which allows the user to automatically "type" a disk file, the clipboard or the contents of Smartcom Editor in either plain text (ASCII) or ANSI.SYS format to a remote computer. Pacing, send lines and await character echo options are provided. If necessary, character set mapping translates between different code pages and systems (Macintosh or Windows text files, for example).

AUTOVON The AUTOmatic VOice Network. The principle long-haul voice communications network within the Defense Communications System. The system includes conferencing and secure voice communications (scrambling), among other features.

Auxiliary Equipment See also Peripheral Device or Applications Processor.

Auxiliary Equipment Access The ability of a telephone system to interface with (i.e. talk to) auxiliary equipment such as a paging system or dial dictation system.

Auxiliary Line A telephone trunk in addition to the main number you rent from the phone company. Phone systems are often equipped for calls to hunt from a busy main number to one or more auxiliary lines (Incoming Service Group, or ISG).

For example, the publisher's main office main number is 212-691-8215. But it also has 8216, 8217, 8218 and several unmarked or coded trunks. These are auxiliary lines and they don't receive their own billing or listing from the phone company. Sometimes, people have single line private lines which "appear" on their phone and no one else's. Sometimes they call these auxiliary lines. Sometimes these are called private lines. Sometimes they are also called terminal numbers.

Auxiliary Network Address In IBM parlance, in ACF/VTM, any network address except the main network address, assigned to a logical unit which is capable of having parallel sessions.

Auxiliary Power An alternate source of electric power, serving as backup for the primary power at the station main bus or prescribed sub-bus. An off-line unit provides electrical isolation between the primary power and the critical technical load; an on-line unit does not. These are government definitions: A Class A power source is primary power source; i.e., a source that assures an essentially continuous supply of power. Types of auxiliary power service include: Class B: a standby power plant to cover extended outages (days); Class C: a quick-start (10 to 60 seconds) unit to cover short-term outages (hours); Class D: an uninterruptible (no-break) unit using stored energy to provide continuous power within specified voltage and frequency tolerances.

Auxiliary Ringer This is a separate external telephone ringer or bell. It can be programmed to ring when a line or a telephone, or both ring; or, when Night Service is turned on.

Auxiliary Service Trunk Groups A category of trunk groups that provides selected services for customer or operators and terminates on announcement systems, switchboards, or desks. Examples include Directory Assistance, Intercept, Public Announcement, Repair Service, Time, and Weather.

Auxiliary Storage A mass storage device capable of holding a larger amount of information than the main memory (i.e. RAM) of the computer or telephone system, but with slower access time. For example — magnetic tape, floppy disks, etc.

Auxiliary Work State A call center term. An agent work state that is typically not associated with handling telephone calls. When agents are in an auxiliary mode, they will not receive inbound calls.

AUU ATM User-to-User

Available In automatic call distribution language, an agent state, between calls, when an agent, having finished the previous transaction, returns to accept the next inbound caller. See also Availability.

Available Channel In the CDPD cellular mobile system, a Radio Frequency (RF) channel is available if it is an allocated channel that is not currently in use for either Cellular Digital Packet Data (CDPD) or non-CDPD-based wireless packet data service.

Available State According to Bellcore, an available circuit state occurs when all of the following are true:
1. The Bit Error Ratio (BER) is better than 1 in 10 to the nth power for a specific number of consecutive observation periods of fixed duration.
2. Block Error Ratio (BLER) is better than 1 in 10 to the nth power under the same conditions.
3. There are a specific number of consecutive observation periods of fixed duration without a severely errored unit time.

Availability The amount of time a computer or a telephone system is available for processing transactions or telephone calls. Here's a more technical definition: The ratio of the total time a functional unit is capable of being used during a given interval to the length of the interval; e.g., if the unit is capable of being used for 100 hours in a week, the availability is 100/168. Contrast this with the term Reliability, which is different.

In SONET, the basic Bellcore reliability criterion is an end to end two way availability of 99.98% for interoffice applications (0.02% unavailability or 105 minuters/year down time). The objective for loop transport between the central office and the customer premises is 99.99%. For interoffice transport the objective refers to a two way broadband channel, e.g. SONET OC-N, over a 250 mile path. For loop applications the objective refers to a two way narrowband channel, e.g. DS0 or equivalent. See RELIABILITY.

Availability Reports Availability reports show how often and for how long nodes, links or paths were unavailable due to outages between specified dates. They can be used to monitor network reliability and to calculate rebates for users.

Avalanche Photo Diode APD. A fiber optic transmission device. A light detector that generates an output current many times the light energy striking its face. A photodiode that shows gain in its output power compared to the optical power that it receives through avalanche multiplication (signal gain) of the current that flows through a photosensitive device. This type of diode is used in receivers requiring high light sensitivity. See APD.

Avalanching The process by which an electrical signal is multiplied within a device by electron impact ionization.

Avatar 1. A graphical icon that represents a real person in a virtual reality system. When you enter the system, you can choose from a number of fanciful avatars. Sophisticated 3D avatars even change shape depending on what they are doing (e.g., walking, sitting, etc.).
2. A common name for the superuser account on UNIX systems. The other common name is root.

AVC Automatic Volume Control. In radio it maintains constant sound level despite undesired differences in strength of incoming signal.

AVD Alternative Voice Data or Alternating Voice Data. AVD used to mean a voice circuit that would also handle data. See AVD Circuits. Now AVD means it's a normal switched analog phone line connected to which is a AVD modem, which lets you transfer bursts of fax, data and images between your voice conversation. The protocol mostly used is VoiceView. See AVD CIRCUITS and DSVD.

AVD Circuits Alternate Voice Data Circuits. Telephone lines which have been electrically treated to handle both voice and data signals. Typically used on leased overseas circuits to save money. See AVD.

Average The average and the mean are the same. What's different is the median. See Mean for a full explanation.

Average Business Day ABD. The sum of the busy hour data (usage, peg count, or overflow) recorded for each of the five busiest days of the week and divided by the total number of days being reported during any given basic data service month. Traditionally, Monday through Friday are the five busiest days of the week, however, traffic characteristics in a particular central office may produce high loads during the weekend. This average (5 days) figure is used in selecting the busy season months and the average busy season value.

Average Busy Season ABS. Average Busy Season is the month's (normally three but not necessarily consecutive) with the highest average busy hour CCS per Network Access Line (NAL). Research supports these calling volumes are highly stable and thereby extremely predictable.

Average Call Duration Divide the total number of minutes of conversation by the number of conversations. Bingo, that's your average call duration.

Average Customer Arrival Rate Represents the number of entities (humans, packets, calls, etc.) reaching a queuing system in a unit of time. This average is denoted by the Greek letter lambda. One would prefer to know, if possible, the full distribution of the calls arriving.

Average Delay The delay between the time a call is answered by the ACD and the time it is answered by a person. This typically includes time for an initial recorded announcement plus time spent waiting in queue. Average delay can be chosen as the criterion for measuring service quality.

Average Handle Time AHT. The period of time an employee is occupied with an incoming call. This is the sum of talk time and after-call-work time.

Average Holding Time The sum of the lengths (in minutes or seconds) of all phone calls during the busiest hour of the day divided by the number of calls. There are two definitions. The one above refers to average speaking time (it's the more common one). There's a second definition for "average holding time." This refers to how long each call was on hold, and thus not speaking. This second definition is typically found in the automatic call distribution business (ACD). Check before you do your calculations.

Average Latency The time required for a disk to rotate one-half revolution.

Average Pulse Density In T-1 bipolar transmissions, refers to the number of "1" pulses per "0" conditions and is usually tied to a maximum number of "0"s in a row (i.e., FCC Part 68 requires 12.5% pulse density and no more than 80 consecutive "0"s where as AT&T Pub 62411 uses a formula and no more than 15 consecutive "0"s).

Average Speed Of Answer ASA. How many seconds it takes an operator on average to answer a call.

Average 10-High Day ATHD. A mathematical average of the data generated during the 10-high day busy hour. This is the same hour for all ten days and generally will occur during the busy period. However, predictable recurring heavy traffic days which occur outside the busy period should be included.

Average Transfer Delay Average time between the arrival of a packet to a station interface and its complete delivery to the destination station.

AVI Audio Video Interleaved. File format for digital video and audio under Windows. Use the "Media Player," which comes with Windows, to play AVI files. The AVI file format is cross-platform compatible, allowing AVI video files to be played under Windows and other operating systems.

AVK Audio Video Kernel. DVI (Digital Video) system software designed to play motion video and audio across hardware and operating system environments.

AVL Automatic Vehicle Location — not an elegant name, but an umbrella description for the fleet management version of mobiletelematics, which involves integrating wireless communications and (usually)location tracking devices (generally GPS) into automobiles. The best known example of mobile telematics is GM's OnStar system, which automatically calls for assistance if the vehicle is in an accident. These systems can also perform such functions as remote engine diagnostics, tracking stolen vehicles, provide roadside assistance, etc. www.onstar.com. Best known operators of AVL services (mobile telematics for fleets) are Qualcomm's OmniTRACS division and Teletrac. They generally involve integrating wireless communications and location sensing technology (frequently GPS) into commercial vehicles, to allow mobile communications, automated dispatching, cargo tracking, etc.

Avoidable Costs A wonderful concept used by the regulated telephone industry. It refers to those costs which would be avoided (i.e. not incurred) if the service were not offered. Examples of costs to avoid are maintenance, taxes, labor, and other direct costs. The concept of Avoidable Costs is to allow the phone industry the justification to price a competitive service very low.

AVRS Automated Voice Response System. See IVR and VOICE RESPONSE SYSTEM.

AVSS Audio-Video Support System. DVI (Digital Video) system software for DOS. It plays motion video and audio.

AWC Area-Wide Centrex.

AWG American Wire Gauge. Originally known as the Brown and Sharpe (B&S) Wire Gauge. The U.S. standard measuring gauge for non-ferrous conductors (i.e., non-iron and non-steel), AWG covers copper, aluminum, and other conductors. Gauge is a measure of the diameter of the conductor. The AWG numbering system is retrogressive (i.e., backwards): The higher the AWG number the thinner the wire. This is due to the fact that the AWG number originally indicated the number of times the copper wire was drawn through the wire machine to reduce its diameter. For example, a 24-gauge wire was drawn through the wire machine 24 times; therefore, it is thinner than a 22-gauge wire, which was drawn through the wire machine only 22 times. A 24-gauge (AWG) wire has a diameter of .0201 in. (.511mm), a weight of 1.22 lbs./ft. (1.82 kg./km.), a maximum break strength of 12.69 lbs. (5.756 kg.) and D.C. resistance ohms of 25.7/1000ft. (84.2/km.). For example, heavy industrial electrical wiring may be No. 2; homes are typically wired with No. 12 or No. 14. Telephone systems typically use No. 22, No. 24 or No. 26. The thicker the wire, the more current it can carry farther without creating heat and without suffering attenuation (signal or power loss) due to resistance. You need thicker phone cabling when your phones are farther away. Also, you need thicker wire when transmitting at higher frequencies such as would be the case in a data application (e.g., 10/100Base-T); high-frequency signals attenuate to a greater extent than do low-frequency signals. Some vendors save money by installing systems with thin wire. Make sure you specify.

AX.25 An amateur radio implementation of the X.25 protocol. Used by some private VANs (Value Added Networks) to avoid PTT (post, Telephone, and Telegraph Administration) monopolies (and thus high prices) on X.25 transmission and switching.

Axial Ratio Of a wave having elliptical polarization, the ratio of the major axis to the minor axis of the ellipse described by the tip of the electric field vector.

Axial Ray A ray that travels along the axis of an optical fiber.

Axial Slab Interferometry Synonym for Slab Interferometry.

Axis The center of an optical fiber.

Azimuth The horizontal angle which the radiating lobe of an antenna makes in angular degrees, in a clockwise direction, from a north-south line in the northern hemisphere. In the southern hemisphere, the reference is the south-north line. Azimuth actually involves a lot more than antennas. For example, it covers the alignment of a recording head in a tape recorder.

B 1. A capital B stands for Byte. A small b stands for bit. Typically a byte is eight bits (but it could be more or fewer). Virtually all telecommunications transmission — data, voice or video, etc. — is stated in bits per second. Virtually all transmission inside a computer (like movement between a disk and the computer's central micropocessor is written in bytes per second. Thus Mbps would be million bits per second, while MBbps would be million bytes per second. For a much longer explanation, see the Introduction to the printed version of this dictionary.
2. The local wireline cellular carrier. In one of its less intelligent decisions, the Federal Communications Commission decided to issue two cellular franchises in each city of the United States. They gave one to the local phone company (the B carrier) and one to a competitor (the A carrier). This duopoly has naturally meant little real price competition. At least it did until the FCC issued PCS (Personal Communications Services. licences and competition started. See PCS.
3. Beta. As in beta test. A beta test is a test of a product in a real production environment with a real, live customer. Beta tests come after alpha tests, and before the general release of the product.

B Battery A section of a phone system power supply that provides unfiltered Direct Current for operating relays and various other components. Typically 20 volts. See A Battery.

B Channel A "bearer" channel is a fundamental component of ISDN interfaces. It carries 64,000 bits per seconds in both directions, is circuit switched and is able to carry either voice or data. Whether it does or not depends on how your local telephone company has tariffed its ISDN service. See BASIC RATE INTERFACE (BRI) and ISDN.

B Connector A commonly-used wire-splicing device consisting of a flexible plastic sleeve over a toothed metal cylinder that bites through insulation when crimped with pliers or a special crimping tool. It is about one inch long and can hold three or four wires. A gel-filled version (water-retardant jelly) is available for installation in damp or humid areas. B connectors are also known as chiclets, beans, beanies, and rodent rubbers.

B Frame Bi-directional or B frames (often called B pictures) refer to part of the MPEG video compression process whereby both past and future pictures / frames are used as references. B frames typically produce the most compression.

B Links A SS7 term. Bridge links assigned in a quad arrangement. Bridge Link. A CCS/SS7 signaling link used to connect STP (Signal Transfer Point) pairs that perform work at the same functional level. These links are arranged in sets of four (called quads). A minimum of three-way path diversity is recommended to allow three completely separate paths. The B-Links are assigned so that local Signal Transfer Points (STPs) can communicate with other local STPs or mated pairs of STPs at the same hierarchical level; local to local, regional to regional, or mated pair to mated pair.

B Number A cellular term for the number of the called party. The originating switch analyzes the A Number (the number of the calling party) in order to route the call to the B Number.

B Port The port which connects the outgoing primary ring and the incoming secondary ring of the FDDI dual ring. This port is part of a dual attached station or concentrator. See FDDI.

B-911 Basic 911. There is also an E-911, which stands for Enhanced 911. B-911 is a centralized emergency reporting system which may have many features but which does NOT provide ALI (Automatic Location Information) to the 911 operator. In most cases, it does not provide ANI (Automatic Number Identification) either. B-911 provides a common emergency response number and relies on Emergency Hold and Forced Disconnect to maintain effective service. See 911 for a full explanation.

B-Block Cellular licenses received from the FCC with an initial association to a telephone company. Also referred to as wireline.

B-CDMA Broadband Code Division Multiple Access.

B-CRYPT A symmetric cryptographic algorithm designed by British Telecom.

B-DCS Broadband Digital Cross-connect System. B-DCS is a generic term for an electronic digital cross-connect system capable of cross-connecting signals at or above the DS3 rate.

B-ICI B-ISDN Inter-Carrier Interface: An ATM Forum defined specification for the interface between public ATM networks to support user services across multiple public carriers.

B-ICI SAAL B-ICI Signaling ATM Adaptation Layer: A signaling layer that permits the transfer of connection control signaling and ensures reliable delivery of the protocol message. The SAAL is divided into a Service Specific part and a Common part (AAL5).

B-ISDN Broadband ISDN. A very vague term that defines a communications channel as anything larger than a single voice channel of 64 Kbps. Under this terminology, "broadband" can be as little as two voice channels. The official ITU-T recommendation I.113 [45] defines Broadband ISDN as "a service or system requiring transmission channels capable of supporting (transmission) rates greater than the primary rate [DS1]." Thus, broadband ISDN is a new concept in information transfer, although exactly what it is isn't clear yet. There is some discussion that broadband ISDN begins at 155 Mbps (the OC-3 version of SONET/SDH). In another ITU-T recommendation (I.121 [47]), the ITU-T presents an overview of what it sees as B-ISDN capabilities:
"B-ISDN supports switched, semipermanent and permanent point-to-point and point-to-multipoint connections and provides on demand, reserved and permanent services. Connections in B-ISDN support both circuit mode and packet mode services of a mono- and/or multi-media type and of connectionless or connection-oriented nature and in a bidirectional or unidirectional configuration. A B-ISDN will contain intelligent capabilities for the purpose of providing advanced service characteristics, supporting powerful operation and maintenance tools, network control and management." The ITU-T (nee ITU) was to decide on an international standard for B-ISDN by 1996, although it has not yet been released. The ITU-T has defined two types of services in the context of B-ISDN, Interactive and Distribution. For a defini-

tion of those services, SEE INTERACTIVE SERVICES and DISTRIBUTION SERVICES.

Bellcore says that "National and international standards bodies have made the Asynchronous Transfer Mode (ATM) the target solution for providing the flexibility required by B-ISDN. ATM provides a common platform capable of supporting both broadband and narrowband services ... The physical layer-transmission standard for B-ISDN is the Synchronous Optical Network (SONET), also known as the Synchronous Digital Hierarchy (SDH). See N-ISDN and SONET."

B-ISDN Inter-Carrier Interface An ATM Forum specification for a public UNI (User Network Interface) providing the interface via PVCs between public ATM networks to support user services across multiple public carriers. See UNI.

B-LLI Broadband Lower Layer Information: This is a Q.2931 information element that identifies a layer 2 and a layer 3 protocol used by the application.

B-PCS Broadband Personal Communications Services. The FCC designated 140 MHz in the 1850-1990 MHz range for PCS services. The B-PCS auctions ended in 1995, yielding over $7 billion in revenues to the U.S. government. B-PCS is intended to support features such as "follow-me." See also 500 Service and PCS.

B-picture Bi-directionally predictive-coded; an MPEG term for picture a that is coded using compensated prediction from a past and/or future refernce picture.

B-TA Broadband Terminal Adapter. A form of DCE which provides the interface into a B-ISDN network from B-TE2, which is Terminal Equipment not compatible with the B-ISDN network.

B-TE Broadband Terminal Equipment: An equipment category for B-ISDN which includes terminal adapters and terminals.

B/I Busy/Idle bits

B3ZS Bipolar with 3-Zero Substitution. The line coding technique used in the SONET STS-1 (Synchronous Transport Signal-Level 1) electrical signal, which is then converted to an optical signal for transmission over the SONET optical fiber transmission system. B3ZS looks for a series of 3 consecutive zeros, removes them, and replaces them with either B0V or 00V. The choice between B0V and 00V is such that the number of B pulses between consecutive V pulses is odd, where B represents the normal bipolar pulse and V represents a bipolar violation. See also B8ZS, Bipolar Violation, SONET and STS.

B8ZS Bipolar 8 Zero Substitution. A technique used to accommodate the ones density requirement for digital T-carrier facilities in the public network, while allowing 64 Kbps clear data per channel. Rather than inserting a one for every seven consecutive zeros, B8ZS inserts two violations of the bipolar line encoding technique for digital transmission links.

Babble Just what it sounds like — crosstalk from several interfering communications circuits or channels.

Babbling Tributary "A station that continuously transmits meaningless messages," as defined by John McNamara, of DEC and author of "Local Area Networks, an introduction to the technology." Some people might argue this was another word for Harry Newton, the author who didn't know when to stop and wanted to make this dictionary the most comprehensive telecom dictionary ever.

BABT British Approvals Board for Telecommunications. You need their approval before you can sell telecom equipment in Great Britain.

Baby Bells A term for the RBOCs (Regional Bell Operating Companies), also known as RHCs (Regional Holding Companies). The seven RBOCs were spun off from AT&T (the

Bell System) in 1984, as a result of the MFJ (Modified Final Judgement), also known as The Divestiture Agreement. The original seven Baby Bells were Ameritech, Bell Atlantic, BellSouth, NYNEX (since acquired by Bell Atlantic), Pacific Telesis (since acquired by SBC), Southwestern Bell (now SBC) and US West. See also Divestiture.

Baby Bills A term for the numerous companies formed by ex-employees of Microsoft. A play on the "Baby Bell," the reference is to Bill Gates, co-founder and chairman of Microsoft.

Babyphone Feature allowing calls to an off-hook telephone to listen to room noises, for example, to check if a baby is crying.

Back Board A piece of plywood mounted on a wall. Phone equipment is mounted on the plywood. It is more efficient to first mount phone equipment on plywood in the service bay, test it out while it's convenient and diagnostic tools are handy. Then take the phone equipment and the back board (which typically consists of the KSU, power supply and 66-blocks) and install them on the customer's premises. This "pre-installation" makes enormous sense — economically and reliably. Sadly, few installation companies do it.

Back Door A way of getting into a password-protected systems without using the password. Usually a carefully guarded secret to prevent abuse and misuse.

Back End Database server functions and procedures for manipulating data on a network.

Back End Processor A server is often called a back end and a workstation is often called the front end. On a LAN (Local Area Network), A back-end processor runs on a server. It is responsible for preserving data integrity and handles most of the processor-intensive work, such as data storage and manipulation.

Back End Results A call center/marketing term. Used to describe the number of trial or risk free orders that actually pay in terms of customer orders. In a telephone sales context, a bad back-end can be the result of the reps selling the trial and not the product. See Front-End Results.

Back Feed Pull Used in tight locations where it's difficult to take large cable pulling equipment. The cable is fed in two parts from the mid-point. The first section of cable is fed in one direction. After this is fed, the remaining cable is unreeled and fed through the opposite direction to the other end point.

Back Haul Back haul is a verb. A communications channel is back hauling when it takes traffic beyond its destination and back. There are many reasons it might do this. The first is that it may be cheaper to go that route instead of going directly. You might, for example, have a full-time private line from New York Dallas. You might find it cheaper to reach Nashville by going to Dallas first, then dialing back to Nashville. The economics of backhauling may change from one moment to another as the line to Dallas is empty, close to full or full. Another reason for backhauling is that you may do it to accommodate changes in your calling or staffing patterns. You may have an automatic call distributor in Omaha and one in Chicago. A call from New York may come into your Omaha ACD, but when it gets there you may discover that there are no agents available to handle the call. So it may now make sense to back haul the call to the Chicago ACD, where an agent is available.

Back Office Operations Management and support tasks that can be performed away from a company's headquarters, such as telemarketing, credit card processing, data file maintenance and many clerical and accounting functions. Back-office operations are an economic development opportunity for small communities that have the appropriate infrastructure (e.g., advanced telecommunications, reliable express mail services.) Back-office operations are helping share a new definition of place, one in which, for example, geographic remoteness is no longer a liability — because of telecommunications and other linkages to the "outside" world.

Back Hoe Fade When buried fiber optic cable is cut. Called fade because sometimes not all communications are cut off. Also, when they are all cut off, the term becomes a euphemism. Better to report a back hoe fade to your boss than to say, "We're just lost 158,000 circuits between new York and Washington. Our customers are not pleased."

Back Porch The portion of a video signal that occurs during blanking from the end of horizontal sync to the beginning of active video. The blanking signal portion which lies between the trailing edge of a horizontal sync pulse and the trailing edge of the corresponding blanking pulse. Color burst is located on the back porch.

Back Projection When the projection is placed behind a screen (as it is in television and various video conferencing applications where the image is displayed on a monitor or a fabric screen) it is described as a back projection system. In these systems the viewer sees the image via the transmission of light as opposed to reflection used in front projection systems. Audiences generally prefer back projection systems since they seem brighter.

Back To Back Channel Bank The connection of voice frequency and signaling leads between channel banks to allow dropping (i.e. removing) and inserting (i.e. adding) of channels.

Back To Back Connection A connection between the output of a transmitting device and the input of an associated receiving device. When used for equipment measurements or testing purposes, this eliminates the effects of the transmission channel or medium.

Back Up Server A program or device that copies files so at least two up-to-date copies always exist.

Backbone The backbone is the part of the communications network which carries the heaviest traffic. The backbone is also that part of a network which joins LANs together — either inside a building or across a city or the country. LANs are connected to the backbone via bridges and/or routers and the backbone serves as a communications highway for LAN-to-LAN traffic. The backbone is one basis for design of the overall network service. The backbone may be the more permanent part of the network. A backbone in a LAN, a WAN, or a combination of both dedicated to providing connectivity between subnetworks in an enterprise-wide network.

Backbone Bonding Conductor A copper conductor extending from the telecommunications main grounding busbar to the farthest floor telecommunications grounding busbar.

Backbone Closet The closet in a building where the backbone cable is terminated and cross connected to either horizontal distribution cable or other backbone cable.

Backbone Network The part of a communications facility that connects primary nodes; a primary shared communications path that serves multiple users via multiplexing at designated jumping-off points. A transmission facility, or arrangement of such facilities, designed to connect lower speed channels or clusters of dispersed users or devices.

Backbone Subsystem See Riser Subsystem.

Backbone To Horizontal Cross-Connect BHC. Point of interconnection between backbone wiring and horizontal wiring.

Backbone Wiring The physical/electrical interconnec-

tions between telecommunications closets and equipment rooms. Cross-Connect hardware and cabling in the Main and Intermediate Cross-Connects are considered part of the backbone wiring.

Backcharging A phone fraud term. Backcharging is starting the clock on a phone call at the time a customer contacts the long-distant phone service provider — not when the person being called answers the phone — which is what it should be.

Backfeed Pull A method used to pull cable into a conduit or a duct liner when the cable is long or when placing cable into controlled environmental vaults, central offices, or under streets. With this method, the cable pays off its reel at an intermediate manhole and is first pulled in one direction. The remaining cable is then removed from the reel, laid on the ground, and then pulled in the opposite direction.

Backfile Conversion The process of scanning in, indexing and storing a large backlog of paper or microform documents in preparation of an imaging system. Because of the time-consuming and specialized nature of the task, it is generally performed by a service bureau.

Backfilling To designate memory on an expanded memory card and make it available for use as conventional memory.

Background See Background Processing.

Background Communication Data communication, such as downloading a file from a bulletin board, that takes place in the background while the user concentrates on another application (e.g. a spreadsheet) in the foreground.

Background Music This feature allows music to be played through speakers in the ceiling and/or through speakers in each telephone, throughout the office, or office-by-office, or selectively. Background music is typically played through paging speakers, but it can also be played through the speakers of speakerphones. In fact, the two — paging and background music — often go hand-in-hand. When you want to page someone, the music turns off automatically and comes back on when the paging is over. The same thing happens on airplanes. Background music is said to motivate workers, often into shutting it off.

Background Noise The noise you hear when nothing else is being transmitted. On digital circuits it's so quiet, they have actually have to inject some form of White Noise otherwise people begin to suspect that the circuit they're speaking on has gone dead. See also White Noise.

Background Processing The automatic execution of lower priority computer programs when higher priority programs are not using the computer's resources. A higher priority task would be completing calls. A lower priority task would be running diagnostics. Some PBXs have this feature. Some insist on running their diagnostics even though they are choked with calls. The smarter ones tone down their diagnostics when they get busier, which makes sense.

Background Program A low priority program operating automatically when a higher priority (foreground) program is not using the computer system's resources.

Background Task A secondary job performed while the user is performing a primary task. For example, many network servers will carry out the duties of the network (like controlling who is talking to whom) in the background, while at the same time the user is running his own foreground application (like word processing). See also Background Processing.

Backhaul In fiber networks, backhauling a cumbersome traffic management technique used to reduce the expense of multiplexing/demultiplexing. See Back Haul.

Backhoe Fade The degradation in service experienced when a backhoe cuts your cable.

Backoff When a device attempts to transmit data and it finds trouble, the sending device must try again. It may not try again immediately. It may "back off" for a little time so the trouble on the line can be cleared. This happens with LANs. For example, an earlier attempt to transmit may have resulted in a collision in a CSMA/CD (Carrier Sense Multiple Access/Collision Detection) Local Area Network (LAN). So the device "backs off," waits a little and then tries again. How long it waits is determined by preset protocols.

Backoff Algorithm The formula built into a contention local area network used after collision by the media access controller to determine when to try again to get back onto the LAN. See also Backoff.

Backplane The physical area, usually at the rear of an electronics frame, where modules and cables plug into the system. The high-speed communications line to which individual components, especially slide-in cards, are connected. For example, all the extensions of a PBX are connected to line cards (circuit boards) which slide into the PBX's cage. At the rear of the PBX cage, there are several connectors. Each of these connectors is wired to the PBX's backplane, also called a backplane bus. This backplane bus is typically running at a very high speed, since it carries many conversations, address information and considerable signaling. These days, the backplane bus is typically a time division multiplexed line — somewhat like a train with many cars, each representing a time slice of another conversation (data, voice, video or image). The backplane's capacity determines the overall capacity of the switch. See Passive Backplane.

Backplane Bus See Backplane.

Backpressure Propagation effects in a communications network of hop-by-hop flow control to upstream nodes.

Backscattering 1. In fiber optics, the scattering of light into a direction opposite to the original one.
2. Radio wave propagation in which the direction of the incident and scattered waves resolved along a reference direction (usually horizontal) are oppositely directed. A signal received by backscattering is often referred to as "backscatter."

Backslash Also called a virgule, the backslash key achieved fame because Microsoft used it to bring distinguish between subdirectories in MS-DOS. This is a backslash \

Backup A copy of computer data on an external storage medium, such as a floppy disk or tape. Computers and telephone systems (which are computers) are unreliable. They glitch and lose data for all sorts of unusual and impossible-to-predict reasons. Thus the necessity for backups. The theory is when (not "if") a glitch will occur and the PBX's database will disappear off the face of the earth. If this happens, you have a backup and you simply retrieve the back-up file, load it up and, presto, you're back live.

Only information changed (since the backup was made) is lost. Backups save time in restoring the system after a loss. Most modern PBXs work with a database and other extensive customized instructions the user loads in. Most PBX users forget to make and keep backups of their PBX data. They expect their vendor to make backups, but he rarely does. This carelessness costs weeks of aggravation, as the PBX's database and instruction set is manually (and painfully) put back together.

The method by which backups are maintained is also important. The medium should clearly be reliable, i.e. the best quality magnetic medium. The method of backing up is also important. For example, a streaming tape backup is less reli-

able than a file-by-file backup. In a streaming backup, the backup medium simply captures the original data one bit after another in one long stream. In a file-by-file backup, the data moves over in logical segments — command files, data files, etc. Streaming backups will work if their data is placed back on the same precise device from which they were originally taken. But, if they are placed on a different device (even though the same model number, etc.), they may barf because the tape assumes bad sectors are in the same place. This will probably not be true. Streaming tape backup devices are less expensive to buy and much faster to use. Avoid them.

Backup Domain Controller BDC. A server in a network domain that keeps and uses a copy by a computer without interrupting its current or primary task. For Windows NT Server domains, refers to a computer that receives a copy of the domain's security policy and domain database, and authenticates network logons.

Backup Link A resilient (fault tolerant) link which is not used until the primary link fails.

Backup Ring The token ring cabling between MAUs or CAUs consists of the main ring and the backup ring. The data is normally transmitted on the main ring, but if an error occurs, the data can be transmitted on the backup ring until the main ring is repaired. In MAU networks the switching is done manually while in CAU (or CAM) networks it is done automatically.

Backward Channel In data transmission, a secondary channel whose direction of transmission is constrained to be opposite to that of the primary (or forward) channel. The direction of transmission in the backward channel is restricted by the control interchange circuit that controls the direction of transmission in the primary channel. The channel of a data circuit that passes data in a direction opposite to that of its associated forward channel. The backward channel is usually used for transmission of supervisory, acknowledgement, or error-control signals. The direction of flow of these signals is opposite to that in which information is being transferred. The bandwidth of this channel is usually less than that of the forward channel; i.e., the information channel.

Backward Compatible A general term applied to new technologies, products or services which represent a more or less graceful upgrade of those existing. As such, they can be provisioned without requiring dramatic and complete (read prohibitively expensive) changes in the current order of things. For instance, a new and enhanced PBX system technology can be implemented through an upgrade of the existing system rather than a full replacement. Although such an upgrade can be quite expensive, the investment in the current system is protected to a large extent from technical and functional obsolescence. Another example of backward compatibility is that of Frame Relay, which is based on the existing network infrastructure, including LAP-D, the link access protocol used in ISDN. Frame Relay services, therefore, works over existing access links and make use of network standards already in place.

Backward Learning Routing algorithm based on assumed symmetric network conditions. Source node assumes best route to given destination is via neighbor node that was on best route from destination to source.

Backward Recovery The reconstruction of an earlier version of a file by using a newer version of data recorded in a journal.

Backward Signal A signal sent in the direction from the called to the calling station, or from the original communications sink to the original communications source. The backward signal is usually sent in the backward channel and consists of supervisory, acknowledgement, or error control elements.

Backward Supervision The use of supervisory sequences from a secondary to a primary station.

Bad Block A defective unit on a storage medium that software cannot read or write.

Bad Line Button A button on the console that lets the attendant remove a line from service when there's noise or static on it, or when it's just simply busted. Some bad line buttons simply mark the line to separate it from the rest so the phone company can quickly identify the problem. See Bad Line Key.

Bad Line Key When the PBX attendant encounters a bad trunk, he/she pushes this bad line button on the console, automatically flagging the trunk for later checking and repair. See Bad Line Button.

Bad Line Reporting Automatically reports a poor connection without interrupting the current call.

Bad Sectors Defective areas on a hard or soft disk. The MS-DOS FORMAT command locks out bad sectors so they are never used. Other operating systems have similar commands. See also Hot Fix.

BAF Billing format of the 0122 structure code defined by the Bellcore Automatic Message Accounting Format (BAF) Requirements TR#030#NWT-001100. This format identifies paths according to the resource they terminate on.

Baffle A partition used with a loud speaker to prevent air vibrations from the back of the diaphragm from cancelling out the vibrations from the front of the diaphragm. Particularly valuable in the reproduction of bass notes.

Bag Phone A slang expression for a transportable cellular phone whose characteristics are 3 WATT output, heavy weight (for a portable), and a bag with a handle. Bag phones are not designed for carrying around. They are designed to carry from one place to another and used at that place for serious conversations, including possibly faxing and modemming. Bag phones' big "plus" is that they give off more power as a handheld cellular phone. This makes them useful for semipermanent "installation" in places like construction sites, etc. Bag phones are as powerful as a car phone, which also have 3 WATT output. That compares with handheld cellular phones which are typically 0.6 WATT.

According to my friends at Motorola, the main distinction that constitutes a Bag Phone is the lack of a self-contained battery. A Bag Phone, depending on the actual manufacturer, or more often the garage-shop assembler, was a mobile transceiver and handset that was placed in a soft-sided bag and powered via a cigarette lighter connected cord. At times, the dealer would sell a camcorder battery outfitted with a female cigarette lighter receptacle that was stuffed into the bag. Often a bag phone had a clip- or suction-mounted antenna that was affixed temporarily to the vehicle.

The formal name for bag phone was "Transmobile," a phone that could be moved from car to car, thus avoiding a fixed installation. A Transportable was a distinctly different category. Transportables had their own integral battery, and antenna, and therefore could be operated anywhere, independent of 12 volt DC power supply.

BAIC Barring of All Incoming Calls. A wireless telecommunications term. A supplementary service provided under GSM (Global System for Mobile Communications).

Bakelite Bakelite, invented in (c.) 1920 by a Mr. L.W. Bakele is a flour-filled (yes, wheat flour..!) phenol-formalde-

hyde resin, which superseded flimsy celluloid plastic. One important distinction of Bakelite is that it is a CAST material, as opposed to Catalin plastic (another phenolic plastic invented by George Catalin) which is a molded material. Most 'Bakelite' jewelry is actually Catalin plastic, which is an inferior material due to its tendency to separate and flake apart, as seen in so many (valuable!) old radios of this composition (see the Fada Bullet, Emerson red tabletop etc.). Genuine bakelite is only available in two colors: Black or Brown, whereas Catalin comes in a rainbow of hues. Both Catalin and Bakelite can be identified, and distinguished from their modern look-alike imitations by this simple test: Rub the sample plastic hard with your thumb until it feels rather hot, then quickly smell the heated area. The distinct smell of phenol will be evident if the piece is authentic! Bill Layer, who works for Viking Electronics, Inc. in Hudson, Wisconsin contributed this definition. He says we should not ask why remembers the above. b.layer@vikingelectronics.com

Balance To equalize load or current between parts or elements of a telephone line or circuit. Balancing helps get the best out of a phone line. In more technical terms, balancing a line is to adjust the impedance of circuits and balance networks to achieve specified return loss objectives at junctions of two-wire and four-wire circuits.

Balanced Circuit Telephone circuit in which the two conductors are electrically balanced to each other and to the ground. A balanced electrical interface generally allows data to be transmitted over longer distances than does an unbalanced circuit. See BALANCE.

Balanced Configuration Point-to-point network configuration in HDLC with two combined stations.

Balanced Electrical Interface An electrical interface on which each circuit consists of a separate pair of wires. A balanced electrical interface generally allows data to be transmitted over longer distances than does an unbalanced electrical interface.

Balanced Line A transmission line which has two conductors and a ground. When the voltages of the two conductors are equal in strength but opposite in direction, then you have a balanced line.

Balanced Mode Transmission Data transmission with information conveyed by differences in voltages on two circuits to minimize effects of induced voltages.

Balanced Modulator An amplitude modulating circuit that suppresses the carrier signal, producing an output consisting only of upper and lower sidebands.

Balanced Return Loss A measure of the effectiveness with which a balanced network simulates the impedance of the two-wire circuit at a hybrid coil. More generally, a measure of the degree of balance between two impedances connected to two conjugate sides of a hybrid set, network, or junction.

Balanced To Ground In a two-conductor circuit, a balanced-to-ground condition exists where the impedance-to-ground on one wire equals the impedance-to-ground on the other. This is the preferred condition for decent data communications.

Balanced Transmission Line A line having conductors with equal resistance per unit length and equal capacitance and inductance between each conductor and ground. Coaxial cable, for example, is configured easily as a balanced transmission system by the use of resistance-to-ground terminators.

Balancing Network A network used in a set ending a four-wire circuit to match the impedance of the two-wire circuit. 2. Sometimes employed as a synonym for balun.

Balcony A little platform up a telephone pole where people can work or sleep safely.

Ballast Device that modifies incoming voltage and current to provide the circuit conditions necessary to start and operate electric discharge lamps, e.g. fluorescent bulbs.

Balloon Help 1. Place your cursor on a button or a command, in a couple of seconds a balloon pops up explaining in lesser or greater detail what clicking on the button or command will do.

2. From Wired Magazine: When someone insists on explaining every obvious detail and function of an electronic device. Refers to the rarely used Balloon Help feature on Macs. "Um, I don't really need balloon help. Just give me the domain address."

Ballot A release form that authorizes a customer's long-distance phone service to be switched to (another) long-distance carrier or reseller.

Balun BALanced/UNbalanced. An impedance matching transformer. Baluns are small, passive devices that convert the impedance of coaxial cable so that its signal can run on twisted-pair wiring. They are used often so that IBM 3270-type terminals, which traditionally require coaxial cable connection to their host computer, can run off twisted-pair. Works for some types of protocols and not for others. There is often some performance degradation with baluns. And the signal cannot run as far on twisted wire as it can on coaxial cable.

BAN Billing Account Number. Used by telephone companies to designate a customer or customer location that will be billed. A single customer may have multiple billing accounts.

Banana A telecommunications tool. A banana is an induction probe, usually yellow, size of a banana. w/ metal clip, (ears) for clipping a test set onto.

Band 1. Originally referred to AT&T's WATS Bands. AT&T WATS service was organized into circles of increasing distance from the caller. Each circle or BAND (also called SERVICE AREA), cost more per minute. But within each service area, each call costs the same per minute, even though the distances the calls travel might be different. There were typically six interstate bands covering the US and several intrastate bands (depending on how large the state is) which a customer can buy. The word "band" was invented by AT&T Communications (originally known as AT&T Long Lines) when it introduced WATS service. Recently it changed the word "band" to "Service Area." Nobody knows why. See Postalized and WATS.

2. Band can also to the range of frequencies between two defined limits. For example, the band of frequencies able to be heard by the human ear ranges between 30 to 25,000 hertz. The ear can hear a band (or more correctly, a bandwidth) of about 25,000 hertz.

Band Elimination Filter BEF. A filter that has a single continuous attenuation band, with neither the upper nor lower cut-off frequencies being zero or infinite.

Band Marking A label placed on an insulated wire or fiber during installation or manufacture to identify it.

Band Splitter A multiplexer designed to split the available frequency band into several smaller channels. A band splitter can use time division or frequency division multiplexing.

Band Stop Filter A device that bars passage of frequencies within its designed range(s), and allows passage of higher or lower frequencies, or both.

Band, Citizens One of two bands used for low power radio transmissions in the United States — either 26.965 to 27.225 megahertz or 462.55 to 469.95 megahertz. Citizens band radio is not allowed in many countries, even some civi-

lized countries. In some countries they use different frequencies. CB radios, in the United States, are limited by FCC rule to four WATTS of power, which gives each CB radio a range of several miles. Some naughty people boost their CBs with external power. The author of this dictionary has actually spoken to Australia while driving on the Santa Monica Freeway in Los Angeles. See also CB.

Band, Frequency The frequencies between the upper and lower bands. See also BAND. Here is the accepted explanation of "bands:"

Below 300 Hertz	—	ELF —	Extremely low frequency
300—3,000 Hertz	—	ILF —	Infra Low Frequency
3—30 kHz	—	VLF —	Very Low Frequency
30—300 kHz	—	LF —	Low Frequency
300—3,000 kHz	—	MF —	Medium Frequency
3—30 MHz	—	HF —	High Frequency
30—300 MHz	—	VHF —	Very High Frequency
300—3,000 MHz	—	UHF —	Ultra High Frequency
3—30GHz	—	SHF —	Super High Frequency
30—300GHz	—	EHF —	Extremely High Frequency
300—3,000 GHz	—	THF —	Tremendously High Frequency

Band	American	European
P	0.2-1.0 Ghz	0.2-0.375 Ghz
L	1-2 Ghz	0.375-1.5 Ghz
S	2-4 Ghz	1.5-3.75 Ghz
C	4-8 Ghz	3.75-6 Ghz
X	8-12.5 Ghz	6-11.5 Ghz
J		11.5-18 Ghz
Ku	12.5-18 Ghz	-
K	18-26.5 Ghz	18-30 Ghz
Ka	26.5-40 Ghz	-
Q	-	30-47 Ghz

Banded Memory In a PostScript printer, virtual printer memory is a part of memory that stores font information. The memory in PostScript printers is divided into banded memory and virtual memory. Banded memory contains graphics and page-layout information needed to print your documents. Virtual memory contains any font information that is sent to your printer either when you print a document or when you download fonts.

Banded Rate A price range for regulated telephone service that has a minimum floor and maximum ceiling. The minimum covers the cost of service; the maximum is the rate filed in the price list.

Bandit Mobile A mobile subscriber that is revealed in the toll-ticketing records as having an invalid ESN, invalid telephone number, or other problem that warrants denial of service to that mobile.

Banjo Also called beaver tail. Used to connect devices to modular jack wiring for testing. See Modular Breakout Adapter.

Banjo Clip See Modular Breakout Adapter.

Bandmarking A continuous circumferential band applied to an insulated conductor at regular intervals for identification.

Bandpass The range of frequencies that a channel will transmit (i.e. pass through) without excessive attenuation.

Bandpass Filter A device which transmits a band of frequencies and blocks or absorbs all other frequencies not in the specified band. Often used in frequency division multiplexing to separate one conversation from many.

Bandpass Limiter A device that imposes hard limiting on a signal and contains a filter that suppresses the unwanted products of the limiting process.

Bandwidth 1. In telecommunications, bandwidth is the

width of a communications channel. In analog communications, bandwidth is typically measured in Hertz — cycles per second. In digital communications, bandwidth is typically measured in bits per second (bps). A voice conversation in analog format is typically 3,000 Hertz, carried in a 4,000 Hertz analog channel. In digital communications, encoded in PCM, it's 64,000 bits per second. Do not confuse bandwidth with band. Let's say we're running a communications device in the 12 GHz band. What's its bandwidth? That's the space it's occupying. Let's say it's occupying from 12 GHz to 12.1 GHz. This means that it's occupying the space from 12,000,000,000 Hz to 12,100,000,000 Hz. This means its bandwidth is one hundred million cycles or one hundred megahertz (1 MHz). Affiliated terms are narrowband, wideband and broadband. While these are not precise terms, narrowband generally refers to some number of 64 Kbps channels (Nx64) providing aggregate bandwidth less then 1.544 Mbps (24x64 Kbps, or T-1), wideband is 1.544 Mbps-45 Mbps (T-1 to T-3) and broadband provides 45 Mbps (T-3) or better.

2. The capacity to move information. A person who can master hardware, software, manufacturing and marketing — and plays the oboe or some other musical instrument — is "high bandwidth." The term is believed to have originated in Redmond, WA in the headquarters of Microsoft. People there (e.g., Bill Gates) who are super-intelligent and have generally broad capabilities, are said to have "high bandwidth."

3. Microsoft jargon for schedule. For example, "I have a bandwidth problem" means that I have an overloaded schedule.

4. The combined girth of a rock band. By way of example, the band "Meatloaf" is broadband, largely due to the individual girth of the singer by the same name. On the other hand, the "Rolling Stones" are narrowband due largely to the svelte Mick Jagger. While the "Rolling Stones" are older, they are also richer than is "Meatloaf." So, bandwidth is not everything!

Bandwidth Augmentation Bandwidth augmentation is the ability to add another communications channel to an already existing communications channel.

Bandwidth Compression A technique to reduce the bandwidth needed to transmit a given amount of information. Bandwidth compression is used typically in "picture type" transmissions — such as facsimile, imaging or video-conferencing. For example, early facsimile machines scanned each bit of the document to be sent and sent a YES or NO (if there was material in that spot or not). More modern machines simply skip over all the blank spaces and transmit a message to the receiving facsimile machine when to start printing dots again. A facsimile "picture" is made up of tiny dots, similar to printing photos in a magazine. Today, bandwidth compression is used to transmit voice, video and data. There are many techniques, few of which are standard. The key, of course, is that if you're going to compress a "conversation" at one end, you must "de-compress" it at the other end. Thus, in every bandwidth compressed conversation there must be two sets of equipment, one at each end. And they better be compatible.

Bandwidth Junkie One who worships brute speed when it comes to Internet connections. He's the type of person who has a T-1 line in his bedroom.

Bandwidth Limited Operation The condition prevailing when the system bandwidth, rather than the amplitude (or power) of the received signal, limits performance. The condition is reached when the system distorts the shape of the signal waveform beyond specified limits. For linear systems, bandwidth-limited operation is equivalent to distortion-limited operation.

Bandwidth On Demand Just what it sounds like. You want two 56 Kbps circuits this moment for a videoconference. No problem. Use one of the newer pieces of telecommunications equipment and "dial up" the bandwidth you need. An example of such a piece of equipment is an inverse multiplexer. Uses for bandwidth on demand include video conferencing, LAN interconnection and disaster recovery. Bandwidth on demand is typically done only with digital circuits (they're easier to combine). Bandwidth on demand is typically carved out of a T-1 circuit, which is permanently connected to the customer's premises from a long distance carrier's central office, also called a POP — Point of Presence.

Bang An exclamation point (!) used in a Unix-to-Unix Copy Program (UUCP) electronic mail address. People who are on AT&T Mail often give you their mail address as "Bang Their Name." My AT&T Mail address used to be Bang HarryNewton, i.e. !HarryNewton.

Bang Path A series of UUCP nodes mail will pass through to reach a remote user. Node names are separated by exclamation marks nicknamed "bangs." The first node in the path must be on the local system, the second node must be linked to the first, and so on. To reach user 1 on sys2 if your computer's address is sys1 you would use the following address: sys1! sys2! sys3! user1

Bank A row of similar components used as a single device, like a bank of memory. Banks must be installed or removed together. See BANK SWITCHING.

Bank Switching A way of expanding memory beyond an operating system's or microprocessor's address limitations by switching rapidly between two banks of memory. In MS-DOS, a 64K bank of memory between 640K and one megabyte is set aside. When more money is needed, the bank, or page, is switched with a 64K page of free memory. This is repeated with additional 64K pages of memory. When the computer requires data or program instructions not now in memory, expanded memory software finds the bank containing the data and switches it with the current bank of memory. Although effective, bank switching results in memory access times that are slower than true, extended memory.

Bantam Connector Plug A plug and jack used to connect test equipment with digital circuits such as (DS1, DS3, STS1). Wired to DSX patch panels.

BAOC Barring of All Outgoing Calls. A wireless telecommunications term. A supplementary service provided under GSM (Global System for Mobile Communications).

Bar Code A bunch of lines of varying width printed on something. The bar code is designed to be read optically by some data capturing device. Bar codes are turning up on letters. They are read by image scanning devices in the post office and allegedly help the mails move faster. Bar codes are on most things you buy now in supermarkets. By scanning those bar codes at the checkout counter, the supermarket knows what's being sold and not being sold. And presumably the supermarket, or its computer, can order supplies to keep the supermarket stocked with goods that are selling and not re-order those which aren't.

Barbie, Jane The electronic "Voice With A Smile" on many telephone company intercept recordings. Ms. Barbie does her work for Electronic Communications, Inc. of Atlanta, GA.

Bare Wire An electrical conductor having no covering or insulation. See also Hard Cable.

Barge-In Interrupting a call in progress.

Barge-Out Leaving a call in progress without notice.

Barium Ferrite A type of magnetic particle used in some

recording media including floptical diskettes. See also FER-RITE, HARD FERRITE and SOFT FERRITE.

Barrel An imaging term. Distortion that swells an image in the middle, narrows it at the top and the bottom.

Barrel Connector This connector is a cylindrical (barrel-shaped) connector used to splice together two lengths of thick Ethernet coaxial cable.

Barrel Contact A term in cabling. A barrel contact is an insulation displacement type contact consisting of a slotted tube that cuts the insulation when the wire is inserted.

Barrel Distortion When a screen is distorted — with the top, bottom and sides pushing outwards (like a beer barrel) — the screen is said to be suffering barrel distortion.

Barton Enos Barton once said he was "disgusted" when told that it would be possible to send conversation along a wire. He later co-founded (with Elisha Gray) the Western Electric Company, which became AT&T's manufacturing subsidiary and was once the largest electrical equipment manufacturer in the US. In addition to phones, the company made sewing machines, typewriters, movie sound equipment, radio station gear, radar systems and guided missile parts. See also GRAYBAR.

Base Address The first address in a series of addresses in memory, often used to describe the beginning of a network interface card's I/O space.

Base Amount A call center term. One of the historical patterns; what the monthly call volume would be if there were no long-term trend or seasonal fluctuation- in other words, the average number of calls per month.

Base-Controlled Hand-off Cell transfers managed by and initiated from the cellular radio network, typical of Advanced Mobile Phone Systems (AMPS).

Base Load In trunk forecasting, an amount of telephone traffic measuring during a certain defined time. See BASE PERIOD.

Base Memory What many people refer to as the first 640 kilobytes of memory in an MS-DOS PC.

Base Period In trunk forecasting, a time span of consecutive study during which a base load is determined.

Base Schedules A call center term. A fixed set of pre-existing schedules that you can use as a starting point in scheduling. And new schedules created are in addition to the base schedules.

Base Staff A call center term. The minimum number of people, or "bodies in chairs," required to handle the workload in a given period. The actual required number of staff is always greater than base the staff, because of various human factors such as the need for breaks and time off. Therefore, schedules need to add in extra people to accommodate breaks, absenteeism and other factors that will keep agents from the phones. See Rostered Staff Factor.

Base Station A wireless term. A base station is the fixed device a mobile radio transceiver (transmitter/receiver) talks to, to talk to a person or to get to the landline phone network, public or private.

Base Transceiver Station BTS. The electronic equipment housed in cabinets that together with antennas comprises a PCS facility or "site". The cabinets include an air-conditioning unit, heating unit, electrical supply, telephone hook-up, and back-up and back-up power supply.

base64 A standard algorithm for encoding and decoding non-ASCII data for attachment to an e-mail message, base 64 is the foundation for MIME (Multipurpose Internet Mail Extensions). MIME was standardized by the Internet Engineering Task Force (IETF) in RFC 1521 — Appendix G — Canonical Encoding Model. Base64 uses a 65-character subset of US-ASCII to represent all 26 characters of the English language, in upper and lower case; all 10 digits in the numbering scheme (i.e., 1-9); "+" and "/." The 65th character is "=," which is used to signify special processing functions. Each of these 64 (actually 65) values can be represented by 6 bits contained within an 8-bit byte; the 1st and 8th bits, known respectively as high-order and low-order bits, are used to frame the data. The encoding process starts with 24-bit input groups, which are formed by concatenating (linking together) three 8-bit groups. These 24-bit input groups are then treated as 4 concatenated 6-bit groups, each of which is translated into a single digit in the base64 alphabet. Each of those single digits is framed with a high-order and a low-order bit. The resulting "characters" are formed into an output stream of "lines" of no more than 76 characters each. The specific type of content (e.g., audio, image or video) is identified by a content header, which precedes the attached data, which then is transmitted across the IP network, usually the Internet. The receiving device decodes the data by reversing the process. As base64 yields a data stream which is approximately 33% greater than the original content, it is not a compression algorithm. It does, however, comprise a standard means of transmitting non-ASCII data over an IP network, and is unaffected by gateways between networks and systems. UUencode and binhex are alternative, non-standard methods of accomplishing the same thing, although their content may be affected by gateway intervention. See also binhex, MIME and UUencode.

Baseband A form of modulation in which signals are pulsed directly on the transmission medium without frequency division. Local area networks as a rule, fall into two categories — broadband and baseband. The simpler, cheaper and less sophisticated of the two is baseband. In baseband LANs, the entire bandwidth (capacity) of the LAN cable is used to transmit a single digital signal. In broadband networks, the capacity of the cable is divided into many channels, which can transmit many simultaneous signals. While a baseband channel can only transmit one signal and that signal is usually digital, a broadband LAN can transmit video, voice and data simultaneously by splitting the signals on that cable using frequency division multiplexing. The electronics of a baseband LAN are simpler than a broadband LAN. The digital signals from the sending devices are put directly onto the cable without modulation of any kind. Only one signal is transmitted at a time. Multiple "simultaneous" transmissions can be achieved by a technique called time division multiplexing (see multiplexing). In contrast, broadband networks (which typically run on coaxial cable) need more complex electronics to decipher and pick off the various signals they transmit. Attached devices on a broadband network require modems to transmit. Attached devices to baseband networks do not.

Baseband LANs typically work with one high speed channel, which all the attached devices — printers, computers, databases — share. They share it by using it in turns — for example, passing a "token" to the next device. That token entitles the device with the token to transmit. IBM's LAN is a token ring passing local area network. Another way of sharing the baseband LAN is that each device, when it is ready to transmit, simply transmits into the channel and waits for a reply. If it doesn't receive a reply, it retransmits. Thus there are two main network or baseband access control schemes — Token Ring Passing and CSMA/CD. See also CSMA/CD, BROADBAND, ETHERNET and LOCAL AREA NETWORKS.

Baseband Modem A modem which does not apply a

complex modulation scheme to the data before transmission, but which applies the digital input (or a simple transformation of it) to the transmission channel. This technique can only be used when a very wide bandwidth is available. It and only operates over short distances where signal amplification is not necessary. Sometimes called a limited distance or short-haul modem.

Baseband Signaling Transmission of a digital or analog signal at its original frequencies, i.e., a signal in its original form, not changed by modulation.

Baseboard A term used in voice processing/computer telephony to mean a printed circuit board without any daughterboards attached.

Baseboard Raceway A floor distribution method in which metal or wood channels, containing cables, run along the baseboards of the building. The front panel of the baseboard channel is removable, and outlets may be placed at any point along the channel.

Baseline 1. The line from which a graph is drawn. The base line is the X axis on vertically oriented graphs, the Y axis on horizontal bar graphs, or the line representing zero if the data contains both positive and negative numbers.

2. The imaginary line extending through a font and representing the line on which characters are aligned for printing. In conventional, alphanumeric fonts, the baseline is usually defined as the imaginary line touching the bottom of upper-case characters.

Baseline Sequential JPEG The most popular of the JPEG modes which employs the lossy DCT (Discrete Cosine Transform) to compress image data as well as lossless processes based on variations of DPCM (Differential Pulse Code Modulation). The "baseline" system represents a minimum capability that must be present in all Sequential JPEG decoder systems. In this mode image components are compressed either individually or in groups. A single scan pass completely codes a component or group of components.

BASIC Beginners All-purpose Symbolic Instruction Code. A programming language. BASIC is an easy language to learn. It's worth learning. Not all "Basics" are the same.

Basic Budget Service An inexpensive local phone service often restricted to people with limited incomes. It may or may not include any outgoing calls. It may include only a few outgoing calls.

Basic Cable Television Service A tier of cable television service which is available to all subscribers, is specifically identified as "basic" service, includes all television broadcast stations listed on the Must-Carry Station List in the system's PIF, includes all television broadcast stations which the system offers to any subscriber pursuant to any retransmission-consent agreement, includes all public, educational, and government access channels which are designated by franchise for carriage on the basic tier and may include, at the cable operator's option, any other service.

Basic Call A call between two users that does not require Advanced Intelligent Network Release 1 features (e.g. a POTS call). Definition from Bellcore. See AIN.

Basic Call State Model BCSM. An abstraction of the ASC call processing activities for a basic two-party call. The BCSM is split into Originating and Terminating BCSMs.

Basic Carrier Platform Common Carrier Platform. A common carriage transmission service coupled with the means by which consumers can access any or all video provides making use of the platform. Video dialtone service at the basic platform level differs from the "channel service " that

Local Exchange Carriers (LECs) currently may provide cable television operators in that the Commission will require LECs to provide sufficient transmission capacity to serve multiple video programmers.

Basic Exchange Telecommunications Radio Service BETRS. A service that can extend telephone service to rural areas by replacing the local loop with radio communications, sharing the UHF and VHF common carrier and private radio frequencies.

Basic Mode Link Control Control of data links by use of the control characters of the 7-bit character set for information processing interchange as given in ISO Standard 646-1983 and ITU-T Recommendation V.3-1972.

Basic Rate The default rate paid by a telephone subscriber who does not elect any other calling plan. Basic rate can apply to phone service from either a local phone company or a long distance phone company. Basic rate for local service can be a flat monthly rate or a monthly rate plus charges for each call — on length of call, time of day and distance sensitive charging. A basic rate for long distance or local toll service will be length of call, time of day and distance sensitive charging.

Basic Rate Interface BRI. There are two "interfaces" in ISDN: BRI and PRI. In BRI, you get two bearer B-channels at 64 kilobits per second and a data D-channel at 16 kilobits per second. The bearer B-channels are designed for PCM voice, video conferencing, group 4 facsimile machines, or whatever you can squeeze into 64,000 bits per second full duplex. The data D-channel is for bringing in information about incoming calls and taking out information about outgoing calls. It is also for access to slow-speed data networks, like videotex, packet switched networks, etc. One BRI standard is the "U" interface, which uses two wires. Another BRI standard is the "T" interface which uses four wires. See ISDN for a much fuller explanation.

Basic Service A telephone company service limited to providing local switching and transmission. Basic Service does not include equipment. The term Basic Service is unclear and varies between telephone companies and data communications service providers.

Basic Service Elements BSEs. An Open Network Architecture term. BSEs are services which value-added companies could get from their phone company. BSEs are optional basic network functions that are not required for an ESP to have a BSA, but when combined with BSEs can offer additional features and services. Most BSEs allow an ESP to offer enhanced services to their customers. BSEs fall into four general categories: Switching, where call routing, call management and processing are required; Signaling, for applications like remote alarm monitoring and meter reading; Transmission, where dedicated bandwidth or bit rate is allocated to a customer application; and Network Management, where a customer is given the ability to monitor network performance and reallocate certain capabilities. The selection of available BSEs is an ongoing process, with new arrangements being developed. ANI, Audiotext "Dial-It" Services, and Message Waiting Notification are all examples of BSEs. See AIN and Open Network Architecture.

Basic Serving Arrangement BSA. An old term defining the relationship of an enhanced service provider (value added provider) to the phone company providing the line/s. Under ONA (Open Network Architecture), a BSA is the basic interconnection access arrangement which offers a customer access to the public network (i.e. the normal switch phone service) and provides for the selection of available Basic Service

Elements. It includes an ESP (Enhanced Service Provider) access link, the features and functions associated with that access link at the central office serving the ESP and/or other offices, and the transport (dedicated or switched) within the network that completes the connection from the ESP to the central office serving its customers or to capabilities associated with the customer's complementary network services. Each component may have a number of categories of network characteristics. Within these categories of network characteristics are alternatives from among which the customer must choose. Examples of BSA components are ESP access link, transport and/or usage. See Open Network Architecture.

Basic Telephony The lowest level of service in Windows Telephony Services is called Basic Telephony and provides a guaranteed set of functions that corresponds to "Plain Old Telephone Service" (POTS - only make calls and receive calls). See Windows Telephony Services.

BASR Buffered Automatic Send/Receive.

Bastion Host An Internet term. A computer placed outside your firewall to provide public services (such as WWW and ftp) to other Internet sites. This term is sometimes generalized to refer to any host which is critical to the defense of a local network. Generally, a bastion host is running some form of general purpose operating system (e.g., UNIX, VMS, WNT, etc.) rather than a ROM-based or firmware operating system.

Batch Processing There are two basic types of data processing. One is batch processing. Also called deferred time processing and off-line processing. Batch processing occurs where everything relating to one complete job — such as preparing this week's payroll — is bunched together and transmitted for processing (locally, in the same building or long distance, across the country), usually by the same computer and under the same application program. Batch processing does not permit interaction between the program and the user once the program has been read (i.e. fed into) the computer. In batch processing with telecommunications (i.e. sending the task to be done over the phone line), network response time is not critical, since no one is sitting in front of a screen waiting for a response. On the other hand, accuracy of communications is very critical, since no one is sitting in front of a screen checking entries and responses.

The second type of processing is called interactive or real time processing. Under this method of processing, a user sends in transactions and awaits a response from the distant computer before continuing. In this case, response time on the data communications facility is critical. Seconds count, especially if a customer is sitting at the other end of a voice call awaiting information on whether there's space on that airline flight, for example. See Batch File.

Batmobiling Batmobiling is putting up emotional shields from the retracting armor that covers the batmobile as in "she started talking marriage and he started batmobiling."

Batteries A, B and C In ancient times, when vacuum tubes were still in use, the A battery (often about 6 volts) heated the filament or cathode to boil off electrons, the B battery (usually several hundred volts), positive with respect to cathode, sucked the negative electrons to the plate, while the C battery, a few volts negative with respect to the cathode, tended to repel the electrons back toward the cathode. By varying a small signal voltage riding on the C battery, the flow of electrons to the plate could be controlled to generate a much larger voltage.

Battery All telephone systems work on DC (direct current). DC power is what you use to talk on. Often the DC power is called "talking battery." Most key systems and many PBXs plug directly into an AC on the wall, but that AC power is converted by a built-in power supply to the DC power the phone system needs. All central offices (public exchanges) used rechargeable lead acid batteries to drive them. These batteries perform several functions: 1. They provide the necessary power. 2. They serve as a filter to smooth out fluctuations in the commercial power and remove the "noise" that power often carries. 3. They provide necessary backup power should commercial power stop, as in a "blackout" or should it get very weak, as in a "brownout."

In short, "battery" is the term used to reference the DC power source of a telephone system. Often called "Talking Battery." 2. Storage battery used with central office switching systems and PBXs serving locations which cannot tolerate outages. Batteries serve the following purposes: Act as a filter across the generator or power rectifier output to smooth out the current and reduce noise; provide a cushion against periodic overloads exceeding the generator/rectifier capacity; supply emergency power for a limited time in event of commercial power failure. See also AC, AC Power, Battery Reserve, Central Office Battery and and NICAD Battery.

Battery Backup A battery which provides power to your phone system when the main AC power fails especially during blackouts and brownouts. Hospitals, brokerage companies, airlines and hotel reservation services must have battery backup because of the integral importance of their phone systems to their business.

Battery Eliminator A device which has a rectifier and (hopefully) a filter. This device will convert AC power into the correct DC voltages necessary to drive a telephone system. Such a battery eliminator, or power supply, should deliver "clean" power, i.e. with little "noise" and of low impedance.

Battery Management An umbrella term used by various UPS manufacturers to describe a suite of functions related to charging, testing, and maximizing the life of a UPS (Uninterruptible Power Supply) battery. Battery management may include imminent battery failure diagnosis and indication (first introduced in 1989 by APC corp), scheduled battery testing, hot swappable user replacable batteries, high speed battery charging, output regulation to reduce unnecessary battery usage, and/or special battery charging techniques.

Battery Reserve The capability of the fully charged battery cells to carry the central office power load imposed when commercial power fails and the primary power source (generators/rectifiers) is out of service. Properly described in terms of the number of hours the batteries can furnish operating power for dependent CO apparatus for a demand equal to that on the CO during its busy hour. A busy-hour reserve of eight hours is typical for a telephone office battery plant.

Battle Faxes A cycle of vitriolic faxes exchanged between clients and lawyers, fighting lovers, etc. "Here's the latest round of battle faxes with my record company." Wired Magazine defined this term.

Baud Rate A measure of transmission speed over an analog phone line — i.e. a common POTS line. (POTS stands for Plain Old Telephone Service). Imagine that you want to send digital information (say from your computer) over a POTS phone line. You buy a modem. A modem is a device for converting digital on-off signals, which your computer speaks, to the analog, sine-wave signals your phone line "speaks." For your modem to put data on your phone line means it must send out an analog sine wave (called the carrier signal) and change that carrier signal in concert with the data it's sending.

Baud rate measures the number of number of changes per second in that analog sine wave signal. According to Bell Labs, the most changes you can get out of a 3 KHz (3000 cycles per second) voice channel (which is what all voice channels are) is theoretically twice the bandwidth, or 6,000 baud.

Baud rate is often confused with bits per second, which is a transfer rate measuring exactly how many bits of computer data per second can be sent over a telephone line. You can get more data per second — i.e. more bits per second — on a voice channel than you can change the signal. You do this through the magic of coding techniques, such as phase shift keying. Advanced coding techniques mean that more than one bit can be placed on a baud, so to speak. To take a common example, a 9,600 bit per second modem is, in reality, a 2,400 baud modem with advanced coding such that four bits are impressed on each baud. The continuing development of newer and newer modems point to increasingly advanced coding techniques, bringing higher and higher bit per second speeds. My latest modem, for example, is 56,000 bits per second. Baud is named after Jean-Maurice Emile Baudot. See Baudot Code.

Baudot Code The code set used in Telex Transmission, named for French telegrapher Jean-Maurice Emile Baudot (1845-1903) who invented it. Also known by the ITU approved name, International Telegraph Alphabet 2. The Baudot code has only five bits, meaning that only 32 separate and distinct characters are possible from this code, i.e. 2 x 2 x 2 x 2 x 2 equals 32. By having one character called Letters (usually marked LTRS on the keyboard) which means "all the characters that follow are alphabetic characters," and having one other key called Figures (marked FIGS), meaning "all characters that follow are numerals or punctuation characters," the Baudot character set can represent 52 (26 x 2) printing characters. The characters "space," "carriage return," "line feed" and "blank" mean the same in either FIGS or LTRS. TDD devices (Telecommunications Devices for the Deaf) use the Baudot method of communications to communicate with distant TDD devices over phone lines. See also ASCII, EBCDIC, MORSE CODE and UNICODE, which are other ways of encoding characters into the ones or zeros needed by computers.

Bay A telephone industry term for the space between the vertical panels or mounting strips ("rails") of the rack. One rack may contain several bays. A bay is another place you put equipment.

BBG Basic Business Group.

BBG-I ISDN Basic Business Group.

BBIN BroadBand Intelligent Network. See Broadband and IN.

BBS Bulletin Board System. Another term for an electronic bulletin board. Typically a PC, modem/s and communications bulletin board software attached to the end of one or more phone lines. Callers can call the BBS, read messages and download public domain software. The person who operates a BBS is called a system operator, commonly shortened to SYSOP (pronounced "sis-op"). See also Electronic Bulletin Board.

BCC 1. Bellcore Client Company. What Bellcore called its original owners — the seven regional Bell operating companies and their operating phone company subsidiaries. This definition is something of a historical footnote, as Bellcore was sold to SAIC and there no longer are seven RBOCs. See Bellcore and RBOC.

2. Block Check Character. A control character appended to blocks in character-oriented protocols and used for figuring if the block was received in error. BCC is especially used in longitudinal and cyclic redundancy checking. As a packet (or in IBM jargon, a frame) is assembled for transmission, the bits

are passed through an algorithm to come up with a BCC. When the packet is received at the other end, the receiving computer also runs the same algorithm. Both machines should come up with the same BCC. If they do, the transmission is correct and the receiving computer sends an ACK — a positive acknowledgement. If they don't, an error has occurred during transmission, and they don't have the same bits in the packet. The receiver transmits a signal (a NAK, for negative acknowledgement) that an error has occurred, and the sender retransmits the packet. This process goes on until the BCC checks.

BCCH BroadCast CHannel. A wireless term for the logical channel used in certain cellular networks to broadcast signaling and control information to all cellular phones. BCCH is a logical channel of the FDCCH (Forward Digital Control CHannel), defined by IS-136 for use in digital cellular networks employing TDMA (Time Division Multiple Access). The BCCH comprises the E-BCCH, F-BCCH and S-BCCH. The E-BCCH (Extended-BCCH) contains information which is not of high priority, such as the identification of neighboring cell sites. The F-BCCH (Fast-BCCH) contains critical information which must be transmitted immediately; examples include system information and registration parameters. S-BCCH (System message-BCCH), which has not yet been fully defined, will contain messages for system broadcast. See also IS-136 and TDMA.

BCD Binary Coded Decimal. A system of binary numbers where each digit of a number is represented by four bits. See BINARY. In ATM, binary code decimal is a form of coding of each octet within a cell where each bit has one of two allowable states, 1 or 0.

BCH Bose, Chaudhuri and Hocquenghem error correction code. Named after the three guys who invented it.

BCHO Base-Controlled Hand-Off. A cellular radio term.

BCM 1. Bit Compression Multiplexer.

2. Basic Call Model (a term from the Bellcore discussion of Advanced Intelligent Networks). See AIN.

BCOB An ATM term. Broadband Connection Oriented Bearer: Information in the SETUP message that indicates the type of service requested by the calling user.

BCOB-A Bearer Class A: Indicated by ATM end user in SETUP message for connection-oriented, constant bit rate service. The network may perform internetworking based on AAL information element (IE).

BCOB-C Bearer Class C: Indicated by ATM end user in SETUP message for connection-oriented, variable bit rate service. The network may perform internetworking based on AAI information element (IE).

BCOB-X Bearer Class X: Indicated by ATM end user in SETUP message for ATM transport service where AAL, traffic type and timing requirements are transparent to the network.

BCR Business Communications Review, a good magazine based in Hinsdale, IL. To get a sub, call 1-800-BCR-1234.

BCS Batch Change Supplement.

BDC See Backup Domain Controller

BDLC Burroughs Data Link Control, a bit oriented protocol.

BDN Bell Data Network. The predecessor to ACS (Advanced Communication System). See ACS.

BDSL Broadband Digital Subscriber Line. Same as VDSL.

Beacon A token ring frame sent by an adapter indicating that it has detected a serious ring problem, such as a broken cable or a MAU. An adapter sending such frames is said to be beaconing. See BEACONING.

Beacon Initiative A "plan and rationale to help bring the information highway to Canadian businesses and consumers. The central theme of the initiative is an open collaborative effort with all interested players to bring enhanced interactive data, image and video services to Canadians." That's what the press release of April 5, 1994 said. Companies in the Beacon Initiative include BC Telephone, AGT Limited, SaskTel, Manitoba Telephone System, Bell Canada. NBTel, Maritime Tel & Tel. Island Telephone, Newfoundland Telephone and Stentor Communications.

Beaconing Token Ring process to recover the network when any attached station has sensed that the ring is inoperable due to a hard error. Stations can withdraw themselves from the ring if necessary. A station detecting a ring failure upstream transmits (beacons) a special MAC frame used to isolate the location of the error using beacon transmit and beacon repeat modes. See BEACON.

Beam A way to exchange information between Newton users. Beaming uses the small built-in infrared unit at the top of the Newton MessagePad to send anything that's on one MessagePad to another MessagePad, or to a Sharp Wizard. This can be done across a conference table.

Beam Diameter The distance between the diametrically opposed points on a plane perpendicular to the beam axis at which the irradiance is a specified fraction of the beam's peak irradiance. The term is most commonly applied to beams that are circular or nearly circular in cross-section.

Beam Divergence As through the air telecom signal go further, they diverge. Beam divergence measures that divergence.

Beamsplitter A device for dividing an optical beam into two or more separate beams, often a partially reflecting mirror.

Beans 1. Also called a B Connector or Plain B Wire Connector. A twisted-pair splicing connector that looks like a one-inch drinking straw. They have metal teeth inside them to pierce the vinyl insulation of the wire to make a good connection. Sometimes water-retardant jelly is sometimes placed inside. See B Connector.
2. Beans are also filled with nutrients essential to a healthy diet. My children love them because of their after-effects.

Bearer Channel In its most basic definition, a bearer channel is a basic communication channel with no enhanced or value-added services included other than the bandwidth transmission capability.

In private line international telecommunications, a bearer channel can consist of any number of DS-Os between two countries that have an agreement to pass traffic between them. This type of bearer is usually an E-1 link between two countries but can be 384 Kbps or any number of 64 Kbps channels combined together. Bearer is usually referred to as a half channel between the two countries that have an agreement to pass traffic, and is referred to as the international half circuit. The middle of the ocean or satellite up-link is the demarcation point for these bearers.

In a more elevated (and more recent) definition, a bearer channel is 64,000 bits per second full duplex. It's the basic building block of the digital signaling hierarchy. An ISDN BRI channel consists of two bearer channels of 64,000 bits per second and one data signaling channel of 16,000 bits per second. It is thus called 2B (two Bearer) + 1D (one D, for data channel. An ISDN PRI (Primary Rate Interface) is 23 B + 1D, the D or Data channel for PRI is 64 Kbps. See ISDN.

Beat Frequency An old radio term: The frequency resulting when an oscillation of one frequency is "beat" or heterodyned against an oscillation of different frequency. The figure given is normally in cycles per second.

Beat Reception An old radio term. The resultant audible frequency when two sources of unequal undamped electrical oscillations of constant amplitude act simultaneously in the same circuit. See BEAT FREQUENCY.

Beating The phenomenon in which two or more periodic quantities having slightly different frequencies produce a resultant having periodic variations in amplitude.

Beaver tail Also called Banjo. Used to connect devices to modular jack wiring for testing.

BECKON See BECN.

BECN Backward Explicit Congestion Notification. A Frame Relay term defining a 1-bit field in the frame Address Field for use in congestion management. During periods of severe congestion, the network will advise the (transmitting) device in the backward direction of this fact. This bit notifies the user that congestion-avoidance procedures should be initiated for traffic in the opposite direction of the received frame. It indicates that the frames that the user transmits on this logical connection may encounter congested resources. In other words, slow down, you move too fast. Your bits may not get through or get delayed. In reality, few devices are capable of "throttling back" based on such advice. Consider a router which connects multiple workstations transmitting across multiple LANs or LAN segments to a Frame Relay network. It is highly unlikely that the router can advise those workstations to pause or slow down the rate of transmission. It is possible that the router can buffer some small number of frames for a short period of time, but that's about it. BECN is a wonderful concept, but one which cannot effectively be implemented.

Bed Of Nails Cord A description of a type of alligator clip that attaches to the end of a craft test set, also called a butt set.

Beeper A colloquial term for a mobile pager — one you carry on your belt or in your purse. In the very beginning, when pagers went off, they made beeping sounds. Thus, they became known as beepers. Not very exciting.

Beeper Sitting To take on responsibility for writing down incoming pages on behalf of a vacationing or otherwise out-of-range beeper-owning friend. "Harry is beeper-sitting for Gerry while he's in Aspen."

Beepilepsy A serious disease afflicting those of us with vibrating pagers, characterized by sudden spasms, goofy facial expressions and loss of speech.

BEF Band Elimination Filter. A filter that has a single continuous attenuation band, with neither the upper nor lower cut-off frequencies being zero or infinite.

Bel A relative measurement, denoting a factor of ten change. Rarely used in practice; most measurements are in decibels (0.1 bel).

Belden A major manufacturer of communications cable. Belden has set many cabling standards. Many other cable manufacturers follow their specs.

Bell A bell in a telephone instrument rings when a 20 Hz signal of about 90 volts AC is applied to the subscriber loop. In contrast, the normal voltage applied to a subscriber loop and used for speaking and listening is 48 volts DC.

Bell, Alexander Graham The following biography of Mr. Bell is courtesy The Bell Homestead Museum Complex, Brantford, Ontario, Canada. www.bfree.on.ca./comdir/festivals/sesqui/bell.html. Alexander Graham Bell was born in Scotland on March 3, 1847 and was the second son of Alexander Melville Bell and the former Eliza Grace Symonds. He was educated at the University of Edinburgh and the University of London. In the 1860s, while they were still living in Scotland, disaster struck the Bell family. Aleck's younger brother Edward Charles died of tuberculosis and, soon afterwards, his elder brother Melville James died from the same disease. Doctors warned his parents that Aleck, too, was threatened by the disease. His father promptly sacrificed his career as a noted teacher of speech and researcher into the problems of the deaf and, in July, 1870, sailed with his family for Canada in search of a better climate. Upon arrival, the Bells bought the property at Brantford now known as the Bell Homestead. Alexander Graham Bell was, at this time, 23 years old. In this wondrous new climate, Bell's health was quickly restored. Like his father, Bell was a teacher of speech, and in the spring of 1871, he accepted an invitation to teach in Boston and moved there to pursue his career. In 1872, he opened a school of his own for teachers of the deaf and, the next year, became Professor of Vocal Physiology at Boston University. The deafness of his mother no doubt provided an inspiration for Bell to follow in his father's footsteps with work in speech studies. It was this work that led Bell to both his bride, Mabel Hubbard — left totally and permanently deaf from Scarlet Fever when she was five years old - and to his ideas for the telephone. At Christmas, during the summer vacations, and at every opportunity, Bell came home to Brantford. On July 26, 1874, while visiting the Homestead, he talked far into the night while disclosing the telephone idea to his father. After young Bell returned to Boston, he began to work in earnest on his new invention and, on June 3, 1875, succeeded in transmitting speech sounds. During a visit home to Brantford in September of that same year, he wrote the specifications for the telephone. On March 10, 1876, at Boston, through the Liquid Transmtter he had designed, Alexander Graham Bell uttered the first words to be carried over a wire - "Mr. Watson, come here, I want you!" In the summer of 1876, back in Brantford again, young Bell conducted three great tests of the telephone. In the first of these three tests, Bell received the first successful telephone call carried between two communities on August 3, 1876 from Brantford to Mount Pleasant. The second great test was made on August 4, 1876, when a large dinner party at the Bell Homestead heard speech recitations, songs, and instrumental music from the telegraph office in Brantford over a line three and one-half miles long. The third great test is hailed as the first long-distance phone call in the world made on August 10, 1876 from Brantford to Paris, Ontario. At last! The invention of the telephone was complete. In his later years, Bell interested himself in the problem of mechanical flight, carrying out experiments with man-lifting kites, working with these at his summer home at Baddeck, Nova Scotia. Bell died at his beloved home at Baddeck on August 2, 1922, and is buried there.

Bell 103 AT&T specification for a modem providing asynchronous originate/answer full duplex transmission at speeds up to 300 bits per second (300 baud) using FSK modulation on dialup lines. This is the most common standard (but not the most common speed) for modems running with personal computers. Every dial service in the U.S. adheres to this standard, and obviously a whole bunch of much faster standards. See the next few definitions and the V.xx standards.

Bell 201 An AT&T standard for synchronous 2400-bps full-duplex modems using DPSK modulation. Bell 201B was originally designed for dialup lines and later leased lines. Bell 201C was designed for half-duplex operation over dialup lines.

Bell 202 An AT&T standard for asynchronous 1800-bps full-duplex modems using DPSK modulation over four-wire leased lines as well as 1200-bps half- duplex operation over dialup lines.

Bell 208 An AT&T standard for synchronous 4800-bps modems. Bell 208A is a full-duplex modem using DPSK modulation over four-wire leased lines. Bell 208B was designed for half-duplex operation over dialup lines.

Bell 209
An AT&T standard for synchronous 9600-bps full-duplex modems using
QAM modulation over four-wire leased lines or half-duplex operation over dialup lines.

Bell 212 AT&T specification for a modem providing full duplex asynchronous or synchronous data transmission at speeds up to 1,200 bits per second on the voice dial-up phone network.

Bell 43401 Bell Publication defining requirements for transmission over telco-supplied circuits that have dc continuity (i.e. circuits that are metallic).

Bell Atlantic Formed as a holding company after the AT&T Divestiture. Includes Bell of Pennsylvania, C&P Telephone Companies of D.C., Maryland, Virginia, and West Virginia, Diamond State Telephone (MD), New Jersey Bell and several Bell Atlantic business activities. In April, 1997, Bell Atlantic bought NYNEX (the holding company for New York Telephone and New England Telephone) for $22.1 billion. Nynex has customers in New York, Massachusetts, New Hampshire, Vermont, Maine and a small part of Connecticut. Check out Bell Atlantic's logo. It has a wonderful, small wave in the "A" in Atlantic. Very charming. Very subtle. www.bellatlantic.com. See Bell Operating Companies.

Bell Communications Research Bellcore. Formed at Divestiture to provide certain centralized services to the seven Regional Holding Companies (RHCs) and their operating company subsidiaries. Also serves as a coordinating point for national security and emergency preparedness communications matters of the federal government. Bellcore does not work on customer premise equipment or other areas of potential competition among its owners — the seven (now five) Regional Bell Operating companies. Bellcore also works on standardizing methods by which customers of long distance companies will reach their favorite long distance companies. At time of this writing, around 8,000 people worked at Bellcore, mostly in northern New Jersey. Bellcore had the unenviable task of trying to service the needs of the seven competitors (now five), which owned it. They agreed to sell it to SAIC in 1997. Bellcore's annual budget is around $1 billion, paid for by the seven (now five) Bell regional operating companies. It does work on ISDN and common channel signaling standards. See also Bellcore for a longer explanation.

Bell Compatible A term sometimes applied to modems. A modem is said to be "Bell compatible" if it conforms to the technical specifications set forth by AT&T for the various devices, such as Bell 212.

Bell Customer Code A three-digit numeric code, appended to the end of the Main Billing Telephone Number. Used by Local Exchange Carriers to provide unique identification of customers.

Bell Operating Company BOC. A BOC is one the 22 regulated telephone companies of the former Bell System, which was broken apart (the Divestiture of the Bell System) at midnight on December 31, 1983. At Divestiture, the Bell operating companies were grouped into seven Regional Holding Companies (RHCs). According to the terms of the Divestiture Agreement between the Federal Courts, the Federal Government and AT&T, the divested companies must limit their activities to local telephone services, directory service, customer premise equipment, cellular radio and any other ventures as the Federal Court may approve from time to time. BOCs are specifically limited from manufacturing equipment and from providing long distance service. See also Regional Bell Operating Company.

The following definition of a Bell Operating Company comes from the Telecommunications Act of 1996: The term `Bell operating company'

(A) means any of the following companies: Bell Telephone Company of Nevada, Illinois Bell Telephone Company, Indiana Bell Telephone Company, Incorporated, Michigan Bell Telephone Company, New England Telephone and Telegraph Company, New Jersey Bell Telephone Company, New York Telephone Company, U S West Communications Company, South Central Bell Telephone Company, Southern Bell Telephone and Telegraph Company, Southwestern Bell Telephone Company, The Bell Telephone Company of Pennsylvania, The Chesapeake and Potomac Telephone Company, The Chesapeake and Potomac Telephone Company of Maryland, The Chesapeake and Potomac Telephone Company of Virginia, The Chesapeake and Potomac Telephone Company of West Virginia, The Diamond State Telephone Company, The Ohio Bell Telephone Company, The Pacific Telephone and Telegraph Company, or Wisconsin Telephone Company; and

(B) includes any successor or assign of any such company that provides wireline telephone exchange service; but

(C) does not include an affiliate of any such company, other than an affiliate described in subparagraph (A) or (B).

Bell Speak A term coined by Michael Marcus for insider jargon spoken by "real" telephone people — those who practiced pre-divestiture. Such old jargon is usually incomprehensible to anyone in today's telephone industry who is younger than 46.

Bell System The entire AT&T organization prior to when it was broken up — at the end of 1984. The Bell System included Bell Labs, Long Lines, Western Electric and the 23 Bell operating companies.

Bell System Practices BSPs. The book that explains everything MA Bell did before divestiture in 1984. No longer used by regional Bells or AT&T. Manufacturers now issue their own instructions. Each RBOC has its own way of doing things. And many BOCs inside the RBOCs also do things their own way. This is not designed to confuse the poor business customer, who may have circuits in many states and thus deals with many of them — but just to encourage diversity, or something.

Bellboy A public paging system run by local Bell phone companies. The name survived for many years and only drew criticism in the middle to late seventies with the rise of the Women's Liberation movement. It's now not used. Shucks.

Bellcore Bell Communications Research. Bellcore was formed by federal mandate coincident with the Divestiture of AT&T in 1984 to provide certain centralized research and development services for its client/owners, the seven (now five) Regional Bell Operating Companies (RBOCs), and their operating company subsidiaries, the BOCs (Bell Operating Companies). The formation of Bellcore was deemed critical at the time, as the MFJ (Modified Final Judgment) called for Bell Telephone Laboratories (Bell Labs) to remain with AT&T. This action left the RBOCs without R&D support. Bellcore was initially staffed with researchers who worked at Bell Labs and who didn't move over to AT&T. Bellcore's focus has been on standards, procedures, software development and all manner of R&D of common interest to the RBOCs, with the exception

of the physical sciences. The MFJ precluded the RBOCs from equipment manufacturing, and even from close involvement in the equipment design process. Traditionally, much of Bellcore's efforts have been in the area of development and support for Operations Support Systems (OSSs) such as Centrex management, line testing, order negotiation and processing, and billing systems. Bellcore also has been a key player in the standards design and systems development work for Advanced Intelligent Networks (AINs).

Bellcore has served as a coordinating point for national security and emergency preparedness, and communications matters of the federal government; this was natural, as the RBOCs and their operating units were (and still are) the dominant carriers responsible for such implementing and supporting such things. Bellcore also was responsible for administering the North American Numbering Plan (NANP), although in 1995 this responsibility was shifted to a more impartial entity, the NANC (North American Numbering Council), in the face of a deregulated and competitive telecommunications landscape. You can acquire Bellcore documents from Bellcore — Document Registrar, 445 South Street, Room 2J-125, P.O. Box 1910, Morristown, NJ 07962-1910. Fax 201-829-5982. www.bellcore.com.

In early 1995, Bellcore's Board of directors announced that the owners (read RBOCs) intended to sell the company, as their interests largely were no longer common in anticipation of deregulation and full competition. Bellcore's 5,600 employees and $1 billion+ budget clearly were not supportable in this environment. On November 21, 1996 it was announced that SAIC (Science Applications International Corporation) had agreed to purchase Bellcore. According to the SAIC press release, "Employee-owned SAIC provides high technology products and services to government and private industry, systems integration, national security, energy, transportation, telecommunications, health care and environmental science and engineering." SAIC revenues are over $2.2 billion; the company employs over 22,000 in more than 475 locations worldwide. SAIC also owned Network Solutions, Inc. (NSI), which runs certain functions for the Internet; NSI was spun off as a public company in early 1998. www.saic.com. See also BCC.

Bellcore AMA Format BAF. The standard data format for AMA (Automatic Message Accounting) data to be delivered to the Revenue Accounting Office (RAO) in Advanced Intelligent Network Release 1.

Bellcore Multi-Vendor Interactions MVI. The process for coordinating the efforts of Bellcore, the Bell Operating telephone companies and vendors to address technical issues associated with Advanced Intelligent Network.

Bellman-Ford-Algorithm Shortest-path routing algorithm that figures on number of hops in a route to find shortest-path spanning tree.

BellSouth Corporation The largest regional Bell holding company formed at the Divestiture of AT&T. Includes Southern Bell and South Central Bell and several other BellSouth businesses. See Regional Bell Operating Company.

Bellwether One that takes the lead or initiative; an indicator of trends. The state of California, for instance, long has been a bellwether with respect to the regulation of utilities through its PUC (Public Utilities Commission). The term "bellwether" originated in the 15th century, when the medieval English began the practice of putting a bell around the neck of a male sheep which had been castrated before reaching sexual maturity. As a result of neutering, the ram was much easier to handle, and to train as the leader of the flock. The bell made it much easier to find. Not surprisingly, not every state is anxious to be known as a bellwether.

Below-the-line Expenses incurred that are charged to shareholders of regulated operating telephone companies, not ratepayers.

Benchmark A standardized task to test the capabilities of various devices against each other for such measures as speed.

Bend Loss A form of increased attenuation caused by allowing high-order signals to radiate from the side of the fiber. The two common types of bend losses are those occurring when the fiber is curved and microbends caused by small distortions of the fiber imposed by poor cabling techniques.

Bend Radius The amount of bend that can occur before a cable may sustain damage or increased attenuation.

Bending Radius The smallest bend which may be put into a cable under a stated pulling force. Bending radius affects size of bends in conduit, trays or ducts. It affects pulley size. It affects the size of openings at pull boxes where loops may form. Bend radius is very critical in all aspects of cable laying, especially with under-carpet cabling and fiber optic cable. If the bend radius is too severe, the cable or fiber can crack or break, which is not a good thing. Coax and shielded copper cables can suffer cracks in the outer conductors, which also is not a good thing.

Bent Pipe A signal relay scheme used in some satellite networks, e.g., Globalstar. The bent pipe signal originates in a wireless user terminal which connects directly with a satellite. The satellite establishes the connection to the target device through a groundstation (satellite dish), or gateway, which completes the call over the existing terrestrial network, which can be either wireline or wireless. See also Globalstar.

Bentogram A syntax for encoding Versit's vCard in a binary encoding. Bentograms are based on Benton, the OpenDoc standard interchange format.

BEP Back End Processor.

BER 1. Bit Error Rate. The ratio of error bits to the total number of bits transmitted. If the BER gets too high, it might be worth while to go to a slower baud rate. Otherwise, you would spend more time retransmitting bad packets than getting good ones through. The theory is that the faster the speed of data transmission the more likelihood of error. This is not always so. But if you are getting lots of errors, the first — and easiest — step is to drop the transmission speed. Bit Error Rate is thus a measure of transmission quality. It is generally shown as a negative exponent, (e.g., 10 to the minus 7 which means one out of 10,000,000 bits are in error.) See also Bit Error Rate. 2. A LAN term, which means Basic Encoding Rules. Rules for encoding data units described in ASN.1. (Abstract Syntax Notation One. LAN "grammar," with rules and symbols, that is used to describe and define protocols and programming languages. ASN.1 is the OSI standard language to describe data types.) BER is sometimes incorrectly lumped under the term ASN.1, which properly refers only to the abstract syntax description language, not the encoding technique.

BERT Bit Error Rate Test, or Tester. A known pattern of bits is transmitted, and errors received are counted to figure the BER. The idea is to measure the quality of data transmission. The bit error rate is the ratio of received bits that are in error, relative to the number of bits received. Usually expressed as a power of 10. Sometimes called Block Error Rate Tester.

Bernoulli Daniel Bernoulli was the 18th century Swiss mathematician who first expressed the principle of fluid dynamics — the basis of Bernoulli "boxes" also called disk

cartridges. A Bernoulli box, which is a mass storage device, uses both floppy and hard disk technologies. Bernoulli disks are removable. They physically look like large floppy disks.

Best Effort A term for a Quality of Service (QoS) class with no specified parameters and with no assurances that the traffic will be delivered across the network to the target device. ATM's ABR (Available Bit Rate) and UBR (Unspecified Bit Rate) are both best-effort service examples.

Best Path An internetworking term: The optimal route between source and destination end stations through a wide area network. Determined through routing protocols such as RIP and OSPF, best path can be based on lowest delay, cost or other criteria.

Beta 1. Refers to the final stages of development and testing before a product is released to market. "Alpha" is the term used when a product is in preliminary development. "Her baby is in beta," according to Peter Lewis of the New York Times, means she is expecting soon. In the software industry, beta has been known to last a year or more. Microsoft's Bill Gates has given the word "beta" a whole new meaning by having as many as 500,000 "beta testers" for his Windows 95 operating system. At this level, beta testing is no longer testing, it's marketing. And it's positively brilliant. See Beta Test. 2. Business Equipment Trade Association (UK). 3. Informal name for Betacam, a professional color difference videotape recording format that uses the Y, R-Y, and B-Y color difference components. Also the name of a consumer videotape recording format that is completely different from the professional Betacam format.

Beta Site A place a beta test is conducted. See Beta and Beta Test.

Beta Test Typically the last step in the testing of a product before it is officially released. A beta test is often conducted with customers in their offices. Some customers pay for the equipment or software they get under a beta test; some don't. Some beta tests stay in (if they work). Some don't. Most products don't work when they're first introduced. So beta tests are a good idea. Unfortunately, most manufacturers don't do sufficient beta testing. They want to get their product to market before the competition does. This often means we now have two or three new products on the market, none of which work reliably or do exactly what they're meant to do. Our rule: always wait several months after a product is introduced before buying it. By then the major bugs will have been fixed. The test before the beta test is called the Alpha. It isn't that common. See Beta.

Betacam Portable camera/recorder system using 1/2-inch tape originally developed by Sony. The name may also refer just to the recorder or the interconnect format; Betacam uses a version of the Y, R-Y, B-Y color difference signal set. Betacam is a registered trademark of the Sony Corporation.

Betacam SP A superior performance version of Betacam. SP uses metal particle tape and a wider bandwidth recording system.

Betamax 1. The noun. A format for video tape which Sony introduced too expensively. VHS (Video Home System), using half-inch tape introduced by Matsushita/JVC in 1975, effectively killed Sony's attempt to make Betamax the leading video tape standard. 2. The verb. When a technology is overtaken in the market by inferior but better marketed competition as in "Microsoft betamaxed Apple right out of the market." See VHS.

Betazed A planet in the second Star Trek TV series, inhabited by Betazoids, beings with great powers of empathy and telepathy.

BETRS Basic Exchange Telecommunications Radio Service. A service that can extend telephone service to rural areas by replacing the local loop with radio communications, sharing the UHF and VHF common carrier and private radio frequencies.

Bezel The metal or plastic part — in short, the frame — that surrounds a cathode ray tube — a "boob" tube.

Bezeq The name of the erstwhile-monopoly Israeli local and long distance phone company. Its full name is the Israel Telecommunications Corp. Ltd.

BEZS Bandwidth Efficient Zero Suppression. N.E.T.'s patented T-1 zero suppression technique; maintains Bell specifications for T-1 pulse density without creating errors in end-user data; uses a 32 Kbps overhead channel.

BFT Binary File Transfer. BFT is a method of routing digital files using facsimile protocols instead of traditional modem file transfer protocols. See BINARY FILE TRANSFER for a fuller explanation.

BFV Bipolar violations: The digital data format consists of pulses of opposite polarity. No two consecutive pulses should be the same polarity; if two are detected in a row, the term is a violation, which is also a warning flag.

BGE-I ISDN Business Group Elements.

BGID Business Group ID.

BGP Border Gateway Protocol is a Gateway Protocol which routers employ in order to exchange appropriate levels of information. In an intradomain routing environment between Autonomous Systems (ASs), IBGP (Internal BGP) is run, allowing the free exchange of information between trusted systems. IBGP is in a class of protocols known as IGPs, or Internal Gateway Protocols. In an interdomain environment, EBGP (External BGP) is run, allowing the routers to exchange only prespecified information with other prespecified routers in other domains in order to ensure that their integrity is maintained. EBGP is in a class known as EGPs, or External Gateway Protocols. When BGP peer routers first establish contact, they exchange full routing tables; subsequent contacts involve the transmission of changes, only.

BH Bandwidth Hog. A term defined by Philip Elmer-DeWitt, technology editor of Time Magazine in 1994, who spearheaded the launch of Time Online, the first fully electronic national magazine. He defined BH "as a person who uses the online medium like a bullhorn and attracts like-minded people who then rove in a pack, filling them with up with screeds." (Screed is a long discourse or essay.)

BHANG Broadband High Layer Information: This is a Q.2931 information element that identifies an application (or session layer protocol of an application).

BHC Backbone to Horizontal Cross-connect. Point of interconnection between backbone wiring and horizontal wiring.

BHCA Busy Hour Call Attempts. A traffic engineering term. The number of call attempts made during the busiest hour of the day.

Bi A Latin prefix meaning twice.

Bias 1. A systemic deviation of a value from a reference value.

2. The amount by which the average of a set of values departs from a reference value.

3. An electrical, mechanical, magnetic, or other force field applied to a device to establish a reference level to operate the device.

4. Effect on telegraph signals produced by the electrical characteristics of the terminal equipment.

Bias Distortion Distortion affecting a two-condition (binary) coding in which all the significant intervals corresponding to one of the two significant conditions have uniformly longer or shorter durations than the corresponding theoretical durations. The magnitude of the distortion is expressed in percent of a perfect unit pulse length.

Bias Generator A CBX printed circuit card that generates a signal that reduces idle channel noise for all coders installed in the CBX.

Bias Potential The potential impressed on the grid of a vacuum tube to cause it to operate at the desired part of its characteristic curve.

Bib Signaling ID assigned by Exchange B.

BICEP An ATM term. Bit Interleaved Parity: A method used at the PHY layer to monitor the error performance of the link. A check bit or word is sent in the link overhead covering the previous block or frame. Bit errors in the payload will be detected and may be reported as maintenance information.

BICI Broadband Inter-Carrier Interfaces. This is also the spanish colloquial word for bicycle. See also B-ICI.

Biconic Fiber Optic Connector developed by Lucent.

Biconical Antenna An antenna consisting of two conical conductors having a common axis and vertex. Excitation occurs at the common vertex. If one of the cones is flattened into a plane, the antenna is called a discone.

BICSI Building Industry Consulting Service International, a professional organization. For those who acquire certain requisite education and experience by BICSI, the association makes them a RCDD, Registered Communication Distribution Designer.

Bicycle Networking The practice among cable access TV shows of distributing programming from one local cable access station to another. They use local messengers on bicycles to transport tape.

Bidirectional Bus A bus that may carry information in either direction but not in both simultaneously.

Bidirectional Couplers Fiber optic couplers that operate in the same way regardless of the direction light passes through them.

Bidirectional Line Switched Ring Commonly referred to as BLSR, bi-directional line switched ring is a method of SONET transport in which half of the working network is sent counter-clockwise over one fiber and the other half is sent clockwise over another fiber. BLSR offers bandwidth use advantages for distributed traffic in single-ring architectures. See also Line Switched Ring, Path Switched ring and SONET.

Bidirectional Printing A typewriter always prints from left to right. So did the early computer printers. The newer computer printers will print from left to right, drop down a line, then print from right to left. This increases the printer's speed.

Bifurcated Routing Routing that may split one traffic flow among multiple routes.

Big-endian A networking term. A format for storage or transmission of binary data in which the most significant bit (or byte) comes first. The reverse convention is called little-endian.

Bigamy The only crime in which two rites make a wrong.

Bikini Transmitter The Bikini Transmitter is a body wire developed for a special surveillance project. Law Enforcement professionals needed to secretly record a conversation between a suspect and a female agent. The suspect insisted the meeting take place at a topless beach. An audio transmitter was sewn into a string bikini with the antenna threaded through the string. The largest component, the battery, was carried, uh...internally. It is not known whether the transmitter was waterproof.

Bilateral Synchronization A synchronization control system between exchanges A and B in which the clock at exchange A controls the received data at exchange B and the clock at exchange B controls the received data at exchange A. Normally implemented by deriving the receive timing from the incoming bit stream.

Bildschirmtext German word for interactive videotex. The German Bundespost likes this service. But the German version isn't as successful as the French because the French gave away the videotex terminals. And the Germans didn't. Also, the French really encourage videotex entrepreneurs by giving a real piece of the action — 60% of the collected revenues.

Bill An itemized list or statement of charges. If you can't bill for a product or service, you're engaged in a hobby rather than a business.

Bill and Keep Imagine a phone call from New York to Los Angeles. It may start with the customer of a new phone company, then proceed to a local phone company (let's say New York Telephone, now called Bell Atlantic). Then it may proceed to a long distance company before ending in Los Angeles and going through another one or two local phone companies before reaching the person dialed. Under the existing rules, all the companies carrying these phone calls have to be paid in some way for their transmission and switching services. There are programs in place such that the company doing the billing and collecting the money pays over some of those monies to the other phone companies in the chain. One such program is called "reciprocal compensation." The opposite of reciprocal compensation is called "Bill and Keep." Under this program, the company billing the call gets to keep all the money. The others in the chain (or most of the others in the chain) get nothing. The concept of "bill and keep" has its roots in the international postal service where "bill and keep" has been in place for many years.

Bill Of Materials A list of specific types and amounts of direct materials expected to be used to produce a given job or quantity of output.

Bill To Room A billing option associated with Operator Assisted calls that allows the calling party to bill a call to their hotel room. With this option, the carrier is required to notify the hotel, upon completion of the call, of the time and charges.

Billboard Electronic sales pitches that come up on your computer screen at any time.

Billboard Antenna A broadside antenna array with flat reflectors.

Billed Number Screening You (at home or your business) establish who can and cannot charge a call to your phone by making an agreement with your local telephone company to screen your calls. (e.g. Refusal of all collect call requests.)

Billed Telephone Number BTN. The primary telephone number used for billing regardless of the number of telephone lines associated with that number. Apparently, the term "billing telephone number" is more accurate than billed telephone number. For a longer explanation, see Billing Telephone Number.

Billibit Someone's absolutely awful term for one billion bits. Also (and better) called a gigabit.

Billing Account Number BAN. Used by telephone companies to designate a customer or customer location that will be billed. A single customer may have multiple billing accounts. See Billing Telephone Number.

Billing Increment The increments of time in which the phone company (long distance or local) bills. Some services are measured and billed in one minute increments. Others are measured and billed in six or ten second increments. The billing increment is a major competitive weapon between long distance companies. Short billing increments become important to you, as a user, when your average calls are very short — for example, if you're making a lot of very short data calls (say for credit card authorizations). Being billed for a lot of six second calls is a lot cheaper than being billed for a lot of one minute calls.

Billing Media Converter A Billing Media Converter, as made by the Cook division of Northern Telecom, provides a means of transporting Automatic Message Accounting (AMA) data from DMS-10 central offices to regional accounting offices with the physical transfer of magnetic tapes. The BMC is polled.

Billing Telephone Number BTN. The primary telephone number used for billing regardless of the number of telephone lines associated with that number. Multiple WTNs (Working Telephone Numbers), also known as ETNs (Earning Telephone Numbers) can be associated with a single BTN. A Billing Telephone Number is the number to which calls to given location are billed. It is the seven-digit number with the area code followed by an alphanumeric code assigned by the local telephone company (e.g. NPA-NXX-XXXX).

Billing Validation Service See BVA and BVS

Billion In North America, a billion is a thousand million. In many countries overseas, a billion is a million million.

Binaries Binary, machine readable forms of programs which have been compiled or assembled. As opposed to source language forms of programs.

Binary Where only two values or states are possible for a particular condition, such as "ON" or "OFF" or "One" or "Zero." Binary is the way digital computers function because they can only represent things as "ON" or "OFF." This binary system contrasts with the "normal" way we write numbers — i.e. decimal. In decimal, every time you push the number one position to the left, it means you increase it by ten. For example, 100 is ten times the number 10. Computers don't work this way. They work with binary notation. Every time you push the number one position to the right it means you double it. In binary, only two digits are used — the "0" (zero) and the "1" (the one). If you write the number 10101 in binary, and you want to figure it in decimal as we know it, here's how you do it. 1 is one thing; Zero x 2 = zero; 1 times 2 x 2 = 4; 0 x 2 x 2 x 2 = 0; 1 x 2 x 2 x 2 x 2 = 16. Therefore the total 10101 in binary = 1 + 0 + 4 + 0 + 16 = 21 in decimal.

Binary notation differs slightly from notation used in ASCII or EBCDIC. In ASCII and EBCDIC, the binary values are used for coding of individual characters or keys or symbols on keyboards or in computers. So each string of seven (as in ASCII) or eight (as in EBCDIC) ones and zeros is a unique value — but not a mathematical one.

ASCII uses a seven bit coding scheme. Thus, the maximum number of different things you can code using seven bits is 128, i.e. 2 x 2 x 2 x 2 x 2 x 2 x 2 = 128. The maximum number represented by a byte (8 bits) or the IBM EBCDIC coding system is 256. i.e. 2 x 2 x 2 x 2 x 2 x 2 x 2 x 2 = 256. See Binary Coded Decimal, Binary File and Binary Transfer.

Binary Code A code in which every element has only one of two possible values, which may be the presence or absence of a pulse, a pulse, a one or a zero, or high or a low condition for a voltage or current.

Binary Coded Decimal BCD. A system of binary numbering that uses a 4-bit code to represent each decimal digit from 0 to 9 and multiple 4-bit patterns for higher numbers.

The decimal numbers 0 to 9 are represented by the four-bit binary numbers from 0000 to 1010.

Binary Digit A number in the binary system of notation.

Binary File A file containing information that is in machine-readable form. It is an application. Or it can only be read by an application. See Binary Transfer.

Binary File Transfer BFT. The transmission of binary files, software, documents, images, video and electronic data exchange information between communicating devices, including PCs, fax devices, etc. When binary files are transferred via telecommunications, they must not be changed in any way during the transfer otherwise they will be destroyed. Many electronic mail services, such as CompuServe or MCI Mail, have ways of "attaching" binary files to electronic mail, such that the binary file will not be affected or changed in any way during transmission. But that only works between MCI users or between CompuServe users. The Internet uses a different system. It's called MIME, which stands for Multipurpose Internet Mail Extension. What it does is it uses software to encode the binary file into an ASCII file. It transmits the ASCII file. At the other end, that file is decoded into its original binary format. MIME uses many software techniques for doing this. But all have the same effect of not changing the original binary file. See MIME.

Binary Logarithmic Access Method BLAM. A proposed alternative to the IEEE 802.3 backoff algorithm.

Binary Notation Any notation that uses two different characters, usually the binary digits 0 and 1.

Binary Number System A number system that uses two characters (0 and 1) with two as its base, just as the decimal number system uses ten characters (0 through 9) with ten as its base.

Binary Symmetric Channel A channel designed so that the probability of changing binary bits in one direction is the same as the probability of changing them back to the correct state.

Binary Synchronous Communications BISYNC or BSC. 1. In data transmission the synchronization of the transmitted characters by timing signals. The timing elements at the sending and receiving terminal define where one character ends and another begins. There are no start or stop elements in this form of transmission. For a more detailed explanation, see BSC.
2. Also a uniform discipline or protocol for synchronized transmission of binary coded data using a set of control characters and control character sequences.

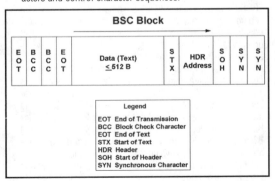

BSC Block

E O T	B C C	B C C	E O T	Data (Text) ≤ 512 B	S T X	HDR Address	S O H	S Y N	S Y N

Legend

EOT End of Transmission
BCC Block Check Character
EOT End of Text
STX Start of Text
HDR Header
SOH Start of Header
SYN Synchronous Character

Binary To Decimal Conversion Conversion from base 2 to base 10. See BINARY.

Binary Transfer See BINARY FILE TRANSFER and MIME.

BIND Berkeley Internet Name Daemon. BIND is the software that allows us to type site names like www.yahoo.com instead of a string of numbers. BIND is the implementation of a DNS server originally developed and distributed by the University of California at Berkeley. Many Internet Hosts run BIND, and it is the ancestor of many commercial BIND implementations.

Bind A request to activate a session between two logical units (LUs). See BIND.

Binder A way to separate groups of 25 pairs in a twisted-pair cable with more than 25. Color plastic ribbon binds separate each group of 25. The first group (1-25) is white/blue, the second (26-50) is white/orange, the third (51-75) is white/green, the fourth (76-100) is white/brown, etc. Binders are helically applied colored thread, yarn, or plastic ribbon. They're used to confine, separate and identify groups of fibers or wires in a cable. Binders are usually used for holding assembled cable components in place.

Binder Group A group of wires within a large cable that can be distinguished from other groups because it is wrapped with colored threads. Normal telephone color-coding provides for only 25 pairs of wire, so binder groups allow multiple pairs of the same color wire to be in one cable. A 50-pair cable has blue and orange binder groups; a 75-pair cable has blue, orange, and green groups. Since several wire pairs have the same color markings, installers must be careful when stripping cable insulation so they do not destroy the binder threads.

Bindery A Novell NetWare database containing definitions for entities such as users, groups, and workgroups. The bindery contains three components: objects, properties, and property data sets. Objects represent any physical or logical entity, including users, user groups, file servers, print servers, or any other entity given a name. Properties are the characteristics of each bindery object, including passwords, account restrictions, account balances, internetwork addresses, list of authorized clients, workgroups, and group members. Property data sets are values assigned to entities bindery properties.

Binding A Windows 95 definition. Binding is a process that establishes the communication channel between a protocol driver and a network adapter driver.

Binding Post 1. A screw with a small nut. You take your wires and join them together on a binding post by wrapping them together around the screw and then tightening the nut on them. You'll find binding posts on huge wall-mounted things called terminal boxes. We have several in the basement of the building in Manhattan, New York, in which I live. Binding posts are numbered on the terminal box and those numbers are entered into the phone company's cable management system. That way, when the technician comes to check out a trouble report, he can quickly find the offending pairs. These days, other more modern termination devices (such as Krone Boxes, 110-connecting blocks, and even 66-blocks) are replacing binding posts. Our building's binding posts are probably 60 years old. See also House Box.
2. A post witches were attached to just before they were burned at the stake. Rude switchboard operators and ACDs with long queues should also suffer this fate.

BinHex When sending files which aren't plain ASCII across a network — dial-up, leased line or the Internet — you basically have two options. First, you can attach them as a binary file (i.e. non-ASCII file). Or second, you can encode them into ASCII characters and send the file as part of your message. The first method is preferable. But you can typically only send binary files from one account to another on the same network, or between two networks that have agreed between themselves to a method of transferring files. But that's a rarity. And it certainly doesn't work in and around the Internet. Thus something called, MIME was created. MIME stands Multipurpose Internet Mail extensions. It is the name for encoding binary files into ASCII characters for transfer across the Internet and on-line services. Under MIME, there are a number of methods of encoding binary files, one of which is called BinHex, which is a popular encoding algorithm that uses Run-Length Encoding (RLE). To send a binary file, you encode it using BinHex encoding software. You then include the ASCII encoded file in your message. The recipient of your message then decodes the ASCII using BinHex decoding software, which may or may not be built into his electronic mail or browser package.

Binomial Distribution The binomial is a two-parameter distribution. The parameters being n, the number of trials, and p, the probability of a particular outcome of a single trial.

Biometric Access Control Any means of controlling access through human measurements, such as fingerprinting and voiceprinting.

Biometric Device A device used in authenticating access to a system. A biometric device authenticates a user by measuring some hard-to-forge physical characteristic, such as a fingerprint or retinal scan. You see a lot of biometric devices in Hollywood thrillers, including Mission Impossible, a movie which was released in the summer of 1996.

Bionic code A problem-solving routine for human behavior as it is exercised in the realm of networks and cyberspace. The first bionic codes were developed by Ebon Fisher based on a series of his theatrical experiments involving communication systems amongst audience members. Fisher's bionic codes have been formalized as a series of diagrams and statements which "float" in the infosphere in a variety of media.

BIOS Basic Input/Output System of desktop computers. The BIOS contains the buffers for sending information from a program to the actual hardware device the information should go to.

BIOS Enumerator A Windows 95 term. Bios enumerator is responsible, in a Plug and Play system, for identifying all hardware devices on the motherboard of the computer. The BIOS supports an API that allows all Plug and Play computers to be queried in a common manner.

Biosensors Devices such as fingerprint readers and signature recognition systems.

BIP-8 Bit Interleaved Parity 8. A method of error monitoring used in SONET optical fiber transmission systems. A SONET frame of data comprises a large number of bits organized into bytes, or 8-bit values. The bytes, of course, are interleaved into a byte stream. The BIP-8 method looks across all matching bit positions for those distinct bytes in a frame, and calculates parity, which should be even, rather than odd. If an odd parity is calculated at the receiving end, a bit error is indicated. The number of such errors in frames over a period of time constitutes the Bit Error Rate (BER). The specific period of time involved is known as a "sliding time window," as it varies according to the maximum detection time, which is sensitive to the transmission rate, or level of the Optical Carrier (e.g., OC-1, OC-3, OC-24, OC-48, OC-192). The higher the OC-N level, the higher the rate of transmission, and the shorter the sliding time window. In other words, the faster the rate of transmission, the more quickly the error detection process must take place in order to establish the BER and to correct the problem.

BIP-N Bit Interleaved Parity N. A method of error monitoring. With even parity, an N bit code is generated by the transmitting equipment over a specified portion of the signal in such a manner that the first bit of the code provides even parity over the first bit of all N-bit sequences within the specified portion, etc. Even parity is generated by setting the BIP-N bits so that there are an even number of 1s in each of all N-bit sequences including the BIP-N. See BIP-8 for a concrete example.

Bipolar The predominant signaling method used for digital transmission services, such as DDS and T-1. The signal carrying the binary value alternates between positive and negative. Zero and one values are represented by the signal amplitude at either polarity, while no-value "spaces" are at zero amplitude.

Bipolar Coding The T carrier line coding system that inverts the polarity of alternate "one" bits.

Bipolar Signal A signal having two polarities, both of which are not zero. It must have two-state or three-state binary coding scheme. It is usually symmetrical with respect to zero amplitude.

Bipolar Violation The presence of two consecutive "one" bits of the same polarity on the T carrier line. See also BIPOLAR CODING.

Bird A satellite.

Birdie A birdie is a lightweight device that you blow through underground cement pipes through which you want to pull cable. Here's how it typically works: First, you use a bore to make an underground hole. Then you fill that hole with hollow concrete cement pipes joined together to form one long underground conduit (i.e. tunnel). Then you go to one end of the tunnel and use a air compressed device to blow a very lightweight "birdie" attached to a lightweight string through the tunnel. Someone at the other end catches the birdie and pulls gently on the string. Attached to the end of the string is strong mule tape. He keeps pulling on it. Attached to the end of the mule tape is the telecommunications cable — fiber or wire — that you really want to install in the underground conduit. The whole point of this elaborate procedure is that it's far better for the cable to lay it after the pipes are laid than it is during the installation process when the cable could be damaged.

Birthdays January 4, 1847 Thomas Edison born
February 16, 1982 Michael Allen Newton born
February 20, 1980 Claire Elizabeth Newton born
February 23, 1965 Michael Dell (Dell Computer) born
March 3, 1847 Alexander Graham Bell born
March 11, 1933 Ben Rosen (Compaq, SRX, etc.) born
March 15, 1949 Gerry Friesen born
April 6, 1939 John Sculley born
April 27, 1791 Samuel Morse born
May 11, 1948 International Communications Association born
June 10, 1942 Harry Newton born in Sydney, Australia
June 13, 1961 TeleCommunications Association born
June 19, 1924 Ray Noorda (Novell) born
June 27, 1968 Carterfone decision handed down by FCC
July 4, 1943 Susan Newton born in Perth, Australia
August 31, 1962 With President Kennedy's signature the Communications Satellite Act is born
September 2, 1936 Andy Grove (Intel)
September 12, 1948 Communications Managers Association
October 28, 1955 Bill Gates (Microsoft)
November 13, 1954 Scott McNealy (Sun Microsystems)
December 10, 1928 William G. McGowan, founder of MCI

Bis 1. The French term for "second" or "encore." It is used by the ITU/ITU to designate the second in a family of related standards. "Ter" designates the third in a family. See V SERIES.
2. Border Intermediate System

BISDN Also spelled B-ISDN. Broadband ISDN. This is a vaguely defined term. It basically means any circuit capable of transmitting more than one Basic rate ISDN, i.e. 144 Kbps. One definition I read recently suggested that BISDN is "a set of public network services that are delivered over ATM, including data, voice, and video. BISDN will provide services such as high-definition television (HDTV), multi-lingual TV, voice and image storage and retrieval, video conferencing. high-speed LANs, and multimedia." See B-ISDN for a longer explanation. See also ISDN.

Bistable Trigger Circuit A trigger circuit that has two stable states.

BISUP Broadband ISDN User's Part: A SS7 protocol which defines the signaling messages to control connections and services. See Signaling System 7.

BISYNC (pronounced bye-sink). BISYNChronous Transmission. A half-duplex, character-oriented, synchronous data communications transmission method originated by IBM in 1964. See Synchronous.

Bisynchronous Transmission Also called BISYNC. A data character-oriented communications protocol developed by IBM for synchronous transmission of binary-coded data between two devices. BISYNC uses a specific set of control characters to synchronize the transmission of that binary coded data. See also Binary Synchronous Communication.

Bit Bit is a contraction of the term BInary digiT. It is the smallest unit of information (data) a computer can process, representing either high or low, yes or no, or 1 or 0. It is the basic unit in data communications. A bit can have a value of zero (a space) or one (a mark).

Bit Bucket Slang for throwing out bits — into a wastepaper bucket.

Bit Buffer A section of memory capable of temporarily storing a single BInary uniT (bit) of information. Used to make data transmission accurate or consistent.

Bit Check A bit added to a digital signal and used for error checking, i.e. a parity bit. See also Parity.

Bit Depth The number of bits used to represent the color of each pixel in a graphic file or the sound in an audio file. The higher the number, the more information included in the file and the higher the quality of the data. Common graphic bit depths are 4-bit, 8-bit, 16-bit, 24-bit and 32-bit. Common sound bit depths are 8-bit and 16-bit.

Bit Duration The time it takes one bit to pass a point on the transmission medium. Used to measure delay times, especially in high speed communications.

Bit Error The value of an encoded bit can be changed due to a transmission problem (e.g. noise on the line) and then interpreted incorrectly by the receiver.

Bit Error Rate BER. The percentage of received bits in error compared to the total number of bits received. Usually expressed as a number to the power of 10. For example, 10 to the fifth power means that one in every 100,000 bits transmitted will be wrong. In transmitting data a high error rate on the transmission medium (i.e. some noise), doesn't mean there'll be lots of problems with the final transmission. It just means there'll have to be lots of re-transmissions — "until one gets it right." These re-transmissions reduce the amount of data transmitted in a unit of time and therefore, increase the time needed to send that information. If the BER gets too high, it might be worth while to go to a slower transmission rate. Otherwise, you would spend more time retransmitting bad packets than getting good ones through. The theory is that the faster the speed of data transmission the more likelihood of

error. This is not always so. But if you are getting lots of errors, the first — and easiest — step is to drop the transmission speed.

Bit Flipper A person who flips bits for a living. In other words, an industrial strength member of the digiterati, which I guess would be a digiteratus. These people are way kewl (that's a NetHead term for cool, or k00l) when it comes to really technical data protocol stuff. SEE ALSO DIGITERATI, NETHEAD and GEARHEAD.

Bit Interleaving A form of TDM for synchronous protocols, including HDLC, SDLC, BiSync and X.25. Bit interleaving retains the sequence and number of bits, so that correct synchronization is achieved between both ends. See Bit Interleaving/Multiplexing.

Bit Interleaving/Multiplexing In multiplexing, individual bits from different lower speed channel sources are combined one bit at a time/one channel at a time into one continuous higher speed bit stream. Compare with byte interleaving/ multiplexing.

Bit Oriented Used to describe communications protocols in which control information may be coded in fields as small as a single bit.

Bit Oriented Protocol BOP. A data link control protocol that uses specific bit patterns to transfer controlling information. Examples are IBM's Synchronous Data Link Control (SDLC) and the ITU-T High-Level Data Link Control (HDLC). Bit-oriented protocols are normally used for synchronous transmission only. Bit-oriented protocols are code transparent (meaning they work regardless of the character encoding method used), since no encoded characters are used in the control sequence.

Bit Oriented Transmission An efficient transmission protocol that encodes communications control information in fields of bits rather than characters or bytes.

Bit Parity A binary bit appended to an array of bits to make the sum of all the bits always odd or always even. See PARITY.

Bit Pattern A group of bits arranged in specified ways to represent numbers, letters or symbols, forming a unique binary number for each character. For example, the 7-bit ASCII code produces 128 different characters, i.e. $2 \times 2 \times 2 \times 2 \times 2 \times 2 \times 2 = 128$.

Bit Pump A device which pumps out bits at a high rate of speed. Slang for high-speed carrier electronics such as ADSL terminating units, which can achieve speeds of multiple Mbps over standard twisted-pair local loops. "Bit pumps" possess no particular intelligence (e.g., protocol conversion, or error detection and correction), they just pump bits.

Bit Rate The number of bits of data transmitted over a phone line per second. You can usually figure how many characters per second you will be transmitting — in asynchronous communications — if you divide the bit rate by ten. For example, if you are transmitting at 1200 bits per second, you will be transmitting 120 characters per second. In real life, it's never this simple, however. The total bits transmitted will depend on re-transmissions, which depends on the noise of the line, etc. See BAUD RATE.

Bit Robbing A technique to signal in-band in digital facilities, which typically use out of band signaling, e.g. Signaling System 7. In bit robbing, we steal bits from the speech path a few line-signal bits. The remaining bits are adequate to recreate the original electrical analog signal (and ultimately, the original sound). Bit robbing typically uses the least significant bit per channel in every sixth frame for signaling. See BIT STUFFING and ROBBED BIT SIGNALING.

Bit Specifications Number of colors or levels of gray that can be displayed at one time. Controlled by the amount of memory in the computer's graphics controller card. An 8-bit controller can display 256 colors or levels of gray; a 16-bit controller, 64,000 colors: a 24-bit controller, 16.8 million colors.

Bit Stream A continuous stream of data bits transmitted over a communications line with no break or separators between the characters.

Bit Stuffing Process in some data communications protocols where a string of "one" bits is broken by an inserted "zero." This inserted zero is added by the sender and stripped by the receiver. The idea of inserting the zero is to avoid confusing the receiver into thinking the series of one bits mean something else, like a flag control character. See ZERO STUFFING.

Bit Synchronous A SONET term describing the manner in which information streams are mapped into the SONET frame format for unchannelized VT (Virtual Tributary) transport. For instance, multiple VT1.5s can be mapped into a SONET frame, with each VT1.5 carrying a single T-1 signal within the STS-1 SPE (Synchronous Payload Envelope). In the LOH (Line OverHead) of the SONET are included VT pointers which identify the established portions of the SPE in which the beginning byte of each VT1.5 is located. As each VT1.5 works its way through the SONET network, its actual position may change within the various SPEs, thereby requiring that the pointer be reset; this process is very complex and expensive, and results in a small level of delay. Regardless of issues of complexity, cost and (slight) processing delay, it is essential that the timing of the various T-carrier and SONET data be maintained in order that the data arrive at an identifiable and consistent point in time, otherwise it would be unrecognizable. As is true with many things in life, timing can be everything. Compare and contrast with Byte Synchronous. See also SONET and VT.

Bit Transfer Rate The number of bits transferred per unit time. Usually expressed in Bits Per Second (BPS).

Bit Twiddler A technical person. Twiddle, according to Random House Dictionary, means "to play or trifle idly with something; fiddle." The expression bit twiddler is used thus, "I'm not a bit twiddler. You'll have to ask Joe in Engineering if you want the answer to that."

BITBLT BIT BLock Transfer. Microsoft Windows relies intensively on a type of operation called bit block transfer (BitBLT) to redraw rectangular areas of the image on the computer's screen. Generally, BitBLT operations are accomplished by software routines in the video driver, a cheap, but slow method that uses many of your CPU's clock cycles. If you add a separate video controller with a special processor to handle BitBLT, you will be able to offload video tasks from your main CPU and make your computer run faster.

Bitmap Representation of characters or graphics by individual pixels arranged in row (horizontal) and column (vertical) order. Each pixel can be represented by one bit (simple black and white) or up to 32 bits (high-definition color). Bitmapped images can be displayed on screens or printed. See Bit Specifications.

Bitmapped Graphics Images which are created with matrices of pixels, or dots. Also called raster graphics. See BITMAP.

BITNET Because It's Time NETwork. An academic computer network based originally on IBM mainframes connected with leased 9600 bps lines. BITNET has recently merged with CSNET, The Computer+Science Network (another academic computer network) to form CREN: The Corporation for Research and Educational Networking. The network connects more than 200 institutions and has more than 900 computational nodes.

Bitnik A person who uses a coin-operated computer terminal installed in a coffee house to log into cyberspace.

Bitronix Hewlett Packard's term for its bidirectional parallel port communications "standard." It introduced this standard with its 600 dps LaserJet 4 plain paper printer in the fall of 1992. It is hoping other manufacturers will adopt the standard. The big plus of the standard is that it allows a printer to tell a connected computer that it (the printer) has run out of paper, or the paper has jammed, etc. Having that communications back and forth will allow the user to clear the problem and get the printer and up and running faster. It will also stop the computer locking up.

BITS Building Integrated Timing Supply. A single building master timing supply. BITS generally supplies DS1 and DS0 level timing throughout an office. The BITS concept minimizes the number of synchronization links entering an office, since only the BITS will receive timing from outside the office. In North America, BITS are thus the clocks that provide and distribute timing to a wireline network's lower levels. Known in the rest of the world as a SSU (Synchronization Supply Unit).

Bits Clock The bits clock provides a pulse that synchronizes the entire network. The pulse is a 1-0-1-0-1-0-1-0 stream. Used extensively in SONET network

Bits Per Second The number of bits passing a specific point per second. See Bps.

A KILObit per second is one thousand bits per second.

A MEGAbit per second is one million bits per second (thousands of kilos).

A GIGAbit per second is one billion bits per second (thousands of millions).

A TERAbit per second is one trillion bits per second (thousands of billions).

A PETAbit per second is equal to 10 to the 15th or 1,000 terabits per second.

Bitslag All the useless rubble on the Net one has to plow through to get to the rich information core.

Bitspit To transmit. "Did you bitspit the file to Harry?"

Bix A Northern Telecom trade name for an in-building termination and cross-connect system for unshielded twisted pair cables.

BL Business Line

Black Body A totally absorbing body that does not reflect radiation (i.e. light). In thermal equilibrium, a black body absorbs and radiates at the same rate; the radiation will just equal absorption when thermal equilibrium is maintained.

Black Box An electronic device that you don't want to take the time to understand. As in, "We'll put the data through a black box that will put it into X.25 format." The term has recently come also to mean PBX switches. While "Black Box" is a generic term, The Black Box Corporation of Pittsburgh, PA, has had the audacity (and brilliance) to register the term as a trademark. The Phone Phreak community has used the term black box to describe a device that's put on phone lines in electromechanical central office areas (they don't work under ESS offices). To the phone phreak community, a black box made up of a resistor, a capacitor and a toggle switch would "fool" the central office into thinking the phone had not been picked up when receiving a long distance call. Since the call was not "answered," the call could not be billed. Clever, eh? (Illegal, too.) See Blue Box, Red Box and White Box.

Black Box Corporation A leading direct marketer of connectivity solutions — everything from cables to routers. It publishes and distributes the award-wining Black Box Catalog. Black Box is based in Pittsburgh. They kindly provided the pinout diagrams for the back of this dictionary. www.blackbox.com

Black Facsimile Transmission 1. In facsimile systems using amplitude modulation, that form of transmission in which the maximum transmitted power corresponds to the maximum density of the subject copy.

2. In facsimile systems using frequency modulation, that form of transmission in which the lowest transmitted frequency corresponds to the maximum density of the subject copy.

Black Level The lowest luminance level that can occur in video or television transmission and which, when viewed on a monitor, appears as the color 'black.'

Black Matrix Picture tube in which the color phosphors are surrounded by black for increased contrast.

Black Recording 1. In facsimile systems using amplitude modulation, that form of recording in which the maximum received power corresponds to the maximum received power corresponds to the maximum density of the record medium.

2. In a facsimile system using frequency modulation, that form of recording in which the lowest received frequency corresponds to the maximum density of the record medium.

Black Signal 1. In facsimile, the signal resulting from the scanning of a maximum-density area of the subject copy.

2. In cryptographic systems, a signal containing only unclassified or encrypted information.

Black Thursday The day that began the Great Depression. It was October 24, 1929.

Blackbird Blackbird is a multifunction system, providing home users with a single box that can work as a game system, network computer, home broadband router, and set-top box. Primary backer behind Blackbird is Motorola. Much of Blackbird's "middleware" software was developed with partners. According to David Lammers, of EE Times, Spyglass built a browser and other networking software that will let companies customize their own offerings and get to the retail channel by next year. Another key partner is VM Labs, a Silicon Valley-based game company headed up by former Atari president Richard Miller. VM Labs worked with Motorola to create a media processor, Nuon, which, at different times, has gone under the code names of Merlin and Project X, Burgess said. Nuon is a 128-bit VLIW engine that will work in tandem with the CPU in Blackbird, a PowerPC 860 core.

Blackout A total loss of commercial electric power. A blackout has a decidedly negative affect on your ability to compute and communicate, assuming that your systems are wired, rather than wireless. UPS (Uninterruptible Power Supply) systems provide battery backup protection from a blackout. Carriers use diesel power generators to keep their central offices operating during a blackout-that way, the phones still work when the lights go out.

Blacksburg Electronic Village Blacksburg is a small town in the mountains of Western Virginia. It has a branch of Virginia Tech. University. They have wired Virginia Tech students and thousands their town's citizens with email, internet access and bulletin boards showing happening in town. The official explanation is: The Blacksburg Electronic Village (BEV) is a cooperative project of Virginia Tech, Bell Atlantic of Virginia and the Town of Blacksburg. It links the town's citizens, via computers lines to each other and to the Internet. Citizens gain access to the Internet from their home or office through a high-speed modem pool or by using Ethernet LANS which are available in some offices and hundreds of apart-

ment units in town. Blacksburg residents may take advantage of a full spectrum of services including the World Wide Web, Gopher, electronic mail, electronic mailing list and thousands of Usenet newsgroups. In addition to full Internet access, citizens enjoy the benefits of extensive online local resources. The Blacksburg Electronic Village is famous because it really is the first town to aggressively ensure that the bulk of citizenry could and would have access to electronic mail and to the various resources on the Internet. It has apparently made a major difference to how people live and communicate in Blacksburg. www.bev.net

BLAM Binary Logarithmic Access Method. A proposed alternative to the IEEE 802.3 backoff algorithm.

Blamestorming Blamestorming occurs when people sit around in a group and discuss why a deadline was missed or a project failed, and most importantly, who was responsible.

Blank A character on teletype terminals that does not punch holes in paper tape (except for feed holes to push the paper through). Also the character between words, usually called a "Space" is referred to in IBM jargon as a Blank.

Blank and Burst On the AMPS cellular telephone network, certain administrative messages are sent on the voice channel by blocking the voice signal (blanking) and sending a short high speed data message (burst). The blank and burst technique is one that causes a momentary dropout of the audio connection (and sometimes disconnection of cellular modem connections) when a power level message is transmitted to the cellular phone.

Blank Cell The hollow space of a cellular metal or cellular concrete floor unit without factory installed fittings.

Blanking The suppression of the display of one or more display elements or display segments.

Blanking Interval Period during the television picture formation when the electron gun returns from right to left after each line (horizontal blanking) or from top to bottom after each field (vertical blanking) during which the picture is suppressed.

Blanking Pulses The process of transmitting pulses that extinguish or blank the reproducing spot during the horizontal and vertical retrace intervals.

Blast BLocked ASynchronous Transmission. A proprietary technology.

Blatherer A Internet user who takes four screens to say something where four words would work a lot better.

Blend To have outbound and inbound phone calls answered by the same agents. See the next two definitions.

Blended Agent A call center person who answers both incoming and makes outgoing calls. This idea of a blended agent is a new concept in a Call Center. In the past call centers have typically kept their inbound and outbound agents separate. The reason? Management felt that the necessary skills were very different and no one could master both.

Blended Call Center A telephone call center whose agents both receive and make calls. In other words, a call center whose phone system acts both as an automatic call distributor and a predictive dialer.

Blended Floor System A combination of cellular floor units with raceway capability and other floor units with raceway capability systematically arranged in a modular pattern.

BLER Block Error Ratio. The ratio of the blocks in error received in a specified time period to the total number of blocks received in the same period.

BLERT BLock Error Rate Test.

BLF The Busy Lamp Field is a visual display of the status of all or some of your phones. Your BLF tells you if a phone is busy or on hold. Your Busy Lamp Field is typically attached to or part of your operator's phone. See BUSY LAMP FIELD.

Blind Bore Imagine you want to lay fiber cable along the side of a highway. You know from blue staking and from the city maps that there are other utility cables along the highway you want. The first thing we know is that we can't trust the maps or the blue staking. The second thing we know is that we don't want our activities to cut someone else's cable. There are expensive implications to doing this. So what we do is we hand dig pot holes every so often along what's known as the running line — where the utilities are meant to be buried. The idea is that our pot holes will locate the existing underground cables and thus make it safer to bore our own cable. Blind boring occurs when we bore underground without digging pot holes. The reason we might do this? Some states and some cities simply don't allow pot holing. They trade the risk of hitting a utility line against creating a hole in the middle of street. They don't want pot holes in their street, since an asphalt patch has never the same integrity as a total overlay and they don't want their streets messed up. See BLUE STAKE.

Blind Dialing All modems come from the factory programmed to "listen" for a dial tone before dialing their connection. However, there are some phone lines which don't have dial tones or, more often, strange dial tones, which your modem doesn't recognize. A "strange" dial tone might be one you find in a strange place, usually not in North America. In this case, you have to tell your modem to start dialing when you want it to. This is called "blind dialing." The old way you did this was to insert an X1 in the dialing stream. The new way, in Windows 95/98, is to go into Control Panel / Modems / Properties / Connection and remove the check mark from "Wait for dial tone before dialing." Actually, you can leave this unchecked. Your modem will work just fine.

Blind Transfer Someone transfers a call to someone else without telling the person who's calling. Also called Unsupervised or Cold Transfer. Contrast with Screened Transfer.

Blinking An intentional periodic change in the intensity of one or more display elements or display segments.

Blister Pack A pocketed polyvinyl chloride shipping container with a snap-on cover.

Blitz A call center/marketing term. Used to describe telephone sales or prospecting activity of intense, high volume accomplished in a short period of time.

Bloatware Ever-fatter packages of "upgraded" software that, with each upgrade, come with dozens and dozens of new features. With each upgrade, the customer has less need to look elsewhere. At least that's the theory. See also Hyperware and Vaporware.

BLOB Binary Large OBjects. When a database includes not only the traditional character, numeric, and memo fields but also pictures or other stuff consuming of large space, a database is said to include BLOBs — binary large objects.

Block In data communications, a group of bits transmitted as a unit and treated as a unit of information. Usually consists of its own starting and ending control deliminators, a header, the text to be transmitted and check characters at the end used for error correction. Sometimes called a Packet.

Block Character Check BCC. The result of transmission verification algorithm accumulated over a transmission block, and normally appended at the end, e.g. CRC, LRC.

Block Cipher A digital encryption method which ciphers long messages by segmenting them into blocks of fixed length, prior to encryption. Each block, which typically is 64 bits in length, is encrypted individually. The blocks may be sent as individual units, or they may be linked in a method knows as Cipher-Block-Chaining. See also Encryption.

Block Diagram A graphic way to show different elements of a program or process by the use of squares, rectangles, diamonds and various shapes connected by lines to show what must be done, when it must be done and what happens if it's done this way or that. In short, it shows how all the small decision points add up to the whole process.

Block Error Ratio BLER. The ratio of the blocks in error received in a specified time period to the total number of blocks received in the same period.

Block Misdelivery Probability The ratio of the number of misdelivered blocks to the total number of block transfer attempts during a specified period.

Block Mode Terminal Interface BMTI. A device used to create (and break down) packets to be transmitted through a ITU-T X.25 network. This device is needed if block-mode terminals (such as IBM bisync devices) are to be connected to the network without an intermediate computer.

Block Multiplexer Channel An IBM mainframe input/output channel that allows interleaving of data blocks.

Block Pair BP. The telephone wires that run from the terminal box to the customer's premises.

Block Parity The designation of one or more bits in a block as parity bits whose purpose is to ensure a designated parity, either odd or even. Used to assist in error detection or correction, or both.

Block the Blocker Call Block. A feature that lets you automatically reject calls from parties that have blocked the transmission of their calling telephone number in order that you are unable to determine who is calling you. These features have meaning only if you subscribe to CLID (Calling Line ID) from your LEC, and you have call display equipment. See also Call Block and CLID.

Block Transfer The process of sending and receiving one or more blocks of data.

Block Transfer Attempt A coordinated sequence of user and telecommunication system activities undertaken to effect transfer of an individual block from a source user to a destination user. A block transfer attempt begins when the first bit of the block crosses the functional interface between the source user and the telecommunication system. A block transfer attempt ends either in successful block transfer or in block transfer failure.

Block Transfer Efficiency The average ratio of user information bits to total bits in successfully transferred blocks.

Block Transfer Failure Failure to deliver a block successfully. Normally the principal block transfer failure outcomes are: lost block, misdelivered block, and added block.

Block Transfer Rate The number of successful block transfers made during a period of time.

Block Transfer Time The average value of the duration of a successful block transfer attempt. A block transfer attempt is successful if 1. The transmitted block is delivered to the intended destination user within the maximum allowable performance period and 2. The contents of the delivered block are correct.

Blocked Calls The fraction of calls failing to be served immediately are called "blocked calls." Blocking can occur in two ways: All facilities are occupied when a demand is originated, and/or a matching of idle facilities cannot be made even though certain facilities are idle in each group.

Blocked Calls Delayed A variable in queuing theory to describe what happens when the user is held in queue because his call is blocked and he can't complete it instantly.

Blocked Calls Held A variable in queuing theory to describe what happens when the user redials the moment he encounters blockage.

Blocked Calls Released A variable in queuing theory to describe what happens when the user, after being blocked, waits a little while before redialing.

Blocking When a telephone call cannot be completed it is said that the call is "blocked." Blocking is a fancy way to say that the caller is "receiving a busy." There are many places a call can be blocked: at the user's own telephone switch — PBX or key system, at the user's local central office or in the long distance network. Blocking happens because switching or transmission capacity is not available at that precise time. The number of calls you try compared to the number of times you get blocked measures "the grade of service" on that network. Blocked calls are different from calls that are not completed because the called number is busy. This is because numbers that are busy are not the fault of the telephone switching and transmission network. One might think the fewer blocked calls, the better. From the user's point of view, the answer is obviously YES, it is better. Less blockage, fewer busies and less frustration. But as one designs a switching and transmission network for less and less blocking, the network becomes more and more expensive. Logarithmically so. We keep adding extra circuits and extra equipment. Thus, in any telecommunications network design there is always a trade-off: What are you prepared to pay, compared to what can you tolerate?

Everyone designs their network with different trade-offs depending on what they and their users or customers, can tolerate and/or are willing to pay. Most companies are willing to pay more for better service if someone explains the logic of telephone design to them. Many network salesmen, however, don't believe this. They practice the sales "theory" of selling better service for less money. This doesn't work in business, and especially not in telephony.

The "Grade of Service" is a measurement of blocking. It varies from almost zero (best, but most expensive case, no calls blocked) to one (worst case, all calls blocked). Grade of Service is written as P.05 (five percent blocking). "Blocking" used to be a technical term but has now become a sales tool especially among PBX manufacturers, who increasingly claim their switch to be "non-blocking." This means it will not, they claim, block a call in the switch.

There are several flaws in this logic: First, it's not logical or useful to buy a non-blocking PBX if the chances of being blocked elsewhere — the local lines, the local exchange or the long distance network — are very high. Second, a true non-blocking PBX can be very expensive, perhaps too much power and too much money for most peoples' needs. Third, most manufacturers define "non-blocking" differently. One defines it strictly in terms of switching capability and ignores the fact that his PBX might not have sufficient other "things," like devices which ring bells on phones (to indicate an incoming call) or devices which deliver dial tone to a phone (to indicate the PBX is ready to receive instructions).

Blocking Factor The number of records in a block; the number is computed by dividing the size of the block by the size of each record contained therein. Each record in the block must be the same size.

Blocking Formulas Specific probability distribution functions that closely approximate the call pattern of telephone users probable behavior in failing to find idle lines.

Blower A microphone.

Blowfish The name of the scrambling algorithm behind Philip Zimmerman's powerful encryption scheme called Pretty Good Privacy, or PGP, which lets you converse in total privacy over normal phone lines. See PGP for a fuller explanation.

Blowing Your Buffer Losing one's train of thought. This happens when the person you're speaking with won't let you get a word in edgewise or just said something so astonishing that your brain gets derailed. "Damn, I just blew my buffer!"

Blown Fiber Blown fiber is an installation method where a housing with lots of smaller tubes is installed without fiber. Once the housing is checked out, they take an air compressor and use air pressure to float the fiber down the tube. This greatly lowers the rate of failures during underground burying of fiber since the housing is empty during the installation. The air flow moves in the same direction as the fiber is installed and helps lower friction and any tugging on the fiber itself. One advantage to blown fiber is that the tubes can be installed with lots of joins (as needed to accommodate the installation problems, such as tight corners), but the fiber strands are later installed with no splices required. Also with blown fiber is that you can upgrade your fiber when you need it.

Blown Fuse A broken fuse.

BLSR See Bidirectional Line Switched Ring.

Blue Alarm Used in T-1 transmission. Also known as the AIS (Alarm Indication Signal). The blue alarm is turned on when two consecutive frames have fewer than three zeros in the data bit stream. A blue alarms sends 1's (ones) in all bits of all time slots on the span. See T-1 and AIS.

Blue Books The CCITT 1988 recommendations were published in books with blue covers, hence the term "Blue Books." The CCITT is now called the ITU-T.

Blue Box A device used to steal long distance phone calls. The classic blue box was slightly larger than a cigarette container. It had a touchtone pad on the front and a single button on top. Typically, you went to a coin phone and dialed an 800 number. While the distant number was ringing, you punched the single button on the top of the blue box. That button caused the blue box's speaker to emit a 2600 Hz tone. This disconnected the ringing at the other end but left the user inside the long distance network. The user then punched in a series of digits on the touchtone pad. The phone network heard those tones and sent the call according to the instructions in the tones. The tones duplicated the tones which the touchtone pads of long distance operators emitted. They are different from those emitted by normal telephones. The first blue box was "discovered" at MIT in a small utility box that was painted blue, thus the term blue box. When they were young, Steve Jobs and Steve Wozniak, founders of Apple Computer, sold blue boxes, which Wozniak built. People who used blue boxes in their salad days included characters with adopted pseudonyms like Dr. No, The Snark and Captain Crunch, who got his name from the free 2600 Hz whistle included as a promotion in boxes of Captain Crunch breakfast cereal. With the advent of CCIS, Common Channel Interoffice Signaling (i.e. out-of-band signaling), blue boxes no longer work.

Blue Collar Computer A colloquial term for a handheld computer which is used by "blue collar" workers for tasks such taking inventory, tracking goods, etc. Such computer may have a pen, a large pen-sensitive or touch-sensitive screens, a bar code scanner and a modem. It may be able to capture signatures — useful for confirmation of the delivery of goods.

Blue Grommet The rubber collar over the joint between

the handset and the armored cable on a pay phone. Blue identifies a "hearing aid compatible" handset.

Blue Pages Section of phone directory commonly used for government phone numbers, as distinct from white and yellow pages.

Blue stake A verb that means to mark an area. Here's how it works. Let's say you're a phone company and you want to lay a cable alongside a highway. You go to the highway authority and ask for permission to lay your cable. They give it to you. Then you call for "blue staking." This means that one organization comes out and marks on the ground using blue paint where all the stuff is located. This one organization represents all the various utilities who have stuff buried in the proposed running line (the path that you intend to lay your cable in). There is a legality here. If you then dig and hit someone's buried cable, you are absolved from legal responsibility — so long as you called for blue staking. If you hit something that was not located (i.e. not blue staked and not located on any map), it's called "off locate."

Bluetooth A code name for a proposed open specification to standardize data synchronization between disparate PC and handheld PC devices. Intel and Microsoft, in April, 1998 formed a consortium between themselves IBM, Toshiba, Nokia, Ericsson and Puma Technology. It was code-named Bluetooth for the 10th century Danish king who unified Denmark. The idea of Bluetooth is to create a single digital wireless protocol to address end-user problems arising from the proliferation of various mobile devices- including smart phones, smart pagers, handheld PCs and notebooks-that need to keep data consistent from one device to another. According to an article in Telecommunications Magazine, December, 1998, "The standard's proponents (Ericsson, Nokia, IBM, Intel, Toshiba) talk of a world where equipment from different vendors works seamlessly together using Bluetooth as a sort of virtual cable, where a laptop can automatically use a mobile phone to pick up e-mail or a PDA can send data wirelessly to a fax machine. Bluetooth operates using FM modulation combined with frequency hopping (1,600 hops per second) to lesson interference. The nominal link range is 10 metres, and the gross data rate is 1 Mbps, although there are plans to double the data rate later." Bluetooth radios will operate in a picocell topology in the 2.4 GHz range of the unlicensed ISM (Industrial, Scientific and Medical) spectrum; the gross data rate is 1 Mbps. The Bluetooth baseband technology will support both SCO (Synchronous Connection Oriented) links for voice and AC (Asynchronous Connectionless) links for packet data. Bluetooth can support

1) an asynchronous data channel in asymmetric mode of maximally 721 Kbps in either direction and 57.6 Kbps in the reverse direction; alternatively, the data channel can be supported in symmetric mode of maximally 432.6 Kbps; 2) up to three simultaneous synchronous packet voice channels; or 3) a channel which simultaneously supports both asynchronous data and synchronous voice. Full-duplex communications will be supported using TDD (Time Division Duplex) as the access technique. Voice coding will be accomplished using the CVSD (Continuously Variable Slope Delta) modulation technique. Security will be provided through encryption and authentication, using the challenge-response mechanism. Frequency hopping, a spread spectrum technique, is used to improve performance in the unlicensed and heavily-used ISM band. www.bluetooth.com See also CVSD, ISM, Picocell, Spread Spectrum and TDD.

BM Burst Modem.

Bmp A Windows BitMaP format. The images you see when Windows starts up and closes, and the wallpaper that adorns the Windows desktop, are all in BMP format.

BMTI Block Mode Terminal Interface. A device used to create (and break down) packets to be transmitted through a ITU-T X.25 network. This device is needed if block-mode terminals (such as IBM bisync devices) are to be connected to the network without an intermediate computer.

BN 1. Bridge Number: A locally administered bridge ID used in Source Route Bridging to uniquely identify a route between two LANs.

2. An ATM term. BECN Cell: A Resource Management (RM) cell type indicator. A Backwards Explicit Congestion Notification (BECN) Rm-cell may be generated by the network or the destination. To do so, BN=1 is set, to indicate the cell is not source-generated, and DIR=1 to indicate the backward flow. Source generated RM-cells are initialized with BN=0.

BNA Burroughs Network Architecture. Communications architecture of Burroughs, now Unisys.

BNC A bayonet-locking connector for slim coaxial cables, like those used with Ethernet. BNC is an acronym for Bayonet-Neill-Concelman. I don't know ask who Neill and Concelman are. Angela McCormack told me it also stands for British Naval Connector, because it was originally developed by the British Navy as a trustworthy connection technique for harsh environments such as onboard ships. Researchers at the College of Engineering, California State Polytechnic University, Pomona, say it was a "Baby Nevel Connector" named after a man called Nevel who invented the large size of connector that resembles a regular BNC connector.

BNC Barrel Connectors These connectors join two lengths of thin Ethernet coaxial cable together. See also BNC.

BNC Connectors The connectors for thin Ethernet coaxial cable are BNC connectors. The BNC connectors on each end of thin Ethernet cable connect to T-connectors, barrel connectors, and other network hardware. A BNC Connector has a twist-lock mechanism to hold it onto its round jack. The connector is the male side of the single-conductor connection, the jack the female side. See BNC.

BNC Female To N-Series Female Adapter The BNC female to N-Series female adapter is a connector which enables you to connect thin coaxial cable to thick coaxial cable. The BNC female connector attaches to the thin cable and the N-Series male connector attaches to the thick Ethernet cable.

BNC T-connectors The top of the T in a BNC T-connector functions as a barrel connector and links two lengths of thin Ethernet coaxial cable; the third end connects to the SpeedLink/PC 16.

BNC Terminators 50-ohm terminators are used to block electrical interference on a Ethernet coaxial cable network and to terminate the network at certain spots. You attach a BNC terminator to one plug on a T-connector if you will not be attaching a length of cable to that plug. You may also need to use a BNC terminator with a grounding wire to ground the network. See BNC.

BNR Bell-Northern Research. Northern Telecom's research arm. Northern Telecom is now called NorTel.

BnZS Code A bipolar line code with n zero substitution.

Board 1. Short for printed circuit board. Phone systems have boards for all sorts of purposes — from boards to serve trunk lines, to boards to serve proprietary phone sets, to boards that serve T-1 lines, etc. Computers also have boards — ones for SCSI ports, for floppy and hard disks, for CD-

ROMs, etc. In display/monitor terminology a board refers to the adapter (or controller) that serves as an interface between the computer and monitor.

2. An SCSA term. Any hardware module that controls its own physical interface to the SCbus or SCxbus. From a programming point of view, a board is an addressable system component that contains resources.

BOB 1. Software from Microsoft, which describes it as a "superapplication" for Windows designed for consumers intimidated by computer technology. According to Microsoft, users will use Bob to write letters, send e-mail, manage their household finances, keep addresses and dates and launch full-blown Windows applications, all under the guidance of cartoon characters. Bob hasn't done well and has effectively died.

2. BreakOut Box.

BOC Bell Operating Company. The local Bell operating telephone company. These days there are 22 Bell Operating Companies. They are organized into (i.e. owned by) seven Regional Bell holding companies, also called RBOCs, pronounced "R-bocks," or RHCs. See BELL OPERATING COMPANY.

Body The main informational part of a message. Body is the information, not the address nor the addressing information. There can be single or multiple parts to a body. For example, a single part could be text, or multiple parts could include text and graphics, or voice and graphics, etc. See Body Part 14.

Body Belt Used to attach telephone workers to poles and structures. Also called safety belt and climbing belt. You got to be gutsy, fit and well trained to use one of these things to go up a telephone pole. It would be good if you had spikes on your boots. That way your feet could dig into the pole.

Body Part 14 BP 14. An X.400 messaging term. A non-specific, identifying body part. A binary attachment with identifying header to explain the nature of the content such as a particular spreadsheet, word processor, etc,

Body Worn Transmitter A body worn transmitter is an audio transmitter secretly worn by an agent for surveillance purposes. Body wires must be carefully designed to be rugged in everyday use and transmit a strong signal regardless of how the antenna and wearer are positioned.

BOE Buffer Overflow Error

BOF 1. Business Operations Framework. A wireless telecommunications term. A document compiled to describe the operations of a telecommunications business entity in a specific area.

2. Birds of a Feather. A group of people with similar interests. If interest in a subject is strong enough, a BOF may develop into a SIG (Special Interest Group).

Bogon Something that is stupid or nonfunctional.

BOIC Barring of Outgoing International Calls. A wireless telecommunications term. A supplementary service provided under GSM (Global System for Mobile Communications).

Bolt Pattern The pattern that the bolts on the back of the device make. The idea is that if you attach something to the device, you don't want to mess up its warranty by opening the box and putting your stuff inside. So you design your stuff to fit the bolt pattern. This does not mess up the warranty. His product is designed to fit the bolt pattern, but not to intrude into the box.

BOM An ATM term. Beginning of Message: An indicator contained in the first cell of an ATM segmented packet.

Bond The electrical connection between two metallic surfaces established to provide a low resistance path between them.

Bonding 1. In ISDN BRI transmissions, bonding refers to

joining the two 64 Kbps B channels together to get one channel of 128 Kbps. Also known as dial-in channel aggregation.

2. Bonding is also the name of a group known as the Bandwidth ON Demand INteroperability Group (BONDING). The group's charter is to develop common control and synchronization standards needed to manage high speed data as it travels through the public network. This will allow equipment from vendors to interoperate over existing Switched 56 and ISDN services. Version 1.0 of the standard, approved on August 17, 1992, describes four modes of inverse multiplexer (I-Mux) interoperability. It allows inverse multiplexers from different manufacturers to subdivide a wideband signal into multiple 56- or 64- Kbps channels, pass these individual channels over a switched digital network, and recombine them into a single high-speed signal at the receiving end.

3. In electrical engineering, bonding is the process of connecting together metal parts so that they make low resistance electrical contact for direct current and lower frequency alternating currents.

Bonding Conductor for Telecommunications A conductor that interconnects the telecommunications bonding infrastructure to the buildings service equipment (power) ground.

Bong A tone that long distance carriers and value added carriers use in order to signal you that they now require additional action on your part — usually dialing more digits in order to provide billing information. For example, you hear a bong (or boing) to prompt you to enter a calling card number. The bong tone consists of a short burst of the # touch tone, followed by a rapidly decaying dial tone. See also VOICE MODEM.

BONTs Broadband Optical Network Terminations.

Booking Factor Booking factor is a percentage of the frame relay links used by frame relay paths, based on the sum of the Committed Information Rates (CIRs) of all the frame relay paths (FRPs) on the frame relay link (FRL).

Bookmark A gopher or Web file that lets you quickly connect to your favorite pre-selected page. Appropriately named. The way it works: You connect to a home page. You decide you'd like to return at some other time. So you command your internet surfing software to mark this web site with a "bookmark." Next time you want to return to that web site, you simply go to your bookmarks, click on which one you want. And bingo, you're there. A bookmark is also known as a hot list. Most Web browsers have bookmarks or hot lists. Microsoft's calls a bookmark a favorite. Netscape calls it a bookmark.

Boolean Expression An expression composed of one or more relational expressions; a string of symbols that specifies a condition that is either true or false.

Boolean Logic Boolean Logic is named after the 19th century mathematician, George Boole. Boolean logic is algebra reduced to either TRUE or FALSE, YES or NO, ON or OFF. Boolean logic is important for computer logic because computers work in binary — TRUE or FALSE, YES or NO, ON or OFF.

Boolean Operators See BOOLEAN LOGIC. Boolean operators are AND, OR, XOR, NOR, NOT. The result of an equation with one or more of the boolean operators is that the result will either be true or false.

Boolean Valued Expression An expression that will return a "true" or "false" evaluation.

Boot Abbreviation for the verb to bootstrap. A technique or device designed to bring itself into a desired state where it can operate on its own. For example, one type of boot is a routine whose first few instructions are sufficient to bring the rest of itself into memory from an input device. See BOOTSTRAP.

Boot Loader A Windows 2000 term. Defines the information needed for system startup, such as the location for the operating system's files. Windows NT automatically creates the correct configuration and checks this information whenever you start your system.

Boot Partition A Windows 2000 term. The volume, formatted for either an NTFS, FAT, or HPFS file system, that contains the Windows NT operating system and its support files. The boot partition can be (but does not have to be) the same as the system partition.

Boot Priority Which disk drive the computer looks to first for the files it needs to get started. Modern PCs start their boot cycle with the hard disk and then move to the floppy disk drive. Older PCs started their boot cycle with the floppy disk drive.

BootP Bootstrap Protocol. A TCP/IP protocol, which allows an Internet node to discover certain startup information such as its IP address.

Boot ROM A read-only memory chip allowing a workstation to communicate with the file server and to read a DOS boot program from the server. Workstations can thus operate on the network without having a disk drive. These are commonly called diskless PCs or diskless workstations.

Bootstrap The process of starting up a computer. Think about the following explanation in regard to your desktop MS-DOS machine. Usually, when the computer is turned on, it goes to a location of permanent Read Only Memory (See ROM) for instructions. These instructions, in turn, load the first instructions from the disk telling the computer what tasks to start performing. The name of this process comes from the expression "pulling oneself up by one's own bootstraps." The typical personal computer BOOT (startup) throws a message on the screen instructing the user to "insert a disk."

To confuse matters, there are WARM boots and COLD boots. Cold boots occur when the ac power switch on the computer is turned on. Warm boots occur when you hit the reset button (or Ctrl/Alt/Del) while the ac power switch stays on. A warm boot — reset — is done when you're changing disks or programs, or have done something dumb, like tried to access a drive that didn't access, or tried to print without connecting up to a printer.

You do a cold boot when the machine locks up rock hard and a warm boot doesn't work. To do a cold boot with your computer, turn the ac power off count to ten and then turn it on. Remember: never leave disks in your computer when you're turning it on and off. The surge of electricity might destroy the disks. When modems give trouble, do a cold boot on them. In fact, when phone systems give trouble, do a cold boot on them also. See also Boot RAM, Device Driver and MS-DOS.

BOP Bit Oriented Protocol. See Bit Oriented Transmission.

Border Gateway Protocol BGP is a Transmission Control Protocol/Internet Protocol (TCP/IP) routing protocol for interdomain routing in large networks. It is used in the Internet and is an alternative to EGP (Exterior Gateway Protocol).

Border Note An ATM term. A logical node that is in a specified peer group, and has at least one link that crosses the peer group boundary.

Borscht A group of functions provided in Line Circuits (LC). It stands for

B: Battery supply to subscriber line.

O: Overvoltage protection.

R: Ringing current supply.

S: Supervision of subscriber terminal.

C: Coder and decoder.

H: Hybrid (2 wire to 4 wire conversion).

T: Test.

Borscht is a group of functions provided to an analog line from a line circuit of a digital central office switch. An analog electronic switch can omit C and possibly H. A line circuit on a switch with a metallic matrix (SXS, Xbar, 1,2,3ESS) only detects call originations and disconnects itself.

Bot Shortened word for Robot. A bot is a program that runs on a computer 24 hours a day, 7 days a week automating mundane tasks for the owner. Bots are used on the Internet in many ways. Most popular is its use in IRC and Web search engines. IRC bots are programs that connect to the Internet and interact with Internet in very much the same way a normal users does (in fact, Internet (IRC) servers treat bots as regular users). Most IRC bots are used for channel control. Bots have also been called automatons. In the world of Web searching, bots are also called spiders and crawlers. They explore the World Wide Web by retrieving a document and following all the hyperlinks in it. Then they generate catalogs that can be accessed by search engines.

Bottle A water-tight device shaped like a glass bottle which contains amplifiers, regenerators and other equipment and used at regular distances along an underwater cable.

Bottom Line A phrase that can mean net profit, the lowest possible price that someone will take or the basic meaning with all the frills and nonsense cut away.

Bottom Lining From Wired's Jargon Watch column. What phone and cable companies consider when picking areas for trials and early deployment of interactive services. They look for areas full of upper- and middle-class households with enough money to pay for these services and generally ignore areas with lower incomes.

Bounce The return of an email message to the sender when the message is undeliverable. This usually means that you have gotten the address wrong, the destination address has been changed or the destination server has died. The bounce often includes information from the email system that explains the nature of the problem.

Bounce Board A large microwave reflector resembling an outdoor movie screen which is use to redirect (bounce) microwave telephone signals between two remote transceivers.

Bounced Mail Mail that is returned to the originator due to an incorrect e-mail address or a downed mail server.

Bound Mode In an optical fiber, a mode whose field decays monotonically in the transverse direction everywhere external to the core and which does not lose power to radiation. Except in a single-mode fiber, the power in bound modes is predominantly contained in the core of the fiber.

Boundary Conditions Boundary conditions are those that are found at the cusp of valid and invalid inputs and parameters. Many faults are found in a computer telephony system's ability to handle boundary conditions, especially when the computer telephony system is under load. For example, for a network that expects a switch to reset a trunk port within two seconds, the associated boundary conditions would be found at 1.9 to 2.1 seconds.

Boundary Function Capability in SNA sub-area node to handle some functions that nearby peripheral nodes are not capable of handling.

Boundary Node In IBM's SNA, a sub-area node that can provide certain protocol support for adjacent sub-area nodes, including transforming network addresses to local addresses, and vice versa, and performing session level sequencing and flow control and less intelligent peripheral nodes.

Boundary Routing A 3Com proprietary name for a method of accessing remote networked locations, such as a bank branch office. Effectively a form of bridging, the idea is to reduce the need for technical expertise locally and the cost of equipment at the remote site and manage the communications from head office.

Bounding Box Traditionally, computer programs have dealt with onscreen objects, such as images, by placing them in an invisible rectangle called a bounding box. You can see an example of a bounding box by clicking an image inside a word processor such as Word. The outline that appears around the image is the bounding box.

Bozo Filter Imagine that you're receiving zillions of emails from MotherInLaw@aol.com. You don't want to receive. Simple. You set up a "bozo filter." This piece of software automatically deletes any incoming emails from MotherInLaw@aol.com. Bozo filters are best set up by your email provider at this site. You don't want to set them up on your machine. See MAIL BOMB.

BP Block Pair. See Block Pair.

BP 14 Body Part 14, an X.400 electronic messaging term referring to a nonspecific body part, commonly used to transfer binary attachments.

BPAD Bisynchronous Packet Assembler/Disassembler.

BPDU An ATM term. Bridge Protocol Data Unit: A message type used by bridges to exchange management and control information.

BPI Bytes Per Inch. How many bytes are recorded per inch of recording surface. Typically used in conjunction with magnetic tape.

BPON Broadband Passive Optical Network.

BPP 1. Brokered Private Peering, an evolving industry plan designed to revamp the way providers exchange traffic.
2. Bits Per Pixel. The number of bits used to represent the color value of each pixel in a digitized image.

Bps Bps is confusing. Is it bits per second or bytes per second? In telecommunications, bps always means bits per second. In computing, BPs (note the capital "B") often means bytes per second. But don't trust people to always be correct — using the correct upper or lower case "B." You have to figure what context you're working in. The "Rule of Thumb" is that outside the computer, in the telecom world — and that means from the computer to the world, on the USB, on the LAN, on the local loop, on the WAN, across the country, across the ocean — it's bits per second. Raw bits per second. In telecom you don't always get the speed you pay for. All telecom circuits require signaling and timing and that requires bits. You need to know if your signaling is "inband" or "out of band." For example, a 64 kbps circuit might use 8 Kbps for inband signaling. This means you only get 56 Kbps (64 minus 8) for sending your precious material. On the other hand, a 64 Kbps ISDN BRI B channel circuit is actually a full 64 Kbps. The signaling for that ISDN channel is handled on a side channel of 16 Kbps, called the D channel.

Inside the computer, Bps is bytes per second. More commonly, it's KBps — kilobytes per second, or MBps — million bytes per second. Virtually all hard disk drive transfer rates (between the hard disk and the main microprocessor, or KSU) are in megabytes per second, Mbps. Sometimes the computer industry refers to KBps (kilobytes per second) when it's talking about transferring files from distant places to your machine — you see the number when you download files over the Internet. You wonder why the number is so much smaller than the alleged speed of your modem, which is measured in bits per

second. You can translate between the two by knowing that the computer industry is referring to serial data communications in which each byte is actually ten bits — eight bits for the letter, number, or character of the information you're receiving and two bits for start and stop information.

Virtually all telecom transmission is full duplex and symmetrical. This means that if you read that T-1 is 1,544,000 bits per second (1,544,000 bps or 1.544 Mbps), it's full duplex (both ways simultaneously) and symmetrical (both directions the same speed). That means it's 1,544,000 bits per second in both directions simultaneously. If the circuit is not full duplex or not symmetrical, this dictionary points that out. For now, the major asymmetrical (but still full duplex) circuit is the xDSL family, starting with ADSL, which stands for asymmetric, which means unbalanced. The DSL "family" no longer starts with "A," but most of it is still asymmetrical. Our definitions point out which is which.

There's one more complication. Inside computers, they measure storage in bytes. Your hard disk contains this many bytes, let's say eight gigabytes. That's fine. But they're not bytes the way we think of them in internal
or external computer transmission terms. They're different and they have to do with a way computer stores material — on hard disks or in RAM. They're what I call "storage bytes." When we talk 1 Kb of storage bytes, we really mean 1,024 bytes. Which comes from the way storage is actually handled inside a computer, and calculated thus: two raised to the power of ten, thus $2 \times 2 \times 2 \times 2 \times 2 \times 2 \times 2 \times 2 \times 2 \times 2 = 1,024$. Ditto for one million, two raised to the power of twenty, thus 1,048,576 bytes.

Finally, in telecom, when talking about transmission speed, the rule to be aware of is that the speed of a circuit is determined by the slowest part of the circuit. If one part of your circuit can only transmit at 9600 bps, then that's going to be the speed of your circuit — irrespective of the fact that other parts can go much faster. When measuring speed, you also have to factor in accuracy. All data communications schemes have error-checking systems, some better than others. Typically such systems force a re-transmission of data if a mistake is detected. You might have a fast, but "dirty" (i.e. lots of errors) transmission medium, which may need lots of re-transmissions. Thus, the "effective" bps (transmission speed) of that communications network is likely to be lower than what it's billed as. See also Baud and Mbps.

BPSK Binary Phase Shift Key.

BQM Business Quality Messaging. An initiative intended to facilitate the collaboration of vendors of e-mail and other messaging-enabled applications toward business systems that run reliably on both corporate networks and the Internet. The BQM SIG (Special Interest Group) was formed in April 1997. Founding members include AT&T, Hewlett Packard, IBM, Intel and Microsoft. www.bqm.com

BRA Basic Rate Access. A Canadian term for the ISDN 2B+D standard, which is called BRI in the U.S. — Basic Rate Interface. See ISDN.

BRB Be Right Back. Used in online chat to tell other participants in the session that you'll be away from the keyboard for awhile (and that your silence shouldn't be misinterpreted). Often used for a bathroom break.

Bragg Reflector A device designed to finely focus a semiconductor laser beam. Dennis Hall, a professor at the University of Rochester's Institute of Optics in New York, told the Economist Magazine in the Spring of 1993 that he and his colleague Gary Wicks have etched into the surface of his gallium-arsenide laser a grating of 600 concentric grooves, each a quarter of a millionth of a meter apart. The grating acts as what is known as a Bragg reflector. As the waves of laser light pass through each of its ridges, they are reflected by each of its ridges, a process which causes them to come together into an even, circular beam.

Braid A fibrous or metallic group of filaments interwoven cylindrically to form a covering over one or more insulated conductors.

Brainerd, Paul S. Founder of Aldus Corporation of 1984, Mr. Brainerd is reputed to be "the father of desktop publishing." His program Aldus FageMaker allowed the average PC user to produce professional-looking documents.

Branch A path in the program which is selected from two or more paths by a program instruction. "To branch" means to choose one of the available paths.

Branch Feeder A cable between the distribution cable and the main feeder cable to connect phone users to the central office. An outside plant term.

Branching Filter A device placed in a waveguide to separate or combine different microwave frequency bands.

Branding A term for identifying the Operator Service Provider (OSP) to the caller. Picture calling from your hotel room. You dial long distance. You have no idea which carrier you're using. But a message comes on: "Thanks for using MCI." Now you know. That's called branding.

BRCS Business and Residence Customer Services. An approach that the AT&T 5ESS switch employs to provision revenue-generating services.

Breadboard A circuit board made by hand, usually in building a prototype. No one knows where the word breadboard came from. Maybe because it looks a little like a breadboard?

Break An interruption. As in "Make and Break." Make means contacts which are usually open, but which close during an operation. "Make and Break" accurately describes rotary dialing.

Break In The attendant can interrupt conversations and announce an emergency or an important call.

Break Key A Break Key is found on some PCs. It is usually used to interrupt the current task running on a remote host. Break is not an ASCII character, it is simply a period of start (space) polarity.

Break Optimization A call center term. The automatic adjustment of break start times for schedules in the Daily Workfile so as to more closely match staff to workload in each period of the day. The program can thus improve upon the originally scheduled break arrangement because it now has information about schedule exceptions, newly added schedules, and additional call volume in AHT (Average Handle Time) history. See BREAK PARAMETERS.

Break Out Box A testing device that permits a user to cross-connect and tie individual leads of an interface cable using jumper wires to monitor, switch, or patch the electrical output of the cable. The most common break out box in our industry is probably the RS-232 box. Some of these boxes have LEDs (Light Emitting Diodes), which allow you to see which lead is "live." See also BREAKOUT BOX.

Break Parameters A call center term. A group of scenario assumptions you set to govern the placement of breaks in employee scheduling. These are typically:
- Earliest allowable break start time
- Latest allowable break start time
- Duration of the break
- Whether the break is paid or unpaid

Break Test Access Method of disconnecting a circuit, which has been electrically bridged, to allow testing on either side of the circuit. Devices that provide break test access include: bridge clips, plug-on protection modules, and plug-on patching devices. Break test access also provides a demarcation point.

Breakage When a prepaid phone card is never used. The distributor / manufacturer gets to keep the money. This is called breakage.

Breakdown Set A device that attaches to copper telephone pairs and sends current down the pairs. The current causes the wire to heat slightly, thus slowly drying out the cable. The device is used by telephone companies to dry out pairs of cables which have become wet.

Breakdown Voltage The voltage at which the insulation between two conductors breaks down.

Breaking Strength The amount of force needed to break a wire or fiber.

Breakout A wire or group of wires in a multi-conductor configuration which terminates somewhere other than at the end of the configuration.

Breakout Box A device that is plugged in between a computer terminal and its connecting cable to re-configure the way the cable is wired. When hooking up a terminal that is wired as if it were a computer itself (such as a VT-100), a break out box is used to break out, or fan out the 25 connections in the RS-232 cable. Each wire in the break out box goes through a switch that can be turned off, and a wire jumper is provided to connect each pin on one side to one or the other pin on the other side. This allows you, for example, to switch pins 2 & 3, thus fooling two computer devices into thinking one is talking to a terminal. (Now you have the essence of a null modem cable.) Break out boxes are necessary because there is no such thing as "standard" pinning on an RS-232 cable. To connect one computer to a printer one minute and to another computer the next minute, usually requires totally different wiring in the RS-232 cable, i.e. two sets of cables. This lack of standardization is why you'll always see dozens of RS-232 cables lying around where computers are used.

BRI Basic Rate Interface. There are two subscriber "interfaces" in ISDN. This one and PRI (Primary Rate Interface). In BRI, you get two bearer B-channels at 64 kilobits per second and a data D-channel at 16 kilobits per second. The bearer B-channels are designed for PCM voice, slow-scan video conferencing, group 4 facsimile machines, or whatever you can squeeze into 64,000 bits per second full duplex. The data (or D) channel is for bringing in information about incoming calls and taking out information about outgoing calls. It is also for access to slow-speed data networks, like videotex, packet switched networks, etc. See BASIC RATE INTERFACE and ISDN.

Brick A large hand-held cellular phone or handheld two-way radio. In more technical language, a "brick" is a station in the mobile service consisting of a hand-held radiotelephone unit licensed under a site authorization. Each unit can work while being hand-carried.

Bridge 1. In classic terms, a bridge is a data communications device that connects two or more network segments and forwards packets between them. Such bridges operate at Layer 1 (Physical Layer) of the OSI Reference Model. At this level, a bridge simply serves as a physical connector between segments, also amplifying the carrier signal in order to compensate for the loss of signal strength incurred as the signal is split across the bridged segments. In other words, the bridge is used to connect multiple segments of a single logical circuit. Classic bridges are relatively dumb devices, which are fast and inexpensive; they simply accept data packets, perhaps buffering them during periods of network congestion, and forward them. Bridges are protocol-specific, e.g., Ethernet or Token Ring in the LAN domain. Bridges also are used in the creation of multipoint circuits in the WAN domain, e.g., DDS (Dataphone Digital Service).

Bridges also can operate at Layer 2 (Link Layer) of the OSI Reference Model. At this level, a bridge connects disparate LANs (e.g., Ethernet and Token Ring) at the Medium Access Control (MAC) sub-layer of Layer 2. In order to accomplish this feat, the MAC Bridge may be of two types, encapsulating or translating.

Encapsulating bridges accept a data packet from one network and in its native format; they then encapsulate, or envelope, that entire packet in a format acceptable to the target network. For instance, an Ethernet frame is encapsulated in a Token Ring packet in order that the Token Ring network can deliver it to the target device, which must strip away several layers of overhead information in order to get to the data payload, or content. In order to accomplish this process, a table lookup must take place in order to change basic MAC-level addressing information.

Translating bridges go a step further. Rather than simply encapsulating the original data packet, they actually translate the data packet into the native format of the target network and attached device. While this level of translation adds a small amount of delay to the packet traffic and while the cost of such a bridge is slightly greater, the level of processing required at the workstation level is much reduced.

Bridges also can serve to reduce LAN congestion through a process of filtering. A filtering bridge reads the destination address of a data packet and performs a quick table lookup in order to determine whether it should forward that packet through a port to a particular physical LAN segment. A four-port bridge, for instance, would accept a packet from an incoming port and forward it only to the LAN segment on which the target device is connected; thereby, the traffic on the other two segments is reduced and the level of traffic on the those segments is reduced accordingly. Filtering bridges may be either programmed by the LAN administrator or may be self-learning. Self-learning bridges "learn" the addresses of the attached devices on each segment by initiating broadcast query packets, and then remembering the originating addresses of the devices which respond. Self-learning bridges perform this process at regular intervals in order to repeat the "learning" process and, thereby, to adjust to the physical relocation of devices, the replacement of NICs (Network Interface Cards), and other changes in the notoriously dynamic LAN environment.

While bridges are relatively simple devices, in the overall scheme of things, they can get quite complex as we move up the bridge food chain. (Please don't blame me. I didn't invent this stuff!) Bridges also can be classified as Spanning Tree Protocol (STP), Source Routing Protocol (SRP), and Source Routing Transparent (SRT).

Spanning Tree Protocol (STP) bridges, defined in the IEEE 802.1 standard, are self-learning, filtering bridges. Some STP bridges also have built-in security mechanisms which can deny access to certain resources on the basis of user and terminal ID. STP bridges can automatically reconfigure themselves for alternate paths should a network segment fail.

Source Routing Protocol (SRP) bridges are programmed with specific routes for each data packet. Routing considerations

include physical node location and the number of hops (intermediate bridges) involved. This IBM bridge protocol provides for a maximum of 13 hops.

Source Routing Transparent (SRT) bridges, defined in IEEE 802.1, are a combination of STP and SRP. SRT bridges can act in either mode, as programmed.

2. In the context of either audioconferencing (voice) or videoconferencing, a bridge connects three or more telecommunications channels so that they can all communicate together. In either case, compensation is made for signal loss (called balancing) in order to maintain consistent quality, thus allowing all participants to hear and see each other with equal ease. In video conferencing, bridges are often called MCUs — Multipoint Conferencing Units. One feature of some video bridges is their ability to figure who's speaking and turn on the camera which is on that person and have that person's face be on everyone's screen.

3. Finally, we'll explain bridge as a verb, as in "to bridge." Imagine a phone line. It winds from your central office through the streets and over the poles to your phone. Now imagine you want to connect another phone to that line. A phone works on two wires, tip and ring (positive and negative). You simply clamp each one of the phone's wires to the cable coming in. That's called bridging. Imagine bridging as connecting a phone at a right angle. When you do that, you've made what's known as a "bridged tap." The first thing to know about bridging is that bridging causes the electrical current coming down the line to lose power. How much? That typically depends on the distance from the bridged tap to the phone. A few feet, and there's no significant loss. But that bridged tap can also be thousands of feet. For example, the phone company could have a bridged tap on your local loop, which joined to another long-defunct subscriber. The phone company technicians simply saved a little time by not disconnecting that tap. If you want the cleanest, loudest phone line, the local loop to your phone should not be bridged. Instead it should be a direct "home run" from your central office to your phone.

Bridging can be a real problem with digital circuits. Circuits above 1 Mbps (e.g., T-1) should never, ever be bridged. Because of the power loss, they simply won't work or will work so poorly they won't be worth having. ISDN BRI channels are also digital. But they were specifically designed to work with the existing telephone cable plant, which has a huge number of bridged circuits. Telephone companies typically will install ISDN BRI circuits with up to six bridged taps and about 6,000 feet of bridged cabling. But that's a rule of thumb. And frankly, if I were getting an ISDN line, I'd ask for a line that had no taps and no bridges.

See INTERNETWORKING, LOADING COIL, ROUTERS, SOURCE ROUTING, and TRANSPARENT ROUTING.

Bridge Amplifier An amplifier installed on a CATV trunk cable to feed branching cables.

Bridge Battery A small supplementary battery on a laptop which holds the contents of the memory and the system status for a few minutes while you replace a drained battery. NEC uses the term on its UltraLite Versa laptops.

Bridge Clip A small metal clip that used to electrically connect together two sides of a 50 pair block. Removing the bridging clips breaks the circuit. You might remove the clips when you want to insert a piece of test gear and check to see which side the trouble is on.

Bridge Equipment Equipment which connects different LANs, allowing communication between devices. As in "to bridge" several LANs. Bridges are protocol-independent but hardware-specific. They will connect LANs with different hardware and different protocols. An example would be a device that connects an Ethernet network to a StarLAN network. With this bridge it is possible to send signals between the two networks, and only these two networks.

These signals will be understood only if the protocols used on each LAN are the same, e.g. XNS or TCP/IP, but they don't have to be the same for the bridge to do its job for the signals to move on either LAN. They just won't be understood. This differs from gateways and routers. Routers connect LANs with the same protocols but different hardware. The best examples are the file servers that accommodate different hardware LANs. Gateways connect two LANs with different protocols by translating between them, enabling them to talk to each other. The bridge does no translation. Bridges are best used to keep networks small by connecting many of them rather than making a large one. This reduces the traffic faced by individual computers and improves network performance.

Bridge Group Virtual LAN terminology for a group of switch interfaces assigned to a singular bridge unit and network interface. Each bridge group runs a separate Spanning Tree and is addressable using a unique IP address.

Bridge Lifter A device that removes, either electrically or physically, bridged telephone pairs. Relays, saturable inductors, and semiconductors are used as bridge lifters.

Bridge Tap An undetermined length of wire attached between the normal endpoints of a circuit that introduces unwanted impedance imbalances for data transmission. Also called bridging trap or bridged tap. See BRIDGED TAP.

Bridged Jack A dual position modular female jack where all pins of one jack are permanently bridged to the other jack in the same order.

Bridged Ringing A system where ringers on a phone line are connected across that line.

Bridged Tap A bridged tap is multiple appearances of the same cable pair at several distribution points. A bridged tap is any section of a cable pair not on the direct electrical path between the central office and the user's offices. A bridged tap increases the electrical loss on the pair — because a signal traveling down the pair will split its signal between the bridge and main pairs. You can't run high-speed digital circuits, e.g. T-1, over cable that has bridged taps in it. But you can run ISDN circuits over cable with a limited number of bridged taps. See BRIDGE and LOADING COIL.

Bridger Bridger Amplifier. An amplifier which is connected directly into the main trunk of a CATV system, providing isolation between the main trunk and multiple (high level) outputs.

Bridging Bridging across a circuit is done by placing one test lead from a test set or a conductor from another circuit and placing it on one conductor of another circuit. And then doing the same thing to the second conductor. You bridge across a circuit to test the circuit by listening in on it, by dialing on it, by running tests on the line, etc. You can bridge across a circuit by going across the pair in wire, by stripping it, etc. You can bridge across a pair (also called a circuit path) by installing external devices across quick clips on a connecting block.

Bridging Adapter A box containing several male and female electrical connectors that allows various phones and accessories to be connected to one cable. Bridging adapters work well with 1A2 key systems and single line phones, but usually not with electronic or digital key systems and electronic or digital telephones behind PBXs.

Bridging Clip A small piece of metal with a U-shape cross-section which is used to connect adjacent terminals on 66-type connecting blocks.

Bridging Connection A parallel connection by means of which some of the signal energy in a circuit may be extracted, usually with negligible effect on the normal operation of the circuit. Most modern phone systems don't encourage bridging connections, since the negligible is rarely negligible.

Bridging Loss The loss at a given frequency resulting from connecting an impedance across a transmission line. Expressed as the ratio (in decibels) of the signal power delivered to that part of the system following the bridging point before bridging, to the signal power delivered to that same part after the bridging.

Bridle Cards Proprietary Basic Rate ISDN Dual Loop Extension lets ISDN service be provided up to 28,000 feet away. See ISDN.

BRIDS Bellcore Rating Input Database System.

Briefcase A Windows 95 feature that allows you to keep multiple versions of a file in different computers in sync with each other.

Brightness An attribute of visual reception in which a source appears to emit more or less light. Since the eye is not equally sensitive to all colors, brightness cannot be a quantitative term.

BRISC Bell-Northern Research Reduced Instruction Set Computing.

Brite Cards And Services Basic Rate Interface Transmission Extension lets telephone companies extend service from ISDN-equipped central offices to conventional central offices. See ISDN.

Brittle Easily broken without much stretching.

Broadband 1. A WAN term. A transmission facility providing bandwidth greater than 45 Mbps (T3). Broadband systems generally are fiber optic in nature. See also Bandwidth and SONET. Contrast with Narrowband and Wideband.

2. A LAN term. A multichannel, analog, coax-based LAN. It almost defies the imagination that one would use an analog LAN for connectivity of digital computers, yet they exist. 10Broad36 is a standard for such a LAN. The real, and only, value of such an approach is that it will support multiple, simultaneous communications channels through Frequency Division Multiplexing (FDM). Some CATV (Community Antenna TeleVision) providers have upgraded their old coax systems to support broadband LAN communications. The coax systems were put in place to support multiple, downstream FDM analog TV channels. The upgrade supports bidirectional data channels for applications such as Internet access, LAN networking, and even POTS (Plain Old Telephone Service). Colleges and universities have upgraded their old CATV networks to broadband LANs, which were put in place to provide entertainment TV to the dormitories. Some theme parks have put them in place to support simultaneous audio, paging, closed-circuit TV and transaction processing. Contrast with Baseband. See also 10Broad36, CATV, FDM, and LAN.

Broadband Bearer Capability A bearer class field that is part of the initial address message.

Broadband Personal Communications Standards BPCS. Consists of 120 MHz of new spectrum available for new cellular networks. Also known as wideband PCS.

Broadband Switching System See BSS

Broadcast 1. To send information to two or more receiving devices simultaneously — over a data communications network, a voice mail, electronic mail system, a local TV or radio station or a satellite system. Broadcast involves sending a transmission simultaneously to all members of a group. In the context of an intelligent communications network, such devices could be host computers, routers, workstations, voice mail systems, or just about anything else. In the less intelligent world of "broadcast media," a local TV or radio station might use a terrestrial antenna or a satellite system to transmit information from a single source to any TV set or radio capable of receiving the signal within the area of coverage. See also NARROWCASTING and POINTCASTING. Contrast with UNICAST, ANYCAST and MULTICAST.

2. As the term applies to cable television, broadcasting is the process of transmitting a signal over a broadcast station pursuant to Parts 73 and 74 of the FCC rules. This definition is deliberately restrictive: it does not include satellite transmission, and it does not include point-to-multipoint transmission over a wired or fiber network. In spite of the fact that the broadcast industry and the cable television industry are forever bound together in a symbiotic relationship, they are frequently at odds over policy issues. See Broadcast Station. Compare with Cablecast.

Broadcast Channel BCCH. A wireless term for the logical channel used in certain cellular networks to broadcast signaling and control information to all cellular phones. BCCH is a logical channel of the FDCCH (Forward Digital Control CHannel), defined by IS-136 for use in digital cellular networks employing TDMA (Time Division Multiple Access). The BCCH comprises the E-BCCH, F-BCCH and S-BCCH. The E-BCCH (Extended-BCCH) contains information which is not of high priority, such as the identification of neighboring cell sites. The F-BCCH (Fast-BCCH) contains critical information which must be transmitted immediately; examples include system information and registration parameters. S-BCCH (System message-BCCH), which has not yet been fully defined, will contain messages for system broadcast. See also IS-136 and TDMA.

Broadcast List A list of two or more system users to whom messages are sent simultaneously. Master Broadcast Lists are shared by all system users and are set up by the System Administrator. Personal Lists are set up by individual subscribers.

Broadcast Message A message from one user sent to all users. Just like a TV station signal. On LANs, all workstations and devices receive the message. Broadcast messages are used for many reasons, including acknowledging receipt of information and locating certain devices. On voice mail systems, broadcast messages are important announcement messages from the system administrator that provide information and instructions regarding the voice processing system. Broadcast messages play before standard Voice Mail or Automated Attendant messages.

Broadcast Net A British Telecom turret feature that allows each trader single key access to a group of outgoing lines. This is designed primarily for sending short messages to multiple destinations. The "net" function allows the user to set up and amend his broadcast group.

Broadcast Quality A specific term applied to pickup tubes of any type — vidicon, plumbicon, etc. — which are without flaws and meet broadcast standards. Also an ambiguous term for equipment and programming that meets the highest technical standardes of the TV industry, such as high-band recorders.

Broadcast station An over-the-air radio or television station licensed by the FCC pursuant to Parts 73 or 74 of the FCC Rules, or an equivalent foreign (Canadian or Mexican) station. Cable television systems are authorized by FCC rules to retransmit broadcast stations; however, such retransmission is subject to a number of restrictions:

• The cable television operator is liable for copyright royalty fees collected by the Copyright Office.

• Under certain conditions, certain broadcast stations are eligible for mandatory carriage.

• Under certain conditions, the cable operator must obtain the permission of the licensee of the broadcast station. This term includes satellite-delivered broadcast "superstations" such as WGN-TV and WWOR, but it does not include:

• Satellite-delivered non-broadcast programming services (HBO, ESPN, C-SPAN, QVC, etc.).

• Video services delivered by terrestrial microwave systems such as MDS, MMDS, or ITFS, unless the actual signal being delivered was originally picked up from a broadcast station.

• Cablecasting programming originated by the cable operator or an access organization.

Broadcast Storm A pathological condition that may occur in a TCP/IP network that can cause large number of broadcast packets to be propagated unnecessarily across an enterprise-wide network, thereby causing network overload. Broadcast storms happen when users mix old TCP/IP routers with routers supporting the new releases of TCP/IP protocol. Routers use broadcast packets to resolve IP addressing requests from stations on LANs. If a station running an old version of TCP/IP sends such a request, TCP/IP routers in an enterprise-wide network misunderstand it and send multiple broadcasts to their brother and sister routers. In turn, these broadcasts cause each router to send more broadcasts, and so on. This chain reaction can produce so many broadcast messages that the network can shut down. It should be noted that this is extremely rare and it happens only in TCP/IP networks that use two specific TCP/IP protocol releases.

Broadcast Transmission A fax machine feature that allows automatic transmission of a document to several locations.

Brochureware A pejorative term for what companies can pull off with a with a clever copy writer, some nice graphics, and a bit of an advertising budget. Ever read a brochure and compared it to the product? You get the idea. See WEBWARE.

Broken Link A link to a file that does not exist or is not at the location indicated by the URL. In short, you click on a hyperlink on a hyperlink on a Web page you're viewing, but nothing happens or you get an error message. Bingo, broken link. You've been sent somewhere that doesn't exist. This is neither exciting, nor good programming.

Broken Pipe This term is usually seen in an error message by browser programs to let the user know that the stream of information which was downloading at the time has been forcibly cut. This can occur for many reasons, most commonly because you are on a very crowded network or your access provider is experiencing heavy traffic.

Broker A company (or person) that buys and sells equipment often without taking ownership. A broker does not test or refurbish the equipment. Often, it never sees the equipment it buys and sells. Instead, it has the equipment shipped from the supplier to the customer, relying on the supplier to have tested and refurbished the equipment. Its specialty is knowing who has what equipment nationwide and selling it, possibly, at below-market price. See SECONDARY EQUIPMENT.

Brokernet A virtual private dedicated network offering

from New York Telephone and provided within Manhattan aimed at brokerage, banking and message industries. It uses digital switching to provide virtual private lines, specifically "hot line" service.

Bronze Alloy of copper and tin, widely used and known since ancient times. Copper content in bronze varies between 89% and 96%.

Brouter Concatenation of "bridge" and "router." Used to refer to devices which perform both bridging and routing functions. In local area networking, a brouter is a device that combines the dynamic routing capability of an internetwork router with the ability of a bridge connect dissimilar local area networks (LANs). It has the ability to route one or more protocols, such as TCP/IP and XNS, and bridge all other traffic.

Brown and Sharpe Wire Gauge An older name for American Wire Gauge, the U.S. standard measuring gauge for non-ferrous conductors (i.e., non-iron and non-steel), AWG covers copper, aluminum, and other conductors. Gauge is a measure of the diameter of the conductor. See AWG for a full explanation.

Brownfield The opposite of greenfield. Brownfield is the sum of all legacy material (equipment, architectures, procedures, etc.) in any given network project. Greenfield refers to the material being developed anew. See GREENFIELD and LEGACY.

Brownout 1. When you lose all your electricity, it's called a blackout. When your voltage drops more than 10% below what it's meant to be, it's a brownout. If a brownout lasts less than about a second, it is called a SAG. Brownouts are sometimes caused by overloaded circuits and are sometimes caused intentionally by the AC utility company in order to reduce the power drawn by users during peak demand periods (like during hot summers when everyone is using their air conditioners). Studies have shown that brownouts of all durations make up the vast majority of power problems that affect telephone systems and computers.
2. In Internet terms, when a system is overloaded by requests that it slows down to the point of near unusability, it is suffering a "brownout."

Browser 1. Software that translates digital bits into pictures and text so you can look at them. A browser displays documents on the Internet and the World Wide Web to your computer. A Web Browser is software which allows a computer user (like you and me) to "surf" the Internet. It lets us move easily from one World Wide Web site to another. Every time we alight on a Web Page, our Web Browser moves a copy of documents on the Web to your computer. A Web Browser uses HTTP — the HyperText Transfer Protocol. Invisible to the user of a Web Browser, HTTP is the actual protocol used by the Web Server and the Client Browser to communicate over the Internet. The most famous Web Browsers are Netscape and Microsoft's Internet Explorer. But there are others, like Opera. See BROWSING, INTERNET and SURF. www.netscape and www.microsoft.
2. A developed tool used to inspect a class hierarchy in an object-oriented software system.

Browsing The act of searching through automated information system storage to locate or acquire information without necessarily knowing of the existence or the format of the information being sought.

Brush A computer imaging term. A paint package's most basic image-creation tool. Most packages let you select a variety of sizes and shapes. Many let you customize shapes.

Brute Force Attack A cracker term. Brute force attack means hurling passwords at a system until it cracks.

BSA See both BASIC SWITCHING ARRANGEMENT and OPEN NETWORK ARCHITECTURE.

BSC 1. Binary Synchronous Communication. A set of IBM operating procedures for synchronous transmission used in teleprocessing networks, BSC has become a de facto standard protocol. BSC is a character-oriented protocol which involves the communication of data in blocks of up to 512 characters. Each block of TeXT (TXT) data is preceded by a header which includes SYNchronizing bits (SYN) in order that the receiving device might synchronize on the rate of transmission, a Start Of Header (SOH), a HeaDeR (HDR) containing application address information, and a Start of Text (STX). Each block is succeeded by a trailer which includes End Of Text (EOT), Block Checking Characters (BCC) for error detection and correction, and End Of Transmission (EOT). BSC is a polling protocol which operates in a half duplex (HDX) mode, generally over the analog PSTN (Public Switched Telecommunications Network). As each block of data is transmitted from the polled device, the receiving computer system responds with either an ACKnowledgement (ACK) indicating successful and error-free receipt of the subject data, or a Negative AcKnowledgement (NAK) indicating detection of an error created in transmission. An ACK prompts the polled device to transmit the next block of data. A NAK prompts the device to retransmit the subject block. The process continues, block-by-block, until such time as all data have been transmitted and received. By way of example, BSC is commonly used in a Call Accounting application to transfer Call Detail Recording (CDR) information to a centralized processor from a pollable buffer attached to a PBX system.
2. Base Station Controller. A wireless telecommunications term. The BSC is a device that manages radio resources in GSM (Global System for Mobile Communications), including the BTS (Base Transceiver Station), for specified cells within the PLMN (Public Land Mobile Network).

BSD Berkeley Software Distribution. Term used when describing different versions of the Berkeley UNIX software, as in "4.3BSD UNIX".

BSE 1. Basic Switching Element. See OPEN NETWORK ARCHITECTURE.
2. Basic Service Elements. A term used in voice processing to describe technical telephone system features such as ANI, DID trunks, call forwarding, stutter dial tone, suppressed ringing, and directory database access.

BSGL Branch Systems General Licence. A British term. A licence that must be obtained by any organization seeking to link a private network to the British PSTN (Public Switched Telephone Network). A separate licence must be held on each site.

BSI British Standards Institution. The body responsible development of UK standards across a wide range, including telecommunications. BSI also is responsible for input to European standards bodies like CEN and CENELEC, as well as international standards bodies such as the ISO and the ITU-T. BSI claims to be the oldest of the over 100 national standards bodies in the world. www.bsi.org.uk. For the U.S. counterpart see also ANSI.

BSIC Base Station Identity Code. An attribute of a GSM (Global System for Mobile Communications) cell which is a code allowing a distinction between local cells having the same radio frequency.

BSP Bell System Practice. A very defined way of writing and presenting instruction and installation manuals. BSPs also establish standards for splicing cable, for installing phones, answering phones, collecting debts, finding phone taps, climbing poles. They are (or once were) the instruction manuals that dictated how to do everything. Divestiture has changed the rules. BSPs are not as important as they were when AT&T handed down all the BSPs.

BSS 1. Base Station System. A wireless telecommunications term. A GSM (Global System for Mobile Communications) device charged with managing radio frequency resources and radio frequency transmission for a group of BTSs. See also GSM.

2. Business Support System. The system used by network operators to manage business operations such as billing, sales management, customer-service management and customer databases. A type of Operations Support System (OSS).

3. Broadband Switching System. A carrier (e.g., LEC or IXC) switch for broadband communications. Such switches are capable of switching frames (Frame Relay) or cells (SMDS and ATM) at a very high rate of speed. They contain multi-Gigabit busses which may be stacked or chained. For example, an "edge" switch, which is located at the edge of the network much as is a Class 5 Central Office in the voice world, may involve a 20 Gbps bus. "Core" switches, which are located in the core of the network and which are the equivalent of a tandem switch in the voice world, may involve eight busses of 20 Gbps each for a total capacity of 160 Gbps.

BT 1. British Telecom. See 1981.

2. Burst Tolerance: BT applies to ATM connections supporting VBR services and is the limit parameter of the GCRA.

BTA 1. Basic Trading Area. A wireless telecommunications term. The United States is broken down into 493 major trading areas for economic purposes. These boundaries were used for licensing PCS wireless phone systems. Several BTAs make up Metropolitan Trading Area, an area defined by the FCC for the purpose of issuing licenses for PCS. Thus, each MTA consists of several Basic Trading Areas (BTAs).

2. Broadband Telecommunications Architecture, an architecture introduced by General Instrument's Broadband Communications Division at the Western Cable Television Show on December 1, 1993. General Instrument said the plant is built to 750 MHz and can support reduced node size and add services such as video-on-demand, telephony, interactivity, data services, etc.

3. Business Technology Association. Previously NOMDA/LANDA, BTA was formed by the merger of the National Office Machine Dealers Association and the Local Area Network Dealers Association. BTA holds several conferences a year. Members include manufacturers, distributors, retailers and consultants. www.btanet.org.

BTAM Basic Telecommunications Access Method. One of IBM's early host-based software programs for controlling remote data communications interface to host applications, supporting pre-SNA protocols. See IBM.

BTag An ATM term. Beginning Tag: A one octet field of the CPCS_PDU used in conjunction with the Etag octet to form an association between the beginning of message and end of message. See ATM.

BTI British Telecom International.

BTL Bell Telephone Laboratories.

BTN Billing Telephone Number. The primary telephone number used for billing, regardless of the number of phone lines associated with that number. According to Bellcore, BTN is sometimes known as a screening telephone number, the BTN is a telephone number used by the AMA process as the calling-party number for recording purposes. See Billing Telephone Number.

BTOS A UNIX program which translates binary files into ASCII.

BTRL British Telecom Research Laboratories.

Btrieve Btrieve is a key-indexed database record management system. You can retrieve, insert, update, or delete records by key value, using sequential or random access methods. First introduced in 1983, Btrieve was one of the first databases designed for LANs. Novell bought the company in the late 1980s and then later sold it. It's now called Pervasive Software.

BTS Base Transceiver Station. A wireless telecommunications term. A GSM (Global System for Mobile Communications) device used to transmit radio frequency over the air interface.

BTSM BTS (Base Transceiver Station) Management. A wireless telecommunications term. Devices configured to manage BTS functions and equipment.

BTSOOM Beats The S*** Out Of Me. Acronym used in e-mail, during online chat sessions, and in newsgroup postings..

BTU 1. Basic Transmission Unit.

2. British Thermal Unit. A measure of thermal energy often used in designing building heating and cooling systems. The heat output of computer equipment is often specified and must be taken into account when sizing building climate control systems. Computer equipment heat output is expressed in BTU per hour. 3.7 BTU per hour is equivalent to 1 Watt of dissipation.

BTV See Business Television.

BTW By The Way. An acronym used in electronic mail on the Internet to save words or to be hip, or whatever.

Bucket RMON terminology for a discrete sample of data. The RMON History group specifically uses buckets in its sampling functions of the different data sources. See Buckets.

Bucket o' Dial Tone Once upon a time, every new central office technician was sent to another central office for a "bucket o' dial tone," when the CO was overloaded. There is no such thing, of course, but the old hands had a big laugh over it. It was sort of a "right of passage." It worked only once. See also Fiber Exhaust and Frequency Grease.

Buckets When competition in cell phones heated up in North America in the mid-1990s, some of the newer competitors pushed increased coverage patterns, small phones, and huge packages of monthly minutes so large that their salespeople started calling the deals "buckets."

Buffer 1. In data transmission, a buffer is a temporary storage location for information being sent or received. Usually located between two different devices that have different abilities or speeds for handling the data. The buffer acts like a dam, capturing the data and then trickling it out at speeds the lower river can handle without, hopefully, flooding or overflowing the banks. A buffer can be located in the sending or receiving device.

2. A coating material used to cover and protect the fiber. The buffer can be constructed using either a tight jacket or loose tube technique.

3. A circuit or component which isolates one electrical circuit from another.

Buffer Coating Protective material applied to fibers. Increases apparent fiber size. May be more than one layer. Stated in microns. Usually thicker or multi-coated on tight-buffer cables.

Buffer Memory Electronic circuitry where data is kept during buffering. See BUFFER.

Buffer Storage Electronic circuitry where data is kept during buffering. See BUFFER.

Buffered Repeater A device that amplifies and regenerates signals so they can travel farther along a local area cable. This type of repeater also controls the flow of messages to prevent collisions.

Bug 1. A concealed microphone or listening device or other audio surveillance device.

2. To install the means for audio surveillance.

3. A semiautomatic telegraph key.

4. A problem in software or hardware. The original computer bug, a moth, is enshrined at the Washington Navy Yard. It was the cause of a hardware failure in an early computer in 1945. The story goes like this: a team of top Navy scientists was developing one of the world's first electronic computers. Suddenly, in the middle of a calculation, the computer ground to a halt, dead. Engineers poured over every wire and every inch of the massive machine. Finally, one of the technicians discovered the cause of the problem. Buried deep insides its electronic innards, crushed between two electric relays, lay the body of a moth. These days, "bugs" in telecom or computer systems are not insects. They're indescribable glitches that adversely affect smooth operations. Bugs usually originate in software. Some programmers call bugs "undocumented features." And they are, indeed. All the above is the story the navy likes to put out. In actual fact, the word "bug" for problem in design has been around for eons. It's mentioned in a 1910 book called, "Edison, His Life and Inventions" by Frank Lewis Dyer. The book talks about Edison harassing his employees to "get all the bugs out."

Bug Mix A silly term for the precise collection of bugs in a particular piece of software.

Bug Rate The frequency with which new bugs are found during the testing cycle is referred to as the bug rate. As these bugs are fixed, and as time passes, bugs will become more and more difficult to find, and new bugs will be found less frequently, i.e., the bug rate decreases. Usually, when the bug rate drops to zero and all major bugs (and most of the minor bugs) have been fixed, a product is ready for the next stage in its lifecycle.

Building Core A three-dimensional space permeating one or more floors of the building and used for the extension and distribution of utility services (e.g., elevators, washrooms, stairwells, mechanical and electrical systems, and telecommunications) throughout the building.

Building Distribution Frame Somewhere in the building where all telephony wiring for the building is. All the cables from the outside the building would first come here and be punched down on the building distribution frame. Additional cables would take the telephone circuits upstairs.

Building Entrance Agreement A piece of paper that lets phone companies and utilities enter a building, gives them permission to build facilities and store equipment and lets them access their equipment. In great demand by Competitive Local Exchange Carriers (CLECS) who give residents a choice of providers and often force existing carriers to improve services or reduce rates. Often the CLECS are charged fees, which they agree to, to avoid paying the RBOCs' access charges.

Building Entrance Area The area inside a building where cables enter the building and are connected to riser cables within the building and where electrical protection is provided. The network interface as well as the protectors and other distribution components for the campus subsystem may

be located here. Typically this area is the end of the local telephone company's responsibility. From here on it's your responsibility. You should protect your equipment inside the building from spikes and surges and other electrical nonsenses which the phone company's cables might bring in. For the best disaster protection, it's wise to have two building entrances by which your telecommunications cables can enter. And they should enter from separate telephone central offices. Some telephone companies, e.g. New York Telephone, are now tariffing such services. See Building Entrance Agreement.

Building Footing The concrete base under the foundation of a building in which copper wire may be laid to form an electrical ground.

Building Integrated Timing Supply BITS. A clock, or a clock with an adjunct, in a building that supplies DS1 and/or composite clock timing reference to all other clocks in that building. See BITS.

Building Module The standard selected as the dimensional coordination for the design of the building, e.g., a multiple of 100 mm (4 in), since the international standards have established a 100 mm (4 in) basic module.

Building Out The process of adding a combination of inductance, capacitance,and resistance to a cable pair so that its electrical length may be increased by a desired amount to control impedance characteristics.

Building Steel The structural steel beams that make up the frame of a building. If the steel frame is buried deep in the earth, it can be used as an electrical ground. But you'd better be careful: Unbalanced three-phase power is also probably using the frame as a ground, and you may pick up huge quantities of 60 hz hum.

Bulk Billing A method of billing for long distance telephone services where no detail of calls made is provided. WATS is a bulk billed service. Therein lies the problem for the cost conscious user. There's no verification of calls made. See CALL DETAIL RECORDING.

Bulk Encryption Simultaneous encryption of all channels of a multichannel telecommunications trunk.

Bulk Storage Lots of storage. Usually reels of magnetic tape or hard disks.

Bulletin Board System A fancy name for an electronic message system running on a microcomputer. Call up, leave messages, read messages. The system is like a physical bulletin board. That's where the name comes from. Some people call bulletin board systems electronic mail systems. See also BBS.

Bumper Beeper Radio beacon transmitter, hidden in or on a vehicle for use with radio tailing equipment.

Bunched Frame-Alignment Signal A frame-alignment signal in which the signal elements occupy consecutive digit positions.

Bundle 1. A group of fibers or wires within a cable sharing a common color-code.

2. In T-1, specifically M44 Multiplexing, a bundle consists of 12 nibbles (4 bits) and may represent 11 channels of 32 Kbps compressed information plus a delta channel. A bundle is typically a subset of a DSI and treated as an entity with its own signaling delta channel.

Bundle Fodder Junk software included on CD-ROMs or packaged with peripherals such as modems and designed to bulk up the presumed value of the total package to entice an unsuspecting consumer to buy. See SHOVELWARE.

Bundled Combining several services under one telephone tariff item at a single charge.

Bundled Rates Several service combined into one offering for one charge.

Burden Test A semi-legitimate test used in regulation to determine if the offering of a new or continued service will cause consumers of other services to pay prices no higher than if the service were not offered. In other words, the question is "Who carries the Burden?" It's sometimes called the "avoidable cost test."

Burford Courier An MCI definition. A communications software package developed for use with Wang VS mainframes enabling users to communicate directly with MCI International's Telex messaging services.

Buried Cable A cable installed under the surface of the ground in such a manner that it cannot be removed without disturbing the soil.

Buried Service Wire Splice A watertight splice filled with an encapsulant.

Burn In To run new devices and printed circuits cards, often at high temperatures, in order to pinpoint early failures. The theory is all semiconductor devices show their defects — if any — in the first few weeks of operation. If they pass this "burn-in" period, they will work for a long time, so the theory goes. "Burn-in" should probably be 30-days under full power and working load. Burn-in should also take place in a room with lots of heat and at least 50% humidity, since this will simulate the poorly-ventilated places most people install telephone systems.

Burn Out A condition, where stress causes agents to be apathetic and lethargic, caused by intensity of calling, lack of variety and poor working conditions. It is particularly associated with outbound cold calling and inbound complaint handling, both of which are stressful for agents if not carefully managed.

Burn Rate The speed at which a new company using up its cash en route to developing a product before it turns cash positive.

Burning A Pole Slang expression to describe when an installer accidentally slides down a telephone pole. This usually happens on an old pole full of gaff holes when his gaff breaks out of the pole. Burning a pole results in painful chemical burns because the installer usually winds up with a chest and legs full of splinters coated with creosote, a coal tar distillate used to preserve the wooden pole. Installers are taught to "kick out," rather than hug the pole and suffer burns.

Burrus Diode A surface-emitting LED with a hole etched to accommodate a light-collecting fiber. Named after its inventor, Charles Burrus.

Burst 1. In data communication, a sequence of signals, noise, or interface counted as a unit in accordance with some specific criterion or measure.

2. To separate continuous-form or multipart paper into discrete sheets.

3. A small reference packet of the subcarrier sine wave, typically 8 or 9 cycles, which is sent on every line of video. Since the carrier is suppressed, this phase and frequency reference is required for synchronous demodulation of the color information in the receiver.

Burst Isochronous Isochronous burst transmission. See ISOCHRONOUS.

Burst Mode A way of doing data transmission in which a continuous block of data is transferred between main memory and an input/output device without interruption until the transfer has been completed.

Burst Switching In a packet-switched network, a switching capability in which each network switch extracts routing instructions from an incoming packet header to establish and maintain the appropriate switch connection for the duration of the packet, following which the connection is automatically released. In concept, burst switching is similar to connectionless mode transmission, but it differs from the latter in that burst switching implies an intent to establish the switch connection in near real time so that only minimum buffering is required at the node switch.

Burst Traffic Burst traffic is many phone calls (usually incoming telephone calls) that simultaneously arrive at a computer telephony system. Burst traffic tests are usually performed as part of a systems load and stress testing. Burst traffic tests are particularly important for computer telephony systems as usually they must perform a lot of processing to set up to handle a call. The arrival of many calls simultaneously, such as in response to a television sales offer (or when Oprah asks you to call 1-800-xxxxxxx to vote on whether OJ is guilty on innocent), can place significant strain on a computer telephony system.

Burst Transmission 1. A method of transmission that combines a very high data signaling rate with very short transmission times.

2. A method of operating a data network by interrupting, at intervals, the data being transmitted. The method enables communication between data terminal equipment and a data network operating at dissimilar data signaling rates.

Bursty Refers to data transmitted in short, uneven spurts.

Bursty Information Information that flows in short bursts with relatively long, silent intervals between. LAN traffic is characterized as "bursty" in nature, as devices tend to transmit substantial amounts of data at irregular intervals.

Bursty Seconds Bursty seconds is a measure of the amount of time spent at maximum data transfer rate.

Bursty Traffic Communications traffic characterized by short periods of high intensity separated by fairly long intervals of little or no utilization. Data traffic and certain kinds of video traffic are inherently bursty.

Bus 1. An electrical connection which allows two or more wires or lines to be connected together. Typically, all circuit cards receive the same information that is put on the Bus. Only the card the information is "addressed" to will use that data. This is convenient so that a circuit card may be plugged in "anywhere on the Bus." There are two common buses inside a PC — the older ISA bus, capable of only five megabytes per second and the newer PCI bus, capable of transmitting up to 132 megabytes per second. All computers and most telephone systems use buses of some type. Computer buses are typically open. Telephone system buses are typically closed. See also BACKPLANE and BUS NETWORK.

2. An ATM term. BUS. Broadcast and Unknown Server: This server handles data sent by an LE Client to the broadcast MAC address ('FFFFFFFFFFFF'), all multicast traffic, and initial unicast frames which are sent by a LAN Emulation Client. The BUS works in conjunction with a LES (LAN Emulation

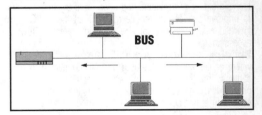

Server), which automatically registers and resolves differences between LAN MAC addresses and ATM addresses. This is accomplished by labeling each device transmission with both addresses.

Bus Card An expansion board that plugs into the computer's expansion bus.

Bus Enumerator A Windows 95 term. A new type of driver required for each specific bus type, responsible for building ("enumerating") the hardware tree on a Plug and Play system.

Bus Extender A device that extends the physical distance of a bus and increases the number of expansion slots.

Bus Hog A device connected to a transmission bus which, after gaining access to the transmission medium, transmits a large number of messages regardless of whether other devices are waiting.

Bus Master A VME board (usually a CPU) that can contend for, seize and control the VME bus for the purpose of accessing bus resources such as voice boards or even other CPUs. See VME.

Bus Mastering A bus design that enables add-in boards to process independently of the CPU and access the computer's memory and peripherals on their own. Bus mastering is a way of transferring data through a bus in which the device takes over the bus and directly controls the transfer of data to the computer's memory. Bus mastering is a method of Direct Memory Access (DMA) transfer.

Bus Mouse Mouse that takes up an expansion slot in a PC, rather that a serial port (those are called "serial mice"). There are generally more expansion slots available than serial ports.

Bus Network All communications devices share a common path. Typically in a bus network, a "conversation" from each device is sampled quickly and interleaved using time division multiplexing. Bus networks are very high-speed — millions of bits per second — forms of transmission and switching. They often form the major switching and transmission backbone of a modern PBX. The printed circuit cards which connect to each trunk and each line are plugged into the PBX's high-speed "backbone" — i.e. the bus network. Similarly, Broadband Switching Systems (BSSs) make use of internal buses, which run at Gbps rates.

In the LAN world, bus networks include Ethernet, which is by far the most common LAN standard. The Ethernet bus specification provides for a common electrical highway which can be shared by as many as 1,024 attached devices per physical Ethernet or Ethernet segment. The Ethernet bus runs at 10 Mbps, although throughput is typically much less. See also BUS.

Bus Slave A VME board(usually a subsystem or I/O board) which can only respond to VME bus accesses mapped to its address. Slaves can usually nterrupt the VME bus on one of 7 levels. See VME.

Bus Speed 1. The speed at which the computer's CPU (central processing unit) communicates with other elements of the computer. For example, the speed at which data moves between the CPU and its various bus-attached devices, such as video controller, disk controllers, voice cards, etc. See AT BUS and PCI.

2. The speed at which the internal bus operates in a network switch.

For example, an ATM Broadband switch may provide multiple, redundant buses, each of which might operate at 20 Gbps, yielding a total rate of 160 Gbps for an 8-bus switch.

3. The speed at which a network bus operates. For instance, an Ethernet bus provides a raw transmission rate of 10 Mbps, although the rate of data throughput typically is much less.

Bus Topology A network topology in which nodes are connected to a single cable with terminators at each end. One form of Ethernet LANs used bus topology, joining the various PCs along one piece of coaxial cable.

Business Audio At one stage business audio was thought to involve voice annotating spreadsheets. Now it means making phone calls through and by your PC.

Business ID An MCI definition. A five-digit numeric code identifying the business to which a customer is assigned. The first two digits indicate division number, the third indicate service type, and the fourth and fifth indicate billing cycle. These are assigned during order entry and passed to MCI A/R with the customer install/ transaction.

Business Service Service used primarily for any purpose other than that of a domestic or family nature. A telephone industry definition.

Business Support System See BSS.

Business Technology Association An association formerly known as Nomda/Landa.

Business Television BTV. Point-to-multipoint videoconferencing. Often refers to the corporate use of video for the transmission of company meetings, training and other one-to-many broadcasts. Typically uses satellite transmission methods and is migrating from analog to digital modulation techniques.

Businessview A Rolm term. Businessview for Call Centers is a centralized system for collecting, integrating, analyzing and reporting call center data.

Busy In use. "Off-hook". There are slow busies and fast busies. Slow busies are when the phone at the other end is busy or off-hook. They happen 60 times a minute. Fast busies (120 times a minute) occur when the network is congested with too many calls. Your distant party may or may not be busy, but you'll never know because you never got that far.

Busy Back A busy signal.

Busy Call Forwarding When you call a busy phone extension, your call is automatically sent to another predetermined telephone extension.

Busy Hour The hour of the day (or the week, or the month, or the year) during which a telephone system carries the most traffic. For many offices, it is 10:30 A.M. to 11:30 A.M. The "busy hour" is perhaps the most important concept in traffic engineering — the science of figuring what telephone switching and transmission capacities one needs. Since the "busy hour" represents the most traffic carried in a hour, the idea is if you create enough capacity to carry that "busy hour" traffic, you will be able to carry all the other traffic during all the other hours. In actuality, one never designs capacity sufficient to carry 100% of the busy hour traffic. That would be too wasteful and too expensive. So, the argument then comes down to, "What percentage of my peak busy or busy hour traffic am I prepared to block?" This percentage might be as low as half of one percent or as high as 10%. Typically, it's between 2% and 5%, depending on what business you're in and the cost to you — in lost sales, etc. — of blocking calls.

PSTN busy hours during the past 20 years or so were from 2:00 to 3:00 P.M., when children arrived home from school and called their working parents to let them know that they were O.K. During the past year or so that busy hour has shifted to 7:00 to 11:00 P.M., reflecting the fact that working parents have put the children to bed and started to "surf the 'Net." See BUSY HOUR USAGE PROFILE.

Busy Hour (Average Busy Season) A telephone company definition. A time-consistent hour, not necessarily a clock hour, having the highest average business day load throughout the busy season. This must be the same hour for the entire busy season.

Busy Hour Call Attempts BHCA. A traffic engineering term. The number of call attempts made during the busiest hour of the day.

Busy Hour Usage Profile The busy hour usage profile identifies how a system will normally be used (i.e., who the users are and what type of transactions they are performing) during the busy hour. Different things that users of CT systems do stress the CT system in different manners. For example, sending a broadcast message in voice mail (where one message is automatically sent to many recipients) may force the system to perform 10 times as much disk I/O as sending message to single recipients. Or, updating an account balance in an IVR transaction forces strain on the IVR to mainframe link. Or, setting up conference calls across a switch or network stresses the use of the network database resources. Understanding and characterizing the type and mix of calls that will take place during the busy hour is key to designing and placing a real-world load on the CT system. Load testing a system should always incorporate tests using the busy hour call profile. See BUSY HOUR.

Busy Lamp A light on a telephone showing a certain line or phone is busy. See BUSY LAMP FIELD.

Busy Lamp Field A device with rows of tiny lights that shows which phones in a telephone system have conversations on them, which phones are ringing, which phones are on hold. Each light corresponds to a telephone extension on the system. The busy lamp field usually sits attached to the attendant's console, telling the attendant if an extension is busy, free, on hold, etc. The benefit of having a busy lamp field is that the operator doesn't have to dial the number to find out what's happening with the extension. This saves the attendant time in handling incoming calls and gives the caller better service. A busy lamp field is often combined with DSS (Direct Station Select.) Next to each light on the busy lamp field there is a button which the operator can push which will dial the corresponding extension (i.e. directly select it) and will typically transfer the call automatically. This button is like an autodial button. This saves the time of dialing the two, three or more numbers of the extension. These days, busy lamp fields are often built into phones on a key system, and everyone, not just the operator, can have one. This gives everyone information on what's happening in the system. It makes transferring calls, etc. easier. See also DIRECT STATION SELECT.

Busy Out To cause a line to return a busy signal to a caller. Busying out lines going into a computer is useful when the computer is not available, i.e. during maintenance periods. This way callers do not get connected to modems with no computers to talk to. This is also known as "taking the phone off the hook." In a voice phone system with trunks that rotary (or hunt) on, sometimes busying one or more broken trunks out helps calls rotary on to trunks that are still working. This way, someone doesn't end up on your third trunk with endless ringing, while your 4th, 5th, 6th etc. trunks are free, leaving them wondering where you are and wondering why you're not getting any calls.

Busy Override A feature of some PBXs which allows the attendant or other high priority user to barge in on a telephone conversation. A warning tone is usually thrown into the conversation to alert the parties of an override. The feature is also called "Barge-In." Sometimes when conversations are over-

ridden, only the person within the organization can hear the barge-in.

Busy Period A telephone company definition. A three to eight month period within which the three busy season months will occur. Example: November through April.

Busy Season An annual recurring and reasonably predictable period of maximum busy hour requirements — normally three months of the year, and typically the three months preceding Christmas.

Busy Season Prior To Exhaust A telephone company definition. The busy season prior to exhaust of an addition is defined as the latest busy season for which an addition will provide objective levels of service.

Busy Signal A signal indicating the line called is busy. The busy signal is generated by the central office. There are two types of busy signals. See BUSY.

Busy Test A method of figuring whether something which can carry traffic is actually doing so or whether it's broken or free and available for use.

Busy Verification Of Station Lines 1. An attendant can confirm that a line is actually in use by establishing a connection (dialing in and listening) to that apparently busy line. 2. In the public switched telephone network, a switching system service feature that permits an attendant to verify the busy or idle state of station lines and to break into the conversation. An alternating tone of 440 Hz is applied to the line for 2 seconds, followed by a 0.5-second burst every 10 seconds to alert both parties that the attendant is connect to the circuit.

Busy/Idle Flag A cellular radio term. A busy/idle flag is an indicator that is transmitted by the Mobile Data Base Station (MDBS) periodically to indicate whether the reverse channel is currently in the busy state or the idle state.

Butcher Colloquial term for a telephone technician who tends to use cutters without verifying whether the cabling is active or inactive.

Butt Set See BUTTINSKY.

Butt Splice Connecting the ends of two wires with a "butt splice connector," from such manufacturers as AT&T and 3M.

Buttinsky Or Butt Set. The one-piece telephone carried on the hips of telephone technicians. It's called buttinsky or butt set because it allows technicians to "butt in" on phone calls, not because the device is worn on their butts. Butt sets used to be essentially telephones without ringers. But now they are much more sophisticated. They will pulse out in rotary or dial out in touchtone and allow you to talk or to monitor a call. They will run computerized tests on the line. Some even have the equivalent of an asynchronous computer terminal built in, which can be used to talk to a distant computer over a phone line. This distant computer could assign them their next jobs, allow them to check and assign features (touchtone, rotary, hunt), report the time spent on this job, etc. In short, the terminal and the computer could replace a raft of clerks and a deluge of paperwork.

The derivation of the term "buttinsky" has long been lost in history. Butt can also mean to attach the end of something to the end of something else. If the clips of a butt set are attached to a pair of wires, it is a "butt" connection. There are those people who also think that the term buttinsky came from some middle European language — Polish or Yiddish — and is slang for someone who butts in a lot, which is what you can do with a Butt Set. I shall keep researching this one. If you can help, let me know, please.

Button Caps Interchangeable plastic squares fit over the buttons of electronic telephones, and are used to label the features programmed onto each programmable button location. Button caps can be either pre-printed or have clear windows which allow features, lines, and Autodial numbers to be labelled on the button.

Buzz 1. To check the continuity of a cable pair by putting an audible buzzer on one end and then checking with a "buttin-sky" to see if you can hear the buzz and thus identify the correct cable pair.

2. A feature of a Rolm CBX which lets the user signal one Rolm desktop product without picking up the handset. Only one buzz per extension is permitted.

Buzzer An electromechanical device that makes a buzzing noise when power is applied, often used to signal someone to answer an intercom call. Battery-powered buzzers were once used to help trace phone circuits. TONE GENERATORS are more common today, but old terminology is still used, as in "buzzing out a line."

Buzzer Leads The wires inside a telephone intended for the connection of a buzzer, usually as part of an intercom system.

BVA AT&T's exclusively-owned Billing Validation Application database. Today, BVA contains all the Regional Bell Operating Companies (RBOCs) calling cards, and other billing information such as billed number screening and payphone numbers. The RBOCs and AT&T access that database today. Prior to Divestiture three market players, the RBOCs, AT&T and most Independent Companies, dominated the "0" Operator Services business which provided alternate billing arrangements such as collect calls, bill to third number and charging calls to calling cards. The three market players still exclusively employ BVA which allows them to validate or authorize alternate billing arrangements. No other long distance carrier or a company needing access to the data for billing validation can use the system. The database is owned by AT&T and is updated daily by the RBOCs and Independent Companies with local exchange information, billing number screening and calling card information. The scenario is further complicated by the 1984 AT&T Plan of Reorganization's exclusive BVA access restrictions. In other words, the three original market players (AT&T, the RBOCs and Independent Companies) have exclusive access to BVA for a predetermined contract length. In most cases these arrangements run into the 1990's. See BVS.

BVS Business Validation Service. US West Service Link was the first in the nation to develop and make available a nationwide Billing Validation Service open to any company that needs to verify the legitimacy of their callers' requests to place charge calls to their local telephone calling cards. US West Service Link developed BVS in 1987 and turned up the system for "on-line" customers in early 1988. Today, the US West BVS system is a national validation source containing calling card data of customers served by the RBOCs, GTE, Southern New England Telephone, United Telecommunications, Cincinnati Bell, Rochester Telephone and Telecomm Canada, a consortium representing all of Canada's local telephone operating companies. In all, more than 60 million records are stored. Sprint, MCI and ITI are among the carriers using BVS. BVS uses X.25 and SS7 protocol. See BVA.

BW Bandwidth: a numerical measurement of throughput of a system or network.

BX.25 AT&T's rules for establishing the sequences of events and the specific types and forms of signals required to transfer data between computers. BX.25 includes the international rules known as X.25 and more.

By-Wire In days gone by, when you turned the steering wheel in a car or an airplane, it moved some wires which moved the wheels or the rudder. Ditto for the brakes. Nowadays, when you do something, the motion you make is detected by some electronic sensors, which is relayed to some computer chips. These chips then signal motors to turn the wheels or make the brakes work. Such a system is said to be more reliable and to allow the injection of computer "intelligence" into the process. For example, the computer might determine that you shouldn't be pushing the brakes because that will make the car skid and it will not apply the brakes.

Bypass 1. A term coming from the idea of using a method to bypass the local exchange network of the Local Exchange Carrier (LEC). Also known as Facilities Bypass, this approach is employed for several reasons:

1. Because the phone company is too expensive, or 2. Because the phone company can't provide the desired levels of bandwidth, quality, or responsiveness.

Bypass means you might be transmitting between two of your offices in the same city, perhaps between your office and your factory, using private microwave or lines you lease from the local power company. You might bypass both the LEC and IXC to link two or more company sites through microwave or satellite transmission systems. You also might be bypassing the LEC to connect directly to your friendly long distance carrier, which then carries your calls to distant cities. In the last case, fiber optic networks deployed by AAVs (Alternative Access Vendors), also known as CAPs (Competitive Access Providers) typically are used, although private microwave may be used as well. Bypass vendors increasingly include CATV providers, some of which have upgraded their coaxial cable networks to support two-way, interactive, switched voice and data communications.

Bypass is a word created by the local telephone industry to sound very threatening. The theory is major users will bypass their local phone company, depriving their phone company of needed revenues. This will drive the local phone company close to imminent bankruptcy, or at the very least, to the state regulators for huge rate increases, hurting the remaining customers, who would, presumably, be customers in areas where bypass is not available (i.e. the poor and the disadvantaged). Additionally, the thought is that the Universal Service Fund would be threatened, thereby depriving users in high cost areas (read "remote and rural") of the right to affordable telephone service. The reality of this threat has not been proven. Nevertheless, the rhetoric frightens sufficient regulators to look at the evils of bypass and to outlaw it, or at least severely restrict it — as several states have done. The Telecommunications Act of 1996 changed the rules on bypass. Now it will happen big-time in the United States. See TELECOMMUNICATIONS ACT OF 1996.

2. An AC power path around one or more functional units of a UPS. An automatic bypass is controlled by UPS control logic and activated when some part of the UPS malfunctions or intentionally shuts down due to overload or other abnormal condition in order to maintain power to the protected load. A manual bypass is a user controlled switch on a UPS that allows a complete electrical bypass of the unit and may be engaged when there is a total UPS failure or when performing certain types of diagnostics or repair. A service bypass is a manual bypass that allows complete maintenance, or even removal, of the UPS without shutting off the load. A true service bypass is commonly a separate device from the UPS. This definition courtesy APC, a maker of UPSes.

Bypass Cabling Bypass cabling or relays are wired connections in a local area ring network that permit traffic between two nodes that are not normally wired next to each other. Such bypass cabling might be used in an emergency or while other parts of the system are being serviced. Usually such bypass relays are arranged so that any node can be removed from the ring and the two nodes on either side of the removed node can then talk.

Bypass Relay A relay used to bypass the normal electrical route in the event of power, signal, or equipment failure.

Bypass Switcher An audio-follow-video switcher usually associated with a master control switcher. Used to bypass the master control switcher output during emergencies, failures, or off line maintenance.

Bypass Trunk Group A trunk group circumvents one or more tandems in its routing ladder.

Byte A set of bits (ones and zeros) of a specific length represent a value in a computer coding scheme. Such a "value" might represent a letter, number, punctuation mark, symbol (e.g., $, @ or &), or control character (e.g., carriage return, line feed, beginning/ending flag, or error check). The term generally is thought as designating a computer value consisting of eight bits. By way of example, ASCII code makes use of an 8-bit byte, comprising 7 information bits and 1 parity bit for error control. EBCDIC makes use of an 8-bit byte, with all bits being information bits. A byte is to a bit what a word is to a character. In some circles a byte is called an octet; this is the case in the world of broadband networking. A logical byte may comprise fewer than 8 bits, or more. A 4-bit byte is often referred to as a "nibble," proving that humor is pervasive, even in the world of computer code. A byte can consist of many more than 8 bits, as is the case with Unicode, which involves 16-bit bytes. Unicode is a standard coding scheme used to accommodate complex alphabets such as Chinese and Japanese. See Bps and Byte Count.

Byte Bonding From Wired Magazine. Byte Bonding occurs when computer users get together and discuss things that noncomputer users don't understand. When byte-bonded people start playing on a computer during a noncomputer-related social situation, they are "geeking out."

Byte Count The number of 8-bit bytes in a message. Since ASCII characters typically have 8-bits, the byte count is also called the character count.

Byte Count Protocol A class of data link protocols in which each frame has a header containing a count of the total number of data characters in the body of the frame.

Byte Multiplexer Channel An IBM mainframe input/output channel that allows for the interleaving, or multiplexing, of data in bytes. Compare with block multiplexer channel.

Byte Multiplexing A byte (or value, or character) from one channel is sent as a unit, with bytes from different channels following in successive time slots. In this manner the bytes are interleaved. See MULTIPLEX.

Byte Stuffing The process whereby dummy bytes are inserted into a transmission stream so that the net data transmission rate will be lower than the actual channel data rate. The dummy bytes are identified by a single controlling bit within the byte.

Byte Synchronous A SONET term describing the manner in which channelized DS-0 level data is mapped into the SONET frame. There are two approaches: Floating Mode and Locked Mode. Floating Mode requires that each frame in the originating T-1 frame be identified in order that the constituent DS-0 bytes can be mapped into the SONET frame effectively through alignment with SONET bytes. The individual T-1 frames, which are converted into VT1.5s as they are mapped into the SONET frame, can float within the SONET frame; this approach is highly effective, but complex and expensive. Locked Mode mapping involves the direct mapping of a T-1 frame into a VT1.5 established at a fixed and inflexible location within the SONET SPE (Synchronous Payload Envelope). This latter approach is much simpler and more efficient than is Floating Mode, but not as effective in maximizing the load balance of the SONET network. In any case, the timing of incoming DS-0 signals must be maintained as they enter the SONET network, are transported through it, and exit it to re-enter the T-carrier network digital hierarchy to be presented to the target device. Loss of timing renders the data useless. Timing is everything. Compare and contrast with Bit Synchronous. See also SONET and VT.

Byte Timing Circuit Optional X.21 circuit used to maintain byte or character synchronization.

Bytecodes A Java compiler creates platform-independent bytecodes that run inside of a Java Virtual Machine (VM). That means that a Java applet can execute on any machine that supports a Java VM.

BZT Bundesamt fur Zulassungen in der Telekommunikation. The name of the German telecom approval authority. It was established in 1982 under the name of the Central Approvals Office for Telecommunications. The name was changed to BZT on March 10, 1992. It is currently based in Saabrucken.

C The programming language AT&T uses for several of its central office switches. It is also used as the programming language of choice for interactive voice response (voice processing) systems. C operates under UNIX, MS-DOS, Windows and other operating systems. It is very powerful and is becoming somewhat of a standard for programming telecom switches.

C Band Portion of the electromagnetic spectrum used heavily for satellite transmission. The uplink frequency is at 6 GHz and the downlink is at 4 GHz. Traditional applications include voice communications, videoconferencing, and broadcast TV and radio. The large dish size and associated high cost of such dishes have contributed to their lack of popularity for TV reception by individuals; Ku band dishes largely have replaced them in support of DBS (Direct Broadcast Satellite) TV reception. Contrast with KU BAND and KA BAND.

C Battery A source of low potential used in the grid circuit of a vacuum tube to cause operation to take place at the desired point on the characteristic curve.

C Conditioning A type of line conditioning which controls attenuation, distortion and delay so that they lie within specified limits. See Conditioning.

C Connector Also called a female amp connector or 25-pair female connector. The male version is called a P connector.

C Drop Clamp Clamp used to fasten aerial wire to buildings.

C Lead The third of three wires which make up trunk lines between central office switches. There are three wires — positive, negative, and the "c lead." The purpose of the "c lead" is to control the grounding, holding and releasing of trunks.

C Links An SS7 term. Cross links used between mated pairs of Signal Transfer Points (STPs). They are primarily used for STP to STP communications or for network management messages. If there is congestion or a failure in the network, this is the link that the STPs use to communicate with each other. See A, B and D Links.

C Message Weighting This definition from James Harry Green, author of the excellent Dow Jones Handbook of Telecommunications. C Message Weighting is a factor in noise measurements to approximate the lesser annoying effect on the human ear of high and low-frequency noise compared to mid-range noise.

C Plane The control plane within the ISDN protocol architecture; these protocols provide the transfer of information for the control of user connections and the allocation/deallocation of network resources; C Plane functions include call establishment, call termination, modifying service characteristics during the call (e.g. alternate speech/unrestricted 64 kbps data), and requesting supplementary services.

C Wire C Wire is what the phone company calls the last piece of its wire that comes into your house or office. It is typically the piece of underground cable that comes in from its pedestal on the street to your network interface box on the side of your house or building.

C&C Computers and Communications. An NEC slogan which focused on the deployment of computer and telephony elements to create an integrated environment. Later on, NEC changed it to "Computing and Communicating" and expanded it into a "Fusion" strategy. See FUSION.

C++ A high-level programming language developed by Bjarne Stroustrup at AT&T's Bell laboratories. Combining all the advantages of the C language with those of object-oriented programming, C++ has been adopted as the standard house programming language by several major software vendors.

C-DTE Character mode Data Terminal Equipment. A term to describe most PCs (personal computers) and printer-terminals that use asynchronous signals for data communications.

C-Link A signaling link used to connect mated pairs of Signal Transfer Points (STPs). An Ericsson term.

C-Message Weighting A type of telephone weighting network that allows for equal attenuation of all frequencies within the voice band in the same manner as it appears to be attenuated by the media.

C.E.R.T. Computer Emergency Response Team. Founded in 1988 by the Pentagon's Advanced Research Projects Agency, after a program written by a graduate student jammed more than 6,000 Internet-connected computers nationwide. C.E.R.T. is based at the Software Engineering Institute of the Carnegie Mellon University in Pittsburgh. C.E.R.T. is Internet's "the fire department." It counsels computer operators on how to keep viruses and intruders out of their computers.

C/A Code The standard Clear/Acquisition GPS (Global Positioning Code) — a sequence of 1023 pseudo-random, binary biphase modulations on the GPS carrier at the chip rate of 1,023 MHz. Also known as the "civilian code." See GPS.

C/R Command Response. A Frame Relay term defining a 1-bit portion of the frame Address Field. Reserved for the use of FRADs, the C/R is applied to the transport of data involving polled protocols such as SNA. Polled protocols require a command/response process for signaling and control during the communications process.

C2 Command and Control. The exercise of authority and direction by a properly designated commander over assigned forces in the accomplishment of the mission. Command and control functions are performed through an arrangement of personnel equipment, communications, facilities, and procedures employed by a commander in planning, directing, coordinating, and controlling forces and operations in the accomplishment of the mission. See C3.

C3 Command, Control and Communications. The capabilities required by military commanders to exercise command and control of their forces. See C2.

C7 European equivalent of the North American System Signaling 7. C7 is not 100% compatible with North American System Signaling 7 and that's where gateway switches come in. These switches convert the signaling between one and the other and do it in real time.

CA 1. Call Appearance

2. Canada, as in a Web address, i.e. www.Corel.ca. Do not type Corel.com, thinking the CA is a mistake. It's not. See URL and Web address.

CAAGR Compound Annual Average Growth Rate.

Cabinet 1. A container that may enclose connection

devices, terminations, apparatus, wiring, and equipment.

2. In telecommunications, an enclosure used for terminating telecommunications cables, wiring and connection devices that has a hinged cover, usually flush mounted in the wall.

Cable May refer to a number of different types of wires or groups of wires capable of carrying voice or data transmissions. The most common interior telephone cable has been two pair. It's typically called quad wiring. It consists of four separate wires each covered with plastic insulation and with all four wires wrapped in an outer plastic covering. Quad wiring is falling into disrepute as it is increasingly obvious that it does not have the capacity to carry data at high speeds. The wire and cable business is immense. The assortment of stuff it produces each year is mind-boggling. In telecommunications, there is one rule: The quality of a circuit is only as good as its weakest link. Often that "weak link" is the quality of the wiring or cabling (we used the words interchangeably) that the user himself puts in. Please put in decent quality wiring. Don't skimp. See CATEGORY OF PERFORMANCE.

Cable Act Of 1984 An Act passed by Congress that deregulated most of the CATV industry including subscriber rates, required programming and fees to municipalities. The FCC was left with virtually no jurisdiction over cable television except among the following areas: (1) registration of each community system prior to the commencement of operations; (2) ensuring subscribers had access to an A-B switch to permit the receipt of off-the-air broadcasts as well as cable programs; (3) carriage of television broadcast programs in full without alteration or deletion; (4) non-duplication of network programs; (5) fines or imprisonment for carrying obscene material; and (6) licensing for receive-only earth stations for satellite-delivered via pay cable. The FCC could impose fines on CATV systems violating the rules. This Act was superseded by the Cable Reregulation Act of 1992.

Cable Assembly A completed cable that typically is terminated with connectors and plugs. It is ready to install.

Cable Bays Lots of cable arranged like bays in a harbor.

Cable Binder In the telephone network, multiple insulated copper pairs are bundled together into a cable called a cable binder.

Cable Budget A local area network term. The overall length of cable allowed between the DTEs located farthest apart within a common collision domain.

Cable Business A magazine on cabling run by Steve Paulov and family in Mesquite (Dallas), TX. A great magazine. 214-270-0860.

Cable Channel The number assigned to a television channel carried by a cable television system. Cable channels 2 through 13 are assigned to the same frequencies as broadcast channels 2 through 13; cable channels above Channel 13 are not. Cable channel assignments are specified in EIA Interim Standard EIA/IS-132, and are incorporated by reference into the FCC's cable television rules.

Cable Cut Service outage caused by cutting or damaging a cable.

Cable Cutoff Wavelength For a cabled single-mode optical fiber, Cable Cutoff Wavelength specifies a complex inter-relation of specified length, bend, and deployment conditions. It is the wavelength at which the fiber's second order mode is attenuated a measurable amount when compared to a multimode reference fiber or to a tightly bent single-mode fiber.

Cable Diameter Expressed in millimeters or inches. Affects space occupied, allowable bend radius, reel size,

length on a reel and reel weight. Also affects selection of pulling grips.

Cable Dog Slang expression. In the West, lifelong cable installer who seeks no upward mobility. In the East, worker who deals with underground cable.

Cable Gland A type of connector. I cannot describe it. Go to this web site. There's a photo of one type of cable gland connector. www.josef-schlemmer.com/1PRODUCT.html

Cable Head The point where a marine cable connects to terrestrial facilities.

Cable Information Technology Convergence Forum This forum is a new organization of the cable television, telephone, computer and switching network industries. The Forum was created to promote greater communication between vendors in the information technology industry, cable television companies and CableLabs, the research and development consortium serving most of the cable operators in North America. The Convergence Forum, based in Louisville, CO, was conceived and sponsored by CableLabs. Companies that have agreed to join the Forum include Apple Computer, Bay Networks, Cisco Systems, Compaq, Digital Equipment Corporation, Fore Systems and LANCity. See also CABLELABS.

Cable Jacket See Sheath.

Cable Loss The amount of radio frequency (RF) signal attenuated (lost) while it travels on a cable. There are many reasons for cable loss, including the cable's shape, its type, its size, its length and what it's made of. For coaxial cable, higher frequencies have greater loss than lower frequencies and follow a logarithmic function. Cable losses are usually calculated for the highest frequency carried on the cable. See ATTENUATION.

Cable Management Companies have oodles of telephone cables in and around their buildings. Cable management is the science and art of managing those cables. Typically, cable management covers keeping track of where the cables are (maps are useful), what type and quality the cable is, and what is attached to either end of it. Cable management for corporations is critical, since stringing, laying and snaking of new cable can be inordinately expensive.

Cable Mapping Cable mapping is the task of trying to track every single pair of wire or circuit from beginning to end. You will need to know where all cables reside, not just the circuits that are in use. Cable mapping is critical for any organization — from company to university — which has a lot of cables floating around. Installing more of it — when there are plenty of spare pairs — is stupid and expensive. Thus, the need for cable mapping.

Cable Modem A cable modem is a small box that connects your PC to the Internet via your local cable TV provider. A cable modem will typically have three connections, one to the coaxial cable wall outlet coming in from your friendly local CATV provider and the other two to connect to your PC and your TV. That PC connection will typically be a standard 10Base-T RJ-45 Ethernet connector, the same connection as you have in your office on your LAN. In essence, a cable modem is a 10 megabit per second Ethernet local area network, in which all the other local cable subscribers are essentially on the same LAN, one end of which is connected to the Internet. Because a cable modem creates a LAN, you have to be careful when using it since all the other local subscribers can "see" your hard disk. As a result, it's important not to allow sharing on your hard disk. Cable modems allow PC users to download information from on-line services at speeds one hundred times faster than today's fastest telephone modems. What actual speed you'll get with your cable modem depends, of course, on a lot of things: how many people are on your "cable LAN" and are presently transmitting or receiving; how fast the connection is from your cable LAN to the Internet; how fast the connections are along the way; how fast the distant server is that you're attached to, etc. etc.

In short, a cable modem is a modem designed for use on a TV coaxial cable circuit. The appeal of a cable modem to a cable TV operator is simple: Put the cable modem on a home cable TV line, provide the subscriber a high-speed circuit to the Internet and the Web, and charge extra for the privilege. The appeal of a cable modem to a user is higher speed. Friends who have such cable modems are in love with them. After using a cable modem to surf the Internet, they all tell me they would never, ever go back to a dial up connection.

Because the cable modem system effectively turns the cable TV system into a very large LAN, users have to share the available bandwidth with any other active users on the same node, which will result in a reduction in data throughput as the number of users increases. This characteristic is the key difference between cable modem systems and the xDSL solutions which telcos are deploying, where the bandwidth to the user is dedicated. See also 10Base-T, DOCSIS, MODEM, NIC and SNMP. www.cable-modems.com and www.ecst.csuchico.edu/~dranch/CABLEMO-DEM/cablemodems.html

Cable Normal Switch A mechanism incorporated into a consumer television receiver which allows the user to select the channel assignment plan. In older receivers, this mechanism is usually a physical switch; in newer receivers, it is usually incorporated as option in the setup menu. All cable/normal switches allow two choices: "standard" (sometimes called "normal" or "off air") which tunes to the channel assignments used by broadcast stations for over-the-air transmission; and "cable" (sometimes called "CATV" or "STD") which tunes to cable channels. Many receivers also include a third option called HRC or Harmonically-Related Carriers.

Cable Plant A term which refers to the physical connection media (optical fiber, copper wiring, connectors, splicers, etc.) in a local area network. It is a term also used less frequently by the telephone to mean all its outside cables — those going from the central office to the subscribers' offices.

Cable Programming Service Any tier of cable television programming except: - The basic tier (see Basic Cable Television Service). - Any programming offered on a per-channel or per-program basis.

Cable Protection There are three basic types of protection in addition to standard plastic cladding:

• ElectroMagnetic (EM) Shielding: Prevents passive coupling. EM shielding can be a metallic conduit or metal wrapping-with appropriate grounding-on the wires.

• Penetration-Resistant Conduit: Used to secure the cable from cutting or tapping. Note, however, not all penetration-resistant conduits provide EM shielding.

• Pressurized Conduit: Detects intrusion by monitoring for pressure loss. Fiber optic cable is extremely difficult to tap and if tapped, the intrusion can be detected through signal attenuation. But since fiber optic cable can be cut, penetration-resistant conduit is recommended to protect the cable.

Cable Racking Framework fastened to bays to support cabling between them.

Cable Reregulation Act Of 1992 Cable Reregulation Bill 1515 passed Congress in October 1992, forcing the FCC to reregulate cable television and cable television rates (after the Cable Act of 1984 effectively de-regulated the cable TV industry). After the Act was passed, the FCC forced the industry to reduce its rates by 10% in 1993 and then again by 7% in 1994.

Cable Riser Cable running vertically in a multi-story building to serve the upper floors.

Cable Run Conduit used to run cables through a building. Also, path taken by a cable or group of cables.

Cable Scanner A device which tests coaxial, twisted-pair, and fiber-optic cable. It measures the length of a cable segment, tests for opens and shorts, and can report on the distance to the problem so the problem can be found and fixed. Many scanners also indicate if a cable segment has RFI or EMI.

Cable Sheath A covering over the conductor assembly that may include one or more metallic members, strength members, or jackets. See Cable Shield.

Cable Shield A metallic component of the cable sheath which prevents outside electrical interference and drains off current induced by lightning.

Cable Stripper 1. Tool used to strip the jackets off ALPETH and lead-jacketed telephone cable. Cable strippers include cable knives and snips.
2. A professional or amateur stripper who appears on X-rated, or community access channels. Quality varies widely. Pay is often non-existent.

Cable Telephony Cable Telephony is transmitting anything other than TV pictures over a cable TV system. That "anything" might be anything from a data connection to the Internet to simple, standard, analog voice phone calls — local, long distance and international. Typically transmitting anything other than TV over the standard coaxial cable CATV providers install at your house requires a cable modem. See Cable Modem.

Cable Television Relay Station CARS. A fixed or mobile station used for the transmission of television and related audio signals, signals of standard and FM broadcast stations, signals of instructional television fixed stations, and cablecasting from the point of reception to a terminal point from which the signals are distributed to the public.

Cable Type The type of cable used. Also called the media. Examples are coaxial. UTP (Unshielded Twisted Pair), STP (Shielded Twisted Pair) and fiber. Factors including cost, connectivity and bandwidth are important in determining cable type. Choosing cable is getting more and more complex. Our tip: Choose and buy well in advance of when you'll need it. The cable you want will not always be in stock.

Cable Vault Room under the main distribution frame in a central office building. Cables from the subscribers lines come into the building through the cable vault. From here they snake their way up to the main distribution frame. The cable vault looks like a bad B-movie portrayal of Hell, replete with thousands of dangerous black snakes. Cable vaults are prime targets for the spontaneous starting of fires. They should be protected with Halon gas, but usually aren't because some parts of the phone industry think Halon is too expensive.

Cable Weight Expressed in lbs. per 1000 (without reel weight included). Affects sag, span and size of messenger in aerial applications.

Cablecast Cablecasting. Non-broadcast radio or television programming transmitted by a cable television system to its subscribers. Cablecast programming may be originated by the cable operator itself ("origination cablecasting" or "local origination") or by an access organization.

Cablegram Service An MCI definition. An MCI

International service which provides cablegram communication to International destinations through the use of a computerized message switching center in New York City.

Cablehead The point where a marine cable connects to terrestrial facilities.

CableLabs Cable Television Laboratories, Inc. A research and development consortium of cable television system operators established in 1988. CableLabs plans and funds research and development projects to help member companies and the cable industry take advantage of opportunities and meet challenges in the telecommunications industry. A good deal of emphasis is placed on digital cable and cable modem technologies. www.cablelabs.com

Cableport Intel Corporation's new technology, which brings high speed multimedia-rich interactive services to personal computers in the home via cable.

Cablespan A Tellabs Operations, Inc. product which deliver two-way voice and data services over coaxial cable used by cable TV operators. Tellabs is in Lisle, Illinois.

Cableway An opening in a work surface that allows access to cords or cables from below, or mounting of an electrical receptacle or telephone jack. Cableways typically come with removable plastic grommets.

Cabling The combination of all cables, wire, cords, and connecting hardware installed. A term used to refer collectively to the installed wiring in a given space.

CABS Carrier Access Billing Specifications

CABS BOS Carrier Access Billing Specifications - Billing Output Specifications

CAC 1. Carrier Access Code. The digits you must dial in North America to reach the long distance carrier of your choice. Those digits fit the following format 101XXXX.
2. Customer Administration Center. A type of terminal used by a PBX user to maintain and troubleshoot his PBX.
3. Connection Admission Control is defined as the set of actions taken by the network during the call setup phase (or during call re-negotiation phase) in order to determine whether a connection request can be accepted or should be rejected (or whether a request for re-allocation can be accomplished).

CACH Call Appearance Call Handling.

Cache From the French "cacher," which translates "to press or hide," especially in terms of tools or provisions. In the context of computer systems and networks, information is cached by placing it closer to the user or user application in order to make it more readily and speedily accessible, and transparently so. At the same time, information which is cached places less stain on limited computer I/O (Input/Output) resources and limited network resources. Let's consider two specific definitions, the first of which relates to computer systems and the second of which relates to computer networks. Let's also consider a combination of the first two, in the context of the Internet.

1. In the context of a computer system, cache memory generally is a partition of SRAM (Static Random Access Memory). Since much of computing is highly repetitive or predictable in nature, and since solid state components (silicon chips) are much faster than mechanical disk drives, the speed of information access can be enhanced if certain information can be stored in RAM. That information typically is in the form of program information, memory addresses, or data. Thereby, the information can be stored in anticipation of your need for it, and can be presented to you faster than if the computer needed to access the hard drive through the execution of an I/O function. The cache memory sits (logically and, perhaps physically) between the CPU and the main memory (RAM).

Caching works because of a phenomenon known as the locality principle which states that a von Neumann CPU (i.e., one that performs instructions and makes database calls sequentially, one after another) tends to access the same memory locations over and over again. A cache works like this. When the CPU needs data from memory, the system checks to see if the information is already in the cache. If it is, it grabs that information; this is called a cache hit. If it isn't, it's called a cache miss and the computer has to fetch the information by accessing the main memory or hard disk, which is slower. Data retrieved during a cache miss is often written into cache in anticipation of further need for it. Let's assume that you open a CD-ROM application with hyperlinks. As the system can reasonably assume that you will exercise the hyperlink options, the information associated with them can be stored in cache memory. If you do, indeed, exercise those options, it's a cache hit and the data is there waiting for you. The cache also will hold information that you recently accessed, in anticipation of your wanting to back up, or access it again. Caching can take place through partitioned or segmented cache memory, which can be in the form of L1 (Level 1) primary cache and L2 (Level 2) secondary cache. L1 cache memory is accessed first, L2 second, the main memory (RAM) and then hard drive last. Also, one cache might hold program instructions and the other might hold data. Generally when the cache is exhausted, it is flushed and the data is written back to main memory, to be replaced with next cache according to a replacement algorithm. Cache freshing and flushing mechanism is designed differently by different vendors. It behaves slightly different. However it mainly depends on main memory type, like write back or WB, write through WT, write protected or WT, write combining or WC and uncached or UC. See also Cache Memory.

2. In the context of a computer network such as a LAN, or the combination of the Internet and World Wide Web, data can be cached in a server which is close to you. In anticipation of your imminent request for that data in a logical sequence of data access, it will be transmitted from the main server to the remote server. Thereby, the data is accessible to you more quickly than if it had to be transmitted across the entire network each time you had a need for it. Should you access a certain set of data frequently, it might be permanently stored on a server in proximity, and refreshed by the main server from time to time in order to ensure its currency (i.e. that it remains up to date).

3. In the context of an Internet client/server application, caching really shows its stuff. First, the network uses distributed cache servers to house the WWW information that users in your region use frequently. As you access a Web site, your speed of access and response is improved because the data is housed on a server closer to you. The data then is loaded into cache memory on your client computer workstation. As you move forward, from page to page and link to link, your client caches the information provided by the cache server, with all of this happening in anticipation of your next move. As you move backward, the same thing happens, in anticipation of that next move, as well. Just in case you don't believe the client side of this story, go to Internet Explorer or Netscape, and click on cache. (The fastest way to regain space on your hard disk is to flush the cache which these programs dump to your hard disk.)

Cache Coherency Managing a cache so that data is not lost or overwritten. See also Cache.

Cache Controller A chip, such as the Intel 82385, that

manages the retrieval, storage and delivery of data to and from memory or the hard disk. Cache controllers may reside in either clients or servers. See also Cache.

Cache Engine A cache engine is a carrier-class, high-speed dedicated Internet appliance that performs Web content caching and retrieval. When a user accesses a Web page, the cache engine locally stores the page's graphics and HTML text. When another user later requests the same Web page, the content is pulled from the cache engine. This process improves download time for the user and reduces bandwidth use on the network. Here is a an explanation of a cache engine from Cisco, which makes one. How does the cache engine work? The cache engine communicates with a Cisco router, which redirects Web requests to the cache engine using the Web Cache Control Protocol (WCCP), a new standard feature of Cisco IOS software. The WCCP also enables load balancing of traffic across multiple cache engines and ensures fault-tolerant, fail-safe operation. What are the benefits of Web caching? By reducing the amount of traffic on WAN links and on overburdened Web servers, caching provides significant benefits to ISPs, enterprise networks and end users. Those benefits include cost savings due to a reduction on WAN usage and dramatic improvements in response times for end users. The cache engine also provides network administrators with a simple method to enforce a site-wide access policy through URL filtering. See also Cache.

Cache Hit When the data you want is actually in cache. Thus you don't have to access your hard disk and your computing is faster. See Cache, Cache Miss and Cache Memory.

Cache Memory Available RAM (Random Access Memory) or SRAM (Static RAM) that you set up to allow your computer to "remember" stuff — so the next time your computer wants that information, it can find it fast from RAM, instead of searching through a slower hard disk I/O (Input/Output) process. This high speed cache memory eliminates the CPU wait state. When the CPU reads data from main memory, a copy of this data is stored in the cache memory. The next time the CPU reads the same address, the data is transferred from the cache memory instead of from main memory. Novell's NetWare, for example, uses cache memory to improve file server access time. In NetWare, cache memory contains the directory and file caches, along with the FAT (File Allocation Table), the turbo FAT, the Hash table, and an open space for other functions. See also Cache.

Cache Miss When the caching software guesses wrongly and you have to read your data off your hard disk rather than reading it from the cache in memory. See also Cache, Cache Hit and Cache Memory.

Caching A process by which information is stored in memory or server in anticipation of next request for information. See Cache for a full explanation.

CAD 1. Computer Aided Dispatch.
2. Computer Aided Design. A computer and its related software and terminals used to design things. A CAD system might be as simple as computerized drafting tools or as complex as detailed layouts of integrated circuits. CAD systems often have terminals on peoples' desks and a central maxi-computer in the company's main computer room. CAD terminals are often run over LANs (local area networks) or through telephone systems. The terminals are often moved, thus having universal wiring and a universal switching system — a LAN or a phone system — is extremely useful.

CAD/CAM Computer Aided Design/Computer Aided Manufacturing. See CAD.

CADB Calling Area Data Base. An MCI definition. An MCI System that stores reference data for various MCI Systems and reconciles MCI Calling Areas with those of Bell.

Caddy The shell of an optical disc. Protects it from grubby fingerprints, and includes write protection devices. AKA case.

Cadence In voice processing, cadence is used to refer to the pattern of tones and silence intervals generated by a given audio signal. Examples are busy and ringing tones. A typical cadence pattern is the US ringing tone, which is one second of tone followed by three seconds of silence. Some other countries, such as the UK, use a double ring, which is two short tones within about a second, followed by a little over two seconds of silence.

CADS Code Abuse Detection System.

CAE Computer Aided Engineering.

Cage Antenna An antenna having conductors arranged cylindrically.

CAGR Compound Annual Growth Rate.

CAI 1. Computer Assisted Instruction. Commonly known as CBT (Computer Based Training). See CBT. See also CAD for a discussion on telecom needs.
2. Common Air Interface. A standard for the interface between a radio network and equipment. A CAI allows multiple vendors to develop equipment, such as radio terminal devices (e.g., cordless phones, cellular phones and PCS terminals) and base stations (e.g., cellular antenna sites), which will interoperate. The yield is a competitive (read less expensive) market for equipment. The British CT2/Telepoint system incorporated one of the first CAI standards. See also CT2.

CALC Customer Access Line Charge. Also known variously as Access Charge, EUCL (End User Line Charge), and SLC (Subscriber Line Charge). See Access Charge.

CALEA Communications Assistance for Law Enforcement Act 1994.

Calendar Routing A call center term for directing calls according to the day of the week and time of day. See also SOURCE/DESTINATION ROUTING, SKILLS BASED ROUTING and END-OF-SHIFT ROUTING.

Calibrate To test and reset a measuring or timing device against a standard to make sure it is functioning correctly.

Call Everyone has a different definition for "call." My definition is simplest: Two people or two machines are on a phone line speaking to each other. That's a call. Bellcore's definition of a call: An arrangement providing for a relation between two or more simultaneously present users for the purpose of exchanging information. The ATM Forum's definition: A call is an association between two or more users or between a user and a network entity that is established by the use of network capabilities. This association may have zero or more connections. Here are some more formal definitions:
1. In communications, any demand to set up a connection.
2. A unit of traffic measurement.
3. The actions performed by a call originator.
4. The operations required to establish, maintain, and release a connection.
5. To use a connection between two stations.
6. The action of bringing a computer program, a routine, or a subroutine into effect, usually by specifying the entry conditions and the entry point.

Call Abandons Also called ABANDONED CALLS. Call Abandons are calls that are dropped by the calling party before their intended transaction is completed. The call may be dropped at various points in the process. The point in the call at which the call is abandoned will have varying impacts on a computer telephony system. Many callers upon hearing an automated system will hang up. For systems that expended significant energy in setting up to answer a call, a large percentage of call abandons can negatively impact the call capacity of the system.

Call Accepted Signal A control signal transmitted by the called equipment to indicate that it accepts an incoming call.

Call Accounting System A computer, a magnetic storage device (floppy or hard disk), software and some mechanical method of attaching itself to a telephone system. A call accounting system is used to record information about telephone calls, organize that information and upon being asked, prepare reports — printed or to disk. The information which it records (or "captures") about telephone calls typically includes from which extension the call is coming, which number it is calling (local or long distance), which circuit is used for the call (WATS, MCI, etc.), when the call started, how long it lasted, for what purpose the call was made (which client? which project?). A call accounting system might also include information on incoming calls — which trunk was used, where the call came from (if ANI or interactive voice response was used), which extension took the call, if it was transferred and to where and how long it took.

There are eleven basic uses for call accounting systems:

1. Controlling Telephone Abuse. It's the 90-10 rule. 10% of your people sit on long distance calls all day to their friends and family. The others work. Some people still think WATS calls are free. Knowing who's calling where and how much they're spending is useful. Often they appreciate being told they're spending money. Big money...and they stop.

2. Controlling Telephone Misuse. I figured once you could call between two major cities for five cents a minute and $1 a minute. That's a 20-fold difference! Often you need different lines. Often a company has different lines. Sometimes the phone system makes the dialing decision. Sometimes the person makes the dialing decision. Whoever's doing it can be wrong. A call accounting system is a good check to see if you're spending money needlessly.

3. Allocating telephone calling costs among departments and divisions. Telephones — voice, data, video and imaging — are some of your biggest expenses. They're a cost that should be allocated to the products you're making, or the departments or divisions in your company. Telephone costs can determine which product is profitable. Which isn't. Item: A software company recently dropped one of its three "big" software packages because phone calls for support got too expensive.

4. Billing Clients and Projects back for telephone charges incurred on their behalf. Every lawyer, government contractor, etc. does it. Makes sense.

5. Sharing and Resale of long distance and local phone calls, as in a hotel/motel, hospital, shared condominium, etc. Someone's got to send out the bills. And it's not the phone company. In fact, with a call accounting system you can be your own phone company!

6. Motivation of Salespeople. The more phone calls they make the more they sell. This rule is as obvious as the nose on your face. You WANT salespeople to make more calls? Hang a list of all their calls on the wall. Give prizes to those make the most! Or those who make more than last week. Or those who set a new record.

7. Personnel Evaluation. Which employees are doing better at being productive on the phone (however you define "produc-

tive"). You want them to get on and off the phone fast? Or you want them to stay on and coddle your customers? You can now correlate phone calls with income — from service or just straight sales.

8. Network Optimization. Two fancy words for figuring which is the best combination of MCI, AT&T, MCI, Sprint, Wiltel, etc. lines. And which is the best combination of all the various services each offer. A rule of thumb: There's a 20-fold difference in per minute telephone calling costs between any two major cities in the US. And — amazing — you won't hear any difference in quality, despite the huge difference in price. I think it's the biggest price difference in any product anywhere. It's amazing.

9. Phone System Diagnostics. Is the phone system working as well as it should? Are all the lines working? Are all the circuit packs (circuit cards) working? Call accounting systems can tell you which lines you're getting no traffic on. Or which line carried the 48 hour call to Germany (it's happened). Either way, you can figure quickly which lines are working and which aren't.

10. Long Distance Bill Verification. Was the bill accurate we received from our chosen long distance phone company? Mostly it isn't. In fact, there's no such thing as an accurate phone bill. That's an oxymoron. Using your call accounting systems to check your long distance gives you some peace of mind. It's cheap peace of mind. Everyone should have one.

11. Tracing Calls. True story: Every third or fourth Friday afternoon a large factory in the south received bomb threats. They'd clear the factory, search the factory and not find anything. By the time they'd checked, it was too late to start up production. One day they checked their call accounting records. The calls were coming from a phone on the factory floor. The whole thing was a ruse to get an afternoon off...And now that many phones give you the number of who's calling, call accounting systems are turning out to be great for checking the effectiveness of regional ad campaigns, figuring the profitability of direct mailings and even figuring the profitability of individual customers.

Call Announcement A telephone operator or person acting as a telephone operator can announce a call to the called party before putting the call through. All modern phone systems have this feature.

Call Answering The name for a central office based answering service, provided by your local phone company. The major advantage of this service is that if you're speaking on your line and another call comes through, it won't receive a busy, it will hear your melodic voice asking if you'd like to leave a message and it will take your message. Once the machine takes the message, it will put some sound on your phone line, which you'll hear next time you pick up. That sound will alert you to the fact that you have a message in your mailbox.

Call Attempt A try at making a telephone call to someone. Tally up call attempts and compare them to completions and you'll have some idea of corporate frustration and thus, the need for more lines or more phone equipment. The measures in this in call attempts,calls answered, calls overflowed, and calls abandoned.

Call Back A security procedure in which a user, dials into access a system and requests service. The system then disconnects and calls the user back at a preauthorized number to establish the access connection. Same as dial-back. See also INTERNATIONAL CALL BACK.

Call Barring The ability to prevent all or certain calls from reaching to or from a phone.

Call Before Dig A preventive maintenance measure in which signs are posted near buried cables advising people to phone before digging in the area.

Call Blending A phone system has a bunch of people answering and making calls. The calls are coming (say in response to an ad). The calls are going out, courtesy of a dialing machine (perhaps a predictive dialer). The idea is to keep the calls at a constant level. The idea is to blend incoming with outgoing calls. Some predictive dialers have call blending. Others don't. They need a dedicated workforce. Call blending automatically transfers staff members between outbound and inbound programs as call volumes change. Some predictive dialers let you choose which workstations will be used for call blending, to avoid training of every staff member.

Call Block 1. A name for an enhanced custom calling service, one of several known as CLASS services. Call Block helps you avoid unwanted calls by rejecting calls from a list of numbers you specify. Depending on the specifics of the LEC offering, the caller may get a message indicating that you are not accepting calls at the present time. Call Block does not work either for numbers outside your local calling area or for calls connected through an operator. See also CLASS.
2.   A feature that allows the calling party to prevent the calling number from being transmitted and displayed on the Caller ID equipment of the called party. Call Block can be provided, as a matter of course, on all your outgoing calls, although you can override Call Block on a call-by-call basis in order to transmit your number. You might choose to do this in order to receive better service from an incoming call center. You can request Call Block from your LEC. You also can invoke Call Block on a call-by-call basis, typically by pressing *67 before placing an outgoing call—this feature also is known as "Cancel Calling Number Delivery."

Call Blocking 1. Check into a hotel. Dial a 0+ call. You're connected to an Alternate Operator Service company. But you know their rates may be high. You ask to be connected to AT&T or MCI, or whoever is the carrier of your choice. Sadly, the AOS cannot connect and neither can (nor will) your hotel's operator. This is called "Call Blocking." The FCC has barred the practice. But it continues. See also CALL SPLASHING.
2. An AIN (Advanced Intelligent Network) service allowing the user to block calls to specific numbers or country codes. Also known as Call Control Service. Content Blocking, a variation on the theme, allows the blocking of calls to either all or specified 900 numbers. While contemporary PBXs and Electronic Key Telephone Systems have such capabilities, not all users enjoy the benefits of working behind such a system. Additionally, PBX systems typically provide the ability to block only all 900 numbers. although some such numbers might have legitimate application.

Call Card A British term. A paper record, used in manual telebusiness systems, to record the results of a call.

Call Center A place where calls are answered and calls are made. A call center will typically have lots of people (also called agents), an automatic call distributor, a computer for order-entry and lookup on customers' orders. A Call Center could also have a predictive dialer for making lots of calls quickly. The term "call center" is broadening. It now includes help desks and service lines. For more information on Call Centers, please read CALL CENTER Magazine. For subscriptions call 215-355-2886. See CALL CENTRE.

Call Centre A British term. An area in an organization where business is conducted by phone in a methodical and organized manner. Call centres are typically based on the integration of a computerized database and an automatic call distribution system. (Note that this definition contains British spellings.) See CALL CENTER.

Call Clearing The process by which a call connection is released from use.

Call Clear Packet An information packet that ends an X.25 communications session, performing the equivalent of hanging up the phone.

Call Collision 1. Contention that occurs when a terminal and a DCE simultaneously transfer a call request and an incoming call specifying the same logical channel. The DCE will proceed with the call request and cancel the incoming call. 2.That arising when a trunk or channel is seized at both ends simultaneously, thereby blocking a call.

Call Completion Rate The ratio of successfully completed calls to the total number of attempted calls. This ratio is typically expressed as either a percentage or a decimal fraction.

Call Completion Service An AIN (Advanced Intelligent Network) service which provides the Directory Assistance operator with the ability to automatically to extend the call to the listed party. In a typical landline application, there is an additional charge to the calling party for such a service. (Frankly, I'd rather dial the number myself and save the 50 cents.) Cellular telephone providers often provide the service free of charge as a value-added service, in order to minimize "driving and dialing" traffic accidents. (I take advantage of this service—It just makes sense.)

Call Control Call control is the term used by the telephone industry to describe the setting up, monitoring, and tearing down of telephone calls. There are two ways of doing call control. A person or a computer can do it via the desktop telephone or a computer attached to that telephone, or the computer attached to the desktop phone line (i.e. without the actual phone being there). That's called First Party Call Control. Third-party call control controls the call through a connection directly to the switch (PBX). Generally third-party call control also refers to the control of other functions that relate to the switch at large, such as ACD queuing, etc.

Call Control Procedure Group of interactive signals required to establish, maintain and release a communication.

Call Control Signal Any one of the entire set of interactive signals necessary to establish, maintain, and release a call.

Call Data Call data refers to any data about a phone call that is passed by a switch to an attached computer telephony system. Call data is usually used by the computer telephony application to process the call more intelligently. Call data may be passed In-Band, over the same physical or logical link as the call — usually via tones, or Out-Of-Band, over a separate link — usually a serial link. Call data may also be passed as part of the data designed to control telephone networks, such as SS7 (Signaling System 7) links. In addition to information about the call, status about the call and even control over the call, can be available as part of the call data link services. Call data almost always includes what number dialed the call (ANI) and/or what number called (DNIS). More complex call data links used for "PBX integration" may also indicate why the call was presented (such as forwarded on busy), tell what trunk the call is coming in on, or to pass message waiting on or off indications, and other functions. Full blown computer telephony links, such as are now being offered by many switching vendors, enhance the call data path, providing additional status information about calls and can even provide a level of call control to the attached computer telephony system. The above definition courtesy of

Steve Gladstone, author of a great book called Testing Computer Telephony Systems, available from 212-691-8215.

Call Delay The delay encountered when a call reaches busy switching equipment. In normal POTS telephone service, the delay is considered OK if no more than one and a half percent of the calls are delayed by three seconds during the busy hour.

Call Detail Recording CDR. A feature of a telephone system which allows the system to collect and record information on outgoing and incoming phone calls — who made/received them, where they went/where they came from, what time of day they happened, how long they took, etc. Sometimes the data is collected by the phone system; sometimes it is pumped out of the phone system as the calls are made. Which ever way, the information must be recorded elsewhere — dumped right into a printer or into a PC with call accounting software. See also CALL ACCOUNTING SYSTEM.

Call Diverter 1. A device which when connected to a called telephone number intercepts calls to that number and connects them to a telephone operator or prerecorded message.
2. An ancillary device which is connected to a telephone line. The device will, when the called telephone rings, initiate a telephone call on another line to a different telephone number. The calling party may or may not be aware that his call has been diverted to another telephone.

Call Duration The time from when the call is actually begun (i.e. answered) to the instant either party hangs up. Call Duration is an important concept for traffic engineering.

Call Establishment The process by which a call connection is created.

Call Forward Busy When your phone is busy, an incoming call is transferred to another number. That number might be one appearing on your phone system. It might be one at your home in the same city. It could even be in another city. Call Forward Busy can perform the same as Rollover Lines. I use Call Forward Busy to move calls from the first line of my residence to my second line, because my local phone company charges too much for Rollover Lines. (Don't ask why. They don't know either.) You can get Call Forward Busy from central offices, as well as PBXs and some key systems. See also Rollover Lines.

Call Forwarding A service available in many central offices, and a feature of many PBXs and some hybrid PBX/key systems, which allows an incoming call to be sent elsewhere. There are many variations on call forwarding: Call forwarding busy. Call forwarding don't answer. Call forwarding all calls, etc.
Call forward is a useful feature. For example, you're going to a meeting but you're expecting an important call. Pick up your phone, punch in some digits and all your calls will go to the new number — perhaps the phone outside the meeting room. The big disadvantage is that many people return to their offices but forget they forwarded their calls elsewhere. As a result, they usually miss a whole bunch of important calls. Some electronic phones now have a reminder light or message on them saying "all calls are being forwarded." Some people program their PBXs to cancel all call forwards at noon and at midnight every day. This makes sense.
Call forwarding is used to send calls to voice mail systems. For example, tell your PBX that if your phone isn't answered in four rings, send that call to your voice mail.
If you are getting call forwarding service from a central office in North America, the code to begin call forwarding is 72# and the number you want to be forwarded to. To cancel it, you punch in 73#.

Call Frame Harris' PBX to computer link. Harris' protocol for linking its PBX to an external computer and having that computer control the movement of calls within a Harris PBX. See also OPEN APPLICATION INTERFACE.

Call Guide A paper or screen "cheat sheet" that provides bullet points or actual copy for call center agents to use while they are on the telephone making marketing calls. They provide responses to commonly asked questions or objections in the most effective way. Call guides are excellent training tools as well as monitoring aids for coaches. See Call Guide Routing.

Call Guide Routing The process by which a call center agent navigates through a call guide. The routing may be driven from a computer-based menu or function key or automatically by the computer system based on responses entered into a field.

Call Handoff A cellular phone term. Call handoff happens when a wireless call is transferred to another cell site in mid-conversation.

Call Hold If you hang the phone up, you lose the caller. Call hold — a feature of most phone systems — allows you to "hold" the call, so the other person can't hear you. You can then return to the conversation by pushing a button on your phone, typically the button flashing which shows which line the person is sitting on hold. Call hold is useful when you have someone on another line calling you.

Call Identifier A network utility that is an identifying name assigned by the originating network for each established or partially established virtual call and, when used in conjunction with the calling DTE address, uniquely identifies the virtual call over a period of time.

Call In Absence Horn Alert A cellular car phone feature that sounds your car's horn when you are receiving a call.

Call In Absence Indicator A cellular car phone feature that ensures that power to the cellular phone is not lost if the car's ignition is turned off.

Call In Progress Override A cellular car phone feature that keeps power to the phone during a call even though you've turned off the car's ignition.

Call Letters Certain combinations of letters assigned to radio stations by the FCC. The group of letters assigned the U.S. by the International Radiotelegraph Convention are all three and four letter combinations beginning with N and/ or W and all combinations of KDA to KZZ inclusive.

Call Manager A Northern Telecom product which allows a PC to process calls on an ISDN BRI access line. See Call Processing.

Call Me Card A special AT&T Card number which permits others to call you, and only you, and have the call charged to your telephone. Give an AT&T CALL ME Card to your very best customers to encourage them to call you. A Call Me Card is like having a private unlisted 800 toll-free number.

Call Me Message A Rolm/IBM feature on their CBX that allows internal users to leave other users a message showing the time, date and the caller's extension number.

Call Mix Call mix is the pattern of call types (each call type defines what the caller will do for that call) that goes into creating a busy hour call profile or other call profile. A voice mail system's busy hour call profile call mix may consist of 10% call abandons, 20% login and send one message, 30% login and listen to one message, and so on. Varying the call mix can often be useful to stress particular parts of a system. For example, a call mix of 100% call abandons is frequently used to stress a computer telephony system's ability to handle high traffic call setup scenarios. This definition courtesy of Steve Gladstone, author of the book Testing Computer Telephony Systems, available from 212-691-8215.

Call Model An abstraction of the call processing functionality of the architecture and the relationship that exists between the functionality of the Service Switching FE in an ASC and the Service Logic and Control FE in a SLEE (Service Logic Execution Environment). The call model consists of two components: Connection View and Basic Call State Model. Definition from Bellcore in reference to Advanced Intelligent Network.

Call Not Accepted Signal A call control signal sent by the called terminal to indicate that it does not accept the incoming call.

Call Notification Service See CALL PICKUP SERVICE.

Call Packet A block of data carrying addressing and other information that is needed to establish an X.25 switched virtual circuit (SVC).

Call Park The phone call is not for you. Or maybe it is, but you don't want to answer it on your phone. Put it into CALL PARK, then you or anyone else can answer it from any other phone. Call Park is similar to placing a call on hold, but your retrieve the call by dialing a code, rather than by pressing a line button. The attendant may have a call for you, but you're not there. So he places the call in Call Park, pages you and tells you the call is in Call Park. You pick up the nearest phone, dial one or two digits (the code for grabbing the call out of Call Park) and you have the call. It's faster than looking for you, then telling you to hang up while she transfers the call.

Call Pickup A phone is ringing but it's not yours. With call pickup, you can punch in a button or two on your phone and answer that person's ringing phone. Saves time. See CALL PICKUP SERVICE.

Call Pickup Group CPUG. All the phones in an area that can be answered by each other by simply punching in a couple of digits. See CALL PICKUP.

Call Pickup Service An AIN (Advanced Intelligent Network) service similar to those offered by many premise-based voice processors but residing on a network-based platform. In the event of an unanswered call, the voice processor records the call and notifies the called party of the message via pager, fax or some other technique.

Call Processing The system and process that sets up the intended connection in a switching system. The system scans the trunk and/or station ports for any "requests" for service. Upon detecting a request, the system checks the stored instructions and look-up tables and sets the connection up accordingly.

Call Progress The status of the telephone line; ringing, busy ring/no answer, voice mail answering, telephone company intercept, etc. See CALL PROGRESS ANALYSIS and CALL PROGRESS TONE.

Call Progress Analysis As the call progresses several things happen. Someone dial or touchtones digits. The phone rings. There might be a busy or operator intercept. An answering machine may answer. A fax machine may answer. Call progress analysis is figuring out which is occurring as the call progresses. This analysis is critical if you're trying to build an automated system, like an interactive voice response system.

Call Progress Signaling All telephone switches use the same three general types of signals: + Event Signaling initiates an event, such as ringing. + Call Progress Signaling denotes the progress (or state) of a call, such as a busy tone, a ringback tone, or an error tone. + Data Packet Signaling communicates certain information about a call, for example, the identify of the calling extension, or the identity of the extension being called.

Call Progress Tone A tone sent from the telephone switch to tell the caller of the progress of the call. Examples of the common ones are dial tone, busy tone, ringback tone, error tone, re-order, etc. Some phone systems provide additional tones, such as confirmation, splash tone, or a reminder tone to indicate that a feature is in use, such as confirmation, hold reminder, hold, intercept tones.

Call Queuing Incoming or outgoing calls may be queued pending an answer. The idea of call queuing is to save money. See also CALLBACK QUEUING.

Call Rate CR. The number of calls within a span of time, such as within an hour, or within a day, etc. It may be confined to a narrow usage, such as the busy hour (BH) originating call rate per main station, or to a broader usage. Hence, its usage should include enough modifying words to assure that it will be properly understood. A telephone company definition.

Call Record The data record of a call transaction. The record is made up of event details that typically include date, time, trunk(s) used, station(s) used and duration. In an ACD these events may also include time in queue, call route used, system disposition flag, inbound or outdialed digits and wrap-up data entered.

Call Reference Information element that identifies to which call a Layer 3 message pertains.

Call Reference Value CRV. A number carried in all Q.931 (I.451) messages, providing a local identifier for a given ISDN call.

Call Release Time The time it takes from sending equipment a signal to close down the call to the time a "free condition" appears and the system is ready for another call.

Call Reorigination Caller reorigination allows a caller with a telephone debit card account number or a telephone credit card account to make unlimited calls without hanging up and redialing their access and their account numbers. At the end the first call, the caller remains on the line. The caller then presses the pound key (the # key) for a prescribed number of seconds and receives a confirmation tone (which sounds like a high-pitched dial tone). After receiving the tone, the caller immediately dials their next phone number. And so on. Sometimes you can hit the # button after the person you were talking to has hung up. My experience has been that it's better to hit the # before the person has hung up. Just tell them what you're doing. All this allows the card account holder to make a series of calls without ever hanging up and redialing the often lengthy access card account numbers. This saves a lot of time for callers with a long list of calls to make. It also saves money for callers from hotels, which charge for each connection to the long distance provider, but don't charge based on the length of the call (especially if it's a local call or toll-free 800 call). The prescribed number of seconds that the user needs to hold down the pound key is configurable. Depending on the application the time may range from 1 to 5 seconds. With most carriers, it's two to three seconds. The above definition was kindly provided by Karen Shelton, Systems Engineer, IEX Corporation, Richardson, TX.

Call Request Packet In packet data switching, a call request packet carries information, such as sender and recipient identification, that is needed to establish an X.25 circuit. In more technical terms, a call request packet is sent by the originating data terminal equipment (DTE) showing requested network terminal number (NTN), network facilities and either X.29 control information or call user data.

Call Restrictor Equipment inserted in a telephone line or trunk which restricts outgoing calls in some way. Usually from making a toll call.

Call Return An enhanced custom calling feature included in what are known as CLASS services and which are offered by local exchange carriers, courtesy of SS7. Call Return service which allows you to automatically dial the number of the last caller, even if you did not answer the telephone. It's a great idea, assuming that you are not blindly making long distance calls to long distance salespeople. Enhanced Call Return allows the called party to access a network-based voice system which announces the date, time and telephone number of the last incoming call. Should you chose to do so, you can launch Call Return by pushing a button on the telephone keypad.

Call Routing Tree A graphical display of complex call routing decision logic.

Call Screening There are several definitions. Here are two.
1. A PBX feature that looks at the digits dialed by the caller to figure whether the call should be completed.
2. A receptionist or secretary answers the executive's phone and checks that the person calling is important enough be put through to the almighty executive whose calls are being screened.

Call Second A unit for measuring communications traffic. Defined as one user making one second of a phone call. One hundred call seconds are called "ccs," as in Centum call seconds. "ccs" is the U.S. standard of telephone traffic. 3600 call-seconds = 1 call hour. 3600 call-seconds per hour = 36 CCS per hour = 1 call-hour = 1 erlang = 1 traffic unit. See also ERLANG and TRAFFIC ENGINEERING.

Call Selector A local phone company service which alerts the subscriber with a distinctive ring that one of the six numbers your pre-selected is calling.

Call Sequencer A call sequencer, also called an Automatic Call Sequencer, is a piece of equipment which attaches to a key system or a PBX. The Call Sequencer's main function is to direct incoming calls to the next available person to answer that phone. It typically does this by causing lights on telephones to flash at different rates. The light with the fastest flashing is the one whose call has been waiting longest. This call is answered first. Call Sequencers also might answer the phone, deliver a message and put the person on hold. They might keep statistical tabs of incoming calls, how fast they were answered, how long the people waited, how many people abandoned (hung up while they were on hold waiting for their call to be answered by a human being), etc. Call Sequencers are usually simple and inexpensive. Better, but much more expensive devices for answering incoming phone calls are Automatic Call Distributors. These are the devices which typically answer when you call an airline. See AUTOMATIC CALL DISTRIBUTOR.

Call Setup The first six PICs (Point In Call) of the Originating BCSM (Basic Call State Model), or the first four PICs of the Terminating BCSM. Definition from Bellcore in reference to Advanced Intelligent Network.

Call Setup Time The amount of time it takes for a circuit-switched call to be established between two people or two data devices. Call set-up includes dialing, wait time and time to move through central offices and long distance services. You don't pay for call set-up, but you will need extra lines to take care of it. See also ANSWER SUPERVISION and TRAFFIC ENGINEERING.

Call Shedding In many states, laws require that a real-life breathing person be available for a phone call being outdialed with an automated device, e.g. a predictive dialer. The reason these laws were enacted is because a lot of times call-center managers have their predictive dialers going so crazy in search of real-life people answering the phone that, when they fluke it and actually hit a person, they don't have an agent ready. Most systems simply hang up the connection when this occurs and they mark it down for later calling. This is called "call shedding" and is illegal in many states.

Call Spill-over In common-channel signaling, the effect on a traffic circuit of the arrival at a switching center of an abnormally delayed call control signal relating to a previous call, while a subsequent call is being set up on the circuit.

Call Splashing A "splash" happens when an Alternate Operator Service (AOS) company, located in a city different to the one you're calling from, connects your call to the long distance carrier of your choice in the city the AOS operator is in. Let's say you're calling from a Hotel in Chicago. You ask AT&T handle your call. The AOS, located in Atlanta, "splashes" your call over to AT&T in Atlanta. But you're calling Los Angeles. Bingo. Your AT&T call to LA is now more expensive than it would be — if you had been connected to AT&T in Chicago.

Call Splitting A feature allowing a phone user to speak privately with either party of a conference call by alternating between the two. Call splitting by an attendant allows the attendant to speak to the called person privately while effectively putting the calling person on hold, or vice versa.

Call Stalker An AT&T PC-based product which gives the 911 attendant the phone number and address of the person calling.

Call Store The temporary memory used in a stored program control switch (SPC) to hold records of calls in progress. These records are then transferred to permanent memory.

Call Stream British Telecom's premium rate service.

Call Tag A term used in the secondary telecom equipment marketplace. A ticket directing a freight carrier (e.g., UPS) to pick up equipment at another site. The company issuing the ticket pays the freight charge. Normally used to return defective equipment, it ensures the dealer a quick return and an accurate tracking mechanism.

Call Trace A name for local telephone company service which permits the tracing of the last call received and holds the results for later use by an authorized law enforcement agency. (Results of the trace are not available to the customer.)

Call Transfer Allows you to transfer a call from your phone to someone else's. On some phones you do this by punching in a bunch of numbers. Some you do it by hitting the "transfer" button and then the number you want to send the call to. The fewer buttons and numbers you have to punch, the easier it will be for your people.

If you're choosing a telephone system, check how easy it is to transfer a call. It is the most commonly used (and misused) feature on a phone system. How many times have you been told, "I'll transfer you to Mr. Smith, but if we get disconnected, please call back on extension 234." If your people are saying this to your customers or prospects, you are giving the outside world the wrong impression of your business. And since 97% of your prospects' contact with your company is first through your phone system, you could be losing precious business.

Call Type A call center term used in Rockwell ACDs. A portion of your call center traffic corresponding to one or more ACD gates or splits. This division of the total ACD traffic is the level at which forecasting and scheduling are done. At setup time, each Call Type is defined in the ACD by a unique three-letter code and specific gate or split number(s) that identifies the corresponding ACD report data.

Call User Data In packet data networking technology, user information transmitted in a call request packet to the destination data terminal equipment (DTE).

Call Waiting Call Waiting is a feature of phone systems that lets you know someone is trying to call you. You're speaking on the phone. A call comes in for you. You might hear a beep in your ear or see a light on your phone turn on. Or you might hear a beep and see a message come across the screen of your phone. When you hear the beep, you can, if you wish, put the present call on hold and answer the new one. You do this typically by hitting the touchhook on your phone (the cradle that sits under your handset.) Or you can ignore the new one, hoping it will go away, and perhaps send it to your attendant/operator, or voice mail. Call Waiting can be done manually by your telephone system operator. Or it can be a service which you buy as a monthly from your local phone company.

A major problem with call waiting is if you're on a data call from your PC, the call waiting "beep" will often cause your modem to hang up, thus destroying your data call. There are two solutions to this, the obvious one being turn off call waiting. Some phone systems will allow you to turn it off. The way to turn it off is to include *70 in your modem dial string before you dial the phone number. That will tell your phone company (in most cases) to turn off the call waiting sound while you're on that (and only that) phone call. Another way is modify your modem's initialization string. Here's how. In all Hayes and Hayes-compatible modems, there's a S10 register. It tells the modem how long before it hangs up after losing carrier. In Hayes modems, the S10 register is set for 1.4 seconds. The typical call waiting tone is 1.5 seconds. Solution, increase the S10 register to six seconds (to be sure). Use your communications software. Go into terminal mode, then type: ATS10=60. You must put this command in every time you power up, because the Hayes 1200 modem (and others) have volatile memory. But the Hayes 2400 and higher speed asynchronous modems have non-volatile memory. They remember the six seconds after they've been switched off. The command to write this to memory is ATS10=60&W. The "&W" means write it to memory.

Callback 1. A feature of some voice and data telephone systems. You dial someone. Their phone or computer is busy. You hit a button or code for "callback." When their phone becomes free, the phone system will call you and them simultaneously. You can only use this callback feature on things internally in your phone system — calling other people, calling long distance lines (which might be busy), calling the dictation pool, etc. See CALL WAITING, CALLBACK MODEM and CALLBACK QUEUING.

2. A quick way of referring to international callback, which works thus: Calling the United States from many countries abroad is far more expensive than calling those countries from the United States. A new business called International Callback has started. It works like this. You're overseas. You dial a number in the United States. You let it ring once. It won't answer. You hang up. You wait a few seconds. The number you dialed in the U.S. knows it was you calling. There is a piece of equipment on that number that "hears" it ring and knows it's you since no one else has that number. (Typically it's done with Centrex service.) That was your special signal that you want to make a call. A switch attached to that line then calls you instantly. When you answer (overseas, obviously) it conferences you with another phone line in the United States and gives you U.S. dial tone. You can then touchtone from overseas your American number, just as if you

would, were you physically in the U.S. There are huge savings. U.S. international callback operators can offer as high as 50% savings on calls from South America, where international calling rates are very high. The process of international callback is being automated with software and dialing devices. International callback is also helping to bring down the high cost of calling the U.S. from overseas. In recent years, deregulation has caused the price of international calls in many countries to fall dramatically. And now international callback or just callback is being done from other countries, including and especially Israel. A company called Kallback in Seattle, WA. has received a service mark from the U.S. Patent and Trademark Office for the words "callback" and "kallback" and sends letters to and threatens law suits against companies who use "their" words. See CALLBACK.

3. A security procedure in which a user, dials into access a system and requests service. The system then disconnects and calls the user back at a pre-authorized number to establish the access connection. This capability is often implemented into a modem, known as a Callback Modem, surprisingly enough. Same as dial-back.

Callback Modem A modem that calls you back. Here's how it works. You dial into a network. A modem answers. You put your password in. It accepts the password. It says "Please hang up. I will now call you back." You hang up. It calls you back. There are two reasons for doing this instead of allowing you to just go straight into the network. 1. It's better security. You have to be at a pre-determined place — an authorized phone number. 2. It may save on phone calls. The modem uses the company's communications network, which is probably cheaper than what the person calling in can use.

Callback Queue The queue used to hold callers who have requested a busy pool or extension. See CALLBACK QUEUE.

Callback Queuing An option on a telephone system which allows outgoing calls to be put in line for one or several trunks. When a trunk becomes available, the phone system calls the user, his phone rings and then the phone system dials the distant party on the trunk it grabbed before calling the user. Phone systems typically have two types of queuing. The first is called Hold-On Queuing. With this, the user dials his long distance number, the phone system searches for the correct trunk, finds it's not available and tells the user with a beep or message. The user then elects to stay on the line and wait. The instant the trunk becomes free, the phone system connects the user to it. The second type of queuing is called Callback Queuing. The user hangs up and the phone system calls you back, as we explained above.

There are tradeoffs between the two types of queuing. Callback queuing obviously can tolerate longer queues. The longer you wait, the more chance you have of reaching a very low-cost trunk. But users don't like waiting so long for a trunk. And when the call does come, it may likely reach a phone, newly-deserted by a user who's gone to the bathroom. In contrast, hold-on queuing is more efficient of the user's time, but less efficient of the user's trunks. The less time you wait, the less chance you have of reaching a low-cost trunk. Life is a trade-off. Queuing is no exception. See also QUEUING.

Callbridge Rolm (now Siemens) open architecture interface. A method of connecting a Rolm CBX (Siemens telephone system) to an outside computer, so that the computer may "talk" to the PBX and make certain things happen, e.g. moving a screen of client information around simultaneously with the phone call from the client. This feature is especially useful in customer service and customer order-entry environments — for example with direct mail order catalog companies, etc. See OPEN APPLICATION INTERFACE.

Called DTE A DTE which receives a call from another DTE.

Called Line Identification Facility A service feature provided by a network (private or public), which enables a calling terminal to be notified by the network of the address to which the call has been connected. See CALLER ID.

Called Line Identification Signal A sequence of characters transmitted to the calling terminal to permit identification of the called line.

Called Party Subaddress Information element that is passed transparently by the SPCS (if certain conditions are met) and can be used to further identify the destination party.

Called Party Camp-on A communication system service feature that enables the system to complete an access attempt in spite of issuance of a user blocking signal. Systems that provide this feature monitor the busy user until the user blocking signal ends, and then proceed to complete the requested access. This feature permits holding an incoming call until the called party is free.

Caller ID Your phone rings. A name pops upon on your phone's screen. It's the name and number of the person calling you. Actually, it's the originating telephone number and the name the phone company thinks is the subscriber. The originating telephone is stored in the originating central office equipment register, which is a database. That number supports a further database lookup, which associates the directory listing, assuming that the originating number is listed (i.e., not unlisted, or "nonpub" for nonpublished). The name and number information is passed through the local and long distance networks, and appears on your Caller ID box or your display telephone between the first and second rings. The delivery of Call ID information assumes several things. First, the entire network of switches must be supported by SS7 (Signaling System System #7). Second, the calling party must originate the call from a single-channel line, rather than a multichannel trunk (e.g., T-1). Third, the originating line/caller must not block the transmission of the information. If all of these criteria are not met, your Caller ID box will display "ANONYMOUS" or "NOT AVAILABLE." Caller ID is one of several CLASS (Custom Local Area Signaling Services) provided by your LEC (Local Exchange Carrier). There generally is both a small installation charge and a monthly charge for Caller ID. Caller ID lets you amaze your parents and scare your technophobic friends, when you answer the phone with something like "Hi, Harry! Great Dictionary!" Caller ID also helps you avoid those dinnertime calls from telemarketers. They always block their numbers. By the way, Caller ID is not the same as ANI, although they often are confused. See also ANI, Caller ID Message Format (for a very detailed explaination), and CLASS.

Caller ID Message Format The following Caller ID message format information is excerpted from "Caller ID", http://testmark.com/callerid.html, by Michael W. Slawson of Intertek Testing Services, TestMark Laboratories. Caller ID (CID), or Calling Number Delivery (CND) as it is sometimes called, came about as an extension of Automatic Number Identification (ANI). ANI is a method that is used by telephone companies to identify the billing account for a toll call. Although ANI is not the service that provides the information for CID, it was the first to offer caller information to authorized parties. The CID service became possible with the implementation of Signaling System 7 (SS7). The CID information is transmitted on the subscriber loop using frequency shift keyed (FSK) modem tones. These FSK modem tones are used to transmit the display message in American Standard Code for Information Interchange (ASCII) character code form. The transmission of the display message takes place between the first and second ring. The information sent includes the date, time, and calling number. The name associated with the calling number is sometimes included also. Since the time CID was first made available, it has been expanded to offer CID on Call Waiting (CIDCW) as well. With CIDCW, the call waiting tone is heard and the identification of the second call is seen. Caller ID is the identification of the originating subscriber line. Although Caller ID is the most common name for the service, a more appropriate name would be Calling Number Identification since it is the directory number of the originating line that is used as the identification. In cases where the name and the number are sent, the name that is used is that which is associated with the number in the directory listing. The telephone company has no way of knowing who the actual caller is. The CID information is sent on the destination subscriber loop between the first and second ring by means of two modem tones. The information is transmitted serially in FSK mode using one of the tones to represent a logic 1 (mark) and the other to represent a logic 0 (space). Caller ID uses the same frequencies, modulation type, and data format as the Bell type 202 modems. The message consists of a Channel Seizure string followed by a Mark string and then the caller information. The information is sent in one of two formats. The Single Data Message Format (SDMF) contains the date, time, and calling number. The Multiple Data Message Format (MDMF) contains the date, time, calling number, and the name associated with that number. Optionally, the number and name fields may contain data indicating that the information has been blocked by the caller or is unavailable. The method for sending and displaying the CID information during a silent interval between rings was invented by Carolyn A. Doughty. This invention was filed for a United States Patent on July 12, 1983. It was assigned patent number 4,582,956 on April 15, 1986 with AT&T Bell Laboratories (now Lucent Technologies Incorporated) listed as the assignee. Caller ID was first offered in New Jersey in 1987 by New Jersey Bell. The telephone company was interested in trying to earn additional revenue from its investments in new high-speed network signaling systems. These new systems use a separate call data circuit based on the SS7 standard to handle the setup, termination, supervisory signals, and other data concerning a call. This separate call data circuit can handle the processing of multiple calls very quickly and eliminates certain types of toll fraud. Prior to SS7 telephone companies used ANI for call billing purposes. With only a few exceptions, the billing information about the caller was not sent beyond the central office that provided that caller's service. The exceptions to this were authorized parties such as 911 services, law enforcement agencies, and more recently 800 and 900 number subscribers. Even today, ANI is still used by these parties. Since ANI is completely independent of CID it will deliver the originating number even when the originator has blocked their number for CID purposes. With the installation of the new SS7 systems it became practical to forward the calling party's identification through the telephone network to the central office serving the called party. This aspect of SS7 is known as Calling Party Number Message (CPNM). The CPNM includes the calling party's telephone number, and whether or not the calling party wants their number blocked from being seen by the called party. Note that the CPNM is sent out on all telephone calls that are made, regardless of whether or not the calling party wants their number blocked. A major issue that came up with the offering of CID was that of privacy. In several states the implementation of CID was slowed by privacy advocate groups. These groups claimed that CID was an invasion of privacy. In April of 1994, the Federal Communications Commission (FCC) issued a ruling that established a national standard for CID and the delivery of the caller's number into the switched network. Local and long distance telephone companies that were equipped to do so had to exchange CPNMs. Along with this, the telephone companies had to offer blocking on a per call basis so that the caller could block their number from being delivered to the called party. The CPNM is still delivered to the final central office serving the end subscriber, but the number is then blocked and the CID message marked as private.

Figure 1 - Sequence Layout and Timing: The checksum word is a twos complement of the modulo 256 sum of each bit in the other words of the message. The Channel Seizure and Mark Signals are not included in this checksum. When the message is received by the CPE it checks for errors by taking the received checksum word and adding the modulo 256 sum of all of the other words received in the message. The addition done by the CPE does not include the Channel Seizure and Mark Signals, nor does it include the received checksum word. The result of this addition should be zero to indicate that no errors have been detected.

Character Description	Decimal Value	ASCII Value	Actual Bits (LSB)
Message Type (SDMF)	4		00000100
Message Length (9)	18		00010010
Month (December)	9	1	00110001
	50	2	00110010
Day (25)	50	2	00110010
	53	5	00110101
Hour (3pm)	49	1	00110001
	53	5	00110101
Minutes (30)	51	3	00110011
	48	0	00110000
Number (6061234567)	54	6	00110110
	48	0	00110000
	54	6	00110110
	49	1	00110001
	50	2	00110010
	51	3	00110011
	52	4	00110100
	53	5	00110101
	54	6	00110110
	55	7	00110111
Checksum	79		01001111

Figure 2 (A Caller ID message in SDMF) shows a CID message in SDMF. For ease in describing the process of determining the checksum, the decimal values will be used for the calculations. The first step is to add up the values of all of the fields (not including the checksum). In this example the total would be 945. This total is then divided by 256. The quotient is discarded and the remainder (177) is the modulo 256 sum. The binary equivalent of 177 is 10110001. To get the twos compliment start with the ones compliment (01001110), which is obtained by inverting each bit, and add 1. The twos compliment of a binary 10110001 is 01001111 (decimal 79). This is the checksum that is sent at the end of the CID information. When the CPE receives the CID message it also does

a modulo 256 sum of the fields, however it does not do a twos complement. If the twos complement of the modulo 256 sum (01001111) is added to just the modulo 256 sum (10110001) the result will be zero. If the result is not zero then the message is discarded. It is important to note that there is no error correction in this method. Even if the CPE were to notify the central office of errors, the central office will not retransmit the information. If an error is detected, the CPE receiving the message should display an error message or nothing at all. Although Bellcore SR-TSV-002476 recommends that the CPE display an error message if erroneous data is received, most CPE manufacturers have elected to just ignore the errored message. The content of the CID message itself depends on whether it is in SDMF or MDMF. A message in SDMF includes a Message Type word, a Message Length word, and the actual Message words. A message in MDMF also includes a Message Type word, a Message Length word, and the actual Message words, but additionally includes Parameter Type and Parameter Length words. There are certain points within these messages where up to 10 Mark bits may be inserted to allow for equipment delays in the central office. These Stuffed Mark

bits are generally not necessary. The Message Type word defines whether the message is in SDMF or MDMF. It will be a binary 00000100 (decimal 4) for SDMF or a binary 10000000 (decimal 128) for MDMF. The Message Length will include the number of characters in the message. This length does not include the checksum at the end of the message. For SDMF the minimum length will be 9 characters. The minimum length for MDMF will depend on whether the customer has subscribed to CNAM service as well as CND. In the case of CND only the minimum length will be 13 characters. If the customer also has CNAM then the minimum will be 16 characters. In all three of the minimums mentioned there will be no actual number or name delivered. The field will be marked either "O" (Out of area) or "P" (Private).

Figure 3 shows an example of a minimum message layout for SDMF. The number will not be delivered because it has been blocked by the calling party. The CPE will receive the date, time, and a "P" to indicate that the caller's identification has been blocked at the caller's request.

Character Description	Decimal Value	ASCII Value	Actual Bits (LSB)
Message Type (SDMF)	4		00000100
Message Length (9)	9		00001001
Month (December)	49	1	00110001
	50	2	00110010
Day (25)	50	2	00110010
	53	5	00110101
Hour (3pm)	49	1	00110001
	53	5	00110101
Minutes (30)	51	3	00110011
	48	0	00110000
Private	80	P	01010000
Checksum	16		00010000

Figure 4 shows an example of a message in MDMF that contains both a number and a name.

Character Description	Decimal Value	ASCII Value	Actual Bits (LSB)
Message Type (MDMF)	128		10000000
Message Length (33)	33		00100001
Parameter Type (Date/Time)	1		00000001
Parameter Length (8)	8		00001000

Month (November)	49	1	00110001
	49	1	00110001
Day (28)	50	2	00110010
	56	8	00111000
Hour (3pm)	49	1	00110001
	53	5	00110101
Minutes (43)	52	4	00110100
	51	3	00110011
Parameter Type (Number)	2		00000010
Parameter Length (10	10		00001010
Number (6062241359)	54	6	00110110
	48	0	00110000
	54	6	00110110
	50	2	00110010
	50	2	00110010
	52	4	00110100
	49	1	00110001
	51	3	00110011
	53	5	00110101
	57	9	00111001
Parameter Type (Name)	7		00000111
Parameter Length (9)	9		00001001
Name (Joe Smith)	74	J	01001010
	11	o	01101111
	101	e	01100101
	32		00100000
	83	S	01010011
	109	m	01101101
	105	i	01101001
	116	t	01110100
	104	h	01101000
Checksum	88		01011000

In *Figure 4*, if the number and name had not been included then the parameter types for those fields would be different. These alternate parameter types are used to signify that the data contained in that parameter is the reason for its absence. The parameter type for the number section would have been a binary 00000100 (decimal 4) and the parameter type for the name section would have been a binary 00001000 (decimal 8). When the parameter type signifies that the data contained is the reason for that fields absence, the parameter length is always a binary 00000001 (decimal 1). If the reason for absence is that the calling party does not want their number/name displayed then the parameter data would be a binary 01010000 (ASCII "P") for Private. If the reason for absence is that the information is just not available then the parameter data would be a binary 01001111 (ASCII "O") for Out of area. The number/name may not be available if the calling party is not served by a central office capable of relaying the information on through the network.

In carrying CID a step further and combining it with Call Waiting, a service commonly called Caller ID on Call Waiting has been developed. This feature is also sometimes called Calling Identity Delivery on Call Waiting. With CIDCW the identification of the party calling can now be seen without putting the current call on hold. CIDCW is done the same way as CID with only a few exceptions. One of those exceptions is that CIDCW only uses MDMF so the message type will always be a binary 10000000 (decimal 128). Since CIDCW will only take place if a call is already in progress there will never be a ring signal preceding the message. The channel seizure signal is not used with CIDCW either. The central office gets the CPE's attention by sending it a CPE Alerting Signal (CAS), a two-tone signal described later. If the customer's equipment

is ready to receive the CID information, it responds to the CAS by sending the central office an acknowledgment (ACK) signal. Note that the CPE will only respond to the CAS signal if it is the only extension that is off-hook. If someone else is involved in the call on another extension, the CPE will not send an ACK signal to the central office because it cannot mute the second device. Most manufacturers detect whether an extension is off-hook by monitoring the line voltage. If the line voltage drops below a certain point then it is assumed that an extension is off-hook. This can lead to problems in situations where the CPE is placed on a very long loop. The long loop reduces the voltage at the CPE and causes its detector to continuously register an extension off-hook. At least one manufacturer is getting around this by designing the equipment to learn the line voltage that it should see during normal operation. This is done by taking the device off-hook two or three times so that it can register the normal off-hook voltage. If the voltage ever drops below this setting, it assumes that there is another extension off-hook. Once the central office sees the ACK signal it sends the FSK data. When the CPE sees the CAS from the central office it mutes the telephone's handset before sending the ACK and receiving the FSK data. This act of muting serves two purposes.

First, it keeps the telephone user from hearing the FSK signal. But, more importantly it avoids interference from speech and other noises that would be picked up by the telephone's microphone. The muting only has to take place on the end receiving the FSK data since before sending the CAS, the central office momentarily removes audio to and from the far end of the current call. The audio is reconnected after the FSK data has been transmitted. The sequence of events that displays the information about the caller begins when the central office temporarily removes the far end party of the current call and sends a Subscriber Alerting Signal (SAS) to the near end party. The SAS is a single frequency of 440 Hz that is applied for approximately 300 ms. This is the tone that is heard when a call is in progress and call waiting beeps to indicate a second call. The SAS tone is mainly for the user and is not required for the CPE to receive the CID information. The SAS tone is followed by a CAS to alert the CPE that it has CID information to send. The CAS is a dual tone signal combination of 2130 Hz and 2750 Hz that is 80 ms in length. Once the CPE hears the CAS it mutes the handset of the telephone and returns an ACK signal to the central office. This ACK signal has a nominal tone duration of 60 ms and is either a Dual-Tone Multifrequency (DTMF) "A" or a DTMF "D". The DTMF "D" is the most common ACK signal and it consists of the frequencies 941 Hz and 1633 Hz. A DTMF "A" consists of the frequencies 697 Hz and 1633 Hz. Once the central office receives the ACK signal it in turn sends the CID information. The CPE un-mutes the handset as soon as it finishes receiving the FSK signal. A problem that is sometimes experienced with CIDCW is Talkoff. Talkoff occurs when the CPE falsely detects a CAS. Speech can sometimes be interpreted by the CPE's detector as a CAS from the central office. When talkoff occurs both parties will hear a brief interruption in the conversation and the party on the far end will hear the ACK signal that is sent to the central office by the CPE. When the handset is muted by a talkoff, the CPE will timeout and unmute the handset after waiting a set period of time for an FSK signal. Bellcore SR-3004 specifies that this timeout period should be no more than 501 ms from the end of the ACK signal. A CPE's detector will behave differently with different people because of the variations in each persons voice and speech patterns. Speech can also cause a valid CAS signal to not be recognized by the CPE. When this occurs, it is called a Talkdown. The last difference between CID and CIDCW is that instead of being preceded by the 300 bits of Channel Seizure Signal and 180 bits of Mark Signal, as with standard CID, the Caller ID message for CIDCW is only preceded by 80 bits of Mark Signal. CIDCW still uses the checksum just like CID.

Conclusion: Although CID can be very useful in screening calls, it is not limited to this application alone. For example, a number of business contact-manager software packages have made it possible to use CID to automatically bring up client information from a database and display it on the screen of a personal computer (PC) before the call is answered. Implementation of this requires a CID hardware device with a data connection to the PC. As time goes on we will see more and more uses for CID. Even though an "Out-of-Area" message may still seen occasionally, it will be less and less often as the remaining telephone offices which are not yet setup for CID install the necessary equipment. Even as there are advances being made in the way Caller ID can be utilized, there are also advances being made in the customer premise equipment. Manufacturers are now coming out with units that allow multiple extensions to be off-hook and still receive CIDCW. Although the same manufacturer's equipment must generally be used on all extensions, this does allow more than one person to be on the same call and still see the identification of a waiting call. Even the small office key telephone systems are now offering CID and CIDCW capabilities so that caller information coming in on a line can be displayed at all stations associated with that line. Implementation does not always mean replacing a currently installed telephone system. Some manufacturers are offering upgrades to existing hardware by simply replacing specific printed wiring cards.

Caller Independent Voice Recognition Having a voice response unit recognize the voice of a caller without having been trained on the caller's voice.

Caller Name An enhancement of Caller ID. Prior to sending the originating telephone number to your display, the carrier associates that number with an electronic white pages, thereby transmitting both the originating number and the associated directory listing. Assuming that the number is listed and that the Caller ID number is not blocked, this provides a much better indication of the identity of the caller.

Calling A procedure which consists of transmitting address signals in order to establish a link between devices that want to talk to each other.

Calling Card A credit card issued by Bell operating companies, AT&T, MCI, Sprint and other phone companies (local and long distance) and used for charging local and long distance calls. Typically, the number on your calling card is the phone number at which you receive bills (home or business phone) plus a four digit Personal Identification Number (PIN number). Increasingly often it's not. I prefer to carry a calling card with digits completely different to my phone number since this provides me with greater security. It's harder for someone to figure out my calling card. Some phone companies — local and long distance — charge more for a call made with a Calling Card. Some don't. Bell Canada claims they trademarked the term "Calling Card" in Canada. If they did, good luck protecting it, since the term "calling card" is generic. See BREAKAGE, DEBIT CARD and PREPAID CALLING CARD.

Calling DTE A DTE (Data Terminal Equipment) which places a call to another DTE.

Calling Line ID Also called Caller ID. You are called. As the call comes in, you receive the phone number of the person calling you. See Caller ID for a full explanation. See also ANI, Calling Number Display, ISDN and Signaling System 7.

Calling Line Identification Facility A service feature, provided by a network, that enables a called terminal to be notified by the network of the address from which the call has originated. See CALLING LINE ID.

Calling List A call center term. A collection of records from a database that is used for a specific telemarketing campaign. See Calling List Penetration.

Calling List Penetration A call center term referring to outbound calls. The percentage of a call list for which the decision makers have been reached after a given number of attempts. See Calling List.

Calling Number An international term for what Americans call "ANI" — or automatic number identification. In other words, calling number simply tells you, the receiver of the call, who's calling. It tells you that by displaying the caller's number on the screen of your phone or the screen of your PC. And sometimes it might tell you who's actually calling. It can do that because some central offices (also called public exchanges) have the ability to dip into a database and replace the calling number with the name of the person who owns that phone number.

Calling Number Display Your phone has a LCD (Liquid Crystal Display) or LED (Light Emitting Diode) display. When your phone rings, it will show which telephone number (internal or external) is calling you. Some phone systems allow you to add the person's name to the calling number display. See also ANI and CALLER ID.

Calling Party The person who makes (originates) the phone call.

Calling Party Camp-on A feature that enables the system to complete an access attempt in spite of temporary unavailability of transmission or switching facilities. Systems that provide this feature monitor the system facilities until the necessary facilities become available, and then proceed to complete the requested access. Such systems may or may not issue a system blocking signal to let the caller know of the access delay.

Calling Party Identification A telephone company service which tells the person being called the number and sometimes the name of the person calling them. They can then decide to answer or not answer it. See ANI, which stands for Automatic Number Identification.

Calling Party Number CPN. When a call is set up over an ISDN network, SS7 sends an IAM (Initial Address Message) as part of the ISUP (ISDN User Part) protocol. Included in the IAM is the Calling Party Number subfield, which contains the number of the calling party. Also included is a two-bit Presentation Indicator (PI), which indicates whether the terminating switch (cellular or PSTN) should pass the CPN to the called party. If the PI says "yes," the originating number is passed to the called party, who can see that number displayed on the telephone set or on an adjunct display device. The end result is Caller ID, perhaps also with the Caller Name, assuming that the called party has subscribed to those features. Based on recognition of the calling party, or the lack of it, Caller ID can prompt the called party to either accept or reject the call. See also CALLER ID, CALLER NAME, IAM, ISUP, and SS7.

Calling Party Pays In the United States, cell phone users pay for incoming as well as outgoing calls. In Europe and in most countries elsewhere, the calling party pays. According to many people, paying for incoming calls retards the industry's growth. As a result, there are many people in the U.S. cell industry, who would like this changed to "Calling Party Pays."

Calling Party Subdividers Information element that is passed transparently by the SPCS (if certain conditions are met) and can be used to further identify the originating party. An AIN term.

Calling Pattern Telecommunications managers are great at looking at phone bills, seeing patterns and smelling out calls that don't fit into those calling patterns. A simple example: zillions of calls over a weekend to Pakistan. The company doesn't do business with Pakistan. Nor does anyone work on the weekend.

Calling Sequence A sequence of instructions together with any associated data necessary to perform a call.

Calling Tone See CNG.

CallPath IBM's telephone system link to IBM's computers. See CALLPATH COORDINATOR, CALLPATH SERVICES ARCHITECTURE, CALLPATH CICS, and CALLPATH HOST.

CallPath CallCoordinator CallCoordinator is IBM's integrated call management application that uses CallPath Services APIs to integrate data processing applications with telephone systems. IBM has versions of CallCoordinator for MVS CICS, OS/2 and Windows workstations. CallCoordinator provides features such as Intelligent Answering (based on ANI, DNIS, or Calling Line ID), Coordinated Voice and Data Transfer, Consultation (both voice and data), Conferencing (both voice and data), Transfer Load balancing between a single or multiple telephone systems, Outbound dialing, Event logging for Management Information Reporting, Personal Dialing Directory (Windows Only), Personal telephony facilities (answer phone, disconnect, transfer, etc.), Integration with CallPath DirectTalk/2 and CallPath DirectTalk/6000, Customizable Application Programming Interfaces. CallCoordinator integrates with existing 3270 or 5250 applications, and on the workstation versions, has the ability to communicate with existing applications via Dynamic Data Exchange or standard LAN communications protocols (such as TCP/IP). See CALLPATH SERVICES ARCHITECTURE.

CallPath CICS Enabling software that connects your telephone systems with your IBM 370 or 390 (i.e. the mainframe version of CallPath/400, which works on the AS/400 platform). See CallPath CallCoordinator.

CallPath Services Architecture CSA is IBM's architecture that defines the protocols for communication between computers and telephone switches. CallPath Services Architecture, announced in 1991, provides an Application Programming Interface (API) that enables a call management application to interact with telephone systems, with little regard to the protocols or communications interface provided by the telephone system. The idea is that with CallPath a call will arrive at a computer terminal simultaneously with the database record of the caller. And such call and database record can be transferred simultaneously to an expert, a supervisor, etc. CallPath has especial value in telephone call centers. As of writing, IBM provided connectivity to PBXs (Lucent Definity Generic 3, Nortel Meridian 1, Siemens/ROLM 9751 and Hicom, Bosch, Alcatel, SDX, Ericsson, Philips, Deutsche Telecom, Cortelco, and GPT), central office switches (AT&T 5ESS and Northern Telecom DMS-100), and ACDs (Aspect and Rockwell). IBM's CallPath products provide support for locally attached applications and

client/server applications. IBM has CallPath APIs available for mainframes, minicomputers and workstations, in particular IBM System 390 and ES9000, AS/400, RISC System/6000, OS/2 workstations, Windows workstations, Sun Solaris, HP UX, and SCO UNIX workstations. See OPEN APPLICATION INTERFACE and DIRECTTALK.

Callpower A Rockwell ACD term. An integrated voice and data workstation for use in combining ACD capabilities with host computer database management.

CallWare CallWare is a company in Salt Lake City, UT, which makes computer telephony software that runs on the Novell NetWare operating system. CallWare software includes voice mail, autoattendant, IVR database lookup, etc.

CALNET The California Network (CALNET) implemented service in September 1991 with the objective of providing cost-effective telecommunications services to state and local government in California by reducing costs through consolidation of user service requirements. In the summer of 1998, CALNET provided services to 300,000 government customers statewide. The Department of General Services' (DGS) Telecommunications Division oversees CALNET.

CAM 1. Call Applications Manager. The name of the Tandem software interface which provides the link between a call center switch telephone switch (either a PBX or an ACD) and all Tandem NonStop (fault tolerant) computers. CAM supports most major PBXs and automatic call distributors.

2. Computer-Aided Manufacture. The actual production of goods implemented and controlled by computers and robots. Often used in conjunction with CAD. Only a few factories are completely automated. Usually, there is some human intervention in the actual construction of the product, often to make sure a part is placed in the robot correctly.

3. Controlled Attachment Module. Intelligent Token-Ring hub.

CAMA Centralized Automatic Message Accounting. See CAMA/LAMA.

CAMA/LAMA Centralized Automatic Message Accounting/Local Automatic Message Accounting. Specific versions of AMA in which the ticketing of toll calls is done automatically at a central location for several COs (CAMA) or only at the local office for that office's subscribers.

Camcorder A camera and a video recording system packaged as a whole.

CAMEL Customized Application of Mobile Enhanced Logic. A new standard proposed by ETSI for GSM (Global System for Mobile Communications). CAMEL enhances GSM for the provisioning of international IN (Intelligent Network) services. In order to effect CAMEL, the GSM operator installs a CSE (CAMEL Service Environment), similar to the wired IN equivalent. The CSE comprises a SSP (Service Switching Point), IPs (Intelligent Peripherals), a SCP (Service Control Point), the SCE (Service Creation Environment) and some additional SS7 (Signaling System 7) software. Once CAMEL is approved, IN services will be available internationally, across GSM networks. Initial services will include voice mail, call waiting, call forwarding, and freephone (toll-free) access. While only approximately 10% of GSM users currently roam internationally, that number is expected to increase significantly in the future. See also GSM and IN.

Cameo Macintosh-based personal videoconferencing system, announced by Compression Labs in January of 1992. Developed jointly with AT&T and designed to work over ISDN lines and, most recently, Ethernet LANs. The Cameo transmits 15 fps of video and needs an external handset for audio.

Camp-on You're calling a telephone an extension or you want to transfer a call to a phone but it's busy. This telephone system feature will allow you to lock the call you're trying to transfer onto the line that's busy. When it becomes free, the phone will ring and the "camped-on" call will be connected automatically.

Campus The buildings and grounds having legal contiguous interconnection.

Campus Backbone Wiring between buildings.

Campus Environment An environment in which users — voice, video and data — are spread out over a broad geographic area, as in a university, hospital, medical center, prison. There may be several telephone systems. There may be several LANs on a campus. They will be connected with bridges and/or routers communicating over telephone, microwave or fiber optic cable.

Campus Subsystem The part of a premises distribution system which connects buildings together. The cable, interbuilding distribution facilities, protectors, and connectors that enable communication among multiple buildings on a premises.

CAN Abbreviation for cancel. The binary code is 100001 and the HEX is 81.

Cancel By touching the "cancel" button on a phone system you're telling the phone system to ignore the last command you gave it. That command might have been transfer, hold, park, etc. The "cancel" button is often mistakenly confused with the "release" button. The "release" button acts the same as hitting "Enter" on a computer system, i.e. it tells the system to go ahead and do what you just told it to do, no matter how stupid your command. In short, "Cancel" means kill the last command. You use when you make a mistake. "Release" means "Enter" — Do it and do it now.

Cancel Call Waiting On a touchtone phone in North America, you typically can cancel the feature, Call Waiting, by touchtoning *70.

Cancelmoose A Newsgroup/Usenet Term. An individual who wages war against spamming.

Cannibalize To devour a phone system by stripping parts from it to repair another system. A common technique for maintaining equipment whose original manufacturer no longer supplies parts. Before you cannibalize, check out the monthly publication Telecom Gear. That publication lists sources of secondary telecom equipment. Good stuff, too.

Canonical Address A method for storing unique telephone numbers. Canonical addressing is used by Windows Telephony TAPI (Telephony API) for making telephone calls from a database of numbers. A canonical address describes all possible aspects of a telephone number. You can call a telephone number using canonical addressing independent of calling location or access method. A canonical address is stored in a database and preceded by an ASCII Hex (2B) to indicate its address type. It includes delimiters and strings for Country Code, Area Code, Subscriber Number, Subaddress and Name.

CAP 1. Competitive Access Provider. Also known as AAV (Alternative Access Provider). CAPs provide an alternative means of establishing a connection between a user organization and an IXC (IntereXchange Carrier), completely bypassing the ILEC (Incumbent Local Exchange Carrier). CAPs typically deploy high-capacity SONET fiber optic transmission systems in a ring topology around geographic areas in which are found a high density of large businesses. Drops from the fiber optic rings are terminated at both the customer locations and the IXC POPs (Points of Presence). Thereby, end user organizations with substantial levels of InterLATA voice and data traffic can bypass the ILEC facilities, which often are made up of poor quality UTP (Unshielded Twisted Pair) and who may take months to provision a T-1 circuit. In addition to providing superior performance and much reduced provisioning time, such fiber optic transmission facilities offer incredible levels of bandwidth, which quickly can be increased, and generally are provided at much lower cost than leased-line ILEC circuits. CAPs also offer the inherent advantage of loop diversity. In the event that the ILEC local loop suffers a catastrophic failure, the CAP loop likely will not be affected, unless both loops follow the same physical path and are destroyed by the same post-hole digger (or other catastrophe). In the unlikely event that the redundant CAP loop fails, the user organization can still access the IXC through the ILEC on a circuit-switched, 1+ dial-up basis. Since the Telecom Act of 1996 and various state initiatives have relaxed regulatory constraints and opened the local exchange to competition, many CAPs have become CLECs (Competitive Local Exchange Carriers). As CLECs, they are free to offer switched voice and data services within the local exchange area. Where they do not provide their own fiber optic local loops, they lease UTP local loops from the ILECs for resale, with those loops terminating in collocated termination equipment in the ILEC central offices and with the traffic then being directed to the CLEC fiber optic transmission facilities. As facilities-based carriers, the traffic then is transported to the CLEC's own switching centers and wire centers for local and long-haul service access. See also CLEC, ILEC, IXC and SONET.

2. Cellular Array Processor.

3. Carrierless Amplitude and Phase modulation is a bandwidth-efficient line coding technique. CAP is a variant of Quadrature Amplitude Modulation (QAM), which is used in today's rate-adaptive voice band V.32/V.34 dial modems. AT&T Bell Laboratories first began development of CAP in the mid-1970's for more efficient implementation of a digital signal processor (DSP), while providing the same high level of performance. Used in conjunction with advanced error correcting codes and channel equalization, CAP modulation provides robust performance and excellent loop reach in the presence of bridge-taps, cross-talk, and other interferers. Carrierless Amplitude & Phase Modulation is now a transmission technology for implementing a Digital Subscriber Line (DSL). The transmit and receive signals are modulated into two wide-frequency bands using passband modulation techniques. CAP is bandwidth-efficient and supports ADSL, HDSL, RADSL, and SDSL line coding.

4. Client Access Protocol. See iCalendar.

Cap'N Crunch see CAPTAIN CRUNCH.

Capacitance The capacity of a medium (wire, cable, resistor, bus) to store an electrical charge. Capacitance is measured in farads.

Capacitive Coupling The transfer of energy from one circuit to another by virtue of the mutual capacitance between the circuits. The coupling may be deliberate or inadvertent. Capacitive coupling favors transfer of higher frequency components, whereas inductive coupling favors transfer of lower frequency components.

Capacitor Capacitors provide a means of storing electric charge so that it can be released at a specific time or rate. The simplest type of capacitor is a parallel plate capacitor and consists of two closely spaced plates of conductive material with an insulating material known as a dielectric sandwiched between them. Dielectrics are chosen for their ability to enhance a capacitor's performance. Capacitors are rated for their capacitance which is measured in farads and voltage.

The rated voltage is usually the breakdown voltage, I.e. The maximum voltage the dielectric can insulate against before the voltage discharges between the two plates of the capacitor as an electric spark.

Because capacitors store charge, they can be used in electronic circuits in place of a battery. However, a battery generates electricity through a chemical reaction. A capacitor will generate an electric current only after it has been charged by another current source. When working on electronic equipment — even equipment unplugged from a power supply — make sure you are well insulated against shock. Capacitors in a circuit can hold a charge, sufficient to cause injury or death, sometimes for many hours after an appliance or device has been turned off. Capacitors in TV sets, for example, can store up to 100,000 volts.

Capacity 1. The information carrying ability of a telecommunications facility. What the "facility" is determines the measurement. You might measure a date line's capacity in bits per second. You might measure a switch's capacity in the maximum number of calls it can switch in one hour, or the maximum number of calls it can keep in conversation simultaneously. You might measure a coaxial cable's capacity in bandwidth.

2. The measure of the amount of electrical energy a condenser can store up. The unit of capacity is the farad.

Capacity Study A local document issued at least once a year for each entity within the telephone company. The capacity study includes information relative to the network access line/trunk capacity of each item of switching equipment as well as the network access line capacity of lines and numbers.

Capacity Transfer Control A Northern Telecom term for a feature which permits single allocation of capacity to be shared among members in a digital switched broadcast connection. For teleconferencing, for instance, a conference leader can transfer transmission capacity among the digital ports in the circuits. 95% of such transfers will take place within 10 seconds.

Capcode A capcode is a four or seven digit number on either side or rear of the casing of a pager, the type you wear on your belt. This number is a paging system necessity to know how to generate the right sequence of tones to alert the pager.

CAPI Cryptography Application Program Interface. The first API developed (by Microsoft) for encryption programs.

CAPS 1. Code Abuse Prevention System.

2. Competitive Access Providers to the local telephone network i.e., Teleport or Metropolitan Fiber System.

Capsizing When downsizing, rightsizing and upsizing fail. Contributed by Fred Schindler of IBM.

Capstan 1. A flangeless pulley used to control speed and motion of magnetic tape through a recorder or playback unit.

2. A rotating drum or cylinder used for pulling cables by exerting traction upon a rope or pull line passing around the drum.

CAPTAIN Character And Pattern Telephone Access Information Network System. A form of videotext developed in Japan and operated through the public switched telephone network. Displays are on a TV set. It's interactive.

Captain Crunch At one point in the 1960s, a breakfast cereal had a promotion. It was a toy bosun's whistle. When you blew the whistle, it let out a nearly precise 2,600 Hz tone. If you blew that whistle into the mouthpiece of a telephone after dialing any long distance number, it terminated the call as far as the AT&T long distance phone system knew, while still allowing the long distance connection to the distant city

to remain open. If you dialed an 800 number, blew the whistle and then pressed in a series of tones (called multi-frequency or MF tones) on your "Blue Box", you could make long distance and international calls for free, since the only the thing the local billing machine knew about was the original toll-free call to the 800 number.

The man who discovered the whistle was John Draper and he picked up the handle of Cap'n Crunch (not to be confused with the spelling of the Quaker Oats breakfast cereal, Cap'n Crunch) in the nether world of the late 1960s phone phreaks. A marvelous account of the exploits of phone phreaks was published in the October 1971 issue of Esquire Magazine. That article described how the Cap'n would call himself (he needed two lines) — choosing to route the connection through Tokyo, India, Greece, South Africa, South America, London, New York and California — to make his second phone next to him ring. He'd have a wonderful time talking to himself, albeit with a round-the-world delay (despite the speed of light) of as long as 20 seconds. Later, AT&T closed the loophole Cap'n Crunch had discovered. AT&T turned from in-band signaling to out-of-band signaling. Cap'n Crunch's legacy (he got put in jail four times during the 1970s) is System Signaling 7, a system of immense benefit to us all. See 2600 TONE, MULTI-FREQUENCY SIGNALING and SIGNALING SYSTEM 7.

Captive Effect An effect associated with the reception of frequency-modulated signals in which, if two signals are received on or near the same frequency, only the stronger of the two will appear in the output. The complete suppression of the weaker carrier occurs at the receiver limiter, where it is treated as noise and rejected. Under conditions where both signals are fading randomly, the receiver may switch from one to the other.

Captive Screw Let's say you have a couple of screws at the front of an industrial grade computer. You unscrew the two screws and a panel pops down. You can then get access to something inside the computer. The screw, however, doesn't leave the computer. No matter how much you unscrew it, you cannot remove the screw. It's "captive."

Capture Division Packet Access CDPA. Capture Division Packet Access is a new cellular access, packet-oriented architecture, which is a packet-oriented architecture able to support the constant bit rate traffic and variable bandwidth on demand for multimedia traffic. According to the September 1996 issue of the IEEE Communications Magazine, the approach integrates multiple access and channel reuse issues to achieve a high rate of spectral efficiency, and presents general advantages even if used for delay-constrained circuit-oriented traffic. Unlike CDMA and TDMA, wherein the effective data rate of each connection is typically a small fraction of the total radio channel allocated for PCN, the CDPA approach allows each user to access the entire channel, if necessary, for brief periods of time (packet access). Spectrum sharing is accomplished by exploiting the different path losses suffered by the various signals as they appear at the base stations (the capture effect), with co-channel interference abated through time diversity (colliding users do not successively retry in the same time interval). Results suggest that abating co-channel interference by random transmission may be more effective than spatial isolation at cells using the same channel, as is usual in FDMA/TDMA systems.

Capture Effect An effect associated with the reception of frequency-modulated signals in which, if two signals are received on or near the same frequency, only the stronger of the two will appear in the output. The complete suppression of the weaker carrier occurs at the receiver limiter, where it is treated as noise and rejected. Under conditions where both signals are fading randomly, the receiver may switch from one to the other.

Capture Ratio The ability of a tuner or receiver to select the stronger of two signals at or near the same frequency. Expressed in decibels, the lower the figure, the better.

Car Phone The type of cellular phone that's installed in a vehicle. There are four types of cellular phones being sold today — mobile, transportable, portable and handheld. A car phone (also called a mobile unit) is attached to the vehicle, its power comes from the vehicle's alternator (or battery if the car if not running) and the car phone has an external antenna, which works best if it's mounted in the middle of the highest point of the car and wired directly with no breaks in the wire. Many window-mounted antennas have a break in their wiring. The wiring ends at the inside. There is no electrical connection between the inside of the window and the antenna glued onto the outside of the window. The "connection" is done through signal radiation. In North America, the car phone transmits with a standard three watts of power.

Carbon Block A device for protecting cable from lightning strikes. The carbon block consists of two electrodes spaced so that any voltage above the design level is arced from line to ground. Carbon block protectors are used commonly in both local customer offices and central offices. They are effective, but can be destroyed if high voltage is directly applied — as in a direct strike by lightning. A more expensive, but more effective method of protection is the gas tube. These are glass capsules that are connected between the circuit and the ground. When a voltage higher than the design voltage strikes the line, the gas ionizes and conducts the excess voltage to ground. When the voltage is gone, the protector restores itself to normal. Gas tubes, however, take a tiny time to ionize. This may not be fast enough for very sensitive things, like PBX circuit cards. So gas tube protectors are often equipped with diodes, which clamp the interfering voltage to a safe level until the gas tube ionizes.

Carbon Fiber A strong synthetic material that is low in mass with excellent damping characteristics, used in the manufacture of tonearms.

Carbon Rheostat A rheostat using carbon as the resistance material. See RHEOSTAT.

Carbon Transmitter The microphone of an telephone set from yesteryear which uses carbon granules and a diaphragm. The diaphragm responds to our voice and varies the pressure on the granules and hence, their resistance. If your carbon mike isn't working well, the humidity has got to it, tap it lightly on your desk. The carbon granules will line up and it will work much better. Carbon microphones are very reliable but are being increasingly replaced with more sensitive electret microphones.

Card Authorization Center CAC. A computer directly linked to MCI switches for authorization and determination of billing center ID for MCI card calls.

CardBus Laptops typically come with slots for what are now known as PC cards — little credit card size devices who do various things — like become a modem, become a network interface card, become a video conferencing card, become an ISDN card, etc. These cards were originally called PCMCIA cards. (For a full explanation see PCMCIA). The original PCMCIA spec was 16-bit. The new spec, called CardBus, which combines the PCI bus, has a 32-bit interface and supports 132 Mbps. The CardBus specification is the significantly improved successor to the previous PC Card standard. But

the two standards are not compatible. You cannot run a laptop with both PCMCIA and CardBus cards. You must run them with cards of the same standard. And these days, the best standard to go with is CardBus. According to 3Com, which makes some excellent CardBus cards, CardBus delivers 33 MHz, 32-bit performance based on PCI bus architecture; Low 3.3-volt power consumption; Bus mastering for sharply improved CPU efficiency; Built-in multifunction capabilities; Zoomed video to handle multimedia applications and (so 3Com says, but I haven't found) Backward compatibility with 16-bit PC Card (PCMCIA) devices. According to 3Com, CardBus provides notebook users with:

Card Cage A frame in a telephone system or computer for mounting circuit cards, power supply, backplane and other equipment.

Card Dialer A device attached to a telephone which accepts a special plastic card and then automatically dials the number on the card as indicated by the holes punched in it. A card dialer is now obsolete except for unusual applications, like systems whereby you carry your card with you and use it as a security device.

Card Issuer Identifier Code CIID - (pronounced "sid") A code issued with certain calling cards. AT&T's CIID cards cannot be used by other interexchange carriers but can be used by LECs.

Card Services The software layer above Socket Services that coordinates access to PCMCIA cards, sockets and system resources. Card Services is a software management interface that allows the allocation of system resources (such as memory and interrupts) automatically once the Socket Services detects that a PC Card has been inserted. This is called "hot swapping." The idea is that you can slide PCMCIA cards in and out of PC at will and your Socket and Card services will recognize them and respond accordingly. It's a great theory. In practice, it doesn't work because certain cards, like network cards, simply can't be connected and disconnected at will. Socket Services is a BIOS level software interface that provides a method for accessing the PCMCIA slots of a computer. Card Services is a software management interface that allows the allocation of system resources (such as memory and interrupts) automatically once the Socket Services detects that a PC Card has been inserted. Both of these specifications are contained in the PCMCIA Standards document. You do not need either Socket or Card Services to successfully use PCMCIA cards in your desktop or laptop. You simply need the correct device drivers and the proper memory exclusions. See PCMCIA, SOCKET SERVICES and SLOT SIZES.

Card Slot A place inside a phone system or computer into which you slide a printed circuit board. See BOARD.

• 20 times the throughput of conventional 16-bit PC Card slots. The 32-bit CardBus interface can transmit data at 400-600 Mbps, compared to 16-bit PC Card's 20-30 Mbps. Users must have that higher bandwidth for linking to a 100 Mbps Fast Ethernet network, for quickly moving data to and from SCSI-2 storage devices (such as Zip) and for handling bandwidth-hungry applications like video conferencing.

• Better systems performance under Windows 95 and Windows NT. Bus mastering lets a CardBus device transfer data to computer memory directly, without intervention from the notebook's processor. This boosts overall computer performance multitasking Windows 95 and Windows NT (soon to be known as Windows 2000) operating environments. Plus, it increases throughput when the notebook is connected to a Fast Ethernet LAN.

• Lower power consumption. CardBus devices run at 3.3 volts, instead of 16-bit PC Card's 5 volts. That means CardBus devices use less power than conventional PC Card devices, and generate less heat inside the computer. The bottom line: batteries last longer, and the computer functions more reliably.

• Easier installation of multifunctional devices. The CardBus specifications enables sharing of multiple resources on a single card with no need for special drivers. As a result, CardBus multifunction cards will be simpler to install than their 16-bit PC Card equivalents, and they will have fewer interoperability and compatibility problems.

• Optimized video performance. The CardBus Zoomed Video feature handles streamed video transmissions more efficiently by transferring the data directly to the PC's video controller over a dedicated bus. That way, video doesn't have to compete for bandwidth on the computer's PCI bus. See Card Services and PCMCIA.

CARE Customer Account Record Exchange. A system developed to facilitate the exchange of customer account information between the IXC and the LEC in order to facilitate the provisioning of telecommunications services. CARE generically identifies data elements that might be exchanged between the IXC and LEC in an industry format. It is intended to provide a consistent definition and data format for the exchange of common data elements.

Caret The symbol ^ which is found above 6 on most keyboards. Also used to indicate the "Ctrl" key in some instruction manuals.

CAROT Centralized Automatic Reporting On Trunks. A test and maintenance facility associated primarily with electronic toll switching systems like the AT&T Communication's #4-ESS. CAROT is a computerized system that automatically accesses and tests trunks for a maximum of fourteen offices simultaneously. It enables rapid routine testing of all trunks to ensure quick identification of faults and potential failures.

Carpal Tunnel Syndrome Carpal tunnel syndrome is a serious disorder of the arm caused by fast, repetitive work, such as typing without support for your wrists or with insufficient time for rest. In carpal tunnel syndrome, the tendons passing through the wrist bones swell and press on the median nerve. Surgery to take pressure off the nerve can relieve numbness and pain, but it's not always effective and many victims remain permanently disabled. The best prevention is using a wrist rest and undertaking specific exercises. There is a good book on the subject — Conquering Carpal Tunnel Syndrome by Sharon J. Butler, New Harbinger Publications, Oakland, CA.

Carriage Return By hitting this key, the printing head or the cursor on your screen will return to the left hand margin. Usually hitting a Carriage Return or the "Enter" key includes a line feed, i.e. the paper will move up one line or the cursor will drop down one line. "Usually" does not mean always. So check. You can usually correct the problem of not having a line feed with a carriage return by moving a dip switch on the printer, changing one of the parameters of the telecommunications software program (the part where it says something about auto linefeed) or changing the computer's operating system (by doing a "Config" or the like). In most microcomputers, a Carriage Return is equivalent to a "Control M," or ASCII 13. A line feed is a "Control J".

Carried Load 1. A telephone industry definition. Carried load is the usage measured on a circuit group. A circuit has a potential carried load capacity of 36 CCS per hour which is rarely approached because of the idle time between calls.
2. A data networking definition. The traffic that occupies a group of servers on a LAN.

Carried Traffic The part of the traffic offered to a group of servers that successfully seizes a server on a LAN.

Carrier 1. A company which provides communications circuits. Carriers are split into "private" and "common." A private carrier can refuse you service. A "common" carrier can't. Most of the carriers in our industry — your local phone company, AT&T, MCI, Sprint, etc. — are common carriers. Common carriers are regulated. Private carriers are not.
2. An electrical signal at a continuous frequency capable of being modified to carry information. For analog systems, the carrier is usually a sine wave of a particular frequency, such as 1800 Hz. It is the modifications or the changes from the carrier's basic frequency that become the information carried. Modifications are made via amplitude, frequency or phase. The process of modifying a carrier signal is called modulation. A carrier is modulated and demodulated (the signal extracted at the other end) according to fixed protocols. Some of the wideband (i.e. multi-frequency) circuits are also called "carriers." T1, which typically has 24-channel PCM voice circuits, is known as a carrier system.

Carrier Access Code CAC. A dialing code used to select a carrier.

Carrier Band The range of frequencies that can be modulated to carry information on a specific transmission system. See also CARRIERBAND.

Carrier Bypass A long distance phone company provides a direct link between its own switching office and a customer's office, thus bypassing the local phone company. Bypass is done to save the customer or the long distance company money. Bypass is also done to get service faster. Sometimes the local phone company simply can't deliver fast enough.

Carrier Circuit A higher level circuit (DS-1, DS-3, Transmission System, etc.) that has been designed to carry lower-level circuits (DS-0, DS-1).

Carrier Class See Carrier Class IP Switch.

Carrier Class IP Switch A Carrier Class IP Switch is a high volume, high reliability hybrid device for routing IP packets. It separates out high priority packets that must all arrive together, like voice and video, and delivers them immediately. All other packets are delivered through normal routing. It adds the timing precision of a switch to the low cost, speed and efficiency of a router. This definition contributed by Ron Acher.

Carrier Common Line Charge CCL. The charge which IXCs (IntereXchange Carriers) pay to LECs (Local Exchange Carriers) for the privilege of connecting to the end user through LEC local loop facilities. The CCL is a charge to cover a portion of the costs associated with the local loop, which is used for origination of local, intraLATA long distance (also known as "local toll"), and interLATA long distance calls. In combination, the CCL, the CALC (Customer Access Line Charge), and the monthly tariff charge for the local loop are intended to cover the costs of provisioning and maintenance of the loop, as well as to provide the LEC with a reasonable rate of return (i.e., profit) on its investment. That they do. They also encourage bypass and may, in the long term, be self-defeating. See also Access Charge.

Carrier Detect CD. The little red LED light on most modems. When this light is on, your modem is connected to another modem or communications device.

Carrier Detect Circuitry Electronic components which

detect the presence of a carrier signal and thus determine if a transmission is about to happen. Used in modems.

Carrier Extension A proposal for modifying the CSMA/CD access mechanism for Gigabit Ethernet. Under a carrier extension, when a device in the network transmits, the signal stays active for a longer time before another device can attempt to transmit. This lets an Ethernet frame travel a longer distance, and thereby increases the potential network diameter.

Carrier Failure Alarm CFA. An alarm telling you that timing has been lost in your digital transmission because there are too many zeros in the message. When this happens, all the calls are lost until the equipment regains timing.

Carrier Frequency The frequency of a carrier wave. The frequency of an unmodulated wave capable of being modulated or impressed with a second (information-carrying) signal. In frequency modulation, the carrier frequency is also referred to as the "center frequency."

Carrier Identification Codes CIC. Four digit numbers used by end-user customers to reach the services of interexchange carriers through equal access arrangements. The primary carrier of choice is reached by dialing "1" plus the area code and called party number. Secondary IX carriers can be reached by dialing 101 plus the CIC assigned to the carrier desired. CIC numbers are used to dial around the carrier presubscribed to the calling telephone number. See 101XXXX.

Carrier Leak The unwanted carrier remaining after carrier suppression in a suppressed carrier transmission system.

Carrier Liaison Committee CLC. A committee formed to help industry participants work together to resolve the issues of implementing 800 Portability. CLC is sponsored by the Exchange Carriers Standards Association (ECSA) and is comprised of the LECs (local exchange carriers), long distance carriers and users of 800 service.

Carrier Loss In T-1, carrier loss means too many zeros. A carrier loss in T-1 is said to occur when 32 consecutive zeros appear on the network. Carrier is said to return when the next 1 is detected.

Carrier Noise Level The noise level resulting from undesired variations of a carrier in the absence of any intended modulation.

Carrier Power (of a Radio Transmitter) The average power supplied to the antenna transmission line by a transmitter during one radio frequency cycle taken under the condition of no modulation. Does not apply to pulse modulation or frequency-shift keying.

Carrier Provided Loop A local phone line owned by a long distance company that is resold as part of a WAN service. This is generally separated from your long distance service, the same way local calls are.

Carrier Select Keys Buttons at the bottom of a payphone used to choose a long distance carrier.

Carrier Selection As a result of Judge Greene's Modified Final Judgment which lead to the breakup of the Bell System, most local phone companies must offer their customers (business and home) the opportunity to select which long distance company they would like to be use on a "primary" basis. That means when you dial 1+ (one plus) you get that carrier. To use any other long distance company you have to dial more digits, e.g. 1-0288 (for AT&T). See NANP.

Carrier Sense In a local area network, a PC or workstation uses its network card to detect if another station is transmitting. See CSMA.

Carrier Sense Multiple Access CSMA. In local area networking, CSMA is a way of getting onto the LAN. Before

starting to transmit, personal computers on the LAN "listen" to make sure no other PC is transmitting. Once the PC figures out that no other PC is transmitting, it sends a packet and then frees the line for other PCs to transmit. With CSMA, though stations do not transmit until the medium is clear, collisions still occur. Two alternative versions (CSMA/CA and CSMA/CD) attempt to reduce both the number of collisions and the severity of their impact. See CSMA/CA and CSMA/CD.

Carrier Sense Multiple Access/Collision Avoidance CSMA/CA. A protocol that requires the PC to sense if another PC is transmitting. If not, it begins transmitting. Under CSMA/CA, a data station that intends to transmit sends a jam signal; after waiting a sufficient time for all stations to pick up the jam signal, it sends a transmission frame; if while transmitting, it detects another station's jam signal, it stops transmitting for a designated time and then tries again.

Carrier Sense Multiple Access/Collision Detection A network control scheme. It is a contention access control scheme. It "listens" for conflicting traffic to avoid data collisions. The Ethernet LAN uses CSMA/CD, then waits a small amount of time and then tries again. See CSMA/CD and ETHERNET.

Carrier Shift 1. A method of keying a radio carrier for transmitting binary data or teletypewriter signals, which consists of shifting the carrier frequency in one direction for a marking signal and in the opposite direction for a spacing signal.

2. In amplitude modulation, a condition resulting from imperfect modulation whereby the positive and negative excursions of the envelope pattern are unequal, thus effecting a change in the power associated with the carrier. There can be positive or negative carrier shift.

Carrier Signal A continuous waveform (usually electrical) whose properties are capable of being modulated or impressed with a second information-carrying signal. The carrier itself conveys no information until altered in some fashion, such as having its amplitude changed (amplitude modulation), its frequency changed (frequency modulation) or its phase changed (phase modulation). These changes convey the information.

Carrier Synchronization In a radio receiver, the generation of a reference carrier with a phase closely matching that of a received signal.

Carrier System A system where several different signals can be combined onto one carrier by changing some feature of the signals transmitting them (modulation) and then converting the signals back to their original form (demodulation). Many information channels can be carried by one broadband carrier system. Common types of carrier systems are frequency division, in which each information channel occupies an assigned portion of the frequency spectrum; and time division, in which each information channel uses the transmission medium for periodic assigned time intervals.

Carrier Terminal The modulation, demodulation and multiplex equipment used to combine and separate individual channels at the ends of a transmission system.

Carrier To Noise Ratio CNR. In radio receivers, the ratio, expressed in decibels, of the level of the carrier to that of the noise in the receiver bandwidth before any nonlinear process such as amplitude limiting and detection takes place.

Carrier Wave The radio frequency wave generated at a transmitting station for the purpose of carrying the modulated or audio frequency wave.

Carrierband Same as single-channel broadband. See also

CARRIER BAND.

CARS Cable Television Relay Service, a microwave service authorized by Part 78 of the FCC Rules for the purpose of transmitting signals intended for carriage over a cable television system. Back in the days when "CATV" stood for Community Antenna Television, "CARS" stood for Community Antenna Relay Service. Over the years, "CATV" has evolved to mean Cable Television, but CARS remains CARS.

Carson's Rule A radiocommunications term. Carson's Rule is a method of estimating the bandwidth of an FM (Frequency Modulation) subcarrier system. It is commonly used in satellite systems in order to ensure that a high-fidelity, sharp TV picture will be delivered over a subcarrier TV channel. Violation of Carson's rule results in a higher video signal-to-noise ratio, at the expense of streaking in fast-moving scenes, sharpness of picture, and loss of audio fidelity. Carson's rule states that $B = 2x(Df+fmax)(A-3)$, where B is the bandwidth, Df is the peak deviation of the carrier frequency, and fmax is the highest (maximum) frequency in the modulating subcarrier signal.

Carterfone A device for connecting a two-way mobile radio system to the telephone network invented by Thomas Carter. It was electrically connected to the base station of the mobile radio system. Its electrical parts were encased in bakelite. When someone on the radio wanted to speak on a "landline" (the phone system), the base station operator would dial the number on a separate phone then place the telephone handset on the Carterfone device. The handset was acoustically, not electrically, connected to the phone system. No more than 4,000 Carterfones were ever installed, yet the Bell System thought they were the most dangerous device ever invented. Tom Carter died in Gun Barrel, TX where he lived, in the early part of 1991. He died a poor man. See Carterfone Decision.

Carterfone Decision In the summer of 1968 the FCC determined that the Carterfone and other customer phone devices could be connected to the nation's phone network — if they were "privately beneficial, but not publicly harmful." The Carterfone decision was a landmark. It allowed the connection of non-telephone company equipment to the public telephone network. This decision marked the beginning of the telephone interconnect business as we know it today. The Carterfone decision made a lot of lawyers rich before all the rules on connection to the network got cleared up, and finally codified in something called Part 68 of the FCC's Rules. See CARTERFONE, NATA and NETWORK HARM.

Cartridge 1. A device which holds magnetic tape of some kind.

2. A device to translate (transduce) stylus motion to electrical energy in a phonograph. It comes in three basic types — moving magnetic coil, induced magnet and ceramic. A phono cartridge is also call a pickup. Most record players use ceramic cartridges because they have higher output than the three magnetic types and can work with a less powerful (i.e. cheaper) amplifier.

CAS 1. Centralized Attendant Service. One group of switchboard operators answers all the incoming calls for several telephone systems located throughout one city. CAS is used by customers with several locations in the same geographic area, i.e. retail stores, banks.

2. Communicating Applications Specification. A high-level API (Application Programming Interface) developed by Intel and DCA that was introduced in 1988 to define a standard software API for fax modems. CAS enables software developers to integrate fax capability and other communication func-

tions into their applications. See CAS 2.0

3. CAS is a generic acronym for Channel Associated Signaling or Call-path Associated Signaling. CAS is in-band signaling used to provide emergency signaling information along with a wireless 911 call to the Public Safety Answering Point (PSAP). This signaling information includes the phone number of the wireless phone and coding used to derive a general location of the caller, and meets the Enhanced 911 Phase 1 FCC requirements. This coding can be either a p-ANI or an ESRD. This in-band signal is made up of tones which pass within the voice frequency band and are carried along the same circuit as the talk/call path that is being established by the signals. NCAS is a generic acronym for Non Call-path Associated Signaling. This definition contributed by Glenda Drizos and Doug Puckett of Sprint PCS, Overland Park, Kansas.

CAS 2.0 Communicating Applications Specification. Fax standard for both fax and voice applications. Developed by Instant Information, Inc. (I3), CAS 2.0 offers a simple, yet highly flexible and scaleable model that has allowed vendors such as Brooktrout, Dialogic and FaxBack to add greater and more sophisticated functionality to their products. Originally designed and developed by Intel and Digital Communications Associates in 1988, CAS is an API (Application Programming Interface) specification that provides support for programs sending data to other devices and computers. It is one of the world's most popular software interfaces to a fax board and is a universal standard embraced by hundreds of developers. Since its inception, more fax ports have shipped supporting CAS standards than any other communications protocol.

New features in CAS 2.0 include:

• Full 32-bit Windows 95, NT 3.51, NT 4.0, support

• Full support for asynchronous real-time fax applications

• Class 1 and Class 2 support

• C++ class library support

• An intuitive redesigned setup user interface

• Full Brooktrout, Gammalink and WildCard co-processed hardware support

Cascade 1. To connect the output of a device into the input of another device, which then may in turn be connected to another device. Image the organization of a company. At the top is the president. Reporting to him are three vice presidents. Reporting to each of these vice presidents are five directors. Reporting to each of these directors are five managers. If each of these positions were a piece of network gear, you'd have a classic cascaded topology. For example, a long-haul circuit may involve multiple, cascading repeaters; multiple, cascading virtual circuits may be involved in a Frame Relay network.

2. A Windows term. When windows cascade, they are arranged in an overlapping pattern so that the title bar of each window remains visible.

3. A series of reply posts to a USENET message, each adding a trivial or nonsense theme to the collection of previous replies. Some consider this art; there is a USENET newsgroup devoted to propagating this art form (alt.cascade).

Cascade Amplification Successively using two or more amplification systems. Most radio and audio products have more than one stage of amplification.

Cascaded Amplifier Two or more amplifiers coupled together. Most radio and audio products have more than one stage of amplification.

Cascaded Stars Local area network topology in which a centralized multiport repeater serves as the focal point for many other multiport repeaters.

Cascaded Topology See CASCADE.

Cascading Faults Faults that cause other faults. Typically faults in a network causing other faults.

Cascading Menu A Windows term. A menu that is a submenu of a menu item. Also known as a hierarchical menu.

Cascading Notification A feature of some sophisticated voice mail systems. Let's say someone leaves a message for you in your voice mail box. Your voice mail system then automatically goes out to find you, i.e. to notify you. It may start by lighting your message light, calling your home phone number, calling your cellular phone, calling your beeper, etc. I like this feature because when I want you, I want you. And a little mechanized help is much appreciated. I first saw the feature in Macrotel's MVX voice mail series.

CASE Computer Aided Software Engineering. CASE is a new, faster, more efficient way of writing software for some applications. The idea with CASE is to sketch out relations between databases, events, and options and then have the computer write the code.

CASE Tools These tools provide automated methods for designing and documenting software programming. Computer-aided software engineering (CASE) sketches relations between databases, events, and options. It then provides a language in which the computer writes the code, letting programmers develop applications faster. It's new and seems to work in limited instances.

Case Method A traditional way of load testing computer telephony systems, the case method involves gathering many individuals together in a room full of telephones along with several cases of an appropriate libation (frequently beer), and using these individuals to simulate real users calling into (or being called by) the computer telephony system. Case method testing usually continues until all the cases have been consumed, the testing is completed, it becomes too late in the evening to continue, or the perspective of the gathered individuals becomes too subjective to be of use any longer to those conducting the test. This definition courtesy of Steve Gladstone, author of the book Testing Computer Telephony Systems, available from 212-691-8215.

Case Sensitive This means that uppercase letters must be typed in uppercase on your keyboard, and that lowercase letters must be typed in lowercase. It is important to key in your data in the exact combination of upper or lower case characters. Inputting in the wrong case could make your entry invalid for some fields (for example, password). DOS and Windows are much less case sensitive than Unix, for example.

Cassegrain Antenna An antenna in which the feed radiator is mounted at or near the surface of a concave main reflector and is aimed at a convex secondary reflector slightly inside the focus of the main reflector. Energy from the feed unit illuminates the secondary, reflects it back to the main reflector, which then forms the desired forward beam. This technique is adapted from optical telescope technology and allows the feed monitor radiator to be more easily supported.

Cassette Tape A slow, inefficient method of storing and retrieving data which uses the same technology as audio cassettes — like the Sony Walkman. Some PBXs use cassette tape to backup their user programming and database.

Castellation A series of ribs and metallized indentations that defines edge contact regions.

Casual Calling 1. A class action lawsuit has been filed in the United States District Court for the District of Columbia, in Washington, D.C., on behalf of all MCI subscribers who were charged MCI's "casual calling" rates (in some cases as high as $2.87 for a one-minute call) instead of the lower rates which MCI advertises and which subscribers expected to be charged. The lawsuit alleges that this practice of MCI violates the Communications Act of 1934. MCI has one set of rates for subscriber direct dialing, including its widely promoted "Five Cents Sunday" Calling Plan. MCI also has another, much higher set of rates it charges for non-subscriber or casual calling: $0.38 per minute, plus a surcharge of $2.49 for every call. The lawsuit alleges, however, that MCI has charged many of its own subscribers the higher non-subscriber, casual calling rates.

2. A phrase referring to dialing 101XXXX in America to place a call over an alternate carrier that may not have the capability to bill the call. This is usually associated with unauthorized calls where the carrier receives Automatic Number Identification but lacks a means to direct a bill to a physical customer location.

CAT 1. An AT&T Merlin term. Call Accounting Terminal. A stand-alone unit with a built-in microprocessor and data buffer that provides simple call accounting at a relatively low cost.

2. Shortened way of saying "Category," as in Cat 1 cabling. Say Cat 5 and you mean Category 5 wiring. See CATEGORY 1 through 5 and CATEGORY OF PERFORMANCE.

CAT1 See Category 1 and Category of Performance.

CAT2 See Category 2 and Category of Performance.

CAT3 See Category 3 and Category of Performance.

CAT4 See Category 4 and Category of Performance.

CAT5 See Category 5 and Category of Performance.

CAT5E See Category 5 Enhanced and Category of Performance.

CAT6 See Category 6 and Category of Performance.

Category 1 CAT1. An unspecified Category of Performance for inside wire and cable systems. CAT1 cables can be of various gauges, and are useful in support of applications requiring a carrier frequency less than 1 MHz, which roughly translates into 1 Mbps, depending on the compression scheme employed. Example applications include analog voice and ISDN BRI. See also Category of Performance.

Category 2 CAT2. A Category of Performance for inside wire and cable systems. CAT2 cables can be of either 22 or 24 gauge, and are useful in support of applications requiring a carrier frequency of up to 1 MHz; the transmission rate achievable depends on the compression scheme employed. Example applications include 4 Mbps Token Ring LANs. See also Category of Performance.

Category 3 CAT3. A Category of Performance for inside wire and cable systems. CAT3 cables can be of either 22 or 24 gauge, and are useful in support of applications requiring a carrier frequency of up to 16 MHz; the transmission rate achievable depends on the compression scheme employed. Example applications include POTS, ISDN, T-1, 4/16 Mbps Token Ring, and 10Base-T. Category 3 technical specifications are defined by FCC Part 68, ANSI/EIA/TIA-568, TIA TSB-36 and TIA TSB-40. Category 3 safety requirements are defined by UL 1459 (Telephone), UL 1863 (Wire and Jacks) and NEC 1993, Article 800-4. See also Category of Performance.

Category 4 CAT4. A Category of Performance for inside wire and cable systems. CAT4 cables can be of various gauges, and are useful in support of applications requiring a carrier frequency of up to 20 MHz; the transmission rate achievable depends on the compression scheme employed. Example applications include 4/16 Mbps Token Ring. Category 4 technical specifications are defined by FCC Part 68, EIA/TIA-568, TIA TSB-36, and TIA TSB-40. Category 4 safety requirements are defined by UL 1459 (Telephone), UL 1863 (Wire and Jacks) and NEC 1993, Article 800-4. See also Category of Performance.

Category 5 CAT5. A Category of Performance for inside wire and cable systems. CAT5 cables can be of various gauges, and are useful in support of applications requiring a carrier frequency of up to 100 MHz; the transmission rate achievable depends on the compression scheme employed. Example applications include 4/16 Mbps Token Ring, 10/100Base-T and 100VG-AnyLAN. Cat 5 is now the most common cabling being installed for LAN connectivity. Increasingly, Cat 5 cabling is being installed for both data and voice use. Increasingly, Cat 5 is the cabling of choice for forward-looking offices. Category 5 technical specifications are defined by FCC Part 68, EIA/TIA-568, TIA TSB-36, and TIA TSB-40. Category 4 safety requirements are defined by UL 1459 (Telephone), UL 1863 (Wire and Jacks) and NEC 1993, Article 800-4. See also Category of Performance.

Category 5 Enhanced CAT5E. A developing, non-standard cabling system. CAT5E is intended to be manufactured according to tight specifications in support of signaling rates of up to 200 MHz over distances of up to 100 meters. Specifications call for a tighter twist, electrical balancing between pairs, and fewer cable anomalies, such as inconsistencies in the core diameter. CAT5E is intended to support of 100Base-T, ATM, and Gigabit Ethernet. See also Category of Performance.

Category 6 CAT6. A developing cabling standard for SFTP (Shielded Foil Twisted Pair) intended to support signaling rates up to 200 MHz. CAT6 comprises 4 twisted pairs separately wrapped in foil insulators and twisted around one another. The group of 4 pairs are contained in an extra insulating shield which is then contained within a flame-retardant polymer jacket. Applications will include 100Base-T, ATM and Gigabit Ethernet. and wiring under development. Categories 1 through 5 are based on the EIA/TIA-568 standard. See Category of Performance.

Category 7 CAT 7. A developing cabling standard for STP intended to support signaling rates up to 600 MHz in support of ATM and Gigabit Ethernet. See Category of Performance.

Category Of Performance As we try and push more and data faster and faster down a pair or two of wires, so the quality of the wires and the components they connect to has become increasingly important. You can't push 100 million bit per second down junky phone lines. As a result, the telecommunications industry has defined cabling and cabling component standards. The idea is that if your stuff conforms to the standard, users will be able to achieve the data rates and reliability they want. Some standards specify physical characteristics, such as thickness of cable, plastic material used in the outer jacket, etc. These "Category of Performance" standards, which the telecommunications industry has adopted, in the main do not specify materials. They specify tests which the cabling and cabling components must pass. There were originally five categories of tests. But now there are effectively only two categories that anyone buys — Category 3 and Category 5. In simple terms, if you want voice and data to 10 megabits per second, i.e. standard Ethernet, then buy Category 3. If you are looking to the future when you may be transmitting 100 million bits per second on your local area network, then you should buy Category 5. Today the difference in price is about 20% to 30%. In other words, Cat 5 is 20% to 30% more expensive than Cat 3. In my unhumble opinion, all LAN cable going in today should be Cat 5, while all voice cable going in should be Cat 3, which will happily support ISDN BRI, when

that happens in the not too distant future.

Let's talk about how they measure Cat 3 and Cat 5. First, you should understand that all the tests are self-certifying, which means that while there are standards, each manufacturer is himself responsible for conforming. No one will put a manufacturer in jail if his Cat 5 stuff doesn't perform to Cat 5 standards. The only thing likely to happen to the manufacturer is that the world will find out his stuff is garbage and he'll go broke.

The concept of the test is simple. The test for cable and components is a swept frequency test. Cat 3 must pass all signals from one through 16 megahertz. (Note megahertz, not megabits). In Cat 5, it must pass signals from one through 100 megahertz. Cat 3 and Cat 5 cables and components are designed to support any applications intended to operate over those frequencies. That includes all forms of modulation of any carrier waves within those frequencies and all forms of digital pulses. Handling digital pulses is more difficult since digital pulsing typically uses a much broader range of frequencies (than simply modulating one or two carrier waves as in AM or FM modulation). Digital pulsing needs a broader range of frequencies in order to get its edges more square. Interestingly, both Cat 3 and Cat 5 cabling are the same thickness, namely 24 gauge. The twist structure on Cat 5 is tighter. Insulation is better. Connecting hardware is definitely different. Below is Harry's quick rule of thumb for what you should buy based on what you want to transmit, as against the swept frequency test:

Category	Cable Type	Application & Speed Supported
1	UTP	Analog Voice
2	UTP	Digital Voice, 1 Mbps Data
3	UTP-STP	16 Mbps Data
4	UTP,ST	20 Mbps Data
5	UTP,STP	100 Mbps Data
6	Coax	100 Mbps+Data
7.	Fiberoptic	100 Mbps+Data

Harry's Rule of thumb: If you want the most flexibly wired office — the Office of The Future — install only cat 5 cabling to everyone's desk. And put in twice as much cabling as you ever dreamed you will need. When installing cable, never use staples; never use tie wraps (some idiot will tighten them too tightly); never untwist wire before you punch it down; (the twists should be right up to the termination); never strip more jacket off the wire than is needed to terminate it; never pull too hard on any cable - especially around a corner; never pull on a cable to straighten out a kink or loop (always go back and untwist it) and never stuff too many cables in too small a conduit. See Commercial Building Telecommunications Wiring Standard.

Catenet A network in which hosts are connected to networks are interconnected by gateways (routers). The Internet is an example of a catenet.

Cathode The heated element which emits electrons in a vacuum tube. It may be a filament, or may be a separate element, heated by proximity to a filament. It is maintained at a negative potential in respect to the anode or plate. Cathodes have other applications, also.

Cathode Ray The beam of electrons emitted by a cathode. See CATHODE RAY TUBE.

Cathode Ray Tube CRT. A TV screen. A CRT is a tube of glass, used in television, oscilloscope and computer terminals, from which air has been removed (i.e. vacuum tube). At the back of the CRT is an electron gun which directs an electron beam to the front of the tube. The inside front of the tube has been coated with fluorescent material which reacts to and lights up once the electron beams hit. CRTs are very reliable if they are vented, since the electron gun gets hot. CRTs have a "memory." They will memorize what's been left on their screen for a while, i.e. the image is burned into the screen. And you'll see it even though the screen is turned off. In short, turn your screen off when you're not using it. Or run a "CRT-saving" program which varies the image on the screen.

Cathodic Protection A form of galvanic corrosion protection in which a conductor with a negative charge will repel chlorine ions, rather than attract them, as would be the case with a positive charge. Cathodic protection is the reason that twisted-pair, copper local loops use -48 volts, rather than +48 volts, for "Ring." At least it was in the olde days of uninsulated local loops. It still works. The same principle works to protect bridges, pipelines, and other metallic structures.

CATI Computer Assisted Telephone Interviewing, a market research term for a call center based on the use of a computerized database.

CATLAS AT&T software standing for Centralized Automatic Trouble Locating and Analysis System. CATLAS is used as a maintenance tool for locating and diagnosing problems in AT&T electronic central offices.

CATNIP Common Architecture for Next Generation Internet Protocol. One of the 3 IPng candidates.

CATS Consortium for Audiographics Teleconferencing Standards, San Ramon, CA. 510-831-4760. CATS describes itself as a non-profit corporation dedicated to promoting standards for this technology, which it describes as enhancing audioconferencing by allowing people at different sites to work together in real time to create, manipulate, edit, annotate and reference still images. Now called IMTC.

CATV CAble TeleVision. This term originally stood for "community antenna television," reflecting the fact that the original cable systems carried only broadcast stations received off the air; however, as cable systems began to originate their own programming, the term evolved to mean Cable Television. CATV is a broadband transmission facility. It generally uses a 75-ohm coaxial cable which simultaneously carries many frequency-divided TV channels. Each channel is separated by guard channels. See Addressable Programming and Broadband.

CAU 1. Northern Telecom term for Connection Arrangement Unit.
2. Controlled Access Unit. CAU. An intelligent hub from IBM for Token Ring networks in conjunction with IBM LAN Network Manager software.

CAVE 1. Cave Automatic Virtual Environment. A sophisticated virtual reality facility developed by the Electronic Visualization Lab at the University of Illinois at Chicago. CAVE is much like the HoloDeck of "Star Trek: The Next Generation," although a headset and a wand are required to create the illusion. CAVE allows scientists to see, touch, hear and manipulate data in order to do such things as create and test new models of various machines, or to manipulate atoms of a molecule. See also Virtual Reality.
2. Cellular Authentication and Voice Encryption. An encryption algorithm specified in ANSI-41 (formerly TIA standard IS-41C) to initiate the system of authentication challenges in order to prove the identity of a mobile phone. CAVE is used in conjunction with the A-key (Authentication key), MIN (Mobile Identification Number) and ESN (Electronic Serial Number) in order to prevent cellular fraud artists from capturing the MIN and ESN data. Such data typically is transmitted "in the clear." The A-key is intended to be known only to the phone and the AC (Authentication Center). Once the A-key is transmitted, the phone and the AC generate a second

secret key, known as a SSD (Shared Secret Data), the value of which can be updated in the event that the IS-41 network suspects that it has been compromised.

Cavity A volume defined by conductor-dielectric or dielectric-dielectric reflective boundaries, or a combination of both, and having dimensions designed to produce specific interference effects (constructive or destructive) when excited by an electromagnetic wave.

Cavity Wall A wall built of solid masonry units arranged to provide air space within the wall.

CB Why 10-4, good buddy, that stands for Citizens Band. Also known as Children's Band, not because of Radio Shack's toy walkie talkies, but for the inane chatter that sometimes goes on in these channels. In short, CB is low-power (up to four WATTS permitted) public radio. You do not need permission from the FCC to transmit or receive at these frequencies. Thus CB's great popularity. CB went through a boom (perhaps a craze?), then it ran out of radio frequencies and public enthusiasm. Its original frequencies were 26.965 to 27.225 Mhz. Now the FCC's given it new frequencies — 462.55 to 469.95 MHz. These new frequencies are much better, clearer and less congested. If you buy a CB set, make sure you get one that operates in these higher frequencies. In some countries they use different frequencies. CB radio is not allowed in many countries, even some civilized countries, though it will obviously work there.

CBDS Connectionless Broadband Data Service: A connectionless service similar to Bellcore's SMDS defined by European Telecommunications Standards Institute (ETSI). In short, the European version of SMDS.

CBEMA Computer Business Equipment Manufacturers Association. A lobbying group created to protect the interests of its members. CBEMA is the author of the AC voltage disturbance tolerance specification to which all computing and business equipment is designed. Specifies overvoltage and undervoltage events that computing equipment must withstand. This standard, for example, provides that all computing and business equipment must withstand a power loss or transfer time of 12ms.

CBF Computer Based Fax.

CBH See Component Busy Hour.

CBK Change BacK.

CBQ Class-Based Queuing. A queuing algorithm used in routers to manage congestion. Through user-definable class definitions, incoming packet traffic is divided into classes. A class might include all traffic from a given interface, all traffic associated with a particular application, all traffic intended for a particular network or device destination, or all traffic of a specific priority classification. Each class of traffic is assigned to a specific FIFO (First In First Out) queue, each of which is guaranteed some portion of the total bandwidth of the router. Should some class(es) of traffic not make full use of their allocated bandwidth, CBQ portions out that available bandwidth to other class-specific FIFO queues on a proportionate basis. See also FIFO, RED, ROUTER and WFQ.

CBR 1. Constant Bit Rate. CBR refers to processes such as voice and video that require a constant, repetitive or uniform transfer of information. The ATM Forum defines it as "an ATM service which supports a constant or guaranteed rate to transport services such as video or voice as well as circuit emulation which requires rigorous timing control and performance parameters."
2. Committed Bit Rate.

CBT Computer Based Training. Also known as CAI

(Computer Assisted Instruction), CBT commonly supports self-paced learning through the use of a CD-ROM storage technology. You plug the CD-ROM into the multi-speed CD-ROM drive of your high-performance PC, and begin the course. CD-ROM storage technology provides enough memory to support great graphics, and is fast enough to provide hyperlinks so you can move around the material quickly. At the end of each section, the CBT system will offer you a quiz to test your knowledge, and will present you with your score. CBT was crude prior to the development of CD-ROM and high-performance PCs. See also CAI and CD-ROM.

CBTA Canadian Business Telecommunications Alliance. The largest organization of business telecom users in Canada. According to the CBTA's own literature, the CBTA is a national non-profit organization that has been working on behalf of a business telecommunications customers for over 34 years. As a major voice in the Canadian telecommunications industry, the CBTA represents about 400 organizations from all sectors of the Canadian economy including industry, commerce, education, health care, and government. CBTA Member organizations have combined annual expenditures of over $4 billion (Canadian dollars). The CBTA strives to facilitate a competitive advantage in Canadian business (where business is understood to include all for-profit, not-for-profit, governmental, educational, and medical institutions), through the strategic application of telecommunications by proactively advocating the common interests of its members and by promoting information exchange and professional development. The main goals of the CBTA are to encourage innovation, quality, and choice in the telecommunications marketplace; to build a strong, influential, and representative national organization; to provide professional development opportunities and relevant services to our members; to enhance the Alliance's visibility as a leading industry authority. CBTA is at the Canadian Trust Tower, 161 Bay Street, Suite 3650, Toronto, Ontario, Canada M5J 2S1. Telephone 416-865-9993 and fax 416-865-0859. www.cbta.ca

CBUD Call Before U Dig. Operational management system for protection of fiber facilities. May have electronic geographic maps of states, counties and city streets where the carrier has buried facilities, upon which reported construction activities are automatically mapped. Human technicians verify that the activities do not pose a danger to the facilities, or dispatch on-site technicians when facilities may be at risk.

CC 1. Call Control. A wireless telecommunications term. A term used to refer to circuit communications management.
2. Country Code. The portion of an international telephone number used to identify the country of the called party. That country code may be one, two or three digits. An international phone number consists of a Country Code (CC) and a NSN (National Significant Number). Until December 31, 1996, the CC can be up to three digits and the NSN up to 11 digits — for a total of no more than 12 digits. After December 31, 1996, the CC stays at up to 3 digits and the NSN goes to up to 14 digits, for a total of no more than 15 digits. See the Appendix for a full list of International Country Codes.
3. Carbon Copy. That's where it comes from. It refers to the person who received one of the copies of a memorandum or document. Now CC is used electronic mail to mean that someone will receive a copy of the electronic mail.
4. Company Code. Also called OCN (Operating Company Number). A unique four-place alphanumeric code (NNXX) assigned to all U.S. domestic telecommunications service providers by NECA (National Exchange Carrier Association).

See Company Code for a complete explanation.

CCB Common Carrier Bureau. One of the largest divisions of the FCC, this Bureau regulates interstate telephone systems — licensing them, monitoring their charges, the conditions they offer service under, etc. The Common Carrier Bureau opened long distance communications to competition in America.

CCBM Came Clear By Magic. This term created (but not contributed) by Nynex. This is pseudo-technical lingo used by Nynex when they repair something that they broke (i.e. the problem was their fault), but they are unwilling to admit it was their fault. They say this "CCBM" to interconnect companies who are trying to get their customers' phone lines fixed.

CCC 1. Clear Coded Channel. A 64 Kbps channel in which all 64 Kbps is available for data.
2. Clear Channel Capability. The bandwidth of a data transmission path available to end users after control and signaling bits are accounted for.
3. Communications Competition Coalition. Lobbying organization established to encourage competition in telecommunications in Canada.

CCD 1. Charge Coupled Device. The "eyes" of a scanner or "digital camera." CCDs are small electronic devices with arrays of light-sensitive elements. The number of these elements and the width determine the scanner's or camera's resolution. Light is bounced off the image onto the CCD, which translates the varying intensities of the reflected light into digital data. CCD technology is used also in "digital still cameras" such as the Sony Mavica. The small size of the array itself — approximately microchip size — and the high resolution — around 1,000 by 1,018 elements — of these cameras have greatly improved "image acquisition" capabilities and opened up new applications in manufacturing quality control and in medicine.
2. Change Coupled Device.

CCDN Corporate Consolidated Data Network. It is the name for IBM's main internal data communications network. It used to be managed by IBM. It's now managed by Advantis, an IBM spin-off company, which is majority-owned by IBM and the rest by Sears.

CCE See Call Carrying Equipment.

CCFL Cold Cathode Fluorescent Lamp. A technology several laptop computer manufacturers use to light their LCD screens.

CCH Connections per Circuit per Hour. How many phone calls one circuit was able to complete in one hour. Compare to ACH, which is call Attempts per Circuit per Hour. See ACH. CCH and ACH are terms useful in traffic engineering, i.e. figuring out how many trunks you need for your incoming and/or outgoing telephone calls.

CCIA Computer and Communications Industry Association. A trade organization of computer, data communications and specialized common carrier services companies headquartered in Arlington VA. It runs seminars, does lobbying and generally tries to take care of the common interests of its members. See CBEMA.

CCIE Cisco Certified Internetwork Expert. Cisco's equivalent of Novell's CNE or Microsoft's MCSE. Attend a class. Study. Become a CCIE. A way of giving some certification as to your ability to manage a network.

CCIR Comite Consultatif International des Radiocommunications. The agency responsible for the international use of the radio spectrum. Effective in 1993, the CCIR is now known as the International Telecommunications Union — Radio. ITU-R and ITU-T form the International Telecommunications Union.

CCIR 601 An internationally agreed-upon standard for the digital encoding of component color television that was derived from the SMPTE RP125 and the EBU 324E standards. It uses the 4:2:2 sampling scheme for Y, U and V with luminance sampled at 13.5 MHz and chrominance (U and V components) sampled at 6.75 MHz. After sampling, 8-bit digitizing is used for each channel. These frequencies are used because they work for both 525/60 (NTSC) and 625/50 (SECAM and PAL) television systems. The system specifies that 720 pixels be displayed on each line of video. The D1 digital videotape format conforms to CCIR 601. See CCIR 656.

CCIR 656 The international standard defining the electrical and mechanical interfaces for digital TV operating under the CCIR 601 standard. It defines the serial and parallel interfaces in terms of connector pin outs as well as synchronization, blanking and multiplexing schemes used in these interfaces.

CCIRN Coordinating Committee for Intercontinental Research Networks. A committee that provides a forum for North American and European network research organizations to cooperate and plan.

CCIS Common Channel Interoffice Signaling. A way of carrying telephone signaling information along a path different from the path used to carry voice. CCIS occurs over a separate packet switched digital network. CCIS is separate from the talk path. A special version of CCIS is called Signaling System #7. SS#7 is integral to ISDN. CCIS offers basically two benefits: first, it dramatically speeds up the setting up and tearing down of phone calls. Second, it allows much more information to be carried about the phone call than what is carried on in-band (old-fashioned) signaling. That information can include the calling number, a message, etc.

Signaling for a group of voice telephone circuits is done on CCIS by encoding the information digitally on one of the voice circuits. In the previous method of signaling — the one replaced by CCIS — multi-frequency tones were sent down the same talkpath and the conversation would eventually travel. By taking the signaling information out of the talk path, the "phone phreak" community could no longer get free calls by using so-called "blue boxes" which duplicated the multi-frequency tones used by switching machines. CCIS is a much more efficient method of signaling, since it doesn't require a full voice grade channel just to check if the called party in LA is free and whether the call coming in from New York should be put through. See also COMMON CHANNEL SIGNALING, SYSTEM SIGNALING 7, ISDN and COMMON CHANNEL INTEROFFICE SIGNALING.

CCITT Comite Consultatif Internationale de Telegraphique et Telephonique, which, in English, means the Consultative Committee on International Telegraphy and Telephony. The ITU is now known as the ITU-T (International Telecommunications Union-Telecommunications Services Sector), based in Geneva Switzerland. The scope of its work is now much broader than just telegraphy and telephony. It now also includes telematics, data, new services, systems and networks (like ISDN). The ITU is a United Nations Agency and all UN members may also belong to the ITU, represented by their governments. In most cases the governments give their rights on their national telecom standards to their telecommunications administrations (PTTs, or TOs). But other national bodies (in the US, for example, the State Department) may additionally authorize Recognized Private Operating Agencies (RPOAs) to participate in the work of the ITU. After approval from their relevant national governmental body, manufacturers and scientific organizations may also be admitted, as well as other international organizations. This means, says the ITU, that participants are drawn from the broad arena. The activities of the ITU-T divide into three areas: Study Groups (at present 15) to set up standard ("recommendations") for telecommunications equipment, systems, networks and services.

Plan Committees (World Plan Committee and Regional Plan Committee) for developing general plans for a harmonized evolution of networks and services.

Specialized Autonomous Groups (GAS, at present three) to produce handbooks, strategies and case studies for support mainly of developing countries.

Each of the 15 Study Groups draws up standards for a certain area - for example, Study Group XVIII specializes in digital networks, including ISDN. Members of Study Groups are experts from administrations, RPOAs, manufacturing companies, scientific or other international organizations - at times there are as many as 500 to 600 delegates per Study Group. They develop standards which have to be agreed upon by consensus. This, says the ITU, can sometimes be rather time-consuming, yet it is a democratic process, permitting active participation from all ITU member organizations.

The long-standing term for such standards is "ITU (ITU-T) recommendations." As the name implies, recommendations have a non-binding status and they are not treaty obligations. Therefore, everyone is free to use ITU-T recommendations without being forced to do so. However, there is increasing awareness of the fact that using such recommendations facilitates interconnection and interoperability in the interest of network providers, manufacturers and customers. This is the reason why ITU-T recommendations are now being increasingly applied — not by force, but because the advantages of standardized equipment are obvious. ISDN is a good example of this. NOTE: ISDN and other standards recommendations include options which allow for multiple "standards," in recognition of differing national and regional legacy "standards;" as a result, international standards recommendations do not necessarily yield evenly applied standards options.

The ITU-T has no power of enforcement, except moral persuasion. Sometimes, manufacturers adopt the ITU-T specs. Sometimes they don't. Mostly they do. The ITU-T standardization process runs in a four-year cycle ending in a Plenary Session. Every four years a series of standards known as Recommendations are published in the form of books. These books are color-coded to represent different four cycles. In 1980 the ITU published the Orange Books, in 1984 the Red Books and, in 1988, the Blue Books. See ITU STUDY GROUPS and ITU V.XX below.

The ITU has now been incorporated into its parent organization, the International Telecommunication Union (ITU). Telecommunication standards are now covered under Telecommunications Standards Sector (TSS). ITU-T (ITU-Telecommunications) replaces ITU. For example, the Bell 212A standard for 1200 bps communication in North America was referred to as ITU V.22. It is now referred to as ITU-T V.22. See ITU.

CCL Configuration Control Link, works with ACD.

CCMA Call Centre Management Association, an association based in the United Kingdom. The CCMA is run by a Committee made up of elected representatives who have an interest in Call Centres. The annual subscription is stlg95.00 and a joining fee of 30 English pounds sterling. One of CCMA's primary goals is to set up an Educational programme where members can attend courses and have those attributed to a recognised qualification.

CCR 1. Customer Controlled Reconfiguration. An AT&T service that lets users make changes in their digital access and cross connect (DACS) network configurations at a DSO Level in either real time or according to a preplanned schedule.

2. Customer Control Routing, works with ACD.

3. An ATM term. Current Cell Rate: The Current Cell Rate is an RM-cell field set by the source to its current ACR when it generates a forward RM-cell. This field may be used to facilitate the calculation of ER, and may not be changed by network elements. CCR is formatted as a rate.

CCS 1. Centi Call Seconds. One hundred call seconds or one hundred seconds of telephone conversation. One hour of telephone traffic is equal to 36 CCS (60 x 60 = 3600 divided by 100 = 36) which is equal to one erlang. CCS are used in network optimization. See also Erlang and Traffic Engineering.

2. Common Channel Signaling: A form of signaling in which a group of circuits share a signaling channel. Refer to SS7.

CCS/SS7 A Bellcore term for Common Channel Signal/System Signaling 7. See SYSTEM SIGNALING 7.

CCS7 COMMON CHANNEL SIGNALING 7. See ISDN.

CCSA Common Control Switching Arrangement. A private network set up by AT&T for very large users and using parts of the public switched network. One important feature of a CCSA is that any user anywhere in a CCSA network can reach any other user by dialing only seven digits. Only very large customers subscribe to this service. It's expensive. AT&T has fewer than 100 customers.

CCSA ACCESS A PBX feature which allows a PBX user to get into a CCSA network. See CCSA.

CCSD Command Communications Service Designator; control communications service designator JP 1-02. It's a circuit designator. Also used as a circuit order number. A U.S. military term somewhat like USOC code in the civilian world.

CCT CONTINUITY CHECK TONE.

CCTV Closed Circuit TV.

CCU 1. Communication Control Unit. A processor, often a minicomputer, associated with a host mainframe computer that performs a number of communications-related functions. Compare with cluster control unit.

2. Camera Control Unit.

CD 1. Carrier Detect. CD is a signal generated by dial-up modem. CD indicates its connection status. If your CD light is on, then your modem is speaking to another modem.

2. Compact Disc. A 12 centimeter diameter (around 4 3/4 inches) disk containing digital audio or digital computer information, which can be played back and (now recorded) on a laser-equipped player. It was introduced by Sony and Philips in 1982. Philips (the inventor) chose the diameter of a Dutch 10-cent coin for the diameter of the hole in the CD. A compact disc originally came in only one flavor — read only. And most music tapes can only be listened to, not recorded to. For music it was a major breakthrough. It recorded music digitally (that is, coded as the zeros and ones of computer-speak) instead of trying to make an electrical copy of the sound waves themselves as devices like audio cassettes and LP records had. The Economist described the CD well. It said, "Instead of using a needle, the sound was plucked from the CD's surface by a tiny beam of laser light and then processed by a microcomputer. To the ear, the leap in performance between a compact disc and a long-playing record was even greater than the difference between color and black and white TV was to this eye."

A CD can typically hold up to 650 megabytes of information.

That is the equivalent of 1,500 floppies or 250,000 pages of print. Most computer CD-ROM drives can play audio CD disks — if they have the software and the speakers. Audio CD players, though, cannot play computer CD-ROM discs. But most computer CD-ROM players can play CD audio discs. As CD-ROM have become more popular and their makers have tried to do more and more with them, so CD-ROM formats have proliferated and some are not compatible with each other. See CD-R, CD-ROM, CD-I, CD-V, DVD and WORM.

CD I See CD-I below.

CD UDF This format defines specifications for a unified logical file format for CD-Recordable. The new format, known as CD UDF, defines a common scheme of "packet writing" to assure interchangeability of CD-R (CD Recordable) discs and to greatly increase performance and flexibility when storing large and small files on recordable disc media. As a result, CD-Recordable drives can be integrated into computer systems to behave much like other removable disk storage products. www.osta.org

CD-Audio Sometimes called "Redbook audio," is the digital sound representation used by CD-ROMs. CD audio is converted to analog sound output within the CD-ROM drive. The sampling frequency for CD audio is 44.1 KHz.

CD-I Compact Disc Interactive. Geared toward home entertainment, the drive connects to a television.

CD-Plus A format for CDs created by Sony and PHilips Electronics which makes the multimedia track on CDs invisible to CD players. The problem: there are few enhanced CDs — discs for both CD player and CD-ROM drives. The reason: the format prior to CD-Plus puts the multimedia data on track one, which listeners must skip over on their CD players. Thus the new format.

CD-R CD Recordable. A standard and technology allowing you to write to and read from a Compact Disc, but not erase or change what you record. This technology is compatible with existing CDs, i.e. you are able to read these discs in existing CD-players, and often in both PC and Macintosh machines. See CD-ROM and MULTI-SESSION.

CD-Recordable See CD-R

CD-ReWritable CD-ReWritable discs are erasable and can be used again and again, such an audio cassette.

CD-ROM Compact Disc Read Only Memory. Also called CD or CD-ROM. The familiar 4.7 inch Compact Disc which you see in the audio stores, but now made for computers. These discs hold huge amounts of data — as much as 660 megabytes per disk, or 330 pages of ASCII test. Put into a computer drive, time to retrieve information off the CD-ROM is much slower than from a hard disc. But CD-ROMs are catching up. Most of today's newer CD-ROMs can now be used in multimedia applications and virtually all CD-ROM drives available today support the Multimedia PC or MPC standard. As of writing, the first desktop device to record CD-ROM disks were beginning to appear. They cost several thousand dollars, but are, of course, coming down in price. 1993 was the first year where sales of encyclopedias on CD-ROM exceeded the number of sales of encyclopedias on paper. See CD, CD-ROM XA, CD-V, CD-WO and SHOVELWARE. See also DVD for an update on what CD-ROMs are becoming.

CD-ROM XA Stands for Compact Disc - Read Only Memory eXtended Architecture. Microsoft's extensions to CD-ROM that let you interleave audio with data. Though it is not a video specification, limited video can be included on disc. Demand for multimedia applications is increasing use of CD-ROM XA. To use it, you must have a drive that reads the audio portions of the disc and audio card in your computer that translates the digital into sound. Not all drives can recognize the extensions. See CD-WO.

CD-V Compact Disc Video. A format for putting 5 minutes of video on a 3-inch disc. This format has come and gone. Video is shifting towards CD-ROM XA.

CD-WO Compact Disc Write Once. A CD-ROM version of the WORM (Write Once Read Many) technology. For companies performing all CD-ROM publishing in-house, this format is useful for creating test discs before sending the master for duplication. CD-WO discs conform to ISO 9660 standards and can be played in CD-ROM drives.

CDA See COMMUNICATIONS DECENT ACT.

CDAR Customer Dialed Account Recording.

CDC Customer Data Change.

CDCS Continuous Dynamic Channel Selection.

CDDI Copper Distributed Data Interface is a version of FDDI (Fiber Distributed Data Interface— a 100 million bit per second local area network) that runs on unshielded twisted-pair cabling rather than optical fiber.

CDE Common Desktop Environment. A graphical user interface which is common and consistent across UNIX platforms. CDE is designed to make UNIX systems easier to use, simpler to support and more cost-effective to target for application development. CDE is defined to converge the major components of the desktop (X Window System, OSF/Motif, and CDE), adding basic features such as printing APIs (Application Programming Interfaces), and providing a competitive, common approach to delivering on-line information access and on-line publishing technology. A major initiative of The Open Group, CDE sponsors include Digital Equipment Corporation, Fujitsu Limited, Hewlett-Packard, Hitachi, IBM, Novell, and Sun.

CDEV Control panel DEVice. An Apple Macintosh term.

CDF 1. An ATM term. Cutoff Decrease Factor: CDF controls the decrease in ACR (Allowed Cell Rate) associated with CRM.
2. Channel Definition Format. A term used in Internet/Intranet push technology.

CDFP Centrex Data Facility Pooling.

CDFS Compact Disc File System, which controls access to the contents of CD-ROM drives in PCs.

CDG The CDMA Development Group (CDG), according to its Web site, is an industry consortium of companies who have come together to develop the products and services necessary to lead the adoption of CDMA wireless systems around the world. In working together, the 100-member companies will help ensure interoperability among systems, while expediting the availability of CDMA technology to consumers. The CDMA Development Group is committed to the definition of CDMA features, services, technical requirements and other activities that promote the availability and evolution of CDMA (IS-95 based) wireless systems worldwide. Specific objectives include:

• Leading the adoption of CDMA based systems around the world

• Maintaining a forum to address issues impacting manufacturers and carriers actively involved in CDMA deployments

• Developing next-generation CDMA systems

• Minimizing the time required to implement CDMA services and features. www.cdg.org

CDH Interface An interface once required by the Bell System to protect their phone lines from "foreign" (i.e. non-AT&T) phone equipment. CDH devices were eventually ruled a total waste of money and the phone companies refunded the

money — at least to the subscribers who asked. If you still have the stuff installed, you may be due a huge refund. Watch out for the statute of limitations.

CDL Coded Digital Locator

CDLC Cellular Data Link Control. A public domain data communications protocol used in cellular telephone systems. In other words, you can attach a data terminal to a cellular telephone and send and receive information. There are more 5,000 modems using CDLC on the Vodaphone Cellular System in the UK, where it is the de facto standard for cellular data communications. Features like improved synchronization field, forward error correction, bit interleaving, and selective retransmission make CDLC ideal for cellular transmissions, according to Millidyne who makes the CDLC modems in the US.

CDLRD Confirming Design Layout Report Date. The date a common carrier accepts the facility designed proposed by the Telco.

CDMA 1. Call Division Multiple Access.
2. Code Division Multiple Access, also called Spread Spectrum, is a name for a new form of digital cellular phone service. Motorola, a leading cellular manufacturer, says CDMA is a spread spectrum technology that assigns a code to all speech bits, sends a scrambled transmission of the encoded speech over the air and reassembles the speech to its original format. The major benefits of CDMA is increased capacity (up to 10 times analog) and more efficient use of spectrum. More importantly, CDMA technology provides three features that improve system quality: 1) The "soft hands-off" feature ensures that a call is connected before handoff is completed, reducing the probability of a dropped call. 2) Variable rate vocoding allows speech bits to be transmitted at only the rates necessary for high quality which conserves the battery power of the subscriber unit.
3) Multipath signal processing techniques combines power for increased signal integrity. Additional benefits to the subscriber include increased talk times for handportable units, more secure transmissions and special service options such as data, integrated voice and data, fax and tiered services.
CDMA works by combining each phone call with a code which only one cellular phone plucks from the air. Business Week said CDMA works "by spreading all signals across the same broad frequency spectrum and assigning a unique code to each. The dispersed signals are pulled out of the background noise by a receiver which knows the code. This method, developed by a San Diego company called Qualcomm Inc. is very new. According to the Wall Street Journal, CDMA systems are said to offer up to 20 times more call handling capacity than the conventional cellular systems by assigning a special electronic code to each call signal, allowing more calls to occupy the same space and be spread over an entire frequency band. Much of the equipment to support CMDA, like cellular switches, however, have not yet been developed."
CDMA is about to be used in inside-building wireless PBX conversations by companies including SpectraLink of Boulder, CO. SpectraLink's explanation: "One of several technologies used to separate multiple transmissions over a finite frequency allocation. CDMA operates in conjunction with spread spectrum transmission. Spread spectrum takes the original information signal and combines it with a correlating code, resulting in a signal which occupies a much greater bandwidth than the original. By assigning a unique correlating code to each transmitter, several simultaneous conversations can share the same frequency allocation. The process of

using spread spectrum in conjunction with individual correlating codes is known as Code Division Multiple Access." See also CODE DIVISION MULTIPLE ACCESS.

cdmaOne I received this press released explaining what cdmaOne is. Don't expect to be overwhelmed by this press release. "REPUBLIC OF SINGAPORE, June 3, 1997 — Facilitating the worldwide acceptance of CDMA technology, the CDMA Development Group (CDG) today announced the creation of a new universal term for IS-95 based CDMA specifications — cdmaOne. This new term will serve as an umbrella for the worldwide family of products that are based on the IS-95 air-interface, namely wireless local loop, PCS and cellular technology. cdmaOne was formally unveiled today by Perry LaForge, executive director for the CDG, at the Second Annual CDMA World Congress at the Raffles City Convention Centre in the Westin Stamford & Westin Plaza, Singapore.

"Our goal in developing this universal, simple industry term is to use a common name when referring to CDMA systems around the world, as

we are now in the process of building a worldwide integrated network based on cdmaOne technology," explained LaForge. "cdmaOne is the only true worldwide CDMA standard being deployed today, and is also the only wireless standard offering global operators superior system economics, improved network performance, increased capacity and enhanced voice quality — the necessary elements required to succeed in an increasingly competitive mobile marketplace."

In describing the selection of the name, LaForge added, "When reflecting on the new cdmaOne logo, the word One acknowledges two pertinent industry issues — first, cdmaOne is the only true worldwide standard for CDMA, and secondly, under this name, the industry is embracing a

family of standards and products that are all based on IS-95 technology. cdmaOne is rich in meaning, simple to communicate and eventually will become an important technology designator, even for the consumer market."

The name cdmaOne and its logo will be for the use of CDG member companies, and will serve to underscore the worldwide support for IS-95 based systems around the world. The new name encompasses all

CDMA standards terminology including IS-95, IS-96, IS-98, IS-99, IS-634 and IS-41."

CDMP Cellular Digital Messaging Protocol.

CDN Control Directory Number

CDO Community Dial Office. A small automatic central office switching system that is completely unattended. Routine maintenance is provided by a traveling technician once or twice each year, or as troubles develop. Such an office usually serves a small community with a few hundred lines in a rural area.

CDP Customized Dial Plan.

CDPA See CAPTURE DIVISION PACKET ACCESS.

CDPD Cellular Digital Packet Data. A radio technology that supports the transmission of packet data at speeds of up to 19.2 Kbps over the existing analog AMPS (Advanced Mobile Phone Service) cellular network, with appropriate CDPD upgrades. The data is structured in packets that are transmitted during pauses in cellular phone conversations, thereby avoiding issues of developing an overlay cellular network for data communications. Estimates suggest that as much as 20%-30% of an AMPS network is idle, even during periods of peak usage. This idle capacity is due to short pauses between the point in time at which you disconnect your circuit-switched cellular telephone conversation and the time

when someone else seizes that same radio channel to place a call. Idle capacity also is created when you are "handed-off" from one cell to another as you travel through the area of coverage in your vehicle. While 19.2 Kbps transmission rates are possible, throughput commonly drops to 2.4 Kbps or so during periods of peak usage.

Connectionless protocols such as IP (Internet Protocol) and the OSI CLNP (Connectionless Network Protocol) are employed to accomplish this minor miracle. The contention method employed is DSMA/CD (Digital Sense Multiple Access/Collision Detect), which is much like CSMA/CD used in Ethernet LANs. CDPD offers a number of advantages over competing wireless mobile data technologies. Among those advantages is the fact that CDPD packets use forward error correction (FEC) to reduce the impacts of noise and interference over the air link. Additionally, CDPD incorporates authentication and encryption to yield much improved security.

CDPD uses a full 30-KHz voice channel, but it can move your connection from one channel to another to avoid congesting voice communications. The drawbacks to CDPD are that it requires you to have a CDPD modem to access your upgraded cellular network, and not all cellular service providers are willing to upgrade their cellular equipment to CDPD. On July 21, 1993, the group of cellular carriers that supports the Cellular Digital Packet Data (CDPD) project released the complete version 1.0 of its open specification designed to enable customers to send computer data over an enhanced cellular network. The group said the packet data approach is ideally suited to those applications that require the transmission of short bursts of data, for example, authorizing a credit card number, exchanging e-mail messages or making database queries. The cellular networks deploying CDPD will enable mobile workers to use a single device to handle all of their voice and packet data needs. This version of the specification includes input from parties that reviewed the earlier release (0.8 and 0.9). The new version provides details of the CDPD architecture, airlink, external network interfaces, encryption and authentication, network support services, network applications services, network management, radio resource management and radio media access control.

CDPD was originally developed by IBM and is backed by the CDPD Forum. Carriers offering CDPD in the U.S. include Ameritech, AT&T Wireless, Bell Atlantic NYNEX Mobile, and GTE Mobilnet. Those carriers offer the service in over 60 markets. Limited CDPD coverage also is provided in Canada, Ecuador, Indonesia and Mexico. Pricing generally is on the basis of a monthly fee, plus a usage charge calculated on the number of kilobytes of data transmitted. Terminal equipment can be in the form of a laptop which plugs into your cellular phone through an adapter, or which has its own CDPD modem and antenna. Recently developed are CDPD terminals which look much like a cellular telephone, although with enhanced display capabilities. The original scope of the CDPD specifications were expanded in July 1995 to include CS-CDPD (Circuit Switched CDPD), which operates on a connection-oriented basis, much like cellular voice call. This approach supports longer file transmissions and yields improved throughput, as a radio channel is seized and maintained for the duration of the transmission. CDPD has application beyond AMPS, as it can be run over TDMA and CDMA networks, as well. See CDPD FORUM.

CDPD Cell Boundary The locus of points at which a Mobile End System (M-ES) should no longer access service by using the transmission of a particular cell. See CDPD.

CDPD Forum The CDPD Forum Inc. is an not-for-profit special interest group formed in 1994 to promote the development, deployment and use of CDPD. The forum comprises companies that develop, deliver or use CDPD products or services. companies with an interest in CDPD. www.cdpd.org See CDPD.

CDPD SNDCP Cellular Digital Packet Data (CDPD) SubNetwork Department Convergence Protocol. See CDPD.

CDR Call Detail Recording (as in Call Accounting) or Call Detail Record, as a record generated by customer traffic later used to bill the customer for service. See CALL ACCOUNTING.

CDR Exclude Table A table listing local central office codes which are not monitored (i.e. ignored) by a call accounting system.

CDSA Common Data Security Architecture. A security framework for developing security and authentication application programs.

CDSL Consumer Digital Subscriber Line Service, introduced by Rockwell in the fall of 1997. CDSL is a one megabit modem technology. The key difference between it and ADSL is that ADSL requires a splitter installed at each home to divide voice and data traffic onto separate lines. CDSL doesn't. According to Rockwell, customers would simply buy a CDSL modem at the local electronics store and call the phone company to have the service turned on. See ADSL.

CDV 1. Compression Labs Compressed Digital Video, a compression technique used in satellite broadcast systems. CDV is the compression technique used in CLI's SpectrumSaver system to digitize and compress a full-motion NTSC or PAL analog TV signal so that it can be transmitted via satellite in as little as 2 MHz of bandwidth. (A normal NTSC signal takes 6 Mhz.)
2. An ATM term. Cell Delay Variation: CDV is a component of cell transfer delay, induced by buffering and cell scheduling. Peak-to-peak CDV is a QoS delay parameter associated with CBR and VBR services. The peak-to-peak CDV is the ((1a) quantile of the CTD) minus the fixed CTD that could be experienced by any delivered cell on a connection during the entire connection holding time. The parameter "a" is the probability of a cell arriving late. See CDVT.
3. CD Video. A small videodisc (5" diameter) that provides five minutes of video with digital sound plus an additional 20 minutes of audio.

CDVT Cell Delay Variation Tolerance-ATM layer functions may alter the traffic characteristics of ATM connections by introducing Cell Delay Variation. When cells from two or more ATM connections are multiplexed, cells of a given ATM connection may be delayed while cells of another ATM connection are being inserted at the output of the multiplexer. Similarly, some cells may be delayed while physical layer overhead or OAM cells are inserted. Consequently, some randomness may affect the inter-arrival time between consecutive cells of a connection as monitored at the UNI. The upper bound on the "clumping" measure is the CDVT.

CE 1. An ATM term. Connection endpoint: A terminator at one end of a layer connection w thin a SAP. 2. Circuit emulation.

CE Mark Conformite Europeene (French) Mark; the English translation is European Conformity Mark. CE Mark is a type of pan-European equipment approval which indicates that the manufactured product complies with all legislated requirements for regulated products. Obtaining the CE Mark allows a product to be sold into 18 European countries without any further in-country testing. Several country regulatory bodies are set up to do the testing and thus the certification. These

"notified bodies" comprise CEN, CENELEC and ETSI. CEN is responsible for European standardization in all fields except electrotechnical (CENELEC) and telecommunications (ETSI). The CE Mark is now a requirement for all telecommunications terminal equipment (TTE) products sold into the European Union (EU), effective January 1, 1996. Countries covered are Austria, Belgium, Denmark, Finland, France, Germany, Greece, Ireland, Italy, Luxembourg, the Netherlands, Portugal, Spain, Sweden, and the United Kingdom. CE marking confirms that a product has been tested and meets the essential requirements of the European Telecom Directive to market it throughout the EU. The European TTE Directive 91/263/EEC specifies approved products to meet appropriate telecommunications technical standards. These include personal safety, protection of public networks, interoperation with public network equipment, and electromagnetic compatibility.
Several country regulatory bodies are set up to do the testing and thus the certification. These bodies are called "notified bodies." When one company, Larscom, reported one of its NDSUs had been granted a CE Mark, it said that the European "Notified Body" issuing the approval for its product was the British Approval Board for Telecommunications (BABT).

CeBIT CeBIT is the world's largest computer and office automation show. It attracts 600,000 or so people to Hannover, Germany in March or so of each year. It is also called the Hannover Fair. It is about five times the size of Comdex, which is North America's largest computer show. Many of the "booths" at Hannover are really small buildings, which are used year-round. May of the "booths" are three stories high, with an open air restaurant on the top floor. Space at the show is sometimes rented for four years.

CEBus Consumer Electronics Bus. EIA IS-60 (Electronics Industry Association Interim Standard-60), known as the home automation standard, includes specification of the CEBus. Essentially a Home Area Network (HAN) standard, CEBus allows connectionless, peer-to-peer communications over a common electrical bus, using standard electrical wiring rather than special voice/data cabling. CEBus employs a CSMA/CD contention protocol similar to that used in Ethernet LANs. Each CEBus has two channels. One channel is for real-time, short-packet, control-oriented functions; the other channel is dedicated to intensive data transfer. The standard includes error detection, automatic retry, end-to-end acknowledgement, and duplicate packet rejection, as well as authentication to ensure the identity of the user and encryption to provide for data security. For more information, contact the CEBus Industry Council in Indianapolis, Indiana.

CED 1. CallEd station iDentification. A 2100 Hz tone with which a fax machine answers a call. See CNG.
2. Capacitance Electronic Disc. System of video recording a grooved disc, employing a groove-guided capacitance pickup.

CED Compression A method of compression used in faxing.

CEI 1. Comparable Efficient Interface. The idea is that the telephone industry will let all its information providers have this interface — defined by technical specs and pricing — and, if it does, then the phone companies can themselves use this information to become information providers themselves. The concept has merit. Implementation has been agonizingly slow.
2. An ATM term. Connection endpoint Identifier: Identifier of a CE that can be used to identify the connection at a SAP.

Ceiling Distribution Systems Cable distribution system that use the space between a suspended or false ceiling and the structural floor for running cable. Methods used in ceiling distribution systems include zone, home-run, raceway, and poke-through.

Ceiling Feed A method of routing communications and / or power cabling vertically from the ceiling / plenum to a cluster of workstations.

CEKS Centrex Electronic Key Set.

Cell 1. The basic geographic unit of a cellular system. It derived its name "cell" from the honeycomb pattern of cell site installations. Cell is the basis for the generic industry term "cellular." A city or county is divided into smaller "cells," each of which is equipped with a low-powered radio transmitter/receiver. The cells can vary in size depending upon terrain, capacity demands, etc. By controlling the transmission power, the radio frequencies assigned to one cell can be limited to the boundaries of that cell. When a cellular phone moves from one cell toward another, a computer at the Mobile Telephone Switching Office (MTSO) monitors the movement and at the proper time, transfers or hands off the phone call to the new cell and another radio frequency. The handoff is performed so quickly that it's not noticeable to the callers. For a longer explanation, see CMTS.
2. The basic unit of a battery, consisting of plates, electrolyte and a container. A chemical device that produces electricity through electrolysis.
3. A unit of transmission in ATM and SMDS. A fixed-size packet consisting of a 48-octet payload and 5 octets of control overhead in the form of a header in the case of ATM, and a header and trailer in the instance of SMDS. SEE ALSO CELL SWITCHING, ATM and SMDS.
4. The smallest component of a table. In a table, a row contains one or more cells.

Cell Dragging An AMPS (Advanced Mobile Phone System) phenomenon in which a mobile cellular terminal moving away from the current serving cell penetrates deeply into or passes through a neighboring cell before the signal weakens to the point of reaching a predetermined signal level threshold, thereby becoming a candidate for handoff. The phenomenon is caused by irregularities of terrain and various other factors affecting radio propagation.

Cell Header A cell header precedes payload data (user information) in an ATM or SMDS cell. The header contains various control data specific to the cell switching protocol.

Cell Interarrival Variation CIV. An ATM term. "Jitter" in common parlance, CIV measures how consistently ATM cells arrive at the receiving end-station. Cell interarrival time is specified by the source application and should vary as little as possible. For constant bit rate (CBR) traffic, the interval between cells should be the same at the destination and the source. If it remains constant, the latency of the ATM switch or the network itself (also known as cell delay) will not affect the cell interarrival interval. But if latency varies, so will the interarrival interval. Any variation could affect the quality of voice or video applications.

Cell Group Color Code A cellular radio term. A color code assigned to a set of cells. Each member of the set is adjacent to at least one other member of the set and no two members of the set are allocated the same Radio Frequency (RF) channel for Cellular Digital Packet Data (CDPD) use. Each cell is assigned exactly one Cell Group Color Code.

Cell Loss Priority A bit in an Asynchronous Transfer Mode header, used to indicate a cell's priority level. A cell set with a CLP bit to zero has higher priority than a cell with the CLP bit set to one. In the case of congestion, cells with a CLP bit set to one may be discarded. See Cell Loss Priority Field.

Cell Loss Priority Field CLP. A single priority bit in the ATM cell header; when set, it indicates that the cell may be discarded should the network suffer congestion. Voice and video data do not tolerate such loss; therefore, such cells would not carry a set CLP bit.

Cell Loss Ratio A negotiated Quality of Service parameter in an Asynchronous Transfer Mode network. This parameter indicates a ratio of lost cells to total transmitted cells.

Cell Padding The space between the contents and inside edges of a table cell.

Cell Relay A form of packet switching using fixed length packets which results in lower processing and higher speeds. Cell relay is a generic term for a protocol based on small fixed packet sizes capable of supporting voice video and data at very high speeds. Information is handled in fixed length cells of 53 octets (bytes). A cell has 48 bytes of information and 5 bytes of address. The objective of cell relay is to develop a single high-speed network based on a switching and multiplexing scheme that works for all data types. Small cells (like 53 bytes) favor low-delay, a requirement of isochronous service. The downside to small cells is that the address information is almost 10 percent of the total packet. That equates to high overhead and raw inefficiency. See ATM and SMDS.

Cell Reversal The reversal of the polarity of the terminals of a battery cell as the result of discharging.

Cell Site A transmitter/receiver location, operated by the WSP (Wireless Service Provider), through which radio links are established between the wireless system and the wireless unit. The area served by a cell site is referred to as a "cell". A cell site consists of an antenna tower, transmission radios and radio controllers. The cell site of an analog cellular radio system handles up to 5,000 users (but not all at once).

Cell Site Controller The cellular radio unit which manages the radio channels within a cell.

Cell Spacing The amount of space between cells in a table. Cell spacing is the thickness, in pixels, of the walls of each cell.

Cell Splitting A means of increasing the capacity of a cellular system by subdividing or splitting cells into two or more smaller cells.

Cell Switching A term that refers to how cellular calls are switched. Cellular systems are built to accommodate moving phones — ones in cars, buses, etc. These phones are low-powered. The "moving-ness" and the low power of the phones poses major design constraints on the design of cellular switching offices, which are called Mobile Telephone Switching Office (or MTSO). First, you have to build many cellular switching sites. That way each phone is close to a cell site. Thus there's always a cell site which can pick up the transmission. Second, because of the closeness of the cell sites, any phone conversation may be simultaneously heard by several MTSOs. As a result, the MTSO constantly monitors signal strength of both the caller and the receiver. When signal strength begins to fade, the MTSO locates the next best cell site and re-routes the conversation to maintain the communications link. The switch from cell site to another takes about 300 milliseconds and is not noticeable to the user. All switching is handled by computer, with the control channels telling each cellular unit when and where to switch.
The Cellular Mobile Telephone System is a low-powered, duplex, radio/telephone which operates between 800 and 900 MHZ, using multiple transceiver sites linked to a central computer for coordination. The sites, or "cells", named for their honeycomb shape, cover a range of three to six, or more, miles in each direction. Their range is limited only by certain natural or man-made objects.

The cells overlap one another and operate at different transmitting and receiving frequencies in order to eliminate cross-talk when transmitting from cell to cell. Each cell can accommodate up to 45 different voice channel transceivers. When a cellular phone is activated, it searches available channels for the strongest signal and locks onto it. While in motion, if signal strength begins to fade, the telephone will automatically switch signal frequencies or cells as necessary without operator assistance If it fails to find an acceptable signal, it will display an "out of service" or "no service" message, indicating that it has reached the limit of its range and is unable to communicate.

Each mobile telephone has a unique identification number which allows the Mobile Telephone Switching Office (MTSO) to track and coordinate all mobile phones in its service area. This ID number is known as the Electronic Security Number (ESN). The ESN and Telephone Number are NOT the same. The ESN is a permanent number engraved into a memory chip called a PROM or EPROM, located in the telephone chassis. This number cannot be changed through programming as the telephone number can, although it can be replaced. Each time the telephone is used, it transmits its ESN to the MTSO by means of DTMF tones during the dialing sequence. The MTSO can determine which ESN's are good or bad, thus individual numbers can be banned from use within the system. See CELLULAR RADIO.

Cell Tax A reference to the demands that ATM (Asynchronous Transfer Mode) places on bandwidth. ATM is a cell-switching technology which segments a data stream into cells, each of which comprises 48 octets of payload (data) and 5 octets of overhead (control information). Therefore, each cell is approximately 10.4% overhead. ATM signaling adds another 5%-10% in overhead. While ATM is wasteful of bandwidth, its benefits are significant. See also ATM and Cell.

Cell Trace Box Wireless Integration/Interface Device. Also refered to as a "Proctor" box (name of the vendor), Cell Trace Box (US West's name for it), and protocol converter.

Cell Transfer A cellular radio term. Cell Transfer is the procedure of changing the channel stream in use to a channel stream originating at a different cell.

Cellpadding The syntax used to control the "padding" or area around the contents of a table's cell. In HTML tables this is used as a layout tool which allows an HTML author to render text and graphics on a Web page in columns and rows. There are many options available with tables and cellpadding is one of them. The syntax looks something like this: <table border=0 cellpadding=5 cellspacing=10>. This syntax would produce a table with a cellpadding of 5 pixels in width.

Cellphone A British term for a cellular telephone — whether car-based or handheld.

Cellular Data Link Control CDLC is a public domain data communications protocol used in cellular telephone systems. In other words, you can attach a data terminal to a cellular telephone and send and receive information. There are more 5,000 modems using CDLC on the Vodaphone Cellular System in the UK, where it is the de facto standard for cellular data communications. Features like improved synchronization field, forward error correction, bit interleaving, and selective retransmission make CDLC ideal for cellular transmissions, according to Millidyne who makes the CDLC modems in the US.

Cellular Digital Packet Data CDPD is an open standard developed by a group of cellular carriers led by McCaw.

The specification provides a standard for using existing cellular networks for wireless data transmission. Packets of data are sent along channels of the cellular network. See CDPD for a much more detailed explanation.

Cellular Digital Packet Data Group In the summer of 1993, a group of cellular carriers that supports the Cellular Digital Packet Data (CDPD) project has released the complete - version 1.0 - of its open specification designed to enable customers to send computer data over an enhanced cellular network. According to the group, the packet data approach is ideally suited to those applications that require the transmission of short bursts of data, for example, authorizing a credit card number, exchanging e-mail messages or making databases queries. The cellular networks deploying CDPD will enable mobile workers to use a single device to handle all of their voice and packet data needs. The 1.0 specification provides network and customer equipment manufacturers the parameters for building to this nationwide approach that sends packets of data over existing cellular networks. This version of the specification includes input from parties that reviewed the earlier release (0.8 and 0.9) and provides details of the CDPD architecture, airlink, external network interfaces, encryption and authentication, network support services, network applications services, network management, radio resource management and radio media access control. Copies of the specification can had from CDPD Project Coordinator Tom Solazzo at Pittiglio Rabin Todd & McGrath, 714-545-9400 extension 235.

Cellular Floor Method A floor distribution method in which cables pass through floor cells, constructed of steel or concrete, that provide a ready-made raceway for distributing power and communication cables.

Cellular Geographic Service Area CGSA. The geographic area served by the wireless (cellular) system within which a WSP is authorized to provide service.

Cellular Mobile Telephone Service See CMTS.

Cellular Modems A device that combines data modem and cellular telephone transceiver technologies in a single unit. This allows a user to transfer data on the cellular network without the use of a separate cellular telephone.

Cellular Phone Service See CMTS.

Cellular Protocols Conventions and procedures which relate to the format and timing of device communications. In data transmission communications, there are currently three major protocols, which are converging into a de facto standard: MNP, SPCL, and PEP.

Cellular Radio A mobile radio system. In the old days, there was one central antenna and everything homed in on that and emanated from it. With cellular radio, a city is broken up into "cells," each maybe no more than several city blocks. Every cell is handled by one transceiver (receiver/transmitter). As a cellular mobile radio moves from one cell to another, it is "handed" off to the next cell by a master computer, which determines from which cell the strength is strongest. Cellular mobile radio has several advantages:

1. You can handle many simultaneous conversations on the same frequencies. One frequency is used in one cell and then re-used in another cell. You can't do this on a normal mobile radio system.

2. Because one cellular system can accommodate many more subscribers than a normal mobile radio system, and therefore because it can achieve certain economies of scale, it has the potential of achieving much lower transmission costs.

3. Because the transceiver is always closer to the user than in

a normal mobile system, and the user's radio device thus needs less power, the device can be cheaper and smaller. Cellular radios started at over $5,000 and are now well under $500. From the first portable units, weight has already dropped to under one pound. There are several units that will fit in your breast pocket and not overly stretch your suit.

The following are specific cellular radio terms, or general telecom terms that mean something special in cellular radio:

A/B Switch Permits user to select either the wireline (B system) or the nonwireline (A system) carrier when roaming.

Alphanumeric memory Capability to store names with phone numbers.

Call-in-absence horn alert User-activated feature that sounds car horn upon receiving a call.

Call-in-absence indicator Feature that displays what calls came in while user was absent.

Call-in-progress override Insures that power to the phone is not lost if the car's ignition is turned off.

Call restriction Security feature that limits phone's use without completely locking it. Variations might include dial from memory only, dial last number only, seven-digit dial only, no memory access, etc.

Call timer Displays information on call duration and quantity. Variations might include present call, last call, total number of calls, or total accumulated time since last reset. Call-timer beep serves as a reminder to help keep calls brief. It might be set to go off once a minute, ten seconds before the minute, for example.

Continuous DTMF (touch-tones) Sends DTMF (dual-tone, multi-frequency) tones — also called touchtones — allowing access to voice mail and answering machines that require long-duration tones. "Continuous" means you get the tone so long as your finger is on the button. This may seem obvious to you and me, except that some "modern" phones just give a short tone no matter how long you keep your finger on the touchtone button.

Dual NAM Allows user to have two phone numbers with separate carriers (see multi-NAM).

Electronic lock Provides security by completely locking phone so it can't be used by unauthorized persons.

Expanded spectrum Full 832-channel analog cellular spectrum currently available to users.

Hands-free operation Allows user to receive calls and converse while leaving handset in cradle (similar to office speakerphone).

Hands-free answering Phone automatically answers incoming call after a fixed number of rings andgoes to hands-free operation.

Memory linkage Allows programming specific memory locations to dial a sequence of other memory locations.

M ulti-NAM A cellular telephone term to allow a phone to have more than two phone numbers, each of which can be on a different cellular system if desired. This lets the user register with both carriers in home city, expanding available geographic coverage.

Mute Silences the telephone's microphone to allow private conversations without discontinuing the phone call. Audio mute turns off the car stereo automatically when the phone is in use, and turns it back on when the call is completed.

NAM Numerical Assignment Module. Basically, your cellular phone number, although it refers specifically to the component or module in the phone where the number is stored.

On-hook dialing Allows dialing with the handset in the cradle.

Roaming Using any cellular system outside your home system. Roaming usually incurs extra charges.

Scratch pad Allows storage of phone numbers in temporary memory during a call. Silent scratch pads allows number entry into scratch pad without making beep tones.

Signal strength indicator Displays strength of cellular signal to let user know if a call is likely to be dropped.

Speed dialing Dialing phone number from memory by pressing a single button.

Standby time Maximum time cellular phone operating on battery power can be left on to receive incoming calls.

Talk time Maximum time cellular phone operating on battery power can transmit.

Voice-activated dialing Your cellular phone recognizes your words and dial accordingly. You say "Dial Mom" and it dials mom.

Cellular Radio Switching Office The electronic switching office which switches calls between cellular (mobile) phones and wireline (i.e. normal wired) phones. The switch controls the "handoff" between cells and monitors usage. Different manufacturers call their equipment different things, as usual.

Cellular Switching See CELL SWITCHING.

Celluplan II A proposed national standard to place packets of data between idle spaces on a cellular voice network.

CELP Code Excited Linear Prediction. An analog-to-digital voice coding scheme. Sun is proposing the use of CELP so that a user could send realtime voice communications over local area or wide are network — bypassing the phone system!

CEMH Controlled Environment ManHole. Environmental control of the CEMH is maintained by a heat pump (a fancy name for an airconditioner — cooler and heater).

CEN Comite European de Normalisation (French); the English translation is European Committee for Standardization. CEN is responsible for European standardization in all fields except electrotechnical (CENELEC) and telecommunications (ETSI). Certified products are awarded the CE Mark, signifying that a company has met the applicable essential health and safety requirements and the specific conformity assessment requirements to market its product in the European Union under the "New Approach" directives. CEN membership includes all EU countries, as well as affiliate members including Turkey, Cyprus, and many countries which formerly were members of the Soviet Bloc. Technical committees address standards in the areas of medical informatics, geographic information systems, character set technology for multilingual information infrastructure, and advanced manufacturing technologies.

CENELEC Comite European de Normalisation ELECtrotechnique (French); the English translation is European Committee for Electrotechnical Standardization. CENELEC is responsible for European standardization in the electrotechnical field, working closely with ETSI (telecommunications) and CEN (all other fields). Certified products are awarded the CE Mark, signifying that a company has met the applicable essential health and safety requirements and the specific conformity assessment requirements to market its product in the European Union under the "New Approach" directives. CENELEC is the European technical organization responsible for coordination of standards for safety and electromagnetic emissions for electrical equipment in the European Economic Community (EEC). The EEC is working toward having a uniform set of standards that will apply for all EEC countries. Membership includes all EU countries, as well as affiliate members including Turkey, Cyprus, and many countries which formerly were members of the Soviet Bloc.

Central Office CO. (pronounced See-Oh). Central office is an ambiguous term in North America. It can mean a telephone company building where subscribers' lines are joined to switching equipment for connecting other subscribers to each other, locally and long distance. Sometimes, that central office means a wire center in which there might be several switching exchanges. That means they'll be switches, cable distribution frames, batteries, air conditioning and heating systems, etc. But a central office is sometimes simply a single telephone switch, what Europeans call a public exchange. In short, you have to figure out by the context if central office means a building or a switch, or a collection of switches. Simple, eh?

Central Office Battery A group of wet cells joined in series to provide 48 volts DC. Central office batteries are typically charged off the main 120 volts AC. The batteries have two basic functions. 1. To provide a constant source of DC power for eight hours or so after AC powers drops, and 2. To isolate the central office from glitches on the AC line.

Central Office Code Part of the national numbering plan, the central office code is the second three digits of a subscriber's telephone number, which identifies the local switching office. Here is Bellcore's definition: A 3-digit identification under which up to 10,000 station numbers are subgrouped. Exchange area boundaries are associated with the central office code that generally have billing significance. Note that multiple central office codes may be served by a central office. Also called NXX code or end office code.
Several central office codes in North America are kept for special purposes
555 — Directory Assistance
950 — Feature Group B Access
958 — Local Plant Test
959 — Local Plant Test
976 — Information Delivery Service

Central Office Equipment Reports COER. A telephone company definition. A large scale computer software package which accepts Central office Engineering Data properly formatted by a Data Collection System (DCS), subjects these data to a series of validation tests, and produces final summarized reports designed to meet both administrative and engineering requirements.

Central Office Override A third party may interrupt or join in your conversation.

Central Office Trunk 1. A trunk between central offices. It may be between major switches or between a major and a minor switch.
2. A trunk between public and private switches.

Central Processing Unit CPU. The part of a computer which performs the logic, computational and decision-making functions. It interprets and executes instructions as it receives them. Personal computers have one CPU, typically a single chip. It is the so-called "computer on a chip." That chip identifies them as an 8-bit, 16-bit or 32-bit machine.
Telephone systems, especially smaller ones, are not that different. Typically they have one main CPU — a chip — which controls the various functions in the telephone. Today's telephone systems are in reality nothing more than special purpose computers. As phone systems get bigger, the question of CPUs - - central processing units — becomes harder to figure. The design of phone systems has, of late, tended away from single processor-controlled telephone systems (as in single processor controlled PCs). There are several reasons for this move. First, it's more economical for growth. Make

modules of "little" switches and join little ones together to make big ones. Second, it's more reliable. It's obviously better not to rely on one big CPU, but to have several. In short, the issue of Central Processing Units — CPUs — is blurring. But the concept is still important because by understanding how your telephone switch works (its architecture), you will understand its strengths and weaknesses.

Centralized Attendant Service Calls to remote (typically branch) locations are automatically directed to operators at a central location. Imagine four retail stores in a town. There are three branch stores and one main, downtown store, each having their own local phone numbers, which customers call. It's clearly inefficient to put operators at each of the stores — when one group is busy, the other will be free, etc. What this feature does is to direct all the calls coming into each of the stores into one bank of operators, who then send those calls back to the outlying stores.
Despite the extra schlepping of calls around town, having one large group of operators is cheaper than maintaining many small groups. Each store has its own local Listed Directory Number (LDN) Service. Special Release Link Trunk circuits connect each unattended location (each store) to the main attendant location. These trunks are only temporarily used during call processing. An incoming call to an unattended store seizes such a trunk circuit for completion of the call to the centralized attendant, who then uses the same trunk circuit to process the call to the remote location's internal extension. (After all if the caller was calling that store, they obviously want to talk to someone in that store.) The circuit is then released and is available for other calls. Since such special trunk circuits are only used during that part of a call that requires connection between locations, such trunks are more efficient than normal tie trunk circuits.

Centralized Automatic Message Accounting CAMA. The recording of toll calls at a centralized point.

Centralized Network Administration See CNA

Centralized Ordering Group COG. An organization provided by some communications service providers (like a local phone company) to coordinate services between the companies and vendors.

Centrex Centrex is a business telephone service offered by a local telephone company from a local central office (also called a public exchange). Centrex is basically normal single line telephone service with "bells and whistles," added. Those "bells and whistles" include intercom, call forwarding, call transfer, toll restrict, least cost routing and call hold (on single line phones).
Think about your home phone. You can often get "Custom Calling" features. These features are typically fourfold: Call forwarding, Call Waiting, Call Conferencing and Speed Calling. Centrex is basically Custom Calling, but instead of four features, it has 19 features. Like Custom Calling, Centrex features are provided by the local phone company's central office.
Phone companies peddle Centrex is leased to businesses as a substitute for that business buying or leasing its own on-premises telephone system — its own PBX, key system or ACD. Before Divestiture in 1984, Centrex was presumed dead. AT&T was, at that time, intent on becoming a major PBX and key system supplier. Then Divestiture came, and the operating phone companies recognized they were no longer part of AT&T, no longer had factories to support, but did have a huge number of Centrex installations providing large monthly revenues. As a result, the local operating companies have injected new life into Centrex, making the service more

attractive in features, price, service and attitude. Here are the main reasons businesses go with Centrex as opposed to going with a stand-alone telephone system:

1. Money. Centrex is typically cheaper to get into (the central office already exists). Installation charges can be low. Commitment can also be low, since most Centrex service is leased on a month-to-month basis. So it's perfect for companies planning an early move. There may be some economies of scale, also. Some phone companies are now offering low cost, large size packages.

2. Multiple locations. Companies with multiple locations in the same city are often cheaper with Centrex than with multiple private phone systems and tie lines, or with one private phone system and OPX lines. (An OPX line is an Off Premise eXtension, a line going from a telephone system in one place to a phone in another. It might be used for an extension to the boss' home.)

3. Growth. It's theoretically easier to grow Centrex than a standalone PBX or key system, which usually has a finite limit. With Centrex, because it's provided by a huge central office switch, it's hard, theoretically, to run out of paths, memory, intercom lines, phones, tie lines, CO lines, etc. The limit on the growth of a Centrex is your central office, which may be many thousands of lines.

4. Footprint Space Savings. You don't have to put any switching equipment in your office. All Centrex switching equipment is at the central office. All you need at your office are phones.

5. Fewer Operators because of Centrex's DID features. Fewer operator positions saves money on people and space.

6. Give better service to your customers. With Centrex, each person has their own direct inward dial number. Many people prefer to dial whomever they want directly rather than going through a central operator. Saves time.

7. Better Reliability. When was the last time a central office crashed? Here are some of the features built into modern central offices: redundancy, load-sharing circuitry, power back-up, on-line diagnostics, 24-hour on-site personnel, mirror image architecture, 100% power failure phones, complete DC battery backup and battery power. Engineered to suffer fewer than three hours down time in every 40 years.

8. Non-blocking. Trunking constraints are largely eliminated with Centrex, since a central office is so large.

9. Minimal Service Costs. Repair is cheap. Service time is immediate. People are right next to the machine 24-hours a day. Phones and wires are the only things that require repair on the customers' premises. You can easily plug new phones in, plug them out yourself. All other equipment is in the central office. You need not hold inventory or test equipment.

10. No technological obsolescence. Renting Centrex means a user has the ultimate flexibility — ability to jump quickly into new technology. Central offices are moving quickly into new technologies, such as ISDN.

11. Ability to manage it yourself. You can now get two important features previously available only on privately-owned self-contained phone systems (like PBXs): 1. The ability for you, the user, to make changes to the programming of your own Centrex installation without having to personally call a phone company representative. 2. The ability to get call detail accounting by extension and then have reports printed by a computer in your office. The phone company does this call accounting by installing a separate data line which carries Centrex call records back to the customer as those calls are made.

The above arguments are pro-Centrex. There are also anti-Centrex arguments. Central offices often run out of capacity.

Centrex is also cable-intensive. A PBX with 20 trunks and 100 phones only needs 20 cable pairs from the user's office to the telephone company. A Centrex installation with the same configuration needs 100 pairs. Every time someone new joins your company, the phone company needs to install another cable pair from the central office to your new employee's desk. Sometimes, they have it. Sometimes, they don't. Delays can get extensive. What with the explosion of telecom demand in recent years — individual fax machines, the Internet, etc. —there just isn't enough copper in the ground, and a typical telco won't plow in the cable unless they receive a pay-off in three years.

The "big" key to Centrex traditionally comes down to price. In some cities the price of Centrex lines is lower than "normal" PBX lines. Of course, you can buy Centrex lines and attach your own PBX or key system to those Centrex lines. The big disadvantage of Centrex is that there are very few specialized Centrex phones able to take better advantage of Centrex central office features the way electronic PBX phones take advantage of PBX features.

Centrex is known by many names among operating phone companies, including Centron and Cenpac. Centrex comes in two variations — CO and CU. CO means the Centrex service is provided by the Central Office. CU means the central office is on the customer's premises. See the following CENTREX definitions.

Centrex Call Management A Centrex feature that provides detailed cost and usage information on toll calls from each Centrex extension, so you can better manage your telephone expenses.

Centrex CCRS Centrex Customer Rearrangement System. Computer software from New York Telephone that allows their Centrex customers to make certain changes in their own line and features arrangements. Other phone companies have similar services under different names.

Centrex CO Indicates that all equipment except the attendant's position and station equipment is located in the central office. See CENTREX.

Centrex CU Indicates that all equipment including the dial switching equipment, is located on the customer's premises. See CENTREX.

Centrex Extend Service The name of a Bell Atlantic service. If you maintain offices in multiple locations — or have work-at-home employees — this service allows you to tie all your locations into one phone system. So everyone in your company can take advantage of Centrex features and services on a cost-efficient, call-by-call sharing basis. With Centrex Extended Service, you can even tie non-Centrex locations into your Centrex system.

Centrex LAN Service Put a modem on a dedicated central office line. Connect that line to a switch. Bingo you have a switched, relatively high-speed service that can connect synchronous and asynchronous terminals and other equipment at speeds up to 19.2 Kbps. Centrex LAN service works with existing wiring Centrex central office to customer premise wiring. Centrex LAN is a name given this service by Nynex. Other phone companies have similar services under different names. It's not a very successful service, since it's very slow. Compare 19,200 bits per second to Ethernet, which is 10 million bits per second!

Centrex SMDI Have you ever called someone and been forwarded to their voice mail, and then had to enter their extension again. This is because their voice mail does not know where the call originated. The voice mail system does not know who you just called, All it knows is that it just received another call. SMDI is simply a modem link back to the central office supplying your Centrex system. This modem link will feed a computer at your location the information about incoming calls as they are forwarded through your system. It feeds the originating number and why the call is transferred, so you know whether the user didn't answer their phone, or that it was busy.

To indicate the health of the SMDI link, it generates a heartbeat message every few seconds. If you don't receive a heartbeat within the time window, you know there is a problem with the SMDI link and know to restart the link. The one good thing about SMDI is the option that your voice mail system can control the status of message waiting indicators on the user's phone. They can either have a message waiting lamp on their phone, or use the "stutter dial tone" which causes a broken dial tone when the user picks up their phone.

Problems to watch out for.. Since SMDI is a communications link, it can be broken. If the SMDI link goes down, make sure you build in the old two step method of finding out what extension the caller was attempting to call. Make sure you also offer a user directory in case the caller does not know the extension number. When the link goes down, make sure the system does not continue to spend the message waiting status commands down an inactive link. SMDI stands for Standard Message Desk Interface or Simplified Message Disk Interface. See also SMDI.

Centronics The name of the printer manufacturer whose method of data transmission between a computer and a parallel printer has become an industry standard. See CENTRONICS PRINTER STANDARD.

Centronics Printer Standard The Centronics standard was developed by the Centronics company which makes computer printers. The Centronics standard is a 36-pin single plug/connector with eight of the 36-pins carrying their respective bits in parallel (eight bits to one character), which means it's much faster than serial transmission which sends only one bit at a time. There are several types of Centronics male and female plugs and receptacles. So know which you want before you buy. The pinring — the location of and function of each of the 36-individual wires, is standard from one Centronics cable to another.

The Centronics printer standard has been adopted by many printer and PC companies, including IBM. It is a narrower standard than the RS-232-C standard. The Centronics works only between a computer and a printer. It won't work over phone lines, unless conversion is done at either end. However, it is standard and has none of the dumb interface problems the RS-232-C standard does.

Centum Call Second 1/36th of an erlang. The formula for a centum call second is the number of calls per hour multiplied by their average duration in seconds, all divided by 100.

CEO Chief Executive Officer.

CEPT Conference des administrations Europeenes des Postes et Telecommunications (European Conference of Postal and Telecommunications Administrations). Standards-setting body whose membership includes European Post, Telephone, and Telegraphy Authorities (PTTs). It in turn participates in relevant areas of the work of CEN/CENELEC. It was originally responsible for the NET standards, but these have subsequently been passed on to ETSI.

CEPT Format Defines how the bits of a PCM carrier system of the 32 channel European type T-1/E-1 will be used and in what sequence. To correctly receive the transmitted intelligence, the receiving end equipment must know exactly what

each bit is used for. CEPT format uses 30 VF channels plus one channel for supervision/control (signaling) and one channel for framing (synchronizing). All 8 bits per channel are used to code the waveshape sample. For a much better explanation, see T-1.

CER An ATM term. Cell Error Ratio: The ratio of errored cells in a transmission in relation to total cells sent in a transmission. The measurement is taken over a time interval and is desired to be measured on an in-service circuit.

CERB Centralized Emergency Reporting Bureau. A Canadian term similar to PSAP — Public Safety Answering Position. See PSAP.

Cerf, Vinton Founding President of the Internet Society (ISOC) from 1992 to 1995 and co-creator of the transmission control protocol/Internet Protocol (TCP/IP), which enables computers to talk to each other over the Internet. Cerf proved that a network can reconfigure itself so that no communications are lost. He did this by breaking apart the Defense Department's Arpanet network artificially, and showed that it could be reconnected by way of flying packet radios in Strategic Air Command jets. Cerf used the airborne radios to link to ground radios which were, in turn, linked to internet gateways (today's routers) to effectively interconnect pieces of ARPANET artificially separated.

CERN European Laboratory for Particle Physics Research in Geneva, Switzerland.

CERT Computer Emergency Response Team. A group of computer experts at Carnegie-Mellon University who are responsible for dealing with Internet security issues. The CERT is chartered to work with the Internet community to facilitate its response to computer security events involving Internet hosts, to take proactive steps to raise the community's awareness of computer security issues, and to conduct research targeted at improving the security of existing systems. The CERT was formed by DARPA in November 1988 in response to the Internet worm incident. CERT exists to facilitate Internet-wide response to computer security events involving Internet hosts and to conduct research targeted at improving the security of existing systems. They maintain an archive of security-related issues on their FTP server at "cert.org." Their email address is "cert@cert.org" and their 24-hour telephone Hotline for reporting Internet security issues is 412-268-7090.

Certificate A cryptography term. Also known as a digital certificate, a "certificate" is a password-protected, encrypted data file which includes the name and other data which serves to identify the transmitting entity. The certificate also includes a public key which serves to verify the digital signature of the sender, which is signed with a matching, private key, unique to the sender. Through the use of keys and certificates, the entities exchanging data can authenticate each other.

Certificate Authority CA. A trusted third-party organization or company that issues digital certificates used to create digital signatures and public-private key pairs. These pairs allow all system users to verify the legitimacy of all other system users with assigned certificates. The role of the certificate authority is to guarantee that the individual granted the unique certificate is, in fact, who he or she claims to be. Usually, this means that the certificate authority has an arrangement with a financial institution, such as a credit card company, which provides it with information to confirm an individual's claimed identity. Certificate authorities are a critical component in data security and electronic commerce because they guarantee that the two parties exchanging information are really who they claim to be.

Certified Several companies in the "secondary" industry test used equipment, parts and/or systems. They have various ways of testing them. Typically they test with working phones operating for extended periods at different temperatures. The idea is to check that this used equipment works the way it's meant to work — to the original manufacturer's design specification. Once these tests have been completed a secondary dealer will "certify" such equipment, usually in writing. Such certification carries the assurance that the used equipment works as it's meant to. Sometimes certified equipment is upgraded to the most current revision level of hardware and software. Sometimes it's not. You, the buyer, must check. Certified equipment typically carries a guarantee — that guarantee being as good, obviously, as the company that backs it.

Certified Equipment A term used in the secondary telecom equipment business. Equipment carrying the written assurance that it will perform up to the manufacturer's specifications. It qualifies for addition to existing maintenance contracts.

CES An ATM term. Circuit Emulation Service: The ATM Forum circuit emulation service interoperability specification specifies interoperability agreements for supporting Constant Bit Rate (CBR) traffic over ATM networks that comply with the other ATM Forum interoperability agreements. Specifically, this specification supports emulation of existing TDM circuits over ATM networks.

CESID Caller Emergency Service ID. Several states in the United States require PBXs to send a telephone extension number to a PSAP (E-911 emergency answering point). This is helpful in cases where the caller is not located near the address listed for the business' Listed Directory Number. A "campus" environment of separate buildings comes to mind. The location information displayed at the PSAP usually comes from the ALI (Automatic Location Identification) database. Customers with PBXs are normally required to provide initial telephone location information as well as updates to the ALI database maintainer (usually the LEC). The CESID information is transmitted in one of two formats: ISDN-PRI ANI (around Chicago and in some areas of New York), or CAMA. CAMA uses R2MF signaling to send the CESID to the PSAP, and, like ISDN-PRI, has to be ordered as a specific type of trunk from the LEC. Unlike ISDN-PRI, however, it appears to have no other use aside from E-911 calls. Until or unless PSAPs in the rest of the country switch over to ISDN-PRI trunks for call receipt, CAMA trunks will be around for awhile. See CAMA, E-911, ISDN PRI, LEC and PSAP.

Cesium Clock A clock containing a cesium standard as a frequency-determining element. It's a very accurate clock. See CESIUM STANDARD.

Cesium Standard A primary frequency standard in which a specified hyperfine transition of cesium-133 atoms is used to control the output frequency. Its accuracy is intrinsic and achieved without calibration.

CEU Commercial End User. See SU, service user.

CEV Controlled Environmental Vault. A below ground room that houses electronic and/or optical equipment under controlled temperature and humidity.

CFA 1. Carrier Facility Assignment. CAPs/CLECs give RBOCs/LECs a slot or channel assignment where their T-1s or T-3s will be connecting. A CFA is the identifier or location where an IXC, CAP, CLEC, or LEC will interconnect with the incumbent Telco. It will come in one of three forms: ACTL/CLLI, APOT, or tie/cable pair. ACTL/CLLI looks like

1001/T3/18/WASHDCAB123/WASHDCXY789, where 1001 is the DS-3 off a SONET OC ring, while T3 signifies that connection will be made at the T-3 slot 18 (the T-1 will connect to this slot). The first location CLLI is the ACTL WASHDC-CAB123. The second location CLLI is the central office CLLI WASHDCXY789 If the CFA is in APOT form, it will be like 1.9.13.23.18 (i.e. floor.aisle.bay.panel.jack). Tie/cable pair looks like 10011/t3/T1TIE/WASHDCAB123/WASHDCAB where the last 1 in 10011 signifies slot 1. The Telco tech will then do the interconnect or x-connect at that facility assignment. See CAP, CLEC and RBOC.

2. Carrier Failure Alarm. The alarm which results from an out-of-frame or loss of carrier condition and which is combined with trunk conditioning to create a CGA.

3. Connecting Facilities Arangement. Identifies a complete communications channel between two places.

CFAC Call Forward All Calls.

CFAMN Call Forwarding Address Modified Notification.

CFB 1. Call Forward Busy.

2. Call Forwarding on mobile subscriber Busy. A wireless telecommunications term. A supplementary service provided under GSM (Global System for Mobile Communications).

CFDA Call Forward Don't Answer.

CFGDA Call Forward Group Don't Answer.

CFM Cubic Feet per Minute. A measure of how much air you move through the fan of a PC.

CFNRc Call Forwarding on mobile subscriber Not Reachable. A wireless telecommunications term. A supplementary service provided under GSM (Global System for Mobile Communications).

CFNRy Call Forwarding on No Reply. A wireless telecommunications term. A supplementary service provided under GSM (Global System for Mobile Communications).

CFO Chief Financial Officer.

CFP Channel Frame Processor.

CFR Confirmation to Receive frame.

CFRP Carbon Fiber Reinforced Plastic. A light and durable material, which has been used (for the wings of advanced fighter jets) in the defense business and which Toshiba introduced in 1991 as casing for a line of notebook sized computer laptops.

CFUC Call Forwarding UnConditional. A wireless telecommunications term. A supplementary service provided under GSM (Global System for Mobile Communications).

CFV Call For Votes. Begins the voting period for a Usenet newsgroup. At least one (occasionally two or more) email addresses is customarily included as a repository for the votes.

CFW Call Forward.

CGA 1. Carrier Group Alarm. A service alarm generated by a channel bank when an out-of-frame (OOF) condition exists for some predetermined length of time (generally 300 milliseconds to 2.5 seconds). The alarm causes the calls using a trunk to be dropped and trunk conditioning to be applied.

2. Color Graphics Adapter. An obsolete IBM standard for displaying material on personal computer screens. The simplest (and conventional) CGA displays 320 horizontal picture elements, known as pels or pixels, by 200 pels vertically. There is also an Enhanced CGA, which is 640 x 400, or 128,000 pixels per screen. Older portables may use CGA monochrome mode. CGA has essentially been obsoleted by VGA. See MONITOR and VGA.

CGI Common Gateway Interface. An Internet term. Programs or Scripts, usually executed on the Web server, that perform actions (like searching or running applications) when the user clicks on certain buttons or parts of the Web Screen. CGI actually refers to the pre-defined way in which these programs communicate with the Web Server but has lately come to refer to the programs themselves. The preferred programming language for CGI is PERL. See also CGI-Bin and PERL.

CGI-Bin The most common name of a directory on a web server in which CGI programs are stored. The "bin" part of "cgi-bin" is a shorthand version of "binary" because once upon a time, most programs were referred to as "binaries." In real life, most programs found in cgi-bin directories are text files — scripts that are executed by binaries located elsewhere on the same machine.

CGI Joe A Wired Magazine definition. A hardcore CGI script programmer with all the social skills and charisma of a plastic action figure.

CGM Computer Graphics Metafile. A standard format that allows for the interchanging of graphics images.

CGSA Cellular Geographic Service Area. The actual area in which a cellular company provides cellular service. CGSAs are usually made up of multiple counties and often cross state lines.

CGSA Restriction If you own a cellular phone, you are prevented from making calls outside your own local Cellular Geographic Service Area. This restriction is an option that is available to subscribers in most cellular cities.

Chad 1. The little solid round dots of paper made when paper tape is punched with information.
2. CHAnge Display.

Chad Tape Punched tape used in telegraphy/teletypewriter operation. The perforations, called "chad," are severed from the tape, making holes representing the characters.

Chadless Tape 1. Punched tape that has been punched in such a way that chad is not formed.
2. A punched tape wherein only partial perforation is completed and the chad remains attached to the tape. This is a deliberate process and should not be confused with imperfect chadding. See CHAD.

Chain Mailboxes Mailboxes that are connected together to provide a service or a number of messages (e.g. Directory, Product Information, etc.).

Chaingang A group of Web homepages which merely link to each other.

Chaining A programming technique linking one activity to another, as in a chain. Each link in the chain may contain a pointer to the next link, or there may be a master control or program instructing the programs to link together.

Chainsaw Consultant An outside "expert" brought in to reduce the employee headcount, leaving the top brass with clean hands and a clean conscience.

Challenge-Handshake Authentication Protocol CHAP. An authentication method that can be used when connecting to an Internet Service Provider. CHAP allows you to log in to your provider automatically, without the need for a terminal screen. It is more secure than the Password Authentication Protocol (another widely used authentication method) since it does not send passwords in text format. An Internet term.

Challenge-Response A type of authentication procedure into a system in which a user must respond correctly to a challenge, usually a secret key code, to gain access.

Channel 1. Typically what you rent from the telephone company. A voice-grade transmission facility with defined frequency response, gain and bandwidth. Also, a path of communication, either electrical or electromagnetic, between two or more points. Also called a circuit, facility, line, link or path.
2. An SCSA term. A transmission path on the SCbus or SCxbus Data Bus that transmits data between two end points.
3. A channel of a GPS (Global Positioning System) receiver consists of the circuitry necessary to tune the signal from a single GPS satellite.
4. A shortened way of saying "distribution channel." Let's say you make a product — hardware or software. You need to have some way of selling it. You can sell it yourself with your own salespeople. Or you can give it to distributors to sell. Such distributors could be wholesalers, small retailers, large retrail chains, direct mail catalogs, etc. Each one of these categories is called a "channel." See also channel ready.
5. A Fibre Channel term. A point-to-point link, the main task of which is to transport data from one point to another.

Channel Aggregator Also known as inverse multiplexors. Devices that allow very large amounts of data to be sent down the narrow band channels of ISDN. The aggregator effectively pulls together ISDN channels at one end to form a higher bandwidth and then re-synchronizes the information at the other end. Re-synchronization is necessary because during transmission the ISDN channels may send the information along different routes, so it arrives at its destination at fractionally different times.

Channel Associated Signaling CAS. A form of circuit state signaling in which the circuit state is indicated by one or more bits of signaling status sent repetitively and associated with that specific circuit.

Channel Attached Describing the attachment of devices directly to the input/output channels of a (mainframe) computer. Devices attached to a controlling unit by cables rather than by telecommunications circuits. Same as locally attached (IBM).

Channel Bank A multiplexer. A device which puts many slow speed voice or data conversations onto one high-speed link and controls the flow of those "conversations." Typically the device that sits between a digital circuit — say a T-1 — and a couple of dozen voice grade lines coming out of a PBX. One side of the channel bank will be connections for terminating two pairs of wires or a coaxial cable — those bringing the T-1 carrier in. On the other side are connections for terminating multiple tip and ring single line analog phone lines or several digital data streams. Sometimes you need channel banks. Sometimes, you don't. For example, if you're shipping a bundle of voice conversations from one digital PBX to another across town in a T-1 format — and both PBXs recognize the signal — then you will probably not need a channel bank. You'll need a Channel Service Unit (CSU). If one, or both, of the PBXs is analog, then you will need a channel bank at the end of the transmission path whose PBX won't take a digital signal. See CHANNEL SERVICE UNIT and T-1.

Channel Capacity A measure of the maximum possible bit rate through a channel, subject to specified constraints.

Channel Capture A condition that occurs when the Ethernet MAC layer temporarily becomes biased toward one workstation on a loaded network, thereby making that one station the contention winner more times than would randomly occur.

Channel Definition Format An open standard announced by Microsoft in March 1997 to be presented to the World Wide Web Consortium (W3C) as a suggested future open standard for "push" technology.

Channel Efficiency In a LAN environment, a measure of the total information that can be communicated in a channel in a unit of time, accounting for noise, collisions and other disruptions.

Channel Gate A device for connecting a channel to a

highway, or a highway to a channel, at specified times.

Channel Hopping A cellular radio term. Cellular Hopping is the process in CDPD of changing the Radio Frequency (RF) channel supporting a CDPD channel stream to a different RF channel on the same cell. This is typically used to avoid collisions with voice traffic use of the RF channel.

Channel Identification Information element that requests or identifies the channel to be used for a call. An AIN term.

Channel Loopback In network management systems, diagnostic test that forms a loop at the multiplexer's channel interface that returns transmitted signals to their source. See also LOOPBACK.

Channel Mastering A term invented by a New York firm called Snickelways.

Channel Mode An AT&T term for a method of communications whereby a fixed bandwidth is established between two or more points on a network as a semi-permanent connection and is rearranged only occasionally.

Channel Modem That portion of multiplexing equipment required to derive a desired subscriber channel from the local facility.

Channel Packing A technique for maximizing the use of voice frequency channels used for data transmission by multiplexing a number of lower data rate signals into a single higher speed data stream for transmission on a single voice frequency channel.

Channel Queue Limit Limit on number of transmit buffers used by a station to guarantee that some receive buffers are always available.

Channel Rate-Adaptation Protocols These protocols tell the ISDN terminal adapter (TA) how to change its transmission/ reception speeds to match those of the connecting device. Europe and Japan primarily use the V.110 protocol. The U.S. uses V.120. See ISDN and PROTOCOL.

Channel Ready A channel is a shortened way of saying "distribution channel." Let's say you make a product — hardware or software. You need to have some way of selling it. You can sell it yourself with your own salespeople. Or you can give it to distributors to sell. Such distributors could be wholesalers, small retailers, large retail chains, direct mail catalogs, etc. Each one of these categories is called a "channel." Channel Ready means that your product is in a form your chosen channel can handle. Typically this means that your software and/or hardware comes and can be delivered in a shrinkwrapped box and that you have set an organization which can service, support, train and otherwise satisfactorily deal with customers contacting you.

Channel Seized The time when a connection is established between the cellular user's mobile equipment and the mobile telephone switching office (MTSO). Channel seizure occurs before the number dialed begins to ring.

Channel Service Unit CSU. A device used to connect a digital phone line (T-1 or Switched 56 line) coming in from the phone company to either a multiplexer, channel bank or directly to another device producing a digital signal, e.g. a digital PBX, a PC, or data communications device. A CSU performs certain line-conditioning, and equalization functions, and responds to loopback commands sent from the central office. A CSU regenerates digital signals. It monitors them for problems. And it provides a way of testing your digital circuit. You can buy your own CSU or rent one from your local or long distance phone company. See also CSU and DSU.

Channel Stream A cellular radio term. Channel Stream is a shared digital communications channel between a Mobile Data Base Station (MDBS) and a set of Mobile End Systems (M-ESs) considered as a logical concept, separate from the frequency of the Radio Frequency (RF) channel used to implement the channel at any given time.

Channel Surfing Flipping channels on a TV set. A person who channel surfs is called a MOUSE POTATO.

Channel Tier An AT&T term for the tier within the Universal Information Services network that partitions transmission capacity into channels and offers the channels to the nodes' higher tiers.

Channel Time Slot A time slot starting at a particular instant in a frame and allocated to a channel for transmitting a character, in-slot signal, or other data. Where appropriate a modifier may be added.

Channel Translator Device used in broadband LANS to increase carrier frequency, converting upstream (toward the head-end) signals into downstream signals (away from the head-end).

Channel Virtual Area Where Internet Relay Chat (IRC) users communicate in real time. There are thousands of channels on the Internet.

Channelization The process of subdividing the bandwidth of a circuit into smaller increments called channels. Typically, each channel carries an individual transmission, e.g., a voice conversation or a data conversation — a computer-to-computer session. Multi-channel circuits always are four-wire in nature, whether physical four-wire or logical four-wire, and the process of channelization is always accomplished through some form of multiplexer (MUX), which can be in the form of either a Frequency Division Multiplexer (FDM) or a Time Division Multiplexer (TDM). The most basic type of MUX is known as a channel bank. Traditional T-1 service is a channelized service, for instance. Through a Statistical Time Division Multiplexer (STDM), a T-1 circuit of 1.544 Mbps is subdivided into 24 channels of 64 Kbps each, with an additional 8 Kbps needed for framing overhead (do the math and you'll see). Each channel (at least for contemporary multiplexers employing ESF, or Extended SuperFrame) provides a full 64 Kbps of bandwidth in support of a single "conversation" (data or voice), which requirement was established for transmission of digitized voice according to the original PCM (Pulse Code Modulation) encoding algorithm. Older multiplexers provided only 56 Kbps of usable channel capacity due to the fact that they employed an intrusive form of signaling and control. The STDM subdivides the digital bandwidth of the circuit into 24 smaller units of bandwidth known as "time slots" through a sampling process. Picture a 24 lane highway leading up to one toll booth. On the other side of the toll booth is a one lane highway. The vehicles in the 24 lanes are allowed through the toll booth (each of the 24 input devices is allotted a piece of time in which it is allowed to send data) if they are ready to pay the toll. If they do not have the toll fee available the attendant will ask the next vehicle to go through (the TDM polls the next device during the next, (fixed) time slot). In TDM devices are polled (asked) if they have anything to transmit during their preordained time slot. If they do not have anything to send, that time slot cannot be used by another device which may have something to send. This is wasted bandwidth. This is traditional TDM. In contrast, Statistical Time Division Multiplexing (STDM) asks why waste the time slot if some other device is ready to send? STDM is a non-channelized way of using the available bandwidth. It allocates time slots to those devices who are ready to send. This principle is called

"Bandwidth on Demand". FDMs subdivide bandwidth through a process of frequency separation, with each conversation occupying a separate and distinct frequency within a larger range of frequencies supported by the circuit. While FDM is seldom employed in the contemporary world of electrically-based (read copper wire) circuits, it is widely used in the wireless world (e.g., cellular and PCS). FDM also is widely employed in the fiber optic world, where it is known as WDM (Wavelength Division Multiplexing) or DWDM (Dense WDM). WDM supports multiple, very high-capacity virtual circuits over a single optical fiber, with each virtual circuit being subdivided into a very large number of TDM channels. See also BANDWIDTH, CHANNEL, CHANNEL BANK, DWDM, FDM, MUX, PCM, STDM, T-1, TDM, and WDM.

Channelize See Channelization.

Chaos Theory Developed by Edward Lorenz in the 1960s, chaos theory states that simple systems may produce complex behavior. It also has been proven that complex systems possess a simple underlying order. The emerging scientific discipline of chaos theory deals with systems with boundaries that are not clearly defined, with a "system" being defined as a set of things which interact. A computer system, for instance, is a set of elements including perhaps hardware, firmware, application software, CPU, hard drive, I/O devices, terminal devices, peripheral devices, drivers, and so on. Ideally, these system elements work together to consistently yield a predictable, desired result. In the context of a complex computer system, chaos theory describes a condition in which the system behaves in a nonlinear, inherently unpredictable manner. Chaos theory commonly is known as the "butterfly effect," with the analogy being that a butterfly flapping its wings in the Amazon Valley may create slight disturbances in the air which may ultimately affect worldwide weather patterns, perhaps resulting in the creation of a cyclone in Southeast Asia some years later. In this analogy, the complex weather system possesses a simple underlying order, although the interaction of the various elements clearly is nonlinear. But I digress. In the context of an information system, the addition of line of computer or the installation of a new application software package may cause the system to crash or to corrupt data residing in a seemingly unrelated application. That is chaos theory. The theory is fascinating. The practical impact is frightening, especially as systems and networks (comprised of systems) grow ever more complex.

CHAP Challenge-Handshake Authentication Protocol. An authentication method that can be used when connecting to an Internet Service Provider. CHAP allows you to log in to your provider automatically, without the need for a terminal screen. It is more secure than the Password Authentication Protocol (another widely used authentication method) since it does not send passwords in text format. An Internet term.

Chapter 7 See Chapter 11.

Chapter 11 The Chapter 11 process is started when a company files a reorganization petition with the federal Bankruptcy Court. From that moment on, creditors are prevented from suing the company, and any creditor lawsuits in process are halted, pending the outcome of the Chapter 11 reorganization. Creditors of the company file claims with the Bankruptcy Court. A creditors committee, usually made up of the seven creditors who have filed the largest claims against the company, represents the interest of all creditors. Under Chapter 11 protection, the company's management usually continues to manage the company's business, subject to judicial review. In rare circumstances, such as fraud, a party may

ask the court to appoint a trustee to manage the company during reorganization. The ultimate objective of a Chapter 11 reorganization is to restructure creditors' claims so that the company can move ahead with its business. Company management includes a negotiated partial payment to creditors. The plan also can include exchanges of debt for equity, a moratorium on repayment or a combination of these actions. In some cases, more than one plan may be proposed. For example, a creditor, or group of creditors, may develop its own plan. The complex process of reaching a consensual plan entails extensive negotiations among the company, its creditors and its shareholders.

Once developed, the company's reorganization plan — or one of the competing plans — must be accepted by specified margins of creditors and shareholders. Creditors representing two-thirds of the total dollar amount of bankruptcy claims against the company and 51 percent of the total number of those voting must accept the plan, and two-thirds of the amount of shares represented by shareholders voting on the plan must approve it for a plan to be accepted. Once accepted, the Bankruptcy Court reviews the plan to ensure that it conforms to certain additional statutory requirements before confirming it. With a restructured balance sheet, the company then emerges from Chapter 11 protection to implement the plan. Some companies emerge from Chapter 11 and become normal operating companies again. Some don't and move into Chapter 7 bankruptcy, which is complete and relatively immediate liquidation of the company (i.e. sale of all the company's assets).

Character A letter, a number or a symbol. A character is sometimes described by the digit represented by the bit pattern that makes up the Character. i.e., the letter A is ASCII code 65, a carriage return is ASCII code 13.

Character Cell In text mode on a PC, each pel is called a character cell. Character cells are arranged in rows and columns. A typical PC will support two text modes — 80 columns by 25 rows and 40 columns by 25 rows. The default text mode on virtually all PCs is 80 x 25.

Character Code One of several standard sets of binary representations for the alphabet, numerals and common symbols, such as ASCII, EBCDIC, BCD.

Character Distortion In telegraphy, the distortion caused by transients that, as a result of previous modulation, are present in the transmission channel. It effects are not consistent. Its influence upon a given transition is to some degree dependent upon the remnants of transients affecting previous signal elements.

Character Generator CG. A computer used to generate text and sometimes graphics for video titles.

Character Impedance The impedance termination of an electrically uniform (approximately) transmission line that minimizes reflections from the end of the line.

Character Interleaving A form of TDM used for asynchronous protocols. A 20% saving can be obtained by omitting the start and stop bits. This can be used either with extra channels or by carrying RS232-C control signals.

Character Interval The total number of unit intervals (including synchronizing, information, error checking, or control bits) required to transmit any given character in any given communication system. Extra signals that are not associated with individual characters are not included. An example of an extra signal that is excluded in the above definition is any additional time added between the end of the stop element and the beginning of the next start element as a result of a speed change, buffering, etc. This additional time is defined as a part of the intercharacter interval.

Character Oriented Protocol A communications protocol in which the beginning of the message and the end of a block of data are flagged with special characters. A good example is IBM Corp's. Binary Synchronous Communications (BSC) protocol. Character-oriented protocols are used in both synchronous and asynchronous transmission.

Character Oriented Windows Interface COW. An SAA-compatible user interface for OS/2 applications.

Character Printer A device which prints a single character at a time. Contrast with a line printer, which prints blocks of characters and is much faster.

Character Set All the letters, numbers and characters which a computer can use. The symbols used to represent data. The ASCII standard has 256 characters, each represented by a binary number from 1 to 256. This set includes all the letters in the alphabet, numbers, most punctuation marks, some mathematical symbols and some other characters typically used by computers. See ASCII.

Character Stuffing A technique used to ensure that transmitted control information is not misinterpreted as data by the receiver during character-based transmission. Special characters are inserted by the transmitter and then removed by the receiver.

Character Terminal A computer terminal that cannot show graphics, only text.

Characteristic Frequency A frequency that can be easily identified and measured in a given emission. A carrier frequency may, for example, be designated as the characteristic frequency.

Characteristic Impedance The impedance of a circuit that, when connected to the output terminals of a uniform transmission line of arbitrary length, causes the line to appear infinitely long. A line terminated in its characteristic impedance will have no standing waves, no reflections from the end, and a constant ratio of voltage to current at a given frequency at every point on the line.

Characters Per Second CPS. A data transfer rate generally estimated from the bit rate and the character length. For example, at 2400 bps, 8-bit characters with Start and Stop bits (for a total of ten bits per character) will be transmitted at a rate of approximately 240 characters per second (cps). Some protocols, such as USR-HST and MNP, employ advanced techniques such as longer transmission frames and compression to increase characters per second.

Chargeback The process of allocating network and telecommunications line, equipment, and usage costs to departments or to individuals. Companies charge back to departments or to users. Colleges and Universities charge costs back to departments and to students.

Charged Coupled Device CCD. The full name of the term is Interline Transfer Charge-Coupled Device or IT CCD. CCD are used as image sensors in an array of elements in which charges are produced by light focused on a surface. They consist of a rectangular array of hundreds of thousands of light-sensitive photo diodes. Light from a lens is focused onto the photo diodes. This frees up electrons (charges) which accumulate in the photo diodes. The charges are periodically released into vertical shift registers which move them along by charge-transfer to be amplified.

Charlie-Foxtrot Slang. Seriously beyond all hope. Very badly broken.

Chat A common name for a type of messaging done over a network, involving short, messages sent from one node to another. Chatting sometimes happens in real-time, sometimes in just short messages, replied to quickly. Sometimes, chatting software is RAM-resident, meaning it can be "popped up" inside an application program. Users are usually notified of an incoming chat by a beep and a message at the bottom of their screens.

Chat Room Real-time chat services offered by many Internet Information Service Providers such as America Online. Supporting a dozen or so participants, they act much like a teleconference, although on a text basis. Private rooms are those that can be entered by invitation. Public rooms allow anyone to participate.

Cheapernet A slang name for the thin wire coaxial cable (0.2-inch, RG58A/U 50-ohm) that uses a smaller diameter coaxial cable than standard thick Ethernet. Thin Ethernet is also called "Cheapernet" due to the lower cabling cost. Thin Ethernet systems tend to have transceivers on the network interface card, rather than in external boxes. PCs connect to the Thin Ethernet bus via a coaxial "T" connector. Thin Ethernet is now the most common Ethernet coaxial cable, though twisted pair is gaining. Thin Ethernet is also referred to as ThinNet or ThinWire. See also 10BASE-T.

Check Bit A bit added to a unit of data, say a byte or a word, and used for performing an accuracy check. See also PARITY.

Check Characters Characters added to the end of a block of data which is determined by an algorithm using the data bits which are sent. The receiving device computes its own check characters. It compares them with those sent by the transmitter. If they do not match, the receiver requests the sender to send the block again. If the check characters match, then all the bits used to compute the check characters have been received properly.

Check-In Mailbox The Centigram VoiceMemo II mailbox used to assign names and passcodes for guests checking into a hotel.

Check-Out Mailbox The Centigram VoiceMemo II mailbox used to clear out guest mailboxes when the guest checks out of the hotel.

Checkpoint Cycle HDLC error recovery cycle formed by pairing an F bit with a previous P bit or vice versa.

Checkpointing HDLC error recovery based on pairing of P and F bits and giving the equivalent of a negative acknowledgment without using either REJ or SREJ.

Checksum The sum of a group of data items used for error checking. Checksum is computed by the sending computer based upon an algorithm that counts the bits going out in a packet. The check digit is then sent to the other end as the tail, or trailer of the packet. As the packet is being received, the receiving computer goes through the same algorithm, and if the check digit it comes up with is the same as the one received, all is well. Otherwise, it requests the packet be sent again.

Cheese The content of a commercial site that mainly consists of pictures of the products or other equally useless information.

Cheesing When a buffered fiber cable appears to stretch during stripping and then cheeses (creeps) back into the outer jacket of the cable, to resume its original place.

Chemical Rectifier A chemical device for changing alternating current to pulsating direct, usually used to storage battery charging.

Chemical Stripping Soaking an optical fiber in a chemical to remove its coating.

Chemical Vapor Deposition Technique CVD. In optical fiber manufacturing, a process in which deposits are produced by heterogeneous gas-solid and gas-liquid chemical reactions at the surface of a substrate. The CVD method is often used in fabricating optical fiber preforms by causing gaseous materials to react and deposit glass oxides. The preform may be processed further in preparation for pulling into an optical fiber.

Cherry Picking A call center term. Calls come in and are identified in some way — by ANI (automatic number identification), Caller ID, or caller touchtone input. The identity of the callers is known to the agents in the call center, who can now answer the callers they wish. They decide to answer those callers who they think will buy the most and presumably give them the highest commission or best reward. Thus the expression "cherry picking.'

Chernobyl Packet A network packet that induces a broadcast storm and/or network meltdown. Named after the April, 1986 nuclear accident at Chernobyl in Ukraine.

Chiclet 1. Another term for a B Connector. See B CONNECTOR.
2. IBM once came out with a PC that had small keys. The press said the PC had a chiclet keyboard, after the chewing gum.

Chief Information Officer The person responsible for planning, choosing, buying, installing — and ultimately taking the blame for — a company's computer and information processing operation. Originally, CIOs were called data processing managers. Then they became Management Information System (MIS) managers. Then, CIOs. The idea of calling them CIOs was to reflect a new idea that the information they controlled was a critical corporate advantage and one that could give the company a competitive edge over its competitors — if played correctly.

Child Group In some systems, a new group of users created under a parent group is called a child group. Child groups sometimes have more properties than their parent groups.

Child Node An ATM term. A node at the next lower level of the hierarchy which is contained in the peer group represented by the logical group node currently referenced. This could be a logical group node, or a physical node.

Child Peer Group An ATM term. A child peer group of a peer group is any one containing a child node of a logical group node in that peer group. A child peer group of a logical group node is the one containing the child node of that logical group node.

Children's programming Cable Television programming originally produced and broadcast primarily for an audience of children 12 years old and younger (reference: FCC Rules, 47 CFR 76.225). This rule also requires:
• Commercial matter in children's programming carried on Origination Cablecasting channels must not exceed specified time limits.
• The cable system must maintain, in its PIF, records ("certifications") to verify compliance with this requirement. Certifications for satellite-delivered programming must be obtained from the programmer.
These requirements apply only to Origination Cablecasting. These requirements do not apply to:
• Broadcast stations.
• Access channel capacity designated by franchise for public, educational, or governmental use.

CHILL ITU HIgh Level Language. A computer language developed by the ITU for the standardization of software in telecommunications switches. Not widely adopted. C is more widely adopted.

Chime An electromechanical or electronic substitute for the conventional telephone bell, that sounds like a musical chime being struck, typically in a "bing-bong" sequence.

Chimney Effect Picture a phone system. We have an upright, rectangular cabinet full of printed circuit cards and all getting hot. How to cool them? Simple, raise the machine a little off the ground, put holes in the bottom of the cabinet and holes in the top of the cabinet. Hot air rises. Bingo, air will rise through the top of the cabinet and cool air will get sucked in the bottom of the cabinet. And bingo, you don't need a fan. This natural cooling technique is called the Chimney Effect and many modern phone systems now use it.

Chip 1. An integrated circuit. The physical structure upon which integrated circuits are fabricated as components of telephone systems, computers, memory systems, etc.
2. The transition time for individual bits in the pseudo-random sequence transmitted by the GPS satellite.

Chip Head Anyone whose education, entertainment and employment is primarily derived from computer-based devices. Also called a BIT HEAD.

Chip Jewelry A Wired Magazine definition: Chip Jewelry is a euphemism for old computers destined to be scrapped or turned into decorative ornaments. "I paid three grand for that Mac SE, and now it's nothing but chip jewelry."

Chip Rate Also known as the spreading rate. The rate at which radio signals are spread across a range of frequencies in a spread spectrum transmission system. See also Spread Spectrum.

Chirping 1. A rapid change (as opposed to a long-term drift) of the wavelength of an electromagnetic wave. Chirping is most often observed in pulsed operation of a source.
2. A pulse compression technique that uses (usually linear) frequency modulation during the pulse.

Choice Chip Your new TV will come with electronics that will allow you to program it not to receive certain programs, e.g. violent ones, you choose not to receive. The idea is that shows will be rated. Before they start, the show will broadcast a digital signal containing its rating. The "choice" chip in your TV will recognize the rating, check it against your instructions and block it or allow it through. The "choice" chip is so named as to give parents a choice of programs they and their children will watch. The provision for a choice chip was contained in telecommunications reform legislation passed by the Senate in mid-1995.

Choke An obsolete term: An inductance with either an air or iron core, designed to retard certain frequencies; as a radio frequency choke or an audio frequency choke.

Choke Coil A coil so wound as to offer a retarding or self inductance effect to an alternating current.

Choke Packet Packet used for flow control. Node detecting congestion generates choke packet and sends it toward source of congestion, which is required to reduce input rate.

Chooser A desk accessory on the Apple Macintosh that allows a user to choose items such as a printer or file server by clicking on an icon of the device.

Chopper A device for rapidly opening and closing a circuit. An ancient radio term.

Christmas Tree Lights The first electric Christmas lights were created by a telephone company PBX installer. Back in the old days, candles were used to decorate Christmas trees. This was obviously very dangerous. Telephone employees are trained to be safety conscious. The installer took the lights from an old switchboard, connected them together, strung them on the tree, and hooked them to a battery. Then he spent the next 40 years looking for the one burnt bulb...

Chroma The level of saturation or intensity of a color. Name sometimes applied to color intensity control in a receiver.

Chroma Key Method of electronically inserting the image from one video source into the picture from another video source using color for discrimination. A selected "key color" is replaced by the background image.

Chromatic Dispersion Chromatic dispersion is one of the mechanisms that limits the bandwidth of optical fibers by producing pulse spreading because of the various colors of light traveling in the fiber. Different wavelengths of light travel at different speeds. Since most optical sources emit light containing a range of wavelengths, each of these wavelengths arrive at different times and thereby cause the transmitted pulse to spread as it travels down the fiber.

Chromatic dispersion is the sum of material and waveguide dispersion. Dispersion can be positive or negative because it measures the change in the refractive index with wavelength. Thus, the total chromatic dispersion can actually be zero (really close to zero). For example, step-index single-mode fibers have zero dispersion at 1300nm, almost exactly at the same wavelength where the optical loss of the fiber is at a minimum. This is what allows single-mode fibers to have low loss and high bandwidth. See also PMD (Polarization Mode Dispersion).

Chrominance The color portion of the video signal. Chrominance includes hue and saturation information but not brightness. Low chroma means the color picture looks pale or washed out; high chroma means the color is too intense, with a tendency to bleed into surrounding areas. Black, gray and white have a chrominance value of O. Brightness is referred to as luminance.

Chromium Dioxide Tape whose coating is of chromium dioxide particles. Noted for its superior frequency output.

CHS Cylinder-Head Sector. The method of identifying a given location on a hard drive used by the original PC-AT BIOS (INT 13) and original IDE specification. Differences between details of the two methods resulted in the 528 MB limit on IDE drives. Enhanced IDE-compliant BIOSes can translate between the two methods, allowing drive sizes up to 8.4 GB. See ENHANCED IDE and IDE.

Chuck Hole Also known as Pot Hole. Slang for when your system hangs up on-line.

Churn Cellular phone and beeper users drop their monthly subscriptions often. Long distance users change their preferred carrier as often as some of them change their udnerwear. The industry calls This phenomenon "churn." Churn is defined as the level of disconnects from service relative to the total subscriber base of the system. Often referred to on a percentage basis monthly, quarterly or annually. Sometimes it's as high as 2% or 3% a month. It drives the cellular, beeper and long distance business mad. It's very expensive to sign up a new customer. Many cell, beeper and long distance companies offer incentives to prospective customers to switch their service. Some users have found ways to switch their long distance service often enough so that they never pay for a long distance phone call.

Chutzpah A Yiddish word which means unmitigated gall (audacity). The word is best exemplified by the story of the 15-year old who goes into court having killed his father and mother and falls on the mercy of the court that he's now an orphan. Now that's Chutzpah! Telecom companies without competition show they understand the meaning of chutzpah very well in the way they typically treat their customers.

CI 1. Customer Interface.

2. Certified Integrator.

3. An ATM term. Congestion Indicator: This is a field in a RM-cell, and is used to cause the source to decrease its ACR. The source sets CI=0 when it sends an RM-cell. Setting CI=1 is typically how destinations indicate that EFCI has been received on a previous data cell.

CIC See Carrier Identification Code.

CICS Customer Information Control System. An IBM program environment designed to allow transactions entered at remote computers to be processed concurrently by a mainframe host. Also, IBM's Customer Information Control System software.

CID 1. A generic term in Britain to identify a customer identity, client identity or contract identity. It is a single record and all the fields of information associated with it; for example, name, address, phone number, contact history and so on.

2. Compatibility ID. Motorola definition.

3. CIrcuit Designator.

CIDB Calling Line Identification Delivery Blocking. A "feature" of central offices which lets you block the sending of your phone number to the person you're calling.

CIDCW CID on Call Waiting. See Caller ID Message Format.

CIDR Common InterDomain Routing. A protocol which allows the assignment of Class C IP addresses in contiguous blocks. CIDR solved a major problem with IP address assignment. Specifically, IPv4 addresses in the Class C block were limited to 256 addresses. If a user required more than 256 addresses, the next step up the IP food chain was Class B, with 64,000 addresses. Clearly, this was wasteful, as only a few more addresses required a huge chunk of precious addresses. Although this was not an issue for the first two decades of IP, the recent popularity of the Internet (and other IP networks) quickly strained the existing IP addressing scheme. CIDR came to the rescue...and will continue to do so, even with the advent of IPv6. The problem is similar to, but much more complex than, that of 800 numbers, which was relieved with the introduction of 888 numbers. See also IP, IPV6, and TCP/IP.

CIF 1. Common Intermediate Format. An option of the ITU-T's H.261/Px64 standard for videoconferencing codecs. It produces a color image of 288 non-interlaced luminance lines, each containing 352 pixels to be sent at a rate of 30 frames per second. The format uses two B channels, with voice taking 32 Kbps and the rest for video. QCIF (Quarter CIF), is a variation on the theme, requiring approximately 1/4 the bandwidth of CIF and delivering approximately 1/4 the resolution. CIF works well for large-screen videoconferencing, due to its greater resolution; QCIF works well for small-screen displays, such as videophones. QCIF is mandatory for ITU-T H.261-compliant codecs, while CIF is optional. See QCIF.

2. Cost, Insurance and Freight are included. That means the seller pays the freight. The opposite of CIF is FOB, which stands for Free On Board. What this means is that you buy something, F.O.B. The seller puts it on a truck or railroad, plane, i.e. some carrier. He's responsible for getting it on the carrier. You — the buyer — are responsible for paying for the cost of the freight of getting you the goods you ordered.

3. Cells In Flight: An ATM term for an ABR (Available Bit Rate) service parameter, CIF is the negotiated number of cells that the network would like to limit the source to sending during idle startup period, before the first RM-cell returns.

4. Cells In Frames. Referring to ATM over Ethernet (or Token Ring), CIF involves the insertion of one or more ATM cells into Ethernet frames for transport over an Ethernet LAN. CIF

allows the user organization to maintain the existing Ethernet wiring, NICs (Network Interface Cards) and other hardware to support ATM applications. The drawback is that CIF must be used in a switched Ethernet environment in order to maintain ATM QoS (Quality of Service) commitments. Additionally, the SAR (Segmentation and Reassembly) process must be accomplished in software at the workstation, which is slow unless the workstation is really fast. CIF is something of a band-aid approach to bridge the gap until such time as ATM really takes hold in the LAN world.

CIF-AD Cells In Frames-Attachment Device. The device which attaches the CIF-ES to the ATM network.

CIF-ES Cells In Frames-End Station. An Ethernet- or Token Ring-attached workstation which supports CIF.

CIFS Common Internet File System. A remote collaborative file sharing technology that, according to Microsoft, dramatically reduces the time it takes to open and work with remote files. Data General, Digital Equipment, Intel, Intergraph Corp, Network Alliance and Microsoft announced support for CIFS n June 13, 1996. According to Microsoft, the Common Internet File System is an enhanced version of the native file-sharing technology used in the Microsoft MS-DOS, Windows and Windows NT operating systems and IBM OS/2, and widely available on leading UNIX systems. It enables millions of computer users to open and share remote files directly on the Internet, expanding the Internet's ability to support interactive computing. CIFS technology provides reliable direct read and write access to files stored on remote computers without first requiring users to download or copy the files to a local machine, as done previously on the Internet. According to Microsoft's June 13 release, CIFS can improve the performance of many types of file access. Because CIFS is based on existing standards, users will be able to use thousands of existing applications over the Internet as well as integrate them with browser applications designed for the World Wide Web. The proposed Common Internet File System protocol runs over TCP/IP and is an enhanced version of the open, cross-platform protocol for distributed file sharing called Server Message Block (SMB). The SMB protocol is the standard way that millions of PC users already share files across corporate intranets and is the native file-sharing protocol in Windows 95, Windows NT and OS/2.

The proposed Common Internet File System protocol has been enhanced over previous versions of the SMB protocol in ways that make it well suited for use on the Internet. CIFS, for example, supports the Internet's Domain Name Service (DNS) for address resolution. The protocol runs optimally over slow-speed dial-up lines, helping improve performance for the vast numbers of users today who access the Internet using a modem. In addition to remote file sharing, CIFS has mechanisms to support remote printer sharing.

CIG Calling Subscriber Identification. A frame that gives the caller's telephone number. See CALLER ID.

CIGOS Canadian Interest Group on Open Systems. Canadian organization which promotes OSI.

CIGRR Common Interest Group on Rating and Routing.

CIIG Canadian ISDN Interest Group. Canadian organization which promotes ISDN.

CIM 1. Computer-Integrated Manufacturing.
2. Common Information Model. CIM is the DMTF's model for describing management information to work with disparate systems.

CIP An ATM term. Carrier Identification Parameter: A 3 or 4 digit code in the initial address message identifying the carri-

er to be used for the connection.

Cipher A means of transforming, or encrypting, data in order to disguise its meaning. Block ciphers, such as DES, are encryption algorithms which encrypt specific blocks of data. Stream ciphers, such as the RC4 algorithm from RSA Data Security, encrypt a steady flow of data. See also Encryption.

Ciphertext The result of processing plaintext (unencrypted information) through an encryption algorithm. Ciphertext is thus the content of an encrypted message. See CLIPPER CHIP.

CIR 1. Committed Information Rate. It is used in the Frame Relay arena. CIR refers to the average maximum transmission speed of a user over a link to the Frame Relay network. The customer is always free to "burst" up to the maximum circuit and port, although any amount of data over the CIR can be surcharged. Additionally, excess bursts can be marked by the carrier as DE or Discard Eligible, and subsequently discarded in the event of network congestion. CIR comes in increments of 16 Kbps (16-32-48-64) with 16 being the lowest. An example would be: A 56k frame relay connection with a CIR of 16k (very common) would allow the customer to transmit or burst data up to the full 56 Kbps. If they bursted up to the full 56k they would only be guaranteed delivery of 16k. Depending on the pricing algorithm employed by the carrier, the customer pays for the Access Rate of the circuit, the Access Port, the CIR, and any usage surcharges for transmission in excess of the CIR. See also committed information rate, access rate, offered load, committed burst size, measurement interval and discard eligibility.
2. An ATM term. Committed Information Range: CIR is the information transfer rate which a network offering frame relay services (FRS) is committed to transfer under normal conditions. The rate is averaged over a minimum increment of time.

Circuit The physical connection (or path) of channels, conductors and equipment between two given points through which an electric current may be established. Includes both sending and receiving capabilities. A circuit can also be a network of circuit elements, such as resistors, inductors, capacitors, semiconductors, etc., that performs a specific function. A circuit can also be a closed path through which current can flow.

Circuit Board Same as a Printed Circuit Board, namely a board with microprocessors, transistors and other small electronics components. Such a board slides into a the slot in a telephone system or personal computer. Also called a circuit card.

Circuit Breaker A special type of switch arranged to open a circuit when overloaded, without injury to itself. A circuit breaker is basically a re-usable fuse. According to APC, a circuit breaker is a protective device that interrupts the flow of current when the current exceeds a specified value. Circuit breakers are calibrated when manufactured to a specific overcurrent value. Building or equipment wiring may overheat and become a fire hazard if excessive current is passed through such wiring. Circuit breakers or fuses are installed and coordinated with wiring by selecting the appropriate trip value so that if equipment malfunction or user error causes too much current to flow through a wire, the circuit breaker will trip to prevent the wire from overheating. For building wiring and power distribution, the values of circuit panel breakers are specified in America by the National Electrical Code.

Circuit Card Same as a Circuit Board. See CIRCUIT BOARD.

Circuit Emulation A connection over a virtual channel-based network providing service to the end user that is indistinguishable from a real, point-to-point, fixed bandwidth circuit.

Circuit Emulation Switching CES. Part of the ATM Forum's proposed Service Aspects and Applications (SAA) standard.

Circuit Grooming The practice of directing selected circuit-switched DS-0s (64 kbit/s channels) from many T-1 trunks into a single T-1 (typical application is voice leased lines from a T-1 access line being 'groomed' in a DACS onto a dozen or more T-1s going to other central offices where those channels may again be groomed with other circuits onto T-1 access lines at those sites). Also used to separate voice circuits from data circuits, and for combining then for delivery to service-specific switches in the CO.

Circuit Identification Code CIC. The part of CCS/SS7 signaling message used to identify the circuit that is being established between two signaling points (14 bits in the ISDNUP).

Circuit Level Gateway A circuit level gateway ensures that a trusted client and an untrusted host have no direct contact. A circuit level gateway accepts a trusted client's requests for specific services and, after verifying the legitimacy of a requested session, establishes a connection with an untrusted host. After the connection is established, a circuit-level gateway copies packets back and forth-without further filtering them.

Circuit Mode 1. An AT&T term for the method of communications in which a fixed bandwidth circuit is established from point to point through a network and held for the duration of a telephone call.

2. An AIN term for a type of switching that causes a one-to-one correspondence between a call and a circuit. That is, a circuit or path is assigned for a call between each switching node, and the circuit or path is not shared with other calls.

Circuit Noise Level At any point in a transmission system, the ratio of the circuit noise at that point to some arbitrary amount of circuit noise chosen as a reference.

Circuit Order Management System COMS. An automated processing system of MCI circuit- and service-related information. Processes hardwire service circuit orders from order entry through scheduling and completion. COMS also provides circuit order data, hardwire customer data, and circuit inventory data to other MCI systems in Finance, Engineering, and Operations.

Circuit Order Record COR. Report generated by the COR Tracking System within NOBIS, indicating circuit installations, changes, and disconnects.

Circuit Provisioning The telephone operating company process that somehow organizes to get you a trunk or other special service circuit.

Circuit Segregation Differentiating between services that are maintained by separate technicians or departments. Can be accomplished through visual and/or mechanical means.

Circuit Switched Digital Capability CSDC. A service implemented by some regional Bell Operating Companies that offers users a 56 Kbps digital service on a user-switchable basis.

Circuit Switching Imagine making a phone call to Grandma. You pick up the phone and dial Grandma. When you finish dialing, the various telephone company switches along the way pick a path for your call and move your call along its way to Grandma. When Grandma answered, you and she are now able to speak. Both of you now have the exclusive and full use of the circuit that was set up between you. You have that circuit until you (or she) hang up, at which time

it goes idle until the system of switches grabs it for another "call." That call might be voice, data and video. Circuit switching has one big advantage: You get the full circuit for the full amount of the time you're using it. And for the most part, it's full duplex. Circuit switching has one big disadvantage. Because you get the full circuit for the full amount of the time, you pay for the privilege of tying up that circuit. Which means it's expensive.

There are basically three types of switching — circuit, packet and message:

Circuit Switching, which I just explained, is like having your own railroad track for your conversation to travel on. It's yours for as long as you keep the connection open. No one else can use it. Once you hang up, the next caller gets to use that track. Virtually all voice telephone calls are circuit switched. All dial up modem calls are circuit switched also.

Packet Switching is like having your own railroad cars which you're sharing with other railroad cars on a railroad track. You slice the information you want to send so it fits into the cars, which join other cars to travel on the railroad track to the other end. You get pretty well as many railroad cars as you need. They will travel on different railroad tracks until they reach the station at the other end, where they'll be assembled in the order you sent them and then dropped off at your destination. Packet switching was originally created for sending data, since it's very efficient (and therefore cheap). One railroad track gets to carry a lot of "conversations." (In circuit switching, it only carries one conversation.) It does have the problem that it takes a little time to break up the data "conversations" into many packets, send them on their different ways and then reassemble them at the other end. But for data, that delay's barely noticeable. It is noticeable, however, in a voice conversation, which is why packet switching hasn't been used much for voice — until recently. See IP Telephony. In packet switching, the addresses on your packets are read by the switches as they approach, and are switched down the tracks. The next packet is read to throw the switches to send that packet where it needs to go. The data conversation is sent in packets. Each packet can be sent along different tracks as they are open. The packets are assembled at the other end — typically in the last switching office before the packets reach the distant computer or distant user.

Message Switching sends a message from one end to the other. But it's not interactive, as in packet or circuit switching. Of course, you can reply. But it's not like having a "conversation." In message switching, the message is typically received in one block, stored in one central place, then retrieved or sent in one clump to the other end. Message switching is like the post office, or like email. It can be slow. But it can also be cheap. Message can use a combination of circuit switching and packet switching to get its message through.

Circuit, Four Wire A path in which four wires are presented to the terminal equipment (phone or data), thus allowing for simultaneous transmission and reception. Two wires are used for transmission in one direction and two in the other direction.

Circular Extension Network Permits two or more single-line phones connected to a PBX, each with its own extension, to operate like a "square" key telephone system. An incoming call directed to any non-busy phone in the group will ring at all of the non-busy phones. The first extension to answer will be connected to the incoming call. At any time, a non-busy extension can make or receive calls.

Circular Hunting When calling a phone, the switching system makes a complete search of all numbers within the hunting group, regardless of the location within that group of the called number. For example, the hunt group is 231, 232, 233 and 234, the call is directed to 233. If it is busy, the equipment will search 234, 231, and 232 to find a non-busy phone or line. Essentially it goes around the ring, remembering where it last connected and then goes to the next line or phone in the circle. See also HUNT GROUP and TERMINATED HUNT GROUP.

Circular Mil The measure of sectional area of a wire.

Circular Polarization In electromagnetic wave propagation, polarization such that the tip of the electric field vector describes a circle in any fixed plane intersecting, and normal to, the direction of propagation. The magnitude of the electric field vector is constant. A circularly polarized wave may be resolved into two linearly polarized waves in phase quadrature with their planes of polarization at right angles to each other.

Circulator 1. In networking, a passive junction of three or more ports in which the ports can be accessed in such an order that when power is fed into any port it is transferred to the next port, the port counted as following the last in order. 2. In radar, a device that switches the antenna alternately between the transmitter and receiver.

Circumnaural A type of headphone that almost totally isolates the listener from room sounds.

CIS Contact Image Sensor. A type of scanner technology in which the photodetectors come in contact with the original document.

CISC Complex Instruction Set Computing. PC Magazine defines CISC as a microprocessor architecture that favors robustness of the instruction set over the speed with which individual instructions are executed. The Intel 486 and Pentium are both examples of CISC microprocessors. See also RISC — Reduced Instruction Set Computing. See RISC.

CISPR 22 This is a European Community standard specifying the limits of radio frequency emissions which appliances and other electrical equipment are allowed. The standard indicates the maximum allowable emissions either radiated or conducted via the power cord at various frequencies. Some countries still use the older VDE 0871 emission standards, which are nearly identical. In the USA, the FCC has a similar standard.

CIT Computer Integrated Telephone is Digital Equipment Company's program, announced in October 1987, that provides a framework for functionally integrating voice and data in an applications environment so that the telephone and terminal on the desktop can be synchronized, the call arriving as the terminal's screen on the caller arrives. CIT uses the DEC VAX line of computers. According to DEC, CIT supports both inbound and outbound telecommunications applications. In an inbound scenario, the application may recognize the caller's originating phone number through Automatic Number Identification (ANI) and/or the dialed number through Dialed Number Identification Service (DNIS), match the information to corresponding data base records and automatically deliver the call and the data to the call center agent. In an outbound application, dialing can be automated, increasing the number of connected calls. In either scenario, the telephone calls and associated data can be simultaneously transferred to alternate locations within an organization, adding a new level of customer service to call center applications. Digital made its first CIT announcements at Telecom '87 in Geneva, Switzerland. The CIT product set, consisting of client and server software implementing a variety of switch-to-computer link protocols, and providing a robust applications interface, was first shipped in 1989. The company announced its latest release, CIT Version 2.1, in January 1991. See also

OPEN APPLICATION INTERFACE.

Citizens Band One of two bands used for low power radio transmissions in the United States — either 26.965 to 27.225 megahertz or 462.55 to 469.95 megahertz. Citizens band radio is not allowed in many countries, even some civilized countries. In some countries they use different frequencies. CB radios, in the United States, are limited by FCC rule to four WATTS of power, which gives each CB radio a range of several miles. Some naughty people boost their CBs with external power. The author of this dictionary has actually spoken to Australia while driving on the Santa Monica Freeway in Los Angeles. See also CB.

CIV Cell Interarrival Variation. An ATM term. See CELL INTERARRIVAL VARIATION

CIVDL Collaboration for Interactive Visual Distance Learning. A collaborative effort by 10 US universities that uses dial-up videoconferencing technology for the delivery of engineering programs.

CIX 1. Commercial Internet Exchange. Pronounced "kicks." As the Internet began to be commercialized, an agreement was reached among a number of commercial network providers that allowed them to exchange traffic. The first CIX router was installed in the Wiltel equipment room in Santa Clara, California. The Santa Clara CIX and that in Herndon, Virginia remain operational, although the CIX concept later gave way to that of the NAP (Network Access Point). See also FIX and NAP.

2. Commercial Internet Exchange Association. The Commercial Internet eXchange Association is a non-profit trade association of Public Data Internetwork service providers that promotes and encourages development of the public data communications internetworking services industry, nationally and internationally.

CL Connectionless service: A service which allows the transfer of information among service subscribers without the need for end-to-end establishment procedures.

Cladding 1. The transparent material, usually glass, that surrounds the core of an optical fiber. Cladding glass has a lower refractive index than core glass. As the light signal travels down the central core transmission path, it naturally spreads out due to a phenomenon known as "modal dispersion." The cladding causes the light to be reflected back into the central core, thereby serving to maintain the signal strength over a long distance. See Cladding Diameter.

2. When referring to a metallic cable, a process of covering with a metal (usually achieved by pressure rolling, extruding, drawing, or swaging) until a bond is achieved.

Cladding Diameter The diameter of the circle that includes the cladding layer in an optical fiber.

Cladding Mode In an optical fiber, a transmission mode supported by the cladding; i.e., a mode in addition to the modes supported by the core material.

Cladding Mode Stripper A device for converting optical fiber cladding modes to radiation modes; as a result, the cladding modes are removed from the fiber. Often a material such as the fiber coating or jacket having a refractive index equal to or greater than that of the fiber cladding will perform this function.

Cladding Ray In an optical fiber, a ray that is confined to the core and cladding by virtue of reflection from the outer surface of the cladding. Cladding rays correspond to cladding modes in the terminology of mode descriptors.

Claim Process A technique used to determine which station will initialize an FDDI ring.

Claim token A token ring frame that initiates an election process for a new active monitor station. Claim tokening can result from expired timers, or from any other condition that causes any station to suspect a problem with the current active monitor.

Claire Harry Newton's favorite daughter. Why "favorite?" Simple. She's his only daughter. Her full name is Claire Elizabeth Newton. "Claire" is a form of "Clare," which means "clear" or "bright"; from Latin "clarus." Actually, she's both clear and bright. Brilliant naming, if I might say so. In early 1999, she was 19 going on 35. She is currently in her first year at college and her father (i.e. me) is facing annual $35,000 bills for the next four, five or six years. Around the Newton household, Claire is known as Princess Claire, a title she has earned by developing fine culinary and sophisticated telecommunications skills. She is, for example, adept at ordering in Chinese food via the telephone. To her credit, she is a born leader. Her unbiased, totally objective father (i.e. me) is convinced that she will become the first woman president of the United States. She has promised me "first sleep" in the Lincoln bedroom, so long as I contribute handsomely to her campaign funds. Look closely. That's me selling Newton's Telecom Dictionaries out the White House's side door. Actually she's a great kid. I couldn't be happier.

CLAMN Called Line Address Modification Notification

Clamper An electronic circuit which sets the level of a signal before the scanning of each line begins to insure that no spurious electronic noise is introduced into the picture signal from the electronics of the video equipment.

Clamping Voltage The voltage at which a surge protector begins to stop electricity from getting through. A good surge protector in a 120 volt circuit (the one common in North America) has a clamping voltage of about 135 volts. Damage to computer equipment can occur as low as 160 volts.

CLAS Centrex Line Assignment Service. A service from Nynex and other Bell operating companies, which allows Centrex subscribers to change their class of service by dialing in on a personal computer, reaching the phone company's computer and then changing things themselves — without phone company personnel assisting or hindering.

Load your PC with communications software. Dial your local central office. Change your Centrex phone numbers. Turn on, turn off features. Change pickup groups. Add numbers to speed dialing, etc. Your on-line changes are checked by the phone company's computers. If they make sense (i.e. one change doesn't conflict with another), they take effect by early the following day — at which time you can call up and get a report on which took, which didn't and who's got what. Saves calling in person. Is more accurate. And, best of all, saves money. Typically just one flat monthly charge. No charge for any of your changes.

CLASS 1. Custom Local Area Signaling Services. It is based on the availability of channel interoffice signaling. Class consists of number-translation services, such as call-forwarding and caller identification, available within a local exchange of Local Access and Transport Area (LATA). CLASS is a service mark of Bellcore. Some of the phone services which Bellcore promotes for CLASS are Automatic Callback, Automatic Recall, Calling Number Delivery, Customer Originated Trace, Distinctive Ringing/Call Waiting, Selective Call Forwarding and Selective Call Rejection. See also Calling Line Identification.

2. In an object-oriented programming environment, a class defines the data content of a specific type of object, the code

that manipulates it, and the public and private programming interfaces to that code. See ANI and ISDN.

3. See also CLASS 1 and 2, below.

Class 1 Also called Class 1/EIA-578. It's an American standard used between facsimile application programs and facsimile modems for sending and receiving Class 1 faxes. The Class 1 interface is an extension of the EIA/TIA's (Electronics Industry Association and the Telecommunications Industry Association) specification for fax communication, known as Group III. Class 1 is a series of Hayes AT commands that can be used by software to control fax boards. In Class 1, both the T.30 (the data packet creation and decision making necessary for call setup) and ECM/BFT (error-correction mode/binary file transfer) are done by the computer. A specification being developed (fall of 1991) Class 2, will allow the modem to handle these functions in hardware. Industry analysts believe Class 2 will be the standard for the long haul, but approval is slow. Even so, some modem makers will shortly deliver data/fax modems. See also CLASS 1 OFFICE.

Class 1 Office A regional toll telephone switching center. The highest level toll office in AT&T's long distance switching hierarchy. There are essentially five levels in the hierarchy, with the lowest level — Class 5 — being those central offices owned by the local telephone companies. Each of the classes can complete calls between themselves. But, if the routes are busy, then calls automatically climb the hierarchy. A Class 1 office is the office of "last resort."

Class 2 Also known as Class 2.0/EIA-592. An American standard used between facsimile application programs and facsimile modems for sending and receiving Class 2.0 faxes. This class places more of the task of establishing the fax connection onto the fax modem, while continuing to rely on the host's processor to send and receive the image data. The Class 2 standard (known as PN-2388) is still under study by the EIA's (Electronic Industries Association) TR.29 committee, with further revisions expected. See CLASS 1.

Class 2 Office The second level in AT&T's long distance toll switching hierarchy.

Class 3 Office The third level in AT&T's long distance toll switching hierarchy.

Class 4 Office The fourth level in AT&T's long distance toll switching hierarchy — the major switching center to which toll calls from Class 5 offices are sent. In U.S. common carrier telephony service, a toll center designated "Class 4C" is an office where assistance in completing incoming calls is provided in addition to other traffic. A toll center designated "Class 4P" is an office where operators handle only outbound calls, or where switching is performed without operator assistance.

Class 5 Office An end office. Your local central office. The lowest level in the hierarchy of local and long distance switching which AT&T set up when it was "The Bell System." A class 5 office is a local Central Office that serves as a network entry point for station loops and certain special-service lines. Also called an End Office. Classes 1, 2, 3, and 4 are toll offices in the telephone network.

Class A Certification A Federal Communications Commission (FCC) certification that a given make and model of computer meets the FCC's Class A limits for radio frequency emissions, which are designed for commercial and industrial environments. See CLASS B CERTIFICATION.

Class B Certification A Federal Communications Commission (FCC) certification that a given make and model of computer meets the FCC's Class b limits for radio frequency emissions, which are designed to protect radio and television reception to residential neighborhoods from excessive radio frequency interference (RFI) generated by computer usage. Class B computers also are shielded more efficiently from external interface. Computers used at home are more likely to be surrounded by radio and television equipment. If you plan to use your computer at home, avoid computers that have only Class A certification (that is, they failed Class B).

Class n Office The way a telephone company defines its switching facilities. Class 5 is an end office (local exchange), Class 4 is a toll center, Class 3 is a primary switching center, Class 2 is a sectional switching center, and Class 1 is a regional switching center. See CLASS 1, CLASS 2, CLASS 3, CLASS 4 and CLASS 5.

Class Of Emission The set of characteristics of an emission, designated by standard symbols, e.g., type of modulation of the main carrier, modulating signal, type of information to be transmitted, and also f appropriate, any additional signal characteristics.

Class Of Office A ranking assigned to switching points in the telephone network, determined by function, interfaces and transmission needs.

Class Of Service 1. Each phone in a system may have a different collection of privileges and features assigned to it, such as access to WATS lines. Class of Service assignments if properly organized, can become an important tool in controlling telephone abuse.
2. A subgrouping of telephone users for the sake of rate distinction. This may distinguish between individual and party lines, between Government lines and others, between those permitted to make unrestricted international dialed calls and others, between business or residence and coin, between flat rate and message rate, and between restricted and extended area service.
3. A category of data transmission provided in a public data network in which the data signaling rate, the terminal operating mode, and the code structure (if any) are standardized. This is defined within ITU-T Recommendations X.1.

Classical IP A set of specifications developed by the Internet Engineering Task Force (IETF) for the operation of LAN-to-LAN IP connectivity over an ATM network.

Classified Ad Log Records required by Section 76.221(f) of the FCC rules which relate to origination cablecasts or classified advertisements sponsored by individuals. This rule provides that the sponsor of such programming need not be identified within the content of the advertisement or program itself provided that two conditions are met:
• The true sponsor must be an individual offering services which he or she personally provides (examples: yard work; babysitting).
• The system must maintain a written record of the name, address, and telephone number of the individual.

Classless Inter-Domain Routing CIDR. An internetworking term. A method for using the existing 32-bit Internet Address Space more efficiently.

Classmark A designator used to describe the service feature privileges, restrictions, and circuit characteristics for lines or trunks accessing a switch; e.g., precedence level, conference privilege, security level, zone restriction. See CLASS OF SERVICE.

CLC Carrier Liaison Committee. A committee formed to help industry participants work together to resolve the issues of implementing 800 Portability. CLC is sponsored by the Exchange Carriers Standards Association (ECSA) and is comprised of the LECs (local exchange carriers), long distance carriers and users of 800 service.

CLD Coalition Inc. Competitive Long Distance Coalition, a Washington, DC-based lobbying group.

Clear To cause one or more storage locations to be in a prescribed state, usually that corresponding to a zero or to the space character.

Clear Channel 1. A digital circuit where no framing or control bits (i.e. for signaling) are required, thus making the full bandwidth available for communications. For example, a 56 Kbps circuit is typically a 64 Kbps digital circuit with 8 Kbps used for signaling. Sometimes called Switched 56, DDS or ADN. Each of the carriers have their own name for clear channel service. The phone companies are obsoleting the 56 Kbps service in favor of the more modern ISDN BRI, which has two 64 Kbps circuits and one 16 Kbps packet service, part of which is used for signaling on the 64 Kbps channels.
2. An SCSA term. A channel which is used exclusively for data transmission, with no bandwidth required for administrative messages such as signaling or synchronization. All SCbus data channels are clear.

Clear Collision Contention that occurs when a DTE and a DCE simultaneously transfer a clear request packet and a clear indication packet specifying the same logical channel. The DCE will consider that the clearing is completed and will not transfer a DCE clear confirmation packet.

Clear Confirmation Signal A call control signal to acknowledge reception of the DTE clear request by the DCE or the reception of the DCE clear indication by the DTE.

Clear To Send CTS. One of the standard attributes of a modem in which the receiving modem indicates to the calling modem that it is now ready to accept data. One of the standard pins used by the RS-232-C standard. In ITU-T V.24, the corresponding pin is called Ready For Sending.

Clearinghouse A service company that collects and processes roaming and billing information from a number of carriers. It then transfers the compiled data to the proper carriers for credits and billing.

Clearline 1.5 A Sprint name for T-1 service. It is an all digital 1.544 Mbps private line service that connects two customer sites via dedicated T-1 access lines.

Clearline 45 A Sprint name for DS-3 service. This high-capacity point-to-point private line service transmits voice, data, and video at 44.736 Mbps.

Clearline Fractional 1.5 A Sprint name for all digital private line service which transmits voice, data, and video at speeds from 112/123 Kbps up to 672/768 Kbps - a fraction of a T-1, also called a DS-1. The service may be ordered in 56/64 Kbps increments from two channels (112/128 Kbps) to 12 channels (672/768 Kbps). Point-to-point service connects customer sites via dedicated T-1 access lines.

Cleaving To cut the end of fiber at 90 degrees with as few rough edges as possible for a fusion splice. With mechanical splices the ends are hand-smoothed with a polishing puck before splicing.

CLEC Competitive Local Exchange Carrier or Certified Local Exchange Carrier. A term coined for the deregulated, competitive telecommunications environment envisioned by the Telecommunications Act of 1996. While The Act is under legal challenge at the time of this writing, many of the state regulatory authorities have moved forward. The CLECs compete on a selective basis for local exchange service, as well as long distance, international, Internet access, and entertainment (e.g., Cable TV and Video on Demand). They build or rebuild their own local loops, wired or wireless. They also lease local loops from the ILECs (Incumbent LEC) at wholesale rates for resale to end users. CLECs include cellular/PCS providers, ISPs, IXCs, CATV providers, CAPs, LMDS operators, and power utilities.

CLEI Codes Comon Language Equipment Identifier codes, that are assigned by Bellcore to provide a standard method of identifying telecommunications equipment in a uniform, feature-oriented language. It's a text/barcode label on the front of all equipment installed at RBOC facilities et. al. that facilitates inventory, maintenence, planning, investment tracking, and circuit maintenence processes. Suppliers of telecommunication equipment give Bellcore technical data on their equipment, and Bellcore assigns a CLEI code to that specific product. Bellcore's GR-485-CORE specification contains the generic guidelines for Common Language Equipment Coding Processes and Guidelines.

CLEOS Conference of the Lasers and Electro-Optics Society.

CLI 1. Command line interface
2. Cumulative Leakage Index. As used in the FCC Rules (in Section 76.611(a)(1)), this term identifies the results of a ground-based measurement of the signal-leakage performance of a cable television distribution system. Under the procedure specified in this rule, each leak is measured on the ground and the CLI is then calculated from measurement data (this term does not include the results of airspace measurements specified in Section 76.611(a)(2)). The calculated CLI value must be reported to the FCC by July 1 of each year on FCC Form 320.

Clifton Clifton is the best educator at Bellcore. I know because he told me so. He also said he would recommend my dictionary even more strongly to his people than he does. Powell to the people. Right on!

Climbers What personnel wear to climb wooden poles. Officially called linesman's climbers, but are often known as spurs, hooks and gaffs. They consist of a steel shanks that strap to a person's leg. The inside has a spike used to stab the pole.

Climbing Belt A belt that
communications/power/construction personnel to attach themselves to poles or
tower structures. Also called safety belt and body belt.

Clipped Frame Transmit packet lost at the interface because no buffer space was available to the host transmit driver for outgoing data.

CLLI Code Common Language Location Identifier. Pronounced silly code. An 11-digit alphanumeric code, CLLI was developed by Bellcore as a method of identifying physical locations and equipment such as buildings, central offices, poles, and antennae. Consider the real-life example of NYCMNY18DSO. The first four characters identify the place name (NYCM is New York City Manhattan). The following two characters identify the state, region or territory (NY is New York). The remaining five characters identify the specific item at that place (18DSO is the AT&T 5E Digital Serving Office on West 18th Street). Phone companies use CLLI Codes for a variety of purposes, including identifying and ordering private lines and trapping and tracing of annoying or threatening calls. See ANNOYANCE CALL BUREAU, TRAP and TRACE, and WIRE TAP.

Click Tones A particular progress tone injected onto the forward voice channel (mobile unit receive, base station transmit) to indicate to the subscriber that the call has not been abandoned by the system. Basically click tones indicates acknowledgment by the cellular system that the cellular system's computer is processing the call.

Clickstreams The paths a user takes as he or she navi-

gates cyberspace. Advertisers and online media providers are developing software that can accurately track user's clickstreams.

CLID Calling Line IDentification. Also called Caller ID. See Caller ID for a full explanation. See also ANI and CLASS.

Client Clients are devices and software that request information. Clients are objects that uses the resources of another object. A client is a fancy name for a PC on a local area network. It used to be called a workstation. Now it is the "client" of the server. See also CLIENT SERVER, CLIENT SERVER MODEL, FAT CLIENT, MAINFRAME SERVER, MEDIA SERVER and THIN CLIENT.

Client Access Protocol CAP. See iCalendar.

Client Application Any computer program making use of the processing resources of another program.

Client Operating System Operating System running on the client platform. See CLIENT.

Client Pull See Meta Tag.

Client Server A computer on a local area network that you can request information or applications from. The idea is that you — the user — are the client and it — the slave — is the server. That was the original meaning of the term. Over time, client server began to refer to a computing system that splits the workload between desktop PCs (called "workstations") and one or more larger computers (called "servers") joined on a local area network (LAN). The splitting of tasks allow the use of desktop graphic user interfaces, like Microsoft's Windows or Apple Macintosh's operating system, which are easier to use (for most people) than the host/terminal world of mainframe computing, which placed a "dumb terminal" on a user's desk. That dumb terminal could only send and receive simple text-based material. And the less it sent, the faster it worked (lines were slow), so some of the "human interfaces" were very cryptic. You often were forced to spend weeks at school learning simple mainframe programs.

A good analogy of client-server computing, according to Peter Lewis of the New York Times is to think of client server as a restaurant where the waiter takes your order for a hamburger, goes to the kitchen and comes back with some raw meat and a bun. You get to cook the burger at your table and add your favorite condiments. In computerese, this is client/server, distributed computing, where some processing work is done by the customer at his or her table, instead of entirely in the kitchen (centralized computing in the old mainframe days). It sounds like more work, but it has many advantages. The service is faster. The food is cooked exactly to your liking, and the giant, expensive stove in the kitchen can be replaced by lots of cheap little grills. See CLIENT SERVER MODEL, DOWNSIZING, REENGINEERING and SERVER.

Client Server Computer Telephony Client server computer telephony delivers ten benefits:
1. Synchronized data screen and phone call pop. Your phone rings. The call comes with the calling number attached (via Caller ID or ANI). Your PBX or ACD passes that number (via Telephony Services) to your server, which does a quick database look up to see if it can find a name and database entry. Bingo, it finds an entry. It passes the call and the database entry simultaneously to whoever is going to answer the phone: The attendant. The boss. The sales agent. The customer service desk. The help desk. All this saves asking a lot of questions. Makes customers happier.
2. Integrated messaging. Also called Unified Messaging. Voice, fax, electronic mail, image and video. All on the one screen. Here's the scenario. You arrive in the morning. Turn on

your PC. Your PC logs onto your LAN and its various servers. In seconds, it gives you a screen listing all your messages — voice mail, electronic mail, fax mail, reports, compound documents .... Anything and everything that came in for you. Each is one line. Each line tells you whom it's from. What it is. How big it is. How urgent. Skip down. Click. Your PC loads up the application. Your LAN hunts down the message. Bingo, it's on screen. If it contains voice — maybe it's a voice mail or compound document with voice in it — it rings your phone (or your headset) and plays the voice to you. Or, if you have a sound card in your PC, it can play the voice through your own PC. If it's an image, it will hunt down (also called launch) imaging software which can open the image you have received, letting you see it. Ditto, if it's a video message.

Messages are deluging us. To stop them is to stop progress. But to run your eye down the list, one line per entry. Pick the key ones. Junk the junk ones. Postpone the others. That's what integrated messaging is all about. Putting some order back into your life.

3. Database transactions. Customer look ups. There are bank account balances, ticket buys, airline reservations, catalog requests, movie times, etc. Doing business over the phone is exploding. Today, the caller inputs his request by touchtone or by recognized speech. The system responds with speech and/or fax. Today's systems are limited in size and flexibility. The voice processing application and the database typically share the same processor, often a PC. Split them. Spread the processing and database access burden. Join them on a LAN (for the data) and on new, broader voice processing "LANs," like SCSA or MVIP. You've suddenly got a computer telephony system that knows no growth constraints. You could also get the system to front-end an operator or an agent. Once the caller has punched in all his information, then the call and the screen can be simultaneously passed to the agent.

4. Telephony work groups. Sales groups. Collections groups. Help desks. R&D. We work in groups. But traditional telephony doesn't. Telephony today is BIG. Telephony today is one giant phone system for the building, for the campus. Everyone shares the same automated attendant, the same voice mail, the same ubiquitous, universal, generic telephone features. But they shouldn't. The sellers need phones that grab the caller's phone number, do a look-up on what the customer bought last and quickly route the call to the appropriate (or available) salesperson. The one who sold the customer last time. The company's help desk needs a front end voice response system that asks for the customer's serial number, some indication of the problem and tries to solve the problem by instantly sending a fax or encouraging the caller to punch his way to one of many canned solutions. "The 10 biggest problems our customers have." When all else fails, the caller can be transferred to a live human, expert at diagnosing and solving his pressing problem. A development group might need e-mails and faxes of meeting agendas sent, meeting reminder notices phoned and scheduled video conferences set up. All automatically. The accounts receivable department needs a predictive dialer to dial all our deadbeats. The telemarketing department also needs a predictive dialer, but different programming.

5. Desktop telephony. There are two important aspects. Call control and media processing services. Call control (also called call processing) is a fancy name for using your PC to get to all your phone system's features — especially those you have difficulty getting to with the forgettable commands phone makers foist on us. *39 to transfer? Or it is *79. With

attractive PC screens, you point and click to easy conferencing, transferring, listening to voice mail messages, forwarding, etc. There are enormous personal productivity benefits to running your office phone from your PC: You can dial by name, not by number you can't remember. You can set up conference calls by clicking on names and have your PC call the participants and call you only when they're all on the phone. You can transfer easily. You can work your voice mail more easily on screen, instead of having to remember "Dial 3 for rewind," "Dial 2 to save," and other obscure commands. Here's a wonderful quote from Marshall R. Goldberg, Developer Relations Group at Microsoft. He says "Voice mail systems that could benefit through integration with the personal computer largely remain isolated, difficult to use, and inflexible. Browsing, storing messages in hierarchical folders, and integration of address books — functions just about everyone could use — are either unavailable or unusable."

The second benefit is media control. Media control is a fancy name for affecting the content of the call. You may wish to record the phone call you're on. You may wish to have all or part of your phone call clipped and sent to someone else — as you often today with voice mail messages. You may wish to simply file your conversations away in appropriate folders. You may wish to be able to call your PC and get it to read you back any e-mails or faxes you received in the last day or so.

6. Applying intelligence. A PC is programmable. The typical office phone isn't. A PC can be programmed to act as your personal secretary, handling different calls differently. It can be programmed to include commands, such as "If Joe calls, break into my conversation and tell me." "If Robert calls, send him to voice mail." etc.

7. The Compound Document. The typed document lacks life. But add voice, image and video clips to it and it gets life. The LAN makes the compound document easier to achieve. The Compound Document gets attention.

8. Management of phone networks. Today, phone networks are very difficult to manage. Often the PBX is managed separately from the voice mail, which is managed separately from the call accounting, etc. It's a rare day in any corporate life when the whole system is up to date, with extensions, bills and voice mail mailboxes reflecting the reality of what's actually happening. The latest generations of LAN software — NetWare 4.1 and Windows NT — have solid enterprise-wide directories and far easier management tools. Integrate these LAN management tools with telecommunications management, and potentially all you need is to make one entry (for a new employee, a change, etc.) and the whole system — telecom and computing — could update itself automatically, including even issue change orders to the MIS and telecom departments and vendors.

9. No dedicated hardware in the PC. With only one link — from the switch to the LAN — there's no need to open the desktop PC and place specialized telephony hardware in each PC that wants to take advantage of the new LAN-based telephony features.

10. Switch elimination. The ultimate potential advantage of LAN-based telephony is to eliminate the connection to the switch (PBX or ACD) by simply populating the LAN server (now called a telephony server) with specialized computer telephony cards and run the company's or department's phones off the telephony server directly.

Client Server Model In most cases, the "client" is a desktop computing device or program "served" by another networked computing device. Computers are integrated over the network by an application, which provides a single system image. The server can be a minicomputer, workstation, or microcomputer with attached storage devices. A client can be served by multiple servers. See CLIENT SERVER.

Client Software An Internet access term. Multiprotocol PPP client software allowing dial-in access to the public Internet or corporate LANs via a dialup switch or a remote access server. The client software dialer is responsible for establishing/terminating the dial-in connection. The client software PPP driver manages the traffic sent/received across the network link.

Client Telescript A General Magic term. Telescript that is integrated into the various client platforms including Macintosh, DOS/Windows, UNIX, PenPoint, and Newton.

CLIP Calling Line Identification Presentation. Also known as CNIP (Calling Number Identification Presentation). A supplementary service provided under by certain wireless "cellular" services, such as GSM (Global System for Mobile Communications) and PCS (Personal Communications Services). Assuming that your subscriber profile includes this service, the number of the calling party displays on your wireless telephone. If the calling party uses CNIR (Calling Number Identification Restriction) however, your display says "PRIVATE." You can choose to have PRIVATE CALLS automatically transferred to your voice mailbox, while retaining the option to override that instruction with a simple keypad command. This feature is known variously as Caller ID and Calling Line ID (CLID) in the wired world. Whether in the wireless or the wired world, this feature is made available courtesy of SS7. See also GSM, PCS, and SS7.

Clip On Toll Fraud Clip on toll fraud occurs when someone connects a phone between someone else's phone (typically a coin phone) and the central office and makes unlawful toll calls. The term "clip on" comes because the telephone service thief "clips on" to the line. Clip on toll fraud is often done on COCOT (Customer Owned Coin Operated Telephone) phone lines because these lines do not enjoy the same protection from toll fraud which is afforded to coin phone lines which local telcos provide to their own coin phones.

Clipboard A generic term for a place in software which holds text, pictures or images that you are copying or moving between applications. The clipboard is a temporary holding place only. When you cut or copy a new item it will replace the current clipboard contents. In other words, there's usually only one clipboard. Windows uses a clipboard and so does the Apple Macintosh.

Clipbook A Windows NT term. Windows NT is soon to be known as Windows 2000. Permanent storage of information you want to save and share with others. This differs from the Clipboard which temporarily stores information. You can save the current contents of the clipboard, which temporarily stores information. You can then share that information, allowing others to connect to the Clipboard. See Clipbook Page.

Clipbook Page A Windows NT term. Windows NT is soon to be known as Windows 2000. A unit of information pasted into a local ClipBook. The ClipBook page is permanently saved. Information on a ClipBook page can be copied back onto the Clipboard and then pasted into documents. You can share ClipBook pages on the notework.

Clipbook Service A Windows NT term. Supports the ClipBook Viewer application, allowing pages to be seen by remote ClipBooks.

Clipper A circuit or device that limits the instantaneous output signal amplitude to a predetermined maximum value, regardless of the amplitude of the input signal. See CLIPPER CHIP.

Clipper Chip A microprocessor chip, officially known as the MYK-78, which the Federal Government wants to add to phones and data communications equipment. The chip would ensure that conversations in Clipper-equipped communicating equipment would be private — from everybody except the Government. With a court-approved wiretap, an agency like the FBI, could listen in, since the Government would have the key to Clipper. On February 4, 194, the Clinton White House announced its approval of the Clipper chip and the "Crypto War" broke out — with many companies and individuals urging a stop to Clipper. In late Spring, 1994 an AT&T Bell Labs researcher revealed that he had found a serious flaw in the Clipper technology. As of writing it wasn't clear what would happen to the Clipper Chip. See NSA.

Clipping Clipping has two basic meanings. The first refers to the effect caused by a simplex (one way at a time) speakerphone. Here the conversation goes one way. When the other person wants to talk, the voice path has to reverse (to "flip"). While the flipping takes place, a few sounds are "clipped" from that person's conversation. This phenomenon happens on some long distance and many overseas channels. These channels are so expensive, they are simultaneously shared by many conversations. Gaps in your conversation are filled by other people's conversation. But when you start talking, the equipment has to recognize you're now talking, find some capacity for your conversation, and send it. In the process of doing this, your first word or part of your first word might be "clipped" and the conversation will sound "broken."

The second way your voice is clipped is what happens every day on the telephone. You're squeezing your own voice which typically spans 10,000 Hertz into a voice channel which is only 3,000 Hertz. This clips the extremes of your conversation — the higher sounds. As a result, your voice sounds flatter over the phone. As you become more economical and try to squeeze your voice into smaller capacity channels, so it becomes increasingly clipped.

CLIR Calling Line Identification Restriction. A wireless telecommunications term. A supplementary service provided under GSM (Global System for Mobile Communications). See GSM.

CLLI Code Common Language Location Identifier. Pronounced silly code. When a phone company institutes trapping and tracing on a phone line, one of the pieces of information it can receive on incoming calls is the CLLI Code, a 11 character code that might look like "nycmny18dso." That says the call is coming in from New York City, Manhattan from a central office called 18DSO (which I happen to know is an AT&T 5E located on West 18th Street. See ANNOYANCE CALL BUREAU, TRAP and TRACE, and WIRE TAP.

CLM Career Limiting Move. An ill-advised activity. Trashing your boss while he or she is within earshot is a serious CLM.

CLNP Connectionless Network Protocol. An OSI network layer protocol that does not require a circuit to be established before data is transmitted. The OSI protocol for providing the OSI Connectionless Network Service (data gram service). CLNP is the OSI equivalent to Internet IP, and is sometimes called ISO IP.

CLNS Connectionless Network Service. CLNS. A Network Layer methodology which does not require a receiver's immediate acknowledgment of communications. See Connectionless Network Service.

Clock Exactly as it sounds. An oscillator-generated signal that provides a timing reference for a transmission link. A clock provides providing signals used in a transmission system to control the timing of certain functions such as the duration of signal elements or the sampling rate. It also generates periodic, accurately spaced signals used for such purposes as timing, regulation of the operations of a processor, or generation of interrupts. In short, a clock has two functions: 1. To generate periodic signals for synchronization on a transmission facility. 2. To provice a time base for the sampling of signal elements. Used in computers, a clock synchronizes certain procedures, such as communication with other devices. It simply keeps track of time, which allows computers to do the same things at the same time so they don't "bump into each other."

Here are other definitions of clock:

A clock is an internal timing device. A clock is a computer chip that uses a quartz crystal to generate a uniform electrical frequency from which digital pulses are created. A clock keeps track of hours, minutes, and seconds and makes this data available to computer programs. A clock is also a timer set to interrupt a CPU at regular intervals in order to provide equal time to all the users of the computer. A clock also maintains the uniform transmission of data between the sending and receiving terminals and computers. In short a clock has many functions.

Clock Bias The difference between the GPS clock's indicated time and true universal time. GPS is Global Positioning System. See GPS.

Clock Cycle The time that elapses from one read or write operation to another in the main memory of a computer's central processing unit (CPU) The more tasks that can be accomplished per cycle, the more efficient the chip. Some chips like the i860 chip can execute two instructions and three operations per clock cycle.

Clock Difference A measure of the separation between the respective time marks of two clocks. Clock differences must be reported as algebraic quantities measured on the same time scale. The date of the measurement should be given.

Clock Doubling Refers a computer whose internal CPU clocks runs twice as fast as the clock for the rest of computer. This has the effect of increasing the computer's speed without the expense of high-speed hardware.

Clock Speed Each CPU contains a special clock circuitry which is connected to a quartz crystal (same as the one in your watch). The quartz crystal's vibrations, which are very fast, coordinate the CPU's operation, keeping everything in step. CPU clock speeds are measured in megahertz, or MHz, which stands for "million cycles per second." The clock speeds of today's computer range from a slow of 4.77 MHz (the original IBM PC) to 300 MHz with some Intel Pentium based PCs. Clock speed is a misleading term. It is only one way of measuring the speed of a computer. One other critical way is how fast you can read and write information to the hard disk. How important that is depends on whether you're running a program with lots of access to your hard disk (like a database program) or running a program which uses a lot of calculations in RAM, e.g. a spreadsheet.

Clock Tolerance The maximum permissible departure of a clock indication from a designated time reference such as Coordinated Universal Time.

Clocking In synchronous communication, a periodic signal used to synchronize transmission and reception of data and control characters.

Clockwise Polarized Wave An elliptically or circularly polarized electromagnetic wave in which the direction of rotation of the electric vector is clockwise as seen by an observer looking in the direction of propagation of the wave.

Clone A mobile device, typically a cellular telephone, that claims to possess the same address identifier as another mobile device. See CLONE FRAUD.

Clone Fraud A way of using cellular phones to steal phone calls. In clone fraud, a legitimate serial number is programmed into an imposter's cellular telephone. This allows unauthorized calling to go on until a huge bill appears on the mailbox of the bewildered subscriber to whom the serial number actually belongs. Crooks get the numbers because the numbers are broadcast with every cellular call and can be picked up by ordinary radio scanners, which you can often buy at your local electronics store. According to the Wall Street Journal, Cellular thieves take advantage of the fact that when a cellular phone call is placed, the phone's unique electronic code is transmitted over the airwaves to update the cellular network on the user's location. Service thieves wait near the busy areas - highways, financial districts - and use scanners to lift the codes from legitimate users. Thousands of such hijacked numbers can be later downloaded through a computer into "clone" phones. Thus, a clone phone lets a user place potentially hundreds of call that are billed to the legitimate owner of the original number. See CLONED PHONE and TUMBLING.

Cloned Phone A cellular phone has two basic ways it identifies itself to the cellular phone company it wants to use — its own telephone number (which can be changed) and a special secret number that's embedded into silicon inside the phone. That number is called an Electronic Serial Number, or ESN. When the phone wants to make a call, it sends those numbers and the cellular carrier uses them to check if the call is authorized. But because the information is traveling through the air, anyone with a scanner can pick up the information and retransmit it later, thus creating a "cloned phone" and pretending that he's authorized to make the call. Of course, the owner of the cloned phone ultimately gets the bill and a nasty shock.

CLONES Central Location On-Line Entry System. Definition courtesy Bellcore.

Close Coupling The condition in which two coils are placed in close magnetic relation to each other, thus establishing a high degree of mutual induction.

Close Talk A voice recognition term. An arrangement where a microphone is fewer than four inches from the speaker's mouth.

Closed Architecture Proprietary design that is compatible only with hardware and software from a single vendor of single product family. Contrast with OPEN ARCHITECTURE.

Closed Captioning A service, designed for people with hearing disabilities, that provides a simultaneous visual presentation of the sound associated with a television program. Closed captioning is not visible except on a TV receiver designed to display it or by use of a specially installed decoder.

Closed End The end of a Foreign Exchange — FX — line which ends on a PBX, a key system or a telephone. The closed end is the end of the circuit beyond which a call cannot progress further. The other end of the FX circuit is called the "open end," because calls can progress further.

Closed Loop System A closed electrical circuit into which a standard signal is feed and received instantly. A measure of the difference between the input signal and the output signal is a measure of the error, and potentially what's causing it.

Closed User Group A group of specified users of a data

network that is assigned a facility that permits them to communicate with each other but precludes communications with other users of the service or services.

Closet Telecommunications closet. An enclosed space for housing telecommunications equipment, cable terminations, and cross-connect cabling that is the recognized location of the cross-connect between the backbone and horizontal facilities.

Closure A cabinet, pedestal, or case used to enclose cable sheath openings necessary for splicing or terminating fibers.

Cloud Beginning with AT&T's BDN (Bell Data Network) and ACS (Advanced Communication System) offerings, the data network was depicted as a "cloud." The user data was presented at one end of the carrier cloud, and was delivered at the other end. What went on inside the cloud was obscured from view. The thinking behind this unsuccessful conceptual sell was that the user needn't be concerned with what went on inside the carrier network; rather, that was the concern and responsibility of the carrier. While BDN, later known as ACS and later still as Net 1000, was unsuccessful, the concept of the cloud was a huge success.

Some of the newer high-speed data, phone company-offered services resemble a local area network. You connect to them directly. To make a call, you don't actually dial a number as you do on a circuit-switched service, you just transmit, putting an address at the front of your transmission. The service reads the address and sends it where you want. Like a LAN, everything is connected and on line. The concept is get stuff sent from one place to another much faster than would be possible if you had to wait to dial, for the circuit to be set up, for the machine at the other end to answer, etc. In these high-speed services, the circuit is "always set up." The provider (the phone company) refers to its network as a "cloud." And when you see diagrams of these newer high-speed services, like SMDS and frame relay, you see the carrier portion drawn as a cloud (like the one you see in the sky). Services with "clouds" are also called "connectionless." See the various definitions below beginning with CONNECTIONLESS.

CLP Cell Loss Priority: This bit in the ATM cell header indicates two levels of priority for ATM cells. CLP=0 cells are higher priority than CLP=1 cells. CLP=1 cells may be discarded during periods of congestion to preserve the CLR of CLP=0 cells. See also Cell-loss Priority Field

CLR Cell Loss Ratio: CLR is a negotiated QoS (Quality of Service) parameter and acceptable values are network specific. The objective is to minimize CLr provided the end-system adapts the traffic to the changing ATM layer transfer characteristics. The Cell Loss Ratio is defined for a connection as: Lost Cells/Total Transmitted Cells. The CLR parameter is the value of CLR that the network agrees to offer as an objective over the lifetime of the connection. It is expressed as an order of magnitude, having a range of 10-1 to 10-15 and unspecified.

CLS Control Line Setting.

CLTP Connectionless Transport Protocol. Provides for end-to-end Transport data addressing (via Transport selector) and error control (via checksum), but cannot guarantee delivery or provide flow control. The OSI equivalent of UDP.

CLTS ConnectionLess Transport Service.

Cluster 1. Collection of terminals or other devices in a single location. A cluster control unit and a cluster controller in IBM 3270 systems are devices that control the input/output operations of a group (cluster) of display stations. See also CLUSTERING.

2. Unit of storage allocation used by MS-DOS usually consisting of four or more 512-byte sectors.

3. Physical grouping of workstations that share one or more panel runs.

4. A mini-network of PCs that work in a fault-resilient manner.

5. A cluster is a group of computers and storage devices that function as a single system.

Cluster Controller A device that can control input/output operations of more than one device connected to it (e.g. a terminal). An interface between several bisynchronous devices and a PAD, NC or communication facility. The cluster controller handles remote communications processing for its attached devices. Most common types are IBM 327X.

Cluster Size An operating function or term describing the number of sectors that the operating system allocates each time disk space is needed.

Clustering A client/server term describing the collection of servers or data in a central location for reasons of increased effectiveness and efficiency of security, administration and performance. Clustering is to help servers become fault resilient. Clustering helps to overcome problems associated with the client/server paradigm in general. In the old days of "heavy iron" (mainframes), user access to applications and files was carefully controlled. The target applications and files were resident on a highly redundant, carefully administered and tightly secured mainframe computer. As client/server has taken hold, the processors tend to be distributed in order that they are located in closer proximity to users, thereby relieving the strain on network resources, improving response times, and so on. The downside is that the resources are more difficult to manage, secure and control. Clustering the servers in a centralized location places them back in the hands of centralized MIS management, relieving these problems. There are two forms of clustering. The first form involves segmenting and spreading the database across multiple servers, with each segment of the database residing on multiple servers in order to achieve some level of redundancy. The second form positions the servers in a communications role, with each providing access for a group of users to data housed in a central repository, generally in the form of a minicomputer or mainframe.

CLUT An imaging term. Color Look-Up Table. The palette used in an indexed color system. Usually consists of 256 colors.

CM 1. Computing Module.

2. Configuration Management. A wireless telecommunications term. The tracking, coordination, and administration of software and hardware related to telecommunication or information systems. Versions are controlled and tracked.

CMA Communications Managers Association. An independent, not-for-profit users group formed in 1948 and serving the New York/New Jersey area. CMA provides a forum for peer-to-peer discussion of common issues, evaluation of technologies and their business applications, and the fostering of constructive relationships between suppliers and users. In addition to end users, CMA welcomes non-voting "Partners" in the form of vendors, consultants and associations. www.cma.org

CMC Common Messaging Calls. A messaging standard defined by the X.400 API Association. CMC 1.0 defines a basic set of calls to inject and extract messages and files and access address information. CMC is intended to define a useful common denominator across a wide variety of messaging systems. The idea is that an electronic mail system, no matter how crude, should be able to support a CMC front end. CMC's major "competition" is MAPI — Messaging Application Programming Interface — from Microsoft, though simple MAPI is almost identical to CMC.

CMEA Cellular Message Encryption Algorithm. See ECMEA.

CMCI Cable Modem to CPE Interface. An element of DOCSIS (Data Over Cable Service Interface Specification), a project intended to develop a set of specifications for high-speed data transfer over cable television systems. CMCI is the interface between the cable modem and the CPE (Customer Premise Equipment), which typically would be in the form of a PC. See also DOCSIS.

CMIP Common Management Information Protocol. CMIP is the network management standard for OSI networks. It has some features that are lacking in SNMP and SNMP-2, and is more complex. CMIP has a far smaller mind share and market share than SNMP in North America, though support for this standard is sometimes mandated, especially in Europe. CMIP is an ITU-TSS standard for the message formats and procedures used to exchange management information in order to operate, administer maintain and provision a network. In short, CMIP is the protocol used for exchanging network management information. Typically, this information is exchanged between two management stations. CMIP can, however, be used to exchange information between an application and a management station. CMIP has been designed for OSI networks, but it is transport independent. Theoretically, it could run across a variety of transports, including, for example, IBM's Systems Network Architecture. See CMIP/CMIS, MIP and SNMP.

CMIP/CMIS Common Management Information Protocol/Common Management Information Services. An OSI network management protocol/service interface created and standardized by ISO for managing heterogeneous networks.

CMISE Common Management Information Service Element. A wireless telecommunications term. The functionality provided by CMIP in transporting network management information.

CMOL Short for "CMIP Over Logical Link Control". An implementation of the CMIP protocol over the second layer of the OSI protocol stack, to be proposed as a standard by 3 Com Corp. and IBM. The goal of CMOL is to create agents that require significantly less memory than CMIP implemented over OSI, or SNMP implemented over UDP.

CMOS Complementary Metal Oxide Semiconductor. A technology for making integrated circuits known for requiring less electricity. See also COMPLEMENTARY METAL OXIDE SEMICONDUCTOR.

CMOS RAM Complementary Metal Oxide Semiconductor Random Access Memory. Memory which contains a personal computer's configuration information. CMOS RAM must have continuous power to preserve its memory. This power is typically supplied by a lithium battery.

CMOS Setup A program which prepares the system to work. CMOS setup records your PC's hardware configuration information into CMOS RAM. It must be modified when you add, change or remove hardware.

CMOT CMIP Over TCP/IP. More correctly, Common Management Information Protocol over TCP/IP. The original CMOT was described by RFC 1095, which is now obsolete. The new RFC 1189 "defines the means for implementing the IS version of CMIS/CMIP on top of both IP-based and OSI-based Internet transport protocols...", an expanded charter. The portion that is CMIS/CMIP over TCP/IP is still referred to as "CMOT". Someone referred to the new RFC as "CMIP over RFC 1066". In short, CMOT is an Internet standard defining the use of CMIP for managing the TCP/IP-based Internet and other attached networks. While the OSI-based CMIP is viewed as the most elegant long-term network management solution for such

networks, it has not received the widespread acceptance of SNMP (Simple Network Management Protocol), which is much simpler (hence the name) and much more easily implemented. See also CMIP, OSI Reference Model and SNMP.

CMP Communications Plenum Cable.

CMR An ATM term. Cell Misinsertion Rate: The ratio of cells received at an endpoint that were not originally transmitted by the source end in relation to the total number of cells properly transmitted.

CMRFI Cable Modem to Radio Frequency Interface. An element of DOCSIS (Data Over Cable Service Interface Specification), a project intended to develop a set of specifications for high-speed data transfer over cable television systems. At the CPE (Customer Premise Equipment) end of the network, the CMRFI provides the interface between the cable modem and the cable system coax drop. The CMRFI specification will include all physical, link and network level aspects of the communications interface, including RF levels, modulation techniques, coding schemes, and multiplexing. See also DOCSIS.

CMRS Commercial Mobile Radio Service. A FCC term for ESMR (Enhanced Specialized Mobile Radio), as well as cellular and PCS.

CMS 1. Call Management System. This is the AT&T label for an inbound call distribution management reporting package. CMS is found on the Horizon, the Merlin, S/75 and S/85 PBX/ACDs. CMS is the successor product to AEMIS/PRO 150/500.

2. Call Management Services. Canadian term for local calling features based on CLID (Calling Line Identification).

CMS 8800 Cellular Mobile Telephone Service (North American version).

CMTRI Cable Modem Telco Return Interface. An element of DOCSIS (Data Over Cable Service Interface Specification), a project intended to develop a set of specifications for high-speed data transfer over cable television systems. At the head-end of the network, the CMTRI provides the interface between the cable modem system and PSTN. See also DOCSIS.

CMTS 1. CMTS stands for the Cellular Mobile Telephone System. The original and still, most common CMTS is a low-powered, duplex, radio/telephone which operates between 800 and 900 MHZ, using multiple transceiver sites linked to a central computer for coordination. The sites, or "cells,", named for their honeycomb shape, cover a range of one to six, or more, miles in each direction. The cells overlap one another and operate at different transmitting and receiving frequencies in order to eliminate crosstalk when transmitting from cell to cell. Each cell can accommodate up to 45 different voice channel transceivers. When a cellular phone is activated, it searches available channels for the strongest signal and locks onto it. While in motion, if the signal strength begins to fade, the telephone will automatically switch signal frequencies or cells as necessary without operator assistance. If it fails to find an acceptable signal, it will display an "out of service" or "no service" message, indicating that it has reached the limit of its range and is unable to communicate. Each cellular telephone has a unique identification number which allows the Mobile Telephone Switching Office (MTSO) to track and coordinate all mobile phones in its service area. This ID number is known as the Electronic Security Number (ESN). The ESN and cellular phone's telephone Number are NOT the same. The ESN is a permanent number engraved into a memory chip called a PROM or EPROM, located in the telephone chassis. This number cannot be changed through programming as the telephone number can, although it can be replaced. Each time the telephone is used, it transmits its ESN to the MTSO by means of DTMF tones during the dialing sequence. The MTSO may be able to determine which ESNs are good or bad, thus individual numbers can be banned from use within the system. See also CELL and CELLULAR.

2. Cable Modem Termination System. An element of DOCSIS (Data Over Cable Service Interface Specification), a project intended to develop a set of specifications for high-speed data transfer over cable television systems. CMTS comprises CMTS-DRFI (CMTS-Downstream RF Interface), CMTS-NSI (CMTS-Network Side Interface), and CMTS-URFI (CMTS-Upstream RF Interface) in order to provide two-day communications. See also CMTS-DRFI, CMTS-NSI, CMTS-URFI and DOCSIS.

CMTS-DRFI Cable Modem Termination System-Downstream RF Interface. An element of DOCSIS (Data Over Cable Service Interface Specification), a project intended to develop a set of specifications for high-speed data transfer over cable television systems. At the head-end of the network, the CMTS-DRFI provides the interface between the cable modem system and the downstream RF (Radio Frequency) path, which terminates in the cable modem at the customer premise. It works in conjunction with the CMTS-URFI (CMTS-Upstream RF Interface) in order to provide two-day communications. See also CMTS-URFI and DOCSIS.

CMTS-NSI Cable Modem Termination System-Network Side Interface. An element of DOCSIS (Data Over Cable Service Interface Specification), a project intended to develop a set of specifications for high-speed data transfer over cable television systems. CMTS-NSI is the interface of the cable modem system at the head-end of the network; i.e., at the CATV provider's premise. CMTS-NSI provides the interface between the backbone cable system and the CATV provider's server complex. See also DOCSIS.

CMTS-URFI Cable Modem Termination System-Upstream RF Interface. An element of DOCSIS (Data Over Cable Service Interface Specification), a project intended to develop a set of specifications for high-speed data transfer over cable television systems. At the head-end end of the network, the CMTS-URFI provides the interface between the cable modem system and the upstream RF (Radio Frequency) path, which terminates in the cable modem at the customer premise. It works in conjunction with the CMTS-DRFI (CMTS-Downstream RF Interface) in order to provide two-day communications. See also CMTS-DRFI and DOCSIS.

CMTRI Cable Modem Telco Return Interface. An element of DOCSIS (Data Over Cable Service Interface Specification), a project intended to develop a set of specifications for high-speed data transfer over cable television systems. At the head-end of the network, the CMTRI provides the interface between the cable modem system and PSTN. See also DOCSIS.

CMY A computer imaging term. A color model used by the printing industry that is based on mixing cyan, magenta, and yellow. It's also referred to as CMYK, with the K denoting black. The K was added after printers discovered they could obtain a darker black using special black colorants rather than by combining cyan, magenta, and yellow alone. See also CMYK.

CMYK A computer imaging term. A color model used by the printing industry that is based on mixing cyan, magenta, yellow and black (called "K.") It used to be called CMY. The K was added after printers discovered they could obtain a darker black using special black colorants (i.e. black ink) rather than by combining cyan, magenta, and yellow. CMYK is the basis of what's known now as "four-color" printing.

But there is also five, six, seven and eight color printing, etc. Each of these "extra" colors are basically "colors" which are better printed as their own color rather than by printing a combination the basic three. Silver, copper, gold, aluminum, etc. are all printed traditionally as extra colors. They cannot be created by combining CMYK. Sometimes people are picky about the way their colors come out — i.e. Coca Cola red — so they may be printed with that color ink rather than combining CMYK. Typically a full color printing job requires passes under four printing presses — each laying down C,M,Y and K. These extra colors — silver, gold, copper, special colors — will need additional printings by additional printing presses. Thus a color print job could easily become five, six or seven color print job. The most common full color print job we hear about is four color. But you see an awful lot of jobs that contain silver and copper and may be six color print jobs.

CN Complementary network.

CNA 1. Cooperative Network Architecture.

2. Centralized Network Administration is an AMP-defined architecture that consolidates all network electronics into a single closet instead of distributing them throughout the building. According to AMP, CNA saves money over the long haul because of reduced administration costs. Centralized Network Administration can be executed with optical fiber for runs up to 300 meters, with Category 5 unshielded twisted pair (UTP) if no user is more than 90 meters from the central cross-connected and equipment room. Compare with DISTRIBUTED NETWORK ADMINISTRATION.

CNAM Caller ID with NAMe. See LNP (as in Local Number Portability).

CNC Complementary Network Service. See OPEN NETWORK ARCHITECTURE.

CND 1. Calling Number Delivery. See Caller ID Message Format and Call Block.

2. Calling Number Display. See also CLID.

CNE Certified (local area) Network Engineer. When you graduate from Novell's third level class, you become a certified network engineer. CNEs are an elite group in the LAN industry. See also CCIE and MCSE.

CNET Centre National d'Etudes de Telecommunication. The French organization that approves telecommunications products for sale in France.

CNG Also called Auto Fax Tone, or Calling Tone. This tone is the sound produced by virtually all fax machines when they dial another fax machine. CNG is a medium pitch tone (1100 Hz) that lasts 1/2 second and repeats every 3 1/2 seconds. A fax machine will produce CNG for about 45 seconds after it dials. The CNG tone is useful for owners of fax/phone/modem switches. Such switches answer an incoming call. If they hear a CNG tone, they will transfer the call to a fax machine. If they don't, they'll transfer the call to a phone, answering machine or perhaps a modem. Depends on how they're set up. Some fax machines do not transmit a CNG tone with manually-dialed transmissions — i.e. where the caller picked up the handset on the fax machine, dialed and waited for high-pitched squeal before pushing his fax machine's "start" button. A manual dialed fax transmission will "fool" fax/voice switches. See CED and FACSIMILE.

CNIP Calling Number Identification Presentation. See CLIP.

CNIR Calling Number Identification Restriction. A wireless "cellular" term (GSM and PCS) for Call Block, also known as Calling Number Delivery (CND), in the wired world. See CALL BLOCK.

CNIS Calling Number Identification Services.

CNM An ATM an d SMDS (Switched Megabit Data Service) term. Customer Network Management. All activities that customers perform to manage their communications networks. SMDS CNM service enables customers to directly manage many aspects of the SMDS service provided by telecommunications carriers. See CUSTOMER NETWORK MANAGEMENT.

CNO Corporate Networking Officer, a term invented by William Y. O'Connor, CEO of Ascom Timeplex.

CNR 1. Telephone company term for re-scheduling a telephone installation appointment because the "Customer is Not Ready."

2. An ATM term. Complex Node Representation: A collection of nodal state parameters that provide detailed state information associated with a logical node.

CNRI The Corporation for National Research Initiatives, a Reston, VA-based not-for-profit organization works with industry, academia, and government on national-level initiatives in information technology. It will host the initial operations of IOPS.ORG. "IOPS.ORG will play a key role in the healthy technical and operational evolution of the Internet as an increasingly important component of the economy," said CNRI President Robert Kahn. www.cnri.reston.va.us

CNS Complementary Network Service. CNSs are basic services associated with end user's lines that make it easier for Enhanced Service Providers (ESPs) to offer them enhanced services. Some examples of CNSs include Call Forwarding Busy/Don't Answer, Three Way Calling, and Virtual Dial Tone. See OPEN NETWORK ARCHITECTURE.

CO Central Office. In North America, a CO is that location which houses a switch to serve local telephone subscribers. Sometimes the words "central office" are confused with the switch itself. In Europe and abroad, the words "central office" are not known. The more common words are "public exchange." But those words tend to refer more to the switch itself, rather than the site, as in North America. See also CENTRAL OFFICE or PUBLIC EXCHANGE.

CO Lines These are the lines connecting your office to your local telephone company"s Central Office which in turn connects you to the nationwide telephone system.

CO Location See COLOCATION.

CO Simulator A desktop device which pretends to act like a mini-central office. The smallest version will consist of two lines and two REJ-11 jacks. Plug a phone into both jacks. Pick up one phone. You hear dial tone. Dial or touchtone two or three digits. Bingo, the second phone rings. You pick up the second phone. You can have a conversation with yourself or with a machine — like a voice processing system. Most central office simulators can simulate normal on-hook, off-hook, dialing, answering, speaking, etc. Some now can simulate caller ID features — including number of person calling.

Co-channel Interference Interference between signals transmitted in a given Radio Frequency (RF) channel in a particular cell and signals transmitted on the same RF channel in a different cell. A receiver that is in a position to receive from both cannot filter out the undesired signal. and consequently the noise level at the receiver increases.

Co-location The ability of a someone who is not the local phone company to put their equipment in the phone company's offices and join their equipment to the phone company's equipment. That "someone" who might co-locate their equipment on the phone company's premises might be an end-user or it might be another local or long distance telecommunica-

tions company. It might even be a competitor of the local phone company, i.e. another local phone company. The idea of co-location is to save money, give better service, ensure better interconnection, get technical problems solved faster, etc. Not all local phone companies offer their customers co-location. New York Telephone and New England Telephone (now Nynex) are two that now do. Also spelled collocation. See also VIRTUAL CO-LOCATION.

COAM Customer Owned And Maintained equipment.

Coasters Unsolicited floppy disks or CD-ROMs, such as the ubiquitous American On Line software, that arrive in one's mailbox too often.

Coasting Mode In timing-dependent systems, a free-running operational timing mode in which continuous or periodic measurement of timing error is not available. In some systems, operation in this mode can be enhanced for a period of time by using clock or timing error (or correction) information obtained during a prior tracking mode to estimate clock or timing corrections to be made in the free-running mode.

Coated Filament A vacuum tube filament coated with a metallic oxide to provide greater electron emission and longer life.

Coating A protective material (usually plastic) applied to the optical fiber immediately after drawing to preserve its mechanical strength and cushion it from external forces that can induce microbending losses.

Coaxial Cable A cable composed of an insulated central conducting wire wrapped in another cylindrical conducting wire. The whole thing is usually wrapped in another insulating layer and an outer protective layer. A coaxial cable has great capacity to carry great quantities of information. It is typically used to carry high-speed data (as in connections of 327X terminals to computer hosts) and in CATV (multiplexed TV stations).

COB Close Of Business.

COBOL Common Business Oriented Language. A very popular computer programming language for business applications.

COBRA Common Object BRoker Architecture.

Cobweb Site A World Wide Web site that hasn't been updated for a long time. A dead Web page.

COC Central Office Connection. Separately tariffed part of T-1 circuit.

Cochannel interference C/I. Cochannel interference refers to the interference caused between two cells transmitting on the same frequency within a network. Since cochannel interference is caused by another cell transmitting the same frequency, you can't simply filter out the interference. You can only minimize the cochannel interference through proper cellular network design. A cellular network must be designed to maximize the C/I ratio. The C/I ratio is the carrier-to-cochannel interference ratio. One of the ways to maximize the C/I ratio is to increase the frequency re-use distance, i.e. increase the distance between cells using the same set of transmission frequencies. The C/I ratio in part determines the frequency re-use distance of a cellular network.

Cockpit Effect An acoustics phenomenon describing the difficulty we have in modulating our speech if there is background noise significant to impair our ability to hear ourselves. Under such circumstances, the lack of feedback causes us to speak loudly, and to alter our speech patterns. Also known as Lombard Speech, the cockpit effect particularly is a problem when using a cellular phone in an automobile to communicate with a voice processor employing speech recognition technology.

COCOT Customer Owned Coin Operated Telephone. See also CLIP ON TOLL FRAUD.

COD An ATM term. Connection Oriented Data: Data requiring sequential delivery of its component PDUs to assure correct functioning of its supported application, (e.g., voice or video).

Code 1. As a verb, it means to write instructions in computer language. As a noun, it means software.

2. In telecommunications, code is the system of dots and dashes used to represent the letters of the alphabet, numerals, punctuation and other symbols.

Code Bit The smallest signaling element used by the Physical Layer for transmission on the fiber cable.

Code Blocking A switch's ability to block calls to a specified area code, central office code or phone number.

Code Blue A PBX feature for hospital application. If a patient is in distress, he can simply knock the telephone handset off of the cradle. After a brief period, the PBX recognizes the lack of dialing activity and sends a "code blue" alarm to the nurses station. Nurses and doctors come running. People turn blue when they can't breathe — hence the term.

Code Breaker See KEY and KEY HOLDER.

Code Call Access A very useful PBX feature. It allows attendants and extension users to activate, by dialing an access code followed by a two or three digit called code, customer-provided signaling devices throughout the premises. The signaling devices then issue a series of tones or visual coded signals corresponding to the called code. The called or paged party responds by dialing a meet-me answering code from any phone and is then connected to the paging party.

Code Coverage Modern computer telephony systems are composed largely of software, or "code." Invariably this code has many different logic paths and options. During normal system usage, many code paths are used only infrequently, if at all, meaning that normal usage will really only test a small portion of the total system. Code coverage refers to the amount of the system code that has been accessed during the testing of the system. The greater the code coverage, the more code that has been tested. 100% code coverage means that all the code has been tested. The amount of code coverage that has been achieved is usually determined through the use of a code coverage tool. Code coverage tools are available for most computer operating systems. Code coverage is especially important for computer telephony applications because many features are only infrequently used or are turned on or off based on the user's class of service. Also, many features interact with other features and the use of one feature often turns off another. Functional anomalies frequently exist in little-used paths and feature interactions. Tests should be designed to make sure that all code paths have been accessed and are adequately exercised. This definition courtesy of Steve Gladstone, author of the book Testing Computer Telephony Systems, available from 212-691-8215.

Code Conversion A process which converts the codes coming in from one network into codes that can be recognized on another network, such as converting from the Baudot code in a telex network to the ASCII code on the TWX network. Usually, the hardware will convert differences in transmission speed.

Code Division Multiple Access CDMA, also called Spread Spectrum, is a name for a new form of digital cellular phone service. The idea is that each phone call is combined with a code which only one cellular phone plucks from the air. Business Week said CDMA works "by spreading all signals across the same broad frequency spectrum and assigning a unique code — the company says one of 42 billion — to

each. The dispersed signals are pulled out of the background noise by a receiver which knows the code. This method, developed by a San Diego company called Qualcomm Inc. is very new. Much of the equipment to support it — like the cellular switches have not yet been developed." CDMA is also being used by wireless PBXs. See CDMA for a longer and better explanation.

Code Excited Linear Prediction CELP. An analog-to-digital voice coding scheme.

Code Independent Data Communication Data communication mode using a link procedure associated with the character and not dependent on the set of characters or the code used by the data source.

Code Level Number of bits used to represent a character.

Code Of Federal Regulations CFR. CFR is a codification of the general and permanent rules published in the Federal Register. It is divided into 50 titles that represent broad areas subject to federal regulation. Title 47 of the CFR pertains to telecommunications and contains the rules covering Part 22 Common Carriers and Part 90 Private Carriers.

Code Violation Violation of a coding rule; for example, the AMI coding rule is corrupted by a bipolar violation.

Code Word When used in the context of the Reed-Solomon encoding, it refers to the 63, 6-bit symbols (378 bits) resulting from the encoding of 47 6-bit (282 bits) information symbols. This is done by appending 16 6-bit parity symbols.

CODEC Originally CODEC stood for CODer-DECoder, i.e. microprocessor chip. Now the PC industry thinks it stands for COmpression/DEcompression, i.e. an overall term for the technology used in digital video and stereo audio. The original CODEC (still in big use in today's telephony industry) converts voice signals from their analog form to digital signals acceptable to modern digital PBXs and digital transmission systems. It then converts those digital signals back to analog so that you may hear and understand what the other person is saying. In some phone systems, the CODEC is in the PBX and shared by many analog phone extensions. In other phone systems, the CODEC is actually in the phone. Thus the phone itself sends out a digital signal and can, as a result, be more easily designed to accept a digital RS-232-C signal.

CODEC Conversion The back-to-back transfer of an analog signal from one codec into another codec in order to convert from one proprietary coding scheme (for instance, that used by CLI) to one used by another codec manufacturer (PictureTel, VTEL, GPT, BT, NEC, etc). The analog signal, instead of being displayed to a monitor, is delivered to the dissimilar codec where it is redigitized, compressed and passed to the receiving end. This is obviously a bi-directional process. Conversion service is offered by carriers such as AT&T, MCI and Sprint.

Coded Character Set A set of unambiguous rules that establish a character set and the one-to-one relationships between the characters of the set and their coded representations.

Coded Image A representation of an image in a form suitable for storage and processing.

Coded Trunks You buy several trunks. They hunt on. The main number is 555-3000. If the main number is busy, the call goes to the next line. There are two types of "next lines." One type can have an actual number, like 555-3001, which you can call directly. The other can be a coded trunk with no actual number and which you can't call directly. It's better to have no coded trunks because it's hard to test coded trunks. You can't dial them directly. Actual dial-able numbers are better.

Coder An analog-to-digital converter that changes analog voice signals to their digital equivalents. See CODEC.

Codial Office CDO. A small central office designed for unattended operation in a distant community. Usually a community dial office is fairly small, rarely more than 10,000 lines.

Coding Theory Mathematical theory describing how to encode data into streams of digital symbols at transmitter and decode it at receiver to maximize accuracy of data presented to user.

COE 1. Central - Office - Equipment as in Central office Equipment (COE) engineer.
2. Central Office Engineer.

Coefficient Of Variation A telephone company definition. Relates the standard deviation of a distribution to the mean of the distribution, usually as a percent. Example: If all of the busy hour, busy season loads for an office have a mean of 10,000 CCS and a standard deviation of 1000 CCS, the coefficient of variation is 10%.

COER Central Office Equipment Reports. A telephone company definition. A large scale computer software package which accepts Central office Engineering Data properly formatted by a Data Collection System (DCS), subjects these data to a series of validation tests, and produces final summarized reports designed to meet both administrative and engineering requirements.

Coherence Area In optical communications, the area in a plane perpendicular to the direction of propagation over which light may be considered highly coherent.

Coherence Length The propagation distance over which a light beam may be considered coherent. See COHERENT LIGHT.

Coherent Light Light emitted from laser and some light-emitting-diodes. It is made up of light of a single frequency in which all the light waves are in phase (i.e., their wave peaks and troughs are all in alignment so the waves reinforce or amplify each other.)

Coil 1. In electronics, a number of turns of wire, so wound as to afford inductance.
2. In telecommunications a coil refers to a load coil. It's a voice-amplifying device for twisted-pair wire. A load coil is usually placed every 3000 feet past the CO. They are usually placed in vaults with twisted-pair splices. They should be removed for high speed data communications.

Coil Antenna One consisting of one or more complete turns of wire. See LOOP ANTENNA.

Coin Acceptor/Rejector A mechanical or electro-mechanical device that checks and validates the coins deposited in a coin pay phone. They measure the coin's size and weight and steel content. These coin acceptor/rejector units transmit the value of the coin deposits to the processing part of a smart payphone or they signal the information to the telephone company central office via coded tones.

Coin Supervisory Trunk Group A trunk group that lets a switchboard operator collect overtime monies due on coin phones and check for stuck coins.

Coin Telephone A pay telephone that takes coins. The coin telephone was invented by William Gray, an American whose previous inventions included the inflatable chest protector for baseball players. Mr. Gray's first phone lacked a dial. Its instructions read:
"Call Central in the usual manner. When told by the operator, drop coin in proper channel and push plunger down."
In today's nomenclature, Mr. Gray's original phone is known as a post-pay coin phone. See also PAYPHONE and several entries following it.

COLD Computer Output to Laser Disk. A computer storage management term referring hardware and software solutions which store, index and retrieve formatted computer output on various media, including optical disks. Large scale COLD systems can be used to manage and archive storage-intensive image files such as those associated with credit card bills, telephone bills, brokerage statements, or tax returns. COLD systems also are used to store compressed graphics and image data on high-powered Web servers for Internet access.

Cold Docking Docking is to insert a portable computer into a base unit. Cold docking means the computer must begin from a power-off state and restart before docking. Hot docking means the computer can be docked while running at full power. See COLD START.

Cold Start Everything starts from scratch. The power to the computer or telephone system is turned off. Everything in the system's volatile memory is erased. A cold start may be needed on a microcomputer when something has happened to "lock up" the keyboard and the Reset button (if there is one) doesn't clear the problem completely. A Cold Start is also needed when you want to load a new operating system. When your phone system gives troubles you find hard to diagnose, turn it off, count to ten and turn it on. This cold boot to your phone system will often fix the problem, as it will typically do on a computer system.

Cold Transfer An incoming phone call transferred without notice or explanation from the transferring party. "Someone in customer service cold transferred the call to me. By that point the guy was ready to crawl through the wires and kill somebody." I got this definition from Wired Magazine. In the telephone industry, they call the same thing "blind transfer."

Collaboration A multimedia term. Collaboration involves two or more people working together in real-time, or in a 'store-and-forward" mode. Applications will enable a group of people to collaborate in real-time over the network using shared screens, shared whiteboards, and video conferencing. Collaboration can range from two people reviewing a slide set on line to a conference of doctors at different locations sharing patient files and discussing treatment options.

Collaboration Software Software that lets two or more people do a task together. See COLLABORATION.

Collaborating Two or more people working together on a project to share information and ideas, view suggestions, and make modifications. Computers can enable users to collaborate in real-time over a network or phone line using tools such as shared documents, shared whiteboards, and video conferencing, or time-efficient workflow such as document forwarding.

Collapsed Backbone The backbone network connecting all network segments is contained within a hub, allowing for the possibility of speeds far exceeding the standard backbone media.

Collapsed Ring A SONET term. SONET optical fiber systems are deployed in a ring architecture, with two or four fibers for redundancy and, therefore, network resiliency. A collapsed ring topology is one in which the ring fibers are laid in the same fiber bundle. If the fiber bundle is cut, and all fibers in the ring are cut, the ring collapses. A collapsed ring is a very bad thing. See also SONET.

Collateral Duties A call center term. Non-phone tasks (e.g., data entry) that are flexible, and can be scheduled for periods when call load is slow.

Collect Call A telephone call in which the called person pays for the call. The person calling calls a number and asks that the call be made "collect." Sometimes collect calls are handled by live operators, sometimes by machines. In a collect call, the phone company has to get some authorization from the person receiving the call that he will pay for it. This may be done by saying "Yes" or hitting a button a touchtone phone.

Collector Ring Metallic ring generally on the armature of a generator in contact with brushes for completing the circuit to a rotating member.

Collimate The condition of parallel light rays.

Collimation The process by which a divergent or convergent beam of electromagnetic radiation is converted into a beam with the minimum divergence or convergence possible for that system (ideally, a parallel bundle of rays).

Collinear Antenna A cellular car antenna which looks like a pigtail, because it has a little curlicue in the middle. The curlicue is not a spring, but a clever bit of electro-mechanical magic known as a phasing network, which allows the antenna to boost the effective power of the transmitter's signal. Typically a collinear cellular car antenna is 13 inches high.

Collision The result of two workstations trying to use a shared transmission medium (cable) simultaneously. The electrical signals, which carry the information they are sending bump into each other. This ruins both signals and both will have to re-transmit their information. In most systems, a built in delay will make sure the collision does not occur again. The whole process takes fractions of a second. Collisions in LANs make no sound. Collisions do, however, slow a LAN down. See ALOHA, COLLISION DETECTION and CONTENTION.

Collision Detection The process of detecting that simultaneous (and therefore damaging) transmission has taken place. Typically, each transmitting workstation that detects the collision will wait some period of time and try again. Collision detection is an essential part of the CSMA/CD access method. Workstations can tell that a collision has taken place if they do not receive an acknowledgement from the receiving station within a certain amount of time (fractions of a second). See ALOHA, CONTENTION and ETHERNET.

Collision Domain A local area network term. A single CSMA/CD network. If two or more Media Access Controllers (MAC) are within the same collision domain and both transmit at the same time a collision will occur. MACs that are separated by a repeater are in the same collision domain. MACs that are separated by a bridge are within different collision domains.

Collision Window The time it takes for a data pulse to travel the length of the network. During this interval, the network is vulnerable to collision.

Collocation A competing local phone company can locate its switches within a local exchange company's (LEC) central office. That's what the word means. OK, now to the real stuff — how to spell it. Several readers have complained that in previous editions of this dictionary it was spelled "colocation." But their non-technical English language dictionaries spell it collocation — with two "l"s. Random House Dictionary says that back in 1505-15 the word collocation appeared and was based on the latin collocatus, which derives from collocare. But Random House also includes a spelling from the era of 1965-1970 which it spells colocate and defines as to locate or be located in jointly or together, as two or more groups, military units, or the like; share or designate to share the same place. My preference is colocation, since it seems to me a logical shortening of co-location. But I'm not arguing. Choose

which spelling you'd like. See Adjacent Collocation, Physical Collocation and Virtual Collocation.

Colocation A competing local phone company can locate its switches within a local exchange company's (LEC) central office. That's what the word means. OK, now to the real stuff — how to spell it. Several readers have complained that their non-technical English language dictionaries spell it collocation — with two "l"s. Random House Dictionary says that back in 1505-15 the word collocation appeared and was based on the latin collocatus, which derives from collocare. But Random House also includes a spelling from the era of 1965-1970 which it spells colocate and defines as to locate or be located in jointly or together, as two or more groups, military units, or the like; share or designate to share the same place. My preference is colocation, since it seems to me a logical shortening of co-location. But I'm not arguing. Choose which spelling you'd like. See also CO LOCATION.

Colophon Did you ever notice a paragraph at the end of a book describing the typefaces used, the production methods, and so forth? That little paragraph is called a colophon.

Color See CMYK.

Color Code A color system for circuit identification by use of solid colors, contrasting stripes, tracers, braids, surface marking, etc.

Color Difference Signal The first step in encoding the color television signal. The color difference signals are formed by subtracting the luminance information from each primary color: red, green or blue. Color difference conventions include the Betacam format, the SMPTE format, the EBU-N10 format and the MII format.

Color Model A technique for describing a color (see CMY, HSL, HSV, and RGB).

Color Picture Signal The electrical signal which represents complete color picture information excluding all the synchronizing signals.

Color Space Inversion A video compression technique which reduces the amount of color information in each of a series of still images. It is based on the fact that the human eye is not highly sensitive to variations in color.

Color Subcarrier The 3.579545 MHz subcarrier that carries the chrominance information of the television signal. This signal is superimposed on the luminance level. Amplitude of the color subcarrier represents saturation and phase angle represents hue.

Color Temperature Selecting that determines the overall color cast of a display. 9,300 degrees Kelvin is good in environments lit by fluorescent lights, 6,500 degrees Kelvin is preferable under incandescent light.

COLT Cellular On Light Trucks. A temporary, mobile installation. I first heard the expression when I heard that BellSouth was using them to help expand its communications capacity in Atlanta for the 1996 Summer Olympics.

Column A database definition: The logical equivalent of a field, a column contains an individual data item within a row or record.

COM 1. Continuation of Message: An indicator used by the ATM Adaptation Layer to indicate that a particular ATM cell is a continuation of a higher layer information packet which has been segmented.
2. Component Object Model. COM is Microsoft's cornerstone of the ActiveX platform. COM is a language independent component architecture (not a programming language). It is meant to be a general purpose, object-oriented means to encapsulate commonly used functions and services. The COM architecture

provides a platform independent and distributed platform for multi-threaded applications. COM also encompasses everything previously known as OLE Automation (Object Linking and Embedding). OLE Automation was originally for letting higher level programming languages access COM objects. An object is a set of functions collected into interfaces. Each object has data associated with it. The source of the data itself is called the data object. With CCM, the transfer of the data itself is separated from the transfer protocol. See ActiveX and WINDOWS TELEPHONY.

3. A type of Internet domain assigned to URLs which are business or commercial entities (for example, www.bidworld.com). There is also .edu, .gov, .net, and .org. See Domain.

COM Port The communications port on a PC, a workstation, server, or other DTE (Data Terminal Equipment). This port is sometimes referred to as the serial, RS-232, DB-9 or DB-25 port (depending on if it has nine or 25 pins).

Combat Net Radio CNR. A radio operating in a network, providing a half-duplex circuit employing a single radio frequency or a discrete set of radio frequencies (frequency hopping). Combat net radios are primarily used for command and control of combat, combat support, and combat service support operations between and among ground, naval, and airborne forces.

Combination System An alternative to upgrading older telephone equipment, combination systems makes it possible to add network-based features to an equipment-based telephone system.

Combination Trunk A central office trunk circuit which is available as either an incoming or outgoing circuit to the attendant and also available through dial access to internal phone users for outgoing calls.

Combined Distribution Frame CDF. A distribution frame that combines the functions of main and intermediate distribution frames. The frame contains both vertical and horizontal terminating blocks. The vertical blocks are used to terminate the permanent outside lines entering the station. Horizontal blocks are used to terminate inside plant equipment. This arrangement permits the association of any outside line with any desired terminal equipment. These connections are made with equipment. These connections are made with twisted pair wire, normally referred to as jumper wire, or with optical fiber cables, normally referred to as jumper cables. In technical control facilities, the vertical side may be used to terminate equipment as well as outside lines. The horizontal side is then used for jackfields and battery terminations.

Combined Station MDLC station containing both a primary and a secondary and used in asynchronous balanced mode.

Comcode AT&T's numbering system for telecom equipment, replacing older KS-prefix numbers, and supplements standard industry part designations. Comcode No. 102092848 is touchtone Princess phone with a transparent plastic housing. How many would you like to order? See also KS NUMBER.

ComForum The Network Reliability ComForum is a gathering of senior telecommunication industry executives and government officials that serve on the Network Reliability Council. The ComForum meets to report on the Council's findings. See USITA.

Comfort Tone Michael Boom heard this one at an engineering meeting at Ascend Corporation, where an engineer was describing digital phone lines. It turns out that many phone connections are so clean now that there is no background noise at all. The phone customer, hearing absolutely nothing, believes the connection is broken and hangs up. The solution? A "comfort tone." A comfort tone is very low-level synthesized white noise deliberately added to a digital line to give a comforting "hiss" to the connection. It assures customers that there is indeed a connection and gives the voices on the other end that slightly distant quality the person making the telephone call expects to hear.

COMINT COMunications INTelligence.

Comite Consultatif International des Radiocommunications CCIR. The agency responsible for the international use of the radio spectrum. Effective in 1993, the CCIR is now known as the International Telecommunications Union — Radio. ITU-R and ITU-T form the International Telecommunications Union. See ITU for a much longer explanation.

Comma-Free Code A code constructed such that any partial code word, beginning at the start of a code word but terminating prior to the end of that code word, is not a valid code word. The comma-free property permits the proper framing of transmitted code words, provided that: (a) external synchronization is provided to identify the start of the first code word in a sequence of code words, and (b) no uncorrected errors occur in the symbol stream. Huffman codes (variable length) are examples of comma-free codes.

Command See COMMAND SET.

Command And Control C2. The exercise of authority and direction by a properly designated commander over assigned forces in the accomplishment of the mission. Command and control functions are performed through an arrangement of personnel equipment, communications, facilities, and procedures employed by a commander in planning, directing, coordinating, and controlling forces and operations in the accomplishment of the mission.

Command And Control System The facilities, equipment, communications, procedures, and personnel essential to a commander for planning, directing and controlling operations of assigned forces pursuant to the missions assigned. See COMMAND, CONTROL and COMMUNICATIONS.

Command Buffer A segment of memory used to temporarily store commands. The command buffer only holds a copy of the last command issued.

Command Conference System A conference calling arrangement in a Northern Telecom PBX which allows a designated phone to originate a conference to and between a group of PBX extensions. Any phone that is busy when the conference begins is automatically connected to the conference as soon as that phone becomes free.

Command, Control And Communications C3. The capabilities required by military commanders to exercise command and control of their forces.

Command Interpreter The operating system that controls a computer's shell. The command interpreter for MS-DOS is COMMAND.COM. The command interpreter for Windows is WIN.COM.

Command Line the line on the screen, in MS-DOS, where the cursor is. The command line is where you enter MS-DOS commands.

Command Line Interpreter CLI. A Rolm user interface to the CBX software and used for things like testing.

Command Net A communications network which connects an echelon of command with some or all of its subordinate echelons for the purpose of command control. See C2 and C3.

Command Path The list of path names that tells MS-DOS where to look for files that aren't in the current directory.

Command Port In network management systems an interface used to monitor and control the system.

Command Processor The MS-DOS program, COMMAND.COM, that contains all DOS's internal commands, like DIR, ERASE and REName. Once you have your hard disk set up, it's a good idea to made your COMMAND.COM "read only." This way it will be difficult for anyone to erase the file. If you don't have COMMAND.COM on your disk, or don't have it in a place where MS-DOS can find it, your disk will not boot (i.e. start) your computer. The command to make COMMAND.COM read only is

ATTRIB +R COMMAND.COM

A mistake many novices make is to "open" COMMAND.COM using their word processor, find that it's full of "junk," then save the file. Saving COMMAND.COM with a word processor destroys COMMAND.COM. The next time you start your computer it will "hang." The solution: Boot from a floppy. Erase COMMAND.COM and REName COMMAND.BAK to COMMAND.COM. Reboot your machine again. This time it should work. If it doesn't, copy COMMAND.COM from your original MS-DOS disk.

Command Prompt The MS-DOS command prompt appears on the screen as the default drive letter followed by a greater than > sign. The command prompt lets you know MS-DOS is ready to receive a command.

Command Response C/R. A Frame Relay term defining a 1-bit portion of the frame Address Field. Reserved for the use of FRADs, the C/R is applied to the transport of data involving polled protocols such as SNA. Polled protocols require a command/response process for signaling and control during the communications process.

Command Save A Rockwell ACD term. The introduction of a new demand command defines up to 10 commands per terminal position to enhance the productivity of both IST and non-IST supervisors.

Command Set In computer telephony, a command set is a collection of special software instructions that do special jobs. These software instructions are often called function calls. For example, the command M_Make_Call (plus parameters) tells Northern's Norstar phone system to have telephone set number 21 dial a phone number. Northern's Norstar and other open phone systems (those that can be commanded by an external computer which you and I can program) all have their own command sets. Each command set is made up of function calls with funny words like M_Make_Call. Typically those function calls work in C, a common software language. A function call will reach into the specialized driver that controls the phone system (an exact analogy is the driver that drives a laser printer) and get the phone system to do something. A programmer must use these function calls if he/she wants to control the phone system from software. Exactly how M_Make_Call works is typically not revealed to the programmer. That keeps the manufacturer's technology proprietary and secret. It also saves the programmer the time and expense of writing the driver.

Command.com The program that carries out MS-DOS commands. The generic term for this program is command interpreter.

COMMDesk Banker A communications software package offered by MCI International that provides all the capabilities of COMMDesk, plus a security feature essential for financial transactions.

COMMDesk Manager An MCI definition. A communications software package designed to run on an IBM PC/XT/AT or compatible that gives the user full access to all MCI International and MCI communications services.

Commercenet A not-for-profit industry association that works to accelerate the development of electronic commerce. In conjunction with the World Wide Web Consortium (W3C), CommerceNet developed the Joint Electronic Payments Initiative (JEPI), an E-Commerce standard. The approximately 500 members of CommerceNet include leading banks, telecommunications companies, VANs, ISPs, online service, software and services companies, as well as major end users. See also Electronic Commerce and JEPI. www.commerce.net

Commercial Building Telecommunications Wiring Standard In 1985, the Electronic Industries Association undertook the task of developing a standard for commercial and industrial building wiring. Approved and published on July 9, 1991, the EIA/TIA-568 "Commercial Building Telecommunications Wiring Standard" defines a generic wiring system which will support a multiproduct, multivendor environment and which will have a useful life of over 10 years. The EIA/TIA standard is based on star topology in which each workstation is connected to a telecommunications closet situated within 90 meters of the work area. Backbone wiring between the communications closets and the main cross-connect is also organized in a star topology. However, direct connections between closets are allowed to accommodate bus and ring configurations. Distances between closets and the main cross-connect are dependent on backbone cable types and applications. Each workstation is to provided with a minimum of two communications outlets (which may be on the same faceplate). One outlet is supported by a four-pair, 100-ohm unshielded twisted-pair (UTP) cable. The other may be supported by (a) an additional four-pair UTP cable, (b) a two-pair, 150 ohm shielded twisted pair (STP) cable or (c) a two-fiber 62.5 /125 micron fiber optic cable. For more on cabling and cabling components, see CATEGORY OF PERFORMANCE.

Commercial Internet The part of the Internet provided by commercial services. Allows business usage of the Internet without violating the appropriate usage clause of the National Science Foundation NETwork (NSFNET), who actually runs the Internet.

Commercial Internet Exchange Association This is a non-profit trade association for public data internetworking service providers. www.cix.org/cixhome.html

Committed Burst Size A Frame Relay term defining the maximum data rate that the carrier agrees to handle over a subscriber link under normal conditions. SEE OFFERED LOAD, COMMITTED INFORMATION RATE, DISCARD ELIGIBILITY, EXCESS BURST SIZE.

Committed Information Rate CIR. A Frame Relay term identifying the user's commitment to a certain average maximum data transmission rate. The monthly bill a customer receives may include at least two elements, depending on the pricing algorithm employed by the carrier; those charges can include a charge for the CIR and a surcharge for usage above the CIR. Usage above the CIR may be measured in terms of the Burst Size and Burst Interval. Usage above the CIR may be subject to discard in the event of network congestion. In the early days of frame relay when few people were using the service, customers were opting for a low Committed Information Rate, thus keeping their bills low, but knowing that they could always get their transmissions through — because there were few other people on the service. As the service got more popular, many customers found they had to hike their Committed Information Rate if they wanted to get their information through. For a more technical explanation, see CIR.

Common Audible The same as Common Bell. Ringer wiring is such that ringing occurs on more than one CO or PBX line.

Common Audible Ringer A loud ringer connected to a phone line in a noisy area. When the phone rings, the loud ringer also rings.

Common Battery A battery (or several batteries) that acts as a central source of energy for many pieces of equipment. A common battery provides 48 volts of power to a central office switch and to all the phones connected downstream.

Common Battery Signaling A system in which the signaling power of a telephone is supplied by the battery at the servicing switchboard. Switchboards may be manual or automatic, and "talking power" may be supplied by common or local battery.

Common Bell A bell or ringer which sounds when any of the lines terminating on that phone rings. A term harking back to 1A2 key system days.

Common Business Line CBL. An option with 800 Service that has been replaced by 800 Business Line.

Common Carrier A company that furnishes communications services to the general public. It is typically licensed by a state or federal government agency. A common carrier cannot refuse to carry you, your information or your freight as long as you conform to the rules and regulations as filed with the state or federal authorities. See OTHER COMMON CARRIER.

Common Carrier Bureau A department of the Federal Communications Commission responsible for recommending and implementing regulatory policies on interstate and international common carrier (voice, video, data) activities.

Common Channel Interoffice Signaling CCIS. A way of transmitting all signaling information for a group of trunks by encoding that information and transmitting it over a separate channel using time-division digital methods. By transmitting that signaling information over a separate channel, CCIS saves huge long distance bandwidth, which in the past was used to switch calls across the country only to find a busy signal and then come all the way back again to signal the calling party a busy. For the biggest explanation of common channel signaling, see SIGNALING SYSTEM 7. See also MTP, SCCP, ISUP, ISDN and TCAP.

Common Channel Signaling This is a Bellcore definition: A network architecture which uses Signaling System 7 (SS7) protocol for the exchange of information between telecommunications nodes and networks on an out-of-band basis. It performs three major functions: 1. It allows the exchange of signaling information for interoffice circuit connections. 2. It allows the exchange of additional information services and features, e.g. CLASS, database query/response, etc. 3. It provides improved operations procedures for network management and administration of the telecommunications network. For the biggest explanation of common channel signaling, see SIGNALING SYSTEM 7. See also ISDN, ISUP, MTP, SCCP, STP and TCAP.

Common Channel Transit Exchange An intermediate exchange where networking of common channel signaling systems occurs.

Common Control A method of telephone switching in which the central logic system (or control equipment) is responsible for routing calls through the network. The control equipment is connected with a given call only for the period required to accomplish the routing function. In other words, the common control equipment is associated with a given call only during the periods required to accomplish the control functions. All crossbar and electronic switching systems have common control.

Common Control Equipment Certain components in a common control switching system that are available in common to all phones and are used only during the setting up of connections. See Common Control Switching Arrangement

Common Control Switching Arrangement CCSA. An AT&T offering for very big companies. Those big companies can create their own private networks and dial anywhere on them by dialing a standard seven digit number, similar to a local phone number. The corporate subscriber rents private, dedicated lines and then shares central office switches. CCSA uses special CCSA software at the central office.

Common Costs Costs of the provision of some group of services that cannot be directly attributed to any one of those services.

Common Equipment In telephone systems Common Equipment are items that are used by several or all phones for processing calls. On a key system, the device that permits a light on any instrument to flash on and off may be Common Equipment when used to control all lights on all instruments.

Common Intermediate Format A videophone ISDN standard which is part of the ITU-T's H.261. It produces a color image of 352 by 288 pixels. The format uses two B channels, with voice taking 32 Kbps and the rest for video.

Common Language Code Codes used to ensure uniform abbreviation of equipment and facility names, place names, etc.

Common Language Location Identification Code CLLI. The CLLI code is an 11 character mnemonic code used to uniquely identify a location in the United States, Canada or other countries. These codes are known as CLLI or 'Location Codes' and may be used in either a manual or mechanized record keeping system. For a bigger explanation, see CLLI Code.

Common Mail Calls New APIs (Application Programming Interfaces) from Microsoft which allow you to move information around your various mail services — the ones on your LAN, on your wireless pager, etc.

Common Mode The potential or voltage that exists between neutral and ground. Electronic equipment requires this to be as close to 0 volts as possible or not to exceed 1/2 volt. For AC power systems, the term common mode may refer to either noise or surge voltage disturbances. Common mode disturbances are those that occur between the power neutral (white wire) and the grounding conductor (green wire) Ideally, no common mode disturbances should exist since the neutral and grounding wires are always connected at the service distribution panel in most countries. However, unwanted common mode disturbances exist as a result of noise injection into the neutral or grounding wires, wiring faults, or overloaded power circuits. Modern computers are quite immune from common mode noise. Common mode noise is frequently mistakenly confused with inter-system ground noise, a distinct problem which frequently causes computer damage and data errors. See Common Mode Interference.

Common Mode Interference 1. Interference that appears between signal leads, or the terminals of a measuring circuit and ground.

2. A form of coherent interference that affects two or more elements of a network in a similar manner (i.e., highly coupled) as distinct from locally generated noise or interference that is statistically independent between pairs of network elements.

Common Mode Rejection Ratio CMRR. The ratio of the common mode interference voltage at the input of a circuit to the corresponding interference voltage at the output.

Common Mode Voltage 1. The voltage common to both input terminals of a device.
2. In a differential amplifier, the unwanted part of the voltage between each input connection point and ground that is added to the voltage of each original signal.

Common Peer Group An ATM term. The lowest level peer group in which a set of nodes is represented. A node is represented in a peer group either directly or through one of its ancestors.

Common Return A return path that is common to two or more circuits and that serves to return currents to their source or to ground.

Common Return Offset The dc common return potential difference of a line.

Common Trunk In telephone systems having a grading arrangement, a trunk accessible to all groups of the grading.

Common User Circuit A circuit designated to furnish a communication service to a number of users.

Common User Network A system of circuits or channels allocated to furnish communication paths between switching centers to provide communication service on a common basis to all connected stations or subscribers.

Commonality 1. A quality that applies to material or systems: (a) possessing like and interchangeable characteristics enabling each to be utilized, or operated and maintained by personnel trained on the others without additional specialized training; (b) having interchangeable repair parts and/or components; (c) applying to consumable items interchangeably equivalent without adjustment.
2. A term applied to equipment or systems that have the quality of one entity possessing like and interchangeable parts with another equipment or system entity.

Communicating Applications A General Magic term. An application whose design presupposes the user's desire to send and receive messages. For a Personal Intelligent Communicator to be effective, it needs to be equipped with a suite of communicating applications. All Magic Cap applications are built to communicate.

Communicating Applications Platform A General Magic term. The Cap in Magic Cap. Software on which Personal Intelligent Communicators are based. It is designed to make it easy for developers to create communicating applications and services. Magic Cap can run on dedicated devices as well as other computer operating systems.

Communicating Applications Specification A facsimile specification. See CAS 2.0

Communicating Objects A term created in the fall of 1992 by Mitel's VP Tony Bawcutt for a new Mitel division which specializes in making PC printed cards for and software drivers and developer tools for those cards. Those cards are designed to be the building blocks of what Mitel calls multimedia applications — but what are more properly called PC-based voice and call processing telecom developer building blocks. One of the first cards Mitel introduced was an ISDN S-access card which converts PCs into ISDN telephones, also called voice and data workstations.

Communicating Word Processor A dedicated word processor that includes software for sending word processed files over phone lines. Communicating word processors have now largely been replaced by PCs (Personal Computers) running word processing programs and asynchronous communications software programs.

Communication Channel A two-way path for transmitting voice and/or data signals. See also CIRCUIT.

Communication Controller Another name for a Front End Processor, a specialized computer which was common in 3270 data communications networks. The FEP acted as a data communications "traffic cop," removing the communications traffic routing and controlling burden from the mainframe computer which lay behind the FEP. In short, the FEP designates a device placed between the network and an input/output channel of a processing system (i.e. the computer).

Communication Endpoint An ATM term. An object associated with a set of attributes which are specified at the communication creation time.

Communication Server A dedicated, standalone system that manages communications activities for other computers.

Communication Workers Of America CWA. A national union of telephone industry employees, currently very worried about its future membership growth given the phone industry's propensity to let its surplus workers go.

Communications Act Of 1934 Federal legislation which established national telecommunications goals and created the Federal Communications Commission to regulate all interstate and international communications.

Communications Act of 1996 It is really called the Telecommunications Act of 1996. It is a federal bill signed into law on February 8, 1996 "to promote competition and reduce regulation in order to secure lower prices and higher quality services for American telecommunications consumers and encourage rapid deployment of new telecommunications technologies." The Act, amongst other things, allowed the local regional Bell operating phone companies into long distance once they had met certain conditions about allowing competition in their local monopoly areas. You can download a copy of this Act (all 391,861 bytes) from http://thomas.loc.gov/cgi-bin/query/1?c104:./ temp/~c104iLoA:: See also COMMUNICATIONS DECENCY ACT OF 1996.

Communications Decency Act of 1996 CDA. An element of the Telecommunications Act of 1996, the CDA provided for penalties of as much as $250,000 and 2 years in jail for U.S. citizens who transmit indecent material that minors could access by computer. Targeted at those who would make such material available over the Internet through Web sites, this element of the act was blocked in June 1996 by a panel of federal judges on the basis of successful arguments that the provision violated the right of free speech, as guaranteed in the Constitution. The Supreme Court in mid-1997 Supreme Court said the Communications Decency Act was unconstitutional. The Supreme Court handed down a seven-to-two decision that upheld a lower-court ruling against the CDA. The case, Reno vs. ACLU, was the first time that the Supreme Court has dealt with issues involving the Internet. See TELECOMMUNICATIONS ACT OF 1996.

Communications Adapter Device attached to an IBM System 3X computer or an IBM PC that allows communications over RS-232 lines.

Communications Control Character A character intended to control or help transmission over data networks. There are ten control characters specified in ASCII which form the basis for character-oriented communications control procedures.

Communications Protocol Procedures which are employed to ensure the orderly transfer of data between devices on a communications link, over a communications network, or within a system. The major functions of a protocol are those of Handshaking and Line Discipline. Handshaking is a specific sequence of data exchange between devices over a circuit. This initial step establishes the fact that the circuit is operational, establishes the level of device compatibility, determines speed of transmission, and so on. Line discipline is a sequence of operations which includes transmission and receipt of data, error control, and sequencing of message sets (e.g., characters, blocks, packets, frames and cells). Line discipline also includes error detection and correction processes, providing for the confirmation or validation of data received, and by implication, the failure to receive data sets.

Communications Satellite A satellite circling the earth, usually at a distance of about 22,000 miles, with electronic equipment for relaying signals received from the earth back to other points on the earth. See GEOSTATIONARY SATELLITE.

Communications Server Also called an asynchronous server or asynchronous gateway. A communications server is a type of gateway that translates the packetized signals of a LAN to asynchronous signals, usually used on telephone lines or on direct connections to DEC and other minicomputers and mainframes. It handles different asynchronous protocols and allows nodes on a LAN to share modems or host connections. Usually one machine on a LAN will act as a gateway, sharing its serial ports or an RS-232 connection to a minicomputer. All devices on the LAN can use this machine to get to the modems and the minicomputer. See also the UnPBX.

Communications Settings Settings that specify how information is transferred from your computer to a device (usually a printer or modem).

Communications System Engineering The translation of user requirements for the exchange of information into cost-effective technical solutions of equipment and subsystems.

Communications Toolbox An extension of he Apple Macintosh operating system that provides protocol conversion and the drivers needed for communications tasks.

Communications Workers of America CWA. The main labor union of the RBOCs. www.cwa.org.

Communications Zone A military term: Rear part of theater of operations (behind but contiguous to the combat zone), which contains the lines of communications, establishments for supply and evacuation, and other agencies required for the immediate support and maintenance of the field forces.

Communicator A British term. An alternative, and probably more meaningful, name for a telebusiness agent. A communicator is called a telemarketer in North America.

Community For the purposes of the FCC's cable television rules, when this term is used to specify the location of a transmitter or an antenna structure, it generally includes any named, urbanized area, without regard to size.

Community Antenna Television CATV. Signals from distant TV stations are picked up by a large antenna, typically located on a hill, then amplified and piped all over the community below on coaxial cable. That's the original definition. See CATV for a more up-to-date definition.

Community Name Community Name is a password

shared by a Network Agent and the the Network Management Station so their communications cannot be easily intercepted by an unauthorized workstation or device.

Community Of Interest A grouping of telephone users that call each other with a high degree of frequency. Often several Communities of Interest exist within an organization. This phenomenon can influence design for service when new switches are planned.

Community of License The community to which a broadcast station is licensed, as specified on the station license. The transmitter may or may not be located within the community (indeed, it is frequently located at some distance, and may even be in a different state). A broadcast station is required to provide a specified field intensity over the entire Community of License.

Community String A password used with the SNMP protocol. ANMP community strings are used for both read only and read/write privileges. A community string is case sensitive, and may include some punctuation characters.

Community Unit A discrete geographic area served by a single cable television system, to which a single Community Unit Identification Number has been assigned by the FCC. At a minimum, each franchise area served by one cable operator constitutes one Community Unit; however, a single franchise area may include two or more Community Units in the following situations:

• If a cable television operator serves a community from two different headends, the portion of the community served by each headend constitutes a separate community unit.

• If two cable operators hold separate franchises for the same geographic area, each operator is assigned a separate Community Unit Identification Number.

• If the system serves discrete, unincorporated areas within a township or county pursuant to a township-wide or county-wide franchise, each area may constitute a separate Community Unit.

• If the system serves an incorporated municipality which overlaps two or more counties, the portion of the municipality within each county is considered a separate Community Unit.

Community Unit Identification Number An identification number assigned by the FCC to each Community Unit. The format of the number is: SSNNNN, where:

• SS is the U.S. Postal Service two-letter abbreviation for the state or territory in which the Community Unit is located.

• NNNN = A four-digit serial number assigned by the FCC.

Commutator A device used on a dynamo to reverse the connection periodically in order to cause the current flow in one direction, i.e., to produce direct current.

Compact Disc A standard medium for storage of digital audio data, accessible with a laser-based reader. CDs are 12 centimeters (about 4 3/4") in diameter. CDs are faster and more accurate than magnetic tape for audio. Faster, because even though data is generally written on a CD contiguously within each track, the tracks themselves are directly accessible. This means the tracks can be accessed and played back in any order. More accurate, because data is recorded directly into binary code; mag tape requires data to be translated into analog form. Also, extraneous noise (tape hiss) associated with mag tape is absent from CDs. See CD-ROM and CVD.

Compact Disc Interactive A compact disc format, developed by Philips and Sony, which provides audio, digital data, still graphics and limited motion video. See CD-I.

CompactPCI CompactPCI is a ruggedized variation of the

PCI bus, which a bunch of industrial grade PC makers have designed for two reasons: First, to be able to put more PCI cards into one PC (8 versus 4). Second, to make the resulting PC more rugged, i.e. better able to withstand shaking, etc. The physical configuration of the hardware conforms to the Eurocard (VMR-style) standard. The cards, are identical to VME cards in size. They differ, however, in that they use a high density 2mm (contact spacing) pin-and-socket connector for interface to a passive backplane. CompactPCI typically comes in a rugged 3U or 6U Eurocard form factor and has a 32/64 data bus with transfer rates up to 528 megabytes per second. The PC makers told me that "CompactPCI is an adaptation of the Peripheral Component Interconnect (PCI) Specification for industrial and/or embedded applications requiring a more robust mechanism form factor than desktop PCI. CompactPCI uses industry standard mechanical components and high-performance connector technologies to provide a system optimized for rugged applications. CompactPCI is electrically compatible with the PCI Specification, allowing low-cost PCI chipsets to be used in a mechanical form factor suited for rugged environments."

Compander See COMPANDING.

Companding The word is a contraction of the words "compressing" and "expanding." Companding is the process of compressing the amplitude range of a signal for economical transmission and then expanding them back to their original form at the receiving end.

Company Code A unique four-place alphanumeric code which must be assigned by NECA (National Exchange Carrier Association) to all U.S. domestic local exchange telecommunications providers. The alphanumeric code is expressed as NNXX, where N=0-9 and X=A-Z. Company Codes are assigned for each type of service provided by a company. Additionally, separate and distinct codes are required for ILECs (Incumbent Local Exchange Carriers), certified facilities-based CLECs (Competitive LECs), local exchange resellers, and wireless carriers. In 1996, NECA assumed this responsibility. Previously, the codes were assigned unofficially from a series of numbers, sensitive to either the type of service provided or the nature of the carrier entity. See also NECA.

Compartmentation A military/government term: A method employed to segregate information of different desired accessibilities from each other. It may be used for communications security purposes.

Compatible A widely misused word. In the computer world, two computers are said to be compatible when they will produce the identical result if they run identical programs. Another meaning is whether equipment — peripherals and components — can be used interchangeably with each other, from one computer to another. Compatibility used to be regarded as a useful trait in 1A2 key telephones. But then the electronics revolution came along and now there are no compatible electronic telephones from different manufacturers. This means you can't take an electronic Mitel PBX phone, example, and have it work behind a Northern Telecom PBX. And a Northern Telecom PBX phone won't work behind an AT&T PBX. The major compatibility in our industry is at the lowest common denominator — the tip and ring analog phone. See interoperability testing.

Compatible Sideband Transmission That method of independent sideband transmission wherein the carrier is deliberately reinserted at a lower level after its normal suppression to permit reception by conventional AM receivers. The normal method of transmitting compatible SSB (AME) is

the emission of the carrier plus the upper sideband.

Compelled Signaling A signaling method in which the transmission of each signal in the forward direction is inhibited until an acknowledgement of the satisfactory receipt of the previous signal has been sent back from the receiver terminal.

Competitive Access Provider See CAP.

Competitive Long Distance Coalition A Washington, D.C. lobbying group.

Compilation The translation of programs written in a language understandable to programmers into instructions understandable to the computer. Think of programmers writing in every language but Greek and computers understanding only Greek. In this case, Greek is called machine language. The other languages (the programmer languages) are called things like COBOL, FORTRAN, Pascal, dBASE. A compiler is a special program that translates from all these other languages into machine language.

Compile To translate a program written in a higher language into machine language so it can be executed by a computer.

Compiler A program that takes the source code a programmer has written and translates it into object code the computer can understand. A computer program used to convert symbols meaningful to a human into codes meaningful to a computer. For example, a compiler takes instructions written in a "higher" level language such as BASIC, COBOL or ALGOL and converts them into machine code which can be read and acted upon by a computer. Compilers converts large sections of code at one time, compared to an interpreter which translates commands one at a time.

Complementary Metal Oxide Semiconductor CMOS. A method of building chips which produces a logic circuit family which uses very little power.

Complementary Network Service CNS. CNSs are basic services associated with end user's lines that make it easier for ESPs to offer them enhanced services. Some examples of CNSs include Call Forwarding Busy/Don't Answer, Three Way Calling, and Virtual Dial Tone. See OPEN NETWORK ARCHITECTURE.

Complete Document Recognition The ability to perform recognition on documents, retaining as much information as possible about the features and formatting of the original, and including the ability to capture images as well as text.

Completed Call Careful with this one. In telephone dialect, a Completed Call is one that has been switched to its destination and conversation has begun but has not yet ended.

Completion Ratio The proportion of the number of attempted calls to the number of completed calls.

Complimentary Network Services CNS. The means for an enhanced-service provider's customer to connect to the network and to the enhanced service provider. Complimentary network services usually consist of the customer's local service (e.g.,business or residence line) and several associated service options, e.g., call-forwarding service.

Component An element of equipment which unto itself does not form a system.

Component Busy Hour CBH. A telephone company definition. The busy hour of an individual component of a switching system. Often, Component Busy Hours coincide with the overall office busy hour. Each component or group will have its own time consistent busy hour during the busy season. While the hour may or may not vary from one busy season year to another, only one hour may be used during a busy season year. It is upon the data collected during this component busy hour that trends are established, projections made, capacities set and future requirements derived. The component busy hour is used to determine the high day (HDCBH), ten highest days (10HDCBH), average of the ten highest days (ATHD), average busy day (ABDCBH) and average busy season (ABS CBH) values.

Component Object Model COM is Microsoft's cornerstone of the ActiveX platform. COM is a language independent component architecture (not a programming language). It is meant to be a general purpose, object-oriented means to encapsulate commonly used functions and services. The COM architecture provides a platform independent and distributed platform for multi-threaded applications. COM also encompasses everything previously known as OLE Automation (Object Linking and Embedding). OLE Automation was originally for letting higher level programming languages access COM objects. An object is a set of functions collected into interfaces. Each object has data associated with it. The source of the data itself is called the data object. With COM, the transfer of the data itself is separated from the transfer protocol.

Component Software Component software is software constructed from reusable components. It was popularized by Microsoft Visual Basic and its successful custom control architecture. This architecture allows third party software components to "plug" into and extend the Visual Basic development environment. Hundreds of third party components, or custom controls, exist — for everything from accessing a mainframe database to programming a computer telephony board. Component-based software development is a productive way to build software. System developers benefit from being able to tailor their development environment for a specific need. Consider the development of an IVR system that allows callers to access their account balance stored on an IBM mainframe. To build this system, Visual Basic developers extend their development environment with a custom control for telephony and another that provides access to an IBM mainframe. There's no need to learn a new and proprietary language for telephony development. Plus, every control is accessed through a common interface of actions, properties, and events.

Component Video Transmission and recording of color television that stores separate channels of red, green and blue.

Composite 1. Output of a multiplexer that includes all data from the multiplexed channels. Contrast with AGGREGATE.
2. Refers to a type of color monitor in which the color signals all come in on the same line and are separated electronically inside the monitor. Compare this type of monitor to RGB, where the colors come in on different cables.

Composite Link The datastream composed of all the input channels and control and signaling information in a multiplexed circuit. The Composite Link Speed is the transmission speed of the circuit.

Composite Materials Composite materials consist of two or more components. They make it possible to combine the best properties of different materials; for example, the compression strength and low price of concrete with the tensile strength of reinforcing rods. Composite materials include: Reinforced concrete, fiber-reinforced plastic, fiber-reinforced metals, plywood, chipboard and ceramics. The composites mainly considered for antennas are fiber-reinforced plastics. They combine the low weight and protective properties of plastics with the stiffness and strength of fiber.

Composite Signaling A direct current signaling system

that separates the signals from the voice band by filters. Two pairs (a quad) provide talking paths and full-duplex signaling for three channels. Also called CX Signaling.

Composite Video Composite video is a mixed signal comprised oof the luminance (black and white), chrominance (color), blanking pulses, sync pulses and color burst. Composite video is a television signal where the chrominance (color) signal is a sine wave that is modulated onto the luminance (black and white) signal which acts as a subcarrier. This is used in NTSC and PAL TV systems. Composite video is the visual wave form representation used in color television. Composite video is analog and must be converted to digital to be used in multimedia computing. See also composite video signal, digial video and NTSC.

Composite Video Signal The completed video signal that is the combined result of the primary colors of red, greeen and blue (RGB) producing all the necessary picture information, such as with the NTSC or PAL formats. A composite video signal of standard amplitude, form correct sync to reference white level, should be presented between -40 and +100 units of the IRE scale on a waveform monitor.

Composited Circuit A circuit that can be used simultaneously for telephony and dc telegraphy, or signaling; separation between the two being accomplished by frequency discrimination.

Compound A term used to designate an insulating or jacketing material made by mixing two or more ingredients.

Compound Document The simple explanation: A compound document contains information created by using more than one application. It is a document often composed of a variety of data types and formats. Each data type is linked to the application that created it. A compound document might include audio, video, images, text, and graphics. Compound documents first became possible to the world of PCs with the introduction of Windows 3.1, which included OLE (Object Linking and Embedding). OLE allows you to write a letter in your favorite Windows word processor, embed a small voice icon in your document, send your letter to someone else, have them open your letter, place their mouse on the voice icon and hear whatever comments you recorded. To make this possible, both you (the creator) and your recipient would need access to programs that could read both the text and the voice. Ideally, you would both be on a LAN (Local Area Network) and would both get access to the identical applications software, resident, presumably, on the LAN's file server. See COMPOUND MAILBOX.

Compound Document Mail See COMPOUND DOCUMENT.

Compound Mailbox A mailbox for mail from all sources — fax, voice mail, e-mail, pager, etc. See COMPOUND DOCUMENT.

Compressed Video Television signals transmitted with much less than the usual bit rate. Full standard coding of broadcast quality television typically requires 45 to 90 megabits per second. Compressed video includes signals from 3 mb/s down to 56 Kbps. The lower bit rates typically involve some compromise in picture quality, particularly when there's rapid motion on the screen. See MPEG.

Compression Reducing the representation of the information, but not the information itself. Reducing the bandwidth or number of bits needed to encode information or encode a signal, typically by eliminating long strings of identical bits or bits that do not change in successive sampling intervals (e.g., video frames). Compression saves transmission time or capacity. It also saves storage space on storage devices such as hard disks, tape drives and floppy disks.

Compression Algorithm The arithmetic formulae which convert a signal into smaller bandwidth or fewer bits.

Compression Artifacts Compression artifacts are introduced by filtering, conversion transformation, quantization and transmission compression. Loss of resolution, block errors, quantization noise and block errors are typically observed as a result of these processes.

Compressor See COMPANDING.

Compromise Equalizer Equalizer set for best overall operation for a given range of line conditions. This is often fixed but may be manually adjustable.

COMPSURF COMPrehensive SURFace Analysis. A Novell program that checks the surface of a hard disk, marks off sections that are lousy and therefore shouldn't be written to, and then low level formats the disk. The program is slow, but thorough and rigorous. No hard disk should ever be used on a file server on a Novell local area network without being subjected to this wonderful program. Don't believe Novell when it says that you don't need to subject new disks to COMPSURF. You should submit ALL disks.

CompTel Competitive Telecommunications Association. A national U.S. organization of competitive local and long distance carriers, most of which are facilities-based. www.comptel.org

CompuCALL Northern Telecom DMS central office link to computer interface. An open architecture specification. Northern Telecom spells it as CompuCALL. According to Northern Telecom's own words, CompuCALL employs the Switch Computer Application Interface (SCAI) open architecture standard to connect the central office with customers' general-purpose business computers. CompuCALL consists of:

• The CompuCALL base software (NTXJ59AA) in the Northern Telecom DMS-100 switch, or the Meridian Automatic Call Distribution (ACD) Server, that sends and receives SCAI messages; and

• The CompuCALL transport mechanism that physically links the switch to the computer or other external processor and carries the SCAI message.

Northern Telecom also supplies the Meridian ACD CompuCALL Options (NTXJ39AA) and other applications software which rely on the CompuCALL base. Computers and software vendors provide application programming interface (API) software as well as business application software. The API, which resides in the business computer, converts SCAI messages into information that can be used by computer-based business application software.

The first applications of CompuCALL integrate computer databases with voice telephony:

• CompuCALL Coordinated Voice and Data provides an agent a screen of information about a caller concurrently with receipt of a call.

• CompuCALL Voice Processing Integration uses Interactive Voice Response (IVR) systems and Voice Response Units (VRU) to obtain additional information about callers and direct them to the appropriate agent.

• CompuCALL Third-Party Call Control lets the customer's computer place outgoing calls.

Successful implementation of CompuCALL requires interaction with application software on a business computer. CompuCALL consists of:

• The CompuCALL base software residing in the Northern Telecom DMS-100 central office switch that sends and receives SCAI messages; and

• The CompuCALL transport mechanism that physically links the switch to the computer or other external processor and carries the SCAI message.

Computer and software vendors provide application programming interface (API) software as well as business application software. The API, which resides in the business computer, converts SCAI messages into information that can be used by computer-based business application software. CompuCALL is based on the seven-layer protocol defined by the International Telegraph and Telephone Consultative Committee's (ITU) Open System Interconnection (OSI) reference model. The product is transport independent. With BCS33, CompuCALL is scheduled to be available for Verifications Office (VO) on X.25. See also OPEN APPLICATION INTERFACE.

Compunications A recent creation meaning the combination of telephones, computers, television and data systems.

CompuServe An on-line, dial-up service — one of the largest worldwide. CompuServe has everything from electronic mail to manufacturer-sponsored forums where you can download files for updated drivers, etc. CompuServe is one of the hardest on line services to find your way around. See the following definitions. CompuServe is now owned by American Online.

CompuServe Electronic Mail You can send electronic mail to CompuServe addresses. Here's the formula: All CompuServe addresses are either of the form 7xxxx,xxx or 1xxxxx,xxx. (where each "x" signifies a digit from 0 to 7). There can be from 2 to 4 digits following the comma. To send mail to such an address from the Internet, change the comma to a period and attach "@CompuServe.com" as is shown in the following examples:

74906.1610@compuserve.com or
100906.1610@compuserve.com

CompuServe B+ File Transfer This file transfer protocol is used by the CompuServe information service and no one else. Recovery of interrupted transfers is supported. In CompuServe B+, the host initiates the transfer. In contrast, in XMODEM, the receiver initiates the transfer, i.e. tells the distant computer to begin sending the file.

CompuServe Mail Hub A facility of CompuServe which enables users on a local area network operating Novell Message Handling Service (MHS) software to exchange electronic messages with other MHS users, CompuServe Mail subscribers and users of other E-mail services that can be reached via a CompuServe gateway.

Compute Servers Very powerful computers that sit on networks and are dedicated to heavy mathematical calculations. Brokerage firms use them for complex yield calculations, mathematical modeling, derivatives analysis, etc. Such servers often have as much as a gigabyte in RAM.

Computer This is a definition straight from AT&T Bell Laboratories. "An electronic device that accepts and processes information mathematically according to previous instructions. It provides the result of this processing via visual displays, printed summaries or in an audible form." When it works, it's wonderful. When it doesn't, it's a disaster. The major lessons every computer user should learn: Save your work regularly. Back it up regularly. back it up to many different media. The value of your work on your computer exceeds the value of your computer many, many, manyfold.

Computer Aided Dialing A newer (and allegedly less offensive) term for predictive dialing. See also PREDICTIVE DIALING.

Computer Aided Professional Publishing CAP. The computerization of professional publishing (as opposed to desktop operations), including true color representation of the layout on the workstation screen.

Computer And Business Equipment Manufacturers Association CBEMA. Association active before Congress and the FCC promoting the interests of the competitive terminal, computer and peripheral equipment industries.

Computer And Communications Industry Association CCIA. Organization of data processing and communications companies which promotes their interests before Congress and the FCC.

Computer Emergency Response Team CERT. A group of computer experts at Carnegie-Mellon University who are responsible for dealing with Internet security issues. The CERT is chartered to work with the Internet community to facilitate its response to computer security events involving Internet hosts, to take proactive steps to raise the community's awareness of computer security issues, and to conduct research targeted at improving the security of existing systems. The CERT was formed by DARPA in November 1988 in response to the Internet worm incident. CERT exists to facilitate Internet-wide response to computer security events involving Internet hosts and to conduct research targeted at improving the security of existing systems. They maintain an archive of security-related issues on their FTP server at "cert.org." Their email address is "cert@cert.org" and their 24-hour telephone Hotline for reporting Internet security issues is 412-268-7090.

Computer Fraud Deliberate misrepresentation, alteration or disclosure of computer-based data to obtain something of value.

Computer Inquiry A series of ongoing FCC proceedings examining the distinctions between communications and information processing to determine which services are subject to common carrier regulation. The FCC decision in 1980 resulting from the second inquiry was to limit common carrier regulation to basic services. Enhanced services and customer premises equipment are not to be regulated. This meant the Bell operating companies had to set up separate subsidiaries if they were to offer non-regulated services. Computer Inquiry III, adopted by the FCC in May, 1986, removed the structural separation requirement between basic and enhanced services for the BOCs and for AT&T. CI III replaced that requirement with "nonstructural safeguards." This action resulted in the imposition of such concepts as "comparably efficient interconnection" (CEI) and Open Network Architecture (ONA). The FCC's jurisdiction regarding Computer Inquiry I, II and III has now been usurped by Judge Greene, who insists on fairly tight control over the non-basic telephone company activities of the Bell operating companies. Sometimes he gives dispensations (waivers). Sometimes he doesn't. His word these days is final law on what the Bell operating companies can and can't do.

Computer Security Service An AIN (Advanced Intelligent Network) service providing for additional computer access security to be embedded in the network. Based on Caller ID or ANI (Automatic Number Identification), plus password protection and other authorization schemes, callers would be afforded or denied access to a networked computer on a customer premise. The network security service also would maintain an audit of all access attempts; the user organization could access that audit data in order to reconfigure access privileges, plug holes in network security, identify access anomalies, and so on. In this fashion, organizations supporting such applications as remote LAN access would realize an extra measure of security through a security system physically and logically separate from the premise.

Computer Support Telephony See CST.

Computer Telephony Computer telephony adds computer intelligence to the making, receiving, and managing of telephone calls. Harry Newton coined the term in 1992. Computer telephony has two basic goals: to make making and receiving phone calls easier, i.e. to enhance one's personal productivity and second, to please corporate customers who call in or who are called for information, service, help, etc. Computer telephony encompasses six broad elements:
1. Messaging.
Voice, fax and electronic mail, fax blasters, fax servers and fax routers, paging and unified messaging (also called integrated messaging) and Internet Web-vectored phones, fax and video messaging.
2. Real-time Connectivity.
Inbound and outbound call handling, "predictive" and "preview" dialing, automated attendants, LAN / screen-based call routing, one number calling / "follow me" numbers, video, audio and text-based conferencing, "PBX in a PC," collaborative computing.
3. Transaction Processing and Information Access via the Phone.
Interactive voice response, audiotex, customer access to enterprise data, "giving data a voice," fax on demand and shopping on the World Wide Web.
4. Adding Intelligence (and thus value) to Phone Calls.
Screen pops of customer records coincident with inbound and outbound phone calls, mirrored Web page "pops," smart agents, skills-based call routing, virtual (geographically distributed) call centers, computer telephony groupware, intelligent help desks and "AIN" network-based computer telephony services.
5. Core Technologies.
Voice recognition, text-to-speech, digital signal processing, applications generators (of all varieties — GUI to forms-based to script-based), VoiceView, DSVD, computer-based fax routing, USB (Universal Serial Bus), GeoPort, video and audio compression, call progress, dial pulse recognition, caller ID and ANI, digital network interfaces (T-1, E-1, ISDN BRI and PRI, SS7, frame relay and ATM), voice modems, client-server telephony, logical modem interfaces, multi-PC telephony synchronization and coordination software, the communicating PC, the Internet, the Web and the "Intranet."
6. New Core Standards.
The ITU-T's T.120 (document conferencing) and H.320 (video conferencing), Microsoft's TAPI — an integral part of Windows 95 and NT, Novell's TSAPI — a phone switch control NLM running under NetWare, Intel's USB and InstantON, Natural MicroSystems / Mitel's MVIP and H-MVIP, Dialogic has SCSA. And the industry has ECTF.
That's today. But what really excites is the potential. It's huge. Despite the above, phone calls today are dumb, seriously bereft of common sense. Few phones have "backspace erase." 75% of business calls end in voice mail! Often in voice mail jail. Every call not completed is an irritated customer and a lost sale. Computer telephony addresses the waste. Computer telephony adds intelligence to the making and receiving of phone calls. Bingo, happier customers and more completed transactions.

The best news: We now have the technology, the resources, the computer power, the new standards and the muscle to back our hype. We also have many new players who are, thankfully, not burdened by the assumptions of yesteryear's telecommunications industry. We also have legions of developers and systems and integrators who are grabbing these computer telephony tools and are cranking out hundreds of customer-pleasing, productivity-enhancing solutions for your business. Computer telephony delivers. And fortunately, industry now wants it.

Once a year, typically in March, industry leaders meet at a trade show called Computer Telephony Conference and Exposition. There is also a monthly magazine covering the industry called Computer Telephony Magazine. See also Telephony, Telephony Services, and Windows Telephony.

Computer-Like Transport An AT&T term for the carrying of digital information with the potential for acting on that information at any network node as appropriate.

Computername The name by which the lan identifies a server or a workstation in lan Manager terminology. Each computername must be unique on the network.

COMSAT The COMmunications SATellite corporation was created by Congress as the exclusive provider to the U.S. of satellite channels for international communications. COMSAT is also the U.S. representative to Intelsat and Inmarsat, two international groups responsible for satellite and maritime communications. In 1996 its revenues were $1 billion. www.comsat.com

COMSTAR A domestic communications satellite system from Comsat.

CON Circuit Order Number

Concatenation A SONET/SDH term. Concatenation is a mechanism for allocating contiguous bandwidth for transport of a payload associated with a "superrate service." The set of bits in the payload is treated as a single entity, as opposed to being treated as separate bits or bytes or time slots. The payload, therefore, is accepted, multiplexed, switched, transported and delivered as a single, contiguous "chunk" of payload data. Certain data protocols (e.g., ESCON) require huge chunks of bandwidth-far more than can be provided by the STS-1 data rate. In SONET, STS-1 is the electrical equivalent OC-1 (Optical Carrier Level 1), which is 51.84 Mbps-T-3 at 44.736 Mbps, plus SONET overhead. Though the use of Concatenation Pointers, multiple OC-1s can be linked together, end-to-end-hence the term "concatenation"-to provide contiguous bandwidth through the network, from end-to-end. The same approach is used at higher SONET OC-N levels, as well. Concatenation applications include bandwidth-intensive video (e.g., HDTV) and high-speed data. See also SDH, SONET, and STS.

Concentration A fundamental concept to telephony. Applies to a Switching Network (or portion of one) that has more inputs than outputs. For example, communications from a number of phones are sent out on a smaller number of outgoing lines. The theory is that, since not all the phones are being used at any one time, fewer trunks than phones are needed. Some phone system designs assume that only 5% of the phones will be in use at any one time. Some phone systems design assume 10%. In some phone-intensive industries, you can't make any assumptions about concentration. You have to assume one line per phone. No concentration. See Concentration Ratio.

Concentration Ratio The ratio between lines and trunks in a concentrated carrier system or line concentrator. See Concentrator.

Concentrator 1. A device which allows a relatively large number of devices or circuits (typically slow speed ones) to share either a single circuit or a relatively small number of circuits. In other words, the traffic is concentrated through a process of multiplexing, in which many relatively low capacity inputs from devices or circuits are folded together in order that they might share a single and typically higher capacity circuit which connects to a device or network of a higher order. Assuming that the capacity of the shared facility is sufficient to support all the lower order inputs in a satisfactory manner (i.e., transmission time and response time are not compromised to an unreasonable extent), the benefit of the concentrator is that communications costs are typically lowered through the process of sharing. An further, underlying assumption is that not all of the devices will be active at the same time. While a concentrator is akin to a multiplexer, it is limited to a single type of information stream and it is not capable of accomplishing some of the more sophisticated processes of the latter . For instance, an ATM concentrator might simply concentrate traffic from downstream ATM switches in a backbone LAN environment in order to share a very expensive high-capacity access circuit to an ATM WAN. The ATM adaptation process would have been accomplished previously either in the ATM LAN switch or in the workstation. In the realm of xDSL (generic Digital Subscriber Line) technology, a DSLAM (DSL Access Multiplexer) can be considered as a concentrator, as it simple concentrates traffic from xDSL circuits through a multiplexing process in order that the traffic might share a high-capacity circuit to the Internet backbone. See also DSLAM, Hub and Multiplexer.

2. A Multistation Access Unit (MAU). Token Ring LANs make use of MAUs to concentrate traffic from multiple nodes (e.g., workstations) to the LAN backbone, which may consist of nothing more than multiple interconnected MAUs. Through these central points of connection, the nodes are attached in a physical star configuration, also called home run, typically using Unshielded Twisted Pair (UTP) for reasons of lower cost of connectivity. Assuming that the concentrator is able to support all attached nodes without seriously degrading their access to the larger network, MAUs offer the advantages of lower overall cost of LAN connectivity and increased manageability.

3. A LAN hub. See also Hub.

Concentricity In a wire or cable, the measurement of the location of the center of the conductor with respect to the geometric center of the circular insulation.

Concentricity Error The amount by which a fiber's core is not centered in its cladding. The distance between the center of the two concentric circles specifying the cladding diameter and the center of the two concentric circles specifying the core diameter.

Concrete Fill A minimal-depth concrete pour to encase single-level underfloor duct.

Concurrency The shared use of resources by multiple interactive users or applications at the same time. Concurrency often means that a company need only buy as many licenses to a program as it has people using the program at one time — concurrent users, in other words. See SOFTWARE METERING.

Concurrency Control A feature that allows multiple users to execute database transactions simultaneously without interfering with each other.

Concurrent Site License Companies that buy software for multiple computers typically buy one copy of the program and a license to reproduce it up to a certain number of times. This is called a site license, though it may apply to its use throughout an organization. Site licenses vary. Some require that a copy be bought for each potential user — the only purpose being to indicate the volume discount and keep tabs. Others allow for a copy to be placed on a network server but limit the number of users who can gain simultaneous access. This is called a Concurrent Site License. And many network administrators prefer this concurrent license, since it gives them greater control. For example, if the software is customized, it need be customized only once, namely on the server.

Condenser A device for storing up electrical energy and consisting of two or more conducting surfaces or electrodes separated by an insulating medium called a dialectic.

Condenser Antenna An antenna consisting of two capacity areas.

Condenser Microphone Microphone which operates through changes in capacitance caused by vibrations of its conductive diaphragm.

Conditioned Circuit A circuit that has conditioning equipment to obtain the desired characteristics for voice or data transmission. See CONDITIONING.

Conditioned Loop A loop that has conditioning equipment to obtain the desired line characteristics for voice or data transmission. See CONDITIONING.

Conditioning The adjustment of the electrical characteristics of transmission lines to improve their performance for specific uses. Conditioning involves the "tuning" of the line or addition/deletion of equipment to improve its transmission characteristics. Conditioning may involve the insertion of components such as equalizers, resistors, capacitors, transformers or inductors. Long voice-grade, twisted-pair local loops, for instance, often have inductors, or "loading coils," installed every 6,000 feet or so in order to amplify the analog signal. Such "loaded" circuits, however, have a decidedly negative impact on data communications, especially at high transmission rates. Therefore, it is necessary to condition the circuit by removing all such electronics, thereby yielding what is known as a "dry copper" circuit. It often is required that the circuit be further conditioned by removing all bridged taps, ensuring that all pairs in the circuit are of consistent gauge, that all cross-connects are mechanically sound, and so on. Carriers provide two types of conditioning for leased lines. C conditioning controls attenuation, distortion, and delay distortion. D conditioning controls harmonic distortion and signal-to-noise ratio.

Conditions Busy. Voice Mail. Out of service. All the situations that a phone line is likely to find itself in.

Condofiber A shared tenancy cable or shared ownership facility such as a transatlantic fiber cable. Multiple vendors such as Sprint, MCI and AT&T may all own a group of fibers with responsibility for maintaining their own operation while at the same time paying an overall "association" fee for the common maintenance of the overall cable.

Conductance The opposite of resistance; a measure of the ability of a conductor to carry an electrical charge. Conductance is a ratio of the current flow to the potential difference causing the current flow. The unit of conductance is Mho (a reversed spelling of Ohm).

Conducting Materials Substances which offer relatively little resistance to the passage of an electric current.

Conductivity A term used to describe the ability of a material to carry an electrical charge. The opposite of specific resistance. Usually expressed as a percentage of copper conductivity - copper being one hundred percent. Conductivity is

expressed for a standard configuration of conductor.

Conductor Some atoms do not hold their electrons tightly, and in materials made of these atoms, the electrons can drift randomly from one atom to the next very easily. These materials make good electrical conductors. Most metals have electrons that can move easily this way and are generally good conductors. The best conductors are silver, copper and aluminum. Another type of good conductor is an electrolyte. An electrolyte is composed of charged ions that are free to move, carrying charge from one location to another. One example of an electrolyte is a solution of table salt in water. The positively charged sodium ions and the negatively charged chloride ions are capable of carrying charge from one part of the solution to another.

Conduit A pipe, usually metal but often plastic, that runs either from floor to floor or along a floor or ceiling to protect cables. A conduit protects the cable and prevents burning cable from spreading flames or smoke. Many fire codes in large cities thus require that cable be placed in metal conduit. In the riser subsystem when riser closets are not aligned, conduit is used to protect cable as well as to provide the means for pulling cable from floor to floor. In the horizontal wiring subsystem, conduit may be used between a riser or satellite closet and an information outlet in an office or other room. Conduit is also used for in-conduit campus distribution, where it is run underground between buildings and intermediate manholes and encased in concrete. Multiduct, clay tile conduit may also be used.

Conduit Run The path taken by a conduit or group of conduits.

Conduit System Any combination of ducts, conduits, maintenance holes, handholes and vaults joined to form an integrated whole.

Conferee Participant in a conference call who is not the call controller. This definition courtesy Hayes. According to Hayes, a "controller" is the person who sets up the conference call.

Conference Bridge A telecommunications facility or service which permits callers from several diverse locations to be connected together for a conference call. The conference bridge contains electronics for amplifying and balancing the conference call so everyone can hear each other and speak to each other. The conference call's progress is monitored through the bridge in order to produce a high quality voice conference and to maintain decent quality as people enter or leave the conference.

Conference Call Connecting three or more people into one phone conversation. You used to have to place conference calls through an AT&T operator (you still can). But now you can also organize conference calls with most modern phone systems or a conference bridge. If conferencing is important to you, make sure your conferencing device has amplification and balancing. If not, it will simply electrically join the various conversations together and people at either end won't be able to hear each other. There are different types of conference devices you can buy, including special teleconferencing devices that sit on conference tables and perform the function of a speakerphone, albeit a lot better. There are also dial-in devices called conference bridges. But, however, you use these devices, they will requires lines (and/or trunks). If you install one inside your phone system, be careful have the extra spare extensions. For a conference of 10 people, you'll typically need 10 extensions connected to your conference bridge. See CONFERENCE BRIDGE.

Conference, Meet-Me A conference call in which each of the people wishing to join the conference simply dials a special "Meet-Me" Conference phone number, which automatically connects them into the conference. It is a feature of some PBXs and also some special Conferencing Equipment. See CONFERENCE BRIDGE.

Conferencing Several parties can be added to a phone conversation through Conferencing.

Confidence Interval A confidence interval is the range of values within which the true value is assured to lie. Confidence level must be two figures.

Confidencer A noise-cancelling microphone for use on a telephone in noisy places. A confidencer is not an easy device to use.

Confidential Reception The ability to receive a facsimile transmission directly into memory which can be printed out or viewed at a later time.

Confidential Transmission A facsimile message that is sent confidentially into memory or a private mailbox, to be retrieved by the receiver at a later time. It's usually retrieved by using a confidential passcode or password.

Configuration 1. The hardware and software arrangements that define a computer or telecommunications system and thus determine what the system will do and how well it will do it. This information can be entered in the CMOS and EEPROM setup programs.
2. An ATM term. The phase in which the LE Client discovers the LE Service.

Configuration Databases Rolm/IBM words for those databases which represent unique user specifications relating to system and phone features. These databases can be entered on-site and are not part of the generic software which runs the phone system.

Configuration File An unformatted ASCII file that stores initialization information for an application.

Configuration Management One of five categories of network management defined by the ISO (International Standards Organization). Configuration management is the process of adding, deleting and modifying connections, addresses and topologies within a network.

Configuration Manager 1. A SCSA system service which manages configuration information and controls system startup.
2. An Intel Plug'n Play term. A driver, such as the ISA Configuration Utility, that configures devices and informs other device drivers of the resource requirements of all devices installed in a computer system. The Windows 95 Resource Kit defined configuration manager as the central component of a Plug and Play system that drives the process of locating devices, setting up their nodes in the hardware tree, and running the resource allocation process. Each of the three phases of configuration management-boot time (BIOS), real mode, and protected mode-has its own configuration manager.

Configuration Registry A database repository for information about a computer's configuration.

Confirming Design Layout Report Date CDLRD. The date a common carrier accepts the facility designed proposed by the Telco.

Conformance Test A test performed by an independent body to determine if a particular piece of equipment or system satisfies the criteria of a particular standard, sometimes a contract to buy the equipment.

Conforming End Office Central office with the ability to provide originating and terminating feature group D local access and transport area access service.

Congestion Control The process whereby packets are dis-

carded to clear buffer congestion in a packet-switched network.

Congestion Management The ability of a network to effectively deal with heavy traffic volumes; solutions include traffic scheduling and enabling output ports to control the traffic flow. See BECN.

Connect Time Measure of computer and telecommunications system usage. The interval during which the user was on-line for a session.

Connected 1. On line.
2. A voice recognition term for words spoken clearly in succession without pauses. For recognition to occur, words or utterances must be separated by at least 50 milliseconds (1/20th of a second). Generally refers to digit recognition and sometimes used to describe fast discrete recognition.

Connected State A state in which a device is actively participating in a call. This state includes logical participation in a call as well as physical participation (i.e., a Connected device cannot be on Hold).

Connected Time The length of time a path between two objects is active.

Connected User A Windows NT term. A user accessing a computer or a resource across the network.

Connecting Arrangement The manner in which the facilities of a common carrier (phone company) and the customer are interconnected.

Connecting Block A plastic block containing metal wiring terminals to establish connections. from one group of wires to another. Usually each wire can be connected to several other wires in a bus or common arrangement. A 66-type block is the most common type of connecting block. It was invented by Western Electric. Northern Telecom has one called a Bix block. There are others. These two are probably the most common. A connecting block is also called a terminal block, a punch-down block, a quick-connect block, a cross-connect block. A connecting block will include insulation displacement connections (IDC). In other words, with a connecting block, you don't have to remove the plastic shielding from around your wire conductor before you "punch it down."

Connecting Hardware A device providing mechanical cable terminations.

Connection 1. A path between telephones that allows the transmission of speech and other signals.
2. An electrical continuity of circuit between two wires or two units, in a piece of apparatus.
3. An SCSA term which means a TDM data path between two Resources or two Groups. It connects the inputs and outputs of the two Resources, and may be unidirectional (simplex) if either of the Resources has only an input or an output. Otherwise it is bi-directional (dual simplex). It usually has a bandwidth that is a multiple of a DS0 (64kbit) channel. Intergroup connections are made between the Primary Resource of each Resource Group.
4. An ATM connection consists of concatenation of ATM Layer links in order to provide an end-to-end information transfer capability to access points.

Connection Master Software from Mitel, which brings the Connection Control Standard to an even higher level for the MVIP developer. Connection Master interacts with circuit switches on multiple MVIP cards to make connections and resolve switching contention. It also interfaces between applications and makes connections in such a way that simple one-chassis applications become networked applications. Connection Master fully supports MC-MVIP, Multi-Chassis MVIP. See also MVIP.

Connection Oriented The model of interconnection in which communication proceeds through three well-defined phases: connection establishment (call setup), information transfer (call maintenance), connection release (call teardown). Examples include ordinary circuit-switched voice and data calls, ISDN calls, X.25 packet switching, Frame Relay, and ATM. See CONNECTION SERVICE and CONNECTIONLESS MODE TRANSMISSION.

Connection Oriented Network Service CONS. An OSI protocol for packet-switched networks that exchange information over a virtual circuit (a logical circuit where connection methods and protocols are pre-established); address information is exchanged only once. CONS must detect a virtual circuit between the sending and receiving systems before it can send packets.

Connection Oriented Operation A communications protocol in which a logical connection is established between communicating devices. Connection-oriented service is also referred to as virtual-circuit service.

Connection Orientated Protocol A protocol in which a connection is established prior to initiation of data transmission, maintained during transmission, and effectively terminated on completion of transmission. (Examples: SPX. TCP)

Connection Oriented Transmission Data transmission technique involving setting up connection before transmission and disconnecting it afterward. A type of service in which information always traverses the same pre-established path or link between two points. See CONNECTIONLESS SERVICE.

Connection Protocol A protocol in which it is not necessary to establish, maintain, and terminate a connection between source and destination prior to transmission. (Example: IPX, IP)

Connection Service A circuit-switching service whereby a connection is switched into place at the beginning of a session and held in place until the session is completed. Also referred to as circuit switching. The circuit switched in place may be real or virtual. See CIRCUIT SWITCHING.

Connectionless The model of interconnection in which communication takes place without first establishing a connection and without immediate acknowledgment of receipt. Sometimes (imprecisely) called datagram. Examples: Internet IP and OSI CLNP, UDP.

Connectionless Communication A form of communication between applications in which all data is exchanged during a single connection.

Connectionless Mode Transmission A mode of data transmission in which the transmitting device accesses the network and begins transmission without the establishment of a logical connection to the receiving device. In other words, the transmitter simply begins "blasting" data. Connectionless mode is very much unlike "Connection Oriented Transmission," wherein communications involves a process of call set-up, call maintenance and call teardown. Connectionless mode is limited to LAN communications and SMDS, which essentially is a MAN extension of the LAN concept.

In connectionless transmission, each packet is prepended with a header containing destination address information sufficient to permit the independent delivery of the packet. In other words, each packet within a stream of packets is independently survivable. While this approach is characteristic of connectionless mode, it also is characteristic of connection-oriented protocols such as X.25 (packet switching), Frame Relay and ATM. See also CLOUD, CONNECTIONLESS NETWORK, CONNECTIONLESS SERVICE, CONNECTION ORIENTED, SMDS, X.25 and ATM.

Connectionless Network A type of communications network in which no logical connection (i.e. no leased line or dialed-up channel) is required between sending and receiving stations. Each data unit (datagram) is sent and addressed independently, and, thereby, is independently survivable. IEEE 802 LAN standards specify connectionless networks. SMDS also is a connectionless network, as an extension of the LAN concept for broadband data communications over a metropolitan area. Connectionless networks are becoming more common in broadband city networks now increasingly offered by phone companies.

Connectionless Network Service CLNS. Packet-switched network where each packet of data is independent and contains complete address and control information; can minimize the effect of individual line failures and distribute the load more efficiently across the network.

Connectionless Service A networking mode in which individual data packets in a network (local or long distance) traveling from one point to another are directed from one intermediate node to the next until they reach their ultimate destination. Because packets may take different routes, they must be reassembled at their destination. The receipt of a transmission is typically acknowledged from the ultimate destination to the point of origin. A connectionless packet is frequently called a datagram. A connectionless service is inherently unreliable in the sense that the service provider usually cannot provide assurance against the loss, error insertion, misdelivery, duplication, or out-of-sequence delivery of a connectionless packet.

Connectionless Transmission Data transmission without prior establishment of a connection.

Connections Per Circuit Hour CCH. A unit of traffic measurement; the number of connections established at a switching point per hour.

Connectivity Property of a network that allows dissimilar devices to communicate with each other.

Connectoid Connectoid is the icon you create for a connection in the Dial-up Networking window in Windows 95.

Connector A device that electrically connects wires or fibers in cable to equipment, or other wires or fibers. Wire and optical connectors most often join transmission media to equipment (host computers and terminal devices) or cross connects. A Connector at the end of a telephone cable or wire is used to join that cable to another cable with a matching Connector or to some other telecommunications device. Residential telephones use the REJ-11C connector. Computer terminals with an RS-232-C interface, use the DB-25 connector. The RS-232-C standard is actually the electrical method of using the pins on a DB-25. See RS-232-C.

CONS Connection Oriented Network Service. See CONNECTION ORIENTED NETWORK SERVICE.

Consent Decree 1982 The agreement which divested the Bell Operating Companies from AT&T. It took effect at midnight on December 31, 1983. Also known as the MFJ (Modified Final Judgment), as it modified the 1956 Consent Decree.

Conservation Of Radiance A basic principle stating that no passive optical system can increase the quantity L/n2, where L is the radiance of a beam and n is the local refractive index. Formerly called conservation of brightness, or the brightness theorem.

Console 1. A large telephone which a PBX attendant uses to answer incoming calls and transfer them around the organization. Before you buy a PBX for your company, make sure your operator has checked out its console. Some are very difficult to use. Some are easy. Some operators hate some consoles. Some consoles hate some operators. You can measure the efficiency of consoles by comparing keystrokes to do simple jobs and comparing them — e.g. answer an incoming line, dial an extension and transfer the call. How many keystrokes does your PBX take?

2. The device which allows communications between a computer operator and a computer.

3. The console is the Novell NetWare name for the monitor and keyboard of the file server. Here you can view and control the file server or router activity. At the console, you can enter commands to control disk drives, send messages, set the file server or router clock, shut down the file server, and view file server information. NetWare commands you can enter only from the console (for example, MONITOR) are called console commands. Keep your file server locked up and away from prying eyes. It's clearly not just a case of changing passwords and getting in and mucking around. There have been examples of thieves simply removing the file server's hard disk, putting it in their briefcase and walking off with it.

Consoleless Operation Some PBXs can work without a console. Some must have a console. It's good to check. Consoles are expensive. If you don't want one — because your company is small — you don't want to be forced to buy one, only to have it sit idly by.

Consolidated Carrier Carriers that provide connection both as interexchange carriers as well as international carriers.

Consolidation Point A location for interconnection between horizontal cables extending from building pathways and horizontal cables extending into furniture pathways.

Constant Bit Rate CBR. A data service where the bits are conveyed regularly in time and at a constant rate, carefully timed between source (transmitter) and sink (receiver), i.e., following a timing source or clock just as members of a marching band follow the beat of the drummer. Examples include uncompressed voice and video traffic, which have to be transported at constant bit rate because they are sensitive to variable delay and, as such, have to be transported without any interruptions in the flow of data.

Constant Carrier Physical line specification selection indicating full duplex line in bisync network.

Constant Holding Time A telephone company definition. Certain devices used in dial equipment for setting up calls may well have practically constant holding times. For estimating the probabilities of congesting, the result of substituting a constant holding time equal to the average of a varying holding time seems to be of negligible moment from a theoretical standpoint. (See Holding Time).

Constellation The assemblage of satellites in a LEO (Low Earth Orbiting) or MEO (Middle Earth Orbiting) system. See LEO and MEO.

Construction Budget A detailed plan of placement, removal, and rearrangement of facilities to modernize and expand the capacity of the facilities network. A telephone company term.

Construction Zone The building of the new information infrastructure by telecommunications and cable companies.

Consult 1. To ask or seek the advice of another.

2. To seek another's endorsement of a decision you've already made.

Consultant See CONSULT and CONSULTANT LIAISON PROGRAMS.

Consultant Liaison Programs Large users often use communications consultants to help them choose systems and long distance phone lines. In recognition of the important role consultants play, many suppliers have consultant liaison programs. Such programs typically consist of a toll-free number and somebody on the other end to answer technical and pricing questions, a three-ring containing information on all the company's products and services, occasional seminars and, for those extra-privileged consultants, trips to all expense paid trips to exotic places and "something" else. With MCI that "something else" is a dial-up, toll-free, bulletin board. Dial it up with your PC, you can download MCI's latest prices and services. It's truly splendid as most of the paperwork others issue is obsolete the moment it's issued.

Consultation See CONSULTATION HOLD.

Consultation Hold PBX feature which allows an extension to place a call on hold while speaking with another call. The idea is "consulting with" someone while you have someone else on the phone.

Contact A strip or piece of metal which makes an electrical contact when some electromechanical device like a relay or a magnet operates. Contacts are often plated with precious metal to prevent them from oxidizing (i.e. rusting) and thus messing up the switch.

Contact Card See Smart Card.

Contact History A log of all the contacts, either by phone or letter, made with a prospect or customer. This is an important factor in building up a marketing database which can be used to accurately target prospects.

Contact Image Sensor Uses a flat bar of light-emitting diode that directly touches the original. It eliminates the step of having the diodes move through the lens, which causes poorer resolution. This method is a more sophisticated than the charged-coupled device scanning method.

Contact Management A business has customers and prospects. In computerese, they're called "contacts." Software to "manage" your customers and prospects is called contact management software. It has three elements: First, a screen or two of information about that contact (address, phone number, notes about your conversations, etc.) Second, the ability to print lists, and mailers, etc. And third, often a tie-in with your phone system to let your computer dial your clients and fax them stuff. With many newer phone systems, you have one extra benefit — namely when your phone rings, your contact management software will receive the calling phone number and pop up the screen or two about your contact. This way you'll be a little prepared before you answer the phone. See also ANI and CLID.

Contact Region The section of the jack wire inside the plug opening as shown in Subpart F of FCC rule 6B, figures 6B.500 (a) (3) and 6B.500 (b) (3).

Contended Access In local area networking technology it's the shared access method that allows stations to use the medium on a first-come, first-served basis.

Contending Port A programmable port type which can initiate a connection only to a preprogrammed port or group of ports.

Content In today's information rich and hyped society, "carriage" is the new name for transmission. And "content" is the new name for what we carry. Content is a more than just phone calls, of course. It's movies, music, games, on-line books, information, etc. Content used to be called information. Now it's called content. You figure. If words, pictures, sound or video are used as part of buying or selling, they are "transactive content," says Stanley Dolberg, an analyst at Forrester Research Inc. In the computer world, according to William Safire, content means "information on a Web site." Companies who provide content are called content suppliers, or O.S.Ps (on-line service providers). A content provider was once called an information provider. See also Content Supplier.

Content Processing Voice processing is the broad term made up of two narrower terms — call processing and content processing. Call processing consists of physically moving the call around. Think of call processing as switching. Content consists of actually doing something to the call's content, like digitizing it and storing it on a hard disk, or editing it, or recognizing it (voice recognition) for some purpose (e.g. using it as input into a computer program.

Content Provider 1. In the worlds of Convergence, the Internet and the World Wide Web (WWW), the Content Provider is the company which provides the material (content), rather than the network. See CONTENT SUPPLIER.
2. A fancy name for a writer, also called a language therapist by William Safire in the Sunday New York Times Magazine of January 28, 1996.

Content Supplier Content is a new fancy name for what telecommunications facilities carry. It includes movies, music, games, on-line books, information, etc. Content suppliers are thus movie studios, publishers, and music companies.

Contextual Ecommerce Imagine you receive a email from your friendly CD supplier. In it, he talks about the latest from Madonna. The email mentions the name of the CD. You notice its title is in blue and underlined — like a hot link to a Web site. You click on it. Instantly you've bought the CD. You receive it the next day by Fedex. Bingo, we now have contextual ecommerce.

Contention Contention occurs when several devices (phones, PCs, workstations, etc.) are vying for access to a line and only one of them can get it at one time. Some method is usually established for selecting the winner (first in, first out, camp on, etc.) and accommodating the loser(s) (.e.g. giving them a busy tone, giving them another shot at the line). When you cannot get an outside line from your extension you have been in contention and lost. See also ETHERNET.

Context Corporation An independent consultancy headed by Ray Horak, Consulting Editor of Newton's Telecom Dictionary. In fact, and according to Horak's mother, wife and children, The Context Corporation is the world's greatest consultancy, headed by the world's greatest, sweetest and most handsome man. Note: Horak's ex-wives (and others) have offered differing and understandably unprintable definitions.

Context Dependent Soft Keys Many telephones now have an LCD screen. Sometimes such screens have unmarked keys underneath them and/or at their side. What these keys do depends on the "labels" appearing on the screen. They are called "context dependent" because what those keys do depends on where the call is at that time. The first context dependent soft keys were on the Mitel SuperSet 4 phones. When the handset was resting on the phone, only three of the six context sensitive keys had meaning. One said "Program," one said "Msg" and one said "Redial." When you picked the phone up, three buttons would now be alive. One would say "Page," one would say "Redial" and one would say "Hangup." If the phone rang and you picked it up, one button would now say "trans/conf" (meaning transfer/conference. When another phone was ringing, one button would say

"Pickup," letting you push that button and answer someone else's phone. And so on. The neatest implementation of context sensitive keys was probably on the Telenova (now no longer manufactured). At one point when you were in voice mail, this phone's six buttons looked exactly like a cassette recorder — record, play, fast forward, fast reverse, etc. It was brilliant. No one has ever made using voice mail so easy.

Context Keys Buttons on a phone or device that have a display next to them. The buttons perform different functions depending on the what the screen shows when you press the button. See Context Dependent Soft Keys.

Context Sensitive A term from the computer industry which means that "Help" is only a keystroke away. Hit F1 and Help information will flash on the screen. That information will be relevant to what you're doing now, i.e. that help is within the context of what's going on right this moment. See also CONTEXT DEPENDENT SOFT KEYS.

Context Switch The technique with which an Intel 80x86 microprocessor handles multitasking is called a context switch. The CPU performs a context switch when it transfers control from one task to another. In the process, it saves the processor state (including registers) of one task, then loads the values for the task that is taking control. Context switching is the kind of multitasking that is done in standard mode Windows, where the CPU switches from one task to another, rather than allocating time to each task in turn, as in timeslicing.

Contiguous Port Ports occurring in unbroken numeric sequence.

Contiguous Slotting This term refers to the process of selecting individual DS-0 circuits, within a DS-1 circuit or DS-3 circuit, which are adjacent to one another. Due to the timing difference which can result when non-adjacent channels are selected, contiguously slotted channels are preferable when the end equipment is designed to multiplex the individual low-speed channels into a single, higher speed connection.

Contiguous United States The area within the boundaries of the District of Columbia and the 48 contiguous states as well as the offshore areas outside the boundaries of the coastal states of the 48 contiguous states, (including artificial islands, anchored vessels and fixed structures erected in such offshore areas for the purpose of exploring for, developing, removing and transporting resources therefrom) to the extent that such areas appertain to and are subject to the jurisdiction and control of the United States within the meaning of the Outer Continental Shelf Land Act, 43 U.S.C. Section 1331, et seq.

Continental Telecom Inc. CONTEL. A telephone company made up of more than 600 small phone companies. In 1990 it merged with GTE in a tax-free swap of shares. Contel was formed and grown by Charles Wohlstetter, an ex-stockbroker, who became financially comfortable (to say the least) in the process of growing Contel. In late 1990, Contel merged with GTE, which is a euphemism for GTE buying Contel.

Continental Morse Code See Morse Code

Continuity An uninterrupted electrical path.

Continuity Check A check to determine whether electrical current flows continuously throughout the length of a single wire on individual wires in a cable.

Continuity Check Tone CCT. A single frequency of 2000 Hz which is transmitted by the sending exchange and looped back by the receiving exchange. Reception of the returned indicates availability of the channel. See ITU-T Recommendation.271.

Continuous A word used in voice recognition to mean a type of recognition that requires no pause between utterances.

Continuous DTMF This is a feature of some phones (especially cellular phones) that sends touchtone sounds for as long as the key is held down, allowing access to services such as voice mail and answering machines that need long-duration tones. Some phones automatically have continuous DTMF; some don't. It's worth checking. Continuous DTMF makes a lot more sense.

Continuous Information Environment A term for the world we live in — in which information (text, voice, video, images, etc.) is flowing at us continuously. And our job is, somehow, to manage the information. The idea is to use the new computer telephony terms to manage the information.

Continuous Waves A series of wave or cycle of current all of which have a constant or unvarying amplitude.

Continuously Variable Capable of having one of an infinite number of values, differing from each other by an arbitrarily small amount. Usually used to describe analog signals or analog transmission.

Contract For the purposes developing applications in the telecommunications industry, there are two types of contracts: Active and Passive. An active contract is one you must sign. A passive contract is the type of contract you find in a software package. By opening the shrink wrapped package, you are committing yourself to the terms of the contract inside the package — the terms of which mostly consist of not duplicating the software in an authorized way.

Contributing Whore A title

Control Cable A multiconductor cable made for operation in control or signal circuits.

Control Channel Within a cellular telephone system, several of the channels are assigned as 'control' channels. Instead of supporting voice communications, these channels allow the base station to broadcast information to the cellular phones in its area. Cellular phones continuously monitor this broadcast information, selecting the base station that provides the best signal.

Control Character A non-printing ASCII character which controls the flow of communications or a device. Control characters are entered from computer terminal keyboards by holding down the Control key (marked CTRL on most keyboards) while the letter is pressed. To ring a bell at remote telex terminal, an operator could hold down the CTRL key, and tap the "G" key, since Control-G is the BELL character. Most computers display Control as the "^" character in front of the designated letter. For example, ^M is the Carriage Return character.

Control Circuit X.21 interface circuit used to send control information from DTE to DCE.

Control Connections A Control VCC links the LEC to the LECS. Control VCCs also link the LEC to the LES and carry LE_ARP traffic and control frames. The control VCCs never carry data frames.

Control Equipment 1. The central "brains" of a telephone system. That part which controls the signaling and switching to the attached telephones. Known as the KSU (or key service unit) in a key system.
2. Equipment used to transmit orders from an alarm center to remote site to enable you to do things by remote control.

Control Field Field in frame containing control information.

Control Flag A cellular phone term. A 6-bit flag transmitted in the forward channel data stream, comprised of a 5-bit busy/idle flag and one bit of the 5-bit decode status flag.

Control Head Roam Lights Indicates that the cellular phone is outside the "home" system.

Control Of Electromagnetic Radiation 1. Measures taken to minimize electromagnetic radiation emanating from a system or component, or to minimize electromagnetic interference. Such measures are taken for purposes of security and/or the reduction of interference, especially on ships and aircraft.
2. A national operational plan to minimize the use of electromagnetic radiation in the United States and its possessions and the Panama Canal Zone in the event of attack of imminent threat thereof, as an aid to the navigation of hostile aircraft, guided missiles, or other devices.

Control Of Flow Language Programming-like constructs (IF, ELSE, WHILE, GOTO, and so on) provided by Transact-SQL so that the user can control the flow of execution of SQL Server queries, stored procedures, and triggers. This definition from Microsoft SQL server.

Control Panel The control panel on the Apple Macintosh is for general hardware and software settings. Icons allow a user to customize the system or application, or select a particular service, such as a specific printer, set the sound level, the date and time and choose an Ethernet connection through the network control panel.

Control Plane The ATM protocol includes a Control Plane which addresses all aspects of network signaling and control, through all 4 layers of the model.

Control Point A program that manages an APPN network node and its resources, enabling communications to other control points in the network.

Control Segment A worldwide network of Global Positioning System monitoring and control installations that ensure the accuracy of satellite positions and their clocks.

Control Signal 1. In the public network, control signals are used for auxiliary functions in both customer loop signaling and interoffice trunk signaling. Control signals are used in the customer loop for Coin Collect and Coin Return and Party Identification. Control signals used in interoffice trunk signaling include Start Dial (Wink or Delay Dial) signals, Keypulse (KP) signals or Start Pulse (ST) signals.
2. In modem communications, control signals are modem interface signals used to announce, start, stop or modify a function. Here's a table showing common RS-232-C and ITU-T V.24 control signals

Pin	Control Signal	From	To
4	Request-To-Send (RTS)	DTE	DCE
5	Clear-To-Send (CTS)	DCE	DTE
6	Data Set Ready (DSR)	DCE	DTE
8	Carrier Detect (CD)	DCE	DTE
20	Data Terminal Ready (DTR)	DTE	DCE
22	Ring Indicator (RI)	DCE	DTE

Control Station On a multi-access link, a station that is in charge of such functions as selection and polling.

Control Tier An AT&T term for the tier within the Universal Information Services network node that provides the transport network's connection control function.

Control Unit An architectural component of a processor chip which orchestrates processor activity and handles timing to make sure the processor doesn't overlap functions.

Controlled Access When access to a system is limited to authorized programs, processes or other systems (as in a network).

Controlled Environment Vault CEV. It is a low maintenance, water-tight concrete or fiberglass container typically buried in the ground which provides permanent housing for remote switches, remote line concentrators, pair gain and fiber transmission systems. Because it is buried, it can often be installed in utility easements or other places where local building laws may be a problem. This below ground room that houses electronic and/or optical equipment is under controlled temperature and humidity conditions.

Controller 1. In the truest sense, a device which controls the operation of another piece of equipment. In its more common data communications sense, a device between a host and terminals that relays information between them. It administers their communication. Controllers may be housed in the host, can be stand-alone, or can be located in a file server. Typically one controller will be connected to several terminals. The most common controller is the IBM Cluster Controller for their 370 family of mainframes. In an automated radio, a controller is a device that commands the radio transmitter and receiver, and that performs processes, such as automatic link establishment, channel scanning and selection, link quality analysis, polling, sounding, message store and forward, address protection, and anti-spoofing.
2. Participant in a conference call who sets up the conference call.

Controller Card Also called a hard disk/diskette drive controller. It's an add-in card which controls how data is written to and retrieved from your PC's various floppy and hard drives. Controller cards come in various flavors, including MFN and SCSI. Controller cards are the devices used to format hard drives. Controller cards are not hard drive specific (except within categories). Controller cards will format many drives. But once you have a hard drive that has been formatted by that one controller card, it tends to prefer talking to that controller card forever. If you switch your hard disk to another machine, switch the controller card along with it. If you switch your hard disk to another machine, but not the controller card, then format the hard disk. That's not a "100% Do It Or Else You'll Be Disappointed" rule. But just a "Play It Safe and Switch Them" rule.

Conturing In digital facsimile, density step lines in received copy resulting from analog-to-digital conversion when the original image has observable gray shadings between the smallest density steps of the digital system.

CONUS A military term for CONtiguous United States (lower 48 states). See CONTIGUOUS UNITED STATES.

Convection Cooling Design techniques used in switching system construction to permit safe heat dissipation from the equipment without the need for cooling fans.

Convector The device which covers the steam heating radiator in buildings and typically sits underneath a window. Also called a weathermaster.

Convector Area An area allocated for heat circulation and distribution. Convector areas, typically built into a wall, can be used as a satellite location only if a more suitable area is unavailable.

Convergence 1. A measure of the clarity of a color monitor. A measure of how closely the red, green and blue guns in a color monitor track each other when drawing a color image. The other measures are focus and dot pitch.
2. A LAN term. The point at which all the internetworking devices share a common understanding of the routing topology. The slower the convergence time, the slower the recovery from link failure.
3. The word to describe a trend, now that most media can be represented digitally, for the traditional distinctions between industries to blur and for companies from consumer electronics, computer and telecommunications industries to form

alliances, partnerships and other relationships, as well as to raid each others markets.
4. The word "convergence" was set in motion in 1992 when Tele-Communications Inc. chairman John C. Malone told a cable-show audience that his vision of all-digital, fiber-optic networks would enable TCI and other cable operators to offer 500 TV channels, interactive programming, electronic mail, and telephony. According to Business Week of June 23, 1997 that picture of digital convergence was so compelling that cable, media and phone companies promptly hopped on the bandwagon. Business Week continued, "Several billion dollars later, it become clear that convergence was a bust. Cable companies, perennially strapped for cash, scaled back on their plans to upgrade their networks to handle huge amounts of interactive data. Phone companies that had hoped to offer television service — on their own wires or in joint ventures with cable companies — went back to their core businesses." However, such concerns should never let a good word die. In the May 17, 1998 issue of the The New York Times, Richard C. Notebaert, chairman and chief executive of the Ameritech Corporation, wrote, "Conventional wisdom holds that convergence — the gradual blurring of telecommunications, computers and the Internet — is primarily about technology and the inevitable clash of voice and data networks. But that narrow viewpoint misses the bigger picture...Convergence is about fundamental changes in the way we work — even behave." Now the latest concept of convergence is that all communications — the Internet and the PSTN (the public switched telephone network) — shall run over one network. As Notebaert says, "Our public voice network will become the public multimedia network and the Internet as we know it will cease to exist. With such a robust and ubiquitous network, we'll never have to go to the time and trouble of dialing into a private network when we want to surf the Web. In essence, we'll always be on line. And that will let us develop applications we can't even dream of today."

Convergence Billing Also known as convergence or composite billing. This is a fancy name for one phone company — local or long distance — providing a total communications bill to the customer. That total bill would include everything the customer buys in telecommunications services — from local, long distance, Internet access, cell phones, paging, etc. In late 1996, the belief developed in the telecom industry that if you "controlled" the bill to the customer, you would be in far better shape to sell the customer more services. The concept has some validity, especially if you also believe in fairies.

Convergent Convergent billing software is software which allows telecom companies (such as local and long distance companies) to bundle services, such as long distance, cellular, paging and cable, together onto a single monthly invoice. Bundling helps service providers offer competitive rates, boost revenue per customer and reduce customer turnover. Customers love the simplicity and convenience of one bill for all their telecom services. One company calls itself a "one stop shop with an integrated bill."

Conversation Path The route from originating port to terminating port for a two-way call. A conversation thus typically requires two ports on most PBXs.

Convergence Sublayer CS. An ATM term. SEE CS.

Conversation Time The time spent on a conversation from the time the person at the other end picks up to the time you or him hang up. Conversation time plus dialing, searching and ringing time equal the time your circuit will be used during a call.

Conversational Mode Telex An MCI International product providing real time exchange between Telex terminals or other compatible devices that allows instantaneous, two-way conversations in writing.

Converter 1. A vacuum tube which combines the functions of oscillator and mixed tube.

2. A device for changing AC to DC and vice versa. An ancient radio term.

3. An adapter, such as one that allows a modular phone to be plugged into a 4-hole jack.

4. A British term. A repeater that also converts from one media type to another, such as from fibre (British spelling) to copper. Often called a media adaptor.

5. A device used in RF distribution systems to convert from one frequency to another. May also control channel access.

Convolutional Code Error protection code encoding data bits in a continuous stream. An error-correction code in which each m-bit information symbol to be encoded is transformed into an n-bit symbol (n>m) where the transformation is a function of the last k information symbols, and k is referred to as the constraint length of the code. Convolutional codes are often used to improve the performance of radio and satellite links.

Cookie A cookie is an Internet mechanism that lets site developers place information on the client's (your) computer for later use. For example, some shopping cart technologies allow you to return to shopping at a later time. What this means is that a "cookie" containing your order is placed on your computer for the site's computer to retrieve when you return. Cookie is a basically a mechanism which is an feature of the HTTP (Hypertext Transport Protocol) protocol used in the Internet and WWW. In this client/server environment, a cookie allows the server side of the connections to both store and retrieve information on the client side. When connecting to a WWW computer in the form of a server, that server can store information on your client PC. This process takes place when the server returns a HTTP object such as graphic or screen to the client computer; at that point, the server also may send a set of "state" information which is stored on the client hard drive. This information is persistent, meaning that it remains in memory for subsequent use by the specified URL (Uniform Resource Locator, which is the address of the Web site) or group of URLs. As future client requests are made against the server, the cookie is automatically passed from client to server.

Cookies are used widely in the client/server environments of the WWW element of the Internet, as well as in Intranets. Their advantage is that they can automatically identify the client to the server, thereby shortening or eliminating the user identification element of the log-in process. For example, an electronic shopping application can use a cookie to identify the shopper during subsequent access sessions, storing information about shopping preferences. Further, the service provider can alter the content of the accessed Web site to appeal more to the client user based on that specific user's profile. The downside is that cookies are placed on the client computer without the knowledge of the user, giving rise to concerns about privacy through electronic trespass. Some Web browsers such as Netscape Navigator will alert the user to the desire of the server to apply a cookie, thereby providing the user with the option of rejecting that request. Cookie blocking also can be accomplished through the use of "Cookie.Cutter," developed by Phil Zimmerman, who gained fame with his development of PGP (Pretty Good Privacy) encryption software.

So, why are people so concerned about cookies? Let's consider an example. I recently was cruising the Web and tapped into the Web site of a book distributor. I had the opportunity to develop a profile of my interests. I chose not to do that for the same reason that I choose to shop for clothes without the assistance of a salesperson. Basically, I don't want to be slotted into a particular style, price range, etc. Rather, I prefer to scan the options, quickly, unassisted and without pressure. When I need the assistance of a salesperson, I'll ask for it. Further, I prefer that my identity be a private matter. I like Caller ID (when I am the calling party), but there may be times when I want to make a conscious decision to block it. I especially don't like cookies downloaded to MY computer, and I really especially don't want someone else taking the liberty of putting stuff into my computer without my OK. It's sort of like putting something in my car or in my house without my approval. It may be harmless and well-intentioned, but it's a "privacy" thing. See CLIENT/SERVER, COOKIE FILE, HTTP, URL and Caller ID.

According to Microsoft, "a cookie is a very tiny piece of text we're asking permission to place on your computer's hard drive. If you agree, then your browser adds the text in a small file. Its purpose is to let us know when you visit microsoft.com. This text, by itself, only tells us that a previous microsoft.com visitor has returned. It doesn't tell us who you are, or your email address or anything else personal. If you want to give us that information later, that's your choice. So why do we offer cookies? Cookies help us evaluate visitors' use of our site, such as what customers want to see and what they never read. That information allows us to better focus our online product, to concentrate on information people are reading and products they are using. And guess what? A cookie can help you. If you accept a cookie, nothing affects you immediately. But you know what happens whenever you want to download software, access a premium site or even request permission to use a Microsoft logo on your Web page? You get asked questions like who you are and your email address. And that happens every time you want to download stuff. If you have accepted a cookie, however, those questions eventually will be asked just once, no matter how often you download software or how many Microsoft sites you visit. In the future, a cookie will allow you to tell us what information you prefer to read and what you don't. If you're a gamer, for example, we can advise you on content specific to games. Why are we telling you all this? Because we want you to know why we ask you to accept a cookie. We want to be sure you understand that accepting a cookie in no way gives us access to your computer or any personal information about you. Cookies are harmless, occupying just a few bytes on your hard drive. They also can be a Web site browser's very good friend.

That having been said, consider the concept of a "third party cookie," also known as a DoubleClick cookie, after DoubleClick Inc. (www.doubleclick.net, but don't touch that website unless you want to run the risk of eating a DoubleClick cookie). After being planted by a participating Web site, a third-party cookie will follow you around the Web, recording your movements and interests.

Cookie File The file (usually in your browser's directory structure) where cookies are kept. The file name is cookie.txt, just in case you'd like to delete it. See COOKIE.

Cooperative Processing Mainframe and intelligent workstations dividing application code between them.

Coopetition A made-up word which means that you partner with your competition. You might be a wholesaler of PCS services who is now partnering with one of your retailers to create a new service which might compete with you, at some point. That's coopetition. The people at Nextwave Telecom, Inc. a wholesale PCS provider, loves this word.

Coordinates "Thank you for all your coordinates." A fancy way of saying "Thank you for your address, phone and fax numbers and email addresses."

Copia Latin for abundance. Steve Hersee called his Wheaton, Illinois all-things-fax company, Copia International in the hope that he would become rich as a horn of plenty. See Cornucopia in your normal dictionary.

Copolymer Compound resulting from the polymerization of two different monomers.

Coppertone Bob Metcalfe of InfoWorld Magazine coined coppertone for bare copper wire which you can rent from your local phone company. By "bare" he means that the copper you rent will contain no electronics on it anywhere, will be unloaded, unconditioned and unpowered. How might you use such copper wire? Let's say you have an office in New York City on on 12 West 21 Street and home on 215 West 19 Street. Let's say the office LAN had a T-1 connection to the Internet. Let's say you wanted to extend the LAN with its T-1 to your home. Simple, add networking equipment on your LAN and add similar equipment at your home. Bingo, your home is now on your LAN. Can it be done easily? Yes, but only if your phone company will rent you coppertone, i.e. bare copper wire. If they gave you something else — like one of their tariffed items — e.g. a 56 Kbps circuit — you would find yourself with a slower more inconvenient, less flexible and most likely, more expensive solution. The telephone companies don't like to rent you coppertone do this, for obvious reasons; in fact, several now refuse to lease coppertone except to legitimate burglar alarm companies. See also ADSL, HDSL, and DRY COPPER.

Coprocessor An additional processor which takes care of specific tasks, the objective being to reduce the load on the main CPU. Many IBM PCs and IBM clones have the capacity to install a coprocessor chip which does only arithmetic functions. This significantly speeds up your computer if you do a lot of calculations. See MATH COPROCESSOR.

COPW Customer Owned Premises Wire. You own the telephone wiring in your office.

Copy A nice new telephone system programming feature. We found it on Northern Telecom's Norstar phone. With this button, certain programmed settings can be copied from one line to another, or from one telephone to another. Line programmable settings that can be copied on the Norstar are Line Data, Restrictions, Overrides, and Night Service. Telephone settings that can be copied are Line Access, Restrictions, Overrides, and Permissions.

Copyright A copyright protects the original author of a story, software program, song, movie, piece of sculpture, or other original work from direct copying. Copying may be inferred where alleged copyist had access to the copyrighted work. The copyright notice (the @ symbol, or the word "Copyright", the year of creation, and the name of the copyright owner) should be provided on each copy of the work. Copyrights may also be registered with the Library of Congress, but this is not necessary in all cases. Copyright protects the expression of an idea, not the idea itself. (In appropriate cases, patents can be used to protect the idea.) Where the idea is so simple that there is only one way to express it, the idea and its expression may merge, preventing copyrightability. This logic was used successfully in defense

of several suits involving "clean room" reverse engineering of microcode: A first group hacked out the code, and prepared a complete functional specification defining the function of each instruction. A second group then wrote new code implementing these functions. Since there had been no copying, there could be no copyright infringement; the fact that both versions of the code for some instructions were identical merely showed that the idea and expression had merged. There are only so many ways to code an ADD instruction, after all. See INTELLECTUAL PROPERTY, PATENT and TRADE SECRET.

CORBA Common Object Request Broker Architecture. An ORB (Object Request Broker) standard developed by the OMG (Object Management Group). CORBA provides for standard object-oriented interfaces between ORBs, as well as to external applications and application platforms. The yield is that of interoperability of object-oriented software systems residing on disparate platforms. Additionally, CORBA provides for portability of such systems across platforms. See also Object Request Broker and OMG.

Cord 1. A small, flexible insulated wire.
2. The Cibernet On-Line Roaming Database. CORD is an online database that acts as a repository for information that wireless carriers need to exchange in order to support roaming in their territories.

Cord Board The earliest manual PBX. Usually an elegant wooden device consisting of lots of cords with plugs on them. These cords sat horizontally sticking up, like missiles in a silo. Each cord corresponded to an extension. Whenever the phone rang, the cord board attendant would answer it. Each incoming line was a vertical hole. When the operator had figured for whom the call was, he/she would simply plug the cord corresponding to the desired extension into the hole corresponding to the incoming trunk. The operator would reverse the process if the internal user wanted to make an external call. Either the operator would dial the call first, or simply plug in the user's extension and thus allow the user to dial the call directly. The tip of the plug and the circular ring on the plug gave the term "tip and ring" to telephony. In electronics, it's known as positive and negative. See CORD CIRCUIT.

Cord Circuit A switchboard circuit, terminated in two plug-ended cords, used to establish connections manually between user lines or between trunks and user lines. A number of cord circuits are furnished as part of the manual switchboard position equipment. The cords may be referred to as front cord and rear cord or trunk cord and station cord. In modern cordless switchboards, the cord circuit is switch operated. See CORD BOARD.

Cord Lamp The lamp associated with a cord circuit that indicates supervisory conditions for the respective part of the connection. See CORD BOARD.

Cordboard See CORD BOARD.

Cordless Telephone A telephone with no cord between handset and base. Each piece contains a radio transmitter, receiver, and antenna. The handset contains a rechargeable battery; the base must be plugged into an AC outlet. Depending on product design, radio frequency, environmental conditions, and national law, range between handset and base can be 10 feet to several miles. Cordless phones were once all analog. Now a breed of digital ones is out. They work much better in electrically noisy environments — like the typical office.

Cordless Switchboard A telephone switchboard in which manually operated keys are used to make connections. See CORD BOARD.

Core The central glass element of a fiber optic cable through which the light is transmitted (typically 8-12 microns in diameter for single mode fiber and 50-100 microns in diameter for multimode fiber). This light conducting portion of the fiber is defined by the high refraction index region. The core is normally in the center of the fiber, bounded by the cladding material. See also CORE

CORE Council Of REgistrars. An organization proposed to be charged with the responsibility for establishing and maintaining a new set of gTLDs (generic Top Level Domains) for the Internet. Effective March 1998, those gTLDs were to comprise the following: .arts (entities emphasizing cultural and entertainment activities); .firm (businesses, or firms); .info (entities providing information services); .nom (individual or personal nomenclature, i.e., a personal nom de plume, or pen name); .rec (entities emphasizing recreation/entertainment activities); .shop (businesses offering goods to purchase); and .web (entities emphasizing activities related to the World Wide Web).

The administration of the new gTLDs was contracted by CORE to Emergent Corporation, which was to develop, maintain, and operate the Shared Registry System (SRS). SRS is a neutral, shared, and centralized database of the new gTLGs. As many as 90 independent entities, known as "registrars," were authorized to register domain names, or URLs (Uniform Resource Locators), with each relying on the SRS. A URL, such as www.happypaintings.arts (this is not a real URL, at least not at the time of this writing), is translated into an IP address by a Domain Name Server (DNS), also known as a resolver. At the time of this writing, the proposal for new gTLDs has been forestalled. See also DNS, IP Address, SRS, TLD, and URL.

Core Gateway The primary routers in the Internet. Historically, one of a set of gateways (routers) operated by the Internet Network Operations Center at BBN. The core gateway system formed a central part of Internet routing in that all groups would advertise paths to their networks from a core gateway, using the Exterior Gateway Protocol (EGP).

Core Network A combination of high-capacity switches and transmission facilities which form the backbone of a carrier network. End users gain access to the core of the network from the Edge Network.

Core Non-Circularity The percent that the shape of the core's cross section deviates from a circle. Sometimes referred to as core ovality.

Core Processing Unit CPU. The card or shelf that controls the system or part of the system. It's called the CPU because all the RAM, subprocessors, buffers, clocking circuitry and ROM are included in this part of the system.

Core Size Primary description of a fiber. Stated in microns. Does not include cladding. Determines end surface area which accepts and transmits light.

Core Switch A Broadband Switching System (BSS) which is located in the core of the network. Conceptually equivalent to a Tandem Office in the voice world, a core switch serves to interconnect "Edge Switches," which provide user access to the broadband network much as do Central Offices in the circuit switched voice world.

Core Wall A wall that runs between structural floor and structural ceiling to separate stairwells, elevators, etc. from the rest of the building.

Cornea Gumbo A visually noisy, overdesigned Photoshopped mess. "We've got to redesign that page, it's become total cornea gumbo."

Corner Reflector 1. A device, normally consisting of three metallic surfaces or screens perpendicular to one another, designed to act as a radar target or marker.

2. In radar interpretation, an object that, by means of multiple reflections from smooth surfaces, produces a radar return of greater magnitude than might be expected from the physical size of the object.

3. A reflected electromagnetic wave to its point of origin. Such reflectors are often used as radar targets.

4. Passive optical mirror, that consists of three mutually perpendicular flat, intersecting reflecting surfaces, which returns an incident light beam in the opposite direction. 5. A reflector consisting of two mutually intersecting conducting flat surfaces.

Cornet A Siemens protocol for PBX-to-PBX signaling over a Primary Rate connection.

Corporate Account Service An MCI specific service involving a single, unified reporting system for multiple business that the customer owns, franchises, manages, or directs.

Corporate ID Number The MCI term for the number which identifies a customer on a corporate level. (Not all MCI customers have a corporate ID number.)

Corporate Network Also called an internetwork or a wide area network. A network of networks (the mother of all networks) that connects most or all of a corporation's voice, data, and video resources using various methods, including the phone system, LANs, private data networks, leased telecommunications lines, and public data networks. Connections between networks are made with bridges and routers.

Corporate networks come in many shapes and sizes. Often, they will consist of networks within the same building or facility. Here, networks are combined using bridges and routers. Corporate networks may also span great distances. Such internetworks require different types of connections than single-facility internetworks, though the fundamentals are similar. Internetworks that connect remote facilities usually rely on some type of public or leased data communications network provided by the phone company or a data network service company. Bridges and routers are still required to connect networks to the long-distance data service, whether it's an X.25 packet switched network, a T-1 line, or even a regular phone line. See also BRIDGE and ROUTER.

Correlation The AMA (Automatic Message Accounting) function that permits the association of AMA data generated at the same network system or at physically separate network systems. There are three levels of correlation that affect Advanced Intelligent Network Release 1: record level, service level, and customer level. Definition from Bellcore in reference to Advanced Intelligent Network.

Corresponding Entities Peer entities with a lower layer connection among them.

Corridor Optional Calling Plan Nynex offers a discounted way for subscribers in the 212 and 718 area codes to call five northern New Jersey counties — Bergen, Essex, Hudson, Passaic and Union. See CORRIDOR SERVICE.

Corridor Service A term that Bell Atlantic and Nynex are using for calls to and from the New York City area to and from Northern New Jersey, or between Philadelphia and Southern New Jersey.

Corrosion The destruction of the surface of a metal by chemical reaction.

COS 1. See CLASS OF SERVICE.

2. Compatible for Open Systems.

3. Corporation for Open Systems international. A Federal Government blessed organization which aims towards standardizing OSI and ISDN. COS members includes everyone from end-users to manufacturers. COS deals with private and public networking issues.

COSINE Cooperation for Open Systems Interconnection Networking in Europe. A program sponsored by the European Commission aimed at using OSI to tie together European research networks.

CoSN Consortium for School Networking A non-profit organization that promotes the use of telecommunications in Kindergarten to 12th grade education to improve learning. Members represent state and local education agencies, as well as hardware and software vendors, Internet Service Providers (ISPs) and interested individuals. www.cosn.org

COSName Identifies class of service SNA.

Cost Of Service Pricing A procedure, rationale or methodology for pricing services strictly on the basis of the cost to provide those services.

Cost per Phone Hour A call center term. Basic unit of resource measurement. Total costs (fixed, variable and semi-variable) divided by the number of workstation call hours that are projected or actually achieved.

COT 1. Continuity Check Message. The second of the ISUP call set-up messages. Indicates success or failure of continuity check if one is needed. See ISUP and COMMON CHANNEL SIGNALING.
2. Central Office Terminal or Termination. The termination of a local loop facility at the central office. See Digital Loop Carrier.

COTS 1. COnnection Transport Service.
2. Commercial Off The Shelf.

Couch Commando A couch potato who insists on taking charge of what he and the rest of the couch potatoes are watching on the TV.

Couch Potato A person who spends their life sitting on a couch surfing TV channels with remote control TV device. See MOUSE POTATO.

Coulomb The quantity of electricity transferred by a current of one ampere in one second. One unit of quantity in measuring electricity.

Council of Registrars CORE. An organization charged with the responsibility for development, implementation, and maintenance of a set of new Top Level Domains (TLDs) for the Internet. See CORE for a longer explanation.

Counter Rotating Ring An arrangement whereby two signal paths, the directions of which are opposite, exist in a physical ring topology. Such rings typically are described as "Dual Counter Rotating Rings," such as described in SONET and FDDI standards. In such a physical configuration, one or more transmission paths operate in a clockwise manner, while one or more other paths operate in counter-clockwise, or anti-clockwise. Should the primary path suffer catastrophic failure, the secondary path come on line. It does this to ensure virtually uninterrupted communications. See also FDDI and SONET.

Counterpoise A system of electrical conductors used to complete the antenna system in place of the usual ground connection.

Country Code 1. The one, two or three digit number that in the world numbering plan, identifies each country or integrated numbering plan in the world. In short, the one, two or three digits that precede the national number in an international phone call. This code is assigned in and taken from Recommendation E.163 (Numbering Plan for International Service) adopted by the ITU-T. There's a list of country codes and key country area codes in the Appendix at the back of this book. See also www.the-acr.com and www.sprint.com/ssi/intl_codes.html
2. In international record carrier transmissions, the country code is a two or three alpha or numeric abbreviation of the country name following the geographical place name.

County For the purposes of the FCC's cable television rules, this term includes: • Borough (in Alaska).
• District (in District of Columbia).
• Independent City (in Alaska, Maryland, Missouri, Nevada, and Virginia).
• Municipio (in Puerto Rico).
• Parish (in Louisiana).

Coupled Modes 1. In fiber optics, a condition wherein energy is transferred among modes. The energy share of each mode does not differ after the equilibrium length has been reached
2. In microwave transmission, a condition where energy is transferred from the fundamental mode to higher order modes. Energy transferred to coupled modes is undesirable in usual microwave transmission in a waveguide. The frequency is kept low enough so that propagation in the waveguide is only in the fundamental mode.

Coupling Any means by which energy is transferred from one conductive or dielectric medium (e.g., optical waveguide) to another, including fortuitous occurrences. Types of electrical coupling include capacitive (electrostatic) coupling, inductive coupling, and conductive (hard wire) coupling. Coupling may occur between optical fibers unless specific action is taken to prevent it. Coupling between fibers is very effectively prevented by the polymer overcoat, which also prevents the propagation of cladding modes, and provides some degree of physical protection.

Coupling Loss The power loss suffered when coupling light from one optical device to another.

Coupon From Britain: A tear-off slip to encourage response to advertisements or to a promotion on packaging. The information is keyed into a telebusiness system which automatically handles the follow-up. This may be a phone call, acknowledgement letter, brochure, distribution of lead to a distributor and so on.

Courier Dispatch The Courier Dispatch service offered by MCI International allows customers to generate and send high-priority messages from their own Telex terminals to any destination in the Continental U.S. and Hawaii.

Courseware A combination of Web pages, E-mail, threaded discussions, chat rooms, listservs and distance learning tools used to provide online educational services or supplement regular classroom instruction.

COV Control Over Voice. Mitel's proprietary signaling protocol which they use between their PBX and their proprietary analog phones.

Cover Page The first page of a fax message. It generally includes a header, typically the sender company's logo; the recipient's name and fax telephone number; the sender's fax and voice telephone numbers; the system's date and time; a message; a footer.

Coverage The percent of completeness with which a metal braid covers the underlying surface.

Coverage Area The geographic area served by a cellular system; that is, the area in which service is available to users of the system. Once the mobile telephone number has traveled outside the coverage area, the mobile telephone will show "NO SERVICE."

COW Cellsite On Wheels. A trailer with antenna and transmitting/receiving hardware used to provide temporary service in emergencies, special events, remote testing and repair, or until a normal tower can be erected. Comes with climate control, diesel generator, and self-supporting wind-resistant 84' antenna mast. Both full-size COWs and mini-COWs are available.

COW Interface Character-Oriented Windows Interface. An SAA-compatible user interface for OS/2 applications.

CP Connection Point in Northern Telecom parlance.

CP/M Control Program for Microcomputers. An erstwhile popular operating system for primarily 8-bit microcomputer systems based on the family of Intel 8080 family of microprocessor chips. CP/M system was originally written by Gary Kidall a programmer and consultant who later formed a company called Intergalatic Digital Research (later just Digital Research). Sadly, that company never upgraded CP/M to 16-bit machines. Thus it left the way open for Bill Gates and the company he formed, Microsoft, to create MS-DOS, which, in its initial form, bore a remarkable resemblance to CP/M.

CPC 1. Calling Party Control.

2. Calling Party Connected.

CPCS An ATM term. Common Part Convergence Sublayer: The portion of the convergence sublayer of an AAL that remains the same regardless of the traffic type.

CPCS-SDU An ATM term. Common Part Convergence Sublayer-Service Data Unit: Protocol data unit to be delivered to the receiving AAL layer by the destination CP convergence sublayer.

CPE Customer Provided Equipment, or Customer Premises Equipment. Originally it referred to equipment on the customer's premises which had been bought from a vendor who was not the local phone company. Now it simply refers to telephone equipment — key systems, PBXs, answering machines, etc. — which reside on the customer's premises. "Premises" might be anything from an office to a factory to a home. GTE once used CPE to refer to "Company Provided Equipment." It doesn't any longer. What the Americans call CPE, the Europeans now call CTE, which stands for Connected Telecommunications Equipment. See CTE Directive.

CPI Computer to PBX Interface. This proprietary hardware/software interface provides direct connectivity between a PBX's switching network and a host computer to allow switched access between the host computer and data terminal equipment connected with the PBX. The interface is based on the North American Standard T-Carrier specification (24 multiplexed 64 Kbps channels operating at a combined speed of 1,544 Mbps). Developed by Northern Telecom, Inc. this interface uses in-band signaling and provides bidirectional data transmission at speeds up to 56 Kbps synchronous per channel. See OPEN APPLICATION INTERFACE.

CPI-C IBM SAA Common Programming Interface-Communication between SNA and OSI environments.

CPM 1. Customer Premise Management.

2. Critical Path Method. See also CP/M.

3. Cable Plant Management. See CABLE MANAGEMENT.

CPN 1. Computer PBX Network. 2. Customer premises network. 3. Calling Party Number. See CALLING PARTY NUMBER.

CPNI Customer Proprietary Network Information. Information which is available to a telephone company by virtue of the telephone company's basic service customer relationship. This information may include the quantity, location, type and amount of use of local telephone service subscribed to, and information contained on telephone company bills. This is the definition of CPNI that the independent voice mail and live telephone answering industry uses.

CPODA Compression Priority Demand Assignment. Another protocol for converting voice into data bits. See also PCM.

CPP Calling Party Pays

CPS Characters per second, or cycles per second. In asynchronous communications, there are typically ten bits per characters — 8 bits for the character and one stop and one start bit.

CPU The Central Processing Unit. The computing part of a computer. The "brain" of the computer. It manipulates data and processes instructions coming from software or a human operator. See CENTRAL PROCESSING UNIT.

CPUG Call Pickup Group. All the phones in an area that can be answered by each other by simply punching in a couple of digits. See CALL PICKUP.

CR 1. Carriage Return. The key on a computer called Carriage Return or sometimes "ENTER." Touching this key usually signals the computer that the entry has been completed and is now ready for processing by the computer. See CARRIAGE RETURN.

2. Critical (alarm status). Indicates a failure affecting more than 96 customers. An AT&T definition.

3. Call Reference.

Cracker A person who "cracks" computer and telephone systems by gaining access to passwords, or by "cracking" the copy protection of computer software. A cracker usually does illegal acts. A Cracker is a "Hacker" whose hacks are beyond the bounds of propriety, and usually beyond the law. The term "cracker" is said to derive from the word "safecracker." See HACKER.

Cradle Cams On-line cameras attached to computers attached to the Internet that allow parents to monitor their children from their desks at their offices. Cradle cams (cameras) — also known as Kiddie cams — are often installed in daycare centers and grade schools.

CRAFT 1. Cooperative Research Action For Technology.

2. Craft. Nonmanagement RBOC staff. Many craft employees are members of the Communications Workers of America (CWA).

Craft Terminal A PCS wireless term. A craft terminal is a device built specifically to provide a man-machine interface that is otherwise not available. The interface is customized to provide a view into a particular device's operation such as a proprietary switch or BSS, which is a Base Station Sub-system charged with managing radio frequency resources and radio frequency transmission for a group of BTSs, which is a Base Transceiver Station, used to transmit radio frequency over the air.

Craft Test Set Also called Goat or Butt-Set. Portable telephone used to test analog phone lines.

Craftsperson In the phone industry, a craftsperson has two distinct meanings. First, it is the person who toils to install phones, repair outside plant and fix problems inside central offices. This person typically carries tools and dresses in jeans. Second, craftspeople are at the bottom of the management hierarchy in most phone companies. They typically belong to a union. Craftspeople are not "in management." See LEVEL.

Cramming A practice in which customers are billed for unexpected telephone charges, which they typically didn't order, authorize or use. "Cramming" refers to the fact that the charges are crammed onto the telephone bill in an inconspicuous place in order that they will go unnoticed. Most of us quickly review (at best) our telephone bills, and write a check for the total-we really don't examine them in great detail. Cramming is a practice of only a very few of the most unethical carriers.

Crankback An ATM term. A mechanism for partially releasing a connection setup in progress which has encountered a failure.

This mechanism allows PNNI to perform alternate routing.

Crapplet A poorly written or totally useless Java applet. "I just wasted 30 minutes downloading this awful crapplet!"

Crash The complete failure of a hardware device or a software operation. Usually used to mean a "fatal" crash in which the device or software must be started from a "power up" condition. See BOOT.

CRC Cyclic Redundancy Check. A process used to check the integrity of a block of data. A CRC character is generated at the transmission end. Its value depends on the hexadecimal value of the number of ones in the data block. The transmitting device calculates the value and appends it to the data block. The receiving end makes a similar calculation and compares its results with the added character. If there is a difference, the recipient requests retransmission. CRC is a common method of establishing that data was correctly received in data communications. See CRC CHARACTER and CYCLIC REDUNDANCY CHECK.

CRC Character A character used to check the integrity of a block of data. The character is generated at the transmission end. Its value depends on the hexadecimal value of the number of ones in the data block. And it is added to the data block. The receiving end makes a similar calculation and compares its results with the added character. If there's a difference, there's been a mistake in transmission. So, please, re-send the data.

CRD Contention Resolution Device.

Cream Skimming Selecting only the most profitable markets or services to sell into. Choosing the cream of the market. An erstwhile popular economic theory to deny new entrants into the telephone industry.

Credentials A way of establishing, via a trusted third party, that you are who you claim to be.

CREDFACS Conduit, Risers, Equipment space, Ducts and FACilitieS. Collective term for pathway elements used in communications cabling.

Credit Card Phone A pay telephone that accepts credit cards with magnetic strips on them instead of coins.

CREN Corporation for Research and Educational Networking. An organization formed in October 1989, when Bitnet and CSNET were combined. CSNET is no longer around, but CREN still operates Bitnet.

Crest Factor The crest factor is the ratio of the crest (peak, maximum) value of a current to the root-mean-square (RMS) value. A square wave of current has a crest factor of 1. A sine wave has a crest factor of 1.414. The current drawn by a typical computer power supply when powered from a typical wall outlet has a crest factor of 4. The crest factor in this case results from a complex interaction between the power supply and the utility power sine wave. The crest factor of a computer or telephone system power supply is usually reduced when it is operated from a UPS. The reduction in crest factor when operating from a UPS does not adversely affect a computer or telephone power supply, and in fact actually makes it run cooler. Crest factor is always a property of the interaction between a load and a source, so it is meaningless to attribute to either a load or source independently. Factors which generally affect the ability of a UPS to supply high crest factors are: output impedance at harmonic frequencies, output distortion, and current limit. Although a high crest factor rating of a UPS has been considered to be a measure of UPS output stability and quality, differences in measurement techniques make product comparisons on this basis useless. A preferred method is to specify the output voltage response to a step load or output

voltage distortion under load. Definition supplied by APC.

CRF 1. An ATM term. Cell Relay Function: This is the basic function that an ATM network performs in order to provide a cell relay service to ATM end-stations.

2. An ATM term. Connection Related Function: A term used by Traffic Management to reference a point in a network or a network element where per connection functions are occurring. This is the point where policing at the VCC or VPC level may occur.

Crimp Die These are the part of the crimp tool that actually come in contact with the connector that is being crimped. They slide into the jaws of the crimp tool. Crimp dies are interchangeable and many types are available. See Crimp Tool.

Crimp Tool Crimp tools form connectors onto cables. They are used for BNC, F-Type and RJ-11, RJ-45 connectors, among others. They have a padded handle and jaws where the crimp dies are inserted. Crimps are installed by inserting the cable into the crimp, then the crimp into the crimp die, and squeezing the handles of the crimp tool. See Crimp Die.

CRIS Cryptography & Information Security Research Laboratory.

Critical Angle The smallest angle at which a ray will be totally reflected within a fiber.

Critical Mess An unstable stage in a software project's life when any single change or bug fix can result in two or more new bugs. Continued development at this stage leads to an exponential increase in the number of bugs.

Critical Technical Load That part of the total technical power load required for synchronous communications and automatic switching equipment.

CRM An ATM term. Cell Rate Margin: This is a measure of the difference between the effective bandwidth allocation and the allocation for sustainable rate in cells per second.

Cross Assembler An assembler that can run symbolic-language on one type of computer and produce machine-language output for another type of computer.

Cross Border Digital Data Service CDDS. An MCI International digital, private-line service that provides customers with service that provides customers with 56 Kbps dedicated terrestrial channels between the U.S. and Canada.

Cross Border Terrestrial Digital Data Service CTDDS. An MCI International point-to-point dedicated, leased channel service enabling customers to transmit traffic between the U.S. and Canada over digital terrestrial facilities at a transmission speed of 1.544 Mbps.

Cross Compiler A compiler that runs on one computer but produces object code for a different type of computer. Cross compilers are used to generate software that can run on computers with a new architecture or on special-purpose devices that cannot host their own compilers.

Cross Connect I've read 15 definitions of cross connect. They're all awful. Let's try. Let's imagine you have an office that you need to wire up for voice and data. So you wire every desk with a bunch of wires. You punch one end of the wires into various plugs at the desk. You punch the other onto some form of punchdown block, for example a 66-block. Then you bring the wires in from your telecom suppliers. The T-1s, the ATM, the frame relay, the local lines, the analog lines, the digital lines, etc. You punch them down on another punchdown block, for example a 66-block. Now you have two sets of blocks (they can be any form of punchdown block) — one for those going to the office and those coming in from the outside world. You now have to join them. Joining them is called "cross connecting" in the telecom world. You simply run wires from one 66-block (or other punchdown device) to the other one. The reason you use cross connect wires rather than just punching down an incoming phone line, for example, directly to your phone system is that moves, adds and changes would, over time, horribly confuse things, screw connections up, and eventually become a total mess. Easier to simply have all the changes accomplished through the cross connect wires and wiring. Follow the short wires. Easy to see what's connected to what. Easier for labeling, documentation, etc. In short, cross connect is a connection scheme between cabling runs, subsystems, and equipment using patch cords or jumpers that attach to connecting hardware on each end. Cross-connection is the attachment of one wire to another usually by anchoring each wire to a connecting block and then placing a third wire between them so that an electrical connection is made. The TIA/EIA-568-A standard specifies that cross connect cables (also called patch cords) are to be made out of stranded cable. See also Cross Connect Equipment and Cross Connect Field.

Cross Connect Equipment Distribution system equipment used to terminate and administer communication circuits. In a wire cross connect, jumper wires or patch cords are used to make circuit connections. In an optical cross connect, fiber patch cords are used. The cross connect is located in an equipment room, riser closet, or satellite closet.

Cross Connect Field Wire terminations grouped to provide cross connect capability. The groups are identified by color-coded sections of backboards mounted on the wall in equipment rooms, riser closets, or satellite closets, or by designation strips placed on the wiring block or unit. The color coding identifies the type of circuit that terminates at the field.

Cross Connection See Cross Connect.

Cross Coupling The coupling of a signal from one channel, circuit, or conductor to another, where it becomes an undesired signal. See Cross Connect.

Cross Extension Cable When you make an REJ-11 extension cable, the wiring crosses over. Conductor 1 becomes 4. Conductor 2 becomes 3. Conductor 3 becomes 2. And conductor 4 becomes one. Next time you have an REJ-11 extension cable in your hand, hold the REJ-11s next to each other and compare them. You'll notice the cross-over of the conductors.

Cross Modulation Distortion The amount of modulation impressed on an unmodulated carrier when a signal is simultaneously applied to the RF port of a mixer under specified operating conditions. The tendency of a mixer to produce cross modulation is decreased with an increase in conversion compression point and intercept point.

Cross Over Cable See Crossover Cable

Cross Pinned See Crossover Cable.

Cross Plan Termination The conversion of ten-digit telephone numbers to seven digits, or vice versa.

Cross Polarization The relationship between two radio waves where one is polarized vertically and the other horizontally.

Cross Selling You buy a shirt from me. I sell you a tie. You buy a car from me. I sell you a mobile phone for your car. There is another term. it's called "up selling." That's when I sell you a more expensive shirt or a more expensive car.

Cross Subsidization Supporting one area of a business from revenues generated by another area. Local phone companies in the U.S. have long argued that if they are required by government or regulatory decree to provide "universal service" to households, they should be allowed to price business

service higher. This way they can cross subsidize low-priced residential with high-priced business service. At least that's the theory. There are, however, many other cross-subsidies in the telephone business — people who stay longer with one phone line cross-subsidize those who move frequently; international service in most countries is priced high and the profits used to provide other services. The problem with cross-subsidies is that everyone knows they exist, but no one knows the actual financial extent of them. The problem is of allocation. A phone company runs with one plant — switches and wires. Those switches and wires provide everything from local to international calling. How to figure how to allocate how much is used for what? It's a question that has provided millions of dollars in consulting fees for thousands of economists over the years. Despite the money, there are no conclusive answers. See also Tariff Rebalancing.

Cross Wye A cable used at the host system, or network interface equipment that changes pin/signal assignment in order to conform to a given wiring standard (USOC, AT&T PDS, DEC MMJ, etc).

Cross-connection Non-permanent wire connections that run between terminals of a cross-connect field. See Cross Connect.

Crossbar Xbar. A switching system that uses a centrally-controlled matrix switching network of electromagnetic switches which worked with magnets and which connect horizontal and vertical paths to establish a path through the network. Crossbar switches are circuit switches, typically in the form of voice PBXs and Central Offices (COs). Crossbar was known for its reliability, at least in comparison to earlier Step-by-Step (SxS) electromechanical switches, but is now largely obsolete because it takes up a lot of space and isn't programmable. The first crossbar switch was a central ofice installed in Brooklyn, NY in 1937.

Crossbar Tandem A 2-wire common-control switching system with a space-division network used as local tandem, toll tandem, and CAMA switching. While originally designed to switch trunks, some systems have been locally modified to accept loop-start or ground-start lines.

Crossed Pinning Configuration that allows two DTE devices or two DCE devices to communicate. See also CROSSOVER CABLE.

Crosslink An X.25 link connecting two XTX NCs on the same level.

Crossover Cable Another word for a null modem cable or a cross-pinned cable. Such a cable is a RS-232 cable that enables two DTE devices or two DCE devices to be connected through serial ports and transmit and receive information across the cable. The sending wire on one end is joined to the receiving wire on the other. In an RS-232 cable, this typically means that conductors 2 and 3 are reversed. See RS-232-C.

Crosspinned Cable See Crossover Cable.

Crosspoint A single element in an array of elements that comprise a switch. It is a set of physical or logical contacts that operate together to extend the speech and signal channels in a switching network.

Crosspolarized Operation The use of two transmitters operating on the same frequency, with one transmitter-receiver pair being vertically polarized and the other horizontally polarized (orthogonal polarization).

Crossposting Crossposting is putting one copy of an electronic file up on the Internet in such a way that it can be viewed from any of several newsgroups (discussion areas). Today's Internet software lets readers avoid seeing a widely crossposted article more than once. They see it in the first group they find it. Crossposting is frowned upon in the Internet when it becomes excessive and off-topic. Crossposting is a less serious offense than spamming, which is seriously frowned upon. See SPAMMING.

Crosstalk Crosstalk occurs when you can hear someone you did not call talking on your telephone line to another person you did not call. You may also only hear half the other conversation. Just one person speaking. There are several technical causes for crosstalk. They relate to wire placement, shielding and transmission techniques. CROSSTALK is also the name of once popular telecommunications software program for 8- and 16-bit microcomputers.

Crosstalk Attenuation The extent to which a communications system resists crosstalk.

CRP Command repeat

CRS An ATM term. Cell Relay Service: A carrier service which supports the receipt and transmission of ATM cells between end users in compliance with ATM standards and implementation specifications.

CRT Cathode Ray Tube. The glass display device found in television sets and video computer terminals. See CATHODE RAY TUBE.

CRTC Canadian Radio Television and Telecommunications Commission. Canada's federal telecom regulator. It's based in Ottawa.

Cryptanalysis 1. The steps and operations performed in converting encrypted messages into plain text without initial knowledge of the key employed in the encryption.
2. The study of encrypted texts. The steps or processes involved in converting encrypted text into plain text without initial knowledge of the key employed in the encryption.

Crypto A term used to describe encrypted information. The use of encryption on data communications circuits lessens the chance that the information will be successfully copied by eavesdroppers.

Cryptochannel A complete system of crypto-communications between two or more holders. The basic unit for naval cryptographic communication. It includes: (a) the cryptographic aids prescribed; (b) the holders thereof; (c) the indicators or other means of identification; (d) the area or areas in which effective; (e) the special purpose, if any, for which provided; and (f) pertinent notes as to distribution, usage, etc. A cryptochannel is analogous to a radio circuit.

Cryptography The process of concealing the contents of a message from all except those who know the key. Cryptography is unregulated in the United States. See CLIPPER CHIP.

Crystal Microphone A microphone, the diaphragm of which is attached to a piezo-electric crystal, which generates electrical currents when torque is applied, due to the vibration of the diaphragm. The earliest form of microphone, now obsolete. See also CONDENSER and ELECTRET MICROPHONE.

CS Convergence Sublayer. The upper portion of BISDN Layer 3. As an ATM term, it covers the general procedures and functions that convert between ATM and non-ATM formats. It describes the functions of the upper half of the AAL layer. It is also used to describe the conversion functions between non-ATM protocols such as Frame Relay or SMDS and ATM protocols above the AAL layer. The exact functions of the CS are dictated by the particular AAL (1, 2, 3/4, or 5) in support of the specific Service Class (A, B, C, or D). SEE AAL.

CS-1 Capability Set 1. Term used by ITU-T to refer to their initial set of Advanced Intelligent Network (AIN) standards. Contains 18 trigger detection points. Bellcore (Bell

Communications Research) plans to adopt the CS-1 terminology for its own AIN.

CS-CDPD Circuit Switched-Cellular Digital Packet Data. A variation on the CDPD theme. Developed in July 1995, the expanded specification provides for packet radio transmission on a circuit-switched basis over the analog AMPS cellular network. See also CDPD.

CSA 1. CallPath Services Architecture. IBM's computer host to PBX interface. It links computer and telephone systems. See CALLPATH SERVICES ARCHITECTURE for detail. See also CALLBRIDGE and OPEN APPLICATION INTERFACE.
2. Canadian Standards Association. A non-profit, independent organization which operates a listing service for electrical and electronic materials and equipment. It is the body that establishes telephone equipment (and other) standards for use in Canada. At least in part, CSA is the Canadian counterpart of the Underwriters Laboratories. CSA also, by way of example, is heavily involved in the development of the ISO 9000 series of standards on quality and the ISO 14000 series on Environmental Management.
3. Carrier Serving Area. A concept which categorizes local loops by length, gauge and subscriber distribution in order to determine how a specific geographic area can best be served. The concept is critical when LECs evaluate the potential for deployment of services which challenge the capabilities of the embedded voice-grade, twisted-pair cable plant. Such services include xDSL, e.g., ADSL and IDSL.

CSA T527-94 Canadian guidelines for Grounding and Bonding for Telecommunications in Commercial Buildings

CSA T528-92 Canadian Design Guidelines for Administration of Telecommunications Infrastructure in Commercial Buildings.

CSA T529-M91 Canadian Design Guidelines for Telecommunications Wiring Systems in Commercial Buildings

CSA T530-M90 Canadian Building Facilities Design Guidelines for Telecommunications CEC Canadian Electrical Code, Part I - 1994

CSA T-530 Canadian equivalent of EIA-569 standard, harmonized. defined. Also has Rcv Clock, and both Xmit Clocks for synchronous systems.

CSC 1. Customer Service Center
2. Customer Service Consultant
3. Customer Service Coordinator
4. Customer Support Center
5. Customer Support Consultant.

CSCD Circuit Switched Cellular Data. A developing alternative to CDPD (Cellular Digital Packet Data) for transmitting data over analog AMPS (Advanced Mobile Phone System) networks. The problems with CDPD are that it is optimized for short messages (smaller than 1KB), it is expensive (approximately $.10 per KB), and it is not universally available. CSCD is optimized for large files, typical of contemporary e-mail transmissions. It also is billed at the same rate as a voice call. Unlike CDPC, CSCD does not require a TCP/IP interface on both ends of the connection.

CSDC Circuit Switched Digital Capability. AT&T defines it as a technique for making end-to-end digital connections. Customers can place telephone calls normally, then use the same private connection to transmit high-speed data. CSDC is a circuit-switched, 56 Kbps, full-duplex data service that provides high-speed data communications over regular telephone lines.

CSFI IBM Communications Subsystem For Interconnection:

networking software.

CSI 1. Called Subscriber Identification. This is an identifier whose coding format contains a number, usually a phone number from the remote terminal used in fax.

2. Capability Set I. A set of service-independent building blocks for the creation of IN services developed by the European Telecommunications Standards Institute and the ITU-T.

CSID Calling Station ID. When you receive a fax from someone, you'll see on the top of the page the phone number of the fax machine that sent the fax to you. That's called the Calling Station ID. Most people think that that number is the phone number from which they're receiving the fax. In fact, it's not. It's the number you enter yourself into your fax machine when you first set it up. You could happily put in a completely different phone number to the number you're sending from. And no one would be any the wiser.

CSMA Carrier Sense Multiple Access. In local area networking, CSMA is a way of getting onto the LAN. Before starting to transmit, personal computers on the LAN "listen" to make sure no other PC is transmitting. Once the PC figures out that no other PC is transmitting, it sends a packet and then frees the line for other PCs to transmit. With CSMA, though stations do not transmit until the medium is clear, collisions still occur. Two alternative versions (CSMA/CA and CSMA/CD) attempt to reduce both the number of collisions and the severity of their impact. See CSMA/CA and CSMA/CD.

CSMA/CA Carrier Sense Multiple Access (CSMA) with Collision Avoidance. In local area networking, CSMA technique that combines slotted time-division multiplexing (TDM) with carrier sense multiple access/collision detection (CSMA/C) to avoid having collisions occur a second time. CSMA/CA works best if the time allocated is short compared to packet length and if the number of stations (these days PCs) is small. See CARRIER SENSE MULTIPLE ACCESS/COLLISION AVOIDANCE and CSMA.

CSMA/CD Carrier Sense Multiple Access with Collision Detection. In local area networking technology, CSMA technique that also listens while transmitting to detect collisions. CSMA/CD is a leading control technique for getting onto and off a local area network. All devices attached to the network listen for transmissions in progress (i.e. carrier sense) before starting to transmit (multiple access). If two or more begin transmitting at the same time and their transmissions crash into each other, each backs off (collision detection) for a different amount of time (determined by an algorithm) before again attempting to transmit.

If you didn't understand the above definition, try this one: CSMA/CD: Abbreviation for Carrier Sense Multiple Access with Collision Detection, a method of having multiple workstations access a transmission medium (multiple access) by listening until no signals are detected (carrier sense), then transmitting and checking to see if more than one signal is present (collision detection). Each workstation attempts to transmit when they "believe" the network to be free. If there is a collision, each workstation attempts to retransmit after a preset delay, which is different for each workstation. It is one of the most popular access methods for PC-based LANs. Think of it as entering a highway from an access road, except that you can crash and still try again. Or think of it as two polite people who start to talk at the same time. Each politely backs off and waits a random amount of time before starting to speak again. Ethernet-based LANs use CSMA/CD. See Ethernet and IEEE 802.3.

CSMDR Centralized Station Message Detail Recording.

CSMI Call Screening, Monitoring and Intercept.

CSO Central Services Organization; an Internet service that makes it easy to find user names and addresses.

CSP Certified Service Provider. Initially developed for the automotive industry, a CSP is an ISP which has met the mission requirements of performance, reliability, security, and manageability for the big three auto manufacturers and their trading partners (suppliers) to exchange critical transaction and planning documents over the web.

CSPDN Circuit-Switched Public Data Networks.

CSQP Customer/Supplier Quality Process is a program designed to help suppliers improve the quality of their products and services and strengthen customer relationships. I first heard about CSQP from Newbridge Networks, which told me that it had been nominated for the program by several regional Bell operating companies because of the volume of products it was selling to these customers. These RBOCs funded BellCore to work with Newbridge to assist in improving product and service quality to attain CSQP registration. The CSQP program is built upon ISO9000 standards, Malcolm Baldrige National Quality criteria and additional Bellcore criteria. According to Newbridge, CSQP establishes clear, concise requirements for each element of the requirements, including the ISO elements. Evidence of compliance must be given to the CSQP Management Team for all elements or action items opened in order to track the problem area to closure.

CSR 1. Customer Station Rearrangement (as in Centrex).

2. Customer Service Representative. A customer care agent that provides direct customer support.

3. Customer Service Record. Computer printout that details the fixed monthly charges billed by your local telephone company. The CSR is composed of computer codes called USOCs, which in turn correspond to a particular tariffed service. USOCs tell the telephone company's billing system what tariff rate should be billed for a particular service. In order to ensure your telephone bill is correct you must request and review this document. No telecom manager should be without this important document.

4. Cell Switch Router. A technology which is proposed in the form of an IETF submission to fill gaps in ATM (Asynchronous Transfer Mode) standards. The objective of CSR is to provide a standard means of building enterprise backbones or carrier-level infrastructures which support high throughput, multicast, and QoS (Quality of Service) capabilities. CSR offers ATM-level connectivity from edge-to-edge of the network via cut-through paths in support of legacy networks such as Ethernet, bypassing packet-level switch processing at the intermediate and core switches.

CSS Cascading Style Sheets. According to CNET, Cascading style sheets are a big breakthrough in Web design because they allow developers to control the style and layout of multiple Web pages all at once. Before cascading style sheets, changing an element that appeared on many pages required changing it on each individual page. Cascading style sheets work just like a template, allowing Web developers to define a style for an HTML element and then apply it to as many Web pages as they'd like. With CSS, when you want to make a change, you simply change the style, and that element is updated automatically wherever it appears within the site. Both Navigator 4.0 and Internet Explorer 4.0 support cascading style sheets. If you needed any more proof of the problem-solving nature of CSS, the World Wide Web Consortium

(W3C) has recommended cascading style sheets (level 1) as an industry standard. See also: DHTML, HTML

CST Computer Supported Telephony, a term coined by Siemens. Here is an explanation from Dr. Peter Pawlita of Siemens. "More people communicate by telephone than by any other means. The reason is simple: The telephone bridges any distance, saves travel time and can be used spontaneously and is universally available. Unfortunately telephone usage is often associated with annoying delays and frayed nerves resulting from such things as time wasted in finding a number, dialing errors, and the absence of the dialed party. Added to this the person to whom you are speaking does not have the knowledge you require, or has to spend a long time looking or documents. What could be more obvious, therefore, than to turn these problems over to the computer — to implement Computer Supported Telephony (CST). CST denotes the functional connection of a computer system to a PBX at the application level. CST applications can automatically initiate calls, receive incoming calls, and provide "just-in-time" business data, documents and notes on the screen. All this makes telephony more convenient, time-saving, efficient and largely error-free."

CSTA Computer Supported Telephony Application. A standard from the European Computer Manufacturers Association (ECMA) for linking computers to telephone systems. Basic CSTA is a set of API call agreed upon by the ECMA. See also CST and OPEN APPLICATION INTERFACE.

CSTP Customer Specific Term Plan. See CUSTOMER SPECIFIC TERM PLAN.

CSU 1. Channel Service Unit. Also called a Channel Service Unit/Data Service Unit or CSU/DSU because it contains a built-in DSU device. A device to terminate a digital channel on a customer's premises. It performs certain line coding, line-conditioning and equalization functions, and responds to loopback commands sent from the central office. A CSU sits between the digital line coming in from the central office and devices such as channel banks or data communications devices. A Channel Service Unit is found on every digital link and allows the transfer of data at a range greater than 56 Kbps. A 56 Kbps circuit would need a 56 Kbps DSU on both ends to transfer data from one end to the other. A CSU looks like your basic "modem," except it can pass data at rates much greater and does not permit dial-up functions (unless it has an asynch dial-backup feature).

2. Channel Sharing Unit. Line bridging device that allows several inputs to share one output. CSUs exist to handle any input/output combination of sync or asynch terminals, computer ports, or modems and thus these units are variously called modem sharing units, digital bridges, port sharing units, digital sharing devices, modem contention units, multiple access units, control signal activated electronic switches or data-activated electronic switches.

CSU/DSU See CSU.

CSUA Canadian Satellite Users Association. Trade association of satellite users.

CT 1. Call Type.
2. Cordless Telephone.
3. Computer Telephony.

CT Connect A computer telephony call control server software that connects a wide range of telephone switches (PBXs and ACDs) to a variety of data processing environments. By bridging the PBX and IT infrastructure, CTI applications such as screen pops and intelligent call routing are easily implemented in call centers. CT Connect runs on Windows NT and SCO UnixWare and supports standard programming interfaces such as TAPI, TSAPI and DDE. See Computer Telephony.

CT-2 Interim Standard for the Telepoint service favored by UK Memorandum of Understanding with other network operators.

CT1 Cordless Telephony Generation 1. A new type of low-cost public cordless telephone system getting popular in Europe. You carry a cheap handset. You go to within several hundred yards of a local antenna and you make your phone call. You can't receive calls as you can on a cellular radio. You can't make calls unless you're close to the antenna. The service helps overcome the serious lack of street-side coin and public phones in Europe. CT1 is the analog version of the interface specification. See CT2, CT2+, CT3 and DECT.

CT2 Cordless Telephony Generation 2, interface specification for digital technology, currently in use in the U.K. for telepoint (payphone) applications. Think of telepoint phones as cellular phones but using micro-cells. By having smaller cells than normal cellular cells, CT2 phones can be smaller, cheaper and lighter. The first generation of these phones didn't do well, since they weren't smaller and lighter; there weren't many micro-cells and you couldn't receive an incoming call. See CT1.

CT2+ An expansion of the CT2 interface specification that would extend network capabilities and allow backwards compatibility with CT2 handsets. See CT1 and CT2.

CT3 Ericsson's proprietary cordless phone system.

CTA Competitive Telecommunications Association. Trade association of alternate long distance carriers (resellers) in Canada.

CTCA Canadian Telecommunications Consultants Association. Professional organization of telecommunications consultants.

CTD 1. Continuity Tone Detector.
2. An ATM term. Cell Transfer Delay: This is defined as the elapsed time between a cell exit event at the measurement point 1 (e.g., at the source UNI) and the corresponding cell entry event at measurement point 2 (e.g., the destination UNI) for a particular connection. The cell transfer delay between two measurement points is the sum of the total inter-ATM node transmission delay and the total ATM node processing delay.

CTE 1. Connected Telecommunications Equipment. The European term for what the Americans call CPE — Customer Premise Equipment. See CTE Directive.
2. Channel Translation Equipment.
3. Coefficient of Thermal Expansion.

CTE Directive CTE stands Connected Telecommunications Equipment. The European term for what the Americans call CPE — Customer Premise Equipment. The CTE Directive refers to a paper on the proposed European-wide regulation of telecommunications terminals. That paper was published in the summer of 1997 by the European Commission. The proposed title is "European Parliament and Council Directive connected telecommunications equipment and the mutual recognition of the conformity of equipment". The timetable indicated by the EC is for a common position to be agreed by the end of 1997 with formal adoption by the Parliament and Council by mid 1998 and the legislation coming into force one year later ie, July 1999. The new Directive is designed to complement other relevant "horizontal" legislation such as that on electrical safety, EMC (electro-Magnetic Compatibility) and ONP (Open Network Provision);
Conformity assessment will be based upon the principle of manufacturers' declarations and the principle that products

reaching the market which do not conform to the applicable essential requirements will be considered to be defective, with the possibility of heavy penalties — equipment using radio comms techniques is included;

CTRs and ACTE disappear with the repeal of Directive 91/263 but a new Telecommunications Conformity Assessment and Market surveillance committee (TCAM) will advise the Commission and Notified Bodies still have a role. CTRs remain applicable until replaced;

Operators of all networks will be required to publish, and regularly update, accurate and adequate technical technical specifications of the available network termination points and the terminal types supported.

Flexibility is achieved by means whereby the essential requirements applicable to new network termination types can be determined in a timely manner. The essential requirements are restricted to:

(a) Prevention of misuse of public network resources causing a degradation of service to third parties.

(b) Interworking via the public network(s) and Community-wide portability between ONTPs specifying a basic level of interworking, e.g. simple voice telephony but excluding supplementary services.

(c) Effective use of spectrum allocated to terrestrial/space radio communication and used for radio services recognizing that trades-off will be necessary between the quality, capacity, and availability.

For each type of Connected Terminal Equipment (CTE) formerly defined as Telecommunications Terminal Equipment, the essential requirements applicable are to be selected from a master list contained in the Directive. The technical requirements will be defined in appropriate technical specifications. These will be harmonized European standards or, in cases where such standards do not yet exist, other appropriate technical specifications. The specifications of essential requirements will take into account the following additional requirements for the common good:

(a) Protection of health, e.g. minimizing the health hazards of radio frequency radiation.

(b) Features for users with disabilities.

(c) Features for emergency and security services.

(d) Protection of individual privacy.

CTI Computer Telephone Integration. A term for connecting a computer (single workstation or file server on a local area network) to a telephone switch (a PBX or an ACD) and have the computer issue the telephone switch commands to move calls around. The classic application for CTI is in call centers. Picture this: A call comes in. That call carries some form of caller ID — either ANI or Caller ID. The switch "hears" the calling number, strips it off, sends it to the computer. The computer then does a lookup for the numbers in a database, sends the switch back instructions on what to do with the call. The switch follows orders. It might send the call to a specialized agent or maybe just to the agent the caller dealt with last time. Meantime, the agent sees a screen pop of information about the caller — such information having been pulled up out of the server's database, using the caller ID information. CTI and CT (computer telephony) are often confused. In fact, CTI is the older and smaller term. CTI has been the dismal part of computer telephony — the difficult integration of reluctant, closed phone systems with outside computers they were never meant to talk. Computer telephony (or CT) is more exciting because it's building new phone systems with fantastic new features based on open standards, open hardware and open software. CTI covers integration with switches. CT covers that AND a lot more — like callback, the UnPBX (communications server), the central office in a PC, IP telephony, one number find me, predictive dialing, unified messaging, interactive voice response, fax blasting and serving, etc. See also Computer Telephony, TAPI, TAPI 3.0, TSAPI and Windows Telephony.

CTIA 1. Cellular Telecommunications Industry Association. Washington, D.C.-based industry association. The Washington, D.C.-based trade association representing the interests of the wireless telecommunications industry.
2. Computer Technology Industry Association, a Lombard, Illinois trade association.

CTIP Computer Telephony Interface Products. Adapters that allow telephones to work with computers. An example is the Konexx connector, which fits between the handset and the phone, allows a connection to a PC modem or fax machine. This definition contributed by Larry Kettler of San Diego.

Ctrl Control. The label on the control key on your computer.

CTS 1. Clear To Send. Pin 5 on the 25-conductor RS-232-C interface or an RS-232-C signal used in the exchange of data between the computer and a serial device. In short, Clear to send is one of the nine wires in a serial port used in modem communications, CTS carries a signal from the modem to the computer saying, "I'm ready to start when you are."
2. Communication Transport System. CTS is The Siemon Company's proprietary structured wiring system. It consists of the methodology and the connecting hardware products to plan, design, and implement the communications wiring infrastructure for commercial buildings (for more information see the company's CTS Design Workbook and CTS Training Videotape). The Siemon Company is based in Watertown, CT.
3. Conformance Testing Services.

CTTS Coax To The Curb. An approach that provisions a multiline remote terminal to deliver voice and data to concentrated residential applications.

CTTU Centralized Trunk Test Unit. An operational support system providing centralized trunk maintenance through a data link on a switch.

CTX Centrex.

CUA Common User Access. The policy of using the same command for a given function in all software. This makes the software easier to learn and use because you only have to learn one set of commands. Windows has a set of CUA guidelines which many Windows programs follow. For example, Alt+F4 always means close this window.

Cube Farm An office filled with cubicles. See also Prairie Dogging.

Cuckoo-Clock Telecom slang for some 6- and 10-button models of AT&T 1A2 wall phones shaped vaguely like traditional cuckoo-clocks. These were probably the first multi-line phones to come with handsets that plugged into the base with Trimline-style 5-pin plugs, before the current modular connectors were adopted. Often seen in hospitals on TV shows.

CUG Closed User Group. Selected collection of terminal users that do not accept calls from sources not in their group and also often restricted from sending messages outside the group.

Curie Point The temperature at which certain elements (usually so-called "rare earth" elements) relax their resistance to magnetic changes. In a magneto-optic disk drive the surface to be marked is heated briefly by a laser light to its Curie point. Magnetism is then applied in the proper polarity to make the spot a "1" or a "0." It cools, and is locked in that position, until it re-heated and changed again. This is how magneto-optic drives can be erasable.

Current A measure of how much electricity passes a point on a wire in a given time frame. Current is measured in amperes, or amps. The abbreviation for current is I. See OHM'S LAW.

Current Carrying Capacity The maximum current an insulated conductor can safely carry without exceeding its insulation and jacket temperature limitations.

Current Limit The function of a circuit or system that maintains a current within its prescribed limits. A circuit breaker terminates current flow when current exceeds the trip limit. Most UPS systems have an electrical subcycle current limit that regulates the output current to a value within the UPS design limits. This subcycle current limit may activate when a load demanding high inrush current (like a computer or phone system) is switched on. The activation of the subcycle current limit protects the UPS from damage but allows the output voltage to become distorted or even collapse momentarily. Most on-line UPS systems will have the subcycle current limit activated by computer load switching and use a bypass in order to maintain load continuity when the current limit activates. Standby and line-interactive UPS systems can draw on the utility grid directly to supply load switching current transients and therefore do not activate the subcycle current limit or need to use the automatic bypass feature. This definition from APC.

Current Loop Transmission technique that recognizes current flows, rather than voltage levels. It has traditionally been used in teletypewriter networks incorporating batteries as the transmission power source. In this serial transmission system, a pair of wires connecting the receiving and sending devices transmit binary 0 (zero) when no current flows and a binary 1 (one) when current is flowing.

Cursor A symbol on a screen indicating where the next character may be typed. Cursors may be solid, blinking, underlines, etc. Many programs, computers and phone systems allow you to reprogram the cursor to what you like. One author of this dictionary, Harry Newton, likes a non-blinking solid block, which came standard with his original CP/M version of WordStar, but doesn't any longer.

Cursor Submarining A liquid crystal display on a computer laptop screen doesn't write to screen very fast. When you move a cursor across your screen or move your mouse quickly across the screen, the cursor disappears. This phenomenon is known as cursor submarining. Cute.

Curves And Arcs A computer imaging term. Paint packages handle curves and arcs in a variety of ways. Examples include spline curves, where-in you specify a series of points and the package draws a curve that smoothly approaches those points, and "three point" curves, in which the first two points anchor the ends of the curve and third selects the apex.

CUSEEME An Internet videoconferencing system that enables up to eight users to see and hear each other on their computer screens. Pronounced "See You, See me."

Custom Calling A group of special services available from the central office switching system which the telco can offer its subscribers without the need for any special terminal equipment on their premises. Basic custom calling features now available include call waiting, 3-way calling, abbreviated dialing (speed calling), call forwarding, series completing (busy or no answer) and wake up or reminder service.

Custom Controls Controls are software objects that you embed in a Visual Basic or other Windows development tool. In the old days you would compile your DOS program with a

"library" of some precompiled subprograms and functions. Controls take the idea a step further and give you tremendous power, all within the Windows Graphical User Interface (GUI). The original Visual Basic "custom controls" were programmed by third parties and behave identically to controls shipped with Visual Basic:

• They appear in the Visual Basic toolbox.
• You control their behavior from your software.
• They generate events that your program can respond to.
• And they have properties that your program can change.

There are hundreds of controls out there for Visual Basic for database management, multimedia presentations, imaging, host connectivity, etc. The ones that concern us do computer telephony stuff (though anything can be leveraged, like host connectivity for IVR, etc.):

Custom ISDN A version of ISDN BRI (Basic Rate Interface) provided off an AT&T 5ESS central office. It actually offers more features and is easier to install than a National ISDN-1 BRI line. We are all awaiting the specifications on National ISDN-2, which is meant to be "standard." Meantime, Custom ISDN is the most popular, most versatile and most understood ISDN service in North America. See ISDN.

Custom Local Area Signaling Services CLASS. A generic term (like WATS) describing several enhanced local service offerings such as incoming-call identification, call trace, call blocking, automatic return of the most recent incoming call, call redial, and selective forwarding and programming to permit distinctive ringing for incoming calls. See CLASS.

Customer Access Line Charge CALC. Also known variously as Access Charge, EUCL (End User Line Charge), and SLC (Subscriber Line Charge). See Access Charge.

Customer Care Center A term created by Alex Szlam, the president of Melita International, Norcross, GA to describe a telephone call center with three basic elements: First, the database technology and the marketing savvy to fill that database with individual customer preference information. Second, the ability to intelligently handle inbound phone calls. Third, the ability to intelligently make outbound calls. See also CUSTOMER SENSITIVITY KNOWLEDGE BASE.

Customer Contact Zone A term invented by Keith Dawson, editor of Call Center Magazine. It refers to all the information a customer requires which is delivered through multiple media, including manned call centers, interactive voice response machines, fax back devices, etc.

Customer Control An AT&T term for the ability for an end user to monitor, choose, modify, redesign and/or program the type of service received from a network.

Customer Information Manager CIM. An MCI definition. A component of the NCS which supports the creation and maintenance of customer databases for Vnet customers. Customers have remote access to and control over their portion of the NCS database via a terminal at the customer's location.

Customer Interaction Software Software that handles your entire

Customer Network Management CNM. An ATM term. CNM allows users of ATM public networks to monitor and manage their portion of the carrier's circuits. Thus far, the ATM forum has agreed that the CNM interface will give users the ability to monitor physical ports, virtual paths, usage parameters, and quality of service parameters.

Customer Premises Equipment CPE. Terminal equipment — telephones, key systems, PBXs, modems, video conferencing devices, etc. — connected to the telephone network and residing on the customer's premises. What North America calls CPE, Europe calls CTE — for Connected Telecommunications Equipment.

Customer Proprietary Network Information CPNI. Information which is available to a telephone company by virtue of the telephone company's basic service customer relationship. This information may include the quantity, location, type and amount of use of local telephone service subscribed to, and information contained on telephone company bills. This is the definition of CPNI that the independent voice mail and live telephone answering industry uses.

Customer Provided Loop The customer assumes responsibility for ordering, coordinating, maintaining,and billing for the local loop.

Customer Provided Terminal Equipment Or just Customer Provided Equipment (CPE). Terminal equipment connected to the telephone network which is owned by the user or leased from a supplier other than the local telephone operating company.

Customer Sensitivity Knowledge Base A term created by Alex Szlam of Melita International, Norcross, GA to describe a complex database that would keep track of your customers' preferences. Such database would be updated almost automatically based on every contact you had with the customer. The database would probably be object-oriented since the idea is define customer preferences based on individual preferences, not on a statistical analysis of conglomerate preferences such as those typically gleaned from existing character databases.

Customer Service Center CSC. MCI organization responsible for installing, verifying, and maintaining MCI customers and customer service.

Customer Service Record CSR. Computer printout that details the fixed monthly charges billed by your local telephone company. The CSR is composed of computer codes called USOCs, which in turn correspond to a particular tariffed service. USOCs tell the telephone company's billing system what tariff rate should be billed for a particular service. In order to ensure your telephone bill is correct you must request and review this document. No telecom manager should be without this important document.

Customer Service Unit CSU. A device that provides an accessing arrangement at a user location to either switched or point-to-point, digital circuits. A CSU provides local loop equalization, transient protection, isolation, and central office loop-back testing capability. See also CSU/DSU.

Customer Specific Term Plan A Customer Specific Term Plan is an option offered by AT&T on the purchase of its 800 services whereby customers can earn additional discounts by committing to a multi year contract. This also is one of two plans used by aggregators to resell 800 services. The other is the Revenue Volume Pricing Plan.

Cut To transfer a service from one facility to another.

Cut Back Technique A technique for measuring optical fiber attenuation or distortion by performing two transmission measurements. One is at the output end of the full length of the fiber. The other is within 1 to 3 meters of the input end. Without disturbing the source-to-fiber coupling, access to the short length output is accomplished by "cutting back" the test fiber.

Cut Down A method of securing a wire to a wiring terminal. The insulated wire is placed in the terminal groove and pushed down with a special tool. As the wire is seated, the terminal cuts through the insulation to make an electrical con-

nection, and the tool's spring-loaded blade trims the wire flush with the terminal. Also called punch down.

Cut Through 1. Cut-through, in voice processing, is what stops voice prompt playback when a touchtone key is pressed. Some of the speech recognition solutions also add cut-through that will stop voice prompt playback as soon as you start talking. Only voice cards that support continuous speech recognition are able to provide cut-through. Cut-though can be a problem in some cases. Imagine yourself at the airport trying to make a call using a speech recognition system. At the start of a new prompt, the airport public address system blares out a last boarding call for a flight. If cut-through is active, it would stop playing the prompt and wait on your response. Now what do you do?
2. The act of connecting one circuit to another, or a phone to a circuit. This is when a user dials the access code for the circuit and is immediately "cut through" to the tie line. The user controls the call. It is a tie line operation.
3. See also CUT THROUGH SWITCH.

Cut Through Resistance A measure of an insulation's ability to withstand penetration by sharp edges.

Cut Through Switch A type of switch algorithm in which the destination address of a packet is read and the packet immediately forwarded to the switch port where the destination MAC address device is attached.

Cutoff Attenuator A waveguide of adjustable length that varies the attenuation of signals passing through the waveguide.

Cutoff Frequency 1. The frequency above which, or below which, the output current in a circuit, such as a line or a filter, is reduced to a specified level.
2. The frequency below which a radio wave fails to penetrate a layer of the ionosphere at the angle of incidence required for transmission between two specified points by reflection from the layer.

Cutoff Mode The highest order mode that will propagate in a given waveguide at a given frequency.

Cutoff Wavelength In fiber optic systems, the cutoff wavelength is the shortest wavelength at which only the fundamental node on optical waveguide is capable of propagation. For single mode fibers, the cutoff wavelength must be smaller than the wavelength of the light to be transmitted.

Cutover The physical changing of lines from one phone system to another, or the installation of a new system. It's usually done over the weekend, accompanied by heavy praying that everything will go right. There are two types of cutovers — flash cuts and parallel cuts. Parallel cuts occur when the old phone system is left functioning and the new one, central switching equipment and phones, is installed around it. This means that for some weeks there are two sets of phones, two sets of wires, two switches, two sets of phone lines, etc. The parallel cut is a far more reliable method of cutting over a new switch. But it's also more expensive.

A "flash cut" occurs in a flash. On Friday, everyone is using the old switch. When everyone comes to work on Monday, the old switch and its phones have disappeared. In its place, there's a brand new system. Sometimes it works. More often than not, there are remaining nagging problems. With any Cutover, it's a good idea to set up a Complaint or Cutover Number. Thus, if anyone's having trouble with their phone, they can call this number and get their problems taken care of. How well these problems are taken care of will determine how well the cutover went and how well the employees perceive the new switch is working. Perception, not reality, is what's at stake here.

CV 1. Old Bell-Speak for single-line phone. It stands for Combined Voice. In old Bell-Speak it meant that the two parts of the phone that dealt with voices were combined into one unit (the handset). Before this, there were phones like the HH (Hand-Held) where there was a piece you spoke into and another piece you put to your ear. From CV, you get CVW (CV Wall phone) and later on, KV (Key Voice) and KVW (Key Wall phone). Later on all this crept into the USOC codes — the Universal Service Order Code numbering systems the local Bell operating phone companies used to identify products and services. See USOC.
2. Checksum Value
3. Code Violation. A violation in the coding of a signal over a digital circuit. See CODE VIOLATION.

CVD Chemical Vapor Deposition.

CVF Compressed Volume File. A Microsoft term which refers to a file on a compressed disk. The term was first introduced in MS-DOS 6.0, which first had double-your-disk-space technology. That technology was later removed when Stac Electronics, originator of Stacker disk doubling technology, took Microsoft to court and won.

CVP 1. Certified Vertical Partner.
2. A British term: Co-operative Voice Processing, gives the caller the ability to move seamlessly between an Interactive Voice Processing device and a live agent.

CVSD Continuously Variable Slope Delta modulation. A method for coding analog voice signals into digital signals that uses 16,000 to 64,000 bps bandwidth, depending on the sampling rate.

CW 1. Call Waiting (as in Custom Calling Service).
2. Continuous Wave.

CWA Communications Workers of America. A national union of telephone industry employees, currently very worried about its future membership growth given the phone industry's propensity to let surplus workers go. www.cwa-union.org

CX Signaling A direct current (DC) signaling system that separates the signal from the voice band by filters. Also called Composite Signaling.

CXR Carrier.

Cyber Five letters which can, seemingly, be attached to a word and made into a noun or a verb. The first Cyber word was Cyberspace, a term coined by science fiction writer William Gibson in his 1984 fantasy novel "Neuromancer" to describe the "world" of connected computers and the society that gathers around them. The idea of Cyberspace is that this world of computer networks can be explored with the proper addresses and codes. People who use the system for hours on end are said to be lost in cyberspace. Today, many people say that world has arrived in the form of the Internet. And, so with projections that there will be 100 million users of the Internet by the year 2000, the word Cyber has become popular. There's "The Cyberbrary of Congress" (books Congress has on on-line). According to William Safire writing in the New York Times Magazine of December 11, 1994, "cyber is the hot combining form of our time. If you don't have cyberphobia, you are a cyberpiliac." The US News & World Report labels its election night on-line forum a cybercast. The Washington Post wrote that "battlefield valor belongs not to the brawny soldier but to the astrophysics major who invented smart bombs," somebody who's called a cyberwonk. See all the following definitions which begin with CYBER.

Cyberbusiness A company that does most of its business on the Internet is called a Cyberbusiness.

Cybercad An Internet term. The electronic equivalent of a lounge lizard.

Cybercafe Establishment with both coffee and Internet access. Trendy in some place, unknown in others. Often used as a retail store to sign up customers to Internet service by a local ISP.

Cybercash Cybercash An electronic payment system integrated into E-Commerce (Electronic Commerce) servers, which typically make use of the Internet. Also called digital cash, the term "Cybercash" was coined by CyberCash Inc. to describe its systems for verification of credit cards and processing of payments. See also E-Commerce.

Cybernoir An Internet term. Used to describe dark, trippy, weird "cyber" films and shows like "Wild Palms," "Tank Girl," and "VR.5."

Cyberia Electronic stuff for the cyberpeople. An advertisement in the November 12, 1995 issue of the New York Times (Sunday) Magazine showed Cyberia covering everything from modern chairs to laptop computers, to cellular phones to an Apple Newton PDA.

Cybermall A Web site designed for online shopping, shared by two or more commercial organizations.

Cybernetics A term invented in 1948 by Norbert Wiener, the automation genius, who declared "We have decided to call the entire field of control and communications theory, whether in the machine or the animal, by the same term Cybernetics." From the Greek "kybernetes," meaning "pilot" or "governor." The science of communication and control theory which is concerned most especially with the comparative study of automatic control systems. Examples include the brain and nervous system, and mechanical/electrical/electronic communication systems.

Cyberpork Government money flowing to well-connected information superhighway contractors.

Cyberpunk A work coined by a book called "Cyberpunk: Outlaws and Hackers on the Computer Frontier" by Katie Hafner and John Markoff. The book defines Cyberpunk as what you and I know as a computer hacker — a person who manages to get into other people's computer systems. He does this usually through telephone lines. In most cases, hackers see themselves as harmless electronic joyriders. But they occasionally steal data, inject viruses and misleading information and disrupt legitimate business and research. Sometimes they get caught.

Cybersex Adult-oriented computer games, images and chat lines. A place where people can discuss their sex lives and wanton desires with total strangers in online (over phone line) forums, even falling in love without having ever met face to face.

Cyberskating Browsing the Internet. See CYBERSPACE.

Cyberspace A term coined by science fiction writer William Gibson in his fantasy novel Neuromancer to describe the 'world" of connected computers and the society that gathers around them. The idea of Cyberspace is that this world of computer networks can be explored with the proper addresses and codes. People who use the system for hours on end are said to be lost in cyberspace. Today, many people say that world has arrived in the form of Internet. John Perry Barlow, a rock-'n'-roll lyricist turned computer activist, defined cyberspace in Time magazine as "that place you are in when you are talking on the phone." Thus by Barlow's definition, just about everybody has already been to cyberspace. I prefer Gibson's definition.

Cybertechnology A term I first saw in the mid-December, 1997 injunction from Judge Thomas Penfield Jackson of the

U.S. District Court in Washington. Judge Jackson ruled that Microsoft could not force PC makers to load a Windows operating system bundled with Microsoft's browser, Internet Explorer. In making the ruling, Judge Jackson appointed a "special master," a Harvard Law School Professor

Cybored State one quickly gets in while waiting for the screen to change on busy (or just plain slow) sites.

Cyborg A contraction of CYBERnetics and ORGanism. A human being who is linked to one or more devices on which he is dependent for survival in a hostile environment. See CYBERNETICS.

Cybrarian A person who makes a living doing online research and information retrieval (comes from cyberspace librarian). According to one definition I read, a cybrarian is a futurist librarian who swims in the electronic ocean or cyberspace. The term is alleged to have been coined by Michel Bauwens of BP Nutrition. A cybrarian is also known as a data surfer or a super searcher.

Cycle One complete sequence of an event or activity. Often refers to electrical phenomena. One electrical cycle is a complete sine wave. (A complete set of one positive and one negative alternation of current.) In the battery business, a cycle is the process of one complete battery discharge and recharge. See CYCLE LIFE.

Cycle Life In the battery business, cycle life is the useful life of a rechargeable battery, expressed as the total number of discharges and recharges.

Cycle Manager Extraction An MCI system which selects processable calls from Distribution and forwards them to the appropriate MCI Reference System for billing.

Cycle Master Part of the bus management scheme used in the IEEE 1394 connection technology. The cycle master broadcasts cycle start packets, which are required for isochronous operation. An isochronous resource manager, for DV and DA applications, is also included for those nodes that support isochronous operation. Also included is an optional bus master.

Cycle Pools Where dial-up call records are stored in MCI's Revenue System until extracted for billing.

Cycle Slip A discontinuity in the measured carrier beat phase resulting from a temporary loss-of-lock in the carrier tracking loop of a Global Positioning System receiver.

Cycle Time The time to complete a cycle. In microcomputers, it's the time between successive RAM read or write operations.

Cyclic Distortion In telegraphy, distortion that is neither characteristic, bias, nor fortuitous, and which in general has a periodic character. Its causes are, for example, irregularities in the duration of contact time of the brushes of a transmitter distributor or interference by distributing alternating currents.

Cyclic Redundancy Check CRC. A check performed on data to see if an error has occurred in the transmitting, reading or writing of the data. A CRC is performed by reading the data, calculating the CRC character and comparing its value to the CRC character already present in the data. If they are equal, the new data is presumed to be the same as the old data. Otherwise, it's wrong. Re-send the data. A CRC character is figured by treating a block of data as a string of bits that are equal to a binary number. Then we divide that number by another predetermined binary number and append the remainder from the division to the data. We call that appended number the CRC character.

Cylinder A hard disk drive contains a number of platters, which are divided into tracks. A cylinder is a collection of all corresponding tracks on all sides of the platters in a disk drive. Think of a hard disk consisting of dozens of concentric cylinders, each of slightly different diameters. These distinct concentric storage areas on the hard disk roughly correspond to the tracks on a floppy diskette. Generally, the more cylinders a hard disk has, the greater its storage capacity.

D Bank Also called Channel Bank. It breaks down a T-1 circuit to its 24 channels.

D Channel In an ISDN interface, the "D" channel (the Data channel) is used to carry control signals and customer call data in a packet switched mode. In the BRI (Basic Rate Interface, i.e. the lowest ISDN service) the "D" channel run at 16,000 bits per second, part of which will carry setup, teardown, ANI and other characteristics of the call. 9,600 bps will be free for a separate "conversation" by the user. That "conversation" will typically be data. And many phone companies are now selling it as an "on the Internet all the time" channel, allowing you to receive and send email continuously. In the PRI (Primary Rate Interface, i.e. ISDN equivalent of T-1), the "D" channel runs at 64,000 bits per second. The D channel provides the signaling information for each of the 23 voice channels (referred to as "B channels"). The actual data which travels on the D channel is much like that of a common serial port. Bytes are loaded from the network and shifted out to the customer site in a serial bit stream. The customer site of course responds with its serial bit stream, too. An example of a data packet sent from the network to indicate a new call has the following components:

- Customer Site ID
- Type of Channel Required (Usually a B channel)
- Call Handle (Not unlike a file handle)
- ANI and DNIS information
- Channel Number Requested
- A Request for a Response

This packet is responded to by the customer site with a format similar to:

- Network ID
- Channel Type is OK
- Call Handle

The packets change as the state of the call changes, and finally ends with one side or the other sending a disconnect notice. The important concept here is the fact the information on the D channel could actually be anything — any kind of serial data. It could just as well be sports scores! So with that in mind, consider the Channel Number Requested packet above. This is the networks' selected channel for the customer site to use. Normally, this number is between 1 and 23, but could be a higher number if needed. This is what NFAS is all about. NFAS (Non Facility Associated signaling, pronounced N-FAST without the T) allows a D channel to carry call information regarding channels which may not even exist in the same PBX or PC system. See also ISDN and DS-0.

D Conditioning A type of line conditioning which controls harmonic distortion and signal-to-noise ratio so that they lie within specified limits. See also Conditioning.

D Link Diagonal Link. A SS7 signaling link used to connect STP pairs that perform work at different functional levels. These links are arranged in sets of four (called quads). Connected mated pairs of Signal Transfer Points (STPs) at different hierarchical levels. For example, from a local pair of STPs to a regional pair of STPs. Like the B-Links, D-links are assigned in a quad arrangement. Once again, the recommendation is for three way path diversity. See A, B and C Links.

D Region That portion of the ionosphere existing approximately 50 to 90 kilometers above the surface of the Earth. Attenuation of radio waves, caused by ionospheric free-electron density generated by cosmic rays from the sun, is pronounced during daylight hours.

D Type The standard connector used for RS-232-C, RS-423 and RS-422 communications. D-type connectors are typically seen in nine, 15 and 25 pin configurations.

D-AMPS Digital Advanced Mobile Phone Service. A term for digital cellular radio in North America. In Europe, it's called GSM and in Japan PDC.

D-Bank Another name for channel bank. A device that multiplexes groups of 24 channels of digitized voice input at 64 Kbps into T-1 aggregate outputs of 1,544 million bits per second.

D-Bit Also called DBIT. The delivery confirmation bit in an X.25 packet used to indicate whether or not the DTE wishes to receive an end-to-end acknowledgment of delivery. In short, a bit in the X.25 packet header that assures data integrity between the TPAD and the HPAD.

D-Inside Wire Direct-Inside Wire. Made of 24-gauge, annealed-copper conductors with color-added PVC, which allows it to be pulled in conduit without the aid of lubricants. Generally used in the horizontal subsystem.

D-Marc see Demarc.

D/A Digital to Analog conversion.

D/A Converter Digital to Analog converter. A device which converts digital pulses, i.e. data, into analog signals so that the signal can be used by analog device such as amplifier, speaker, phone, or meter.

D/I Drop and Insert. See BIT STUFFING and BIT ROBBING.

D1, D1D, D2, D3, D4 and D5 T-1 framing formats developed for channel banks. All formats contain a framing bit in every 193rd bit position. The Superframe (introduced in D2 channel banks) is made up of 12 193-bit frames, with the 193rd bit sequence being repeated every 12 frames. D2 framing also introduced robbed bit signaling, where the eighth bit in frames 6 and 12 were "robbed" for signaling information (like dial pulses). D1D was introduced after D2 to allow backwards compatibility of Superframe concepts to D1 banks.

D2-MAC One of two European formats for analog HDTV.

D3 Format 24 data channels on one standard (North American standard) T-1/D3 span line. Each data channel is 8-bits wide and has a bandwidth of 8KHz. See also DS-1.

D3/D4 Refers to compliance with AT&T TR (Technical Reference) 62411 definitions for coding, supervision and alarm support. D3/D4 compatibility ensures support of digital PBXs, M24 services, Megacom services and Mode 3 D3/D4 channel banks at a DS-1 level.

D4 In T-1 digital transmission technology, D4 is the fourth-generation channel bank. A channel bank is the interface between the T-1 carrier system and a analog premises device such as an analog PBX (private branch exchange).

D4 Channelization Refers to compliance with AT&T TR (Technical Reference) 62411 in regards to the DSI frame layout (the sequential assignment of channels and time slot numbers within the DSI).

D4 Framing First read T-1 FRAMING. The most popular framing format in the T-1 environment is D-4 framing. The name stems from the way framing is performed in the D-series of channel banks from AT&T. There are 12 separate 193-bit frames in a super-frame. The D-4 framing bit is used to identify both the channel and the signaling frame. In voice communications, signaling is an important function that is simulated and carried by all the equipment in the transmission path. In D-4 framing, signaling for voice channels is carried "in-band" by every channel, along with the encoded voice. "Robbed-bit-signaling" is a technique used in D-4 channel banks to convey signaling information. With this technique, the eighth bit (least significant bit) of each of the 24 8-bit time slots is "robbed" every sixth frame to convey voice related signaling information (on-hook, off-hook, etc.) for each voice channel. See also EXTENDED SUPER-FRAME FORMAT.

DA 1. Doesn't answer, as in "The phone rang DA."
2. Directory Assistance.
3. Demand Assignment.
4. Discontinued Availability. Meaning a circuit that was once available is now no longer.
5. Destination Address, a field in FDDI, Ethernet and Token Ring packets which identifies the unique MAC (Media Access Control, the lower part of ISO layer two) address of the recipient. A six octet value uniquely identifying an endpoint and which is sent in IEEE LAN frame headers to indicate frame destination.
6. Desk Accessory. Standard desk accessories on the Apple Macintosh include a calculator, alarm clock and the chooser. Desk accessories are available to the user regardless of the application currently in use, networked or non-networked. Desk accessories are installed in the Apple menu and accessed from there.

DAA Data Access Arrangement. A device required before the FCC registration program if a customer was going to hook up CPE (Customer Provided Equipment), usually modems and other data equipment, to the telephone network. Today, equipment is FCC registered (under Part 68) meaning that the device itself is approved for connection to the phone network. DAAs can still be found in old DP (data processing) installations.

DAB 1. Dynamically Allocable Bandwidth.
2. Digital Audio Broadcasting. Radio broadcasting using digital modulation and digital source coding techniques.
3. Digital Audio Broadcast. The international term for DARS (Digital Audio Radio System), which are proposed satellite-delivered audio/radio systems. See DARS.

DAC 1. Digital to Analog Converter. A device which converts digital pulses, i.e. data, into analog signals so that the signal can be used by analog device such as amplifier, speaker, phone, or meter. In the imaging field, a DAC is a chip that converts the binary numbers that represent particular colors to analog red, green and blue signals that a color monitor displays.
2. Dual Attachment Connector. See DUAL ATTACHMENT CONNECTOR.

DACC 1. Digital Access Cross-Connect.
2. Directory Assistance Call Completion.

DACD Digital Automatic Call Distributor. Nynex's name for its central office provided ACD.

DACOMNET A packet-switched network in South Korea.

DACS Digital Access and Cross-connect System. A digital switching device for routing and switching T-1 lines, and DS-0 portions of lines, among multiple T-1 ports. It performs all the functions of a normal "switch," except connections are typically set up in advance of the call, not together with the call, as in most, normal low bandwidth communications systems (e.g. voice- band voice and data). A DACS is in essence a manual T-1 switch.

DACS/CCR Digital Access Crosscontrol System/Customer Controlled Reconfiguration is a feature of AT&T Accunet T1.5 service. DACS/CCR allows Accunet subscribers to redirect T-1 trunk or DS-0 data traffic over the public network from a terminal on their own premises.

Daemon 1. An agent program which continuously operates on a UNIX server and which provides resources to client systems on the network. Daemon is a background process used for handling low-level operating system tasks. In Greek mythology, "Daemon" was a supernatural being acting as an intermediary between the gods and man.
2. Disk And Execution MONitor. A harmless UNIX program that waits in the background and runs when a request is made on the port that it is watching. It normally works out of sight of the user. On the Internet, it is most likely encountered only when e-mail is not delivered to the recipient. You'll receive your original message plus a message from a "mailer daemon."

DAF 1. Destination Address Field.
2. Decrement All Frame. Motorola definition.

Daisy Chain A method of connecting devices in a series, much as one might interweave daisies to make a lovely floral wreath, or so the story goes. Signals are passed through the chain from one device to the next. Jack 1 is connected to jack 2, which is connected to jack 3 and so on. The last jack in the chain is not connected to jack 1. A SCSI adapter, for instance, is a daisy chain, supporting a daisy chain of up to seven devices. Intel's Universal Serial Bus also is a daisy chain. Stackable hubs, switches and other devices are daisy-chained. While this approach yields the lovely advantage of scalability, interconnection of such devices in this manner also yields some less-than-lovely level of performance degradation, as each device in the chain becomes a point of contention and, therefore, a point of potential congestion. See also SCALABLE and STACKABLE.

DAL Dedicated Access Line. A private tie line from you to your long distance or local phone company. The line may be analog or digital, e.g. a T-1 circuit.

DAMA Demand Assigned Multiple Access. A way of sharing a channel's capacity by assigning capacity on demand to an idle channel or an unused time slot.

Damped Wave A wave consisting of a series of oscillations or cycles of current gradually decreasing amplitude.

Damping 1. The decreasing of the amplitude of oscillations caused by resistance in the circuit. 2 The progressive diminution with time of certain quantities characteristic of a phenomenon. 3. The progressive decay with time in the amplitude of the free oscillations in a circuit. 4. More generally, decreasing some dimension of a phenomenon, such as its power.

Dancing Baloney Gratuitous animated GIFs files and other Web special effects that are used to impress people. "This page is kinda dull. Maybe a little dancing baloney will help?" This definition courtesy Wired Magazine.

DAP Directory Access Protocol. The protocol used between a Directory User Agent (DUA) and Directory System Agent (DSA) in an X.500 directory system. See X.500 and LDAP.

DAP Information Distributor DID. An MCI definition. Software which directs the distribution of network information from the Operational Data Integrator to the Data Access Points.

Dark Current The flow of electricity through the diode in a photodiode when no light is present. These are used as light-sensitive switches. When light hits them they turn on.

Dark Fiber Unused fiber through which no light is transmitted or installed fiber optic cable not carrying a signal. Sometimes dark fiber is sold by a carrier without the (usually) accompanying transmission service. It's "dark" because it's sold without light communications transmission. The customer is expected to put his own electronics and signals on the fiber and make it light.

A Dense Wavelength Division Multiplexing (DWDM) term. Dark wavelength refers to a virtual channel in a fiber optic system utilizing DWDM. Each virtual channel is supported through a specific wavelength of light, with many such channels riding over the same fiber. Once the fiber system is deployed and the DWDM equipment is activated, some of the wavelengths may be activated immediately and others may be left dark for future needs. When the need arises, those dark wavelengths are lit up. See also DWDM, Fiber, Optical Fiber and SONET.

DARPA Defense Advanced Research Projects Agency. Formerly called ARPA, it is a US government agency that funded research and experimentation with the ARPANET and later the Internet. The group within DARPA responsible for the ARPANET is ISTO (Information Systems Techniques Office) formerly IPTO (Information Processing Techniques Office). DARPA had sponsored research in the 1960s and the 1970s to a create a computer network that could survive a nuclear detonation. See also DARPA INTERNET, IAB, IETF and Internet.

DARPA Internet World's largest internetwork, linking together thousands of networks around world. Sponsored by U.S. Defense Advanced Research Projects Agency. Now called DARPANET. See next definition.

DARPANET Defense ARPANET. Also known as DARPA Internet. In 1983 the ARPANET was officially split into DARPANET and MILNET. World's largest internetwork, linking together thousands of networks around world. Sponsored by U.S. Defense Advanced Research Projects Agency.

DARS Digital Audio Radio System. Also known as DAB (Digital Audio Broadcasting) outside the U.S. Proposed satellite-delivered audio/radio systems, similar to DBS (Direct Broadcast System) TV systems, which have been enormously successful in competition with CATV. DARS has been debated by the FCC and the ITU-R since the initial application by CD Radio Inc. in 1990. Assuming that the FCC and ITU-R eventually agree on frequency assignments, you may want to make room for one more satellite dish on your rooftop.

DAS Tape A cellular term. The magnetic tape that is used at the MTSO to record traffic statistics and call billing information. This tape is sent to a third-party 'billing-house' where the actual billing of the subscribers is done.

DASD Direct Access Storage Device. Any on-line data storage device. Usually refers to a magnetic disk drive, because optical drives and tape are considered too slow to be direct access devices. Pronounced DAZ-dee. The term is said to have been invented by IBM.

DASS Direct Access Secondary Storage. Same as near-line: storage on pretty-fast storage devices (e.g., rewritable optical) that are less expensive than hard drives but faster than off-line devices.

DASS1 Digital Access Signaling. A British term. The original British Telecom (BT) ISDN signalling developed for both single line and multi-line Integrated Digital Access but used in the BT ISDN pilot service for single line IDA only.

DASS2 Digital Access Signaling System No. 2. A British Term. A message-based signalling system following the ISO-

based model developed by British Telecom to provide multi-line IDA interconnection to the BT network.

DAT Digital Audio Tape used to identify a type of digital tape recorder and player as well as the tape cassette. DAT tape machines record music that is much crisper, and free of the hisses and pops that mar traditional analog recordings. The drawback with DAT tape machines is they require considerable tape to store music digitally. In a DAT machine, the music is recorded by sampling the music 48,000 each second. Each of those samples is represented by a number that is written as a 16-digit string of zeros and ones. There are two such signals, once for each stereo channel, meaning that storing a single second of music requires about 1.5 million bits. On top of that, extra bits are added to allow the system to mathematically correct errors and help the machine automatically find a particular song on the tape. All together, according to Andrew Pollack writing in the New York Times, a single second of music on a digital audio tape requires 2.8 million bits. But compression techniques are cutting down the amount of information required to be recorded.

Data This is AT&T Bell Labs' definition: "A representation of facts, concepts or instructions in a formalized manner, suitable for communication, interpretation or processing." Typically anything other than voice.

Data Abstraction A term in object-oriented programming. An object is sometimes referred to as an instance of an abstract data type or class. Abstract data types are constructed using the built-in data types supported by the underlying programming language, such as integer and date. The common characteristics (both attributes and methods) of a group of similar objects are collected to create a new data type or class. Not only is this a natural way to think about the problem domain, it is a very efficient way to write programs. Instead of individually describing several dozen instances, the programmer describes the class once. Once identified, each instance is complete with the exception of its instance variables. The instance variables are associated with each instance, i.e., each object; methods exist only with the classes. See OBJECT ORIENTED PROGRAMMING.

Data Access Arrangement DAA. Equipment that allows you to attach your data equipment to the nation's phone system. At one stage, DAAs were required by FCC "law." Now, their limited functions are built into directly attached devices, such as terminals, computers, etc.

Data Access Point DAP. MCI computer that holds the number translation and call-routing information for 800 and Vnet services. These computers respond to inquiries from MCI switches on how to handle these calls.

Data Arrangement In the public switched telephone networks, a single item or group of items present at the customer side of the purposes, including all equipment that may affect the characteristics of the interface.

Data Attribute A characteristic of a data element such as length, value, or method of representation.

Data Bank A collection of data in one place. The data is not necessarily logically related, nor is it necessarily consistently maintained. See DATABASE.

Data Base See DATABASE, which is our preferred spelling.

Data Base Administrator DBA. A computer at MCI that maintains the master file of Vnet translation information. The master file is created when a customer begins service and can be changed at anytime through CIM. The updated copies of the database are downloaded each night to the DAPs.

Data Base Lookup A software program which allows

telephone users to find information on someone calling via the LCD window on their phone. This information comes to the user via CLID (Calling Line IDentification) or ANI (Automatic Number Identification). See also CLASS.

Data Broadcasting A method of high speed data distribution for text and graphics which uses the spare capacity in the broadcasting television, cable and satellite transmission systems.

Data Bubble A new organization within BellSouth to provide high-speed digital services. No one seems to know why it's called "Data Bubble," except that someone inside BellSouth clearly thinks the term is cute.

Data Burst Burst transmission.

Data Bus A bus transmits and receives data signals throughout the computer or telephone system. See BUS.

Data Circuit Terminating Equipment See DATA COMMUNICATIONS EQUIPMENT and DCE.

Data Circuit Transparency The capability of a circuit to transmit all data without changing its content or structure.

Data Cleansing And Scrubbing A process of removing redundancies and inconsistencies in operational data.

Data Communications The transfer of data between points. This includes all manual and machine operations necessary for this transfer. In short, the movement of encoded information by means of electrical transmission systems. See DATA COMMUNICATIONS EQUIPMENT.

Data Communications Channel A three-byte, 192 Kbps portion of the SONET signal that contains alarm, surveillance and performance information. It can be used for internally or externally generated messages, or for manufacturer specific messages.

Data Communications Equipment DCE. A definition of an interface standard between computers and printers. A device typically comes configured as a DCE or DTE. See DCE and DTE. Which way it comes determines how you connect it to another device.

Data Compression Reducing the size of a file of data by eliminating unnecessary information, such as blanks and redundant data. The idea of reducing the size is to save money on transmission or to save money on storing the data. The file or program which has been compressed is useless in its compressed form and must be "decompressed," i.e. brought back to normal before use. One method of data compression replaces a string of repeated characters by a character count. Another method uses fewer bits to represent the characters that occur more frequently. See also COMPRESSION.

Here's another definition, courtesy the US Department of Commerce: 1. The process of reducing (a) bandwidth, (b) cost, and (c) time for the generation, transmission, and storage of data by employing techniques designed to remove data redundancy. 2. The use of techniques such as null suppression, bit mapping, and pattern substitution for purposes of reducing the amount of space required for storage of textual files and data records. Some data compaction methods employ fixed tolerance bands, variable tolerance bands, slope-keypoints, sample changes, curve patterns, curve fitting, floating- point coding, variable precision coding, frequency analysis, and probability analysis. Simply squeezing noncompacted data into a smaller space, e.g., by transferring data on punched cards onto magnetic tape, is not considered data compression. See DATA COMPRESSION TABLE.

Data Compression Protocols All current high-speed dial-up modems also support data compression protocols. This means the sending modem will compress the data on

the-fly (as it transmits) and the receiving modem will decompress the data (as it receives it) to its original form. There are two standards for data compression protocols, MNP-5 and ITU-T V.42 bis. Some modems also use proprietary data compression protocols. A modem cannot support data compression without using an error control protocol, although it is possible to have a modem that only supports an error control protocol but not any data compression protocol. A MNP-5 modem requires MNP-4 error control protocol and a V.42 bis modem requires V.42 error control protocol. Note that although V.42 include MNP-4, V.42 bis does not include MNP-5. However, virtually all high-speed modems that support ITU-T V.42 bis also incorporate MNP-5. The maximum compression ratio that a MNP-5 modem can achieve is 2:1. That is to say, a 9,600 bps MNP-5 modem can transfer data up to 19,200 bps. The maximum compression ratio for a V.42 bis modem is 4:1. That is why all those V.32 (9,600 bps) modem manufacturers claim that their modems provide throughput up to 38,400 bps.

Are MNP-5 and V.42 bis useful? Don't be fooled by the claim. It is extremely rare, if ever, that you will be able to transfer files at 38,400 or 57,600 bps. In fact, V.42 bis and MNP-5 are not very useful when you are downloading files from online services. Why? How well the modem compression works depends on what kind of files are being transferred. In general, you will be able to achieve twice the speed for transferring a standard text file (like the one you are reading right now). V.42 bis and MNP-5 modem cannot compress a file which is already compressed by software. In the case of MNP-5, it will even try to compress a precompressed file and actually expand it, thus slow down the file transfer! The above information courtesy modem expert, Patrick Chen.

Data Compression Table A term from US Robotics, makers of fine modems. A data compression table is a table of values assigned for each character during a call under data compression. Default values in the table are continually altered and built during each call. The longer the table, the more efficient the throughput gained.

Data Compressors Also called compactors. These devices take over where high speed modems and statistical multiplexers leave off. They save phone lines by a doubling of data throughput by further compressing async or sync data streams.

Data Concentrator A device which permits the use of a transmission media by a number of data sources greater than the number of channels currently available for transmission.

Data Connections An ATM term. Data VCCs connect the LECs to each other and to the Broadcast and Unknown Server. These carry Ethernet/IEEE 802.3 or IEEE 802.5 data frames as well as flush messages.

Data Contamination Data corruption.

Data Control Block A data block usually at the beginning of a file containing descriptive information about the file.

Data Conversion Converting data from one format to another. Conversion typically falls into three basic categories. 1. To convert to a form usable by the equipment you have, e.g. you convert some data from tape to disk (because you don't have a tape drive). Or you may convert from one method of encoding data to another, say from EBCDIC to ASCII, because you don't have software which can understand IBM's EBCDIC method of coding. 3. Or you may convert from one format to another, e.g. from the dBASE method of encoding databases to the Paradox method, or from WordStar to WordPerfect. There are many service bureaus whose job is to convert computer data from one form to another and there are now many programs out to do the conversion. Our favorite programs are Word-for-Word for converting word processing formats and Data Junction for converting database formats. See also DATA COMPRESSION.

Data Country Code A 3-digit numerical country identifier that is part of the 14-digit network terminal number plan. This prescribed numerical designation further constitutes a segment of the overall 14-digit X.121 numbering plan for a ITU-T X.25 network.

Data Dialtone Networking as widespread as the telephone. The Internet is widely thought to contain the beginnings of Data Dialtone. See INTERNET, METCALFE's LAW, and WORLD WIDE WEB.

Data Dictionary 1. A part of a database management system that provides a centralized meaning, relationship to other data, origin, usage, and format.
2. An inventory that describes, defines, and lists all of the data elements that are stored in a database.

Data Diddling Unauthorized altering of data before, during or after it is input into a computer system.

Data Directory An inventory that specifies the source, location, ownership, usage, and destination of all of the data elements that are stored in a database.

Data Element A basic unit of information having a unique meaning and subcategories (data items) of distinct units or values. Examples of data elements are military personnel grade, sex, race, geographic location, and military unit.

Data Encryption Standard DES. A 56-bit, private key, symmetric cryptographic algorithm for the protection of unclassified computer data issued as Federal Information Processing Standard Publication. DES, which was developed by IBM in 1977, was promulgated by the National Institute of Standards and Technology (NIST) — formerly the National Bureau of Standards (NBS) — for public and Government use. It was thought to be uncrackable until 1997, when a nationwide network of computer users broke a DES key in 140 days. Triple DES, a later version, encodes the data three times for additional security. As the cost of computer equipment has dropped and as computer power has increased, DES no longer is considered to be totally secure. It has been said that anyone who can afford a BMW can afford a DEScracker. As a result, NIST is searching for a replacement encryption standard to be known as AES (Advanced Encryption Standard). See also Encryption and AES.

Data Entry Using an I/O device (input/output device), such as a keyboard on a terminal, to enter data into a computer.

Data File A database typically contains multiple files of information. Each file contains multiple records. Each record is made up of one or more fields. Each field contains one or more bytes of data. The terms file, record and field find their roots in manual office filing systems.

Data Fill One name for the specifications your ISDN phone lines. Ask for your ISDN Data Fill. It will give you useful information, such as how your lines are set up — voice, data, data/voice, etc.

Data Frame An SCSA term. A set of time slots which are grouped together for synchronization purposes. The number of time slots in each frame depends on the SCbus or SCxbus Data Bus data rate. Each frame has a fixed period of 125us. Frames are delineated by the timing signal FSYNC.

Data Freight The long-haul transport of bits in bulk. Any corporation with cross-country rights-of-way can get into the datafreight biz by laying fat fiber. Qwest is in the datafreight business.

Data Grade Circuit A circuit which is suitable for transmitting data. High speed data needs better quality phone lines than normal dial-up phone circuits. You can acquire such circuits from many telephone companies. To upgrade voice phone lines to high-speed data circuits, you must sometimes "condition" the phone line. See CONDITIONING.

Data Group A number of data lines providing access to the same resource.

Data Hunt Group An AT&T Merlin term. A group of analog or digital data stations that share a common extension number. Calls are connected in a round-robin fashion to the first available data station in the group.

Data Integrity The data you receive is exactly what was sent you. Typically the concept of data integrity relates to data transmission. Data integrity is also a performance measure based on the rate of undetected errors. A measure of how consistent and accurate computer data is. In data transmission, error correcting protocols, such a LAP-M, MNP and X-modem — provide methods of ensuring that data arrives at the destination in its full integrity. This is done in many ways, including retransmitting messed-up blocks.

Data integrity can be threatened by hardware problems, power failures and disk crashes, but most often by application software. In a database system, data integrity can be threatened if two users are allowed to update the same item or record of data at the same time. Record locking, where only a single user at a time is allowed access to a given data record, is a method of insuring data integrity.

Data Line Interface The point at which a data line is connected to a telephone system.

Data Line Monitor A measuring device that bridges a data line and looks at how clean the data is, whether the addressing is accurate, the protocol, etc. Being only a monitor, it does not in any way affect the information traveling on the line. See DATA MONITOR.

Data Line Privacy Prohibits activities which would insert tones on a data station line used by a facsimile machine, a computer terminal or some other device sensitive to extraneous noise.

Data Link A term used to describe the communications link used for data transmission from a source to a destination. In short, a phone line for data transmission. Or, A fiber optic transmitter, cable, and receiver that transmits digital data between two points.

Data Link Control DLC. Characters used in data communications that control transmission by performing various error checking and housekeeping functions — connect, initiate, terminate, etc.

Data Link Control Protocol A Microsoft driver used to connect Windows 95 and NT workstations to IBM mainframes when TCP/IP is not available. However, the protocol is not routable and it is not designed for peer-to-peer communications between workstations. It is designed only for connectivity to mainframes and minicomputers.

Data Link Escape The first control character of two-character sequence used exclusively to provide supplementary line-control signals.

Data Link Layer The second layer of the Open Systems Interconnection data communications model of the International Standards Organization. It puts messages together and coordinates their flow. Also used to refer to a connection between two computers over a phone line. See OSI MODEL.

Data Mart A small, single-subject warehouse used by individual departments or groups of users.

Data Message A message included in the GPS (Global Positioning System) signal which reports the satellite's location, clock corrections and health. Included is rough information on the other satellites in the constellation.

Data Mining Data mining refers to using sophisticated data search capabilities that use statistical algorithms to discover patterns and correlations in data. A comparison to traditional gold or coal mining, data mining is defined as a way to find buried knowledge ("data nuggets" — no kidding!) in a corporate data warehouse and to improve business users' understanding of this data. The data mining approach is complementary to other data analysis techniques such as statistics, on-line analytical processing (OLAP), spreadsheets, and basic data access. In a nutshell, data mining is another way to find meaning in data. See DATA WAREHOUSE.

Data Monitor A device used to look at a bit stream as it travels on a circuit. It will show the user what is going down both sides of a data channel. It will show what the user at his terminal is typing, and what the computer is responding with. Extremely useful for troubleshooting data communications problems. Also called a Data Scope.

Data Multiplexer A device allowing several data sources to simultaneously use a common transmission medium while guaranteeing each source its own independent channel. See MULTIPLEXER.

Data Network Identification Code DNIC. In the ITU-T International X.121 format, the first four digits of the 14-digit international data number; the set of digits that may comprise the three digits of the data country code (DCC) and the 1-digit network code (which is called the "network digit"). See DNIC.

Data Numbering Plan Area DNPA. In the U.S. implementation of the ITU-T X.25 network, the first three digits of a network terminal number (NTN). The 10-digit NTN is the specified addressing information for an end-point terminal in an X.25 network.

Data Object An individually addressable unit of information, specified by a data template and its content, that can persist independently of the invocation of a service.

Data Over Voice A device that takes a voice grade line and multiplexes it so it can carry a voice and a data signal. Typically the data is carried in analog form. Thus, to put data on this type of circuit, you need a modem. It is called Data Over Voice, because the data streams (transmission and reception) travel at a higher frequency than the voice conversations using a technique called frequency division multiplexing (FDM).

Data Overrun Also called UART Overrun. This definition from Derrick Moore of JDR Microdevices (1-800-538-5000). Data overrun occurs when your PC is unable to accept interrupts as fast as needed from your serial port chip. Because your PC may be busy reading from a disk or refreshing the screen, it may not handle a serial interrupt before the next byte of data overfills the receiver buffer on the chip. Early PCs used the 8250 which only had a one byte buffer. It was enough because modems were slow and PCs only did one thing at a time. Later, when modems became faster, the serial chips like the 16550 were upgraded to 16 byte buffers. Startech's 16C650 chip has a 32 byte buffer and allows more time for the operating system to service interrupts as well as reducing the number of interrupts that occur. If you run Windows 3.1, 95, NT, or OS/2, and use a high speed modem, you have probably experienced data overrun. Since most communications programs merely request that data be re-sent, the problem is

masked. In short, make sure your PC has at least a 16-byte UART. See UART and UART OVERRUN.

Data Packet Although a computer and modem can send data one character at a time, when you're surfing the Internet, downloading files, or sending email, it's much more efficient to send information in larger blocks called data packets. Modems generally send packets of around 64 characters along with some extras for error checking. When downloading files using a protocol like Xmodem, however, the packets are larger. And when using Internet protocols such as TCP/IP, the packets are larger still — around 1,500 characters. Such packets of data contain the information you're sending or receiving, an address (i.e. where it's going) and some start and stop information. See also asynchronous communication, Ethernet, TCP/IP and Xmodem.

Data Packet Signaling All telephone switches use the same three general types of signals: + Event Signaling initiates an event, such as ringing. + Call Progress Signaling denotes the progress (or state) of a call, such as a busy tone, a ringback tone, or an error tone. + Data Packet Signaling communicates certain information about a call, for example, the identify of the calling extension, or the identity of the extension being called.

Data Packet Switch System-common equipment that electronically distributes information among data terminal equipment connected to a data transmission network. The switch distributes information by means of information packets addressed to specific terminal devices.

Data PBX A PBX for switching lots of low-speed asynchronous data. A switch that allows a user on an attached circuit to select from among other circuits, usually one at a time and on a contention basis, for the purpose of establishing a through connection. Distinguished from a PBX in that only digital transmissions, and not analog voice, is supported. Like a telecommunications PBX that makes and breaks phone connections, a data PBX makes and breaks connections between computers and peripherals. In response to dynamic demand, it establishes communications paths between devices attached to its input/output ports by receiving, transmitting and processing electrical signals. Usually, data PBXs work off PCs' serial ports rather than through cable attached to a network interface card. For that reason, they are restricted to serial speeds, topping out at about 19.2K bits per second. For switching lots of low-speed asynchronous data, a data PBX (also called a line selector) can be better than a LAN. Total throughput can actually be higher. See also LINE SELECTOR.

Data Phase A phase of a data call during which data signals may be transferred between DTEs that are interconnected via the network.

Data Port Point of access to a computer that uses trunks or lines for transmitting or receiving data.

Data Protection A means of ensuring that data on the network is safe. Novell's NetWare protects data primarily by maintaining duplicate file directories and by redirecting data from bad blocks to reliable blocks on the NetWare server's hard disk. A hard disk's Directory Entry Table (DET) and File Allocation Table (FAT) contain address information that tells the operating system where data can be stored or retrieved from. If the blocks containing these tables are damaged, some or all of the data may be irretrievable. NetWare reduces the possibility of losing this information by maintaining duplicate copies of the DET and FAT on separate areas of the disk. If one of the blocks in the original tables is damaged, the operating system switches to the duplicate tables to get the location data it needs. Data protection within standard NetWare also involves such features as read-after-write verification, Hot Fix, and disk mirroring or duplexing. See DISK MIRRORING and HOT FIX.

Data Rate The rate at which a channel carries data, measured in bits per second, also known as data signaling rate. If there are restrictions on the pattern of bits, the information capacity of the channel could be less than the data rate. In short, data rate is the measurement of how quickly data is transmitted — but it may be very different (i.e. less) than what the channel is theoretically capable of.

Data Rate Mismatch A condition that occurs when a packet's transmission frequency (data rate) does not match the local transmit frequency.

Data Record See DATA FILE and DATABASE MANAGEMENT SYSTEM.

Data Scope See DATA MONITOR.

Data Scrambler A device used in digital transmission systems to convert an input digital signal into a pseudo random sequence free from long runs of marks, spaces, or other simple repetitive patterns.

Data Secure Line A single tip and ring line off the PBX which is protected against any tones (like call waiting) or break-ins that would otherwise mess up any ongoing data transmission call.

Data Security The protection of data from unauthorized (accidental or intentional) modification, destruction, disclosure, or delay.

Data Segment A pre-defined set of data elements ordered sequentially within a set, beginning and ending with unique segment identifies and terminators. Data segments combine to form a message. Their relation to the message is specified by a Data Segment Requirement Designator and a Data Segment Sequence. Data Segment is an EDI (Electronic Data Interchange) term.

Data Segment Requirement Designator An EDI (Electronic Data Interchange) requirement designator determines that if and when the data segment will occur in a message: - MANDATORY. The segment must appear in the message. - CONDITIONAL. The segment will occur in the message depending on agreement conditions. The relevant conditions must be given as part of the message definition. - OPTIONAL. The segment may or may not occur.

Data Segment Sequence In Electronic Data Interchanges, each data segment has a specific place within a message. The data segment sequence determines exactly where a segment will occur in a message:

• HEADING AREA. A segment occurring in this area refers to the entire message.

• DETAIL AREA. A segment occurring here is detail information only will override any similar specification in the header area.

• SUMMARY AREA. Only segments containing total or control information may occur in this area (e.g., invoice total, etc.)

Data Service Unit DSU. Device designed to connect a DTE (Data Terminal Equipment like a PC or a LAN) to a digital phone line to allow fully-digital communications. A DSU is sort of the digital equivalent of a modem. In more technical terms, a DSU is a type of short haul, synchronous data line driver, normally installed at a user location that connects a user's synchronous equipment over a 4-wire circuit to a serving dial-central-office. This service can be for a point-to-point or multipoint operation in a digital data network. DSUs are typically used for leased lines. For switched digital services, you need a CSU/DSU also called a DSU/CSU. See CSU/DSU and DSU/CSU.

Data Set In AT&T jargon, a data set is a modem, i.e. a device which performs the modulation/demodulation and control functions necessary to provide compatibility between business machines which work in digital (on-off) signals and voice telephone lines. In IBM jargon however, a data set is a collection of data, usually in a file on a disk. See also MODEM.

Data Set Ready One of the control signals on a standard RS-232-C connector. It indicates whether the data communications equipment is connected and ready to start handshaking control signals so that transmission can start. See RS-232-C and the Appendix.

Data Sheets What Business Communications Review calls its statistical and descriptive material comparing PBXs. BCR includes these data sheets in its excellent BCR Manual of PBXs. This manual is the most extensive write-up of larger (more than 200 line) PBXs in the world. It is available from Telecom Library on 1-800-LIBRARY.

Data Signaling Rate The total of the number of bits per second in the transmission path of a data transmission system. A measurement of how quickly data is transmitted, expressed in bps, bits-per-second.

Data Sink Part of a terminal in which data is received from a data link.

Data Source 1. The originating device in a data communications link.

2. An object identifier in RMON that represents a particular interface.

Data Span Any digital service, T-1, 56K, ISDN or data carrying service.

Data Steward A new role of data caretaker emerging in business units. Individual takes responsibilities for the data content and quality.

Data Stream 1. Collection of characters and data bits transmitted through a channel.

2. An SCSA term. A continuous flow of call processing data.

Data Surfer A person who makes a living doing online research and information retrieval. Also known as a Cybrarian (comes from cyberspace librarian) or a super searcher. See CYBRARIAN.

Data Switching Exchange DSE. The equipment installed at a single location to perform switching functions such as circuit switching, message switching, and packet switching.

Data Synchronization The process of keeping database data timely and relevant by sending and receiving information between laptops, between desktops in the field and between bigger computers at headquarters. See also SYNCHRONIZATION and REPLICATION.

Data Terminal Equipment DTE. A definition of hardware specifications that provides for data communications. There are two basic specs your hardware can conform to, DTE (Data Terminal Equipment) or DCE (Data Communications Equipment). See DCE and DTE.

Data Terminal Ready One of the control signals on a standard RS-232-C connector. It indicates if the data terminal equipment is present, connected and ready and has had handshaking signals verified. See RS-232-C and the Appendix.

Data Transfer Rate The average number of bits, characters, or blocks per unit time passing in a data transmission system.

Data Transfer Request Signal A call control signal transmitted by a DCE to a DTE to indicate that a distant DTE wants to exchange data.

Data Transfer Time The time that elapses between the initial offering of a unit of user data to a network by transmitting data terminal equipment and the complete delivery of that unit to receiving data terminal equipment.

Data Typing When converting a database from one format to another, several conversion programs will convert the data to a common format before converting it to the final version. During the conversion process a program may check through the data in the database to determine what it is and arbitrarily make one field numeric, one field character, one field memo, etc.

Data User Part DUP. Higher layer application protocol in SS7 for the exchange of circuit switched data; not supported by ISDNs.

Data Warehouse A database warehouse consolidates information from many departments within a company. This data can either be accessed quickly by users or put on an OLAP server for more thorough analysis. Data warehouses often use OLAP servers. OLAP stands for On Line Analytical Processing, also called a multidimensional database. According to PC Week, these databases can slice and dice reams of data to produce meaningful results that go far beyond what can be produced using the traditional two-dimensional query and report tools that work with most relational databases. OLAP data servers are best suited to work with data warehouses. See DATA WAREHOUSING.

Imaging Magazine, one of our publications, wrote a story on data warehouses. The writer, Joni Blecher found "defining a data warehouse to be puzzling at best." She said these definitions seem to make the most sense.

A collection of physical data stores designed to concisely present a historical perspective of the events that occur in an enterprise. Data warehousing is a set of activities some of which are optional and some mandatory that create, operate and evolve the collection of data stores that make up the data warehouse.

• Actium. An extremely comprehensive solution that includes hardware, software, middleware, partner products as well as their own professional services focused on solving business problems through the enterprise level.

• NCR. The place where business managers can access information for managerial processes. They're built for decision making purposes. It's an elaborate process that consists of a solution made up of many products.

• Oracle. A group of individuals, processes, methodologies — all the things that deal with and manage data including cleansing, enhancing, standardizing, consolidating and disseminating it.

• Acxiom. A data store that companies build where they're storing their information assets so they can extract knowledge and understanding to the operation and performance of their business. — Logic Works.

Data Warehousing A software strategy in which data are extracted from large transactional databases and other sources and stored in smaller databases, making analysis of the data somewhat easier. See DATA WAREHOUSE.

Database A collection of data structured and organized in a disciplined fashion so that access is possible quickly to information of interest. There are many ways of organizing databases. Most corporate databases are not one single, huge file. They are multiple databases related to each other by some common thread, e.g. an employee identification number. Databases are made up of two elements, a record and a field. A record is one complete entry in a database, e.g. Gerry Friesen, 12 West 21 Street, New York, NY 10010, 212-691-

8215. A field would be the street address field, namely 12 West 21 Street.

Databases are stored on computers in different ways. Some are comma delineated. They differentiate between their fields with commas — like Gerry's record above. A more common way of storing databases is with fixed length records. Here, all the fields and all the records are of the same length. The computer finds fields by index and by counting. For example, Gerry's first name might occupy the first 15 characters. Gerry's last name might be the next 20 characters, etc. Where Gerry's names are too short to fill the full 15 or 20 characters, their fields are "padded" with specially-chosen characters which the computer recognizes as padded characters to be ignored. The most important thing to remember about databases is that all the common database programs, like dBASE, Paradox, Rbase, etc. don't automatically make backups of their files like word processing programs do. Therefore, before you muck with a database file — sort it, index it, restructure it, etc. Please make sure you make a backup of the main database file.

Database Management System DBMS. Computer software used to create, store, retrieve, change, manipulate, sort, format and print the information in a database. Database management systems are probably the fastest growing part of the computer industry. Increasingly, databases are being organized so they can be accessible from places remote to the computer they're kept on. The "classic" database management system is probably an airline reservation system.

Database Object One of the components of a database: a table, view, index, procedure, trigger, column, default, or rule.

Database Server A specialized computer that doles out database data to PCs on a LAN the way a file server doles out files. Where a traditional DBMS runs both a database application and the DBMS program on each PC on the LAN, a database server splits up the two processes. The application you wrote with your DBMS runs on your local PC, while the DBMS program runs on the database server computer. With a regular file server setup, all the database data has to be downloaded over the LAN to your PC, so that the DBMS can pick out what information your application wants. With a database server, the server itself does the picking, sending only the data you need over the network to your PC. So a database server means vastly less network traffic in a multi-user database system. It also provides for better data integrity since one computer handles all the record and file locking. See SERVER.

Datablade Datablades are components for particular types of data that plug into a central database, similar to razor blades snap into a razor. Informix is rewriting some of its databases so other companies can produce datablades.

Datagram A transmission method in which sections of a message are transmitted in scattered order and the correct order is re-established by the receiving workstation. Used on packet-switching networks. The Dow Jones Handbook of Telecommunications defines it as, "A single unacknowledged packet of information that is sent over a network as an individual packet without regard to previous or subsequent packets." Here's another definition I found. A finite-length packet with sufficient information to be independently routed from source to destination. In packet switching, a self-contained packet, independent of other packets, that carriers information sufficient for routing from the originating data terminal equipment to the destination data terminal equipment, without relying on earlier exchanges between the equipment and the network. Unlike virtual call service, there are no call establishment or clearing procedures, and the network does not generally provide protection against loss, duplication, or misdelivery. Datagram transmission typically does not involve end-to-end session establishment and may or may not entail delivery confirmation acknowledgment. A datagram is the basic unit of information passed across the Internet. It contains a source and destination address along with data. Large messages are broken down into a sequence of IP datagrams. See CONNECTIONLESS MODE TRANSMISSION.

Datagram Packet Network The type of packet-switched network in which each packet is individually routed. This may result in a loss of sequence within a message because of alternate routing, or a loss of portions of a message because of packet elimination for congestion control. See DATAGRAM.

Datakit An AT&T (now Lucent) wide area data switching system. See Datakit VCS.

Datakit VCS The Datakit VCS is an AT&T packet switch that can switch over 44,000 packets per second and support up to 3,500 simultaneous virtual circuits. According to AT&T, the Datakit VCS is a digital virtual circuit switch with an architecture that combines the advantages of LANs, PBXs, data circuits and X.25 packet switches.

DATAP The programmer and system house in Atlanta which provides several long distance carriers with value-added services — what Sprint calls its SCADA system manufacturer.

Datapak A packet-switched network run in Denmark, Sweden, Finland and Norway and operated by their respective governments.

Datapath A name that's becoming generic for a data service that provides digital, full-duplex data transmission at speeds of 300 bps through 19.2 Bps asynchronous and 1,200 through 64 Kbps synchronous. Datapath has built in auto-baud and hand-shaking protocols.

Dataphone A service mark of AT&T for various data hardware products such as modems and printers and services. See also DDS.

Datascope A diagnostic tool for monitoring and capturing data transmissions which displays real-time transmissions of raw data in hexadecimal, binary, or character-oriented displays.

DataSPAN DataSPAN is generally characterized as a "fast packet" service and is based on frame relay standards recommended by the International Consultative Committee for Telephone & Telegraph (ITU)(now called the ITU-T) and the American National Standards Institute (ANSI). Northern Telecom has introduced DataSPAN as a new DMS SuperNode value-added, data communications service that is targeted toward connecting high-speed Local Area Networks. Northern Telecom asserts that DataSPAN's rapid and efficient data transport assures reliable delivery and substantial performance improvement over current LAN interconnect solutions. DataSpan switching and transmission delay is less than 3 ms per node; X.25 switching and transmission delay can be up to 50 ms per node. Using Frame Relay, wide-area packet switching can be accomplished with the same level of performance that is traditionally limited to complex, dedicated private-line networks. DataSPAN is accessed through standard DS-0 or DS-1 links.

Date And Time Stamp Many voice mail systems will append the date and time of receipt of voice messages for their users/subscribers.

Datel Services An MCI International service that carries data for medium-volume users. This service provides two-way data and voice transmission for simplex, half-duplex, and full duplex operations.

Datex P An Austrian packet-switched network in Austria and Germany and run by their respective governments.

Dating Format The format employed to express the time of an event. The time of an event on the UTC time scale is given in the following sequence; hour, day, month, year; e.g., 0917 UT, 30 August 1997. The hour is designated by the 24-hour system. UTC stands for Coordinated Universal Time, believe it or not.

DATU See Direct-Access Test Unit.

DAU Dumb Ass User.A term meaning user induced error in any technical application: computing, telecom or whereever mankind meets the mechanical. It's a stupid acronym. I don't make this stuff up.

Daughterboard First there is the motherboard. That's the main circuit board of a computer system. The motherboard contains edge connectors or sockets so other PC (printed circuit) boards can be plugged into it. Those PC boards are called Fatherboards. Some fatherboards have pins on them into which you can plug smaller boards. Those boards are called Daughterboards. In a voice processing system, you might have a Fatherboard to do faxing. And you might have a range of Daughterboards, which allow you to connect different types of phone connections. Different boards exist for standard analog tip and ring, digital switched 56, etc. See also MOTHERBOARD.

Daughtercard Same as DAUGHTERBOARD.

DAVIC Digital Audio Video Council. A voluntary SIG (Special Interest Group) organized to "favour the success of emerging digital audio-visual applications and services, by the timely availability of internationally agreed specifications of open interfaces and protocols that maximize interoperability across countries and applications/services." The current set of DAVIC 1.0 specifications are to further the deployment of broadcast and interactive systems that support initial applications such as TV distribution, near video on demand, video on demand, and basic forms of teleshopping. DAVIC is one of the organizations examining the developing specifications for VDSL (Very-high-bit-rate Digital Subscriber Line), which is intended for video applications. www.davic.org

Day-Of-Week Factors A call center term. A historical pattern consisting of seven factors, one for each day of the week, that defines the typical pattern of call arrival throughout the week. Each factor measures how far call volume on that day deviates from the average daily call volume. This is a historical pattern.

db 1. Decibel. A unit of measure of signal strength, usually relationship between a transmitted signal and a standard signal source. db stands for decibels, which is a way to represent logarithmic ratios. Since a db number is a ratio, it must always be a ratio between two things. The db number is equal to 20 times the log of the ratio between two numbers. A ratio of 10 is 20db, a ratio of 100 is 40db, a ratio of 1000 is 60 db, etc. For example, if a filter has -40db of noise reduction, that means that the ratio between the output noise and the input noise is 40db or 100. The letters "db" are always used in lower case. A 6 dB of loss would mean that there is 6 dB difference between what arrives down a communications circuit and what was transmitted by a standard signal generator. See decibel.

2. Database

3. Short for Data Bus connector. Usually shown with a number that represents the number of wire conductors in the connector, e.g. DB-9 or DB-25, both of which are very common connectors to plug into serial ports on PCs. See DB-9, DB-15 and DB-25.

db Loss Budget Should be written as dB. dB Loss Budget is the amount of light available to overcome the attenuation in the optical link and still maintain specifications.

DB-9 This is the standard nine-pin RS-232 serial port on all laptop PC computers and most desktop computers today. The term DB-9 is used to describe both the male and female plug. So be careful when you order. See also the Appendix in this Dictionary for more information on the pinning of RS-232-C plugs. The above is contemporary usage, i.e. the way it's used today. In fact, it's wrong historically. The "D" originally described the shape of the housing. The second letter: A, B, C, D or E originally specified the size of the housing where the "E" is used somewhat like it's used as a drawing size (i.e., smaller that a "D"). There are connectors made with, e.g., size "B" housing with other than 9, 15 or 25 pins. Sometimes coax "pins" are included, which use up several little pin locations. See the Appendix in this dictionary for more information on the pinning of RS-232 plugs.

DB-15 A standardized connector with 15 pins. It can be used in Ethernet transceivers. It can also be used for connecting VGA monitors. Careful when ordering a DB-15 that order a female or male connector. DB-15 can refer to either. See also DB-9 for a longer explanation.

DB-25 The standard 25-pin connector used for RS-232 serial data communications. In a DB-25 there are 25-pins, with 13 pins in one row and 12 in the other row. DB-25 is used to describe both the male and female plug. So be careful when you order. See also DB-9 and the Appendix in this dictionary for more information on the pinning of RS-232 plugs.

DB-60 A connector with 60 pins in four rows (15 pins in each row).

DB2 IBM's relational database system that runs on System 370-compatible mainframes under the MVS operating system.

DBA 1. Dynamic Bandwidth Allocation.

2. Database administrator. The individual in the organization with the responsibility for the design and control of databases.

3. dbA. dbA is a measure of sound level. The ratio for determining db in this case is the ratio between the sound level measured with a microphone and an implicit reference sound level, namely 0db, which is defined to be approximately equal to the threshold of human hearing. 45dbA is a very faint whisper, 75 dbA is typical conversation, 100 dbA is about how loud a Walkman can get, and 120dbA is a jet plane taking off at 20 feet away from the engine. The "A" represents a special filter which is used to take into account the fact that people are less sensitive to very low and very high frequencies. The dbA system is used for measuring background sounds like computer fans or office noise. A system with a different filter called dbC is used when very loud sounds are being measured. This definition courtesy APC.

DBIT Also called D-BIT. The delivery confirmation bit in an X.25 packet that is used to indicate whether or not the DTE wishes to receive an end-to-end acknowledgment of delivery. In short, a bit in the X.25 packet header that assures data integrity between the TPAD and the HPAD.

DBM Decibels below 1mW. This should be written as dBm. Output power of a signal referenced to an input signal of 1mW (Milliwatt). Similarly, dBm0 refers to output power, expressed in dBm, with no input signal. (0 dBM = 1 milliwatt and -30 dBm = 0.001 milliwatt). See DECIBEL.

DBMS Database management system. A computer program that manages data by providing the services of centralized control, data independence, and complex physical structures. Advantages include efficient access, integrity, recovery, con-

currency control, privacy, and security. A DBMS enables users to perform a variety of operations on data, including retrieving, appending, editing, updating, and generating reports.

bBrn DeciBels above Reference Noise. A ratio of power level in dB relative to a noise reference. dBrnC uses a noise reference of -90 dBrn, as measured with a noise meter, weighted by a frequency function known as C-message weighting which expresses average subjective reaction to interference as a function of frequency.

DBS Direct Broadcast Satellite. A term for a satellite which sends relatively powerful signals to small (typically 18-inch diameter) dishes installed at homes. See C Band, 1994 and Direct Broadcast Satellite.

DBT Deutsche Bundespost Telecom.

DBU Decibels below 1uW.

DC Direct Current. The flow of free electrons in one direction within an electrical conductor, such as a wire. DC also stands for Delayed Call.

DC Signaling A collection of ways of transmitting communications signals using direct current — the type of current produced by a dry cell household "D" cell battery. DC signaling is only used on cable. It's an out-of-band signal.

DCA 1. Defense Communication Agency. US government agency responsible for installation of Defense Data Networks, including the ARPANET and MILNET, and PSNs. The DCA writes contracts for operation of the DDN (Defense Data Network) and pays for network services.

2. Document Content Architecture. The IBM approach to storing documents as two types of document group: draft documents and final form documents. For presentation, the draft document is transformed into a final document through an office system.

DCC 1. Data Communications Channel. Channels contained within section and line overhead used as embedded operations channels to communicate to each network element. An AT&T SONET term.

2. An ATM term. Data Country Code: This specifies the country in which an address is registered. The codes are given in ISO 3166. The length of this field is two octets. The digits of the data country code are encoded in Binary Coded Decimal (BCD) syntax. The codes will be left justified and padded on the right with the hexadecimal value "F" to fill the two octets. An ATM term.

3. Digital Compact Cassette. A digital version of the familiar analog audio cassette. A DCC recorder can play and record both analog and digital cassettes. But the digital ones will sound a lot better.

DCCH Digital Control CHannel. A channel used in most newer digital cellular and PCS systems for signal and control purposes between the mobile terminal device and the radio base station. See also CELLULAR and PCS.

DCD 1. Data Carrier Detect. Signal from the DCE (modem or printer) to the DTE (typically your PC), indicating it (the modem) is receiving a carrier signal from the DCE (modem) at the other end of the telephone circuit.

2. Dynamically Configurable Device. A dynamically configurable device is a fancy name for a Plug and Play device, so-called because you don't have to reboot the system after installing one.

DCE 1. Data Communications Equipment. The classic definition of DCE is that it resolves issues of interface between Data Terminal Equipment (DTE) and the network. Examples include Network Interface Cards (NICs), CSUs and DSUs, modems, and routers. DCE may accomplish such functions as changes in electrical coding schemes, electro-optical conversion, and data formatting. In the RS-232 "standard" developed by the Electronic Industries Association (EIA), there are DCE devices (typically modems) and DTE (Data Terminal Equipment) devices, which are typically personal computers, data terminals, or peripherals such as printers. The main difference between a DCE and DTE in RS-232 is the wiring of pins two and three. But there is, of course, no standardization. When wiring one RS-232 device to another, it's good to know which device is wired as a DCE and which as a DTE. But it's actually best to go straight to the wiring diagram in the appendix of the device's instruction manual. Then you compare the wiring diagram of the device you want to connect and build yourself a cable that takes into account the peculiar (i.e. strange) vagaries of the engineers who designed each product. In short, with an RS-232 connection, the modem is usually regarded as DCE, while the user device (terminal or computer) is DTE. In a X.25 connection, the network access and packet switching node is viewed as the DCE. DCE devices typically transmit on pin 3 and receive on pin 2. DTE (Data Terminal Equipment) devices typically transmit on pin 2 and receive on pin 3. See also the Appendix.

2. Distributed Computing Environment. An industry-standard, vendor-neutral set of distributed computing technologies developed by the Open Software Foundation (OSF). According to The Open Group, successor to the OSF, DCE provides security services to protect and control access to data, name services that make it easy to find distributed resources, and a highly scalable model for organizing widely scattered users, services and data. DCE runs on all major computing platforms, supporting distributed applications in heterogeneous hardware and software environments.

DCG Dispersion Compensation Grating. DCG overcomes the distortion of optical signals as they are transmitted through a network. Instead of trying to compensate for large amounts of signal dispersion at the end of a network, DCG periodically removes the distortion where needed along the transmission line. See Solitons.

DCH D-Channel Handler.

DCM Digital Circuit Multiplication. A means of increasing the effective capacity of primary rate and higher level PCM hierarchies, based upon speech coding at 64 Kbit/s. Also a Digital Carrier Module.

DCN Disconnect frame. Indicates the fax call is done. The sender transmits before hanging up. It does not wait for a response.

DCOM Distributed Component Object Model. Microsoft's distributed version of its COM, a language-independent component architecture (not a programming language) meant to be a general purpose, object-oriented means to encapsulate commonly used functions and services. COM encompasses everything previously known as OLE Automation (Object Linking and Embedding). Microsoft describes DCOM as "COM with a longer wire." See also COM.

DCP Digital Communications Protocol.

DCP Telephone A digital voice telephone of the AT&T model 7400 series, operated with the Digital Communications Protocol (DCP) used in System 75/85 digital PBXs.

DCS 1. Distributed Communications System. See DISTRIBUTED SWITCHING.
2. Digital Crossconnect System. A device for switching and rearranging private line voice, private line analog data and T-1 lines. A DCS performs all the functions of a normal "switch," except that connections are typically set up in

advance of when the circuits are to be switched — not together with the call. You make those "connections" by calling an attendant who makes them manually, or dialing in on a computer terminal — one similar to an airline agent's. See also NETWORK RECONFIGURATION SERVICE.
3. Digital Cellular System.
4. Digital Communications System.
5. DCS. A packet-switched network implemented in Belgium and operated by the Belgian government.

DCS 1800 Digital Cellular System at 1800 MHz. A GSM (Global System for Mobile Communications) standard for cellular mobile telephony established by ETSI (European Telecommunications Standards Institute) for operation at 1800 MHz. In short, DCS 1800 is GSM adopted to the 1800Mhz frequency band. This means that existing GSM phones won't be able to talk on the DCS 1800.

DCT 1. Digital Carrier Termination.
2. Discrete Cosine Transform. A compression algorithm used in most of the current image compression systems for bit rate reduction, including the ITU-T Px64 standard for video conferencing. DCT represents a discrete signal or image as a sum of sinusoidal wave forms.

DCTI Desktop Computer Telephone Integration. Basically, providing a way for your computer to control your telephone set.

DCTU Digital Cordless Phone US.

DCV Digital Compressed Video.

DD Dotted Decimal Notation See DOTTED QUAD.

DDA Domain Defined Attribute. A way of adding additional information to the address of your electronic mail in order to avoid confusion between people of the same or similar name.

DDB Digital Data Bank.

DDC Direct Department Calling.

DDCMP Digital Data Communications Message Protocol. A byte-oriented, link-layer protocol from Digital Equipment Corp., used to transmit messages between stations over a communications line. DDCMP supports half- or full-duplex modes, and either point-to-point or multipoint lines in a DNA (Digital Network Architecture) network.

DDD Direct Distance Dialing. The "brand name" the Bell System used to call its Message Toll Telephone Network. It used the words "Direct Distance Dialing" to convince the public to dial their own long distance calls directly without the help of an operator.

DDE See DYNAMIC DATA EXCHANGE

DDEML Dynamic Data Exchange Management Library. A feature of Microsoft Windows.

DDI Direct Dialing Inward. A British term. It is a service where a call made to a DDI number arrives direct, without the intervention of an organization's operator, at an extension or, if routed via an ACD, a group of extensions. A specific DDI number is assigned to each campaign. When an agent answers a call made to one of these DDI numbers, the relevant script and screen are instantly displayed, so the agent can give an appropriate response.

DDM Distributed Data Management Architecture. An IBM SNA LU 6.2 transaction providing users with facilities to locate and access data in the network. It involves two structures: DDM Source, and DDM Target. The DDM Source works with a transaction application to retrieve distributed data and transmits commands to the DDM Target program on another system where the data that has been requested is stored. The DDM Target interprets the DDM commands, retrieves the data and sends it back to the DDM Source that originated the request.

DDN Defense Data Network. A network that provides long

haul and area data communications and interconnectivity for DoD systems, and supports the DoD suite of protocols (especially TCP and IP). All equipment attached to the DDN by military subscribers must incorporate, or be compatible with, the DoD Internet and transport protocols. DDN was split off from ARPANET to handle U.S. military needs. It is also called MILNET. See also DDN NIC.

DDN NIC Defense Data Network Network Information Center. Also called the "NIC," by those in the defense domain, the DDN NIC is responsible for the assignment of Internet addresses and Autonomous System numbers, the administration of the root domain, and the provision of information and support services to the DDN. See also DDN, INTERNIC and NIC.

DDP Distributed Data Processing. See DISTRIBUTED SWITCHING.

DDS 1. Dataphone Digital Service, also called Digital Data System. DDS is private line digital service, typically with data rates at 2,400; 4,800; 9,600 and 56,000 bits per second. DDS is offered on an inter-LATA basis by AT&T and on an intra-LATA basis by the Bell operating companies. AT&T has now incorporated it into their Accunet family of offerings. But, by the time your read this, they may have a new name for the service, like Accunet.
2. Digital Data Storage. A DAT format for storing data. It is sequential — all data that is recorded to the tape falls after the previous block of data.

DDSD Delay Dial Start Dial. A start-stop protocol for dialing into a switch.

DE 1. Discard Eligible. The frame relay standard specifies that data sent across a virtual circuit in excess of that connection's Committed Information Rate (CIR) will be marked by the user as being eligible for discard in the event of network congestion. DE data is the first to be discarded by the network when congestion occurs, thus providing protection for data sent within the parameters of the CIR. It is the responsibility of the intelligent end equipment and/or protocol to recognize the discard and respond by resending the information. As a practical matter, few users are likely to volunteer to have frames discarded, and few manufacturers provide for the setting of the DE bit. SEE OVERSUBSCRIPTION.
2. Designated Entities

De Facto Used to describe a standard reflecting current or actual practice, but not having approval or sanction by any official standards-setting organization. In other words, a de facto is an unofficial standard that exists because sufficient companies adhere to it or because of market acceptance. Usually created by an individual manufacturer or developer — hence, often used as a synonym for proprietary. The Hayes AT auto-dial modem command language is an example of a de facto standard. Very often, de facto standards form the basis for de jure standards. Ethernet was the basis of Institute of Electrical and Electronic Engineers or IEEE standard 802.3. Contrast with de jure.

De Forest, Lee Lee de Forest invented the vacuum tube in 1907. Until the invention of the transistor after World War II, all radio, long distance telephony and complicated electronics, including electronic computers, were derived from de Forest's invention.

De Jure Used to describe standards approved or sanctioned by an official standards setting organization. ITU-T Recommendation X.25 for packet data networks is an example of a de jure standard. Contrast with DE FACTO.

De-encapsulation The process of extracting a data packet from the user data field of the encapsulating data packet.

De-regulation See deregulation.

De-spreading The process used by a correlator to recover narrowband information from a spread-spectrum signal.

DEA Digital Exchange Access This is a service that Bell Canada offers. It gives the customer a local number and access to their digital equipment where the number resides.

Dead Sector In facsimile, the elapsed time between the end of scanning of one line and the start of scanning of the following line.

Dead Spot A cellular radio term. It denotes an area within a cell where service is not available. A dead spot is usually caused by hilly terrain which blocks the signal to or from the cell tower. It can also occur in tunnels and indoor parking garages. Excessive foliage or electronic interference can also cause dead spots. If you encounter a dead spot, tell your cellular radio supplier. They often will do something.

Dead Tree Edition The paper version of a publication available in both paper and electronic forms. If you are reading my dictionary on paper, you are reading the dead tree edition.

Deadline During the Civil War, captured troops were placed in a field. They were told they could move wherever they wanted in the field except beyond a line that was drawn around the field. If they did, they would be shot. Thus the origin of the word deadline.

Deadlock See DEADLY EMBRACE.

Deadly Embrace Stalemate that occurs when two elements in a process are each waiting for the other to respond. For example, in a network, if one user is working on file A and needs file B to continue, but another user is working on file B and needs file A to continue, each one waits for the other. Both are temporarily locked out. The software must be able to deal with this.

Dealer Dealer is simply a person who sells equipment — hardware or software — made by someone else. That dealer may be a distributor of Novell LAN software. Or that dealer may be in the secondary telecom equipment business, i.e. a company that buys and sells secondary market telephone equipment. Generally in the secondary equipment business, a dealer takes ownership of the equipment (see BROKER). He tests and refurbishes it before remarketing. An authorized dealer has a contract with the manufacturer to buy equipment at a preset price. He also has the added support of a manufacturer's warranty. An independent dealer has no formal agreement with the manufacturer. He uses a variety of sources to obtain equipment and his warranty is backed only by the company's internal resources.

Dealer Board British term meaning "trading turret."

Dealing Room British term meaning "trading room."

Death Star Villages Suburbs around New Jersey where many AT&T employees live. Makes reference to the AT&T logo, which employees have dubbed "The Death Star" (from the Star Wars films).

Debbie Warren A remarkable woman of great stamina, great perseverance and great tolerance. How do we know this? Anyone married to Stuart desperately needs these character traits. Favorite pastimes: Rachael and Jennifer.

Debit Card The term telephone "debit card" covers three categories of a new type of telephone calling card, variously called "calling card" and "prepaid telephone card." The definitions are in flux. Here's the best shot at debit card: A telephone debit card is a piece of credit-card size plastic with some technology on it or embedded into which represents the value of the money remaining on the debit card. Such money can be used to make phone calls. The technology on the card is most typically an integrated circuit, a magnetic strip, bar codes which can be read by an optical reader. See BREAKAGE, CALLING CARD and PREPAID TELEPHONE CARD.

Debitel A big provider of mobile telephone services in Germany. It is owned mainly by Daimler-Benz and Metro, a big retail group.

Debug A MS-DOS program to examine or alter memory, load and look at sectors of data from disk and create simple assembly-language programs. MS-DOS DEBUG.COM lets you write some small programs. You can use DEBUG to correct problems in some programs.

DEC Digital Equipment Corporation. A leading manufacturer of minicomputers. The Unix operating system, developed at Bell Labs, runs on DEC computers. DEC, with its headquarters in Massachusetts and having sold so many computers to Western Electric for inclusion in central office switches and toll switches, is sometimes referred to as "Eastern Electric." Kenneth H. Olsen founded DEC in 1957 with three employees in 8,500 square feet of leased space in a corner of an old woolen mill. DEC is now part of Compaq.

DEC LAT A proprietary data communications protocol developed by DEC.

Decibel dB. A unit for measuring the power of a sound or the strength of a signal. It is expressed as the ratio of two values. Here is an explanation from James Harry Green's excellent book The Dow Jones-Irwin Handbook of Telecommunications. (Get a copy from 1-800-LIBRARY or 212-691-8215.) Mr. Green writes, "The power in telecommunications circuits is so low that it is normally measured in milliwatts. However, the milliwatt is not a convenient way to express differences in power level between circuits. Voice frequency circuits are designed around the human ear, which has a logarithmic response to changes in power. (Likewise, the human eye has a logarithmic response to changes in light.) Therefore in telephony, the decibel, which is a logarithmic rather than a linear measurement, is used as a measure of relative power between circuits or transmission level points. A change in level of 1 dB is barely perceptible under ideal conditions...Increases or reductions of 3 dB result in doubling or halving the power in a circuit. This ratio is handy to remember when evaluating power differences. The corresponding figure for doubling or halving voltage is 6 dB. See DB.

Decide A deadly computer virus.

Decimal Our normal numbering system. It is to the base 10.

Decision Circuit A circuit that measures the probable value of a signal element and makes an output signal decision based on the value of the input signal and a predetermined criterion or criteria.

Decision Instant In the reception of a digital signal, the instant at which a decision is made by a receiving device as to the probable value of a signal condition.

Decision Support Systems DSS. Computerized systems for transforming data into useful information, such as statistical models or predictions of trends, which is used by management in the decision-making process. There are several aspects of the best decision support systems: First, they are connected to mainframe databases. Second, they are accessible by executives from their desktops. Third, there are usually lots of programs for producing graphs, charts and writing simple reports, i.e. for the executives to extract and portray information in forms that are most useful to them.

Decision Tree The organization of the call flow for a particular application, expressed in a tree-like logic structure.

DECNET DEC's proprietary Ethernet LAN that works across all of the company's machines, endowed with a peer-to-peer methodology. See DEC.

Decoder A device that converts information from one form to another — typically from analog to digital and vice versa. See DECODING.

Decode Status Flag A cellular industry term. A 5-bit flag used by the CDPD Mobile Data Base Station (MDBS) in the forward channel transmission to indicate the decoding status of Reed-Solomon blocks received on the reverse channel from the CDPD Mobile End System (M-ES).

Decoding Changing a digital signal into analog form or into another type of digital signal. The opposite of Encoding. See also MODEM. Decoding and coding should not be confused with deciphering and ciphering. See DES.

Decollimation In optics, that effect wherein a beam of parallel light rays is caused to diverge or converge from parallelism. Any of a large number of factors may cause this effect, e.g., refractive index inhomogeneities, occlusions, scattering, deflection, diffraction, reflection, refraction.

Decolumnization The process of reformatting multi-column documents into a signal column. Generally, when you are processing a document for use in a word processing program, a single column of text is preferable to multiple columns.

Decompression Decompression is the process of expanding a compressed image or file so it can be viewed, printed, faxed or otherwise processed.

Decrypt To convert encrypted text into its equivalent plain text by means of a cryptosystem. This does not include solution by cryptanalysis. The term decrypt covers the meanings of decipher and decode.

DECT Digital European Cordless Telecommunication. The pan-European standard based on time division multiple access used for limited-range wireless services. Based on advanced TDMA technology, and used primarily for wireless PBX systems, telepoint and residential cordless telephony today, potential uses for DECT include paging and cordless LANs. DECT frequency is 1800-1900 MHz.

Dedicated Access A connection between a phone or phone system (like a PBX) and an IntereXchange Carrier (IXC) through a dedicated line. All calls over the line are automatically routed to a particular IXC.

Dedicated Access Line Service A type of service often used by large companies which have a direct telephone line going directly to the long distance companies' "Point of Presence" (POP), thereby bypassing the local telephone company and reducing the cost per minute. Often referred to as "T-1" service.

Dedicated Array Processor A microprocessor on a hardware-based RAID array that controls the execution of RAID array-specific functions, such as rebuilding. See RAID.

Dedicated Attendant Link Assures that there will always be an intercom link available for your attendant or receptionist or operator to announce incoming calls.

Dedicated Bypass A connection between a phone or phone system (like a PBX) and an IntereXchange Carrier (IXC) through a dedicated line that is not provided by the dominant local provider of local phone service. For example, I live in New York. I might order a leased T-1 to MCI. If that line is not provided by Nynex, it is "dedicated bypass." Such bypass circuits are often cheaper and better quality than what the dominant local carrier can provide.

Dedicated Channel Or Circuit A channel leased from a common carrier by an end user used exclusively by that end user. The channel is available for use 24 hours a day, seven days a week, 52 weeks of the year, assuming it works that efficiently.

Dedicated Circuit A circuit designated for exclusive use by specified users. See also DEDICATED CHANNEL.

Dedicated Feature Buttons The imprinted feature buttons on a telephone: Conf or Conference, Drop, HFAI (Hands Free Answerer on Intercom), Hold, Mute or Microphone, Speaker or Speakerphone, Transfer, Message, and Recall.

Dedicated Inside Plant DIP. Inside plant is the portion of a LEC's (Local Exchange Carrier's) plant, or physical facilities, that are located inside its buildings. Such inside plant comprises a wide variety of equipment such as channel banks, multiplexers, switching systems, Main Distribution Frames (MDFs) and Intermediate Distribution Frames (IDFs). Dedicated Inside Plant, most commonly, is a term describing an IDF which is dedicated to the purpose of providing a CLEC (Competitive LEC) with a point of interconnection between the local loops it has leased from the ILEC (Incumbent LEC) for purposes of customer access, and the facilities which the CLEC uses to serve those customers. The dedicated IDF also is known as a Single Point of Termination (SPOT) frame. From the SPOT frame, the circuit commonly is directed to a secure enclosure in which the CLEC has collocated in the ILEC building a concentrator or multiplexer which is connected to a high-speed transmission link which hauls traffic to the CLEC's own facilities-based network. A common SPOT is a shared dedicated IDF for use by multiple CLECs which are unable to cost justify a SPOT of their own — the cost of the SPOT, plus a reasonable profit margin — is passed on to the CLEC by the ILEC. See also Collocation and Dedicated Outside Plant.

Dedicated Line Another name for a private leased line or dedicated channel. A dedicated line provides the ability to have constant transmission path from point A to point B. A dedicated line may be leased or owned. It may be assigned a single purpose, such as monitoring a distant building. It may be part of a network, with the ability for many to dial into it. It may be a tie-line between your offices or it may be a line to a long distance carrier. In this case, you do not have to dial a local connection number or put in an authorization code. A WATS line is in effect a Dedicated Line to AT&T or whomever you purchased WATS from.

Dedicated Machine A computer designed to run only one program or do one thing. This machine cannot easily be re-programmed to do another task, as, for example, a general-purpose machine can. A general-purpose machine, however, can be dedicated to running only one task if programmed to do so.

Dedicated Mode When a file server or router on a local area network is set up to work only as a file server or router, it runs in dedicated mode.

Dedicated Outside Plant DOP. Outside plant is the local loop facilities which connect the customer premises to the LEC (Local Exchange Carrier) network. Typically, the connection is to the LEC central office (CO) exchange. Dedicated outside plant, at the extreme, means that each customer premises has one or more local loops which connect directly to the "wire center" in which the CO is housed. While this approach is copper-intensive, it allows local loops to be activated remotely, as it is not necessary to "roll a truck" in order to make cross connections between various trunk and feeder facilities in the outside plant network in order to effect the connection. Once the initial investment is made, therefore, the ongoing installation and maintenance expenses are much

reduced. In the context of a competitive local exchange environment, the CLEC (Competitive LEC) commonly desires to lease from the ILEC (Incumbent LEC) a dedicated outside plant in the form of a "dry copper" circuit. A dry copper circuit is one which is has no electronics (e.g., load coils, repeaters or subscriber carrier systems) between the wire center and the customer premises. Such electronics interfere with the provisioning and support of most data services, including high-speed DSL (Digital Subscriber Line) services, which many of the CLECs are interested in providing. See also CLEC, Dedicated Inside Plant, Dry Copper, DSL, ILEC.

Dedicated Server A computer on a network that performs specialized network tasks, such as storing files. The word "dedicated" means that the computer is used exclusively as a server. It is not used as a workstation, which means no one is sitting in front of it, using it. A dedicated server sits all alone, attached to its network, working happily all by itself.

Dedicated Service A communication network devoted to a single purpose or group of users, e.g., AUTOVON, FTS. It may also be a subset of a larger network; e.g., AUTOVON, FTS.

Dedicated Trunk A trunk which bypasses the Attendant Console and rings through to a particular phone, hunt group or distribution group.

Deep Space Space at distances from the Earth approximately equal to or greater than the distance between the Earth and the Moon.

Deep UV Printed circuits are made by optical lithography, by shining light through a negative, or mask. In the beginning, it was visible light, but lately it's been made with more precise ultraviolet light, which can print lines as thin as 0.15 micron wide. They call this light "deep uv" and it is ultraviolet light emitted by an exciter or pulsing laser.

Defacto Standards Standards, widely accepted and used, but lacking formal approval by a recognized standards organization. The Hayes AT command set, MS-DOS, Windows, TCP/IP etc. are defacto standards.

Default The default is a factory-set hardware of software setting or configuration. It is the preset value that the program or equipment comes with. It will work with default values in the absence of any other command from the user. For example, communications software programs, such as Crosstalk, Blast, etc., have as their default settings 300 baud, 8 bit, one stop bit, no parity. If you want to run at 1,200 baud, you have to change that "default" setting.

Default Carrier Generic name given to the long distance carrier which will carry the traffic of customers who haven't pre subscribed to a long distance carrier.

Default Node Representation An ATM term. A single value for each nodal state parameter giving the presumed value between any entry or exit to the logical node and the nucleus.

Default Route A routing table entry that is used to direct any data addressed to any network numbers not explicitly listed in the routing table.

Defense Data Network DDN. The Department of Defense integrated packet switching network capable of worldwide multilevel secure and non-secure data transmission.

Deferred Processing Performing operations as a group or batch, all at once. Using batch mode, you can quickly prescan your documents, capturing just the image of each page, then perform recognition on these images later, freeing your computer for interactive work.

Definition A figure of merit for image quality. For video-type displays, it is normally expressed in terms of the smallest resolvable element of the reproduced received image.

Deflection Some automatic call distributors can be programmed to give callers a busy signal if the waiting time is past a certain threshold for callers to your 800 numbers (since you pay for all the waiting time). This is called "deflection" by Intecom and may have another name for other manufacturers. Of course the financial benefit of this must be weighed against the impression it gives your callers.

Degauss To demagnetize. To degauss a magnetic tape means to erase it. See Degaussing Coi. and Degausser.

Degaussing Coil Degaussing is to demagnetize. A degaussing coil is a long piece of wire bent into the shape of a circle. When a CRT becomes magnetized an area of the screen becomes discolored If you wave the degaussing coil around this area, the screen becomes demagnetized and the picture quality improves.

Degausser Device to demagnetize color picture tube for color purity. See Degaussing Coil.

Degradation In communications, that condition in which one or more of the established performance parameters fall outside predetermined limits, resulting in a lower quality of service.

Degraded Service State The condition wherein degradation prevails in a communication link. For some applications e.g., automatic switching to a non degraded standby link, degradation must persist for a specified period of time before a degraded service state is considered to exist.

Degree Of Coherence A measure of the coherence of a light source.

Degree Of Isochronous Distortion In data transmission, the ratio of (a) the absolute value of the maximum measured difference between the actual and the theoretical intervals separating any two significant instants of modulation (or demodulation) to (b) the unit interval. These instants are not necessarily consecutive. The degree of isochronous distortion is usually expressed as a percentage. The result of the measurement should be completed by an indication of the period, usually limited, of the observation.

Degree Of Start-Stop Distortion 1. In asynchronous data transmission, the ratio of (a) the absolute value of the maximum measured difference between the actual and theoretical intervals separating any significant instant of modulation (or demodulation) from the significant instant of the start element immediately preceding it to (b) the unit interval.
2. The highest absolute value of individual distortion affecting the significant instants of a start-stop modulation. The degree of distortion of a start-stop modulation (or demodulation) is usually expressed as a percentage.

Deinstallation A term used in the secondary telecom equipment business. The shutoff and disconnect of machine power and the disassembly of a PBX switch to prepare for its removal from a switch room facility. A properly executed deinstallation will include all necessary parts for the reassembly, operation, maintenance and acceptance of the switch and any of its components at the next location.

Dejitterizer A device for reducing jitter in a digital signal, consisting essentially of an elastic buffer into which the signal is written and from which it is read at a rate determined by the average rate of the incoming signal. Such a device is largely ineffective in dealing with low-frequency impairments such as waiting-time jitter.

DEK Data Encryption Key. The key by which one can unlock an encrypted message. There are "private" and "public" keys. See ENCRYPTION.

Delay The wait time between two events, such as the time from when a signal is sent to the time it is received. There are all sorts of reasons for delays, such as propagation delays, satellite delays, the additional time introduced by the network in delivering a packet's worth of data compared to the time the same information would take on a full-period, dedicated point-to-point circuit, etc. See also Latency.

Delay Announcements These are pre-recorded announcements to incoming callers that they are being delayed and being placed in an ACD queue. Sample: "Please wait. All our agents are permanently busy. You are being placed on Eternity Hold. Don't go away or you'll never be allowed back." Some announcements are giving callers sales pitches and some idea of how long they'll have to stay on line until someone helps them.

Delay Distortion The difference, expressed in time, for signals of different frequencies to pass through a phone line. Some frequencies travel slower than others in a given transmission medium and therefore arrive at the destination at different times. Delay distortion is measured in microseconds of delay relative to the delay at 1700 Hz. Also called Envelope Delay.

Delay Encoding A method of encoding binary data to form a two-level signal. A binary zero caused no change of signal level unless it is followed by another zero, in which case a transition takes place at the end of the first bit period. A binary "1" causes a transition from one level to the other in the middle of the bit period. Used primarily for encoding of radio signals since the spectrum of the encoding of signal contains less low frequency energy than an NRZ signal and less high frequency energy than a biphase signal.

Delay Equalizer A corrective piece of electronic circuitry designed to make communications circuit delays constant over a desired frequency range. A delay equalizer is a device that adds a delay to analog signals, which will travel through a medium faster than other frequencies used to transmit portions of the same data. The objective is to create a medium that transfers the information on all the used frequencies in the same time over the same distance, thus eliminating transmission delay distortion.

Delay Length Call Factor DLCF. An intermediate factor found in an Erlang C table that defines service level as a delay factor. It measures the delay of a call compared to the average handle time (AHT) of a call. It is used to determine the number of staff necessary to meet service level goals. For example, if the desired delay time is 20 seconds and is divided by a AHT of 150 seconds, the corresponding service level to locate in the Erlang table is 0.133. The table will give a staffing level.

Delay Line A transmission line, or equivalent device, designed to introduce delay.

Delay Modulation A modulation scheme that uses different forms of delay in a signal element. Frequently used in radio, microwave and fiberoptic systems.

Delay Skew The difference in timing of the transmission of signals between pairs in a cable. Delay skew is an issue when multiple pairs in a cable are used to support a single transmission. Examples include 100Base-T, which uses as many as four pairs to support a connection between a LAN hub and an attached device. By splitting the transmission across multiple pairs, the distance between the hub and the attached device can be increased considerably, as each element of the split signal runs at a lower carrier frequency than would the native signal over a single pair. This enhanced performance is due to the fact that lower frequency signals suffer less from the effects of attenuation (loss of signal strength). Delay skew, however, can have significantly adverse effects on this approach. See also 100Base-T and ATTENUATION.

Delay, Absolute The time elapsed between transmission of a signal and reception of the same signal.

Delayed Call Forwarding A phone system feaure in which you have your incoming calls forwarded to another number only after several rings.

Delayed Calls The fraction of calls delayed longer than a given time for service are called delayed calls. A telephone company definition.

Delayed Delivery Hold a message for delivery later. Just as the words say.

Delayed Delivery Facility A facility that employs storage within the data network whereby data destination for delivery to one or more addresses may be held for subsequent delivery at a later time.

Delayed Ring Transfer An optional KTU facility that provides for automatic transfer to the ringing signal from a principal telephone set to the attendant telephone station after an adjustable number of rings.

Delayed Sending A feature of fax machines which allows the machine to be programmed to send its transmissions at a later time — to take advantage of lower phone rates, for example.

Delayed Transmission A fax machine feature that allows a document to be transmitted automatically at a specific time.

Delimiter A character that separates the parts of a DOS command. For example, a backslash is the delimiter between subdirectory names. Also, the character (typically a comma or tab) that separates field items in a database.

Delivered Block A successfully transferred block.

Delivered Overhead Bit A bit transferred to a destination user, but having its primary functional effect within the telecommunication system.

Delivered Overhead Block A successfully transferred block that contains no user information bits.

Delivery Confirmation Information returned to the originator indicating that a given unit of information has been delivered to the intended addresses.

Delivery Envelope An X.400 term.

DELNI Digital Ethernet Local Network Interconnect. An industry adopted term indicating a multiport transceiver, also known as a fan-out, and generally limited to AUI connections in Ethernet. Originally, this was a product introduced by Digital Equipment corporation as the DEC Local Network Interconnect.

Delphi Forecasting One of the silliest methods of forecasting the future. Namely, to ask a bunch of alleged experts (often academic eggheads) what they think might happen and then averaging out their opinions, sort of.

Delta 1. The mathematical term for a finite increment in a variable. Once a value has been established, the delta is the difference between that value and the succeeding value. The delta can be either positive or negative. The benefit of delta modulation, for instance, is that the difference between the subject value and a previous value can be expressed in fewer bits than can the absolute subject value — it is a form of compression, in effect, yielding better utilization of expensive network bandwidth. See DELTA MODULATION.

2. A wiring system for distributing and utilizing three phase electrical power. In this system, three power carrying conductors are used, possibly with a fourth safety ground wire. The voltage between any two of the three power wires is the rated distribution voltage, which is most commonly 380 to 415 VAC in most countries or 208 VAC in North America. The other type of three phase power distribution is called the WYE style.

Delta Channel In T-carrier/ISDN communications, a delta channel/"D channel" contains signaling and status information.

Delta Frame Also called Difference Frame. Contains only the pixels different from the preceding Key Frame. Delta frames reduce the overall size of the video clip to be stored on disk or transmitted on phone lines.

Delta Modulation A method for converting analog voice to digital form for transmission. It is the second most common method of digitizing voice after Pulse Code Modulation, PCM. Sampling is done in all conversion of analog voice to digital signals. The method of sampling is what distinguishes the various methods of digitization (Delta vs.' PCM, etc.). In delta modulation, the voice signal is scanned 32,000 times a second, and a reading is taken to see if the latest value is greater or less than it was at the previous scan. If it's greater, a "1" is sent. If it's smaller, a "0" is sent.

Delta modulation's sampling rate of 32,000 times a second is four times faster than PCM. But Delta records its samples as a zero (0) or a one (1), while PCM takes an 8-bit sample. Thus PCM encodes voice into 64,000 bits per second, while Delta codes it into 32,000. Because delta has fewer bits, it could theoretically produce a poorer representation of the voice. In actual fact, the human ear can't hear the difference between a PCM and a Delta encoded voice conversation.

Delta modulation has much to recommend it, especially its use of fewer bits. Unfortunately no two delta modulation schemes are compatible with each other. So to get one delta-mod digital PBX to speak to another, you have to convert the voice signals back to analog. With AT&T making T-1 a de facto digital encoding scheme, PCM has become the de facto standard for digitally encoding voice. And although there are three types of PCM in general use, they can be made compatible on direct digital basis (i.e. without having to go back to analog voice). One problem with PCM is that American manufacturers typically put twenty four 64,000 bit per second voice conversations on a channel and call it T-1. The Europeans put 30 conversations on their equivalent transmission path. Thus, you can't directly interface the American and the European systems. But there are "black boxes" available...(In this business, there are always black boxes available.)

Deltas The changes.

Delta Sigma Modulation A variant of delta modulation in which the integral of the input signal is encoded rather than the signal itself. A normal delta modulation encoder by an integrating network. See DELTA MODULATION.

Delta Technology An Internet access tern. Delt technology is specialized remote adaptive routing protocols for optimizing bandwidth. It prevents unnecessary traffic from being sent over slow WAN connections by only sending the changes (deltas).

Delurking Coming out of online "lurking mode," usually motivated by an irresistible need to flame about something. "I just had to delurk and add my two cents to that conversation about a woman's right to abortion."

Deluxe Queuing A feature that allows incoming calls from phone users, tie trunks and attendants to be placed in a queue when all routes for completing a particular call are busy. The queue can be either a Ringback Queue (RBQ)— the user hangs up and is called back when a trunk becomes available — or an Off-Hook Queue (OHQ) — the user waits off-hook and is connected to the next available trunk. Deluxe Queuing is a term used mainly by AT&T. Most modern PBXs have this feature. Most have simpler names, however.

Demand And Facility Chart D&F. A telephone company definition. A chart designed to: a. Record an up-to-date picture of working network access lines, actual usage rates, future gains in working network access lines, and future usage rates.

b. Record the capacity of existing equipment and the current picture of the planned capacity additions.

c. Provide a recording vehicle to report consistent data (using standardized terminology and definitions) for planning, and budget review evaluation purposes.

Demand Assigned Multiple Access See DAMA.

Demand Assignment A technique where users share a communications channel. A user needing to communicate with another user on the network activates the required circuit. Upon completion of the call, the circuit is deactivated and the capacity is available for other users.

Demand Factor The ratio of the maximum demand on a power system to the total connected load of the system.

Demand Load In general, the total power required by a facility. The demand load is the sum of the operational load (including any tactical load) and non operational demand loads. It is determined by applying the proper demand factor to each of the connected loads and a diversity factor to the sum total.

Demand Paging The common implementation in a PC of virtual memory, where pages of data are read into memory from storage in response to page faults.

Demand Priority Access method providing support for time sensitive applications such as video and multimedia as part of the proposed 100BaseVG standard offering 100Mbit/s over voice grade UTP (Unshielded Twisted Pair) cable. By managing and allocating access to the network centrally, at a hub rather than from individual workstations, sufficient bandwidth for the particular application is guaranteed on demand. Users, say its proponents, can be assured of reliable, continuous transmission of information.

Demand Publishing The production of just the number of printed documents you need at the present time, as in "just in time." In short, the immediate production of printed documents which have been created and stored electronically.

Demand Service In ISDN applications, a telecommunications service that establishes an immediate communication path in response to a user request made through user-network signaling.

Demarc or Demark (Pronounced D-Mark.) The demarcation point between the wiring that comes in from your local telephone company and the wiring you install to hook up your telephone system — your CPE (Customer Provided Equipment) wiring. A Demarc might be anything as simple as an RJ-11C jack (one trunk) or an RJ-14C (two trunks) or an RJ-21X (up to 25 trunks) or a 66-block. A 66-block is a punchdown block on one side of which the telephone company may punch down its trunks and on the other, you may punch down your connections into your phone system. On a 66 block, there are little metal clips called "bridging clips" between you and the phone company. Lifting these clips off cuts your equipment from the phone company's trunks. This way you can quickly see whose fault it is. Often it's yours. But, if you think it's theirs and they come and find it's yours, they'll bill you. See DEMARCATION POINT or DEMARCATION STRIP.

Demarcation Point The point of a demarcation and/or interconnection between telephone company communications facilities and terminal equipment, protective apparatus,

or wiring at a subscriber's premises. Carrier-installed facilities at or constituting the demarcation point consist of a wire or a jack conforming to Subpart F of Part 68 of the FCC Rules.

Demarcation Strip The terminal strip or block (typically a 66 block) which is the physical interface between the phone company's lines and the lines going directly to your own phone system. See also DEMARC.

Demilitarized Zone See Screened Subnet.

DEMKO Denmark Elektriske MaterielKOntrol (Denmark Testing Laboratory).

Democratically Synchronized Network A mutually synchronized network in which all clocks in the network are of equal status and exert equal amounts of control on the others.

Demodulation The process of retrieving an electrical signal from a carrier signal or wave. The reverse of modulation. See MODEM.

Demodulator In general, this term refers to any device which recovers the original signal after it has modulated a high frequency carrier. In television, it may refer to an instrument which takes video in its transmitted form (modulated picture carrier) and converts it to baseband.

Demultiplex DEMUX. To separate two or more signals previously combined by compatible multiplexing equipment. As an ATM term, it is a function performed by a layer entity that identifies and separates SDUs from a single connection to more than one connection.

Demultiplexer A device that pulls several streams of data out of a bigger, fatter or faster stream of data.

Demultiplexing A process applied to a multiplex signal for recovering signals combined within it and for restoring the distinct individual channels of the signals.

Demux Jargon for demultiplexer.

DEN Directory Enabled Network. DEN is designed to integrate network hardware with directory services, such as Novell's Directory Services or Microsoft's Active Directory. DEN is designed as an extension to the Desktop Management Task Force's (DMTF's) Common Information Model (CIM). The DEN specification is under development by the Desktop Management Task Force (DMTF). It includes a standard approach to defining schema for integrating network equipment, such as switches and routers, with a directory service.

DENet The Danish Ethernet Network which consists of many Ethernet networks in universities connected together by bridges.

Denial Of Service You're no longer allowed to use a service. That service might be anything from normal phone service (you didn't pay your bills) or not being allowed into the company's email because you were just fired.

Denis The Little The 6th century monk who decided that history should be split between B.C. and A.D.

Density 1. The number of bits (or bytes) in a defined length on a magnetic medium. Density describes the amount of data that can be stored.

2. The number of circuits that can be packed into an integrated circuit.

3. In a facsimile system, a measure of the light transmission or reflection properties of an area, expressed by the logarithm of the ratio of incident to transmitted or reflected light flux.

Denizen A low life citizen of the Internet. Not a complimentary term.

Denwa Japanese for telephone. An aka denwa is a red telephone. Some coin phones in Japan are red and are often known as aka denwas.

Departmental Firewall NEC PrivateNet Systems

Group issued a White Paper called Connecting Safely to the Internet — A study in Proxy-Based Firewall Technology. In that White Paper, they defined an Departmental Firewall:
A departmental firewall is identical to an Internet firewall except that it controls access to and from a single department in a larger organization. It is used to protect sensitive corporate data, such as financial information and personnel records, from access by unauthorized people. A departmental firewall tends to be more generous in the access it allows, but if insecure services, such as NFS (Network File System), are allowed through a departmental firewall, the purpose of installing the firewall in the first place might be defeated. See FIREWALL and NFS.

Departure Angle The angle between the axis of the main lobe of an antenna pattern and the horizontal plane at the transmitting antenna.

Depersonalization In 1879 a flu epidemic in Lowell, MA made it likely that all four of the telephone operators would get sick simultaneously. To help substitute operators, management numbered each of the exchange's two hundred plus customers. No problem. The customers accepted the change easily.

Deplaning Getting off a plane. See DETRAINING.

Depolarization 1. In electromagnetic wave propagation, that condition wherein a polarized transmission being transmitted through a nonhomogeneous medium has its polarization reduced or randomized by the effects of the medium being traversed.
2. Prevention of polarization in an electric cell or battery.

Depopulate A technique to reduce the traffic load on a switch by removing devices from the shelf or cabinet. Depopulating reduces the effective device capacity of a switch but can increase switching capacity. This is a ploy used to give older PBX systems traffic capacity nearer true ACD systems.

Depressed Cladding Fiber An optical fiber construction, usually single mode, that has double cladding, the outer cladding having an index of refraction intermediate between the core and the inner cladding.

Deregistration A cellular radio term. In the CDPD network, the process of dissociating an Network Entity Identifier (NEI) from the CDPD network.

Deregulation The removal of regulatory authority to control certain activities of entrenched telephone companies. An attempt by federal authorities to make the telephone industry more competitive. Deregulation is meant to benefit the consumer. Sometimes it does. Sometimes it doesn't. Often, it's a scapegoat for whatever subsequently goes wrong. But it means different things in different countries. In some it means "a new company can come in and bash the living hell out of the local supplier," according to Philip Khoo, of Miller Freeman, Singapore. In other countries, it simply means means giving the new comers 5% of the market, while the old-timer keeps 95%.

Derivation Equipment Produces narrow band facilities from a wider band facility. Such equipment can, for instance, derive telegraph grade lines from the unused portion of a voice circuit.

Derived MAC PDU DMPDU. A Connectionless Broadband Data Service (CBDS) term that corresponds to the 1.2 PDU in Switched Multimegabit Data Service (SMDS). CBDC is the European equivalent of SMDS.

DES 1. Data Encryption Standard. A block cipher algorithm for encrypting (coding) data designed by the National Bureau of Standards so it is impossible for anyone without the decryption key to get the data back in unscrambled form. The DES standard enciphers and deciphers data using a 64-bit key specified in the Federal Information Processing Standard Publication 46, dated January 15, 1977. DES is not the most advanced system in computer security and there are possible problems with its use. Proprietary encryption schemes are also available. Some of these are more modern and more secure. The quality of your data security is typically a function of how much money you spend. See BLOCK CIPHER, CLIPPER CHIP and NSA.
2. Destination End Station: An ATM termination point which is the destination for ATM messages of a connection and is used as a reference point for ABR services. See SES.

Descenders Those parts (or tails) of the letters p, y, j and g which descend below the base line. This type style is much easier to read than one in which the tails rest on the base line. Watch out for true (i.e. real) descenders when you're buying a system. Most have them these days, but many telephone screens don't. Careful.

Descrambler A device which corrects a signal (often video) that has been intentionally distorted to prevent unauthorized viewing. Used with satellite TV systems. See DES.

DESI Strip A slang term for Designation Strip, the small printed piece of paper or card that slides into or attaches onto a telephone and tells you which button answers which line or which button does what in the way of features or intercom.

Designation Strip Also called Desi Strip. A designation Strip is the small printed piece of paper that slides into or attaches onto a telephone and tells you which button answers which line or which button does what in the way of features or intercom.

Designing Around Designing around is a legal process which deals with designing a new patent around an existing patent. This process often yields a better product than the patented device or method. Designing around is a perfectly legal operation. It has only been recognized and sanctioned by the courts for about 5 or 6 years. Most engineers and business owners are completely unaware of this option and think it sounds illegal when described to them. Designing around can be used to protect an existing patent or to file for a new one.

Design Layout Report DLR. A record containing the technical information that describes the facilities and terminations provided by a local telephone company to a long distance telephone company. The technical information is needed by the long distance carrier to design the overall service and includes such items as cable makeup (gauge, loading, length, etc.), carrier channel bank type and system mileage, signaling termination compatibility, etc. The DLR is sent to the designated Carrier representative via the local Telephone Company's Engineering Department.

Deskewing An imaging term. Adjusting — straightening — an image in software to compensate for a crooked scan.

Desktop The computer's working environment. The screen layout, the menu bar, and the program icons associated with the machine's operating environment. Apple's Macintosh (introduced on January 24, 1984) really started the idea that the computer's screen was a desktop. With the introduction of Windows, IBM PCs and clones now also have desktops.

Desktop Collaboration Using ISDN lines or analog lines with high speed lines, you can link your desktop computers so teleworkers, suppliers and clients can share documents and work together no matter where they are.

Desktop Connection AT&T's code name for hardware which includes an AT&T serial port adapter, DSS cable and a

9 to 25 pin connector and some software to make it work. The new name for desktop connection is PassageWay. It attaches to the back of the AT&T phone. The cable connects the adapter to your PC's serial port. In AT&T's words, the AT&T PassageWay integrates telephone functions with a Microsoft Windows 3.1 or greater application, facilitating outdialing using the Hayes Command Set used by modems, linking caller identification to PC business applications, and paving the way to an open interface between the PC software and the telephone. As a development platform, AT&T PassageWay offers an application layer access to features of the telephone including receiving information on visual/audible alerts and telephone displays, activating button presses, turning the speakerphone on or off, adjusting the volume of the telephone and dialing, including some call state progress. In addition, operating with Windows provides a unique way of accessing and sharing information between applications that co-reside on a PC. Windows provides a Graphic User Interface (GUI) that allows software packages to be easily developed. Windows also offers the Dynamic Data Exchange (DDE) interface between co-resident applications that is a standard for sharing information on DOS PCs. AT&T PassageWay also passes information using Microsoft's DDE standard. See also AT&T APPLICATIONS PARTNER.

Desktop Management Task Force A consortium of vendors working toward a set of standards for network management software which will ease the management of desktop systems, their components, and peripherals. Activities address "standard groups" such as software, PC systems, servers, monitors, network interface cards (NICs), printers, mass storage devices, and mobile technologies. The most notable of its accomplishments is the development of the Desktop Management Interface (DMI). www.dmtf.org See also DMI.

Desktop Metaphor A desktop metaphor is the conceptual way a workstation screen area is used to emulate a user's physical desktop through graphic icon images. The icon maps directly to its real life function. For example, a trash can icon will allow a user to "throw out" a document. Gives an application a "user friendly" feel. Desktop metaphors are consistent throughout all Windows and windows-like applications, like Sparc's OPEN LOOK.

Desktop Pattern A Windows 3.1 term. A design that appears across the desktop. You can use Control Panel to create your own pattern or choose a pattern provided by Windows.

Desktop Video Communications that rely either on video phones or personal computers offering a video window.

Desktop Videoconferencing By combining ISDN technology and individual PCs, people can meet "face-to-face" without leaving their offices. It's a way to reduce costly and time-consuming travel, maybe.

Despun Antenna Of a rotating communications satellite, an antenna, the direction of whose main beam with respect to the satellite is continually adjusted so that it illuminates a given area on the surface of the Earth, i.e., the footprint does not move.

Destination Address That part of a message which indicates for whom the message is intended. Usually a collection of characters or bits. Just like putting a destination address on an envelope. On a token ring network this is a 48-bit sequence that uniquely defines the physical name of the computer to which a LAN data packet is being sent. The IEEE assures that in the world of LANs no two devices have the same physical address. It does so by assigning certain num-

bers to vendors of token ring adapters, the devices that connect computers to a token ring network.

Destination Address Filtering A feature of bridges that allows only those messages intended for the extended LAN to be forwarded.

Destination Code See DESTINATION FIELD.

Destination Document The document into which an object is linked or embedded via OLE. The destination document is sometimes also called the container document.

Destination Field 1. A telephone company definition. A combination of digits that provides a complete address to reach a destination in the message network. Most destination codes are made up of some of the following components: Access Code, Area (NPA) Code, End Office Code (NNX), Main Station Number, Service Code, Toll Center Code.
2. A networking term. The field in a message header that contains the network address of the individual for whom the message is meant and who will (with luck and good management) receive the message.

Destination Node Those system nodes which receive messages over the control packet network from the source or transmitting node.

Destination Point Code DPC. The part of a routing label that identifies where the SS7 signaling message should be sent. See also POINT CODE.

Destination Service Access Point DSAP. The logical address of the specific service entity to which data must be delivered when it arrives at its destination. This information may be built into the data field of an IEEE 802.3 transmission frame.

Destuffing The controlled deletion of stuffing bits from a stuffed digital signal, to recover the original signal.

Detector 1. In a radio receiver, a circuit or device which converts or rectifies high frequency oscillations into a pulsating direct current or which translates radio frequency power into a form suitable for the operation of an indicator. This is most frequently a vacuum tube, less commonly a crystal. Coherers and delicate chemical rectifiers were used in former years.
2. In an optical communications receiver, a device that converts the received optical signal to another form. Currently, this conversion is from optical to electrical power; however, optical-to-optical techniques are under development.

Detem An opto-electronic transducer that combines the function of an optical detector and emitter in a single device or module. Do not confuse with DTERM, a name for one of NEC's telephones that work on its NEAX 2000 and 2400 PBXs.

Detent Tuner "Click" type of TV tuner.

Detour Difficulty in gaining access to the information highway. Often involves a Highway Construction Supervisor solving your access problem.

Detraining Getting off a train. An absolutely ghastly word invented by the railroad industry to keep them on a par with new, awful language invented by the airline industry. See also DEPLANING.

Device Contention 1. Occurs when more than one application is trying to use the same device, such as a modem or printer. Some of the newer operating systems do a better job handling device contention.
2. The way Windows 95 allocates access to peripheral devices, such as a modem or a printer, when more than one application is trying to use the same device.

Device Control A multimedia definition. Device control enables you to control different media devices over the network through software. The media devices include VCRs, laser disc players, video cameras, CD players, and so on.

Control capabilities are available on the workstation through a graphical user interface. They are similar to the controls on the device itself, such as play, record, reverse, eject, and fast forward. Device control is important because it enables you to control video and audio remotely — without requiring physical access.

Device Driver 1. A special type of software (which may or may not embedded in firmware) that controls devices attached to the computer, such as a printer, a scanner, a voice card, a diskette drive, a CD-ROM, a hard disk, monitor or mouse. Device drivers are typically loaded low into the memory of PCs in the MS-DOS CONFIG.SYS command, or they're loaded high using the Devicehigh command. A device driver is software that expands an operating system's ability to work with peripherals. A device driver controls the software routines that make the peripherals work.
2. A program that enables a specific piece of hardware (device) to communicate with Windows 95. Although a device may be installed on your computer, Windows 95 cannot recognize the device until you have installed and configured the appropriate driver.

Device ID A Plug and Play term. A code in a device's Plug and Play extension that indicates the type of device it is. The device ID and the vendor ID create a unique identifier for each PnP (Plug N Play) device.

DFA Doped Fiber Amplifier. An amplifier used in fiber optic systems to amplify light pulses. Such amplifiers typically are known as EDFAs (Erbium-Doped Fiber Amplifiers) as they are doped with erbium, a rare-earth element. DFAs are more effective than regenerative repeaters in many applications, as they simply amplify the light pulses through a chemo-optical process. Regenerative repeaters, on the other hand, require that the light signal be converted to an electrical signal, amplified, and then re-converted to an optical signal. Additionally, DFAs can simultaneously amplify multiple wavelengths of light in a Wavelength Division Multiplex (WDM) fiber system. See also EDFA, SONET and WDM.

DFB Distributed Feedback Laser. A type of laser used in fiber-optic transmission systems, at the distribution level of the local loop. DFBs are point-to-point lasers distributed among nodes in a geographic area such a neighborhood. They transmit and receive optical signals between the distributed nodes and the centralized node, where the signals are multiplexed over a higher-speed fiber link to the head-end (point of signal origin). DFBs can be more effective than the traditional approach of using a single laser which serves multiple nodes through a broadcast approach, as the available bandwidth can be segmented. DFBs have application in a FTTN (Fiber-To-The-Neighborhood) local loop scenario. See also FTTN and SONET.

DFI Digital Facility Interface. An 5ESS switch circuitry in a DTLU responsible for terminating a single digital facility and generating one PIDB (Peripheral Interface Data Bus).

DFT Direct Facility Termination. A telephone company trunk that terminates directly on one or more telephones.

DG Directorate General (CEC).

DGPS Differential Global Positioning Service — a new venture of the US Coast Guard, which it hopes to have ready by 1996. It will use an existing network of radio beacons throughout the US to create a fixed grid of known reference points in order to improve the accuracy of the Defense Department's GPS signal. The Coast Guard hopes to achieve an accuracy of about 10 meters.

DGT Direccion General de Telecommunicaciones (Spanish General Directorate of Telecommunications).

DHACP Dynamic Host Automatic Configuration Protocol. See DHCP.

DHCP Dynamic Host Configuration Protocol. A developing IETF (Internet Engineering Task Force) protocol which allows a server to dynamically assign IP addresses to nodes (workstations) on the fly. Like its predecessor, The Bootstrap Protocol (Bootp), DHCP allows supports manual, automatic and dynamic address assignment; provides client information including the subnetwork mask, gateway address, and DNS (Domain Address Server) addresses; and is routable. DHCP offers the advantage of automatic configuration, whereas Bootp must be configured manually. A DHCP server, generally in the form of a dedicated server, verifies the device's identity, "leases" it the IP address for a predetermined period of time, and reclaims the address for reassignment at the expiration of that period. DHCP relieves the pressure on the current IPv4 numbering scheme; the emergence of IPv6, with its expanded numbering scheme, may obviate DHCP. See also BOOTP and IPv6.

Dhrystones Benchmark program for testing the speed of a computer. It tests a general mix of instructions. The results in Dhrystones per second are the number of times the program can be executed in one second. The Dhrystone benchmark program is used as a standard indicating aspects of a computer system's performance in areas other than its floating-point performance, for instance, integer processes per second, enumeration, record and pointer manipulation. Since the program does not use any floating-point operations, performs no I/O, and makes no operating system calls, it is most applicable to measuring the performance of systems programming applications. The program was developed in 1984 and was originally written in Ada, although the C and PASCAL versions became more popular by 1989. See WHETSTONES.

DHT Direct to Home satellite Tv.

DIA/DCA Document Interchange Architecture/Document Content Architecture. IBM promulgated architectures, part of SNA, for transmission and storage of documents over networks, whether text, data, voice or video. Becoming industry standards by default.

Diagnostic Programs Programs run by the computer portion of a PBX to detect faults in the system. Such programs may run automatically at regular intervals or continuously. The goal of diagnostic programs is to detect faults before they become serious and to alert someone — typically the attendant — to do fix it. Some diagnostic programs stop running when the switch gets too busy. Some don't. You can dial into some diagnostic programs from afar. And you can't in some. Remote diagnostic programs are probably the greatest boons to improved reliability of telephone systems.

Diagnostics A term used in the secondary marketplace. Original Equipment's Manufacturer (O.E.M) prescribed test procedure whose successful completion is normally required for maintenance acceptance of a switch, cabinet, or peripheral piece of equipment. Comment: A new maintenance contract will not go into effect until the maintenance company accepts the results of the diagnostics.

DIALAN DMS Integrated Access Local Area Network

Dial A Prayer A sarcastic name for the local 611 number run by the local telephone companies as their centralized number for repair.

Dial Around A method used by callers to purposely bypass a payphone company's local or long distance carrier services. Such methods include calling cards and alternative carrier's collect services, such as 1-800-COLLECT or 1-800-

CALL-ATT. Payphone operators receive little or no revenue from such calls.

Dialback Security Dialback security is a telecom security feature. If a person calls in wanting remote access, the system asks for a password. Once it receives a correct password, it hangs up on the caller and dials back a pre-defined remote number, only then giving the caller access. Unless the hacker has you tied up in your living room, it makes things very secure. It can be made even more secure with multiple passwords and features like voice recognition.

Dial Backup 1. A network scheme using dial-up phone lines as a replacement for failed leased data lines. In one typical case, two dial-up lines can be used. One dial-up link is used to transmit data and the other to receive data, thus giving us full-duplex data transmission.

2. A security feature that ensures people do not log into modems that they shouldn't have access to. When a connection is requested, the system checks the username presented for validity, then "dials back" the number associated with that username.

Dial By Name You can dial someone by spelling their name out on the touchtone pad. Typically, the system plays a recorded announcement giving directions for using the Dial by Name feature: the caller then inputs the appropriate digits/letters. When the system recognizes a match, a recorded announcement states the name of the dialed party for confirmation by the caller before automatically completing the call. If the input digits are not uniquely associated with a particular station the system may ask the caller to pick a name from a menu of choices. Dial by name is getting cheaper. Automated attendants are being programmed to have the feature. And you shouldn't buy an auto attendant unless it has this feature.

Dial Call Pickup A phone user on a PBX or hybrid can dial a special code and answer calls ringing on any other phone within his own predefined pickup group.

Dial Dictation Access A service feature available with some switching systems that permits dialing a special number to access centralized dictation equipment.

Dial In Banner An Internet Access Term. Optional pop-up window for dial-in connections. Allows network managers to display information or warning messages when users dial into remote networks.

Dial In Channel Aggregation The ability to use more than one communications channel per connection. By aggregating both 64 kbit/s ISDN B channels, users can take advantage of 128 kilobits per second dial-in connections. Fast 128 kbit/s data transfer rates reduce large file transfer times. The same as BONDING.

Dial In Tie Trunk A Dial In Tie trunk is a trunk that may be accessed by dialing an access code and then seizing a dedicated transmission path to a distant PBX (or another PBX a short distance away). Once the trunk is seized in the distant PBX, the caller may then use the features of that PBX, depending on the class-of-service and restrictions assigned to the trunk.

Dial IT 900 Service A special one-way mass calling service that allows prospects, customers and others to reach you from anywhere across the country. In contrast to 800-service, the caller pays the 900 charge, generally one charge for the first minute, with a lesser charge for each additional minute. DIAL-IT 900 Service is a great way to involve your customers and prospects in a promotion! Premium Billing lets you select a rate above standard DIAL-IT 900 rates. The long distance carriers (through their deals with local phone companies)

handle the billing. You, the information provider, split the revenues with the long distance provider. International DIAL-IT 900 service is currently available from a growing number of countries.

Dial Level The selection of stations or services associated with a PBX, based on the first digit(s) dialed.

Dial Pick-Up PBX feature. A phone on a PBX can answer another ringing phone by dialing a few digits. Also called an access code.

Dial Pulse Signaling A type of address signaling in which dial pulse is implemented to signal the distant equipment. See DIAL PULSING.

Dial Pulsing A means of signaling consisting of regular momentary interruptions of a direct or alternating current at the sending end in which the number of interruptions corresponds to the value of the digit or character. In short, the old style of rotary dialing. Dial the number "five" and you'll hear five "clicks." See DIAL SPEED, DIAL TRAIN and DTMF.

Dial Repeating Trunks PBX tie trunks used with terminating PBX equipment capable of handling telephone signaling without attendant involvement.

Dial Service Assistance DSA. A service feature associated with the switching center equipment to provide operator services, such as information, intercepting, random conferencing, and precedence calling assistance.

Dial Speed The number of pulses a rotary dial can send in a given period of time, typically 10 per second. A Hayes modem with a communications package, like Crosstalk, can send 20 pulses per second.

Dial String A Dial String is the sequence of characters sent to a device which can dial a phone number. Such a device might be a modem or a voice processing card. Here are some "digits" in a dial string: ! — flashhook (TAPI standard); & — flashhook (Dialogic); T — use tone dialing; , — pause (typically of half a second to two seconds); W — wait for dial tone.

Dial String/Command String A sequence of characters and digits used for dial-in access; ATDT5107861000,,,,,,,,,123456, H<CR> for example.

Dial Through A technique, applicable to access circuits, that permits an outgoing routine call to be dialed by the PBX user after the PBX attendant has established the initial connection.

Dial Tone The sound you hear when you pick up a telephone. Dial tone in North America is a unbroken signal (350 + 440 Hz) from your local telephone company that it is alive and ready to receive the number you dial. If you have a PBX, dial tone will typically be provided by the PBX. Dial tone does not come from God or the telephone instrument on your desk. It comes from the switch to which your phone is connected to. Outside North America, dial tone often sounds very different. And modems PCs, which are set up to "wait for dial tone" often don't recognize these unusual dial tones. The key then is a disable the modem's property that says "wait for dial tone," and have it begin dialing the second it goes off hook.

Dial Tone Delay 1. The specific time that transpires between a subscriber's going off-hook and the receipt of dial tone from a servicing telephone central office. It's a measure of the time needed to provide dial tone to customers. Many of the local public service commissions in the United States say that 90% of customers should receive dial tone in fewer than three seconds.

2. A telephone company definition. Percent Dial Tone Delay (% DTD) over three seconds is a measurement of calls that did not receive dial tone within three seconds. The average busy season objective for an entity in the busy season of

exhaust is a maximum of 1.5 percent and is the engineering objective ceiling for all types of equipment. In addition" the following maximum DTD engineering ceilings for an entity to be included in the equipment design:

1. Highest Annually Recurring Day - Not over 20%. types of offices. This maximum ceiling is to be applied to all types of offices.

2. Average 10 High Day — not over 8%. This maximum ceiling is to be applied in analog ESS offices.

Dial Tone First Coin Service A type of pay phone service in which dial tone is received when the caller goes off-hook and coins must be inserted only after the call is connected.

Dial Tone Speed DTS. A telephone company definition. The length of time required for switching equipment to provide dial tone to a subscriber originating a call. Usually expressed as the percentage of attempts that delayed over three seconds.

Dial Train The series of pulses or tones sent from the phone that's calling and the switching system it's attached to in order to signify the call's destination.

Dial Up The use of a dial or push button telephone to create a telephone or data call. Dial-up calls are usually billed by time of day, duration of call and distance traveled. A connection to the Internet, or any network, where a modem and a standard telephone are used to make a connection between computers. See DIAL UP LINE.

Dial Up Account You want to access the Internet. You dial your local Internet Service Provider (ISP) via a local phone number. That's called a Dial Up Account. You can also have a dedicated account, which means that physically a piece of wire (and other electronics) connects you (i.e. your computer) to your ISP 24-hours a day, seven days a week. Mostly, dedicated is more expensive than dial up (depending on rates, etc.). But there are big advantages to dedicated — including, faster access, getting your mail the instant it comes in, etc. See xDSL and Cable Modem.

Dial Up Line A telephone line which is part of the switched nationwide telephone system. Typically a "dial up line" is a standard analog POTS line. These days, ISDN lines are dial up, also. So this definition is changing also.

Dial Up Modem A modem that works on the public switched telephone network (PSTN) and connects to a remote computer resource for the duration of an individual call. See Modem.

Dial Up Networking A Windows 95 and 98 definition. It is a service that provides remote networking for telecommuters, mobile workers, and system administrators who monitor and manage servers at multiple branch offices. Users with Dial-Up Networking on a computer running Windows 95 can dial in for remote access to their networks for services such as file and printer sharing, electronic mail, scheduling, and SQL database access.

Dial Zero Phone A telephone on a Northern Telecom Norstar phone system which is assigned to ring when someone dials 0 (zero) from another Norstar telephone.

Dialed Number Identification Service See DNIS and 800 SERVICE.

Dialed Number Recorder Also called a Pen Register. An instrument that records telephone dial pulses as inked dashes on paper tape. A touchtone decoder performs the same thing for a touchtone telephone.

Dialing Parity Dialing parity is a technological capability that enables a telephone customer to route a call over the network of the customer's preselected local or long distance phone company without having to dial an access code of extra digits. Here's a more technical definition from the Telecommunications Act of 1996: The term `dialing parity' means that a person that is not an affiliate of a local exchange carrier is able to provide telecommunications services in such a manner that customers have the ability to route automatically, without the use of any access code, their telecommunications to the telecommunications services provider of the customer's designation from among 2 or more telecommunications services providers (including such local exchange carrier).

Dialing Pattern Dialing pattern refers to the digits you need to dial to place local, long distance, collect calls, or other phone calls. Dialing patterns will vary due to different types of telephone carrier switching equipment, computer software and the host carrier's credit policies (e.g. automatic roaming versus credit card roaming when using a cellular phone).

Dialing Plan A description of the dialing arrangements for customer use on a network.

Dialing Report A call center term. A report that summarizes the results of all numbers dialed for one or more telemarketing campaigns over a specified period of time.

Dialog Box A dialog box is a temporary window which prompts you to input information or make selections necessary for a task to continue.

Dialup Switch An Internet Access Term. Category of switching equipment designed to manages the dialup connections between the PSTN and either the Internet or a corporate LAN internetwork, providing security, accounting, and service management capabilities.

Diaphragm The thin flexible sheet which vibrates in response to sound waves (as in a microphone) or in response to electrical signals (as in a speaker or the receiver of telephone handset).

Dibit A group of two bits which can be represented by a single change of modulation of the carrier signal. On phase modulation, one of four phases in four-phase modulation is used to represent 00, 01, 10 or 11.

Dichroic Filter An optical filter designed to transmit light selectively according to wavelength (most often, a high-pass or low-pass filter).

Dichroic Mirror A mirror designed to reflect light selectively according to wavelength.

Dictation Access and Control A telephone system feature which allows a user to dial a dictation machine and use that machine (giving it instructions by push button) as if it were in his office. Typically, the material on that dictation machine is taken off by one or several typists of a centralized pool and word processed into letters, reports, legal briefs, etc. Telephone suppliers usually don't supply the dictation equipment. Newer telephone dictation machinery is, in reality, a specialized application of voice processing equipment. See VOICE PROCESSING.

Dictation Tank A recording gadget which receives messages dictated through the telephone system. This tank contains tape which can then be transcribed into letters or documents. See DICTATION ACCESS AND CONTROL.

Dictionary Attack Someone is trying to figure out a password to your network. So they throw millions of possible passwords at your system one word after another after another. This is called a dictionary attack. The simple solution to such an attack is to configure your network so that it takes three attempts at typing in the correct password and then hangs up on the caller.

DID Direct Inward Dialing. You can dial inside a company directly without going through the attendant. This feature used to be an exclusive feature of Centrex but it can now be provided by virtually all modern PBXs and some modern hybrids, but you must connect via specially configured DID lines from your local central office. A DID (Direct Inward Dial) trunk is a trunk from the Central office which passes the last two to four digits of the Listed Directory Number to the PBX or hybrid phone system, and the digits may then be used verbatim or modified by phone system programming to be the equivalent of an internal extension. Therefore, an external caller may reach an internal extension by dialing a 7-digit central office number. Notice: DID is different from a DIL (Direct-In-Line) where a standard, both-way central office trunk is programmed to always ring a specific extension or hunt group. DID lines cannot be used for outdial operation, since there is no dialtone offered.

DID director A standalone box which interpret DID data from trunks and route incoming calls to the appropriate fax mailboxes on your fax server. The boxes strip DID, read it, then tell the server who the fax is for by sending out DTMF. In a real fax, this means the answering fax sends its capabilities in HDLC (High-Level Data-Link Control)-encoded data frames: a Digital Identification Signal (DIS) frame spouts off the standard feature set (defined by the TSS, or ITU, the international fax standards body) the fax has. An NSF (Non-Standard Facilities) frame about what vendor-specific features the fax has comes next; a CSI (Called Subscriber Identification) frame gives the calling fax's telephone number. The sending fax responds with its Digital Command Signal (DCS) frames, informing the answering fax of modem speed, image width, image encoding and page length. The sender's phone number then comes across in a Transmitter Subscriber Information (TSI) frame, as well as a response to the answering fax's non-standard facilities frame.

Synchronous, real-time communication lets senders know if their fax stopped going through while they're still faxing. Computer-based fax servers, blasters, etc., used to use the store-and-forward paradigm (longer-term storage, so faxes could be sent using off-peak phone rates). This gave away the alerting advantage. Dialogic is now making boards for fax modems that let you know if your fax went through (sending faxes real-time over local phone lines and the Internet, an even bigger trick). Traditional faxing is losing ground here.

THESE ARE INDIVIDUAL BOXES

• Your fax is connected on a two-wire Tip-and-Ring line circuit, running from either an analog card in your PBX, or straight from your telco's local CO. Tip is the (-) wire, and ring is the (+) wire.

In analog signaling, sound and fax data are represented by varying current on the line. Digital signaling involves faster and more accurate data transfer of ones and zeroes, on/off signaling, plus some packet framing information.

Not every switch can retransmit this information. As a result, digital fax signaling can involve ISDN, more wires, proprietary protocols. All of this would destroy the fax's universality as a communications medium, it's biggest strength. At present, scanned digital image data gets converted back and forth from/to analog signals by modems, generally slowing things down and introducing errors.

• All faxes have modems. A modem (MOdulator/DEModulator) ✎
hooked to the analog line's tip and ring wires and converts digital ones and zeroes to sine-wave shaped variations in the voltage on the lines. The modulates the frequency / amplitude

/ phase of this sine-wave voltage signal to send ones and zeros, and demodulates current at the far end back into "one" bits and "zero" bits. Few faxes can swap their modems out with any ease, but computer-based fax servers / blasters / routers can readily replace a modem. Software drivers can make a data modem double as a fax.

• CNG tone is the key in fax routing. Line managers look for the 1100 Hz sine-wave CNG calling tone (use AC sine-waves, not DC square-waves, for a good one), route calls with CNG tone to a fax machine and calls without tone to a phone. They can also work with distinctive ringing (one ring for fax, two for phone) provided by the phone company. Great for SOHOs. To route your fax to a live fax machine prior to CNG, phone companies offer a service which lets you terminate different phone numbers on the same set of lines. This is a hunt group; when an incoming call arrives, the switch searches (hunts) for an available, non-busy line and sends the fax there. If your equipment responds with a wink signal (reversing polarity +/- on the line circuit), the CO switch can send you the digits dialed as DTMF. This is DNIS, Dialed Number Identification Service. You can have faxes routed to fax machines by number. Analog DNIS is called DID, for Direct Inward Dial. This lets you call numbers on your company's PBX or with a three- or four-digit extension. Computer-based fax systems can use OCR (Optical Character Recognition), in addition to DTMF. Using OCR, a computer-based system looks for keywords, like names or what follows "TO:" or an extension number and routes the fax that way.

The little sending fax number at the top line of a fax comes from the Transmitter Subscriber Information (TSI) frame, transmitted by the sending fax machine during the capability-exchange handshake phase. It doesn't depend on Caller ID or the CO switch; you put the information on the sending fax machine.

DIEL Advisory committee on telecommunications for DIsabled and Elderly People (UK).

Dielectric A nonconducting or insulating substance which resists passage of electric current, allowing electrostatic induction to act across it, as in the insulating medium between the plates of a condenser. Also an insulating material otherwise used (e.g. a Bakelite panel, or the cambric covering of a wire is a dielectric material). See also SEMICONDUCTOR.

Dielectric Absorption The penetration of a dielectric by the electric strain during a period of time.

Dielectric Constant The ratio of the capacity of a condenser with a given dielectric to the capacity of the same condenser with air as the dielectric.

Dielectric Lens A lens made of dielectric material that refracts radio waves in the same material that an optical lens refracts light waves.

Dielectric Process A printing process that uses a specially treated, charge-sensitive paper. Paper is roller-fed past an electrode array where an electrical charge is applied on line-by-line to form a latent image, then passed through a toner. The toner adheres to the charged image and heat fuses the toner to the paper to create the printed document.

Dielectric Sheath Or Cable A sheath or cable that contains no electrically conducting materials such as metals. Dielectric cables are sometimes used in areas subject to high lightning or electro-magnetic interference. Synonym for non-metallic cable.

Dielectric Strength The property of material which resists the passage of an electric current. It is measured in terms of voltage required to break down this resistance (such as volts per mil.).

Dielectric Test A test in which a voltage higher than the rated voltage is applied for a specified time to determine the adequacy of the insulation under normal conditions.

Differential Manchester Encoding A digital signaling technique in which there is a transition in the middle of each bit time to provide clocking. The encoding of a zero or one is represented by the presence (absence) of a transition at the beginning of the bit period.

Differential Mode For AC power systems, the term differential mode may refer to either noise or surge voltage disturbances. The terms normal mode and differential mode are interchangeable. Differential mode disturbances are those that occur between the power hot (black wire) and the neutral conductor (white wire). Most differential mode disturbances result from load switching within a building, with motor type loads being the biggest contributor. Surge voltages that come from outside of the building, such as surges caused by lightning, enter the building on the hot (black) wire and are therefore primarily differential mode in nature since the neutral (white) wire is nominally at ground voltage. Surge suppressors sometimes divert differential mode noise and surges into the neutral wire, resulting in voltages on the neutral wire called common modem noise or surge voltages. This definition courtesy APC.

Differential Mode Termination A type of cable termination where a pair of wires is terminated by a resistance matching the cable impedance, but there is no termination resistance between that pair and any adjacent pairs. For low-frequency signals this is often acceptable, but for a high-frequency environment (whether due to high-speed network protocols, or due to transmission towers nearby), this allows large voltages to exist between one pair and an adjacent pair.

Differential Positioning Precise measurements of the relative positions of two receivers tracking the same GPS (Global Positioning System) signals.

Diffie-Hellman Key A technique of changing encryption techniques on the fly. In a landmark 1976 paper, called New Directions in Cryptograph, IEEE Transactions on Information Theory, W. Diffie and M. Hellman describe a method by which a secret key can be exchanged using messages that do not need to be kept secret. This type of "public" key management provides a significant cost advantage by eliminating the need for a courier service. In addition, security can be considerably enhanced by permitting more frequent key changes and eliminating the need for any individual to have access to the key's actual value.

Diffraction The deviation of a wavefront from the path predicted by geometric optics when a wavefront is restricted by an opening or an edge. Diffraction is usually most noticeable for openings of the order of a wavelength.

Diffraction Grating An array of fine, parallel, equally spaced reflecting or transmitting lines that mutually enhance the effects of diffraction at the edges of each so as to concentrate the diffracted light very close to a few directions depending on the spacing of the lines and the wavelength of the diffracted light.

Digicash A name for electronic money transmitted in and around the Internet.

Digiterati The digital version of literati. That vague collection of people who seem to be hip and know something about the digital revolution.

Digroup 24 channels.

Digirepeater Digital Repeater

Digit Any whole number from 0 to 9.

Digit Deletion It's nice to make it easy for people to dial their desired numbers. Part of making it "nice" is to keep their pattern of dialing consistent. The charm of our ten-digit numbering system in North America — the three digit area code and seven digit local number — is its consistency, making for easy use and easy remembering. Some corporate networks, however, don't use a common numbering scheme. They might use tie trunks to get to Chicago, and insist on the user dialing 69, instead of the more common 312 area code. They might insist on the user dialing 73 when he wants to go to Los Angeles. But if he wants to reach the LA office, he might dial 235. This can be awfully confusing. So some switches — central office and PBXs — have the ability to insert or delete digits. That is, they will recognize the number dialed and change it as it progresses through the network. The user, however, knows nothing of this. He simply dials a normal phone number and listens as his call progresses normally. Digit insertion and digit deletion are components of a PBX feature called common number dialing.

Digit Insertion See DIGIT DELETION.

Digital 1. In displays, the use of digits for direct readout. 2. In telecommunications, in recording or in computing, digital is the use of a binary code to represent information. See PCM (as in Pulse Code Modulation.) Analog signals — like voice or music — are encoded digitally by sampling the voice or music analog signal many times a second and assigning a number to each sample. Recording or transmitting information digitally has two major benefits. First, the signal can be reproduced precisely. In a long telecommunications transmission circuit, the signal will progressively lose its strength and progressively pick up distortions, static and other electrical interference "noises."

In analog transmission, the signal, along with all the garbage it picked up, is simply amplified. In digital transmission, the signal is first regenerated. It's put through a little "Yes-No" question. Is this signal a "one" or a "zero?" The signal is reconstructed (i.e. squared off) to what it was identically. Then it is amplified and sent along its way. So digital transmission is much "cleaner" than analog transmission. The second major benefit of digital is that the electronic circuitry to handle digital is getting cheaper and more powerful. It's the stuff of computers. Analog transmission equipment doesn't lend itself to the technical breakthroughs of recent years in digital. See also PCM, as in Pulse Code Modulation.

Digital Access And Cross-Connect System See DACS and DIGITAL CROSS-CONNECT SYSTEM.

Digital Audio The storage and processing of audio signals digitally. It usually requires at least 16 bits of linear coding to represent each digital sample.

Digital Audio Radio Service See DARS.

Digital Audio Tape See DAT.

Digital Cash Once there were only stores. To buy something you needed money. Then they invented checks. And storekeepers took them. Then they invented credit cards, which were sort of checks that you paid later. Once we had credit cards, we could invent direct mail catalogs and 800 lines and call centers that took your orders via the 800 lines and you paid for what you bought by giving them your credit card number. Then came the Internet and vendors started to put catalogs on the Internet in the hope that somebody would buy from them. But they needed a way to get paid. Credit cards worked but many journalists wrote about how the Internet was "insecure" and anyone could steal your number and go on a spending binge. No one asked the journalists to cite instances of spending binges. Nor did they ask the journalists about the Federal Government legislation which limits credit card liability in the case of fraud to $50. But scaring people was a good story. Meantime, some entrepreneurs thought there was an opportunity to solve people's paranoia by creating "digital cash." No one exactly knows what digital cash is, yet. Lots of people are working on variations of it. But the idea is that it will be some form of encoded information transfer that contains instructions to take money from one person and pay it to another. We'll see how it evolves.

Digital Cellular The state of the art in cellular communications technology. Implementation will result in substantial increases in capacity (up to 15 times that of analog technology). In addition, digital will virtually eliminate three major problems encountered by users of analog cellular: static, loss/interruption of signal when passing between cells (during handoff), and failure to get a connection because of congested relays. Specifications for TDMA digital systems have been developed in North America (D-AMPS), in Europe (GSM) and in Japan (PDC).

Digital Certificate A fancy term for buying goods and services on-line over the Internet using your credit card, possibly in conjunction with some verification of who you are from an independent certification authority. See CERTIFICATE.

Digital Circuit Multiplication DCM is a variation of analog TASI — Time Assigned Speech Interpolation. In DCM, speech is encoded digitally and advanced voice band coding algorithms are applied to TASI's old speech interpolation techniques. DCM delivers a four to fivefold increase in the effective capacity of normal pulse code modulation (PCM) T-1 links operating at 1.544 megabits per second. DCM equipment is used on the TAT-8 transatlantic optical fiber submarine cable. Most DCM equipment has three operating elements: a speech activity detector. An assignment mapping and message unit, and a speech reconstitution unit. See February, 1987 issue of Data Communications for more.

Digital Coast The City of Los Angeles. In March of 1998, the mayor of Los Angeles, Richard J. Riordan, announced that Los Angeles would now be known as Digital Coast.

Digital Command Signal Signal sent by a fax machine or card when the caller is transmitting, which tells the answerer how to receive the fax. Modem speed, image width, image encoding and page length are all included in this frame.

Digital Communications Manager DCM. An MCI monitoring system that maintains communications through the network with the Site Controllers, the Extended Superframe Monitoring Units, and the I/O DXCs. The DCM issues requests for data and collects alarm and performance information, which is processed and stored in real-time for further computation and display.

Digital Compact Cassette DCC. A digital version of the familiar analog audio cassette. A DCC recorder can play and record both analog and digital cassettes. But the digital ones will sound a lot better.

Digital Convergence A Microsoft term for getting all the digital devices of the office and the home together working in a seamless architecture. See AT WORK.

Digital Cross-Connect System DACS (Digital Automatic Cross-Connect System) or DCS (Digital Cross-Connect System). A specialized type of high-speed data channel switch. It differs from a normal voice switch, which switches transmission paths in response to dialing instructions. In a digital cross-connect system, you give it separate and specific instructions to connect this line to that. These

instructions are given independently of any calls that might flow over the system. This contrasts with a normal voice switching in which switching instructions and conversations go together. Commands to a digital cross-connect system can be given by an operator at a console or can be programmed to switch at certain times. For example, you might want to change the T-1 24-voice conversation circuit to Chicago at 11 A.M. each day to allow for the president's 30 minute video conference call.

Digital Echo Canceller A Digital Echo Canceller is an echo canceller as opposed to an echo suppressor. An echo canceller filters out unwanted echoes among incoming signals, while the echo suppressor shut-offs the entire signal, by using an analog voice switch. The Digital Echo Canceller is one application of a digital transversal filter.

Digital Enveloping Digital enveloping is an application in which someone "seals" a message m in such a way that no one other than the intended recipient, say "Bob," can "open" the sealed message. The typical implementation of digital enveloping involves a secret-key algorithm for encrypting the message (i.e., a content-encryption algorithm) and a public-key algorithm for encrypting the secret key (i.e., a key-encryption algorithm)

Digital Ethernet Local Network Interconnect DELNI. The product offered by Digital Equipment Corp. that allows up to eight active devices to be connected to a single Ethernet transceiver. A similar device is manufactured by many other suppliers under various names. The DELNI can be thought of as "Ethernet in a box."

Digital European Cordless Telecommunication See DECT.

Digital Facilities Management System A Northern Telecom software which integrates the maintenance of all types of digital facilities from T-1 to the high-bit fiber. Largely used by telephone companies.

Digital Facsimile Equipment Facsimile equipment that employs digital techniques to encode the image detected by the scanner. The output signal may be either digital or analog. Examples of digital facsimile equipment are ITU-T Group 3, ITU-T Group 4, STANAG 5000 Type I and STANAG 5000 Type II.

Digital Frequency Modulation The transmission of digital data by frequency modulation of a carrier, as in binary frequency-shift keying.

Digital Hierarchy The standardized increments for multiplexing digital channels. For twenty four 64 Kbps DS-O channels are multiplexed into one 1,544 million bits per second DS-1 channel. See also SONET.

Digital LAT Protocol The LAT protocol, announced by Digital in the mid '80s, is today one of the industry's most widely used protocols for supporting character terminals over Ethernet networks. See LAT.

Digital Line Protection Many extensions behind PBXs deliver greater voltage to the desk than do normal tip and ring analog lines. This higher voltage can damage a PCMCIA modem inside a laptop. In fact, it can destroy the modem. Newer PCMCIA cards (now called PC cards) have Digital Line Protection, which protects against that higher voltage — what one manufacturer called "innovative isolation circuitry."

Digital Loop Carrier See DLC.

Digital Loopback A diagnostic feature on a modem, a short haul microwave or some other digital transmission equipment which allows the user to loop a signal back from one part of the system to another to test the circuit or the equipment. Digital loopbacks can be as long or as short as are necessary to isolate the problem. By looping a signal back and measuring it at both ends of the loop (at the beginning and at the end), you can see if the device carried the message cleanly and is thus, operating correctly.

Digital Microwave A microwave system in which the modulation of the radio frequency carrier is digital. The carrier is still a standard microwave radio wave. The digital modulation may be frequency or phase shift, but the control of that modulation is the digital bit stream.

Digital Modem Originally, a digital modem was a term used to describe a piece of hardware that can support ISDN lines. It was also called an ISDN terminal adapter. But as

Digital Modulation A method of decoding information for transmission. Information, e.g. a voice conversation, is turned into a series of digital bits - the 0s and 1s of computer binary language. At the receiving end, the information is reconverted back into its analog form.

Digital Monitor Receives discrete binary signals at two levels; one level corresponds to Logic 1 (true) while the other corresponds to Logic 0 (false). Monitors generally were of this type before VGA models appeared. Digital monitors do not have as wide a range of color choices as analog types; digital EGA monitors, for example, can display just 16 colors out of a palette of 64.

Digital Multiplex Hierarchy An ordered scheme for the combining of digital signals by the repeated application of digital multiplexing. Digital multiplexing schemes may be implemented in many different configurations depending upon the number of channels desired, the signaling system to be used, and the bit rate allowed by the communication medium. Some currently available multiplexers have been designated as D1-, DS-, or M-series, all of which operate at T-carrier rates. Extreme care must be exercised when selecting equipment for a specific system to ensure interoperability, because there are incompatibilities among manufacturers' designs (and various nations' standards).

Digital Multiplexed Interface A ISDN PRI-like connection between a PBX and a computer, developed by AT&T.

Digital Multiplexer A device for combining digital signals. Usually implemented by interleaving bits, in rotation, from several digital bit streams either with or without the addition of extra framing, control, or error detection bits. In short, equipment that combines by time division multiplexing several signals into a single composite digital signal.

Digital Network A network in which the information is encoded as a series of ones and zeros rather than as a continuously varying wave — as in traditional analog networks. Digital networks have several major pluses over analog ones. First, they're "cleaner." They have far less noise, static, etc. Second, they're easier to monitor because you can measure them more easily. Third, you can typically pump more digital information down a communications line than you can analog information.

Digital Phase-Locked Loop A phase-locked loop in which the reference signal, the controlled signal, or the controlling signal, or any combination of these, is in digital form.

Digital Phase Modulation The process whereby the instantaneous phase of the modulated wave is shifted between a set of predetermined discrete values in accordance with significant conditions of the modulating digital signal.

Digital Plastic A fancy term for buying goods and services on-line over the Internet using your credit card, possibly in conjunction with some verification of who you are from an independent certification authority.

Digital Port Adapter DPA. A device which provides conversion from the RS449/422 interface to the more common interfaces of RS-232-C, v.35, WE-306 and others.

Digital Private Network Signaling System See DPNSS.

Digital Recording A system of recording by conversion of musical information into a series of pulses that are translated into a binary code intelligible to computer circuits and stored on magnetic tape or magnetic discs. Also called PCM - Pulse Code Modulation.

Digital Reference Signal DRS. A digital reference signal is a sequence of bits that represents a 1004-Hz to 1020-Hz signal.

Digital Selective Calling DSC. A synchronous system developed by the International Radio Consultative Committee (CCIR), used to establish contact with a station or group of stations automatically by radio. The operational and technical characteristics of this system are contained in CCIR Recommendation 493.

Digital Service Cross-Connect DSX. A termination/patch panel that lets DS1 and DS3 circuits be monitored by test equipment.

Digital Signal A discontinuous signal. One whose state consists of discrete elements, representing very specific information. When viewed on an oscilloscope, a digital signal is "squared." This compares with an analog signal which typically looks more like a sine wave, i.e. curvy. Usually amplitude is represented at discrete time intervals with a digital value.

Digital Signal Processor A specialized digital microprocessor that performs calculations on digitized signals that were originally analog (e.g. voice) and then sends the results on. There are two main advantages of DSPs — first, they have powerful mathematical computational abilities, much more than normal computer microprocessors. DSPs need to have heavy mathematical computation skills because manipulating analog signals requires it. For example, DSPs are often called upon to compress video signals. Each sample must be examined and processed. And all done in very little time. The second advantage of a DSP lies in the programmability of digital microprocessors. Just as digital microprocessors have operating systems, so DSPs are now acquiring their very own operating systems. DSPs are used extensively in telecommunications for tasks such as echo cancellation, call progress monitoring, voice processing and for the compression of voice and video signals. They are also used in devices from fetal monitors, to anti-skid brakes, seismic and vibration sensing gadgets, super-sensitive hearing aids, multimedia presentations and low cost desktop fax machines. DSPs are replacing the dedicated chipsets in modems and fax machines with programmable modules — which, from one minute to another, can become a fax machine, a modem, a teleconferencing device, an answering machine, a voice digitizer and device to store voice on a hard disk, to a proprietary electronic phone. DSPs will do (and are already doing) for the telecom industry what the general purpose microprocessor (e.g. Intel's 80286 or 80386) did for the personal computer industry. DSPs are made by Analog Devices, AT&T, Motorola, NEC and Texas Instruments, among others.

Digital Signature A digial signature is the network equivalent of signing a message so that you cannot deny that you sent it and that the recipient knows it must have come from you. In short, a digital signature is an electronic signature which cannot be forged. It verifies that the document originated from the individual whose signature is attached to it and that it has not been altered since it was signed. There are two types of digital signatures. Ones you encrypt yourself and are the result of an ongoing relationship between you and the other party. Second, there are encrypted certificates issued by a company that is not affiliated with you. That company basically certifies that you are who you say you are. It does this because it's sent you a code. And it has retained a code for you, too. Join the two codes together mathematically, come up with the correct answer, and bingo, it's you. Utah has a Digital Signature Program whose goal is to develop, implement and manage a reliable means of secure electronic messaging over open, unsecured computer networks, minimize the incidence of forged digital signatures and possible fraud in electronic commerce and establish standards and develop uniform rules regarding verification and reliability of electronic messages. According to Utah, "digial signatures will enable us to determine who sent a document, identify what document was sent, and determine whether the document had been altered in route. It reasonably ensures the recipient that the message came from an identifiable sender and contains a specific, unaltered message. It may be used where there needs to be sufficient confidence in the source, content and integrity of a message." www.commerce.state.ut.us/web/commerce/digsig/dsmain.htm

Digital Speech Interpolation DSI. A type of multiplexing. A way of sharing bandwidth among a larger number of users than we really have circuits for. DSI allocates the silent periods in human speech to active users. At least 50% of a voice conversation is always quiet. The technique was originally called TASI (Time Assigned Speech Interpolation). TASI and DSI are lousy for data because they "clip" the first little bit of every new snippet of conversation — unless you hog the channel the whole time by talking incessantly or transmitting continuously. If you pause, you'll get clipping as the system drops you and then reconnects you. Clipping can ruin the meaning of the beginning of data conversations, unless the header knows that TASI or DSI is coming up or the data transmission is following some reasonable protocol and can resend the data. Unfortunately, this slows transmission. DSI and TASI are somewhat similar techniques to STATISTICAL MULTIPLEXING.

Digital Storage Media Command And Control DSMCC. Network protocols specified in Part 6 of MPEG-2 (ISO 13818) standards dealing with user-to-network and user-to-user signaling and communications.

Digital Subscriber Line A generic name for a family of evolving digital services to be provided by local telephone companies to their local subscribers. Such services go by different names and acronyms — ADSL (Asymmetric Digital Subscriber Line), HDSL (High Bit Rate Digital Subscriber Line) and SDSL (Single Pair Symmetrical Services). Such services propose to give the subscriber up to eight million bits per second one way, downstream to the customer and somewhat fewer bits per second upstream to the phone company.

A Digital Subscriber Line is also a fancy name for an ISDN BRI channel. Here's AT&T's definition: "A three-channel digital line that links the ISDN customer's terminal to the telephone company switch when four ordinary copper telephone wires. Operated at the Basic Rate Interface (with two 64-kilobit per second circuit switched channels and one 16-kilobit packet switched channel) the DSL can carry both voice and data signals at the same time, in both directions, as well as the signaling data used for call information and

customer data. With the introduction of the AT&T 5E5 generic, up to eight different users can be served by a single DSL." See also ISDN.

Digital Switching A connection in which binary encoded information is routed between an input and an output port by means of time division multiplexing rather than by a dedicated circuit.

Digital Telephony A digital telephone system transmits specific voltage values of "1" and "0" to transmit information. An analog system uses a continuous signal that uses the entire range of voltages. The human voice is an analog signal. To transmit it digitally, it must be converted into a digital signal. This is accomplished by sampling the value of the analog signal several times a second and converting each value into an binary number. These binary numbers are transmitted as a series of voltage levels representing ones and zeros. See PCM.

Digital To Analog Conversion A circuit that accepts digital signals and converts them into analog signals. A modem typically has such a circuit. It also has other circuits, such as those doing with signaling. See MODEM.

Digital Traffic Channel DTC. A digital cellular term. Defined in IS-136, the DTC is the portion of the air interface which carries the actual data transmitted. The DTC operates over frequencies separate from the DCCH (Digital Control CHannel), which is used for signaling and control purposes. See also DCCH and IS-136.

Digital Versatile Video See DVD.

Digital Video Digital video is video recorded and played digitally, i.e. in on-on bits. Traditional analog video — the one we have seen in our homes for eons — is recorded and played back in analog format, i.e. in analog wave forms. Why Digital Video? Due to its versatility, digital video has several advantages over analog video. You can edit it, store it, and transmit it easily. Digital video may be taken from analog source — such as standard over-the-air National Television Systems Committee (NTSC) analog source. Or it me be taken from an analog video camera and a VCR. To convert analog video into digital video typically requires a board inside a PC. Analog video is typically recorded on tape, such as a VCR. Digital video is typically recorded on a hard disk (magnetic or optical) or on a CD-ROM. Two common digital video technologies are Intel's Digital Video Interactive (DVI) and Microsoft's Audio/Video Interleaved (AVI). See ANALOG, DIGITAL VIDEO INTERACTIVE and INDEO VIDEO.

Digital Video Interactive DVI. A compression and playback technology originally developed by RCA's Sarnoff Research Institute and eventually acquired by Intel Corp. DVI is not a compression technique per se but a brand name for a set of processor chips that Intel is developing to compress video onto disk and to de-compress it for playback in real time at the U.S. standard motion video rate of 30 frames per second. The chip set includes both a pixel processor, which performs most of the decompression and also handles special video effects, and a display processor, which performs the rest of the decompression and produces the video output. DVI's greatest long-term advantage, according to Nick Arnett writing in PC Magazine, is that its microprocessors are programmable, so DVI can be adapted to a variety of compression and decompression schemes.

Digital Voice Coding Technology by which linear audio (voice) samples are collected and then compressed using an encoding algorithm. Typically used to store voice data for future decoding.

Digital Wireless Standards

Standards	Japan	US	Europe
Cordless	PHS	PACS	DECT
Cellular	PDS	IS54,IS136,IS95	GSM
PCS	PHS	IS136,IS95 PCS1900 (GSM) PACS Omnipoint BB CDMA	DCS1800 (GSM)
Data	PDC RAM PHS	CDPD (AMPS) RAM, ARDIS IS136,IS95 PACS Omnipoint 802.11 (Wireless Lan) BBCDMA	GSM RAM DCS11800 HyperLan DECT

Digitize Converting an analog or continuous signal into a series of ones and zeros, i.e. into a digital format. See DELTA MODULATION and PULSE CODE MODULATION. See also the Appendix.

Digitized Voice Analog voice signals represented in digital form. There are many ways of digitizing voice. See PULSE CODE MODULATION for the most common.

Digitizer 1. A device that converts an analog signal into a digital representation of that signal. Usually implemented by sampling the analog signal at a regular rate and encoding each sample into a numeric representation of the amplitude value of the sample.
2. A device that converts the position of a point on a surface into digital coordinate data.

Digroup Two groups of 12 digital channels combined to form one single 24-channel T-1 system.

Dijkstra's Algorithm An algorithm that is sometimes used to calculate routes given a link and nodal state topology database.

Dikes A wire-cutter.

DIL Direct-In-Line. A standard, both-way central office trunk is programmed to always ring a specific extension or hunt group within the PBX. This contrasts with Direct Inward Dialing, which allows an external caller to reach an internal extension by dialing a 7-digit central office number. A DID (Direct Inward Dial) trunk is a trunk from the Central office which passes the last two to four digits of the Listed Directory Number into the PBX, thus allowing the PBX to switch the call to and thus ring the correct extension.

Dilution Of Precision The multiplicative factor that modifies the ranging error. It is caused solely by the geometry between the user and his set of GPS (Global Positioning System) satellites. Known as DOP or GDOP.

DIM Document Image Management. The electronic access to and manipulation of documents stored in image format, accomplished through the use of automated methods such as high-powered graphical workstations, sophisticated database management techniques and networking.

Dimension An analog PBX that used PAM techniques, first introduced in the late 1970s by Western Electric (now AT&T Technologies) for AT&T. Now effectively discontinued except for the hotel/motel version. Some claim Archie McGill was responsible for Dimension. Others claim it was Bob Hawke.

DIMM Dual In-line Memory Module. A DIMM has a lot more bandwidth than a single in-line memory module (SIMM). It's a small circuit board filled with RAM chips, and its data path is 128 bits wide, making it up to 10 percent faster than a SIMM. DIMMs are prevalent on the Power Mac platform but are also creeping into high-performance systems.

Dimmed In Windows, dimmed means unavailable, disabled, or grayed. A dimmed button or command is displayed in light gray instead of black, and it cannot be chosen.

DIMS Document Image Management System.

DIN Deutsche Institut fur Normung (German Institute for Standardization). DIN specifications are issued under the control of the German government. Some are used on a worldwide basis to specify, for example, the dimensions of cable connectors, often called DIN connectors. See also VME.

DINA Distributed Intelligence Network Architecture.

DIOCES Distributed Interoperable and Operable Computing Environments and Systems.

Diode Diodes are devices that conduct electricity in one direction only. The simplest semiconductor devices are diodes. They are sometimes referred to as PN devices because they are made of a single semiconductive crystal with a P-type region and an N-type region.
This wonderful explanation comes from George Gilder's book Microcosm: "Named from the Greek words meaning two roads, an ordinary diode is one of the simplest and most useful of tools. It is a tiny block of silicon made positive on one side and negative on the other. At each end it has a terminal or electrode (route for electrons). In the middle of the silicon block, the positive side meets the negative side in an electrically complex zone called a positive-negative, or p-n, junction. Because a diode is positive on one side and negative on the other, it normally conducts current only in one direction. Thus diodes play an indispensable role as rectifiers. That is, they can take alternating current (AC) from your wall and convert it into direct current (DC) to run your computer.
"In this role, diodes demonstrate a prime law of electrons. Negatively charged, electrons flow only toward a positive voltage. They cannot flow back against the grain. Like water pressure, which impels current only in the direction of the pressure, voltage impels electrical current only in the direction of the voltage. To attempt to run current against a voltage is a little like attaching a gushing hose to a running faucet.
"It had long been known, however, that if you apply a strong enough voltage against the grain of a diode, the p-n wall or junction will burst. Under this contrary pressure, or reverse bias, the diode will eventually suffer what is called avalanche breakdown. Negative electrons will overcome the p-n barrier by brute force of numbers and flood "uphill" from the positive side to the negative side. In erasable programmable read-only memories (EPROMs), this effect is used in programming computer chips used to store permanent software, such as the Microsoft operating system in your personal computer (MS-DOS). Avalanche breakdown is also used in Zener diodes to provide a stable source of voltage unaffected by changes in current."

Diodea A semiconductor device which allows electricity to pass through it in only one direction, restricting flow the other way.

Diode Matrix Ringing A method of connecting a common audible line to a system so that all stations do not ring on all lines. See also MATRIX RINGING.

DIP 1. The act of consulting a database for information. Much like dipping into a bucket of water to extract a drink, carrier switches must dip into centralized databases in order to access various types of information. The database is housed in one or more SCPs (Service Control Points), which are centralized in the networks in order that many switches can share access to them, generally via SS7 links. Dips are made into such databases in order to accomplish tasks like calling card verification. 800/800 number routing requires that a dip be made in order to determine the serving IXC or LEC. LNP (Local Number Portability), mandated by the Telecommunications Act of 1996, requires the deployment of SCPs in order that the call can be terminated by the serving LEC. See also LNP and Number Portability.

2. See DIP Switch.

3. Document Image Processing. A term for converting paperwork into electronic images manipulable by a computer. Components of include input via scanner, storage on optical media and output via video display terminal, printer, fax, micrographics, etc.

4. See Dedicated Inside Plant.

DIP Switch Dual In-Line Package. A teeny tiny switch usually attached to a printed circuit board. It may peek through an opening in a piece of equipment. It may not. It usually requires a ball point pen or small screwdriver to change. There are only two settings — on or off. Or 1 or 0. But printed circuit boards often have many DIP switches. They're used to configure the board in a semipermanent way. The DIP switches are similar to integrated circuit chips which have two rows (dual) of pins in a row (in-line) that fit into holes on a printed circuit board. If something doesn't work when you first install it, check the dip switches first. Then check the cable connecting it to something else.

DIP/DOP Dedicated Inside Plant/Dedicated Outside Plant.

Diphones Speech segment beginning in the middle of one phoneme and concluding in the middle of another. See PHONEME.

Diplexer A device that permits parallel feeding of one antenna from two transmitters at the same or different frequencies without the transmitters interfering with each other. Diplexers couple transmitter and receiver to the same antenna for use in mobile communications.

Dipole Antenna fed from the center. Name often applied to "rabbit ear" antenna.

DIR An ATM term. This is a field in an RM-cell which indicates the direction of the RM-cell with respect to the data flow with which it is associated. The source sets DIR=0 and the destination sets DIR=1.

Direct Access Test Unit DATU. Also called Mechanized Loop Test (MLT) added or built into a central office switch. With DATU a technician can execute tests for shorts, opens and grounds remotely. The technician gets a digital voice, enters a password and is given a series of options. The technician can get results as a digital recording or through an alpha-numeric pager. DATU units can send a locating tone as TIP, RING or a combination of both. The unit can short lines and remove battery voltage for testing.

Direct Bond An electrical connection using continuous metal-to-metal contact between the things being joined.

Direct Broadcast Satellite DBS. A digital satellite system transmitting TV programs which can be received by small and relatively inexpensive dish antennas typically mounted on either the roofs or sides of houses. The receiving dish is stationary, being locked in on the position of the DBS service provider with which you have a subscription agreement. As a result, it can receive only those channels broadcast by that specific provider. DBS satellites operate in the Ku-band spectrum, and at fairly high power levels; hence the small size of the receiving dish, which commonly is as small as 19.7 inch-

es in diameter. DBS has virtually eliminated the old C-band satellite dishes — huge things about 3 meters (118.1 inches) across which you mounted in your back yard. The C-band systems were tunable, however; that is to say that they could be adjusted to pick up programming from just about any TV broadcast satellite which didn't encrypt its signals. Once DBS was introduced in the U.S. in 1994, it very quickly took significant market share away from the CATV providers.

Direct Connect A term describing a customer hooking directly into a long-distance telephone company's switching office, bypassing the local phone company. Such "direct connect" could be via a leased copper pair, a specially-run copper pair, a fiber optic or a private microwave system. See DIRECT ELECTRICAL CONNECTION, which is different. See DIRECT CONNECT MODEM.

Direct Connect Modem A modem connected to telephone lines using a modular plug or wired directly to the outside phone line. It thus transfers electrical signals directly to the phone network without any intermediary protective device. Direct connect modems must be certified by the FCC. Direct connect modems are more much reliable and more accurate than acoustically coupled modems. Virtually all modems these days are directly connected. One day pay phones will even come with RJ-11 jacks into which you can plug the modem of your portable laptop computer.

Direct Connection Connection of terminal equipment to the telephone network by means other than acoustic and/or inductive coupling.

Direct Control Switching The switching path is set up directly through the network by dial pulses without the use of central control. The telex network is an example of direct control switching. A step by step central office also uses direct control switching.

Direct Current DC. A flow of electricity always in the same direction. Contrast with alternate current (AC).

Direct Current Signaling DX. A method whereby the signaling circuit E & M leads use the same cable pair as the voice circuit and no filter is required to separate the control signals from the voice transmission.

Direct Department Calling DDC. A telephone service that routes incoming calls on a specific trunk or group of trunks to specific phones or groups of phones.

Direct Distance Dialing DDD. A telephone service which lets a user dial long distance calls directly to telephones outside the user's local service area without operator assistance.

Direct Electrical Connection A metallic connection between two things. The normal electrical way of connecting two things. This dumb definition is included in this dictionary because there was a time back in the early 1970s and before, when you couldn't (i.e. weren't allowed to) directly electrically connect your own phone or phone system to the nation's phone network. Those were the "good old days" when they (the Bell System) were trying to convince the world that electrically connecting anyone else's phones could harm the network. They never did prove this, and so today we have direct electrical connection of FCC-certified phone equipment. It's certified so it won't cause any harm to the network. See PART 68.

Direct In Line See DIL.

Direct In Termination Incoming calls on a PBX may be programmed to route directly to pre selected telephones without the attendant intervening. DIT features may be assigned to trunk circuits on a day, night or full time basis. Direct In Termination is slightly different from DIRECT

INWARD DIALING, though how different depends on whose PBX you're using.

Direct InterLATA Connecting Trunk Groups Those trunk groups used for switched LATA access that interconnect Interexchange Carriers (IXCs) used to connect that Point of Presence (POPs) directly with the Bell Operating Company (BOC) end office switching system.

Direct Inward Dialing DID. The ability for a caller outside a company to call an internal extension without having to pass through an operator or attendant. In large PBX systems, the dialed digits are passed down the line from the CO (central office). The PBX then completes the call. Direct Inward Dialing is often proposed as Centrex's major feature. See also DIRECT INWARD SYSTEM ACCESS (DISA) for another approach to DID.

Direct Inward System Access DISA. This feature of a telephone system allows an outside caller to dial directly into the telephone system and to access all the system's features and facilities. DISA is typically used for making long distance calls from home using the company's less expensive long distance lines, like WATS or tie lines. It's also used for leaving dictation for the typing pool. With DISA, you can dial individual extensions without the aid (or hindrance) of an operator. To use DISA, one must punch in from your touchtone phone a short string of numbers as a password code.

The problem with DISA is that "phone phreakers" (i.e. unauthorized people) often acquire that number or figure it out and run up expensive long distance phone calls. It's best to restrict DISA to trusted people and check the numbers called and bills generated. Changing the password code from time to time can help prevent this. DISA is acquiring a whole new life. It's becoming something called AUTOMATED ATTENDANT. An additional piece of equipment, called an automated attendant, is placed next to the phone system. You dial a special phone number (as you do with DISA). You're answered by a recording that says "Dial the extension you want." In DISA, the response is typically just a tone. An automated attendant is designed to save on operators and speed up outside people getting to talk to your inside people. Automated attendant is being suggested as a lower cost alternative to Centrex.

The following is excerpted from a document Northern Telecom sent to its PBX users. Read it. It's well-done:

PBX features that are vulnerable to unauthorized access include call forwarding, call prompting and call processing features. But the most common ways hackers enter a company's PBX is through Direct Inward System Access (DISA) and voice mail systems. They often search a company's trash for directories or call detail reports that contain 800 numbers and codes. They have also posed as systems administrators and conned employees into telling them PBX authorization codes. More "sophisticated" hackers use personal computers and modems to break into databases containing customer records showing phone numbers and voice mail access codes, or simply dial 800 numbers with the help of sequential number generators and computers until they find one code that gives access to a phone system. Once these thieves have the numbers and codes, they can call into the PBX and place calls out to other locations. In many cases, the PBX is only the first point of entry for such criminals. They can also use the PBX to access the company's data system. Call-sell operators can even hide their activities from law enforcement officials by using "PBX-looping" - using one PBX to place calls out through another switch in another state.

To minimize the vulnerability of the Meridian 1 system to unauthorized access through DISA, the following safeguards are suggested:

1) Assign restricted Class of Service, TGAR and NCOS to the DISA DN.
2) Require users to enter a security code upon reaching the DISA DN.
3) In addition to a security code, require users to enter an authorization code. The calling privileges provided will be associated with the specific authorization code.
4) Use Call Detail Recording (CDR) to identify calling activity associated with individual authorization codes. As a further precaution, you may choose to limit printed copies of these records.
5) Change security codes frequently.
6) Limit access to administration of authorization codes to a few, carefully selected employees.

Direct Line Terminations The term refers to central office/PBX lines which terminate directly on telephones, and are generally common to all instruments within the system. In a square configuration on a Key Telephone System, these lines must appear at the same button location on each phone.

Direct Memory Access DMA is a technique in which on adapter bypasses a computer's CPU, and handles the transfer of data between itself and the system's memory directly.

Direct Outward Dialing DOD. The ability to dial directly from an extension without having to go through an operator or attendant. In PBX and hybrid phone systems, you dial 9, listen for a dial tone, and then dial the number you want to reach. In some phone systems, you don't have to listen for the second dial tone. You can dial straight through. All phone systems now have DOD. The older ones didn't, especially cordboard PBXs. Some Club Meds and lots of cheap hotels (especially the ones Harry — the editor — stays in) do not have DOD.

Direct Set An ATM term. A set of host interfaces which can establish direct layer two communications for unicast (not needed in MPOA).

Direct Show see TAPI 3.0

Direct Station Select DSS. A piece of key system equipment usually attached to an operator's phone set. When the operator needs to call a particular extension he/she simply touches the corresponding button on the Direct Station Select equipment. Typically DSS equipment/feature is part of a Busy Lamp Field (BLF), which shows with lights what's happening at each extension. Is it busy? Is it on hold? Is it ringing? See BUSY LAMP FIELD.

Direct Station Select Intercom DSS. An interoffice caller can punch one button on his or her phone and dial his desired person, instead of dialing the full intercom number. Direct station select is like having an auto dial or speed dial button for everyone in the office. DSS saves time, but adds more buttons to the phone — one button for each extension the user wants to dial.

Direct Termination Overflow DTO. An optional MCI Vnet and 800 Service feature, which allows a call to "overflow" to shared lines for completion by the local telephone company if the dedicated line is busy.

Direct Trunk A trunk between two class 5 central offices.

Direct Trunk Access A PBX feature. By dialing some digits, the attendant can directly access any specific trunk. You'd do this if you want to check the trunk for problems, etc.

Direct Trunk Select Permits you, the user, the attendant, or an attached computer telephony system to access an individual outgoing trunk instead of one chosen by the PBX from a group of trunks. You may want to grab a special trunk to get access to a specially conditioned data line, for example. Direct trunk select is particularly important in testing computer telephony systems and facilities as it is usually the best method to use to address specific ports on a VRU (Voice Response Unit) or switching system.

Directed Call Pickup A telephone system feature. An extension user on a phone system user can answer calls — ringing or holding — on any other phone by dialing a unique answer code. If the call has already been answered by the called phone, the user who dials the answer code will join the connection in conference. Some tones will alert the conversing parties to the intrusion.

Directed Pickup A PBX feature. Directed pickup is when you pick up a call ringing at another, specific, known extension. You dial the code for directed pickup and the other extension number and you now answer the ringing call at your own phone. Also called Directed Call Pickup. In contrast, Undirected Pickup picks up any call ringing at any extension in the pickup group in which your extension is a member. The pickup groups are pre-programmed in the switch.

Directed Push The server interacts with the push client only occasionally, providing directions (agents, modules, and so on) for how content should be handled or where content is located. The client then gets the information directly from a variety of services and processes it locally. Lanacom Headliner is a good example of this. See ACTIVE PUSH and PUSH.

Directional Antenna An antenna which impels electrical waves with more energy in one direction than in another, or which receives electrical waves more readily from one direction than from another.

Directional Coupler 1. A device put in a microwave system's waveguide to couple a transmitter and receiver to the same antenna.
2. A transmission coupling device for separately sampling (through a known coupling loss) either the forward (incident) or the backward (reflected) wave in a transmission line. A directional coupler may be used to sample either a forward or backward wave in a transmission line. A unidirectional coupler has available terminals or connections for sampling only one direction of transmission; a bidirectional coupler has available terminals for sampling both directions. For optical fiber applications.

Directories Are places within a hard disk volume where you can store files or subdirectories. The term subdirectory is relative. A directory is a subdirectory only in relation to the directory above it. To a directory below it, the same directory is a parent directory.

Directory 1. A list of all the files on a floppy diskette or hard disk. A directory may also contain other information such as the size of the files and the amount of free space remaining.
2. Also a telephone directory.

Directory Assistance DA. Formerly known as "Information", but changed to "Directory Assistance" because the operators were getting too many stupid questions like "Who was the third president of the United States?" DA allows you to get telephone numbers from an operator. It comes in real handy when you don't have access to a phone book, or when the number is a new one not included in the book. DA once was provided exclusively by the local telephone company. It also once was free — even long distance DA was free. In most states of the United States, the local phone company now charges for this service, or will begin to very shortly, perhaps even from payphones.

Directory Assistance has changed a lot in the past few years. Not only have the local telephone companies begun to charge for the service, but also the service has become competitive. The local phone companies offered to sell their Directory Assistance databases to their competitors, although at a cost that was claimed to be unreasonably high. The competitors generally refused those offers, opting instead to build their own databases from whatever sources they could find. The end result is that Directory Assistance information often is grossly inaccurate. It also is provided by operators who have no knowledge of the area in question. For instance, the New York Telephone Company (later NYNEX; now Bell Atlantic) operators once could provide you directions to Flatiron Publishing; the operator you talk to today may live in Omaha and may never have been to New York City. Most local phone companies will give you the person's address as well as his phone number if you ask for it.

Directory Caching A method of decreasing the time it takes to find a file's location on a PC's disk. The FAT (File Allocation Table) and directory entry table are written into the file server's memory. The area holding all directory entries is called the directory cache. The file server can find a file's address (from directory cache) and the file data (from the file cache) much faster than retrieving the information from disk.

Directory Date A telephone company term. The issue date of the telephone directory. Improvements which will require changes in dialing procedures, directory numbers, etc., are usually timed to occur coincidental to directory issue dates.

Directory Information Base DIB. Made up of information about objects. The collective information held in the directory. An X.500 term.

Directory Information Tree DIT. Information that outlines the structure of an X.500 directory.

Directory Name As defined in the X.500 Recommendations, the Directory Name is an ITU term for the name of a directory entry. For example, used to retrieve an O/R (Originator/Recipient) Name for message submission.

Directory Number 1. The full complement of digits associated with the name of a subscriber in a telephone directory. This is a very long way of saying the obvious, namely your phone number.
2. A Northern Telecom definition: A unique phone number which is automatically assigned to each telephone during System Startup. The DN, also referred to as an intercom number, is often used to identify a telephone when settings are assigned during programming. A DN may be changed during programming.

Directory Package The process of adding, deleting and moving people attached to PBX or Centrex phones is more than simply programming in new extensions. Or should be. First, there are the changes necessary for the call accounting system. Second, there are changes necessary for "The Corporate Phone Directory". In the past, the directory bore little relation to the telephone system and it was often months, and sometimes years behind the actual phone system. Now some phone systems are incorporating various Directory Software Packages into the PBXs.
Some features included in these systems are the ability to dial someone by name — both for the attendant and for the individual phone user, i.e. dial HARRY, or HAR, instead of 3245. Some also include the ability to interface to the call accounting system. So that who's in what corporate department corresponds to which department's bill. There's also that important thing called a Telephone Directory. It would be useful if

you could hook a laser printer to the telephone system and tell it to print, in neat, photo-ready columns, an alphabetical (by last name), departmental or any other sorted telephone directory. Some of the newer PBXs have "directory package" features which include some or all of the above. Most users find today's necessity of at least three different systems to be a pain in the behind. Rightly so. The three different systems are CDR, phone directory and extension dialing in the PBX. The three are often out of synch.

Directory Replication A Windows NT term. The copying of a master set of directories from a server (called an export server) to specified servers or workstations (called an import computers) in the same or other domains. Replication simplifies the task of maintaining identical sets of directories and files on multiple computers, because only a single master copy of the data must be maintained. Files are replicated when they are added to an exported directory and every time a change is saved to the file.

Directory Replicator Service A Windows NT term. Replicates directories, and the files in those directories, between computers. See DIRECTORY REPLICATION.

Directory Service 1. A simple term for the information service which the telephone company runs on 411 or 555-1212.

2. A computer networking term. The facility within networking software that provides information on resource available on the network, including files, users, printers, data sources, applications, and so on. The directory service provides users with access to resources and information on extended networks.

Directory Services A service that provides information about network objects. DNS (Domain Name System) provides node address information. An X.500 Directory service provides any appropriate information an enterprise wishes to include in the X.500 directory itself.

Directory Synchronization The reconciliation of user directories from two electronic mail post offices. Many gateways and messaging switches have software to automate reconciling these directories.

Directory Tree A list of directories. A directory tree looks like an organizational chart and shows how your directories and subdirectories are related. Our favorite program that shows the directory tree in your hard disk is called CTREE. You can pick it up from many electronic bulletin boards for free, including from TELECONNECT Magazine's on 212-989-4675.

DirectShow See TAPI 3.0.

DIRTBAGS Digitally Initiated Resale of Telecommunications Bypass Applications by Scumbags. A term created by Ron Adams of Marylhurst College in Washington. There is some confusion between another definition — Digitally Initiated Reorigination of Telecommunications Bypass Access Generated by Scumbags. It is ascribed to Karen Corcoran of MCI's Atlanta office to describe as a group the various hackers, phreakers, and others who invade our networks and/or steal long distance service.

Dirty Power Dirty power typically refers to alternating current that is not a perfect sine wave and not perfectly 120 volt. There are all sorts of ways electricity can be made "dirty." It can be affected by spikes. Spikes are transient impulses (sometimes called glitches) of relatively high amplitude but very short duration. Spikes so short that a very high-speed oscilloscope is needed to observe them can often cause problems. Many spikes can occur in a fraction of a cycle.

Power can also be affected by sags and surges. Sags and surges are rapid changes in the amplitude of an AC voltage.

These are generally caused by abrupt changes in the load on a power source or circuit (such as when an air conditioner starts up), and can range from a fraction of a cycle to several complete cycles. Power can also be delivered consistently beyond its rating. In New York Con Edison guarantees 120 volts plus or minus 10%. Any level below 108 volts or above 132 volts Con Edison would consider dirty power. These are called low or high average variations. And they occur when the average voltage is above or below a desired level for significant periods of time, usually measured in seconds, minutes or longer.

Other kinds of dirty power include: Blackouts or brownouts. They occur when the power is switched off or lost completely (blackout), or when the voltage feeding a load is deliberately or inadvertently reduced significantly for a sustained period (brownout). Common mode noise is a small (+1V-2V) signal that appears between a neutral line and ground (earth) where there should be no signal. High/low frequency variations occur when the instantaneous frequency of an AC power source differs from its normal frequency, e.g., 60 Hz, by 0.5 Hz or more. Phase angle variations can be observed in three-phase systems whenever the phase relationships vary from their normal 120-degrees.

DIS Draft International Standard. As specified by ISO, a development step representing near final status on a specification. Once a specification has reached DIS status, companies are encouraged to develop actual products based on it.

DISA 1. Direct Inward System Access. DISA is a way of dialing into a phone system. It can be used to access one's voice mail, often over 800 lines. It can be used to make long distance calls, often on the theory that it's cheaper for employees to dial over the company's leased line phone network than to dial direct from their homes. DISA has been the major way crooks have dialed into PBXs and stolen toll calls. See DIRECT INWARD SYSTEM ACCESS.

2. Data Interchange Standards Association, Inc. DISA is the Secretariat and administrative arm of the Accredited Standards Committee (ASC) X12 which has responsibility for developing Electronic Data Interchange (EDI) standards. DISA OnLine is an electronic messaging and information system designed for use by DISA's member constituency. www.disa.org

Disable You figured this one. It means to prevent a hardware device from working. Unplugging it is the easiest way. It also refers to a tone or other signal which you send over a phone line to disable the equipment at the other end.

Disaster Recovery A generic term for all the tools and planning you need to bring your telecommunications facilities back to where they were before the disaster hit you.

Disc An older method of spelling DISK, as in Floppy DISK. Disc (spelled with a C) is now more commonly used to refer to optical storage devices, like CD (Compact Discs) and MO (Magneto Optical) discs. See also MS-DOS.

Discard Eligibility DE. A bit set by the user in a Frame Relay network, to indicate that a frame may be discarded in the event of congestion.

Discharge Block A protective device through which unwanted voltages discharge to ground.

Disco 1. Some phone and cable companies call orders to disconnect service "disco orders." See Disco Tech.

2. A club where annoying loud music is played, often similar to the "music on hold" offered my many PBXs.

Disco Tech Slang expression for a technician who only handles disconnections.

Disconnect 1. The breaking or release of a circuit connecting two telephones or data devices.

2. The occasional April Fools issue of TELECONNECT Magazine, New York City. Sometimes funny. Sometimes not.

Disconnect Frame Indicates in a fax call that the call is done. The sending fax machine sends the disconnect frame before hanging it. It does not wait for a response.

Disconnect Signal The signal sent from one end to indicate to the other to shut down the connection.

Disconnect Supervision The change in electrical state from off-hook to on-hook. This indicates that the transmission connection is no longer needed.

Disconnect Switch In a power system, a switch used for closing, opening, or changing the connections in a circuit or system or for purposes of isolation. It has no interrupting rating and is intended to be operated only after the circuit has been opened by some other means, such as by a circuit breaker or variable transformer.

Discontinuity An interruption or drop out of the optical signal.

Discounted Payback Period The number of years in which a stream of cash flows, discounted at an organization's cost of money, repays an initial investment.

Discounting The process of computing the present worth of a future cash flow by reducing it by a factor equivalent to the organization's cost of money (or some other measure of the value of money as measured by an interest rate) and the time until the cash flow occurs.

Discrete In voice recognition, refers to an isolated word. A discrete word is preceded and followed by silence, hence isolated in speech. Discrete words need to be separated by about half a second of silence when spoken to a discrete recognizer.

Discrete Cosine Transform DCT. A pixel-block based process of formatting video data where it is converted from a three dimensional form to a two dimensional form suitable for further compression. In the process the average luminance of each block or tile is evaluated using the DC coefficient. Used in the ITU-T's Px64 videoconferencing compression standard and in the ISO/ITU-T's MPEG and JPEG image compression recommendations.

Discrete Multi-Tone See DMT.

Discretionary Preview Dialing A single button dialing technique where the agent initiates a call with a single key stroke. Often used in association with a CRT tied to a database. Upon hitting a single button the system selects the phone number field from the screen and dials the number. Contrast this with Forced Preview Dialing. When the call ends, the computer brings up the next screen and starts dialing the call without the agent helping or hindering.

Discriminating Ringing See DISTINCTIVE RINGING.

Discussion A method of confirming others in their errors.

Disengagement Denial Disengagement failure due to excessive delay by the telecommunication system.

Disengagement Failure Failure of a disengagement attempt to return a communication system to the idle state, for a given user, within a specified maximum disengagement time.

Dish Typically a parabolic microwave antenna — used for receiving line-of-sight terrestrial signals or signals from satellites.

Disintermediation The decline of middlemen companies that today operate between the buyer and maker of goods. Pundits predict this will happen with the rise of commerce on the Internet. For example: the insurance, auto, mortgage, news delivery, and stock brokerage industries may change dramatically over the next few years. The Internet allows many industries such as these to do business directly with their customers.

Disk A piece of plastic or metal upon which a coating has been applied and which can thus, record computer information magnetically. The present convention is that a "disk" with a K refers to magnetic storage, while "disc" with a C refers to optical storage. See also MS-DOS.

Disk Array Also called a Drive Array. Although any set of disk drives put into a common enclosure could be called an array, in terms of RAID technology disk arrays tend to be those drives which are subject to a hardware or software-based controller that makes them appear to be a single drive to the host CPU or operating system. In short, a disk array is a disk subsystem combined with management software which controls the operation of the physical disks and presents them as one or more virtual disks to the host computer.

Disk Cache On a PC, a disk cache is the part of RAM that is set aside to temporarily hold data read from disk. A disk cache doesn't have to hold an entire file, as a RAM disk does, but can hold parts of running application software or parts of a data file. Disk-caching software manages the process of swapping data to and from the disk cache. See DISK CACHING.

Disk Caching A technique used to speed up processing. Each time your application retrieves data from the disk, a special program, called a disk caching program, stores data read from the disk in an area of RAM. When the application next requests more data, some of it may already be in RAM, thereby dramatically speeding the retrieval of data. See DISK CACHE and CACHING.

Disk Controller A hardware device that controls how data is written to and retrieved from the disk drive. The disk controller sends signals to the disk drive's logic board to regulate the movement of the head as it reads data from or writes data from or writes data to the disk. Gateway Computers defines disk controller thus: " Circuitry that manages the physical activity of a disk drive, such as moving the drive heads and creating the actual signals recorded on the disks. The controller is usually a card on the underside of the drive itself. The expansion card that connects to the drive by a ribbon cable enables communication between the disk drive and the computer and is called an adapter or host adapter. Sometimes the adapter circuitry is directly on the system board. Older hard drives had two ribbon cables to the expansion card because one part of the card held controller circuitry and another part held the adapter circuitry.

Disk Dancers Teenagers who uses the AOL disks given away in magazines and via direct mail to hop from one free account to another.

Disk Drive A device containing motors, electronics and other gadgetry for storing (writing) and retrieving (reading) data on a disk. See DISK DUPLEXING, DISK MIRRORING, DISK OPERATING SYSTEM and other definitions below starting with the work "Disk." See also IDE and ENHANCED IDE.

Disk Drive Performance Three basic things affect your perception of the speed of your hard disk drive: 1. The disk transfer rate — the speed at which data can be read from or written to the drive's media. This speed is governed mainly by drive mechanics — especially rotational speed, latency and seek time. 2. The controller transfer rate — the speed at which the drive's controller electronics can move data across the interface. This is governed by the design of the drive controller. 3. The host transfer rate — the speed at which the computer can transfer data across the interface. This is a matter of both

CPU and bus speeds as well as the hardware that provides the bridge between the host bus and the disk interface.

Disk Duplexing A method of failsafe protection, used on file servers on local area networks. Disk duplexing involves copying data onto two duplicate hard disks, each using a separate controllers and a separate disk channel. Disk duplexing protects data against the failure of a hard disk or of the hard disk channel between the disk and the file server. The hard disk "channel" includes the disk controller and interface cable. If any component on one channel fails, the other channel continues to operate normally. (You hope.) The operating system sends a warning message to the workstations to indicate the failure. It's a good idea then to fix the problem fast. See also DISK MIRRORING.

Disk Mirroring A technique for protecting the information on your hard disk. Disk mirroring writes data simultaneously to two identical hard disks using the same hard disk controller. Here's how it works: You have a special hard disk controller card. That's the card which organizes getting information into and off your hard disk. When you come to write information to your hard disk, your hard disk controller writes first to the first hard disk, called the primary hard disk. The controller retains that information in its memory and then writes it to the second hard disk. This causes a 50% degradation in performance since it now takes twice as long to write to the disk. When the controller comes to read, it reads only from the primary disk. Thus there is no performance degradation in reading. Mirroring is designed to protect against mechanical problems with one of your hard disks. If one of the hard disks break, the other one will take over instantly. You will get a warning message. You will be told to repair the broken disk and you will be told to designate the other disk now as your primary disk (it may already be). That primary disk will now become your bootable disk, the one you boot your computer from. Mirroring does not protect against viruses, or corrupt data or losing data. Any idiocy you can perform on one hard disk you can now happily perform on two. Mirroring does not protect you against a lightning strike which could knock out both your hard disks. Mirroring does not protect against the loss of a controller card since you're only using one. Another protection technique called disk duplexing uses two separate controllers to drive two separate hard disks. See also DISK DUPLEXING.

Disk Operating System See DOS.

Disk Pack A series of disks mounted horizontally and arranged as a single unit. A disk pack contains more space for storing and retrieving information than one single disk.

Disk Sector Magnetic diskettes are typically divided into tracks, each of which contains a number of sectors. A sector typically contains a predetermined amount of data, such as 256 bytes.

Disk Server A device equipped with disks and a program that permits users to create and store files on the disks. Each user has access to their own section of the disk. It gives users disk space which they would not normally have at their own personal computers. The disk server is linked to the PCs via a LAN. The next level of sophistication would be a file server, which would allow users to share files.

Disk Spanning When you want to save a single big file to several floppy disks, it's called disk spanning. Many of the zipping programs allow you to do this. See PKZIP.

Disk Striping Writing data in stripes across a volume that has been created form areas of free space on from 2 to 32 disks. A Windows NT definition.

Disk Subsystem A collection of physical disks and the hardware required to connect them to a host computer or network.

Disk/File Server A mass storage device that can be accessed by several computers, usually through a local area network (LAN).

Diskless PC Just what it says: a PC without a disk drive. Used on a LAN, a diskless PC runs by booting DOS from the file server. It does this via a read-only memory chip on its network interface card called a remote boot ROM. Diskless PCs are cheaper than PCs with disks, they're more compact and they offer better security since users can't make off with floppy disks of important and sensitive data or add their own virus-ridden programs to file servers. Diskless PCs appeal primarily to users interested in security. One system of diskless PCs allow the system operator to disable and physically lock the machine's various ports and the computer case itself. Should something go wrong with the machine, the ports can be restored to operating condition by letting the system allow a technician to attach a laptop computer and run diagnostic programs.

Diskless Workstation See DISKLESS PC.

DISN Defense Information Systems Network.

DISOSS IBM's Distributed Office Support System.

Disparity Control A function of the 8B/10B transmission encoding scheme, in which the two remaining bits are used for error detection and correction. 8B/10B transmits 8 bits as a 10-bit group, thereby leaving two additional bits for disparity control.

Dispatch A radio communications technique where one communicates to many through short bursts of communication. Users of dispatch services include taxis, trucking companies and service personnel.

Dispersion More correctly known as chromatic dispersion. A term used to describe how an electromagnetic signal is distorted because the various frequency components of that signal have different propagation characteristics. Each frequency is chromatically distinct; i.e., has a different color. Dispersion is the degree of scattering taking place in the light beam as it travels along the fiber optic. Or the overlapping of a light signal on one wavelength to different wavelengths because of reflected rays and the different refractive index of the core fiber material. The problem is that each pulse of light comprises multiple wavelengths, each with different propagation characteristics-i.e., each travels at a slightly different speed. In a high-speed, long-haul fiber optic system, particularly one which employs WDM (Wavelength Division Multiplexing), portions of the pulses can overtake portions of the preceding pulses. Dispersion is a particularly significant problem in high-speed networks, which operate at Gbps speeds, and in long-haul networks which involve links exceeding several hundred kilometers. In order to defeat dispersion, the network must be optimized through the placement of regenerative repeaters at appropriate intervals, or through the use of Dispersion-Shifted Fiber. See also Dispersion-Shifted Single-Mode Fibers, SONET, and WDM.

Dispersion Compensation Grating DCG. DCG overcomes the distortion of optical signals as they are transmitted through a network. Instead of trying to compensate for large amounts of signal dispersion at the end of a network, DCG periodically removes the distortion where needed along the transmission line. See Solitons.

Dispersion-Shifted Single-Mode Fibers These types of fibers have a different internal configuration. This changes the zero total chromatic dispersion point to 1550nm. This is important because the attenuation at this wavelength is only about half as much as at 1300nm.

Display The visual presentation of information, usually on a TV-like screen or an array of illuminated digits.

Display Driver A piece of software which translate instructions from the software you are running into thousands of colored dots, or pixels, that appear on your video monitor. A display driver is also called a Video Driver. Symptoms of a video driver giving trouble can range from colors that don't look right, to horizontal flashing lines to simply a black screen. In the Macintosh world, Apple rigidly defined video drivers. Windows, in contrast, is a free-for-all. Windows 3.1 defined the lowest common denominator of displays — namely 16 colors at 640 x 480 pixels. But most multimedia programs and many games won't run with only 16 colors. They require at least 256 colors.

Display Phone A telephone that has a LED display. Also called an Executive Phone. Display phones are usually difficult to read because the displays are not backlit. Phone lines and PBX extension lines don't have enough power to run the display without an external power adapter and most telephone equipment makers don't want to sell phones with an external power adapter. It limits where phones can be placed and therefore how many can be sold. Display phones typically allow both "programmable keys", buttons that can be programmed for speed dial, conference, etc. and "soft keys", buttons that change function as different features are used, and also can show the number that is being dialed, and internal callers' names and extensions.

Displayboards Also called Readerboards or Wall Displays. Readerboards are typically found in call centers. They are electronic displays, sort of like giant TVs. They are typically hooked into the ACD or PC monitoring the machine and they throw up information about how many people are waiting in line, how long the longest person has been in line, how well the agents are doing and, often, whose birthday it is today. the idea is that all the agents in the call center can see the Readerboards and change their behavior accordingly.

Distance Learning Video and audio technologies into education so students can attend classes in a location distant from where the course is being presented.

Distance Sensitive Pricing Product pricing based on the distance (airline mileage) between the originating and terminating locations of a call/data transmission.

Distance Vector An approach, or algorithm, used by network equipment in selecting the best available network path by calculating the total distance over which a packet would travel on each alternative route. The shortest distance is usually the most preferred, however, it is possible to consider other factors in the decision as well.

Distance Vector Protocol A protocol run in LAN routers in order to minimize the number of hops (link-level connections) which a data packet must travel from the originating device to the terminating device. Such a protocol causes each router to regularly broadcast the entire contents of its routing tables to all neighboring routers. As this method is bandwidth-intensive, it is best used in relatively small router networks with relatively few inter-router connections. Distance vector protocols include IP RIP, IPX RIP, and AppleTalk RTMP. See also LINK-STATE PROTOCOL, POLICY ROUTING PROTOCOL, and ROUTER.

Distant Learning A Pacific Bell term for students sitting in front of TVs and phones and participating in classes that are being held and delivered elsewhere. In one of PacBell's trials, they used a T-1 signal so the distant lecturer could see and hear his distant students using full-color video.

Distinctive Dial Tones In some phone systems, dial tones sound different. An internal dial tone sounds different to an external dial tone. The logical reason for this is simply to alert the user as to whether he or she is making an intercom or an outside local or long distance call.

Distinctive Ringing First, distinctive ringing is a feature that offers extra numbers which cause different ringing patterns on a line. When the main number is called, the called party will receive the normal ringing pattern. If one of the extra numbers is dialed, that line would ring with a different cadence. In North America, the normal ringing pattern is a single ring every six seconds. The distinctive ring patterns are 1) two short rings every six seconds, or 2) a short-long-short ring.

Different ringing patterns are also used in conjunction with such features as busy call return, to indicate a freed line. One test done by Bell Canada set up a special ringing pattern (different from any of the featured distinctive rings) to indicate an incoming long distance call.

Each telephone company has its own name for this feature: Ident-a-Call, Teen Ring, Feature Ring, etc. In any case, different ringing patterns allow for calls to certain people, or to sort out different call purposes such as for voice, fax, modem, or answering machine.

Distort To change some characteristic of a signal during its transmission. See Distortion.

Distortion 1. The difference in values between two measurements of a signal — for example, between the transmitted and received signal. "Distortion" typically refers to analog signals.

2. In imaging, distortion is any deformation of the on-screen image. Two common types of distortion are pincushion and barrel.

3. When used in relation to AC power distribution, this refers to deviations between the actual AC voltage waveform delivered to the user and the ideal sine wave of voltage. Total distortion is usually expressed as a percentage of desired sine wave, for example, a square wave has approximately 33% distortion. Distortion in AC power systems can also be resolved into a series of harmonics. In this case percentages for each harmonic, such as the third, fifth, seventh, etc. are provided. The square root of the sum of the squares of the individual harmonics is equal to the total distortion. This definition courtesy APC.

Distributed Capacity The capacity in a coil due to the proximity of the turns.

Distributed Common Control There are two elements of telephone switching: The switching itself and the control of that switching. The earliest step-by-step telephone switches had their "Control" built into them. The dialing information at the beginning of the call physically moved switches. You could say, as a result, that control was distributed throughout the switching system. Then came the 1940s and crossbar exchanges, and the economics pointed to centralizing control. Then came computerized or stored program control (SPC) switches in which large computers were used centrally to perform virtually all the functions of the erstwhile electro-mechanical senders, registers, markers, etc. — those things which affect the setting up and tearing down of the call. As computers got smaller and as microprocessors appeared (the so-called computer on a chip), it became economical and efficient to place inexpensive microprocessors in the telephone circuits themselves, in essence getting much of the processing done before it hits the central processing unit.

Increasingly, as special microprocessors (so-called "computers on a chip") for telecommunications evolve, we will see more and more of the processing being distributed to further and further away from the central point and closer and closer to the originating telephone instrument. It will be rare in coming years for telephones to come without microprocessors. One day, each phone will have its own switch and the rest of the system will just be one gigantic loop of cable — not unlike today's local area networks.

Distributed Computing Environment DCE. A comprehensive integrated set of services that supports the development, use, and maintenance of distributed applications. Digital Equipment Corporation's DCE is an implementation of the Open Software Foundation's DCE (OSF DCE). In response to OSF's request for distributed computing technology, Digital submitted for consideration four of Digital's established distributed computing technologies:
Remote Procedure Call (RPC), a joint effort with HP/Apollo; Threads Service, based on Digital DECthreads; Cell Directory Service (CDS), based on the Digital Distributed Name Service (DECdns); Distributed Time Service (DTS), based on the Digital Distributed Time Service (DECdts). See DCE for more detail.

Distributed Data Processing DDP. A data processing arrangement in which the computers are decentralized — i.e. scattered in various places. Hence, processing occurs in a number of distributed locations and only semi-processed information is communicated on data communications lines from remote points to the central computers. The object of DDP is to save telecommunications charges and to improve network response time.

Distributed Environment Refers to a network environment, or topology, in which decision making, file storage and other network functions are not centralized but instead are found through the network. This type of environment is typical for client-server applications and peer-to-peer architectures.

Distributed Feedback Laser DFB. A type of laser used in fiber-optic transmission systems, at the distribution level of the local loop. DFBs are point-to-point lasers distributed among nodes in a geographic area such a neighborhood. They transmit and receive optical signals between the distributed nodes and the centralized node, where the signals are multiplexed over a higher-speed fiber link to the head-end (point of signal origin). DFBs can be more effective than the traditional approach of using a single laser which serves multiple nodes through a broadcast approach, as the available bandwidth can be segmented. DFBs have application in a FTTN (Fiber-To-The-Neighborhood) local loop scenario. See also FTTN and SONET.

Distributed File System A type of file system in which the file system itself manages and transparently locates pieces of information from remote files and distributes files across a network. It can recognize multiple servers and be accessed independently of where it physically resides on the network.

Distributed Management Environment A compilation of technologies now being selected by the Open Software Foundation to create a unified network and systems management framework, as well as applications. Those technologies will complement OSF's own Unix implementation, OSF/1, as well as other operating systems.

Distributed Microprocessor Common Control This means that the system employs many individual microprocessors to control system and telephone phone functions. The microprocessors may be located in central processing equipment or in the telephones themselves.

Distributed Name Service A technique for storing network node names so that the information is stored throughout the network (either one LAN or many joined together), and can be requested from, and supplied by, any node.

Distributed Network Service Introduced in March 1991, AT&T's Distributed Network Service was designed expressly for the switchless resale community unlike SDN. It allows resellers to purchase large volumes of services and receive progressive discounts on all direct dial domestic and international calls. Resellers may designate any number of locations to participate in the plan with the flexibility of adding locations.

Distributed Nodes PBX and its "slave" switches which are physically in separate buildings, in separate areas of the campus, in separate parts of the town.

Distributed Processing A network of computers such that the processing of information is initiated in local computers, and the resultant data is sent to a central computer for further processing with the data from other local systems. The term also covers computing jobs "farmed out" from a central site to remote processors where faster processing or specialized databases are available. Distributed Processing is often a more efficient use of computer processing power since each CPU can be devoted to a certain task. A LAN is the perfect example of distributed processing. See also DISTRIBUTED DATA PROCESSING.

Distributed Queue Dual Bus DQDB. A connectionless packet-switched protocol, normally residing in the Medium-Access Control sublayer of the data link layer. Definition from Bellcore in reference to Switched Multimegabit Data Service (SMDS). See also DQDB and SMDS.

Distributed Switching When electronics and computers were expensive it made sense to centralize them and run individual lines out for miles to subscribers. Then the economies changed. Electronics and computers became cheaper and running phone lines for miles became very expensive. So switching companies started building small switches which they could put closer to subscribers. Thus, individual local loops would be shorter and the long lines going back to the larger central office would be more efficiently used — namely by more people. The remote, or distributed switches, are called everything from remote switches to slave switches (because they slave off the main one which is distant). Usually these remote switches are unattended.

Distribution 1. The portion of a switching system in which a number of inputs is given access to an equal number of outputs. 2. Refers to the arrangement of premises wiring runs and their associated hardware required to implement the planned customer premises wiring system extending from the network interface jack to each communications outlet at the desktop.

Distribution Cable Part of the outside cable plant connecting feeder or subfeeder cables to drop wires or buried service wires that connect to the customer's premises. In simpler language, it's the cable from the serving area interface — a box on a pole, in the ground, etc. — to the lightning protection at the entrance to the customer's premises. See also FEEDER PLANT and DROP WIRE.

Distribution Cable, Inside Plant Cables usually running horizontally from a closet on a given floor within a building. Distribution cables may be under carpet, simplex, duplex, quad, or higher fiber count cables.

Distribution Cable, Outside Plant The cable running from a central office or remote terminal to the side of a subscriber's lot.

Distribution Duct A raceway of rectangular cross-section placed within or just below the finished floor and used to extend the wires or cables to a specific work area.

Distribution Frame Cables coming in from thousands of subscribers need to connect to the correct ports on a central office. Similarly, cables coming in from many PBX extensions need to connect to the PBX. The cables could be directly wired to the CO or to the PBX. This would be inflexible. It would make future moves and changes a nightmare. So the solution is something called a Distribution Frame. Basically it's a giant wire connecting devices made of metal. There are no electronics in it whatsoever. On one side we punch down the wires coming in from the outside world. On the other side, we punch down the wires coming in from the CO or PBX. Both sides are connected with wire that's called "jumper" wire. By pulling off one end of the jumper wire and moving it to another location we can quickly change phone numbers, add or subtract cabling (one, two or three pairs for normal or electronic phones, etc.). In big central offices, distribution frames can span whole city blocks and the "jumper" wires can be several hundred yards long. Designing distribution frames and their layout in advance is critical, otherwise it becomes a mess and tracing where jumper wires go becomes an enormously time consuming job.

Distribution Frequency The number of times used in the Internet for translating names of host computers into addresses. It is a network database system that provides translation between host names and addresses.

Distribution Group 1. A group made of phone extensions on a PBX arranged to share the load. In the Rolm PBX, each group is assigned a dummy extension number called a pilot number.

2. A group of telephone extensions on an automatic call distributor (ACD). The ACD answers the incoming calls then checks to see if any agents' phones are free. If none are free, it delivers the caller a message and then puts the caller on hold. Which line the call has come in on may determine which group of agents should handle that call. They would be called a Distribution Group. Once the call is released from hold, it may be sent to a member of that Distribution Group following some pre-determined mathematical formula — for example, so that everyone's workload is kept constant, or a group of people are kept busy.

Distribution Rack A device used to mount communications equipment and cables.

Distribution Service In ISDN applications, a telecommunications service that allows one-way of information from one point in the network to other points in the network with or without user individual presentation control. See DISTRIBUTION SERVICES.

Distribution Services In the world of B-ISDN applications, distribution services are communications services that emphasize one-way, bandwidth-intensive transfer of information from one point in the network to other point(s) in the network. There are 2 classes of Distribution Services defined within that context, revolving around the issue of "Presentation Control:" Services requiring no presentation control include what we normally think of as broadcast services. Such services include such data as TV, Video On Demand (VOD), audio and multicast data. The data is broadcast or multicast across the network, with no requirement that the receiving device exercise any form of control over the transmission or presentation of the data. Services requiring presentation control include Interactive TV. While engaged in an Interactive TV session, a viewer might wish to control a TV broadcast in much the same way as he would control a videotape through a VCR remote control. While watching the Super Bowl, for instance, Ray Horak might wish to rewind and replay the touchdown scored by The Dallas Cowboys' Emmit Smith against the San Francisco 49'ers. Further, the viewer might wish to view that play from a multiple angles covered by cameras positioned around the stadium. While the viewer is exercising these options. the live broadcast is buffered in large-scale temporary memory in the TV set of the future. Once the viewer has sufficiently relished the play, he can play rejoin the live broadcast, begin the program where he left off, or exercise other options. Harry Newton, on the other hand, might wish to exercise the same control over a tennis match broadcast from Australia.

Distribution Voltage Drop The voltage drop between any two defined points of interest in a power distribution system.

Distributor A company with a contractual relationship with a manufacturer to buy equipment at a preset price. The manufacturer provides training, advertising and warranty support. Often called an authorized dealer, although a dealer may be one step lower in the distribution chain. A distributor is often used as a generic term for any supplier. Therefore you should clarify whether a distributor is an authorized distributor.

DIT Directory Information Tree. The global tree of entries corresponding to information objects in the OSI X.500 Directory.

Dithering Dithering is an imaging term with at least two meanings. One meaning that it's the processing of an image containing more colors than a system can handle to an image containing exactly the right number of colors that the system can handle. For example, some of the color images on my laptop contain 16 million colors. But my laptop (the way I have it set up) will only handle 256 colors. If I ask my image display software to display that image, it will "dither" it to 256 colors. This means it will give its best shot guess at what the image should look like.

In another meaning, dithering is patterning black and white dots to approximate shades of grey on a scanned image.

Diurnal Phase Shift The phase shift of electromagnetic signals associated with daily changes in the ionosphere. The major changes usually occur during the period of time when sunrise or sunset is present at critical points along the path. Significant phase shifts may occur on paths wherein a reflection area of the path is subject to a large tidal range. In cable systems, significant phase shifts can be occasioned by diurnal temperature variance. SEE ALSO DIURNAL WANDER and WANDER.

Diurnal Wander A loss of signal synchronization in digital cable systems caused by temperature variations over the course of 24 hours. (Diurnal means "daily cycle.") As the ambient temperature varies from the heat of the day to the cool of the night, the cable stretches and contracts, with the overall length of the cable changing, if only ever so slightly. As the length of the medium changes, the speed of signal propagation (the time it takes for the signal to transverse the cable) is affected. As a result, the number of digital pulses effectively stored in the medium changes. The end result is that the network elements (e.g., repeaters and multiplexers) can get out of synch. Diurnal wander affects all types of cable systems — twisted pair, coax cable, and optical fiber; it especially affects cables hung from poles, rather than buried, as such cables are more exposed to temperature variations and as the weight of the cable adds to the problem. The impacts of diurnal wander are particularly great in very high-speed transmission systems. See also DIURNAL PHASE SHIFT and WANDER.

Diverse Entry You have a building with phone service. You are concerned about the reliability of your phone service. You are concerned that the wires coming in from your phone company might be cut. So you organize to have service coming in from the phone company along different routes and entering your building from opposite sides of your building. Thus the term diverse entry.

Diversity 1. In microwave communications, the strength of a microwave signal can decrease for many reasons — heat, rain, fog, etc. This is not good if the objective is to get reliable communications. One solution is to simultaneously send and receive two microwave signals at slightly different frequencies. Since different frequencies respond differently to weather problems, the likelihood is that at least one will get through well. This is called diversity.
2. A means of effecting redundancy in a network, with the result being protection from catastrophic failure. Consider the typical end user — one cable entrance to one group of wire pairs housed in one cable connected to one central office provided by one local exchange carrier and connecting to one interexchange carrier. That is a catastrophe waiting to happen. Consider the alternative of full diversity. The fully redundant end user has several cable entrances into the building — this is Entry Diversity. The local loop connection is provided through multiple, non-adjacent pairs in multiple cables — this is Pair and Cable Diversity. The several cables follow different routes to the CO — this is Route Diversity. The local loops terminate in multiple COs — this is Central Office Diversity. The carrier connection is to multiple LECs and IXCs — this is Carrier Diversity. In this scenario of full diversity, no single point of failure can totally isolate the user organization from the network. Diversity is good — perhaps expensive and complex, but good.

Diversity Combiner A circuit or device for combining two or more signals carrying the same information received via separate paths or channels with the objective or providing a single resultant signal that is superior in quality to any of the contributing signals.

Diversity Receive A method commonly employed by cellular manufacturers to improve the signal strength of received signals. Uses two independent antennas that receive signals which differ in phase and amplitude resulting from the slight difference in antennas position. These two signals are either summed or the strongest is accepted by voting. The most popular methods include dual-antenna phase switching, dual-receiver audio switching and "ratio diversity" audio combining. The most effective method is ratio diversity combining.

Divestiture On January 8, 1982 AT&T signed a Consent Decree with the U.S. Department of Justice, stipulating that on midnight December 30, 1983, AT&T would divest itself of its 22 telephone operating companies. According to the terms of the Divestiture, those 22 operating Bell telephone companies would be formed into seven regional holding companies of roughly equal size. Terms of the Divestiture placed business restrictions on AT&T and the BOCs. Those restrictions were threefold: The BOCs weren't allowed into long distance, equipment manufacturing, or information services. AT&T wasn't allowed into local telecommunications (i.e. to compete with the BOCs). But it was allowed into computers. The federal Judge overseeing Divestiture, Judge Harold Greene, is slowing the lifting the restrictions against the BOCs being allowed into information services. He has stayed firm on the other two — equipment manufacturing and long distance.

Divx A new DVD (Digital Video Format) format that supports encryption and timed rentals. Divx is essentially a more expensive version of DVD. The foremost difference between it and standard DVD is the disk's encryption technology. The second difference is the way the Divx system allows for an on-demand viewing experience. Divx disks require a Divx player, which will sell for about $100 more than the price of a standard DVD player, and which will also play standard DVD disks and audio CDs. Users buy a Divx disk for about $5, and can watch it as many times as they like for a two-day period; the period begins when they first hit "play" on the Divx player. After the initial two-day viewing period, customers can play the disk at a much later time by just paying a fee. Moviephiles can also buy a special password to gain unlimited playback of the movie. Future viewings are billed in two-day periods, just like the initial viewing. The player keeps track of the number of periods the consumer has used and transmits this information over a household phone line to the consumer's account; the consumer then receives a bill in the mail. This system includes the key characteristics of the video rental model, like convenience, variety and low cost, but with important advantages. For example, the user never has to return the disks so, there are never any late fees. "It basically moves DVD technology into the rental domain," said one observer at the time of the announcement of the Divx technology in the fall of 1997.

DIW Type D Inside Wire. Originated as a specific AT&T cable. Now commonly used to describe any 22, 24, or 26 gauge PVC jacketed twisted-pair cable used primarily for inside telephony wiring.

DIX Digital/Intel/Xerox. The early 1980s consortium of manufacturers that promoted the Ethernet Version 1 and Ethernet Version 2 variations of a CSMA/CD media access protocol. This DIX "standard" was then submitted to the IEEE, where after some modifications it was released as IEEE Standard 802.3. The DIX version did not include specifications for UTP or Fiber Optic cable.

DIX Connectors A local area network connector. DIX connectors on the transceiver local area network cable link it to the network; the male DIX connector plugs into the SpeedLink/PC16 and the female DIX connector attaches to an external transceiver.

DIX Ethernet The DEC, Intel, Xerox Ethernet standard, also known as Version 1 or Bluebook Ethernet. There are subtle differences between IEEE 802.3 and the DIX Ethernet.

DL 1. Distribution List.
2. Distance Learning.

DLC 1. Digital Loop Carrier. Network transmission equipment used to provide pair gain on a local loop. The digital loop carrier system derives multiple channels, typically 64 Kbps voice-grade, from a single four-wire distribution cable running from the central office to a remote site. In the traditional deployment, Central Office Termination (COT) comprises multiplexing equipment in the central office (CO). A four-wire, twisted-pair circuit is deployed from the CO to the remote location at the point of Remote Termination (RT), where it terminates in matching DLC electronics. From the remote node, the interface is joined to individual voice-grade local loops which extend to the individual customer premises. Effectively, traditional DLCs are channel banks — devices which multiplex and demultiplex multiple channels over a high-bandwidth, electrical distribution facility. Such DLCs are used in situations in which the cost of the equipment is more than offset by the savings in distribution facilities through eliminating the need for a large number of individual copper pairs. Traditional DLCs also are known as SLCs (Subscriber Loop Carrier systems).

In a more contemporary scenario, the carriers deploy high-bandwidth fiber optic facilities from the COT to the RT. The carrier electronics at each end accomplish the optoelectric conversion process, as well as that of multiplexing/demultiplexing. The final leg of the local loop remains embedded twisted pair. This type of system can be characterized as a hybrid local loop system. Given the high cost of such conversion process, DLCs offer clear advantages in comparison to FTTC (Fiber-To-The Curb) and, certainly, to FTTH (Fiber-To-The-Home). Additionally, the deployment of fiber optic distribution facilities yields much greater aggregate bandwidth — typically a minimum of 51.84 million bits per second, which is the optical equivalent of 45 million bits per second (T3) in the electrical world. See also SLC, CHANNEL BANK, and MULTIPLEXER.

2. See DATA LINK CONTROL.

3. Direct Line Console. An AT&T Merlin term. An answering position used by system operators to answer calls, transfer calls, make calls, set up conference calls, and monitor system operations. Calls can ring on any of the line buttons, and several calls can ring simultaneously (unlike the QCC where calls are sent to a common QCC queue and wait until a QCC is available to receive a call).

DLCI Data Link Connection Identifier. A Frame Relay term defining a 10-bit field of the Address Field. The DLCI identifies the data link and its service parameters, including frame size, Committed Information Rate (CIR), Committed Burst Size (Bc), Burst Excess Size (Be) and Committed Rate Measurement Interval (Tc). See TUPLE ADDRESS.

DLE Data Link Escape. A control character used exclusively to provide supplementary line control signals, control character sequences or DLE sequences. In packet switching, Data Link Escape is a name applied to the Control P non-print character which is used to swap the PAD from the data mode to the command mode in packet switched networks.

DLEC Data CLEC. Companies that deliver high-speed access to the Internet, and not voice.

DLL 1. Dynamic Link Library. A feature of OS/2 and Windows that allow executable code modules to be loaded on demand and linked at run time. This lets library code be field-updated — transparent to applications — and then unloaded when they are no longer needed. Unlike a standard programming library, whose functions are linked into an application when the application's code is compiled, an application that uses DLL links with those DLL functions at runtime. Hence, the term dynamic.

2. Data Link Layer driver. A driver specification developed by DEC primarily to work with DECnet PCSA for DOS. DLL is a shared driver specification, allowing multiple protocol stacks to share a single network interface card.

DLPI An ATM term. UNIX International, Data Link Provider Interface (DLPI) Specification: Revision 2.0.0, OSI Work Group, August 1991.

DLR Design Layout Report. A description of how a circuit is engineered. Often used between LECs and CLECs. See LEC and CLEC.

DLS Data Link Switching. IBM's method for carrying SNA and NetBIOS over TCP/IP operating at the Data Link layer. DLS, now an open Internet spec, can be used with OSPF or PPP.

DLSE Dial Line Service Evaluation.

DLSw DataLink Switching Workgroup. This workgroup has issued a new interoperability standard for integrating SNA and NetBIOS over the TCP/IP protocol. According to Cisco, the new DLSw standard provides interoperability and functionality not currently offered by Informational RFC 1434 or existing DLSw implementations.

DLTU Digital Link Trunk Unit. An AT&T term for a device which provides the interface to digital trunks and lines such as T-1, EDSL, and remote line units.

DM Delta Modulation.

DMA 1. Direct Memory Access. A fast method of moving data from a storage device or LAN interface card directly to RAM which speeds processing. In essence, DMA is direct access to memory by a peripheral device that bypasses the CPU.

2. Direct Marketing Association.

3. Document Management Alliance.

4. Designated Market Area. This term specifies the geographic area established by Nielsen Media Research for the purpose of rating the viewership of commercial television stations. DMAs represent the geographic areas covered by groups of competing commercial television broadcast stations. The boundaries of these areas are of considerable financial importance to the stations involved because they determine the number of viewers each station can claim, and, hence, the dollar amount the station can charge for advertising time. Section 76.55(e) of the FCC Rules provides that, for the purpose of the cable television must-carry rules, the local market area of a commercial television broadcast station is the Nielsen DMA. A map showing DMA boundaries may be obtained from:

Nielsen Media Research
1290 Avenue of the Americas
New York, NY 10104-0061
212-708-7500

DMA Channel A channel for direct memory access that does not involve the microprocessor, providing data transfer directly between memory and a disk drive. See DMA.

DMB Digital Multipoint Bridge.

DMD Differential Mode Delay. A fiber optic term. DMD refers to the fact that multimode and monomode (single mode) fiber optic cabling systems differ considerably in their performance characteristics. Specifically and in a LAN environment, some multimode systems are not capable of supporting signals from high-performance laser diodes, which operate at very high speeds and which emit narrowly-defined pulses of light. Traditionally, LEDs (Light-Emitting Diodes), which operate at lower speeds and which emit light pulses of a broader range of frequencies, have been use in conjunction with multimode fiber systems. In a LAN environment, such a combination has proved very satisfactory, even at high speeds and over relatively long distances. The combination of LEDs and multimode fiber also is much less costly. Ethernet at 10/100 million bits per second, Token Ring at 4/16 million bits per second, and ATM at 25/155/622 million bits per second all have made use of this traditional combination. However, the emerging Gigabit Ethernet standard requires the use of laser diodes to achieve such speeds. Multimode fiber, at least in some cases, appears to underperform in this environment, even over short distances. The problem is DMD, which yields unacceptably unmanageable levels of signal delay.

DMI 1. Digital Multiplexed Interface. AT&T's Digital Multiplexed Interface. A PBX to computer interface that divides the T-1 trunk into 23 user channels and one signaling channel. Also used as a T-1 PBX to computer interface. See OPEN APPLICATION INTERFACE.

2. Desktop Management Interface, a protocol independent management interface developed by the DMTF, the Desktop Management Task Force, a group working on improving network printing. Making network printers easier to use — more standard and easier to use — requires three basic elements: standardized definitions of printer objects, a protocol for communicating with those objects and an application interface.

Question: What is DMI?

Answer: The Desktop Management Interface (DMI) is a the result of a cooperative, industry-wide effort to make PC systems easier to manage, use and control. Specifically, DMI is a specification developed by the Desktop Management Task Force (DMTF), a consortium of hardware, software, and peripheral vendors. DMI describes hardware and software components in a format that can be easily accessed over a phone line by PC management applications and technical support personal. Thanks to DMI, technical support personnel and PC management applications can:

• Access an inventory of hardware and software components
• Access and change parameter values and settings
• View data generated by software agents and diagnostics routines

With DMI, technical support applications and PC management applications use a Management Interface (MI) to access data stored in Management Information Files (MIFs).

DMI-BOS Digital Multiplexed Interface-Bit Oriented Signaling. A form of signaling, which uses the 24th channel of each DSI to carry signaling information, allowing clear channel 64 Kbps functionality.

DMO Digital Modification Order.

DMS Digital Multiplex System. Also the name of a line of digital central office switches from Northern Telecom. There are DMS-10s, DMS-100s, DMS-100Fs and DMS-200s and, by the time you read this, probably more. See also DMS SUPERNODE.

DMS Integrated Access Local Area Network Shortened to DIALAN. A central office provided local area network offering completely digital, full duplex data transmission at speeds of 300 bps through 19.2 asynchronous and 1,200 bps through 64 Kbps synchronous. DIALAN users use existing telephone sets and an Integrated Voice and Data Multiplexer (IVDM) that plugs into a telephone jack.

DMS Supernode The SuperNode is a very flexible central office switch from Northern Telecom which can be configured as a high-capacity local or tandem switch with Common Channel Signaling 7 (CCS7) Service Switching Point (SSP) function, a Signaling Transfer Point (STP), a Service Control Point (SCP), a Digital Network CrossConnect (DNX) system or a network service node with custom programming applications. In 1988, the DMS SuperNode offered twice the capacity of the DMS-100 switch (the older Northern Telecom central office) based on the NT40 processor.

DMS-100 A digital central office (public exchange) manufactured by Northern Telecom (Nortel).

DMS-250 A digital central office (public exchange) manufactured by Northern Telecom (Nortel).

DMS-INode DMS Integrated STP/SSP Node.

DMT Discrete Multi-Tone. A new technology using digital signal processors to pump more than 6 megabits per second of video, data, image and voice signals over today's existing one pair copper wiring. DMT technology, according to Northern Telecom, provides the following:

• Four "A" channels at 1.5 million bits per second. Each "A" channel may carry a "VCR"- quality video signal, or two channels may be merged to carry a "sports"- quality real-time video signal. In the future, all four channels operating together will be able to transport an Extended Definition TV signal with significantly improved quality over anything available today. ("A" channels are asymmetric — carrying information only from the telephone company to the subscriber's residence. All other channels within ADSL are symmetric or bi-directional.)

• One ISDN "H zero" channel at 384 Kbps (kilobits per second). This channel is compatible with Northern Telecom's multi rate ISDN Dialable Wideband Service or equivalent services. This channel could also be used for fast, efficient access to corporate LANs for work-at-home applications, using Northern Telecom's DataSPAN or other frame-relay services.

• One ISDN Basic Rate channel, containing two "B" channels (64 Kbps) and one "D" channel (16 Kbps). Basic Rate access allows the home user to access the wide range of emerging ISDN services without requiring a dedicated copper pair or the expense of a dedicated NT1 unit at the home. It also permits the extension of Northern Telecom's VISIT personal video conferencing to the home at fractional-T-1 rates (Px64).

• One signaling/control channel, operating at 16 Kbps giving the home user VCR-type controls over movies and other services provided on the "A" channel including fast-forward, reverse, search, and pause.

• Embedded operations channels for internal system maintenance, audits, and telephone company administration.

• Finally, the home user can place or receive telephone calls over the same copper pair without affecting the digital transmission channels listed above. And since ADSL is passively coupled to the POTS line, the subscriber's POTS capability is unimpaired in the event of a system failure.

DMTE Desktop Management Task Force

DMTF Desktop Management Task Force. A consortium of vendors working toward a set of standards for network management software which will ease the management of desktop systems, their components, and peripherals. Activities address "standard groups" such as software, PC systems, servers, monitors, network interface cards (NICs), printers, mass storage devices, and mobile technologies. The most notable of its accomplishments is the development of the Desktop Management Interface (DMI). www.dmtf.org See also DMI.

DN Directory Number or subscriber number or telephone number entries. DN is typically the phone number associated with a telephone line. The international format for a phone number consists of four items:

1. A + plus, which indicates that the caller should dial whatever digits are necessary to get him internationally. In the U.S., the prefix is 011.

2. The country code. The country code for the U.S. and Canada is 1. For Australia, it's 61. For England, it's 44.

3. The routing code, which is also called an area code.

4. The local number. In North America it's 7-digits. But in other countries it varies. At one stage in Sydney, Australia, some local numbers were five digits; some were six, some were seven and some were eight.

See DIRECTORY NUMBER.

DNA 1. Digital Network Architecture. The framework within which Digital Equipment Corporation (DEC) designs and develops all of its communications products. DNA includes many standards of the OSI Model. Some of these standards will be adopted into ISDN. Acronym also by Network Development Corporation for their network offering.

2. Dynamic Node Access. A high-speed bus invented by Dialogic to join together multiple voice processing PEB-based systems. PEB stands for PCM Expansion Bus. See PEB.

3. Distributed Network Administration is an AMP wiring architecture that decentralizes network electronics, into closets on every floor. It saves money on initial installation and works especially well in multi-tenant building and small workgroup situations where you don't share want to share facilities, according to AMP. Distributed Network Administration can be executed with optical fiber or with Category 5 UTP. Since closets will be placed throughout the building, distance is usually not an issue in the choice of the medium.

DNAR Directory Number Analysis Reporting.

DNC Dynamic Network Controller.

DNIC Data Network Identification Code. An address to reach a host computer system residing on a different packet switched network than the one you're on. The data equivalent of a telephone number with country code and area code. Typically the DNIC is a four digit number. The first three digits of a DNIC specify a country. The fourth digit specifies a public data network within that country. See also DATA NETWORK IDENTIFICATION CCDE.

DNIS Dialed Number Identification Service. DNIS is a feature of 800 and 900 lines that provides the number the caller dialed to reach the attached computer telephony system (manual or automatic). Using DNIS capabilities, one trunk group can be used to serve multiple applications. The DNIS number can be provided in a number of ways, inband or out-of-band, ISDN or via a separate data channel. Generally, a DNIS number will be used to identify to the answering computer telephony system the "application" the caller dialed. For example, a 401K status program may be offered by a service provider to a number of different companies. The employees of each company are provided their own 800 number to call to access their account status. When the computer telephony system sees the incoming DNIS number, it will know to which company the call was directed, and can so answer the phone correctly with a customized "you have reached the 401K line for xyz company. Please enter your personal account code and password..."

Here's another application: You use one 800 phone number for testing your advertisements on TV stations in Phoenix; another number for testing your ads on TV stations in Chicago; and yet another for Milwaukee. The DNIS information can be used in a multitude of ways — from playing different messages to different people, to routing those people to different operators, to routing those people to the same operators, but flashing different messages on their screens, so the operators answer the phone differently. In Ireland, incoming toll free phone calls from the rest of Europe arrive with DNIS. As a result a phone call arriving from Germany is routed to a computer telephony system playing messages in German. The advantage of DNIS is basically economic: You simply need fewer phone lines. Without DNIS you would need at least one phone line for every different 800 or 900 number you gave out to your callers. Make sure you understand the difference between DNIS and ANI and Caller ID. DNIS tells you the number your caller called. ANI or Caller ID is the number your caller called from. See 800 SERVICE and 900 SERVICE.

DNPA Data Numbering Plan Area. In the U.S. implementation of the ITU-T X.25 network, the first three digits of a net-

work terminal number (NTN). See also DATA NUMBERING PLAN AREA.

DNPIC Directory Number Primary InterLATA Carrier.

DNR 1. Dialed Number Recorder. Also called a Pen Register. An instrument that records telephone dial pulses as inked dashes on paper tape. A touchtone decoder performs the same thing for a touchtone telephone.

2. Dynamic Network Reconfiguration. Allows IBM networks to change addresses without reloading and bringing the network down.

DNS 1. The Domain Naming System is a mechanism used in the Internet for translating names of host computers into addresses. The DNS also allows host computers not directly on the Internet to have registered names in the same style. The DNS is a distributed database system for translating computer names (like ruby.or-a.com) and vice-versa. DNS allows you to use the Internet without remembering long lists of numbers. On TCP/IP networks (like the Internet), the Domain Naming System provides IP address translation for a given computer's domain name. DNS would change a computer name such as harry.newton.com to the machine's actual numeric IP address, which is in the format xxx.xxx.xxx.xxx. The DNS makes it easier to remember where you want to go. See Domain Naming System for more a more detailed explanation.

2. Domain Name Server. Domain Name Servers, also known as resolvers, are a system of computers which convert domain names into IP addresses, which consist of a string of four numbers up to three digits each. Each applicant for a domain name (e.g., www.harrynewton.com) must provide both a primary and a secondary DNS server; a domain name which fails to provide both primary and secondary DNS servers is known as a "lame delegation." See also the first definition above.

3. See DISTRIBUTED NETWORK SERVICE.

DO A word in a high-level language program which comes before a collection of things to be done, i.e. statements to be executed.

Do-Not-Call A legal requirement to remove from a calling list the telephone numbers of people who have asked that they not receive unsolicited telephone calls. See Telephone Consumer Protection Act.

Do-Not-Disturb Makes a telephone appear busy to any incoming calls. May be used on intercom-only, by extension line-only or both.

Do-While A programming statement used to perform instructions in a loop while a certain condition exists — i.e. do something while the variable Y is less than 20.

DOA Dead On Arrival. A term several manufacturers use to refer to equipment which arrives at the customer's premises not working. A person who receives a DOA machine will ask the company for a NPR number — New Product Return number. This allows them to return the product and have the factory replace it with another new one.

DOC 1. Department of Communications. Canadian government department.

2. See DYNAMIC OVERLOAD CONTROL.

DOC-IT Okidata's name for a combination scanner, printer, copier, fax machine. A truly wonderful machine.

Dock To insert a portable computer into a base unit. Cold docking means the computer must begin from a power-off state and restart before docking. Hot docking means the computer can be docked while running at full power.

Docket Formal FCC/State regulatory commission proceeding, also referred to as a case.

Docking Station Base station for a laptop that includes a power supply, expansion slots, monitor, keyboard connectors, CD-ROM and extra hard disk connectors. A user slides his laptop into a base station and in effect, gets the equivalent of a desktop machine.

Docobjects See DOCUMENT OBJECTS.

DOCSIS Data Over Cable Service Interface Specifications. A project with the objective of developing on behalf of the North American cable industry a set of necessary communications and operations support interface specifications for cable modems and associated equipment. The project is hoped to lead to the rapid deployment of HFC (Hybrid Fiber-Coax) cable television systems in support of high-speed data transfer, as well as entertainment TV. The project is being led by the MCNS (Multimedia Cable Network System Partners Limited), which consists four leading CATV operators. Activities are in phases, with the interfaces to be addressed including Cable Modem to CPE Interface (CMCI), Cable Modem Termination System-Network Side Interface (CMTS-NSI), Operations Support System Interface (OSSI), Cable Modem Telco Return Interface (CMTRI), Cable Modem to RF Interface (CMRFI), Cable Modem Termination System-Downstream RF Interface (CMTS-DRFI), Cable Modem Termination System-Upstream RF Interface (CMTS-URFI), and Data Over Cable Security System (DOCSS). For more detail see CMCI, CMTS-NSI, OSSI, CMTRI, CMRFI, CMTS-DRFI, CMTS-URFI, and DOCSS. See also MCNS.

Document A Windows 95 file that contains the information you are working on.

Document Camera A specialized camera on a long neck that is used for taking pictures of still images — pictures, graphics, pages of text and objects which can then be sent stand alone or as part of a video conference.

Document Commenting A Microsoft Web-based feature in which users will be able to add comments to a document, which are then saved in a database, not as HTML. Once the document is posted on a Web site, only users on a special discussion list will be able to view the comments.

Document Database An organized collection of related documents.

Document Image Management DIM. The electronic access to and manipulation of documents stored in image format, accomplished through the use of automated methods such as high-powered graphical workstations, sophisticated database management techniques and networking.

Document Objects DocObjects for short. Microsoft Office for Windows95 has an app called Office Binder which is a "document container that enables a user to manipulate text files, spreadsheets, graphics presentations, and other documents as a single entity." DocObjects is the "core technology that makes Office Binder work." This technology can support document containers other than Office Binder and document server as well. "The DocObjects technology consists of a set of extensions to OLE Documents. One application for this technology is in Internet browsers; a user would only need one navigation tool to browse and view all documents, whether local or network-based. A document provided from a DocObjects server is essentially a full-scale, conventional document that is embedded as an object within another DocObjects container (binder, browser, etc.)."

Document Recognition The ability to capture all the information on a page (text and images) and perform not only character recognition, but page structure analysis as well.

Documentation Written text describing the system, how it works and how to work it. In most cases of high technology products, documentation is awful. Better documentation helps sell equipment and software. Please write your instruction manuals better. Please.

DOD 1. Direct Outward Dial

2. Department of Defense

DOD Master Clock The U.S. Naval Observatory master clock, which has been designated as the DOD Master Clock to which DoD time and frequency measurements are referenced. This clock is one of two standard time references for the U.S. Government in accordance with Federal Standard 1002; the other standard time reference is the National Institute for Standards and Technology (NIST) master clock.

Doghouse 1. The closure containing the cellular PCS equipment. Some have heaters and airconditioning units for environmental control. The doghouse is located in a hut near the antenna.

2. Doghouse is where your customers will put you if they have problems with their cellular or fixed phone service.

Dolby A system of noise/hiss reduction invented by Ray Dolby, widely used in consumer, professional and broadcast audio applications.

Domain In the broadest sense, a "domain" is a sphere of influence or activity. In the vernacular, domain equals turf. In the MIS world, a domain is "the part of a computer network in which the data processing resources are under common control." In the Internet, a domain is a place you can visit with your browser — i.e. a World Wide Web site. In reality, "a place you can visit" might be a single computer. It might be a group of computers masquerading as a single computer. Or it might even be a logical/physical partition of a computer or group of computers. On the Internet, the domain is the address that gets you there, and consists of a hierarchical sequence of names (labels) separated by periods (dots). Examples of the 100 million plus Internet domain addresses include computertelephony.com (Computer Telephony magazine), harrynewton.com, teleconnect.com (Teleconnect magazine), or ctexpo.com (CT Expo). A Top Level Domain (TLG) is defined as the alphabetical address suffix, which identifies the nature of the organization, e.g. com, edu, org. Currently, the central naming registry on the Internet is administered by the Internet Assigned Numbers Authority, (IANA), which includes the National Science Foundation (NSF), InterNIC, Network Solutions Inc., and the International Ad Hoc Committee (IAHC). Only a select few Top Level Domains currently can be registered, including .com (commercial), .edu (educational institution), .gov (government), .org (non-profit organization), .net (network provider) and .mil (military). In this case and for example, they will register the domain name "teleconnect.com" or "computertelephony.com." (They are different sub-domains, also known as secondary domains under the Top Level Domain.) What you put in front of your domain name is your concern. Everything that is SomethingInFrontOf@harrynewton.com or Something.In.Front.Of@HarryNewton.com will be routed to the domain "harrynewton.com" for resolution (further routing, processing, etc.). There was a proposal that beginning in March 1998, new TLDs (Top Level Domain names) (e.g. firm, arts) could be registered under the authority of CORE (Council of Registers), which contracted Emergent Corporation to build and operate the new, expanded Internet Domain Name Shared Registry System (SRS). As many as 90 new registrars were proposed to be authorized to assign those TLDs through the SRS, which is a neutral, shared repository of URLs. New TLDs were to include the following: .arts: entities emphasizing cultural and entertainment activities; .firm: businesses, or firms; .info: entities providing information services; .nom: individual or personal nomenclature, i.e., a personal nom de plume, or pen name; .rec: entities emphasizing recreation/entertainment activities; .shop: businesses offering goods to purchase; and .web: entities emphasizing activities related to the World Wide Web. This effort has been forestalled, at the time of this writing. Preceding both the secondary and Top Level Domain, increasingly can be found tertiary domains. For instance, a large organization might have hosts "www.company.com" and "mail.company.com." The tertiary domain might be called "support.company.com", with hosts "www.support.company.com" and "ftp.support.company.com." The domain may also include an geographical suffix, indicating that the target organization is physically located outside the U.S. For example, .au is for Australia, .nz for New Zealand, and .po for Poland. Without a geographical suffix, it is assumed that the address identifies a device physically located in the U.S., where the Internet originated. In order to route the user or the user's mail to the correct location (correct domain) a "dip" is made into a database housed on a Domain Name Server (DNS). The DNS translates the alphabetical address into an IP (Internet Protocol) address, which corresponds to a logical address which is tied to a physical device in the form of a server, which is located at physical address, which is connected to the Internet.

Electronic mail vendors define domain in various ways. PC Magazine's networking editor, Frank Derfler defines a domain as referring to a set of hosts on a single LAN that needs only one intermediary post office to move mail from one host to another. A domain may consist of only one host, depending on its design and implementation.

In IBM's SNA, a domain is a host-based systems services control point (SSCP), the physical units (PUs), logical units (LUs), links, link stations and all the affiliated resources that the host (SSCP) can control. In Microsoft networking, a domain is a collection of computers that share a common domain database and security policy that is stored on a Windows NT Server domain controller. Each domain has a unique name. See also Domain Controller, Domain Defined Attribute, Domain Name, Domain Name Server, Domain Naming System, gTLD, URL and Workgroup.

Domain Controller For a Windows NT Server domain, the primary or backup domain controller that authenticates domain logons and maintains the security policy and the master database for a domain

Domain Defined Attribute DDA. In X.400 addressing, the DDA is a special field that may be required to assist a receiving E-mail system in delivering a message to the intended recipient. Up to four DDAs are allowed per address, with each DDA address entry made up of two parts, a type and a value. For example, if I were a subscriber to MCI Mail and I wanted to send a message to Harry Newton, this is how I would address it:

TO: Harry Newton
EMS: CompuServe / 592-7515
MBX: P=CSMAIL
MBX: DDA=ID=70600,2451

If I were a subscriber to CompuServe, the only addressing information I would need would be the number 70600,2451.

Domain Name In networks using the TCP/IP (Transfer Control Protocol/Internet Protocol), the full domain name consists of a sequence of names (labels) separated by periods (dots), e.g.,"pictures.computertelephony.com." See DOMAIN and TCP/IP. All domain names have extensions. The common Top Level Domain extensions include the following:
.arpa (Old style Arpanet)
.com (Commercial (mainly USA))
.edu (USA Educational)
.gov (USA Government)
.org (Non-Profit Making Organizations)
.mil (USA Military)
.net (Network (network provider, such as an ISP))
International extensions also are included where national boundaries are crossed. International extensions succeed the organizational extension. Examples include the following:

.ae (United Arab Emirates)	.int (International)
.ag (Antigua and Barbuda)	.is (Iceland)
.ar (Argentina)	.jm (Jamaica)
.at (Austria)	.kr (South Korea)
.au (Australia)	.kz (Kazakhstan)
.be (Belgium)	.lk (Sri Lanka)
.bm (Bermuda)	.lt (Lithuania)
.bn (Brunei Darussalam)	.lu (Luxembourg)
.bo (Bolivia)	.my (Malaysia)
.ch (Switzerland)	.nl (Netherlands)
.co (Colombia)	.nz (New Zealand)
.cr (Costa Rica)	.ph (Philippines)
.cz (Czech Republic)	.pt (Portugal)
.de (Germany)	.ro (Romania)
.dk (Denmark)	.ru (Russian Federation)
.do (Dominican Republic)	.sg (Singapore)
.ec (Ecuador)	.si (Slovenia)
.ee (Estonia)	.su (Former USSR)
.fi (Finland)	.th (Thailand)
.gt (Guatemala)	.uk (United Kingdom)
.hk (Hong Kong)	.us (United States)
.hr (Croatia)	.uy (Uruguay)
.hu (Hungary)	.ve (Venezuela)
.id (Indonesia)	.za (South Africa)
.ie (Ireland)	

New top level domain names are allegedly about to be augmented with seven new names: .arts, .firm, info, .nom, .rec, .shop, and .web. There is also talk of "personalized second level domains," i.e. words that fall before the @ sign. See also Web Address and URL.

Domain Name Server Domain Name Server (DNSs) is a computer on the Internet which contains the programs and files which make up a domain's name database. Using a name server is much like placing a call to a 800/888 voice telephone number. The 800/888 number requires a "dip" into a database (on the DNS) in order to translate the name (e.g., harry@ctexpo.com) into a telephone number (IP address), which you then use to establish connection with the person (host computer). In other words, the DNS translates the logical alphanumeric address into a logical IP address associated with an applications server (or perhaps, a logical partition of a server), which is connected to the Internet. The telephone network (Internet) addresses by telephone number (IP address), not really by the person's (domain's) name. See Domain and Domain Naming System.

Domain Naming System DNS. In Unix-based networks, of which the Internet is the largest, a domain naming system is the commonly accepted way of giving attached computers names. Domain naming system is sometimes referred to as the BIND (berkeley internet name domain) service in BSD Unix, a static, hierarchical naming service for tcp/ip hosts. A DNS server computer maintains a database for resolving host names and IP addresses, allowing users of computers configured to query the DNS to specify remote computers by host names (in words) rather than IP addresses (which are only numbers). DNS domains should not be confused with Windows NT networking domains, although Windows NT does support and can use the Internet's DNS scheme. See DOMAIN.

Domestic Arc The portion of the geostationary orbit allowing a satellite to return a footprint that almost covers the continental United States.

Dominant Carrier The long distance service provider which dominates a particular market and is subject to tougher regulation than its competitors. An FCC term. Essentially it means AT&T.

Domino Server The Internet and Intranet server packaged with Lotus Notes. Domino is the server, Lotus Notes is the client.

DOMSAT Domestic Communications Satellite.

Donald Elliptical Projection Named after Jay K. Donald of AT&T, who invented in 1956 the basis for the V&H system used to rate calls in North America and many other countries. The concept is one of flattening the Earth, as it is much easier to calculate the distance between two points using a flat plane. The more difficult process would be that of performing a trigonometric function using degrees, minutes and seconds of latitude and longitude. See also V&H.

Don't Answer Recall Allows an extension user on a PBX to automatically retry a call by dialing a special digit code.

Done Deal A term used in the secondary telecom equipment business. Term used between seller and buyer to signify that a sale has been agreed to and an oral contract is now in effect, binding both parties to the agreed-to-sale as if a written contract has been signed. A written contract submitted later that does not conform to the original oral agreement is not justification for dissolving the original agreement unless both parties agree to the new written contract.

Dongle 1. A device to prevent copies made of software programs. A dongle is a small device supplied with software that plugs into a computer port. The software interrogates the device's serial number during execution to verify its presence. If it's not there, the software won't work. A dongle is also called a hardware key.

2. A small cable with a connector at one end that attaches to a PCMCIA (also called a PC Card) inside your laptop and has a female RJ-11 at the other end, into which you plug a phone line. See also Xjack — which is another alternative to a PC Card modem dongle.

3. An issue of SunExpert Magazine defines dongle as a 15-pin to 13-pin adapter. It is primarily used for large color monitors that require the 13W3 13-pin adapter.

Door A software program that allows access to files and programs not built into an electronic bulletin board system, thus letting users run them on-line.

DOP See Dedicated Outside Plant.

Doped Fiber Fiber optical cable treated with erbium. Such cable can carry signals three times farther than untreated fiber. Doped fiber is now the fiber of choice in long distance networks.

Doped Fiber Amplifier DFA. An amplifier used in fiber optic systems to amplify light pulses. Such amplifiers typically are known as EDFAs (Erbium-Doped Fiber Amplifiers) as they are doped with erbium, a rare-earth element. DFAs are more effective than regenerative repeaters in many applications, as they simply amplify the light pulses through a chemo-optical process. Regenerative repeaters, on the other hand, require that the light signal be converted to an electrical signal, amplified, and then re-converted to an optical signal. Additionally, DFAs can simultaneously amplify multiple wavelengths of light in a Wavelength Division Multiplex (WDM) fiber system. See also EDFA, SONET and WDM.

Doppler Aiding A signal processing strategy that uses a measured doppler shift to help the GPS Global Positioning System) receiver smoothly track the GPS signal. It allows more precise velocity and position measurement.

Doppler Shift The apparent shift in the frequency of a signal caused by the relative motion of the transmitter and the receiver.

DOS Disk Operating System, as in MS-DOS, which stands for MicroSoft Disk Operating System. A disk operating system is software that organizes how a computer reads, writes and reacts with its disks — floppy or hard — and talks to its various input/output devices, including keyboards, screens, serial and parallel ports, printers, modems, etc. Until the introduction of Windows, the most popular operating system for PCs was MS-DOS from Microsoft, Bellevue, WA.

Dot A dot is an integral part of an email address or a web site. You say "dot," not "period," since for most people a period means the end of a sentence. The English say "full stop," which is even more specific. My email address is "harry underscore newton (my mailbox) at msn (the site domain) dot com (commercial domain)." Keyboard translation: "Harry_Newton@msn.com" Dots are now replacing dashes in phone numbers. My phone number used to be 212-206-7140. Now many people believe it's 212.206.7140. Artists prefer the dots in phone numbers since dots save space. Some people claim it's the Swiss minimalist influence. See also Dot Address.

Dot Address Also known as "Dotted Quad" and "Dotted Decimal Notation." A set of four numbers connected with periods that make up an Internet address; for example, 147.31.254.130.

Dot Addressable Graphics Refers to the mode of operation on a dot matrix printer which allows you to control each element in the dot matrix printhead. With this feature, you may produce complex graphics drawings.

Dot Pitch A measure of the clarity of a color monitor. Dot pitch measures the vertical distance between the centers of like-colored phosphors on your screen. The smaller the distance, the sharper the monitor. Dot pitch is the major determinant in the clarity of an image on screen. And you can't do anything about it. When you buy a monitor, you buy it with a certain dot pitch and you're stuck with that dot pitch. You may be able, however, to do something about improving convergence and focus — the other measures of the clarity of a color monitor.

Dot Zero When new software is issued, it often bears the number .0 (i.e. dot zero) as in MS-DOS 5.0. The theory among software gurus is that you should always avoid a "Dot Zero" revision, since it will likely contain bugs and that one should wait for 4.01 or 5.01 etc. This theory has some validity, although MS-DOS 5.0 came out very clean and was not revised until 6.0.

Dotted Decimal Notation Also known as "Dotted Quad" and "Dot Address." A set of four numbers connected with periods that make up an Internet address; for example, 147.31.254.130.

Dotted Quad Also known as "Dot Address" and "Dotted

Decimal Notation." A set of four numbers connected with periods that make up an Internet address; for example, 147.31.254.130.

Dotting Sequence A cellular radio term. An alternating series of 38 bits used for the purpose of symbol (bit) timing recovery at the CDPD Mobile Data BAse Station (MDBS) for reverse channel transmissions by the CDPD Mobile End System.

Double Buffering The use of two buffers rather than one to temporarily hold data being moved to and from an I/O device. Double buffering increases data transfer speed because one buffer can be filled while the other is being emptied.

Double Camp-On Indication A PBX feature. A phone attempting to camp on to another phone which is already being "camped on" shall receive a distinctive audible signal and may be denied the ability to camp-on.

Double Click With a Mac and an IBM-compatible PC running Windows, double-clicking carries out an action, such as beginning a new program. Press and release the mouse button twice in rapid succession to double-click. If you don't double click fast enough, it won't work. Some mouse software allow you to assign the left hand button on the mouse to a single click and the right hand button to a double click. This is very useful.

Double Crucible Method A method of fabricating optical fiber by melting core and clad glasses into two suitably joined concentric crucibles and then drawing a fiber from the combined melted glass.

Double Density Refers to a diskette which can contain twice the amount of data in the same amount of space as a single-density diskette. For example, a double-density 360k diskette has a 720k storage capacity. These days double density is an obsolete term, since there are now disks that are "double double" density. Most 3 1/2 inch disks will now hold twice 720K — or 1,440,000 bytes. These are called high density disks. Toshiba has introduced a "double double double" floppy, which will hold 2,880,000 bytes. But it hasn't caught on, yet.

Double Ended Synchronization A synchronization control system between two exchanges, in which the phase error signals used to control the clock at one exchange are derived from comparison w th the phase of the internal clock at both exchanges.

Double Interrupted Ring Two quick rings followed by a period of silence indicating the arrival of an outside call in some systems.

Double Modulation Modulation of a carrier wave of one frequency by a signal wave, this carrier then being used to modulate another carrier wave of different frequency.

Double Pole A double pole switch is one which opens and closes both sides of the same circuit simultaneously. Most electrical circuits open and close with only one side being broken.

Double Pull A method for pulling cable into conduit or duct liner that is similar to backfeed pulling except that it eliminates the need to lay the backfeed cable on the ground.

Double Sideband Carrier Transmission DSBTC. That method of transmission in which frequencies produced by the process of Amplitude Modulation (AM) are symmetrically spaced above and below the carrier. The carrier level is reduced for transmission at a fixed level below that which is provided to the modulator. Carrier is usually transmitted at a level suitable for use as a reference by the receiver except in those cases where it is reduced to the minimum practical level

(suppressed carrier). See also Amplitude Modulation, DSBSC, SSB and VSB.

Doublewide Two trailer homes stapled together with a modest gabled roof.

Doubler A device which doubles the distance of certain types of circuits. HDSL (High bit-rate Digital Subscriber Line) is a repeaterless means of provisioning a T-1 access circuit over a standard UTP (Unshielded Twisted Pair) local loop. Signal attenuation issues limit the range, however, to 12,000 feet on 24 gauge wire, and 9,000 feet on 26 gauge wire. A doubler can be added at that point in the outside plant to double the distance range, and a second doubler can be added to triple the range. As the doublers are line-powered, no local power supply is required. See also Attenuation, HDSL, Outside Plant, Repeater, T-1 and UTP.

Doubly Clad Fiber An optical fiber, usually single mode, that has a core surrounded by an inner cladding of lower refractive index, which is in turn surrounded by an outer cladding, which has a higher refractive index than the inner cladding. This type of construction is often employed in single-mode fibers to reduce bending losses.

DOV Data Over Voice. A technology used primarily with local Centrex services or special customer premises PBXs for transmitting data and voice simultaneously over twisted-pair copper wiring. Typical data rates for Centrex operation are 9.6 Kbps and 19.2 Kbps.

Down Sampling A sampling technique used in conjunction with certain compression algorithms, such as Wavelet Transform. Down-sampling computation involves disregarding certain samples; for instance, "down-sampling by two" considers only every other sample. This approach is an operation fundamental to the Fast Pyramid Algorithm used in wavelet transform, which is commonly applied to compression of image information. Up-sampling effectively is the reverse process of decompression on the receiving end of the data transfer. See Wavelet Transform, Fast Pyramid Algorithm and Up-Sampling.

Downconverter Integrated assembly of components required to convert microwave signals to an intermediate frequency range for further processing. Generally consists of an input filter, local oscillator filter, IF filter, mixer and frequently an LO frequency multiplier and one or more stages of IF amplification. May also incorporate the local oscillator, AGC/gain compensation components and RF preamplifier.

Downline Loading A system in which programs are loaded into the memory of a computer system, such as a LAN bridge, router or server, via the same communication line(s) the system normally uses to communicate with the rest of a network. As opposed to systems in which all programs are loaded into the computer from a disk or tape associated with the computer. A PC connected to a LAN may use this type of loading when it is first turned on in the morning to get the information it needs from a file server. Diskless PCs always work this way.

Downlink 1. The part of a transmission link reaching from a satellite to the ground. Some satellite transmission circuits, especially international ones, are priced and billed separately for the uplink and the downlink. This is because their transmissions are provided by different carriers.

2. In packet data communications, a downlink is a link from an NC or PAD to another NC or PAD on a different level. The defining of downlinks and uplinks depends on the network configuration of PADs, their relationships to each other and the direction of data transmission.

Download 1. To receive data from another computer (often called a host computer or host system or just plain host) into your computer. It's also called to RECEIVE. The opposite is UPLOAD or TRANSMIT. You have to be very careful distinguishing between the two. Choosing the "Download" option in some communications programs automatically erases a file of the same name that was meant for transmission. If that happens, stop everything. Grab your file unerase program and use it. Don't wait. If you wait, you may write over your erased file and never get it back. See DOWNLOADED FONTS.

2. The Vermonter's Guide to Computer Lingo defines download as to take the firewood off the pickup.

Downloaded Fonts Fonts that you send to a printer either before or during the printing of a document. When you send a font to a printer, it is stored in printer memory until it is needed. Downloaded fonts are one reason for loading lots of RAM memory into your printer.

Downsampling See DOWN SAMPLING

Downsizing Downsizing is what happens when companies move from large computer systems to smaller systems. There are four major reasons companies downsize from mainframe-based computer to local area network-based computing: 1. They save money. There are several reasons: a. Mainframe computers cost lots each month in maintenance. They require costly maintenance agreements with the supplier, e.g. IBM. Servers are usually bought without maintenance agreements. When they break, the managers or the workers simply replace the broken parts themselves. b. Mainframe computers cost lots to program. There are comparatively few programs available for mainframes, compared to the plethora of off-the-shelf programs available for workstations. c. Servers require a far less costly home to live. You don't need air-conditioning, special buildings with raised floors, etc. 2. Servers today have the power of mainframes 10 years ago. In fact, servers are now beginning to acquire more power than mainframes of ten years ago. And as servers increasingly acquire several processors, they will leap in power beyond what mainframes have. 3. Servers are typically manufactured from off-the-shelf, standard components that are usually available from several manufacturers. As a result, there is constant competition and constant improvement in quality and features. 4. Servers are much more flexible tools to design networks. You can start with one baby network containing one server and several workstations (a.k.a. clients) and confined to one floor of one small building and grow to a huge, complex network containing thousands of workstations, dozens of servers and spanning the globe.

To most people, downsizing is not only swapping out the "big iron" (the mainframe) and bringing in servers and local area networks. It's also a new way of thinking about the way corporations are organized. Downsizing is often accompanied by re-engineering, which is basically re-organizing for a greater focus on the customer — a focus which means responding faster to customer needs. See SERVERS.

Downstream 1. Refers to the relative position of two stations in a local area network ring topology. A station is downstream if it receives a token after the previous station. See also DOWNSTREAM CHANNEL.

2. In a communications circuit, there are directions of transmission — coming to you and going away from you. Downstream is another term for the transmission coming towards you. Downstream is used in cable TV — for the signal flowing to you from the cable head end. Downstream is used in modem connection for information coming at you. See UPSTREAM CHANNEL.

Downstream Channel The frequency multiplexed band

in a CATV channel which distributes signals from the headend to the users. Compare with UPSTREAM CHANNEL, the band of frequencies on a CATV channel reserved for transmission from the user to the CATV company's headend.

Downtime The total time a telephone system is not working due to some software or hardware failure. Downtime is also defined as a time interval when a system is not in use either because of equipment failure (unplanned downtime) or scheduled maintenance (planned downtime).

DP 1. Dial Pulse (as in dialing a phone) or Data Processing. Also called EDP for Electronic Data Processing. Now more commonly called Management Information Systems — or MIS.

2. Demarcation Point. The point of a demarcation and/or interconnection between telephone company communications facilities and terminal equipment, protective apparatus, or wiring at a subscriber's premises. Carrier-installed facilities at or constituting the demarcation point consist of a wire or a jack conforming to Subpart F of Part 68 of the FCC Rules.

DPA 1. Digital Port Adapter. A Northern Telecom word.

2. Demand Protocol Architecture. A technique for loading protocol stacks dynamically as they are required. It is associated with adapter cards in workstations and servers. Only the protocol stacks that are required for a particular communications sessions are loaded. Examples of such stacks include TCP/IP, XNS, SPX/IPX and NetBios.

DPBX Digital PBX. Not a common term. Most PBXs these days are digital. And they are simply called PBXs.

DPCM Differential Pulse Code Modulation

dpi Dots Per Inch. A measure of a scanner's ability to scan. The higher the number, the sharper the image and the greater the potential size of the image. See SCANNER ACCURACY.

DPLB Digital Private Line Billing.

DPMI An acronym for DOS Protected Mode Interface. DPMI is an industry standard that allows MS-DOS applications to execute code in the protected operating mode of the 80286 or 80386 processor. The DPMI specification is available from Intel Corporation. It is a superset of the VCPI (Virtual Control Program Interface) specification for controlling multiple programs inside a PC, as well as programs that use protected mode.

DPMS Display Power Manager Signaling is a power reduction feature that places a computer monitor in reduced power when the monitor is still on but has been idle for some time.

DPNSS Digital Private Network Signaling System. A standard in Britain which enables PBXs from different manufacturers to be tied together with E-1 lines and pass calls transparently between each — as easily as if the phones were extensions off the same PBX and were simply making intercom calls. The international version of DPNSS is called Q.SIG, which is now becoming Q.931, which is Euro-ISDN.

DPO Dedicated Pair Out (DPO) and Dedicated Pair In (DPI). If you move out of a residential apartment, the pair that you were using is dedicated to that Apt. so when the next occupant moves in, it is a flick of a switch to get phone service turned on. That's called Dedicated Pair Out. The DPI pair is the OE (Office Equipment) or the Switch pair. Now the pair in the C.O. (Central Office), when it's joined to the DPO pair, is often called a MATED PAIR. The DPI pair is mated to the DPO pair.

DPP Distributed Processing Peripheral.

DPRAM Dual port RAM.

DPSK See Differential Phase Shift Keying.

DPU Dynamic Path Update. Allows IBM networks to add new network nodes or change backup routing paths while the front-end is still operating.

DPX DataPath loop eXtension.

DQDB Distributed Queue Dual Bus. The Metropolitan Area Network (MAN) access technique defined by the IEEE 802.6 standard for Switched Multimegabit Data Service (SMDS). Based on QPSX (Queued Packet Synchronous Exchange), developed at the University of Western Australia (of which fact Harry Newton is ever so proud), DQDB operates by maintaining a queue at each station to determine when the station may access its dual buses. The dual buses provide bidirectional transmission between originating and terminating stations with excellent congestion control ensured through the network, from end-to-end. DQDB consists of two uni-directional buses connected as an open ring (i.e., each bus is not directly interconnected). It provides for full duplex operation between any two nodes. Each SMDS cell consists of 5 bytes of control information, involving a header and trailer, plus 48 bytes of payload. See also SMDS and ATM.

DQPSK Differential Quadrature Phase Shift Keying. An improvement on QPSK, this compression technique transmits only the differences between the values of the phase of the sine wave, rather than the full absolute value. QPSK makes use of two carrier signals, separated by 90 degrees. See also PSK and QPSK.

Draft Proposal An ISO standards document that has been registered and numbered but not yet approved.

Drag Dragging is a way of moving an item on the screen using your mouse. To drag a window in Windows 3.1, for example, move the mouse pointer onto a window's title bar, then hold down the mouse button while moving the mouse across your desktop. When you release the mouse button, the window will remain in its new location. Apply this technique to drag any data object, such as icons or list box items.

Drag And Drop The "drag and drop" definition defines how objects from one desktop application can be "dragged" out of that application, through clicking on the object with a mouse, across the desktop and "dropped" on another application. Most of the graphics operating systems, like Windows, Apple's Macintosh and Sun Sparc use Drag and Drop.

Drag Line A length of rope or string used to pull wire and cable through conduit or inaccessible spaces. Drag lines are often inserted in wall and ceilings during construction to ease future wire installation.

Drain Wire In a cable, an uninsulated wire laid over the component or components and used as a ground connection.

DRAM 1. Dynamic Random Access Memory chip. Pronounced "dee-ram." The readable/writable memory used to store data in PCs. DRAM stores each bit of information in a "cell" composed of a capacitor and a transistor. Because the capacitor in a DRAM cell can hold a charge for only a few milliseconds, DRAM must be continually refreshed to retain its data. In contrast, static RAM, or SRAM, requires no refresh and delivers better performance, but it is more expensive to manufacture. See also EDO RAM, SRAM, MICROPROCESSOR (for a full explanation). See also DYNAMIC RANDOM ACCESS MEMORY and SDRAM, which is slowly replacing DRAM.
2. Digital Recorder Announcer Module.

Drawing In the manufacture of wire, pulling the metal through a die or series of dies in order to reduce the diameter to a specified size.

Drawing Tools A computer imaging term. The means of creating freehand lines or basic geometric shapes. Paint packages often provide an ellipse-drawing function as a variation of the circle (or vice versa) and a square drawing function as a variation of the rectangle. Virtually all packages offer

filled geometric figures, the fill item being either a solid color or a pattern.

Dressing Cable You "dress" cable by taking multiple cables and joining them together neatly with cable ties. A nicely-dressed, clean installation is a sign of telephony professionalism. It still exists.

Drift 1. When a carrier frequency changes due to a transmitter problem. It can be caused by bad connections or defective components, temperature changes or diffraction. Crystal oscillators are the most drift-reliable circuits.
2. When customers leave and go to another carrier. See also Churn.

Drill Down Microsoft jargon which means to learn more about a subject.

Drive Icon A Windows NT term. An icon in a directory window in File Manager that represents a disk drive on your system. Different icons depict floppy disk drives, CD-ROM drives, network drives, etc.

Drive Mappings A Novell NetWare term. Drive mappings provide direct access to particular locations in the directory structure. They are a "shorthand" method for accessing directories on a disk. Instead of typing in the complete path name of a directory that you want to access, you can simply enter a drive letter that has been assigned to that directory. NetWare recognizes two types of drives (physical drives and logical drives) and three types of drive mappings (local, network, and search drive mappings).

Drive Type A number representing a standard configuration of physical parameters (cylinders, heads and sectors) of a particular type of hard disk drive. You need to know your drive's drive type; otherwise the BIOS of your machine will not recognize your drive on boot up and your PC will not work. Normally this information of your drive's type resides in memory kept alive by a small lithium battery. However, should the lithium battery die, your PC will "forget" which drive it has and it will ask you. If you don't know, you're in big trouble. Your PC simply won't work. My recommendation: write the drive type on two labels — one to stick on the drive and the other to stick on the bottom of your machine. This way you'll be able to find it easily. You will usually find the drive type's number on paperwork sent originally with your machine.

Drivebar A Windows NT term. Allows you to change drives by selecting one of the drive icons.

Driver A driver (which is always software) provides instructions for reformatting or interpreting software commands for transfer to and from peripheral devices and the central processor unit (CPU). Many printed circuit boards which you drop into a PC require a software driver in order for the other parts of the computer and the software you're running to work correctly. In other words, the driver is a software module that "drives" the data out of a specific hardware port. The port in question will usually have another device connected, such as a printer or modem, and the driver will be organized in software (i.e. configured) to communicate with the device.

Drop 1. A wire or cable from a pole or cable terminal to a building.
2. That portion of a device that looks toward the internal station facilities, e.g., toward an AUTOVON 4-wire switch, toward a switchboard, or toward a switching center.
3. Single channel attachment to the horizontal wiring grid (wall plate, coupling, MOD-MOD adapter).
4. The central office side of test jacks.
5. To delete, intentionally or unintentionally, part of a signal for some reason, e.g., dropping bits.

Drop And Insert That process wherein a part of the information carried in a transmission system is demodulated (dropped) at an intermediate point and different information is entered (inserted) for subsequent transmission.

Drop Cable 1. The outside wire pair which connects your house or office to the transmission line coming from the phone company's central office. See also DROP WIRE, which is different.

2. In local area networks, a cable that connects a network device such as a computer to a physical medium such as an Ethernet network. Drop cable is also called transceiver cable because it runs from a network node to a transceiver (a transmit / receiver) attached to the trunk cable.

Drop Channel Operation A type of operation where one or more channels of a multichannel system are terminated (dropped) at any intermediate point between the end terminals of the system.

Drop Clamp A piece of equipment used to attach aerial wire to a J hook or Ram's Horn on a building or pole.

Drop Loop The segment of wire from the nearest telephone pole to your home or business.

Drop Outs Drop outs are one major cause of errors in data communications circuits. The technical definition is that the signal level drops more than 12 dB (decibels) for more than 4 milliseconds. It means some of your data will not arrive. A four millisecond drop out in a transmission at 2,400 bits per second will lose about ten bits. A "drop out" is similar to a person's voice fades away in a telephone conversation. To correct the problem of drop out,we will ask the person (or computer) at the other end to repeat what they just said. "Huh?" This is called retransmission of data. In telephony, drop outs are defined as incidents when signal level unexpectedly drops at least 12 dB for more than 4 milliseconds. (Bell standard allows no more than two drop outs per 15 minute period.)

Drop Reel A reel used to transport and distribute drop wire.

Drop Repeater A repeater that is provided with the necessary equipment for local termination (dropping) of one or more channels of a multichannel system.

Drop Set All parts needed to complete connection from the drop (wall plate, coupling, MOD-MOD) to the terminal equipment. This would typically include a modular line cord and interface adapter.

Drop Shipment Equipment shipped to a buyer from a location separate from the seller's premise. If this third location is a different company, then this third-party supplier bills the seller and the seller bills the buyer. This saves time, but the seller loses control over the equipment's condition.

Drop Side Defines all cabling and connectors from the terminal equipment to the patch panel or punch down block designated for terminal equipment at the distribution frame.

Drop Wire Wires going from your phone company to the 66 Block or protector in your building. See also DISTRIBUTION CABLE.

Dropout A short period of time during which a transmission service looses the ability to transmit data. Bell System specifications define a dropout as any such loss which lasts for more than four milliseconds. See DROP OUTS for a longer explanation.

Dropped Call A call in which the radio link between the cellular customer and the cell site is broken. Dropped calls can happen often, and for many reasons, including terrain, equipment problems, atmospheric interference, and traveling out of range. In short, a dropped call is a call terminated by other than the calling or called party.

DRP Distribution and Replication Protocol. A proposal for an alleged improvement to HTTP. DRP is intended to improve on HTTP in two ways: by enabling multiple files to be downloaded over a single connection and by limiting the amount of data that needs to be downloaded each time a user returns to a Web site or downloads an update of an application. Currently HTTP requires a separate connection for each item being downloaded, which can slow the download process if many connections need to be opened. HTTP is also stateless, meaning that no information about a Web site is maintained on the client machine once the user moves on to another side, with the exception of a temporary cache and cookies. See HTTP. DRP is currently being studied by The World Wide Web Consortium (W3C).

DRU 1. DACS Remote Unit.

2. Digital Remote Unit. An NEC term. It's a multiplexer used to distribute NEC Dterm digital telephones and analog sets throughout the user's communications network, whether that network is local or geographically dispersed. The multiplexing technology used is North American Standard T-1 and European Standard E-1.

Drum Factor In facsimile systems, the ratio of drum length to drum diameter. Where drums are not used, it is the ratio of the equivalent dimensions.

Drum Speed The angular speed of the facsimile transmitter or recorder drum, measured in revolutions per minute.

Drunken Swede A way of describing the sound of a computer doing text-to-speech conversion. "Why, he sounds like a drunken Swede." This great definition from Stuart Segal of Phone Base Systems, Inc. in Vienna, VA. Says Stuart, "Our people think that a drunken Swede has recorded this message." It is possible to have a computer generate speech that doesn't sound like a drunken Swede if you throw sufficient horsepower (MIPS and memory) at it. Throwing sufficient horsepower, however, has been expensive, until recently. Drunken Swedes are going to get less and less common as horsepower gets cheaper and cheaper.

Dry Cable with no electronics and with no telecommunications transmission on it. In short, raw copper pair. You rent raw copper pair when you want to put your own electronics on it.

Dry Cell A type of primary cell in which the electrolyte is in the form of a paste rather than that of a liquid.

Dry Circuit A circuit over which voice signals are transmitted and which carries no direct current.

Dry Contact A dry contact refers to a circuit with an energy level such that a spark is not created in a mechanical relay or switch contracts when the circuit is opened. As a result, no cleaning of the contacts takes place (sparking vaporizes contact materials thereby continually exposing fresh contact material). Sometimes general purpose contacts are goldflashed so the relay or switch can be used on dry circuits; when used on higher energy circuits the gold coating is destroyed, but no damage is done except that the contact should not be used to carry dry circuit signal levels again. A dry contact might operate a relay which might turn something of higher power on or off. For example, a low voltage signal in a key system might cause a dry contact to close, thus causing much higher voltage to flow to a bell, a klaxon, a strobe light. And, yes, there is a "wet" contact. The term "mercury-wetted relay" refers to a relay or switch in which the movable contact of the device makes contact with a pool of mercury. In fact, before solid-state devices, this was a common technology for switching dry circuits.

Dry Copper Pair A dry copper pair or circuit is one which is has no electronics (e.g., load coils, repeaters or subscriber carrier systems) between the wire center and the customer premises. Such electronics interfere with the provisioning and support of high-speed data lines, such as DSL (Digital Subscriber Line) services, which many CLECs are interested in selling. See also CLEC, Dedicated Inside Plant, Dry Copper, DSL, ILEC and See DRY TWISTED PAIR.

Dry Electrolytic Condenser An electrolytic condenser in which the electrolyte is in the form of a paste or jelly rather than that of a liquid.

Dry Loop Powering Refers to local (not span) powering, and a transmission medium other than copper wire (microwave/fiber optic).

Dry Pair A "dry pair" is a simple pair of wires with no voltage, signals or protocols, just two wires. These are sometimes used for alarm systems but can conceivably be used for other communications if the pair is "loop qualified" before implementation.

Dry T-1 A T-1 line with an unpowered interface. A T-1 line with a power is called "Wet.'

Dry Twisted Pair Also known as "dry copper," a dedicated twisted-pair circuit without loading coils or any sort of electronics, whatsoever. Dry twisted pair circuits are commonly leased from the LEC for burglar alarm circuits, with one end terminating at the customer's premise in an alarm box and with the other end terminating at the premise of the alarm company. Recently, a number of CAPs (Competitive Access Providers), CLECs (Competitive Local Exchange Carriers) and ISPs (Internet Service Providers) have been leasing dry copper circuits from telephone companies for purposes of provisioning high-speed xDSL services. Since such xDSL services compete with LEC services and since xDSL can cause crosstalk problems with adjacent pairs in the same cable system and since it deprives them of revenue, several of the LECs recently have refused to lease dry copper. See Coppertone.

DS 1. Digital Signal. See DS-, DS-0, DS-1
2. Danske Standardiseringsrad (Danish Standards Institution).
3. An ATM term. Distributed Single Layer Test Method: An abstract test method in which the upper tester is located within the system under test and the point of control and observation (PCO) is located at the upper service boundary of the Implementation Under Test (IUT) - for testing one protocol layer. Test events are specified in terms of the abstract service primitives (ASP) at the upper tester above the IUT and ASPs and/or protocol data units (PDU) at the lower tester PCO.

DS- A hierarchy of digital signal (that's where the DS comes from) speeds used to classify capacities of lines and trunks. The fundamental speed level is DS-0 (64-kilobits per second). It's the speed you need when you use PCM to sample a voice call 8,000 times a second and encode it in an 8-bit code, thus 8 x 8,000 = 64,000 bits per second. The highest in the digital signal (DS) hierarchy is DS-4 (about 274 million bits per second. Here they are: DS-1, DS-1C, DS-2, DS-3, DS-4. They correspond to 1.544, 3.152, 6.312, 44.736, and 274.176 million bits per second. DS-1 is also called T-1.

DS-0 Digital Service, level 0. It is 64,000 bps, the worldwide standard speed for digitizing one voice conversation using pulse code modulation (PCM) and sampling the voice 8,000 times a second and encoding the result in an 8-bit code (thus 8 x 8,000 = 64,000). There are 24 DS-0 channels in a DS-1.

DS-1 Digital Service, level 1. It is 1.544 million bits per sec-

ond in North America, 2.048 million bits per second elsewhere where it's called E-1. Why there's no consistency is one of those wonderful, unanswered, questions. The 1.544 Mbps standard is an old Bell System standard. The 2.048 Mbps standard is a ITU-T standard. Standard for 1.544 million bits per second is 24 voice conversations each encoded at 64 Kbps. Standard for 2.048 megabits is 30 conversations plus control/signaling channels, for a total of 32 channels, each of 64,000 bits per second.

DS-1C Digital Service, level 1C. It is 3.152 million bits per second in North America and is carried on T-1. See DS-1.

DS-2 Digital Service, level 2. It is 6.312 million bits per second in North America and is carried on T-2. See DS-0, DS-1, T-1 and T-2.

DS-3 Digital Service, level 3. Equivalent of 28 T-1 channels, and operating at 44.736 million bits per second. Also called T-3. See DS-0, DS-1, T-1 and T-2. See DS-0, DS-1, T-1, T-2 and T-3.

DS-4 Digital Signal, level 4. 274,176,000 bits per second. 168 T-1s. 168 DS-1s. 4032 standard voice channels. See DS-0, DS-1, T-1, T-2 and T-3.

DS0 Usually written DS-0 (and pronounced D S zero). Digital Signal, level Zero. DS0 is 64,000 bits per second. It is equal to one voice conversation digitized under PCM. Twenty-four DS0s (24x64 Kbps) equal one DS1, which is T-1 or 1.544 million bits per second. See DS-0.

DS0-A Refers to a process where a subrate signal (2.4, 4.8, or 9.6 Kbps) is repeated 20, 10 or 5 times, respectively to make a 64 Kbps DS-0 channel.

DS0-B Refers to a process performed by a subrate multiplexer where twenty 2.4 Kbps, ten 4.8 Kbps, or five 9.6 Kbps signals are bundled into one 64 Kbps DS-0 channel.

DS1 Digital Signal, level One. A 1.544 million bits per second digital signal carried on a T-1 transmission facility. See DS-1.

DS2 Digital Signal, level Two. 6,312,000 bits per second. Four T-1s. Four DS1s. 96 standard voice channels.

DS3 Digital Signal, level Three. 44,736,000 bits per second. 28 T-1s. 28 DS1s. 672 standard voice channels.

DS3 PLCP An ATM term. Physical Layer Convergence Protocol: An alternate method used by older T carrier equipment to locate ATM cell boundaries. This method has recently been moved to an informative appendix of the ATM DS3 specification and has been replaced by the HEC method.

DS4 Digital Signal, level Four. 274,176,000 bits per second. 168 T1s. 168 DS1s. 4032 standard voice channels.

DSA 1. Distributed Systems Architecture, the network architecture developed by Honeywell.
2. Directory System Agent. The software that provides the X.500 Directory Service for a portion of the directory information base. Generally, each DSA is responsible for the directory information for a single organization or organizational unit.
3. Data Service Adapter

DSAP Destination Service Access Point. The logical address of the specific service entity to which data must be delivered when it arrives at its destination. This information may be built into the data field of an IEEE 802.3 transmission frame.

DSAT Digital Supervisory Audio Tones. A supervisory signaling scheme used in NAMPS — a new form of digital cellular radial called Narrow-band Advanced Mobile Phone service. See also NAMP.

DSBSC Double SideBand Suppressed Carrier. A variation on the theme of DSBTC (Double SideBand Transmitted Carrier), DSBSC uses less power since the carrier is not sent. The same amount of bandwidth is used, and carrier synchronization is lost. See also Amplitude Modulation, DSBTC, SSB and VSB.

DSBTC Double SideBand Transmitted Carrier. A form of Amplitude Modulation (AM) used to encode analog signals for transmission over a digital facility. DSBTC multiplies the carrier signal by modulating the amplitude (volume) of the carrier signal, plus adding a dc (direct current) component. The resulting output has some level of redundancy as the output is the sum of the carrier signal plus symmetric components in sidebands, which are some frequency separation from the carrier frequency. See also Amplitude Modulation, DSBSC, SSB and VSB.

DSC Digital Selecting Calling. A synchronous system developed by the International Radio Consultative Committee (CCIR), used to establish contact with a station or group of stations automatically by radio. The operational and technical characteristics of this system are contained in CCIR Recommendation 493.

DSCWID Call Waiting Display with Disposition.

DSDC Direct Service Dialing Capability. Network services provided by local switches interacting with remote data bases via CCIS.

DSE 1. Data Switching Equipment.

2. An ATM term. Distributed Single-Layer Embedded (Test Method): An abstract test method in which the upper tester is located within the system under test and there is a point of control and observation at the upper service boundary of the Implementation Under Test (IUT) for testing a protocol layer, or sublayer, which is part of a multi-protocol IUT.

DSI Digital Speed Interpolation. A technique for squeezing more voice conversations onto a line. DSI digitizes speech so it can be cut into slices, such that no bits are transmitted when no one is speaking. As soon as speech begins, bits flow again. See DIGITAL SPEECH INTERPOLATION.

DSL See DIGITAL SUBSCRIBER LINE.

DSLAM Digital Subscriber Line Access Multiplexer. A new technology being developed to concentrate traffic in ADSL implementations through TDM (Time Division Multiplexing) at the CO (Central Office) or remote line shelf. See Digital Subscriber Line, ADSL and TDM.

DSMCC Digital Storage Media Command And Control. Network protocols specified in Part 6 of MPEG-2 (ISO 13818) standards dealing with user-to-network and user-to-user signaling and communications.

DSN 1. Distributed Systems Network, the network architecture developed by Hewlett-Packard.

2. Double Shelf Network.

DSO See DS-0.

DSOB An AT&T Digital Data Service standard that specifies a means of multiplexing several subrate data channels within one DSO. Five 9.6, ten 4.8 or 20 2.4 Kbps subrate channels may be multiplexed within one DSO.

DSP 1. Display System Protocol

2. Digital Signal Processor. A Digital Signal Processor is a specialized computer chip designed to perform speedy and complex operations on digitized waveforms. Useful in processing sound and video. See DIGITAL SIGNAL PROCESSOR for a much better explanation.

DSP Modem Digital Signaling Processor chip set used for analog modem emulation. The software is programmable for easy upgrades.

DSR Data Set Ready. This signal is on pin 6 of the RS-232-C connector. It means the modem (which some telephone companies call a "data set") is ready to send data from the terminal. Some modems use Data Set Ready. Some don't. Modems that

are snooty enough to give you the DSR signal, are obnoxious enough to not work until they receive the DTR (Data Terminal Ready) signal from the terminal on pin 20. By bridging pins 6 and 20 on the connector at the modem, you can usually get it to work. If it doesn't, bridge in pin 8 (carrier detect) as well.

DSRR Digital Short Range Radio.

DSS Direct Station Select. A piece of key telephone equipment usually attached to an operator's phone set. When the operator needs to call a particular extension he/she simply touches the corresponding button on the Direct Station Select equipment. Typically DSS equipment/feature is part of a Busy Lamp Field (BLF), which shows with lights what's happening at each extension. Is it busy? Is it on hold? Is it ringing?

DSS is also Decision Support Systems. Computerized systems for transforming data into useful information, such as statistical models or predictions of trends, usually in a graphical format, which is used by management in the decision-making process. See DECISION SUPPORT SYSTEMS.

DSS2 Setup An ATM term. Digital Subscriber Signaling #2: ATM Broadband signaling.

DSSS Direct Sequence Spread Spectrum. A technique used in spread spectrum radio transmission systems, such as Wireless LANs and some PCS cellular systems. DSSS involves the conversation of a datastream into a stream of packets, each of which is prepended by an ID contained in the packet header. The stream of packets then is transmitted over a wide range of frequencies, using an approach known as "scattering." A large number of other transmissions also may share the same range of frequencies at the same time, with the potential for overlapping of packets. The receiving device is able to distinguish each packet in a packet stream by reading the various IDs, treating competing signals as noise. In a wireless LAN environment, DSSS typically operates in the 2.4 GHz frequency band, which is one of the ISM (Industrial Scientific Medical) bands defined by the FCC for unlicensed use. Although some manufacturers continue to use DHSS, Frequency Hopping Spread Spectrum (FHSS) is the preferred approach. See also CDMA, FHSS, ISM and SPREAD SPECTRUM.

DSTN Double Super Twisted Nematic. A display technology which uses two layers of crystal to correct distortions caused by the first layer and so improve readability.

DSU Digital Service Unit, also called Data Service Unit. Converts RS-232-C or other terminal interface to DSX-1 interface. See DATA SERVICE UNIT.

DSU/CSU The devices used to access digital data channels are called DSU/CSUs (Data Service Unit/Channel Service Units). At the customer's end of the telephone connection, these devices perform much the same function for digital circuits that modems provide for analog connections. For example, DSU/CSUs take data from terminals and computers, encode it, and transmit it down the link. At the receive end, another DSU/CSU equalizes the received signal, filters it, and decodes it for interpretation by the end-user.

DSVD Digital Simultaneous Voice and Data. Technology in a modem that allows you to send and receive voice and data (fax, images, files, etc.) on the same "conversation" on one analog phone line. DSVD allows the simultaneous transmission of data and digitally-encoded voice signals over a single dial-up analog phone line. DSVD modems use V.34 modulation (up to 33.6 kilobits per second), but may also use V.32 bis modulation (14,400 kilobits per second). DSVD modems reserve eight kilobits per second for voice transmissions. The remaining bandwidth is available for data transmission. The

DSVD voice coder is a modified version of an existing specification and is defined as G.729 Annex A. The DSVD voice/data multiplexing scheme is an extension of the V.42 error correction protocol widely used in modems today. DSVD also specifies fallbacks that enable DSVD modems to communicated with standard data modems (i.e. V.34, V.32 bis, V.32 and V.22).

What Can You Do With DSVD? (Below is good manufacturer promotion on DSVD. The words came from Rockwell.)

Have you ever promised someone on the telephone that you'd fax a document to them, then forget to send it? Have you ever traveled cross-country or halfway around the world to make a presentation? Have you considered expensive video conferencing hardware and applications as a way to communicate? Ever been frustrated waiting online, hoping to get some technical support?

When you need to work on a project, spreadsheet or presentation with someone from another location, a DSVD modem lets you do it without leaving your office or requiring costly video conferencing equipment. Both parties can simultaneously collaborate on a shared document, each viewing and discussing the additions and modifications made. This can slash time and money previously wasted exchanging faxes and relying on overnight express mail deliveries. Computer and software demonstrations can be easily conducted through a DSVD modem. This is a great way to highlight or test market new features and get immediate customer feedback. For the computer company, DSVD means fewer support follow-up calls and lower telephone bills. With a DSVD modem, you will be able to dial your favorite online catalog showrooms, browse the aisles, ask a salesperson a question or two, then place an order. Unlike other online "malls", with DSVD the human interface is still intact. While you're interactively shopping you can talk with salespeople or even other shoppers, just like the real mall. See V.70.

DSX Digital System Cross-connect frame. A bay or panel to which T-1 lines and DS1 circuit packs are wired and that permits cross-connections by patch cords and plugs. A DSX panel is used in small office applications where only a few digital trunks are installed. See also DACS.

DSX-1 Digital Signal Cross-connect Level 1. The set of parameters for cross connecting DS-1 lines.

DSX-3 The designation for the DS3 point of interface (cross-connect).

DTC 1. Digital Trunk Controller.
2. Digital Transmit Command.
3. Digital Traffic Channel. A digital cellular term. Defined in IS-136, the DTC is the portion of the air interface which carries the actual data transmitted. The DTC operates over frequencies separate from the DCCH (Digital Control CHannel), which is used for signaling and control purposes. See also DCCH and IS-136.

DTE 1. Data Terminal Equipment. In the RS-232-C standard specification, the RS-232-C is connected between the DCE (Data Communications Equipment) and a DTE. The main difference between a DCE and a DTE is that pins two and three are reversed. See also DCE and the Appendix.
2. Defense Technology Enterprise.

DTE-DCE Rate Data terminal equipment/data communications equipment rate. A designation for the maximum rate at which a modem and a PC can exchange information, expressed in kilobits per second (kbps). For maximum performance, a modem must support a DTE-DCE rate in excess of its maximum theoretical throughput.

Dterm A line of proprietary electronic phones made by NEC for use with its PBXs. The Dterm terminal derives its intelligence from its own microprocessor, which detects events and accepts direction from the PBX.

DTH Direct To Home. Intended as a replacement for C-band satellite systems, DTH was proposed to operate on medium-powered FSS (Fixed Satellite Systems) in the Ku-band. DTH was superseded by DBS (Direct Broadcast Satellite), which allows the use of even smaller receive antennae than possible with DTH. See also DIRECT BROADCAST SATELLITE and KU BAND.

DTL An ATM term. Designated Transit List: A list of nodes and optional link IDs that completely specify a path across a single PNNI peer group.

DTL Originator An ATM term. The first switching system within the entire PNNI routing domain to build the initial DTL stack for a given connection.

DTL Terminator An ATM term. The last switching system within the entire PNNI routing domain to process the connection and thus the connection's DTL.

DTLU Digital Trunk and Line Unit. Provides system access for T1-carrier lines used for inter office trunks or remote switching module umbilicals.

DTMF Dual Tone Multi-Frequency. A fancy term describing push button or Touchtone dialing. (Touchtone is a not registered trademark of AT&T, though until 1984 it was.) In DTMF, when you touch a button on a push button pad, it makes a tone, actually a combination of two tones, one high frequency and one low frequency. Thus the name Dual Tone Multi Frequency. In U.S. telephony, there are actually two types of "tone" signaling, one used on normal business or home pushbutton/touchtone phones, and one used for signaling within the telephone network itself. When you go into a central office, look for the test board. There you'll see what looks like a standard touchtone pad. Next to the pad there'll be a small toggle switch that allows you to choose the sounds the touchtone pad will make — either normal touchtone dialing (DTMF) or the network version (MF).

The eight possible tones that comprise the DTMF signaling system were specially selected to easily pass through the telephone network without attenuation and with minimum interaction with each other. Since these tones fall within the frequency range of the human voice, additional considerations were added to prevent the human voice from inadvertently imitating or "falsing" DTMF signaling digits. One way this was done was to break the tones into two groups, a high frequency group and a low frequency group. A valid DTMF tone has one tone from each group. In other words, each DTMF tone has two tones. Here is a table of the DTMF digits with their respective frequencies. One Hertz (abbreviated Hz.) is one cycle per second of frequency.

How each touchtone button makes two tones

				Low tones
1	2	3	A	697Hz
4	5	6	B	770Hz
7	8	9	C	852Hz
*	0	#	D	941Hz

High Tones 1209Hz 1336Hz 1477Hz 1633Hz

Normal telephones (yours and mine) have 12 buttons, thus 12 combinations. Government Autovon (Automatic Voice Network) telephones have 16 combinations, the extra four (those above) being used for "precedence," which in Federal government parlance is a designation assigned to a phone call by the caller to indicate to communications personnel the relative urgency (therefore the order of handling) of the call and to the called person the order in which the message is to be noted. See also LONG TONES and the four following definitions.

DTMF Automatic Routing This is a term relating to a fax server operating on a Novell file server. In this system, the fax software assigns a four-digit number to each user. A fax sender dials the fax line, and after the fax server answers, it sends a special auto routing request signal. The sender dials the four-digit number for the correct user, and the fax is auto-matically sent to the user's workstation on the LAN.

DTMF Cut-Through The capability of a voice response system to receive DTMF tones while the voice synthesizer is delivering information, i.e. during speech playback. This capability of DTMF cut-through saves the user waiting until the machine has played the whole message (which typically is a menu with options). The user can simply touchtone his response anytime during the message — when he first hears his selection number, when the message first starts, etc. When the voice processor hears the touchtoned selection (i.e. the DTMF cut-through), it stops speaking and jumps to the chosen selection. For example, the machine starts to say, "If you know the person you're calling, touchtone his extension in now." But before you hear the "If you know" you push but-ton in 230, which you know is Joe's extension. Bingo, the message stops and Joe's extension starts ringing. DTMF Cut-Through is also known as touchtone type-ahead.

DTMF Register A printed circuit card in a switch that con-verts the DTMF signals coming from the phone into signals which can be used by the switch's stored program control, central computer to do its switching, etc.

DTMF To Dial Pulse Conversion A PBX feature. DTMF (push button) phones are very popular. But sometimes you install a PBX with push button phones in an area which does-n't have a central office which will respond to push button tones. It's old. In this case, anyone dialing on a push button phone will find that the PBX converts that dialing to rotary pulsing when the PBX accesses a trunk which can't handle push button dialing. All this doesn't speed up the time the call takes to get through. It just speeds up the user's dialing and makes him or her feel she is dealing with a more modern phone system.

DTP DeskTop Publishing.

DTR Data Terminal Ready. A control signal sent from the DTE to the DCE that indicates that the DTE is powered on and ready to communicate. DTR can also be used for hardware flow control.

DTS Digital Termination Systems. Microwave based trans-mission technology designed for bypass functions for short-hop, line-of-sight applications. It never converts to analog. Is useful in high-volume, pure-data applications in urban set-tings where line costs are high. It requires FCC license and is referred to formally by FCC as Digital Electronic Message Service, or DEMS.

DTSR Dial Tone Speed Recording.

DTT Digital Trunk Testing.

DTU Digital Test Unit.

DTV Digital TV. The generic term for TV broadcast systems employing digital, rather than analog, transmission. See ATV.

DTX Battery-saving feature on a cellular phone that cuts back the output power when you stop speaking.

DU Fiber Optic Connector developed by Nippon Electric Group.

DUA Directory User Agent. The software that accesses the X.500 Directory Service on behalf of the directory user. The directory user may be a person or another software element.

DUAL Distributed Update Algorithm. A routing algorithm that provides fast rerouting (convergence) with minimal con-sumption of resources.

Dual Attachment Concentrator DAC. A concentrator that offers two S ports for connections to the FDDI network and multiple M ports for attachment of devices such as worksta-tions and other concentrators. A DAC takes maximum advan-tage of the resiliency of the FDDI dual counter-rotating ring architecture through connection to both rings, unlike a SAC (Single Attachment Concentrator). Connection of devices through the concentrator obviates the need for each device to attach directly, thereby reducing the number of optoelectric conversions and resulting in reduced cost of attachment. The concentrator also serves as a point of contention for network access. See also Single Attachment Concentrator, Single Attachment Station, Dual Attachment Station and FDDI.

A concentrator that offers two connections to the FDDI net-work capable of accommodating the FDDI dual ring, and additional ports for connection of other FDDI devices.

Dual Attachment Station DAS. A device such as a workstation which connects directly to both FDDI dual counter-rotating rings, rather than gaining ring access through a Dual Attachment Concentrator (DAC). A DAS allows access to two separate cable systems at the same time, providing protection against cable failure or damage. See also Single Attachment Concentrator, Single Attachment Station, Dual Attachment Concentrator and FDDI

Dual Band Dual band describes a cellular phone handset which works on both the 800 Mhz (in AMPS — analog) and 1900 Mhz (in Digital) bands for North America. This allows the phone to switch between the two bands. The reason you'd want this is simple: Digital service is often cheaper better in areas you can get it. But you can't get it everywhere. If you travel you need a cell phone you can use everywhere. Thus the idea of carrying a dual band cell phone and subscribing to a service that gets you access to both. See also Tri-Mode.

Dual Cable A two-cable system in broadband LANs in which coaxial cable provides two physical paths for transmis-sion, one for transmit and one for receive, as against dividing the capacity of a single cable.

Dual Coat An optical fiber coating structure consisting of a soft inner coating and a hard outer coating.

Dual Duplex A term sometimes used to describe HDSL (High bit rate Digital Subscriber Line), a relatively new tech-nology which commonly is used to provide T-1 and E-1 local loops. HDSL makes use of 2 twisted pair loops, with each operating in full duplex. The traditional approach to T-1 pro-visioning involves 2 twisted pairs, each operating in simplex. See also HDSL.

Dual Fiber Cable A type of optical fiber cable that has two single-fiber cables enclosed in an extruded overjacket of polyvinyl chloride with a rip cord for peeling back the over-jacket to access the fibers.

Dual Headset Also known as an integrated headset. A special type of headset for the blind. One jack plugs into a telephone and another jack plugs into a telephone and anoth-er jack plugs into a specially configured PC. This PC provides voice synthesized output. The dual headset allows a visually impaired TSR (Telephone Sales Representative) hands-free capability. Example: The Social Security Administration has numerous blind TSRs handling incoming public calls. Dual headsets allow these blind TSRs to perform their duties with no deterioration in public service. This definition provided by

Matt Gottlieb, telecommunications specialist for the Social Security Administration.

Dual Homed Gateway A dual homed gateway is a computer that runs firewall software and has two network interface boards: One board is attached to an untrusted network, and the other board is attached to a trusted network. A dual-homed gateway relays information between the two networks and prevents any direct contact between them. Both circuit-level gateways and application-level gateways are dual homed gateways.

Dual Homing 1. The process of using two geographically diverse frame relay port connections, each with its own set of virtual circuits, to support a network location running critical business applications which cannot afford network down time. 2. A method of cabling FDD concentrators and stations that permits a alternative path to the dual ring. Can be used in a tree or dual ring of trees topology.

Dual Line Registration The ability to have two cellular telephone numbers in a single cellular telephone. This allows the user to have service on two cellular systems without "roaming" in a second city (and paying higher toll charges) or to have one number on the wireline and one number on the non-wireline system. This assures the cellular user of back-up. I know a salesman who lives in Los Angeles, but spends much of his week serving customers in Phoenix. He has a cellular phone with Los Angeles number and a Phoenix number.

Dual Line Service Telephone service where two pairs of wires are connected to the premises. One or both could be in service.

Dual Mode Dual mode is the cellular industry's term for a cellular phone which will work for both analog and digital cellular phone systems. The cellular phone industry is going digital. Today's analog phones won't work on tomorrow's digital systems. But some phones — dual mode — will work on both. You may need to buy them over today's analog phones. You also need to be careful that the digital mode which you get in your dual mode cellular phone will work with the digital technology of your local carrier. That's not as standard, as yet, as today's analog technology, which works universally in North America. In short, there are many variations of digital cell phones service in North America, many of which are incompatible.

Dual NAM Allows a cellular phone user to have two phone numbers with the same or separate carriers. Very useful for someone who spends half his life in one place and half in another. For example, a friend of mine lives in LA, but works weekdays in Phoenix. His handheld cellular has phone numbers from LA and Phoenix carriers.

Dual Processing An SFT II configuration under Novell's NetWare that assigns parts of the operating system to separate processors. Because SFT II is split into two engines (the IOEngine and the MSEngine), it is possible to install each engine on a separate CPU, creating dual processing system. However, unless such a system is extremely busy, the extra CPU will not help network performance. Dual processing improves performance only when the servers are being used at near-maximum capacity.

Dual Ring Of Trees Topology An FDDI network topology of concentrators and nodes that cascades from concentrators on a dual ring.

Dual Tone Multi-Frequency DTMF. A way of signaling consisting of a push button or touchtone dial that sends out a sound which consists of two discrete tones, picked up and interpreted by telephone switches — either PBXs or central offices. See DTMF for a bigger explanation.

Dual Universal Asynchronous Receiver

Transmitter Dual Universal Asynchronous Receiver Transmitter. A DUART provides hardware support for two serial communications ports.

DUART Dual Universal Asynchronous Receiver Transmitter. A DUART provides hardware support for two serial communications ports.

Duct A pipe, tube or conduit through which cables or wires are passed. Duct space is always at a premium. If you ever install a duct, make sure it's twice the diameter you think you need. If you're lucky, it will last a couple of years. The cost of putting in thicker or extra ducts is peanuts compared to the cost of having to install additional ones later. Digging up places is getting very expensive, despite Ditch Witch, a company that makes the greatest backhoe trenching equipment. And also has the greatest name.

Duct Cycle The relationship between the time a device or facility is used and the time it is idle.

Duct Liner A small diameter pipe or tubing placed inside conventional underground conduit so you can install fiber optic or cables. Its main purpose is to provide a clean, continuous path with known frictional characteristics.

Ductbank An arrangement of ducts, for wires or cables, arranged in tiers.

Ductile Capable of being drawn out, hammered thin or being flexed or bent without failure.

Due Date The date an event is to occur, i.e. an installation, a change or a connection. Some vendors give accurate due dates. Penalty clauses work most effectively in ensuring due dates are met.

Dumb Switch A slang word for a telecommunications switch that contains only basic switching software and relies on instructions sent it by an outside computer. Those instructions are typically fed the "dumb" switch through a cable from the computer to one or more RS-232 serial ports which the dumb switch sports. The switch makes no demands on what type of computer it talks to, but simply insists that it be able to feed the computer questions and promptly receive responses in a form that it (the switch) can understand. Plain ASCII is OK. For example, the dumb switch might signal the computer, "A call is coming in on port 23, what do I do now?" The computer might reply "Answer it and transfer it to extension 23." Or it might say "answer it and put it on hold," or "answer it, put it on hold and play it recording number three." In essence, a dumb switch is anything but. It is in reality an empty cage containing whatever network interface cards the user has chosen. Each of these network interface cards is designed to "talk" to one type of telephone line. That line might be a T-1 line. It might be a normal tip and ring loop start line. It might be a tie trunk with E&M signaling. The card may handle one or many lines, but always of the same type. The card knows how to answer a call or pulse out a call on that particular type of line. It has all the telephony smarts. What it lacks is the intelligence of what to do with the calls. That is provided by the outside computer. Well, almost. Most "dumb" switches do contain rudimentary intelligence — a small computer and some memory. That computer is usually programmed to handle "default" calls — and to handle calls should the link to the outside computer fail, or the outside computer itself fail. Dumb switches come in flavors all the way from residing in their own cabinet to being printed circuit cards which reside in one or more of the personal computer's slots. Dumb switches are programmed to do "specialized" telecom applications, for example emergency 911, added value 800 services, cellular switching, automatic call distributors, predictive dialers, etc. They can, of course, be programmed to be "normal" PBXs. The question increasingly being asked is "If I want to program a specialized telecom application should I use a dumb switch or should I use an open PBX?" And the answer is "It depends." Depends on what you want to do. Depends on what software is available, etc. See also OAI.

Dumb Terminal A computer terminal with no processing or no programming capabilities. Hence, it derives all its power from the computer it is attached to — typically over a local hardwire or a phone line. A dumb terminal does not employ a data transmission protocol and only sends or receives data one character at a time, sequentially. There are many reasons for "dumb" terminals. They're cheap. They're foolproof. Operators don't have to mess with floppy disks, etc. The require minimal training. Dumb terminals are typically used for simple data entry and data retrieval tasks. Their disadvantage is that everything must come from the central computer — not only the information (data record) but also the form in which to put it. This has led to the creation of "intelligent" terminals, which have a modicum of capabilities — such as an inbuilt (with software) form, some smart function keys and perhaps, a modicum of processing power, etc.

Dumb Terminal Access An Internet Access Term. Telnet shell command that allows remote terminals to connect to a LAN host. Basic terminal server support over the same modems and phone lines used for remote access.

Dummy Load A dissipative impedance-matched network, used at the end of a transmission line to absorb all incident power, usually converted to heat.

Dump To copy the entire contents of something — memory, a file on a disk, a complete disk — to a printer or another magnetic storage medium. A dump is often called a "core dump," which is a bigger dump than dump.

Dumpster Diving Searching for access codes or other sensitive information in the trash. In North America, large trash receptacles — those found outside buildings — are often known as dumpsters.

Duobinary Signal A pseudobinary-coded signal in which a "0" ("zero") bit is represented by a zero-level electric current or voltage; a "1" ("one") bit is represented by a positive-level current or voltage if the quality of "0" bits since the last "1" bit is even. and by a negative-level current or voltage if the quantity of "0" bits since the last "1" bit is odd. Duobinary signals require less bandwidth than NRZ. Duobinary signaling also permits addition of error-checking bits.

Duopoly Similar to monopoly except there are two licensed competitors per market instead of one. The cellular business in the United States is a duopoly. Each major city in the United States has two licensed cellular providers — a feat accomplished by the FCC. As a result, prices in most cities as between the carriers are identical.

DUP Data User Part. Higher-layer application protocol in SS7 for the exchange of circuit switched data; not supported by ISDNs.

Duplex 1. Simultaneous two-way transmission in both directions. A data communications term.
2. Two-sided printing.
3. Two hard disks that have separate disk controllers and are mirror copies of each other. Data is written simultaneously to both. See Mirroring.

Duplex Circuit A telephone line or circuit used to transmit in both directions at the same time. Also referred to as full duplex as opposed to half duplex which allows transmission in only one direction at one time.

Duplex Operation The simultaneous transmission and reception of signals in both directions.

Duplex Signaling DX. A direct current signaling system that transmits signals directly on the cable pair. Duplex signaling is a facility signaling system and range extension technique that used bridge type detection of small dc changes. Duplex signaling is typically used on long metallic trunks.

Duplex Transmission The simultaneous transmission of two series of signals by a single operating communicating device. A data communications term.

Duplexer 1. A device which splits a higher speed source data stream into two separate streams for transmission over two data channels. Another duplexer at the other end puts the two slower speed streams back together into one higher-spread stream.
2. A waveguide device designed to allow an antenna to be used both transmission and reception simultaneously.

Dustbuster A phone call or e-mail message you send to someone after a long silence — just to "shake off the dust" and see if the connection still works.

Duty Cycle The ratio of operating time to total elapsed time for a device that operates intermittently. Usually expressed as a percentage.

DV-1 As in Northern Telecom's PBX/computer product called the Meridian DV-1.

DVB The Digital Video Broadcasting Group is a European organization which publishes its work through ETSI that has authored many specifications for satellite and cable broadcasting of digital signals. Part of the DVB work has been focused on conditional access and has produced draft recommendations for a common CA module and the DVB Superscrambling algorithm.

DVD Digital Video Disc. Also called Digital Versatile Disc. A specification announced in early December, 1995 by nine companies for a new type of digital disk, similar to CD-ROMs but able to store far more music, video or data. DVDs are 5 inches in diameter (same diameter as today's CDs) and will be able to store as much as 17 gigabytes, compared to the 650 megabytes on today's CD-ROMs. Current CD-ROMs are 1.2 mm thick. The DVD disc uses two bonded 0.66 mm substrates, reducing the amount of distance between the surface of the disc and the physical pits on the disc that hold information. The laser doesn't have to penetrate as much plastic so it can be focused on a smaller area. Therefore, the pits are smaller and packed more tightly together, increasing capacity. At the same time, two different lens apertures can be included in the pick-ups, so information can be read by both DVDs and standard CDs. Not only can the disc hold more information, but having two bonded sides strengthens it and prevents warping. Here is a comparison of DVD storage:
Single-sided, single layer DVD5 = 4.7 gigabytes
Single-sided, dual layer DVD9 = 8.5 gigabytes
Double-sided, single layer DVD10 = 9.4 gigabytes (i.e. 2 x 4.7 gigabytes)
Double-sided, double layer DVD18 = 17 gigabytes (i.e. 2 x 8.5 gigabytes)
Write Once DVD-R (Recordable) = 3.8 gigabyte per side
Overwrite DVD RAM "More than 2.6 gigabyte per side
According to Toshiba, a single DVD can play up to 133 minutes of a full-featured Hollywood film. That covers nearly 92% of all movies ever made. A DVD can offer up to eight different sound tracks and 32 subtitles. DVD technology produces spectacular motion pictures. The standard VHS resolution is

210 horizontal lines, laser disc comes in second at 425 lines but DVD wins with 540 lines of resolution. That's not all, DVD technology allows you to chose from different viewing formats. You can watch your movies in the:
Windscreen format 20:99
Letterbox format 16:9
Panned format 4:3
All DVD players allow you to listen to audio CDs. The sampling rate is higher than with audio CDs giving a much richer sound. DVD brings Dolby AC3 sound to your own house, you can experience the same audio performance as the best movie theaters. Six audio channels envelop you in the most realistic sound you will have ever experienced.

DVD-R Digital Video Disc (also called Digital Versatile Disc) — Record. A write once 5" doubled-sided disc with a capacity of 3.8 gigabytes per side. See DVD.

DVD-RAM Digital Video Disc (also called Digital Versatile Disc) — Rewritable. Both single and double-sided discs can be used in this rerecordable format. The music capacity is 296 minutes (for one sided) and 592 minutes (for two sided), compared with up to 75 minutes for a normal CD. Matsushita was the first major company to support the DVD-RAM format. And the DVD Forum, a working group of 10 DVD disc drive makers, officially supports Matsushita's rewritable format, called DVD-RAM. Two Forum members, Sony and Philips, however, have introduced their own rewritable format called DVD+RW, which can record 3.0 GB per side. Another standard is Pioneer's, called DVD-R/W, which can record 3.95 GB per side. Another format is NEC's, which it calls Multi Media Video File (MMVF), and which can record 5.6 GB per side. DVD-RAM, DVD+RW and MMVD are incompatible.

DVD-ROM Digital Video Disc (also called Digital Versatile Disc) Read Only Memory. A misnamed term for a DVD disc player, which plays DVD discs, but which does NOT record DVD discs. See DVD.

DVD-R/W Another rewritable DVD format. See DVD-RAM.

DVD+RW Digital Video Disc (also called Digital Versatile Disc) — Rewritable. The six companies introducing a DVD+RW format at Comdex 1997 endorse a specification called Phase-Change Rewritable. See DVD-RAM.

DVI Digital Video Interactive. A name for including still and moving video pictures in material shown on a PC's screen. DVI is part of multimedia. DVI is also Intel's old name for a scheme for digitizing and compressing video and audio for storage, editing, playback and integration into PC applications. The name has been replaced on the software side with Indeo video technology, on the retail side with Smart Video Recorder and on the hardware side with i750 Processors. See DIGITAL VIDEO INTERACTIVE for a much bigger explanation.

Dvorak Keyboard A keyboard, invented mainly by August Dvorak, on which letters and characters are arranged for faster and easier typing than on the standard QWERTY keyboard. The QWERTY keyboard was actually designed to be difficult to use, to slow down typists so they wouldn't jam the old typewriters' mechanisms! In that respect the QWERTY keyboard resembles the present touchtone in that it also was designed to be slow and difficult to use so that it wouldn't confuse the early and slow central offices of the time.

DVTS Pronounced "Divitz." Stands for Desktop Video conferencing Telecommunications System.

DWDM Dense Wavelength Division Multiplexing. DWDM is the higher-capacity version of WDM (Wavelength Division Multiplexing), which is a means of increasing the capacity of fiber-optic data transmission systems through the multiplex-

ing of multiple wavelengths of light. WDM systems support the multiplexing of as many as four wavelengths; commercially available DWDM systems support from 8 to 40 wavelengths. The capacity is steadily increasing, both by ever-expanding channel counts and faster supported TDM rates of the individual wavelengths. The top numbers are expected to continue to increase for the next few years, at least. Each wavelength operates as though it were a separate light pipe, with each currently supporting as much as OC-192 transmission (9.953 Gbps). Generally, existing systems trade-off channel count against maximum supported rate: the current maximum channel count of 40 (for the present — we are writing this in December, 1998) is limited to OC-48 (almost 100 Gbps net throughput). Systems which support higher signal rate (i.e., OC-192) support fewer than half as many channels, at most. At OC-192, a 40-channel system would yield an incredible 400 Gbps, rounded up. While such a system is not yet available, it may be little more than a year away. Within the next 2-3 years it is reasonable to expect to see systems supporting on the order of 100 wavelengths of OC-192 each, providing almost 1 Terabit per second transport! See WDM (Wavelength Division Multiplexing) for much more detailed explanation.

Dwell Time The period of time that a satellite is over the desired area of coverage. The term has commercial significance in LEO (Low Earth Orbiting) and MEO (Middle Earth Orbiting) satellite systems. In such systems, the individual satellites in a constellation travel in elliptical orbits, rather than in traditional geosynchronous orbits. See LEO and MEO.

DWI Data Warehousing Institute. A Special Interest Group dedicated to "helping organizations increase their understanding and use of business intelligence by educating decision makers and I/S professionals on the proper deployment of data warehousing strategies and technologies." DWI has over 3,000 members. www.dw-institute.com

DWS Dialable Wideband Service.

DX Signaling A form of DC (direct current) signaling in which the differences in voltage on two pairs of a four-wire trunk indicates the supervision information, i.e. the call's beginning, its end, etc. See DUPLEX SIGNALING.

DXI Data eXchange Interface. A specification developed by the SMDS (Switched Megabit (or Multi-megabit) Data Services) Interest Group to define the interaction between internetworking devices and CSUs/DSUs that are transmitting over an SMDS access line. SMDS is a way for a corporate network to dial up switched data services as fast as 45 megabits per second. The ATM Forum defines DXI as "a variable length frame-based ATM interface between a DTE and a special ATM CSU/DSU. The ATM CSU/DSU converts between the variable-length DXI frames and the fixed-length ATM cells."

Dye Sublimation a spectacular printing process where exactly measured temperatures control the amount of ink transferred from colored ribbons to paper. Under high temperature and pressures, the ink is not melted, but is transformed directly to gas, which hardens on the paper after passing through a porous coating. Dye sub printers create very nearly continuous tones, making them great for natural images. Because the gas makes "fuzzy" dots, dye sub is not recommended for sharp-edged "computer-y" graphics or type. But it does turn out gorgeous photo-like images.

Dynamic In English, dynamic means that things are changing. In telecomese, it tends to mean that our equipment — hardware and/or software — can respond instantly to changes as they occur. For example, dynamic routing in the call center world means that when a machine can switch the incoming calls from moment to moment. We may want to this because we want calls from the east to go to our call center in the east. Our eastern call center may presently be busy. So we may want to flip the calls over

Dynamic Answer This a term typically used in Automatic Call Distributors. The ability to dynamically assign the number of ring cycles (interrupt, more or less) to the queue period when agents are unavailable. The implication of being able to assign this number allows return supervision to the calling in person to be delayed and thus not allow billing on 800 INWATS lines to begin. This is a money saving feature. But it can cost you some customers if they get bored waiting for your phones to pick up.

Dynamic Backup A backup made while the database is active.

Dynamic Bandwidth Allocation The capability of subdividing large, high-capacity network transmission resources among multiple applications almost instantaneously, and providing each application with only that share of the bandwidth that the application needs at that moment. Dynamic bandwidth allocation is a feature available on certain high-end T-1 multiplexers that allows the total bit rate of the multiplexer's circuits to exceed the bandwidth of the network trunk. This works because the multiplexer only assigns channels on the network trunk to circuits that are transmitting.

Dynamic Beam Focusing When you have a curved cathode ray tube, the distance between the gun which shoots the electrons and all the parts of the screen are equal. When you have a flat screen, the distance varies slightly. Some beams have to travel further. When some have to travel not so far, Dynamic beam focusing, a term I first heard used by NEC, focuses each electron to the precise distance it must travel, thus ensuring edge-to-edge clarity on the screen.

Dynamic Capacity Allocation The process of determining and changing the amount of shared communications capacity assigned to nodes in the network based on current need.

Dynamic Configuration Registry A part of Chicago (Windows 4.0) which contains a list of all the various hardware bits and pieces that make up your computer. The dynamic configuration registry is a vital element of what Microsoft calls "Plug and Play," which is the ability to remove and add bits and pieces of hardware while the machine is running and have the machine automatically recognize those hardwares and alert applications accordingly.

Dynamic Data Exchange DDE. A form of InterProcess Communication (IPC) in Microsoft Windows and OS/2. When two or more programs that support DDE are running simultaneously, they can exchange information, data and commands. In Windows 3.xx this capability is enhanced with Object Linking and Embedding (OLE). See OLE.

Dynamic Host Configuration Protocol DHCP. A protocol for automatic TCP/IP configuration that provides static and dynamic address allocation and management. See DHCP for a longer explanation.

Dynamic HTML Dynamic HTML combines HTML, scripts and style sheets to bring animation to the Web. With Dynamic HTML, you can program your Web site such that a visitor surfing it alights on a button or some object and instantly a "help" or "explanation" balloon pops up. This balloon explains in greater detail what will happen if the visitor clicks on the button. Or the type may change and suddenly become bigger. Dynamic HTML is being incorporated into both Netscape and Microsoft Internet Explorer. The World Wide Web Consortium is considering the various flavors of dynam-

ic HTML as part of its DOM (Document Object Model) specification. See also www.astound.com.

Dynamic IP addressing An Internet Access Term. Allows dial-in client to update a variety of TCP/IP protocol stacks with a dynamically acquired address. Removes the burden of assigning per-user IP addresses and addresses the issue of dialing into different subnets. Virtually all the 30 million or so people who regularly surf the Internet do so through dynamic IP addressing.

Dynamic Link Library DLL. An executable code module for Microsoft Windows that can be loaded on demand and linked at run time, and then unloaded when the code is no longer needed.

Dynamic Load & Stress Testing An advanced and accurate form of load testing, dynamic load and stress testing more accurately presents the variety of stimulus of external callers, systems and networks to a computer telephony system. It will test your system using real-world user actions and a realistic call mix and busy hour usage profile. Most modern computer telephony network and systems provide a number of services to their users. Each of those services may stress the system in a different fashion. As the usage patterns of those users is varied and what each user will do can vary significantly, it is critical to load test systems with traffic patterns as dynamically and as close as possible to the way the system will be used in the real world. This definition courtesy of Steve Gladstone, author of the book Testing Computer Telephony Systems, available from 212-691-8215.

Dynamic Load Balancing A technique where a switching system, particularly multiple connected ACDs, apportion incoming calls (the load) to balance the workload. This is done dynamically in real time.

Dynamic Loud Speaker A loud speaker in which the diaphragm is driven by means of a small "voice coil" suspended in a powerful magnetic field.

Dynamic Memory The most common form of memory, used for RAM, with an access speed ranging from about 60 to 150 nanoseconds. (A nanosecond is a billionith of a second.) Dynamic memory is an inexpensive but relatively complicated form of semiconductor memory with two states: presence and absence of electrical charge. Dynamic memory requires a continuous electrical current. All data is lost when the power is cut. Frequent saving files to disk helps preserve your data.

Dynamic Microphone A microphone, the coil of which is moved in a strong magnetic field by vibrations striking the diaphragm to which it is attached. Electrical currents are thus generated in the moving coil.

Dynamic Node Access A high-speed bus invented by Dialogic to join together multiple voice processing PEB-based systems. PEB stands for PCM Expansion Bus. See PEB.

Dynamic Overload Control DOC. The feature of a switch which uses its translation tables and intelligence to allow the switch to adapt to changes in traffic loads by re-routing and blocking call attempts.

Dynamic Port Allocation In a voice processing system running multiple applications, dynamic port allocation is automatic allocation of ports based on the traffic being used by each application.

Dynamic RAM RAM memory that requires data to be refreshed periodically to prevent its loss in memory.

Dynamic Random Access Memory RAM which requires electronic refresh cycles every few milliseconds to preserve its data. See also RANDOM ACCESS MEMORY.

Dynamic Range In a transmission system, the ratio of the overload level to the noise level of the system, usually expressed in decibels. The ratio of the specified maximum level of a parameter (e.g., power, voltage, frequency, or floating point number representation) to its minimum detectable or positive value, usually expressed in decibels.

Dynamic Resource Allocation The assignment of network capacity to specific users and specific services as required on a moment-to-moment basis.

Dynamic Routing Routing that adjusts automatically to changes in network topology or traffic. Dynamic routing automatically accomplishes load balancing, therefore optimizing the performance of the network "on the fly." Static routing, on the other hand, involves the selection of a route for data traffic on the basis of predetermined routing options preset by the network administrator. Dynamic routing is more effective, but the routers are more costly and the more complex decision-making process imposes additional delays on the subject packet traffic. See also Router.

Dynamic Storage Allocation The allocation of memory space while a program is running. The memory is released when the program is complete.

Dynamic Variation A short time variation outside of steady-state conditions in the characteristics of power delivered to communication equipment.

Dynamically Adaptive Routing An algorithm, used for route determination in packet-switched networks that automatically routes traffic around congested, damaged, or destroyed switches and trunks and allows the system to continue to function over the remaining portions of the network.

Dynamically Assigned IP Address See IP ADDRESSING.

Dynamo An electrical machine which generates a direct current.

Dynamotor A direct current machine having two windings on its armature: one acting as a motor, the other as a generator.

E An abreviation for EMF — Electro Motive Force, a synonym for voltage.

E & M Ear & Mouth, Earth and Magneto, rEceive and transMit (take your choice). In telephony, a trunking arrangement that is generally used for two way (either side may initiate actions) switch-to-switch or switch-to-network connections. It also frequently used for computer telephony system to switch connections. See E & M LEADS.

E & M Leads The pair of wires carrying signals between trunk equipment and a separate signaling equipment unit. The "M" lead transmits a ground or battery conditions to the signaling equipment. The "E" lead receives open or ground signals from the signaling equipment. These leads are also known as Ear and Mouth Leads. The Ear lead typically means to receive and the Mouth lead typically means to transmit. Changes of voltage on these leads convey such information as seizure of circuit, recognition of seizure, release of circuit, dialed digits, etc. In the old days it was the PBX operators who originated trunk calls by asking the long distance carrier for free trunks using their mouth or M lead. If the carrier had a free trunk, the PBX heard about it through its ear or E lead. See also E & M and E & M SIGNALING.

E & M Signaling In telephony, an arrangement that uses separate leads, called respectively the "E" lead and "M" lead, for signaling and supervisory purposes. The near end signals the far end by applying -48 volts dc (vdc) to the "M" lead, which results in a ground being applied to the far end's "E" lead. When -48 vdc is applied to the far end "M" lead, the near-end "E" lead is grounded. The "E" originally stood for "ear," i.e., when the near-end "E" lead was grounded, the far end was calling and "wanted your ear." The "M" originally stood for "mouth," because when the near-end wanted to call (i.e., speak to) the far end, -48 vdc was applied to that lead.

When a PBX wishes to connect to another PBX directly or to a remote PBX or extension telephone over a leased voice grade line, a channel on T-1, the PBX uses a special line interface which is quite different from that which it uses to interface to the phones it's attached directly to (i.e. with in-building wires). The basic reason for the difference between a normal extension interface and the long distance interface is that the signaling requirements differ — even if the voice signal parameters such as level and two-wire, 4-wire remain the same. When dealing with tie lines or trunks it is costly, inefficient and too slow for a PBX to do what an extension telephone would do, i.e. go off hook, wait for dial tone, dial, wait for ringing to stop, etc. The E&M tie trunk interface device is the closest thing there is to a standard that exists in the PBX, T-1 multiplexer, voice digitizer telco world. But even then it comes in at least five different flavors. E&M signaling is the most common interface signaling method used to interconnect switching signaling systems with transmission signaling systems. See E & M and E & M LEADS.

E Block Carrier A 10 MHz PCS carrier serving a Basic Trading Area in the frequency block 1885 — 1890 MHz paired with 1965 — 1970 MHz.

E Channel E stands for echo. It is the 16 Kbps ISDN basic rate channel echoing contents of DCEs to DTEs. Used in bidding for access to multipoint link.

E Link Extended Link. A Signaling System 7 (SS7) connection. This protocol controls all transfers between COs in North America. A SS7 signaling link used to connect a Signaling End Point (SEP) to an STP pair not considered its home STP pair.

E Mail Electronic Mail.

E Mail Gateway A LAN application that fetches messages from one electronic mail system, translates them to the format of another electronic mail system, and then sends them to the "post office" of that other system. The post office is the public entry point — the place you put mail you want the other system to receive.

E Port A expansion port on a switch. It is used to link multiple switches together into a Fibre Channel fabric. See Fibre Channel.

E Purse Electronic purse. An electronic monetary transaction card being proposed by several government agencies.

E Rate From Network World, December 15, 1997: "E Rate is the program President Clinton and Vice President Gore are referring to when they say they want every classroom connected to the Internet by the year 2000. It grants elementary and secondary schools a discount on carrier services, including not only Internet access but also a raft of other offerings."

E-1 The European equivalent of the North American 1.544 million bits per second (Mbps) T-1,except that E-1 carries information at the rate of 2.048 million bits per second. This is the rate used by European CEPT carriers to transmit 30 64 Kbps digital channels for voice or data calls, plus a 64 kilobits per second (Kbps) channel for signaling, and a 64 Kbps channel for framing (synchronization) and maintenance. CEPT stands for the Conference of European Postal and Telecommunication Administrations. Since robbed-bit signaling is not used (as it is for T-1 in North America) all 8 bits per channel are used to code the waveshape sample. See E1, E2, E3, and T-1.

E-2 Interim data signal that carries four multiplexed E-1 signals. Effective data rate is 8.448 million bits per second (Mbps).

E-3 CEPT signal which carries 16 CEPT E-1s and overhead. Effective data rate is 34.368 million bits per second (Mbps).

E-911 Service Enhanced 911 service. Dial 911 in most major cities and you'll be connected with an emergency service run typically by a combination of the local police and local fire departments. 911 service becomes enhanced 911 emergency reporting service when there is a minimum of two special features added to it. E-911 provides ANI (Automatic Number Identification) and ALI (Automatic Location Information) to the 911 operator. Picture: A call comes in. Someone is dying. The 911 operator's screen comes alive as his phone rings. The number calling is on the screen. The caller is dying and needs an ambulance. The operator punches a button or two and his screen immediately indicates the location of the ambulance dispatch center nearest the caller. The operator contacts the dispatch center, another button may dispatch a fax of a map of how to get there to the ambulance and an ambulance gets there in short order and saves a live. (Remember, this is a book, not the real world.)

E-BCCH Extended-Broadcast Control CHannel. A logical channel element of the BCCH signaling and control channel used in digital cellular networks employing TDMA (Time Division Multiple Access), as defined by IS-136. See also BCCH, IS-136 and TDMA.

E-Bend A smooth change in the direction of the axis of a waveguide, throughout which the axis remains in a plane parallel to the direction of electric E-field (transverse) polarization.

E-Check An E-commerce term for an electronic check. The E-check is a demand for payment which is sent electronically over a network from the buyer to the seller. The e-check subsequently is sent from the seller to the seller's bank, and then to the buyer's bank. See also E-Commerce.

E-Commerce Electronic Commerce. Some experts think using electronic commerce to place orders could slash the cost of a typical company purchase order from $150 to $25.

E-IDE Enhanced IDE. An enhancement to the original IDE disk drive found on many PCs. E-IDE raises the storage capacity limit from 504 megabytes to 8,033 megabytes and the data transfer rate from up to 3 megabytes per second to up to 16.6 megabytes per second. See also ENHANCED IDE.

E-Interface The network interface between the Cellular Digital Packet Data (CDPD) networks and other external networks.

E-Mail Electronic Mail.

E-Stamp Electronic Stamp. Developed by E-Stamp Inc., and planned for trial by the USPS (U.S. Postal Service), the E-Stamp is a means for buying postage over the Internet. In support of a corporate Intranet post office, the system comprises PC software, a small security device which attaches to a user's printer port, and 1,024-bit encryption software for purposes of security. Think of it as a PC-based postage meter which can be refreshed over the Internet.

E-TDMA Extended Time Division Multiple Access. A proposed, new, standard for cellular. Other standards are TDMA (Time Division Multiple Access), CDMA (Code Division Multiple Access) and NAMPS (Narrow Advanced Mobile Phone Service). Refers to the extended (digital cellular) transmission technology developed by Hughes Network Systems. E-TDMA is alleged to have 15 times the capacity of today's analog cellular phone systems — in other words to allow the simultaneous use of 15 times as many cellular phones as today's analog cellular phone system.

E-Zine Magazines that are published (i.e. made public) on the World Wide Web. Typically an e-zine is available for anyone to read who wants to visit the site the electronic magazine is located at. An e-zine is also called a Webzine.

E.164 1. ITU-T recommendation for international telecommunication numbering, especially ISDN, B-ISDN (Broadband ISDN), and SMDS. An evolution of normal telephone numbers. In short, a scheme to assign numbers to phone lines.
2. An ATM term. A public network addressing standard utilizing up to a maximum of 15 digits. ATM uses E.164 addressing for public network addressing.

E.169 An ITU-T recommendation which will allow International Freephone Service customers to be allocated a unique Universal International Freephone Number (UIFN) which will remain the same throughout the world, regardless of country or telecommunications carrier. "Freephone" is a service which permits the cost of a telephone call to be charged to the called party, rather than the calling party. In North America, 800 and 888 numbers are "freephone" numbers.

E1 Name given to the CEPT (Conference of European Postal

and Telecommunication Administration) digital telephony format devised by the ITU-T that carries data at the rate of 2.048 million bits per second (DS-1 level). It's designed to carry 32 (thirty two) 64 Kbps digital channels. E1 is the rate used by European CEPT carriers to transmit 30 64 Kbps digital channels for voice or data calls, plus a 64 Kbps channel for signaling, and a 64 Kbps channel for framing and maintenance. Plesiochronous means "almost synchronous." In the network sense, when two networks operate with clocks of sufficiently high quality such that the signals in the two networks are nearly synchronous, the networks are plesiochronous. In the synchronization hierarchy, stratum I clocks are required for plesiochronous operation. See E-1 and T-1, which is the North American equivalent.

E2 Data signal that carries four multiplexed E-1 signals. Effective data rate is 8.448 million bits per second or 128 simultaneous conversations.

E3 Signal which carries 16 CEPT E-1 circuits and overhead. Effective data rate is 34.368 million bits per second or 512 simultaneous voice conversations. Also known as CEPT3.

E4 Signal which carries four E3 channels — or 139.264 million bits per second, or 1,920 simultaneous voice conversations.

E5 Signal which carries four E4 channels — or 565.148 million bits per second, or 7,680 simultaneous voice conversations.

EA See EQUAL ACCESS and ADDRESS FIELD EXTENSION.

EACEM The European Association of Consumer Electronics Manufacturers.

EADAS Engineering and Administrative Data Acquisition System.

EADAS/NWM EADAS NetWork Management.

Early Packet Discard EPD. A congestion control technique that selectively drops all but the last ATM cell in a Classical IP over ATM packet. When congestion occurs, EPD discards cells at the beginning of an IP packet, leaving the rest intact. The last cell is preserved because it alerts the switch and the destination station of the beginning of a new packet. Because IP packets from cells have been discarded receive no acknowledgment from the source. Most vendors expect EPD to be used in conjunction with unspecified bit rate (UBR) service. Switches simply junk UBR cells when congestion occurs, without regard for application traffic. By discarding ceiling selectively, so that whole IP packets are resent, EPD makes UBR a safer option.

Early Token Release This is a method of token passing which allows for two tokens to exist on the network simultaneously. It is used primarily in 16 million bit per second (Mbps) token ring LANs. On a regular four million bit per second token ring LAN, the token is passed on only after the sending computer receives its message back from the destination computer. With early token release, the sending computer does not wait for its message to return before passing the token. This means there are two tokens on the network at the same time. This is done to take advantage of the idle time created on the faster token ring. While the message is moving to its destination and back the sending computer is idle. On the four Mbps token ring, this is not much of a problem since most of the time the message is on the ring. On the 16 Mbps token ring less of this idle time is transmission time and more is taken by copying the message to the token ring card. That is, on the faster ring, there is more of a window for a second token. Early token release is especially helpful when traffic is heavy.

EARN European Academic Research Network. A network

using BITNET technology connecting universities and research labs in Europe.

EAROM An acronym for Electrically Alterable Read-Only Memory. A type of ROM chip which can be erased and reprogrammed without having to be removed from the circuit board. An EAROM is reprogrammed electrically faster and more conveniently than an EPROM (Erasable Programmable Read-Only Memory). An EAROM chip does not lose its memory when power is turned off.

Earth The term the English use for what Americans call ground. See Grounding.

Earth Ground The connection of an electrical system to earth. This connection is necessary to provide lightning and static protection as well as to establish the zero-voltage reference for the system. See EARTH GROUNDING.

Earth Ground Connection The conductor which connects directly to earth ground usually via a water pipe or possibly via a copper rod driven into the earth. This ground is different than the logic ground used in electronic circuits.

Earth Ground Electrode The conducting body in contact with the earth. The grounding electrode may be a metallic cold water pipe when used in conjunction with a driven rod, a mat, a grid, etc. Earth should never be used as the sole equipment grounding conductor.

Earth Grounding The purpose of earth grounding is essentially threefold: 1. Lightning protection; 2. Static protection; and 3. Establish a zero voltage reference. See GROUND and GROUNDING.

Earth Station A ground-based antenna and associated equipment used to receive and/or transmit telecommunications signals via satellite. Earth stations come in all sizes, shapes and purposes. For example,M most earth stations used by cable television operators are receive-only. Other terms used synonymously with "earth station" include downlink, ground station, and TVRO ("Television Receive Only").

Earthing There are two distinct and unique categories in the broad area called grounding. One is earthing, which is designed to guard against the adverse effects of lightning, assist in the reduction of static and bring a zero-voltage reference to system components in order that logic circuits can communicate from a known reference. The other category of grounding is known as equipment grounding. This is the primary means of protecting personnel from electrocution. According to the Electric Power Research Institute, "electrical wiring and grounding defects are the source of 90% of all equipment failures." Many telephone system installer/contractors have found that checking for and repairing grounding problems can solve many telephone system problems, especially intermittent "no trouble found" problems. As electrical connections age, they loosen, corrode and become subject to thermal stress that can increase the impedance of the ground path or increase the resistance of the connection to earth. Equipment is available to test for proper grounding. One of our favorite devices is made by Ecos Electronics in Oak Park, IL. Before you attach any equipment (computer, telephone, hi-fi set, etc.) to an improperly grounded electrical outlet, you should have the problems corrected. See also GROUND and GROUNDING.

EAS 1. Emergency Alert System. A system for radio and TV that is designed to provide warnings inthe event of emergencies. The system was originally designed during the Cold War to provide the President with a meas to address the American people in the event of a national emergency. It has never been used for that purpose (thank God), but it has been used to warn of natural disasters. The current use of the Emergency Alert System is voluntary and involves participation by three groups — 1. Radio, TV and cable broadcasters, 2. The National Weather Service and 3. State and local emergency management agencies. The FCC reports that 85% of the messages have dealt with weather emergenccies.

2. Extended Area Service. A novel name for a larger than normal local telephone calling area. The local phone company extends its subscribers the option of paying less per month for a small calling area and paying extra per individual call outside that area (i.e. the extended area), or paying more per month flat rate but having a larger calling area (i.e. having extended area service).

EASE A voice processing applications generator from Expert Systems, Inc. in Atlanta, GA.

Easter Egg A feature hidden in a software program. According to Bill Gates, many commercial programs contain secret screen credits listing the people involved its creation. To find the "Easter egg" in Excel 95, for instance, select row 95 in a new workbook, press the Tab key, choose the Help menu's "About Microsoft Excel" command, holding down the Shift, Ctrl and Alt keys simultaneously while clicking on the Tech Support button. If you can follow Bill Gates' instructions (this guy either is a contortionist, or has a roomful of people to press keys for him) you may find yourself in a roomlike environment which you can navigate with the arrow keys, leading up stairways to see scrolling lines of credits. You can turn away from the stairway and type "excelkfa" to find a zigzag walkway leading to photos of some of the program creators.

EAX Electronic Automatic eXchange. Term used throughout the non-Bell telephone industry to refer to an electronic central office. Similar to ESS (Electronic Switching System), the term used by AT&T and the Bell operating telephone companies.

EB Exabyte. See Exabyte.

EBCDIC (Pronounced Eb-si-dick.) Extended Binary Coded Decimal Interexchange Code. It is the way IBM codes characters, letters and numbers into a digital binary stream for use in its larger computers. EBCDIC codes characters into eight bits. This gives it 256 possible characters, $2 \times 2 \times 2 \times 2 \times 2 \times 2 \times 2 \times 2 = 256$. See also EXTENDED CHARACTER SET.

EBCDIC is mainly used in IBM mainframes and minicomputers, while ASCII is used in IBM and non-IBM desktop microcomputers. EBCDIC is not compatible with ASCII, meaning that a computer which understands EBCDIC will not understand ASCII. But there are many real-time and non-real time translation programs that will convert text files back and forth. A good program is Word-For-Word from MasterSoft. See ASCII.

EBDI Electronic Business Data Interchange. Term for EDI (Electronic Data Interchange). See EDI.

EBITDA Earnings Before Interest, Taxes, Depreciation and Amortization. Otherwise referred to as cash flow. Most commonly used for the valuation of wireless telecom stocks because of the heavy fixed-cost nature of the business.

EBONE European Backbone. A pan-European network backbone service.

EBPP Electronic Bill Presentation and Payment. Billing and payment over the Internet. Developing standards include Open Financial Exchange (OFX) and GOLD. See also Electronic Commerce, GOLD and OFX.

EBS 1. Electronic Business Set.

2. Enhanced Business Service (also known as P-Phone) is an analog Centrex offering provided by Northern Telecom. It operates over a single-pair subscriber loop., providing nor-

mal full duplex audio conversations and a secondary 8 KHz half-duplex amplitude shift-keyed signal, which is used to transmit signaling information to and from the Northern Telecom-equipped central office.

3. Emergency Broadcast System. Some local radio stations have volunteered their services to be part of a group of radio stations which would broadcast information should there be a public emergency. They operate the EBS under the aegis of the Federal Communications Commission. Such emergency would be a natural disaster, a technological disaster or a war. In 1994, Emergency Broadcast System changed. For a full explanation, see EMERGENCY BROADCAST SYSTEM.

4. End System Bye packet. Part of CDPD Mobile End System (M-ES) registration procedures.

EBU European Broadcasting Union. Formed in 1950 and headquartered in Geneva, Switzerland, the EBU was formed to address technical and legal problems in Western European broadcasting. The EBU now includes former Soviet Bloc broadcasters, having merged with the former International Radio and Television Organization (OIRT) in 1993. Associate membership includes 30 countries in Africa, the Americas, and Asia. The EBU runs EUROVISION, a permanent network of 13 channels on a Eutelsat satellite, plus 5,500 kilometers of permanently leased terrestrial circuits; the network serves as a vehicle for daily news and program exchanges. www.ebu.ch

EC European Community. Also known as the Common Market; an organization of 15 nations in Western Europe that has its own institutional structure and decision-making framework. The intent of the organization is to promote trade and reduce barriers. Member nations are Austria, Belgium, Denmark, Finland, France, Germany, Greece, Ireland, Italy, Luxembourg, the Netherlands, Portugal, Spain, Sweden and the United Kingdom.

Ecash Developed by DigiCash and the Mark Twain Bank, ecash is the ability to use real money in a electronic purchasing system over the World Wide Web. The process involves you sending a check to Mark Twain Bank which in turn sends you software which gives you access to the Ecash Mint where you draw funds to your hard drive for use when purchasing goods and services on the Internet. www.marktwain.com

ECC Elliptic Curve Cryptography. The Wireless Application Protocol (WAP) Forum has added ECC into Version 1.0 of its specification for wireless security that is bandwidth, memory and power efficient. More details www.certicom.com

ECC RAM Error-correcting code memory, often used in local area network servers. ECC memory tests for and corrects errors on the fly, using circuitry that generates checksums to correct errors greater than one hit.

Eccentricity Like concentricity a measure of the center of a conductor's location with respect to the circular cross section of the insulation. Expressed as a percentage of center displacement of one circle within the other.

ECCKT Exchange Company Circuit. A circuit ID for a trunk line. Such term is often used between telephone companies to identify lines that one is leasing to the other. See also FOC.

ECH Enhanced Call Handling. ECH systems are those in which a telephone call is handled "intelligently" b a variety of network, human, computer and telecommunications resources. ECH systems cover those from voice mail, to interactive voice response to computer telephone integration to fax-on-demand to complex telephone networks.

Echelon The name of a startup company in Palo Alto that is making a microprocessor chip destined for mundane house-

hold appliances like toasters, air conditioners, ovens, etc. The idea is that chip will be used by these devices to talk to other devices and thus coordinate their coming on and going off and doing things. The chips are destined to be networked together. Early uses for the chip includes smoke detectors which call you when they detect smoke and wall switches that detect when you come into a room and turn the lights on.

ECHO 1. European Commission Host Organization.

2. Exactly what you expect it to mean. You hear yourself speak. Echoes happen in both voice and data conversation. Echoes are good and bad. In voice, an echo happens when the equipment meant to amplify the voice of the party at one end, picks up the signals from the party at the other end, and amplifies them back to that party. Some echo is acceptable (in fact, almost a necessity) in voice conversations. It's called SIDETONE. When the speaker can hear himself speak through the receiver, sidetone gives the speaker some feeling his conversation is actually going through. But too much (i.e. too loud) echo is unacceptable. There are devices called echo suppressors, which do exactly that.

In low speed or on-line data conversations, an echo is positively vital. An echo in a data conversation is where I send my words to the distant computer which "echoes" them back to me and my screen displays them. This way I can visually check if the distant computer received my words/data accurately. In some software programs there's a command called ECHO or ECHOPLEX. Switch it one way, the distant computer echoes the words to my screen. Switch it the other, the words I'm typing on my keyboard are put on my screen.

There are two basic transmission modes in data communications — full duplex and half duplex. In full duplex, I have simultaneous two-way data flowing. In full duplex, I therefore have the capacity to "echo" back the data I am sending and have it displayed on my screen. But this "echoing" depends on the capability and/or programming of the computer at the other end. All dial-up services — Tymnet, GTE Telenet, MCI Mail, etc. — will echo my data back to me so I can check it. Some computers, such as the extremely dumb Compugraphic typesetter which typeset the first edition of this book, won't send an echo. In this case, if I want to see the data I am sending, I can change the parameters on my communications software to "half duplex." This way I will see what I am sending, but I will not see what the computer is receiving. Which may be very different. (And with our Compugraphic often is.) Of course, watching characters being echoed across your screen is a very poor method of data transmission and only useful in on-line transmissions. Some form of error-checking protocol is much better.

An echo is also a public discussion group that extends over more than one BBS (bulletin board system) via echomail.

Echo Attenuation In a communications circuit (4- or 2-wire) in which the two directions of transmission can be separated from each other, the attenuation of echo signals that return to the input of the circuit under consideration. Echo attenuation is expressed as the ratio of the transmitted power to the received echo power in decibels. See also ATTENUATION.

Echo Cancellation Technique that allows for the isolation and filtering of unwanted signals caused by echoes from the main transmitted signal. An echo cancellation device puts a signal on the return transmission path which is equal and opposite to the echo signal. Echo cancellation allows full duplex modems to send and receive on the same frequency. Network-based echo cancellation can interfere with modems which perform their own echo cancellation. To avoid cancellation problems, modems capable of echo cancellation (such as V.32 modems) send a unique answer tone with a phase reversal ever half second. The network echo cancellers detect the phase reversal in the answer tone and disable themselves. Contrast with echo suppression. See echo canceller.

Echo Canceller Device that allows for the isolation and filtering of unwanted signal caused by echoes from the main transmitted signal. In data communications networks, echo cancelers are used in the same way as PADS are in the network, but some brands of echo cancelers have the ability of being disabled by a 2100 Hertz tone transmitted by the data device prior to the exchange of the data device's handshaking protocol. If the echo canceler cannot be disabled by the data device, it will block the data call from completing. See also echo cancellation.

Echo Check A technique for verifying data sent to another location by returning the received data (echoing it back) to the sending end.

Echo Modeling A mathematical process where an echo is conceptually created from an audio waveform and subtracted from that form. The process involves sampling the acoustical properties of a room and guessing what form an echo might take, then removing that information from the audio signal.

Echo Return Loss ERL. The difference between a frequency signal and the echo on that signal as it reaches its destination. For example, a way of measuring echo return loss is to say it's the frequency-weighted measure of return loss over the middle of the voiceband (appox. 560-1965 Hz) where talker echo is most annoying.

Echo Suppression The process of turning off reverse transmissions on a telephone line to reduce the annoying effects of echoes in telephone connections, especially on satellite circuits. An active echo suppressor impedes full-duplex data transmission. Contrast with echo cancellation.

Echo Suppressor Used to reduce the annoying effects of echoes in telephone connections. The worst echoes occur on satellite circuits. An echo suppressor works by turning off transmission in the reverse direction while a person is talking, thus effectively making the circuit one way. An echo suppressor obviously impedes full-duplex data — data flowing both ways simultaneously. Echo suppressors are turned off by the high-pitched tone (typically 2025 Hz) in the answering modem, which it uses to signal it's answered the phone and is ready for a data conversation.

Echo Suppressor Disabler An echo suppressor disabler is a device which causes an echo suppressor to be disabled (i.e. turned off). Echo suppressors are turned off by the high-pitched tone (typically 2025 Hz) in the answering modem, which it uses to signal it's answered the phone and is ready for a data conversation. A disabled echo suppressor stays disabled until the circuit is disconnected and restored to its "ready" connection. Because an echo suppressor hinders full duplex transmission in data communications, it is necessary to disable the echo suppressor.

Echomail A public message area or conference on a bulletin board system (BBS) that is "echoed" to other systems in a BBS network. EchoMail is organized into different groups, each with a different topic and the term normally references communications on a FidoNet network. Also a term referring to the electronic transfer of messages between bulletin board systems.

Echoplex A way of checking the accuracy of data transmitted whereby the data received are returned to the sender for comparison with the original data. Somewhat time consuming. Used typically in slow speed transmissions. See ECHO.

ECITC European Committee for Information Technology testing and Certification.

ECL Emitter Coupled Logic.

ECM Error Correction Mode. An enhancement to Group 3 fax machines. Encapsulated data within HDLC frames providing the received with an opportunity to check for, and request retransmission of garbled data. See FACSIMILE and V.17.

ECMA European Computer Manufacturers Association. An international, Europe-based industry association founded in 1961 and dedicated to the standardization of information and communications systems. ECMA addresses standards in areas such as software engineering and interfaces, APIs and languages, data presentation, character sets and coding schemes, file structures, LAN protocols, IT security, and optical disks. www.ecma.ch See also CEPT.

ECMEA Enhanced Cellular Messaging Encryption Algorithm. The Telecommunications Industry Association's CMEA is used for confidentiality of the control channel in the most recent American digital telephony systems. It is a variable-length block cipher. A paper called "Cryptanalysis of the Cellular Message Encryption Algorithm" by Wagner, Schieier, and Kelsey) describes an attack on CMEA that "demonstrates that CMEA is deeply flawed." ECMEA (Enhanced CMEA) corrects the flaw.

ECOC European Conference on Optical Communication.

Economic Bandwidth An AT&T term for the maximum bandwidth that a physical medium can support without a significant increase in its cost.

Economic CCS ECCS. The load that should be carried on the last trunk of a high usage trunk group to minimize the total cost of routing the offered traffic, assuming that overflow from the high usage route is offered to an alternate route engineered to meet an objective blocking probability.

Economist The standing joke is that an economist is someone who didn't have the personality to make accountant. There is a theory floating around in some academic circles that God invented economists to make weather forecasters and astrologers look good. See ACCOUNTANT.

Ecommerce Electronic Commerce.

Economy Of Scale As throughput gets bigger, so the per unit cost comes down. This is the argument used by economists to justify monopolies — namely that the per unit costs of one supplier are far lower than having two suppliers. The economy of scale argument is used to justify having only one water company in town. It makes more sense in that industry than in the telephone industry.

It was once used in the telephone industry to justify one combined local phone company/long distance phone company/one supplier of terminal equipment. This argument does not really apply to telephony as technological breakthroughs have brought down the cost of getting into the telephone industry and have allowed smaller, competitive companies to become cost effective. Some large telecommunications monopolies, in fact, are experiencing diseconomies of scale. In this case, their cost of per unit business starts to rise as they get very large. Diseconomies of scale are caused by bloated bureaucracies and inertia in management decision making.

Econopath Calling Plans Discount plans for New York City businesses to make calls in their regional calling area. For example, Econopath offers a Manhattan business a discount for calls to the East Suffolk region.

ECP 1. Enhanced Call Processing. An Octel term for an interactive customized menu in its voice mail system which provides levels of call routing. See ENHANCED CALL PROCESSING.

2. Extended Capabilities (Parallel) Port. An upgrade to the original parallel port on a PC, which gives you: Transfer rates of more than two million bytes per second; bidirectional 8-bit operation)a standard parallel port has only 4 input bits); support for CD-ROM and scanner connections; 16-byte FIFO buffer; support for run length coding data compression. See also USB.

ECPA Electronic Communications Privacy Act.

ECSA Exchange Carriers Standards Association. See EXCHANGE CARRIERS STANDARDS ASSOCIATION.

ECTEL The European Telecommunications and Professional Electronics Industry.

ECTF The ECTF (Enterprise Computer Telephony Forum) is an industry organization formed to foster an open, competitive market for Computer Telephony technology. Participants include industry suppliers, developers, system integrators and users working to achieve agreement on multi-vendor implementations of Computer Telephony technology based on international defacto and dejure standards.

According to its own words, "the ECTF facilities the development, implementation and acceptance of Computer Telephony (CT) solutions by bringing together suppliers, developers, systems integrators and users. The Forum discusses, develops and tests interoperability techniques for dealing with the diverse technical approaches currently available. The ECTF incorporates and augments existing industry standards and publishes CT interoperability agreements."

Principal members of the ECTF are: Aculab plc; Amarex Technology, Inc; Amtelco; Amteva, Inc.; Aspect Telecommunications; AT&T; Brite Voice Systems; Brooktrout Technology; CallScan Limited; Centigram Communications; Cintech Tele-Management; COM2001 Technologies; CSELT; Deutsche Telekom AG; Dialogic Corporation; Digital Equipment Corporation; Ericsson Business Networks AB; Excel Inc.; Fujitsu Limited; Hewlett-Packard; IBM Corporation; InterVoice; Lernout & Hauspie Speech Products; Linkon Corporation; Lucent Technologies; Mitel Corporation; Natural MicroSystems Corp.; NEC America Inc.; Northern Telecom; Periphonics; Rhetorex; Rockwell Telecommunications; Siemens AG; Sun Microsystems; Tandem Computers; Texas Instruments; Trident Data Systems; Unimax Systems Corporation and Voicetek Corporation.

The ECTF has a Web site at www.ectf.org. Membership inquiries — 510-608-5915 cr ectf@ectf.org.

ECTF defintions Enterprise Computer Telephony Forum M.100 Revision 1.0 Administration Services "C" Language Application Programming Interfaces See M.100 and ECTF.

ECTUA European Council of Telecommunications Users Association.

ED 1. CallED party. The ED receives a call from the ING, or callING party, to set up a data transfer. ING and ED apply to any type of data transfer, including both voice and data.
2. Ending Delimiter.

EDA 1. Electronic Directory Assistance. A method by which companies can get access to telephone directories electronically using an X.25 packet switched connection.
2. Electronic Design Automation.

EDAC Error Detection And Correction.

EDCH Enhanced D-Channel Handler.

EDDA The European Digital Dealers Association. An European association of DEC resellers. EDDA members include VARs, systems integrators, leasing companies and service organizations.

Eddy Current Losses Losses in electrical devices using iron, due to the currents set up in it by magnetic action.

EDF Erbium-Doped Fiber.

EDFA Erbium-Doped Fiber Amplifier. A form of fiber optical amplification in which the transmitted light signal passes through a section of erbium-doped fiber and is amplified by means of a laser pump diode. EDFA is used in transmitter booster amplifiers, in-line repeating amplifiers, and receiver preamplifiers. EDFA involves doping a section of glass fiber approximately one meter in length with erbium, a rare earth element. When the light pulse hits that section of the fiber, it excites the erbium which, in turn, amplifies the light pulse. The advantage is the elimination of regenerative repeaters, which are costly, power hungry, and generally problematic. Repeaters must convert the optical signal to an electrical signal, boost it, and then regenerate an optical signal — that can cause errors and create latency. EDFA recently has been demonstrated to support a mix of four 2.5-Gbps digital video streams in delivery of an 80-channel digital AM (Amplitude Modulation) cable television network over distances of 100 kilometers; one EDFA was located at the output node and one at midspan. The signals were multiplexed using DWDM (Dense Wave Division Multiplexing) over standard single-mode fiber. See also DWDM and SONET.

Edgar Electronic Data Gathering Archiving and Retrieval. A database of corporate disclosure, transaction, and financial status data maintained by the U.S. Securities and Exchange Commission (SEC).

Edge Connector A connector made of strips of brass or other conductive metal found at the edge of a printed circuit board. The connector plus into a socket of another circuit board to exchange electronic signals.

Edge Device 1. A physical device which is capable of forwarding packets between legacy interworking interfaces (e.g., Ethernet, Token Ring, etc.) and ATM interfaces based on data-link and network layer information but which does not participate in the running of any network layer routing protocol. An Edge Device obtains forwarding descriptions using the route distribution protocol.
2. A physical device which sits on the edges of the Internet under the control of an ISP (Internet Service Provider) and allows the ISP to provide its customers with ancillary services, such as voice mail, fax forwarding, video downloading, etc. For an example of an edge device maker, see www.mediagate.com

Edge Effect A video term. The overemphasizing of well defined objects from the addition of black or white outlines to the vertical edges of the objects. Examples of this phenomena are, trailing white, leading black around the outline of a figure in movement within a scene.

Edge Router A new device for Internet Service Providers which will forward packets at high speed, do tunneling, authentication, filtering, packet accounting, trafffic shaping and address translation. Such a device, when it's out, service providers will be able to save money and create different services — combining features in different ways.

Edge Site A remote network site. A site at the edge of the network.

Edge Switch A Broadband Switching System (BSS) which is located at the edge of the network. Conceptually equivalent to a Central Office in the voice world, an edge switch is the first point of user access (and the final point of exit) for a broadband network (i.e., Frame Relay, SMDS and ATM). Edge switches are interconnected by "Core Switches," which are the functional equivalent of Tandem switches in the circuit-switched voice world.

EDH Electronic Document Handling.

EDI Electronic Data Interchange. A series of standards which provide computer-to-computer exchange of business documents between different companies' computers over phone lines and the Internet. These standards allow for the transmission of purchase orders, shipping documents, invoices, invoice payments, etc. between an enterprise and its "trading partners." A trading partner in EDI parlance is a supplier, customer, subsidiary, or any other organization with which an enterprise conducts business. EDI is used for placing orders, for billing and paying for goods and services via private electronic networks or via the Internet. According to studies, it costs about $50 a process paper-based purchase order and about $2.50 to process the same order with EDI. Internet-based EDI can lower the cost to less than $1.25. EDI software translates fixed field or "flat" files that are extracted from applications into a standard format and hands off the translated data to communications software for transmission. EDI standards are supported (i.e. have been adopted) by virtually every computer company in the country and increasingly, by every packet switched data communications company. The formats used to convert the documents into EDI data are defined by international standards bodies and by specific industry bodies. See also Electronic Data Interchange and IES. and http://www.edi-info-center.com/html/hotline.html

Edison Battery A type of storage battery in which the elements are nickel and iron and the electrolyte is potassium hydroxide. An old type of battery.

Edison Effect The phenomenon attributed to Edison, that when a filament is incandescent a current will flow between it and another electrode in the tube. In other words, the light bulb.

Editing Editing is a familiar process of changing the content of files to achieve more effective communication by cutting, pasting, cropping, resizing, or copying. Multimedia editing can be done on all types of media: voice annotations, music, still images, motion video, graphics and text. Tools for editing vary from simple tools for email voice annotations to more sophisticated tools for video manipulation. See also ELECTRONIC MAIL.

Editor A software program used to modify programs or files while they are being prepared or after they are (allegedly) complete. You can use an editor to write in. I wrote this dictionary using an editor called The Semware Editor. This program is much faster and far more flexible than a word processing program like Microsoft Word. But it can't do all the fancy layout, bolding, underlining, etc. that Word can do. An editor produces only ASCII text. The editor that comes with Windows95 is called Notepad. I use my editor to write in, because when I'm finished with this dictionary, I send it to sophisticated desktop publishing software called QuarkXpress. And that's where the layout for this book is done. I believe that the two functions — of writing and desktop publishing — are different and shouldn't be combined in one tool.

EDLIN The MS-DOS line editor that came originally with DOS and which you can use to create and edit batch files and other small text files. It's not very good. There are far better editors around, including The Semware Editor.

EDM Electronic Document Management. EDM unites the disparate workflow, document management and imaging.

EDMS An imaging term. Engineering Document Management System.

EDO RAM Extended Data-Out Random Access Memory. A form of DRAM (Dynamic RAM memory) that, according to PC Magazine, speeds accesses to consecutive locations in memory by (1.) assuming that the access to next memory will target an address in the same transistor row as the previous one and (2.) latching data at the output of the chip so it can be read even as the inputs are being changed for the next memory location. EDO RAM reduces memory access times by an average of 10 percent over with standard DRAM chips and costs only a little more to manufacture. EDO RAM has already replaced DRAM in many computers, and the trend is expected to continue. See also DRAM and EDRAM.

EDP Electronic Data Processing. Also DP, as in Data Processing. Basically, a machine (also called a computer) that receives, stores, operates on, records and outputs data. The word "electronic" was added to Data Processing when the industry moved away from tab cards — the 80 column "do not spindle," etc. — and was able to accept data electronically, instead of electromechanically as with the tab cards. People in the industry used to be called EDPers. Now the term MIS — Management Information System — is more common.

EDRAM 1. Enhanced Digital Announcement Machine
2. Enhanced Dynamic Random Access Memory. A form of DRAM (Dynamic RAM) that boosts performance by placing a small complement of static RAM (SRAM) in each DRAM chip and using the SRAM as a cache. Also known as cached DRAM, or CDRAM.

EDS See 1962.

EDSL Extended Digital Subscriber Line. The ISDN EDSL combines 23 B-channels and one 64-Kbps D-channel on a single line. Also called the Primary Access Rate.

EDTV See EXTENDED DEFINITION TV.

Edu An Internet address domain name for an educational organization.

Edutainment The answer to the question "What do you get when you cross educational material with interactive video?" A term coined by "someone who obviously knows nothing about either education or entertainment," says Laura Buddine, president of multimedia games maker Tiger Media. But it is becoming popular in residences and it's typically played on PCs with CD-ROM players. See also MMX.

EE End to End signaling. Punch DTMF are sent through the lines to signal the end of a conversation. A tone code is also used to access long distance carriers, to signal your answering machine, or to access your voice mail.

EEC European Economic Community.

EEHLLAPI Entry Emulator High Level Language Applications Programming Interface. An IBM API subset of HLLAPI.

EEHO Either End Hop Off. In private networks, a switch program that allows a call destined for an off-net location to be placed into the public network at either the closest switch to the origination or the closest switch to the destination. The choice is usually by time of day and is usually done to take advantage of cheaper rates.

EEMA European Electronic Messaging Association. See also ELECTRONIC MESSAGING ASSOCIATION.

EEPROM Electronically Erasable Programmable Read Only Memory. A read only memory device which can be erased and reprogrammed. Typically, it is programmed electronically (not electromagnetically) with ultraviolet light. EEPROMs don't lose their memory when you lose power. EEPROM used to be often used in PBXs and were the way manufacturers of older style PBXs upgraded their software. In other words, every time they sent you a software upgrade, they'd send you a bunch of chips. You'd pull out a bunch of chips on one of the main boards in the your PBX. And you'd replace them with the new chips. When you don't have a disk drive (and in the olden days disk drives were very expensive), EEPROMs were the only way to go. EEPROM Setup in a computer allows it to recognize certain system board configurations during initialization. You, the user, can then choose options such as the type of memory chips installed and base memory size without changing jumpers on the system board.

EES Escrow Encryption Standard. A security system proposed by the U.S. Department of Justice for the U.S. government's data communication. It involves inserting into all new federal computers a special encryption chip whose output would be reasonably secure but could be tapped by law enforcement agencies.

EETDN End-to-End Transit Delay Negotiation.

EF Entrance Facility (also called TEF or Telecommunications Entrance Facility).

EF&I Engineer, Furnish and Install.

EFCI Explicit Forward Congestion Indication: EFCI is an indication in the ATM cell header. A network element in an impending-congested state or a congested state may set EFCI so that this indication may be examined by the destination end-system. For example, the end-system may use this indication to implement a protocol that adaptively lowers the cell rate of the connection during congestion or impending congestion. A network element that is not in a congestion state or an impending congestion state will not modify the value of this indication. Impending congestion is the state when a network equipment is operating around its engineered capacity level.

EFD Event Forwarding Discriminator. A wireless telecommunications term. Software that contains a discriminator that determines if a notification should be forwarded on to a particular destination.

Efficiency Factor In data communications, the ratio of the time to transmit a text automatically and at a specified modulation rate, to the time actually required to receive the same text at a specified maximum error rate.

EFOC European Fiber Optics and Communications conference.

EFS Error Free Seconds: A unit used to specify the error performance of T carrier systems, usually expressed as EFS per hour, day, or week. This method gives a better indication of the distribution of bit errors than a simple bit error rate (BER). Also refer to SES.

EFT Electronic Funds Transfer. The moving of bits of data from one bank to another. Done in place of moving little green pieces of paper, called money.

EFTA European Free Trade Association.

EFTPOS Electronic Funds Transfer Point Of Sale.

EGA Enhanced Graphics Adapter. Second color video interface standard established for IBM PCs. Maximum resolution is 640 x 350 pixels. See MONITOR.

Ego Surfing Scanning the Net, databases, print media, or research papers looking for mentions of your own name.

EGP Exterior Gateway Protocol. An Internet protocol for exchanging routing information between autonomous systems.

Egress 1. The exit point. This typically refers to information being sent out of, as opposed to being sent in to a frame relay port connection or other network element.

2. In video it's often called "signal leakage" and it's a condition in which signals carried by the distribution system leak into the air.

EHz Exahertz (10 to the 18th power hertz). See also SPECTRUM DESIGNATION OF FREQUENCY.

EIA Electronic Industries Alliance, previously Electronic Industries Association. A trade organization of manufacturers which sets standards for use of its member companies, conducts educational programs, and lobbies in Washington for its members' collective prosperity. Founded in 1924 as the Radio Manufacturers Association, the EIA is organized along specific electronic product and market lines. Each group or division has its own board of directors, and sets its own specific agenda, designed to enhance the competitiveness of its own business sector. Under the umbrella of EIA, and representative of the range of interests of the organization, are the Consumer Electronics Manufacturers Association (CEMA); Electronic Components, Assemblies, Equipment & Supplies Association (ECA); Electronic Industries Foundation (EIF); Electronic Information Group (EIG); Government Electronics & Information Technology Association (GEIA); JEDEC Solid State Products Technology Division; and Telecommunications Industry Association (TIA). Membership is open to companies and individuals. www.eia.org See EIA INTERFACE.

EIA 232-D New version of RS-232-C physical layer interface adopted in 1987. See also RS-232.

EIA 530 Interface using DB-25 connector, but for higher speeds than EIA-232. Has balanced signals (like EIA-422) except for three maintenance signals which are EIA-423. See also RS-530.

EIA 561 EIA-232E interface on DIN-8 connector (like Macs use). See also RS-561.

EIA 568 An EIA standard for commercial building wiring. The standard covers four general areas; the medium, the topology of he medium, terminations and connections, and general administration. See EIA/TIA 568 below.

EIA 569 Commercial Building Standard for Telecommunications Pathways & Spaces - EIA/TIA, 1991. Lays out guidelines for sizing telecom closets, equipment rooms, conduit, etc . Every architect doing commercial buildings should have to memorize it. Few have even heard of it!

EIA 574 EIA 232E interface on DE-9 connector. (OK, DE is a bit pedantic. Most call it a DB-9, though you'll never find one in a parts book).

EIA 606 Telecommunications Administration Standard for Commercial Buildings. Guidelines covering design and identification of two- level backbone cabling for individual buildings and for campuses.

EIA Interface A set of signal characteristics (time, duration, voltage and current) set up by the Electronic Industries Association to standardize the transfer of information between different electronic devices, like computers, modems, printers, etc. The most famous EIA interface is the RS-232-C (now called the RS-232-E.) EIA-232 specifies three things: the functions of the interchange circuits, the electrical characteristics AND the connector (EIA-232-E includes two different connectors).

In contrast, the ITU-T's V.24 specifies ONLY the interchange circuit FUNCTIONS. V.28 specifies electrical characteristics compatible with EIA 232. ISO 2110 is the internal standard that defines the 25-pole D-shell connector compatible with EIA-232. Following the merger of EIA and the ITG part of EIA, all formed EIA telecommunications standards are now EIA/TIA publications and the standard referred to is now known as EIA/TIE-232-D, edition D, being the most recent. See EIA, EIA/TIA-RS-232-E and RS-232-E.

EIA/IS-132 An EIA Standard, "Cable Television Channel

Identification Plan", which specifies the cable-television channel-numbering plan proposed by the EIA and the NCTA. The cable television channel assignments specified in this document are incorporated by reference into the FCC Rules.

EIA/TIA-232-E The latest version of the familiar RS-232-C serial data transfer standard for communicating between Data Terminal Equipment (DTE) and Data Circuit Terminating equipment employing serial binary data interchange. (EIA/TIE's exact words.) This standard defines the serial ports on computers, which communicate with such things as external modems, serial printers, data PBXs, etc. See also EIA and RS-232.

EIA/TIA 568 EIA/TIA 568 Commercial Building Wiring Standard. This telecommunications standard in early 1991 was out for industry review under draft specification SP-1907B. Its purpose is to define a generic telecommunications wiring system for commercial buildings that will support a multi-product, multi-vendor environment. It covers topics such as:

- Recognized Media
- Topology
- Cable Lengths/Performance
- Interface Standards
- Wiring Practices
- Hardware Practices
- Administration

EID Equipment Identifier. A cellular radio term.

EIDE Enhanced IDE. See IDE and ENHANCED IDE.

Eiffel Tower The famous Paris landmark built in 1889 for the 1890 World's Fair. On October 21, 1915 the first wireless transAtlantic telephone call took place between H.R. Shreeve, a Bell Telephone engineer listening at the Eiffel Tower and using borrowed French equipment, and B.B. Webb in Arlington, Virginia. The transmission was " Hello, Shreeve! Hello, Shreeve! And now, Shreeve, good night!" During the period following W.W.I, it was often proposed that the Eiffel Tower be dismantled and sold for scrap. Literally the only thing that stalled those plans was the usefulness of the Eiffel Tower as the world's largest radio transmitter/receiver.

Eight Hundred Service 800-Service. A generic and common (and not trademarked) term for AT&T's, MCI's, Sprint's and the Bell operating companies' IN-WATS service. All these IN-WATS services have "800," "888" or "887" as their "area code." Dialing an 800, 888 or 887-number is free to the person making the call. The call is billed to the person or company being called.

800 Service works like this: You're somewhere in North America. You dial 1-800 and seven digits. Your local central office sees the "1" and recognizes the call as long distance. It ships that call to a bigger central office (or perhaps processes the call itself). At that central office it's processed, a machine will recognize the 800 "area code" and examine the next three digits. Those three digits will tell which long distance carrier to ship the call to.

Until 800 portability, happened in May, 1993 each 800 provider (local and long distance company) was assigned specific 800 three digit "exchanges." For example, MCI had the exchange 999. AT&T had the exchange 542. If you wanted a phone number beginning with 800-999, then you had to subscribe to MCI 800 service. If you wanted a phone number beginning with 800-542, you had to subscribe to AT&T 800 service. With 800 Portability that is no longer the case. Here is a history of what the phone industry calls 800 Data Base Access Service. It comes courtesy Bellcore:

"After divestiture (1984), the seven regional telecommunications companies began to provide limited 800 Service on their own as well as in conjunction with interexchange carriers. The regional companies transported 800 calls only within their own calling areas. The 800 number — containing 10 digits in accordance with the North American Numbering Plan (NANP) — was routed onto the long distance carrier's networks."

"The Common Carrier Bureau of the Federal Communications Commission (FCC) endorsed an incremental approach that would ultimately give the seven companies the right to create their own 800 Service architecture and eliminate reliance on the only existing signaling system (AT&T's). That approach involved assigning to 800 service providers one or more special numbers from the NANP. These numbers, known as "NXX codes," allowed carriers (MCI, Sprint, NY Telephone, etc.) to identify their own 800 numbers and offer their customers 800 numbers."

"Bellcore — Bell Communications Research — began to develop a new network architecture that would allow an 800 Service subscriber to change to another carrier without changing their existing 800 number (full number portability), in accordance with a September 1991 FCC order. That order declared that 800 data base service should be implemented by March 1993 (later extended to May, 1993) and that the old NXX plan be eliminated as long as access times met certain FCC standards."

"In September 1991, the FCC endorsed the plan initially set forth by the Bell operating companies, which provided that the administration of the Number Administration and Service Center (NASC) be transferred from Bellcore to an independent third party outside the telecommunications industry. Lockheed Information Management Systems Company (IMS) was selected by competitive bid to succeed Bellcore as NASC administrator. NANPA (North American Numbering Plan Administration) since has been shifted to NANC (North American Numbering Council.)"

"How 800 data base service works: The telecommunications network architecture that supports 800 Data Base Access Service is considered "intelligent" because data bases within the network supplement the call processing function performed by network switches. The Service uses a Common Channel Signaling (CCS) network and a collection of computers that accept message queries and provide responses. When a caller dials an 800 number, a Service Switching Point (SSP) recognizes from the digits "8-0-0" that the call requires special treatment and processes that call according to routing instructions it receives from a centralized database. This database, called the Service Control Point (SCP), can store millions of customer records."

"Although each regional company maintains whatever number of SCPs it needs to provide 800 Data Bases Access Service, information about how an 800 call should be handled is entered into the SCP through the off-line Bellcore support system called the Service Management System (SMS). SMS is a national computer system which administers assignment of 800 numbers to 800 service providers. It is located in Kansas City, maintained by Southwestern Bell Telephone Company, and administered by Bellcore with information received from 800 Number Administration and Service Center (NASC). The NASC provides user support and system administration for all 800 Service providers who access the SMS/800."

Because 800 service is essentially a data base lookup service, there are an endless "800 services" you can create. Here are the variables that can be used to influence how an 800 phone call

is handled and where an 800 phone call ultimately gets sent:
• The number calling. Virtually all 800 calls in North America (excepting Mexico) now come with the information as to from which number the call came.
• The number being called.
• The time of day, week, month etc.
• The instructions given at that particular moment. A computer might say "Sorry, our phone system is busy. We can't take any more calls in New York. Please send this one and all subsequent ones — until informed otherwise — to our phone system in Kansas City.

Here are a few examples of the services 800 providers have created using the above variables:
• TIME OF DAY ROUTING: Allows you to route incoming calls to alternate, predetermined locations at specified days of the week and times of the day.
• PERCENTAGE ALLOCATION ROUTING: Allows you to route pre-selected percentages of calls from each Originating Routing Group (ORG) to two or more answering locations. Allocation percentages can be defined for each ORG (typically an area code), for each day type and for each time slot.
• SINGLE NUMBER: The same 800 number is used for intrastate and interstate calling.
• CALL BLOCKAGE: You can block calling areas by state or area code. The caller from a blocked area hears the message: "Your 800 call cannot be completed as dialed. Please check the number and dial again or call 1- 800-XXX-XXXX for assistance." (You may want to block callers from areas which didn't see your special commercial, for example.)
• POINT OF CALL ROUTING: Allows a customer to route calls made to a single 800 number to different terminating locations based on the call's point of origin (state or area code.) You establish Originating Routing Groups (ORGs) and designate a specific answering location for each ORG's call.
• CALL ATTEMPT PROFILE: A special service that allows subscribers to purchase a record of the number of attempts that are made to an 800 number. The attempts are captured at the Network Control Point, and from this data a report is produced for the subscriber.
• ALTERNATE ROUTING: Allows a customer to create alternate routing plans that can be activated by the 800 carrier upon command in the event of an emergency. Several alternate plans can be set up using any features previously subscribed to in the main 800 routing plan. Each alternate plan must specify termination in a location previously set up during the order entry process.
• DIALED NUMBER IDENTIFICATION SERVICE: DNIS. Allows a customer to terminate two or more 800 numbers to a single service group and to receive pulsed digits to identify the specific 800 number called. DNIS is only available on dedicated access lines with four-wire E & M type signaling or a digital interface. The customer's equipment must be configured to process the DNIS digits.
• ANI: The carrier will deliver to you the incoming 800 call plus the phone number of the calling party. See also ANI, COMMON CHANNEL INTEROFFICE SIGNALING and ISDN.
• COMMAND ROUTING: Allows the customer to route calls differently on command at any time his business requires it.
• FOLLOW ME 800: Allows the customer to change his call routing whenever he wants to.

Now to the question of how to complete an 800 call. There are essentially two ways to terminate an 800 call. You can end the call on your normal phone line — business or residence. This is the phone line you use for normal in and out calling.

That's called not having a dedicated local loop. Or you can end the call on a dedicated phone line. By "dedicated," we mean there's a leased line between your office and the local office of your 800 provider, local or long distance carrier. There are several ways this dedicated "line" might be installed. It could be part of a T-1 circuit. It could be one circuit on one single copper pair. It could even be a phone number dedicated to your 800 number — a phone number you can't make an outgoing call on.

There is one major problem with 800 lines. They're hard to test. You may have bought an 800 number to cover the country, but you may be unreachable from certain parts of the country for weeks on end and not know it. That's part of the problem of a service which uses multiple databases lookup tables and relies on many exchanges to carry the calls. Many companies — like the airlines — recruit their distant employees to call their 800 number regularly. The only part of your 800 IN-WATS line you can test is your local loop (assuming you have one) from the local central office to your office. If you have a dedicated, leased line, you may have local Plant Test Numbers — standard seven digit numbers. You can call these numbers. If they work, you know that the end parts of your lines are working. One of our WATS lines is 1-800-LIBRARY. When it had a dedicated local phone number, it had a plant test number of 212-206-6870. So we could call this and all the subsequent hunt-on numbers every day first thing. Just to check. And when we go traveling, we call our own numbers. Just to check. Now we don't have any dedicated local phone numbers, we rely on prayer. The most common problem we have with our 800 numbers is at our local central office. Seems that it crashes every so often for very short amounts of time. When it starts up, it's meant to load all the tables to give us the features we're paying for — like hunting. Sadly, it doesn't always do this. We then report the trouble to our local phone company. It's usually fixed within an hour or so. Depends on how busy they are. See also 800 at the front of this dictionary. 800 Service now includes 888 (April 1996) numbers and 877 (April 1998) numbers.

EIGRP Enhanced Interior Gateway Routing Protocol; also known as Enhanced Internet Gateway Routing Protocol. Cisco System's newest version of its proprietary routing algorithm, IGRP, EIGRP provides link-to-link protocol-level security to avoid unauthorized access to routing tables.

Eight Way Server A motherboard with up to eight processors, e.g. eight Intel Pentium chips. Eight-way servers were introduced in 1997 to bring more power to the Windows NT operating system.

EIP Early Implementers Program. A term Novell coined to refer to those companies who had early on committed to adopt its Telephony Services architecture. See TELEPHONY SERVICES.

EIR Equipment Identity Register. A database repository used to verify the validity of equipment used in mobile telephone service. It can provide security features such as blocking calls from stolen mobile stations and preventing unauthorized access to the network. Black-listed equipment prevents call completion for a user.

EIRP Equivalent Isotropic Radiated Power.

EIRPAC A ITU-T X.25 packet-switched network operating in Ireland under the control of the Irish government.

EISA In a computer a "bus" is an electrical channel for getting information and commands in and around the computer. It is the way the central microprocessor running the computer gets its information and commands to the various peripheral devices or device controllers, such as video controllers, hard disk controllers, etc. The original IBM PC was "balanced" in that the microprocessor matched the speed of the bus that came in the machine. But the microprocessor got faster and more powerful and the bus lagged behind. So there has been much effort to speed the bus up, including EISA, which stands for Extended Industry Standard Architecture. EISA is the independent computer industry's alternate to IBM's Micro-Channel data bus architecture which IBM uses in some of its high end PS/2 line of desktop computers. EISA, like Micro-Channel (also called MCA), is a 32-bit channel. But, unlike IBM's Micro-Channel, plug-in boards which work inside the XT and AT-series of IBM and IBM clone desktop computers will work within EISA machines. They won't work in Micro-Channel machines. EISA expands the 16-bit ISA (Industry Standard Architecture) to 32-bit. EISA technology is useful in computing environments where multiple high performance peripherals are operating in parallel. The intelligent bus master can share the burden on the main CPU by performing direct data transfers into and out of memory. EISA capabilities are valuable when the system is being used as a server on a local area network or is running a multi-user operating system such as UNIX or OS/2. As of writing, over 200 manufacturers had endorsed EISA. Broader, wider buses than EISA are now available. 64-bit is not uncommon, especially among servers. See LOCAL BUS, PCI and VESA for examples of newer, faster buses.

Either End Hop Off EEHO. In private networks, a switch program that allows a call destined for an off-net location to be placed into the public network at either the closest switch to the origination or the closest switch to the destination. The choice is usually by time of day and is usually done to take advantage of cheaper rates.

Either Way Operation Same as half-duplex.

EIU Ethernet Interface Unit

EKE Electronic key exchange.

EKTS Electronic Key Telephone System.

ELAN Emulated Local Area Network: A logical network initiated by using the mechanisms defined by LANE (LAN Emulation). This could include ATM and legacy attached end stations. See also LANE.

Elastic Buffer A variable storage device having adjustable capacity and/or delay, in which a signal can be temporarily stored.

Elasticity Of Demand The relationship between price and the quantity sold. The theory is the lower the price, the more you'll sell. In telecommunications, this has traditionally been true, though sometimes it has taken time for demand to catch up with dramatic price cuts.

Elastomeric Firestop A firestopping material resembling rubber. See also firestopping.

ELDAP The beginnings of an Internet directory protocol.

ELEC Enterprise Local Exchange Carrier. Generally, a larger corporation operating as their own LEC as a means of obtaining better carrier rates for themselves, possibly selling services to others for a profit to enhance revenue in the process. ELECs are a new breed of LEC. The ILECs (Incumbent Local Exchange Carriers) for around 100 years have had the exclusive right and responsibility for providing local telephone service. During the recent past, many state PUCs (Public Utility Commissions) have allowed competition for local exchange service; hence, the origin of the CLECs (Competitive LECs). ELECs actually are a subset of the CLEC concept, although they actually preceded it by a few years. It works like this:

Let's say that you are the telecommunications manager for a large college or university, or for a large corporation with theme parks and hotels around the country. In other words, your enterprise owns a piece of property on which sit a lot of buildings. You provide a wide range of voice, data and video services to management, staff, and guests in dorms or hotels. In a very significant sense, you are providing communications services to what, in effect, is a self-contained town or small city. You go to the state PUC and file for certification as a LEC. You also file local exchange tariffs, defining available services, the terms under which they are offered, and all associated costs. You arrange to interconnect your PBX with an IXC, just like an ILEC would connect a CO to an IXC POP. The IXC pays you access charges for all interLATA traffic you hand off to it, just as they would pay an ILEC for that traffic. You have just become the manager of a telephone company — perhaps a very profitable telephone company. You also have become the manager of a facilities-based long distance resale company. Take this scenario one step further. Perhaps you become your own facilities-based IXC, with leased-line connections between your properties. Take it still one step further. In addition to hauling your own interLATA traffic, you market your long distance service and your Internet access service to other companies close to your ELEC properties, connecting those companies to your switch via dedicated circuits leased from the LEC. Once built, you can run this ELEC as a separate profit center, or you can sell it to a traditional LEC or CLEC. See also CLEC and LEC.

Electret Microphone Electret is a combination of ELECTRicity and magnET. An electret microphone operates on the basis of a dielectric (non-electrically conducting) material in which a permanent state of polarization, or electrical bias, has been established. The dielectric material is then spread over a conductive metal backplate. A back-electret microphone involves a diaphragm of dielectric material which is not electrically biased, but which is spread over a conductive metal back-plate which is electrically biased. Variants on standard condenser microphones, electret microphones are very inexpensive and require very little electrical energy to operate. Electret microphones are preferred for today's telephone handsets, since they are more sensitive (and cheaper) than the older carbon microphones, which many old-fashioned people (like me) still prefer because (a) I think it sounds better, (b) It's far less susceptible to interference from external sources, such as neon signs and (c) It works reliably with hearing aids (i.e. people who wear hearing aids have no trouble with carbon mics.).

Electric Banana Telecom installers' slang for TONE PROBE.

Electric Lock A cellular phone feature that provides security by locking a cellular phone so it can't be used by unauthorized persons.

Electrical Closet Floor-serving facility for housing electrical equipment, panelboards, and controls.

Electrical Service Equipment That portion of the electrical power installation, the service enclosure or its equivalent, up to and including the point at which the supply authority makes connection.

Electrically Powered Telephone A telephone in which the operating power is obtained either from batteries located at the telephone (local battery) or from a telephone central office (common battery).

Electrician's Scissors Used to cut cables. They have flat blades and look like very heavy scissors except they have notches on the side of one blade that are used to strip cable.

Electricity Electricity is the flow of electric charge. Normally this is thought of as electrons flowing through wire but it can also be protons or electrically charged ions flowing through a fluid.

Electrodeposition The deposition of a conductive material from a plating solution by the application of electric current.

Electroluminescence The direct conversion of electrical energy into light.

Electrolysis The production of chemical changes by passage of current through an electrolyte.

Electrolyte A chemical solution used in batteries, chemical rectifiers, and certain types of fixed condensers.

Electrolytic Process A printing process where paper is treated with an electrolyte and a stylus passes the signal current through the paper to produce an image. Paper is roll-fed past the stylus and changes color depending on the intensity of current passing through the stylus.

Electromagnetic Compatibility EMC. The ability of equipment or systems to be used in their intended environment within designed efficiency levels without causing or receiving degradation due to unintentional EMI (Electromagnetic Interference). EMI is reduced by, amongst other things, copper shielding.

Electromagnetic Emission Control The control of electromagnetic emissions. e.g., radio, radar, and sonar transmissions, for the purpose of preventing or minimizing their use by unintended recipients. A military term. Electromagnetic emission is reduced by, amongst other things, copper shielding.

Electromagnetic Force EMF. Also called voltage.

Electromagnetic Interface EMI. interference in signal transmission or reception caused by the radiation of electrical and magnetic fields. That's the easy explanation. Here's a more comprehensive explanation: Any electrical or electromagnetic phenomenon, manmade or natural, either radiated or conducted, that results in unintentional and undesirable responses from, or performance degradation or malfunction of, electronic equipment.

Electromagnetic Interference See ELECTROMAGNETIC INTERFACE.

Electromagnetic Lines Of Force The lines of force existing about an electromagnet or a current carrying conductor.

Electromagnetic Wave The electric wave propagated by an electrostatic and magnetic field of varying intensity. Its velocity is 186,300 miles per second.

Electromechanical Ringing The traditional bell or buzzer in a telephone which announces incoming calls.

Electromigration A phenomenon in which metal migrates along a current path in current rails, which are power-carrying circuits. Eventually, this phenomenon causes an open in the circuit or a shorting of an adjacent circuit. It is caused as the metal ions move in the direction of the current flow through metal wires comprising the circuit. Electromigration is particularly likely to affect the very thin and very tightly-spaced power distribution lines of sub-micron chip designs. As the phenomenon occurs only after months or years of use, it cannot be prevented by production testing, but only through careful circuit design. (Just in case you don't have enough things to worry about, Newton's Telecom Dictionary is pleased to bring you this source of consternation.)

Electron An electron is a light, subatomic particle that carries a negative charge. Electrons are found in atoms where they balance out the positive charge of the protons in the nucleus. Electrons in an atom are arranged in layers or shells around the nucleus. All atoms follow the pattern where the shells fill in from the inside to the outside. Each layer must fill to capacity before the next layer can be started. See electricity.

Electron Gun Device in a television picture tube from which electrons are emitted toward screen.

Electron Tube Rectifier A device for rectifying an alternating current by utilizing the flow of electrons between a hot cathode and a relatively cold anode.

Electronic Blackboard This is a teleconferencing tool. At one end there's a large "whiteboard." Write on this board and electronics behind the board pick up your writing and transmit it over phone lines to a remote TV set. The idea is that remote viewers can hear your voice on the phone and see the presentation on the electronic blackboard. The product has not done well because it is expensive — typically several hundred dollars a month just for rent, plus extra hundreds for transmission costs. In Japan, there are similar boards called OABoards — Office Automation Boards. They do one thing differently — they will print a copy on normal letter-size paper of what's written on the board. This takes about 20 seconds. Some of these Japanese OABoards will also transmit their contents over phone lines. So far, neither the OABoards nor the electronic blackboards have found a sizable market in the United States.

Electronic Bonding EB. A term for the exchange of information between carriers' Operations Support Systems (OSSs). Through secure gateways, the carriers can exchange information such as trouble tickets, which is very important in a multivendor network. The specific technique generally is either EDI (Electronic Data Interchange) or Telecommunications Management Network (TMN). See also EDI and TMN.

Electronic Bracelet A device attached to the legs of criminals who have been sentenced to confinement in their homes. The device allows them to move around a confined area. In most iterations, the device emits a regular signal to a nearby receiving station. If the criminal leaves the permitted area or tampers with the device, the in-home receiving device will dial local police authorities and effectively say, "the crim has flown his coop." An electronic bracelet is not the same as an electronic leash, which is a beeper.

Electronic Bulletin Board A computer, a modem, a phone line and a piece of software. Load communications software in your computer, dial the distant electronic bulletin board. The system will answer and present you with a menu of options. Typically those options will include leave messages, pick up messages, find out information, fill in a survey and upload and download a file.

Electronic Business Card Also known as vCard. See VERSIT and VERSITCARD.

Electronic Call Distribution Another term for Automatic Call Distribution. See AUTOMATIC CALL DISTRIBUTOR.

Electronic Commerce Using electronic information technologies to conduct business between trading partners, using or not using EDI (Electronic Data Interchange), using or not using the Internet. See EDI and OFX.

Electronic Commerce Services ECS. A set of e-mail authentication and certification services announced by the U.S. Postal Service in October 1996 and scheduled for early 1997 rollout. The service will provide an electronic postmark aimed at making e-mail authentic and traceable and, therefore, legally binding in support of electronic commerce. E-mail fraud effectively will become mail fraud, as a result. The service initially will be priced at 22 cents. ECS seems like a great idea! Hopefully, it will be more successful than the USPS' failed 1980s attempt at offering e-mail services. Critical to its success is the support of leading Internet e-mail vendors such as Microsoft and Netscape, both of whom are planning such support.

Electronic Custom Telephone Service Provides deluxe key telephone features and simplified access to certain AT&T Dimension PBX phones.

Electronic Data Interchange EDI. The process whereby standardized forms of electronic commerce documents are transferred between computer systems often run by different companies and without human intervention. EDI is used for placing orders, for billing and paying for goods and services via private electronic networks or via the Internet. According to studies, it costs about $50 a process paper-based purchase order and about $2.50 to process the same order with EDI. Internet-based EDI can lower the cost to less than $1.25. The form and format of EDI documents may be defined by vendor specifications, ITU-T standards, the ANSI X.12 standard, or the United Nations EDIFACT standard. See EDI for a fuller explanation. See also the next definition. See also http://www.edi-info-center.com/html/hotline.html and http://www.ecworld.org/Members/edi-uk.html

Electronic Data Interchange Association EDIA. An organization which works to provide a common platform to communicate global EDI activity, bypassing language conventions and national boundaries.

Electronic Data Processing See EDP.

Electronic Frontier Foundation EFF. A foundation established in July 1990 by Mitch Kapor, founder of Lotus, to "ensure that the principles embodied in the Constitution and Bill of Rights are protected as new communications technologies emerge." The EFF addresses a wide range social and legal issues arising from the impact on society of the increasingly pervasive use of computers as the means of communication and information distribution. Efforts include working to defeat the Communications Decency Act, in order to protect the right to free speech over the Internet. EFF also works to support both legal and technical means of enhancing privacy in communications, specifically focusing on the unfettered use of encryption algorithms. See appendix for address. www.eff.org

Electronic Funds Transfer EFT. A system which transfers money electronically between accounts or organizations without moving the actual money.

Electronic Image Mail The transmission of slow scan TV or facsimile via "Store and Forward." Not a common term.

Electronic Key Exchange A security procedure by which two entities establish secret keys used to encrypt and decrypt data exchanged between them. The procedure used in CDPD is based on a form of public key cryptology developed by Diffe and Hellman.

Electronic Key System A key telephone system in which the electromechanical relays and switches have been replaced by electronic devices — often in the phone and in the central cabinet. The innards of the central cabinet of an electronic key system more resemble a computer than a conventional key system. These days, virtually all key systems are electronic. Production of electromechanical key systems (such as 1A2) has been severely curtailed and most manufacturers have ceased making it.

Electronic Leash Pagers or beepers are often called "electronic leashes" because they allow your boss to contact you, to control you, to keep you on a leash. At one stage, beepers were carried by doctors and technicians who could never be "out of touch." As a result, beepers got a bad reputation and "real" people, i.e. bosses, wouldn't carry them. But that's getting better now and real people are now carrying them. See also ELECTRONIC BRACELET.

Electronic Lock Lets you lock your cellular phone so no one can use it. If you use Electronic Lock, you'll have to punch in some extra digits — like a password — to unlock the lock.

Electronic Mail A term which usually means Electronic Text Mail, as opposed to Electronic Voice Mail or Electronic Image Mail. Sometimes electronic mail is written as E-Mail. Sometimes as email. These days electronic mail is everything from simple messages flowing over a local area network from one cubicle to another, to messages flowing across the globe on an X.400 network. Such messages may be simple text messages containing only ASCII or they may be complex messages containing embedded voice messages, spreadsheets and images. See ELECTRONIC TEXT MAIL, ELECTRONIC VOICE MAIL, ELECTRONIC IMAGE MAIL and WINDOWS, WINDOWS TELEPHONY.

Electronic Mail Gateways A collection of hardware and software that allows users on an E-mail system to communicate and exchange messages with other mail systems that use a different protocol.

Electronic Mall A virtual shopping mall where you can browse and buy products and services online.

Electronic Message Registration A system to detect and count a phone user's completed local calls and then tell the central office the number of message units used. Also used in hotels.

Electronic Messaging Association EMA. The trade association for electronic messaging and information exchange. Formerly known as the Electronic Mail Association, EMA is a membership forum that seeks to enable users to work in partnership with providers of the technologies. Vendor members offer a wide range of services, including electronic mail, network, directories, computer facsimile, electronic data interchange (EDI), paging, groupware, and voice mail. EMAs technology programs aim to remove barriers to global interconnectivity and interoperability through assisting in the definition, endorsement, development, demonstration, and implementation of all messaging standards, operating conventions, and practices for use in electronic commerce. EMA lobbies governments, standards bodies, and consumer groups in advocacy of favorable public policies. www.ema.org

Electronic Order Exchange EOE. Inter-company transactions between buyers and sellers handled electronically via standard data communications protocols. EOE can be employed to send purchase orders, price and product listings and order-related information.

Electronic Phone General description for most phones designed after about 1980, where many mechanical and electrical parts are replaced by smaller, lighter, and cheaper electronic parts. Features such as mute, redial and memory became popular with these phones, which range in price from $5 to hundreds of dollars.

Electronic Publishing Electronic Publishing is synonymous with Desktop Publishing. Electronic Publishing software packages give the user the ability to perform page composition, insert images and manipulate text on the computer screen and display the document on the screen exactly as it will look when it is printed.

Electronic Receptionist A fancy name for a voice processing automated attendant, except that in addition to all the normal auto attendant features, it also sends messages to personal PCs on LANs telling the owner who's calling and giving the owner (the called party) the choice of doing something with the call — like answering it or putting it into voice mail.

Electronic Redlining A term for disenfranchising people and institutions because of their lack of telecommunications services and apparatus. In December, 1993, Vice President of the United States, Al Gore, told the National Press Club, When it comes to ensuring universal service, our schools are the most impoverished institution in society. Only 14% of our public schools used educational networks in even one classroom last year. Only 22% possess even one modem. Video-on-demand will be a great thing. It will be a far greater thing to demand that our efforts give every child access to the educational riches we have in such abundance. The recent article in the Washington Post on the proposed video communication network in the D.C. area is a wake-up call to all of us concerned about "electronic redlining."

Electronic Ringer A substitute for the conventional telephone bell, that uses music synthesizer circuitry to generate an attention-getting signal played through a speaker. Typical sounds include warbles, chirps, beeps, squawks, and chimes. The writer of this entry, Michael Marcus, once installed a phone with a chirp sound. A few days later, the customer complained that she had not been receiving any calls, and the birds in her yard were chirping much more than usual.

Electronic Serial Number ESN. A 32-bit binary number which uniquely identifies each cellular phone. The ESN consists of three parts: the manufacturer code, a reserved area, and a manufacturer-assigned serial number. The ESN, which represents the terminal, is hard-coded, fixed and supposedly cannot be changed. Paired with a MIN (Mobile Identification Number), the ESN and MIN are automatically transmitted to the mobile base station every time a cellular call is placed. The Mobile Telephone Switching office checks the ESN/MIN to make sure the pair are valid, that the phone has not been reported stolen, that the user's monthly bill has been paid, etc., before permitting the call to go through. At least that's the theory. It doesn't always work this way on calls made from roaming cellular phones. And some cellular phones have been known to have their ESNs tampered with (it's called fraud) which tends to mess up the billing mechanisms. See MIN.

Electronic Sweep Variation in the frequency of a signal over a whole band as a means of checking the response of equipment under the test.

Electronic Switching System A telephone switch which uses electronics or computers to control the switching of calls, their billing and other functions. The term is now vaguely defined, with each manufacturer defining it as something somewhat different. In fact, every telephone switch sold today is electronic. The term originally came about because early telephone switches were entirely electro-mechanical. The switch consisted entirely of a moving switch. Devices like relays physically moved in order to send the call through the exchange and on its way. These things moved in direct response to the digits dialed by the telephone subscriber. These switches contained no "intelligence" — i.e. no ability to deviate from a set number of very simple tasks which could be accomplished by electromechanical relays.

Then someone said: it would be more efficient if the "instruc-

tion part" of the process were divorced the switching mechanism. This lead to the creation of the "electronic" switch in which the "brains" of the switch are separated from the switching mechanism itself. Thus the "brains" can do simple things like collect the dialed digits as they are slowly dialed and pulse them out quickly to the switch — as fast as it can handle them. Now, the "brains" are typically a digital computer.

Electronic Tandem Network 1. Two or more switching systems operating in parallel as part of providing network services (usually voice) to large users.

2. A telephone company switching device used to connect telephone company toll offices located in the same geographic area.

Electronic Telephone Directory Service A PBX feature which stores and produces, on demand, a directory of all extension phone numbers. The directory may include all users in a network. A CRT with keyboard and/or printer is usually required for input and retrieval. In some systems, the CRT or another type of alphanumeric display is part of the Attendant Console. In some systems, the directory may also include names and telephone numbers of frequently called outside people, especially those in the speed calling system. The directory may also be enhanced to include SMDR data such as client codes, account codes and client telephone numbers.

Electronic Text Mail A "Store and Forward" service for the transmission of textual messages transmitted in machine readable form from a computer terminal or computer system. A message sent from one computer user to another is stored in the recipient's "mailbox" until that person next logs onto the system. The system then can deliver the message. Telex, in which a machine readable form of message transmission takes place, is also considered an Electronic Text Mail medium, albeit a very slow one. For an example of electronic mail, please dial our electronic mail system on 212-989-4675. It's free. Parameters are 300, 1200 or 2400 baud, 8 data bits, one stop bit, and no parity.

Electronic Voice Mail A system which stores messages usually spoken over a telephone. These messages can be retrieved by the intended recipient when that person next calls into the system. Also called Voice Mail, it operates just like a touch-tone controlled answering machine.

Electronic Warfare See EW.

Electrophotographic Printing A printing method that uses light to modify electrostatic charges on a photoconductive substrate.

Electrostatic Charge An electric charge at rest.

Electrostatic Discharge ESD. Discharge of a static charge on a surface or body through a conductive path to ground. Can be damaging to integrated circuits.

Electrostatic Printing A method of printing, very common in photocopying, in which charges are beamed onto the surface of paper. The charges attract particles of a very fine (typically black powder) which sticks to the charges. The black powder is fused permanently on the paper by great heat. "XEROXing" is electrostatic printing. In xeroxing, the black powder is called toner.

Elegant An elegant program is one that is efficiently written to use the smallest possible amount of main memory and the fewest instructions.

Elektrosvyaz A Russian phone company.

Element 1. Network Element (NE). A constituent part of a network. An element might be in the form of a modem, a multiplexer, a switch, or some other basic unit of a network.

2. Elements are the structural building blocks of HTML docu-

ments. Blocks of text in HTML documents are contained in elements, according to their function in the document, for example, headings, lists, paragraphs of text and links are all surrounded by specific elements. See HTML.

Element Management Layer See EML.

Elevator Eyes A term used in sexual harassment to mean viewing someone up and down.

Elevator Seeking Organizes the way data is read from hard disks and logically organizes disk operations as they arrive at the Novell NetWare local area network server for processing. A queue is maintained for each disk driver operating within the server. As disk read and write requests are queued for a specific drive, the operating system sorts incoming requests into a priority based on the drive's current head position. As the disk driver services the queue, subsequent requests are located either in the vicinity of the last request or in the opposite direction. Thus, the drive heads operate in a sweeping fashion, from the outside to the inside of the disk. Elevator seeking improves disk channel performance by significantly reducing disk head thrashing (rapid back-and-forth movements of the disk head) and by minimizing head seek times. Imagine how inefficient an elevator would be if the people using it had to get off the elevator in the order they got on.

ELIU Electrical Line Interface Unit.

Elongation The fractional increase in length of a material stressed in tension.

ELOT Hellenic Organization for Standardization (Greece).

ELSU Ethernet LAN Service Unit. An ELSU provides 12 independent virtual Ethernet bridges for running over ATM networks. ELSUs are designed for flexible deployment, either local to an ATM switch or at a remote site. ELSUs are designed for LAN internetworking services over ATM networks.

Elvis Year The peak year of something's popularity.

EM 1. Element Manager. Software and hardware used to manage and monitor components of a telecommunications network at their lowest level.

2. Abbreviation for End of Medium. The binary code is 1001001, the Hex is 91.

EMA Electronic Messaging Association, Arlington, VA. www.ema.org

EMACS A standard Unix text editor preferred by Unix types that beginners tend to hate.

EMAG ETSI MIS Advisory Group.

Email A colloquial term for electronic mail. See EMAIL address.

Email Address The UUCP or domain-based address by which a user is referred to. My email address is HARRYNEWTON@MCIMAIL.COM.

Email Gateway An email gateway is typically a PC on LAN. The PC has one or more modem and/or fax/modem cards. Its job is to send and receive e-mails and/or send and receive faxes for everyone on the LAN. To pick up emails, it might dial once an hour into various mail systems, like MCI Mail, CompuServe, and download all the messages for all the people on the LAN. Once it has those messages, it brings them onto its hard disk and then alerts the recipients that they now have an e-mail. See SERVER.

Email Reflector An Internet electronic mail address which automatically sends you back a reply (i.e. reflects mail to you) if you include certain key words in your message to it. Such key words might be "subscribe" or "lists help."

Email Server See EMAIL GATEWAY.

E-Mail Shorthand Acronyms for commonly used phrases that one would otherwise type. Some of the most popular

ones are: IMHO: In My Humble Opinion; BTW: By The Way; RTM: Read The Manual; LOL: Laughing Out Loud; FWIW: For What It's Worth; and ROFL: Rolling On The Floor Laughing.

Embarc Motorola's company which does wireless electronic mail to people carrying laptops and palmtops. Embarc, according to Motorola, stands for Electronic Mail Broadcast to A Roaming Computer. Actually Embarc does more than mail. It also broadcasts snippets of news.

Embed To insert information (an object) that was created in one document into another document (most often the two documents were created with different applications). The embedded object can be edited directly from within the document. To embed under Windows 3.1, you must be using applications that support object linking and embedding.

Embedded Base Equipment All customer-premises equipment that has been provided by the Bell Operating Companies (BOCs) prior to January 1, 1984, that was ordered transferred from the BOCs to AT&T by court order.

Embedded Code Formatting ECF. A NetWare definition. This is something of a programming language, in which faxing commands or other program that automatically generates information, formats it, and faxes it without user intervention.

Embedded Customer-Premises Equipment Telephone-company-provided premises equipment in use or in inventory of a regulated telephone utility as a December 31, 1982.

Embedded Hyperlink A hyperlink that is in a line of text. A hotspot is the place in a document that contains an embedded hyperlink.

Embedded Object A Windows term. An embedded object is information in a document that is a copy of information created in another application. By choosing an embedded object, you can start the application that was used to create it, while remaining in the document you're working in.

Embedded SCSI A hard disk that has a SCSI (Small Computer System Interface) and a hard disk controller built into the hard disk unit. See also SCSI.

Embedded SQL SQL statements embedded within a source program and prepared before the program is executed.

Embedded System Processors National Semiconductor's line of high-performance microprocessors used in dedicated systems, such as fax machines and laser printers.

Embossing A means of marker identification by thermal indentation leaving raised lettering on a cable's sheath material.

EMC ElectroMagnetic Compatibility.

Emergency Access An alarm system built into some PBXs. In an emergency it rings all phones.

Emergency Alert System See Emergency Broadcast System and EAS.

Emergency Broadcast System EBS. The EBS is composed of AM, FM, and TV broadcast stations; low-power TV stations; and non-Government industry entities operating on a voluntary, organized basis during emergencies at national, state, or operational (local) area levels. "This is a test of the Emergency Broadcast System — this is only a test." That warning, a remnant of the cold war, is about to disappear. The high-pitched tone is to be replaced by a few short buzzes, and the "this is a test" warning may be dropped altogether. The buzzes are generated by new computer technology. The new system, approved by the FCC in 1994, is expected to be fully operational by 1998 as the Emergency Alert System. The current test lasts ca. 35 or 40 seconds; the new test will be shorter, although the duration is not yet decided upon. The system has never been used for a nuclear emergency, but is used regularly for civil emergencies and severe weather alerts. The current emergency broadcast system is serial, that is it works on a daisy chain where one station receives the warning and sends it on to the next. That means that, if one station's equipment fails, the warning may not get further down the line. The new system looks more like a 'web' in which a station does not rely on one sole source for the signal, but will receive digital signals that will activate computers at broadcast facilities and download emergency messages.

Emergency Dialing A variation on speed calling to call numbers for police, fire department, ambulance, etc. Typically found as special buttons on an electronic phone.

Emergency Hold "Emergi-hold" allows a 911 caller's line to be held open in the event that a caller attempts to hang up. This gives the PSAP (Public Service Answering Position) agent full control of the call. It will not be released until the agent finishes the call.

Emergency Power A stand-alone secondary electrical supply source not dependent upon the primary electrical source.

Emergency Ringback This feature enables the 911 PSAP (Public Service Answering Position) attendant to signal a caller who has either hung up or left the phone off hook. Emergency Ringback enables the PSAP agent to ring a phone which has been hung up or issue a loud "howling" sound from the customer's phone if it has been left off hook.

Emergency Stand Alone Service A feature of a central office switch which allows it to keep working — switching and transferring calls — even though some of its connections to other central offices switches have been broken.

Emergency Telephone A single line telephone that becomes active when there is no commercial AC power to the Key Service Unit.

EMF Electromagnetic Force, a synonym for voltage. See also OHM's Law.

EMI Electromagnetic Interference, (EMI) happens when one device leaks so much energy that it adversely affects the operation of another device. EMI is reduced by copper shielding. National and international regulatory agencies (FCC, CISPR, etc.) set limits for these emissions. Class A is for industrial use and Class B is for residential use.

Here's a definition from APC: EMI usually refers to unwanted electrical noise present on a power line. This noise may "leak" from the power lines and affect equipment that is not even connected to the power line. Such "leakage" is called a magnetic field. Magnetic fields are formed when unwanted noise voltages give rise to noise currents. Such noise signals may adversely affect electronic equipment and cause intermittent data problems. EMI protection is provided by noise filters placed on the AC power line. The filter reduces the noise voltage on the protected line, and by doing so also eliminates the magnetic fields of noise generated by the protected line. Noise signals that act over a significant distance are called RFI (Radio Frequency Interference). Equipment power cords and building wiring often act as antennas to receive RFI and convert it to EMI.

EMI Segregation Isolation of the telecommunications signal from electromagnetic interference.

EMI/RFI Filter A circuit or device containing series inductive (load bearing) and parallel capacitive (non-load bearing) components, which provide a low impedance path for high-frequency noise around a protected circuit.

Emission 1. Electromagnetic energy propagated from a source by radiation or conduction. The energy thus propagat-

ed may be either desired or undesired and may occur anywhere in the electromagnetic spectrum.
2. Radiation produced, or the production of radiation, by a radio transmitting station. For example, the energy radiated by the local oscillator of a radio receiver would not be an emission but a radiation.

Emissivity Ratio of flux radiated by a substance to the flux radiated by black body at the same temperature. Emissivity is usually a function of wavelength.

Emitter The source of optical power.

EML Element Management Layer. A layer representing the management and monitoring of components, at their lowest level, in a telecommunications network. In short, an abstraction of the functions provided by systems that manage each network element on an individual basis.

Emotags Mock HTML tags (<smile>, <smirk>) used in WWW-related e-mail and newsgroups in place of ASCII emoticons, for example: "<flames> Someone tell that jerk to shut up, I'm sick of his vapid whining! </flame>." Definition from Wired Magazine. See EMOTICON.

Emoticon From Emotional Icon, one of a growing number of typographical cartoons used on BBSs (Bulletin Board Systems) to portray the mood of the sender, or indicate physical appearance. They are meant to be looked at sideways. Some examples:

:-D writer talks too much
:-# writer's lips are sealed
:-o writer is surprised
:-& writer is tongue-tied
ALL CAPS writer is shouting
:) is a smiley face
;) is a smile with a wink
;(is a frown with a wink
(:(is very sad
;? is a bad guy
[:0 is a wide-open mouth and a crewcut
(:{>X is bald with a handlebar mustache and bow tie
:-I is Wayne Newton
{8<)# is Michael Marcus, the writer of this entry: balding, glasses, mustache, smiling, beard.
Here's another collection of emoticons which I found on the Internet:

:-)	Smile	:-D	Laughing
:)	Smile	:-}	Grin
:-]	Smirk	:-(	Frown
;-)	Wink	:-X	Close-mouthed
8-)	Wide-eyed	:-O	Open-mouthed
:-I	I wear a	:-Q	But I don't inhale
	moustache	:-o	Oh, no!
<g>	Grin		
<ggg>	Wide Grin		
<g....g>	Very wide grin		

EMP A large and fast-moving electromagnetic pulse caused by lightning.

Emphasis In FM transmission, the intentional alteration of the amplitude-versus-frequency characteristics of the signal to reduce adverse effects of noise in a communication system. The higher frequency signals are emphasized to produce a more equal modulation index for the transmitted frequency spectrum, and therefore a better signal-to-noise ratio for the entire frequency range.

Empty Slot Ring In LAN technology an empty slot ring is a ring LAN in which a free packet circulates through every workstation. A bit in the packet's header indicates whether it contains any messages for the workstation. If it contains messages, it also contains source and destination addresses.

EMR Exchange Message Record. Bellcore standard format of messages used for the interchange of telecommunications message information among telephone companies. Telephone companies use EMR to exchange billable, non-billable, sample, settlement and study data. EMR formatted data is provided to all interdepartmental applications and to large customers (users) who request reproduced message records for control and allocation of their communication costs. Bellcore BR-010-200-010 Issue 15, Oct 96. In November of 1998, I heard that EMR was being replaced by something called EMI so that it applies to IXCs as well as LECs.

EMS Enterprise Messaging Server. A Microsoft concept which allows users to transparently access the messaging engine from within desktop applications to route messages, share files, or retrieve reference data. According to Microsoft, corporate developers will be able to add capabilities using Visual Basic and access EMS by writing either to the X.400 Application Program Interface Association's (XAPIA's) Common Mail Calls (CMC) or to Microsoft's Messaging API (MAPI). See MAPI.

EMT Electrical Metal Tubing. In many towns you must run your electrical AC wire inside metal tubing. In other towns you can run normal plastic insulated wiring. Theoretically, EMT is a safer fire hazard. What you are allowed to run depends on local laws and regulations. Tip: Dimmers for incandescent lights raise havoc with LAN data. Solution: Put the plastic electrical wires inside EMT (Electrical Metal Tubing) and ground the conduit.

Emulate To duplicate one system or network element with another. For instance, to imitate a computer or computer system by a combination of hardware and software that allows programs written for one computer or terminal to run on another. The most common data terminal is a DEC VT-100. Our communications program, Crosstalk, allows us to "emulate" a DEC-VT100 on our IBM PCs and PC clones.
Circuit emulation, an ATM term, refers to the ability of an ATM network to emulate a circuit over a channel in a T-carrier electrical environment or the over a Virtual Channel in a SONET/SDH fiber optic transmission system. LANE (LAN Emulation) allows an ATM network to emulate a LAN, offering LAN functionality over an ATM network. See LANE for more detail.

Emulation What happens one gadget emulates another. See EMULATE.

Emulator A device or computer program which can act as if it is a different device or program, that is Emulate (i.e. pretend to be) another device. Certain computer terminals are necessary in specific systems and a terminal that is not that type may be able to act as if it was. If it can, it is an Emulator. This is not a common term. See also the verb EMULATE.

EN50-091 A European test standard for UPS system safety. Supercedes and is a superset of the IEC950 standard formerly used for UPS testing. In addition to the typical safety tests found in the IEC950 standard, this standard includes special sections on batteries and other safety concerns specific to UPS systems. UPS products are normally certified to this standard by VDE, TUV, SEMKO or other authorized certification body.

ENA Enterprise Network Accounting. "Enterprise Network Accounting is software that allows end users to collect call data from routers and generate communications management reports. ENA software tracks and allocates the costs of using a corporate network or the Internet, which allows network administrators to bill users for time spent on the network. ENA software also generates traffic statistics reports that show traffic patterns, potential misuse/abuse, and network inefficiencies. As voice traffic moves to the net, communications managers need tools to track and account for network usage. ENA represents the next phase in call accounting products. "Network World" coined the term concerning an announcement by Cisco Systems regarding a partnership with Telco Research, which is developing ENA products for Cisco Systems.

Enable To make something happen. Or, in more complex language, to set various hardware and software parameters so that the central computer will recognize those parameters and start doing what you want.

Enabler An "enabler" is a strange name for a piece of software.

Enabling Signal A signal that permits the occurrence of an event.

Encapsulated Postscript File EPS. A file that prints at the highest possible resolution for your printer. An EPS file may print faster than other graphical representations. Some Windows NT and non-Windows NT graphical applications can import EPS files.

Encapsulating Bridge A LAN/WAN term. A special bridge type usually associated with backbone/subnetwork architectures. Encapsulating bridges place forwarded packets in a backbone-specific envelope — FDDI, for example - and send them out onto the backbone LAN as broadcast packets. The receiving bridges remove the envelope, check the destination address and, if it is local, send the packet to the destination device. For a much longer explanation, see Bridge.

Encapsulation 1. Encasing a splice or closure in a protective material to make it watertight.
2. In object-oriented programming, the grouping of data and the code that manipulates it into a single entity or object. Encapsulation refers to the hiding of most of the details of the object. Both the attributes (data structure) and the methods (procedures) are hidden. Associated with the object is a set of operations that it can perform. These are not hidden. They constitute a well-defined interface — that aspect of the object that is externally visible. The point of encapsulation is to isolate the internal workings of the object so that, if they must be modified, those changes will also be isolated and not affect any part of the program. See OBJECT-ORIENTED PROGRAMMING.
3. Component lingo. Encapsulation is the isolation of a component's attributes and behaviors from surrounding structures. The technique protects components from outside interference and protects other components from relying on information that may change over time. Components are often encapsulated.
4. An electronic messaging term. The technique used by layered protocols in which a layer adds header information to the PDU (Protocol Data Unit) form the layer above. As an example, in Internet terminology, a packet would contain a header from the physical layer, followed by a header from the network layer (IP), followed by a header from the transport layer (TCP), followed by the application protocol data.
5. A networking term. It means carrying frames of one protocol as the data in another. Often the encapsulating protocol will be TCP/IP.
6. See also ENCAPSULATION BRIDGING.

Encapsulation Bridging Method of bringing dissimilar networks where the entire frame from one network is simply enclosed in the header used by the link-layer protocol of the other network.

Enclosure Reverberation A phenomenon of acoustics in which sound is reflected, or echoed, within an enclosure. This effect is particularly troublesome in automobile hands-free cell phone applications, as the coupling of the speakerphone, cell phone microphone, and the enclosed automobile chamber can degrade the quality of the transmitted signal.

Encoding The process of converting data into code or analog voice into a digital signal. See also PCM and ADPCM.

Encrippling Encrippling is the name of a technology which Hyperlock Technologies (www.hyperlock.com) has created which allows CD owners to unlock premium content stored on music compact discs. According to Hyperlock, instead of typical encryption approaches that wrap the equivalent of a digital security envelope around a complete piece of content, Hyperlock's system removes key pieces of data from content stored on the compact disc. The content can only be played them, by retrieving the missing data from a preselected Web site, e.g. the publisher of the compact disc.

Encryption The transformation of data into a form unreadable by anyone without a secret decryption key. Its purpose is to ensure privacy by keeping the information hidden from anyone for whom it is not intended. In security, encryption is the ciphering of data by applying an algorithm to plain text to convert it to ciphertext. "Private key" demands that the key to both encoding and decoding be kept secret. "Public key" encryption utilizes the RSA encryption key, although the key for decoding is kept secret. See PGP for detailed discussion of a classic "public key."

Encryption key A unique, secret data block used to encrypt e-mail. See ENCRYPTION.

End Access End Office EAEO. An end office that provides Feature Group D.

End Delimiter ED. Sequence of bits used by IEEE 802 MAC to indicate the end of a frame. Used in token bus and ring networks, with nondata bits making ED easy to recognize.

End Distortion In start-stop teletypewriter operations, the shifting of the end of all marking pulses except the stop pulse from their proper positions in relation to the beginning of the next pulse. Shifting of the end of the stop pulse would constitute a deviation in character time and rate rather than being an end distortion. Spacing end distortion is the termination of marking pulses before the proper time. Marking end distortion is the continuation of marking pulses past the proper time. Magnitude of the distortion is expressed in percent of a perfect unit pulse length.

End Finish Surface condition at the optical fiber face.

End Instrument A communication device that is connected to the terminals of a circuit.

End Node A node such as a PC that can only send and receive information for its own use. It cannot route and forward information to another node.

End Of File EOF. A control character or byte used in data communications that indicates the last character of the last record of a file has been read.

End Of Medium EM. A control character used to denote the end of the used (or useful) portion of a storage medium.

End Of Message EOM. A control character used in data communications to indicate the end of a message.

End Of Shift Routing A call center term for a process that calls won't be left in limbo when a shift ends. See also SOURCE/DESTINATION ROUTING, SKILLS-BASED ROUTING and CALENDAR ROUTING.

End Of Text Message ETX. A control character used in data communications to indicate the end of a text message. See ETX.

End Of Transmission Block A communications control character indicating the end of a block of Bisync data for communication purposes.

End Of Transmission Block Character A control character used in data communications to indicate the end of a block where data are divided into blocks for transmission purposes.

End Office A central office to which a telephone subscriber is connected. Frequently referred to as a Class 5 office. The last central office before the subscriber's phone equipment. The central office which actually delivers dial tone to the subscriber. It establishes line to line, line to trunk, and trunk to line connections. See End Office Code.

End Office Code That part of a destination code consisting of the first three digits of a customer's seven digit directory number. It is usually expressed as an "NXX Code" where N represents digits 2 through 9 and X represents digits zero through 9.

End Point A network element (component) at the end of the network. In other words, a transmitter or receiver, or an originating or terminating device.

End Station An ATM term. These devices (e.g., hosts or PCs) enable the communication between ATM end stations and end stations on "legacy" LAN or among ATM end stations.

End System A host computer, in the context of the Internet.

End to End Communications Data delivered between a source and destination endpoint.

End to End Confidentiality The provision of data confidentiality between the sender and receiver of a communication.

End to End Connection Connections between the source system and the destination system.

End To End Loss The loss of an installed transmission path. The loss consists of the loss of the transmission cable or fiber, splices and connectors.

End To End Signaling A signaling system capable of generating and transmitting signals directly from the originating station to the terminating end after the connection is established, without disturbing the connection. Touchtone dialing is such a system, allowing the user to send tones to a remote computer for data or other access. See POINT TO POINT.

End User A highfalutin' term for a user. It's actually the occupant of the premises who uses and pays for the telephone service received and does not resell it to others. Bellcore's definition: A user who uses a loop-start, ground-start, or ISDN access signaling arrangement. Definition from Bellcore is part of its concept of the Advanced Intelligent Network.

Endurability The property of a system, subsystem, equipment, or process that enables it to continue to function within specified performance limits for an extended period of time, usually months, despite a potentially severe natural or man-made disturbance, e.g., nuclear attack, and a subsequent loss of external logistic or utility support.

Energy Communications EC. A PBX feature which communicates with energy consuming and monitoring devices and perform functions like dimming the lights or turning down the heat in a vacant hotel room. See also ENERGY CONTROL.

Energy Control Indicates that phone system has software and hardware necessary to control and regulate the energy consuming devices in a user's facility (heating, ventilating, air conditioning, electrical machinery etc.). The system's processor transmits control signals, over existing telephone wiring where possible, to control units at each power-consuming device. This feature always includes user reconfiguration of the system's control parameters in response to operational and/or environmental changes. At one stage, AT&T and some other telephone equipment manufacturers sold energy control as a integral feature of their phone systems. The idea didn't take off for a lot of reasons.

Energy Density A beam's energy per unit area, expressed in joules per square meter. Equivalent to the radiometric term "irradiance."

Energy Star A U.S. Government program that mandates strict limits on power consumption on electronic equipment, like computers and monitors, to the Federal Government. Products that comply often carry the symbol of a green star.

ENET Enhanced Network.

ENFIA Exchange Network Facilities for Interstate Access. A tariff providing a series of options for connecting long distance carriers with local exchange facilities of the local telephone company.

Engineer Furnish and Install EF&I. A way to buy a product. If you buy a PBX (or anything else) the company will ask you if you want to buy the equipment and install it yourself, or get them to engineer, furnish and install it.

Engineered Capacity A telephone company term. The highest possible load level for a trunk group or a switching system at which service objectives are met. In general, for a switching-system, carried-load is equal to offered-load below engineered capacity, but is less than offered load above engineered capacity. Engineered capacity does not include equipment provided for maintenance or service protection.

Engineering Administration Data Acquisition System A telephone company term. EAD. The system is composed of traffic measuring and indicating devices, data converters, data accumulators, an EADAS central control unit (CCU), and a general purpose computer. The downstream general purpose computer provides data to the data management system which in turn provides the raw data, properly formatted and for the measurement intervals requested, to other downstream programs.

Engineering Judgment A telephone company definition. A term used by Network Engineering Managers and in various system publications to describe a behavior; expected of engineers when factual data and calculations are unavailable to justify engineering decisions.

Engineering Orderwire EOW. A communication path for voice or data, or both, that is provided to facilitate the installation, maintenance, restoral, or deactivation of segments of a communication system by equipment operators, attendants, and controllers.

Engineering Period A telephone company definition. Usually a one to four year period starting with the required service date of a new office or addition and concluding at the planned exhaust date of the switching equipment.

Enhanced 800 Services A name MCI uses for a family of 800 services with additional features added to them. It includes time of day and day of week routing.

Enhanced 911 Enhanced 911 is an advanced form of 911 service. With E-911, the telephone number of the caller is transmitted to the Public Safety Answering Point (PSAP) where it is cross-referenced with an address database to determine the caller's location. That information is then displayed on a video-monitor for the emergency dispatcher to direct public safety personnel responding to the emergency.

This enables police, fire departments and ambulances to find callers who cannot orally provide their precise location. See also E-911.

Enhanced Call Processing An Octel term for the interactive voice response option in its voice mail system. Here's how Octel defines the term: "Companies and departments that receive a heavy volume of calls can use ECP to create menus that are presented to callers. When the system answers a call, a recorded voice instructs the caller how to use a touch-tone telephone to send call routing instructions to the system. Depending on which option is chosen, ECP's customized call routing feature allows a caller to press a single key to reach a predetermined extension, a voice messaging mailbox where he can leave a message, an Information Center Mailbox where he can listen to a series of recordings giving frequently requested information or additional levels of ECP menus. ECP menus are easily custom-built by the customer to meet its specific needs. Each menu can offer as many as ten options."

Enhanced Call Routing An AIN (Advanced Intelligent Network) service which is an enhancement to 800 / 888 services. The calling party is voice prompted through a set of menu options which serve to define the specifics of the request and the particular needs of the caller. Based on that input, the caller is directed to the most appropriate incoming call center and agent. By way of example, language preference might be a cause for changing call routing.

Enhanced Dialing Features allow for speed dialing, preview dialing, and manual dialing from a host or workstation application.

Enhanced DNIS Enhanced DNIS is a combination of ANI and DNIS delivered before the first ring on a T-1 span. The number of digits delivered is configurable on a per span basis.

Enhanced IDE An improved interface to the IDE hard disk interface. Enhanced IDE allows you to attach hard disks of larger than 528 megabytes (the largest normal IDE will handle) up to a maximum of 8.4 gigabytes. Enhanced IDE has a data transfer rate of between 11 and 13 megabytes per second, compared to the 2 to 3 megabytes per second, which normal IDE drives sport. See IDE.

Enhanced Mode The Intel 8088 and 8086 microprocessors, used in the earliest PCs, run DOS programs using real mode. Real mode causes problems when you try to run more than one program at a time because nothing prevents a poorly designed program from invading another program's memory space, resulting in a system crash, i.e. the PC seizing up. The Intel 80386 microprocessor introduced several technical improvements. For compatibility, an 80386 can run in real mode, but also offers protected mode. In protected mode, the 80386 can address up to four gigabytes of RAM, far more than you can install in any PC. The chip (and later versions such as the 486 and the Pentium) also can simulate more than one 8086 machines, protected from one another, thus preventing memory conflicts.

Running a DOS program in the protected mode of an 80386 computer (and later versions such as the 486 and the Pentium) requires software to manage the memory. Like a traffic cop, this software — called memory-management or the Windows operating system — puts DOS programs into their own 640K virtual machines, where they work away without interfering with other programs. The most popular memory-management available for 80386 and 80486 is Windows. In 386 Enhanced mode, Windows takes advantage of the 80386/80486/Pentium's virtual memory capabilities. Virtual memory is a way of extending RAM. Most DOS applications swap program instructions and data back and fourth from disk rather than keep them in memory. See EXPANDED MEMORY and EXTENDED MEMORY.

Enhanced Parallel Port EPP. A new hardware and software innovation (and now a standard) which allows computers so equipped to send data out their parallel port at twice the speed of present parallel ports. There's no difference in the shape of the plug or the number of conductor. See EPP for a fuller explanation.

Enhanced Private Switched Communications Service EPSCS (pronounced EP-SIS). A private line networking offering from AT&T which provides functions similar to CCSA. Big companies are its customers.

Enhanced Serial Interface ESI. A new, broader serial interface announced by Hayes Microcomputer Products, Norcross, GA, and placed in the public domain. The ESI is an extension of the familiar COM card used in personal computers. ESI includes the definition of I/O, control registers, buffer control, Direct Memory Access (DMA) to the system and interaction with attached modem devices. ESI specification is available from Hayes Customer Service at no charge. Combined with Hayes' announcement of ESI was their announcement of new Enhanced Serial Port hardware products for the IBM microchannel and IBM XT/AT or EISA bus personal computers. According to Hayes, the ESI spec and the supporting ESP hardware provide a "cost-effective" communication coprocessor to manager the flow of data between an external high speed modem and PC. This technology prevents loss of data resulting from buffer overflow errors and provides maximum data throughput for high speed modems. Hayes says that the combination of ESP and ESI will allow through-the-phone modem speeds of up to 38.4 Kbps.

Enhanced Serial Port See ENHANCED SERIAL INTERFACE.

Enhanced Service Provider ESP. An ESP is a company that provides enhanced or value-added services to end users. An ESP typically adds value to telephone lines using his own software and hardware. Also called an IP, or Information Provider. An example of an ESP is a public voice mail box provider or a database provider, for example, one giving the latest airline fares. An ESP is an American term, unknown in Europe, where they're most called VANs, or Value Added Networks. See also OPEN NETWORK ARCHITECTURE and INFORMATION PROVIDER.

Enhanced Services Services offered over transmission facilities which may be provided without filing a tariff. These services usually involve some computer related feature such as formatting data or restructuring the information. Most Bell operating companies (BOCs) are prohibited from offering enhanced services at present. But the restrictions are disappearing.

The FCC defines enhanced services as "services offered over common carrier transmission facilities used in interstate communications, which employ computer processing applications that act on the format, content, code, protocol or similar aspects of the subscriber's transmitted information; provide the subscriber additional, different or restructured information; or involve subscriber interaction with stored information." In other words, an enhanced service is a computer processing application that messes in some way with the information transmitted over the phone lines. Value-Added Networks, Transaction Services, Videotex, Alarm Monitoring and Telemetry, Voice Mail Services and E-Mail are all examples of enhanced services.

Enhanced Small Device Interface An interface which improves the rate of data transfer for hard disk drives and increases the drive's storage capacity.

Enhanced Unshielded Twisted Pair EUTP. UTP (Unshielded Twisted Pair) cables that have enhanced transmission characteristics. Cables that fall under this classification include Category 4 and above.

ENIAC Electronic Numerical Integrator and Computor (spelled with an O). Early computer, built in 1944.

ENOS A Sun Microsystems term, Enterprise Network Operating Systems. Part of Sun's Networking Solutions. This provides the foundation for Sun's networking environment. It uses NFS (de facto standard for global file sharing), and TCP/IP protocol. TCP/IP is Sun Microsystems' vehicle of choice for transferring data between database clients and server. Sun uses WebNFS here. It's the de facto standard TCP/IP protocol for remote file access.

ENQ ENQuiry character. A control character (Control E in ASCII) used as a request to obtain identification or status. Abbreviation for enquiry. The binary code is 0101000 and the hex is 50.

ENQ/ACK Protocol Hewlett-Packard communications protocol in which the HP3000 computer follows each transmission block with ENQ to determine if the destination terminal is ready to receive more data. The terminal indicates its readiness by responding with ACK.

Enriched Services Providers Those third-party service providers (other than Network Providers) who provide value-added services that are accessed through telecommunications networks.

ENS Emergency Number Services.

ENSO ETSI National Standardization Organizations (ETSI).

ENTELEC ENergy TELECommunications and electrical association, the oldest nationwide user group in telecommunications. It is an association of communications managers and engineers in the oil, gas, pipeline and utility industries. ENTELEC played an important role in the early opening of competition in the telecommunications industry, including the famous "Above 890" decision, which allowed private companies to build their own long distance microwave system. The decision was called "Above 890" because electromagnetic waves in the radio frequency spectrum above 890 Megahertz (million cycles per second) and below 20 Gigahertz (billion cycles per second) are typically called microwave. Microwave used to be a common method of transmitting telephone conversations and was used by common carriers as well as by private networks. Now fiber is far more common. Microwave signals only travel in straight lines. In terrestrial microwave systems, a single transmission is typically good for 30 miles, at which point you need another repeater tower. Microwave is the frequency for communicating to and from satellites. ENTELEC was formerly known as the Petroleum Industry Electrical Association.

Entering Distribution A call center term. In this mode of the alerting state, a call is being presented to an ACD group or hunt group in preparation for distribution to a device associated with that group. This mode is indicated by a Delivered event with a cause code of Entering Distribution.

Enterprise Enterprise means the whole corporation. It tends to refer to corporations with more than one location. See Enterprise Computing.

Enterprise Calendaring iCal

Enterprise Computing Enterprise means the whole corporation. Enterprise computing refers to the computing applications on which a company's life depends: order entry, accounts receivable, payroll, inventory, etc. It is also known by the phrase "mission critical." See also Enterprise Network.

Enterprise Network The word Enterprise was invented by IBM. It means the whole corporation. An enterprise-wide network is one covering the whole corporation. Local PBXs. Local area networks. Internetworking bridges. Wide area networks, etc, etc. See also CORPORATE NETWORK and ENTERPRISE COMPUTING.

Enterprise Number A service provided by AT&T and the Bell operating companies (a.k.a. the Bell System) years ago which allowed people to make collect calls and have their calls automatically accepted by the company at the other end. It was very expensive. It has largely been replaced with 800 IN-WATS service, which is much more successful.

Enterprise RMON A proprietary extension of RMON and RMON-2, Enterprise RMON was developed by NetScout Systems (formerly Frontier Software Development) and is supported by several other vendors, including Cisco Systems. Enterprise RMON's extensions monitor FDDI and switched LANS. See RMON, RMON-2.

Enterprise Server A Sun Microsystems term, Part of Solaris' Server Suite. Used to develop and deploy mission critical applications on large server systems. Provides distributed computing. Comes with Solstice DiskSuite and Networker products for on-line backup and recovery.

Enterprise Solution Software that enables individuals and groups (either within an organization or part of a virtual organization beyond one company) to use computers in a networked environment to access information from a wide range of sources, collaborate on projects, and communicate easily with text, graphics, video, or sound.

Entity 1. An active element within an OSI layer or sublayer. 2. A telephone company definition. A group of lines served by common originating equipment.

Entity Coordination Management The portion of connection management which controls bypass relays and signals connection management that the medium is available.

Entity. Nongrowth A telephone company term. Also referred to in some areas as 'capped' or 'floating.' The term non-growth entity will be used to identify those entities where we do not intend to add capacity. However, we must always insure that these entities continue to provide objective levels of service.

Entrance And Exit Ramps The companies who control access to the internet and other networks of the information superhighway, whatever that is.

Entrance Bridge A terminal strip that is an optional component in a network interface device and is provided for the connection of ADO cable.

Entrance Facility Point of interconnection between the Network Demarcation Point and/or campus backbone and intra building wiring. The Entrance Facility includes overvoltage protection and connecting hardware for the transition between outdoor and indoor cable.

Entrance Point/Telecommunications The point of emergence of telecommunications conductors through an exterior wall, a concrete floor slab, or from a rigid metal conduit or intermediate metal conduit.

Entrance Room/Telecommunications A space in which the joining of inter or intra building telecommunications backbone facilities takes place.

Entrapment The deliberate planting of apparent flaws in a system for the purpose of detecting attempted penetrations.

Entrenched Transactors Banking industry jargon for people who refuse to use cost-saving ATMs, preferring to deal only with more expensive human bank tellers.

Entry Border Node An ATM term. The node which receives a call over an outside link. This is the first node within a peer group to see this call.

Enumerator A Windows 95 term. A Plug and Play device driver that detects devices below its own device node, creates unique device IDs, and reports to Configuration Manager during startup. For example, a SCSI adapter provides a SCSI enumerator that detects devices on the SCSI bus.

Envelope 1. The boundary of the family of curves obtained by varying a parameter of a wave.
2. The part of messaging that varies in composition from one transmittal step to another. It identifies the message originator and potential recipients, documents its past, directs its subsequent movement by the MTS (Message Transfer System) and characterizes its content.

Envelope Delay The difference, expressed in time, for signals of different frequencies to pass through a phone line. Some frequencies travel slower than others in a given transmission medium and therefore arrive at the destination at different times. Delay distortion is measured in microseconds of delay relative to the delay at 1700 Hz. Also called Delay Distortion.

Envelope Delay Distortion The distortion that results when the rate of change of phase shift with frequency over the bandwidth of interest is not constant. It is usually stated as one-half the difference between the delays of the two frequency extremes of the band of interest. See Envelope Delay.

Envelope Distortion Distortion of the transmitted signal which results from the different transmission speed characteristics of different frequency components to the signal. Mathematically it is the derivative of the phase shift with respect to frequency.

Environment The place your telephone system's main cabinet and main electronics live. While most PBX vendors will specify the room's characteristics, the ultimate responsibility for the room is yours, the user. Not designing your telephone system's environment correctly is tantamount to jinxing your telephone system from the start.
Here are some things to watch out for (your vendor has a more comprehensive list): 1. Sufficient air conditioning? Telephone systems give off heat. You need some way of getting rid of the heat. If you don't, you will blow some of your phone system's delicate electronic circuitry. 2. Sufficient space? Is there room for technicians to get in and around your telephone system so they can repair it? Will you have room for additional cabinets when you want to grow your phone system? 3. Sufficient and correct power? Will you have sufficient clean commercial AC power? Will you require isolation regulators? Or you will require extensive wet cell batteries? Will you have space? 4. Will you have a solid electrical ground? Can you find somewhere solid to ground your telephone system to — other than the third wire on the AC power, which is not suitable for most telephone systems? Beware of cold water pipes which end in PVC plastic pipes.

Environment Variable A Windows 95 and NT definition. A string consisting of environment information, such as a drive, path, or filename, associated with a symbolic name that can be used by Windows 95 or Windows NT. You use the System option in Control Panel or the set command from the Windows NT command prompt to define environment variables.

Envoy 1. A palmtop communicator introduced by Motorola in March of 1994. The device lets its users receive and transmit messages via Ardis, a network owned by Motorola and IBM. Envoy contains software from General Magic.
2. Spectrum Envoy is a DSP-based PC-board used for "telephone management" from a company called Spectrum Signal Processing, Burnaby, BC. Telephone management includes voice mail, contact manager, upgradable fax/modem, business audio, etc.

EO 1. End Office. Typically your own central office.
2. Erasable Optical drive. EO drives act like hard drives yet offer virtually unlimited storage because their cartridges are removable. Each cartridge sports at least 650 MB. Some sport 1 gigabyte.
3. EO was a startup in Mountain View, CA which did wireless data. It made a device called EO Personal Communicator 440 and 880. It uses GO's PenPoint operating system and the Hobbit microprocessor made by AT&T, which is "optimized" for telecommunications. In fall of 1994, AT&T closed EO down and stopped the sale of EO devices. It was too expensive and wasn't selling. An excellent book was written about EO. It is called "Startup; A Silicon Valley Adventure Story." It was written by Jerry Kaplan, one of EO's founders. The book is published by Houghton Mifflin.

EOA End Of Address. A header code.

EOB End Of Block. A control character or code that marks the end of a block of data.

EOC Embedded Operations Channel.

EOE See ELECTRONIC ORDER EXCHANGE.

EOF The abbreviation for End Of File. MS-DOS files and some programs often mark the end of their files with a Ctrl Z — or ASCII 26.

EOM End of Message (indicator). In ATM network, EOM is an indicator used in the AAL that identifies the last ATM cell containing information from a data packet that has been segmented.

EOP End of Procedure frame. A frame indicating that the sender wants to end the call.

EOT End of Transmission, End of Tape.

EOTC European Organization for Testing and Certification.

EOW Engineered OrderWire.

EPA Energy Star Monitors that comply with this standard consume less electricity by powering down when not in use.

EPABX Electronic Private Automatic Branch eXchange. A fancy name for a modern PBX. Other fancy names include CBX, Computerized Branch Exchange.

EPD Early Packet Discard. See EARLY PACKET DISCARD.

Ephemeris The predictions of current satellite position that are transmitted to the user in the data message of a GPS (Global Positioning System) satellite message.

EPLANS Engineering, PLanning and ANalysis Systems. Software offered by Western Electric (now called AT&T Technologies) to help operating telephone company people run their business better.

Epoxy A liquid material that solidifies upon heat curing, ultraviolet light curing, or mixing with another material. Epoxy is sometimes used for fastening fibers to other fibers or for fastening fibers to joining hardware.

EPP Enhanced Parallel Port. A new hardware and software innovation (and now a standard) which allows computers so equipped to send data out their parallel port at twice the speed of older parallel ports, i.e. those that came on the original IBM PC. The EPP conforms to the EPP standard developed by the IEEE (Institute of Electrical and Electronics Engineers) 1284 standards committee. The EPP specification transforms a parallel port into an expansion bus that theoretically can handle up to 64 disk drives, tape drives, CD-ROM drives, and other mass-storage devices. EPPs are rapidly gaining acceptance as inexpensive means to connect portable drives to notebook computers. There's no difference in the shape of the ordinary, 25-pin D-connector plug/connector or the number of conductors. The Enhanced Parallel Port (EPP) was developed by Intel Corp., Xircom Inc., Zenith, and other companies that planned to exploit two-way communications to external devices. Many laptops built since mid-1991 have EPP ports. See also ECP.

EPROM Erasable Programmable Read Only Memory. A read only memory device which can be erased and reprogrammed. Typically, it is programmed electronically (not electromagnetically) with ultraviolet light. EPROMS are typically returned to the vendor or factory for reprogramming. An Eprom on a graphics card might contain the default or ROM character set. EPROM chips normally contain UV-permeable quartz windows exposing the chips' internals. See also ROM and EEPROM.

EPS An extension of the PostScript graphics file format developed by Adobe Systems. EPS lets PostScript graphics files be incorporated into other documents. FrontPage supports importing EPS files.

EPSCS (Pronounced Ep-Sis.) Enhanced Private Switched Communications Service. An AT&T offering for large businesses with offices scattered all over the country. This service allows such businesses to rent space on AT&T electronic switches and join that switching capacity to leased lines. EPSCS customers get a network control center in their offices which gives them information on the continuing operation of their network and allows them some limited options for changing their services.

EPSN Enhanced Private Switched Network.

EQ See Equalization, Equalizer.

Equal Access All long distance carriers must be accessible by dialing 1 — and not a string of long dialing codes. This is laid down in Judge Green's Modified Final Judgment (MFJ), which spelled out the terms of the Divestiture of the Bell Operating phone Companies (BOCs) from their parent, AT&T. Under the terms of this Divestiture, all long distance common carriers must have Equal Access for their long distance caller customers. City by city telephone subscribers are being asked to choose their primary carrier who they will reach by dialing 1 before their long distance number. All other carriers (including AT&T, if not chosen as primary) can be reached by dialing a five digit code (10XXX), thus providing Equal Access for all carriers. Not all long distance companies will opt for full equal access since this involves considerable expense to the local phone companies. See also FEATURE GROUP A, B, C and D.

Equal Access End Office A central office capable of providing equal access. See also EQUAL ACCESS.

Equal Gain Combiner A diversity combiner in which the signals on each channel are added together. The channel gains are all equal and can be made to vary equally so that the resultant signal is approximately constant.

Equalizer A device inserted in a transmission line or amplifier circuit to improve its frequency response. An equalizer adds loss or delay to specific frequencies to produce a flat frequency response. The signal may then be amplified to restore its original form.

Equalization The process of reducing distortion over transmission paths by putting in compensating devices. The telephone network is equalized by the spacing and operation of amplifiers along the way. In recording, equalization is frequency manipulation to meet the requirements of recording; also the inverse manipulation in playback to achieve uniform or "flat" response. Also called Compensation. See equalizer and equalization circuit.

Equalization Circuit A compensation circuit designed into modems to counteract certain distortions introduced by the telephone channel. Two types are used: fixed (compromise) equalizers and those that adapt to channel conditions. U.S. Robotics high speed modems use adaptive equalization.

Equatorial Orbit An orbit with a zero degree inclination angle, i.e. the orbital plane and the Earths' equatorial plane are coincident.

Equipment Cabinet The metal box which houses relays, circuit boards or other phone apparatus. Usually also contains the power supply, which converts the 120 volt AC current into the low voltage direct current necessary to run the telephone system.

Equipment Compatibility One computer system will successfully do the same thing that another computer will do with the same data. There are many levels of "equipment compatibility." The only true compatibility, however, is identical machinery. And identical means "identical" down to the very last chip and very last integrated circuit. We have found that some computers — even those consecutively numbered — do not always perform the same. We have empirically proven this for both IBM and AT&T computers.

Equipment Identity Register See EIR.

Equipment Wiring Subsystem The cable and distribution components in an equipment room that interconnect system-common equipment, other associated equipment, and cross connects.

Equipped For Capacity The maximum number of lines and trunks that can be supported by the available hardware. It is not a totally effective measure of the size of a PBX. See WIRED-FOR-CAPACITY.

Equivalent Four-Wire System Transmission using frequency division to get full duplex transmission over only one pair of wires. Normally two pairs are needed for full duplex.

Equivalent Network 1. A network that may replace another network without altering the performance of that portion of the system external to the network.

2. A theoretical representation of an actual network.

Equivalent PCM Noise Through comparative tests, the amount of thermal noise power on an FDM or wire channel necessary to approximate the same judgment of speech quality created by quantizing noise in a PCM channel.

ER 1. Explicit Rate. The current mechanism for flow control in ATM networks. ATM RM (Resource Management) cells are circulated by the transmitting device, indicating both the current and the desired rates of transmission. Assuming that the receiving device is able to accommodate that desired rate without overflowing its buffers, the request is granted and is honored by all intermediate switches in the network.

2. Equipment Room.

Erasable Programmable Read-Only Memory See EPROM.

Erasable Storage A storage device whose contents can be changed, i.e. random access memory, or RAM. Compare with read-only storage.

Erase Head On a magnetic tape recorder — voice or video — this is the "head" which erases the tape by demagnetizing it immediately before a new recording is placed on the tape by the adjacent record head.

Erbium A rare earth element that when added to fiber optic

cabling could obviate the need for repeaters every 20 miles on undersea cables and expand fiber optic cabling to capacities of trillions of bits a second. See ERBIUM-DOPED FIBER AMPLIFIER.

Erbium-Doped Fiber Amplifier EDFA. Erbium-Doped Fiber Amplifiers have become the dominant method for signal amplification in long-haul lightwave transmission systems. EDFAs differ from the normal method of regenerative or electro-optic repeaters in that light does not have be converted to an electrical signal, amplified, and then converted back to light. Optical amplifiers contain a length of erbium-doped (a rare earth) that provides the gain medium, an energy source or "pump" from a laser source as the correct frequency, and a coupler to couple the pump laser to the doped fiber. Both the signal to be amplified and the pump energy are coupled into the doped fiber section of the transmission system. The pump laser puts the erbium-doped fiber into an excited state where it is able to provide optical gain through emission stimulated by a passing signal photon. One of the most important features, after the fact that EDFAs are amazingly simple, is that they are not frequency dependent, and therefore allow bandwidth upgrades (within limits) without replacing the entire transmission systems. Undersea transmission systems, such as Americas 1, TAT-12/13, and TCP-5 use EDFA technology.

Erector Set Telecom In North America, there's a children's game of building blocks called Lego. The game comes with hundreds of small plastic blocks, which can be assembled into all sorts of wonderful designs, from castles to gas stations. In England, Lego sets are also called Mecano sets. The generic term for Lego and Mecano sets is erector sets. The term "erector set telecom" is a concept created by Harry Newton as a way of explaining "the new open" telecommunications equipment, namely that you build your own computer telephony system from freely-available, non-proprietary hardware and software components. In short, a telecom industry along the same open hardware and software lines as the PC industry. See COMPUTER TELEPHONY.

Ergonomics The science of determining proper relations between mechanical and computerized devices and personal comfort and convenience; e.g., how a telephone handset should be shaped, how a keyboard should be laid out.

Erlang A measurement of telephone traffic. One Erlang is equal to one full hour of use (e.g. conversation), or 60 x 60 = 3,600 seconds of phone conversation. You convert CCS (hundred call seconds) into Erlangs by multiplying by 100 and then dividing by 3,600 (i.e. dividing by 36). Numerically, traffic on a trunk group, when measured in erlangs, is equal to the average number of trunks in use during the hour in question. Thus, if a group of trunks carries 12.35 erlangs during an hour, a little more than 12 trunks were busy, on the average.

Erlang gets its name from the father of queuing theory, A. K. Erlang, a Danish telephone engineer, who, in 1908, began to study congestion in the telephone service of the Copenhagen Telephone Company. A few years later he arrived at a mathematical approach to assist in designing the size of telephone switches. Central to queuing theory are basic facts of queuing life. First, traffic varies widely. Second, anyone who designs a telephone switch to completely handle all peak traffic will find the switch idle for most of the time. He will also find he's built a very expensive switch. Third, it is possible, with varying degrees of certainty to predict upcoming "busy" periods. See also ERLANG, A.K., ERLANG B, ERLANG C and POISSON.

Erlang, A. K. In 1918, A. K. Erlang, a Danish telephone engineer, published his work on blocking in "The Post Office

Electrical Engineers' Journal," a British publication. Like E.C. Molina, an AT&T engineer, Erlang assumed a Poisson distribution of calls arriving in a given time. Molina had assumed a constant holding time for all calls, whereas Erlang assumed an exponential distribution for holding times. That means that longer calls occur less frequently than shorter calls. Erlang assumed that blocked calls are immediately cleared and lost and do not return. A formula that Erlang worked out based on these assumptions (Erlang B) is still in use in telephone engineering. See ERLANG, ERLANG B, ERLANG C and POISSON.

Erlang B A probability distribution developed by A.K. Erlang to estimate the number of telephone trunks needed to carry a given amount of traffic. Erlang B assumes that, when a call arriving at random finds all trunks busy, it vanishes (the blocked calls cleared condition). Erlang B is also known as "Lost Calls Cleared." Erlang B is used when traffic is random and there is no queuing. Calls which cannot get through, go away and do not return. This is the primary assumption behind Erlang B. Erlang B is easier to program than Poisson or Erlang C. This convenience is one of its main recommendations. Using Erlang B will produce a phone network with fewer trunks than one using Poisson formulae. See also ERLANG, ERLANG A. K., ERLANG C, and TRAFFIC ENGINEERING.

Erlang C A formula for designing telephone traffic handling for PBXs and networks. Used when traffic is random and there is queuing. It assumes that all callers will wait indefinitely to get through. Therefore offered traffic (see ERLANG) cannot be bigger than the number of trunks available (if it is, more traffic will come in than goes out, and queue delay will become infinite). Erlang C is not a perfect traffic engineering formula. There are none that are.

Erlang Formula A mathematical way of making predictions about randomly arriving work-load (such as telephone calls) based on known information (such as average call duration). Although traditionally used in telephone traffic engineering (to determine the required number of trunks), Erlang formulas have applications in call center staffing as well. See ERLANG.

ERMES 1. European Radio MEssaging System.
2. One of the communications protocols used between paging towers and the mobile pagers/receivers/beepers themselves. Other protocols are POCSAG, ERMES, FLEX, GOLAY and REFLEX. The same paging tower equipment can transmit messages one moment in GOLAY and the next moment in ERMES, or any of the other protocols.

ERP 1. Effective Radiated Power.
2. Enterprise Resource Planning. Software which links together back-office computer systems such as manufacturing, financial, human resources, sales force automation, supply-chain management, data warehousing, document management, and after-sales service and support. Such systems typically run on networks of PCs. These often replace older mainframe-based systems. ERP software typically makes heavy use of telecommunications.

Error Burst A sequence of transmitted signals containing one or more errors but regarded as a unit in error in accordance with a predefined measure. Enough consecutive transmitted bits in error to cause a loss of synchronization between sending and receiving stations and to necessitate resynchronization.

Error Checking And Correction Error checking is the process of checking a "packet" being transmitted over a network to determine if the package, or the data content within the package, has been damaged. If checked and found wanting, damaged packets are discarded. Error correction is the

process of correcting the damage by resending a copy of the original packet. In public frame relay services, the network performs the function of error checking, but not error correction. That function is left to the intelligent end equipment (at the user's site).

Error Control Various techniques which check the reliability and accuracy of characters (parity) or blocks of data sent over telecommunications lines. V.42, MNP and HST error control protocols (three common dial-up phone line modem protocols) use error detection (CRC) and retransmission of errored frames (ARQ). See ERROR CONTROL PROTOCOLS.

Error Control Protocols Besides high-speed modulation protocols, all current models of high-speed dial-up modems also support error control and data compression protocols. There are two standards for error control protocols: MNP-4 and V.42. The Microcom Networking Protocol, MNP, was developed by Microcom. MNP 2 to 4 are error correction protocols. V.42 was established by ITU-T. V.42 actually incorporates two error control schemes. V.42 uses LAP-M (Link Access Procedure for Modems) as the primary scheme and includes MNP-4 as the alternate scheme. V.42 and MNP-4 can provide error-free connections. Modems without error control protocols, such as most 2400 bps Hayes-compatible modems, cannot provide error-free data communications. The noise and other phone line anomalies are beyond the capabilities of any standard modem to deliver error-free data. V.42 (and MNP 2-4) copes with phone line impairments by filtering out the line noise and automatically retransmitting corrupted data. The filtering process used by V.42 (and MNP 2-4) is similar to the error correction scheme used by file transfer protocols (such as XMODEM). The two modems use a sophisticated algorithm to make sure that the data received match with the data sent. If there is a discrepancy, the data is re-sent.

What is the difference between error control protocols (such as V.42) and file transfer protocols (such as XMODEM)? For one thing, file transfer protocols provide error detection and correction only during file transfers. File transfer protocols do not provide any error control when you are reading e-mail messages or chatting on line. Even though an error control protocol is "on" all the time, we still need file transfer protocols when two modems establish a reliable link. A modem works with bit streams, timing and tones. It does not understand what a file is. When you download or upload a file, your communications software needs to take care of the details related to the file: the filename, file size, etc. This is handled by the file transfer protocol which does more than error-checking.

The other benefit of V.42 (or MNP 4) is that it can improve throughput. Before sending the data to a remote system, a modem with V.42 (or MNP 4) assembles the data into packets and during that process it is able to reduce the size of the data by stripping out the start and stop bits. A character typically takes up 1 start bit, 8 data bits and 1 stop bit for a total of 10 bits. When two modems establish a reliable link using V.42 or MNP 4, the sending modem strips the start and stop bits (which subtracts 20% of the data) and sends the data to the other end. The receiving modem then reinserts the start and stop bits and passes the data to the computer.

Therefore, even without compressing the data you can expect to see as much as 1150 characters per second on a 9600 bps connection. Although the modem subtracts 20% of the data, the speed increase is less than 20% due to the overhead incurred by the error control protocol.

The above definition with great thanks to modem expert Patrick Chen.

Error Correcting Protocol A method of transmitting bit streams in a mathematical way such that the receiving computer verifies to the sending computer that all bits have been received properly. SNA and XMODEM protocols, in the mainframe and microcomputer environments respectively, are Error Correcting Protocols. See ERROR CONTROL PROTOCOL.

Error Correction Code In computers, rules of code construction that facilitate reconstruction of part or all of a message received with errors.

Error Correction Mode A method of transmitting and receiving data that eliminates errors.

Error Free Second A Bellcore definition. An error-free second is, surprise, surprise, a one second time interval of digital signal transmission during which no error occurs. That's it.

Error Level A numeric value set by some programs that you can test with the errorlevel option of the "If" batch command. It works as follows. Some programs set the DOS errorlevel to a certain number depending on a certain input or response to an event. Let's say when you type the letter "Y" in response to a question the errorlevel is set to 32. Once this is done, you may condition other events based upon this number using an If command in a batch file. You can say "IF ERRORLEVEL = 32 THEN GOTO END." That way, when you type "Y" you will get whatever is at END. This can be very helpful in batch files and other programs for providing "branching" from one event to another based on certain inputs.

Error Logical An error in the binary content of a signal, for example, bit error.

Error Rate In data transmission, the ratio of the number of incorrect elements transmitted to the total number of elements transmitted.

Error Suspense An MCI definition. An automated process which allows billable MCI calls on switch tapes to be processed for billing, while calls with errors are held in the Error Suspense File (a separate file for each switch).

Error Trapping In software programming, an exception is an interruption to the normal flow of a program. Common exceptions are division-by-zero, stack overflow, disk full errors and I/O (input/output) problems with a file that isn't open. The quality of a software program depends on how completely it checks for possible errors and deals with them. Code used for trapping errors can be excessive. Some programming languages have error trapping built in. Others don't and you have to program it in.

Erstwhile An English word meaning previous. I define this word because I use it in this dictionary and lots of readers have told me they don't know it.

ES 1. Errored-Second. A count of the number of seconds in which at least one code violation (CV) was detected on a digital circuit. See also CODE VIOLATION).
2. End System: A system where an ATM connection is terminated or initiated. An originating end system initiates the ATM connection, and terminating end system terminates the ATM connection. OAM cells may be generated and received.

ES-IS End system to Intermediate systems protocol. The OSI protocol used for router detection and address resolution.

ES/9000 IBM Enterprise System/9000: mainframe computer family.

ESA Emergency Stand Alone.

ESC The ESC key on the keyboard. Often used to leave (escape) a program. Appears on the upper left of some keyboards on the IBM or compatibles but moves around with IBM's latest keyboard redesign whim. See also ESCAPE.

Escalation A formal word for taking your trouble up through the levels of management at the vendor — until you get your problem resolved. Some users have formal Escalation Charts, which detail action to be taken depending on how many hours the problem persists, etc. Escalation sometimes works and sometimes doesn't, depending on the vendor. Usually it does. The rule in telecommunications (and we guess most other industries) is that "the squeaky wheel gets the most attention." Escalation works well with honey, flowers, plants and chocolates.

Escape 1. The button on many computer keyboards which allows you to "escape" the present program. ESCape is the ASCII control character — code 27. It is often used to mark the beginning of a series of characters that represent a command rather than data. So called "ESCAPE" because it escapes from the usual meaning of the ASCII code and allows commands to be interspersed in a file of data, especially for data transmission to peripheral devices such as printers and modems. See ESCAPE SEQUENCE.
2. A means of aborting the task currently in progress.
3. A code used to force a smart modem back to the command state from the on-line state.

Escape Guard Time An idle period of time before and after the escape code sent to a smart modem, which distinguishes between data and escapes that are intended as a command to the modem.

Escape Sequence A series of characters, usually beginning with the escape character, that is to be interpreted as a command, not as data. Escape sequences are used with ANSI.SYS to change the color of a screen. They are mostly used to send print commands to printers. The name Escape is due to the fact that it "escapes" from the usual meaning of the ASCII code, letting characters be commands instead of data, yet interspersed with data in a transmission.

ESCON Enterprise Systems Connection. A 10 to 17 megabytes per second fiber optic LAN for linking IBM mainframes to disk drives and other mainframes. Not very popular with smaller users.

Escrow Bucket A hopper at the outlet of a coin phone's acceptor/rejector that is tipped electrically to return money through the Coin Return or to send the money to the Cash Box as a collection for a completed call.

ESD 1. Electrostatic Discharge.
2. Electronic Software Delivery. A technique whereby software (both initial installations and upgrades) on computers can be accomplished electronically without the need for floppy disks or CDs. One ESD scenario might be for a bunch of PCs attached to a LAN to be upgraded by a file server on the LAN. Another ESD scenario might be for a single user

ESDI 1. Enhanced Small Device Interface. An interface which improves the rate of data transfer for hard disk drives and increases the drive's storage capacity.
2. Northern Telecom term for Enhanced Serial Data Interface.

ESF Extended Super Frame or Extended Superframe Format. A T-1 format that uses the framing bit for non-intrusive signaling and control. A T-1 frame is sent 8,000 times a second, with each frame consisting of a payload of 192 bits, and with each frame preceded by a framing bit. Therefore, there are 8,000 framing bits per second. Previous generations of channel banks (D1, D2, D3, and D4) used the framing bit exclusively for network synchronization purposes. As ESF requires only 2,000 framing bits for synchronization, the remaining 6,000 framing bits can be used for error detection, using cyclic redundancy checking, and data link monitoring and maintenance. As a result, the channel banks need not rob bits from the data payload for such purposes. The ultimate yield is that the payload is not compromised, and the full 1.536 Mbps is available for user data In a channelized T-1 application, such as traditional PCM-encoded voice, each channel is a reliable 64 Kbps, rather than the 56 Kbps realized through use of older channel banks. While voice is unaffected by the process of bit robbing, data suffers greatly. The impact of intrusive signaling and control is the reason that most carriers limit data communications to 56 Kbps. They have a lot of old channel banks still in use. Note that channel banks must be matched throughout the entire network. In other words, buying an ESF channel bank won't do you any good unless the carrier also has them in place. In networking, as in life in general, the lowest common denominator rules. See also Bit Robbing, Channel, Channel Bank and T-1.

ESH End System Hello packet. Part of CDPD Mobile End System (M-ES) registration sequence.

ESI 1. See ENHANCED SERIAL INTERFACE.

2. End System Identifier: This identifier distinguishes multiple nodes at the same level in case the lower level peer group is partitioned.

ESM Extended subscriber module.

ESMA Expanded Subscriber Module-100A.

ESMR Enhanced Specialized Mobile Radio. An enhancement of SMR technology, allowing two-way radio service with the capability to provide wireless voice telephone service to compete against cellular. It uses TDMA technology to put six voice conversations into one 25 kilohertz UHF radio channel in the 806-821 MHz band. ESMR can be deployed on a cellular basis, and supports hand-off, like cellular radio. ESMR was developed by Nextel and Geotek Communications. Nextel acquired a large number of SMR frequencies, applied ESMR technology, and began to offer what is planned to be a nationwide radio service in direct competition with cellular and PCS. ESMR will support low-speed data, as well as voice, with digital technology yielding both improved efficiency of bandwidth utilization and improved security. The network remains to be fully deployed. See CDMA.

ESMTP Extended Simple Mail Transport Protocol (see SMTP), described in RFC 1651. Seldom used outside the Unix community.

ESMU Expanded Subscriber Carrier Module-100 URBAN.

ESN 1. Emergency Service Number. An ESN is a "list" of emergency numbers that corresponds to a particular ESZ (Emergency Service Zone). This list has to do with 911 service. Usually this ESN list is unique and contains a listing of the corresponding police, figure and ambulance dispatch centers for the caller's area. This "list" is used for selective routing and one button transfer to secondary PSAPs — Public Safety Answering Positions. The ESN/ESZ concept is especially useful in fringe areas.

2. Electronic Serial Number. A 32-bit binary number which uniquely identifies each cellular phone. The ESN consists of three parts: the manufacturer code, a reserved area, and a manufacturer-assigned serial number. The ESN, which represents the terminal, is hard-coded, fixed and supposedly cannot be changed. Paired with a MIN (Mobile Identification Number), the ESN and MIN are automatically transmitted to the mobile base station every time a cellular call is placed. The Mobile Telephone Switching office checks the ESN/MIN to make sure the pair are valid, that the phone has not been reported stolen, that the user's monthly bill has been paid, etc., before permitting the call to go through. At least that's the theory. It doesn't always work this way on calls made from roaming cellular phones. And some cellular phones have been known to have their ESNs tampered with (it's called fraud) which tends to mess up the billing mechanisms. See MIN.

3. Electronic Switched Network.

ESOs European Standardization Organizations.

ESP 1. Enhanced Serial Port. The Hayes Enhanced Serial Port (ESP) adapter, introduced in late 1990, replaces and extends the traditional COM1/COM2 serial port adapter. The ESP combines dual 16550 UARTS with an on-board communications coprocessor. The ESP has two distinct modes of operation to provide both old and new standards in the same package: Compatibility Mode and Enhanced Mode. Each ESP port can be independently operated in either mode. Default modes are configured via DIP switches and can be modified by ESP commands. The MCA-bus version of the ESP uses Programmable Option Selection (POS) rather than DIP switches. See ENHANCED SERIAL INTERFACE.

2. Enhanced Service Provider — a vendor who adds value to telephone lines using his own software and hardware. Also called an IP, or Information Provider. An example of an ESP is a public voice mail box provider or a database provider, say one giving the latest airline fares. An ESP is an American term, unknown in Europe, where they're most called VANs, or Value Added Networks. See also OPEN NETWORK ARCHITECTURE and INFORMATION PROVIDER.

3. EncapSulated Postscript File.

ESPA European Selective Paging Association.

ESPRIT European Strategic Program for Research and development in Information Technology. A $1.7 billion research and development program funded by the European Community.

ESQ End System Query packet. Part of CDPD Mobile End System (M-ES) registration procedures.

ESS 1. Electronic Switching System. ESS was originally a designation for the switching equipment in Bell System central offices but has slightly more general use now. In the independent telephone company industry, the abbreviation for the same thing is EAX. The first 1ESS switch went into service in May 1965 in Succasunna, New Jersey.

2. European Standardization System.

ESS No 4 AT&T's large toll telephone switch. It will handle over 100,000 trunks and over 500,000 attempts at making a call each hour. It's large and sophisticated and can probably be configured to be the largest telephone switch in the world.

ESS No 5 AT&T's Class 5 digital central office. See also END OFFICE.

Essential Lines A telephone company definition. In order to guarantee to certain customers the ability to make outgoing calls during an emergency, the telephone company's Customer Services Department designates these customers as "essential." Examples of essential lines are: police and fire departments, ambulance companies, hospitals, etc. Whenever Line Load Control is activated, outgoing service may be selectively denied to nonessential customers in order to preserve originating calling capacity for those customers having a documented priority. Also see - Class A lines.

Essential Service 1. A service provided by a telecommunications provider, such as an operating telephone company or a carrier, for delivery of priority dial tone. Generally, only up to 10 percent of the customers may request this type of service. See Essential Lines.

2. A service that is recommended for use in conjunction with NS/EP (national emergency) telecommunications services.

ESSX ESSX (pronounced essex) is some local phone companies' name for Centrex. See CENTREX.

Established Connection A telephone company term. A connection on which all necessary switching or operating steps have been taken to connect the calling and called lines. Generally speaking, it is somewhat broader than the term "completed call," in that it includes established connections to tones or announcements, as well as completed calls. A completed call is a connection between two telephones.

ESZ Emergency Service Zone. This term is used in conjunction with 911 emergency service. An ESZ is a geographic area that is served by a unique mix of emergency services. Each ESZ has a corresponding ESN (a list of Emergency Service Numbers) which enables 911 service to properly route incoming calls.

ET Exchange Termination. Refers to the central office link with the ISDN user.

ETACS Extended TACS. The cellular technology used in the United Kingdom and other countries. It is developed from the U.S. AMPS technology. See also AMPS, TACS, NTACS and NAMPS.

Etailer A retailer who conducts his business by electronic commerce — i.e. over the Internet and the world wide web.

ETB Abbreviation for end-of-transmission block. The binary code is 0111001, the hex is 71.

ETC Enhanced Throughput Cellular is an error correction cellular communications protocol, which helps prevent disruptive signal fading and thus reduces the number of dropped calls.

Etched Antiglare treatment that prevents glare but also reduces screen sharpness and clarity on monitors. Generally considered an obsolete technology.

Eternity Hold Our own creation for what happens when someone puts you on long-term hold. Governmental agencies, airlines and police departments (especially when you need them) tend to be firm believers in placing their callers on Eternity Hold. A new service adjunct to Eternity Hold is Conference Hold. Here everyone on Eternity Hold can speak to each other. We made this up. It doesn't exist, but we think it would be great if it did.

Ether The medium which, according to one theory, permeates all space and matter and which transmits all electromagnetic waves.

EtherRing ADC Telecommunications' name for a "revolutionary new idea allowing transport of native mode Ethernet and Fast Ethernet (100 million bits per second) data packets over a wide area network (WAN). Unlike typical Ethernet transport solutions," according to ADC Telecommunications (www.adc.com), "EtherRing has no distance limitations."

Ethernet A local area network used for connecting computers, printers, workstations, terminals, servers, etc., within the same building or campus. Ethernet operates over twisted wire and over coaxial cable at speeds up to 10 million bits per second (Mbps). For LAN interconnection, Ethernet is a physical link and data link protocol reflecting the two lowest layers of the DNA/OSI model. The theoretical limit of Ethernet, measured in 64 byte packets, is 14,800 packets per second (PPS). By comparison, Token Ring is 30,000 and FDDI is 170,000. Ethernet specifies a CSMA/CD (Carrier Sense Multiple Access with Collision Detection). CSMA/CD is a technique of sharing a common medium (wire, coaxial cable) among several devices. As Byte Magazine explained in its January, 1991 issue, Ethernet is based on the same etiquette that makes for a polite conversation: "Listen before talking." Of course, even when people are trying not to interrupt each other, there are those embarrassing moment when two people accidentally start talking at the same time. This is essentially what happens in Ethernet networks, where such a situation is called a collision. If a node on the network detects a collision, it alerts the other nodes by jamming the network. Then, after a random pause, the sending nodes try again. The messages are called frames (see the diagram).

The first personal computer Ethernet LAN adapter was

AN ETHERNET FRAME					
Preamble	Destination address	Source address	Type	Data up to 1500 bytes	Frame check sequence
8 bytes	6 bytes	6 bytes	2 bytes	bytes	4 bytes (contains CRC check)

shipped by 3Com on September 29, 1982 using the first Ethernet silicon from SEEQ Technology. Bob Metcalfe created the original Ethernet specification at Xerox PARC and later went on to found 3Com. In the October 31, 1994 issue of the magazine InfoWorld, Bob Metcalfe explained that Ethernet got its name "when I was writing a memo at the Xerox Palo Alto Research Center on May 22, 1973. Until then I had been calling our proposed multimegabit LAN the Alto Aloha Network. The purpose of the Alto Aloha Network was to connect experimental personal computers called Altos. And it used randomized retransmission ideas from the University of Hawaii's Aloha System packet radio network, circa 1970. The word ether came from luminiferous ether — the omnipresent passive medium once theorized to carry electromagnetic waves through space, in particular light from the Sun to the Earth. Around the time of Einstein's Theory of Relativity, the light-bearing ether was proven not to exist. So, in naming our LAN's omnipresent passive medium, then a coaxial cable, which would propagate electromagnetic waves, namely data packets, I chose to recycle ether. Hence, Ethernet."

According to Metcalfe, "Ethernet has been renamed repeatedly since 1973. In 1976, when Xerox began turning Ethernet into a product at 20 million bits per second (Mbps), we called it The Xerox Wire. When Digital, Intel, and Xerox decided in 1979 to make it a LAN standard at 10 Mbps, they went back to Ethernet. IEEE tried calling its Ethernet standard 802.3 CSMA/CD — carrier sense multiple access with collision detection. And as the 802.3 standard evolved, it picked up such names as Thick Ethernet (IEEE 10Base-5), Thin Ethernet (10Base-2), Twisted Ethernet (10Base-T), and now Fast Ethernet (100Base-T)."

Ethernet PC cards now come in a couple of basic varieties — for connecting to an Ethernet LAN via coaxial cable or via two twisted pairs of phone wires, called 10Base-T. See also 10Base-T, Ethernet Controller, Ethernet Identification Number, Ethernet Switch, Ethertalk, Frame, Thinnet and Token Ring.

Ethernet Controller The unit that connects a device to the Ethernet cable. An Ethernet controller typically consists of part of the physical layer and much or all of the data link layer and the appropriate electronics.

Ethernet Identification Number This is a unique, hexadecimal Ethernet number that identifies a device, such as a PC/AT with a Speed_ink/PC16 network interface card installed, on an Ethernet network.

Ethernet II (DIX) Defined by Digital, Intel and Xerox. The frame format for Ethernet II differs from that of IEEE 802.3 in that the header specifies a packet type instead of the packet length.

Ethernet Switch A new device that connects local area networks. Here's a definition of the capabilities which an Ethernet switch must have from a company called Kalpana Inc. in Santa Clara CA. "What capabilities must a device have to be an Ethernet Switch? Ethernet switching is being embraced as the next milestone solution for bandwidth-constrained Ethernets. To qualify as an Ethernet Switch, the device must: Be capable of switching packets from one Ethernet segment to another "on-the-fly;" Avoid using slower store-and-forward technologies to route packets from one segment to another; Exhibit very low port-to-port latency (the elapsed time between receiving and transmitting a LAN packet is measured in 10s of microseconds, not 100s); Offer a busless, scaleable architecture that increases network carrying capacity as switched connections are added; Support all higher-level products; Provide the technology for creating a massively parallel system with a tens-of-gigabits per second capacity."

Ethernet Transceiver A device used in an Ethernet local area network that couples data terminal equipment to other transmission media.

EtherTalk An Ethernet protocol used by Apple computers. AppleTalk protocol governing Ethernet local area network transmissions. Also the Apple Computer Ethernet adapter and drivers. Apple's implementation of Ethernet is compliant with IEEE specification 802.3.

Ethertype A two-byte code indicating protocol type in an Ethernet local area network packet.

ETI Electronic Telephone Interface.

ETISALAT Emirates Telecommunications Corporation is the sole provider of telecommunications services throughout the United Arab Emirates. The head office is in Abu Dhabi.

ETM Electronic Ticketing Machine. A machine that looks like a banking Automated Teller Machine (ATM), except that it will dispense airline tickets and possibly, hotel reservations, car rental agreements, etc.

ETN 1. Electronic Tandem Network. An ETN is a large private network which comprises dedicated leased lines interconnecting electronic tandem switches. ETNs were deployed in the 1970s and 1980s by very large user organizations such as state and federal governments.
2. Earning Telephone Number. A billing term which is synonymous with WTN (Working Telephone Number). Multiple ETNs can be associated with a single BTN (Billing Telephone Number).

ETNO The European Public Telecommunications Network Operations Association (ETNO).

ETS 1. European Telecommunications Standard. A standard defined by the European Telecommunications Standards Institute (ETSI).
2. Electronic Tandem Switching. See ELECTRONIC TANDEM SWITCHING.

ETS 300 211 Metropolitan Area Network (MAN) Principles and Architecture.

ERS 300 212 Metropolitan Area Network (MAN) Media Access Control Layer and Physical Layer Specification.

ETS 300 217 Connectionless Broadband Data Service (CBDS).

ETS Set A Northern Telecom term for an electronic Telephone Set.

ETSI European Telecommunications Standards Institute, is the European counterpart to ANSI, the American National Standards Institute. ETSI is based in Sophia-Antipolis, near Nice, France. ETSI's task is to pave the way for telecommunications integration in the European community as part of the single European market program. ETSI was founded in 1988 as a result of an in initiative of the European Commission. It was established to produce telecommunications standards by democratic means, for users, manufacturers, suppliers, administrations, and PTTs. ETSI's main aim is the unrestricted communication between all the member states by the provision of essential European standards. It is now an independent, self-funding organization, which works closely with both CEPT (European Conference of Posts and Telecommunications Administrations) and EBU (European Broadcasting Union).

ETSI also works closely with CENELEC, which is responsible for electrotechnical standards, and CEN, which is responsible for European standardization in all remaining fields. Certified products are awarded the CE Mark, signifying that a company has met the applicable essential health and safety requirements and the specific conformity assessment requirements to market its product in the European Union under the "New Approach" directives. www.etsi.org. See also ANSI, CE MARK, CEN, CENELEC, CEPT, and EBU.

ETSI V5 European Telecommunications Standards Institute's open standard interface between an Access Node (AN) and a Local Exchange (LE) for supporting PSTN (Public Switched Telephone Network) and ISDN (Integrated Services Digital Network). Examples of Access Nodes include Digital Loop Carrier (DLC) systems, wireless loop carrier system, and Hybrid Fiber Coax (HFC) systems.

ETX End of Text. Indicates the end of a message. If multiple transmission blocks are contained in a message in Bisynch systems, ETX terminates the last block of the message. ETB is used to terminate preceding blocks. The block check character is sent immediately following ETX. ETX requires a reply indicating the receiving station's status. The binary code is 0011000, the hex is 30. See PACKET.

EUCL End User Common Line charge. A FCC tariff term defined in FCC Rules 69.104 as follows: "A charge that is expressed in dollars and cents per line per month shall be assessed upon end users that subscribe to local exchange telephone service, Centrex or semi-public coin telephone service to the extent they do not pay carrier common line charges. Such charge EUCL shall be assessed for each line between the premises of an end user and a Class 5 office that is or may be used for local exchange service transmissions. Each Single Line Service is charged one CALC or EUCL. The amount varies by state." The intent is that the EUCL, in combination with the CCL (Carrier Common Line Charge) compensate the LEC for the use of the local loop for the purposes of originating/terminating interLATA long distance calls. The LEC receives no direct benefit from such traffic, although its investment in, and maintenance cost associated with, the local loop is considerable. The EUCL is known variously as the Access Charge, CALC, and SLC. See also Access Charge and CCL.

Eudora An electronic mail sending and receiving software program that runs on Macs and under Windows and is probably the most common e-mail service used by people on the Internet. It is manufactured and distributed by Qualcomm Enterprises, San Diego, California. Eudora Pro is the commercial version (i.e. the one that costs money). It includes features that are not in Eudora Light, the freeware version and one in heaviest use on the Internet.

EUnet European UNIX Network. (Original name). Now a major European Internet Service provider.

EURESCOM EUropean institute for REsearch and Strategic studies in TeleCOMmunication.

Euro New European single currency introduced on January 1, 1999 by 11 European countries, who agreed to lock their exchange rates to the euro, which was, at the time of its birth, worth around $1.17.

EUROBIT European Association of Business Machines Manufacturers and Information Technology Industry.

Euro-ISDN The European implementation of ISDN. It differs from North American National ISDN-1 in that Euro-ISDN is very limited in the options it offers. In the United States, ISDN comes with many options, including two call appearances, conference calling, call forwarding variable, call forwarding — busy, call forwarding — no answer, voice mail with indicator, two secondary directory numbers, etc. That makes North American ISDN more full-featured, but much less easy to order. Users, can however, call from the United States to Europe and complete ISDN calls. They can not carry their end-user ISDN equipment from the United States and use it in most places in Europe.

European Commission The administrative body of the European Union, and a central source of policy, legislation and funding for pan-European research and development in ICT applications.

European Space Agency The European equivalent of the U.S.'s NASA — National Aeronautics and Space Administration.

EUROSINET-EUROTOP International ISDN pilot project for travel agents.

EUROTELDEV EUROpean TELecommunications DEVelopment. An organization involved in telecommunications standardization.

EUTELSAT EUropean TELecommunications SATellite organization. Inter-governmental organization that aims to provide and operate a communications satellite for public intra-European international telecommunications services. The segment is also used to meet domestic needs by offering leased capacity, primarily for television. U.K. and France are the largest shareholders, with about 25 member countries in total.

EUTP Enhanced Unshielded Twisted Pair. UTP Cables that have enhanced transmission characteristics. Cables that fall under this classification include Category 4 and above.

EV European Videotelephony.

EVA Economic Value Added. A financial measure of whether you're making more money with your plant, factory, assets, etc. in your present business than you would be if you sold everything and stuck the proceeds in a investment. The common assumption is that you can get a 10% or 11% return. If you earn more than that, you're EVA positive. EVA is a term you hear a lot around AT&T. You can become EVA positive in any ways — writing down the value of your capital is one way. Sacking people works too. And, so does selling more (presumably, at a decent price).

Even Parity In data communications there's something called a PARITY BIT that's used for error checking. The transmitting device adds that parity bit to a data word to make the sum of all the "1" ("one") bits either odd or even. If the sum is odd, the result is called ODD parity. If it's even, it's called EVEN PARITY. See also PARITY.

Event An unsolicited communication from a hardware device to a computer operating system, application, or driver. Events are generally attention-getting messages, allowing a process to know when a task is complete or when an external event occurs.

Event Code A code that an agent in a call center enters at the conclusion of a call. Event codes can trigger a variety of follow-up activities such as an acknowledgement letter, or inclusion in a list for a subsequent campaign.

Event Driven A style of programming under which programs wait for messages to be sent to them and react to those messages. See EVENT DRIVEN ALARMS/TRIGGERS/TICKLERS.

Event Driven Alarms/Triggers/Ticklers In a parallel process, an event trigger can be set to move the processing forward when a set of criteria is met (ex. the last piece of documentation is added to the file). Alarms can also be time-driven, as when a folder is automatically routed to exception processing if no action is taken within a specified time frame. This term is often found in workflow management.

Event History A history of the activities that have been carried out on a record, (customer or prospect). For example, phone calls and mailers, offers and so on.

Event Mask The set of events that the SLEE (Service Logic Execution Environment) designates the ASC (AIN Switch Capabilities) to report for a particular connection segment, and an indication for each event if the ASC should suspend processing events for that connection segment until the SLEE sends a message back. Definition from Bellcore in reference to its concept of the Advanced Intelligent Network.

Event Message A message provided by the switching domain to the computing domain to indicate one of the following: 1) A change in the state of a telephone call by reporting state transitions of each connection in the call (Call Control Events); 2) A physical or logical device change that has taken place (such as Do Not Disturb being set or a device's microphone being muted) at a device physical and Logical Device Feature Events); 3) A change in call-associated information (Call Associated Feature Events); 4) A change in switching domain specific information that is associated either a device or call (Private Data / Information Events). See Call-Progress Event Message.

Event Report Synonymous with Event Message.

Event Signaling All telephone switches use the same three general types of signals: + Event Signaling initiates an event, such as ringing. + Call Progress Signaling denotes the progress (or state) of a call, such as a busy tone, a ringback tone, or an error tone. + Data Packet Signaling communicates certain information about a call, for example, the identify of the calling extension, or the identity of the extension being called.

EW Electronic Warfare. The military use of radar, electronic counter measures and electronic counter-counter measures to keep an enemy from finding invading forces, on land or in the air. It covers such methods as sending planes equipped with equipment which transmit thousands of signals purporting to be signals that an enemy radar might see on locating an incoming plane. By sending thousands of such signals, the enemy's radar becomes a myriad of "radar" signals, of bright spots. Thus it's impossible for the enemy to read any intelligent information. There are also anti-radiation missiles which home in on and destroy air-defense radar facilities. the only defense against such anti-radiation missiles is to turn off the radar. Electronic warfare also covers such techniques as jamming radio frequencies, anti-jamming. It is not the state of Judge Greene's courtroom or the boardroom at the FCC.

EWOS European Workshop for Open Systems.

EWP Electronic White Pages.

EWSD Name of digital central office switches from Siemens Stromberg-Carlson. Model numbers start with APS.

Exabyte EB. A unit of measurement for physical data storage on some form of storage device — hard disk, optical disk, RAM memory etc. and equal to two raised to the 60th power, i.e. 1,152,921,504,606,800,000 bytes.

MB = Megabyte (2 to the 20th power)
GB = Gigabyte (2 to the 30th power)
TB = Terabyte (2 to the 40th power)
PB = Petabyte (2 to the 50th power)
EB = Exabyte (2 to the 60th power)
ZB = Zettabyte (2 to the 70th power)
YB = Yottabyte (2 to the 80th power)

One googolbyte equals 2 to the 100th power

Exalted Carrier Reception A method of receiving either amplitude- or phase-modulated signals in which the carrier is separated from the sidebands, filtered and amplified, and then combined with the sidebands again at a higher level prior to demodulation.

EXCA Exchangeable Card Architecture. ExCA is a hardware and software architectural implementation of PCMCIA 2.0 from Intel that allows card interoperability and exchangeability from system to system, regardless of manufacturer. See PCMCIA.

Exception 1. In telecom, when something happens that's "unusual," it's an exception. The key is to define what's "unusual." For example, you might define that every phone call of longer than 15 minutes is an "exception." Now you have defined an "exception," the question is how to use that information. You might ask the phone system to print out each "exception" call on a printer next to your desk immediately after the call is over. Or you might ask the machine to print "Exceptions" reports at the end of the month listing all the calls over 15 minutes. These reports might be by perpetrator. Or in chronological order, or order of phone number called, etc. In short, any event you define by certain strict parameters can be an "exception." Management reports printed in full are almost useless because they contain so much information, so much paper. Management reports which list only previously-defined "exceptions" are more useful. They show you where to focus your attention so as to improve your or your company's performance. 2. In software programming, an exception is an interruption to the normal flow of a program. Common exceptions are division-by-zero, stack overflow, disk full errors and I/O (input/output) problems with a file that isn't open. The quality of a software program depends on how completely it checks for possible errors and deals with them. Code used for trapping errors can be excessive. Some programming languages have error trapping built in. Others don't and you have to program it in.

Exception Condition In data transmission, the condition assumed by a device when it receives a command that it cannot execute.

Exception Reports Reports generated by "exceptions," often detailing extra long calls or indications of bad circuits. See EXCEPTION.

Excess Burst Size A Frame Relay term defining the maximum data rate that the carrier network will attempt to transport over a specified period of time, known as the Measurement Interval. Beyond reasonable excesses, the carrier may mark the excess frames as Discard Eligible. SEE ALSO DISCARD ELIGIBILITY, MEASUREMENT INTERVAL, COMMITTED BURST SIZE, AND COMMITTED INFORMATION RATE.

Excess Insertion Loss In a optical fiber coupler, the optical loss associated with that portion of the light which does not emerge from the operational ports of the device.

Excessive Zeros More consecutive zeros received than are permitted for the selected coding scheme. For AMI-encoded T-1 signals, 16 or more zeros are excessive. For B8ZS encoded serial data, 8 or more zeros are excessive.

Exchange 1. Sometimes used to refer to a telephone switching center — a physical room or building. Outside North America, telephone central offices are called "Public Exchanges."

2. A geographic area established by a common communications carrier for the administration and pricing of telecommunications services in a specific area that usually includes a city, town or village. An exchange consists of one or more central offices and their associated facilities. An exchange is not the same as a LATA. A LATA consists of several adjacent exchanges.

3. A term that refers to one of the Fibre Channel "building blocks," composed of one or more non-concurrent sequences for a single operation. See Fibre Channel.

Exchange Access In the telephone networks, the provision of exchange services for the purpose of originating or terminating interexchange telecommunications. Such services are provided by facilities in an exchange area for the transmission, switching, or routing of interexchange telecommunications originating or terminating within the exchange area.
The Telecommunications Act of 1996 defined it as follows: The term `exchange access' means the offering of access to telephone exchange services or facilities for the purpose of the origination or termination of telephone toll services.

Exchange Area Geographic area in which telephone services and prices are the same. The concept of exchange is based on geography and regulation, not equipment. An exchange might have one or several central offices. Anyone in that exchange area could get service from any one of those central offices. It's good to ask which central offices could serve your home or office and take service from the most modern. There will be no difference in price between being served by a one-year old central office, or a 50-year old step-by-step central office.

Exchange Carrier A Bellcore definition. A company that provides telecommunication within a franchised territory.

Exchange Carriers Association An organization of long distance telephone companies with specific administrative duties relative to tariffs, access charges and payments. See EXCHANGE CARRIERS STANDARDS ASSOCIATION.

Exchange Carriers Standards Association ECSA. According to their literature ECSA is "the national problem-solving and standards-setting organization where local exchange carriers, interexchange carriers, manufacturers, vendors and users rationally resolve significant operating and technical issues such as network interconnection standards and 800 database trouble reporting guidelines. The Association was created in 1983. The major committees sponsored by ECSA are The Carrier Liaison Committee (to coordinate and resolve national issues related to provision of exchange access); the Telecommunications Industry Forum (TCIF) (to respond to the growing need for voluntary guidelines to facilitate the use of new technology that offers cost savings throughout the telecommunications industry — e.g. EDI, bar coding, automatic number identification); and the

Information Industry Liaison Committee (IILC) (an inter industry forum for discussion and voluntary resolution of industry wide concerns about the provision of Open Network Architecture (ONA) services and related matters and Committee T1-Telecommunications (an accredited standards group under ANSI to develop technical standards and reports for US telecommunications networks. In October, 1993, The Exchange Carriers Standards Association changed its name to the Alliance for Telecommunications Industry Solutions (ATIS). It is based in Washington, D.C. See ATIS. www.atis.org

Exchange Facilities Those facilities included within a local access and transport area.

Exchange Network Facilities For Interstate Access See ENFIA.

Exchange Message Record EMR. Bellcore standard format of messages used for the interchange of telecommunications message information among telephone companies. Telephone companies use EMR to exchange billable, non-billable, sample, settlement and study data. EMR formatted data is provided to all interdepartmental applications and to large customers (users) who request reproduced message records for control and allocation of their communication costs. Bellcore BR-010-200-010 Issue 15, Oct 96

Exchange, Private Automatic Branch (pabx) A private telephone exchange which transmits calls internally and to and from the public telephone network.

Exchange Service A name that BellSouth gives to its local phone services, which it also calls Plain Old Telephone Service (POTS).

Exchange Termination ET. In Integrated Services Digital Network (ISDN) nomenclature, ET refers to the central office link with the end user.

Exciting The most boring and the most over-used word in the whole high-tech world. If I read another press release describing their shiny new product as "exciting," I'll puke. If it excites, I'll get excited. But I don't need (or want) to be told I'm about to be excited.

Exclude A memory management command-line option that tells the memory manager in an MS-DOS machine not to use a certain segment of memory. For example, you may exclude upper memory locations D200 through D800 (hexadecimal) because your network adapter card uses that space. The reciprocal term — include — specifically directs the memory manager to use an area of memory.

Exclusion A PBX feature that prevents the attendant from silently monitoring a call once he/she has extended it.

Exclusive Hold Only the telephone putting the call on hold can take it off. This feature assures that the call on hold will not be picked up by someone at another telephone who can then listen to your call.

Exclusive Hold Recall When a call is placed on "Exclusive Hold" and is not picked up after a predetermined amount of time, you will hear a beeping at that phone, which indicates the call is still on hold.

Exclusive Or Private Unit A circuit card installed in each key telephone set sharing the same line or intercom path that causes the first caller on the line to lock out (exclude) all other stations from using or listening in, until the line is released (or privacy feature is defeated by the active caller).

Execunet An intercity switched telephone service introduced by MCI in 1975. Execunet was the first dial-up switched service introduced by a long distance phone company in competition with AT&T. At that time, all of AT&T's competitors, including MCI, were selling full-time private lines and shared private lines. The service was named by Carl Vorder-Bruegge, MCI's VP marketing at that time and introduced and made successful by Jerry Taylor, who was MCI's regional manager in Texas and is now president of MCI. The service was the forerunner of what is today a $20 billion plus per year industry — the non-AT&T provided switched long distance business. MCI no longer uses the word Execunet to describe its switched long distance service. It's just plain long distance. Jerry Taylor started Execunet using a 104-port Action WATSBOX in Dallas, Texas. He deserves a place in the history books, not just a dictionary.

Executable File A computer program that is ready to run. Application programs, such as spreadsheets and word processors, are examples of executable files. Such files in PCs running MS-DOS and Windows usually end with the BAT, COM or EXE extension.

Execute To complete a task.

Execution Time The time needed to complete a task.

Executive Barge-In See EXECUTIVE OVERRIDE.

Executive Busy Override See EXECUTIVE OVERRIDE.

Executive Camp-On A feature for use by executives or other privileged people. When they call a someone lowly, that low person hears a special distinctive tone or sees a special light or sees a special signal that their phone has been camped on by someone significant. These days many PBXs let you know who's calling — even though you're on the phone. So executive camp-on is not that useful.

Executive Override A feature of some telephone systems which permits certain users to intrude on conversations on other extensions. In some systems, executive barging-in will not be heard by the person outside the office, only the one inside the office. In some systems, such as the Mitel SX series with the Mitel Superset 4 phones, this feature activates the hands-free speakerphone of the called party, who is using his other line to speak on a normal phone conversation.

Exhaust A telephone company definition. Equipment is said to exhaust when it has reached its most limiting network access line capacity level.

Exhaust Date A telephone company definition. The exhaust date for an entity(s) refers to the calendar date on which the entity(s) will have reached the most limiting network access line capacity level.

Exit Border Node An ATM term. The node that will progress a call over an outside link. This is the last node within a peer group to see this call.

Exit Event An event occurring in an ASC (AIN Switch Capabilities) that causes call processing to leave a PIC (Point in Call). Definition from Bellcore in reference to its concept of the Advanced Intelligent Network.

EXM Exit Message. The seventh ISUP message. It's a message sent in the backward direction from the access tandem to the end office indicating that call setup information has successfully proceeded to the adjacent network. See ISUP and COMMON CHANNEL SIGNALING.

EXOS Abbreviation for EXtension OutSide; a phone connected to a key system based in another building. The wiring belongs to the telephone company, even though the phone equipment may not. Unlike an OPX, the circuit between the two locations does not pass through a central office.

Exosphere This region lies beyond an altitude of about 400km from the surface of the earth. The density is such that an air molecule moving directly outwards has an even chance of colliding with another molecule or escaping into space.

Expanded Memory MS-DOS running on the Intel 80286, 80386, 80486 and Pentium family of microprocessors can only address one megabyte at one time. Expanded memory is memory located between base memory (either 512K or 640K) and one megabyte. Expanded memory is reserved by MS-DOS for "housekeeping" tasks such as managing output to the screen. As programs got larger and more hungry for memory (640K was no longer enough), people started jealously eyeing the memory between 640K and 1024K (one megabyte). The first technique was a standard called LIM-EMS, named after the three companies which developed it — Lotus, Intel and Microsoft. Essentially LIM grabs 64K of the 640-1024 memory and uses it to swap pages of other memory in and out quickly. This fools DOS into thinking that it has actually more memory. LIM-EMS lets you work on bigger spreadsheets, and do other jobs faster.

There are many ways of using expanded memory, including special memory management application program or DOS 5.0 or higher. 80386, 80386SX, 80486 and Pentium computers can create expanded memory readily by using the EMS (expanded memory specification) driver provided with DOS, through

Windows 3.xx or Windows 95, or through a memory manager such as Quarterdeck QEMM or Qualitas 386. To use expanded memory, a program must be EMS-aware or run under an environment such as Microsoft Windows. See also EXTENDED MEMORY, which is memory above 1MB.

Expanded Spectrum A cellular telephone term for having the full 832-channel analog cellular spectrum currently available to you, the user of the cellular phone.

Expander That device in a transmission facility which expands the amplitude of received compressed signals to their approximate normal range. The receiving side of a compandor.

Expandor See EXPANDER.

Expansion The switching of a number of input channels, such as telephone lines onto a larger number of output channels.

Expansion Carrier An AT&T Merlin term. A carrier added to the control unit when the basic carrier cannot house all the modules needed. An expansion carrier houses a power supply module and up to six additional modules.

Expansion Slots In a computer there are card slots for adding accessories such as internal modems, extra drivers, hard disks, monitor adapters, hard disk drivers, etc. Most modern PBXs are actually cabinets with nothing but expansion slots inside. Into these slots we fit trunk cards, line cards, console cards, etc. Some phone systems have "universal" slots, meaning you can put any card in any slot. Some phone systems have dedicated expansion slots, meaning that they expect only a certain card in that slot. In the PC industry, many manufacturers make cards for IBM and IBM compatible slots. In the phone industry, nobody makes cards for expansion slots in anyone else's phone system. See also EISA and MCA.

Expert What we all become when we're away from the office.

Expert System A very sophisticated computer program consisting of three parts. 1. A stock of rules or general statements, e.g. Some long distance phone calls are free. These rules are generally based on the collective wisdom of human "experts" who are interviewed. 2. A set of particular facts, e.g. Three companies provide the bulk of long distance service in the United States. 3. Most importantly, a "logical engine" which can apply facts to rules to reach all the conclusions that can be drawn from them — one of which might be "Three

companies give away long distance phone calls." (Which would be wrong.) The idea of expert systems is to help people solve problems. For example, Compaq is trying to improve its customer service by installing automated assistants that work on the principle that reasoning is often just a matter of remembering the best precedent. The simplest expert systems, according to the Economist Magazine, assume that their rules and facts tell them everything there is to know. Any statement that cannot be deduced from the system's rules and facts is assumed to be false. This can lead machines to answer "YES" or "NO," when they should say "I don't know." Slowly we are beginning to find ways of dealing with the inflexibility of machines. One such gadget is a "truth maintenance machine" invented by Dr. Jon Doyle of MIT. As each fact is fed into the system, Dr. Doyle's program checks to see if it (or the deductions derived from it) contradict any of the facts or deductions already in the system. If there is a contradiction, the machine works backward along its chain of reasoning to find the source and dispose of that troublesome fact or deduction. So the system maintains one consistent set of beliefs.

Explicit Access In LAn Technology, explicit access is a shared access method that allows workstations to use the transmission medium individually for a specific time period. Every workstation is guaranteed a turn, but every station must also wait for its turn. Contrast with CONTENDED ACCESS. See also CSMA.

Exponential Back-Off Delay The back-off algorithm used in IEEE 802.3 systems by which the delay before retransmission is increased as an exponential function of the number of attempts to transmit a specific frame.

Exponential Holding Time A telephone company term. A great number of calls are assumed to have a relatively short holding time while decreasing numbers of calls are assumed to have longer and longer holding times out to the point where a very small number of calls exhibit exceedingly long holding times.

Export Imagine you have a software program, like a spreadsheet or a database. And you have information in that program. Let's say it's Microsoft Word or Lotus 123. And you want to get it into a different program, say to give it to a workmate who uses WordPerfect or Excel. You have to convert it from one format to another. From Word to WordPerfect or from Lotus to Excel. That process is typically called "exporting." And you'll typically see the word "EXPORT" as a choice on one of your menus. The opposite is called importing. See IMPORT.

Export Script First read my definition of EXPORT. An export script is a series of specifications which control the export process. It contains the fields to be sent, which records to be sent, the name of file to send as well as the name of the import script (if there is one) located at the receiver's end which will control the merge. See EXPORT.

Export Server A Windows NT term. In directory replication, a server from which a master set of directories is exported to specified servers or workstations (called import computers) in the same or other domains.

Express Call Completion Someone calls an information operator. "What is the name?" the operator answers. "Here is the number. Would you like me to get that number for you now? If so, please hit 1." Express Call Completion lets the operator complete the call for you while you're on line. Express Call Completion was begun in September of 1990 by Pacific Bell using a Northern Telecom central office. Express Call Completion is part of Northern Telecom's Automated Directory Assistance Call Completion (ADACC) software and

Traffic Operator Position Systems Multipurpose (TOPS MP).

Express Client Installation An Internet Access Term. A client installation scripting utility that enables network managers to establish defined defaults that make client installation and deployment easier.

Express Orderwire A permanently connected voice circuit between selected stations for technical control purposes.

EXT See EXTENSION.

Extend A verb used by the phone industry to describe an operator transferring a call to a telephone extension. The word is used thus: The operator extended the call to Mr. Smith on extension 200. "Putting a call through" is a clearer way of saying "extending" a call. The word "extend" probably comes from the old days when the operator extended her arm to plug you in on her cordboard.

Extended Addressing In many bit-oriented protocols, extended addressing is a facility allowing larger addresses than normal to be used. In IBM's SNA, the addition of two high-order bits to the basic addressing scheme.

Extended Area Service A geographic area beyond the local service area to which traffic is classified as local for selected customers, i.e., telephone service that allows subscribers in one exchange to call subscribers of another exchange without a toll charge. Sometimes subscribers may be given the option of paying more for the privilege of calling these more distant phone companies. Sometimes, they have no option. The local public service commission deems EAS a "good idea." And, typically, everyone's monthly rate for basic telephone service goes up.

Extended Binary Coded Decimal Interchange Code EBCDIC. (Pronounced Eb-Si-Dick.) An IBM standard of coding characters. It's an 8-bit code and can represent up to 256 characters. A ninth bit is used as a parity bit. See PARITY and EBCDIC.

Extended BIOS Data Area In PCs, extended BIOS data area is 1KB of RAM located at 639KB. It is used to support extended BIOS functions including support for PS/2.

Extended Call Management A Northern Telecom term for a collection of features being added to its DMS Meridian central office Automatic Call Distribution (ACD) service. Using Switch-to-Computer Applications Interface (SCAI), ECM will work with user-provided computer equipment to integrate call processing, voice processing (recorded announcements, voice mail and voice response) and data processing. For example, ECM will allow an outboard computer device to coordinate the presentation of customer data on the ACD agent's computer screen with an incoming call. The D channel of an ISDN Basic Rate Interface (BRI) serves as the transport mechanism from the DMS-100 central office switch to an outboard computing device. Communication is peer-to-peer, meaning that neither the switch nor the computer is in a "slave" relationship to the other. The application layer messaging — i.e. layer 7 messaging as defined by the Open Systems Interconnection (OSI) reference model — is in the Q.932 format and is designed to conform to the T1S1 SCAI message protocol.

Extended Character Set The characters assigned to ASCII codes 128 through 255 on IBM and IBM-compatible microcomputers. These characters are not defined by the ASCII standard and are therefore not "standard." See EXTENDED GRAPHICS CHARACTER SET.

Extended Definition Television EDTV. Television that includes improvements to the standard NTSC television system, which improvements are receiver-compatible with the NTSC standard, but modify the NTSC emission standards. Such improvements may include (a) a wider aspect ratio, (b) higher picture definition than distribution-quality definition but lower than HDTV, and/or (c)any of the improvements used in improved-definition television. When EDTV is transmitted in the 4:3 aspect ratio, it is referred to simply as "EDTV." When transmitted in a wider aspect ratio, it is referred to as "EDTV-Wide."

Extended Digital Subscriber Line The ISDN EDSL combines 24 B-channels and one 64-Kbps D-channel on a single line, ISDN primary rate interface.

Extended Graphics Character Set The characters assigned to ASCII codes 128 through 255 on IBM and IBM-compatible microcomputers. These characters are not defined by the ASCII standard and are therefore not "standard." The original ASCII code used a seven bit one-or-zero code. There are two to the seventh power, or 128 possible combinations. The IBM PC uses a 16-bit CPU with an eight bit data bus and thus transmits data internally in eight big bytes. Instead of using the seven bit ASCII code, the PC uses the equivalent eight bit code, by simply making the left most digit, a zero. In seven bit code, an R is 1010010. In 8-bit, it's 01010010. The only difference between the first 128 characters and the second 128 characters is that in the second, the first bit is a 1.

Extended Life Battery Toshiba's wonderful name for a battery for a laptop that's physically larger than the normal one, costs more than the normal one and lasts longer than the normal one. Extended life batteries make lightweight laptops into heavyweight laptops.

Extended Key Code The two digit code that represents pressing a key outside the typewriter portion of the keyboard, such as a function key, cursor-control key or combinations of CTRL (control) and ALT keys with another key. The first number is always 0 (zero) and is separated from the second number by a semicolon.

Extended LAN A collection of local area networks connected by protocol independent devices such as bridges or routers.

Extended Memory Memory beyond 1 megabyte in 80286, 80386, 80486 and Pentium computers. Windows uses extended memory to manage and run applications. Extended memory can be used for RAM disks, disk caches, or Microsoft Windows, but requires the processor to operate in a special mode (protected mode or virtual real mode). With a special driver, you can use extended memory to create expanded memory. Extended memory typically is not available to non-Windows applications or MS-DOS. See also EXPANDED MEMORY.

Extended Superframe Format ESF. A new T-1 framing standard used in Wide Area Networks (WANs). With this format 24 frames — instead of 12 — are grouped together. In this grouping, the 8,000 bps frame is redefined as follows:
• 2,000 bps for framing and signaling to provide the functions generally defined in the D-4 format.
• 2,000 bps are CRC-6 (Cyclic Redundancy Check-code 6) to detect logic errors caused by line equipment, noise, lightning and other interference. Performance checking is done by both the carrier and the customer without causing any interference with the T-1 traffic.
• 4,000 bps are used as a data link. This link is to perform functions such as enhanced end-to-end diagnostics, networking reporting and control, channel or equipment switching, and/or optional functions or services. See also T-1 FRAMING and D-4 FRAMING.

Extended Superframe Monitoring Unit ESFMU. An MCI definition. Placed on customer data circuits to provide performance monitoring throughout MCI's Digital Data Network.

Extended Telephony Level The lowest level of service in Windows Telephony Services is called Basic Telephony and provides a guaranteed set of functions that corresponds to "Plain Old Telephone Service" (POTS - only make calls and receive calls). The next service level is Supplementary Telephone Service providing advanced switch features such as hold, transfer, etc. All supplementary services are optional. Finally, there is the Extended Telephony level. This API level provides numerous and well-defined API extension mechanisms that enable application developers to access service provider-specific functions not directly defined by the Telephony API. See WINDOWS TELEPHONY SERVICES.

Extended Text Mode Standard text mode is 80 columns wide. So-called extended text mode is 132 columns wide. This mode allows you to view more text on-screen when using such applications as Lotus 1-2-3.

Extensible In strictest terms, the word means "capable of being extended." When Microsoft introduced its At Work operating system on June 9, 1993 it said that one of the operating system's key features was that it was "extensible." Microsoft's explanation: The software is designed to allow both manufacturers and customers to add new features. For example, local area network connectivity will be able to be added easily by installing an optional LAN hardware module and a software driver. Additional memory will be able to be added to the system, and the system will automatically make use of this memory. New image-processing software and communications protocols will be able to be added on the premises, and it will even be able to be done over the phone line, allowing manufacturers to create basic models that can be enhanced in many different ways to fit the needs of different user groups.

Extensibility This means it's easy to add new technologies without reinventing the wheel.

Extension 1. An additional telephone connected to a line. Allows two or more locations to be served by the same telephone line or line group. May also refer to an intercom phone number in an office.

2. The optional second part of an PC computer filename. Extensions begin with a period and contain from one to three characters. Most application programs supply extensions for files they create. Checking a file's extension often tells you what the file does or contains. For example, most BASIC files use a filename extension of .BAS. Most backup files have an extension of .BAK. MS-DOS programs have .EXE or .COM. dBASE database files have the extension .DBF and .DBT. Paradox files have the extension .DB. Files of sounds have their own extensions. Here are the typical extensions on sound files of various computers:

Microsoft Windows — .wav
Apple — .aif
NeXT — .snd
MIDI — .mid and .nni
Sound Blaster — .voc

Here are the typical extensions on graphics formats:
.TIFF, .EPS, .CGM, .PCX, .DRW, .WMF, and .BMP

Extension Cord A multi-conductor, male/female modular line cord generally used to permit greater separation between the Communications Outlet and the telephone equipment. Available in various lengths up to 25 feet. May be of tinsel or stranded wire construction.

Extensions to Unix Extensions to UNIX are additional features or functions not found in the standard UNIX implementation. The extension are classified either as "open extensions" or "proprietary extensions." An "open extension" usually consists of a surface addition, such as a driver for peripheral or a software patch for a new mode of I/O. The "open extension" is transparent to standard UNIX and its application programs. The "proprietary extension" is for the implementation of custom hardware or software. It results in a version of UNIX that is not transparent to UNIX and its applications.

Exterior An ATM term. Denotes that an item (e.g., link, node, or reachable address) is outside of a PNNI routing domain.

Exterior Link An ATM term. A link which crosses the boundary of the PNNI routing domain. The PNNI protocol does not run over an exterior link.

Exterior Reachable Address An ATM term. An address that can be reached through a PNNI routing domain, but which is not located in that PNNI routing domain.

Exterior Route An ATM term. A route which traverses an exterior link.

External F-ES A Fixed End System (F-ES) connected to the CDPD network outside the administrative domain of the service provider.

External Interface A cellular radio term. The Cellular Digital Packet Data (CDPD)-based wireless packet data service provider's interface to existing external networks. The external application service providers communicate with CDPD subscribers through this external interface.

External Memory Storage devices, such as magnetic disks, drums or tapes which are outside (externally attached) to the main telephone or computer system.

External Modem A modem external to the computer, it sits in its own little box connected to a computer through the computer's serial port. Compare with an internal modem, which typically comes on one printed circuit card and is placed into one of the computer's expansion slots and thus connects to the computer through the computer's "backplane." Internal modems cost less because they don't need any external housing and separate power supply. But because they're mounted inside the computer, it's harder to see what they're doing. You can't see the various status lights, like OH (for Off-Hook) and CD (for Carrier Detect). They also take up valuable slots instead of a serial port.

External Photoeffect In fiber optics, an external photoeffect consists of photon-excited electrons that are emitted after overcoming the energy barrier at the surface of a photoemissive surface.

External Storage See also EXTERNAL MEMORY.

External Timing Reference A timing reference obtained from a source external to the communications system such as one of the navigation systems. Many of which are referenced to Coordinated Universal Time (UTC).

External Viewer This is the program that is launched or used by Web browsers for presenting graphics, audio, video, VRML, and other multimedia found on the Internet. Sometimes referred to as helper applications. Usually when you initially set up your browser you configure what external viewers you want to use by associating a program with a file type or extension. This way the browser knows what to do when these files are "clicked on" by the user.

EXTN Extension.

Extranet An extranet, coined by Bob Metcalfe in the April 8, 1996 issue of InfoWorld, is a Internet-like network which a company runs to conduct business with its employees, its customers and/or its suppliers. Extranets typically include Web sites that provide information to internal employees and also have secure areas to provide information to customers and external partners like suppliers, manufacturers and distributors. A company might place a call for product on its extranet's web site. Its suppliers will check the site regularly and bid on the product. It is called an extranet because it typically uses the technology of the public Internet (TCP/IP and browsers) and customers and suppliers often access the extranet through the Internet via their local ISP — Internet Service Provider. But an Extranet is not a public entity. You typically need accounts and passwords, typically issued by the firm running the Extranet. The word Extranet, however, is a term in evolution. In the October 21, 1996 issue of InfoWorld, Bob Metcalfe defined Extranets as IP networks through which companies run Web applications for external use by customers. He explained that ISPs have deployed private networks (i.e. extranets) which operate on the same principles and make use of the same network technologies as the Internet, but are external to it. Access to the Internet can be gained through an Extranet. Extranets, he said, were for electronic commerce. See also INTERNET and INTRANET.

Extremely High Frequency EHF. Frequencies from 30 GHz to 300 GHz.

Extremely Low Frequency ELF. Frequencies from 30 Hz to 300 Hz.

Extrinsic Joint Loss For an optical fiber, that portion of a joint loss that is not intrinsic to the fibers, e.g., loss caused by end separation, angular misalignment, or lateral misalignment.

Extrusion Method of continuously forcing plastic, rubber or elastometer material through an orifice to apply insulation or jacketing over a conductor or cable core.

Eye Pattern An oscilloscope display used to visually determine the quality of an equalized transmission line signal being received. So called because portions of the pattern appearing on the scope resemble the elliptical shape of the human eye.

Eye Phone Several researchers are studying something they call "virtual reality." One version of it, a system developed by a company called VPL Research, Redwood City, CA, is based around three things: a three-dimensional glove worn on the head (called an Eye Phone), an electronic glove (the Data Glove) and a high-speed computer. The whole system cost $250,000 in the fall of 1990.

Eyeball A viewing audience. "There are plenty of new eyeballs available in this time slot."

Eyeball Shot Also called microwave link. The link is made by two radio transceivers equipped with parabolic dish antennas pointed at each other. The transmissions can be carried on many bandwidths including DS1, DS2, DS3, STS1 and OC1. The range varies from 0-50 miles depending on the dish size, weather and transmitter power.

EZTV A software created by Apple which will enable consumers to order movies, go shopping and play games on their TV sets. The product is mostly on the drawing boards.

F Connector A 75 ohm coaxial cable connector commonly found on consumer television and video equipment.

F Link Fully Associated Link. A link used to connect two SS7 signaling points when there is a high community of interest between them and it is economical to link them. Also called associated signaling.

F Port Fabric Port. A Fibre Channel term, referring to the port residing on the fabric (switch) side of the link. It attaches to a N Port (Node Port) at the connected device, across a link. See FIBRE CHANNEL.

F Type Connector A low cost connector used by the TV industry to connect coaxial cable to equipment. See also F-Type Connector (which is the same thing, except spelled with a dash.)

F-BCCH Fast-Broadcast Control CHannel. A logical channel element of the BCCH signaling and control channel used in digital cellular networks employing TDMA (Time Division Multiple Access), as defined by IS-136. See also BCCH, IS-136 and TDMA.

F-Block Carrier A 10 MHz PCS carrier serving a Basic Trading Area in the frequency block `890 - 1895 MHz paired with 1970 - 1975 MHz.

F-ES Fixed End System

F-Type Connector These are used to terminate coaxial cable. This connector is mostly used for video applications. It's a male single-conductor connector and screws into the female jack.

F2F Face to Face. When you actually meet someone with whom you have been corresponding electronically, perhaps through a chat room over the Internet. F2F often is quite a surprise, as your "pen pal" may not be anything like he said he was. F2F also can be very dangerous. Never, ever meet someone F2F unless you have a companion with you and you meet in a well-lit public place. Never, ever give the other person your real name, address or telephone number until you have met him F2F and are confident that he is who he says he is. Tell your children to never, never, ever agree to meet someone F2F unless you approve in advance and you are with them at the meeting. This is a very, very dangerous world full of very, very dangerous people who prey on the unsuspecting.

Fab Factory that makes ("fabricates") IC chips.

Fabric 1. A descriptive term referring to the physical structure of a switch or network. Much like a piece of cloth, physical/logical communications channels (threads) are interwoven from port-to-port (end-to-end). Ideally, data are transferred through this switch or network on a seamless basis. In ATM and Fibre Channel, the switching fabric generally is non-blocking, or virtually so, from port-to-port. In the Internet, data works its way through a complex, and even unpredictable, interwoven network of networks comprising transmission facilities, packet switches and multiple carriers.
2. Multiple Fibre Channel switches interconnected and using Fibre Channel methodology for linking nodes and routing frames in a Fibre Channel network. See Fibre Channel.

FAC See FORCED AUTHORIZATION CODE.

Faceplate A cover that fits around the pushbuttons or rotary dial of a telephone. Hotels and motels put instructions on them. More businesses should also.

Facilities A stupid, imprecisely defined word that means anything and everything. To me it sounds like toilets. But it's not. It can mean the equipment and services which make up a telecom system. It can mean offices, factories, and/or building. It can be anywhere you choose to put telecom things. Oops, I nearly said telecom facilities. So "facilities" means practically anything you want it to mean so long as it covers a sufficiently broad variety of "things" which you haven't got a convenient name for. "Facilities" sounds better than things, especially if you want to sound pompous.

Facilities Administration And Control A PBX feature which allows you, the subscriber, to assign to your users features and privileges like authorization codes, restriction levels and calling privileges.

Facilities Assurance Reports This feature allows a subscriber to get an audit trail of the referrals produced by the automatic circuit assurance feature of some PBXs. The audit trail will identify the trunk circuit, the time of referral, nature of the problem and if a test was performed, the outcome of the test.

Facilities Based Carrier A carrier which owns some of its own facilities (i.e., stuff), such as switching equipment and transmission lines. ILECs (Incumbent Local Exchange Carriers) such as Bell Atlantic, BellSouth, Citizens Communications, GTE and SBC fit this definition. Incumbent IXCs (IntereXchange Carriers) such as AT&T, MCI Worldcom and Sprint also fit this definition. The major facilities-based IXCs have switching offices, or POPs (Points OF Presence) in all service areas of the country and provide both originating and terminating service nationwide. Major facilities-based carriers sell their services to business and residential users and to other carriers which resell those services. Non facilities-based long distance carriers are known as switchless resellers. To be recognized as a CLEC (Competitive Local Exchange Carrier) by most local regulatory authorities and to receive reciprocal compensation from the local ILEC, you must, at minimum, own a central office switch; thus you must be a facilities based carrier. There's probably not one single carrier — local,long distance or international — in the entire North America that is 100% facilities based these days. Everyone seems to be renting some facilities — usually lines — from someone else. The most facilities based would be the ILECs. The least facilities based would be the CLECs (Competitive LECs). They tend to resell local loops from the local ILEC which they terminate in their switching centers. See also CLEC, ILEC, IXC and POP.

Facilities Data Link FDL. An Extended SuperFrame (ESF)term. ESF extends the superframe from 12 to 24 consecutive and repetitive frames of information. The framing overhead of 8 Kbps in previous T-1 versions was used exclusively for purposes of synchronization. ESF takes advantage of newer channel banks and multiplexers which can accomplish this process of synchronization using only 2 Kbps of the framing bits, with the framing bit of only every fourth frame being used for this purpose. As a result, 6 Kbps is freed up for other purposes. This allows 2 Kbps to be used for continuous error checking using a CRC-6 (Cyclic Redundancy Check-6), and 4 Kbps to be used for a FDL which supports the communication of various network information in the form of in-service monitoring and diagnostics. ESF, through the FDL, supports non-intrusive signaling and control, thereby offering the user "clear channel" communications of a full 64 Kbps per channel, rather than the 56 Kbps offered through older versions as a result of "bit robbing." FDL implementations can vary, and the 4Kbps data bandwidth can be allocated among different functions, such as managing line side and equipment side operations. For example, ADC just introduced a new HDSL service to compete with T-1. ADC allocates 2 Kbps of the FDL channel for managing the HDSL interface and 2K bps for remotely managing attached channel banks and data terminals. As you can see, FDL has great potential for reducing the cost of onsite provisioning and maintenance. See also ESF and T-1.

Facilities Management Also called Outsourcing, facilities management is having someone else run your computers or your telecommunications system. The concept is that you're a great bank and you should concentrate on being in the banking business. Your outside facilities manager should concentrate on running your computers or telecom systems. He can do it cheaper, allegedly. Ross Perot's Electronic Data Systems (EDS) probably started facilities management. Mr. Perot incorporated EDS on June 27, 1962. At that time he was a leading IBM salesman. See also OUTSOURCING.

Facilities Restriction Level Which types of calls a PBX user is entitled to make.

Facility A telephone industry term for a phone or data line. Sometimes (but rarely) used to describe equipment. See FACILITIES.

Facility Grounding System The electrically interconnected system of conductors and conductive elements that provides multiple current paths to the earth electrode subsystem. The facility grounding system consists of the earth electrode subsystem, the lightning protection subsystem, the signal reference subsystem, and the fault protection subsystem. Faulty grounding causes more phone and computer problems than any other single factor.

Facility Work Order An order to a phone company to rearrange things.

FACS 1. How they abbreviate the word facsimile in Bermuda. 2. Facilities Access Control Systems. A collection of dozens of interrelated computer applications developed by the former AT&T Bell Operating Companies which manage the local loops connecting customers to the Public Switched Telephone Network.

Facsimile Equipment FAX. Equipment which allows hard copy (written, typed or drawn material) to be sent through the switched telephone system and printed out elsewhere. Think of a fax machine as essentially two machines — one for transmitting and one for receiving. The sending fax machines typically consists of a scanner for converting material to be faxed into digital bits, a digital signal processor (a single chip specialized microprocessor) for reducing those bits (encoding white space into a formula and not an endless series of bits representing white), and a modem for converting the bits into an analog signal for transmission over analog dial-up phone lines. The receiving fax consists of a modem and a printer which converts the incoming bits into black and white images on paper. More modern and more expensive machines also have memory — such that if the machine runs out of paper, it will still continue to receive incoming faxes, storing those faxes into memory until someone fills the machine with paper and it prints the faxes out.

There are six internationally accepted specifications for facsimile equipment. Group 1, Group 2, Group 3, Group 3 Enhanced, Super GE and Group 4. Only 1, 2, 3 and 3 Enhanced will work on "normal" analog dial-up phone lines. Group 4 is designed for digital lines running at 56/64 Kbps, e.g. ISDN lines. Among the analog line fax machines, Group 2 is faster than Group 1. Group 3 is faster than Group 2, etc. Virtually all machines sold today are Group 3, though an increasing percentage are Group 3 enhanced, which has speeded up Group 3's transmission speed from 9,600 bps to 14,400 bps and improved its error correction. Group 3 faxes send an 8-1/2 x 11 inch page over a normal phone line in about 20 seconds. How much time it actually takes depends on how much stuff is actually on the paper. Unlike older machines, Group 3 machines are "intelligent." They only transmit the information that's on the paper. They do not transmit white space, as earlier machines did. Super G3 is a new "standard" for higher speed fax machines, which contain a 33.6 Kbps V.34 modem, V.8 handshaking and the new ITU-T T.85 JBIG image compression. On most phone lines such a machine should get close to double the speed of the highest speed Group 3 fax machines — 14.4 Kbps. But, the JBIG image compression will speed faxing of gray scale images by at as much as five to six times. In short, these machines will send faxes much faster — if they send to a Super G3 machine at the other end. Super G3 is compatible with and can communicate with older fax machines, Group 1, 2, 3 and 3 Enhanced.

When a fax machine calls a phone line, it emits a standard ITU-T-defined, "CNG tone" (calling tone) — 1100 Hz tone every three seconds. When the receiving fax machine hears this tone, it knows it's an incoming fax call and it can automatically connect. With this tone it is possible to insert a "fax switch," which would "listen" for an incoming fax call and switch it to a fax machine if it heard the CNG tones or to something else — like a phone or answering machine — if it didn't. It is not possible to do this with a modem. A calling modem does not issue any tones whatsoever. A modem works backwards — when the receiving modem answers the phone, it emits a tone.

Typically, a Group 3 machine can speak to a Group 2 and a Group 1 machine. A Group 2 can speak to a Group 1. Speaking down means slowing down. Fax machines are dropping in price. "Personal" fax machines are emerging. Most fax machines today at Group 3 or Group 3 enhanced.

Group 4 machines are 100% digital, transmit at 64,000 bps and directly attach to the B (bearer channel) of a digital ISDN line. They will transmit a sheet of 8 1/2 x 11 paper in under six seconds. The author of this dictionary has seen a working Group 4 fax machine. It's mighty impressive.

The latest ITU standards include T.37 and T.38. T.37 is a new ITU standard for transferring of facsimile messages via store and forward over packet-switched IP networks — the Internet, corporate Intranets, etc. T.38 is a new ITU-T standard for sending real-time facsimile messages over packet-switched IP networks — the Internet, corporate Intranets, etc.

Some warnings on fax machines:

1.   All analog Group 3 and enhanced Group 3 fax machines pose a security risk. Anyone can attach a normal audio cassette recorder to a phone line, record the incoming

or outgoing fax "tones" of an analog fax machine. By playing back to another fax machine at a later time, you'll get a perfect reproduction of the fax. There are now fax encryption devices which make the fax transmission unintelligible to any machine other than the one it's intended for — i.e. the one that has a similar un-encryption device.

2. Some plain paper fax machines present a different security risk. Some (not all) use a carbon ribbon the width of their paper. As a result, if you want to read what came in, you simply read the carbon ribbon, which you open like a scroll, which the cleaning lady finds in the trash. These machines are increasingly less common, as plain paper fax machines acquire laser printing engines.

3. Most fax machines record all the digits dialed into them which were used to set up a fax call. If a fax machine is sitting behind a PBX (as many are these days) it will capture all the confidential authorization codes of all the company's employees. To get those codes all you need do is ask the machine to print out a report. There is no easy solution to this problem as at the time of writing this dictionary, except that some fax makers have told me they intend to obscure these numbers on their reports, at some stage. Some may, by the time you read this.

4. Slimy paper fades. How long it takes to fade depends on a bunch of factors — from what's sitting on top of the fax, to the temperature in the room, to whether it's exposed to sunlight, etc. Recommendation: If you want to retain a slimy fax, make a photo copy of it the moment you get it and throw out the original.

5. Poor quality slimy fax paper can abrade the fax machine's drum and cause a costly repair. Don't buy cheap slimy fax paper.

6. Plain paper fax machines cost more to buy, but less to run. You can buy a second tray for some plain paper fax machines which will hold 8 1/2" wide x 14" long paper, which is useful for receiving faxes from outside the US where they use longer paper. This way you save a sheet of paper.

7. It makes sense to have banks of fax machines attached to phones which roll over — also called "hunt." It makes absolutely no sense to have multiple fax machines on separate phone lines that don't hunt, i.e. one for everybody in the office. Two fax machines in rotary can receive and transmit more than twice the number of faxes that two machines on separate, non-hunting phone lines can send and receive. "Personal" fax machines should be out. Banks of fax machines should be in. Egos, though, usually prevail over logic.

8. The paper feed mechanism on plain fax machines has a tendency to jam. Slimy paper fax machines don't jam because their paper typically comes in rolls. And roll paper doesn't jam. The feed mechanism is much simpler.

9. Plain paper fax machines, like laser printers (which many are) use supplies, like toner, which run out. When the supplies run out, such machines usually accept incoming faxes into memory — until that runs out. Then they just ring and ring and ring. Which means that incoming faxes don't get through and don't roll over to the next machine. There is no simple solution since the FCC (Federal Communications Commission) has ruled that fax machines must not return a busy signal to the central office if it runs out of supplies or paper. We have a separate machine that automatically busies out a line if it failed to answer on the fifth ring. But so far, the device is not commercially available. I don't know the answer to this problem except to make sure your fax machine is always stuffed with supplies. Especially check every Friday night. A final note: If your plain paper fax machine is missing supplies, but stuffed with incoming messages in memory, don't turn it off, since you'll lose the messages. Simply

replace the supplies and pray your messages will emerge.

10. Some slimy fax paper rolls are coated on the inside of the paper. Others are coated on the outside. When you put one in a fax machine and images don't appear on the paper, then turn the roll over and feed it from underneath. In short, ignore what the instruction book says.

11. Fax modem switches only work when they're called automatically by a fax machine — not by a person using a fax machine manually and is waiting the sound of the distant fax prior to pushing the "Send" button. Make sure you warn your senders. It's remarkable how many people manually dial their faxes and thus penetrate fax modem switches.

11. Think about putting your fax machine on "fine." You'll transmit better quality faxes and may only cost yourself 10% more in transmission time. But that savings depends on the quality of the fax machine at the other end. If it's an older machine, it may cost you as much as double the transmission time. Here are the numbers: Standard is 203 x 98 dpi. Fine is 203 x 196 dpi. "Fine" faxes obviously look much better.

12. Printed circuit cards which slide into slots of PCs and allow you to transmit and receive faxes work well — when transmitting faxes. They work far less well when receiving faxes — largely because of the difficulty of reading faxes. Faxes conform to one type of digital encoding and PC screens conform to another. Moreover a PC screen is landscape (i.e. horizontal), while a fax message is portrait (i.e. vertical). Viewing vertical images on horizontal screens is difficult. Here is a comparison of how fax machines and how personal computer screens encode their images. Obviously, the more digits or pixels, the clearer the end picture. Notice that the encodings are completely dissimilar:

FAX ENCODING

Standard, Group III	.203 x 98
Fine, Group III	203 x 196
Superfine, Group III	203 x 391
Standard, Group IV	400 x 400

PC SCREEN ENCODING

CGA.	320 x 200
Enhanced CGA	640 x 400
EGA	640 x 350
Hercules	720 x 348
VGA	640 x 480
Super VGA	800 x 600
8514/A (also called XGA)	1,024 x 768

See also 1966, 1978, 1980, demodulation, facsimile converter, FACSIMILE RECORDER, FACSIMILE SIGNAL LEVEL, FACSIMILE SWITCH, FAX, FAX AT WORK, FAX BACK, FAX BOARD, FAX DATA MODEM, FAX DEMODULATION, FAX MAILBOX, FAX MODEM, FAX PUBLISHING, FAX SERVER, FAX SWITCH, FAXBIOS, GROUP 1, 2, 3, 3 BIS AND 4, PHASE A thru E, T.37 and T.38 and WINDOWS TELEPHONY.

Facsimile Converter A facsimile device that changes the type of modulation from frequency shift to amplitude and vice versa.

Facsimile Recorder That part of the facsimile receiver that performs the final conversion of the facsimile picture signal to an image of the original subject copy on the record medium.

Facsimile Signal Level The facsimile signal power or voltage measured at any point in a facsimile system. It is used to establish the operating levels in a facsimile system, and may be expressed in decibels with respect to some standard value such as 1 milliwatt.

Facsimile Switch A new breed of "black box." Its purpose is to avoid having to lease a separate phone line for your facsimile machine, for your phone and for your modem. You buy this box, connect it to an incoming line, connect it to your fax machine, your phone and, possibly, your modem. When a call comes in, the fax switch answers the call, listens if the call coming in is from a fax machine (it can hear the fax machine's CNG calling tone) and switches the call to the fax machine, or switches the call to your modem if a computer is calling. It knows if a computer is calling because the calling computer will, when it hears the fax switch answer, send out some ASCII characters — e.g. 22. (You must put those numbers in your modem dialing stream.) And it knows if a person is calling because it hears neither a CNG tone from a fax machine nor touchtones from the dialing stream of a modem.

The above are the basics of how fax switches work. There are variations on this theme. Some fax switches work automatically. Some work by the incoming caller punching in digits. Some allow you to switch from fax machine to modem to phone and back again. And some fax switches will answer and connect to three modems and one fax or other combinations. The major problem with fax switches is that they typically send a DC ringing tone to whatever device they're trying to connect you (the incoming caller with). Sometimes some devices — for example, high-speed 9600 baud and higher modems — have difficulties responding to low power, DC ringing signals. And they just sit there not answering. Better to buy one that sends standard telephone company AC ringing signals. In short, before you buy a fax/modem/phone switch, test it on your favorite 9,600 or 14,400 bps modem. The more expensive switches tend to work better.

Factoid Factoids are paragraph size pieces of "Gee Whiz" information. They were originally made famous in the newspaper, USA Today.

Factory Refurbished A term used in the secondary telecom equipment business. Equipment that has been returned to the factory and the factory has replaced plastic, repaired, upgraded boards, or otherwise reconditioned.

Fade A reduction in a received signal which is caused by reflecting, refraction or absorption. See also FADING.

Fade Margin The depth of fade, expressed in dB, that a microwave receiver can tolerate while still maintaining acceptable circuit quality.

Fading 1. The reduction in signal intensity of one or all of the components of a radio signal.
2. A video term. A progressive deterioration of picture quality due to increasing losses in an electromagnetic (radio) propagation path. The term "fading" may be illustrated by the following sequence: (a) Noise appears on the porches and tip of the sync pulses. (b) Noise appears in the picture. (c) Loss of picture due to loss of synchronization which in turn is caused by distortion of the sync pulse by noise.

FADS Force Administration Data System. A system which takes basic statistics on telephone traffic and gives hints as to how many operators should be employed to answer the incoming calls and when they should be present.

Fail Safe A specially designed system that continues working after a failure of some component or piece of the system. There are precious few, genuinely fail safe systems. To be genuinely fail safe, a system needs to be completed duplicated. It is prohibitively expensive for most commercial users to duplicate every part of their system. But you can duplicate selectively and bring yourself closer to "fail safe." The extent of the duplications you choose (and thus the cost of your telephone equipment and transmission system) depends on how important it is that your system function as close to 100% as

possible. The idea is to identify those things most likely to break and to duplicate them. Power is clearly the first area to focus on. These days, the words "FAIL SAFE" are increasingly being replaced with "FAULT TOLERANT." Given the number of times your local, friendly airline has told you that its "computer is down," you can understand the reason for the wording change.

Fair Condition A term used in the secondary telecom equipment business. One step up from "as is" condition. Equipment may have been tested; i.e., product is in working order but looks semi-awful.

Fair Market Value See FMV.

Fairness An ATM term. As related to Generic Flow Control (GFC), fairness is defined as meeting all the agreed quality of service (QOS) requirements, by controlling the order of service for all active connections.

Fake Root A subdirectory on the file server of a local area network that functions as a root directory, where you can safely assign rights to users. Fake roots only work with NetWare shells included with NetWare v2.2 and above. If you use older versions of the workstation shell, you will not be able to create fake roots.

Fall time The length of time during which a pulse decreases from 90 to 10 percent of its maximum amplitude.

Fall.com An early virus which made the characters on a screen fall to the bottom.

Fallback Rate A modem speed that is lower than its normal (that is, maximum) speed of operation. May be used when communicating with a slower, compatible modem, or to help transmission over a line that is too noisy for full speed operation.

False Ringing False ringing is a recording of a telephone ringing signal (two seconds on, four seconds off, which is played while a call is transferred or while a switching device listens for modem for facsimile CNG (calling) tones.

Falsing In telecom signaling, DTMF tones are created using specific combinations of frequencies to prevent the possibility of "falsing." Falsing is the condition where a DTMF detector incorrectly believes a DTMF is present when in fact it is actually a combination of voice, noise and/or music.

Fan Antenna An aerial consisting of a number of wires radiating upwards from a common terminal to points on a supporting wire.

Fan Out Equipment that breaks down DS1 or DS3 service to the size demanded by the customer. On a DS3 line it breaks out the 28 DS1 channels. On a DS1 line it breaks them into 24 DS0 channels.

Fanatic Someone who's overly enthusiastic about something in which you have zero interest.

FAP Formats And Protocols. The set of rules that specifies the format, timing, sequencing and/or error checking for communication between clients and servers.

FAQ Either a Frequently Asked Question, or a list of frequently asked questions and their answers. Many Internet USENET news groups, and some non-USENET mailing lists, maintain FAQ lists (FAQs) so that participants won't spend lots of time answering the same set of questions.

Far End Crosstalk Crosstalk which travels along a circuit in the same direction as the signals in the circuit. The terminals of the disturbed channel at which the far-end crosstalk is present and the energized terminals of the disturbing channel are usually remote from each other.

Far Field Pattern Synonym for Far-Field Radiation Pattern.

Far Talk In voice recognition, far talk is an arrangement where a microphone is more than four inches from the speaker's mouth. The opposite is CLOSE TALK, where the microphone is closer than four inches.

Farad The practical unit of capacity. A capacitor which retains a charge of one coulomb with a potential difference of one volt. See FARADAY and FARADAY CAGE.

Faraday As a Faraday shield: refers to the protection a material or container provides to electronic devices to keep them from exposure to electrostatic fields. Named after M. Faraday, the English physicist.

Faraday Cage A structure designed to isolate a sensitive electronic system or device from outside interference, usually constructed of metal screens. Named for 19th century inventor Michael Faraday whose name also gave us the FARAD, the unit of measuring capacitance.

Farms of Mainframes Picture a hall full of mainframe computers, lined one after another. Now you have the concept of a farm of mainframes.

FAS Frame Alignment Signal or Frame Alignment Sequence. See Frame Alignment Signal.

Fast Broadcast Control Channel F-BCCH. A logical channel element of the BCCH signaling and control channel used in digital cellular networks employing TDMA (Time Division Multiple Access) as defined by IS-136. See also BCCH, IS-136 and TDMA.

Fast Busy A busy signal which sounds at twice the normal rate (120 interruptions per minute vs. 60 a minute). A "fast busy" signal indicates all trunks are busy.

Fast Clear Down A call center term. A caller who hangs up immediately when they hear a delay announcement.

Fast Ethernet 100BaseT. Ethernet at 100 Mbps, a tenfold improvement over the original Ethernet speed of 10 Mbps. Fast Ethernet is in the form of an Ethernet hub with an internal bus that runs at 100 Mbps. The interface to the hub is through a port which generally is selectible (i.e., programmable) to run at either 10 Mbps or 100 Mbps, depending on the requirement of the attached device. Connection between the hub and the attached workstation or other device is over data-grade UTP (Unshielded Twisted Pair) in the form of Cat (Category) 5, at a minimum, and over distances of up to 100 meters, at a maximum. The attached device connects to the UTP connection via a 10/100 Mbps NIC (Network Interface Card). 100BaseT hubs interconnect over fiber optic facilities, which can support 100 Mbps over relatively long distances with no loss of performance. Fast Ethernet is no longer all that fast- Gigabit Ethernet switches were standardized in 1998. See also 10BaseT, 100BaseT, Cat 5, Ethernet, Gigabit Ethernet, NIC and UTP.

Fast Ethernet Alliance A group of vendors that participated in writing the 100Base-X technical hub and wiring specifications, which would allow fast Ethernet (100 megabits a second) to run over Category 5, data-grade unshielded twisted pair wiring.

Fast File Transfer FFT. An ISDN term referring to the fact that file transfers can be accomplished "fast." Reason #1: Two B channels at 64 Kbps each are available to be bonded to provide as much as 128 Kbps. Reason #2: Data transfer is accomplished in an "optimistic" streaming mode, rather than a "pessimistic" packet mode. Therefore, there is no delay associated with acknowledgments. This is possible due to the excellent level of error performance inherent in digital services. The end result is FFT.

Fast Fourier Transform See FFT.

Fast Network An AT&T term for a network with low delay relative to the needs of the application.

Fast Packet Multiplexing Multiplexing, from Latin "multi" and "plex" translates as "manyfold." In other words, folding many "conversations" onto a single circuit. You can do this in either of two ways — by splitting the channels sideways into subchannels of narrower frequency. This is called Frequency Division Multiplexing (FDM), which is used in analog networks. Or you could split it by time, through a process of Time Division Multiplexing (TDM), which is used in digital networks. TDM is much like a railroad train. The first car carries "Conversation 1." The second carries "Conversation 2." And split them apart at the other end.

Fast packet multiplexing is a combination of three techniques — time division multiplexing, packetizing of voice and other analog signals, and computer intelligence. Here are the main advantages fast packet multiplexing has over today's industry standard time division multiplexing:

1. Fast packet multiplexing doesn't blindly slot in "information" from devices if there's no information to send. Most other multiplexing techniques, including the most common — time division and frequency division — slot in capacity, whether the device is "talking" or not.

2. The fast packet multiplexer can start sending a packet before it has completely received the packet. This is accomplished by reading the destination address, which is contained in the header portion of the packet. This speed of movement is critical to voice, for example, which must move ultra-fast. Delays are devastating. (No one can afford to replace the phone instruments broken in anger.)

3. Fast packet multiplexing can interrupt the delivery of one packet in favor of sending another. It's OK to delay a packet of data by several milliseconds. It's not OK to delay a packet of voice or video.

Fast Packet Services An umbrella term for ATM, Frame Relay, and SMDS service offerings, all of which operate at broadband speeds and all of which make use of Fast Packet Switches and Multiplexers

Fast Packet Switching A wide area networking technology capable of switching data at a very high rate of speed in the context of a broadband network service such as Frame Relay, SMDS or ATM. The term "packet" is generic, referring to the manner in which data is formatted. "Data" is also generic in this context, referring to voice data, video data, and image data, as well as data data. Should the data be analog in its native form, it is digitized and packetized before being presented to the network for transport and switching. The packets are in the form of short (53 octets), fixed length cells in the case of SMDS and ATM. The packets are in the form of variable (0-4,096 octets) frames in the case of Frame Relay. The underlying switching technology is based on the statistical multiplexing of data contained within the cells or frames. While any of these packets could carry digital voice, video, data or image information, only ATM is specifically intended for other than data use. All the packets travel at Level Two of the OSI Model, and routing is performed on the basis of the Level Two addressing. Fast packet is claimed to be very effective way of make best use of available bandwidth. It is claimed to offer the benefits of conventional multiplexing techniques and circuit switching techniques because of the way it operates. It is one of the transmission technologies being developed for use with B-ISDN (Broadband ISDN), which is based on ATM. The switch used to route packets in a fast packet network is termed a fast packet switch. See ATM, FAST PACKET

MULTIPLEXING, FRAME RELAY, and SMDS.

Fast Pyramid Algorithm Pyramid algorithms are used in image compression to compute the wavelet transform. The algorithm implements a complex mathematical procedure, using far fewer calculations than are nominally required, thereby yielding an approximation of the original data. The algorithm involves a series of linear filtering operations, in combination with down-sampling by two of the output. Up-sampling is employed to reconstruct a highly satisfactory approximation of the original data. See DOWN-SAMPLING, UP-SAMPLING and WAVELET TRANSFORM.

Fast Scan Receiver A cellular term. A piece of equipment that scans all 1,300 channels in an entire cellular network. It is a quick way to determine channel usage and signal strength.

Fast Select In packet switched networks, a calling method which allows the user to send a limited amount of information along with a "call req packet" rather than after the packet. A more technical explanation: An optional user facility in the virtual call service of ITU-T X.25 protocol that allows the inclusion of user data in the call request/connected and clear indication packets. An essential feature of the ITU-T X.25 (1984) protocol.

Fast Stat MUX MICOM's advanced statistical multiplexer that uses data compression, priority echoplex handling and fast packet technology to improve throughput.

Fast Switching Channel A single channel on a GPS (Global Positioning System) which rapidly samples a number of satellite ranges. "Fast" means that the switching time is sufficiently fast (2 to 5 milliseconds) to recover the data message.

Fast-20 A type of SCSI, introduced in the SCSI-3 specification, in which the data rate is quadrupled to 20 MBytes per second for narrow SCSI or 40 MBytes per second for wide SCSI. Also known as Fast-20 or Double Speed SCSI.

Fast-40 A type of SCSI in which the data rate is increased to 40 MBytes per second for narrow SCSI or 80 MBytes per second for wide SCSI. Also known as Fast-40.

FAT File Allocation Table. The FAT is an integral part of the MS-DOS and Windows operating systems. It is like a roadmap (or index) of a hard, floppy disk, magneto optical, zip drive, etc. It keeps track of where the various pieces of each file on a disk are stored. A hard disk's directory and file allocation tables are extremely important because they contain the address and mapping information the operating system needs to figure where to store and where to retrieve our precious data. If any of the data storage blocks containing these tables is damaged, it will be very hard, if not impossible to find the data on the hard disk. As a result, all operating systems keep multiple updated copies of their file allocation tables on your computer's hard disk. The two most common

FAT Client Clients are devices and software that request information. Client is a fancy name for a PC on a local area network. It used to be called a workstation. Now it is the "client" of the server. Clients come in two varieties — Fat and Thin. Here's a definition of Fat Client, courtesy of Oracle Corporation, writing in early 1994: "Since the early 1980s, users have loaded their personal computers with more and more software and data. PCs often are connected to file servers that store information. With each loaded PC costing thousands of dollars, the fat client model has a high cost per machine. Example, a PC or a Macintosh. See also CLIENT, CLIENT SERVER, CLIENT SERVER MODEL, MAINFRAME SERVER, MEDIA SERVER and THIN CLIENT.

FAT File System A FAT is a file allocation table. It is a roadmap to what's on your hard disk, your floppy, your mag-

neto optical disk, your CD — any disk associated with your PC. The FAT file system is a fancy way of saying the whole procedure the FAT uses to organize information on your disks. Various computer operating systems use different FAT file systems.

Fat Pipe A fiber-optic cable used on a network backbone for high-speed communications. It has a wide bandwidth for baseband and broadband high-capacity communications.

FATbits A computer imaging term. Extreme magnification of individual pixels to allow easy pixel-by-pixel editing of images.

Fatfingered Verb, a common term for miss dialing a phone number. Derived from blaming end users who can't dial correctly on their over sized fingers.

Fatherboard First there is the motherboard. That's the main circuit board of a computer system. The motherboard contains edge connectors or sockets so other PC (printed circuit) boards can be plugged into it. Those PC boards are called Fatherboards. Some fatherboards have pins on them into which you can plug smaller boards. Those boards are called Daughterboards. In a voice processing system, you might have a Fatherboard to do faxing. And you might have a range of Daughterboards, which allow you to connect different types of phone connections. Different boards exist for standard analog tip and ring, digital switched 56, t-1, etc.

FAU Fixed Access Unit. A fixed, wireless telephone placed in a user's home or business using cellular or PCS (Personal Communications Service). A new, lower powered, higher-frequency technology. The device provides local telephony service circumventing existing LEC (Local Exchange Carrier) transmission equipment using wired connections.

Fault A hard failure or a performance degradation so serious as to destroy the ability of a network element to function effectively. Opens, short circuits and breaks are examples of common cable faults.

Fault Current The current that can flow in a circuit as a result of a undesired short circuit.

Fault Domain A fault domain defines the boundaries of an isolating soft error on a Token Ring network. The fault domain limits the problem to two stations, their connecting cables, and any equipment (a MAU, for example) between the two stations. The two stations involved are the station reporting the error and its Nearest Active Upstream Neighbor (NAUN).

Fault Isolation The process of determining where a network problem, or fault located.

Fault Management Detects, isolates and corrects network faults. It is also one of five categories of network management defined by the ISO (International Standards Organization). See also Fault.

Fault Resilient A fault resilient computer tends to means that it must be relied on to run 99% of the time. In contrast, a fault tolerant machine must run 100% of the time, which typically means that the design must duplicate the CPU microprocessor. A fault resilient machine will typically be less expensive than a fault tolerant machine. Which you buy depends on what your needs are and the possible cost of losing transactions should your machine go down. See fault tolerant.

Fault Tolerant A method of making a computer or network system resistant to software errors and hardware problems. A fault tolerant LAN system tries to ensure that even in the event of a power failure, a disk crash or a major user error, data isn't lost and the system can keep running. In fact, the general concept is that a fault tolerant machine must be designed with sufficient duplicated parts that it can be relied upon to run 100%. Cabling systems can also be fault toler-ant, using redundant wiring so that even if a cable is cut, the system can keep running. True fault tolerance is very difficult to achieve. See fault resilient.

Faults Conditions that degrade or destroy a cable's ability to transmit data. Opens, short circuits and breaks are examples of common cable faults.

Favorite Microsoft's term for what Netscape calls a book-mark. See also Bookmark.

Fax An abbreviation for facsimile. See facsimile and facsim-ile switches.

Fax Adapter Hook one up between your printer and your phone. Bingo, your printer now becomes a fax machine.

Fax At Work Fax At Work was a subset of Microsoft's office equipment architecture called At Work which was it announced on June 9, 1993. Microsoft's idea was to put a set of software building blocks into both office machines and PC products, including:

- Desktop and network-connected printers.
- Digital monochrome and color copiers.
- Telephones and voice messaging systems.
- Fax machines and PC fax products.
- Handheld systems.
- Hybrid combinations of the above.

Microsoft Fax At Work has not been adopted by the fax and telecommunications industry, since (the story goes) Microsoft has wanted too much in the way of royalties and Fax at Work is now effectively dead. Newer fax standards have been issued by the ITU — including T.37 and T.38 — which have interested the industry far more. They cover send-ing faxes over packet switched networks, such as the Internet and corporate Intranets.

Fax Back You dial a computer using the handset of your fax machine. The distant computer answers. "What documents would you like? Here's a menu." You touchtone in 123. It says "Touch your Start button." You do. Seconds later your fax machine disgorges the document you wanted. Fax-back is the generic term for the process of ordering fax documents from remote machines. Fax-back uses a combination of fax and voice processing technology. Fax-back is also called fax on demand.

Fax Board A specialized synchronous modem for designed to transmit and receive facsimile documents. Many fax boards also allow for binary synchronous file transfer and V.22 bis communication. See also FAX SERVER.

Fax Broadcasting Automatically distributes faxes to pre-selected destinations.

Fax Data Modem See FAX MODEM.

Fax Demodulation A technique for taking a Group III fax signal and converting it back to its original 9.6 Kbps. It works like this: When a sheet of paper is inserted into a fax machine, the fax machine scans that paper into digital bits — a stream of 9600 bps. Then, for transmission over phone lines, that 9.6 Kbps is converted into an analog signal. But if you wish to transmit the fax signal over a digital line, then it makes sense to convert it back to its original 9.6 Kbps. That means you can put several fax transmissions on one 56 Kbps or 64 Kbps line — the capacity you'd normally need if you transmitted one voice conversation, or one erstwhile analog fax transmission. See FAX/DATA modem.

Fax Enhancer A standalone add-on device that connects to legacy fax machines for the purpose of boosting fax trans-mission speeds and enabling faxes to be sent over any net-work (Internet, PSTN and private networks).

Fax Jack A device that connects to a phone line with two jacks, one for a phone and one for a fax machine and one for a phone line. When a call comes in the fax jack answers the call and waits for the mechanical tone from the other fax machine. If it doesn't hear the tone, it rings the phone. The downside is that they block caller-ID (ANI) signals.

Fax Mailbox Companies can send facsimiles of docu-ments to be stored for later retrieval to a fax mailbox. Fax mailbox is like voice mail for faxes. Travelers can check their fax mailboxes and have the faxes sent to convenient locations, like a hotel front desk. You can do fax mailboxes with fax servers. You can also subscribe to a fax mailbox, as you can subscribe to a voice mailbox.

One company with a fax mailbox service is Bell Atlantic. Here's their explanation of their service, which they call FAX Mailbox. They say it gives you the freedom to leave the office and pick-up your faxes on the run. Wherever you need them, they're always as close as the nearest fax machine. With FAX Mailbox, you can control your faxes as though you were in the office. Every fax sent to you is kept strictly confidential in a private mailbox, instead of being left by the machine for any-one to see. You can access the faxes in your mailbox by call-ing from any touchtone phone and entering your personal code. Then you can review, store or delete faxes — or send them to any machine you choose. There's no need to be dependent on office support to receive and send faxes. Or have to delegate or postpone work because you're out of the office. You determine when and where you receive faxes, according to your schedule and itinerary.

Fax Mode The mode in which the fax modem is capable of sending and receiving files in a facsimile format. See FAX MODEM.

Fax Modem A combination facsimile machine/modem. A device which lets you send documents from a computer to a fax machine. It comes in many shapes and sizes. It may come as a card which you slip into a vacant slot in your desktop PC (called an internal fax/modem). It may come as a PCMCIA card which you slip into your laptop. It may come as a small box which you connect by a cable to your computer's serial port. It may also come as a small self-contained package about the size of a cigarette package. The technology of "fax modems" is changing radically. Originally they contained ded-icated fax/modem chipsets, i.e. microprocessors designed as fax modems and good for nothing else. Increasingly, fax modems are now coming with powerful, general purpose dig-ital signal processors (DSPs), instead of dedicated fax modem chipsets. These DSP devices become fax modems when you load the appropriate software. When you load other software they can also become the equivalent of sound blaster cards, or become a Microsoft Sound System, etc.

There are big advantages to sending faxes from a fax modem, as compared to sending it from a fax machine. First, faxes sent are cleaner because they're not scanned but computer generated. Second, sending faxes directly from your comput-er is faster than printing the document, then sliding it in a fax machine, dialing and sending it. Third, a fax modem is typi-cally cheaper than a fax machine. Fourth, because a fax modem uses computer software it may have some neat fea-tures, like the ability to send faxes when phone costs are low, like running the fax software in the background while you're doing something else.

There are two main disadvantages:

1. Viewing incoming faxes on your PC's screen is not easy. A PC screen is horizontal. Most faxes are vertical. It's also not

easy to translate an incoming fax into the pixels on your computer screen. Here's why:

FAX ENCODING

Standard, Group III	203 x 98 pixels
Fine, Group III	.203 x 196
Superfine, Group III	203 x 391
Standard, Group IV	400 x 400

PC SCREEN ENCODING

CGA.	320 x 200 pixels
Enhanced CGA	640 x 400
EGA	640 x 350
Hercules	720 x 348
VGA	.640 x 480
Super VGA	.800 x 600
8514/A (also called XGA)	1,024 x 768

2. For these two reasons, you may still have to print your incoming faxes.

3. Keeping faxes on your hard disk is also pretty consuming of hard disk space. A typical one page fax can easily use between 40,000 and 50,000 bytes. Twenty pages and you've used up a megabyte.
See also FAX DEMODULATION, FAX SERVER AND FAX SWITCH.

Fax On Demand You dial a computer using the handset of your fax machine. The distant computer answers. "What documents would you like? Here's a menu." You touchtone in 123. It says "Touch your Start button." You do. Seconds later your fax machine disgorges the document you wanted. Fax on demand is one term for the process of ordering fax documents from remote machines. Fax on demand uses a combination of fax and voice processing technology. Fax on demand is also called faxback. See FAX SERVER for a more complete explanation.

Fax Publishing Fax publishing allows a caller to have electronically stored information automatically faxed to them via a touchtone telephone. By pressing touchtone keys, callers can have timely information, including product brochures, business forms and benefits information, automatically faxed to them anytime, anywhere. See also FAX SERVER.

Fax Server A fax server sits on a local area network and literally serves faxes to those people using it. Those people may be on the LAN physically, i.e. joined by wires to the server. Or they may be outside, reaching the LAN over phone lines. Basically anything that a live person can do with a fax machine, a fax server can do. It can receive faxes and distribute them to people they're addressed to. It can send faxes for people who are typically sending those faxes from their PCs. It can send faxes to people who call it on the phone and request certain faxes — those stored on its hard disk. A fax is typically a relatively high-powered computer which has one or more PC fax boards in its slots. It can receive faxes. It can send faxes. It can store faxes. It can forward them. If it doesn't know for whom the faxes are meant, it may send the faxes to a printer or alert a supervisor to manually check the incoming faxes and distribute them — electronically or on paper. The fax server also accepts faxes from PCs on the LAN, stores them and gets them ready for sending out over phone lines. It might send the faxes immediately or wait until later, when phone calls are cheaper. It might send the same fax to thousands of people. It might send a personalized fax to thousands of people, grabbing the names from a database on it or on another computer. A fax server can also be an interactive voice response system which you call. When you call it, it answers, reads you a menu of options — including various documents it can send you. You choose which documents you want by

touchtoning in numbers. Then you designate to which fax machine you want the documents sent. The fax machine you designate might be the one you're calling from (i.e. you dialed using your fax machine's handset).
There are two types of interactive voice response fax servers. One is a one-call machine. The caller calls from his own fax machine. When he's chosen his faxes and he's ready to receive a fax, he simply hits the "Start" button on his fax machine and his machine receives the chosen faxes. There is also a two-call machine. The caller will call from a phone and touchtone in the phone number of a fax machine he wants the fax of his desired documents sent. One-call IVR fax servers are the newer breed, harder to build than the older two-call machines. There are obvious advantages to both. The one call machine — in which the user pays the phone bill — will, I suspect, become the more popular type. See also FAX PUBLISHING and other FAX definitions.

Fax Switch A device which allows you to share one phone line with a fax machine, a phone and a modem. Here's how it works. A call comes in. The device answers the call. The switch listens for the distinctive CNG (Calling) tone which a calling fax machine emits (the "cry" of the fax machine). When it hears this sound, it switches the call to the fax machine. If it doesn't and hears nothing (or at least nothing it can recognize) it switches the call to the phone. If it hears some touchtones — e.g. 44, or *6 — it will switch the call to the modem (and therefore the attached computer) or whatever other device you've designated, including a modem-equipped cash register, etc. Some fax switches allow you to have a data conversation with one device (the cash register), then switch to another device (the second cash register) and another, etc. — all on the one conversation.
The advantage of a fax switch is that it saves having to buy several phone lines. Phone lines are expensive compared to fax switches. There are disadvantages to a fax switch — it typically must hear an incoming CNG tone to switch the call to the fax machine. This means if your friend wanting to send you a fax is dialing manually (i.e. not letting his fax machine do it), your fax switch may not ever send the call to your fax machine. Also you have to set up to dial those extra digits for your distant computer to "dial through" your fax switch. And finally, some fax switches don't send the "right" ringing signal to their attached devices. Some 9,600 baud and 14,400 baud modems, for example, are very sensitive and won't answer certain fax switches' ringing signals, especially if the fax switch's ringing signal is a DC square wave, not an AC sine wave. All this can be solved, however, with intelligence, checking and proper programming. I use a fax modem switch every day. It saves me money and is convenient. See CNG.

Fax Waiting Service The name of a Bell Atlantic service. FAX WAITING Service is like Call Waiting for your fax machine. If your fax line is busy, a second incoming fax is electronically stored. When your machine is clear, FAX Waiting service sends it through. So customers, prospects and suppliers can get their faxes through on the first call — without the frustration of busy signals. And employees can use your fax machine without interrupting your important fax communications. In fact, FAX Waiting is the next best thing to a second fax machine. A great idea.

FaxBios The FaxBios Association is an organization of fax printed circuit card manufacturers who have formed an association in order to promulgate a standard applications programming interface (API) which they are calling FaxBios. Phone 801-225-1850; 2625 Alcatraz Avenue, Berkeley CA 94705.

Faxed The past tense of the new verb "to fax," as in "I faxed the document to him."

FB Framing bit.

FBT Fused Biconic Tape.

FBU Functional Business Unit. A fancy name for a group of workers inside a company. An FBU might be your sales department, your accounting department, etc.

FBus Frame Transport Bus.

FC 1. Fiber optic Connector (developed by NTT).
2. Frame Control. On Token Ring networks, this data supplies the frame type.
3. Feedback Control: Feedback controls are defined as the set of actions taken by the network and by the end-systems to regulate the traffic submitted on ATM connections according to the state of network elements.
4. Fibre Channel.

FC and PC Face Contact and Point Contact. Designations for fiber optic connectors designed by Nippon Telegraph and Telephone which feature a movable anti-rotation key allowing good repeatable performance despite numerous matings.

FC-0 Lowest level of the Fibre Channel Physical standard, covering the physical characteristics of the interface and media.

FC-1 Middle level of the FC-PH standard, defining the 8B/10B encoding/decoding and transmission protocol.

FC-2 Highest level of FC-PH, defining the rules for signaling protocol and describing transfer of the frame, sequence, and exchanges.

FC-3 The hierarchical level in the Fibre Channel standard that provides common services, such as striping definition.

FC-4 The hierarchical level in the Fibre Channel standard that specifies the mapping of upper-layer protocols (ULPs) to levels below.

FC-AL Fibre Channel-Arbitrated Loop

FC-EL Fibre Channel-Enhanced Loop

FC-PH Fibre Channel Physical standard, consisting of the three lower levels, FC-0, FC-1, and FC-2.

FCB The abbreviation for File Control Block. FCBs are used by older MS-DOS application programs to create, open, delete, read, and write files. One FCB is set up for each file you open.

FCC Federal Communications Commission. See FEDERAL COMMUNICATIONS COMMISSION.

FCC Registration Number A number assigned to specific telephone equipment registered with the FCC, as set forth in FCC docket 19528, part 68. The presence of this number affixed to a device indicates that the FCC has approved it as being a compatible device for direct connection to telephone line facilities.

FCC Tariff #9 The FCC tariff for private line services including Accunet T-1.5, DDS, Voice Grade circuits, and Accunet T45.

FCC Tariff #11 AT&T's tariff file at the FCC for local private line services.

FCC Tariff #12 AT&T's tariff filed at the FCC tariff for custom-designed integrated services. A special tariff that allows AT&T to develop custom network solutions, including allowing customers to install their networking multiplexers in AT&T central offices and letting AT&T manage the network.

FCC Tariff #15 AT&T's FCC tariff filed at the FCC that allows AT&T to lower rates after all bids are placed to be competitive with other carriers.

FCFS A silly abbreviation for First Come First Served. See FIFO (First In, First Out).

FCKT Facility Circuit ID.

FCN Abbreviation for Function. This button enables your cellular phone or fax machine or other telecom device to access special features, like switching from one cellular phone company to another. See also DUAL NAM.

FCOS Fully programmable classes of service that control user (Feature Class of Service) access to mailbox features, operations and options. Feature Classes of Service (FCOS) are entirely independent of Limits Classes of Service (LCOS).

FCOT See Fiber Control Office Terminal.

FCS 1. Frame Check Sequence. Any mathematical formula which derives a numeric value based on the bit pattern of a transmitted block of information and uses that value at the receiving end to determine the existence of any transmission errors. In bit-oriented protocols, a frame check sequence is typically a 16-bit field that contains transmission error checking information, usually appended to the end of the frame. See FRAME CHECK ERROR and FRAME CHECK SEQUENCE.
2. Federation of Communications Services.
3. An MCI term for Fraud Control System.

FCS Error A Frame Check Sequence error occurs when a packet is involved in a collision or is corrupted by noise.

FDCCH Forward Digital Control CHannel. A digital cellular term defined by IS-136, which addresses cellular standards for networks employing TDMA (Time Division Multiple Access). The FDCCH includes all signaling and control information passed downstream from the cell site to the user terminal equipment. The FDCCH acts in conjunction with the RDCCH (Reverse Digital Control CHannel), which includes all such information sent upstream from the user terminal equipment to the cell site. The FDCCH includes the BCCH, SCF and SPACH. See also BCCH, IS-136, SCF, SPACH and TDMA.

FDD Floppy Disk Drive. A Hard Disk Drive is a HDD.

FDDI Fiber Distributed Data Interface. FDDI is a 100 million bits per second fiber optic LAN. It is an ANSI standard. It uses a "counter-rotating" token ring topology. FDDI is ANSI X3T12 standard for a dual-ring LAN operating at 100 Mbps and using token passing; FDDI rings may use up to 200 km of optical fiber, or may employ twisted copper pairs for short hops. FDDI is compatible with the standards for the physical layer of the OSI model. An FDDI LAN is often known as a "backbone" LAN. It is used to join file servers together and to join LANs together. The theoretical limit of Ethernet, measured in 64 byte packets, is 14,800 packets per second (PPS). By comparison, Token Ring is 30,000 and FDDI is 170,000 pps. See FDDI TERMS and FDDI-II.

FDDI Follow-On LAN A faster FDDI. Said to operate at up to 2.4 gigabits per second.

FDDI Terms DAC Dual Attachment Concentrator DAS Dual Attachment Station ECF Echo Frames ESF Extended Service Frames LER Link Error Rate LLC Logical Link Control MAC Media Access Control MIC Media Interface Connector NIF Neighborhood Information Frame NSA Next Station Addressing PDU Protocol Data Unit PHY Physical Protocol PMD Physical Media Department PMF Parameter Management Frames RAF Resource Allocation Frames RDF Request Denied Frames SAC Single Attachment Concentrator SAS Single Attachment Station SDU Service Data Unit SIF Station Information Frames SMT Station Management SRF Status Report Frame THT Token Holding Timer TRT Token Rotation Timer TTRT Target Token Rotation Timer TVX Valid Transmission Timer UNA Upstream Neighbor Address

FDDI-II Fiber Distributed Data Interface-II is a recently standardized enhancement to FDDI. It still runs at 100 million bits per second on fiber or on twisted copper pairs, but in addition to transporting conventional packet data like other LANs, FDDI-II allows portions of the 100 Mbps bandwidth to carry low delay, constant bit rate, isochronous data like 64 Kbps telephone channels. This means the same LAN that carries computer packet data can carry live voice or live video calls. Some additional terms used with FDDI-II are: I-MAC which stands for Isochronous Media Access Control; P-MAC which stands for Packet Media Access Control; and WBC which stands for Wide Band Channel. See FDDI, FDDI TERMS, ISOCHRONOUS and ISOETHERNET.

FDL Facilities Data Link. A T-1 term, specifically relating to Extended SuperFrame (ESF). ESF extends the superframe from 12 to 24 consecutive and repetitive frames of information. The framing overhead of 8 Kbps in previous T-1 versions was used exclusively for purposes of synchronization. ESF takes advantage of newer channel banks and multiplexers which can accomplish this process of synchronization using only 2 Kbps of the framing bits, with the framing bit of only every fourth frame being used for this purpose. As a result, 6 Kbps is freed up for other purposes. This allows 2 Kbps to be used for continuous error checking using a CRC-6 (Cyclic Redundancy Check-6), and 4 Kbps to be used for a FDL which supports the communication of various network information in the form of in-service monitoring and diagnostics. ESF, through the FDL, supports non-intrusive signaling and control, thereby offering the user "clear channel" communications of a full 64 Kbps per channel, rather than the 56 Kbps offered through older versions as a result of "bit robbing." See also ESF and T-1. The FDL is embedded in the framing bits, using half the bits or 4000 bit/s. It is over this channel that two schemes operate:

1. In the original Bell System scheme, the repair station in the CO queries the CSU at the customer site, which responds with error statistics for the last 24 hours (in 15-min increments). The repairman uses this info to diagnose line condition.

2. The more modern ANSI method has the CSU broadcast the error statistics for the last three seconds, every second (with overlap). Automatic monitoring equipment in the CO can tell when the line is going bad. Both systems can co-exist and operate on the same link, but that's unlikely in reality.

FDM Frequency Division Multiplexing. A technique in which the available transmission bandwidth of a circuit is divided by frequency into narrower bands, each used for a separate voice or data transmission channel. This means you can carry many conversations on one circuit. The conversations are separated by "guard channels." At one point, FDM was the most used method of multiplexing long haul conversations when they were transmitted in analog microwave signals. No more. Fiber optic transmission (today's preferred method) uses TDM — Time Division Multiplexing.

FDMA Frequency Division Multiple Access. One of several technologies used to separate multiple transmissions over a finite frequency allocation. FDMA refers to the method of allocating a discrete amount of frequency bandwidth to each user to permit many simultaneous conversations. In cellular telephony, for example, each caller occupies approximately 25 kHz of frequency spectrum. The cellular telephone frequency band, allocated from 824 MHz to 849 MHz and 869 MHz, consists of 416 total channels, or frequency slots, available for conversations. Within each cell, approximately 48 channels are available for mobile users. Different channels are allocated for neighboring cell sites, allowing for re-use of frequencies with a minimum of interference. This technique of assigning individual frequency slots, and re-using these fre-

quency slots throughout the system, is known as FDMA. See CDMA, TDMA.

FDP Fiber Optic Distribution Panel.

FDS Frequency Division Switching. Seldom used for voice switching. Primarily used for radio and TV broadcasting.

FDX See FULL DUPLEX.

FE Extended Framing ("F sub E"). An old name for ESF, also known as Extended SuperFrame, a T-1 carrier framing format that provides a 64 Kbps clear channel, error checking, 16 state signaling and some other nice data transmission features.

FE D4 Superframe Extended Another designation for AT&T's ESF (Extended Super Frame).

Feather An imaging term. An effect in which the edges of a pasted selection or paint tool fade progressively at the edges for a seamless blend with the background.

Feature Buttons Think of a feature button on a telephone as a collection of numbers stored in a bin. When you hit the button, the bin quickly disgorges all the numbers one after another. Feature buttons are fast ways of doing things. You have a feature button labelled "Conference." Hit the button, set up a conference call. Without a feature button, you'd probably have to hit the switch hook and some numbers on your touchtone pad. In computer terms, a feature button on a phone is the same as a macro — an easy way of doing something. On most phones with feature buttons, the feature buttons are "programmable." This means you can assign different features to different buttons, i.e. the ones you want. For example, I always assign "Last Number Redial," "Saved Number Redial" and "Conference Call" to the buttons of any phone I'm programming. Some phones have many feature buttons. Some don't.

Feature Cartridge A replaceable software cartridge containing software features. The Feature Cartridge is inserted into the central cabinet, or Key Service Unit (if it's a key system). Several small phone systems (under 100 lines) use cartridges to upgrade their software. The manufacturers find cartridges are cheaper than equipping their phone systems with a floppy drive and the associated electronics.

Feature Code This is a number that is used to activate a particular feature on a phone system.

Feature Creep Occupational hazard The enemy of the good is the better. A term to show how features tend to get added to telecom equipment as time passes and new models appear. The term "feature creep" makes no judgments about whether the new features are actually useful. In book called "Startup; A Silicon Valley Adventure Story," Jerry Kaplan, the author, describes "Feature creep as "the irresistible temptation for engineers to load a product down with their favorite special features."

Feature Function Testing Feature/function testing is designed to assure that everything a system is supposed to do is done correctly, e.g. calls are switched to the correct destination, messages are left and deleted, billing records collected accurately, and so on. Feature/function testing is the most detailed portion of the test process. The people who perform functional testing must be extremely detail oriented and have the discipline to test every feature to their written functional requirement. No function of the system should be overlooked. Definition courtesy Steve Gladstone, from his book "Testing Computer Telephony Systems."

Feature Group A, B, C, D FGA, FGB, FGC, FGD, are four separate switching arrangements available from local exchange carrier (LEC) end central offices to interexchange (long distance) carriers. These switching arrangements allow the LECs' end-users to make toll calls via their favorite long distance carrier. Feature groups are described in a tariff filed by the National Exchange Carrier Association with the FCC. The feature group used by each IX (IntereXchange, also called long distance) carrier together with any special access surcharge determines the service they can provide their customers and the carrier common line access fee they will pay to the local exchange carrier involved. The most common Feature Group now is D. See the next four definitions. See FEATURE GROUP A, FEATURE GROUP B, FEATURE GROUP C, FEATURE GROUP D

Feature Group A Offers access to the local exchange carrier's network through a subscriber-type line connection rather than a trunk. It is a continuation of the ENFIA arrangement used in the early days of OCCs, until equal access using an access tandem central office is available. Remember, without equal access the IX carrier had to require its customers to dial a local number to reach their long distance facilities, then dial an identification number, then dial long distance numbers of the called party desired. This service handicap, compared to AT&T's superior connections, qualifies the OCC for a discount off the FGA rate until access is equal. The IX carrier is billed by the LEC based upon actual monthly use rather than the ENFIA method of projected "minutes of use" rate.

Feature Group B Is similar to FGA, but provides a higher quality trunk line connection from end CO to the IX carrier's facilities, instead of the subscriber-type line. The IX customer can originate a call from anywhere within the LATA, while FGA requires customers to initiate the call from within the local exchange of the exchange carrier connecting to the IXC. FGB billings to the IX are on a flat usage basis, and a discount is applicable.

Feature Group C Is the traditional toll service arrangement offered by LECs to AT&T prior to breakup of the Bell System. Quality is superior, and the service includes automatic number identification of the calling party, answerback, and disconnection supervision, and the subscribers can use either a dial or touchtone pad. This FGC service is offered only to AT&T without a discount.

Feature Group D FGD. The class of service associated with equal access arrangements. All facilities-based IXCs (IntereXchange Carriers) and resellers of significance pay extra for Feature Group D terminations (connections), which is a trunk-side connection provided by the ILECs (Incumbent Local Exchange Carriers). Feature Group D is required for equal access, which allows phone users in the United States to pick up the telephone and dial 1+ to place a long distance call, with the call being handled by the IXC they have preselected. Without FGD, the user must first dial a 7- or 10-digit number, a calling card number and PIN number, and then the desired telephone number. FGD also is required for an end user organization desiring ANI (Automatic Number Identification) information. See also 1+, 101-XXXX, ANI, Equal Access, FGD, ILEC and IXC.

Feature Keys Same as FEATURE BUTTONS. A key is to a telephone man what a switch is to an electrical man.

Feature Phone A generic name for a telephone that has extra features (often speed dial buttons) designed to simplify and speed making and receiving phone calls.

Feature/Function Testing Feature/function testing is designed to assure that everything a system is supposed to do is done correctly, e.g. calls are switched to the correct destination, messages are left and deleted, billing records collected accurately, and so on. Feature/function testing is the

most detailed portion of the test process. The people who perform functional testing must be extremely detail oriented and have the discipline to test every feature to their written functional requirement. No function of the system should be overlooked. Definition courtesy Steve Gladstone, from his book "Testing Computer Telephony Systems."

FEBE Far End Block Error: A maintenance signal transmitted in the PHY overhead that a bit error(s) has been detected at the PHY layer at the far end of the link. This is used to monitor bit error performance of the link.

FEC Forward Error Correction. A technique used by a receiver for correcting errors incurred in transmission over a communications channel without requiring retransmission of any information by the transmitter. Typically involves a convolution of the transmitter using a common algorithm and embedding sufficient redundant information in the data block to allow the receiver to correct. While this technique is processor-intensive, it improves the efficiency with which the network is used. See FORWARD ERROR CORRECTION.

FECN Forward Explicit Congestion Notification. A Frame Relay term. This bit contained within the Address Field notifies the receiving device that the network is experiencing congestion. Thereby, the target device is advised that frames may be delayed, discarded, or damaged in transit. It is the responsibility of the target device to adjust to that condition. In conjunction with BECN, devices in both the forward and backward directions are advised. See BECN.

FED Field Emission Display. A new way of making TV and computer screen displays. FED screens are flat and potentially cheap. Like conventional glass screens, they emit light. LCDs, by comparison, don't. A typical FED screen packs millions of tiny individual emitters between two ultra-thin glass layers. Each emitter fires electrons simultaneously across a minuscule vacuum gap onto a phosphor coating very much like a CRT's. See also FIELD EMISSION DISPLAYS.

FED-STD A system of standards numbered FED-STD-1001 to 1008 which set modulation specifications for data transmission.

Federal Communications Commission FCC. The federal organization in Washington D.C. set up by the Communications Act of 1934. It has the authority to regulate all interstate (but not intrastate) communications originating in the United States. The FCC is the U.S. federal regulatory agency responsible for the regulation of interstate and international communications by radio, television, wire, satellite and cable. Established by the Communications Act of 1934, it is responsible directly to Congress and is directed by five Commissioners appointed by the President and confirmed by the Senate for 5-year terms. The President designates one of the Commissioners to serve as Chairman. The Chairman's tenure is at the pleasure of the President. No more than three Commissioners may be members of the same political party. None can have a financial interest in any Commission-related business.

Stripped of all the extensive regulatory and legal mumbo jumbo, the FCC does three things: 1. It sets the prices for interstate phone, data and video service. 2. It determines who can or cannot get into the business of providing telecommunications service or equipment in the United States. 3. It determines the electrical and physical standards for telecommunications equipment and services. The FCC's powers, although strong, are tempered (limited) by the Federal Courts. Anyone who disagrees with FCC rulings can appeal them to a Federal Court. The FCC's power and rulings are also affected by the Justice Department (The Justice Department changed the industry with Divestiture), Congress and The 50 state public service commissions. The FCC changed with the passage of the Telecommunications Act of 1996 — the first telecom act passed by Congress since the Communications Act of 1934.

How is the FCC organized? Most items considered by the Commission are developed by one of seven operating bureaus and offices organized by substantive area:

• The Common Carrier Bureau handles domestic wireline telephony.

• The Mass Media Bureau regulates television and radio broadcasts.

• The Wireless Bureau oversees wireless services such as private radio, cellular telephone, personal communications service (PCS), and pagers.

• The Cable Services Bureau regulates cable television and related services.

• The International Bureau regulates international and satellite communications.

• The Compliance & Information Bureau investigates violations and answers questions.

• The Office of Engineering & Technology evaluates technologies and equipment.

In addition, the FCC includes the following other offices:

• The Office of Plans and Policy develops and analyzes policy proposals.

• The Office of the General Counsel reviews legal issues and defends FCC actions in court.

• The Office of the Secretary oversees the filing of documents in FCC proceedings.

• The Office of Public Affairs distributes information to the public and the media.

• The Office of the Managing Director manages the internal administration of the FCC.

• The Office of Legislative and Intergovernmental Affairs coordinates FCC activities with other branches of government.

• The Office of the Inspector General reviews FCC activities.

• The Office of Communications Business Opportunities provides assistance to small businesses in the communications industry.

• The Office of Administrative Law Judges adjudicates disputes.

• The Office of Workplace Diversity ensures equal employment opportunities within the FCC.

Federal Information Processing Standards FIPS. The identifier attached to standards developed to support the U.S. government computer standardization program. The FIPS effort is carried out by the U.S. Department of Commerce, Springfield, VA. See the next definition.

Federal Telecommunications Standards Committee FTSC. A U.S. government agency established in 1973 to promote standardization of communications and network interfaces. FTSC standards are identified by the designator FED-STD. The FTSC's address is General Services Administration, Specification Service Administration, Bldg 197, Washington Navy Yard, Washington DC 20407.

Federal Telecommunications System FTS. The private network used primarily by the civilian agencies of the federal government to call other government locations and to place calls to phones connected to the public network.

Federal-State Joint Board An organization with representatives from the FCC and the state public service commissions which tries to resolve Federal and State conflicts on telecommunications regulatory issues. Sometimes successfully and sometimes not successfully.

Feed A television signal source.

Feedback The return of part of an output signal back to the input side of the device. Think of the high-pitched squeal you hear when someone brings a microphone too close to the loudspeaker. Not all feedback is as obvious or as irritating. Some feedback is good. See SIDETONE, which is what happens when you hear a little in the receiver of you're saying in the transmitter of a phone.

Feeder Cable A group of wires, usually 25-pair or multiples of 25-pair, that supports multiple phones in a single cable sheath. These cables may or may not be terminated with a connector on one or both ends. Feeder cable typically connects an intermediate distribution frame (IDF) to a main distribution frame (MDF). But the term "feeder cable" is also used in backbone wiring. And Bellcore defines the term slightly differently: A large pair-size loop cable emanating from a central office and usually placed in an underground conduit system with access available at periodically place manholes.

Feedholes Holes punched in paper or papertape which allow the paper or paper tape to be driven by sprocket wheels.

Feedware Software designed to get demand for a product or a new market segment started. Feedware is typically a less-full featured piece of software than the software you're really trying to sell. Feedware typically costs very little. It may even be free. See also SEEDWARE.

FEFO First Ended, First Out. A rule for dealing with things in a queue. For example, higher priority messages will be sent before lower priority messages.

Female Amp Connector Also called a C Connector or 25-pair female connector. The male version is called a P connector.

Femtosecond One-millionth of a billionth of a second. Femtoseconds are used in laser transmission and in other measures of very small happenings. It's 10 to the minus 15. There are as many femtoseconds in one second as there are seconds in thirty million years. There are 1,000,000,000,000,000 femtoseconds in one second. How small is a femtosecond? In a little more than a second, light can travel from the moon to the earth, but in a femtosecond it only travels one hundredth the width of a human hair.

FEP 1. Front End Processor. The "traffic cop" of the mainframe data communications world. Typically sits in front of a mainframe computer and is designed to handle the telecommunications burden, so the mainframe computer can concentrate on handling the processing burden. Here's a more technical definition: A dedicated communications system that intercepts and handles activity for the host. Can perform line control, message handling, code conversion, error control, and such applications functions as control and operation of special-purpose terminals. Designed to offload from the host computer all or most of its data communications functions. Front end processors are not used in the client/server world. 2. Fluorinated Ethylene Propylene. Also known by the trade name Teflon, a registered trademark of Dupont. FEP is the insulation of choice for high performance cable and wire systems installed in return air plenums. As FEP is really slick, it makes the wire really easy to pull through conduits, around corners, and so on — the same property that makes it so wonderful in the kitchen. FEP's fire retardant properties have led many countries to require its use, particularly in plenum ceilings. See plenum.

FER Frame Error Rate. A computation determined based on the number of frames received with errors compared to frames received without errors.

FERF Far-End Remote Failure. An alarm indicating a failure at the far end of an ATM network, identifying the specific circuit in a failure condition.

Fernsprechvermittlungsstelle A central office in German. In Europe, they call a central office a "public exchange," or just plain "exchange." They look at you kinda strange when you say the North American word, namely "central office."

Ferreed Assembly A glass enclosed reed relay switch in which the reeds are made of some metal which can be opened or closed by an external magnetic field.

Ferri Chrome A coating used on tape comprising a layer of ferric oxide particles and a layer of chromium dioxide particles and combining the attributes of both.

Ferric Oxide A coating used on tape comprised of red iron oxide, the original material used for magnetic recording tapes.

Ferrite A type of ceramic material having magnetic properties and consisting of a crystalline structure of ferric oxide and one or more metallic oxides, such as those of nickel or zinc. See BARIUM FERRITE HARD FERRITE and SOFT FERRITE.

Ferrosesonant Transformer A special transformer which puts out regulated AC voltage even when the input voltage is variable. A ferroresonant transformer may be used by itself to correct brownouts or it may be built into a UPS. A ferroresonant transformer has an undesirable characteristic called "high output impedance" which can prevent protective devices such as circuit breakers on equipment plugged into it from functioning, resulting is possible safety hazard. Another problem is that computer loads applied to ferro based line conditioners or UPS systems cause the voltage waveform applied to the computer to be very distorted, which may result in undervoltage conditions within the computer. Ferro based UPS systems are becoming obsolete because they can become unstable and oscillate when supplying modern power factor corrected power supplies. This definition courtesy APC.

Ferrule A component of a fiber optic connection that holds a fiber in place and aids in its alignment.

FET Field Effect Transistor. Very thin and small transistors are used to control pixels in a TFT (Thin Film Transistor) display.

FEXT Far-End CrossTalk. A type of crosstalk which occurs when signals on one twisted pair are coupled to another pair as they arrive at the far end of multi-pair cable system. FEXT is an issue on short loops supporting high-bandwidth services such as VDSL (Very-high-bit-rate Digital Subscriber Line), given the relatively high carrier frequencies involved. Services such as ADSL and HDSL, are not affected to the same extent, as the loops are longer and as such interference tends to be attenuated (weakened) on longer loops. See also CROSSTALK. Compare with NEXT.

Fewer A smaller number. The word "fewer" is always confused with the word "less." According to the Oxford American Dictionary, the word "less" s used of things that are measured by amount (for example, eat less butter, use less fuel). Its use with things measured by numbers is regarded as incorrect (for example in "we need less workers"; correct usage is "fewer workers").

FF Form Feed. A printer function used to skip to the top of the next page or form.

FFDI Fast Fiber Data Interface. A proprietary 100 megabit per second local area network that uses fiber optic, coax, shielded twisted pair or unshielded twisted pair. It is manufactured by PlusNet, Phoenix, Arizona.

FFOL Fiber Follow On LAN. Emerging LAN technology.

FFT 1. Fast File Transfer. An ISDN term referring to the fact that file transfers can be accomplished "fast." Reason #1: Two B channels at 64 Kbps each and a D channel at up to 16 Kbps are available to be bonded to provide as much as 144 Kbps. Reason #2: Data transfer is accomplished in an "optimistic" streaming mode, rather than a "pessimistic" packet mode. Therefore, there is no delay associated with acknowledgements. This is possible due to the excellent level of error performance inherent in digital services. The end result is FFT.
2. Fast Fourier Transform. A signal processing term for a common computer implementation of Fourier Transforms. The FFT, as a practical implementation, will always result in a finite series of sine and cosine waves as an extremely close approximation of the possibly infinite series described by the purely mathematical application of the Fourier Transform. See FOURIER'S THEOREM.

FG An ATM term. Functional Group: A collection of functions related in such a way that they will be provided by a single logical component. Examples include the Route Server Functional Group (RSFG), the IASG (Internetwork Address Sub-Group), Coordination Functional Group (CFG), the Edge Device Functional Group (EDFG) and the ATM attached host Behavior Functional Group (AHFG).

FGB See FEATURE GROUP B.

FGC See FEATURE GROUP C.

FGC-EA See FEATURE GROUP C and EQUAL ACCESS.

FGD See FEATURE GROUP D.

FGD-EA Feature Group D - Equal Access. See FEATURE GROUP D, also See EQUAL ACCESS.

FHSS Frequency Hopping Spread Spectrum. A technique used in spread spectrum radio transmission systems, such as Wireless LANs and some PCS cellular systems. FHSS involves the conversion of a datastream into a stream of packets, each of which is prepended (prepend means added to the front of) by an ID contained in the packet header. Short bursts of packets then are transmitted over a range of 75 or more frequencies, with the transmitter and receiver hopping from one frequency to another in a carefully choreographed "hop sequence." FCC regulations specify that each transmission can dwell on a particular frequency no more than 400 milliseconds. A large number of other transmissions also may share the same range of frequencies at the same time, with each using a different hop sequence. The potential remains, however, for the overlapping of packets. The receiving device is able to distinguish each packet in a packet stream by reading the various IDs, treating competing signals as noise. In a wireless LAN environment, DSSS typically operates in the 2.4 GHz frequency band, which is one of the ISM (Industrial Scientific Medical) bands defined by the FCC for unlicensed use. Although most manufacturers prefer FHSS, Direct Sequence Spread Spectrum (DSSS) is used by some. See also CDMA, DSSS, ISM and SPREAD SPECTRUM.

FIAT Fix It Again Tony.

Fiber A shortened way of saying "fiber optic." Fiber is made of very pure glass. In Bill Gates' book called "the Road Ahead," he says that optical fiber is so clear and pure that if you looked through a wall of it 70 miles thick, you'd be able to see a candle burning on the other side. Digital signals, in the form of modulated light, travel on strands of fiber for long distances. The big advantage that fiber has over copper is that it can carry far, far more information over much, much longer distances. The short history of fiber optics for communications is that scientists keep discovering more and more ways of putting more and more information down one single strand of fiber. Based on my own personal researches, no one has any idea what the eventual capacity limit of a strand of fiber optic might be. I have personally asked many scientists (including one Nobel Physics prize winner) and all seem to think there must be a theoretical limit. But they don't know what it is. And they believe we have many, many years of breakthroughs in fiber still to go. As of the time of this writing, SONET OC-192 (Synchronous Optical NETwork Optical Carrier Level 192) systems are being deployed fairly routinely by a number of major long distance carriers. Each OC-192 strand supports approximately 10 Gbps. With DWDM (Dense Wavelength Division Multiplexing), as many as 32 "windows," or wavelengths of light, can be overlaid into a single strand at OC-912, yielding a total of approximately 320 Gbps. Fiber is the American spelling. The spelling in England, Europe, Canada, Australia and New Zealand is fibre. See also the following definitions beginning with fiber. See Optical Fiber for an essay on the advantages of fiber as a communications medium. See also OC-192 and SONET.

Fiber Axis In an optical fiber, the line connecting the centers of line circles that circumscribe the core, as defined under "tolerance field."

Fiber Buffer The material surrounding and immediately adjacent to an optical fiber that provides mechanical isolation and protection. Buffers are generally softer than jackets.

Fiber Bundle An assembly of parallel unbuffered optical fibers, in intimate contact with one another and secured, usually with an epoxy or other adhesive. Each endface of the bundle is typically finished to a flat or other optical surface, usually at right angles to the axis of the bundle. Such bundles are used to transmit optical power or images. Bundles used to transmit images must maintain spatial coherence amongst the relative positions of the respective fibers at each end (aligned bundles). There is no requirement for this if the bundle is used to transmit optical power only. Fiber bundles were employed in early, short-distance communication applications, but have become obsolete in modern telecommunications.

Fiber Channel There is no such thing as Fiber Channel. See Fibre Channel.

Fiber Control Office Terminal FCOT is a generic term for a fiber terminal that can be configured for full digital, full analog or mixed communications.

Fiber Distributed Data Interface FDDI. A set of ANSI/ISO standards that, when taken together, define a 100 million bits per second (Mbps), timed-token protocol, Local Area Network that uses fiber optic cable as the transmission medium. The standards define Physical Layer Medium Dependent, Physical Layer, Media Access Control, and Station Management entities. The standard specifies: multi-mode fiber, 50/125, 62.5/125, or 85/125 core-cladding specification; and LED or laser light source; and 2 kilometers for unrepeatered data transmission at 40 million bits per second (Mbps).

Fiber Exhaust A fiber optic term. Fiber exhaust comprises the noxious light emissions from a fiber optic engine, the device which generates the light signals signals in an optical fiber transmission system. Much as an internal combustion engine creates noxious emissions, fiber optic engines (e.g., LEDs and Laser Diodes) create noxious light emissions, which can be extremely hazardous to your health. Actually, the preceding is a joke, like "frequency grease," which is used to overcome static noise in a radio system. Now for the truth: Fiber exhaust simply means that the capacity of a fiber optic transmission has been exhausted. The solution is 1) to lay

more fiber, 2) to increase the speed of the system through an upgrade of light sources and detectors, or 3) to use WDM (Wavelength Division Multiplexing) or DWDM (Dense WDM) to increase capacity through the support of multiple wavelengths of light. See also Bucket o' Dial Tone, DWDM, Frequency Grease, SONET and WDM.

Fiber Identifier A test instrument that can differentiate between live and dead fibers in a working cable and can identify a preselected fiber to which a special transmitter has been attached.

Fiber Loss The attenuation (deterioration) of the light signal in optical fiber transmission.

Fiber Mile Let's say that you have two sheaths of fiber, each of which contains ten fibers and runs for one mile. That is one route mile (total distance of all fibers), two sheath miles (two sheaths running one mile), and twenty fiber miles (20 fibers running one mile).

Fiber Optic See FIBER.

Fiber Optic Amplifier As light, like electricity or any other form of electromagnetic energy, travels through a physical medium, it attenuates, or loses intensity. At some point in a communications transmission system, you must take your increasingly weak signal and boost it back to its original strength. In analog systems, you simply amplify the weak incoming signal and send it on its way. In addition to amplifying the information signal, any accumulated noise is also amplified. In digital signals, you first regenerate the signal, then amplify it, then send it on its way, with no recognition of or regard for any noise present. Most fiber transmission systems accomplish this process by converting the original light signal on the fiber to electrical impulses, regenerate the signal, then amplify it, then convert it back to light pulses, then send it on its way. This takes significant energy and equipment — not altogether convenient for an underwater cable of several thousand miles. With new fiber optic amplifiers you no longer need to convert the light signal to electrical impulses. A fiber optic amplifier uses special fiber doped with erbium to act as the amplifier. Light comes into this special fiber, is pumped with the correct frequency laser and is amplified with extremely high gain and very low noise through a process of chemical light amplification. It's truly amazing technology. See ERBIUM-DOPED FIBER AMPLIFIER for a more technical explanation.

Fiber Optic Attenuator A small device with two connectors. It reduces the amount of light passing through it, similar to the way sunglasses reduce the amount of light entering your eyes so you can see better.

Fiber Optic Buffer Plastic coating on individual fibers. There are 12 colors to distinguish them from each other.

Fiber Optic Cable See FIBER.

Fiber Optic Connector Three types: SC, ST, FT.

Fiber Optic Distribution Panel A termination device and organizer. It houses splice trays where connector plugs (called pigtails) are attached to the ends of fiber-optic cables.

Fiber Optic Inter Repeater Link An 802.3 Ethernet standard for connecting two repeater devices at 10 million bits per second.

Fiber Optic Transmission System See FOTS.

Fiber Optic Waveguide A relatively long thin strand of transparent substance, usually glass, capable of conducting an electromagnetic wave of optical wavelength (visible region of the frequency spectrum) with some ability to confine longitudinally directed, or near-longitudinally directed, lightwaves to its interior by means of internal reflection.

Fiber Optics A technology in which light is used to transport information from one point to another. More specifically, fiber optics are thin filaments of glass through which light beams are transmitted over long distances carrying enormous amounts of data. Modulating light on thin strands of glass produces major benefits in high bandwidth, relatively low cost, low power consumption, small space needs, total insensitivity to electromagnetic interference and great insensitivity to being bugged. All these benefits have great attraction to anyone who needs vast, clean transmission capacity, to the military and to anyone who runs a factory with lots of electronic machinery. The first field trial of an AT&T lightwave system took place in Chicago in 1977. There has been a rapid improvement in cost effectiveness of fiber systems, expressed as cost per bit per kilometer. A one hundredfold increase in cost performance in one five-year period — from 1980 to 1985. Some versions of fiber optics now carry 10 Gbps (ten billion bits per second) to carry more than 129,000 voice conversations. See SONET.

Fiber Pigtail A short length of optical fiber, permanently fixed to a component, used to couple power between the component and the transmission fiber.

Fiber Remote Fiber Remote extends what Northern Telecom calls Intelligent Peripheral Equipment using dark single or multimode fiber cable. Fiber Remote operates over a range of typical campus distances — from thousands of feet to several miles. Distance is site specific, determined by variables such as the type and quality of the fiber installed and the number of connectors and splices. The signal attenuation between the local PBX system and the remote IPE shelves should not exceed a 13 dB loss. Fiber Remote is used when users prefer a single switch; when users have a campus with right of way for running fiber where the distance between the local and remote site is within 6 miles; where there's limited riser space in a large high rise building; where a user needs to alleviate switch room congestion and where security is of utmost importance (e.g. in military bases).

Fiber Spudger Shaped like a pencil, it's a gadget technicians use to move around and find their way through fiber optic cables on their hunt for one single fiber optic.

Fiberoptichead Slang expression to describe a customer who thinks he knows everything about cable. Usage: "That fiberoptichead wouldn't know a drop from a fish job."

Fiberphone A battery-powered device that connects to both ends of a fiber optic cable allowing people (typically craftspeople) to talk over the cable. The complete device (two ends) costs $1,000 to $2,000 and often comes with a headset.

Fiberworld In simultaneous media events in Washington and Montreal on October 12, 1989, Northern Telecom and Bell Northern Research (BNR) unveiled "FiberWorld," which they referred to "as a vision and commitment to deliver the world's first completely family of fiber-optic access, transport, and switching products."

Fibre The European, Australian, Canadian, British and New Zealand spelling. The American spelling is fiber. Whoops, except in the case of Fibre Channel, which is correct, even in American English. See FIBER, FIBER OPTICS and FIBRE CHANNEL.

Fibre Channel An set of standards developed by ANSI (American National Standards Institute). Fibre Channel (FC) is intended to provide a practical and inexpensive means of rapidly "transferring data between workstations, mainframes, supercomputers, desktop computers, storage devices, displays and other peripherals," according to the Fibre Channel Association

(FCA). The mysterious (to Americans) spelling of "fibre," so says Datapro Information Systems, refers to the generic underlying transmission mechanism. As that mechanism includes, but is not limited to, fiber optics, the term "fibre" avoids confusion with respect to the transmission medium.

Fibre Channel standards support a number of speeds, including 133 million bits per second (Mbps), 266 Mbps, 530 Mbps, and 1 gigabit per second (Gbps). The transmission media can include coaxial cable, as well as either monomode or multimode fiber. Monomode fiber with longwave laser light is preferred, as this combination supports transmission distances of 10 km at 1 Gbps with a BER (Bit-Error Ratio) in the range of 10 to the minus 12th power (i.e., an errored bit in every 1,000,000,000,000. In the aggregate, dual, unidirectional links provide bidirectional communications.

FC's speed of data transmission is due not only to the fundamental nature of the transmission system, but also to the fact that FC is a serial link technology. In other words, FC is a new I/O (Input/Output) interface over which data is streamed in serial fashion across an established link. FC provides a channel connection for dedicated or switched point-to-point connection between devices. "Channel connections" are hardware-intensive, low in overhead, and high in speed; "network connections" typically are software-intensive, high in overhead, and therefore slower. The downside is that channel connections tend to be limited to a relatively small number of devices with pre-defined addresses. Actually, FC will support, through separate ports, both channel and network connections. It also will support not only its own protocol, but also higher level protocols such as FDDI, SCSI, HIPPI and IPI. The physical topology of Fibre Channel can be point-to-point, ring or star. In a star configuration, the interconnecting switching device is known as a "Fabric," which can be a circuit switch (star), an active hub (star) or a loop (ring).

Fibre Channel supports the transfer of data in frames, with a payload of 2,048 bytes. A CRC (Cyclic Redundancy Check) mechanism is employed for purposes of detection and correction of transmission errors. Flow control is supported through switch buffers, in a Fabric implementation.

Three service classes are supported. Class 1 provides the equivalent of a dedicated physical connection; it is the highest quality of service, and is most effective for very high-speed transfers of large amounts of data. Class 2 is a connectionless grade of service, making use of multiplexed frame switching, with multiple sources sharing the same channels; Class 2 supports confirmation of frame delivery. Class 3 is identical to Class 2, minus confirmation of frame delivery.

The applications for Fibre Channel initially were for high-speed data and image transfer. In the recent past, much attention has been focused on the real-time transfer of audio and video, as well. As a result, Fibre Channel is being implemented for video file transfer and video playback applications in post-production digital video and movie studios, and in the broadcast backbone. www.FibreChannel.com

Fibre Channel — Arbitrated Loop FC-AL. A Fibre Channel application for Storage Area Networks (SANs), FC-AL supports high-speed access to storage arrays over loops as long as 10 kilometers over a single Fibre Channel link, non-amplified, and at data rates as high as 100 MBps (MegaBytes per second). In a dual loop architecture, data rates are doubled to as much as 200 MBps; rates of 400 MBps are anticipated in the near future. Logically, FC-AL operates as a full-duplex, point-to-point, serial data channel. As many as 126 hosts can be connected to a given storage device; intermediate and cascading FC-AL hubs and concentrators can serve to improve costs, although there are corresponding performance degradations. FC-AL earns the tag "arbitrated" by virtue of the fact that access to the storage system is arbitrated on the basis of level of privilege, with fractional bandwidth services supported. The next generation of FC-AL, designated FC-EL (Fibre Channel-Enhanced Loop) is under development at ANSI. See also Fibre Channel and SAN.

Fibre Channel — Enhanced Loop FC-EL. The next generation of FC-AL (Fibre Channel-Arbitrated Loop), FC-EL is under development at ANSI. See also FC-AL and Fibre Channel.

Fibre Channel Association FCA. Formed in January 1993, the FCA works to encourage the utilization of Fibre Channel, complementing the standards development efforts of the ANSI T11 committee. The mission of the FCA is "to provide a support structure for system integrators, peripheral manufacturers, software developers, component manufacturers, communications companies and computer service providers." www.FibreChannel.com

FID Field Identifier, an ISDN SPID term.

Fidonet An electronic bulletin board technology for transfer and receipt of messages. According to PC Magazine, the origins of FidoNet date back to the early 1980s, when the two authors of the BBS software Fido, who lived on opposite coasts, needed an easy way to exchange modifications they made to the source code. They designed a system where, as a nightly event, the board would shut down and run utilities that automatically transferred the changed files between the author's BBSs. The logical next step was to permit the exchange of private mail messages called NetMail, between the sysops. The author found these capabilities so useful that they include them as part of the Fido BBS (Bulletin Board Software) package. It didn't take long for an informal network of Fido nodes to come into existence, all running the Fido software and exchanging various utility and program files and NetMail among sysops. Like other BBSs, the FidoNet BBSs had their own SIGs, or Special Interest Groups, where users with similar interests could exchange messages in a way similar to what on-line services call conferences or forums. By 1986 a Fido sysop had extended the NetMail concept to allow SIGs to share public messages among the BBSs, and EchoMail was born. In the years since, BBS authors and FidoNet users and sysops extended these capabilities to other BBS packages, and FidoNet grew. It currently has over 11,000 nodes covering most of the world. Many of the existing public and private networks go through FidoNet gateways into the Internet Mail system, which carries e-mail over a group of interconnected networks to universities, government agencies, military branches, and corporations. FidoNet technology uses store-and-forward messaging and is based on point-to-point communications between nodes.

Field 1. One half (every other line) of a complete television picture "frame", consisting of every other analog scan line. There are 60 fields per second in American television.

2. A place with no phones or other communications capability where an important person inevitably is when you need some vital information, service or device that only he or she can provide. "I'm sorry, the chief technician is in the field today, and can't be reached." Few "fields" are actually fields. They're usually downtown office buildings.

3. The specific location of data within a record. In the jargon of database management systems, many fields make up one record Many records make up one file. A field is one of the

basic subdivisions of a data record. The record on you in your company's database might include your name, your address, your salary, etc. A field is simply one of these — e.g. your salary, your last name, or your street address. All the records of all the employees in your company make up a file, also called a database.

4. The name given to that part of an electrical system in which electromagnetic lines of force are established.

5. In Windows, the field is the empty line in a dialog box where you enter data.

6. In call center jargon. A field is a single piece of data, such as an employee ID, stored in a record. The fields are organized under column headings.

Field Effect Transistor A field effect transistor (FET) is composed of a single piece or channel of either P or N-type semiconductor surrounded by a ring or collar of opposite semiconductor. The collar is called the gate and the ends of the channel are called the source and the drain. As the voltage across the gate and source is varied, the resistance between the source and drain changes. A large current between the source and drain can be controlled by a small gate-source voltage.

Field Emission Displays FED. Another way of making thin, flat, lightweight computer displays for laptops, planes, etc. The other way is called "active matrix liquid crystal display." In field emission displays, a tiny color cathode ray tube sits behind each of the many pixels in the screen. This results in a brighter picture that uses less energy than the active matrix LCD displays. See also FED.

Field Intensity The irradiance of an electromagnetic beam under specified conditions. Usually specified in terms of power per unit area, e.g., watts per square meter, milliwatts per square centimeter.

Field Interlacing In television, field interlacing is the process of creating a complete video frame by dividing the picture into two halves with one containing the odd lines and the other containing the even lines. This is done to eliminate flicker.

Field Measurement Refers to both signal strength and qualitative field tests of wireless networks.

Field Programmable The ability of a system to have changes made in its program while it is installed — without having to be returned to the factory.

Field Repairable A characteristic of an unfortunately-decreasing number of electronic devices, that allows users or technicians to fix them where they are used ("in the field"), instead of having to send them to a centralized repair facility where esoteric parts and tools are available.

Field Rheostat A variable resistance used in the field circuit of a generator or motor to control the field current and consequently the strength of the electromagnetic field, thereby regulating the speed or power of the motor, or the output of the generator.

Field Sequential System Field sequential system was the first broadcast color television system, approved by the FCC in 1950. It was later changed to the NTSC standard for color broadcasting.

Field Strength The intensity of an electric, magnetic, or electromagnetic field at a given point. Normally used to refer to the rms value of the electric field, expressed in volts per meter, or of the magnetic field, expressed in amperes per meter.

Field Strength Meter Electronic instrument that measures the intensity of the magnetic field.

Field Upgradable A desirable characteristic of telecom equipment, computers, etc., that allows new features to be added and other improvements to be made, where the device is used, rather than having to return it to the manufacturer or a repair facility.

Field Wire A flexible insulated wire used in field telephone and telegraph systems. WD-1 and WF-16 are types of field wire. Usually contains a strength member. See Field Wiring.

Field Wiring An electrical connection intended to be made at the time of installation, in the field, as opposed to factory wired.

FIFO First In, First Out. All telephone networks are a trade-off. It's simply too expensive to build a phone network which will be ready to give everyone dial tone and a circuit — if everyone picked up the phone simultaneously and tried to make a call. There are basically two ways of handling calls which cannot be sent on their way — i.e. for which there's no present available capacity. First, you can "block" the call. This means giving the caller a busy or a "nothing" (also called "high and dry"). Second, you can put the call into a queue. Now you have people waiting in queue, how do you handle them? The most equitable — the way most queues work — is to handle the calls on the basis of First In, First Out. (First call to come in is handled first.) There are other ways of handling calls in a queue — including First In, Last Out, by priority (e.g. which line you came in on and how much it cost, or how high you are in the corporation, etc.)

FIFO queuing also is used in some routers, although it has disadvantages in TCP/IP application. For example, if the buffer memory is full and data packets are dropped from the tail of the queue, the originating devices will assume that they are sending to rapidly and will slow down their rate of transmission. As multiple devices subsequently probe to seek the capacity of the network by sending data at higher rates, they can create another congestion condition.

FIFO also is a term used in data communications. It is a buffering scheme in which the first byte of data that enters the buffer is also the first byte retrieved by the CPU. This scheme is used in the 16550 (the UART chip which controls the serial port on most PCs and most other serial-buffering designs, because it closely mimics the way serial data is actually transmitted; that is, one bit at a time.

Fifth Generation Fifth generation computers and telephone systems will be based on artificial intelligence. A fifth generation phone system may make far more sophisticated decisions about routing calls across networks. Those decisions may be made on how many calls have already happened so far that month, the choice of carrier by the likely quality of his connection, etc.

FIGS FIGure Shift. A physical shift in a terminal using Baudot Code that enables the printing of numbers, symbols and upper-case letters.

Filament 1. An electrically heated wire in an evacuated glass bulb, forming one element (the cathode) of a vacuum tube.
2. The part of an incandescent light bulb that heats and lights. Filaments are often made from Tungsten.

File 1. A set of similarly structured data records (such as personnel records using a standardized form). See FIELD.
2. A call center term. A logical division of the data stored on a disk or diskette; for example, employee information vs. supervisor information vs. call volume history. Files generally consist of one or more records of a certain structure.

File Allocation Table A file allocation table is essentially a road map of the location of files on a hard disk. See FAT.

File Caching A Novell local area network NetWare file server can service requests from workstations up to 100 times faster when it reads from and writes to the file server's cache

memory (in RAM) rather than executing direct reads from and writes to the file server's hard disks.

File Extensions MS-DOS files can have an 8-character filename followed by a period and a three-character file extension. Windows 95, Windows 98 and Windows NT files can have an up-to-256 letter filename followed by a period and then a three character extension. While most extensions are arbitrarily assigned by users or companies, some extensions are reserved for special purposes, e.g. exe, com and bat. Windows has a built in program called Explorer. When you click on that file, Explorer uses the file extension to launch an application it has associated with that extension. When you install a new application, it usually tells Explorer which file extensions it creates. Thus when you click on a file with that extension, it will launch the correct application.

Here are some common extensions:

EXE DOS executable file
BAT DOS executable batch file
DAT ASCII text file (usually)
COM DOS executable command file
ERR Error log file
OVL Overlay file
HLP Help screens which appear by pressing F1
INI Initialization file
SYS Operating system file
BAS Basic language file
GIF Graphic Interchange Format
PS Postscript file
TAR UNIX tape archive format
Z UNIX compressed file
ZIP DOS compressed file

You'll find a definition of the extensions you're using in your Windows WIN.INI file. Here are some of mine

txt=notepad.exe ^.txt
ini=notepad.exe ^.ini
pcx=pbrush.exe ^.pcx
bmp=pbrush.exe ^.bmp
wri=write.exe ^.wri
doc=C:\WINWORD\winword.exe ^.doc
rtf=C:\WINWORD\winword.exe ^.rtf
ppt=C:\powerpnt\powerpnt.exe ^.ppt

File Format The way in which data is stored. The file's format is indicated by the three or four letter extension after its name. For example, Word documents end in .doc and Excel documents in .xls. An industry standard interchange file formats (IF/IFF) example is .gif for graphics. See File Extensions.

File Gap A short length of blank tape used to separate files stored on linear magnetic tape.

File Locking Picture a cabinet of file folders. Now I remove a folder to work on it. I make a photocopy of the folder in the cabinet and leave the original. You come along and remove the original because you want to work on it. You make changes and replace the changed copy in the cabinet. Ten minutes later I pull your file out and replace it with mine. Bingo, all your changes are lost. But let's say when I remove the file to work on it, I staple the remaining folder shut. That's a message to anyone else — including you that you shouldn't mess with the file. When I return, I unstaple the file, and add my changes. Now it's ready for you to do your thing. File locking ensures that a file will be updated correctly before another user, application, or process will be allowed to write to the file. When a file is locked, no one else can write to it. Without file locking, one user could overwrite the file update of another user. In contrast to file locking, record locking allows many users to access the same file at once, but have only one access the record. See also ATTRIB and RECORD LOCKING.

File Maintenance The job of keeping your data base files up to date by adding, changing or deleting data.

File Management The system of rules and policies for maintaining a set of files — including how files can be created, accessed, retrieved and deleted.

File Server A file server is a device on a local area which "serves" files to everyone on that local area network (LAN). It allows everyone on the network to get to files in a single place, on one computer. It typically is a combination computer, data management software, and large hard disk drive. A file server directs all movement of files and data on a multi-user communications network, namely the LAN. It allows the user to store information, leave electronic mail messages for other users on the system and access application software on the file server — e.g. word processing, spreadsheet. In computer telephony applications, potentially many users or voice channels need to access data on a file server. The file server may therefore present a significant bottleneck for computer telephony especially if it is used to store large files, such as voice prompts. The ability for the file server to handle the transaction load planned for your computer telephony application is therefore a key design consideration and issue to test.

File Server Console Operator A user or a member of a group to whom a Novell NetWare SUPERVISOR delegates certain rights in managing the file server. A file server console operator has rights to use FCONSOLE to broadcast messages to users, to change file servers, to access connection information, to monitor file/lock activity, to check LAN driver configurations and to purge all salvageable files.

File Sharing A topology-independent feature of Apple Macintosh's System 7 operating system which allows users to share files and folders on their disks with other users across the LAN. File sharing is slow but acceptable for sharing small numbers of files among small groups. For larger networking, the user must consider AppleShare, Netware, Vines, etc.

File Transfer Protocol FTP. A service that supports file transfer between local and remote computers, including the Internet. FTP supports several commands that allow bidirectional transfer of binary and ASCII files between computers. The FTP client is installed with the TCP/IP connectivity utilities. See FTP and FILE TRANSFER PROTOCOLS.

File Transfer Protocols One problem with transmitting information over phone lines is the noise on the phone line. One way to overcome the problem of noise is a file transfer protocol. The idea is simple: send your information in bundles (called packets). Accompany those packets with a special number derived in some way from the information in the packet. Send it all to the other end. Have the computer at the other check the number and see if corresponds to the packet. If not, send a signal back, saying "Something went wrong. Please send the packet of information back again." Most asynchronous file transfer protocols use some form of error detection, typically checksum or cyclic redundancy check (CRC). Both the checksum and the CRC are values derived from the data being sent (or received) according to mathematical algorithms. The protocol sends the value long with the information (the bits) in the packet. The receiving program compares with the check values with the values it calculates. If the check values do not match, the receiver asks the sending computer to retransmit the packet. Older protocols required a positive acknowledgement (an ACK) before they sent another packet.

But newer protocols allow transmission of several packets before they receive an acknowledgement. This is particularly useful for circuits with long delays, especially satellites. See also XMODEM.

File Transfer Software Software to transmit files between computers, over phone lines or over a direct cable connection between the two computers.

Fill Bit Stuffing

Fills A computer imaging term. Designated areas that are flooded with a particular color. Most paint packages let you create geometric shapes in filled form. All packages also let you fill irregular closed regions. Two types of such fills exist: A seed fill floods all connected regions with the color specified by the mouse or stylus pointer; a boundary fill floods a color until the algorithm encounters a specified boundary color.

Filter 1. A device which transmits a selected range of energy. An electrical filter transmits a selected range of frequencies, while stopping (attenuating) all others. It is used to suppress unwanted frequencies or noise, or to separate channels in communications circuits. Such a filter might be called a BANDPASS filter. You can also use a filter to remove certain characters you might be receiving over a data communications channel, for example control characters or higher-order nonstandard ASCII bits.

2.An operating parameter used in LAN bridges and routers that when set will cause these devices to block the transfer of packets from one LAN to another. Filters can be set to prevent the internetworking of several types of messages. They may be set to block all packets originating from a specific destination, called source address filtering, or all packets heading for a particular destination, called destination address filtering. Filters may also be set to exclude packet of a particular protocol or any particular filed in a LAN packet.

3. See TAPI 3.0

Filtering 1. A process used in both analog and digital image processing to reduce bandwidth. Filters can be designed to remove information content such as high or low frequencies, or example, or to average adjacent pixels, creating a new value from two or more pixels. "Tap" refers to the number of adjacent lines or pixels considered in this process. MPEG, for instance, makes use of a 7-tap filter.

2. Bridges can reduce LAN congestion through a process of filtering. A filtering bridge reads the destination address of a data packet and performs a quick table lookup in order to determine whether it should forward that packet through a port to a particular physical LAN segment. A four-port bridge, for instance, would accept a packet from an incoming port and forward it only to the LAN segment on which the target device is connected; thereby, the traffic on the other two segments is reduced and the level of traffic on those segments is reduced accordingly. Filtering bridges may be either programmed by the LAN administrator or may be self-learning. Self-learning bridges "learn" the addresses of the attached devices on each segment by initiating broadcast query packets, and then remembering the originating addresses of the devices which respond. Self-learning bridges perform this process at regular intervals in order to repeat the "learning" process and, thereby, to adjust to the physical relocation of devices, the replacement of NICs (Network Interface Cards), and other changes in the notoriously dynamic LAN environment.

Filtering Agent A new form of smart agent whose basic job is to keep away all the stuff you don't want and find the stuff you do want — such as information gleaned from the Internet. See also V-CHIP.

Filtering Bridge See Bridge.

Filtering Traffic This is the process of selecting which traffic will be allowed into a certain portion of a network, such as the wide area network. It is also the process of determining which traffic is transmitted first, then next, and so on. The traffic is compared to a filter, or a set of specifications, to determine if it can pass through or not.

Final Trunk Group A last-choice trunk group that receives overflow traffic and which may receive first-route traffic for which there is no alternate route.

Find Me Service An AIN version of call forwarding, allowing the forward numbers to be programmed or re-programmed from any location. Additionally, priority access can be extended to specific callers based on password privilege. For instance, only highly privileged callers would be forwarded to your cell phone, in consideration of the high cost of airtime.

Finder The user interface portion of the Apple Macintosh operating system. Unlike running Windows on top of DOS, tight integration of the finder and system requires both to be running.

Finger 1.A standard protocol specified in RFC-742. A program implementing this protocol lists who is currently logged in on another host. In short, finger is a computer command that displays information about people using a particular computer, such as their names and their identification numbers.
2. Also known as a tine. An individual digital channel of a wireless rake receiver. A rake receiver can support a number of tines, which can be combined to form a stronger received signal.

Finite State Machine A computer system with a defined set of possible states and defined transitions form state to state. Given the same inputs, two identical state machines will change states identically.

FIPS Federal Information Processing Standard. See also FIPS PUBS nn.

FIPS PUBS nn Various standards for data communications.

Fire To discharge someone. The story goes that in the early part of the 20th century, if an NCR salesman lost an order, when he returned to his office, they put his desk out on the front lawn and burned the desk. Then they "fired" the salesman.

Fire Break A material, device, or assembly of parts installed along a cable, other than at a cable penetration of a fire barrier, to prevent the spread of fire along a cable.

Firestop A material, device, or assembly of parts installed in a cable pathway at a fire-rated wall or floor to prevent passage of flame, smoke or gases through the rated barrier, (e.g., between cubicles or separated rooms or spaces).

Firestop System A specific construction consisting of the material(s) (firestop penetration seals) that fill the opening in the wall or floor assembly and any items that penetrate the wall or floor, such as cables, cable trays, conduit, ducts, pipes, and any termination devices, such as electrical outlet boxes, along with their means of support.

Firestopping The process of installing specialty materials into penetrations in fire-rated barriers to reestablish the integrity of the barrier.

Firewall A combination of hardware and software which limits the exposure of a computer or group of computers to an attack from outside. The most common use of a firewall is on a local area network (LAN) connected to the Internet. Without a firewall, anyone on the Internet could theoretically jump onto the corporate LAN and pick up any information on or dump anything to any of the computers on the LAN. A firewall is a system or combination of systems that enforce a boundary between two or more networks. There are several types of firewalls — packet filter, circuit gateway, application gateway or trusted gateway. A network-level firewall, or packet filter, examines traffic at the network protocol packet level. An application-level firewall examines traffic at the application level — for example, FTP, E-mail, or Telenet. An application-level firewall also often readdresses outgoing traffic so it appears to have originated from the firewall rather than the internal host. NEC PrivateNet Systems Group issued a White Paper called Connecting Safely to the Internet — A study in Proxy-Based Firewall Technology. In that White Paper, they defined an Internet firewall:

The primary purpose of an Internet firewall is to provide a single point of entry where a defense can be implemented, allowing access to resources on the Internet from within the organization, and providing controlled access from the Internet to hosts inside the organization's internal networks. The firewall must provide a method for a security or system administrator to configure access control lists to establish the rules for access according to local security policies. All access should be logged to ensure adequate information for detailed security audit.

A traditional firewall is implemented through a combination of hosts and routers. A router can control traffic at the packet level, allowing or denying packets based on the source/destination address of the port number. This technique is called packet filtering. A host, on the other hand, can control traffic at the application level, allowing access control based on a more detailed and protocol-dependent examination of the traffic. The process that examines and forwards packet traffic is known as a proxy.

A firewall based on packet filtering must permit at least some level of direct packet traffic between the Internet and the hosts on the protected networks. A firewall based on proxy technology does not have this characteristic and can therefore provide a higher level of security, albeit at the cost of somewhat lower performance and the need for a dedicated proxy for each type of connectivity.

Each organization needs to choose one of these basic types of technologies. The right choice depends on the organization's access and protection requirements.

FireWire It's a 100 Mbps serial bus, also known as IEEE 1394. It is geared to become a digital interface for consumer video electronics and hard-disk drives. It's designed for up to 4.5 meters per segment and features six pins per connector. See IEEE 1394, USB, UNIVERSAL SERIAL BUS.

Firm Order Confirmation FOC. The form a local phone company submits to another phone company indicating the date when the circuits ordered by the other company will be installed. See FOC for a longer explanation.

Firmware Software kept in semipermanent memory. Firmware is used in conjunction with hardware and software. It also shares the characteristics of both. Firmware is usually stored on PROMS (Programmable Read Only Memory) or EPROMs (Electrical PROMS). Firmware contains software which is so constantly called upon by a computer or phone system that it is "burned" into a chip, thereby becoming firmware. The computer program is written into the PROM electrically at higher than usual voltage, causing the bits to "retain" the pattern as it is "burned in." Firmware is nonvolatile. It will not be "forgotten" when the power is shut off. Handheld calculators contain firmware with the instructions for doing their various mathematical operations. Firmware programs can be altered. An EPROM is typically erased using intense ultraviolet light.

First In, First Out See FIFO.

First Office Application The first office to have the guts to try a new system in a real, live production mode. The same thing as a beta test. See also Beta Test.

First Party Call Control A call comes into your desktop phone. You can transfer that call. When the phone call has left your desk, you can no longer control it. That is called First Party Call Control. If you were still able to control the call (and let's say, switch it elsewhere) that would be called Third Party Call Control. First party call control is mostly done at your desk with your telephone or with a card in your PC, which emulates a telephone. Third party call control is usually done via a computer (often a server on a LAN) attached to a special link directly into your PBX. There are some evolving standards in call control — chiefly Microsoft's Windows Telephony and Novell's TSAPI (Telephony Services API). There is no such animal as Second Party Call Control. See CALL CONTROL, TELEPHONY SERVICES and WINDOWS TELEPHONY.

First Ring Suppression Caller ID in North America comes in just after the first ring. You don't want to answer the call before the second ring otherwise you will mess up your receiving of Caller ID information. You can simply not answer until the second ring. Or you can get a trunk-based gadget which will turn off the first ring so you or your voice processing equipment won't hear it.

FIS Forms Interchange Standard.

Fish 1. To push a stiff steel wire or tape through a conduit or interior wall. Pull through wires, cable or a heavier pulling-in is then attached to one end of the steel wire. The other end is then pulled until the wire or cable appears.

2. First In Still Here. A non-standard term used in inventory accounting. Roughly equivalent to FILO (First In Last Out), but suggesting that the inventory is not moving because you aren't selling any of it.

Fish Food Internet Webmasters who want to draw attention to new on-line content call the tidbits "fish food." The morsels are posted in "What's new" buttons or "Click here" icons on the home page, so, like the flakes that feed your guppies, they float at the top and attract hungry users. Contributed by Judy Ehrenreich, PaciCare Health Systems, Santa Ana, CA.

Fish Job Running cables inside walls. Usage: "That fish job is too tough for a rookie."

Fish Tape Non-conductive tape with a reinforced fiberglass core and slippery outer nylon coating which slides easily through conduit without jamming. The idea is to push the tape through, attach it to the cable and pull the cable back. You also might use wire pulling lubricants to make the job even easier. They come in various formulations — for use in different temperatures, for pulling different cable, etc.

FISK Fax a dISK. Method of sending information on 3 1/2" disks painlessly across phone lines. Plug your disk into a fisk machine, choose which files you want to send, dial the number you want to send your files to and walk away.

FITC Fiber To The Curb. See FTTP.

FITH FIber To The Home. (I kid you not. That's what it stands for.) See FTTP.

FITL Fiber In The Loop. A local loop transmission system which includes fiber optics. Generally, the fiber is used for the main distribution facilities from the central office or POP (Point Of Presence) of the carrier to a remote node. At the remote node, twisted pair is used to extend the loop the last mile or so. See also Fiber Optics, FTTC, FTTH, Local Loop and POP.

Five By Five Slang expression meaning satisfactory transmission in both directions.

FIX Federal Internet Exchange. A connection point largely serving to interconnect network traffic from MILNET (MILitary NETwork), NASA Science Net and other federal government networks, as well as providing those network users with access to the Internet. FIX-EAST is located at the University of Maryland in College Park, Maryland; FIX-WEST is located at the NASA Ames Research Center at Moffet Field between Sunnyvale and Mountain View, California. See also CIX, MAE and NAP.

Fixed Condenser A condenser, the plates of which are stationary and the capacity of which cannot be changed.

Fixed Disk Old name for a hard disk.

Fixed End System F-ES. A non-mobile end system. A host system that supports or provides access to data and applications.

Fixed Format A way of communicating in which everything to be sent follows a predetermined sequence, i.e. it fits into a specific length and format. The idea is to allow you to predict message length, the location of the message, where the control characters are, etc.

Fixed IP Address See IP ADDRESSING.

Fixed Length Records A set of data records all having the same number of characters in them. Think of a database of name, address, city, state, zip. Clearly, not everyone's record will be the same length. In order to make a fixed length record, the computer will pad the record with "padding characters" which the computer will ignore when it reads the record. But by including the padding characters it has effectively given everyone the same fixed length record.

Fixed Loop A services feature available in some switching systems that permits an attendant on an assisted call to retain connection through the attendant position for the duration of the call. The attendant will normally receive a disconnect signal when the call has been completed.

Fixed Priority-Oriented-Demand Assignment FPODA. Medium access technique in which one station acts as master and controls channel based on requests from stations.

Fixed Rate A fixed monthly price. See also FLAT RATE.

Fixed Satellite Service A radiocommunication service between Earth stations as specified fixed points when one or more satellites are used; in some cases this service includes satellite-to-satellite links, which may also be effected in the inter-satellite service, the fixed-satellite service may also include feeder linker for other space radiocommunication services.

Fixed Satellite System A system of geosynchronous satellites, which always are in equatorial orbit. See also GEO. Contrast with LEO and MEO.

Fixed Wireless Local Loop Imagine a community of 100 people spread out in a huge area in one of the Western states in the United States — e.g. Airzona or Wyoming.

FL Fault Locating.

Flag 1. A variable in a program to inform the program later on that a condition has been met.

2. In synchronous transmission, a flag is a specific bit pattern (usually 01111110) used to mark the beginning and end of a "frame" of data. Frame Relay and lots of other protocol use this approach in order to delineate one frame from another, and to allow the devices in the network to synchronize on the rate of transmission for purposes of improved bandwidth efficiency. See Frame Relay, Synchronous and Zero Stuffing.

3. Fiberoptic Link Around The Globe. A consortium of phone companies owning an underwater submarine cable made of multiple strands of fiber, each strand carrying information at five gigabits per seconds. Each strand of fiber is unidirectional, i.e. you need two to make a conversation, one for going

and one for coming.

Flag Sequence (HDLC, SDLC, ADCCP, Frame Relay). The unique sequence of eight bits (01111110) employed to delimit the opening and closing of a frame.

Flame An outpouring of verbal abuse that network users write about other users who break the rules. A wonderful term for getting mad via electronic mail. People who frequently write flames are known as "flamers." You can flame by simply sending messages ALL IN CAPS!!!!!!!!! See FLAME FEST, FLAME WAR and MAIL.

Flame Bait An intentionally inflammatory posting in a newsgroup or discussion group designed to elicit a strong reaction thereby creating a flame war.

Flame Fest Massive flaming. See FLAME.

Flame Mail Slang term for rude electronic mail. Bill Gates, Microsoft chairman, is said to be famous for the flame mail he sends to employees who don't perform according to his likings. Mr. Gates is famous for flame mail sent by him between midnight and 2:00 AM.

Flame Resistant Insulated wire which has been chemically treated so it will not aid the spread of flames.

Flame Retardant Constructed or treated so as not be able to convey flame.

Flame War What happens when people send too much flame mail at each other. The online discussion degenerates into a series of personal attacks against the debaters, rather than discussion of their positions. A heated exchange.

Flaming To send an insulting message, usually in the form of a tirade, sent via online postings but also as personal. Flaming is the verb. Flame is the noun. And too much of flaming can cause a nasty flamewar.

Flammability Measure of a material's ability to support combustion.

Flash Quickly depressing and releasing the plunger in or the actual handset-cradle to create a signal to a PBX or Centrex that special instructions will follow such as transferring the call to another extension.

Flash Button A button on a phone which performs the same thing as quickly pressing the switch hook on a phone. See FLASH, FLASH HOOK and FLASHPHONE.

Flash Cut The conversion from an old to a new phone system occurs instantly as one is removed from the circuit and the other is brought in. There are advantages and disadvantages to Flash Cuts. For one, they're likely to be much more dangerous than the opposite view, known as a Parallel Cut, in which the two phone systems run side by side for a month or so. Also known as Hot Cut.

Flash EPROM A type of EPROM that can be electronically erased. It differs from EEPROM in that generally the entire memory must be erased at once.

Flash Hook Another name for Switch Hook. The little button on the telephone that you place your receiver into. It obviously hangs the phone up, releasing that line to receive another call. If you push the flash hook quickly, you can signal the switch at the other end (central office or PBX) to do something, such as place a call on hold and switch to the incoming one (call waiting), or transfer the call to another phone. See FLASH and FLASH BUTTON above.

Flash Memory A technology developed by Intel and licensed to other semiconductor companies, Flash memory is nonvolatile storage that can be electrically erased in the circuit and reprogrammed. Flash memory occupies little space and doesn't need continuous power to retain its memory. Some laptop companies, like Toshiba, are using flash memo-

ry as nonvolatile storage for the BIOS (Basic Input/Output System) and the instructions that start the computer (the bootstrap loader). See also Flash Rom and Memory Cards.

Flash ROM Flash Read Only Memory. Read Only Memory that can be erased and reprogrammed, but stays on when power to your computer is turned off. Flash Rom is used in modems, for example, to hold software known as firmware. When a later software release comes out, you dial a distant computer which downloads new software into your Flash Rom, updating it. See FLASH MEMORY.

Flashpix FlashPix is a new file format designed to optimize the electronic display, manipulation, and distribution of high-resolution images. Developed cooperatively by Eastman Kodak Company, Hewlett-Packard, Live Picture, and Microsoft Corporation, FlashPix has quickly gained recognition as a key enabling technology for companies seeking to sell and license images for reuse over the World Wide Web.

FlashROM See FLASH ROM.

Flat Network A term used primarily in the LAN (Local Area Network) domain to describe a network employing bridges, hubs, or OSI Layer 2 switches. As all such devices are protocol-specific, they are relatively inexpensive and are very fast. Such devices read the address of the target device, which address is contained in the packet header, and forward the packet. Some such devices also can filter packets or encapsulate them to resolve protocol differences (e.g., between Ethernet and Token Ring), although this latter function is accomplished at a relatively low level in order not to compromise speed of packet transfer. Flat networks are fairly easy to configure, but tend to be limited in terms of scalability (i.e.size they can grow to); they also can suffer from congestion caused by broadcast traffic. Flat networks are distinguished from hierarchical networks, which employ more complex router technology, operating at OSI Layer 3. See also HIERARCHICAL NETWORK, BRIDGE, HUB, and ROUTER.

Flat Rate Service FR or FRS. An erstwhile common method of pricing local phone calls in the United States. The concept was that for a fixed amount of money — say $10 a month — you received a plain old desk telephone and an unlimited number of local calls. For years, most residential and most business phones were on a flat rate service. The first thing to go was the phone instrument. You had to pay a dollar or so a month to continue renting it, or you could send it back and buy your own. Second to go was the size of the local calling area you could call. It got smaller.

Third to go were the phone calls themselves. This happened first with businesses and now increasingly with residential service. Under this new "pay-per-call" you get charged a "message unit" for each local calls. A message unit is typically eight to ten cents. But psychologically, "message units" sound better than dimes. Fourth to go was the definition of local calls. What was now a "local" call got smaller, i.e. you could call less far for the price of a local call. And what was now a "local" long distance call changed. Calls which, years ago, were free (i.e. on flat rate service) have now become long distance calls. You can witness this phenomenon of changing local pricing in California, New York and Jersey. In other states, it's taking a little longer. There are cities where flat rate service still exists. Treasure them. They're disappearing, too.

Flat Top Antenna An aerial consisting of one or more parallel horizontal wires supported between masts. The "T" type and the inverted "L" type belong in this class.

Flat Topping Flat topping is where the frame relay carrier limits the ability of the customer to burst above the CIR (committed information rate), thereby flat topping the customer to only the CIR. This is especially important when the customer expects to be able to burst above the CIR. In practice, many customers burst all the time above the CIR to their port speed.

Flatpack In general microwave usage, a miniature hermetic package for MIC components, designed for a minimum height, with pins for RF and DC connections existing through the sides (narrowest dimensions), and designed to be surface mounted or "dropped in" to a cutout in a micro-strip printed circuit board. The leads and the largest surface of the package are in parallel planes.

Flattery The art of telling someone exactly what he thinks of himself. Flattery is the most powerful sales tool.

Flattopping See Flat Topping

FLEC Forward Looking Economic Cost. The general pricing methodology adopted by the Federal Communications Commission in its implementation of the Local Competition Provisions in the Telecommunications Act of 1996, August 8, 1996, FCC 96-325. FLEC consists of two parts: the total element long-run incremental cost of the network element and an allocation of common costs. The total element long-run incremental cost is the cost of providing the total quantity demanded in the future of a network element based upon an efficient network configuration, projected values of the cost of capital and economic depreciation rates. The allocation of common costs is required to be forward-looking as well. Appropriate common costs are those that are realized by an efficient company that cannot be attributed directly to a network element or set of network elements. FLEC has been contested in Federal Court. The U.S. Court of Appeals, Eighth Circuit, vacated the specific FCC rule requiring that state commissions use FLEC in arbitration proceedings. However, state commissions have adopted versions of FLEC as an acceptable methodology to be used in arbitration proceedings. This definition courtesy Douglas Meredith.

Flex One of the communications protocols used between paging towers and the mobile pagers/receivers/beepers themselves. Other protocols are POCSAG, ERMES, GOLAY and REFLEX. The same paging tower equipment can transmit messages one moment in GOLAY and the next moment in ERMES, or any of the other protocols. In mid-February, 1997 Motorola announced tht its Products Sector was now shipping its 68175 FLEX chip paging protocol IC (integrated circuit) in volume to customers worldwide. FLEX protocol, an open paging standard developed by Motorola, offers product developers, according to Motorola, a common set of rules that ensure applications work across different service providers' equipment. Currently the FLEX protocol has been adopted by service providers around the world, including providers in China, Southern Asia, India, Japan and the Middle East, along with North America and Latin America. According to Motorola, the FLEX chip IC processes information that has been received and demodulated from a FLEX radio paging channel, selects messages addressed to the paging device, and communicates the message information to the host.

Flex ANI Flexible Automatic Number Identification. Additional two-digit ANI identifiers for PSPs (Payphone Service Providers). Flex ANI provides a means of identifying the specific class of service associated with the originating telephone number in order that the PSP can be compensated properly for originating long distance calls. Flex ANI also provides for enhanced routing, call screening. Carriers can order Flex ANI to identify calls originating from 1) dumb payphones with switch-generated coin signaling, 2) smart payphones with coin signaling resident in the phone, and 3) inmate/detention facility payphones. A FCC mandate (March 1998) requires that facilities-based LECs (Local Exchange Carriers) deploy Flex ANI. Flex ANI is provided per end office (central office) on a CIC (Carrier Identification Code) basis; FGD (Feature Group D) is required. See also ANI and CIC.

Flex Life The measurement of the ability of a conductor or cable to withstand repeated bending.

Flexible Dialing Pattern A PBX dialing pattern that allows you to set your PBX so it can have one, two, three or four digit numbers for its extensions. See also FLEXIBLE NUMBERING OF STATIONS.

Flexible Drill Bit A long drill bit that bends and is used for pulling cable and wire through walls in one operation. This means you drill the hole and then reverse the drill and it pulls the cable through the hole, while the drill bit is still inserted. Diversified Manufacturing of Graham, NC, makes such a marvelous product.

Flexible Intercept Allows you to assign "operator intercept" service to those extensions you wish for whatever reason, unassigned number, temporary disconnect, etc.

Flexible Line Ringing A PBX feature which allows different phones to have different ringing for incoming calls from inside the building and from outside. Different ringing for intercom calls, different for inter-net calls, different ringing for outside calls, etc.

Flexible Numbering of Stations A PBX feature which allows you some flexibility in the way you number the extensions off your PBX. How much flexibility depends on the particular PBX and the number of extensions you have. Hotels like giving their hotel phones the same number as the rooms. Makes sense. See FLEXIBLE STATION NUMBERING.

Flexible Pricing Tariffs A regulatory procedure which permits rates for certain services to be changed quickly to meet market conditions, i.e. competition.

Flexible Release The ability of the switching system to release a connection when either party hangs up.

Flexible Ringing Also called Distinctive Ringing. A PBX feature that lets phones ring differently. Useful to separate inside and outside calls. Also good when phones are close to each other because people can recognize their phone instantly.

Flexible Routing The ability to choose different physical paths through a network for different calls as circumstances warrant.

Flexible Station Numbering A feature that allows telephone extensions to be numbered according to their physical location or departmental location, etc. No rewiring is required for in-place telephones. It's all done in software. See FLEXIBLE NUMBERING OF STATIONS.

Flexibility The quality of a cable or cable component which allows for bending under the influence of outside force, as opposed to limpness which is bending due to the cable's own weight..

Flicker The wavering or unsteady image sometimes seen on monitors. A major cause is a refresh rate that's too low. Above 60 Hz, flicker disappears completely. See MONITOR.

FLINK A FLash and a wINK makes a flink signal.

Flip Flop A device or circuit which can assume either of two stable states. Flip flop devices are used to store one bit of information.

Float Charging The battery charging technique for which sealed lead acid batteries are designed. A float charger maintains a voltage on the battery known as the "float voltage". The float voltage is the ideal maintenance voltage for the battery which maximizes battery life. When the float voltage is applied to a battery a current known as the "float current" flows into the battery, exactly cancelling the batteries' own internal self discharge current. Sealed lead acid batteries require float charging at least occasionally or they will become permanently degraded by a process called "sulfation". Maximum lifetime is obtained when a sealed lead acid battery is permanently float charged. This definition courtesy APC.

Floating Batteries The normal technique for powering telephone equipment in which batteries are simultaneously charged from a commercial source or generator and discharged to operate the telephone equipment.

Floating Point Using an exponent with numbers to indicate the location of the decimal point. It's more precise than integer but slower. See FLOATING POINT ARITHMETIC.

Floating Point Arithmetic Calculations performed on floating point, or exponential numbers. These numbers have two parts, a mantissa and an exponent. The mantissa designates the digits in the number, and exponent designates the position of the decimal point. Essentially floating point arithmetic allows the representation of very large numbers using a small number of bits. The speed of scientific computers is often rated in the Millions of FLoating Operations Per Second (MFLOPS) they can perform.

Floating Selection An imaging term. A selected area that is conceptually floating above the image, allowing it to be manipulated without affecting the background (for example, the contents of the Clipboard).

Floating Virtual Connection FVC. The ability to resume an on-demand connection on a port other than the port on which the original on-demand connection was established.

Flood Projection In facsimile, the optical method of scanning in which the original is floodlighted and the scanning spot is defined by a masked portion of the illuminated area.

Flood Search Routing A routing method that employs an algorithm that determines the optimum route for traffic within a network, avoiding failed and congested links.

Floodgaters From Wired's Jargon Watch column. Individuals who send you email inquiries and, after receiving only a slightly favorable response, begin flooding you with multiple messages of little or no interest to you.

Flooding A packet-switched network routing method whereby identical packets are sent in all directions to ensure that they reach their intended destination.

Floor Feed An access point in a raised or cellular floor used for the exit of communications or power cables. Floor feeds can be fixed or drilled as required, depending upon the floor type.

Flop FLoating point Operation. Performing an operation on a floating point number. One measure of microprocessor speed is FLOPs per second, or MFLOPS (million flops per second).

Floppy Disk A thin, flexible plastic disk resembling a phonograph record upon which computer data is stored magnetically. Called a floppy disk because it is flexible and can (and will) flop inside a drive as it is being turned. And it may sound as though it is flopping. Floppy disks were never designed as the permanent storage many people are using them for at present. Floppy disks were designed by IBM as a way of having its sellers and engineers carry programs and program updates to its customers. Floppy disks were lighter and less cumbersome than carrying heavy spools of magnetic tape. IBM designed its floppy disks to be thrown away once their information was loaded into the mainframe computer. The moral of this story is that floppy disks are NOT permanent reliable storage. Anything stored on floppy disks should be backed up at least once and, if possible, twice. Floppy disks come in three diameter sizes — 3 1/2, 5 1/4 and 8 inches. Floppy disks can be now safely put through X-ray machines at US airports.

Floppy Mini A floppy disk smaller than the traditional 5 1/4 inch diameter floppy disk. Now most commonly the 3 1/2 inch size invented by Sony, and used by the Apple Macintosh, among others. All MS-DOS laptop computers have 3 1/2 inch disks.

Floptical Technology The combination of optical servo track positioning and magnetic read-and-write technologies used in 3 1/2-inch Very High Density floppy disk drives. Floptical is a registered trademark of Insite Peripherals.

Flow Control The hardware, software and procedure for controlling the transfer of messages or characters between two points in a data network — such as between a protocol converter and a printer — to prevent loss of data when the receiving device's buffer begins to reach its capacity. In flow control, you can also deny access to additional traffic that would further add to congestion. (Think about flow control and the airlines.) See FLOW CONTROL PROCEDURE, RATE-BASED FLOW CONTROL, QFC and ER.

Flow Control Parameter Facility X.25 facility that allows the negotiation of packet and window sizes in both directions of transmission.

Flow Control Procedure The procedure for controlling the rate of transfer of data among elements of a network, e.g., between a DTE and a data switching exchange network, to prevent overload.

Flowchart A graphic or diagram which shows how a complex operation, such as programming, takes place. The flowchart breaks that operation down into its smallest, and easiest-to-understand events.

Flush Jack 1. A telephone or data-connection jack mounted on and recessed in a wall. Each flush jack can have up to six connections on its face.
2. A toilet in a casino that works even after you have lost all your money (you still have to tip the attendant).

Flush Protocol An ATM term. The flush protocol is provided to ensure the correct order of delivery of unicast data frames.

Flushing Out the Queue A call center term. Changing system thresholds so that calls waiting for an agent group are redirected to another group with a shorter queue or available agents.

Fluorinated Ethylene Propylene FEP. Also known by the trade name Teflon, a registered trademark of Dupont. FEP is the insulation of choice for high performance cable and wire systems installed in return air plenums. As FEP is really slick, it makes the wire really easy to pull through conduits, around corners, and so on — the same property that makes it so wonderful in the kitchen. FEP's fire retardant properties have led many countries to require its use, particularly in plenum ceilings. See plenum.

Flushofone A cordless headset with a noise cancelling microphone so you can't hear a toilet flush when on the phone in the bathroom.

Flutter A rapid change in an electrical signal. The change may be in strength, frequency or phase. Distortion due to variation in loss resulting from the simultaneous transmission of a signal to another frequency.

Flux In soldering, a substance used to remove oxides from metal so the metal can be wet with molten solder for soldering.

Fly-By-Wire In traditional airplanes, the controls pilots moved were attached to heavy cables and hydraulic systems which themselves physically moved the rudder or the flaps, etc. Fly-by-wire replaced these wires and the hydraulic systems with computers and thin electrical wires. There are two main advantages to fly-by-wire. The computers can continuously adjust the aircraft's controls without the input of the pilot, trimming control surfaces so that the plane slides through the air with a minimum of air drag. Second, by eliminating heavy control cables and cutting down on hydraulic lines you can cut several hundred pounds off the weight of the plane, thus saving huge amounts of fuel over the life of the plane.

Flying Lead A grounding lead that exits the back of the connector hook on the outside of the cable jacket. It's normally attached to the drain wire or shield and then connected to the chassis of the switch, modem, etc.

Flywheel A flywheel is a large heavy wheel used in electrical power generation. It's connected to an electrical power generator and will keep the generator spinning after the power source (a waterfall, or whatever) is unavailable.

FM 1. Fault Management. A network management function designed to receive fault information into a centralized management function. The faults are monitored, tracked, and resolved.
2. Frequency Modulation. See FREQUENCY MODULATION.

FM Blanketing That form of interference to the reception of other broadcast stations, which is caused by the presence of an FM broadcast signal of 115 dBu (562 mV/m) or greater signal strength in the area adjacent to the antenna of the transmitting station. The 115-dBu contour is referred to as the "blanking area."

FM Capture A cellular radio term. In cases of extreme co-channel interference, a receiver may experience "FM capture", which is a co-channel interference condition where is selected. Cellular users often experience FM capture as a momentary burst of someone else's conversation.

FM Stereo Separation A measure of a radio tuner's ability to separate the left and right hand channels of a stereo broadcast. The higher the number, the greater the separation. The unit of measure is the Decibel (dB), a logarithmic unit which expresses the ratio between two voltage, current or power levels, usually relating to a standard reference level, or a background noise level.

FM Subcarrier One-way data transmission using the modulation of an unwanted portion of an FM broadcast station's frequency band.

FMAS Facility Maintenance and Administration System.

FMIC Flexible MVIP Interface Circuit. The FMIC provides a complete MVIP compliant interface between the MVIP bus and a variety of processors, telephony interfaces and other circuits. A built-in digital time slot switch provides Enhanced-Compliant MVIP switching between the full MVIP bus and any combination of up to 128 full duplex local channels of 64 kbps each. An 8-bit microprocessor port allows real-time control of switching and programmable device configuration. On board clock circuitry, including both analog and digital phase-locked loops, supports all MVIP clock modes. The local interface supports ST_BUS (Mitel), PCM Highway (Siemens), CHI (AT&T) signal formats at programmable rates of 2.048 MHz, 4.096 MHz, and 8.192 MHz as well as parallel DMA through the microprocessor port. See MVIP.

FMV Fair Market Value. A special lease for IRS purposes. Be careful. With Fair Market Value (FMV) leases, there is a catch to having the lessor guarantee the dollar amount or the percentage of your buyout. In order to be a FMV lease there must

be a risk. That is why you are paying a lower rate of interest. If you agree on an amount up front, make sure it is not in writing, otherwise it does not meet the IRS test for a FMV lease. With A FMV lease, the lessor owned the asset and depreciates it; lessee expenses monthly payments and deducts them for tax purposes. If the buyout is determined in writing and the IRS can prove it, then it is a financing lease and lessee owns asset and depreciates it. Beware of this. This advice from Jane A Blank, telecom consultant, Westerville OH.

FNA A Brussels-based strategic alliance, which exists to facilitate global communications connections for companies in the financial services sector. The 12 founding FNA companies are Stentor of Canada, AOTC of Australia, RTT-Belgacom of Belgium, France Telecom, Deutsche Bundespost Telekom of Germany, Hong Kong Telecom, Italcable of Italy, KDD of Japan, Singapore Telecom, Telefonica of Spain, Mercury Communications of the United Kingdom, and MCI of the United States.

FNC Federal Networking Council. The body responsible for coordinating networking needs among U.S. Federal Agencies. A US group of representatives from those federal agencies involved in the development and use of federal networking, especially those networks using TCP/IP, and the connected Internet. The FNC coordinates research and engineering. Current members include representatives from the DoD, DOE, DARPA, NSF, NASA and HHS.

FNR Fixed Network Reconfiguration

FNS Fiber Network Systems

FO Fiber Optics.

FOA 1. First Office Application. A telephone company term for what you and I know as beta testing.
2. Fiber Optic Amplifier. See FIBER OPTIC AMPLIFIER for a full definition.

FOB Free On Board. Term indicating where the seller's responsibility ends and the buyer's begins. You buy something, F.O.B. The seller puts it on a truck or railroad, plane, i.e. some carrier. He's responsible for getting it on the carrier. It's FOB, the truck. You — the buyer — are responsible for paying for the cost of the freight of getting you the goods you ordered. The opposite of F.O.B. is C.I.F. That stands for Cost, Insurance and Freight are included. That means the seller pays the freight. See FOB, FOB DESTINATION, FOB PLACE OF DELIVERY and FOB SHIPPING POINT.

FOB Destination Seller retains ownership until delivered to buyer. See FOB.

FOB Place of Delivery Seller retains ownership until delivered to buyer. See FOB.

FOB Shipping Point Seller responsibility ends when item is turned over to carrier. The buyer is responsible for payment if goods are damaged in transit. The buyer also handles any insurance claim. See FOB.

FOC Firm Order Confirmation. A Service Center response to an ASR such as a circuit order. An ASR is an Access Service Request. It is a request that a telephone company gives to another telephone company for any of many kinds of interconnectivity or data sharing needs. These requests can be between Local carriers or long distance carriers and can originate with either an incumbent or an alternative company.

Focal Point An IBM Network management term, it consolidates the functions needed to manage centrally all parts of a network. It provides an end-to-end network view and receives information from entry points and service points. NetView is IBM's key implementation of the focal point.

Focus A measure of the clarity of a color monitor. Focus relates to the sharpness of a monitor's electron beam as it paints the face of a Cathode Ray Tube (CRT). The other measures are convergence and dot pitch.

FMC Fixed Mobile Convergence. It means one phone, one telephone number (instead of our current private, business, mobile and fixed telephone numbers). With FMC, subscribers get to have one handset and one phone number, using that single handset and number to make and receive calls in the home, the know who made this semi-silly term up. It sure sounds to me like FMC is fancy word for having a cell phone, carrying it with you everywhere and using it as your main phone — wherever you are. Good luck. I've never found cell phones to be that reliable, that good quality or their serviceso cheap that I'd give up my wired phones.

FOD Fax On Demand. See FAX BACK.

Foil A slang term for an overhead transparency. The expression "he gives good foil" reflects an executive's ability to make great presentations using overhead transparencies. In the 1970s and early 1980s, so many managers at IBM made presentations that some senior executives actually got overhead projectors built into their desks.

Foilware Foil is a slang term for an overhead transparency. There are various iterations in the development of a product. One of the first is a description of the product on overhead transparencies. Such overheads are often used to convince investors to put money into the company or to convince distributors to sell the product. This often happens long before the product actually exists. Sometimes the company will pretend with its foilware that its products actually exist. In this case, the products then become the foilware. See Foil.

FOIRL Fiber Optic Inter Repeater Link. Defined in IEEE 802.3 and implemented over two fiber links, transmit and receive, this medium may be up to one kilometer in length, depending on the number of repeaters in the network. A FOIRL is the perfect transmission medium to join a local area network on the eleventh floor to the fourth floor of the same building.

Folder A subdirectory or a file folder. The Apple Macintosh was the first to use them. Microsoft picked up on the idea when it introduced Windows 95.

Follow Me 800 Service Basically, Follow Me 800 Service is call forwarding of your personal 800 line. MCI announced this service in the Spring of 1991. It differs from local call forwarding in that you can dial into MCI from anywhere in the world and change the number your 800 line will send its calls to. Your 800 number always stays the same. What changes is the number it calls. A simple explanation: We buy a personal 800 line from MCI. The number is 800-555-6534. When someone calls that number, MCI looks up a database, checks where to send the number and sends it to my office at 212-691-8215. However, one day I go traveling. So I call another MCI 800 number, punch in my identification number and then give it the new number I will be at — namely 212-206-6660. From then on, MCI will send all my calls to that number — until I call and change the number again.

Follow Me Call Forwarding Progressive Call Forwarding. Allows a previously forwarded call to be forwarded from that to another phone extension.

Follow Me Roaming The ability for the cellular system to automatically forward calls to a roaming mobile that has left it's primary service area. Without this feature, the calling party must know the location of the roamer and place a call to that area.

Follow Me Services Also called One Number Services. Follow me systems and services are based on the premise that people are mobile (e.g., they move around a lot in and out of the office), and have many phone numbers or places they might be. A person could have an office number, a cellular number, a voice mail number, a home number, and a pager number. Which phone number will a caller be at? Follow me systems will "track-down" the user being called no matter where they are and connect the caller to the user. The caller need only dial a single phone number. Usually, network or local switch provided call data (or data gathered by a voice response unit) is used to identify each call as being intended for a specific user. Based on options the user has selected, the caller will hear an answering prompt customized to that user, and the one number system will then automatically attempt to locate the user at one of several locations. The tracking-down process varies considerably between follow me systems. Some systems try multiple locations at once, others will try the possible destination locations sequentially. Almost all follow me services provide the caller an exit to voice mail at various points of the call.

Follow The Sun Dialing A technique used in call centers whereby the agents call those parts of the country where it's convenient to call and move the calling across the country as the sun moves. Our agents might call New York households between 6 P.M. and 9 P.M. When the time hits 9 P.M., the agents stop calling New York households and start focusing calls on households in the central time zone. To accomplish Follow The Sun Dialing, a call center needs software which knows in which phone numbers are in which time zones.

Fonline 800 Sprint's inbound service for small to medium-sized customers with applications up to 500 hours per month.

Font Alphanumeric and other characters in a distinctively shaped type style or type face. Common fonts are Helvetica, Times Roman, Century.

Font Family A group designation that describes the general look of a font.

Font Size See point size.

Foo An Internet term. A place-holder for nearly anything — a variable, function, procedure, or even person. "A given user foo has the address 'foo@bar.com'."

Football Also called Aerial Service Wire Splice. A device used to splice aerial service wire shaped a bit like a football when it's installed.

Footprint 1. The area on the earth's surface where the signals from a specific satellite can be received. A footprint is shown as a series of concentric contour lines that show the area covered and the decreasing power of the signal as it spreads out from the center.
2. The area on a desk a device occupies, i.e. the computer's footprint.

Force Administration Data System See FADS.

Force Feed An arrangement in an outbound telebusiness unit where agents are force fed with a new call which is automatically dialed, a pre-determined time after finishing the previous call.

Forced Account Code Billing A telephone feature which prevents call from being completed if the user does not pushbutton in a billing code. That billing code may correspond to the department within the company. Or it may conform to the client and to the client's matter number the call must be billed to.

Forced Authorization Code FAC. A PBX feature which requires all or certain users to enter a code before dialing an outside number.

Forced Hop A channel hop made by the Mobile Data Base Station (MDSB) because non-Cellular Digital Packet Data (CDPD) activity is detected on the channel that is currently in use.

Forced Perfect Terminator A type of terminator containing a sophisticated circuit that can compensate for variations in the power supplied by the host adapter, as well as variations in bus impedance of complex SCSI systems.

Forced Release/Disconnect The switching center's automatic hang-up if the calling party fails to do so at the end of a conversation.

Forced Route Override Allows a PBX user to automatically redirect an outgoing call to a different trunk if the first trunk is busy or the connection is poor.

FORD 1. Fixed Or Repaired Daily.
2. Found On the Road Dead.
3. First On Race Day.
4. Fast is a FORD Letter Word

Forecasting Taking historical data (what happened in the past) from your ACD and using that information to predict what might happen in the future. Has your call volume always doubled on Tuesday? It will probably double next Tuesday too. A very important function of call center management software.

Foreground Processing Automatic execution of computer programs designed to preempt the use of the computing facilities. Usually a real time, urgent program. Contrast this with Background Processing, which might be something less urgent, for example, diagnostics of the system.

Foreign Address An ATM term. An address that does not match any of a given node's summary addresses.

Foreign Agent A Mobile IP term. A service which enables mobile nodes associated with nomadic users to register their presence at a remote location. The foreign agent communicates with the home agent in order that data packets can be forwarded to the remote subnet. See also Mobile IP.

Foreign Area Translation Translating the office codes of a distant (foreign) area to codes that make sense to a PBX which has more than one way of completing the call to that area.

Foreign Central Office Service Getting telephone service in a multi-office exchange from a central office other than the one you are normally served by. Not a common term any longer. Foreign central office service is the same price as normal local central office service. It typically just involves asking for service off another central office. For example, our main number in New York City is 212-691-8215. Our 691- central office is in the 18th Street Exchange, a tall building on 18th Street. There is another central office in the same building. It is 206- and it is a more modern central office. When we ordered additional lines, we ordered them from this central office. You can now also call us on 212-206-6660. Don't trust my definition, however. Ask you local telephone company. See also FOREIGN EXCHANGE SERVICE.

Foreign EMF Any unwanted voltage on a telecommunications circuit.

Foreign Exchange Service FX. Provides local telephone service from a central office which is outside (foreign to) the subscriber's exchange area. In its simplest form, a user picks up the phone in one city and receives a dial tone in the foreign city. He will also receive calls dialed to the phone in the foreign city. This means that people located in the foreign city can place a local call to get the user. The airlines use a lot of foreign exchange service. Many times, the seven digit local phone number for the airline you just called will be answered in another city, hundreds of miles away. See also FOREIGN CENTRAL OFFICE SERVICE and FOREIGN EXCHANGE TRUNK.

Foreign Exchange Trunk A Foreign EXchange (FEX) trunk provides a direct connection between a PBX switch and a remote central office other than the central office that serves the location of the PBX.

Foreign Numbering Plan Area FNPA. Any other NPA (Numbering Plan Area) outside the geographic NPA where the customer's number is located.

Foreign Prefix Service Getting dial tone in a multi wire center exchange from a foreign wire center other than the one you are normally served by. Similar to Foreign Central Office Service, except that you may get charged extra for Foreign Prefix Service. Don't trust my definition, however. Ask you local telephone company. See also FOREIGN CENTRAL OFFICE.

Forklift Upgrades A forklift is a self-propelled machine used to lift and transport heavy objects by means of steel fingers inserted under the load. An upgrade is an improvement or advancement in size or functionality. A forklift upgrade has its roots in the days of "heavy iron," when it literally took a forklift to upgrade the PBX or mainframe computer technology. One drove a forklift into the switchroom or computer room, picked up the system, transported it out the door, and brought in an improved system. The old system, which had little use, was made into a boat anchor or artificial reef, or so the story goes. Such upgrades typically cost an arm and a leg. Such upgrades are increasingly uncommon today. Rather, much gear is upgradable by simply changing the generic software load, possibly inserting a card or microchip or two, and perhaps swapping out a power supply. Most switch manufacturers are trying to figure ways to avoid forcing their customers into forklift upgrades when their requirements outgrow the existing system. Examples include stackable hubs and distributed switches, which are scalable to one degree or another. See SCALABLE and STACKABLE.

Form A group of graphical controls in an HTML document: text boxes, radio buttons, drop-down lists, check boxes, etc. A user on the Web browsing the document can

Form 230 Form 730 Application Guide is a collection of literature you'll need to register your telephone/telecom equipment under Part 68 of Title 47 at the Federal Communications Commissions. To get this material (it's free) drop a line or call the Federal Communications Commission, Washington DC 20554. As I write this edition, the person at the FCC in charge is William H. Von Alven, who also puts out a newsletter for Part 68 applicants. See PART 68 for a much larger explanation.

Form Effectors FEs. Control characters intended for the layout and format of data on an output device such as a printer or CRT. Examples are CR (carriage return) and LF (line feed).

Form Factor Fancy way of saying shape and size (width, depth, height).

Formal Call Centre A British term. A telebusiness unit in which all of the staff are dedicated to telephone based work. See also Informal Call Centre.

Formal Standards Specifications which are approved by vendor-independent standards bodies, such as ANSI (American National Standards Institute), ISO (International Standards Organization), IEEE (Institute of Electrical and Electronic Engineers) and NIST (National Institute of Standards and Technology).

Formant A point of excitation, or high energy, in a speech waveform caused by resonance in the human vocal tract. Formants are responsible for the unique timbre of each individual's voice.

Formant Synthesis A form of synthesized speech in which the computer creates the voice. The result is smooth but sometimes artificial-sounding. Formant synthesis is used in text-to-speech (TTS) technology in which the computer "reads" text as voice. Another technology used in TTS is called concatenation synthesis, which uses actual samples of human voice, chopped up and put back together. Concatenation synthesis sounds choppy.

Format 1. Arrangement of bits or characters within a group, such as a word, message, or language.
2. Shape, size and general makeup of a document. As a verb, its most common usage is in "to format this disk."

FORTEZZA A cryptology mechanism developed by Mykotronx, Inc., a subsidiary of Rainbow Technologies, in conjunction with the NSA, which holds the registered trademark. The family of FORTEZZA security products includes PCMCIA-based client cards, and server boards; compatible implementations are available variously in hardware and software. All FORTEZZA Crypto implementations support data privacy, user ID authentication, data integrity, non-repudiation, and time-stamping. FORTEZZA is the crypto token chosen to secure the Defense Messaging System (DMS), including both the MILNET and the Internet. Applications include e-mail, voice communications and file transfer. Depending on the application, the encryption keys are either 80 or 160 bits in length, thereby providing excellent security for "Sensitive But Unclassified" (SBU) government data, as well as for commercial applications. FORTEZZA opponents suggest that the NSA is attempting to force the mechanism on the private sector as a replacement for the rejected Clipper Chip technology. The fear is that the NSA holds the keys to the secret encryption algorithm, and that the agency, therefore, can gain access to your data even more easily that it could have through the "backdoor" built into the Clipper Chip. See also Clipper Chip and MISSI.

FORTRAN FORmula TRANslating system. A computer programming language.

Fortuitous Conductor Any conductor that may provide an unintended path for intel igible signals, e.g., water pipes, wire or cable, metal structural members.

Fortune Cookie An inane/witty/profound comment that can be found around the Internet.

Forum A section within an online service (such as CompuServe, America Online, etc.) where you can find out information on a specific subject — computers made by Toshiba or printers made by Hewlett Packard. Forums may include a library from which you can download various files (programs, bug fixes, printer drivers, text, press releases of new products and so on). Many forums also include one or more "conference rooms" which users may "enter" for conversations (on-line or off-line) with representatives of companies or the person running the forum, who is typically called the "sysop," as in system operator. Most manufacturers run forums as a relatively cheap and painless way of getting help information to their customers.

Forward A switch feature that temporarily redirects incoming calls. The incoming calls are redirected from the forwarding telephone to another destination by the person associated with the telephone or by the computing domain. The other destination has previously been defined to the switch by the device associated with the telephone.

Forward Busying That feature of a telecommunications system wherein supervisory signals are forwarded in advance of address signals to seize assets of the system before attempting to establish a call.

Forward Channel The communications path carrying data or voice from the person who made the call. The Forward Channel is the opposite of the Reverse Channel.

Forward Direction The forward direction of data away from the head-end in a broadband LAN.

Forward Echo An echo propagating in the same direction as the original wave in a transmission line, and formed by energy reflected back from one irregularity and then onward again by a second. Forward echoes can occur at all irregularities in a length of cable, and, when they add systematically, can impair its performance as a transmission medium.

Forward Error Correction FEC. A technique of error detection and correction in which the transmitting host computer includes some number of redundant bits in the payload (data field) of a block or frame of data. The receiving device uses those bits to detect, isolate and correct any errors created in transmission. The idea of forward error correction is to avoid having to retransmit information which incurred errors in network transit. The additional bits add a small amount of overhead to the block or frame. Therefore, they create some level of inefficiency in transmission. The alternative is retransmission of the block or frame of data, which can be much more inefficient where large numbers of errors occur during transmission. This inefficiency is compounded when the retransmitted block or frame is errored, as well. From the standpoint of network throughput, FEC can be much more effective, particularly when bandwidth is expensive or limited. On the other hand, FEC is processor-intensive, as it places a load on the computational capabilities of the receiving computer. The simple idea of forward error correction is to avoid having to retransmit information sent incorrectly. The technique is consuming of bandwidth and can make the transmission take longer.

I asked Ray Horak, how can a few redundant bits of information significantly reduce error rates on transmission. Here's his reply: The process is extraordinarily complex. Explaining it would take pages and pages and would do no one but a mathematician any good. Essentially, a few redundant data bits are added at strategic places in the data field (for example, just suppose that every 50th bit were repeated — the exact repeated bits vary according to the specific algorithm used). The very few redundant bits significantly lower the potential for an individual data bit to be transmitted in error and go undetected, given the complex sampling technique and complex algorithms used to develop a description of the data field. The receiving host computer is intelligent enough and has enough computational horsepower at its disposal to figure it out, unlike most of us real human types. The issue and the tradeoff is one of the cost of processing power vs. the cost of retransmission across the network. As the cost of computers comes down and the cost of bandwidth comes down, the best solution remains specific to the specifics of the user and the application.

Forward Prediction A technique used in video compression, specifically compression techniques based on motion compensation, where a compressed frame of video is reconstructed by working with the differences between successive video frames.

Forwarding Description An ATM term. The resolved mapping of an MPOA Target to a set of parameters used to set up an ATM connection on which to forward packets.

FOSSIL Fido/Opus/Seadog Standard Interface Layer. This is the interface used as an add-on to mailer software packages to connect them to PCs that are not 100% IBM-compatible.

FOTS Fiber Optic Transmission System. Not the same as POTS. But a neat acronym, nevertheless.

Foundation Graphics A set of graphics libraries or imaging models that form the lowest level graphics programmer's interface in Sun's OpenWindows. Examples: a graphics sub-routine library that a program could call to draw graphics primitives like arcs, circles, rectangles, etc.

Four Horsemen of the Apocalypse War, Plague, Famine and Death.

Four Pair UTP Cable There are four pairs of conductors in this cable for a total of eight conductors. The cable jacket (also called the cable sheath) holds all four pairs together. Many manufacturers also include a ripcord, used for cutting the cable jacket. Pull on it, the sheath opens, allowing you to get to your conductors to attach them to things. Sometimes, the ripcord works. Sometimes, it doesn't. See UTP Cable.

Four Wavelength Wave Division Multiplexing 4WL-WDM, also called Quad-WDM. MCI announced this technology in the Spring of 1996 as a method of allowing a single fiber to accommodate four light signals instead of one, by routing them at different wavelengths through the use of narrow-band wave division multiplexing equipment. The technology allowed MCI to transmit four times the amount of traffic along existing fiber. At that time MCI's backbone network operated at 2.5 Gbps (2.5 billion bits per second) over a single strand of fiber optic glass. Using Quad-WDM the same fiber's capacity rose to 10 gigabits — enough capacity to carry approximately 130,000 simultaneous voice transmissions over one single strand of fiber. Since then, a number of carriers have deployed OC-192 (Optical Carrier Level 192) fiber, running at 10 Gbps. They are opening four "windows," or wavelengths, each running at 10 Gbps. That's 40 Gbps over a single strand through DWDM (Dense Wavelength Division Multiplexing). They are pulling as many as 620 strands at a time. While most carriers have elected to implement WDM/DWDM in their networks by purchasing equipment from vendors, MCI has concentrated on developing their own WDM capability internally. Generally speaking, the intense competition among the various DWDM vendors is pushing capacity upwards faster than MCI's own internal development. See also WDM and DWDM.

Four-wire See Four-wire Circuit.

Four-wire Adapter A device which allows the connection of two-wire telephone equipment to a four-wire line. See FOUR-WIRE CIRCUITS.

Four-wire Circuit A high-performance circuit, which offers lots of bandwidth and which is capable of multi-channel communications. Four-wire circuits are of two types: physical and logical. Physical four-wire was the original approach. In other words, they all comprised four wires, which were organized into two copper pairs of UTP (Unshielded Twisted Pair). These original four-wire circuits were analog, and used amplifiers to overcome the effects of signal attenuation, which is a significant problem at high frequencies. As the amplifiers worked in only one direction, two pairs of wires and two sets of amplifiers were needed: one for transmission in one direction and another in the reverse direction. A lot of physical four-wire circuits remain in use, and more are being deployed every day. Even though such circuits mostly are digital today, it generally still requires four physical UTP wires to provide four-wire service such as T-1. Logical four-wire performs like physical four-wire, but with fewer wires. ISDN BRI (Basic Rate Interface) is an example of logical four-wire, as it usually uses only two wires to achieve

relatively lots of bandwidth (144 Kbps), and multiple channels (2B+D, or 2 Bearer channels plus 1 Data channel). HDSL2 (High bit-rate Digital Subscriber Line, version 2), an emerging local loop technology, provides T-1 service over only 2 UTP wires. SONET fiber optic technology provides incredible amounts of bandwidth and supports hundreds of thousands of channels using only 2, or even 1, physical wires (glass fibers). Microwave, satellite and infrared transmission systems support four-wire service without any wires at all. See also ISDN, T-1 and SONET.

Four-wire Repeater See Four-wire Circuit.

Four-wire Terminating Set An electrical device which takes a four-wire circuit — one pair coming and one pair going — and turns it into the "normal" tip and ring circuit you need for a typical telephone, key system or PBX. See FOUR-WIRE CIRCUITS.

Fourier's Theorem In the early 1800s, the French mathematician Emile Fourier proved that a repeating, time-varying function may be expressed as the sum of a (possibly infinite) series of sine and cosine waves. Digital data is a bit stream, which can be sent as a sequence of square waves. Fourier's Theorem shows that to send a square wave (digital signal), a series of sine waves (analog signals) are actually summed together. If 1,000 square waves are to be sent every second, for example, the frequency components of the sine waves that are summed together are 1 kHz, 3 kHz, 5 kHz, 7 kHz, etc. The point of this analysis is to show that high frequency signals are required to form a stable, recognizable square wave.

As the bit rate increases, the square wave frequency increases and the width of the square waves decrease. Thus, narrower square waves require sine waves of even higher frequencies to form the digital signal. Note, then, that there is insufficient bandwidth in the 3 kHz voiceband to send square waves due to the absence of frequency components above 3,300 Hz. Even low frequency square waves cannot be sent because sine waves below 300 Hz are also absent. Thus, the local loop, according to Fourier's Theorem, cannot be used for the transmission of digital signals! The last paragraph is, in fact, no longer totally correct, as the increasingly successful ISDN trials are proving.

Fourth Estate The press. In May 1789, Louis XVI, King of France, summoned to Versailles a full meeting of the "Estates General." The First Estate consisted of 300 nobles; the Second Estate, 300 clergy; the Third Estate, 600 commoners. Some years later, and well after the French Revolution, Edmund Burke, looking up at the press gallery of the British House of Commons, said "Yonder sits the Fourth Estate, and they (i.e. the press) are more important than them all."

Fourth Utility The non-vendor specific communications premise wiring system which you use for integrated information distribution (voice, data, video, etc.) Leviton in Bothell, Washington has trademarked the term Fourth Utility. They make a broad range of premise wiring products.

Fox Message A standard sentence for testing teletypewriter circuits because it uses most of the letters on the keyboard. That sentence is "The quick brown fox jumped over the lazy sleeping dog, 1234567890"

FP 1. Feature Package. A software release for a telephone system. Originated with AT&T's Dimension PBX, now manufacturer discontinued. 2. File Processor.

FPDL Foreign Processor Data Link. A link from a Rockwell ACD to an external computer.

FPG Feature Planning Guide.

FPGA Field Programmable Gate Array. An FPGA is a specialized microprocessor that has no physical connections between its logic gates when it leaves the factory. But it has a huge number of potential connections, which can be firmed up in the field by a programmer with the right tools. FPGAs are a competitor to the cheaper ASICs — Application Specific Integrated Circuits. See VIRTUAL COMPUTING.

FPI Formal Public Identifier. A string expression that represents a public identifier for an object. FPI syntax is defined by ISO 9070.

FPLMTS Future Public Land Mobile Telecommunication Systems. A subject under discussion among the world's standards bodies. FPLMTS's objective is global terminal mobility.

FPM DRAM Fast Page Mode Dynamic Random Access Memory.

FPP Fiber Optic Patch Panel.

FPS 1. Fast Packet Switching.
2. Frames Per Second. A measure of the quality of a video signal. NTSC TV — the standard in North America — uses 30 fps. Film is 24 FPS. PAL/SECAM (European) is 25 FPS.

FPT Forced Perfect Terminator. A high-quality type of single-ended SCSI terminator, developed by IBM, with special circuitry that compensates not only for variations in terminator power but also for variations in bus impedance. See also Active Terminator and Passive Terminator.

FPU Floating Point Unit. A formal term for the math coprocessors (also called numeric data processors, or NDPs) found in many PCs. The Intel 80387 is an example of an FPU. FPUs perform certain calculations faster than CPUs because they specialize in floating-point math, whereas CPUs are geared for integer math. Today, most FPUs are integrated with the CPU rather than sold separately. See also CPU and DSP.

FQDN Fully Qualified Domain Name. An Internet term. The FQDN is the full site name of an Internet computer system, rather than just its hostname. For example, the system lisa at Widener University has a FQDN of lisa.cs.widener.edu.

FR See Flat Rate Service.

FR-1 A flammability rating established by Underwriters Laboratories for wires and cables that pass a specially designed vertical flame test. This designation has been replaced by VW-1.

Fractal A word coined in 1975 by Benoit B. Mandlebrot from the Latin fractus ("to break"). One fractal creator called fractals a shape with the property of "self-similarity."

Fractal Compression An asymmetrical compression technique that shrinks an image into extremely small resolution-independent files by storing it as a mathematical equation as opposed to storing it as pixels. The process starts with the identification of patterns within an image and results in collection of shapes that resemble each other but that have different sizes and locations within an image. Each shape-pattern is summarized and reproduced by a formula that starts with the largest shape and repeatedly displaces and shrinks it. These patterns are stored as equations and the image is reconstructed by iterating the mathematical model. Fractal compression can store as many as 60,000 images on one CD-ROM. One disadvantage of fractal compression is that it is time consuming, taking as long as four minutes to convert a 1.3 MB TIFF file to a 228 KB file. See FRACTALS.

Fractal Geometry The underlying mathematics behind fractal image compression, discovered by two Georgia Tech mathematicians, Michael Barneley and Alan Sloan.

Fractal Image Format FIF. A compression technique that uses on-board ASIC chips to look for patterns. Exact matches are rare and the process works on finding close matches using a function known as an affine map.

Fractals Along with raster and vector graphics, fractals are a way of defining graphics in a computer. Fractal graphics translate the natural curves of an object into mathematical formulas, from which the image can later be constructed. See FRACTAL COMPRESSION.

Fractional Services A British term. Bandwidth available from carriers in increments of 64Kbit/s such as Mercury's Switchband. See FRACTIONAL T-1 for the North American definition.

Fractional T-1 FT-1. Fractional T-1 refers to any data transmission rate between 56/64 Kbps (DSO rate) and 1.544 Mbps (T-1). Fractional T-1 is a four-wire (two copper pairs) digital circuit that's not as fast as a T-1. Fractional T-1 is popular because it's typically provided by a LEC (Local Exchange Carrier) or IXC (IntereXchange Carrier) at less cost than a full T-1, and in support of applications that don't require the level of bandwidth provided by a full T-1. While FT-1 is less costly than a full T-1, it is more costly on a channel-by-channel basis, as you would expect. Users love FT-1, but carriers hate it. FT-1 costs the carriers just as much to provision as does as full T-1, they just turn down some of the channels. FT-1 is typically used for LAN interconnection, videoconferencing, high-speed mainframe connection and computer imaging.

Fractional T-3 A telephone company service in which portions of a T-3 (44.7364 Mbps) transmission service are leased to provide a service similar to a T-1 (1.544 Mbps) or T-2 (3.152 Mbps) channel, but normally at a lower cost.

FRAD Frame Relay Access Device, also sometimes referred to as a Frame Relay Assembler/Disassembler. Analogous to a PAD (Packet Assembler/Disassembler) in the X.25 world, a FRAD is responsible for framing data with header and trailer information prior to presentation of the frame to a Frame Relay switch. On the receiving end of the communication, the FRAD serves to strip away the Frame Relay control information in order that the target device is presented with the data packaged in its original form. On the receiving end, the FRAD also generally is responsible for detecting errors in the payload data created during the process of network switching and transmission; error correction generally is accomplished through a process of retransmission. A FRAD may be a stand-alone device, although the function generally is embedded in a router.

Fragment The pieces of a frame left on an FDDI ring, caused by a station stripping a frame from the ring.

Fragmentation 1. In messaging it is the process in which an IP (Internet Protocol) datagram is broken into smaller pieces to fit the requirements of a given physical network. The reverse process is termed "reassembly."
2. ATM and SMDS networks routinely perform a process of Segmentation and Reassembly (SAR), segmenting the native PDU into 48-octet payloads which are carried in 53-octet cells. The process is reversed on the receiving end.
3. A condition that affects data stored on a disk. Adding and deleting records in a file, creates what is sometimes called the Swiss cheese effect. The operating system stores the data for an individual file in many different physical locations on the disk, leaving large holes between records. Fragmented files slow system performance because it takes time to locate all parts of a file.

Frame 1. A generic term specific to a number of data communications protocols. A frame of data is a logical unit of data, which commonly is a fragment of a much larger set of data, such as a file of text or image information. As the larger file is prepared for transmission, it is fragmented into smaller data units. Each fragment of data is packaged into a frame format, which comprises a header, payload, and trailer. The header prepends (prepend means added to the front of) the payload and includes a beginning flag, or set of framing bits, which are used for purposes of both frame delineation (beginning of the frame) and synchronization of the receiving device with the speed of transmission across the transmission link. Also included in the header are control information (frame number), and address information (e.g., originating and terminating addresses). Following the header is the payload, which is the data unit (fragment) being transmitted. Appending the payload is the trailer, which comprises data bits used for error detection and correction, and a final set of framing bits, or ending flag, for purposes of frame delineation (ending of the frame). This frame format, in the broader generic sense, also is known as a data packet. Frame, therefore, is a term specific to certain bit-oriented data transmission protocols such as SDLC (Synchronous Data Link Control) and HDLC (High-level Data Link Control), with the latter being a generic derivative of SDLC. In the case of SDLC, a frame is very similar to a block, which would be employed in a character-oriented protocol such as IBM's BSC (Binary Synchronous Communications), also known as Bisync. See also BSC, HDLC, Packet, and SDLC.

2. In TV video, a frame is a single, complete picture in video or film recording. A video frame consists of two interlaced fields of either 525 lines (NTSC) or 625 lines (PAL/SECAM), running at 30 frames per second (NTSC) or 25 frames per second (PAL/SEACAM). 24 frames are sent in moving picture films and a variable number, typically between 8 and 30, sent in videoconferencing systems, depending on the transmission bandwidth available. Up to about 12 frames a second looks "jerky."

3. One complete cycle of events in time division multiplexing. The frame usually includes a sequence of time slots for the various sub channels as well as extra bits for control, calibration, etc. T-Carrier makes use of such a framing convention for packaging data. Channelized T-1, for instance, frames 24 time slots with a framing bit which precedes each set of sampled data.

4. A unit of data in a Frame Relay environment. The frame includes a payload of variable length, plus header and trailer information specific to the operation of a Frame Relay network service.

5. A metal framework, such as a relay rack, on which equipment is mounted. A distribution frame. A rectangular steel bar framework having "verticals and horizontals" which is used to place semipermanent wire cross connections to permanent equipment. Found in telephone rooms and central offices. See Distribution Frame.

Frame Alignment The extent to which the frame of the receiving equipment is correctly phased (synchronized) with respect to that of the received signal.

Frame Alignment Errors A frame alignment error occurs when a packet is received but not properly framed (that is, not a multiple of 8 bits).

Frame Alignment Sequence See Frame Alignment Signal

Frame Alignment Signal FAS. Frame Alignment Signal or Frame Alignment Sequence.
The distinctive signal inserted in every frame or once in n frames that always occupies the same relative position within the frame and is used to establish and maintain frame align-

ment, i.e. synchronization. See FRAME ALIGNMENT ERRORS.

Frame Buffer A section of memory used to store an image to be displayed on screen as well as parts of the image that lie outside the limits of the display. Some systems have frame buffers that will hold several frames, in which case they should be called "frames buffers." But they're not.

Frame Check Sequence Bits added to the end of a frame for error detection. Similar to a block check character (BCC). In bit-oriented protocols, a frame check sequence is a 16-bit field added to the end of a frame that contains transmission error-checking information. In a token ring LAN, the FCS is a 32-bit field which follows the data field in every token ring packet. This field contains a value which is calculated by the source computer. The receiving computer performs the same calculation. If the receiving computer's calculation does not match the result sent by the source computer, the packet is judged corrupt and discarded. An FCS calculation is made for each packet. This calculation is done by plugging the numbers (1's and 0's) from three fields in the packet (destination address, source address, and data) into a polynomial equation. The result is a 32-bit number (again 1's and 0's) that can be checked at the destination computer. This corruption detection method is accurate to one packet in 4 billion. See FRAME CHECK SEQUENCE ERRORS.

Frame Check Sequence Errors Errors that occur when a packet is involved in a collision or a corrupted by noise.

Frame Dropping The process of dropping video frames to accommodate the transmission speed available.

Frame Duration The sum of all the unit time intervals of a frame. The time from the start of one frame until the start of the next frame.

Frame DS1 The DS1 frame comprises 193 bit positions. The first bit is the frame overhead bit, while the remaining 192 bits are available for data (payload) and are divided into 24 blocks (channels) of 8 bits each.

Frame Error An invalid frame identified by the Frame Check Sum (FCS). See also FRAME ERRORS.

Frame Errors In the 12-bit, D4 frame word, an error is counted when the 12-bit frame word received does not conform to the standard 12-bit frame word pattern.

Frame Flag Sequence The unique bit pattern "01111110" used as the opening and closing delimiter for the link layer frames.

Frame Frequency A video term. The number of times per second a frame is scanned.

Frame Grab To capture a video frame and temporarily store it for later manipulation by a graphics input device.

Frame Grabber A PC board used to capture and digitize a single frame of NTSC video and store it on a hard disk. Also known as Frame Storer. See VIDEO CAPTURE BOARD.

Frame Ground FGD. Frame Ground is connected to the equipment chassis and thus provides a protective ground. Frame Ground is usually connected to an external ground such as the ground pin of an AC power plug.

Frame Header Address information required for transmission of a packet across a communications link.

Frame Multiplexing The process of handling traffic from multiple simultaneous inputs by sending the frames out one at a time in accordance with a specific set of rules. Instead of multiplexing traffic from a lower-speed connection into a higher speed connection based on a specific time duration for each low-speed channel, frame multiplexing using the length of a given frame as the measurement.

Frame Rate The number of images displayed per second in a video or animation file. The Frame Rate is highly significant is determining the quality of the image, with a high frame rate creating the illusion of full fluidity of motion. 30 frames per second (30 fps) is considered to be full-motion, broadcast quality. On the other end of the scale, 2fps is most annoying. At 30 fps, the brain processes the images, filling in the blanks due to the "Phi Phenomenon." See PHI PHENOMENON.

Frame Relay Frame relay, technically speaking, is an access standard defined by the ITU-T in the I.122 recommendation, "Framework for Providing Additional Packet Mode Bearer Services." Frame relay services, as delivered by the telecommunications carriers, employ a form of packet switching analogous to a streamlined version of X.25 networks. The packets are in the form of "frames," which are variable in length, with the payload being anywhere between 0 and 4,096 octets. The key advantage to this approach is that a frame relay network can accommodate data packets of various sizes associated with virtually any native data protocol. In other words, a X.25 packet of 128 bytes or 256 bytes can be switched and transported over the network just as can an Ethernet frame of 1,500 bytes. The native Protocol Data Unit (PDU) is encapsulated in a Frame Relay frame, which involves header and trailer information specific to the operation of the Frame Relay network.

Further, a Frame Relay network is completely protocol independent. Not only can any set of data be accepted, switched and transported across the network, but the specific control data associated with the payload is undisturbed in the process of encapsulation. Additionally, and unlike a X.25 network, a Frame Relay network assumes no responsibility for protocol conversion; rather, such conversions are the responsibility of the user. While this may seem like a step down from X.25, the data neither requires segmentation into fixed length packets nor does the network have to undertake processor-intensive and time-consuming protocol conversion. The yield is faster and less expensive switching.

A Frame Relay network also assumes no responsibility for errors created in the processes of transport and switching. Rather, the user also must accept full responsibility for the detection and correction of such errors. The user also must accept responsibility for the detection of lost packets (frames), as well for the recovery of them through retransmission. Again, this may seem like a step down from X.25 networks, which correct for errors at each network node, but which detect and recover from lost packets. Once again, however, the yield is faster and less expensive switching. In fact, it is unlikely that frames will be damaged, as the switches and transmission facilities are fully digital and offer excellent error performance.

Much like X.25, Frame Relay employs the concept of a shared network. In other words, the network switches accept frames of data, buffer them as required, read the target address and forward them one-by-one as the next transmission link becomes available. In this fashion, the efficiency of transmission bandwidth is maximized, yielding much improved cost of service. The downside is that some level of congestion is ensured during times of peak usage. The level of congestion will vary from time-to-time and frame-to-frame, resulting in latency (delay) which is unpredictable and variable in length. This is especially true in a Frame Relay network (as opposed to X.25), as the length of the frames is variable—the switches never quite know what to expect.

Access to a Frame Relay is over a dedicated, digital circuit which typically is 56/64 Kbps, Nx56/64 Kbps, T-1 or T-3. The device which interfaces the user to the network is in the form of a Frame Relay Access Device (FRAD) which serves to encapsulate the native PDU before presenting it to the network. The FRAD at the destination address unframes the data before presenting it to the target device, with the two FRADs working together much as do PADs in a X.25 environment. Further, it generally is the responsibility of the FRAD to accomplish the error detection and correction process, although this responsibility may be that of the eventual target device. Across the digital local loop, the FRADs connect functionally to Frame Relay Network Devices (FRNDs, pronounced "friends"), proving once again that the carriers want to be your friends (especially as Frame Relay users tend to be large organizations with lots of $$$ to spend).

Frame Relay is intended for data communications applications, most especially LAN-to-LAN internetworking, which is bursty in nature. Frame Relay is very good at efficiently handling high-speed, bursty data over wide area networks. It offers lower costs and higher performance for those applications in contrast to the traditional point-to-point services (leased lines). Additionally, Frame Relay offers a highly cost-effective alternative to meshed private line networks. As the Frame Relay network is a shared, switched network, there is no need for dedicated private lines, although special-purpose local loops connect each customer location to a frame switch. Transmission of frames between the user sites is on the basis of Permanent Virtual Circuits (PVCs), which are pre-determined paths specifically defined in the Frame Relay routing logic. All frames transmitted between any two sites always follow the same PVC path, ensuring that the frames will not arrive out of sequence. Backup PVCs, generally offered by the carrier at trivial cost, provide redundancy and, therefore, network resiliency in the event of a catastrophic network failure. With frame relay, a pool of bandwidth is made instantly available to any of the concurrent data sessions sharing the access circuit whenever a burst of data occurs. An addressed frame is sent into the network, which in turn interprets the address and sends the information to its destination over broadband facilities. Those facilities may be as "slow" as 45 Mbps, but more often are SONET fiber optics in nature and operating at much higher speeds. Like traditional X.25 packet networks, frame relay networks use bandwidth only when there is traffic to send.

Frame Relay, while intended for data communications, also supports compressed and packetized voice and video. While such isochronous data is highly sensitive to the variable latency characteristic of packet networks, improved compression algorithms such as ACELP provide quite acceptable support for voice over Frame Relay, subject to the level of con-

---

Frame Relay Frame

1	2	n =0-4096B	2	1	octets
Flag (0111110)	Address Field	Information Field	Frame Check Sequence	Flag (0111110)	

DLCI (high order)	C/R	EA	DLCI (low Order)	FECN	BECN	DE	EA

DLCI:	Data Link Connection Identifier
C/R:	Command/Response Field
FECN:	Forward Explicit Congestion Notification
BECN:	Backward Explicit Congestion Notification
DE:	Discard Eligibility
EA:	Address Field Extension

gestion in the network. For voice to be supported satisfactorily in a packet network, the receiving end compensates for delay and delay variation.

In addition to public network services, Frame Relay can also be implemented in a private network environment consisting of unchannelized T-Carrier circuits. Such an implementation offers exceptional data communications performance over an existing leased line network. Additionally, framed voice and video can ride over such a network, essentially for "free" when the circuits are not being used for data communications purposes. Thereby, the usage of the circuits is maximized, with little concern for poor quality due to network congestion.

A Frame Relay frame consists of a header, information field, and trailer. The header comprises a Flag denoting the beginning of the frame, and an Address Field used for routing of the frame, as well as for purposes of congestion notification. The Information Field is of variable length, from 0 to 4,096 Bytes. The trailer consists of a Frame Check Sequence (FCS) for detection and correction of errors in the Address Field, and an ending Flag denoting the end of the frame.

The American National Standards Institute (ANSI) describes frame relay service in the following documents:

ANSI T1.602 — Telecommunications — ISDN — Data Link Layer Signaling Specification for Application at the User Network Interface.

ANSI T1.606 — Frame Relaying Bearer Service — Architectural Framework and Service Description.

ANSI T1S1/90 - 175 - Addendum to T1.606 - Frame Relaying Bearer Service — Architectural Framework and Service Description.

T1.607-1990 ISDN Layer 3 Signaling Specification for Circuit-Switched Bearer Service for DSS-1

T1.618 DSS-1 Core aspects of Frame Protocol for use with frame relay bearer service, ANSI, 1991

ANSI T1.617a, Signaling specification for Frame Relay bearer service for DSS-1, 1994

Frame relay access makes use of the LAP-D signaling protocol developed for ISDN. Frame relay, technically speaking again, does not address the operation of the network switches, multiplexers or other elements. Both the ITU-T and ANSI were highly active in the development of Frame Relay standards, as was ETSI in Europe. See the next three definitions.

Frame Relay Access Device Required for connection into a frame relay network.

Frame Relay Forum Organization of frame-relay equipment vendors, carriers, end users and consultants working to speed the development and deployment of frame relay products, as well as interfaces with other broadband technologies, such as ATM. The Frame Relay Forum is based in Foster City, CA. 415-578-6980. It was formed in May 1991 as a non-profit mutual corporation. It has over 300 members. See also FRAME RELAY IMPLEMENTORS FORUM and ATM. www.frforum.com.

Frame Relay Implementors Forum A group of companies which have announced their support for a common specification for frame relay connections to link customers premises equipment to networking equipment. The common specification was originally announced on September 4, 1990. The common specification is based on the standard frame relay interface proposed by the American National Standards Institute (ANSI). The common specification supports the proposed ANSI standard and defines the extensions to that standard, including a local management interface that allows the exchange of control information between the user

device and the frame relay network equipment. The specification is available for review from Cisco Systems, Digital Equipment Corporation, Northern Telecom and StrataCom. See FRAME RELAY and FRAME RELAY FORUM.

Frame Relay Modem A data communications device which connects to a PC's COM (serial) port and emulates a dial tone while actually establishing a dedicated 56Kbps frame relay connection.

Frame Slip That condition in a TDM network under which a receiver of a digital signal experiences starvation or overflow in its receive buffer due to a small difference in the speeds of clocks and the clock (transmission rate) at the transmitter. The receiver will drop or repeat of a full TDM frame (193 bits on a T-1 line) in order to maintain synchronization.

Frame Store A system capable of storing complete frames of video information in digital form. This system is used for television standards conversion, computer applications incorporating graphics, video walls and video production and editing systems.

Frame Switch A device similar to a bridge that forwards frames based on the frames' layer 2 address. Frame switches are generally of two basic forms, cut-through switch (on-the-fly-switching) or store and forward switch. LAN switches such as Ethernet, Token Ring, and FDDI switches are all examples of frame switches.

Frame Synchronization The process whereby a given digital channel (time slot) at the receiving end is aligned with the corresponding channel (time slot) of the transmitting end as it occurs in the received signal. Usually extra bits (frame synchronization bits) are inserted at regular intervals to indicate the beginning of a frame and for use in frame synchronization.

Frame UNI Frame-based User-Network Interface, a frame format for access to ATM networks. Defined by the Frame Relay Forum, Frame UNI is a derivative of the DXI standard. For low-speed access application, it provides for a router to send frames (much like Frame Relay frames) to an ATM Edge Switch, where the conversion to cell format takes place.

Frames A term used to describe a viewing and layout style of a World Wide Web site, it refers to the simultaneous loading of 2 or more web pages at the same time within the same screen. Originally developed by Netscape and implemented in their Navigator 2.0 browser, today many other popular Web browsers support this feature. Some Web sites come in two versions; a "frames" and "no frames" version. The frames version usually takes a longer to load and may contain other "enhanced" features such as Java and Animation.

Frames Received OK The number of frames received without error. See FRAMES RECEIVED TOO LONG.

Frames Too Long An Ethernet statistic that indicates the number of frames that are longer than the maximum length of a proper Ethernet frame, but not as long as frames resulting from jabbering.

Framework A Taligent definition. A set of prefabricated software building blocks that programmers can use, extend, or customize for specific computing solutions. With frameworks, software developers don't have to start from scratch each time they write an application. Frameworks are built from a collection of objects, so both the design and code of a framework may be reused.

Framing An error control procedure with multiplexed digital channels, such as T-1, where bits are inserted so that the receiver can identify the time slots that are allocated to each subchannel. Framing bits may also carry alarm signals indicating specific alarms. In TDM reception, framing is the process of adjusting the timing of the receiver to coincide with that of the received framing signals. In video reception, the process of adjusting the timing of the receiving to coincide with the received video sync pulse. In facsimile the adjustment of the facsimile picture to a desired position in the direction of line progression

Framing Bit 1. A bit used for frame synchronization purposes. A bit at a specific interval in a bit stream used in determining the beginning or end of a frame. Framing bits are non-information-carrying bits used to make possible the separation of characters in a bit stream into lines, paragraphs, pages, channels etc. Framing in a digital signal is usually repetitive.

Framing Error An error occurring when a receiver improperly interprets the set of bits within a frame.

Franchise The exclusive right to operate telephone service in a community. This right — also called the franchise — is granted by some government agency. Some phone companies existed before the appropriate regulatory authority, so they're "grandfathered" in their exclusivity. Some phone companies have an exclusive area to serve more because of their presence than because of the legal right conferred on them. The question of who has a franchise to serve what community with what service is becoming increasingly unclear as competition penetrates all aspects of the phone industry.

Franchise Authority The contractual agreement between a cable operator and a local governmental body, that defines the rights and responsibilities of each in the construction and operation of a cable system with n a specified geographical area.

FRD Fire RetarDant. A rating used for cable within duPont's Teflon or equivalent fluorpolymer material. FRD cable is used when local fire codes call for low flame and low smoke cable. FRD cable is typically run in forced air plenums as an alternative to metal conduits.

FRED A system for searching the international X.500 user directory.

Free Address Office Office arrangement in which all personal spaces are eliminated in favor of employees picking up their supplies at a front desk upon arriving and then choosing a temporary work area each day.

Free On Board FOB. Board. Term indicating where the seller's responsibility ends and the buyer's begins. You buy something, F.O.B. The seller puts it on a truck or railroad, plane, i.e. some carrier. He's responsible for getting it on the carrier. It's FOB, the truck. You — the buyer — are responsible for paying for the cost of the freight of getting you the goods you ordered. The term FOB is typically enhanced (or made clear), thus:

FOB Destination: Seller retains ownership until delivered to buyer.

FOB Place of Delivery. Seller retains ownership until delivered to buyer.

FOB Shipping Point. Seller responsibility ends when item is turned over to carrier. The buyer is responsible for payment if goods are damaged in transit. The buyer also handles any insurance claim.

The opposite of F.O.B. is C.I.F. That stands for Cost, Insurance and Freight are included. That means the seller pays the freight.

Free Space Communications Any form of telecommunications that doesn't a conductor (e.g., copper wire, or glass or plastic fiber). In other words, free space communications is accomplished using "space," rather than a conductor, as a medium. Radio (e.g., radio, microwave, satellite, wireless LANs, cellular) and optical (i.e., infrared) transmission systems communicate through space, rather than through a conductor. See also Airwave.

Free Space Loss This is simply the power loss of the signal as a result of the signal spreading out as it travels through space. As a wave travels, it spreads out its power over space, i.e. as the wave front spreads, so does its power.

FREENET An organization to provide free Internet access to people in a certain area, usually through public libraries.

Freephone A service which permits the cost of the call to be charged to the called party, rather than the calling party. Pioneered in the US in 1966, the freephone service now carries around 100 millions calls per day in the US alone. US companies currently hold around 90% of the world's 9 million freephone numbers.

Freephone has proves particularly popular with business subscribers, who are often willing to bear the cost of a telephone call in order to promote their services or to encourage customers to order their products by phone. Recent estimates by AT&T indicated some $100 billion is currently traded over the freephone service every year. Until the recent approval of the new ITU-T Recommendation E-169, however, companies have been restricted by only being able to use their freephone number in one country. Those organizations wishing to offer products or services to customers on an international basis have had no choice but to register a separate number in each country, which has proved unwieldy and often inefficient. The new standard for international freephone will greatly free up companies' abilities to operate across international markets, and will benefit consumers by allowing them to obtain information or to shop around for goods and services at no personal expense. It is hoped the new standard might also stimulate the market for freephone services in Europe and Asia-Pacific, regions that until now have been slow to take up the service.

The potential market for the new international freephone service is expected to be considerable. The globalization of markets via new technologies such as the Internet means that many companies are now able to offer their products and services to users in different countries, and will benefit from being able to advertise a single toll-free number to potential customers all over the world. Calls to the new global market number can also be routed to different destinations, allowing companies to direct their incoming calls to the most appropriate location for efficient handling.

Free Space Communications Radio communications including microwave, satellite and cellular.

Freeware Software that doesn't cost anything, but may work just as well as the software you pay for.

Freeze In digital picture manipulators, the ability to stop or hold a frame of video so that the picture is frozen like a snapshot.

Freeze Frame The transmission of discrete video picture frames at a data rate which is too slow to provide the perception of natural motion, referred to as "full-motion." An uncompressed, digitized full-motion video signal is typically transmitted at many millions of bits per second. Freeze frame can be carried on anything from a simple voice grade phone line running at 9.6 Kbps (the same speed as a Group 3 facsimile machine).

Freq A term or abbreviation for a "file request" for a file from another node in a network. In FidoNet a node user usually freq's a file through mailer software which sends an appropriate request to a distant node that has the desired file. Freqing is the ability in FidoNet to transfer files back and forth between BBSs (bulletin board systems) automatically.

Equivalent to file transfer in PCRelay.

Freqing See FREQ.

Frequency The rate at which an electromagnetic waveform (e.g., electrical current) alternates, usually measured in Hertz. Hertz is a unit of measure which means "cycles per second." So, frequency equals the number of complete cycles of energy (e.g., current) occurring in one second. See Bandwidth, Bandwidth, Frequency and Hertz.

Frequency Agile Modem A modem used on some broadband LANs (Local Area Networks). A frequency agile model can search the frequencies on the LAN to find one available in order to communicate with other attached devices.

Frequency Agility The ability of a cellular mobile telephone system to shift automatically between frequencies.

Frequency Band The portion of the electromagnetic spectrum within a specified upper- and lower-frequency limit. Also known as Frequency Range. See also BAND, FREQUENCY for a complete list of all the frequencies.

Frequency Domain Waveforms, such as speech signals, are typically viewed in the time domain, i.e. as power levels or voltages varying over time. The 19th century French mathematician Fourier demonstrated an algorithm called "Fast Fourier Transform", or FFT, which can express any complex waveform over a fixed interval as the sum of a series of sine waves of different energy levels. Analyzing signals in the frequency domain has proven an extremely powerful technique with diverse applications, including filtering, recognition and speech modeling.

Frequency Diversity A way of protecting a radio signal by providing a second, continuously operating radio signal on a different frequency, which will assume the load when the regular channel fails. Here's another way of saying the same thing: Frequency diversity is a any method of diversity transmission and reception wherein the same information signal is transmitted and received simultaneously on two or more independently fading carrier frequencies.

Frequency Division Multiple Access A technique for sharing a single transmission channel (such as a satellite transponder) among two or more users by assigning each to an exclusive frequency band within the channel.

Frequency Division Multiplexing FDM. An older technique in which the available transmission bandwidth of a circuit is divided by frequency into narrower bands, each used for a separate voice or data transmission channel. This means you can carry many conversations on one circuit.

Frequency Frogging The interchanging of the frequency allocations of carrier channels to prevent singing, reduce crosstalk, and to correct for a transmission line frequency-response slope. It is accomplished by having the modulators in a repeater translate a low-frequency group to a high-frequency group, and vice versa. Because of this frequency inversion process, a channel will appear in the low group for one repeater section and will then be translated to the high group for the next section. This results in nearly constant attenuation with frequency over two successive repeater sections, and eliminates the need for large slope equalization and adjustment. Also, singing and crosstalk are minimized because the high-level output of a repeater is at a different frequency from the low-level input to other repeaters.

Frequency Grease A special kind of radio lubricant that is used to overcome problems of static in radio transmissions. Actually, there is no such thing, but every new radio technician falls prey to the joke. It's much like a "pot stretcher." Ray Horak, my Contributing Editor, was a Mess Sergeant

in the US Army. He would send the privates on KP (Kitchen Patrol) to another mess hall to get a pot stretcher if the pot was too small, or to get a screen door for the refrigerator during the summer. His buddies in Communications would send the new radio technicians to get some radio grease. It was a lot of fun during the Vietnam War, which was not a lot of fun. It worked only one time per private (usually). See also Bucket o' Dial Tone.

Frequency Hopping Another name for spread spectrum transmission. A technique developed by Hedy Lamarr, the actress, in the early part of the second world war to prevent the enemy from jamming or eavesdropping on conversations and on commands to steer torpedoes, etc. The idea is to hop from one frequency to another in split-second intervals as you transmit information. Attempts to jam the signal succeed only in knocking out a few small bits of it. So effective is the concept that it is now the principal antijamming device in the US military. Ms. Lamarr never got paid for the invention. But it was definitely hers. She invented it because of her patriotism for the United States. She had fled Austria in 1937. She received a U.S. patent in 1940. See also SPREAD SPECTRUM.

Frequency Modulation A modulation technique in which the carrier frequency is shifted by an amount propor-

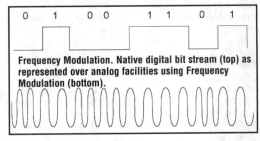

Frequency Modulation. Native digital bit stream (top) as represented over analog facilities using Frequency Modulation (bottom).

tional to the value of the modulating signal. The amplitude of the carrier signals remains constant. The deviation of the carrier frequency determines the signal content of the message. Commercial TV and FM radio use this technique, which is much less sensitive to noise and interference than is amplitude modulation (AM). In the world of modems, digital bit streams can be transmitted over analog facilities through this same technique, whereby a 0 bit might be represented by a high-frequency sine wave (or set of sine waves) and a 1 bit by a low-frequency sine wave (or set of sine waves). Contrast with AMPLITUDE MODULATION and PHASE SHIFT KEYING.

Frequency Offset Non-linear distortion that causes a shift in the frequency of a received signal.

Frequency Response The variation (dB) in relative strength between frequencies in a given frequency band, usually the voice frequency band of an analog telephone line.

Frequency Reuse The ability to use the same frequencies repeatedly within a single system, made possible by the basic design approach used in cellular. Since each cell is designed to use radio frequencies only within its boundaries, the same frequencies can be reused in other cells not far away with little potential for interference. The reuse of frequencies is what allows a cellular system to handle a huge number of calls with a limited number of channels.

Frequency Shift Keying FSK. A modulation technique for data transmission. It shifts the frequency above the carrier for a 1 and below the carrier for a 0 (zero). See also FSK.

Frequency Tolerance The maximum permissible departure by the center frequency of the band occupied by an emission from the assigned frequency or by the characteristic frequency of an emission from the reference frequency. By international agreement, frequency tolerance is expressed in parts per 10 (6) or in hertz. This includes both the initial setting tolerance and excursions related to short- and long-term instability and aging. In the United States, frequency tolerance is expressed in parts per 10(n), in hertz, or in percentages.

Frequency Translator In a split broadband cable system, a frequency translator is an analog device at the headend that converts a block of inbound frequencies to a block of outbound frequencies.

Fresnel Loss The loss at a joint that is caused by a portion of the light being reflected.

Fresnel Reflection In optical physics, fresnel reflection is the reflection of a portion of incident light at a planar interface between two homogeneous media having different reflective indices. Fresnel reflection occurs at the air-glass interfaces at entrances at entrance and exit ends of an optical fiber. Resultant transmission losses (on the order of 4 percent per interface) can be virtually eliminated by using antireflection coatings or index-matching materials. Fresnel reflection depends upon the index difference and the angle of incidence. In optical elements, a thin transparent film is sometimes used to give an additional Fresnel reflection that cancels the original one by interference. This is called an antireflection coating.

Fresnel Reflective Losses For optical fiber communication, the losses incurred at the terminus interface that are due to refractive index differences.

Fresnel Region In radio communications, the region between the near field of an antenna and the Fraunhofer region. The boundary between the two is generally considered to be at a radius equal to twice the square of antenna length divided by wavelength.

Fresnel Zone Fresnel zone is the line-of-sight path between two microwave antennas. It is an elliptical zone between the two antennas where the total path distance varies by more than half of the operating wavelength. The concept is extended to describe the distance by which the direct wave clears any intervening obstacle such as a mountain peak. If the total path distance between transmitter, peak and receiver, is 1 wavelength greater than the direct distance, then the clearance is said to be two Fresnel zones.

FRF11 Basically, there are two main standards regarding voice transmission over data networks: H.323 and "Voice Over Frame Relay Implementation Agreement" (FRF.11). Both specify that the following coders should be used: G.711, G.728, G.729, and G.723.1. The H.323 adds the G.722, and the VoFR (Voice over Frame Relay) adds the G.726/7 coders. The G.711 is a PCM coder that uses 64ks/s and two companding techniques: A-law and Mu-law. Recommendation G.722 describes 7 kHz audio-coding within 64 kbit/s. The G.723.1 describes a Dual rate speech coder for multimedia communications transmitting at 5.3 and 6.3 kbit/s and is based on Multi Pulse Maximum Likelihood Quantizer (MP-MLQ) (Voice frame duration of 30mSec). The G.726 describes 40, 32, 24, 16 kbit/s Adaptive Differential Pulse Code Modulation (ADPCM). The G.727 describes 5-, 4-, 3- and 2-bits sample embedded adaptive differential pulse code modulation (ADPCM). The G.728 describes Coding of speech at 16 kbit/s using Low-Delay Code Excited Linear Prediction (LD CELP). The G.729 describes Coding of speech at 8 kbit/s using Conjugate-Structure Algebraic-Code- Excited Linear-Prediction (CS-ACELP) (Voice frame duration of 10mSec).

FRF92.02 Multiprotocol Interconnect over Frame Relay.

FRF92.07 Frame Relay Multicast Draft Service Description.

FRF92.08 Frame Relay Network-to-Network Interface Implementation Agreement Draft.

Frictional Electricity Static electricity produced by friction (e.g., by rubbing a hard rubber rod with a silk cloth.)

Friendly Name A name, typically identifying a network user or a device, intended to be familiar, meaningful, and easily identifiable. A friendly name for a printer might indicate the printer's physical location (e.g. "Sales Department Printer").

Friesen, Gerry One of the smartest people you'll meet in a long time. Gerry was, and remains my partner in most everything businessy I co. Originally, he was half the operation which published this dictionary (I was the other half). Now Miller Freeman publishes the dictionary, though I still write it. And Gerry has retired to his own personal paradise. On January 1, 1999 he sent me the following email: HAPPY NEW YEAR PARTNER!!! Its been a pleasure working with you over the years. I hope 1999 brings you everything you hope for. My New Year's resolution is to do something fun every day. Hope yours is as rewarding, whatever it is, as I plan mine to be."

FRL Facility Restriction Level. A term created by AT&T for its Dimension PBX. These levels define the calling privileges associated with a line; for example, intragroup calling only in the warehouse, but unrestricted calling from the boardroom.

FRND Frame Relay Network Device. Pronounced "friend." A device that sits at the edge of a public frame relay network. The FRND is the point of ingress into the cloud of the network on the inbound side of the network and the point of egress on the outbound side. The FRND connects to the user's FRAD (Frame Relay Access Device), which often is in the form of a router, over a digital access link, which usually is some form of T-carrier circuit (e.g., T-3, T-1, or Fractional T-1). A FRND can be in the form of either a switch or a router, although it usually is a router. See Frame Relay, FRAD and T-1.

Frogging Frogging is the process of inverting line frequencies of a carrier system so that incoming high-frequency channels leave at low frequencies and vice versa. Frogging equalizes the transmission loss between high and low frequency channels.

Front End The client part of a client/server application that requests services across a network from a server, which is known sometimes as the back end. It typically provides an interactive interface to the user, for example, a data entry front end, allowing database to be entered into a database server. The term "front end" is, of course, a contradiction in terms. But, who said language had to be consistent, or logical? See FRONT END CONTROLLER, FRONT END DEVELOPMENT TOOLS and FRONT END PROCESSOR.

Front End Controller See FRONT END PROCESSOR.

Front End Development Tools These tools let a programmer control the design and manipulation of applications using visual techniques. Front-end development creates a graphical user interface, providing more flexibility and making it easier to link users with data accessed from database servers.

Front End Equipment The equipment positioned between a computer and the communications line(s). Its purpose is to organize data being sent and received.

Front End Mailer A program that operates on a bulletin board system and determines if a caller is another computer that wants to exchange mail or a human that wants to exchange mail or a human that wants to access the BBS resources. Usually the mailer transmits the prompt "Press ESC" and upon receiving an ESC character, or the passing of a timeout period, considers the caller to be human and gives it the resources of the BBS. Also known as a mailer.

Front End Processor FEP. An FEP is a computer under the control of another, larger computer (typically a mainframe) in a network. The FEP does simple, basic "housekeeping" operations on the data streams as they arrive to be processed by the bigger computer. The FEP acts as a sort of intelligent traffic cop. It relieve the bigger, host computer of some of its telecommunications Input/Output burden, so that the host computer can concentrate on handling the processing burden. Depending on its sophistication, the front end processor might also perform serial to parallel conversion, protocol conversion, block or message assembly, etc. Here's a more technical definition: A dedicated communications system that intercepts and handles activity for the host. Can perform line control, message handling, code conversion, error control, and such applications functions as control and operation of special-purpose terminals. Designed to offload from the host computer all or most of its data communications functions. IBM 3705, 3725 and 3745 are Front End Processors. Front end processors are not used in client/server networks.

Front End Results A call center/marketing term. Used to describe the rate of expected or tentative, rather than actual, orders generated, usually as a result of a trial or free offer.

Front Porch The blanking signal portion which lies between the end of the active picture information and the leading edge of horizontal sync.

FRS Frame-Relay Service: A connection oriented service that is capable of carrying up to 4096 bytes per frame.

FRTT Fixed Round-Trip Time: This is the sum of the fixed and propagation delays from the source to the furthest destination and back.

FRU Field Replaceable Unit.

FSA Foreign Serving Arrangement.

FSK Frequency Shift Keying. A modulation technique for translating 1's and 0's into something that can be carried over telephone lines, like sounds. A "1" will be assigned a certain frequency of tone, and a "0" will be assigned to another tone. The transmission of the bits keys the sounds to shift from one frequency to the other. See also FREQUENCY SHIFT KEYING.

FSN Full Service Network. A term introduced by Time Warner Cable, a large cable television operator, in late 1994. According to Time Warner, FSN is a prototype for future interactive games being played over cable TV lines.

FSP 1. File Service Protocol. It's a file-transfer protocol, similar to FTP except that it doesn't create much load on a server. 2. Fiber Optic Splice Panel.

FSS See Fixed Satellite System.

FT-1 Fractional T-1. Any part of a T-1 circuit that's smaller than a full T-1 circuit. Fractional T-1 circuits are cheaper than full T-1 circuits. That's their reason for existing. See FRACTIONAL T-1 for a bigger explanation.

FT-3 Fractional T-3. typically fractional T-3 delivers between four megabits per second, all the way to the full T-3 capacity of 45 megabits per second. See T-3.

FT1 Fractional T-1. See FT-1 and T-1.

FT3 Fractional T-3. See FT-3 and T-3.

FTAM File Transfer and Access Management. The OSI (Open Systems Interconnection) standard for file transfer (i.e., the communication of an entire file between systems), file access (i.e, the ability to remotely access one or more records in a file) and management (e.g., the ability to create/delete, name/rename a file). FTAM is also an international standard.

FTE Full Time Equivalent. A call center term. A scenario

assumption used in budget forecasting and scheduling that defines the number of hours per week full-time employees are normally suppose to work. As a measure of staffing level, an FTE is equivalent to a full-time position, even though the hours may actually be filled by part-time schedules.

FTIP Fiber Transport Inside Plant.

FTP 1. File Transfer Protocol and File Transfer Program. FTP lets users quickly transfer text and binary files to and from a distant or local PC, list directories, delete and rename files on the foreign host, and perform wildcard transfers between hosts. That distant or local PC (also called an FTP host) might be on your local area network, or a phone line across the world or connected to the Internet. FTP, the program, is actually a MS-DOS program. It comes with Windows. Here's how to use it: Make sure you're on the Internet. Then go into MS-DOS. Then type ftp and the name of your ftp host you want to reach. For example, that might be ftp.harry.com. So you'd type ftp ftp.harry.com. If it finds ftp.harry.com, it will come back and ask you for your user name. Give it. Let's say "harry." (Read below for anonymous logins.) Then it will ask you for your password. Then it will say you're logged in. Then you type send d:\work\jokes. And it will send your file called jokes. Once you're through, simply type quit. If you're uncertain of ftp.exe's command, load ftp by typing ftp. You'll get an ftp prompt. Then type ?.

On the Internet, the FTP, or File Transfer Protocol, is an extension of the TCP/IP protocol suite. FTP is best known as an Internet tool for accessing file archives around the world that are linked to the Internet. If you have a modem with a terminal emulation program or are running Windows 95 or 98, you have all the software you need to visit ftp sites. Here's what it's like to visit the a ftp site. The site address may be ftp.pht.com. Most public ftp sites accept anonymous login. That is, when the site asks who you are, you answer "anonymous." Then when they ask for your password, you give them your Internet email address. FTP is file-sharing protocol that operates at layers 5 through 7 of the Open Systems Interconnection (OSI) model. See also Anonymous FTP and TCP/IP.

2. Foil Twisted Pair. A type of STP (Shielded Twisted Pair) cable which is employed to protect the signal-carrying conductors from EMI (ElectroMagnetic Interference). FTP uses a thin metallic foil; ScTP (Screened Twisted Pair) uses a heavy braided mesh for this purpose. See also STP.

FTP Mail Server A server which permits the retrieval of files via e-mail. See FTP.

FTS Federal Telecommunications System is a private telephone network sometimes shared enthusiastically by all federal government agencies. And sometimes not. See FTS2000.

FTS2000 The U.S. General Services Administration in Washington, D.C. describes the FTS2000 as "the state of the art, digital, long distance telecommunications program that provides voice, data and video transmission services to federal government agencies."

FTTB Fiber To The Building. See FTTN and FTTP.

FTTC Fiber To The Curb. A hybrid transmission system which involves fiber optics to the curb, and either twisted pair or coaxial cable to the premises. FTTC is less extreme than FTTH, but more so than FTTN. See also FTTH, FTTN and HFC.

FTTCab Fiber To The Cabinet. Also known as FTTN (Fiber To The Neighborhood). A Hybrid Fiber Coax (HFC) network architecture involving an optical fiber which terminates in either a street-side or neighborhood cabinet which converts the signal from optical to electrical. The subscriber connection is over either UTP (Unshielded Twisted Pair) or coaxial cable. FTTCab can be either FTTC or FTTN. See HFC.

FTTH Fiber To The House. See FTTN and FTTP.

FTTN Fiber To The Neighborhood. Also known as FTTCab (Fiber To The Cabinet). A hybrid network architecture involving optical fiber from the carrier network, terminating in a neighborhood cabinet which converts the signal from optical to electrical. The connection from the cabinet to the user premises is over UTP (Unshielded Twisted Pair) or coaxial cable. ILECs (Incumbent Local Exchange Carriers), i.e., local telephone companies, use the embedded UTP for this purpose; CATV providers use the embedded coaxial cable. The advantages of the fiber include incredible levels of bandwidth, outstanding error performance, and transmission over long distances without the requirement for expensive and troublesome repeaters. The advantage of the UTP and coax is simply it is already there. The advantage of the neighborhood cabinet is that the expensive optoelectric conversion process takes place at a single location per neighborhood of perhaps 100 or 200 users. The cabinet also serves as a sophisticated multiplexer, allowing all the users to share the single, high-capacity fiber optic system for connection to the carrier network. FTTN is the preferred local loop architecture in a full convergence scenario, which involves the delivery of voice, Internet access, and entertainment TV over the same hybrid cable plant. While FTTH (Fiber To The House) is preferable in terms of overall performance, it currently is too expensive for serious consideration. Further, ADSL (Asymmetric Digital Subscriber Line) technologies will support the necessary levels of bandwidth over considerable distances, assuming the UTP is of good quality. See also ADSL, Fiber Optics, FTTH, FTTN, FTTP, HFC, SONET.

FTTP Fiber To The Premise. Also know as FTTB (Fiber To The Building) and FTTH (Fiber To The House). FTTP is the most extreme implementation of fiber optic transmission systems. FTTP literally involves a fiber optic system which connects directly from the carrier network to the user premises. The advantages of the fiber include incredible levels of bandwidth, outstanding error performance, and transmission over long distances without the requirement for expensive and troublesome repeaters. Taking the fiber optic transmission system to the premise extends all of these advantages to the user's front door. FTTP is used extensively by CAPs (Competitive Access Providers) and CLECs (Competitive Local Exchange Carriers). It also is used increasingly by ILECs (Incumbent LECs) and IXCs (IntereXchange Carriers) to provide the optimum level of service to medium and large businesses, which can make full use of the unparalleled performance of the fiber local loop. FTTP also was highly touted as the ultimate solution for residential application (FTTH) in a full convergence scenario. Full convergence involves the support of voice, Internet access, high-speed data communications, entertainment TV, videoconferencing, and other forms of communications-all over a single local loop. While FTTH certainly is attractive, it's just too expensive at the moment. FTTN (Fiber To The Neighborhood) is a much more cost-effective approach. See also FTTN.

FUBAR F...d Up Beyond All Recognition. In short, a mess. A term often used in electronic mail messages.

FUD Fear, Uncertainty, Doubt. A marketing tactic which a dominant player (once IBM) has used to discourage its customers from buying from its competitors.

Fugitive Glue Glue used by printers to affix stuff into magazines. The glue is designed to stick until the magazine is delivered. At that point, the stuck-in thing becomes easier to remove and/or falls into your lap. This definition contributed by Rich Kubik.

Fugitive Odor A smell that leaks out of a composting plant or landfill.

Full Availability Idealized condition which exists when your phone system can provide connections for every telephone connected to it. Also called NON-BLOCKING.

Full Duplex Transmission in two directions simultaneously, or, more technically, bidirectional, simultaneous two-way communications. The best two-direction phone conversations take place on four-wire circuits, two for transmission in one direction and two for transmission in the other. All long distance circuits are four wire. Most local lines are two wire, which means they're a compromise. It's important to contrast full duplex with symmetrical and asymmetrical. Full duplex means simultaneous transmission in both directions. Asymmetric means more bandwidth (or speed) in one direction than in the other. Symmetric means the same bandwidth (or speed) in both directions. Most speakerphones (except the newer more expensive ones) are half-duplex, meaning they only transmit in one direction at one time. The speakerphone flips its direction based on who's talking, or, more precisely, who's talking the loudest. Full duplex speakerphones are the best. See Speakerphones and Four-Wire CIRCUITS.

Full Duplex Audio Audio that allows remote sites to speak simultaneously without losing audio contact (two-way simultaneous audio). See Full Duplex.

Full Echo Suppressor An echo suppressor in which the speech signals on each path are used to control the suppression loss in the other path of a 4-wire circuit. Used for long-distance communications. Compare with split echo compressor.

Full Motion Video Television transmission where images are sent and displayed in real-time and motion is continuous. Video reproduction at 30 frames per second (NTSC-original signals) or 25 frames per second. (PAL-original signals). Compare with freeze frame. See FREEZE FRAME VIDEO.

Full System Battery Backup This means there's sufficient battery power backing the phone system so that during a power outage, the telephone system will continue to work, i.e. you won't even know the commercial power has gone out. All programming will be intact. Calls will get through, etc. Full System Battery Backup is critical to many businesses, especially those in the "life or death" business, such as hospitals, police, fire departments, etc. Other businesses who depend heavily on the phone for their revenues — airlines, brokerage companies, hotel/motels, etc. — often also use full system battery backup.

Full Time Equivalent FTE. A call center term. A scenario assumption used in budget forecasting and scheduling that defines the number of hours per week full-time employees are normally suppose to work. As a measure of staffing level, an FTE is equivalent to a full-time position, even though the hours may actually be filled by part-time schedules.

Fully Connected Network A network topology in which each node is directly connected by branches to all other nodes. This architecture becomes impractical as the number of nodes in the network increases in complexity. Such networks normally go to distributed nodes.

Fully Perforated Paper tape on which information is represented by the holes punched through the paper.

Fully Qualified Domain Name FQDN. In Internet terms, the full name of a system, rather than just a host name. For example, if the host name is Harry, the full name is harry.company.com).

Fully Restricted Stations In a PBX, fully restricted stations (also called phones) can't place any outside calls. They can make intercom calls as well as receive incoming calls.

Function Key 1. One of up to 12 keys on a PC keyboard labeled with the letter F followed by a number. The effect (if any) of pressing a particular function key depends on which program you are running at the time.

2. An undefined key on a computer or telephone that can be defined to perform one function, which would normally require the user hitting one or several keys in succession.

Functional Entity FE. A set of functions that provides one or more specified capabilities. Seven FEs have been identified for the Advanced Intelligent Network Release 1 architecture: Network Access, Service Switching, Service Logic and Control, Information Management, Service Assistance, Automatic Message Accounting and Operations. Definition from Bellcore in reference to its concept of the Advanced Intelligent Network. See AIN.

Functional Group A collection of FEs (Functional Entities) that reside together in a system.

Functional Management Layer A communications layer in SNA that formats presentations.

Functional Profile A defined stack of ISO OSI-Layer elements, such as GOSIP, MAP or TOP. Functional profiles were developed in order to ensure that, when defined, ISO OSL stacks could interoperate. Due to the number of different protocol elements at each OSI layer, it was possible to define stacks that were syntactically correct, but would not be able to exchange information due to differences at particular layers. A functional profile that has been defined as a standard is termed a standardized profile. Likewise, an International Standard Profile is an ISO OSI functional profile.

Functional Resource An abstraction of physical entities (e.g., voice synthesizers) that the Service Assistance FE (Functional Entities) can manipulate.

Functional Signaling In an ISDN circuit, function signaling provides messages with unambiguous, defined meanings known to both the sender and receiver of the messages. Signaling is generated by the terminal.

Functional Specification A description of a system from a working point of view. It differs from a precise technical description which includes each piece of equipment precisely spelled out. A system can often work the same using different hardware and software configurations. By functionally describing a system, a user allows sellers to use their imagination to solve the problem in the most creative, cost-effective way. Most sellers prefer functional descriptions.

Functional Split A division within an automatic call distributor (ACD) which allows incoming calls to be directed from a specific group of trunks to a specific group of agents.

Functional Test A test carried out under normal working conditions to verify that a circuit or particular part of the equipment works properly.

Functional Transparency The ability of a network to carry any user information regardless of its form, so that user applications can operate through the network.

Functional User An entity external to the functional architecture that uses the functional architecture capabilities to exchange information with other functional users. Definition from Bellcore in reference to its concept of the Advanced Intelligent Network.

FUNI Frame-based User-Network Interface, a frame format for access to ATM networks, very much like Frame Relay but with a few additional bits reserved for mapping into the ATM control bits in the cell format. The Frame Relay format and FUNI both pass through a frame switch. See FRAME RELAY and FRAME UNI.

Fuse Verb: To blend together through melting. Noun: An electrical device typically consisting of a wire or strip of fusible metal that melts to interrupt an electrical circuit when current exceeds the rated level of the fuse. The idea is that in any electrical circuit, the fuse should be the weakest point — thus the point that heats up when things go wrong and melts. Better the fuse melts than your expensive PBX. See also CIRCUIT BREAKER.

Fuse Alarm Panel FAP. A distribution panel at the top of the rack. Each device gets its power from the rack. To protect the rectifier from an over-current condition, each device has its own fuse.

Fused Quartz The precise term for glass made by melting natural quartz crystals.

Fusible Links Short lengths (about 25 feet) of fine-gauge wire pairs inside metallic sheath cable that melt to interrupt an electrical circuit and to prevent overheating in building wiring and equipment.

Fusion In 1995 NEC America invented the term "Fusion" for its program to bring a combination of computers and communications into the office. The Fusion strategy focuses on delivery of "multimedia information (text, diagrams, graphics, voice, images and video) to the information worker's desktop." Fusion has three elements — the NEC PBX, various application softwares, and integration links, which consist of Fusion Netlinks (trunk level interfaces), Desklinks (station level interfaces), CPULinks (Processor level interfaces such as OAI and other Ethernet interfaces) and Commandlinks (data stream level interfaces, such as property management system (PMS), Message Center Interface (MCI) and other RS-232 interfaces. See also C&C.

Fusion Splicing In optical transmission systems using solid transmission media, the joining together of two media by butting them, forming an interface between them, and then removing the common surfaces so that there be no interface between them. Thus, no reflection or refraction at the former interface occurs.

Future Proof A term used to describe a phone system (or any technology) that supposedly won't become technologically outdated (at least anytime soon). There's no such thing.

Fuzzy Logic Fuzzy logic is the newest wrinkle in the ancient science of controlling processes that involve constantly changing variables. Contrary to its name, fuzzy logic is a very precise sub discipline in mathematics. It was invented in the 1960s by University of Berkeley's Russian-born Iranian computer science professor Lotfi Zadeh. It enables mathematicians and engineers to simulate human thinking by quantifying concepts such as hot, cold, very far, pretty close, quite true, most usually, almost impossible, etc. It does this by recognizing that measurements are much more useful when they are characterized in linguistic terms that when taken to the fourth decimal point. Fuzzy logic reduces a spectrum of numbers into a few categories called membership groups. Within five years virtually all consumer goods will come with fuzzy logic. Already fuzzy logic is inside video camcorders (to reduce the motion of the camera), in washing machines (to figure the optimum mix of washing conditions for that weight and filth).

FVR Flexible Vocabulary Recognition.

FWA Fixed Wireless Access. Fixed wireless consists of a radio link to the home or the office from a cell site or base station. This "fixed" wireless link replaces the traditional wireless local loop. According to Northern Telecom, FWA is the solution of choice in sparsely-developed areas where potential subscribers have been on lengthy waiting lists, in dense urban areas where rapid expansion is desirable and in suburban settings where new neighborhood developments can be provisioned quickly with FWA.

FWIW Abbreviation for "For What It's Worth;" commonly used on E-mail and BBSs (Bulletin Board Systems).

FWLL Fixed Wireless Local Loop.

FYI An Internet term. An abbreviation for the phrase "for your information." There is also a series of RFCs put only by the Network Information Center called FYIs. They address common questions of new users and many other useful things.

FX Foreign Exchange. A Central Office trunk which has access to a distant central office. Dial Tone is returned from that distant Central Office, and a location can be reached in the area of the foreign Central Office by dialing a local number. This will provide easier access for customers in that area and calls may be made anywhere in the foreign exchange area for a flat rate. See also FOREIGN EXCHANGE and FXO.

FXO Foreign Exchange Office. Foreign exchange (FX) service is a service that can be ordered from the telephone company that provides local telephone service from a central office which is outside (foreign to) the subscriber's exchange area. In its simplest form, a user can pick up the phone in one city and receive a dial tone in the foreign city. This type of connection is provided by a type of trunk called foreign exchange (FX) trunks. FX trunk signaling can be provided over analog or T-1 links. Connecting POTS telephones to a computer telephony system via T-1 links requires a channel bank configured with FX type connections. To generate a call from the POTS set to the computer telephony system, you will need a FXO (foreign exchange office) connection configured. To generate a call from the computer telephony system to the POTS set, you will need a FXS connection configured. See FX.

FXS Foreign Exchange Station. See FXO.

FZA Fernmeldetechnisches Zentralamt. Telecom approval authority Austria. literally translated "long distance communications technical central office." All that in two words. Not bad.

G 1. G stands for giga, which means a billion or one thousand million. In telecommunications, a gig is actually 1,000,000,000. In computers it is ten to the ninth power, which is actually 1,073,741,824. One thousand gigas are a tera. One thousand teras are one peta, which is equal to 10 to the 15th.
2. Abbreviation of "Grin," commonly typed within pointy brackets as <G>, at the end of an item uploaded to a BBS (Bulletin Board System), where the sender wants to make sure that readers realize that the message was meant to be humorous or sarcastic, and not to be taken literally. Example: "If my wife makes meat loaf one more time, I'm going to cut her fingers off <G>." Usage is similar to appending Wayne's-World usage of "Not" to reverse the meaning of a sentence.

G Recommendations A series of standards defined by the ITU-T covering transmission facilities. Namely: G.703 transmission facilities running at 2.048 megabits/second (E1) and 64 kilobits per second. G.703 is the ITU-T standard 1984 current version for the physical and logical traits of transmission over digital circuits. G.703 now includes specifications for the US 1.544 megabits per second as well as the European 2.048 megabits/second, and circuits with larger bandwidths on both continents. G.703 is still generally used to refer to the standard for 2.048 megabits per second; G.821 is the ITU-T Recommendation that specifies performance criteria for digital circuits for ISDN. The G.990 series covers xDSL technology. See G.990.

G-Lite A new ITU standard for xDSL high-speed local loop access to the Internet. For a full explanation, see G.990.

G-Style Handset A G-style handset is a standard round screw-in, screw-out handset, as compared to the K-style handset, which is the newer square handset with the two screws in the middle.

G.5 See G5.

G.703 ITU-T Recommendation G.703, "Physical/Electrical Characteristics of Hierarchical Digital Interfaces". See G RECOMMENDATIONS. 64 Kbps PCM and used for E-1 and T-1.

G.704 ITU-T Recommendation G.704, "Synchronous Frame Structures Used at Primary and Secondary Hierarchy Levels".

G.707 ITU-T Recommendation: Standard - ATM.

G.708 ITU-T Recommendation: Standard - ATM.

G.709 ITU-T Recommendation: Standard - ATM.

G.711 ITU-T Recommendation for an algorithm designed to transmit and receive A-law and Ê-law PCM voice at digital bit rates of 48, 56, and 64 Kbps. It is used for digital telephone sets on digital PBX and ISDN channels. Support for this algorithm is required for ITU-T compliant videoconferencing (the H.320/H.323 standard).

G.721 ITU-T Recommendation: This algorithm is used for digital audio at 64 Kbps.

G.722 ITU-T Recommendation: This algorithm produces digital audio at a rate of 7KHz to 32 Kbps. G.722 is optional for H.320 compliance. See H.320.

G.723 ITU-T Recommendation: This algorithm is for compressed digital audio over POTS lines — Plain Old Telephone Lines. It is the voice part of H.324. This algorithm runs at 6.3 or 5.4 kbps and uses linear predictive coding and dictionaries which help provide smoothing. The smoothing process is CPU-intensive, however (30 MIPs on an Intel Pentium), so don't expect a PC-based implementation to work well for lots of real-time activity. See H.324.

G.723.1 ITU-T Recommendation: Speech encoding/decoding with a low bit rate, 5.3 kbps or 6.3 Kbps output quality. This is what the Voice over IP would use over the Internet. This is the default encoder required for H.323 compliance. G.723.1 is a subset of G.723.

G.728 ITU-T Recommendation: Encoding/decoding of speech at 16 kbps using low-delay code excited linear predictive methods. Like G.722, it is optional for H.320 compliance.

G.729 The International Telecommunications Union's standard voice algorithm — CS-ACELP (Conjugate Structure Algebraic Code Excited Linear Predictive) voice algorithm for the coding of encoding/decoding of speech at 8 Kbps using conjugate-structure, algebraic-code excited linear predictive methods. G.729 is supported by, inter alia, AT&T, France Telecom and Japan's NTT. See V.70, the specification for DSVD, which uses G.729.

G.804 ITU-T Recommendation G.804, "ATM Cell Mapping into Plesiochronous Digital Hierarchy (PDH)".

G.990 On October 29. 1998, the International Telecommunications Union issued the following press release. (My edits in brackets.) Geneva - The International Telecommunication Union today closed a vital link in the high capacity Information Highway by reaching agreement on a set of new technical system specifications for Multi-Megabit/s network access, and initiating the formal approval process. The new specifications, designated as the G.990 series of Recommendations, specify several techniques to provide megabit per second network access on existing telephone subscriber lines (i.e. copper local loops) simultaneously with the regular voice communication. Main applications are high-speed Internet access, video and other on-line data communications such as electronic commerce, home office, distance learning.

"These new specifications for multi-megabit network access link well into the already existing ITU-T fiber- and coax-based standards on Gigabit/s transport systems for the core network, enabling network providers to offer on-demand, high capacity digital services over the last mile — another major step towards building the information society", said Peter Wery, Chairman of ITU-T Study Group 15.

The new access systems are industry's response to the yearning of subscribers for quicker network access without long waiting times and at high bit rates. Commercially very important, industry analysts foresee a market potential of several billion dollars world-wide. The new access network specifications provide for:

Symmetrical bi-directional access at bit rates of up to 2 million bits per second (New Recommendation G.991.1). Asymmetrical bi-directional access bit rates of up to 640 kilobits per second in the upstream (subscriber to network) and up to 6 million bits per second in the downstream (network to subscriber) direction, depending on the subscriber line length

(Draft new Recommendation G.992.1). Splitterless, asymmetrical bi-directional access (Draft new Recommendation G.992.2, previously known as G.lite). This is a simpler, splitterless asymmetrical system which can be installed by the user. Depending on the subscriber line length, the system provides upstream access up to 512 kilobits per second and enables the subscriber to download data and video at speeds of up to 1.5 million bits per second. The standard eliminates the need for a piece of equipment called "splitter" at the consumer's premises. New G.992.2 compliant modems will simply plug into the back of the PC as current modems do. Industry analysts expect that the adoption of the standard will speed up the rollout of high-speed Internet access to consumers over existing phone lines. It is also expected that this type of Megabit per second system to become a 'best seller' in the network access arena, with transmission speed of Internet data 25 times faster than today's 56k analog modems and close to speeds achieved on cable modems. Today's agreement on a single open standard also means that consumers can choose freely from any supplier providing G.992.2-compliant products as all DSL modems will be able to interoperate. "One of the keys to the mass deployment is standardization, which allows a situation where an end-subscriber can comfortably buy a modem and be reasonably assured that they can move to a different location and have it work.

In addition to the system specifications above, a number of complementary technical specifications have also been agreed upon, addressing test procedures, system management, and 'handshaking' procedures.

The ITU, a United Nations agency, coordinates the development of global communications standards. Study Group 15 of the ITU Telecommunication Standardization Sector (ITU-T), where the work on these specifications has been carried out, is responsible for the standards development in the area of transport networks, systems and equipment.

Recommendations G.991.1 and G.992.1 have been approved and have taken effect. In respect of Recommendation G.992.2 (previously G-Lite), the Study Group has agreed to apply the approval procedure under which the draft text is circulated to all ITU-T members to determine whether the Study Group is to be assigned the authority to give it final approval ("decision") at its next meeting. After unanimous approval by the Study Group, the standard takes effect. For the G.992.2 draft standard, the "decision" step is scheduled for end of June 1999 (June 21 to July 2). The agreement by the Study Group covers the key technical specifications, thus providing the technical stability required by manufacturers and service providers to bring compatible products to the market. The next step in June/July is the formal approval of the standard before it can take effect.

G.991 See G.990

G.992 See G.990

G.992.2 This spec was previously known as G.Lite.

G.DSVD Voice digitizer for V.DSVD.

G.lite See ADSL Lite

G.O.O.D. Job A "Get-Out-Of-Debt" job. A well-paying position you take to pay off your debts, and one you'll quit as soon as you're solvent.

G3 Third Generation Mobile System. An ITU-T discussion over a proposed worldwide worldwide cellular phone GSM standard.

G5 Messaging Forum This comes from a press release I received from the G5 Messaging Forum. In May 1996 a meeting of representatives of major fax industry players was called in Anaheim, California. They jointly agreed to create the G5 Messaging Forum, a non-profit organization, to develop and promote a new messaging service. Later, the group established the G5 Messaging Forum, with 25 member organizations, between them accounting for over 50% of inter-company messaging stations worldwide. A Group 5 Messaging Interoperability Agreement (G5 Messaging) which defines a new 5th generation electronic messaging service is being developed by members of the Forum. The G5 Messaging specification is based on MIME enframing, a technique that is well established in Internet e-mail. This immediately provides for transmission and identification of any file type registered with the Internet MIME registry (IANA). G5 messaging reduces the high cost of fax transmissions by allowing Internet, internal network or carrier transfer and least cost routing, irrespective of file format. G5 Messaging also reduces the handling costs associated with document transfer and retention, variously estimated at between 10 and 100 times the communications costs. The eight steps typically required to send and receive electronic documents today can be reduced in G5 Messaging to three — create, index and record. These processes can in themselves be fully automated. G5 Messaging is a new messaging service designed and agreed by the G5 Messaging Forum. Its target is to be the 5th major inter-organization messaging service after post, telex, fax and e-mail. It brings together five currently disparate messaging application areas: Image/fax, Text/e-mail, Voice Messaging, Video and Electronic Commerce. It adds to the functionality of all these current messaging applications. Examples are: Carrier independent delivery. End delivery to person, application, peripheral or department. Transmission of document referencing information. Electronic postmarking. Delivery confirmation. Its design caters for communications of any data type e.g. an information file, instruction (control command) or software. It addresses the costs associated with transmission, handling and retention storage, the complete document life cycle burden. Electronic commerce transactions are possible and acceptable without currently required agreement, (subject to individual country legislation or accepted practice). It interworks with existing Group 3 Facsimile and with Internet SMTP E-mail It is designed to meet the identified and evolving user needs.

Whilst fax has become ubiquitous as a means of inter-organization messaging there has been much change in the organizations served so effectively since its introduction. In particular over 70% of all business documents are generated on computers.

To provide integration with the existing user base of inter-organizational messaging, G5 Messaging is designed to support interworking with Group 3 fax and Internet SMTP e-mail. This means that with a single keystroke, a message may be sent to multiple recipients who use any mix of Group 3 fax, Internet e-mail and full G5 Messaging. G5 Messaging thus sits above existing application specific messaging services.

The G5 Messaging architecture is based on MIME enframing, which allows for transmission and identification of any file type registered with the Internet MIME registry (IANA). The Interoperability Agreement defines a G5 Message Header, which is being registered as a specific MIME file type, and which provides control functions, such as: Sender identification, Recipient identification Title, subject description and keywords. An electronic postmark. Because this header (which provides a number of functions) is itself a MIME data file, the G5 Messaging service is transport independent allowing it to operate over a variety of underlying data transport mechanisms including: PSTN: V.34, T.30 and T.434 (Group 3 fax standards) Internet: SMTP E-mail or direct network connection TCP/IP Intranet: SMTP E-mail or direct network connection. Other transport mechanisms such as cable or mobile networks can be accommodated in future developments.

The sending system generates an electronic postmark at time of transmission consisting of a unique message ID, sender, recipient, date and time stamp and cryptographic checksum). The receiving system automatically generates a confirmation containing the postmark from the original message and constructs its own confirmation postmark. The original and confirmation postmarks can be stored at the point of sending and point of receipt, providing mutual non-repudiation of the message and its contents.

Carrier independence allows flexible selective routing by carrier, Internet, or an internal network. This provides least cost routing opportunities, with major cash savings, and automatic fallback.

Within a G5 Messaging system all messages can be archived at the point of sending (even if they are subsequently transmitted as Group 3 fax or Internet E-mail). This provides an indexed, searchable unified message store and retention of Group 3 Fax, Internet e-mail and G5 messages. G5 Messages carry usable indexing information for the recipient. This saves high re-indexing costs and provides pre-use archiving, searchable message store, and retention.

G5 Messaging combined with optional optical storage conforms to existing and new codes of Practice for legal admissibility. These new Standards Body approved codes will be issued in USA, Japan, and Europe in early 1998. This provides legal status for the electronic version of the document, and a means of avoiding high paper archiving costs.

The G5 Messaging addressing scheme allows people, organizational unit, applications or devices to be directly addressed using the recipient's own choice of inbound addressing. This allows a full range of existing schemes to be used. From any G5 compliant system, messages can be automatically directed to G5, Group 3 Fax and Internet e-mail users on carrier, Internet, or internal networks (universal outbox). People, applications, or peripherals can be addressed. Open and closed copies can be made, personal receipt confirmation requested and sender authentication added. Inboxes can combine G5, Group 3 fax, and Internet e-mail. G5 Messages can be sorted as required using inbuilt indexing, postmark, and media type data (universal inbox).

The G5 Messaging service is designed to integrate with an X.500 style distributed directory service using an access protocol based on LDAP. A local directory can be self-built from the headers of incoming message, confirmations of outgoing messages and calls to external directory services, thus minimizing the cost and effort of individual directory building.

In addition to the basic electronic postmarking, digital signature/encryptographic checksum and message confirmation G5 Messaging provides optional additional security, including file encryption, smart card, one time session encryption, file negotiation and directory referencing.

Current business electronic invoicing is limited in reach. G5 Messaging allows for the first time "total electronic invoicing" between all companies using image, text, or MIME registered EDI formats (e.g. EDIFACT). Version 1.0 of the G5 messaging specification is now available on www.group5forum.org

GA 1. Generally Available or General Availability. A vague term manufacturers use to refer to when their new product will be generally available.

2. Abbreviation for "Go Ahead," used in real-time computer communications to indicate that you have finished a sentence and are awaiting a reply.

GAAP Prounced gap. Generally Accepted Accounting Principles. American companies traditionally report financials according to GAAP. The problem with GAAP is that it's neither "generally accepted," nor are its principles real principles — as in the sense of principles (or laws) of physics. How you report financials under GAAP is open to much interpretation and depends on such factors as your accountant, your auditor and/or the rules and regulations of the Internal Revenue Service. See Operating Income.

GAB Group Access Bridging. A service for bridging of multiple calls to create a conference call.

Gaff Equipment worn by telecommunications and power company staff when they climb poles. The official name is linesman's climbers. They consist of a steel shank that can be strapped to a person's leg. The inside of the shank has a spike that can stab the pole. It's the spike that most people mean when they refer to a gaff.

Gain 1. The increase in signaling power that occurs as the signal is boosted by an electronic device. It's measured in decibels (dB).

2. A radio term. Formally, and according to Bell Telephone Laboratories, "gain" is the ratio of the maximum radiation intensity in a given direction to the maximum radiation intensity in the same direction from an isotropic radiator (an antenna radiating equally in all directions). In other words, "gain" is a measure of the relative efficiency of a directional (focused) radio antenna systems, as compared to an omnidirectional (broadcast) system.

Gain Hits A cause of errors in data transmission over phone lines. Usually the signal surges more than 3dB and lasts for more than four milliseconds. AT&T's standard calls for eight or fewer gain hits in a 15-minute period.

Gallium Arsenide A substance from which microprocessor and memory chips are made. Compared with silicon, GaAs is three to ten times faster or, depending on its speed, uses as little as one-tenth the power; GaAs can detect, emit and convert light into electrical signals, opening the possibility of providing optoelectronic properties on a single chip; GaAs can resist up to 10,000 times the radiation; GaAs can withstand operating temperatures of 200 degrees Centigrade, and GaAs have a higher electron mobility.

Galvanic Isolation A characteristic of a UPS or transformer in which the output is completely electrically disconnected from the input. Power is coupled from input to output by magnetic fields in a transformer within the UPS. A galvanically isolated output is considered to be a separately derived source according to the US National Electirical Code and is required to be grounded, that is, the output grounding wire must be directly bonded to the input grounding wire. It is commonly but falsely believed that galvanic isolation eliminates ground loops. This definition courtesy American Power Conversion Corp.

Galvanometer A delicate instrument used for measuring minute currents.

Galvo Man Telco-talk for Galvanometer Man, a technician who uses a galvanometer to find and repair circuit faults. It's common for the phone company rep to tell you that your new lines can't be installed when promised because the galvo man hasn't finished his work.

GAMMA Distribution A telephone company term. A particular type of rightskewed probability distribution which closely resembles COE load distribution. GAMMA extends farther to the right which means that "for a given average busy season load" the GAMMA will predict higher peak day loads than will the normal distribution.

Gammic Ferric Oxide The type of magnetic particle used in conventional floppy disks.

GAN Global Area Network.

Gap 1. GAP. Generic Access Profile. A wireless term.
2. Gap. An open space in a circuit through which a condenser discharges for producing electric oscillations.
See GAP LOSS.

Gap Loss That optical power loss caused by a space between axially aligned fibers. For waveguide-to-waveguide coupling, it is commonly called "longitudinal offset loss."

Garage Silicon Valley, according to contemporary lore, started in a garage in Palo Alto in 1939. In that year Bill Hewlett and Dave Packard started Hewlett Packard with $538. Hewlett was the inventor and Packard the manager. Their first product, an audio oscillator, was an immediate success. Walt Disney used it in making Fantasia. Microsoft also was started by Bill Gates and Paul Allen in a garage. Even more recently, Excite, the Web browser company, was started in a garage. Excite no longer works out of a garage, but there is a conference room in their new headquarters, that looks just like a garage, from the outside.

Garbage Band A pejorative term for the ISM (Industrial Scientific Medical) radio frequency bands, also known as Part 15.247 of FCC regulations. ISM operates in the 902-928 MHz, 2.4-2.483 GHz, and 5.725-5.875 GHz ranges. Traditionally used for in-building and system applications such as bar code scanners, industrial microwave ovens and wireless monitoring of patient sensors, ISM also is used in many Wireless LANs. As the ISM band is unlicensed, anyone can use it for anything, anywhere in the U.S. Some garage door openers use it-hopefully not the garage door openers at the same hospital that's using it to monitor your pulse rate in the ICU. It's a catch-all, hence the term "garbage band."

Garbage Collection Routine that searches memory for program segments or data that are no longer active in order to reclaim that space.

Garbage In, Garbage Out GIGO. If the input data is wrong or inaccurate, the output data will be inaccurate or wrong. GIGO is problem with data entered by hand into computer systems. Ask yourself how many times you've received "junk" mail with the wrong spelling of your name? That's called Garbage In, Garbage Out.

Garbitrage Sending garbage from one city to another, usually organized by garbitrageurs on the phone.

Gas Carbon Used for lightning protection by phone companies. In the telephone's early days lightning often struck telephone lines, electrocuted people or burned their houses down. Early lightning protectors were made of carbon. When hit they took phone out of action and needed to be replaced by a technician. Newer lightning protectors are made with a gas. When hit by lightning they temporarily short, then re-enable the phone line. This invention has greatly reduced the number of bad lines a phone company has after a storm. Despite their name, there is no carbon in them. Gas carbons are the same size and shape as the older carbon protectors so they fit easily into the old slots.

Gas Pressurization A method for preventing water from entering openings in splice closures or cable sheaths by keeping the cables under pressure with dry gas.

Gas Tube A method of protecting phone lines and phone equipment from high voltage caused by lightning strikes. See CARBON BLOCK (another protection technology) for a more detailed explanation. Here is a definition from American Power Conversion Corp. Gas tube is a surge suppression device that clamps a surge voltage to a limited value. Also called a "spark gap", a gas tube is simply two electrodes that are held at a close distance so that high voltages between the electrodes simply arc through the air or other gas within the tube, thereby effectively clamping the voltage. Gas tubes are very slow, but can handle very large surges. The main problem with the use of gas tubes in AC power circuits is that when they clamp the surge they momentarily short out the utility line which usually trips the circuit breaker feeding the circuit which the tube is connected to. In this case the operation of surge clamping leads directly to power interruption. They are well suited to use in data line surge suppression, but have protective clamping voltages that are too high to provide effective protection for most modems or computer ports.

Gaseous Conductors The gases which, when ionized by an electric field, permit the passage of an electric current.

Gate This term is typically used in Automatic Call Distributors, devices used for handling many incoming telephone calls. Gate refers to a telephone trunk or business transaction grouping that may be handled by one group of telephone answerers (called attendants, operators, agents or telemarketers). That one group of telephone answerers is called "the gate." All calls coming into that gate can, theoretically, be handled by any of the telephone answerers. A telephone call is homogeneous throughout the gate. An automatic call distributor may have one gate — all calls coming in can be handled by everyone. Or it may have many gates, each one consisting of the line (or lines) bringing the call in — e.g. Band 5 WATS, New York City foreign exchange line. Or it may have two gates — one for orders and one for service. ACDs with multiple gates will establish rules for moving the calls between the gates, should one gate become overloaded.

Gate Array A circuit consisting of an array of logic gates aligned on a substrate (a piece of silicon) in a regular pattern.

Gate Assignments Used in context of ACD (Automatic Call Distribution) equipment. Gates are made up of trunks that require similar agent processing. Individual agents can be reassigned from one gate to another gate by the customer via the supervisory control and display station. Also called splits.

Gate D Gateway Daemon. A popular routing software package which supports multiple routing protocols. Developed and maintained by the GateDaemon Consortium at Cornell University.

Gatekeeper In the classic sense of the word, a gatekeeper is someone who is in charge of a gate. His or her job is to identify, control, count, supervise the traffic or flow through it. A network gatekeeper provides the same functions, including terminal and gateway registration, address resolution, bandwidth control, admission control, etc. A gatekeeper is a fancy name for a network administrator.

Gateway 1. A gateway is what it sounds like. It's an entrance and exit into a communications network. That "communications network" may be huge, for example, at the point where AT&T Communications ends and Comsat begins — for taking my satellite call overseas. Gateways may be small — between one LAN and another LAN. Technically, a gateway is an electronic repeater device that intercepts and steers electrical signals from one network to another. Generally, the gateway includes a signal conditioner which filters out unwanted noise and controls characters. In data networks, gateways are typically a node on both two networks that connects two otherwise incompatible networks. For example, PC users on a local area network may need a gateway to gain access to a mainframe computer since the mainframe does not speak the same language (protocols) as the PCs on the LAN. Thus, gateways on data networks often perform code and protocol conversion processes. Gateways also eliminate duplicate wiring by giving all users on the network access to the mainframe without each having a direct, hard-wired connection. Gateways also connect compatible networks owned by different entities, such as X.25 networks linked by X.75 gateways. Gateways are commonly used to connect people on one network, say a token ring network, with those on a long distance network. According to the OSI model, a gateway is a device that provides mapping at all seven layers of the model. A gateway may be used to interface between two incompatible electronic mail systems or for transferring data files from one system to another. Electronic mail systems that sit on local area networks often have gateways into bigger e-mail systems, like Internet or MCI Mail. For example, I might use MCI Mail to send a e-mail to someone's internal LAN e-mail. It might travel from MCI Mail to Internet via a gateway and then from Internet via another gateway to the company's e-mail to its own LAN.

2. A Gateway is an optional element in an H.323 conference. Gateways bridge H.323 conferences to other networks, communications protocols, and multimedia formats. Gateways are not required if connections to other networks or non-H.323 compliant terminals are not needed. Gatekeepers perform two important functions which help maintain the robustness of the network — address translation and bandwidth management. Gatekeepers map LAN aliases to IP addresses and provide address lookups when needed. Gatekeepers also exercise call control functions to limit the number of H.323 connections, and the total bandwidth used by these connections, in an H.323 "zone." A Gatekeeper is not required in an H.323 system-however, if a Gatekeeper is present, terminals must make use of its services. See TAPI 3.0.

Gateway City A city where international calls must be routed. New York, Washington, DC, Miami, New Orleans, and San Francisco are the five gateway cities in the United States.

Gateway Protocol Converter GPC. An application-specific node that connects otherwise incompatible networks or networked devices. Converts data codes and transmission protocols to enable interoperability. Routers are capable of running gateway protocols — we used to call routers "gateways." Contrast to Bridge.

Gating 1. Enabling or disabling a signal through applied logic. If it's turned on, the signal gets through. If not, the signal doesn't get through.

2. The process of selecting only those portions of a wave between specified time intervals or between specified amplitude limits.

Gauge A term for specifying the thickness (diameter) of cables. Thicker cables have a lower number in the American Wire Gauge (AWG) scale. Thicker gauge cables can carry phone conversations further and more cleanly than thinner gauge cable. But thicker cables cost more and take up more room, especially when you bundle them together and put them in a duct. When buying a phone system it is good to specify the thickness of the cables that will be installed — especially if some of your extensions will be a great distance

from the central telephone switch, if you intend to carry high-speed data on them or you intend to live with your cabling scheme for more than a few months. You should, of course, not only specify the cable's thickness, but also whether it's stranded or solid core, coax, etc. Gauge is but one part of a cable description. See AWG for a fuller explanation.

Gauge, Wire The method of specifying the thickness and size of wire. The two important American gauges are the American Wire Gauge (AWG), previously known as Brown & Sharpe, and the Steel Wire Gauge. See AWG for a fuller explanation.

Gauss The unit of magnetic field intensity in terms of the lines of force per square centimeter.

Gaussian Noise Gaussian noise is white noise uniform across the whole range of frequencies. "Gaussian" refers to the measurement of the noise. It is essentially the assumption you make when working on the problem of noise mathematically.

Gazillion An extremely large, indeterminate amount. See Gigabyte.

GB Gigabyte. See Gigabyte.

GBH Group Busy Hour.

Gbps Gigabits per second. Gig is one thousand million bits per second.

GCAC An ATM term. Generic Connection Admission Control: This is a process to determine if a link has potentially enough resources to support a connection.

GCRA An ATM term. Generic Cell Rate Algorithm: The GCRA is used to define conformance with respect to the traffic contract of the connection. For each cell arrival the GCRA determines whether the cell conforms to the traffic contract. The UPC function may implement the GCRA, or one or more equivalent algorithms to enforce conformance. The GCRA is defined with two parameters: the Increment (I) and the Limit (L).

GCT Greenwich Civil Time.

GDDM An SNA definition: Graphical Data Display Manager (GDDM) system software used for graphics display and printer devices and performs the same functions as QuickDraw in Macintosh computers.

GDF Group Distribution Frame.

GDI Graphics Device Interface. The part of Windows that allows applications to draw on screens, printers, and other output devices. The GDI provides hundreds of convenient functions for drawing lines, circles, and polygons; rendering fonts; querying devices for their output capabilities; and more.

GDMO Guidelines for the Definition of Managed Objects.

GE Gigabit Ethernet. See Gigabit Ethernet.

Geekosphere A definition courtesy Wired Magazine: The area surrounding one's computer where trinkets, personal mementos, toys and "monitor pets" are displayed. A place where computer geeks show their colors.

Gender Connectors, plugs and receptacles are assigned a gender to describe their physical type. Ones with pins are male, and those with holes into which the male pins slide are female. See GENDER BENDER.

Gender Bender A device which changes the gender of a connector, plug or receptacle. A gender bender is typically a small plug with all male pins on one side and all male pins on the other. By plugging a female connector into one side of a gender bender, you've effectively changed the female gender of the cable to male. Alternatively, a gender bender could be female on either side. But a gender bender must be the same on both sides.

Gender Changer Another name for a gender bender. See GENDER BENDER.

Genderless Connector Also called data connector or hermaphroditic connector. Invented by IBM. The connector doesn't require male and female plugs to make a connection. It was designed for token-ring applications. It was too big and clunky for my taste.

General Availability How a product gets to market varies from one company to another. But typically, along the way, there's something called an alpha — the first version of hardware or software. It typically has so many bugs you only let your employees play with it. A beta is the next version. It's a pre-release version and selected customers (and the press) become your guinea pigs. They give you feedback. After beta, and when the bugs are removed and the features have been fine-honed, comes "general availability." That's when the product is finally available for buying by the general public.

General Call The letters CQ in the international code and used as a general inquiry call.

General Packet Radio Service GPRS. General Packet Radio Service is the data service for GSM, the European standard digital cellular service. GPRS is a packet-switched service that is widely expected to be the next major step forward in the evolution of GSM technology. GPRS, further enhancing GSM networks to carry datacom, is also an important component in the GSM evolution entitled GSM+. GPRS enables high-speed mobile datacom usage. It is most useful for "bursty" data applications such as mobile Internet browsing, e-mail and push technologies. GPRS has been demonstrated as fast as 115 Kbps. See HSCSD.

General Protection Fault GPF. A General Protection Fault is an indication that Windows 3.xx has tried to assign two or more programs to the same area in memory. Obviously that's not possible, since two things can't occupy the same area in memory. As a result, your screen stops and says "General Protection Fault." If you can save what you're doing, do it. If you have other programs open, try and save the material in them. Close Windows and then do a cold reboot. Do not continue to work after you have received a General Protection Fault. You must reboot. Better do a cold reboot, too.

General Purpose Network An AT&T term for a network suitable for carrying many forms of communication — voice and data, circuit and packet, image, sensor or signaling, for example.

General Release When software is finally finished and ready to be sold to the general public, it said to be in "General Release." Before it is in general release, it is still in beta. Beta software is not alleged to be bug-free. General release software is meant to be bug-free. Sometimes it is.

General Telemetry Processor GTP. A device that receives and processes telecommunications equipment alarming protocols such as TBOS (Telemetry Bit-Oriented Serial).

General Trunk Forecast A telephone company term. GTF. A forecast of future trunk circuit requirements. This forecast covers the current year and the future four years.

Generational Loss The reduction in picture quality resulting in the copying of analog images for editing and distribution.

Generations, Computer As computers have improved, so the industry's pundits have assigned "generations" to those improvements. The concept of generations is not perfect nor finite. Here's our best shot on generations in computers:
• First generation: 1951-1958, core memory 8 Kbytes to 32 Kbytes.
• Second generation: 1958-1964, transistor technology, memory 32 Kbytes to 64 Kbytes.
• Third generation: 1964-1975, integrated circuitry.
• Fourth generation: 1975-date, non procedural languages, software driven.
• Fifth generation: into the 1990s, natural language programming, parallel processing and super computing.
• Sixth generation: in the 2000, will process knowledge rather than data.

Generations, PBX As PBXs have improved, so the industry's pundits have assigned "generations" to those improvements, as they did in computers. The concept of generations is not perfect nor finite. Here's our best shot on generations in PBXs:

First generation: 1920s to the late 1960s. Step-by-step mechanical equipment. The first and the last of the step-by-step Bell PBXs switches was called a 701. Lee Goeller says the 701 "was the best PBX ever built. It was infinitely flexible. It was just too BIG. In fact, it was usually bigger than the office it served. This era of the stepper will be remembered as the era the Bell System was intact and had the gaul to rent operator chairs."

Second Generation: Late 1960s: Bell 801 reed relay switch. Stromberg Carlson 800 series reed relay switch. GTE had a series, also. Reed relay switches were not very popular.

Third Generation: 1974 and 1975: Rolm introduces its first CBX, an electronic, solid-state PBX. AT&T introduces Dimension. Digital Telephone Systems introduced the D1200. Northern introduced its first stored-program controlled SL-1. Some of these PBXs switched voice digitally, though they used different techniques, including PCM, PAM, and Delta Modulation. The codecs were in the switch, not in the instruments.

Fourth generation: early 1980s. Distributed processing. Northern, Rolm and NEC and others introduced remote modules — slaves to the master switch at headquarters. These switches also added the capability of handling data without using modems. Switches like Lexar and InteCom were designed from the beginning to handle data without modems, thus requiring digital capability out to the set.

Fifth generation: CXC, Anderson Jacobson and Ztel and others called themselves "fourth generation." When they started to fail, some people called them the "fifth generation." It wasn't clear exactly what that generation was. But they all got lots of publicity and the PBXs from CXC, Anderson Jacobson and Ztel ultimately failed.

Sixth generation: Networked PBXs. Sit in New York. Operate your national network as if it were in the same building. Bingo, you can transfer calls across the country. All your messaging is the same wherever you sit on the network. Lee Goeller, however, says you can network stepper PBXs. In fact, in 1971 he says he managed one of the biggest integrated voice and data networks in the US using step-by-step electromechanical PBXs (701s made by AT&T). It was the world's largest dial tandem network. He had 63 different locations and three hubs.

Seventh generation: Open Architecture. You can now program your own PBX. For more, see the NORSTAR COMMAND SET.

Eighth generation: Dumb switches. You can now buy completely dumb phone systems which are just basically switches. To get them to do anything they require an external computer (and software programming that computer) to drive the dumb switch. Often the "driving" is done through one or more serial ports.

In reality, the concept of generations amongst PBXs is very flimsy. But it's the stuff dictionaries are made of.

Generator A machine which converts mechanical energy, such as the power from a piston engine into electrical energy.

Generic Cell Rate Algorithm An ATM function that is carried out at the user-to-network interface (UNI) level. It guarantees that traffic matches the negotiated connection that has been established between the user and the network.

Generic Flow Control Field GFC. A 4-bit value in an ATM header for purposes of flow control between the user equipment and the carrier ATM Edge Switch across the User Network Interface (UNI). The GFC field tells the target end-station that the switch may implement some form of congestion control.

Generic Program A set of instructions for an ESS central office or electronic PBX that is the same for all installations of that particular equipment. Detailed differences for each individual installation are listed in a separate parameter table. Here's a more formal definition, from Bellcore. A generic program is a set of instructions for an electronic switching system or operations system that is the same for all central offices using that exact type of system. Detailed differences for each individual office are usually listed in a separated parameter table.

Generic Requirements GR. See GR.

Generic Services Framework GSF. Generic Services Framework is a set of software designs being implemented by Bell Northern Research (a subsidiary of Northern Telecom) to accelerate the development and testing of new features and provide a platform for the later stages of DMS SuperNode interworking with Advanced Intelligent Networking (AIN). GSF applies the principles of object-oriented programming to DMS SuperNode software, and delivers enhancements that simplify feature development and reduce testing needs. The GSF has three distinct elements:

1. Call Separation - The new architecture uses separate software data and processes to handle the two "halves" of a call (originating and terminating).

2. GSF Agent Interworking Protocol (AIP) - The AIP uses a standardized set of instructions to allows the two call halves to communicate.

3. Event-Driven Call Processing (EDCP) - Just as the AIP mediates communications between call halves, EDCP handles communications within each call half. EDCP also uses a standardized set of instructions to simplify messaging.

Generic Top Level Domain see gTLD and Domain.

Genlock Circuitry that synchronizes video signals for mixing. The video circuitry determines the exact moment at which a video frame begins. Genlock allows multiple devices (video recorders cameras, etc.,) to be used together with precise timing so that they capture a scene in unison. A genlock display adapter converts screen output into an NTSC video signal, which it synchronizes with an external video source.

GEO Geosynchronous Earth Orbit. A term for a satellite which is placed in a geosynchronous, or geostationary, orbital slot. Such orbits are always equatorial; the satellites are placed at altitudes of approximately 22,300 miles. As a result, they are synchronized with the rotation of the earth, maintaining their relative position to the earth's surface. In other words, they always appear to be in the same spot. See also GEOSTATIONARY SATELLITE. Contrast with LEO and MEO.

Geographic Information Systems GIS. Computer applications involving the storage and manipulation of electronic maps and related data. Applications include resource planning, commercial development, military mapping, etc.

Geographic Interface First there was the ASCII interface - a screen containing nothing but ASCII letters. You saw them on airline reservation terminals. Just plain boring green type

against a dark, unlit, black background. Then you got the GUI - the Graphical User Interface. Windows is the most famous GUI. It used icons to represent actions. An eraser to indicate you could remove something. A calendar to indicate that you could enter your day's schedules, etc. Now there is the "Geographic Interface." It attempts to depict objects from the real world — or at least some circumscribed part of it, like your office. The first of these products was Magic Cap from General Magic. Another was the opening screen for Apple's eWorld on-line service. Another is Novell's Corsair technology. Click on a filing cabinet to get to its database contents. Click on a bloodhound find something. One idea behind these geographic interfaces is that there should be a libraries of objects. A set useful for an auto mechanic. Another set useful for a bond trader. The idea is that everyone gets to choose the bits and pieces of the geographic interface that he or she is most comfortable with. Some of these new geographic interfaces are endearing. Some are tiresome. We'll see if they endure.

Geographical Portability The ability to take your New York phone number to Boston, including the area code. In short, geographical portability is the ability to take all your phone number (i.e. all its 10-digits) with you wherever you go — anywhere in North America. No timetable has been set for geographical portability in North America. See LNP (Local Number Portability) and 800 Service.

Geometric Dilution of Precision GDOP or DOP (Dilution of Precision). In systems employing position determination technology such as Enhanced 911, GDOP refers to the phenomenon that causes the calculated position's accuracy to be a factor of the relative geometry of the involved units. For example, in an angle of arrival system, two fixed units measure the angle to a mobile unit. The resulting lines intersect in a point defining the mobile's position. If each measured angle has some uncertainty, the intersection is no longer a point, but say, a quadrilateral. In this case, the mobile is known only to be within the area of the quadrilateral. Depending on where the mobile is in relation to the fixed measurement units, the area and shape of the quadrilateral (and therefore the accuracy of the calculated position) may be larger or smaller. GDOP also affects time difference of arrival systems.

Geostationary Orbit An orbit, any point on which has a period equal to the average rotational period of the Earth, is called a synchronous orbit. If the orbit is also circular and equatorial, it is called a stationary or geostationary orbit.

Geostationary Satellite A satellite in geostationary orbit, also called geosynchronous orbit. A satellite placed in an geosynchronous orbit — 22,300 miles (or 42,164 kilometers)) directly over the earth's equator — will appear to be stationary in the sky, turning synchronously with the earth. This means you can plant a satellite receiving/transmitting antenna on the ground, and point it at that one place in the sky to receive signals from and transmit signals to "the bird," as satellites are sometimes called. Most communications satellites are in geostationary or geosynchronous orbit. The Russians have some satellites that orbit the earth and require antennae which move. These satellites are used to transmit to far northern communities which are difficult to reach with normal geosynchronous satellites.

Geosynchronous Orbit Synchronous with the Earth. An orbit 22,300 miles above the earth's equator where satellites circle at the same rate as the earth's rotation, thereby appearing stationary to an earth-bound observer. See also GEOSTATIONARY ORBIT.

Get A Life What your kids say to when you start talking too much about the information superhighway, or the Internet, of fiber optic, or something that doesn't interest them.

GETS Government Emergency Telecommunications Service. GETS is a new service offered by the Office of the Manager, National Communications System (OMNCS).

GFC Generic Flow Control. GFC is a field in the ATM header which can be used to provide local functions (e.g., flow control). It has local significance only and the value encoded in the field is not carried end-to-end.

GFCI Ground Fault Circuit Interrupter. A device intended to interrupt the electrical circuit when the fault current to ground exceeds a predetermined value (usually 4 to 6 milliamps) that is less than required to operate the overcurrent protection (fuse or breaker) for the circuit. This device is intended to protect people against electrocution. It does not protect against fire from circuit overload. GFCI outlets are typically installed in bathrooms, kitchens and garages because the presence of water in these area increases the possibility of electric shock. Sometime GFCI circuits are incorrectly wired. The way to find out if your GFCI is wired correctly is to press the test button on its face. This should shut off power to the GFCI outlet and to those outlets connected to it.

GFI Group Format Identifier. In packet switching, refers to the first four bits in a packet header. Contains the Q bit, D bit and modulus value.

GFLOPS One billion FLoating point Operations Per Second. (G stands for GIGA, meaning billion). Today's fastest supercomputers are able to maintain a sustained throughput of over one billion floating point operations per second (GFLOPS) while performing real-world applications. By contrast, a 25-MHz 486 personal computer can sustain about one million floating operations per second (one MFLOP), or about one-thousandth the throughput of a supercomputer. See also G.

GFXO Ground start FXO.

GFXS Ground start FXS.

GGP Gateway to Gateway Protocol. The protocol that core gateways use to exchange routing information, GGP implements a distributed shortest path routing computation.

Ghost 1. A secondary image resulting from echo or envelope delay distortion.

2. A term coined by Fluke to mean energy (noise) detected on the cable that bears similarities to a real frame, but does not include a valid start frame delimiter. To qualify for this category of error, the event must be a minimum 72 octets. Ghosts are a strong indication of a physical problem on the local segment.

GHz One billion, or one thousand million, hertz, or cycles per second. See also BANDWIDTH.

GIF pronounced JIF, as in peanut butter. GIF stands for Graphics Interface Format, which is pronounced "Jiff." It is a format for encoding images (pictures, drawings, etc.) into bits so that a computer can "read" the GIF file and throw the picture up on a computer screen. GIF can only handle 256 colors. The major advantage of GIF is that it compresses the image (photograph, drawing, etc.), thus making it faster to transmit across phone lines. CompuServe, the on-line service, invented GIF in 1987. There are thousands of GIF images on CompuServe, which its subscribers can download. GIF is based on a mathematical algorithm called LZW, which is a set of mathematical formulae used to compress images into GIF files. Unisys has a patent on LZW. The GIF format has become a standard in the electronic bulletin board world, and on Web sites on the Internet. Pictures and graphics you see on Web pages are usually in GIF format because the files are small and download quickly. Web page authors often use GIF images because GIFs can be interlaced, which produces a melting effect on the client screen as the image is loading. Most of the graphics you see on the Internet are stored in GIF format. You can download

GIF is a popular format also for photographs of scantily-clad women. According to Jack Rickard, original publisher of the excellent Boardwatch Magazine, "some of the photographs are reasonably good, but most feature strikingly plain women rather artlessly photographed by those whose higher calling is probably more aptly found in the building trades or automotive repair." Other graphics formats include BMP, PCX, TIFF, etc. See also INTERNET.

Giga Prefix meaning one billion, which is one thousand million. 1,000,000,000. Giga is the reciprocal of nano. See also GIGABIT, GIGABYTE, GFLOPS.

Gigabit 1. In transmission terms, exactly one billion bits, or one thousand million bits. In the world of transmission systems, we speak of the number of bits which can be transmitted in a period of time - -specifically, one second. Hence, 1 Gbps is one billion bits per second. Let me illustrate. In eight seconds at a transmission rate of 1 Gbps, I could send you 200 copies of my dictionary.

2. In computer terms, a Gigabit is 1024 times mega; in other words, actually 1,073,741,824. One thousand gigas are a tera. One thousand teras one peta, which is equal to 10 to the 15th. Think of it another way, just to put it in personal perspective. If you are a Gigasecond (one billion seconds) old, you have lived to the ripe old age of 31 years, 8 months, 18 days, 18 hours, 50 minutes and 24 seconds. (Feel free to check my math.) At a rate of 1 Gbps, your entire life could flash across your network in a single second's time. Think about SONET fiber optic transmission facilities, which can operate at 2.5 Gbps or more.

Gigabit Ethernet Gigabit Ethernet (GE) was finalized and formally approved on June 29, 1998, as IEEE 802.3z. Although basically compatible with both 10 Mbps Ethernet and 100 Mbps Ethernet, most equipment will have to be upgraded to support the higher transmission level. Gigabit Ethernet addresses the bandwidth problem in 10/100 Mbps Ethernet networks, which are beginning to max out due to bandwidth-intensive, multimedia-based Internet and Intranet applications. As a backbone LAN technology, however, Gigabit Ethernet faces strong opposition from ATM. GE is available in two flavors: shared and switched. Shared GE is essentially a much higher speed of 10BaseT and 100BaseT. In other works, shared GE is a high-speed hub, which effectively is a brute force attack on congestion; CSMA/CD Medium Access Control is exercised in the hub, rather than in one the wire in the collapsed backbone approach. Switched GE addresses the congestion problem through Logical Link Control (LLC), with the more substantial Switched GE products offering non-blocking switching through a crossbar switching matrix. While GE is much like traditional Ethernet, differences include frame size. As the clock speed of GE is one or two orders of magnitude greater than its predecessors, issues of round-trip propagation delay affect error detection. As a result, the minimum frame size has been increased from 64 bytes to 512 bytes. This larger minimum frame size provides more time for the transmitting device to receive a collision notification in the event of a congestion condition. The maximum frame size has been increased from 1,514 bytes to a jumbo frame size of 9,000 bytes; this frame size improves the frame throughput of a GE switch as each frame requires switch processing of header information, the fewer frames presented to the switch, the more data the switch can process, switch and deliver in a given period of time. Physical transmission media currently is limited to fiber optics: multimode fiber will support gigabit transmission at distances up to 550 meters, and single mode fiber up to 5 km; in each case, there is a minimum distance of 2 meters dues to issues of signal reflection (echo). While UTP, STP and other electrically-based media are anticipated in the near future, distances will be limited once technical problems (e.g., signal reflection, or echo) are resolved. Cat 5 UTP is anticipated to support full-duplex transmission over distances up to 25 meters, with each of 4 pair carrying a 125-MHz signal. Clearly, the application for Gigabit Ethernet largely will be in the backbone, for interconnecting lesser Ethernet hubs, Ethernet switches and high-performance servers, rather than connecting individual nodes. GE hubs and switches, however, commonly offer 10/100/1000 Mbps ports. Both half-duplex and full-duplex interfaces will be supported, with full-duplex offering the advantage of virtual elimination of issues of data collisions. Half-/full-duplex declarations will be made on a port-by-port basis. QoS (Quality of Service) guarantees are not an element of GE, by the way-ATM does that. As a practical matter, GE involves a backbone upgrade at the level of the wiring closet/data center switch. Early Gigabit Ethernet products range from $6,000-$160,000, depending on the number of Gbps- and 10/100-Mbps ports supported, the level of stackability, etc. Chassis-based systems, as opposed to standalone systems, offer advantages including hot-swappable power supplies, switching matrixes, and switching modules. Gigabit Ethernet products are anticipated in the near future to achieve estimated costs of $900-$1,400 per port for shared GE, and $1,500-$3,000 per port for switched GE. The market for GE is anticipated to reach as much as $1 billion by the year 2000. Network Interface Cards (NICs), of course, will not be the same as those employed in either the 10 Mbps or 100 Mbps versions. The final result is that complete upgrades will be required at the NIC, cable and switch levels. Many users, faced with such a complete upgrade, will consider ATM as an alternative, given the Quality of Service (QoS) guarantees which ATM provides and which Ethernet does not. Additionally, ATM is more flexible, supporting voice, video and multimedia, as well as data. On the other hand, Ethernet is relatively simple and inexpensive, well understood, highly standardized, and very mature. This definition courtesy of Ray Horak's book, Communications Systems & Networks, which is a wonderful companion to this book, and intentionally so. Ray, you see, is my Contributing Editor. See also ATM, Ethernet and Gigabit Ethernet Alliance.

Gigabit Ethernet Alliance A multi-vendor forum comprised of 86 members committed to driving the industry's adoption of networking standards at up to 1,000 megabits per second. The forum supports the CSMA/CD (Carrier Sense Multiple Access with Collision Detection) protocol of the original Ethernet standard. According to the Alliance, gigabit Ethernet will initially be deployed in backbone environments as the preferred interconnection for switches which support multiple transmission speeds between Ethernet segments (10 Mbps and 100 Mbps). Technical proposals developed by Alliance members have been submitted to the IEEE 802.3z standards committee, furthering efforts to standardize 1,000 Mbps Ethernet technology. See Gigabit Ethernet Alliance Interoperability Consortium. www.gigabit-ethernet.org

Gigabit Ethernet Alliance Interoperability Consortium An organization of 15 vendor members

formed within the Gigabit Ethernet Alliance for the purpose of testing the interoperability of 1Gbps products. The group will work with the University of New Hampshire Lab to conduct its interoperability tests.

Gigabits One thousand million bits. In the U.S., that's the same as one billion bits.

Gigabyte GB. A unit of measurement for physical data storage on some form of storage device — hard disk, optical disk, RAM memory etc. and equal to two raised to the 30th, i.e. 1,073,741,824 bytes.

MB = Megabyte (2 to the 20th power)
    GB = Gigabyte (2 to the 30th power)
    TB = Terabyte (2 to the 40th power)
    PB = Petabyte (2 to the 50th power)
    EB = Exabyte (2 to the 60th power)
    ZB = Zettabyte (2 to the 70th power)
    YB = Yottabyte (2 to the 80th power)
One googolbyte equals 2 to the 100th power

Gigahertz GHz. A measurement of the frequency of a signal equivalent to one billion cycles per second, or one thousand million cycles per second.

Gigaplane A Sun Microsystems term, Center plane bus used in Sun's Ultra Enterprise Server line. Uses separate paths of address, data, and control lines. Communicates with several subsystems concurrently. With a 167MHz UltraSPARC CPU, the Gigaplane can do rates of 2.5 Gbytes per second.

GIGAPOP A POP (Point Of Presence) with a throughput in the range of a billion (giga) packets per second. GIGAPOPs are being implemented in support of Internet2, a high-speed Internet supported by the National Science Foundation and a project of the University Corporation for Advanced Internet Development (UCAID). Internet2 uses the MCI vBNS (very high-speed Backbone Network Services) fiber optic network for transport between the GIGAPOPs, and for access to the GIGAPOPs from member universities and NSF-funded supercomputing centers. A GIGIPOP differs from a NAP (Network Access Point) in that it is a value adding, OSI layer 3 (Network Layer) meet point between customers and network providers. A NAP is a neutral, OSI layer 2 (Link Layer) meet point for ISPs (Internet Service Providers) to exchange traffic and routes.

GIGO Garbage In, Garbage Out. Regardless of the capability of the computer, bad data yields bad results.

GII Global Information Infrastructure, a term first advocated in March, 1994 and since used as a concept around which to form many international committees. For more, see the IEEE Communications Magazine, June 1996.

GIM Group Identification Mark. The Group ID mark is a two digit number used by cellular sites other than your home system to determine if your cellular phone should be allowed to make phone calls, i.e. access on "roam" status. This feature is not yet fully implemented. As cellular systems are upgraded, the GIM will be on line real time, requiring all NAM information, including the Mobile Identification Number (MIN), to be validated before a subscriber is allowed to call outside of their home area.

GIP See GLOBAL INTERNET PROJECT.

GIS Geographic Information Services. Computer applications involving the storage and manipulation of electronic maps and related data. Applications include resource planning, commercial development, military mapping, etc. Raw input comes often from satellite photographs.

GIX Global Internet eXchange. A common routing exchange point which allows pairs of networks to implement agreed upon routing policies. The GIX is intended to allow maximum connectivity to the Internet for networks all over the world.

GL Graphics Library.

Glare Glare occurs when both ends of a telephone line or trunk are seized at the same time for different purposes or by different users. Most embarrassing — glaringly so, in fact. Blame Ray Horak for this awful pun. See Glare Hold and Glare Release and Glare Resolution. See also Ring Splash.

Glare Hold and Glare Release A method of glare resolution. Glare occurs when both the local and distant end of a trunk are seized at the same instant; this usually results in deadlock of the trunk. To prevent this, one end of the trunk is assigned a glare hold status and the other a glare release status. In the event of glare, the glare hold end holds the trunk and the glare release end releases the trunk and attempts to seize another. This approach is used in cellular systems between the MTSO (Mobile Traffic Switching Office) and the connecting cell sites. See also Ring Splash.

Glare Resolution Ability of a system to ensure that if a trunk is seized by both ends simultaneously, one caller is given priority and the other is switched to another trunk. See Glare, Glare Hold and Glare Release.

Glare Window The period of time in a trunk is susceptible to glare, a situation in which a trunk simultaneously is seized by the switches (e.g., a CO and either a PBX or ACD) at both ends. The size of the window can be reduced through ring splash and other techniques. See also Glare Hold and Glare Release, and Ring Splash.

Glass House 1. A colloquial word for a mainframe computer. It derives from the fact that all mainframe computers were once housed in a separate, locked room, with glass windows. You typically needed to pass through heavy security to gain admittance to the glass room.

2. A room, closet, department, floor, or entire building in which special equipment and/or procedures are implemented which allow the data processing within to proceed without interruption even if power or other services are cut off. Creating a "glass house" environment may be a simple as installing a UPS in a wiring closet or may involve providing backup power, heat, light, and telecommunications services for an entire building.

Glass Insulators Glass insulators were widely used in the 1800s to fasten open wire to telephone poles and to protect insulator pins from moisture so they couldn't conduct electricity. This was a technique developed by the telegraph industry over a 40-year period of experimentation and was one of the few basic telegraph practices carried over into telephone line construction. Insulators are found in a variety of different shapes and colors depending on the time period they were developed and on their application. Most have a greenish color from traces of iron oxide from the sand used to make the glass. When insulator design was in its heyday in the mid-to-late 1800s, hundreds of patents to improve the product were issued. For example, the double petticoat, a second lip on the bottom of the insulator, was added to reduce the amount of moisture that could travel up the inside of the cap. The above explanation courtesy Tellabs of Lisle, IL.

Glass Mount Antenna A type of car phone antenna used in cellular service. A glass mount antenna is glued to a car's rear window. Many window-mounted glass antennas have a break in their wiring. The wiring ends at the inside. There is no electrical connection between the inside of the window and the antenna glued onto the outside of the window. The "connection" is done through signal radiation. This type of antenna is not as efficient as one in which the wire goes unbroken from the radio to the antenna.

Glass Terminal A keyboard and screen that conveys data generated by the user directly to a computer or network without buffering or otherwise acting upon the data, and also returns data unchanged from the computer to the user. This terminal type does not provide for cursor addressing or escape sequences.

Glazing Corporate-speak for sleeping with your eyes open. A popular pastime at conferences and early-morning meetings. "Didn't he notice that half the room was glazing by the second session?"

GLines GSF Lines.

Glitch A jargon term used in data communications to describe an extraneous bit that has been introduced into a bit stream usually by a noise source. It can also be a problem or a delay. Can be a noun. "What's the glitch?" Or a verb: "Who glitched this thing up?" Glitch is also a momentary interruption in electrical power.

Global Universal. An adjective meaning the whole world. See GLOBAL SEARCH.

Global 800 International toll-free numbers. See UIFN for more detail.

Global Access A new service of MCI Mail. It allows you, an MCI Mail user, to use your computer and its modem to dial a local number in a foreign (i.e. non-North American city) and reach a port of a packet switched operation called InfoNet. When you reach InfoNet you will then punch in a few letters and reach MCI Mail in the U.S. You can then leave MCI Mail messages, send telexes, send faxes and send paper mail, i.e. do all the normal services MCI Mail allows you to do. The advantage of Global Access is that you don't have to dial back to the U.S. (which usually doesn't work because of all the garbage on the line) or subscribe to a foreign packet switched operation (they have them in all industrialized countries). Sadly, it usually takes weeks to subscribe to a foreign packet switched operator.

Global Directory Imagine a bunch of local area networks all connected together. Today you log onto one server on one LAN, tell it who you are and what your password is. You want to connect to another server? You have to tell it who you are and what your password is, which may be different from the first time. Global Directory, a feature of NetWare 4.x, gives you a central directory. You establish your user name once and associated with your name is your authorized service on all the connected LANs. This way you don't have to sign on again for another server. If telephone systems can attach to NetWare (see Computer Telephony), then they should also be able to benefit from this central directory — for phone bill allocation, people location, phone moves and changes, etc. The reason NetWare never had a central directory is historical. Novell introduced NetWare in 1983 as software to allow a handful of personal computers to share a single hard disk, which at that stage was a costly and scarce resource. As hard disks became more available, the product evolved to allow the sharing of printers and file servers. It was always designed as a departmental computing solution. It's only recently, with more powerful desktop machines, that the Client-Server LAN concept has become more corporate-wide in concept.

Global Directory Service Allows desktop clients to transparently access data on servers across a network.

Global Internet Project GIP. Comprised from a group of senior executives, representing sixteen leading Internet software, telecommunications and digital commerce companies worldwide. Its purpose is to promote the growth of the Internet across geographic boundaries worldwide. Explores present and future impact of the Internet upon commerce and society.

Global Mobile Personal Communications Services GMPCS. A term coined by the ITU-T to refer to satellite telephony to be provided by the proposed Big LEO (Low Earth-Orbiting Satellite) systems such as Iridium and Globalstar, and MEO (Middle Earth-Orbiting Satellite) systems such as ICO and Odyssey.

Global Network Navigator GNN. An application developed at CERN in Switzerland which provides information about new services available on the Internet, articles about existing services, and an online version of Internet related books. The GNN is a World Wide Web (WWW) based information service.

Global One A joint venture of Deutsche Telekom, France Telecom, and Sprint. Launched in January 1996, Global One provides Virtual Network Services (VNS) in more than 65 countries. See also VNS.

Global Positioning System See GPS.

Global Roaming When you go traveling, you want access to the Internet, preferably by making a local phone call. This capability is called global roaming.

Global Search A word processing term meaning to automatically find a character or group of characters wherever they appear in a document.

Global Search and Replace A word processing term meaning to automatically find a character or group of characters wherever they appear in a document and replace them with something else.

Global Security Service Provides networkwide security functions, including single log-in to multiple systems.

Global Software Defined Network AT&T's international virtual private network. It provides business customers with point-to-point, two-way voice and voiceband data communications between the US and various overseas countries. Features include:
- International two-way on-net calling.
- Abbreviated user defined 7-digit dialing.
- Usage sensitivity.
- SDN call screening.

Global Title GT. An address such as customer-dialed digits that does not explicitly contain information that would allow routing in the SS7 signaling network, that is, the GTT translation function is required. See GTT.

Global Title Translation GTT. The process of translating a Global Title from dialed digits to a point code (network node) address and application address (subsystem number). This process is accomplished by the STP (Signal Transfer Point) in the SS7 network. See also Global Title, SS7 and STP.

Global Transaction Network See GTN.

Global Village A term coined in the 1960s by Marshall McLuhan, who wrote a number of very popular books, including "The Medium is the Massage," "War and Peace in the Global Village" and "The Gutenberg Galaxy; The Making of Typographic Man."

GlobalNet A free, electronic amateur bulletin board system network which operates based upon FidoNet technical standards. GlobalNet nodes are located in North America and Europe.

Globalstar Imagine a hand-held, light, low-cost telephone that looks like a cell phone, but works by talking to a satellite. Several companies have proposed a collection of low-orbiting satellites. The idea is that the closer the satellite, the stronger

the signal on the ground, and the smaller the size of the telephone. The low earth orbit also reduces propagation delay, which plagues communications using GEOs (Geosynchronous Earth Orbiting Satellites), which are in equatorial orbits at altitudes of approximately 22,300 miles. Companies proposing such systems include Iridium, Teledesic and Globalstar.

Globalstar is a Low Earth-Orbiting (LEO) satellite-based digital telecommunications system that will offer wireless voice, fax, low-speed data, messaging and position location services worldwide. Globalstar service will be delivered through a 48-satellite LEO constellation (plus 4 backup satellites) at an orbital altitude of 1,414 kilometers (877 miles). In total, the constellation will provide wireless telephone service in virtually every populated area of the world where Globalstar service is authorized by the local telecommunications regulatory authorities. (Globalstar service providers will be required to obtain such approvals before beginning to offer Globalstar service in their territories.) The system will work on a "bent-pipe" signal relay scheme in which the call is launched from a terrestrial wireless device (e.g., wireless phone or data terminal) directly to the satellite. The satellite, with minimal processing, will relay the call to a gateway groundstation (i.e., authorized service provider's satellite dish) for connection to the destination device over the existing local terrestrial wireline or wireless network.

Users of Globalstar will make or receive calls using hand-held or vehicle-mounted terminals similar to today's cellular telephones; fixed wireless terminals (e.g., wireless pay phones) also will be supported. Because Globalstar will be fully integrated with existing fixed and cellular telephone networks, Globalstar's dual-mode handsets units will be able to switch from conventional cellular telephony to satellite telephony as required. In remote areas with little or no existing wireline telephony, users will make or receive calls through fixed-site telephones, similar either to phone booths or ordinary wireline telephones.

Globalstar planned to begin launching satellites in the second half of 1997, and to commence initial commercial operations via a 24-satellite constellation in 1998. That didn't happen. The first eight satellites were launched in 1998. The second launch of 12 satellites was a failure. An additional 24 satellites are expected to be launched by May 1999, and commercial service will be initialized in Fall 1999. Full 48-satellite coverage will occur sometime in the future. Based in San Jose, California, Globalstar is a limited partnership founded by Loral Corporation of New York City, and QUALCOMM Inc., of San Diego, California. Strategic partners represent the world's leading telecommunications service providers and equipment manufacturers. See also GEO, LEO and Propagation Delay.

GLONASS The Russian Global Navigation Satellite System is similar in operation and may prove complimentary to the American NAVSTAR system. Launched in 1996, it is a 24 satellite constellation 19,100 Km above the earth in three orbital planes.

Glueware The trend of joining software applications to physical networks through the deal AT&T and Novell have struck to adapt Novell local area networking software to communicate over AT&T's long-distance network. Intel and Microsoft are considering similar arrangements, according to The Wall Street Journal.

GMD Gesellschaft fur Mathematik und Datenverarbeitung: a German government computer science research institute.

GMDSS Global Maritime Distress and Safety System.

GMPCS Global Mobile Personal Communications Services.

A term coined by the ITU-T to refer to satellite telephony to be provided by the proposed Big LEO (Low Earth-Orbiting Satellite) systems such as Iridium and Globalstar, and MEO (Middle Earth-Orbiting Satellite) systems such as ICO and Odyssey. See GLOBALSTAR.

GMSC Gateway Mobile services Switching Center. A wireless telecommunications term. A means to route a mobile station call to the MSC (Mobile Switching Center) containing the called party's HLR (Home Location Register).

GMSK Gaussian Minimum Shift Keying. A wireless telecommunications term. A means of radio wave modulation used specifically in the GSM (Global System for Mobile Communications) air interface.

GNE Gateway Network Element. A SONET Network Element (NE) that provides a direct OS/NE interface. The GNE provides an indirect OS/NE interface. The GNE provides an indirect OS/NE interface for other NEs in its own management sub-network.

GNN See GLOBAL NETWORK NAVIGATOR.

GNU A non-USENET news group hierarchy devoted to the discussion of things related to the GNU Project of the Free Software Foundation.

GNX Development Tools A set of fax software development tools offered by National Semiconductor.

Goat 1. Also called Craft test Set or Butt Set. A telephone used to test analog phone lines.

2. A hairy animal that can become nasty when upset, like telephone customers who have a problem with their phone lines.

Goats People in our population whose voices cannot — under any circumstances — be recognized by voice recognition machines. No one seems to know where this term came from. I first heard it from Ed Tagg, VP Engineering of Voice Control Systems, Dallas, TX.

Go Local A command typically given in a microcomputer asynchronous data communications program to tell the computer that it will connect to something without a modem over a null modem cable. The command "Go Local" also refers to modem connections and can tell one modem to overlook some of the handshaking and assume it's already taken place.

GO-MVIP Global Organization for MVIP. GO-MVIP is a non-profit trade association established in 1993 to move the Multi-Vendor Integration Protocol standards forward. The stated goals of GO-MVIP are to 1) Develop and establish design specifications for further enhancements of MVIP, 2) Drive MVIP to an official industry standard, 3) Establish a testing laboratory and quality assurance program for current and future MVIP products, and 4) Ensure the continued growth and long-term success of MVIP. As of July 1996, GO-MVIP comprises 170 members; over 300 MVIP-compliant products exist. www.mvip.org. See MVIP.

Going Cyrillic Going cyrillic is when a graphical display (LED panel, bit-mapped text and graphics) starts to display garbage. "The thing just went cyrillic on me."

GOLAY One of the communications protocols used between paging towers and the mobile pagers/receivers/beepers themselves. Other protocols are POCSAG, ERMES, FLEX and REFLEX. The same paging tower equipment can transmit messages one moment in GOLAY and the next moment in ERMES, or any of the other protocols.

GOLD 1. Global OnLine Directory. VocalTec's centralized directory of people who are presently on-line and ready to make an Internet phone call. Here's how it works: You want to make a phone call on the Internet. You dial up your local ISP (Internet Service Provider). As you log on, your service provider assigns you an address-the Internet equivalent of a phone number. As you load your VocalTec Internet phone, it tells VocalTec's GOLD that you're on-line and ready to talk. You can then check that listing of all the other people who are on-line and ready to talk. And you can then choose someone to talk to, or they can choose you. The reason you need an on-line directory is that every time you log onto the Internet, you are assigned a different address. Thus, there's no way for someone to call you. (It would like having a phone number that changed every time you picked up the phone. It would be very hard for someone to find you.) The GOLD is like a central directory of who's on-line and ready to talk. It is not like a paper telephone directory. If you are not on-line and not logged into GOLD, you cannot talk to anyone. Every manufacturer of Internet phones has an equivalent on-line directory. They're all called something different. Eventually there will be a central directory of all people on all Internet phones wishing to make phone calls.

2. An EBPP (Electronic Bill Presentation and Payment) specification developed for billing and payment over the Internet. GOLD was developed by Integrion Financial Network, which is owned by VISA USA, IBM and a number of banks. GOLD was designed to support the display and manipulation of financial data such as bank account information and stock holding, and funds transfer. GOLD also supports transactional Web sites. The competing specification is OFX (Open Financial Exchange). See also Electronic Commerce and OFX.

Gold Codes Named after Robert Gold, Gold codes are used in direct-sequence spread spectrum transmission. Each transmitted signal is assigned to unique Gold code, which correlates the original information signal into a pseudo random sequence. This sequence is then modulated and transmitted as a spread spectrum signal. The receiver, which uses the same Gold code, is able to de-correlate the spread spectrum signal and recover the original information. Gold codes possess two very desirable qualities which are important in a high quality communications system. The first quality is called "auto-correlation." When a receiver is subjected to several spread spectrum signals, it must extract the desired information and reject the remainder. Auto-correlation allows for an excellent signal recovery when the transmitted code matches the reference code in the receiver.

The second quality is called "cross-correlation." Cross-correlation simply means that an undesired transmitted code cannot produce a false match at the receiver. The advantage of Gold codes is that they consistently exhibit superior cross-correlation performance, which is critical in an environment with multiple transmitters, each representing a potential interfering source. See SPREAD SPECTRUM.

Gold Disk You have finally finished your new software. And you're now ready to go into production, to have your software reproduced onto disks you can sell. That final, completed version of software, from which you reproduce commercial production disks, is called your "Gold Disk." From what I can see, it's a "Gold Disk," though there may be more than one disk in your package.

Gold Number Also called vanity number. A service of Nynex. It's a phone number that's easy for your customers to remember, e.g. 555-LIMO. But occasionally hard for them to dial. Tip: If you buy a vanity number, make sure it doesn't have numbers in it. 555-LIMO is harder to remember than CAR-RENT.

Golden Rolodex, The The small handful of experts who are always quoted in news stories and asked to be quests on discussion shows. Example: Henry Kissinger appears to be in The Golden Rolodex under foreign policy.

Gonk On-line jargon. It means To prevaricate or to embellish the truth beyond any reasonable recognition. "You're gonking me. That story you just told me is a bunch of gonk."

Good Condition A term used in the secondary telecom equipment business. One step up from fair condition. Product is in working condition and looks good.

Googolbyte Ten raised to the power 100 bytes, i.e.10 with 100 zeros after it. It's a very large number.

Gopher Programmers at the University of Minnesota — home of the Golden Gophers — developed a kind of menu to "go for" items on Internet, bypassing complicated addresses and commands. If you want to connect to the State Library in Albany you select that option off the menu. Time Magazine once described Gopher as a tool used for "tunneling quickly from one place on the Internet to another." Hence the term Gopher. See also GOPHERSPACE.

Gopherspace The vast number of servers and areas of interest accessible through the Internet gopher. See GOPHER.

GORIZONT The Russian geostationary telecommunications satellite.

GOS Grade of Service. Telecom traffic term. The probability that a random call will be delayed, or receive a busy signal, under a given traffic load.

GOSIP Government Open Systems Interconnection Profile. The U.S. government's version of the OSI protocols. GOSIP compliance is typically a requirement in government networking purchases. GOSIP addresses communication and inter operation among end systems and intermediate systems. It provides specific peer-level, process-to-process and terminal access functionality between computer system users within and across government agencies.

Gotcha Law, The This law comes from "Got You." It's my favorite law. Think about life. You buy a beautiful laptop. You fall in love with it. You want to buy one for your wife. When you go back to the store all hot to do your wife a wonderful favor, they tell you, "Sorry, it's been discontinued." Gotcha! Compare with MURPHY's LAW.

Government Radio Publications Publications on radio subjects by the Bureau of Standards and Signal Corps and sold by the superintendent of Documents, Government Printing Office, Washington D.C.

GPA General Purpose Adapter. An AT&T Merlin device that connects an analog multi line telephone to optional equipment such as an answering machine or a FAX machine.

GPC Gateway Protocol Converter. An application-specific node that connects otherwise incompatible networks. Converts data codes and transmission protocols to enable interoperability. Contrast to Bridge.

GPF General Protection Fault. A problem that happens too often under Windows 3.xx. A General Protection Fault is an indication that Windows 3.xx has tried to assign two or more programs to the same area in memory. Obviously that's not possible, since two things can't occupy the same area in memory. As a result, your screen stops and says "General Protection Fault." If you can save what you're doing, do it. If you have other programs open, try and save the material in them. Close Windows and then do a cold reboot. Do not continue to work after you have received a General Protection Fault. You must reboot. Better do a cold reboot, too.

GPI GammaFax Programmers Interface. C-level programming language. Real-time applications for fax switched and gateways.

GPIB An interconnection bus and protocol that allows connection of multiple instruments in a network under the direction of a controller. Also known as the IEEE 488 bus, it allows test engineers to configure complete systems from off-the-shelf instruments and control those systems with a single, proven interface.

GPRS General Packet Radio Service is the data service for GSM, the European standard digital cellular service. GPRS is a packet-switched mobile datacom service that is widely expected to be the next major step forward in the evolution of GSM technology. GPRS, further enhancing GSM networks to carry datacom, is also an important component in the GSM evolution entitled GSM+. GPRS enables high-speed mobile datacom usage. It is most useful for "bursty" data applications such as mobile Internet browsing, e-mail and push technologies. GPRS has been demonstrated as fast as 115 Kbps. See HSCSD.

GPS Global Positioning System. A system to allow us all to figure out precisely where we are anywhere on earth. The GPS will eventually consist of a constellation of 21 satellites orbiting the earth at 10,900 miles — they circle the earth twice a day. In a way, you can think of them as "man-made stars" to replace the stars that we've traditionally used for navigation. The US Government is investing over $10 billion to build and maintain the system. Applications are almost limitless: Delivery vehicles will be able to pinpoint destinations. Emergency vehicles will be more prompt. Cars will have electronic maps that will instantly show us the way to any destination. Planes will be able to land in zero visibility. GPS is based on satellite ranging. That means we figure our position on earth by measuring our distance from a group of satellites in space. The satellites act as precise reference points. To get your precise position, latitude, longitude and altitude (and in some applications velocity), you need to get a "fix" off at least four satellites. That's the main reason for the 21 satellites. Each GPS satellite transmits on two frequencies: 1575.42 MHz referred to as L1, and 1227.60 MHz referred to as L2. It transmits a host of somewhat complicated data occupying about 20 MHz of the spectrum on each channel. But basically, it boils down to three items. The satellite transmits its own position, its time, and a long pseudo random noise code (PRN). The noise code is used by the receiver to calculate range. If we know precisely where a satellite is located and our precise range from it, and if we can obtain similar readings from other satellites, we can calculate precisely our own location and even altitude by triangulation. Satellite position and time are derived from on-board celestial navigation equipment and atomic clocks accurate to one second in 300,000 years. But the ranging is the heart of GPS. Both in the receiver, and in the satellite, a very long sequence of apparently random bits are generated. By comparing internal stream of bits in the receiver to the precisely duplicate received bits from the satellite, and "aligning" the two streams, a shift error or displacement can be calculated representing the precise travel time from satellite to receiver. Since the receiver also knows the precise position of the satellite, and its range from the receiver, a simple triangulation calculation can give two dimensional position (lat/long) from three satellites and additional elevation information from a fourth.

There are actually two PRN strings transmitted: a course acquisition code (C/A code) and a precision code (P code). The coarse code sequence consists of 1023 bits repeated every 266 days. But each satellite transmits a seven day segment re-initialized at midnight Saturday/Sunday of each week. By using both codes a very accurate position can be calculated. By transmitting them at different frequencies, even the signal attenuating effects of the ionosphere can to some degree be factored out. At the present time, civilian users are only authorized to use the coarse acquisition code and this is referred to as the GPS Standard Positioning Service (SPS), a best accuracy of about 5 meters. Military users use both the coarse acquisition code and the precision code in what is referred to as the Precise Positioning Service or PPS accuracy to centimeters. The Department of Defense (DOD) can at any time encrypt the precision code with another secret code on demand. This is referred to as "anti-spoofing" and ensures that no hostile military forces can also use the GPS service. The DOD can also purposely degrade the accuracy of the coarse acquisition code referred to as "selective availability" to about a 100 meter accuracy. But other than during brief test periods and national emergencies, the service is generally available to all. New techniques now make the civilian use of GPS almost as accurate as the military use.

GPS receivers are now accurate to within 70 meters virtually anywhere in the world. How accurate GPS receivers are depends on how many GPS satellite signals they pick up. Some low-cost receivers can pick up five satellites and are thus accurate to within ten meters.

GR Generic Requirement. A Bellcore document type replacing the Framework Technical Advisory (FA), Technical Advisory (TA) and Technical Refrence (TR) document types. FA, TA and TR documents previously reflected the maturity level of the proposed requirements. In contrast, a Generic Requirements (GR) is a living document that represents Bellcore's preliminary and current view of a technology, equipment, service or interface. It does not necessarily reflect the views of any other company. See Bellcore and GR-303.

GR-1209 Bellcore generic requirements for fiber optic branching components.

GR-1221 Bellcore generic reliability assurance requirements for fiber optic branching.

GR-196 Bellcore generic requirements for optical time domain reflectometer (OTDR) type equiment. See GR.

GR-20 Bellcore generic requirements for optical fiber and fiber optic cable. See GR.

GR-303 GR-303 is a set of technical specifications from Bellcore to help define what the next generation of the world's telecommunications network (i.e. the new PSTN) might look like. The following words are from Bellcore:
"What is GR-303? Network providers are looking to deploy Next Generation Integrated Digital Loop Carrier (NG-IDLC) systems that take advantage of leading edge technology to help reduce operating and capital equipment costs while delivering a full range of telecommunications services. Bellcore's GR-303 family of requirements specifies a set of NG-IDLC generic criteria that creates an Integrated Access System, supporting multiple distribution technologies and architectures (e.g., xDSL, HFC, Fiber-to-the-Curb, etc.), and a wide range of services (narrowband and broadband) on a single access platform. The GR-303 family of generic criteria defines a set of requirements for Integrated Access Systems that includes open interfaces for mix-and-match of (1) Local Digital Switches (LDSs) with Remote Digital Terminals (RDTs) as well as (2) RDTs and Element Management Systems (EMSs). Facilities connecting to the narrowband digital switch (i.e., LDS) are efficiently assigned and managed through the Time-Slot Management Channel (TMC), with remote operations functions supported over the Embedded Operations Channel (EOC) of the GR-303-based Integrated Access System. GR-303-based Integrated Access Systems promote increased net-

work architecture flexibility by providing a consistent approach to deploying a wide range of access system technologies in a consistent manner. Many vendors are developing NG-IDLC products that, although they use different distribution technologies and architectures (e.g., hybrid fiber coax and fiber in the loop), meet the open interfaces described in the GR-303 requirements. This allows network providers to tailor the access system technology deployed area-by-area while utilizing core network features such as the LDS interface and Telecommunications Management Network (TMN) operations capabilities. GR-303-based Integrated Access Systems are intended to reduce capital costs through supplier competition and operating costs through a standards-based, Telecommunications Management Network (TMN) compatible operations environment that provides remote operations capabilities. Integrated Access System products will help to reduce capital costs by enabling mix-and-match among LDS, RDT and EMS products from a wide variety of vendors. The open interfaces described in the GR-303 requirements will help enable the network providers to pursue competitive bids from multiple suppliers for Integrated Access Systems products, thereby potentially obtaining better prices." For more: Bellcore Project Manager - GR-303 Integrated Access Systems 331 Newman Springs Road NVC-1F453 Red Bank NJ 07701-5699 732-758-4001. rrentko@notes.cc.bellcore.com and www.bellcore.com/GR/gr303.html

According to a company called Zarak Systems Corporation, a maker of bulk call simulator GR-303 test equipment, "GR-303 is a specification for a digital loop carrier system (DLC) that operates on T-1 circuits. The GR-303 specification encompasses all aspects of the functionality of the DLC system. Thus, the term GR-303 is commonly used to describe a system or the framing on a set of T-1 circuits. GR-303 is used by telephone operating companies to concentrate telephone traffic and provide better maintainability. The system provides:

a. T-1 circuits exiting a switch (referred to as the IDT), and going directly to the remote digital terminal (RDT) equipment, without the need for additional equipment in the central office (CO)

b. concentration from 1:1 to 44:1

c. a timeslot management channel (TMC) data link that uses messages for call setup and tear down

d. the use of signalling bits to indicate call control

e. a separate embedded operations channel (EOC) data link

f. redundancy on the circuits that carry the data links

g. expandability from two to 28 T-1 circuits that can carry up to 668 channels simultaneously

h. expandability from 1 to 2048 subscriber channels

i. ability to handle ISDN circuits (both BRI and PRI) for the subscriber

j. multiple interface groups (IGs), so that the remote equipment can simultaneously interface to multiple switches

The T-1 circuits are configured for ESF framing, and usually have B8ZS enabled. The first two T-1 circuits each carry the TMC and EOC for redundancy. The EOC is carried in timeslots 12 of the first and second T-1 circuits, and the TMC is carried in timeslots 24 of the first and second T-1 circuits.

According to Zarak, GR-303 has its foundation on SLC-96 mode 2. The two specifications differ in many aspects:

a. GR-303 is expandable, whereas SLC-96 is fixed at 2 T-1 circuits and 96 subscriber channels,

b. GR-303 has continual redundancy, whereas SLC-96 has an optional back up scheme

c. the GR-303 protocols emanate directly from the switch, whereas SLC-96 requires equipment in the CO that is separate from the switch

d. GR-303 has a comprehensive EOC which allows an operating company to do OAM&P remotely, whereas SLC-96 is limited in its capabilities.

Zarak highlights these disadvantages of GR-303:

1. The EOC is enormously complex in its implementation. This had led many manufacturers to implement a minimum number of its features, known as "EOC-light."

2. There is a combination of hybrid signalling, using messages to set up and tear down the allocation of a timeslot, and then robbed bit signalling is used to indicate the call control.

3. There is only 56 kb/s data path because of the robbed bit signalling.

4. The TMC uses messages based on ISDN and Q.931 in particular. However, because the objectives are different, the messages are not standard.

5. To add BRI, a separate channel must be allocated for the D-channel (call set up and tear down). Four BRI D-channels can be merged into one GR-303 channel, called a QDS0 (quad DS0).

6. There is no scheme to handle concentrated PRI, and a whole T-1 circuit must be permanently dedicated. www.zarak.com

GR-485-CORE GR-485-CORE specification contains the generic guidelines for Common Language Equipment Coding Processes and Guidelines (CLEI codes)m which are a standard method of identifying telecommunications equipment in a uniform, feature-oriented language. It's a text/barcode label on the front of all equipment installed at RBOC facilities et. al. that facilitates inventory, maintenance, planning, investment tracking, and circuit maintenance processes. Suppliers of telecommunication equipment give Bellcore technical data on their equipment, and Bellcore assigns a CLEI code to that specific product. See CLEI.

Graceful Close Method terminating a connection at the transport layer with no loss of data.

Graceful Degradation A condition in which a system continues to operate, providing services in a degraded mode rather than failing completely. See ABEND.

Grade 1 Cable Twisted pair cables specifically designed for analog voice circuits and data transmissions up to 1 Mbps. Applications — Key systems, analog and digital PBX, low speed data, RS-232, etc.

Grade 2 Cable Twisted pair cables designed to meet the IBM Type 3 specification. These cables are capable of data transmissions at 4 Mbps, IBM 3270, STAR-LAN I, IBM PC Network, ISDN, etc.

Grade 3 Cable Twisted pair grade 3 LAN cables have performance characteristics that permit data transmissions at 10 Mbps. Each have been tested to insure they meet the EIA/TIA 568 emerging standard. Applications — 802.3 10BASE-T at 10 Mbps, STARLAN 10 and 802.5 token ring at 4 Mbps.

Grade 4 Cable The highest quality twisted pair cables available. Super grade cables have been tested up to speeds of 20 Mbps, 802.5 token ring at 4 Mbps and 802.3 10Base-T at 10 Mbps.

Grade 5 Cable These are the IBM-type individually shielded 2 pair twisted data cables. They're currently being tested for data rates at 100 Mbps. Applications — IBM Cabling System, 802.5 token ring at 16 Mbps and FDDI at 100 Mbps. Grade 5 cable is not the same as CAT 5 cable. See Category of Performance.

Grade Of Service GOS. A term associated with telephone service indicating the probability that a call attempted will receive a busy signal, expressed as a decimal fraction. Grade of service may be applied to the busy hour or to some other specified period. A P.01 Grade of Service means the user has a 1% chance of reaching a busy signal. See TRAFFIC ENGINEERING.

Graded Index A characteristic of fiber optic cable in which the core refractive index is varied so that it is high at the center and matches the refractive index of the cladding at the core-cladding boundary. This reduces dispersion, which is fiber's equivalent of fading.

Graded Index Fiber Graded index fiber is a multimode fiber optic cable that is made with progressively lower refractive index fiber toward the outer core. This reduces dispersion, which is fiber's equivalent of fading.

Gradient In graphics, having an area smoothly blend from one color to another, or from black to white, or vice versa. See GRADIENT FILL.

Gradient Fill A computer imaging term. A fill composed of a smooth blend from a starting color to an ending color. There are many variations on this theme. Most programs let you apply textures, and others have "smart" gradient fill routines that lend a three-dimensional appearance.

Grand Alliance Also known as HDTV Grand Alliance. Comprises AT&T, General Instrument Corporation, Massachusetts Institute of Technology, Philips Electronics North America Corporation, Thomson Consumer Electronics, The David Sarnoff Research Center, and Zenith Electronics Corporation. These organizations had developed and promoted competing digital standards for HDTV. In May 1993, and under pressure from the FCC, they joined together in a "Grand Alliance" to develop a final digital standard for HDTV, which then became known as ATV (Advanced TV). The resulting single standard was documented in the ATSC (Advanced Television Systems Committee) DTV (Digital TV) Standard, which was accepted in large part by the FCC in December 1996. See ATV and HDTV.

Grandfather Clause See GRANDFATHERED.

Grandfather Tape The first backup of a program or a data record, saved so that you can always go back to step one if something goes wrong.

Grandfathered Something that has a right to be a thing or own a thing by reason of it being or owning that thing before laws or rules were introduced to formalize the process. The derivation of term goes back to the Civil War. Grandfathering was a provision in several southern state constitutions designed to enfranchise poor whites and disfranchise blacks by waiving voting requirements for descendants of men who voted before 1867. The word derives from a "grandfather clause." Grandfather clauses stated that the right to vote was only available to those Americans whose grandfathers had been eligible to vote. These clauses were used, primarily in the South, to discriminate against blacks and immigrants shortly after Lincoln's issuance of the Emancipation Proclamation and congressional ratification of the Fourteenth Amendment. As a result, "grandfather" has come to mean something allowable because it was allowable before prohibitive legislation. See also GRANDFATHERED EQUIPMENT.

Grandfathered Equipment Non-FCC registered telephone equipment that was directly connected to the telecommunications network without a phone company-provided protective connecting arrangement (PCA) prior to the formalized FCC registration program. See GRANDFATHERED.

Granularity 1. Microsoft jargon for complexity. For example, when you "achieve granularity," you grasp the complexity of the issue or problem.

2. Scalable in the most agreeable terms. A granular technolo-

gy is scalable in very small increments, like grains of sand. In other words, it can be upscaled in small increments, matching the small, incremental requirements of the user while avoiding disproportionately large increases in cost. It's an overused, overly optimistic, and misleading term which finds its application primarily in sales presentations and brochures. See Brochureware.

Graphic Character A character, other than a character representing a control function (like Ctrl G being, in WordStar and dBASE nomenclature, to delete the character on the right) that has a visual representation normally handwritten, printed, or displayed, and that has a coded representation consisting of one or more bit combinations.

Graphic Equalizer A device which adjusts the tone by changing specific frequencies. The tone control on a radio is a type of equalizer. A radio transmitter may amplify low-end signals better than high-end signals. An equalizer can reduce or increase the amplification of the broadcast for an even and accurate reproduction of the input.

Graphic Violator Picture the home page of a typical Web site. Somewhere on the page is a moving graphic — perhaps an animated GIF — that screams at you and violates the visual integrity and consistency of the page. Most often, such graphic is designed to deliberately violate the integrity of the page. It is often a paid-for advertisement. And the advertiser wants to draw your attention to his graphic ad. After all, he paid big money for the ad.

Graphical Browser A graphical browser is another, more commonly used, term for a World Wide Web(WWW) client program. A graphical browser can display inline graphics and allows the user to choose hyperlinks to move between hypertext documents. All browsers are graphical these days. The two leading browsers are Netscape and Microsoft Internet Explorer.

Graphical User Interface GUI. A fancy name probably originated by Microsoft which lets users get into and out of programs and manipulate the commands in those programs by using a pointing device (often a mouse). Microsoft's own definition is more elaborate. Namely that GUI puts visual metaphor that uses icons representing actual desktop objects that the user can access and manipulate with a pointing device.

Graphics Coprocessor A programmable chip that speeds video performance by carrying out graphics processing independently of the computer's CPU. Among the coprocessor's common abilities are drawing graphics primitives and converting vectors to bitmaps.

Graphics Engine The print component that provides WYSIWYG (What You See Is What You Get) support across devices.

Graphics File In terms of the World Wide Web (WWW), a graphics file is a file in graphics format that can be retrieved through a Web browser. The Web browser may need an add-on or file viewer in order to be able to display the file.

Graphics Interface Format Graphics Interface Format (GIF) is pronounced "Jiff." GIF is a format for encoding images (pictures, drawings, etc.) into bits so that a computer can "read" the GIF file and throw the picture up on a computer screen. CompuServe pioneered the bit-graphics format titled Graphics Interface Format. The format has become a standard in the electronic bulletin board world and it is used primarily to carry photographs of women in semi-clad and naked poses. According to Jack Rickard, publisher of Boardwatch Magazine, "some of the photographs are reasonably good, but most feature strikingly plain women rather artlessly photographed by those whose higher calling is probably more aptly found in the building trades or automotive repair."

Graphics Mode PCs work in two modes — text and graphics. In graphics mode, the pixels are individually addressable. In text mode, the graphics card inside your PC throws a type font on your screen. And that's it. You can't change it. In graphics mode, you can. Graphics and text modes are mutually exclusive.

Grasshopper Fuse A fuse that indicates that it has been blown by the movement of a piece of springy metal.

Grating Lobes Secondary main lobes.

Gravity Cell A closed circuit cell used where a continuous flow of current is desired. This type consists of copper and zinc electrodes with copper sulfate and zinc sulfate electrolyte. These are separated because of difference in their specific gravity.

Gray Elisha Gray was an inventor who filed for a patent on his own telephone design a few hours after Alexander Graham Bell. Gray was involved in a number of lawsuits with the young AT&T, and ultimately co-founded (with Enos BARTON) electrical equipment maker Western Electric, which was later sold to American Bell, which became AT&T. In 1984, Western Electric was renamed AT&T Technologies, and was spun off from AT&T in 1996 as part of Lucent Technologies. Stay tuned for another 10 years or so to see what becomes of Mr. Gray's company.

Gray Code A code used for translating certain analog representations into binary representations, such as the representation of angle. Gray code may also be called cyclic binary code or reflective code.

Gray Scale The spectrum, or range, of shades of black that an image has. An optical pattern consisting of discrete steps or shades of gray between black and white. Early facsimile machines could only receive black and white images and print them in black or white. Now they can print 16 shades of gray. This way if they receive a photo, it will look like a photo.

Gray Whale One of the first non-Bell key systems sold in North America. It was a 1A2 electromechanical key system from TIE/communications. TIE was a shortening of the words Telephone Interconnect Equipment. It was a wonderfully reliable phone system. A gray whale is now a prized possession.

Graybar Probably the oldest distributor of phone equipment, as well as various electrical products. Named for Elisha GRAY and Enos BARTON, who formed Western Electric. The company is headquartered in St. Louis, though there is a famous Graybar building in Manhattan, attached to Grand Central Terminal. A plaque in the terminal near stairs leading into the Graybar building shows Gray and Barton (if it hasn't been defaced or stolen).

Grayscale Monitor Any monitor capable of showing levels of gray and not just black or white.

Great Circle A circle defined by the intersection of the surface of the Earth and any plane that passes through the center of the Earth. The shortest distance, over the idealized surface of the Earth, between two points, lies along a great circle.

Greek Prefixes Remember the word "chronous?" It's used to mean the process of adjusting intervals or events of two signals to get the desired relationship between them. Here are the Greek prefixes that describe different timing conditions:

asyn - not with
hetero - different
homo - the same
iso - equal
meso - middle
piesio - near
syn - together

Green Green is a term being applied by manufacturers to mean their equipment uses less electricity than other equipment. Classic PC equipment include screens that shut themselves almost down if they haven't been accessed for a minute or two. Much of the technology of "green" has already been used in laptops. When it moves to the desktop, there are occasionally compromises — such as moving from a 3/4rds off to full-on might take a moment or two. Are you willing to live with it?

Greenfield "Greenfield" is essentially the opposite of "legacy". A greenfield network is one that is being designed and built from scratch, with no need to accommodate legacy (i.e., old) equipment or architectures. Usage extends to almost any aspect of describing a new venture: greenfield companies, greenfield evaluation, greenfield opportunities, greenfield factors, etc. An existing network being expanded has both greenfield and "brownfield" components. A variant is to describe a greenfield effort as a "greenstart." See BROWNFIELD and LEGACY.

Greenwasher A business that uses the fact that it recycles to promote itself.

Greenwich Mean Time Also called Zulu Time. Coordinated Universal Time. Another term for Greenwich Mean Time (GMT). Greenwich is a borough in SE London, England, which is located on the prime meridian from which geographic longitude is measured. Greenwich was formerly the site of the Greenwich Observatory. And for these historic reasons, Greenwich is the place from which world time starts. For example, GMT time is five hours earlier Eastern Standard Time — i.e. the time in the northern hemisphere Summer. GMT (Zulu Time) is always the same worldwide. Communication network switches are typically coordinated on GMT.

Greetings Only Mailboxes Mailboxes that deliver a message to incoming callers but do not allow a message to be left. The Greeting Only Mailbox may transfer a caller to a designated telephone number.

Grep Generalized Regular Expression Parser. A really powerful UNIX utility which can search a text file or program, finding and displaying or printing lines of computer code which contain specific character strings.

GRIC Global Reach Internet Connection. An alliance of ISPs (Internet Service Providers) and IAPs (Internet Access Providers) to provide roaming capabilities for travelers. Based on proprietary standards, roamers are authenticated before being afforded Internet access. Usage is cross-billed through the GRIC clearinghouse, with fees being set by each ISP for use of its facilities by roamers. GRIC includes over 100 member ISPs in approximately 30 countries, and includes over 1,000 POPs (Points of Presence). Members include Prodigy in the U.S., and Telstra in Australia. GRIC competes with I-PASS. The IETF's Roamops working group is developing a standard for roaming, as well. See also I-PASS and ROAMOPS.

Grid 1. That element in a vacuum tube having the appearance of a grid and which controls the flow of electrons from the filament to the plate. "Grid" generally refers to the control grid.
2. Global Resource Information Database: part of the United Nations environment program.

Gritch A computer complaint.

Groom/Fill In telephony, terms associated with more efficient use of T-1 trunks by combining partially filled input T-1 trunks into fully filled outgoing T-1 trunks.

Grooming Managing bandwidth on a wide area, public or private network to use the long haul transmission facilities as effectively as possible.

Gross Additions A cellular industry term. The amount of new subscribers signing up for the service before adjusting for disconnects (churn).

Ground 1. A problem that exists when a circuit is accidentally crossed with a grounded conductor.
2. Connecting equipment by some conductor (wire) to a route that winds up in the earth (Ground) for electrical purposes. One purpose of a "ground" wire is to carry spurious voltage (e.g. lightning strikes) away from the electrical and electronic circuits it can cause harm to. Incorrect grounding is probably the major cause of telephone systems problems. See GROUNDING (the major explanation), GROUND RETURN and GROUND START.

Ground Absorption The loss of energy in transmission of radio waves due to dissipation in the ground.

Ground Button A button needed on phones used for power failure transfer behind a PBX. You need the button because many trunks behind a PBX are ground start (as compared to loop start).

Ground Clamp A clamp or strap used to provide make a secure connection to a water pipe or grounding rod. It connects a wire to earth ground.

Ground Constants The electrical constants of the earth, such as conductivity and dielectric constant. The values vary with frequency, and also with local moisture content and chemical composition of the earth.

Ground Fault In AC electricity, a ground fault is any unintended connection between a supply conductor and ground (i.e.: hot conductor in contact with the metal case of a piece of equipment). A ground fault will cause a high current flow and should operate the overcurrent protection (fuse or breaker provided such devices are functionally adequate) only if the ground path impedance is sufficiently low — but under no circumstances greater than two ohms.

Ground Fault Circuit Interrupter GFCI or GFI. A device intended to interrupt the electrical circuit when the fault current to ground exceeds a predetermined value (usually 4 to 6 milliamps) that is less than required to operate the overcurrent protection (fuse or breaker) for the circuit. This device is intended to protect people against electrocution. It does not protect against fire from circuit overload.

Ground Fault Protector GFP. A device designed to protect electrical service equipment from arcing ground faults. A GFP does not provide protection for people.

Ground Lead The conductor leading to the ground. Connection.

Ground Loop This occurs when a circuit is grounded at one or more points. It can cause telephone system problems. Here's an explanation from American Power Conversion Corp.: Common wiring conditions where a ground current may take more than one path to return to the grounding electrode at the service panel. AC powered computers all connected to each other through the ground wire in common building wiring. Computers may also be connected by data communications cables. Computers are therefore frequently connected to each other through more than one path. When a multi-path connection between computer circuits exists, the resulting arrangement is known as a "ground loop". Whenever a ground loop exists, there is a potential for damage from inter system ground noise.

Ground Plane The surface existing or provided, that serves as the near-field reflection point for an antenna.

Ground Potential The electrical potential of the earth with respect to another body or region. The ground potential of the earth will vary with locality and also as a function of certain phenomena such as meteorological disturbances.

Ground Return If a battery is connected to a closed electrical circuit, an electric current will flow in the circuit. In the early days of the telegraph, the circuit consisted of a long wire, the telegraph key, the electromagnet of a telegraph sounder and a return path through the ground, which served as a conductor. Thus the current flowed from one terminal of the battery through the wire, through the electromagnet, to a metal stake driven into the ground (a "ground" electrode), back through hundreds of miles of earth to the distant stake at the distant telegraph office and then to the other terminal of the battery. In later telecommunications, the ground return path was replaced by a second wire.

Ground Return Circuit A circuit in which the earth serves as one conductor. A circuit in which there is a common return path, whether or not connected to earth ground.

Ground Start A way of signaling on subscriber trunks in which one side of the two wire trunk (typically the "Ring" conductor of the Tip and Ring) is momentarily grounded (often to a cold water pipe) to get dialtone. There are two types of switched trunks typically for lease by a local phone company — ground start and loop start. PBXs work best on ground start trunks, though many will work — albeit intermittently — on both types Normal single line phones and key systems typically work on loop start lines. You must be careful to order the correct type of trunk from your local phone company and correctly install your telephone system at your end — so that they both match. In technical language, a ground start trunk initiates an outgoing trunk seizure by applying a maximum local resistance of 550 ohms to the tip conductor. See LOOP START.

Ground Start Supervision Telephone circuitry developed to prevent Glare.

Ground Start Trunk A phone line that uses a ground instead of a short to signal the CO for a dial tone. required by some PBXs. See Ground Start

Ground Station A cluster of communications equipment, usually including signal generator, transmitter, receiver and antenna that receives and/or transmits to and from a communications satellite. Also called a satellite earth station.

Ground Wave In radio transmission, a surface wave that propagates close to the surface of the Earth. The Earth has one refractive index and the atmosphere has another, thus constituting an interface. These refractive indices are subject to spatial and temporal changes. Ground waves do not include ionospheric and tropospheric waves.

Grounding There are two distinct and unique categories in this broad area called grounding. One is earth grounding. The purpose of an earth grounding system is essentially threefold: 1. To guard against the adverse effects of lightning, 2. To assist in the reduction of static and 3. To bring a zero-voltage reference to system components in order that logic circuits can communicate from a known reference. The other category of grounding is known as equipment grounding. The purpose of equipment grounding is threefold: 1. To maintain "zero volts" on all metal enclosures under normal operating conditions This provides protection from shock or electrocution to personnel in contact with the enclosure. This is the safety aspect. This is the primary means of protecting personnel from electrocution. 2. To provide an intentional path of high current carrying capacity and low impedance to carry fault current under

ground fault conditions; and 3. To establish a zero voltage reference for the reliable operation of sensitive electronic equipment. Effective equipment grounding is defined in the National Electrical Code, Article 250-51 and the Canadian Electrical Code Article 10-500. These Codes read almost identically. They say, "The path to ground from circuits, equipment and metallic enclosures for conductors shall;

1. Be permanent and continuous.

2. Have the capacity to conduct safely any fault current likely to be imposed on it, and

3. Have sufficiently low impedance to limit the voltage to ground and to facilitate the operation of the circuit protective devices in the circuit.

The Earth shall not be used as the sole equipment grounding conductor."

According to the Electric Power Research Institute, "electrical wiring and grounding defects are the source of 90% of all equipment failures." Many telephone system installer/contractors have found that checking for and repairing grounding problems can solve many telephone system problems, especially intermittent "no trouble found" problems. As electrical connections age, they loosen, corrode and become subject to thermal stress that can increase the impedance of the ground path or increase the resistance of the connection to earth. Equipment is available to test for proper grounding. (Ecos Electronics of Oak Park, IL makes some.) Before you install power conditioning equipment such as voltage regulators, surge arresters, etc. you should test for and correct any problems you have with grounding and wiring.

This story was related by Pat Routledge of Winnipeg, Ontario, Canada about an unusual telephone service call he handled while living in England. It is common practice in England to signal a telephone subscriber by signaling with 90 volts across one side of the two wire circuit and ground (earth in England). When the subscriber answers the phone, it switches to the two wire circuit for the conversation. This method allows two parties on the same line to be signalled without disturbing each other. This particular subscriber, an elderly lady with several pets called to say that her telephone failed to ring when her friends called and that on the few occasions when it did manage to ring, her dog always barked first. Torn between curiosity to see this psychic dog and a realization that standard service techniques might not suffice in this case, Pat proceeded to the scene. Climbing a nearby telephone pole and hooking in his test set, he dialed the subscriber's house. The phone didn't ring. He tried again. The dog barked loudly, followed by a ringing telephone. Climbing down from the pole, Pat found: 1. Dog was tied to the telephone system's ground post via an iron chain and collar 2. Dog was receiving 90 volts of signalling current 3. After several jolts, the dog was urinating on ground and barking 4. Wet ground now conducted and the phone rang.

See also AC, AC POWER and BATTERY.

Grounding Field Grounding rods placed in the ground and connected together around an antenna site or central office site. This provides the best possible ground for electronic equipment.

Grounding Strap A device worn on the wrist or on the shoe when handling a static-sensitive component to prevent static shocks (sparks) which could damage the component. Don't even think of touching a printed circuit card without wearing a grounding strap. See GROUNDING.

Group 1. In call centers or in automatic call distributors, a group is the same as GATE or SPLIT. A group is an ACD routing division that allows calls arriving on certain telephone trunks or calls of certain transaction types to be answered by specific groups of employees.

2. A group is a collection of voice channels, typically 12. In AT&T jargon, a group is 12 channels. A supergroup is 60 channels. A mastergroup is 10 supergroups or 600 voice channels.

3. An SCSA definition. A group is an associated set of one or more Resource Objects. Groups encapsulate the functionality of the Resource Objects that are associated with them. Resource Objects within a Group have defined connectivity. The Group provides three services to the application: implicit management of connectivity between group members; representation of a single entity to the applications (group ID); and reservation of all physical resources (CPU, memory, time slots) required to provide the application with exclusive use of configured resources.

Group 1, 2, 3, 3 bis & 4 These relate to the facsimile machine business. They are essentially standards of speed and sophistication. They were created by the ITU-T in Geneva, Switzerland to make sure facsimile machines from one maker could speak to facsimile machines of another maker.

Group 1 transmits an 8 1/2 by 11-inch page in around six minutes. It conforms to ITU-T Recommendation T.2.

Group 2 transmits an 8 1/2 by 11-inch page in around three minutes. It conforms to ITU-T Recommendation T.3.

Group 3 — the most common fax in the world today — transmits an 8 1/2 by 11-inch page (also called A4) in as little as 20 seconds. It is a digital machine and includes a 9,600 baud modem. It transmits over dial up phone lines. Group 3 standards for facsimile devices were developed by ITU-T adopted in 1980 and modified in 1984 and 1988. Group 3 defines a resolution of 203 x 98 dots per inch and 203 x 196 for "fine." Group 3 uses modified Huffman code to compress fax data for transmission. For example, a white line with no text, called a run, extending across an 8.5" page equals 1728 bits. Modified Huffman Code compresses the 1728 bits into a 17-bit code word. The lengths for all possible white runs are grouped together into 92 binary codes that will handle any white run length from 0 to 1728.

Group 3 bis. This is an update to Group 3. It includes an image resolution of 406 x 196 dpi and a transfer rate of 14,400 bits per second. Fax machines that are Group 3 bis can transmit 50% faster to fax machines that are also Group 3 bis, which is a big speed improvement. Group 3 bis can drop to Group 3 if there's a Group 3 on the other end. Most of the modern plain paper fax machines and most of the today's computer fax modems are Group 3 bis. That means they transmit and receive at up to 14,400 bps.

Group 4 Fax. The latest and fastest international standard for facsimile machines. It specifies a machine which operates at 64 Kbps, which can only work on a digital channel and which takes six seconds to transmit a 8 1/2 x 11 inch page. The Group 4 standard was promulgated in January, 1987. Group 4 fax machines are designed to use one of the 64,000 bit per second B (Bearer) channels on ISDN. The main difference between Group 3 and Group 4 fax machines is that Group 4 fax machines do not convert the scanned information into an analog format before transmitting it down phone lines. Group 4 fax machines simply send the digitally scanned information down ISDN lines. The advantages of Group 4 fax are that quality is much higher, and call costs are much lower due to the increase in speed of transmission.

Most Group 3 machines will transmit and receive from Group 1, Group 2 and Group 3 machines (but at their slower

speeds). Group 1, 2, 3 & 4 are international standards. Group 3 is now by far and away the most common. All Group 3 machines will transmit and receive from each other. Some manufacturers have improved on the standards by offering Group 3 "fine," for example. These "fines" can talk to the same machines. But often can't talk to other "fines." If you're buying a facsimile machine and want super-quality transmission, check its compatibility with other machines. Or, easier, buy all identical machines. See FACSIMILE.

Group Address A single address that refers to multiple network devices. Synonymous with multicast address.

Group Addressing In transmission, the use of an address that is common to two or more stations. On a multipoint line, where all stations recognize addressing characters but only one station responds.

Group Busy Hour GBH. The busy hour offered to a given trunk group.

Group Call A special type of station (i.e. extension) hunting that requires a special access number to permit a call to the special access number and ring the first available phone in that group.

Group Distribution Frame GDF. In frequency-division multiplexing, a distribution frame that provides terminating and interconnecting facilities for the modulator output and demodulator input circuits of the channel transmitting equipment and modular input and demodulator output circuits for the group translating equipment operating in the basic spectrum of 60 kHz to 108 kHz.

Group Hunting Automatically finds free telephones in a designated group. See HUNT

Group Scheduling Software Software designed to coordinate and manage both worker schedules and office resources such as equipment and conference rooms.

Group Velocity 1. The velocity of propagation of an envelope produced when an electromagnetic wave is modulated by, or mixed with, other waves of different frequencies. The group velocity is the velocity of information propagation. In optical fiber transmission, for a particular mode, the reciprocal of the rate of change of the phase constant with respect to angular frequency.

Grouping A facsimile term for periodic error in the spacing of recorded lines.

GRoupIPC Ohio, January 8, 1997, GRoupIPC - North America (GRoupIPC-NA) has announced their official incorporation as a non-profit organization. Susan M. Chicoine, of Systran Corp., Dayton, Ohio, who served as a trustee during the incorporation process, was elected president of the group. The parent organization, GRoupIPC, was established in Europe in 1994 to provide a worldwide forum for the exchange of information on Industry Pack (IP) and PCI Mezzanine Card (PMC) technologies. GRoupIPC-NA promotes the embedded system industry's trend towards open, internationally recognized, standards solutions reinforcing the movement away from sole-vendor, proprietary solutions that limit the flexibility and upgradeability of embedded system designs. Member organizations support open-system solutions based upon the internationally recognized PMC and IP mezzanine board standards, and will promote their use to bring the benefits of stable, multi-vendor standards to embedded designs. GRoupIPC-NA will: (1) Promote market acceptance and the use of PMC and IP mezzanine board technology, (2) Disseminate information about products, applications and technical requirements, using or affecting IP and PMC mezzanine board technology and (3) Provide market and technical support to users, distributors, and manufacturers of IP and PMC technology and products. GRoupIPC-NA is headquartered in Dayton, Ohio. www.GRoupIPC.com, email GRoupIPCNA@aol.com or phone 937-427-9735.

Groupware A term for software which runs on a local area network and which allows people on the network (typically a team) to participate in a joint (often complex) project. According to Fortune Magazine, March 23, 1992, using groupware, "Boeing has cut the time needed to complete a wide range of team projects by an average of 91%, or to one-tenth of what similar work took in the past." Groupware can be used in a meeting, with everyone sitting around a conference table and typing their ideas into the PC in front of them. Groupware can also be used off-line, with members of the "team" in different cities adding their comments. The "bell-wether" of groupware software is Lotus' program called notes.

Growth Addition A telephone company definition. Any equipment addition that increases the limiting capacity of an entity. Hence, a trunk relay addition will not generally be considered a growth addition since trunk relay equipments are not considered as limiting. However, if a trunk relay addition requires other equipment, such as trunk frames, or other common control equipment, then the addition should be considered a growth addition.

Growth Entity A telephone company definition. The growth entity in a multi-entity-wire center is: that entity where all future network access line growth is engineered to take place. Non growth entities are normally loaded at or near capacity and excess demand is served via the growth entity. This distribution of demand is accomplished via the loading plan.

Growth Factor A telephone company definition. A ratio derived by trending network access lines, traffic or loads and relating the future levels to current levels. Growth factors may be combined to develop a projection ratio or used individually as a projection ratio.

GRSU Generic Remote Switch Unit.

GS Abbreviation for ground separator. The binary code is 1101001, the hex is D1.

GS Trunk Ground Start Trunk. A trunk on which the communications system, after verifying that the trunk is idle (no ground on tip), transmits a request for service (puts ground on ring) to a telephone company. The other and more common type of trunk is called a Loop Start Trunk.

GSA General Services Administration.

GSDN See GLOBAL SOFTWARE DEFINED NETWORK.

GSF See GENERIC SERVICES FRAMEWORK.

GSM GSM stands for Groupe Speciale Mobile, now known as Global System for Mobile Communications, is the standard digital cellular phone service you will find in Europe, Japan, Australia and elsewhere — a total of 85 countries around the world. GSM actually is a set of ETSI standards specifying the infrastructure for a digital cellular service. To ensure interoperability between countries, these standards address much of the network wireless infrastructure, including the radio interface (900 MHz), switching, signaling, and intelligent network. An 1,800 MHz version, DCS1800, has been defined to facilitate implementation in some countries, particularly the UK. Since GSM is limited to technical standards, an association of GSM operators called the Memorandum of Understanding (MoU) ensures service interoperability, allowing subscribers to roam across networks. GSM has gained widespread acceptance in several parts of the world, most notably Europe, with deployment in 52 countries by mid year '94. GSM subscriber data is carried on a

Subscriber Identity Module (SIM) or "smartcard" which is inserted into the phone to get it going. As a result, the subscriber potentially has the option of either SIM card mobility or terminal mobility across multiple networks.

One of GSM's major problems is that people wearing hearing aids can't use GSM cell phones, because GSM phones produce unpleasant high-pitched squealing the closer they get to the cell phone.

GSM technical Characteristics: Receiver frequency: 935.2 — 959.8 MHz. Transmitter frequency: 890.2 — 914.8 MHz. Access method: mixed TDMA & FDMA with optional frequency hopping. Security: Optional radio interface encryption. Carrier frequency division: 200 KHz. Users per carrier frequency: 8. Speech bit rate (transfer rate): full rate (13 kbps) or half rate. Total bit rate: 21 Kbps. Bandwidth per channel: 25 KHz." A good book on GSM is "The GSM System for Mobile Communications" by Michael Mouly and Marie-Bernadette Pautet, both of France. The authors, in fact, contributed to the development of GSM. It's really the definitive document on GSM. See also BSS, which stands for BASE STATION SYSTEM. See also SIM CARD.

The audio encoding subset of the GSM standard is best known to computer users because its data compression and decompression techniques are also being user for Web phone communication and encoding .wav and .aiff files.

GSM-900 Global System for Mobile Communications at 900 MHz. A wireless telecommunications term. A GSM (Global System for Mobile Communications) standard for cellular phone systems operating at 900 Megahertz. See GSM.

GSN Gigabyte System Network. See HIPPI.

GSTN General Switched Telephone Network. Same as public telephone network.

GT Global Title. An address such as customer-dialed digits that does not explicitly contain information that would allow routing in the SS7 signaling network, that is, the GTT translation function is required. See GTT.

GTA 1. Government Telecommunications Association. An association of local, state and federal telecommunications professionals in Washington, D.C.

2. Government Telecommunications Agency. Specialized agency which provides telecom service to Canadian federal government departments.

GTE General Telephone and Electronics. A major telecommunications company, whose main business is owning and operating "independent" (i.e. non Bell) local telephone companies. It used to own part of Sprint, the long distance company, but no longer owns any part of it.

GTE Sprint A long distance service once provided by GTE Sprint, then a 50-50 joint venture of GTE and United Telecom. In 1989 it became majority owned (80.1%) by United Telecom and 19.9% GTE. In 1991 United bought the remaining 19.9% of Sprint from GTE. Now, GTE Sprint is just called Sprint. And Sprint is no longer owned at all by GTE.

GTE TELENET A public data network which operates by using the ITU-T approved X.25 packet switching protocol. The new name for GTE TELENET is SPRINTNET. It's owned and run by Sprint Corporation.

gTLD Generic Top Level Domain. An Internet term. The idea of gTLD is to allow web sites with creative, descriptive names such as www.RayHorakConsulting.firm and www.MargaretHorakAstrology.web (these are not real). gTLD names were slated to become operational in March, 1998. CORE, the Council Of REgistrars, was to be the organization charged with the responsibility for establishing and maintaining a new set of gTLDs (generic Top Level Domains) for the Internet. Effective March 1998, those proposed gTLDs were meant to comprise the following: .arts (entities emphasizing cultural and entertainment activities) .firm (businesses, or firms); .info (entities providing information services); .nom (individual or personal nomenclature, i.e., a personal nom de plume, or pen name); .rec (entities emphasizing recreation/entertainment activities); .shop (businesses offering goods to purchase) and .web (entities emphasizing activities related to the World Wide Web). The administration of the new gTLDs has now been contracted by CORE to Emergent Corporation, which was to develop, maintain, and operate the Shared Registry System (SRS). SRS is a neutral, shared, and centralized database of the new gTLGs. As many as 90 independent entities, known as "registrants," were authorized to register domain names, or URLs (Uniform Resource Locators), with each relying on the SRS. A URL, such as www.HarryNewtonsGhastlyPaintings.arts (this is not a real URL, at least not at the time of this writing), is translated into an IP address by a Domain Name Server (DNS), also known as a resolver. At the time of this writing, the proposal for new gTLDs has been forestalled. See also CORE, DNS, IP Address, SRS, TLD, and URL.

GTN Global Transaction Network. An AT&T service which adds smarts to the routing of inbound 800 calls. It offers six call processing services: Next available agent routing, call recognition routing, transfer connect service (allows agents to transfer calls to distant ACDs), network queuing, 800 select again service and multiple number database (allows multiple 800 numbers to be assigned to a single routing plan in the network, rather than each 800 number having its own unique routing plan).

GTP General Telemetry Processor.

GTT Global Title Translation. The process of translating a Global Title from dialed digits to a point code (network node) address and application address (subsystem number). This process is accomplished by the STP (Signal Transfer Point) in the SS7 network. See also Global Title, SS7 and STP.

Guardband A narrow bandwidth between adjacent channels which serves to reduce interference between those adjacent channels.   That interference might be crosstalk. Guardbands are typically used in frequency division multiplexing. They are not used in time division multiplexing, because the technology is completely different.

Guardian Agent A Guardian Agent is similar to an Intelligent Agent (which hunts for and grabs information off of the web that you specify) only the Guardian Agent prevents certain sites, such as pornographic pages, gambling sites, or other areas you don't want a child to see, from being accessed.

Guarding The process of holding a circuit busy for a certain interval after its release to assure that a necessary minimum disconnect interval will occur between calls.

Guest Mailbox A mailbox used by a hotel or motel to set up temporary mailboxes for their guests. At least that was the original definition. Now it seems every voice mail system comes with guest mailboxes that could be used for visitors, employees from out of town, etc. Same application as a hotel — temporary use.

GUI Graphical User Interface. A generic name for any computer interface that substitutes graphics for characters. GUIs usually work with a mouse or trackball. Windows 3.0 and Windows 3.1 are the most famous GUIs.   Second most famous is the Apple Macintosh. GUI is pronounced "GOO-

ey." See GRAPHICAL USER INTERFACE.

GUID A Versit term. Globally Unique IDentifier.

Guided Ray In an optical waveguide, a ray that is completely confined to the core.

Guided Wave A wave whose energy is concentrated near a boundary or between substantially parallel boundaries separating materials of different properties and whose direction of propagation is effectively parallel to these boundaries.

Gutta-Percha A latex substance first discovered in 1847 and derived from the sap of Malayan evergreen trees. It first found use circa 1851 as the insulation in the first international telegraph cable which ran between England and France. Gutta-Percha was also the first insulator to survive in under seas applications (particularly for submarine cables) and was still the insulator material of choice for golf balls and telephone receivers until about 1947, when polyethylene finally began to gain acceptance.

Guy Hook A hook bolted to telephone poles and used to attach guy wires.

Guy Thimble A device used to attach a guy wire to a bolt which is attached to an anchor in the ground.

Guy Wire A wire used to support a radio mast.

Guyed A type of wireless transmission tower that is supported by thin guy wires.

Gzip A free compression software program available for Unix and MS-DOS. Appends either a .z or a .gz to the file name. Compressing a file makes it smaller and therefore it takes less time to transmit. The most popular programs are Winzip and PKZIP.

H PAD Host Packet Assembler/Dissembler. See HPAD.

H Schedule A separate and distinct list in the translations area of program store in No. 1 or IA ESS central office. The H (hourly) Schedule is normally used for collecting counts on items required for day-to-day administration of the central office equipment and for the engineering of general growth jobs. They include such items as call processing registers, service circuits and miscellaneous trunks, intraoffice trunks and junctors networks.

H-Channel The packet-switched channel on an ISDN PRI (Primary Rate Interface) which is designed to carry user bandwidth-intensive videoconferencing information streams at varying rates, depending on type: H0 — 384 Kbps; H11 — 1,536 Kbps; and H12 — 1,920 Kbps. In short, H-channels are ISDN bearer services that have pre-defined speeds, starting and stopping at locations on an ISDN PRI circuit. They are contiguously transported from one PRI site through networks to another ISDN PRI site. H-channels are accomplished by aggregation of multiple individual 64Kbps B channels; in the carrier domain, this aggregation is accomplished through a process known as "bonding." See ISDN and PRI.

H-MVIP The original MVIP standard, now called MVIP-90, was developed in 1989-90 and first deployed in 1990. MVIP-90 supports up to 512 telephony channels of 64 Kbps each between circuit boards within a single computer chassis. MVIP-90 has been widely adopted for voice, FAX, data and video services as well as for telephony switching applications. Other MVIP standards have since been developed to address multi-chassis MVIP systems and higher level software APIs. Beginning in 1993, there was interest in higher capacities within individual single-chassis MVIP nodes. Specific applications include large audio conferences, multimedia servers, and the termination and switching of all traffic on a dual FDDI-II fiber ring (as used in MC2 standard Multi-Chassis MVIP) as well as the termination and switching of traffic from T-3/E-3 telephone trunks and SONET/SDH links at OC-3 (155 Mbps) rates (as used in MC3 standard Multi-Chassis MVIP). H-MVIP addresses this need for higher telephony traffic capacity in individual computer chassis.

H-MVIP defines three major items that together make a useful digital telephony transport and switching environment. These are the H-MVIP digital telephony bus with up to 3072 "time-slots" of 64 Kbps each; a bus interface with digital switching that allows a group of H-MVIP interfaced circuit boards to provide distributed telephony switching and a logical device driver model and standard software interface to that logical model. The bus definition includes the mechanical, electrical and timing requirements for a high capacity telephony bus that is a super-set of the existing MVIP-90 standard for computer-based telephony. Several levels of capacity expansion are defined to support a range of system implementations. Among them, H-MVIP "24/2" is a wider version of MVIP-90, while the full H-MVIP bus is both wider and faster. See ECTF, SCSA and TAO.

H.100 H.100 is a hardware specification that provides all the necessary information to implement a CT bus interface at the physical layer for the PCI computer chassis card slot independent of software applications. It is the first card-level definition of the overall CT Bus single-communications bus specification. CT Bus defines a single isochronous communications bus, often called a mezzanine bus, across newer PC chassis card slots (PCI, and the emerging compact PCI). H.100 CT Bus will be compatible with the most popular existing implementations, SCBus, HMVIP and MVIP (as well as ANSI VITA 6) implemented in ISA/EISA card slots. A CT Bus specification for compact PCI, H.110, is also under development for a later release.

Adoption of the single-bus specification, CT Bus, will allow a fluid inter-operation of components to provide an unprecedented level of flexibility for product design and operation. CT Bus provides more capacity to allow development of a new class of applications as well as to increase the capabilities of existing applications. Its addition of greater fault-tolerance will increase the reliability of applications, and its provision for implementing a subset of the specification will provide for many lower cost applications. H.100 offers the following features:

• A PCI card slot form factor to accommodate the growing popularity of PCI slots in computer chassis

• 4,096 bi-directional time slots (permitting up to 2,048 full duplex calls) for larger communications capacity

• An eight megabit data rate and 128 channels per stream for greater bandwidth

• Redundant clocking scheme for increased fault tolerance

• Backwards compatibility and interoperability with SCBus, HMVIP and MVIP

The H.100 is part of a complementary suite of Interoperability Agreements sponsored by the ECTF (Enterprise Computer Telephony Forum). Each specification is fully self contained, yet designed to be complementary with all of the others in the suite. Other ratified specifications include S.100 and S.200. Software developers are creating applications with the S.100 and S.200 specifications, and hardware manufacturers will introduce H.100-based communications cards.

S.100, published in March, 1996 with an addendum last month, specifies a set of software interfaces that provide an effective way to develop CT applications in an open environment, independent of underlying hardware. It defines a client-server model in which applications use a collection of services to allocate, configure and operate hardware resources. S.100 enables multiple vendors' applications to operate on any S.100-compliant platform.

H.110 H.110 defines H.100 on the CompactPCI (cPCI) bus. The biggest difference between H.100 and H.110 is that H.110 supports CompactPCI Hot Swap (the removal and insertion of cards in a live system). www.ectf.org/ectf/home.html. See also H.100

H.221 A framing recommendation which is part of the ITU-T's H.320 family of video interoperability Recommendations. The Recommendation specifies synchronous operation where the coder and decoder handshake and agree upon timing. Synchronization is arranged for individual B channels or bonded 384 Kbps (H0) connections.

H.222.0 Defines the general form of elementary stream multiplexing as the Moving Picture Experts Group 2 (MPEG-2) system part. See H.222.1.

H.222.1 Specifies the parameters of H.222.0 for communication use.

H.230 A multiplexing recommendation which is part of the ITU-T's H.320 family of video interoperability recommendations. The recommendation specifies how individual frames of audiovisual information are to be multiplexed onto a digital channel.

H.231 A recommendation, formally added to the ITU-T's H.320 family of recommendations in March, 1993, which specifies the multipoint control unit used to bridge three or more H.320-compliant codecs together in a multipoint conference.

H.233 An Recommendation, part of the ITU-T's H.320 family, which specifies the encryption method to be used for protecting the confidentiality of video data in H.320-compliant exchanges. Also called H.KEY.

H.235 An ITU-T standard (February, 1998) for securing H.323. voice and videoconference information streams over IP networks (e.g., the Internet, Intranets and LANs. H.235 provides authentication, integrity and privacy services. Authentication serves to establish as genuine the identity of all endpoints in the conference in order that unauthorized users or machines cannot participate. Integrity validates the payload of data packets, thereby ensuring that the data was neither corrupted nor altered in transit. Privacy, accomplished through an encryption mechanism, ensures that the data payload cannot be read by users or machines not authorized. Non-repudiation, planned for inclusion in future releases of H.235, protects against an endpoint's denial of participation in the conference. See also H.323.

H.242 Part of the ITU-T's H.320 family of video interoperability Recommendations. This Recommendation specifies the protocol for establishing an audio session and taking it down after the communication has terminated.

H.244 Recommendation on a channel aggregation method for audiovisual communications. This enables several ISDN B-channels to behave as a single higher-rate channel.

H.245 H.245 specifies the in-band signaling protocol necessary to actually establish a call, determine capabilities, and issue the commands necessary to open and close the media channels. The H.245 control channel is responsible for control messages governing operation of the H.323 terminal, including capability exchanges, commands, and indications. See H.323 and TAPI 3.0.

H.246 See H.245.

H.261 The ITU-T's H.261 is the standards watershed in videoconferencing. Also known as p x 64, H.261 specifies the video coding algorithms, the picture format, and forward error correction techniques to make it possible for video codecs from different manufacturers to successfully communicate. H.261 is an ITU-standard video codec designed to transmit compressed video at a rate of 64 Kbps and at a resolution of 176x44 pixels (QCIF). Announced in November 1990, it relates to the decoding process used when decompressing video conferencing pictures, providing a uniform process for codecs to read the incoming signals. Any H.323 client is guaranteed to support the following standards: H.261 and G.711. Other important standards are H.221: communications framing; H.230 control and indication signals and H.242d: call set-up and disconnect. Encryption, still-frame graphics coding and data transmission standards are still being developed. See H.320.

H.263 H.263 is an ITU-standard video codec based on and compatible with H.261. It offers improved compression over H.261 and transmits video at a resolution of 176 x 44 pixels (QCIF).

H.310 Recommendations for a videoconferencing terminal in an ATM environment.

H.320 The most common family of ITU-T videoconferencing standards. These standards allow ISDN BRI videoconferencing systems and videophones to communicate with each other. I've personally had several H.320 compatible videoconferencing systems and videophones on my desk and have received from and made videoconferencing calls to many different H.320 compatible video phones. The quality is not brilliant. But you can recognize the person at the other end. And they can recognize you. Most H.320 systems allow you to bond together the two B channels of a 2B+D ISDN BRI channel and thus get better video. See all the H.2NN and H.3NN explanations above and below. See also G.711 and V.80.

H.321 The adaptation to the ATM environment of H.320 videoconferencing standards. See H.320 and V.80.

H.322 The adaption to a guaranteed quality of service LAN of H.320 terminals. See H.320 and V.80.

H.323 H.323, a standard from the International Telecommunications Union (ITU-T), serves as the "umbrella" for a set of standards defining real-time multimedia communications for packet-based networks — what are now called IP telephony. Much of the excitement surrounding the H.323 standards involves the use of H.323 entities to communicate over the Internet or managed Internet Protocol (IP) networks. The standards under the H.323 umbrella define how components that are built in compliance with H.323 can set up calls, exchange compressed audio and/or video, participate in conferences, and interoperate with non-H.323 endpoints.

H.323 is an ITU-T standard, which defines a set of call control, channel setup and codec specifications for transmitting real-time voice and video over networks that don't offer guaranteed service or quality of service — such as packet networks, and in particular Internet, LANs, WANs and Intranets. This ITU-T standard (ratified initially in March of 1996) defines the negotiation and adaptation layer for video and audio over packet switched networks. "Negotiation" means that this layer defines the way the devices on either end of the data conversation will figure out what is the fastest speed they can accommodate. H.323 doesn't mean you get good videoconferencing over lousy circuits. But it does mean that you should get some videoconferencing. H.323 is comprised of the following standards:

H.225 Middleware which specifies a message set for call signaling registration and admissions, supporting call negotiations — i.e. synchronization.

• G.711: Pulse Code Modulation (PCM) (64 Kbps) encoder/decoder specification for voice.

• G.722: 7 kHz audio-coding.

• G.723.1: Speech encoding/decoding with a low bit rate, high output quality. This is the default encoder required for H.323 compliance.

• G.728: Encoding/decoding of speech at 16 kbps using low-delay code excited linear predictive methods.

• G.729: Encoding/decoding of speech at 8 kbps using conjugate-structure, algebraic-code excited linear predictive methods.

The H.323 standard has the endorsement of several key client vendors such as Netscape, for use within their Cool Talk application; Microsoft, for use in NetMeeting, now part of

Internet Explorer; and Intel, for their Internet Phone product. With Netscape and Microsoft representing approximately 95% of the Internet browser market, and Intel and Microsoft dominating the current platforms, this collective support makes H.323 a defacto standard. See H.320, H.324 and V.80. In September, 1997, Microsoft issued White Paper on "IP Telephony with TAPI." In that paper, Microsoft said:

H.323 is a comprehensive International Telecommunications Union (ITU) standard for multimedia communications (voice, video, and data) over connectionless networks that do not provide a guaranteed quality of service, such as IP-based networks and the Internet. It provides for call control, multimedia management, and bandwidth management for point-to-point and multipoint conferences. H.323 mandates support for standard audio and video codecs and supports data sharing via the T.120 standard. Furthermore, the H.323 standard is network, platform and application independent, allowing any H.323 compliant terminal to interoperate with any other. H.323 allows multimedia streaming over current packet-switched networks. To counter the effects of LAN latency, H.323 uses as a transport the Real-time Transport Protocol (RTP), an IETF standard designed to handle the requirements of streaming real-time audio and video over the Internet.

The H.323 standard specifies three command and control protocols: + H.245 for call control
• Q.931 for call signaling
• The RAS (Registration, Admissions, and Status) signaling function

The H.245 control channel is responsible for control messages governing operation of the H.323 terminal, including capability exchanges, commands, and indications. Q.931 is used to set up a connection between two terminals, while RAS governs registration, admission, and bandwidth functions between endpoints and gatekeepers (RAS is not used if a gatekeeper is not present). See below for more information on gatekeepers.

H.323 defines four major components for an H.323-based communications system: + Terminals
• Gateways
• Gatekeepers
• Multipoint Control Units (MCUs)

Terminals are the client endpoints on the network. All terminals must support voice communications; video and data support is optional. A Gateway is an optional element in an H.323 conference. Gateways bridge H.323 conferences to other networks, communications protocols, and multimedia formats. Gateways are not required if connections to other networks or non-H.323 compliant terminals are not needed. Gatekeepers perform two important functions which help maintain the robustness of the network - address translation and bandwidth management. Gatekeepers map LAN aliases to IP addresses and provide address lookups when needed. Gatekeepers also exercise call control functions to limit the number of H.323 connections, and the total bandwidth used by these connections, in an H.323 "zone." A Gatekeeper is not required in an H.323 system-however, if a Gatekeeper is present, terminals must make use of its services.

H.323 Components: Multipoint Control Units (MCU) support conferences between three or more endpoints. An MCU consists of a required Multipoint Controller (MC) and zero or more Multipoint Processors (MPs). The MC performs H.245 negotiations between all terminals to determine common audio and video processing capabilities, while the Multipoint Processor (MP) routes audio, video, and data streams between

terminal endpoints. Any H.323 client is guaranteed to support the following standards: H.261 and G.711. H.261 is an ITU-standard video codec designed to transmit compressed video at a rate of 64 Kbps and at a resolution of 176x44 pixels (QCIF). G.711 is an ITU-standard audio codec designed to transmit A-law and E-law PCM audio at bit rates of 48, 56, and 64 Kbps. Optionally, an H.323 client may support additional codecs: H.263 and G.723. H.263 is an ITU-standard video codec based on and compatible with H.261. It offers improved compression over H.261 and transmits video at a resolution of 176 x 44 pixels (QCIF). G.723 is an ITU-standard audio codec designed to operate at very low bit rates. The TAPI 3.0 H.323 Telephony Service Provider The H.323 Telephony Service Provider (along with its associated Media Stream Provider) allows TAPI-enabled applications to engage in multimedia sessions with any H.323-compliant terminal on the local area network. Specifically, the H.323 Telephony Service Provider (TSP) implements the H.323 signaling stack. The TSP accepts a number of different address formats, including name, machine name, and e-mail address. The H.323 MSP is responsible for constructing the DirectShow filter graph for an H.323 connection (including the RTP, RTP payload handler, codec, sink, and renderer filters).

H.323 telephony is complicated by the reality that a user's network address (in this case, a user's IP address) is highly volatile and cannot be counted on to remain unchanged between H.323 sessions. The TAPI H.323 TSP uses the services of the Windows NT Active Directory to perform user-to-IP address resolution. Specifically, user-to-IP mapping information is stored and continually refreshed using the Internet Locator Service (ILS) Dynamic Directory, a real-time server component of the Active Directory. See TAPI 3.0.

The official ITU definition of H.323 is as follows: H.323 describes terminals, equipment and services for multimedia communication over Local Area Networks (LAN) which do not provide a guaranteed quality of service. H.323 terminals and equipment may carry real-time voice, data and video, or any combination, including videotelephony.The LAN over which H.323 terminals communicate, may be a single segment or ring, or it may be multiple segments with complex topologies. It should be noted that operation of H.323 terminals over the multiple LAN segments (including the Internet) may result in poor performance. The possible means by which quality of service might be assured on such types of LANs/internetworks is beyond the scope of this Recommendation.H.323 terminals may be integrated into personal computers or implemented in stand-alone devices such as videotelephones. Support for voice is mandatory, while data and video are optional, but if supported, the ability to use a specified common mode of operation is required, so that all terminals supporting that media type can interwork. This Recommendation allows more than one channel of each type to be in use. Other Recommendations in the H.323-Series include H.225.0 packet and synchronization, H.245 control, H.261 and H.263 video codecs, G.711, G.722, G.728, G.729, and G.723 audio codecs, and the T.120-Series of multimedia communications protocols.This Recommendation makes use of the logical channel signalling procedures of Recommendation H.245, in which the content of each logical channel is described when the channel is opened. Procedures are provided for expression of receiver and transmitter capabilities, so transmissions are limited to what receivers can decode, and so that receivers may request a particular desired mode from transmitters. Since the procedures of Recommendation H.245 are also used by Recommendation H.310 for ATM networks, Recommendation H.324 for GSTN, and V.70, interworking with these systems should not require H.242 to H.245 translation as would be the case for H.320 systems.H.323 terminals may be used in multipoint configurations, and may interwork with H.310 terminals on B-ISDN, H.320 terminals on N-ISDN, H.321 terminals on B-ISDN, H.322 terminals on Guaranteed Quality of Service LANs, H.324 terminals on GSTN and wireless networks, and V.70 terminals on GSTN.

H.324 Standard for analog POTS telephone line based videoconferencing via modems. H.324 is an interoperability standard, meaning that if a vendor's videoconferencing product conforms to H.324 it should communicate with all the other vendors who say their products conform to H.324. H.324 contains several standards for videoconferencing. They are H.263 for real time video compression/decompression, G.723 for real time audio compression/decompression, H245/H.223 control protocol and multiplexing and V.80 application interface for modems. See H.320 and V.80.

H.gcp A proposed new ITU-T standard being added to the H.323 family of ITU-T recommendations, which have been widely adopted by industry as the main standards for multimedia communications over the Internet. H.gcp will permit control of gateway devices that pass voice, video, facsimile and data traffic between conventional telephony networks, i.e. the Public Switched Telephone Network and packet based data networks such as the Internet. Connections through such gateway devices allow callers from a normal telephone to make long distance voice calls over the Internet. According to the ITU-T, the H.323 family of standards already provides an extensive framework for the provision of new services. The new recommendation (i.e. H.gcp) will permit low-cost Internet gateway devices for the first time to be interfaced in a standard way with the signaling systems found in conventional telephony networks. • H.245: Adds the ability to open and close logical channels on the network, i.e. transmission control.

H.R.nnnn A proposed law introduced into the House of Representatives by a Congressman. Typically, four digits follow the H.R., signifying the proposed bill's number. The reason for including this definition in this dictionary is that every few months since divestiture some Congressman has attempted to introduce a bill into the House of Representatives changing the Communications Act of 1934. Such a bill is generally supported by a bevy of Bell telephone companies trying to use the proposed bill to remove those restrictions placed on them by Divestiture — manufacturing, creating information content and getting into long distance.

H0 Channel An H zero channel is a 384 kbps channel that consists of six contiguous DS0s (64 kbps) of a T-1 line.

H10 Channel The H ten channel is the North American 1,472 Kbps channel from a T-1 or primary rate carrier. It is equivalent to twenty-three (23) 64 kbps channels.

H11 Channel The North American primary rate used as a single 1,536 Kbps channel. This channel uses 24 contiguous DS0s (DS zeros) or the entire T-1 line except for the 8 Kbps framing pattern.

H12 The European primary rate used as a single 1,920 kbps channel (30 64 kbps channels) or the entire E-1 line except for the 64 kbps framing and maintenance channel.

HA Horn Alert. A cellular car phone feature that automatically blows the car's horn if a call is coming in.

Hack The output of a hacker. Usually good programs, but sometimes just something clever of no discernible use. Just a "good hack", or something done for the "hack value."

Hacker A person who "hacks" away at a computer until his program works. The term has been wrongly used by the press to mean people who break into computer systems. The word hacker has gone through many meanings. In the late 1950s MIT students who loved to tinker with the university's gigantic early computers started calling themselves "hackers." At one stage being a hacker was a badge of honor conferred on an elite programmer or computer hardware designer. But in 1983 the movie "War Games" presented another view of the hacker mentality — someone who tries to break into computer systems for fun and sport. Today the term tends to have positive meanings, while the word "cracker" is reserved for individuals who willfully break into computer systems seeking to wreak damage. See CRACKER, Hacker Ethic and Hacker Tourism.

Hacker Ethic A set of moral principles common to the first generation hacker community. According to hacker ethic, all technical information should, in principle, be freely available to all. However, destroying, altering, or moving data in a way that could cause injury or expense to others is always unethical.

Hacker Tourism From Wired's Jargon Watch column. Travel to exotic locations in search of sights and sensations that only a technogeek could love. The term was coined by Neal Stephenson in his colossal article for Wired on FLAG, a fiber-optic cable now being built from England to Japan.

Hairpinning Hairpinning is a term for information/data going into a central office based switch and turning around and going back out to another device before leaving the central office site. The term is fairly descriptive, if you visualize a hairpin and its u-shaped bend.

Here's an answer from Lucent to the question, "What is hair pinning, and why would I want to use it?" Answer: Hairpinning is bringing traffic in on a tributary and instead of putting it on the high speed OC-N line you direct it out another low speed tributary port. You might want to do this if you had interfaces to two IXCs on different nodes. If one of your IXCs went down you could hairpin the other to pick the traffic, assuming the spare capacity existed on the tributary. Hairpin cross-connections allow local drop of signals, ring extensions supported by a ring host node, and allow passing traffic between two ring interfaces on a single host node. In this case, no high speed channel is involved and the cross-connections are entirely within the interfaces.

HAL 1. The computer from the movie 2001: A Space Odyssey. HAL is an acronym for "Heuristically programmed ALgorithmic computer." The one-letter-shift transposition to "IBM" (The I became H. The B became A. The M became L) was noted shortly after the film's release and was widely accepted as a subtle Kubrick/Clarke joke directed at the computer giant. Clarke himself has pointed out that when he named HAL, he didn't catch the one-letter-shift bit, and if he had, he would have changed the name. IBM had been a huge supporter of the film project, and he wouldn't have dreamed of poking fun at them, however subtly. (Incidentally, according to Clarke, the one intentional joke in the film was the scene where a knuckle-chewing Heywood Floyd read the instructions for the Zero-Gravity Toilet.)

Half-Bridge Apple Computer term for a device linking LANS over a low-speed link such as a telephone line or X.25 link. It is termed a half-bridge as one is required at each end of the link.

Half-duplex A circuit designed for data transmission in both directions, but not at the same time. Telex is an example of a half duplex system, as is speaking on with most speakerphones. (The best speakerphones are full duplex. They're rare and expensive.)

Half-duplex M-ES A cellular radio term. A Mobile End System (M-ES) that can either transmit or receive, but cannot do both simultaneously, for example, an M-ES that has a single transceiver (radio).

Half-Life In science, the time it takes for half the radioactivity of a substance to disappear. Among techies, it is a gauge of an individual's usefulness. "He may have a short half-life here."

Half-Repeater A device which extends the distance a LAN can cover by joining two lengths of cable over another communication medium.

Half-tapping The action of making an analog trunk appear in two places for simultaneous service. Half-tapping refers to the duplication of service on the customer side of the demarcation point and back-tapping is the description used by the telephone company when the duplicate service originates from their side of the demarcation point. Half-tapping is useful when new telephone equipment is being installed in the same location as the current equipment because the new system can be tested while the old system is still in use.

Half-Tone Any photomechanical printing surface or the impression therefrom in which detail and tone values are represented by a series of evenly spaced dots in varying size and shape, varying in direct proportion to the intensity of tones they represent.

Half-Wave Antenna An antenna which is half as long as the wave being received.

Halo Effect, The Websters defines a halo as a conventional, geometric shape, usually in the form of a disk, circle, ring or rayed structure representing a radiant light around or above the head of a divine or sacred personage. A company or person acquiring "the Halo Effect" suggests that the company is doing that most outside observers really view as "right." In short, the company or person is on a major roll.

HAM HAM or HAM-radio. Home AMateur radio. A person who operates a HAM radio. www.ham.org and www.fcc.gov

HAM Operator A person who operates a HAM radio. See HAM.

Hamming Code An error correcting code named after R. W. Hamming of Bell Labs. The code has four information bits and three check bits per character.

HAN Home Area Network. A residential network for data communication and control based on the same concepts as a LAN, but using standard electrical wiring. For more detail, see CEBus.

Han Characters Han characters are Chinese language symbols which are used to represent whole words or concepts in Chinese, Japanese and Korean. See UNICODE.

Hand Off 1. To connect a phone call or service from one telephone company to another. These usually occur in a place called co-location.

2. The process of transferring cellular-based calls from one cell site to another as the mobile or portable moves through the service area.

Handhole A buried box whose lid is even with the surface of the ground. It provides a space for splicing and terminating cables.

Handle 1. In the Windows 95 user interface, an interface added to the object that facilitates moving, sizing, reshaping, or other functions pertaining to an object.

2. In programming, a pointer to a pointer- that is, a token that lets a program access a resource identified in another variable. See also object handle.

3. The name you use in an online computer service. It's typically not your own name. You adopt this name to give yourself anonymity for whatever reasons you find convenient. A handle is called a "Nom de ligne" in some circles.

Handled Call A call center term. A call that is answered by an employee, as opposed to being blocked or abandoned.

Handoff 1. The process by which the Mobile Telephone Switching Office (MTSO) passes a cellular phone conversation from cell to another. There are two forms of handoff: hard and soft. A hard handoff is performed on a "break and make" basis, requiring the connection to be broken in the original cell before it is made in the successor cell. Hard handoffs are required in cellular systems using FDMA (Frequency Division Multiple Access), such as the analog AMPS (Advanced Mobile Phone System), and those using TDMA (Time Division Multiple Access), such as GSM (Global System for Mobile Communications). As AMPS and GSM employ different frequencies in adjacent cells, hard handoffs are required — you don't notice the difference in a voice conversation, as the process takes only 250 milliseconds or so, but data communications are affected very adversely. A soft handoff, on the other hand, employs a "make and break" handoff algorithm. Some emerging PCS (Personal Communications Services) systems employ CDMA (Code Division Multiple Access), which does not require the use of different frequencies in adjacent cells. Those systems, therefore, take advantage of soft handoffs. See also AMPS, CDMA, FDMA, GSM, MTSO, PCS, and TDMA.

2. An SCSA definition. The change of ownership of a Group (and therefore, typically, a call) from one session to another. For example, if a call center application discovers that a caller wishes to access a technical support audiotex database, it hands off the call to an application servicing that database.

Handover Word The word in the GPS (Global Positioning System) message that contains synchronization information for the transfer of tracking from C/A to P code.

Handset The part of a phone held in the hand to speak and listen, it contains a transmitter and receiver. In the old days, the transmitter was a carbon mike. Now it's mostly electronic. Some electronic mikes are awful. Some phone makers are going back to carbon mikes. There are two basic types of telephone handsets in North America: the G-style handset, which has round, screw-in ear and mouthpiece, and the new K-style handset, which has square ear and mouthpieces and has the two screws in the middle. I prefer the older G-style one. I think it's sturdier. It also has the advantage that you can unscrew it and quickly remove the transmitter — very useful if you want someone to listen in on your conversation, but you don't want the other party to hear his breathing and coughing.

Handset Management Imagine you have a phone attached to your computer through a telephony board inside your computer. Now imagine that you pick up the phone and dial a number. If the company knows you have dialed a number and knows which number you have dialed, that feature is called handset management. It is the ability of the computer to be aware of every button pushed on the phone. The advantage of this is obvious: You really want the PC to collect those digits, so it can, for example, add a price to each call and use them for monthly billing (lawyer, accountant, etc.). You also want to be able re-dial those numbers by simply clicking on the number one you want, hitting Enter and bingo, you're redialing that number, without having to key it in again. This term, handset management, has now been replaced by a more meaningful term which we're now calling "Telephone Set Management."

Handsfree This a term with different meanings in the telephone business. It can mean that you have a telephone with a speakerphone and thus you are able to talk on it "handsfree," i.e. without your hands touching the handset, but you still must dial manually. In the car cell phone business, it means the same thing, i.e. the ability to use your phone without lifting or holding the handset to your ear, plus it may mean that your car phone has voice recognition and you can also dial handsfree by talking to your care phone, e.g. "Call mother." See also HANDSFREE DIALING.

Handsfree Answerback This feature, when activated, automatically turns ON the microphone at a telephone receiving a Call so that the person receiving the call can respond without lifting the receiver. Handsfree answerback is typically used on intercom calls.

Handsfree Dialing A telephone feature which allows the user to place outside calls and listen to the progress of those calls without lifting the handset of his telephone. This feature is unbelievably useful when calling airlines which inevitably put you on "eternity hold." (We made that term up.)

Handsfree Monitoring You can dial an outside call and hear the call's progress without having to lift your handset. Similar to hands-free dialing. With hands-free monitoring, you can only listen. To speak, you must pick up the handset. To be able to speak, you need a full speakerphone. Be careful of the distinction. Many people have been caught.

Handsfree Telephone Could be another word for a speakerphone or for a phone that does hands-free dialing.

Handshake Two modems trying to connect. Two modems trying to agree on how to transfer data. The series of signals between a computer and another peripheral device (for example, a modem) that establishes the parameters required for passing data.

Handshake In HIN. A general purpose control signal sent from the DTE to DCE in a Newbridge Networks RS-232-C connection. HIN can be used in place of Request to Send (RTS), Carrier Detect (CD) or Ring Indicator (RI).

Handshake Out HOUT. A general purpose control signal sent from the DCE to the DTE. For example, in the case of a Newbridge Networks Mainstreet Data Controller with ports configured as DCE, HOUT is sent from the Data Controller to an attached device. HOUT can be used in place of Clear to Send (CTS), Carrier Detect (CD) or Ring Indicator (RI).

Handshaking The initial exchange between two data communications systems prior to and during data transmission to ensure proper data transmission. A handshake method is part of the complete transmission protocol. A serial (asynchronous) transmission protocol might include the handshake method (XON/XOFF), baud rate, parity setting, number of data bits and number of stop bits. Just as people shake hands, and go through a perfunctory "Hi, how are ya?", computers must go through a procedure of greeting the opposite party, verifying the identity of the other party, and other functions that can be described by this "humanizing term." As with human contacts, once the Handshaking is complete, the business of communications begins.

Handwriting Recognition A system for taking handwritten generated with a stylus on a computer pad or directly onto the computer screen, and converting then into machine-readable text.

Hang Up Hang up lets you disconnect from an ISDN call. To hang up from the phone set you must depress and hold the receiver button for a specified amount of time. By default, the time is set for 0.8 of a second.

Hard Cable Coaxial cable commonly used in the cable television industry for trunk and feeder. At a minimum, hard cable consists of a copper (or copper-clad aluminum) center conductor, a plastic foam dielectric and a solid-aluminum sheath. The solid aluminum sheath is quite stiff; hence the name. Hard cable is available in several configurations; the most common are:

• Bare: There is no protective cover over the aluminum sheath. Bare cable is used in aerial installations where it's lashed to a steel supporting strand; the bare aluminum sheath

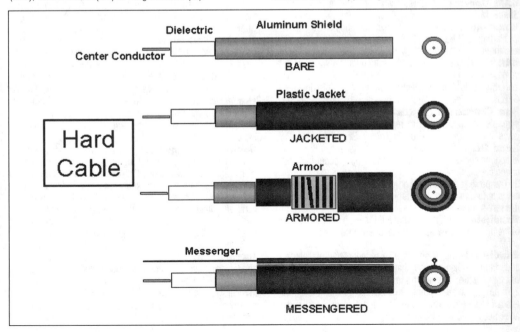

is readily visible from the ground because of its dull silver color. See Strand.

• Jacketed: The sheath is covered with a vinyl jacket, typically black. Jacketed cable is often used in aerial installations located in areas subject to corrosion from industrial pollution or salt; however, many cable companies use jacketed cable in all aerial locations.

• Armored: The sheath is covered with three protective layers: a vinyl jacket, a steel wrap, and another vinyl jacket. Armored cable is intended for use in direct-burial applications; the steel armor protects the sheath from damage during installation.

• Messengered: Similar to jacketed cable, but includes an integral steel "messenger wire" to provide mechanical support. Messengered cable is intended for aerial installation without strand.

Jacketed and armored cables are also available in "flooded" configurations. Flooded cable contains a sticky, viscous substance called "flooding compound" between the sheath and the jacket. Like a self-sealing tire, flooding compound seals microscopic holes in the jacket to prevent water intrusion. Flooded cables are intended for underground use, but they are not recommended for aerial use (flooding compound is a sticky mess — literally and politically — if it drips onto a parked car). Hard cables are identified by "trade size"; i.e., the outside diameter of the sheath. The standard trade sizes are 0.412", 0.500" ("half-inch cable"), 0.625", 0.750", 0.875", and 1.000" ("one inch cable").

Hard Copy Anything on paper. It is all well and good to have information flash by on your CRT or video display terminal, but there are times when you want to take a Hard Copy with you. This dictionary was written on a computer screen. Now you have a hard copy in your hands. In this case, that's a lot more useful than having a disk.

Hard Disk Assembly A sealed mass storage unit used for storing large amounts of data. Now available on personal computers.

Hard Drawn Copper Wire Copper wire that has been drawn to size and not annealed.

Hard Drive 1. A sealed hard disk. Originally the hard drive was called the Winchester magnetic storage device. It was pioneered by IBM for use in its 3030 disk system. It was called Winchester because "Winchester" was IBM's code name for the secret research project that led to its invention. A Winchester hard disk drive consists of several "platters" of metal stacked on top of each other. Each of the platter's surfaces is coated with magnetic material and is "read" to and "written" from by "heads" which float across (but don't touch) the surface. The whole system works roughly like the old-style Wurlitzer jukebox. See Winchester.
2. The Vermonter's Guide to Computer Lingo defines hard drive as getting home during mud season, i.e. it was a hard drive.

Hard Ferrite Ferrite that remains permanently magnetized. Used to make magnets.

Hard Ground When the

Hard Handoff A hand off that occurs when the subscriber is already using the TPC (three-party-conference) card, and the TPC card cannot be used to aid in a smooth handoff.

Hard Problem A type of calculation which is easy to perform in one direction, but difficult and even impractical to perform in the other direction. In the context of cryptography, hard problems provide extreme levels of security for encrypted data. See also DES and AES.

Hard RAM Carve some memory out of a computer's RAM;

power it continuously and bingo you have Hard RAM, also called a Virtual disk. Setting up a RAM disk lets you use your computer's conventional, extended or expanded memory to simulate a disk drive (or drives). The primary advantages of a RAM disk are its very fast access speed and its battery power-saving properties. It has no mechanical element to slow it down or to use additional power.

Hard Rubber A hard insulating material made of rubber, and having a dielectric constant of from two to four.

Hard Sectoring Physically marking the sector boundaries of a magnetic disk by punching holes in the disk where there's space available to store data. Hard sectored disks are not very common these days. Most disks — like those used on the IBM PC — are soft-sectored.

Hard Tubes Vacuum tubes having a high vacuum.

Hard Wired 1. Describes a circuit designed to do one task (e.g. a leased line).
2. A person with a very narrow and rigid view of his or her job. "That security guard is really hard-wired."

Hardened Resistant to disaster. Facilities with protective features that have been designed to withstand an explosion, a natural disaster, or ionizing radiation.

Hardware The actual physical computing machinery, as opposed to Software which is the list of instructions to operate the hardware, or the Firmware which is combination hardware/software that is "burned into" a Programmable Read Only Memory chip or chips. See FIRMWARE and SOFTWARE.

Hardware Address Also called physical address or MAC-layer address, a data-link layer address associated with a particular network device. Contrasts with network or protocol address, which is a network layer address.

Hardware Flow Control Hardware flow control is the method used by the UART chip (that chip controls the serial port) to modulate the flow of data. It does this by controlling the Clear to Send/Ready to Send (CTS/RTS) lines of the serial port's interface. For example, it can turn off or re-enable the flow of data. Most high-speed sessions require hardware flow control due to their need for precise, instantaneous control over the flow of incoming and outgoing data.

Hardware Interrupt See INTERRUPT.

Hardware Tree A Windows 95 term. A record in RAM of the current system configuration information for all devices in the hardware branch of the Registry. The hardware tree is created each time the computer is started or whenever a dynamic change occurs to the system configuration.

Hardwire To permanently connect by wire two or more devices rather than to connect them temporarily through connectors or switches. Hardwire is a term also used to represent a leased line.

Hardwire Services An MCI definition. Services providing intercity communications facilities dedicated to the use of a specific customer, and provided through a dedicated access line from the customer to the MCI switch.

Hardwire Terminating City City of circuit termination for hardwire services.

Harm See NETWORK HARM.

Harmonic A frequency which is an exact multiple of a fundamental frequency.

Harmonic Distortion A problem caused when the non-linearities in communication channels cause the harmonics of the input frequencies to appear in the output channel.

Harmonic Ringing A way of stopping users on a party line from hearing other than their own ring. We do this by tuning the ringer in their phone to a given ringing frequency, so it only rings when their frequency comes down the line.

Harmonic Signals Signals which are coherently related to the output frequency. In general, these signals are integer multiples of the output frequency.

Harmonica A device attached to the end of a connectorized feeder cable that converts the 25 pair into individual 4, 6 or 8 wire modular channels.

Harmonica Adapter An adapter that connects a 25-pair cable plug into 12 four-conductor RJ11 plugs or two 24 connector RJ11 plugs. These are often used as an alternative connectivity to hardwired blocks or temporary installations of key of PBX phone systems.

Harmonica Bug See Infinity Transmitter.

Harmonics This definition courtesy American Power Conversion Corp. In an AC power system, distortion of voltage or current waveforms may be expressed as a series of harmonics. Harmonics are voltage or current signals that are not at the desired 50 or 60 Hz fundamental frequency, but rather at some multiple frequency. For example, the fifth harmonic of 60 Hz is 300Hz. It is a characteristic of AC signals that any distortion will have components only at integer multiples of the fundamental frequency. In AC power distribution, these distortion components only occur at odd multiples of the fundamental frequency. The third harmonic voltage distortion at a typical wall outlet in the U.S. is about 3%. Harmonic voltages have virtually no effect on modern computers, but can cause overheating in some equipment.

Hash Total Adding up one or more information fields in order to provide a check number for error control. The addition is not intended to have any meaning other than for checking.

Hashing A cryptographic term for a small mathmetical summary or digest of an original clear-text data file or message. A hash algorithm ensures data integrity through the detection of changes to the data caused either by communications errors occuring in transit, or by tampering. In combination, hashing and the use of a digital signature (digital certificate) prevent the forging of an altered message. See also Certificate and Encryption.

HASP Houston Automatic Spooling Program. A control protocol adopted by IBM for transmitting data processing files and jobs to IBM 360 and 370 computers. An early job control language still in limited use.

Hayes At Command Set Before 1981, the modem was a dumb device. It had no memory or ability to recognize commands. It simply modulated and demodulated signals between the telephone line and the computer or terminal. In 1981, Hayes Microcomputer Products, Inc. in Norcross, GA produced the first "smart" modem, appropriately named the Smartmodem 300. It was "smart" because it understood commands, such as "ATD" which means "ATtention, Dial the phone." The Hayes Standard AT Command Set (its full name) — a language for modems — has been accepted as a standard by the modem industry. And now many modems claim to be 100% Hayes compatible, which may mean they are and may mean they aren't. As in all cases of claimed compatibility, one should check. You'll find the complete Hayes AT Command Set spelled out in virtually every manual of every modem which purports to be "100% Hayes Compatible." See also CLASS 1.

HBA Host Bus Adapter. A printed circuit board that acts as an interface between the host microprocessor and the disk controller. The HBA relieves the host microprocessor of data storage and retrieval tasks, usually increasing the computer's performance time. A host bus adapter (or host adapter) and its disk subsystems make up a disk channel.

HBFG Host Behavior Functional Group: The group of functions performed by an ATM-attached host that is participating in the MPOA service.

HBS Home Base Station. A wireless PCS term. Supports the PCS 1900 air interface in combination with the PCS 1900 handset.

HCI Host Command Interface. Mitel SX-2000 PBX to computer link. HCI is designed to work with Digital Equipment Corporation computers. See OPEN APPLICATION INTERFACE.

HCL Hardware Compatibility List. See HCL.

HCO Hearing Carry Over. A reduced form of Telecommunication Relay Service (TRS) where the person with the speech disability is able to listen to the other end user and, in reply, the Communications Assistant speaks the text as typed by the person with the speech disability. The Communications Assistant does not type any conversation.

HCS 1. Hundred Call Seconds. One hundred seconds of telephone conversation. See CCS.

2. Hard Clad Silica.

HCT Hardware Compatibility Test. Microsoft came up with this definition and concept when it found several computers didn't run its software as well as they should. Basically the Microsoft Hardware Compatibility Test (HCT) is a series of tests for verifying the compatibility of hardware systems with Windows NT. The TCT Test Manager is an application that provides a way to launch the tests, keep track of test results and return the results to Microsoft. Microsoft maintains a Windows NT Hardware Compatibility List (HCL). If you send back the test results and your stuff passes, your hardware will be included on Microsoft's list of hardware that works with NT.

HD Half Duplex circuit.

HDB3 High Density Bipolar 3. A bipolar coding method that does not allow more than 3 consecutive zeros.

HDD Hard Disk Drive.

HDLC High level Data Link Control. An ITU-TSS link layer protocol standard for point-to-point and multi-point communications. In HDLC, control information is always placed in the same position. And specific bit patterns used for control differ dramatically from those used in representing data, so that errors are less likely to occur. SDLC and ADCCP are similar protocols. See also HIGH LEVEL DATA LINK CONTROL.

HDMAC Another potential high definition TV standard. HDMAC was spawned by Britain's Independent Broadcasting Authority. Unlike Japan's Hi-Vision, HDMAC has the attraction of being compatible with existing TV sets, i.e. those in Europe.

HDML Handheld Device Markup Language, which is Unwired Planet of Redwood Shores, CA's modification of standard HTML for use on mobile phones. HDML is a text-based markup language which uses HyperText Transfer Protocol (HTTP) and is compatible with all Web servers. HDML is designed to display on a smaller screen such as one might find on a cellular phone, PDA, pager, or PCS device. The basic structural unit for HDML is a "card," while that of HTML is a "page." HDML allows the mobile user to access the Internet, and send, receive and redirect e-mail. As PCS devices are graphics-challenged, a Web site must be HDML-enabled in order to allow access by such devices. www.wapforum.org

HDSL High-bit-rate Digital Subscriber Line. The most mature of the xDSL technologies, HDSL allows the provisioning of T-1/E-1 local loop circuits much more quickly and at much lower cost than through conventional means. In the U.S., HDSL delivers T-1 (1.536 Mbps usable bandwidth) over a four-wire loop of two pairs. E-1 capacity of 2.048 Mbps requires three pairs. Unlike ADSL, HDSL bandwidth is symmetric, as equal bandwidth is provided in each direction.

The traditional approach of provisioning T-1/E-1 access loops on copper wires requires specially-conditioned UTP (Unshielded Twisted Pair), with repeaters spaced every 6,000 feet in order to compensate for signal attenuation at the high carrier frequencies required. Each pair supports simplex (one-way) transmission at 1.544 Mbps, of which 1.536 Mbps is usable for data transmission; in combination, the two simplex circuits yield a full-duplex circuit.

HDSL, which involves special electronics at both the CO and the customer premise, delivers the same transmission capacity over standard UTP at distances up to 12,000 feet without the requirement for repeaters.

The UTP loop may be bridged, although loading coils are not tolerated. This is accomplished through full-duplex transmission at 784 Kbps over each pair of the four-wire circuit; 768 Kbps is usable for data transmission, with the remaining 16 Kbps being required for signaling and control. In the aggregate, the yield is 1.536 Mbps (T1). The lower transmission rate on each pair implies a much lower carrier frequency. As lower frequency signals can travel much longer distances without experiencing unacceptable levels of attenuation (loss of signal strength), the requirement for repeaters is obviated for distances up to 12,000 feet. Note: T-1 requires 1.5 MHz over each pair, while HDSL operates at frequencies ranging from 80 KHz to 240 KHz, depending on the specific techniques employed.

HDSL has been deployed aggressively by LECs for some years. Well over 300,000 systems reportedly are in service (as of summer, 1997). Although both the COT (Central Office Termination) and the RT (Remote Termination) require the placement of HDSL electronics, the overall carrier costs of provisioning are much reduced. No special circuit engineering, no physical inspection of cable plant, and no repeater acquisition and placement is required. Additionally, the circuit can be provisioned much more quickly, which fact results in much happier customers and much faster revenue generation. In fact, several LECs have lowered their T-1 rates in consideration of the lower costs.

At the time of this writing, a proposal for a new variation on the HDSL theme recently was proposed as a standard. HDSL2, based on technology from Adtran Inc., provides the same capability over a single pair, although the local loop length is limited to about 10,000 feet. This technology also is known as SDSL (Single line DSL). S-HDSL (Single-line HDSL) is a variation on this non-standard variation (It gets confusing, doesn't it? Remember that this is an emerging technology.) run at speeds of 768 and 384 Kbps for loop lengths of 12,000 feet and 18,000 feet, respectively. See also DSL, ADSL, HDSL2, IDSL, RADSL, SDSL, T-1 and VDSL. www.adtran.com and www.adsl.com

HDSL2 January 6, 1998, Level One Communications, ADC Telecommunications, ADTRAn, PairGain Technologies and the Siemens Semiconductor Group today announced agreement within the American National Standards Institute (ANSI) TIE1.4 committee on the basis of an HDSL2 standard. A provisional agreement, T1/E1 contribution number 41/97-471, has been approved marking a milestone within the ANSI HDSL2 standards effort. The elements agreed upon were line code, spectral shaping, system performance and forward error correction. These elements make up the core of the HDSL2 standard. The agreement reached is expected to accelerate the development HDSL2 technology and promote industry interoperability. The HDSL2 standard proposal will enable service providers to deliver full T-1, and potentially E-1, performance over a single twisted pair cable, with the same reach, robustness and spectral compatibility of today's two pair HDSL. This will permit local exchange carriers and telecom service providers to meet rapidly increasing demands for business and Internet access services, according to the press release I received on January 6.

HDT Host Digital Terminal.

HDTP 1. Handheld Device Transport Protocol - a wireless-optimized protocol sitting between the UP.Browser's HDML interpreter and a datagram transport (typically UDP). HDTP provides security and reliability for the transport in a way that is significantly more efficient (optimized for wireless and communication with a minimal number of IP addresses) than is TCP. HDTP is the protocol in use on UP's UP.Link Platform V2.x. Version 3 sees UP migrating towards WAP - the new Wireless Application Protocol, from the WAP Forum. www.wapforum.org.

2. Hoofddirectie Telecommunicatie en Post (Directorate for Telecommunications and Posts, The Netherlands).

HDTV High Definition TeleVision. Today's typical TV set in North America contains 336,000 pixels. HDTV will offer approximately twice the vertical and horizontal resolution of current NTSC analog television broadcasting, which is a picture quality approaching 35mm film. Further, it will support sound quality approaching that of a CD (Compact Disc). The ideal HDTV set would be flat screen, cheap, reliable and require very little electrical power. In December 1996, the FCC established standards for ATV (Advanced TV), the successor to HDTV, based on the recommendation of the ATSC (Advanced Television Systems Committee). See ATV, HIGH DEFINITION TELEVISION and NTSC.

HDX Half DupleX.

HE See HEAD END.

Head 1. A device that reads, writes, or erases data on a storage medium. The device which comes in contact with or comes very close to the magnetic storage device (disk, diskette, drum, tape) and reads and/or writes to the medium. In computer devices, it performs the same function as the head on a home cassette tape recorder.

2. A sub-component accessory which serves a translator type function for signal being input to some larger system or instrument. Many pieces of test equipment have various available plug-in "sampling heads" which allow the equipment to be used with a variety of different signal types. The most common type is probably optical-to-electrical converters which convert an optical signal for display on an oscilloscope.

Head End 1. The originating point of a signal in cable TV systems. At the head end, you'll often find large satellite receiving antennae. Now increasingly spelled headend.

2. A central control device required within some LAN/MAN systems to provide such centralized functions as remodulation, re-timing, message accountability, contention control, diagnostic control, and access.

Head Landing Zone In older hard drives, the head landing zone is an area of the hard disk set aside for take off and landing of the heads when the drive is turned on and off. In newer drives, the heads are retracted.

Head Slap Similar to head crash but occurs while the drive is turned off. It usually occurs during mishandling or shipping. Head slap can cause permanent damage to a hard disk drive.

Head Thrashing A term for rapid back and forth movements of the disk head of a hard drive.

Headend 1. The originating point of a signal in cable TV systems. At the head end, you'll often find large satellite receiving antennae.

Header 1. Protocol control information located at the beginning of a protocol data unit.

2. The portion of a message that contains information that will guide the message to the correct destination. This information contains such things as the sender's and receiver's addresses, precedence level, routing instructions, and synchronization pulses.

Header Area The area containing preliminary information for the entire document, such as the data, company name, address, purchase order, terms, etc. An EDI (Electronic Data Interchange) term.

Header Error Control HEC. An 8-bit CRC code contained within the header of an ATM data cell. The HEC is used for checking the integrity of the cell header at the various cell switches.

Headset A telephone transmitter and receiver assembly worn on the head. Headsets are becoming very light and very comfortable and are no longer worn only by switchboard attendants and airline clerks. Northern Telecom says the following are traditional applications for headset use: Receptionists. Console attendants. Telemarketers. Customer service reps. Order entry reps. Financial Service professionals. Stockbrokers. Sales reps and reservation agents.

Headset Jack A place on a phone or console into which you can plug a headset.

Hearing Aid Compatible A hearing aid compatible phone may be used with inductively coupled hearing aid devices. You can find hearing aid compatible coin phones by looking for the blue grommet between the handset and the cord.

Heartbeat Ethernet-defined SQE signal quality test function, defined in IEEE 802.3. Heartbeat is created by a circuit (normally part of the transceiver) that generates a collision signal at the end of a transmission. This signal is used by the controller interface for self-testing.

Heartbeat Support A function that generates a frame periodically, even if no data is sent, for network management purposes.

Heat Electromagnetic waves of a frequency between that of light waves and radio waves. A form of energy.

Heat Coil An electrical protection device used to prevent equipment from overheating as a result of foreign voltages that do not trigger voltage limiting devices. It typically consists of a coil of fine wire around a brass tube that encloses a pin soldered with a low-melting alloy. When abnormal currents occur, the coil heats the brass to soften the solder, allowing the spring-loaded pin to move against a ground plate directing currents to ground.

Heavy Iron Hardware, really BIG hardware. Contemporary hardware hardly qualifies. For example, the ENIAC (Electronic Numerical Integrator And Computer), built in 1946 at a cost of about $400,000, was the first large-scale electronic digital computer built. The ENIAC contained about 18,000 vacuum tubes, weighed 30 tons and occupied a footprint of 30x50 feet. In 1949, Popular Mechanics magazine forecast that "Computers in the future may...perhaps only weigh 1.5 tons." Technology marched on. Your laptop provides more horsepower than the ENIAC, has more memory, and more functionality.

HEC Header Error Control - a CRC code located in the last byte of an ATM (Asynchronous Transfer Mode) cell header used for checking integrity only. Using the fifth octet in the ATM cell header, ATM equipment may check for an error and

corrects the contents of the header. The check character is calculated using a CRC algorithm allowing a single bit error in the header to be corrected or multiple errors to be detected.

HEHO Head-End Hop Off. You have a private network. You overflow a long distance call to WATS or DDD at the originating end (the end the call is coming from). This HEHO (Head-End Hop Off) is done because it's usually cheaper than carrying the call part way through the network, then jumping off the network at that point (because the network is busy or it won't reach the end point). The opposite of HEHO is TEHO — Tail-End Hop Off. In TEHO, you carry the call as far as possible through the network, then pass it off to WATS or DDD as close to its destination as possible. The decision to go HEHO or TEHO has to do with economics, primarily which is cheaper.

Heisenberg's Realization The mere act of observing affects what is being observed.

Held Call A held call is a call to which you are connected but which is on hold.

Held Orders A telephone company term for requests for telephone lines which the phone company cannot fill. Thus it is "holding" the orders. The reasons for holding customer orders might range from lack of capacity at the serving central office to a lack of local cable plant.

Helical Antenna An antenna that has the form of a helix. When the helix circumference is much smaller than one wavelength, the antenna radiates at right angles to the axis of the helix. When the helix circumference is one wavelength, maximum radiation is along the helix axis.

Helical Scan Storage method that increases media capacity by laying data out in diagonal strips. Used in video tape recorders, etc.

Helical Strand A process of twisting conductors of a cable together in a helix, or spiral fashion, in order to improve the break strength of the conductors. See Helix, Stranded Copper and Stranded Fiber.

Helical Stripe A continuous, colored, spiral stripe applied over the outer perimeter of an insulated conductor for circuit identification purposes.

Helix A spiral. The shape of screw.

Hello When the phone was first invented, no one was sure how to begin the conversation. Thomas Edison saw the telephone as being used by businesses with permanently open lines. How would anyone know that the other party wanted to speak? A letter was found from Thomas Edison, dated August 15, 1877 to TBA, David, president of the Central District and Printing Telegraph Co. in Pittsburgh. "Friend David, I don't think we need a call bell as Hello! can be heard 10 to 20 feet away. What do you think? " Edison. At that time Alexander Graham Bell insisted on answering the telephone with "Ahoy." Hello! became the standard as the first telephone exchanges were set up across the country. Hello first appeared in the Oxford English Dictionary in 1883. In September of 1880, the first National Convention of Telephone Companies was held in Niagara Falls. "Hello" was used on everyone's name tag for the first time. Besides electricity, the phonograph and hundreds of other inventions, we can thank Edison for the "Hello" greeting. The above from New Pueblo Communications in Tucson, AZ.

Hello Packet A type of PNNI Routing packet that is exchanged between neighboring logical nodes.

Help Desk A centralized location where queries about product usage, installation, problems or services are answered. Sometimes help-desks are provided by the manufacturer of the product. Sometimes help desks are provided by outside companies — systems integrators and integrators, independent software developers and third party companies.

Helper Applications Programs that can be linked to various file types and commands. Helper apps will launch automatically when linked files are accessed through a browser.

Henry The inductance in a circuit in which the electromotive force induced is one volt when the inducing current varies at the rate of one ampere per second. It is 1,000,000.000 electromagnetic units, and is the unit of inductance.

HEPNET An Internet term. A non-USENET set of newsgroups devoted to discussing the topic of high-energy nuclear physics.

HER High Energy Radio Frequency Gun. Shoots a high-powered radio signal at an electronic target (such as a computer) and puts it out of commission.

Hercules Graphics Hercules graphics adheres to the Hercules standard of monochrome graphics on a monochrome PC monitor. That standard is 720 x 348 pixel resolution and 64K screen memory. This encoding was never adopted as a color standard and is now pretty well obsolete. See also MONITOR and FACSIMILE.

HERF gun A High Energy Radio Frequency gun capable of destroying magnetic data storage.

Hermaphroditic connector A loopback or self-shorting connector typically used with Type 1 (STP) Token Ring cable.

Hermaticity Test A fine and gross leak test of a hermetically sealed IC to see if there are any leaks in the seal. The gross leak test uses a fluorocarbon fluid, and the fine leak uses a light gas such as helium.

Hermetic Coating A coating applied over the cladding of a fiber that retards the permeation of moisture and hydrogen into the fiber.

Heroinware Doom played on corporate networks.

Hertz Abbreviated Hz. A measurement of frequency in cycles per second. A hertz is one cycle per second, and is the basic measurement for bandwidth in analog terms. "Hertz" is named after Heinrich Rudolf Hertz, the physicist who discovered the presence of electromagnetic radio waves.

Hertzian Wave A name sometimes given to electromagnetic waves.

Hetero The Greek prefix meaning different.

Heterodyne To generate new frequencies by mixing two or more signals in a nonlinear device such as a vacuum tube, transistor, or diode mixer. A superheterodyne receiver converts any selected incoming frequency by heterodyne action to a common intermediate frequency where amplification and selectivity (filtering) are provided. A frequency produced by mixing two or more signals in a nonlinear device. See HETERODYNING.

Heterodyne Repeater A repeater for a radio system in which the received signals are converted to an intermediate frequency, amplified, and reconverted to a new frequency band for transmission over the next repeater section.

Heterodyning Here is an explanation from James Harry Green's book, the Dow Jones Handbook of Telecommunications: Analog microwave repeaters use either of two techniques to amplify the received signal for retransmission: HETERODYNING or BASEBAND. In a baseband repeater, the signal is demodulated to the multiplex (or video) signal at every repeater point. In heterodyne repeaters the signal is demodulated to an intermediate frequency, typically 70 MHz, and modulated or heterodyned to the transmitter output frequency. Heterodyne radio is reduced to baseband only at main repeater stations where the baseband signal is required to drop off voice channels. The primary advantage of base-

band radio is that some carrier channel groups can be dropped off at repeater stations. Heterodyne radio has the advantage of avoiding the distortions caused by multiple modulation/demodulation and amplification of a baseband signal. Therefore, heterodyne radio is employed for transcontinental use with drop-off points only at major junctions.

Heterogeneous Networks Networks composed of hardware and software from multiple vendors usually implementing multiple protocols.

Heterojunction A junction between semiconductors that differ in their doping level conductivities, and also in their atomic or alloy compositions.

Heuristic Using much trial and error to arrive at a solution to a problem.

Hexadecimal A numbering system of 16 characters, ten digits and six letters. It is used to condense the long strings of zeros and ones in large binary numbers. This base-16 numeric notation system is frequently used to specify addresses in computer memory. It makes life simpler for programmers. In hexadecimal notation, the decimal number numbers 0 through 15 are represented by the decimal digits 0 through 9 and the alphabet "digits" A through F (A=decimal 10, B=decimal 11, and so forth).

HF Hands Free.

HFAI Hands Free Answer on Intercom. A desirable feature of several phone systems.

HFC Hybrid Fiber Coax. An outside plant distribution cabling concept employing both fiber optic and coaxial cable. Fiber is deployed as the backbone distribution medium, terminating in a remote unit where optoelectric conversion takes place. At that remote unit, the signal then is passed on to coax cables which carry the data the last leg to the individual business, residence, dormitory room, etc. HFC systems provide substantial bandwidth at lower cost than a system based exclusively on fiber. Given the embedded base of coaxial cable in college and university campuses, HFC is an effective means of delivering combined voice, data, video and CATV to dormitory rooms and classrooms. HFC also is planned for extensive use in CATV networks for the same reasons. US West, for instance, plans to deploy HFC as a means of upgrading the distribution plant of Continental Cablevision, which it acquired in 1996.

HFU Hands Free Unit

HGC Hercules Graphic Card; long the standard monochrome graphics adapter for PCs and compatibles. Maximum resolution is 720 x 348 pixels.

Hi-Lo Tariff A long distance private line tariff filed by AT&T whereby private lines between major cities were priced lower than private lines between smaller cities. In effect, those "larger" cities were those MCI operated in and those "smaller" cities were those MCI didn't operate it. Eventually the tariff was thrown out by the FCC and it figured in anti-trust suits by MCI and the Federal Government against AT&T.

HI8 Video The high-quality extension of the Video 8 (or 8mm) format, which features higher luminance resolution.

Hibernation You turn your computer off in the middle of a program. It grows to sleep. When you turn it on, it returns to exactly where you left it, without the need for rebooting. Compaq invented the term after copying the feature from Toshiba, who called it Resume. These days most laptops have some form of hibernation or resume.

Hidden Markov Method HMM. A common algorithm in voice recognition which uses probabilistic techniques for recognizing discrete and continuous speech.

Hierarchal File System A system of arranging files in directories and subdirectories to maintain hierarchical relationships (one file ranked above the other) between the files and make them easier to find and retrieve. See RELATIONSHIP DATABASE.

Hierarchical Network 1. A network that includes two or more different classes of switching systems in a defined homing arrangement, meaning to home in on the telephone you wish to be connected to. This is a fancy Bell System (Oops, I mean AT&T - Old habits die hard) term meaning that when direct circuits between two switches are busy or too far apart to be directly connected, the machinery will seek a higher level of switches to route the call through.

2. A LAN (Local Area Network) term describing a network employing OSI Layer 3 routers. Flat networks, on the other hand, make use of Layer 2 bridges, hubs and switches for LAN interconnection and segmentation. Routers make intelligent decisions about routing data packets, with such decisions taking into consideration the condition of the entire network; Layer 2 devices forward packets on a link basis. Hierarchical router networks are more complex and expensive, and are slower. However, they also can add value through protocol conversion and flow control. Routers, by the way, typically serve to interconnect bridges, hubs and switched, as well as to provide access to the WAN (Wide Area Network). See also FLAT NETWORK, BRIDGE, HUB, SWITCH and ROUTER.

Hierarchical Routing The process of establishing a network data path to a destination based on addresses with some kind of ranking criteria.

Hierarchically Complete Source Route An ATM term. A stack of DTLs representing a route across a PNNI routing domain such that a DTL is included for each hierarchical level between and including the current level and the lowest visible level in which the source and destination are reachable.

Hierarchy A hierarchy is a group of things arranged in order of rank. It is a set of transmission speeds arranged to multiplex successively higher numbers of circuits. See also HIERARCHICAL NETWORK. It is also an Internet USENET newsgroup hierarchy which refers to the set of all newsgroups contained within a specific broad subject category.

High and Dry What happens when you dial into a long distance network and nothing happens. You don't hear anything. Your call doesn't go anywhere. You're simply left High and Dry!

High ASCII ASCII characters whose values exceed 127. In most bulletin board networks, The use of high-ASCII in messages is prohibited since some types of personal computers cannot correctly interpret those characters.

High Bandwidth A person who is super intelligent is said to have "high bandwidth." The term is believed to have originated at Microsoft in Seattle, Washington. See also BANDWIDTH.

High Capacity Service Generally refers to tariffed, digital-data transmission service equal to, or in excess of T-1 data rates (1,544 Mbits.)

High Cost Area A term describing a serving area of a LEC (Local Exchange Carrier) in which the cost of providing local telephone service is at least 115% of the national average. Through the "settlements" process, administered by NECA (National Exchange Carrier Association) under the direction of the FCC (Federal Communications Commission), high cost LECs are compensated for this extraordinary cost through the Universal Service Fund. See also NECA, SEPARATIONS AND SETTLEMENTS, and UNIVERSAL SERVICE FUND.

High Definition TV HDTV. A system standard for transmitting a TV signal with far greater resolution than specified by the current NTSC standard. NTSC was developed by the National Television System Committee and is the standard for analog TV broadcast and reception in North America. HDTV employs a wide aspect ratio of 16:9 (horizontal:vertical), offering approximately twice the horizontal and twice the vertical emitted resolution specified by the NTSC standard. The total number of pixels, therefore, is more than four times that of NTSC. In December 1996, the FCC adopted standards for digital ATV (Advanced TV), the successor to the HDTV concept. Those standards, which are voluntary, largely follow the recommendations of the ATSC (Advanced Television Systems Committee). See ATV and HDTV.

High Dome Synonym for "egghead." A scientist.

High Energy Radio Frequency Gun HER. Shoots a high-powered radio signal at an electronic target (such as a computer) and puts it out of commission.

High Fidelity Systems of radio transmission and reception which permit a wide band of audio frequencies to be transmitted and/or reproduced.

High Frequency Noise A signal frequency more than 1,000 times the normal AC power line frequency of 60 cycles. The frequency will lie between 3 and 30MHz.

High Level Data Link Control HDLC. A communications protocol that is bit oriented in which control codes differ according to their bit positions and patterns.

High Level Languages Essentially any of the computer languages whose code is not unique to the hardware or architecture of a particular computer. High level languages are more like human language than the machine language which computers talk. High level languages translate human instructions into the machine language computers can understand, but which humans don't have to (in order to tell the computer what to do). Computer languages such as BASIC, FORTRAN, COBOL and Pascal are high level languages. They are a number of levels (at a High Level) away from the actual bit manipulation (machine language, also called "bit twiddling" by the Hackers). Compare with LOW LEVEL.

High Low Tariff A tariff in which two prices are given for something — a high price and a low price. The first high/low tariff from AT&T was for leased voice lines where a lower charge was made per mile for connections between routes that have much traffic (High Density) and greater charges per mile are made for all other (Low Density) routes. The High/Low tariff was significant because it was AT&T's response to competition from long distance carriers like MCI and it was one of the first moves away from nationwide rate averaging, which was the way things were done under monopoly.

High Memory Area HMA. High Memory Area is the first 64KB of extended memory. If you're using MS-DOS 5.0 or 6.0, you can save some conventional memory (i.e. below 640K memory) by loading the operating system into HMA. Add the line DOS=HIGH to your CONFIG.SYS to use HMA for the operating system.

High Pass Filter A filter which passes frequencies above a certain frequency and stops (attenuates) those below.

High Performance Computing Act An Act passed by Congress in 1991 to foster the creation of computer "superhighways" linking computers at universities, national laboratories and industrial organizations. One objective of the High Performance Computing Program is the establishment of a gigabit/second National Research and Education Network (NREN) that will link the government, industrial and higher education communities involved in general research activities. Such a gigabit network would provide a significant increase in bandwidth compared with the existing National Science Foundation network, which is evolving from a 1.5 megabit per second (T-1) backbone to 45 megabit per second (T-3).

High Performance Routing HPR. A local area networking term. HPR is the next-generation APPN — referred to in the past as APPN+ — that adds IP-like dynamic networking — e.g., dynamic alternate routing in the event of path failure — features to APPN, and uses a routing mechanism that works at Layer 2 using a RIF concept similar to that found in SRB.

High Power Amplifier HPA. A device which provides the high power needed to shoot signals 22,000 miles plus from an earth station to a satellite.

High Rejection The ability of a voice recognition system active vocabulary words and reject those sounds that do not match closely with those words.

High Resolution TV Television with over 1000 lines per screen, about double the resolution of present systems. Sometimes called HDTV, for high-definition television. We're still awaiting standards for high resolution TV.

High Sierra Format A standard format for placing files and directories on CD-ROM , revised and adopted by the International Standards Organization as ISO 9660.

High Speed Digital Subscriber Loop See HDSL.

High Speed Local Network HSLN. A local network designed to provide high throughput between expensive, high-speed devices, such as mainframes and mass storage devices.

High Speed Printer Any printer which can print at over 100 lines a minute. Like many definitions, this one is arbitrary. Some people claim a dot matrix is "high speed" and a letter quality, daisy wheel is a "low speed" printer. Laser printers could be classed as high speed printers, maybe.

High Speed Register Set Registers are storage locations within the CPU that are used to hold both the data to be operated on and the instructions to accomplish the operations.

High Speed Signal An AT&T definition for a signal traveling at the DS3 rate of 44.736 mbps (million bits per second) or at either 90 mbps or at 180 mbps (Optical mode).

High Split 1. A broadband cable system in which the bandwidth used to send toward the head-end (reverse direction) is approximately 6 MHz to 180 MHz, and the bandwidth used to send away from head-end (forward direction) is approximately 200 MHz to 400 MHz. The guard band between the forward and reverse directions (180 MHz to 220 MHz) provides isolation from interference. High split requires a frequency translator which transfers the originating signals to other frequency ranges at the head-end in either direction. Historically, CATV systems used the spectrum below Channel 2 for inbound transmissions from the user premise to the head-end; that frequency range is 5-30/40 MHz.
2. A term used in radio communications, including paging and cellular, for several ranges of frequency used to connect a remote site to a main site. For instance, the low-split might be 806.0125 MHz and the high-split 851.0125-869.9875 MHz. Frequency translators are used to transfer the signal to another frequency range from that point forward.

High Usage Groups Trunk groups established between two central office switching machines to serve as the first choice path between the machines and thus, handle the bulk of the traffic. See HIGH USAGE TRUNK GROUP.

High Usage Trunk Group A Bellcore definition. A trunk group that is designed to overflow a portion of its offered traffic to an alternate route.

Highway 1. Another word for BUS. A common path or set of paths over which many channels of information are transmitted. The channels of the highway are separated by some electrical technique.
2. The Information Superhighway. In 1995, a consulting firm called Ovum defined the superhighway as a mechanism for providing access to electronic information and content held on network servers. It has four key features, according to Ovum: A. It supports two way communications. B. It offers more than just simple voice telephony. C. It is interactive and provides real-time, cooperative communications, and D. It supports electronic screen-based applications.

Highway Construction Supervisor A consultant to provide assistance in specification, installation and/ or operation of systems and software for accessing the information highway.

Highway Patrol A slang term for the U.S. Congress.

HiPeR-LAN High Performance Radio LAN

HIPPI High Performance Parallel Interface. In 1989, researchers at the Los Alamos National Laboratories began work on a standard for high-speed, point-to-point data transport between supercomputers. The result of that effort was HIPPI, which later became known as HIPPI-800 and which was standardized by the ANSI X3T9.3 committee as X.3.183-1991. HIPPI also is used to move data between supercomputers or high-end workstations and peripherals (e.g., disk arrays and frame buffers) through high-capacity, non-blocking crossbar-type circuit switches. HIPPI provides for transfer rates of 800 Mbps over 32 shielded twisted pair (STP) copper wires (single HIPPI) and 1600 Mbps over 64 pairs (double HIPPI). HIPPI connections are limited to 25 meters over STP and 10 kilometers over fiber. HIPPI is currently the most common interface in supercomputing environments. Work is in progress on HIPPI-6400, which supports transmission rates as high as 6400 Mbps, or 800 MBps (MegaBytes per second), in each direction, an 8-fold increase over the original version. Also known as SuperHIPPI and GSN (Gigabyte System Network), HIPPI-6400 is compatible with HIPPI-800. Distances of 50 meters can be bridged with parallel copper cables, and 200 meters with parallel fiber-optic cables. The connection is devised in four virtual circuits in each direction, capable of supporting various combinations of traffic such as 10Base-T, 100Base-T, Fibre Channel, ATM and HIPPI-800.

Histogram An imaging term. A display plotting the density of the various colors and/or values in an image.

HIT 1. Electrical interference that causes the loss or introduction of spurious bits into a data stream.
2. The unit of measure most commonly cited by companies that have set up shop on the Internet's World Wide Web. Hits is a measure of file openings done on a Web site. It is not a measure of how many people have visited your Web site. Since one file is needed for every chunk of text and every graphic element on a Web page, one mouse click may count as a dozen or more hits, depending on the complexity of the page. Also, because most Web sites contain more than one page, hits can rapidly multiply. So, using "hits" as a measure of the popularity of a Web site is not accurate. The industry is working on a measure of people visits that's a little more accurate.

HITS In the language of the Internet, hits has two meanings. The first and more common, is the number of times your Web site or a file within your Web site is accessed by people visiting it. "Hits" is often used as a measure of how popular your site is. But the measurement of "hits" is not very scientific. For example, if someone visits a home page, then jumps to another page, then comes back to the home page, that is registered as two "hits." Still, people who are selling advertising on Web pages use "hits" as a measure of how many people visit the site and therefore how much to charge for advertising. The second meaning of "hits" is how many matches you might find in a search; e.g., a Veronica search for the word "NASA" will return a long list of hits for your query.

HIVR Host Interactive Voice Response. tying a voice response unit into a mainframe computer which has lots of data. Applications which can be produced include bank-by-phone, reservations-by-phone, etc. See INTERACTIVE VOICE RESPONSE.

Hizzoner His Honor. Hizzoner is the way the New York City tabloids refer to the mayor of New York. They call him Hizzoner. Say it fast.

HKSW Abbreviation for HOOK SWITCH, the actual electrical switch inside a phone that is controlled by the motion of the SWITCH HOOK.

HLC High Level Committee of ITU (International Telecommunication Union).

HLD High Level Domain. See Web Address.

HLF High Level Function.

HLLAPI High Level Language Applications Programming Interface. An IBM API.

HLR Home Location Register. A wireless telecommunications term. A permanent SS7 database used in cellular networks, including AMPS (Advanced Mobile Phone System), GSM (Global System for Mobile Communications), and PCS. The HLR is located on the SCP (Signal Control Point) of the cellular provider of record, and is used to identify/verify a subscriber; it also contains subscriber data related to features and services. The HLR is used not only when you are making a call within the area of coverage supported by your cellular provider of record. It also is used to verify your legitimacy and to support the features to which you subscribe when you are roaming outside that home area. In a roaming scenario, the local service provider queries the HLR via a SS7 link. Once verified, your data is transferred via SS7 to the VLR (Visitor Location Register), where it is maintained during your period of roaming activity within the coverage area of that provider. HLR is a key element of IS-41, the predominant wireless standard in North America. See also AMPS, GSM, IS-41, PCS, SCP, SS7 and VLR.

HMA See HIGH MEMORY AREA.

HMI 1. Novell's Hub Management Interface. See HMI DRIVER. 2. Human-to-Machine Interface.

HMI Driver A Hub Management Interface (HMI) driver is an ODI driver running on a NetWare server that is compliant with the Novell HMI specification. A node may emulate an HMI driver by supporting the Novell NWHUB.MIB and IPX autodiscovery.

HMM Hidden Markov Method. A common algorithm in voice recognition which uses probabilistic techniques for recognizing discrete and continuous speech.

HO Tone A cellular term. Handoff Tone. 50ms of signaling tone sent by the mobile on the REVC to indicate leaving the source cell site during handoff.

Hobbit A microprocessor chip developed by AT&T's Bell Labs and used in the EO handheld devices. As of this writing, EO was 50% owned by AT&T.

HOBIC HOtel Billing Information Center used by hotels for getting immediate charges for long distance calls placed by their guests. A service of AT&T and the local telephone operating company delivered through a distinct and separate trunk and usually terminating on a telex machine, which prints guest long distance charges.

HOBIS HOtel Billing Information System.

Hobo The word hobo comes from the time after the Civil War, when men who had lost their farms and homesteads would ride the rails looking for temporary work at the various farms that were along the railroad tracks. Because they usually carried their own hoes, the farmers called them "hoe boys."

Hold To temporarily leave a phone call without disconnecting it. You can return to the call at any time, sometimes from other extensions. There are several types of "HOLD" on a telephone system. How they work and what lamping they put on instruments varies from phone system to phone system.
Exclusive Hold: Prevents every other telephone from picking up the call. Only the telephone instrument that put the call on hold can retrieve it.
I-Hold: Effectively the same as Exclusive Hold.
Line Hold: The call is on hold. Anyone with a phone with the held line appearing on it can pick up the phone.

Hold Recall A telephone system feature which reminds you periodically that you've put someone on hold.

Holding Tank A queue in which a call is held until it can either use its assigned route or overflow into the next available route.

Holding Time The total time from the instant you pick up the handset, to dialing a call, to waiting for it to answer, to speaking on the phone, to hanging up and replacing the handset in its cradle. You are never billed for holding time. You are always billed for conversation time which is shorter than holding time. But holding time is an important figure to know when you're trying to determine how many circuits you need. For you will need sufficient circuits to take care of dialing, etc. — even though you're not being billed for that time.

Holdup Time The amount of time that a power supply can continue to supply the load after input power is terminated. The duration of a blackout or transfer time that a power supply can accept without any disturbance of the output. Holdup time is specified by CBEMA to be a minimum of 8 milliseconds for business and computer equipment. The typical value specifed for commercial computer power supplies is 25ms. Holdup time is increased when a power supply is lightly loaded. Therefore typical computers have holdup times in the range of 100ms. This definition courtesy American Power Conversion Corp.

Holiday Factor A call center term. A historical factor associated with a specific date and multiplied by the forecast call volume for that data in order to take into account an expected increase or decrease in the call volume. For example, if on a given day only half the usual number of calls occur for that day of the week and that time of year, the holiday factor for that date would be .5.

Hollerith Card A punched-hole 80 column card used for storing information for input into a computer. Remember the cards you got telling you "not to fold, bend, punch, spindle, etc."? They were Hollerith Cards. They're now falling into disfavor as other, less tamper-proof methods appear.

Hollerith Code Twelve level punched card code.

Hollow Pipeline Jargon for a broad bandwidth circuit that has no framing. A private out-of-band signaled (CCC, Clear Coded Channel) DS1. There is no timing, framing or error connection. You input a bit stream into one end and it comes out the other end in the same order. The maximum speed for DS-1 is 1.536 Mbps. This is 1.544 Mbps less than the framing overhead of 8 Kbps.

Hologram A three-dimensional image produced by a system that uses lasers instead of lenses.

Holographic Data Storage A technology still in the

labs. It uses lasers and crystals rather than magnetic medium to store bits and bytes in holograms. Holographic data storage portends major reduction in physical storage space and much shorter seek time.

Holy War Arguments that involve basic tenets of faith, about which one cannot disagree without setting one of these off. For example: PCs are superior to Macintoshes.

Home The beginning place of a cursor on a CRT screen. Usually it's the top left hand corner. The function key on an IBM PC or clone marked "Home" will take the cursor to the home position, namely the top left hand corner.

Home Agent A Mobile IP term. Mobile nodes associated with nomadic users register their presence at a remote location through a foreign agent. The foreign agent communicates with the home agent in order that data packets can be forwarded to the remote subnet. See also Mobile Agent and Mobile IP.

Home Carrier The cellular operating company which a subscriber is registered with and pays the monthly service charge and usage charges to.

Home Location Register HLR. A wireless telecommunications term. A permanent SS7 database used in cellular networks, including AMPS (Advanced Mobile Phone System), GSM (Global System for Mobile Communications), and PCS. The HLR is located on the SCP (Signal Control Point) of the cellular provider of record, and is used to identify/verify a subscriber; it also contains subscriber data related to features and services. The HLR is used not only when you are making a call within the area of coverage supported by your cellular provider of record. It also is used to verify your legitimacy and to support the features to which you subscribe when you are roaming outside that home area. In a roaming scenario, the local service provider queries the HLR via a SS7 link. Once verified, your data is transferred via SS7 to the VLR (Visitor Location Register), where it is maintained during your period of roaming activity within the coverage area of that provider. HLR is a key element of IS-41, the predominant wireless standard in North America. See also AMPS, GSM, IS-41, PCS, SCP, SS7 and VLR.

Home Page The classic definition: The front page of an "online brochure" about an individual or organization. The Internet definition: The first page browsers see of the information you have posted or your computer attached to the World Wide Web is your "home page." It's a "welcome" page. It says "Welcome to my site, my home." It typically contains some sort of table of contents to more information which a visitor (browser, surfer, etc.) will find at your site by clicking onto hypertext links you've created. In a Web site, a home page is usually called index.htm, index.html or index.asp. The biggest mistake made by people creating Web sites is that they fail to call their home page index.* (that star depends on the operating system which the Web hoster is using). See HTML, INTERNET, STREAMING and WORLD WIDE WEB.

Home Run Phone system wiring where the individual cables run from each phone directly back to the central switching equipment. Home run cabling can be thought of as "star" cabling. Every cable radiates out from the central equipment. All PBXs and virtually all key systems work on home run cabling. Some local area networks work on home run wiring. See LOOP THROUGH.

HomePNA Home Phoneline Networking Alliance. An association of companies working toward the adoption of a single, unified phoneline networking standard and bringing to market a range of interoperable home networking solutions using in-place phone wiring. HomePNA solutions are intended to be plug-and-play for networking of multiple PCs, peripherals (e.g., printers, scanners and video cameras), multi-player network games, home automation devices (e.g., environmental control and security systems), digital televisions and digital telephones. An all-purpose Home Area Network (HAN) using existing telephone wiring, the HomePNA solution also is intended as a means of shared access to IP voice and video networks, the IP-based Internet, and the conventional circuit-switched Wide Area Network (WAN). Network access technologies are intended to include analog, ISDN and xDSL local loops. Initial efforts are directed at a technology that will support spatial separation of nodes by as much as 500 feet, which represents a home of up to 10,000 square feet (which is bigger than my home, and probably bigger than yours, unless you are Bill Gates and live in a monstrosity of a castle, in which case you probably already have an ATM-based LAN with SONET fiber optics pipes running at 10 Gbps, but I digress), and running at data rates of 1 Mbps. Frequency Division Multiplexing (FDM) is intended to support simultaneous voice and data traffic; frequency ranges are intended to avoid interference from devices (e.g., refrigerators and air conditioners) found in the home. HomePNA solutions are based on an Ethernet derivative, running at 1 Mbps at frequencies above 2 MHz using a proprietary compression technique from Tut Systems, and using the CSMA/CD protocol native to Ethernet; speeds of 10 Mbps are planned into the future, with the theoretical potential being as much as 100 Mbps. Members include 2Com, AT&T, Compaq, Hewlett-Packard, IBM, Intel, Lucent and Tut Systems. www. homepna.org. See also Ethernet, FDM, ISDN, SONET and xDSL. www.homepna.org

Homeostasis The state of a system in which the input and output are exactly balanced, so there is no change.

Homes Passed An expressed of the number of dwellings that a CATV provider's distribution facilities pass by in a given cable service area and an expression of the market potential of the area.

Homing When you dial a long distance number, your central office will choose a special set of trunks to send your call onto the next switching center for movement through the nationwide toll system. Those trunks are said to be the homing trunks for your central office. In other words, your central office is said to home on these trunks. If you're consistently encountering lousy long distance lines (and so are others on your central office), then ask your telephone company to check these trunks out.

Homo 1. The Greek prefix meaning the same.
2. Home Office Mobile Office. See also SOHO, which stands for Small Office Home Office.

Homogeneous Networks Composed of similar hardware from the same manufacturer.

Homologation Conformity of a product or specification to international telephony connection standards. What this means in simple language is that you have submitted your product to a regulatory agency or a government testing agency in a foreign country and they have said that your product is OK for use and sale in that country and it is allowed to be connected to the local phone system. In short, your product has now been homologated in that country.

Honeycomb Coil A type of inductance in which the turns do not lie adjacent to each other.

Honeymoon It was the accepted practice in Babylon 4,000 years ago that for a month after the wedding, the bride's father would supply his son-in-law with all the mead he could drink.

Mead is a honey beer, and because their calendar was lunar based, this period was called the "honey month" or what we know today as the honeymoon.

Hook-up Wire A wire used for low current, low voltage (under 1000 volts) applications within enclosed electronic equipment.

Hookemware Free software that contains a limited number of features designed to entice the user into purchasing the more comprehensive version. See also HYPERWARE, MEATWARE, SHOVELWARE, SLIDEWARE and VAPORWARE.

Hookflash Momentarily depressing the hookswitch (up to 0.8 of a second) can signal various services such as calling the attendant, conferencing or transferring calls. In ISDN a hookflash signals the System Adapter to perform an operation, such as placing such as placing a call on hold. To hookflash, simply depress and release the receiver button. By default, the Hayes ISDN System Adapter recognizes a hookflash when the receiver button is depressed less than 0.8 of a second. You can change the default.

Hooking Signal An on-hook signal of 0.1 to 0.2 seconds duration used to indicate that a subscriber intends to initiate a new process such as "add-on."

Hookswitch Also called SWITCHHOOK or switch hook. The place on your telephone instrument where you lay your handset. A hookswitch was originally an electrical "switch" connected to the "hook" on which the handset (or receiver) was placed when the telephone was not in use. The hookswitch is now the little plunger at the top of most telephones which is pushed down when the handset is resting in its cradle (on-hook). When the handset is raised, the plunger pops up and the phone goes off-hook. Momentarily depressing the hookswitch (up to 0.8 of a second) can signal various services such as calling the attendant, conferencing or transferring calls. See HOOKING SIGNAL.

Hookswitch Dialing You can make phone calls by depressing the hookswitch carefully. If you push it five times, you dial five. Push it ten times you dial 0. Some coin phones discourage hookswitch dialing. Some don't.

Hoot'n'Holler Hoot'n'Holler are special 24-hour a day phone circuits which stay open 24-hours a day, seven days a week. Anyone picking up a phone on the circuit can listen and talk to whoever's on the line, or whoever might be within earshot. The idea is that to get someone to speak to you, you "hoot and holler," i.e. make a noise. A hoot'n'holler circuit is also called a Junkyard Circuit, Holler Down, Shout Down, Open Speech Circuit, Squawk Box System, FP or Full Period (as in FP 123456 circuit #). Hoot'n'Holler is a circuit consisting of 4 wires (2 pair — a transmit pair and a receive pair). Technically, this is how it works: Audio energy present on the transmit pair at any location will appear on the receive of all the other locations, usually a multipoint circuit. The transmitted audio will not return on the receive pair of the originating location. Hoot'n'Holler circuits are voice conferencing oriented party lines and are non private by nature. There is no signaling on a hoot and holler circuit except when one "shouts down" to open speakers and "listening" at the distant out points. A Hoot'n'Holler circuit is a dedicated full time voice network. Individual four wire "drops" are connected via various bridging mechanisms. These "bridges" can be analog or digital and act as mix minus devices (mixes everyone else minus yourself). They provide all drops with connectivity with each other.

Hop 1. Each short, individual trip that packets (or e-mail messages) make many times over, from router to router, on their way to their destinations.

2. A change of Radio Frequency (RF) channel used to carry the cellular Digital Packet Data (CDPD) data for a channel stream.

Hop by Hop Route A route that is created by having each switch along the path use its own routing knowledge to determine the next hop of the route, with the expectation that all switches will choose consistent hops such that the call will reach the desired destination. PNNI (Private Network-Network Interface) does not use hop-by-hop routing.

Hop Channel A Radio Frequency (RF) channel that has been declared a candidate for carrying a Cellular Digital Packet Data (CDPD) channel stream after a channel hop.

Hop Count The number of hops it will take for a packet to make it from a source to a destination. In short, the number of nodes (routers or other devices) between a source and a destination. In TCP/IP networks, hop count is recorded in a special field in the IP packet header and packets are discarded when the hop count reaches a specified maximum value.

Hop Off When you make a phone call on the Internet you can call from one phone attached to the Internet to another phone attached to the Internet or you can call from one phone attached to the Internet and, at the other end, go into a PC stuffed with voice and switching cards and which is attached to local phone lines. The process of leaving the Internet is called "Hop Off."

Hop Sequence The carefully coordinated sequence by which radio transmitters and receivers hop from on frequency to another, hop sequence is used in FHSS (Frequency Hopping Spread Spectrum) systems. FHSS is used extensively in Wireless LANs and certain PCS (Personal Communications Systems) cellular systems. See also FHSS.

Hops Term describing the number of times a message traverses different nodes.

Horizontal H. In television signals, the horizontal line of video information which is controlled by a horizontal synch pulse.

Horizontal Blanking Interval The period of time during which an electron gun shuts off to return from the right side of a monitor or TV screen to the left side in order to start painting a new line of video.

Horizontal Cable Defines the cable used to link the communications closet / room with individual end use devices.

Horizontal Cross-Connect A cross-connect in the telecommunications closet or equipment room to the horizontal distribution cabling.

Horizontal Distribution Frame Located on the floor of a building. Consists of the active, passive, and support components that provide the connection between inter-building cabling (i.e. cabling coming from outside the building) and the intra-building cabling for a building.

Horizontal Interval The sum of Horizontal Retrace.

Horizontal Link An ATM term. A link between two logical nodes that belong to the same peer group.

Horizontal Output The amplifier that amplifies the horizontal output sync signal in a TV or monitor. The output runs through a deflection yolk. This creates magnetic fields that control tracing of the CRT beam sideways. A vertical amplifier does the same for the up and down tracing of the CRT beam. A TV's horizontal output frequency is 15.73425 kHz. On some TVs you can hear the high-pitched sound of the horizontal output circuitry when you turn on the equipment.

Horizontal Retrace A video term. The return of the electron beam from the right to the left side of the raster after the scanning of one line.

Horizontal Resolution Detail expressed in pixels that provide chrominance and luminance information across a line of video information.

Horizontal Scan Rate The frequency in Hz (hertz) at which the monitor is scanned in a horizontal direction; high horizontal scan rates produce higher resolution and less flicker. Thus, the EGA horizontal scan rate is 21.5Hz, while the VGA standard scan rate is 31.4Hz. Some displays now offer even higher scan rates, as much as 70 Khz. See MONITOR.

Horizontal Sync A video term. Horizontal sync is the -40 IRE pulse occurring at the beginning of each line. This pulse signals the picture monitor to back to the left side of the screen and trace another horizontal line of picture information. See interlace.

Horizontal Wiring The portion of the wiring system extending from the workstation's outlet to the BHC (Backbone to Horizontal Cross-Connect) in the telecommunications closet. The outlet and cross-connect facilities in the telecommunications closet are considered part of the horizontal wiring. See HORIZONTAL WIRING SUBSYSTEM.

Horizontal Wiring Subsystem The part of a premises distribution system installed on one floor that includes the cabling and distribution components connecting the riser subsystem and equipment wiring subsystem to the information outlet via cross connects, components of the administration subsystem.

Horn In radio transmitting, a waveguide section of increasing cross-sectional area used to radiate directly in the desired direction or to feed into a reflector that forms the desired beam.

Horn Alert HA. A cellular car phone feature that automatically blows the car's horn if a call is coming in.

Horsepower A unit of power equivalent to 550 foot pounds per second or to 746 watts.

Horton A software program which provides an automatic method for creating a directory of e-mail addresses. Users can look up electronic addresses via a search key which can be a fragment of a person's name.

Hose And Close A pattern of behavior exhibited by phone tech-support people who spout a bunch of jargon you don't understand, ask you to perform a bunch of procedures you don't follow, and then abruptly hang up. This definition from Wired Magazine.

Host 1. An intelligent device attached to a network.
2. A mainframe computer.
3. A computer with full two-way access to other computers on the Internet. A host can use virtually any Internet tool, such as WAIS, Mosaic and Netscape.

Host Apparent Address A set of internetwork layer addresses which a host will directly resolve to lower layer addresses.

Host Based Firewall A hostbased firewall is a firewall system that includes a bastion host (a general-purpose computer running firewall software). A host based firewall usually includes a circuit-level gateway, an application level gateway, a hybrid of both gateways, or a stateful inspection firewall.

Host Bus Adapter HBA. A printed circuit board that acts as an interface between the host microprocessor and the disk controller. The HBA relieves the host microprocessor of data storage and retrieval tasks, usually increasing the computer's performance time. A host bus adapter (or host adapter) and its disk subsystems make up a disk channel.

Host Carrier The cellular operating company a subscriber from another cellular system would be billed roamer charges.

Host Computer A computer attached to a network providing primarily services such as computation, database access or specific programs of special programming languages.

Host Interactive Voice Response A voice response system that can communicate with a host computer, typically a mainframe. Applications which can be produced include bank-by-phone, reservations-by-phone, etc.

Host Name The name given to a mainframe computer.

Host Name Resolution A mechanism that provides static and dynamic mechanisms for resolving host names into numeric addresses. The Internet Name Server Protocol accesses an Internet name server that provides dynamic name-to-number translation (this process is specified in IEN 116). The Domain Name Protocol accesses a Domain Name Server that provides dynamic name-to-number translation (this process is specified in RFC-1034 and RFC-1035). A static local host table can also be accessed for name-to-number translation.

Host Number The part of an internet address that designates which node on the (sub)network is being addressed. See DOMAIN.

Host Processor Same as HOST COMPUTER.

Host Server A device which connects to a LAN and then allows a computer, which cannot directly support the LAN protocols, to connect to it, providing all necessary LAN support.

Host Site In the transfer of files, the host site is the location receiving a file. When two individuals are exchanging files, the one who receives the file first would be the host, the other would be considered the remote.

Host Switch A central office switching system which provides certain functions to a smaller switch located remotely.

Host Table 1. A list of TCP/ IP hosts along with their IP addresses.
2. In Windows 95, the host table is HOSTS or LMHOST file that contains lists of known IP addresses mapped to host names or NetBIOS computer names.
3. An ASCII text file where each line is an entry consisting of one numeric address and one or more names associated with that address. Host tables are used to resolve hostnames into numeric addresses.

Hot Live wire. A conductor carrying a signal is said to be a hot conductor, i.e. the wire carrying the signal or the ground as opposed to the neutral or ground wire.

Hot And Ground Reversed In AC electrical power, the correct connection of the Hot and Ground wires is reversed. This is an extremely dangerous condition because the GROUND path will rise to 129 Volts and can present a lethal shock hazard to anyone in contact with equipment powered from this outlet or any outlet using the same ground path.

Hot And Neutral Reversed Also called reversed Polarity. A symptom of poor AC electrical wiring. In this case the correct connection of the Hot and Neutral conductors is reversed. Dangers include increased leakage current, and damage to electronic equipment or motors and appliances requiring correct polarity.

Hot Chat An Internet term. Sex talk, in real time, online, usually between two or more consenting people (through not necessarily adults).

Hot Cut The conversion from an old to a new phone system which occurs instantly as one is removed from the circuit and the other is brought in. There are advantages and disadvantages to Hot Cuts. For one, they're likely to be much more dangerous than a Parallel Cut, in which the two phone systems run side by side for a month or so. Also known as Flash Cut.

Hot Desk An employee of a company no longer has a permanent office. He works out of their home, visits customers and communicates with the office through fax and electronic mail. Occasionally that person finds it necessary to visit an office of the company. He is allocated a desk and perhaps an office for his stay. That stay might be as short as an hour or as long as several weeks. Once he checks out, someone else gets the desk. This arrangement is called a "Hot Desk."

Hot Docking Docking is to insert a portable computer into a base unit. Cold docking means the computer must begin from a power-off state and restart before docking. Hot docking means the computer can be docked while running at full power.

Hot Fix A feature of Novell's NetWare LAN (local area network) operating system in which a small portion of the hard disk's storage area is set aside as a "Hot Fix Redirection Area." This area is set up as a table to hold data that are "redirected" there from faulty blocks in the main storage area of the disk. It's a safety feature.

Hot Key Combination A combination of keys on the keyboard that are pressed down simultaneously to make the computer perform a function. For example, the Ctrl, Alt, Del hot-key combination will warm boot an MS-DOS computer.

Hot Line A private line dedicated between two phones. When you pick up either phone or do some act of signaling (like push a button), the other phone rings instantly. Hot lines are useful in emergencies and other areas where time is of the essence — e.g. trading currencies.

Hot Line Service When you pick up the phone, you're automatically connected with a phone number. Such Hot-Line Service on a PBX typically gets you emergency service, etc. See also HOT-LINE.

Hot Links A methodology that references and can connect information from one document to another, regardless of the type of application used.

Hot List A gopher or Web file that lets you quickly connect to your favorite pre-selected page. Appropriately named. The way it works: You connect to a home page. You decide you'd like to return at some other time. So you command your internet surfing software to mark this web site on your hot list (also called marking it with a "bookmark." Next time you want to return to that web site, you simply go to your hot list or your bookmarks, click on which one you want. And bingo, you're there. A hot list is also known as a bookmark. Most Web browsers have book marks or hot lists.

Hot On Neutral, Hot Unwired In AC electrical power, the HOT wire is connected to the NEUTRAL terminal of the outlet and the HOT terminal in UNWIRED. Dangers include shock hazard from excessive leakage current and fire hazard. Depending on other conditions, equipment may or may not operate.

Hot Plug When a system component (e.g., a computer disk drive) fails, it may be replaced without turning the system off. During this period, the system's activity is suspended, however. Also known as a Warm Swap, it is unlike a Hot Swap, during which the system remains active. See HOT SWAP.

Hot Plugging The ability to add and remove devices to a computer while the computer is running and have the operating system automatically recognize the change. Two new external bus standards — Universal Serial Bus (USB) and FireWire support hot plugging. This is also a theoretical feature of PCMCIA cards.

Hot Potato Routing A packet-switching routing technique that retransmits a packet as soon as possible after receiving a packet, even if it means making a poor routing choice.

Hot Racking A Navy term referring to the practice in sub-marines of having sailors sleep in the same bunk at different times. This occurs because of the shortage of bunks on sub-marines. Hot racking is the reason the US Navy gave in May of 1995 for vetoing the idea of having women serve on crowded submarines.

Hot Redundancy A term used in conjunction with very critical telecom and computing systems, such as 911 service. With Hot Redundancy, the component or the system runs in parallel with an identical "twin." Should one twin fail, the other is already running and provides full service without interruption.

Hot Restart Imagine a corporate telephone system, perhaps a PBX or an ACD. It's handling phone calls to and from customers every second of the day. It's mission critical. You can't allow it to crash for even a second a day. But phone systems are nothing more than specialized computers with specialized software. They will crash or lock up just like your PC, though perhaps not as often. But they will crash and lock up. What happens when your PC crashes and locks up? You will lose data. You will probably then reboot your PC, probably from its hard disk. This will cost you anywhere from 30 to 60 seconds. You can tolerate this on your PC. You can't tolerate this on your phone system. An integral part of most phone systems' software is a feature called "Hot Restart." When the phone system's software crashes or locks up, there is "Hot Restart" software that enables the phone system to restart itself (another word for reboot) without losing the phone calls in progress or without taking 30 to 60 seconds to load a new operating system from hard disk or tape. How this exactly is done seems to vary from one telephone system maker to another. Some may keep an operating system in RAM. Some may keep two identical processors chugging away simultaneously and switch from one to another, when things go awry. How manufacturers do "Hot Restarts" is something of a trade secret.

Hot Spot See HOTSPOT.

Hot Standby Backup equipment kept turned on and running in case some equipment fails. Also known as a Hot Spare.

Hot Swap The process of replacing a failed component — e.g. a RAID drive — while the rest of the system (in this case, the disks) continues working and continues to provide function normally, i.e. providing data to the network users and providing a place for them to store their data. See HOT SWAPPABLE and RAID.

Hot Swappable The ability of a component (such as a redundant power supply) to be added to or removed from a device (for instance, a repeater) without powering down the device, thus providing a maximum uptime.

Hotel/Motel Console A specialized PBX console or a normal console programmed to work specifically in hotels and motels. The console will often show room status information.

Hotfix A Novell program that dynamically marks defective blocks on the hard disk so they will not be used. See HOT FIX.

Hotjava Java is a programming language from Sun Microsystems designed primarily for writing software to leave on World Wide Web sites and downloadable over the Internet to a PC owned by you or me. HotJava, its brother, is another piece of software installed on a Web browser at your desktop. HotJava enables Java programs delivered over the Web to run on your desktop PC. In short, Java is the programming language the programs on the Web are written in. HotJava is the software that will sit on your PC. Java is basically a new virtual machine and interpretive dynamic language and environment. It abstracts the data on bytecodes so that when you develop an app, the same code runs on whatever operating system you choose to port the Java compiler/interpreter to. What's a Java application? According to Wired Magazine, point to Ford Motor Company's website today, for instance, and all you'll get are words and pictures of the latest cars and trucks. Using Java, however, Ford could relay a small application (called an applet) to a customer's computer (the one on your desk which are using the surf the Internet). The customer could then customize options on an F-series pickup while calculating the monthly tab on various loan rates offered by a finance company or local bank. Add animation to these applications and you could get to "drive" the truck.

Hotline There are several definitions. In the cellular business, a hotline is a system restriction that allows a cellular customer to call only one prearranged number. In the landline phone business, a hotline is often a dedicated line. But it may also be a phone which only dials one number. In this case it's often call a virtual private line. See HOTLINE VIRTUAL PRIVATE LINE SERVICE.

Hotline Virtual Private Line Service A Nynex offering that simulates private line service. HotLine lets you automatically dial a predetermined line within the HotLine network, just by picking up the handset. Hotline uses facilities, switches and programmed intelligence of the public network to create a closed network of simulated private lines.

Hotshot dialer A piece of equipment used to create a hotline or ring-down circuit. A hotline is when one phone rings another without dialing (thing of the Batphone to Commissioner Gordon's office, or the hotline between Washington and Moscow). The dialer automatically dials the number when the handset is lifted.

Hotspot An embedded hyperlink is a hyperlink that is in a line of text. A hotspot is the place in a document that contains an embedded hyperlink. A hotspot is a graphically defined area in an image that contains a hyperlink. An image with hotspots is called an image map. In browsers, hotspots are invisible. Users can tell that a hotspot is present by the changing appearance of the pointer.

House Cable Communication cable within a building or a complex of buildings and owned by the local phone company. House cable comes from the terminal box in the basement or the nearby outside pedestal box and goes straight to the apartment or house. Often it's not terminated. Thus, a technician installing a phone line will often have to break into a house cable and search around for an unused cable pair. In a multi-storey building, a house cable is called a riser cable. Thanks to John Arias of Bell Atlantic for help on this definition. House cable owned before divestiture by the Bell System and after divestiture by the Regional Bell Operating Companies will eventually be fully depreciated and will then belong to the customer. See also Binding Post, Block Cable, Block Pair, Feeder Box, House Box, Krone Block, Riser Cable, Terminal Box,

Howler A device which produces a loud sound to a subscriber's phone or private branch exchange (PBX) extension to indicate that the handset is off-hook and it ought to be put back on hook.

Howler Tone A tone which gets increasingly louder over a short period of time. It is used to notify a user that his phone handset is off its hook.

Howling Howling is typically heard in a speakerphone or conferencing unit when there is "Acoustic coupling" between the microphone and the speaker. This is due to putting the microphone too near the speaker. New circuits called acoustic echo cancelers allow you to operate the microphone and the speaker simultaneously and much closer to each other.

HP Hewlett-Packard Company. HP was formed in 1939 by Bill Hewlett and David Packard. David lost the toss to Bill, which is why the company is called Hewlett-Packard, not Packard-Hewlett. They started the company in a small garage in Addison Avenue in Palo Alto, CA. Their first sale was an audio oscillator used in a Disney film, "Fantasia." See also MBWA.

HP Openview Hewlett-Packard's Openview network management products allow network administrators to monitor and control network devices from an MS-DOS PC or UNIX workstation.

HPA See HIGH POWER AMPLIFIER.

HPAD Host Packet Assembler/Disassembler. The HPAD can link to a host or FEP with native protocol data, or if the host can accept it, with X.25 input. The 4400 PAD functions as either an HPAD or a TPAD See TPAD.

HPFS High-performance file system (HPFS); primarily used with the OS/2 operating system version 1.2 or later. It supports long filenames but does not provide security. OS/2 can use any file system it wants, thanks to its installable file system (IFS) architecture. Two choices available are the FAT file system, used by MS-DOS, and the High Performance File System (HPFS). You can mix and match each and select one at boot time, thanks to OS/2's Dual Boot option. IBM, which created OS/2, claims HPFS is much more efficient than FAT. It tries to store all files on disk contiguously and uses its own built-in cache. However, HPFS' most notable attribute is the long 254-character file names and case preservation. OS/2 remembers file names as upper and lower case (though it's not case-sensitive to commands).

HPO High Performance Option. A way of improving equipment transmission characteristics. For instance, the upgrading of a voice-grade line to meet standards for data transmission.

HRC See Cable Normal Switch.

HSCS High Speed Circuit Switched.

HSCSD High Speed Circuit Switched Data. HSCSD has relevance in the GSM wireless world. According to Ericsson, "Today's data transfer rate of 9,600 bits per second (supporting fax, e-mail, voice/fax mail, PC file transfer and short message service) will be expanded to 19.2, 28.8, and even 64 kbit/s in the near future. The first step will be to introduce high-speed circuit-switched data (HSCSD) solutions which enable users to access two time slots instead of one — thus doubling the data capability. The second step will be to introduce bandwidth-on-demand (as a built-in capability of HSCSD). By dynamically allocating up to eight time slots for each single data call (64 kbit/s; the full PCS bandwidth), new services can be offered, such as high-speed multimedia access, videoconferencing and CD-quality sound. With the HSCSD high-speed data capacity, graphics-heavy World Wide Web pages can in principle be downloaded as easily and quickly as via a terrestrial connection. See www.ericsson.se/Review/ According to www.telecoms-mag.com, "GSM already meets many of the requirements for UMTS (Universal Mobile Telecommunications Systems), with the key exception of wideband radio access. However, two new service classes under development for GSM will expand the current user data rate of 9.6 kbps to 100 kbps and beyond: high-speed circuit switched data (HSCSD) and general packet radio service (GPRS). Both techniques are designed to integrate with current GSM infrastructure. HSCSD bearer services up to 64 kbps in GSM using multi-slot transmission have already been demonstrated. This technique bundles up to eight TDMA slots within the 200 kHz GSM carrier to create a higher bandwidth channel. HSCSD is also being developed to provide bandwidth-on-demand at variable data rates.

HSD Home Satellite Dish.

HSDA High Speed Data Access.

HSDL High-speed Subscriber Data Line. A Bellcore idea for a two pair phone line coming into a house or business that is a full-duplex T-1 line. See also ADSL.

HSDU High Speed Data Unit.

HSL A computer imaging term. A color model based on hue, saturation, and luminance. Hue is the attribute that gives a color its name (e.g., red, blue, yellow, or green). In this model, saturation refers to the strength, or purity, of the color. If you mix watercolors, saturation would specify how much pigment you added to a given amount of water. Luminance identifies the brightness of a color. For example, full luminance yields white, while no luminance yields black. See also HSV.

HSLN High-Speed Local Network. A local network designed to provide high throughput between expensive, high-speed devices, such as mainframes and mass storage devices.

HSRP Hot Standby Routing Protocol, a proprietary routing protocol from Cisco for fault-tolerant IP routing. According to Cisco, "HSRP enables a set of routers to work together to present the appearance of a single virtual router or default gateway to the hosts on a LAN. HSRP is particularly useful in environments where critical applications are running and fault-tolerant networks have been designed. By sharing an IP address and a MAC address two or more routers acting as one virtual router are able to seamlessly assume the routing responsibility in the case of a defined event or the unexpected failure. This enables hosts on a LAN to continue to forward IP packets to a consistent IP and MAC address enabling the changeover of devices doing the routing to be transparent to them and their sessions."

HSSI High Speed Serial Interface. A serial data communication interface optimized for high speeds up to 52 Mbps. Used for connecting an ATM switch to a T-3 DSU/CSU, for example.

HSSP High Speed Switched Ports. The HSSP is a module inserted into the lower right bay of a Compaq Netelligent 5000 Ethernet SWitch. It provides one connection to a 100Mb/s network - either Fast Ethernet or FDDI.

HST High Speed Technology, a U.S. Robotics proprietary signaling scheme, design and error control protocol for high-speed modems. HST incorporates trellis-coded modulation, for greater immunity from variable phone line conditions, and asymmetrical modulation for more efficient use of the phone channel at speeds of 4,800 bps and above. HST also incorporates MNP-compatible error control procedures adapted to asymmetrical modulation.

HSTR High Speed Token Ring. Proposals have been made (1997) by IBM to the IEEE 802.5 working group for high-speed versions of the Token Ring LAN standard. HSTR proposals specify operating speeds of 100, 128 and 155 Mbps. Current and traditional versions of Token Ring operate at 4 and 16 Mbps, putting Token Ring at a decided disadvantage in comparison to Fast Ethernet and ATM. IBM plans to move forward with product development while the standards process works its magic, i.e. takes its long slow time. See also ATM, FAST ETHERNET and TOKEN RING.

HSV A computer imaging term. A color model based on hue, saturation, and value. Hue specifies the color, as in the HSL model. In this model, saturation specifies the amount of black pigment added to or subtracted from the hue. Value identifies the addition or subtraction of white pigment from the hue.

HTML HyperText Markup Language. This is the authoring software language used on the Internet's World Wide Web. HTML is used for creating World Wide Web pages. HTML is basically ASCII text surrounded by HTML commands in angle brackets, which your browser interprets whichever way it feels. That means (and I reiterate this) different browsers will display the exact same HTML code differently.) Here is an example of a simple line of HTML code:

<H2>Call Center Magazine Editorial Calendar</H2> This line says to the browser: Display those words as a type 2 headline. You can also include an image in an HTML page.

<img src="photos/harry.gif">

Your browser would go find the picture "harry.gif" in the subdirectory called "photos" on the computer which had the URL and the Web page you were visiting. An HTML document has three types of content: tags (which define type styles, like the one above), comments (words to tell the HTML author what he's doing, but which aren't displayed) and text (i.e. the words Call Center Magazine Editorial Calendar. Tags also let you do hyperlinking, which lets a person browsing click your HTML document, click on that hyperlink and go elsewhere -- either to another page which you wrote, or to another Web page (i.e. URL) across the world.

Here's an example of text and tags you might put in your HTML document:

<h2>For more information on computer telephony, please visit</h2> <a href="http://www.computertelephony.com"></a>

When someone browsing your page paces their mouse on the words "www.computertelephony.com," their cursor changes to a hand. They click on it. A few seconds later, they see the computer telephony home page begin to download. See HTML 1.0, HTML 2.0, HTML 3.0 and HTML 3.2.

HTML 1.0 This original specification was drafted in 1990. It was designed primarily for publishing scientific papers to the Web. The spec contained features such as six levels of headings, simple character attributes, quotations, source code listings, list, and hyperlinks to other documents and images. It is no longer used. See HTML.

HTML 2.0 This revised specification for HTML arrived in 1994. It added forms and eliminated many seldom-used tags from the original spec. It also included support for pop-up and pull-down menus, buttons, and text-entry boxes that could be used for filling in forms. As of this writing (Winter of 1996-97), this is the specification that all products support in full. See HTML.

HTML 3.0 The proposal for this HTML specification was published in 1995. The spec called for the inclusion of coding for tables, text flow around figures, and mathematical equations. The spec was too progressive and garnered only piecemeal product support. It has since been dropped in favor of HTML 3.2. See HTML.

HTML 3.2 This new specification for HTML was developed by vendors that include IBM, Microsoft, Netscape, Novell, SoftQuad, Spyglass, and Sun Microsystems. It proposed the inclusion of support for tables, applets, text flow around images, superscripts, and subscripts. Both Microsoft and Netscape have added extensions to HTML that in some cases have been included in the subsequent standard. Netscape first introduced Java, tables, and frames, and Microsoft introduced ActiveX controls. See HTML.

HTML tag A symbol used in HTML to identify a page element's type, format, and structure. The FrontPage Editor automatically creates HTML tags to represent each element on the page.

HTR Hard-To-Reach.

HTTP HyperText Transfer Protocol. Invisible to the user, HTTP is the actual protocol used by the Web Server and the Client Browser to communicate over the "wire". In short, the protocol used for moving documents around the Internet. See DOMAIN, DRP, INTERNET, SURF, URL, Web Address and WEB BROWSER.

HTTPS Hypertext Transfer Protocol Secure. A type of server software which provides the ability for "secure" transactions to take place on the World Wide Web. If a Web site is running off a HTTPS server you can type in HTTPS instead of HTTP in the URL section of your browser to enter into the "secured mode". Windows NT HTTPS and Netscape Commerce server software support this protocol.

Hub The point on a network where a bunch of circuits are connected. Also, a switching node. In Local Area Networks, a hub is the core of a star as in ARCNET, StarLAN, Ethernet, and Token Ring. Hub hardware can be either active or passive. Wiring hubs are useful for their centralized management capabilities and for their ability to isolate nodes from disruption.

HUB Expansion Port An older local area network term for two ports located to the right of a 2008, 2016 or 2116 repeater used to interconnect these repeaters in a stack. The interconnect cable is standard Category 5 UTP cable. These ports are now called Repeater Expansion Ports (REP).

Hub Junction Box A box used to connect a Hub Interface when a node is placed at a remote location.

Hub Management Interface A network management protocol developed by Novell to allow network managers to manage hubs anywhere on a NetWare LAN.

Hub Polling A polling system in which a polled station sends its traffic and passes the polling message (after it's sent its message) to the next station.

Hub Site The location(s) on a network where many circuits are brought together to be multiplexed into a single higher speed connection.

Hublet A mini-hub, submitted by Tracy Meyer, Network Technician, Pacific Bell Mobile Services. The term was created by Mark Alexander an Engineer at Pacific Bell Mobile Services who came up with it. "It is growing in popularity here," according to Tracy.

Hue The attribute by which a color may be identified within the visible spectrum. Hue refers to the spectral colors of red, orange, yellow, green blue and violet.

Huffman Encoding Developed by D.A. Huffman, a popular loss-less data compression algorithm that replaces frequently occurring characters and data strings with shorter codes. Dynamic Huffman encoding reads the information twice, the first time to determine the frequency with which each data character appears in the text and the second time to accomplish the actual encoding process. Huffman encoding suffers from the fact that it does not recognize characters between 0 and 1; for instance, 4.6 would be rounded to either 4 or 5. Huffman encoding is often used in image compression. It also is used in PKZIP, along with other compression algorithms. See also COMPRESSION.

Huge Pipes From Wired's Jargon Watch column. A high-bandwidth Internet connection. "CU-SeeMe doesn't look half-bad_if you've got huge pipes."

Hum Hum on phone lines sounds awful and can severely cut your data throughput. Hum on a phone line may have many sources — grounded carbon, lightning-damaged protection, left in drops or jumpers, "half-tap," or even a wet cable. A wet cable pair "usually" manifests itself with a "frying" sound or crosstalk with other pairs. The solution to hum

on the line is to replace the pair or remove the offending section from your circuit.

Hum Bucker A circuit (often a coil) that introduces a small amount of voltage at power line frequency into the video path to cancel unwanted AC hum.

Humidification The process of adding moisture to the air within a critical space. Without humidity, you get static electricity. With static electricity comes strong shocks of computers equipment. With strong shocks comes loss of data. Lack of humidity is particularly bad in the middle of winter.

Hundred Call Seconds Known by the initials CCS where C is the roman numeral for Hundred. One CCS is 36 times the traffic expressed in Erlangs. See CCS.

Hunt Refers to the progress of a call reaching a group of lines. The call will try the first line of the group. If that line is busy, it will try the second line, then it will hunt to the third, etc. See also HUNT GROUP.

Hunt Group A series of telephone lines organized in such a way that if the first line is busy the next line is hunted and so on until a free line is found. Often this arrangement is used on a group of incoming lines. Hunt groups may start with one trunk and hunt downwards. They may start randomly and hunt in clockwise circles. They may start randomly and hunt in counter-clockwise circles. Inter-Tel uses the terms "Linear, Distributed and Terminal" to refer to different types of hunt groups. In data communications, a hunt group is a set of links which provides a common resource and which is assigned a single hunt group designation. A user requesting that designation may then be connected to any member of the hunt group. Hunt group members may also receive calls by station address. See also TERMINAL HUNT GROUP.

Hunt group helpers Most intelligent businesses run their faxes in hunt groups. Five fax machines in a hunt group can process as many calls as 45 faxes on individual lines. The problem with hunt groups is that when the first fax machine runs out of paper, it typically won't answer the phone. This means calls won't roll over. They just keep landing on the "sleeping" fax machine. A hunt-group helper, which is a piece of hardware installed before the fax machine — between it and the central office phone line — listens to the number of rings on each line of a hunt-group set and rolls the call over to the next line if a given machine lets the line ring too many times (i.e. it's probably out of paper). A simple fix, works great.

Hunting See ROLLOVER LINES.

HVAC Heating, Ventilating, and Air-Conditioning systems. High voltage stuff. Keep your telecommunications cables away from the motors in HVAC systems, please.

HVQ Hierarchical Vector Quantization — a method of video compression introduced by PictureTel in 1988 which reduced the bandwidth necessary to transmit acceptable color video picture quality to 112 Kbps.

HW Email abbreviation for hardware.

Hybrid A device used for converting a conversation coming in on two pairs (one pair for each direction of the conversation) onto one pair and vice versa. This is necessary because all long distance circuits are two pairs, while most local circuits are one pair. Here is a longer explanation from "Signals, The Science of Telecommunications," by John Pierce and Mike Noll:

The telephone instrument in your home is connected to a single pair of wires called the subscriber loop or local loop, which carries both the outgoing voice signal and the incoming one. This pair of wires creates an electrical circuit for each of the two signals. A device in your phone called a hybrid or hybrid coil keeps the two signals separate, more or less, so that what you say into your phone's transmitter doesn't blast into your ear from the receiver.

In contrast, all multiplex systems provide separate talking paths in two directions. Separate paths are necessary because the amplifiers placed along the lines between terminals amplify signals traveling in one direction only. When two people talk between New York and San Francisco the call goes from one phone through a local two-wire voice circuit to a multiplex terminal. There the call is transferred to a four-wire long distance circuit that consists of two separate one-way circuits. At the end of the system, a hybrid reconverts each four-wire circuit into a two-wire circuit.

Hybrid Backbone Two or more types of facilities in a corporate telecom WAN — e.g. ATM and SONET.

Hybrid Cable A communication cable that contains two or more types of conductors that bear electrical signals, a mixture of signal-bearing electrical conductors and optical fibers, and/or two or more different types of optical fibers. A communication cable containing signal-bearing media and electric power conductors.

Hybrid CDPD Circuit-switched CDPD (Cellular Digital Packet Data), a technology known as hybrid CDPD, is the system architecture developed by the CDPD Forum for interconnecting circuit-switched data, including cellular and landline, with the CDPD network.

Hybrid Coil Transformer used in a balancing network to connect a 4-wire line to a 2-wire line.

Hybrid Communication Network A communication system that uses a combination of trunks, loops, or links, some of which are capable of transmitting (and receiving) only analog or quasi-analog signals and some of which are capable of transmitting (and receiving) only digital signals.

Hybrid Connector A connector containing both optical fiber and electrical conductors.

Hybrid Coupler In antenna work, a hybrid junction forming a directional coupler. The coupling factor is normally 3 dB.

Hybrid Disk 1. A CD-ROM term. Under the Orange Book standard for recordable CD, a hybrid disc is a recordable disc on which one or more sessions are already recorded, but the disc is not closed, leaving space open for future recording. However, in popular use the term "hybrid" often refers to a disc containing both DOS/Windows platform is seen as a ISO 9660 disc, while on a Mac it appears as an HFS disc.
2. A CD-ROM disc which works in both a Macintosh and an MS-DOS/Windows. Most these days do.

Hybrid Integrated Circuits The combination of thin-film or thick-film circuitry deposited on substrates with chip transistors, capacitors and other components. Thin-film construction is used for microwave integrated circuits (MICs).

Hybrid Junction A waveguide or transmission line arrangement having four ports that, when terminated in their characteristic impedance, have the property that energy entering any one port is transferred (usually equally) to two of the remaining three ports. Widely used as a mixing or dividing device.

Hybrid Key System Term used to describe a system which has attributes of both Key Telephone Systems and PBXs. The one distinguishing feature these days is that a hybrid key system can use normal single line phones in addition to the normal electronic key phones. A single line phone behind a hybrid works very much like a single line phone behind a PBX. The second distinguishing feature of a hybrid is that it's "non-squared." This means that not every trunk appears as a button on every phone in the system — as occurs on virtually every electronic key system manufactured today.

Hybrid Local Network An integrated local network consisting of more than one type of local network (e.g. LAN, HSLN, digital PBX).

Hybrid Mode A mode possessing components of both electrical and magnetic field vectors in the direction of propagation.

Hybrid Network 1. A communications network which has some links capable of sending and receiving only analog signals and other links capable of handling only digital signals. The current public switched telephone network is Hybrid. A Hybrid Network is also a network with a combination of dissimilar network services, such as frame relay, private lines and/or X.25.
2. An amalgam of public and private network transmission facilities.

Hybrid Set Two or more transformers interconnected to form a network having four pairs of accessible terminals to which may be connected four impedances so that the branches containing them may be made interchangeable.

Hydra A 25-pair cable that at one end has an Amphenol connector (typical of what 1A2 phone systems were connected with) and at the other has many individual 2, 4, 6 and 8 wire connectors, typically male RJ-11s. A hydra cable is named for a mythological multi-headed monster. It's more commonly called an octopus cable. The reason it's called an octopus is that it looks a bit like an octopus — one body and many arms.

Hydrogen Loss Increases in optical fiber attenuation that occur when hydrogen diffuses into the glass matrix and absorbs some light.

Hydrometer Instrument for determining the density of liquids. Formerly in wide use for testing radio storage "A" batteries.

Hygroscopic Capable of absorbing moisture from the air.

Hyperband A cellular wireless term meaning the ability of a radio to handle calls originating in different frequencies. On Feb. 10, 1997, Ericsson conducted the first public hyperband hand-off between the 1900 MHz and 850 MHz systems at the Universal Wireless Communications (UWC) conference in Orlando. The hyperband call, which used the TDMA IS-136 digital wireless technology, was connected and carried across the United States through AT&T Wireless Services' network in five cities. This demonstrates the ability for operators to establish networks in either band, and connect them in a multi-vendor environment, enabling nationwide and international roaming. Advanced digital wireless applications deploying public and private networks were demonstrated including: Location and Caller ID, Short Message, Message Waiting Indicator, Voice Mail, four digit extension dialing and intelligent roaming. These advanced capabilities were demonstrated using Ericsson's 1900/850 MHz dual-band phones. These phones are equipped with the new Enhanced Full Rate ACELP vocoder delivering enhanced digital voice quality.

Hypercard The first desktop program that allowed hypertext creation. It ran on the Mac. Not long after its introduction, Hypercard "stacks" became available, especially in the artistic and educational communities. Stacks are a collection of documents within one package through which the user can jump using hypertext links. Stacks were often made available free through online services.

Hyperchannel An SCSA term. A data path on the SCbus or SCxbus Data Bus made up of more than one time slot. By bundling time slots into a hyperchannel, data paths with a bandwidth greater than 64 Kbps can be created.

Hyperfiction According to the New York Times, hyperfiction is a new narrative art form, readable only on the computer, and made possible by the developing technology of hypertext and hypermedia. Not all adults have familiarized themselves with hypertext, but most children have, for it is the basis of many of their computer games and is fast becoming the dominant pedagogical tool of our digitalized times. See Hypertext.

Hyperlink A link from one part of a page on the Internet to another page, either on the same size or a distant site. For example, a restaurant's home page may have a hyperlink or link to its menu. A retailer or laptops might have a link to the site of the laptop's manufacturer. A hyperlink is a way to connect two Internet resources via a simple word or phrase on which a user can click to start the connection. A user can access a Web site and exercise the option to hyperlink to another, related Web site by clicking on that option. You'll recognize hyperlinks on Web pages because the links look different. Typically they're in blue and underlined. Sometimes, you'll see a button saying "For more," "Full specs," or "Our biography," etc. During the linking process, the user remains connected in a Web session through a process known as spoofing. A hyperlink is also called an anchor.

Hypermedia A way of delivering information that provides multiple connected pathways through a body of information. Hypermedia allows the user to jump easily from one topic to related or supplementary material found in various forms, such as text, graphics, audio or video.

Another definition we found is: Non-linear media, of which multimedia can be a form. Just as hypertext is a non-sequential, random-access arrangement of text, hypermedia is a non-sequential, random-access arrangement of multiple media such as video, sound and computer data.

A third definition: Hypermedia is a type of authoring and playback software through which you can access multiple layers of multimedia information related to a specific topic. The information can be in the form of text, graphics, images, audio, or video. For example, suppose you received a hypermedia document about the Sun file system. You could click on a hotspot (such as the words file system) and then read a description. You could then click an icon to see an illustration of a file structure, and then click the file icon to see and hear information in a video explaining the file system.

Hyperstream A name for frame relay service from Stentor and MCI.

Hypertext The term "hypertext" was coined by Ted Nelson in a paper delivered to the Association for Computing Machinery at its national conference in 1965. Nelson envisioned a nonsequential writing tool which included a feature he called "zippered lists," which allowed textual elements in documents to be linked. We currently associate "hypertext" with the World Wide Web and the HTML language. Imagine you're reading something. You come to a word that's in a different color or perhaps is underlined. You click on the word with your mouse. Suddenly you're transported to another sentence, to another paragraph, to another section somewhere else. That new sentence, paragraph or section may explain the original word. It may take you to another thought. It may take you to another part of the story. The New York Times defines Hypertext as "nonsequential writing made up of text blocks that can be linked by the readers in multiple ways." Hypertext is not only the words and the links. It's also the software that allows users to explore and create their own paths through written, visual, and audio information. Capabilities include being able to jump from topic to topic at any time and follow cross-references easily. Hypertext is often used for Help files. It's being used for "hyperfiction," a new narrative art form, readable only on the computer, made possible by the developing technology of hypertext and hypermedia. See HTTP and Hypertext Transfer Protocol.

Hypertext Transfer Protocol HTTP is the transport protocol in transmitting hypertext documents around the Internet. See Hypertext.

Hyperware New hardware that has been announced and perhaps even publicly demonstrated, but is not being shipped to commercial customers. Vaporware is software which has been announced, but is not yet shipping to commercial customers. Years can pass between public announcement and actual commercial shipment. Be wary.

Hysteresis In pure physics, hysteresis is the lag in response exhibited by a body in reacting to changes in the forces, esp. magnetic forces, affecting it. In our industry it has come to acquire a couple of meanings. See also hysteresis loss.
1. An uninterruptible power supply (UPS) definition. The voltage output from the wall will continually shift within a certain range, causing some UPSs to constantly switch back and forth from AC to battery power.
2. A buffering approach used in digital cellular networks to prevent the "ping pong" effect which occurs when a telephone repeatedly reselects two cell sites of approximately equal strength.

Hysteresis loss A physics term. The loss of energy by conversion to heat in a system exhibiting hysteresis.

Hytelnet A hypertext system which contains information about the Internet, such as accessible library catalogs, Freenets, Gophers, bulletin boards, etc.

HZ See Hertz.

I Used on switches to mean "ON." The "OFF" setting is "O," (the letter coming after N.)

I&M Abbreviation for Installation and Maintenance.

I&R Installation and Repair. The telephone company department responsible for these jobs. I&R refers to a person's job area, a department, or tools and test gear made for I&R

I-CF ISDN Call Forwarding.

I-CFDA ISDN Call Forwarding Don't Answer.

I-CFDAIO ISDN Call Forwarding Don't Answer Incoming Only.

I-CFIB ISDN Call Forwarding Interface Busy.

I-CFIBIO ISDN Call Forwarding Interface Busy Incoming Only.

I-CFIG ISDN Call Forwarding IntraGroup only.

I-CFIO ISDN Call Forwarding Incoming Only.

I-CFPF ISDN Call Forwarding over Private Facilities.

I-CFV ISDN Call Forwarding Variable.

I-CFVCG ISDN Call Forwarding Variable facilities for Customer Groups.

I-CNIS ISDN Calling Number Information Services.

I-EDI Internet-based Electronic Data Interchange.

I-ETS Interim European Telecommunications Standard.

I-HC ISDN Hold Capability.

I-Hold Indication A telephone system feature. If I put someone on hold at my phone, all the other phones which have the same line appearing on them will start flashing — indicating that the call is on hold.

I-MAC Isochronous Media Access Control. An FDDI-II term. See FDDI-II.

I-MUX Inverse Multiplexer. See INVERSE MULTIPLEXER.

I-PASS An alliance of ISPs (Internet Service Providers) and IAPs (Internet Access Providers) to provide roaming capabilities for travelers. Based on proprietary standards, roamers are authenticated before being afforded Internet access. Usage is cross-billed through the I-PASS clearinghouse, with fees being set by each ISP for use of its facilities by roamers. I-PASS includes over 100 member ISPs in approximately 150 countries, and includes over 1,000 POPs (Points of Presence). I-PASS competes with GRIC (Global Reach Internet Connection). The IETF's Roamops working group is developing a standard for roaming, as well. See also GRIC and ROAMOPS.

I-Series Recommendations ITU-T recommendations on standards for ISDN services, ISDN networks, user-network interfaces, and internetwork and maintenance principals.

I-TV Interactive TV.

I-Use Shows a user which line the phone is connected to when the receiver is off-hook. It does this by illuminating a small light below that line button. Most key sets and PBX sets have I-Use buttons. Most of the newer two-line phones do also.

I-Way An acronym for the Information SuperHighway, that nebulous concept which refers to interconnected telecommunications channels snaking their way into every household, every company, every college, every university in the world. Essentially, the Information SuperHighway is a fancy term for the nation's phone network overlaid with heavy data communications ability.

I.122 ITU-T description of the general "bearer" services offered by ISDN networks, including both packet-switched and frame relay data services.

I.356 ITU-T Specifications for Traffic Measurement.

I.361 B-ISDN ATM Layer Specification.

I.362 B-ISDN ATM Layer (AAL) Functional Description.

I.363 B-ISDN ATM Layer (AAL) Specification.

I.430 Basic rate physical layer interface defined for ISDN. The ITU-T Layer 1 specification for the ISDN BRI S/T-interface, which consists of four wires. I.430 specifies ASI line coding. The beginning and end of each frame is marked with deliberate bi-polar violations. Each BRI frame is forty-eight bits in length including the bi-polar violations; and repeated 4,000 times per second for a total line rate of 192 Kb/s in each direction. The point-to-point limit is one kilometer. the passive bus is limited to about 10 meters.

I.431 Primary rate physical layer interface defined for ISDN. The ITU-T Recommendation for Layer 1 of the ISDN PRI. Specifies operation on North American T-1 at 1.544 megabits per second (23B+D) or European E-1 at 2.048 megabits per second (30B+2D).

I.432 ITU-T Recommendation for B-ISDN User-network Interface.

I.440 The ITU-T specification, commonly known as Q.920, which describes the general network aspects of the LAPD protocol (also known as DSS1).

I.450 (Q.930) The ITU-T specification describing the general network aspects of the ISDN D channel Layer 3 protocol.

I.451 See Q.931

I.452 See Q.932

I.R.D.A. Infrared Data Association. See INFRARED.

I/G Bit Bit in IEEE 802 MAC address field distinguishing between individual and group addresses.

I/O Input/Output. See the following definitions.

I/O Bound When a computer systems spend much of its time waiting for peripherals like the hard disk or video display, it is said to be I/O bound. If your computer is I/O Bound, going to a faster CPU (like a 386 or 486) might make little perceived difference. What you need is a faster hard disk or faster video card, etc.

I/O Channel Equipment forming part of the input/output system of a computer.

I/O Controller Provides communications between the central processor and the I/O devices.

I/O Device An input/output device, which is a piece of hardware used for providing information to and receiving information from the computer, for example, a disk drive, which transfers information in one of two directions, depending on the situation. Some input devices, such as keyboards, can be used only for input; some output devices (such as a printer or a monitor) can be used for output. Most of these devices require installation of device drivers.

I/O Request Packet IRP. Data structures that drivers use to communicate with each other.

IX An IntereXchange carrier, i.e. a long distance phone company.

iXXX The little i stands for Intel and the numbers that follow refer to the particular microprocessor chip. For example, i386 refers to the 80386 chip.

I2 See Internet2.

i386SL One version of Intel's '386 family of microprocessors. The i386SL's special feature is that it can be slowed to 0 megahertz and still maintain register integrity (memory) practically indefinitely. This results in significant power savings for computers (especially laptops) that advantage of this feature.

i750 Name of the programmable video processor family from Intel.

IA5 International Alphabet No. 5

IAB Internet Architecture Board, formed in 1981 by the Defense Department's Advanced Research Projects Agency (DARPA). A policy setting and decision-review board for the TCP/IP-based Internet. The IAB supervises the Internet Engineering Task Force (IETF) and Internet Research Task Force (IRTF). The IAB included researchers such as Vint Cerf and Robert Kahn, who had created the TCP/IP protocol that became the universal language of the Internet. See IETF.

IAD 1. Integrated Access Device. A device which supports voice, data and video information streams over a single, high-capacity circuit.

2. Internet Addiction Disorder. I found this on the Internet (where else?): "A growing number of men are losing friends, family, and jobs, and sometimes all touch with reality through an addiction to the Internet, an Italian psychiatrist warned recently. Professor Tonino Cantelmi, University of Rome, told a conference he has studied 24 cases of certifiable "Internet Addiction Disorder" (IAD), a condition with symptoms of spending up to 10 hours online and a physical fallout of uncontrollable shaking hands and memory loss." Only 10 hours?

IAHC See International Ad Hoc Committee

IAL Intel Architecture Labs, home of the ISA Bus, Plug and Play, Universal Serial Bus and other PC advances designed to sell more Intel products.

IAM Initial Address Message. In SS7 networks, a message sent in the forward direction as part of the ISUP (ISDN User Part) call set-up protocol. The IAM is a mandatory message which initiates seizure of an outgoing circuit and which transmits address and other information relating to the routing and handling of a call. Included in the IAM is Calling Number Identification (CNI), also known as Calling Line Identification (CLI). CNI is the telephone number of the calling party, which is sent to the called party for identification purposes. Many carriers also support Caller Name, which transmits the name of the calling party along with the originating telephone number. It's interesting that part of the IAM is my identification, as in "I am. See also CNI, COMMON CHANNEL SIGNALING, ISDN, ISUP, and SS7.

IANA Internet Assigned Numbers Authority. IANA is responsible for assignment of unique Internet parameters (e.g., TCP port numbers, and ARP hardware types), and managing domain names. It also was responsible for administration and assignment of IP (Internet Protocol) numbers within the geographic areas of North America, South America, the Caribbean and sub-Saharan Africa; on December 22, 1997, that responsibility was shifted to ARIN (American Registry for Internet Numbers). For full details see Internet Assigned Numbers Authority. See also ARIN.

IAO IntrAOffice SONET Signal. Standard SONET signal used within an Operating Company central office, remote site, or similar location.

IAP Internet Access Provider. An IAP is provides companies and individuals with a link to the Internet. That link typically is a high-speed one —— T-1 or T-3. An IAP may also do Web hosting, which involves renting a customer a computer and communications links to the Internet and charging by the size of the computer and the use of the phone lines. An IAP is not the same as an Internet Service Provider, which also provides dial-in Internet service. But frankly, the distinction between ISPs and IAPs and exactly what they do is blurry. See ISP.

IASG Internetwork Address Sub-Group: A range of internetwork layer addresses summarized in an internetwork layer routing protocol.

IBDN The Integrated Building Distribution Network (IBDN) is an unshielded twisted pair/fiber optic based structured wiring system based on the EIA\TIA 568 wiring standard. IBDN is a creation of Northern Telecom. IBDN is an open wiring system meaning that it can support any standards based data or voice application available today on unshielded twisted pair horizontal wiring.

IBERPAK A ITU-T X.25 packet-switched network operated in Spain by the Spanish government.

IBM International Business Machines. Also known affectionately as I've Been Moved, International Big Mother, Itty Bitty Machines, It's Better Manually, along with others not suitable for printing (but used in the back of better computer rooms all across the North American continent). IBM is powerful in some parts of the computer industry. This is why many of the IBM terms in this book are described as how IBM defines them.

IBM 8514/A Graphics standard introduced by IBM with 1,024 x 768 resolution. Many current monitors are 8514/A-compatible. 8514/A is also called XGA, or eXtended Graphics Array, which is IBM's high-resolution extension to its VGA adapter. It provides a resolution of 1,024 horizontally x 768 vertically, yielding 786,432 possible bits of information on one screen, more than two and a half times what is possible with VGA.

IBM Cabling System IBM's specification for the kind of cable to be used in connection its products.

IBM PC Introduced in the summer of 1981 using the 16-bit Intel 8088 processor.

IBM Token Ring A local area network using star wiring architecture of two pair cabling to each location — one pair from the hub to the workstation and one pair from the workstation back to the hub to continue the ring. The IBM 8228 Multiple Access Unit (MAU) will support communications for eight PCs (workstations). Up to 33 IBM 8228 MAUs may be connected together into a single ring, supporting up to 260 data devices. MAU to MAU connection is accomplished with data connectors equipped with Type 1 cables from a MAU's RO (Ring Out) connection to the next MAU's RI (Ring In). The final MAU's RO connects back to the initial MAU's RI to complete the ring. See also TOKEN PASSING and TOKEN RING.

IBND Interim Billed Number Database

IBR Bellcore spec 54019, which covers specs on delivering fractional T-1.

IBS Intelligent Battery System. A conventional battery system interfaces to its host product and charger through a power and perhaps a simple sensor port. An intelligent battery system has state sequential intelligence, memory, and a data communications protocol to the conventional battery sensor package. Thee IBS additions allow sensor data, events and memory access to take place between the battery pack and the host device and charger.

IBX Integrated Business eXchange. Another name for a PBX.

This is the name InteCom uses for their PBX family. InteCom is now owned by Matra, a French company.

IC 1. Intercom.

2. Integrated Circuit.

3. Intermediate Cross-connect. An interconnect point within backbone wiring. for example, the interconnection between the main cross-connect and telecommunications closet or between the building entrance facility and the main cross-connect.

IC DRAM Integrated Circuit Dynamic Random Access Memory.

ICA International Communications Association. ICA is the biggest trade association of the largest corporate telecommunications users — the people whose companies spend the most. ICA was founded in 1949 as the National Committee of Communications Supervisors. On behalf of its members, the ICA works to influence the FCC, Congress, and other regulatory and law-making bodies on issues of national telecom and information distribution issues. www.icanet.com

iCal See iCalendar.

iCalendar iCal. Internet Calendaring and Scheduling Core Object Specification is a specification from the Internet Engineering Task Force designed to allow people to share and coordinate their appointment calendar over the Internet. At the heart of iCalendar is the Time Zone Calendar Component, a protocol that accounts for different time zones. As a result, users of a typical day planner can schedule appointments with users in other time zones and with other calendaring programs. The iCalendar spec is the foundation for four new specifications. iTIP is the iCalendar Transport Independent Interoperability Protocol, which details how calendaring systems use iCalendar objects to interoperate and defines a message protocol for finding free time or searching to-do lists. iMIP is the iCalendar Message Based Interoperability Protocol which addresses defines interoperability among calendaring systems piggybacked on Internet email using MIME (Multipurpose Internet Mail Extension). iRIP is the iCalendar Real Time Interoperability Protocol that addresses the how diverse scheduling systems query each other in real time. Client Access Protocol (CAP) allows any calendaring client to access information from heterogeneous back-end systems. iCalendar is replacing an older specification called vCalendar. See vCalendar. www.ietf.org

ICANN The Internet Corporation for Assigned Names and Numbers, a new non-profit, international organization based in California. ICANN is slated to have broad authority to reform the present system of issuing Internet addresses, including adding new top-level domains, like shop, as in www.mycompany.shop.

ICAPI International Call Control API.

ICB Individual Case Basis Nonstandard situations where special arrangements are required to satisfy unusual requirements. General tariffs do not apply.

ICD International Code Designator. A 2-byte field of the 20-byte NSAP (Network Service Access Point) address, the ICD is used to identify an international organization. NSAP and ICD are used in a variety of networks, including ATM. The British Standards Institution is the registration authority for the International Code Designator. See also BSI.

ICE Information Content and Exchange. An emerging protocol in the form of a XML (eXtensible Markup Language) application, ICE is designed to facilitate the automation of content syndication on the World Wide Web. The ICE architecture is intended to define business rules that would support data

(e.g., user profile) exchange among partner sites. Such data can then be processed, loaded into a repository and resold under the user interface of the licensee. For example, ICE would allow a review of a book, movie or theater production to be licensed to multiple Web site hosts. A kill date would automatically kill the content across all sites on a specified date.

ICEA Insulated Cable Engineers Association.

ICFA International Computer Facsimile Association. The mission of this new organization is to create awareness of the benefits and uses of computer fax to increase worldwide market size.

ICI 1. Interexchange Carrier Interface. The interface between carrier networks that support SMDS.

ICIT International Center for Information Technologies. A part of MCI.

ICM See INTEGRATED CALL MANAGEMENT.

ICMP Internet Control Message Protocol. As specified in RFC-792, ICMP provides a number of diagnostic functions and can send error packets to hosts. ICMP uses the basic support of IP and is an integral part of IP.

iCOMP Intel Comparative Microprocessor Performance Index. A test for measuring how fast microprocessors are.

Icon An icon is a picture or symbol representing an object, task, command or choice you can select from a piece of software (e.g. a trash can for a deletion command).

Iconography The science of icons. A fancy name for talking about icons — those little visual representations of objects, tasks or commands you find in Windows and other GUI (Graphical User Interface) programs and operating systems.

ICP 1. Intelligent Call Processing. The ability of the latest ACDs to intelligently route calls based on information provided by the caller, a database on callers and system parameters within the ACD such as volumes within agent groups and number of agents available.

2. Instituto das Communicacoes de Portugal (The Portuguese Institute of Communications).

3. Intelligent Call Processor. The name of an AT&T service. ICP allows users to directly link their customer premise equipment to its network for individual call processing based on customer-specific information.

4. Independent Communications Provider. An ICP is a switchless CLEC.

ICR 1. Initial Cell Rate: An ABR service parameter, in cells/sec, that is the rate at which a source should send initially and after an idle period.

2. Internet Call Routing node. A Bellcore proposed device that would communicate with both the voice and data networks through the Signaling System 7. After a local telephone company's central office detects an incoming data call, the ICR would instruct the voice switch to reroute the connection to a data network through remote access gear, which would then send the data to the Internet.

ICS 1. See INTERACTIVE CALL SETUP.

2. Integrated Communications System. A Northern Telecom definition: A telecommunications based platform with advanced processing power and capacity that enables integration and orchestration of typical business equipment (telephones, fax, etc.) through open architecture interfaces, as in the Norstar-PLUS Modular ICS.

ICSA International Computer Security Association. Formerly the National Computer Security Association, founded in 1989. An independent organization which strives to improve security and confidence in global computing through awareness and certification of products, systems and people. www.ncsa.com

ICT The Information and Communications Technologies Standards Board.

ICTA The International Computer-Telephony Association (ICTA), based at Campus Box 350, University of Colorado, Boulder, CO 80302.

ICV Integrity Check Value is a digest of a message which provides a high level of assurance that the message has not been tampered with. Also referred to as Message Authentication Code.

ID Identifier.

IDA Integrated Data Access or Integrated Digital Access.

IDC Insulation Displacement Connection. A type of wire connection device in which the wire is "punched down" into a double metal holder and as it is the metal holders strip the insulation away from the wire, thus causing the electrical connection to be made. The alternate method of connecting wires is with a screw-down post. There are advantages and disadvantages to both systems. The IDC system, obviously, is faster and uses less space. But it requires a special tool. The screw system takes more time, but may produce a longer-lasting and stronger, more thorough (more of the wire exposed) electrical connection. The most common IDC wiring scheme is the 66-block, originally invented by Western Electric. See Punchdown Tool.

IDC Clip IDC Clips are a method of jack termination. They look like a modular jack with a mini 66-block attached to the back. They are usually more expensive than 110-type blocks but are easier to install. See IDC and Punch-Down Tool.

IDCMA Independent Data Communications Manufacturers Association, a lobbying and education group based near Washington, DC.

IDDD International Direct Distance Dialing. The capability to directly dial telephones in foreign countries from your own home or office telephone.

IDDS Installable Device Driver Server. A Dialogic term.

IDE 1. Integrated Drive Electronics. IDE is a hard disk drive standard interface for PCs. It appeared in 1989 as a low-cost answer to two other standard hard disk interfaces, ESDI and SCSI. The distinguishing feature of the IDE interface is that it incorporates the drive controller functions right on the drive. Instead of connecting to a controller card, an IDE drive attaches directly to the motherboard with a 40-pin connector. IDE drives offer a data transfer rate of three megabytes per second, which is not very fast. Several methods of data encryption can be used with the IDE interface, including MFM and RLL. Many laptops use IDE drives. IDE has a limit of 528 megabytes. Enhanced IDE drives, which appeared around 1994 to solve the problem that computers had gotten much faster and IDE wasn't keeping up, have a data transfer rate of between 11 and 13 megabytes per second and can handle drives of up 8.4 gigabyte. See ENHANCED IDE.
2. Integrated Development Environment. A term for products such as Microsoft's Visual C++ and Borland's Delphi that combine a program editor, a compiler, a debugger, and other software development tools into one integrated software package. The first of the IDEs, Borland's Turbo Pascal changed the way programmers write code by allowing programs to be edited and compiled within the same application.

IDEA 1. International Data Encryptions Algorithm. A secret key encryption algorithm developed by Dr. X. Lai and Professor J. Massey in Switzerland to replace DES.
2. Internet Development & Exchange Association, formed in

1995 and developed as part of a capstone MBA strategy project at West Virginia University to address the ever increasing competitive nature of the ISP market. Today, IDEA claims to be the largest trade association of independent Internet Service Providers (ISPs) in the world. www.auidea.org

iDEN Integrated Dispatch Enhanced Network. A wireless technology developed by Motorola, iDEN operates in the 800 MHz, 900 MHz and 1.5 GHz radio bands; the 900 MHz development is aimed at operators of digital Commercial Mobile Radio Service (CMRS), also known as ESMR (Enhanced Specialized Mobile Radio). iDEN is a digital technology using M16QAM (Quadrature Amplitude Modulation) for compression, and TDMA (Time Division Multiple Access). Through a single proprietary handset, iDEN supports voice in the form of both dispatch radio and PSTN interconnection, numeric paging, SMS (Short Message Service) for text, data, and fax transmission. See also ESMR, QAM, SMS and TDMA.

Identification Failure Automatic Number Identification (ANI) equipment in the originating office failed to identify the calling number. See ANI.

Identified Outward Dialing Same as AIOD. It's a PBX feature which provides identification of the PBX extension making the outward toll calls. This identification may be provided by automatic equipment or by attendant identification of the extension.

Identifier The name of a database object (table, view, index, procedure, trigger, column, default, or rule). An identifier can be from 1 to 30 characters long.

IDF Intermediate Distribution Frame.

IDLC See Integrated Digital Loop Carrier.

Idle 1. Not being used but ready.
2. An SCSA term. A state of the SCbus or SCxbus Message Bus where no information is being transmitted and the bus line is pulled high.

Idle Cell An ATM cell used for cell stuffing where rate adaption is required. As Physical Layer cells, idle cells are required and cannot, therefore, be replaced by assigned cells during the process of cell multiplexing; this is unlike Unassigned Cells, which are not necessary at a network level and which can be replaced, therefore. See also ATM Reference Model, Physical Layer, Rate Adaption and Unassigned Cell.

Idle Channel Code A repetitive pattern (code) that identifies an idle channel.

Idle Channel Noise Noise which exists in a communications channel when no signals are present.

Idle Line Termination An electronic network which is switch controlled to maintain a desired impedance at a trunk or line terminal when that terminal is in an idle state.

Idling Signal Any signal that indicates no data is being sent.

IDN Integrated Digital Network.

IDP InterDigital Pause.

IDSL A developing xDSL technology which uses ISDN technology to deliver transmission speeds of 128 Kbps on copper loops as long as 18,000 feet. IDSL is a dedicated service for data communications applications, only; whereas ISDN is a circuit-switched service technology for voice, data, video and multimedia applications. IDSL terminates at the user premise on a standard ISDN TA (Terminal Adapter). At the LEC CO, the loop terminates in collocated ISP electronics in the form of either an IDSL access switch or a IDSL modem bank connected to a router. The connection is then made to the ISP POP via a high-bandwidth dedicated circuit. See also xDSL, ADSL, ISDN, HDSL, RADSL, SDSL and VDSL.

IDSU Intelligent Data Service Unit from ADC Kentrox.

IDT Inter-DXC Trunk

IDTV Improved Definition TeleVision. See IMPROVED DEFINITION TELEVISION.

IDU Interface Data Unit: The unit of information transferred to/from the upper layer in a single interaction across the SAP. Each IDU contains interface control information and may also contain the whole or part of the SDU.

IEC 1. InterExchange Carrier. Also called an IXC (as in IntereXchange Carrier). In practice, an IEC or IXC is any common carrier authorized by the FCC to carry customer transmissions between LATAs. In practice this means anyone and his brother who print up stationery, rent a few lines and proclaim themselves to be in the long distance phone business. Except for AT&T, regulation of long distance carriers by the FCC is perfunctory. It is less perfunctory by the local state authorities, some of whom still think competition in telecommunications is a mild form of insanity.
2. International Electrotechnical Commission. The international standards and conformity assessment body for all fields of electrotechnology, including electricity and electronics. The IEC publishes a number of international standards and technical reports on a wide variety of subjects including telecommunications (LANs, MANs and WANs), video cameras, electrical cables, communications protocols (e.g., HDLC), Open Systems Interconnection (OSI), optical fiber cables and connectors, and diagnostic X-ray imaging equipment. www.iec.ch

IEEE Institute of Electrical and Electronics Engineers, Inc. IEEE, founded in 1884, says it's the world's largest technical professional society, consisting of over 320,000 members in 147 countries. The IEEE's technical objectives "focus on advancing the theory and practice of electrical, electronics and computer engineering and computer science." The IEEE sponsors technical symposia, conferences and local meetings, and publishes technical papers. It also is a significant standards-making body responsible for many telecom and computing standards, including those used in LANs — e.g. the 802 series. www.ieee.org.

IEEE 1394 Also called Firewire. An IEEE data transport bus that supports up to 63 nodes per bus, and up to 1023 buses. The bus can be tree, daisy chained or any combination. It supports both asynchronous and isochronous data. 1394 is a complimentary technology with higher bandwidth (and associated cost) than Universal Serial Bus. Intel told me in summer of 1996 that it was supporting USB for most devices that attach to PC up through audio and video conferencing. Intel told that they are "supporting IEEE 1394 as the preferred interface for higher bandwidth applications such as high quality digital video editing, and connection to new digital consumer electronics equipment. We expect that 1394 will show up in low volumes in 97, and ramp into high volumes in 99." See USB.

IEEE 488 IEEE 488 is the most widely-used international standard for computer-to-electronic instrument communication. It is also known as GPIB and HPIB.

IEEE 802 The main IEEE standard for local area networking (LAN) and metropolitan area networking (MAN), including an overview of networking architecture. It was approved in 1990. See IEEE 802 standards.

IEEE 802.1 This IEEE committee defines the LAN Management and bridging standards.

IEEE 802.1d An algorithm, the original version of which was invented by Digital Equipment Corporation, that is used to prevent bridging loops by creating a spanning tree. The algorithm is now documented in the IEEE 802.1d specification, although the Digital algorithm and the IEEE 802.2 handles errors, framing, flow control, and the Layer 3 service interface.

IEEE 802.2 A data link layer standard used with the IEEE 802.3, 802.4 and 802.5 standards. 802.2 is the more modern form of 802.3. Novell recommends that it be used on its NetWare networks in preference to the 802.3 (which will still work). Most Ethernet networks support both 802.2 and 802.3. For more on the 802 series, see the numbers definitions at the front of this dictionary.

IEEE 802.3 A Local Area Network protocol suite commonly known as Ethernet. Ethernet has either a 10 Mbps or 100 Mbps throughput and uses Carrier Sense Multiple Access bus with Collision Detection CSMA/CD. This method allows users to share the network cable. However, only one station can use the cable at a time. A variety of physical medium dependent protocols are supported. This is the most common local area network specification. See 802.2.

IEEE 802.3 10 BASE-T This is the standard for Ethernet over twisted pair cabling using home runs. See 802.3

IEEE 802.3 10Broad36 This IEEE standard describes a long-distance type of Ethernet cabling with a 10-megabit-per-second signaling rate, a broadband signaling technique, and a maximum cable-segment distance of 3,600 meters.

IEEE 802.4 A physical layer standard specifying a LAN with a token-passing access method on a bus topology. Used with Manufacturing Automation Protocol (MAP) LANs. Arcnet can work this way. Typical transmission speed is 10 megabits per second.

IEEE 802.5 A physical layer standard specifying a LAN with a token-passing access method on a ring topology. Used by IBM's Token Ring hardware. Typical transmission speed is 4 or 16 megabits per second. Typical topology is star.

IEEE 802.6 This IEEE standard for metropolitan area networks (MANs) describes what is called a Distributed Queue Dual Bus (DQDB). The DQDB topology includes two parallel runs of cable — typically fiber-optic cable — linking each node (typically a router for a LAN segment) using signaling rates in the range of 100 megabits per second.

IEEE 802.9 This IEEE committee deals with integrated voice and data LANs, i.e. Isochronous Ethernet.

IEEE 802.10 This committee deals with LAN security.

IEN Internet Experimental Note. A standards document similar to an RFC, and is available from the Network Information Center (NIC). IENs contain suggestions and proposals for Internet implementations or specifications.

IES 1. Inter-Enterprise Systems is EDI (Electronic Data Interchange) and inter-company electronic mail, fax, electronic funds transfer, videotex/online databases and the exchange of CAD/CAM graphics. See also EDI. 2. Information Exchange Services.

IETF Internet Engineering Task Force. Formed in 1986, the IETF is one of two technical working bodies of the Internet Activities Board. The little-known all volunteer group, the IETF, meets three times a year to set the technical standards that run the Internet. The IETF is the primary working body developing new TCP/IP (Transmission Control Protocol/Internet Protocol) standards for the Internet. The IETF was formed in 1986 when the Internet was evolving from a Defense Department experiment into an academic network. Of late, the IETF has been the forum where engineers and programmers have cooperated to solve the succession of crises caused by the Internet's phenomenal growth. The IETF's home page is: www.letf.cnrl.reston.va.us/home.html

IF Intermediate frequency.

IFG The minimum idle time between the end of one frame transmission and the beginning of another. On Ethernet 802.3 LANs the minimum interframe gap is 9.6 micro-seconds.

IFIP International Federation for Information Processing. A research organization that performs substantive pre-standardization work for OSI. IFIP is noted for having formalized the original Message Handling System (MHS) model. www.dit.upm.es/~cdk/Iflp.html

IFRB International Frequency Registration Board.

IG 1. AT&T's ISDN Gateway. A set of specs for hooking up an outside computer to an AT&T switch. Under IG, information travels in one direction — from the switch to the host. See also ITG (which is two-directional), ASAI and OPEN APPLICATION INTERFACE. 2. Isolated Ground. In AC electricity, an isolated ground is a type of outlet characterized by the following features and uses:

• It may be orange and must have a Greek "delta" on the front of the outlet. (A delta looks like a triangle.)

• It must be grounded by an insulated green wire.

• It must have insulation between the ground terminal and the mounting bracket.

• It is used primarily to power electronic equipment because it reduces the incidence of electrical "noise" on the ground path.

IGMP Internet Group Management Protocol. A protocol used by IP hosts to report their multicast group memberships to an adjacent multicast router.

Ignition Key A rod arranged to strike the arc in an arc generator of high frequency currents.

IGP Interior Gateway Protocol. The protocol used to exchange routing information between collaborating routers in the Internet. RIP and OSPF are examples of IGPs.

IGRP Interior Gateway Routing Protocol. A distance-vector routing protocol developed by Cisco Systems for use in large, heterogeneous networks.

IGT Ispettorato Generale delle Telcomunicazioni (General Inspectorate of Telecommunications, Italy).

IILC Information Industry Liaison Committee (part of ATIS). See EXCHANGE CARRIERS STANDARDS ASSOCIATION.

IIR Interactive Information Response.

IIS Microsoft Windows NT's Internet Information Server, which is similar to Netscape's Webserver. IIS lets you set up a web site and control and manage it remotely though the Internet, assuming you have a necessary privileges and a Web browser.

IISP 1. The Information Infrastructure Standards Panel formed by ANSI in July 1994 to accelerate the development and acceptance of standards critical to the establishment and deployment of the information superhighway. See ANSI. 2. Interim Interface Signaling Protocol. A call routing scheme used in STM networks. Formerly known as PNNI Phase 0. IISP is an interim technology meant to be used pending completion of PNNI Phase 1. IISP uses static routing tables established by the network administrator to route connections around link failures.

IITF Information Infrastructure Task Force. This task force of high level representatives from federal agencies was formed by the Clinton Administration to identify and address the issues of creating a National Information Infrastructure. The Task Force relies on the members of the Industry Advisory Council as it assesses the requirements of individuals and businesses that will shape future networks. See also NATION-AL INFORMATION INFRASTRUCTURE.

IIW The IPv6 Implementors Workshop, a group within the North American IPv6 Forum.

iKP Internet Keyed Payments Protocol. An architecture for secure payments over the Internet in the general context of Electronic Commerce. iKP is a public-key cryptography which defines transactions of a credit card nature, where a buyer and seller interact with a third party, such as a credit card company, in order to authorize transactions on a secure basis. SEPP (Secure Electronic Payment Protocol) is a standard implementation of iKP. iKP specifies RSA as the public-key encryption and signature algorithm. See also Electronic Commerce, RSA and SEPP.

ILAN A protocol independent router for token ring and Ethernet networks from CrossComm Corporation, Marlboro, MA.

ILEC Incumbent Local Exchange Carrier. Not to be confused with "Independent Telco," a traditional, old, incumbent LEC which was never part of the Bell System. See CLEC and ELEC.

ILLP Inter Link-to-Link Protocol.

ILMI Interim Link Management Interface: An ATM Forum-defined interim specification for network management functions between an end user and a public or private network and between a public network and a private network. ILMI is based on a limited subset of SNMP capabilities.

ILS 1. Input buffer Limiting Scheme. A flow control scheme used in data communications that blocks overload by limiting the number of blocks arriving at a buffer.

2. Internet Locator Service. In a September 1997 White Paper on "IP Telephony," Microsoft wrote, "User-to-IP mapping information is stored and continually refreshed using the Internet Locator Service (ILS) Dynamic Directory, a real-time server component of the Active Directory, which is part of Windows NT."

ILSR IPX Link State Router. Novell's improvement on its RIP distance vector-based routing protocol.

IMA 1. Interactive Multimedia Association. Formed in 1991 (rooted in IVIA, Interactive Video Industry Association), an industry association chartered with creating and maintaining standard specifications for multimedia systems. www.ima.org 2. Inverse Multiplexing over ATM. You have a high speed data stream. But not a high speed tranmssion link. You have several low speed links. Inverse multiplexing lets you join several slow speed links together and pretend that they're one high speed link. Here's a formal definition. IMA is an access specification approved in 1997 by the ATM Forum. This User Network Interface (UNI) standard allows a single ATM cell stream to be split across multiple access circuits from the user site to the edge of the carrier's ATM network. In an ATM LAN application, for instance, the ATM switch deployed in the enterprise backbone typically operates at 155 million bits per second or 622 Mbps. In this example, ATM traffic from the enterprise to the public ATM carrier-based network, requires 6 Mbps — well more than the 1.544 million bits per second provided by a T-1, but less than a full T-3 (which is 45 million bits per second). Rather than subscribing to a T-3, which requires a fiber optic access circuit and is very expensive, IMA is used. Thereby, the ATM datastream is split across four T-1 circuits by an access concentrator which possesses IMA capability. The IMA process works in a round-robin fashion, with cell number 1 traveling over T-1 number 1, cell #2 traveling over T-1 #2, and so on. Each of the four T-1 circuits is relatively inexpensive, can be provisioned over twisted-pair, and is readily available. At the edge of the carrier network, the ATM switch receives each of the four separate datastreams, and reverses the IMA process to put the original datastream back together, which is then switched and transported through the network to the far edge. At that far edge, the IMA process may take place again, from the edge of the carrier network, over four T-1s, and to the IMA-capable ATM concentrator on the user premises. IMA is specified in the ATM forum specification AF-PHY-0086.000 Inverse Multiplexing for ATM Specification Version 1.0 and dated July 1997. Go to ftp://ftp.atmforum.com/pub/approved-specs/af-phy-0086.000.pdf to read and download a copy of the 140-page document.

The definition of IMA is not easy. One reader, Rosario Brinquis of Madrid Spain, has contributed the following definition, which he thinks really captivates the essence of IMA: A methodology is described which provides a modular bandwidth for user access to ATM networks and for connection between ATM network elements at rates between the traditional order multiplex levels, for instance, between the DS1/E1 and DS3/E3 levels in the asynchronous digital hierarchies. DS3/E3 links are not necessarily readily available throughout a given network and therefore the introduction of ATM Inverse Multiplexers provides an effective method of combining the transport bandwidths of multiple links (e.g., DS1/E1 links) grouped to collectively provide higher intermediate rates. See also ATM, Inverse Multiplexer and TDM.

Image Antenna A hypothetical, mirror-image antenna considered to be located as far below ground as the actual antenna is above ground.

Image map A Web term. An image containing one or more invisible regions, called hotspots, which are assigned hyperlinks. Typically, an image map gives users visual cues about the information made available by clicking on each part of the image. For example, a geographical map could be made into an image map by assigning hotspots to each region of interest on the map.

Image Resolution The fineness or coarseness of an image as it was digitized, measured in Dots Per Inch (DPI), typically from 200 to 400 DPI.

IMAP Internet Messaging Access Protocol (1993), originally Interactive Mail Access Protocol (1986). A next-generation e-mail protocol which is likely to replace POP (Post Office Protocol) for Internet mail servers. IMAP allows users to create and manage mail folders over the WAN, as well as to scan message headers and then download only selected messages. IMAP4, the current version, is specified in RFC 1730. See also MIME and POP.

IMAP4 Internet Messaging Access Protocol 4, an emerging Internet email standard that vendors promise will make electronic messaging management easier and safer. The original version of IMAP4 was originally written in 1987. See IMAP

IMAS Intelligent Maintenance Administration System. Northern Telecom software which is a menu-driven PC-based program that provides enhanced maintenance and administrative capabilities for DMS-10 central offices.

iMIP iCalendar Message based Interoperability Protocol. See iCalendar.

IMC See INTERNET MAIL CONSORTIUM.

IMEI International Mobile station Equipment Identity. A wireless telecommunications term. An equipment identification number, similar to a serial number, used to identify a mobile station.

IMHO Abbreviation for "In My Humble Opinion;" commonly used on E-mail, on the Internet and BBSs (Bulletin Board Systems).

Immediate Ringing A PBX feature which makes the called telephone begin ringing the instant the phone has been dialed. Normally there's a small wait between dialing the number and having the phone ring.

Immunity From Suit A term I first saw in licensing agreements with Microsoft. The provision says that the company signing the agreement with Microsoft agrees not to sue Microsoft or Microsoft's customers and OEMs for infringement of said company's own patents. Some observers are claiming that signing an agreement with this provision would give Microsoft a royalty-free license to an outside company's patents.

IMNSHO In My Not So Humble Opinion. An acronym used in electronic mail on the Internet to save words or to be hip, or whatever. See IMHO.

IMO Abbreviation for "In My Opinion;" commonly used on E-mail and BBSs (Bulletin Board Systems). See IMHO.

IMPACS An MCI International packets switching service that is useful to firms with overseas remote computing needs, and to scientific, educational or commercial organizations that need periodic access to U.S. database facilities. IMPACS also provides overseas users with communications links to their own computers in the USA for applications such as order entry, inventory control, billing, payroll, and sales statistics.

Impact Strength A test designed to ascertain the abuse a cable configuration can absorb, without physical or electrical breakdown. Done by impacting with a given weight, dropped from a given height, in a controlled environment.

Impact Tool Also called a "punch down" tool. See PUNCH DOWN TOOL.

Impaired When an individual circuit exceeds the transmission limits or its signaling functions (e.g., seizure, disconnect, ANI) are experiencing failures.

IMPDU Initial MAC Protocol Data Unit. A Connectionless Broadband Data Service (CBDS) term that corresponds to the L3 PDU in Switched Multimegabit Data Service (SMDS). CBDS is the European equivalent of SMDS.

Impedance The total opposition (i.e. resistance and reactance) a circuit offers to the flow of alternating current. It is measured in ohms and the lower the ohmic value, the better the quality of the conductor. Low impedance will help provide safety and fire protection and a reduction in the severity of common and normal mode electrical noise and transient voltages. For telecommunications, impedance varies at different frequencies. Ohm's law says that voltage equals the product of current and impedance at any single frequency. The unit of impedance is the ohm.

Impedance Matching The connection of additional impedance to existing impedance one in order to improve the performance of an electrical circuit. Impedance Matching is done to minimize distortion, especially to data circuits.

Implementors' Agreement An agreement about the specifics of implementing as a standard, reached by vendors who are developing products for the standard. Compare with De Facto Standard and De Jure.

Implied Acknowledgment Implied acknowledgment is a process whereby negative acknowledgment of a specific packet of information implies that all previously transmitted packets have been received correctly. See also PIPELINING.

Import Imagine you have a software program, like a spreadsheet or a database. And you have information in that program. Let's say it's Microsoft Word or Lotus 123. And you want to get it into a different program, say to give it to a workmate who uses WordPerfect or Excel. You have to convert it from one format to another. From Word to WordPerfect or from Lotus to Excel. That process is typically called "exporting" and the process of your workmate getting it into his computer is called "importing." And you'll typically see the words "EXPORT" and "IMPORT" as choices on one of your menus.

Import Computers A Windows NT. In directory replication, the servers or workstations that receive copies of the master set of directories from an export server.

Import Script First read my definition of IMPORT. An import script is a series of specifications which control the merging processes. It contains a series of merge rules which specify how the fields are to be merged and a record precedence rule which governs which records to merge of the ones received.

Important Call Waiting Notifies you with a special ring that someone you want to hear from is calling you.

Improved Definition Television IDTV. Television that includes improvements to the standard NTSC television system, which improvements remain within the general parameters of NTSC television emission standards. These improvements may be made at the transmitter and/or receiver and may include enhancements in parameters such as encoding, digital filtering, scan interpolation, interlaced scan lines, and ghost cancellation. Such improvements must permit the signal to be transmitted and received in the historical 4:3 aspect ratio.

Improved Mobile Telephone Service IMTS. In the beginning, there was dispatch mobile service. The base operator broadcast a message to you. Everyone could hear it. You responded. Then they had mobile telephone service. You picked up the phone in your car, the operator responded. You asked for the number you wanted and she/he dialed it and connected you. You had the channel to yourself but others could still tune in. Then came Improved Mobile Telephone Service (IMTS). Now you could dial from your car without using an operator with some assurance of privacy. The latest development is cellular mobile telephone service. See CELLULAR.

Impulse A surge of electrical energy usually of short duration, of a non repetitive nature.

Impulse Hits Errors in telephone line data transmission are caused by voltage surges lasting from 1/3 to 4 milliseconds and at a level within 6 dB of the normal signal level (Bell standard allows no more than 15 impulse hits per 15 minute period).

Impulse Noise High level, short duration noise that comes on a circuit. You can get impulse noise from electromechanical relays. These noise "spikes" have little effect on voice transmission but can be devastating to data. You can get a piece of test equipment called an impulse noise measuring set. Such a machine establishes a threshold and counts the number of impulses (hits) above that threshold.

Impurity Level An energy level outside the normal energy band of the material, caused by he presence of impurity atoms. Such levels are capable of making an insulator semiconductor.

IMS/VS Information Management System/Virtual Storage. An IBM host operating environment.

IMSI International Mobile Subscriber Identity. An ITU-T specification used to uniquely identify a subscriber to mobile telephone service. It is used internally to a GSM (Global System for Mobile Communications) network, and has been adopted for future use in all cellular networks. The IMSI is a 50-bit field which identifies the phone's home country and carrier.

IMT 1. InterMachine Trunk. A circuit which connects two automatic switching centers, both owned by the same company.
2. International Mobile Telecommunications

IMT-2000 International Mobile Telecommunications for the year 2000. IMT-2000 is an ITU-T initiative for a process to develop a 21st century architecture to meet the needs of emerging nations. A G3 (Generation 3) wireless concept, specifications include 128 Kbps for high mobility and ISDN applications, 384 Kbps for pedestrian speed and full motion compressed video, and 2 Mbps for fixed E-1/T-1 access and wireless LANs.

IMTC International Multimedia Teleconferencing Consortium. A non-profit corporation with the mission of promoting, encouraging and facilitating the development and implementation of interoperable multimedia teleconferencing solutions based on open international standards. Emphasis is on ITU-T standards such as T.120, H.320, H.323, and H.324. IMTC sponsors and conducts interoperability test sessions between suppliers of conferencing products and services based on those standards. It also focuses on market education. IMTC comprises over 140 members, including 3Com, Alcatel, BellSouth, Cisco, Compaq, Dialogic, IBM and Motorola — manufacturers, carriers, end users and others committed to open standards are welcome. www.imtc.org

IMTS See IMPROVED MOBILE TELEPHONE SERVICE.

IMUIMG ISDN Memorandum of Understanding Implementation Management Group. Formed in 1992, the IMUIMG is intended to ease ISDN implementation in Europe. The organization's stated goal is to ensure consistency when ordering or using ISDN services, regardless of provider or country. Carriers in the U.S., Canada and the Asia-Pacific have been invited to join.

IMUX Inverse Multiplexer

IN Intelligent Network. Ericsson has done focus groups on Mobile Intelligent Network Services. Among the new IN (Intelligent Network) services, Ericsson identified: * Enhanced number translation services functions * Enhanced screening services, i.e. selective call diversion * Selective forwarding of calls * Location-dependent call forwarding * Improvements to voice announcements * Services to support fixed and mobile integration, i.e. personal communications services, PCS and universal personal telecommunications, UPT, and * Enhanced billing. See AIN.

In The Clear A cellular term referring to the fact that certain signaling and control information is transmitted between a cell phone and a cell site in an insecure manner. A cellular phone is equipped with 2 identification numbers, a MIN and an ESN. The Mobile Identification Number (MIN) is a changeable number assigned to the terminal by the retailer activating the service. The Electronic Serial Number (ESN) is hard-coded into each terminal at the time of manufacture. In combination, the MIN and ESN are intended to identify both the terminal and subscriber to the network for purposes of authentication and billing. Not only are the MIN and ESN transmitted as the user seeks access to a channel for purposes of initiating a call, they also are transmitted frequently by the terminal to the cell sites in order that the cellular network can keep track of the terminal for purposes of terminating incoming calls. Further, and especially in the case of analog networks, those numbers are transmitted "in the clear"; in other words, they are not encrypted. Even most digital cellular standards do not provide for encryption of such numbers. As a result, it is relatively easy for criminals to gain access to the numbers through the use of a low-cost radio scanner. Standing on a freeway overpass, for instance, it is a simple matter for the criminal to capture a number of MIN/ESN numbers and to clone them into a number of other terminals. In fact, multiple terminals may be cloned with the same MIN/ESN numbers in different cities across the nation, with the information being posted to BBSs accessible through the Internet. This definition courtesy of an article on Voice Network Fraud that Ray Horak wrote for Datapro Information Services. See also Clone.

In-band Control Control information that is provided in the same channel as data.

In-band Signaling Signaling made up of tones which pass within the voice frequency band and are carried along the same circuit as the talk path that is being established by the signals. Virtually all signaling — request for service, dialing, disconnect, etc. — is in-band signaling. Most of that signaling is MF — multi-frequency dialing. The more modern form of signaling is out-of-band. Several local and long distance companies provide ANI (Automatic Number Identification) via in-band signaling. Some long distance companies provide it out-of-band, using the D-channel in a PRI ISDN loop. In cellular networks, In-band Signaling is known as CAS (Callpath Associated Signaling). See also CAS, ISDN, Out of Band Signaling and SS7 (ITU Signaling System Number 7).

In-Collect A CLEC term for the process of collecting long distance calling records from IXCs for purposes of subscriber billing. See CLEC.

In-line Device Hardware that is physically attached between two communications lines.

In-Safe An inbound store-and-forward MCI International Telex service that automatically answers a subscriber's incoming calls, provides an answerback, and accepts the messages.

INA Information Networking Architecture. Bellcore developed INA to facilitate the interoperation of proprietary software components through open interfaces based on voluntary international standards. INAsoft is the set of guidelines used to design interoperable, vendor-independent solutions, allowing rapid and successful product development.

Inactivity Time-outs Dial-in users can be disconnected after specific periods of inactivity. By eliminating idle connections, you reduce the number of ports required for remote access to a network.

INAsoft See INA

Inbound Path On a broadband LAN, the transmission path used by stations to transmit packets toward the headend.

incAlliance incAlliance stands for the Isochronous Network Communication Alliance. It was announced publicly on June 13, 1995. It was formed by high-technology businesses, including several Fortune 500 companies, to promote the use of isochronous Ethernet (also called isoEthernet) to provide interactive multimedia applications to the enterprise desktop over existing cable infrastructures. isoEthernet defines ISDN channels with standard ethernet packet traffic to facilitate real-time interactive voice/video/data transmission over 10Base-T local area networks. Isochronous Ethernet is a draft standard which is expected to be ratified by the IEEE 802.9a multimedia committee by June 1995. The incAlliance will act as a plenary liaison and resource to industry groups such as the IEEE, ITU, Electronic Computer Telephony Forum (ECTF), International Multimedia Teleconferencing Consortium, ATM Forum and the Multimedia Communications Forum. The incAlliance's mission is to promote the value of high quality, real-time, interactive multimedia including new isochronous LAN technologies such as isoEthernet; educate the industry on differences and requirements between voice/video and data for LAN/WAN implementation; foster industry growth through joint applications development and interoperability testing; provide a vision and roadmap for upgrading existing data networks. Two key players in the formation of this group are National Semiconductor's Interactive Multimedia Group and Dialogic Corporation.

INCC Internal Network Control Center

Incentive Regulation Prices of services provided by the local regulated phone company are fixed or capped but incentives are provided to improve earnings through cost savings. Earning levels are flexible, within a range of rates, allowing opportunity for earnings improvement.

Incident Angle The angle between an incident ray and a line perpendicular to an optical surface.

Inclination Angle The angle at which a satellite orbit is tilted relative to the equator.

Inclined Orbit Any nonequatorial orbit of a satellite. Inclined orbits may be circular or elliptical and may be synchronous or nonsynchronous. Inclined orbits are used for many reasons — for photographing, for reaching places in the extreme north and south which normal geosynchronous satellites can't reach.

Incoherent Light A random form of light whereby the phase of the light is unpredictable. LEDs emit incoherent light.

Incoming Call Identification ICI. Some way of telling the user who's calling. It might be the caller's extension number on an LCD screen or even the caller's name spelled out, e.g. "KATE BRODIE-DAVID CALLING." Today, most "incoming call identification" is done totally within one PBX. However, the days of ISDN and ITU Signaling System 7 are arriving. They promise to deliver to us the phone number of everyone calling us — from within the PBX or key system and from the outside world.

Incoming Calls Barred An interface configuration option that blocks call delivery attempts. Only outgoing calls are allowed.

Incoming First Failure To Match IFFM. A telephone company term. The multi-line recycle feature permits a second incoming attempt to complete to hunting lines. This feature may substantially lower the total office incoming matching loss in offices with high multiline development. At the same time, individual lines may be incurring a high %IML. IFFM registers will count all incoming first failures to match. The system objective for this measurement is 2.3%.

Incoming Matching Loss IML. A telephone company term. Percent Incoming Matching Loss (% IML) is a measurement of incoming calls unable to complete to a line equipment because of the lack of an available path between the incoming trunk (or junctor) and the called line. The engineering objective ceiling to be included in the equipment design is 2% in the busy season of exhaust.

Incoming Server Group See ISG.

Incoming WATS An incoming WATS (INWATS) trunk is used exclusively for received incoming calls from a defined geographical area to a customer's PBX. An incoming WATS trunk can only be used to receive calls via the dial-up telephone network. Originally WATS lines came in only lines that could receive calls or only lines that could make calls. Now, you can buy a WATS line that handles both incoming and outgoing lines. See WATS.

Increment A small change in the value of a quantity.

Incremental Cursor Control The user-controlled function that moves the focus in increments dictated by the application. In character-based text editing, the increment is typically one character in the horizontal direction and one line in the vertical direction.

Indefeasible Right of Use IRU. (or INdefeasible Right of User). A term used in the underseas cable and fiber optic carrier business. Someone owning an IRU means he has the right to use the circuit for the time and bandwidth the IRU applies to. An IRU is to a submarine or fiber optic cable what a lease is to a building.

Independence If two or more events occur in nature with no influence on each other they are said to be independent. The probability of these two or more events occurring simultaneously is the product of their individual probabilities. This definition of great relevance to the network planning in the telephone industry.

Independent Clocks A communication network timing subsystem using precise free running clocks at the nodes for synchronization purposes. Variable storage buffers installed to accommodate variations in transmission delay between nodes are made large enough to accommodate small time (phase) departures among the nodal clocks that control transmission. Traffic is occasionally interrupted to reset the buffers.

Independent Sideband Transmission ISB. That method of double sideband transmission in which the information carried by each sideband is different. The carrier may be suppressed.

Independent Software Vendor ISV. Typically a company which writes and sells software, but not hardware. Manufacturers of hardware and operating systems, i.e. IBM or Northern Telecom, often contract with ISVs to produce specialized software to make their hardware and operating system more attractive.

Independent Telephone Company A telephone company not affiliated with one of the "Bell" telephone companies. There are about 1,400 independent phone companies. They serve more than half the geographic area of the United States, but only around 15% of its telephones. The independent phone companies used to be represented by the United States Independent Telephone Association (USITA). But once Divestiture happened, the association dropped the word "Independent" from its name, accepted membership of the Bell operating companies (but not AT&T) and became USTA.

Index Think of a filing cabinet. It contains oodles of information. Think of a computer hard disk. Same thing as a filing cabinet. Oodles of information. Now think of putting everything in the filing cabinet into filing folders and putting them in alphabetical order. Makes finding things a lot easier. Now think of a computer. You ask it to find you the name of a file folder. Nothing sophisticated here. Except it's dumber than you. It starts at the top and searches down. Of course, it searches fast. But it still searches from the top down. The fastest way for it to search is to give it less stuff to search through. Thus you make an index. Just as you do in a book. Only, compared to you and me, a computer is willing to do more stupid work. It will index in alphabetical order. It will index in date order. It will index in order of how much you sold the guy recently. It will index in any order you ask it to. And many database software programs will let you keep several indexes concurrently, thus allowing you to find things quickly. The rules of database are simple: The more indexes you keep concurrently, the more time your computer will take to update its indexes every time you enter a new record or update an old record.

Index Dip In an optical fiber, a decrease in the refractive index at the center of the core, caused by certain manufacturing techniques.

Index Field The field to be used when indexing a database.

Index File An (optional) file used for indexing the data in a database. Index files are usually given extensions which identify them as index files. For example, when using dBASE III+, the index files are given the NDX extension.

Index Matching Material In fiber optics, a material (liquid, gel, or cement) whose refractive index is nearly equal to the fiber core index. It is used to reduce Fresnel reflections from a fiber end face.

Index Of Cooperation In facsimile, the product of the total line length in millimeters times the lines per millimeter divided by r. For rotating devices, the index of cooperation is the product of the drum diameter times the number of lines per unit length.

Index Of Refraction How light bends as it travels through a specific substance. When light travels through water it refracts and makes everything appear wavy and distorted. Fiber optic cable has different types of glass with different refractive indexes. As light travels from the core to the outer edge it's bent back inward because of the increasing refractive indexes of the glass it passes through.

Index Profile In an optical fiber, the refractive index as a function of radial distance from the optical axis.

Indexed Database A database indexed on a key field. Indexing allows for rapid retrieval of records through an index field. Microsoft's Outlook is not indexed, which is why it is so slow.

Indication Circuit X.21 circuit used to send control information from DCE to DTE.

Indication Of Lights, bells and buzzers indicating that something has or is about to happen. For example, indication of camp-on to a station: short bursts of tone are periodically transmitted to the busy phone to indicate that another call is camped on and waiting.

Indirect Control In digital data transmission, the use of a clock at a higher standard modulation rate, e.g., 4, 8, 128 times the modulation rate, rather than twice the data modulation rate, as is done in direct control.

Indirect Tapping A current in a conductor gives rise to a magnetic field around the conductor. When the current varies, the magnetic field changes. Conversely, if a conductor is immersed in a magnetic field, changes in the magnetic field will induce currents in the conductor. A coil of wire attached to a telephone or clamped to a telephone line can pick up conversations on the telephone. This type of wiretap, where there is no physical connection between the tap and the target line, is called Indirect Tapping.

Indium Gallium Arsenide InGaAs, a semiconductor material used in lasers, LEDs, and detectors.

Individual Load Cycling Feature This is one feature of AT&T's Dimension Energy Communications Service Adjunct. Individual Load Cycling reduces energy consumption by turning devices on and off (e.g. Air-conditioning) on an hourly basis.

Individual Speed Calling A key system or PBX feature by which a user can dial a longer number by punching one or two buttons on his phone. Sometimes this speed dial ability is programmed into the phone. Sometimes it's programmed into the system. Whichever it is, each user has a bunch of numbers he/she can speed dial. These are his/her own. No one else can speed dial them.

Induced Uplink An ATM term. An uplink "A" that is created due to the existence of an uplink "B" in the child peer group represented by the node that created uplink "A". Both "A" and "B" share the same upnode, which is higher in the PNNI hierarchy than the peer group in which nk "A" is seen.

Inductance is the property of an electric force field built up around a conductor. Inductance allows a circuit to store up electrical energy in electromagnetic form. When current flows through a wire, lines of force are built up around the wire. The field created by DC current is steady. When AC flows through a wire, the lines of force are constantly building and collapsing. An inductor is formed by winding a conductor into a coil. In long local loops, conversation gets difficult because the long wires encounter capacitive resistance. To counter this, inductors known as load coils are connected in series, increasing the inductance. When load coils, or inductors are connected in parallel, they reduce the inductance. See also INDUCTIVE CONNECTION.

Induction Electromagnetic transfer of energy from one coil to another.

Induction Coil A coil having a high turn ratio used for raising the voltage. A step-up transformer.

Inductive Amplifier An inductive amplified is a handheld device used by telephone installers which amplifies inductive signals and plays them over a speaker built into one end. The other end of the tool has a metal probe to touch the wire or connector to pick up tones coming from a tone generator. The two tools are used together to locate and / or test cable. You attach tone generator at one end. You go hunting for it with an Inductive Amplifier tool. If the tone comes through to a connection point (see IDF, Termination Block, Patch Panel), the cable is OK to there, there are no breaks. Tone Generators and Inductive Amplifiers are adequate for testing for voice quality circuits, but this testing method doesn't tell you if the cable is good enough for data. See Inductive Connection.

Inductive Connection A connection between a telephone instrument and another device by means of the electromagnetic field generated by the telephone instrument. No direct electrical connection is established between the two. See INDUCTANCE.

Inductive Coupling The transfer of energy from one circuit to another by means of the mutual inductance between the circuits, i.e. energy jumping from one circuit to another without actually touching it copper wire to copper wire. The coupling may be deliberate and desired as in an antenna coupler or may be undesired as in powerline inductive coupling into telephone lines. See also INDUCTANCE.

Inductive Pickup A coil used to tap phone lines without direct connection.

Inductive Tap Wiretap that is not physically connected to the telephone wires. A voltage proportional to the varying line current is induced into a coil.

Inductively Coupled Receiver A radio receiver in which the energy in the antenna circuit is transferred to the secondary circuit by induction.

Inductivity A term sometimes used to denote the dielectric constant or the specific inductive capacity.

Inductor See INDUCTANCE and INDUCTIVE CONNECTION.

INE Intelligent Network Element. Network equipment which contains autonomously intelligent computing capabilities.

INET Institutional Network. Generally dedicated to linking government and other public buildings for such uses as training, meetings, data and voice. Such INETs are provided by cable TV operators pursuant to 47 U.S.C. 531 for public, educational and governmental use. It is not available for consumers.

INF File A file that provides Windows 95 Setup with the information required to set up a device, such as a list of valid logical configurations for the device, the names of driver files associated with the device, and so on. An INF file is typically provided by the device manufacturer on a disk.

Inference Engine The AI (Artificial Intelligence) heart of a knowledge base system. The inference engine is the technology which directs the reasoning process. The inference engine contains the general problem-solving knowledge such as how to interact with the user and how to make the best use of the domain information.

Infinite Loop A state in which specific steps of a program are executed repeatedly, not allowing the program execution to advance further.

Infinity Transmitter Also called Harmonic Bug. Infinity transmitters got their name because the original manufacturers claimed they could pick up conversations from an infinite distance. They are sometimes called harmonica bugs because original versions were activated by a 440 Hz tone created by a harmonica. An infinity transmitter is a room listening device that uses the telephone lines to send audio back to the surveillance operator. An infinity transmitter is attached to a telephone line as if it were an extension or it can be installed on a telephone instrument. To use an infinity transmitter, the target telephone is dialed and the tone or signal is sent down the telephone line before the target phone has a chance to ring. The infinity transmitter closes the telephone circuit and activates its microphone to pick up room conversations and transmit the audio down the telephone line. The disadvantage of the infinity transmitter is that it is easily detected. Telephone calls cannot be made by the target phone when the device is operating. Some telephone systems may not send audio to the target telephone line until after the phone is picked up.

Inflight Packet See Mobile IP.

Info Look Gateway Service A Nynex service. Use your PC, modem and communications software to dial Info-Look and check on news, weather, health, finance, sports. Do some shopping. Info-Look connects you to many computer services and databases, none of which are provided by Nynex or New England Telephone. That's why Info-Look is called a "gateway." It's a gateway to a myriad of services. As of writing this there was talk Nynex was closing Info-Look down. And it eventually did.

Infobahn A new term for the Information Superhighway.

Infomercial A short segment shown on the video system of a plane purporting to be informational/newsy and educational. In fact, the segment is a commercial paid for by the company whose products and/or services are featured. According to research, infomercials create three types of phone calls — order calls, inquiry/incomplete calls and customer service calls.

InfoNet A consortium of about 10 European telephone companies and MCI. InfoNet has built a worldwide packet switched network that is great for data communications. It offers one major advantage over packet switched networks owned by the individual phone companies — You don't have to separately sign up to use InfoNet in all the countries it operates. You just dial Infonet up, punch in and communicate. Not having to sign up is a big savings in time, since it can take weeks to sign up for European packet switched networks and cost minimum monthly fees. For MCI Mail subscribers, using Infonet means you only get one bill — directly from MCI. See also MCI Mail and MCI FAX. Infonet is based in El Segundo, CA 90254.

Information At Your Fingertips At Fall Comdex 1990 Bill Gates, Microsoft chairman, suggested the idea. With Information at your fingertips, he said, PC users can easily access company wide information "anywhere at anytime" through an icon-based graphical user interface. In the speech, Gates demonstrated applications that used Object Linking and Embedding (OLE), Dynamic Data Exchange (DDE), handwriting recognition, cellular communications and multimedia. See OLE.

Information Bearer Channel 1. A channel provided for data transmission that is capable of carrying all the necessary information to permit communication, including user's data, synchronizing sequences, control signals, etc. It may therefore, operate at a greater signaling rate than that required solely for the user data.

2. A basic communication channel made available by the circuit provider with no enhanced or value-added services included other than the bandwidth transmission capability.

Information Center Mailboxes An Octel term for a voice bulletin board on a voice mail system. Here's their explanation: Multiple callers can access, directly or indirectly, recorded announcements containing information that would otherwise have been given live by employees. Callers are frequently "outside" users of the system. One type of "listen only" mailbox simply plays the messages to the callers. This technology, sometimes known as audiotex, makes it possible to create a verbal database so callers can select which information they want to hear. Another type of Information Center Mailbox prompts callers to reply to announcements. Callers wanting further information can be given the opportunity to leave their names and phone numbers after listening to a product description. They can also be transferred to a designated employee who can immediately take an order. If desired, a password can be required before confidential or controlled access information can be heard.

Information Digits CDR call type options. Two digit codes which precede the 7-or-10 digit destination number and inform exchange carriers and IECs about the type of line that originated the call, any special characteristics of the billing number, or certain service classes. These codes plus the destination number are part of the signalling protocol of equal access offices. These codes are defined by Bellcore. Examples: 00 - POTS, 01 - Multiparty, 02 - ANI Failure, 06 - Hotel/Motel, 07 - Special Operator Handling, 20 - AIOD, 24 - 800, 27 - Coin, 30 - Unassigned DN, 31- Trouble/Busy, 32 - Recent change or disconnect, 34 - Telco Operator, 52 - Outward WATS, 61- Cellular 1, 62 - Cellular 2, 63 - Roaming, 70 - Private Pay Phone, 93 - Private virtual Network

Information Element The name for the data fields within an ISDN Layer 3 message.

Information Engineering Coined by James Martin, the most prolific writer in data processing, the term refers to systems within data processing and their impact on giving the corporation a greater competitive edge. In short, a fancy term for Management Information Systems (MIS), which itself was a fancy term for DP, namely Data Processing.

Information Field A Frame Relay term referring to the variable length field of data, which can include either user payload or internetwork control data to be passed between routers or other intelligent end user devices. The information field may be 0-4,096B, although ANSI recommendations are that the field be 1600B, which accommodates most LAN packets.

Information Frame Frame in HDLC, DDCMP, or related protocols containing user data.

Information Highway A term coined by Al Gore. This fact affirmed by Dan Lynch, the man who started the trade show, InterOp and who was very heavily involved with Internet from the very beginning. As the term got developed and people got turned on by the idea, it became known as The Information Superhighway. See also INFORMATION SUPERHIGHWAY.

Information Outlet IO. Sort of like an AC power outlet, but a little more cerebral. A connecting device designed for a fixed location (usually a wall in the office) on which horizontal wiring subsystem cable pairs terminate and which receives an inserted plug; it is an administration point located between the horizontal wiring subsystem and work location wiring subsystem. Although such devices are also referred to as jacks, the term information outlets encompasses the integration of voice, data, and other communication services that can be supported via a premises distribution system.

Information Packet A bundle of data sent over a network. The protocol used determines the size and makeup of the packet.

Information Page Mapping See ADSI.

Information Provider A business or person providing information to the public for money. The information is typically selected by the caller through touchtones, delivered using voice processing equipment and transmitted over tariffed phone lines, e.g., 900, 976, 970. Typically, billing for information providers' services is done by a local or long distance phone company. Sometimes the revenues for the service are split by the information provider and the phone company. Sometimes the phone company simply bills a per minute or flat charge. A typical "information provider" is American Express, which provides a service — 1-900-WEATHER. By dialing that number you can touchtone in city names and find out temperatures, weather forecasts, etc. Calling 1-900-WEATHER costs several dollars a minute.

Information Service The Telecommunications Act of 1996 defined Information Service as: The term `information service' means the offering of a capability for generating, acquiring, storing, transforming, processing, retrieving, utilizing, or making available information via telecommunications, and includes electronic publishing, but does not include any use of any such capability for the management, control, or operation of a telecommunications system or the management of a telecommunications service.

Information Signals A Bellcore definition. Information signals inform the customer or operator about the progress of the call. They are generally in the form of universally understood audible tone (for example, dial tone, busy, ringing) or recorded announcements (for example, intercept, all circuits busy).

Information Superhighway A very vague concept which Senator Al Gore created in the early 1990s and which gained great popularity when he became vice president and the Clinton/Gore administration started pushing the concept. The Information Superhighway is a term sufficiently vague that it can mean anything to anyone. It can mean a gigantic Internet reaching everybody in North America, or the planet (if you're that expansive). It could just as easily mean a combination 500-channel interactive cable TV system with full video on demand to every household in North America. Somewhere in all this is the idea that easy access to large amounts of information will enrich our lives immeasurably. Who's going to get first access to it all, what the precise technical details will be, and who's going to pay for it are, naturally, minor details to be worked out. We can be assured that the details will be worked out, since the idea originated in Washington, DC., home of so many practical ideas.

Information Technology IT. A fancy name for data processing, which became management information systems (MIS), which became information technology. All the same thing, essentially. See also IT.

Information Technology District The Information Technology District (ITD) is New York City's fastest growing totally-wired community. Anchored by the New York Information Technology Center @ 55 Broad Street and sharing the Downtown Business Improvement District's boundaries of City Hall to the southern tip of Manhattan, the ITD serves as the headquarters for Silicon Alley. The ITD is home to more than 250 IT companies, from web page developers to financial modeling firms. According to promotion from ITD, these companies are quickly emerging as the City's prime economic generators, creating jobs and innovative products, and serving as pioneers in the ongoing revitalization of Downtown into a 24-hour, 21st Century global community.

Information To Go A term coined by Digital Equipment Corporation to refer to the transmission of data over airwaves instead of fixed wires.

InfoSpace A service that helps surfers locate listings of people, businesses, government offices, toll-free numbers, fax numbers, e-mail addresses, maps and URLs, all on one Web site. InfoSpace has developed a patent pending technology that integrates all of these services. www.infospace.com

Infostrada SpA A new phone company which started service in July 1998, competing against Italy's erstwhile monopoly, Telecom Italia. Infostrada is controlled by Olivetti SpA and the German company, Mannesmann AG.

Infrared The band of electromagnetic wavelengths between the extreme of the visible part of the spectrum (about 0.75 um) and the shortest microwaves (about 100 um). This portion of the electromagnetic spectrum is used in some fiber-optic transmission systems, but more commonly for airwave communications. In such application, the system typically consists of a two transmitter/receivers. The infrared light signal is transmitted through a focused lens to a collecting lens in the receiving device. Transmission rates of as much as 6.312 Mbps can be achieved over distances of as much as several miles. Typically deployed in campus environments or other very short-haul applications where cabled systems are not possible or practical, infrared offers the advantage of no FCC licensing requirements, thereby sometimes making it preferable to microwave.

Infrared also is commonly used for short haul (up to 20 feet) through-the-air data transmission. With the adoption of new infrared standards at a meeting of over 50 manufacturers in June 1994, many PC devices will begin sporting something called the "Infrared Serial Data Link" (IRDA) with speeds up to 1.5 Mbps. This standard is designed to ensure that products sporting this link will work together and interchangeably.

Infrared Data Association See IrDA.

Infrared Fiber Optical fibers with best transmission at wavelengths of 2 um or longer, made of materials other than silica glass.

Infrared Serial Data Link As a result of a meeting at Microsoft of over 50 manufacturers in June 1994, many PC devices will begin sporting something called the "Infrared Serial Data Link," an infrared through-the-air (up to 20 feet) link with speeds up to 1.5 million bytes per second. This standard is designed to insure that products sporting this link will work together and interchangeably. There is now an organization called I.R.D.A., the Infrared Data Association, representing over 80 manufacturers.

Infrastructure/Telecommunications A collection of those telecommunications components, excluding equipment, that together provide the basic support for the distribution of all information within a building or campus.

ING CallING party. The ING calls the ED, or callED party, to set up a data transfer. ING and ED apply to any type of data transfer, including both voice and data.

Ingredient Technology See INDEO VIDEO.

Ingress Ingress is a cable TV term. Ingress occurs when strong outside signals leak into a CATV coaxial cable and interfere with the signal quality inside the home and nearby homes. Picture a car driving along outside a house. The car has a strong CB radio. It sends the signal out. It is picked up by the coaxial CATV cable in the house, which then sends it to nearby houses. The primary cause of ingress is cheap wiring and/or loose connectors. But the interfering signal is caused by radio transmitters of all types (including short wave transmitters), electrical appliances, motors with brushes, light dimmers or speed controls on toys. Leakage is really a shielding problem. The number of houses that can be affected by ingress depends on the strength of the signal and the number of service areas around a CATV node, which could be as many as 1,000. Companies like Trilithic in Indianapolis are expert in measuring ingress. See also LEAKAGE.

Inheritance A term from object oriented programming. Data abstraction can be carried up several levels. Classes can have super-classes and subclasses. In moving to a level of greater specificity, the application developer has the option to retain some attributes and methods of the super-class, while dropping or adding new attributes or methods. This allows greater flexibility in class definition. It is even possible in some languages to inherit from more than one parent. This is referred to as multiple inheritance. See OBJECT ORIENTED PROGRAMMING.

INIC ISDN Network Identification Code.

INIM ISDN Network Interface Module (INIM) is both hardware and software. It does the job of an NT-1, so the physical network interface is ISDN-U. When calls arrive, the INIM collects the number dialed and Caller ID. This data is passed on to the Call Processing Module, which does the actual call handling. When you go off hook to place an outbound call, the INIM assigns an available ISDN B-channel to the call. For 'Find Me' scenarios, the INIM lets Front Desk place multiple, simultaneous out-bound calls. During data calls, the INIM constantly monitors the data transmission rate on the ISDN line. The INIM will automatically build up or tear down the second ISDN B channel from a data call to match bandwidth requirements. Because telcos charge for usage per B-channel, the INIM uses both B channels only when necessary. If both B channels are doing data when a new voice or fax call arrives, the INIM instantly tears down one of the B channels to let the new call through. Ditto for when you make an outbound voice or fax call.

INIT An INIT is the Macintosh System 7 equivalent of a terminate and stay resident (TSR) program . An init might load to initialize a fax modem, screen saver, etc. Similar to the DOS environment, some inits conflict. When troubleshooting operating system problems, remove inits first.

Initial Address Message. IAM. A SS7 signaling message that contains the address and routing information required to establish a point-to-point telephone connection.

Initial Answer Initial answer refers to the point in time at which a computer telephony system answers an incoming call. Many computer telephony systems require significant processing to set up to answer incoming calls. For example, the system may examine the incoming ANI, DNIS, or PBX integration data to determine how to answer (which prompt to use), or where to switch the call. This can involve significant database access and processing time. Therefore, the ability to handle large number of incoming calls (especially in burst mode) may delay the initial answer. The delay from when a call reaches a computer telephony system until the computer telephony system answers the call (the initial answer) is usually a key response time to understand when testing a computer telephony system.

Initial MAC Protocol Data Unit IMPDU. A Connectionless Broadband Data Service (CBDS) term that corresponds to the L3 PDU in Switched Multimegabit Data Service (SMDS). CBDS is the European equivalent of SMDS.

Initial Period The minimum billing period on a call. For interstate or inter-LATA AT&T calls, the initial period is one-minute. Some non-AT&T long distance companies have initial periods under one-minute. This also applies to local calls in Measured areas.

Initial Program Load The initial loading of generic and/or configuration software into a PBX or other phone system. The Initial Program Load is a pain in the rear end. But an even bigger pain is what happens when you lose your programming and you've forgotten to back it all up.

Initial Sequence Number ISN. Generated at each end of TCP connection to help to uniquely identify that connection.

Initialization String A group of commands sent to the modem by a communications program at start-up — before the number has been dialed. Such a string tells the modem to set itself up in a way that will make it easy to correctly communicate with a distant modem.

Initializing Terminals An ISDN term. These devices, sometimes called self-initializing terminals, are basically ISDN terminals that can generate their own terminal identification number. This makes it easier for the network and the terminal to agree on the number to use.

Initiating Event An event that causes the ASC (AIN Switch Capabilities) to assign an ID to a connection segment for a certain user ID and to communicate with a SLEE (Service Logic Execution Environment) for the first time with respect to the combination of the specific user ID and connection segment ID.

Initiator A SCSI device, usually a host system, that requests an I/O process to be performed by another SCSI device.

Injection Laser Another name for a semiconductor or diode laser. See INJECTION LASER DIODE.

Injection Laser Diode ILD. A solid-state device that works on the laser principle to produce a light source for optical fiber.

INL See INTERNODE LINK.

Inline Image A built-in graphic that is displayed by a Web browser as part of an HTML document and is retrieved along with it.

Inline Plug-in An application that, when inserted into a Web browser that supports it (through an installation procedure), allows greater functionality and flexibility of the browser to view multimedia that would otherwise require an outside application.

INMAC International Network Management Center

INMARSAT The INternational MARitime SATellite service that has satellites and provides mobile communications to ships at sea, aircraft in flight, vehicles on the road and to small stationary satellite antennas which people carry with them. Inmarsat provides dial-up telephone, telex, fax, electronic mail, Internet access, data connections and fleet management. Typically you pay a per minute charge for the use of Inmarsat's communications services. Inmarsat describes itself as the international mobile satellite organization. It is based in London and has 84 member countries. www.inmarsat.org

Inmate Call Management A typical inmate call management application package serves the special requirements of correctional institutions. By assigning Personal identification Numbers (PINs), inmates may be allowed debit calling only, collect calling only, or a combination of both. Some packages can also access external commissary databases, allowing immediate debit of an inmate's account in real time. They may also do:
- billed number screening, reducing billing fraud
- blocking of calls after commissary funds are exhausted, minimizing lost revenue on commissary funds
- identifying announcement to called party
- time-of-day restrictions to specified destination numbers
- temporary call prohibition for specific PINs.
- restricted destination numbers for specific PINs
- system-wide deny numbers

Inmate Calling Dennis Squires from Bell Atlantic says, "Prison inmates can get very creative when they have a lot of time on their hands, like 20 years." So, to avoid stealing of phone calls by prisoners, some phone companies have a service called "Inmate Calling." The inmate enters his authorization code, which allows him to make phone calls — but only to authorized phone numbers, i.e. mothers, lawyers, bail bondsmen, etc.

INN See InterNode Network.

Inner Duct A flexible plastic conduit that's placed in larger conduits. This is used when different companies lease space in the same conduit.

Inner Wires For a standard four wires connection, the wires fastened to the inner two pin positions in a jack or connector. For example, in a six position connector, the inner wires are pins 3 and 4.

Innerduct A nonmetallic raceway, usually circular, placed within a larger raceway.

INode Integrated Node (SSP and STP).

INOS Intelligent Network Operations System. An operations system designated to manage, monitor, and control elements of an intelligent telecommunications network.

iNOW! iNOW! is a standards-based, multi-vendor initiative announced in December 1998 to provide interoperability among IP telephony platforms. Ascend, Cisco, Clarent, Dialogic, Natural MicroSystems and Siemens will be working with the iNOW! Profile to make their gateways and gatekeepers interoperable with each other's products and with those from Lucent and VocalTec. The specs on the standard are due to be published in January 1998. According to the January, 1998 release, the iNOW! Interoperability Profile will detail how to achieve interoperability between gateways from different vendors and interoperability between gatekeepers from different vendors. Up until recently, carriers and callers were limited in the destinations they could reach since calls had to be terminated on the exact same platform from which they originated. Internet telephony service providers had to choose between dependence on a single vendor or operating multiple parallel networks of incompatible gateways. According to the release, Lucent and VocalTec are responsible for the development of the programming and engineering for the iNow! Interoperability Profile. The interoperability guidelines are based on the International Telecommunications Union (ITU) H.323 standard, and the upcoming H.225.0 Annex G standard.

INP Board Intelligent Network Processor board for synchronous transmission on an Hewlett Packard 3000 system.

INPA Interchangeable Numbering Plan Area. An area code that looks like an office code. There is no particular name for an office code that looks like an area code, according to Lee Goeller.

INPUT A signal fed into a circuit.

Input Buffer Limiting Buffering strategy that divides buffer at a mode into two classes, both available to transit packets but only one available to packets input at the node.

Input Circuit The grid circuit of an electron tube.

Input/Output I/O. Input and output are two of the three functions that computers perform (the other is processing). Input/Output describes the interrelated tasks of providing information to the computer and providing the results of processing to the user. I/O devices include keyboards (input) and printers (output). A disk drive is both an input and an output device, since it can both provide information to the computer and receive information from the computer.

Inquiry A request for specific information.

INS 1500 A term for a digital T-1 line transmitting at 1.544 million bits per second.

INS 64 A term for a digital ISDN BRI line transmitting at 144,000 bits per second.

Insertion Gain The gain resulting from the insertion of a device in a transmission line, expressed as the ratio of the power delivered to that part of the line following the device to the power delivered to that same part before insertion. If more than one component is involved in the input or output, the particular component used must be specified. If the resulting number is negative, an "insertion loss" is indicated. This ratio is usually expressed in decibels. See INSERTION LOSS.

Insertion Loss The difference in the amount of power received before and after something is inserted into the circuit (viz. another telephone instrument) or a call is connected. In an optical fiber, insertion loss is the optical power loss due to all causes, usually expressed as decibel/kilometer. Causes of insertion loss may be absorption, scattering, diffusion, dispersion, microbending, or methods of coupling power outside the fiber. In lightwave transmission systems, the power lost at the entrance to a waveguide due to causes, such as fresnel reflection, packing fraction, limited numerical aperture, axial misalignment, lateral displacement, initial scattering, or diffusion. See INSERTION GAIN.

Inside Dial Tone The dial tone provided by a PBX. This lets you dial an internal number. When you dial 9 (internal extensions never start with 9), you get the dial tone provided by your local RBOC. Outside North America, you often dial 0 to get an outside line.

Inside Link An ATM term. Synonymous with horizontal link.

Inside Plant Everything inside a telephone company central office. Thus, electronic equipment in buildings. Includes central office switches, PBX switches, broadband equipment, distribution frame, power supply equipment, etc. It doesn't include telephone poles, cable, terminals, cross boxes, cable vaults or equipment found outdoors. See also INSIDE WIRING.

Inside Wiring That telephone wiring located inside your premises or building. Inside Wiring starts at the telephone company's Demarcation Point and extends to the individual phone extensions. Traditionally, Inside Wiring was installed and owned by the telephone company. But now you can install your own wiring. And most companies installing new phone systems are installing their own new wiring because of potential problems with reusing the old telephone company cable. See also INSIDE PLANT.

Inspect Screen An AT&T Merlin term. A display screen on digital telephones that allows users to preview incoming calls and see a list of the features programmed on line buttons.

Instabus1080 and Instabus1480 Trademarks for MICOM's direct host attachment products.

Instalink 1. An MCI International service that allows access to a host computer in the U.S.A. from a Telex machine anywhere overseas. This allows easier retrieval of information from a U.S. database.

2. A trademark for MICOM's data-over-voice products.

Installation The physical hook-up and diagnostic testing of a PBX switch, cabinet, or peripheral item prior to a cutover and maintenance acceptance by the maintaining vendor.

Installed Base How many of whatever are in and working. Installed base is often confused with annual shipments. They're very different. Shipments is what goes out the factory. Installed base is what's out there. The equation is: Installed base at beginning of year plus annual shipments less equipment taken out of service during the year is equal to the installed base at the end of the year.

Installer's Tone Also called test tone. A small box that runs on batteries and puts an RF tone on a pair of wires. If the technician can't find a pair of wires by color or binding post, they attach a tone at one end and use an inductive amplifier (also called a banana or probe) at the other end to find a beeping tone.

Instance ID An ATM term. A subset of an object's attributes which serve to uniquely identify a MIB instance.

Instanet Trademark for MICOM's family of local data distribution and data private automatic branch exchange (PABX) products.

Instant On Buy a PC (Personal Computer). Turn it on. Bingo, it's already loaded with Windows or OS/2. Instant On is a new term for preloading software onto hard disks of new computers and shipping those computers already pre-loaded with that software.

Instantaneous Override Energy Function IOEF. A feature of the AT&T PBX Dimension Energy Communications Service Adjunct (ECSA), which allows the user to turn all the ECSA energy functions ON or OFF. IOEF is most often used for periodic maintenance, or to adjust to sudden changes in weather.

Institute for Telecommunications Sciences ITS is the research and engineering branch of the National Telecommunications and Information Administration (NTIA), which is part of the U.S. Department of Commerce (DoC). www.ntia.doc.gov See NTIA

Instruction Register The register which contains the instruction to be executed and functions as the source for the subsequent operations of the arithmetic unit.

Instructional Television Fixed Service ITFS. A service provided by one or more fixed microwave stations operated by an educational organization and used mainly to transmit instructional, cultural and other educational information to fixed receiving stations.

Insulated Wire Wire which has a nonconducting covering.

Insulating Materials Those substances which oppose the passage of an electric current through them.

Insulation A material which does not conduct electricity but is suitable for surrounding conductors to prevent the loss of current.

Insulation Displacement Connection IDC. The IDC has replaced wire wrap and solder and screw post terminations as the way for connecting conductors (i.e. wires carrying telecom) to jacks, patch panels and blocks. Insulation Displacement Connections are typically two sharp pieces of metal in a slight V. As the plastic-covered wire is pushed into these metal teeth, the teeth pierce the plastic jacket (the insulation) and make connection with the inside metal conductor.

This saves the installer having to strip off the conductor's insulation. This saves time. Since IDCs are very small, they can be placed very close together. This reduces the size of jacks, patch panels and blocks. IDCs are the best termination for high speed data cabling since a gas-tight, uniform connection is made. The alternate method of connecting wires is with a screw-down post. There are advantages and disadvantages to both systems. The IDC system, obviously, is faster and uses less space. But it requires a special tool. The screw system takes more time, but may produce a longer-lasting and stronger, more thorough (more of the wire exposed) electrical connection. The most common IDC wiring scheme is the 66-block, invented by Western Electric, now Lucent. But there are other systems — from other telecom manufacturers. See Punchdown Tool.

Insulation Resistance That property of an insulating material which resists electrical current flow through the insulating material when a potential difference is applied.

Insulators Some atoms hold onto their electrons tightly. Since electrons cannot move freely these material can't easily conduct electricity and are know as non-conductors or insulators. Common insulators include glass, ceramic, plastics, paper and air. Insulators are also called dielectrics.

INT14 A software interrupt designed to communicate with the com (serial) port in a PC. Communications programs use interrupt 14h to talk to a modem physically attached to another computer on the network.

Integer A computing procedure for solving or finding the optimum solution for complex problems in which the variables are based on integers. Integers include all the natural numbers, the negatives of these numbers, or zeros.

Integrated Access An AT&T term for the provision of access for multiple services such as voice and data through a single system built on common principles and providing similar service features for the different classes of services.

Integrated Call Management ICM. A family of Rolm networking products and services for single-site and multi-site CBX installations.

Integrated Circuit IC. After the transistor and other solid state devices were invented, electronic circuits were designed that were more complex than ever. It became a real problem wiring all the components together. In 1958-1959, Jack Kilby and Robert Noyce independently invented the integrated circuit. An integrated circuit is a piece of silicon or other semiconductor called a chip on which is etched or imprinted a network of electronic components such as transistors, diodes, resistors, etc. and their interconnections.

Integrated Development Environment IDE. A Windows program within which a developer may perform all the essential tasks of development including editing, compiling and debugging.

Integrated Digital Loop Carrier IDLC. Access equipment which extends Central Office services; it connects to a SONET ring on the network side while providing telephony services on the subscriber side (POTS, ISDN, leased lines, etc.).

Integrated Dispatch Enhanced Network. iDEN. A wireless technology developed by Motorola, iDEN operates in the 800 MHz, 900 MHz and 1.5 GHz radio bands; the 900 MHz development is aimed at operators of digital Commercial Mobile Radio Service (CMRS), also known as ESMR (Enhanced Specialized Mobile Radio). iDEN is a digital technology using M16QAM (Quadrature Amplitude Modulation) for compression, and TDMA (Time Division Multiple Access). Through a single proprietary handset, iDEN supports voice in the form of both dispatch radio and PSTN interconnection, numeric paging, SMS (Short Message Service) for text, data, and fax transmission. See also ESMR, QAM, SMS and TDMA.

Integrated IS-IS Formerly Dual IS-IS. Routing protocol based on the OSI routing protocol IS-IS, but with support for IP or other networks. Integrated IS-IS implementations send only one set of routing updates, regardless of protocol type, making it more efficient than two separate implementations.

Integrated Messaging Also called Unified Messaging. Integrated messaging is one of many benefits of running your telephony via a local area network. Here's the scenario: Voice, fax, electronic mail, image and video. All on the one screen. You arrive in the morning. Turn on your PC. It logs onto your LAN and its various servers. In seconds, it gives you a screen listing all your messages — voice mail, electronic mail, fax mail, reports, compound documents .... Anything and everything that came in for you. Each is one line. Each line tells you whom it's from. What it is. How big it is. How urgent. Skip down. Click. Your PC loads up the application. Your LAN hunts down the message. Bingo, it's on screen. If it contains voice — maybe it's a voice mail or compound document with voice in it — it rings your phone and plays the voice to you. Or, if you have a sound card, it can play the voice through your own PC. If it's an image it may hunt down (also called launch) an imaging application which can open the image you have received, letting you see it. Ditto, if it's a video message. Messages are deluging us. To stop them is to stop progress. Run your eye down the list, one line per entry. Pick the key ones. Junk the junk ones. Postpone the others.

It gets better. You're out. Dial in on a gateway with your laptop. Skim your messages. Dial in on a phone. Punch in some buttons. Hear your voice mail messages. Or if you're not on laptop, have your e-mail read to you. Better, have your fax server OCR your faxes and image mail and have it read them to you. A LAN server is the perfect repository for messages. It can search for them, assemble them, process them, store them, convert them, compress them, shape them, shuffle them, interpret them. Integrated messaging essentially applies intelligence and order to the messages deluging you each day. See TELEPHONY SERVICES.

Integrated Personal Computer Interface IPCI. A ROLM-designed communications printed circuit card designed to provide an IBM PC with asynchronous data transmission over two-strand wiring to and from a Rolm CBX PBX.

Integrated Photonics Integrated photonics are devices that include optical waveguides embedded in a semiconductor or ferroelectric substrate, and which perform some type of signal processing function under electrical control. These functions include: routing of light signals in different directions, filtering out one or several wavelengths, emitting light or modulating the intensity and/or phase of an incoming light signals. An optical waveguide consists of a region in which the refractive index is higher than in the surrounding material so that a light signal can propagate without spreading (diffraction). In any applications, only single mode waveguides are useful. This definition courtesy Ericsson.

Integrated Services Digital Network See ISDN and SIGNALING SYSTEM 7.

Integrated Services Digital Network User Part ISDN-UP The part of SS7 (Signaling System Number 7) that encompasses the signaling functions required to provide voice and non-voice services in ISDN and pre-ISDN architectures. The basic service offered by the ISDN-UP is the control of circuit switched connections between subscriber line exchange terminations. Definition from Bellcore in reference to its concept of the Advanced Intelligent Network.

Integrated Voice Data There are many different meanings to this concept. The most common (we'll get arguments on this) is that a workstation or a combination telephone/personal computer on a desk can combine voice and data signals over a single communications channel. That channel might be carried digitally on one pair of wires. That is "the most integrated" voice/data. Less integrated is when you carry voice and data digitally on two pairs — one pair for transmitting and one pair for receiving. Even less integrated are some systems which use three pairs of cabling set up as one voice analog pair, one digital data pair and one power/signaling pair. In short, "integrated voice/data" means different things to different people and depends on the technology. See also ISDN.

Integrated Voice Data Workstation See ISDN, IVDT and INTEGRATED VOICE/DATA.

Integration Testing Integration (or single thread) testing is the phase in the computer telephony lifecycle that begins as individual modules are pulled together to make a complete system. Testing in this phase is related to making sure the interfaces between the various modules function correctly, and is oriented to functional issues. Inter-module functions are be checked for load stability by exposing them to a variety of real-world stimuli. Definition courtesy Steve Gladstone, from his book "Testing Computer Telephony Systems."

Intel Blue Specifications required to provision the ISDN line to meet the needs of Intel's ISDN-based products. When ordering your ISDN phone line and you want to use it for data or video, tell them it's "Intel Blue." That should tell your local phone company the correct technical specifications for your line. And when you come to plug in your ISDN equipment (assuming your chosen manufacturer has made it compatible with Intel Blue), it should work. This is not a guarantee, but a probability. See ISDN.

Intellectual Property Property produced by effort of the mind, as distinct from real or personal property. Intellectual property may or may not enjoy the benefit of legal protection: See Copyright, Patent, Trademark, Trade Secret, and WIPO.

Intelligence The part of a computer which performs the arithmetic and logic functions. Also, the information impressed or modulated on a transmission carrier — either voice or data.

Intelligent Agent Software that has been taught something of your desires or preferences and acts on your behalf to do things for you. It might, for example, search through incoming material on networks (e-mail and news) and find what you're interested in or looking for. It might, for example, monitor your TV viewing habits, accept general instructions about your preferences and then, on its own, browse through huge databases of available videos and make recommendations about programs you might be interested in viewing.

Intelligent Answering A Rolm term, explained thus: "When your customers call — or you call them — the Rolm 9751 CBX system can use automatic number identification (ANI) or dialed number identification service (DNIS) to identify the caller and the reason for that call."

Intelligent Assistance A concept Apple is pushing for its Newton PDA. Newton can anticipate what you want to do and provide a bit of help. This is how Fortune Magazine explained it: For example, scrawl "lunch with John Thursday." My Newton would assume that Thursday means next Thursday and that John is the John I've been meeting with a lot lately, John Sculley, and that I want to eat at 12:30, my

usual lunch hour. Newton updates my calendar, and presto, displays the entry for my approval. I can okay it or change it.

Intelligent Battery System See IBS.

Intelligent Concentrator A concentrator which receives signals from a device on one port and retransmits them to devices on other ports. An intelligent concentrator is one that has software and therefore has programming capabilities.

Intelligent Hub A hub that functions both as a bridge and multiprotocol router.

Intelligent Network IN. A network that allows functionality to be distributed flexibly at a variety of nodes on and off the network and allows the architecture to be modified to control the services; (In North America) an advanced network concept that is envisioned to offer such things as (a) distributed call-processing capabilities across multiple network modules, (b) real-time authorization code verification, (c) one-number services, and (d) flexible private network services (including (1) reconfiguration by subscriber, (2) traffic analyses, (3) service restrictions, (4) routing control, and (5) data on call histories). Levels of IN development are:

• IN/1. A protocol intelligent network targeted toward services that allow increased customer control and that can be provided by centralized switching vehicles serving a large customer base.

• IN1+. A protocol intelligent network targeted toward services that can be provided by centralized switching vehicles, e.g., access tandems, serving a large customer base.

• IN/2. A proposed, advanced intelligent-network concept that extends the distributed IN/1 architecture to accommodate the concept called the "service independence." Traditionally, service logic has been localized at individual switching systems. The IN/2 architecture provides flexibility in the placement of service logic, requiring the use of advanced techniques to manage the distribution of both network data and service logic across multiple IN/2 modules. See AIN, which stands for Advanced Intelligent Network.

Intelligent Peripheral IP. A network system in the Advanced Intelligent Network Release 1 architecture containing an Resource containing an Resource Control Execution Environment (RCEE) functional group that enables flexible information interactions between a user and the network.

Intelligent Phone When the Bell operating companies get bored they occasionally fantasize about applications for the networks they provide. Here are some of their ideas for what intelligent phones could, if motivated, do:
Select entertainment on demand (movies, music, video); Order groceries or other services or products; Record customized news and sports programming; Enroll and participate in education programs from the convenience of subscribers' living rooms; Find up-to-minute medical, legal and encyclopedic information; Pay bills and manage finances; Make airline, rental car and hotel reservations and buy sports and entertainment tickets.

Intelligent Premises Equipment This refers to modern equipment, such as routers and intelligent switches. These devices are often capable of taking on roles traditionally performed by the network service, such as error correction.

Intelligent Routing A voice call comes in. Your voice mail machine recognizes it as being urgent, so it gives the caller a message, "Please hold. Harry is away from his desk. I'll find Harry for you." Meantime, it dials several numbers looking for me. It also beeps me. Eventually I call in. It tells me, "John Smith is calling for you. You want him?" Yes, I say and we're connected. This is a simple form of a broad concept that many are beginning to call intelligent routing. See also AT WORK and WINDOWS TELEPHONY.

Intelligent Terminal A terminal is an input/output device to a distant computer. The terminal may communicate with the computer over a dedicated collection of wires or over phone lines. In the early days, terminals contained no processing power. They simply reflected what the user typed in and what the distant computer responded. As computers became cheaper and with the advent of the "computer on a chip," so it was economically possible to put computing power into a terminal. This reduced the load on the main computer and cut down on communications costs. There are levels of "intelligence" in terminals. An intelligent terminal might perform simple arithmetic functions or it might check the accuracy of input data (does the zip code match the state?). It may perform far more comprehensive processing — as doing virtually all the local processing, and only transmitting summary results to corporate headquarters once a day. A personal computer can be used and act as an Intelligent Terminal. Many personal computer communications software can emulate terminals, the most common being the DEC VT-100.

Intelligent Token A hardware device which generates one-time passwords. In turn, the passwords are verified by a secure server, yielding additional security.

Intelligible Crosstalk Crosstalk from which information can be derived.

INTELSAT INternational TELecommunications SATellite organization. A worldwide consortium of national satellite communications organizations. As of writing Intelsat is owned by 138 governments and Intelsat itself owns 24 satellites worldwide. INTELSAT owns and operates the world's most extensive global communications satellite system. INTELSAT provides international, regional and domestic telephone and television services. It also offers, business services such as international video, teleconferencing, facsimile, data and telex. Comsat acts as the exclusive manager for Intelsat. Comsat is also the exclusive U.S. representative. Comsat stands for Communications Satellite Corporation.

Intensity Modulation IM. In optical communication, a form of modulation in which the optical power output of a source is varied in accordance with some characteristic of the modulating signal. In intensity modulation, there are no discrete upper and lower sidebands in the usually understood sense of these terms, because present optical sources lack sufficient coherence to produce them. The envelope of the modulated optical signal is an analog of the modulating signal in the sense that the instantaneous power of the envelope is an analog of the characteristic of interest in the modulating signal. Recovery of the modulating signal is by direct detection, not heterodyning.

Inter- 1. Means between two things, as opposed to intra, which means inside one thing. Interstate means phone calls and communications between states. Intrastate means communications inside one state. Calling between New York and California is interstate. Calling from Los Angeles to San Francisco is intrastate.
2. There is some argument about whether words should be spelled intra-state or inter-frame. In this dictionary, I spell them all without the dash. It seems to make more sense, since the word intra or inter has become an integral part of so many words.

Interactive The ability of a person or device to talk to or communicate with another device (typically a computer) in real time, i.e. no delays. The term generally is applied in the context of interaction with a computer over a network in a conversational mode. Interactive processing is very time-dependent since a user is sitting there, waiting for the computer to ask him/her questions. The opposite of Interactive processing is batch processing. See BATCH and REAL TIME.

Interactive Call Setup A Rolm definition. An interface between a station user and the Rolm CBX that provides prompts and error messages to process the setting up of data calls, and provides for the displaying of data line parameters.

Interactive CATV Interactive CATV is a two-way cable system from which subscribers can receive and send signals. They will probably do this by punching buttons on their cable TV's remote control, which may look more like a computer keyboard than a traditional cable TV handheld remote signaling device.

Interactive Data Transaction A single (one-way) message, transmitted via a data channel to which a reply is required for work to proceed logically.

Interactive Kiosk Interactive kiosks represent a powerful new product delivery vehicle for "non-store" marketers, to increase sales by offering their products and services in high traffic areas, such as airports. According to research analyst Warren Hersch, financial services kiosks are not confined to full-service bank locations; they are found in supermarkets, shopping malls and auto dealerships. Self-service terminals offering government services reside in discount stores, libraries, outdoor pavilions and subway stations. Kiosks purveying travel-related services occupy office complexes, colleges and universities, pharmacies and other retail outlets.

Interactive Services A B-ISDN term referring to two-way communications in support of three types of services. "Conversational Services" include interactive voice, video and data communications. "Messaging Services" include video mail and compound mail. "Retrieval Services" include retrieval of data, image, video, and compound mail documents. Most of these services are highly bandwidth-intensive, hence their inclusion in the concept of Broadband ISDN.

Interactive Television Association ITV has now changed its name to the Association for Interactive Media (AIM). See AIM.

Interactive Video The fusion of video an computer technology. A video program and a computer program running in tandem under the control of the user. In interactive video, the user's actions, choices, and decisions affect the way in which the program unfolds. See INDEO VIDEO.

Interactive Voice Response IVR. Think of Interactive Voice Response as a voice computer. Where a computer has a keyboard for entering information, an IVR uses remote touchtone telephones. Where a computer has a screen for showing the results, an IVR uses a prerecorded human voice that is stored (digitized) on a hard drive. In addition it can use a synthesized voice (computerized voice) for read back information that is constantly changing. (The synthesized voice is commonly referred to as Text-to-Speech.) Whatever a computer can do, an IVR can too, from looking up train timetables to moving calls around an automatic call distributor (ACD). The only limitation on an IVR is that you can't present as many alternatives on a phone as you can on a screen. The caller's brain simply won't remember more than a few. With IVR, you have to present the menus in smaller chunks.

Interarea Cell Transfer A cellular radio term. A cell transfer between two cells that are controlled by different serving Mobile Data Intermediate Systems (MD-ISs).

Interaxial Spacing Center to center conductor spacing between any two wires.

Interbreak Interval A call center term. A Scenario scheduling assumption specifying the minimum amount of time that must elapse between the end of one break and the beginning of another.

Interbuilding Cable The communications cable that is part of the campus subsystem and runs between buildings. There are four methods of installing interbuilding cable: in-conduit (in underground conduit), direct-buried (in trenches), aerial (on poles), and in-tunnel (in steam tunnels).

Interbuilding Cable Entrance The point at which campus subsystem cables enter a building.

Interbuilding Wiring Consists of underground or aerial telephone wire/cables used on the premises to connect structures remote from the primary building to the premises telephone system.

Intercast New plug-in cards from Intel which will allow your PC to simultaneously receive TV pictures, and in the blank spaces of TV signals, Internet Web pages and text.

Intercept Calls which cannot reach their destination may be intercepted and diverted to a station attendant, a recording or some other place. See INTERCEPT RECORDING and INTERCEPT SERVICE.

Intercept Interval A telephone company term. The intercept interval is the amount of time a changed or disconnected telephone number must remain unassigned in order to insure that after reassignment the new customer does not receive calls intended for the previous subscriber. Intercept intervals vary by customer class of service and are established by the utilities commissions and/or telephone companies.

Intercept Recording You make a phone call. It doesn't go through. The phone company intercepts that call and sends it somewhere. Intercept Recording is a recording telling you your call cannot be completed and has been intercepted on its way to the destination number for some reason that will be explained by the recording. The most common voice you hear on intercept announcements is Jane Barbie's. See BARBIE, JANE and INTERCEPT SERVICE.

Intercept Service A service of the local phone in which a phone call is redirected by an operator or a recording to another phone number or a message.

Interchange Carrier IC. A common carrier that provides services to the public local exchanges on an intra or interLATA basis in compliance with local or Federal regulatory requirements and that is not an end user of the services provided.

Interchangeable NPA Code Code in the NXX format used as a central office code (NNX format), but that can also be used as an NPA code. Interchangeable NPA codes will be introduced on or after January 1, 1995.

Intercom Intercommunication. An internal communication system which allows you to dial another phone in your building, office complex, factory or home. There are three types of intercom: 1. Dial: It allows you to dial or pushbutton another extension; 2. Automatic: One phone goes off hook and automatically dials another; and 3. Manual: The user can manually signal another phone by pushing a button for that phone. An example is a buzzer between a boss and a secretary.

Intercom Blocking A PBX feature by which phones with a particular Class Of Service (COS) are blocked from calling certain phones. A rare feature.

Interconnect A circuit administration point, other than a cross connect or an information outlet, that provides capability for routing and re-routing circuits. It does not use patch cords or jumper wires, and typically is a jack-and-plug device used in smaller distribution arrangements or that connects circuits in large cables to those in smaller cables. See INTERCONNECT COMPANIES.

Interconnect Agreement An agreement between an established local phone company and a new local phone company for both companies to allow their subscribers to dial each other. Such agreement covers issues such as sharing of revenues and if a subscriber, who changes local phone companies, can keep his phone number.

Interconnect Companies Companies which sell, install and maintain telephone systems for end users, typically businesses. AT&T coined the word "interconnect" as a pejorative word — to indicate that these companies "interconnected" to AT&T's telephone network — but didn't really belong there and, if they were there, they were probably unreliable. These "interconnect" companies contrasted with true-blue companies belonging to AT&T which did a sterling job. Anyway, despite the changes in the industry, the term stuck and the nasty associations have pretty well gone away. Now the irony is that the indpendent (i.e. non-Bell, non-AT&T) interconnect companies often deliver better service at a lower price. The industry is looking for a better word. TELECONNECT Magazine once started a campaign to make "TELECONNECT" a replacement for interconnect. But TELECONNECT's lawyers and the lawyers for a manufacturing/interconnect company called Teleconnect told us to lay off and stop trying to make the word generic. Since then we rather like the terms "Telecommunications Systems Integrator," "Telecommunications VAR" or "Telecom Developer." They seem to be catching on.

Interdomain Trust Relationships With Windows NT, Unix and some other operating systems, the user accounts and global groups from one domain can be used in another domain. In the MIS world, a domain is the part of a computer network in which the data processing resources are under common control. In the Internet, a domain is a place you can visit with your browser — i.e. a World Wide Web site. When a domain is configured to allow accounts from another domain to have access to its resources, it effectively trusts the other domain. The trusted domain has made its accounts available to be used in the trusting domain. These trusted accounts are available on Windows NT Server computers and Windows NT Workstation computers participating in the trusting domain.

Hint By using trust relationships in your multidomain network, you reduce the need for duplicate user account information and reduce the risk of problems caused by unsynchronized account information.

The trust relationship is the link between two domains that enables a user with an account in one domain to have access to resources on another domain. The trusting domain is allowing the trusted domain to return to the trusting domain a list of global groups and other information about users who are authenticated in the trusted domain. There is an implicit trust relationship between a Windows NT Workstation participating in a domain and its PDC.

In this example, the following statements are true because the London domain trusts the Topeka domain:

Users defined in the Topeka domain can access resources in the London domain without creating an account within that domain. Topeka appears in the From box at the initial logon screen of Windows NT computers in the London domain. Thus, a user from the Topeka domain can log on at a computer in the London domain. When trust relationships are defined, user accounts and global groups can be given rights and permissions in domains other than the domain where these accounts are located. Administration is then much easier, because you need to create each user account only once on your entire network, and then the user account can be given access to any computer on your network (provided you set up domains and trust relationships to allow it).

Note Trust relationships can be configured only between two Windows NT Server domains. Workgroups and LAN Manager 2.x domains cannot be configured to use trust relationships.

Interdrive The name of the FTP Software client implementation of the Sun NFS protocol.

Interenterprise Communications Communications exchanged between multiple organizations, e.g., between business trading partners, collaborators, affiliates or a business and its customers.

Interexchange Carrier IXC. At one stage an IXC was a telephone company that was allowed to provide long-distance telephone service between LATAs but not within any one LATA. Then some states in the United States started allowing intra-LATA competition. Now an IXC is best defined as a telephone company that is allowed to provide long-distance telephone service between LATAs. See IXC and LATA. Contrast with LEC.

Interexchange Channel IXC. A communications channel or path between two or more telephone exchanges.

Interexchange Customer Service Center ICSC. The Telephone Company's primary point of contact for handling the service needs of all long distance carriers.

Interface 1. A mechanical or electrical link connecting two or more pieces of equipment together.
2. A shared boundary. A physical point of demarcation between two devices where the electrical signals, connectors, timing and handshaking are defined. The procedures, codes and protocols that enable two entities to interact for a meaningful exchange of information.
3. To bring two things or people together to allow them to talk.
4. A poorly-defined word often used when the speaker is incapable of figuring precisely what he means. No one would ever invite a pretty girl out to lunch asking her to "interface" with you. See also INTERFACE DEVICE.
5. According to Steven Johnson's book, "Interface Culture — How new technology transforms the way we create and communicate," the word interface "refers to software that shapes the interaction between user and computer: The interface serves as a kind of translator, mediating between the two parties, making one sensible to the other. In other words, the relationship governed by the interface is a semantic one, characterized by meaning and expression rather than physical force."

Interface Device A device which meets a standard electrical interface on one side and meets some other nonstandard interface on the other. The purpose of the device is to allow a device with a nonstandard interface to connect to a device with a standard interface. See also INTERFACE.

Interface Functionality The characteristic of interfaces that allows them to support transmission, switching, and signaling functions identical to those used in the enhanced services provided by the carrier. As part of its comparably efficient interconnection (CEI) offering, the carrier must make available standardized hardware and software interfaces that are able to support transmission, switching, and signaling functions identical to those used in the enhanced services provided by the carrier.

Interface Manager The original name for Microsoft's Windows. Later called Windows, and finally shipped in its first version in November 1985.

Interface Message Processor IMP. A processor-controlled switch used in packet-switched networks to route packets to their proper destination.

Interface Overhead the interface overhead is the remaining portion of the bit stream after deducting the information payload. The interface overhead may be essential (e.g. framing for an interface shared by users) or ancillary (e.g. performance monitoring).

Interface Payload The portion of the bit stream which can be used for telecommunications services. Any signaling is included in the interface payload. See also INTERFACE OVERHEAD.

Interface Shelves Shelves in a Rolm PBX cabinet containing the printed circuit card groups that connect telephones, terminals, lines and trunks to CBX interface channels. These shelves also contain shared electronics cards.

Interference Energy you receive with a signal. You don't want the energy. You want the signal. Getting rid of the interference may be a pain. The interference may be manmade (e.g. elevator motors) or it may be GOD-made, e.g. lightning, thunderstorms. Some media (fancy word for cabling) may be more immune to interference than other. Media immune to interference, in order
1. Optical fiber
2. Coax
3. Shielded twisted pair
4. Unshielded twisted pair
5. Unshielded untwisted pair

Interference Emission Emission that results in an electrical signal being propagated into and interfering with the proper operation of electrical or electronic equipment. The frequency range of such interference may be taken to include the entire electromagnetic spectrum.

Interferometer An instrument that employs the interference of light waves for measurement.

Interflow The ability to establish a connection to a second ACD and overflow a call from one ACD to the other. This provides a greater level of service to the caller.

Interframe See INTERFRAME CODING.

Interframe Coding A video term. It's a technique to cut down the size of the video to save on transmission costs. It's a way of source coding where the temporal correlation of moving pictures is used for data reduction. Interframe coding use compression techniques which track the differences between frames of video and eliminates redundant information between frames. Interframe coding stores only once those pixels that don't change. Then multiple frames access those pixels during decompression. This results in more compression over a range of frames than intraframe coding, which compresses information within a single frame.

Interframe Encoding A way of video compression that transmits only changed information between successive frames. This saves bandwidth. See interframe coding.

Interframe Gap IFG. The minimum idle time between the end of one frame transmission and the beginning of another. On Ethernet 802.3 LANs the minimum interframe gap is 9.6 micro-seconds.

Interim Interswitch Signal Protocol A call routing scheme used in STM networks. Formerly known as PNNI Phase 0. IISP is an interim technology meant to be used pending completion of PNNI Phase 1. IISP uses static routing tables established by the network administrator to route connections around link failures.

Interior An ATM term. Denotes that an item (e.g., link, node, or reachable address) is inside of a PNNI routing domain.

Interlace In TV, each video frame is divided into two fields with one field composed of the odd- numbered horizontal scan lines and the other composed of the even-numbered horizontal scan lines. Each field is displayed on an alternating basis. This is called interlacing. It is done to avoid flicker. See also INTERLACING.

Interlaced GIF When you're downloading a Web page, which contains images, interlaced GIF images appear first with poor resolution and then improve in resolution until the entire image has arrived, as opposed to arriving linearly from the top row to the bottom row. This lets users get a quick idea of what the entire image will look like while waiting for the rest to load. Your Web browser has to support progressive display. Non-progressive-display Web browsers will still display interlaced GIFs, but only after they have arrived in their entirety.

Interlaced Image A Web term. A GIF image that is displayed full-sized at low resolution while it is being loaded, and at increasingly higher resolutions until it is fully loaded and has a normal appearance. See Interlaced GIFs.

Interlacing Regular TV signals are interlaced. In the US there are 525 scanning lines on the regular TV screen. This is the NTSC standard. Interlaced means the signal refreshes every second line 60 times a second and then jumps to the top and refreshes the other set of lines also 60 times a second. Non-interlaced signals, which are used in the computer industry, means each line on the entire screen is refreshed X times. X times depends on what the video card is outputting to the color monitor. The more expensive the card and the monitor, the more often the monitor will be refreshed. The more it's refreshed, the better and more stable it looks — the less perceived flicker. For example, text on an NTSC United States TV set tends to "flicker." It doesn't on a non-interlaced monitor. Typical non-interlaced computer monitors refresh at 60 to 72 times a second. But good ones refresh at higher rates. Generally, anything over 70 Hz (i.e. 70 times a second) is considered to be flicker-free and therefore preferred, if you can afford it. In short, buy an non-interlaced monitor. You'll like it better.

InterLATA Telecommunications services that originate in one and terminate in another Local Access and Transport Area (LATA). Under provisions of Divestiture, the Bell operating companies cannot provide Inter-LATA service, but can provide Intra-LATA service. Some LATAs are very large. So some "local" phone companies provide the equivalent of long distance service. And some of these phone companies have different pricing packages. Some of these packages are cheap, but not highly-publicized. See also LATA.

InterLATA Call A call that is placed within one LATA (Local Access Transport Area) and received in a different LATA. These calls are currently carried by a long distance company.

InterLATA Competition Originally long distance telephone companies in the United States were now allowed to provide InterLATA telecom services. Later, many Public Utility Commission (PUC) in ARF (Alternative Regulatory Framework) Phase III started to consider it. And many state agencies started to allow long distance phone companies to compete with local monopolies to carry intraLATA toll calls.

Interleave 1. The transmission of pulses from two or more digital sources in time-division sequence over a single path.

2. A data communication technique, used in conjunction with error-correcting codes, to reduce the number of undetected error bursts. In the interleaving process, code symbols are reordered before transmission in such a manner that any two successive code symbols are separated by I-1 symbols in the transmitted sequence, where I is called the degree of interleaving. Upon reception, the interleaved code symbols are reordered into their original sequence, thus effectively spreading or randomizing the errors (in time) to enable more complete correction by a random error-correcting code.

3. Interleaving also refers to the way a computer writes to and reads from a hard disk. Understanding interleaving is critical if you want to get your hard disk to work at its maximum speed (without in any way damaging the disk). Let's look at the way MS-DOS reads information from a hard disk. All hard disks are controlled by a special card called a hard disk controller card. Let's say your computer wants a file. It tells the hard disk controller card it wants the file. The controller searches the disk for the first sector of the requested file, reads that sector (usually 512 bytes) to your computer's RAM and then transfers the information to the CPU to be processed. When this is complete, the controller goes back to the hard disk and searches for the file's second sector. The process continues in this way until the file is completely read. The problem is that while the controller and the CPU are doing their things, the hard disk itself is spinning 3,600 times a minute. By the time the controller reads one sector and it is ready to return to the disk, the next consecutive file sector has spun past the read/write head. If the file is stored in contiguous sectors, the controller must wait for the disk to complete its revolution before it can read the next file sector. To solve this problem, hard drive makers developed a concept called interleave setting, which tells the hard drive controller to skip a certain number of sectors when it writes a file to disk. Thus, when the file is later read back, the appropriate file sectors should fall under the read/write heads at the appropriate time. If the controller reads or writes one sector and then skips a sector, the interleave is 2 (every other sector is used to store logically consecutive blocks). The interleave is sometimes written as 2:1. If the controller writes to one sector and then skips two, the interleave is 3 or 3:1. The interleave factor is usually established by the manufacturer or reseller of the hard disk/controller combination. If someone else assembles the hard disk/controller combination, that person may need to experiment to determine the correct interleave factor — i.e. the one that works fastest without messing up.

Setting the "correct" interleave settings on your hard disk is critical to getting maximum performance out of your hard disk. Here's a test that a writer for PC Resource Magazine did. He copied the same files from one part of his hard disk to another part using different interleave settings:

Interleave Setting	Time to copy file
3	1 min 15 seconds
4	1 min 17 seconds
5	1 min 1 second
6	35 seconds
7	41 seconds
8	1 min 10 seconds

Clearly his best interleave setting is six. There are two ways of choosing the correct interleave setting. You can do it by trial and error as the writer did. His test took three hours. Or buy a program and do it in seconds. The best program is called

Disk Technician. It's from a company called Prime Solutions in San Diego. Sadly, the program doesn't work on certain laptops and on certain controller card/hard disk combinations.

Interleaved Memory An option on some system boards that increases processing speed by assigning memory locations on an alternating basis to two banks of RAM. The computer has to wait one cycle between accesses to a single bank of memory, but it can access a different bank without having to wait.

Intermediate Assist A method for pulling cables into conduits or duct liners in which manual labor or machines are used to assist the pulling at intermediate manholes.

Intermediate Cross-Connect an interconnect point within backbone wiring. for example, the interconnection between the main cross-connect and telecommunications closet or between the building entrance facility and the main cross-connect.

Intermediate Distribution Frame IDF. A metal rack designed to connect cables and located in an equipment room or closet. Consists of bits and pieces that provide the connection between inter-building cabling and the intra-building cabling, i.e. between the Main Distribution Frame (MDF) and individual phone wiring. There's usually a permanent big, fat cable running between the MDF and IDF. The changes in wiring are done at the IDF. This saves confusion in wiring. See also FEEDER CABLE and CONNECTING BLOCK.

Intermediate Frequency Transformer A transformer designed to amplify the intermediate frequencies generated in a superheterodyne radio receiver. These are normally sharply tuned to a single frequency band.

Intermediate High-Usage Trunk Group A Bellcore definition. A high-usage trunk group that receives route-advanced overflow traffic and may receive first-route traffic and/or switched-overflow traffic.

Intermediate Reach Intermediate reach refers to optical sections from a few kilometers (km) to approximately 15 km. An AT&T SONET term.

Intermediate System An OSI term which refers to a system that originates and terminates traffic, as well as forwarding traffic to other systems.

Intermittent Problems Intermittent problems are issues or bugs that come to light only after systems have been running for some time, or certain infrequently performed sequences of events are performed. Often many thousands of calls need to be put through before they are discovered. And bugs may only be seen occasionally, perhaps one of every 100 times something is done. Intermittent problems are among the hardest to find and duplicate.

Intermix A mode of service defined by Fibre Channel that reserves the full Fibre Channel bandwidth for a dedicated (Class 1) connection, but also allows connectionless (Class 2) traffic to share the link if the bandwidth is available.

Intermodulation IM. The production, in a nonlinear element of a system, of frequencies corresponding to the sum and difference frequencies of the fundamentals and integral multiples (harmonics) of the component frequencies that are transmitted through the element.

Intermodulation Distortion IMD. Nonlinear distortion characterized by the appearance of frequencies in the output, equal to the sum and difference frequencies of integral multiples (harmonics) of the component frequencies present in the input. Harmonic components also present in the output are usually not included as part of the intermodulation distortion.

Internal Bus See LOCAL BUS.

Internal F-ES A cellular radio term. A fixed End System within the administrative domain of the CDPD service provider. Typically provides value-added support services such as network management, accounting, directory, and authentication services.

Internal Modem A modem on a printed circuit card which is inserted into one of the slots on a PC (personal computer). The other type of modem for a PC is an external modem — essentially a modem with the same circuitry as an internal modem but with a metal or plastic case. An internal modem costs slightly less than an external one. Internal modems are good if you're short of desk space and afraid your external modem will be stolen. External modems have lights so it's easier to tell what's going on. Everybody has their theories on which type of modem is best. We prefer the external ones — largely for their lights and ease of moving around.

Internal Reachable Address An ATM term. An address of a destination that is directly attached to the logical node advertising the address.

International 800 Service You can now have your customers overseas call you for free on an 800 line, just as your domestic customers do. The service is available from countries including Australia, Brazil, France, Hong Kong, Israel, Italy, Japan, Sweden, Switzerland and the United Kingdom. Overseas 800 service is often known as "Freephone."

International Ad Hoc Committee IAHC. One of the organizations which parcels out Internet domain names. The IAHC has proposed to add a number of TLD (Top Level Domain) names to the existing list of .com (commercial), .edu (education), .gov (government), .mil (military) and .org (not-for-profit organizations). Those proposed TLDs include .firm (businesses), .store (stores), .web (entities emphasizing cultural and entertainment activities), .rec (entities emphasizing recreation/entertainment activities), .info (entities providing information services) and .nom (those wishing individual or personal nomenclature). See DOMAIN, DOMAIN NAME SERVER, AND DOMAIN NAMING SERVICE.

International Alphabet No. 5 IA5. Internationally standardized alphanumeric code with national options. ASCII is United States version.

International Ampere The current which will in one second deposit 0.001118 gram of silver from a neutral solution of silver nitrate.

International Callback Calling the United States from many countries abroad is far more expensive than calling those countries from the United States. A new business called International Callback has started. It works like this. You're overseas. You dial a number in the United States. You let it ring once. It won't answer. You hang up. You wait a few seconds. The number you dialed in the U.S. knows it was you calling. There is a piece of equipment on that number that "hears" it ring and knows it's you since no one else has that number. (Typically it's done with Centrex service.) That was your special signal that you want to make a call. A switch attached to that line then calls you instantly. When you answer (overseas, obviously) it conferences you with another phone line in the United States and gives you U.S. dial tone. You can then touchtone from overseas your American number, just as if you would, were you physically in the U.S. There are huge savings. U.S. international callback operators can offer as high as 50% savings on calls from South America, where international calling rates are very high. The process of international callback is being automated with software and dialing devices. International callback is also helping to bring down the high cost of calling the U.S. from overseas. In recent years, deregulation has caused the price of international calls in many countries to fall dramatically. And now international callback or just callback is being done from other countries, including and especially Israel. A company called Kallback in Seattle, WA. has received a service mark from the U.S. Patent and Trademark Office for the words "callback" and "kallback" and sends letters to and threatens law suits against companies who use "their" words. See CALLBACK.

International Carrier A carrier that generally provides connections between a customer located in World Zone 1 and a customer located outside of World Zone 1, but with option of providing service to World Zone 1 points in North American Numbering Plan area codes outside the U.S.

International Center For Information Technologies A Washington "think tank" whose mission is to bring together discussion on new telecommunications and computer technologies. Targeted at senior executives who are looking for ways to apply new technologies to gaining competitive edges for their company.

International Denial An optional restriction on your cellular phone that prevents the cellular number from marketing international calls. Some carriers place this restriction on all subscribers using their service.

International Direct Distance Dialing IDDD. Being able to automatically dial international long distance calls from your own phone. The direct calling by the originating customer to the distant (international) called customer via automatic switching. IDDD is synonymous with the phrases international direct dialing and international subscriber dialing.

International Engineering Consortium The International Engineering Consortium, established in 1944, is a non-profit organization dedicated to catalyzing positive change in the information industry and its university communities. The Consortium provides educational opportunities for today's information industry professionals and conducts a variety of industry-university programs. The IEC also conducts research and provides publications addressing major opportunities and challenges of the information age. More than 70 leading, high technology universities are currently affiliated with the Consortium. www@iec.org

International Gateways The switches in the various domestic long distance networks (e.g. MCI, AT&T and Sprint) which interface their networks with International telecommunications networks. All US International calls are routed through an international gateway.

International Morse Code See Morse Code

International Organization For Standardization ISO. An organization established to create standards. See ISO.

International Prefix The combination of customer-dialed digits prior to dialing of the country code required to access the automatic outgoing international equipment in the originating country.

International Record Carrier IRC. One of a group of carriers that, until recently, was part of a monopoly of U.S. common carriers certified to carry data and text to locations outside the U.S. In recent years, regulation of this type of service has been markedly relaxed. Most of the IRCs got bought by MCI.

International Computer Security Association. See ICSA

International Standards Organization ISO. An international standards-setting organization.

International Switching Carrier ISC. An exchange whose function is to switch telecommunications traffic between national network and the networks of other countries. Also known as an international gateway office.

International Telecommunication Union-Radiocommunication Sector ITU-R. Formerly the International Radio Consultative Committee (CCIR). The technical study branch of the International Telecommunication Union responsible for the study of technical and operating questions relating specifically to radio communications. See also ITU-T.

International Telecommunications Union ITU. The specialized agency of the United Nations which tries to establish standardized communications procedures and practices. Its most successful work is done in the allocation of radio frequencies worldwide — including satellites, etc. See ITU and ITU-T.

International Telephone Address A four-part code specifying a unique address for any telephone company in the world.

Internet It is very hard to define the Internet in a way that is either meaningful or easy to grasp. To say the Internet is the world's largest computer network is to trivialize it. But it is. And it is undoubtedly the most important happening to the computing and communications industries since the invention of the transistor. The Internet has its roots as a cooperative research effort of the United States Federal Government known as the Advanced Research Project Agency NETwork (ARPAnet), which was established in 1969 by the Defense Department. ARPAnet tied universities and research and development organizations to their military customers, and provided connectivity to a small number of supercomputer centers to support timesharing applications. Much of the funding was provided by NSFNET (National Science Foundation NETwork). More recently, the Internet has been commercialized, extending its use to anyone with a PC, a modem, a telephone line and an access provider — a special company known as an Internet Service Provider or Internet Access Provider, who allows their customers to reach the Internet via dial-up or dedicated line. The Internet has become a major new publishing, research and commerce medium. I believe that its invention is as important to the dissemination of knowledge, to peoples' life styles and to the way we'll be conducting business in coming years as the invention of the Gutenberg Press was in 1453.

At its heart, the Internet is many large computer networks joined together over high-speed backbone data links ranging from 56 Kbps (now rare) to T-1, T-3, OC-1 and OC-3. The Internet, in short, is a network of computer networks. The Internet now reaches worldwide. Depending on the whim of the local government (which typically controls the local phone company and thus access to the Internet for its citizenry) you can pretty well get onto the Internet and roam it unchecked. The governments of Singapore, the People's Republic of China, Burma and few others limit their peoples' access.

The topology of the Internet and its subnetworks changes daily, as do its providers and its content. The bottom line is that the makeup of the Internet — i.e. how it works — is not all that important. It is the applications and information available on it that are important — the most significant of which are e-mail (electronic mail) and the World Wide Web. Commercial networks from AT&T, MCI, SPRINT, Worldcom and many others now carry the bulk of the traffic. As NSFNET no longer funds the Internet, it has been commercialized, with money changing hands in complex ways between users, companies with Web sites, Internet Access Providers, long distance providers, government, universities and others. The Internet remains supported by some level of public funding, although it is less direct that in the past. Increasingly, businesses are joining their computers to the Internet. According to Network Wizards (www.nw.com), as of year end 1996 the Internet linked over 60,000 networks, 9.5 million computers and 35 million users in 150 countries. The commercialization of the Net has led to exponential growth in both the number of connected hosts and the overall volume of traffic, creating bottlenecks. As a result, a large number of research universities have begun the development of Internet II, also known as Internet2, which effectively is a separate Internet for colleges, universities and government organizations. In other words, the organizations which founded the original Internet are getting off it in favor of building their own Internet.

The Internet's networking technology is very smart. Every time someone hooks a new computer to the Internet, the Internet adopts that hookup as its own and begins to route Internet traffic over that hookup and through that new computer. Thus as more computers are hooked to the Internet, its network (and its value) grows exponentially. The Internet is basically a packet switched network based on a family of protocols called TCP/IP, which stands for Transmission Control Protocol/Internet Protocol (TCP/IP), a family of networking protocols providing communication across interconnected networks, between computers with diverse hardware architectures and between various computer operating systems. Most PCs, including Windows-based machines and Macintoshes, will happily communicate using TCP/IP.

How TCP Works: TCP is a reliable, connection-oriented protocol. Connection-oriented implies that TCP first establishes a connection between the two computer systems that intend to exchange data (e.g. your PC and the host computer you're trying to reach, which may be thousands of miles away). Since most networks are built on shared media (for example, several systems sharing the same cabling), it is necessary to break chunks of data into manageable pieces so that no two communicating computers monopolize the network. These pieces are called packets. When an application sends a message to TCP for transmission, TCP breaks the message into packets, sized appropriately for the network, and sends them over the network. Because a single message is often broken into many packets, TCP marks these packets with sequence numbers before sending them. The sequence numbers allow the receiving system to properly reassemble the packets into the original original order, i.e. the original message. TCP checks for errors. And finally, TCP uses port IDs to specify which application running on the system is sending or receiving the data. The port ID, checksum, and sequence number are inserted into the TCP packet in a special section called the header. The header is at the beginning of the packet containing this and other "control" information for TCP.

How IP Works: IP is the messenger protocol of TCP/IP. The IP protocol, much simpler than TCP, basically addresses and sends packets. IP relies on three pieces of information, which you provide, to receive and deliver packets successfully: IP address, subnet mask, and default gateway. The IP address identifies your system on the TCP/IP network. IP addresses are 32-bit addresses that are globally unique on a network. There's much more on TCP/IP in my definition on TCP/IP and on Internet Addresses in that definition.

Here's how the Internet is used: As a computer network joining two (or more) computers together in a session, it is basically transparent to what it carries. It doesn't care if it carries electronic mail, research material, shopping requests, video, images, voice phone calls, requests for information, faxes ... or anything that can be digitized, placed in a packet of information and sent. A packet-switched network like the Internet injects short delays into its communications as it disassembles and assembles the packets of information it sends. And while these short delays are not a problem for non-real time communications, like email, they present a problem for "real-time" information such as voice and video. The Internet can inject a delay of as much as half a second between speaking and being heard at the other end. This makes conversation difficult. Internet telephony, as it's called when it runs on the Internet, is getting better, however, as the Internet improves and voice coding and compression techniques improve. I've enjoyed some relatively decent conversations to distant places in recent months.

See various Internet definitions following. See also Domain, Domain Naming System, gTLD, Internet2, Internet Appliance, Internet Protocol, Internet Telephony, Intranet, IP Telephony, Surf, TCP/IP and Web Browser.

Internet2 The next generation Internet, replacing the current Internet exclusively for the use of member universities, Internet2 is a project of the University Corporation for Advanced Internet Development (UCAID). As a result of the deteriorating performance of the Internet, 34 U.S. universities announced in October 1996 the formation of Internet2. Subsequently, the central goals of the project were adopted as part of the Clinton administration's Next Generation Initiative (NGI). This second version of the Internet is a collaboration of the National Science Foundation (NSF), the U.S. Department of Energy, over 110 research universities, and a small number of private businesses. Each participating university has committed at least $500,000 to fund the project. Intended to serve as a private Internet for the exclusive use of its member organizations, it will be separate from the traditional Internet. The network eventually will operate over fiber optic transmission facilities at speeds of up to 2.4 Gbps (SONET OC-48), although current speeds of connection are at 155 Mbps (OC-3) and 622 Mbps (OC-12). Internet2 will connect through gigiPOPs, switches with throughput in the range of billions of packets per second, and will run the IPv6 protocol. www.internet2.edu. See also Internet.

Internet Address A unique, 32-bit identifier for a specific TCP/IP host computer on a network. Also called an Internet Protocol (IP) address. IP addresses are in dotted decimal form, such as 128.127.50.224, with each of the four address fields assigned as many as 255 values. Internet addresses are organized into hierarchical "classes," as follows:

Class A Networks: 0.x.x.x to 127.x.x.x (1 network number, 3 host numbers). Only a few Class A networks exist. Examples include General Electric Company, IBM Corporation, AT&T, Hewlett-Packard Company, Ford Motor Company, and the Defense Information Systems Agency. They all are huge organizations, and require the highest possible categorization.

Class B Networks: 128.0.x.x to 191.255.x.x (2 network numbers, 2 host numbers)

Class C Networks: 192.0.0.x to 255.255.255.x (3 network numbers, 1 host number). Most applicants are assigned Class C addresses in blocks of 255 IP addresses.

Internet Appliance A sub-$500 machine specially designed for Internet browsing and first proposed in the late Fall of 1995 by Larry Ellison, head of database software company Oracle. Part of its appeal to people outside Microsoft and Intel is that the Internet Appliance would not have to be based on standard PC technology. It need have an Intel chip and need not run Windows. This device is also called an Internet Terminal, a Network Computer or an IPC, an Interpersonal computer. The original description of the Internet Appliance was that it would come with 4mb of RAM, 4mb of flash memory, processor, monitor, keyboard and mouse — all for under $500.

Internet Architectures Board The Internet Architectures Board oversees Internet protocols and procedures and the creation of Internet standards. According to Computerworld Magazine, it has 16 members. www.iab.org/lab/

Internet Assigned Numbers Authority IANA. This group is responsible for the assignment of unique Internet parameters (e.g., TCP port numbers, and ARP hardware types), and managing domain names. It also was responsible for administration and assignment of IP (Internet Protocol) numbers within the geographic areas of North America, South America, the Caribbean and sub-Saharan Africa; on December 22, 1997, that responsibility was shifted to ARIN (American Registry for Internet Numbers). www.arin.net. The IANA has well-established working relationships with the US Government, the Internet Society (ISOC), and the InterNIC. ISOC provides coordination of IANA activities with the Internet Engineering Task Force (IETF) through the participation of IANA in the Internet Architecture Board (IAB). IANA responsibility was assigned by DARPA (Defense Advanced Research Project Agency) to the Information Sciences Institute (ISI) of the University of Southern California. ISI has discretionary authority to delegate portions of its functions to an Internet Registry (IR), previously performed by SRI International and currently performed by Network Solutions Inc. (NSI), a subsidiary of SAIC. Beginning March 1998, that function is shared with the Council of Registrars (CORE). CORE contracted (November 1997) with Emergent Corporation to build and operate the new Internet Name Shared Registry System (SRS), which is a neutral, shared database repository that coordinates registrations from CORE and propagates those names to the global Internet Domain Name System (DNS). www.isi.edu/div7/iana/ See also ARIN, CORE, DNS, Internet, InterNIC, and SRS.

Internet Cable Access A general term used to describe accessing the Internet using the cable TV coaxial cable for inbound Internet access (i.e. downstream) and the phone line for up sending commands and requests (i.e. upstream information). The cable TV is very fast — as much as six million bits per second. The phone is relatively slow — no more than fifty thousand bits per second. But it works because most information from the Internet flows at you, not away from you. The cable and telecom industry is working on standards to make disparate cable systems and TV set-top boxes work with each other. The industry has developed Data Over Cable Service Interface Specification (DOCSIS), which sets standards for both two-way and cable-plus-phone specifications. See DOCSIS.

Internet Control Message Protocol ICMP. The protocol used to handle errors and control messages at the IP player. ICMP is actually part of the IP protocol.

Internet Engineering Steering Group IESG. The executive committee of the IETF (Internet Engineering Task Force).

Internet Engineering Task Force IETF. One of two technical working bodies of the Internet Activities Board. The IETF is the primary working body developing new TCP/IP (Transmission Control Protocol/Internet Protocol) standards for the Internet. It has more than one thousand active participants. www.ietf.org

Internet Fax Internet fax is, as it sounds, sending faxes over the Internet. There are a whole bunch of manual ways to send faxes over the Internet — most of which are akin to sending a fax over the PSTN, as we do it today. Dial up, etc. There are movements, however, to automate this process and get Internet faxing more along the lines of Internet email. Internet Fax is coming in two parts. The first is a store and forward model that is essentially based on the MIME attachment of TIFF files to standard E-Mail messages delivered by SMTP. The standards for this model are found in the IETF - ITU agreements of January 1998. The second part is an Internet draft that extends SMTP itself. The draft turns a fax machine into a virtual SMTP server so that transmission of the fax from point-to-point happens in real time. The protocol would extend SMTP beyond its function of a simple mail transport protocol to the point where, when a transport session is established, the user can exchange capabilities between devices - something that cannot be done with store and forward mail. Implementing these will be a series of hybrid "stupid-smart" devices that bridge faxes between the PSTN and the Internet. The Panasonic FO-770I, which is already on the market, is one such device with almost all the capabilities of the new standard . Load your fax, toggle "send" in one direction to transmit via the PSTN, toggle "send" in the other direction to go via the Internet. Other manufacturers are working on the introduction of inexpensive "black boxes" to connect standard G3 faxes in small-office, home-office (SOHO) environments directly to one's PC and from there to the Internet.

Internet Gateway Internet gateways are devices which typically sit on a local area network and handle all the translations between IPX traffic on your LAN (IPX is the NetWare protocol) and the TCP/IP traffic on the Internet. TCP/IP is the protocol used on the Internet. See also INTERNET SERVERS and other definitions beginning with INTERNET.

Internet Group Name In Microsoft networking, a name registered by the domain controller that contains a list of the specific addresses of computers that have registered the name. The name has a 16th character ending in 0x1C.

Internet Intellectual Infrastructure Fund A fund created in 1995 to offset government funding for the preservation and enhancement of the intellectual infrastructure of the Internet. The fund was funded by 30% of the Internet domain registration fee, which was set at $50 per year at that time. On March 16, 1998, the funding for the Intellectual Infrastructure was completed, and the InterNIC ceased to collect that portion of the annual fee, thereby reducing it to $35 for new registrations. Proceeds of the fund are to be used to build Internet2, which will be a separate Internet for institutions of higher learning. See also CORE, DNS and InterNIC.

Internet Mail Consortium IMC. A technical trade association which pursues cooperative promotion and enhancement of electronic mail and messaging on the Internet. Activities cover promotion of Internet mail and the products and services which serve to implement it. IMC is involved in formative efforts for IETF (Internet Engineering Task Force) mail standards, with a focus on implementation guidelines. (www.imc.org)

Internet MIB Subtree A tree-shaped data structure in which network devices on a local area network and their attributes can be identified within the confines of a network management scheme. The name of an object or attribute is derived from its location on this tree.

For example, an object in MIB-I might be named 1.2.1.1.1.0. the first 1 indicates the object is on the Internet. The 2 denotes that it falls within the Management category. The second 1 shows the object is part of the first fully defined MIB, known as MIB-I. The third 1 indicates which of the eight object groups is being referenced. And the fourth 1 is a textual description of the network component. The 0 indicates there is only one object instance. An object instance links a particular object to a specific node on the network. The numbering system is infinitely extendible to accommodate additions to this base identification scheme. This common naming structure permits equipment from a variety of vendors to be managed by a single management station that uses SNMP. The four main categories of the tree are Directory, Management, Experimental and Private/Enterprises.

Internet Number The dotted-quad address used to specify a certain system. The Internet number for cs.widener.edu is 147.31.130. A resolver is used to translate between hostnames and Internet addresses.

Internet Numbers Registry IR. The officially designated organization responsible for the assignment of IP addresses, the IR assigns unique UFLs (Uniform Resource Locators), which are translated into IP addresses through a resolver. IR is a responsibility of the IANA (Internet Assigned Numbers Authority), a function assigned to the Information Sciences Institute (ISI) of the University of Southern California. In accordance with its discretionary authority, ISI initially delegated that responsibility to SRI International and, subsequently, to Network Solutions Inc. (NSI). Beginning March 1998, NSI shares that responsibility with CORE and Emergent Corporation, which administers the Shared Numbers Registry (SRS). See also CORE, IANA, SRS, and URL.

Internet Packet Exchange IPX. Novell NetWare's native LAN communications protocol, used to move data between server and/or workstation programs running on different network nodes.

Internet Protocol IP. Part of the TCP/IP family of protocols describing software that tracks the Internet address of nodes, routes outgoing messages, and recognizes incoming messages. Used in gateways to connect networks at OSI network Level 3 and above. See Internet Protocol Address and TCP/IP.

Internet Protocol (IP) Address A unique, 32-bit number for a specific TCP/IP host on the Internet. IP addresses are normally printed in dotted decimal form, such as 128.127.50.224. Once your domain is assigned a group of numbers by the Internet's central registry, it can house one or several domains and/or hosts, i.e. computertelephony.com and teleconnect.com. People looking for those domains will be pointed to that server where they will find all information in the domain — perhaps a home page, or a place to leave e-mail, etc. There are three classes of IP address A, B, and C — the most common of which is a class "C" address block. A class "C" address block can address about 256 hosts (e.g., 128.10.10.*). a class "B" address block can contain about 256*256 (e.g., 128.10.*.*) hosts. Some ip addresses are reserved for broadcasts in respective domains. See DOMAIN and INTERNET.

Internet Protocol Datagram The fundamental unit of information passed across the Internet. Contains source and destination addresses along with data and a number of fields which define such things as the length of the datagram, the header checksum, and flags to say whether the datagram can

be (or has been) fragmented, This is a self-contained packet, independent of other packets.

Internet Protocol Suite A suite of network protocols which have been adopted as the main de facto protocols for LANs.

Internet Registry Activities involved in the administration of generic Top Level Domains (gTLDs) in the CORE (Council or REgistrars) Domain System. Such activities comprise all the services needed for assignment and maintenance of Internet domain names. As many as 90 registrars will be authorized by CORE as registrars to administer and maintain the new gTLDs: .arts, .firm, .info, . nom, .rec, .shop and .web. InterNIC historically has been primarily responsible for the assignment, administration and maintenance of a subset of the traditional gTLDs, specifically, .com, .edu and .org. Future responsibility for those traditional gTLDs is uncertain. See also CORE, DNS, gTLD and InterNIC.

Internet Relay Chat IRC. Sort of like CB radio, but run on the Internet, and far more confusing than CB radio.

Internet Router see router.

Internet Security Information traveling on the Internet usually takes a circuitous route through several intermediary computers to reach any destination computer. The actual route your information takes to reach its destination is not under your control. As your information travels on Internet computers, any intermediary computer has the potential to eavesdrop and make copies. An intermediary computer could even deceive you and exchange information with you by misrepresenting itself as your intended destination. These possibilities make the transfer of confidential information such as passwords or credit card numbers susceptible to abuse. This is where Internet security comes in and why it has become a rapidly growing concern for all who use the Internet. See the Internet and Secure Channel.

Internet Server 1. An Internet server is a device which users on the Internet access to get services. Such services might be electronic mail, news, a Web page, etc. A company will have one or more Internet servers attached to the Internet when it wants to deliver services to people on the Internet. Such Internet servers could be called e-mail servers, FTP servers, News servers and World Wide Web servers. Internet servers most commonly run on Unix. But Microsoft Windows NT is increasingly gaining popularity.

2. A Sun Microsystems term, Part of Solaris' Server Suite. Provides secure, scalable workgroup-based Internet computing.

Internet Server API See ISAPI.

Internet Service Provider ISP. A vendor who provides access for customers (companies and private individuals) to the Internet and the World Wide Web. The ISP also typically provides a core group of internet utilities and services like E-mail, News Group Readers and sometimes weather reports and local restaurant reviews. The user typically reaches his SP by either dialing-up with their own computer, modem and phone line, or over a dedicated line installed by a telephone company. An ISP is also called a TSP, for Telecommunications Service Provider, and a ITSP, for Internet Telephony Service Provider.

Internet Society ISOC. A non-profit organization that fosters the voluntary interconnection of computer networks into a global communications and information infrastructure. According to Computerworld, the Internet Society is concerned with the evolution of the Internet and its social, political and technical issues. The ISOC is the umbrella organization for the IAB, IETF and IRTF. The ISOC has 5,000 members in 120 countries. Its Web site is www.isoc.org. See also IOPS and NANOG.

Internet Suite Of Protocols The combination of TCP (Transmission Control Protocol) and IP (Internet Protocol). See TCP/IP.

Internet Telephony In the very beginning, Internet telephony simply meant the technology and the techniques to let you make voice phone calls — local, long distance and international — over the Internet using your PC. To make these calls, both people on the phone need appropriate hardware and software. The hardware is typically a sound card or voice modem in a PC. There are almost as many ways of making phone calls on the Internet as there are software packages. The key is to figure a way that your PC can dial and reach someone else's distant PC — which must be turned on, plugged in and connected to some place that my PC can find you at. In short, making voice phone calls was the first definition of internet telephony. But then people started thinking of other things internet telephony could become. For example, internet telephony could let you talk to someone while the two of you worked on making perfect a document that was on both your screens. If the Internet could send email, people started thinking of sending fax, voice, video and imaging mail / messages. And maybe, as you cruise the Internet and find a product you'd like to buy, you might see a button that says "I'd like to know more. Have an operator call me." So you click the button, and 15 seconds later your phone rings. The operator is calling, wanting to know how he can help? In short, the definition of Internet telephony is broadening day by day to include all forms of media (voice, video, image), all forms of messaging and all variations of speed from real-time to time-delayed. See Gold, Packet Switching, Tier 1 and, for the best explanation, TAPI 3.0.

Internet Terminal A sub-$500 machine specially designed for Internet browsing and first proposed in the late Fall of 1995 by Larry Ellison, head of database software company Oracle. Part of its appeal to people outside Microsoft and Intel is that the Internet Appliance would not have to be based on standard PC technology. It need have an Intel chip and need not run Windows. This device is also called an Internet Terminal, a Network Computer or an IPC, an Interpersonal computer. The original description of the Internet Appliance was that it would come with 4mb of RAM, 4mb of flash memory, processor, monitor, keyboard and mouse — all for under $500. Also called a NC, or Network Computer.

Internet Worm This software program caused a major part of the Internet network to crash by replicating and generating spurious data.

Internetwork See INTERNETWORKING.

Internetwork Management A generic term used to describe the actions that help maintain, a complex network.

Internetwork Operating System IOS. Cisco's massive operating system that runs most routers on the Internet.

Internetwork Packet Exchange IPX. A network layer protocol developed by Novell Inc. and used in NetWare implementations. See IPX.

Internetwork Router In local area networking technology, an internetwork router is a device used for communications between networks. Messages for the connected network are addressed to the internetwork router, which chooses the best path to the selected destination via dynamic routing. Internetwork routers function at the network layer of the Open Systems Interconnection (OSI) model. Also known as a network router or simply as a router.

Internetworking 1. Communication between two networks or two types of networks or end equipment. This may or may not involve a difference in signaling or protocol elements supported. And, in the narrower sense — to join local area networks together. This way users can get access to other files, databases and applications. Bridges and routers are the devices which typically accomplish the task of joining LANs. Internetworking may be done with cables — joining LANs together in the same building, for example. Or it may be done with telecommunications circuits — joining LANs together across the globe.

InterNIC Internet Network Information Center. The InterNIC registry is where you always used to go to register your domain name. Registration was free until 1995; then it changed to $50 a year; now it's $35 a year for domain names with anniversary dates on or after April 1, 1998. The InterNIC Registration Services Host computer contains information on Internet Networks, ASNs, Domains, and POCs. The InterNIC was established in January 1993 as a collaborative project between AT&T, General Atomics (no longer involved) and Network Solutions, Inc., and was supported through a 5-year cooperative agreement with the NSF (National Science Foundation). InterNIC participates in Internet forums to promote Internet services, explore new tools and technologies, and contribute to the Internet community. InterNIC currently is operated by Network Solutions Inc., a subsidiary of SAIC (Science Applications International Corporation), the private company which also has acquired Bellcore. On December 22, 1997, responsibility for assignment of IP numbers was shifted to ARIN (American Registry of Internet Numbers) for the geographic areas of North American, South America, the Caribbean and sub-Saharan Africa. About the same time, the decision was made to shift responsibility for domain name registration to the Council of Registrars (CORE). CORE is empowered to authorize as many as 90 independent registrars, including InterNIC, to register domain names, or URLs (Uniform Resource Locators). www.internic.net. See also ARIN, CORE, Domain, and Domain Name.

Internode Communication paths which originate in one node and terminate in another.

Internode Link A data line for high-bandwidth connections between PBXs.

Interoffice Between two telephone company switching offices.

Interoffice Channel A transmission path between two AT&T serving offices.

Interoffice Trunk A trunk circuit connecting two local telephone company central offices.

Interoperate The ability of multi vendor computers to work together using a common set of protocols. With interoperability, PCs, Macs, Suns, DEC VAXes, IBM mainframes, etc. all work together allowing one host computer to communicate with and take advantage of the resources of another.

Interoperability The ability to operate software and exchange information in a heterogeneous network, i.e. one large network made up of several different local area networks.

Interoperability Testing In our industry there are several levels of testing of new products and services. Clearly, the manufacturer does one level of testing. His job is to ensure that his product meets the claims that he promotes for it. Then there is a interoperability testing. This is testing to ensure that his product works with other products that allegedly conform to the same standard/s. Clearly it is one thing for a bunch of engineers to create a new standard on paper. And it is another thing to have engineers build product to that standard and have their products communicate with other products from other manufacturers who also allegedly conform to the standard. The most public example of interoperability testing in the telecom industry is that which occurred so successfully in the ATM Forum. That organization was established by manufacturers who make ATM products. Their objective was not to establish new standards (they already existed), but to make sure that their products worked successfully with each other without hardware modification. Plug the equipment together. Configure the software. Bingo it would work.

Interoperator Modular hardware or software that implements part of the OSI model and can work with components implementing the other parts of the model.

Interpacket Gap IPG. A delay or time gap between CSMA/CD packets intended to provide interface recovery time for other CSMA/CD sublayers and for the Physical Medium. For 10Base-T, the IPG is 9.6 us (96 bit times). For 100Base-T, the IPG is 0.96 us (96 bit times).

InterPBX Calls coming into one PBX can be transferred to extensions on another PBX using direct tie lines between the two PBXs.

Interpersonal Message IPM. The term used in the 1984 X.400 recommendations to refer to a message in the Interpersonal Messaging System. The 1988 X.400 recommendations use the term "interpersonal message" (IPM).

Interpersonal Message (IPM) User Agent A class of cooperating user Agents capable of processing Interpersonal (IP) messages. An x.400 term.

Interpersonal Messaging IPM. Electronic exchange of information between two or more persons.

Interpolation 1. The process of estimating values of (a function) between two known values.
2. A video technique used in motion compensation where a current frame of video is reconstructed by using the differences between it and past and future frames. This technique is also known as forward and backward prediction. Intel (originator of Indeo Video) defines interpolation slightly differently, namely: The process of averaging pixel information when scaling an image. When the size of an image is reduced, pixels are averaged to create a single new pixel; when an image is scaled up in size, additional pixels are created by averaging pixels of the smaller image.

Interposition Calling One operator in a multi-position system calling another.

Interposition Transfer Transfer of a call from one operator to another.

Interposition Trunk 1. A connection between two positions of a large switchboard so that a line on one position can be connected to a line on another position.
2. Connections terminated at test positions for testing and patching between testboards and patch bays within a technical control facility.

Interpret Interpret means that the computer will translate a stored program expressed in pseudocode into machine language and will perform the indicated operations as they are translated.

Interprocess Communications this is the capability of programs to share information. At the most basic level, it consists of cutting and pasting information between two programs. Above that ranks the "live" paste, in which information shared between two documents is updated whenever one of the documents is modified. This is referred to as Dynamic Data Exchange (DDE). In advanced DDE, programs can send messages as well as data to other programs running locally or remotely. Beyond DDE is Object Linking and Embedding (OLE), which lets one program borrow the specialized capabilities of another program loaded on the machine (say, advanced chart creation) rather than having to implement that capability redundantly.

InterRepeater Link IRL. A networking term. A mechanism for connecting two and only two repeater sets.

Interrogate To determine the state of a device or unit.

Interrupt A temporary suspension of a process caused by an event outside of that process. More specifically, an interrupt is a signal or call to a specific routine. An interrupt setting allows the hardware in a file server, router, workstation or PC to send an interrupt signal to the processor. The interrupt signal temporarily suspends the other station tasks while the processor performs the task requested by the interrupting device. After the routine is completed, the processor then continues with the original tasks. Each piece of hardware (serial and parallel ports and network boards) installed in the same computer needs a unique interrupt. Interrupts are divided into two general types, hardware and software. A hardware interrupt is caused by a signal from a hardware device, such as a printer. A software interrupt is created by instructions from within a software program.
TIP: When you slide a new card into one of the empty slots on your PC and things go awry, check that the new card's Interrupt is not the same as one of the other cards in your bus. An interrupt is also called a hardware interrupt or an InterRupt reQuest (IRQ). For a listing of normal IRQs see IRQs. See also INTERRUPT REQUEST and POLLING.

Interrupt Driven Someone who moves through a workday responding to interruptions rather than the work goals as originally set.

Interrupt Flag In a PC, there is a configuration control (addressed as bit IF of the processor flag register). This control process manages the CPU's ability to receive and process interrupt requests. The flag is often set to zero (which means interrupts disabled) by device drivers or other I/O privilege-level code that needs exclusive access to the CPU during critical operations. See also INTERRUPT, INTERRUPT HANDLING ROUTINE and UART.

Interrupt Handling Routine This program, which is often part of a device driver, handles all requests from a particular interrupt line. Interrupt-handling routines are defined in the CPU's Interrupt Descriptor Table (IDT). When the CPU (the Central Processing Unit of your PC) receives an interrupt request, it looks up the matching interrupt-handling routine in the IDT, then transfers control to the routine until it (the CPU) gives an interrupt return call (IRET), indicating the task is complete.

Interrupt Latency The delay in servicing an interrupt request is known as interrupt latency. It is not a problem with devices that are not sensitive to timing inconsistencies (such as hard-disk controllers or video boards). But it is a problem with high-speed, asynchronous communications (9,600 bps and above), which are highly time-sensitive operations.

Interrupt Overhead The cumulative demand on your computer's central microprocessor by peripheral devices that generate interrupt requests is referred to as interrupt overhead. Such devices include hard-disk controllers, network interface cards, parallel and serial ports.

Interrupt Request IRQ. This is the communications channel through which devices issue interrupts to the interrupt handler of an IBM PC or IBM compatible PC's microprocessor. It's the channel through which these devices get the microprocessor's attention. Different IRQs are assigned to different devices. This assignment pattern differs from PC to PC. Many LAN interface cards use an IRQ to get to the microprocessor. You must be sure that your LAN interface card is not trying to use the IRQ assigned to another peripheral, like the hard disk controller or EGA card. See also IRQ for a different and longer explanation.

Interrupt Request Lines Hardware lines over which devices can send signals to get the attention of the processor when the device is ready to accept or send information. Typically, each device connected to the computer uses a separate IRQ.

Interrupter An automatically operated electromechanical device used to turn lights, bells or other signals on and off in timed sequences. An interrupter makes lights wink on and off on a key system. Or did, when everything was electromechanical. It was used on 1A key telephone systems.

Interrupting Equipment Motor-driven mechanical devices used to break the ringing generator's output into ringing and silent periods, creating the busy and ringback tone pulses.

InterSpan The full name is InterSpan Frame Relay and it's AT&T's new frame relay data communications service, announced in the late fall of 1991.

Intersputnik A Russian satellite system similar in concept to the West's Intelsat, except that it's set up by Russia and the Eastern bloc countries. Two US carriers, AT&T and IDB Communications, once used Intersputnik to alleviate their shortage of US-Russia circuits. See INTELSTAT.

Interstate Between states (crossing a state line).

Interstices In cable construction, the spaces, valleys or voids between or around the cable's components.

Interswitch Trunk A circuit between two switching machines.

Intersymbol Interference Distortion of signals due to preceding or following pulses affecting desired pulse amplitude at time of sampling.

Intertandem Trunk Groups A category of trunk groups that interconnects tandems.

Intertoll Trunks Trunks connecting Class 4 and higher switching machines in the AT&T long distance network.

Interval Time. Pulse interval, for example, means the time from the start of one pulse to the start of the next.

Interworking The ability to seamless communicate between devices supporting dissimilar protocols, such as frame relay and ATM, by translating between the protocols, not through encapsulation. Many carriers are planning to implement the necessary equipment and conversion algorithms to allow the network itself to transparently convert from frame relay to ATM, and vice versa.

INTFC Interface.

Intra Intra means inside. Intrastate means inside the state. Interstate means between states.

Intra-Area Cell Transfer A cellular radio term. A cell transfer between two cells that are controlled by the same serving Mobile Data Intermediate System (MD-IS).

Intracalling This is an outside plant term. Intracalling refers to the ability of a remote line concentrator to interconnect users served by the same concentrator without providing two trunks directly back to the central office.

Intraday Distribution A call center term. A historical pattern consisting of factors for each intra-day period of the week that define the typical distribution of call arrival or average handle time throughout each day. Each factor measures how far call volume or average handle time in that half hour or quarter hour deviates from the average half-hourly or quarter-hourly figure for that day. This information enables the program to forecast intra-day call volumes and staffing requirements.

IntraEnterprise Communications Communications that are exchanged within a single organization (including multiple sites of the organization).

Intraflow This is an automatic call distribution term. It refers to the ability to select a second or subsequent group of agents to backup the primary agent group. This is designed to allow the caller to be serviced more efficiently and less expensively.

Intraframe Coding A way of video compression that compresses information within a single frame. Compare to INTERFRAME CODING.

IntraLATA Telecommunications services that originate and terminate in the same Local Access and Transport Area. See also LOCAL ACCESS AND TRANSPORT AREA.

Intramodal Distortion In an optical fiber, the distortion resulting from dispersion of group velocity of a propagating mode. It is the only form of multi mode distortion occurring in single-mode fibers.

Intranet A private network that uses Internet software and Internet standards. In essence, an Intranet is a private Internet reserved for use by people who have been given the authority and passwords necessary to use that network. Those people are typically employees and often customers of a company. An Intranet might use circuits also used by the Internet or it might not. Companies are increasingly using Intranets — internal Web servers — to give their employees easy access to corporate information.

According to my friends at Strategic Networks Consulting, Boiled down to its simplest, an Intranet is: a private network environment built around Internet technologies and standards — predominantly the World Wide Web. The primary user interface, called a Web browser, accesses Web servers located locally, remotely or on the Internet. The Web server is the heart of an Intranet, making selection of Web server software a crucial decision, even though much fanfare has focused on browsers (Netscape's Navigator vs. Microsoft's Explorer).

At its core, a Web server handles two arcane languages (HTML and CGI) that are the meat and potatoes of generating Web pages dynamically, making connections and responding to user requests. But in the rush to dominate the potentially lucrative Intranet market, these simple Web functions are being bundled into operating systems and vendors are now touting pricey "Intranet suites" which encompass everything from database and application interfaces, to e-mail and newsgroups, to the kitchen sink.

Most medium- or larger-sized companies will need more than just a handful of simple Web servers to deploy a reasonably robust Intranet. To help a company post current job openings, or make up-to-date product specs and available inventory accessible by traveling sales reps, an Intranet needs the following capabilities:

• Database access. Getting at critical data housed in corporate databases can be accomplished via generic, universal ODBC linking or based on "native" links directly to Sybase, Oracle et al. allowing use of all the database's features.

• Application hooks. Used by developers, a standard programming interface (API) allows outside applications like Lotus Notes to interact with Web data and vice versa. In addition, proprietary APIs exist — most notably Microsoft's ISAPI (for "Internet Server API") which lets developers link directly to Microsoft applications.

• User publishing. In addition to dialogues via chat/newsgroup/bulletin board features, users will want to post their own content on Web servers without having to attain Webmaster status.

• Search vehicles. How does an engineer find the current specs on Project #686-2 among thousands of pages spread across a bunch of Web servers? The answer: an indexing and search engine that creates an internal Yahoo! for your own Web sites.

• Admin/management. A catch-all for loads of important, but still ill-conceived features for managing access, users, content and the servers themselves. Intranet administrators are currently fascinated with analyzing Web server logs which contain data of some sort, including user connections and page activity.

According to a white paper released by Sun Microsystems in the summer of 1996, the basic infrastructure for an intranet consists of an internal TCP/IP network connecting servers and desktops, which may or may not be connected to the Internet through a firewall. The intranet provides services to desktops via standard open Internet protocols. In addition to TCP/IP for basic network communication, these also include protocols for:

Browsing	HTTP
File Service	NFS
Mail Service	IMAP4/SMTP
Naming Service	DNS/NIS+
Directory Services	DNS/LDAP
Booting Services	Bootp/DHCP
Network Administration	SNMP
Object Services	IIOP (CORBA)

See also EXTRANET and INTRANET.

Intranodal Service Intranodal service is a feature of some central office switches and smaller remote switches. It means that it will continue to switch in which

Intranode Communications path which originates and terminates in the same node.

Intraoffice Call A call involving only one switching system.

Intraoffice Trunk A telephone channel between two pieces of equipment within the same central office.

Intrapreneur An entrepreneur who works inside a big company. Hence, intra, as in inside. It's hard to imagine it actually happening. But the word has became popular as a way for large companies to motivate their employees to take personal career risks and introduce new products.

Intrastate Remaining entirely within the boundaries of a single state and, therefore, if related to telephone, falling under the jurisdiction of that state's telephone regulatory procedures.

Intrastructure A term coined by "Data Communications" and referring to the software, hardware, and Internet services underlying a corporate Intranet.

Intrinsic Joint Loss That loss in optical power transmission, intrinsic to the optical fiber, caused by fiber parameters, e.g., dimensions, profile parameter, mode field diameter, mismatches when two non identical fibers are joined.

Intrinsics Intrinsics are a component of many windows toolkits. The windows toolkit intrinsics definition has been developed by the MIT X Consortium. The intrinsics define the function of specific graphical user interface and window objects. They do not define any particular look or feel, just the function. Example: A pull down menu intrinsic would define

the function of a pull down menu within a toolkit but not the appearance of it.

Intumescent Firestop A firestopping material that expands under the influence of heat.

Inverse Fourier Transform Inversion of Fourier transform to convert frequency representation of signal to time representation.

Inverse Multiplexer I-Mux. An inverse multiplexer performs the inverse function of a multiplexer. "Multiplexer" translates to "many fold." For example, a TDM Mux (Time Division Multiplexing Multiplexer) accepts many (typically 24) low-capacity inputs in the form of information streams, and folds them together through a process know as byte interleaving in order to send them over a single, high-capacity, shared digital circuit. In this example, each of the 24 voice-grade channels supports a transmission of 64 Kbps; the total capacity of the T-1 circuit is 1.536 Mbps. The advantage of this approach is that of economy of scale-a single, high-capacity T-1 circuit is far less expensive that are 24 individual voice-grade circuits. An inverse multiplexer does just the inverse. In other words, it accepts a single, high-capacity information stream and splits it up into multiple information streams, each of which is sent over a separate and lower-capacity circuit; the process is reversed on the receiving end. Videoconferencing, for example, may make use of inverse multiplexers. A full-motion videoconference requires a full T-1. While a user organization may have multiple T-1s at a given site, a full T-1 may not be available at the moment it is required. Therefore, an inverse mux might split that video datastream into four datastreams of 384 Kbps, and send each over 6 channels of four separate T-1s. At the receiving end of the video datastream, the four datastreams are received, demultiplexed and resynchronized in order to reconstitute the original datastream. Resynchronization is critical, as each of the four circuits may impose different levels of propagation delay on the signal due to reasons such as differing route lengths. Synchronization prevents your head (which traveled over T-1 #1) from appearing at your knees (which traveled over T-1 #3) on the receiving TV monitor. Inverse Multiplexing over ATM (IMA) is an access specification approved in 1997 by the ATM Forum. This User-Network Interface (UNI) standard allows a single ATM cell stream to be split across multiple access circuits from the user site to the edge of the carrier's ATM network. In an ATM LAN application, for instance, the ATM switch deployed in the enterprise backbone typically operates at 155 Mbps or 622 Mbps. In this example, ATM traffic from the enterprise to the public ATM carrier-based network, requires 6 Mbps-well more than the level supported by a T-1, but less than a full T-3. Rather than subscribing to a T-3, which requires a fiber optic access circuit and which is very expensive, IMA is used. Thereby, the ATM datastream is split across four T-1 circuits by an access concentrator which possesses IMA capability. The IMA process works in a round-robin fashion, with cell #1 traveling over T-1 #1, cell #2 traveling over T-1 #2, and so on. Each of the four T-1 circuits is relatively inexpensive, can be provisioned over twisted-pair, and is readily available. At the edge of the carrier network, the ATM switch receives each of the four separate datastreams, and reverses the IMA process in order to reconstitute the original datastream, which then is switched and transported through the network to the far edge. At that far edge, the IMA process may take place again, from the edge of the carrier network, over four T-1s, and to the IMA-capable ATM concentrator on the user premises. See also ATM and TDM.

Inverse Multiplexing The process of splitting a single high-speed channel into multiple signals, transmitting each of the multiple signals over a separate facility operating at a lower rate than the original signal, and then recombining the separately-transmitted portions into the original signal at the original rate. See Inverse Multiplexer for a full description of the process.

Inverter A device which converts direct current electricity to alternating current electricity, often used to power AC devices in a car.

Invitation To Send A character or sequence of characters which calls for a station to begin transmission. Usually this is part of a polling arrangement.

Inward Restriction A Centrex service feature which stops Centrex lines from receiving certain incoming calls.

Inward Trunk Used only for incoming calls, these trunks cannot dial out. "800" lines, for example, can only be used to receive calls.

INWATS INward Wide Area Telephone Service. A service of interexchange carriers (e.g., AT&T, MCI, and Sprint), local exchange carriers, the Bell operating companies and the independent phone companies and long distance resellers in North America which allows subscribers to receive calls from specified areas (depending on the rate band chosen) with no charge to the person who's calling. Rather, the charges are billed to the called party. See 800 SERVICE for a much bigger explanation.

IOC 1. Inter-Office Channel

2. Independent Operating Carrier

IOD Identified Outward Dialing. See also AIOD and CALL ACCOUNTING SYSTEM.

IOEngine Input/Output Engine. The part of the Novell SFT III operating system that handles physical processes, such as network and disk I/O, hardware interrupts, device drivers, timing, and routing. SFT III is split into two parts: the IOEngine and the MSEngine (Mirrored Server Engine). The IOEngine routes packets between the network and the MSEngine. To network workstations (i.e. PCs on the LAN), the IOEngine appears as a standard NetWare router or bridge. The primary server and the secondary server each have an IOEngine, but they share the same MSEngine. Because the IOEngines are not mirrored, NetWare Loadable Modules (NLMs) and applications that directly interface with hardware, such as backup NLMs may be installed in the IOEngines on both the primary and the secondary server. See MSENGINE.

IOF Inter Office Facility

Ion Exchange Technique A method of fabricating a graded-index optical fiber by an ion exchange process.

Ionization The process of breaking up molecules into positively and negatively charged carriers of electricity called ions.

IONL Internal Organization of the Network Layer. The OSI standard for the detailed architecture of the Network layer. Basically, it partitions the Network layer into subnetworks interconnected by convergence protocols (equivalent to inter-net working protocols), creating what the Internet community calls a catenet or internet.

Ionosphere That part of the atmosphere in which reflection and/or refraction of electromagnetic waves occurs. It extends from about 70 to 500 kilometers. At that point, ions and free electrons exist in sufficient quantities to reflect electromagnetic waves.

Ionospheric Absorption Attenuation of the energy in a radio wave due to the interaction between it and gas molecules. Deviative absorption describes the appreciable bend-ing that occurs in an ionospheric layer at close to critical frequency. Non-deviative absorption describes the condition where little or no bending occurs as the wave passes through an ionized layer. See IONOSPHERE.

Ionospheric Cross Modulation Nonlinearities within the medium can produce nonlinear absorption. This can lead to the modulation on a strong signal being transferred to a weaker carrier. Sometimes described as the Luxembourg effect.

Ionospheric Disturbance An increase in the ionization of the D region of the ionosphere, caused by solar activity, which results in greatly increased radio wave absorption. See IONOSPHERE.

Ionospheric Focusing A variation in the curvature of the ionospheric layers can give rise to a focusing/defocusing effect at a receiving antenna. This may produce either an enhancement or attenuation in the received field strength due to signal phase variations.

Ionospheric Refraction The change in the propagation speed of a signal as it passes through the ionosphere.

IOP 1. Input/Output Processor.

2. Interoperability: The ability of equipment from different manufacturers (or different implementations) to operate together.

IOPS Internet OPerators Group. On May 20, 1997 Nine of the nation's major Internet service providers announced the formation of IOPS.ORG, a group of Internet service providers (ISPs) dedicated to making the commercial Internet more robust and reliable. IOPS.ORG will focus primarily on resolving and preventing network integrity problems, addressing issues that require technical coordination and technical information-sharing across and among ISPs. These issues include joint problem resolution, technology assessment, and global Internet scaling and integrity. IOPS.ORG will provide a point-of-contact for these industry-wide technical issues.

The founding members of IOPS.ORG are ANS Communications, AT&T, BBN Corporation, EarthLink Network, GTE, MCI, NETCOM, PSINet, and UUNET, and it is expected that additional national and international Internet operators will join. IOPS.ORG will work with other Internet organizations, with Internet equipment vendors, and with businesses that rely on the Internet. IOPS.ORG members individually will continue to support other Internet organizations such as the Internet Engineering Task Force (IETF), the North American Network Operators Group (NANOG), and the Internet Society.

The Corporation for National Research Initiatives (CNRI), a Reston, VA-based not-for-profit organization which works with industry, academia, and government on national-level initiatives in information technology, will host the initial operations of IOPS.ORG. "IOPS.ORG will play a key role in the healthy technical and operational evolution of the Internet as an increasingly important component of the economy," said CNRI President Robert Kahn. www.iops.org

IOS 1. The International Organization for Standardization based in Geneva, Switzerland. The IOS develops and publishes hundreds of international ISO standards such as ISO-9000 for quality assurance or ISO 14000 for environmental performance. The IOS is comprised of more than 90 member standards bodies worldwide plus other international associations, government and non-government bodies. The U.S. member of the IOS is ANSI. See ANSI and ISO.

2. Internetwork Operating System from Cisco. This operating system runs the vast majority of routers now deployed in the core of the Internet. See also Junos Code.

IP 1. The Internet Protocol. IP is the most important of the protocols on which the Internet is based. The IP Protocol is a standard describing software that keeps track of the Internetwork addresses for different nodes, routes outgoing messages, and recognizes incoming messages. It allows a packet to traverse multiple networks on the way to its final destination. Originally developed by the Department of Defense to support interworking of dissimilar computers across a network. While its roots are in the ARPAnet development, IP was first standardized in RFC 791, published in 1981, and updated in RFC 1349. This protocol works in conjunction with TCP and is usually identified as TCP/IP. It is a connectionless protocol that operates at the network layer (layer 3) of the OSI model. See IP ADDRESS, IPv4, IPv5, IPv6, the INTERNET, and TAPI 3.0.
2. Internal Protocol.
3. Intelligent Peripheral. A device in an IN (Intelligent Network) or AIN (Advanced IN) that provides capabilities such as voice announcements, voice recognition, voice printing and help guidance. By way of example, MCI's 1-800-COLLECT makes use of IPs, which are specialized voice processing systems. The IP prompts the caller to enter the target telephone number and speak his or her name. The system then instructs the network to connect the call. Based on a spoken acceptance of the call by the called party, the system authorizes call completion.

IP Address See Internet Address.

IP Address Mask Internet Protocol address mask. A range of IP addresses defined so that only machines with IP addresses within the range are allowed access to an Internet service. To mask a portion of the IP address, replace it with the asterisk wild card character (*), For example, 192.44.*.* represents every computer on the Internet with an IP address beginning with 192.44. See IP Addressing.

IP Addressing A networking term. IP (Internet Protocol) addressing is a system for assigning numbers to network subdivisions, domains, and nodes in TCP/IP networks. IP addresses are figured as 32-bit (four-byte) numbers. The high bytes constitute the "Class A" and "Class B" portions of the address, which denote network and subnetwork. The low bytes ("Class C" address segments) identify unique nodes — individual machines or (in the case of multi-addressing) individual node processes. The Class C address segment (two bytes) can represent 65,536 unique values — enough so that in most conventional TCP/IP LANs, sufficient values are available to afford each machine its own "fixed" IP address. In public internet-access, however, the number of fixed addresses available to a provider may not be sufficient to provide each dialup client with a permanent IP address. In such scenarios, available Class C addresses can be assigned dynamically, as machines log into network access ports — on the presumption that no more than N clients will attempt to log on, simultaneously (where N denotes the number of absolute addresses in the pool).
Thus:
"Fixed" or "Static" IP address: a four-byte TCP/IP network address permanently assigned to an individual machine or account.
"Dynamically-assigned" IP address: a four-byte TCP/IP network address assigned to a machine or account for the duration of a single session.

IP Datagram The fundamental unit of information passed across the Internet. Contains source and destination addresses along with data and a number of fields which define such things

as the length of the datagram. the header checksum, and flags to say whether the datagram can be (or has been) fragmented.

IP Device Control IPDC. See Simple Gateway Control Protocol.

IP Multicasting The transmission of an IP datagram to a host group, a set of zero or more hosts identified by a single IP destination address. A multicast datagram delivered to all members of its destination host group with the same "best efforts" reliability as regular unicast IP datagrams, i.e., the datagram is not guaranteed to arrive intact at all members of the destination group or in the same order relative to other datagrams.

IP Router A computer connected to a multiple physical TCP/IP networks that can route or deliver IP packets between networks. See also GATEWAY.

IP Spoofing An attack whereby a system attempts to illicitly impersonate another system by using its IP network address. See IP, IP ADDRESS and IP ROUTER.

IP Subnet All devices which share the same network address. Routers are boundaries between subnets so each connection to a router has a different network address.

IP Switching A term coined by Ipsilon Networks to describe a new class of switch it developed, combining intelligent IP routing with high-speed ATM switching hardware in a single, scalable platform. The IP switch implements the IP protocol stack on ATM hardware, allowing the device to dynamically shift between store-and-forward and cut-through switching based on the flow requirements of the traffic as defined in the packet header. Data flows of long duration, thereby, can be optimized by cut-through switching, with the balance of the traffic afforded the default ATM treatment, which is hop-by-hop, store-and-forward routing. Ipsilon suggests that first-generation IP Switches can achieve rates of up to 5.3 million PPS (packets per second) by avoiding ATM cell segmentation and reassembly, ATM overhead, and ATM switch processing of each cell header. Clearly, one of the advantages of IP Switching is the use of IP (Internet Protocol), which protocol is mature, well-understood, and widely deployed across a wide range of networks. Contrast with TAG SWITCHING.

IP Telephony Here is Microsoft's definition, excerpted from their white paper on TAPI 3.0: IP Telephony is an emerging set of technologies that enables voice, data, and video collaboration over existing IP-based LANs, WANs, and the Internet. Specifically, IP Telephony uses open IETF and ITU standards to move multimedia traffic over any network that uses IP (the Internet Protocol). This offers users both flexibility in physical media (for example, POTS lines, ADSL, ISDN, leased lines, coaxial cable, satellite, and twisted pair) and flexibility of physical location. As a result, the same ubiquitous networks that carry Web, e-mail and data traffic can be used to connect to individuals, businesses, schools and governments worldwide.
What are the benefits of IP Telephony? IP Telephony allows organizations and individuals to lower the costs of existing services, such as voice and broadcast video, while at the same time broadening their means of communication to include modern video conferencing, application sharing, and whiteboarding tools. In the past, organizations have deployed separate networks to handle traditional voice, data, and video traffic. Each with different transport requirements, these networks were expensive to install, maintain, and reconfigure. Furthermore, since these networks were physically distinct, integration was difficult if not impossible, limiting their potential usefulness.
IP Telephony blends voice, video and data by specifying a common transport. IP, for each, effectively collapsing three

networks into one. The result is increased manageability, lower support costs, a new breed of collaboration tools, and increased productivity. Possible applications for IP Telephony include telecommuting, real-time document collaboration, distance learning, employee training, video conferencing, video mail, and video on demand. See the Internet, IP Telephony Algorithms, TAPI, TAPI 3.0 and TCP/IP.

IP Telephony Algorithms The major IP Telephony Algorithms in the market today (fall of 1997), according to a white paper, called "IP Telephony powered by Fusion" from Natural MicroSystems (www.nmss.com), include:

• MS-GSM: This algorithm, marketed by Microsoft, runs at 13kbps and is a derivative of the ITU (International Telecommunications Union) standard GSM work. GSM is used in 85 countries around the world as the standard for digital cellular communications. Microsoft's implementation varies from the standard in several ways including how the encoded data is represented and what aspects of the encoder are supported. Natural MicroSystems provides an MS-GSM encoder that is compatible with Microsoft's Win95/WinNT embedded product.

• ITU G.723.1: This algorithm runs at 6.3 or 5.4 kbps and uses linear predictive coding and dictionaries which help provide smoothing. The smoothing process is CPU-intensive, however (30Mips on an Intel Pentium), so don't expect a PC-based implementation to work well for lots of real-time activity.

• VoxWare: This is a proprietary encoder that has been bundled by Netscape with their Browser. It delivers 53:1 compression and very low jitter. VoxWare presents very low network bandwidth requirements; however, it also has lower speech quality.

Most speech encoder algorithms have a set of rules concerning packet delivery and disposition management. This is often called jitter buffer management. "Jitter" in this case refers to when the signal is put into frames. The decoding algorithm must decompress and sequence data and make "smoothing" decisions (when to discard packets versus waiting for an out-of-sequence packet to arrive). Given the real-time nature of a live connection, jitter buffer management policies have a large affect on voice quality. Actual sound losses range from a syllable to a word, depending on how much data is in a given packet. The first buffer size is often a quarter-second, large enough to be a piece of a word or a short word — similar to drop-outs on a cellular connection in a poor coverage area.

IPA Intellectual Property Attorney.

IPARS The International Passenger Airline Reservation System. An IBM-originated term.

I-PASS An alliance of ISPs (Internet Service Providers) and IAPs (Internet Access Providers) to provide roaming capabilities for travelers. Based on proprietary standards, roamers are authenticated before being afforded Internet access. Usage is cross-billed through the I-PASS clearinghouse, with fees being set by each ISP for use of its facilities by roamers. I-PASS includes over 100 member ISPs in approximately 150 countries, and includes over 1,000 POPs (Points of Presence). I-PASS competes with GRIC (Global Reach Internet Connection). The IETF's Roamops working group is developing a standard for roaming, as well. See also GRIC and ROAMOPS.

IPC Interprocess Communications. A system that lets threads and processes transfer data and messages among themselves; used to offer services to and receive services from other programs. Supported IPC mechanisms under MS OS/2 are semaphores, signals, pipes, queues, shared memory, and dynamic data exchange.

IPCH Initial Paging CHannel is the channel number used by your cellular provider to "page" the phones on the system. The term "paging" refers to notifying a particular phone that it has an incoming call. All idle, turned-on phones on a system monitor the data stream on the IPCM. Non-wireline cellular carriers use channel 0333 as the IPCH, while wireline providers (those operated by a telephone company use channel 0334.

IPCI See INTEGRATED PERSONAL COMPUTER INTERFACE.

IPCP IP Control Protocol; protocol for transporting IP traffic over a PPP connection.

IPDC IP Device Control. See Simple Gateway Control Protocol.

IPDS Intelligent Printer Data Stream. It's IBM's host-to-printer page description protocol for printing. You can now buy kits which let you use your present printer to emulate an IBM printer.

IPE Intelligent Peripheral Equipment. Northern Telecom's term for being able to extend all the features of its PBX over distances longer than a normal extension in a building. See FIBER REMOTE.

IPEI International Portable Equipment Identities. A wireless term.

IPEM If the Product Ever Materalizes.

IPL Initial Program Load.

IPLC International Private Leased Circuit.

IPM Interruptions Per Minute or Impulses Per Minute.

IPNG IPng. IP Next Generation. Collective term used to describe the efforts of the Internet Engineering Task force to define the next generation of the Internet Protocol (IP) which includes security measures, as well as larger IP addresses to cope with the explosive growth of the Internet. The were three candidate protocols for IPng (CATNIP, TUBA and SIPP), were blended into IPv6, which is in trial stages at the time of this writing. See IPv6.

IPNS International Private Network Service. It actually international private line service and it's typically a circuit from 9.6 Kbit/sec up to T-1 or E-1. Domestically you would simply call it "Private line data service."

IPP IPP is the Internet Print Protocol, a collection of IETF standards developed through the Printer Work Group, www.pwg.org, that will make it as easy to print over the Internet as it is to print from your PC. IPP uses the HTTP protocols to "POST" a supported MIME Page Discription Language file to a printer. Printers are given Internet addresses such as, www.mydomain.com/ipp/my_printer, so they can be located on the Internet. IPP has the support of all the major printer companies including, Xerox, HP, Lexmark, IBM as well as Novell and Microsoft. Since fax, at a sufficient level of abstraction, is "remote printing," work is under way to create a Fax Profile for IPP as well, so that IPP can duplicate the legal as well as common practices of fax transmissions. Richard Shockey. Rshockey@ix.netcom.com contributed this definition. Thank you.

IPO Initial Public Offering. Start a company. Some years later, take it public. Come out at $12. A year later, your stock is at $24. You're a success, and rich. IPOs are critical in saying "Thank You" to all your hardworking employees.

IPS Internet Protocol Suite.

IPSEC A secure version of the Internet Protocol that provides optional authentication and encryption at the packet level. IPSEC is required for the next release of the IP, IPv6 (also called IPng) and optional for the current version of Ip (IPv4).

IPT IP Telephony.

IPT Gateway IP Telephony Gateway. Imagine you and I work for a company which has a PBX — a telephone system. You dial 234 to reach Harry. You dial 9 and a long distance number to dial your biggest client in Los Angeles. Now imagine you want to call your the company's branch office in London. You dial 22. You hear a dial tone. You then punch in 689. You hear another dial tone. Then you punch 123. Bingo, the boss of the London office answers. Here's what all those numbers mean. Dialing 22 dials you into a PC called the IP Telephony Gateway, which, on the one side, is connected to your PBX and on the other side is connected to a data line your company has between your office and your London office. Dialing 689 is you telling the IPT Gateway that you want to speak to the PBX in your London office. Dialing 123 tells the London PBX to dial extension 123.

That connection between your PBX and your London office's PBX might be anything from a dedicated private data line (e.g. part of your company's Intranet), to a virtual circuit on a Virtual Private Network (VPN) or it might be the public Internet. The IPT Gateway's major function is to convert the analog voice coming out of your PBX into VoIP (voice over Internet Protocol) and then send it on a packet switched data circuit which conforms to the IP. In short, an IPT Gateway allows users to use the Internet (or most likely an Intranet or Virtual Private Network) to talk with remote sites using (Voice over Internet Protocol).

IPTC On April 30, 1998, Ericsson Inc. released a press release which contained, inter alia, "Ericsson Inc. has developed a new IP telephony platform called Internet Telephony Solution for Carriers (IPTC) that raises the standard for IP telephony systems. IPTC offers phone-to-phone, fax-to-fax and PC-to-phone services over a TCP/IP network. It provides a superior operations and management (O&M) facility that moves IP telephony to a true carrier-class communications system. IPTC works by taking phone and fax calls that originate in the public switched telephony network (PSTN) and passing them to the IPTC platform, which carries them over the TCP/IP network to their destination where they are fed back to the PSTN network. PC-to-phone calls are taken directly from the TCP/IP network and carried to their destination in the same way...IPTC software runs on industry standard platforms that are based on Intel Pentium processors and Microsoft Windows NT...IPTC uses a Web-based management program to update and control multiple gateways. No longer is it necessary to change the parameters in individual gateways when IPTC can update all gateways within a network through one "netkeeper" applications program. The call and traffic control for individual gateways in a network is handled by sitekeepers. The sitekeepers connect to the netkeeper, which acts as a single point of control for the O&M functions of the whole IPTC platform. The netkeeper is not involved in the processing of calls but stores the platform topology information, routing configuration and alarm information. Other features included in the IPTC platform are least-cost routing, dynamic route allocation, multiple IP networks support, and the ability to handle validated and un-validated traffic. Real-time billing with fraud prevention and call duration advice with integrated voice response software is also provided."

IPv4 Internet Protocol Version 4. The current version of the Internet Protocol, which is the fundamental protocol on which the Internet is based. Although its roots are in the initial development work for ARPAnet, IPv4 was first formalized as a standard in 1981. Since that time, t has been widely deployed in all variety of data networks, including LANs and LAN internetworks. While IPv4 served its purpose for some 25 years, it

has lately proved to be inadequate, largely in terms of security and limitations of the address field. The address field is limited to 32 bits; although 2 to the 32nd power is a very large number, we are running out of IP addresses just as we have run out of 800 numbers and traditional area codes. Hence, the development of IPv6. See IP.

IPv5 Internet Protocol Version 5. IPv5 is not exactly a missing link, although it might appear so. Rather, IPv5 was assigned to ST2, Internet Stream Protocol Version 2, which is documented in RFC 1819. ST2 is an experimental protocol developed as an adjunct to IP for support of real-time transport of multimedia data. See IP and IPv6.

IPv6 Internet Protocol Version 6. The new proposed Internet Protocol designed to replace and enhance the present protocol which is called TCP/IP, or officially IPv4. IPv6 has 128-bit addressing, auto configuration, new security features and supports real-time communications and multicasting. IPv6 is described in RFC 1752, The Recommendation for IP Next Generation Protocol, including the strengths and weaknesses of each of the proposed protocols which were blended to form the final proposed solution. At the time of this writing, IPv6 is in the trial stages and is expected to be available in mid-1997. IPv6 offers 128-bit addressing, auto configuration, new security features and supports real-time communications and multicasting. The 128-bit addressing scheme will relieve pressure on the current 32-bit scheme, which is nearly exhausted due to the widespread use of IP in the Internet and a wide variety of LAN, MAN and WAN networks. Clearly, 2 to the 128th power is a huge number, yielding a staggering number of IP addresses. According to Mark Miller of Diginet Corporation, it equates to approximately 1,500 addresses per square angstrom, with an angstrom being one ten-billionth of the earth's surface. Another way of looking at this is that IPv6 conceivably yields literally billions of addresses per square meter of the earth's surface; in practice, however, a variety of factors serve to reduce the number of addresses available. According to Miller, even the most pessimistic estimates suggest that there will be approximately 50,000 addresses per square inch—in other words, we are not likely to run out of IPv6 addresses. (Don't be surprised to see your telephone assigned an IP address in the future.)

Autoconfiguration Protocol, an intrinsic part of IPv6, allows a device to assign itself a unique IP address without the intervention of a server. The self-assigned address is based in part on the unique LAN MAC (Media Access Control) address of the device, which might be in the form of laptop computer. This feature allows the user the same full IPv6 capability when on the road as he might enjoy in the office when the laptop is inserted into a LAN-attached docking station. IPv6 security is provided in several ways. Data integrity and user authentication are provided by any of a number of authentication schemes. Second, the Encapsulating Security Payload feature provides for confidentiality of data through encryption algorithms such as DES (Data Encryption Standard). Several different types of IPv6 addresses support various types of communications. Unicast supports point-to-point transmission, Anycast allows communications with the closest member of a device group, and Multicast supports communications with multiple members of a device group.

IPX Internet Packet eXchange. Novell NetWare's native LAN communications protocol, used to move data between server and/or workstation programs running on different network nodes. IPX packets are encapsulated and carried by the packets used in Ethernet and the similar frames used in Token-

Ring networks. IPX supports packet sizes up to 64 bytes. Novell's NCP and SPX both use IPX. See also IPX.COM.

IPX Autodiscovery The ability of a network manager to discover the node address and functionality of network devices.

IPX.COM The Novell IPX/SPX (Internetwork Packet eXchange/Sequenced Packet eXchange) communication protocol that creates, maintains, and terminates connections between network devices (workstations, file servers, routers, etc.). IPX.COM uses a LAN driver routine to control the station's network board and address and to route outgoing data packets for delivery on the network. IPX/SPX reads the assigned addresses of returning data and directs the data to the proper area within a workstation's shell or the file server's operating system. See also NETWARE.

IPX/SPX Internetwork Packet Exchange/Sequenced Packet Exchange. Two network protocols. IPX is NetWare protocol for moving information across the network; SPX works on top of IPX and adds extra commands. In the OSI model, IPX conforms to the network layer and SPX is the transport layer.

IPXCP IPX Control Protocol; protocol for transporting IPX traffic over a PPP connection.

IPXWAN A Novell specification describing the protocol to be used for exchanging router-to-router information to enable the transmission of Novell IPX data traffic across WAN (Wide Area Network) links.

IR 1. Infrared. The band of electromagnetic wavelengths between the extreme of the visible part of the spectrum (about 0.75 um) and the shortest microwaves (about 100 um).

2. Internet Registry. See also Internet Assigned Numbers Authority.

IRAM Intelligent RAM. The idea is to put a microprocessor into a memory chip — a move that dramatically improve computer performance.

IRC 1. International Record (i.e. non-voice) Carrier. One of a group of common carriers that, until a few years ago, exclusively carried data and text traffic from gateway cities in the U.S. to other countries. The distinction between international companies providing "record" and data has eroded and now both types of companies provide voice and data services internationally.

2. Internet Relay Chat. IRC is another Internet-based technology, like FTP, Telnet, Gopher, and the Web. It is live text communication between two or groups of people that uses special IRC software and ASCII commands. Each IRC is delegated to a single channel and each channel is dedicated to a different area of interest. IRC requires special software, use of complicated ASCII-based commands and it doesn't have a graphical interface, so people more generally use World Wide Web-based chat rooms instead. See Internet.

3. Interference Rejection Combining. A cellular term.

IRD Integrated Receiver/Descrambler. A receiver for satellite signals that also decodes encrypted or scrambled signals. Especially used in the cable TV business.

IrDA InfraRed Data Association. A not-for-profit organization formed in 1993 to set and support hardware and software standards for infrared data transmission, IrDA membership now exceeds 160 corporations worldwide. Standards activities are across hardware, software, systems, components, peripherals, communications, and consumer markets. Specific standards have been set for Serial Infrared Link (SIR), Infrared Link Access Protocol (IrLAP), and Infrared Link Management Protocol (IrLMP). IrLAP explains how link initialization, device address discovery, connection start-up (including link data rate negotiation), information exchange,

disconnection, link shutdown, and device address conflict resolution occur on an IR connection. IrLAP implements the high-level data-link control (HDLC) communications protocol for infrared environments; the rules for discovery and address-conflict resolution are IrLAP's most significant departure from HDLC. Transmission speeds included in the specifications range from 1.152 Mbps to 4.0 Mbps. Imagine that you're carrying around a small portable laptop, PDA or other device and you want to exchange data with your desktop. You simply aim the device at your desktop PC and transmit information back and forth, courtesy of IrDA's work. See also Infrared, IrLAP and IrLMP.

IREQ The Interrupt Request signal between a PCMCIA Card and a socket when the I/O interface is active.

Iridium The name for Motorola's ambitious satellite project "to bring personal communications to every square inch of the earth." According to Motorola, "for the first time, anyone, anywhere, at any time can communicate via voice, fax, or data." Motorola has targeted the band 1610 to 1626.5 MHz. Motorola estimates the service costing $3 a minute. The idea is that we all carry an Iridium handset — a device larger than today's cellular phone — and that we talk directly from the phone to the satellite (one of 66) circling the Earth at 480 miles up and then down to the satellite closest to the called person, then down to an Iridium phone on the ground or to a satellite dish, through landlines to the phone of the called person. The big benefit is that the system knows who you are and where you are the moment you turn on your phone. This way it can always complete calls from somewhere calling you who doesn't know — or doesn't need to know — where you precisely are. It's called Iridium after the element called Iridium which has 77 electrons, which used to be the number of satellites needed. In 1994, the number got cut to 66. But the name stuck. In November of 1992, Business Week estimated that putting up the full Iridium system would cost $3.4 billion. Iridium started launching satellites in November of 1997 and started service in late September, 1998. See also TELEDISC.

iRIP See iCalendar.

IRL 1. Inter-Repeater Link. A networking term. A mechanism for connecting two and only two repeater sets.

2. In Real Life. An Internet term (sort of) — and (sort of) the opposite of URL (Uniform Resource Locator). People who meet "In Real Life" don't meet in chat rooms over the Internet. Real life is good! The Internet is good, too. It's just not "real life."

IrLAP InfraRed Link Access Protocol from IRDA, the InfraRed Data Association. IrLAP defines a link protocol for serial infrared links. IrLAP explains how link initialization, device address discovery, connection start-up (including link data rate negotiation), information exchange, disconnection, link shutdown, and device address conflict resolution occur on an IR (Infrared) connection. IRLAP implements the high-level data-link control (HDLC) communications protocol for infrared environments and adds procedures for infrared-based link initialization and shutdown plus connection start-up, disconnection, and information transfer. The rules for discovery and address-conflict resolution are IrLAP's most significant departure from HDLC. Until recently, IRDA's standards characterized infrared ports as serial links operating at speeds up to 115 Kbps. IRDA's latest standards allow transmission rates as high as 4 Mbps and provide for LAN access via a new IRLAN protocol. The 4 Mbps mode uses pulse-position-modulation data encoding with four possible chip or time-slice positions per data symbol. The system can recognize and prevent interference with UART-based systems by including a Serial Infrared physical-layer link Interaction Pulse (SIP) at least every 500 milliseconds. IRDA has developed APIs for accessing the infrared port. The first, IRCOMM, emulates existing communications device drivers to handle legacy serial and parallel-port connectivity. There's also a native API that infrared-aware programs can use to locate and communicate with each other.

IrLMP InfraRed Link Management Protocol. See IrDA.
IROB 1. In Range of Building. An Underwriters Laboratories term to define where the protection of UL 1459 will apply. See UL 1459.

2. An Lucent Merlin term. In-Range Out-of-Building Protector. A surge protection device for off-premises telephones at a location within 1000 ft. (305 m) of cable distance

8 Bit XT Bus		16 Bit AT Bus
IRQ0	TIMER SERVICES	IRQ0 TIMER SERVICES
IRQ1	KEYBOARD	IRQ1 KEYBOARD
IRQ2	UNUSED	IRQ2 SLAVE INTERRUPT
IRQ3	COM2 & COM4	IRQ3 COM2 & COM4 *
IRQ4	COM1 & COM3	IRQ4 COM1 & COM3
IRQ5	LPT2	IRQ5 LPT2 *
IRQ6	FLOPPY DISK	IRQ6 FLOPPY DISK
IRQ7	LPT1	IRQ7 LPT1
		IRQ8 REAL TIME CLOCK
		IRQ9 IRQ2 VECTOR
*Available for assigning to new devices, such as network and video cards.		IRQ10 Available *
		IRQ11 Available *
		IRQ12 Available *
		IRQ13 MATH COPROCESSOR

IRQs in alphabetical order:		
Device	**IRQ**	
ARCnet card	2	
Bus Mouse	2	
Cascade	2	
CD-ROM drive	5	This list is not set in concrete.
COM1	4	These are suggestions and
COM2	3	ideas. Experimentation is the
COM3	4	best solution.
COM4	3	
Diskette Controller	6	
Ethernet card (old ones)	5	
Ethernet card (new ones)	10 or 15	
Floppy Drive	6	
Hard disk drive	14	
Keyboard	1	
LPT1 (PARALLEL)	7	
LPT2 (PARALLEL)	5	
Math Coprocessor	13	
PC Timer	0	
Printer 1	7	
Printer 2	5	
PS/2 Mouse	12	
Real time	8	
Scanner	7	
Sound Card	10	

from the communications system control unit.

Iron Hardware, as opposed to firmware and software. See also HEAVY IRON.

IRP I/O Request Packet. Data structures that drivers use to communicate with each other. See INTERRUPT and IRQs.

IRQs Interrupt ReQuests. IRQs are found in PCs. IRQs are also called hardware interrupts. They are the way a device signals the data bus and the CPU that it needs attention. In more technical terms, an IRQ is a signal sent to the central processing unit (CPU) to temporarily suspend normal processing and transfer control to an interrupt handling routine. Interrupts may be generated by conditions such as completion of an I/O process, detection of hardware failure, power failures, etc. Devices that use hardware interrupts include the serial and parallel ports, mouse interface cards, modems, game ports, and even the hard disk on XTs. The original IBM PC and PC-XT had only seven hardware interrupts. The bigger AT bus extended that to 15. Until the advent of the 32-bit PS/2 micro-channel and the 32-bit EISA buses, hardware interrupts could not be shared by two or more devices within the PC. Thus if one device had a specific hardware interrupt, even though you weren't using it that time, nothing else could use it. When you start filling your PC with devices (and remember most PCs still use the old AT bus) — like serial ports, modems and mice, you may suddenly find your modem no longer works. There are two solutions — change the interrupts (either in software or using jumpers), making sure no two devices are trying to share the same interrupt — or simply remove one of the printed circuit devices you're not using from the bus. (That's typically my solution.) These are the "normal" IRQs used by current hardware devices in PCs. Below is a list of 16-bit IRQs as they have become used. See also INTERRUPT and INTERRUPT REQUEST.

Irradiation In insulations, the exposure of the material to high energy emissions for the purpose of favorably altering the molecular structure by crosslinking.

Irritainment Annoying entertainment and media spectacles you're unable to stop watching. The O.J. trial is a prime example.

IRSG Internet Research Steering Group. See IRTF.

IRTF Internet Research Task Force. The IRTF is a community of network researchers, generally with an Internet focus. The work of the IRTF is governed by its Internet Research Steering Group (IRSG).

IRTU Integrated Remote Test Unit.

IRU Indefeasible Right of Use (or User). A term used in the underseas cable and fiber optic carrier business. Someone owning an IRU means he has the right to use the circuit for the time and bandwidth the IRU applies to. An IRU is to a submarine or fiber optic cable what a lease is to a building.

IS 1. Information Separator. A type of control character used to separate and qualify data logically. Its specific meaning has to be defined for each application.

2. Interim Standard. EIA/TIA terminology for a "standard" before it becomes a standard. See EIA and TIA.

3. Intermediate System: OSI terminology for a router, which functionally sits between devices on the originating and terminating ends of a session. Such a system provides forwarding functions or relaying functions or both for a specific service such as Frame Relay or ATM.

IS-136 The EIA/TIA Interim Standard which succeeded IS-54, and which addresses digital cellular systems employing TDMA (Time Division Multiple Access). IS-136 also specifies a DCCH (Digital Control CHannel) in support of new features

controlled by a signaling and control channel between the cell site and the terminal equipment. IS-136 gave rise to a high-tier standard for PCS, developed by a Joint Technical Committee (JTC) comprising representatives from ATIS and the TIA. High-Tier PCS supports fast-moving vehicular traffic, much like traditional cellular. See also DCCH, PCS and TDMA.

IS-41 Interim Standard 41. A signaling protocol used in the North American standard cellular system. IS-41 includes pre-call validation of the ESN/MIN combination in order the ensure the legitimacy of the originating device. The signaling protocol has been effective in countering "Tumbling" fraud. See ESN, MIN and TUMBLING.

IS-410 Interim Standard 41 Zero. The initial version of IS-41, which was released in February 1988. IS-410 defined the process for intersystem call hand-off.

IS-41A Interim Standard 41a. A version of IS-41 which supports automatic roaming.

IS-41B Interim Standard 41B. A version of IS-41 which defined Global Title Translation (GTT), the process of translating Global Titles (telephone numbers) into Point Codes. Point Codes are unique addresses of SS7 network nodes. See also GLOBAL TITLE, GLOBAL TITLE TRANSLATION, POINT CODE, and SS7.

IS-41C Interim Standard 41C. A version of IS-41 which supports PCS SMS (Short Messaging Service), defining message formats and authentication standards. See also IS-41, SMS and AUTHENTICATION.

IS-41D Interim Standard 41D. A version of IS-41 which addresses Calling Number ID (also known as Calling Line ID, or CLID), Enhanced 911, and Law Enforcement Intercept. See also IS-41.

IS-54 Interim Standard 54. It is the dual mode (analog and digital) standard for cellular phone service in North America. In its analog form, it conforms to the AMPS standard. IS-54 is an EIA/TIA, developed with the involvement of the CTIA. Since 1995, IS-54 enhancements fall under IS-136. See IS-54B.

IS-54B Interim Standard 54B, the second version of IS-54. IS-54B defined TDMA (Time Division Multiple Access), an access technique used in digital cellular networks. See also IS-54.

IS-55 Interim Standard 55. Standard for TDMA digital cellular service, which is three times the capacity of today's analog cellular service. IS-55 is a fully digital cellular system.

IS-634 The Interim Standard for the interface between cellular base stations and Mobile Traffic Switching Offices (MTSOs). Issued by the TIA subcommittee TR45.4, IS-634 standardizes the functionality of the A-interface associated with the handling of call processing in order that terminal equipment and MTSOs of disparate origin can interoperate in a predictable fashion. The interim standard is intended to support AMPS, N-AMPS, CDMA and TDMA. The first release of IS-634 employs SS7 and 64-Kbps PCM encoding.

IS-661 The Interim Standard for a hybrid CDMA/TDMA wireless system.

IS-95 Digital CDMA standard for U.S. cellular radio systems.

IS-III An AT&T Merlin term. Integrated Solution. One or more UNIX-based applications for improving voice and data communications and automating office operations.

IS-IS Intermediate System to Intermediate System. OSI link-state hierarchical routing protocol, based on DECnet Phase V routing, whereby intermediate systems (routers) exchange routing information based on a single metric to determine network topology.

ISA 1. Interactive Services Association. http:/policy.net/isa

2. Industry Standard Architecture. The most common bus architecture on the motherboard of MS-DOS computers. The ISA bus was originally pioneered by IBM on its PC, then its XT and then its AT. ISA is also called classic bus. It comes in an 8-bit and 16-bit version. Most references to ISA mean the 16-bit version (which carries data at up to 5 megabytes per second). Many machines claiming ISA compatibility will have both 8- and 16-bit connectors on the motherboard. In 1987 IBM introduced a 32-bit bus which it called MCA for Micro Channel Architecture, which is the internal bus inside some of IBM's line of PS/2 MCA machines. But MCA isn't popular because it is incompatible with ISA, so the industry (excluding IBM) invented a 32-bit bus called EISA which stands for Extended Industry Standard Architecture, which is compatible with ISA. EISA, however, suffered from some of the same problems as the MCA bus, namely it was complicated to program to get the card's full benefit. As a result, other buses have been invented, including the VL bus from VESA (the Video Electronics Standards Association) and Intel's PCI bus. PCI stands for Peripheral Component Interconnect. Both the VL and PCI buses claim to be more than 20 times faster than the ISA bus. The PCI bus, according to Intel, can transfer data at up to 132 megabytes per second. Until the advent of complex Windows graphics and imaging programs, the speed of the ISA was not a gating factor in a computer. As of writing, a VL bus could only handle three drop-in printed cards, while a PCI bus could handle 10. See also EISA and MICROCHANNEL.

ISA Configuration Utility The Intel Architecture Lab has been co-developing the Plug and Play specifications with industry partners to ensure long-term compatibility across cards, systems and software. AL has openly licensed the necessary BIOS software to PC manufacturers so they can add Plug and Play capabilities to their systems. Intel Architecture Labs also designed the ISA Configuration Utility for system and add-in card manufacturers to include with their products. This software utility makes it easier for users to install existing ISA cards in their PCs. The software tells the user which resources are available, but configuration is still done manually. The utility also allows the user to optimize the way resources are assigned, which is particularly important for memory addresses.

ISAM Indexed Sequential Access Method. It is a procedure for storing and retrieving data from a disk file. When the programmer designs the format of the file, a set of indexes is created which describes where the records of the file are located on the disk. This provides a quick method of retrieving the data, and eliminates the need to read all the data from the beginning to find the desired information. The indexes can be stored as part of the data file or in a separate index file.

ISAPI Internet Server Application Program Interface. This API was created by Process Software Corp. and Microsoft and announced by Microsoft at the 1995 Fall Interop show. It is tailored to Internet servers and uses Windows' dynamic link libraries (DLLs) to make processes faster than under other APIs. It allows Internet browsers supporting it to access remote server applications from Microsoft and others. See ISAPI Filter.

ISAPI Filter A World Wide Web term. An ISAPI Filter A replaceable DLL which the server calls whenever there is an HTTP request. When the filter is first loaded, it communicates to the server what sort of notifications will be accepted. After that, whenever a selected event occurs, the filter is called to process the event. Example applications of ISAPI filters include custom authentication schemes, compression,

encryption, logging, traffic analysis or other request analyses.

Isarithmic Flow Control Approach to flow control in which transmission permits circulate throughout network. Node wishing to transmit must first capture permit and destroy it, then recreate permit after transmission finished.

ISB InterShelf Bus.

ISCP Integrated Services Control Point. Bellcore's ISCP software system manages and distributes the information needed to run intelligent networks that provide inexpensive, rapidly deployed customized service, according to Bellcore.

ISD Incremental Service Delivery.

ISDL Integrated Services Digital Line. Part of the family of xDSL technologies, ISDL is a means of provisioning ISDN BRI capacity of 128 Kbps to the premise without the need for upgrade of existing cable pairs. Such existing copper pairs are often in such poor condition as to deny the provisioning of conventional ISDN. See ISDN, xDSL, and ADSL.

ISDN Integrated Services Digital Network. ISDN comes today in two basic flavors — BRI, which is 144,000 bits per second and designed for the desktop and, in North America PRI, which is 1,544,000 bits per second and in Europe PRI, which is 2,048,000 bits per second. PRI, which can be made into as many as 24 and 32 phone calls (respectively). PRI is designed for telephone switches, computer telephony and voice processing systems. ISDN BRI is a wonderful service in your home or office because it can give you videoconferencing, and ultrafaster data communications. But it is not an easy service to get up and running. The best advice I can give you is: 1. Figure out what you want to do with your ISDN. 2. Find which equipment you're going to need that will do the best job for you. 3. Call the manufacturer of that equipment, tell him where you're located and ask him which ISDN service to order. 4. After he tells you, order your ISDN service from your local phone company. 5. Then buy the equipment. 6. Allow yourself at least a month to get up and running. 7. Any ISDN equipment you install in a PC will cause major interrupt problems. Make sure you know which interrupts your PC is using for what. See IRQs.

ISDN is essentially a totally new concept of what the world's telephone system should be. According to AT&T, today's public switched phone network has the following limitations: 1. Each voice line is only 4 KHz, which is very narrow, which limits also the speed you can send data across. 2. Most signaling is in-band signaling, which is very consuming of

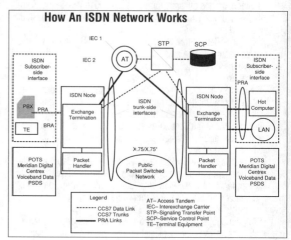

How An ISDN Network Works

bandwidth (i.e. it's expensive and inefficient). 3. The little out-of-band signaling that exists today runs on lines separate to the network. This includes signaling for PBX attendants, hotel/motel, Centrex and PBX calling information. 4. Most users have separate voice and data networks, which is inefficient, expensive and limiting. 5. Premises telephone and data equipment must be separately administered from the network it runs on. 6. There is a wide and growing variety of voice, data and digital interface standards, many of which are incompatible.

ISDN's "vision" is to overcome these deficiencies in four ways: 1. By providing an internationally accepted standard for voice, data and signaling. That standard has pretty well achieved, though don't try and take North American ISDN equipment to Europe. 2. By making all transmission circuits end-to-end digital. 3. By adopting a standard out-of-band signaling system. 4. By bringing significantly more bandwidth to the desktop.

One of the best features of ISDN is the speed of dialing. Instead of 20 seconds for a call to go through on today's old analog network, with ISDN it takes less than a second. It's beautiful. Here are some sample ISDN services:

Call waiting: A line is busy. A call comes in. The user knows who is calling. He can then accept, reject, ignore, transfer the call.

Citywide Centrex: A myriad of services: Specialized numbering and dialing plans. Central management of all ISDN terminals, including PBXs, key systems, etc.

Credit card calling: Automatic billing of certain or all calls into accounts independent of the calling line/s.

Calling line identification presentation: Provides the calling party the ISDN "phone" number, possibly with additional address information, of the called party. Such information may flash across the screen of an ISDN phone or be announced by a synthesized voice. The called party can then accept, reject or transfer the call. If the called party is not there, then his/her phone will automatically record the incoming call's phone number and allow automatic callbacks when he/she returns or calls back in from elsewhere.

Calling line identification restriction: Restricts presentation of the calling party's ISDN "phone" number, possibly with additional address information, to the called party.

Closed user group: Restricts conversations to or among a select group of phone numbers, local, long distance or international.

Collaborative Computing. Work on the same document or drawing or design with someone 10,000 miles away. With ISDN, it doesn't really matter where members of the design team live.

Desktop videoconferencing. I have an ISDN desktop videoconferencing device on my desk. It's wonderful to see the person at the other end. It makes for a far more meaningful conversation.

E-Mail (a.k.a. Personal mailbox): ISDN can carry information to and from unattended phones as long as they're equipped with proper hardware and software.

Internet Access: It's much nicer to browse the Internet at 128 Kbps than at 28.8 Kbps which is the fastest I can do it today with the fastest analog modem I can buy.

Shared Screen — Switched data services provided via ISDN lets two people in remote locations, both equipped with a computer terminal, view the same information on their screens and discuss its contents while making changes — all over one telephone line.

Simultaneous Data Calls: Two users can talk and exchange information over the D packet and/or the B circuit or packet

switched channel.

There are two major problems to the widespread acceptance of ISDN: First, the cost of ISDN terminal equipment is too high. Second, the cost of upgrading central office hardware and software to ISDN is too high. Both costs are coming down. In early, 1995 Pacific Bell announced that it would install one million lines of ISDN BRI by the end of 1998. And, to make this happen, it dropped its ISDN monthly prices to an affordable $24.95 a month for residences and $26.50 a month for businesses.

There are three basic configurations you can get ISDN:

1. The 2B+D "S" interface (also called the "T" interface). The 2B+D is called the Basic Rate Interface (BRI). The "S" interface uses four unshielded normal telephone wires (two twisted wire pairs) to deliver two "Bearer" 64,000 bits per second channels and one "data" signaling channel of 16,000 bits per second. An S-interfaced phone can be located up to one kilometer from the central office switch driving it. Each of the two 64 kpbs "bearer" or B channels can be used to carry a voice conversation, or one high speed data or several data channels, which are multiplexed into zone 64 kbps high speed data line. The "D" channel of 16 kbps will carry control and signaling information to set up and break down the voice and data calls. The "D" channel can also carry data up to 9600 bits per second in addition to the control and signaling information. Signaling and control on the D channel conforms to a protocol (LAPD) and a messaging structure (Q.931). These two allow intelligent endpoints and switching nodes from different vendors to talk a common language and thus be able to transfer features across a network, from one switch to another, e.g. to transfer a Centrex call across town through several switches and to have it arrive at the end phone with the calling party's name.

2. The 2B+D "U" interface. This "U" interface delivers the same two 64 kbps bearer channels and one 16 kbps data channel, except that it uses 2-wires (one pair) and can work at 5-10 kilometers from the central office switch driving it. The "U" interface is the most common ISDN interface. It carries 160,000 bits per second from the central office to your home or office. Of those 160,000 bits, two are used for 64,000 bps Bearer (B) channels and one is used by the subscriber for 16,000 bps of data (the D channel). The other 16,000 bps is used by the network for signaling between the black box on the subscriber premises and the central office. The idea is to get the ISDN "U" interface working to 18,000 feet — the average length of a North American subscriber local loop. You connect the two "U" wires (local loop pair) coming in from your local ISDN CO into a black box about the size of desk printing calculator, called an NT-1. Out the side of the black box comes four wires, which are called the "S Bus." Onto these four wires you can attach, in a loop configuration (also called single bus), as many as eight ISDN terminals — telephones, fax machines, etc. See ISDN.

3. The 23B+D or 30B+D. This is called the Primary Rate Interface (PRI). At 23B+D, it is 1.544 megabits per second. At 30B+D, it is 2.048 megabits per second. The first, 23B+D is the standard T-1 line in the U.S. which operates on two pairs. The second 30B+D is the standard E-1 line in Europe, which also operates on two pairs or wires (i.e. four conductors).

Integral to ISDN's ability to produce new customer services is ITU Signaling System 7. This is a ITU-T recommendation which does two basic things: First, it removes all phone signaling from the present network onto a separate packet switched data network, thus providing enormous economies of bandwidth. Second, it broadens the information that is generated by a call, or call attempt. This information — like the phone number of the person who's calling — will significantly broaden the number of useful new services the ISDN telephone network of tomorrow will be able to deliver.

ISDN has "enjoyed" many "meanings," including I Still Don't Know to It Still Does Nothing its most recent, I Smell Dollars Now.

For more on ISDN, see euro-ISDN, Intel blue, ISDN 2, ISDN 30, ISDN standards, ISDN telephone, ISUP, ISDN network termination 1, personal computer terminal adapter, Proshare, q.931, robbed bit signaling, signaling system 7, spid and tcap.

ISDN 2 What the Americans call ISDN BRI, the British call ISDN 2, which is ISDN with two BRI channels and one D channel.

ISDN 30 The name of an ISDN service which delivers 30 ISDN BRI lines over a single line. ISDN 30 is a fancy name for ISDN on an E-1 line. You find it in countries outside North America, especially Europe. ISDN 2, in the UK, is their name for what Americans call ISDN BRI. See ISDN.

ISDN Basic Link Facility Here is a Nynex definition: The ISDN Basic Link Facility consists of a local transmission facility terminated in the local central office and in a suitable network interface device that is capable of supporting ANSI standard ISDN Basic Rate 2 Binary 1 Quaternary U interface line coding scheme. The standard ISDN Basic Link Facility is 18,000 cable feet or less from the central office termination or served via appropriate electronic equipment, to the Network Termination One device located on the customer's premises.

ISDN BRI Service 2B+D - Two bearer channels and one D channel to your desktop. There are many varieties of ISDN BRI service. The three most common are National ISDN-1 compliant, AT&T 5ESS Custom (an older form of ISDN BRI) and Northern Telecom DMS 100 ISDN. With ISDN BRI, you choose your ISDN equipment first, figure out what it needs, check whether what it needs is available from your local telephone company, then get your line installed, then buy your equipment. Changing your line specs later is expensive and slow. ISDN is still a very first generation product, though when you get it working, it usually works thereafter relatively flawlessly. See ISDN.

ISDN Forum See Vendors ISDN Association.

ISDN Modem A special kind of ISDN Terminal Adapter (see ISDN Terminal Adapter) with enhanced functionality (software, hardware, etc.) to allow it to connect and exchange data with conventional analog modems. Ordinary ISDN terminal adapters can only support data communications with other ISDN devices. They lack the ability to communicate with conventional modems. Because of this, the many on-line services and other analog modem-based applications are not available to users of normal ISDN terminal adapters. The extra functionality in ISDN modems allow their users to make use of these and other existing analog modem-based services. This definition supplied by Paul D. Cook or Ameritech, who once wrote a White Paper describing the obvious need for such a device.

ISDN Network Termination Device You can't plug your ISDN phone directly into an ISDN line like you can with today's analog lines. You need a black box, called a Network Termination device, called an NT1, as in Network Termination 1. In North America you can pick one of these devices up for under $250. The NT1 provides an interface between the ISDN loop and an S or T interface terminal, such as an ISDN phone, or the PCTA (Personal Computer Terminal Adapter). The PCTA is the device which turns a PC into an ISDN termi-

nal/phone. The NT1 is the classic ISDN "black box." It sits on the subscriber's premises at the end of the subscriber loop coming in from the phone company. It talks to the ISDN central office. And, in turn, all ISDN terminals, phones and other devices on the subscriber premises are plugged into this black box. The basic NT1 functions are:
• Line transmission termination.
• Layer 1 line maintenance functions and performance monitoring.
• Layer 1 multiplexing, and
• Interface termination, including multi drop termination employing layer 1 contention resolution.
Some ISDN devices — such as LAN hubs — now come with NT1s built in. See ISDN.

ISDN Overflow/Diversion A feature of Rockwell Galaxy ACDs. ISDN Overflow/Diversion allows Galaxy ACD users to overflow calls between multiple switches using PRI D-channels and B-channels through the public network. This gives the user a virtual private network without the cost of dedicated trunks. By using ISDN messages to overflow a call, specific information associated with the call can be passed to the destination switch, such as ANI, DNIS, and delay time in queue at the originating switch.

ISDN PRI PRI stands for Primary Rate Interface. In North America, ISDN PRI can be thought of as "enhanced T-1." And some long distance carriers are only delivering T-1 in this format. In this ISDN-PRI format, it has major benefits — chiefly the extra bandwidth and benefits derived from the much richer and much faster out-of-band signaling. For example, ANI (Automatic Number Identification, DNIS (Dialed Number Identification Service), etc. are delivered much better this way. ISDN PRI is 24 B (bearer) channels, each of which is a full 64,000 bits per second. One of these channels is typically used to carry signaling information for the 23 other channels. If you're running voice on a single ISDN-PRI, you get 23 voice channels, compared to 24 if you run voice on a T-1 line. However, if you get multiple ISDN-PRI lines, you can often carry the signaling for those voice lines on that one B channel and thus get 24 on each of the others. For example, some voice processing cards will let you support up to eight ISDN PRI channels on one B channel — i.e. the first PRI gives you 23 voice channels. All seven others will give you 24 channels. In Europe, ISDN PRI is 30 bearer channels of 64 Kbps and two signaling channels, each of 64 KBps.

ISDN Repeaters ISDN repeaters let telephone companies extend ISDN lines up to 48,000 feet from their central office. This means they can send ISDN to people further away.

ISDN Router A device which combines the functionality of an ISDN Terminal Adapter (TA) and a TCP/IP router for access to an ISDN network in a small office environment. The TA function accomplishes the interface between the ISDN network and devices which are not ISDN compatible. The router function allows multiple workstations to access the ISDN network, as well each other, through what essentially is a LAN hub with TCP/IP routing capabilities. See also ISDN Terminal Adapter.

ISDN Standards The path of a call in an ISDN network is based on standards:
1. ISDN User A signals the public network over a standard interface. The 2B1Q protocol is used by the terminal as well as the line card in the telephone company central office. The 2B1Q arranges the bits of a digital ISDN signal in a standard manner over the twisted pair connecting the user and the central office. Bellcore TR268 establishes protocols for signaling between the caller and the network.

2. TR444 and TR448 define the standard protocols that allow ISDN services to be carried by the SS7 network.

3. TR317 defines the protocols for standard SS7 networking between the LEC's intraLATA switches that, when combined with TR444 and TR448, can deliver ISDN services.

4. TR394 defines the protocols for standard SS7 networking of the interLATA switches that, when combined with TR444 and TR448, provide the interface for ISDN services and the interexchange network.

5. ISDN UserB, equipped with the standard network interface and using the 2B1Q and TR268 protocols, is prepared to receive the ISDN call from User A. The network delivers the ISDN call information carried over SS7 from User A to User B in the call set-up message.

ISDN Teladapter A Northern Telecom term for a device which connects a national ISDN 1 telephone and a Macintosh to a Northern Telecom switch via an ISDN line card.

ISDN Telephone An ISDN phone can attach to an ISDN basic rate interface. It typically has one digital voice (at 64 kbps) channel and two data options — one for packet switched services (up to 9600 bps) and another for circuit switched data (up to 64 kbps). It will also have an RS-232-C connector on its back and a two line, 48-character LCD adjustable display. It will also have a bunch of dedicated buttons for standard stuff — last number redial, speed dial, on-hook dialing, listen-on-hold, etc. Some ISDN phones work behind most ISDN central offices — e.g. the Telrad phones. Most don't. They have to work behind the central office they were designed for. Or did have to. In February, 1991, Bellcore issued a technical specification for a standard ISDN phone line. The idea of National ISDN-1 is that it be a set of standards which every manufacturer can conform to. A consumer can buy an ISDN phone (one conforming to National ISDN-1) at his local Radio Shack (or other store) take it home, plug it in and know it will work, irrespective of whose central office he's connected to. At time of writing most ISDN phones cost over $600. Some cost nearly $1,000. Some cost more than a personal computer. In late fall of 1992 there was increasing talk that PCs will soon come with telecom ports — able to accept the ISDN signal directly from the central office, without the need for a separate (and expensive ISDN) phone instrument. The PC and the software within it, will then become the phone (presumably with a handset, headset or earset attached to the back of the PC). See ISDN.

ISDN Terminal Adapters ISDN terminal adapters are devices that typically allow analog devices to speak on digital ISDN lines. Terminal adapters are essentially similar to modems, with the following difference. Modems connect terminals to the traditional analog network, and terminal adapters connect those analog-network (non-ISDN) terminals to the digital ISDN network.

ISDN.ISO 8877 ITU-T is the standards group responsible for ISDN-Integrated Services Digital Network. ISDN is an international standard that provides end-to-end digital connectivity to support a wide range of voice, data, and video services. It uses a single communications channel for all forms of information transfer. The ISDN interface connector is specified as an 8-position connector (plug and jack). The four center contacts are assigned to transmit and receive pairs. The four outer contacts are for power.

ISDN/AP Northern Telecom's host to SL1 PBX protocol, which supports NT's Meridian Link. See MERIDIAN LINK and ISDN.

ISE Integrated Switching Element.

ISG Incoming Service Group or Grouping. A fancy name for hunting or rollover. You receive many incoming calls. You don't want to miss a call, so you ask your phone company to set your phone lines up to roll over, also called hunt, also called ISG (Incoming Service Group) in telephonese. You order five lines in hunt. The calls come into the first. If the first one is busy, the second rings. If it's busy, the third rings. If they're all busy, then the caller receives a busy. The commonest types of hunting are sequential and circular hunting. Sequential hunting starts at the number dialed, keeps trying one number after another in number order and ends at the last number in the group. It's typically descending. For example, it starts at 691-8215, goes to 691-8216, then 691-8217, etc. But it can also be ascending — from 691-8217 up. Circular hunting hunts all the lines in the hunting group, regardless of the starting point. Circular hunting, according to our understanding, circles only once (though your phone company may be able to program it circle a couple of times). The differences between sequential and circular are subtle. Circular seems to work better for large groups of numbers. You don't need consecutive phone numbers to do rollovers. Nowadays you can roll lines forwards, backwards and jump around, for example most idle, least idle. Rollovers are now done in software. This also has its downside, since software fails. For example, theoretically if a rollover strikes a dead trunk, it should bounce to the next live trunk. But sometimes it hangs on the dead trunk and many of your incoming calls never get answered. They might ring and ring. They might hit a busy. My recommendation: Test your rollovers at least twice a day. In particular, test that your callers ultimately get a busy if all your lines are busy. Nothing worse your customer should receive a ring-no-answer or a constant busy when calling your company. See also Terminal Number.

ISL 1. Northern Telecom term for ISDN Signaling Link.

2. InterSatellite Link. A relatively new development in satellite technology, ISLs allow LEOs (Low Earth Orbiting satellites) and MEOs (Middle Earth Orbiting satellites) to communicate directly, rather than through earth stations. ISL functions include selection of the shortest circuit path between originating and terminating device, selection of lowest cost terrestrial route, and complete bypass of terrestrial carriers. In the case of this last function, the following scenario best explains: A user of the Iridium LEO system, for instance, places a call from his satellite phone to another user of a satellite phone on the same system. The caller connects directly to a LEO, which finds the other user through querying a GEO (Geosynchronous Earth Orbiting satellite) which then queries the other 65 LEOs in the satellite constellation over ISL signaling links. As the target user answers the call, the connection is established and maintained over another ISL, perhaps established directly between the LEOs in best positions to communicate with the users. The terrestrial networks, both local and long-haul, are bypassed completely, and any charges associated with those terrestrial networks are avoided. As the satellites communicate directly, no earth stations are involved, and precious bandwidth is conserved. Also, issues of propagation delay are minimized, as only one uplink/downlink combination is required, as are the number of systems and processes which must act on the call. This process effectively is the same as in a cellular network, although it is much more complex — the satellites are whizzing around, the users may be mobile and, therefore, multiple handoffs may be required from satellite to satellite on both ends of the connection. While cellular users also typi-

cally are mobile, at least the cell sites are stationary. While at the time of this writing there are no ISL standards, either radio or laser light frequencies are appropriate.

ISM Band Industrial Scientific Medical. A term describing several frequency bands in the radio spectrum, also referred to as Part 15.247 of FCC regulations. Specifically, ISM bands include 902-928 MHz, 2.4-2.483 GHz, and 5.725-5.875 GHz. ISM frequencies are unlicensed; in other words, they can be used for any variety of applications without the requirement for FCC permission. Traditionally used for in-building and system applications such as bar code scanners, industrial microwave ovens and wireless monitoring of patient sensors, ISM also is used in many Wireless LANs. As there is no licensing requirement, there exists the potential for interference from other applications in close physical proximity. Therefore, spread spectrum technology is often used to protect the data transmission integrity of a Wireless LAN.

ISN 1. Intelligent Services Node
2. AT&T's Information Systems Network.

ISO 1. The Greek prefix which means equal or symmetrical, as in isometric. See also ISOCHRONOUS.
2. Most people believe that ISO stands for The International Standards Organization in Paris, which it doesn't. Actually the organization is strictly called the International Organization for Standardization (IOS) and is based in Geneva. The U.S. representative to the ISO is ANSI. IOS or ISO (it's known as both) is a voluntary, non-treaty organization chartered by the United Nations. It began to function officially on February 23, 1947 and in 1951 published its first standard, entitled "Standard reference temperature for industrial length measurement." Its role is to define international standards covering all fields other than electrical and electronic engineering, which is the responsibility of the IEC (International Electrotechnical Commission). In the world of communications, the ISO is best known for the 7-layer OSI (Open Systems Interconnection) Reference Model. See ANSI and IEC. www.iso.ch.
3. ISO also stands for Independent Service Organizations or Independent Sales Organizations in the computer sales community.

ISO 11172 ISO 11172 MPEG-1 and ISO 13818 MPEG-2 Specifications define audio compression algorithms at bit rates from 32 kbps to 384 kbps. The MPEG-1 define three similar compression techniques which are referred to as layer I, II and III. In progressing from layer I to layer III, improvements in compression efficiency are achieved at the expense of additional complexity and algorithmic delay. All layers support two audio channels at sample rates of 32, 44.1 or 48kHz at bit-rats from 16 to 384 kbps. The MPEG-2 standard extends the number of audio channels to five plus a low frequency effects channel. MPEG-2 also provides the additional sample rate options of 16, 22.05 and 24kHz. See MPEG.

ISO 13818 See ISO-11172 above.

ISO 8877 Information Processing Systems Q Interface Connector and Contact Assignment for ISDN Basic access interface located at reference points S and T - International Organization for Standardization. Part of this standard describes pin/pair assignments for 8-line modular connectors. The assignments are the same as EIA's T-568A.

ISO 9000 Series The ISC 9000 series, published in 1987, outlines the requirements for the quality system of an organization. It is a set of generic standards that provide quality assurance requirements and quality management guidance. It is now evolving into a mandatory requirement, especially for

manufacturers of regulated products such as medical and telecommunications equipment. ISO 9001, the most comprehensive of three compliance standards — 9001, 9002, 9003 — is a model for quality assurance for companies involved with design, test, manufacture, delivery and service of products. ISO 9002 covers manufacturing and installation only. ISO 9003 covers product testing and final inspection of standards. ISO is the International Standards Organization based in Paris. See ISO 9001.

ISO 9001 ISO 9001 is a rigorous international quality standard covering a company's research and development, design, production, installation and service procedures. Compliance with the standard is of increasing significance for vendors trading in international markets, in particular in Europe where ISO 9001 registration is widely recognized as an indication of the integrity of a supplier's quality processes. ISO is the International Standards Organization. ISO 9001 is the most rigorous of the three standards. See ISO 9000 SERIES.

ISO 9002 ISO 9002 covers manufacturing and installation only. See ISO 9000 Series.

ISO 9003 ISO 9003 covers testing and final inspections of manufactured products. See ISO 9000 Series.

ISO 9660 The CD-ROM logical file format standard adopted by ISO in 1987. Describes a table of contents but not the format of the actual data. This has led to incompatibilities between different computers. Based on a specification developed by the High Sierra Group (HSG) which included Apple, Microsoft, 3M, Philips, Hitachi, DEC. Also know as Yellow Book and High Sierra.

ISO/IEC International Standards Organization international and the International ElectroTechnical Commission. www.iso.ch

ISOC The Internet Society: a membership organization whose members support a worldwide information network. It is also the governing body to which the IAB reports. See INTERNET.

Isochronous Isochronous transmission means "two-way without delay." Normal everyday voice conversations are isochronous. They have always been isochronous. We could not tolerate delays. We just never called them isochronous. The word isochronous appeared when we started digitizing voice, then joining it with data on a single channel. The data guys suddenly woke up to the fact that users would only tolerate joining voice and data — if the voice went through without delays. So they came up with this fancy new term "isochronous." By accepting this realization, they then could design buses — e.g. Universal Serial Bus — where other flows of data (e.g. printing, keyboard entry, data communications from the Internet) could be delayed minute amounts of time.

Isochronous comes from the Greek "iso" (equal) and "chronous" (time). Isochronous transmission is used to move stuff which must get to its destination with absolutely no delays. Voice and video need isochronous transmission. Let me explain. In the beginning, the phone network switched a call from A to B. It kept the circuit open. Whatever you said at one end went to the other end at the speed of light, effectively instantly. Then they invented other methods of transmission, where the circuit isn't open 100% from end to end during the "conversation." One example is packet switching, used widely for sending data. If you're sending an electronic mail, it clearly doesn't matter if your electronic letter arrives half a second faster or slower. It does matter with voice and video.

In isochronous data transmission, timing is derived from the signal carrying the data. No timing or clock lead is provided at the customer interface. In isochronous data transmission, data has no embedded timing - send it slower and it is still valid, just late. Voice and video are intimately tied to timing. Send voice slower and it sounds very different. With TDM, there is a direct relationship between the signal rate used to digitize the voice samples and the bearer channel rate, allowing accurate reconstruction of the voice (or other signals) at the far end. With packet technologies no such relationship exists. See ISOETHERNET and UNIVERSAL SERIAL BUS.

Isochronous Ethernet A 10 Mbps LAN topology that sets aside 96 ISDN channel to carry voice, data and video. See ISOCHRONOUS and ISOETHERNET.

ISODE ISO Development Environment. An implementation of OSI's upper layers on a TPC/IP protocol stack. Pronounced "eye-so-dee-eee".

ISODE Consortium X.500 directories www.isode.com

IsoENET Also called ISOETHERNET. Developed by National Semiconductor in cooperation with IBM, which helped establish the call set-up procedure, IsoENET adds isochronous services to established Ethernet LANs without affecting, i.e. degrading, normal Ethernet's data traffic. Isoenet allocates a six megabit isochronous channel which can be divided into 97 standard 64 kbps ISDN-like B-channels, 96 of which can be used for real-time interactive multimedia connections, such as voice and video, and one, the D channel, is used for signaling. Isoenet may also be used in combination with Ethernet's standard 10 Mbps to provide faster access to local area network servers. See ISOCHRONOUS.

IsoEthernet An extension to the Ethernet LAN standard proposed by IBM and National Semiconductor and first demonstrated at Fall Comdex 1992. IsoEthernet adds 6 Mbps of capacity to regular Ethernet, specifically to carry low delay, constant bit rate, isochronous data, especially voice and video. This isochronous capacity appears as up to 97 telephony channels of 64 Kbps each — 96 for transmission of information (voice, video, data, etc.) and one (called the D channel) for signaling. Like FDDI-II, IsoEthernet has the potential to carry both live voice or video calls together with LAN packet data on the same cable. See ETHERNET, ISOCHRONOUS, FDDI-II.

Isolated Ground IG. In AC electricity, an isolated ground is a type of outlet characterized by the following features and uses:
• It may be orange and must have a Greek "delta" on the front of the outlet. (A delta looks like a triangle.)
• It must be grounded by an insulated green wire.
• It must have an insulator between the ground terminal and the mounting bracket.
• It is used primarily to power electronic equipment because it reduces the incidence of electrical "noise" on the ground path.

Isolation See POWER CONDITIONING.

Isolator A device that permits microwave energy to pass in one direction while providing high isolation to reflected energy in the reverse direction. Used primarily at the input of communications-band microwave amplifiers to provide good reverse isolation and minimize VSWR. Consists of microwave circulator with one port (port 3) terminated in the characteristic impedance.

Isotropic Radiator A completely non-directional antenna (one which radiates equally well in all directions.) This antenna exists only as a mathematical concept and is used as a known reference to measure antenna gain.

ISP 1. Internet Service Provider. A vendor who provides access for customers (companies and private individuals) to the Internet and the World Wide Web. The ISP also typically provides a core group of internet utilities and services like E-mail, News Group Readers and sometimes weather reports and local restaurant reviews. The user typically reaches his ISP by either dialing-up with their own computer, modem and phone line, or over a dedicated line installed by a telephone company. An ISP is also called a TSP, for Telecommunications Service Provider, and a ITSP, for Internet Telephony Service Provider.
2. ISDN Signal Processor.
3. Information Services Platform.

ISPBX Integrated Services Private Branch eXchange.

ISPT Instituto Superiore delle Poste e delle Telcomunicazioni. (Superior Institute for Posts and Telecommunications, Italy).

ISR International Simple Resale. A system which allows international carriers to buy transmission capacity in bulk, to plug it into the public network at each end, and to resell it, one call at a time. This eliminates the need for settlements between international carriers.

ISS Intelligent Services Switch.

ISSI InterSwitching Interface. An interface between two SMDS switching systems within a LATA.

ISSN Integrated Special Services Network.

Issue A euphemism for "problem." Margaret Horak, wonderful wife Ray Horak, my Contributing Editor, is a top-notch consultant involved in customer service and certain related OSSs (Operations Support Systems). Seems as though her clients lately refer to "problems" as "issues." It's less scary that way.

ISTF Integrated Services Test Facility.

ISUP Integrated Services Digital Network User Part. The call control part of the SS7 protocol. ISUP determines the procedures for setting up, coordinating, and taking down trunk calls on the SS7 network. ISUP is defined by ITU-T recommendations Q.761 and Q.764. ISUP also provides:
• Calling party number information (including privacy indicator).
• Call status checking, to keep trunks in consistent states at both ends.
• Trunk management, and
• Relates of trunks and the application of tones and/or announcements in the originating switch upon encountering error, blockage or busy conditions.
There are seven ISUP Messages: Initial Address Message (IAM), Continuity Check Message (COT), Address Complete Message (ACM), Answer Message (ANM), Release Message (REL), Release Complete Message (RLC) and Exit Message (EXM). For you to benefit from these capabilities, your phone equipment must first be able to access the CCS7 network. One suggested way (but not the only way) is through the ISDN primary rate access (PRA) standard, which supports Q.931 protocol. See IAM, ISDN and COMMON CHANNEL SIGNALING.

ISV Independent Software Vendor. Typically a company which writes and sells software, but not hardware. Manufacturers of hardware and operating systems, i.e. IBM or Northern Telecom, often contract with ISVs to produce specialized software to make their hardware and operating system more attractive.

IT Information Technology. A fancy name for data processing, which became management information systems (MIS), which became information technology. All the same thing, essentially. IT may have come from Europe. I heard it first from Siemens and Nixdorf who merged in 1989.

IT&T Information Technology and Telecommunications.

It's A shortened form of "It is." Not to be confused with Its, which is the possessive of it. Its house. It's a house.

ITAA Information Technology Association of America. ITAA was founded in 1961 as ADAPSO (Automated Data Processing and Services Association). ITAA says its 9,000+ members are IT companies who create and market products and services associated with computers, communications and data. ITAA divisions are Software, IT Services, Information Services and E-Commerce, and Enterprise Solutions. www.itaa.org.

ITAR International Traffic in Arms Regulations. A U.S. government document which established the rules for import and export of goods and services which have significance in terms of national security. Such goods and services are assigned to the U.S. Munitions List, and cannot be exported. The most complex, (read "effective) encryption technologies are included in this list. For example, PGP and DES are on the list. Don't take your laptop out of the country if you have these encryption algorithms loaded on it, unless you want to spend a few years in a federal prison.

ITB Intermediate Block Character. A transmission control character that terminates an intermediate block. A Block Check Character (BCC) usually follows. Using ITBs allows for error checking of smaller blocks in data communications.

ITC 1. International Teletraffic Congress.
2. Japan's Telecommunications Technology Committee.

ITCA International TeleConferencing Association. A professional association organized to promote the use of teleconferencing, including audio, videographics, video, business TV, and distance education. Membership is open to service and product providers, consultants and users. www.itca.org

Iterative Development An approach to application development in which prototypes are continually refined into increasingly complete and correct systems. Similar to prototyping.

Iterative Process The process of repeatedly processing a bunch of instructions. Each repetition, theoretically, comes progressively closer to the desired result, the "correct" answer, etc.

ITESF Internet Traffic Engineering Solutions Forum. An initiative of Bellcore formed in 2Q 1997 at the request of several Incumbent LECs (ILECs) to address common issues of Internet congestion of the Public Switched Telephone Network (PSTN). The ITESF seeks to develop generic requirements for products and features designed to off-load Internet traffic from the PSTN-such traffic is characterized by very long holding times during which relatively bursty traffic is supported. In other words, sometimes the circuit-switched network is used to full capacity while, at other times, little or no data is transmitted-regardless, the network is committed to supporting the traffic, whether or not it is present and whether or not the capacity of the network is required. The ITESF's solution, in abbreviated form, is to recognize Internet traffic for what it is by virtue of the dialed number, and to shunt it off to an IP data network. While this seems very obvious and very simple to do, it does require that some entity recognize the issue, take charge, and do something about it (i.e., set standards). Therefore, the ITESF is forming. See also Bellcore.

ITG AT&T's Integrated Telemarketing Gateway. This is a set of specs for hooking up an outside computer to an AT&T switch. Under ITG, information travels in both directions — from the switch to the host computer and from the host computer to the switch. See also IG (one-directional link), ASAI and OPEN APPLICATION INTERFACE.

ITI 1. Idle Trunk Indicator.

2. Information Technology Industry Council. ITI, according to ITI, represents the leading U.S. providers of information technology products and services. Its members had worldwide revenues of $323 billion in 1994. They employ more than one million people in the United States. 202-626-5725.

ITIC Information Technology Industry Council. www.itic.org

iTIP iCalendar Transport Independent Interoperability Protocol. See iCalendar.

ITM See Information Technology Management.

ITFS Instructional Television Fixed Service. A service provided by one or more fixed microwave stations operated by an educational organization and used mainly to transmit instructional, cultural and other educational information to fixed receiving stations.

ITS 1. Institute for Telecommunications Sciences. The research and engineering branch of the National Telecommunications and Information Administration (NTIA), which is part of the U.S. Department of Commerce (DoC). www.its.bldrdoc.gov/its
2. Intelligent Transportation System. A concept for a transportation system using IT (Information Technologies) to reduce highway transit time, provide necessary emergency services and traffic advisories, reduce traffic congestion, and improve travel safety. The concept is being translated into reality through the development of a number of wireless applications. For example, vehicle navigation systems relying on GPS (Global Positioning System) satellites can track your location, with directions to your destination offered through graphic maps displayed on a monitor. The same terminal can be used to display alternate routes as traffic congestion develops. The same terminal also can display emergency messages. Trucks no longer need to stop at weigh-in stations, as the gross weight of the truck can be transmitted on a wireless basis. The Intelligent Transportation Society of America (ITAS) is heavily involved in the promotion of the concept, as well as the underlying technologies and applications. www.itsa.org

ITSEC The European Information Technology Security classification and evaluation initiative.

ITSP Internet Telephony Service Provider.

ITT International Telephone and Telegraph. A company that once was the largest manufacturer of telecommunications equipment outside the U.S.

ITU ITU Comite Consultatif Internationale de Telegraphique et Telephonique, which, in English, means the Consultative Committee on International Telegraphy and Telephony. The ITU is now known as the ITU-T (International Telecommunications Union-Telecommunications Services Sector), based in Geneva, Switzerland. The scope of its work is now much broader than just telegraphy and telephony. It now also includes telematics, data, new services, systems and networks (like ISDN). The ITU is a United Nations Agency and all UN members may also belong to the ITU, represented by their governments. In most cases the governments give their rights on their national telecom standards to their telecommunications administrations (PTTs, or TOs). But other national bodies (in the US, for example, the State Department) may additionally authorize Recognized Private Operating Agencies (RPOAs) to participate in the work of the ITU. After approval from their relevant national governmental body, manufacturers and scientific organizations may also be admitted, as well as other international organizations. This means, says the ITU, that participants are drawn from the broad arena. The activities of the ITU-T divide into three areas: Study Groups (at present 15) to set up standard ("recommendations") for telecommunications equipment, systems, networks and services. Plan Committees (World Plan Committee and Regional Plan Committee) for developing general plans for a harmonized evolution of networks and services. Specialized Autonomous Groups (GAS, at present three) to produce handbooks, strategies and case studies for support mainly of developing countries. Each of the 15 Study Groups draws up standards for a certain area — for example, Study Group XVIII specializes in digital networks, including ISDN. Members of Study Groups are experts from administrations, RPOAs, manufacturing companies, scientific or other international organizations - at times there are as many as 500 to 600 delegates per Study Group. They develop standards which have to be agreed upon by consensus. This, says the ITU, can sometimes be rather time-consuming, yet it is a democratic process, permitting active participation from all ITU member organizations. The long-standing term for such standards is "ITU (ITU-T) recommendations." As the name implies, recommendations have a non-binding status and they are not treaty obligations. Therefore, everyone is free to use ITU-T recommendations without being forced to do so. However, there is increasing awareness of the fact that using such recommendations facilitates interconnection and interoperability which is in the interest of network providers, manufacturers and customers. This is the reason why ITU-T recommendations are now being increasingly applied — not by force, but because the advantages of standardized equipment are obvious. ISDN is a good example of this. ISDN and other standards recommendations include options which allow for multiple "standards," in recognition of differing national and regional legacy "standards;" as a result, international standards recommendations do not necessarily yield evenly applied standards options. The ITU-T has no power of enforcement, except moral persuasion. Sometimes, manufacturers adopt the ITU-T specs. Sometimes they don't. Mostly they do, as for example with modem specifications, including V.90, H.XXX standards. The ITU-T standardization process runs in a four-year cycle ending in a Plenary Session. Every four years a series of standards known as Recommendations are published in the form of books. These books are colorcoded to represent different four cycles. In 1980 the ITU published the Orange Books, in 1984 the Red Books and, in 1988, the Blue Books. See ITU STUDY GROUPS and ITU V.XX below. The ITU has now been incorporated into its parent organization, the International Telecommunication Union (ITU). Telecommunication standards are now covered under Telecommunications Standards Sector (TSS). ITU-T (ITU-Telecommunications) replaces ITU. For example, the Bell 212A standard for 1200 bps communication in North America was referred to as ITU V.22. It is now referred to as ITU-T V.22. ITU itself says that it specializes in three main activities — defining and adopting telecommunications standards, regulating the use of the radio frequency spectrum and furthering telecommunications development around the world, particularly in the developing countries. It also holds a major trade show in Geneva every four years. As satellites have become more important as a method of long distance communications, so the ITU's allocation of scarce satellite frequencies among countries has become a hot bed of controversy. There are many who believe the ITU to be the most important telecommunications organization in the world. The organization owes its origins to Union Telegraphique which was formed in 1865, with the specific aim of developing standards for the telegraph industry. In 1947, under a United Nations charter, it was reformed as the ITU. This body has three main aims:

a. To maintain and extend, international cooperation for the improvement and interconnectivity of equipment and systems, through the establishment of technical standards.

b. To promote the development of the technical and natural facilities (the spectrum) for most efficient applications.

c. To harmonize the actions of national standards bodies to attain these common aims. In particular, to encourage the growth of communications facilities in developing countries. Due to the rapid growth of the telecommunications industry, it was necessary to set up the International Consultative Committees (ITU, CCIR and IFRB) within the ITU's jurisdiction in order to adequately manage this expansion. The aims are achieved by organizing international conferences and meetings, by sponsoring technical cooperation, and by publishing information and promoting world exhibitions. Currently, the ITU has about 170 member nations. I.T.U., Place des Nations, CH-1211 Geneve 20, Switzerland. Tel +41 22 99 51 11. Fax +41 22 33 72 56. International Telecommunication Union. See the following definitions. ITU can be contacted at www.itu.org, www.itu.ch (where CH stands for Switzerland. Do not type com.)

ITU E.169 An ITU-T Recommendation allowing International Freephone customers to be allocated a unique Universal Freephone Number (UIFN) which will remain the same throughout the world, regardless of country or telecommunications carrier.

ITU H.222 An ITU-T Study Group 15 standard that addresses the multiplexing of multimedia data on an ATM network. See H.222 and other H.XXX entries.

ITU Q.XXX See all the definitions beginning with Q.

ITU Study Groups The ITU operates as a series of groups considering specialist areas. Key study groups applicable to networking and communications are: Study Group VII responsible for terminal equipment for telematic services, including fax and higher level OSI standards; Study Group X covering Languages and methods for telecommunications applications; Study Group XI covering ISDN, telephone network including V-series Recommendations; Study Group XVIII covering digital networks including ISDN. See ITU above and ITU-T.

ITU V.XX A set of evolving telecom standards. For more on those standards, see under the letter "V."

ITU-T The Telecommunications Standards Section (TSS) is one of four organs of the ITU. Any specification with an ITU-T or ITU-TSS designation refers to the TSS organ.

ITUSA Information Technology Users' Standards Association.

ITV Interactive TV. Also abbreviated to I-TV.

IU Interface Unit

IUT Implementation Under Test: The particular portion of equipment which is to be studied for testing. The implementation may include one or more protocols.

IUW The ISDN Users Workshop.

IVDM Integrated Voice and Data Multiplexer. A device that Northern Telecom uses to provide DIALAN, a central office provided local area network offering completely digital, full duplex data transmission at speeds of 300 bps through 19.2 asynchronous and 1,200 bps through 64 kbps synchronous. DIALAN users use existing telephone sets and an Integrated Voice and Data Multiplexer (IVDM) that plugs into a telephone jack.

IVDS Interactive Video and Data Service. A service that the FCC figures will be used for new narrow band PCS licenses.

IVDT Integrated Voice/Data Terminal. A device with a terminal keyboard/display and a voice telephone with or without its own processing power. See INTEGRATED VOICE/DATA TERMINAL.

IVR Interactive Voice Response. Think of IVR as a voice computer. Where a computer has a keyboard for entering information, an IVR uses remote touchtone telephones. Where a computer has a screen for showing the results, an IVR uses snippets of recordings of human voice or a synthesized voice (computerized voice). Recordings are used for repetitive messages, "Thanks for calling ABC Company. Push one for our sales department. Push two for our service department." Synthesized voice (also called Text-To-Speech) is used for reading information from files which contain information that can't be put into neat "sound bites," like numbers and dates, e.g. reading my incoming email. Whatever a computer can do, an IVR can too — from looking up train timetables to moving calls around an automatic call distributor (ACD). The only limitation on an IVR is that you can't present as many alternatives on a phone as you can on a screen. The caller's brain simply won't remember more than a few. With IVR, you have to present the menus in smaller, cascading chunks.

The benefits of Interactive Voice Response are obvious. By automating the retrieval and processing of information by phone, you can "give data a voice" and "add intelligence to the phone call." By doing that, you can:

Put information to work. The classic IVR "killer app" takes an existing database (e.g., a magazine's article archives, a freight company's package-tracking system) and makes it available by phone (or other media, such as fax, e-mail, or DSVD — Digital Simultaneous Voice and Data). You can automate telephone-based tasks. From "bank by phone" to "find my package" to "sell me an airline ticket," to "validate my new credit card," IVR gives access to and takes in information; performs record-keeping, and makes sales, 24 hours a day — supplementing or standing in for human personnel. IVR can add value to communications. Any call-handling phone system (e.g. an automatic call distributor used by airlines) can profit from IVR. Used as a front-end for an ACD, an IVR system can ask questions (e.g., "what's your product serial code?") that help routing and enable more intelligent and informed call processing (by people or automatic systems). IVR can add interactive value to what would otherwise be wait-time. The IVR can be used to distribute info, make callers aware of specials — even provide entertainment. The result: fewer callers drop off queue; you make more sales.

Periphonics, a early and leading IVR company, explains it thus:

IVR systems allow individuals to access information in an organization's computer database and to receive that information either verbally, using an ordinary touch-tone phone, or on a PC, via the Internet. In addition, these systems enable customers to execute certain transactions on-line without the intervention of customer service personnel. Typically, 30-60% of the repetitive and/or routine inbound calls are automated, which can maximize the effectiveness of the current customer service staff.

Benefits of IVR:

1. Reduces costs

2. Improves access to information (24 hours a day - 7 days a week)

3. Enhances customer service

4. Improves competitive position with increased customer retention

5. Streamlines operations

6. Generates new revenues

7. Better utilization of telephone and computer systems capabilities

8. Improves productivity of customer support staff and reduces the need to increase staff for peak periods

9. Provides more services in less time and at lower costs

10. Reduces errors in data capture/input, with feedback for valid entries

11. Gives a typical ROI of six to nine months

IVS Interactive Voice Service.

IW 1. Interworking.

2. Information Warfare.

3. Inside Wire.

IWS Intelligent Workstation.

IWV German for pulse dialing. Impulsewahlverfahren.

IX IntereXchange. Any service which crosses exchange boundaries.

IXC IntereXchange Carrier. Also known as IEC (InterExchange Carrier). Long-haul long distance carriers, IXCs include all facilities-based inter-LATA carriers. The largest IXCs are AT&T, MCI, Sprint and Worldcom; a huge number of smaller, regional companies also fit this definition. The term generally applies to voice and data carriers, but not to Internet carriers. IXC is in contrast to LEC (Local Exchange Carrier), a term applied to traditional telephone companies which provide local service and intraLATA toll service. IXCs also provide intraLATA toll service and operate as CLECs (Competitive Local Exchange Carriers) in many states. Once upon a time, the non-AT&T IXCs were called OCCs (Other Common Carriers), a status which they resented for understandable reasons.

J Box See Junction Box)

J Carrier The Japanese version of the T Carrier system of North America. It's different to the North American one in more ways (e.g. signaling and such) than are apparent in the two tables below.

JAPANESE HIERARCHY

DS 1	1.544 Mbps	24 voice channels
DS 2	6.312 Mbps	96 voice channels
DS 3	32.064 Mbps	480 voice channels
DS 4	97.728 Mbps	1440 voice channels
DS 5	400.352 Mbps	5760 voice channels

NORTH AMERICAN DESIGNATOR (DS LEVEL)

T-1 (DS 1)	1.544 Mbps	24 voice channels
T-1C	3.152 Mbps	48 voice channels
T-2 (DS 2)	6.312 Mbps	96 voice channels
T-3 (DS 3)	44.736 Mbps	672 voice channels
T-4 (DS 4)	274.176 Mbps	4032 voice channels

J-Hook 1. In a microwave or satellite antenna, a j-hook is the name for a length of waveguide with one end turned through 180 degrees. This passes through the reflector vertex to illuminate the reflector's surface from an electronics unit mounted behind the microwave or satellite structure. It looks like a J-Hook.

2. J-hooks are also pieces of J-shaped pieces of bent metal used to hold cables in an equipment rack.

Jabber To jabber. In local area networking technology, continuously sending random data (garbage). Normally used to describe the action of a station (whose circuitry or logic has failed) that locks up the network with its incessant transmission.

Jack Common term for communications terminals found at the end of cables. Also known as modular jack.

Jack Contacts Metallic elements of telephone jacks that carry the central office currents/voltages to the CPE plus contacts.

Jack Header A raceway similar to a header duct, usually provided in short lengths to connect a quantity of distribution ducts together.

Jack Pins See JACK CONTACTS.

Jack Positions A numbering scheme to permit consistent identification of the Jack Contact(s) position. Position identification helps assure compatibility between the wiring system and the associated terminal equipment.

Jack Type Different types of jacks (RJ-11, RJ-45, or RJ-48) are used on telephone lines in North America — analog or digital. The RJ-11 is the most common in the world and is most often used for analog phones, modems, and fax machines. The RJ-11 can be wired with two conductors, four conductors and six conductors. If it has two pairs for two separate phone lines, it's called an RJ-14. One of the lines is the "normal" RJ-11 line — the red and green conductors in the center. The second line is the second set of conductors — black and yellow — on the outside. The RJ-14C is surface or flushmounted for use with desk telephone sets while the RJ-14W is for walmounted telephone sets. The RJ-48 and RJ-45 are slightly bigger jacks and are both virtually the same, but

they both have an 8-pin configuration and are often used for high-speed LANs or T-1 lines. An RJ-11 jack can fit into an RJ-45/RJ-48 connector, however, an RJ-45/RJ-48 jack cannot fit into an RJ-11 connector. See RJ for a complete listing of all available jacks.

Jacket The protective and insulating housing of a cable. Not part of the fiber or the fiber buffer. See also Hard Cable.

Jacket Material The material used as the outer insulator of a cable. See also Hard Cable.

Jacks A receptacle used in conjunction with a plug to make electrical contact between communication circuits. Jacks and their associated plugs are used in a variety of connecting hardware applications including cross connects, interconnects, information outlets, and equipment connections. Jacks are used to connect cords or lines to telephone systems. A jack can be female or male.

Jam In an IEEE 802.3 network, the jam signal, which is normally produced by fixing the minimum number of data bytes that must be transmitted, is used to ensure that if a collision is produced, all devices on the network will detect it. See JAM SIGNAL.

Jam Signal A signal generated by a printed circuit card to ensure that other cards know that a packet collision on a local area network has taken place.

Jamming The interference with through-the-air radio transmission, the object being to hinder the receiver's ability to pick up and understand the signal. An example is the Russians' jamming Radio Free Europe.

Jane Barbie The electronic "Voice With A Smile" on most telephone company intercept recordings. Ms. Barbie does her work for the Electronic Telecommunications Inc., Atlanta, GA. See BARBIE, JANE.

JANET Joint Academic Network. A university network in the U.K. In recent years, with the advent of higher speed links, renamed to "Super-JANET." I have a wonderful lady who works for me called Janet Lindsley. I think of her also as Super-Janet.

JATE The Japanese equivalent of the U.S. FCC part 68 certification for equipment to be attached to the Japanese telephone network. It stands for Japan Approvals Institute for Telecommunications Equipment. Getting JATE approval is expensive, complex and immensely time consuming. At least that's what I wrote in the ninth edition of my edition. JATE wrote me and suggested that I correct my definition as follows:

1. Getting JATE approval is not expensive. It is not expensive compared with the same approval in U.S.A. We don't require to give detailed descriptions on a testing machine nor environmental tests, while FCC requires those. Therefore it is easier to get approval in JATE than in FCC. In some cases, FCC registration costs more than JATE approval, with fee for application and for a test laboratory.

2. Getting JATE approval is not complex. Japanese technical conditions only require not to harm network, like condition in U.S.A. does so. It's just simple.

3. Getting JATE approval is not time consuming. The average

period is 23.4 days for documentation examination process, from the date an application is received by JATE to the date it is completed. And 90% were completed within 38.4 days.

Java Java, invented in 1995, is a programming language from Sun Microsystems designed primarily for writing software to leave on World Wide Web sites and downloadable over the Internet to a PC owned by you or me. The initial excitement over Java was over its ability to bring motion to static Web pages — to make animated figures dance and stock tickers flash. But Java has a larger potential. In the past, software programs had always been written for particular computers and had resided on one machine. Java theoretically enables software to run on any machine (to "write once, run anywhere," as Sun puts it), Java would allow programs to reside anywhere on the Web, flowing across the wires of the Internet and working equally well wherever they land, thus rendering Windows irrelevant, if not obsolete (Sun's hope). In a Java-fuelled computing world, the reign of the Wintel PC (Windows/Intel) machine would be challenged by cheap, bare-bones devices known as "thin clients" — the most prominent of which is the network computer, or NC — a stripped down PC that stores and accesses files and programs on a network rather than a hard drive. The reality of Java since 1995 is that, like all new languages and computer "breakthroughs," getting it implemented into the real world of day-to-day programming has proven sticky.

Java is basically a new virtual machine and interpretive dynamic language and environment. It abstracts the data on bytecodes so that when you develop an app, the same code runs on whatever operating system you choose to port the Java compiler/interpreter to. What's a Java application? According to Wired Magazine. point to Ford Motor Company's website today, for instance, and all you'll get are words and pictures of the latest cars and trucks. Using Java, however, Ford could relay a small application (called an applet) to a customer's computer (the one on your desk which are using the surf the Internet). The customer could then customize options on an F-series pickup while calculating the monthly tab on various loan rates offered by a finance company or local bank. Add animation to these applications and you could get to "drive" the truck.

http://java.sun.com

Java Telephony API JTAPI. A set of modularly-designed, application programming interfaces for Java-based computer telephony applications. JTAPI is designed to serve a broad audience, from call control centers to Web-page designers. JTAPI offers telephony interface extensions grouped into building-block "packages." JTAPI consists of one Core package and several extension packages. JTAPI applications are portable across platforms without modification. Applications written to JTAPI are independent of platform or phone system.

There are two configurations for JTAPI: Desktop Computer and Network Computer. In a desktop configuration, the JTAPI application or Java applet runs on the same workstation that houses the telephony resources. In a network configuration, the JTAPI application or Java applet runs on a remote workstation. This workstation can be a network computer with only a display, keyboard, processor, and some memory. It accesses resources off of the network making use of a centralized server that manages telephony resources. JTAPI communicates with this server via a remote access mechanism, such as Java Remote Method Invocation (RMI), JOE, or a telephony protocol. JTAPI interfaces for other computer

telephony applications, such as SunXTL, TAPI, TSAPI and IBM Call Path are being produced.

JavaMail API The JavaMail API provides a set of abstract classes that model a mail system. The API provides a platform independent and protocol independent framework to build Java based mail and messaging applications. The JavaMail API is implemented as a Java standard extension. Sun provides a royalty-free reference implementation, in binary form, that developers will be able to use and ship. See Java.

JavaScript A scripting language for Web pages. Scripts written with JavaScript can be embedded into HTML documents. With JavaScript you have many possibilities for enhancing your Web page with interesting elements. It makes it easy to respond to use initiated events (such as form input). Some effects that are now possible with JavaScript were once only possible with CGI. Some computer languages are compiled, which means that you run your program through a compiler, which performs a one-time translation of the human- readable program into a binary that the computer can execute. JavaScript is an interpreted language, which means that the computer must evaluate the program each time it is run. Java and JavaScript are not the same thing. JavaScript was designed to resemble Java, which in turn looks a lot like C and C++. The difference is that Java was built as a general purpose object language, while JavaScript is intended to provide a quicker and simpler language for enhancing Web pages and servers.

JavaTel Java Technology Toolkit, or JavaTel, a cross-platform product designed to link any telephone, appliance or networked computer to any Java-based application. In October, 1996, IBM, Intel, Lucent Technologies, Nortel and Novell said they'll support the standard. JavaTel will offer software developers and device manufacturers a uniform interface for driving basic telephony functions, such as call setup, disconnect, hold and call transfer. A series of JavaTel Extension Packages will deliver interfaces such as advanced call control, media services, terminal management, call center management and mobile services. See JAVATEL API.

JavaTel API Java Telephony Application Programming Interface. One of many Java Media APIs developed by Sun Microsystems with help from Lucent Technologies. Provides for call set-up, tear-down and media stream control. JavaTel can run on top of the Sun XTL Teleservices architecture. See JAVATEL.

JB7 Jam Bit 7. It is the same zero suppression format found on T-1s as AMI, which is Alternate Mark Inversion. See AMI.

JBIG Joint Bitonal Image Group. Standard for black and white, and grayscale image representation.

JBIG Alliance In late September, 1996 12 companies announced the formation of the JBIG Alliance. The JBIG Alliance is an industry group formed to create a public forum for the dissemination of information encouraging the adoption and use of ISO/IEC Standard 11544:1993 (JBIG compression) for storage and transmission of bitonal and grayscale image data. JBIG (Joint Bitonal Image Group) is an advanced compression scheme originally developed, like the ITU/ITU Group IV standard that it is intended to replace, as an improved facsimile transmission standard. JBIG's exceptional compression is the result of an advanced compression technique known as arithmetic coding. For bitonal images of standard business documents, JBIG provides file size reductions of 20 to 60 percent with the existing Group IV standard (see table). According to the JBIG Alliance, users can use JBIG's efficient compression for either reducing stor-

	GroupIII	GroupIV	JBIG
Invoices	254,187	287,419	122,813
Line Art/text	306,256	166,098	119,060
Photo/magazines	274,883	241,742	119,066

age and transmission costs, or for substituting higher resolution images without incurring substantially higher storage or transmission costs. JBIG can also store many grayscale images of equal or better quality in less space than required for JPEG compressed files.

Here is a chart showing average file sizes in bytes of identical quality Group III, Group IV, and JBIG Files

JBOD Just a Bunch Of Disks. A very simple and inexpensive storage technology used in storage-intensive applications, such as imaging. JBOD is much simpler than even RAID (Redundant Array of Inexpensive Disks). See also RAID.

JCL Job Control Language.

JEDEC Joint Electronic Devices Engineering Council. An organization of the U.S. Semiconductor manufacturers and users that sets package outline dimension standards for packages made in the U.S.

Jeep The name Jeep came from the abbreviation used in the army for the "General Purpose" vehicle, G.P.

Jeep Fishing When you use your Hummer automobile to pull Jeeps and other four-wheeled, all-terrain vehicles out of the mud on weekends. This definition contributed by Jack Rickard, editor rotundus of Boardwatch Magazine.

JEMA Japan Electronic Messaging Association.

Jeopardy A wonderful AT&T word meaning anything occurring during the course of accomplishing scheduled work which might cause the scheduled completion date to slip.

JEPI Joint Electronics Payments Initiative. A specification from the World Wide Web Consortium (W3C) and CommerceNet for a universal payment platform to allow merchants and consumers to transact E-Commerce (Electronic Commerce) over the Internet. JEPI comprises a standard mechanism for web clients and servers to negotiate payment instrument, protocol and transport with one another. JEPI consists of two parts: Protocol Extensions Protocol (PEP) is an extension layer that sits on top of HTTP (HyperText Transfer Protocol), and Universal Payment Preamble (UPP) is the negotiation protocol that identifies appropriate payment methodology. These protocols are intended to make payment negotiations automatic for end users, happening at the moment of purchase, based on browser configurations. See also Electronic Commerce.

JES Job Entry Subsystem. Control protocol and procedure for directing host processing of a task in an IBM host environment. Also the specific IBM software release, host-based, that performs job control functions.

Jitter If a network provides various latency (i.e. different waiting times) for different packets or cells, it introduces jitter, which is particularly disruptive to audio communications because it can cause audible pops and clicks. Jitter is also the tendency towards lack of synchronization caused by mechanical or electrical changes. Technically, jitter is the phase shift of digital pulses over a transmission medium. Three forms of jitter exist: Data Dependent Jitter (DDJ), Duty Cycle Distortion (DCD), and Random Jitter (RJ). Data Dependent Jitter is caused by limited bandwidth characteristics and imperfections in the optical channel components as it relates to the transmitted symbol sequence, according to Information Gatekeepers. This jitter results from less than ideal individual pulse responses and from variation in the

average value of the encoded pulse sequence which may cause baseline wander and may change the sampling threshold level in the receiver. DCD Jitter is caused by propagation delay differences between low-to-high and high-to-low transitions. DCD is manifested as a pulse width distortion of the nominal baud time. RJ is the result of thermal noise.

Jitter Buffer Management Most speech encoder algorithms have a set of rules concerning packet delivery and disposition management. This is often called jitter buffer management. "Jitter" in this case refers to when the signal is put into frames. The decoding algorithm must decompress and sequence data and make "smoothing" decisions (when to discard packets versus waiting for an out-of-sequence packet to arrive). Given the real-time nature of a live connection, jitter buffer management policies have a large affect on voice quality. Actual sound losses range from a syllable to a word, depending on how much data is in a given packet. The first buffer size is often a quarter-second, large enough to be a piece of a word or a short word — similar to drop-outs on a cellular connection in a poor coverage area. See also IP Telephony Algorithms.

Jitterati What the digital generation becomes after tanking up on too much coffee. This definition courtesy Wired Magazine.

Job A file, typically sent in batch mode. Specifically a set of data, including programs, files and instructions to a computer, that together amount to a unit of work to be done by a computer.

Joel Vietnam was no preparation for Harry and the summer of 1997. Fortunately the green was bucks, not beret. Bucks are mitigating.

Join 1. A basic operation in a relational system that links the rows in two or more tables by comparing the values in specified columns.

2. A service/feature which allows a device to join an existing call, i.e., Conference.

Joining An ATM term. The phase in which the LE Client establishes its control connections to the LE Server.

Joint Costs A regulatory concept. Joint costs are essentially overhead costs. They cover the costs of providing more than one service. Most costs in the telecommunications industry are joint. And being "joint" they give regulators enormous pleasure trying to allocate those costs to various services and therefore trying to figure what prices for those services should be.

Joint Pole A utility pole which supports the facilities of two or more companies. A typical joint pole supports three facilities: electric power, cable television and telephone. In many locations, joint poles also support other devices such as street lights, municipal communications systems, signs, traffic signals, seasonal decorations, fire and police call boxes, and alarm signal wiring. The figure below illustrates the typical allocation of space on utility poles in the United States. The allocation is similar in Canada except that cable television and telephone are sometimes lashed to the same strand. Starting at the top and working down, facilities on this pole are:

• Static wire: a grounded wire at the very top of the pole intended to protect lower conductors from lightning.

• Transmission: three uninsulated conductors which carry 3-phase high voltage (typically 69 to 200 kilovolts) circuits among substations.

• MGN (multi-grounded neutral): a single uninsulated grounded conductor. The currents in the three phases of the transmission line are never quite equal; the MGN carries the

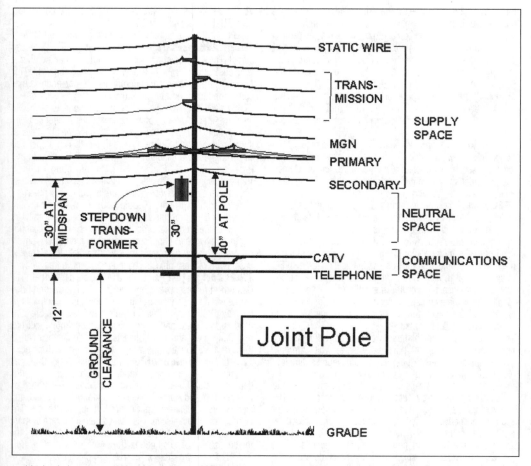

Joint Pole

residual unbalance current. At many poles, the MGN is physically grounded to a groundrod at the base of the pole.
• Primary: one to four uninsulated conductors, frequently supported on a crossarm, which carry power from substations to pole-mounted stepdown transformers. Primary circuits may be single-phase or three-phase, and operate at 4 to 15 kilovolts.
• Secondary: one or two insulated conductors, accompanied by an uninsulated grounded neutral conductor. The secondary circuit (so named because it is fed from the secondary winding of the stepdown transformer) provides the standard 3-wire 115/230-volt electric service for residential and small commercial customers. Secondary conductors are usually twisted together in a bundle called "triplex," although older secondary lines may consist of three separate conductors.
• Stepdown transformer: an oil-cooled transformer which converts the primary voltage to the secondary voltage. Most stepdown transformers are designed for single-phase operation; if a three-phase secondary circuit is required, three physical transformers are sometimes mounted on the same pole.
• Neutral Space: an unused space which separates electric power facilities from communications facilities. This space is specified by the National Electrical Safety Code for safety reasons.
• CATV: cable television facilities supported by steel strand.

An expansion loop at each pole absorbs expansion and contraction caused by temperature variations.
• Telephone: copper telephone cables supported by steel strand. Each telephone cable contains several individual wire pairs; a large cable may contain as many as several hundred pair.

Joint Procurement Consortium An organization formed by Ameritech, BellSouth, Pacific Bell and SBC Communications to help them buy things.

Joint Trench A company wants to lay cable under a busy street. It applies to the city for permission. The city wants to limit the number of times a trench will be dug under that street. So the city announces that this will be a "joint trench" and it tells you that you must contact all the other utilities in town (typically giving them 30 days to respond) and check out those who might wish to locate their cable in that trench. Once you've determined who wants to participate in the joint trench, the trench will probably change in depth and width. Something called the "Western Formula" will be applied. Each of the utilities will pay then less than what they would have, had they built it by themselves.

Joint User Service An arrangement whereby a corporation, association, partnership or individual whose telecommunications needs do not warrant the provision of separate leased service, is permitted to use the service of another customer by mutual agreement. The primary objective of joint user service is to save money by buying circuits in bulk.

Joule The unit of work or energy. The energy expended when a current of one ampere flows through a resistance of one ohm for one second. Joule's Law says the heat produced in a circuit in joules is proportional to the resistance, to the square of the current and to the time.

Journal Printers These are special purpose printers which provide hard copy output for audit trail and demand printing functions associated with hotel/motel management features.

Joy Clicker One who nervously fiddles with a mouse.

Joystick A pointing device for a computer whose upright level is used to manipulate a pointer on a screen. Named after a similar shaped control in airplanes. Joysticks are often used in computer gaming.

JPEG Joint Photographic Experts Group. So called as it was developed jointly by the International Standards Organization (ISO) and the ITU-T, it formally is known as ISO 10918-1 Recommendation T.81. JPEG is a compression technique used primarily in the editing of still images, and in color fax, desktop publishing, graphic arts and medical imaging. JPEG is symmetrical in nature, requiring equal processing power, time and expense on both the transmitting side (compression) and the receiving side (decompression). Its complexity renders it ineffective for real-time video; imaging applications are not so delay-sensitive.

The JPEG compression standard works by converting a color image into rows of pixels (picture elements), which are dots of color image, each with a numerical value representing levels of brightness and color. The picture is then broken down into blocks, each 16 pixels x 16 pixels, and then reduced to 8 pixels by 8 pixels by subtracting every other pixel. The software uses a formula that computes an average value for each block, permitting it to be represented with less data. Further steps subtract even more information from the image. To retrieve the data and thus decompress the image, the process is reversed. A specialized chip decompresses the images hundreds of times faster than is possible on a standard desktop computer. JPEG is a lossy image-compression algorithm that reduces the size of bitmapped images by a factor of 20:1 to 30:1 which compromises the absolute quality of the image in terms of resolution and color fidelity; JPEG can be pushed to yield a 40:1 compression ratio, although the loss in quality is noticeable at this level. JPEG compression works by filtering out an image's high-frequency information to reduce the volume of data and then compressing the resulting data with a lossless compression algorithm. Low-frequency information does more to define the characteristics of an image than does high-frequency information which serves to define sharp edges—losing some high-frequency information doesn't necessarily affect the image quality. In complex images, however, JPEG suffers from an effect known as "tiling," yielding a mosaic-like effect due to the block-oriented compression technique. When you see an image with the .JPG extension, that means it's JPEG image. See also JPEG ++, Motion JPEG, and MPEG.

JPEG++ Storm Technology's proprietary extension of the JPEG algorithm. It lets users determine the degree of compression that the foreground and background of an image receive; for example, in a portrait, you could compress the face in the foreground only slightly, while you could compress it in the background to a much higher degree. See JPEG.

JPG See JPEG.

JT-2 6.312 Mb/s data rate. Same as T-2. Signal compatible with ITU-T document G.704 signal specification.

JTAPI See Java Telephony API and the definitions below.

JTAPI Address Object Part of the JTAPI Core call model. The Address object represents a telephone number. It is an abstraction for the logical endpoint of a phone call. This is distinct from a physical endpoint. In fact, one address may correspond to several physical devices.

JTAPI Call Model The JTAPI Core call model is defined in the Core API package. A call model describes a set of software objects that correspond to physical and conceptual entities in the telephony world. These objects fit together in a specified way to represent a telephone call. The Core API objects are: Provider Object, Call Object, Connection Object, TerminalConnection Object, Terminal Object and Address Object. In the physical view, each Core object represents a tangible property or telephony equipment. From a logical view, the call model represents an abstraction of telephony software entities or the functional properties of the objects. In describing these objects, it is difficult to separate the objects' physical representation from their logical properties, therefore, the description of these objects changes perspective frequently.

JTAPI Call Object Part of the JTAPI Core call model. The Call object represents a telephone call, the information flowing between the service provider and the call participants. A telephone call comprises a Call object and zero or more connections. In a two-party call scenario, a telephone call has one Call object and two connections. A conference call is three or more connections associated with one Call object.

JTAPI Connection Object Part of the JTAPI Core call model. A Connection object models the communication link between a Call object and an Address object. Relationships between Call and Address like connected, disconnected, and alerting are modeled by the Connection object as states. The Connection object also serves as a container for zero or more TerminalConnection objects. Connection objects model the logical aspects of a call connection.

JTAPI Core Package All JTAPI implementations make use of the Core package. Many application developers will only need basic telephony, in which case they will only need to use the Core API package. The Core API package provides basic telephony: placing calls, answering calls, and dropping calls. It defines the basic call model that the extension packages follow in design.

JTAPI Provider Object Part of the JTAPI Core call model. The Provider object is an abstraction of telephony service provider software. The provider might manage a PBX connected to a server, a telephony/fax card in a desktop machine, or a computer networking technology, such as IP. A Provider hides the service-specific aspects of the telephony subsystem and enables Java applications and applets to interact with the telephony subsystem in a device-independent manner.

JTAPI Standard Extension Packages The JTAPI specification defines standard extension packages. The core telephony package, Call Control, Call Center, Private Data and Terminal Set Management extension packages are at version 1.0. The specifications for Media, and Capabilities extension packages are at version 0.3. Mobile, and Synchronous are still under consideration. There are currently eight standard extension packages: Call Control, Call Center, Private Data, Terminal Set Management, Capabilities, Media Services, Mobile Phones and Synchronous.

JTAPI Terminal Object Part of the JTAPI Core call

model. The Terminal object represents a physical device like a telephone and its associated properties. Each Terminal object may have one or more Address Objects (telephone numbers) associated with t, as in the case of some office phones capable of managing multiple call appearances. Additionally, Terminal objects that have more than one phone line may share a telephone number and a single Address object with another Terminal in an adjacent office. However, each Terminal has a unique TerminalConnection associated with a Call even though it may share an Address.

JTAPI TerminalConnection Object Part of the JTAPI Core call model. TerminalConnection objects model the relationship between a call and physical entities, represented by the Terminal object. The TerminalConnection object signals a Terminal when there is an incoming call and monitors the Terminal's activity during the process of a call. This object also closely communicates with the Connection object to receive information on a Call's change in state and to send information on the Terminal's state change. It models the physical aspects of a call connection.

JTC Joint technical Committee

Judge Harold Greene Judge Greene presided over the 1982 AT&T Antitrust settlement, enforcing its provisions and making decisions about requests from the participants to modify or reinterpret the provisions of the settlement. As long as he doesn't allow AT&T to be completely free of regulation, Judge Greene will probably always be involved in figuring the future of the telecommunications industry.

Juggling On-demand Connection An Internet Access Term. The ability to have more suspended on-demand connections than there are ports on the Dialup Switch.

Jughead Jonzy's Universal Gopher Hierarchy Excavation And Display. A database of Gopher links which accepts word searches and allows search results to be used on many remote Gophers. See ARCHIE.

Juice a Brick From Wired's Jargon Watch column. To recharge the big and heavy NiCad batteries used in portable video cameras. "You better start juicin' those bricks, we got a long shoot tomorrow."

Jukebox A jukebox is a piece of hardware that holds storage media, such as optical disks or cartridge tapes. Jukeboxes are typically designed to hold as few as five and as many as 120 devices. Like old-fashioned record playing jukeboxes, media is moved by a robot-like device from the storage slot to the drive reading it. This lets the user share one drive among several cartridges or disk. Jukeboxes are typically used for secondary and archival storage. Access to information is not fast. See JUKEBOX MANAGEMENT.

Jukebox Management In a network, tasks like retrieval and writes to a jukebox come randomly from all the users. These tasks vary in urgency — retrievals are higher priority than writes, for example. Jukebox management software sorts out requests from the network by priority. Management also enhances the performance of a jukebox, by intelligently reordering requests. For example, if there are three requests for images on platter 1 and two from platter 2 and the another from platter 1, jukebox management means the requests from platter 1 will get handled together, then go to platter two. Sometimes it's called "elevator sorting" — responding to requests in logical order, not in the order in which they were made.

Jumbo Frame A Gigabit Ethernet (GE) term. Gigabit Ethernet standards have adjusted the standard Ethernet frame size. The minimum size has been increased from 64 bytes to 512 bytes. The maximum frame size has been increased from

1,518 bytes to a "jumbo frame" of 9,000 bytes. This larger frame size reduces the number of frames that must be processed by a Gigabit Ethernet switch in the process of switching a large data set. As each frame must be processed by the switch, the fewer frames involved, the less the processing demands on the switch, and the less the delay in doing so. Jumbo frames, therefore, increase the throughput of the switch. However, jumbo frames require that multiple standard Ethernet frames be consolidated through a relatively minor process of protocol conversion. See also Gigabit Ethernet.

Jumbo Group A 3,600 channel band of frequencies formed from the inputs of six master groups. See Mastergroup and Supergroup.

Jump Hunting See Nonconsecutive Hunting.

Jumper 1. A wire used to connect equipment and cable on a distributing frame.

2. A patch cable or wire used to establish a circuit, often temporarily, for testing or diagnostics.

3. Jumpers are pairs or sets of small prongs on adapters and motherboards. Jumpers allow the user to instruct the computer to select one of its available operation options. When two pins are covered with a plug, an electrical circuit is completed, When the jumper is uncovered the connection is not made. The computer interprets these electrical connections as configuration information. 4. When errors are found on printed circuit boards, a Jumper cable is sometimes soldered in to correct the problem.

Jumper Wire A short length of wire used to route a circuit by linking two cross connect points. See also JUMPER.

Jumperless No jumpers on the hardware. Settings are accomplished with software — but not by setting jumpers. See Jumper.

Junction Box A metal or plastic box used as an access for cable or wire (coax, fiber, UTP, STP). When companies build a network in a building, building management usually require the J box to be located close to the building's entry point.

Junctor A connection or circuit between inlets and outlets of the same or different switching networks.

JUNET Japan UNIX Network.

Junk Bonds Junk bonds, or high-yield bonds, are rated below investment grade because they are allegedly riskier than higher-rated bonds. As a result, they carry a higher yield than investment grade bonds. I included this definition because in April, 1998, Level 3 Communications, Inc., one of the newest telecommunications transmissions companies sold $2 billion in junk bonds, equaling the largest junk bond deal up to that point in the 1990s.

Junos Code The operating system which Juniper Networks has created for its router. According to Juniper, Junos allows high-speed forwarding across ever-more complex sets of paths. Junos will compete with IOS, the Cisco Internetwork Operating System (IOS) that runs the vast majority of routers now deployed in the core of the Internet.

Jurisdiction A geographic area presided over by the same regulatory body, within the boundary of a single state and an area in which a common carrier is authorized to provide service.

K 1. See K-STYLE HANDSET below.
2. In metric terms it means one thousand (1,000), taken from the Greek word kilo. It is often appended to a measurement such as KHertz or kHz, which means 1,000 Hertz. In data communications, a kilobit means a thousand bits per second (Kbps). In computer storage terms, it means 1,024, which is the figure for two raised to the 10th power, i.e. 2 x 2 x 2 x 2 x 2 x 2 x 2 x 2 x 2 x 2. See also MB.

K Band That portion of the electromagnetic spectrum in the high microwave/millimeter range — from 10.9 GHz to 36 GHz. See also Ka BAND and Ku BAND.

K Plans Also called keysheets. When designing a phone system, you need to assign features and line assignments to each extension. A keysheet or a K Plan is an organized way of figuring and keeping track of those features and assignments for system design and programming. Typically it's one page per extension. These days, keysheets are often done on computer.

K Plant Old Bell System lingo for equipment used in key systems.

K Style Handset A K-style handset is the newer, square telephone handset. The older, round handset is called the G-style handset.

K10 Old Bell-Speak for 10-button (9-lines) key telephone.

K20 Old Bell-Speak for 20-button (19 lines) key telephone.

K30 Old Bell-Speak for 30-button (29 lines) key telephone.

K56flex K56flex was one of two de facto "standard" for running data over dial-up phone lines at up to 56,000 bits per second one way and up to 33.6 Kbps the other way. The standard was developed for use on the Internet, with the 56 Kbps channel flowing to you. The logic is that at 56 Kbps, Web pages fill a lot faster on your screen. 56Kflex was developed by Rockwell Semiconductor and Lucent Technologies, two of the world's leading manufacturers of modem chips. More than 700 modem makers, PC manufacturers, including Compaq and Toshiba, and ISPs like Microsoft Network, supported this de facto standard. The competing 56 Kbps "standard" was called 2x. It was developed by US Robotics. 2x was not compatible with 56Kbps. In other words, a 56flex modem cannot talk at 56Kbps to a 2x modem. On February 6, 1998, the ITU-T ratified a new standard called V.90 for Pulse Code Modulation (PCM) modems running at speeds to 56 Kbps. The idea of this V.90 was to create an international standard so that all 56 Kbps modems could talk to each other. It had been referred to informally as V.PCM until the numeric designation, V.90, was assigned in Geneva on February 6, 1998. It allows speeds of up to 56 Kbps in one direction only, from the central site equipment to the end user. The "back channel" upstream from end user to the central site remains limited to 33.6 kbps (V.34 speeds). Actually, in North American use, the modem is actually limited to only 53 Kbps. The reason for this? The FCC determined that running the modem at 56 Kbps, which it's perfectly capable of, would entail pumping out too much power, which might interfere with adjacent telephone circuit pairs in the same bundle. Some 56Kflex modems can be software upgraded to V.90. Others can't. See 56 Kbps Modem (for a longer technical explanation), V.91 and V.PCM.

Ka Band That portion of the electromagnetic spectrum in the high microwave/millimeter range — approximately 33 GHz to 36 GHz. Ka Band is used primarily in satellites operating at 30 GHz uplink and 20 GHz downlink and is intended in support of such future applications as mobile voice.

KA9Q A popular implementation of TCP/IP (Transmission Control Protocol/Internet Protocol) and associated protocols for amateur packet radio systems.

Kb Kb equals Kilobit. KB is kilobyte. Kilo is one thousand. See Bps for a much more detailed explanation.

KBIT/S KiloBITs per Second. One thousand bits per second. Standard measure of data rate and transmission capacity. See Bps for a much longer explanation.

KBps KBps is kilobytes per second. Kbps is kilobits per second. In short, one thousand bits or bytes per second. For a much better explanation, see Bps.

Kearney System An AT&T numbering scheme for telecom parts. See KS NUMBER.

Keep-Alive Signal A generic term for a signal transmitted when a DTE detects a loss of input from the customer's equipment for a specified period of time (sometimes called a blue signal or AIS). T-carrier systems, for instance, transmit a keep-alive signal during periods of circuit idleness which exceed 150 msec. The purpose of the keep-alive signal is to maintain the circuit during periods of idleness; otherwise, the circuit would time out and the logical connection between devices would be terminated.

Kerberos A security system for client/server computing. Kerberos is a scheme, developed at MIT in the 1980s, to enable secure multiple system access to a client/server computing environment. Named for the three-headed dog, Cerberus, who guarded the gates of Hades in Greek mythology. Kerberos is a UNIX-based distributed database used for user authentication.

Kermit An asynchronous file transfer protocol originally developed at Columbia University in New York City in 1981. The protocol has become popular because of its flexibility. Kermit is found most frequently on DEC VAX computers, IBM mainframes and other minicomputers. One of the clearest advantages of Kermit is its ability to be tailored for virtually any equipment. Protocols break a file into equal parts called blocks or packets, with the data (also known as text or payload) preceded and succeeded by specific control data. The receiving computer checks each arriving packet and sends either an acknowledgement (ACK) or a negative acknowledgement (NAK) back to the sending computer, explicitly indicating the arrival of the packet and its arrival condition. Because modems use phone lines to transfer data, noise or interference on the line will often mess up the block of data. When a block is damaged in transit, an error occurs. The purpose of a protocol is to set up a mathematical way of measuring if the block came through accurately. And if it didn't, ask the distant end to re-transmit the block until it gets it right. Kermit believers say that Kermit is robust, platform-independent, medium-independent, extensible, and highly configurable. XMODEM, YMODEM, and ZMODEM are the file transfer protocols most commonly compared with Kermit, and which are found in numerous shareware and commercial communication software packages. XMODEM and YMODEM are stop-and-wait protocols; XMODEM uses short blocks (128 data bytes), YMODEM uses longer ones (1024 data bytes). ZMODEM is a streaming protocol. In the results of tests of file transfers shown to me, ZMODEM and Kermit are closest in terms of speed and efficiency of transfer, with Kermit edging out ZMODEM for first place. The Kermit Project can be found at www.columbia.edu/kermit/. The Kermit newsgroups are comp.protocols.kermit.misc and comp.protocols.kermit.announce.

2. A suite of software programs for hundreds of different computers and operating systems from the Kermit Project at Columbia University, incorporating Kermit protocol as well as other features including serial communications, automatic dialing, Telnet, Rlogin, LAT, X.25, and other clients and servers, terminal emulation, script programming, and

Kernel The level of an operating system or networking system that contains the system-level commands or all of the functions hidden from the user. In Unix, the kernel is a program that contains the device drivers, the memory management routines, the scheduler and system calls. This program is always running while the system is operating. See KERNEL-BASED WINDOW SYSTEM.

Kernel Based Window System Kernel-based window systems are those in which the software application executes and displays in the same physical machine. Examples include personal computers and Macintoshes. The advantage is speed. The disadvantage is that applications are closely tied to the system environment and are therefore not portable. Kernel-based window systems also do not allow users/developers to use the network as a means of sharing computer resources.

Kernel Driver A Windows NT term. A driver that accesses hardware.

Kerr Effect When polarized light is shone onto a magnetized surface, the light is reflected back at an angle and in a different direction, depending on the polarity of the magnetism. This quirk of nature is called the Kerr Effect and it is the basis of magneto-optical (erasable) discs. The Kerr Effect also affects optical fiber transmission systems. This phenomenon is manifested where the index of refraction of a fiber optic cable varies with the intensity of the transmitted light signal. This nonlinear phenomenon occurs in systems with milliwatt transmitters and very long span lengths, resulting in self phase modulation of the signal, which is not a good thing.

Kevlar A strong synthetic material used in cable strength members. The name is a trademark of the Dupont Company. Kevlar is also used in bulletproof vests worn by police.

Key 1. One or more characters or perhaps a field within a data record used to identify the data and perhaps control its use.
2. The physical button on a telephone set. What normal people call a "Switch," telephone people call a "Key."
3. The physical button on a key telephone set. In a KTS (Key Telephone System environment, the user selects an outside line or intercom line by depressing the appropriate key. The term "Key" originated in the manual switchboard (cordboard) systems, with the operator flipping a key (switch) to set up a talk path. What normal people call a "Switch," telephone people call a "Key."
4. In encryption, a key is a data string which, when combined with the source data according to an algorithm, produces output that is unreadable until decrypted. A key can also be used to decrypt a data string. See KEY HOLDER.

5. The device which unlocks your front door or perhaps your terminal or computer, assuming that you haven't lost it. I have been told that all of my lost keys will be waiting for me on my desk in my next life.

Key Exchange A procedure by which the value of a key is shared between two or more parties. See also ENCRYPTION.

Key Generation The process of creating a key.

Key Holder In encryption, a key is a data string which, when combined with the source data according to an algorithm, produces output that is unreadable until decrypted. A key can also be used to decrypt a data string. In the mid-1990s in the United States, there was great controversy about software that could encrypt electronic communications. Under U.S. laws, such software could not be exported. There was a movement to change the law and create organizations called "key holders." These would be organizations that would be given copies of an individual's decryption key or codebreaker. Such organizations, the theory went, would, under court order, give an individual's decryption key to a law enforcement agency. This might happen with or without the individual's consent.

Key Illumination A lamp under a button (called a "key" in telephony) which flashes at different rates to signal an incoming call, a steady busy and "wink" (fast) hold.

Key Management Digital cryptography systems are based on the use of keys. Before secure transmissions can take place, the appropriate keys must be obtained for use by the sender and the receiver. The total operations and services related to the use and distribution of cryptographic keys is known as key management.

Key Map A MIDI patch-map entry that translates key values for certain MIDI messages, for example, the keys used to play the appropriate percussion instrument or a melodic instrument in the appropriate octave.

Key Pad The touchtone dial pad on a pushbutton phone. Contrary to popular belief, touchtone is not a registered trademark of AT&T. See TOUCHTONE.

Key Pad State An AT&T enhanced fax term. The KEYPAD state can be either NULL or NON-NULL. NULL: The keypad is in use by a feature and is not available for use on a call. NON-NULL: The keypad is available for use in originating a call or for sending DTMF tones on an existing call.

Key Pulse In multi-frequency (MF) tone signaling, a signal used to prepare the distant equipment to receive digits.

Key Pulsing A pulsing system in which digits are transmitted using pushbuttons. Each button corresponds to a digit and generates a unique set of tones.

Key Sequence Number An identifier associated with a key that allows one value of a key to be distinguished from an older or newer value of the key.

Key Service Panel An old 1A2 key telephone term. Wired or unwired connector panel for modular expansion of key system service by allowing the installation of additional Line Cards and/or other KTUs. Typically, Key Service Panels are available in different jack configurations to accommodate 18-, 20-, 36- and 40-pin KTUs. Commonly abbreviated as KSP. Most KSPs are supplied as rack-mount equipment.

Key Service Unit KSU. This is a small metal cabinet which contains all the electronics of a business key telephone system. The KSU fits between the lines coming in from the central office and the lines going to the individual phones. Be careful where you place the Key Service Unit. That place should be well-ventilated as the KSU gets hot. It should be near a power outlet (it needs one). Unless it's a very small phone system, it should be plugged into a power outlet dedicated to it (other devices, such as typewriters, computers, TV sets, and vacuum cleaners, plugged into the same electrical circuit could affect it). And the power outlet should be above the reach of the mops and brooms of the local cleaning people. Otherwise the plug will get knocked out and the phone system won't work the next day. See KEY TELEPHONE SYSTEM.

Key Set Also called Key Telephone Set. A telephone set having several buttons which can be used for call holding, line pickup, auto-dialing, intercom and other features. Ericsson calls a keyset a touchtone telephone.

Key Strip The row(s) of buttons on key telephone sets used for line or extension access and for features like call hold and intercom.

Key System Power Supply The local source for all DC voltages required for talking and lamp signaling within the Key Telephone System. The power supply may or may not also provide an AC voltage output for ringing. If it does, a separate Ringing Generator will normally not be required.

Key Telephone System KTS. A system in which the telephones have multiple buttons permitting (requiring) the user to directly select central office phone lines and intercom lines. According to strict, traditional definition, a KTS is not a switch. A PBX switch allows the sharing of pooled trunks (outside lines), to which the user typically gains access by dialing "9," with software in the switch managing contention for the pooled lines, selecting an available line, and setting up the connection. A KTS system, on the other hand, requires that the user make the selection of an available outside line through the use of "grayware" (brain power).

KTS systems generally and traditionally find most appropriate application in relatively small business environments, typically in the range of 50 telephones and requiring relatively unsophisticated functionality and feature content. PBX systems generally are applied to larger and more demanding situations. Contemporary Electronic Key Telephone Systems (EKTSs), however, often cross the line into the PBX world, providing switching capabilities, as well as impressive functionality and feature content.

Key Telephone Unit A modular 1A2 Key Telephone System building block that plugs into a KSU or KSP. Commonly abbreviated as KTU. Typical KTU examples include 4000 Series Line Cards, 4448 Delayed Ring Transfer Card, 6606 Interrupter, etc.

Key To Disk A method of entering data whereby it's sent directly from the keyboard to a disk, usually a hard disk.

Keyboard 1. A series of switches, arranged somewhat like a standard "QWERTY" typewriter that allows you to send information to a computer. There is no such animal as a "standard" computer keyboard. For speed typists, this is a terrible pity. If you are buying multiple computers for your office, check out your peoples' preference for keyboards. Getting the right one can make a big difference. You can often buy PCs without keyboards and buy third party differently designed keyboards.
2. According to the Vermonter's Guide to Computer Lingo, a keyboard is where you hang your keys. Joke.

Keyboard Buffer A temporary storage area in memory that keeps track of keys that you typed, even if the computer did not immediately respond to the keys when you typed them.

Keyboard Call Setup Allows you to set up a data call using the buttons of a telephone, or it allows you to set up a voice call using the keyboard of your PC. Which definition you choose — setting up a voice or a data call — depends on which manufacturer you're working with.

Keyboard Plaque The disgusting buildup of dirt and crud found on computer keyboards. "Are there any other terminals I can use? This one has a bad case of keyboard plaque."

Keyboarding A really stupid word for typing.

Keyed A term used in data communications whereby the RJ-45 male plug has a small, square bump on its end and the female RJ-45 plug is shaped to accommodate the plug. A keyed RJ-45 plug will not fit into a female, non-keyed (i.e. normal) RJ-45. The purpose of keying a plug is to differentiate it from a "normal" non-keyed plug. Keyed RJ-45 plugs are typically used for data communications. See also RJ-11, RJ-22 and RJ-45.

Keying Modulation of a carrier signal, usually by frequency or phase, to encode binary (digital) information, (as in FSK or Frequency Shift Keying).

Keypad See KEY PAD.

Keystone A monitor distortion where one end of the screen — either side to side, or top to bottom — is larger than the other end.

KHz KiloHertz. One thousand hertz. See K.

Kiddie Cams On-line cameras attached to computers attached to the Internet that allow parents to monitor their children from their desks at their offices. Kiddie cams (cameras) are often installed in daycare centers and grade schools. Also called Cradle Cams. See also Web Cam.

Kill File A file that lets you filter USENET postings to your account to some extent. It excludes messages on certain topics or from certain people.

Kill Message A recorded message played at the beginning of a call to a 900 (or other pay-per-call) number that warns the caller of the charges and gives him the option to hang up before it starts.

Killer App Killer Application. The high-tech industry's life-long dream. That dream is to discover a new application that is so useful and so persuasive that millions of customers will rush in, and throw money at you to buy your killer app. The term derives from the PC industry where a killer app was so powerful that it alone justifies the purchase of a computer. The first PBX killer app was probably being able to dial out without an operator (i.e. dial 9). The second was probably least cost routing. The third was probably call accounting. The next may be some form of hookup to PCs, with desktop and LAN connectivity. In the PC industry, killer apps have been spreadsheets, word processing and databases. Finding that one killer app that will make them wealthy beyond their wildest dreams is what drives many software programmers and entrepreneurs. See WINDOWS TELEPHONY.

Kilo One thousand. See kilobyte.

Kilobyte KB. Two to the power ten, which equals 1,024 bytes. Roughly the amount of information in half a typewritten page.

Kilocharacter One thousand characters. Used as a measure of billing for data communications by some overseas phone companies. See also KILOSEGMENT.

Kilosegment 64,000 characters. Used as a measure of billing for data communications by some overseas phone companies. See also KILOCHARACTER.

Kingsbury Commitment December 13, 1913: A letter from Nathan C. Kingsbury, VP AT&T, to the Attorney General of the United States committed AT&T to dispose of its stock in Western Union Telegraph Company. It also promised to provide long distance connection of Bell System lines to independent phone companies (where there was no local competition) and further agreed not to purchase any more independent telephone companies, except as approved by the

Interstate Commerce Commission which regulated the phone industry at that time. See also DIVESTITURE.

Kittyhawk A trade name for a line of very small hard disks manufactured by Hewlett Packard.

Kludge A hardware solution that has been improvised from various mismatched parts. A slang word meaning makeshift. A kludge can also be in software. It may not be elegant and is probably only a temporary fix. As in, "That patch to the software is a real kludge."

Kluge Another way of spelling kludge. We think spelling it kludge is correct. See KLUDGE.

KM Knowledge Management. See Knowledge Management.

KNET Kangaroo Network Hardware/software product (Spartacus/Fibronics) that lets IBM mainframes communicate over networks using the TCP/IP protocol.

Knowbots Intelligent computer programs that automate the search and gathering of data from distributed databases. The creation of knowbots is part of a research project headed up by the Corporation for National Research Initiatives, Reston, VA. Two knowbot-based databases for the medical field are expected to be available in 1991. Knowbots could become more widespread for general use, according to networking experts.

Knowledge Base System In its most simple term, it means knowledge that is known by the system. Software in which application specific information is programmed into something called the "knowledge base" in the form of rules. The system uses artificial intelligence (AI) procedures to mimic human problem solving. It applies the rules stored in the knowledge base and the facts supplied to the system to solve a particular business problem.

Knowledge Management The big buzzword at the 1998 March Internet World in Los Angeles was "knowledge management." Defining precisely what it means is difficult. Some people define knowledge management as the ability to get the right information to the right people at the right time. Yun Wang, wrote in InfoWorld, "In a way, this is what IS should have been about from the beginning...What has kept these networks from serving as true knowledge management systems in the past is the haphazard way in which they tend to grow. New technology routinely is added on to old without systemwide reengineering; new data sources are hooked up to the network without recategorization of the entire information base."

Wang wrote, "In the past, this mish-mash quality of much corporate technology would have posed a large obstacle for anyone trying to turn the network into an easy-to-use knowledge management system. But today, more tools exist that allow you to integrate disparate data sources and interfaces into one coherent whole. This integration ability is one key to taking full advantage of your company's information repositories, both formal and informal. The first step ... is clearly identifying your strategic business goals. Converting to such a system means going beyond such needs as email or Internet access — clear needs, but with little strategy attached to them — and defining how your new system will have a direct and positive impact on productivity, revenue and your company's competitive position..."

When planning your knowledge management system, says Wang, you should start by thinking about which information is most critical to move to which people, and begin planning accordingly. Thinking about the purpose of each kind of information can help you devise a scheme for information distribution and storage that best fits your company's business needs. The difference between a typical company network and a knowledge management system has to do with the deliberate engineering of an information structure. If you want to empower your employees to use your company's formal and informal information base to its full potential, you have to begin thinking of disparate data sources, applications, and interfaces as parts to a greater whole. From this vantage point, you can begin engineering an information retrieval system that makes the most sense for your company's purposes — the integration of multiple sources and interfaces in a logical fashion is what causes "information" to magically transform into "knowledge."

Knowledge Worker In its simplest use, a knowledge worker is a person who uses a computer. But that's not the end of it. Some people take this term real seriously. It's as though they've discovered a new religion. A reader sent me this definition: "An organizational employee who, whenever he/she performs knowledge work, adds intellectual value to the organization's memory. A knowledge worker is an empowered person who both knows (has access to) and affects (measurably change) the organizational memory in a profitable sense. Profitable sense assumes the business-process being aligned to organization strategy, and the value (outcome) of the individual's work effort being measurable." John Perry Barlow, who is a cattle rancher, computer hacker, poet, and a lyricist for the rock band, The Grateful Dead, thinks the expression was created by the "droids" who run Microsoft and Apple.

KS Number Abbreviation for Kearney System number. AT&T's Western Electric division had a major manufacturing and distribution facility in Kearney, New Jersey. Thousands of items were assigned part numbers with KS prefixes. KS numbers still appear on certain basic telecom hardware items made by various manufacturers. AT&T now uses both a Code number which reflects standard industry numbering and a different "Comcode" number on most products. The Code No. 259C modular-to-Amphenol adapter is Comcode No. 103339396, and KS No. 21997L15. Very confusing. Kearney, by the way, is pronounced "carny."

KSR Keyboard Send Receive. A combination teleprinter/transmitter/receiver with transmission capability from the keyboard only, i.e. there is no punch paper tape device and no magnetic memory device, such as a floppy disk.

KSU Key Service Unit. The main cabinet containing all the equipment, switching and electronics necessary to run a key telephone system. See also KEY SERVICE UNIT.

KSU-Less Phone System KSU stands for Key Service Unit. It's a funny term for the main cabinet which contains all the equipment and electronics necessary to run a key telephone system. When you pick up your key telephone's handset and punch a button for an outside side, the KSU connects you to an outside line. When you pick up the handset and punch a button to make an intercom call, the KSU gives you intercom dial tone and receives your dialing instructions, rings the correct extension and gives you a talk path. When you have a KSU key system, you plug outside phone lines into the KSU and run lines in a star configuration to each phone. A KSU-less phone system, on the other hand, has all its electronics in the phone sets themselves. You run the outside phone lines into each KSU-less phone and typically one pair of wires looping for intercom to each one of the KSU-less phones. A KSU-less phone system is typically a very small system, consisting of usually no more than six phones. A KSU-less phone can be very easy to install. And that's its primary charm. It may also be cheaper than having a small KSU phone system. I always think of a KSU-less phones as multiple single line phones joined together in one plastic multi-line phone with one intercom path to other KSU-less phones. See also KEY. What normal people call a "Switch," telephone people call a "Key." That's were the term Key Service Unit gets it derivation from.

KTA Key Telephone Adapter. A Rolm multiplexing unit which connects a standard 1A2 key telephone to a three-pair cable coming in from the Rolm CBX.

KTI Key Telephone Interface.

KTILA Development Centre for Telecommunications (Greece).

KTS Key Telephone System.

KTU Key Telephone Unit. The circuit cards found in a KSU that control telephone sets and their features in a key system.

Ku Band Portion of the electromagnetic spectrum in the 12 GHz to 14 GHz range. Used for satellites, employing 14 GHz on the uplink and 11 GHz on the downlink in support of such applications as broadcast TV for man-on-the-street interviews and other situations requiring a small, portable dish. Ku also is used in Direct Broadcast Satellite (DBS) systems, also know as Direct Satellite System (DSS), such as DirectTV.

KV Old Bell-Speak for key telephone. K stands for Key; V stands for Voice. CV stood for Combined Voice, a single line, simple telephone. All this had to do with letters included in USOC (Universal Service Order Code) codes — the alphanumeric naming convention the local Bell operating phone companies used to identify products and services. See USOC.

KVM Switch Keyboard/Video/Mouse Switch. A switchbox used to control multiple computers from a single keyboard, monitor and mouse.

KVW Old Bell-Speak for a key telephone that's designed specifically for wall mounting. Most modern key telephones can be used on a desk or wall.

KWH KiloWatt Hour. One thousand WATTS of electricity used for one hour.

L Band Portion of the electromagnetic spectrum commonly used in satellite and microwave applications, with frequencies in the 390 MHz to 1550 Mhz range. The GPS (Global Positioning System) frequencies are in the L-Band. GPS uses 1227.6 MHz and 1575.42 MHz.

L Carrier A long haul frequency division multiplexed coax-cabled long haul carrier system. It was first introduced just before the second World War. Eventually it grew to a capacity of 13,300 voice channels over a pair of coaxial tubes. L Carrier systems are still used today. They are the most widely-used analog long distance transmission system.

L Multiplex A system of analog multiplexers built up through groups, supergroups, master groups and jumbo groups of circuits. See L CARRIER.

L-to-T Connector A device that mates two FDM (Frequency Division Multiplexed) groups with one TDM (Time Division Multiplexed) digigroup to allow 24 voice conversations in analog form to talk to (tie into) a DS-1 line — a T-1 line.

L2 Cache Refers to "level 2 or secondary" cache. A type of cache that sits on the motherboard of a PC except when referring to a Pentium Pro machine, where it lives in the processor.

L2TP Layer 2 Tunneling Protocol. A proposal before the IETF for a means of providing secure, high-priority, temporary paths through the Internet. See TUNNELING.

L3 An acronymym for layer 3 switching. See Layer 3 and Layer 3 Switching.

L8R Later. Shorthand added to an email.

Label A set of symbols used to identify or describe an item, record, message or file. It can also be the same as the address in storage.

Labeling Algorithm Algorithm for shortest path routing or similar problems which labels individual nodes, updating labels as appropriate to reach a solution.

LAD LATA Architecture Database

LADT Local Area Data Transport. A method by which customers will send and receive digital data over existing customer loop wiring.

LAI Location Area Identity. A wireless term. A LAI is part of GSM (Global System for Mobile Communications), used in the radio interface. An LAI is part of the GSM Temporary Mobile Subscriber Identity (TMSI). The TMSI is allocated by the GSM network on a location basis. More simply put, it identifies the cell that a mobile telephone user is in.

LAMA Local Automatic Message Accounting. A process using equipment in the central office which records the information necessary to bill your local phone calls by your local phone company.

Lambda The 11th letter of the Greek alphabet. Lambda is used as the symbol for wavelength in lightwave systems. Fiber optic systems may use multiple wavelengths of light through a process known as WDM (Wavelength Division Multiplexing) or DWDM (Dense WDM), with each range of wavelengths appearing in a "window," roughly corresponding to a color in the visible light spectrum. Light wavelengths are measured in nanometers, with a nanometer being one billionth of a meter. See also Lambda Switch.

Lambda Switch A type of switch which is capable of switching light signals. Such a switch is capable of identifying different wavelengths (frequencies) of light, which roughly correspond to visible colors in the light spectrum. In a fiber optic transmission systems employing DWDM (Dense Wavelength Division Multiplexing), the lambda switch would identify those separate wavelengths of light in an incoming fiber and perhaps switch each over a separate fiber on the outgoing side. Or perhaps a lambda switch would identify various datastreams in multiple wavelength light streams, select those intended to travel in a particular direction, pluck them out and redirect them to a particular fiber going in the right direction, multiplex them, and shift them to a different wavelength. If this definition seems a bit fuzzy, it's because there is no such thing as a lambda switch at the time of this writing...at least not one that's commercially available. But, that is expected to change by 2005 or so. See also Lambda, DWDM, SONET, and WDM.

Lame A user who behaves in a stupid or uneducated manner.

Lame Delegation On the Internet, there are Domain Name Servers, also known as resolvers, which are a system of computers which convert domain names into IP addresses, which consist of a string of four numbers up to three digits each. Each applicant for a domain name (e.g., www.harrynewton.com) must provide both a primary and a secondary DNS server. A domain name which fails to provide both primary and secondary DNS servers is known as a "lame delegation." See also DNS.

Laminate The whole structure in the mold, consisting of several piles. A layer in a composite is called a "ply."

Laminations Thin sheets of steel used as the magnetic core in electrical apparatus, (e.g., the core of an audio frequency transformer is normally composed of laminations).

Lamp The technically correct term for light bulb, which nontechnical folks put into their lamps.

Lamp Battery A steady (unpulsing) 10 volt AC source of power to operate the lamps in key telephone sets; usually one of the outputs of the local Key System Power Supply.

Lamp Flash A pulsed 10 VAC source of lamp power sent to a key telephone set to indicate a CO or PBX line is ringing in. Pulse repetition rate is normally 60 Hertz with a duty cycle of .5 sec on and .5 sec off. This signal is usually provided by the local Interrupter KTU.

Lamp Leads Lamp and Lamp Ground (L&LG) wires connected to all lamps in the key telephone set over which steady and pulsed 10 VAC signals from the Line Card KTU are sent.

Lamp Steady A steady (unpulsed) 10 VAC source of lamp power sent to a key telephone set to indicate that the line is in use. See also Lamp Battery.

Lamp Wink A pulsed 10 VAC source of lamp power sent to a key telephone set to indicate that the line is on Hold status; pulse repetition rate is normally 120 Hertz with a duty cycle of .4 sec on and .1 sec off. This signal is usually provided by the local Interrupter KTU.

LAMs Line Adapter Modules.

LAN Local Area Network. A short distance data communications network (typically within a building or campus) used to link together computers and peripheral devices (such as printers) under some form of standard control. For a longer, more detailed explanation, See LOCAL AREA NETWORK.

LAN Adapter A PC-compatible circuit card that provides the PC-to-LAN hardware connection. In addition, LAN software drivers and LAN operating systems need to be run on the PC for it to function as a LAN station.

LAN Aware Applications that have file and record locking for use on a network.

LAN Emulation Also known as LAN-E, it is a set of specifications developed by the ATM Forum for the operation of LAN-to-LAN bridged connectivity over an ATM network, allowing ATM to be deployed on a legacy LAN or with legacy LAN applications.

LAN Ignorant Applications written for single users only. These are not recommended for use on LANs (local area networks).

LAN Intrinsic Applications written for client-server networks.

LAN Manager 1. A person who manages a LAN. Duties can includes adding new users, installing new hardware and software, diagnosing network problems, helping users, performing backup and setting up a security system. Unlike MIS managers, LAN managers are rarely formally trained in LAN management. Sometimes they're called LAN Network Managers.

2. The multi-user network operating system co-developed by Microsoft and 3Com. LAN Manager offers a wide range of network-management and control capabilities. It has been superseded by Windows NT Advanced Server.

LAN Network Manager An IBM-developed network management tool. It is a software program that runs under OS/2 and which provides management and diagnostics tools needed to manage a Token Ring LAN. A PS/2 running LAN Network Manager collects vital statistics and special management data packets on the ring to which it is connected. When multiple rings are involved, the LAN Network Manager relies on the token ring bridges and routers to help in managing those token ring LANs that are not directly connected to the LAN Network Manager station. IBM has installed software in its bridges called the LAN Network Manager Agent. The agent software acts as the eyes and ears for the LAN Network Manager station so that the station can manage the remote rings as if it were connected directly to them. If there were no such agents, managers of networks would be blind to what's going on these LANs. Remote management with LAN Network Manager includes the ability to perform ring testing, analyze traffic and error statistics, and force adapters off the network.

LAN Server IBM's implementation of LAN manager, now largely superseded by OS/2 2.1.

Land Line A telephone circuit that travels over terrestrial circuits, be they wire, fiber or microwave. A call may originate from a source not connected to the "terrestrial" network, as from a car telephone or a ship to shore radio, and the call be completed via a Landline.

LANDA Local Area Network Dealers Association. It runs a number of excellent trade shows each year. Its members are LAN resellers, distributors, manufacturers, and consultants. In 1993 it merged with NOMDA, the National Office Machine Dealers Association. And shortly, thereafter, NOMDA/LANDA changed its name to the Business Technology Association, headquartered in Kansas City. www.btanet.org. See BTA.

Landscape Most computer screens are horizontal, i.e. they are wider than they are high. In the new language of computer screens, such screens are called "landscape." When a computer screen is higher than it is wide, it's called "portrait." Some computer screens can actually work both ways. Some even have a small mercury switch in them that determines which way the screen is standing (portrait or landscape) and will adjust their image accordingly.

LANE An ATM term. LAN Emulation: The set of services, functional groups and protocols which provide for the emulation of Ethernet and Token Ring LANs over an ATM backbone. Operating at the Link Layer (Layer 2 of the OSI Reference Model), LAN Emulation takes over the MAC (Medium Access Control) layer function found on Ethernet and Token Ring NICs (Network Interface Cards). LANE supports connectionless service in either a broadcast or multicast mode. The network addresses of the LECs (LAN Emulation Clients are resolved through a LES (LAN Emulation Server), by virtue of a LUNI (LAN emulation User-to-Network Interface), as defined by the ATM Forum. The LES maintains a table of MAC-to-ATM addresses in order that the native MAC addresses of the LAN-attached devices (e.g., workstations) can be mapped into ATM addresses, with the process being reversed on the destination end of the transmission. Another server, known as a BUS (Broadcast and Unknown Server) handles data addressed to the MAC broadcast address, all multicast and unicast traffic sent by a LEC prior to the establishment of an ATM address for the destination LEC. As a Layer 2 ATM service, LANE functions only within a single ELAN (Emulated LAN) environment, but offers significant advantage in the establishment of a VLAN (Virtual LAN) consisting of multiple physical LAN segments interconnected over the WAN via an ATM VC (Virtual Channel). LANE is not simple and is prone to bottlenecks, but is a cool way to internetwork LANs over ATM. MPOA (MultiProtocol Over ATM) is even more cool, as it operates at the Network Layer (Layer 3 of the OSI Reference Model). Thereby, MPOA overcomes the limitations of LANE by supporting multiple network protocols such as IP, IPX, and AppleTalk. See also MPOA.

Lane A single interactive multimedia network.

Language Computer software that allows you to write programs.

Language Interpreter Any processor, assembler or software that accepts statements in one software language and then produces equivalent statements in another language.

LAP Link Access Procedure.

LAP-B LAPB. Link Access Procedure Balanced, the most common data-link control protocol used to interface X.25 DTEs with X.25 also specifies a LAP or link access procedure (not balanced). Both LAP and LAP-B are full-duplex, point-to-point bit-synchronous protocols. The unit of data transmission is called a frame. Frames may contain one or more X.25 packets. LAP-B is the data link level of X.25 in a packet switched network. Same as a subset of the asynchronous balanced mode of HDLC. It is the link initialization procedure that establishes and maintains communications between the data terminal equipment (DTE) and data communications equipment (DCE). All public packet data networks (PDNs) support LAPB.

LAP-D Or LAPD. Link Access Procedure-D. Also called Link Access Protocol for the D channel. Link-level protocol devised for ISDN connections, differing from LAPB (LAP-Balanced) in its framing sequence. Likely to be used as basis for LAPM, the proposed ITU-T modem error-control standard.

Laplink The word is originally the name of a DOS program that transferred files between laptop and desktop computers.

It came with a special cable that attached from one computer's serial port to the other. You plugged the cable in, loaded the LapLink software and transferred files back and forth. You could do the transferring from either machine. It was a nifty program, the brainchild of Mark Eppley and he formed a company to sell it, called Traveling Software in Bothell, WA. Eventually the program became so successful that the word "to laplink" became a common verb to connote the transferring of files between computers, as in "I'll go laplink these files over to Mary's machine." Eppley's program is now in its sixth revision. It's in Windows format. You can transfer across a parallel port cable (which is faster than the serial port). You can also transfer files across phone lines. You don't even have to be in the same city to transfer files. And Eppley has a new feature called SpeedSync that really cuts transfer time down by sending only the changed parts of the file. We use LapLink every day. It's one of those "must have" pieces of software.

LAPM Link Access Procedure for Modems. A type of error control used in V.42 and V.42bis modems. LAPM uses the Automatic Repeat Request (ARR) method, whereby a request for retransmission of an errored data frame is automatically requested by the receiving device. As I was writing this entry, a reader asked "if I can transmit HDLC format data thru the GSM network. The data sheet for one of the GSM engines states it can handle V42 bis data transfer." According to Ray Horak, LAPM and both V.42 and V.42bis are specifically linked. If GSM supports v.42, then it supports LAPM, by definition. Such modems also use HDLC frames, which and are synchronous modems, by definition. Therefore, he should be able to accomplish all of this via GSM, as best I can determine.

Laptop Computer A portable computer you can use on your lap. Usually weighing fewer than than ten pounds. Laptops are probably the most useful gadget to come along in years. I wrote much of this dictionary on a laptop in planes, trains, airports, etc. My laptop has a data modem and an Ethernet network card. I don't own a desktop. I do all my work on my laptop.

Large Squaring Capability A feature on some key systems which permits all lines to appear on all telephone sets.

Laser An acronym for Light Amplification by Stimulated Emission of Radiation. It is a device which produces a single frequency light. By turning on and off the laser light signal quickly, you can transmit the ones and zeros of a digital communications channel. Lasers carried through glass fiber are ideal for telecommunications transmission for two major reasons. 1. Glass fiber of such purity has now been developed that only a very minute portion of the laser light traveling through it is lost. In telecom terms, this means very little of the laser signal is attenuated, or loses power. The signal maintains its strength and thus reduces the need for frequent and expensive repeaters (the digital word for "amplifiers.") Laser transmission systems can now carry thousands of voice conversations for hundreds of miles without repeaters on two fibers no thicker than a human hair. (You need two fibers — one for the conversations coming and one for them going.) 2. Laser fiber optic telecom systems are totally immune to electromagnetic interference of any kind. There's no humming from electrical motors. You can't pick up the local TV station in the background. You can't pick up any interference from adjacent cables.

In recent years, laser fiber optic transmission has been getting cheaper, more reliable and more powerful at roughly the same rate as computers, and like computers, nobody believes there is an end in sight. As we were writing the first edition of this dictionary, Russell Dewitt of Contel (since acquired by GTE) delivered a paper entitled Evolution of fiber optics in rural telephone networks. In it, he talked about "Fiber Optics Progress" and said:

"In 1860, the Pony Express could deliver a letter from St. Louis to San Francisco in ten days. For three typed pages the data transmission rate was about three bits per minute. In comparison, today in Contel we are transmitting at the rate of 565 Mbps (million bits per second) over single mode fiber. This is a capacity of 8064 voice channels. Gigabit per second systems will be available for use this year (a gigabit is a thousand million bits) and a 20 Gbps system has been demonstrated in the laboratory. For the future, the ultimate potential of a single mode fiber has been estimated. It is about 25,000 Gbps (25,000,000,000,000,000 bits per second). At that rate you could transmit all the knowledge recorded since the beginning of time in 20 seconds. See also LASER DIODE, LASER FAX, LCD, LED and FIBER OPTICS.

Laser Diode Conceptually similar to LEDs, Laser Diodes are the light sources in high-speed fiber optic systems. While LEDs are limited to transmission rates of 500 Mbps or so, Laser Diodes operate at speeds of many Gbps.

Laser Fax A conventional laser printer that is also capable of being used as a FAX machine when combined with an optional plug-in cartridge and used with a personal computer.

Laser Optical System of recording on grooveless discs using a laser-optical-tracking pickup. Originally, the technology was WORM — Write Once (i.e. not erasable) Read Many. It's now erasable.

Laser Printer A high speed non impact dot matrix printer which uses a laser beam to electrostatically form characters on paper. The printer then heats the paper which melts a metallic dust attracted to the electrostatic areas which form the inked images on the paper. Laser printers are fast and the quality of their printing beautiful, rivaling that produced by conventional photo typeset (the way this book was produced).

Lashing Attachment of a cable to a support strand by wrapping steel wire or dielectric filament around the cable.

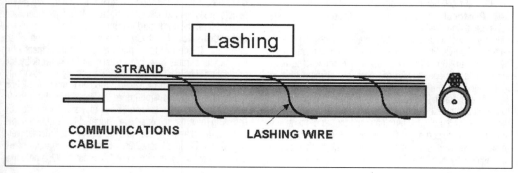

Lashing

STRAND

COMMUNICATIONS CABLE

LASHING WIRE

LASS See LOCAL AREA SIGNALING SERVICES

LAST Local Area Systems Technology. A Digital (DEC) protocol

Last Digit Dialed Signal Allows the use of the # sign on Touchtone telephones to indicate that the last digit has been dialed on outgoing calls. This signal enables the PBX to process calls more rapidly, since some PBXs count the time after a digit was dialed. If nothing else is dialed within a certain time, it assumes that the dialing is complete and then pulses out the call.

Last Extension Called Same as LAST NUMBER REDIAL.

Last In First Out LIFO. The last phone call (or data) arriving is the first call (or data) to leave — to be processed, to be saved, whatever. The term LIFO comes from accounting. It's one of several ways to value an inventory.

Last Mile Not to be taken literally. "Last mile" is an imprecise term that typically means the link — usually twisted pair — between an end-user and the telephone company central office — local, long distance or Internet. Of course, it doesn't mean a "mile," since that "mile" could be less than a mile or several miles. The term has entered the language referring to the problems of your communications making it that last mile. Often that last mile runs over old, limited bandwidth copper wire that has been in the ground for eons and is supplied by a sleepy phone company who doesn't have any competition and not much incentive to perform and hasn't improved the quality of the cable in the loop. The vast majority of local loops are less than 12,000 feet in length — a little over two miles. Generally provisioned with twisted-pair cable plant intended to support voice-grade analog service, the "last mile" is the source of much difficulty for high-speed data services.

Last Number Redial Most modern electronic phones have a button on them called "Last Number Redial." When you touch this button, your phone will automatically dial the last number you dialed. If you also have speed dial numbers on your phone, any number you dialed with a Speed Dial button will not appear in Last Number Redial. Most Last Number Redial buttons on electronic phones attached to a PBX will only recognize completely-dialed numbers. Last Number Redial buttons are useful. Also useful — but less common — is a stored number dial button. Dial a number, punch in "save," then that number will be saved to that button, ready to be dialed later, even though you might dial some other numbers in the meantime.

LAT 1. Local Area Transport. A proprietary communications protocol developed by DEC for terminal-to-host communications. LAT allows terminal emulators to access VAX and VMS systems over Ethernet. See LAT PROTOCOL.

2. A proprietary protocol used in Digital Equipment Corp. terminal servers, providing communication for terminals across an Ethernet LAN. See LAT Protocol.

LAT Protocol The LAT protocol, announced by Digital Equipment Corporation in the mid-80s, is today one of the industry's most widely used protocols for supporting character terminals over Ethernet networks. LAT is currently licensed by more than 40 third party hardware and software developers, and is compatible with the products of more than 30 major system vendors, from Apollo, and Apple to IBM, Tandem and Wang. The basic function of the LAT protocol is to permit a terminal server to connect multiple asynchronous devices — video display terminals, printers or plotters— to a host timeshare computer. To do this, LAT (or any other terminal server protocol) puts data into packets that can be understood by both the asynchronous device and the host.

Essentially, a terminal server protocol is responsible for establishing lower level communications connections, and for routing appropriate transmissions to their destinations.

LATA Local Access and Transport Area, also called Service Areas by some Bell Operating Companies. One of 196 local geographical areas in the US within which a local telephone company may offer telecommunications services — local or long distance. At one stage, AT&T was expressly prohibited from offering intraLATA calls by the terms of the Divestiture. But it is now allowed to offer intraLATA phone calls. Other competitors, such as MCI and Sprint, though rules vary by state, have always been allowed to offer intraLATA phone calls and do so in many states. LATAs serve basically two purposes. First, they provide a method for delineating the area within which the Bell Operating Companies may offer service. Second, they provided a basis for determining how the assets of the former Bell System were to be divided between the BOCs and AT&T.

While writing this edition of the dictionary, Ray and I got into an argument about LATAs. I thought there were fewer than 196 LATAs. But Ray researched the subject to death and affirms 196 is the correct number. In the midst of his research, he sent me the following memo:

Dear Harry,

I've got the story on LATAs...and it is very strange, indeed! There originally were 161 LATAs established by the MFJ. Those LATAs were identified by three-digit codes, as follows: 1xx designated NYNEX (now Bell Atlantic) LATAs; 2xx, Bell Atlantic;, 3xx, Ameritech; 4xx, BellSouth; 5xx US West; 6xx, Southwestern Bell; and 7xx, Pacific Telesis (now Southwestern Bell). 8xx was assigned by Bellcore to areas such as the Commonwealth of North Mariana Islands, Midway/Wake, Guam, and other Caribbean islands. 9xx was assigned by Bellcore to areas covered by Southern New England Telephone (SNET), Cincinnati Bell, and the Navajo Nation (one LATA in Arizona and one in Utah). Since the initial designation of LATAs, a number of subLATAs have been identified, for a variety of reasons.

There are 17 subLATAs in Florida, mandated in 1984 by the FL PUC for equal access purposes. There are 23 "900" LATAs set aside for places like SNET (CT), Cincinnati Bell (OH), and the Navajo nation (1 in AZ and 1 in UT). Add to that the "800" LATAs for Puerto Rico, the Virgin Islands, the Bahamas, Jamaica, AK, HI, and various other the Caribbean, Atlantic and Pacific Islands and it gets very strange. Most of these are pseudo-LATAs set up by Bellcore for toll routing purposes, especially since many of the island nations were only recently added to the NANP.

Equally interesting, if not more so, is the story on the 8xx area codes. There are 8 NPAs in the 8xx range—with 1,298 NXXs— set aside for "non-dialable toll points." These are remote areas served by a cordboard. The subscribers are reached from the cordboard via ring-down circuits.

Late Collision A networking term. Late Collision is an Ethernet collision that takes place on the local segment after 64 bytes of a frame have been placed on the network by the originating device. Late collisions are usually detected only on coax networks, because the 10Base-T monitor station would have to be transmitting at the same time in order to detect a late collision. Late collisions may also be inferred by detecting the presence of a "jam" signal at the end of a frame that is larger than 64 bytes. Note that traditional Ethernet (versus Gigabit Ethernet) specifies a frame size minimum of 64 octets (bytes) and a maximum of 1,514 octets. Also note that a single logical Ethernet may comprise multiple physical segments, with the segments being connected by bridges, hubs, switches or routers. If all of this seems a trite confusing, it's because it is. At some level, however, it's really pretty simple. First, it takes a certain amount of time for a data bit, and certainly a frame of data bits, to propagate (move) across a wire and through all of the intermediate devices that might be involved. The original Ethernet standard specifies big, thick coaxial cable that will support LAN (Local Area Network) communications over a maximum reach of 2.5 kilometers, from one extreme end of the cable to the other. As many as 1,024 devices may be attached, each with a minimum spatial separation of one meter (due to issues of echo, or signal reflection) and with a maximum spatial separation of 500 meters (due to issues of signal attenuation, or power loss). In the most extreme case, therefore, as many as 1,022 devices might be positioned between transmitting device and receiving device. Each device must read the incoming frame of data, determine if it is intended for it, and, if not, pass it on. This process takes some time, and it takes some time for the frame of data to work its way across the wire to the next device, where the process is repeated. Second, Medium Access Control (MAC), or collision control, technique used in Ethernet is CSMA/CD (Carrier Sense Multiple Access with Collision Detection). CSMA/CD allows multiple devices to sense the status of the carrier frequency to determine whether it is "clear" to send a frame of data. Assuming that they sense that it is "clear," they can access the wire at their own option, and at their own risk. If multiple devices access the wire at about the same time, a collision is likely. All attached devices constantly monitor the wire. If a collision is detected, they broadcast a "jam" signal, or collision detection, over a separate subcarrier frequency. All attached devices also monitor the subcarrier frequency, and adjust as necessary, backing off and re-accessing the wire when it becomes available again. Third, a transmitting device is assumed to be transmitting a series of frames of data. The series of frames is assumed to be associated with the transmission of a set of information, which is organized into frames of certain minimum and maximum sizes, 46 bytes and 1,500 bytes, respectively. Small sets of information, such as a query, are very small. Some sets of information, such as file transfers, are potentially very large, and are fragmented into data frames of as much as 1,500 bytes, plus overhead. The maximum size of 1,500 is set so that no single transmitting device can lay claim to all of the capacity of the network, thereby giving other devices a chance. The minimum size is mandated so that a device can be advised of a collision, and have a chance to adjust in that event, and before it assumes that the first frame was received without collision. When you extend the traditional Ethernet with intermediate hubs and switches, you mess with the original concept, and with the underlying physics of signal propagation, which is tuned to network length and link length and number of attached devices and the processes, all of which is tuned to minimum and maximum frame sizes, in consideration of assumptions about the supported applications. It's all very confusing at some level, but relatively simple at another. Read this lucid explanation by Ray Horak several times. You'll get it. See also Ethernet and CSMA/CD.

Late Target Channel Keyup A cellular term. A condition when the target cell does not receive the execute target order in time for the arriving mobile, caused by link delays between MTSO and target cell site. After the mobile retunes to the target cell, noise will be heard on the downlink audio from

the target cell, as the assigned voice channel is not on the air (yet). This results in noise during the handoff.

Latency A fancy term for waiting time or time delay. The time it takes to get information through a network. Real-time, interactive applications such as voice and desktop conferencing are sensitive to accumulated delay, which is referred to as latency. For example, telephone networks are engineered to provide less than 400 milliseconds (ms) round-trip latency. You can get latency in several ways:
1. From propagation delay — the length of time it takes information to travel the distant of the line. This period is mostly determined by the speed of light; therefore, the propagation delay factor is not affected by the networking technology in use.
2. From transmission delay — the length of time it takes to send the packet across the given media. Transmission delay is determined by the speed of the media and the size of the packet.
3. From processing delay - the time required by a networking device for route lookup, changing the header, and other switching tasks. In some cases, the packet also must be manipulated; for instance, changing the encapsulation type, changing the hop count, and so on. Each of these steps can contribute to the processing delay.
4. Rotation delay. The delay in accessing data which comes from waiting for a disk to rotate to the currant location.
In a bridge or a router, latency is the amount of time elapsed between receiving and retransmitting the LAN packet. The length of time the packet is stuck in a bridge or router. See INTERRUPT LATENCY.

Latent Cooling Capacity An air conditioner's capability to remove moisture from the air.

Launch A new term for starting a program from within another. Typically what might happen is you're working in a messaging program, which has individual lines showing you've just received several faxes, voice mails, electronic mail documents. You click on one of the lines. Your program recognizes that it's an electronic mail message and says "quickly open the electronic messaging software and get it to read the message." So it "launches" the messaging software.

Laurus Unit A rack for T-1 interface units.

Lay A term used in cable manufacturing to denote the distance of advance of one member of a group of spirally twisted members, in one turn, measured axially.

Layer See LAYERING.

Layer 1 In the widely-adopted OSI (Open Standards Interconnection) model, there are seven levels of interconnection. Layer 1 is the the Physical layer. It deals with the physical means of sending data over lines (i.e. the electrical, mechanical and functional control of data circuits). See next definitions.

Layer 2 In the widely-adopted OSI (Open Standards Interconnection) model, there are seven levels of interconnection. Layer 2 is the Data Link layer. It is concerned with procedures and protocols for operating the communications lines. It also has a way of detecting and correcting message errors. See next definitions.

Layer 3 In the widely-adopted OSI (Open Standards Interconnection) model, there are seven levels defined of interconnection. Layer 3 is the Network layer. It determines how data is transferred between computers. It also addresses routing within and between individual networks. See next definitions.

Layer 3 Switching Layer 3 switching is simply a combination of Layer 2 switching and Layer 3 routing. But there is some confusion — a tribute to the fact that vendors of these popular switches tend to confuse things. According to

Kristina B. Sullivan, writing in the February 16, 1998 issue of PC Week, Layer 3 switches integrate routing and switching to provide high-speed performance without the drawbacks of a of a flat layer 2 network — broadcast storms, address limitations and spanning-tree loops. In addition, Layer 3 switches can define items, such as traffic prioritization, security and intranet allocation. Layer 3 switches also can control larger network segments than Layer 2 switches, eliminating the need to create subnets and isolate them with routers. There are two main types of Layer 3 switches. Packet-by-packet Layer 3 switches examine each packet and forward it to its destination, just like a router. Cut-through Layer 3 switches examine the first packets, determine their destination and make a connection using Layer 2 switching, thus gaining a high throughput rate. Layer 4 switching, sometimes called Layer 3-plus switches, involves a switch's ability to examine applications to determine priorities and allocate bandwidth accordingly.

Layer 4 In the widely-adopted OSI (Open Standards Interconnection) model, there are seven levels defined of interconnection. Layer 4 is the Transport layer, which defines the rules for information exchange and manages end-to-end delivery of information within and between networks, including error recovery and flow control. See next definitions.

Layer 4 Switching See Layer 3 Switching.

Layer 5 In the widely-adopted OSI (Open Standards Interconnection) model, there are seven levels defined of interconnection. Layer 5 is the Session layer. It is concerned with dialog management. It controls the use of the basic communications facility provided by the Transport layer. See next definitions.

Layer 6 In the widely-adopted OSI (Open Standards Interconnection) model, there are seven levels defined of interconnection. Layer 6 is the Presentation layer. It provides transparent communications services by masking the differences of varying data formats (character codes, for example) between dissimilar systems. See next definitions.

Layer 7 In the widely-adopted OSI (Open Standards Interconnection) model, there are seven levels defined of interconnection. Layer 7 is the Applications layer. It contains functions for particular applications services, such as file transfer, remote file access and virtual terminals. See next definitions.

Layer Entity An active element within a layer.

Layer Function A part of the activity of the layer entities.

Layer Service A capability of a layer and the layers beneath it that is provided to the upper layer entities at the boundary between that layer and the next higher layer.

Layer Service Provider Each layer of the OSI Reference Model is a Layer Service Provider. Examples of layers are Presentation, Session, Transport, Network, and Link. The Layer Service is made available through Service Access Points.

Layer User Data Data transferred between corresponding entities on behalf of the upper layer or layer management entities for which they are providing services.

Layered Network Architectures Currently the basis of all telecommunication network architecture standards, with functions allocated to different layers and standardized interfaces between layers. The OSI Model is a layered network architecture. See LAYERING.

Layered Network Sales Strategy An AT&T sales strategy which involves segmenting customer network needs into four usage levels or layers, then pursuing sales opportunities at these various entry points. The layers are Premises Networks, Office Networks, "metropolitan" or " campus" networks and Wide Area Networks.

Layering Layering is a technique to write complex software faster and more easily. Layering is often used with public, open software. The idea is to have layers of software on top of other layers. Each performs a specific task. The idea is that if your software works at one layer — i.e. conforms to the rules of that layer — it should be compatible (i.e. work with) the layers of software above and below it. The most famous layered software is the seven-layer OSI (Open Systems Interconnection) model. It breaks each step of a transmission between two devices into a discrete set of functions. These functions are grouped within a layer according to what they are meant to accomplish. The data link layer, for example, is concerned with the transmission of frames of data between devices and covers protocols that are aimed at packaging raw data characters into frames, detecting and correcting errors when frames get lost or mutilated, arranging for retransmission and adding flags and headers so that DTE can recognize the beginning and end of a frame. Other layers serve other purposes. Each layer communicates with its counterpart through header records. The flexibility offered through the layering approach allows products and services to evolve. Accommodating changes are made at the layer level rather than having to rework the entire OSI model. Another layered software architecture is Microsoft/Intel's Windows Telephony. It has three layers. At the lowest is SAPI, which is the Service providers' API. In the center is the actual Windows Telephony code. At the top is the TAPI — Telephony applications API. See OSI and Windows Telephony.

Lays The twists in twisted pair cable. Two single wires are twisted together to form a pair; by varying the length of the twists, or lays, the potential for signal interference between pairs is reduced.

LB An ATM term. Leaky Bucket: Leaky Bucket is the term used as an analogous description of the algorithm used for conformance checking of cell flows from a user or network. See GCRA, UPC and NPC. The "leaking hole in the bucket" applies to the sustained rate at which cells can be accommodated, while the "bucket depth" applies to the tolerance to cell bursting over a given time period.

LBA An abbreviation for Logical Block Address. See Logical Block Address.

LBO Electrical Line Build Out.

LBRV Low Bit Rate Voice. Digitized voice that requires a bandwidth of fewer than 32 Kbps. LBRV digitizing techniques include packetized voice, APV, DSI, and LPC.

LCD Liquid Crystal Display. An alphanumeric display using liquid crystal sealed between two pieces of glass. The display is divided into hundreds or thousands of individual dots, which are charged or not charged, reflecting or not reflecting external light to form characters, letters and numbers. LCD displays have certain advantages. They use little electricity and react reasonably quickly — though not nearly as quickly as a glass cathode ray tube or a gas plasma screen. They are reasonably legible. They require external light to reflect their information to the user. The newer so-called "supertwist" LCDs are much more readable. You see LCDs on computer laptops and telephone screens. The reason computer laptop LCD screens are brighter than phone screens is that laptops use fluorescent or other light sources to illuminate the LCD (typically from the back or the side). Phones typically don't have the power. The only way for a phone to get the power is to plug the phone and its screen into an AC outlet. Most users, however, don't want to have to plug their phone into both AC and phone outlets. It's cumbersome. In newer LCDs, called active matrix displays, the circuit board contains individual transistors for each pixel, or dot on the screen. The enables the crystals to shift quickly, resulting in a higher quality image and the ability to display full-motion video. Active matrix displays in color are hard to manufacture. Low production yields are common, though improving.

LCF-PMD The ANSI X3T9.5 standard which defines the requirements for the transmission of data over low cost fiber in an FDDI topology. Also refers to the ANSI working group responsible for the development and perpetuation of the standard.

LCM 1. Line Control Module.
2. Line Concentrating Module is a cabinetized peripheral which contains two duplicated Enhanced Line Concentrating Modules (LCME) to interface analog circuits.

LCN Logical Channel Number. An ISDN term which applies when multiple users share a single B or D channel for X.25 data communications. Each user is assigned a LCN. For instance, as many as eight users can share a single access to a single B channel through an 8-port Terminal Adapter in a BRI application. Effectively eight logical virtual channels are derived from a single B channel.

LCR Least Cost Routing. A telephone system feature that automatically chooses the lowest cost phone line to the destination. What actually is the "lowest cost" is determined by algorithms, equations and decision trees programmed into the PBX. Least Cost Routing typically works with "look-up" tables in the PBX's memory. These tables are put into the PBX by the user. The PBX does not automatically know how to route each call. It must be told so by the user. That "telling" might be as simple as saying "all 312 area codes will go via the AT&T FX line."

Or it might be as complex as actually listing which exchanges in the 312 area code go by which method. LEAST COST ROUTING tells the calls to go over the lines which are perceived by the user to be the least cost way of getting the call from point A to point B. There are typically two types of "least cost routing" translation — that which examines the first three digits of the phone number (i.e. just the area code) and the first six digits of the phone number (i.e. the area code and the three digits of the local central office). Six digit translation is preferred because it allows you more flexibility in routing — particularly to big area codes, like 213 in LA, where there are long distance calls within the area code. See also AUTOMATIC ROUTE SELECTION, ALTERNATE ROUTING and SIX DIGIT TRANSLATION.

LCS Live Call Screening. This is a wonderful phone/voice mail feature, at this stage confined to Panasonic phone systems. Here's how it works. Someone calls you and ends up in your voice mail. You can hit a LCS button on your phone (or it will happen automatically) and you will hear the message being left in your voice mail. If you choose to speak to this person (they're important), you press a button or simply pick up your handset. There are various options with this feature. You can listen through your speakerphone or through your handset. You can have it happen on every voice mail call or just the ones you choose. You'll know if someone is leaving you a voice mail because a special light lights on your phone. There's an allied feature, called Remote Live Screening, which lets you plug a single line cordless phone into your Panasonic phone and monitor incoming voice mails — even though you're in the bathroom. This way you can always be assured you will always receive the important call you're waiting for. These features are presently only available on the Panasonic Super Hybrid Digital PBX with its integrated digital Panasonic voice mail system. But they should be on every phone system.

LD 1. Long Distance.
2. Loop Disconnect.

LDA Long Distance Alerting.

LDAP Lightweight Directory Access Protocol. An emerging standard which is being touted as an Internet-based solution to the intricacies of DAP. LDAP also is based on the ITU-T X.500 standard, but will be installed on disparate, legacy e-mail directories, network operating system directories and databases. In effect, LDAP will be a front-end solution, tying into X.500 directory gateways. With initial development work having been accomplished at the University of Michigan in 1991, Netscape and a number of vendor have thrown their weight behind it. At the time of this writing, LDAPv2 is specified in RFC 1777 and LDAPv3 in an Internet Draft; Commercial products are expected in 1997 or 1998. See DAP and X.500.

LDBS Local Data Base Services.

LDM See LIMITED DISTANCE MODEM.

LDN Listed Directory Number. Your main phone number. The one you list in the telephone directory and Directory Assistance.

LDT Line appearance on a Digital Trunk.

LDTV Low Definition TeleVision (e.g. VHS).

LE LAN Emulation. Refer to LANE.

LE_ARP An ATM term. LAN Emulation Address Resolution Protocol: A message issued by a LE client to solicit the ATM address of another function.

Lead Agent The first agent in an ACD group. See also AUTOMATIC CALL DISTRIBUTOR.

Lead Cable Before plastic (polyethylene) was invented telephone cable was insulated with paper and covered in lead. Much of this cable is still used by the RBOCs. This cable is heavy and nonbuoyant in underwater applications. It is now being removed due to the effects of lead on the environment.

Lead In 1. Wire or cable from antenna to TV set.
2. The conductor from the antenna to the radio receiver.

Leader The section at the beginning of a roll of magnetic tape which holds no data and often is not even magnetic tape. A leader is used to feed the magnetic tape through the tape mechanism and secure it onto the roll.

Leadership Priority An ATM term. The priority with which a logical node wishes to be elected peer group leader of its peer group. Generally, of all nodes in a peer group, the one with the highest leadership priority will be elected as peer group leader.

Leading Current The phrase difference in a capacitive alternating current where the current leads the E. M. F.

LEAF Law Enforcement Access Field. See CLIPPER CHIP and the NSA.

Leakage 1. Leakage is a cable TV term. Leakage occurs when certain radio frequencies ooze out of the CATV's coaxial cable in such strength that they are evident outside the home. They might be sufficiently strong to interfere with aircraft navigation. Leakage is really a shielding problem. See INGRESS.
2. See LEAKING MEMORY.

Leaking Memory Under Windows (3.xx or Windows 95), when you close a program, Windows sometimes fails to release all the memory that it's used. This is called "leakage." It is a cumulative problem. The more programs you open and close, the more memory you lose. This can create problems.

The simplest solution is to get out of Windows regularly and reboot your computer.

Leaky Bridge A type of LAN bridge that forwards packets from one LAN to another even though the packet should not be forwarded. Usually due to poor engineering.

Leaky Bucket An ATM term. An informal term for the Generic Cell Rate Algorithm.

Leaky Coax A device to assist wireless transmission. A leaky coax is a coaxial cable that has the tops of the corrugated shield milled of to make a series of holes on one side of the cable. Instead of preventing signal loss, the cable will now leak the signal the entire length. This provides a much easier method of evenly covering tunnels, underpasses, stairwells, elevator shafts and basements. Any building constructed with steel, or re-enforced concrete will have dead areas (called nulls) that are in the shadows of traditional transmitters. Instead of boosting the power output and overpowering nearby receivers, a length of leaky coax run high in the ceiling will cover the same areas evenly. The outer insulation is applied after the holes have been cut to provide a weather tight cable. You must orient the cable so the holes are pointed where you want the coverage. Many cellular carriers are working with subways to install leaky coax in subway stations and tunnels. The English Channel Tunnel has leaky coax installed in its tunnels.

Leaky PBX One of those really silly terms the phone industry is famous for. Picture this. You dial a distant PBX. If you can dial on — either locally or make a long distance call, that distant PBX is referred to as "leaky." All PBXs are, of course, "leaky," or at least capable of being made leaky.

Leaky Roof Syndrome You don't notice that your roof is leaky when the sun is shining. When it rains and the roof leaks, it's too late to fix it.

LEAS LATA Equal Access System.

Leased Circuit Same as LEASED LINE or PRIVATE LINE.

Leased Line Same as a LEASED or DEDICATED CIRCUIT, PRIVATE LINE, LEASED CHANNEL. A telephone line rented for exclusive use of the customer 24-hour a day, seven days a week from a telephone company — a local phone company (like Nynex) or a long distance company like AT&T or MCI.

Least Cost Routing LCR. A telephone system feature which automatically chooses the "least cost" long distance line to send out a long distance call. The user typically dials "9" and then his 10-digit long distance number which is routed over the least costly service. See also LCR.

Leatherneck The first U.S. Marines wore high leather collars to protect their necks from sabres. Hence the name "leathernecks."

Leave Word Calling For AT&T System 75 and 85 users for internal messages. The caller's name, time of call and extension number are taken and can be retrieved on a Digital Display Monitor or BCT. The service is accompanied by an Integrated Directory — a simple electronic listing of employee names and extension numbers, access by a terminal equipped with a DDM or via the attendant console.

LEC 1. Local Exchange Carrier. The local phone companies, which can be either a Bell Operating Company (BOC) or an independent (e.g., GTE) which traditionally had the exclusive, franchised right and responsibility to provide local transmission and switching services. Prior to divestiture, the LECs were called telephone companies or telcos. With the advent of deregulation and competition, LECs now are known as ILECs (Incumbent LECs). This terminology delineates them from CLECs (Competitive LECs).

2. LAN Emulation Client: An ATM term for a router capable of supporting LANE (LAN Emulation). It works like this: A LAN-attached device, typically in the form of a PC, addresses another LAN-attached device-of course, the originating device hasn't a clue where the other device is physically located. When the router receives the data packets, it exercises its LEC capability, establishing a connection to an edge ATM switch, mapping the native LAN MAC addresses to ATM addresses. Through the ATM network, a matching connection is established to a matching LEC. MPOA (MultiProtocol Over ATM) is much better, but less mature. See also LANE and MPOA.

LECID An ATM term. LAN Emulation Client Identifier: This identifier, contained in the LAN Emulation header, indicates the ID of the ATM host or ATM-LAN bridge. It is unique for every ATM Client.

LECS An ATM term. LAN Emulation Configuration Server: This implements the policy controlled assignment of individual LE clients to different emulated LANs by providing the LES ATM addresses.

LED Light Emitting Diode. A semiconductor diode which emits light when a current is passed through it. In lightwave transmission systems, light emitting diodes or lasers are used as sources of light. These devices are fabricated from multi layered structures of compound semiconductors epitaxially grown on a single-crystal substrate. LEDs are used as sources for optical data link applications in which the data rates are less than about 500 megabits per second and the transmission distances do not exceed a few kilometers. LEDs are also used in alphanumeric displays on calculators and computer devices. LEDs use less power than normal incandescent light bulbs, but more power than LCDs (Liquid Crystal Displays). Contrast with Laser Diodes.

Leg My definition: A segment of a multipoint circuit which lies between any two of the points. Bellcore's definition: An object within a connection view that represents a communication path toward some addressable entity.

Leg Iron 1. Also called spurs and climbers. What personnel wear to climb wooden poles. Officially linesman's climbers. They consist of a steel shanks that strap to a person's leg. The inside has a spike used to stab the pole.

2. Worn by prisoners to prevent them running away. Many customers want their telephone technicians to wear them until their system is up-and-running 100%.

Legacy 1. All the stuff you have on hand — equipment, software, files and paperwork. In short, everything of a data processing/telecommunications nature in your business today. The use of the word "legacy" suggests that you've inherited all this stuff from previous generations of obsolete management. Most English dictionaries define the word legacy as "Anything handed down from the past. The idea is that you're forced to update it, without junking it altogether — which is expensive and potentially problematical. All you have is the legacy of previous generations. Preserve it, because no one can afford to junk it. Or so the theory goes."

2. Microsoft defines legacy in its Windows 95 Resource Kit as hardware and devices cards that don't conform to its Plug and Play standard.

Legacy Technology Outdated stuff that is basically obsolete but still too expensive to trash. Also called "heritage system."

Legal Holiday A call center term. Any holiday for which special wages are paid to employees who work on that day.

Lego Lego is a Danish-created child's game of many standard plastic pieces which fit together to make wonderful objects. I include this term in this dictionary because many people see computer telephony as "lego telephony." They see computer telephony comprised of standard building blocks — hardware and software — which fit together to make wonderful objects, just like LEGO. See LEGOWARE TELECOM.

Legoware Telecom The British term for what North Americans know as Lego Set or the British term Erector Set Telecom.

Lempel-Ziv-Welsh LZW. The Lempel-Ziv-Welsh compression algorithm is way of reducing the number of bits to transfer. See LZW for a full explanation.

LEN Line Equipment Number.

Length The number of bits or bytes in a computer word, a field, a record, etc.

LEO Low Earth Orbit satellites. Also called LEOS, as in Low Earth Orbiting Satellites. Low Earth Orbit satellites move a few hundred miles in orbit around the earth in various orbits like electrons whizzing around the nucleus of an atom (remember your high school physics). A group of such satellites is known as a "constellation." To establish a connection, you gain access to one of these satellites much as you gain access to a cell site in close proximity in the case of cellular telephony. When that satellite moves out of range, you are handed off to another satellite which has come into view. During a lengthy conversation, this process may take place many times. LEOs are being promoted for functions as diverse as worldwide paging with acknowledgement, worldwide hand-held telephone service and tracking cargo (with the truck sending up a continuous stream of info about its whereabouts). A primary advantage of LEOs is that the transmitting terminal — the one on earth — doesn't have to be very powerful, because the LEO satellite is so much closer than traditional geostationary satellites, which are satellites placed in an geosynchronous orbit — 22,300 miles (or 42,164 kilometers) directly over the earth's equator. The close proximity of the satellites also minimizes propagation delay, thereby avoiding that aggravating CB-radio like problem of conversation delay and clipping. LEOs are divided into two groups, Little LEOs and Big LEOs, with each group having been assigned specific radio frequencies by international agreement. Little LEOs support data services, while Big LEOs support both voice and data communications. See IRIDIUM and GLOBALSTAR. Contrast with GEO and MEO.

LEOS Low Earth Orbiting Satellites. Motorola's proposed Iridium is one of these. See LEO and IRIDIUM.

LEP Large Electron Positron Collider.

LERG Local Exchange Routing Guide. A Bellcore document which lists all North American Class 5 offices (Central Offices, or end offices) and which describes their relationships to Class 4 offices (Tandem Offices). Carriers use the LERG in the network design process. See also BELLCORE, CENTRAL OFFICE and TANDEM OFFICE.

LES An ATM term. LAN Emulation Server: This implements the control coordination function for the emulated LAN, in order to perform LAN Emulation (LANE) over an ATM network. Examples are enabling a LEC or IXC to extend a LAN, resolving issues of addressing between LAN MAC addresses and ATM addresses. The LES works in conjunction with a BUS (Broadcast and Unknown Server), which distributes the broadcast and multicast packets. See Emulation and LANE.

Less Than Of smaller quantity. Of less importance. The word "less" is always confused with the word "fewer." According to the Oxford American Dictionary, the word "less" is used of things that are measured by amount (for example,

eat less butter, use less fuel). Its use with things measured by numbers is regarded as incorrect (for example in "we need less workers;" correct usage is fewer workers.)

Letter Of Agency A letter sent by an end user to a telephone company — local or long distance — authorizing the end user's equipment vendor to deal on the end user's behalf with the phone company. A letter of agency is actually a specialized Power of Attorney.

Letterbox The black bars above and below a picture on a TV screen. Explanation: A TV screen does not have the same aspect ratio that Hollywood typically uses. Hollywood's much wider, typically at least twice as wide as it is high. When you take a Hollywood movie and put it on a TV screen, you can run it with letterboxes and get the whole Hollywood image. Or you can run it full TV screen and lose parts of the left and right hand side of the picture. See LETTERBOXING. Contrast with PAN-AND-SCAN.

Letterboxing A TV term referring to the technique in which the aspect ratio (width:height) of an original film is preserved by blacking out portions of the TV screen, typically at the top and bottom. No material is cut out, however. You've noticed this when viewing classic films like Ben Hur. Contrast with PAN-AND-SCAN. See LETTERBOX.

Letters Shift A physical shift in a teletypewriter, specifically Telex, which enables the printing of alphabetic characters.

Level 1. The power of a signal measured at a certain point in the circuit.

2. Your management position (i.e. "level" in the management structure) in a telephone company. In AT&T and members of the operating Bell telephone companies, employees are identified by their "Levels." At the bottom of the totem pole are crafts people, the installers, the repair people, the trench diggers, etc. They do not have a level. They are often unionized. Management begins one level above the union. They are called first level. They are often called supervisors. Above them are second level managers. Above them are third level managers. They are called district managers. Above them are fourth level managers. They are division managers. Fifth level managers are assistant vice presidents. Sixth level managers are vice presidents. Above vice presidents, levels get fairly vague. Salary is contingent upon level. There are several levels with different salary levels within each level. It is not uncommon for AT&T or for a Bell operating company to have as many as 16 different management levels. At one stage, there was talk about eliminating the fourth level altogether.

3. As wiring got to carry faster and faster data flows, so the quality of wiring has become increasingly important. Thus more and more companies have started specifying cabling standards. Here is a series of standards, which Anixter has promoted:

• Level 1 VOICE

Level 1 cables are MADE to meet minimum telecommunication cable requirements. Typical uses include analog and digital voice plus low speed data (20 Kbps). Plenum constructions are available in shielded and unshielded designs while PVC constructions are available in shielded designs only.

• Level 2 ISDN & LOW SPEED DATA

Cables support the IBM Type 3 Media requirement. Most uses are defined through the IBM Cabling System guidelines. This specification defines electrical requirements through 1 MHz. These products are available in both plenum and PVC UTP (unshielded twisted pair) constructions. There are no shielded options in Level 2.

• Level 3 LAN & MEDIUM SPEED DATA

These products support the ANSI/EIA/TIA-568 Commercial Building Telecommunications Wiring Standard specification horizontal cable (also known as Category 3). This standard defines cable performance through 16 MHz and thus supports high speed LAN applications. Shielded constructions are available.

• Level 4 EXTENDED DISTANCE LAN

Level 4 identifies the first 100 ohm premises cables specifically designed for LAN applications. Most UTP LANs require a higher degree of performance than the standard telecommunications design offers. Level 4 cables require performance testing through 20 MHz and provide outstanding crosstalk isolation and attenuation. They are ideal for extended distance 10Base-T and 16 Mbps Token Ring. The specification for Level 4 is referenced from TIA TR41.8.1 Category 4 and NEMA "Low Loss."

• Level 5 HIGH SPEED LAN 100 OHM

This level requires the ultimate design for 100 ohm UTP cable. TIA TR4 and the NEMA Premises Wiring Task Force have recently defined this new specification for 100 ohm cable tested through 100 MHz. These cables are intended to be used up to and including 100 Mbps CDDI applications.

• HIGH SPEED LAN 150 OHM - DGM

The 150 ohm shielded twisted pair (STP) data grade media is the cornerstone of the IBM Cabling System. In addition to the many IBM applications, this cable is now supported by a consortium of five system vendors for 100 Mbps twisted pair transmission until the ANSI X3T9.5 standard is complete.

• Level 6

Increases UTP cable performance by requiring 10dB of ACR (Attenuation-to-Crosstalk Ratio) at 155 MHz. Level 6 cable also must meet more stringent four-pair NEXT (power sum) requirements than must Level 5.

• Level 7

Meets at least twice the Category 5 (Level 5) bandwidth requirement. Level 7 UTP achieves 10 dB ACR at 200 MHz, and is power sum-tested to higher NEXT values than is Level 6. Level 7 can support multiple applications at different frequencies with a single cable jacket, and will support Gigabit Ethernet at distances up to 100 meters.

Level 1 Relay Another name for a repeater. Level 1 indicates that the device operates at the lowest layer (physical layer), as defined by the Open Systems Interconnect (OSI) architecture.

Level 2 Relay Another name for a bridge. Level 2 indicates that the device operates at the second layer (data link layer), as defined by the Open Systems Interconnect (OSI) architecture.

Level 3 Relay Another name for a router. Level 3 indicates that the device operates at the third layer (network layer), as defined by the Open Systems Interconnect (OSI) architecture.

Level 7 Relay Another name for a gateway. Level 7 indicates that the device operates at the seventh layer (application layer) of the Open Systems Interconnect (OSI) architecture. There are other standards. See also CATEGORY OF PERFORMANCE.

Level Playing Field An area of business competition where all the players enjoy the same rights and privileges. None has special privileges, such as conferred by government regulation. The term has special meaning in the telecom industry where regulation is so pervasive. At one point, the aspiring competitors to the long distance carriers argued for a level playing field. Then when these new competitors got rights and privileges and the older long distance carriers still

had the remnants of regulation, the older ones complained. They now wanted a "level playing field." There really is no definition of "level playing field." Everyone defines it the way they want.

Level Mode Interrupt A method of transmitting an Interrupt Request from a PCMCIA Card to a socket using the IREQ signal. In this mode, the IREQ signal is asserted when the Card initiates an interrupt and is negated when the Host acknowledges to the PC Card that the interrupt has been serviced. The method of acknowledgment is specific to devices on the PCMCIA Card.

Level Sensitive Interrupt A host system interrupt which causes repeated interrupts as long as the interrupt request signal is in the asserted state and the interrupt request is not disabled. Used in Micro Channel Architecture bus hosts and available in EISA hosts.

Level Sensitive Interrupt Trigger These adjustable triggers are the key to the operation of the new UART chip, called the 16550. They determine both the amount of data (in bytes) that the UART can receive before generating an interrupt request and the remaining buffer space available to store additional, incoming data. See 16550 and UART.

LF Line Feed. ASCII character 10. This character is now identified in the ASCII code set as New Line. See NEW LINE.

LGC Line Group Controller.

LGCI ISDN Line Group Controller.

LGN An ATM term. Logical Group Node: LGN is a single node that represents the lowest level peer groups in the respective higher level peer group.

Liberation A Northern Telecom line of headsets designed for use with Meridian telephones and attendant consoles. The Liberation product line includes monaural and binaural styles. Northern Telecom says the following are traditional applications for headset use:

Receptionists. Console attendants. Telemarketers. Customer service reps. Order entry reps. Financial Service professionals. Stockbrokers. Sales reps and reservation agents.

Library A file that stores related modules of compiled code.

Library of violators Picture the home page of a typical Web site. Somewhere on the page is a moving graphic — perhaps an animated GIF — that screams at you and violates the visual integrity and consistency of the page. Most often, such graphic is designed to deliberately violate the integrity of the page. It is often a paid-for advertisement. And the advertiser wants to draw your attention to his graphic ad. After all, he paid big money for the ad. Such graphic is called a graphic violator. A collection of graphics violators is called a library.

LIDB Line Information Data Bases which are being developed by the Regional Bell Operating Companies and all the local phone companies will include such services as Originating Line number Screening, Calling Card Validation, Billing Number Screening, Calling Card Fraud and Public Telephone Check. The LIDB systems contain all valid telephone and calling card numbers in their regions, and have the necessary information to perform billing validation. The LIDB systems operational. A national system connecting them all together started working at the beginning of 1992.

Life Cycle A test performed on a material or configuration to determine the length of time before failure in a controlled, usually accelerated, environment.

Lifeline Service A minimal telephone service designed for the poor and elderly to assure they can be reached by phone and have a "Lifeline" to the world in case of emergency. Typically, Lifeline Service entitles you to a phone line, a listing in the directory and a minimal number of outgoing local calls, e.g. 10. Some people who are neither poor nor elderly, subscribe to Lifeline Service and use it for incoming calls — for an answering machine or a computer electronic mail or bulletin boards. There's no difference in the quality of service provided by Lifeline Rates and normal phone lines. The cost of providing lifeline service is subsidized at the national level through the settlements pool administered by NECA (National Exchange Carrier Association) under the supervision of the FCC (Federal Communications Commission). NECA Lifeline Assistance Programs include SLC Waiver, which waives either the entire Subscriber Line Charge up to $3.50, or a portion of it; and Link-Up America, which offsets half the initial installation fee, up to $30, and defrays interest expenses. See also NECA, SEPARATIONS AND SETTLEMENTS, and UNIVERSAL SERVICE.

LIFO Last In First Out. A method of organizing queues. See LAST IN FIRST OUT.

Light Technically, light is electromagnetic radiation visible to the human eye. The term is also applied to electromagnetic radiation with properties similar to visible light, including the invisible near-infrared "light" (or more technically correct, radiation) that carries signals in most fiber optic communication systems. Light consists of electromagnetic waves ordinarily applied to those having a wave length of from .000075 cm. (the red ray) to .000038 cm. (the violet ray).

Light Amplification of Stimulated Emission of Radiation LASER. A device which transmits a narrow beam of electromagnetic energy in the visible light spectrum. The light waves are in phase with one another, or coherent, rather than jumbled as in normal light. See laser.

Light Emitting Diode See LED.

Light Pen A video terminal input device which is a light sensitive stylus connected by a cable to the video terminal. The user brings it to the desired point on the screen surface and presses a button. A light pen is used to select options from a menu on the screen or to draw images by dragging the cursor around the screen on a graphics terminal.

Light Piping Use of optical fibers to illuminate.

Light Year The distance that light travels in a pure vacuum (e.g., outer space) during a year. It's a big number. Do the math: 186,000 miles per second x 60 seconds per minute x 60 minutes per hour x 24 hours per day x 365 days per year = 5,865,696,000,000 miles (that's almost six trillion miles). All wave forms in the electromagnetic spectrum propagate at roughly this speed, assuming they are unimpaired by physical matter such as copper wires, earth's atmosphere or glass fibers. Such physical matter not only slows the rate of travel as a result of resistance, but also creates distortion in the signal. See Fiber Optics and Loss.

Lightgate Service A BellSouth INtraLATA optical fiber-based private line services that allows high-volume customers integrated voice, data and video transmission. Lightgate is the equivalent of 672 voice or data, private line or dial up circuits.

Lightwave Communications Fiber Optic communications using light to carry information.

Lightwave Transmission This term now means laser communications systems shot through the air (as opposed to glass fiber). Also called "free space lightwave communications." Typically, a signal is radiated directly from a light transmitter to a receiver less than a mile away. Advantages of lightwave transmission: easy to install, no digging of cables, wide bandwidth, reliable, cheap, no FCC frequency clearance approvals required and the receiving and transmitting equipment occupy little space. Disadvantages: only works for a mile or so and is subject to attenuation (fading) from fog and dust. It's perfect for between downtown buildings, where installing cables is too expensive, too cumbersome, too slow, etc. See LASER.

LIJP An ATM term. Leaf Initiated Joint Parameter: Root screening options and Information Element (IE) instructions carried in SETUP message.

Like New A term used in the secondary telecom equipment business. It means in excellent condition. Under normal conditions, the like new equipment could pass as new (i.e., not used, but not necessarily in the O.E.M. packaging). See LIKE NEW REPAIR AND UPDATE.

Like New Repair And Update LNRU. A term in the industry which repairs telecom equipment. It means all equipment is repaired and updated to the current manufacturer's specifications. New plastic is used to refurbish to a "like new" status. Also added are a new coil cord, line cord and address tray. Included is a full diagnostic test with a burn-in (if required) and an operational system test. Definition courtesy Nitsuko America. See also REPAIR AND QUICK CLEAN and REPAIR, UPDATE AND REFURBISH.

LIM Link Interface Module.

LIM-EMS The abbreviation for Lotus Intel Microsoft-Expanded Memory Specification. A software technique that allows MS-DOS to access memory beyond one megabyte by mapping the memory into a window in an area that MS-DOS can access. LIM-EMS is one of the greatest techniques for speeding up getting in and out of programs. For example, when my calendar program called Maxi-Calendar is not running, it occupies only 7K of normal RAM and 350K of expanded RAM. When I need it, it swaps itself quickly out of expanded RAM into normal RAM, taking less than half a second. If I didn't have expanded memory, it would take as long as 15 seconds to swap the program onto and off my hard disk, which is the other alternative. LIM stands for Lotus/Intel/Microsoft, the founding organizations that developed the Expanded Memory Specification. AST Research was also part of the driving force behind EMS, though its name doesn't appear in the acronym.

Limited Distance Modem LDM. A special purpose conversion device designed to connect two DTEs (data communications devices) over a relatively short distance, typically up to several miles. An LDM is not really a modem since it does not perform a digital-to-analog conversion, but transmits a special type of digital signal to the other LDM on the circuit. Also called a line driver, local dataset or short-haul modem.

Limiter A circuit which shapes a signal sent through it to conform to certain preset tolerances, used in both audio and video to regulate signal flow and prevent overloading, which would lead to distortion and the introduction of spurious noise.

Limiting Amplifier Relating to analog signals and their processing. Also refers to the operating range of an amplifier where little or no distortion occurs.

Line The word line is confusing. In traditional telecom, a line is an electrical path (two wires) between a phone company central office and a subscriber, usually with an individual phone number that can be used for incoming and outgoing calls. A line, in this definition, is the most common type of loop. In carrier systems, a line is the portion of a transmission system that extends between two terminal locations. The line includes the transmission media and associated line repeaters. A line is also used to indicate the side of a piece of central office equipment that connects to or toward the outside plant. The other side of the equipment is called the drop side. And finally, a line is a family of equipment or apparatus designed to provide a variety of styles, a range of sizes, or a choice of service features. The confusion over the word line starts with an office phone system. Some people believe a line to be the same animal as a trunk — i.e. the line coming in from the central office to the PBX. Other people think a line is an extension, i.e. the line from the PBX to the phone on the user's desk.

Line Build Out Because T-1 circuits require the last span to lose 15-22.5 dB, a selectable out put attenuation is generally required of DTE equipment (typical selections include 0.0, 7.5, and 15 dB of loss at 772 KHz).

Line Capacity A telephone company definition. The maximum number of network access lines that can be working on installed lines at the entity's derived objective percent line fill.

Line Card A plug-in electronic Printed Circuit (PC) card that operates lamps, ringing, holding and other features associated with one or several telephone lines or telephones in a telephone system.

Line Circuit The sensor in the CO which detects and advises the switching system that one of its subscribers has gone off-hook and wishes to make a call. One line circuit is dedicated to each line of each subscriber.

Line Conditioning A service offered by telephone companies to reduce envelope delay, noise and amplitude distortion. By doing this, you allow for transmission of higher speed data than over a traditional dial-up phone line.

Line Cord The connecting cord between the phone and the jack in the wall.

Line Current A telephone's average off-hook current is about 35 milliamps (mA) or 0.035A. Line current is electrical current measured on an idle telephone line. Typical range is 20 - 100 Ma DC, with 40 - 50 mA considered optimum for proper operation of the phone.

Line Discipline Archaic term for communications protocol.

Line Disturbance Analyzer A tool used in analyzing problems in a facility's incoming power. The line disturbance analyzer is connected at the power input to measure and record incoming power, then left in place long enough to gather data typical of the site.

Line Driver A short haul communications device used when cable lengths between RS-232 devices begin to the alleged 50-foot RS-232 limit. A line driver is a signal converter that conditions the digital signal transmitted by an RS-232 interface to ensure reliable transmission beyond the 50-foot RS-232 limit and often up to several miles; it is a baseband transmission device. Also called baseband modem, limited distance modem, or short-haul modem. See also LIMITED DISTANCE MODEM.

Line Equipment Equipment in a central office which is there to serve a phone line. That line equipment includes a line relay or equivalent which starts to work when the customer's telephone goes off-hook.

Line Equipment Number A line equipment number identifies the physical central office line equipment for each subscriber.

Line Error A Token Ring error reported by any ring station that detects an FCS failure, or some type of protocol code violation in a received frame.

Line Feed The act of moving a cursor or the head or a printer or telex machine down one line. These days, on the key-

boards of most equipment — Personal Computers, etc. — there's no single key that says "Line Feed." There's usually a key that's labelled "Enter" or "Return." This key does two functions — a line feed and a carriage return (i.e. sending the print head or cursor to the left hand side of the carriage or the screen). In many (but not all) programs, a line feed is control J or ASCII character 10. The name for this character has been changed to New Line. See NEW LINE.

Line Finder The first switching element of a step-by-step phone system which recognizes a calling party is waiting for dial tone to make a call, identifies the party, and connects that party to the switching system so that the processing of the call may begin. Normally line finders serve 100 or 200 subscriber lines.

Line Hit Electrical interference that causes a hit, that is, a loss or introduction of spurious bits into a data stream.

Line Hold Provides a winking, blinking flash on the line lamp at every telephone which has the line appearing on it.

Line Insulation Test LIT. A test performed from the central office, which measures resistance and voltages on local lines to find faults.

Line Link Frame LLF. An arrangement that permits a crossbar office to transmit dial pulse information over a line to a PBX for switching Direct Inward Dial (DID) calls to the indicated phone. See CROSSBAR SWITCHING

Line Lockout When a phone stays off-hook for longer than a predetermined time, line lockout provides some loud noise and then puts the phone line out of service — until someone puts the phone back on hook again.

Line Loop Back LLB. A troubleshooting function of CSU/DSU equipment. The receive pair of a circuit is connected directly back to the transmitter so the line can be tested. If a clear signal is "looped back" to it the line is OK. If there's a problem, it's inside or beyond the receiving equipment.

Line Noise Spurious signals introduced into a line by static, or other imbalances in the circuit. Line Noise is the most common cause of "Hits" or problems in data calls.

Line Of Sight Some through the air transmission media — such as microwave, infrared, and laser — operate at a frequency which transmit in a perfectly straight line. Or in "line of sight." In other words, the area between a transmitter and a receiver must be clear of obstructions.

Line Pool A Line Pool is a specific group of lines in certain key systems used for making outside calls. In Northern Telecom's Norstar, three Line Pools give phone access to outside lines without taking up too many Line buttons on each phone instrument.

Line Powered Telephone equipment that is powered solely by the CO talk battery supplied in a standard phone line.

Line Preference User selects the line to be used simply by pressing the button associated with that line.

Line Printer A type of printer which prints an entire line of text at one time. This printer is obviously a high speed printer. It is used, for example, to print TELECONNECT Magazine's monthly mailing list

Line Protocol Rules for controlling transmission on a synchronous data transmission line. Includes rules for bidding for the line, for positive and negative acknowledgements, requests for retransmission, and transmitter time-outs.

Line Queuing Dial an outside line (typically a long distance line). It's busy. Your phone system will put you in queue for that line. The queue might involve your waiting a few seconds on hold; or it might involve your hanging up and having the phone system call you back. There are thus two types of Line Queuing — hold-on and callback queuing.

Line Relay A telephone company term. Relay in a subscriber's line which operates on the calling-in signal.

Line Ringing Provides the user with an audible indication of a call on a specific line that appears on his/her telephone.

Line Side Connection A carrier term. A local loop, which connects the customer premise to the carrier network. The carrier community uses this term to describe the customer side of the network, regardless of whether it is specifically in the form of a line or a trunk. In this context, the term "trunk" refers to a local loop which connects a network switch (e.g., a central office circuit switch, a frame relay switch or router, or an ATM switch) to a customer switch (e.g., a PBX circuit switch, a frame relay router, or an ATM switch). Also in this context, a "line" connects a network switch to a non-switch (e.g., a telephone set, a computer modem, or a traditional key system). In other words, a trunk connects one switching device to another switching device, while a line connects a non-switching device to another device, which can be in the form of either a switch or a non-switch. Compare with Trunk Side Connection.

Line Speed The maximum number of bits you can transmit over a line in a certain defined time, say one second.

Line Status Indication Provides a visual indication on an ECTS (electronic telephone set) telephone of the idle, busy, ringing or held state for each line appearing on the telephone.

Line Switched Ring A technique for providing redundancy in a SONET network. Line-switched rings use either 2 or 4 fibers per ring. The primary ring transmits in one direction (e.g., clockwise), with the other transmitting in the reverse direction. Through this technique, a failure in a SONET ring will not prevent devices from communicating, as they can transmit and receive at least one direction, assuming that there is no more than 1 break in a fiber. See Path-Switched Ring.

Line Switching Another term for circuit switching. See CIRCUIT SWITCHING and SWITCHING.

Line Termination LT. An ISDN term. Electronics at the ISDN network side of the user-network interface, that complements the electronics equipment. The classic line termination devices is the NT1 at the user side of the interface.

Line Transfers A telephone company definition. Line transfers consist of physically removing a customer line from one line equipment and moving it to another. Line transfers are used as an important corrective tool to improve load balance.

Line Turnaround Time The delay in a circuit as the direction of communications changes, usually in half duplex communications. When one side of the communications stops sending, there is a delay before the other party stars sending in return. This is the Line Turnaround delay.

Line Voltage Voltage measured on a telephone circuit; typically 48 volts DC when phone is idle. Voltage may be lower at great distance from the central office, or when carrier equipment is used to multiplex several phone lines on one pair of wires. Line voltage on PBX systems is typically 24 volts.

Linear Distortion Amplitude distortion wherein the output signal envelope is not proportional to the input signal envelope. This distortion is often caused by part of the signal being bounced off something, while part arrives free and clear. Thus the receiver hears the same signal but bits of it arrive earlier and later than other bits, causing distortion. Anyone coming across a more understandable definition, please send it to the editors for the next edition.

Linear Predictive Coding A speech coding method that analyzes a speech waveform to produce a time-varying filter

as a model of the human vocal tract. See also Digital Signal Processing.

Linear Programming Techniques in Operations Research (OR) to find an optimum solution to a linear function, given certain restrictions and typically expressed in many equations. A typical linear programming problem might be to find the least expensive, most efficient route between various pick-up and drop-off points in a transportation route.

Linearity The straightness of a frequency response curve as an indication of true or accurate sound reproduction. In an A/D (Analog to Digital) or D/A (Digital to Analog), linearity measures the precision with which the digital output/input tracks the analog input/output.

Linearly Polarized A mode of operation of fiber optics for which the field components in the direction of propagation are small compared to components perpendicular to that direction.

Lineman A person who fixes the telephone company's outside aerial plant — typically the wires hanging from poles dotted across the country-side. See also LINEMAN'S CLIMBERS.

Lineman's Climbers Telephone pole climbing irons which are strapped to the telephone lineman's legs, allowing him or her to climb a wooden telephone line. You can tell when a pole has been climbed by the holes left in it by the lineman's climbers.

Lines A computer imaging term. The line tool draws straight lines, typically from point to point. Most paint packages let you continue lines in a fashion that permits rapid creation of polygons.

Lines Of Force The directional lines of magnetic or static field which represent the stresses.

Lines Per Minute The way of measuring the speed of a line printer. Like any measure of speed, the speed you will get from your printer may be different than what the manufacturer says. Your speed will depend on how fast you feed the printer from your computer — a function of how fast you're transmitting, what software you're running, how fast that software can get the information to be printed off your disk, etc.

Link 1. Another name for a communications channel or circuit. The ATM Forum defines link as an entity that defines a topological relationship (including available transport capacity) between two nodes in different subnetworks. Multiple links may exist between a pair of subnetworks. Synonymous with logical link.

2. A Windows command that takes several programs and subprograms that were meant to be used together, but were written separately, and combines them into one. Usually used to create an executable program out of modules that were not themselves directly executable.

3. An element in an HTML document that points to a document or to a specific location in a document, using a URL. When the document is displayed in a browser, clicking on a link causes the browser to display the document and/or location that it points to. Links usually appear on-screen as underlined text and are usually in blue, although Web page designers can change how they look.

Link Aggregation Token See AGGREGATION TOKEN.

Link Attached Describing devices that are connected to a network, a communications data link, or telecommunications circuit; compare with channel-attached.

Link Attribute A link state parameter that is considered individually to determine whether a given link is acceptable and/or desirable for carrying a given connection.

Link Connection An ATM term. A link connection (e.g., at the VP-level) is a connection capable of transferring information transparently across a link without adding any overhead, such as cells for purposes for monitoring. It is delineated by connection points at the boundary of the subnetwork.

Link Constraint A restriction on the use of links for path selection for a specific connection.

Link Control Facility A Fibre Channel hardware facility which attaches to the end of a link and manages the transmission and reception of data. The LCF is contained within each F Port (Fabric Port, i.e., switch port) and N Port (Node Port, i.e., device port). See FIBRE CHANNEL.

Link Converter A device for an InteCom S/80 which connects distributed switching modules to the centralized switching equipment through a coaxial cable or a fiber optic cable.

Link Encapsulation LE. A function of the HiPPI Framing Protocol (FP) layer. LE encapsulates IEEE 802.2 Logical Link PDUs (Protocol Data Units) inside of HiPPI packets, thereby allowing IP traffic to travel over a HiPPI connection.

Link Layer Access Method The algorithm that determines when any given network interface in a PC/TCP local area network is allowed to transmit. It is also known as the access method. CSMA/CD is the access method for the Ethernet.

Link Metric An ATM term. A link parameter that requires the values of the parameter for all links along a given path to be combined to determine whether the path is acceptable and/or desirable for carrying a given connection.

Link Optimization ISDN feature that prevents administration packet from opening the communications link and allows only user data to open the line. Link optimization ensures that remote connections are not kept open unnecessarily, which saves usage costs.

Link Protocol The set of rules by which a logical data link is set up and by which data transfers across the link. It includes formatting of the data.

Link Pulse A communication mechanism used in 10Base-T networks to indicate link status and, in auto-negotiation equipped devices, to communicate information about abilities and negotiate communication methods. 10Base-T uses Normal Link Pulses (NLPs) which indicate link status only. These are transmitted periodically while not transmitting packets. 10Base-T and 100Base-T nodes equipped with auto-negotiation exchange information using a Fast Link Pulse (FLP) mechanism which is compatible with NLP.

Link Rot The process by which links on a Web page become obsolete by virtue of changes in location or expiration of the sites to which they are connected. Link rot happens quickly.

Link Set A group of signaling links directly connecting two signaling points.

Link State Protocol A type of LAN routing protocol in which updates to routing tables are exchanged between neighboring routers only when modifications are made to the table. Distance Vector Protocols, on the other hand, exchange such data on a highly regular and predetermined basis, regardless of whether updates are required. Therefore, Link State protocols consume less networking resources and reduce network congestion by only providing updates when needed, and only sending the changes. Periodically, the entire route table will be sent as a precautionary procedure. Examples include OSPF (Open Shortest Path First, ISO's IS-IS (Intermediate System-to-Intermediate System), and Novell's NLSP (NetWare Link Services Protocol). See also

DISTANCE VECTOR PROTOCOL, POLICY ROUTING PROTOCOL, and ROUTER.

Link State Parameter Information that captures an aspect or property of a link.

Link Status Signal Unit LSSU. A packet sent between MTPs (Message Transfer Part of the SS7 Protocol) to provide SS7 information about the sending node and its links. This information is sent during the initial alignment of the links, when there is an associated processor outrage, or when link congestion is detected.

Link Test A test that is performed by the hardware to ensure the integrity of the cable in a local area network. The link test can be disabled to allow old style NICs incapable of performing a link test to connect to the repeater.

Link Time This is a specific time delay that allows access to PBX or Centrex features through a telephone system. Link Time is also referred to as a Hookswitch Flash or Recall.

Linked Object A representation or place holder for an object that is inserted into a destination document. The object still exists in the source file and, when it is changed, the linked object is updated to reflect these changes.

Links 1. The transmission portion of the local loop.
2. An affectionate name for Apple's electronic mail system, called AppleLink.

Linux Linux (Officially pronounced Lih-nucks) is the creation of Linus Torvalds, who started the Linux kernel as a research project at the University of Helsinki, Finland. Since then, Linux has evolved into a full-featured, powerful and robust Unix operating system. Linux is among the most powerful and feature-packed Unix operating systems available for the PC, offering a large base of hardware peripheral compatibility. Linux hosts an impressive array of compilers and development environments, including C/C++, Perl, Pascal, SmallTalk, and complete X-windows system that rivals many commercial offerings. For most of its life, Linux has been shareware. Linux has many script languages and parsers such as Awk, Sed, Yacc as well as all popular shells (Borne, Korn, C, BASH, etc.). The Linux kernel was written for the Intel 386/486/Pentium processor. In many ways, Linux is a better performing UNIX than many of the commercial versions of UNIX. Ray Noorda, one of the geniuses behind Novell, has recently become involved with Linux and is commercializing it. For more information, contact Linux Journal, Seattle, Washington.

Liquid Crystal Display LCD. A low power display that aligns material suspended in a liquid under the influence of a low voltage so it reflects ambient light and displays alphanumeric characters. LCD displays are finding great use as methods of displaying information on new electronic telephones, especially those positioned behind PBXs. The advantage of putting such displays on telephones is that the power to drive the display is very small. The display can be line powered — i.e. powered by the one or two pairs coming from the PBX. This avoids the necessity and cost of a transformer/rectifier — the little black box you plug into the wall to run your answering machine or to power up your rechargeable calculator/laptop computer. Such LCD displays on electronic phones can perform many functions. The most useful is that of "walking" the user through the phone call — showing him/her how to transfer a call, to make a conference call, to split a conference, etc. An LCD can also alert you as to who's calling you.

LIS Link Interface Shelf.

List Box A Windows term. In a dialog box, a box that lists available choices, for example, a list of all files in a directory.

If all the choices do not fit in the list box, there is a scroll bar.

List Server An automated mailing list distribution system.

Listed Directory Number LDN. Incoming exchange network calls to the PBX via assigned listed local telephone directory number are directed to the attendant.

Listen Before Talk LBT. Same as carrier sense multiple access (CSMA). Compare with LISTEN WHILE TALK.

Listen While Talk LWT. Same as carrier sense multiple access with collision detection (CSMA /CD). Compare with LISTEN BEFORE TALK.

LIT See LINE INSULATION TEST.

Lithium Ion Type of highly efficient rechargeable battery, often used in computer laptops and cellular portables. Here's an explanation from 1-800-BATTERIES, a seller of recharge-able batteries:

NiCad: Nickel Cadmium is the most popular and durable type of rechargeable battery. It is quick to charge, lasts about 700 charge and discharge cycles, and works well in extreme tem-peratures. Unfortunately, NiCads suffer from "memory effect" if they are not completely discharged during each cycle. The memory effect reduces the overall capacity and run time of the battery.

Nickel Metal Hydride (NiMH) batteries do not suffer from memory effect. Compared to a NiCad battery of equal size, NiMH batteries run for 30% longer on each charge. They are also made from non-toxic metals so they are environmentally friendly. The downside to NiMH technology is overall battery life. These batteries last for 400 charge and discharge cycles. Lithium Ion (LiON) is the latest development in portable bat-tery technology. These batteries do not suffer from memory effect. Compared to a NiMH of equal size, a LiON will deliver twice the run time from each charge. Unfortunately, these bat-teries are only available for a limited number of models and are expensive. Similar to NiMH technology, LiON batteries have a life expectancy of 400 charge and discharge cycles.

Little Endian A format for storage or transmission of bina-ry data in which the least significant byte (bit) comes first.

Little LEO Relatively small and inexpensive low earth orbit-ing satellites that provide low-cost, low-data rate, two-way digital communications, and location positioning to small handheld terminals. The frequency allocations are in the VHF band below 400 MHz. Systems include Leosat, Orbcomm, Starnet, and Vitasat. For example, the Orbcomm system requires 34 satellites for reliable full-world coverage.

LIU Line Interface Unit.

LIU7 Line Interface Unit for CCS7.

Live Bug Colloquialism used to refer to a leaded integrated circuit package when the leads are down, like a bug that is alive and standing upright.

Live Call Screening See LCS.

Livelock A request for an exclusive lock that is repeatedly denied because a series of overlapping shared locks in a shared database keeps interfering. A SQL server will detect the situation after several denials, and refuse further shared locks.

Liveware People.

LLB Line LoopBack. A maintenance and/or diagnostic mode of operation whereby a CSU regenerates a signal received from a span line and retransmits that signal back onto the span towards its point of origin.

LLC Logical Link Control. A protocol developed by the IEEE 802.2 committee for data-link-level transmission control. It is the upper sublayer of the IEEE Layer 2 (OSI) protocol that complements the MAC protocol. IEEE standard 802.2 includes end-system addressing and error checking. It also

provides a common access control standard and governs the assembly of data packets and their exchange between data stations independent of how the packets are transmitted on the LAN. See 802 STANDARDS.

LLC2 Logical Link Control 2. The frame format used to carry 3270 traffic on Token Ring LANs.

LLDP Local Loop Demarcation Point. See MPOE.

LLF 1. Line Link Frame. See CROSSBAR SWITCHING.

2. Low Layer Functions.

LM Long distance Marketer.

LMCS Local Multipoint Communication Systems. Using fre-quencies above 25 GHz, MCS produces a wireless broadband digital network capable of delivering high-bandwidth signals over the air. Industry Canada has allocated the frequency band 25.35 to 28.35 GHz for LMCS networks. Some observers believes LMCS may represent fiber to the curb.

LMDS Local Multipoint Distribution System. Developed by Bellcore for Wireless Local Loop (WLL) applications, LMDS systems initially were trialed commercially in New York for point-to-multipoint broadcast TV, on the basis of an experi-mental FCC license granted to CellularVision. In that applica-tion, broadcast microwave signals operating at 28 GHz trans-mit to small receiver dishes, typically installed on the top of apartment buildings. Each of 12 transmitters in the boroughs of Manhattan, Brooklyn, Queens and portions of The Bronx covers an area of 28 square miles. At that high frequency, line-of-sight is required for maximum signal performance. This necessity for line-of-sight is the reason it wasn't installed in Manhattan (too many tall buildings). The received LMDS signal is often then distributed through the building's central CATV system. It also can be used to broadcast directly to a subscriber's home via an 18" flat antenna sitting in the sub-criber's window. There are actually all sorts of variations on the LMDS theme. In one trial, the service was used for high-speed Internet downloads to LMDS subscribers — the Internet downloads coming from LMDS, the command to ini-tiate those downloads being sent from the subscriber's PC over his local phone line. There is R&D going on at present to enable LMDS to carry two-way voice conversations. In Brazil, CellularVision uses LMDS technology, transmitting in the FM range, which means the signal has the ability to bounce, and to reflect off virtually any surface, thereby avoid-ing issues of line-of-sight and increasing the coverage area significantly. Two-way or interactive communication may be inserted between video channels for transmission back on the opposite polarity. This reverse polarization, or interweaving, theoretically allows simultaneous use of signals at the same frequency for two applications. LMDS is competitive with conventional cable-based CATV. In March 1997, the FCC set aside total LMDS bandwidth of 1.15 GHz in the 28-GHz, 30-GHz and 31-GHz frequency bands. The intent is to use LMDS for its original intended purpose of WLL (Wireless Local Loop). LMDS is competitive with conventional cable-based CATV. See also ADML, MMDL, MMDS and Wireless Local Loop (WLL).

LMEI Layer Management Entity Identifier.

LMI 1. Local Management Interface. A specification for the use of frame-relay products that define a method of exchang-ing status information between devices such as routers.

2. Logical Modem Interface. The core of the Microsoft Fax interface. LMI lets third-party licensed vendors write plug-in modules to provide instant and transparent access to diverse underlying systems. An easy analogy for the LMI is to consid-er the Windows print manager. To the user, simply installing

the printer driver suited to their printer is all that is required. According to Microsoft, "The LMI interface provides a similar layer between the internal fax components of Windows 95 and the fax hardware or, in our case, the fax server."

LMOS Loop Maintenance Operations System.

LMS Local Message Switch.

LMSS Land Mobile Satellite Service.

LMU Line Monitor Unit.

LNA Low Noise Amplifier.

LND Last Number dialed

LNNI LANE Network-to-Network Interface. An ATM term for the standardized interface protocol between LANE (LAN Emulation) servers (LES-LES, BUS-BUS, LECS-LECS and LECS-LES). See also LANE and LUNI.

LNP Local Number Portability. Similar in concept to 800/888 number portability, LNP was mandated by the Telecommunications Act of 1996 in order to level the playing field between the ILECs (Incumbent Local Exchange Carriers) and the CLECs (Competitive Local Exchange Carriers). In July 1996, the FCC issued a ruling that LNP must be in place nationwide by January 1, 1998. Since each state is responsible for implementation of LNP, timetables vary; the specifics of the implementations vary, as well.

In some states, the implementation approach is exactly like that for 800 number portability. In other words, the originating CO "dips" into a centralized database of numbers via an SS7 link. The database, known as a SCP (Service Control Point) in IN (Intelligent Network) terms, identifies the LEC providing service to the target telephone number in order that the originating carrier can hand the call off to the terminating carrier.

In other states, such as Illinois, which is the first to implement LNP, a totally different approach is taken. This implementation involves the use of a new 10-digit telephone number, known as a LRN (Local Routing Number). When the originating CO switch consults the SCP, the new 10-digit number is provided along with the identification of the CLEC to which the service has been ported. The originating carrier then hands off the call to the CLEC. While this approach is claimed to be faster, clearly two telephone numbers are required, thereby placing additional pressure on the North American Numbering Plan (NANP).

To implement LNP, the FCC has mandated a system of regional databases, which will store master copies of all porting information. These databases will be maintaineed by regional Number Portability Administration Centers (NPACs) that will serve as number portability clearinghouses for all local operators. Originally, the deal was that Lockheed Martin would maintain the databases in four regions and Perot Systems in three regions. After Perot had some problems getting going on time, Lockheed Martin is running all seven.

In either case, LNP will require the SCPs be established by LECs, CLECs, IXCs and wireless carriers. Further, the SCPs must be synchronized in order that the databases are consistent across them all. The concept is simple, but its implementation is complex and expensive. See also NUMBER PORTABILITY.

LNPA Local Number Portability Administration. See also LNP and NANC.

LNRU Like New Repair and Update. A term in the industry which repairs telecom equipment. It means all equipment is repaired and updated to the current manufacturer's specifications. New plastic is used to refurbish to a "like new" status. Also added are a new coil cord, line cord and address tray. Included is a full diagnostic test with a burn-in (if required)

and an operational system test. Definition courtesy Nitsuko America. See also REPAIR AND QUICK CLEAN and REPAIR, UPDATE AND REFURBISH.

LO Local Operator. In the PCS sense, a functional entity providing local wireless service to customers in a geographical region. The LO (Local Operator) is serviced by the National Services Organization in the PCN (Personal Communications Network) for long-distance communications and for marketing/sales.

LOA Letter Of Agency. A letter that you give to someone whom you allow to represent you and act on your behalf. For example, a letter of agency is used when your interconnect company orders lines from your local phone company on your behalf. Letters of Agency are also used when companies switch their long distance service from one carrier to another.

Load 1. The act of taking a program or data from external storage — a cassette, a floppy or hard disk, etc. and storing it in the computer's main RAM memory.

2. The load is any electric or electronic appliance or gadget plugged into an AC electrical outlet. It completes the circuit from the transformer through the hot conductor, to the load, through the neutral conductor and back to the utility transformer. See AC, AC POWER, GROUND and GROUNDING.

Load Balance A telephone company term. Load Balance is the even distribution of customer traffic volume across all loading units in a switching entity. Load Balance is not related to the absolute level of load, but only to how well the existing load is distributed. See also Load Balancing.

Load Balance Index LBI. A telephone company term. Indicates trends, identifies superior performances and points up opportunities for improvement in load balance administration of dial Central Office line equipment.

Load Balancing The practice of splitting communication into two (or more) routes. By balancing the traffic on each route, communication is made faster and more reliable. In telephone systems, you can change phone and trunk terminations in order to even out traffic on the network. An example: You have a PBX of three separate cabinets, each of which are joined by tie lines. Instead of having each cabinet serve anyone in the building, you might figure which groups talk to each other the most and concentrate them into specific cabinets. The objective is to maximize the number of calls that can be handled inside each cabinet and reduce the number of calls that need to travel between the cabinets. This makes the calls go faster and reduces the need for inter- cabinet lines.

In data internetworking, bridges and routers perform load balancing by splitting LAN-to-LAN traffic among two or more WAN links. This allows for the combination of several lower speed lines to transmit higher speed LAN data simultaneously. In local area networking, load balancing is a function performed by token ring routers. In data networking, load balancing can also be a form of inverse multiplexing where data packets are alternated over all available circuits. At the receiving end, the packets are reassembled in their proper order.

In disk arrays, load balancing means using multiple power supplies within a disk array so that power usage is spread equally across all the power supplies. The failure of one supply will not cause the entire array to fail.

Load Coil A bridged tap is multiple appearances of the same cable pair at several distribution points. A bridged tap is any section of a cable pair not on the direct electrical path between the central office and the user's offices. A bridged tap increases the electrical loss on the pair — because a signal traveling down the pair will split its signal between the bridge and main pairs. Since most existing telephone company cable pair is bridged, the phone company puts loading coils in the circuit. The effect of load coils is to modify the loss versus frequency response of the pair so it is nearly constant across the voice band. This works for voice. However the loss above the voice band due to load coils increases rapidly. ISDN and other digital circuits operates above the voice band. So, when the phone company installs digital circuits, it must remove the load coils. Bridged taps, on the other hand, are acceptable to some degree, especially with slower speed digital circuits, such as ISDN. But you can't run high-speed digital circuits, e.g. T-1, over cable that has bridged taps in it. See BRIDGE and LOADING COIL.

Load Coil Detector A device use to detect unseen load coils on a wire pair.

Load Leveling Distributing traffic over more than one route.

Load Number Load number is the Canadian equivalent of the U.S. concept of Ringer Equivalence. The idea is that each phone or "phone thing" you buy (e.g. answering machine) comes with a number. You add the numbers together and if you get above a certain number, you are drawing too much current and none of the bells on the phones will ring. In Canada, single line phones are typically rated at 10 for the newer ones with electronic "bells" or 20 for the older electro-mechanical ones with real metal bells. In Canada, the rule is not more than 100 points on a line. In the U.S., phones are typically one and the rule is not more than five points on a line.

Load Service Curves The output from load and stress testing on a computer telephony system is a set of load service curves. Load service curves identify how individual areas of the system respond under various load amounts. Traffic is provided to the computer telephony system at defined steps (perhaps at 1,000 call per hour increments) until the system design threshold is reached. Measurements are taken at each step, and usually shown graphically, in a "curve". Most computer telephony systems are designed to handle up to a specified number of busy hour calls with specific response times. For example, the time that passes between the point in time a caller enters a DTMF digit and the point the computer telephony system speaks a response should usually be no more than 1 second 97% of the time, nor more than 3 seconds 99% of the time. A load service curve would be used to illustrate the response time at each step of increasing load. When the load curve shows the response time is slower than the above parameters, the system has reached its capacity. Of course, the load placed on the system must accurately mimic the real world load the system will experience or it is largely meaningless. This definition from Steve Gladstone, president, Hammer Technologies, makers of fine computer telephony testing systems, 508-694-9959.

Load Sharing In data processing, load sharing is the technique of using two computers to balance the processing normally assigned to one of them. In local area networking, load sharing is performed by token ring routers when connecting remote LANs. It allows combining Ethernet and Token Ring traffic over a common WAN (Wide Area Network) link such as T-1 or 56 Kbps circuit. Loads sharing eliminates the need for duplicate WAN links (and bridges or routers) each serving a different type of LAN.

Load Testing Also known as stress testing, the goal of load testing is to make sure the system will meet or exceed its busy hour load capacity objectives under all operating conditions. This requires stressing the system in incremental steps until

it breaks and understanding what happens when the system is operating under its full rated transaction load? Beyond its load? Does it slow down? How? Does it fail? Where? How is service restored after an outage? Is service restoration graceful or must the system reboot? Is restart manual or will the system reset itself? Individual load tests may be performed to understand the impact of load on specific system bottlenecks. Most significant architectural problems will come to light under load testing. It is critical that any load placed on a computer telephony system be dynamic, and mimic the load characteristics the system will experience under real-world usage and conditions. See also DYNAMIC LOAD TESTING and LOAD SERVICE CURVES.

Loaded Line A telephone line equipped with loading coils to add inductance in order to minimize amplitude distortion. See LOADING and LOADING COIL.

Loading A method of improving the voice quality of a phone line. Telephone companies put load coils on local lines. What this loading does is to insert inductance in a local loop circuit to offset the effect of capacitance in the cable. Loading "tunes" the circuit to the voice frequency band (500 to 2500 Hz) and thus improves the quality at the expense of overall bandwidth. You usually have to ask that the loading coils be removed if you're planning to transmit high-speed data exclusively on that circuit. See LOADING COIL.

Loading Coil An induction device employed in local loops exceeding 18,000 feet in length, that compensates for wire capacitance and boosts voice grade frequencies. Loading coils are often removed for higher speed data services, as distortion will occur at frequencies higher than those used for voice. See LOADING.

Loading Division A telephone company term. A group of the same type of equipment designed to be loaded similarly by both usage and classes of service.

Loading High A memory management verb for loading a device driver or TSR (Terminate and Stay Resident) program into upper memory, out of conventional memory. Under DOS, the loading high commands are DEVICEHIGH for device drivers and LOADHIGH (or LH) for TSRs. Third party memory managers use their own routines to load high, though they can sometimes borrow DOS commands.

Loading Plan A telephone company term. A Loading Plan is a systematic scheme for fully utilizing all existing capacity in a given switching entity; Utilizing and coordinating the capabilities and capacity limitations of various entities in a multi-entity wire center and maintaining objective service levels at all times. A Loading Plan is the basis for achieving and retaining good Load Balance.

LOC An ATM term. Loss of Cell Delineation: A condition at the receiver or a maintenance signal transmitted in the PHY overhead indicating that the receiving equipment has lost cell delineation. Used to monitor the performance of the PHY layer.

Local Pertaining to a system or device that resides within a subject device's switching domain.

Local Access and Transport Area LATA. The MFJ (Modified Final Judgement), which broke up the Bell System, also defined 196 distinct geographical areas known as LATAs. The LATA boundaries generally were drawn in consideration of SMSAs (Standard Metropolitan Statistical Areas), which were defined by the Census Bureau to identify "communities of interest" in economic terms. Generally speaking, the LATA boundaries also were coterminous with state lines and existing area code boundaries, and generally included the territory served by only a single RBOC. The basic purpose of the

LATA concept was to delineate the serving areas reserved for LEC (Local Exchange Carrier) activity. In other words, IntraLATA traffic (i.e., local and local long distance) became the sole right and responsibility of the LECs. InterLATA traffic, on the other hand, became the sole right and responsibility of the IXCs. Over time, a number of state PUCs allowed the IXCs to compete for IntraLATA long distance; they also allowed CAPs (Competitive Access Providers) to provided limited local service in competition with the LECs. The Telecommunications Act of 1996 (The Act) opened the floodgates for competition with the LATA boundaries. The Act also allows the RBOCs to provide InterLATA service outside the states in which they provide local service. Additionally, The Act contains provisions for the RBOCs to offer InterLATA service within the state in which they provide local service, once they have satisfied a 14-point checklist, the most significant conditions of which relate to significant, demonstrated levels of competition within their respective local exchange serving areas. California is divided into 10 LATAs. Sparsely populated states such as South Dakota comprise only a single LATA.

Local Airtime Detail This cellular telephone carrier option (which means it costs money) provides a line-itemized, detailed billing of all calls, including call attempts and incoming calls to the mobile. What you get for free is generally a non-detailed, total summary of all calls.

Local Area And Transport Area See LATA.

Local Area Data Transport LADT. A service of your local phone company which provides you, the user, with synchronous data communications.

Local Area Network LAN. A short distance data communications network (typically within a building or campus) used to link computers and peripheral devices (such as printers, CD-ROMs, modems) under some form of standard control. Older data communications networks used dumb terminals (devices with no computing power) to talk to distant computers. But the economics of computing changed with the invention of the personal computer which had "intelligence" and which was cheap. LANs were invented as an afterthought — after PCs — and were originally designed to let cheap PCs share peripherals — like laser printers — which were too expensive to dedicate to individual PCs. And as time went on, what LANs were used for got broader and broader. Today, LANs have four main advantages: 1. Anyone on the LAN can use any of the peripheral devices connected to the LAN. 2. Anyone on the LAN can access databases and programs running on client servers (super powerful PCs) attached to the LAN; and 3. Anyone on the LAN can send messages to and work jointly with others on the LAN. 4. While a LAN does not use common carrier circuits, it may have gateways and/or bridges to public telecommunications networks. See LAN MANAGER, TOKEN RING and ETHERNET.

Local Area Signaling Services LASS is a group of central office features provided now by virtually all central office switch makers that uses existing customer lines to provide some extra features to the end user (typically a business user). They are based on delivery of calling party number via the local signaling network. LASS can be implemented on a standalone single central office basis for intra office calls or on a multiple central office grouping in a LATA (what the local phone companies are allowed to serve) for interoffice calls. Local CCS7 (Common Channel Signaling Seven) is required for all configurations. The following features typically make up LASS:

Automatic Callback: Lets the customer automatically call the last incoming call directory number associated with the customer's phone when both phones become idle. This feature gives the customer the ability to camp-on to a line.

Automatic Recall: Lets the customer automatically call the last outgoing call currently associated with the customer's station when both stations become idle. This feature gives the customer the ability to camp-on to a line.

Customer-Originated Trace: Lets the terminating party request an automatic trace of the last call received. The trace includes the calling line directory number and time and date of the call. This information is transmitted via an AM IOP channel to a designated agency, such as the telephone company or law enforcement agency.

Individual Calling Line Identification: Consists of two distinct features:
1. Calling Number Delivery which transmits data on an incoming call to the terminating phone. 1. Directory Number Privacy which prevents delivery of the directory number to the terminating phone.

Also, LASS has some selective features:

Selective Call Acceptance: Allows users to restrict which incoming voice calls can terminate, based on the identity attribute of the calling party. Only calls from parties identified on a screening lists are allowed to terminate. Calls from parties not specified on a screening list are rerouted to an appropriate announcement or forwarded to an alternate directory number.

Selective Call Forwarding: Allows a customer to pre-select which calls are forwarded based on the identity attribute of the calling party.

Selective Call Rejection: Allows a customer to reject incoming voice calls from identity attributes which are on the customer's rejection list. Call attempts from parties specified on the rejection list are prevented from terminating to the customer and are routed to an announcement which informs the caller that his/her call is not presently being accepted by the called party.

Selective Distinctive Alert: Allows a customer to pre-select which voice calls are to be provided distinctive alerting treatment based on the identify attributes of the calling party.

Users can, at their convenience, activate or modify any of these features by sending commands to the central switch from their existing touchtone telephones.

Local Automatic Message Accounting LAMA. A combination of automatic message accounting equipment and automatic number identification equipment in your telephone company's central office and used by them to bill your local phone calls.

Local Battery Having "local battery" means the telecom equipment — the telephone, the PBX, the key system, etc. — has its own source of power and does not draw from the power coming down the phone line. The term came from telegraphy and was used to distinguish the battery which provided power to the telegraphic station as against the power that went to drive the line and the signal traveling down it. See BATTERY.

Local Bridge A bridge between two or more similar networks on a local site (within same building).

Local Bus A microprocessor inside a PC must communicate with certain integral devices, including memory, video controllers, hard disks. This is typically called an internal bus. That is to distinguish it from the "external" bus, such as the AT, ISA, EISA, MCA buses, which define the communications between the motherboard and the various peripheral devices, such as the I/O cards like those handling modems and LAN connections. As microprocessors have gotten faster, so they have begun to outpace the speed of their computer's internal bus, which has tended to narrow the stream of data in and out of the CPU, slowing the computer. A Local Bus is a new type of internal bus. It is a faster bus. The idea is to get a broader path between your critical components — memory, video and disk controller — and your microprocessor. The idea is to get the data in and out of the microprocessor at the same speed as the microprocessor's system clock. Local Bus is an emerging standard. See also EISA, PCI and VESA.

Local Call Any call within the local service area of the calling phone. Individual local calls may or may not cost money. In many parts of the US, the phone company bills its local service as a "flat" monthly fee. This means you can make as many local calls per month as you wish and not pay extra. Increasingly this luxury is dying and local calls are costing money.

Local Call Accounting Computes the dollar amount for local calls based on the total message units stored for each phone.

Local Call Billing Computes the dollar amount for local calls placed by guests based on total message units.

Local Central Office Switching office in which a subscriber's lines terminate.

Local Channel Controller An AT&T name for its family of 3270 compatible cluster controllers.

Local Composite Loopback In network management systems. Composite loopback test that forms the loop at the output of the local multiplexer that returns transmitted signals to their source. See loopback.

Local Dataset Signal converter that conditions the digital signal transmitted by an RS-232 interface to ensure reliable transmission over a dc continuous metallic circuit without interfering with adjacent pairs in the same telephone cable. Normally conforms with Bell 43401. Also called baseband modem, limited distance modem, local modem, or short-haul modem. See line driver.

Local Distribution Frame LDF. Another word for an Intermediate Distribution Frame. It's a device for cross connecting cables — from one thing to another. On one side of the LDF are the pairs from individual phones in that part of the building or area. On the other side are trunks coming in from a central office or cables coming in from the central, larger PBX. LDFs typically help with the organization of cables in a building or area. See also Intermediate Distribution Frame.

Local Echo A modem feature that enables the modem to send copies of keyboard commands and transmitted data to the screen. When the modem is in Command mode (not online to another system) the local echo is invoked through the ATE1 command. The command causes the modem to display your typed commands. When the modem is online to another system, the local echo is invoked through the ATF0 command. This command causes the modem to display the data it transmits to the remote system.

Local Exchange The telephone company exchange where subscribers lines are terminated. Also called an "End Office."

Local Exchange Carrier A local phone company. See also LEC.

Local Heap A memory storage area limited to 64K in size.

Local IP A telephone company AIN term. The Internet Protocol (IP) indicated when an SCP or Adjunct requests a local AIN switch to make a connection to an IP to which the SSP or ASC switch has a direct ISDN connection.

Local Long Distance IntraLATA long distance. A marketing term invented by the LECs (local exchange carriers) to distinguish intraLATA from interLATA toll calling. Specifically, the term was invented by the RHCs (Regional Holding Companies), which currently are limited to providing "long distance" calls only within the intraLATA toll market in their home states.

Local Loop The physical connection from the subscriber's premise to the carrier's POP (Point of Presence). The local loop can be provided over any suitable transmission medium, including twisted pair, fiber optic, coax, or microwave. Traditionally and most commonly, the local loop comprises twisted pair or pairs between the telephone set, PBX or key telephone system, and the LEC (Local Exchange Carrier) CO (Central Office). As a result of the deregulation of inside wire and cable in the United States, the local loop typically goes from the demarc (demarcation point) in the phone room closet, in the basement or garage, or on the outside of the house, to the CO. The subscriber or building owner is responsible for extending the connection from the demarc to the phone, PBX, key system, router, or other CPE device. See also Demarc.

Local Management Interface LMI. The specification for a polling protocol for use in Frame Relay networks between the user equipment in the form of a FRAD (Frame Relay Access Device) and the network equipment in the form of a FRND (Frame Relay Network Device). The LMI verifies the existence of the UNI (User Network Interface) and the Permanent Virtual Circuit (PVC).

Local Measured Service LMS. Years ago virtually all phone lines in the United States were FLAT RATE. That meant that for a fixed amount of money each month, you, the customer (a.k.a. subscriber) were allowed to make as many local calls as you wanted. For many reasons, the U.S. phone industry has progressively moved to LOCAL MEASURED SERVICE for local calls. Typically this means that for a fixed amount of money each month, you, the customer, can receive as many calls as you want and can make a finite number of outgoing local calls — typically 50. Each additional call beyond the 50 (or whatever the number is) costs extra. How much that call costs depends on the distant the call travels, the time of day, the day of the week, and the local company's tariffs.

Local Multipoint Distribution System LMDS. A method of distributing TV signals to households in a local community. LDMS uses broadcast microwave (or sometimes FM) signals operating at 28 GHz to contact local dishes, typically installed on the top of apartment buildings. The received signal is then distributed through the building's central CATV system. Sometimes it broadcasts directly to a subscriber's home. There are actually all sorts of variations on the LMDS theme. In Brazil, CellularVision uses LMDS technology, transmitting in the FM range, which means the signal has the ability to bounce, and to reflect off virtually any surface, thus increasing the coverage area significantly. Two-way or interactive communication may be inserted between video channels for transmission back on the opposite polarity. This reverse polarization, or interweaving, allows simultaneous use of signals at the same frequency for two applications. LMDS is competitive with conventional cable-based CATV. See also MMDS.

Local Net The broadband architecture used in Sytek's work. Also the product name of their network. Sytek is in Sunnyvale, CA.

Local Number Portability LNP. Imagine a town in which there are many local phone companies. You have service from one company. But another comes along offering better service, lower price and more features. You want to switch. But you don't want to change your phone number. That's what LNP is all about — the ability to change your phone company and still keep your phone number.

Regardless of the local provider selected, consumers will continue to have access to Emergency 911 service; operator and directory assistance services; advanced services such as voice mail, Caller ID and Call Forwarding; and other customized local area and signaling capabilities, including equal access to all 800 and 888 toll-free telephone numbers.

On November 15, 1997 I received a press release from Lockheed Martin saying that they had successfully developed and tested Local Number Portability (LNP) in the Midwest, the FCC's mandated national test region. The implementation was mandated by the Federal government as part of the Telecommunications Act of 1996, which required this process be completed by October, 1997. In the press release, Lockheed Martin explain that in 1994, amidst concern over the need for fair, open telecommunications markets, the MCI Telecommunications Corporation initiated a Gallup Poll to investigate the demand for a Local Number Portability system. The poll randomly surveyed approximately 2000 businesses and consumers across the country in September and October 1994. The study assessed whether business or consumers would switch local telephone service providers under various market scenarios. The results of the poll indicated that 83 percent of business customers and 80 percent of residential customers would not change local service providers if changing service providers meant changing phone numbers. The study helped articulate the need for LNP in order to facilitate fair and open market competition in the local telecommunications industry. The LNP system developed in response to that survey and completed by Lockheed Martin in October, 1997, represents a substantial benefit to consumers: They now can choose local service from a variety of providers without changing their phone numbers. The system works for voice, data, and video lines, residential and business lines.

Here is what Lockheed wrote about local number portability technology in that press release. "The database that facilitates Local Number Portability is a technological marvel in its complexity and the speed with which it operates. Each phone number has a network address. The LNP database keeps track of these addresses. When a customer places a call, the database records the caller's network address, locates the dialed number's network address, and notifies all telecommunications companies involved where to route the call and which companies to credit for the call.

Simplified, the process works like this:
- Call placed
- Network address of caller identified
- Network address of call recipient identified
- Telecommunications companies told how to route the call and which companies to credit.

All of this happens within nano-seconds of the customer placing the call — an imperceptible lapse of time. The LNP system also records the appropriate information whenever a customer changes local carriers, updating account information and ensuring that no interruption in service occurs.

In developing LNP, Lockheed Martin IMS drew on two types of existing technology. In 1993, Lockheed Martin IMS developed a portability system for 800 numbers that was used by 140 telecommunications companies in the United States at the request of Bellcore, the research and engineering division of the Regional Bell Operating Companies. This database allowed customers to handpick 800 numbers (such as 1-800-FLOWERS) and keep those numbers regardless of the long-distance carrier used. That experience laid the groundwork for developing LNP.

Lockheed Martin IMS also used existing infrastructure to run LNP. Twenty years ago, each local phone company installed computerized databases for their own internal use. LNP is an incremental application of this network, the Advanced Intelligence Network (AIN), and was made possible by the investment that local service providers made years ago. See also LNP and Location Portability.

Local Order Wire A communications circuit between a technical control center and selected terminal or repeater locations.

Local Phone A phone attached to your computer. See also HANDSET MANAGEMENT.

Local Phone Service When I dial the pizza store on the corner, I'm making a local phone call. But when I'm calling the pizza store 50 blocks away, is that a local phone call? Answer, it could be. It depends. What's local phone service? What do you charge for it? Once upon a time, most Americans didn't pay for local phone service. They paid a flat monthly fee. Then the phone companies needed money, so they started charging for local service. A few cents per call. Then the phone companies timed the call and charged more the longer you talked. Then they started charging for longer local calls — maybe for calls of ten miles and further. In short, the definition and pricing of local phone calls is changing. Now local calls are looking increasingly like long distance calls — charged by time, distance, and day of the week.

Local Printer A printer that is directly connected to one of the ports on your computer.

Local Redirector A local redirector is a shim that redirects HTTP requests to a local proxy server. A local redirector is also known as a load balancer, a local redirector is a piece of software that receives a server request (e.g., HTTP, FTP or NFS) and reroutes it to one of a cluster of Web servers to be actioned. The distribution function may be based on which machine in the cluster has the lowest current level of utilization, the proximity of the server to the client, or which machine has the resources necessary to carry out the request. This definition courtesy of Mark Gibbs.

Local Service Area The geographic area that telephones may call without incurring toll charges. A flat rate calling area. Increasingly rare.

Local Switch The term local switch refers to the switch (PBX, ACD, dumb) to which the computer telephony system is directly connected. Usually, the local switch will provide better integration with the computer telephony system (more comprehensive call data) than connections that take place over the network. Additionally, the local switch will have both line and trunk side connections and will also support connection to whatever agents or desktop users that may use the computer telephony application.

Local Tandem A central office, usually in large metropolitan areas, serving as a transit switch between noncontiguous class 5 exchanges. It connects end office trunks.

Local Test Desk A testing system that is used to test local loops and central office subscriber line equipment from a central point, typically a central office.

Local Trunk Trunks between Class 5 offices (local central offices)

Locality A measure of how close commonly-accessed files are to one another on a hard disk. "High locality" means the files reside on sectors or tracks which are close to each other. When this is the case, seek times during operation are shorter than average.

LocalTalk Apple Computer's proprietary local area network for linking Macintosh computers and peripherals, especially LaserWriter printers is called Appletalk. AppleTalk's LAN hardware is called LocalTalk. Appletalk is a CSMA/CD network that runs at 230.4 kilo bits per second and is therefore, incompatible with any other local area network. It is also a lot slower than the present top speeds of Ethernet (10 Mbps) and Token Ring (16 Mbps). Outside manufacturers, however, make gateways which will connect an Appletalk LAN to other local area and telecommunications networks — LANs, WANs and MANs. See also APPLETALK.

Location One definition is the place where a telephone jack is located. This location is given a number. The wire going to that location is given a number. All this in hope of being organized for installation, moves, changes and maintenance.

Location ID A feature of the IS-136 standards for digital cellular networks employing TDMA (Time Division Multiple Access). A capable telephone set will display the name of the cellular carrier providing service. In a wireless office system application, the phone can display the name of the company. When you are at home, connected to your PBS (Personal Base Station), the phone can display "cordless."

Location Portability The ability of an end user to retain the same geographic or non-geographic telephone number (NANP numbers) as he/she moves from one permanent physical location to another. Location Portability will involve either of the following scenarios: 1) new location is within the same central office area, or 2) new location is within a different central office area. See also Local Number Portability.

Location Transparency More professionals are working from home, customer sites or from the road. Location transparency means that your communications system — faxing, email, voice mail, etc — works as well for you, the user, whether you're in the office or in the field, or where in the field you are.

Locator A term used in the secondary telecom equipment business. A locator is a company that assists both a buyer and seller to quickly find each other. A locator contracts with dealers to provide them with daily lists of potential customers. The list develops from phone calls to an 800-number asking for a specific component.

Lock And Load Original y a military term. Then it became software speak for freezing code on a program in development. Then it became "Let's make a decision and get on with it."

Lock Code The lock code locks a cellular telephone to prevent unauthorized use. The lock code is programmed into the NAM (Numerical Assignment Module) and is frequently factory set to either 1234 or 00004.

Lock On The process by which an earth station initially acquires the signal from a satellite.

Lock Out In a satellite telephone circuit controlled by an echo suppressor, one or both subscribers can't get through because of excessive noise at one end. You get this also with speakerphones. The person with the speakerphone can simply hog the conversation because his speakerphone keeps transmitting his voice. There are weird variations on this. Sometimes you might call someone on your speakerphone and wait for them to pick up. They do. They shout into the phone "I'm here." All they can hear is you at the other end talking or typing. The sound at your end is hogging the channel and thus locking out the person at the other end. The solu-

tion? Turn the "mute" button on your speakerphone. This will stop your end transmitting and allow the other end to say "Hello, I'm here." See also LOCKOUT.

Locked Mode In SONET networks, a mode of operation for a Virtual Tributary (VT) group. A VT group can function in either locked or floating mode. While floating mode minimizes delays in distributed VT switching, locked mode is used to enhance the efficiency of the network devices performing the switching.

Locked Resources An Intel Plug and Play term. Resources that must be used by the same card each time the system is booted. The configuration manager cannot assign these resources to any other card.

Locker Telco-speak for a storage area, often an urban storefront, where phone company installers and repairman can pick up and drop off tools and installation material such as phones, wire and hardware. These places are prime targets for burglary and robbery, and often have no identification to show their valuable contents.

Locking Preventing several people getting to and changing the same data in a shared database simultaneously. Locks may be permanent and prevent access completely, or they may be "advisory." A user is warned the data is being used by someone else and that the data is not presently available. Locks prevent the destruction of data that can occur if two people access a file at the same time. In any data base or other computer system, there are typically two types of "locks" — record and file locks. A record lock occurs when an airline agent pulls up your travel plans. No other travel agent can access those records at that time. A file lock occurs when the whole file is locked up. This might occur in a centralized word processing program. The whole document will be locked when it is being used by someone.

Lockout A PBX feature. Denies the attendant the ability to re-enter an incoming central office connection directly terminated or held on her position, unless specifically recalled by the phone user.

Locutorio In Buenos Aires, they have places (normally privately owned) where there are a lot of telephone booths, fax machines, that you walk in, make a call, pay and walk out. The word in Spanish in Buenos Aires is "locutorio", which was the name given to place in convents or prisons where, nuns or prisoners could speak to visitors from behind bars (in both cases). The word comes from Latin and is mostly used in Spain (Telefonica de Espaa purchase part of the old Argentine telco). This definition contributed by Jorge E. Corbalan, head translator in the Arthur Anderson office in Buenos Aires.

LOF Loss of Frame. LOF is a generic term with specific variations of meaning, depending on the signal standards domain in which it is being used. In the OSI/ATM world for instance, LOF is a condition at the receiver or a maintenance signal transmitted in the PHY overhead indicating that the receiving equipment has lost frame delineation. This is used to monitor the performance of the PHY layer. In the SONET world also, LOF is a condition detected in the signal overhead at the receiver, indicating that a valid framing pattern could not be obtained. There is however no "LOF indication" per se transmitted in the SONET overhead: SONET uses the AIS-L overhead indication to inform downstream equipment that the receiving equipment upstream has experienced a failure. A hierarchy of LOF defect and alarm notifications is implemented at the LOF-detecting equipment based on the duration the condition persists. See also AIS and SEF.

Log In The process of identifying and authenticating oneself to a computer system. Used to control access to computer systems. See LOGIN SCRIPT.

Log Off 1. To type in the needed keystrokes for ending a session that's on-line with a computer. Often those keystrokes are "Logoff." Usually it's very easy to Log Off. It's more difficult to Log On.

2. Employees wear their photo-ID badges in little plastic holders attached by a clip to their clothing. It's a good idea to "log off" when you're in public. This means to turn your badge around so nobody can see whom you work for.

Log On To enter the needed keystrokes to start an on-line session with a computer. "Logging On" may be done with a computer that's local or one that's long distance and your work is done over communications lines.

According to the Vermonter's Guide to Computer Lingo, log on is making the wood stove hotter.

Logic Logic is the application of mathematical analysis and deductive reasoning to propositions that may or may not be true or false. The logic we're interested in owes much to the work of George Boole in the mid-19th century. He formulated a system that could be applied to the relationships between propositions to which only a binary choice of truth existed, i.e. yes or no. The first application of the Boolean Algebra that derived from this was Shannon's research into the analysis of relay switching circuits in 1938.

Logic Bomb Program routine that destroys data. For example, a logic bomb may reformat the hard disk or insert random bits into data files. It may be brought into a personal computer by downloading a corrupt public domain program. Once executed, it does its damage right away and then stops, whereas a virus keeps on destroying. Another definition of a logic bomb is that it is a resident computer program that lies dormant for a period, and then triggers an unauthorized act when a certain event, such as a date, occurs.

Logical Address See Physical Address

Logical Block Address A logical block address is a sequential address for accessing blocks on storage media. The first block of the media is addressed as block O and succeeding blocks are numbered sequentially until the last block is encountered. This is the traditional method for accessing peripherals on a SCSI bus.

Logical Topology The logical layout of a network, as opposed to the Physical Topology. In the LAN world, for instance, a 10Base-T network is a Star from the standpoint of Physical Topology (the way it looks). Yet the network operates as a logical Bus. In other words, the devices arrayed around the 10Base-T hub connect through ports into the hub chassis which houses a collapsed bus which supports communications over the shared physical path just as does a traditional Ethernet bus network, which is a bus in both physical and logical terms.

Logical Bus A LAN topology, such as Ethernet, which shares a common communications channel.

Logical Channel A software based connection through which data is sent. The channel is assigned by the switch. In X.25 talk, a logical channel refers to a virtual connection operated over a physical connection that can support one or more virtual connections simultaneously.

Logical Channel Number Virtual circuit identified at the packet level of X.25. See LOGICAL CHANNEL.

Logical Client Refers to one component in a pair of communicating components which is obtaining access to CTI functionality through the other component. This term is used to differentiate between two components which are communicating

across an inter-component boundary. See Logical Server.

Logical Drive A disk drive recognized by the operating system. A computer's logical drives may differ from its physical drives. For example, a single hard disk drive may be partitioned into two or more logical drives. Before MS-DOS 5.0, the biggest logical "drive" that DOS could address was 32 megabytes. If you had a larger hard disk, you partitioned the bigger driver into logical drives of 32 megabytes.

Logical Formatting The third step in structuring a data medium so that data may be written to it. Logical formatting must follow physical formatting (also called low-level formatting) and partitioning (figuring into how many drives you wanted to slice the one drive into).

Logical Group Node An ATM term. A logical node that represents a lower level peer group as a single point for purposes of operating at one level of the PNNI routing hierarchy.

Logical ID An AT&T Merlin term. A numbering sequence used to identify station and trunk locations on the communications system control unit.

Logical Link An abstract representation of the connectivity between two logical nodes. This includes individual physical links, individual virtual path connections, and parallel physical links and/or virtual path connections.

Logical Link Control LLC; A protocol developed by the IEEE 802 committee, common to all of its LAN standards, for data link-level transmission control; the upper sublayer of the IEEE Layer 2 (OSI) protocol that complements the MAC protocol; IEEE standard 802.2; includes end-system addressing and error checking.

Logical Modem Interfaces LMIs are to Microsoft's MAPI what SPIs are to TAPI. LMIs serve network-based fax servers and multi-port fax boards. MAPI sits on top of Microsoft's Exchange and other MAPI-compliant messaging systems.

Logical Node An abstract representation of a peer group or a switching system as a single point.

Logical Node ID A string of bits that unambiguously identifies a logical node within a routing domain.

Logical Provisioning An AT&T term for the establishment of network services by changing software controls, rather than by physically installing or rearranging hardware.

Logical Ring A network which is treated logically as a ring even though it maybe cabled as a physical star topology.

Logical Server Refers to one component in a pair of communicating components which is providing CTI functionality to the other component. This term is used to differentiate between two components which are communicating across an inter-component boundary. See Logical Client.

Logical Unit Interface See LU 6.2.

Logiciel French for software.

Login Script When users log into a local area network, they may wish to do many things or the network supervisor may wish to do several things. These commands are part of something called a "login script." In computerize (Novell's words), a login script contains commands that initialize environmental variables, map network drives, and/or control the user's program execution. Login scripts are similar to batch files. The familiar AUTOEXEC.BAT can be thought of as an MS-DOS login script.

LOH Line Overhead. 18 octets in a SONET frame for purposes of controlling the reliable transport of payload data between SONET network elements such as repeaters. LOH and SOH comprise Transport Overhead (TOH).

LOI Letter of Intent

Long Distance Any telephone call to a location outside the local service area. Also called toll call or trunk call.

Long Haul Communications That type of phone call which reaches outside a local exchange or serving area.

Long Key A long key is a character held down for a prescribed period of time. The time period is generally longer than the time for other keys. In call reorigination, the # sound is defined as a "long key." This definition was kindly provided by Karen Shelton, Systems Engineer, IEX Corporation, Richardson, TX.

Long Lines AT&T Long Lines. The department of AT&T which operates long distance toll service. It is no longer called Long Lines. It is called AT&T Communications.

Long Reach Long reach refers to optical sections of approximately 25 kilometers or more in length and is applicable to all SONET rates.

Long Tones First, we invented touchtone, also called DTMF, Dual Tone Multi Frequency tones. You'd punch your number with tones, instead of dialing them. Then someone thought you could control telephone response gadgets, like voice mail, interactive voice response, etc. with touch tones. For these gadgets to work, they had to "hear" the tones you sent. No one really set standards as to the minimum length tone they would hear. But it was generally conceded that they were to be 120 milliseconds. So some manufacturers of telephone equipment started to make phone equipment that, if you pushed a touchtone button, the machine would only sent a touchtone of 120 millisecond duration. That was called a short tone. It wasn't very useful because the manufacturers quickly discovered that many pieces of equipment couldn't respond that quickly. And the manufacturers got complains that their customers couldn't call their voice mail, their bank, etc. As a result, some manufacturers of equipment brought out new hardware (replacing the old) to allow you to send "long tones," which are now defined as touchtones that last for as long as you hold down the button — just as it is (and has always been) on a normal single line, non-electronic, non-digital telephone. Isn't progress wonderful? See also DTMF for a much longer explanation of tone dialing.

Long Wavelength Light whose wavelength is greater than about 1 micrometer.

Longest Available This is a method of distributing incoming calls to a bunch of people. This method selects an agent based on the amount of time that each agent has been on the phone. This allows for an equitable distribution of calls to each agent. See also TOP DOWN and ROUND ROBIN.

Longevity Testing When you're building a computer telephony system, many problems do not come to light until it's been running under high traffic for a long term. According to Steve Gladstone, author of the book "Testing Computer Telephony Systems," (available from 212-691-8215), there can be slow memory leaks, counter overflows, or disk fragmentation which slows down data access, system resets that cause calls to be dropped inadvertently, and so on. Frequently these issues are discovered only after thousands of calls are placed or after some random call pattern has occurred. Longevity tests, according to Gladstone, provide a high load to a system over an extended period of time. The most effective way to do this is to run a load test based on your busy hour usage profile over several days and track failures over the entire testing period. If failure rates increase, or specific failures occur, the events leading up to the fault can be relatively easily duplicated. If the system generating the load provides good error tracking and event logging, it may be possible to immediately identify the sequence that caused the fault, the corresponding reactions of the computer telephony system, and even immediately duplicate the fault causing scenario.

Longitudinal Balance A measure of the electrical balance between the two conductors (tip and ring) of a telephone circuit; specifically, the difference between the tip-to-ground and ring-to-ground AC signal voltages, expressed in decibels.

Longitudinal Redundancy Check LRC. An error checking technique based on an accumulated collection of transmitted characters. An LRC character is accumulated at both the sending and receiving stations during the transmission of a block of data. This accumulation is called the Block Check Character (BCC) and is transmitted in the last character in the block. The transmitted BCC is compared with the accumulated BCC character at the receiving station for an equal condition. When they're equal, you know your transmission of that block has been fine. LRC commonly is used in combination with VRC (Vertical Redundancy Checking) to improve the reliability or error control in asynchronous transmission and in support of the ASCII coding scheme.

With VRC, a check bit, or parity bit added to each ASCII character in a message such that the number of bits in each character, including the parity bit, is odd (odd parity), or even (even parity). The term comes from the fact that the bits representing each character of data logically is viewed in a vertical fashion. When LRC is used in combination with VRC, the parity of a block (set) of data characters is checked for parity longitudinally (along the horizontal plane) of characters, as though they were laid out logically in a matrix format. For instance, the word "CONTEXT" consists of 7 letters, each of which consists of 7 bits, viewed in a block matrix format as follows:

BIT/VALUE	C	O	N	T	E	X	T	P**
1*	1	1	0	0	1	0	0	0
2*	1	1	0	0	0	0	0	0
3*	0	1	1	1	1	0	1	0
4*	0	1	1	0	0	1	0	0
5*	0	0	1	0	1	0	1	0
6*	0	0	0	0	0	0	0	1
7*	1	1	1	1	1	1	1	0
8**	0	0	1	0	0	0	0	

* INFORMATION BIT
** PARITY BIT

The transmitting machine sums the bit values for each character, beginning with "nothing," which is an even value in mathematical terms. In the case of the letter "C," for instance, the next bit is a "1" bit, which creates an odd value. The next bit is a "1" bit, which creates an even value. The next four bits are "0" bits, which do not change the even value. The seventh bit is a "1" bit, which creates an odd value, once again. Assuming that the device is set for odd parity, which is the default, it will insert a "0" bit in the eighth bit position, retaining the odd value. (Should the device be set for even parity, a "1" bit would have been inserted in the eighth bit position.) Should the value of the 7 information bits be an even value, the device appends a "1" bit in order to create an odd value. Across the longitudinal plane, the transmitting device accomplishes exactly the same process, appending "0" to retain odd values or "1" bits in order to create odd values.

After the data, character-by-character, has been formatted in this fashion, each bit sequence is transmitted across the network to the target device, which also is set for odd (or even) parity. The receiving device goes through exactly the same process, examining each character for parity. If the parity does not match the expectation of the receiving device, the subject character is flagged as errored, although no remedial action is taken. As VRC exposes the transmission to reasonable likelihood that two bits in a given character can be errored in the process of transmission of each character, that the parity of the character therefore would not be affected, and that the receiving device would not detect the fact that the character was errored, this technique is known as "send and pray." LRC substantially improves the likelihood that errors in a block of data will be detected, although there remains potential for compensating errors to affect the data, without detection. Any remedial action must be accomplished on a man-to-machine basis. See also VERTICAL REDUNDANCY CHECKING and PARITY.

Longitudinal Transmission Check LTC. An even or odd parity check at fixed intervals during data transmission.

Longitudinal Wrap A tape applied longitudinally with the axis of the core being covered, a opposed to a helical, or spiral, tape wrapped core.

Look Ahead Routing A Common Channel Signaling System 7 (SS7) technique that determines the availability of a communications channel before the call is sent over the network. The technique maximizes efficiency and use of the public switched network. see Signaling System 7.

Look-Up Table 1. A translation table. You dial a certain number. But the number is meaningless. For the phone system to complete the call, it needs routing instructions. It gets that by "looking up" that number in a table, which translates that number to another number that is now meaningful to the switching network. There are lots of applications for Look-up Tables. Least Cost Routing tables are essentially look up tables. IN-WATS dialing works by looking up the 800 number and finding its real ten digit normal number. Most private networks use look up tables to translate the number dialed by the internal user into a number that the network can recognize.
2. A set of addresses (source and destination) used by a bridge or router to determine what should be done with a packet. As the packet comes in, its address information is read and compared with the information in the look-up table. Depending on the information, the bridge may forward the packet, or discard it, leaving it for the local LAN. Many bridges and routers can build their look-up tables as they operate. See also 800 SERVICE, EIGHTHUNDRED SERVICE, BRIDGE, PERSONAL 800 NUMBER and ROUTER.

Loop 1. Typically a complete electrical circuit.
2. The loop is also the pair of wires that winds its way from the central office to the telephone set or system at the customer's office, home or factory, i.e. "premises" in telephones.
3. In computer software. A loop repeats a series of instructions many times until some prestated event has happened or until some test has been passed.

Loop Antenna An antenna consisting of one or more complete turns of wire, both ends of which are to be connected to the input circuit of the radio receiver.

Loop Back A diagnostic test in which a signal is transmitted across a medium while the sending device waits for its return. See LOOKBACK.

Loop Checking A method of checking the accuracy of transmission of data in which the received data are returned to the sending end for comparison with the original data.

Loop Circuit Generally refers to the circuit connecting the subscriber's set with the local switching equipment.

Loop Current Detection When a modem, telephone or fax card (etc.) seizes the line (i.e. completes the connection between tip and ring terminals of the telephone cable) current flows from the positive battery supply in the telephone central

office, through the twisted pair in the loop, through the card (or phone) and back to the central office negative terminal where it is detected, showing that this telephone or telephone device is off hook. The fax card or modem can detect problems such as disconnects, shutting down the connection or a busy signal.

Loop Extender Device in the central office that supplies augmented voltage out to subscribers who are at considerable distances. It provides satisfactory signaling and speech for such subscribers.

Loop Plant Telco-talk for all the wires and hardware and poles and manholes used to connect their central offices to their customers.

Loop Qualification Test done by the phone company to make sure the customer is within the maximum distance of 18,000 feet from the central office that services that customer. 18,000 is the maximum distance an ISDN-BRI phone line will work.

Loop Reverse-battery A method of signaling over interoffice trunks in which changes associated with battery reversal are used for supervisory states. This technique provides 2-way signaling on 2-wire trunks; however, a trunk can be seized at only one end. It cannot be seized at the office at which battery is applied. It is also called reverse-battery signaling.

Loop Signaling A method of signaling over circuit paths that uses the metallic loop formed by the line or trunk conductors and terminating circuits.

Loop Signaling Systems Any of three types of signaling which transmit signaling information over the metallic loop formed by the trunk conductors and the terminating equipment bridges.

Loop Start LS. You "start" (seize) a phone line or trunk by giving it a supervisory signal. That signal is typically taking your phone off hook. There are two ways you can do that — ground start or loop start. With loop start, you seize a line by bridging through a resistance the tip and ring (both wires) of your telephone line. The Loop Start trunk is the most common type of trunk found in residential installations. The ring lead is connected to -48V and the tip lead is connected to 0V (ground). To initiate a call, you form a "loop" ring through the telephone to the tip. Your central office rings a telephone by sending an AC voltage to the ringer within the telephone. When the telephone goes off-hook, the DC loop is formed. The central office detects the loop and the fact that it is drawing DC current and stops sending the ringing voltage. In ground start trunks, ground Starting is a handshaking routine that is performed by the central office and the PBX prior to making a phone call. The central office and the PBX agree to dedicate a path so incoming and outgoing calls cannot conflict, so "glare" cannot occur. See GLARE. Here are two questions that help in understanding:
How does a PBX check to see if a CO Ground Start trunk has been dedicated?
To see if the trunk has been dedicated, the PBX checks to see if the TIP lead is grounded. An undedicated Ground Start Trunk has an open relay between 0V (ground) and the TIP lead connected to the PBX. If the trunk has been dedicated the CO will close the relay and ground the TIP lead.
How does a PBX indicate to the CO that it requires the trunk?
A CO ground start trunk is called by the PBX CO Caller circuit. This circuit briefly grounds the ring lead causing DC current to flow. The CO detects the current flow and interprets it as a request for service from the PBX. See also POTS.

Loop Test A way of testing a circuit to find a fault in it by completing a loop and sending a signal around that loop. See LOOPBACK.

Loop Through A type of phone system wiring that allows phones to connect to one cable in parallel going to the common central switching equipment. The most common type of Loop Through wiring is that which you have in your home. You have one cable with two conductors — a red and a green — winding through your home. Whenever you want to connect a phone, you simply attach it to the red and green conductors. The other way of connecting phones is called HOME RUN. In that system, every phone has its own one, two or three pairs of conductors which wind their lonely way back to the central PBX or key system cabinet. In Loop Through wiring, many phones share one set of cables. In Home Run Cabling, only one phone sits on that line.

Loop Timing A way of synchronizing a circuit that works by taking a synchronizing clock signal from incoming digital pulses.

Loop Up/Loop Down In T-1, there are generally two loopback types, LLB (line loopback) and TLB or DLB (terminal or DTE loopback). Loop Up refers to activating one of these loop backs, where as Loop Down refers to deactivating one of these loopbacks.

Loopback Type of diagnostic test in which the transmitted signal is returned to the sending device after passing through a data communications link or network. This allows a technician (or built-in diagnostic circuit) to compare the returned signal with the transmitted signal and get some sense of what's wrong. Loopbacks are often done by excluding one piece of equipment after another. This allows you to figure out logically what's wrong. (It's called Sherlock Holmes deductive reasoning.) See LOOPBACK TEST.

Loopback Test A test typically run on a four-wire circuit. You take the two transmit leads and join them to the two receive leads. Then you put a signal around the loop and see what happens. Measuring differences between the sent and the received signal is the essence of a loopback test. See LOOPBACK.

Looping Problem encountered in distributed datagram routing in which packets return to a previously visited node.

Loopstart Circuit The standard world-wide telephone circuit. For the phone to signal the phone system that it wants to make a call, it applies a DC termination across the phone line. See LOOP START for a longer explanation.

Loose Tube Buffer A cable construction in which the optical fiber is placed in a plastic tube having an inner diameter much larger than the fiber itself. The loose tube isolates the fiber from the exterior mechanical forces acting on the cable. The space between the tube and the fiber is often filled with a gel to cushion the fiber.

Loosely Coupled A computer system architecture consisting of multiple computer systems, each with its own dedicated memory and its own copy of the operating system, connected over a communications link. See also TIGHTLY COUPLED.

LOP Loss of Pointer. LOP is a generic term with specific variations of meaning, depending on the signal standards domain in which it is being used. In the OSI/ATM world for instance, LOP is a condition at the receiver or a maintenance signal transmitted in the PHY overhead indicating that the receiving equipment has lost the pointer to the start of cell in the payload. This is used to monitor the performance of the PHY layer. In the SONET world also, LOP is a condition detected in the signal overhead at the receiver, indicating that a payload position pointer could not be obtained. There is however no "LOP indication" per se transmitted in the SONET overhead: SONET uses the AIS-L overhead indication to inform downstream equipment that the receiving equipment upstream has experienced a failure. A hierarchy of LOP defect and alarm notifications is implemented at the LOP-detecting equipment based on the duration the condition persists. See also AIS.

LORAN LOng Range Aid to Navigation. A radio-navigation system which helps you find where you are. It works by timing the difference in reception of pulses from one or more fixed transmitters, usually on land. It's a radio based systems that's pretty good for coastal waters where there are LORAN transmitters. The maximum range is about 1,400 miles at night (about a half that during the day), in virtually any sort of weather. LORAN doesn't cover much of the rest of the earth and its accuracy varies depending on electronic interference and geographic variations. The first LORAN transmitters were put into operation by the U.S. Navy in 1944. LORAN range limitations were overcome in 1973, when the U.S. Navy first installed the "Transit" SatNav (Satellite Navigation) system, which set the stage for a much better system known as GPS (Global Positioning Satellite System). Over time and as costs have decreased, even small pleasure craft have replaced their LORAN receivers with GPS. See also GPS.

LOS Loss of Signal. LOS is a generic term with specific variations of meaning, depending on the signal standards domain in which it is being used. In the OSI/ATM world for instance, LOS is a condition at the receiver or a maintenance signal transmitted in the PHY overhead indicating that the receiving equipment has lost the received signal. This is used to monitor the performance of the PHY layer. In the SONET world also, LOS is a condition directly detected at the physical level (photonic or electronic) at the receiver. There is however no "LOS indication" per se transmitted in the SONET overhead: SONET uses the AIS-L overhead indication to inform downstream equipment that the receiving equipment upstream has experienced a failure. A hierarchy of LOS defect and alarm notifications is implemented at the LOS-detecting equipment based on the duration the condition persists. See also AIS.

Loss The drop in signal level between two points on a network. It is important to distinguish between LOSS and LEVEL. Level is measured at a finite point. Loss is the difference between levels. Loss occurs constantly throughout telephony — from long distance circuits to switches. Loss is usually measured in dB — decibels. Loss is cumulative. Add two circuits each with a loss of 10 dB. You will have 20 dB loss in the total circuit. The human ear can detect a 3 dB loss.

Loss Budget A loss budget is the maximum amount of signal degradation a data communications network can withstand before it becomes susceptible to errors and/or loss of data. The idea is a establish a "loss budget" by consulting your equipment vendors for recommended wire types and maximum allocable lengths of cable before you build the network. We first heard the idea of a loss budget from The Siemon Company, Watertown CT.

Loss of Frame See LOF

Loss of Pointer See LOP

Lossless Compression Image- and data-compression applications and algorithms, such as Huffman Encoding, that reduce the number of bits a picture would normally take up without losing any data. In this way, no information is lost or altered in the compression and/or transmission process. See LOSSY.

Lossy Compression Methods of image compression, such as JPEG, that reduce the size of an image by disregarding some pictorial information.

Lost Call Attempt A call attempt that cannot be further advanced to its destination due to an equipment shortage or failure in the network.

Lost Calls Cleared Traffic engineering assumption used in Erlang C that calls not satisfied on the first attempt are held (delayed) in the system until satisfied.

Lost Calls Held Traffic engineering assumption used in Poisson that calls not satisfied on the first attempt are held in the phone system for a period not exceeding the average holding time of all calls.

Lottery A method, authorized by the Congress, designed to provide the FCC with an alternative or option to comparative hearings for allocating spectrum space to competing applicants in various services.

Lotus Express An e-mail communications software program for PC users of MCI Mail that allows the user to automatically send and receive messages, as well as binary files, such as spreadsheets or documents.

Loudspeaker See SOUND.

Loudspeaker Paging Access Interface to customer-provided paging equipment.

Lovelace, Ada Augusta Ada Augusta Lovelace was the daughter of Lord Byron, the English poet. Miss Lovelace is regarded as the first computer programmer because she worked for the computer pioneer Charles Babbage. A computer language was named after her. That language is called ADA.

Low Battery Cutoff An uninterruptible power supply (UPS) definition. This UPS feature automatically switches off battery power before the batteries discharge beyond safe limits. Without this feature, batteries can be taken into deep discharge, making them useless.

Low Entry Networking LEN. A peer-oriented extension to SNA, first implemented on IBM's System/36, that allows networks to be more easily built and managed by such techniques as topology database exchange and dynamic route selection.

Low Frequency The band of frequencies between 30 and 300 kilohertz.

Low Level Formatting The first step in preparing a drive to store information after physical installation in complete. The process sets up the "handshake" between the drive and the controller. Most drivers are now low level formatted at the factory. See PHYSICAL FORMATTING.

Low Level Language A programming language that uses symbols — one step away from the machine language of a computer. Low level computer languages, such as Assembler and C, actually manipulate the bits in computer registers. Higher level languages such as Basic and Fortran will take care of the piddling details of doing specific functions when you give it a broad command like "PRINT". In a lower level language, you must provide all the details of instruction necessary in the code (program) to perform the operation. It is possible to do this by calling standard routines, but still takes up the programmers' time in deciding which routines, and keeping the registers straight as he designs the program.

Low Noise Amplifier LNA. Typically a parametric amplifier in a satellite earth station.

Low Pass Filter A device that cuts frequencies off above a certain point and allow all other frequencies to pass. Opposite of high pass.

Low Power Television Service LPTV. A broadcast service that permits program origination or subscription service or both via low powered television translators. LPTV operates

secondarily to regular television stations. Transmitter output is limited to a 1000 watts for a UHF station, 10 watts for a VHF station, except when VHF operation is on an allocated channel when 100 watts may be used.

Low Speed Loopback A closed circuit feature useful for maintenance or testing.

Low Speed Signal Signal traveling at the DS1 rate of 1,544 Mb/s or at the DS1C rate of 3,152 Mb/s.

Low Voltage A low voltage condition exist when fewer than 105 volts AC is present at a 120 VAC outlet or HOT conductor. This figure was chosen by many manufacturers of electronic and telephone equipment. It is also the test of "low voltage" tested by a wonderful AC electric outlet-testing product called the Accu-Test II made by Ecos Electronics Corporations of Oak Park, Illinois. Below 105 volts, motors deteriorate and electronic circuits overheat. Long-term damage can occur to most gadgets plugged into an electrical outlet which consistently delivers below 105 volts. Also 105 volts is below the stated tolerance levels of all North American power utilities who state that their acceptable power is 120 volts plus or minus 10 percent. If your power is consistently below 105 volts, you should contact your local power utility.

LPC 1. Linear Predictive Coding. Low bit rate voice (LBRV) digitizing technique that requires a bandwidth of only 2.4 or 4.8 Kbps. This technique may result in poor quality voice signals.

2. Late Payment Charge.

LPD Line Printer Daemon, a process on Berkeley spooler implementations that provides LPR systems.

LPDA-2 IBM's protocol under NetView for monitoring of dial-up modems for error correction.

LPF Low Pass Filter: In an MPEG-2 clock recovery circuit, it is a technique for smoothing or averaging changes to the system clock.

LPI Lines Per Inch. The number of lines both horizontal and vertical, that a facsimile machine will print in a square inch.

LPP Link Peripheral Processor or Link Peripheral Processing.

LPR The LPR command is used to queue print jobs on Berkeley queuing systems.

LPS Line Profile System.

LPT Port A logical designation for a series of I/O (Input/Output) addresses that allows the computer to communicate with a parallel printer.

LPT1 The first or primary parallel printer port on the IBM PC or clone. LPT2 is the second parallel port. COM1 is the first serial port. LPT1 is usually the default printer port, i.e. the one your computer will print to, if you don't tell it something else.

LPTV Low Power Television Service. A broadcast service that permits program origination or subscription service or both via low powered television translators. LPTV operates secondarily to regular television stations. Transmitter output is limited to a 1000 watts for a UHF station, 10 watts for a VHF station, except when VHF operation is on an allocated channel when 100 watts may be used.

LQA Link Quality Analysis. See AUTOMATIC LINK ESTABLISHMENT.

LRC Longitudinal Redundancy Check.

LRN Local Routing Number. A 10-digit telephone number. The term is used in the context of LNP (Local Number Portability). See LNP.

LRU Least Recently Used. Refers to an algorithm that sorts items according to time last accessed and then discards the oldest items in the list to free up needed space.

LS Trunk Loop-Start Trunk. A trunk on which a closure between the tip and ring leads is used to originate or answer a call. High-voltage 20-Hz AC ringing from the telephone company signals an incoming call. See LOOP START.

LS1A-A Single Mode, Single-Fiber Interconnection Cable. An AT&T definition.

LSAP An ATM term. Link Service Access Point: Logical address of boundary between layer 3 and LLC sublayer 2.

LSB Least Significant Bit and Least Significant Byte. That portion of a number, address or field which occurs right most when its value is written as a single number in conventional hexadecimal or binary notation. The portion of the number having the least weight in a mathematical calculation using the value.

LSCIE Lightguide Stranded-Cable Interconnect Equipment.

LSCIM Lightguide Stranded-Cable Interconnect Module.

LSCIT Lightguide Stranded-Cable Interconnect Terminal.

LSDU Link layer Service Data Unit.

LSI Large Scale Integration. Refers to micro electronic components which combine many hundreds of transistors on an integrated circuit. See CHIP.

LSL Link Support Layer. A layer within the Novell Open Data-Link Driver specification. This layer lets multiple protocol stacks access a network card simultaneously.

LSO Local Service Office. Defined for North America as a six digit number consisting of the area code and the first three digits of the exchange code (i.e. 410-638 (area code for northern Maryland, with exchange code for Belair Maryland). Used to identify a geographical area and local service provider for that circuit. An LSO is important when ordering circuits from a telephone company, especially a long distance one.

LSOA Local Service Order Administration system. An OSS (Operations Support System) used by LECs (Local Exchange Carriers) to administer service orders such as orders for new service, service rearrangements, and changes of carrier. See also LNP, NPAC and OSS.

LSR An ATM term. Leaf Setup Request: A setup message type used when a leaf node requests connection to existing point-to-multipoint connection or requests creation of a new multipoint connection.

LSSGR LATA Switching System Generic Requirements.

LT 1. Logical Terminal
2. An ATM term. Lower Tester: The representation in ISO/IEC 9646 of the means of providing, during test execution, indirect control and observation of the lower service boundary of the IUT using the underlying service provider.

LTB Last Trunk Busy.

LTC Line Trunk Controller.

LTCI ISDN Line Trunk Controller.

LTE SONET Lite Terminating Equipment: ATM equipment terminating a communications facility using a SONET Lite Transmission Convergence (TC) layer. This is usually reserved for end user or LAN equipment. The SONET Lite TC does not implement some of the maintenance functions used in long haul networks such as termination of path, line and section overhead. In short, line terminating equipment includes network elements which originate and/or terminate line (OC-N) signals. LTEs originate, access, modify, and/or terminate the transport overhead.

LTO See Linear Tape Open Architecture.

LTRS Letters Shift. 1. Physical shift in a terminal using Baudot Code that enables the printing of alphabetic characters. 2. Character that causes the shift.

LTS Loop Testing System.

LU 1. Line Unit.
2. Logical Unit, access port for users in SNA. In a bisync network, a port through which the user gains access to the network services. A LU can support sessions with the host-based System Services Control Point (SSCP) and other LUs.
3. Local Use flag. Occasionally used to initialized approval for local cellular calls. The Cellular carrier insures that local users are registered with a local system.

LU 6.2 Logical Unit Interface. Version 6.2. An IBM SNA protocol that allows for peer-to-peer or program-to-program communications. The LU 6.2 protocol standard frees application programs from network specific details. On an IBM PC, a LU 6.2 program accepts commands and passes them on to an SDLC card to communicate directly with the mainframe or a token ring handler. LU 6.2 enables users to develop applications programs for peer-to-peer communications between PC's and IBM host systems. It increases the processing power of the PC user without the constraints of mainframe-based slave devices, i.e. 3274/3276 controllers. It creates a transparent environment for application-to-application communications, regardless of the types of systems used or their relative locations. Also referred to as Advanced Program-to-Program Communication (APPC).

LU Type 1 LU 1 is the SNA protocol that describes generic input/output devices (e.g. line printer).

LU Type 3 LU 3 is the SNA protocol that describes a print output device that uses 3270 data streams.

Lucent Technologies On September 30, 1996, AT&T split its manufacturing operations (Western Electric) and its Bell Labs off from its company and floated those operations as a separate company. It called that combined operation Lucent Technologies. Lucent now comprises Bell Laboratories (R&D), Network Systems (develops and manufactures switches, and related systems and software for the carrier market), Business Communications Systems (develops, manufactures, markets and services advanced communications products for business customers), Microelectronics Group (designs and manufactures high-performance integrated circuits, optoelectronic components and power systems), and Consumer Products (designs, manufactures, sells, services and leases both wired and wireless communications products for consumers and small businesses). One reason for the split was that Western Electric's major customers (the RBOCs) were unlikely to purchase equipment and services from their strongest likely competitor (i.e. AT&T) in a deregulated environment, i.e. AT&T. www.lucent.com

Lug Something which sticks out and onto which a wire may be connected by wrapping or soldering.

Luma The brightness signal in a video transmission.

Luminance The measurable, luminous intensity of a video signal. Differentiated from brightness in that the latter is non-measurable and sensory. The color video picture information contains two components: luminance (brightness and contrast) and chrominance (hue and saturation). The photometric quantity of light radiation. Luminance is that part of the video signal which carries the information on how bright the TV signal is to be.

LUNI LANE User Network Interface. Pronounced "looney," as in "Sometimes all these acronyms make me looney." An ATM term for standardized network interface protocol between a LANE (Local Area Network Emulation) client and a LANE Server. See also LANE and LNNI.

LUNS Logical Unit Numbers. An identification number given to devices connected to a SCSI adapter. Each SCSI ID can have eight LUNs. Normally, there is only one device with LUN 0. See DAISY CHAIN and SCSI.

Lurker A person who "hangs around" online bulletin boards and forums, browsing through the messages and, if moved, replying to some of them. That's one definition. Here's another: A visitor to a newsgroup or online service who only reads other people's posts but never posts his or her own messages, thus remaining anonymous.

Lux A contraction of luminance and flux and a basic unit for measuring light intensity. A Lux is approximately 10 foot candles.

LUXPAC A ITU-T X.25 packet switched network operated in Luxembourg by the Luxembourg government.

LZS Lempel-Ziv-Stac. A data compression algorithm developed by Stac Electronics and sometimes used by routers.

LZW The Lempel-Ziv-Welsh compression algorithm is way of reducing the number of bits to transfer. Northern Telecom uses this compression algorithm for its Distributed Processing Peripheral (DPP) — the Automatic Message Accounting Transmitter (AMAT) for the DMS-100 family of central office switches. Northern Telecom selected this non proprietary protocol and helped promote it as an industry standard. The nominal compression ration is 2.8:1, without considering field suppression. Transmitting data compressed at a ratio of 2.8:1 at 9600 bps is equivalent to transmitting non-compressed data at 27 Kbps. Compatible compression collectors can poll the DPP in compressed or non-compressed mode. DPPs equipped with the Data Compression feature can transmit in either compressed or non-compressed mode, based on the collector's polling request for a specific polling session. To preserve data integrity, AMA data are still stored in non-compressed form on the DPP disks. LZW is a lossless data-compression algorithm. See LOSSLESS and LOSSY.

m 1. (small letter) Milli. One-thousandth. M (big letter) Mega. One million, e.g. Mbps or Mbit/s, one million bits per second. 2. Meter. The fundamental metric unit of length, a meter is equivalent to 39.37 inches.

M Bit The More Data mark in an X.25 packet that allows the DTE or DCE to indicate a sequence of more than one packet.

M Hop The transmission of satellite signals through an uplink to a satellite that downlinks to a receiving station halfway or between the final receiving station. The intermediate receiving station uplinks the signal again to a satellite for delivery to the final, targeted receiving station. As an example, data from a company in the UK for a business in Japan may be sent to a satellite covering North America and Europe. The signal is downlinked to an earth station in Colorado. The signal is then uplinked to a satellite covering Asia and North America for final delivery to an earth station in Japan. The signal flow forms an 'M', thus the name.

M Patch Bay A patching facility designed for patching and monitoring of digital data circuits at rates from 1 Mbps to 3 Mbps.

M Port The port in an FDDI topology which connects a concentrator to a single attachment station, dual attachment station, or implemented in a concentrator. This port is only implemented in a concentrator.

M VTS Marconi Video Telephone Standard sends color pictures over regular phone lines at up to 10 frames per second at 14.4 Kbps. The MCI Video Phone, which conforms to this standard, has a resolution of 128 by 96 pixels.

M-ES Mobile End System.

M.100 M.100, a specific API for S.100 Server Configuration and Control, now available from the Enterprise Computer Telephony Forum (ECTF). By managing the configuration, startup and shutdown of S.100-conforming computer telephony servers, M.100 allows Call Centers to provide more consistent service through better-configured, more stable servers and orderly shutdown when problems occur. In its July issue, Computer Telephony Magazine said M.100 "addresses (a) configuration management, (b) performance management, (c) statistic management and (d) fault management across the system/server and application level. M.100 was announced on April 22, 1998. At that time, the ECTF said M.100 allows users to ensure that all of the resources and other elements of a server are brought up in an orderly manner during startup to produce a stable environment. If problems occur during operation, M.100 allows the server to be shut down without loss of data, a critically important benefit to large Call Centers. During a shutdown, M.100 allows all current calls to be handled and prevents additional calls to be taken or created until shutdown has been completed. This enables Call Centers to continue servicing customers without catastrophic interruptions even when problems occur. M.100 employs session and event management, symbols and data types from S.100, and it defines functions that allow administrators and developers to create customized administration applications. These include management of configuration data, management of services, startup and shutdown, information about service providers

and handling of generic administration commands. M.100 also contains the infrastructure hooks that enable inclusion of vendor-supplied diagnostic tools. This feature allows users to easily configure a system with those diagnostic routines needed to create a highly stable environment. Because M.100 is a specific API for S.100 servers, it addresses several key areas not covered by the network management APIs. First, the startup and shutdown of an S.100 server is specific to its operating system environment. If the server were running under Windows NT, it would be started as an NT service, and M.100 would control the startup and the shutdown of the various services. M.100 also supports making persistent information available on the server that can be accessed whenever new applications or services are installed or removed. This data is referred to as a profile that is similar to an INI file. An S.100 server contains multiple profiles that describe all the components that make up the server. M.100 allows for manipulation of the profiles making for much cleaner server configurations and re-configurations. Also, M.100 makes it much easier to handle KVSets, the preferred mechanism for using data within an S.100 server. M.100 is available from the ECTF Web Site at www.ectf.org.

M1 1. Multiplexer in the U.S. digital signal hierarchy. See M1, which is listed as if it were M-ONE in this dictionary. 2. Management Interface 1: The management of ATM end devices.

M2 Management Interface 2: The management of Private ATM networks or switches.

M3 Management Interface 3: The management of links between public and private networks.

M4 Management Interface 4: The management of public ATM networks.

M5 Management Interface 5: The management of links between two public networks.

M12 A designation for a multiplex which interfaces between four DSIs and one DS2 circuit.

M13 The multiplexer equivalent of T-1. In the U.S. digital hierarchy, multiplexers are called by the digital signal levels they interface with. For example, a multiplexer, which joins DS-1 channels to DS-3 is called a M1-3. A M1-3 takes 28 DS-1 inputs and combines them into a single 45 megabit per second stream. (The bit stream is actually 44.736 megabits.)

M24 A T-1 service that allows a user to multiplex up to 24 voice or data channels into a single T-1 link, compatible with AT&T central office based channel banks (M24 compatibility generally refers to compliance with the channelization and coding techniques specified by AT&T TR62411).

M28 The telephone company multiplexing scheme that multiplexes 28 T-1 data streams onto a single carrier system, the T-3.

M34 A designation for a multiplex which interfaces between six DS3s and one DS4 circuit.

M44 A T-1 service that allows up to 44 voice channels (48 without signaling) to operate over a single T-1 link by using ADPCM, and is compatible with AT&T central office based equipment. MJ44 compatibility generally refers to the ability to accept the 44 channel T-1 aggregate and break out one of the channels individually for routing purposes.

MA Abbreviation for MILLIAMP or MILLIAMPERES, unit of electric current.

Ma Bell A term used to refer affectionately to the old AT&T and the old Bell System. Several Women's Lib organizations objected to it some years ago on the basis that there were no women in the higher corporate structure of AT&T, and that women, as over supervised operators, were the downtrodden majority within the Bell System. There was a movement afoot to change it Pa Bell. But then came Divestiture and the breakup of the Bell System. The term, MA BELL, now largely belongs in the history books. And there are now a handful of women in senior management in the Bell operating companies.

MAC 1. Moves, Adds and Changes. When you first install a phone system it will cost money to run wires and install phones all over the building. Very quickly you will notice that you'll need to move people and their phones, add phones for new people and change phones around. This will cost money, often lots of it. How much it will cost you depends on the arrangement you have negotiated with the vendor of your phone system. It's a good idea to get a good deal on Moves, Adds and Changes later on BEFORE you sign your original deal to buy the phone system.
2. IEEE specifications for the lower half of the data link layer (layer 2) that defines topology dependent access control protocols for IEEE LAN specifications. MAC is a media-specific access control protocol within IEEE 802 specifications. It currently includes variations for the token ring, token bus and CSMA/CD. The lower sublayer of the IEEE's link layer (OSI) which complements the Logical Link Control (LLC).

MAC Address The address for a device as it is identified at the Media Access Control layer in the network architecture.

MAC Layer That layer of a distributed communications system concerned with the control of access to a medium that is shared between two or more entities.

MAC Name A MAC name (also called a MAC address) is a 48-bit number, unique to each local area network card, that is programmed into the card, usually at the time of manufacture. Unlike Network Layer Addresses, MAC names are location-independent. Destination and source MAC names are contained in the LAN packet and are used by bridges to filter and forward packets. See also NETWORK LAYER ADDRESS.

MAC Protocol The procedures used to control access to a medium that is shared between two or more entities.

MACs Moves, Adds and Changes.

Machine Code Same as Machine Language. See MACHINE LANGUAGE.

Machine Dependent Software which will only run (i.e. is dependent) on a certain computer.

Machine Directory A Windows 95 definition. For shared installations, the directory that contains the required configuration files for a particular computer. The machine directory contains WIN.COM, the Registry, and startup configuration files.

Machine Language A computer language composed of machine instructions that can be executed directly by a computer without further compilation. Instructions and data coded in binary code. Machine language is the native language of computer hardware. Machine language is the only language recognized by the microprocessor that controls all the operations in your PC. All programs and all data to be processed by your computer (PC, mini or mainframe) have to be translated into machine language at some stage.

Macro 1. MAChine ROutine. An instruction in a source lan-

guage (e.g., FORTRAN) that is equivalent to a specified sequence of machine or assembler instructions. See Macro Language and Machine and Assembler.
2. Macro means very big. Random House's dictionary says it means "very large in scale, scope, or capability." Macro also refers to macroeconomics — the study of bigger things in economics, like the factors that affect the wealth and growth of countries. In contrast, microeconomics focuses on factors that affect the wealth and growth of organizations, such as corporations.

Macro Language A collection of instructions by which any kind of information in the system can be located and manipulated and by which new information types can be added to the system.

Macrobending In an optical fiber, all macroscopic deviations of the fiber's axis from a straight line; distinguished from microbending.

Macrocell macrocell is a new word for what we used to call a "cell," as in cellular radio. This is what some researchers at Bell Northern Research (BNR) wrote: "Today's cellular networks employ macrocells and are optimized to serve users in automobiles, moving at relatively high speeds. Yet, a growing proportion of cellular traffic is originating from users who are not driving in vehicles, but are on foot... If a portion of the radio frequencies in a geographic area were transferred from macrocell (optimized for cars) to microcell technology (optimized for pedestrians), cellular traffic would increase up to a hundredfold. Microcell networks will entail the deployment of many more transceivers than today's macrocell systems. However, microcell equipment will be less costly because it is low power, simpler and smaller, say the researchers at Bell Northern Research.

Macros A common name for a microcomputer software program which lets you alter the definitions of what the keys on your computer keyboard are. With a "macro" software program, you could change the letter "M" on your keyboard to type "Michael" every time you hit it. But this would be stupid. Better to hit a combination of letters to get "Michael." Most personal computers have extra non-alphabetic keys, like Control and Alternate. For example, type Ctrl-Alt M and bingo, the machine types "Michael Newton is a good son."

Macstar An old AT&T 3B2 computer-based software system that interfaces with the Remote Memory Administration System (RMAS) to effect customer moves and rearrangements on the 5ESS switch. The user needs only a terminal and printer.

Mad as a hatter In the 19th century, workmen who used mercury, a poison, to cure beaver skins for top hats over time developed nervous twitches, drooled and spoke incoherently. Thus the expression, "mad as a hatter."

MADN Multiple Appearance Directory Number.

MAE Metropolitan Area Exchange or Ethernet. Both Exchange and Ethernet are correct, but there's an ongoing Internet industry argument over which is more appropriate. A MAE is a Network Access Point (NAP) where Internet Service Providers (ISPs) connect with each other. The original MAE was originally called MERIT Access Exchange. MERIT was a backbone Internet Access Provider which was acquired by MFS, which then was acquired by MCI, which then was acquired by Worldcom to become MCI Worldcom. The original MAE — namely MAE-East — is based in suburban Washington, D.C. (Vienna, Virginia). MAE-West was MFS's second MAE, located in San Jose, CA. Although MAE refers really only to the NAPs from MFS, the two terms are often used interchangeably. MAE West is made up of two networks: an ATM network that can switch a billion bits per second, and an FDDI ring that runs at 100 Mbits per second. Companies connect to these networks by Ethernet, FDDI, or ATM over OC3. Physically a MAE is a building with zillions of wires, gigantic switches and computers with routing tables containing the location of ISPs and how to get to them. MAEs are run by networking companies and carriers, and ISPs pay a fee to these networking companies to locate equipment there and for the transmission, switching and interconnect services. For more on MAEs and charts of the traffic they carry, see www.nap.net/where/w_mae-east.shtml and www.mae.net/east.html See also MAE-East. For a good explanation of how all the Internet infrastructure fell (and was hammered) into place. www.ieng.com/documents/one.isp-con.81096/index.html

MAE-East MAE stands for Metropitan Area Exchange or Ethernet. Both Exchange and Ethernet are correct. MAE is a huge interconnection point for Internet Service Providers (ISPs). They use MAEs for routing their internet traffic from customers on their network to Internet sites on other peoples' Internet networks. There are two MAEs — MAE-East in Vienna, Virginia and MAE-West in San Jose, California. Physically a MAE is a building with zillions of wires, gigantic switches and computers with routing tables containing the location of Internet sites and how to get to them. MAEs are run by networking companies (MAE-East and MAE-West are run by MFS) and ISPs pay a fee to these networking companies to locate equipment there and for the transmission, switching and interconnect services. See also MAE-East++ and MERIT.

MAE-East++ An expansion of MAE-East, located in a nearby building. i++ (plus plus) gets its term from the C programming language where it means "use the current variable i and add 1." Plus plus has come to mean an expansion, an improvement, an upgrading etc. See MAE-East.

MAE-West See MAE-East.

MAE-West++ An expansion of MAE-West, located in a nearby building. i++ (plus plus) gets its term from the C programming language where it means "use the current variable i and add whatever one to it." Plus plus has come to mean an expansion, an improvement, an upgrading, etc. See MAE-East.

Magalog A mail order catalog disguised as a magazine in the hopes of confusing its recipients.

Magazine Hardware unit, mounted in a frame, containing printed board assemblies.

MAGIC Multidimensional Applications and Gigabit Internetwork Consortium One of the Information Superhighway projects funded by the Information Technology Office (ITO) of the Defense Advanced Research Projects Agency (DARPA) and the National Science Foundation (NSF). MAGIC is a gigabit-per-second ATM-based network connecting various high-tech research and development sites in Minneapolis (MN), Sioux Falls (SD), Lawrence (KS), Kansas City (KS) and Ft. Leavenworth (KS).

Magic Wand One of the devices used by con men when pretending to do TSCM (Technical Surveillance CounterMeasures). A magic wand is typically a field strength meter or box with many fancy lights.

Maggots Telco slang for B-connectors, which are about the size and shape of fly larvae.

Magnetic Bubble A device in which information is stored in a magnetic film as a pattern of oppositely directed magnetic fields. Magnetic bubble devices hold their memory even if you lose power.

Magnetic Disk A computer storage device that records data bits as tiny spots on magnetic-coated disk platters. Hard disks

and floppy disks are variations on magnetic disks. See also MAGNETO OPTICAL and HOLOGRAPHIC DATA STORAGE.

Magnetic Ink An ink that contains particles of a magnetic substance whose presence can be detected by magnetic sensors. Typical is the ink on your checks, which carry your name, your account number and the check's number.

Magnetic Medium Any data-storage medium and related technology including diskettes and tapes, in which different patterns of magnetization are used to represent the values of stored bits and bytes.

Magnetic Storage Any medium (generally tape or disk) upon which information is encoded as variations in magnetic polarity. The hard disk on your computer is magnetic storage.

Magnetic Stripe A strip of magnetic material, usually tape, attached to a credit card containing data relating to the card holder. You have a magnetic stripe now on the back of most of your credit cards. That stripe tells a computer who you are, what your account number is, etc.

Magnetic Tape A tape made of magnetic material upon which data may be stored for later retrieval by a computer.

Magneto A small hand-cranked AC generator which uses permanent field magnets and can make electricity to ring telephone bells.

Magneto Optic Relating to the change in a material's refractive index under the influence of a magnetic field. Magneto-optic materials generally are used to rotate the plane of the polarization. This phenomenon is how magneto optical disk drives work.

Magneto Optical Drive A computer data storage device that writes data using magnetism (in the form of a magnetic field called the bias field) and light (a laser beam) to write to a disk that resembles a CD-ROM disk. A magneto optical drive holds huge amounts of information — as much as 500 megabytes on a single disk. Introduced in 1988, the magneto optical drive (also spelled with a dash between the magneto and optical), the drive provides the convenience of the removability of floppies and the Bernoulli Box, the random access convenience of hard disk, the reliability of CD-ROMs and the promise of DAT-like capacity. But, according to PC Magazine, before you rush out and buy this ultimate storage solution, note that they're expensive, can't provide as much storage on one side of a disk as the largest hard disks and they're slower than today's hard disks. An explanation of how magneto optical drives work, courtesy PC Magazine: The recording layer on the disk stores the equivalent of binary 1s and 0s in the magnetic domains. The disk is designed so that the bias field by itself is to too weak to change the polarity of the magnetic domains. But when a spot on the disk is heated by a high-powered laser beam, its resistance to changing polarity drops. The bias field can now change the disk area's polarity. To read the disk, the drive uses a laser beam that is not hot enough to allow the bias field to change the disk area's polarity.

Magneto Phone A magneto phone has a crank handle on the side and dial pad. It is a very old phone — typically manufactured around the turn of the century. It is totally manual. You crank the handle which turns a bunch of permanent magnets on the inside which in turns generates electricity which in turn is used to ring a bell at the central office. When that bell rings, an operator answers, asks where the person wants to dial, dials and connects the call.

MAHO Mobile Assisted Hand-Off. A process in which the wreless mobile station assists the base station in assigning a voice channel by reporting its surrounding F signal strengths to the base station.

MAHR Abbreviation for MILLIAMPERES HOUR, 1/1000th of an ampere hour. Term is commonly used with small rechargeable battery packs, such as those used by portable phones and laptop PCs.

Mail Bomb The flooding of an e-mail address with frequent messages, often done as an act of protest or harassment. See also BOZO FILTER.

Mail Enable Applications Applications that use mail as a way of addressing and transporting information to and from users on a network.

Mail Exploder Part of an electronic mail delivery system which allows a message to be delivered to a list of addresses. Mail exploders are used to implement mailing lists. Users send messages to a single address (e.g., smith@somehost.edu) and the mail exploder at somehost.edu takes care of delivery to the individual mailboxes on the lists, including Smith.

Mail Filter A piece of software which lets a user sort his/her email messages according to information in the header.

Mail Gateway A machine that connects two or more electronic mail systems (especially dissimilar mail systems on two different networks) and transfers messages between them.

Mail Path A series of machine names used to direct electronic mail from one user to another.

Mail Reader Software which enables a user to select unread electronic mail and unread conferences messages and have them downloaded for reading off-line. Most mail readers also permit users to create responses off-line and upload them at their convenience.

Mail Reflector An Internet term. A special mail address; electronic mail sent to this address is automatically forwarded to a set of other addresses. Typically, used to implement a mail discussion group.

Mail Server Mail Server is the "post office" of a messaging network. Mail server is a computer host and its associated software that offer electronic mail reception and (optionally) forwarding service. Users may send messages to, and receive messages from, any other user in the system.

Mailbot An email server that automatically responds to requests for information and sends that information by return email.

Mailbox Messages belonging to a single owner in a voice mail system. Today, these will be recorded voice messages, but increasingly mailboxes will include E-mail and fax documents. See Unified Messaging.

Mailer Also known as the front end; a program that allows BBSs (bulletin board systems) using different software to "talk" to each other. The mailer acts as a transfer layer for the messages passed. Some BBSs require mailers to talk to some netmail systems (for example, Spitfire for FidoNet). Seadog is generally acknowledged to be the market leader, according to PC Magazine.

Mailer Daemon The Mailer-Daemon is not a person, but rather a computer program on Internet which runs automatically to perform the service of telling you why it could not deliver your message.

Mailgram An overnight electronic mail service of Western Union. The letter is phoned in or sent by computer to a central Western Union computer, from where it is sent to teleprinter machines located in post offices in major cities. When it's printed at the post office, the Mailgram is placed in an envelope and hand delivered by your friendly, local mailman in next day's mail.

Mailto A piece of HTML code that sits on a Web site and, through your browser, displays an email address underlined and in blue. When you click on this email address, this mailto HTML code automatically launches your email program and automatically inserts the email address. Mailto HTML code looks like this <AHREF=mailto:Harry@HarryNewton.com>Harry@HarryNewton.com</A>. In this line, the second email address is what the person browsing your Web site will set in blue and underlined. When he clicks on it, the first email address will be what's dropped into the email as the address to send to. Typically the two are the same. But you could easily write it as: <AHREF=mailto:Harry@HarryNewton.com>email to Harry Newton</A>. In which case, "email to Harry Newton" in blue and underlined would appear when viewed through a browser by someone visiting my Web site.

Main PBX or Centrex switch into which other PBXs or remote concentration of switching modules are homed. A PBX or Centrex connected directly to an electronic tandem switch (ETS). Also, a power source.

Main Cross-Connect The interconnect point where wiring from the Entrance Facility and from the Workstation is connected to telecom equipment.

Main Distribution Frame MDF. A wiring arrangement which connects the telephone lines coming from outside on one side and the internal lines on the other. A main distribution frame may also carry protective devices as well as function as a central testing point. See MAIN DISTRIBUTION FRAME FILL, DISTRIBUTION FRAME and FRAME.

Main Distribution Frame Fill The central office mainframe is the termination point for outside plant cables. The "fill" is the percentage of pairs used by customers of the total number of pairs on the frame. Optimum fills vary based on the size of the central office and the amount of growth in the area. A low fill means idle lines and wasted investment in outside plant. A high fill, plus unexpected growth, forces budget busting and crisis construction projects.

Main Feeder Feeder cable that transports pairs from the central office to branching or taper points.

Main Lobe The main lobe is the area with the maximum intensity in the pattern of radiation produced by an antenna. One presumes it's called "lobe" because the pattern in a microwave signal of the main lobe typically looks like a ear lobe.

Main Memory The principal random storage area inside the computer. Used for storing data and programs and under the direct control of the CPU — the main processor. Also called RAM memory.

Main Network Address In IBM's SNA, the logical unit (LU) network address within ACF/VTAM used for SSCP-to-LU sessions for certain LU-to-LU sessions. Compare with auxiliary network address.

Main PBX A main PBX is one which has a Directory Number (DN) and can connect PBX stations to the public network for both incoming and outgoing calls. A main PBX can have an associated satellite PBX, and can be part of a tandem tie trunk network (TTTN). If the main PBX provides tandem switching for tie trunks, it is called a tandem PBX. In the context of ESN (Electronic Switched Network), a main PBX has tie trunks to only one node. See PBX.

Main Satellite Service A PBX feature that allows multi-location customers to concentrate their attendant positions at one location referred to as the Main. Other unattended locations are referred to as Satellites.

Main Service Entrance In AC electricity, the main service entrance is the necessary equipment, usually consisting of main circuit breakers or fuses, a switch and branch circuit breakers or fuses, in a grounded enclosure (panel) connected directly to earth. Located in the building at the point of entrance of the supply conductors from the power utility. Other panels in the building are referred to as branch, service or supply panels.

Main Station A subscriber's telephone instrument, terminal or workstation used to originate and receive calls. Very often if two instruments have the same extension number (are bridged), one becomes the Main Station and the other is a bridged station for inventory purposes. See Main Station to Line Ratio.

Main Station To Line Ratio A telephone company term. The ratio of main stations to lines. This ratio will normally be greater than 1.0 because of 2 and 4 party service, etc.

Main Terminal Room The location of the cross-connect point between the incoming cables from the telecommunications external network and the premises cabling system.

Mainframe A powerful computer, almost always linked to a large set of peripheral devices (disk storage, printers, and so forth), and used in a multipurpose environment at the corporate or major divisional level. A mainframe is a large-scale computer typically containing hundreds of megabytes of main memory and hundreds of gigabytes of disk storage. It is capable of "serving" thousands of "on-line" terminals. The term — main frame — derives from the racks that typically hold a large computer and its memory.

Mainframe Chiller System Water-cooled mainframe computers rely on mainframe chillers for a continuous supply of liquid coolant to maintain processor temperature within a specified range. Exceeding the temperature specifications or an interruption of coolant flow can cause a sudden shutdown, interrupting of computer operations, and possible hardware damage, requiring costly repairs.

Mainframe Gateway A hardware/software system that allows PCs on a LAN (Local Area Network) to communicate with a mainframe. A single, usually dedicated, PC acts as the gateway. PCs on the LAN share its hardware and its communication link, communicating with it over the LAN cable. The most common mainframe gateway is an SNA gateway, which hooks a LAN into an IBM mainframe.

Mainframe Server Clients are devices and software that request information. Client is a fancy name for a PC on a local area network. It used to be called a workstation. Now it is the "client" of the server. A mainframe server is a large computer that stores lots of information and manages libraries of information. Here's a definition of Thin Client, courtesy of Oracle Corporation, writing in early 1994: "Mainframe systems store lots of data, but they're expensive, slow and difficult to use. Because all the processing happens on one large computer, they can't move large amounts of multimedia information to large numbers of users. Example, the IBM ES/9000, Amdahl's 5995-1400 or any plug compatible mainframe." See also CLIENT, CLIENT SERVER, CLIENT SERVER MODEL, FAT CLIENT and MEDIA SERVER.

Mains Some countries call their normal commercial power outlets — "mains." In Europe the frequency of commercial power is 50 Hz. In the United States, its frequency is 60 Hz. It's hard to convert the frequency of commercial power. It's easier to convert voltage. In Europe and Australia, normal voltage is 240 volts. In the U.S., it's 120 volts.

Mains Modem A modem which is part of a system called remote metering which monitors electricity usage and allows electric companies to offer such services as electronic mail, burglar alarms and energy management. The idea of energy management is that if the electric companies could turn off unnecessary appliances for a few hours during peak times, they might not have to build expensive new power stations. In exchange for that favor, they undoubtedly would be prepared to offer their customers price reductions.

Maintenance 1. All work needed to keep the telephone system operating properly, including periodic testing, repairs, etc. See PREVENTIVE MAINTENANCE.

2. All work needed to keep a software program operating properly, operating on new machinery and operating with new management needs. Often, software maintenance means substantially rewriting the original software program. Most of the work done by data processing departments in large companies involves maintaining old programs. This is not a put-down.

Maintenance Acceptance A term used in the secondary telecom equipment business. The point at which a maintenance company has tested a system, component, or peripheral device and determined that it meets manufacturer's specifications. The product can now be added to a maintenance contract. Once under contract, the maintenance company is responsible for repairing or replacing any defective components.

Maintenance Contract Contract guaranteeing the repair of a PBX switch to support it at operational levels for a predetermined fixed term and fixed price.

Maintenance Control Center MCC. A central place in a stored program control central office from which system configuration and trouble testing are controlled.

Maintenance Control Circuit MCC. A voice circuit used by maintenance personnel over microwave links for coordination. This is not available to operations or technical control personnel.

Maintenance Hole A vault located in the ground or earth as part of an underground duct system and used to facilitate placing, connectorization, and maintenance of cables as well as the placing of associated equipment, in which it is expected that a person will enter to perform work. Also called a manhole.

Maintenance Release An euphemism for a new piece of software that fixes a buggy piece of old software. When software is released, it's usually buggy. A maintenance releases attempts to fix the bugs. A more correct term for the new software would be a "bug fix." A maintenance release often carries the number one as the second digit after the decimal point, e.g. 3.51 or 4.01.

Maintenance Services In IBM's SNA, network services performed between a host SSCP and remote physical units (PUs) that test links and collect and record error information. Related facilities include configuration services, management services and session services.

Maintenance Termination Unit MTU. A MTU is an electronic circuit that is owned and deployed by a telephone company to aid in fault sectionalization and is installed at the network interface. The MTU should meet the requirements of Bellcore Technical Advisory TSY-000324 and be testable with the Mechanized Loop Test System (MLT). The MTU is designed to work on single line residence or business service.

Maintenance Update A euphemism for a piece of software which fixes bugs in a previously-released version of the software. A maintenance update rarely has any new features and rarely costs anything. Software companies send them because they find bad bugs in their software and want to fix those bugs asap, or because they can't stand the heat from complaining customers.

Maintenance Usage A telephone company term. The amount of time, measured in CCS that equipment components are removed from service. Can be caused by equipment malfunction, routine maintenance, transitions, etc.

Majordomo A Majordomo is a mail-processing software program. It maintains multiple mailing lists. It automatically interprets (i.e. understands) commands in emails sent to it by individuals who wish to subscribe to, unsubscribe from, receive periodic summaries of, or otherwise become associated with a mailing list. It also handles the mass mailing of messages to members of the list. Most MajorDomo applications are unix based. You find Majordomo all over on the Internet. For example of how Majordomo mailing lists work, here's a note on Subscription Information for an electronic newsletter called WinNews published by Microsoft. "If you know someone who might be interested in WinNews, please instruct them to:

1. Send Internet e-mail to: ENEWS99@ MICROSOFT.NWNET.COM

2. Send the message from the account that you wish to subscribe (some people use more than one e-mail account).

3. Subject line should be blank.

4. Body of message should ONLY have in the text: SUBSCRIBE WINNEWS If you wish to stop receiving WinNews, send mail to

enews@microsoft.nwnet.com with a blank subject line and the body of the message should only save in the text: UNSUBSCRIBE WINNEWS."

Make Busy To make a communication circuit unavailable for connection. The technical term for taking the phone off the hook and leaving it off hook.

Malicious Call Tracing An ISDN service which enables to User to Network message to be sent while the call is in progress, ensuring that origination details are captured at the local exchange.

MAN Metropolitan Area Network. A high-speed data intracity network that links multiple locations within a campus, city, or LATA. Typically extends as far as 50-kilometers, operates at speeds from 1 Mbit/s to 200 Mbps and provides an integrated set of services for real-time data, voice and image transmission. The IEEE 802.6 standard defines MAN standards and SMDS is the MAN service offered by local phone companies. Private and public MANs may use the ANSI FDDI standard.

Man Machine Interface A term coined by James Martin to designate the ease (or lack of ease) of a person working with a computer.

Man Page Manual page. On-line documentation that commonly comes bundled with computers running Unix.

Managed Object A telephone company AIN term. If you understand this definition, you're a better person than I am. Given a Common Management Information Services Element (CMISE) interface between an Operations System and a network element or network system, a managed object is an abstract representation of a physical or logical network element or network system resource. The managed object constitutes an Operations Systems' view of that resource from the CMISE operations system and a network element or network system interfere. It can also be called a Managed Object Instance to emphasize the distinction between Managed Objects and Managed Object Classes.

Managed Object Class A telephone company AIN term. A group of managed objects that all have the same types of attributes, same permissible ranges of attribute values, same semantics and pragmatics for interpretation of Common Management Information Services Element (CMISE) requests, and the same capabilities for issuing event reports.

Managed System An entity that is managed by one or more management systems, which can be either Element Management Systems, Subnetwork or Network Management Systems, or any other management systems.

Management Domain MD. An X.400 term describing a set of messaging systems. At least one system contains, or realizes, an MTA (Messaging Transfer Agent) managed by a single organization. It is a primary building block in the organizational construction of an MHS (Messaging Handling System), referring to an organizational area for the provision of messaging services. MD is used in a similar manner in X.500, consisting of at least one DSA (Directory System Agent). A management domain may or may not be identical with a geographical area. See DOMAIN.

Management Extranet A management extranet uses the Web to electronically join entire supply chains, so the IS department can tie itself via the Web to its suppliers and service providers. See also Management Intranet.

Management Information Base See MIB.

Management Information System MIS. Management information provided by computer data processing. Once upon a time called data processing.

Management Intranet A management intranet uses World Wide Web technologies and techniques to integrate disparate management tools and databases, provide universal access to documentation and promote distributed collaboration among far-flung IS support people. See also Management Extranet.

Management Plane An element of the ATM Protocol Reference Model, the Management Plane addresses the management of the ATM switches and hubs, cutting through all 4 layers of the model. Included in Management Plane functions are Operation, Administration and Maintenance (OA&M) functions.

Management Services In IBM's SNA, network services performed between a host SSCP and remote physical units (PUs) that include the request and retrieval of network statistics.

Management System An entity that manages a set of managed systems, which can be either NEs, subnetworks or other management systems.

Manchester Encoding A digital encoding technique in which each bit period is divided into two complementary halves. A negative-to-positive (voltage) transition in the middle of the bit period designates a binary "1" while a positive-to-negative transition represents a "0". This encoding technique is self clocking (the receiving device can recover transmitted clock from the data stream).

Mandatory Dialing When permissive dialing is over after an area code change, it becomes mandatory dialing. See Permissive Dialing.

Mandrel Wrapping A mandrel is a cylindrical shaft or bar. In machining, a mandrel is inserted into the piece you are working on. This holds it in place during machining. In multimode fiber optics, mandrel wrapping is a technique used to modify the modal distribution of a propagating optical signal. Basically you wrap a specified number turns of fiber on a mandrel of specified size, depending on the fiber characteristics and the desired modal distribution. It has application in optical transmission performance tests, to simulate, i.e., establish, equilibrium mode distribution in a launch fiber (a fiber used to inject a test signal in another fiber that is under test). If the launch fiber is fully filled ahead of the mandrel wrap, the higher-order modes will be stripped off, leaving only lower-order modes. If the launch fiber is underfilled, e.g., as a consequence of being energized by a laser diode or edge-emitting LED, there will be a redistribution to higher-order modes until modal equilibrium is reached.

Manhole An underground concrete vault in which cables may be spliced, and transmission equipment (repeaters, etc.) may be located. A manhole is used in conjunction with an underground cable running in conduits.

Manual Exclusion A PBX extension user, by entering a certain code, can block all other phones on that line from entering the call. Assures privacy on the line.

Manual Gain Control MGC. There are two electronic ways you can control the recording of something — Manual or Automatic Gain Control (AGC). AGC is an electronic circuit in tape recorders, speakerphones and other voice devices which is used to maintain volume. AGC is not always a brilliant idea since it attempts to produce a constant volume level. This means it will try to equalize all sounds — the volume of your voice and, when you stop talking, the circuit static and/or general room noise which you undoubtedly do not want amplified. Sometimes it's better to have quiet, when you want quiet. Manual Gain Control is preferred in professional applications. Manual Gain Control is simply an elegant way of saying there's a record volume control. Never record a seminar or speech using AGC. The end result will be decidedly amateurish.

Manual Hold The method of placing a line circuit on "hold' by activating a non-locking "hold" button on the phone, usually one colored red.

Manual Intercom A crude, single-path communications link between telephones without the ability to signal the receiving party.

Manual Modem Adapter An external device for the Merlin key system from AT&T. It allows connection of single line accessories to any Merl n telephone. The device, in effect, draws a standard tip and ring line out of the Merlin proprietary cabling/signaling scheme. Some other key systems have similar devices. Comdial calls theirs a "data port" and their phones contain extra RJ-11 jacks.

Manual Originating Line Service The attendant must complete all outgoing calls. All other calls are blocked. This "feature" is used to cut down on long distance phone abuse. There's a wonderful story. When many of the PBXs in Europe went from manual originating line service to automatic dial "9" long distance, the number of long distance calls doubled within two months. Some of these calls were legitimate. Some were not. How much abuse there was varied from company to company. Typically, those companies with employees who were more bored suffered (or enjoyed?) more abuse.

Manual PBXs Refers to PBXs which are not automatic and which require that all calls, including intercom calls, be placed through the attendant. Such PBXs are still used today, though in limited applications. You can still find manual PBXs in vacation hotels, nursing homes and in the data communications departments of some firms, who use manual PBXs as manual dataPBXs These are especially useful in places where long data calls and sold metal-to-metal connections are an advantage.

Manual Ring Down Line Two phones connected by a pair of wires and a battery. Signaling is performed manually by flipping a switch on and off which connects and disconnects the battery. This causes a weak ringing. It's used by rescue teams in caves and mines because radio range is often limited.

Manual Ring Down Tie Trunk A direct talk path between two distant phones Signaling must be done manually from either phone. Contrast this with Automatic

Ringdown Tie Trunk, in which the signaling occurs the moment one of the phones is lifted off hook.

Manual Signaling Pushing a button on a telephone sends an audible signal to a predetermined phone. Manual signaling can be used for secretary/boss communications.

Manual Telephone A telephone without a dial. Taking the receiver off hook automatically rings a predetermined number. A courtesy phone.

Manual Terminating Line Service Provides extension lines that require all calls be completed by the attendant. For a better explanation see MANUAL ORIGINATING LINE SERVICE.

Manufacturing And Automation Protocol MAP. A protocol initially developed as an internal specification for its own factory floor equipment and now championed by General Motors as the industry standard to facilitate communications among the diverse automation devices found in production environments. AT&T, IBM and DEC have endorsed this standard and have already or will introduce MAP-compatible products. TOP (Technical and Office Protocol) was initiated by Boeing Computer Services (one of the nine companies that helped form the MAP Users Group in 1984) and is designed for use in the engineering and office environment and to move information from the factory floor to other parts of the company. Implementation of these protocols would lead to GM's factory of the future concept.

Manufacturing Message Format Standard An Application Layer protocol developed as a part of MAP to provide a syntax for exchanging messages in the manufacturing environment.

Manufacturing Message Specification MMS. An International Standards Organization (ISO) application layer protocol that defines the framework for distributing manufacturing messages within a network. This specification is used in MAP 3.n.

MAP 1. A new term for multiplexing, implying more visibility inside the resultant multiplexed bit stream than available with conventional asynchronous techniques.

2. Mobile Application Part. As defined by IS-41 (Interim Standard 41) a User Part of the SS7 protocol used in wireless mobile telephony. MAP standards address registration of roamers and intersystem hand-off procedures. As a query-and-response procedure, MAP makes use of TCAP (Transaction Capabilities Application Part) over the SS7 network. See also IS-41, SS7 and TCAP.

3. Maintenance and Administration Position. See MAP/MAAP below.

4. Manufacturing Automation Protocol. A protocol initially developed as an internal specification for its own factory floor equipment and now championed by General Motors as the industry standard to facilitate communications among the diverse automation devices found in production environments. AT&T, IBM and DEC have endorsed this standard and have already or will introduce MAP-compatible products. TOP (Technical and Office Protocol) was initiated by Boeing Computer Services (one of the nine companies that helped form the MAP Users Group in 1984) and is designed for use in the engineering and office environment and to move information from the factory floor to other parts of the company. Implementation of these protocols would lead to GM's factory of the future concept.

MAP/MAAP Maintenance and Administration Panel. A device attached to a PBX to allow you to maintain and administer the system — to change phone features, etc.

MAP/TOP Manufacturing Automation Protocol/Technical Office Protocol.

MAPI Microsoft's Windows Messaging Application Programming Interface, which is part of WOSA (Windows Open Services Architecture). MAPI is a set of API functions and as OLE interface that lets messaging clients, such as Microsoft Exchange, interact with various message service providers, such as Microsoft Mail, Microsoft Exchange Server, Microsoft Fax and various computer telephony servers running under Windows NT server. Overall, MAPI helps Exchange manage stored messages and defines the purpose and content of messages — with the objective that most end users will never know or care about it. A friend of mine, who's a great programmer, Pete MacLean, explained MAPI as: MAPI is Microsoft's new foundation for a modular mail system. You can pick and choose among various email clients, address books, message stores (foldering systems), and transports (the message-service specific pieces) and build your own custom mail system. See also AT WORK, MICROSOFT EXCHANGE, WINDOWS 95, WINDOWS TELEPHONY and WOSA. The biggest explanation of MAPI is in the definition for WINDOWS 95.

Mapping 1. In network operations, the logical association of one set of values, such as addresses on one network, with quantities or values of another set, such as devices on another network (e.g. name-address mapping, internetwork-route mapping).
2. A Novell NetWare term. To assign a drive letter to a chosen directory path on a particular volume of a particular file server. For example, if you map drive F to the directory SYS:ACCTS\RECEIVE, you will access that directory every time you enter "F:" at the DOS prompt. See also DRIVE MAPPINGS.
3. In EDI (Electronic Data Interchange), mapping defines the translation between a company's unique data layout and an EDI formal structure.

Marathon A family of products that are combination fast packet multiplexer, data compression, voice compression and fax de-modulation devices that fit many, voice, data, fax and LAN "conversations" onto one leased circuit — analog or digital. The idea of Marathon is to save money on long distance telecommunications charges. The Marathon family of products is made by Micom Communications Corporation, Simi Valley CA, now a subsidiary of Northern Telecom (Nortel).

Marconi, Guglielmo Guglielmo Marconi, born in Bologna, Italy in 1874, was on a holiday when he read of the electromagnetic wave experiments of Hertz. This article established the thought in Guglielmo's mind that electromagnetic waves could free telegraphy from the wires and submarine cables, which at that time constrained its use. Finding out if electromagnetic waves could be used to communicate at a distance became an obsession for Marconi. His mother allowed him to use two large rooms on the top floor of their house as a laboratory. She also helped persuade Guglielmo's father to provide (albeit grudgingly) the money necessary for the batteries, wire and other equipment Guglielmo needed. Marconi started by repeating Hertz's experiments. His oscillator was an induction coil equipped with four spheres for the spark discharge. The frequency of the oscillations was in what we, today, call the VHF range. The detector he used with his receiving coil was a Branly coherer, similar to that used by Oliver Lodge. The coherer provided much greater sensitivity than the spark-gap equipped loop of wire Hertz had used. Marconi placed a curved metal detector behind his oscillator to direct the waves toward the detecting circuit. Soon, Marconi was able to cause a bell, located thirty feet away, to ring

when the oscillator was keyed. Through trial-and-error experimentation, he was able to increase the sensitivity of the coherer significantly over what others had achieved. The following spring, Marconi took his experiments outdoors. Connecting metal plates to the oscillator's spark gap lowered the frequency and strengthened the intensity of the oscillations produced. Similar plates were connected to each side of the coherer. By chance, Marconi found that if one of the metal plates was elevated high in the air and the other was laid on the ground, the range at which oscillations could be detected increased to over one-half mile. Soon, the elevated plates at the oscillator and detector were replaced by long vertical wires. The plates which had lain on top of the ground now were buried. This arrangement increased the distance at which signals could be received to one and one-quarter miles. An intervening hill was found to be no barrier to the reception of the signals. The combination of using lower-frequency oscillations and using the Earth as an element in his antenna system were crucially important achievements. Another demonstration was held in March of 1897. This time longer wavelengths were used in conjunction with wire antennas raised some 120 feet above the ground by means of kites and balloons. This arrangement resulted in signals being received over a distance of four and one-half miles. In May of 1887, Marconi demonstrated that wireless signals could span significant lengths across water by sending signals between the shore and an island in the Bristol Channel, a distance of 8.7 miles. This was a crucial test because the submarine cable that normally provided communications to the island had failed several times in recent months. Repairing the cable was costly both in time and in money so Marconi's system must have appeared as an excellent alternative. Marconi established the Wireless Telegraph and Signa. Ltd. in July of 1897. In 1899, he changed the name of his company to The Marconi Wireless Telegraph Co. Ltd. A major goal Marconi had in mind was to show the value of wireless for communicating with ships. In 1897, he returned home to Italy to convincingly demonstrate that wireless could communicate between naval warships. The Italian Navy soon adopted the Marconi wireless system. In 1896, in England, Marconi obtained the first patent on the wireless. In 1901 he succeeded in transmitting signals across the Atlantic. In 1909 he received jointly with C. F. Braun the Nobel Prize in Physics. Marconi was made a Marchese and a member of the Italian senate. He died in Rome on July 20, 1937. See also TESLA, Nikola.

Margaret From the Greek "margaron," translating to "pearl" in English. Margaret Hinano Horak is the lovely wife of Ray Horak, my Contributing Editor. Margaret is one-quarter Hawaiian, which makes the "pearl" thing especially appropriate. "Hinano" is the name her mother gave her. Margaret's mother was half Hawaiian, with some Swedish and Seneca Indian mixed in, but leaned very much toward the Hawaiian side. The meaning of "Hinano" is a bit unclear. It seems to refer to a native Hawaiian flower or plant, which apparently is extinct. Neither "Margaret" nor "Hinano" can adequately describe this loveliest of women. She has the luminescence of the finest pearl and the delicacy of the loveliest of flowers. Margaret is one of a kind, and Ray is the luckiest man ever to walk the face of the earth. (You can bet who wrote this objective definition.) See Ray.

Marginal Cost The cost of supplying an extra unit of output. The telecommunications transport business is the only one in the world where the marginal cost of providing an extra unit of product (i.e. a phone call) is zero. This makes for wonderful economics once your network is in place.

Marine Telephone Marine telephones operate on assigned

radiotelephone frequencies much as a radio broadcast does. Marine telephones can be used to contact other marine telephones or to reach land-based telephones through an operator.

MARISAT A satellite for marine use. Conversations on MARISAT are crystal clear. Call Comsat and ask them for a demo call to a ship somewhere in the world. It's very exciting.

Marine Broadcast Station A coast station which makes scheduled broadcasts of time, meteorological, and hydrographic information.

Marine Utility Station A station in the maritime mobile service consisting of one or more hand-held radiotelephone units licensed under a single authorization. Each unit is capable of operation while being hand-carried by an individual.

Maritime Air Communications Communications systems, procedures, operations, and equipment that are used for message traffic between aircraft stations and ship stations in the maritime service. Commercial, private, naval, and other ships are included in maritime air communications.

Maritime Broadcast Communications Net A communications net that is used for international distress calling, including international lifeboat, lifecraft, and survival-craft high-frequency (HF); aeronautical emergency very high-frequency (VHF); survival ultra high-frequency (UHF); international calling and safety very high-frequency (VHF); combined scene-of-search-and-rescue; and other similar and related purposes. Basic international distress calling is performed at either medium frequency (MF) or at high frequency (HF).

Maritime Mobile Satellite Service A mobile satellite service in which mobile earth stations are located on board ships; survival craft stations and emergency position-indicating radiobeacon stations may also participate in this service.

Maritime Mobile Service A mobile service between coast stations and ship stations, or between ship stations, or between associated on-board communication stations; survival craft stations and emergency position-indicating radiobeacon stations may also participate in this service.

Maritime Radio Navigation Satellite Service A radionavigation-satellite service in which earth stations are located on board ships.

Maritime Radio Navigation Service A radio navigation service intended for the benefit and for the safe operation of ships.

Mark 1. A term that originated with the telegraph. It currently indicates the binary digit "1" (one) in most coding schemes. A space is zero in most coding schemes.

2. A call center term. To flag a record in a browse listing for some special purpose. Typically you mark a record that you want to copy information from. As long as the record is marked, you can continue copying the information to other records.

Marker The logic circuitry in a crossbar central office that controls call processing functions.

Marker Beacon A transmitter in the aeronautical radio navigation service which radiates vertically a distinctive pattern for providing position information to aircraft.

Marker Tape A tape laid parallel to the conductors under the sheath in a cable imprinted with the manufacturer's name and the specification to which the cable is made. Other information such as date of manufacture may also be included.

Marker Thread A colored thread lain parallel and adjacent to the strands of an insulated conductor which identifies the cable manufacturer. It may also denote a temperature rating or the specification to which the cable is made.

Market Price Prices set at market rates but, in most cases, are not permitted to be less than cost.

Marking Bias The uniform lengthening of all marking signal pulses at the expense of all spacing pulses.

Mark-Hold The normal no-traffic line condition where a steady mark is transmitted.

Markup Special codes in a document that specify how parts of it are to be processed by an application. In a word-processor file, markup specifies how the text is to be formatted; in an HTML document, the markup specifies the text's structural function (heading, title, paragraph, and so on.)

Marquee A region on a Web home page that displays a horizontally scrolling message.

Martian Packets that turn up unexpectedly on the wrong network because of bogus routing entries. Also used as a name for a packet which has an altogether bogus (non-registered or ill-formed) Internet address.

Martian Mail An email message that arrives months after it was sent (as if it had been routed via Mars).

Martian Packet Strange fragments (data packets) of electronic mail that turn up unexpectedly on the wrong computer network because of bogus routing. Also used for a fragment that has an altogether unregistered or ill-formed Internet address.

MAS 1. Multiple Address System. A microwave point-to-multipoint communications system, either one-way or two-way, serving a minimum of four remote stations. The private radio MAS channels are not suitable for providing a communications service to a larger sector of the general public, such as channels the commission has allocated for cellular paging or specialized mobile radio services. (SMR).

2. Mobile Application Subsystem. An MAS is application software that is independent of the Cellular Digital Packet Data (CDPD) network. A cellular radio term.

MASER Microwave Amplification by Stimulated Emission of Radiation. A device that generates electromagnetic signals in the microwave range, known for relatively low noise.

Mask 1. A field made up of letters or numbers and wildcard characters, used to filter data. For example, a mask 800xxxxxxx may be applied to the dialed digits field of a call record to identify toll-free calls.

2. A computer imaging term. The electronic equivalent of placing transparent tape over selected regions of an image, a mask marks pixels that remain unchanged by subsequent painting operations. For example, you might mask out a mountain range and add background clouds to the sky. In the final image, the clouds will appear between the peaks.

Maskable Interrupt An interrupt on a computer that can be interrupted by another interrupt.

Masquerade To pretend to be someone else by using another person's password or Token.

MASS860 An organization of computer system vendors formed to promote open system standards and the writing of applications software for the Intel i860 microprocessor. Members include Intel, Oki Electric Industry Co., Ltd., IBM, Stratus, Olivetti, Alliant, Samsung and Stardent Computer.

Massively Parallel Systems Tightly coupled multi processing computers that house 100 or more CPUs, each with its own memory.

Mast Clamp A piece of equipment used to attach a ram hook to a pole. Aerial service wire is attached with a ram hook (sometimes called a ram horn) with a drop clamp.

Master Term applied to the data communications equipment at one end of a synchronous digital transmission network that supplies the clock timing signal that determines the rate of transmission in both directions.

Master Address Street Guide MSAG. In the emergency services telephone network in the United States, the Master Address Street Guide is a database containing the mapping of street addresses to Emergency Service Numbers within a given community. This allows the derivation of call routing information from a call's Automatic Location Identification. See Automatic Location Identification.

Master Clock An electronic timing circuit which synchronizes the entire data communications network. The source of timing signals, or the signals themselves, that all network stations use for synchronization.

Master Control Unit MCU. An InteCom word for the device which controls the main operating functions of the system.

Master Frequency Generator In FDM, equipment used to provide system end-to-end carrier frequency synchronization and frequency accuracy of tones over the system.

Master Gateway Control Protocol MGCP. See Single Gateway Control Protocol.

Master Group MG. In frequency division multiplexing (the old way of putting many voice conversations onto on communications line) a master group consists of 300 voice-grade (4 kHz) channels.

Master Number Hunting When a call is directed to the pilot number of a hunt group, it will hunt to the first non-busy station in that group. If a call is directed to a specific station in that hunt group it will go directly to that station and not hunt to another station in the group.

Master Slave Switching System A configuration consisting of a central switch and one or more remote switches. The master switch typically controls all I/O (input/output) information. The slave system performs tasks as directed by the master, including switching calls between phones attached to that remote module — without sending those calls back to the central switch. There are enormous savings in wiring since not every remote phone has to have a pair back to the central switch.

Master Slave Timing In a communication system, a timing subsystem wherein one station or node supplies the timing reference for all other interconnected stations or nodes.

Master Station 1. The main phone or station in a group. The one controlling the transmission of the others.

2. The unit which controls all the workstations on a LAN, usually through some type of polling. The master station on a token-passing ring allows recovery from error conditions, such as lost, busy or duplicate tokens, usually by generating a new token. Sometimes servers are referred to as master stations.

3. In navigation systems employing precise time dissemination, a station whose clock is used to synchronize the clocks of subordinate stations.

4. In basic mode link control, a data station that has accepted an invitation to ensure a data transfer to one or more slave stations. At a given instant, there can be only one master station on a data link.

Mastergroup A mastergroup consists of 600 voice channels, or 10 supergroups. Six mastergroups are equal to one jumbogroup.

MAT Meridian Administration Tools.

Matched Junction A waveguide component having four or more ports, and so arranged that if all ports except one are terminated in the correct impedance, there will be no reflection of energy from the junction.

Matching Loss ML. A telephone company definition. The inability to find an idle path between two idle equipment components. Usually expressed as a percentage. Example: Incoming Matching Loss, Originating Matching Loss, etc.

Mated Pair 1. A pair of devices which are perfectly matched. In other words, they perform identical functions. Modems, for instance, are perfectly mated as they perform the same functions of modulating and demodulating signals, depending on the direction of the transmission. Otherwise, they can't communicate. Like most things in network technology (and most things in life) communications is best accomplished between entities that are balanced and symmetrical. By the way, two mated pairs of things are known as "quads."

2. If you move out of a residential apartment, the pair that you were using is dedicated to that Apt. so when the next occupant moves in, it is a flick of a switch to get phone service turned on. That's called Dedicated Pair Out (DPO). The DPI pair is the OE (Office Equipment) or the Switch pair. Now the pair in the C.O. (Central Office), when it's joined to the DPO pair, is often called a mated pair. The DPI pair is mated to the DPO pair.

MATEL A Multiplex Automatic TELephone system. Picture one long, up to two miles wire (any decent quality works). You roll it out, then you clip phones anywhere into the wire. Then you have, in effect, a seven channel bus PBX. You have two digit extension dialing between the phones (up to 60). You also have conferencing, broadcast, call back, DID and connection to one central office line and one radio channel. The uses? String it around a rioting prison, a siege, an airport hijack, an emergency in New York City subway, etc. Three advantages: instant communications, communications where radio is bad and radio silence — you keep the press and the bad guys in the dark. This definition from John McCann, general manager of Racal Acoustics Limited, Frederick, MD, which makes the MATEL.

Material Dispersion Material dispersion occurs because of pulse of light in a fiber includes more than one wavelength. Because of the refractive index of a material varies with wavelength (check out a prism in the sun!) different wavelengths travel down the fiber at different paths. See also CHROMATIC DISPERSION.

Material Scattering In an optical fiber, that part of the total scattering attributable to the properties of the materials used for fiber fabrication.

Math Coprocessor A coprocessor is a special purpose microprocessor which assists the computer's main microprocessor in doing special tasks. A math coprocessor performs mathematical calculations, especially floating point operations. Math coprocessors are also called numeric and floating point coprocessors. If you do a lot of mathematical tasks on your PC, like recalculating large spreadsheets, then installing a math coprocessor makes huge sense. Intel included a math coprocessor with its 486DX chip, but removed it for the 486SX. No other Intel chip has a math coprocessor built in. When you buy a math coprocessor make sure it's the same speed as your existing processor.

MATR Minimum Average Time Requirement.

Matrix 1. A switch. A device for moving calls from one input to the desired output. There are many types of switching matrices — from simple step-by-step matrices to complex digital pulse code modulated matrices. Most switching matrices are "blocking." They do not have sufficient capacity to switch every call. There are some switches that are "non-blocking." These have the ability to switch every call simultaneously. By definition, non-blocking matrices are more expensive. They are only needed in special situations of high traffic.

2. The encompassing material in the composite (i.e. the plastic).

Matrix Ringing Two key system phones picking up the same extensions with different lines ringing on different phones. These days, with electronic phones, matrix ringing is easy. In the old days, with 1A2 phones, you needed to do considerable wiring.

Matrix Switch Device that allows multiple channels connected via serial interfaces (typically RS-232C) to connect under operator control to designated remote or local analog circuits or other serial interfaces.

MATV Master Antenna System, such as used in apartment buildings and motels.

MAU 1. Math Acceleration Unit.

2. Multistation Access Unit. A MAU is a wiring concentrator used in Local Area Networks. In token ring networks it's called a Multi-station Access Unit. In Ethernet networks it's called a Medium Attachment Unit. Basically a MAU is a device that allows terminals, PCs, printers, and other devices to be connected in a star-based configuration to Token Ring or Ethernet LANs. MAU hardware can be either active or passive. Each computer is wired directly to the MAU which then provides the connection to all other computers. MAUs themselves can be connected to expand the network. The MAU is a small box with eight or sixteen connectors and an arrangement of relays that function as bypass switches. When only one MAU is used, its relays and internal wiring arrange themselves so that the MAU and the connected computers form a complete electrical ring. MAUs can be cascaded to create bigger rings. The MAU listens for the "I'm here" signal sent by a computer when the computer is attached to the MAU. If the token ring adapter card in a computer is not working properly or the computer is turned off, the MAU no longer hears the "I'm here" signal and automatically disconnects the computer from the ring using the bypass relay.

A managed MAU contains on-board intelligence which enables it to communicate with network management software. This software can then be used to control the MAU. This type of MAU provides the network administrator with an "out-of-band" (a separate line) method of port control, monitoring, and diagnostics. For example, in the event that a token ring network is down, conventional network management may not be possible as it relies on the network itself to communicate with the various components on the token ring LAN. The out-of-band management provided by the managed MAU gives the administrator the ability to diagnose and reconfigure the network without physically inspecting and reconfiguring each MAU in the network.

MaxCR Maximum Cell Rate: This is the maximum capacity usable by connections belonging to the specified service category.

Maximize Button The maximize button in Windows is the up-arrow button at the far right of a title bar in the Windows operating system. Click on the maximize button to enlarge the IMARA Lite window to full size. See also, Minimize button and Restore button.

Maximum Access Time Maximum allowable waiting time between initiation of an access attempt and successful access.

Maximum Block Transfer Time Maximum allowable waiting time initiation of a block transfer attempt and completion of a successful block transfer.

Maximum Calling Area Geographic calling limits permitted to a particular access line based on requirements for a particular line.

Maximum Keying Frequency In facsimile systems, the frequency in hertz numerically equal to the spot speed

divided by twice the X-dimension of the scanning spot.

Maximum Modulating Frequency The highest picture frequency required for a given facsimile transmission system. The maximum modulating frequency and the maximum keying frequency are not necessarily equal.

Maximum Power Level Maximum power output limit for Mobile end System (M-ES).

Maximum Stuffing Rate The maximum rate at which bits can be inserted or deleted.

Maximum Transmission Unit MTU. The largest possible unit of data that can be sent on a given physical medium. Example: The MTU of Ethernet is 1500 bytes.

Maximum Usable Frequency MUF. The upper limit of the frequencies that can be used at a specified time for radio transmission between two points and involving propagation by reflection from the regular ionized layers of the ionosphere. MUF is a median frequency applicable to 50 percent of the days of the month, as opposed to 90 percent cited for the lowest usable high frequency (LUF) and the optimum traffic frequency (OTF).

Maximum User Signaling Rate The maximum rate, in bits per second at which binary information can be transferred (in a given direction) between users over the telecommunication system facilities dedicated to a particular information transfer transaction, under conditions of continuous transmission and no overhead information.

MAYPAC A X.25 packet-switched network operated in Malaysia by the Malaysian government.

MB Megabyte. A unit of measurement for physical data storage on some form of storage device — hard disk, optical disk, RAM memory etc. and equal to two raised to the 20th, i.e. 1,048,576 bytes. Here is a summary of sizes:

MB = Megabyte (2 to the 20th power)
GB = Gigabyte (2 to the 30th power)
TB = Terabyte (2 to the 40th power)
PB = Petabyte (2 to the 50th power)
EB = Exabyte (2 to the 60th power)
ZB = Zettabyte (2 to the 70th power)
YB = Yottabyte (2 to the 80th power)

One googolbyte equals 2 to the 100th power

MBE Micom Business Exchange. Trademark for MICOM's series of communications processors.

MBG Multilocation Business Group.

Mbits Million bits. See Mbps.

MBONE Multicast Backbone. A collection of Internet routers that support IP multicasting. The MBONE is used as a "broadcast (actually multicast) channel" on which various public and private audio and video programs are sent. Circa 1992 IETF (Internet Engineering Task Force) effort. Came out of earlier ARPA DARTnet experiments. Supports multicast audio and video across the Internet. Provides one-to-many and many-to-many network delivery services for apps like videoconferencing and audio. Supports simultaneous communication between several hosts. At present, the Internet MBONE is the largest demonstration of the capabilities of IP Multicast. The MBONE is an experimental, global, volunteer effort, and topographically is layered on top of portions of the physical Internet. (IP multicast packet routing is not supported by many installed production routers.) The network is linked by virtual point-to-point links called "tunnels". The tunnel endpoints are typically workstation-class machines having operating system support for IP multicast and run the "mrouted" multicast routing daemon. It presently carries IETF meetings, NASA space shuttle launches, music, concerts, and many other live meetings and performances. www.mbone.com

Mbps This one is confusing. When you see Mbps as the speed of a telecommunications, networking or local area networking transmission facility (i.e. something that moves information), Mbps means million bits per second — exactly one million. No more. No less.

When you see Mbps or MBps referred to the context of computing, it means million bytes per second, which is the same as one million bytes per second. How many bits that is depends on how many bits there are in a byte. Typically it's eight (but it could be more or fewer). So, in this case, one Mbps would be eight million bits per second.

To be correct, Mbps is million bits per second and MBps is million bytes per second. You will also see it written as Mb/s. That usually means million bits per second. For a much longer explanation, see Bps.

MBS 1. Maximum Burst Size. An ATM term for a traffic parameter which specifies the maximum number of cells which can be transmitted at the Peak Cell Rate (PCR). In the signaling message used for call setup, the Burst Tolerance (BT) is conveyed through the MBS. The BT, together with the SCR and the GCRA, determine the MBS that may be transmitted at the peak rate and still be in conformance with the GCRA.
2. Meridian Business Set.

MBus A Sun Microsystems definition: An open specification for connecting multiple CPUs (such as those in SPARC modules) with a 64-bit, 320-MB/second data path. Designed by Sun Microsystems; available from SPARC International.

MBWA Management By Walking Around. A technique pioneered by David Packard, one of the two founders of Hewlett-Packard which he and Bill Hewlett started in 1939 in a small garage in Addison Avenue in Palo Alto, CA.

MC 1. Main Cross-connect. The interconnect point where wiring from the entrance facility and from the workstation is connected to telecom equipment.
2. Multi-Carrier.

MC1 Cable The inter-PC chassis MVIP bus cable, which can support up to 20 PCs. It allows a developer to distribute MVIP's resources across all the connected computers. MC-MVIP type MC1 media provides 1536 x 64 Kbps of inter-chassis connectivity using twisted-pair copper cables. See MC2 and MVIP.

MC2 Cable The advanced inter-PC chassis MVIP bus cable, which leverages FDDI-II to provide up to 3072 x 64 Kbps of inter-chassis connectivity on fiber or copper. MC3 leverages SONET/SDH fiber technology at 155 Mbps to provide 2400/4800 x 64 Kbps of inter-chassis telephony. See MC1 and MVIP.

MC-MVIP Multi-Chassis MVIP. See MVIP.

MCA Micro Channel Architecture. The internal 32-bit bus inside some of IBM's PS/2 machines. It was originally introduced by IBM as a proprietary bus which manufacturers of IBM clone PCs would have to pay IBM large royalties if they wanted to include the bus in their machines. Sadly, this strategy backfired and few people wanted IBM PCs with the MCA bus. ISA remains the most popular PC bus. See ISA and EISA.

MCC 1. Mobile Country Code. A portion of the LAI and the IMSI (International Mobile Subscriber Identity).
2. Mobile Control Channel. See CONTROL CHANNEL.

MCCS Mechanized Calling Card Service was formerly known as ABC Service. MCCS is a CO switch facility that automatically bills credit card calls made on DDD without the involvement of an operator.

MCDV An ATM term. Maximum Cell Delay Variance: This is

the maximum two-point CDV objective across a link or node for the specified service category.

MCF Message Confirmation Frame. Confirmation by the receiver in a fax transmission that the receiver is ready to receive the next page.

MCHO Mobile Controlled Hand-off.

MCI 1. Once upon a time it was called Microwave Communications Inc. Then it became just MCI, which stands for nothing. MCI was the largest long distance phone company in the US after AT&T. In MCI's early days, the initials were said to stand for "Money Coming In." MCI was a full-service long distance company offering every service from switched single channel voice to leased T-1. In 1996, it announced that it had accepted a takeover offer from British Telecom (BT), the leading phone company in England. BT had held a 20% stake in MCI for a number of years, and viewed MCI as a vehicle to gain a significant position in the highly lucrative U.S. market. Then GTE offered more. Then upstart Worldcom offered even more. The MCI/Worldcom merger was completed on September 20, 1998 for approximately $40 billion (not cash, but shares). MCI Worldcom, at the time of the merger, boasted annual revenues of $30 billion, and a presence in over 65 countries. In order to gain regulatory approvals in the U.S. and Europe, MCI sold its Internet backbone to Cable & Wireless for $1.75 billion. See also Worldcom.
2. Media Control Interface. A standard control interface for multimedia devices and files. Using MCI, a multimedia application can control a variety of multimedia devices and files. Windows provides two MCI drivers; one controls the MIDI sequencer, and one controls sound for .WAV files.
3. Message Center Interface. An interface in some PBXs which allows you to connect an external PC and do voice mail/IVR (interactive voice response).

MCI Worldcom The company formed on September 20, 1998, when Worldcom merged with (read acquired) MCI for approximately $40 billion. MCI Worldcom, at the time of the merger, boasted annual revenues of $30 billion, and a presence in over 65 countries. In order to gain regulatory approvals in the U.S. and Europe, MCI sold its Internet backbone to Cable & Wireless for $1.75 billion. See also MCI and Worldcom.

MCL Mercury Communications Limited (UK). The second long distance company in England. It is competitor of British Telecom, the erstwhile monopoly local and long distance company in the U.K.

MCLR An ATM term. Maximum Cell Loss Ratio: This is the maximum ratio of the number of cells that do not make it across the link or node to the total number of cells arriving at the link or node.

MCNC Microelectronic Center of North Carolina.

MCNS Multimedia Cable Network System Partners Ltd. An organization which is leading the development of DOCSIS (Data Over Cable Service Interface Specification), a set of interface specifications for the delivery of high-speed data services over cable television systems. MCNS consists of Comcast Cable Communications Inc., Cox Communications, Tele-Communications Inc. (TCI), and Time Warner Cable. MCNS has partnered with a number of other companies in this project, as well. See also DOCSIS.

McNutt, Emma M. The first woman telephone operator, Ms. Emma M. McNutt was hired in September 1878 by the New England Telephone company in Boston, Massachusetts. Her hiring caused quite a stir, as "proper ladies" didn't work outside the home in those days. Within a few short years male operators were extinct-for about 100 years, at least. Ms. McNutt worked for the Bell System until her retirement in 1911.

MCR An ATM term. Minimum Cell Rate: An ABR service traffic descriptor, in cells/sec, that is the rate at which the source is always allowed to send.

MCSE Microsoft Certified Systems Engineer

MCSP Microsoft Certified Solutions Provider

MCT Algorithm A compression algorithm introduced in 1986 by PictureTel. MCT reduced the bandwidth necessary to transmit acceptable picture quality from 768 kbps to 224 kbps making two-way videoconferencing convenient and economical at relatively low data rates (for those times).

MCTD An ATM term. Maximum Cell Transfer Delay: This is the sum of the fixed delay component across the link or node and MCDV.

MCU Multipoint Control Unit. A bridging or switching device used in support of multipoint videoconferencing and supporting as many as 28 conferenced sites. The devices may be in the form of CPE or may be embedded in the WAN in support of carrier-based videoconferencing services. MCU standards are defined in ITU-T H.231, with T.120 describing generic conference control functions.

MCVF Multi-Channel Voice Frequency.

MD 1. Mediation Device. A SONET device that performs mediation functions between network elements and OSs. Potential mediation functions include protocol conversion, concentration of NE to OS links, conversion of languages, and message processing.

2. Manufacturer Discontinued. A product that the manufacturer no longer makes is called "manufacturer discontinued." Some people think it's a nice way of saying obsolete. But there are many "obsolete" products that do just fine, often for less money.

MD-IS Mobile Data Intermediate System.

MD-IS Serving Area A cellular radio term. The set of cells controlled by a single serving Mobile Data Intermediate System.

MDA Monochrome Display Adapter.

MDBS Mobile Data Base Station.

MDC Meridian Digital Centrex. A Northern Telecom abbreviation.

MDDB Multi-Drop Data Bridge. MDDB. A technique for combining data circuits in a computer environment in which the host machine polls other equipment.

MDF See MAIN DISTRIBUTION FRAME.

MDI Medium Dependent Interface.

MDK Modem Developer's Kit. Definition invented, I believe, by Microsoft.

MDLP Mobile Data Link Protocol. The Link Layer protocol defined in CDPD networks.

MDMF Multiple Data Message Format. See Caller ID Message Format.

MDN Mobile Dialing Number. The originating telephone number of the cellular caller.

MDQ Market Driven Quality. An IBM term of the mid-1980s.

MDRAM Multibank Dynamic Random Access Memory. Memory normally used in video boards that boasts extended performance with high bandwidth and short access times. The MDRAM chip can access several memory banks at a time.

MDT Mean Down Time.

MDU Message Display Unit. See Readerboard.

MDUs Multiple Dwelling Units. A telephony jargon word for high-rise apartment buildings.

MDVC Mobile Digital Voice Channel. The channel between a mobile phone and a cell site antenna in a digital cellular or PCS environment. The MDVC supports both voice and data transmission, although the allocated bandwidth is designed primarily to support voice. Signaling and control functions take place over separate channels set aside specifically for that purpose.

MEA Metropolitan Economic Area. See Metropolitan Statistical Area and MSA.

Meaconing A system for receiving radio beacon signals and retransmitting them on the same frequency to confuse navigation and cause inaccurate bearings to be obtained by aircraft or ground stations.

Mean The sum of all items divided by the number of items, e.g., for the five numbers 7, 7, 8, 10, and 11, the mean is (7 + 7 + 8 + 10 + 11)/5 = 43/5 = 8.6. The average (also the arithmetic average) and the mean are the same. What's different is the median. The median of a set is the number that divides the set in half, so that as many numbers are larger than the median as are smaller. The median of 7, 7, 8, 10, and 11 is 8. If the set has an even number of elements, the median is the number halfway between the middle pair. the median is widely used as a measure of central tendency. See several Mean definitions below.

Mean Busy Hour For a telephone line or group of lines or a switch the Mean Busy hour is the 60 minute period where traffic is the greatest.

Mean Deviation An average of all deviations, plus or minus from the mean. It is occasionally used as a measure of dispersion.

Mean Power Of A Radio Transmitter The average power supplied to the antenna transmission line by a transmitter during an interval of time sufficiently long compared with the lowest frequency encountered in the modulation taken under normal operating conditions. Normally, a time of 0.1 second, during which the mean power is greatest, will be selected.

Mean Time Between Failure MTBF. The average time a manufacturer estimates before a failure occurs in a component, a printed circuit board or a complete telephone system. One must check, since MTBFs are cumulative.

Mean Time Between Outages MTBO. The mean time between equipment failures or significant outages which essentially render transmission useless. See MEAN TIME BETWEEN FAILURE.

Mean Time To Repair MTTR. The vendor's estimated average time required to do repairs on equipment.

Mean Time To Service Restoral MTSR. The mean time to restore service following system failures that result in a service outage. The time to restore includes all time from the occurrence of the failure until the restoral of service.

Measured Rate A message rate structure in which the monthly phone line rental includes a specified number of calls within a defined area, plus a charge for additional calls. See LOCAL MEASURED SERVICE.

Measured Service Also known as USAGE SENSITIVE PRICING (USP). A local phone company method of pricing used to bill local phone calls. Measured service is often charged on the number of calls, the time of day, the distance traveled and the length of the call. See LOCAL MEASURED SERVICE.

Measurement Interval A Frame Relay term defining the interval of time which the carrier uses to measure burst rates which exceed the CIR, as well as the length of the bursts.

Meatware People. See also VAPORWARE.

MECCA Multiplex Engineering Control Center Activity

Mechanic A programmer.

Mechanical Equipment Room A room serving the space needs for HVAC and other building systems other than telecommunications equipment. These are often special-purpose rooms.

Mechanical Hold A very basic line-holding mechanism used on simple two- and three-line phones that operated by placing a short circuit or a resistor across one phone line while talking on another. Chief disadvantage was that a call put on hold at one phone could not be taken off hold at another phone. Inexpensive multi-line phones with electronic holds largely replaced mechanical holds in the 1980s.

Mechanical Loop Test MLT. Also called Direct-Access Test Unit (DATU) added or built into a central office switch. With MLT a technician can execute tests for shorts, opens and grounds remotely. The technician gets a digital voice, enters a password and is given a series of options. The technician can get results as a digital recording or through an alphanumeric pager. MLT units can send a locating tone as TIP, RING or a combination of both. The unit can short lines and remove battery voltage for testing.

Mechanical Splice A splice in which optical glass fibers are joined mechanically (e.g., glued or crimped in place) but not fused (i.e. melted) together.

Mechanical Stripping Removing the coating from a fiber using a tool similar to those used for removing insulation from wires.

Mechanized Calling Card Service MCCS was formerly known as ABC Service. MCCS is a central office switch feature that automatically bills credit card calls made on DDD (direct distance dial) rates without the involvement of an operator.

Mechanized Line Testing MLT. The system provides computer control of accurate and extensive loop testing functions in the customer contact, screening, testing, dispatch and closeout phases of trouble report handling. It also provides full diagnostic outputs instead of just pass/fail indications.

Media In the context of telecommunications, media is most often the conduit or link that carries transmissions. Transport media include coaxial cable, copper wire, radio waves, waveguide and fiber.

Media Access Control MAC. The real term is Medium Access Control. But some naughty people call it, incorrectly, Media Access Control. See Medium Access Control for a full explanation.

Media Compatible Usually used to refer to floppy disk media. Even though two different computers (e.g. an AT&T PC 6300 and an Apple IIe) both use 5 1/4 inch floppy disks, the information recorded on them is recorded in a different format and thus, they are not media compatible. You can put one disk in another's machine. But it won't work. You'll get a dumb error message.

Media Independent Interface MII. A part of the Fast Ethernet specification. The MII replaces 10Base-T Ethernet's Attachment Unit Interface (AUI), and is used to connect the MAC layer to the physical layer. The MII establishes a single interface for the three 100Base-T media specifications (100Base-TX, 100Base-T4, and 100Base-FX).

Media Interface Connector An optical fiber connector which links the fiber media to the FDDI node or another cable.

Media Path Same as wire run. The means by which telephone signals are conveyed from the Network Interface Jack to the Communications Outlet.

Media Processor A special microprocessor whose job is to perform processing for multimedia devices, e.g. videophones, audio, computer telephony devices, voice recognition and 3-D, while the computer's main microprocessor (e.g. a Pentium) handled the basic processing and input/output (I/O) processing. Such media processors might have all the characteristics of digital signal processors and then some. See also MMX.

Media Processing The processing of transactions during a telephone call; these transaction may include fax operations, speech recognition and synthesis, Touch Tone recognition, voice and fax store-and-forward messaging, and the conversion of messages from one format to another (such as from text to voice, or fax to text).

Media Server A new term for a file server on a local area network which contains files containing voice, images, pictures, video, etc. In short, a media server is a repository for media of all types. Media servers are also called file servers. Here's a definition of Media Server, courtesy of Oracle Corporation, writing in early 1994: "The media server harnesses the power of the fastest computers ever built — massively parallel computers. Media servers support diverse clients so that users don't have to discard existing systems. They provide storage, network interfaces and memory — plus support for all forms of multimedia information. Because they use thousands of low-cost microprocessors, media servers offer astonishing performance at low cost."

Media Service Instance A logical server providing access to media services (e.g., Accessing a DataStream, Sending & Receiving Faxes, Playing & Recording Sounds, Engaging other VRU services) that can be associated with a call through a Media Access Device. See MediaStreamID.

Media Stream The information content carried on a call- that is, what actually is transmitted and received over the line, and can, with the necessary hardware, be read and written by a media stream API.

Media StreamID Allows an association to be established between a given call and Media Services available on a Media Service Instance that can be associated with the call through a Media Access Device. See Media Service Instance.

Media Type A call's media type describes what type of information the call is carrying, such as data or voice. A computing domain can use this information, for example, to route the call to a more appropriate computing domain, such as a data computing domain for an incoming data call.

Median The average and the mean are the same. What's different is the median. See Mean for a full explanation.

Mediation System A wireless telecommunications term. A mediation system provides for three functions for transporting data from one device to another. These include protocol conversion, message routing, and store-and-forward processing.

Medium Any material substance that can be used for the telecommunications transmission of signals. "Mediums'" (or media) include optical fiber, cable, wire, dielectric slab, water, air or free space.

Medium Access Control MAC. The IEEE sublayer in a LAN (Local Area Network) which controls access to the shared medium by LAN-attached devices. In the context of the OSI Reference Model, the MAC layer extends above to the Data Link Layer (Layer 2), and below to the Physical Layer (Layer 1). Within the MAC sublayer are defined Data Link Layer options which specify the basis on which devices access the shared medium, and the basis on which congestion control is exercised. Defined at the Physical Layer are media options such as UTP (Unshielded Twisted Pair) CAT 3, 4, and 5 (Categories of wire); STP (Shielded Twisted Pair); fiber optic cable; and wireless radio and infrared. Specific

IEEE MAC standards are defined for LANs such as CSMA/CD (802.3), Token Passing Bus (802.4), Token Passing Ring (802.5), Metropolitan Area Networks (802.6), and Wireless LANs (802.11). The MAC sublayer works in conjunction with the Logical Link Control Layer of the IEEE model; at this higher level are defined specific LLC conventions such as frame format and addressing.

Medium Attachment Unit A device used in a data station to couple the data terminal equipment (DTE) to the transmission medium.

Medium Dependent Interface MDI-X. The physical components of a network interface which handle the electrical or optical connection to a cable. This includes the connector, transceivers, and other physical layer components. MDI-X refers to a physical connection which includes an internal crossover of the transmit and receive signals. All standard repeater ports are MDI-X and are often marked with just an X by the port. Some repeater ports are changeable to a DTE port. In this case, the port is changed to a MDI port for connection to a MDI-X port on another repeater. An example of a DTE port is the connection on a NIC.

Medium Frequency MF. Radio frequencies from 300 KHz to 3000 KHz.

Medium Interface Connector MIC. In LAN/MAN systems, the connector at the interface point between the bus interface unit and the terminal, termed the medium interface point.

Meet-Me Conference A teleconferencing term. Meet-Me Conferencing is an arrangement by which you can dial a specific, pre-determined telephone number and security access code to join a conference with other participants. You are automatically connected to the conference through a conference bridge. Conference participants may call in at a preset time or may be directed to do so by a conference coordinator. Meet-Me Conferences may be set up through a teleconferencing service provider, generally with the capability to conference thousands of participants. It also can be provided through a phone system, such as a PBX, key system or hybrid. Some phone systems restrict this to intercom circuits only. In almost all phone systems there is a maximum number of parties that can be connected in such conference at one time.

Meet-Me Intercom Conference Dial a special number ("access code") and any telephone can join an intercom conference call.

Meet-Me Page A feature which allows a person to answer an intercom page from any phone in the system.

Meet-Point Billing This is a billing arrangement that applies when two Local Exchange Carriers in the same LATA are used to complete a private line circuit.

Mega A prefix meaning one million, also represented as an M. MEGABIT = one million bits. MEGABYTE = one million bytes. MEGAHERTZ = one million cycles per second. See also MEGABYTE.

Megabyte Megabyte. A unit of measurement for data storage equal to 1,048,576 bytes. This is calculated by raising two to the twentieth power. 1,048,576 is the power of 2 closest to a million, i.e. 1024 X 1024.
Here is a summary of sizes:

MB = Megabyte (10 to the 6th power)
GB = Gigabyte (10 to the 9th power)
TB = Terabyte (10 to the 12th power)
PB = Petabyte (10 to the 15th power)
EB = Exabyte (10 to the 18th power)
ZB = Zettabyte (10 to the 21th power)
YB = Yottabyte (10 to the 24th power)

One googolbyte = 10 to the 100th power

Megaflops Million Floating point Operations Per Second. A measure of computing power usually associated with large computers. Mega means million. Also known as MFLOPS.

Megahertz 1. MHz. A unit of frequency denoting one million Hz or one million cycles per second. See BANDWIDTH and HERTZ.
2. What the Vermonter's Guide to Computer Lingo says you get when you're not careful downloading, which is defines as taking firewood off the pickup truck.

Megalink Name for BellSouth's leased T-1 service.

Megastream British Telecom's brand name for a service of 30 64-Kbps channels (i.e. E-1).

Megohm A resistance of 1,000,000 ohms.

Memo 1. A telephone feature that enables the user to store a phone number for calling in the future. For example, while speaking to a Directory Assistance operator, you can put the number she gives you into memory, and then call that number by pushing one or two buttons.
2. A call center term. A free form field used to store descriptive text or comments. The information in a memo field can be of any length and type.

Memory The part of a computer or sophisticated phone system which stores information or instructions for use. Memory comes in many variations. There is memory which is lost when the power is switched off. There is memory which is retained when power is turned off.

Memory Administration MA. A set of functions that provide network system database updates, network system database integrity, network system database security and network system database backup and restoration. Definition from Bellcore in reference to its concept of the Advanced Intelligent Network.

Memory Board An add-on board designed to increase a computer's amount of RAM.

Memory Caching A technology for increasing hardware performance by storing frequently used sequences of instructions in a memory cache separate from the computer's main memory where they can be more quickly accessed by the CPU.

Memory Call Service A family of central office based voice messaging services from BellSouth.

Memory Cards The memory card is a bunch of memory chips crammed into a small plastic cartridge about the size of a credit card and about three times the thickness. It is used in several palmtop computers. As this dictionary was being written, we were awaiting the release of a 16 megabyte memory card. In contrast to flash memory, a memory card requires small batteries, typically the same ones as used in watches. See Smart Card.

Memory Effect The gradual shorting of a battery's useful life, caused by recharging before the battery is completely discharged. This is a real problem with nickel cadmium batteries, less of a problem with Nickel Hydride and even less of a problem with Lithium ion batteries.

Memory Interface A PCMCIA definition. The memory interface is the default interface after power up, PCMCIA Hard Reset and PCMCIA Soft Reset for both PCMCIA cards and sockets. This interface supports memory operations as defined in PCMCIA Release 1.0 and later and is used by both Memory Cards and I/O Cards.

Memory Map An indication of what type of data is stored where in a computer's RAM memory.

Memory Protection The structuring of memory resources in Novell's NetWare 4.0 that guards the NetWare server memory from corruption by NLMs. Memory protection allows you to run NLMs in a separate memory domain called the OS PROTECTED domain. Once you determine the NLM to be safe, you can load it into the OS domain, where it can run most efficiently.

Memory Reserve Power The operating voltage, generally provided by a battery, which supplies power to the memory modules when your commercial power fails. You should check your memory reserve power before it's too late. You should test it even when you don't need it.

Memory Technology Driver A PCMCIA card definition. A memory technology driver is a memory device specific software that interfaces to Card Services to mask the details of accessing different memory technologies.

Menu Options displayed on a computer terminal screen or spoken by a voice processing system. The user can choose what he wants done by simply choosing a menu option — either typing it on the computer keyboard, hitting a touchtone on his phone or speaking a word or two. There are basically two ways of organizing computer or voice processing software — menu-driven and non-menu driven. Menu-driven programs are easier to use but they can only present as many options as can be reasonably crammed on a screen or spoken in a few seconds. Non-menu driven screens allow more alternatives but are much more complex and frightening. It's the difference between receiving a bland "A" or "C" prompt on the screen — as in MS-DOS and receiving a menu of "Press A if you want Word Processing," "Press B if you want Spread Sheet," etc. It's very easy to write menus in MS-DOS using BATch files. See also AUDIO MENUS and PROMPTS.

Menu Bar call center term. The part of the menu system visible as a single row across the top of the display.

MEO Middle Earth Orbiting satellite. MEOs operate much like LEOs (Low Earth Orbiting satellites systems), although in slightly higher orbits. MEOs are capable of supporting both voice and data services. Contrast with GEO and LEO.

Merced Merced is a 64-bit multichip module jointly developed by Intel with Hewlett-Packard Co. It extends the Intel architecture in both raw speed (one version will reach 600-MHz) and overall performance. Merced will be followed by the two-chip processor called Flagstaff in the year 2000. Flagstaff chips will be the first built using a process that creates much smaller 0.18-micron-wide circuits, which will enable Intel to build smaller and faster chips in greater volume. There will be two versions of Flagstaff, with a choice of 4 Mbytes or 8 Mbytes of secondary cache, according to writer, Tom Davey.

Merchant Silicon Merchant Silicon is a special microprocessor chip, which is a non-ASIC, commercially available semiconductor chip.

Mercury-Wetted Relay A relay or switch in which the movable contact of the device makes contact with a pool of mercury. Before solid-state devices, this was common technology for switching dry circuits. See also DRY CONTACT.

Merge A Sun Microsystems term, it is a Solaris utility for Intel platforms. Gives Sun users access to MS-DOS and Microsoft Windows applications.

Merging Traffic The telecommunications, cable, consumer electronic and media conglomerates all vying for access to the same markets. See SILIWOOD.

Meridian A Northern Telecom name for its family of PBXs. In recent years, Northern's PBXs have evolved from the SL-1 family to the Meridian SL-1 family to the Meridian 1 family, which Northern says will support from 30 to 60,000 lines.

Meridian Link Meridian Link from Northern Telecom enables Meridian 1 Communication Systems (i.e. PBXs and ACDs) to exchange information with host computers and the application software that resides on those computers. This means that the application software can use information such as Automatic Number Identification (ANI) and Dialed Number Identification System (DNIS) to automatically perform a series of routines, such as record look-up. According to Northern, its customer installations have proven that these capabilities increase employee productivity, reduce call lengths and result in higher revenues. Meridian Link offers these capabilities to application software:

• Intelligent Answering. Intelligent Answering provides agents with the capability of knowing who is calling (ANI) and why that person is calling (DNIS). Intelligent answering improves agent productivity by saving the record look-up time.

• Computer Assisted Dialing. This provides a productive means of passing dial requests from a computer to Meridian 1. Computer assisted dialing eliminates errors and improves productivity.

• Intelligent Dialing. With predictive dialing, the computer makes calls and monitors call progress, transferring active (or answered) calls to available agents.

• Coordinated Transfer of Voice and Data. A telephone call and a screen of information are simultaneously transferred. A customer does not have to repeat information.

• Telephone Operations from a Terminal. Capabilities include making a call, disconnecting, transferring, or conferencing by a keyboard command.

Meridian Link Interface A Northern Telecom definition. The Meridian Link interface is the link between the host computer and the Meridian, through the Meridian Link Module. It uses LAB protocol between the MLM and Meridian 1, and either X.25 or LAB protocol between the MLM and host computer. See MERIDIAN LINK.

Meridian Link Module MLM. A Northern Telecom definition. The MLM supports the Meridian Link interface to host processors. It is packaged within the Application Equipment Module. See MERIDIAN LINK.

Meridian Mail Northern Telecom's voice messaging system. Meridian Mail provides voice processing capability to Meridian Link applications.

Meridian Teladapter TCM Connects the digital telephone and Macintosh to the switch. The TelAdapter connects to a QPC578 or NT8D02AB digital line card.

Meridional Ray In fiber optics, a ray that passes through the optical axis of an optical fiber. This contrasts with a skew ray, which does not.

MERIT The successor to NSFNET, MERIT originally was a statewide IP network operated by the University of Michigan. It also was a substantial regional subnetwork (subnet) of the NSFNET and the Internet. MERIT provides access into the Internet through MAEs (Merit Access Exchanges) located in San Jose (MAE West) and Vienna, Virginia (MAE East); those points of access actually are provided in partnership with MFS Datanet.

Merlin AT&T's first electronic key system, distinguished by futuristic styling, horrible membrane line keys, nonstandard 8-conductor wiring scheme, and expensive accessories. Later versions were much better. See below.

Merlin Classic Term applied to original series of AT&T's MERLIN 206, 410 and 820 electronic key telephone systems.

Merlin II Third generation of AT&T's Merlin electronic key telephone system. Uses digital technology, and either older analog Merlin phones, dedicated single-line phones or newer digital multi-line phones. Programming is menu-driven. Features include hospitality functions and automatic route selection. Accepts Legend circuit boards and phones.

Merlin Legend Fourth Generation of AT&T's Merlin phone system, actually considered an enhancement of previous Merlin II. Maximum size is 90 lines and 144 phones (but not both maximums in same system). Operates in either key system, PBX or "behind switch" mode. T-1 compatible. Works with either older Merlin phones or new line of MLX digital phones, with more conventional styling than older Merlin phones.

Merlin Plus Second Generation of AT&T's Merlin electronic key telephone system. It introduced such features as call forwarding, automated attendant, remote system access, auto-busy-redial, and direct station access.

MERS Most Economical Route Selection. A term used by GTE and some other PBX manufacturers to mean Least Cost Routing. See LEAST COST ROUTING.

MESA Architecture Centigram's trademarked brand name for its architecture that stands for Modularly Expandable System Architecture:
1. The capability to expand and/or enhance a VoiceMemo II system hardware and software in a modular fashion.
2. The capability to expand a single module VoiceMemo II into a multi-module system with a single database and centralized control.

Mesa-Flex Centigram's service design utility that allows a VoiceMemo II system to be fully customized through individually designed Feature Class of Service, independent Limits Classes of Service, Group Classes of Service and Network Classes of Service.

Mesalink Centigram's registered trademark for its interprocessor high speed bus that carries control information to the distributed processors.

Mesh Network architecture in which each node has a dedicated connection to all other nodes.

Mesh Connectivity A Wide Area Network (WAN) term for connectivity over a mesh network, wherein each site is directly connected with every other site. See MESH NETWORK for a much fuller explanation.

Mesh Network A leased line network that provides a direct connection between each site and every other site. Through the use of intelligent internetworking devices, such as Nodal Multiplexers in a T-Carrier network, each transmission might be routed over an alternative path should the primary (direct) path between the two sites be either congested or in a state of failure. The advantages of mesh networking include high availability of efficient transmission links of high quality. Through the use of intelligent internetworking devices, network redundancy offers the advantage of network resiliency. The disadvantages of a full mesh include high cost, difficulty of configuration and reconfiguration, and long lead times associated with carrier provisioning. Mesh networks are relatively easy and inexpensive to configure where four or fewer sites must be interconnected; the cost and complexity increase significantly where more than four sites are involved. As a result, large organizations increasingly tend to favor alternative solutions for voice and data. Such alternatives as Virtual Private Networks (VPNs) Frame Relay, SMDS and ATM offer performance comparable to leased mesh networks, but at lower cost, with greater flexibility of configuration, and without the long lead times.

Meso The Greek prefix meaning the middle.

Mesochronous The relationship between two signals such that their corresponding significant instants occur at the same average rate.

Message 1. A sequence of characters used to convey information or data. In data communications, messages are usually in an agreed format with a heading which establishes the address to which the message will be sent and the text which is the actual message and maybe some information to signify the end of the message. A Northern Telecom Norstar definition: A message, which appears on the telephone display that informs the recipient to call the person who sent the message. Messages can only be sent within the Norstar system.
2. The Layer 3 information in the OSI model that is passed between the CPE and SPCS for signaling.
3. A SCSA definition. The transport container for SCSA requests, replies and events. Assumes a set of conventions for directing the delivery of the message to the proper entity, either a client or service provider. See also SCSA Message Protocol.

Message Alert A cellular phone term, also called "call-in-absence" indicator. A light or other indicator announcing that a phone call came in, an especially important feature if the cellular subscriber has VOICE MAIL.

Message Alignment Indicator In a signal message, data transmitted between the user part and the message transfer part to identify the boundaries of the signal message.

Message Backbone A single format message transport system designed for the electronic mail and messaging needs of an entire corporation or enterprise.

Message Center A centralized place within the corporation where messages are taken and (occasionally) delivered. Message centers are good if they are staffed with competent, motivated people and the various phones in the place have message waiting lights (like they do in hotels). If staffed by talented people, message centers work a lot better than the amateur message takers called secretaries and their part-time short-term replacements.

Message Detail Recording See CALL DETAIL RECORDING and CALL ACCOUNTING SYSTEMS.

Messaging Enabled Applications MEA. This term defined by the Electronic Messaging Association, Arlington, VA. Applications that directly access the messaging service as a way of addressing and transporting information to and from objects on a network. Messaging Enabled Applications differ from Mail Enabled Application because they do not require electronic mail to successfully navigate the network. They may or may not have their own directory and also have the ability to directly access Directory Services. For example: A student requests a class via electronic mail to the education system. The education system automatically schedules the student (this an electronic mail enabled application).

Message Format The rules for placing information necessary for an electronic message. The format includes where the heading is and how long it will be as well as other control information.

Message Frame A SCSA term. A data link layer frame the encapsulates control and signaling data transmitted on the SCbus or SCxbus Message Bus. The form of a Message Bus frame is fully compliant with ISO HDLC UI (Unnumbered Information) Frame specifications. See SCSA.

Message Handling Services MHS. Software whose primary function is the movement of messages between application programs. MHS was developed by Action Technologies and Novell acquired full marketing and development rights to MHS for Netware based Lans. Under MHS each application sends messages to the server's \ mhs \ mail \ snd directory. MHS delivers messages to an application by placing them in the application's assigned directory. MHS is most commonly used for e-mail (electronic mail). Various e-mail programs use the MHS format for exchanging e-mail messages.

MHS also means the standard defined by ITU-T as X.400 and by ISO as Message-Oriented Text Interchange Standard (MOTIS). MHS is the X.400 family of services and protocols that provides the functions for global electronic-mail transfer among local mail systems.

Message Header The header before a string of data containing information regarding the destination of the data, usually in a packet in X.25 format.

Message Management A new term for managing all your voice mail, fax mail and electronic mail. The major concept is to join the control of the devices that produce voice, fax and e-mail through your desktop PC connected over your LAN. For example, you come into work in the morning, turn your PC on, and immediately see a screenful of messages — one line per message. By clicking on that line with your mouse, your PC would pull up the application that will then let you read or hear your message. The benefit is that you can handle your messages faster and be more discriminating about which ones you pay most and least attention to.

Message Packet A unit of information used in network communication. Messages sent between network devices (workstations, file servers, etc.) are formed into packets at the source device. The packets are reassembled, if necessary, into complete messages when they reach their destination. A message packet might contain a request for service, information on how to handle the request, and the data that will be serviced. An individual packet consists of headers and a data portion. Additional headers are appended to the data portion as the packet travels through the layers of the communication protocol. Any message that exceeds the maximum size is partitioned and carried as several packets. When the packet arrives at its destination, the headers are stripped off, the message delivered and the request serviced.

Message Rate A method of billing local phone calls that varies from one place to another. Phone calls are billed as "message units." Message units are a combination of length of call and distance of the call. In a city like New York you might buy "basic" phone service and be entitled to 50 "message units." That may mean you can call the local pizza house for under 5 minutes 50 times. Or it may mean that you can call from Manhattan to the Bronx 25 times — assuming each call is two message units.

Message Register Leads Terminal equipment leads at the interface used solely for receiving dc message register pulses from a central office at a PBX so that message unit information normally recorded at the central office only is also recorded at the PBX.

Message Registration A phone system feature that records the number of message units incurred by each phone. Useful in hotels which bill local calls by message units.

Message Retrieval The ability of a fax machine to store material already transmitted so it can be retransmitted.

Message Signal Unit Signal unit of CCS that carries a message corresponding to the information part or packet of the HDLC frame plus a message transfer part corresponding to the HDLC frame header.

Message Store An X.400 electronic mail term: A staging point, similar to a post office, in which messages are temporarily held for later transmission to one or more recipients.

Message Switch A message switch is another term for an electronic mail gateway, which is a LAN application that fetches messages from one electronic mail system, translates them to the format of another electronic mail system, and then sends them to the "post office" of that other system. The post office is the public entry point — the place you put mail you want the other system to receive.

Message Switching A technique for receiving a message, storing it for a while and then sending it on. Message switching is normally used when the desired recipient is not there. The message switch will keep attempting delivery, freeing the calling party to handle other work. Unlike voice phone calls no direct connection is made in message switching between the incoming and outgoing messages. Each message, like a Western Union telegram, contains a destination address and is recipient through intermediate nodes. Each node along the way receives the message, stores it briefly, and then passes it on to the next node.

Message Switching Network A public data communications network over which subscribers send primarily text messages to one another.

Message Telecommunications Service MTS. The regulatory term for long distance or message toll voice service. Misnamed. Actually there's nothing "message-y" about this service. This is a 100% switched telephone service (which can, obviously, be used for voice, data, video or fax).

Message Telephone Service MTS. Official designation for tariffed long-distance, or toll, telephone service. See MESSAGE TELECOMMUNICATIONS SERVICE.

Message Toll Voice Service See Message Telecommunications Service.

Message Transfer MT. The carriage of information between parties using computers as intermediaries. It is one aspect of message handling.

Message Transfer Agent An X.400 electronic mail term: Software usually residing in a LAN server or host computer that moves messages between senders and recipients.

Message Transfer Part MTP. The part of SS7 signaling node that is used to place formatted signaling messages into packets, strip formatted signaling messages from packets, and send or receive packets.

Message Unit 1. The charge for one unit of local telephone service. How many message units you will find on your bill is a function of how many calls you made, how far the calls traveled, what time of the day or night you called and for how long you talked.

2. In IBM Corp's Systems Network Architecture, the portion of the data within a message that is passed to, and processed by, a specific software layer.

Message Unit Detail MUD. A service offered by local telephone companies in which they give you a report listing the phone number of all local calls made from each of your billing numbers. The billing number may be the main number for a PBX or the individual extensions if Centrex service is used. MUD reports are usually available to a telephone company customer at additional cost. MUD reports generally have to be requested in advance. Some telephone company central offices cannot generate MUD reports. When available, MUD reports may not be in machine processable form.

Message Waiting A light on the phone or some letters or characters on the phone's display indicating there's a message waiting somewhere for the owner of the phone. That message might be with a special message center (as in a hotel or a larger company), with the operator or with a computer attached to the phone system or with someone else in the company. Message waiting lights are incredibly useful at hotels. It's amazing more companies don't use them also.

Message-To-Slave Directory Propagation In electronic mail on a local area network, message-to-slave directory propagation is a way of updating user addresses where changes in the master post office are sent to the slaves, but changes in the slave post offices are not sent to the master.

Messages Roses French term for pay telephone pornographic messages. What the English call phone sex lines.

Messaging Application Programming Interface MAPI. A set of calls used to add mail-enabled features to other Windows-based applications. See MAPI.

Messaging Middleware Lets applications communicate and exchange data via asynchronous messages and queues.

Messenger A piece of heavy metal cabling attached to a pole line to support aerial phone cable. See also Hard Cable and Messengered Cable.

Messengered Cable Messengered cable is similar to jacketed cable, but includes an integral steel "messenger wire" to provide mechanical support. Messengered cable is intended for aerial installation without strand. See Hard Cable for a much bigger explanation.

MET Multibutton Electronic Telephone for an old phone system called AT&T Horizon.

Meta Tag An optional HTML tag that is used to specify information about a Web document. It appears in the <head> portion of a Web page. Search engines use "spiders" to index Web pages by reading the information contained within the tag's code. A HTML or Web page author uses these tags to help his or her page get noticed or "come up" when an Internet surfer queries a search engine for a particular keyword or topic. The META tag can also be used to specify an HTTP or URL address for the page to "jump" to after a certain amount of time. This is known as Client-Pull. This means a Web page author can control the amount of time a Web page is up on the screen as well as where the browser will go next. Here's the HTML syntax for search engine indexing: <HTML><HEAD><TITLE></TITLE><META NAME="keywords" CONTENT="description"></HEAD></HTML>
Here's the HTML syntax for Client Pull: <HTML><HEAD><TITLE></TITLE> <META HTTP-QUIV= "REFRESH" CONTENT="30; URL=meta2.html"></HEAD> </HTML>. This code makes the Web page "refresh" or change to the URL specified in 30 seconds.

Metal Oxide Varistor Metal Oxide Varistor. A voltage dependent resistor which absorbs voltage and current surges and spikes. This low-cost, effective device can sustain large surges and switch in 1 to 5 nanoseconds. It is used as a surge protector and suppressor. It often the first electronic component that electrons coming in on an incoming phone line hit. Many trunk boards inside PBX are protected by MOVs. If the voltage or current is high, it will blow the MOV, thus protecting the remaining the far more valuable devices on the board.

Metal Tape Recording tape coated with iron particles and noted for its wide dynamic range and wide frequency response.

Metallic Circuit A circuit completely provided by metallic wire conductors, and not containing any carrier, radio or fiber and in which the ground or earth forms no part.

Metallic Voltage A potential difference between metallic conductors, as opposed to a potential difference between metallic conductor and ground.

Metallization Metallization is necessary if composites are to be used in antenna applications. Antennas are mainly metallized with copper or gold. Composite systems that lend themselves to metallization are those epoxy and thermoplastic composites with Kevlar, glass or carbon fiber reinforcement.

Metaphor See DESKTOP METAPHOR.

Metasignaling ATM Layer Management (LM) process that manages different types of signaling and possibly semi-permanent virtual channels (VCs), including the assignment, removal and checking of VCs.

Metasignaling VCs An ATM term. The standardized VCs that convey metasignaling information across a User-Network Interface (UNI).

Metcalfe's Law Named after Robert N. Metcalfe, co-inventor of Ethernet. Metcalfe's Law states that the value of a network — defined as its utility to a population — is V=A*N*N+B*N+C where V is the value and N the number of users of a network. The value of a network grows with the square of the number of its users. Often, for small N, the COST of a network exceeds its VALUE. This means there is a critical mass phenomenon in networks. Therefore, small pilots of some networks might fail where a larger operational network might succeed wildly. It's tough getting them started, but then BAM! off they go. A simple example: One telephone is useless. Two telephones are better, but not much. Only when a good proportion of the population has a telephone does the network gain the power to change the society. Some people estimate that a totalitarian government cannot survive even one telephone per hundred people. According to many observers, the Internet owes its extraordinary growth and impact to its ability to harness both Metcalfe's Law and Moore's Law (which says that computing power and capacity double every 18 months). See also DATA DIALTONE.

Meteor Burst Communications Communications by radio signals reflected by ionized meteor trails.

Meter The metric unit of length, equivalent to 39.37 inches. An instrument for measuring quantities of length.

Metering Pulses In virtually all foreign countries, periodic pulses are returned from the distant exchange to the exchange (central office) of the calling number. These pulses determine the cost of the call, local or long distance. Typically, all pulses cost the same. However, the farther you call, the quicker the pulses come. This system contrasts to the North American long distance billing scheme which typically charges a certain amount for the first minute — no matter how much of the minute is actually used in conversation. After the first minute in the U.S., conversations are billed in one minute increments. Overseas pulses can be as short as three or four seconds (especially for international calls).

Method An SCSA definition. The specific implementation of an operation for a class; code that can be executed in response to a request.

Metric A measurement. A benchmark. A metric has nothing to do with a 100. See Service Level Agreement.

Metropolitan Area Network MAN. A loosely defined term generally understood to describe a data network covering an area larger than a local area network (LAN), but less than a wide area network (WAN). A MAN typically interconnects two or more local area networks, may operate at a higher speed, may cross administrative boundaries, and may use multiple access methods. While MAN is a data term, a MAN may carry data, voice, video, image, and multimedia data. The only true MAN technology is SMDS, which, in fact, is limited to the MAN.

Metropolitan Dial The common rotary dial or touchtone

pad that contains both numbers and letters. Dials and pads are also available without the letters. Presumably metropolitan areas required the letters because of multiple central office exchanges, but rural areas with few subscribers and only one CO, required just a few digits and no letters.

Metropolitan Fiber Ring A metropolitan fiber ring is an advanced, high-speed local network that can also be used to connect businesses and residences directly to a long distance carrier's network, and provide alternatives to the local telecommunications services they have today. This definition, courtesy MCI.

Metropolitan Statistical Area MSA. Sometimes known as SMSA, MSAs are areas based on countries as defined by the U.S. Census Bureau that contain cities of 50,000 or more population and the surrounding countries. Using data from the 1980 census, the FCC allocated two cellular licenses to each of the 305 MSAs in the United States.

MF Multi-Frequency.

MFD Abbreviation for Microfarad; one thousandth of a farad, the unit of measuring capacitance. The capacitor is a common electrical device that can store electric charges, and pass AC but not DC. Most phones use capacitors to disconnect the bell during conversations.

MFJ The Modified Final Judgment is the federal court ruling that set up the rules and regulations concerning deregulation and divestiture of AT&T and the Bell system. See Modified Final Judgment for a bigger explanation.

MFLOPS Million Floating point Operations Per Second. A measure of computing power usually associated with large computers. Also known as MEGAFLOPS.

MFM 1. Modified Frequency Modulation. An encoding scheme used to record data on the magnetic surfaces of hard disks. It is the oldest and slowest of the Winchester hard disk interface standards. RLL (Run Length Limited encoding) is a newer standard, for example.

2. Multi Function Module. A term in the AT&T Merlin phone system. MLM is an adapter that has a tip/ring mode for answering machines, modems, FAX machines, and tip/ring alerts, and a Supplemental Alert Adapter mode for 48VDC alerts. It supplies the connection of optional equipment such as answering machines, external alerts, and FAX machines to a Merlin MLX telephone. The MFM is installed inside the MLX telephone.

MFOS MultiFunction Operations System. An AT&T term.

MFP See MULTI-FUNCTION PERIPHERAL.

MFS Multifunction Peripherals. A gadget you connect to your computer that can print, photocopy, fax and scan.

MFSK Multiple Frequency Shift Keying.

MFV German for tone dialing. Mehrfrequenzverfahren

MGCP Master Gateway Control Protocol. See Single Gateway Control Protocol.

MGN Multi Grounded Neutral. A single uninsulated grounded conductor. The currents in the three phases of the transmission line are never quite equal; the MGN carries the residual unbalance current. At many poles, the MGN is physically grounded to a groundrod at the base of the pole. See Joint Pole.

MH Modified Huffman data compression method.

MHF Mobile Home Function.

MHO The unit of conductivity.

MHS Message Handling Service. A program developed by Action Technologies (and others) and marketed by those firms and Novell to exchange files with other programs and send files out through gateways to other computers and mail networks. It is used particularly to link dissimilar electronic-mail systems.

A company running e-mail on their internal LAN will dedicate one computer on the network to be a MHS machine. Every hour or so it will call MCI Mail, CompuServe, etc. and download e-mail messages for people and upload messages from people on the network. Once it has the messages downloaded it will distribute them to the people on the LAN the messages are destined for. See MHS MESSAGE HANDLING SYSTEM.

MHS Enterprise A messaging installation either on a local or corporate-wide level that uses MHS as its backbone between several messaging applications such as E-mail, scheduling, fax, workflow and more. Gateways are used to connect to X.400 systems, public carriers and mainframe systems.

MHS Message Handling System An ISO standard Application Layer protocol that defines a framework for distributing data from one network to several others. It transfers relatively small messages in a store-and-forward manner (defined by ITU-T as X.400 and by ISO as MOTIS/Message-Oriented Text Interchange Standard). See MHS.

MHTML MIME encapsulation of aggregate HTML documents is a proposed standard that would, if deployed on the Internet, allow the easy attachment of complex Web pages — or entire sites — to an email viewer. According to InfoWorld, that means that all components of a Web site could be sent as attachments on a single e-mail and then reassembled with full integrity to produce a functional site for the end viewer, even if that viewer does not have Web access. For example, companies could "push" via email Web or intranet content — as well as applications and software downloads— to employees without giving those employees carte blanche Web access.

MHz An abbreviation for Megahertz. One million Hertz. One million cycles per second. Used to measure band and bandwidth. See BAND and BANDWIDTH. Megahertz is also used by the computer industry to mean millions of clock cycles per second, a measure usually applied to the computer's main microprocessor. Everything that happens in a computer is timed according to a clock which ticks millions of times every second. Higher MHz computers work faster than lower MHz computers. But megahertz is not an accurate measure of a microprocessor's speed. Other factors, such as wider data paths and the ability to execute more than one instruction per clock cycle, affect the actual speed of a microprocessor. Which is why a 100 MHz Pentium chip outpaces a 100 MHz Intel DX4 chip. When comparing the speed of one PC to another, there are other factors also, such as the amount and speed of the system's random access memory (RAM).

MI/MIC Mode Indicate/Mode Indicate Common, also called Forced or Manual Originate. Provided for installations where other equipment, rather than the modem, does the dialing. In such installations, the modem operates in Dumb mode (no Auto Dial capability), yet must go off hook in Originate mode to connect with answering modems.

MIB Management Information Base. MIB is a database of network performance information that is stored on a Network Agent for access by a Network Management Station. MIB consists of a repository of characteristics and parameters managed in a network device such as a NIC, hub, switch, or router. Each managed device knows how to respond to standard queries issued by network management protocols. To be compatible with CMIP, SNMP, SNMP-2, RMON, or RMON-2, devices gather statistics and respond to queries in the manner specified by those specific standards. Within the Internet MIB employed for SNMP (Simple Network Management Protocol)-based management, ASN.1 (Abstract Syntax Notation One) is used to describe network management variables. These variables, which include such information as error counts or on/off status of a device, are assigned a place on a tree data structure. MIB is used in X.400 electronic mail. Many managed devices also have "private" MIB extensions. These extensions make it possible to report additional information to a particular vendor's proprietary management software or to other management software that's aware of the extensions. See CMIP, INTERNET MIB TREE, MIB-2, RMON, RMON-2, SNMP.

MIB Attribute An ATM term. A single piece of configuration, management, or statistical information which pertains to a specific part of the PNNI protocol operation.

MIB-1 The initial collection of objects and attributes defined by the TCP/IP (Transmission Control Protocol/Internet Protocol) standards community. MIB-I was elevated to Internet standard status in May 1990.

MIB-2 The expression MIB refers to the original SNMB MIB definition in IETF RFC 1157. The broader MIB-2 (RFC 1213) adds to the number of monitoring objects supported and is included in SNMP-2's MIB. However, SNMP-2s MIB (RFC 1907) is a superset of MIB-2.

MIB Instance An ATM term. An incarnation of a MIS object that applies to a specific part, piece, or aspect of the PNNI protocol's operation.

MIB Object An ATM term. A collection of attributes that can be used to configure, manage, or analyze an aspect of the PNNI protocol's operation.

MIC 1. Microphone.

2. Medium Interface Connector. FDDI de facto standard connector.

3. A designation for a multiplex which interfaces between two DSIs and one DSIC circuit.

4. Microwave Integrated Circuit - In the microwave industry, a hybrid using thin- or thick-film conductors and passive components on a ceramic substrate combined with chip-form active and passive components.

Michael Harry's favorite son. His only one. Once Michael couldn't read what was in this dictionary. Now he's disagreeing with the definitions, especially the wrong ones. At 16 he became the youngest owner of a licensed, New York State Public Service Commission approved reseller (but not CLEC) telephone company. His company is called OnTheNetAllTheTime. His idea is to wire tony buildings in Manhattan for high-speed Internet access, a market totally ignored by NYNEX and Manhattan Cable. He quickly hit the complexities of the local technologies, the finances and Bell Atlantic's intransigence ... and decided that school was less challenging (i.e. easier). So, now while he tops his bio tests, his father languishes on glacial, analog dial-up Bell Atlantic Internet access. For now I'll take a bright kid and dumb dial-up. But the thought still lurks: for the cost of Michael's private schooling, the Newton household could more than afford a full-time T-1 (I kid you not).

Mickey Unit of space that a mouse moves, measured at 1/200th of an inch.

Micom A manufacturer of data and voice equipment located in Simi Valley, CA.

MICR Magnetic Ink Character Recognition. A process of character recognition where printed characters containing particles of magnetic material, are read by a scanner and converted into a computer-readable digital format.

Micro One-millionth.

Micro Cassette Miniaturized version of the standard audio cassette.

Micro Channel A proprietary bus developed by IBM for its PS/2 family of computers' internal expansion cards. Also

offered by NCR, Tandy and other vendors. See ISA and EISA.

Micro Components Miniaturized audio components that provide the benefits of traditional sized components in far less space.

Micro Farad One millionth of a farad. This is the common unit for designation capacitance in electronics and communications.

Micro To Mainframe Link The telecommunications path over which data between a microcomputer and a mainframe computer travels.

Micro To Mainframe Software Software which provides the logic by which data can be transferred back and forth between a microcomputer and a mainframe computer.

Microbend Loss In an optical fiber, that loss attributable to microbending.

Microcell 1. A PacTel technology to improve cellular calling coverage in areas where high quality cellular service has never been available. PacTel MicroCell, according to PacTel, will reach into canyons, freeway underpasses and high-rise areas where it has sometimes been impossible to provide good cellular coverage. MicroCell uses very small antennas that can be mounted on utility poles or billboards or in other inconspicuous places. The microcell technology can direct the cellular signal right into an isolated trouble spot, leaving broader coverage to conventional cellular sites.

2. A cellular radio cell aimed at serving pedestrians, not automobile drivers. Microcells are smaller, lower-powered, simpler and smaller than macrocells. See MACROCELL.

Microchannel A proprietary bus developed by IBM for its PS/2 family of computers, which it introduced in 1987. The bus was one of IBM's "weapons" against the disturbing increase in clone PCs (disturbing to IBM). IBM figured that introducing a proprietary bus like the microchannel (called MCA for the MicroChannel Architecture), it could slay the clones or control them by charging large amounts of money to license the microchannel architecture. The microchannel, however, engendered a big yawn from consumers, who didn't at the time need its speed, nor its greater complexity, nor its higher cost. As the need to process large amounts of Windows' video information became more urgent in the late 1980s and early 1990s, so faster buses have become more critical. And two new buses, VL and PCI, have been created. For a fuller explanation, see ISA, EISA, PCI and VESA.

Microchips What's left in the bag when the big chips are gone.

Microcode Programmed instructions that typically are unalterable. Usually synonymous with firmware and programmable read-only-memory (PROM).

Microcom Networking Protocol The Microcom Networking Protocol, MNP, is a defacto standard protocol that provides error correction and data compression in dial-up modems. The protocol's design allows for a broad range of services to be implemented, while maintaining compatibility among modems with different levels of MNP capabilities. For example, a modem capable of MNP Class 5 and Class 7 data compression can talk to a modem that lacks MNP data compression.

According to Microcom, MNP is an error correction protocol accepted by international standards authorities (ITU-T Rec. V.42). MNP offers a reliable and widely accepted method of correcting errors in transmissions over dial-up communications lines. MNP incorporates three different data compression methods, including the ITU-T recommendation, V.42bis. Since its original definition, MNP has evolved through nine classes of enhancements. Of those nine classes, the first four provide error control and are in the public domain. Classes 5 through 7 may be licensed from Microcom. Currently MNP error control (Classes 2,3 and 4) has been adopted, along with the LAPM protocol, as mandatory elements of the Consultative Committee on International Telegraphy and Telephony (ITU-T) V.42 recommendation for modem error control.

An Overview Of MNP Service Classes

Class 1
This is the first level of MNP performance. MNP Class 1 uses an asynchronous byte-oriented half-duplex method of exchanging data. MNP Class 1 implementations make minimal demands of processor speeds and memory storage. MNP Class 1 makes it possible for devices with few hardware resources to communicate error-free. Class 1 implementations are no longer included in modems.

Class 2
MNP Class 2 uses asynchronous byte-oriented full-duplex data exchange (i.e., data goes in both directions at once). All microprocessor-based modems are capable of supporting MNP Class 2 performance.

Class 3
This class uses synchronous bit-oriented full-duplex data exchange, eliminating the overhead of start and stop bits used in byte-oriented asynchronous communications. The user still sends data asynchronously to the modem while communications between modems is synchronous.

Class 4
This class introduces two new concepts Adaptive Packet Assembly and Data Phase Optimization, both of which further enhance performance. Adaptive Packet Assembly means that the size of the packets in which data is sent between modems is altered according to the quality of the physical link. The higher the line quality, the larger the packets. Larger packets, while more efficient (the ratio of user data to control data is higher), are also more susceptible to errors. Data Phase Optimization means that repetitive control information is removed from the data stream to make packets more efficient. Both techniques, when combined with Class 3, yield a protocol efficiency of about 120 percent (A V.22bis 2400 bps modem will realize approximately a 2900 bps throughput).

Class 5
This class implements MNP basic data compression to realize a net throughput efficiency of 200 percent on average. (A 2400 bps modem will realize 4800 bps). Class 5 uses a real-time adaptive algorithm to compress data. The real-time aspects of the algorithm allow the data compression to operate on interactive terminal data as well as file transfer data. The adaptive nature of the algorithm means data compression is always optimized for the user's data. The compression algorithm continuously analyzes the user data and adjusts the compression parameters to maximize data throughput.

Class 6
This class implements Universal Link Negotiation and Statistical Duplexing. The first feature allows a single modem to operate at a full range of speeds between 300 and 9600 bps, depending on the maximum speed of the modem on the other end of the link. Modems begin operation at a common slower speed and negotiate the use of an alternative high speed modulation technique. The Microcom AX/9624c modem is an example of a modem that uses Universal Link Negotiation, starting with 2400 bps V.22bis technology and shifting to 9600 bps V.29 fast train technology, if the other modem has that technology too. Statistical Duplexing allows the modem to simulate full-duplex service on the half-duplex

V.29 modem connection.

Class 7
This class implements a more efficient data compression method than the one used in Class 5. The difference between the two classes is that Class 5 realizes an average 200 percent speed improvement over a non-MNP modem, versus an average 300 percent improvement for Class 7. Class 7 data compression uses Huffman encoding with a predictive algorithm to represent user data in the shortest possible Huffman codes. In addition to Class 5 and Class 7 data compression, MNP also supports V.42bis data compression. Based on the Lempel-Ziv-Welsh data compression model, V.42bis supports an average 400 percent efficiency improvement.

Class 8
Not defined.

Class 9
This class reduces the amount of time required for the modem to perform two frequently occurring administrative activities: to acknowledge that a message was received and to retransmit information following an error. Message acknowledgment is streamlined by "piggy-backing" the acknowledgment in its own dedicated packet. Retransmission is streamlined by indicating in the error or Negative Acknowledgment Packet (NAK) the order sequence number of each of the failed messages. Rather than sending all the messages over again (even the good ones) from the point of the error, as is usually done with error correcting protocols, only the failed messages are resent.

Class 10
MNP Class 10 consists of Adverse Channel Enhancements that optimize performance in environments with poor or varying line quality, such as cellular telephones, international telephone calls, and rural telephone service. These enhancements fall into four categories:
1. Multiple aggressive attempts at link setup
2. Adapting packet size to accommodate varying levels of interference
3. Negotiating transmission speed shifts to achieve the maximum acceptable line speed
4. Dynamically shifting to the modem speed most suitable to transmission line conditions
See ERROR CONTROL PROTOCOLS and LZW.

Microcomputer The combination of CPU (Central Processing Unit) and other peripherals (I/O, memory, etc.) that form a basic computer system. See MICROPROCESSOR.

Microfloppies The latest generation of floppy disks at 3 1/2 inches diameter, invented by Sony. The microfloppy is used in the Apple Macintosh and most MS-DOS laptop computers. Used in an MS-DOS machine, a 3 1/2 inch microfloppy diskette will currently format to carry 1.44 million bytes of data — equivalent to about 500 pages of double spaced text.

Microform Microform means Microfiche and Microfilm.

Micrographics Conversion of information into or from microfilm or microfiche.

Micron One thousandth of a millimeter. Or one millionth of a meter. A unit of measurement corresponding to 1/25,000 of an inch. A micron can be used to specify the core diameter of fiber-optic network cabling. This diameter should match your hardware vendor's requirements; but if you install fiber before you buy the equipment, specify the 62.5-micron size.

Micropayment An on-line payment of a dime or less. Touted as the key catalyst for Internet commerce, micropayments were conceived as a means of generating revenues which would be significant for vendors, in the aggregate, while being so trivial to the individual users that they would not hesitate make micropayments freely. While still rhetorical, micropayments were to apply to such services as custom newsfeeds, processing applets and data queries.

Microphone A transducer that changes the air pressure of sound waves into an electrical signal that can be recorded, amplified and/or transmitted to another location.

Microprocessor An electronic circuit, usually on a single chip, which performs arithmetic, logic and control operations, with the assistance of internal memory. The microprocessor is the fabled "computer on a chip," the "brains" behind all desktop personal computers. Typically, the microprocessor contains read only memory — ROM — (permanently stored instructions), read and write memory — RAM, and a control decoder for breaking down the instructions stored in ROM into detailed steps for action by the arithmetic logic unit — ALU — which actually carries out the numerical calculations. There's also a clock circuitry which connects the chip to an exterior quartz crystal whose vibrations coordinate the chip's operations, keeping everything in step. And finally, the input/output section directs communications with devices on the outside of the chip, such as the keyboard, the screen and the various disk drives.

The Fortune Magazine issue of May 6, 1991 contained a very good explanation of chips and microprocessors (usually used interchangeably). Here is the article, slightly condensed:

Chips today can store and retrieve data, perform a simple mathematical calculation, or compare two numbers or words in a few billionths of a second. And they can carry out tens of thousands of such tasks in the blink of an eye. Today's chips contain millions of transistors, capacitors, diodes, and other electronic components, all connected by metallic threads a fraction of the diameter of a human hair. A single chip the size of a fingernail can store dozens of pages of text or combine circuits that can perform scores of tasks simultaneously.

Most chips fall into one of two categories - memory chips and logic chips. Memory chips have the easier job: They merely store information that will be manipulated by the logic chips, the ones with the smarts. Today's biggest-selling memory chip (mid-1991) is the one-megabit dynamic random access memory, or DRAM. Each DRAM is a slice of silicon embedded with a lattice of 1,000 vertical and 1,000 horizontal aluminum wires that circumscribe one million data cells. The densest DRAM designed so far has 64 million cells.

Think of those wires as streets and those cells as blocks. Each block contains a transistor that can be turned on or off — to signify 1 or 0 — and that can be identified by it's unique "address" in the wire grid, much like a house in a suburban subdivision. Each digit, letter, or punctuation mark is represented by 1's or 0's stored in eight-cell strings. (See ASCII.) The word "chip" takes up 32 cells in a memory chip. Most PCs sold today have at least eight one-megabit DRAMs.

It's the job of the logic chips to turn those transistors in the DRAMs on or off, and to retrieve and manipulate that information once it's stored. The most important and complex logic chips are microprocessors like Intel's 80386DX, the brains of the more powerful IBM-compatible PCs sold today. If the structure of a memory chip is a suburban subdivision, the layout of a microprocessor is more like an entire metropolitan area, with distinct neighborhoods devoted to different activities. A typical microprocessor contains among other things:

• A timing system that synchronizes the flow of information to and from memory and throughout the rest of the chip.

• An address directory that keeps track of where data and program instructions are stored in the DRAMs.

• An arithmetic logic unit with all the circuits needed to crunch numbers.

• On-board instructions that control the sequence of microprocessor operations.

Other logic chips in a computer take their cues from the microprocessor millions of times each second to draw images on the screen, to feed instructions from a spreadsheet program, say, out of the disk drives into DRAMs, or to dispatch data to a modem or a printer. Perhaps most amazing of all, memory and logic chips can accomplish all this with just a trickle of electricity - far less than it takes to light a flashlight bulb.

Ted Hoff at Intel invented the microprocessor in 1971. See also 1971 in the beginning of this dictionary.

Microprocessor Controls A control system that uses computer logic to operate and monitor an air conditioning system. Microprocessor controls are commonly used on modem precision air conditioning systems to maintain precise control of temperature and humidity and to monitor the unit's operation.

Microsecond One millionth of a second. A microsecond is ten to the minus six. One microsecond — a millionth of a second — is the duration of the light from a camera's electronic flash. Light that short freezes motion, making a pitched ball or a bullet appear stationary. See ATTO, NANOSECOND, FEMTO and PICO.

Microsegmenting The process of configuring Ethernet and other LANs with a single workstation per segment. The objective is to remove contention from Ethernet segments. With each segment having access to a full 10 Mbps of Ethernet bandwidth, users can do things involving significant bandwidth, such as imaging, video and multimedia.

Microsegmentation Division of a network into smaller segments, usually with the intention of increasing aggregate bandwidth to devices.

Microslot The time between two consecutive busy/idle flags (60 bits, or 3.125 milliseconds at 19.2 kbps). It is used in CDPD only. A cellular radio term.

Microsoft Founded in 1975 by Bill Gates and Paul Allen, Microsoft is (or was at the time of writing this edition of this dictionary) one of the largest software companies in the world. It is the originator of At Work, MICROSOFT AT WORK, MS-DOS, Windows, Windows NT and Windows Telephony. See AT WORK, MS-DOS, WINDOWS, WINDOWS NT and WINDOWS TELEPHONY.

Microsoft At Work A new architecture announced by Microsoft on June 9, 1993 and then put into retirement a couple of years later. Many of its features and ideas surfaced in Windows 95. It consisted of a set of software building blocks that will sit in both office machines and PC products, including:

• Desktop and network-connected printers.

• Digital monochrome and color copiers.

• Telephones and voice messaging systems.

• Fax machines and PC fax products.

• Handheld systems.

• Hybrid combinations of the above.

According to Microsoft, the Microsoft At Work architecture focuses on creating digital connections between machines (i.e. the ones above) to allow information to flow freely throughout the workplace. The Microsoft At Work software architecture consists of several technology components that serve as building blocks to enable these connections. Only one of the components, desktop software, will reside on PCs. The rest will be incorporated into other types of office devices (the ones above), making these products easier to use, compatible with one another and compatible with Microsoft Windows-based PCs. The components, according to Microsoft, are:

• Microsoft At Work operating system. A real-time, preemptive, multi tasking operating system that is designed to specifically address the requirements of the office automation and communication industries. The new operating system supports Windows compatible application programming interfaces (APIs) where appropriate for the device.

• Microsoft At Work communications. Will provide the connectivity between Microsoft At Work-based devices and PCs. It will support the secure transmission of original digital documents, and it is compatible with the Windows Messaging API and the Windows Telephony API of the Windows Open Services Architecture (WOSA).

• Microsoft At Work rendering. Will make the transmission of digital documents, with formatting and fonts intact, very fast and, consequently, cost-effective; will ensure that a document sent to any of these devices will produce high-quality output, referred to as "What You Print Is What You Fax Is What You Copy Is What You See."

• Microsoft At Work graphical user interface. Will make all devices very easy to use and will make sophisticated features accessible; will provide useful feedback to users. Leveraging Microsoft's experience in the Windows user interface, Microsoft At Work-based products will use very simple graphical user interfaces designed for people who are not computer users.

• Microsoft At Work desktop software for Windows-based PCs. Will provide Windows-based PC applications the ability to control, access and exchange information with any product based on Microsoft At Work. Desktop software is the one piece of the Microsoft At Work architecture that will reside on PCs. See also FAX AT WORK, VOICE SERVER, WINDOWS, WINDOWS CE, WINDOWS 95, WINDOWS TELEPHONY and WOSA.

Microsoft Exchange A family of products that offers enterprise computing and information sharing. According to the Windows 95 Resource Kit, Windows 95 includes the Microsoft Exchange client, an advanced messaging application that retrieves messages into one inbox from many kinds of messaging service providers, including Microsoft Mail, The Microsoft Network and Microsoft Fax. Its integration with Microsoft Fax software allows you to send rich-text documents as faxes or mail messages. With Microsoft Exchange client, you can do the following:

• Send or receive electronic mail in a Win 95 workgroup

• Include files and objects created in other applications as part of messages

• Use multiple fonts, font sizes and colors, and text alignments in messages

• Create a Personal Address Book or use address books from multiple service providers

• Create folders for storing related messages, files, and other items

• Organize and sort messages in a variety of ways

• Send and receive messages to and from the following service providers Microsoft Mail, the Microsoft Network (online service), Microsoft Fax and other messaging services that use MAPI service providers.

Microsoft Fax Microsoft Fax is an integral part of Windows 95. It is the first Microsoft operating system to include built-in faxing capabilities. I plucked the following explanation from the Windows 95 Resource Development Kit: With Microsoft Fax, users with modems can exchange faxes

and editable files as easily as printing a document or sending an electronic mail message. Microsoft Fax is compatible with the millions of traditional Group 3 fax machines worldwide, yet it provides advanced security and binary file transfer (BFT) features that make sharing information by means of a fax easier and more powerful. To use Microsoft Fax, you must install Microsoft Exchange. Microsoft Fax has been integrated into Microsoft Exchange as a Messaging Application Programming Interface (MAPI) service provider. All faxes sent to Microsoft Fax are received in the Microsoft Exchange universal inbox. You can send a fax by composing a Microsoft Exchange message, or by using the Send option on the File menu of a MAPI-compatible application (such as Microsoft Excel or Microsoft Word).

Microsoft Solution Provider This is Microsoft's definition: Microsoft Solution Providers are independent organizations that have teamed with Microsoft to use technology to solve business problems for companies of all sizes and industries. SPs use the Microsoft Solutions Platform of products as building blocks and offer various value-added services, such as integration, consulting, software customization, developing turnkey applications and technical training and support. All Solution Providers have at least one Microsoft Certified Professional on staff who has demonstrated expertise in developing, implementing, and supporting Microsoft solutions.

Microsoft Speech API The Microsoft Speech API is a set of applications programming interfaces (APIs) which allow applications to incorporate speech recognition and text-to-speech in their Win32 applications (for Windows 95 or Windows NT). All of the APIs are accessed through the OLE Component Object Model and derive benefits from this object-oriented approach. Like other Windows Open Services Architecture (WOSA) services (e.g. TAPI), the Microsoft Speech API provides a set of standard APIs which allow Windows developers to add speech capabilities to their applications without being tied directly to a specific speech engine technology. Here are some specific areas where speech recognition and text-to-speech can improve a user-interface in any application, and a few uses to avoid. This list of pros and cons is courtesy Mike Rozak, Microsoft software design engineer. What speech recognition is good for...

• Fast access to complex features. Some applications have features that are frequently used but which are difficult to present/control with a GUI. Often times the features can be more easily accessed by speech recognition.

• Magic keystrokes. Many applications have overloaded the keyboard with not only text entry features, but also commands. Some applications have so many keyboard accelerators that they distinguish between ctrl-f2, shift-f2, alt-f2, etc. These are difficult for the end-user to memorize. Using voice commands to replace these makes life easier for the user because voice commands are easier to memorize.

• Macros. Macros are related to magic keystrokes. Many applications allow users to create macros that speed up frequently-done or difficult tasks. To activate these macros, users often have to invent and memorize a magic keystroke. The keystroke can be replaced with speech, whose commands are easier to memorize and are less likely to cause an accelerator conflict.

• Global commands. At times, an application wants to have an input hook that is always active, even when the application doesn't have keyboard focus. If the application uses a keyboard hook then it's likely that the chosen key will already be used by another application, causing conflicts. Because there are so many different possible sentences, an application can provide a global command that is always active and not worry about conflict. Example: A PC-based phone can use global commands for "Call <name>", so that a user can make a phone call even when the phone isn't the active application.

• Anthropomorphize the computer. If you want the user to perceive the computer as a person, then the application should use speech (synthesized or recorded) to talk to the user information, and speech recognition to get information from the user. The computer can listen for commands, or answers to questions, using speech recognition.

• Interaction with characters. Applications that have characters, especially adventure games, have a lot to gain from speech. Simulating characters is a specialized anthropomorphizing of the system.

• Form entry - Applications can use speech recognition for form entry, both for selecting elements from a list, and entering numbers. Because speech can have so many commands active at once, it can be listening for all of the possible values for all of the fields at once, ard infer which field the user wanted filled in. Example: If a user spoke, "Male, sixteen, one hundred and fifty pounds," the application could correctly identify which field was being referred to by "male", "sixteen", and "one hundred and fifty pounds", set the focus to the proper field for each command, and place in the proper data. The user doesn't have to worry about tabbing around.

• Sit back and relax. If you watch users sitting in front of a computer, you'll notice that whenever they type or use the mouse they hunch forward. If you want the user to relax and sit back while using your application, you should provide a speech recognition interface. Example: A consumer title might want to use speech recognition in order to make the computer seem less work-oriented because the user can sit back in his chair without touching the keyboard and mouse.

• Dictation for poor typists. Current dictation technology (which requires that users leave pauses between words) is good for people who cannot type or who are poor typists. The only reason not to include dictation is that dictation systems require about 8 megabytes of extra RAM.

• Access over the telephone. With the arrival of voice-modems in more and more PCs, applications will start providing "phone-based" UIs. Speech recognition is a much better interface than maneuvering through menus with touch-tones. especially since most Europeans don't have touch-tone phones. Example: A user will call up his computer and ask to speak to his PIM application, which then allows him to look up his schedule, address book, etc., all over the phone. The user can merely say, "Give me the phone number for John Smith."

• Accessibility. Some people cannot use the keyboard or mouse effectively. The most common disability is carpel tunnel syndrome or equivalent. Adding speech recognition enables them to use the application.

• Mouse overloaded. Some applications (especially CAD systems) drag, drop, and select objects with the mouse. If the user wants to do any more complex actions, he either has to memorize a magic keystroke (which may not have the information bandwidth), or move the mouse to a menu/toolbar. Instead, the user could just speak, "Rotate this sixty two degrees."

• No keyboard/mouse available. Sometimes a keyboard or mouse is not available. This happens a lot in kiosk situations, where the kiosk has the keyboard and mouse hidden so that users don't feel intimidated, won't break the devices, and won't steal the devices. In order to get input from the users,

current systems use a touch-screen. Why not use speech recognition?

• Hands busy. If a user's hands are busy, then speech is useful. Example: Some dentists' offices use speech recognition to enter charts about the patient's teeth because their hands are busy probing around the patient's mouth.

• Multiple people in front of machine. Speech recognition is good for applications that have several people sitting in front of the same computer, each participating in the application. Normally, an application which is to be used by several users would require that they pass the mouse and/or keyboard around. Since speech recognition is effective up to several feet, users would just have to sit around the computer and talk to it. Where not to use speech recognition...

• Selecting from a large list of words. Most recognition systems break down when more than 100 words/commands are active, so keep lists below this number. For example, allowing the user to address electronic mail to any of 10,000 employees will not work well; instead, it would be better to allow any name from the 100 (or 75, or 50) employees to whom the user frequently sends mail.

• Spelling. Asking a user to spell a word does not work well because many letters such as "m" and "n" sound the same. Instead, it would be better to ask the user to type the word or offer a list of possibilities. Spelling with communications code words_"alpha," "bravo," "charlie," and the like_may be appropriate for certain vertical markets but not for general-purpose applications.

• Entering long sequences of numbers. Most engines will have a very high error rate for long series of digits that are spoken continuously. For phone numbers or other long series, either break the number into groups of four or fewer digits or have the user speak each digit as an isolated word.

• Pointing device. Do not use speech as a pointing device. Speaking "up" five times in a row is very annoying to the user.

• Action games. Because of the background noise (music and sound effects) that is typically present, speech recognition does not work well for action games. The speech-recognition engine spends time processing audio from the computer's speakers and may even recognize it as commands. Speech-recognition works for action games only if the user wears a close-talk microphone (for example, a headset). Additionally, by the time that the user finishes saying, "Fire," it's probably too late.

Microsoft SQL Server A Microsoft retail product that provides distributed database management. Multiple workstations manipulate data stored on a server, where the server coordinates operations and performs resource intensive calculations.

Microsoft TAPI See TAPI and Windows Telephony.

Microspeak A term coined by James Gleick in The New York Times Magazine of June 18, 1997 to refer to the language of euphemisms Microsoft Corporation often indulges in. For example, Mr. Gleick referred to Microsoft's seeming unwillingness to use the word "bug" and use words such as "known issue," "intermittent issue", "design side effect," "undocumented behavior," or "technical glitch."

Microtransaction A small electronic ecommerce transaction — under ten cents. The significance of the term "micro-transaction" is that there is a real need for such small transactions on the Internet — but no one has figured a way to bill economically in such small amounts. Ultimately, someone will. You might need a microtransaction, for example, if you were renting software for small amounts of time.

Microwave Electromagnetic waves in the radio frequency

spectrum above 890 Megahertz (million cycles per second) and below 20 Gigahertz (billion cycles per second). (Some people say microwave refers to frequencies between 1 GHz and 30 GHz.) Microwave is a common form of transmitting telephone, facsimile, video and data conversations used by common carriers as well as by private networks. Microwave signals only travel in straight lines. In terrestrial microwave systems, they're typically good for 30 miles, at which point you need another repeater tower. Microwave is the frequency for communicating to and from satellites.

Microwave Band Loosely defined as those frequencies from about 1 gigahertz upward, Services that use microwave frequencies for point-to-point and point-to-multipoint communications include common carrier, cable TV operators broadcasters, and private operational fixed users.

Microwave Multi-Point Distribution System MMDS. Microwave Multi-point Distribution System. A means of distributing cable television signals, through microwave, from a single transmission point to multiple receiving points. Often used as an alternative to cable-based cable TV. According to an April, 1995 press release from Pacific Telesis, which was starting an MMDS service, "in digital form, it will provide more than 100 channels to a radius of approximately 40 miles from the transmitter. The MMDS transmitter delivers video to homes that are in its 'line of sight.' MMDS transmissions are limited by the terrain and foliage of a given market. The microwave signal is received by an antenna on the subscriber's home, then sent down coaxial cable to a box atop the customer's TV set. The box decodes and decompresses the digital signal."

Microwave Pulse Generator MPG. A device that generates pulses at microwave frequencies.

MID An ATM term. Message Identifier: The message identifier is used to associate ATM cells that carry segments from the same higher layer packet.

Mid-air Passenger Exchange Grim air traffic controller speak for a head-on collision. Midair passenger exchanges are quickly followed by "aluminum rain." Definition from Wired Magazine.

Mid-Span A phone service that runs from a pole to a hook attached to a cable strand before it reaches the building.

Mid-span Meet Sonet's ability to mix the terminal, multiplexing and cross-connect equipment from different vendors. A major accomplishment for standardization. Wish more telecom systems could meet mid-span. Sadly, most can't.

Mid-split A broadband cable system in which the cable bandwidth is divided between transmit and receive functions.

Midband Microsoft's word for telecom speeds that are faster than phone lines but slower than broadband networks. ISDN BRI is midband.

Middleware Middleware is software which sits between layers of software to make the layers below and on the side work with each other. On that broad definition, middleware could be almost any software in a layered software stack. And, to be sure, middleware is not a 100% cast-in-concrete term. It is an evolving term, with many people giving the software the name "middleware" — if they see it fashionable that week. In computer telephony, middleware tends to be software that sits right above that part of the operating system dealing with telephony — TSAPI in NetWare or TAPI in Windows — but below the computer telephony application above it which the user sees on their desktop. In short, middleware is software invisible to the user which takes two or more different applications and makes them work seamlessly together.

MIDI Musical Instrument Digital Interface, a standard for connecting musical instruments, synthesizers and computers. The MIDI standard provides a way to translate music into computer data, and vice versa. A file with a MIDI extension means that it's a file containing music. In contrast, a file with the extension WAV typically means it's a sound file, which means it may contain a voice recording.

Midrange System Medium-scale computer that functions as a workstation or as a multiuser system handling several hundred terminals.

Mid Span Repeater A device that amplifies the signal coming or going to the central office. This device is necessary for ISDN service if you are outside the 18,000 feet distance requirement from the central office.

Midsplit A broadband cable system in which the cable bandwidth is divided between transmit and receive frequencies. The bandwidth used to send toward the head-end (reverse direction) is about 5 MHz to 116 Mhz, and the bandwidth used to send away from the head-end (forward direction) is about 168 MHz to 400 Mhz. The guard band between the forward and reverse directions (100 MHz to 160 MHz) provides isolation from interference. Requires a frequency translator.

MIF 1. Management Information File. MIF is a file format for DMI that describes components within a PC. See WFM.
2. Minimum Internetworking Functionality. A general principle within the ISO that calls for minimum local area network station complexity when interconnecting with resources outside the local area network.

Migration An AT&T marketing strategy designed to encourage all phone equipment month-to-month renters into long-term contracts for AT&T's "flagship products", i.e., System 75, 85 and Definity PBXs.

MII See Media Independent Interface.

Mike Microphone.

MIL 1/1000 of an inch.

Mileage See AIRLINE MILEAGE.

Mileage Sensitive rates Mileage sensitive rates (also called "banded rates") are rates that increase with physical distance within the United States. For example, if you live in New York, a call to (or from) New Jersey will cost less than a call to (or from) California. Things are changing, however. Most calls now in the United States are distance insensitive, and are only time sensitive. They charge by time, not far you call.

Millennium Bug The year 2000. The millennium bug. What happens when your computer hits the year 2000. There are three happenings. First, it will move its date to 1900 and stay there. In this case programs that use date calculations — like figuring how much money you have in your insurance policy will screw up. For example, because many mainframe programs used 95, instead of 1995 in the date field. That means that when it hits 00 or 01, the computer won't be able to figure that the difference between 01 and 95 is 6. (Figure it. 2001 minus 1995 = 6). Second, it will move its date to 1900 and you'll be able to manually change it back. Third, it will move its date to 2000 flawlessly. Most PCs manufactured after 1995 will handle the Y2K problem flawlessly. Many mainframe and minicomputers manufactured before that won't. Some phone systems will also mess up. They may stop working altogether. They may work in strange ways, like producing the wrong information on calls being made or being received. For more info: www.ibm.com/IBM/year2000; www.microsoft.com/CIO/articles/YEAR2000faq.htm; www.year2000.com. See 2000.

Miller Freeman Miller Freeman is a San Francisco-based subsidiary of United News & Media of the U.K. Having sold our publishing company to them in September 1997, they now have the great distinction of publishing this wonderful dictionary.

Milli One thousandth. Millisecond equals one thousand of a second.

Milliwatt One thousandth of a WATT. Used as a reference point for signal levels at a given point in a circuit.

MILNET MILitary NETwork. Along with DARPANET, MILNET was created in 1983 as successor to the ARPANET. One of the DDN networks that make up the Internet; devoted to non-classified U.S. military communications. SEE ARPANET.

Milspec Military Specification. Milspecs are very demanding.

MIM Metal-Insulator-Metal. A display technology which uses active matrix technology that uses diodes behind each pixel to produce images. It is an improvement on passive displays but a step behind TFT (Thin Film Transistor) technology. See also LCD.

MIMD Multiple Instruction Multiple Data is a type of parallel processing computer, which includes dozens of processors. Each processor can run different parts of the same program and execute those instructions on different data. This makes it more flexible, though more expensive than a computer running SIMD — single instruction multiple data.

MIME Multipurpose Internet Mail Extensions. Developed and adopted by the Internet Engineering Task Force (IETF) in RFC 1521, MIME is the standard format for including non-text information in Internet mail, thereby supporting the transmission of mixed-media messages across TCP/IP networks. The MIME protocol, which is actually an extension to SMTP, covers binary, audio and video data. MIME also is the standard for transmitting foreign language text (e.g., Russian or Chinese) which cannot be represented in plain ASCII code. Here's an explanation: When sending files which aren't plain US-ASCII across a network — dial-up, leased or the Internet — you basically have two options. First, you can attach them as a binary file (i.e., non-ASCII file). Or second, you can encode them into ASCII characters and send the file as part of your message. The first method is preferable. But you can typically only send binary files from one account to another on the same network, or between two networks that have agreed between themselves to a method of transferring files. That's a rarity. And it certainly doesn't work in and around the Internet. Therefore, MIME was created, employing a base64 coding scheme, which involves relatively simple encoding and decoding algorithms. See also base64, BinHex, Multipurpose Internet Mail Extensions and UUencode.

MIMJ Modified Modular Jack. These are the 6-pin connectors used to connect serial terminal lines to terminal devices. MIMJ jacks can be distinguished from the familiar RJ11 jacks by having a side-looking tab, rather than a center-mounted one.

MIN Mark Can be 0 or 1. Your home cellular station sends extended address data upon origination and page response. See MIN1.

MIN1 Mobile Identification Number. A 24-bit number corresponding to the actual 7-digit telephone number assigned by the cellular carrier exclusively to your phone, used for both billing and for receiving calls. The MIN is meant to be changeable, as the ownership of the device may change hands, the owner may change telephone numbers, or the owner may change cities. The MIN is paired with the Electronic Serial Number (ESN). Theoretically, both numbers are verified, and in combination, every time a call is placed in order to verify the legitimacy of the device and the call. See MIN2 and ESN.

MIN2 The area code of your cellular phone number.

Mini-COW See CCW.

Mini-floppy A floppy disk that is 3 1/2 inches in diameter. Also called a microfloppy. See MICROFLOPPY.

Mini-Link A generic name for an Ericsson family of compact radio links of one or more 2 megabit per second channels over distances of up to 30 km per hop.

Mini-MAP Mini-Manufacturing Automation Protocol. A version of MAP consisting of only physical, link and application layers intended for lower-cost process-control networks. With mini-MAP, a device with a token can request a response from an addressed device. Unlike a standard MAP protocol, the addressed Mini-MAP device need not wait for the token to respond.

MiniCLAS A service from Nynex which allows a small (up to 199 lines) Centrex III user to make certain changes to his own Centrex features and lines with an easy-to-use computer program or a touchtone phone. See also CLAS.

Minimal Regulation Regulated local telephone company under limited state regulation has the ability to file price lists for services and those price lists are usually effective on 10-days' notice to local state commission and customers. This means there are no extensive hearings.

Minimize Button The minimize button is the down-arrow button at the immediate right of a title bar in Windows software. Clicking the minimize button will shrink a window to its icon.

Minimum Internetworking Functionality MIF. A general principle within the ISO that calls for minimum local network station complexity when interconnecting with resources outside the local network.

Minimum Point of entry The closest practical point to where the carrier facilities cross the property line or the closest practical point to where the carrier cabling enters a multi-unit building or buildings.

Minitel French name for videotex. Very popular. See VIDEOTEX and packet switched networks.

MIPS Millions of Instructions Per Second. A measure of computer speed that refers to the average number of machine language instructions performed by the CPU in one second. A typical Intel 80386-based PC is a 3 to 5 MIPS machine, whereas an IBM System 370 mainframe typically delivers between 5 and 40 MIPS. MIPS measures raw CPU performance, but not overall system performance.

MIR Maximum Information Rate: See also PCR.

Mirror A term used to reference Internet FTP sites that copy files from other archives every day or so. By accessing a mirror site close to your location you reduce transmission over the Internet.

Mirror Server Imagine you're a big international software company with branches in every country and in every town in the entire universe. You have zillions of users. Every day your users need to download software from you — fixes, upgrades, etc. If you have only one server (big computer to serve multiple users), that server will bog down in heavy traffic. Also, all your customers overseas will receive their software very very slowly because they're farther away. So the solution is to install a bunch of "mirror servers" all around the world. Each server would be a mirror image of the main one — the one at your headquarters. That means its content and structure would be identical. The only difference is that servers all around the world would be closer to the users. And when the users come first on-line to your main server and find software they need, they would then be given the choice of having their material downloaded to them from a server that's closer to them. Go to www.microsoft.com. They often allow you to download some new software. As you are about to download, they ask you which server in which country you'd like.

Mirrored Server Link MSL. A dedicated, high-speed, point-to-point connection between the primary server and the secondary server. The MSL can either be a coaxial or fiber-optic cable (with maximum distances of 100 feet and 2.5 miles, respectively). See MSL.

Mirroring A fault tolerance method in which a backup data storage device maintains data identical to that on the primary device and can replace the primary if it fails. Mirroring will typically cost you a 50% performance degradation when your write to disk and 0% performance degradation when you are reading. For a full explanation, see DISK MIRRORING.

MIS 1. Management Information System. A fancy name for Data Processing. MIS are also the first three letters of the word MISanthrope and MISguided and lots of other MIS words. MIS departments are taking over corporate telecommunications departments which is why there's little love lost between the two.
2. IBM-speak for Management Initiated Separation. Translation: You're fired.

Miscellaneous Common Carrier A communications common carrier (typically one using microwave) which is not offering switched service to the public or to companies. A miscellaneous common carrier usually provides video and radio leased line transmission services to TV and radio networks.

Miscellaneous Trunk Restrictions Denies preselected lines access to preselected trunk groups (e.g., FX or WATS trunks). A call attempt over a restricted group routes to an intercept tone.

MISSI Multilevel Information Systems Security Initiative. Developed under the leadership of the NSA (National Security Administration), MISSI is a framework for the development and evolution of interoperable, complementary security products intended to provide flexible, modular security for networked information systems. MISSI encompasses a suite of security technologies developed to support national defense operations across the Defense Information Intrastructure (DII) and the National Information Infrastructure (NII), and with application in secure corporate environments. Included in the MISSI Security Solutions suite are a set of best security practices, as well as endorsement of compliant products which implement elements of the architecture. A fully compliant MISSI architecture controls system access by "level" (e.g., unclassified, sensitive, confidential, secret and top secret) and "compartment" (i.e., topical area of interest). Authentication and encryption are provided courtesy of the Fortezza family of chips developed by Mykotronx Inc. See also FORTEZZA.

Mission Critical Systems Systems on which the future success of an organization depends.

Mitel Mitel is a PBX and computer telephony component maker. The word Mitel actually stands for MIke and TErry's electric Lawn mower. Here's how it happened. In the early 1970s, Mike Cowpland and Terry Matthews were engineers working in the semi-conductor factory of Northern Telecom in Kanata, Ontario, Canada. They decided they wanted to go out on their own. Their first idea was cordless, electric lawn mowers. Apparently they thought there was a demand such devices. They formed Mitel and ordered lawn mowers from a manufacturer in England. The mowers arrived late, after the onset of the Canadian winter. No one wanted lawn mowers. So they decided to produce telephone systems. They went more successfully. Mitel went public. Most of the company was later sold to British Telecom, who lost their shirt on it and eventually sold it on a fire sale to a venture capital company. Terry moved on and founded the eminently successful Newbridge Networks. Mike got involved with Corel.

Mixed Cable Cables have characteristic impedances which vary according to the cables' physical parameters. When cables of different characteristic impedance are mixed, an impedance imbalance will occur. It is thus bad practice to mix different wire gauges, and twisted or non-twisted pair cables. However, it is accepted practice to combine long runs of twisted pair cable with short lengths of modular patch cords and line cords. Baluns are an exception to this rule.

Mixed Mode An imprecise term which suggests that one digital bit stream can carry voice, data, facsimile and video signals.

Mixed Mode Night Service After-hours answering of incoming calls in which Assigned Night Answer (ANA) is specified for some trunks, and Universal Night Answer (UNA) specified for others.

Mixed Station Dialing A telephone system feature which allows you to install both rotary and pushbutton phones on your phone system. Most modern PBXs have this feature.

MJ Modular Jack. A jack used for connecting voice cables to a faceplate, as for a telephone. See RJ (Registered Jack).

MK2 MK2 is the code name for the first product resulting from the best-of-breed combination of SCO's SCO OpenServer Release 5 operating system with Chorus Systems' open microkernel technology. MK2, a binary product initially targeted at the Intel ix86 platform, is specifically designed for telecommunications manufacturers. According to its maker, it provides the modularity and scalability needed to meet a wide range of telecommunications products ranging from switches to telephony servers. Scheduled to ship in early 1997, MK2, says its maker, enables telecommunications vendors to benefit from the open systems software market while protecting their current investment in legacy telecommunications-specific software.

MLHG MultiLine Hunt Group

MLID Multiple Link Interface Driver. A layer of the Novell Open DataLink Interface specification. The MLID layer controls a specific network interface, and works below the Link Support Layer.

MLL Monthly Leased Lines.

MLM Meridian Link Module. A Northern Telecom term for an Application Module that provides a link to a host processor through the Meridian Link interface.

MLPP See MULTILEVEL PRECEDENCE PREEMPTION.

MLT Mechanized Loop Testing.

MLT-3 Multi Level Transmit - 3 Levels. The ANSI approved modulation scheme used for the transmission of data on an FDDI network over shielded and unshielded copper twisted pair media (see TP-PMD).

MM Mobility Management. A wireless industry term.

MMCF Multimedia Communications Forum. Formed in June 1993, the MMCF is an non-profit research and development organization of telecommunications service providers, multimedia application and equipment developers, and end users who realize the revolutionary potential of multimedia communications. Forum members are dedicated to accelerating market acceptance and multivendor interoperability of multimedia communications worldwide. MMCF acts as a central clearinghouse for all multimedia communications-related standards, specifications and recommendations.

MMDS Microwave Multi-point Distribution System or

Multipoint Multichannel Distribution Service. Nobody seems to know which MMDS means. But, irrespective of what the acronym means, the definition is the same. MMDS is a way of distributing cable television signals, through microwave, from a single transmission point to multiple receiving points. Often used as an alternative to cable-based cable TV. According to an April, 1995 press release from Pacific Telesis, which was starting an MMDS service, "in digital form, it will provide more than 100 channels to a radius of approximately 40 miles from the transmitter. The MMDS transmitter delivers video to homes that are in its 'line of sight.' MMDS transmissions are limited by the terrain and foliage of a given market. The microwave signal is received by an antenna on the subscriber's home, then sent down coaxial cable to a box atop the customer's TV set. The box decodes and decompresses the digital signal." MMDS is increasingly being called "Wireless Cable." See WIRELESS CABLE.

MME Mobility Management Entity.

MMF Multimode Fiber optic Cable: Fiberoptic cable in which the signal or light propagates in multiple modes or paths. Since these paths may have varying lengths, a transmitted pulse of light may be received at different times and smeared to the point that pulses may interfere with surrounding pulses. This may cause the signal to be difficult or impossible to receive. This pulse dispersion sometimes limits the distance over which a MMF link can operate.

MMFD Abbreviation for micromicrofarad; one millionth of a farad, the unit of measuring capacitance.

MMI Machine-to-Machine Interface.

MMIC Microwave Monolithic Integrated Circuit.

MMITS Modular Multifunction Information Transfer System. Hiding behind something call the MMITS Forums is a group of people working to define software programmable radios. There are two areas of emphasis:

Handheld — working with being able to download software into cellular handsets so they can work with a variety of air interfaces.

Mobile — concentrating on military requirements initially, but looking at the needs of public safety (Police, Fire, etc.) for the future.

The basic idea is to bring PC concepts to the radio world by moving the Digital/Analog — Analog/Digital function very close to the antenna, and do all of the tuning, spread/despread, modulation/demodulation, etc with DSPs. www.mmitsforum.org

MMJ A six wire modular jack with the locking tab shifted off to the right hand side. Used in the DEC wiring system.

MML Man Machine Language.

MMR Modified Modified Read data compression method used in newer Group 3 facsimile machines.

MMSU Modular Metallic Service Unit.

MMTA MultiMedia Telecommunications Association. The successor organization to NATA. MMTA is organized around five divisions — computer telephony integration, conferencing/collaboration and messaging, LAN/WAN internetworking, Voice/Multimedia and Wireless Communications. MMTA is on 202-296-9800. In November, 1996 MMTA announced its intent to merge with TIA, another Washington organization called Telecommunications Industry Association. On December 15, 1997 MMTA announced that it had been officially combined into the Telecommunications Industry Association. www.mmta.org and www.tiaonline.org. See ACTAS and NATA.

MMU Memory Management Unit. Circuitry that manages the swapping of blocks of memory.

MMVF A format of rewritable DVD disc proposed by NEC. It stands for Multi Media Video File. For a longer explanation, see DVD-RAM and DVD.

MNA Multimedia Network Applications.

MNA7 Multiple CCS7 Network Addresses

MNC Mobile Network Code. A part of the IMSI (International Mobile Subscriber Identity) or LAI.

Mnemonic From the Greek mnemonikos, a shorthand label or term that is easy to remember. A mnemonic is a symbolic representation of an address (e.g., ATL for Atlanta, or DLS for Dallas) or operation code (e.g., JMP for jump). Acronyms are a type of mnemonic; LASER, for instance, is shorthand for Light Amplification by Stimulated Emission of Radiation. See also Acronym.

Mnemonic Dial Plan Pronounced "nemonic." A way of dialing using characters typed on the keyboard of a terminal. The word Mnemonic comes from the same roots as memory. It's a memory jogging way of remembering something, like a way to dial. See MNEMONIC PROMPTS.

Mnemonic Prompts System commands represented by the appropriate alphabet letter rather than by a number, (for example, "P" to "Play", "A" to Answer"). See MNEMONIC DIAL PLAN.

MNLP Mobile Network Location Protocol.

MNP10-EC Error correction protocol for awful communications environments, like cellular networks. Use of MNP10-EC helps prevent disruptive signal fading and reduces the number of dropped calls that occur when you're trying to send data over cell networks. See MICROCOM NETWORKING PROTOCOL for a greater explanation.

MNRP Mobile Network Registration Protocol.

Mobile Application Subsystems MAS. That portion of a Mobile End Systems (M-ES) concerned with the provision of application services. The MAS contains the applications software that is independent of the CDPD network. In most cases, the includes network software.

Mobile Attenuation The power of the mobile phone can be adjusted (or attenuated) dynamically to one of seven discrete power levels (analog cellular). This is done so that when a mobile comes closer to a base receiver its power is reduced to prevent the chance of interfering with other mobiles operating on the same voice channel in another cell (co-channel interference). Additionally, this is even more important to portable units to keep the transmit power at a minimum to increase the talk usage time before the batteries expire.

Mobile Cellular Phone The cellular handset unit permanently mounted in a vehicle. See CAR PHONE.

Mobile Cellular Service A fancy name for cellular phone service or cell phone service. See CELL, CELL SWITCHING, CELLULAR RADIO and IN.

Mobile Controlled Handoff This means the decision to initiate a transfer or handoff from one cell to another cell is under the control of the mobile device. Used in CDPD.

Mobile Data Mobile data is a generic term used to describe data communications through the air from and to field workers — from package deliverers, to car rental companies (to track cars), to field service personnel, to law enforcement officials checking license plates.

Mobile Data Base Station MDSB. Component of the CDPD network that provides data link relay functions for a set of radio channels serving a cell. An MDBS is located in each cell site, and its primary role is to relay data between Mobile End System (M-ES) and the Mobile Data Intermediate System (MD-IS). it is the stationary network component responsible for managing interactions across the airlink interface.

Mobile Data Intermediate System MD-IS. The CDPD network element that performs routing functions based in knowledge of the current location of the M-ES. Responsible for CDPD mobility management. A cellular radio term.

Mobile Data Link Protocol MDLP. The Link Layer protocol used in Cellular Digital Packet Data (CDPD). Provides Temporary Equipment Identifier (TEI) management, multiple frame operation, unidata transfers, exception condition detection with selective reject recovery, etc.

Mobile Digital Voice Channel MDVC. The channel between a mobile phone and a cell site antenna in a digital cellular or PCS environment. The MDVC supports both voice and data transmission, although the allocated bandwidth is designed primarily to support voice. Signaling and control functions take place over separate channels set aside specifically for that purpose.

Mobile End System M-ES. An end system that accesses the CDPD network through the airlink interface. The device that allows mobile users to work in an untethered fashion while remaining connected to a data network. The system's physical position may change during data transmission. A cellular radio term.

Mobile Home Function A Mobile Data Intermediate System, that (1) maintains an information database of the current serving area of each of its homed Mobile End Systems (M-ESs), and (2) operates a packet forwarding service for its homed M-ESs. A cellular radio term.

Mobile Identification Number When the "SEND" key on a cellular phone is pressed, the phone transmits an origination message to the base station. This message includes the dialed digits and the identity of the calling cellular phone. The calling cellular phone is identified by its Mobile Identification Number (MIN), which is usually the same as its ten-digit phone number. See also ESN.

Mobile IP An emerging set of extensions to the Internet Protocol for packet data transmission, Mobile IP is intended to serve nomadic users connecting on a wireline, rather than a wireless, basis. Mobile IP is being developed by the IETF (Internet Engineering Task Force) to operate much like a highly secure and dynamic packet data communications version of a postal service forwarding address. The benefit is that the nomadic user will not have to continually change IP addresses and reinitialize sessions. It will work like this:

The mobile node will have one permanent address and another for location purposes and another for identifying it to other network nodes. Data will be transmitted to the permanent address, associated with the "home agent." When the nomadic node is traveling, the "home agent" will forward the data in care of the "foreign agent," the IP server serving the foreign subnet, through a process of encapsulating that data with another IP address contained in a data header preceding the original packet. Once the data packets are received by the foreign agent, the additional header will be removed through a process known as decapsulation. Should the node relocate yet another time, both the "home agent" and the previous foreign agent will be advised of that fact; thereby, inflight packets can be forwarded by the previous foreign agent to the new foreign agent through a process known as "smooth handoff." While there currently is no Mobile IP standard being developed for wireless mobility, Mobile IP promises to make life easier for users that roam from location to location within a multisite corporate enterprise. See also IP.

Mobile Mounting Kit An optional cellular phone acces-

sory that allows a transportable or portable to be connected to a vehicle's power supply and antenna lead, thereby boosting power and improving reception. Sometimes referred to as a car kit or car mounting kit. Some of these kits are very expensive. Check the price of the kit before you buy your phone.

Mobile Network Location Protocol MNLP. A cellular radio term. In the CDPD network, the MNLP is the protocol used between the Home Mobile Data Intermediate System (MD-IS) and the Serving MD-IS and it used to keep the Home MD-IS updated on the location of a Mobile End System (M-ES) (i.e. the location of the cell phone).

Mobile Network Registration Protocol MNRP. In the CDPD cellular radio network, protocol used between the Mobile End System (M-ES) and the Serving Mobile Data Intermediate System (MD-IS) to announce the M-ES's Network Entity Identifier (NEI) and to confirm the service provider's willingness to provide service.

Mobile Phone One term for a cellular phone. There are four main types of cellular phones — mobile (also called car phone), transportable, portable and personal. A mobile phone is attached to the vehicle, the vehicle's battery and has an external antenna. The mobile phone (the car phone) transmits with a standard three watts of power. Mobile telephone service is provided from a broadcast point located within range of the moving vehicle. That range is called a "cell." The broadcast point in turn is connected to the public network so that calls can be completed to or from any stationary telephone, i.e. one connected to a land line. See CELLULAR and CAR PHONE.

Mobile Serving Function MSF. A Mobile Data Intermediate System (MD-IS) function that (1) maintains an information database of the Mobile End Systems (M-ES) currently registered in the serving area, and (2) de-encapsulates forwarded packets from the MHF and routes them to the correct channel stream in a cell where the destined M-ES is located.

Mobile Switching Center MSC. The location of the Digital Access and Cross-connect System (DACS) in a cellular telephone network.

Mobile Telematics Sometimes just called telematics.It involves integrating wireless communications and (usually) location trackingdevices (generally GPS) into automobiles. The best known example is GM's OnStar system, which automatically calls for assistance if the vehicle is in an accident. These systems can also perform such functions as remote engine diagnostics, tracking stolen vehicles, provide roadside assistance, etc. www.onstar.com

Mobile Terminated The term used to describe a call where the destination of the call is a mobile (i.e. cellular) telephone.

Mobile Unit The cellular telephone equipment installed in a vehicle. It consists of a transceiver, control head, handset and antenna.

Mobilink A "unified" cellular phone service covering 83% of North America's population. It is a consortium of six Bell cellular operators and some Canadian cellular operators. The idea is simple. Anyone calling a subscriber of one of these companies would have his call automatically routed to the subscriber, no matter where in Mobilink that subscriber was. Before Mobilink, you had to know where the person was you wanted and then dial a bunch of complex codes to get to him.

Modal Dispersion Modal dispersion can be thought of as the blurring of the input signal in a fiber by the several modes that may propagate down a fiber. Each mode may take a separate path down the fiber, and thus that signal may arrive at slightly different times. The dispersion depends on the fiber's internal characteristics and its length. In short, modal dispersion is pulse rounding in lightwave communications that takes place because of the slightly different paths followed by the laser light rays as they arrive at the detector slightly out of phase.

Modal Distribution In an optical fiber operating at a single wavelength, the number of modes supported by the fiber, and their propagation time differences. In an optical fiber operating at multiple wavelengths simultaneously, the separation in wavelengths among the modes being supported by the fiber.

Modal Loss In an open waveguide, such as an optical fiber, a loss of energy on the part of an electromagnetic wave due to obstacles outside the waveguide, abrupt changes in direction of the waveguide, or other anomalies, that cause changes in the propagation mode of the wave in the waveguide.

Mode 1. A mode is a stable propagation of state of an electromagnetic wave in an optical fiber. In rough terms a mode can be thought of bundles of light rays, of the same wavelength, that enter a fiber at the same angle. See MODAL DISPERSION.

2. Mode is essentially a switch inside a computer that makes it run like another computer, usually an older one.

Modem 1. Acronym for MOdulator/DEModulator. Conventional modems comprise equipment which converts digital signals to analog signals and vice versa. Modems are used to send digital data signals over the analog PSTN (Public Switched Telephone Network. Although the carrier switches (e.g., central offices and tandem offices) are typically digital, as is the backbone transmission network (e.g., T-carrier), the local loop always is analog unless the user orders a more costly digital loop (e.g., ISDN or T-1). Therefore, the PSTN is analog as far as most people are concerned.

Conventional modems work like this. Your PC outputs data in the form of "1's" and "0's" which are represented by varying levels of voltage. The modem converts the digital signal into variations of the analog sine wave so the data can be transmitted over the PSTN. A matching modem on the other end reverses the process in order to present the target device with a digital bit stream. The modulation techniques include some combination of Amplitude Modulation (AM), Frequency Modulation (FM) and Phase Modulation (PM), also known as Phase Shift Keying (PSK). Used in combination, these techniques allow multiple bits to be represented with a single (or single set) of sine waves. In this fashion, compression is accomplished, which allows more data to be transmitted in the same period of time and which therefore reduces the connect time and the associated cost of the data transfer. Contemporary, conventional modems are standardized by the ITU-T as part of the "V" series of standards. Such modems are characterized by error detection and correction mechanisms, adaptive equalization, internal dialing, and numerous other sophisticated capabilities. 56 Kbps modems are the latest development in the world of conventional modems; they remain to be standardized. The term "modem" also is applied (and correctly so, in the purely technical sense) to ISDN TAs (Terminal Adapters), ADSL TUs (Terminating Units), line drivers and short-haul modems. The last two, in fact, are voltage converters.

See also LINE DRIVER, MODEM ELIMINATOR, MODEM POOL, MODEM STANDARDS, MODULATION PROTOCOLS, SERIAL PORT, SHORT-HAUL MODEM and 56 Kbps MODEM.

2. According to the Vermonter's Guide to Computer Lingo, modem is what landscapers do to dem lawns. (This is a joke.)

Modem Bonding A term which describes the bonding, or linking, of two 56 Kbps modems over two phone lines to double the performance. This process is accomplished through matching devices, one on each end of the connection; each modem operates at its maximum achievable rate, with the aggregate rate being roughly double that of each individual modem. Theoretically, modem bonding can yield speeds of as much as 128 Kbps downstream and 67.2 Kbps upstream, although the FCC currently limits maximum performance through each modem to 53 Kbps, for a total of 106 Kbps. Modem bonding technologies are proprietary. See 56 KBPS MODEM.

Modem Cowboy A slang term for someone who typically lives in the mountain states of Western America and does most of his work via modem.

Modem Eliminator A wiring device designed to replace two modems; it connects equipment over a distance of up to several hundred feet. In asynchronous systems, this is a simple cable. Here is a specific application using a Modem Eliminator: You can connect a PC to a printer, or a PC to another printer using a cable. But you can only go a certain distance — maybe 100 feet. After that, the traditional solution has been to use a modem and go over traditional phone lines. Instead, you can connect the two devices directly by wire using a Modem Eliminator. There are two advantages of a Modem Eliminator over a normal modem. The eliminator is cheaper and it can often transmit faster. According to Glasgal Communications, there are many cases where it is either unnecessary, cumbersome or too expensive to interconnect terminals using modems or line drivers in an experimental or a very short-haul environment. A modem eliminator functionally resembles two modems back-to-back on a leased line and therefore saves the cost of two modems and a line in many situations.

Modem Fallback When the telephone line quality is not good enough to accommodate the top rated speed of a modem — for example 14,400 bps — the modem drops down to lower speeds — initially to 9,600 bps, then if necessary, to 4,800 bps, or even down to 2,400 bps.

Modem Pool A collection of modems which a user can dial up from his terminal, access one and use that one to make a data call over the switched telephone network. Modem pools are obviously designed to allow many users to share few modems, thus saving on modems. Now that modems have become less expensive, the advantages of modem pooling are no longer as great. There are also some advantages in having a modem right next to your terminal or computer — namely you can see how it is functioning. And modems have lights to indicate what they're doing. One of the most useful lights on most modems indicates whether the line is "off hook" or not. It is possible for your computer to instruct your modem to hang up the line and for your modem to forget to do it, leaving you with a huge phone bill. One problem with giving people their own modems is they (the modems) have a tendency to get pinched. It's hard to screw down modems. Harder, anyway, than computers and disk drives.

Modem Server A networked computer with a modem or group of modems attached to it that allows network users to share the modems for outbound calls.

Modem Standards Definitions of electrical and telecommunications characteristics which enable modems from dissimilar manufacturers to speak to each other. Bell 103...US standard for 300 bps; ITU-T V.21...International standard for 300 bps; Bell 212A...US standard for 1200 bps; ITU-T V.22...International standard for 1200 bps; ITU-T V.22

bis...US and international standard for 2400 bps; ITU-T V.23...International videotex standard (1200/75 bps or 75/1200 bps). See also HAYES COMMAND SET and V.SERIES recommendations, i.e. V.34.

Modem Turnaround Time The time needed for a half-duplex modem (an old-fashioned one) to reverse its transmission direction.

Moderated Mailing List A mailing list where messages are first sent to the list owner before they are distributed to the subscribers.

Moderator A moderator is a person who controls what gets posted to a particular Internet newsgroup, A moderator is used to ensure that a newsgroup's article stick to the agreed upon subject matter. A newsgroup may or may not have a moderator.

Modified Final Judgment MFJ. The agreement reached on January 8, 1982 between the United States Department of Justice and AT&T approved by the courts on August 24, 1982 that settled the 1974 antitrust case of the United States versus AT&T. The MFJ divested AT&T of the local regulated exchange business and created seven regional holding companies — Ameritech, Bell Atlantic, Bell South, NYNEX, Pacific Telesis, Southwestern Bell and US West. The MFJ placed restrictions on the local exchange carriers, namely that they couldn't get into long distance communications. The Modified Final Judgment did not prohibit AT&T from providing local telecommunications, it prohibited AT&T from purchasing the stock of the divested RBOCs.

Modified Finite Queue A traffic engineering term. Erlang C assumes an infinite queue, that is, callers will wait indefinitely to have their call answered. Since this is obviously not the case, some parties have suggested that a different algorithm should be used in order to produce more accurate forecast. In practice, while Erlang C will produce some degree of overstaffing based on its assumption of infinite queuing, alternatives that assume finite queues result in understaffing. Since most call centers would prefer slight overstaffing, and a greater likelihood of meeting grade of service, to under-staffing, with a greater degree of customer dissatisfaction, Erlang C continues to be the preferred and recommended algorithm.

Modified Frequency Modulation MFM. An encoding scheme used to record data on the magnetic surfaces of hard disks. It is the oldest and slowest of the Winchester hard disk interface standards. RLL (Run Length Limited encoding) is a newer standard, for example.

Modified Huffman A one-dimensional data-compression scheme that compresses data in a horizontal direction only. Allows no transmission of redundant data.

Modified Read A two-dimensional coding scheme for facsimile machines that handles the data compression of the vertical line and that concentrates on space between the lines and within given characters.

Modular Equipment is said to be modular when it is made of "plug-in units" which can be added together to make the system larger, improve its capabilities or expand its size. There are very few phone systems that are truly modular.

Modular Breakout Adapter Allows the technician to access each individual conductor of a cable. Sometimes called a "banjo clip." It's a rectangular plastic box, with conductors on the sides and a modular plug at the long end.

Modular Cord A cord containing four twisted pairs of wires with a modular plug on one or both ends.

Modular Jack A device that conforms to the Code of Federations, Title 47, part 68, which defines size and configuration of all units that are permitted for connection to the public telephone network.

Modular Plug Connecting devices adopted by the FCC as the standard interface for telephone and data equipment to the public network. These are the plastic "ends" you see on cables. They come in two conductor, four, eight and six. Two, four and six conductor plugs are the same physical size, and are usually used for telephone voice and low speed data communications. Eight conductor (four pair) plugs are wider, and most often used for data, e.g. Ethernet LAN connections. There are several wiring configurations for modular plugs. The most common are T568A and T568B. See UTP Cable. It's important to match the modular plugs to the type of cable you are using. Plugs made for stranded cable will not work with solid conductor wire because they're designed to pierce the cable in-between the strands. Used with a solid conductor cable, they don't pierce the cable and just get smashed. Plugs made for solid conductors usually work with stranded cable. But I wouldn't recommend trying.

Modulated Waves Alternating current waves which have their amplitude varied periodically. The signals transmitted by a radio station are examples of a modulated wave.

Modulation The process of varying some characteristic of the electrical carrier wave as the information to be transmitted on that carrier wave varies. Three types of modulation are commonly used for communications, Amplitude Modulation, Frequency Modulation and Phase Modulation. And there are variations on these themes called Phase Shift Keying (PSK) and Quadrature Amplitude Modulation (QAM).

Modulation Index In angle modulation, the ratio of the frequency deviation of the modulated signal to the frequency of a sinusoidal modulating signal. The modulation index is numerically equal to the phase deviation in radians.

Modulation Protocols Modem stands for MOdulator/DEModulator. A modem converts digital signals

INTERNATIONALLY ACCEPTED MODEM MODULATION PROTOCOLS					
Standard	Speed bps	Modulation	Duplex	Symbol Rate	Bits per symbol
Bell 103	300	FSK	Full	300	1
v.21	300	FSK	Full	300	1
v.22	1200	DPSK	Full	600	2
v.23	1200/75	FSK	Half	1200	1
Bell 202	1200/75	FSK	Half	1200	1
Bell 212A	1200	DPSK	Full	600	2
v.22bis	2400	QAM	Full	600	4
v.32	9600	QAM	Full (EC)	2400	4
v.32bis	14400	TCM	Full (EC)	2400	6
v.32ter	19200	TCM	Full (EC)	2400	8

generated by the computer into analog signals which can be transmitted over an analog telephone line. It also transforms incoming analog signals into their digital signals for inputting into a computer. The specific techniques used to encode the digital bits into analog signals are called modulation protocols. The various modulation protocols define the exact methods of encoding and the data transfer speed. In fact, you cannot have a modem without modulation protocols. A modem typically supports more than one modulation protocol. The raw speed (the speed without data compression) of a modem is determined by the modulation protocols. Here are the main internationally accepted modulation protocols:
Two modems can establish a connection only when they support the same modulation protocol. A modem with a proprietary modulation protocol can only establish a connection with another modem which also supports that modulation protocol. That protocol is typically from the same manufacturer, or from one of several manufacturers that say they are supporting it. For example, there once was a modulation protocol called V.FAST, which delivered 28.8 Kbps over normal analog phone lines. Several manufacturers supported it. Later, the ITU-T came out with V.34. Every modem maker adopted it and V.FAST went away, leaving some modem owners with modems that only worked at 28.8 Kbps with other proprietary modems.

Modulation Rate The reciprocal of the measure of the shortest nominal time interval between successive significant instants of the modulated signal.

Modulation Suppression In the reception of an amplitude-modulated signal, an apparent reduction in the depth of modulation of a wanted signal, caused by presence, at the detector, of a stronger unwanted signal.

Modulator A device which converts a voice or data signal into a form that can be transmitted.

Modulo Term used to express the maximum number of states for a counter. Used to describe several packet-switched network parameters, such as packet number (usually set to modulo 8 — counted from 0 to 7). When the maximum count is exceeded, the counter is reset to 0.

Modulo N In communications, refers to a quantity, such as the number of frames or packets to be counted before the counter resets to zero. Relates to the number of frames or packets that can be outstanding from a transmitter before an acknowledgement is required from the receiver. Also indicates the maximum number of frames or packets stored, in case a retransmission is required (i.e., Modulo 8 or Modulo 128).

MOH Music On Hold.

Moire In a video image, a wavy pattern caused by the combination of excessively high frequency signals. Mixing of these signals causes a visible low frequency that looks a bit like French watered silk, after which it is named.

Moire pattern Wavy distortions, most obvious in image areas filled with solid color, that result from interference between the screen's phosphor layer and image signal.

Moisture Barrier Bag MBB. A three-ply bag with characteristics that allow minimal moisture transmission, thereby preserving plastic surface-mount packages, which are packed into the bag, in a dry state.

Moisture Resistance The ability of a material to resist absorbing moisture from the air or when immersed in water.

Molding Raceway Method A cable-distribution method in which hollow metal or wood moldings support cables. Small sleeves of pipe are placed in the wall behind the molding to allow cable to pass through the wall.

MOM Message Oriented Middleware. See MOMA

MOMA 1. Museum of Modern Art. www.moma.org
2. Message Oriented Middleware Association. According to MOMA, an international not-for-profit association of vendors, users and consultants focusing on the promotion of the use of messaging middleware to provide multi-platform, multi-tier message passing and queuing services for distributed computing architectures. MOMA serves as a point of interchange for experiences and ideas related to the development of MOM, as well as a point of concentration for interoperability and technology requirements toward influencing appropriate standards bodies. MOMA also directs its attention toward promotion of functional interoperability between applications built using disparate message-passing tools and mechanisms. See MIDDLEWARE and MOM. http://198.93.24.24 (I'm sure that they would have preferred www.moma.org, but the Museum of Modern Art got there first. Hence the use of an IP address for MOMA's Web site, rather than a URL.) See IP and URL.

MONET Multiwavelength Optical Network. A high-end fiber optic testbed network on the US East Coast. The $100 million project, funded by ARPA (Advanced Research Projects Agency), is intended to test DWDM (Dense Wavelength Division Multiplexing), a means of increasing the capacity of SONET fiber optic transmission systems through the multiplexing of multiple wavelengths of light. MONET participants include AT&T, Bell Atlantic, Bellcore, BellSouth, Lucent Technologies, NRL (Naval Research Lab), NSA (National Security Agency), Pacific Telesis, and Southwestern Bell. See also DWDM, WDM, and SONET.

Money Suck "Net Guide" Magazine before CMP closed it down was referred to internally as a "money suck," i.e. it was consuming vastly more money than it was bringing it.

Monitor 1. To listen in on a conversation for the purpose of determining the quality of the attendant's or agent's response and politeness to customers.
2. Video monitor. TV screen and surrounding electronics. IBM personal computer monitors come in monochrome and color. The more you pay the better quality the monitor is. The quality of monitors is typically measured by how monitor's screen resolution, which is measured by pixels, or picture elements. In monochrome monitors, the "standard" is Hercules Graphics, which offers a resolution of 720 pixels horizontally by 348 vertically. The worst color monitor is called CGA, which stands for Color Graphics Adapter. It offers 320 x 200. Most low-end laptops have this resolution screen. Next better is EGA, Enhanced Graphics Adapter. It offers 640 x 350 pixels. Next better is VGA, or Video Graphics Array, which offers 640 x 480. Next up is 8514/A also called XGA, or eXtended Graphics Array, which is IBM's high-resolution extension to its VGA adapter. It provides a resolution of 1,024 x 768 vertically, yielding 786,432 possible bits of information on one screen, more than two and a half times what is possible with VGA. It will only work on IBM machines with the Micro Channel Architecture (MCA). Here is a chart showing the pixel coding of various PC standard screens.

PC SCREEN ENCODING

CGA	320 x	...200
Enhanced CGA	640 x	...400
EGA	640 x	...350
Hercules	720 x	...348
VGA	640 x	...480
Super VGA	800 x	...600
8514/A (also called XGA)	1,024 x	...768
I don't know the name	1,600 x	1,200

I don't know the name   1,800 x 1,440
I don't know the name   2,040 x 1,664

Monitor On Hold A telephone feature. If the person you're speaking with puts you "on hold," you can turn your speaker on your phone and hang up your handset but keep listening until the other person comes back to the phone.

Monochromatic Consisting of one color or wavelength. Although light in practice is never perfectly monochromatic, it can have a narrow wavelength.

Monochrome Monitor A monitor with 720 x 348 pixel resolution in a single color. Most monochrome monitors display in paper white, green or amber.

Monopole A slender self-supporting tower on which wireless antennas can be placed.

Monopoly Leveraging Monopoly leveraging is one of the main charges brought against Microsoft by Department of Justice in its 1998 antitrust suit. Monopoly leveraging is using a monopoly in one area to gain a monopoly in another. According to the Justice Department, Microsoft was using its 90% or so market share in desktop and laptop operating systems to gain a monopoly in the browser market. One wonders if the Federal Government is so uncreative it can't think of anything better to do with its time or money.

Monospaced Font A font in which all characters have uniform widths. See also PROPORTIONAL FONT.

Month To Month The standard way of paying for telephone service. Some services now come in "rate stability" packages, which means if you commit to keeping the service for a while — typically three or five years — it's cheaper each month.

Monthly Factors A call center term. A historical pattern consisting of 12 factors, one of each month, that tells the program how much that much that month can be expected to deviate from the average monthly traffic year after year. For example, a monthly factor of .75 means that the month will be 25% slower than average, while a factor of 1.15 means that the month will be 15% busier than average.

MOON Magneto Optical On Network.

Moore's Law Gordon Moore, one of the cofounders of Intel Corporation, in 1965 in an Article in Electronics Magazine forecasted that computer chip complexity would double every twelve months for the next ten years. Ten years later his forecast proved true. He then forecasted that the doubling would occur every two years for the next ten years. Again history demonstrated his accuracy. The average of the two estimates is often stated as doubling every 18 months. He has again estimated this doubling will continue for another decade. His estimating is often referred to as "Moore's Law." There are many interpretations of what Moore actually said. For example, the July 4, 1994 Business Week said that in 1964, Moore observed that, by shrinking the size of the tiny lines that form transistor circuits in silicon by roughly 10% a year, chip makers unleash a new generation of chips every three years — with four times as many transistors. In memory chips, according to Business Week, this has quadrupled the capacity of dynamic random memories (DRAMs) every three years. In microprocessors, the addition of new circuits — and the speed boost that comes from reducing the distance between them — has improved performance by four or five times every three years since Intel launched its X86 family in 1979 (the family that has powered IBM-compatible PCs).

MOP 1. Method of Procedure. A formal, written procedure detailing a job that will take place. Required for any work on the network. Includes detailed instructions for completing the work and for backing out of trouble or mistakes that may occur. For example, when network engineers extend a network by adding a new node to a SONET ring (or by doing something else) they write a MOP. The MOP tells technicians which circuits to reroute, which cards to swap and when to turn on the new node. These MOPs are essential communication tools for engineers and technicians.
2. Maintenance Operations Protocol, a DEC protocol used for remote communications between hosts and servers. **MOPS** Millions of operations per second. Refers to a processor's performance. In the case of DVI (Digital Video Interactive) technology, more MOPS translates to better video quality. Intel's video processor can perform multiple video operations per instruction, thus the MOPS rating is usually greater than the MIPS rating.

MQA Multiple Queue Assignment. lets ACD system agents log into multiple queues.

Morph Computer animation technique that allows figures to change from one shape to another in increments you choose.

Morphology A cellular radio term. Morphology describes population density. Higher population densities cover more POPs per cell, leading to economies of scale. Lower densities imply improved propagation characteristics and a greater coverage area.

Morse Code There are (or were) two Morse Codes. One called American Morse Code and one called Continental or International Morse Code. The first one, American Morse code was invented by Samuel F. B. Morse, an American born in Boston, in 1837 for use on the electric telegraph. When the electric telegraph was adopted later in Europe, the code Morse invented was not used. The so-called "Continental" code was devised. This code, among other changes, eliminated the spaced dots for C, O, R, Y and Z and the long dash for L. These changes were needed for the satisfactory use of the early visual "needle telegraph" instruments in Europe wherein a needle swung slowly between right and left positions to indicate dots and dashes. The Continental (or International) Code was adopted as a worldwide standard at the Telegraph Conference in Berlin in 1851. American landline telegraphists, however, steadfastly refused to abandon their Morse Code. According to the history books, there was more to their refusal than just plain American stubbornness. When skilfully handled, the American Morse Code actually transmits information somewhat faster because of spaced dots being used in place of dash combinations for some letters. However, careless sending of American Morse Code will produce more errors. Morse code is referred to as International Telegraph Alphabet 1. Mr. Morse also invented the telegraph and first demonstrated it in 1844. Morse Code was used in landlines and in radio telegraphy to ships at sea. The United States Coast Guard abandoned Morse Code in 1996 and member nations of the International Maritime Organization agreed to officially stop its use by February 1, 1999. The French maritime radio authorities sent their last Morse code message on February 1, 1996. Governments are abandoning Morse code in favor of faster, better radio and satellite voice and data communications. International Morse Code is still widely used by U.S. and foreign amateur (Ham) radio operators.
Morse Code represents letters by combinations of long and short signals. Morse Code can be written in dots and dashes or signaled with flashlights and radio bleeps, or taped and scratched between cells in prison. In 1912, the easily-memo-

rized letters SOS were chosen as the international distress signal. "Save Our Souls" was the catch phrase devised later. You'll notice that in Morse Code, more commonly-used letters, such as vowels, have fewer dots and dashes. You'll also notice that some letters are represented by one dot or dash and some by as many as five. This was not a data transmission code which you can easily apply error checking and correction to. It relied heavily on the skill of the operators for error-checking.

American Morse Code

```
A .-      B -...    C .. .
D -..     E .       F .-.
G —.      H ....    I ..
J -.-.    K -.-     L — (L is long dash, not double
                          dash)
M —       N -.      O . .
P .....   Q ..-.    R . ..
S ...     T -       U ..-
V ...-    W .—      X .-..
Y .. ..   Z ... .
```

Continental (also called International) Morse Code

```
A .-      B -...    C -.-.
D -..     E .       F ..-.
G —.      H ....    I ..
J .—      K -.-     L .-..
M —       N -.      O —
P .—.     Q —.-     R .-.
S ...     T -       U ..-
V ...-    W .—      X -..-
Y -.—     Z —..
```

See also Inmarsat, Marisat and Morse, Samuel.

Morse, Samuel Samuel Finley Breese Morse was born in Charlestown (now a part of Boston) on April 27, 1791. He entered Yale University at the age of 14 and graduated in 1810. Although he attended lectures on electricity while at Yale, after graduation he went to England to study art. He returned to the United States in 1815 and became a well-known painter. In 1832, while returning from Europe, he and a fellow passenger discussed the electromagnet and Morse conceived the idea of his telegraph. He made a working model about 1835 and filed for a patent in 1838 and in 1844 inaugurated public service between Washington and Baltimore with his famous message, "What hath God wrought?" He died in 1872. See MORSE CODE.

MORT AT&T's database of its dead employees.

MOS Metal Oxide Semiconductor. Technology describing a transistor composed of a semiconductor layer including "source" and "drain" regions separated by a channel. Above the channel is a thin layer of oxide and over that a metal electrode called a gate. A voltage applied to this gate controls the current between the source and drain regions, or in another format, stops a flow between the two areas. (Definition courtesy of Bell Labs)

Mosaic The first graphical Web browser, developed by National Center for Supercomputing Applications, which greatly popularized the Web in the last few years, and by extension the Internet, as it made the multimedia capabilities of the Net accessible via mouse clicks. Mosaic let you surf the Internet's Worldwide Web. Mosaic lets you see hypertext documents with embedded graphics and occasionally sound, movie clips and animation. Mosaic was the first popular software that allowed people to browse around the Web by pointing and clicking. In short, Mosaic is an interface to the WWW (World Wide Web) distributed-information system. Like Gopher, the WWW is functionally split into two parts, the server and the client. Using a GUI (Graphical User Interface) interface (like Mosaic), it has the ability to display:

• Hypertext and hypermedia (sounds, movies, extended character sets, and interactive graphics) documents.

• Electronic text in an enormous variety of fonts.

• Text in bold and italic

• Layout elements such as paragraphs, bulleted lists, and quoted paragraphs. It is mostly distinguishable by its support of multiple hardware platforms, and the WWW HTML (Hypertext Markup Language) document format. Mosaic used to be the most popular Web browser. At the time I wrote this, the most popular one was Netscape and browsers based on Netscape. See NETSCAPE.

MOSS MIME Object Security Services. An Internet mail security standard which was introduced in 1995 as the successor to PEM (Privacy Enhanced Mail). PEM didn't address MIME (Multipurpose Internet Mail Extension) attachments. MOSS failed to secure widespread support. S/MIME (Secure/MIME), introduced in 1996 has become the de facto standard. See also MIME, PEM and S/MIME.

Most Economical Route Selection MERS. Used by several phone companies and several manufacturers to mean Least Cost Routing — the feature of a telephone system which automatically chooses the least cost route for a long distance call. See LEAST COST ROUTING.

Most Favored Nation Clause A clause added to a purchase contract with a vendor saying that for a certain period after signing the contract, if the buyer finds out that the product has been bought for less, then the seller will refund the difference. The idea is to give the purchaser the assurance of the least expensive price.

Most Limiting Capacity A telephone company term. The arithmetic minimum of (1) Line Capacity (2) Number Capacity or (3) Switching Equipment Capacity.

MOTD Message Of The Day.

Mother Of The largest, greatest, grandest, of something. An expression coined in 1990 by Saddam Hussein, Iraqi dictator. The Mother Of all telephone switches would be the largest, most powerful, most elaborate etc. Such a device doesn't exist, as yet — as, in fact, Saddam Hussein's Mother of all Battles (the one against the Allies in 1990-1991) didn't exist. He lost in the Mother Of all defeats.

Motherboard The main circuit board of a computer system. The motherboard contains edge connectors or sockets so other PC (printed circuit) boards can be plugged into it. (Those boards are typically called Fatherboards, because they plug into the Motherboard.) On IBM's new OS series of personal computers, the motherboard is called the PLANAR BOARD. Motherboards are common in key systems and hybrid key/PBXs. They are not common in PBXs, where all the electronics are typically on printed circuit cards which slide into the PBX's cage and which attach to a backplane, which is typically a wiring scheme connecting the PBX's printed circuit cards. See also DAUGHTERBOARD.

Motif Motif is the name given to the Open Software Foundation's (OSF) toolkit (Application Programming Interface) and look and feel. Standardized as IEEE 1295, OSF/Motif has become the major GUI (Graphical User Interface) for open computer systems, as defined by The Open Group, a consolidation of the OSP and X/Open industry consortia. Now in Version 2.0, Motif is the basis of the Common Desktop Environment (CDE) developed jointly by HP, IBM, Novell and SunSoft.

MOTIS Message Oriented Text Interchange System. Original name for the ISO (International Organization for Standardization) standard now being changed to MHS (Messaging Handling System).

MOTO Mail Order Telephone Order. A credit/debit card classification by the banking and finance industry reflecting what the banking industry thinks as its highest risk transaction type.

Motorola According to the family which founded Motorola in the 1920s, the name Motorola was chosen to mean "Music in motion" to signify one of the company's first products — a car radio.

MOU Memorandum Of Understanding. In the context of GSM, the memorandum of understanding signed my a potential GSM PLMN operators.

Mount The method in NFS and other networks by which modes access network resources. The word "mount" is often used as a verb, as in my workstation "mounts" the file server, called DALLAS2.

Mouse A device that generates the coordinates of a cursor or position indicator on your computer screen (e.g. a hand, an arrow) as you move it around on a flat surface, generally in the form of a "mouse pad." The term "mouse" comes from the appearance of the device, as it generally is connected to the mouse input port by a wire which is reminiscent of the tail of a mouse (there also now are wireless mice.) The body of the mouse has one or more buttons which allow you to select objects, icons or text for the performance of certain functions, depending on the application running at the time. This "point-and-click" mode is a critical element of a Graphical User Interface (GUI), such as Windows and its variations; such GUIs were first popularized by Apple Computer.

On the bottom of the mechanical mouse is a ball that rolls on the surface of the mouse pad. In contrast, a trackball is a stationary device with a ball that you move with your finger — essentially an upside-down mechanical mouse. The mechanical mouse, which is what most of us use, was invented by Douglas Engelbart of Stanford Research Center in 1963; it was commercialized by Xerox in the 1970s.

An optical mouse makes use of a laser to detect the movement of the mouse in relationship to a grid on a special mouse pad. While optical mice are very precise, the also are relatively expensive. Optomechanical mice combine the technologies, without the requirement for the grid pad.

Mouse Blur Move your mouse quickly across your screen and if you're running an LCD (for example on a laptop), the mouse's pointer will blur — due to the screen's inability to change as fast as you can move the mouse. Another term for mouse blur is Cursor Submarining.

Mouse Potato A person who uses his mouse to view educational or entertainment on his computer. Museums are afraid, for example, that if they sell the electronic rights of the art hanging on their walls, every one will stay at home, become mouse potatoes and never visit the museums. The concept of a mouse potato derives from a couch potato — namely someone who sits on his couch and changes channels on his TV set using a remote control.

MOV 1. Metal Oxide Varistor. A voltage dependent resistor which absorbs voltage and current surges and spikes. This low-cost, effective device can sustain large surges and switch in 1 to 5 nanoseconds. It is used as a surge protector and suppressor. It often the first electronic component that electrons coming in on an incoming phone line hit. Many trunk boards inside PBX are protected by MOVs. If the voltage or current is high, it will blow the MOV, thus protecting the remaining the far more valuable devices on the board.

2. A Macintosh-based audio/video (multimedia) file. A MOV file has a file extension of .mov and can be played on a Windows operating system if you have the QuickTime Movie Player application installed.

Moves, Adds and Changes MACs. Any of the above ancillary work performed on a PBX switch, cabinet, or peripheral item after installation. See MAC for a fuller explanation.

MP1 MPEG Layer-1. An extension of the MPEG (Moving Picture Experts Group) standards for compressed digital video, Layer 1 supports CD-quality audio using 4:1 (4-to-1) compression, which reduces the required bandwidth from approximately 1.411 Mbps to 384 Kbps. See MP3 for more detail.

MP2 MPEG Layer-2. An extension of the MPEG (Moving Picture Experts Group) standards for compressed digital video, Layer 2 supports CD-quality audio using compression of 6:1 to 8:1, thereby reducing the bandwidth requirement from 1.411 Mbps to 256-192 Kbps. See MP3 for more detail.

MP3 MPEG Layer-3. MP3 is the most popular audio-compression format on the Internet. MP3 provides an efficient audio-coding scheme, which allows compression of audio files by a factor of up to 12, with little loss in quality from the original CD. For example, a five minute CD song takes about 50 megabytes of storage space on your computer's hard drive. In MP3 format the same song occupies only about 5 megabytes and delivers the original digital quality sound. MP3 is also an extension of the MPEG (Moving Picture Experts Group) standards for compressed digital video. To play MP3 music, you'll need the music. You get download it from the Internet or convert it on your PC from your favorite CD. You play your MP3 music on your PC using a software player you downloaded (for free or for pay) from the Internet (www.mp3.com, www.musicmatch.com or www.trellian.com). Or you can play MP3 music on portable devices that look like Sony Walkmans, but have no moving parts. The first such device was Diamond Multimedia's Rio.

MPC 1. MPOA Client. An ATM term. A protocol entity that implements the client side of the MPOA (MultiProtocol Over ATM) architecture. A MPOA client, typically in the form of a host computer, establishes a VCC (Virtual Channel Connection) with a MPOA server in order either to forward data packets to a destination MPOA client, or to request information so that the originating MPOA client can establish a more direct path on the basis of a cut-through SVC (Switched Virtual Circuit). In this latter case, the server acts as a virtual router. The MPOA client implements the Next Hop Client (NHC) functionality of the Next Hop Resolution Protocol (NHRP). See MPOA.

2. Multimedia PC. See the following definitions for MPC1, MPC2 and MPC3.

MPC1 Published in 1991, this original Multimedia PC (MPC) specification was adopted worldwide as the basic multimedia extension of the PC standard. MPC standards are established by the MPC Working Group of the SPA (Software Publishers Association). In 1993, MPC1 was followed by MPC2. MPC3, the latest, does not replace MPC2, but takes it one step further. See MPC2 and MPC3.

MPC2 Published in 1993 as the successor to MPC1, MPC2 standards specify elements including: 1. 4MB RAM; 2. 485 SX or equivalent microprocessor of 25 MHz; 3. hard drive of 160 MB; 4. CD-ROM drive supporting a sustained transfer rate of 300 KBps, and a maximum average seek time of 400 ms; and 5. Windows 3.0 plus multimedia extensions, or binary compatibility. No video playback standards were included. See MPC3 for the latest standards.

MPC3 MPC3 is the latest specification (Release 1.3, February 26, 1996) for multimedia PCs as defined by the Multimedia PC Working Group, an independent special interest group of the Software Publishers Association (SPA). Minimum requirements for MPC3 machines include: 1.8 MB RAM; 2. CPU which can pass the MPC Test Suite, which is benchmarked on a 75 MHz Pentium; 3. hard drive of 540 MB; 4. CD-ROM Drive supporting a sustained transfer rate of 600 KBps and an average access time of 250 ms; and 5. Windows 3.11 and DOS 6.0 or binary compatibility. MPC3 also adds the requirement for video playback capability compatible with MPEG-1 (hardware or software). See also MPEG.

MPEG MPEG is commonly known as a series of hardware and software standards designed to reduce the storage requirements of digital video, i.e. video recorded digitally or converted into digital bits. MPEG is most commonly known as an compression scheme for full motion video. The word MPEG is actually the acronym for the Moving Pictures Experts Group, a joint committee of the International Standards Organization (ISO) and the International Electrotechnical Commission (EG). The first MPEG specification, known as MPEG-1, was introduced by this committee in 1991. The common goal of all MPEG compression is to convert the equivalent of about 7.7 meg down to under 150 Kb, which represents a compression ratio of about 52 to one. The two requirements of MPEG-1 are 30 frames per second of Standard Image Format (SIG) of 352 pixels x 240 pixels and CD-quality sound at 44.1 Khz, 16 bit stereo. MPEG image scheme offers more compression than the other poplar JPEG image compression scheme, which is largely for still images. MPEG takes advantage of the fact that full motion video is made up of many successive frames consisting of large areas that are not changed — like blue sky background. While JPEG compresses each still frame in a video sequence as much as possible, MPEG performs "differencing," noting differences between consecutive frames. If two consecutive frames are identical, the second can be stored in remarkably few bits. MPEG condenses moving images about three times more tightly than JPEG. See also JPEG.

There are two types of MPEG Playback: Software and Hardware. Software MPEG playback is the decompression of MPEG video and audio files using the processing power of the CPU. Hardware MPEG Playback uses an add-in card to deliver full-screen, full-motion, full-color video and CD-quality audio at the full NTSC video frame rate of 30 frames per second, with no dropped frames. The card plays the video from a computer file that has been compressed using the MPEG video standard. Hardware playback is typically much better quality than software playback.

There are actually two MPEG standards: MPEG-1 and MPEG-2. A third, MPEG-4, is currently under development. MPEG-1 is a small-picture mode of MPEG geared to a resolution of 352 by 240 pixels at 30 frames per second (U.S.), with full CD-quality audio. MPEG-1 was originally designed to handle much larger picture sizes than 352 by 240 through interpolation or scaling, but MPEG-2 is more efficient. MPEG-2 offers a "main profile at main level" resolution of 720 by 480 pixels at 30 frames per second (U.S.), with full CD-quality audio. This picture size enables full-screen playback on PCs or TVs. MPEG-2 can incorporate a range of compression ratios, which trade off economies of storage and transmission bandwidth against picture quality. At compression ratios of 30:1 and

smaller, MPEG-2 offers the perception of broadcast-quality TV. For greater economy, MPEG-2 supports up to 200:1 compression. MPEG-2 decodes such as the IBM decoder chip can also recognize and decode MPEG-1 bitstreams, enabling the IBM chip to support both compression standards.

MPEG-3 has been dropped. It was focused on HDTV with sampling dimensions up to 1,920 by 1,080 at 30 frames per second. The standard was to address bit rates between 20 and 40 Mbit/sec. Nevertheless, it was discovered that with a little tweaking, MPEG-2 and MPEG-1 work extremely well at the HDTV rate. HDTV is now part of the MPEG-2 High-1440 Level specification.

MPEG-4 is currently in the application identification phase, with a target of November 1998 for the official sanction of the proposed standard. Intended for very narrow bandwidths, MPEG-4 is exploring ideas in frame reconstruction. Much like MIDI music creates realistic sound from a narrow bandwidth command string, using pre-existing sound components, MPEG-4 is considering speech and video synthesis, fractal geometry, computer visualization and artificial intelligence to build accurate pictures from minimal data.

If you want to find out even more gory detail about MPEG, hyperlink over to the Moving Pictures Experts Group Web site in in Italy at

http://www.crs4.it/~luigi/MPEG/mpegfaq1.html

MPEG-1 See MPEG.

MPEG-2 MPEG-2 is one of the most important standards developed by the Moving Pictures Expert Group, an International Standards Organization (ISO) group responsible for the standardization of coded representations of video and audio signals. MPEG-2 has been chosen as a leading digital video compression for a broad range of future video and broadcast applications. See MPEG and MPEG-2 Audio.

MPEG-2 Audio MPEG-2 audio is a compatible extension of the MPEG-1 audio coding which enables the transfer of mono, steroid, or multichannel audio in a single bitstream. It can operate at data rates from 32 kbps up to more than 1 Mbps, and supports sampling rates of 32, 44.1 and 48 kHz. For stereo, a typical application would operate at an average data rate of 128-256 kbps. A multichannel movie soundtrack requires an average bit rate of 320-640 kbps, depending on the number of channels (5 to 7, plus a sub woofer channel) and the complexity of the encoded audio.

MPEG-3 See MPC3.

MPEG-4 See MPEG.

MPG Microwave Pulse Generator. A device that generates electrical pulses at microwave frequencies.

MPI 1. Multi-Path Interface. Between a transmitter and receiver, the radio wave can take a direct path and one or more reflected paths. The direct radio wave always arrives prior to the reflected waves. If the reflected waves are of sufficient amplitude, they will interfere with the direct wave. The relationship of the amplitude and time delay between the direct and reflected waves create peaks and nulls at the receiver, causing momentary signal fading or loss. In a digital system, this can result in very significant degradation, as the receiver loses signal acquisition and frame synchronization during each fade. The net effect is an increase in the residual bit error rate.

2. Media Platform Interface libraries. Part of Sun Microsystems' XTL Teleservices architecture. MPIs provide a layer of abstraction between details of the system services, applications and providers. The system services include a message passing "server", a data stream multiplexor streams driver, a provider configuration database and a database

administration tool.

MPLS MultiProtocol Label Switching. An evolving IETF standard intended for Internet application. MPLS is a widely supported method of speeding up data communication over combined IP/ATM networks. As IP and ATM come together, the concept is that of "route at the edge and switch in the core." In other words, routers are used at the ingress and egress edges of the network, where their high levels of intelligence can be best utilized and where their inherent slowness can be tolerated. Switches are used in the core of the network, where they can take advantage of the intelligent routing instructions provided by the routers, and where their inherent speed offers great advantage. MPLS takes this concept to new heights in an IP (Internet Protocol) WAN (Wide Area Network) such as the Internet, much as does Cisco's proprietary Tag Switching in the LAN (Local Area Network) domain. MPLS works like this: As an IP datastream enters the edge of the network, the ingress router reads the full address of the first data packet and attaches a small "label" in the packet header, which precedes the packet. The ATM switches in the core of the network examine the much-abbreviated label, and switch the packet with much greater speed than if they were forced to consult programmed routing tables associate with the full IP address. All subsequent packets in a datastream are automatically labeled in this fashion...and very quickly, as they have been anticipated. See also ATM, IP and Tag Switching.

MPOA MultiProtocol Over Asynchronous Transfer Mode. A developing set of architectural specifications defined by the ATM Forum. Working at Layer 3 (Network Layer) MPOA specifies standards for Layer 2 (Link Layer) switching through a Layer 3 router — i.e., switched routing — over an ATM fabric. MPOA allows companies to build scalable, enterprise-wide LAN internetworks that seamlessly interwork ATM with LAN protocols such as Ethernet, Token Ring, FDDI and Fast Ethernet. In effect, MPOA provides for inter-LAN cut-through, for the deployment of a WAN VLAN (Virtual Local Area Network) over an ATM backbone. MPOA accomplishes this by separating the route calculation function from the Network Layer forwarding function. In support of Network Layer packets such as IP and IPX, the edge routers will recognize the beginning of a data transfer and respond with an ATM network destination address. At that point, the router network will establish a cut-through SVC (Switched Virtual Circuit) which will eliminate router-by-router delays, thereby considerably increasing the speed of associated data transfer. This is accomplished by distributing the connection intelligence through the network to the edge devices; the traditional approach involves each router's acting independently on each packet in an effort coordinated by a centralized router, which can become overloaded. MPOA draws on existing standards, including the Layer 2 LANE (Local Area Network Emulation) from the ATM Forum, and the Layer 3 NHRP (Next Hop Resolution Protocol) from the IETF. MPOA also draws on IP extensions such as RSVP (Resource ReSerVation Protocol), which is used in support of isochronous data such as streaming video over IP networks. See the following four definitions. See also Classical IP over ATM, IP, LANE, NHRP, RSVP and VLAN.

MPOA Client MPC. An ATM term. A protocol entity that implements the client side of the MPOA architecture. An MPOA client implements the Next Hop Client (NHC) functionality of the Next Hop Resolution Protocol (NHRP). See MPOA.

MPOA Server MPS. An ATM term. A protocol entity that implements the server side of the MPOA architecture. An MPOA Server implements Next Hop Server (NHS) functional-

ity of the NHRP. See MPOA.

MPOA Service Area An ATM term. The collection of server functions and their clients. A collection of physical devices consisting of an MPOA server plus the set of clients served by that server. See the three definitions above and one below.

MPOA Target An ATM term. A set of protocol address, path attributes, (e.g., internetwork layer QoS, other information derivable from received packet) describing the intended destination and its path attributes. See the four definitions immediately above.

MPOE Minimum Point Of Entry, pronounced em-poe. Also known as MPOP (Minimum Point Of Presence), as defined by the FCC, and the LLDP (Local Loop Demarcation Point). The MPOE is the main point of physical and logical demarcation between the LEC (Local Exchange Carrier) and the customer premises. Up to the point of the MPOE, the telco is fully responsible for deployment and maintenance of the local loop connection. Beyond the MPOE, the user organization or building owner is responsible for the extension of the connection to the PBX, Centrex telephone sets, etc. In a campus environment comprising multiple buildings, there may be multiple points of demarcation, in which case one is designated by mutual agreement as the MPOE. Here's a working explanation from Ty Osborn, who works for the best CLEC in California (he says), tosborn@email.pacwest.com, "I was first intoduced to MPOE when I had a (telco) tech out on prem (an apartment building) and could not find the DS-1 Bell had delivered earlier that day. He called and asked "where the heck is the MPOE?" (pronounced em-poe). In short, a MPOE might be a phone room with some punchdown blocks. An MPOE might be a box on the outside of the building. An MPOE is the last piece of equipment a phone company installs in a building for a customer it is providing service to. The MPOE is basically the phone company's last point of responsibility for that circuit. After that it's the customer's or the equipment or service company who is servicing the customer with on-premise equipment — for example, a PBX.

MPOP 1. Minimum Point Of Presence. See MPOE
2. Metropolitan Point Of Presence. The point of presence of a carrier within a metropolitan area. See also POP.

MPPP Multilink PPP. A technique which combines two ISDN BRI B channels into a single, high-speed data path.

MPRII A Swedish PC monitor standard that specified a limit of 2.5 milligauss of ELF emissions when measured at about 20 inches from the screen.

MPS 1. Multi Page Signal. A frame sent in fax transmission if the sender has more pages to transmit.
2. Mobile Positioning Service. This technology uses the idle time of a GSM cellular system to figure where you (or more precisely, your cell phone) is. An alternative to GPS. Such MPS is designed for applications like emergencies, in which the cell phone can transmit to emergency authorities — fire, police — where the person in trouble actually is at that moment.

MPSK Mobile switching centre.

MPTN Multi Protocol Transport Networking. IBM scheme addressing multi protocol network support including TCP over SNA and SNA over IP.

MPTy Multiparty. A wireless telecommunications term. A supplementary service provided under GSM (Global System for Mobile Communications).

MPX Multiplex.

MR 1. Modem Ready. An ASCII signal and visible "on" light that tells you that your modem is on and ready.
2. Modified Read. Relative element address differentiation

code. A two-dimensional compression technique for fax machines that handles the data compression of the vertical line and that concentrates on space between the lines and within given characters. See MMR.

Mrm An ATM term. An ABR service parameter that controls allocation of bandwidth between forward RM-cells, backward RM-cells, and data cells.

MS 1. Mobile Station. A wireless telephone allowing mobility so that calls may be placed locally or in another geographic region.

2. Message Switch.

3. Microsoft.

MS-DOS MicroSoft Disk Operating System, now buried under Windows

MSA 1. Metropolitan Statistical Area. Sometimes known as SMSAs (Standard Metropolitan Statistical Areas) and MEAs (Metropolitan Economic Areas) and as defined by the U.S. Census Bureau, MSAs are geographic areas that contain cities of 50,000 or more population, and which include the surrounding counties. Such areas are characterized by the "community of interests." Using data from the 1980 census, the FCC allocated two cellular licenses to each of the 305 MSAs in the United States. The FCC developed LATA boundaries based largely on the SMSA concept.

2. Message Service Application.

MSAT Mobile SATellite. Technology for transmitting and receiving satellite transmissions in moving vehicles.

MSAU Multi-Station Access Unit.

MSB Abbreviation for Most Significant Bit and Most Significant Byte. That portion of a number, address or field which occurs leftmost when its value is written as a single number in conventional hexadecimal or binary notation. The portion of the number having the most weight in a mathematical calculation using the value.

MSC 1. Mobile Switching Center. A switch providing services and coordination between mobile users in a network and external networks.

2. Malaysian Multimedia Super Corridor. The MSC is an ambitious development concept of the Malaysia government to develop an area of Malaysia with high-tech (telecom and computer) companies. The area is scheduled to house government offices, moving them out of traffic-congested Kuala Lumpur, the capital of the country. The government is installing a 4,300 kilometer fiber optic network in the MSC. If a company agrees to invest in the area, the Malaysian government extends tax breaks and other incentives. Overseeing the MSC is something called The Multimedia Development Corporation.

MSEngine Mirrored Server Engine. The part of the Novell SFT III operating system that handles nonphysical processes, such as the NetWare file system, queue management, and the bindery. SFT III is split into two parts: the IOEngine (Input/Output Engine) and the MSEngine) and the MSEngine. The primary server and the secondary server each have a separate IOEngine, but they share the same MSEngine. The file system, receive buffers, and queue management system all reside in the MSEngine. Applications and NetWare Loadable Modules (NLMs) that do not address hardware directly can be mirrored by loading them in the MSEngine. If one server fails, applications and NLMs in the MSEngine continue to run. The MSEngine keeps track of active network processes; it provides uninterrupted network service when the primary server fails and the secondary server takes over. See IOENGINE.

MSF Mobile Serving Function.

MSI Modular Station Interface. A Dialogic board that interfaces analog phones to SCbus and PEB-based products.

MSISDN Mobile Station ISDN number. An ISDN number provisioned to a mobile station subscriber and used to place a call.

MSIX Metered Services Information Exchange. This protocol, first announced in June of 1997, provides a common interface for Internet applications to easily exchange detailed usage information with network billing and information systems. This specification includes a way for Internet Service Providers (ISPs) to effectively meter usage and charge for Internet services. Here are words from the June press release. Compaq and NetCentric announced a New Internet Metering Technology called MSIX, which is intended to accelerate the growth of new billable Internet services such as telephony, fax, video conferencing, content distribution and gaming. The commercial deployment of such applications has been hindered by the immaturity of the management and billing tools inside the Internet infrastructure. MSIX addresses many of these issues by providing a common mechanism to effectively meter application usage by a subscriber. It provides software developers and ISPs a common accounting framework that significantly simplifies network and billing integration. MSIX will make value-added services available to a much larger audience for the first time. MSIX complements other emerging protocols such as Reservation Protocol (RSVP) that allow network resources to be reserved and different quality-of-service levels to be offered. www.compaq.com and www.netcentric.com

MSL Mirror Server Link. A dedicated, high-speed connection between Novell NetWare SFT III primary and secondary servers. The mirrored server link is essentially a bus extension from the primary to the secondary server. It requires similar boards in each server, directly connected by fiber-optic or other cables.

MSN MSN 1. The Microsoft Network. Microsoft's version of American On Line (AOL). It has three main features. You can use it to access the Internet. You can send and receive email. You can use some of its own information services, none of which have impressed me sufficiently to use them. I use MSN as my primary email vendor because they do a good, reliable job with email and they have high-speed (56 Kbps) local phone numbers in all the domestic places I visit. They are weak overseas. IBM.Net is much better overseas.

2. Mobile Station Number. Also known as MIN (Mobile Identification Number) The telephone number of a cellular or PCS telephone. The unique MSN is paired with a unique ESN (Electronic Serial Number) for reasons of security. See also MIN and ESN.

MSNF Multisystem Networking Facility. An optional feature of certain IBM telecommunications access methods that allows more than one host running ACF/TCAM or VTAM to jointly control an ACF/NCP program.

MSMQ MicroSoft Message Queueing. MSMQ provides fault-tolerant support for distributed applications which work in conjuction with Message Transaction Server, MTS, which is a component manager for Windows NT, soon to be known as Windows 2000.

MSO Multiple System Operator. A company that operates more than one cable TV system.

MSP The name of a general purpose programmable switch made by Redcom Laboratories.

MSS Mobile Satellite System. MSSs are satellite systems which support mobile voice and/or data services. Constellations of such satellites are launched in various non-equatorial paths so that they whiz around the earth much like electrons whiz around the nucleus of an atom. As a result, one or more of the satellites always is in "view" of a small, low-power terminal. With orbital paths over all major land masses, one can conduct a cellular-like voice or data conversation from the jungles of New Guinea to another person in the Sahara desert.

MSX Mobile Switching eXchange. In a cellular environment, a MSX is akin to a central office (CO) in the wired world. More commonly, a MSX is known as a MSC (Mobile Switching Center) or a MTSO (Mobile Telephone Switching Office). See MTSO.

MT An ATM term. Message Type: Message type is the field containing the bit flags of a RM-cell. These flags are as follows: DIR = 0 for forward RM-cells = 1 for backward; RM-cells BN = 1 for Non-Source Generated (BECN), RM-cells = 0 for Source Generated RM-cells CI = 1 to indicate congestion = 0 otherwise NI = 1 to indicate no additive increase allowed = 0 otherwise RA - Not used for ATM Forum ABR.

MTA 1. Message Transfer Agent. An OSI application process used to store and forward messages in the X.400 Message Handling System. Equivalent to Internet mail agent.
2. Apple Computer's Macintosh Telephony Architecture.
3. Metropolitan Trading Area. An area defined by the FCC for the purpose of issuing licenses for PCS. Each MTA consists of several Basic Trading Areas (BTAs). The United States is broken down into 51 metropolitan trading areas for economic purposes. These boundaries were used for licensing PCS.

MTBF Mean Time Between Failure. The length of time a user may reasonably expect a device or system to work before an incapacitating fault occurs.

MTD Memory Technology Drive.

MTM Maintenance Trunk Monitor.

MTP Message Transfer Part of the SS7 Protocol. It provides functions for basic routing of signaling messages between signaling points. It is Level 1 through 3 protocols of the SS7 protocol stack. MTP 3 (Level 3) is used to support BISUP.

MTS 1. Message Telecommunications Service. AT&T's name for standard switched telephone service. Also called DDD, for Direct Distance Dial.
2. Member of the Technical Staff. A common term at AT&T Bell Labs, Bellcore and other R&D labs.

MTSO Mobile Telephone Switching Office. This central office houses the field monitoring and relay stations for switching calls between the cellular and wire-based (land-line) central office. The MTSO controls the entire operation of a cellular system. It is a sophisticated computer that monitors all cellular calls, keeps track of the location of all cellular-equipped vehicles traveling in the system, arranges handoffs, keeps track of billing information, etc.

MTTR Mean Time to Repair. The average time required to return a failed device or system to service.

MTU Maximum Transmission Unit. The largest possible unit of data that can be sent on a given physical medium. Example: The MTU of Ethernet is 1500 bytes.

MU Monitoring Unit. A wireless telecommunications term. Devices added to circuit configurations that use sophisticated trending rules with fault and topology information to determine potential outages.

Mu Law The PCM voice coding and companding standard used in Japan and North America. A PCM encoding algorithm where the analog voice signal is sampled eight thousand times per second, with each sample being represented by an eight bit value, thus yielding a raw 64 Kbps transmission rate. A sample consists of a sign bit, a three bit segment specify-

ing a logarithmic range, and a four bit step offset into the range. All bits of the sample are inverted before transmission. See A Law and PCM.

MUA An acronym for Mail User Agent, is the end user's mail program, like Eudora.

MUD Multi-User Dungeons. A term that Time Magazine in its 9/13/1993 issue called "the latest twist in the already somewhat twisted world of computer communications." Time called it "a sort of poor man's virtual reality" — created by using words, not expensive head-mounted displays. The first MUD apparently was invented in 1979 as a way for British university students to play the fantasy game Dungeons & Dragons by networked computers. MUD are basically now online games environments that use a great deal of network bandwidth.

Mudbox An unsheltered item of equipment that is sufficiently rugged to withstand adverse environments. It is expected to work perfectly though it sits outdoors in good and bad weather.

MUDS Multi-User Dungeons. A cyberspace term. MUDS are elaborate fictional gathering places that users create one room at a time. All these "spaces" have one thing in common, according to cyberspace wisdom, they are egalitarian. Anybody can enter the rooms (provided he has the correct equipment) and everybody is afforded the same level of respect. A significant feature of most MUDs is that users can create interactive objects that remain in the program after they leave. MUD worlds can be built gradually and collectively. See also USENET.

MULDEM A contraction for Multiplexer Demultiplexer, referring to a piece of equipment which performs both functions and generally operates between two of the AT&T digital hierarchy rates (i.e., DSI to DS3).

Mule Tape Mule tape is very strong, flat tape which is used to pull cable through underground conduit. Here's how it typically works: First, you use a bore to make an underground hole. Then you fill that hole with hollow concrete cement pipes joined together to form one long underground conduit (i.e. tunnel). Then you go to one end of the tunnel and use a air compressed device to blow a very lightweight "birdie" attached to a lightweight string through the tunnel. Someone at the other end catches the birdie and pulls gently on the string. Attached to the end of the string is strong mule tape. He keeps pulling on it. Attached to the end of the mule tape is the telecommunications cable — fiber or wire — that you really want to instal in the underground conduit. The whole point of this elaborate procedure is that it's far better for the cable to lay it after the pipes are laid than it is during the installation process when the cable could be damaged.

Multi Channel Aggregation A feature under Windows NT which gives remote users the option of using two phone lines for the same remote session. This way you double bandwidth, thus making their session go twice as fast.

Multi Vendor Integration Protocol See MVIP.

Multi- In this dictionary, I include a dash between words beginning with "multi" and ending with something else. See below for examples.

Multi-Access The ability of several users to communicate with a computer at the same time with each working independently on their own job.

Multi-Address Calling Facility A system service feature that permits a user to nominate more than one addressee for the same data. The network may accomplish this sequentially or simultaneously.

Multi-Alternating Routing Alternate routing with provision for advancing a call to more than one alternate route, each of which is tested in sequence in the process of seeking an idle path. A Bellcore definition.

Multi-Carrier Modulation MCM. A technique of transmitting data by dividing the data into several interleaved bit streams and using these to modulate several carriers. MCM is a form of frequency division multiplexing.

Multi-Cast The broadcast of messages to a selected group of workstations on a LAN, WAN or the Internet. Multicast is communication between a single device and multiple members of a device group. For example, an IPv6 router might address a series of packets associated with a routing table update to a number of other routers in a LAN internetwork. Similarly, a LAN-attached workstation might address a transmission to a number of other LAN-attached devices. Companies are discovering they can distribute material to large numbers of employees and others on their intranets more efficiently using multicast than they can by sending such material in separate bursts to each user. In multicast mode, routers distribute a given file to all hosts that have signaled they want to receive the material, using the Class D addresses of the IP addressing hierarchy. See See MULTI-CAST PACKETS AND IPV6. CONTRAST WITH UNICAST, ANYCAST AND BROADCAST.

Multi-Cast Packets Multi-cast packets are addressed to multiple devices within a group of devices. For example, LAN stations use multi-cast packets to deliver information to a specific set of devices such as routers, file servers, and hosts. See MULTI-CAST.

Multi-Cast User Message A user message generated at the source node and distributed to two or more destination nodes.

Multi-Casting The ability of one network node to send identical data to a number of end points - known as broadcast in other circles; one example is if new software or addressing updates need to be distributed to all users; also, a point-to-multipoint video transmission is a multi-cast operation.

Multi-Channel The use of a common channel to make two or more channels either by splitting the frequency band of the common channel into several narrower bands (called frequency division multiplexing) or by allocating time slots in the entire channel (time division multiplexing).

Multi-Conductor More than one conductor within a single cable complex.

Multi-Domain Network In IBM Systems Network Architecture technology, a network that contains more than one host based System Services Control Point (SSCP).

Multi-Drop A multi-drop private line or data line is a communications path between two or more locations requiring two or more LECs, but there are multiple 'drops' per LEC. For example, a hospital in Detroit has a data line going to NY, NY. But in New York, NY there are four hospitals in a several block area. Therefore, one data line with four drops. Then you can have a multipoint - multidrop line, which is a combination of both. This explanation courtesy of Robert.Chatters@MCI.COM

Multi-Drop Line A communications channel that services many data terminals at different geographical locations and in which a computer (node) controls utilization of the channel by polling one distant terminal after another and asking it, in effect, "Do you have anything for me?"

Multi-Fiber A fiber that supports propagation of more than one of a given wavelength. See MULTI-MODE.

Multi-Frame In PCM systems, a set of consecutive frames in which the position of each frame can be identified by reference to a multi-frame alignment signal. The multi-frame alignment signal does not necessarily occur, in whole or in part, in each multi-frame.

Multi-Frequency Monitors Also known as multisync or multiscan monitors. They can show images in several resolution standards. Such versatility makes them more expensive than single-resolution monitors (e.g. a standard VGA) but also less prone to instant obsolescence. A multisync monitor showing a VGA may or may not look better than VGA monitor showing a VGA image. That depends on the screen's other attributes.

Multi-Frequency Pulsing An in-band address signaling method in which ten decimal digits (the numbers on the touchtone pad) and five auxiliary signals are each represented by selecting two frequencies and combining them into one "musical" sound.. The frequencies are selected from six separate frequencies — 700, 900, 1100, 1300, 1500 and 1700 Hz. See also CAPTAIN CRUNCH.

Multi-Function Peripherals MFS. These are devices which take on two or more functions generally associated with individual peripherals and combine these into one product or in a linker series of modules. A multi-function peripheral might combine a fax machine with a photocopier with a computer printer with a scanner. The term is not very precise, but it tends increasingly to mean a computer device that will print, photocopy, fax and/or scan.

Multi-Homed Host A computer connected to more than one physical datalink. The data links may or may not be attached to the same network.

Multi-Hop An example of a single hop system is a microwave system between one building (let's say downtown San Francisco) and another across town (let's say uptown San Francisco). Each with one microwave antenna on its roof. Let's say we wanted to extend that system to Oakland. We'd put a second antenna on the uptown San Francisco building and shoot across to an antenna in Oakland. That building would now have a multi-hop transmission system.

Multi-Hosting The ability of a Web server to support more than one Internet address and more than one home page on a single server. Also called Multi-Homing.

Multi-Leaving In communications, the transmission (usually via bisync facilities and using bisync protocols) of a variable number of data streams between user devices and a computer.

Multi-Level Precedence Preemption MLPP. A system in which selected customers may exercise preemption capabilities to seize facilities being used for calls with lower precedence levels.

Multi-Line Hunt The ability of switching equipment to connect calls to another phone in the group when other numbers in the group are busy.

Multi-Line Telephone Any telephone set with buttons which can answer or originate calls on one or more central office lines or trunks. Originally all multi-line telephones were 1A2 and they came in sizes of 2, 3, 5, 9, 11, 17, 19, 29, and 60 lines. Now, skinny wire electronic key systems come in all sizes. See KEY TELEPHONE SYSTEM.

Multi-Line Terminating System Premises switching equipment and key telephone type systems which are capable of terminating more than one local exchange service line, WATS access line, FX circuit, etc.

Multi-Link Procedure A procedure (defined in ITU-T Recommendations X.25 and X.75) that permits multiple links connecting a single pair of nodes to provide service to the network layer on a shared basis. Such sharing provides greater effective throughput capacity and availability than a single link.

Multi-Link PPP MPPP or MLPPP. This technique combines two ISDN BRI B channels into a single, high-speed data path.

Multi-Location Billing Multilocation Billing is an option whereby a long distance carrier bills separate locations and applies volume discounts pro-rated to each site based on usage, or fixed percentage with pro-rated discounts to sites based on usage.

Multi-Location Extension Dialing An AIN (Advanced Intelligent Network) service providing network-based dialing between multiple company locations on the basis of an abbreviated dialing plan. Working much like coordinated dialing plans in the PBX world, the user need only dial an access code (e.g., "7") and a 4-digit extension number. The network connects the call to the target extension at the correct location.

Multi-Master An inherent mode of the VME bus which allows the controlled sharing of the bus by multiple CPUs under a flexible priority structure.

Multi-Media Multi-media is the combination of multiple forms of media in the communication of information. Multimedia enables people to communicate using integrated media: audio, video, text, graphics, fax, and telephony. The benefit is more powerful communication. The combination of several media often provides richer, more effective communication of information or ideas than a single media such as traditional text-based communication can accomplish. Multi-media communication formats vary, but they usually include voice communications (vocoding, speech recognition, speaker verification and text-to-speech), audio processing (music synthesis, CD-ROMs), data communications, image processing and telecommunications using LANs, MANs and WANs in ISDN and POTS networks. Multimedia technology will ultimately take the disparate technologies of the computer, the telephone, the fax machine, the CD player, and the video camera and combine them into one powerful communication center. Technologies that were once analog — video, audio, telephony — are now digital. The power of multimedia is the integration of these digital technologies. To many people, "multimedia" (as defined above) is a disparate collection of technologies in search of a purpose. And it's true: most of the merger of media (as above) is taking place in business communications in the moving around of compound documents. Meantime, multimedia has moved into training and in the home for education and entertainment. See AUTHORING, COMPOUND DOCUMENTS, HYPERMEDIA, OLE, SHARED SCREENS, SHARED WHITEBOARDS, and SYNCHRONIZATION.

Multi-Media Capabilities The ability to run simultaneous voice, image, data and video applications on a computer. A technology that requires enormous bandwidth and processing power. See MULTI-MEDIA.

Multi-Media Network A network capable of carrying multiple forms of user information such as voice text, sounds, etc.

Multi-Media PC The Multimedia PC Council now defines a multimedia PC as a PC having a minimum of two megabytes of memory, a 30 megabyte hard drive, a CD ROM drive, digital sound support and Microsoft's Multimedia Extensions for Windows. See MULTI-MEDIA.

Multi-Media Protocol A protocol suitable for handling multiple forms of information such as voice, text, pictures, numbers, etc.

Multi-Mode In fiber-optics, an optical fiber designed to allow light to carry multiple carrier signals distinguished by frequency or phase at the same time (contrasts with single-

mode). (Also spelled multimode.)

Multi-Mode Distortion In an optical fiber, a result of different values of the group delay for each individual mode at a single wavelength. It isn't the same as "multi-mode dispersion."

Multi-Mode Dispersion Dispersion fattens or smears the transmitted signal, making it difficult to identify the original information sent. Dispersion limits the amount of information that can be transmitted. Dispersion is much worse on the information carried on multi-mode fibers, because there's simple more going on. As a result, dispersion limits the rate and how you can send the information. Single mode is the preferred mode for long distance.

Multi-Mode Fiber Multi-mode fiber allows many modes of light to propagate down the fiber-optic path. Multimode fibers are generally used for short-distance data links, as they provide limited bandwidth due to modal dispersion. The relatively large core of a multimode fiber allows good coupling from inexpensive LEDs, and the use of inexpensive couplers and connectors. Multi-mode fiber typically has a core diameter of 25 to 200 microns. The core is much larger than single-mode fiber and allows several modes of light to be passed through it. Multimode fiber was the original medium specified for FDDI. See also SINGLE MODE FIBER.

Multi-Mode Optical Fiber An optical fiber that will allow many modes to propagate. The fiber may be either graded-index or step index fiber. Multimode optical fibers have a much larger core diameter than a single-mode fiber. See MULTI-MODE FIBER for a better explanation. See also SINGLE MODE FIBER.

Multi-NAM A cellular telephone term to allow a cellular phone to have two (or more) phone numbers, each of which can be on a different cellular system. This lets you get service from many cellular phone companies. For example, you could subscribe to one carrier in your home city and another in a distant city — perhaps one you travel to often. This saves you paying high roaming charges.

Multi-Path Multiple routes taken by RF energy between the transmitter and the receiver. Signal can cancel or reinforce. Varying multipath (at sunset or sunrise) causes varying signal strength that sounds like a train's steam engine starting up.

Multi-Path Error Errors caused by the interference of a signal that has reached the receiver antenna by two or more different paths. Usually caused by one path being bounced or reflected.

Multi-Point A configuration or topology, designed to transmit data between a central site and a number of remote terminals on the same circuit. Individual terminals will generally be able to transmit to the central site but not to each other.

Multi-Point Circuit A circuit connecting three or more locations. It is often called a multidrop circuit. See also MULTI-POINT and MULTI-DROP LINE.

Multi-Point Distribution Service A one-way domestic public radio service rendered on microwave frequencies from a fixed station transmitting (usually in an omnidirectional pattern) to multiple receiving facilities located at fixed points.

Multi-Point Grounding System A system of equipment bonded together and also bonded to the facility ground at the nearest location of the facility ground.

Multi-Point Private Line 1. A multi-point private line or data line is a communications path between two or more locations requiring two or more LECs. For example.... A location in Detroit, MI has a data line going to NY, NY, then down to Washington, DC. See also Multi-drop private line. 2. A single communications link for two or more devices shared by one computer and or more computers or terminals. Use of this line requires a polling mechanism. It is also called a multidrop line.

Multi-Processing A type of computing characterized by systems that use more than one CPU to execute applications. Multi-processing is not multi-tasking, which is the ability to have more one application running on a system at the same time. The technique is not associated with multi-processing, nor does it require multi processing to take place. Multi-tasking typically uses a computer with one CPU (e.g. your desktop or laptop). Multi-processing uses a computer with several CPUs, often a server. See MULTI-TASKING.

Multi-Processor Kernel Software that enables a computer operating system to use more than one CPU chip simultaneously. Such software is typically used on servers. It is used to speed up a server.

Multi-Programming Computer system operation whereby a number of independent jobs are processed together.

Multi-Protocol Message Routers A device which converts different electronic mail formats. Such a router would be used to move electronic mail from a cc mail equipped-LAN to a Davinci e-mail LAN to a Wang mini-computer based system.

Multi-Purpose Internet Mail Extension See MIME.

Multi-Rate ISDN "Nx64" or switched-fractional T-1 lets users combine multiple ISDN PRI "B" channels on a call-by-call basis. Since each "B" channel operates at 64 kbps, users could get from 128 kbps to 1.544 Mbps of bandwidth in 64 kbps increments.

Multi-Server Network A single local area network (one cabling system) that has two or more file servers attached. Network addresses assigned to the LAN drivers are the same in each file server because the network boards are attached to the same cabling system. On a multi-server network, users can access files from any file server to which they are attached (if they have access rights). A multi-server network should not be confused with an inter network (two or more networks linked together through an internal or external router).

Multi-Session When pictures are placed on Kodak Photo CDs the first time, that's called the first session. And the result is a single session CD. The next time more photos are put on the disk, the disk is now called a multi-session Photo CD. To read such a disk, you need a CD-ROM player, which is specifically called a multi-session CD-ROM player.

Multi-Stage Dialing The device needs to break the dialing sequence up into a number of stages in order to execute and complete the call. This is referred to as "multi-stage" or "incremental" dialing. This type of dialing is needed in cases where the switching domain prompts the device for more digits (by sending dial tone again or some other tone).

Multi-Stage Queuing This is a term typically used in Automatic Call Distributors. It is the ability to array a number of agent groups in a routing table. The notion of multiple agent groups being addressed may mean that the system may be able to "look-back" and "look forward" as it searches for a free agent in the right group to take the call presently holding.

Multi-Station Any network of stations capable of communicating with each other on one circuit or through a switching center.

Multi-Tasking The concurrent management of two or more distinct tasks by a computer. Although a computer with a single processing unit (as virtually all PCs are) can only execute one application's code at a given moment, a multi tasking operating system can load and manage the execution of multiple

applications, allocating processing cycles to each in sequence. Because of the processing speed of computers, the apparent result is the simultaneous processing of multiple tasks. Standard mode Windows performs multi tasking only in the form of context switching. 386 enhanced mode allows multi tasking in the form of time slicing. See MULTI-PROCESSING.

Multi-Tenant Sharing The capability of a PBX to serve more than one tenant in a building. This process is a new option for building owners. They now become the Telephone Company for tenants in the building. There is money in this business, but chiefly on the resale of long distance phone calls.

Multi-Tester Usually an alternate name for VOLT-OHM-MILLIMETER, but may also apply to other MULTI-FUNCTION testing devices.

Multi-Threading Concurrent processing of more than one message by an application program. One of OS/2's advantages over Windows is that IBM designed it as a multi-threaded operating system. Each program in OS/2 can start two or more threads, which carry out various interrelated tasks with less overhead than two separate programs would require. For example, a communications program could have three threads running: one that waits for characters to be received, another that monitors the keyboard, and a third that displays information. This is more efficient than running multiple tasks because it doesn't require the overhead of an operating system context switch.

In short, a new and complex method of programming. Here is a definition from Sun Microsystems. A traditional UNIX process has a single thread of control. A thread of control, or more simply a thread, is a sequence of instructions being executed in a program. A thread has a program counter (PC) and a stack to keep track of local variables and return addresses. A multi-threaded UNIX process is no longer a thread of control of itself; instead, it is associated with one or more threads. Threads execute independently. There is in general no way to predict how the instructions of different threads are interleaved, though they have execution priorities that can influence the relative speed of execution. In general, the number or identities of threads that an application process chooses to apply to a problem are invisible from outside the process. Threads can be viewed as execution resources that may be applied to solving the problem at hand.

Threads share the process instructions and most of its data. A change in shared data by one thread can be seen by the other threads in the process. Threads also share the operating system. Each sees the same open files. For example, if one thread opens a file, another thread can read it. Because threads share so much of the process state, threads can affect each other in sometimes surprising ways. Programming with threads requires more care and discipline than ordinary programming because there is no system enforced protection between threads. See WINDOWS NT.

Multi-Tier Tariffs A way of paying for something (i.e. equipment) from your local phone company. The idea is that one tier of your monthly payments is to pay off the equipment, and after a finite period, this tier payment drops to zero. The next tier is to pay for your monthly service and it is ongoing. Other tiers are for other reasons. As this technique was practiced by the Bell System, it was called "two tier." You will no longer find two tier tariffs in common use.

Multi-User PC A microcomputer that has several terminals attached to it, so that multiple users can simultaneously use its resources. Multi-user PCs can either slice up the time of a single microprocessor or can give each terminal-based user his own microprocessor. Multi-user PCs are an alternative to LANs and are typically used in specialized, one-application solutions, such as a doctor's office billing system.

Multi-User Software An application designed for simultaneous access by two or more network nodes, i.e. two or more users on a network. It typically employs file and/or record locking. It is not associated with multi processing, nor does it require multi processing to implement.

Multi-User Telecommunications Outlet Assembly A grouping in one location of several telecommunications outlets/connectors.

Multi-Vendor Integration Protocol See MVIP.

Multi-Way Communication A multimedia definition. Multi-way communication goes between two people, or between groups of people in all directions. Multi-way communication can be in real-time, or in store-and-forward mode. Examples of multi-way communication include a video conference, where one individual is giving a presentation to a group of people who listen and ask questions from their workstations; and group conferencing, where several people collaborate, supported by audio, video, and graphics on their workstation screens.

Multicast 1. The broadcast of messages to a selected group of workstations on a LAN, WAN or the Internet. Multicast is communication between a single device and multiple members of a device group. For example, an IPv6 router might address a series of packets associated with a routing table update to a number of other routers in a LAN internetwork. Similarly, a LAN-attached workstation might address a transmission to a number of other LAN-attached devices. Companies are discovering they can distribute material to large numbers of employees and others on their intranets more efficiently using multicast than they can by sending such material in separate bursts to each user. In multicast mode, routers distribute a given file to all hosts that have signaled they want to receive the material, using the Class D addresses of the IP addressing hierarchy. See See MULTI-CAST PACKETS AND IPV6. CONTRAST WITH UNICAST, ANYCAST AND BROADCAST.

2. A TV term that simply means more channels will be available for viewers. It's often used to refer to the explosion of special interest channels that will be around when digital TV hits the scene and uses the CATV network to transmit its programs. I suspect that the explosion of channels will lend new meaning to Johnny Carson quips about channels for "one-eyed, one-legged transvestites."

Multifunction Peripheral MFS. A single computer device that prints, photocopies, faxes and scans.

Multimedia See MULTI-MEDIA.

Multimedia ACD Also called multimedia queueing. Automated Contact Distribution system that is enabled to process multiple media types, such as voice calls, e-mails, incoming fax documents, web chat requests and Internet voice/video interactions using queuing/hold strategies. For example, a sales queue could be configured and staffed by sales associates that could evenly distribute all sorts of incoming customer sales requests (fax, phone calls, emails, etc.) to sales associates by blending computer and phone capabilities into a common sales response system. This makes sales associates more efficient and also provides a more consistent service level to all customers. See also Multimedia Queuing.

Multimedia Communications Forum See MMCF.

Multimedia Extensions See MMX

Multimedia Queuing The term was originally invented by Interactive Intelligence of Indianapolis, who makes a communications server — a telephone and messaging system based in a PC. The idea is that the communications server should do way more than servicing mere phone calls. It should handle all media streams — email, fax mail, chats, etc. — with the same discipline as phone systems (particularly automatic call distributors) handle plain ordinary phone calls. All "calls" of whatever media should be treated by the system as of equal importance. See also Multimedia ACD.

Multimedia Super Corridor The MSC is an ambitious development concept of the Malayasia government to develop an area of Malayasia with high-tech (telecom and computer) companies. The area is scheduled to house government offices, moving them out of traffic-congested Kuala Lumpur, the capital of the country. The government is installing a 4,300 kilometer fiber optic network in the MSC. If a company agrees to invest in the area, the Malayasian government extends tax breaks and other incentives. Overseeing the MSC is something called The Multimedia Development Corporation.

Multimode See MULTI-MODE above.

Multiparty Line A single telephone line which serves two or more subscribers (network access lines). The term usually means either a "two party" or a "four party" line. Lines serving more than four parties are called "rural lines".

Multipath Fading The signal degradation in a cellular radio system that occurs when multiple copies of the same radio signal arrive at the receiver through different reflected paths. The interference of these signals, each having traveled a different distance, result in phase and amplitude variations. The radio signal processing in both the base station and mobile units have to be designed to tolerate a certain level of multipath fading.

Multiplayer Gaming See TELEGAMING.

Multiple Access The ability of several personal computers connected to a Local Area Network to access one another through a common addressing scheme and protocol.

Multiple Address Message A message to be delivered to more than one destination.

Multiple Address Systems MAS. A microwave point-to-multipoint communications system, either one-way or two-way, serving a minimum of four remote stations. The private radio MAS channels are not suitable for providing a communications service to a larger sector of the general public, such as channels the commission has allocated for cellular paging or specialized mobile radio services. (SMR).

Multiple Console Operation A phone system with this feature can use more than one attendant console. It's good to know the maximum number of consoles your chosen PBX can use.

Multiple Customer Group Operation A PBX shared by several different companies, each having separate consoles and trunks.

Multiple Domains A set of domains on a single LAN, each of which has its own domain-wide post office. Hosts within each domain can exchange mail by going through one domain post office. Hosts in different domains must generally send mail though two intermediary post offices: the sender's domain post office and the receiver's domain post office.

Multiple Frequency-Shift Keying MFSK. A form of frequency-shift keying in which multiple codes are used in the transmission of digital signals. The coding systems may use multiple frequencies transmitted concurrently or sequentially.

Multiple Homing Connecting your phone so it can be served by one or several switching centers. This service may use a single directory number. It may also use several directory numbers (another term for phone numbers). It all depends on how you set the service up with your local phone company. The idea is to give you more ways of reaching the switched network — in case one or more of your local loops breaks.

Multiple Listed Directory Number Service Permits more than one listed directory number to be associated with a single PBX.

Multiple Master Domain Consists of two or more Single Master Domains connected through two-way trust relationships. See Trust Relationship.

Multiple Name Spaces The association of several names or other pieces of information with the same file. This allows renaming files and designating them for dissimilar computer systems such as the PC and the Mac.

Multiple Parallel Processing A method of fault tolerance used with host computers. Several CPUs cooperate to process data. It one CPU fails, its processing tasks are automatically assigned to other processors.

Multiple Protocol Router A communications device designed to make decisions about which path a packet of information will take. The packets are routed according to address information contained within, and can route across different protocols.

Multiple Routing The process of sending a message to more than one recipient, usually when all destinations are specified in the header of the message.

Multiple Spot Scanning In facsimile systems, the method in which scanning is carried on simultaneously by two or more scanning spots, each one analyzing its fraction of the total scanned area of the subject copy.

Multiple Token Operation Variant of token passing for rings in which a free token on a LAN is transmitted immediately after the last bit of the data packet, allowing multiple tokens on ring (but only one free token) simultaneously.

Multiple Tuned Antenna An antenna with connections through inductances to ground at more than one point and so determined that the total reactances in parallel are equal to those necessary to give the antenna the desired natural frequency.

Multiplex 1. To transmit two or more signals over a single channel.
2. In the world of CATV and the explosion of choices that digital TV is bringing, to multiplex means to offer subscribers a choice of various starting times for movies and events.

Multiplex Aggregate Bit Rate The bit rate in a time division multiplexer that is equal to the sum of the input channel data signaling rates available to the user plus the rate of the overhead bits required.

Multiplex Baseband In frequency division multiplexing, the frequency band occupied by the aggregate of the signals in the line interconnecting the multiplexing and radio or line equipment.

Multiplex Hierarchy In the U.S. frequency division multiplex hierarchy,
12 channels = 1 group
5 groups (60 channels) = 1 supergroup
10 supergroups (600 channels) = 1 mastergroup
6 mastergroups = 1 jumbo group
In contrast, the ITU-T standard says 5 supergroups (i.e. 300 channels) = 1 mastergroup.

Multiplexed Channel A communications channel capable of carrying the telecommunications transmissions of a number of devices or users at one time.

Multiplexer Electronic equipment which allows two or more signals to pass over one communications circuit. That "circuit' may be a phone line, a microwave circuit, a through-the-air TV signal. That circuit may be analog or digital. There are many multiplexing techniques to accommodate both.

Multiplexing Efficiency Figure of merit for multiplexers. The ratio of the aggregate channel input data rate to the composite output data rate. Many statistical multiplexers achieve a multiplexing efficiency of 8 or more.

Multipoint Access User access in which more than one terminal equipment (TE) is supported by a single network termination.

Multipoint Grounding System A system of equipment bonded together and also bonded to the facility ground.

Multipoint-to-Multipoint Connection A Multipoint-to-Multipoint Connection is a collection of associated ATM VC or VP links, and their associated nodes, with the following properties:
1. All Nodes in the connection, called endpoints, serve as a Root Node in a Point-to-Multipoint connection to all of the (N-1) remaining endpoints.
2. Each of the endpoints on the connection can send information directly to any other endpoint, but the receiving endpoint cannot distinguish which of the endpoints is sending information without additional (e.g., higher layer) information.

Multipoint-to-Point Connection A Point-to-Multipoint Connection may have zero bandwidth from the Root node to the Leaf Nodes, and non-zero return bandwidth from the Leaf Nodes to the Root Node. Such a connection is also known as a Multipoint-to-Point Connection. Note that UNI 4.0 does not support this connection type.

Multiport Card A circuit board with two or more ports for modems or other devices. Useful for enabling one PC to handle multiple incoming or outgoing calls at one time.

Multiport Repeater A repeater, either standalone or connected to standard Ethernet cable, for interconnecting up to eight ThinWire Ethernet segments.

Multiport Switch A local area network term. A device which allows packets to switch from one cable to another.

Multiprocessing A type of computing characterized by systems that use more than one CPU to execute applications. Multi processing is not multi tasking, which is the ability to have more one application running on a system at the same time. The technique is not associated with multi processing, nor does it require multi processing to take place. Multi tasking typically uses a computer with one CPU (e.g. your desktop or laptop). Multi processing uses a computer with several CPUs, often a server. See MULTI-PROCESSING, MULTI-TASKING and MULTI-THREADED.

Multiprotocol Over ATM MPOA. A proposed ATM Forum spec that defines how ATM traffic is routed from one virtual LAN to another. MPOA is key to making LAN emulation, Classical IP over ATM, and proprietary virtual LAN schemes interoperate in a multiprotocol environment. At this point, its unclear how MPOA will deal with conventional routers, distributed ATM edge routers (which shunt LAN traffic across an ATM cloud, while also performing conventional routing functions between non-ATM networks), and route servers (which centralize lookup tables on a dedicated network server in a switched LAN).

Multipurpose Internet Mail Extension MIME. An extension to electronic mail (Internet and other e-mail systems) which provides the ability to transfer non-ASCII data,

Multipurpose Internet Mail Extension (MIME)

Uuencoded File Attachment: ANTARESN.DOC

```
begin 660 ANTARESN.DOC
MVZ4M`````"00``````````````````````@`$```$%```![" P`````````*
M```````($$# 
M``!_.````*``!!_ `'\*```*`````\ `````\*```````\*```..  ( T*`````
M_____
M_____
9_____T16
end
```

such as graphics, software, audio, video, binary and fax. MIME was developed and adopted by the Internet Engineering Task Force. It was designed for transmitting mixed-media files across TCP/IP networks. The MIME protocol, which is actually an extension to SMTP, covers binary, audio and video data. Essentially what happens with MIME is that you pick a binary file (any file that isn't ASCII) to send along with your e-mail. Your e-mail software then converts your binary file to ASCII text which can easily be transmitted across one or more e-mail systems. All e-mail networks will transmit ASCII (i.e. ASCII 128 and below). The technique that makes MIME encoding possible is called UUencoding. Here is an example. Notice that every line begins with a M. Every line is the same length. At one end you UUencode the file (and it looks as below). When you receive the file, you must UUdecode it. Mostly the coding and encoding is done automatically by your e-mail program. Sometimes you must decode it manually. Don't try and UUdecode the following file. It won't work. I chopped the middle of the file out. I'm running it below purely as an example. See UUENCODE and MIME.

Multisession An incrementally updated Kodak Photo CD. See MULTI-SESSION for a fuller explanation.

Multitasking See MULTI-TASKING.

Multithreaded See MULTI-THREADED.

Muriel Muriel Fullam worked for me for ten years before she worked with me on this dictionary. That had to be the longest apprenticeship ever. It shows. She is married to Jerry Fullam who is absolutely the best husband and will love it when he sees his name here.

Murphy's Law When something can go wrong, it will. Compare with the GOTCHA LAW.

MUSE MUltiple Sub-Nyquist Samplying Encoding. The Japanese bandwidth compression algorithm for analog HDTV transmission. The Japanese began work on analog HDTV in 1968, and demonstrated it in Washington, D.C. in 1987. The Japanese system ultimately was rejected for use in the U.S., in favor of a digital standard proposed by The Grand Alliance. The Japanese government invested over US$1 billion in the project. See also THE GRAND ALLIANCE.

Mushroom Board Also called a white board or peg board. This is placed between termination blocks to support route crossing wire.

Music Source An external music source such as a radio, can be connected to the Key Service Unit for Music on Hold, Background Music, or both.

Music-On-Camp Audio source input for use with attendant camp-on. See also MUSIC-ON-HOLD.

Music-On-Hold Background music heard when someone is put on hold, letting them know they are still connected. Some modern phone systems generate their own electronic synthesized music. Most phone systems have the ability to connect any sound-producing device, e.g. a radio or a cassette player. Most companies, unfortunately, devote little attention to the sound source they select. Sometimes competitors will deliberately advertise on the radio station that callers will hear on hold. Thus, Macys is now selling Gimbels. It's better to use pre-recorded music. Better yet are tapes of "specials" and other happenings around your firm. Use the "Music-on-Hold" feature as another method of selling. "Ask the operator about our special on ladies' underwear." "Ask our operator to send you a copy of our latest annual report."

Musical Chair Stocks When the music stops, the last one holding the stock is bankrupt. This Aphorism, which refers to the phenomenon of forever-rising Internet stocks, which happened in the fall of 1998, came from the best stockbroker in North America, Todd Kingsley of Smith Barney in Washington, D.C. I know he's the best, because he, modestly, told me so.

Musical Instrument Digital Interface (MIDI) A standardized communications protocol. MIDI files contain musical note information and other performance data that can "play" a MIDI instrument or sound module to produce music.

Must Carry Signals Mandatory signal carriage of both commercial and noncommercial television broadcast stations that are "local" to the area served by the cable system.

Mute A feature which disconnects the handset microphone or speakerphone microphone so that side conversations won't be heard.

Mutual Capacitance The capacitance between two conductors when all other conductors, including the shield, are short circuited to ground.

Mutual Synchronization A timing subsystem not employing directed control, by which the frequency of the clock at a particular node is controlled by some weighted average of the timing on all signals received from neighboring nodes.

Mutually Synchronized Network A network-synchronized arrangement in which each clock in the network exerts a degree of control on all others.

Mutually Exclusive If there are two or more ways in which an event can occur then each way is mutually exclusive of the other way. The probability of the event occurring is the sum of the probabilities of the two or more ways.

MUX See MULTIPLEXER.

MVIP Multi-Vendor Integration Protocol. Pronounced M-

VIP. MVIP is a family of standards designed to let computer telephony products from different vendors inter-operate within a computer or group of computers. MVIP started in 1990 with a computer telephony bus for use inside a single computer. Picture a printed circuit card that fits into an empty slot in a personal computer. The slot carries information to and from the computer. This is called the data bus. Printed circuit cards that do voice processing typically have a second "bus" — the voice bus. That "bus" is actually a ribbon cable which connects one voice processing card to another. The ribbon cable is typically connected to the top of the printed circuit card, while the data bus is at the bottom. As of this writing, there are at least four such buses defined. Several (AEB, PEB, SCSA) have been defined by Dialogic Corporation, Parsippany, NJ for connecting devices with Dialogic products and Dialogic compatible devices. There is also an organization called the ECTF which helps promulgate buses. The MVIP Bus was defined by Natural MicroSystems, Natick, MA with assistance from Mitel, Promptus Communications and Rhetorex as a vendor-independent means of connecting telephony devices within a computer chassis. MVIP was introduced by a seven-company group in 1990 and has been distributed by Natural MicroSystems, Mitel and NTT International (part of the Japanese telephone company). By July 1996, MVIP had over 170 companies manufacturing over 300 board-level MVIP products including telephone line interfaces, voice boards, FAX boards, video codecs, data multiplexers and LAN/WAN interfaces. MVIP now has its own trade association the Global Organization for MVIP (see GO-MVIP), which develops extensions such as higher-level APIs and multi-chassis switching. Here is a write-up on the original MVIP Bus from Mitel's Communicating Objects Division:

"The MVIP Bus consists of communications hardware and software that allows printed circuit cards from multiple vendors to exchange information in a standardized digital format. The MVIP bus consists of eight 2 megabyte serial highways and clock signals that are routed from one card to another over a ribbon cable. Each of these highways is partitioned into 32 channels for a total capacity of 256 full-duplex voice channels on the MVIP bus. These serial link from one card to another. They are electronically compatible with Mitel's ST_BUS specification for inter-chip communications. By letting expansion cards exchange data directly, the MVIP bus opens the PC architecture to voice/data applications that would otherwise overburden the PC processor with data transfers. The MVIP bus is equivalent to an extra backplane that is capable of routing circuit switched data.

"MVIP systems generally have two types of cards ; network cards and resource cards. They differ by the switching they provide and in the way they are wired to the bus. Network cards almost always provide more flexible switching and can drive either the input or the output side of the bus, although they usually drive the output side of the bus. Resource cards usually provide very little switching and are only able to drive the input side of the bus. Resource cards usually rely on the network cards to do most or all of the switching on the MVIP bus."

According to Brough Turner, chairman of the GO-MVIP Technical Committee, MVIP Evangelist and vice president of Natural MicroSystems, "The original MVIP bus distributes switching elements but only requires switches on telephone network interface cards. This has simplified access to MVIP for many communications service providers and has permitted easy inter-connection of pre-existing and proprietary technology to the common bus. More recently, the Flexible MVIP Interface Circuit (see FMIC) has reduced the complete MVIP interface, including clocks and switching, to one chip, further simplifying MVIP connections.

"MVIP's distributed switching architecture provides software controlled digital switching within the PC chassis. MVIP software driver standards simplify access for application developers and support the integration of components from multiple vendors. As a result of work by more than a dozen companies active in the GO-MVIP Technical Committees, MVIP now addresses multi-chassis connections and higher-level APIs. Multi-Chassis MVIP (MC-MVIP) provides telephony connections and distributed digital switching between MVIP-based PCs and supports interconnection of MVIP PCs with proprietary telephony equipment including PBXs, ACDs, and voice processing systems. MC-MVIP includes redundant clocks and specifically permits plugging and unplugging connections while a multi-chassis system is operating. This allows application developers to construct fault-tolerant systems."

Since the original MVIP standard (now known as MVIP-90), the GO-MVIP Technical Committee has developed several standards which offer significant improvement. H-MVIP (High-Capacity MVIP) was approved in 1995 as a single-chassis standard which increases the call capacity of MVIP from 512 to 3,072 time slots. Additionally, four draft standards have been developed for MC-MVIP (Multi-Chassis MVIP). Those standards address operation and synchronization of MVIP in a multi-chassis environment over twisted pair and fiber (FDDI-II and SDH/SONET), with MC4 addressing ATM networking involving both SVCs and PVCs. See FMIC, GO-MVIP, NMS, NTT, PEB and SCSA.

MVPD Multichannel Video Program Distribution. It is provided by cable and satellite. See IB docket No. 95-168 or PP docket No. 93-253 In the Matter of Revision of Rules and Policies for the Direct Broadcast Satellite Service before the FCC.

MVS Multiple Virtual Storage.

MW Abbreviation for MILLIWATT.

MX3 A designation for a multiplex which interfaces between any of the following circuit combinations: 28 DSIs to one DS3 (M13), 14 DSICs to one DS3 (MC3), or 7 DS2s to one DS3 (M23).

MXR Multiplexer

MZI Mach Zehnder Interferometer.

N Any digit 2 through 9. X is any digit 0 through 9. This is telephony nomenclature, not computer nomenclature. Accept it, whether it's logical or not.

N Port Node Port. A Fibre Channel term, referring to the link control facility which connects across a link to the F Port (Fabric Port) at the Fabric (switch). The node can be a mainframe, storage device, workstation, peripheral, or any other attached device. See FIBRE CHANNEL.

N Series Terminators Most local area networks are bus configurations. This means one long piece of cable (coaxial or fiber) with workstations connected along the way, typically with "T" connectors. For a network to work properly, you need to place resistance at the end of the bus, to terminate it. A thin wire Ethernet typically requires a 50 ohm resistance at either end of the bus. Thin wire Ethernet terminators are commonly called N series terminators. They may also be used with grounding wires to ground the network. Networks don't work well without resistance at the end of their buses.

N-1 Pronounced N minus One. This is a term used in central office (also called exchange) switching. It refers to the central office switch just before the last one, i.e. the penultimate switch. It's an important term because that second last switch through which your incoming call passed will determine the signaling your central office switch received. See also N-1 Network.

N-1 Network A telephone company AIN term. The telecommunications network in the call path just prior to the network containing the ported-from switch. If there are only two networks in the call path, then the N-1 network is the originating network. See also N-1.

N-ary Code A code used to encode real-world characters that use n different code elements. Examples are binary (two states) and tertiary (three states).

N-ISDN Narrowband ISDN (Integrated Services Digital Network). Narrowband ISDN is an unkind name for the present form of ISDN presently implemented. In short, anything under an ISDN PRI. According to Professor Michael L. Dertouzos of MIT, narrowband ISDN suffers from the same constraints as classical voice telephony. For example, N-ISDN can carry reasonably be altered to accommodate the great information only in fixed chunks of 64 kilobits per second. This would be like a road system closed to everything except motorcycle. An 18-wheel truck carrying produce or a heating fuel tanker would be "welcome" to use this road as long as it could repackage its cargo into (and out of) motorcycle size chunks. See also B-ISDN.

N1 The first short-haul multiplex carrier in the U.S. by the Bell System was Western Electric's N1. The N1 transmitted 12 voice frequency channels over separate transmit and receive pairs in a single cable. N carriers have been progressively improving.

N2 In packet data networking technology, parameter used to specify the allowable number of retransmissions before disconnection.

NA See NIGHT ANSWER.

NAB National Association of Broadcasters. A U.S.-based association which fosters and promotes radio and television broadcasting, represent those industries before the government, works to strengthen the abilities of its members to serve the public, and assists its members in operational matters. www.nab.org

NACN The North American Cellular Network. Various operators nationwide have linked up to offer seamless roaming services so that roaming subscribers can make calls anywhere nationally without speaking to an operator. See North American Cellular Network for much more detail.

NADF North American Directory Plan. An association of electronic mail providers who are figuring standards and ways of sending mail between their subscribers. They plan to use X.500 Recommendations.

Nailed Connection A permanent circuit of a previously switched circuit/s. Defined by Fujitsu, their Omni SI PBX provides up to eight dedicated (constant) trunk-to-trunk connections for special applications. See also NAILED DATA CONNECTION and NAILED-UP CIRCUIT.

Nailed Up Circuit A private line. A circuit permanently established through a circuit switching facility for point-to-point connectivity. Originally, private lines were, in fact, dedicated circuits which literally could be physically traced through the network. They also were known as "nailed-up circuits," as telephone company technicians hung the circuits on nails driven into the walls of the central offices. Today, private lines are actually dedicated channel capacity typically now provided over high-capacity, multi-channel transmission facilities. See PRIVATE LINE.

NAK Negative AcKnowledgement. NAK is a control character in ASCII that means a packet arrived with the check digits in error. It is sent from the computer receiving the packets to the sender, implying that the packet should be retransmitted so that all bits will arrive intact in the next go-round. The binary code is 0101001. The hex code is 51. See CHECK DIGITS and ACK.

Naked Call An incoming call that is routed into an ACD queue without getting call menus or flexible routing.

NAM Number Assignment Module. An electronic or module in a cellular phone which associates the MIN (Mobile Identification Number) with the ESN (Electronic Serial Number). Phones with dual or Multi-NAM features offer the user the option of having a cellular phone with more than one phone number.

NAR Network Access Register. Centrex term describing a Central Office register which is required in order to complete a call involving access to the network outside the confines of that Centrex CO. NARs may be defined in support of local, intraLATA or interLATA traffic. The specifics of NAR implementation vary by Centrex provider.

NAMAS NAtional Measurement Accreditation Service.

Name A name, as opposed to an address, is a location. Independent description of an end-station or node on a network (LAN or WAN) that contains no information about where the name entity is located. Certain protocols, such as IBM NetBIOS, make extensive use of a naming scheme.

Name Registration A Windows 95 definition. The

method by which a computer registers its unique name with a name server on the network. In a Microsoft network, a WINS server can provide name registration services.

Name Resolution The process of mapping a name to the corresponding address. The process used on the network for resolving a computer address as a computer name, to support the process of finding and connecting to other computers on the network.

Name Server 1. An AIN (Advanced Intelligent Network) term. A directory service located in the SLEE (Service Logic Execution Environment) that provides a mapping between a resource's global name and its physical location in the network. 2. An electronic messaging term. A program which provides information about network objects, such as domains and hosts within a domain, by answering queries.

Named Pipe A connection used to transfer data between separate processes, usually on separate computers. Named pipes are the foundation of interprocess communications (IPC). An administrator can set permissions on named pipes, but only LAN Manager and network applications can create them. See also NAMED PIPES.

Named Pipes A technique used for communications between applications operating on the same computer or across the local area network. It includes an applications programming interface, providing application programmers with a way to create interprogram communications using routines similar to disk-file opening, reading, and writing. In Microsoft's words, named pipes allow two or more processes to communicate with each other. Any process that knows the name of a named pipe can access it (subject to security checks).

Naming Authority An authority responsible for the allocation of names.

NAMPS Narrowband Analog Mobile Phone Service. A proposed new standard for cellular radio. NAMPS combines current voice processing with digital signaling. According to Motorola, NAMPS triples the capacity of today's cellular AMPS system, reduces the number of dropped calls and offers a range of new performance enhancements and digital messaging services. The other cellular standards include E-TDMA (Extended Time Division Multiple Access), TDMA (Time Division Multiple Access) and CDMA (Code Division Multiple Access).

NANC North American Numbering Council, pronounced "nancy." An industry council chartered by the FCC in October 1995 to assume administration of the NANP (North American Numbering Plan) from Bellcore, as well as to select LNP (Local Number Portability) administrators. The impartial council comprises 32 voting members from the carrier, manufacturer and end user communities. Another four non-voting members were selected, including representatives from Bellcore, ATIS, the U.S. NTIA, and the U.S. State Department. Ex-officio participants are selected from Canada, the Caribbean, and Bermuda. In October 1997, Lockheed Martin was selected as the primary administrator of NANP, formally replacing Bellcore. See also BELLCORE, NANP and LNP.

NANOG North American Network Operators Group. www.nanog.org. See also IOPS, Internet Society (ISOC) and IOPS.

Nanometer One billionth of a meter. Written nm. The nanometer is a convenient unit for describing the wavelength of light. The light spectrum extends from 750 nm (near infrared) to 390 nm (lowest energy ultraviolet). A nanometer is equal to 10 angstroms. A nanometer is also a millimicron.

Nanosecond One billionth of a second. Written nsec. It's

ten to the minus 9. One nanosecond — a billionth of a second — is the speed at which transistors in today's computers turn on and off to represent the ones and zeros of binary logic and arithmetic. It is a time-duration so short that light, which can speed seven times around Earth in the second between our heartbeats, travels only one foot. See picosecond.

NANP The North American Numbering Plan. It assigns area codes and sets rules for calls to be routed across North America (i.e. the US and Canada). The new one, put into effect in January, 1995 has one major change: The middle number in a North American area code no longer is required to be a 1 or a 0 (one or zero); rather, it can range between 0 and 9. NANP was administered by Bellcore since its formation in 1986. Responsibility was shifted to NANC (North American Numbering Council) in 1995; it was shifted again in 1997 to Lockheed Martin. See North American Area Codes, North American Numbering Plan and Nanc.

NANPA North American Numbering Plan Administration. See also NANP and NANC.

NAP 1. Network Action Point. An AT&T term describing the switching point through which a call is processed. The NAP switches the call based on routing instructions received from the NCP.

2. Network Access Point. A point of access into the Internet used by ISPs and providers of Internet regional and local subnets. NAPs operate at Layer 2 (Link Layer) of the OSI Reference Model, providing meet points where ISPs exchange traffic and routes. Similar to the original concept of the CIX (Commercial Internet eXchange), the NAPs provide a means of direct connection to the Internet, rather than serving solely as an intermediate point of exchanging commercial traffic. The initial NAPs were located in San Francisco under the operation of PacBell; Chicago, Bellcore and Ameritech, and New York (actually, Pennsauken, New Jersey), SprintLink. A fourth was awarded for MAE-East (MERIT Access Exchange) in Washington, DC, and is operated by MFS (Metropolitan Fiber Systems), which now is a business unit of Worldcom. On April 30, 1995, the NSFNet backbone was essentially shut down, and the NAP architecture effectively became the Internet. See also GigaPOP, FIX and MAE.

3. Network Access Point, an AIN term. See Network Acccess Point.

4. Network Access Provider. The NAP provides a transit network service permitting connection of service subscribers to NSPs. The NAP is typically the network provider that has access to the copper twisted pairs over which the DSL-based service operates.

NAPI Numbering/Addressing Plan Identifier.

NAPLPS North American Presentation-Level Protocol Syntax. A protocol for videotex text graphics and screen formats, developed by AT&T and since standardized within ANSI, based on Canada's Telidon videographics protocol.

NAR 1. Nothing Added Reseller. In contrast to a VAR, which is a Value Added Reseller.

2. Network Access Register. Centrex / ESSX term required for a voice conversation. Each conversation requires a LINE and a NAR to complete a connection.

Narrative Traffic Messages normally prepared in accordance with standardized procedures for transmission via optical character recognition equipment or teletypewriter. In contrast to data pattern traffic, narrative messages must contain additional message format lines.

Narrowband 1. An imprecise term. Some people think it's sub-voice grade channels capable of only carrying 100 to 200 bits per second. Others think it means lines or circuits able to carry data up to 2400 bits per second. So as lines get broader, narrowband gets broader. The latest definition of narrowband is up to and including T-1 — or 1.544 megabits per second. See also BANDWIDTH, WIDEBAND, BROADBAND, N-ISDN and B-ISDN.

2. In cellular radio terminology, narrowband refers to the methodology of gaining more channels (and hence more capacity) by splitting FM channels into channels that are narrower in bandwidth. See NAMPS and NTACS.

3. PCS. Mobile or portable radio services which can be used to provide services to both individuals and businesses such as acknowledgement and voice paging and data services.

Narrowband FM Narrowband FM is an FM signal with a bandwidth approximately equal to that of an AM signal modulated with the same audio information. Narrowband FM is used on many emergency bands because it conserves bandwidth while being clear and free from static.

Narrowband ISDN Any ISDN speed up to 1.544 Mbps, which is called PRI or PRA. But this definition is imprecise. And as speeds get faster, so the definition of narrowband ISDN means faster and faster. See N-ISDN and B-ISDN.

Narrowband Signal Any analog signal or analog representation of a digital signal whose essential spectral content is limited to that which can be contained within a voice channel of nominal 4-kHz bandwidth.

Narrowcasting First, there was broadcasting. One signal went to many people. Radio and TV are the classic concepts of broadcasting. One signal — the same signal — to many people. Then came the idea of narrowcasting. One signal to a select number of people — maybe only those people who subscribed to the service and had the equipment to receive it. Then there came pointcasting. This is a fancy name for sending someone a collection of customized information — snippets of stuff that they chose from a palette of information offerings.

NARTE National Association of Radio and Telecommunications Engineers. A worldwide, non-profit, professional organization which certifies engineers and technicians in the areas of telecommunications and electromagnetic compatibility (EMC). NARTE was founded in 1983 to address the professional testing and certification void created when the FCC reduced its role in that regard. www.narte.org

NARUC National Association of Regulatory Utility Commissioners. Members are commissioners of utility regulatory agencies of the states, the federal government and U.S. territories (i.e., the District of Columbia, Puerto Rico, and the Virgin Islands). Objectives are the advancement of uniform regulation, coordinated action, and protection of the common interests of the people with respect to utility regulation. www.erols.com/naruc

NAS NetWare Access Server. See Remote Access Server.

NASA National Aeronautics and Space Administration.

NASC Number Administration and Service Center. Provides centralized administration of the Service Management System (SMS) database of 800 numbers. The NASC keeps track of the 800 numbers that are in use, or available for use, by new 800 users.

NATA North American Telecommunications Association. A trade association of manufacturers and distributors of telephone equipment that was formed in 1970, two years after the FCC's Carterfone decision, which said that the Carterfone and other customer phone devices could be connected to the nation's phone network — if they were "privately beneficial, but not publicly harmful." The Carterfone decision was a land-

mark. It allowed the connection of non-telephone company equipment to the public telephone network. This decision marked the beginning of the telephone interconnect business as we know it today. And NATA's mission was to fight all the restrictive rules and regulations which the telephone companies subsequently threw up to make connection of customer-owned equipment difficult and expensive. At that time the phone companies (especially AT&T and GTE) were among the largest manufacturers of phone equipment. Eventually NATA won the legal fight and lost its mission, its reason for being. In the early 1990s, NATA collapsed. The remaining bits of it were re-assembled into a new trade association called the MMTA, which stood for MultiMedia Telecommunications Association. MMTA is organized around five divisions — computer telephony integration, conferencing/collaboration and messaging, LAN/WAN internetworking, Voice/Multimedia and Wireless Communications. www.mmta.org.

National Association Of Regulatory Utility Commissioners NARUC. An organization supporting the needs of the commissioners of U.S. federal and state regulatory agencies. www.naruc.org

National Bureau Of Standards NBS. The U.S. government organization that helps prepare non-Department of Defense communications standards and operates a testing service to indicate conformity to existing standards. Address: Institute for Computer Sciences and Technology, National Bureau of Standards, Gaithersburg, MD 20899.

National Cable Television Association NCTA. An organization representing the major U.S. cable television operators in the United States. Founded in 1952, the NCTA's mission is to advance the public policies of the CATV industry before the legislative, judicial and regulatory bodies, as well as before the American public. www.ncta.com

National Communications System NCS. The organization established by Section 1(a) of Executive Order No. 12472 to assist the President, the National Security Council, the Director of the Office of Science and Technology Policy, and the Director of the Office of Management and Budget, in the discharge of their national security emergency preparedness telecommunications functions.

National Coordinating Center NCC. The joint telecommunications industry-Federal Government operation established by the National Communications System to assist in the initiation, coordination, restoration, and reconstitution of NS/EP telecommunication services or facilities.

National Electrical Code NEC. A nationally recognized safety standard for the design, construction, and maintenance of electrical circuits. The NEC also gives rules for the installation of electrical and telephone cabling. The NEC is developed by the NEC Committee of the American National Standards Institute (ANSI), sponsored by the National Fire Protection Association (NFPA, Boston) and identified by the description ANSI/NFPA 70-1990. This code has been adopted and enforced by many states and municipalities as law.

National Exchange Carrier Association NECA. An association of local exchange carriers mandated by the U.S. Federal Communications Commission in 1983, in anticipation of the breakup of the Bell System. NECA's primary responsibility was to file new interstate access tariffs and to perform the "settlement" functions previously performed by AT&T and the BOCs (Bell Operating Companies). Under the direction of the FCC, NECA administers approximately $2 billion in annual revenues, based on pooled tariff rates. Through the pooling process, the LECs (Local Exchange Carriers) bill the IXCs (IntereXchange Carriers) for the use of their local exchange networks in originating and terminating long distance calls. Individual revenues and costs then are submitted to NECA, which distributes them to the member companies on an averaged basis. NECA also administers the Universal Service Fund (USF), the Lifeline Assistance Programs, and the Telecommunications Relay Services Fund. See also DIVISION OF REVENUES, LIFELINE ASSISTANCE PROGRAMS, SEPARATIONS AND SETTLEMENTS, TELECOMMUNICATIONS RELAY SERVICES FUND, and UNIVERSAL SERVICE FUND. www.neca.org

National Information Infrastructure NII. What the Clinton Administration prefers to call the Information Superhighway — basically a switched, broadband network that could, in theory, deliver everything from switched video to high-speed access to the Internet. The NII is intended to connect people, businesses, institutions and governments with one another. Its purpose is to expand the availability of a wide variety of information and communications services.

National Institute For Standards And Technology NIST. The U.S. government agency that oversees the operation of the U.S. National Bureau of Standards. The NIST is based in Gaithersburg, MD. See NIST.

National ISDN Council NIC. A council of carriers and manufacturers which determines the generic guidelines for National ISDN. It also publishes those guidelines, as they are developed. NIC is a voluntary council. Membership includes Ameritech, Bell Atlantic, BellSouth, Cincinnati Bell, Lucent, Nortel and SBC. Administration and project management are the responsibility of Bellcore. See also National ISDN-1.

National ISDN-1 NI-1. National ISDN-1 is a set of specifications for a "standard" national implementation of ISDN (BRI and PRI). Based on international ISDN standards recommendations from the ITU-T, and technical references (TRs) specified by Bellcore for the U.S., NI-1 lays the groundwork for a national ISDN infrastructure. Bellcore issued the National ISDN-1 document SR-NWT-001937, Issue 1, in February 1991. Currently, National ISDN is the responsibility of the National ISDN Council (NIC), comprising certain carriers and manufacturers, and administered by Bellcore. The idea of National ISDN-1 is that it be a set of standards to which every manufacturer and carrier can conform. Thereby, all manufacturers can build ISDN equipment (i.e., terminal equipment, terminal adapters, PBXs and COs) which can interoperate effectively, both through and across the various carrier ISDN networks. Similarly, all carriers can interconnect through a standard set of interfaces in order to support seamless connectivity, and transparency of feature content and access. As a result, a consumer can buy an ISDN phone (one conforming to National ISDN-1) at his local Radio Shack (or other store) take it home, plug it in and know it will work, irrespective of carrier and central office manufacturer. This was not always the case. Sadly, National ISDN-1 has not been a rousing success, as ISDN generally has been a failure; also, NI-1 addressed a set of only 17 features. National ISDN-2 documents were first published in 1992, and National ISDN-3 documents in 1993. See also Bellcore Custom ISDN, ISDN and National ISDN Council.

National ISDN-2 NI-2. A set of specifications published in 1992 and building on NI-1. NI-2 defines a set of features and functional capabilities to be provided in the NI offering, and composed of those directly related to serving user applications. NI-2 also defines operations support and billing capabilities directed toward service providers. See NATIONAL ISDN-1.

National ISDN-3 NI-3. A set of specifications published in 1993, building on NI-1 and NI-2. NI-3 defines a marketable and feasible set of features in consideration of market needs, difficulty of implementation and ease of use. See National ISDN-1.

National Park A government-provided repository of on-line information and entertainment, such as is provided by the Smithsonian.

National Science Foundation National Science Foundation. An independent agency of the U.S. government established by the National Science Foundation Act of 1950 with the mission "to promote the progress of science; to advance the national health, prosperity, and welfare; and to secure the national defense." Among the NSF activities are the fostering of the interchange of scientific information among scientists and engineers in the U.S. and foreign countries, and the fostering of the development and use of computers. This mission and charter gave rise to NSFNET, which was in large part the impetus behind the development of the Internet. See NSFNET.

National Security Agency See NSA.

National Technical Information Service. NTIS. An non-appropriated agency of the U.S. Department of Commerce's Technology Administration, the NTIS serves as the official resource for government-sponsored U.S. and worldwide scientific, technical, engineering and business-related information. Information is acquired from over 200 U.S. government agencies, numerous international governmental departments and other international organizations, and through contracts or cooperative agreements with the private sector and other organizations. Information is available in a variety of formats. www.nist.com.

National Technology Grid An effort of the NSF (National Science Foundation) to develop a nationwide computational infrastructure. The NSF will fund the National Computational Science Alliance with up to $170 million over a period of five years, beginning in 1997. The aim of the Alliance is to enable the science and engineering community to take full advantage of rapidly improving high-performance computing and communications technologies. The alliance comprises more than 50 research partners, and is led by the National Center for Supercomputing Applications (NCSA). See also NCSA.

National Television Standards Committee NTSC. The North American standard for the generation, transmission, and reception of television communication wherein the 525-line picture is the standard. The picture information is transmitted in AM and the sound information is transmitted in FM. Compatible with CCIR Standard M. This standard is used also in Central America, a number of South American countries, and some Asian countries, including Japan. See NTSC.

Nationwide/Statewide cost Averaging A method of averaging costs to establish uniform prices for telephone service so that subscribers using more costly-to-serve, lightly trafficked routes — such as those between small communities — receive the same service for the same price as subscribers on lower-cost highly trafficked routes. The idea is that people in rural areas shouldn't pay more for phone service than those living in cities. Theoretically, they would, since theoretically it costs more to provide phone service to rural communities.

Native Address An address that matches one of a given node's summary addresses.

Native Applications 1. Software that runs directly on the

computer's operating system — without requiring an emulation or other program to sit between the software and the operating system. The term "native" acquired new meaning when Apple moved from its Motorola 680x0 microprocessor line of computers to the Power Macintosh line. See NATIVE MODE.
2. An ATM term. A term given to any application written to use any communications protocol prior to ATM.

Native LAN Services Carrier-provided services which interconnect LANs at their full speed, i.e., the native speed of the LANs.

Native Mode 1. Uncompressed.
2. Able to run software directly in the computer's indigenous operating system without intervening emulation software. The term became important when Apple moved from its Motorola 680x0 microprocessor line of computers (called Quadra, Macintosh IIX, Performa, etc.) to the Power Macintosh line of computers powered by the RISC microprocessors called PowerPC 601. This new line of computers runs an operating system different to what Apple had been running on its Macintoshes. The new Power Macintoshes can run software for the Macintosh 680x0 machines — but only in emulation. This makes the software run slowly. So, for the software to run at speeds the new faster machines are capable at running, application software (word processing, desktop publishing, etc.) has to be rewritten to run in native mode.

Native Signal Processing Analog signals, such as voice and video, often are converted into digital signals — for transmission, for computer storage, for compression, etc. Once an analog signal becomes digital, it often has to be manipulated — compressed (to save storage space or transmission time), to be edited, to be conferenced, etc. These speciality tasks of working on digital signals that originally were analog have largely been done by a specialized microprocessor, called a digital signal processor (DSP). DSPs are used extensively in telecommunications for tasks such as echo cancellation, call progress monitoring, voice processing and for the compression of voice and video signals. Lately, as the PC's general purpose microprocessor has become more powerful, the makers of these general purpose microprocessors (especially Intel) have started talking about using some of their spare MIPS (spare processing power) to do some of the tasks previously done by DSPs. They call this new idea — Native Signal Processing.

Native X.400 A term describing a messaging system or service developed using the protocol specifications and service definitions in the X.400 recommendations. Typically used to describe full X.400 services native to a user's home mail environment.

NATOA The National Association of Telecommunications Officers and Advisors, an affiliate of the National League of Cities. NATOA is involved in telecommunications issues that affect state and local governments. Such issues include rights-of-way, radio frequency emissions, placement of radio towers, and universal service. www.natoa.org

Natural A voice recognition term for a language as in normal spoken conversational sentences. The vocabulary would include words like fifty, sixty, hundred etc. and be used in digit recognition.

Natural Computing As computers get more and more powerful, they get easier to use as they are endowed with human-like senses — fluid speech, a good ear (speech recognition) and keen vision. IBM coined the term "natural computing" to describe this.

Natural Frequency The natural frequency of an antenna,

the lowest frequency at which the antenna resonates without the addition of any inductance or capacitance.

Natural Language Query A query written in natural language (for example, plain English) seeking information from a database.

Natural Monopoly A term used by economists to justify regulation. The idea is that one company can provide certain services (such as gas, water, or telecommunications) considerably cheaper than two or three. Therefore, let one company have the monopoly on the service. But substitute government regulation for free competition and this way keep prices down. How well the theory and the practice of government regulation has worked is the subject of acres of learned prose. Suffice, the theory of "natural monopoly" has evaporated in most areas it was practiced — from airlines to telephones, local and long distance. When regulation is removed, prices have usually fallen.

NAU Network Addressable Unit. SNA term for LU, PU and SSCP. Each unit in SNA has a unique address.

NAUCS Network Access Usage and Cost System.

Nautical Mile 6,076 feet. 15% longer than a normal mile, which is 5,280 feet. A measure of distance equal of one minute of arc on the Earth. An international nautical mile is equal to 1,852 meters or 6,076.11549 feet.

NAV Network Applications Vehicle.

Navigate An Internet term. It means simply to move around by the World Wide Web by following hypertext links / paths from document to document on different computers, typically in different places.

Navigation The ability to route telecommunications traffic over diverse circuit options to achieve communications continuity between the desired end customer stations.

Nax Network fAXing; to send faxes over the internet and any other network. Can also be used as a noun. Naxing.

NBFCP NetBios Framing Control Protocol; protocol for transporting NetBIOS traffic over a PPP connection.

NBI Some people think this once famous company's name stood for "Nothing But Initials." Others thought it stood for Nectum Bilinium Inc. It actually stood for "Nothing But Initials."

NBIF No Basis In Fact. Acronym used as a greeting in e-mail, during online chat sessions, and in newsgroup postings to save keystrokes and time.

NBMA NonBroadcast MultiAccess. A term for networks which provide access for multiple devices and which do not broadcast data. Ethernet LANs broadcast data; ATM and X.25 do not.

NBP Name Binding Protocol. AppleTalk protocol for translating device names to addresses.

NBS National Bureau of Standards. A US government agency that produces Federal Information Processing Standards (FIPS) for all other agencies except the Department of Defense (DoD).

NBS/ICST National Bureau of Standards/Institute for Computer Sciences and Technology. The NBS directorate, based in Gaithersburg, MD, is concerned with developing computer and data communications.

NC Network Computer. Larry Ellison of Oracle's idea of a $500 (or so) PC that lacks a hard disk and may lack a monitor but can be used to browse the Internet and run applications on a server on the Internet or corporate intranet. Ellison, who is Oracle's chairman, sees the NC as a "universal digital appliance."
The New Yorker of September 8, 1997 discussed the implications of the network computer thus: Microsoft's worries about

Ellison and NCs are not trivial. After a prolonged period of being in denial about the rise of the Internet, Gates and his team now understand that it is the central fact of the next phase of computing, and that it poses a real threat to Microsoft's power. In 1995, Sun Microsystems introduced an Internet-centric programming language called Java, which creates programs that can run on any operating system and is fast becoming the standard lingo of the Net. In a Java-fuelled future, the reign of the PC might be challenged by the NC which would let users "borrow" programs from the Net and would have no need for Microsoft's Windows — developments that would create enormous upheaval in many of the software markets that Gate's firm now dominates. See also Internet Terminal, NetPC, and NetStation.

NCA Number of Calls Abandoned. The number of calls accepted into the ACD on the trunks only but lost before being connected to a person.

NCAS NCAS is a generic acronym for Non Call-path Associated Signaling. NCAS is out-of-band signaling used to provide emergency signaling information separate from the wireless 911 call to the Public Safety Answering Point (PSAP). This signaling information includes the phone number of the wireless phone and coding to derive a general location of the caller, and meets the Enhanced 911 Phase 1 FCC requirements. This coding can be either a p-ANI or an ESRD. Out-of-band signaling is separated from the channel carrying the voice. Several solutions are available to accomplish this in the wireless 911 calling environment. CAS is a generic acronym for Call-path Associated Signaling. This definition contributed by Glenda Drizos and Doug Puckett of Sprint PCS, Overland Park, Kansas

NCC 1. Network Control Center. A central location on a network where remote diagnostics and network management are controlled.

2. National Computer Conference, once upon a time the largest annual conference of the computer industry.

NCCF Network Communication Control Facility. An IBM term. See NETVIEW.

NCCS Network Control Center System.

NCEO NonCompliant End Office.

NCL 1. Network Control Language. A command line interface language used by Digital Equipment Corp.'s Digital Network Architecture (DNA).

2. Non-Computing Module Load.

NCO Network Control Office

NCOP Network Code Of Practice.

NCP 1. In AT&T language, Network Control Point. A routing billing, and call control data base system for DSDC which uses the AT&T 3B2OD computer as the feature processor.

2. In IBM language. Network Control Program, which is a program that controls the operations of the communication controllers, 3704 and 3705 in an IBM SNA network.

3. In Northern Telecom language, it means Network Configuration Process.

4. In Novell language, it means NetWare Core Protocol. It is NetWare's format for requesting and replying to requests for file and print services.

NCR Originally National Cash Register, then called NCR Corp. AT&T acquired the company in 1991, and changed the name to AT&T Global Information Solutions (GIS) in January 1994. AT&T had grand plans to become a voice and data powerhouse, in both the equipment and network domains. It didn't work out that way. AT&T split into three companies on January 1, 1997. AT&T GIS was spun off and once again

became NCR Corp. NCR is trying to regain its status as a leading manufacturer of data processing systems, ATMs and electronic cash registers.

NCRA National Cellular Resellers Association. A Washington-based trade association and lobbying organization which became the NWRA (National Wireless Resellers Association), which merged into TRA (Telecommunications Resellers Association) in 1997. See TRA.

NCSA National Center for Supercomputer Applications. A supercomputer operations and research center located at the University of Illinois at Urbana-Champaign. the NCSA was one of the national supercomputer centers funded by the NSF (National Science Foundation). The six centers were interconnected over a network known as the NSFNet, which originally operated at 56 Kbps, and which was upgraded in 1987 to 1.544 Mbps (T-1) and to 45 Mbps (T-3) in 1989. The NSFNet was essentially disbanded in 1995, in favor of commercialization of the backbone network.

NCSANet National Center for Supercomputer Applications Network. A regional TCP/IP network which connected users in Illinois, Wisconsin and Indiana. NCSANet was also a mid-level network in NSFNET, which has been abandoned. The NSF (National Science Foundation) decided to get out of the business of being an Internet backbone provider, in favor of allowing the commercialization of the backbone.

NCTA National Cable Television Association. A trade organization representing U.S. cable television carriers.

NCTE Network Channel Terminating Equipment. The general name for equipment that provides line transmission termination and layer-1 maintenance and multiplexing, terminating a 2-wire U-interface. Another name for a CSU, a Channel Service Unit. Also called a Data Service Unit. A device to terminate a digital channel on a customer's premises. It performs certain line conditioning and equalization functions, and responds to loop back commands sent from the central office. A CSU sits between the digital line coming in from the central office and devices such as channel banks or data communications devices. FCC decisions have established that most NCTE is customer premises equipment (CPE) and may therefore, by supplied by third party vendors as well as the telephone company.

NCUG National Centrex Users Group

NDA 1. Non Disclosure Agreement. Check carefully before you sign one. There are precious few new ideas in the history of the world.

2. National Directory Assistance.

NDC National Destination Code. A wireless telecommunications term. The second part of an MSISDN (Mobile Station ISDN) number. It is used to identify a PLMN (Public Land Mobile Network) within a country.

NDD Network Descriptive Database.

NDIS Network Driver Interface Specification. A device driver specification co-developed by Microsoft and 3Com and used by LAN Manager and Vines and supported by many network card vendors. Besides providing hardware and protocol independence for network drivers, NDIS supports both MS-DOS and OS/2. It offers protocol multiplexing so that multiple protocol stacks can coexist in the same host. NDIS is conceptually similar to ODI.

NDM Network Data Mover, a delayed interface using the "Connect: Direct" transmission method. I don't know much about this term, but apparently it is a one-way street, may be somewhat error prone, and is being moved away from by the long distance industry as a means of transmission.

NDO Network Design Order. See Telephone Equipment Order.

NDS Novell Directory Services, although I've also seen it written as NetWare Directory Services. A new feature of Novell's NetWare 4.0 and higher. Log onto NetWare, you're part of a group. That group gives you various "directories," i.e. access to files. Those directories may be on one or many servers.

NDT No Dial Tone. Abbreviation often used on phone company repair orders by staff.

NDX NetWare Directory Services. A feature of Novell's NetWare operating system which allows users to sign on to multiple NetWare servers through one simple sign-on. The ultimate idea is that NDS will become somewhat of a front-end sign-on to other networked devices, e.g. telephone systems, voice mail, electronic mail, Lotus Notes, etc. Sign on in one place, get into many. A big user time-saving benefit. A competitor to NDS is Microsoft's ODSI Open Directory Services Interface.

NE Network Element. A single piece of telecommunications equipment used to perform a function or service integral to the underlying bearer network. In ATM networks, NE is a system that supports at least NEFs and may also support Operation System Functions/Mediation Functions. An ATM NE may be realized as either a standalone device or a geographically distributed system. It cannot be further decomposed into managed elements in the context of a given management function.

NE&I Network Engineering and Implementation. An area of responsibility in a telecommunications organization that identifies business processes and conducts activities that identify, plan, design, and construct network and system resources.

Near End Crosstalk Interference from an adjacent channel. If crosstalk at the near end is great enough, it may interfere with signals received across the circuit.

Near Instantaneous Companding NIC. The very fast quantizing of an analog signal into digital representation...and converting it back. See also companding.

Near-end Crosstalk interference from an adjacent channel. In wires packed together within a cable, the signals generated at one end of the link can drown out the weaker signals coming back from the recipient.

Near-line A storage term typically used as "near-line storage." A digital audio tape (DAT) might be considered near-line storage. A DAT would typically store a database of information sequentially. To find a record might take anywhere from three to 23 seconds. That's an eternity in most computer applications, but it may be adequate for finding something you rarely need to find — maybe several times a year. Thus the concept "near-line" storage.

NEARNET A commercial Internet access service run by Bolt Beranek and Newman, Inc., Cambridge, MA.

NEBS 1. New Equipment Building System.
2. Network Equipment Building Standards. NEBS defines a rigid and extensive set of performance, quality, environmental and safety requirements developed by Bellcore, the R&D and standards organization once owned by the seven regional Bell operating companies (RBOCs). NEBS compliance is often required by telecommunications service providers such BOCs (Bell Operating Companies) and Interexchange Carriers (IEC) for equipment installed in their switching offices. NEBS defines everything from fire spread and extinguish ability test to Zone-4 earthquake tests to thermal shock, cyclic temperature, mechanical shock, and electro-static discharge.

Conforming to NEBS is not inexpensive.
Here is a more detailed explanation of NEBS: NEBS, Generic Equipment Requirement, is a subset of the Family of Requirements as published by Bellcore known as LSSGR, LATA Switching System Generic Requirement. Bellcore has published a technical reference (TR), TR-NWT-000063, Issue 5, 1994, which outlines the NEBS "standard." Bellcore states that the intent of this TR is to "inform the industry of Bellcore's view of the proposed minimum generic requirements that appear to be appropriate for all new telecommunications equipment systems used in central offices (CO) and other telephone buildings of a typical BCC (Bellcore Client Company)." The requirements defined in this TR are broken into two major categories:
1) spacial requirements and 2) environmental requirements. Spacial requirements apply to the equipment systems' cable distributing and interconnecting frames, power equipment, operations support systems and cable entrance facilities. Compliance with these requirements is intended to improve the use of space in the CO, simplify building-equipment interfaces and help make the planning and engineering of central offices simpler and more economical. In general, the spacial requirements relate to the size and location of the equipment going into a CO, how to power it and how to cable it all together.
The second category relates to environmental requirements that define the conditions under which the equipment could potentially be exposed to and still operate reliably or not cause any catastrophic situation such as fire spread. These environmental requirements include temperature and humidity, fire resistance (which seems to be at the top of many BCCs' lists), shock and vibration (which covers transportation, office and earthquake), electrostatic discharge (susceptibility and immunity) and electromagnetic compatibility (emission and immunity). Bellcore has also defined test methods in this TR which provide procedures on how to test the equipment to ensure that it meets the requirement. See Bellcore.

NEC See NATIONAL ELECTRIC CODE.

NEC Requirements The National Electrical Code (NEC) is written and administered by the National Fire Protection Agency (NFPA). The latest 1990 version states that any equipment connected to the telecommunications networks must be listed for that purpose. Listing is acquired through Underwriters Laboratories (UL) or a similar approved lab. Listing requirements for premises wiring between the Network Interface Device and the modular jack at the work area took effect on October 1, 1990.

NECA National Exchange Carrier Association. Nonprofit organization established by the FCC in 1983 to implement the access charge objectives being introduced replacing interstate division of revenue procedures. Membership includes all U.S. local exchange carriers. NECA files interstate access charges with the FCC, and after approval, pools such collections from the LECs and distributes these revenues equitably to the members based on each one's contribution in terms of expenses and capital investment for interstate toll.

NECTAR Nectar is a collaboration among Bell Atlantic/Bell of PA, Carnegie-Mellon University, the Pittsburgh Supercomputing Center and Bellcore.

NEF Network Element Function: A function within an ATM entity that supports the ATM based network transport services, (e.g., multiplexing, cross-connection).

Negated A signal is negated when it is in the state opposed to that which is indicated by the name of the signal. Opposite of Asserted.

Negative Absorption Amplification. The positive difference between stimulated and absorbed radiation.

Negative Acknowledgement NAK. A communications control character sent by a receiving station to indicate that the last message or block received was not received correctly.

Neighbor Node An ATM term. A node that is directly connected to a particular node via a logical link.

NEL Network Element Layer: An abstraction of functions related specifically to the technology, vendor, and the network resources or network elements that provide basic communications services.

Nelson, Lorraine Routh The lady who records the messages on Lucent's Audix voice mail system. She records her timeless messages at the Lucent Bell Labs in Denver. She lives in Oregon.

Nelson, Ted The software visionary who coined the term "hypertext" in a 1965 paper delivered to the Association for Computing Machinery. Extending the concept, Nelson proposed "zippered lists," whereby elements in one textual document would be linked to identical or related elements in other texts. Nelson also started Xanadu, a yet unsuccessful venture for the development of a system which would support the network sale of documents, or snippets of documents, with an automatic royalty on every byte. Nelson since has gone on to other things. See also Hypertext and Xanadu.

NEMA National Electrical Manufacturers Association.

NEMKO Norwegian Board for Testing and Approval of Electrical Equipment.

NEP 1. Ted Rose works in a NEP for a large telephone company. His employer calls NEP a Network Entry Point. Mr. Rose explains, "We are not unlike the hub of a wagon wheel where data comes in from the two computer centers at DS1 and DS3 speeds. The equipment here breaks down those high speeds to both DS0 and DS1 speeds and then we feed that low speed data to Pacific Bell offices and garages to the people who are the users of the data that are customer facing. They would be the spokes of the wagon wheel."

2. See noise equivalent power.

NEPA The National Environment Policy Act of 1969. An Act of Congress which requires federal agencies to take into consideration the potential environmental effects of a particular proposal such as construction of a radio station.

Neper Np. A method of expressing the ratio between two quantities (Q1, Q2). Np=log(to the base e)xQ1/Q2. One Neper equals 8.686 decibels.

Nest 1. To embed a set of instructions or a block data within another.

2. Novell Embedded Systems Technology. Software that Novell wants to put into cars, office machines, telephones and other noncomputer products. It is a variant of NetWare.

Nested list A list that is contained within a member of another list. Nesting is indicated by indentation in most Web browsers. When you create one list element within another list element in FrontPage, the new list element is automatically nested.

Net 1. The Internet. See INTERNET.

2. NETwork.

3. Normes Europeenes de Telecommunication (European Telecommunications Standards).

Net 1000 A data communications network and processing service of AT&T that never got off the ground and was closed down at a cost of about $1 billion. At one stage it was called Advanced Communications Service.

Net Additions A cellular term. The amount of new subscribers signing up for service after adjusting for disconnects (churn).

Net Address The location on a network where an addressee's mail is held, usually in storage until the user logs into the system. The system delivers the message at that time. The Net Address in the header of a message gives the information required by an automated message processing or message switch, to deliver a message to an intended addressee. A net address of RA3Y#SAIL would designate that the message was addressed to the user with the "Net Name" of "RA3Y" (pronounced "Ray". The 3 is silent) located at (signified by the symbol "#") the system in the network called "SAIL", in this case, the Stanford (University) Artificial Intelligence Laboratory.

Net Card Detection Library NCD. Part of Microsoft's Windows for Workgroups. Its purpose is to determine the network adapted installed in the workstation and minimize mistakes made by the user.

Net Citizen An inhabitant of Cyberspace. One usually tries to be a good net.citizen, lest one be flamed, i.e. insulted through e-mail.

Net Data Throughout NDT. The rate at which data is transferred on a communications channel, normally specified in bits per second.

Net Income See Operating Income.

Net Logon Service A Windows NT term. For Window NT Server, performs authentication of domain logons, and keeps the domain's database synchronized between the primary domain controller and the other backup domain controllers of the Windows NT Server domain.

Net Loss The signal loss encountered in a transmission facility or network, or the sum of all the losses the signal encounters on its way to its final destination.

Net Nanny A Net Nanny program lets parents, guardians and teachers keep children from accessing pornographic material on the 'Net, while preventing the childrens' personal information — names, addresses, telephone numbers, etc. — from being accessed.

Net Personality Somebody sufficiently opinionated/flaky/with plenty of time on his hands to regularly post in dozens of different Usenet newsgroups, whose presence is known to thousands of people. See INTERNET.

Net Police Derogatory term for those who would impose their standards on other users of the Internet. Often used in vigorous flame wars (in which it occasionally mutates to net.nazis).

Net Staffing A call center term. The actual number of staff minus the required number of staff in a given period. Net staffing that is positive indicates overstaffing; net staffing that is negative, understaffing. See NET STAFFING MATRIX.

Net Staffing Matrix A call center term. A report that shows the actual number of staff, required number of staff, and net staffing for each period of a given day. See NET STAFFING.

NETBEUI NetBIOS Extended User Interface. The transport layer driver frequently used by Microsoft's LAN Manager, Windows for Workgroups and Windows NT. NetBEUI implements the OSI LLC2 protocol.

NETBIOS Network Basic Input/Output System. A layer of software originally developed by IBM and Sytek to link a network operating system with specific hardware. Originally designed as the network controller for IBM's PC Network LAN, NetBIOS has now been extended to allow programs written using the NetBIOS interface to operate on the IBM Token ring. NetBIOS has been adopted as something of an industry standard and now it's common to refer to Netbios-compatible LANs. It offers LAN applications, a variety of "hooks" to carry out inter-application communications and data transfer. Essentially, NetBIOS is a way for application programs to talk to the network. Other applications interfaces are also being used these days, such as IBM's APPC. To run an application that works with NetBIOS, a non-IBM network operating system or network interface card must offer a NetBIOS emulator. More and more hardware and software vendors offer these emulators. They aren't always perfectly compatible, though. Today, many vendors either provide a version of NetBIOS to interface with their hardware or emulate its transport layer communications services in their network products.

NETBIOS Emulator An emulator program provided with NetWare that allows workstations to run applications that support IBM's NetBIOS calls.

NetBT NBT. NetBios over TCP/IP. A protocol supporting NetBIOS services in a TCP/IP environment, defined by RFCs 1001 and 1002.

NETDDE Network Dynamic Data Exchange, a feature of Microsoft Windows.

Netfind A white pages service which enables a person to query one service and have that service search other databases for addresses matching the originally entered query.

NetFX Consortium NetFX consortium is a group of companies that have come together to provide solutions for delivering real-time multimedia over the Internet's World Wide Web. The members are Diamond Multimedia Systems www.diamondmm.com, NETCOM www.netcom.com, SAIC http://merkury.saic.com/demo), Template Graphics Software www.sd.tgs.com/~template), and Xing Technology Corp www.xingtech.com).

Netgod Also spelled NET.GOD. A person very visible on a network and who played an important role in the development of the network.

NetHead Also spelled net.head. A person who like really loves the Internet.

Nethopper A wonderfully cute name for a device that organizes data to hop from one local area network to another. Nethoppers are also called bridges and routers.

NetHub The home of the RIME network located in Bethesda, MD. All Super-Regional Hubs in the RIME network call the NetHub to exchange mail packets.

NETI Network Information Table.

Netiquette A pun on "etiquette". Contraction for "Network Etiquette". It means proper behavior on Internet. Because the recipients of text-only messages cannot see your face or hear the inflections of your voice, special care must be taken to avoid misunderstanding and to convey the "flavor" that you intended your messages to have. Some common techniques are the use of smileys (also called emoticons): punctuation that suggests sideways faces, bracketed abbreviations like <g> for grin and <G> for Big Grin, and careful capitalization as in I AM NOW SHOUTING AT YOU!

Netizen A citizen of the Internet.

Netkeeper See IPTC.

Netmail A private message transmission capability on FidoNet in which nodes directly communicate with one another on a point-to point (in FidoNet) or hub-routed (in PCRelay). NetMail was originally developed for use by SYSOPS to communicate with one another and is available on some BBs for regular users. See ECHO.

NetMeeting Microsoft's NetMeeting is a conferencing and collaboration tool designed for the Internet or intranet. I'm

giving a lot of space to it in this edition of my dictionary, because I believe NetMeeting is an important product that will have a major impact on the way we communicate (voice, video, data, images, etc.) over the PSTN, over private corporate networks and over IP networks, such as the Internet and private IP networks, known as intranets.

NetMeeting, according to Microsoft, has the following features:

• H.323 standards-based voice and video conferencing. Real-time, point-to- point audio conferencing over the Internet or corporate intranet enables a user to make voice calls to associates and organizations around the world. NetMeeting voice conferencing offers many features, including half-duplex and full-duplex audio support for real-time conversations, automatic microphone sensitivity level setting to ensure that meeting participants hear each other clearly, and microphone muting, which lets users control the audio signal sent during a call. This voice conferencing supports network TCP/IP connections.

• Support for the H.323 protocol enables interoperability between NetMeeting and other H.323-compatible voice clients. The H.323 protocol supports the ITU G.711 and G.723 audio standards and Internet Engineering Task Force (IETF) RTP and RTCP specifications for controlling audio flow to improve voice quality. On MMX-enabled computers, NetMeeting uses the MMX-enabled voice codecs to improve performance for voice compression and decompression algorithms. This will result in lower CPU use and improved voice quality during a call.

With NetMeeting, a user can send and receive real-time visual images with another conference participant using any video for Windows-compatible equipment. They can share ideas and information face-to-face, and use the camera to instantly view items, such as hardware or devices, that the user chooses to display in front of the lens. Combined with the video and data capabilities of NetMeeting, a user can both see and hear the other conference participant, as well as share information and applications. This H.323 standards-based video technology is also compliant with the H.261 and H.263 video codecs. Multipoint data conferencing using T.120. Two or more users can communicate and collaborate as a group in real time. Participants can share applications, exchange information through a shared clipboard, transfer files, collaborate on a shared whiteboard, and use a text-based chat feature. Also, support for the T.120 data conferencing standard enables interoperability with other T.120-based products and services. The following features comprise multipoint data conferencing: Application sharing: A user can share a program running on one computer with other participants in the conference. Participants can review the same data or information, and see the actions as the person sharing the application works on the program (for example, editing content or scrolling through information.) Participants can share Windows-based applications transparently without any special knowledge of the application capabilities. The person sharing the application can choose to collaborate with other conference participants, and they can take turns editing or controlling the application. Only the person sharing the program needs to have the given application installed on their computer.

Shared Clipboard: The shared clipboard enables a user to exchange its contents with other participants in a conference using familiar cut, copy, and paste operations. For example, a participant can copy information from a local document and paste the contents into a shared application as part of a group collaboration. File Transfer: With the file transfer capability, a user can send a file in the background to one or all of the conference participants. When one user drags a file into the main window, the file is automatically sent to each person in the conference; they can then accept or decline receipt. This file transfer capability is fully compliant with the T.127 standard. Whiteboard: Multiple users can simultaneously collaborate using the whiteboard to review, create, and update graphic information. The whiteboard is object-oriented (versus pixel-oriented), enabling participants to manipulate the contents by clicking and dragging with the mouse. In addition, they can use a remote pointer or highlighting tool to point out specific contents or sections of shared pages.

Chat: A user can type text messages to share common ideas or topics with other conference participants, or record meeting notes and action items as part of a collaborative process. Also, participants in a conference can use chat to communicate in the absence of audio support. A "whisper" feature lets a user have a separate, private conversation with another person during a group chat session.

TAPI 3.0 and NetMeeting both support core IP Telephony capabilities. Each platform offers unique benefits: TAPI 3.0 integrates traditional telephony with IP Telephony, providing a COM-based, protocol-independent call-control and data streaming infrastructure. NetMeeting SDK supports T.120 conferencing and application sharing in addition to IP Telephony. Applications using TAPI 3.0 and the NetMeeting API interoperate using H.323 audio and video conferencing.

Because TAPI 3.0 and NetMeeting both support core IP Telephony capabilities (including support for H.323), developers may want to consider the following guidelines when choosing an API for their IP Telephony applications: TAPI 3.0. This is the API to use if you are doing IP Telephony in your application. TAPI 3.0 is especially valuable in the world of client/server computer telephony integration, for combining IP Telephony with traditional telephony, and for IP multicast of voice and video.

NetMeeting API. This is the API to use if you are doing real-time collaboration and want to integrate voice, video, and data conferencing into your application. The NetMeeting API is useful for applications that want to integrate application sharing, whiteboard functionality, and multipoint file transfer with voice and video sessions.

www.microsoft.com/netmeeting.

NetPC See Network Computer and Thin Client.

Netscape The most famous Web Browser for now is Netscape. It owes its genesis to a software program called Mosaic, which Marc Andreessen and a small team of student programmers working for $6.85 an hour in 1993 wrote at the National Center for Supercomputing Applications at the University of Illinois at Urbana-Champaign. A year later, after becoming annoyed at the way the University had taken over his Mosaic creation, Mr. Andreessen proposed a "Mosaic Killer" — a new and improved version of his own creation. The team was back at work by April 1994 in a company called Netscape. And by October, they had created a new version of Mosaic, called Netscape Navigator. Netscape went public in August of 1995 in one of the most successful IPOs (Initial Public Offerings) ever. when Microsoft started giving away its Internet Explorer Browser for free, Netscape fell on hard times. And in late 1998, America OnLine (AOL) bought it.

Netserv A file server used for distributing files directly related to the BITNET network.

Netsite The term Netscape Navigator uses to refer to a URL or WWW address.

Netsploitation Flick Any one of the Hollywood films about the Internet.

Netstation In an Internet scenario, thin clients are known as NetPCs or Netstations. The NetStation is reliant on the server, which is provided by your company or a service provider (e.g., America OnLine, CompuServe, or your ISP). In addition to providing some combination of content and Internet access, the service provider's server will provide your client NetPC with access to all necessary applications (e.g., word processing and spreadsheet), will store all your personal files, will provide all significant processing power, and so on. In this Internet example, the Netstation differs from the standard thin client by virtue of the fact that it does contain a modem, a communications port and communications software, all of which are required for Internet access. See also Client, Client/Server, Client/Server Model, Fat Client, Mainframe Server, Media Server, and Thin Client.

Netview An IBM product for management of heterogeneous networks that integrates the functions of three formerly separate Communications Network Management (CNM) software programs: 1. NCCF. Network Communication Control Facility. 2. NLDM. Network Logical Data Manager, which uses functions from NCCF and helps pinpoint problems along the logical connection/path of an SNA session. 3. NPDA. Network Problem Determination Application, which displays various alerts using IBM equipment located at strategic points in the network and allows diagnostic information to be displayed. Also, NetView incorporates some of the functions from two other programs: VNCA (Virtual Telecommunications Access Method/Node Control Application) which monitors the status and current activity of all resources in a domain, and NMPF (Network Management Productivity Facility) which helps the network operator to install, learn and use many network management products. See also NETWORK and NETWORK MANAGEMENT.

NetWare NetWare is an extremely popular and extremely good operating system for a local area network from Novell, Orem, UT. NetWare is actually its own operating system. This means it is the link between machine hardware (file servers, printers, modems, etc.) and people who want it use that hardware. NetWare is neither DOS, nor OS/2 nor Windows though it can be made to look and act like them. That's part (a small part) of its popularity. See NETWARE MHS, NETWARE WORKSTATION FILES, NETx.COM and NOVELL.

NetWare Bindery Centralized authentication database for NetWare 3.xx LANs.

NetWare Directory Services. See NDS.

NetWare Global MHS Novell's implementation of MHS as a NetWare Loadable Module (NLM), providing powerful integration with NetWare services. This supports additional modules to connect to X.400, SNA and SMTP systems.

NetWare Loadable Module NLM. An driver that runs in a server on a local area network under Novell's NetWare operating system and can be loaded or unloaded on the fly as it's needed. In other networks, such applications could require dedicated PCs. A telephony NLM might allow a workstation on a LAN to control a PBX attached to a NetWare file server. It might also allow the workstation to control one or more voice processing cards sitting on in a NetWare server. In early 1993, AT&T became the first PBX maker to ink a deal with Novell, the creator of NetWare, to put telephony onto Novell LANs. AT&T created a PC-card resident in a Novell File server. The card connects to the ASAI (Adjunct Switch Applications Interface) BRI port on the AT&T Definity PBX. Anyone with a PC on the Novell network and an AT&T phone on their desk can use telephone features, such as auto-dialing, conference calling and message management (a new term for integrating voice, fax and e-mail on your desktop PC via your LAN). The Novell/AT&T deal intends to create open Application Programming Interfaces (APIs) that third party developers can work with. A Novell/AT&T example of what could be developed: A user could select names from a directory on his PC. He could tell the Definity PBX through the PC over the LAN to place a conference call to those names. At the same time, a program running under NetWare would automatically send an e-mail to the people, alerting them to the conference call and giving them the agenda. All participants would have access to both the document and the conference call simultaneously. See TELEPHONY SERVICES.

NetWare MHS Netware MHS, which is software that provides store-and-forward capability. Fax and E-mail systems that support MHS format their message transmissions according to MHS specifications. MHS reads compatible transmissions, determines the intended recipient and his location, and then sends the message to that location, regardless of the type of fax or E-mail system at the different ends. See MHS.

NetWare Telephony Services See TELEPHONY SERVICES.

Network Networks are common in our lives. Think about trains and phones. A networks ties things together. Computer networks connect all types of computers and computer related things — terminals, printers, modems, door entry sensors, temperature monitors, etc. The networks we're most familiar with are long distance ones, like phones and trains. But there are also Local Area Networks (LANs) which exist within a limited geographic area — like the few hundred feet of a small office, an entire building or even a "campus," such as a university or industrial park. There are also Metropolitan Area Networks (MANs). See also LAN and MAN.

Network Access Control Electronic circuitry that determines which workstation may transmit next or when a particular workstation may transmit.

Network Access Point NAP. A telephone company AIN term. Software within a switch capable of recognizing a call that requires processing by AIN logic which, upon recognizing such a call, routes the call to an SSP or ASC switch.

Network ACD Network ACD allows ACD agent groups, at different locations (nodes), to service calls over the network independent of where the call first entered the network. NACD uses ISDN D-channel messaging to exchange information between nodes.

Network Address Every node on an Ethernet network has one or more addresses associated with it, including at least one fixed hardware address such as "ae-34-2c-1d-69-f1" assigned by the device's manufacturer. Most nodes also have protocol specific addresses assigned by a network manager.

Network Addressable Unit NAU. In IBM's SNA, a logical unit (LU), physical unit (PU) or system services control point (SSCP), which is host-based, that is the origin or destination of information transmitted by the path control portion of an SNA network.

Network Agent A network agent is a device, such as a workstation or a router, that is equipped to gather network performance information to send to the network management agent. See network management agent.

Network Analyzer A microwave test system that characterizes devices in terms of their complex small-signal scattering parameters (S-parameters). Measurements involve determining the ratio of magnitude and phase of input and output signals at the various ports of a network with the other ports terminated in the specified characteristic impedance (generally 50 ohms).

Network Application Architecture A generalized architecture allowing interoperability at the application level. Examples are Digital Equipment Corp.'s Network Application Support (NAS) and IBM Corp.'s Systems Application Architecture (SAA).

Network Application Support Digital Equipment Corporation's set of open software which allegedly allows its customers to integrate, port and distribute applications across different computer systems, including VMS, UNIX, MS-DOS, OS/2 and Apple MacIntosh.

Network Architecture The structure and protocols of a computer network. See ARCHITECTURE.

Network Balancing 1. Lumped circuit elements (inductances, capacitances and resistances) connected so as to simulate the impedance of a uniform cable or open-wire circuit over a band of frequencies.
2. Moving circuits around in a multi-node switching network such the switching loads on each of similar switching modules are roughly equal.

Network Basic Input/Output System NETBIOS. Within the context of the MS-DOS operating system, the software or software and firmware services that implement the interface between applications and a network adaptor, such as a CSMA\CD or token-ring adaptor.

Network Board 1. A circuit board installed in each network station to allow stations to communicate with each other and with the file server.
2. An SCSA term. A board device designed to act as an interface between a computer-based signal processing system and a telephone network.

Network Byte Order The Internet standard way of ordering of bytes corresponding to numeric values.

Network Channel Terminating Equipment NCTE. A device or devices at the user's premises used to amplify, match impedance or match network signaling to the customer's equipment connected to the network. Basically, network channel terminating equipment is a general name for equipment linking the network to a customer's premises. When NCTE connects to digital circuits, it typically consists of DSUs and CSUs. They are used for balancing of signals and providing for loop-back testing.

Network Computer NC. Larry Ellison of Oracle's idea of a $500 (or so) PC that lacks a hard disk and may lack a monitor but can be used to browse the Internet and run applications on a server on the Internet or corporate intranet. Ellison, who is Oracle's chairman, sees the NC as a "universal digital appliance."
The New Yorker of September 8, 1997 discussed the implications of the network computer thus: Microsoft's worries about Ellison and NCs are not trivial. After a prolonged period of being in denial about the rise of the Internet, Gates and his team now understand that it is the central fact of the next phase of computing, and that it poses a real threat to Microsoft's power. In 1995, Sun Microsystems introduced an Internet-centric programming language called Java, which creates programs that can run on any operating system and is fast becoming the standard lingo of the Net. In a Java-fuelled future, the reign of the PC might be challenged by the NC which would let users "borrow" programs from the Net and would have no need for Microsoft's Windows — developments that would create enormous upheaval in many of the

software markets that Gate's firm now dominates. See also Internet Terminal, NetPC and NetStation.

Network Control Center A physical point within a network where various management and control functions are implemented.

Network Control Program NCP. An IBM Systems Network Architecture (SNA) term. This is the program that switches the virtual circuit connections into place, implements path control, and operates the Synchronous Data Link Control (SDLC) link. The Network Control Program is normally resident in the communications controller or the host processor.

Network Control Signaling The transmission of signals used in the telecommunications system which perform functions such as supervision, address signaling and audible tone signals to control the operation of switching machines in the telecommunication system.

Network Control Signaling Unit A telephone set that controls the transmission of signals into the telephone system which will perform supervision, number identification and control of the switching machines.

Network Controller A powerful microprocessor device designed to perform communications protocol translations between various terminals and computers and an X.25 packet switching network.

Network DDE Service A Windows NT definition. The Network DDE (dynamic data exchange) service manages shared DDE conversations. It is used by the Network DDE service.

Network Demarcation Point The network demarcation point is the point of interconnection between the local exchange carrier's facilities and the wiring and equipment at the end user's facilities. The demarcation point is located on the subscriber's side of the telephone company's protector.

Network Design and Optimization Network design and optimization is a process which balances network performance (availability) against cost. There are two fundamental tools in network design and optimization: a traffic usage recorder and software to interpret the results and make recommendations. A traffic usage recorder (TUR) is a device which connects to a network element in order to capture and record traffic statistics. Most network elements (e.g., PBXs, ACDs, data switches and routers) have special ports to which such a device can connect, usually via an RS-232 cable. As traffic flows through the network element, various information about that traffic is sent to the TUR in real time. The TUR holds that raw data in buffer memory until such time as it is polled by a centralized computer and the data is downloaded to that centralized computer. Later, the data is processed and reports are generated by traffic analysis software. That software will help you figure out which circuits you need, what speeds, to where, etc.

Network Design Order NDO. See Telephone Equipment Order.

Network Device Driver Software that coordinates communication between the network adapter card and the computer's hardware and other software, controlling the physical function of the local area network adapter cards.

Network Diversity A simply concept that says if you have a network which is important you need to have multiple ways of moving information around that network. Some of those ways should be provided by circuits from one vendor and some should be provided by circuits from several vendors. In other words, there should be diversity in both circuits and vendors. As an example of the necessity of taking net-

work diversity very seriously, consider the year 1991. In 1991, nearly 30 million residential and business customers, more than one hundred thousand airline passengers and an alphabet soup of state and federal agencies were temporarily crippled by nine major telephone network outages. Also in a 12-month period, the FAA (Federal Aviation Commission) reported a total of 114 phone service failures that disrupted air traffic facilities around the nation. Between June 10 and July 1, 1991, six major outages in three states comprised the largest series of network outages in history.

Network Drive A disk drive that is available to multiple users and computers on a network. Network drives often store data files for many people in a work group.

Network Elements NE. Processor controlled entities of the telecommunications network that primarily provide switching and transport network functions and contain network operations functions. Examples are: Non-AIN switching systems, digital cross-connect systems, AIN switching systems, and Signaling Transfer Points (STPs). In SONET, five basic network elements are: add/drop multiplexer; broadband digital cross-connect; wideband digital cross-connect; digital loop carrier; and switch interface.

Network Entity Identifier NEI. A cellular radio term. An Internet protocol address, or ConnectionLess Networking Protocol (CLNP) address, or any other protocol addressing used by the provider in transmission and receipt of Cellular Digital Packet Data (CDPD) services. The address that identifies that a user is authorized to use the service.

Network Equalizing A device connected to a transmission path in order to alter the characteristics of that path in a specified way. Often used to equalize the frequency response characteristic of a circuit for data transmission.

Network Extension Unit NEU. An AT&T thing which sits in a telephone satellite closet (their words) which links up to 11 Starlan daisy chained clusters of PCs, etc. in a star configuration through standard telephone wiring, modular cords and plugs.

Network Externalities First seen in an article by John Cassidy in The New Yorker Jan 12, 1998 issue. In plain English, "network externalities" means that the value of a product increases along with the number of other people using it. Few people care about how many people are buying corn flakes or pizzas. But it usually applies to high-tech goods for two reasons. They have to be compatible with one and other (how useless is a Betamax videocassette today?). Second, the buyers are often linked to a network. The more people on the network, the more valuable the product becomes. Clearly, one person and one phone has no value. Two people and two phones have more value. 10,000 people and 10,000 phones have even more value, etc.

Network Fax Server An network fax server is typically a PC on LAN. The PC has one or more fax/modem cards. Its job is to send and receive faxes for everyone on the LAN. As a result it is connected to dial out phone lines. See SERVER.

Network File System NFS. A method developed by Sun Microsystems Inc. for distributing files within a heterogeneous network. NFS has become a de facto standard. It requires special software drivers to suit specific vendor hardware and operating system software. NFS hardware and software gateway products are also available.

Network Harm The reasoning behind the Bell System's insistence that the public could not hook its own equipment to the telephone line for fear it would produce "Irreparable Harm To The Network". A Bell System publication quoted the National

Academy of Sciences as identifying four areas of potential harm. They were: Excessive signal power, hazardous voltage, improper network control signaling and line imbalance. Since the Carterfone Decision in the summer of 1968, the FCC has seen fit to allow devices, which pass an FCC Registration program, to be connected to the network, and while the quality of the stuff is occasionally poor, the fears of Network Harm have proven groundless. The fears were, of course, raised to preserve the Bell System's erstwhile almost monopoly on the manufacture of telephone equipment in the U.S.

Network Information Center NIC. Centers providing user assistance, document service, training on the Internet.

Network Interface The point of interconnection between Telephone Company communications facilities and terminal equipment, protective apparatus or wiring at a subscriber's premises. The network interface or demarcation point shall be located on the subscriber's side of the Telephone Company's protector, or the equivalent thereof in cases where a protector is not employed, as provided under the local telephone company's reasonable and nondiscriminatory standard operating practices.

Network Interface Card Also called a NIC card. A printed circuit board comprising electronic circuitry for the purpose of connecting a workstation to a LAN. A NIC usually is in the form of a card that fits into one of the expansion slots inside a PC. Alternatively, it can fit into a slot of a MAU (Multistation Access Unit), which serves multiple LAN-attached devices, such as workstations and printers. In the context of IEEE standards, NICs operate at the MAC (Medium Access Control) layer. In the context of the OSI Reference Model, NICs operate at Layers 1 (Physical Layer) and 2 (Data Link Layer). The basic job of the NIC is to take data from the transmitting workstation, form it into the specific packet format demanded by the LAN protocol you are running (e.g., Ethernet or Token Ring), and present it to the shared medium (usually a cable). On the receiving end, the process is reversed, of course. Hard-coded into the NIC at the time of manufacture is a MAC address, unique in all the world to that NIC; the MAC address effectively identifies the LAN-attached device with which it is associated. A NIC works with the network software and computer operating system to transmit and receive messages on the network.

Network Interface Controller Same as a Network Interface Card. See above definition.

Network Interface Device 1. NID. A device wired between a telephone protector and the inside wiring to isolate the customer's equipment from the network.
2. A device that performs, functions such as code and protocol conversion, and buffering required for communications to and from a network.
3. A device used primarily within a local area network to allow a number of independent devices, with varying protocols, to communicate with each other. This communication is accomplished by converting each device protocol into a common transmission protocol.

Network Interface Module Electronic circuitry connecting a system (typically a PC) to the telephone network. Network interface modules come in as many versions as there are ways of connecting to the telephone network. They range from a simple loop start telephone line to complex ISDN PRI circuits. Usually the network interface modules slides into one of the expansion slots inside a PC. The card transmits and receives messages from the resource modules and provides access to the telephone network.

Network Interface Unit 1. A Network Interface Card (NIC) or Multistation Access Unit (MAU) that attaches to a LAN. It implements the local network protocols and provides an interface for device attachment. See also NIC and MAU for much more detail.
2. NIU. A semi-intelligent device that serves as the point of physical and logical demarcation between the LEC (Local Exchange Carrier) and your residential or small business premise. The NIU includes a silicon-based protector that trips the circuit in the event of a lightening strike or some other form of aberrant voltage that otherwise would fry your telephone and burn down your house. The NIU also has enough intelligence to allow the carrier to conduct an automated loopback test, which tests the integrity of the electrically-based, twisted-pair, local loop from the central office (CO) to your premise...and back. See also Loopback and NIU.

Network Interworking A Frame Relay/ATM term. Network internetworking is a method of connecting two frame relay devices over an ATM backbone network. This approach leaves the entire frame relay frame intact, including header, payload and trailer; the ATM devices act to set up the connection on a cut-through basis through the use of a tunneling protocol. Network internetworking is defined in the FRF.5 specification from the Frame Relay Forum, and is recognized by the ATM Forum. Contrast with SERVICE INTERWORKING.

Network Inward Dialing NID. A service feature of an automatically switched telephone network that allows a calling user to dial directly to an extension number at the called user facility without operator intervention.

Network Layer Third layer of the OSI model of data communications, sometimes called the packet layer. Involves routing data messages through the network on alternate routes. See OSI STANDARDS.

Network Layer Address NLA. An address appended to the LAN packet that, unlike a MAC name, indicates exactly where a computing device is located within an internetwork. TCP/IP, DECNet and IPX support network layer addressing and each has its own unique NLA format. Protocol dependent routers use the NLA to make routing decisions.

Network Location In a URL, the unique name that identifies an Internet server. A network location has two or more parts, separated by periods, as in my.network.location. Also called host name and Internet address.

Network Management A set of procedures, software, equipment and operations designed to keep a network operating near maximum efficiency. Network management generally falls into five areas:
1. Configuration Management deals with installing, initializing, "boot" loading, modifying and tracking the configuration parameters of network hardware and software.
2. Fault Location and Repair Management Tools let you find out what's going wrong with what equipment or lines and give you the ability to fix those resources — by re-routing traffic on different lines or reporting problems to the carrier, or suggesting to whoever that certain equipments should be replaced. Fault location and management tools have strong error and alarm characteristics.
3. Security Management Tools allow the network manager to restrict access to various resources in the network. There are devices such as password protection schemes, giving users different levels of access to different network resources.
4. Performance Management Tools provide real-time and historical statistical information about the network's operation. Such tools show, for example, how many packets are being

transmitted at any given moment, the number of users logged into a specific server and use of network lines.

5. Accounting Management applications help users allocate the costs of the various network resources — from lines to PBXs, from access to a mainframe to time used on printers. See NETVIEW.

Network Layer Protocol Identifier NLPI. An identifier allowing entities providing different Network Layer protocols to be distinguished from each other.

Network Management Control Center NMCC. A central place from which the network is maintained and changed and from where statistical information is collected.

Network Management Forum NMF. Now known as the TeleManagement Forum (TMF). An international consortium of service providers and suppliers working toward the development of "practical solutions for cost-effective integration of support systems for improved management of services and networks...through a common, service-based approach." NMF is not a standards organization. Rather, it picks up where the standards bodies leave off through highly-focused activities, oriented toward standards implementation. www.nmf.org. See Network Management and TeleManagement Forum.

Network Management Software The software that manages and controls all network functions within a network.

Network Management Station NMS. A network management station is a dedicated workstation that gathers and stores network performance data, obtaining that data from network nodes (computers) running network agent software that enables them to collect the data. The NMS runs network management software that enables it to compile the information and perform network management functions. See NMS.

Network Management System A comprehensive system of equipment used in monitoring, controlling and managing a data communications network. Usually consists of testing devices, CRT displays and printers, patch panels and circuitry for diagnostics and reconfiguration of channels, generally housed together in an operator console unit. See NMS.

Network Monitor Agent A service that can be installed on an NT workstation, NT server or Windows 95 that lets that computer collect statistics about its network performance, such as the number of packets sent from and received. It gathers the information thorugh its NIC about NetBEUI and NWLink (IPX/SPX) traffic that passes through it.

Network Number The part of an Internet address that designates the network to which the addressed node belongs.

Network Professional Association NPA. A self-regulating, not-for-profit organization with the mission of advancing the network computing profession by educating and providing resources for its members. Membership is approximately 7,000 in 100 chapters, worldwide. NPA runs a program called CNP, which stands for Certified Network Professional. That program is designed for individuals whose career is to design, integrate, manage and maintain networked computing environments. www.npa.org.

Network Protocol Data Unit NPDU. Network Layer Protocol Data Unit. The NPDU comprises the Network Layer Service Data Unit and the Network Layer protocol Control Information.

Network Redundancy In a network, the state of having more connecting links than the minimum required to provide a connecting path between all nodes.

Network Reliability Council A committee comprising senior-level officials from a cross-section of telecommunications service providers and user organizations. Organized under the provisions of the Federal Advisory Committee Act, the council provides recommendations to both the FCC and the telecommunications industry to enhance the reliability of the nations telecom networks.

Network Resource A facility or device that supports call processing.

Network Supervisor A network supervisor is responsible for monitoring network-wide call activity.

Network Synchronization Unit See NSU.

Network Systems A telephone company AIN term. Processor controlled entities of the telecommunications network that provide ancillary network functions and contain network operations functions. Examples are: Service Control Points (SCPs), Adjuncts, Service Nodes (SNs), and Intelligent Peripherals (IPs).

Network Traffic Management NTM. A telephone company AIN term. Functionality that maximizes the traffic throughput of the network during times of overload or failure and minimizes the impact of one service on the performance of others.

Network Vehicle A network vehicle is basically a car equipped with PC-like gadgetry. Such a car, according to the Economist Magazine, "could tap into a regional road system not only to get directions but also to map out a route around rush-hour snags. Drivers and passengers will be able to send and receive email, track the latest sports scores or stock quotes, surf the Web and even play video games. Apparently one of Detroit's largest suppliers of electrical and electronic hardware to the automotive industry, United Technologies Automotive (UTA), has licensed Microsoft's CE operating system. Writing about the deal, the Economist commented, "Any driver familiar with the flakiness of some early releases of Microsoft's products need not fear, however: his car will not crash each time the operating system does. The UTA/Microsoft system will, according to a spokesman, be rigorously tested; and to start with it will only be put in charge of office-type tastes and certain "comfort and convenience" fripperies such as a garage door open, or more extravagantly, temperature-controlled cup-holders. A car thus equipped might not be any safer

Network Virtual Terminal A process whereby a network accommodates terminal devices with different characteristics by converting the varying characteristics into a single format so all terminals appear the same.

Networked Multimedia This definition courtesy Intel: Just as standalone PCs bring you rich audio, video, animation and three dimensional graphics, so networked multimedia brings these same capabilities to PCs connected to networks and the Internet. You could create an electronic letter with an embedded video clip of yourself in the salutation. Include hotlinks to your company's web site for more information on new products. Allow the receiver to have the entire letter read aloud by a real human voice. Hold a virtual conference in shared interactive spaces, with people in different locations virtually meeting and interacting. Videoconferencing and broadcasting over the Internet will become commonplace. With innovations such as real-time video streaming and multicasting, your company could design a live instructional video that covers their problem. If they wanted more help, they might click and initiate an interactive video conference with a customer supper engineer. In the personal sphere, networked multimedia will create many new applications, from real-time multi-player games between people scattered all over the world, to interactive chat sessions complete with voice-enabled and video images of participants, to virtual amusements.

Network Name Automatic Routing A system for getting network-delivered faxes delivered to individual users. The fax network assigns each user a name (usually the user's network login name), and the sender uses this name. The receiving fax system detects this name and routes the fax to the intended recipient.

Network News Transfer Protocol NNTP. A common method by which articles over Internet's Usenet are transferred.

Network Number The part of an Internet address that specifies the network to which the host belongs.

Network Operating System NOS. The software side of a LAN. The program that controls the operation of a network. It allows users to communicate and share files and peripherals. It provides the user interface to the LAN, and it communicates with the LAN hardware or network interface card. It is important to note that a network operating system is different from a network interface card: IBM's Token Ring, for example, is an interface card, not a NOS. IBM's NOS is called the PC LAN Support Program. A variety of other vendors also offer NOSs, including Novell and 3Com. Most have drivers to support a variety of network interface cards.

Network Operations Functions that provide, maintain and administer services supported by the network systems. These functions reside in network systems or network operations applications that interface directly to network systems and include memory administration, surveillance, testing, network traffic management and network data collection. Definition from Bellcore in reference to its concept of the Advanced Intelligent Network.

Network Operations Center A name for the central place which monitors the status of a corporate network and sends out instructions to repair bits and pieces of the network when they break. In more formal terms, monitoring of network status, supervision and coordination of network maintenance, accumulation of accounting and usage data and user support.

Network Outward Dialing NOD. A service feature of an automatically switched telephone network that allows a calling user to dial directly all network user numbers without operator intervention.

Network Printer A printer shared by multiple computers over a network. See also local printer.

Network Problem Determination Application NPDA. A host resident IBM program product that aids a network operator in interactively identifying network problems from a central point.

Network Professional Association An organization that runs a program called CNP, which stands for Certified Network Professional. That program is designed for individuals whose career is to design, integrate, manage and maintain networked computing environments. The Association can be reached at 801-373-7888.

Network Protection A term used to describe an array of strategies to protect your network from crashing around your ears should a disaster happen. The ultimate network protection means duplicating every item in the network — including the people who operate it. Obviously anything else (which is what we are all forced to do) is a compromise. The typical network protection these days tends to focus on alternate routing and duplication of network lines, including local loops and long distance lines.

Network Protection Device NPD. A device which pro-

vides isolation between PBX circuits and CO trunks or tie lines.

Network Reconfiguration Service NRS. A Nynex service that lets you control, rearrange and switch your private line voice, private line analog data and Superpath 1.544 Mbps lines. You might wish to do this because you want to temporarily join one office to another for a high-speed data dump, or for a videoconference. Or in case something happens and you may want to route yourself around the trouble. You can reconfigure you own network in one of two ways — using a computer terminal or PC to dial in your instructions or call an 800 number and have an Nynex attendant do it for you. Rearrangements can be made in near-real time or scheduled for an appointed time.

Network Redundancy Network redundancy means that the network topology has been constructed such that a failure of a network component can be automatically and/or rapidly recovered by using identical components engineered into the network for recovery purposes. For example, IXC network redundancy may take the form of spare network capacity on geographically diverse IXC circuits so that applications can be recovered to an alternative network path in the event of a failure on the primary path.

Network Relay A device which allows interconnection of dissimilar networks.

Network Reliability ComForum The Network Reliability ComForum is a gathering of senior telecommunication industry executives and government officials that serve on the Network Reliability Council. The ComForum meets to report on the Council's findings. See USITA.

Network Server A powerful computer on a LAN designed to serve the needs of all the people on the LAN with everything from database, email, fax, images, voice messages, communications, etc. See also EMAIL SERVER and FAX SERVER.

Network Services In IBM's SNA, the services within network addressable units (NAUs) that control network operations via sessions to and from the SSCP.

Network Synchronization Within a data transmission network, especially one using multiplexers, the need for network synchronization is two-fold: 1. The network must transmit data to the receiving DTE device at the same rate it is being received from the transmitting DTE; and 2. The data originating from the low-speed channels must be capable of being inserted into the composite link (and vice versa) without the loss of any information (channel signaling rate too low) or the creation of unwanted information (channel signaling too high). The need for network synchronization is necessary with networks with time division multiplexing and synchronous data. It does not arise with time division multiplexing in which only asynchronous data is transmitted, because asynchronous data contains "start" and "stop" bits and therefore doesn't need the synchronization of a single master clock. When time division multiplexers are to multiplex synchronous channels, it is essential that the composite links and the low-speed channels are strictly synchronized, i.e. that their clocks operate at precise multiples of the basic clock rate. A single clock, therefore, must be in overall control of the whole network. See SYNCHRONOUS.

Network Terminal Number NTN. The number assigned to a data terminal under the Data Network Identification Code system. In the ITU-T International X.121 format, the sets of digits that comprise the complete address of the data terminal end point. For an NTN that is not part of a national integrated numbering format, the NTN is the 10 digits of the ITU-T X.25 14-digit address that follow the Data Network Identification Code (DNIC). When part of a national integrated numbering format, the NTN is the 11 digits of the ITU-T X.25 14-digit address that follow the DNIC.

Network Terminal Option NTO. An IBM program product that enables an SNA network to accommodate a select group of non-SNA asynchronous and bisynch devices via the NCP-driven communications controller.

Network Terminating Equipment NTE. See NTE and Network Terminating Interface.

Network Terminating Interface 1. NTI. The point where the network service provider's responsibilities for service begin or end.

2. NTI. The interface between DCE and its connected DTE.

Network Terminating Unit NTU. The part of the network equipment which connects directly to the data terminal equipment.

Network Termination Type 1 NT1 or NT-1. An ISDN term. The NT1 is a device which provides the interface between an ISDN local loop and the customer premises. The NT1 provides the functions of physical and electrical connection to the ISDN local loop, thereby operating at Layer 1 of the OSI Reference Model. In a BRI implementation, the NT1 is in the form of a special NIU (Network Interface Unit). Many TAs (Terminal Adapters) also perform NT1 functions; some ISDN-compatible terminal devices (e.g., ISDN phones and PCs) also include NT1s. The basic NT1 functions are:

• Line transmission termination. The ISDN local loop is a multichannel, full duplex loop which may be presented physically in the form of either a two-wire or four-wire connection. In a BRI implementation, this function typically is provided by the NIU; in a PRI implementation, by the DSU (Data Service Unit), which generally is embedded in the intelligent switching or concentrating device.

• Line maintenance functions and performance monitoring. Through the D Channel, the ISDN local loop is supervised and monitored from the central office to the user premises in order to ensure that it performs at a satisfactory level. Loopback testing is included in this functional grouping. In a BRI implementation, this function typically is provided by the NIU; in a PRI implementation, by the DSU (Data Service Unit).

• Multiplexing. This function allows multiple devices to share access to the multichannel digital local loop. In a BRI implementation, this function typically is provided by the TA; in a PRI implementation, by the P3X, ACD, or other CPE switching device.

• Interface termination, including multi drop termination and contention resolution. At the physical level, the devices which access to an ISDN loop may connect through 2, 4 or more wires; the ISDN loop is in the form of a physical circuit of either two or four wires. Physical interconnectivity is resolved through the NT1 functional grouping, generally in the form of a TA for a BRI implementation; a PBX, ACD, router, or other intelligent switching or concentrating device performs this function in a PRI implementation. Contention resolution is the process of resolving access issues as multiple physical devices contend for a limited number of ISDN logical channels. Contention resolution generally is performed by a TA in a BRI implementation; a PBX, ACD, router, or other intelligent switching or concentrating device performs this function in a PRI implementation.

See also ISDN and Network Termination Type 2.

Network Termination Type 2 NT2 or NT-2. An ISDN term for a functional grouping embodied in an intelligent switching device such as a PBX, ACD, router or concentrator. Depending on the requirements of the specific device, NT2 functions fall into Layer 1 (Physical Layer), Layer 2 (Data Link Layer) and Layer 3 (Network Layer) of the OSI Reference Model. A full description of Layer 1 functions is described immediately above in the definition of Network Termination type 1. NT2 functions, again depending on the requirements of the specific device, also include the following:

• Protocol handling at Layers 2 and 3

• Multiplexing at Layers 2 and 3

• Switching at Layer 2

• Concentration at Layer 2

See also ISDN and Network Termination type 1.

Network Time Protocol NTP. A protocol built on top of TCP that assures accurate local time keeping with reference to radio and atomic clocks located on the Internet. This protocol is capable of synchronizing distributed clocks within milliseconds over long time periods.

Network-to-Node Interface A set of ATM Forum-developed specifications for the interface between two ATM nodes in the same network. Two variations are being developed: an interface between nodes in a public network and an interface between nodes in a private network.

Network Topology The geometric arrangement of links and nodes of a network. The geography of a network. Networks are typically of either a star, ring, tree or bus topology, or some hybrid combination.

Network Traffic The number, size, and frequency of packets transmitted on the network in a given amount of time.

Network Traffic Management NTM. Functionality that maximizes the traffic throughput of the network during times of overload or failure and minimizes the impact of one service on the performance of others. This is done through centralized surveillance of maintenance conditions and traffic, and centralized control of traffic volumes being originated in network systems. Definition from Bellcore in reference to its concept of the Advanced Intelligent Network.

Network Virtual Terminal A communications concept wherein a variety of DTEs, with different data rates, protocols, codes and formats, are accommodated in the same network. This is done as a result of network processing, where each device's data is converted into a network standard format and then converted into the format of the receiving device at the destination end.

Networking Blueprint IBM's latest and most ambitious network initiative. The Blueprint is structured to allow applications to work over a wide range of networking schemes, e.g., SNA applications over TCP/IP or OSI, TCP/IP over APPN or OSI, DECnet applications over SNA.

Networking Systems IBM Networking Systems is new name (summer, 1991) for its Communication Systems group, formed in January, 1988.

Netx.com This Novell NetWare workstation shell file provides an interface between the application and DOS, monitoring all data transmissions that move in or out of either one. When a function request needs network service, such as a call to read a file from the file server, the shell begins the process of protocol conversion and network transmission. Netx intercepts requests by taking over software interrupts 21h (used to call standard DOS functions), 24h (DOS's critical error handler vector), and 17th (used to send data to local printer ports). In general, the shell intercepts all interrupt 21h DOS requests and inspects each one. The "x" typically stands for which version of DOS the workstation is running. NET3.COM would run on a workstation running MS-DOS 3.X, i.e. 3.1, 3.2 and 3.3. (There is no 3.4).

Neural Network A neural network is designed to take a pattern of data and generalize from it. Thus, if the data are daily temperatures in New York city over two years, the neural network should emerge with a simple undulating curve that describes the way temperature rises in summer and falls in winter. It does this in effect by a sophisticated form of trial and error, or, in the jargon, varying the strengths of connections between individual processors until the input yields the right output. The two essential features of neural network technology are that it improves its performance on a particular task by trial and error (neural networkers prefer to say that "it learns") and that it can be a "black box." That is, you do not need to know what mathematical equation describes its output, although you could find out if you wanted to. (Thanks for the Economist Magazine for help on this definition.)

Neural Network Computer A very different kind of computer. Neural network computers are built from webs of randomly connected electronic neurons. These machines are designed to be trained, not programmed. In design and in function, they are meant to closely resemble the human brain. As in the brain, the neurons send signals to one another through thousands of adjustable connections, or synapses. As the machine learns, the settings of these controls are automatically turned up or down. And the chaos of connections evolves into a finely tuned machine, one that can read hand-written letters or recognize the spoken words. With a neural network computer, a trainer simply speaks words to the machine, rewarding it if it acts correctly and punishing it when it acts incorrectly. There is hope that neural network computers may perform better than traditional computers in running artificial intelligence software. See NEURAL NETWORKS.

Neutral The ac power system conductor that is intentionally grounded on the supply side of the service disconnect. It is the low potential (white) side of a single phase ac circuit or the low potential fourth wire of a three-phase Wye distribution system. The neutral provides a current return path for ac power currents whereas the safety ground (green) conductor should not, except during fault conditions.

Neutral Ground An intentional ground applied to the neutral conductor or neutral point of a circuit, transformer, machine, apparatus, or system.

Neutral Transmission Unipolar transmission. A form of signaling which employs two distinct states, one of which represents the existence of a space as well as the absence of current.

Nevada Bell The Nevada Bell Operating Company (BOC) which, along with Pacific Bell (California), formed Pacific Telesis (PacTel). PacTel was one of the seven original Regional Bell Operating Companies (RBOCs) formed in 1984 by the Modified Final Judgement (MFJ). As PacTel was acquired by SBC Corporation (Southwestern Bell) in 1996, Nevada Bell is now a subsidiary of SBC.

Never-Busy Fax A fax service offered by LECs and IXCs. Here's how it works. You identify the telephone line dedicated to your fax machine, only. If someone tries to send you a fax, but your fax line is busy, the carrier redirects the fax call to it own fax server, which receives the fax and stores it. The fax server then periodically tries to forward the fax to you. After a predetermined number of tries or after a predetermined period of time has elapsed, an exception report is created so that the carrier personnel can call you on your voice line to advise you of the waiting fax.

New A term used in the secondary telecom equipment business. Generally defined as being sold by authorized vendor of the O.E.M. and carrying the O.E.M.'s standard warranty. See UNUSED.

New Line ASCII character 10, abbreviated "NL". This character replaced the abbreviation "LF" which meant Line Feed. Newer terminals accept the NL character to mean both CR (Carriage Return) and LF functions for "end of line" sequences. The function of transmitting an "end of line" sequence to a printer or remote computer system is to designate to a computer to print what follows on the next line of a print out. This usually consists of transmitting the characters CR, LF (Carriage Return, followed by Line Feed). On older printing terminals, these two characters would be followed by a number of null characters whose job would be to waste time until the print head got back to the left margin before printing the first letter of the next line on the page. Otherwise, that character would be in the middle of the page, and the second character would start the line. Telex machines usually send the sequence CR, LF, CR as a New Line sequence to give the head enough time to get back.

New York Stock Exchange On May 17, 1792, a group of New York brokers who had been buying and selling securities under an old buttonwood tree met to formalize rules and methods for securities trading. A agreement was composed and signed by the local brokers. These were the humble beginnings of the New York Stock Exchange.

New York Telephone A phone company that once was part of the Bell System. When AT&T was broken up in 1984, it was lumped with New England Telephone to become Nynex. Then in January of 1994, Nynex decided to eliminate the names New York Telephone and New England Telephone and call them both Nynex. A public relations spokesman told me the reason for this name change had something to do with consistency of marketing image. Frankly, I never understood the necessity for the name change, nor its cost — which included painting the 16,700 trucks the two companies owned. Then Nynex got bought by Bell Atlantic, which it's now part.

Newbie A newcomer to cyberspace; usually applied condescendingly, the way sophomores talk about freshmen. Somebody new to the Net. Sometimes used derogatorily by net.veterans who have forgotten that, they, too, were once newbies who did not know the answer to everything. "Clueless newbie" is always derogatory.

Newbridge Networks A manufacturer of T-1 muxes, channel banks, ISDN devices and other digital communications devices founded by Terry Matthews, co-founder of Mitel Newbridge is one of our favorite companies.

Newsfeed ISPs (Internet Service Providers) get their newsgroups from different newsfeeds, or news sources, by transferring them over the Internet or other networks. See NEWSGROUP.

Newsgroup A USENET newsgroup is a place on the Internet where people can have conversations about a well-defined topic. Physically it is made of the computer files that contain the conversation elements to the discussions currently in progress about the agreed-upon topic. See NEWSFEED.

Newsgroup Distribution Software A set of computer programs used to manage the distribution of USENET news groups and articles.

Newsgroup Feed One remote computer receiving USENET newsgroups from another remote computer.

Newsgroup Management Software A set of computer programs used to manage USENET newsgroups and articles.

Newsreader Software for reading and posting articles to newsgroups on the Internet. A newsreader is a program that organizes conversations in a sensible and presentable manner. The news reader allows the person using it to read and/or participate in those conversations.

Newton 1. The name of Apple's personal communicator, or PDA (Personal Digital Assistant). In its day, the Newton was truly revolutionary. Sadly, Apple Computer Inc. announced on February 27, 1998, the demise of the Newton operating system and Newton OS-based products, including the MessagePad 2100 and eMate 300.

2. The author of this dictionary, Harry Newton, has reached a pinnacle of achievement, of sorts. According to Susan, his wife of over 21 years, he has become a sex symbol for women who no longer care. The photo on the back cover (in case you hadn't already figured) is years old. Maybe decades old.

NEXT Near End crosstalk. A type of crosstalk which occurs when signals transmitted on one pair of wires are fed back into another pair. Since at this point on the link the transmitted signal is at maximum strength and the receive signal has been attenuated, it may be difficult to maintain an acceptable ACR (Attenuation-to-Crosstalk Ratio). NEXT is particularly troublesome when a number of high-speed transmission services (e.g., ADSL, HDSL and T-carrier) are supported within a single copper cable system. Shielded or screened cable systems are more desirable in addressing this problem than are unshielded varieties. See also ACR and FEXT.

NF Noise Figure.

NFAS Non-Facility-Associated Signaling. Another term for out-of-band signaling, such as SS7 (Signaling System 7), which is a Common Channel Signaling technique. ISDN is dependent on the use of SS7 for non-intrusive signaling and control. In an ISDN PRI (Primary Rate Interface) application, a PRI delivers 23 B (Bearer) channels of 64 Kbps each for information transfer, and a D (Data) channel of 64 Kbps primarily for signaling and control purposes between the CPE (Customer Premises Equipment) switch (e.g., PBX, ACD or router) and the LEC (Local Exchange Carrier) CO (Central Office) switch. However, as many as eight PRIs can be supported with a single D channel, as the signaling and control activities of a B channel aren't excessively bandwidth-intensive. In the cellular domain, NFAS is known as NCAS (Non Callpath Associated Signaling).See also ISDN, NCAS, PRI, and D Channel.

NFPA National Fire Protection Association. www.npfa.org. See NEC REQUIREMENTS.

NFS Network File System. One of many distributed-file-system protocols that allow a computer on a network to use the files and peripherals of another networked computer as if they were local. This protocol was developed by Sun Micro-Systems and adopted by other vendors.

NGDLC Next Generation Digital Loop Carrier. DLC can receive and aggregate large amounts of Bandwidth (higher than T-1).

NGI Next Generation Internet. An initiative of the Clinton administration announced in February 1997, NGI is "a second generation of the Internet so that our leading universities and national laboratories can communicate in speeds 1,000 times faster than today." A large part of NGI is Internet2, a project of the University Corporation for Advanced Internet Development (UCAID). See also Internet2.

NHRP Next Hop Resolution Protocol. A protocol suggested by the IETF (Internet Engineering Task Force) as a means of extending the issue of address resolution beyond the borders of individual IP subnets. The central issue is one of address resolution between IP addresses and ATM addresses when one is attempting to send IP packets over an ATM network, i.e. Classical IP over ATM. NHRP makes use of a NHS (Next Hop Server) to advise routers and other network clients where the next hop toward a destination resides. NHRP is a critical protocol for NBMA (NonBroadcast MultiAccess) networks like ATM and X.25. NHRP is likely to be integrated into the Multiprotocol Over ATM (MPOA) architecture. See also Classical IP over ATM and MPOA.

NHS Next Hop Server. See NHRP.

NI 1. National ISDN.

2. Network Interface. Demarcation point between PSTN and CPE.

NI-1 National ISDN 1. It's a BRI circuit. See ISDN.

NI-2 National ISDN 2. It's a BRI circuit. See ISDN.

NI-3 National ISDN 3. It's a BRI circuit. See ISDN.

Ni-MH Nickel Hydride is a new technology used in batteries for portable devices, such as cell phones and laptops, etc. Allegedly, these batteries do not have a "memory effect." Today, Ni-MH batteries cost about 30% more for a 20% or so gain in capacity. The average life of Ni-MH batteries is about 500 cycles. See NICKEL METAL HYDRIDE.

NIA Network Interface Adapter. An IBM hardware device that with certain software, will allow an SNA device to communicate over a packet switching network.

Nibble Four bits or half a byte. Usually described by one hexadecimal digit.

Nibble Interleaving/Multiplexing A technique where 4 bit nibbles (one at a time) from each lower speed input a channel are used to build the higher speed frame output of a multiplexer.

NIC 1. Network Interface Card. The device that connects a device to a LAN. Usually in the form of a PC expansion board, the NIC executes the code needed by the connected device to share a cable or some other media with other stations. See Network Interface Card for a full explanation.

2. Near Instantaneous Companding. This describes the very fast, essentially real-time, process of quantizing an analog signal into digital symbols..and converting it back into its native form. See Companding.

3. Network Information Center; any organization that's responsible for supplying information about any network.

4. The InterNIC, which plays an important role in overall Internet coordination. See InterNIC.

5. The DDN NIC (Defense Data Network Network Information Center), which provides support for the defense community, much as does the InterNIC for all other Internet user communities.

6. National ISDN Council. A council of carriers and manufacturers which determines the generic guidelines for National ISDN. It also publishes those guidelines, as they are developed. NIC is a voluntary council. Membership includes Ameritech, Bell Atlantic, BellSouth, Cincinnati Bell, Lucent, Nortel and SBC. Administration and project management are the responsibility of Bellcore. See also National ISDN-1.

Nickel Cadmium Battery NiCad: Nickel Cadmium is the most popular and durable type of rechargeable battery. It is quick to charge, lasts about 700 charge and discharge cycles, and works well in extreme temperatures. Unfortunately, NiCads suffer from "memory effect" if they are not completely discharged during each cycle. The memory effect reduces the overall capacity and run time of the battery.

Nickel Metal Hydride (NiMH) batteries do not suffer from memory effect. Compared to a NiCad battery of equal size, NiMH batteries run for 30% longer on each charge. They are also made from non-toxic metals so they are environmentally friendly. The downside to NiMH technology is overall battery life. These batteries last for 400 charge and discharge cycles. Lithium Ion (LiON) is the latest development in portable battery technology. These batteries do not suffer from memory effect. Compared to a NiMH of equal size, a LiON will deliver twice the run time from each charge. Unfortunately, these batteries are only available for a limited number of models and are expensive. Similar to NiMH technology, LiON batteries have a life expectancy of 400 charge and discharge cycles. See NICKEL METAL HYDRIDE

Nickel Metal Hydride Battery Ni-MH or NiMH. A rechargeable battery, now being used in laptop computers. Most of today's rechargeable batteries are nickel cadmium. These suffer from several problems:
1. They have a "memory." Which means they must be fully discharged once a month or they won't deliver their full potential. 2. They are hard to dispose of. Nickel hydride is a new technology with some benefits: It's easier to dispose of. It can hold at least 1.25 times as much power per unit as a standard nickel cadmium battery. It doesn't have a "memory." (Neither do car or telephone batteries, which are typically lead acid.) But there are some downsides: Nickel hydride batteries lose their charge faster than nickel cadmium. Nickel hydride batteries are ABOUT 30% more expensive than nickel cadmium. The Nickel hydride battery also can not put out power as fast as a nickel cadmium one can, so it probably won't be suitable for power tools and appliances that drain batteries quickly. But it's great for "slow-burning" items, like laptops and cellular phones. See NICKEL CADMIUM BATTERY.

Nicname A LAN protocol specified in RFC-812. It requests information about a specific user or hostname from the Network Information Center (or NIC) name database service.

NID 1. Network Inward Dialing. A service feature of an automatically switched telephone network that allows a calling user to dial directly to an extension number without operator intervention.
2. Network Interface Device. An electronic device that connects the telephone line and the POTS splitter to the local loop.

NIF Network Interface Function.

Night Answer Incoming calls to a switchboard during evening and weekend hours are automatically rerouted to ring only at designated night answering phones such as the security desk. See the next few definitions.

Night Answer — Assigned Night answer going to specific, assigned telephones.

Night Answer — Offsite Phones being answered after hours by people and machinery located offsite — someplace else.

Night Answer — Universal Anybody and everybody can answer the incoming calls from any phone. In TELECONNECT, if the phone rings after hours, we simply touch "6" on any phone and we can answer that incoming call — whether it's a local, normal DDD or incoming WATS.

Night Audit This feature provides automatic printout of message registration data for all quest rooms at the front desk console.

Night Chime An auxiliary ringer, usually wall mounted, used to indicate a ringing trunk during night operations or used as a "phantom" extension for overflow applications.

Night Console Position Provides an alternate attendant position which can be used at night in lieu of the regular console. Usually cheaper than buying a second normal console.

Night Patch Assigned Night Answer, Fixed Night Service, Programmable Night Connections. Provides arrangements (which are prewired into the system) to route incoming central office calls normally answered at the attendant position, to preselected stations within the PBX system when the attendant is not on duty.

Night Service Your operator goes home, puts your phone system into "night service." A call now comes in. The night bell rings. You hear it. You go to a phone and hit a code (such as #6) or hit the flashing "Nigh Answer" button. And you answer the call. Not all phone systems need to be placed in Night Service for you to answer an incoming call from any phone. And you can program the night bell to ring all day.

Night Service Automatic Switching Should the attendant neglect to place the console in the night answering mode, after a certain period of timed ringing from an incoming central office call, the entire system will automatically jump to the night service.

Night Service — Expanded Service Routes calls normally directed to the attendant to preselected station lines within the system when it is arranged for night service. Calls to specific exchange trunks can be arranged to route to specific station lines and can be assigned on a flexible basis. Trunk Answer From Any Station capability is provided for calls which are handled by assigned night stations.

Night Service — Fixed Service Provides arrangements to route calls normally directed to the attendant to preselected station lines within the PBX system when regular attendant positions are not in use. In addition, calls to specific trunks can be arranged to ring on specific phones. The receiving phone can then transfer the call.

Night Station The phone assigned to automatically handle incoming calls after the main switchboard has shut down for the day.

NIH 1. Not Invented Here. The tendency of organizations to reject ideas and inventions which they didn't think of. A major and continuing problem in the telephone industry.
2. National Information Highway. A term coined by Al Gore when he was Vice President of the United States under President Clinton. See also NII.

NII National Information Infrastructure. A clumsy name for the Information SuperHighway. NII is a term that came from a paper which the Gore-Clinton Administration released, called "National Information Infrastructure; Agenda for Action." According to one engineer, NII is "multiple, interconnected interoperable networks."

NiMH Nickel Metal Hydride. See NICKEL METAL HYDRIDE.

NIMQ (pronounced "nihm-kyoo") Acronym for "Not in My Queue." Said in response to suggestions to take on more work when you're already overwhelmed. Similar to the more common "It's not my job."

NIOD Network Inward/Outward Dialing.

NIOSH National Institute of Occupational Safety and Health.

NIS 1. Network Information Service. A way of centralizing user configuration files in a big distributed computing environment. NIS was created by Sun Microsystems.
2. Network Imaging Server. A local area network based server largely devoted to storing, retrieving and possibly manipulating images. See also SERVER.

NISDN-1 See National ISDN-1.

NIST The National Institute of Standards and Technology, which was formed in 1901 as the National Bureau of Standards, is part of the U.S. Department of Commerce. NIST works with industry and government to advance measurement science and to develop standards in support of industry, commerce, scientific institutions, and all branches of government. NIST comprises four major programs, according to their literature. The Advanced Technology Program (ATP) provides cost-shared awards to companies and consortia for project geared toward the development of high-risk, enabling technologies during pre-product phases of research and development. The Manufacturing Extension Partnership (MEP) is a network of extension centers, co-funded by state and local governments, to provide small and medium-sized manufacturers with technical assistance toward improved performance and competitiveness. Laboratory research and services focus on development and delivery of measurement techniques, test methods and standards. NIST manages the Baldrige National Quality Award program, including providing U.S. industry with comprehensive guides to quality improvement. www.nist.gov

NIU Network Interface Unit. Also known as a NID (Network Interface Device). The NIU serves as the point of demarcation between the local exchange carrier network and the customer premise. As required by the Modified Final Judgement (MFJ), which broke up the Bell System, the NIU currently is positioned outside the main body of the premise, generally on an exterior wall, inside the garage, or just inside the point of cable entrance. NIUs are multi-function devices, including a protector block which serves to premise equipment and inside wire from high-voltage surges. The NIU also typically includes electronics which allow the carrier to initiate a loopback test from the central office for purposes of testing the integrity of the local loop.

NIUF North American ISDN Users' Forum.

NKT Nederlands Keuringsinstituut voor Telecommunicatieapparatuur (Private laboratory for regulatory testing, The Netherlands).

NL See NEW LINE.

NLA Network Layer Address. An address appended to the LAN packet that, unlike a MAC name, indicates exactly where a computing device is located within an internetwork. TCP/IP, DECNet and IPX support network layer addressing and each has its own unique NLA format. Protocol dependent routers use the NLA to make routing decisions.

NLDM Network Logical Data Manager. An IBM term. See also NETVIEW.

NLM NetWare Loadable Modules. NLMs are applications and drivers that run in a server on a local area network under Novell's NetWare operating system and can be loaded or unloaded on the fly. The category of applications ranges from complex voice mail/auto attendant programs and electronic mail gateways to simple programs like drivers for PCI cards. In some other networks, such applications could require dedicated PCs. See NETWARE LOADABLE MODULE for a longer explanation.

NLPI Network Layer Protocol Identifier.

NLS Name Lookup Service. An electronic directory service which is designed to respond to external queries for general information about a large group of users. Once installed, NLS can be accessed via finger or whois.

NLSP NetWare Link Services Protocol. A link state protocol from Novell to improve performance of IPX traffic in large internetworks.

NM 1. nm. Nanometer. A bi lionith of a meter. Often used as a measure of frequency in a fiber optic link.

2. Network Management. A business process in a telecommunications organization that constantly monitors the state of the network and invokes changes to the network when outages or problems occur. In an ATM network, NM is the body of software in a switching system that provides the ability to manage the PNNI protocol. NM interacts with the PNNI protocol through the MIB.

NM Forum Network Management Forum. International forum for network management.

NMC Network Management Center. A centralized location at the network management layer used to consolidate input form various network elements to monitor, control, and manage the state of a network in a telecommunications organization.

NMF Network Management Forum. An international forum of service providers and suppliers developing network management solutions. Now known as the TeleManagement Forum. See Network Management Forum and TeleManagement Forum. www.nmf.org.

NML Network Management Layer. The layer in the Telecommunications Management Network (TMN) standard addressing network functions including monitoring, management, and control. The network management layer is an abstraction of the functions provided by systems which manage network elements on a collective basis, so as to monitor and control the network end-to-end.

NMOS N-channel Metal Oxide Semiconductor.

NMP Network Management Protocol. An AT&T-developed set of protocols designed to exchange information with and control the devices that govern various components of a network, including modems and T-1 multiplexers.

NMS Network Management Station or Network Management System. The system responsible for managing a portion of a network. The NMS talks to network management agents, which reside in the managed nodes, via a network management protocol. The NMS is the entity that implements functions at the Network Management Layer. It may also include Element Management Layer functions. A Network Management System may manage one or more other Network Management Systems.

NMT Nordic Mobile Telephone. Analog cellular system in Sweden, Norway, Denmark and Finland.

NMVT Network Management Vector Transport. A network management protocol which provides alert problem determination statistics and other network management data within SNA management services.

NN Network Node. In frame relay, a network node is typically the frame relay service port connection and its associated virtual circuits.

NNI 1. Network Node Interface. An Asynchronous Transfer Mode (ATM) term. The interface between two public network pieces of equipment (contrast that to UNI, which stands for User Network Interface).

2. Network to Network Interface. A protocol defined by the Frame Relay Forum and the ATM forum to govern how ATM switches establish connection and how ATM signaling requests are routed through an ATM network. Equivalent to a routing protocol in a router environment.

3. Network to Network Interface. There are two types of network interfaces specified by frame relay standards. The first is a user-to-network (UNI) interface and the second is a network-to-network interface (NNI). The NNI describes the connection between two public frame relay services, and includes elements such as bidirectional polling, to assist the network services providers with gaining information on the status of

the public networks being interconnected.

NNTP Network News Transport Protocol. An extension of the TCP/IP local area network protocol that provides a network news transport service. NNTP is the standard for Internet exchange of Usenet messages, published in RFC-977, Network News Transfer Protocol: A Proposed Standard for the Stream-Based Transmission of News.

NNX The first three digits of a North American local telephone number, NNX once upon a time was used to identify the local central office exchange, or CO prefix. This prefix code breaks down as follows: N is a specific digit (i.e., 2- 9) and X is any digit (i.e., 0-9). Originally, only NNX codes were used to identify and number local central offices. Now all subscribers dial 1+ when making a direct distance dialed long distance call. NNX has been changed to NXX, which allows local central offices to have numbers which look like area codes, e.g. 206, 210, etc. This gives us more central office numbers. We needed more central office prefixes for several reasons. First, CO prefixes are unique within each area code, or NPA, and we assign both NPAs and CO prefixes on a very wasteful basis. Second, the popularity of fax machines, pagers and cell phones has resulted in demand for millions of telephone numbers. See NPA and NXX.

No Answer Transfer A service provided by a cellular carrier that automatically transfers an incoming cellular call to another phone number if the cellular subscriber is unable to answer. Most no-answer transfer systems will automatically transfer an incoming cellular call to another phone number if the cellular subscriber is unable to answer (it's not turned on) or if it's not answered after the third or fourth ring if the phone is turned on.

No Attendant Option CBX systems with Direct Inward Dialing may be designed (configured) without an attendant console.

No Bill Phone A name for a cellular phone from which you can make free phone calls — local or long distance. The name "no bill" was given by crooks, largely in Southern California in the Spring of 1992, to phones they had modified to emulate other legitimate cellular phones. Thus all calls made by "no bill" phones are billed to legitimate users, who happen to have other cellular phones with identical codes and now have huge cellular bills. Sorry about that.

No Busy Test A circuit used to connect to a busy subscriber's line number.

No Hold Conference/transfer In the event that a call to 911 is being transferred to a secondary or what they call a downstream PSAP (Public Service Answering Position), it is important that the caller never be left in disconcerting silence. After all it is an emergency call. "No hold" features allow the conference or transfer to be done while the PSAP 911 agent is in full uninterrupted communication with the caller.

No Incoming Calls A carrier restriction that prevents incoming calls to the assigned cellular number. Only outgoing calls are permitted. This is an optional feature that some cellular phone uses subscribe to because it saves them

No Line Preference Requires the user to manually select (i.e. punch down) a line for each call.

No Op Instruction that does nothing. It is used to hold the place for future insertion of a machine instruction.

NOAA National Oceanic and Atmospheric Administration. A US government agency which runs satellites used, inter alia, to track wildlife movements. Two satellites, called NOAA-11 and NOAA-12, orbit the earth via the poles every 100 minutes. As the Earth revolves beneath them, they are able to scan

every point on its surface. The satellites work by listening for the "Doppler Shift" in the signal being transmitted by a collar round an animal's neck. The shift is a change in the perceived frequency of the radio signal — similar to what a pedestrian hears happen to the pitch of a police siren as the police car speeds by. Using this technique, you can track animals to with an accuracy of a little more than half a mile, or one kilometer.

NOC Network Operations Center, a group which is responsible for the day-to-day care and feeding of a network. Each service provider usually has a separate NOC, so you need to know which one to call when you have problems.

NOD Network Outward Dialing. A service feature of an automatically switched telephone network that allows a calling user to dial directly all user numbers on the network without operator intervention. See also DIRECT OUTWARD DIALING.

Nodal Architectures Also called Hub architectures. Nodal network architectures means that network traffic from many locations are connected into a single network site for consolidation and aggregation to higher speed circuits. The traffic is then transmitted from this hub site to a central network site, or to other hub sites. Nodal architectures save money, but also increase the risk of a single network failure affecting multiple network locations.

Nodal Attribute A nodal state parameter that is considered individually to determine whether a given node is acceptable and/or desirable for carrying a given connection.

Nodal Clock The principal clock or alternate clock located at a particular node that provides the timing reference for all major functions at that node.

Nodal Constraint A restriction on the use of nodes for path selection for a specific connection.

Nodal Metric A nodal parameter that requires the values of the parameter for all nodes along a given path to be combined to determine whether the path is acceptable and/or desirable for carrying a given connection.

Nodal State Parameter Information that captures an aspect or property of a node.

Node 1. A point of connection into a network. In multipoint networks, it means it's a unit that's polled. In LANs, it's a device on the ring. In packet switched networks, it's one of the many packet switches which form the network's backbone.
2. An SCSA term. An independent SCSA unit in a distributed processing SCSA network, consisting of one or more resource and/or network boards, and one or more SCxbus adapter boards. Communication between nodes take place via the SCxbus. From a device programming point of view, a node is simply an addressable system unit which contains boards connected by an SCbus. See S.100.

Node Address The unique identifier used to describe a specific node. See NODE NUMBER.

Node Number A node number identifies a network board on a local area network. Every station on a network must contain at least one network board. Each network board must have a unique node number to distinguish it from all the other network boards in that network. In a file server with more than one network board, the node number in LAN A is designated for all traffic addressed to that server. Node numbers can be set in a variety of ways, depending on which network board you use: (1) with jumpers or switches on boards such as Arcnet, (2) at the factory for Token-Ring and Ethernet boards, or (3) with software.

Node Type In IBM's SNA, the classification of a network device based on the protocols it supports and the network addressable units (NAUs) it can contain. Type 1 and Type 2 nodes are peripheral nodes. Type 4 and 5 nodes are sub-area nodes.

NOI Notice Of Inquiry. The first public notification that the FCC is about to hold a public inquiry into a particular subject.

NOIS Microsoft has coined the term, NOIS, for the group of four Net-centric companies that tend to believe that what happens on the Internet might be so strong as to alleviate Microsoft's hegemony over the PC industry. The four members are Netscape, Oracle, IBM and Sun.

Noise Unwanted electrical signals introduced into telephone lines by circuit components or natural disturbances which tend to degrade the performance of the line. Also known as Line Noise.

Noise Cancelling Headset manufactures have long sought to reduce the background noise transmitted via headsets. One approach is the use of noise cancelling microphones. These microphones consist of two separate microphones, one directed at the headset user's mouth, the other in the opposite direction. The room side element will pick up ambient room noise along with some ambient user sound. The microphone directed at the user will receive the same amount of ambient room noise as the other microphone, but a much greater amplitude of the user's voice. Both signals are then transmitted to the amplifier. At this point, signals common to both microphones are cancelled out. What remains is the extra voice signals received by the user side microphone. This signal is then amplified and transmitted to the party on the receiving end of the call. This approach has one drawback. It demands perfect microphone positioning, because without it, the headset user's voice is cancelled. The technology works well with highly-trained people such as pilots, astronauts, and military personnel, but can be difficult to implement in the office environment where less skilled personnel struggle to properly position sensitive microphones. Headset manufacturers compromised by using noise cancelling microphones with more limited capabilities but that were easier to use.

A second approach to noise reduction is the use of voice switching technology. This technique only allows the microphone to transmit when volume reaches a predetermined level. When the headset user is not talking, or is pausing during the conversation, no sound is transmitted. When the headset user speaks at a normal level, the microphone is "live" and will transmit in a normal fashion. This approach also has it drawbacks. When the microphone is "live" it picks up not only the voice of the person using the headset, but any and all background noise. Voice switching helps the headset user hear what is being said more clearly, but does little to help the person to whom they are talking.

As a solution, some headset manufacturers have merged the two technologies. By using a noise cancelling microphone and voice switching, they achieve near perfect noise reduction. Each manufacturer offers noise cancelling technology on some of their headsets.

Noise cancelling is important in a telephone call center. In a large center, as room noise rises, agents speak louder. For those employees, noise is more than just an inconvenience, or a black spot on a professional image, it directly affects productivity. When conversations must be repeated, call durations increase. Multiply this by enough calls, and staffing and equipment must also be increased. The above information from headset distributor, CommuniTech.

Noise Equivalent Power At a given data-signaling rate or modulation frequency, operating wavelength, and effective noise bandwidth, the radiant power that produces a signal-to-noise ratio of unity at the output of a given optical detector. Information Gatekeepers defines NEP as a measurement in fiber optics that at a given modulation frequency, wavelength, and for a given effective noise bandwidth, the radiant power that produces a signal-to-noise ratio of 1 at the output of a given detector. Some manufacturers and authors, according to Information Gatekeepers, define NEP as the minimum detectable power per root unit bandwidth; when defined in this way, NEP has the units of watts/(hertz) 1/2. Therefore, the term is a misnomer, because the units of power per watts. Some manufacturers define NEP as the radiant power that produces a signal-to-dark current noise ratio of unity. This is misleading when dark-current noise does not dominate, as is often true in fiber systems.

Noise Figure NF. The ratio (in dB) between the signal-to-noise ratio applied to the input of the microwave component and the signal-to-noise ratio measured at its output. It is an indication of the amount of noise added to a signal by the component during normal operation. Lower noise figures mean less degradation and better performance.

Noise Floor The lowest input signal power level which will produce a detectable output signal from a microwave component, determined by the thermal noise generated within the microwave component itself. The noise floor limits the ultimate sensitivity to the weak signals of the microwave system, since any signal below the noise floor will result in an output signal with a signal-to-noise ratio of less than one and will be more difficult to recover.

Noise Measurement Units A series of terms used to express circuit noise. These units include: -dB RN — Decibel rated noise -dBrnC — C message weighting refers to the noise measured at 1000 Hz.

Noise Suppressor Filtering or digital signal-processing circuitry in a receiver or transmitter that automatically eliminates or reduces noise.

Noise Voltage In optical communication, an rms component of the optical detector electrical output voltage which is incoherent with the signal radiant power.

Noise Weighting A method of assigning a specific value to the transmission impairment due to the noise encountered by an average user operating a particular telephone. Noise weightings generally in use have been established by regulatory agencies concerned with public telephone service. They are tightened as technology improves.

Nom De Ligne For those of you who don't speak French, particularly Old French, "Nom de Ligne" translates to "Line Name." It's a pseudonym, or "handle" that you use when on-line-in an Internet Chat Room, for instance. You adopt a Nom de Ligne to give yourself anonymity for whatever reasons you find convenient. Nom de Ligne is called a "handle" in some circles. Other French terms for pseudonyms include "Nom de Guerre" (war name) and "Nom de Plume" (pen name). I don't have a "Nom de Ligne" or a "Nom de Plume." My "Nom de Guerre" is "Harry The Terrible."

Nomadic Computing An enterprise application describing laptop users dialing over the public telephone network, from hotel rooms, from client offices, from home or from airports to access information on the corporate LAN.

Nomadic Node See Mobile IP.

NOMDA National Office Machine Dealers Association. NOMDA merged with LANDA (Local Area Network Dealers Association) and became NOMDA/LANDA. It's now known as BTA (Business Technology Association). See BTA.

Nominal Bit Stuffing Rate The rate at which stuffing

bits are inserted (or deleted) when both the input and output bit rates are at their nominal values.

Nominal Linewidth In facsimile systems, the average separation between centers of adjacent scanning or recording lines.

Non-adaptive Routing A routing method that cannot adapt to or accommodate changes in a network.

Non-blocking A device which can support a full traffic load without experiencing congestion. The term typically is applied to a switch, such as a PBX or an ATM switch. A PBX can be characterized as fully non-blocking if the internal switching matrix is capable of supporting connectivity between all ports, simultaneously. In other words, the total number of transmission paths is equal to the total number of ports. In a purely voice application, this means that every user with a telephone set can be talking at the same time-this is a highly unlikely scenario involving a ridiculously expensive PBX. It is especially expensive if the PBX is designed to provide a non-blocked path between every PBX station set and the wide area network, as trunk cards and trunks both are very expensive-it's one thing to equip a system so that all station users can talk to each other, but quite another to expect that they all would be connected to the outside world at the same time. ACDs, on the other hand, typically are effectively non-blocking, as productivity objectives demand that all agents in a call center be active as much as possible. Non-blocking switch architectures can be very important in the data world, as holding times are very long in comparison to voice calls. ATM backbone switches commonly are virtually non-blocking, as traffic loads are very heavy and as Quality of Service (QoS) requirements can be extreme.

Non-blocking Switch A switching system where a connection path always exists for attached device. See NON-BLOCKING.

Non-busy Season A telephone company definition. The nine months not selected as part of the Busy Season.

Non-concur IBM-speak for to withhold approval, as in I non-concur with this proposal.

Non-consecutive Hunting Often referred to as "jump hunting." NON-CONSECUTIVE lines, trunks or extensions can be accessed or "searched" by the switching equipment upon dialing the initial number in the hunting group to find a connection to the first non-busy phone. NON-CONSECUTIVE hunting can be used on incoming and outgoing lines. For example, you could order four trunks from your local phone company which are to "hunt" on. Should the first be busy, the call will go to the next. If the next is busy, then it will hunt to the next. TELECONNECT Magazine has "consecutive" numbering in its first four phone numbers — 212-691-8215, 8216, 8217 and 8218. But let's say we only started with one number and then, because of growth, we needed three more. It's possible consecutive numbers might be taken. Therefore to get the hunting feature, the phone company might have assigned us NON-CONSECUTIVE hunting numbers like 691-8220, 8256, 8678. Sometimes the phone company will also assign us "coded" trunks as part of our trunk group. These are trunks which have no dialable number associated with them. It's best to get real numbers so you can test them by calling them individually. You cannot test coded trunks.

Non-critical Technical Load That part of the technical load not required for synchronous operation.

Non-data Bit Bit with encoding violating normal format: used for special control purposes.

Non-dedicated Server A node on which user applications are available while network resource maintenance appli-

cations execute in the background.

Non-deterministic A term which refers to the inability of being able to predict the performance or delay on a network where collision is possible.

Non-deterministic Network A network where access to the transmission medium within some specified period, cannot be guaranteed.

Non-directional Antenna An antenna that transmits and receives equally well in all directions, usually on one plane; also called omnidirectional antenna.

Non-dominant carrier Have we run any press on the FCC's August decision to detariff nondominant long distance carriers, and their simultaneous new policy decision to not require those carriers to make public rate information?

Non-duplication Rules Restrictions placed on cable television systems prohibiting them from importing distant programming that is simultaneously available locally.

Non-erasable A switch where a through traffic path always exists for each attached phone. Generically, a switch or switching environment designed never to experience a busy condition due to call volume. See NON-BLOCKING, which is a better term.

Non-impact Printer Refers to printers that do not strike a hammer to a platen as typewriters do. Usually a heat sensitive paper or laser printing technology is involved.

Non-intelligible Crosstalk Crosstalk which is not of sufficient level to be understood by a listener but which is more annoying than other crosstalk because you think it's intelligible. Or think it should be intelligible.

Non-interlaced Non-interlaced refers to monitors whose electron gun scans the entire screen without skipping any scan lines. There's a good definition on interlacing and non-interlacing under INTERLACING.

Non-ionizing Emissions Radio waves, infrared rays and visible light rays, none of which can affect an atom's electrical balance.

Non-ISDN Line Any connection from a CPE to a central office switch that is not served by D-channel signaling.

Non-ISDN Trunk In ISDN language, a non-ISDN trunk is any trunk not served by either SS7 or D-channel signaling.

Non-linear Distortion Amplitude distortion of a signal in which the output signal does not have a linear relationship to the input signal.

Non-loaded Lines Cable pairs or transmission lines with no added inductive loading coils. In short, straight, raw copper pairs. See LOAD COILS.

Non-maskable interrupt An interrupt on a PC that cannot be disabled by another interrupt.

Non-persistent In local area networking technology, describes a carrier sense multiple access (CSMA) local area network (LAN) in which the stations involved in a collision do not try to retransmit immediately — even if the network is quiet. Compare with persistent and p-persistent.

Non-printing Character A character in a transmission code which performs a control function but is not reproduced when the transmission is printed.

Non-progressive Display See Interlaced GIF.

Non-proprietary LAN A Local Area Network that can connect the equipment of many vendors. See PROPRIETARY LAN.

Non-published There are various interpretations of what constitutes an "non-published," "unpublished" and "unlisted" phone number in North America. Some phone companies use these words interchangeably. Some don't. In California, Pacific Bell offers unpublished phone service. Your phone number is not listed in the paper phone directories, but is listed with dial up "Directory Assistance." Pacific Bell also has a more expensive service called "Unlisted Service." Here, your phone number is not included in the paper phone directories or given out to callers to Directory Assistance. Nynex (and I presume other phone companies) has a service whereby you can leave a message for the owner of an unlisted number. "Please call me. You've won the lottery." The owner of the unlisted number then has the choice to return the call or not. He doesn't pay to receive this message. Some telephone companies confuse the definitions and some invent new ones. For example, some phone companies use the term "non-published" number. You won't find the number in a phone book or by calling Directory Assistance. Over 25% of many private phone numbers in major metropolitan areas are unlisted, unpublished or non-published — a "service" their subscribers pay extra for. To my simple brain, it's a lot easier to simply publish your name as "Apple Plumpudding." See UNPUBLISHED.

Non-receipt Notification In X.400, a non-receipt notification is a report prepared by a recipient UA (User Agent) or Access Unit (upon request) and sent to the originating UA or Access Unit when a message is deemed unreceivable by a recipient.

Non-repudiation A mechanism which prevents a user from denying a legitimate, billable charge. For instance, a user engaged in an Internet-based voice or video conference might later repudiate the charge, alleging that he was not a party to the conference. A non-repudiation mechanism would provide for monitoring of all endpoints (connected users or machines) during the course of the conference in order that any applicable charges might be supported in the event that they are challenged.

Non-routable Protocols LAN protocols, such as IBM's NetBIOS, LAN Server and SNA, that use names and not Network Layer Addresses to identify devices and therefore supply no routing information. Internetworking devices must find other ways to route traffic in networks that use non-routable protocols.

Non-sent Paid Utility industry term for calls made as third party billings, reversed charges or with a Calling Card.

Non-session A type of synchronous access mode that can exist between an end user and a host computer in an Amdahl packet network. In non-session access, the calling line (PU) is mapped to a specific offering line (LU). The offering line is reserved for that calling line at all times.

Non-simultaneous Transmission Half duplex transmission. Transmission in one direction at a time. This mode of transmission may be the result of limitations of the transmission channel or of the transmitting/receiving equipment.

Non-subscriber calling See Casual Calling.

Non-synchronous Communications See ASYNCHRONOUS COMMUNICATIONS.

Non-traditional Retailing A fancy word for selling your wares via a Web page on the Internet.

Non-traffic Sensitive Plant Telephone company facilities which are unaffected by changes in volume of telephone activity.

Non-transparent Mode A transmission environment, mainly of bisynch transmission, in which control characters and control-character sequences are recognized through the examination of all transmitted data. Compare with transparent mode. Also called normal mode.

Non-trivial Microsoft word for "hard."

Non-volatile Memory which is not lost (i.e. that does not

"forget") when the power is shut off.

Non-volatile Random Access Memory Electronic circuitry that provides back-up operation of CMOS RAM and/or Flash PROM in case of a power failure.

Non-wireline Also called the block "A" carrier. The "A" originally stood for "alternate." The FCC, in setting up the licensing and systems in each market. It reserved one for the local wireline telephone company, and opened the second system — the Block A system — to other interested applicants. The distinction between Block A and Block B is meaningful only during the licensing phase at the FCC. Once a system is constructed, it can be sold to anyone. Thus in some markets today, both the A and B systems are owned by a telephone company. One happens to be the local phone company, and the other is a phone company that decided to buy a cellular system outside its home territory. Non-Wireline, or Block A systems, operate on the radio frequencies from 824 to 849 Megahertz.

NORC Network Operators Research Committee.

Normal Distribution (Gaussian) Bell shaped curve, with 90% of the values within 1.645 standard deviations of the mean and 98% of the values within 2.33 standard deviations of the mean. In each case, half of the values are greater than the mean and half are less than the mean. See Mean.

Normal Mode The AC voltage that exists between the normal current-carrying wires, that is, between neutral and hot or live.

Normal Response Mode NRM. HDLC mode for use on links with one primary and one or more secondaries. Under NRM, a secondary can transmit only after receiving a poll addressed to it by a primary. It may then send a series of responses. But after it sets the F bit in a response, it cannot transmit any more until it receives another poll.

Normal Tennis Tennis without the emotion, mistakes and frustration. What Patrick Bullot and Harry Newton would prefer to play, but can't.

Normalize A call center term. To change an unusual call statistic reported by the ACD so as to reflect what would have been usual for that period of the day or that day of the week. Normalizing is something you do before updating historical patterns so that the historical patterns will not be distorted by the unusual data.

Normalized Average Transfer Delay Average transfer delay divided by packet transmission time at the clock rate of the medium.

Normalized Network Throughput Network throughput in packets per second divided by maximum throughput possible at clock rate of medium. Less than one.

Normalized Offered Traffic The average number of attempted packet transmissions per second divided by the average number of packet transmissions/second possible at the clock rate of the medium. May exceed one.

Nortel Nortel is a shorthand version of Northern Telecom's corporate name (which remains Northern Telecom Limited). According to the company's annual report, it is designed to reflect the corporation's heritage, while signaling a new direction, presenting a single face that reinforces Northern Telecom's global presence. The new logo, with its Globemark 'O', creates a dynamic visual symbol for a corporation whose business knows no boundaries and whose spirit is one of leadership, innovation, dedication and excellence. In the U.S., Northern Telecom's principal subsidiary is called Northern Telecom Inc. www.nortel.com

North American Area Codes North American Area Codes are a numbering plan (called NANP, North American

Numbering Plan) for the public switched telephone network in the United States and its territories, Canada, Bermuda, and many Caribbean nations, including Anguilla, Antigua & Barbuda, Bahamas, Barbados, British Virgin Islands, Cayman Islands, Dominica, Dominican Republic, Grenada, Jamaica, Montserrat, St. Kitts and Nevis, St. Lucia, St. Vincent and the Grenadines, Trinidad and Tobago, and Turks & Caicos. NANP numbers are ten digits in length, and they are in the format: NXX-NXX-XXXX, where N is any digit 2-9 and X is any digit 0-9. The first three digits are called the numbering plan area (NPA) code, often called simply the area code. The second three digits are called the central office code or prefix. The final four digits are called the telephone line number. For an up-to-date collection, see www.nanpa.com. Here are two lists, one by numerically by area code and one alphabetically by area:

By Number

201 New Jersey
202 District of Columbia
203 Connecticut
204 Manitoba
205 Alabama
206 Washington
207 Maine
208 Idaho
209 California
210 Texas
212 New York
213 California
214 Texas
215 Pennsylvania
216 Ohio
217 Illinois
218 Minnesota
219 Indiana
224 Illinois
225 Louisiana
228 Mississippi
240 Maryland
242 Bahamas
246 Barbados
248 Michigan
250 British Columbia
252 North Carolina
253 Washington
254 Texas
256 Alabama
264 Anguilla
267 Pennsylvania
268 Antigua/Barbuda
270 Kentucky
281 Texas
284 British Virgin Islands
301 Maryland
302 Delaware
303 Colorado
304 West Virginia
305 Florida
306 Saskatchewan
307 Wyoming
308 Nebraska
309 Illinois
310 California
311 Non-Emergency Access
312 Illinois

313 Michigan
314 Missouri
315 New York
316 Kansas
317 Indiana
318 Louisiana
319 Iowa
320 Minnesota
323 California
330 Ohio
334 Alabama
336 North Carolina
340 US Virgin Islands
345 Cayman Islands
352 Florida
360 Washington
401 Rhode Island
402 Nebraska
403 Alberta
404 Georgia
405 Oklahoma
406 Montana
407 Florida
408 California
409 Texas
410 Maryland
411 Local Directory Assistance
412 Pennsylvania
413 Massachusetts
414 Wisconsin
415 California
416 Ontario
417 Missouri
418 Quebec
419 Ohio
423 Tennessee
424 California
425 Washington
435 Utah
440 Ohio
441 Bermuda
443 Maryland
450 Quebec
456 Inbound International
469 Texas
473 Grenada
484 Pennsylvania
500 Personal Communication Services
501 Arkansas
502 Kentucky
503 Oregon
504 Louisiana
505 New Mexico
506 New Brunswick
507 Minnesota
508 Massachusetts
509 Washington
510 California
512 Texas
513 Ohio
514 Quebec
515 Iowa
516 New York
517 Michigan

518 New York
519 Ontario
520 Arizona
530 California
540 Virginia
541 Oregon
559 California
561 Florida
562 California
570 Pennsylvania
573 Missouri
580 Oklahoma
600 Canada (Services)
601 Mississippi
602 Arizona
603 New Hampshire
604 British Columbia
605 South Dakota
606 Kentucky
607 New York
608 Wisconsin
609 New Jersey
610 Pennsylvania
611 Repair Service
612 Minnesota
613 Ontario
614 Ohio
615 Tennessee
616 Michigan
617 Massachusetts
618 Illinois
619 California
626 California
630 Illinois
649 Turks & Caicos Islands
650 California
651 Minnesota
660 Missouri
661 California
664 Montserrat
670 CNMI
671 Guam
678 Georgia
700 IC Services
701 North Dakota
702 Nevada
703 Virginia
704 North Carolina
705 Ontario
706 Georgia
707 California
708 Illinois
709 Newfoundland
710 U.S. Government
711 TRS Access
712 Iowa
713 Texas
714 California
715 Wisconsin
716 New York
717 Pennsylvania
718 New York
719 Colorado
720 Colorado

724 Pennsylvania
727 Florida
732 New Jersey
734 Michigan
740 Ohio
757 Virginia
758 St. Lucia
760 California
765 Indiana
767 Dominica
770 Georgia
773 Illinois
775 Nevada
780 Alberta
781 Massachusetts
784 St. Vincent & Grenada
785 Kansas
786 Florida
787 Puerto Rico
800 800 Service
801 Utah
802 Vermont
803 South Carolina
804 Virginia
805 California
806 Texas
807 Ontario
808 Hawaii
809 Caribbean Islands
810 Michigan
811 Business Office
812 Indiana
813 Florida
814 Pennsylvania
815 Illinois
816 Missouri
817 Texas
818 California
819 Quebec
828 North Carolina
830 Texas
831 California
832 Texas
843 South Carolina
847 Illinois
850 Florida
858 California
860 Connecticut
864 South Carolina
867 Yukon & NW Territories
868 Trinidad and Tobago
869 St. Kitts & Nevis
870 Arkansas
876 Jamaica
877 800 Service (also 888)
880 PAID-800 Service
881 PAID-888 Service
882 PAID-877 Service
888 800 Service (also 877)
900 900 Service
901 Tennessee
902 Nova Scotia
903 Texas
904 Florida

905 Ontario
906 Michigan
907 Alaska
908 New Jersey
909 California
910 North Carolina
911 Emergency
912 Georgia
913 Kansas
914 New York
915 Texas
916 California
917 New York
918 Oklahoma
919 North Carolina
920 Wisconsin
925 California
931 Tennessee
935 California
937 Ohio
940 Texas
941 Florida
949 California
954 Florida
956 Texas
970 Colorado
972 Texas
973 New Jersey
978 Massachusetts

By Place
800 800 Service
888 800 Service Expansion
877 888 Service Expansion
900 900 Service
205 Alabama
256 Alabama
334 Alabama
907 Alaska
403 Alberta
780 Alberta
264 Anguilla
268 Antigua/Barbuda
520 Arizona
602 Arizona
501 Arkansas
870 Arkansas
242 Bahamas
246 Barbados
441 Bermuda
250 British Columbia
604 British Columbia
284 British Virgin Islands
811 Business Office
209 California
213 California
310 California
323 California
408 California
415 California
424 California
510 California
530 California
559 California

562 California
619 California
626 California
650 California
661 California
707 California
714 California
760 California
805 California
818 California
831 California
858 California
909 California
916 California
925 California
935 California
949 California
600 Canada (Services)
809 Caribbean Islands
345 Cayman Islands
670 CNMI
303 Colorado
719 Colorado
720 Colorado
970 Colorado
203 Connecticut
860 Connecticut
302 Delaware
202 Dist. of Columbia
767 Dominica
911 Emergency
305 Florida
352 Florida
407 Florida
561 Florida
727 Florida
786 Florida
813 Florida
850 Florida
904 Florida
941 Florida
954 Florida
404 Georgia
678 Georgia
706 Georgia
770 Georgia
912 Georgia
473 Grenada
671 Guam
808 Hawaii
700 IC Services
208 Idaho
217 Illinois
224 Illinois
309 Illinois
312 Illinois
618 Illinois
630 Illinois
708 Illinois
773 Illinois
815 Illinois
847 Illinois
456 Inbound International
219 Indiana

317 Indiana
765 Indiana
812 Indiana
319 Iowa
515 Iowa
712 Iowa
876 Jamaica
316 Kansas
785 Kansas
913 Kansas
270 Kentucky
502 Kentucky
606 Kentucky
411 Local Directory Assistance
225 Louisiana
318 Louisiana
504 Louisiana
207 Maine
204 Manitoba
240 Maryland
301 Maryland
410 Maryland
443 Maryland
413 Massachusetts
508 Massachusetts
617 Massachusetts
781 Massachusetts
978 Massachusetts
231 Michigan
248 Michigan
313 Michigan
517 Michigan
616 Michigan
734 Michigan
810 Michigan
906 Michigan
218 Minnesota
320 Minnesota
507 Minnesota
612 Minnesota
651 Minnesota
228 Mississippi
601 Mississippi
314 Missouri
417 Missouri
573 Missouri
660 Missouri
816 Missouri
406 Montana
664 Montserrat
308 Nebraska
402 Nebraska
702 Nevada
775 Nevada
506 New Brunswick
603 New Hampshire
201 New Jersey
609 New Jersey
732 New Jersey
908 New Jersey
973 New Jersey
505 New Mexico
212 New York
315 New York

516 New York
518 New York
607 New York
716 New York
718 New York
914 New York
917 New York
709 Newfoundland
311 Non-Emergency Access
252 North Carolina
336 North Carolina
704 North Carolina
828 North Carolina
910 North Carolina
919 North Carolina
701 North Dakota
902 Nova Scotia
216 Ohio
330 Ohio
419 Ohio
440 Ohio
513 Ohio
614 Ohio
740 Ohio
937 Ohio
405 Oklahoma
580 Oklahoma
918 Oklahoma
416 Ontario
519 Ontario
613 Ontario
647 Geographic Relief Code
705 Ontario
807 Ontario
905 Ontario
503 Oregon
541 Oregon
880 PAID-800 Service
882 PAID-877 Service
881 PAID-888 Service
215 Pennsylvania
267 Pennsylvania
412 Pennsylvania
484 Pennsylvania
570 Pennsylvania
610 Pennsylvania
717 Pennsylvania
724 Pennsylvania
814 Pennsylvania
500 Personal Communication Services
787 Puerto Rico
418 Quebec
450 Quebec
514 Quebec
819 Quebec
611 Repair Service
401 Rhode Island
306 Saskatchewan
803 South Carolina
843 South Carolina
864 South Carolina
605 South Dakota
869 St. Kitts & Nevis
758 St. Lucia

784 St. Vincent & Grenada
423 Tennessee
615 Tennessee
901 Tennessee
931 Tennessee
210 Texas
214 Texas
254 Texas
281 Texas
409 Texas
469 Texas
512 Texas
713 Texas
806 Texas
817 Texas
830 Texas
832 Texas
903 Texas
915 Texas
940 Texas
956 Texas
972 Texas
868 Trinidad and Tobago
711 TRS Access
649 Turks & Caicos Islands
710 U.S. Government
340 US Virgin Islands
435 Utah
801 Utah
802 Vermont
540 Virginia
703 Virginia
757 Virginia
804 Virginia
206 Washington
253 Washington
360 Washington
425 Washington
509 Washington
304 West Virginia
414 Wisconsin
608 Wisconsin
715 Wisconsin
920 Wisconsin
307 Wyoming
867 Yukon & NW Territories

North American Cellular Network A Craig McCaw idea to join together a bunch of cellular phone companies providers who would provide a painless, simple way for someone making or receiving a cellular call in their territory. Previously, traveling cellular users — what the industry calls "roamers" — were forced to pay heavy charges for calling from the territory of a cellular company not theirs. And roamers were, in effect, incommunicado from incoming calls. It was impossible to call someone on a cellular phone if they were not in the own territory and you didn't know where they were. Craig McCaw was the founder of McCaw Communications, a large cellular phone company, now owned by AT&T.

North American Directory Forum NADF. An association of electronic mail providers who are figuring standards and ways of sending mail between their subscribers.

North American ISDN Users Forum NIUF. Here is the complete explanation of this important group, as excerpted from

their brochure: The barriers to the widespread use of ISDN nationally and internationally is difficult because the technology is complex and developing rapidly. One underlying problem is the lack of standard implementations of ISDN applications. ISDN standards are currently developed by the International Telecommunications Union (ITU-T) and by accredited standards committee T-1 under the umbrella of the American National Standards Institute (ANSI). But standards designed to meet many requirements offer multiple options which are open to diverse interpretations. As a result, services and products produced by different manufacturers are incompatible.

To solve this problem, NIST (National Institute of Standards and Technology) collaborated with industry in 1988 to establish the North American ISDN Users' Forum (NIUF). NIST's Computer Systems Laboratory (CSL) serves as chair of the forum and hosts the NIUF Secretariat under the terms of a cooperative Research and Development Agreement (CRADA) established with industry in 1991. Through support of the forum, CSL advances new uses of computer and telecommunications technology in government and industry. The objectives of the NIUF are "to provide users the opportunity to influence developing ISDN technology to reflect their needs; to identify ISDN applications, develop implementation requirements, and facilitate their timely, harmonized, and interoperable introduction; and to solicit user, product provider, and service provider participation in the process." www.niuf.nist.gov

North American Numbering Plan NANP. The method of identifying telephone trunks in the public network of North America, called World Numbering Zone 1 by the ITU-T. The Plan has three ways of identifying phone numbers in North America — a three digit area code, a three digit exchange or central office code and four digit subscriber code. Other countries have much more complicated numbering schemes. There are some countries, for example, where the length of the area code actually exceeds the subscriber code. There are many countries where there is no consistency in the length of phone numbers. Some are nine-digit. Some are 12-digit, etc. All these varying number lengths may be within 100 miles of each other.

Under the NAMP format prior to January 1, 1995, the second digit of the area code was always a one or a zero. With this system, approximately one billion telephone numbers, 152 area code combinations and 640 prefixes were available. Numbers started to run out as a result of increased use of fax, modem and cellular lines. With the new Numbering Plan introduced on January 1, 1995, the area code can be any combination of three digits. This will allow more than 6 billion telephone numbers and 792 area code combinations. See NANC.

North American Telecommunications Association NATA. Now known as the MultiMedia Telecommunications Association MMTA). The national trade association for companies providing customer premise telephone equipment. NATA represents the interests of the industry before the Congress, the FCC and in court actions.

NOS Network Operating System, the software for a network that runs in a file server and controls access to files and other resources from multiple users. It can run on top of DOS, and also provides security and administrative tools. Novell's NetWare, Banyan's VINES, Microsoft's Windows NT Server and IBM's OS/2 LAN Server are NOS examples.

Notch An out-of-phase impulse causing spontaneous dip in voltage. This is an under-voltage impulse, similar to a spike, but of reverse polarity to the instantaneous value of the AC sine wave. A notch normally is associated with the power company removing a generator from the power grid. See NOTCHED NOISE.

Notch Filter A filter that blocks or passes a specific band of frequencies. The filters are: lowpass/block, high pass/block, notch pass/block. If the filter is set in a series with a circuit, the desired frequencies pass down the line. If it's passed to ground, the desired frequencies are sent to ground. They can't get through the circuit.

Notched Noise Noise in which a narrow band of frequencies has been removed. Normally used for testing devices or circuits.

Notes on The Network A famous book explaining how the North American public switched network works. It's now called BOC Notes on the LEC Networks.

NOTHS Network Operations Trouble Handling System.

Notwork A network in its nonworking state.

Novell Novell is the reincarnation of NDSI, a computer firm that almost went under in the early 1980s. The then nearly 60-year-old Ray Noorda, who had 20 years of experience in systems automation with General Electric before striking out for Silicon Valley, was called in to help NDSI prepare for a trade show. He spotted that it had some potentially interesting technology, and bought the ailing company in 1983. Their software is called NetWare. See NetWare.

Novell IPX See IPX.

Novell NLSP Novell NetWare Link Services Protocol. A link state protocol under development by Novell to improve performance of IPX traffic in large internetworks.

Novell SAP Novell Service Advertisement Protocol. See SAP.

Novell Telephony Services See TELEPHONY SERVICES.

NP See NUMBER PORTABILITY

NPA 1. In IBM language, it means Network Performance Analyzer, a product for network tuning, determining performance, degradation and determining the affect of network growth.
2. Numbering Plan Area. A fancy way of saying Area Codes. There are over 200 area codes in the United States, Canada, Bermuda, the Caribbean, Northwestern Mexico, Alaska and Hawaii. Within any of these area codes, no two telephone lines may have the same seven digit phone number. The middle number has been either "1" or "0" creating "N 1/0 x" codes. The number of codes available on this basis were nearing depletion. Bellcore has now modified the plan to obtain more area codes. Future area codes will be NXX like the central office numbering scheme. Switching systems in the national network will differentiate between the central office and area codes by recognizing the subscriber always dials 1+ or 0+ preceding an area code when direct dialing such long distance calls.

Here are the special, unassigned and reserved NPAs:
200 Reserved for special services
211 Assigned to local operators
300 Assigned to special services
311 Reserved for special local services
400 Reserved for special services
500 Reserved for special services
511 Reserved for special local services
600 Reserved for special services
700 Assigned special access code for interLATA carriers/resellers

711 Reserved for special local services
See NNX and NORTH AMERICAN AREA CODES.

NPAC Number Portability Administration (also called Access) Center. Regional centers which will be developed to assist in the implementation of Local Number Portability (LNP). The NPAC will interact with the LSOAs (Local Service Order Administration) systems of all carriers in order to effect the transportation of the end user's telephone number from the legacy Local Exchange Carrier (LEC) to the new LEC of choice. LNP effectively decouples customer identification number (Directory Number) from Network Routing Address (Location Routing Number or LRN). The NPACs serve to synchronize the numbering databases, and to coordinate the porting process. As orders are placed to change the serving LEC, the OSSs (Operations Support Systems) of the victorious LEC communicate that request to its Local Service Order Administration (LSOA) system, which passes the request to the NPAC. The NPAC notifies the legacy carrier's LSOA of the request for purposes of confirmation. Once confirmed and ported, the NPAC passes relevant data to to all carrier LSOAs in its region, which, in turn, pass that information to its LSMSs (Local Service Management Systems), which pass the change to the LSOAs, which pass it on to the appropriate OSSs. There are seven NPACs designated in the U.S., with Lockheed Martin serving all seven regions. Originally, Perot Systems served three regions, but it pulled out. Now Lockheed Martin runs all seven. There is a single NPAC in Canada. See also LNP.

NPC 1. Network Processing Card.
2. Network Parameter Control: Network Parameter Control is defined as the set of actions taken by the network to monitor and control traffic from the NNI. Its main purpose is to protect network resources from malicious as well as unintentional misbehavior which can affect the QoS of other already established connections by detecting violations of negotiated parameters and taking appropriate actions. Refer to UPC.

NPD See NETWORK PROTECTION DEVICE.

NPDA Network Problem Determination Application. A program which allows for the monitoring of an entire network from a single location, collection of statistics, and isolation of communication faults. An IBM term. See NETVIEW.

NPDU Network Protocol Data Unit.

NPM Network Process Monitor.

NPR DOA is Dead On Arrival. And it's a term several manufacturers use to refer to equipment which arrives at the customer's premises not working. A person who receives a DOA machine will ask the company for a NPR number — New Product Return number. This allows them to return the product and have the factory replace it with another new one.

NPRM Notice of Proposed Rule Making. A term used in regulatory agencies. The agency runs an idea up the flagpole, then typically hold hearings to find out how people react.

NPSI IBM X.25 NCP Packet Switching Interface. Networking software package that allows Systems Network Architecture (SNA) 3270 traffic to be transmitted over an X.25 packet data network (PDN). See also QLLC. Contrast with DSP.

NRAM Nonvolatile Random Access Memory. NRAM does not lose its memory when you turn off the computer or phone system. Many modems, for example, use nonvolatile memory to store configuration information (in place of the switches used on other modems). The command &W writes instructions to the NRAM in many PC modems.

NRC 1. Non-Recurring Charge.
2. Network Reliability Council.

NREN The National Research and Education Network, an ultragigabit network established by legislation in 1991 by the U.S. House and Senate. Preliminary steps toward deploying this information superhighway (for talking between University computers) include some gigabit network testbeds and the cutover to a 45 megabit per second backbone for the National Science Foundation Network (NSFnet) used by the scientific research and university community nationwide.

NRM 1. NoRMal response of HDLC. See NORMAL RESPONSE MODE.

2. An ATM term. An ASP service parameter, Nrm is the maximum number of cells a source may send for each forward RM-cell.

NRN No Response Necessary. A Proposed e-mail convention to prevent endless back-and-forth acknowledgments: "Thanks for the info." " You're welcome ...hope it helps." "I hope so too. Thanks." By putting NRN at the bottom of your mail, you absolve the receiver from having to reply.

NRS See Network Reconfiguration System

NRUG National Rolm Users Group.

NRZ Non-Return to Zero. A binary encoding scheme in which ones and zeroes are represented by opposite and alternating high and low voltages and where there is no return to a zero (reference) voltage between encoded bits. NRZ is now used as an encryption scheme for getting data onto and off hard disk fast. It eliminates the need for clock pulses and yields up to 18.5 kilobytes per track and high read/write speeds.

NRZI Non-Return to Zero Inverted. A binary encoding scheme that inverts the signal on a "one" and leaves the signal unchanged for a "zero". Where a change in the voltage signals a "one" bit, and the absence of a change denotes a "zero" bit value. Also called transition coding.

NS/EP Telecommunications National Security and Emergency Preparedness Teecommunications. A Federal government definition. Telecommunications services that are used to maintain a state of readiness or to respond to and manage any event or crisis (local, national, or international) that causes or could cause injury or harm to the population, damage to or loss of property, or degrade or threaten the national security or emergency preparedness of the United States.

NSA 1. Network Service Address. NSAs are unique addresses that define physical or logical locations in equipment, such as a residential telephone number (TN).

2. National Security Agency. A super-secret agency of the Federal Government. NSA is the Federal agency responsible for the design and use of nonmilitary encryption technology, developing sophisticated codes to scramble data, voice or video information. In short, it is charged with signals intelligence and is widely assumed to monitor all communications traffic (phone, fax, data, video, etc.) into and out of the United States with foreign countries. It is barred from intercepting domestic communications. NSA grabbed the headlines in 1993 and 1994 when it adopted its most visible attempt to outgun cybervillains with something called the Clipper Chip. The idea is that the Clipper Chip (a microprocessor) would be installed in every phone, computer, and personal digital assistant in America would carry a device identification number or electronic "key" — a family key and unit key unique to each Clipper chip. The device key is split into two numbers that, when combined into what's called a Law Enforcement Access Field number, can unscramble the encrypted messages. The device keys and the corresponding device numbers, according to NSA proposals, would be kept by the US government

through key escrow agents. Under a plan proposed, the attorney general would deposit the two device keys in huge, separate electronic database vaults. One key would be held by the National Institute for Standards and Technology (NIST) and the other by the Automated Systems Division of the U.S. Treasury. Access to these keys would be limited to government officials with legal authorization to conduct a digital wiretap. When a law enforcement agency wants to tap into information encrypted by the Clipper ship, they must obtain a court order and then apply to each of the escrow agents. The agents electronically send their key into to an electronic black box operated by the law enforcement agency. When these keys are electronically inserted, encrypted conversations stream into the black box and come as standard voice transmissions or as ASCII characters in the case of electronic mail. At least that's the theory. See NSA LINE EATER.

NSA Line Eater The more paranoid Internet users believe that the National Security Agency has a super-powerful computer assigned to reading everything posted on the Net. They will jokingly refer to this line eater in their postings. Goes back to the early days of the Internet when the bottom lines of messages would sometimes disappear for no apparent reason.

NSAP Network Service Access Point. The point at which the OSI Network Service is made available to a Transport entity. The NSAPs are identified by OSI Network Addresses. The NSAP is a generic standard for a network address consisting of 20 octets. ATM has specified E.164 for public network addressing and the NSAP address structure for private network addresses.

NSAP-Selector A component of an Network Service Access Point (NSAP)-Address used to select the Network Layer service user. The NSAP-Selector is sometimes referred to as a Transport-Selector; however, a user of the Network Layer need not be a transport service.

NSAPI Netscape Server Application Programming Interface. A Netscape-only Web server application development interface, developed by Netscape Communications Corporation. NSAPI was designed as a more robust and efficient replacement for CGI.

NSC 1. Network Service Center. See NETWORK CONTROL POINT and SDN.

2. Non-Standard facilities Command. A response to the called fax DIS response.

NSCP National Scalable Cluster Project. A consortium of universities and corporate partners dedicated to developing distributed computing and deploying high-speed networking through ATM technology. The project is intended to solve the technical hardware and software problems associated with the support of sophisticated multi-processing at widely separated locations. The meta-cluster will operate over an ATM-based WAN; cluster computing within the LAN domain is emerging as a scalable and cost effective to supercomputing and massively parallel computing. Applications are anticipated to include digital libraries and linguistic data, imaging and virtual reality, and data mining. Initial participants include the University of Pennsylvania; the University of Illinois at Chicago; and the University of Maryland, College Park.

NSEP Telecommunications See NS/EP TELECOMMUNICATIONS.

NSF See NATIONAL SCIENCE FOUNDATION.

NSFNET National Science Foundation's TCP/IP-based NETwork, funded by the U.S. Government. Linking supercomputing centers and over 2500 academic and scientific institutions across the world, largely through the Internet, the

NSFNET was founded in 1985. It was officially retired in 1995, replaced by the MERIT Network. See NREN and MERIT.

NSG National Systems Group. Division of Canadian telcos responsible for multiple accounts — large customers with service in multiple telephone company territories.

NSLOOKUP Name Server Lookup. An interactive query program of the InterNIC DNS (Domain Name Server). NSLOOKUP allows the user to contact servers to request information about a specific host or to print a list of hosts in the domain. See INTERNIC, DNS, WHOIS, RWHOIS and DIG.

NSO National Services Organization. An NSO provides the interconnecting network infrastructure, intelligent network services, network management, and roaming in a national Personal Communications Network (PCN). An NSO has ownership of resources that allow it to provide services to local operators. Operations such as marketing are centralized.

NSP 1. Network Service Provider. Can include a local telephone company, ISP, or CLEC.
2. Native Signal Processing. Intel's idea to use the "spare MIPS" on its Pentium, Pentium Pro and later versions to take over the processing that once dedicated chips, like DSPs, used to do. The idea, clearly, is to sell more Intel chips and fewer chips from other makers.

NSR Non-Source Routed: Frame forwarding through a mechanism other than Source Route Bridging.

NSS 1. Non Standard Facilities Setup command, a response to an NSF frame.
2. Network and Switching Sub-system. A wireless telecommunications term. The part of the GSM (Global System for Mobile Communications) system in charge of the management of calls and the interface with other networks.

NSU Network Synchronization Unit. A timing distribution system that provides synchronization signals to all electronic equipment within a wireline network node or office.

NT 1. Network Termination. Network Termination represents the termination point of a Virtual Channel, Virtual Path, or Virtual Path/Virtual Channel at the UNI. See NT-1.
2. New Technology, usually known as Windows NT. It's a new operating system from Microsoft which will let Windows run on high-end machines, such as file servers and workstations. NT has two sets of goals: to provide true multi-tasking, security, network connectivity and 32-bit power. Secondly, it's to provide a smooth upgrade path from Windows and MS-DOS. For a much larger explanation, see WINDOWS NT and WINDOWS NT ADVANCED SERVER.
3. Night transfer of ringing and station class of service.

NT-1 National ISDN 1. See ISDN.

NI-2 National ISDN 2. See ISDN.

NT-3 National ISDN 3. See ISDN.

NT1 Network Termination type 1. An ISDN term. The NT1 provides functions related to the physical and electrical termination of the local loop between the carrier network and the user premise. These functions, which fall into Layer 1 of the OSI Reference Model, are necessary to support multichannel digital communications over the loop, support loopback testing, and a variety of other functions. NT1 functions are required for both BRI (Basic Rate Interface), also known as 2B+D, and PRI (Primary Rate Interface), also known as 23B+D. See Network Termination type 1 for a more detailed explanation. See also BRI, ISDN, NT2, NT12, OSI Reference Model and PRI.

The NT1 is the first customer premise device on a two-wire ISDN circuit coming in from the ISDN central office. It does several things. It converts the two-wire ISDN circuit (called

"U" interface) to four-wire so you can hook up several terminals — like a voice phone and a videophone. The NT1 typically has several lights on it which indicate if it's working. An ISDN central office can usually "talk" to the NT1 and do testing and maintenance by instructing the NT1 to loop signals back to the central office. An NT1 will support up to eight terminal devices, though I've never seen it work with that many. The basic NT1 functions are:

- Line transmission termination.
- Layer 1 line maintenance functions and performance monitoring.
- Layer 1 multiplexing, and
- Interface termination, including multi drop termination employing layer 1 contention resolution.

Increasingly, NT1s are being incorporated into customer premises equipment, like phones, PC ISDN data cards, etc. See ISDN.

NT12 Network Termination type ´. An ISDN term for a NT2 device which supports both NT1 and NT2 functionality. Such a device performs functions described in Layers 1 and 2, and perhaps Layer 3, of the OSI Reference Model. See also NT1 and NT2.

NT2 In ISDN, the Network Termination type 2 is an intelligent CPE (Customer Premise Equipment) switching or concentrating device (e.g., PBX, ACD, router, or concentrator). A NT2 device typically terminates PRI (Primary Rate Interface), also known as 23B+D (or 30B+D in Europe), access lines from the local ISDN CO switch. Depending on the specific nature of the NT2 device, it performs protocol handling functions described in Layers 1, 2 and 3 of the OSI Reference Model. See also BRI, ISDN, NT1, NT12, OSI Reference Model and PRI.

NTACS Narrow TACS. Cellular radio system deployed in Japan using narrow band technology to increase capacity by splitting TACs channels into two narrow channels.

NTCA National Telephone Cooperative Association. A trade association representing primarily rural telephone cooperatives and other small telephone companies.

NTE Network Termination Equipment. The equipment on both sides of a subscriber line. A term used in the DSL (Digital Subscriber Line) world. See ADSL, ATU-C, ATU-R, and xDSL.

NTFS Windows NT file system; an advanced file within the Windows NT operating system. It supports file system recovery, extremely large storage media, and various features for the POSIX subsystem. It also supports object-oriented applications by treating all files as objects with user-defined and system-defined attributes.

NTI 1. Northern Telecom Inc. Now called NorTel.
2. Network Terminating Interface. A. The point where the network service provider's responsibilities for service begin or end, or B. The interface between DCE and its connected DTE.

NTIA National Telecommunications and Information Administration. An agency of the U.S. Department of Commerce concerned with spectrum management, public safety (e.g., electromagnetic fields), and communications (primarily telephony) standards. The NTIA is responsible for managing the Federal government's use of the spectrum, while the FCC is responsible for managing the use of spectrum by the private sector, and state and local governments. Clearly, the two agencies work closely in such matters. Both the FCC and the NTIA serve to represent the U.S. at the biannual World (Administrative) Radiocommunications Conference (WRC, pronounced "wark" as in "cork"), which are sponsored by the ITU (International Telecommunications Union) every two years. The NTIA also serves as the principal adviser to the President, Vice President and Secretary of Commerce on issues of domestic and international communications and information. The NTIA has been highly instrumental in furthering the concept of the NII (National Information Infrastructure). www.ntia.doc.gov See FCC, ITU, NII and WARC.

NTIS National Technical Information Service. www.ntis.gov. See NATIONAL TECHNICAL INFORMATION SERVICE.

NTL Northern Telecom

NTN See NETWORK TERMINAL NUMBER.

NTO Network Terminal Option. An IBM program product that enables an SNA network to accommodate a select group non-SNA asynchronous and bisynchronous devices via the NCP-driven communications controller.

NTP 1. Network Termination Point.
2. The Network Time Protocol was developed to maintain a common sense of "time" among Internet hosts around the world. Many systems on the Internet run NTP, and have the same time (relative to Greenwich Mean Time), with a maximum difference of about one second.

NTS 1. Network Test System.
2. Non-Traffic-Sensitive commercial line costs levied on the user.
3. Number Translation Services.

NTSC National Television Standards Committee of Electronic Industries Association (EIA) that prepared the standard of specifications approved by the Federal Communications Commissions in 1953 for commercial broadcasting. The initials are used to describe the standard method of television transmission in U.S., Canada, Japan, Central America and half of South America. The North American system uses interlaced scans and 525 horizontal lines per frame at a rate of 30 frames per second. The picture information is transmitted in AM (amplitude modulation) and the sound information is transmitted in FM (frequency modulation). NTSC is compatible with ITU-R (nee CCIR) Standard M. PAL (which stands for Phase Alternate Line) is the name of the format for color TV signals used in West Germany, England, Holland, Australia and several other countries. It uses an interlaced format with 25 frames per second and 625 lines per frame. The two systems are not compatible.   You cannot view an Australian videotape on a U.S. TV.

When TV engineers get together to hoist some brews, however, the initials for the various standards for TV broadcast take on other meanings, such as: NTSC, Never Twice the Same Color. PAL, Peace At Last. FAL-M, Peace At Last — Maybe. SECAM (French system) becomes System Essentially Contrary to the American Method.

NTSC Signal National Television System Committee specified signal. De-facto standard governing the format of television transmission signals in the United States.

NTSI Format A color television format having 525 scan lines; a field frequency of 60 Hz; a broadcast bandwidth of 4 MHz; line frequency of 15.75 KHz; frame frequency of 1/30 of a second; and a color subcarrier frequency of 3.58 MHz.

NTT Nippon Telephone and Telegraph, the major Japanese local and long distance telephone company.

NTU Network Terminating Unit. A device which is placed at the final interconnect point between the PSTN and the customer owned equipment. NTUs allow the Carrier to isolate their facilities from those of their customers for testing, fault detection, and some service feature functionality.

NU Network unit.

NuBus (Pronounced "New Bus.") The name of the bus design for most Apple Macintosh computers. See NUBUS CARD.

NuBus Card An add-on card that fits inside the NuBus slots of a Modular Macintosh. Often used for video cards and modems.

Nucleus An ATM term. The interior reference point of a logical node in the PNNI complex node representation.

NUI Network User Identifier. A unique alphanumeric number provided to dial-up users to identify them to packet switched networks around the world. The number is used to get onto the network and for billing.

Null Having no value. A dummy letter, letter symbol, or code group inserted in an encrypted message to delay or prevent its solution, or to complete encrypted groups for transmission or transmission security purposes. See also NULL CHARACTERS.

Null Call_id AN ISDN term. A null call_id is the call_id used to convey information that does not pertain to a specific call between the ISDN System Adapter and the central office switch. Null call_ids are primarily used for accessing feature buttons that do not relate to a specific call.

Null Characters Characters transmitted to fill space, time or to "pad" something. They add nothing to the meaning of a transmitted message, but the null characters are expected by the system. On older teletype machines, for example, when the type head reaches the end of a line and the New Line sequence is transmitted, it usually includes a number of Null Characters in order to give the mechanical type head enough time to reach the left margin of the page before transmitting the next line to the terminal. In this manner, no characters are lost. MCI Mail uses five null characters at the beginning of every line — unless you tell it otherwise.

Null Modem A null modem is a shortened way of saying "Null Modem Cable." See Null Modem Cable.

Null Modem Cable Crossover or cross-pinned wiring of an RS-232 cable such that a DTE (Data Terminal Equipment) device (such as a PC) can talk to another such device without the use of a modem, hence the term "null", which means "amounting to nothing." A null modem cable allows one PC to connect directly to another PC for file transfer over maximum distances of 50-100 feet (depending on the quality of the cable) without the use of either a modem or a line driver. A null modem cable also can be used to connect one DCE (Data Communications Equipment) device to another, in what is known as a "tail circuit" configuration. Essentially, a null modem cable reverses pins 2 and 3 on an RS-232 cable. But there are no standard null modem cables. And other pins sometimes need changing and jumpering together. Null Modems also are known as Modem Eliminators.

Null State A state in which there is no relationship between a call and device. Synonymous with Idle State.

Null Suppression A data compression technique whereby streams of null characters are identified at a transmission source and replaced by two or more control characters. The first character indicates the null suppression, and more characters indicate the number of null characters removed. The receiver uses this information to replace the removed data.

Null Value A parameter or field position for which no value is specified.

Nulls See LEAKY COAX.

NUMA Non-Uniform Memory Access. A symmetrical multiprocessing (SMP) technology. Proponents claim the technology is an improvement over traditional Intel-based SMP systems that can suffer from traffic jams on their shared-memory buses and typically cannot accommodate more than 16 or 32 processors. With NUMA, each Intel processor has its own local memory and is able to form static or dynamic connections with other chips' memories. NUMA servers can be powered by 64 or more processors.

Number Administration And Service Center
NASC. Provides centralized administration of the Service Management System (SMS) database of 800 numbers. The NASC keeps track of the 800 numbers that are in use, or available for use, by new 800 customers.

Number Capacity A telephone company term. The network access line capacity of numbers or terminals is the maximum number of network access lines that can be working on installed numbers at the entity's derived objective percent number fill.

Number Crunching A repetitive series of mathematical calculations.

Number Portability Number Portability (NP) refers to the ability of end users to retain their geographic or non-geographic telephone number when they change any of:
1. Their service provider. This means that users can retain the same telephone number as they change from one service provider (telephone company) to another.
2. Their location. This means that users can retain the same telephone number as they move from one permanent location to another.
3. Their service. This means that users can keep the same telephone number as they move from one type of service to another (e.g., POTS to ISDN).
Number portability started with 800 toll-free numbers as that industry and then moved to local number portability (LNP) as competition developed for local telephone companies. Here's a longer explanation of how 800 number portability developed. Once upon a time, 800 numbers belonged to the phone companies who supplied 800 service. So if you got 800 service from AT&T and you wanted to keep your number, you were stuck with AT&T. For example, we had (and still have) 1-800-LIBRARY (800-542-7279). The reason? The database of 800 numbers was maintained by Bellcore, who allocated certain "exchanges," like 800-542, to certain carriers. In the case of 542, it was AT&T. If we wanted to go to a different phone company, we had to give up our phone number. Since many companies had invested vast monies in promoting their 800 numbers, this was inconvenient and, in essence, forced companies to stay with the same long distance company. In 1993, things changed. The FCC mandated Number Portability, which allowed you to change your long distance carrier but keep your valued 800 number. The way the whole thing works now is simple to understand, although somewhat complex and expensive for the carriers to implement. You pick up the phone. You dial an 800-number. Your local central office holds the call for a moment, while it" dips" (checks) into an external, central database of 800 numbers for the routing of that call. (A number of such centralized databases are maintained by the carriers. In other words, the centralized databases are distributed throughout the network in order to balance the load on the processors and associated databases, placing them in strategic proximity to concentrations of traffic.) This database dip typically takes place via the packet-switched SS7 (Signaling System 7) network. When it receives a reply from that external database, it simply then sends the call to the appropriate long distance carrier which is providing service on that 800 number. To change the carrier of your 800 number, simply tell the database who the new carrier is. And that's 800 (actually, 800/888) number portability. See also LNP (Local Number Portability)

Numbering To have any sort of network work, everyone on the network has to have a unique number which we can all dial. I only included this definition because I found the following definition in a glossary of LAN terms and I thought, "Maybe someone doesn't know, though if they didn't know, this definition wouldn't help them much." Here's the LAN definition: "The assignment of unique identities to a user-network interface."

Numbering Plan 1. In Wide Area Networks, the method for assigning NNX codes to provide a unique telephone address for each subscriber, special line or trunk destination. 2. In PBXs, the method of assigning extension numbers and trunk designations at the local premises.

Numbering Plan Area NPA. A fancy term the Bell System came up with years ago to mean Area Codes. See NPA.

Numbers Shift A character in the Baudot code which establishes that the characters following in the transmission are to be interpreted as numeric characters. See LETTERS SHIFT.

Numeric Key Pad A separate section of a computer keyboard which contains all the numerals 0 through 9. Sometimes, some special keys are included — a plus sign, a minus sign, a multiplication sign and a division sign. The numeric key pad on a computer is the same as that found on calculators and adding machines. The top row is 789. The second top row is 456. The third top row is 123. The lowest row is typically 0, "." and "+". The numeric key pad is exactly opposite that of the touchtone telephone keypad, which was designed deliberately to be unfamiliar to users, so they may not input digits into the nation's telephone system faster than it could take them. Early touchtone central offices were very slow.

Numeric User Identifier According to the 1988 X.400 recommendations, a numeric user identifier is a standard attribute of an O/R (Originator/Recipient) address that consists of a unique sequence of numbers for identifying a user. (Numeric User Identifier was referred to as Unique Identifier in the 1984 recommendations.)

Numeris The French name for ISDN.

Numeronym A telephone number that spells a word. For example, 1-800-542-7279 is the telephone for the company which distributes this dictionary. That telephone number is advertised as 1-800-LIBRARY. It is a great number to have because it's easy to remember, although it's not so easy to dial. Numeronyms are in great demand, for obvious reasons. With the expansion of the toll-free dialing plan to include 888 (and other prefixes in the near future), it's also important (but increasingly difficult) to protect those numbers from competition.

NVOD Near Video On Demand. Providing a consumer a multimedia item — movie, TV program, etc. — on a rotating schedule, thus giving the appearance of an on-demand system, i.e. VOD (Video On Demand).

NVP 1. Network Voice Protocol. Circa 1973 ARPANET protocol. Used to support real-time voice over the ARPANET. Both LPC and CVSD encoding schemes were successfully implemented by Culler-Harrison, Inc., the Information Sciences Institute, Lincoln Laboratory and Stanford Research Institute. 2. In regards to cable, NVP stands for the Nominal Velocity of Propagation. All communications cable has a spec called NVP. An electrical signal in a vacuum would travel at the speed of light. In the world of land-based communications, electrical signals travel through twisted copper pairs at a percentage of the speed of light, around 72% for good cat 5 cable. When you test cat 5 cable, the testing instrument is supposed to be calibrated to the NVP of the cable so that the device can measure the time it takes to go end to end and back (or end to fault and back).

Without knowing the NVP, the test device cannot accurately locate a fault in the cable. See also Time Domain Reflectometer. This definition contributed kindly by Steven Waxman.

NVRAM Nonvolatile Random Access Memory. RAM that doesn't lose its memory when you shut the electricity off to it.

NVS NonVolatile Storage is a storage device, like a disk or EPROM, that retains data when you turn off.

NWL Non-Wireline. Cellular radio licenses received from the FCC with no initial association to telephone company. Also referred to as A-Block.

NWLink Microsoft's network protocol that simulates Novell's IPX/SPX for Windows 95 and NT communications with Novell NetWare file servers and compatible devices. NWLink is an IPX/SPX-compatible transport stack that gives NetWare-compatible clients access to NT applications services.

NWRA National Wireless Resellers Association. A Washington-based trade association and lobbying organization which formerly was known as the NCRA (National Cellular Resellers Association) NWRA merged into TRA (Telecommunications Resellers Association) in 1997. See TRA.

Nx384 N-by 384. The ITU-T's approach to creating a standard algorithm for video codec interoperability It is based on the ITU-T's HO switched digital network standard, which was expanded into the Px64 or H.261 standard, approved in 1990.

Nx64K An ATM term. This refers to a circuit bandwidth or speed provided by the aggregation of nx64 kbps channels (where n= integer> 1). The 64K or DSO channel is the basic rate provided by the T Carrier systems.

NXX In a seven digit local phone number, the first three digits identify the specific telephone company central office which serves that number. These digits are referred to as the NXX where N can be any number from 2 to 9 and X can be any number. At one stage, many moons ago, it was not permissible to have a 1 or a 0 as the second digit in an NXX and it was called an NNX. But that was before everyone had to dial a "1" before making a direct distance dialed long distance call, whether within their own area code or outside it. This little trick of forcing everyone to dial "1" for long distance allowed us to introduce telephone exchanges with the same three digits as area codes. For example, one of our company's numbers is 212-206-6660. The "206" elsewhere is an area code for Seattle and other parts of Washington state.

NYNEX Corporation (Pronounced "nine X." One of the seven Regional Holding Companies (RHCs) formed at Divestiture. It included New York Telephone and New England Telephone Company and sundry service and cellular radio companies. The company said its name was spelled in all capitals, thus NYNEX. The New York Times spells it Nynex. In any event, NYNEX no longer exists. The company was merged (read "acquired") by Bell Atlantic in 1997 and it now is part of Bell Atlantic. Internally, Bell Atlantic refers to the old NYNEX region as Bell Atlantic North, but we're not supposed to know that. Internally, some people Bell Atlantic people tend to look down on NYNEX people as inferior. After all, Bell Atlantic bought the NYNEX. Ain't snobbery fun?

Nyetscape Nickname for AOL's less-than-full-featured Web browser.

Nyquist Theorem In communications theory, a formula stating that two samples per cycle is sufficient to characterize an analog signal. In other words, the sampling rate must be twice the highest frequency component of the signal (i.e., sample 4 KHz analog voice channels 8000 times per second.)

O Used on switches to mean "OFF." The "ON" setting is "I."

O&M Operations and Maintenance

O.E.M. Original Equipment Manufacturer. See OEM.

O/E Optic to Electric conversion.

O/R Short for Originator/Recipient in the X.400 MHS (Message Handling System).

O/R Name A ITU (International Telegraph and Telephone Consultative Committee) term for the set of user attributes that identifies a specific MHS (Message Handling System) user. Example components are: given name, surname, ADMD (ADministrative Management Domain), country and PRMD (PRivate Management Domain), country and PRMD (PRivate Management Domain). An example of a complete O/R Address is: G=HARRY; S=NEWTON; I=HN; A=MARK400; O=PIPELINE; OUI=COMPUTER TELEPHONY; C=US.

OA Office Automation. Nobody knows what it means. But there are many consultants out there who will tell you for the right amount of money. Actually, the early "office automation" seemed to translate into word processing for the masses. When someone discovered that word processing didn't enhance office productivity, office automation fell into disrepute.

OA&M Operations, Administration and Maintenance. See OAM.

OACSU Off Air Call Set Up. A wireless telecommunications term. A method of establishing a call to a mobile telephone where the radio channel is set up at the last possible moment.

OADM Optical Add/Drop Multiplexer. Another term for ADM, which is a SONET term. See ADM and SONET.

OAI Open Application Interface. Basically one or many openings in a telephone system that lets you link a computer to that phone system and lets the computer command the phone system to answer, delay, switch, hold etc. calls. The term is also called PHI — as in PBX-Host-Interface. The term OAI was first used by PBX makers, NEC and InteCom. And now the term has become somewhat generic. Essentially every manufacturer of phone systems is evolving towards open application interfaces of their own. Many of these open interfaces are TSAPI-compatible and increasingly most are TAPI-compatible. See TAPI and TSAPI. According to Probe Research, there are really two separate "markets" for OAI or PHI:

First, there's Horizontal/Office Automation applications. These are applications that support business functions across organizational groups or industry verticals in inter- or intra-department business settings. Examples include voice mail, electronic mail, message centers, corporate telephone directories, automated screen-based dialing, personal productivity tools, conferencing, PBX feature enhancements, ANI interfaces, time clocks, 911 emergency service and compound image (data, text, image and voice) processing. I see these applications as all those useful, productivity-enhancing things I always wished my telephone systems would do if only they would let me program the thing. (Until OAI, all phone systems were totally closed architecture.)

Second, there's transaction applications. These are applications that support an actual business transaction — customer service, inbound telephone order taking, outbound telemarketing,

market research, data gathering, inventory inquiry, account time billing, credit collections, locator services with or without transfer to the local dealer. These always require access to a computer database. These applications are generally complicated, time-sensitive and customized for each installation.

A sample OAI arrangement: In early 1993, Novell and AT&T inked a deal to put telephony onto Novell LANs. The Telephony Server NetWare Loadable Module (NLM) will be the first product. It is an AT&T PC-card sitting in the Novell File server. The card connects to the ASAI (Adjunct Switch Applications Interface) port on the AT&T Definity PBX. Anyone with a PC on the network and an AT&T phone on their desk can use telephone features, such as auto-dialing, conference calling and message management (a new term for integrating voice, fax and e-mail on your desktop PC via your LAN). The Novell/AT&T deal intends to create open Application Programming Interfaces (APIs) that third party developers can work with. A Novell/AT&T example of what could be developed: A user could select names from a directory on his PC. He could tell the Definity PBX through the PC over the LAN to place a conference call to those names. At the same time, a program running under NetWare would automatically send an e-mail to the people, alerting them to the conference call and giving them the agenda. All participants would have access to both the document and the conference call simultaneously. Here are some of the names which manufacturers of PBXs and computers have coined for their open application interfaces, also called PHIs:

ACL — Applications Connectivity Link — Siemens' protocol

ACT — Applied Computer Telephony — Hewlett Packard's generic application interface to PBXs

Application Bridge — Aspect Telecommunications' ACD to host computer link

ASAI — AT&T's Adjunct Switch Application Interface

CAM — Tandem's Call Applications Manager — the name of the Tandem software interface which provides the link between a call center switch telephone switch (either a PBX or an ACD) and all Tandem NonStop (fault tolerant) computers.

CIT — Digital Equipment Corporation's Computer Integrated Telephony (works with major PBXs)

CSA — Callpath Services Architecture — IBM's Computer to PBX link

Call Frame — Harris' PBX to computer link

Callbridge — Rolm's CBX and Siemens to IBM host or non-IBM host computer link

Callpath — IBM's announced, CICS application link to IBM's CSA, available on the AS400 in April of 1991

Callpath Host — IBM and ROLM's CICS-based integrated voice and data applications platform which links to ROLM's 9751 PBX

CompuCall — Northern Telecom's DMS central office link to computer interface

CPI — Computer to PBX Interface developed by Northern Telecom and DEC

CSTA — Computer Supported Telephony Application, RSL standard from ECMA

DECags — DEC ASAI Gateway Services. Two-directional link to AT&T's Definity

DMI — AT&T's Digital Multiplexed Interface, a T-1 PBX to computer interface

HCI — Host Command Interface. Mitel's digital PBX link to DEC computer

IG — AT&T's ISDN Gateway (one direction from the switch to the host)

ITG — AT&T's Integrated Telemarketing Gateway (two directional)

ISDN/AP — Northern Telecom's host to SL1 PBX protocol, which supports NT's Meridian Link

Meridian Link — Northern Telecom's host to PBX link available on the Meridian PBX

ONA — Open Network Architecture (for telephone central offices)

PACT — Siemens' PBX and Computer Teaming, protocols between Siemens PBXs and computers

PDI — Telenova/Lexar's Predictive Dialing Interface

PHI — PBX Host Interface (a generic term coined by Probe Research)

SAI — Stratus Computer Switch Application Interface

SCAI — Switch to Computer Application Interface, one name given by Northern Telecom to PHI

SCIL — Aristacom's Switch Computer Interface Link Transaction Link

Solid State Applications Interface Bridge — Solid's State's PBX to external computer link.

STEP — Speech and Telephony Environment for Programmers; WANG's link

Transaction Link — Rockwell's link from its Galaxy ACD to an external computer

Teleos IRX-9000 — Teleos' Intelligent Call Distribution platform

VoiceFrame — Harris Digital Telephone Systems Division Platform

OAM Operations, Administration and Maintenance. This usually term refers to the specifics of managing a system or network. It is typically a group of network management functions that provide network fault indication, performance information, and data and diagnosis functions. Some switches have computers devoted to OAM.

OAM&P Operation, Administration, Maintenance and Provisioning.

OBF Ordering and Billing Forum. A forum of the Carrier Liaison Committee (CLC) of ATIS (Alliance for Telecommunications Industry Solutions), originally called the Exchange Carriers Standards Association (ECSA). ATIS is heavily involved in standards issues including interconnection and interoperability. According to ATIS, the OBF "provides a forum for telecommunications customers and providers to identify, discuss and resolve national issues which affect ordering, billing, provisioning and exchange of information about access services, other connectivity and related matters." The six standing committees of the OBF are heavily involved in the development of standard mechanisms by which CLECs (Competitive Local Exchange Carriers) and ILECs (Incumbent LECs) can interface effectively; such interface is required in the competitive environment fostered by the Telecommunications Act of 1996, and by various state initiatives. See also ATIS, CLEC and ILEC. www.atis.org

OBI Open Buying on the Internet. A standard which provides a generic set of requirements, an architecture, and a technical specification for Internet purchasing solutions in the general context of Electronic Commerce. The OBI standard was developed by the OBI Consortium; membership is open to buying and selling organizations, technology companies, financial institutions, and other interested parties. See also Electronic Commerce. www.supplyworks.com/obi

Object 1. In the context of network management, an object is a numeric value that represents some aspect of a managed device. An object identifier is a sequence of numbers separated by periods, which uniquely defines the object within a MIB. See MIB.
2. In its simplest form in computing, an object is a unit of information. It can be used much more broadly, depending on the application. In X.400, an object contains both attributes and method describing how the content is to be interpreted and/or operated on.
3. An entity or component, identifiable by the user, that may be distinguished by its properties, operations, and relationships.
4. Any piece of information, created by using a Windows-based application with OLE capabilities, that can be linked or embedded into another document.
See Object Oriented Programming.

Object Code A term usually applied to the executable, machine-readable, form of a software program. Object code is instruction code in machine language produced as the output of a compiler or a assembler. The original program, or code, is called the source Code. The term is also used to refer to an intermediate state of compilation, which is different from the initial source code and the final binary object code. See Source Code.

Object Definition Alliance A group formed to establish software standards for interactive multimedia applications that will operate uniformly over a variety of hardware, operating systems and networks.

Object Encapsulation Data and procedures may be encapsulated to produce a single object, thereby hiding complexity.

Object Handle Code that includes access control information and a pointer to the object itself.

Object Inheritance The transfer of characteristics down a hierarchy from one to another.

Object Linking And Embedding OLE. An enhancement to DDE protocol that allows you to embed or link data created in one Windows application in a document created in another application, and subsequently edit that data in the original application without leaving the compound document. See DDE and OLE.

Object Oriented File System A file system, based on object-oriented programming, that allows permanent storage of objects and associated links.

Object Oriented Programming OOP. Object oriented programming is a form of software development that models the real world through representation of "objects" or modules that contain data as well as instructions that work upon that data. These objects are the encapsulation of the attributes, relationships, and methods of software-identifiable program components. Object-oriented methodology differs from conventional software programming where functions contained in code are found within an application. Although hype about object-oriented technology is fairly recent, the approach was first introduced in the programming language Simula developed in Norway in the late 1960s. The methodology was based on the way children learn (i.e., object + action = result). The idea of object oriented programming is make the writing of complex computer software much easier. The idea is to simply combine objects together to produce a fully-written software application. The work, goes into producing all the objects. They are the building blocks. Theoretically, libraries of objects will be worth a fortune to those companies who develop them.

Object oriented programming is not easy to understand. Here's a definition of object-oriented programming from Business Week, September 30, 1991. "Software objects are chunks of programming and data that can behave like things in the real world. An object can be a business form, an insurance policy or even an automobile axle. The axle object would include data describing its physical dimensions and programming that describes how it interacts with other parts, such as wheels and struts. A system for a human resources department would have objects called employees, which would have data about each worker and the programming needed to calculate salary raises and vacation pay, sign up dependents for benefits, and make payroll deductions. Because objects have 'intelligence,' they know what they are and what they can and can't do. Thus objects can automatically carry out tasks such as calling into another computer, perhaps to update a file when an employee is promoted. The biggest advantage of objects is they can be reused in different programs. The object in an electronic-mail program that places messages in alphabetical order can also be used to alphabetize invoices. Thus, programs can be built from prefabricated, pretested building blocks in a fraction of the time it would take to build them from scratch. Programs can be upgraded by simply adding new objects."

The key concepts of object-oriented programming, according to ComputerWorld are 1. OBJECTS. The basic building block of a program is an object. Objects are software entities. They may model something physical like a person, or they may model something virtual like a checking account. Normally an object has one or more attributes (fields) that collectively define the state of the object; behavior defined by a set of methods (procedures) that can modify those attributes; and an identity that distinguishes it from all other objects. Some objects may be transient, existing temporarily during the execution of a program, i.e., only during run time. Others may be persistent, existing on some form of permanent storage (file, database, programming library) after the program finishes. 2. ENCAPSULATION. This concept refers to the hiding of most of the details of the object. Both the attributes (data structure) and the methods (procedures) are hidden. Associated with the object is a set of operations it can perform. These are not hidden. They constitute a well-defined interface — that aspect of the object that is externally visible. The point of encapsulation is to isolate the internal workings of the object so that, if they must be modified, those changes will also be isolated and not affect any part of the program. 3. MESSAGING. One object requests another object to perform its operation through messaging. The client object sends a message to the server object consisting of the identity of the server object, the name of the operation and, in some cases, optional parameters. The names of the operations are limited to those defined for that object. For example, the operations for a checking account object may be defined to be OPEN, DEBIT, CREDIT, COMPUTE INTEREST, ISSUE STATEMENT, SCHEDULE AUDIT, AND CLOSE. 4. DATA ABSTRACTION. An object is sometimes referred to as an instance of an abstract data type or class. Abstract data types are constructed using the built-in data types supported by the underlying programming language, such as integer and date.

The common characteristics (both attributes and methods) of a group of similar objects are collected to create a new data type or class. Not only is this a natural way to think about the problem domain, it is a very efficient way to write programs. Instead of individually describing several dozen instances, the programmer describes the class once. Once identified, each instance is complete with the exception of its instance variables. The instance variables are associated with each instance, i.e., each object; methods exist only with the classes. 5. INHERITANCE. Data abstraction can be carried up several levels. Classes can have super-classes and subclasses. In moving to a level of greater specificity, the application developer has the option to retain some attributes and methods of the super-class, while dropping or adding new attributes or methods. This allows greater flexibility in class definition. It is even possible in some languages to inherit from more than one parent. This is referred to as multiple inheritance.

Objective Percent Fill — Lines A telephone company term. The objective percent line fill provides for line equipments that are administratively unusable because of their use for test, assignment restrictions such as class-of-service, and for being out on assignment lists. This percent which is a percent of the total line equipment installed less those required for trunks will vary according to the entity equipment type and its service features. The objective percent line fill should not be one value applied to all entities, but is derived for each individual entity on an empirical basis by the Network Administrator. Local OTC policy may dictate specific values to be used with each type of switching system and the procedures to be followed when changing these values. See also Objective Percent Fill — Numbers.

Objective Percent Fill — Numbers A telephone company term. The objective percent number fill provides for numbers that are administratively unusable because of intercept requirements, PBX growth coin and official series, rate protection, and for other requirements such as being out on assignment lists. Care also needs to be taken, where applicable, to make allowance for CENTREX-CO and CENTREX-CU requirements. As in percent line fill, the quantity of numbers classified as administratively unusable will vary with an entity's characteristics. The objective percent number fill which is a percent of the total numbers installed less those required for trunks is derived on an empirical basis for each individual entity by the Network Administrator. Local OTC policy may dictate specific values to be used with each type of switching system and the procedures to be followed when changing these values. See also Objective Percent Fill — Lines.

Object Request Broker ORB. An object-oriented system consisting of middleware which manages message traffic between application software and computer/software platforms. As an application sends a message to an object, it need only identify the object by name. The ORB keeps track of the actual addresses of all such objects and, therefore, is able to route the request to the specific address space where the object resides in the system. CORBA (Common Object Request Broker Architecture), developed as an ORB standard by the OMG (Object Management Group), provides a standard interface for interoperability between object management systems residing on disparate platforms. SEE CORBA and OMG.

Obscenity The Supreme Court of the United States has defined obscenity as speech lacking in any social value. The Court has said that obscenity is entitled for First Amendment protection. But it may be regulated to a greater or less degree, depending on the medium — TV yes, paper no. Indecency on the Internet is under the jurisdiction of the Communications Decency Act of 1996.

OC 1. Operator Centralization.

2. Optical Carrier. A SONET optical signal. See OC-1.

OC-1 Optical Carrier level-1. OC-1 is 51.840 million bits per second. The optical counterpart of STS-1 (Synchronous Transport Signal-1), which is the fundamental signaling rate of 51.840 Mbps on which the SONET (Synchronous Optical NETwork) hierarchy is based. OC-1 provides for the direct electrical-to-optical mapping of the STS-1 signal with frame synchronous scrambling. The STS-1 signal originates as a DS-3 (T-3) electrical signal, which operates at a raw signaling rate of 44.736 Mbps and which supports 672 voice-grade digital channels of 64 Kbps; the additional bps are attributable to optical processing overhead (signaling and control data). All higher levels are direct multiples of OC-1 (e.g., OC-3 equals three times OC-1). See OC-N and SONET.

Optical Carrier level-1. The optical counterpart of STS-1 (the basic rate, 51.840 Mbps, on which Sonet is based). Direct electrical-to-optical mapping of the STS-1 signal with frame synchronous scrambling. All higher levels are direct multiples of OC-1 (i.e., OC-3 equals three times OC-1, etc.). See OC-N. See O-3, OC-12, OC-48, OC-192 and OC-256.

OC-3 Optical Carrier 3. A SONET carrier equal to three DS-3s, which is equal to 155.52 million bits per second.

OC-12 Optical Carrier level-12. SONET channel of 622.08 million bits per second. See OC-1, OC-N and SONET.

OC-48 Optical Carrier level 48. SONET channel of 2.488 thousand million bits per second (Gbps). How you calculate OC-48 (51.840 million bits per second) by 48. That gives you 2.488 thousand million bits per second. See OC-1, OC-N and SONET.

OC-192 Optical Carrier level 192. SONET channel of 9.953 thousand million bits per second (Gbps). How you calculate the capacity of an OC-192 is to multiply times 192 by 51.840 million bits per second and thus you get 9.953 thousand million bits per second — gigabits per second. See OC-1, OC-N and SONET.

OC-256 Optical Carrier level 256. The Digital Bit Rate of the OC256 is 13.271.04 thousand million (Gbps) bits per second, which will accommodate 172,032 voice circuits, which is equivalent to 7,168 DS1s (T-1s) and equivalent to 256 DS3s (T-3s). See OC-1, OC-N and SONET.

OC-N Optical Carrier — n. The optical interface designed to work with the STS-n signaling rate in a Synchronous Optical Network (SONET). Optical Carrier level N. The optical signal that results from an optical conversion of an STS-N signal. N= 1, 3, 9, 12, 18, 24, 36, 48, 192 or 256.

OCC Other Common Carrier. A long distance carrier other than AT&T. Once upon a time, the only long distance carrier was AT&T Long Lines. In 1984, the Carterphone decision allowed competition. Then there was AT&T Long Lines and the "other" category. Nowadays, these companies (including AT&T) are called IXCs — IntereXchange Carriers.

Occupancy 1. The time a circuit or a switch is in use, i.e. occupied. Occupancy is normally expressed as a percentage, occupancy represents the actual usage versus the maximum amount of time available during a 1-hour period.

2. A call center term. The percentage of the scheduled work time that employees are actually handling calls or after-call wrap-up work, as opposed to waiting for calls.

OCDD On-line Call Detail Data. Actually, its OCDD/RT (OCDD/Real Time). A security feature of AT&T's SDN (Software Defined Network), which is their term for a classic VPN (Virtual Private Network). OCDD is intended for identification and control of SDN toll fraud; in effect, it is a carrier network-based Call Accounting service for toll fraud control for large, multi-site VPN customers. OCDD/RT allows the user organization to access call usage data, including that associated with dedicated, switched or NRA (Network Remote Access). NRA activity is identified by the specific SDN card (much like a calling card or credit card) number. Inbound and outbound calling activity both can be tracked. Outbound activity from a SDN location is recorded, is posted to the SDN database, and is available to the telecommunications manager within two minutes of the event. Access to the data is via a dial-up modem connection, through a VT-100 terminal or a PC with terminal emulation capability. While OCDD relies on ANI (Automatic Number Identification), OCDD is a carrier-based network management service offering which carries an additional feature charge. See also ANI, Call Accounting, SDN and VPN.

OCE See OPEN COLLABORATION ENVIRONMENT.

OCIS On-Line Customer Information System. An MCI definition: OCIS. An MCI customer service and sales system used for on-line access of customer information stored in the MCI Customer Database and for processing information to update the Customer Database.

OCN Operating Company Number. A code used in the telephone industry to identify a telephone company. NECA-assigned company codes may be used as OCNs. NECA stands for the National Exchange Carrier Association. See NECA.

OCR See OPTICAL CHARACTER RECOGNITION or OUTGOING CALL RESTRICTION.

OCR Automatic Routing Implemented by Optus and others, this technique also assigns a number to each user up to four digits long. The sender types this number within double parentheses anywhere on the cover sheet of a fax transmission, and the LAN-networked fax system uses optical scanning technology to read the number and route the fax to the intended recipient's workstation.

OCL Office Code Location.

Octal A numbering system with the base eight.

Octathorp The character on the bottom right of your touchtone keyboard, which is also typically above the 3 on your computer keyboard. It's commonly called the pound sign, but it's also called the number sign, the crosshatch sign, the tic-tack-toe sign, the enter key, the octothorpe (also spelled octathorp) and the hash. On some phones it represents "NO." And on others it represents "YES." MCI and AT&T and some other long distance companies use it as the key for making another long distance credit card call without having to punch in your authorization code again.

Octet An eight-bit byte.

Octets received OK The number of octets (bytes) received without error.

Octopus A 25-pair cable that at one end has an amphenol connector (typical of what 1A2 phone systems were connected with) and at the other has many individual 2, 4, 6 and 8 wire connectors, typically male RJ-11s. The reason it's called an octopus is that it looks a bit like an octopus — one body and many arms. It's also called a Hydra.

Octothorpe The character on the bottom right of your touchtone keyboard, which is also typically above the 3 on your computer keyboard. It's commonly called the pound sign, but it's also called the number sign, the crosshatch sign, the tic-tack-toe sign, the enter key, the octothorpe (also spelled octathorp) and the hash. On some phones it represents "NO." And on others it represents "YES." MCI, AT&T, Premiere and some other long distance companies use it as the key for making another long distance credit card call without having to redial. www.nynews.com/octohome.htm

OCUDP Office Channel Unit Data Port. A channel bank unit used to interface between the channel bank and a customer's DDS CSU or DSU.

OCX Ole Custom Control. According to InfoWorld, OCXes are the core of Microsoft's plans for packaging prewritten code that can be downloaded from Web servers.

ODA Office Document Architecture. ISO's standard 8613-1/8 for document architecture and interchange format adopted by MAP/TOP 3.0, GOSIP, and standardized by ECMA as ECMA-101.

ODBC Open Database Connectivity is a standard put out by Microsoft that allows databases created by various relational and non-relational database programs — such as dBASE, Microsoft Access, Microsoft FoxPro, and Oracle — to be accessed by a common interface. That interface is independent of the database file format. By relying on ODBC, one can write an application that uses the same code to read records from a dBASE file or a FoxPro file. ODBC drivers use a form of SQL to carry out database operations.

Odd Ball Day A telephone company term. A day experiencing an extremely heavy traffic load caused by an event that is not expected to reoccur. The data would be excluded from the engineering historical base.

Odd Parity One of many methods for detecting errors in transmitted data. An extra bit is added to each character sent and that bit is given a value of 0 ("zero") or 1 ("one") such that the total number of ones in the character (including the parity bit) will be odd.

ODI Open Data-link Interface. A device driver standard from Novell. ODI allows you to run multiple protocols on the same network adapter card. Interconnectivity strategy that adds functionality to Novell's NetWare and network computing environments by supporting multiple protocols and drivers. ODI's benefits allow you to:
• Expand your network by using multiple protocols without adding network boards to the workstation. ODI creates a "logical network board" to send different packet types over one board and wire.
• Communicate with a variety of workstations, file servers, and mainframe computers via different protocols without rebooting your workstation.
• Configure the LAP driver for any possible hardware configuration with NET.CFG, instead of using limited SHELL.CFG choices.

ODP Open Distributed Processing.

ODSI Open Directory Services Interface. Announced by Microsoft in the summer of 1995, two years after Novell's NetWare Directory Services (NDS), it is designed to do essentially what NDS does, namely give users a way to sign on to multiple servers through one simple sign-on. The ultimate idea is that NDS or ODSI will become a single place for a networked user (i.e. one on a local area network — a LAN) to sign on to multiple networked devices, e.g. telephone systems, voice mail, electronic mail, Lotus Notes and, of course, multiple file servers (wherever they're located). Sign on in one place, get into many. A big user time-saving benefit.

OEM 1. Original Equipment Manufacturer. This term is confusing. An OEM maker might be a manufacturer of original equipment, typically from close to scratch. Intel is an OEM, though it buys raw silicon from somebody else. Compaq, Dell

or Gateway are not OEMs, since they buy their components from outside and simply assemble the components into PCs and PC devices. They might be considered value-added resellers. But if a manufacturer came to Compaq, Dell and Gateway with a 100 meg disk drive, for example, and asked them to it in in the PCs they sell, Compaq, Dell and Gateway would be considered to be OEMing the disk drive.
2. Operations Enterprise Model. MCI's name for the mass of its internal productivity tools. Covers everything from group calendars to presentation tools.

OES Operations Evaluation System. An MCI internal system, which generates daily, weekly, and monthly switch data reports; used to scan high-level switch degradation problems and to analyze specific switch problems.

OFDM Orthogonal Frequency Division Multiplexing. A next-generation modulation technique being tested for use by CATV companies to provide new services in advance of the deployment of fiber optic cable systems. Much like DMT (Discrete MultiTone), OFDM splits the datastream into multiple RF (Radio Frequency) channels, each of which is sent over a subcarrier frequency. The signal-to-noise ratio of each of those very precisely defined frequencies is carefully monitored to ensure maximum performance.

Off-Hook When the handset is lifted from its cradle it's Off-Hook. When you lift the handset of many phones, the hookswitch is moved by a spring and alerts the central office that the user wants the phone to do something like receive an incoming call or dial an outgoing call. A dial tone is a sign saying "Give me an order." The term "off-hook" originated when the early handsets were actually suspended from a metal hook on the phone. When the handset is removed from its hook or its cradle (in modern phones), it completes the electrical loop, thus signaling the central office that it wishes dial tone. Some leased line circuits work by lifting the handset, signaling the central office at the other end which rings the phone at the other end; such circuits are known as "ring down circuits." Some phones have autodialers in them. Lifting the phone signals the phone to dial that one number. An example is a phone without a dial at an airport, which automatically dials the local taxi company. All this by simply lifting the handset at one end-going "off-hook." See also Ring Down Circuit.

Off-Hook Call Announce A telephone system feature. A telephone has a speaker. If the person is speaking on the phone, another person (inside the building, on the same phone system) can "off-hook voice announce" you and can give you a message or speak with you. You will hear their voice coming through the speaker on your phone. (So may the person on the other end.) Depending on your phone, you may be able to put your hand over your telephone handset and whisper something back to the person who's "off-hook voice announcing" you. Otherwise you'll have to hang up or put the person on hold, and speak to the person on another line.

Off-Hook Queue There are two types of queuing: ON-HOOK and OFF-HOOK. In On-Hook Queuing, the user dials his number, the switch tells him the outgoing trunks are busy. The user then hangs up. The switch calls him back when a trunk becomes available. In Off-Hook dialing, the user waits with his receiver screwed into his ear until a trunk comes free and the PBX connects him to the next available trunk. Off-Hook queues are usually shorter than on-hook queues. If a trunk doesn't come free quickly in off-hook queuing, the call will often flow over onto the more expensive DDD trunks. Off-hook queuing costs more but keeps the user waiting less. On-hook queuing costs less by waiting for a cheaper trunk but can be tiring and frustrating for the workers waiting for their calls to go through.

Off-Hook Routing See OFFHOOK and RINGDOWN.

Off-Hook Voice Announce See OFF-HOOK CALL ANNOUNCE.

Off-Line Any equipment not actively connected to a phone line but which can be activated to work with that system is Off-Line. This concept also applies to computer systems. For example, a modem attached to or built into a microcomputer can be plugged permanently into a phone line. But the microcomputer can be used for word processing most of the time. While it's doing word processing, it is "off-line." When the user loads the communications software and turns on the modem, the microcomputer is now said to be "on line." Off-line computer storage is a place to put stuff which a computer cannot access "on-line," like a hard-disk. Off-line storage might be microfilm or microfiche.

Off-Line Storage Storage that is not under the control of a processing unit.

Off-Net Calling Phone calls which are carried in part on a network but are destined for a phone not on the network, i.e. some part of the conversation's journey will be over the public switched network or over someone else's network. MCI defines off-net calls as "Billable calls to non-tariffed cities. Can be MCI Off-Net or WATS Off-Net. Classified as Tier 2 for tariff purposes."

Off-Network Access Line ONAL. A circuit in a private network which allows the user to go off the private network and complete calls on the public dial network.

Off-Peak The periods of time after the business day has ended during which carriers offer discounted airtime charges. Usually, OFF-PEAK rates are available for cellular calls between 7:00 p.m. and 7:00 a.m. and on weekends and holidays, but times vary among carriers. Among landline carriers, the business day usually ends at 5 p.m., after which time residential calling builds, and that ends at 11 p.m., after which little happens, except rates drop once again until they rise at the beginning of the next business day at 8 a.m. the next morning.

Off-Premises Extension OPX. Now also called OPS for Off-Premises Station. A telephone located in a different office or building from the main phone system. The OPX is connected by a phone line dedicated to it. It acts as if it were in the same place as the main phone system and can use its full capabilities. Here's another explanation a reader sent in. OPX is the appearance of an actual telephone line (such as 212-691-8215) in two physically separate locations. For example, this line (212-691-8215) could appear and ring in my office and at my home without my home phone being a part of the office telephone system. An OPX is commonly used for answering services or for small businesses. Bell operating phone companies are sharply increasing the charge for decicated OPS lines — often by several hundred percent per year.

Off-Site Night Answer A feature of some phone systems that allows phone calls to the main line to be forwarded to another phone line after hours.

Off-The-Grid Euphemism for being off the Net. "Sorry I didn't email you last week; I was off the grid in Europe." Off-the-grid also refers to someone who lives in a rural area without running water, electricity or phone service.

Off-the-shelf When something has already been produced and is available for immediate delivery, it is said to be available "off-the-shelf." It is presumably sitting on the shelf in a warehouse waiting for your order. Sometimes called "shrinkwrapped," referring to the plastic wrap around the box.

It means you don't are not meant to need intervention by a programmer or integrator to make the software work.

Off/On Hook A modem term. Modem operations which are the equivalent of manually lifting a phone receiver (taking it off hook) and replacing it (going on hook). See OFF-HOOK.

Offered Call 1. A call that is presented to a trunk or group of trunks. See TRAFFIC ENGINEERING.

2. A call center term. A call that is received by the ACD. Offered calls are then either answered by an employee (handled) or abandoned.

Offered Load 1. The total traffic load, including load that results from retries, submitted to a system, group of servers, or the network over a circuit. See also OFFERED TRAFFIC.

2. In Frame Relay terms, the total data rate presented to the network. Note that Offered Load does not translate to carried load.

Offered Traffic The total attempts to seize a group of servers.

Office Automation Nobody knows what office automation means, though there are many consultants out there who will tell you what it means for an grand sum of money. In reality, it's a benign, imprecise term for data processing when it applies to self-focused (as opposed to customer-focused) white collar-type activities — accounting, word processing, communications, document management.

Office Characteristics A telephone company definition. The peculiarities that make one switching entity different from others. Some examples: a. Class of Service Mixture b. Trunking Configuration c. Holding Times d. Special Services e. Calling habits of subscribers Located in that entity and f. Size of office

Office Class Functional ranking of a telephone central office switch depending on transmission requirements and hierarchical relationship to other switching centers. (Awful mouthful!) There used to be five classes of switches in the U.S. telephone network hierarchy, with the one closest to the end-subscriber being a class 5 central office. But technology and marketing is changing things and, by distributing intelligence closer to the end user, it is diffusing our traditional definitions of network hierarchies and the class of switches. See CLASS 5 central office. See OFFICE CLASSIFICATION.

Office Classification Prior to divestiture, those numbers that were assigned to offices according to their hierarchical function in the U.S. public switched telephone network. The following class numbers are used:

Class 1: Regional Center (RC)
Class 2: Sectional Center (SC)
Class 3: Primary Center (PC)
Class 4: Toll Center (TC) if operators are present, or else Toll Point (TP)
Class 5: End Office (EO) (local central office)

Any one center handles traffic from one to two or more centers lower in the hierarchy. Since divestiture and with more intelligent software going into telephone switching offices, these designations have become less firm.

Office Code The first three digits of your seven-digit local telephone number. Also called NXX code. See also NXX.

Office Network A network within an office. An older term for a Local Area Network. User concern is with application sharing, file/database sharing, electronic mail, word processing and circuit switching.

Office User Interface A special shell program which sets up windows with menus of available utilities and applications.

Office Window Interface See OFFICE USER INTERFACE.

Offset The offset of a port or a memory location is the difference between the address of the specific port or memory address and the address of the first port or memory address within a contiguous group of ports or a memory window. This term is used when identifying the locations of registers located with respect to the base address of the 16 contiguous I/O ports in a PCMCIA card. It is also used when identifying the location of memory mapped registers with respect to the base address of the memory window. See also OFFSET PARABOLIC ANTENNA and OFFSET GEOMETRY.

Offset Geometry Shadow-free geometry. The feeder in the primary focus is mounted so that the effect of its shadow on the secondary radiation is negligible.

Offset Parabolic Antenna An offset antenna is a new form of satellite antenna that is taller than it is wide. According to the manufacturers, the antenna design makes for more efficient use of the antenna surface. What that means is that it captures more of the satellite signal hitting the antenna. Offset antennas are more expensive than the "normal" parabolic satellite antennas, which are called "prime focus parabolic antennas." They are more expensive because they cost more to make since they typically must be made out of one sheet of metal. Offset antennas are harder to carry around, since you can't make them out of several foldover sheets of metal.

Offsite Night Answer This mode allows incoming after-hours calls to be forwarded automatically to an off site location.

OFTEL The OFfice of TELecommunications in the United Kingdom. OFTEL is the main regulatory body over the U.K. telecommunications industry, which includes phones and cable. OFTEL was created at the time British telecommunications was sort of de-regulated in 1984 by the Telecommunications Act. On its Web site, Oftel describes itself: "OFTEL is the regulator - or "watchdog" - for the UK telecoms industry. Broadcast transmission is also part of OFTEL's remit. Our aim is for customers to get the best possible deal in terms of quality, choice and value for money. OFTEL is a government department but independent of ministerial control. It is headed by the Director General of Telecommunications, who is appointed by the Secretary of State for Trade and Industry." All telecommunications operators in the U.K. — such as BT, Mercury, local cable companies, mobile network operators and the increasing number of new operators — must have an operating licence. These set out what the operators can — or must - do or not do. For example, BT's licence contains the formula (currently RPI — 7.5%) which controls the prices of its main network services. Users of telecom services supplied by the operators also need a licence. In nearly all cases they are covered by a class licence — a licence issued to a group, not an individual, allowing certain activities. For example, the Self Provision Licence (SPL) enables customers to use telephones in their homes. What are OFTEL's functions? Under the Telecommunications Act 1984, OFTEL has a number of functions. Briefly these are:
• to ensure that licensees comply with their licence conditions
• to initiate the modification of licence conditions either by agreement with the licensee or, failing that, by reference to the Monopolies and Mergers Commission (MMC) together with the Director General of Fair Trading to enforce competition legislation - under both the Fair Trading Act 1973 and the Competition Act 1980 - in relation to telecommunications. OFTEL expects to gain wider powers under the new 1998 Competition Bill. This will bring UK law into line with

European law, and is much more flexible.
• to advise the Secretary of State for Trade and Industry on telecommunications matters and the granting of new licences
• to obtain information and arrange for publication where this would help users
• to consider complaints and enquiries made about telecommunications services or apparatus www.oftel.gov.uk

OFX Open Financial eXchange. A technical specification for the exchange of electronic financial data over the Internet for the purpose of Electronic Commerce. OFX was developed jointly by CheckFree, Intuit and Microsoft in concert with financial services and technology companies. The OFX specification is an Internet-oriented client/server system which provides security, features full data synchronization, and offers error recovery mechanisms to simplify and streamline the process by which financial services companies connect to transactional Web sites, thin clients and financial software. OFX supports a range of financial activities including consumer and small business banking, bill presentment, and investments.
According to Open Financial Exchange, Specification 1.0.2 issued on May 30, 1997, Open Financial Exchange is a broad-based framework for exchanging financial data and instructions between customers and their financial institutions. It allows institutions to connect directly to their customers without requiring an intermediary. Open Financial Exchange is an open specification that anyone can implement: any financial institution, transaction processor, software developer, or other party. It uses widely accepted open standards for data formatting (such as SGML), connectivity (such as TCP/IP and HTTP), and security (such as SSL).
Open Financial Exchange defines the request and response messages used by each financial service as well as the common framework and infrastructure to support the communication of those messages. This specification does not describe any specific product implementation.
The following principles were used in designing Open Financial Exchange:
• Broad Range of Financial Activities. Open Financial Exchange provides support for a broad range of financial activities. Open Financial Exchange 1.0.1 specifies the following services:
• Bank statement download
• Credit card statement download
• Funds transfers including recurring transfers
• Consumer payments, including recurring payments
• Business payments, including recurring payments
• Brokerage and mutual fund statement download, including transaction
history, current holdings, and balances.
• Broad Range of Financial Institutions - Open Financial Exchange supports communication with a broad range of financial institutions (FIs), including Banks, Brokerage houses, Merchants, Processors, Financial advisors, Government agencies
• Platform Independent -Open Financial Exchange can be implemented on a wide variety of front-end client devices, including those running Windows 3.1, Windows 95, Windows NT, Macintosh, or UNIX. It also supports a wide variety of Web-based environments, including those using HTML, Java, JavaScript, or ActiveX. Similarly on the back-end, Open Financial Exchange can be implemented on a wide variety of server systems, including those running UNIX, Windows NT, or OS/2.
The design of Open Financial Exchange is as a client and

server system. An end-user uses a client application to communicate with a server at a financial institution. The form of communication is requests from the client to the server and responses from the server back to the client. Open Financial Exchange uses the Internet Protocol (IP) suite to provide the communication channel between a client and a server. IP protocols are the foundation of the public Internet and a private network can also use them.

Clients use the HyperText Transport Protocol (HTTP) to communicate to an Open Financial Exchange server. The World Wide Web throughout uses the same HTTP protocol. In principle, a financial institution can use any off-the-shelf web server to implement its support for Open Financial Exchange. To communicate by means of Open Financial Exchange over the Internet, the client must establish an Internet connection. This connection can be a dial-up Point-to-Point Protocol (PPP) connection to an Internet Service Provider (ISP) or a connection over a local area network that has a gateway to the Internet. Clients use the HTTP POST command to send a request to the previously acquired Uniform Resource Locator (URL) for the desired financial institution. The URL presumably identifies a Common Gateway Interface (CGI) or other process on an FI server that can accept Open Financial Exchange requests and produce a response. www.ofx.net See also Electronic Commerce. See also Electronic Commerce, GOLD and XML. www.ofx.net.

OGM OutGoing Message. The message an answering machine delivers to someone who calls. Sample, "I'm not here. Leave a message after the beep."

OGT OutGoing trunk.

Ohnosecond That minuscule fraction of time in which you realize you've just made a gigantic mistake.

OHD Optical Hard Drive. A term pioneered by Pinnacle Micro, Irvine, CA. OHD technology, according to Pinnacle, combines the advantages of magneto-optical technology with speeds faster than most hard drives.

Ohm The practical unit of resistance. The resistance that will allow one ampere of current to pass at the electrical potential of one volt. Ohm's Law dictates the relation between the current, electromotive force and resistance in a circuit:

Amperes = Volts divided by Ohms
Volts = product of Amperes and Ohms
Ohms = Volts divided by Amperes

Ohm's Law The law that for any circuit, the electric current is directly proportional to the voltage and is inversely proportional to the resistance. The law relates current measured as Amps (I), voltage (V) and resistance measured as Ohms (R). Ohm's Law is $V = I \times R$. It can also be expressed as $I = V/R$, or $R = V/I$. Sometimes E is used instead of volts. E is short for EMF or Electro Motive Force, a synonym for voltage.

Ohms Measures of resistance. A resistance of one Ohm allows one Ampere to flow when a potential difference of one volt is applied to the resistance. See Ohm's LAW.

OHQ Off-Hook Queue. See OFF-HOOK QUEUING.

OHR Optical Handwriting Recognition. Exactly what it says. Machine reading of handwriting.

OHSA Occupational Health and Safety Act. Specifically the Williams-Steiger law passed in 1970 covering all factors relating to safety in places of employment.

OHVA Off-Hook Voice Announce. A phone system feature that permits an intercom announcement to be heard through a speaker at a phone where the handset is in use on an outside call.

OLAP On-Line Analytical Processing, also called a multidi-

mensional database. According to PC Week, these databases can slice and dice reams of data to produce meaningful results that go far beyond what can be produced using the traditional two-dimensional query and report tools that work with most relational databases. OLAP data servers are best suited to work with data warehouses. A database warehouse consolidates information from many departments within a company. This data can either be accessed quickly by users or put on an OLAP server for more thorough analysis.

According to Microsoft, OLAP refers to a class of database-management systems and client software that arranges data in multiple dimensions for high-speed analysis. Microsoft does not currently ship its own OLAP software, but is in the process of adding OLAP functionality to its SQL Server database and some client applications. In October, 1996, Microsoft announced the acquisition of OLAP technology from Panorama Software Systems in Tel Aviv. At the time, Microsoft said it intended to use the technology to add OLAP features to SQL Server and to some of its desktop tools, such as its Excel spreadsheet and the Internet Explorer browser.

OLAP Client End-user applications, that can request slices from OLAP servers and provide two- or multidimensional displays, user modifications, selections, ranking, calculations, etc.,for visualization and navigational purposes. OLAP clients can be as simple as a spreadsheet program retrieving a slice for further work by a spreadsheet-literate user or or as high functioned as a financial-modeling or sales-analysis application.

OLAP Server An OLAP server is a high-capacity, multi-user, data manipulation engine specifically designed to support and operate on multi-dimensional data structures. A multi-dimensional is arranged so that every data item is located and accessed based on intersection of the dimension members that define the item. The design of the server and the structure of the data are optimized for rapid, ad-hoc information retrieval in any orientation, as well as for fast, flexible calculation and transformation of raw data based on formulaic relationships.

OLE Object Linking and Embedding. A Microsoft Corp. software technology that allows Windows programs to exchange information and work together. For instance, a word processing document with OLE capabilities could contain a link to a chart created in a spreadsheet. Version 2.0 of OLE was released in the Windows 95 operating system. OLE means tying one piece of information in one form into a document in another form, such that a change in one piece of information will be automatically reflected in the other document. Here's an explanation from the New York Times: Business reports may contain information in a variety of formats, including text and numbers, charts, tables, images, graphics, sound and video. Typically, these are created in separate applications programs (e.g. spreadsheet, word processing, charting, database, etc.) and are merged into a single document (i.e. the report). But when the numbers used to create a chart are changed the chart must be updated as well. The executive then has to track down all the various components of the report, call up their respective applications, make the changes and stitch everything back together. OLE promises to keep track of those links and update the various components as they change. Here's an explanation from PC Magazine: Ole is a complex specification that describes the interfaces used for such tasks as embedding objects created by one application within documents created by another, performing drag-and-drop data transfers within or between applications, creating automation servers that expose their inner functionality to other programs, extending the Windows 95 shell with custom DLLs, and much more. Version 1.0 of the specification was originally created for placing objects such as Excel spreadsheets inside documents created by other applications such as Microsoft Word for Windows. OLE 2.0 greatly expanded the scope of OLE and made the original name obsolete, but the name had achieved widespread recognition and was retained. See OLE DB.

OLE DB OLE DB is a Microsoft interface for sharing multimedia data among different kinds of servers, including SQL Server, and clients. The OLE DB for OLAP spec and a software developers kit are available for downloading on Microsoft's Web (http://www.microsoft.com/data/oledb).

OLE DB for OLAP, will make it possible for OLAP products from multiple vendors to share data. OLAP refers to a class of database-management systems and client software that arranges data in multiple dimensions for high-speed analysis. Microsoft does not currently ship its own OLAP software, but is in the process of adding OLAP functionality to its SQL Server database and some client applications. In October, 1996, Microsoft announced the acquisition of OLAP technology from Panorama Software Systems in Tel Aviv. At the time, Microsoft said it intended to use the technology to add OLAP features to SQL Server and to some of its desktop tools, such as its Excel spreadsheet and the Internet Explorer browser.

OLIU Optical Line Interface Unit

OLNS Originating Line Number Screening

OLS Originating Line Screening.

OLTP OnLine Transaction Processing. A generic concept in the computer industry to cover everything from issuing airline tickets to dispensing money out of street-corner, automated teller machines.

OM Operational Measurement

OMAT Operational Measurement and Analysis Tool.

OMC Operation and Maintenance Center. Computer hardware and software assigned specifically to monitor and manage one part of a telecommunications network, usually employed in GSM (Global System for Mobile Communications).

OMC-Env Operations and Maintenance Center — Environment. An OMC dedicated to monitoring and managing the physical environment where telecommunications equipment resides.

OMC-IN Operations and Maintenance Center — Intelligent Network. A wireless telecommunications term. An OMC dedicated to monitoring and managing components of the Intelligent Network in GSM (Global System for Mobile Communications). This includes the HLR, AUC, EIR, SMSC, and VMS.

OMC-Misc Operations and Maintenance Center — Miscellaneous. An OMC that manages and monitors non-intelligent devices in a mobile telecommunications network under GSM (Global System for Mobile Communications).

OMC-R Operations and Maintenance Center — Radio. An OMC that manages and monitors the radio interface under GSM (Global System for Mobile Communications) including the BSS, BSC, and BTS.

OMC-S Operations and Maintenance Center — Switching. An OMC dedicated to monitoring and managing switches in a GSM (Global System for Mobile Communications) telecommunications network.

OMC-SS7 Operations and Maintenance Center — SS7. A wireless telecommunications term. An OMC dedicated to monitoring and managing the SS7 signaling network in a GSM (Global System for Mobile Communications) telecom-

munications network.

OMC-T Operations and Maintenance Center — Transmission. A wireless telecommunications term. An OMC dedicated to managing and monitoring transmission activities under a GSM (Global System for Mobile Communications) telecommunications network.

OMC-WAN Operations and Maintenance Center — Wide Area Network. A wireless telecommunications term. An OMC dedicated to managing and monitoring components of the Wide Area Network under GSM. The WAN links all or most devices in a GSM (Global System for Mobile Communications) telecommunications network together.

OMEGA A global radio navigation system that provides position information by measuring phase difference between signals radiated by a network of eight transmitting stations deployed worldwide. The transmitted signals time-share transmission on frequencies of 10.2, 11.05, 11.33, and 13.6 KHz. Since the transmissions are coordinated with UTC (Universal Time Coordinated), they also provide time reference. In the U.S., UTC is the responsibility of the USNO (U.S. Naval Observatory). See also UTC.

OMG Object Management Group. A group of major systems vendors involved in the definition of standards for object management. According the OMG, "a non-profit consortium dedicated to promoting the theory and practice of Object Technology for the development of distributed computing systems. OMG was formed to help reduce the complexity, lower the costs, and hasten the introduction of new software applications." The stated goal of the OMG is to "provide a common architectural framework for object-oriented applications based on widely available interface specifications." OMG membership currently stands at over 600 software vendors, developers and end users. www.omg.org

Omnidirectional A microphone with a pickup pattern essentially uniform in all directions.

Omnidirectional Antenna An antenna whose pattern is nondirectional in azimuth. The vertical pattern may be of any shape.

OMNIPoint A program established by the Network Management Forum (NMF) to speed the implementation of TMN (Telecommunications Management Network). OMNIPoint is a collaborative partnership between NMF and a number of other groups. Each OMNIPoint document release specifies a strategy and a comprehensive set of network management components, such as standards, de-facto standards, software development tools, and implementation and procurement guides. The intent of the document releases is to provide users with sufficient information to prepare and evaluate responses to RFPs and to guide suppliers in implementing network management products. See also Network Management Forum and TMN.

OMR Optical Mark Recognition. Refers to machine recognition of filled-in "bubbles" on reader service bingo cards.

On-Hook See ON-HOOK.

On-Demand Connection An ISDN BRI term. The ability to automatically suspend and resume a physical connection while "spoofing" network protocols, routing and applications. The physical connection is only brought up on-demand. This ensures that users' ISDN holding time charges are proportional to their useful holding time not the total holding time.

On-Demand Dialing ISDN cost-savings feature that sets up, transfers, and closes a call only if the ISDN device detects a data packet that is addressed to a remote network.

On-Going Maintenance A term used in the secondary

telecom equipment business. A manufacturer's guarantee of maintainability from one owner to the next. There are all types of conditions which must be met including: The machine has been under the manufacturer's maintenance contract until its time of deinstallation. There hasn't been any damage in storage or transit. And there have not been any modifications made.

On-Hook When the phone handset is resting in its cradle. The phone is not connected to any particular line. Only the bell is active, i.e. it will ring if a call comes in. On-Hook is thus the normal, inactive condition of a telephone system terminal device. See ON-HOCK DIALING and OFF-HOOK.

On-Hook Dialing Allows a caller to dial a call without lifting his handset. After dialing, the caller can listen to the progress of the call over the phone's built-in speaker. When you hear the called person answer, you can pick up the handset and speak or you can talk hands-free in the direction of your phone, if it's a speakerphone. Critical: Many phones have speakers for hands-free listening. Not all phones with speakers are speakerphones — i.e. have microphones, which allow you to speak, also.

On-Line When a device is actively connected to a PBX or a computer, it is On-Line. Terminals, PCs, modems and phones are often On-Line. More and more people are spelling it online, as one word.

On-Net Telephone calls which stay on a customer's private network, traveling by private line from beginning to end are said to be on-net. Here's MCI's definition: Billable calls to MCI-tariffed cities, including those cities reached via leased lines. Can be MCI On-Net or WATS On-Net. Classified as Tier 1 for tariff purposes.

On-Ramps A way of getting onto The Information Superhighway. Such an on-ramp could be anything from a phone line to a two-way cable TV channel. Many companies are trying to create on-ramps, however they define them.

ONA See OPEN NETWORK ARCHITECTURE.

ONAC Operations Network Administration Center.

ONAL Off Network Access Line.

ONC Open Network Computing. A distributed applications architecture developed by Sun Microsystems. Includes NFS, NIS and RPC. Now part of Solaris OS.

ONC+ Part of Sun Microsystems' ENOS Networking Solutions.

One Armed Cable Locator A backhoe. Ben Kurtzer of Qwest Communications first heard this term at AT&T, but it's used throughout the company here also. It refers to the uncanny ability of the backhoe to locate buried cable, even when you clearly aren't looking for it and, in fact, are trying to avoid it.

One Call Fax-back system that requires you to call from your fax machine to send documents on the same line after picking from menu of verbal prompts. the other fax-back system is called a two-call system. You dial from one line and tell the fax-back machine on the other end that you want it to send your requested fax to second number, i.e. one where your fax machine will receive your requested message.

One Dimensional Coding A data compression scheme for fax machines that considers each scan line as being unique without referencing it to a previous scan line. One dimensional coding operates horizontally only.

One Hop Set A set of hosts which are one hop apart in terms of internetwork protocols.

One Number Calling You give someone a phone number at which they can reach you 24 hours a day. It's typically an 800 number but it could be a local number. Here's how it

works: They call the number. A computer answers. That computer might ask the caller to touchtone in some digits. That will identify whom the caller is trying to call. The computer will then check its memory. What number is the person likely to be at, at this very moment? It will then dial that number. If the number answers. it will connect the caller. If the number doesn't answer, it will call another number, and, if it answers, it will connect the call. How does it know which numbers to call? The subscriber (i.e. the person who wants to be reached) might have given the computer several numbers — office number, cell phone numbers, home number, etc.

One Number Presence This is the consistent use of one telephone number (particularly an 800 number) across all advertising media. The long distance vendors can arrange this for both in-state and national 800 services.

One Number Systems Also called Follow Me Systems. Follow me systems and services are based on the premise that people are mobile (e.g., they move around a lot in and out of the office), and have many phone numbers or places they might be. A person could have an office number, a cellular number, a voice mail number, a home number, and a pager number. Which phone number will a caller be at? Follow me systems will "track-down" the user being called no matter where they are and connect the caller to the user. The caller need only dial a single phone number. Usually, network or local switch provided call data (or data gathered by a voice response unit) is used to identify each call as being intended for a specific user. Based on options the user has selected, the caller will hear an answering prompt customized to that user, and the one number system will then automatically attempt to locate the user at one of several locations. The tracking-down process varies considerably between follow me systems. Some systems try multiple locations at once, others will try the possible destination locations sequentially. Almost all follow me services provide the caller an exit to voice mail at various points of the call.

One Plus Bulk Restriction This is the name for a local service provided by a Northern Telecom DMS central office switch. One-Plus Bulk Restriction allows subscribers to deny or permit all one-plus (i.e. long distance) calls from their phones by dialing a special PIN (Personal Identification Number).

One Plus Per-Call Restriction This is the name for a local service provided by a Northern Telecom DMS central office switch. Subscribers can restrict one-plus toll calls from their phones by requiring that a PIN (Personal Identification Number) be dialed prior to a one-plus call. If the PIN is valid, the caller hears a second dial tone, and the one-plus number can then be dialed. When a one-plus call is attempted without the PIN or with a wrong PIN, the caller is routed to a tone or announcement and is not able to place the one-plus call.

One Plus Restriction A central office service. Telephone subscribers can now limit one-plus toll calls by selecting an authorization code that must be dialed before any one-plus call will be connected. If the code is valid, the caller hears a second dial tone, permitting the number to be dialed and connected.

One Thirtieth One thirtieth of a second is the time it takes human eyes to react to light. Project each frame of a home movie for one thirtieth of a second, and viewers, unable to distinguish separate frames, see continuous motion. Light, during the time one frame is projected, travels 6,200 miles. If you climb aboard a light beam in Chicago, you'll be in Tokyo in the blink of an eye.

One Time Programmable OTP. A term describing

memory that can be programmed to a specific value once, and thereafter cannot be changed (or can only be revised in a limited way.) OTP EPROMs are typically ordinary EPROMs that have been packaged in such as way that ultra-violet light cannot be used to erase the contents of the EPROM. Such packaging is usually less expensive.

One Way Bypass This is a term we are beginning to increasingly see as international telecommunications markets are liberalized, i.e. deregulated and competitors are allowed in. One Way Bypass is an abuse of its dominant position by an incumbent operator in an unliberalised country. The incumbent insists on receiving all traffic into the country via the accounting rate system (and so gets the benefits of very high settlement rates) but takes advantage of the liberalised system in other countries (to avoid paying similarly high outgoing settlement rates) by bypassing the accounting rate system for outgoing traffic and finding other ways to terminate its traffic. Some regulators have devised rules to try to prevent this by insisting that all operators have a proportion of total incoming traffic no larger than their proportion of total outgoing traffic on any given route (known as "proportional return"). See also whipsaw.

One Way Operation See SIMPLEX.

One Way Splitting When the attendant is connected to an outside trunk and an internal phone, pushing a button on the console allows her to speak privately with the internal extension, thus "splitting" her off from the external trunk.

One Way Trade A call center term. A schedule trade in which only one employee is working the other's schedule.

One Way Trunk A trunk between a switch (PBX) and a central office, or between central offices, where traffic originates from only one end. You can, of course, still speak and listen on the trunk. It's just like a normal two-way trunk except that a one-way trunk can only be used for dialing out or only to receive calls.

Ones Density The requirement for digital transmission lines in the public switched telephone network that eight consecutive zeros cannot exist in a digital data transmission. On a T-1 line, 0 means no voltage, no pulse. Too many zeroes and the repeaters lost count because they had no signal pulses to count. Ones density exists because repeaters and clocking devices within the network will lose timing after receiving eight zeros in a row. There are many techniques or algorithms used to insert a one after every seventh-consecutive zero. The question of how many consecutive zeros you can have on a digital line is changing. In the old days, the FCC actually said you could have 15 zeroes in a row. Now the FCC says you can have up to about 40 something (just like you) 0s without 'harming the network." For all practical purposes, seven consecutive zeros is the maximum today. See BIT STUFFING.

ONI 1. Operator Number Identification.
2. Optical Network Interface. A device which converts photons to electrons and vice versa. It's a device which converts an optical signal into an electrical signal that non-optical telecommunications transmission and switching devices can understand and vice versa.

Online Available through the computer. Online may refer to information on the hard disk, such as online documentation or online help, or a connection, through a modem, to another computer.

Online Fallback A modem feature. It allows high speed error-control modems to monitor line quality and fall back to the next lower speed if line quality degrades. Some modems fall forward as line quality improves.

Online Gaming Gambling over the Internet or on a dial-up connection.

Online Service A commercial service that gives computer users (i.e. its customers) access to a variety of online offerings such as shopping, games, and chat rooms, as well as access to the Internet. America Online and Microsoft Network (MSN) are examples of online services.

Online Transaction Processing See OLTP.

ONMS Open Network Management System. Digital Communications Associates architecture for products confirming to ISO's CMIP.

ONP 1. One Night Process.
2. Open Network Provision.

ONTC Optical Networks Technology Consortium. This research consortium was organized and coordinated by Bellcore's Optical Network Research Department with assistance from ARPA. The ONTC's research spans material and device technologies to network design and management. Other ONTC members include Case Western Reserve University, Columbia University, Hughes Research Laboratories, Northern Telecom, Bell Northern Research, Lawrence Livermore National Laboratory, Rockwell Science Center, United Technologies Photonics, and United Technologies Research Center. P ONU Optical Network Unit. A type of Access Node which converts optical signals to electrical signals and vice versa. Hybrid networks make use of ONUs to accomplish the interface between fiber optic feeder cables and metallic cables (e.g., coaxial cable or twisted pair). See also HFC.

OOCM Object-oriented call model.

OOF 1. Out Of Frame. A designation for a condition defined as either the network or the DTE equipment sensing an error in framing bits. It's declared when 2 of 4 or 2 of 5 framing bits are missed (the OOF condition existing for 2.5 seconds generally creates a local Red Alarm). See LOF.
2. Out Of Franchise. Often used by a local phone company to refer to business and other activities that are outside its local franchise boundaries. The concept often is that these OOF activies are often subject to fewer government regulations. Also referred to as OOR, or Out Of Region.

OOP Object-oriented programming (OOP) is simply combining objects to produce a finished software application. The real work is developing the objects that carry out tasks within the program. Programs can be produced by recycling objects from other programs and upgraded by replacing old objects.

OOPS Open Outsourcing Policy Services. A specification from the IETF for policy-based networking. OOPS defines a protocol for exchanging QoS (Quality of Service) policy information and policy-based decisions between a RSVP-capable router and a policy server. The data are transferred between routers and servers via TCP/IP. Proposed for use in large and complex networks, OOPS is intended to allow the prioritization of traffic based on policy-level parameters established by the network administrator. Those parameters are stored in a policy server, which is queried by the client reouters. This centralized approach avoids the requirement for programming each router. See also POLICY-BASED NETWORKING, QoS, and RSVP.

OOR See OOF.

OPAC Outside Plant Access Cabinet.

Open Means the circuit is not complete or that the fiber is broken. There is a break in it. A break does not necessarily mean it's malfunctioning, only that it's been turned off.

Open Air Transmission Referring to a transmission type or associated equipment, that uses no physical communications medium other than air. Most radio communications systems, including microwave, shortwave and FM radio and infrared are open-air (also called "through-the-air") transmission systems. Open air transmission is not to be confused with the OpenAir specification, an ad hoc standard promoted by the Wireless LAN Interoperability Forum (WLI Forum). See also OpenAir.

Open Application Interface See OAI.

Open Architecture See OAI.

Open Circuit A circuit is not complete. There is no complete path for current flow. Electrical current cannot flow in the circuit. In electrical engineering, a loop or path that contains an infinite impedance. In communications, a circuit available for use.

Open Collaboration Environment O.C.E. Apple's Open Collaboration Environment extends the Macintosh operating system to provide a platform for the integration of fax, voicemail, electronic mail, directories, telephony and agents. From a user's perspective, according to Apple, O.C.E.'s functionality will be seen through:
• System-wide directory services, including a desktop directory browser and electronic business cards.
• A compound mailbox for mail from all sources — fax, voice mail, e-mail, pager, etc.
• Application integration, with all applications having the ability to send documents.

Open Ended Access Term used to describe the ability to terminate a call to any public network destination. For example, on the open end of a foreign exchange circuit, the customer may call any number in the local calling area without being charged for long distance service.

Open Financial Exchange See OFX for a full explanation.

Open Ground, Neutral Or Hot In AC electrical power, an "open" is a break, an extremely loose or an unconnected wire in any electrical path. Dangers of an "open" GROUND include serious shock and fire hazard and are life-threatening. Caution: an "open" GROUND will not stop equipment from operating. However it will stop a fuse or circuit breaker from operating should a ground fault occur.

Open Line Dealing This is a term used by British Telecom in its turrets. The system electronically recreates the original "pit" share trading environment in the stock exchange in which everybody could talk to everybody else. In open line dealing, the trader has the ability to program a number of parties onto speakers. Full duplex speech is achieved using the associated microphones and all lines receive a simultaneous broadcast.

Open Loop System A control system which does not use feedback to determine its output.

Open Network Architecture ONA. The "network" refers to the public switched network. The FCC wants to encourage companies to get into the value-added telecom business — voice mail, electronic mail, shopping by phone, etc. These companies may be called "value added providers" or "enhanced service providers." The FCC's idea of encouraging companies to add value to phone lines is a nice idea, except that all these companies will rely on phone lines provided by local phone companies who also want to be in value-added telecommunications business. And the local phone companies would prefer that the business be a monopoly (easier to manager, higher prices, etc.) If the FCC is to allow the Bell operating companies and other local phone companies into the value-added business and encourage others in, then it must figure a way the Bell operating companies don't organize things so they have an unfair advantage. The FCC's

latest idea is called ONA — Open Network Architecture. Under this concept, the telephone companies are obliged to provide a certain class of service to their own internal value-added divisions and the SAME class of service to nonaffiliated (i.e. outside) valued-added companies. The concept is that the phone company's architecture is to be "open" and that everyone and anyone can gain access to it on equal footing. ONA is only a concept at present and still needs some rigorous defining. There is not much pressure from outside entrepreneurial companies for ONA access. Thus ONA at the FCC and elsewhere drags its feet.

The March 18, 1991 issue of Telephony Magazine said that as conceptualized by the FCC, ONA "is the overall design of a carrier's basic network facilities and services to permit all users of the basic network, including the enhanced services operations of a carrier and its competitors, to interconnect to specific basic network functions on an unbundled and "equal access" "basis." Selected regional Bell holding companies ONA services would include

1. Basic Serving arrangements (BSAs):

A BSA is the basic interconnection access arrangement which offers a customer access to the public network and provides for the selection of available Basic Service Elements (BSEs, see below). Basic serving arrangements are:
• Switched, line side connection
• Switched, trunk side connection
• Dedicated, metallic dedicated

2. Basic service elements (BSEs):

Basic Service Elements are optional basic network functions that are not required for an ESP to have a BSA, but when combined with BSEs can offer additional features and services. Most BSEs allow an ESP to offer enhanced services to their customers in a more flexible manner. BSEs fall into four general categories: Switching, where call routing, call management and processing are required; Signaling, for applications like remote alarm monitoring and meter reading; Transmission, where dedicated bandwidth or bit rate is allocated to a customer application; and Network Management, where a customer is given the ability to monitor network performance and reallocate certain capabilities. The selection of available BSEs is an ongoing process, with new arrangements being developed many times in response to customer demands. ANI, Audiotext "Dial-It" Services, and Message Waiting Notification are all examples of BSEs, which also include:
• Multiline hunt group
• Uniform call distribution
• Central office announcements
• Three-way call transfer

3. Complementary network services (CNSs):

CNSs are basic services associated with end user's lines that make it easier for ESPs (Enhanced Service Providers) to offer enhanced services. Some examples of CNSs include
• Call forwarding Busy/Don't Answer
• Three way calling
• Call waiting
• Virtual dial tone
• Message waiting/indicator
• Speed calling
• Warm line

4. Ancillary services.

Ancillary Services. These are options available to an ESP which support and complement the provision of enhanced services. Examples of ancillary services are protocol conversion, and DID with third number billing inhibited.

In June, 1991, according to Communications Week, the FCC established a tariff structure that will determine how much the telcos can charge enhanced-services providers — and ultimately how much end users will have to pay for those services. FCC Chairman Alfred Sikes called the agency's action "one of our most pivotal steps" in the implementation of ONA. The FCC's idea is that ONA tariffs will be filed with the FCC in November, 1991 and will take effect February 1, 1992. They didn't. At the time I wrote this, the FCC had a new chairman, with a different agenda.

See also COMPUCALL, ENHANCED SERVICES, OAI, and OPEN APPLICATION INTERFACE.

Open Office Typically, this layout places the manager's desk in the foreground, within view of all other desks and enclosed spaces. The first open office appeared around 1960 when it was introduced in Germany as the "office landscape."

Open Skies When a government or government agency allows virtually anyone to sell satellite telecommunications service, you have "OPEN SKIES." The United States has an Open Skies satellite policy. Virtually anyone can apply to launch and operate a telecommunications satellite and, with a high degree of certainty, you'll be granted your wish. The European community is just now (summer, 1991) beginning to think of opening its skies. They have gone one small step — namely allowing anyone to buy and operate a receive-only satellite earth station not connected to the public network.

Open Software Foundation OSF. An industry organization founded in 1988 to deliver technology innovations in all areas of open computer systems, including interoperability, scalability, portability and usability. The OSF is an international coalition of vendors and users in industry, government and academia that work to provide technology solutions for a distributed computing environment. In February 1996, the OSF consolidated with X/Open Company Ltd. to form The Open Group. (www.osf.org) See The Open Group (www.opengroup.org)

Open Solutions Developers A Dialogic term for companies which develop and sell end-user voice processing applications. We'd probably call these people value added resellers (VARs).

Open System Interconnect OSI. An ISO publication that defines seven independent layers of communication protocols. Each layer enhances the communication services of the layer just below it and shields the layer above it from the implementation details of the lower layer. In theory, this allows communication systems to be built from independently developed layers. See OSI.

Open Systems Open systems refers to that best of all possible worlds, where everyone would comply with a set of hardware and software standards. You could buy a server from company A, a client from company B, a networking system from companies C, D or E, and applications software from companies F-Z, and everything would work harmoniously. In real life, some "open" systems are more open than others. Some companies talk, without embarrassment, of their "proprietary open systems." Some things are open and closed. PBXs have many "open" ports on which you can attach things. But there are no PBXs I know of which have open backlanes, meaning you (or someone else) can design a board and plug it into the PBX's backplane. That's closed. The concept of open systems has been more popular in the computer industry.

Open Systems Interconnection See OSI.

Open Toolkit Developers A Dialogic term for outside developers (outside of Dialogic) who provide applications generators that simply application development and work in a variety of operating systems — including MS-DOS, UNIX, OS/2, Windows, etc.

Open Wire A transmission facility typically consisting of pairs of bare (uninsulated) conductors supported on insulators which are mounted on poles to form an aerial (above ground) pole line. Most basic of all practical types of transmission media. Open wire may be used in both communication and power.

OpenAir A wireless LAN specification promoted by the Wireless LAN Interoperability Forum (WLI Forum), Open Air is recognized by many as an ad hoc standard. Actually an interface specification, OpenAir describes the physical and MAC layer (Layers 1 and 2 of the CSI Reference Model) interface used by WLI Forum products. Based on the RangeLAN2 protocol developed by Proxim, the specification employs Frequency Hopping Spread Spectrum (FHSS) technology in the 2.4 GHz portion of the unlicensed Industrial, Scientific and Medical (ISM) radio band. The specification includes Access Points (APs) in the form of protocol-independent wireless bridges which connect to a standard wired backbone, such as IEEE 802.3 Ethernet. Client workstations are equipped with compatible client adapters, completing the wireless circuit. Security is provided at several levels. First, the FHSS implementation includes 79 frequencies, each of which can be used in a pseudo-random fashion in any of 15 hopping sequences. Second, every station is programmed with a 20-bit security code number which must be validated by the network during the process of circuit initiation and network synchronization. Third, the security code is encrypted. See also FHSS, ISM and WLI Forum.

OpenView Hewlett-Packard's suite of a network-management application, a server platform, and support services. OpenView is based on HP-UX, which complies with AT&T's Unix system.

Operand That which is being operated on. An operand is usually identified by the address part of an instruction.

Operating Environment Referring to the combination of (usually IBM) host software that includes operating system, telecommunications access method, database software and user applications. Some common operating environments include MVS/CICS and MVS/TSO.

Operating Income There are two ways American telecom companies traditionally report their earnings — operating income and net income. Operating income purports to show how much money the company earned from running its basic business — that of making and selling its primary products and services. Operating income starts with total revenues (i.e. sales from products and services) and then deducts the cost of delivering those sales, i.e. raw materials, components, supplies, packagings, etc. In short, all direct costs. Next deduction is sales and marketing expenses. The final result gives you "operating income." If you then deduct interest expense (or add in interest income), goodwill amortization, "other (unusual) income or losses" (e.g. profit or loss on the sale of a building) and state and federal income taxes, you end up with a number that is referred to as Net Income. See GAAP.

Operating System A software program which manages the basic operations of a computer system. It figures how the computer main memory will be apportioned, how and in what order it will handle tasks assigned to it, how it will manage the flow of information into and out of the main processor, how it will get material to the printer for printing, to the screen for viewing, how it will receive information from the keyboard, etc. In short, the operating system handles the computer's basic housekeeping. MS-DOS, UNIX, PICK, etc. are operating systems.

Operating Time The time required for seizing the line, dialing the call and waiting for the connection to be established.

Operation Code The command part of a machine instruction.

Operational Data Integrator ODI. An MCI term: Combines data from the Customer Information Manager, the MCI Information Manager, and the Management Information Systems to build databases containing network information.

Operational Grammar A voice recognition term. A vocabulary structure where certain word sets activate other word sets.

Operational Load The total power requirements for communication facilities.

Operational Security OPSEC. A thorough on-site examination of an operation or activity to determine if there are vulnerabilities that would permit adversaries and exploitation of critical information during the planning, preparation, execution, and post-execution phases of any operation or activity. A Federal Government definition.

Operational Service Period A performance measurement period, or succession of performance measurement periods, during which a telecommunication service remains in an operational service state. An operational service period begins at the beginning of the performance measurement period in which the telecommunications service enters the operational service state, and ends at the beginning of the performance measurement period in which the telecommunications service leaves the operational service state.

Operations Applications OAs. A telephone company AIN term. A class of functions to provide provisioning, administration, maintenance, and management capabilities for network elements, network systems, software, and services (e.g., assessment of service quality over the group of systems and software that support the service). These functions usually reside in Operations Systems but may be assigned to network elements or network systems.

Operations Domains A telephone company AIN term. The set of operations functions residing in network elements, network systems, Operations Applications, and associated interfaces necessary to accomplish memory administration, network surveillance, network testing, network traffic management, and network data collection.

Operations Evaluation System OES. An MCI internal system, which generates daily, weekly, and monthly switch data reports; used to scan high-level switch degradation problems and to analyze specific switch problems.

Operations Support System See OSS.

Operator 1. Employee of telephone company, or an individual business or institution, who aids in the completion of phone calls. Traditionally a woman's occupation, now increasingly the role of men and machines. In some countries, like Germany, the phone company doesn't have operators.

2. In PCs, an operator is a symbol that represents a mathematical action, such as a +-/ and * (plus, minus, divide and multiply). Operators can also be words like AND, OR and NOT.

Operator Assisted A phone call placed with the assistance of the carrier's operator. You pay more when you use an operator.

Operator Console Same as attendant console. See ATTENDANT CONSOLE.

Operator Services OS. Any of a variety of telephone services which need the assistance of an operator or an automated "operator" (i.e. using interactive voice response technology and speech recognition). Such services include collect calls, third party billed calls and person-to-person calls. The responsibility for operator services used to be very straightforward. The LECs had their own operators and so did the IXCs. That's no longer necessarily true. Many IXCs use an Alternative Operator Services (AOS) service bureau, which essentially is a huge incoming call center. The AOSs typically provide operator services for a large number of small long distance companies, many of which are resellers and aggregators who don't own their own facilities. That is to say that they own no switches or transmission facilities — they simply resell long distance services which they buy at bulk wholesale rates from the larger carriers. Since the MFJ (Modified Final Judgment) broke up the Bell System in 1984, operator services have gotten more confusing still. The RBOCs can, and generally do, provide their own operator services for both local and IntraLATA calls within their home states. The Telecom Act of 1996 added to the confusion. Where the RBOCs operate outside their home states (e.g., NYNEX in Arizona), they may provide operator services directly or they may use an AOS. The larger CLECs (Competitive LECs), many of which also are IXCs (e.g., AT&T, MCIWorldcom, and Sprint), also generally provide their own operator services; the smaller CLECs generally use an AOS. As a rule, operator services are much faster today, thanks to automation. Also as a rule, operator services today are much less personal, much, much more expensive, and much less accurate. Accuracy is a problem particularly in the case of Directory Assistance. See also Directory Assistance.

Operator Workstation OWS. The OWS is an advanced voice and data workstation (typically a PC running a flavor of Windows) that streamlines and automates many of the routine tasks of an operator, thus reducing the amount of time needed for call handling. Color screens, pop-up windows, one-touch commands, and database look-up are some of the features that simplify the operator's tasks and speed call processing.

OPRE Operations Order Review

OPS 1. Operator Services

2. Off-Premises Station. See OFF PREMISES EXTENSIONS.

3. Open Profiling Standard. A recent (1997) privacy initiative intended to provide guidelines for collecting information on users accessing Web sites. OPS software will allow the user to fill out a personal profile, which will be stored in a file residing on a client computer. The OPS file will conform to the vCard specification which is managed by the Internet Mail Consortium. Much like an electronic version of a business card, the OPS file will store information such as name, company, address, telephone number, fax number, and e-mail address. The OPS information will be shared only with the consent of the user — very much unlike a cookie, which is embedded by the host of the Web site in a file on the client computer. It's a privacy issue. See also COOKIE and vCARD.

Optical Amplifier A device to amplify an optical signal without converting the signal from optical to electrical back again to optical energy. The two most common optical amplifiers are erbium-doped fiber amplifiers (EDFAs), which amplify with a laser pump diode and a section of erbium-doped fiber, and semiconductor laser amplifiers.

Optical Attenuator In optical communications, a device used to reduce the intensity of the optical signal. In some optical attenuators used in optical fiber systems, the amount of attenuation depends on the modal distribution of the optical signal.

Optical Blank A casting consisting of an optical material molded into the desired geometry for grinding; polishing; or, in the case of some optical fiber manufacturing processes, drawing to the final optical/mechanical specifications.

Optical Cavity A region bounded by two or more cavity surfaces, referred to as mirrors or cavity mirrors, whose elements are aligned to provide multiple reflections of lightwaves. The resonator in a laser is an optical cavity.

Optical Character Recognition OCR. Reading data using a machine that visually scans the characters in a document and converts that data into standard form which can be stored on conventional magnetic medium, e.g. floppy or hard disk. OCR is not 100% accurate and so requires manual cleanup or fuzzy logic to make your OCRed relatively clean and thus retrievable through a search engine. See OPTICAL SCANNER.

Optical Combiner A passive device in which power from several input fibers is distributed among the smaller number (one or more) of output fibers.

Optical Computer A computer that uses photons, not electrons as in today's old-fashioned computers. Scientists think photon computers and photon switches could be a thousand times faster than present computers and switches, and, of course, totally impervious to electromagnetic interference.

Optical Connectors Connectors designed to terminate and connect either single or multiple optical fibers. Optical connectors are used to connect fiber cable to equipment and interconnect cables.

Optical Cross-Connect Panel A cross-connect unit used for circuit administration and built from modular cabinets. It provides for the connection of individual optical fibers with optical fiber patch cords.

Optical Disk Peripheral storage disk for programs and information. Optical disks are emerging as computer storage devices because of their tremendous storage capacities in comparison to magnetic disk. Optical disks are WORM — which stands for Write Once, Read Many Times, meaning you write once to the disk, and read that information many times. But with today's technology, you can't erase the information on the optical disk. It is not uncommon to be able to store 600 megabytes of information on an optical disk the size and shape of one you currently buy recorded music on. See OPTICAL STORAGE DEVICE.

Optical Fault Finder A device which measures power and distance characteristcs in fiber-optic cable. Optical Fault Finders are more expensive than Optical Power Meters but have the additional ability to make distance measurements and locate tiny fiber breaks called Microbends. See Optical Power Meter.

Optical Fiber Any filament made of dielectric materials, that guides light, whether or not it is used to transmit signals. Optical fiber is an almost ideal transmission medium. It has these advantages:

• Transmission losses are very small.

• Bandwidth is greater than any other transmission medium we know of today. And we have no idea what the theoretical bandwidth of a strand of fiber might be.

• Fiber is immune to electromagnetic interference. This means it can operate in hostile or hazardous environments, like on the factory floor, in elevators shafts, on battleships, etc.

• Fiber does not radiate. You can't place a receiver next to it

and figure out what's going on the fiber, as you can with cable.

• You can put many strands of fiber carrying much information in the same bundle and they won't interfere with each other, i.e. there won't be any significant cross-talk between the adjacent fibers.

• Its basic raw material, silica (sand) is the second most abundant element on earth. Actually, optical fiber is made from a synthetic glass produced by burning two chemicals together; the resulting soot collects on a bait rod or in a bait tube, and is baked until all moisture is removed. this is called the blank which fiber is drawn from.

The insulating properties of the glass fiber produces the only major disadvantage. Metallic conductors need to be included to power repeater amplifiers needed over long fiber runs. However, these can form the strength members that are necessary to aid the laying of fiber cables. Typical fiber applications range from use in local area network sections, where there is a high degree of electrical interference, to trans-oceanic telecommunications cables.

An optical fiber not carrying signals is typically called a "dark fiber." There are two types of optical fiber: Single Mode and Multimode. In single mode fiber, light can only take a single path through a core that measures about 10 microns in diameter. A micron is one millionth of a meter. Multimode fibers have thicker cores — typically 50 to 200 microns. Single mode fiber is more efficient. It offers low dispersion, travels great distances without repeaters and has enormous information-carrying capacity. The relatively large core of multimode fiber lightguide allows light pulses to zig-zag along many different paths. It's also ideal for light sources larger than lasers, such as LEDs (Light Emitting Diodes). Multimode fiber is not the preferred method of optical telecommunications any longer. See also FIBER, OPTICAL FIBER CABLE and OPTICAL SPECTRUM.

Optical Fiber Cable A transmission medium consisting of a core of glass or plastic surrounded by a protective cladding, strengthening material, and outer jacket. Signals are transmitted as light pulses, introduced into the fiber by light transmitter (either a laser or light emitting diode). Low data loss, high-speed transmission, large bandwidth, small physical size, light weight, and freedom from electromagnetic interference and grounding problems are some of the advantages offered by optical fiber cable. There are five common types: single, dual, quad, stranded, and ribbon. See OPTICAL FIBER.

Optical Fiber Duplex Adapter Mechanical media termination device designed to align and join two duplex connectors.

Optical Fiber Duplex Connection Mated assembly of two duplex connectors and a duplex adapter.

Optical Fiber Duplex Connector Mechanical media termination device designed to transfer optical power between two pairs of optical fibers.

Optical Fiber Facility Transmission system which uses glass fibers as the transmission medium. See optical fiber.

Optical Fiber Patch Panel One way to terminate fiber optic cable. Fiber patch panels have a fiber splice tray with pigtails. Pigtails are fiber connectors with a piece of fiber optic connected so fiber from within a cable can be easily spliced to it. Connectors are on the front of the fiber patch panel.

Optical Fiber Preform Optical fiber material from which an optical fiber is made, usually by drawing or rolling.

Optical Fiber Ribbon A cable of optical fibers laminated in a flat plastic strip.

Optical Fiber Splice A permanent joint whose purpose is

to couple optical power between two fibers.

Optical Fibre British spelling. In an optical fibre transmission system, the data is carried by pulses of light along glass fibres. This method of transmission has a much higher bandwidth than copper cables and is less subject to distortion and interference. It is safer than copper because it provides electrical isolation and as it carries no current can be used in flammable areas. See OPTICAL FIBER for a longer explanation.

Optical Interconnection Panel An interconnection unit used for circuit administration and built from modular cabinets. It provides interconnection for individual optical fibers. Unlike the optical cross-connect panel, the interconnection panel does not use patch cords.

Optical Power Meter A device which measures the light signal transmitted through fiber-optic cable.

Optical Receiver An optoelectric circuit that converts an incoming optical signal to an electrical signal.

Optical Repeater In an optical fiber communication system, an optoelectronic device or module that receives a signal, amplifies it (or, in the case of a digital signal, reshapes, retimes, or otherwise reconstructs it) and retransmits it. See also OPTICAL AMPLIFIER.

Optical Scanner A hardware device that recognizes images on paper, film and other media and converts them into digital form which can be stored in a conventional computer readable magnetic medium, such as floppy or hard disk. Optical scanners are getting better and better but are still not perfect. See OCR.

Optical Spectrum Generally, the electromagnetic spectrum within the wavelength region extending from the vacuum ultraviolet at 1 nm to the far infrared at 0.1 mm. The term was originally applied to that region of the electromagnetic spectrum visible to the normal human eye, but is now considered to include all wavelengths between the shortest wavelengths of radio and the longest of X-rays. See OPTICAL FIBER.

Optical Storage Device Optical storage devices use a source of coherent light — usually a semiconductor laser — to read and write the data. There are three big advantages to using a laser — size, safety and portability. Because you can focus a laser into approximately one micron in size — a far smaller area for encoding a bit of data than conventional drives — you can fit more data in.

Optical media are also more stable than metal-oxide disks. They aren't affected by light, normal temperatures or electromagnetic fields. (You can put through as many airport x-ray machines as you wish.) And best, the read/write head doesn't get as close to the recording medium as it does in conventional disk drives. Optical drives are interchangeable, also. You can remove them and store them. That makes them great for archiving.

Optical Time Domain Reflectometer OTDR. A device that measures distance to a reflection surface by measuring the time it takes for a lightwave pulse to reflect from the surface. Reflection surfaces include the ends of cables and breaks in fiber. The reflectometer measures the ratio of incident and reflected light power. By using this device you can figure precisely where a fiber optic link is broken. This device operates like Radar. It sends a light pulse down the cable and waits for it to return. It measures the time taken and calculates the distance based on the speed of light through the fiber optic cable. The reflectometer usually also displays the reflected waves on a time axis for precise reading of, e.g., the leading edges of the transmitted and reflected waves. The reflectometer is also capable of launching a light pulse into the fiber optic transmission medium and measuring the time

required for its reflection to return by backscattering or end reflection, thus indicating the continuity, crack, fracture, break, or other anisotropic features of the medium.

Optical Transmitter An optoelectric circuit that converts an electrical signal to an optical signal.

Optical Virtual Tributary Group OVTG contains four DS-1 signals packaged into an optical SONET virtual tributary, or VT.

Optical Waveguide Technically, any structure that can guide light. Sometimes used as a synonym for optical fiber, it also can apply to planar light waveguides.

Optical Waveguide Connector A device whose purpose is to transfer optical power between two optical waveguides or bundles, and that is designed to be connected and disconnected repeatedly.

Optically Active Material A material that can rotate the polarization of light that passes through it. An optically active material exhibits different refractive indices for left and right circular polarizations (circular birefringence).

Opto-Electric Transducer A device which converts electrical energy to optical energy and vice versa. Used as transmitters and receivers in fiber optic communications systems.

Opto-Electronics The range of materials and devices associated with fiber optic and infrared transmission systems. As there are no practical optical computers, all information originates as an electrical signal. Therefore, opto-electronic light sources convert the electrical signal generated to an optical signal which is transmitted to the receiving light detector for reverse conversion back to an electrical signal. In fiber optic systems, the light sources are in the form of either LEDs (Light-Emitting Diodes) or laser diodes. The light detectors are in the form of either PINs (Photo-INtrinsic diodes or APDs (Avalanche PhotoDiodes). Laser diodes and APDs are matched in high-speed, long-haul networks. LEDs commonly are used in fiber optic LANs and other short-haul, relatively low-speed applications. Optical repeaters, which also are opto-electronic devices, repeat the optical signal. Such repeaters accept the optical signal through a light detector, convert it to electrical energy, boost the signal and clean it up, and reconvert it to an optical signal for insertion to the next link of the optical fiber system. Also, each of these functions requires electrical energy to operate and depends on electronic devices to sense and control this energy. See also Optical Amplifiers, a new breed of amplifiers that doesn't take the light signal back to electrical energy.

Optoelectric Transducer Electronic components that turn light energy into electrical energy and electrical energy into light energy.

Optronics Opt(o-Elect)ronics. See Opto-Electronics.

OPX Off Premise Extension. An extension or phone terminating in a location other than the location of the PBX. The station uses a line circuit out of the PBX. OPX is commonly used to provide a company executive with an extension off the PBX in his home. This way he can pretend to be at the office when he's really at home. He can also make toll calls and have them easily charged to the office.

Orange Book The common name for the U.S. Department of Defense's Trusted Computer System Evaluation Criteria (TSEC).

ORB 1. Office Repeater Bay.
2. Object Request Broker. See OBJECT REQUEST BROKER and CORBA.

Order Entry A voice processing application which allows someone with a touchtone phone to buy something, i.e. enter their order.

Order Wire 1. A circuit used by telephone personnel for fixing, installing and removing phone lines.

2. Equipment and the circuit providing a telephone company with the means to establish voice contact between central office and carrier repeater locations.

3. A SONET/SDH term for a connection request, and consisting of one octet contained within the SOH (Section Overhead).

Order Wire Circuit A voice or data circuit used by telephone company technical control and maintenance personnel for the coordination and control action relating to activation, deactivation, change, rerouting, reporting and maintenance of communication systems and services.

Originate Mode The "originate mode" sets the modem to begin a data phone call — i.e. dial the phone, listen for a carrier tone from a remote modem and connect to that modem. The modem at the receiving end must be set to "Answer" mode. In any asynchronous data conversation, one side must be set to "Originate" and the other to "Answer." Such settings are usually made in software.

Originate/Answer The two modes of operation for a modem. Originate and answer states define the frequencies used to transmit and receive. In a two-way communication system, one modem must be set to originate and the other to answer.

Originating Restriction A phone line with this restriction cannot place calls at any time. Calls directed to the phone, however, will be completed normally.

Origination A call that is placed by the mobile subscriber, calling either a land-line circuit or another mobile subscriber.

Origination Cablecasting Programming over which a cable television system operator exercises editorial control. This term includes programming produced by the operator; Non-broadcast local programming produced by other entities and carried voluntarily by the system. Example: PRISM; regional news channels; Satellite-delivered non-broadcast programming carried voluntarily by the system, such as HBO, ESPN, CNN, C-SPAN, QVC, etc.

This term does not include programming over which the operator does not exercise editorial control, including any broadcast signal, including satellite-delivered broadcast "superstations" (WGN-TV, WWOR, etc.); Any access channel designated by franchise for public, educational, or governmental use; Leased-access channels.

The cable system operator is required by Section 76.225c of the FCC Rules to maintain records, in the PIF, to verify compliance with rules governing commercial matter in children's programming carried on origination-cablecasting channels. See PIF.

Originator The user that is the ultimate source of a message or probe.

ORM Optically Remote Module. A type of switching module made by AT&T which connects directly to the 5ESS switch communications module via optical fibers.

Orphan A Windows NT term. A member of a mirror set or a stripe set with parity that has failed in a severe manner, such as a loss of power or a complete head crash. When this happens, the fault-tolerance driver determines that it can no longer use the orphaned member and directs all new reads and writes to the remaining members of the fault-tolerant volume.

OS 1. Outage Seconds.

2. Operating System, as in MS-DOS (Microsoft Disk Operating System), Windows NT or OS/2. See OPERATING SYSTEM.

3. Operator Services. See OPERATOR SERVICES.

4. Operations System. Includes SCOTS, FMAS, etc.

OS/2 Operating System/2. An operating system originally developed by IBM and Microsoft for use with Intel's microprocessors and for use with IBM personal system/2 personal computers. Now OS/2 is the prime responsibility of IBM and it will run on many PCs, including those using the Intel family of PC microprocessors. OS/2 is a multitasking operating system. This means many programs can run at the same time. See OS/2 EXTENDED EDITION.

OS/2 2.0 A 32-bit version of the IBM's OS/2 operating system. Apple Macintosh's operating system is 32-bit.

Osborne Effect Once there was a personal computer company called Osborne Computer Company. One day, the president announced a revolutionary new computer. It was so good not one of his dealers wanted to (or could) sell the existing product and they sent all their inventory back. Meantime, it was six months before the company could deliver the new product. But without any sales in the meantime, it had no money and Osborne went broke. There is a lesson here for companies who are attempting to manage transition between old and new product lines. Be careful, or suffer the horrible consequences of The Osborne Effect.

Oscillator 1. A device for generating an analog test signal.

2. Electronic circuit that creates a single frequency signal.

Oscilloscope Electronic testing device that can display wave forms and other information on a TV-screen-like cathode ray tube. A basic fixture in sci-fi movies.

OSF Open Software Foundation. An industry organization founded in 1988 to deliver technology innovations in all areas of open computer systems, including interoperability, scalability, portability and usability. The OSF is an international coalition of vendors and users in industry, government and academia that work to provide technology solutions for a distributed computing environment. In February 1996, the OSF consolidated with X/Open Company Ltc. to form The Open Group. (www.osf.org) See The Open Group (www.opengroup.org)

OSF/1 Version 1 of the Open Software Foundation's Unix-based operating system

OSI Open Systems Interconnection. The only internationally accepted framework of standards for communication between different systems made by different vendors. ISO's goal is to create an open systems networking environment where any vendor's computer system, connected to any network, can freely share data with any other computer system on that network or a linked network. OSI was developed by the International Standards Organization. Most of the dominant communications protocols used today have a structure based on the OSI model. Although OSI is a model and not an actively used protocol, and there are still very few pure OSI-based products on the market today, it is still important to understand its structure. The OSI model organizes the communications process into seven different categories and places these categories in a layered sequence based on their relation to the user. Layers 7 through 4 deal with end to end communications between the message source and the message destination, while layers 3 through 1 deal with network access.

Layer 1 — The Physical layer deals with the physical means of sending data over lines (i.e. the electrical, mechanical and functional control of data circuits).

Layer 2 — The Data Link layer is concerned with procedures and protocols for operating the communications lines. It also has a way of detecting and correcting message errors.

Layer 3 — The Network layer determines how data is transferred between computers. It also addresses routing within and between individual networks.

Layer 4 — The Transport layer defines the rules for information exchange and manages end-to-end delivery of information within and between networks, including error recovery and flow control.

Layer 5 — The Session layer is concerned with dialog management. It controls the use of the basic communications facility provided by the Transport layer.

Layer 6 — The Presentation layer provides transparent communications services by masking the differences of varying data formats (character codes, for example) between dissimilar systems.

Layer 7 — The Applications layer contains functions for particular applications services, such as file transfer, remote file access and virtual terminals

See also OSI STANDARDS.

OSI Model Open Systems Interconnection Model. See OSI.

OSI Network Address The address, consisting of up to

OSI Reference Model

Layer 7	Application	Semantics
Layer 6	Presentation	Syntax
Layer 5	Session	Dialog Coordination
Layer 4	Transport	Reliable Data Transfer
Layer 3	Network	Routing & Relaying
Layer 2	Data Link	Technology-Specific Transfer
Layer 1	Physical	Physical Connections

20 octets, used to locate an OSI Transport entity. The address is formatted into an Initial Domain Part which is the responsibility of the addressing authority for that domain and a domain-specific part which is the responsibility of the addressing authority for that domain.

OSI Presentation Address The address used to locate an OSI Application entity. It consists of an OSI Network Address and up to three selectors, one each for use by the Transport, Session, and Presentation entities.

OSI Standards The International Standards Organization (ISO) has established the Open Systems Interconnection (OSI). The idea of OSI is to provide a network design framework to allow equipment from different vendors to be able to communicate. Codex of Mansfield, MA. has published an excellent booklet called "The Basics Booklet of Local Area Networking". Here is a shortened excerpt of what Codex says about standards. 'Standards allow us to buy items such as batteries and bulbs. Many of us have learned "the hard way" that the lack of computer standards can make it impossible for computers from different vendors to talk to each other. Because a major goal of a LAN (Local Area Network) is to connect varied systems, standards are being developed to specify the set of rules networks will follow. The OSI Model is a design in which groups of protocols, or rules for communicating, are arranged in layers. Each layer performs a specific data communications function. The concept of layered protocols is analogous to the steps we follow in making a phone call:

Step 1 — Listen for dial tone.
Step 2 — Dial a phone number.
Step 3 — Wait for a ring.
Step 4 — Exchange greetings.

Step 5 — Communicate message.
Step 6 — Say Good-bye.
Step 7 — Hang up.

Each of these steps, or OSI "layers," builds upon the one below it. Although each step must be performed in preset order, within each layer there are several options. Within the OSI model, there are seven layers. The first three are the PHYSICAL, DATA LINK, and NETWORK layers, all of which are concerned with data transmission and routing. The last three — SESSION, PRESENTATION and APPLICATIONS — focus on user applications. The fourth layer TRANSMISSION provides an interface between the first and last three layers. The X.25 PROTOCOL which created a standard for data transmission and routing is equivalent to the last three layers of the OSI Model. The OSI model is quickly becoming the standard for how LAN products should be built." See also OSI and X.25.

OSINet A test network sponsored by the National Bureau of Standards (NBS) designed to provide vendors of products based on the OSI model a forum for doing interoperability testing.

OSN Operations System Network.

OSP 1. Operator Service Provider. A new breed of long distance phone company. It handles operator-assisted calls, in particular Credit Card, Collect, Third Party Billed and Person-to-Person. Phone calls provided by OSP companies are often more expensive than phone calls provided by "normal" long distance companies, i.e. those which have their own long distance networks and which you see advertised on TV. You normally encounter an OSP only when you're making a phone call from a hotel or hospital phone, or privately-owned payphone. It's a good idea to ask the operator what the cost of your call will be before you make it.

2. Online Service Provider. A company that provides content only to subscribers of their service. This content is not available to regular Web surfers. The idea was to build subscription and other revenues from a closed knit group of people. The problem with this idea was the Internet came along and no one any longer could afford a team to compete with the Web's exploding and varied content. So, some online service providers dropped their attempt at content altogether. Others severely limited it. But all were forced to offer (and do offer) access to the Internet. As a result the term "online service provider" has virtually become obsolete, to be replaced by the term, Internet Service Provider.

OSPF Open Shortest Path First. A link-state routing algorithm that is used to calculate routes based on the number of routers, transmission speed, delays and route cost.

OSPFIGP Open Shortest-Path First Internet Gateway Protocol. An experimental replacement for RIP. It addresses some problems of RIP and is based upon principles that have been well-tested in non-internet protocols. Often referred to simply as OSPF. See CSPF.

OSPS An AT&T word for Operator Services Position System.

OSS Operations Support System. Methods and procedures (mechanized or not) which directly support the daily operation of the telecommunications infrastructure. The average LEC (Local Exchange Carrier) has hundreds of OSSs, including automated systems supporting order negotiation, order processing, line assignment, line testing and billing.

OSS7 Operator Services Signaling System Number 7.

OSSI Operations Support System Interface. An element of DOCSIS (Data Over Cable Service Interface Specification), a project intended to develop a set of specifications for high-speed data transfer over cable television systems. At the head-end of the network, the OSSI provides the interface between the cable modem system and the OSSs. The OSSs, according to the OSI (Open Systems Integration) model, provide for the management of faults, performance, configuration, security and accounting. See also DOCSIS and OSI.

OSTA The Optical Storage Technology Association. An international trade association dedicated to promoting the use of writeable optical technology for storing computer data and images. With a membership of more than 60, OSTA helps the optical storage industry define practical implementations of standards to assure the compatibility of resulting products. www.osta.org

OTC Operating Telephone Company.

OTDR Optical Time Domain Reflectometer, a test and measurement device often used to check the accuracy of fusion splices and the location of fiber optic breakers. See GR.196 and Optical Time Domain Reflectometer.

OTGR Operations Technology Generic Requirements.

Other Common Carriers Providers of long distance telephone service in competition with AT&T. OCCs often (but not always) have lower rates than AT&T. All long distance carriers — including AT&T — are now called interexchange carriers.

OTIA NTIA's Office of Telecommunications and Information Applications (OTIA) assists state and local governments, educations and health care entities, libraries, public service agencies, and other groups in effectively using telecommunications and information technologies to better provide public services and advance other national goals. This is accomplished through the administration of the Telecommunications and Information Infrastructure Assistance Program (TIIAP), the Public Telecommunications Facilities Program (PTFP) and the National Endowment for Children's Educational Television (NECET). The Telecommunications and Information Infrastructure Assistance Program promotes the widespread use of advanced telecommunications and information technologies in the public and non-profit sectors. The program provides matching demonstration grants to state and local governments, health care providers, school districts, libraries, social service organizations, public safety services, and other non-profit entities to help them develop information infrastructures and services that are accessible to all citizens, in rural as well as urban areas. The program was specifically created to support the development of the National Information Infrastructure. The Public Telecommunications Facilities Program supports the expansion and improvement of public telecommunications services by providing matching grants for equipment that disseminate noncommercial educational and cultural programs to the American public. The main objective of the program is to extend the delivery of public radio and television to unserved areas of the United States. Under the program's authority, funds are also allocated to support the Pan-Pacific Educational and Cultural Experiments by Satellite (PEACESAT) project. PEACESAT provides satellite-delivered education, medical, and environmental emergency telecommunications to many small-island nations and territories in the Pacific Ocean. The National Endowment For Children's Educational Television supports the creation and production of television programming that enhances the education of children. The program provides matching grants for television productions, which are designed to supplement the current children's educational program offerings and strengthen the fundamental intellectual skills of children. In addition, a ten-member national Advisory Council on Children's Educational Television provides advice to the Secretary of Commerce on funding criteria for the program and other matters pertaining to its administration. See www.ntia.doc.gov/otiahome/otiahome.html

OTOH Abbreviation for "On The Other Hand;" commonly used on E-mail and BBSs (Bulletin Board Systems).

OTS Operations Technical Support. See also OFFICE TELESYSTEM.

OUI Organizationally Unique Identifier: The OUI is a three-octet field in the IEEE 802.1a defined SubNetwork Attachment Point (SNAP) header, identifying an organization which administers the meaning of the following two octet Protocol Identifier (PID) field in the SNAP header. Together they identify a distinct routed or bridged protocol.

Out-of-Band A LAN term. It refers to the capacity to deliver information via modem or other asynchronous connection.

Out-Of-Band Network Management A method of managing LAN bridges and routers that uses telephone lines for communications between the network management station and the managed devices. This type of management is normally in addition to the conventional method which uses the LANs and WANs that are being connected by these devices. The principal advantage is that in the event of a system failure (which may take a LAN or a WAN down), a network supervisor can bypass the failed system and use a telephone link to reach a bridge/router to diagnose a network problem. Bridges and routers must have built-in telephone modems for this to work.

Out-Of-Band Signaling Signaling that is separated from the channel carrying the information. Also known as NFAS (Non Facilities Associated Signaling). In the cellular domain, it is known as NCAS (Non Callpath Associated Signaling). Out-Of-Band Signaling is non-intrusive, as it is carried over separate facilities or over separate frequency channels or time slots than those used to support the actual information transfer (i.e., the call). Thereby, the signaling and control information does not intrude on the information transfer. SS7 (Signaling System 7) is an example of NFAS. The signaling information includes called number, calling number, and other supervisory signals. See also In-Band Signaling, NCAS, NFAS and SS7.

Out-Of-Frame In T-1 transmission, an OOF (Out Of Frame) error occurs when two or more of four consecutive framing bits are in error. When this condition exists for more than 2.5 seconds a RED alarms is sent by OOF detecting unit. Equipment receiving this RED alarm responds with a YELLOW alarm.

Out-Of-Order Tone A tone which indicates the phone line is broken.

Out-Of-Paper Reception The ability to receive a facsimile transmission into memory when the facsimile machine is out of paper. The facsimile paper will be printed when you put in new paper.

Out-Of-Service Or Used. A term used in the secondary telecom equipment business. Equipment taken from service. Can be in any condition. Expected to work and be complete. May not be.

Out-Tasking Using a vendor to perform specific network management tasks; as opposed to "outsourcing" where the whole operation is turned over to an outside vendor. See also OUTSOURCING.

Outage Service interrupted.

Outage Ratio The sum of all the outage durations divided by the time period of measurement.

Outdoor Jack Closure Closures that protect jacks from

moisture, dirt and the elements.

Outgoing Access A ITU description of the ability of a device in one network to communicate with a device in another network.

Outgoing Calls Barred A switch configuration option that blocks call origination attempts. Only incoming calls are allowed.

Outgoing Line Restriction The ability of the system to selectively restrict any outgoing line to "incoming only."

Outgoing Station Restriction The ability of the system to restrict any given phone from making outside calls.

Outgoing Trunk A line or trunk used to make calls.

Outgoing Trunk Circuit Used to carry traffic to a connecting (distant) office. depending on the traffic in an individual office. The types of outgoing trunks used will vary depending on the traffic in an individual office.

Outgoing Trunk Queuing OTQ. Extensions can dial a busy outgoing trunk group, be automatically placed in a queue and then called back when a trunk in the group is available. This feature allows more efficient use of expensive special lines such as WATS or FX. Instead of having to redial the trunk access code until a line is free, the caller can activate OTQ. See also OFF-HOOK QUEUING.

Outgoing WATS An outgoing WATS (OUTWATS) trunk can only be used for outgoing bulk-rate calls from a customer's phone system to a defined geographical area via the dial-up telephone network. Originally WATS lines came in only lines that could receive calls or lines that could make calls. Now, you can buy a WATS line that handles both incoming and outgoing lines. See WATS.

Outlet A set of openings containing electrical contacts into which an electrical device can be plugged. See OUTLET TELECOMMUNICATIONS.

Outlet Box A metallic or nonmetallic box mounted within a wall, floor, or ceiling and used to hold telecommunications outlets/connectors or transition devices.

Outlet Cable A cable placed in a residential unit extending directly between the telecommunications outlet/connector and the distribution device.

Outlet Connector A connecting device in the work area on which horizontal cable terminates.

Outlet Telecommunications A single-piece cable termination assembly (typically on the floor or in the wall) and containing one or more modular telecom jacks. Such jacks might be RJ-11, RJ-45, coaxial terminators, etc.

Outlier An ATM term. A node whose exclusion from its containing peer group would significantly improve the accuracy and simplicity of the aggregation of the remainder of the peer group topology.

Outline Font Font is the design of printed letters, like the ones you see on this page. The first type was produced with raised metal or wooden blocks. Put ink on the blocks. Put paper on the inked blocks. Lift paper off. Bingo you have type on paper. Blocks came in fonts — styles of type, which has neat names like Times Roman Helvetica, Souvenir, etc. Blocks also came in various sizes — 10 point, 12 point, 14 point, 36 point, etc. "Point" is simply the name for a way of measuring the size of type, like miles measure distance. When computers came along, they simply copied this technique. You picked type and you picked the size. Printers with print cartridges still work this way. They have to. They couldn't simply take one size font and enlarge or contract it because type enlarged or contracted doesn't look "right." Then two men, John Warnock and Martin Newell, said there had to be a better way and they came

up with the idea of an outline font, originally called JaM, then Interpress and now PostScript. In PostScript letters and numbers become mathematical formulas for lines, curves and which parts of the character are to be filled with ink and which parts are not. Because they are mathematical, outline fonts are resolution independent. They can be scaled up or down in size in as fine detail as the printer or typesetter is capable of producing. PostScript outline fonts contain "hints" which control how much detail is given up as the type becomes smaller. This makes smaller type faces much more readable than they otherwise would be. Before outline fonts can be printed, they have to be rasterized. This means that a description of which bits to print where on the page has to be generated. And this is one reason printing outline fonts is so consuming of computer power (whether the power is in the computer or in the printer — usually it's in both). But it's also the reason why outline fonts, of which PostScript is the most successful and the most common, look so great.

Outpulse Dial A pushbutton dial which allows rotary dial users the convenience of "Touch-Tone" dialing. Pushing the buttons makes the phone pretend to be a rotary dial phone. This is necessary because touch-tones are not recognized everywhere.

Outpulsing The process of transmitting address information over a trunk from one switching center to another.

Output Data that flows out of a computer to any device.

Output Device A device by which a computer transfers its information to the outside world. For example, a monitor, a printer and a speaker.

Output Impedance The impedance a device presents to its load. The impedance measured at the output terminals of a transducer with the load disconnected and all impressed driving forces taken as zero.

Output Return Loss A measure of the accuracy of the impedance match between a signal source (such as a cable) and its terminating load. An unequal impedance match causes some of the power from the source to be reflected back to the source, resulting in signal distortion. The ratio of the signal voltage at the load to that voltage reflected back to the source is defined as the return loss. This ratio is generally expressed in decibels (dB).

Output To Output Isolation The ratio of attenuation provided by the output stage to an interfering signal driving one output compared to a second output. The ratio is measured at the second output. A good specification protects output signals against incorrect cabling, such as accidental untermination or double termination.

Outside Link An ATM term. A link to an outside node.

Outside Node An ATM term. A node which is participating in PNNI routing, but which is not a member of a particular peer group.

Outside Plant The part of the LEC (Local Exchange Carrier) telephone network that is physically located outside of telephone company buildings. This includes cables, conduits, poles and other supporting structures, and certain equipment items such as load coils. Microwave towers, antennas, and cable-system repeaters traditionally are not considered outside plant. Outside plant includes the local loops from the LEC's switching centers to the customers' premises, and all facilities which serve to interconnect the various switches (e.g., central office and tandem) in the carrier's internal network. Dedicated outside plant comprises physical local loop facilities which are dedicated from the switching center to the customers' premises. See also

Dedicated Outside Plant and Inside Plant.

Outsourcing An allegedly more modern word for the term "facilities management." What outsourcing means is that a company contracts one of its internal functions out to an outside company. Those functions might include running the company's phone systems and telecom networks and/or running the company's computer system. A company might be motivated to do this because they lack the internal resources (typically people) or feel they can bring their phone costs into line and those phone costs (or at least certain of them) might now become able to be budgeted with some precision. This has appeal to senior management, who are trying to reduce their uncertainties. This usually has no appeal to lower level management who might be fired, especially if the new corporate outsourcing manager felt they were useless.

Here are some questions and answers about outsourcing taken from an information brochure from a company called Cyclix Communications Corporations, which does a lot of outsourcing:

Q: What is outsourcing?

A: Outsourcing is the process by which a company arranges for a third party to implement and manage a specific department or function of the company.

Q: Why would a company outsource?

A: A company would outsource for a number of reasons. In a recent Network World "Critical Issues Survey," two-thirds of the respondents noted lack of in-house resources, staff or expertise, while one-third noted cost effectiveness as reasons for outsourcing. Other reasons for outsourcing may include to cope with new technologies, to minimize capital expenses by having a third party provide, manage and maintain hardware and software, to acquire skilled personnel without having to train them or put them on the payroll and to remain focused on what the company can do best.

Q: What is an advantage of outsourcing?

A: The advantages of outsourcing will largely be dependent upon the objectives which were predetermined before the arrangement begins. For example, if a relatively small MIS department is responsible for both data processing and data networking for a national company, outsourcing one of those functions to an experienced vendor will allow the internal department to concentrate more fully on the other function.

Q: What is a disadvantage of outsourcing?

A: Once you decide to outsource, the transition back to managing those functions in-house can be costly. Companies must realize that outsourcing is a long term financial and competitive strategy. To simply try outsourcing creates false expectations and limits the long-term benefits which can be derived from outsourcing. See also OUT-TASKING.

Outward Restriction Phone lines within the PBX can be denied the ability to access the exchange network without the assistance of the attendant. Restricted calls are routed to intercept tone.

Outward Trunk Queuing A process of holding outgoing long distance calls in queue until the appropriate long distance facility is available. See OFF-HOOK QUEUING.

OutWATS Outward Wide Area Telephone Service. See WATS.

OV Ground. See GROUND START.

Overbuild A term which means just what you think it does: to build more capacity into a network than you really need. Overbuilding has several potential benefits, depending on which side of the equation you find yourself. First, it provides for anticipated growth in traffic requirements. Second, it yields greater revenues for the vendor, although at the expense of the customer. ILECs (Incumbent Local Exchange Carriers) long have been accused of overbuilding their networks, as the total investment goes into the rate base, on which the carrier realizes a rate of return guaranteed by the regulator. Overbuilding also is known as "gold-plating."

Overdrive Processor Intel's name for its line of single-chip performance upgrade chips. Based on Intel486 DX2 "speed doubling" technology, the Intel overdrive processors allow users of Intel486SX systems to double the internal speed of their computer's CPU by adding a single chip, without upgrading or modifying any other system components.

Overfloor Duct Method A distribution method that uses metal or rubber ducts to protect and conceal exposed wiring across floor surfaces.

Overflow Additional traffic beyond the capacity of a specific trunking group which is then offered to another group or line. For example, overflowing calls from WATS lines to (DDD) direct distance dial lines.

Overflow Load The part of an offered load that is not carried. Overflow load equals offered load minus carried load.

Overflow Tie-Line Enhancement A call center term. Using Overflow Tie-Line Enhancement, non-ISDN calls diverted to an overflow call center now convey the city-of-origin announcement prior to being connected to an agent.

Overflow Traffic The part of the offered traffic that is not carried, for example, overflow traffic equals offered traffic minus carried traffic.

Overhead In communications, all information, such as control, routing and error-checking characters, that is in addition to user-transmitted data. Includes information that carries network status or operational instructions, network routing information, as well as retransmissions of user-data messages that are received in error.

Overhead Bit A bit other than one containing information. It may be an error checking bit or a framing bit. See OVERHEAD BITS.

Overhead Bits Overhead bits are bits assigned at the source. They are transmitted with the information payload and are for functions associated with transporting that payload.

Overlay 1. Typically a piece of cut-out cardboard, which you place over several keys on a phone or console. When you punch in a certain code, the buttons become what's written on the programming overlay. Also called a Programming Overlay.

2. The ability to superimpose computer graphics over a live or recorded video signal and store the resulting video image on videotape. It is often used to add titles to videotape.

3. A Northern Telecom definition: Generally used to describe some software that is not always memory resident. It is loaded on request. In the Meridian 1 most configuration, administration and maintenance functions are done from a tty terminal using various overlays. Each overlay is designed for a specific task: For example, Overlay 10 is used to configure PBX (500/2500) sets, Overlay 11 is used to configure proprietary sets. Overlay 17 is used to configure I/O ports, Overlay 15 is used to configure customer data, Overlay 48 is used to configure link maintenance, and so on. The Meridian 1 has some 100 overlays. An overlay is loaded from the tty by typing LD nn where is the overlay number. Overlays are exited by typing ****./

Overlay Area Code A new area code which overlays an existing area code. New York's Manhattan has always had 212 as an area code. When the local phone company needed more numbers, it simply added an area code, 917. It uses that area code for cell phones and beepers. Thus no one had to change their phone number. In contrast, we have something called a "split" area code. The "authorities" simply say the left hand side of town will have a new area code, which means that all the people and companies on the left side of town now get a new area code, and thus a new phone number. This could be classed as "dumb" idea since it forces a huge number of people and companies to reprint all their stationery and all their customers to change their listings. In contrast an overlay area code allows everyone to keep their old phone numbers, forcing only the new area code on new phone numbers. The problem, according to some state public service commissions is that an overlay area code forces some neighbors to dial 1+area code+a number (i.e. eleven digits) when dialing their neighbors. In today's exploding communicating world, this doesn't seem like too high a price to pay for the huge savings and lack of business disruption. Local public service commissions make the split/overlay decision. For example, a few years ago when Dallas, Texas, ran out of numbers, the local phone companies simply wanted to overlay a new area code — 972 — on top of the existing 214 area code. But the Texas PUC refused to allow it for whatever dumb reason. Now, here's an update for you: Pacific Bell announced the 424 area code to overlay the 310 area code effective July 17, 1999. Both area codes will serve the Westside and South Bay areas of Los Angeles County and a very small portion of Ventura County. All existing numbers will retain the 310 area code; new telephone numbers assigned in the same area may receive the new 424 area code. In an overlay area, all calls require 1+10-digit dialing. In other words, you will have to dial 1+10 digits to call your next door neighbor, regardless of whether he has a 424 or a 310 number. The call will remain priced as a local call. While overlay area codes are less disruptive in many ways than are area code splits, dialing 1+10 digits to place a local call is very inconvenient. Overlay area codes are also a dumb idea.

Overlay Cell A cellular/PCS term for an additional cell sector which overlays the underlay cell. Overlay cells are used when traffic regularly exceeds the capacity of an individual cell. The overlay cell handles the excess traffic.

Overlay Network A separate network for a particular service covering most of the same geographical locations as the basic telephone network, but operating independently. Overlay networks typically are deployed on a selective basis to address performance issues in the existing PSTN (Public Switched Telephone Network). For example, developing countries in Asia and Central/Eastern Europe have built digital microwave overlay networks to serve major businesses, educational institutions, and government offices. Those networks overlay the legacy network, bypassing it and its inherent problems resulting from old, overloaded and poorly maintained infrastructure. Such an approach serves as an immediate stop-gap measure, providing state-of-the-technology voice and data communications capabilities. Eventually, such overlay networks will be fully integrated into the PSTN at such time as the PSTN is upgraded.

Overload Control How a system responds to being overstressed is called Overload Control. When a system is overloaded, frequently there are so many extra events being processed that the system's actual capacity or throughput goes down. Even though it may be rated at, say 10,000 busy hour calls, when overloaded, for example, with 11,000 calls, the computer telephony system may be only able to process only 8,000 calls.

Overload Management An AT&T term for handling

peak demands by selectively delaying, degrading, or dropping only those portions of traffic flow that are tolerant of those particular types of impairments.

Overload Protection An uninterruptible power supply (UPS) definition. This feature automatically shuts the unit off when the battery is overloaded in order to protect it against overload damage.

Override When a circuit already in use is seized. For example, when your boss can break into your telephone conversation.

Override Prompts The ability of callers and users to key over system prompts.

Overrun Loss of data because the receiving equipment could not accept the data at the speed at which it was being transmitted.

Oversampling Time division multiplexing (TDM) technique where each bit from each channel is sampled more than once.

Overscan The image fills the screen from bezel to bezel. The bezel is the metal or plastic part — in short, the frame — that surrounds a cathode ray tube — a "boob" tube. NEC's term for overscan is "full scan."

Overspeed Condition in which the transmitting device runs slightly faster than the data presented for transmission. Overspeeds of 0.1% for data PABXs are typical.

Oversubscription In frame relay, this refers to a situation in which a device is transmitting over a circuit at a rate greater than the target device can receive the data over a circuit. For instance, the transmitter may be pumping data at 128 Kbps over a T-1 circuit, while the receiver has only a 64 Kbps circuit connection to the network. In such an instance, the network will impose flow control, adapting to the rate of reception within the limits of its buffering capability. Once that limit is reached, the network is oversubscribed. In order to maintain its ability to support "normal" traffic requirements of users in the aggregate, the network will mark the excess frames as Discard Eligible and spill them onto the switchroom floor for the night shift to clean up. (Just kidding about the switchroom floor part.) As always, it is the responsibility of the end user equipment to detect the fact that this had occurred and to recover through retransmission of the discarded frames. SEE DE.

Overtime Period Those minutes of use of a telephone service beyond the initially defined period for which a basic charge is quoted. The initial period on many calls is one or three minutes. After that, the next minute is overtime.

Overview Proteon's architecture for products conforming to SNMP.

Overwrite A call center term. To replace the contents of something — a file or record, for example — with new data.

OVS Open Video Systems, the successor to Video Dialtone.

OVTG See Optical Virtual Tributary Group.

Owner Microsoft jargon for person in charge.

OWS Operator Workstation. See OPERATOR WORKSTATION.

OWT Operator Work Time.

P Connector Also called a male amp connector or 25-pair male connector. The female version is called a C connector.

p-ANI pseudo-ANI. A cellular term. p-ANI is a fictitious non-dialable telephone number which comprises the seven to ten digits of the cell site or base station identifier. p-ANI information is used by the PSAP (Public Safety Answering Position) in a wireless E-911 (Enhanced-911) application in order that the 911 personnel can locate the caller. Along with p-ANI information are sent the 10 digits of the cellular calling number, also known as the MDN (Mobile Dialing Number), in order that the 911 personnel can callback for more information. The FCC has mandated that, in the future, p-ANI information will be used to identify the specific location of the caller, within a range of 125 meters. See also E-911 Service and PSAP.

P-Code The Precise or Protected Code. A very long sequence of pseudo random binary biphase modulations on the GPS (Global Positioning System) carrier at a chip rate of 10.23 MHz which repeats every 267 days. Each one week segment of this code is unique to one GPS satellite and is reset each week.

P-Frame Predictive framing, specified by the MPEG Recommendation. Pictures are coded through a process of predicting the current frame using a past frame. The picture is broken up into 16x16 pixel blocks and each block is compared to a previous frame's block that occupies the same vertical and horizontal position.

P-MAC Packet Media Access Control. An FDDI-II term. See FDDI-II.

P-Phone Enhanced Business Service (also known as P-Phone) is an analog Centrex offering provided by Northern Telecom. It operates over a single-pair subscriber loop., providing normal full duplex audio conversations and a secondary 8 KHz half-duplex amplitude shift-keyed signal, which is used to transmit signaling information to and from the Northern Telecom-equipped central office.

P/AR Peak to Average Ratio. A standard analog transmission-line test signal of varying frequencies and amplitudes, which is then compared with the received signal. Composite results are a weighted number from 1 to 100 being the maximum. The P/AR is used increasingly as a standard quick test of a telecommunications channel comparative quality. Per Bell standard, the minimal acceptable P/AR rating for medium-speed data transmission is 48.

P.139 Serial Bus The P.139 Serial Bus is promulgated by the IEEE (202-371-0101). Initially it will operate at 100 million bits per second. But it has extensions for 200 and 400 megabits per second. The cable topology allows for branching and daisychaining of the peripherals attache, with a maximum of 32 4.5 meter hops using the baseline copper cable. That cable is a specially-designed combination consisting of two shielded twisted pairs for clock and data, two wires for power and ground and an overall shield that surrounds the twisted pairs and power and ground wires. See also USB, which stands for Universal Serial Bus.

PO Protocol The protocol for messaging headers used for undefined messaging in an X.400 MHS (Message Handling System). Not officially part of the X.400 standards but became the U.S. intercept solution to P35 and X.435.

P01, P.01, Pnn or P.nn The Grade of Service for a telephone system. The digits following the P, i.e. nn, indicate the number of calls per hundred that are or can be blocked by the system. It is a goal or a measure of an event. In this example, P01 (also spelled P.01) means one call in a hundred (i.e. one divided by 100) can be blocked, so the system is designed to meet this criterion. See GRADE OF SERVICE and TRAFFIC ENGINEERING.

P1 Protocol P1 is the protocol defined in the X.400 standard used between MTAs (Message Transfer Agents) for relaying messages. Defined as the envelope, P1 is a portion of an X.400 message the identifies the message originator and potential recipients, records information about the message's path through the MTS (Message Transfer Service), directs the message's subsequent movement through the MTS, and includes characteristics of the message's contents. The composition of an envelope changes as the message is submitted, relayed and delivered.

P2 Format The 1984 defined message format, or protocol, used between cooperating User Agents in the IPM (Interpersonal Messaging System). The P2 heading is a component of an IP-message that indicates such items as the originator, recipients(s) and subject of the message.

P3 Protocol X.400 standard protocol used between a RUA (Remote User Agent) and an MTA (Message Transfer Agent) and an MTA (Message Transfer Agent) or between a remote MS (Message Store) and an MTA. P3 may be used optically between a co-resident UA (User Agent) and MTA. See P3P

P3P Platform for Privacy Preferences Project. P3P is a way of that Web sites you visit could capture a great deal of information about you — so long as you allow that capture. You have control over how much information also. Writing about P3P, the New York Times' John Markoff said, "the new standard establishes a complicated computerized negotiation that dramatically extends both the information-gathering potential and the privacy-protection possibilities inherent in each visit to a Web site." The P3P standard, wrote Markoff, "foreshadows new Internet technology that will make each visit to a Web site a complicated exchange of information in which a personal computer will have the opportunity to automatically disclose a range of information covering every conceivable category, from birth data to shoe size, depending on rules set by the computer user. At the same time, the Web site will be forced to disclose its policy for using the information it is gathering. hopefully giving the computer user the ability to decline to share data about himself."

P5 Protocol The 1984 X.400 protocol used between a MTA (Message Transfer Agent) and a Teletex Unit. This protocol is not popular.

P6 Intel's successor to the Pentium processor. The P6 is now called the Pentium Pro.

P7 Protocol The protocol used between a UA (User Agent) and a MS (Message Store). Defined in the 1988 X.400 standard.

P22 Format An enchanted version of the P2 format that appeared in the 1988 X.400 standard.

P2T Pulse to Tone & Rotary Dial Recognition products.

P35 Format The EDI (Electronic Data Interchange) message header enhancement to X.400 that enables EDI-specific addressing, routing and handling of EDI messaging. Defined in 1990 in the X.435 standard.

P802.11 P802.11 is a series of evolving standards in the wireless Local Area Network arena. The IEEE Program P802.11 Standards Working Group aims to define universal protocols for wireless LANs in the 900 MHz, 2.4 GHz and infrared frequency bands. Main components in the 802.11 standard are twofold: 1. The physical specifications for medium-dependent protocols. There are different physical specifications for each frequency band supported in 802.11. 2. The Medium Access Control (MAC) specifications for ad-hoc wireless networks and wireless network infrastructures. A single medium-independent MAC protocol provides a unified network interface between different wireless PHYs (Physical Specifications) and wired networks.

PA Public Address. Loud speaker system, sometimes used for paging.

PABX Private Automatic Branch eXchange. Originally, PBX was the word for a switch inside a private business (as against one serving the public). PBX means a Private Branch Exchange. Such a "PBX" was typically a manual device, requiring operator assistance to complete a call. Then the PBX went "modern" (i.e. automatic) and no operator was needed any longer to complete outgoing calls. You could dial "9." Thus it became a "PABX." Now all PABXs are modern. And a PABX is now commonly referred to as a "PBX.

PAC Personal Activity Center. A combination IBM PC clone, alarm clock, answering machine, speakerphone, fax machine, modem, compact-disk player and AM/FM radio all rolled into one unit sitting on your desk.

Pac Bell Pacific Bell, the California Bell Operating Company (BOC) which, along with Nevada Bell, formed Pacific Telesis (PacTel). PacTel was one of the seven original Regional Bell Operating Companies (RBOCs) formed in 1984 by the Modified Final Judgement (MFJ). As PacTel was acquired by SBC Corporation (Southwestern Bell) in 1996, Pac Bell is now a subsidiary of SBC.

Pacific Telesis One of the seven, now-independent Regional Holding Companies formed at the Divestiture of AT&T of 1983. It holds Pacific Bell, Nevada Bell and several non-regulated subsidiaries. In 1996, SBC Corporation bought Pacific Telesis.

Pacing Controlled rate of flow dictated by the receiving component, to prevent congestion. A method of flow control in IBM's SNA. See PACING GROUP.

Pacing Algorithm The mathematical rules established to control the rate at which calls are placed by an automatic dialing machine, also called a predictive dialer. See PREDICTIVE DIALER.

Pacing Control SNA term for flow control. See PACING GROUP.

Pacing Group In IBM's SNA, the number of data units (Path Information Units, or PIUs) that can be sent before a response is received. An IBM term for window.

Pack To compress data items so they take up less space. A process used by many database programs to remove records marked for deletion.

Packet 1. Generic term for a bundle of data, usually in binary form, organized in a specific way for transmission. The

Generic Format of a Data Packet STX=Start of Text EXT=End of Text

Header	S T X	Text	E T X	Trailer

specific native protocol of the data network may term the packet as a packet, block, frame or cell. A packet consists of the data to be transmitted and certain control information. The three principal elements of a packet include: 1. Header — control information such as synchronizing bits, address of the destination or target device, address of originating device, length of packet, etc., 2. Text or payload — the data to be transmitted, and 3. Trailer — end of packet, error detection and correction bits. See also Frame.

2. Specific packaging of data in a packet-switched network, such as X.25. A true packet-switched network such as X.25 involves packets of a specific and fixed length. In a public packet switched network, such packet payloads are specified as being either 128B or 256B, where B=Byte. Custom networks, such as those traditionally used in airline reservation systems, may employ a packet payload of 1024B. A X.25 packet prepends the payload with header information including a flag of eight bits; the flag denotes the beginning of the packet and also serves to assist the network nodes (packet switches) in synchronizing on the rate of transmission. An address field of eight bits also prepends the payload, with four bits identifying the target device and four bits identifying the transmitting device. Control data of eight-to-sixteen bits comprise the last element of the header; included in control data is the packet number in order that the network nodes might identify and correct for lost or errored packets. Appending the payload is a trailer consisting of a sixteen-bit CRC, which is used by all packet nodes for purposes of error detection and correction. See Packet ASSEMELER/DISASSEMBLER (PAD) and CRC.

Packet Assembler/Disassembler PAD. A hardware/software combination that forms the interface between a X.25 network such as PDN and an asynchronous device such as a PC. The PAD generates call request, call clear, and other information packets in addition to the ones that contain user data. The PAD is responsible for packetizing the data from the transmitting device before it is forwarded through the packet network. On the receiving end of the transmission, the PAD strips away the control information contained in the header and trailer in order to get at the original text or payload, in effect disassembling the packets and reconstituting the original set of data in its native or original form. The PAD may be in the form of a standalone DCE device on the customer premise and supporting one or more terminals, or in the form of a printed circuit board which fits into an expansion slot of the terminal. Smaller users generally rely on the carrier to provide the PAD, which is embedded in the packet switch.

Packet Buffer Memory set aside for storing a packet awaiting transmission or for storing a received packet. The memory may be located in the network interface controller or in the computer to which the controller is connected. See BUFFER.

Packet Burst Protocol A protocol built on top of IPX that speeds the transfer of NCP data between a workstation and a NetWare server by eliminating the need to sequence and acknowledge each packet. With packet burst, the server sends a whole set (or burst) of packets before it requires an acknowledgement.

Packet Controller The hub of the AT&T ISDN system. It acts as a fast packet switch providing virtual circuit services to the devices hooked to the system.

Packet Driver The specification developed by John Romkey at FTP Software to allow TCP/IP and other transport protocols to share a common network interface card. Packet Drivers have been written for a variety of network interface cards, and in many cases provide NetWare compatibility.

Packet Filter A router-based firewall that can accept or reject packets based on predefined rules. The ability to search a packet to determine its destination and to then route it or block it accordingly. This ability helps to control network traffic. See Packet Filtering.

Packet Filtering The recognition and selective transmission or blocking of individual packets based on destination addresses or other packet contents.

Packet Filtering Firewall A packet filtering firewall is a router or a computer running software that has been configured to block certain types of incoming and outgoing packets. A packet-filtering firewall screens packets based on information contained in the packets' TCP and IP headers, including some or all of the following: Source address; Destination address; Application or protocol; Source port number; and Destination port number.

Packet Forwarding Copying the packet to another node without looking at the destination address.

Packet Handler Function The packet switching function within an ISDN switch, for the packet mode bearer service.

Packet Interleaving Refers to the process of multiplexing multiple incoming packets from multiple channels on to a single outgoing channel by sampling one or more packets from the first channel, then the next, and so on.

Packet Level In packet data networking technology, level 3 of X.25. Defines how user messages are broken into packets, how calls are established and cleared over the packet data network (PDN) and how data flows across the entire PDN. The packet level also handles missing and duplicate packets.

Packet Level Procedure PLP. A full-duplex protocol that defines the means of packet transfer between a X.25 DTE and a X.25 DCE. It supports packet sequencing, flow control (including maintenance of transmission speed), and error detection and recovery.

Packet Mode Bearer Service An ISDN term for X.25 packet data transmission over the D channel in a BRI application. Always a part of the ITU-T (nee CCITT) standards, the service has only recently been made available. The 16-Kbps D channel can accomplish its primary responsibilities for signaling and control while still leaving 9.6 Kbps free for end user transmission of low-speed data. Retailers make extensive use of this service for credit card authorization. Only in the very recent past has the D channel been made available to end users like you and me. See also AO/DI, BRI and ISDN.

Packet Overhead A measure of the ratio of the total packet bits occupied by control information to the number of bits of data, usually expressed as a percent.

Packet Radio Packet Radio is the transmission of data over radio using a version of the international standard X.25 data communications protocol adapted to radio (AX.25). It takes your information, and breaks it up into "packets" which are each sent and acknowledged separately. This assures error-free delivery from sender to receiver. A packet is a

stream of characters consisting of a header, the information the user is sending, and a check sequence. The header gives the destination call sign, the call sign of the sender, and any digipeaters (digital repeater) call signs that will be used for relaying the packet. The check sequence makes certain that the data received is what was sent. AlohaNET, a packet radio network developed for a number of years ago for use at the University of Hawaii, was an early packet radio network for LAN networking among the islands and laying a foundation for subsequent packet networks, both wired and wireless. Packet radio data networks recently have been deployed by a number of carriers serving mobile and fleet applications, with such carriers including ARDIS, RAM Mobile Data and Nextel.

Packet Size The length of a packet, expressed in bytes (B). Packet size is of specified and fixed length in X.25 and other true packet networks. The size of the "packet" in other networks may be variable within limits, as is the case with an Ethernet frame or a Frame Relay frame.

Packet Switching Sending data in packets through a network to some remote location. The data to be sent is assembled by the PAD (Packet Assembler/Disassembler) into individual packets of data, involving a process of segmentation or subdivision of larger sets of data as specified by the native protocol of the transmitting device. Each packet has a unique identification and each packet carries its own destination address. Thereby, each packet is independent, with multiple packets in a stream of packets often traversing the network from originating to destination packet switch by different routes. Since the packets may follow different physical paths of varying lengths, they may experience varying levels of propagation delay, also known as latency. Additionally, they may encounter varying levels of delay as they are held in packet buffers awaiting the availability of a subsequent circuit. Finally, they may be acted upon by varying numbers of packet switches in their journeys through the network, with each switch accomplishing the process of error detection and correction. As a result, the packets may also arrive in a different order than they were presented to the network. The packet sequence number allows the destination node to reassemble the packet data in the proper sequence before presenting it to the target device.

Originally developed to support interactive communications between asynchronous computers for time-share applications, packet switched networks are shared networks, based on the assumption of varying levels of latency and, thereby, yielding a high level of efficiency for digital data networking. Isochronous data such as realtime voice and video, on the other hand, are stream-oriented and highly intolerant of latency. As a result, packet switched networks are considered to be inappropriate for such applications. Recent development of certain software and making use of complex compression algorithms, however, has introduced packetized voice and video to the corporate intranets and the Internet, which was the first public packet-switched data network and remains by far the most heavily used.

Here is another way of explaining packet switching: There are two basic ways of making a call. First, the one everyone's familiar with — the common phone call. You dial. Your local switch finds an unused path to the person you called and joins you. While you are speaking, the circuit is 100% all yours. It's dedicated to the conversation. This is called circuit switched. Packet switching is different. In packet switching, the "conversation" (which may be voice, video, images, data, etc.) is sliced into small packets of information. Each packet is given a unique identification and each packet carries its own destination address — i.e. where it's going. Each packet may go by a different route. The packets may also arrive in a different order than how they were shipped. The identification and sequencing information on each packet lets the data be reassembled in proper sequence. Packet switching is the way the Internet works. Circuit switching is the way the worldwide phone system works, also called the PSTN (Public Switched Telephone Network).

Packet and Circuit Switching each have their own significant advantages. Packet switching for example does a wonderful job getting oodles of data into circuits. Think about a voice conversation. When you are talking, he's listening. Therefore half the circuit is dead. There are pauses between your voice. Packet switching takes advantage of those pauses to send data. Packet switching has been used primarily for data. But with the growth of the Internet it has been used also for voice. Because of the need to re-assemble packets and other reasons, there's up to a half second delay between talking and the person at the other end hearing anything. Packet voice on the Internet is not as clear as circuit switched voice. But that's changing as the packets come faster and the technology improves. See INTERNET, IP Telephony and TAPI 3.0.

Packet Switching Exchange PSE. The part of a packet switching network that receives the data from a PAD Packet Assembly Disassembler) through a modem. The PSE makes and holds copies of each packet before sending them to the PSE they're addressed to. After the far-end PSE acknowledges receipt of the original, the copies are discarded.

Packet Switching Network A network designed to carry data in the form of packets. See PACKET SWITCHING.

Packet Tracing The monitoring and reporting a particular packet addresses or types for diagnostic purposes.

Packet Type Identifier In packet data networking technology, the third octet in the packet header that identifies the packet's function and, if applicable, its sequence number.

Packetized Video First, read the definition of "Packet Switching." Then read the definition of "Packetized Voice" just below. The concept of packetized video is basically the same as that of packetized voice. A video camera feeds the signal into a codec, which converts the native analog signal into a digital format, and segments the data into data packets. The packets are sent across a packet network as a packet stream for reassembly by a codec on the receiving end of the transmission before presentation on a monitor. While packetized video performance is improving in quality through the application of increasingly sophisticated video compression techniques, it suffers from the same intrinsic packet-switching characteristics as does packetized voice. Namely, packet latency and loss. The result often is a video image which is less than pleasing. Note that voice and video are isochronous data, meaning that they are stream-oriented. In other words, the transmitting device must have regular and reliable access to the network. Further, the network must transport and deliver the data on a regular and reliable basis in order that a stream of information reach the presentation device. Such regular and reliable ingress, transport and egress of data results in a image of consistent quality. As packet-switched networks are not designed to support isochronous data communications, they generally are considered unsuitable for voice and video communications. Additionally, video is very bandwidth-intensive, thereby placing additional stress on packet-switched networks such as the Internet, which already is overloaded.

An example might help. Let's say that you are using an inexpensive ($200 or so) videoconferencing package consisting of a camera and software. Your friend has the same package. At a pre-arranged time, you place a call over the Internet to establish a videoconference. At two fps (frames per second) the videoconference goes along pretty smoothly, although both the video and voice quality are a bit rough. At some point, your friend turns his head quickly; at the same time, the Internet bogs down. The packet which contains the image of your friend's nose gets delayed or lost in the network. The video image of your friend now is missing a nose. Funny the first time, aggravating the second, maddening thereafter. The upside is that the videoconference is cheap, if not free, depending on your cost of Internet access. See also PACKET SWITCHING, VIDEOCONFERENCING, INTERNET and ISOCHRONOUS.

Packetized Voice First read the definition of "Packet Switching" just above. The idea is to digitize voice and, compress it, and then slice it up into packets and send those packets from the sender by various routes and assemble them as they get to the receiver. Packet switching for data makes sense. Packet switching for voice has not made sense because the voice is too sensitive to latency, or delay, especially the variable delay which is part and parcel of packet-switched networking. Recently developed software and DSP hardware, which employs sophisticated compression techniques has improved the ability to conduct "reasonable" quality packet voice conversations over the Internet. See Packet Switching, IP Telephony, TAPI 3.0.

PacketNet Sprint's internal X.25 Packet Network.

PACS Personal Communications Access System. PACS is a cellular system providing limited, regional mobility in a given area. It provides mobility between that of a cordless phone and a full-fledged cellular system. Originally developed by Bell Labs in the early 1980s, PACS is a comprehensive framework for the deployment of PCS and applies to both licensed and unlicensed applications. Now it is approved by the TIA and Exchange Carriers Standards Associations. Today's currently implemented versions of PCS are "up-banded" versions of the 900 MHz AMPS and GSM cellular standards.

PACT Siemens' PBX And Computer Teaming. It defines protocols between Siemens PBXs and external computers,

Pad 1. A device inserted into a circuit to introduce loss.
2. Packet Assembler/Disassembler. A device that accepts characters from a terminal or host computer and puts the characters into packets that can be handled by a packet switching network. It also accepts packets from the network, and disassembles them into character streams that can be handled by the terminal or host.

Pad Characters In (primarily) synchronous transmission, characters that are inserted to ensure that the first and last characters of a packet or block are received correctly. Inserted characters that aid in clock synchronization at the receiving end of a synchronous transmission link. Also called fill characters.

Pad Switching A technique of automatically cutting a transmission loss pad into and out of a transmission circuit for different operating conditions.

PAF File A British term. Post Office Address file, a publicly available data file that, when integrated with an application, links postcodes to full addresses. When using a PAF file, an agent can save time by entering only the postcode. The PAF file automatically inserts post town, street and country.

Page A chunk of information, like a document or file, on the

Web. A hypermedia document as viewed through a World Wide Web browser. Pages are the way you make information available on the Web. They can contain text, black and white and color photographs, audio and video.

Page hits A measure of the number of Web pages accessed at a particular site, or of the number of times a single page is accessed.

Page Mirroring You're surfing the Internet. You come upon an Web site and find something you want to buy. It has a button that says, "To speak to an Agent, push here." You do. An agent in a distant office calls you on another phone line. You're now both speaking to each other and your agent is also seeing the screen you're seeing. As you move to different screens (to different pages), the agent sees the same screens you are looking at. This is called page mirroring. At present, the technology won't allow the agent to move the pages which you see. Only you can move the screens.

Page Zone A local area in the office that can receive directed Page announcements independently of the remainder of the office.

Pager A small one-way (typically) wireless receiver you carry with you. When someone wants you, they make your pager receiver alert you via a tone or a vibrator. They can activate your pager in a number of ways, including dialing your pager digits directly into a computer; calling your pager from a telephone; or the old, low-tech approach of giving your name and pager number to an operator who then punches out your numbers. Pagers have become small, cheap and very reliable. Monthly service costs have dropped and the area which you can be paged in most areas has widened dramatically. With most pagers you can be reached in most major metropolitan areas of the US. This minor miracle is accomplished through a combination of satellite and terrestrial radio networks; if the network can't reach you to deliver the page, it will store the message until you can be reached. Some pagers also display small alphanumeric messages-like phone numbers to call and names of babies born; this capability is known as SMS (Short Message Service). Many pagers and pager networks now include an "acknowledgment" feature, which allows you to press a button to acknowledge the receipt of the page through two-way communications capability. All this has made the pager far more useful. Some multi-function devices incorporate pagers into cellular phones. See also Paging and SMS.

Pager Codes Forget the phone. Forget the PC. Pagers are teenagers' new communication tools of choice. Wondering what your kid is receiving? Here are the secret codes, according the September 1997 Seventeen Magazine and Danielle Cioffi.

007 — I've got a secret.
143 — I love you.
07734 — Hello (upside down).
55 — Let's go for a drive.
2468 — Who do we appreciate? You.
13 — I'm having a bad day.
10-4 — Is everything ok?
2-2 — Shall we dance?
666 — He's a creep.
90210 — She's a snob.
9-5 — Time to go home.
121 — I need to talk to you alone.
100-2-1 — Bad odds. That's not likely.

Paging To give a message to someone who is somewhere, but where we don't know. Paging can be done with a little

"beeper" carried in her purse or on his belt. Paging can also be done through speakers in phones or from speakers in the ceiling. Most phone systems offer a paging channel access. You dial that number and page your party. The system comes "live" and your voice is heard everywhere. Similar to the stuff they have at airline terminals. Paging systems as an accessory to phone systems always cost extra. They're one of the most valuable features on a phone system. Don't skimp, however, on the quality of the speakers or the power of the paging system. If you do, your system will sound awful and people will not use it. See PAGER.

Paging Access, Rapid Your attendant and you or anybody else with an extension off your PBX can make a page (access the paging equipment) by pushbuttoning one or several digits. Sometimes you can dial one several numbers for different paging alternatives. One number gives you the tenth floor. One number gives you the fourth floor, etc. And one number to page everyone — also called an "ALL CALL" page. See also RAPID PAGING ACCESS.

Paging By Zone By dialing the appropriate access code, any phone is able to selectively page "groups" of pre designated phones or speakers.

Paging Channel PCH (from Paging CHannel). Specified in IS-136, PCH carries signaling information for set up and delivery of paging messages from the cell site to the user terminal equipment. PCH is a logical subchannel of SPACH (SMS (Short Message Service) point-to-point messaging, Paging, and Access response CHannel), which is a logical channel of the DCCH (Digital Control CHannel), a signaling and control channel which is employed in cellular systems based on TDMA (Time Division Multiple Access). The DCCH operates on a set of frequencies separate from those used to support cellular conversations. See also DCCH, IS-136, PAGING, SPACH and TDMA.

Paging Code Call Access A feature of the ROLM Attendant Console which offers direct, one-touch access to the paging or code call features.

Paging Speakers Speakers in the telephone. Also external units located in ceilings, on walls, etc.

Paging Total System Upon dialing the appropriate special code, any station may make a paging announcement through all the loudspeakers.

Paid Call The usual type of a toll telephone call automatically billed to the calling telephone number.

Paid Hours A call center term. The time that an employee is either on duty-handling calls, doing other work, in meetings, etc., or on a paid schedule exception, such as an excused absence. In TCS's TeleCenter System this is calculated as scheduled hours minus any unpaid schedule exceptions that occur within those scheduled hours. TCS makes manpower scheduling packages. It is based in Nashville, TN.

Paintmonkey Someone with a less-than-glamorous, entry-level computer graphics job. A paintmonkey may spend months on a nanosecond of digitized film footage, painting mattes, or doing monotonous touch-ups.

Pair The two wires of a circuit. Those which make up the subscriber's loop from his office to the central office.

Pair Gain The multiplexing of x phone conversations over a lesser number of physical facilities. Pair gain usually refers to electronic systems used in outside plant — from the central office to the subscriber's premises. In "pair gain" you might do something as simple as take one pair of wires and carry two conversations on it. You might also take two pairs and carry 128 conversations. "Pair gain" is actually the number of conversations you get minus the number of wire pairs used by the system.

Lucent Technologies has various subscriber pair gain devices called "SLC" (pronounced "s ick" and standing for Subscriber Loop Carrier Systems). Other companies, like Rockwell, have comparable systems. The more circuits these devices produce, i.e. the more cable pairs they save you, the more they cost. The cost of subscriber pair gain equipment — like all electronics — has been dropping in recent years, reducing the phone company's need to install outside cable and thus making better use of the presently installed cable. T-1 is a type of subscriber pair gain equipment.

Paired Cable A cable in which all conductors are arranged in twisted pairs. This form of cable is the most common for communications.

PAL 1. Public Access Line. Also called a COCOT, if you live outside of US West territory. A line that is tariffed to be attached to a pay phone. There are two basic flavors, "smart" and "basic" (or dumb). Smart lines may be connected to dumb phones, and offer features such as coin signalling. Basic lines may be connected to smart phones, and the phone does all the work. Between the phone and the line, somebody has to have some intelligence. I first heard about this from one of my customers, a pay phone provider that puts up phones around town and pays commissions to the building owners. The service is tariffed by US West as a Public Access Line. (With multiple providers such as this one, anxious to pay you a percentage on a pay phone installed at your location, it's pretty silly to pay Bell for a pay phone).

2. Programmable Array Logic

2. Proprietary ALgorithm. A designation for a privately designed and owned intelligence-based electronic method for performing a task (such as voice compression).

3. Phase Alternate Line. PAL is the format for color TV signals used in the United Kingdom, West Germany, Holland, much of the rest of western Europe, several South American countries, some Middle East and Asian countries, several African countries, Australia, New Zealand, and other Pacific island countries. PAL inverts the phase of the color signal 180 degrees on alternate lines, hence the term Phase Alternate Line. It was invented in 1961 and is used in England and many other European countries. With its 625-line scan picture delivered at 25 frames/second (primary power 220 volts), it provides a better image and an improved color transmission over the US system called NTSC, which uses interlaced scans and 525 horizontal lines per frames at a rate of 30 frames per second. SECAM is used in France and in a modified form in Russia. SECAM uses an 819-line scan picture which provides better resolution than PAL's 625-line and NTSC's 525. All three systems are not compatible. You cannot view an Australian or English videotape on a US TV. See PAL-M.

PAL-M A modified version of the phase-alternation-by-line (PAL) television signal standard (525 lines, 50 hertz, 220 volts primary power), used in Brazil. See also NTSC, PAL, SECAM.

Palette In some programs, a palette is a collection of drawing tools, brush widths, line widths and colors. In other programs it is the part of the color lookup table that determines the number and type of colors that will be displayed on the screen.

PALS A standard library database interface. An Internet term.

PAM Pulse Amplitude Modulation. Process of representing a continuous analog signal (a voice conversation) with a series of discrete analog samples. This concept is based on the information theory which suggests that the signal can be accurately recreated from a sufficient sample. Why bother? Sampling allows several signals to then be combined on a channel that otherwise would only carry one telephone conversation. PAM was used as part of a method of switching phones calls in several PBXs. It is not a truly "digital" switching system. PAM is the basis of PCM, Pulse Code Modulation. See PCM and T-1.

PAMA Pulse Address Multiple Access. Where carriers are distinguished by their time and space characteristics simultaneously.

Pan-European Across all of Europe. "We'll launch that magazine in a Pan-European edition."

Pan-and-Scan A TV term referring to the translation of widescreen movies for TV broadcast through the introduction of moves and cuts which were never intended in the original. Less than the complete frame is transmitted, and portions of the picture are left out. The technique makes the action visible in a narrower frame such as your TV set. Contrast with LETTERBOXING.

Pancake Coil A type of inductance having flat spiral windings. An old radio definition.

Panel, Patch See PATCH PANEL.

Panel Office A very early type of central office switch.

Panel System Workstation defined by thin panels that provide privacy and insulation from noise.

Panne Fatale An equipment crash in Italian.

PANS Pretty Amazing New Services/Stuff. PANS is a term coined to describe ISDN Capabilities which should eventually replace POTS. Contrast with POTS. In May, 1998, Rodney G. Seiler, Telecommunications Engineer, QUALCOMM Incorporated, wrote me "Harry, I take issue with your definition and origin of 'PANS.' I first heard this as 'Peculiar And Novel Services' in 1978, possible even earlier. I can assure you that ISDN was not in the picture when PANS first came along. Thank you for an almost perfect (this is not faint praise) reference much used in the business."

PAP 1. Packet-Level Procedure. A protocol for the transfer of packets between an X.25 DTE and an X.25 DCE.X.25 PAP is a full-duplex protocol that supports data sequencing, flow control, accountability, and error detection and recovery.

2. Password Authentication Protocol and CHAP are widely-used authentication methods for communicating between routers, both for reaching the Internet and for securing temporary WAN connections such as a dial-backup line. CHAP uses a 3 way handshake process that, in concept, resembles a dial-back routine and uses encrypted passwords. With PAP, one router connects to the other and sends a plaintext login and password.

Paper Sizes US	Europe and Japan
A = 8 1/2" x 11"	A3 = 11.7" x 16.5"
B = 11" by 17"	A4 = 8.3" x 11.7"
C = 18" by 24"	A5 = 5.8" x 8.3"
D = 24" by 36"	B4 = 10.1" x 14.3"
E = 34" by 44"	B5 = 7.2" x 10.1"
B6 = 5.1" x 7.2"	

Paper Tape A long thin paper roll on which data is stored in the form of punched holes. Usually used as input to other systems. Many old-fashioned telex machines still use paper tape as their storage medium. Punch up the message on the paper tape, rewind the paper tape, call the distant telex machine, then start the paper tape containing the message. The primary benefit of paper tape is that you save on transmission line cost. The paper tape will run through at the maximum speed of the line, while a human operator typing manually would be slower. The disadvantage of paper tape is that you can't change the message once you've typed it. Magnetic medium — floppy disks, hard disks, bubble memory — are much more flexible. They are rapidly replacing paper tape, even on telex machines, or on personal computers, which are replacing telex machines as telex data entry devices.

Paper Tape Punch A device to physically punch holes in a roll of paper tape in order to store information.

Paper Tape Reader A device which translates the holes in coded perforated tape into electrical signals suitable for further handling. The reader may be attached to a keyboard-printer or it may be a free standing device.

PAR Positive Acknowledgement Retransmit.

Parabola A shape which can focus a microwave signal into one narrow beam. All satellite and microwave antennae are parabolic, not spherical.

Parabolic Reflector The technical name for a dish antenna shaped like a perfect parabola.

Paradigm An assumption about the ways things work. The word paradigm (pronounced par-a-dime) is typically used by people who want to sound a little more pompous and intellectual than you and I. A yuppie word. If you want to talk about how things are changing you can talk about a "paradigm shift." According to the Economist Magazine, Thomas Kuhn invented the notion of the paradigm shift to explain what happens in scientific revolutions. A revolution happens, his theory goes, not because of startling new facts, but because of a change in the overall way the universe is seen. After this shift, old knowledge suddenly takes on new meaning. A classic paradigm shift is the way we have changed our concepts of computing from mainframe centralized mainframe computing to distributed LAN-based computing.

Parallel Circuit In a parallel circuit there are at least two paths for the electric current to flow through. To find the resistance of a parallel circuit, add up the reciprocal of the resistance of each of the paths the electric current may follow. The reciprocal of the sum is the total resistance of the circuit. As components are added to a circuit in series the total resistance of the circuit increases. As components are added to a circuit in parallel, the total resistance of the circuit decreases. Household wiring is the most common type of parallel circuitry. Every outlet is parallel with every other outlet on the same circuit-breaker. So, if a bulb blows in a light fixture on that circuit, all the other devices on that circuit will still function. Another explanation. Imagine three 1.5 volt batteries. With the batteries connected in parallel, the circuit will deliver 1.5 volts. Connected in series it will deliver 3 x 1.5 volts, or 4.5 volts. By contrast, a string of Christmas tree lights is strung in a series circuit, one long continuous circuit. Should one bulb go out, usually the rest will also. See also PARALLEL DATA.

Parallel Connection A connection in which the current divides, only a part of the total current passing through each device.

Parallel Data The transmission of bits over multiple wires at one time. This is usually accomplished by having one wire for each bit of an eight-bit byte going from a device, usually a computer, to another device, usually a printer. Thus the word "Parallel." Data transmission in parallel is very fast, but usually happens only over short distances (typically under 500 feet) because of the need for huge amounts of cable. In contrast, the other common method of data transmission, serial transmission, takes place over one pair of wires and is usually slower than parallel transmission, but can happen over

much longer distances, especially using phone lines. Parallel data transmission does not happen on phone lines. See SERIAL DATA and see the APPENDIX.

Parallel Interfacing A method of interfacing peripherals to computers, usually printers. Not as common as RS-232-C serial interfacing.

Parallel Networks Parallel networks, or segregated networks, exist when a single network location supports more than one physical wide area network connection for the purpose of supporting one or more applications.

Parallel Port An output receptacle often located on the rear of a computer. Unlike serial, there is no EIA standard for parallel transmission, but most equipment adheres to a quasi-standard called the Centronics Parallel Standard. Almost every PC since the original IBM PC has come with an ordinary, 25-pin D-connector parallel port. These low-speed ports are fine for sending output to a printer (which is usually the slowest device in a computer system). But when transferring data between two PC parallel ports or using the parallel port as a method of getting to and from external hard disks, the speed is too slow. As a result, there have been a number of attempts to speed up and add intelligence to the lowly parallel port. How can you tell what type of parallel port you have? A free utility called PARA14.ZIP is available on the Internet at netlab2.usu.edu/misc, or on CompuServe in the IBMHW forum, Library2. See EPP (Enhanced Parallel Port), ECP (Extended Capabilities Port) and USB (Universal Serial Bus).

Parallel Processing 1. A computer technology in which several or even hundreds of low-cost microprocessors are linked and able to work on different parts of a problem simultaneously.

2. A computer performs two or more tasks simultaneously. This contrasts with multi-tasking in which the computer works fast and gives the impression of performing several tasks at once.

Parallel Sessions In IBM's SNA, two or more concurrently active sessions between the same two logical units (LUs), using different network addresses. Each session can have different transmission parameters.

Parallel Tasking Technology which allows LAN adapters to transmit data to the network before an entire frame has been loaded from the computer to the adapter's buffer and to transmit the data to the computer's main memory before an entire frame has been received from the network. In effect, a frame can reside on the network, the adapter and in the computer memory simultaneously, thus boosting throughput.

Parallel Transmission 1. Method of information transfer in which all bits of a character are sent simultaneously as opposed to serial transmission where the bits are sent one after another.

2. Method of achieving higher system reliability through use of completely redundant transmission facilities.

Parallel Wiretaps A parallel wiretap is connected across the two lines of a telephone line pair in parallel with the telephone instrument. A parallel must have a high resistance, otherwise the telephone line will be closed and the central office will think that the telephone is off hook. Parallel wiretaps can use high value resistors to isolate the tap. These are easy to detect. Serious wiretaps will use capacitors to isolate the tap from the telephone line.

Parameterize A verb which means to control the behavior of a piece of software by supplying required data at time of execution. A basic precept of structured programming is that, wherever possible, data should be kept separate from active program code. If, for example, you're writing a function that dials a phone number, the phone number should never be "hardwired" into the function itself — but should be passed to it, at runtime, as a so-called "argument" or "parameter." This encourages the creation of flexible, bulletproof software components that are easy to re-use. Much application software is written so that its behavior can be controlled by supplying parameters at time of execution. For example, the DOS command 'delete' requires a filename (the name of a file to be deleted) or wildcard expression as a parameter. More complex application programs (e.g., voicemail systems) are configured by supplying parameters on special screens or in dialog boxes, or by modifying external parameter databases.

Parameters The record in a stored program control central office's data base that specifies equipment and software and options and addresses of peripheral equipment for use in call processing. See also PARAMETERIZE.

Parametric Equalizer A device for manipulating sound by boosting and cutting selected frequencies by specific amounts. Basically, a much more elaborate and precise version of the bass and treble controls found on stereo systems. See EQUALIZATION.

PARC Palo Alto Research Center. A laboratory owned by Xerox and populated by it in the 1970s with some of the most creative scientists of the day. It is legendary for having pioneered technologies ranging from the laser printer to the Ethernet local area network and the graphical user interface for PCs. Sadly, Xerox senior management didn't recognize what it had and didn't recognize the value of its inventions, and didn't exploit most of them. Fortunately, other companies did.

Parasite A radio tap that takes its power from the phone line.

Parent Node The logical group node that represents the containing peer group of a specific node at the next higher level of the hierarchy.

Parent Peer Group The parent peer group of a peer group is the one containing the logical group node representing that peer group. The parent peer group of a node is the one containing the parent node of that node.

Parity A process for detecting whether bits of data (parts of characters) have been altered during transmission of that data. Since data is transmitted as a stream of bits with values of one or zero, each character of data composed of, say seven bits has another bit added to it. The value of that bit is chosen so that either the total number of one bits is always even if Even Parity error correction is to be obeyed or always Odd if odd Parity error correction is chosen.

Here's an explanation (better but longer) from The Black Box Corporation in Pittsburgh: Many asynchronous systems append a parity bit following the data bits for error detection. Parity bits trap errors in the following way. When the transmitting device frames a character, it counts either the number of 0s or 1s in the data bits and appends a parity bit that corresponds to whether or not the count in the data bits was even or odd. The receiving end also counts the data bit 0s or 1s as it receives them and then compares the computation to the parity bit. If an error is detected, a flag can be set and retransmission may be requested. When even parity is chosen, the parity bit is set at 0 if the number of 1's in the data bits is even and it is set at 1 if the number of 1's is odd. Conversely, odd parity sets the parity bit at 1 if the number of 1's in the data bits is even, and it is set at 0 if the number of 1's is odd. Other parity selections include mark, space or off. Mark parity always sets parity at 1. Space parity always sets parity at 0, and "off" tells the system to ignore the parity bit.

Parity Bit A binary bit appended to an array of bits to make the sum of all the bits always odd or always even. See PARITY and ASCII. See PARITY.

Parity Check A method of error-detection in binary data transmission whereby an extra bit is added to each group of bits (usually a character of data). If parity is to be odd, then the extra or parity bit is assigned either a one or zero so the total number of ones in the character will be odd. If the parity is even, the parity bit is assigned a value so that the total number of ones in the character is even. This way errors can be detected. See PARITY.

Park 1. A telephone system feature that (like many features) may mean different things depending on who created it. One definition of "park" is that I dial another extension and park the call at that extension. It doesn't ring. Then I go over to that extension and pick up the phone and I'll be speaking with whoever I parked over there. This feature is useful if I have to go to another phone to find some information the caller wants. There's another definition of the telephone system meaning of "park." You have a single line phone. You put that call on a variation of hold. Then you or anyone else can pick up any phone in that pickup group and you will have your parked call. 2. In the language of hard disks, "parking" means moving the read/write head to a safe area of the hard disk when you're ready to turn the disk off. "Parking" places the heads of a hard disk in a locked position so that the storage medium (i.e. the hard disk) will not be damaged during transit. This is useful because it keeps the head from bouncing on data areas of the disk and damaging the disk. Some hard disks have a program called "park" which you run before you turn off the machine. Others do it (self-park) automatically. All hard disks on laptops are self-parking. Most modern disks are. You ought to check. It's very important.

Park Timeout A PBX feature. This is the period of time before an unanswered Call Park call is redirected to the Prime Phone for the line the call is on.

Parse In linguistics it means to divide the language into components that can be analyzed. Parsing a sentence involves dividing it into words and phrases, then identifying and naming each component. Parsing is very common in computer science. Compilers must parse source code to translate it into object code. Applications that processes complex commands must also parse the commands. Parsing is divided into lexical analysis and semantic parsing. Lexical analysis divides strings into components, called tokens, based on punctuation and other keys. Semantic parsing works to define the meaning of the string once it's been broken down into individual components.

Part 68 Requirements Specifications established by the FCC as the minimum acceptable protection communications equipment must provide the telephone network. Meeting these requirements does not certify that equipment performs any task. Part 68 is the section of Title 47 of the Code of Federal Regulations governing the direct connection of telecommunications equipment and premises wiring with the public switched telephone network and certain private line services, e.g., foreign exchange lines (customer premises end), the station end of off-premises stations associated with PBX and Centrex services, trunk-to-station tie lines (trunk end only), and switched service network station lines (common control switching arrangements); and the direct connection of all PBX (or similar) systems to private line services for tie trunk type interfaces, off-premises station lines, automatic identified outward dialing and message registration. These rules provide the technical, procedural and labeling standards under which direct electrical connection of customer-provided telephone equipment, systems, and protective apparatus may be made to the nationwide network without causing harm and without a requirement for protective circuit arrangements in the service provider's network. Form 730 Application Guide is a collection of literature you'll need to register your telephone/telecom equipment under Part 68 of Title 47 at the Federal Communications Commissions. To get this material (it's free) drop a line or call the Federal Communications Commission, Washington DC 20554. FCC's general # is 202-418-0200. The Bureau Chief's # is 202-418-1500. As I write this edition, the person at the FCC in charge is Reed Hunt who also puts out a very useful newsletter for Part 68 applicants. The newsletter is called "The Billboard." You can file Form 730 yourself, but the Form 730 Application Guide also contains a list of Part 68 Certification Laboratories, a list of technical references and a list of reference sources. www.fcc.gov

Part X A reference to Part 64 or 68 of the MFJ (Modified Final Judgement) given to the RBOCS by Judge Harold Green. It specified the separation of customer-owned equipment and telephone owned equipment as well as telephone company demarcation.

Partial Meshed Network A type of wide area network topology in which every remote location is not connected directly to every other remote location, but instead is connected directly to a small subset of locations. It is a topology with more direct connectivity than a star configuration, but less direct connectivity than a fully meshed configuration.

Partially Perforated Tape Same as chadless tape. See CHADLESS TAPE.

Partition As a verb, partition means to divide a network into independent segments or to divide a disk or tape drive into independent volumes. As a noun, a partition is a division of memory or hard disk. For example, the MS-DOS and Windows operating systems can allocate hard disk space for one or more partitions, each of which behaves as a physically distinct hard disk. The largest MS-DOS partition permitted is two gigabytes.

Partitioned Emulation Programming Extension PEP. An IBM special software package that, with the Network Control Program (NCP), allows the same communications controller to operate in split mode, controlling an SNA network while at the same time managing a number of non-SNA communications lines. It was developed by IBM to facilitate migration of users to SNA.

Partitions Sections on a hard disk. You can divide your hard disk into as many as four partitions to run four operating systems. See also 32-MEGABYTE BARRIER.

Party A particularly stupid word for the person making or receiving a phone call, as in the calling party (caller) or the called party (person called). Sometimes the phone industry calls a subscriber to their service a "party," as in four-party phone lines. Party is now now used in the airline business, as in "How many people will there be in your party?" As if traveling were fun any longer.

Party Identification Identifying the person who is placing a call on a party line, i.e. phone line with several people sharing it. Often found in rural locations.

Party Line 1. Saying what your company or boss wants you to say.

2. A telephone line with several subscribers sharing its use.

Party Line Service Telephone service which provides for two or more phones to share the same loop circuit. Party line service, which is becoming less common, is offered in two-

party, four-party and eight-party versions. Interestingly, there is a version of ISDN in which several subscribers do share the same ISDN line — but they would rarely be affected by it because of ISDN's specialized signaling and the two phone lines in its 2B+D bandwidth.

Party Line Stations Two party phone service can be expanded to support to multi-party service.

Pascal A programming language designed for general information processing and noted for its structured design. Pascal originally was specified by Niklaus Wirth, a computer scientist at the Institut fur Informatic in Zurich, Switzerland in 1968. It is named in honor of Blaise Pascal, a 17th century mathematician who developed one of the first calculating machines.

Pass Through The process of accessing one device via another device. The intermediate device that sends backs the transmitted messages for testing.

Pass Window The range of frequencies used in a transmission system to transmit voice or data signals. More often referred to as bandwidth. See BANDWIDTH.

Passband The range of frequencies that can pass through a filter without being attenuated (i.e. stopped).

Passive Backplane A technology where all of the active circuitry that is normally found or an "active" PC motherboard (such as the CPU) is moved onto a plug-in card. The motherboard itself is replaced with a passive backplane that has nothing on it other than connectors and joining, etched-in wires. This is why this technology is sometimes referred to as "slot cards". The chance of a passive backplane failing is very low, since it has essentially no functioning componentry. A passive backplane has several advantages: You can swap cards in and out faster. You can upgrade your processor and change faster and easier. A computer made with a passive backplane will typically have slots — as many as 25 versus only 8 or so in a "normal" PC. Passive backplane computers are increasingly used in critical computer telephony applications.

Passive Bus ISDN feature which allows up to six terminal devices and two voice devices (also called telephones) to simultaneously share the same twisted pair, each being uniquely identifiable to the switched ISDN telephone network. See ISDN.

Passive Contract In the software business, there are two types of contracts. One you sign and one you don't. A passive contract is the one you don't sign. A passive contract typically comes with an over-the-counter, shrink-wrapped software package and you execute it by breaking the seal on the package. The passive contract spells out terms and conditions you agree to — like not copying the software, not selling, etc.

Passive Coupler A coupler that divides entering light among output ports without generating new light.

Passive Device Electronic components that don't require external power to manipulate or react to electronic output. These include include capacitors, resisters and coils (inductors). Active devices include transistors, op amps, diodes, cathode ray tubes and ICs.

Passive Headend A device that connects the two broadband cables of a dual-cable system. It does not provide frequency translation.

Passive Hub A device used in certain network topologies to split a transmission signal, allowing additional workstations to be added. A passive hub cannot amplify the signal, so it must be connected directly to a workstation or an active hub.

Passive Leg A telephone company AIN term. The leg to a terminating access of an SSP or ASC switch. There is no access signaling on a passive leg to directly control the progress of a call.

Passive Optical Network See PON.

Passive Reflector A simple reflector used to change the direction of radiation from a microwave beam. For example, a reflecting surface mounted on a hill top and so positioned as to direct the energy down onto a valley receiving site. See PASSIVE REPEATER.

Passive Repeater A passive reflector system constructed from two reflectors that are simply coupled together with a short length of waveguide. The first reflector acts as a receiver while the second transmits but in a different direction. There's no electronics in the system.

Passive Side When describing a loopback test, the passive side is used to identify the device that sends back the transmitted messages for testing.

Passive Splicing Aligning the two ends of a fiber without monitoring its splice loss.

Passive Star A star-topology local network configuration in which the central switch or node is a passive device. Each station is connected to the central node by two links, one for transit and one for receive. A signal input on one of the transmit links passes through the central node where it is split equally among and output to all of the receive links. Also called a star coupler or a retransmissive star.

Passive Terminator A crude type of single-ended SCSI terminator that can't compensate for variations in terminator power or bus impedance. No longer recommended by ANSI, it's adequate for most simple SCSI-1 Applications. See also Active Terminator and Forced-Perfect Terminator.

Passthrough Gaining access to one network through another element. Also spelled PASS THROUGH.

Password A word or string of characters recognized by automatic means permitting a user access to a place or to protected storage, files or input or output devices.

Password Control Of Changes A feature that makes it impossible to alter the performance of a piece of equipment without first entering a password.

Patch 1. A small addition to the original software code, written to bypass or correct a problem. See Y2K.

2. To connect circuits temporarily with a jack and a cable. Patching is typically done on devices called PATCH BAYS, PATCHBOARDS or PATCH PANELS. See also PATCHING.

Patch Bay A collection of hardware put together in such a way that circuits appear on jacks and can be connected together for transmission, monitoring and testing. See PATCH PANEL.

Patch Cord A short length of wire or fiber cable with connectors on each end used to join communication circuits at a cross connect. For bigger explanation, see Cross Connect.

Patch Panel A device in which temporary connections can be made between incoming lines and outgoing lines. It is used for modifying or reconfiguring a communications system or for connecting devices such as test instruments to specific lines. A patch panel differs from a distribution frame in that the interconnections on a distribution frame are intended to be permanent.

Patchboard Same as a patch bay. See PATCH BAY.

Patching Means of connecting circuits via cords and connectors that can be easily disconnected and reconnected at another point. May be accomplished by using modular cords connected between jack fields or by patch cord assemblies that plug onto connecting blocks.

Patchmaster A patch panel in which multiple pair lines can be interconnected as a group.

Patent A patent is a grant, limited in time and technological extent, of the right to exclude others from making, using or selling the invention. The temporal extent of a U.S. patent (other countries' laws are similar) is usually at least 17 years. The technological extent of protection is defined by the claims of the patent. Claims are allowed only after examination by a technically degreed examiner of the Patent and Trademark Office to ensure that an opinion is defined that is nonobvious with respect to the predecessor technology, or "prior art." Having a patent does not confer the right to manufacture the thing claimed; others may have dominating patents. Independent invention is no defense to a charge of infringement. Rights in an invention made by an employee in performance of his normal duties belong to the employer in most states, even without a written agreement.

The claims made for the invention in applying for the patent may define a "process, machine, manufacture, or composition of matter". The great software patent controversy has largely been settled with the understanding that software is normally best claimed as a process, although sometimes claims usefully intermix hardware and software. The only software that remains per se unpatentable is that defining a process operating on pure numbers, that is, not tied to any particular end use, control process, or the like. A Fast Fourier Transform algorithm would thus not be patentable although an unobvious method of using it to (say) remove harmonic noise from a signal might be.

Patent serve several functions. A patent is a technical disclosure, forming part of the scientific literature, as the invention must be described with sufficient particularity that others can use it "without undue experimentation." The idea is that the inventor is given a limited monopoly in exchange for the benefit to the public of having inventions made and disclosed. A patent is also a legal document, in that the claims define the exact extent of protection. The third function of a patent is to market the invention.

A well-written patent explains the underlying technical problem, the deficiencies of the prior art, and the way in which invention solves these problems, in clear and non-technical language. Only then can a federal judge or jury, or the CEO of a competitor being asked to pay damages, be expected to understand the relation of the claimed invention to the prior art and the allegedly infringing product.

Path The route a telecommunications signal follows through a circuit or through the air.

Path Clearance In through-the-air microwave transmission, you must find a line-of-sight path, free of obstruction of buildings, trees, other microwave towers, etc. In microwave line-of-sight communications, the perpendicular distance from the radio-beam axis to obstructions such as trees, buildings, or terrain. The required path clearance is usually expressed, for a particular k-factor, as some fraction of the first Fresnel zone radius. That's the technical definition of path clearance.

Path Constraint A bound on the combined value of a topology metric along a path for a specific connection.

Path Control IBM Corp.'s implementation of what is normally referred to as the network layer in the International Standards Organization Open Systems Interconnect (OSI) layered network architecture.

Path Control Layer In IBM's SNA, the network processing layer that handles primarily the routing of data units as they travel through the network and manages shared link resources.

Path Switched Ring A technique for providing redundancy in a SONET network. Path switched rings use 2 fibers, with both transmitting simultaneously in both directions. Through this technique, a failure in a SONET ring will not prevent devices from communicating, as they transmit and receive in both clockwise and counter-clockwise directions. All devices monitor both rings, locking in on the better signal, thereby improving on the inherently high quality of fiber optic transmission. See Line Switched Ring.

Pathway A facility for routing communications cables.

Pattern Recognition A small element of human intelligence. The ability to recognize and match visual patterns. (Auditory pattern recognition is the ability to recognize spoken words.) Pattern recognition basically works by having the computer seek out particular attributes of the character (assuming it's pattern recognition for reading words) and then having the computer compare what it finds to what's in its database of patterns. By a process of breaking down letters into curves and lines, and by a process of elimination, the computer can figure out what it's seeing. As Forbes said, "think of pattern recognition as a kind of super detective, a tireless if unimaginative collector of clues, distinguished not by brilliance, but by ceaseless legwork.

Pause This feature on some phone systems which, by hitting #, inserts a 1.5 second delay in a speed dialing sequence. This way you can program your phone to call a main number, wait a few seconds for the machine to answer and then punch out your person's extension.

PAX Private Automatic eXchange. Typically an intercom system not joined to the public telephone system. PAXs are more common in Europe, where is it common for business people to have two phones on their desk — one for internal intercom calls and one for external calls.

Pay Phone See PAYPHONE

Payload 1. From the perspective of a network service provider: of a data field, block or stream being processed or transported, the part that represents information useful to the user, as opposed to system overhead information. Payload includes user information and may include such additional information as user-requested network management and accounting information. In Sonet, the STS-1 signal is divided into a transport overhead section and an information payload section (similar to signaling and data). See SPE (Synchronous Payload Envelope) for a description of what would be found in the payload.
2. The activity carried out by a computer virus when it is activated by a triggering event. Depending on the virus, the payload may be as benign as putting a message on your screen or as destructive as erasing your hard disk or scrambling your data.

Payload Type Indicator Field PTI. A three-bit field in the ATM cell header that indicates the type of information being carried in the payload. The PTI is used to distinguish between cells carrying user data and those carrying service information such as call set-up and call termination.

Payphone Used to be just a public phone that accepted only coins. Now pay phones can be coinless and can read credit cards. Soon they will be acquiring keyboards, computer screens and dataports for plugging in fax machines and portable computers. The payphone was invented by William Gray, an American whose previous inventions included the inflatable chest protector for baseball players. Mr. Gray's first phone lacked a dial. Its instructions read: "Call Central in the usual manner. When told by the operator, drop coin in proper channel and push plunger down." In today's nomenclature, Mr. Gray's original phone is known

as a post-pay coin phone. See other entries below.

Payphone-Postpay Calls are paid for after they are completed, typically with a credit card or calling card, etc.

Payphone-Prepay At a coin phone, calls must be paid for before they can be dialed. Virtually all local calls are prepay.

Payphone-Private Referred to as Customer Owned Coin Operated Telephone Companies (COCOTs). Installed and maintained by companies other than local exchange carriers who are rapidly entering this industry. COCOTs may have access to more than one IXC.

Payphone-Public A coin phone installed in a "public" place. The local operating Company is totally responsible for its installation. The phone company will typically pay someone — the city, the bus station owner — a commission on the calls made from this phone. Also see PAYPHONE-SEMI-PUBLIC.

Payphone-Semi-Public A coin phone installed for public use but installed in a "semi-public" place, such as a restaurant or bar. The proprietor of the establishment is obliged to guarantee that the phone company will receive a minimum amount of money out of the phone. The phone company will typically not pay a commission on this type of phone and takes all the money in the coin box for itself. What is a "public" and what is a "semi-public" phone is a decision made by the local telephone company for whatever reason it chooses. The pay phone business is rapidly deregulating. It is now legal to own your own payphone.

Paystation, Postpay Calls are paid for after they are completed, typically with a credit or calling card, etc.

Paystation, Prepay Calls must be paid for before they can be dialed. Virtually all local calls are prepay.

Paystation, Public A coin phone installed in a "public" place. The phone company is totally responsible for its installation. The phone company will typically pay someone — the city, the bus station owner — a commission on the calls made from this phone. See PAYSTATION, SEMI-PUBLIC.

Paystation, Semi-Public A coin phone installed for public use but installed in a "semi-private" place, such as a restaurant or bar. The proprietor of the establishment is obliged to guarantee that the phone company will receive a minimum amount of money out of the phone. The phone company will typically not pay a commission on this type of phone and takes all the money in the coin box for itself. What is a "Public" and what is a "Semi- Public" phone is a decision made by the local telephone company for whatever reasons it chooses. The pay phone business is rapidly deregulating. So the rules are changing. And it is now legal to own your own payphone.

PB Petabyte. See Petabyte.

PBS Personal Base Station. A PCS (Personal Communications System) term. A PCS subscriber might use a "High-Tier" PCS service, which effectively is cellular service using PCS frequencies. When at home, the PCS set acts as a cordless phone, establishing a wireless link to the PBS. When in close enough proximity to have sufficient signal strength, the PBS takes over from the PCS carrier's cell site. All PBS calls then are routed over the landline PSTN, thereby avoiding cellular usage charges. In a business environment using a PCS wireless office system, the PCS set and the wireless controllers establish the same relationship.

PBX Private Branch eXchange. A private (i.e. you, as against the phone company owns it), branch (meaning it is a small phone company central office), exchange (a central office was originally called a public exchange, or simply an exchange). In other words, a PBX is a small version of the phone company's larger central switching office. A PBX is also called a Private Automatic Branch Exchange, though that has now become an obsolete term. In the very old days, you called the operator to make an external call, except in Europe. Then later someone made a phone system that you simply dialed nine (or another digit — in Europe it's often zero), got a second dial tone and dialed some more digits to dial out, locally or long distance. So, the early name of Private Branch Exchange (which needed an operator) became Private AUTOMATIC Branch Exchange (which didn't need an operator). Now, all PBXs are automatic. And now they're all called PBXs, except overseas where they still have PBXs that are not automatic.

At the time of the Carterfone decision in the summer of 1968, PBXs were electro-mechanical step-by-step monsters. They were 100% the monopoly of the local phone company. AT&T was the major manufacturer with over 90% of all the PBXs in the U.S. GTE was next. But the Carterfone decision allowed anyone to make and sell a PBX. And the resulting inflow of manufacturers and outflow of innovation caused PBXs to go through five, six or seven generations — depending on which guru you listen to. (See my definition for GENERATIONS in this dictionary). Anyway, by the fall of 1991, PBXs were thoroughly digital, very reliable, and very full featured. There wasn't much you couldn't do with them. They had oodles of features. You could combine them and make your company a mini-network. And you could buy electronic phones that made getting to all the features that much easier. Sadly, by the late 1980s the manufacturers seemed to have finished innovating and were into price cutting. As a result, the secondary market in telephone systems was booming. Fortunately, that isn't the end of the story. For some of the manufacturers in the late 1980s figured that if they opened their PBXs' architecture to outside computers, their customers could realize some significant benefits. (You must remember that up until this time, PBXs were one of the last remaining special purpose computers that had totally closed architecture. No one else could program them other than their makers.) Some of the benefits customers could realize from open architecture included:

• Simultaneous voice call and data screen transfer.
• Automated dial-outs from computer databases of phone numbers and automatic transfers to idle operators.
• Transfers to experts based on responses to questions, not on phone numbers.

And a million more benefits. We discuss them at our annual trade show called TELECOM DEVELOPERS held in May each year. Call 212-691-8215 for more information. For more on open architecture, see OAI.

An alternative to getting a PBX is to subscribe to your local telephone company's Centrex service. For a long explanation on Centrex and its benefits, see CENTREX. Here are some of the benefits of a PBX versus Centrex:

1. Ownership. Once you've paid for it, you own it. There are obvious financial and tax benefits.
2. Flexibility. A PBX is a far more flexible than a central office based Centrex. A PBX has more features. You can change them faster. You can expand faster. Drop another card in, plug some phones in, do your programming and bingo you're live.
3. Centrex benefits. You can always put Centrex lines behind a PBX and get the advantages of both. In some towns, Centrex lines are cheaper than PBX lines. So buy Centrex lines and put them behind your PBX. Make sure you don't pay for Centrex features your PBX already has. (It has most.)
4. PBX phones. There are really no Centrex phones — other than a few Centrex consoles. If you want to take advantage of

Centrex features, you have to punch in cumbersome, difficult-to-remember codes on typically single line phones. PBXs have electronic phones, often with screens and dedicated buttons. They're usually a lot easier to work. A lot easier to transfer a call. Conference another, etc. A lot more productive.
5. Footprint savings. Modern PBXs take up room, more than Centrex. But the space they take up is far less than it used to be. PBXs are getting smaller.
6. Voice Processing/Automated Attendants. Centrex's DID (Direct Inward Dialing) feature was always pushed as a big "plus." You saved operators. However, you can now do operator-saving things with PC-based voice processing and automated attendants you couldn't do five years ago. These things work better with on-site standalone PBXs than with distant, central office based Centrex. Moreover, virtually every PBX in existence today supports DID. You can dial directly into PBXs and reach someone at their desk just as easily as you can dial directly using Centrex.
7. Open Architecture. Most PBXs have open architecture. See OAI for the benefits. Central offices don't.
8. Good Reliability. There have been sufficient central office crashes and sufficient improvement in the reliability of PBXs that you could happily argue that the two are on a par with each other today. Both are equally reliable, or unreliable. The only caveat, of course, is that you back your PBX up with sufficient batteries that it will last a decent power outage. Of course, that assumes that your people will be prepared to hang around and answer the phones during a blackout.
9. Expansion. Central offices are big. Allegedly you can grow your lines to whatever size you want. In contrast, PBXs have finite growth. It's true about PBXs. But it's equally true about central offices. I've personally heard too many stories about central office line shortages to believe in the nonsense about "infinite Centrex" growth. Fact is central offices grow out, just like PBXs. Given the tight economy of recent years, local phone companies have not been buying the central offices they should have. And they have been filling central offices up a little too tight for my taste.
10. Technological obsolescence. Allegedly central offices are upgraded faster than PBXs and therefore are always up to date technologically. It's nonsense. The life cycle of a typical central office was 40 years until recently. It's now around 20 years. Think of what's happened to PCs in the past 10 years — the IBM PC debuted only in 1981 — and you can imagine how obsolete many of the nation's central offices are.

PBX Central Office Trunk PBX central office (CO) trunks connect the PBX switch to the central office serving the PBX location. The trunks appear as station lines at the central office equipment.

PBX Driver Profiles Telephony Services (also called TSAPI) is software which AT&T (now Lucent) invented to run on NetWare servers and allow those servers to communicate with PBXs. According to a presentation made by Oliver Tavakoli of Novell in late August, 1994, "The purpose of TSAPI is to make it easier for developers to create telephony applications. Given the broad scope of TSAPI and the number of ways in which PBX vendors can package and deliver solutions, applications being developed against TSAPI are unlikely to work with a large cross-section of PBXs. This is particularly problematic for application developers wishing to add simple telephony features to a product that is not telephony centric. In order to spur the growth of CTI application development, Novell has created 'profiles' that are placed on top of TSAPI. The profiles listed in this document (i.e. the speech)

are based on documents created by other organizations:
"• A NOTA document entitled PABX Driver NLM Conference Meeting and Draft Recommendations dated July 20th, 1994
"• An ECMA document entitled 'Technical report on CSTA Scenarios/Second Draft' dated July 1994.
"Each profile is made up of the following components: Functions (and matching confirmation events) included in the profile; unsolicited events that may be encountered in the event flow scenarios in the profile; event flow scenarios that attempt to describe what should occur in the common scenarios encountered when using the functions in the profile.
"Group A Profile: Using the functions and event flows in the Group A profile, an application should be able to: provide "screen pops" to the end user; make outbound calls originating at an end user's device; provide hands-free operation for answering the phone and making calls (assuming the PBX and end user's device support hands-free operation)
"Group B Profile: Using the functions and event flows in the Group B profile, an application should be able to: perform operations involving two calls on a single device (the connection to one of the calls is in the held state); perform operations involving more than two parties on the same call; obtain information and unsolicited events from a call-centric view (as opposed to a device-centric view).
"Group C Profile: Using the functions and event flows in the Group C profile, an application should be able to: perform the rest of the telephony functions usually done from a phone — pickup, group pickup, call completion and perform more sophisticated monitoring functions.
(There is apparently no Group D profile.)
"Group E Profile: Using the functions in the Group E ('E' for Environmental) profile, an application should be able to: activate and deactivate features of a device, query the state of features on a device, query the generic information about a device."
Once the profiles are finalized, they will be incorporated into Novell's PBX Driver certification process; groups supported by the driver will be included in literature describing all certified drivers. See also NETWARE TELEPHONY SERVICES.

PBX Extension A telephone phone line connected to a PBX.

PBX Fraud Same as TOLL FRAUD.

PBX Generations See GENERATIONS, PBX.

PBX Integration A loose term to mean joining the PBX to any number of outside computer based gadgets and services, from voice mail to call accounting. To make voice mail integrate into PBXs, you minimally need the ability to provide a message waiting indicator (light or stutter dial-tone) at the user's phone when a message is received, and to forward a call to the user's mailbox when a call is sent to the recipient and they are on the phone or do not answer (forward on busy or ring no answer). This requires PBX "integration." Most PBX integrations provide the attached voice mail system call data that includes the calling phone number, the number of the caller, why the call was presented (such as forwarded on busy, or ring no answer), to pass message waiting on or off indications. PBX integration data may be implemented in-band or out-of-band on a separate link, most often a serial link. Some PBXs "integrate" with outside equipment better than others.

PBX Profiles PBXs do things differently. To make a conference call, one PBX's phone may put the caller on hold automatically, while another may insist that you put that person on hold manually and then dial the next person to join the conference call. As Novell in the fall of 1994 attempted to get as

many PBXs to conform to TSAPI, it discovered that PBX features often work very differently. So it decided to categorize PBXs and their features. This, it called, PBX profiles. The idea being that Profile A would contain the most common, easy-to-integrate-to-TSAPI features. B would contain the second most common, etc. Novell also calls PBX Profiles PBX Driver Profiles. See PBX DRIVER PROFILES for a much bigger explanation.

PBX Station Line A transmission path extending from the station (phone instrument) location to the switching equipment.

PBX Tie Line A tie line between two PBX's, permitting extensions in one PBX to be connected to extensions in the other without having to dial through the public switched network. See also OPX and OPS, which are different and are lines between PBXs and distant extensions, not tie lines between PBXs.

PBX Trunk A circuit which connects the PBX to the local telephone company's central office switching center or other switching system center.

PC 1. Personal Computer. See PERSONAL COMPUTER for a bigger explanation.
2. Peg Count.
3. Point Code.
4. Printed Circuit.
5. Product Committee.
6. Protocol Control.
7. Politically Correct.

PC Administration Server A Sun Microsystems term, Part of Solaris' Server Suite. Automates and centralizes PC network administration.

PC As Phone See also HANDSET MANAGEMENT.

PC Card A memory or I/O card compatible with the PCMCIA PC Card Standard. In short, PC Cards are a new name for PCMCIA cards. For a much fuller definition, see PCMCIA, which stands for the Personal Computer Memory Card International Association.

PC Centric There are two ways you can organize a computer to control telephone calls on an office telephone system. One way is to join a file server on a local area network to a phone system. Commands to move calls around are passed from the desktop PC over the LAN to the server and then to the phone system via the cable connection between the server and the system. A second way to get a computer to control phone calls is through a connection at the desktop. This is called PC Centric. There are two ways you can do this. The first is to join the desktop phone to the computer with a cable. This is often done via the PC's serial port connecting via cable to the phone's data communications port (if it has one — if it doesn't, you get one). The second way to be PC Centric is by simply replacing the standalone phone with a board that emulates in a phone and drop it into the PC's bus.

PC Network IBM's first LAN (Local Area Network).

PC Telephony Another term for Computer Telephony. See COMPUTER TELEPHONY.

PC's The plural of the word PC, according to the New York Times. However, every other computer and general magazine spells them PCs. And that's the spelling which this dictionary writer prefers also.

PCA 1. Premises Cabling Association. A association in Great Britain.
2. Protective Connecting Arrangement. A device that AT&T and members of the Bell System insisted be connected between a telecommunications device (like a phone) that was-

n't made and sold by AT&T and a phone line provided by a local Bell operating phone company. Many years later, the PCAs were found by the FCC to be totally unnecessary and AT&T and members of the Bell System were ordered to refund all payments received for rental of PCAs. The Bell System insisted on the PCAs as a way of protecting AT&T's effective monopoly of telecommunications equipment. See also PROTECTIVE CONNECTIVE ARRANGEMENT.

PCB Printed Circuit Board.

PCC Personal Companion Computer. What other companies call a PDA (Personal Digital Assistant), Intel calls a PCC. A PCC or PDA is meant to have significant telecommunications abilities — including wired and wireless. See PDA.

PCCA AT Command Set The new PCCA AT command set for wireless modems contains well-defined commands for obtaining link status information.

PCF Physical Control Fields. The AC (Access Control) and FC (Frame Control) bytes in a Token Ring header.

PCH Paging CHannel. Specified in IS-136, PCH carries signaling information for set up and delivery of paging messages from the cell site to the user terminal equipment. PCH is a logical subchannel of SPACH (SMS (Short Message Service) point-to-point messaging, Paging, and Access response CHannel), which is a logical channel of the DCCH (Digital Control CHannel), a signaling and control channel which is employed in cellular systems based on TDMA (Time Division Multiple Access). The DCCH operates on a set of frequencies separate from those used to support cellular conversations. See also DCCH, IS-136, PAGING, SPACH and TDMA.

PCI 1. Protocol Control Information. The protocol information added by an OSI entity to the service data unit passed down from the layer above, all together forming a Protocol Data Unit (PDU).
2. Peripheral Component Interconnect, a 32 bit local bus inside a PC or a Mac designed by Intel for the PC. According to Intel, it can transfer data between the PC's main microprocessor (its CPU) and peripherals (hard disks, video adapters, etc.) at up to 132 megabytes per second, compared to only five megabytes per second which the original PC's ISA bus is capable of. PCI is one of two widely adopted local-bus standards. The other, the VL-Bus, is primarily used in 486 PCs. See also CompactPCI and VLB.

PCIA Personal Communications Industry Association. The association of the new cellular providers.

PCL 1. Hewlett-Packard's Printer Control Language, developed by HP in 1984 as a way for the then-new PC to communicate with a new breed of laser printers — the HP LaserJet printer. HP's PCL language is now the de facto industry standard for PC printing. Most of the printers in the world today are equipped with PCL or a PCL-compatible language. PCL allows the type of sophisticated page creation generally referred to as "laser quality output." PCL supports such advanced features as fully scalable typefaces and rotation of text. PCL defines a standard set of commands enabling applications to communicate with HP or HP-compatible printers. PCL has become a de facto standard for laser and ink jet printers and is supported by virtually all printer manufacturers. On April 8, 1996 HP announced PCL 6 which it billed as "the next generation" of HP Printer Control Language). HP said that PCL 6 includes font synthesis technology for true what-you-see-is-what-you-get (WYSIWYG) printing and better document fidelity. PCL 6 commands were designed by HP to closely match Microsoft Windows GDI (Graphical Direct Interface) commands.

2. Product Compute-Module Load.

PCLEC Packet Competitive Local Exchange Carrier.Covad invented this term for a CLEC who provides dedicated high-speed digital communications services using DSL technology to Internet Service Providers ("ISPs") and corporate enterprise customers.

PCM Pulse Code Modulation. The most common method of encoding an analog voice signal into a digital bit stream. First, the amplitude of the voice conversation is sampled. This is called PAM, Pulse Amplitude Modulation. This PAM sample is then coded (quantized) into a binary (digital) number. This digital number consists of zeros and ones. The voice signal can then be switched, transmitted and stored digitally. There are three basic advantages to PCM voice. They are the three basic advantages of digital switching and transmission. First, it is less expensive to switch and transmit a digital signal. Second, by making an analog voice signal into a digital signal, you can interleave it with other digital signals — such as those from computers or facsimile machines. Third, a voice signal which is switched and transmitted end-to-end in a digital format will usually come through "cleaner," i.e. have less noise, than one transmitted and switched in analog. The reason is simple: An electrical signal loses strength over a distance. It must then be amplified. In analog transmission, everything is amplified, including the noise and static the signal has collected along the way. In digital transmission, the signal is "regenerated," i.e. put back together again, by comparing the incoming signal to a logical question: Is it a one or a zero? Then, the signal is regenerated, amplified and sent along its way.

PCM refers to a technique of digitization. It does not refer to a universally accepted standard of digitizing voice. The most common PCM method is to sample a voice conversation at 8000 times a seconds. The theory is that if the sampling is at least twice the highest frequency on the channel, then the result sounds OK. (See NYQUIST THEOREM.) Thus, the highest frequency on a voice phone line is 4,000 Hertz. So one must sample it at 8,000 times a second. Many PCM digital voice conversations are typically put on one communications channel. In North America, the most typical channel is called the T-1 (also spelled T1). It places 24 voice conversations on two pairs of copper wires (one for receiving and one for transmitting). It contains 8000 frames each of 8 bits of 24 voice channels plus one framing (synchronizing bit) bit which equals 1.544 Mbps, i.e. 8000 x (8 x 24 + 1) equals 1.544 megabits.

Countries outside of the United States and North America use a different scheme for multiplexing voice conversations. It is

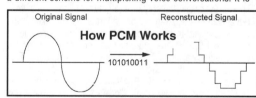

based not on 24 voice channels, but on 32. This scheme keeps two of the 32 channels for control, actually transmitting 30 voice conversations at a data rate of 2.048 Mbps. The European system is calculated as 8 bits x 32 channels x 8000 frames per second. European PCM multiplexing is not compatible with North American multiplexing. The two systems cannot be directly connected. Some PBXs in the U.S. conform to the U.S. standard only. Some (very few) conform to both. Both the European and North American T-1 "standards" have now been accepted as ISDN "standards." In addition to PCM, there are many other ways of digitally encoding voice. PCM remains the most common.

See T-1 and VOICE COMPRESSION.

PCM Voice Transmission Synchronization There are three levels in PCM voice transmission synchronization:

1. Bit Level Synch — operating the transmitter and receiver at the same bit rate so the bits are not lost.

2. Frame Level Synch — phase alignment between the transmitter and receiver so the beginning of the frame can be identified.

3. Time Slot Synch — phase alignment between the transmitter and receiver so the time slots are lined up for information retrieval.

PCM-30 Short name of international 2.048 Mbps T-1 (also known as E1) service derived from the fact that 30 channels are available for 64 Kbps digitized voice each using pulse code modulation (PCM).

PCMCIA The Personal Computer Memory Card International Association (an awful mouthful) standardizes credit-card size packages for memory and input/output (modems, LAN cards, etc.) for computers, laptops, palmtops, etc. There are three physical standards for PCMCIA cards — Type I, II, and III and undefined standard called Type IV, which only Toshiba has at this moment. The cards are 69.2 millimeters (3.37 inches) long x 51.46 millimeters (2.126 inches) wide. All three types use the same 68 female pin edge connector for attachment to the computer, and differ only in thickness. The thickness for Type I, Type II and Type III are 3.3, 5.0, 10.5 millimeters respectively. Toshiba's Type IV is 16 mm. A Type I PC Card is typically used for various types of memory enhancements, including RAM, FLASH memory, one-time programmable (OTP) memory, and electronically erasable programmable read only memory (EEPROM). A Type II PC Card (the most common) is typically used for input/output such as modem, LANs, host communications and SCSI device connection. A Type III PC Card is twice the thickness of the Type II and is typically used for I/O features that require a larger size, such as rotating mass storage devices (removable hard disk drives) and radio communication devices. Since Type I, Type II and Type III Cards all use the same interface, the size of the card chosen for the application is dependent on the miniaturization of the technology to be implemented.

PCMCIA's first standards were issued in September, 1991. The idea is that small computers will use these cards for modems, fax cards, hard disks, LAN connections, Ethernet connections, SCSI device connections, etc. A PCMCIA card is, in most cases, the only way to get to a laptop's bus without attaching a docking station. In terms of performance, the bus is comparable, but not equivalent to the ISA bus on desktops. The PCMCIA bus is 8 bits wide and does not allow for direct memory access (DMA) transfers or bus mastering. This means, for example, that you can't equip a laptop with a sound blaster compatible PCMCIA card, since it requires DMA transfers. A new spec called PCMCIA v3 (now called CardBus) calls for a number of improvements, including 32-bit data paths, DMA, bus mastering and significant improvements in speed with the bus as fast as the PCI bus.

PCMCIA standards exist so others can make these cards. Some computers — like some pre-summer 1991 notebooks and laptops from Compaq and Toshiba — don't comply because these computers were released before the standards were released. The PCMCIA v2.0 standard contains a software specification for XIP, the "eXecute In Place" mechanism that maps application software stored on the PCMCIA card into the system address space. This means application software will run directly from the card, start faster and not require precious RAM from the host computer.

One key element of the PCMCIA software architecture are Socket Services and Card Services. Socket Services is a BIOS level software interface that provides a way to access the PCMCIA sockets (slots) of a computer. Socket Services identifies how many sockets are in your computer system and detects the insertion or removal of a PC Card while the system is powered on. Socket Services is part of the PCMCIA Specification and interfaces with Card Services. Card Services is a software management interface that allows the allocation of system resources (such as memory and interrupts) automatically, once the Socket Services detects that a PC Card has been added. Card Services also releases these resources when the PC Card has been removed. Card Services also provides you with an interface to higher level software to load any needed hardware drivers.

The combination of PC Card hardware, Card Services software and Socket Services software provides an almost "plug-and-play" capability in the portable computing environment. Once the software has been installed, it is theoretically possible to add and remove PC Cards without powering off the system. But this is theory. And in practice, it has worked only intermittently for this writer. It doesn't work for network cards. It's meant to be possible, for example, to insert a modem PCMCIA Card to access another computer system, download information into the portable computer's memory, remove the modem PCMCIA Card, replace it with a flash PCMCIA Card, and store the downloaded information — all while your portable computer is still powered on. Great theory.

The PCMCIA has around 300 members, including manufacturers of semiconductors, connectors, peripherals and systems, as well as BIOS and software developers and related industries. Members include Intel, IBM, Toshiba, Lotus, Epson and Fujitsu. The association is based in Sunnyvale, CA. 408-720-0107. It has an electronic bulletin board — 408-720-9388. Its standards are also recognized by the Japanese Electronic Industry Development Association (JEIDA). The Association publishes a free book listing all the manufacturers making cards which comply to their standards. Advice in buying PCMCIA cards: PCMCIA standards are still being formed and are thus not truly standards. Check that cards you buy work in your machine. The manufacturer should have tested it in your machine. Don't assume compatibility. Check your socket and card services supports that card. It may not. Use "enabler" software instead. Avoid PCMCIA cards with "pigtails." They break. You lose them. Use PCMCIA cards with an "X-JACK." You can plug directly into a PCMCIA card with an X-JACK, or equivalent. If you have to buy a card with a pigtail, buy a second pigtail, just in case.

In 1994, the PCMCIA started calling its specs, PC Card. Some people said it was because many people claimed that PCMCIA stood for "People Can't Memorize Computer Industry Acronyms."

In the latter part of 1995, the PCMCIA came out with a new specification called CardBus which is a 32-bit bus, as against the present 16-bit cards. CardBus is an extension of the PCI bus. This means that new CardBus will support 132 Mbps, much faster than the present 8 Mbps. See CardBus for a full explanation of that new standard. See also Card Services, PCMCIA standards, Socket Services, and Slot Sizes. www.pcmcia.org

PCMCIA Standards The complete set of all of the PCM-

CIA PC Card Standards. It includes the PC Card Standard Release v2.01, Socket Services Specification Release v2.0, Card Services Specification Release v2.0, ATA Specification Release v1.01, AIMS Specification Release v1.0, and the Recommended Extensions Release v1.0. Standard v3.0 has been proposed. See PCMCIA. www.pcmcia.org

PCN Personal Communications Network. A new type of wireless telephone system that would use light, inexpensive handheld handsets and communicate via low-power antennas. When it was originally conceived, PCN was primarily seen as an a city communications system, with far less range than cellular. Subscribers would be able to make and receive calls while they are traveling, as they can do today with cellular radio systems, but at a low price. Now PCN is seen as what Dr. Sorin Cohn of Northern Telecom calls an "enabler of unplanned growth." One idea for PCN is to locate a PCN cell site (transmitter/receiver) in a residential community. When someone wanted a new phone line, they'd simply drop down to their local phone store, pick up a PCN portable phone and, by the time, they got back home, their frequency would be "switched on" and they'd be "live." The original plans for PCN never materialized fully. However, the concept has been implemented in the forms of Personal Communications Service (PCS) and Wireless Local Loop (WLL). See also PCS, Personal Communications Network, Wireless Local Loop.

PCO Point of Control and Observation: A place (point) within a testing environment where the occurrence of test events is to be controlled and observed as defined by the particular abstract test method used.

PCP 1. Post Call Processing.
2. Program Clock Reference: A timestamp that is inserted by the MPEG-2 encoder into the Transport Stream to aid the decoder in the recovering and tracking the encoder clock.

PCR An ATM term. Peak Cell Rate: The Peak Cell Rate, in cells/sec, is the cell rate which the source may never exceed.

PCS 1. the plural of PCs, i.e. PCc.
2. Personal Communications Service. A new, lower powered, higher-frequency competitive technology to cellular. Whereas cellular typically operates in the 800-900 MHz range, PCS operates in the 1.5 to 1.8 Ghz range. The idea with PCS is that the phones are cheaper, have less range, are digital; the cells would be smaller and closer together and the airtime would be cheaper also. Several licenses have been awarded and several systems have started in North America. The concept of PCS is evolving. It is not clear exactly where it will end up. So far, it looks like another cellular system with some digital messaging on the phone's larger screen. this is how the Federal Government awarded PCS licenses:
• "C-Block" Carrier
A 30 MHz PCS carrier serving a Basic Trading Area (BTA) in the frequency
block 1895-1910 MHz paired with 1975-1990 MHz.
• "D-Block" Carrier
A 10 MHz PCS carrier serving a Basic Trading Area (BTA) in the frequency
block 1865-1870 MHz paired with 1945-1950 MHz.
• "E-Block" Carrier
A 10 MHz PCS carrier serving a Basic Trading Area (BTA) in the frequency
block 1885-1890 MHz paired with 1965-1970 MHz.
• "F-Block" Carrier
A 10 MHz PCS carrier serving a Basic Trading Area (BTA) in the frequency
block 1890-1895 MHz paired with 1970-1975 MHz.

See PERSONAL COMMUNICATIONS NETWORKS and PERSONAL COMMUNICATION SERVICES.

PCS Over Cable You run a CATV — cable TV company. You have a wires strung all over the neighborhood. On one of your wires you attach a six foot by four foot by four box of electronics and three two feet attennae. Bingo, you're now a way station — also called a cell site — for a PCS cellular phone system. People who are PCS subscribers will talk and receive calls when they're close to your cell site. Calls come and go via your coax cable, up it to a landline connection point with the PCS carrier. You, the CATV company, get paid money for completing calls. See www.sanders.com/telecomm

PCSA Personal Computing System Architecture. A PC implementation of DECnet, that lets PCs work in a DECnet environment. PCSA is a network architecture defined and supported by Digital Equipment Corporation for the incorporation of personal computers into server-based networks.

PCT Personal Communications Technology. A security protocol developed by Microsoft for online Web commerce and financial transactions. Transparent to the user, PCT provides authentication and encryption routines that complement credit-card based commerce on the World Wide Web. Internet Explorer, Microsoft's Web browser, makes use of PCT. See also AUTHENTICATION and ENCRYPTION.

PCTA Personal Computer Terminal Adapter. A printed circuit card that slips into an IBM PC or PC compatible and allows that PC to be connected to the ISDN T-interface. See PERSONAL COMPUTER TERMINAL ADAPTER.

PCTE Portable Common Tool Environment.

PCTS Public Cordless Telephone Service. A Canadian digital cordless telephone service for residential, business and public use. For other variations of digital cordless telephone service, see CT1, CT2, CT2Plus, CT3, and DECT.

PCWG Personal Conferencing Work Group (www.gogcwg.org/pcwg/)

PCX Server Software PCX server software turns your PC into a graphics terminal front-end for Unix and X applications. Thus, your PC can display application output generated by remote X-based client applications.

PDA Personal Digital Assistant. A consumer electronics gadget that looks like a palmtop computer. Unlike personal computers, PDAs will perform specific tasks — acting like an electronic diary, carry-along personal database, multimedia player, personal communicator, memo taker, calculator, alarm clock. The communications will take place through the phone or through wireless. Apple has announced a PDA, which it has named Newton. When I added this definition in the late fall of 1992, sales of PDAs weren't doing well and some wag in Silicon Valley called them Probably Disappointed Again. IBM prefers to call them Personal Communicators. General Magic prefers to called them PICs, Personal Intelligent Communicators.

PDAU Physical Delivery Access Unit. A gateway device that facilitates the delivery of messages (excluding probes and reports) in physical form. This is an X.400 term.

PDC 1. Personal Digital Cellular. (Digital system used in Japan).
2. See Primary Domain Controller

PDF Portable Document Format. This is the file format for documents viewed and created by Adobe's Acrobat Reader, Capture, Distiller, Exchange and the Acrobat Amber Plug-in for Netscape Navigator. The PDF file format was developed to standardize Internet-based documents. One of the benefits of using Acrobat and the .pdf format is you can deliver business documents others without reauthoring them and read them

PDH Plesiochronous Digital Hierarchy. Developed to carry digitized voice over twisted pair cabling more efficiently. This evolved into the North American, European, and Japanese Digital Hierarchies where only a discrete set of fixed rates is available, namely, nxDS0 (DS0 is a 64 kbps rate) and then the next levels in the respective multiplex hierarchies. See also PLESIOCHRONOUS. See T CARRIER for hierarchy detail.

PDI A Versit term. Personal Data Interchange, a collaborative application area which involves the communication of data between people who have a business or personal relationship, but do not necessarily share a common computing infrastructure.

PDL A page description language (PDL) is a is a clever shortcut for transmitting bit-mapped images from a PC application to a printer. They save processing time, by sending only "instructions" to a printer, rather than the entire bitmapped image. They also allow the printer to print any font, any size. PDL is the generic term. Hewlett-Packard has been the major proponent of the concept of PDLs. And they include something called PCL with all their printers. PCL stands for Printer Command Language. Postscript is also a PDL, but different to PCL. HP includes PCL with all its laser printers, but has only included What we really meant was that HP had never built PostScript into their printers. They still haven't, really. You still have to buy PostScript as either a plug-in cartridge or an add-on "SIM" chip from HP, Adobe or third parties for most of the HP line. The exception is the top-of-the-line LaserJet IIIsi, for which built-in PostScript is an option that you pay extra for. That may change. For now, if you're using Windows applications, Windows' built-in "True Type" scalable fonts — which do work on HP and other non-PostScript printers — will satisfy most of your printing needs.

PDM 1. Pulse Duration Modulation.

2. See PERSONAL DATA MODULE.

PDN 1. An ISDN Term. Primary Directory Number (7 digits for 5ESS switch; 10 digits for DMS-100 switch)

2. Public Data Network. A public network for the transmission of data, particularly a network compatible with X.25 protocol. A public data network is to data what the Public Switched Voice Network is to voice. To access a public data network, you typically dial a local number, receive a carrier tone and then follow very specific instructions. Public data networks send their digital data in packets over high speed channels. The major reason to use a PDN is that it may be cheaper than dialing directly on a switched voice line. Also, they get you into some databases and services which are hard to get into by dialing direct. There are many public data networks in the United States. The two best-known public data networks are Tymnet and Telenet. But there are probably 30 more. Every industrialized foreign country has at least one, usually owned by the local government phone company.

3. Premises Distribution Network. The electronics and cabling system which connect terminal equipment to the NTI (Network Terminating Interface). The PDN may be either point-to-point or multipoint, may be configured as either a bus or star, and may be either active or passive in nature. In an ADSL Asymmetric Digital Subscriber Line) implementation, for example, a PDN system would be used to connect the ATU-R (ADSL Termination Unit-Remote) to Service Modules which perform terminal adaptation functions. See also NTI, ADSL and ATU-R.

PDP 1. See POWER DISTRIBUTION PANEL.

2. Plasma Display Panel.

PDP-11 On the last day of March, 1970 Digital Equipment shipped its first PDP-11/20 minicomputer to a customer in Tennessee. The PDP-11 family was among the first minicomputer that incorporated open standards into its operations. And the PDP-11 software platform has continued through subsequent generations of hardware.

PDS 1. Premise Distribution System. Lucent's proprietary structured cable and wire system for intrabuilding application. Such a cabling system provides a number of options for deploying various combinations of UTP, STP, coax and fiber optic cables contained within a single cable sheath. Thereby, various media can be deployed in appropriate combination, considerably reducing overall deployment costs.

2. Processor Direct Slot. Some non-NuBus Apple Macintosh computers have one PDS that allows for expansion cards.

PDU Protocol Data Unit. OSI (Open Systems Interconnection) terminology for a generic "packet." A PDU is a message of a given protocol comprising payload and protocol-specific control information, typically contained in a header. PDUs pass over the protocol interfaces which exist between the layers of protocols (per OSI model). Basically, PDU is OSI (Open Systems Interconnection) terminology for "packet". A PDU is a data object exchanged by protocol machines (entities) within a given layer. PDUs consist of both data and control (protocol) information that allows the two to coordinate their interactions. The native PDU may be altered through a process of Segmentation and Reassembly (SAR) in order to achieve compatibility with a service offering. For example, an Ethernet PDU in the form of a frame becomes known as an SMDS Service Data Unit (SDU) when considered in the context of a SMDS network service. The SDU then is encapsulated with control information to become a Level 3 (SMDS) PDU, which subsequently is segmented into 48-octet payloads, each of which is preceded with another 5 octets of control data to constitute a 53-octet SMDS cell, also known as a Level 2 PDU. The cells are transported over the SMDS network, with the process of reassembly being performed on the receiving end. As a result, the receiving device is presented with the data in its native PDU format.

PE Processing Element.

Peak That part of the business day in which customers expect to pay full service rates. For cellular customers, peak hours are generally 7:00 a.m. to 7:00 p.m. For business landline customers, peak hours are generally 8:00 a.m. to 5:00 p.m.

Peak Emission Wavelength Of an optical emitter, the spectral line having the greatest power.

Peak Hour When used with an automatic call distributor, the peak hour is when the number of calls coming into your center are at their highest level. ACDs allow you to track and report on calls by hour. Some allow you to also track peak half-hours, or peak delays of the week or months of the year.

Peak Load A higher than average quantity of traffic. Peak Load is usually expressed for a one-hour period, often the busiest hour of the busiest day of the year. See BUSY HOUR.

Peak Position Requirements A call center term. The maximum number of base staff required in any half hour within a given date range.

Peak Rate The per-minute price for using a communications device in the "peak" time period. For cellular, "Peak" time generally includes hours such as evenings Monday through Friday, and all day Saturday, Sunday, and certain holidays. For normal landline phone service, peak rates will include workday days.

Peak To Average Ratio P/AR. An analog test that provides a measure of data circuit quality by sending a pulse into one end of a circuit and measuring its envelope at the distant end of the circuit. See P/AR.

Peak Power Many phone systems use more power when more people are talking on them. You need to know peak power so you don't suddenly blow all your fuses when everybody gets on the phone.

Peaked Load The load that results from peaked traffic.

Peaked Traffic Random traffic that has a variance-to-mean ratio grater than one.

Peakedness Within-the-hour or 'moment-to-moment' variations in traffic.

Peaking Momentary bursts of high volume traffic which occur during the busy hour.

Peanut Tubes A name given to the smaller sizes of vacuum tubes.

PEB PCM Expansion Bus. A digital voice bus for sending voice across different voice processing cards and components in the same PC. PEB is from Dialogic, Parsippany, NJ. It is an open platform. Many companies make voice processing products connecting via PEB. See also DYNAMIC NODE ACCESS.

PEBCAK Tech support shorthand for "Problems Exist Between Chair And Keyboard." A way of indicating there's nothing wrong with the computer — the user is clueless. A Wired Magazine definition.

Ped See Pedestal

Pedestal 1. A small green box sits outside and houses cables coming in, cables going out and cable splices inside the box to join cables. This way a cable coming in from the telephone central office can be joined to one going to someone's house.

2. A pedestal is also a mounting device used in pay telephone installations where the instrument is not attached to a wall.

PEEK Taligent's Partners Early Experience Kit. I guess it's their cute name for their developer's kit.

Peek/Poke Instructions that view and alter a byte of memory by referencing a specific memory address. Peek displays the contents; poke changes it.

Peer Entities Entities within the same layer.

Peer Group A set of logical nodes which are grouped for purposes of creating a routing hierarchy. PTSEs are exchanged among all members of the group.

Peer Group Identifier A string of bits that is used to unambiguously identify a peer group.

Peer Group Leader A node which has been elected to perform some of the functions associated with a logical group node.

Peer Group Level The number of significant bits in the peer group identifier of a particular peer group.

Peer Network Entities The origin and destination of all data transmissions.

Peer Node A node that is a member of the same peer group as a given node.

Peer Protocol The set of rules defining the procedures for communication between like entities. The identical entities may be devices or software modules at specific layers in a layered network architecture implementation.

Peer-Peer Directory Propagation A way of updating user addresses in which charges in any post office on a LAN (local area network) are sent to all other post offices.

Peer-to-Peer Communications Communications between two entities that operate within the same protocol layer of a system.

Peer-To-Peer Network 1. A network (typically a local area network) in which every node has equal access to the network and can send and receive data at any time without having to wait for permission from a control node. While peer-to-peer resource sharing is effective in small networks, security and reliability issues prevent its widespread use in larger networks.

2. A new telephony term describing the relationship between a telephone system and the external computer working with it. Picture a telephone switch acting as an automatic call distributor and an outboard computer processor. The idea is to coordinate the call and the screen at the agent. Communication must take place between the switch and the computer. If that communication is peer-to-peer, as it is, for example, in the DMS Meridian ACD, then neither the switch nor the computer is in a "slave" relationship to the other. See EXTENDED CALL MANAGEMENT.

Peer-to-Peer Protocols Describes the relationship between a telephone system and the external computer.

Peer-To-Peer Resource Sharing An architecture that lets any station contribute resources to the network while still running local application programs. 10-Net Local Area Network allows for peer-to-peer resource sharing.

PEG Public, Educational or Government access. A cable TV term to denote the local public access channel/s.

Peg Board Also called a mushroom board or white board. This is placed between termination blocks to support route crossing wire.

Peg Count A raw count of some event. Because this was originally maintained by moving pegs on a board with units of 1,10s,100, 1000s it became a peg count. A count of the number of calls placed or received at a certain point or over certain lines during a period such as an hour or day or week. A peg count simply tells you how many calls you made or received. It does not tell you how long they were or where they went or anything else. In the old days before we had accurate and relatively inexpensive call accounting equipment, we relied on Peg Counts to figure out how many circuits we needed. No more. The peg count method is too inaccurate.

PEL Picture ELement. A pel is the smallest area on a video screen that can be controlled by software. Pels are arranged on screens in a grid-like fashion. Depending on the screen mode selected by an application, a pel may be a single pixel or several pixels. (A pixel is the smallest visual element on a screen, which can be turned on or off or varied in intensity.) The larger the number of pixels in a pel, the greater the size of the palette of distinct colors. When a pel consists of multiple pixels, the number of colors can be quite large. The pel size determines the clarity of the image — called screen resolution. Larger individual pels reduce the total number of available pels, resulting in lower resolution. Smaller pels increase the number of pels that can fit on the screen, resulting in higher resolution and a clearer picture. See also PIXEL, FAX and VGA.

PEM Privacy Enhanced Mail. An Internet electronic mail capability which provides confidentially and message integrity using various encryption methods. PEM was adopted by the IETF in 1985 as an Internet standard. PEM involves digital certificates which are issued by the Internet Policy Registration Authority (IPRA), which serves as the top-level "trusted authority." The IPRA signs certificates for a second level of trusted bodies, which then sign certificates for Certificate Authorities (CAs), which then issue them to the public. As PEM does not address MIME (Multipurpose Internet Mail Extension), it is seldom used. S/MIME (Secure MIME) is the preferred approach. See also Certificate Authority, Digital Signature, Encryption, MIME, and S/MIME.

Pen Register Also called a Dialed Number Recorder (DNR). An instrument that records telephone dial pulses as inked dashes on paper tape. A touchtone decoder performs the same thing for a touchtone telephone. Pen registers and DNRs are used by law enforcement agencies to gather information about the telephone numbers that suspected criminals are calling. They also are used by toll fraud criminals to steal calling card numbers.

Pen Windows A new Microsoft operating system for note-book size computers that uses a stylus instead of a keyboard.

Penetration Bypassing the security mechanisms of a system.

Penetration Tap A connection method used in installations that allows devices to be connected to cable without interrupting network operation. Penetration taps are most commonly found used with coaxial cable. A sharp, pointed probe is used to penetrate the outer insulation and grounding shield of the coaxial cable and to make direct contact with the inner conductor.

Penetration Testing Attempting to circumvent a system's security features to identify weaknesses.

Pentium In the fall of 1992, Intel adopted the name Pentium for its 80586 chip, its successor to the 80486. It introduced the Pentium formally in early April, 1993. The chip is capable of 112 million instructions per second and is 80% faster than the fastest 80486. It contains more than three million transistors and is said to be a superscalar chip, which means it can execute two instructions at a time.

Pentium Pro Intel's name for the 80686 microprocessor chip, its successor to the Pentium.

PEP 1. Protocol Extensions Protocol. A part of the JEPI (Joint Electronics Payments Initiative) specification from the World Wide Web Consortium (W3C) and CommerceNet for a universal payment platform to allow merchants and consumers to transact E-Commerce (Electronic Commerce) over the Internet. PEP is an extension layer that sits on top of HTTP (HyperText Transfer Protocol). PEP works in conjunction with Universal Payment Preamble (UPP), the negotiation protocol that identifies appropriate payment methodology. These protocols are intended to make payment negotiations automatic for end users, happening at the moment of purchase, based on browser configurations. See also Electronic Commerce and JEPI.
2. Packet Ensemble Protocol. A high-speed modulation method from Telebit Corporation for dial-up modems. Now obsolete.

PEPCI Protocol for Exchange of PoliCy Information. A draft protocol from the IETF intended to support the exchange of policy information between the policy server and its clients. Policy-based networks manage traffic through the establishment of priorities based on parameters such as traffic type, application type, and user. See also POLICY-BASED NETWORKING.

Per Call Calling Identity Delivery Blocking Feature Allows a caller to toggle or override the value of a calling identity item's Permanent Presentation Status (PPS) for a particular call.

Percentage ATB Percentage of All Trunks Busy. Percentage of time during a reporting period that all trunks in a group or split were busy. This may be measured in two ways, actual simultaneous busies and call length per event, or backed into statistically. Neither technique is absolutely accurate as each depends on "snap shots" in a environment of random interleaved call events.

Percentage CA Percentage of Calls Abandoned. Indicates the percentage of calls abandoned by callers after being accepted by the ACD.

Percentage HLD Percentage of total calls HeLD in queue within a reporting group.

Percentage NCO Percentage of total of Number of Calls Offered to a particular reporting group.

Percentage TUT Percentage of Trunk Utilization Time. The percentage of a time during a reporting period that a trunk is in use and not idle.

Perforator An instrument for the manual preparation of a perforated tape, on which telegraph signals are represented by holes punched in accordance with a predetermined code.

Perforator, Paper Tape An electro-mechanical device which converts electrical signals into coded holes in a paper tape. See PAPER TAPE PUNCH.

Performance Management Measures and records resource utilization. It is one of the categories of network management defined by the ISO (international Standards Organization).

Periapsis In satellite systems, the point on a satellite's orbit at which it is closest to the center of the primary body about which it is orbiting. Where Earth-based satellite systems are concerned, the term is synonymous with perigee. See also GEOSTATIONARY ORBIT.

Perigee The point at which a satellite orbit is the least distance from the center of the gravitational field of the Earth. The point in an orbit at which the satellite is farthest from the Earth is known as the apogee. In commercial application, the terms have most significance with respect to LEOs (Low Earth Orbiting) and MEOs (Middle Earth Orbiting) satellite constellations, which travel in elliptical orbits. See also LEO, MEO, and PERIAPSIS.

Perimeter Protection System A field disturbance sensor which uses buried leaky cables installed around a facility to detect any unauthorized entry or exit.

Period Of A Satellite The time elapsing between two consecutive passages of a satellite through a characteristic point on its orbit.

Periodic Postings Articles that are posted periodically to a newsgroup for the benefit of people who are new to the newsgroup. An Internet term.

Peripheral Device See PERIPHERAL EQUIPMENT or APPLICATIONS PROCESSOR.

Peripheral Equipment Equipment not integral to but working with a phone system. An example might be a printer or television screen on which calling traffic statistics are displayed. It might also be a voice mail or an automated attendant system. AT&T once called PBX peripheral equipment "applications processors," because they process specific applications. Some people now call them Adjunct Processor or Outboard Processors.

PERL Practical Extraction and Report Language. An interpreted scripting programming language, created in 1986 by Larry Walls to streamline the administration of a network of Sun and DEC VAX computers. PERL is a highly portable language widely used in writing CGI (Common Gateway Interface) scripts, which are the standard means of performing actions — like searching or running applications when the user clicks on certain buttons or on parts of the Web scree. pages. See CGI.

Permalancer A permanent freelancer. A person hired on a per-project basis who lives a benefits-free existence.

Permanent Presentation Status PPS. A PPS has a

public value of a "public" or "anonymous" and is used as the presentation status of a call if no per-call CIDB (Calling Line Identification Delivery Blocking) feature is active. A PPS should exist for each calling identity item. A Bellcore definition.

Permanent Shift Type A call center term. A shift definition that the program gives priority to in creating schedules but uses only as long as no overstaffing results in any intra-day period (at which point flexible shift types are used). When scheduling is done for more than a week at a time, the permanent schedules are always identical from one week to the next.

Permanent Signal A sustained off-hook supervisory signal originating outside a switching system and not related to a call in progress. Permanent signals can occupy a substantial part of the capacity of a switching system.

Permanent vacation What Gerry, my partner, wants. What Harry won't let him have.

Permanent Virtual Circuit PVC. A virtual circuit that provides the equivalent of a dedicated private line service over a packet switching network between two DTEs. The path between users is fixed. A PVC uses a fixed logical channel to maintain a permanent association between the FTEs. Once a PVC is defined, it requires no setup operation before data is sent and no disconnect operation after data is sent. A new connection between the same users may be routed along a different path. See also VIRTUAL CIRCUIT and SDN for another type of virtual network.

Permissible Interference Observed or predicted interference which complies with quantitative interference and sharing criteria contained in these [Radio] Regulations or in CCIR Recommendations or in special agreements as provided for in these Regulations. (RR) See also accepted interference, interference.

Permission A Windows NT term. A rule tied to an object (usually a directory, file or printer) to regulate which users can have access to the object and in what manner.

Permissions 1. A call center term. Privileges granted to each user with respect to what data that user is allowed to access and what menu options or commands he or she is allowed to use. Permissions are under control of the System Administrator.
2. A Northern Telecom Norstar definition to define specific characteristics that can be assigned to an individual telephone. Permissions includes Full Handsfree, Handsfree Answerback, Pickup Group, Page Zone, Auxiliary Ringer, Receive tones, and Priority Call.

Permissive Dialing There's an area code change. Many peoples' phones now have different area codes. People calling them will have to dial a different area code. There'll be a time, perhaps three months. in which a caller will be able to reach the number by dialing the old area code or the new area code. That period is known as permissive dialing. Once the period of permissive dialing is over, the period of mandatory dialing begins.

Perpetrator Someone who carries out an illegal or malicious act affecting information security.

Persistence 1. In a CRT, the time a phosphor dot remains illuminated after being energized. Long-persistent phosphors reduce flicker, but generate ghost-like images that linger on screen for a fraction of a second.
2. The probability that when a device on a local area network which has the data to transmit senses a free transmission line it will attempt the transmission. 1.0 persistence indicates that there is 100 percent probability that the device will always

attempt to transmit. IEEE 802.3 and Ethernet both use 1.0 persistence. A 0.5 persistence indicates that when a device senses a free line it will only attempt to transmit 50 percent of the time.

Persistency A persistent configured site will attempt to re-establish the connection in the event of an unexpected line drop. A non-persistent configured site will not.

Persistent Communication A dialogue between an ASC (AIN Switch Capabilities) and a SLEE (Service Logic Execution Environment) that may involve a sequence of messages. Definition from Bellcore in reference to its concept of the Advanced Intelligent Network.

Persistent Information Information for which a permanent data object exists. Definition from Bellcore in reference to its concept of the Advanced Intelligent Network.

Person Call There will be two types of calls: person calls and place calls. I make a place call when I call a phone number which ends in one designated, fixed RJ-11 jack attached to the wall or floor. I make a person call when I call a phone number which doesn't necessarily end on a fixed place. A typical person call might be a cellular or wireless phone call. It might also be a service like MCI's Personal 800 Service. The major characteristic of a person call is that I am calling a person and I don't know where that person is. As a result, the person I called might answer my call anywhere — from his car, from vacation home, from his wireless phone, etc.

Person To Person Call The most expensive way to make a long distance call. Call the operator. Say "I want to speak to Harry Newton on 212-691-8215.' The operator dials Mr. Newton's phone number, gets Mr. Newton on the phone. "Are you Mr. Newton?" Mr. Newton replies: "Yes, I am." The operator bows out of the conversation and sends you, the caller, a hefty bill for that personalized service. Until recently, person-to-person service was only offered by AT&T. Now it's offered by most phone companies. But prices, in the main, are not regulated. And some companies charge an arm and a leg. Please be careful.

Personal 800 Number Several long distance companies are now offering Personal 800 numbers, which are basically party line 800 numbers with call routing. The way they work is as follows: You dial a number, e.g. 800-484-1000. A machine answers with a double beep. You punch in four or five digits on your touchtone pad. A voice response unit at the other end hears the digits, says "Thank you for using MCI" and dials out your long distance number which might be 212-691-8215 (mine). The long distance carriers are charging under $5 a month and 15 to 25 cents a minute for the service. The per minute charges are more expensive than normal 800 lines. One company, MCI, is also offering FOLLOW ME 800 which allows you to change the routing of your personal 800 number instantly with one phone line.

Personal Authentication Device P.A.D. A portable or fixed device which allows for precise identification and validation of each user. Used in fraud prevention and unauthorized access. Also called a "token".

Personal Central Office Trunk Line Also known as a "private line." Allows a user behind a PBX to access a central office trunk line dedicated to him; also allows people to call him directly, bypassing the PBX. Such a private line can appear on a line access button on an electronic PBX phone, or can terminate on a separate phone. As the private line bypasses the PBX, the console attendants can't listen to your conversations; neither can the technicians, unless they physically tap the line. Private lines, therefore, are more secure than the typical PBX station line. They also are CO-powered;

as a result, they are not susceptible to power outages which might affect the PBX system. Senior executives use them, as do venture capitalists and others who conduct highly confidential negotiations. See also Private Line.

Personal Communications Industry Association PCIA. The trade association of the new cellular phone providers. PCIA represents a wide range of interests, including Broadband PCS, paging and narrowband PCS, antennae site owners and managers, Specialized Mobile Radio (SMR), suppliers and manufacturers. See also PCS. www.pcia.com

Personal Communications Networks PCN. A new type of wireless telephone system that would use light, inexpensive handheld handsets and communicate via low-power antennas. PCN is primarily seen as a city communications system, with far less range than cellular. Subscribers would be able to make and perhaps receive calls while they are traveling, as they can do today with cellular radio systems, but at a low price. There's talk that they'll put PCN antennas in communities and issue everyone in the community with a PCN phone that would hub off the PCN antenna. This would save the phone company wiring up each house. In this way, the PCN phone would resemble the common household cordless phone, but with a larger range. Dr. Sorin Cohn of Northern Telecom calls the new low-power, wireless, personal communications systems an "enabler of unplanned growth." Nice definition. See PCN and PERSONAL COMMUNICATIONS SERVICES.

In the fall of 1992 MCI broadened the definition of PCN. It proposed a national consortium of local PCNs joined together by a long distance carrier (namely it). In its release, MCI said that "PCNs are the next generation of digital wireless communications technology. PCNs use less power and are less expensive than the current cellular technology and permit the use of inexpensive pocket telephones with much longer battery life than cellular portables. PCN phones will have many more features than today's cellular or conventional telephones, and unlike cellular phones, will be usable in most areas of the world. PCNs will operate in the same frequency band in most countries, (1850-1990 MHz) while cellular is operating in several different frequency bands in various countries, and thus is not portable from country to country." See also PERSONAL COMMUNICATIONS SERVICES.

Personal Communications Services PCS are a broad range of individualized telecommunications services that let people or devices to communicate irrespective of where they are. Some of the services include:

• Personal numbers assigned to individuals rather than telephones

• Call completion regardless of location ("find me")

• Calls to the PCS customer can be paid for by the caller, or by the PCS customer

• Call management services giving the called party much greater control over incoming calls.

PCS can both find and complete a call to a person regardless of location, but give that person the choice of accepting or rejecting the call or sending it somewhere else. PCS will possibly use a new category of wireless, voice and data communications — using low power, lightweight pocket telephones and hand-held computers.

Personal Computer PC. A computer for one person's use. That's why it was originally called a personal computer — to distinguish it from other computers that existed at the time of the PC's invention, or 1981, the date IBM introduced its first PC. Those other computers were mainframes and

mini-computers. Their main use was to be shared by many users. Airline reservations, being an example. Of course, as the PC got more powerful, things got more complex. In 1983, Novell introduced its first local area network software called NetWare. NetWare was originally introduced to allow a handful of personal computers to share a single hard disk, which at that stage was costly and scarce. As hard disks became more available and cheaper, NetWare evolved to allow sharing of printers and file servers. Networking developed in the 1980s and 1990s and gave PCs the shared-user power of mainframes and minicomputers. As a result, today some PCs are personal computers, used by one person standalone. Some PCs are PCs that sit on a local area network (LAN). For some reason, people call these PCs workstations. And there are some PCs which have become servers — i.e. they "serve" many PCs. They may do faxing. They may do voice mail for the company. They may run the company's entire phone system. In short, meanings change as technology changes. Not all PCs are personal computers.

Personal Computer Memory Card International Association See PCMCIA.

Personal Computing Systems Architecture PCSA. A network architecture defined and supported by Digital Equipment Corp. for the incorporation of personal computers into server-based networks.

Personal Computer Terminal Adapter PCTA. A printed circuit card that slips into an IBM PC or PC compatible and connects the PC to the ISDN T-interface. The PCTA basically turns a normal PC into an ISDN phone and ISDN terminal, ready for voice and data communications. According to Northern Telecom, one of the manufacturers of such a device, the PCTA does:

• Functional signaling call setup.

• B channel for circuit switched data.

• X.25 packet data services on the B or D channel.

• Simultaneous operation of the B and D channels.

• M5317T digital telephone to PC messaging for integrated voice and data operation.

• NetBIOS interface to applications software, such as Microsoft Networks.

For more, see Northern Telecom's brochure ISDN PC Terminal Adapter NetBIOS Interface Description (D307-1).

Personal Data Interchange PDI. A format for exchanging information, such as electronic business, cards via wired or wireless connections.

Personal Data Module A removable module, unique to the ROLM Cypress and Cedar, which stores all information entered by the user. Such items as phone numbers, terminal profiles and log-on sequences are stored in the PDM, which also features battery backup to protect the memory in the event of a power failure. The user can take his personal data module with him, plug it into another Cypress or Cedar and get all his speed dial and other personalized programming on his phone he just moved from.

Personal Digital Assistant PDA. A consumer electronics gadget that looks like a palmtop computer. Unlike personal computers, PDAs will perform specific tasks-acting like an electronic diary, carry-along personal database, personal communicator, memo taker, calculator, alarm clock. The communications will take place through the phone, through a cable attached to your PC or through infrared. See also Windows CE.

Personal Identification Number PIN number. 1. An AT&T term meaning the last four digits of your AT&T, MCI

Bell operating company Credit Card — the card you use for making long distance numbers. 2. Some banks and financial institutions issue credit cards for machine, teller-less banking. These machines, called Automated Teller Machines, ask you for a password consisting of several numbers or characters. These are not on your credit card. These numbers or characters, called PIN numbers, are designed to make sure the right person is using your card. It's not a good idea to use your birthday as your PIN number.

Personal Intelligent Communications What General Magic calls the products and services its alliance members will create using General Magic technologies that will help people remember, communicate and know things in new and powerful ways. According to General Magic, "the alliance's shared long-term vision is to bring personal communications to people who may not use a computer today, to people whose personal technology is a car, a television set and a telephone."

Personal Information Assistant Tandy's name for what Apple calls a Personal Digital Assistant. See PDA.

Personal Information Manager PIM. Software application which allow the user to organize personal information. similar to an appointment book but personalized and programmed into the PC. PIMs now support Caller ID. When the phone rings, the phone system passes the calling number to the PC, which does a lookup in the PIM on the phone number and throws information about the caller up on screen.

Personal IVR An Interactive Voice Response system running on your own personal PC and designed to serve the needs of only one person. See IVR.

Personal Line A feature which allows specific key telephones to have their own private Central Office line. Sometimes called an AUXILIARY LINE. You can typically receive and make calls on this line. No one else can answer it, since it does not appear on any other phone instrument in the office. You can give this number to your wife or girlfriend. It is not a good idea to give it to both.

Personal Name An X.400 term for a standard attribute of an O/R (Originator/Recipient) Address form that identifies a person relative to another attribute (e.g., an organization name). The personal name may include surname, given name, initials, and generation qualifier. Initials consists of the first letter of all the user's names except the user's surname.

Personal Productivity Tool Another term for a computer. John Perry Barlow thinks the expression was created by the "druids" who run Microsoft and Apple. Mr. Barlow is a cattle rancher, computer hacker, poet, and a lyricist for the rock band The Grateful Dead.

Personal Speed Dial Simplified ways of dialing. You do them by dialing a couple of digits. Or you punch in a button at your phone. Personal Speed Dial codes are programmed for each telephone, and can only be used at the telephone on which they are programmed. System Speed Dials, in contrast, can be used from every phone in the system.

Personal System/2 PS/2. IBM's current family of microcomputers some of whom sport one major difference from its predecessors, namely the existence of a 32-bit micro-channel bus. This "bus" serves the same purpose as a PBX's backplane — namely to move information from the printed circuit cards and to other printed circuit cards, which may contain their own individual microprocessors (computers on a chip) and which may communicate with the outside world through their own communications ports.

Personal Telephone The category of cellular telephones pioneered by Motorola's Pan American Cellular Subscriber

Group with the introduction of the MicroTAC Digital Personal Communicator Telephone. Weighing less than one pound, they are so compact and lightweight, they fit comfortably into a shirt pocket or purse making them "body friendly."

Personality Module A small motherboard added to a voice board to give it the "personality" of a proprietary electronic PBX telephone. "Personality" means electrical characteristics and the same button configuration and responsiveness, all of which can be recreated on the screen of a PC which has the voice board installed.

Personalization Technology Software that lets suppliers of information to "nonintrusively" (their word, not mine) learn the individual interests of their customers in order to deliver personalized Web content, targeted advertising, and product recommendations.

Personalized Ringing A telephone feature which allows you to select different ringing sounds for your telephone. This feature is useful if you work in a big room with lots of other people and it's hard to tell whose phone is ringing.

PERT Project Evaluation and Review Technique. A variation on the Critical Path Method of organizing the completion of projects. Projects are examined for the their worst, best, average completion times. A critical path is determined and overall standards for completion times are created. The PERT technique was created by the military. It is used for organizing complex tasks.

PES Packetized Elementary Stream: In MPEG-2, after the media stream has been digitized and compressed, it is formatted into packets before it is multiplexed into either a Program Stream or Transport Stream.

Petabyte PB. A unit of measurement for physical data storage on some form of storage device — hard disk, optical disk, RAM memory etc. and equal to two raised to the 50th power, i.e. 1,125,899,906,842,600 bytes.

MB = Megabyte (2 to the 20th power)
GB = Gigabyte (2 to the 30th power)
TB = Terabyte (2 to the 40th power)
PB = Petabyte (2 to the 50th power)
EB = Exabyte (2 to the 60th power)
ZB = Zettabyte (2 to the 70th power)
YB = Yottabyte (2 to the 80th power)
One googolbyte equals 2 to the 100th power

PF Xfer Power Failure Transfer.

PG Peer Group: A set of logical nodes which are grouped for purposes of creating a routing hierarchy. PTSEs are exchanged among all members of the group.

PGA Programmable Gain Amplifier, a part of the Analog Front End.

PGL Peer Group Leader: A single real physical system which has been elected to perform some of the functions associated with a logical group node.

PGP PGP stands for Pretty Good Privacy. It is a cryptography program for computer data, electronic mail and voice conversations. PGP was written in 1991 by Philip R. Zimmermann, who gave the software away. It was posted on a public computer on Internet and thousands have downloaded copies of it. So far, according to the New York Times Magazine, June 12, 1994, no has been able to crack information encoded with PGP. Originally, PGP was just used for the transmission of computer data and electronic mail. Then it got extended to voice conversations on the phone, called PGPfone. The idea is that you use modems to dial. Then the PCs "shake hands" and jointly agree on a complex number that is plugged in a scrambling algorithm equation. This notoriously complex scrambling algorithm, called Blowfish, recalculates the digital ones and zeroes of the sampled voices into a stream of numbers unintelligible even to highly sophisticated eavesdroppers. Finally, PGPfone unscrambles the stream to provide intelligible — though not great quality — sound.

http://web.mit.edu/network/pgpfone/

PGPfone The name of the phone is used for Pretty Good Privacy. See PGP for a full explanation.

PH 1. Packet Handler, or Packet Handling function.
2. The Ph system allows you to look up directory information, usually including e-mail addresses at universities, research institutions, and some governmental agencies throughout the world. You need a program that lets you use Ph. Tell that program which Ph server to use, and then enter a name you would like to search for.

Phantom Circuit A circuit derived from two suitably arranged pairs of wires, called side circuits, with each pair of wires being a circuit in itself and at the same time acting as one conductor of the phantom.

Phantom Directory Number Also called a virtual DN. A directory number with a voice mailbox, but not a phone on it. Calls are then transferred to this number. The mailbox user can dial into the system, enter the extension number, security information and retrieve their messages.

Phase The relationship between a signal and its horizontal axis, also called zero-crossing point. A full cycle describes a 360 degree arc. A sine wave that crosses the zero-point when another has attained its highest point is 90 degrees out of phase with the other. See PHASE SHIFT KEYING.

Phase A Phase A is the first part of a fax machine's call process. It is the call establishment. It occurs when transmitting and receiving units connect over the phone line, recognizing one another as fax machines. This is the start of the handshaking procedure. See PHASE B.

Phase B Phase B is the second part of a fax machine's call process. It is the premessage procedure, where the answering machine identifies itself, describing its capabilities in a burst of digital information packed in frames conforming to the HDLC standard. See PHASE C.

Phase C Phase C is the third part of a fax machine's call process. It is the fax transmission portion of the operation. This step consists of two parts C1 and C2 which take place simultaneously. Phase C1 deals with synchronization, line monitoring and problem detection. Phase C2 includes data transmission. See PHASE D.

Phase D Phase D is the fourth part of a fax machine's call process. This phase begins once a page has been transmitted. Both the sender and receiver revert to using HDLC packets as during Phase B. If the sender has further pages to transmit, it sends an MPS and Phase C recommences for the following page. See PHASE E.

Phase Displacement Antenna An antenna constructed from a driven element and a group of reflectors, the secondary radiation from which produces an antenna with directivity. The Yagi-Uda array is a member of this family.

Phase Delay See ENVELOPE DELAY.

Phase Distortion An unwanted modification of a transmitted signal caused by the non-uniform transmission of the different frequency components of the signal. Same as Delay Distortion.

Phase E Phase E is the fifth part of a fax machine's call process. This phase is the call release portion. The side that transmitted last sends a DCN frame and hangs up without awaiting a response.

Phase Hit In telephony, the unwanted and significant shifting in phase of an analog signal. As defined by AT&T: any case where the phase of a 1004 HZ test signal shifts more than 20 degrees. Also, error-causing events more severe than phase jitter, especially for data transmission equipment using PSK modulation.

Phase inversion The condition whereby the output of a circuit produces a wave of the same shape and frequency but 180 degrees out of phase with the input.

Phase Jitter In telephony, the measurement, in degrees out of phase, that an analog signal deviates from the referenced phase of the main data-carrying signal. Phase jitter is often caused by alternating current components in a network.

Phase lock The phase of a signal follows exactly the phase of a reference signal.

Phase Lock Loop PLL. Phase Lock Loop is a mechanism whereby timing information is transferred within a data stream and the receiver derives the signal element timing by locking its local clock source to the received timing information.

Phase Modulation One of three ways to change a sine wave (or L signal) to let it carry information. In this case, the phase of the sine wave is changed as the information to be carried is changed. See PHASE SHIFT KEYING and MODULATION.

Phase Roll Variations in the phase of a transmitted signal and its echoed back modem verification. Phase roll is encountered most often in international systems.

Phase Shift A change in the time or amplitude that a signal is delayed with respect to a reference signal.

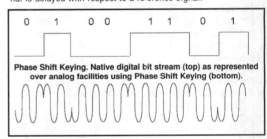

Phase Shift Keying. Native digital bit stream (top) as represented over analog facilities using Phase Shift Keying (bottom).

Phase Shift Keying Also known as PSK and Phase Modulation. Used by relatively sophisticated modems for transmitting digital signals over analog phone lines. Picture an electromagnetic sine wave. In its natural state, the sine wave is a continuous and uninterrupted wave form of a certain amplitude and carrier frequency. If we want to place a digital signal on it, it is necessary that we cause the signal to change in some way to reflect the presence of a "1" or the absence of a one, i.e. a zero. In phase shift keying, we simply change the phase of the carrier signal to reflect a change in value of the adjacent bits. Contrast with AMPLITUDE MODULATION and FREQUENCY MODULATION.

Phased Array A type of radio antenna used in certain satellite and wireless TV applications, and soon to be used for WLL (Wireless Local Loop). Phased Array Antennae are small (e.g., 4.5 inches square), flat antennae which mount on the side of your building or on your rooftop. Inside the thin flat box is an array of chip-based radio receivers which lock in on the desired transmission frequency on a dynamic basis. That is to say that they have sufficient intelligence to direct and redirect their focus in order to maximize the strength of the incoming signal. As neither the device nor its components, are physically dynamic (i.e., they cannot be repositioned physically, in the kinetic sense of the word), this capability is critical. These devices have the appearance of a pizza box, the term "pizza box" sometimes is applied to them. See also WLL.

Phasing The process of ensuring that both sending and receiving facsimile machines start at the same position on a page.

Phasing Orbit A temporary satellite orbit which is used prior to putting the satellite into its final orbit.

Phasor Temporary buffer storage that compensates for slight differences in data rate between TDM I/O ports and devices.

Phenolic Insulating Materials A type of insulating materials, one of which is bakelite. Now no longer used.

PHF Packet Handling Function. The switching capability that processes and routes X.25 virtual calls.

PHI 1. PBX-to-Host Interface. The same as CPI (Computer to PBX Interface) but it puts the PBX first, which is the way many telephone manufacturers prefer to see it. It refers to a connection between a telephone system and a computer, such that the computer can signal the telephone system to switch calls and the telephone system can signal the computer when it has switched them. There are major advantages in joining a telephone system to a computer. For a much greater explanation, see OAI, which stands for OPEN APPLICATION INTERFACE.
2. PHI is also a Northern Telecom term for Packet Handler Interface.

Phi Phenomenon A theory developed by psychologist Hugo Munsterberg, it explains the illusion of motion created by rapid presentation of a series of still images. Munsterberg suggests that the brain hallucinates, effectively filling in the voids between the images. At 30 fps (frames per second), the brain processes the images as fully fluid motion. See FRAME RATE.

Phoenix MCI's ill-fated order entry system that never rose from the dead, i.e. never worked.

Phone A simpler way of saying the word "telephone." An invention of the devil. See TELEPHONE.

Phone Bomb A phone bomb is a bobby-trapped cellular phone. I saw the term first used on the front page of the January 6, 1996 New York Daily News. The story read: "'The engineer,' an extremist Palestinian bomb maker which was No. 1 on Israel's most-wanted list was killed on the Gaza Strip yesterday when his head was blown up by a booby-trapped cellular phone. The phone was rigged with about two ounces of explosives." No one admitted to killing the man.

Phone Centric Users These are knowledge workers for whom the telephone plays a key role in the success of their business — e.g. salesmen, stock brokers.

Phone Freak See Phone Phreaks.

Phone Phreaks Communication hobbyists. People, usually kids, who like to figure out how the telephone network works and sometimes make free calls on the network by figuring a way to bypass billing mechanisms. Phone Phreaks have become Computer Phreaks with the advent of PCs and the advent of out-of-band signaling, making it a lot more difficult to make long distance calls for free. See Phreak.

PhoneMail A ROLM term for Voice Mail. Rolm's PhoneMail is a voice messaging system that provides telephone answering (with the user's own greeting), the capability to store and forward voice messages and the capability to turn on a message waiting light or message on the recipient's phone. PhoneMail can be used positively to speed the flow of information. It can also be used negatively to allow the user to "hide behind" the system and avoid the outside world and anyone in the outside world who might actually want to buy

something. See also VOICE MAIL.

Phoneme A voice recognition term. The minimal significant structural unit in the sound system of any language that can be used to distinguish one word from another. For example, the p of pit and the b of bit are considered two separate phonemes, while the p of spin is not. These minimal sound units comprise words.

PhoneNet Farallon's twist on Apple's local area network called LocalTalk. PhoneNet uses standard one pair UTP (unshielded twisted pair) wiring for networking. PhoneNet is compatible with LocalTalk.

Phonetic Alphabet A list of standard words used to identify letters in a message transmitted by radio or telephone. The following are the authorized words, listed in order, for each letter in the alphabet: Alpha, Bravo, Charlie, Delta, Echo, Foxtrot, Golf, Hotel, India, Juliet, Kilo Lima, Mike, November, Oscar, Papa, Quebec, Romeo, Sierra, Tango, Uniform, Victor, Whiskey, X-ray, Yankee, Zulu.

Phosphor Substance which glows when struck by electrons. The back of a picture tube face is coated with phosphor.

Photo Diode 1. The basic element that responds to light energy in a solid state imaging system. It generates an electric current that is proportional to the intensity of the light falling on it.
2. The light source or detector in a fiber optic transmission system. Light sources and light detectors are paired: LEDs (Light Emitting Diodes) as light sources, and PINs (Photo INtrinsic diodes) as light detectors in the slower systems; Laser Diodes and APDs (Avalanche PhotoDiodes) in the high-bandwidth systems such as SONET. See also APD, Laser Diode, LED, PIN and SONET.

Photo Etch The process of forming a circuit pattern in metal film by light hardening a photo sensitive plastic material through a photo negative of the circuit and etching away the unprotected metal.

Photoconductive Effect Some non-metallic materials exhibit a marked increase in electrical conductivity when they absorb photon or light energy. This is called the photoconductive effect. The conductivity increase is due to the additional free carriers generated when photon energies are absorbed in electronic transitions. The rate at which free carriers are generated and the length of time they persist in conducting states (their lifetime) determines the amount of conductivity change.

Photoconductivity The conductivity increase exhibited by some nonmetallic materials, resulting from the free carriers generated when photon (i.e. light) energy is absorbed in electronic transitions. The rate at which free carriers are generated, the mobility of the carriers, the length of time they persist in conducting states (their lifetime) are some of the factors that determine the extent of conductivity charge. See PHOTOCONDUCTIVE EFFECT.

Photoconductor 1. Any transducer that produces a current which varies in accordance with the incident light energy. A fiber optic communications term.
2. Photoconductor is also material, available in many forms (sheets, belts, and drums), which changes in electrical conductivity when acted upon by light. Electrophotography (a form of facsimile machine printing) relies on the action of light to selectively change the potential of a charged photoconductive surface, creating areas receptive to an oppositely charged toner, thus making the latent charged-image visible.

Photocurrent The current that flows through a photosensitive device (such as a photodiode) as the result of exposure to radiant power. Internal gain, such as that in an avalanche photodiode, may enhance or increase the current flow but is a distinct mechanism.

Photodetector In a lightwave system, a device which turns pulses of light into bursts of electricity.

Photodiode See PHOTO DIODE.

Photoelectric Effect The emission of electrons by a material when it is exposed to light. Albert Einstein received a Nobel Prize for explaining this phenomenon. Amazingly, he never received one for his brilliant theories of relativity.

Photon The Photon is a particle of light. For hundreds of years light was thought of solely as a wave. In 1905 Einstein discovered that under certain circumstances the energy of a light wave only came in specific amounts or quanta. These quanta are called photons.

Photonic Ethernet A high-speed networking technology based on Polymer Optical Fiber (POF) cabling that can deliver gigabit networking speeds for a fraction of the cost of conventional fiber cable, even though the new POF has the same optical characteristics as glass. With this new technology, Gigabit Ethernet can be delivered to the desktop cost-effectively — about $200 per port (Summer of 1998) — making it potentially interesting for bandwidth-constricted workgroups.

Photonic Layer The lowest of four layers of Sonet capability, which specifies the kind of fiber to be used including sensitivity and laser type. See SONET.

Photonic Switch A switch which switches photonics, or light signals. See Lambda Switch.

Photonics The technology that uses light particles (photons) to carry information over hair-thin fibers of very pure glass.

Photophone In 1880 (4 years after he invented the telephone) Alexander Graham Bell invented the photophone, which he felt was his greatest invention. Consisting of a set of specially ground and shaped mirrors, and associated electrical gear, the photophone was capable of voice transmission over short distances, using sunlight. It was, in fact, the first optical transmission system, preceding fiber optics by nearly 100 years. It also was highly impractical, relying on fragile mirrors and sunny days. The Nazis experimented with a variation on the theme for application in W.W.II tank warfare — the results were not positive.

Phototransistor A transistor that detects light and amplifies the resulting electrical signal. Light falling on the base-collector junction generates a current, which is amplified internally.

Photovoltaic Adjective to describe material which develops voltage and electrical current when light shines on it.

Photovoltaic Effect Using light to produce electricity. Shine light on a device, typically a "cell." If the device produces electricity, that's called the photovoltaic effect. It's not very efficient at present. Less than 10% of the light energy emerges as electricity. But it's getting better.

PHP Personal Handy Phone. Japan's standard for digital cordless phones.

Phreak A phone phreak is to the phone network community what the original hackers were to the computer revolution. These were the hobbyists who couldn't get enough information about the telephone from reading published works or taking one apart; rather they learned how the touchtone frequencies worked to route calls. Granted, one aspect of what they did was making illegal phone calls, but the larger picture was a hunger for information that they couldn't get elsewhere.

PHS Personal Handyphone System. PHS, previously known

as PHP (Personal HandyPhone). The Japanese version of the U.S.'s PCS (Personal Communications Service, with three key differences. It's not as powerful as PCS. You can't use a PHS phone in a rapidly moving vehicle, since there is no cell-handoff, i.e. it won't move you from one cell to another. And thus, if you move outside your cell with PHS, you lose connection. PHS is a perfect mobile phone for pedestrians in high density cities like Tokyo. PHS is truly a phenomenon, having grown from zero users to four million plus subscribers in 1996. In the meantime, the (slower) growth of cellular and wireline services have been unabated.

PHY PHYsical, as in physical specifications. OSI Physical Layer: The physical layer provides for transmission of cells over a physical medium connecting two ATM devices. This physical layer is comprised of two sublayers: the PMD Physical Medium Dependent sublayer, and the TC Transmission Convergence sublayer. See PHYSICAL LAYER, PMD and TC.

Physical Address 1. A number of digits which identifies the physical location of a communications channel or port within a system. In the old Rolm CBX, for instance, the number was in the form of xxyyzz, where xx=shelf, yy=slot and zz=channel.

2. The address where something physically resides. A physical addresses is translated from a logical address. Allow me to illustrate. When someone dials your telephone number, they are dialing a logical address; in other words, the series of numbers means nothing until they are translated into a physical address. The physical address is the port to which your local loop is connected to which your telephone is connected. Similarly, your postal address is a logical address. It has meaning only when translated by the post office into the plot of earth on which your house sits. A logical address, on the other hand and just to confuse you, may have no fixed physical address. For example, your e-mail address has no fixed physical address. Rather, it is translated into an IP (Internet Protocol) address which is associated with your e-mail server, which can be moved from place to place. Ultimately, your e-mail address actually is associated with you, and you and your computer can move all over the world without losing access to your e-mail. Rather, you gain access to your e-mail by dialing a telephone number (logical address) which connects you to your e-mail server which has a physical address which can change as the server is moved from one location to another.

Physical Colocation A local exchange carrier (LEC) provides space within the building housing its central office to other phone companies, to interconnect companies and/or to users to place their equipment. Typically it's done to connect circuits — transmission or switching — to the phone company's central office equipment. The interconnector (i.e. the company placing the equipment) installs, maintains, and repairs its own equipment, while the LEC provides power, environmental conditioning, and conduit and riser space for the interconnector's cable.

Physical Connection The full-duplex physical layer association between adjacent PHYs in an FDDI ring.

Physical Delivery Delivery of a message in physical form through a Physical Delivery System; for example, delivery of a letter through the U.S. Postal Service. This term is used in X.400.

Physical Delivery Address Component An X.400 address component that describes how to physically deliver a message. For example, the name and mail stop to hand deliver a message after it is printed. The concept is that the X.400 address would cause a message to be printed on a printer and an individual would complete the hand delivery.

Physical Delivery Office Name Standard attribute of a Postal O/R (Original/Recipient) Address, in the context of physical delivery, specifying the name of the city, town, etc., where the physical delivery is to be accomplished. An X.400 term.

Physical Delivery Office Number Standard attribute in a Postal O/R (Originator/Recipient) Address that distinguishes between more than one physical delivery office within a city, etc. An X.400 term.

Physical Delivery Organization Name A free form name of the addressed entity in the postal address, taking into account the specified limitations in length. An X.400 term.

Physical Delivery Personal Name In a postal address a free form name of the addressed individual containing the family name and optionally the given name(s), the initial(s), title(s) and generation qualifier, taking into account the specified limitations in length. An X.400 term.

Physical Delivery Service The service provided by a Physical Delivery System. An X.400 term.

Physical Delivery Service Name Standard attribute of a Postal O/R (Original/Recipient) Address in the form of the name of the service in the country electronically receiving the message on behalf of the physical delivery service. An X.400 term.

Physical Formatting The second step in structuring a hard drive so that you may write to it. Physical formatting follows partitioning.

Physical Layer 1. The OSI model defines Layer 1 as the Physical Layer and as including all electrical and mechanical aspects relating to the connection of a device to a transmission medium, such as the connection of a workstation to a LAN. Included at this layer are issues specific to the manner in which a device gains physical access to the medium and how it goes about putting bits on the wire or extracting bits from the wire. As the lowest level of network processing, below the Link Layer, the Physical Layer deals with issues such as volts, amps, and pin configurations and handshaking procedures. Communications hardware (e.g., NICs and MAUs) and software drivers are specified at the Physical Layer.
2. The ATM Physical Layer (PHY) loosely corresponds with the OSI version. In the ATM world, Physical Layer functionality is discussed in terms of the Physical Medium sublayer (PM) and the Transmission Convergence (TC) sublayer. The implementation of the ATM Physical Layer is addressed in the ATM Forum's UNI (User Network Interface) specifications.

Physical Layer Connection An association established by the PHY (OSI Physical Layer) between two or more ATM entities. A PHY connection consists of the concatenation of PHY links in order to provide an end-to-end transfer capability to PHY SAPs.

Physical Layer Medium Dependent The Physical Layer sublayer that defines the media dependent portion of the Physical Layer in FDDI. Items defined by PMD include transmit and receive power levels, connector requirements, and fiber optic cable requirements.

Physical Layer Protocol The Physical Layer sublayer that defines the media independent portion of the Physical Layer in FDDI. Items defined by the PMD include transmit and receive power levels, connector requirements, and fiber optic cable requirements.

Physical Link A real link which attaches two switching systems.

Physical Media Any means in the world for transferring

signals between OSI systems. Considered to be outside the OSI Model, and therefore sometimes referred to as "Layer 0." The physical connector to the media can be considered as defining the bottom interface of the Physical Layer, i.e. Layer 1 of the OSI Reference Model.

Physical Medium PM. In ATM terms, the Physical Medium sublayer is the dimension of the Physical Layer (PHY) which specifies the physical and electrical/optical interfaces with the physical media. SEE ALSO PHYSICAL LAYER AND TRANSMISSION CONVERGENCE.

Physical Rendition The transformation of an MHS (Message Handling System) message to a physical message (e.g., by printing the message on paper and enclosing it in a paper envelope). An X.400 term.

Physical Security Ways to stop someone from gaining physical access to your stuff. Methods include locks, security personnel and guard dogs.

Physical Signaling Sublayer PLS. In a LAN or MAN system, that portion of the OSI Physical Layer that interfaces with the medium access control sublayer and performs bit symbol encoding and transmission, bit symbol reception and decoding, and optional isolation functions.

Physical Slots Slots that are available to cards in a card shelf.

Physical Topology The actual arrangement of cables and hardware that comprise a network. In other words, the actual physical appearance of a network. Typical physical topologies in the LAN world, for instance, include Bus, Ring and Star. The Physical Topology may differ significantly from the Logical Topology.

Physical Unit PU. In IBM's SNA, the component that manages and monitors the resources of a node, such as attached links and adjacent link stations. PU types follow the same classification as node types.

Physical Unit Control Point PUCP. In SNA, the component that provides a subset of the system-services control point (SSCP) within a node. Types 1, 2 and 4 nodes contain a PUCP, while a Type 5 (host) contains a SSCP.

PHz Petahertz (10 to the 15th power hertz). See also SPECTRUM DESIGNATION OF FREQUENCY.

PI Presentation Indicator. A two-bit field in the Calling Party Number (CPN) subfield of the Initial Address Message (IAM). In an ISDN network, the IAM is part of the call set-up protocol. The PI indicates to the terminating switch whether it should pass to the called party the telephone number of the calling party, which telephone number is contained in the CPN. See also CPN, IAM, and ISDN.

PIA Personal Information Appliance. A name for a product most people call a Personal Digital Assistant (PDA).

PIC 1. See PRIMARY INTEREXCHANGE CARRIER.
2. Plastic Insulated Conductor. Conductors covered with an extruded coating of plastic.
3. Also an imaging term. Picture Image Compression. Intel-DVI Technology's on-line still image compression algorithm. See DVI.
4. Personal Intelligent Communicator. A General Magic term for a product most other people called a Personal Digital Assistant. See PDA.
5. Point In Call.
6. Programmable Interrupt Controller. A chip or device that prioritizes interrupt requests generated by keyboards, serial ports, and other devices and passes them on to the CPU in PC in order of highest priority. See also IRQ.

PIC Freeze Pre-subscribed interexchange Carrier Freeze. A

PIC Freeze is when a customer makes arrangements with their LEC (Local Exchange Carrier), or telephone company, to prevent "slamming," which is unauthorized changing of their long distance telephone carrier. See also PICC and Slamming.

Picasso Porn The semi-scrambled transmissions from adult cable channels that can sometimes be seen and heard by nonsubscribers. This definition courtesy Wired Magazine. According to Ray Horak, my Contributing Editor, the scrambling technique used by CATV operators over analog coaxial cable systems (which they mostly are) simply involves "twisting" the audio and video frequencies; thereby, you are able to hear and view the adult channel only if you have paid for the service and, therefore, have a converter box that can unscramble the signal. However, and according to Ray, you'll notice that the picture comes in clear when the actors stop moaning and the obnoxious music stops.

PICC Primary Interexchange Carrier Charge, also known as Pre-subscribed Interexchange Carrier Charge. The FCC-mandated (May 1997) flat-rate charge which applies to pre-subscribed IXCs connecting to the end user through LEC facilities. The PICC applies first to primary lines; to the extent that PICC charges, in combination with the SLC (Subscriber Line Charge) and the monthly tariff line charge, are insufficient to provide the LEC with full recovery of the costs of the local loop, a lower PICC also may apply to non-primary (i.e., secondary) residential lines and multi-line business lines. The PICC became effective in 1998, and can either increase or decrease over time. While the LEC bills the end user directly for the LSC, it bills the IXC for the PICC. The IXCs are free to recover the PICC from end users; AT&T, for instance, has imposed a Carrier Line Charge on its end users in order to recover this cost. See also Access Charge and SLC.

Pick The computer operating system of VMark Computer Inc. Pick is a neat operating system that unfortunately never really caught on.

Pick-And-Place The manufacturing operation in which components are selected and placed in the correct position on a substrate for the purpose of interconnection to the substrate. This is most commonly done with a programmable machine equipped with a robot arm.

Pick cable Outside telephone cable with plastic insulated pairs.

PICMG PCI Industrial Computer Manufacturers Group. An international consortium of vendors of industrial computer products, PICMG was formed to develop specifications for PCI (Peripheral Component Interconnect) based systems and boards for use in industrial and telecommunications computing applications. The electrical and mechanical specifications of PICMG enable CPU boards and backplanes from different manufacturers of industrial-grade computers to be interchangeable. PICMG specifications include CompactPCI, rackmount applications and PCI for passive backplane, standard format cards. See CompactPCI and PCI. www.picmg.org

Pickup Means you can answer a call from your phone. There all sorts of "pickups." The most common is GROUP PICKUP. Here you are part of a group and you can answer — from your phone — the call of anybody in that group, usually by punching a digit or a button or two. There's also NIGHT PICKUP, which typically allows anyone to answer an incoming call after hours, again by punching down a digit or a button or two. In TELECONNECT Magazine, we have one GROUP PICKUP. Everybody in the company belongs to that Group PICKUP and everyone can answer everyone else's phone. We believe

this simplifies things. It also allows anyone, anywhere, from any phone, to play telephone attendant, answering any and all incoming calls and transferring them through the system.

Pickup Group Imagine you're on a phone behind a PBX. Imagine you work in accounting, a group of five people. You can program most PBXs such that a call to your phone could be answered by anyone else in your group, and vice versa, you could answer someone else's ringing phone in your group. When you set your PBX up, you need to program which Pickup Groups which phones belong in.

Pickup Pattern A determination of the directions from which a microphone is sensitive to sound waves. It varies with the mike element and mike design. The two most common pickup patterns are omni- and uni-directional.

Pico Prefix meaning one-trillionth, or one-millionth of a millionth of a second. A pico is ten to the minus 12. See ATTO, FEMTO and NANOSECOND.

Picocell A wireless base station with extremely low output power designed to cover an extremely small area, such as one floor of an office building.

Picofarad One-trillionth of a farad. A unit of capacitance usually used to designate capacitance unbalance between pairs and capacitance unbalance of the two wires of a pair to ground.

Picosecond One-millionth of a millionth of a second. A picosecond is one to the minus 12 of a second. One picosecond — a trillionth of a second — is a spot of time from the domain of molecules. Light, traveling for one picosecond, would barely make it across the period at the end of this sentence. Only with a laser that generates picosecond light pulses can scientists freeze the short-duration motion of molecules and produce images of what goes on at the molecular level. Used in this way, the picosecond laser is comparable to a strobe, which can freeze the motion of a sprinter's stride in time-lapse photography. See nanosecond.

PICS 1. Protocol Implementation Conformance Statement: A statement made by the supplier of an implementation or system stating which capabilities have been implemented for a given protocol.

2. Product Inventory Control System.

3. USWest defines the term as Plug-In Inventory Control System. Plug-Ins, according to USWest, are circuit cards which fit into central office equipment and which control various aspects of transmission and carrier circuit functionality and usage.

4. A Macintosh-specific "multimedia" format for exchanging animation sequences (developed in 1988 by Macromind and others). PICS assembles several PICT files (frames) and combines them into one file.

PICT Picture Format. Developed by Apple in 1984 as the standard format for storing and exchanging black-and-white graphics files. PICT2 (1987) supports eight-bit color and gray scale.

Picture Element See PIXEL.

Picturephone AT&T's trademark for a video telephone that permitted the user to see as well as talk with the person at the distant end. AT&T introduced it at the 1964 World's Fair in Flushing Meadow, Queens, New York City. The device had a camera mounted on the top and a 5.25" x 4.75" screen. Audio signals were transmitted separately from video signals and the system could not use the public switched telephone network. It needed a transmission bandwidth of 6.3 Mbps and no one wanted to pay the price for the service. It never got off the ground. AT&T picked up on the name Picturephone and came out with an offering called Picturephone Meeting Service which provided full video teleconferencing. It was available through rented rooms or through equipment sold or rented to corporations. It also didn't do too well since the service was expensive and no one wanted to spend the time traveling to the few and far between conferencing room. We believe, as of writing, AT&T has abandoned Picturephone, the product, and has closed down the Picturephone Meeting Service rooms it rented to corporations.

Picturephone Meeting Service An AT&T service once provided under experimental tariff. It combined TV techniques with voice transmission. PMS is usually only available between telephone company-located picturephone centers. Most (we think all) have now been closed down. The venture was losing too much money.

PID Protocol Identifier. A field in the Call User Data included in the Call Request Packet sent to the ISP host for POS terminal initiated calls.

PIECE Productivity, Information, Education, Creativity, Entertainment. Microsoft's trick for remembering the big five multimedia computing applications.

Piepser The German word for beeper. Also the name of a small beeper made by Swatch and sold by BellSouth.

Piesio The Greek prefix meaning near.

Piezo-Electric Crystal A type of crystal which, when subjected to mechanical stress, generates current; or which, when subjected to varying electrical stresses, generates mechanical movement. Most familiar type is Rochelle Salts crystal. An old radio term.

PIF 1. Personal Communications Services Industry Forum.
2. Program Information File, a binary file, which contains information about how Windows should run an MS-DOS application, such as how much memory it needs, the path to the executable file, and whether the window in which the program is run closes automatically when the program terminates.
3. Public Inspection File. A set of documents which must be maintained by every cable television system at a convenient location in the cable community: at the system's office, or at some other convenient location such as an attorney's office or the local public library. The contents of the PIF are specified in Sections 76.302 and 75.305 of the FCC Rules. Any member of the public:
• Has the right to see the PIF on request.
• Has the right to request accommodations where the PIF can be reviewed without disturbance.
• Has the right to request photocopies of any or all documents in the file at a reasonable cost.
The Public Inspection File should be kept separate from all other files, both physically and operationally. This will reduce the chance of inadvertently releasing to the public any information which is not specifically required.

Piggy Back Data Slurp Imagine a data communications connection between two distant computers. Somewhere along the line, someone has attached a terminal or communicating computer and begun to capture the data as it flows across the line. In short, someone had slurped out data on a piggyback terminal or communicating computer. A piggyback data slurp could continue forever, without either party being aware their data and their conversations were being stolen — so long as the piggyback terminal didn't let its presence be known. The piggyback terminal, however, may insert itself into the conversation, pretending to send back authentic communications, thus misleading one or both of the parties. You could call this active deception or proactive espionage.

Piggyback Board Another name for a daughterboard on a card inside a PC.

Piggybacking A technique used at the data link or transport layer in a layered network architecture that allows for transmission acknowledgments to be carried in transmission frames received from the destination.

Pigtail 1. Multiple pieces of short cable with single circuit connectors connected to a multi-conductor cable. See OCTO-PUS.
2. A short, permanently attached piece of optical fiber used to link the transmitter and receiver to the transmission fiber.

Pigtail Antenna The standard cellular antenna for a car. The term "pigtail" refers to the spring-like section in the lower third of the antenna, the phasing coil.

Pilot Number Identifies a Hunt Group or Distribution Group. See also DISTRIBUTION GROUP.

Pilot-Make-Busy Circuit A circuit arrangement by which trunks provided over a carrier system are made busy to the switching equipment in the event of carrier system failure, or during a fade of the radio system.

PIM 1. Protocol-Independent Multicast. Multicast routing architecture that enables IP multicast routing on existing IP networks.
2. Personal Information Manager. A specialized form of software used by individuals and groups for keeping track of their contacts (and their addresses and phone numbers), their appointments, their lists of Things to Do, their reminder notes, their anniversaries, etc. PIMs are also called contact managers. Examples are Maximizer, TeleMagic, Borland's SideKick and Lotus' Organizer.
3. Plug In ISDN Module. A British term. A device used

PIN 1. Procedure Interrupt Negative. A fax term.
2. Photo INtrinsic diode. Also known as PIN Diode. A type of photodetector used to sense lightwave energy and then to convert it into electrical signals. PINs are matched with LEDs (Light Emitting Diodes) in fiber optic transmission systems of relatively low capacity (i.e., less than 500 Mbps for); currently, such systems are deployed for interconnection of hubs, switches and routers in a LAN environment. See also LED. See APD and Laser Diode for descriptions of the components used in high-bandwidth fiber optic systems.
3. Personal Identification Number. A code used by a mobile telephone subscriber in conjunction with a SIM card to complete a call. A code used by a credit card user. A code used by an ATM card user, etc. In short, a code to protect you against fraud. A code that you remember, which you don't write down anywhere and which theoretically can't fall into the wrong hands.

PIN Code Personal Identity Number code. A code used by a mobile telephone subscriber in conjunction with a SIM card to complete a call. See PIN.

Pin Diode A photodiode made with an intrinsic layer of undoped material between doped P and N layers and used as a lightwave detector.

PIN Number Personal Identification Number. A group of characters entered as a secret code to gain access to a computer system, such as the one that completes long distance calls. See PERSONAL IDENTIFICATION NUMBER.

Pin Photodiode An optical detector that converts light into electricity. This type is the typical diode used in a fiber optic receiver.

Pincushion Distortion When a video screen is distorted — with the top, bottom and sides pushing in — the screen is said to be suffering pincushion distortion.

PING 1. Packet InterNet Groper. Packet Internet Groper

(PING) is a program used to test whether a particular network destination on the Internet is online (i.e. working) by repeatedly bouncing a "signal" (called the Internet Control Message Protocol —ICMP— Echo Request Packet) off a specified address (i.e. the destination) and seeing how long that signal takes to complete the round trip. If you get no return signal, the site is either down or unreachable. If only a portion of the signal is returned, it indicates some trouble with the connection that will slow down performance. The term is often used as verb: "Ping host X to see if it is up!" Ping is useful for testing and debugging networks. In addition, PING reports how many hops are required to connect two Internet hosts. There are many freeware and shareware PING utilities available for personal computers.

2. A graphics file format ending in .png, which was developed to overcome deficiencies of .gif and .jpeg file formats, which are commonly used on the Web. Ping allows 16-, 24-, and 32-bit images, providing for better color depth. The downside is that higher-quality images require more storage and involve longer download times.

Ping Pong 1. A method of getting full duplex data transmission over a two wire circuit by rapidly alternating the transmission direction. See PING PONGING.

2.A disruptive phenomenon that occurs in digital cellular networks when the cell phone repeatedly reselects two cell sites of approximately equal strength. This problem is overcome through the use of a buffer area known as a "hysteresis."

3. A disruptive phenomenon that occurs in digital cellular networks when both transmission and reception take place over the same frequency channel, although in separate time slots. This problem is overcome through the use of separate frequency channels.

Ping Ponging Routing that causes a packet to bounce back and forth between two modes. See Ping Pong.

Pink Adult audiotext, i.e. dirty talking over the phone for money.

Pink Noise Noise in which power distribution is logarithmic through the spectrum, with an equal amount of power in each octave.

Pink Pages In Australia, at one stage, the list of businesses organized by industry, were printed on pink paper and called the Pink Pages. They are the equivalent of the North American "Yellow Pages."

Pinouts Pin configurations for cabling. In other words, which pin connects to which cable. Not all pins are always connected. Not all cables always connected.

Pinosecond One-trillionth of a second. One-millionth of a microsecond.

Pip Tone The tone that notifies you of a call waiting, assuming you subscribe to "call waiting." See CALL WAITING.

Pipe 1. A communications process within the operating system that acts as an interface between a computer's devices (keyboard, disk drives, memory, and so on) and an applications program. A pipe simplifies the development of application programs by "buffering" a program from the intricacies of the hardware or the software that controls the hardware; the application developer writes code to a single pipe, not to several individual devices. A pipe is also used for program-to-program communications and can be a connection between two processes so that the output from one immediately becomes the input for the other. Indicated by the I character.

2. A transmission facility. Pipe usually is used when discussing transmission bandwidth. For instance, fiber optics is a "big pipe," because it offers lots of bandwidth. See also BANDWIDTH.

Pipelining 1. Executing instructions by breaking them into component parts and processing them in parallel on separate processors. This reduces reduce cycle time and increases the computer's performance.

2. In imaging, pipelining also lets an imaging card start compressing and writing the image to disk while it is still being scanned.

3. In networking, pipelining is a technique used at the transport layer or data link layer in a layered network architecture that allows for the transmission of multiple frames without waiting to see if they are acknowledged on an individual basis. Each frame may have to be acknowledged later and in sequence, or a process of implied acknowledgment may be employed. Implied acknowledgment is a process whereby negative acknowledgment of a specific frame implies that all previously transmitted frames have been received correctly.

Piracy Any impersonation, unauthorized browsing, falsification or theft of data or disruption of service or control information in a network.

PITA Abbreviation for "Pain In The Ass;" commonly used on E-mail and BBSs (Bulletin Board Systems).

Pitch Control Variable control for increasing or decreasing the speed of a tape deck or turntable.

PIU Path Information Unit in SNA.

Pixilation A technique used by cinematographers and stage managers to make human performers appear to move as it artificially animated. Using a stop-frame camera, the pixilator can distort and speed up the motion of actors.

Pixel PIcture ELement. The smallest unit of area of a video screen image that can be turned on or off, or varied in intensity. The single point on a CRT display. The single point in a facsimile transmission. The image you see on your screen is the result of some pixels being on and others off. A pixel is the smallest part of the video screen that can be turned on or off or varied in intensity. It is one of the phosphor elements that coat the inside of a CRT tube. Pixels glow when struck by an electron beam. The number of pixels in the most common computer screen, a VGA monitor, is 640 x 480, with the first number (640) being the number of pixels in each horizontal row and the second number (480) the number of rows displayed. VGA stands for Video Graphics Array. Resolution (crispness and clarity of text and images) improves as the number of pixels displayed increases. If it's a color screen, a pixel is really three dots together, or clusters of red, green and blue — the triad of colors that, when energized, add up to white, or when the set is turned off, show as black. The phrase "picture element" was first used in 1927 in the magazine "Wireless World" writing about the mosaic of dots, or picture elements. See PEL.

Pixel Shim A small, usually invisible graphic used in an HTML document to create a page format. "I had to use a pixel shim to get the type to space correctly." A Wired Magazine term.

PIXIT Protocol Implementation eXtra Information for Testing: A statement made by a supplier or implementor of an IUT (Implementation Under Testing) which contains information about the IUT and its testing environment which will enable a test laboratory to run an appropriate test suite against the IUT.

Pixrects Pixrects is the primary graphics programming interface in the SunView Window System from Sun Microsystems. It is replaced in the OpenWindows XView toolkit by the Pixwin interface, which is a thin layer on top of Xlib.

Pixwin Pixwin is the primary graphics programming interface in the XView toolkit from Sun Microsystems. Pixwin is a thin layer on top of Xlib.

Pizza Box A wireless term referring to a phased array antenna. The device is small, square, and flat, resembling a pizza box. If you open the cover of the pizza box, you will find an array of little antenna gizmos where you normally would expect to find the pepperoni slices on a pizza. The pizza box mounts flat against the side of your building in a WLL (Wireless Local Loop) application, for instance. It also can mount on your rooftop in a satellite application. Hence, it is aesthetically pleasing. See also Phased Array and WLL.

PJ-327 A double RCA dipole plug for connecting a headset into a PBX console. When you buy a headset you need to specify — 4-pin modular jack (for connecting to a telephone) or two prong plug (for connecting to a PBX console).

PKI See Public Key Infrastructure.

PL Private Line.

PL/1 Programming Language One (IBM).

Place Call There will be two types of calls: person calls and place calls. I make a place call when I call a phone number which ends in one designated, fixed RJ-11 jack attached to the wall or floor. I make a person call when I call a phone number which doesn't necessarily end on a fixed place. A typical person call might be a cellular or wireless phone call. It might also be a service like MCI's Personal 800 Service. The major characteristic of a person call is that I am calling a person and I don't know where that person is. As a result, the person I called might answer my call anywhere — from his car, from vacation home, from his wireless phone, etc.

Plain B Wire Connector Also called a B Connector or beans. A twisted-pair splicing connector that looks like a one-inch drinking straw. They have metal teeth inside them to pierce the vinyl insulation of the wire to make a good connection. Sometimes water-retardant jelly is sometimes place inside.

Plaintext A message which is not encrypted. A message which is encrypted is called a ciphertext. See CLIPPER CHIP.

Plan A route to an end or objective usually achieved accidentally and often written in hindsight. Derived from the expression: "Most people spend more time planning their annual vacation than they do planning their careers."

Plan File A file that lists anything you want others on the Internet to know about you. You place it in your home directory on your public-access site. Then, anybody who fingers (sees) you, will get to see this file.

Planar Array Antenna A planar array antenna is designed for use at microwave frequencies. It resembles a double-sided printed circuit board. One side of the substrate carries an etched pattern of microstrip whilst the other is left completely metallic to act as a ground plane.

Planar Board IBM's new name for a motherboard in their new series of System/2 Personal Computers. A motherboard is the main board in a PC on which the main CPU, the main memory, the clock and sundry other things like serial and parallel ports are mounted. Other boards, i.e. graphics boards, are plugged into the motherboard. Thus the expression "motherboard." No one knows why IBM dropped the word. Maybe it was too risque? Maybe they included more on their motherboards in the System/2 series that they would no longer function as motherboards? Maybe a feminist group of mothers objected?

Plant A general term for all equipment used by a telephone company to provide telecommunications services. Usually divided into inside and outside plant. See Inside Plant and Outside Plant.

Plant Hump A very friendly term for a craftsperson in a phone company — installer, splicer, underground cable guy — who works hard with outside phone equipment, often under adverse conditions and whose labors tend to be undervalued, particularly by the white collar, pencil pushers back at headquarters.In British slang, "hump" means to exert oneself. There is a badge of honor to being a "plant hump." These are the people who bring phone service to your door. This definition contributed by Steve Marcus of Nynex, now called Bell Atlantic. Mr. Marcus readily acknowledges he is one of the pencil pushers back at headquarters. The word plant hump is occasionally written as one word, i.e. Planthump.

Plant Test Numbers Virtually every 800 IN-WATS number has a plant test number. This is its equivalent seven digit local number. That number looks like a normal local seven digit number, with a standard three-digit central office exchange code and a four-digit extension. The purpose of plant test numbers is to allow the telephone company to test the local part of the incoming 800 number by simply dialing that number. For example, Miller Freeman, which published this dictionary has an 800 number — 800-LIBRARY (or 800-542-7279). The plant test number of the first line of that 800-LIBRARY group is 212-206-6870. The second line is 212-206-6871 and so on. It is valuable to know the plant test numbers of your incoming WATS lines so you can test the local loop part of those lines. The local loop part is the part which typically gives the most problem. It is, unfortunately, the only part of your 800 lines you can test yourself — unless you ask someone (or several people) to call you regularly on your 800 lines, just to test them. You can get plant test numbers out of your local and/or your long distance carrier. When they tell you those numbers are "not available," beg a little. They are available and you are entitled to them. Calling plant test numbers costs exactly what a normal long distance IN-WATS call on that line costs. So keep your test calls short. You should call your plant test numbers once a day.

Planthump A colloquial word for a telephone company craftsperson. It derives from the term "plant," a telephone company word used to describe their "factory" — i.e. everything from their inside plant, their central office switch, to their outside plant, which includes wire strung on telephone poles. In British slang, "hump" means to exert oneself. Planthump is a term of endearment in the telephone industry. Definition courtesy, Steve Marcus, New York Telephone, now called Bell Atlantic. The term is often spelled as two words, i.e. plant hump.

PLAR Private Line, Automatic Ringdown. In telecommunications, leased voice circuit that connects two single instruments together. When either handset is lifted, the other instrument automatically rings.

Plasma An ionized gas with a mixture of positive and negative electrons. See PLASMA DISPLAY.

Plasma Display Type of flat visual display device in which selected electrodes, part of a grid of crisscross electrodes in a gas-filled panel, are energized, causing the gas to be ionized and light to be emitted. Some computers use plasma displays. They're fabulous, and quite expensive. See PLASMA DISPLAY PANELS.

Plasma Display Panels PDPs. Plasma gases composed of helium, neon and xenon are sandwiched into cells between two vertical glass plates. Bursts of electricity are applied between transparent electrodes attached to one pane of glass. These bursts causes the plasma gases to emit ultraviolet rays. This activates red, blue and green phosphor dots, which emit visible light and form pictures on the screen. The more common cathode ray tube technology (e.g. computer monitors) uses an electron gun to direct a beam that lights up phosphors on a screen. Directing that beam requires CRT sets to be deep, heavy and unwieldy.

Plasmatron The name Sony chose for a flat screen display it demoed at Comdex in the fall of 1995. The screen measured 20 inches diagonally. It was as bright as normal CRT screen, but was less than four inches wide. It is the beginning of flat screen entertainment screens for the home. It was spectacular.

Plaster Ring A metal or plastic plate that attaches to wallboard for the purpose of mounting a telecommunications outlet box.

Plastic Fiber Optics See Plastic Optic Fiber.

Plastic Optic Fiber POF. A fiber optic transmission medium made from plastic, rather than glass. Glass clearly (double entendre intended) performs better than plastic, as it offers less attenuation and, therefore, better transmission quality at higher speeds and over longer distances. Plastic, however, is less expensive and less susceptible to breakage. POF uses low-quality light sources and carry data at speeds greater than 10 Mbps over distances up to 100 meters. The ATM Forum is developing specifications for 50- and 155-Mbps transmission over POF. POF is evolving as a replacement for twisted-pair copper wire. see PLASTIC FIBER OPTICS.

Plasticizer A chemical agent added in compounding plastics to make them softer and more flexible.

Plat An imaging term. When a CAD/CAM plotter prints a large drawing, it's called a plat.

Plate The anode in a vacuum tube, which collects the electrons emitted by the filament.

Plate Battery The source of E.M.F. connected in the plate circuit to give the plate element its positive charge.

Plate Voltage The potential applied to the plate of the vacuum tube by the plate voltage supply.

Platen A cylinder in a printer or typewriter around which the paper goes and which the printing mechanism strikes to produce an impression.

Platform A loosely-defined word for a software operating system and/or open hardware, which an outsider could write software for. Windows98 is a platform. So is Windows 2000. If every phone system were a platform, then every owner of that phone system could write software for his phone system or buy outside-produced software and have his phone system work more to his liking. That's the objective of creating a "platform." See also OAI and PLATFORM INDEPENDENCE.

Platform Independence A term from IBM and Metaphor Computer Systems. The idea, they say, is to produce a layer of software that would rest atop any operating system on any piece of hardware. The applications developers would write their software just once, rather than start from scratch each time they wanted get their software working on a different computer. If the whole idea sounds rather daunting, you're right.

Platter The round magnetic disk surfaces used for read/write operations in a hard disk system.

Play Off In voice processing, in response to questions such as "Press one for Harry," the user touchtones buttons on his phone. Those buttons generate DTMF (Dual Tone Multi-Frequency) tones. The system has to figure out what the person "said" with his touchtones. The tricky part of DTMF detection is distinguishing between tones generated from an actual "key press" and "tones" caused by speech. Mistaking a person's speech (as in leaving a message) for DTMF is called "talk-off." Mistaking a person's recorded speech (as in playing back a message) for DTMF is called "play-off."

You can imagine the havoc poor DTMF detection can cause a voice processing system. For example, if touchtoning three means "delete this message" and while playing the message, the system incorrectly detects a portion of the message playback as the touchtones for a key press three, I'll delete the message when I had intended to listen to it. On the other hand, if I'm listening to a message and want to delete it prior to finishing the message, I want the system to detect my key press three as the real thing and go ahead and delete the message.

Playback Retrieval, decoding and transmission of encoded data. It is also a multimedia term. Playback is the process of viewing multimedia materials created by an author. Playback can include a range of activities, from viewing a single video clip to participating in a series of interactive multimedia training modules. Some playback applications (for example many training and presentation applications) are sold separately from their authoring applications. However, many developers are selling authoring and playback capabilities in a single product.

Playback Head The part which converts the magnetic information on the tape or disk into an electrical signal. Moving the magnetic fields on the medium (tape or disk) past the playback head generates a tiny voltage, which is picked up in a conductor (a coil) in the payback head and sent onto the electronic equipment where it is amplified or transmitted.

Player An SCSA definition. A resource object that plays TVM data. The audio data can come from a voice or audio encoded file, or from text that has passed through a text-to-speech service. The output of a player can be analog audio, TDD, ADSI, etc.

PLCP Physical Layer Convergence Protocol. The part of the physical layer that adapts the transmission facility to handle DQDB functions as defined in IEEE 802.6-1990. It is used for DS-3 transmission of ATM. ATM cells are encapsulated in a 125microsecond frame defined by the PLCP which is defined inside the DS3 M-frame.

PLD Programmable logic device.

Plenum In some modern buildings, the ducts carrying the heat return are not metal ducts but actually are part of the ceiling. This is called a plenum ceiling. Most cities now have rules and regulations which say that if you run cabling through these plenum ceilings, you must not use cabling sheathed in PVC (polyvinyl chloride), the standard jacketing of most electrical cable. The reason is that PVC burns and emits toxic smoke ferociously. Plenum cable is low smoking so that if it catches fire it won't circulate toxic smoke through the vent system and suffocate everyone. Plenum cabling is often made of teflon. It's much more expensive than normal cabling. See also FEP.

Plenum Area The space between the drop ceiling and the floor above. Continuous throughout the length and width of each commercial building floor.

Plenum Cable Cable specifically designed for use in a plenum (the space above a suspended ceiling used to circulate air back to the heating or cooling system in a building). Plenum cable has insulated conductors often jacketed with polyvinylidene diflouride (PVDF) material to give them low flame spread and low smoke-producing properties. Plenum cable has fully color coded insulated copper conductors and is available in various pair sizes. It can be either 22 or 24 AWG.

Plesiochronous Plesiochronous, based on Greek and Latin roots, roughly translates as "more together in time." Plesiochronous networks involve multiple digital synchronous circuits running at different clock rates. For instance, a

NYNEX T-1 circuit may meet a MCI T-1 circuit, with each taking making use of a different clocking source. Also for example, multiple MCI T-1 circuits may require multiplexing into a T-3 circuit; with the T-1's and the T-3 running at different clock speeds. In either case, the differences in clock speeds must be resolved through the use of a master clocking source such as a Stratum I clock, which relies on a highly reliable cesium clocking source. T-carrier and E-carrier networks are plesiochronous. Compare to SYNCHRONOUS, ASYNCHRONOUS and ISOCHRONOUS. See also PDH.

Plesiochronous Networks Network elements that derive timing from more than one primary reference source. Network elements accommodate minor frequency differences between nodes.

PLL Phase Lock Loop: Phase Lock Loop is a mechanism whereby timing information is transferred within a data stream and the receiver derives the signal element timing by locking its local clock source to the received timing information.

PLMN Public Land Mobile Network. A mobile telephone communications network established by a provider to facilitate mobile telecommunications services. This includes equipment, operations, and staff. A single provider may have more than one PLMN.

Plotter A type of computer peripheral printer that displays data in two-dimensional graphics form.

PLS Premises Lightwave System.

PLSC Private Line Service Center.

Plug A male element of a plug/jack connector system. In the Premises Wiring System it provides the means for the user to connect his communications devices to the Communications Outlet as well as the means to disconnect his service at the Network Interface Jack when trouble analysis is required.

Plug 'N Play 1. Manufacturers' concept of how easy it is to install their equipment. "Why it's just plug 'n play," says the manufacturer. In reality, nothing, absolutely nothing, is plug 'n play. It's a fantasy concept. See PLUG AND PLAY.
2. Also defined as a new hire who doesn't need any training. "The new guy, Harry, is great. He's 100% plug-and-play."

Plug And Play This explanation comes from an Intel Technology Primer: Since add-in cards first appeared over a decade ago, they've given users a lot of different ways to improve their PCs and given them a lot of installation headaches. In this brief, we'll tell you how Intel, together with industry leaders, has spent years developing Plug and Play technology to make add-in cards both easier to use and install. Never before has the PC had as many capabilities as it does today. That's due in part to the large number of add-in cards available, like those for multimedia and faxmodems. Yet, as more cards are added to a PC, their installation can become quite complex. Installing a card can be a time-consuming and technical process, and there's no guarantee it will even work the first time. Sometimes the user must configure the card manually, which means selecting a variety of system resources for each card. These include Interrupt Requests (IRQ), I/O and memory addresses, and Direct Memory Access (DMA) channels. Every PC has a limited number of these resources available. Each card is designed to use a small group of them. Assigning these resources means opening the computer and physically setting the jumpers and DIP switches. And since no standard has been set to determine which cards can use which resources, numerous conflicts can arise between cards. Often, it's a process of trial and error to determine which resources aren't already being used by other cards. Since the ISA bus was introduced, several new bus architectures have followed to solve the resource allocation problem. For example, the MCA and the EISA bus standards both defined a mechanism where add-in cards were configured somewhat automatically. These bus architectures allocated the resources, but the process wasn't always flexible and still required some manual intervention. And they still left the current ISA cards without a solution. Plug and Play technology, co-developed by Intel and other industry partners, consists of hardware and software components that card, PC, and operating system manufacturers incorporate into their products. With this technology, the user is responsible for simply inserting the card. Plug and Play makes the card capable of identifying itself and the resources it requires. The system's software automatically sets up a suitable configuration for the card. Newly developed PCI and Plug and Play ISA cards are all built to eliminate user intervention during the installation process. See PLUG AND PLAY BIOS EXTENSIONS.

Plug And Play BIOS Extensions Software code added to a PC's bios which purports to automatically recognize which peripherals are in the PC and automatically configure the PC for those peripherals — without the need for fiddling with dip switches or setting interrupts, etc. Plug and Play comes from Intel. And more and more PC cards are coming Plug and Play compatible.

Plug Compatible Devices made by different manufacturers that are totally interchangeable. The word derives from the fact that the devices are so completely interchangeable that you can simply unplug one device and plug in another device made by different manufacturer and it will work the same, or better.

Plug-in A program of data that enhances, or adds to, the operation of a (usually larger) parent program. A paint package, for example, might contain plug-in tools that create special effects, like speckling. A Web browser might have a plug-in that allows you to hear sound or view movies you download over the Internet. There are hundreds of plug-ins. See www.netscape.com/plugins/index.html

Plugboard A telephone switchboard on which connections are made by a jack and an attached cord representing a trunk (the male jack and the cord) and a female plug (the telephone extension). Early plugboards needed an operator to place outside calls and connect incoming calls. All calls were completed by the operator. Plugboards were common in the days of "PBXs." Then came PABXs (Private Automated Branch Exchanges) and you could dial out without the help of an operator. Then electronic PABXs came in and you could dial directly in to many internal extensions, using a feature called DID (Direct Inward Dial). Now PABXs are called PBXs because they're all automatic. Plugboards are rapidly disappearing. They do have two great uses, however. First, operators who grew up with them, still like them. Operators who now live in nursing homes like them. Second, because the jack and plug make a pure metallic connection, they're great for data transmission and occasional data switching.

Plugs Circuit cards which control various aspects of transmission and carrier circuit functionality and usage.

PLV An imaging term. Production Level Video. DVI Technology's highest quality motion video compression algorithm. It's about 120-1 compression. Compression is done "off-line". i.e. non-real time, and playback (decompression) is real time. Independent of the technology in use, off-line compression will produce a better image quality than real time since more time and processing power is used per frame.

PLY One layer in a composite.

PM 1. Physical Medium: Physical Medium refers to the actual physical interfaces. Several interfaces are defined including

STS-1, STS-3c, STS-12c, STM-1, STM-4, DS1, E1, DS2, E3, DS3, E4, FDDI-based, Fiber Channel-based, and STP These range in speeds from 1.544Mbps through 622.08 Mbps.
2. Performance Monitoring. Gives a measure of the quality of service and identifies degrading or marginally operating systems (before an alarm would be generated). Digital signal parameters, including errored seconds and out of frame, measure the integrity of a communication channel as defined in AT&T Compatibility Bulletin 149 (CB 149).
3. Peripheral Module.
4. An ATM term for the Physical Medium sublayer. See PHYSICAL MEDIUM.

PMC Public Mobile Carrier

PMD Physical Medium Dependent. This sublayer defines the parameters at the lowest level, such as speed of the bits on the media. The bottom half of BISDN Layer 1.
2. Polarization Mode Dispersion. A fiber optic term describing distortion created by irregularities in the shape of the fiber optic cable and its core; the problem is exacerbated by splicing, expansion and contraction of the cable due to variations in ambient temperature, and spooling of the cable. At high transmission speeds (e.g., SONET OC-192) digital light pulses can suffer from PMD. As the pulses travel down the fiber, the cable's physical irregularities cause delays on the at the outer edges of the core; the center portion of the pulse which travels through the "sweet spot" is unimpaired and, therefore, travels at a higher rate of speed from end-to-end. The result is that portions of an individual light pulse can arrive at slightly different time, with the delay being measured in picoseconds. The effect is one of distortion as the center portion of subsequent light pulses can overrun the outer portions of the preceding pulses. The impact is a higher bit error rate (BER). PMD is especially a problem at high transmission speeds, and particularly over older fiber optic cables deployed prior to the anticipation of high-bit rate SONET. Confused?...Consider the following explanation: If you drop a perfectly round rock into a pond of water, the waves run at the same speed toward the edges of the pond. If the pond is not perfectly round, some waves will reach the bank before other waves. If you now think of shooting the rock into a water pipe, you can see that some of the resulting compression waves will reach the other end before others. The combination of resistance at the edges of the pipe and irregularities in the shape of the pipe act to compound the problem, which varies unpredictably from pipe to pipe. See CHROMATIC DISPERSION.

PMI Project Management Institute.

PMMU Paged Memory Management Unit. Macintosh computers equipped with a PMMU may use virtual memory with the System 7 operating system.

PMN Indicates loss of ac power at the far-end terminal.

PMP 1. Point-to-MultiPoint. An ATM term describing the connecting circuitry between a single end point (root node) and multiple end points (leaf nodes). The root node can transmit data over a PMP connection to multiple leaf nodes, usually through a switch or router. The leaf nodes can transmit back to the root node, but do not have the ability to transmit data directly to other leaf nodes.
2. Project Management Professional as certified/designated by the Project Management Institute (PMI).

PMR 1. Poor Man's Routing. A technique used in any packet-switched networks to allow a source node to predefine the routing to the destination, bypassing the normal routing algorithm implemented at the network layer.
2. Private Mobile Radio

PMS 1. Picturephone Meeting Service. An AT&T service once provided under experimental tariff. It combined TV techniques with voice transmission. PMS is usually only available between telephone company-located picturephone centers. Most have now been closed down. The venture was losing too much money.
2. Property Management System, a software program and computer that controls all guest billing and guest services functions in a hotel. In short, the guts of a hotel's computer system. Some telephone systems have a PMS Interface, which allows various degrees of integration between the telephone system and the hotel's computer systems. For example, voice mail could be administered through the hotel's Property Management System.
3. The Pantone Matching System, a universal language for solid-color specification and reproduction. Colors defined by PMS receive a unique number and mixing formula. Consequently, when artists specify a PMS number they can be sure that the final printed product will match the chosen color. But, be careful, PMS colors look different when printed on different papers. The biggest perceived difference is when you print on glossy or matte paper.

PMS Interface An interface that allows telephone system functions (like voice mail) to be administered through a hotel's Property Management System.

PNG Portable Network Graphics, pronounced "Ping." A graphics file format ending in .png, which was developed to overcome deficiencies of .gif and .jpeg file formats, which are commonly used on the Web. PNG, under development by the W3C, allows 16-, 24-, and 32-bit images, providing for better color depth. The downside is that higher-quality images require more storage and involve longer download times. PNG overcomes this problem through the use of better image compression technology.

PNI An ATM term. Permit Next Increase: An ABR service parameter, PNI is a flag controlling the increase of ACR upon reception of the next backward RM-cell. PNI=0 inhibits increase. The range is 0 or 1.

PNNI Private Network-Network Interface: A routing information protocol that enables extremely scalable, full function, dynamic multi-vendor ATM switches to be integrated in the same network.

PNNI Protocol Entity An ATM term. The body of software in a switching system that executes the PNNI protocol and provides the routing service.

PNNI Routing Control Channel An ATM term. VCCs used for the exchange of PNNI routing protocol messages.

PNNI Routing Domain An ATM term. A group of topologically contiguous systems which are running one instance of PNNI routing.

PNNI Routing Hierarchy An ATM term. The hierarchy of peer groups used for PNNI routing.

PNNI Topology State Element An ATM term. A collection of PNNI information that is flooded among all logical nodes within a peer group.

PNNI Topology State Packet An ATM term. A type of PNNI Routing packet that is used for flooding PTSEs among logical nodes within a peer group.

PNM Public Network Management.

PNO Public Network Operator. Usually a PTT of some sort. See PTT.

PnP Plug and Play. The technology that lets Windows 95 and soon other operating systems automatically detect and configure most of the adapters and peripherals connected to or sitting inside a PC. A fully Plug and Play-enabled PC requires three PnP pieces: a PnP BIOS, PnP adapters and peripherals, and a PnP operating system. Adding a PnP-compliant CD-ROM drive, hard disk, monitor, printer, scanner, or other device to a PnP PC means little more than making the physical connection. The operating system, together with PnP logic present in the BIOS and in the device itself, handles the IRQ settings, I/O addresses, and other technical aspects of the installation to make sure that the thing will work. The idea of PnP is to make installation of complex gadgets — such as sound cards and modems — easy, taking care of the major bane of everyone's life: That your new device now conflicts with an old device, effectively killing both devices and maybe crashing your PC at the same time. PnP is a great idea. Its success has been slow in coming, because so many devices are not PnP compatible.

PNS Personal Number Service is a new concept in telecommunications that assigns a telephone number to a person, not a location, effectively allowing a subscriber to use one number for all calls and helping them manage their incoming communications. The service does not require the user to change any existing phone numbers. The subscriber simply provides the various numbers — office, cellular, pager, fax and home — and instructions on where and when the calls should be routed, and the PNS directs the calls in the order requested by the subscriber.

PO Point of Origin. It is used in relationship with a Message Transfer Agent (MTA).

POCSAG One of the communications protocols used between paging towers and the mobile pagers/receivers/beepers themselves. Other protocols are GOLAY, ERMES, FLEX and REFLEX. The same paging tower equipment can transmit messages one moment in POCSAG and the next moment in ERMES, or any of the other protocols.

PODP Public Office Dialing Plan.

Podiumware You're presenting a great speech detailing some great new concept in hardware or software. You don't have many precise details, except your vague words. This is called podiumware. When your thinking has become more concrete, and you make slides on your new hardware or software, you have moved to slideware. Eventually when you announce your new hardware or software, you have moved to hypeware or vaporware. I was first introduced to the word "podiumware" by Bob Lewis, a columnist for InfoWorld Magazine. See also HOOKEMWARE, HYPERWARE, MEATWARE, PODIUMWARE, SHOVELWARE, and VAPORWARE.

POF Plastic Optic Fiber. A fiber optic transmission medium made from plastic, rather than glass. Glass clearly (double entendre intended) performs better than plastic, as it offers less attenuation and, therefore, better transmission quality at higher speeds and over longer distances. Plastic, however, is less expensive and less susceptible to breakage. Plastic Optic Fiber (POF) uses low-quality light sources and carry data at speeds greater than 10 Mbps over distances up to 100 meters. POF is evolving as a replacement for twisted-pair copper wire.

POFS Private Operation Fixed Systems. Microwave incumbents in the 2.0 Ghz band. Must be relocated with comparable alternative facilities funded.

POGO Post Office Goes Obsolete. When MCI Mail was originally being planned, its code name was POGO. The idea was obvious. In September of 1994, I asked MCI what "POGO" meant in and they answered: "Pogo" is an internal message format used by MCI for coding purposes.

POH Path OverHead. SONET overhead assigned to and transported with the payload until the payload is demultiplexed. It is used for functions that are necessary to transport the payload; i.e., end-to-end network management. These functions include parity check and trace capability. It is not implemented in SONET Lite.

POI Point Of Interface. The physical telecommunications interface between the LATA access and the interLATA functions. A POI is a demarcation point between LEC and a Wireless Services Provider (WSP). This point establishes the technical interface, the test point(s) and the point(s) for operational division of responsibility. See also POINT OF PRESENCE.

Point Code A SS7 term for a unique code which identifies a network node in order that the SS7 network can route calls properly. When placing a call, you dial a Global Title in the form of dialed digits (i.e., a telephone number). Those digits are translated from the Global Title to a Point Code by the STP (Signal Transfer Point) through a process known as Global Title Translation (GTT). See also GLOBAL TITLE, GLOBAL TITLE TRANSLATION, SS7, and STP.

Point In Call PIC. A representation of a sequence of activities that the ASC (AIN Switch Capabilities) performs in setting up and maintaining a basic two-party call. PICs occur in Originating and Terminating BCSMs (Basic Call State Model).

Point of Demarcation Physical point at which the phone company's responsibility for the wiring of the phone line ends.

Point Of Interface POI. The physical telecommunications interface between the LATA access and the interLATA functions. A POI is a demarcation point between LEC and a Wireless Services Provider (WSP). This point establishes the technical interface, the test point(s) and the point(s) for operational division of responsibility.

Point Of Presence POP. A physical place where a carrier has a presence for network access, a POP generally is in the form of a switch or router. For example, an large IXC will have a great many POPs, at which they interface with the LEC networks to accept originating traffic and deliver terminating long distance traffic. The basis on which the interface is accomplished can include switched and dedicated (leased line) connections. Similarly, providers of X.25, Frame Relay and ATM services have specialized POPs, which may be collocated with the circuit-switched POP for voice traffic. A POP also is a meet point for ISPs (Internet Service Providers), where they exchange traffic and routes. See also GIGAPOP and POP.

Point Of Purchase Politics Politically correct shopping or cause-related marketing, such as that advocated by Benetton or Ben and Jerry's.

Point Of Sale Terminal A special type of computer terminal which is used to collect and store retail sales data. This terminal may be connected to a bar code reader and it may query a central computer for the current price of that item. It may also contain a device for getting authorizations on credit cards.

Point Of Termination POT. The point of demarcation within a customer-designated premises at which the telephone company's responsibility for the provision of access service ends.

Point Size The height of a printed character specified in units called points. A point equals 1/72 inch. Also known as font size.

Point To Multipoint A circuit by which a single signal goes from one origination point to many destination points. The classic example is a TV signal (say a Home Box Office program) being broadcast from one satellite to many CATV subscribers all around the country. Not to be confused with a multi-drop circuit. See POINT TO MULTIPOINT CONNECTION.

Point To Multipoint Connection A Point-to-Multipoint Connection is a collection of associated ATM VC (Virtual Channel) or VP (Virtual Path) links, with associated endpoint nodes, with the following properties:
1. One ATM link, called the Root Link, serves as the root in a simple tree topology. When the Root Node sends information, all of the remaining nodes on the connection, called Leaf Nodes, receive copies of the information.
2. Each of the Leaf Nodes on the connection can send information directly to the Root Node. The Root Node cannot distinguish which Leaf is sending information without additional (higher layer) information. (Note: UNI 4.0 does not support traffic sent from a Leaf to the Root.)
3. The Leaf Nodes cannot communicate directly to each other with this connection type. See ATM.

Point-to-Multipoint Delivery Delivery of data from a single source to several destinations.

Point To Point A private circuit, conversation or teleconference in which there is one person at each end, usually connected by some dedicated transmission line. In short, a connection with only two endpoints. See also POINT TO MULTIPOINT.

Point To Point Connection An uninterrupted connection between one piece of equipment and another.

Point-to-Point Delivery Delivery of data from a single source to a single destination.

Point-To-Point Protocol PPP. An 8-bit serial interconnection protocol which allows a computing device, such as a PC, to connect as a TCP/IP host to a network through an asynchronous port. PPP commonly is used for connection across the PSTN from a PC to an ISP for purposes of Internet access. PPP is the successor to SLIP (Serial Line Internet Protocol). PPP provides router-to-router and host-to-network connections over both synchronous and asynchronous circuits. PPP includes error detection and data protection features, unlike SLIP and other protocols. See also SLIP.

Point To Point Signaling A signaling method where signals must be completely received by an intermediate station before that station can set up a call connection. See END TO END SIGNALING.

Point To Point Topology A network topology where one node connects directly to another node.

Pointcasting First, there was broadcasting. One signal went to many people. Radio and TV are the classic concepts of broadcasting. One signal — the same signal — to many people. Then came the idea of narrowcasting. One signal to a select number of people — maybe only those people who subscribed to the service and had the equipment to receive it. Then there came pointcasting. This is a fancy name for sending someone a collection of customized information — snippets of stuff that they chose from a palette of information offerings.

Pointer Processing Pointer processing accommodates frequency differences by adjusting the starting position of the payload within the frame. A pointer keeps track of the starting position of the payload.

Points Of Failure A simple term to indicate that in a complex network there are many places things can go wrong. Those places need to be identified so that you can anticipate and plan for things to go wrong.

Poisson See Poisson Distribution.

Poisson Distribution A mathematical formula named after the French mathematician S. D. Poisson, which indicates the probability of certain events occurring. It is used in traffic engineering to design telephone networks. It is one method of figuring how many trunks you will need in the future based on measurements of past calls. Poisson distribution describes how calls react when they encounter blockage (see QUEUING THEORY for a detailed explanation of blockage). There are two main formulas used today in traffic engineering: Erlang B and Poisson. The Erlang B formula assumes all blocked calls are cleared. This means they disappear, never to reappear. The Poisson formula assumes no blocked calls disappear. The user simply redials and redials. If you use the Poisson method of prediction, you will buy more trunks than if you use Erlang B. Poisson typically overestimates the number of trunks you will need, while Erlang B typically underestimates the number of trunks you will need. There are other more complex but more accurate ways of figuring trunks — Erlang C (blocked calls delayed or queued) and computer simulation. Poisson has been used extensively by AT&T to recommend to its customers the number of trunks they needed. Since AT&T was selling the circuits and preferred its customers to have excellent service, it made sense to use the Poisson formula. As competition in long distance has heated up, as circuits have become more costly and as companies have become more economically-minded (more aware of their rising phone bills), Poisson has become widely ignored.

After I wrote the above definition, Lee Goeller, a noted traffic engineering expert contributed the following definition of Poisson Distribution: A probability distribution developed by E.C. Molina of AT&T in the early 1900s for use in solving problems in telephone traffic (see TRAFFIC ENGINEERING), although it has many other uses and is widely applied in many fields. When made aware of Poisson's prior effort (circa 1820), Molina gave him full credit and even taught himself French so he could read Poisson in the original. The Poisson distribution assumes a call is in the system for one holding time, whether it is served or not (blocked calls held); the first form of the distribution estimates the probability that exactly X calls will be in the system, while the second estimates the probability that X or more calls will be present. If there are only X trunks to serve the calls, the second form gives the probability of blocking. Although limited tabulations of the Poisson distribution had been made earlier, Molina published an extensive set of tables in 1942. The Poisson distribution slightly overstates the number of trunks needed when compared to the Erlang B distribution (see ERLANG B).

Poisson Process A kind of random process based on simplified mathematical assumptions which makes the development of complex probability functions easier. In traffic theory, the arrival of telephone calls for service is considered a Poisson process. Calls arrive "individually and collectively at random," and the probability of a new call arriving in any time interval is independent of the number of calls already present. A Poisson process should not be confused with the Poisson Distribution, which gives the probability that a certain number of calls will be present if certain additional assumptions are made. See POISSON DISTRIBUTION.

Poke-Through Method A distribution method that involves drilling a hole through the floor and poking cables through to terminal equipment from the ceiling space of the floor below. See also CEILING DISTRIBUTION SYSTEMS and NEWTON.

Poke-Through System Penetrations through the fire-resistive floor structure to permit the installation of horizontal telecommunications cables.

Polar Keying A transmission technique for digital signals in which the current flows in opposite directions for 1s and 0s or marks and spaces. It is used in telegraph signaling. It is also known as polar transmission.

Polar Relay A relay containing a permanent magnet that centers the armature. The direction of movement of the armature is governed by the direction of current flow.

Polarity Which side of an electrical circuit is the positive? Which is the negative? Polarity is the term describing which is which. Knowing polarity is not critical with rotary phones. They will work irrespective of which way the telephone circuit's polarity is. Touchtone phones, however, need correct polarity for their touchtone pads to work. How to tell? If you can receive an incoming call, can speak on the phone clearly, but can't "break" dial tone by touching a digit on your touchtone pad, then the polarity of your line is reversed. Simply reverse the red and green wires. Some electronic phones behind PBXs and key systems are also sensitive to polarity. If in doubt, simply reverse the wires. In video, reversed polarity results in a negative picture.

Polarization Characteristic of electromagnetic radiation (e.g. lightwave, radio or microwave) where the electric-field vector of the wave energy is perpendicular to the main direction, or vector, of the electromagnetic beam.

Pole Attachment Cost to cable TV, cellular provider and other telecom operators (including end users) to rent space to attach cables to telephone company and power company poles. There are charges and often significant restrictions on the attachment of your cable to their pole.

Policy Based Networking A traffic management concept involving the establishment of priorities for network traffic based on parameters such as traffic type, application, and user ID. ATM does a great job of policy-based networking as a result of QoS (Quality of Service) levels. RSVP (Resource ReserVation Protocol) from the IETF is emerging as a solution to managing traffic priorities over the Internet. Policy-based networking can be implemented in capable switches, routers and servers. See also OOPS, RSVP and QoS.

Policy Routing Protocol An extension of Vector Distance Protocols used in router networks. Used primarily in Internet routers, Policy Routing Protocols determine the route of a packet in consideration of "permissions" and reciprocal business contracts between and among backbone carriers, ISPs and Internet Access Providers. In other words, the route is determined on the basis of non-technical policy, rather than the number of hops a packet must travel. Assuming that the intercarrier policy accepts the offered traffic, the packet is routed based on technical considerations according to Vector Distance Protocols. Examples of Policy Routing Protocols include BGP (Border Gateway Protocol) and IDRP (InterDomain Routing Protocol). See also DISTANCE VECTOR PROTOCOL, LINK-STATE PROTOCOL, and ROUTER.

Polishing Preparing a fiber end by moving the end over an abrasive material.

Politeness The most acceptable hypocrisy. Mostly seen before the sale.

Political File Records required by Section 76.207 which relate to origination cablecasts by, or on behalf of, candidates for public office. This rule requires each cable television system to keep a record, in its PIF, of all requests for cablecast time, together with detailed supporting information.

Politician John Maynard Keynes said that politicians are apt to be slaves to the ideas of long-deceased economists. John Kenneth Galbraith defined economists as people who didn't have the personality to become accountants.

Politics A clash of self-interests masquerading as a clash of principles. Also, the technique by which most telephone systems are bought in large corporations.

Poll In data communications, an individual control message from a central controller to an individual station on a multipoint network inviting that station to send if it has any traffic to send. See POLLING.

Poll Cycle The complete sequence in which stations are polled on a polled network.

Poll/Final Bit Bit in HDLC frame control field. If frame is a command, bit is a poll bit asking station to reply. If frame is a response, bit is a final bit identifying last frame in message.

Polling Connecting to another system to check for things like mail or news. A form of data or fax network arrangement whereby a central computer or fax machine asks each remote location in turn (and very quickly) whether they want to send some information. The purpose is to give each user or each remote data terminal an opportunity to transmit and receive information on a circuit or using facilities which are being shared. Polling is typically used on a multipoint or multidrop line. Polling is done to save money on telephone lines.

Polling Delay Communications control procedure where a master station systematically invites tributary stations on a multipoint circuit to transmit data. Polling delay is a measure of the time to transmit and receive on a polled network versus a direct point-to-point circuit.

POLSK POLarization Shift Keying.

Polybutylene Terephthalate PBT. An insulating material used extensively for buffer tubes which surround optical fibers.

Polyethylene A family of insulating (thermoplastic) materials derived from polymerization of ethylene gas. They are basically pure hydrocarbon resins with excellent dielectric properties. Used extensively in cables.

Polymer A material having molecules of high molecular weight formed by polymerization of lower molecular weight molecules.

Polymerization A chemical reaction in which low molecular weight molecules unite with each other to form molecules with higher molecular weights.

Polymorphism The ability of objects to handle different types of information and different requests for actions. Components are not typically polymorphic.

Polyolefin Any of the polymers and copolymers of the ethylene family of hydrocarbons.

Polypropylene A thermoplastic similar to polyethylene but stiffer and having a higher softening point (temperature) and excellent electric properties.

Polyvinylchloride PVC. A thermoplastic material composed of polymers of vinyl chloride. A tough, water and flame-retardant thermoplastic insulation material that is commonly used in the jackets of building cables when fire retardant, but not smoke retardant properties are required. Unfortunately, it burns and gives out noxious gases which kill. PVC can't be run in air return ducts, also called plenum ducts and most towns, therefore, don't allow PVC to be run in their plenum ceilings. See PLENUM.

Polyvinylidene Difluoride PVDF. A fluoropolymer material that is resistant to heat and used in the jackets of plenum cable.

PON Passive Optical Network. A fiber-based network which uses passive splitters to deliver signals to multiple users. TPON means telephony on passive optical network.

Pony Express Out of the summer haze bursts a horse and

rider, swiftly approaching a lonely sod building on the prairie. Arriving in a cloud of dust, the rider leaps from his horse and heads for a water barrel to quench his thirst. Meanwhile, a leather sack filled with mail is whisked off the tired horse and thrown over the saddle of a fresh mount. Within two minutes, the rider is gone, galloping toward the far horizon. This young man in a hurry was one of some 200 Pony Express riders who carried the mail in a giant relay between St. Joseph, Missouri, and Sacramento, California, a distance of 1,966 miles, in ten days or less. Changing horses every ten to fifteen miles at swing stations, and switching riders at home stations after a run of 75 miles or more, the riders averaged 250 miles a day. During the short time the Pony Express was in operation — from April 1860, through October, 1861 — its rider defied hostile Indians, blazing desert heat, and bone-chilling blizzards to travel a total of 650,000 miles with 34,753 pieces of mail. To save weight the letters they carried were written on tissue-thin paper as postage cost $10 an ounce, later cut to $2. The best time ever achieved was in March 1861, when Lincoln's inaugural address was carried from Missouri to California in seven days, 17 hours.

The Pony Express was organized by stagecoach operator William Hepburn Russell, who had been convinced by a group of prominent Californians that an overland mail route to their state was feasible. Russell's business partners opposed the venture because it was not protected by a U.S. mail contract. (They had competition and de-regulation even in those days.) But Russell went ahead, building stations and purchasing 500 top quality Indian horses. In advertising for riders, he hinted at the hazardous nature of the job by asking for "small, daring young men, preferably orphans." The riders received board and keep and were paid $100 to $150 a month. Their average age was 19, but one rider, David Jay, was 13, and William F. Cody, who became famous as "Buffalo Bill," was 15. In a further effort to save weight, a rider usually carried only a pistol and a knife. He was expected to outrun the Indians, not out-fight them.

The Pony Express days of glory ended abruptly in 1861 following completion of the transcontinental telegraph. Russell's firm lost more than $200,000 in the venture, but the daring of the Pony Express riders caught the imagination of every American, and their exploits became an important part of the legend and lore of the nation. The above history copyright 1979 by Panarizon Publishing Corp.

Pool A collection of things available to all for the asking or the dialing. A modem pool is a collection of modems typically attached to a PBX. Dial a special extension and you can use the modem, which answers that extension (or one of the extensions in the hunt group) to make a data call. Pooling is sharing. The purpose of having a "pool" is to avoid buying everybody one of whatever it is you're pooling. Actually, "pooling" is a fancy word for something we've been doing in the telephone business for the past 100 years — sharing. We started sharing lines, then sharing switches, then sharing voice mail devices, now we're sharing equipment, like modems.

POP 1. Point Of Presence. The IXC equivalent of a local phone company's central office. The POP is a long distance carrier's office in your local community (defined as your LATA). A POP is the place your long distance carrier, called an IntereXchange Carrier (IXC), terminates your long distance lines just before those lines are connected to your local phone company's lines or to your own direct hookup. Each IXC can have multiple POPs within one LATA. All long distance phone connections go through the POPs.

2. Point Of Presence at which ISPs (Internet Service Providers) exchange traffic and routes at Layer 2 (Link Layer) of the OSI model.

3. Short for "population." One "pop" equals one person. In the cellular industry, systems are valued financially based on the population of the market served.

4. Post Office Protocol. An e-mail server protocol used in the Internet. You use POP to get your mail and download it to your PC, using SMTP (Simple Mail Transfer Protocol). POP3 is the current version, as defined in RFC 1725. POP is increasingly being replaced by IMAP.

POP3 POP stands for Post Office Protocol. POP is a protocol used on the Internet to retrieve electronic mail from a mail server. You use POP to get your mail from the server it is sitting on and to download it to your PC. Most e-mail software (sometimes called e-mail clients) use the POP protocol. A few use the newer IMAP (Internet Message Access Protocol). There are two versions of POP. The first, called POP2, became a standard in the mid-1980s and requires SMTP (Simple Mail Transfer Protocol) to send messages. The newer version, POP3, can be used with or without SMTP. POP3 servers are a wonderful invention because, if they are attached to the Internet, they are independent of the transport mechanism used to access them. For example, my main email address is Harry_Newton@email.msn.com. MSN stands for the MicroSoft Network. Let's say I'm in Singapore and I want to get my emails. I jump onto the Internet, using any local Internet Service Provider (ISP). I then instruct my email client (in my case Microsoft Outlook) to go find my MSN POP3 email server, whose address happens to be pop3.email.msn.com. It finds that server. The server asks me to identify myself by telling who I am and what my password is. Then it starts sending me my emails. By being a POP3 server, MSN is particularly useful because it means I can pick up my emails from wherever I am in the world — even those cities which MSN doesn't serve (which is most of them). It also means that I can pick up my emails with typically only a local phone call. When you are choosing an email provider, make sure that email provider places your email on a POP3 server and that you can retrieve your email from anywhere. Some companies claim to have "POP3" servers but, for some reason (technical, security or incompetence), they don't allow you grab your mail over the Internet. When I wrote an article on the subject of POP3 servers, many readers wrote me that there are "POP3" and "POP3" servers. Thus my warning to "check, check, check."

Pop-Up A call center term. A button that displays, on demand, several items from which you can choose (by clicking on it, for example). In effect, this is a list box that you don't see until you push button. Pop-ups always have double lines on their right and bottom sides.

Pop-Up Electronic Mail An electronic mail system that runs as a terminate-and-stay-resident program (typically within DOS) and can be popped up inside any application to send or read mail. Our office, we have a TSR electronic mail program that pops up. It is called Noteworks and we really love it.

Pop-Up Program A memory-resident program that is loaded into memory but isn't visible until you press a certain key combination or until a certain event occurs, such as receiving a message. See also TSR.

POPS A cellular industry term for its customers or its potential customers. (It varies with usage.) POPS, short for "population" (well, sort of) refers to members of the population. According to Ron Schneiderman's book on Wireless, "if the coverage area of a cellular carrier include a popular base of one million people, it is said to have one million POPS. The financial community uses the number of potential users as measuring stick to value cellular carriers."

Populated Occupied by chips on printed circuit boards.

Port 1. noun. An entrance to, and/or an exit from a network.
2. noun. The physical or electrical interface through which one gains access. A point in the computer or telephone system where data may be accessed. Peripherals — like call accounting devices — are connected to ports. The two most common ports are the parallel and serial ports.
3. noun. The interface between a process or program and a communications or transmission facility.
4. verb. To move a process, program or subroutine from one processor to controller to another ("port it over").
5. noun. Network access point for data entry or exit. In Internet terms, it is the identifier (16-bit unsigned integer) used by Internet transport protocols to distinguish among multiple simultaneous connections to a single destination host.

Port Aliasing Imagine a switch on a local area network. It switches one port to another via a common backplane. We're having trouble and we're looking to do some diagnosis. So we grab all the information flowing through the switch and mirror it (i.e. forward it, but keep it going elsewhere in the switch) to a special port, which we can hook up equipment to and then monitor and check for problems in the resulting data flow — errors in packets, etc.

Port Connection The point of entry into a public frame relay network service.

Port Group A collection of switch interfaces through which packets can be switched. Port groupings can be distinct for different types of destination addresses: multicast, broadcast, and unicast sprays.

Port Identifier The identifier assigned by a logical node to represent the point of attachment of a link to that node.

Port Level VLAN VLAN based on source port ID. This is a multiple bridge configuration of a switch.

Port Multiplier A local area network interconnect, a concentrator providing connection to a network for multiple devices.

Port Per Pillow A goal set by colleges seeking to install network connections in the bedrooms of every student on campus.

Port Replicators Low-cost docking station substitutes that provide one-step connection to multiple desktop devices.

Port Selector Another name for a dataPBX. Since the advent of LANs (local area networks) these devices have been getting a bad rap. Not fair. These gadgets are really great at transmitting and switching huge number of low-speed asynchronous lines. If you put this sort of traffic on a LAN, you could severely mess up its performance. Some port selectors have data throughput in excess of 20 million bits per second.

Port Sharing 1. A system which connects multiple lines to a single port by means of a manual or automatic line selection method.
2. In frame relay, where multiple virtual connections share the same port connection.

Port Sharing Device A system which connects multiple lines to a single port by means of a manual or automatic line selection method.

Port Switching According to 3Com, port switching is merely an electronic patch panel function, not the genuine switching capability that provides a performance boost. Port switching lets administrators configure their networks, allo-

cating any port to any backplane on their hub. Unlike true switching, it doesn't increase the bandwidth available to the network manager.

Portability 1. The ability of a customer to take his telephone number from place to place and, for 800 numbers, from one long-distance company to another.
2. The ability of software designed for one computer system to be used on other systems. Little software outside MS-DOS software for IBM and IBM clone computers is portable. UNIX software is portable to an extent.

Portable A one-piece, self-contained cellular telephone — easily carried in a brief case or purse. Portables normally have a built-in antenna and rechargeable battery and operate with six-tenths of one watt (0.6 watt) of power. Car cellular phones operate with three watts.

Portable Cellular Phone Also known as a "hand-held phone". Refers to a lightweight, compact cellular handset that incorporates a battery power supply, and can be used without any peripheral power or antenna. See PORTABLE.

Portable Teletransaction Computers PTC. These are typically handheld devices used for retail (inventory), healthcare (tracking supplies), mobile field repair (reporting fixes), insurance (visiting car wrecks and other disasters), etc. The devices typically have telecommunications capabilities, sometimes wireless, sometimes landlines. And they typically include microprocessors, memories, displays, keyboards, touchscreens, character recognition software, barcode readers, printers, modems and local and/or wide area data radios.

Portal Site The classic definition of a portal is a door, gate, or entrance, especially one of imposing appearance, as to a palace. In the Internet business, a portal is a site on the Internet, which the owner positions (through marketing) as an entrance to other sites on the Internet. The concept is that he convinces visitors to the Internet to visit his site first, and savor the advertising on his site (his way of making money). A portal typically has, at minimum, search engines, free email, instant messaging and chat, personalized home pages and Web hosting. In the past, companies like America Online and CompuServe would have been called portals. Many browsers (e.g., Netscape Navigator and Internet Explorer) point you to a Web site — their own Web site, which they are endeavoring to position as a "portal." Once there, you might then want to use a search engine (e.g., AltaVista, Excite, Infoseek, Lycos, Yahoo). You are then taken to their Web site. These sites serve as your secondary portal, or point of entry, into the Web. At each of these Web sites you are assailed with advertisements at every opportunity. Also, cookies are embedded in your computer so they can track your movement and activities on the Web to get a feel for your buying preferences and, ultimately develop a profile on you for purposes of targeted on-line advertising. You can change your initial portal, if you like, perhaps pointing your browser to the Web site of favorite financial information provider (e.g. Bloomberg or Nasdaq), rather than that of Netscape or Internet Explorer. Portals are BIG business. In July 1998, Walt Disney Co. offered approximately $900 million for 43 percent interest in Infoseek, a 4-year old startup with fewer than 200 employees and annual sales of about $35 million. At the time, Infoseek had never made a profit. See also Portal Service.

Portal Service First, read my definition for Portal above. Then add these words. Portal Service is a whole architectural concept surrounding and including a point of entry. Portal Service is a whole architectural concept surrounding and including a point of entry. Portal Service introduces the concept of service on which other Internet services could be built. Such services provide an entry point on which other applications and services can be built, customized and enchanced to suit and envisaged deployment of any new services via a library of building blocks.

Portal Service Portrait Most computer screens are horizontal, i.e. they are wider than they are high. In the new language of computer screens, this is called "landscape." When a computer screen is higher than it is wide, it's called "portrait." Some computer screens can actually work both ways. Some even have a small mercury switch in them that determines which way the screen is standing (portrait or landscape) and will adjust their image accordingly. See also PORTRAIT MODE.

Portrait Mode 1. In facsimile, the mode of scanning lines across the shorter dimension of a rectangular original. ITU-T Group 1, 2 and 3 facsimile machines use portrait mode.
2. In computer graphics, the orientation of a page in which shorter dimension is horizontal. The opposite is called landscape mode. See also PORTRAIT.

POS 1. Point Of Service. Also called Point of Presence. See POINT OF PRESENCE.
2. POS Device. A point of sale device such as a credit card scanner used for authorization when a purchase is made.
3. Packet Over SONET. A high-speed means of transmitting data over a SONET fiber optic transmission system through a direct fiber connection to a data switch or router. POS is a point-to-point, dedicated leased-line approach intended purely for high-speed data applications. Where a point-to-point approach is possible, POS offers significant advantages when compared to ATM's cell-switching approach. Specifically, POS allows a user organization to pass data in its native format, without the addition of any significant level of overhead in the form of signaling and control information. For example, POS allows Ethernet frames to be packed into STS-1 (Synchronous Transport Signal-1) frames of 810 bytes, with only 36 bytes of overhead- an overhead factor of less than 1% — and then sent over an OC-1 (Optical Carrier-1) frame at 51.84 Mbps. For higher capacity applications, SONET supports example rates of approximately 155 Mbps, 622 Mbps, 2.4 Gbps and 10 Gbps-all with the same low level of overhead, or inefficiency. ATM, on the other hand, is a cell-switching approach that segments each data frame into small cells of 53 octets, of which 5 octets are overhead-an inefficiency, or overhead, factor of approximately 11% ATM, on the other hand, will support voice, video, fax and any other form of traffic, as well as true data traffic.

POSI Promoting Conference for OSI. Consists of executives from the six major Japanese computer manufacturers and Nippon Telephone and Telegraph. They set policies and commit resources to promote OSI.

Position A telephone console at a switchboard manned, oops, staffed by an attendant, or operator, or agent, or whatever the latest PC-correct word is.

Positive Action Digit A digit that must be dialed before a PBX will advance a call to a higher-cost route. The WATS lines are busy. Time on the queue is over. It's time to move the call to the more expensive direct distance dial. Before it can go that route, the caller must punch in a positive action digit. This affirms that the user knows he is now making a more expensive call. It causes him to think twice, allegedly.

Position Determination Technology PDT. also known as Geolocation Technology. A technology used to determine the geographic coordinates of a radio-equipped mobile device, e.g., a cellular handset. Generally refers to

technology that allows the device's position to be monitored remotely, such as by an emergency service operator in the case of a wireless E-911 call. "Position" is preferred in some circles to avoid confusion with the use of "location" in reference to cellular system roaming. Three general classes of PDT are as follows.
• Network-based. Fixed-site network infrastructure (e.g., collocated or consolidated with the cellular base station) determines device's position based on received signal measurements. Includes Angle of Arrival and Time Difference of Arrival technologies. See Angle of Arrival.
• Handset-based. The mobile device determines its own position (e.g., from an integral Global Positioning System receiver) and reports it to the network.
• Handset-assisted. A hybrid approach wherein the mobile device collects some measurements from its environment that are reported to the network infrastructure which in turn uses them to derive the mobile's position.

POSIX Portable Operating System Interface uniX. A proposed universal UNIX interface to user-created application programs that would run on all vendor equipment, thereby improving system interoperability.

Post 1. To compose a message for an Internet Usenet newsgroup and then send it out for others to see.
2. Power-On Self-Test

Post Dial Delay PDD. The time from when the last digit is dialed to the moment the phone rings at the receiving location.

Post Office Any part of an e-mail system that directors or delivers mail. But, says PC Magazine's David Stone, to refer to a particular level of mail handling, the term post office needs a modifier. A local post office or host post office is the module on a LAN that users directly interact with to send and receive mail. A domain post office is the module that controls the mail delivery within a domain of multiple hosts on a single network. In Windows 95, the postoffice is a temporary message store, holding the message until the recipient's workstation retrieves it. The postoffice exists as a directory structure on a server and has no programmatic components.

Post Pay A method of coin phone operation characterized by the operation of a lever or button that causes the collection of deposits after the called party answers. This method of "A" and "Buttons" is still used on coin phones overseas, especially in Great Britain.

Post production The editing process after the video footage has been shot.

Post Restante Address A standard attribute in a postal address indicating that physical delivery at the counter is requested. It may also carry a code.

Post Tensioned Concrete A type of reinforced concrete construction in which the steel is put under tension and the concrete under compression, after the concrete has hardened.

Postalize To structure rates or prices so that they are not distance sensitive, but depend on other factors (such as duration of a call, etc.) See also POSTALIZED.

Postalized Long distance phone calls were traditionally billed based on a costing algorithm which considered the distance between the originating and terminating points, call duration, and time of day. U.S. carriers, for the most part, currently no longer consider distance in the costing algorithm for domestic calls. In other words, long distance calls are subject to "postalized" charging — from the fact that Post Office also charges a flat rate irrespective of how far it carries the mail (within the country). Charges for circuit-switched data calls (e.g., Switched 56/64 Kbps), for the most part, also are postalized.

Postalized Rates Refers to the way the post office prices their delivery services, namely one price to anywhere in the United States. See Postalized.

Posting An individual article sent to a USENET news group on Internet; or the act of sending an article to a USENET news group.

Postlink A program used by a RIME bulletin board system node in place of mailer software.

Postmaster A postmaster could be the person responsible for taking care of mail problems, answering queries about users, and performing similar work for a given site. It could also be an alias for a mail server (i.e. computer) for routing and handling of electronic mail within an organization.

Postoffice See POST OFFICE.

Postpay See POST PAY.

Postscript 1. PostScript is a popular format used to create World Wide Web screens and send documents over the Internet.

2. PostScript by Adobe Systems Inc. is the standard page description language for desktop computer systems. It describes type, graphics and halftones as well as the placement of each on the page. The big advantage of PostScript is that it is device independent. Thus if you create a postscript image (text and/or photo and/or drawing), you can print it to a relatively cheap, low quality printer like a laser printer or a magazine quality printer like a Linotronics. PostScript is a printer language, much the same that BASIC is a computer language. By sending your PostScript printer a series of commands, you can make it do almost anything from printing text in a circle to printing foot-high letters to printing halftones. If you need PostScript, buy a printer that has built-in PostScript. If your printer doesn't have built-in PostScript, you may be able to get an external software interpreter but that interpreter will slow down your printer and tie up your computer while printing. See OUTLINE FONT.

POT 1. Techie-slang for potentiometer.

2. Point Of Termination. The point of demarcation within a customer designated premises at which the telephone company's responsibility for the provision of access service ends.

POT Bay The POT Bay, or Point-Of-Termination bay, is a device placed between a competitor's network and the natural point of connection to the local exchange carrier (LEC) network. It is located between a LEC's main distribution frame (MDF), which directs traffic to proper channels for distribution throughout the LEC network and the interconnector's colocated equipment. The POT Bay, according to most new local phone companies (i.e. ones that are not a Bell operating company or GTE), is an unnecessary obstacle that adds to the costs of interconnection, serves no necessary engineering function, can degrade quality, and is nothing more than a latter-day "protective coupling arrangement." As of writing, Nynex BellSouth, U.S. West, Southwestern Bell, and Pacific Bell insist that a LEC-provided POT Bay be installed as a point of demarcation between the LEC and the competitive interconnector. Bell Atlantic and GTE do not require competitive interconnectors to use POT Bays.

Potato Also called Aerial Service Wire Splice. A tool used to splice aerial service wire.

Potemkin Village A village, a showcase of progress thrown up by Catherine the Great's principal adviser, Prince Grigori Aleksandrovich Potemkin, to fool her into believing in the great progress of the New Russia. Catherine the Great, Catherine II, was Empress of Russia from 1762 to 1796.

Potential The difference in voltage between one point and another. One point is usually ground.

Potential Revenue A call center term. The revenue value per call times the number of calls forecast for a given period.

Potentiometer A variable RESISTOR, such as the ubiquitous volume control.

Potting In many European countries, the local regulatory authorities are very strict about voice boards that are attached to phone lines. Because phone lines have high voltage for ringing — 90 volts AC and higher — the authorities feel that the phone lines may caused an electrical short and possibly a fire. As a result, the authorities insist that the area of the voice boards which receives the high voltage be covered with some protective non-flammable material. Covering your board is called potting.

POTS Plain Old Telephone Service. Pronounced POTS, like in pots and pans. The basic service supplying standard single line telephones, telephone lines and access to the public switched network. Nothing fancy. No added features. Just receive and place calls. Nothing like Call Waiting or Call Forwarding. They are not POTS services. All POTS lines work on loop start signaling. See also LOOP START.

POTS Splitter A device that rejects the DSL signal and allows the POTS frequencies to pass through.

POTS-C Plain Old Telephone Service-Centralized. An ADSL term for the functional interface between the PSTN (Public Switched Telephone Network) and the POTS splitter at the Centralized (i.e., Central Office) end of the network. The POTS splitter is a filter that uses FDM (Frequency Division Multiplexing) to separate the low-frequency POTS voice channel from the high-frequency ADSL data channel. See also POTS-R, ADSL, FDM and SPLITTER.

POTS-R Plain Old Telephone Service-Remote. An ADSL term for the functional interface between the POTS splitter at the Remote (i.e., customer premise) end of the network and the individual telephone sets on premise. The POTS splitter is a filter that uses FDM (Frequency Division Multiplexing) to separate the low-frequency POTS voice channel from the high-frequency ADSL data channel. See also POTS-C, ADSL, FDM and SPLITTER.

POTV Plain Old TV. A Microsoft definition. I kid you not.

Pound The # on a pushbutton touchtone key pad is called the pound key. It's also called the number sign, the crosshatch sign, the tic-tack-toe sign, the enter key, the octothorpe (also spelled octathorp) and the hash. Musicians call the # sign a "sharp."

Pound Key See pound

Power The term which describes the amount of work an electric current can do in a unit of time. We measure power in WATTS (note we spell it with two "T"s.) A WATT measures the amount of work done in lifting a quarter-pound weight a distance of one yard in one second. Metric WATTS are a little more powerful. They go the distance of one meter. Power is the product of the current in amperes times the voltage, i.e. P = IV. See OHM's LAW.

Power Budget In fiber optic cable communications, power budget is the difference between the transmitted power and the receiver sensitivity, measured in decibels. It is the minimum transmitter power and receiver sensitivity needed for a signal to be sent and received intact.

Power Conditioner A combination voltage regulating transformer and isolation transformer, providing smooth, regulated, noise-free, AC voltage. See POWER CONDITIONING.

Power Conditioning Power conditioning is a generic concept to encompass all the methods of protecting sensitive hardware against power fluctuations. When electricity leaves a commercial power generating plant, it is very clean. In fact, most power companies make sure the power they put out is a pure sine wave. Unfortunately, nearly all devices connected to power lines — and the worst are things with motors, like elevators, air conditioners, etc. — create disturbances that pollute the sine wave. As power travels through a wire away from the power plant, it picks up more of these interferences. A pure AC power sine wave appears as a smooth wave. The height of the wave is measured in volts. The wave starts at zero volts and moves to the highest point of 120 volts. The wave then cycles through a low point of -120 volts and back to zero. The speed at which it travels through this cycle is the frequency. Normal frequency in North America is 60 cycles per second (Hz). (In other places it's often 50 cycles per second.) Anything that disrupts this wave can cause hardware or data problems and needs to be regulated.

Power disturbances can be categorized in several ways. A transient, sometimes called a spike or surge, is a very short, but extreme, burst of voltage. Noise or static is a smaller change in voltage. Brownouts and blackouts are the temporary drop in or loss of electrical power. Three types of protection against these three events are available: suppression, isolation, and regulation.

Suppression protects against transients. The most common suppression devices are surge protectors that include circuitry to prevent excess voltage. Although manufacturers originally designed surge protectors to prevent large voltage changes, most have also added circuitry to reduce noise on the line. Isolation protects against noise. Ferro-resonant isolation transformers use a transformer within the circuitry to envelop the sine wave at a slightly higher and lower voltage. Any voltage irregularity that extends beyond this envelope is clamped. Isolation transformers are usually expensive.

Regulation protects against brownouts and blackouts. Regulation modifies the power wave to conform to a nearly pure wave form. The Uninterruptible Power Supply (UPS) is the most commonly used form of regulation. A UPS comes in two varieties, on-line and off-line. An on-line UPS actively modifies the power as it moves through the unit. This is closer to true regulation than the off-line variety. If a power outage occurs, the unit is already active and continues to provide power. The on-line UPS is usually more expensive but provides a nearly constant source of energy during power outages. The off-line UPS monitors the AC line. When power drops, the UPS is activated. The drawback to this method is the slight lag before the off-line UPS jumps into action. That lag is getting shorter as electronics improves. So it's rarely a problem any longer.

Because UPS systems are expensive, most companies attach them only to the most critical devices, such as phone systems, network file servers, routers, and hard disk subsystems. Attaching a UPS to a local area network file server enables the server to properly close files and rewrite the system directory to disk. Sadly, most programs run on the workstation and data stored in their RAM is not saved during a power outage unless each workstation has its own UPS. If the UPS doesn't have its own form of surge protection, it is a good idea to install a surge protector to protect the UPS from transients. Proper use of power conditioning devices greatly reduces telephone system and network maintenance costs. Make sure that proper amperage is available for each system and that all outlets are grounded. Power conditioning devices connected to poorly-grounded outlets offer very little protection.

Studies have shown that total local area network maintenance costs are higher with line-surge suppressors and ferro-resonant isolation transformers alone, than with uninterruptible power supplies.

Power Conditioning Systems A broad class of equipment that includes filters, isolation transformers, and voltage regulators. Generally, these types of equipment offer no protection against power outages.

Power Cross A situation in which AC current flows into a telephone circuit, as a result of contact with a power line.

Power Dialer A piece of hardware to which you feed a list of phone numbers you want called. It calls them one after another. It detects busies, no answers, fax machines, answering machines, voice mail machines, etc. When it hears these, it disconnects and dials the next number in the list. If it reaches a live human being, it will pass the call to the owner of the machine, who will presumably try and sell the fellow at the other end something, or try and collect money from him — the two main uses which power dialers are typically put to. The term "power dialer" is fairly new. I first saw it used on a single line device. It seemed to need a prompt from its owner to dial the next call, very much like a preview dialer. See also PREDICTIVE DIALING and PREVIEW DIALER.

Power Distribution Panel A part of the Rolm CBX power distribution system that receives voltages from the main power supply and distributes them to the cabinet shelves.

Power Down The sequence of things you to have to do to turn off a computer or telephone system. Not following the correct power down procedures can cause a loss of data.

Power Factor This is a number between 0 and 1 which represents the portion of the VA (Volt-Amps rating) delivered to the AC load which actually delivers energy to the AC load. With some equipment such as motors or computers, AMPS flow into the equipment without being usefully converted to energy. This happens if the current is distorted (has harmonics) or if the current is not in phase with the voltage applied to the equipment. Computers draw harmonic currents which cause their power factor to be less than 1. Motors draw out of phase or reactive currents that cause their power factor to be less than 1. See also VA. This definition courtesy American Power Company.

Power Factor Corrected Supply PFC. A recently developed type of computer power supply, which exhibits an input power factor equal to one. IEC555 may force most computers to use a power supply of this type at some point in the future.

Power Fail Bypass A feature that allows analog trunks to be answered if your commercial AC power or your telephone system crashes.

Power Failure Backup If your AC power fails, your telephone system can still operate by switching to a backup battery power supply, often called a UPS — Uninterruptible Power Supply.

Power Failure Transfer When the commercial AC power fails and there is no backup power source — such as a battery or a generator — this feature switches some of the trunks connected to the phone system to several single line phones, which don't need external power and can draw their power from the phone lines.

Power Level The measure of signal power at some point. The measure can be referenced to some power level in which case the measurement is expressed in dB (decibels). It may also be referenced to 1 milliwatt in which case the measurement is expressed in dBm.

Power Line Carrier PLC. An AC power line can be made

to carry high frequency radio waves, which can carry "information," which could be a voice or data call. Over the years many companies have tried power line carrier — a seductively easy method of not having to install conventional phone lines. The results have been poor-to-awful. For example, rural telephone companies once made use of PLC in order to serve very remote farms and ranches over power lines run the local power utilities. The power utilities, themselves, still make extensive use of PLC as a backup for more conventional copper and microwave transmission systems. A number of manufacturers over the years have developed residential and small business telephone systems which use in-building AC power. The telephone line is terminated by special equipment at the common electrical bus, with the individual sets simply plugging into electrical outlets like a toaster or any other electrical appliance. (Do not confuse this odd approach with the term "Information Appliance," which refers to multipurpose, multimedia terminals used in a convergence (Information Superhighway) context.) Such systems have not been successful.

Recently, several technical developments have resurrected interest in PLC. First, Spread Spectrum technology has been applied to overcome the inherent noise problems associated with data communications over in-building electrical wiring. Second, the EIA's (Electronic Industries Association) CEBus (Consumer Electronics Bus) was adopted industrywide. CEBus essentially is a Home Area Network (HAN), a residential and small business version of a Local Area Network; CEBus makes use of Spread Spectrum. Current commercial PLC applications using Spread Spectrum include control of building environments and managing utility electrical distribution systems. For instance, heating, ventilation, and air-conditioning systems in a commercial building can be managed by a central controller which polls various temperature and humidity sensors scattered throughout the building with communications taking place over the electrical wiring and through the common electrical bus. See Spread Spectrum and CEBus.

Power On See POWER UP.

Power Open A new operating system which is planned to run on a new super-powerful PC manufactured by a joint IBM-Apple alliance. The idea of the IBM-Apple alliance is make a super-powerful PC that runs virtually every PC operating system imaginable, including MS-DOS, UNIX, Windows, OS/2, Macintosh. The new, all powerful operating system, would be called "Power Open."

Power Product A cellular radio term. A configurable parameter broadcast by the Mobile Data Base Station (MDBS), defining the desired relationship between received signal strength and transmitted power level at any single point.

Power Regulator Equipment that regulates the power delivered to a system. Designed to mitigate transients in the commercial electric power source.

Power Supply Most single line phones are powered by the electricity that comes in over the phone line. That's why they'll work when there's a power outage. Single line phones that have gadgetry associated with them, all ISDN phones and all multi-line phones (like key systems and PBXs) require a power supply. A power supply which is a device which converts the normal 120 or 240 volts AC power to AC and DC at the various voltages and frequencies needed by the components and circuits of the phone or computer system. Power supplies are usually the least reliable part of modern electronic gadgetry. This is because they take the hits form the lousy power the local utility sends in and also because many manufacturers skimp on the quality of their power supplies. A

cheap power supply is not evident immediately. It may take time to break down. Whenever you're having intermittent problems with your phone system or computer, suspect the power supply. And, given a choice, buy the best quality power supply you can.

Power Synthesizer Power synthesizers actually use the incoming utility power as an energy source to create a new sine wave that's free from power disturbances. They can be as much as 99% effective against power disturbances. Types of power synthesizers include magnetic synthesizers (capable of generating a sine wave of the same frequency as the incoming power - 60 Hz), motor generators (which use an electric motor to drive a generator that provides electrical power), and UPSes.

Power Systems A system that provides a conversion of a primary alternating current power to direct current voltages required by telecom equipment, and may generate emergency power when the primary alternating current source is interrupted.

Power Up The sequence of things you have to do in order to turn a computer or telephone system on. You can't cut corners starting up electronic equipment. It must be done carefully and in the correct order. Always count to ten after turning something off before turning it back on again. See also POWER DOWN.

Power Vendor One who has a major chunk of a market. Some users believe that a good IS strategy is to buy from a power vendor in the belief that "you can't go wrong buying from AT&T, IBM, Northern Telecom..."fill in the name of your favorite power vendor.

Power, Peak In a pulsed laser, the maximum power emitted.

PPDN Public Packet Data Network.

PPI Pixels Per Inch. See RESOLUTION.

PPL Pioneer Preference Licensees. A PCS wireless term. Three US companies were awarded licenses before the A and B band auctions began. They all adopted PCS-1900, at least in part.

PPM 1. Pulse Position Modulation. Also known as Pulse Phase Modulation. A form of pulse time modulation in which the position in time of a light pulse is varied; signal amplitude and frequency both remain constant. The typical implementations are known as 4PPM and 16PPM, as a pulse of light is emitted in one of 4 or 16 time slots, respectively. For example, 4PPM is a dibit coding scheme which allows two bits of information to be impressed on a single light pulse, with the exact bit pattern (i.e., 00, 01, 10 or 11) indicated by the specific position in time in which the pulse appears in a synchronized light stream. PPM can be used in both analog and digital transmission systems; it is particularly useful when power requirements must be kept low and when the transmission medium may be pulsed easily. PPM commonly is used in infrared (IR) transmission systems, in both wireless LAN (WLAN) and short-haul networking applications; it also is used in deep-space laser communications systems.
2. Northern Telecom term for Periodic Pulse Metering.

PPP Point-to-Point Protocol; a protocol that allows a computer to connect to the Internet with a standard dial-up telephone line and a high-speed modem and enjoy most of the benefits of a direct connection, including the ability to use graphical front ends such as Mosaic and Netscape. PPP is considered to be better than SLIP, because it features error detection, data compression, and other elements of modern communications protocols which SLIP, the older Internet protocol, lacks. PPP is thus replacing SLIP. See also

PPP/ML/PPP and SLIP.

PPP/ML/PPP Point to point protocol, the commonest way to carry IP frames over a circuit (either a leased line or switched channel). Available in both sync and async versions, also has defined formats to carry IPX and AppleTalk, with frame relay in the works. Multi-Link/PPP is a 'bandwidth on demand' technology that allows one logical PPP connection to add additional channels (as in a second ISDN channel) when the bandwidth is needed (however the vendor defines that situation). Also may be used with leased lines when the total bandwidth needed exceeds the available line speed — a form of inverse muxing. See also PPP.

PPS 1. Pulses Per Second.
2. Precise Positioning Service. The most accurate dynamic positioning possible with GPS (Global Positioning System), based on the dual frequency P-code. 3. Packets Per Second. The theoretical limit of 10 Mbps Ethernet, measured in 64 byte packets, is 14,800 packets per second (PPS). By comparison, Token Ring is 30,000 and FDDI is 170,000 pps. 4. The Path Protection Switched ring defined by Bellcore TA-496. PPS is really a fancy name for a duplicated SONET signal traveling over diverse (i.e. different) routes. When one route of network crashes, the other will take over. This enables it to survive service outages caused by cable cuts, earthquakes, lightning strikes and equipment failures. The PPS ring gives a SONET route a greater degree of survivability than other Sonet transmission paths that don't have route diversity.

PPSN Public Packet Switched Network

PPSS Public Packet Switched Service. A connection-oriented, packet-switched data communication service that permits users to communicate with data terminals of other customers and on other packet networks.

PPTP Point-to-Point Tunneling Protocol, a new protocol that enables virtual private networking — enabling secure remote access to corporate networks over the Internet. The protocol was first demonstrated at InterOp, Spring 1996. According to the press release from U.S. Robotics and Microsoft which demonstrated the protocol, PPTP will help companies deploy remote access to employees more quickly, using fewer resources, by allowing them to take advantage of existing enterprise network infrastructures such as the Internet for remote access. U.S. Robotics developed the Windows NT PPTP driver, which will be integrated into Microsoft's Windows NT Server 4.0. Under an agreement with Microsoft, U.S. Robotics will license a variety of software components for PPTP to Microsoft. PPTP will be added as a standard feature to Microsoft's Windows NT Server and U.S. Robotics' Total Control NETServer remote access server platform. PPTP streamlines access in NT networking environments, and allows NT network clients to take full advantage of the services provided by Microsoft's RAS (Remote Access Service). For remote access, over analog or ISDN lines, PPTP creates a "tunnel" directly to the appropriate departmental NT Server on a network — even if there are hundreds of NT Servers. The PPTP specification builds on standards such as PPP and TCP/IP. PPTP 'tunnels' a remote user's PPP packets from the NETServer to a Windows NT server. By terminating the remote user's PPP connection at the NT server, rather than at the remote access hardware, PPTP allows network administrators to standardize security using the existing services and capabilities built into the Windows NT security domain. Using PPTP, network administrators can extend a virtual private network from their Windows NT server throughout the Internet

and still retain control of their user passwords and accounts. NT provides its own knowledge of enterprise users, databases, allowed access and network addressing integrated into its RAS capabilities. With PPTP, users accessing their NT-based network will use these services, including DHCP and WINS, for access.

PQFP Plastic Quad Flat Pack, a format used in the design of PCMCIA devices. Another format is called TQFP, which stands for Thin Quad Flat Pack.

PRA An ISDN ITU-T term used internationally to mean 23B+D and also 30B+D. The international version of what Americans call PRI. See also ISDN and Primary Rate Interface for longer explanations.

Practice The technical and installation manuals often used by Bell Operating Companies. A poor use of the word. A better word would be procedure.

Prairie Dogging Something loud happens in an office made up of cubicles (also called a cube farm), and people's heads pop up over the walls to see what's going on.

PRBS Pseudo Random Bit Sequence/pattern. A test pattern having the properties of random data (generally 511 or 2047 bits), but generated in such a manner that another circuit, operating independently, can synchronize on the pattern and detect individual transmission bit errors.

Pre-Arbitrated PA. A type of SMDS slot defined in support of isochronous traffic such as voice and video. As such traffic is stream oriented and time-sensitive, isochronous data must have regular and reliable access to time slots in order that it might be presented to, travel across, and exit the network without suffering from delay. Otherwise, the presentation of the data through the target device would be most unpleasant (e.g., "herky-jerky video).

Pre-Fetching Imagine a call center with extensive customer records. Perhaps the company is a utility, a phone company or an insurance company. Now imagine that the phone rings. CallerID picks up the customer's name and pops a screen of information on the customer. Meantime, the database system has sent a signal to what's known as "near line" file storage. That near line file storage might be a magneto optical jukebox. It might be a tape file library. The "normal" time to retrieve a record from such systems might range from 30 seconds to two minutes. But if the signal is sent early, the records could be retrieved and loaded into a cache memory, ready if the customer asks a question, for example, about his old bills. This called pre-fetching. See also PRE-LOADING.

Pre-Loading Or pre-fetching. Documents are manually (by the administrator) retrieved from a jukebox or optical drive on a local area network in anticipation of some special need. Pre-loading may take advantage of the regular cycles of business, such as payroll processing on the 10th and 20th, personnel reviews the first week of each quarter, year-end processing of accounting reports. The network's administrator can take advantage of regular cycles to improve the system's response to users. See also PRE-FETCHING.

Pre-paid Calling Card A credit card size card which you buy at your local store. Each card comes with information that allows you to use a certain dollar amount of local or long distance phone service. Typically the card will give you an 800 (toll-free number) to call. You call it. You touchtone in the authorization number you find on the back of your calling card and the switch at the other end dials your desired number. The switch measures how long you talk and lets you talk as long as the dollar entitlement on the card — $5, $10, $20, $50, etc. Pre-paid calling cards are typically expensive for long calls, but cheap for short calls, based on the alternative — using a credit card with one of the traditional long distance carriers, many of whom apply hefty up-front, per call surcharges.

Pre-Wiring 1. Wiring installed before walls are enclosed or finished.

2. Wiring installed in anticipation of future use or need.

Preamplifier An electronic circuit which maintains or establishes an audio or video signal at a predetermined signal strength, prior to that signal being amplified for reproduction through a monitor or speaker.

Precedence Precedence is Federal government parlance to mean a designation assigned to a phone call by the caller to indicate to communications personnel the relative urgency (therefore the order of handling) of the call and to the called person the order in which the message is to be noted. Autovon phones which have "precedence" have an additional four touchtone buttons. You can find frequencies for those buttons under the definition for DTMF.

Precedence Prosign An introductory character or set of characters which indicate how a message is to be handled by the receiving unit.

Precise Positioning Service The most accurate dynamic positioning possible with GPS (Global Positioning System), based on the dual frequency P-code.

Precision Air Conditioning Precision air conditioning systems are primarily designed for cooling electronic equipment, rather than people. These pre-packaged systems offer excellent reliability and typically have a high ratio of sensible-to-total cooling capacity and a high CFM/ton ratio.

Predictive Coding Method or source coding using prediction. The prediction error resulting from the difference between the prediction value and actual sample value is transmitted.

Predictive Dialer I like predictive dialers because they deliver the benefits they promise. And they do it neatly and painlessly. I've never met a predictive dialer user who didn't realize the productivity improvements he was promised and figured on to justify the system. I have only one problem with predictive dialers. Most people use them for collecting money. Dial up the deadbeats. Dun them for money. B-O-R-I-N-G! I'm a mail order junkie. Everyone is. Mail order catalogs have become our escapist reading. I'm dressed in clothes from L.L. Bean, Eddie Bauer, Nordstrom, etc. I'm typing on a computer from PC Connection. My computer is tied into a server. We bought the wire and the server and the network cards by mail order. But I'm an unloved mail order junkie. No one ever calls me. Bean and Bauer know I like "large tall." They know I'm 11N in shoes. They know what I like and what I don't like. I've been buying from them for years. How come they never call me and tell me of their fantastic tall and narrow new products? I'm Platinum on American Airlines. Their most frequent flyer. So why don't they call me and offer me a special "weekend" surprise? Call me on Wednesday. Fill their planes on Friday and Sunday. Don't call me if their planes are full. To my naive brain, predictive dialers are the best sales tool ever invented. They let you sell when you've got something to sell. And not sell when you got nothing to sell. They let you sell "talls" to tall people and narrow shoes to narrowed-footed people. This is not rocket science. One day, some companies will figure there are millions of people (like me) waiting at the end of their phone line waiting for someone to call them and sell them something. That someone, if they're smart, will be using a predic-

tive dialer. See PREDICTIVE DIALING.

Predictive Dialing An automated method of making many outbound calls without people and then passing answered calls to a person as the calls are answered. Here's the story: Imagine a bunch of operators having to call a bunch of people. Those calls may be for collections. They may be for employee callups. They may be for alumnae fund raising. When it's done manual, here's how it works: Before each call operators spend time reviewing paper records or computer terminal screens, selecting the person to be called, finding the phone number, dialing the numbers, listening to rings, listening to phone company intercepts, busy signals and answering machines. Operators also spend time updating the records after each call. Predictive dialing automates this process, with the computer choosing the person to be called and dialing the number and only passing it to an operator when a real live human being answers. There are enormous productivity gains made by screening out answering machines, busy signals, network busy signals, non-completed calls, operator intercepts etc. The result is productivity increases of 200% to 300%. According to generally accepted industry lore, a well-run manual dialing center can get its people talking on the phone for 25 minutes an hour. With a predictive dialer you can get them on the phone making sales, collecting money, etc. for 55 minutes an hour. It's a major productivity gain.

True predictive dialing should not be confused with automated dialing. True predictive dialing has complex mathematical algorithms that consider, in real time, the number of available telephone lines, the number of available operators, the probability of getting no answer, a busy signal, a disconnected number, operator intercept or an answering machine, the time between calls required for maximum operator efficiency, the length of an average conversation and the average length of time the operators need to enter the relevant data. Some predictive dialing systems constantly adjust the dialing rate by monitoring changes in all these factors.

Some people don't like the term "predictive dialing," since they think it's getting "a bad rap" in Washington, DC by being associated with junk phone calls. As a result some people would prefer to call it Computer Aided Dialing. See also PREVIEW DIALING.

Preemptive Multi-Tasking Operating System A multi-tasking operating system allows more than one task to be active at the same time. Under Windows 3.x, a task is defined as a single program. For instance, if you have ever had a word processor and a spreadsheet program open at the same time, then you have used Win 3.x multi-tasking. Win 3.x is a cooperative multi-tasking environment. In other words, applications must cooperate for multi-tasking to work. The system cannot preempt the program that has control of the CPU. It is the responsibility of each program to share the CPU. Windows 95 and Windows NT is a preemptive multi-tasking environment. This means the CPU is in charge and can seize control from applications when necessary. This environment reduces the risk of your system freezing up.

Preemptive Operating System An operating system scheduling technique that allows the operating system to take control of the processor at any instant, regardless of the state of the currently running application. Preemption guarantees better response to the user and higher data throughput. Most operating systems are not preemptive multitasking, meaning that task-switching occurs asynchronously and only when an executing task relinquishes control of the processor. See PREEMPTIVE,

PREEMPTIVE MULTITASKING, REAL TIME SUPPORT.

Preemptive, Real Time Support When Microsoft released its At Work operating system, it said it had a number of key features, one of which was "pre-emptive, real-time support." Here's Microsoft's description:

• Pre-emptive, real-time support. Communication devices such as fax machines and phones are distinct from personal computers in that they have critical real-time needs. Consequently, the software in these devices must attend to communication hardware such as modems very frequently, so that pieces of the communication are not lost. To support this need, the operating system was designed to be able to put other processes "on hold" temporarily in order to service the communication hardware before continuing other functions. See AT WORK and WINDOWS TELEPHONY.

Preference Setting A set of parameters on software tools, including Web Browsers that allow the user to choose which stuff shows on his screen, which colors he will use to display text, whether a signature file should be attached to his email, etc.

Preferred Call A local phone company services which lets you forward calls from a bunch of numbers you have preselected. The service uses the calling number ID as the basis for choosing which calls to forward.

Prefix One or several digits dialed or touchtoned in front of a phone number, usually to indicate something to the phone system. For example, dialing a zero in front of a long distance number in the United States would indicate to the phone company you wanted operator assistance on the call.

Preform Optical fiber source material. Preform is glass rod formed and used as source material for drawing an optical fiber. The glass structure is a magnified version of the fiber to be drawn from it.

Premise A thesis. a proposition supposed or proved as a basis of argument or inference. Often misused to mean the space occupied by a customer or authorized or joint user in a building or buildings on continuous or contiguous property (except railroad rights of way, etc.) not separated by a public road or highway. See PREMISES.

Premises The space occupied by a customer or authorized or joint user in a building or buildings on continuous or contiguous property (except railroad rights of way, etc.) not separated by a public road or highway.

Premises Distribution System PDS. There are two meanings for premises distribution system — a general one and a specific one, specific to Lucent. Here, first, is the general definition. A PDS is the transmission network inside a building or group of buildings that connects various types of voice and data communications devices, switching equipment, and other information management systems to each other, as well as to outside communications networks. It includes cabling and distribution hardware components and facilities between the point where building wiring connects to the outside network lines and back to the voice and data terminals in your office or other user work location. The system consists of all the transmission media and electronics, administration points, connectors, adapters, plugs, and support hardware between the building's side of the network interface and the terminal equipment required to make the system operational.

Here is the specific definition — A multi-functional distribution system from Lucent to support voice, data, graphics and video communications on premise. PDS includes cables, adapters, electronics, eight pin universal wall jacks and pro-

tective devices, all arranged in a logically coherent and economic fashion. It uses fiber optic cable and twisted pair copper wire and is suitable for single building, multi-tenant high rise or campus environment.

Premises Lightwave System The fiber optic part of the Premises Distribution System (PDS) from AT&T. PDS, which can replace the coaxial cables linking IBM terminals and printers, consists of two fiber optic interface units, one at a controller-end and the optic interface units and one at a terminal-end linked by a fiber optic pair. The fiber optic interface units connect to the terminals through four-pair building wiring and balun adaptors. Balun adaptors also enable direct connections of the terminals to the cluster controller through building wiring.

Premises Wire The twisted-pair, quad or other wire installed at the user's location to provide telephone service. Includes both intra-building and inter-building wiring.

Premises Wiring System The entire wiring system on the user's premises, especially the supporting wiring that connects the communications outlets to the network interface jack.

PReP PowerPC Reference Platform specification. It details the hardware, operating system and software elements necessary to build PowerPC-based systems that meet certain compatibility goals. According to P Magazine, these machines will run 32-bit operating systems and will resemble today's high-end desktop system with lots of memory, CD-ROM drives, stereo audio support, and PCI/PCMCIA expansion buses.

Prepaid Phone Cards With a Prepaid Phone Card, (sometimes called a "Debit phone card") a customer purchases in advance a specific amount of telephone calling time. For example, a typical phone card may offer 30 minutes for about $10.

Prepaid Telephone Card A prepaid telephone card is a piece of credit-card size plastic entitling the owner to make phone calls. Prepaid telephone cards are sold in many denominations — typically $1, $5, $10, $20 and $50. Once the owner of the card makes phone calls, the value of the card decreases. The accounting is typically done in a remote switch which the user dials to make his calls. He often dials via an 800 number. Companies sell prepaid telephone calls as a business, often owning the switch and renting the long distance lines. Sometimes companies buy cards from a supplier, print up their names on the cards and give the cards away as promotion. It allows them to reward clients and customers with what one seller of cards calls "a universally valuable commodity — long distance service." Some people have started collecting prepaid telephone cards the way people collect baseball cards. Surprise, surprise, the value of some telephone cards is increasing. I know someone who has a collection of 30,000 telephone prepaid cards. He tells me he is going to retire on this collection.

A prepaid telephone card doesn't typically have technology on it or embedded into which represents the value of the money remaining on the debit card. Such technology might be an integrated circuit, a magnetic strip, bar codes which can be read by an optical reader. That's a debit card. See BREAKAGE, CALLING CARD and PREPAID TELEPHONE CARD.

Prepay The industry standard for coin phone operation which requires that the full cost of a call be deposited before a connection is attempted through the Central Office.

Prepend The opposite of append. Prepend means added to the front of, whereas append means added to the back of. Prepend is different from precede, which just means comes before, rather than being part of the data frame. See Frame.

Prepended Prepended means preceded as part of the same data unit (generic "packet"), whereas appended means succeeded as part of the same data unit. See SMDS.

Preprocessor A device or information handling system which converts raw data into a form more easily processed with standard equipment.

Prereq Prerequisite

Presentation Address According to the book, "Internetwork Mobility," by Mark Taylor, William Waung and Mohsen Banan, the presentation address is the network name of an Application Entity within a Data Service. The Presentation Address takes the form of: optional Presentation Selector (SSAP-Selector + Session Selector (TSAP-Selector) + Transport Selector (NSAP-Selector) + a required NET. Presentation Addresses are stored in the Directory Service. Application Entity Titles are used to retrieve the Presentation Address from the Directory Service.

Presentation Indicator PI. An ISDN term. See PI.

Presentation Layer The sixth layer of the OSI model of data communications. It controls the formats of screens and files. Control codes, special graphics and character sets work in this layer. See OSI STANDARDS.

Presentation Manager Presentation Manager is a look and feel specification and kernel-based toolkit development environment. It was developed for IBM by Microsoft with input from IBM. Presentation Manager is the standard graphical user interface and toolkit for the OS/2 operating system, which is a multi tasking operating system for personal computers. The screens are similar to those of Microsoft Windows.

Presentation Status For a particular call, an item that indicates if a calling identity item may be presented to the called party. If the presentation status is "public" presentation is allowed. If it is anonymous", presentation is restricted. Presentation status has to do with the presentation or not or calling line identification numbers.

Preset The "programming" of radio station frequencies on a tuner or receiver or musical selections on a tape, for instant recall at the push of a button.

Preset Call Forwarding Incoming calls will be re-routed to a pre-determined secondary number.

Press-To-Talk Telephone circuits are two way. Some circuits, such as mobile dispatch services for taxis, etc., are one-way. They use a microphone or handset with a button you must press-to-talk and release to listen. You can also buy a normal telephone handset with a press-to-talk button. Such a handset is useful in noisy places.

Pressure Cable Telephone cable equipped with air-pressure equipment so the phone company can determine when there's a problem with the line. When a cable is cut, the pressure drops and the company is notified of the problem. Nitrogen is often used instead of air because it's noncorrosive. Nitrogen also prevents water entering the cable when there's a break.

Pressurization Pumping inert gas into a heavy casing in which a couple of thick cables will be joined. The pressure is usually maintained at a few pounds above the surrounding atmospheric pressure. The idea is that the higher pressure inside keeps moisture out of the splice and thus improves the quality of phone service.

Prestel A videotex system used only in Britain.

Prestel Terminal Emulation Prestel is a character-based graphics emulation for communicating with the Viewdata service, particularly popular in the United Kingdom.

Presubscription A local Bell or local independent operat-

ing telephone company service that encourages each subscriber to select one long distance carrier he may use without having to dial a multiple digit access code. If you pre-subscribe to MCI in America, you will simply reach MCI by dialing "1" plus the 10-digit long distance number. AT&T's code is 10-10-288 (as in 1-0-ATT). Sprint's is 10-10-333.

Pretrip A central office malfunction that causes a phone to ring only once, and then stop, as if it had been answered. IT can be very confusing and difficult to get repaired, because most telco repair people have never heard of the problem and will insist your phone is at fault. Often fixed by replacing a faulty heat coil.

Pretty Good Privacy A powerful encryption scheme. See PGP for a full description.

Prevail An office automation UNIX software package which combines functions usually available only in individual programs, such as spreadsheet, a word processor, a database management system, communication capabilities and more.

Preventive Maintenance The periodic inspection, cleaning, adjusting and repair to eliminate problems before they affect service. Usually ignored.

Preview Dialing Preview dialing is a term used to describe an automatic dialer. Preview dialing is also called "screen dialing" or "cursor dialing." Typically the prospect's account information and/or phone number appears on the screen BEFORE the call is made. Thus the agent can "preview" the number, the screen, the customer. If the agent wants to make the call, the agent hits a key, such as "Enter" and the computer dials the number. In some preview dialing equipments, the agent must hit a key if he/she DOESN'T want the number dialed. Contrast preview dialing with Predictive Dialing where the computer makes all the dialing decisions and presents the calls to the agent only after they are connected. Predictive dialing is a lot faster than preview dialing. See PREDICTIVE DIALING.

Prewiring 1. Wiring installed before walls are enclosed or finished.

2. Wiring installed in anticipation of future use or need.

PRI See PRIMARY RATE INTERFACE and ISDN.

PRI-EOP A fax signal. PRocedure Interrupt-End of Page.

PRI-MPS A fax signal. PRocedure Interrupt-MultiPage Signal.

Price Cap The phone industry has always been regulated on the basis of the profits it earns compared to the investment it had. That was called Rate of Return Regulation. How you figure profits — since you also have to figure what are allowable expenses — has been the subject of on-going debate for over 100 years. The latest idea in regulation is to replace the rate of return regulation with something called "Price caps" which allow the price of phone company services to rise by x% at maximum — the so-called price cap.

Primary Agent Group An automatic call distributor term: Primary agent group for which the inbound calls are intended. Intraflow goes to secondary and tertiary groups if the primary group does not have an agent available after the time or after overflow parameters are exceeded. Different ACD systems label this process differently.

Primary Alias A term used in the Sun Solaris Teleservices platform. See PROVIDER.

Primary Buffer A part of a computer's memory where fast incoming or outgoing data is kept until the computer has a chance to process it.

Primary Center A control center connecting toll centers — a Class 3 Central Office. It can also serve as a toll center for its local end offices.

Primary Domain Controller PDC. For a Windows NT Server domain, the computer that authenticates domain logons and maintains the security policy and the master database for a domain. See DOMAIN.

Primary Group A group of basic signals which are combined by multiplexing. The lowest level of the multiplexing hierarchy.

Primary High-Usage Trunk Group A high-usage trunk group that is offered first-route traffic only.

Primary Insulation The first layer of non-conductive material applied over a conductor to act as electrical insulation.

Primary Interexchange Carrier A primary Interexchange Carrier is the long distance company to which traffic from a given location is automatically routed when dialing 1+ in equal access areas. The PIC is identified by a code number which is assigned by the local telephone company to the telephone numbers of all the subscribers to that carrier to ensure the calls are routed over the correct network. When a subscriber switches long distance carriers, it often is referred to as a PIC change.

Primary Link The active LAN connection. When it fails the LAN is switched to the Backup link.

Primary Partition A portion of a physical disk that can be marked for use by an operating system. Under MS-DOS, there can be up to four primary partitions (or up to three, if there is an extended partition) per physical disk. A primary partition cannot be subpartitioned.

Primary Rate Interface The ISDN equivalent of a T-1 circuit. The Primary Rate Interface (that which is delivered to the customer's premises) provides 23B+D (in North America) or 30B+D (in Europe) running at 1.544 megabits per second and 2.048 megabits per second, respectively. There is another ISDN interface. It's called the Basic Rate Interface. It delivers 2B+D over either one or two pairs. In ISDN, the "B" stands for Bearer, which is 64,000 bits per second, which can carry PCM-digitized voice or data. See ISDN for a much better explanation.

Primary Resource An SCSA definition. The main resource around which a Group is constructed. Typically, the primary resource will be an interface to the telephone network, but it may also be a switch port.

Primary Server The SFT III Novell NetWare server that has been operating longer than its partner and is currently servicing the attached workstations. The primary server is the SFT III server that network workstations "see," and the one to which they send requests for network services. Routers on the internetwork see only the primary server and send routing packets to it. The primary server's IOEngine determines the order and type of events that are sent to the MSEngine. Only the primary server sends reply packets to network workstations. The secondary server is the SFT III NetWare server that is activated after the primary server. Either server may function as primary or secondary, depending on the state of the system. You cannot permanently designate which server is primary or secondary. System failure determines each server's role, that is, when the primary server fails, the secondary server becomes the new primary server. When the failed server is restored, it becomes the new secondary server.

Primary Station A network node that controls the flow of information on a communications link. Also, the station that, for some period of time, has control of information flow on a communications link (in this case primary status is temporary).

Primary Storage The main internal storage.

Primary Wire Center A switching center in the

AT&T/Bell system hierarchy of exchange classes. The primary center is a Class 3 exchange. It is used to connect toll offices and less frequently to connect a toll center with a local end office. Primary centers are capable of connecting toll centers through sectional centers and then to local end offices to establish communication connections when simple routing possibilities are busy.

Prime Focus Parabolic Antenna There are two types of satellite antennas — a prime focus parabolic antenna and an offset parabolic antenna. A "normal" satellite antenna is a pure parabola. It's been around forever. A newer antenna, called the offset antenna is taller than it is wide. According to the manufacturers, the offset antenna design makes for more efficient use of the antenna surface than a traditional prime focus parabolic antenna. What that means is that it captures more of the satellite signal hitting the antenna. Offset antennas are more expensive than the "normal" parabolic satellite antennas, which are called "prime focus parabolic antennas." They are more expensive because they cost more to make since they typically must be made out of one sheet of metal. Offset antennas are harder to carry around, since you can't make them out of several foldover sheets of metal.

Prime Line A key system feature. You can program your phone set to automatically select a certain phone line whenever your lift the receiver or press the Handsfree/Mute button. The line that appears is called the Prime Line.

Prime Line Preference When you pick up the handset on your key system or hybrid key, you are automatically connected to your preferred line (central office or intercom), rather than having to punch down an extra line button. Some phone systems tout this as a feature. Some have it set up where you simply leave one of the line buttons depressed and it doesn't pop up when you put your handset back into its cradle. Most 1A2 phone systems have this feature. Not all electronic phone systems do. Walker breaks the features into Prime Line Outgoing and Prime Line Incoming, and allows you to program the phones separately.

Primitive An abstract, implementation independent, interaction between a layer service user and a layer service provider. See PRIMITIVES.

Primitives Abstract representations of interactions across the service access points indicating information is passed between the service user and service provider. There are four types of primitives in the OSI Reference Model — request, indication, response and confirm.

Principal First or highest in importance. An owner or part-owner of a business. A person who authorizes another, as an agent, to represent him. Often confused with principle, not that a principal necessarily has any principles. See Principle.

Principle A general or fundamental truth on which others are based. A rule of conduct. Often confused with principal. See Principal.

Principal Headend of a cable television system According to the FCC's definition:
• If the system has one headend, that headend is the "Principal Headend".
• If the system has two or more headends, the operator may designate the "Principal Headend". However, once designated, it cannot be changed except for "good cause". The location of the Principal Headend is a factor in determining the must-carry status of certain broadcast stations.

Print Control Character A coded control character used to instruct the receiving unit on how a message is to be formatted in hard copy. Print control characters include carriage returns, back spaces, line feeds, tabs, etc.

Print Server A networked computer, usually consisting of fixed-disk storage and a CPU, that controls one or more printers that can be shared by users.

Print Spooler An application that manages print requests or jobs so that one job can be processed while other jobs are placed in a queue until the printer has finished with previous jobs. See PRINT SPOOLING.

Print Spooling A technique used to schedule printing tasks to one printer and to free up computer time from the slow task of feeding a slow printer (Any printer is slow compared to the speed of a computer). A small program or program/machine called the spooler does the scheduling. A user loads the print task to the spooler and when the print task's turn comes, the job is printed. Print spooling is handled several ways: You can allocate part of the computer's main memory to become a print spooler. You can allocate part of the company's disk memory to become a print spooler. You can get an external device called a print spooler. It will have all the storage space and software necessary. There are two primary advantages to Print Spooling: 1. You can use the spooler to save your and your computer's time. Dump the report to a print spooler at thousands of bits per second. Get on with something else on the computer. 2. You can use a print spooler to schedule several users' printing requests. This is particularly good in multi-user environments — for example, where the printer is a laser printer (and therefore expensive) and is attached to a LAN (Local Area Network).

Printed Circuit Board PCB. Flat material (fiberglass/epoxy) on which electronic components are mounted. A PCB also provides electrical pathways called traces, that connect components. Printed circuit boards are what PBXs and computers are made of these days. Be careful when you're replacing PCBs. They're usually very sensitive to static electricity. Handle them only when you're attached to a static electricity strap that is properly grounded. Lay them down only on a surface you're sure is static electricity free. And don't touch the components on PCBs whatever you do.

Printer A device which takes computer information and prints it on paper.

Printer Control Language PCL. See PCL.

Printer Driver A program that controls how your computer and printer interact. A printer driver file supplies information such as the printing interface, description of fonts, and features of the installed printer.

Printer Emulation A fax term for mimicking a printer-generated document. This way, the outgoing fax will look as if it has come from the printer attached to the computer. This can include full formatting, as well as letterhead, signature and graphic images.

Printer Font A font stored in your printer's memory, or soft fonts that are sent to your printer before a document is printed.

Printer Server A computer and/or program providing LAN (Local Area Network) users with access to a centralized printer. A person using the LAN will send a message to the printer server computer. This computer will then assign it a piece of memory or disk space to store its file while it waits to be printed. With a printer server, users can send to the printer any time. Their print jobs are usually handled in the order they are received. But "big bosses" can be given priority and can be bumped to the top of the queue. Print servers allow fewer printers to satisfy more users. Print servers are also especially useful for expensive, laser or high speed printers because they (the print servers) spread the cost of these expensive machines over many users, making them more affordable. See PRINT SPOOLING.

Printer, Wire A matrix printer which prints using a set of wire hammers which strike the page through a carbon ribbon to generate the matrix characters.

Prioritization The process of assigning different values to network users, such that a user with higher priority will be offered access or service before a user with lower priority. Increasingly available as an added option with network operation. Any procedure where different levels of precedence exist.

Prioritization Parameters The hierarchical rules which a network device, such as a router, applies to incoming traffic to determine which traffic should be handled first, next, and so on.

Priority A ranking given to a task which determines when it will be processed.

Priority Bumping The process during a link, trunk or facility failure where lower priority user access to network services is interrupted in order to offer those services or bandwidth to a pre designated higher priority user.

Priority Call 1. Emergency calls to the attendant bypass the normal queue and alert the attendant with some special signal.

2. The name of a Bell Atlantic service. While all callers are important, some are more critical to your business. So give your high-priority callers a ring of their own with Priority Call. Simply use your phone to program up to six callers' numbers. When any of these people call, you'll know right away because you'll hear a different ring. If you have Call Waiting, you'll hear a Priority "beep" when you're on the phone. This way, you'll only have to interrupt your work to take priority calls.

3. This PBX feature allows an urgent intercom voice call to be made when the called telephone is busy or has Do Not Disturb activated. Priority Call should not be made available to everyone, and should be selectively programmed.

Priority Indicator A character or group of characters which determine the position in queue of the message in relation to the urgency of other messages. Priority indicators control the order in which messages are to be delivered.

Priority Ringing A name for a Pacific Bell (and possibly other local telephone companies') service which alerts you to have calls from selected numbers ring at another number.

Priority Transport The capability of a network for certain classes of traffic to have priority over others and thus have lower delay or otherwise better performance.

Priority Trunk Queuing Through user-chosen trunk access level, this PBX feature places any caller with this or higher level in the class of service assignment ahead of callers waiting for the same trunk group (or Agent Group in the case of incoming ACD calls).

Privacy Privacy usually means that once a caller "seizes" a line, no other user can access that same line even though it appears on his/her key set. Privacy can be automatic or selected for each call.

Privacy And Privacy Release All other extensions of a line are unable to enter a conversation in progress unless the initiating telephone releases the feature.

Privacy Enhanced Mail. PEM. An Internet electronic mail capability which provides confidentially and message integrity using various encryption methods.

Privacy Lockout Privacy automatically splits the connection whenever an attendant would otherwise be included on the call, i.e. the attendant can't listen in to a call she's just extended to someone. A tone warning is generated when the attendant bridges into a conversation in progress.

Privacy Override Activation of a special pushbutton allows the phone user to access a given busy line, even though the automatic exclusion facility is being used by the station on that line. This privilege of Privacy Override is usually only given to Big Bosses.

Private ATM Address A twenty-byte address used to identify an ATM connection termination point.

Private Automatic Branch Exchange PABX. A private telephone switch for a business or an organization in which people have to dial "9" to access a local line. In the old days, private branch exchanges were manual, meaning that operators/attendants were needed to manually place calls. Then the systems improved and you were able to dial the outside world from your extension without the help (or hindrance?) of an operator. Thus they became known as private automatic branch exchanges. But then all PBXs became Automatic. So these days, PABXs are all called PBXs, except in some countries outside North America, where they're still called PABXs. See also the next definition and PBX.

Private Branch Exchange PBX. Term used now interchangeably with PABX. PBX is a private telephone switching system, usually located on a customer's premises with an attendant console. It is connected to a common group of lines from one or more central offices to provide service to a number of individual phones, such as in a hotel, business or government office. For the biggest definition, see PBX. See also PABX.

Private Carrier An entity licensed in private services and authorized to provide communications service to others for money.

Private Dial-In Ports A packet network term. For customers who have many calls, the packet network operator provides dedicated, unpublished phone numbers. The idea is to give the preferred user better service.

Private Domain Name A standard attribute of an O/R (Originator/Recipient) Address that identifies a PRMD (Private Management Domain) generally relative to an ADMD (Administrative management Domain). An X.400 term.

Private Exchange PX. A telephone switch serving a particular organization and having no means of connection with a public exchange. In other words, a phone system just for intercom calls.

Private Facility Trunk A telephone company AIN term. A transmission facility that carries non-public switched telephone network (PST) traffic. An example of a private facility trunk is an access arrangement to a switch supporting PBXs, including the switched end of a Foreign Exchange (FX) and an Off Network Access Line (ONAL).

Private Key An encryption technique which requires that the decrypting key be kept secret. Also known as single-key and secret-key. See PUBLIC KEY and ENCRYPTION for more detail.

Private Line 1. A direct channel specifically dedicated to a customer's use between specified points. A line leased from a carrier, local or long distance. A non-switched circuit. One end of the line is directly connected to the other end. Here's the AT&T definition of a private line. "A dedicated, non-switchable link from one or more customer-specified locations to one or more customer-specified locations..."

Private lines offer highly available connectivity, as they are dedicated to the use of a single organization. As private lines are priced solely based on distance, with no usage-sensitive cost element, they can be used constantly and at maximum capacity. Therefore, they offer a highly cost-effective to usage-sensitive, switched services. Private networks comprise numbers of private lines. Originally, private lines were, in fact, dedicated circuits which literally could be physically traced through the network. They also were known as "nailed-up circuits," as telephone company technicians hung the circuits on nails driven into the walls of the central offices. Contemporary private lines actually involve dedicated channel capacity provided over high-capacity, multi-channel transmission facilities. See also PRIVATE NETWORK.

2. An outside telephone line, with a separate telephone number, which is separate from the PBX. The line is a standard business line which goes around the PABX. It connects the user directly with the LEC central office, rather than going through the PBX. Private lines connections are considered to be very "private" by virtue of the fact that it is not possible for a third party (e.g., technician or console attendant) to listen to conversations without placing a physical tap on the circuit. Additionally, private lines are not subject to congestion in the PBX. As private lines also are not susceptible to catastrophic PBX failure, they often are used to provide failsafe communications to individuals with mission-critical responsibilities in data centers, network operations centers, and the like.

Private Line Service An outside telephone number separate from the PBX, can be set up to appear on one of the buttons of a key telephone. Also called an Auxiliary Line. See also PRIVATE LINE.

Private Management Domain PRMD. An X.400 electronic mail term: A private domain to which MTAs (Message Transfer Agents) send mail. PRMDs are connected to ADMDs (Administrative Management Domains) for message routing over wide area links. Under X.400 addressing, the PRMD represents a private electronic messaging system that may be connected to a Administrative Management Domain. The PRMD is usually a corporate or government agency E-Mail system connected to an ADMD.

Private Message A message designation which prevents that message from being given to another mailbox.

Private Network 1. A network built and owned by an end user organization. Some very large organizations build their own private microwave networks, rather than rely on circuits leased from carriers. This generally is the case where a number of remote sites must be networked, especially where substantial bandwidth is required. In such situations, the public carriers may be unable to provide the necessary bandwidth and network performance.

2. A network comprising dedicated circuits leased from one or more public carriers. Such circuits make use of private lines over carrier transmission facilities, bypassing the switches. Many large organizations deployed complex, dedicated T-carrier networks in the 1970s and 1980s. While such networks continue to be supplemented and while such networks continue to be deployed for data communications, VPNs (Virtual Private Networks) generally are preferred for voice communications. A variety of VPN technology alternatives also exist for data communications. See also PRIVATE LINE, PRIVATE VOICEBAND NETWORK, and VPN.

Private Networks Marketing A Northern Telecom term which defines their organization for making and selling all telecom switches, except central offices. These products include the Meridian 1 PBX family, residential and business telephone sets, including Norstar and data communications.

Private Subscriber Network A virtual private network service supported by Public Packet Switched Service (PPSS) and incorporating interLATA transmission facilities owned or leased by the customer for private traffic. A Bellcore definition.

Private Voiceband Network A network that is made up of voice band circuits, and sometimes switching arrangements, for the exclusive use of one customer. These networks can be nationwide in scope and typically serve large corporations or government agencies.

Private Wire A private line. Derives its name from the old telegraph days when messages were carried on wires that strung across the nation.

Privileges The access rights to a directory, file or program over a local area network. Typically read, write, delete, create and execute.

PRMD PRivate Management Domain. An X.400 Message Handling System private organization mail system. Example: NASAmail.

Proactive Taking the initiative. Doing it before someone (most likely your competition) forces you to do it. The word is currently in vogue among those people who believe the telephone companies should do all the positive, forward-looking actions before the competition does them and gets the public kudos. The word has no real meaning, but serves a purpose as a cry to action. The word actually is grammatically incorrect. The real word is "active." It is the opposite of "reactive." The person who told us this is Norm Brust, a wonderful man and one of the more active (not proactive) people in our industry.

Probe 1. A sensing device, typically about the size and shape of a pencil, that is used to sense various physical conditions such as temperature, humidity, current flow, speed. Usually connected to a meter or oscilloscope which displays the condition being monitored.
2. An empty message that is sent to reach a particular address to determine if an address can be reached.

Probe Envelope In X.400, the envelope that encloses a probe in the MTS (Message transfer System). See PROBE.

Process A software application. Any activity or systematic sequence of operations that produces a specified result. Typically, a computer function that consists of, or involves, procedure code, data storage and an interface for communicating with other processes.

Process Manufacturing The making of things. This contrasts with flow manufacturing which is working on something — like oil — that flows through a production process.

Processing Gain In a spread spectrum transmission system, the original information signal is combined with a pseudo random correlating, or spreading code. The more random and the greater the length of the code, the more robust the resulting spread spectrum signal is against interference and interception. A measure of this robustness is referred to as processing gain. The FCC requires a minimum of 10 dB processing gain for non-licensed equipment operating in the Part 15 902-928 MHz, 2400-2483 MHz, and 5725-5850 MHz frequency bands. See also CDMA.

Processing, Batch A method of computer operation in which a number of similar input items are accumulated and sorted for processing. Compare with ON-LINE or INTERACTIVE PROCESSING.

Processor The intelligent central element of a computer or other information handling system. Also called the Central Processing Unit (CPU).

Processor Card See Smart Card.

Processor Occupancy The time the telephone system processor is in use. There are two typical demands on the central processor in a telephone system, moving calls around and running self-diagnostics. Be sure you factor in the second when you're trying to figure out how many calls your telephone system processor will handle before it dies.

Processor Power The number of computations that a computer, microprocessor, or digital signal processor can complete in a fixed time interval. May be measured in MIPS (millions of instructions per second) or MFlops. Typical low-end DSP chips provide up to 10 MFlops; high-end chips 30 or more.

Procurement Lead Time The interval in months between the initiation of procurement action and receipt into the supply system of the production model (excludes prototypes) purchased as the result of such actions, and is composed of two elements, production lead time and administrative lead time.

Procr Processor.

Prod A device that resembles a pencil, but containing a metal tip in an insulated handle with a wire to connect it to a piece of test equipment, such as a VOM (volt-ohm-millimeter). the metal tip is touched to various points in an electrical circuit for measurements and trouble-shooting.

Prodigy Formed in 1984 as a joint venture of IBM and Sears Roebuck & Company, Prodigy was originally called TRINTEX. The name was changed to Prodigy in 1988, and the company was acquired by employees with the help of International Wireless in 1996. Prodigy used to offer on-line computer services. The company was one of the first to offer such services for a largely flat monthly fee. Recently, Prodigy decided to terminate the activities of 50 staffers who develop "content" for its information service, and instead to link its users to the content of Excite, a Web directory and search engine. Prodigy will now become more a pure Internet Service Provider, offering connections to the Internet.

Productize This is a stupid word. But it means to make an idea into a product. What this means is to complete the R&D on it, to finish the customer documentation, to finish the packaging design, to assign a name, model number and stocking number, to pass the information onto product support, etc. Everything necessary to make it a product that can be sold. See also BETA.

Profile A set of parameters defining the way a device acts. In the LAN world, a profile is often used by one or more workstations to determine the connections they will have with other devices and those devices they will offer for use by other devices. Often called a login file. Profiles and login files usually work like batch files, automatically executing a number of commands when you turn on the machine.

PROFS PRofessional OFfice System. Interactive productivity software developed by IBM that is part of the Virtual Machine (VM) Productivity System and runs under the VM/CMS mainframe system. PROFS is frequently used for electronic mail and is said to give a user an edge in productivity in three areas: business communications (including electronic mail), time management and document handling.

Program Instructions given to a computer or automated phone system to perform certain tasks. Most vendors improve (update) their software programs continuously. It's a good idea to ask what the deal is with getting updates.

Program Circuit A voice circuit used for the transmission of radio program materials. It is a telephone circuit which has been equalized to handle a wider range of frequencies than are required for ordinary speech signals.

Program Counter A device inside a computer which keeps track of which instruction in the program is next, etc.

Program Evaluation Reviews An activity many of us

are consigned to spend our aging years doing. In the early days of our careers, we used to do things — actually do tasks hands-on. Then many of us, sadly, got "successful." Our jobs then became telling younger people what to do and checking that they do what we told them, or doing something better (hopefully). To do all this, we sit in meetings. We call these meetings "Program Evaluation Reviews."

Program Evaluation Review Technique PERT. A management tool for graphically displaying projected tasks and milestones, schedules and discrepancies between tasks.

Program File A file that starts an application or program. A program file has an .EXE, .PIF, .COM, or .BAT filename extension. AKA executable.

Program Information File PIF. A file that provides information about how Windows should run a non-Windows NT application. PIFs contain such items as the name of the file, a start-up directory, and multitasking options for applications running in 836 enhanced mode. See PIF.

Program Logic The particular sequence of instructions in a program.

Programmable In telephony, the ability to change a feature or a function or the extension assigned to a telephone without rewiring.

Programmable Call Forwarding This feature of a telephone system allows a user to instruct his phone to send all his calls to another phone. That phone might be another extension in the same phone system or it might be another phone number altogether in a different part of the country. This feature is great. You're going to a meeting but don't want to miss that one special call. You can send your calls to a person close to the meeting and ask them to interrupt you if that "special" call comes in. The problem with this feature is that often people forget their phone is on "forwarding" and when they return to their office, they sit around all afternoon waiting for that special call, which got call forwarded elsewhere. There are two ways to overcome this. Some phones have lights or messages on their screens which indicate all calls are being forwarded. Also, some telecom managers program their total phone systems so that twice a day, all call forwarding shall cease and all calls shall return to their original phone.

Programmable Configuration Select Refers to the EEPROM setup routine which allows jumperless configuration of the system board.

Programmable Memory Memory that can be both read from and be written into by the processor. Synonym for RANDOM ACCESS MEMORY — RAM.

Programmable Terminal A user terminal that has some limited processing power. Also, intelligent terminal.

Programming Language A language used by a programmer to develop instructions for the computer. It is translated into machine language by language software called assemblers, compilers and interpreters. Each programming language has its own grammar and syntax.

Programming Overlay Typically a piece of cutout cardboard, which you place over certain of the keys on a phone or console. When you punch in a certain code, the buttons become what's written on the programming overlay.

Program Sharing The ability of several users or computers to use a program simultaneously.

Program Store Permanent memory in a stored program control central office that contains the machine's generic software program, parameters and translations.

Progressive Conference A PBX feature. Allows the extension user to create conferences of more than three peo-

ple using the consultation hold and add-on conference features. To create a conference, an extension user typically uses the consultation hold, dials the desired internal or external number and effects an add-on conference. The conference may then be progressively expanded, in this same fashion, to the maximum capability of the phone system offering this feature. A good question to ask before you get sold this feature is "does the conferencing have amplification and balancing?" Without these features, the conferencing conversation will simply get more and more difficult to hear on.

Progressive Dialing A form of predictive dialing, progressive dialing is slightly more automated than preview dialing. The customer data is not displayed until the number is dialed, giving the agent less time to review it and a shorter time between calls. See also PREDICTIVE DIALING and PREVIEW DIALING.

Progressive Display See Interlaced GIF.

Progressive Tuning A method of painting pictures on computer monitors or TV screens in which the picture is painted line by line. It is today's most common way of painting a picture or an image on a computer screen.

Project Evaluation Review Technique PERT. A technique for managing a project — say the installation of a PBX — which produces a guess at the project's critical path (longest task to complete) and of project milestone completion dates. See PERT.

PROM Programmable Read Only Memory. A PROM is a programmable semiconductor device in which the contents are not intended to be altered during normal operation. PROM acts like nonvolatile memory. When you install an autoboot PROM on a LAN network board, the workstation can boot up from the network server. This is particularly useful for diskless workstations.

Promiscuous Mode Most Ethernet cards ignore all the packets on the network that aren't destined for them. But in a Remote server — one serving multiple remote users all calling in over modems — the Ethernet LAN card has to get access to all the packets and grab those that are meant for it — so it can pass them over to the remote callers. I assume it's called "promiscuous mode" because it means that the Ethernet card has to have a relationship with all the packets traveling on the local area network. Another application for promiscuous mode is if you want to attach software or hardware to your computer, monitor and analyze all the packets flying around your network. You can set some (but not all) Ethernet cards to promiscuous mode.

Promotion According to various dictionaries, "promotion" means to raise in station, status, rank or honor." Once upon a time, a promotion meant you got a better title, a bigger office and a raise in pay. Today, it means that a software release or your hardware product just made a change in status from alpha test to beta test, or from beta test to general release. See also Alpha Test and Beta Test.

Prompt An audible or visible signal to the system user that some process is complete or some user action is required. Also used to signify a need for further input and/or location of needed input. See also the next three definitions.

Prompt Tagging And Encoding According to Steve Gladstone, author of the book, "Testing Computer Telephony Systems," (available from 212-691-8215) many test applications require the test system "know" which prompt is being played by the computer telephony system. This is often accomplished by prompt tagging, also known as prompt encoding. Prompt tagging has the computer telephony system play-out an audio tone that can be heard by the test system with each prompt that is played. Prompt tagging is relatively inexpensive to perform, and inexpensive to automate. Prompt tagging can be accomplished in two ways, either by inserting tones into actual user prompts ("insert" mode), or by having a special programmatic switchable test mode that may be toggled on/off by the computer telephony system that will play out a tone sequence before or after each prompt is played ("append" mode).

Prompting Visually or audibly indicating to a user of a telephony device that a call has reached (and been accepted by) the device and is capable of being answered. This is typically done by ringing the device, flashing a lamp, or presenting a message on the device display.

Prompts 1. Recorded instructions delivered by voice processing units. Prompts may include MENUS or other information that is played each time you get into the system.
2. Messages from the computer instructing the user on how to use the system. See MENU and AUDIO MENU.

Proof Of Concept You're at a trade show. You go to a booth. You see some great new technology. You can't buy it. It's simply a demonstration of new technology. What's called "proof of concept." t proves that the idea works. It doesn't mean there's a market for it. The idea of "proof of concept" is to excite people — customers or security analysts. Maybe someone will place a big order or buy the company's stock?

Propagation Delay The delay caused by the finite speed at which electronic signals can travel through a transmission medium. Propagation delay is estimated at 160,000,000 meters per second in copper wire. Note: No signal can travel through any medium faster than the speed of light, which is 300,000,000 meters per second, or 186,000 miles per second. Propagation delay is not a huge issue in most communications networks or applications scenarios, with the clear exception being satellite communications. Given the fact that the originating signal must travel from the earth station 22,300 miles up to the satellite and 22,300 miles back down, a roundtrip transmission takes about 1/4 second. This level of propagation delay renders satellite communications ineffective for highly interactive data communications applications, as the users get really bored. Satellite communications also is highly aggravating for voice communications.

Propagation Time Time required for an electrical wave to travel between two points on a transmission line.

Propagation Velocity The speed at which electrons or photons travel through a transmission medium.

Propeller Head An excessively technical person, whose social skills are lacking.

Properties Windows 95 treats all objects, such as windows, icons, applications, disk drives, documents, folders, modems, and printers as self-contained objects. Each object has its own properties, such as the object's name, size position on-screen, and color, among others. You can change an object's properties using the properties dialog box.

Property Management Interface PMI. A telephone system's ability to talk to a hotel's computer system.

Proportional Font A font in which different characters have varying widths. All magazines and newspapers are printed in proportional characters, which make reading easier. By contrast, in a monospaced font, such as one on an old typewriter, all characters have the same widths.

Proprietary If something is proprietary it means it will only work with one vendor's equipment. See the next three definitions.

Proprietary LAN A LAN (Local Area Network) that runs the equipment of only one vendor. A proprietary LAN, for example, cannot join IBM PCs to DEC minicomputers. Some people say such LANs are more "bug-free" because they have only one vendors' wares to deal with. They also tend to be more expensive. They also tend to tie you to one vendor, although some makers are now coming out with bridges which connect proprietary LANs to non-proprietary LANs. Since Ethernet and the Internet, proprietary LANs are effectively dead.

Proprietary Network A network developed by a vendor that is not based on protocols approved by standards body or on standards that are "open." Typically, you won't be able to connect to a network with any equipment other than that made by the manufacturer who created it.

Proprietary Telephone Sets Proprietary telephones are feature phones that are specific to a particular make of PBX, ACD or other switching system. They may be digital or analog. As they are custom designed for that system, they have non-standard electrical interfaces and have non-standard protocols to communicate between the telephone and the switch. This has several implications: 1. You can't take a proprietary phone from one switch and expect it to run on another switch. It won't. 2. Proprietary phones are expensive, and are highly profitable to their makers. Hence the manufacturers' insistence on keeping them proprietary. 3. Signaling between proprietary phones and their switches is richer than signaling between switches and single line analog phones. As a result, it's preferable to integrate voice mail and automated attendants through proprietary phones. Sometimes the manufacturer of the switch will divulge his secret signaling scheme. Other times he won't. Most times he wont. And you, as a voice mail or auto attendant manufacturer have to reverse engineer it, which is sometimes successful. ISDN is actually the first attempt to make "proprietary" sets standard. So far, only a few manufacturers have used SDN-like phones as their proprietary phones.

Prosodics In speech recognition prosodics refers to the parts of the sentence the speaker emphasizes. For example, "I am going to Paris" with emphasis on the "I" means that only one person is going to Paris and therefore only one ticket should be issued. See also PROSODY.

Prosody Intonation. In text to speech, prosody refers to how natural it sounds — the ups and downs of the sentence. See also PROSODICS.

Prospective The opposite of Retrospective or Retroactive. Most regulatory commission rate cases are prospective, which means they relate to prices and things in the future. Some rate cases, however, are retroactive or retrospective, which means they apply to prices and things in the past. Most of these decisions involve forcing the company to return money to its subscribers in the form of a refund. Interestingly — try this one — most retroactive commission decisions are prospectively retroactive. In other words, they only take effect some time in the future, when the decision is voted upon by all the commission members.

Prospero UNIX software which helps you search archives connected to the Internet. Prospero uses a virtual file system which enables users to transparently view directions and retrieve files. In short, Prospero is a distributed directory service and file system that allows users to construct customized views of available resources while taking advantage of the structure imposed by others.

Protected Distribution System PDS. This is a US

Federal Government definition: A wireline or fiber-optics telecommunication system which includes adequate acoustic, electrical, electromagnetic, and physical safeguards to permit its use for the unencrypted transmission of classified information. A complete protected distribution system includes the subscriber and terminal equipment and the interconnecting lines.

Protected Mode A computer's operating mode that is capable of addressing extended memory directly. The operating mode for the Intel 80286 and higher processors (the 80386, 80486 and Pentium) that supports multi tasking, data security, and virtual memory. The 80286 processor can run in either of two modes: real or protected. In real mode, it emulates an 8086 (it accesses a maximum of 640KB of RAM and runs only one software application at a time). Protected mode allows the 80286 processor to access up to 16MB of memory. It uses a 24-bit address bus. Since a bit can have one of two values, raising the base number of 2 to the power of 24 is equal to 16,777,316 unique memory addresses. Each memory address can store one byte of information (16,777,216 bytes equals 16MB). Protected mode operation also makes it possible to run more than one application at once and to handle more processes because more memory is available. Processes can be requests from an operating system or an application to perform disk I/O, memory management, printing, or other functions. Processes are assigned priority numbers in protected mode. The processor gives priority to those with higher numbers. Operating system processes always have higher priority than application processes. See also REAL MODE and VIRTUAL 8086 MODE.

Protective Connecting Arrangement PCA. A device leased from the telephone company and placed between your own (customer-provided) telephone equipment and the lines of the telephone company. The idea was to protect their lines from your junky equipment. No instance/case was ever proven of harm occurring to the network from faulty customer-provided equipment and the PCAs were thrown out and replaced by the FCC's Part 68 Registration Program. Under this program, customer-owned equipment which passes FCC tests can be registered and connected directly to the phone network without these devices. The phone industry eventually refunded most of the fees it charged on the PCAs. NATA and many manufacturers claimed the PCAs were designed to prevent the growth of the interconnect or customer-owned phone industry. They were probably right. The question is now moot, since the charges and the devices no longer exist, except in a museum or attached to very old equipment. See also PROTECTIVE COUPLING ARRANGEMENT and PCA.

Protective Coupling Arrangement PCA. A device placed between the phone company's trunks and your particular telephone gadget. The objective of the PCA is to isolate the telephone company's lines from your equipment and thus protect their lines from your equipment. The device is not needed if your equipment has passed FCC approval — under Part 68 of the FCC's rules. See also PROTECTIVE CONNECTIVE ARRANGEMENT, which is another term for the same thing.

Protector Block A device connected to an exchange access line to protect connected equipment from over-voltage and/or over-current. Hazardous voltages and currents are shunted to ground. In other words, a surge protector limits unwanted surge voltages to values which can be handled safely by the insulation on inside wire and by the electronics in the customer terminal equipment. Protectors are very important in high-lightning areas, where they (theoretically) keep wires and phones from melting, phone systems from being blown off the

wall, and end users from being electrocuted.

The original protectors were based on carbon blocks which effectively blocked aberrant voltage surges. Subsequently, gas tube protectors were used. Solid state protectors were the third generation. Improvements in the speed of of reacting to incoming high voltage and high currents have been at the forefront of the improvements in technology. While all variety of protectors currently are in place, those currently being deployed are either solid-state or hybrids, which incorporate both gas tube and solid-state technology. Protectors often are an element of a multi-function NID (Network Interface Device), also known as a NIU (Network Interface Unit), which acts as the point of demarcation between the local exchange carrier and the customer premise.

PROTEL PRocedure Oriented Type Enforcing Language. Protel is a block-structured, type-enforcing, high level, software language that enables extensive type checking on the source code at compile time. It was developed at Bell Northern Research, a subsidiary of Northern Telecom. Protel is used in the DMS-100, a family of Northern Telecom central office telephone switches. Both the central control CPU and the DMS SuperNode CPU are programmed in Protel.

Protn Protection. See Protector Block.

Protocol A protocol is a set of rules governing the format of messages that are exchanged between computers and people. Imagine making a phone call. You pick up the phone, listen for dial tone, then punch out some buttons on your phone, then listen for ringing and for an answer. The person says "Hello." You say "Hello." Then you talk... What you're doing is following a protocol to make a call. When computers make calls between themselves — to transfer data, for example — they follow a protocol. They aren't smart, like you and I. They can't distinguish between dial tone and fast busies, unless those sounds and signals are specifically defined. A protocol defines the procedure for adding order to the exchange of data (i.e. a "conversation.") A protocol is a specific set of rules, procedures or conventions relating to format and timing of data transmission between two devices. It is a standard procedure that two data devices must accept and use to be able to understand each other. The protocols for data communications cover such things as framing, error handling, transparency and line control. There are three basic types of protocol: character-oriented, byte-oriented and bit-oriented.

Protocols break a file into equal parts called blocks or packets. These packets are sent and the receiving computer checks the arriving packet and sends an acknowledgement (ACK) back to the sending computer. Because modems use phone lines to transfer data, noise or interference on the line will often mess up the block. When a block is damaged in transit, an error occurs. The purpose of a protocol is to set up a mathematical way of measuring if the block came through accurately. And if it didn't, ask the distant end to re-transmit the block until it gets it right. See PROTOCOLS for a list of the more common protocols. See the following protocol definitions.

Protocol Analyzer A specialized computer and/or program that hooks into a LAN and analyzes its traffic. Good protocol analyzers can record and display data on all levels of traffic on a LAN cable, from the lowest media access control packets to NetBIOS commands and application data. They are excellent for diagnosing network problems, but they require some expertise, as their data output can be obscure.

Protocol Control Protocol Control is a mechanism which a given application protocol may employ to determine or control the performance and health of the application. Example, protocol liveness may require that protocol control information be sent at some minimum rate; some applications may become intolerable to users if they are unable to send at least at some minimum rate. See MCR.

Protocol Control Information PCI. the protocol information added by an OSI (Open Systems Interconnection) entity to the service data unit passed down from the layer above, all together forming a PDU (Protocol Data Unit).

Protocol Conversion A data communications procedure which permits computers operating with different protocols to communicate with each other. See PROTOCOL and PROTOCOL CONVERTER.

Protocol Converter A device which does protocol conversion. It's your classic "black box." Glasgal Communications defines a protocol converter as any device which translates a binary data stream from one format to another according to a fixed algorithm. Compare with bridge and gateway, which are different animals and may contain protocol converters...and more.

Protocol Data Unit See PDU.

Protocol Dependent Routing Any routing method in which routing decisions are made on the basis of information provided by the specific LAN protocol used by the communicating devices. TCP/IP and DECnet routers are protocol dependent routers. So are so-called multi protocol routers, because they must support each protocol running in the network. See also PROTOCOL INDEPENDENT ROUTING.

Protocol Filtering A feature available in some network bridges which allows it to be programmed to always forward or reject transmissions associated which specified protocols.

Protocol Independent Router A routing device that provides the functionality of protocol specific routers such as TCP/IP or DECnet routers but is independent of protocols. In addition to routing "routable" protocols like TCP/IP, DECnet or XNS, it routes IBM protocols which are not routable. The protocol independent router combines the latest in computer hardware with the new advanced routing technologies such as SPF (Shortest Path First) and IS-IS (OSI routing standard). It represents an alternative to conventional routers that use old routing technologies and are protocol dependent. Protocol independent routers provide easy-to-install-and-use enterprise-wide networks in a token ring or Ethernet environment.

Protocol Independent Routing A routing method in which routing decisions are made without reference to the protocol being used by the communicating devices. Protocol independent routers provide the functionality of protocol specific routers such as TCP/IP or DECnet routers, but can also route non-routable protocols. See also PROTOCOL INDEPENDENT ROUTER.

Protocol Mapper Protocol Mappers are employed where a logical server is delivering a proprietary CTI protocol and / or providing connections through a proprietary transport protocol. Mappers deliver one of the specified CTI Protocols alter first mapping, or translating, from the non-specified, proprietary protocol.

Protocol Mapper Device Implementation of a Protocol Mapper as device which sits on a data connection and is transparent to the logical client. See Protocol Mapper.

Protocol Mapper Code Implementation of a Protocol Mapper as a software component. Mapper Code components mimic transport protocol stack implementation, and, using an appropriate R/W interface, are layered above any transport protocol stack and are transparent to Client Implementations. See Protocol Mapper.

Protocol Stack First, read the definition on what a protocol is. Understand that it's a specific set of rules, procedures or conventions relating to format and timing of transmission between two devices. Protocol defines the rules for conversations — voice, data, video, etc. — between two computers. A protocol stack is basically a collection of modules of software that together combine to produce the software that enables the protocol to work, i.e. to allow communications between dissimilar computer devices. It is called a stack because the software modules are p led on top of each other. The process of communicating typically starts at the bottom of the pile and works itself up. Each software module typically (not always) needs the one below it. Sometimes one big protocol stack — such as the one for H.323 — might include specific protocol standards further down the stack. Depending on The TCP/IP protocol stack includes such protocols as TCP, IP, FTP, SMTP, telnet, and so on A protocol stack is also called a protocol family or protocol suite. See PROTOCOL.

Protocol Suite A hierarchical set of related protocols.

Protocols For an explanation of protocols, see PROTOCOL. Here are the more common PC protocol types:

MODULATION PROTOCOLS
    Bell 103: Low Speed (300 baud)
    Bell 212: Low Speed (1200bps)
    ITU-T V.22bis: Medium (2400bps)
    ITU-T V.32: Medium Speed (9600bps)
    ITU-T V.32bis: High Speed (14,400bps)
    ITU-T V.34: High Speed (28,800bps)
ERROR CONTROL PROTOCOLS
    Microcom Network Protocol (MNP)
    ITU-T V.42 (Includes LAP-M & MNP)
DATA COMPRESSION PROTOCOLS
    MNP/5
    ITU-T V.42bis
All data compression requires an underlying error control protocol.
FILE TRANSFER PROTOCOLS
    Kermit: 7-bit data path, quotes control characters.
    XMODEM: 8 bit data path, ACK/NAK protocol.
    YMODEM: 8-bit data path, batch capability.
    ZMODEM: 8-bit data path, quotes some control characters.

Proton A Proton is a heavy subatomic particle that carries a positive charge. Protons are found in the nucleus of the atom.

Prototyping The development of a model that displays the appearance and behavior (look and feel) of an application to be built. A prototype may only demonstrate the application's. It may also demonstrate navigation and user controls, or it may even accept input data that can be stored in and retrieved from a simulated database. (See also iterative development.)

Provider A process that represents an interface between the Sun Solaris Teleservices platform and an installed telephone device, such as a telephone line or a fax machine. Multiple instances of a provider can be configured, each based on the same information. Each configuration is identified by a unique primary alias. A primary alias is the primary label used to prefer to a provider configuration. The primary alias is a provider's default and primary name. This definition courtesy Sun.

Provisioning The act of supplying telecommunications service to a user, including all associated transmission, wiring, and equipment. The telephone industry defines provisioning as an engineering term referring to the act of providing sufficient quantities of switching equipment to meet established service standards. In NS/EP telecommunication services, "provisioning" and "initiation" are synonymous and include altering the state of an existing priority service or capability.

Proxy 1. A proxy is an application running on a gateway that relays packets between a trusted client and an untrusted host. A proxy accepts requests from the trusted client for specific Internet services and then acts on behalf of this client (in other words, serves acts as proxy for this client) by establishing a connection for the requested service. The request appears to originate from the gateway running the proxy, rather than from the client. All application-level gateways use application-specific proxies (that is, modified versions of specific TCP/IP services). Most circuit-level gateways use pipe, or generic, proxies that offer the same forwarding service but support most TCP/IP services. See Dual-Homed Gateway and Proxy Server. 2. A software agent that acts on behalf of a user. Also, the mechanism whereby one system "fronts for" another system in responding to protocol requests. Proxy systems are used in network management to avoid having to implement full protocol stacks in simple devices, such as modems. In SNMP, a proxy is a device which performs SNMP functionality for a separate managed device. The amount of responsibility may vary. Proxy ARPing refers to address recognition for another unit with SNMP capability, while a proxy agent provides an external SNMP agent for a managed device which does not have SNMP capability.

Proxy ARP The technique in which one machine, usually a router, answers ARP requests intended for another machine. By "faking" its identity, the router accepts responsibility for routing packets to the "real" destination. Proxy ARP allows a site to use a single IP address with two physical networks. Subnetting would normally be a better solution.

Proxy Server A proxy is an application running on a gateway that relays packets between a trusted client and an untrusted host. A proxy server is software that runs on a PC and is basically a corporate telephone system for the Internet. Here's what I mean: A telephone system's main job is to allow a large number of people access to a few number of phone lines. Example: we have 100 people in our firm. But we have only 30 outside phone lines. To grab an outside phone line you dial 9 and then dial your number. When the 31st person tries to grab an outside line, he gets a busy. If this happens a lot, we install more phone lines. The reason we have fewer phone lines than people is clearly economic. We save money that way. All phone systems work that way. A proxy server performs the same function. Let's say your company is connected to the Internet on a single high-speed digital line, e.g. a T-1. The provider of this line gives you a certain number of distinct and different IP addresses which you can use — just like our phone system gives us 30 distinct and different phone lines. Since most firms will have fewer IP addresses than they have people wanting to use the Internet, they'll need a proxy server to act like a phone system — allocating precious IP addresses as the people want them. This process is called address translation. A proxy server is typically also a firewall — that means it keeps unwanted intrusion from the Internet getting into your corporate network. Thus, the firewall's IP addresses function as a proxy addresses. Proxy servers provide extra security by replacing calls to insecure systems' subroutines. Proxy servers also allow companies to provide World Wide Web access to selected people, restricting some, allowing others through the firewall — just like a phone system restricts some people from making long distance calls, etc. Acting as behind-the-scenes directors, proxy servers can also help distribute processing load and provide an added

layer of security. A proxy server could also cache some of the material from popular Web sites, saving access time and phone monies.

In short, a proxy server lets your employees access the Internet right from their desktop PCs over a shared, managed, and secure connection to the Internet. No more running modems to desktops — a slow, expensive solution. That connection to the Net can be "nailed up," like a T-1 or equivalent, or it can even be an on-demand connection. That is, if there is no traffic moving over the connection for a period of time, a proxy server can turn off the connection so your company isn't wasting dollars on an Internet connection not being used. And the proxy would then re-establish the connection immediately when a user tried to access a web site.

According to Microsoft, a proxy server has the following advantages:

1. It accelerates access to the Internet with intelligent caching — no more World Wide Wait!

2. It protects your Intranet in ways a packet filtering router can not.

3. It blocks access to undesirable sites and provides other easy-to-use management features.

4. It saves money by consolidating and making the most of your Internet connection.

See also Router-based Firewall and Dual-Homey Gateway.

PRS Primary Reference Source. The master clocking source in a network. Other, distributed devices derive their clocking from the PRS in order that the entire network and all associated network elements maintain synchronization. See also CLOCK.

PRSM Post Release Software Manager.

PRX Program.

PS Paging Systems

PS-ACR Power Sum-Attenuation-to-Crosstalk Ratio. A measurement of the strength of the data signal in one pair compared to one or more other pairs in a common cable. PS-ACR is measured in dB (decibels). See also ACR and dB.

PS-NEXT Power Sum-Near End CrossTalk. A measurement of the extent to which one cable pair resists interference generated by one or more other pairs in a common cable. PS-NEXT is measured in dB (deciBels). See also dB and NEXT.

Ps and Qs In English pubs, ale is ordered by pints and quarts. In old England, when customers got unruly, the bartender would yell at them to mind their own pints and quarts and settle down. It's where we get the phrase "mind your Ps and Qs."

PS/2 IBM Personal System/2 personal computer.

PSAI AT&T's Processor-to-Switch Applications Interface. See also ASAI and SCAI.

PSAP Public Safety Answering Point. PSAPs are customarily segmented as "primary," "secondary" and so on. The primary PSAP is the first contact a 911 caller will get. Here, the PSAP operator verifies or obtains the caller's whereabouts (called locational information), determines the nature of the emergency and decides which emergency response teams should be notified. ALI (Automatic Location Information), contained in a database, provides supplemental information for purposes of locating the caller, determining if hazardous materials are located at the subject, and so on. In some instances, the primary PSAP may dispatch aid. In most cases, the caller is then conferenced or transferred to a secondary PSAP from which help will be dispatched. Secondary PSAPs might be located at fire dispatch areas, municipal police force headquarters or ambulance dispatch centers. Often the primary PSAP will answer for an entire region. See also 911, E-911 and ALI.

PSC/PUC Public Service Commission. Also known as Public Utility Commission. It's the state agency charged with regulating the local phone company utility. In reality, there are only two things the PSC can do: 1. Allow the phone company to increase its prices, and 2. Restrict competition to the phone company by creating all sorts of restrictive rules and regulations. As competition in the telecommunications industry grows — chiefly because of Federal rulings — the state PSCs are losing their power. This bothers them.

PSDS Public Switched Digital Service. A BOC service. AT&T Circuit Switched Digital Capability (CSDC), also known commercially as AT&T's Accunet Switched 56 service. It allows a full-duplex, dial up, 56-Kbit/s digital circuits on an end-to-end basis.

Pseudo Code P-CODE. Program code unrelated to the hardware of a particular computer and requiring conversion to the code used by the computer before the program can be executed or acted upon. Here's a more technical explanation. Pseudo Code is a compiled program written for a hypothetical processor and interpreted at runtime by a P-code interpreter written for a native environment. P-code has many objectives in its different implementations, most often portability and space savings.

Pseudo Flaw A loophole planted in an operating system as a trap for intruders.

Pseudo Lite A ground based differential GPS (Global Positioning System) which transmits a signal like that of an actual GPS satellite, and can be used for ranging.

Pseudo Random Bit Pattern A test pattern consisting of 511 or 2,047 bits ensuring that all possible bit combinations can pass through a network without error.

Pseudo Random Test Signal A pseudo random test signal is a signal consisting of a bit sequence that approximates a random signal.

Pseudo Range A distance measurement based on the correlation of a GPS (Global Positioning System) satellite transmitted code and the local receiver's reference code, that has not been corrected for errors in synchronization between the transmitter's clock and the receiver's clock.

Pseudo Ternary A term used in ISDN Basic rate interface data coding. Refers to three encoded signal levels representing two-level binary data (binary "1"s are represented by no line signal, and binary "0"s by alternating positive and negative pulses).

PSI 1. Packet Switching Interface.
2. Pounds per square inch, a unit of air pressure. Telephone cables that are pressurized with nitrogen (because it's not corrosive) are kept at a pressure of around ten to 15 PSI.

PSK Phase Shift Keying. A method of modulating the phase of a signal to carry information. See PHASE MODULATION.

PSN Packet Switch Node. The modern term used for nodes in the ARPANET and MILNET. These used to be called IMPs (Interface Message Processors). PSNs are currently implemented with BBN C30 or C300 minicomputers.

PSP 1. PCS Service Provider.
2. Payphone Service Provider.

PSPDN Packet Switched Public Data Network. A PSPDN is a general purpose data network using packet transmission techniques, as opposed to circuit techniques as used for instance in the PSTN. It is used primarily for communications with or between computers.

Psophometer An instrument arranged to give visual indication corresponding to the aural effect of disturbing voltages of various frequencies. A psophometer usually incorporates a

weighting network, the characteristics of which differ according to the type of circuit under consideration; e.g., high-quality music or commercial speech circuits.

PSTN Public Switched Telephone Network. PSTN is an abbreviation used by the ITU-T. PSTN simply refers to the local, long distance and international phone system which we use every day. In some countries it's only one phone company. In countries with competition, e.g. the United States, PSTN refers to the entire interconnected collection of local, long distance and international phone companies, which could be thousands.

PSU Packet Switch Unit.

Pseudo Cut Through A switching mechanism where a packet is transmitted from its source port to its destination port only after the first 64 bytes of the packet are in the source port and its destination port is determined.

Pseudophone A pay phone that looks like a real Bell telephone company phone but is owned by a smaller phone company that charges exorbitant fees for long-distance calls.

Psychic ANI A term created by Howard Bubb from Dialogic to designate what happens when you call someone on one line while they're calling you on the other.

PT Payload Type: Payload Type is a 3-bit field in the ATM cell header that discriminates between a cell carrying management information or one which is carrying user information.

PTC 1. Portable Teletransaction Computers. These are typically handheld devices used for retail (inventory), healthcare (tracking supplies), mobile field repair (reporting fixes), insurance (visiting car wrecks and other disasters), etc. The devices typically have telecommunications capabilities, sometimes wireless, sometimes landlines. And they typically include microprocessors, memories, displays, keyboards, touchscreens, character recognition software, barcode readers, printers, modems and local and/or wide area data radios.
2. Personal Telecommunications Center. Infocorp's name for a product most people call a PDA, Personal Digital Assistant.
3. Pacific Telecommunications Council. A not-for-profit organization open worldwide to anyone or any entity interested in the Pacific hemisphere and involved with telecommunications, broadcasting, informatics, digital media and associated fields. www.ptc.org

PTE Path Terminating Equipment. SONET network elements that multiplex and demultiplex the payload and that process the path overhead necessary to transport the payload. See also TERMINATING MULTIPLEXERS.

PTI An ATM term. Payload Type Indicator: Payload Type Indicator is the Payload Type field value distinguishing the various management cells and user cells. Example: Resource Management cell has PTI=110, end-to-end OAM F5 Flow cell has PTI=101.

PTMPT Point-To-Multipoint: A main source to many destination connections.

PTN Public Telecommunications Network.

PTO Public Telephone Operator.

PTS Presentation Time Stamp: A timestamp that is inserted by the MPEG-2 encoder into the packetized elementary stream to allow the decoder to synchronize different elementary streams (i.e. lip sync).

PTS Public Telecommunications Systems.

PTSE An ATM term. PNNI Topology State Element: A collection of PNNI information that is flooded among all logical nodes within a peer group.

PTSP An ATM term. PNNI Topology State Packet. A type of PNNI routing packet used to exchange reachability and resource information among ATM switches to ensure that a connection request is routed to the destination along a path that has a high probability of meeting the requested QOS. Typically, PTSPs include bidirectional information about the transit behavior of particular nodes (based on entry and exit ports) and current internal state.

PTT Post Telephone & Telegraph administration. The PTTs, usually controlled by their governments, provide telephone and telecommunications services in most foreign countries. In ITU-T documents, these are the Administrations referred to as Operating Administrations. The term Operating Administrations also refers to "Private Recognized Operating Agencies" which are the private companies that provide communications services in those very few countries that allow private ownership of telecommunications equipment.

PTY Party.

PU Physical Unit. In IBM's SNA, the component that manages and monitors the resources of a node, such as attached links and adjacent link stations. PU types follow the same classification as node types.

PU 2.0 & 2.1 IBM protocols which allow applications written to APCC and interpreted by LU 6.2 to access the mainframe (2.0) and token ring LAN (2.1).

PU Type 2 A physical unit (PU) refers to the management services in SNA node always contains one physical unit (PU), which represents the device and its resources to the network. PU Type 2 is often referred to as a cluster controller.

PU 4 An IBM SNA front end processor.

PU 5 An IBM SNA mainframe, such as a System/370 or System/390. It runs VTAM to handle data communications.

Public Access Terminal A kiosk (SK series), enclosure (TK series) or a system for special enclosures and custom applications (XE Series) which provides the public access to service. Contains a color monitor and keypad for customer interaction.

Public Announcement Trunk Group A trunk group used to provide multiple types of announcements, such as the weather, time and sports results.

Public Asynchronous Dial-In Port A term used in packet switching networks referring to the local phone number of a port into the packet switched network. Some networks provide different numbers for different speeds. Some provide different speeds on the same numbers. See also PRIVATE DIAL-IN PORTS.

Public Data Network A network available to the public for the transmission of data, usually using packet switching under the ITU-T X.25 packet switching protocol. See PACKET SWITCHED NETWORK.

Public Dial Up Port A port on a computer system or on a communications network which is accessible to devices operating over the public switched telephone network.

Public Domain Imagine you write software. You've just written a great program. You now want to sell it. You have two choices. You can take advertisements, sell it to retailers, get distributors to carry it, hire salespeople, etc. In other words, go the commercial route. This is expensive and requires a major marketing / sales budget. The other choice is to go the Public Domain (also called Shareware) route. This involves giving away your software on various bulletin boards, on many Web sites, in "shareware" direct mail catalogs. In short, putting your software in the public domain. People download the software for free and try it. If they like it, they will send you money. They will do this because you offer them an instruction manual, a new version of the software that doesn't blast "unregistered" on the splash screen when you load the soft-

ware, or an upgraded version of the software, with more features, or a chit that assuages your guilt at using unpaid-for software that someone (i.e. you) worked real hard on.

Public Exchange A British word for Central Office. Outside North America, central offices are all called "public exchanges." In the US, a public exchange is typically a local telephone switch. TELECONNECT's phone number in North America is 212-691-8215. The 212 is our area code. The 691 designates the central office or public exchange which serves us. That public exchange belongs to Nynex Company. See also CENTRAL OFFICE and CO.

Public Key Algorithms that encrypt and decrypt using asymmetric (different), yet mathematically-linked keys. Each security module is assigned a pair of keys. The encryption key is "public" and does not require distribution by secure means. The decryption or "private" key cannot be discovered through knowledge of the public key or its underlying algorithm. Public key algorithms can apply to one or more of the following: key distribution, encryption, authentication, or digital signature. See Public Key Encryption.

Public Key Encryption A form of key encryption which uses two keys, a public key (for encrypting messages) and a private key (for decrypting messages) to enable users to verify each other's messages without having to securely exchange secret keys. See Encryption and Public Key.

Public Key Infrastructure PKI. An means by which public keys can be managed on a secure basis for use by widely distributed users or systems. The IETF's X.509 standard is widely accepted as the basis for such an infrastructure, defining data formats and procedures for the distribution of public keys via digital certificates signed by certification authorities (CAs).

Public Network A network operated by common carriers or telecommunications administrations for the provision of circuit switched, packet switched and leased-line circuits to the public. Compare with private network.

Public Room A video conferencing center that is arranged with transmission services. Public rooms can be rented by business customers who wish to have a video conference but do not have facility available at their own offices.

Public Safety Answering Point PSAP. A generic term for the person or group of people who answer 911 emergency phone calls.

Public Service Commission PSC. The state regulatory authority responsible for communications regulation. Also known as Public Utility Commission, Corporate Commission and in some states, the Railway Commission.

Public Switched Digital Service PSDS. A generic name for Bell telephone companies' service offerings that provide customer switches the capability of sending data at 56 Kbps over the public circuit switched network.

Public Switched Network Any common carrier network that provides circuit switching between public users. The term is usually applied to the public telephone network but it could be applied more generally to other switched networks such as Telex, MCI's Execunet, etc.

Public Switched Telephone Network Usually refers to the worldwide voice telephone network accessible to all those with telephones and access privileges (i.e. In the U.S., it was formerly called the Bell System network or the AT&T long distance network).

Public Telephone Station Coin phone. Pay phone.

Public Telephony A new term to describe what might be possible with a new type of public "phone" which could do more than make and receive analog phone calls. Perhaps it could do videoconferencing? Perhaps it could send faxes? Perhaps it could access remote databases? In short, it could do a bunch of multimedia things. Exactly what hasn't been defined, as yet. Having an RJ-11 plug in it so I could plug my laptop into would be a good beginning, however.

Public Utility Commission PUC. State body charged with regulating phone companies. Also called Public Service Commissions. See PUBLIC SERVICE COMMISSION.

Publish To make information public. These days you can "publish" in many ways — from paper to CD-ROM to the World Wide Web (i.e. via the Internet).

Publishing Making resources available to network users.

PUC See PUBLIC UTILITY COMMISSION.

Pull Box A box with a cover inserted in a long conduit run, particularly at a corner. It makes it easier to pull wire or cable through the conduits.

Pull Tension The pulling force that can be applied to a cable without effecting specified characteristics for the cable.

Pullcord or Pullwire or Pullstring A cord or wire placed within a raceway or conduit or ceiling or wall and used to pull wire and cable through.

Pulling Eye A device on the end of a cable to which a pulling line is attached for pulling cable into conduit or duct liner.

Pulling Glass Laying fiber-optic cable.

Pulling Strength Expressed in lbs. The maximum force which may be applied to strength members of a cable. Pulling strength limits are specified for all Belden Fiber Optic cables in the General Line and Fiber Optic catalogs. Affects pulling methods, pulling tension and operation tension.

Pulp A type of older telephone twisted-pair cable whose wood-pulp "paper" insulation is formed on the cable during manufacture.

Pulp Cable Outside telephone cable that uses paper insulation on the twisted copper pairs. Pulp cable is obsolete technology.

Pulpware Books and magazines. They are printed on paper, which is made of wood pulp. Despite the popularity of the World Wide Web, CD-ROM, audio tapes and other media, pulpware never has been so popular, for which fact I am ever so grateful. There is nothing quite so nice as curling up in front of the fire with a good book. Books not only are user-friendly, but they just feel good to the touch. The only problem with pulpware is that it is impossible to fix "bugs" once the book is published. There are some "bugs" in this book. Please send e-mail if you find one; I'll fix it with the next edition.

Pulse 1. Pulse is the dialing mode for outside lines that is traditionally used by rotary dial telephones.
2. A quick change in the current or voltage produced in a circuit used to operate an electrical switch or relay or which can be detected by a logic circuit.

Pulse Address Multiple Access PAMA. The ability of a communication satellite to receive signals from several Earth terminals simultaneously and to amplify, translate, and relay the signals back to Earth, based on the addressing of each station by an assignment of a unique combination of time and frequency slots. This ability may be restricted by allowing only some of the terminals access to the satellite at any given time.

Pulse Amplitude Modulation PAM. A technique for placing binary information on a carrier to transmit that information. PAM is a technique for analog multiplexing. The amplitude of the information being modulated controls the amplitude of the modulated pulses. Samples of each input voltage are placed between voltage samples from other chan-

nels. The cycle is repeated fast enough so the sampling rate of any one channel is more than twice the highest frequency transmitted. See also PAM and PCM.

Pulse Cable A type of coaxial cable constructed to transmit repeated high voltage pulses without degradation.

Pulse Code Modulation PCM. The most common and most important method a telephone system in North America can use to sample a voice signal and convert that sample into an equivalent digital code. PCM is a digital modulation method that encodes a Pulse Amplitude Modulated (PAM) signal into a PCM signal. See PCM and T-1.

Pulse Density Also known as Ones Density. In electrically-based T-Carrier systems, "O"s are represented by zero voltage (i.e., no pulse) and "1"s by alternating positive and negative voltages (pulses). Pulse density refers to the number of no pulse ("O") periods allowed before a pulse ("1") must occur. Typically, no more than 15 no pulse periods ("O"s) are allowed before a pulse ("1") must occur. Pulse density is required by older T-Carrier systems, as repeaters and other network devices rely on it in order to maintain synchronization; some devices also depend on pulse density for power. Not all T-Carrier systems require pulse density; for instance, unchannelized T-Carrier provides clear-channel transmission, without regard for pulse density. See also T-Carrier.

Pulse Density Violation A pulse density violation occurs if a signal contains more than a specified number of zeros, or the percentage of ones in the signal is less than specified.

Pulse Dialing One or two types of dialing that uses rotary pulses to generate the telephone number. See ROTARY DIAL.

Pulse Dispersion The spreading out of pulses as they travel along an optical fiber.

Pulse Distribution Amplifier DA. A device used to replicate an input timing signal, typically providing 6 outputs, each of which is identical to the input signal. May also perform cable equalization or pulse regeneration.

Pulse Duration Modulation PDM. That form of modulation in which the duration of the pulse is varied in accordance with some characteristic of the modulating signal.

Pulse Link Repeater A signaling set that interconnects the E and M leads of two circuits. In E & M signaling, a device that interfaces the signal paths of concatenated trunk circuits. Such a device responds to a ground on the "E" lead of one trunk by applying -48Vdc to the "M" lead of the connecting trunk, and vice versa. This function is a built-in, switch-selectable option in some commercially available carrier channel units.

Pulse Modulation A general method of carrying digital information on any system that uses fixed-frequency pulses of transmit information from the source to the destination. Examples are pulse amplitude modulation and pulse duration modulation.

Pulse Overshoot In T-1, the amount of signal voltage that can remain at the trailing end of a pulse. It can be no more than 10-30% of the pulse amplitude. Also called afterkick.

Pulse Position Modulation PPM. Pulse position modulation is a variation on frequency modulation. In FM the carrier signal's frequency is modulated by the signal. In Pulse position modulation, instead of a continuous carrier signal, Pulses of constant amplitude are transmitted at different frequencies, the frequency of the pulses being modulated by the transmission.

Pulse Repetition Frequency PRF. In radar, the number of pulses that occur each second. Not to be confused with transmission frequency which is determined by the rate at which cycles are repeated within the transmitted pulse.

Pulse Stuffing When timing signals on digital circuits get out of whack, some method of allowing mismatches must be provided. In time division multiplexing, this is called pulse stuffing. One stream of data has bits added to it so its final rate is the same as the master clock.

Pulse To Tone Conversion Most of the world doesn't have touchtone service. They have rotary, make-and-break phone service and phones. Most computer telephony systems — from airline timetable audiotext machines to bank balance dispensers — require the user to punch in touch tone sounds. For people with rotary phones to get access to computer telephony systems, those systems must have a device called a pulse to tone converter — electronic circuitry which counts the clicks made by a rotary phone and converts them into touch tones. This technology is not 100% accurate and should be accompanied by programming which confirms the input. "You just entered 1034. If that is correct, please say YES or dial 1."

Pulse Train The resulting electronic impulses that transmit encoded information.

Pulse Width In T-1, refers to the width (at half amplitude) of the bipolar pulse (typically 324 + or -45 nsec).

Pulse Width Modulation Another but not very common method of modulating a signal, in which an analog input signal's DC level controls the pulse width of the digital output pulses. See PULSE CODE MODULATION and PULSE AMPLITUDE MODULATION.

Pulsenet Alert Transport Service A service from Nynex, PULSENET Alert Transport Service lets monitoring agencies use a business or residence customer's existing POTS telephone line to transport information (burglar, fire, medical, alert, etc.) without interfering with their basic voice telephone service.

Pulsing The method used for transmitting the phone number dialed to a telephone company switching office.

PUMA Product Upgrade Manager.

Punch 1. The process of perforating a paper tape or card in order to code information into machine readable form.

2. The process of connecting jumper interconnection wires on a distribution frame. It is called punching because of the tool which places the wire on the metal post of the frame. It is called a PUNCH and requires a heavy "punch" to make it strip its wires, then connect into the PUNCH-DOWN BLOCK.

Punch Tool Punch tools are used to conduct copper wires to terminations, either jacks, blocks or patch panels. They come in two varieties: non-impact and impact. Non-impact tools are less costly, and they push a conductor into its connector. Impact punch tools have a spring mechanism that delivers a jolt of force to the conductor being punched, which helps ensure the cable is properly seated. Impact punch tools are best for most applications. There are different blades for each kind of termination, including the 110 blade, the 66 blade, the five pair 110 blade, the Krone blade and the BIX blade. See Punch Down Tool.

Punch Down A term used to describe the connection of twisted pair wires to an insulation displacement block. (e.g. a 66 block, or a patch panel). See PUNCH DOWN TOOL.

Punch Down Tool A device used to connect twisted pair copper wires to an insulation displacement block (e.g. a 66 block or a patch panel). the punch down tool consists of a slotted blade attached to a heavy-duty plastic handle. Loop the wire between the prongs of an insulation displacement block, slide the slotted blade over the terminal and give the tool a downward push. A spring loaded mechanism in the handle complete the

job automatically. The blade spreads the prongs just enough, strips the wire, drives it between the prongs, and neatly cuts the wire with a satisfying "ka chunk." Thus you've produced a perfect "gad tight" termination, which is important since the connection might corrode and the corrosion will insulate the connection. Result: Noise (buzz, hum) on the telephone line and bad voice or data transmission. See Punch Tool.

Punchdown Block A device used to connect one group of wires to another. Usually each wire can be connected to several other wires in a bus or common arrangement. A 66-type block is the most common type of punchdown block. It was invented by Western Electric. Northern Telecom has one called a Bix block. There are others. These two are probably the most common. A punchdown block is also called a terminating block, a connecting block, a punch-down block, a quick connect block, a cross connect block. A punchdown block will include insulation displacement connections (IDC). In other words, with a connecting block, you don't have to remove the plastic shielding from around your wire conductor before you "punch it down."

Punchdown Tool A punchdown tool is used to insert cable onto cross-connects, patch panels and jacks. "Punching" a cable means forcing it into an Insulation Displacement Connector (IDC), such as a 66-block. The IDC has replaced Wire Wrap and Solder and Screw Post terminations for connecting conductors to jacks, patch panels and blocks. They pierce the cable jacket to make a connection with the conductor rather than the installer having to strip off the conductor's plastic insulation, saving time. Since IDCs are very small, they can be placed very close together, reducing the size of cross-connects. IDCs are the best termination for high speed data cabling since a gas-tight, uniform connection is made. Most punchdown tools will allow you to install wires on 66-blocks, also known as RJ-21Xs. But there are other IDC blocks. Each of the following types of IDC blocks requires a unique punchdown die for your punch-down tool: (1) 110 Connector; These are very popular connectors for new installations. 110 Patch panels are typically rated Category 5, but always double-check. (2) Krone. Krone cross-connects have a patented 45 degree angled IDC, which typically surpasses standard 66-block and 110 connectors in attainable transmission speed. (3) BIX. BIX IDCs also have superior transmission properties to 66-block and 110 connectors. See Insulation Displacement Connector.

Pure Aloha A random access technique developed by the University of Hawaii in the early 1970s. In this scheme, a user wishing to transmit does so at will. Collisions are resolved by retransmitting after a random period of time. See also Aloha and Alohanet.

Purge Verb. To remove records from a database. To get rid of old voice mail messages.

Push There are essentially two ways of getting information from a Web or Intranet site — push or pull. In the simplest terms, "pull" means that you go into a Web site, and ask for information — typically by clicking on a button. You will see a page of information. You may ask for a file to be sent (downloaded) to you. This is typically called "pull." You are pulling the information from the web site by doing something — i.e. clicking away. "Pushing" is a new technology that involves the web site sending you specific material you had only generally asked for. An example, you're surfing the net. You have told one or more sites that you're on line and ready to receive whenever they are ready to send to you. This information might be sent to you as a bar which scrolls along the bottom of your screen — like the stockmarket quotes that scroll across CNN's financial channels — or it might be sent to you while your screen saver is on. It might replace your screen saver with live information — your stocks, news of your favorite baseball teams, etc. According to contemporary wisdom in the Spring of 1997, there are two types of "push" — Active push and Directed push. Here are two definitions I found of Active and Directed Push: Active: The server on the Web interacts with the client by sending all the content to the client upon the client's request (polling), essentially the way that a client/server application might. PointCast is an example of this. Directed: The server interacts with the push client only occasionally, providing directions (agents, modules, and so on) for how content should be handled or where content is located. The client then gets the information directly from a variety of services and processes it locally. Lanacom Headliner is a good example of this.

Push To Talk 1. In telephone or two-way radio systems, you have to push a button to talk and stop pushing to listen. Typically you say "over" to indicate it's the other person's turn to talk. In terrestrial radio systems such as SMR (Specialized Mobile Radio) and CB (Citizens Band), it is used where the same frequency channel is employed by both transmitters. See also Citizens Band and SMR.
2. A protocol which one must observe to conduct a successful voice conversation over satellite links and IP Telephony networks. Much like SMR and CB radio communications, you must wait a brief moment to make sure that your transmission was received and to give the other person an opportunity to respond before you start to talk again; otherwise, you will overtalk the other person, which phenomenon also is known as "clipping." In GEO (Geosynchronous Earth Orbiting) satellite systems, this protocol is necessary due to propagation delay, which is caused by the fact that the communications satellites are roughly 22,300 miles above the equator; even at the speed of light, it takes a while for the signal to reach the satellite and return to earth. In IP Telephony networks (especially the Internet) this delay, or latency, is the result of the shared nature of the network.
3. A method of payphone operation in which a push button switch is touched by the caller when the called party answers. Once pushed, money is collected and handset microphone turned on. This system deprives the phone company of revenues for calls to 976 numbers, answering machines or answering services.

Push Technology An Internet/WWW (World Wide Web) term describing the reversal of the traditional information gathering paradigm (I hate that word). Instead of seeking out information, information seeks you, based on your demonstrated preferences. Here's how it works: You cruise the Web (Internet or Intranet, seeking certain kinds of information from certain Web sites. You either register your interests or the Web takes note of them. Automatically, you are then presented with notification of changes in that information, as changes take place. You then can access that changed information, or ignore it, as you choose.

Pushbutton Dialing Instead of rotary dialing, buttons are pushed to generate the tones needed to place a phone call. Also called Touchtone and Touch-call. Some pushbutton phones do not produce tones, but generate the dial pulses of rotary dials. Some phones and phone systems will generate both rotary dial pulses and tone signaling. See TOUCHTONE.

Pushbutton Dialing To Stations A special attendant console feature in which the switching system is served by rotary dial central office trunk circuits. A ten-button keyset is provided on the console which allows fast dialing of extension numbers to complete incoming calls.

Pushbutton Originating Register A register used to store information about originating calls with pushbutton signals.

Put-Up Refers to the packaging of wire and cable. The term itself refers to the packaged product that is ready to be stored or shipped.

PVC 1. Premises Visit Charge.
2. PolyVinyl Chloride, a common type of plastic used for cladding telephone cable (except that to be run in plenum ceilings). Or,
3. Permanent Virtual Circuit, a permanent association between two DTEs established by configuration. A PVC uses a fixed logical channel to maintain a permanent association between the DTEs. Once defined and programmed by the carrier into the network routing logic, all data transmitted between any two points across the network follows a predetermined physical path, making use of a Virtual Circuit. PVCs are widely used in X.25 networks, and are the basis on which communications take place in a Frame Relay network. See also SVC.

PVCC Permanent Virtual Channel Connection: A Virtual Channel Connection (VCC) is an ATM connection where switching is performed on the VPI/VCI fields of each cell. A Permanent VCC is one which is provisioned through some network management function and left up indefinitely.

PVDF See POLYVINYLIDENE DIFLUORIDE.

PVN Private Virtual Network.

PVPC Permanent Virtual Path Connection: A Virtual Path Connection (VPC) is an ATM connection where switching is performed on the VPI field only of each cell. A Permanent VPC is one which is provisioned through some network management function and left up indefinitely.

PWB Printed Wire Board.

PWM Pulse Width Modulation. In communications, encoding information based on variations of the duration of carrier pulses. Also called Pulse Duration Modulation or PDM.

PWR Power.

Px64 Pronounced P-times-sixty four. Informal name for the ITU-T family of videoconferencing interoperability standards correctly known as H.261, and addressing codecs and video formats. "P" refers to the value range of 1 through 30, with "Px64" referring to 1 through 30 64Kbps channels for videoconferencing. At a value of 30, the standard address the use of a full E-Carrier facility for video transmission. Px64, or H.261, addresses standards for codecs, as well as video formats. At the upper end, H.261 supports 352 pixels per frame, 288 lines per frame and 30 fps (frames per second). H.261 also addresses much lower levels of capability, offering the advantage of a standard for digital communication between video transmitters and receivers. As a videoconferencing transmitter and receiver go through the process of handshaking, they negotiate the communications protocol, including such issues as compression technique, frame rate, and format. H.261 provides a common standard which disparate devices (not of the same manufacturer) can use for communications. H.261 is part of a family of ITU-T standards known as H.320.

PXE Preboot Execution Environment. See WFM.

Pyramid Configuration A communications network in which the data link(s) of one or more multiplexers are connected to I/O ports of another multiplexer.

Pyramidal Horn A wave guide feed horn (i.e. antenna) in which both opposite faces are tapered.

Q Queue

Q & A Question and Answer. A teleconferencing term. During a lecture style teleconference, typically only the session sponsor can transmit audio; the other participants can listen, only. Q & A allows the other participants to signal via their touchtone pads their desire to ask a question. The session moderator/speaker can accept that request off-line, screen the question, and allow the participant to ask it on-line, as appropriate.

Q Band A range of radio frequencies in the 40 GHz and 50 GHz range, also known as the V Band. The "Q" is a random, arbitrarily-assigned designation with its roots in the context of military security during World War II.

Q Bit The qualifier bit in an X.25 packet that allows the DTE to indicate that it wishes to transmit data on more than one level. It is Bit 8 in the first octet of a packet header. It is used to indicate whether the packet contains control information.

Q.1541 ITU-T Recommendation. UPT Stage 2 for Service Set 1 on IN CS1 Procedures for universal personal telecommunication functional modelling and information flows.

Q.2100 ITU-T Recommendation. B-ISDN signaling ATM Adaptation Layer Overview.

Q.2110 ITU-T Recommendation. B-ISDN Adaptation Layer - Service Specific Connection Oriented Protocol.

Q.2130 ITU-T Recommendation. B-ISDN Adaptation Layer - Service Specific Connection Oriented Function for Support of signaling at the UNI.

Q.2723.6 ITU-T Recommendation. Broadband integrated services digital network (B-ISDN), Extension to the SS7 B-ISDN user Part (B-ISUP): signaling capabilities to support the indication of the statistical bit rate configuration 2.

Q.2725.1 ITU-T Recommendation. B-ISDN user Part - Support of negotiation during connection setup.

Q.2725.4 ITU-T Recommendation. Broadband integrated services digital network (B-ISDN), Extensions to the signaling system No. 7 B-ISDN user Part (B-ISUP) : Modification procedures with negotiation.

Q.2766.1 ITU-T Recommendation. Switched virtual path capability.

Q.2767.1 ITU-T Recommendation. Soft PVC capability.

Q.2931 ITU-T Recommendation. The signaling standard for ATM to support Switched Virtual Connections. This is based on the signaling standard for ISDN.

Q.2934 ITU-T ITU-T Recommendation. Broadband - Integrated services digital network (B-ISDN) digital subscriber signaling system No. 2 (DSS 2) - Switched virtual path capability.

Q.2961.6 ITU-T Recommendation. Additional signaling procedures for the support of the SBR2 and SBR3 ATM transfer capabilities.

Q.2962 ITU-T Recommendation. Digital subscriber signaling system No. 2 - Connection characteristics negotiation during call/connection establishment phase.

Q.2963.3 ITU-T Recommendation. Broadband integrated services digital network (B-ISDN) digital subscriber signaling system No. 2 (DSS

2) connection modification - ATM traffic descriptor modification with negotiation by the connection owner.

Q.2931 ITU-T Recommendation. The signaling standard for ATM to support Switched Virtual Connections. This is based on the signaling standard for ISDN.

Q.699.1 ITU-T Recommendation. Interworking between ISDN access and Non-ISDN access over ISDN user part of signaling system 7 - Support of VPN applications with PSS1 information flows.

Q.700-Q709 ITU-T Recommendation. Messaging Transfer Part (MTP) of SS7.

Q.710 ITU-T Recommendation. PBX Application part of SS7.

Q.711-Q716 ITU-T Recommendation. Signaling Connection Control Part (SCCP) part of SS7.

Q.721-Q.725 ITU-T Recommendation. Telephone User Part (TUP) part of SS7.

Q.730 ITU-T Recommendation. ISDN Supplementary Systems part of SS7.

Q.741 ITU-T Recommendation. Data User Part (DUP) part of SS7.

Q.751.4 ITU-T Recommendation. Network element information model for SCCP accounting and accounting verification.

Q.755.1 ITU-T Recommendation. MTP Protocol tester.

Q.761-Q.766 ISDN User Part (ISUP) part of SS7.

Q.765 ITU-T Recommendation. Signaling System No. 7 application transport mechanism.

Q.765.1 ITU-T Recommendation. Signaling System No. 7 - Application transport mechanism - Support of VPN applications with PSS1 information flows.

Q.771-Q.775 Transaction Capabilities Application Part (TCAP) part of SS7.

Q.791-Q.795 Monitoring, Operations, and Maintenance part of SS7.

Q.780-Q.783 Test Specifications part of SS7.

Q.825 ITU-T Recommendation. Specification of TMN applications at the Q3 interface: Call detail recording.

Q.850 ITU-T Recommendation. Usage of cause and location in the digital subscriber signaling system No. 1 and the signaling system No. 7 ISDN user part.

Q.921 ITU-T Recommendation. Q.921 defines the ISDN frame format at the data link layer of the OSI/ISDN Model. It contains address information. The ITU-T/OSI Layer 2 protocol used in the D channel. It is synonymous with LAPD.

Q.922 Annex A. ITU-T Recommendation defining the structure of Frame Relay frames. All Frame Relay frames entering the network automatically conform to this frame structure.

Q.931 Q.931 is the powerful message-oriented signaling protocol in the PRI ISDN D-channel. It is also referred as ITU-T Recommendation I.451. This protocol describes what goes into a signaling packet and defines the message type and content. Specifically, Q.931 provides:

• call setup and take down.

• called party number, with type of number indication (private or public).

• calling party number information (including privacy and authenticity indicators).

• bearer capability (to distinguish, for example, voice versus data for compatibility check between terminals.

• status checking (for recovery from abnormal events, such as protocol failures or the manual busying of trunks), and

• release of B-channels and the application of tones and/or announcements in the originating switch upon encountering errors.

Q.931 makes it possible to interwork PBX features with features in the public network. In addition to offering users more access to a wider range of services, this interaction, according to Northern Telecom, wi l improve the revenue potential of service providers. Service provided over PRA, using Q.931, include:

• access to the public network, such as equal access, WATS, DDD, international DDD, dial-800 and other special number services and operator assisted calls.

• access to and from such private networks as Northern Telecom's Meridian Switched Network (previously call Electronic Switched Network — ESN), tandem tie networks, and extension dialing network, and

• integration of voice and circuit-switched data traffic (up to 64 Kbps).

The Q.931 protocol also enables corporations to use B-channels — that is voice and data channels — in ways currently not possible. Today, for example, a separate trunk from the PBX to the central office is often required for each different service, such as voice, data, foreign exchange, 800-service. With PRA, one common trunk between the PBX and the central office can carry multiple call types. Moreover each B-channel within the PRA trunk can be assigned dynamically to carry whatever service is needed at the moment.

Q.933 ITU-T Recommendation. The signaling standard for Frame Relay to support SVCs. This is based on the signaling standard for ISDN.

Q-signal In the NTSC color system the Q signal represents the chrominance on the green-magenta axis.

QA Quality Assurance.

QAM, QSAM Quadrature Amplitude Modulation, Quadrature Sideband Amplitude Modulation. A sophisticated modulation technique, using variations in signal amplitude, that allows data-encoded symbols to be represented as any of 16 or 32 different states. Some QAM modems allow dial-up data rates of up to 9600 bits per second.

QBE Query By Example. A database front-end that requests the user to supply an example of the type of data he wants to retrieve. Typically, the user forms a query by filling in a table with examples of the requested information. IBM created QBE in the 1970s to simplify the process of retrieving information from mainframe databases; it was later implemented on the PC platform in such products as dBASE and Paradox. See also SQL.

QBF A test message containing the "Quick Brown Fox" text. Used to test data terminals. The text is "The Quick Brown Fox jump over the lazy dog." It contains every letter of the alphabet. Check it out.

QC Laser Quantum Cascade laser, a new type of semiconductor laser that works like an electronic waterfall. According to Bell Labs who invented the QC laser, it is the world's first laser that can be tailored to emit light at a specific wavelength at nearly any point over a very wide range from the mid- to far-infrared spectrum. This can be done by simply varying the layer thickness of the laser, using the same combination of materials. Conventional semiconductor lasers, widely used in other applications such as lightwave communications and compact disk players, operate at wavelengths from the near

infrared to the visible. When an electric current flows through the QC laser, electrons cascade down an energy staircase. Every time they hit a step they emit an infrared photon, or light pulse. At each step, the electrons make a quantum jump between well defined energy levels. The emitted photons are reflected back and forth between built-in mirrors, stimulating other quantum jumps and the emission of other photons until the amplified pulse escapes the laser cavity. The QC laser was invented by Federico Capasso and Jerome Faist in collaboration with Debbie Sivco, Carlo Sirtori, Al Hutchinson and Al Cho, according to AT&T Bell Labs.

QCIF Quarter Common Intermediate Format, a mandatory part of the ITU-T's H.261 standard which requires that non-interlaced video frames be sent with 144 luminance lines and 176 pixels at a rate of 30 fps (frames per second). QCIF provides approximately one quarter the resolution of CIF, but requires about one quarter the bandwidth. It works quite nicely for small-screen display devices.

QD Queuing Delay: Queuing delay refers to the delay imposed on a cell by its having to be buffered because of unavailability of resources to pass the cell onto the next network function or element. This buffering could be a result of oversubscription of a physical link, or due to a connection of higher priority or tighter service constraints getting the resource of the physical link.

QDOS In 1980 IBM showed up on Bill Gates' doorstep seeking an operating system for its upcoming personal computer. Mr. Gates did not have one. But he knew someone that had one. A little firm down the road (in Seattle) had developed QDOS — the Quick and Dirty Operating System. It looked just right for IBM's PC. Mr. Gates bought QDOS for $100,000 and renamed it MS-DOS — Microsoft Disk Operating System. According to the Economist Magazine of May 22, 1993, some jealous Microsoft rivals claim that MS-DOS now stands for Microsoft Seeks Domination Over Society.

QDU Quantizing Distortion Units. ITU-T Recommendation G.113 defines one QDU as the amount of degradation introduced into a voice channel by a single conversion from analog to PCM and back to analog (analog-PCM-analog). Where several voice channels are connected in tandem, the end-to-end QDU rating for the whole circuit is calculated by adding the number of conversions from analog to PCM and back. For example: analog - PCM - analog - PCM - analog introduces 2QDUs.

QFC Quantum Flow Control. A method of flow control for ATM proposed as an alternative to ER (Explicit Rate), the current rate-based flow control mechanism. QFC is proposed by the QFC Alliance, which consists of a group of vendors including Digital Equipment Corp., Thomson-CSF and Ascom Nexion. QFC is touted as the solution for flow control in long-haul ATM implementations where data traffic is supported in addition to widely varying levels of high-priority traffic such as voice, video and multimedia information. QFC provides assurances that the buffers in the destination switch will not overflow, which would require retransmissions of low priority data (i.e., data), while the higher-priority data (i.e., voice, video and multimedia) flows through the network without difficulty. QFC also establishes limits on the bandwidth available to any individual connection, thereby avoiding the potential for monopolization of the switch buffers. See also FLOW CONTROL. Compare with RATE-BASED FLOW CONTROL and ER.

QLLC Qualified Logical Link Control. Software package that allows Systems Network Architecture (SNA) commands to be transmitted over an X.25 packet data network (PDN). See also NPSI. Contrast with DSP.

QMS Queue Management System.

QNX A UNIX-like operating system that really works well for computer telephony applications.

QoR Query on Release. See LNP (Local Number Portability.)

QoS Quality of Service. See QUALITY OF SERVICE.

QPSK Quaternary Phase Shift Keying. A compression technique used in modems and in wireless networks, such as CDMA. A simple implementation of QPSK allows the transmission of 2 bits per symbol, with each symbol being a phase range of the sine wave. In this fashion, a 2:1 compression ratio is achieved, resulting in a doubling of the efficiency with which a circuit is employed. For instance, 0-90 degrees of phase indicates a 11 bit pattern; 90-180 degrees a 01; 180-270 a 10; and 270-360 a 00. In wireless networks, two carrier signals can be used, each of which is separated by 90 degrees of phase (position). If the phase of the carrier signals were not separated, one would be indistinguishable from the other. A 90-degree phase shift provides maximum phase separation and, therefore, maximum delineation between the carrier signals.

QPSX Queued Packet Synchronous Exchange. Medium Access Control technology developed by the University of Western Australia for use in extending the reach of LANs across a Metropolitan Area Network (MAN). The technology was licensed to QPSX, Ltd. and subsequently was standardized by the IEEE as 802.6. QPSX was commercialized by Bellcore as DQDB, which is the access technology for SMDS networks. See also DQDB and SMDS.

QRSS Quasi-Random Sequence Signals. Signals used for testing digital circuits, n particular DS-1 (i.e. T-1) circuits.

QSAM Quadrature Sideband Amplitude Modulation. A sophisticated modulation technique, using variations in signal amplitude, that allows data-encoded symbols to be represented as any of 16 or 32 different states.

QSIG An emerging signaling and control standard for PINX-to-PINX (Private Integrated Network eXchange) applications; it is intended as a global standard for use in private corporate ISDN networks. "Q" comes from the fact that the standard is an extension of the "Q" logical reference point defined by the ITU-T in its Q.93x series of recommendations for generic functions and basic services of ISDN SIGnaling systems. The early work on QSIG was accomplished by the European Computer Manufacturers Association (ECMA), which built on ITU-T ISDN standards for public networks. As a result, and for obvious reasons, QSIG, therefore, builds on the ITU-T DSS1 (Digital Subscriber Signaling 1) standard. DSS1 defines the logical reference point for ISDN at the user equipment. The impetus for this effort was that of encouraging the harmonization of existing, proprietary private network "standards" toward the reduction of technical trade barriers in the pan-European market. Subsequently, the EC (European Commission) became involved, charging ETSI (European Telecommunications Standards Institute) with the responsibility for further development and promotion of the standard in collaboration with CENELEC (translated from French as European Electrotechnical Standards Committee). QSIG standards are submitted to the JTC1 (Joint Technical Committee 1), which is a collaboration of the ISO (International Standards Organization) and the IEC (International Electrotechnical Commission). The standards also are promoted by the IPNS (ISDN PBX Networking Specification) Forum, which comprises a number of manufacturing companies such as Alcatel, Ascom, Ericsson, Lucent, Nortel, Philips and Siemens.

QSIG is much like the public network DSS1 standard set by the ITU-T, at least at Layers 1 & 2 of the OSI Reference Model. Differences appear at Layer 3, the Network Layer, as QSIG is intended for use in private networks and is symmetrical in nature, with the user side and the network side being identical. Further, QSIG is designed for peer-to-peer operation, although the standard addresses transit node capabilities, as well. QSIG also addresses both connection-oriented and connectionless services, unlike DSS1 standards which address only the former. ECMA currently is working on B-QSIG, which will extend the QSIG protocol stack to B-ISDN (Broadband ISDN). According to the IPNS Forum, QSIG offers user benefits including vendor independence, guaranteed PBX interoperability, free-form network topology, support for an unlimited number of nodes, flexible numbering plan, flexibility of interconnecting transmission technologies (i.e., analog or digital leased lines, radio and satellite links, and public VPN services). Supplementary services offered by QSIG include name identification, call intrusion, do not disturb, path replacement, operator services, mobility services, and call completion on no reply. As a standards recommendation, QSIG provides manufacturers the freedom to develop custom features, with QSIG providing a standard mechanism for transporting such non-standard features. See also CENELEC, EC, ECMA, ETSI, IEC, ISO, and OSI REFERENCE MODEL.

QTC Quick Time Conference. Apple Computer's cross-platform, video-conferencing, collaborative computing and multimedia communications technology.

QTVR See QUICKTIME VR.

Quad A slang term for cable conductor with four single, plastic coated wires not twisted together and contained in a single plastic covering. Quad wiring has been traditionally used inside houses and small offices. Since it will not handle data well, it is no longer being recommended for installation anywhere, except in single-line analog (never data) applications. In the old days, a quad wire would support two analog phone lines. Color coding in quad wire in North America is red-green, yellow-black. When I showed this definition to a professional installer, he told me that quad wire was generally not used anymore except by ignorant do-it-yourselfers, cheap telcos (telephone companies), irresponsible contractors, etc. See QUAD WIRE.

Quad Fiber Cable A cable consisting of four single optical fiber cables placed inside a polyvinyl chloride jacket with a rip cord to peel back the jacket and gain access to each single cable.

Quad Inside Wire Quad IW. Older phone wire. It has four solid core copper conductors — red, green, black, yellow. Line one colors are green and red, line two colors are yellow and black. Since it's often not twisted, it's susceptible to RFI.

Quad Lock Conduit Conduit that's designed to be buried. The four conduits let companies lease space to each other in a way that's easy to track for fiber-optic cable installers/splicers, etc.

Quad Wire A type of wire which contains four untwisted copper conductors in a plastic sheath. These four conductors are not two separate twisted pairs, although the four may have a very "slow" twist to them. Quad wiring is no longer recommended by the telephone industry for installation in other than analog single line applications. In short, quad is dead. See QUAD.

Quadded Cable A cable in which at least some of the conductors are arranged in the form of a quad.

Quads See Mated Pairs.

QUALDIR QUALification DIRective. A wireless term for changes to a VLR (Visitor Location Register), a database which contains information about legitimate roamers and which describes the features to which they have access. The response to the QUALDIR is a "qualdir" (lower case). See also VLR.

Quality Of Service QoS. Quality of Service is a measure of the telephone service quality provided to a subscriber. It's not easy to define "quality" of telephone service. It's very subjective. Is the call easy to hear? Is it "clear?" Is it loud enough, etc.? The state Public Service Commissions (PSCs) have attempted to define the quality of service they want the residents of their states to have. And they have created various measures to which they insist phone companies conform. They tend to be more measurable. They include the longest time someone should wait after picking up the handset before they receive dial tone (three seconds in most states).

Quality of Service is more easy to define in digital circuits, since you can assign specific error conditions and compare them. For example if you were defining QoS with respect to ATM, it would be defined or an end-to-end basis in terms of the attributes of the end-to-end ATM connection, as detailed in ITU-T Recommendation I.350. The ATM Forum extended this standard through the definition of QoS parameters and reference configurations for the User Network Interface (UNI). ATM Performance Parameters include the following:
- Cell Error Ratio (CER)
- Severely Errored Cell Block Ratio (SECBR)
- Cell Loss Ratio (CLR)Cell Misinsertion Rate (CMR)
- Cell Transfer Delay (CTD)
- Mean Cell Transfer Delay (MCTD)
- Cell Delay Variability (CDV)

ATM Quality of Service (QoS) objectives set by the carriers are defined as Class of Service 1, 2, 3, and 4. Here is an explanation of the various classes: Class 1: Equivalent to digital private lines. Class 2: Supports traffic such as audioconferencing, videoconferencing and multimedia Class 3: Addresses connection-oriented protocols such as SDLC and Frame Relay Class. 4: Supports connectionless data protocols such as SMDS.

In the middle 90s, the concept of carrying voice and video over IP (Internet Protocol) networks suddenly became very important. In a White Paper which Microsoft put out in September 1997, it discussed QoS with the following words: "What is Quality of Service? In contrast to traditional data traffic, multimedia streams, such as those used in IP Telephony or videoconferencing, may be extremely bandwidth and delay sensitive, imposing unique quality of service (QoS) demands on the underlying networks that carry them. Unfortunately, IP, with a connectionless, "best-effort" delivery model, does not guarantee delivery of packets in order, in a timely manner, or at all. In order to deploy real-time applications over IP networks with an acceptable level of quality, certain bandwidth, latency, and jitter requirements must be guaranteed, and must be met in a fashion that allows multimedia traffic to coexist with traditional data traffic on the same network.

Quantization The converting of a native analog signal to digital format through a sampling and quantizing process. This process is accomplished in a CODEC and is necessary in order to send analog data (voice or video) over a digital network (e.g., T-carrier or ATM) or through a digital switch (e.g., PBX or central office).

In the case of a voice signal and using PCM (Pulse Code Modulation), for instance, the amplitude of the native analog signal is sampled 8,000 times per second, with the each sampled amplitude value being expressed as an 8-bit digital value (byte) consisting of a specific combination of ones and zeros. At the receiving end of the communication, the process is reversed, with the digital value being translated into an analog amplitude value. The result is an approximation of the original analog signal, as it was sampled rather than digitized exactly. Note that the original analog signal varied continuously in terms of both amplitude and frequency. Clearly, the higher the rate of sampling, the truer the reproduced approximate signal; the lower the rate of sampling, the less accurate the reproduced signal. In other words, a low rate of sampling would yield relatively unpleasant voice or fuzzy video as a result of what is known as "quantizing noise." However, a lower rate of sampling requires less bandwidth over the network or through the switch, yielding obvious cost benefits. As is that case with many things in life, there are tradeoffs between cost and quality.

Quantization Noise Signal errors which result from the process of digitizing (and therefore ascribing finite quantities to) a continuously variable signal. See QUANTIZATION.

Quantize The process of encoding a PAM signal (Pulse Amplitude Signal) into a PCM signal (Pulse Code Modulation). See QUANTIZATION.

Quantizing The second stage of pulse code modulation (PCM), for instance. The waveform samples obtained from each communication channel are measured to obtain a discrete value of amplitude. These quantized values are converted to a binary code and transmitted to a distant location to reconstruct an approximation of the original waveform. See QUANTIZE and QUANTIZATION.

Quantizing Noise Noise caused by the inability of an analog signal to be exactly replicated in digital form. Such noise is the result of the fact that the original signal was sampled, yielding an approximation (but not exact replica) of the original signal as it is reconstructed on the receiving end of the communication.

Quantum In physics, quantum means a very small indivisible piece of energy. This word is widely misused by people who refer to "a quantum leap," meaning a big leap. See Quantum Leap.

Quantum Flow Control An ATM term. See QFC.

Quantum Leap In physics, quantum means a very small parcel or increment of energy. Also in physics, quantum leap or quantum jump refers to the abrupt transition (of something such as an electron, atom or molecule) from one discrete energy state to another. In popular usage, the term refers to an abrupt change, dramatic advance, or sudden increase. For instance, it might be said that major system enhancements which entail "forklift upgrades" involve quantum leaps in cost. Systems which are scalable do not. See also FORKLIFT UPGRADE and SCALABLE.

Quarter Speed An international leased teletype line capable of transmitting one quarter of Telex speed of 16 2/3 words per minute.

Quarter Wave Antenna An antenna, the length of which is 1/4 that of the wave length received.

Quartz Crystal A small piece of quartz which is cut to a precise size. When electricity is applied to the crystal, it vibrates at a specific and precise frequency. Quartz crystals are often used in watches. They vibrate quickly and make the watch far more accurate than a timing device which vibrates far more slowly, like a pendulum, for example or a tick tock watch.

Quasi-Random Signal QR. A pseudo random test signal

that has artificial constraints to limit the maximum number of zeros in the bit sequence.

Quenched Gap A spark gap so arranged that the spark is quenched quickly by a cooling effect. A method used to give impulse excitation. An old radio term.

Query 1. In data communications, it's the process by which a master station (or mainframe or boss computer) asks a slave station to identify itself and tell its status, i.e. is it busy, alive, OK, waiting, etc.?

2. In database, a query is a request for the retrieval of data.

Query Language A programming language designed to make it easier to specify what information a user wants to retrieve from a database.

Queue A stream of tasks waiting to be executed. A series of calls or messages waiting for connection to a line. See QUEUING.

Queue Management In a network, tasks like retrieval and writes to a jukebox come randomly from all the users. These tasks vary in urgency — retrievals are higher priority than writes, for example. Queue management sorts out requests from the network by priority. Queue management also enhances the performance of a jukebox, by intelligently re-ordering requests. For example, if there are three requests for images on platter 1 and two from platter 2 and the another from platter 1, queue management means the requests from platter 1 will get handled together, then go to platter two. Sometimes it's called "elevator sorting" — responding to requests in logical order, not in the order in which they were made.

Queue Service Interval The maximum length of time a queue will go unsampled.

Queued Arbitrated QA. A type of SMDS time slot which supports asynchronous data traffic. As such data is not time-sensitive, it is acceptable for the data transmission to take place over time slots on an "as available" basis.

Queued Call A call that is waiting in a queue of telephone calls to be serviced by a system resource is a queued call. An ACD group is an example.

Queued Mode Calls entering an Automatic Call Distributing system wait in a queue are presented, one at a time, to the first idle trunk in the chosen group.

Queued Packet Synchronous Exchange See QPSX.

Queued Telecommunications Access Method QTAM. A program component in a computer which handles some of the communications processing tasks for an application program. QTAM is employed in data collection, message switching and many other teleprocessing applications.

Queuing The act of "stacking" or holding calls to be handled by a specific person, trunk or trunk group. There are two reasons to queue telephone calls:

1. Because you simply don't have enough trunks.

2. Because you want to save money.

You can queue calls mechanically using your telephone switch or manually using a human operator or attendant. There are two ways you can queue calls — hold-on or callback. In "hold-on" queuing, you dial, you get some queuing tone (or the operator tells you you're being queued), then you wait on-line until a line becomes free and you're connected. In "call-back" queuing, you tell the operator or the machine you want to dial a call. And you hang up. When the line becomes free you are called back and connected. There are advantages and disadvantages to both systems. In "hold-on" queuing, you waste your time but save on phone time. In callback queuing, you waste less of your time, but more phone line time. In call-back queuing, the operator or the phone sys-

tem has to grab the line you want and simultaneously call you. By then, you may have left your desk. The call may be wasted, etc. The line given to you could have been used by someone else, etc. Queuing calls as a method to save money on long distance calling makes sense ONLY:

1. IF you are out of trunks because of a temporary surge in telephone traffic — perhaps at your peak, peak busy time and it's very expensive to buy sufficient "cheap long distance" trunks to handle every conceivable peak, and

2. IF you never plan on having a queue longer than 20 seconds for a hold-on queue and 60 seconds for a call-back queue and

3. IF you are queuing calls into an expensive fixed-cost line. For example a tie line between New York and London. If you queue calls into a variable cost line, like an interstate WATS line, you will save money over throwing the call onto DDD, but the pennies you save usually won't be worth it — considering the aggravation you're going to cause your people. Queuing is a very sensitive subject in corporate telecommunications departments. People don't like to wait for telephone lines. They consider that insulting to them personally, damaging to their "productivity" and to heck with the cost. Queues do, however, make enormous sense. Even a queue as short as ten seconds can save big amounts of money. Queues of a maximum length of ten seconds are rarely noticed. These days some of the more modern PBXs will allow you to offer "selective" queuing, or levels of queuing. Upper management doesn't have to queue for the cheap long distance lines before it's bounced to the expensive ones. While lower management has to wait up to 30 seconds. And the worker bees (non-management) have to wait even longer. Queuing is also used on incoming trunks. See ACD and QUEUING THEORY.

Queuing Theory The study of the behavior of a system that uses queuing, such as a telephone system. Much of queuing theory derives from the science of Operations Research (OR). Dr. Leonard Kleinrock has written the authoritative books on the subject. He is probably a genius. His books are very difficult to understand for laymen. Here is an explanation of Queuing Theory from James Henry Green's Dow Jones-Irwin Handbook of Telecommunications (you can buy a copy from www.amazon.com or www.barnesandnoble.com):

"The most common (telephone) network design method involves modeling the (phone) network according to principles of queuing theory, which describes how customers or users behave in a queue. Three variables are considered in network design. The first is the arrival or input process that describes the way users array themselves as they arrive to request service...The second variable is the service process, which describes the way users are handled when they are taken from queue and admitted into the service providing mechanism. The third method is the queue discipline, which describes the way users behave when they encounter blockage in the network...Three reactions to blockage are possible:

• Blocked calls held (BCH). When users encounter blockage, they immediately redial and reenter the queue.

• Blocked calls cleared (BCC). When users encounter blockage, they wait for some time before redialing.

• Blocked calls delayed (BCD). When users encounter blockage, they are placed in a holding circuit until capacity to serve them is available. See QUEUE.

"Traffic engineers have different formulas or tables to apply, corresponding to the assumption about how users behave when they encounter blockage." See POISSON.

After I wrote the above definition, Lee Goeller, a noted traffic engineering expert contributed the following definition:
The study of systems in which customers wait in line for servers to become available, the "blocked calls delayed" condition in telephony (see TRAFFIC ENGINEERING). Although seldom used in designing voice networks (other techniques are usually more cost-effective), queuing is very important in the design of packet networks where speed of transmission more than offsets the delay of waiting for a transmission facility to become available, and in staffing Automatic Call Distributors.

QUICC Quad Integrated Communications Controller. A Motorola term.

QuickDraw Programming routines that allow an Apple Macintosh computer to display graphics on a screen. QuickDraw is also used for outputting text and images to printers not compatible with PostScript.

Quicksilver Mercury, an extremely poisonous chemical. See "mad as a hatter."

QuickTime A dynamic-data format developed by Apple to be used for animation. QuickTime files can be used in documents created by other applications. For instance, a QuickTime video clip can be pasted into a word-processing document. QuickTime VR is the new Apple standard for Virtual Reality. See QuickTime VR.

QuickTime VR QuickTime VR (QTVR) is the acknowledged standard for creating and viewing photo-realistic environments (panoramas) and real-world objects on Mac OS and Windows computers. Users interact with QuickTime VR content with a complete 360 degree perspective and control their viewpoint through the mouse, keyboard, trackpad or trackball. Using a QTVR-enabled authoring tool, panoramas and objects are automatically 'stitched together' from digitized photographs or 3D renderings to create a realistic visual perspective. The effect is awesome. As I wrote this, over 5,000 web sites were QuickTime enabled. One site I particularly enjoyed showed a hotel room. As a you moved your cursor, it seemed as though I was turning to see all 360 degrees of the hotel room. For more info http://quicktimevr.apple.com.

Quick Connect Block Also called a 66-block or punch-down block. It's a 18" piece of metal and plastic which allows you to connect telephone wiring coming from two remote points. The quick-connect block has multiple metal "jaws" ranging horizontally and vertically. To connect up, you "punch" (or push) a wire between the two metal teeth of the "jaws." This both holds it firm and strips the wire's insulation, thus allowing for a good electrical connection. (There are special "punch-down" tools for punching wires into 66-blocks.) On a 66-block, one horizontal row of "jaws" is always the same conductor. To connect other wires to it, you simply punch those wires down along the row. Some 66-blocks have a gap between one side of the 66-block and the other. To connect one wire on one side to the wire on the other side, you have to use a BRIDGING CLIP. This is a small metal clip about one inch long. The bridging clip has one purpose: you can slip it off easily and thus cut one side of the circuit from the other. For example, if you connected central office trunks on one side of the 66-block and a PBX on the other, by removing the bridging clips, you can tell instantly if the trouble is in the PBX or in the central office. Two conductors on a 66-block makes a circuit — a trunk or a line. Therefore, the trunk 212-691-8215 (our main number) takes up the first two horizontal rows on our 66-block. The second two horizontal rows are taken up with 212-691-8216 and so on.

It is good to learn where your main 66-block is — the one that connects you to the telephone company's central office lines. The 66-block is what the telephone company calls the "demarcation point." And they (the phone company) usually install the 66-block. On one side (the trunk side) of their block, they're responsible. On the other (the PBX, key system or phone side), you're responsible. By knowing how to test your lines at this point, you can know whose fault it is — the phone company's or your equipment's. This can avoid having to wait until the phone company arrives, discovers it's not their problem and then sends you a hefty bill. Or the interconnect company arrives, finds out it's not their problem, and sends you a hefty bill, etc.

Quick Connect Blocks or 66-blocks are found in the Main Distribution Frame — where lines coming out of the PBX are connected to the individual wires going to the phones, or to big cables going to clumps of phones in other parts of the building. They're also found in Satellite Distribution Frames where they take big cable coming in from the main distribution frame and connect it to the individual cable pairs going to the individual phones. See CONNECTING BLOCK.

Quick Format A DOS program which deletes the file allocation table and root directory of a disk but does not scan for disk for bad areas.

Quick Plug A device which adapts a standard four wire telephone cord into a modular connector.

Quicktime Apple Computer's video environment (like Microsoft's Video For Windows). Quicktime video files must be converted to *.AVI format to run under Microsoft's Video For Windows. Indeo video technology is supported under MacOS.

Quiescent A fancy word for quiet. No noise. No activity. Quiescent time is the best time to write this dictionary. Sadly, it wasn't always to be.

QUIPU This term is not an acronym. This public domain X.500 directory service, developed by University College London, demonstrates X.500 feasibility on TCP/IP (Transfer Control Protocol/Internet Protocol). This pioneering software package was developed to study the OSI (Open Systems Interconnections) Directory and provide extensive pilot capabilities. ISSUE (International Organization for Standardization Development Environment) provides commercial version of this software.

Quorum A family of teleconferencing products linked in a system designed to meet a customers teleconferencing needs.

QWERTY The name for a computer or typewriter keyboard. It got its name from the left side, top row of letter keys which spell QWERTY. One theory for the strange design of the QWERTY keyboard has to do with typewriter keyboards which had long metal arms that physically hit the paper. To keep the arms from jamming, they designed the QWERTY keyboard which split commonly used letters — i.e. a, i, o, e — to opposite sides of the keyboard. For years, people have argued that a keyboard called the Dvorak would be a much faster and more efficient. However, the U.S. General Service Administration in the 1950s contradicted the claims made by advocates of the Dvorak keyboard. The chief advocate was the patent owner, August Dvorak. According to September, 1995 Upside Magazine, his "book on the relative merits of QWERTY versus his own keyboard has about as much objectivity as a modern infomercial found on late night TV." With computers, there is no such thing as a "standard" QWERTY keyboard. Computer keyboards typically have 20 to 30 more keys than "standard" typewriter keyboards. Many of the keys are unique

— on some keyboards, not on others. Many of the keys on computer keyboards are called "function" keys. If you hit one of them, they might perform a complete function on the computer, e.g. save a file, move to the end of the file, etc. There is absolutely no such thing as a standard computer keyboard.

QZ Special billing arrangement provided by your local telephone company. Before there was automatic call accounting and before there was Centrex, the phone company would give you "time and charges" on every outgoing call. This service was called "QZ" billing. It was used by engineers, lawyers, accountants, consultants and other service people who had to bill their calls back to their clients.

R Interface An ISDN term. The 2-wire physical interface which is used for termination between a TA (Terminal Adapter) and TE2 (Terminal Equipment type 2), which is non-ISDN compatible terminal equipment. The physical connection generally follows either the RS-232 or the V.35 specification in terms of its electrical, functional and physical characteristics. TE2 can be in the form of a telephone set, a PC or a fax machine; none of these devices are ISDN-compatible unless specially equipped to be so TE1 is ISDN-compatible terminal equipment, generally at significant additional cost. See also ISDN, S Interface, T Interface, TE1, TE2, and U Interface.

R Reference Point An ISDN reference point between non-ISDN terminal equipment (TE2) and a terminal adapter (TA). Non-ISDN (TE2) terminal equipment connects to ISDN at the R-reference point through a terminal adapter.

R&D Research & Development.

R&E Research & Education.

R-Y A designator used to name one of the color signals (red minus luminance) of a color difference video signal. The formula for deriving R-Y from the red, green, and blue component video signals is .70R - .59G - .11B.

R/T Internet-speak for Real Time. R/T means the time it takes to download stuff. Writing in the New York Times, Charles McGrath said it was "customary in Net-speak to make a distinction between r/t, or real time — the time in which all these delays and jam-ups occur and v/t, or virtual time, which is time on the Net: a kind of external present in which it is neither day nor night and the clock never ticks. V/t is time without urgency, without priority."

R1 The ITU-T's name for a particular North American digital trunk protocol that happens to use multi-frequency (MF) pulsing. Some Europeans refer to any North American MF signaling protocol as R1 when distinguishing it from their own R2. See R2, MULTI-FREQUENCY PULSING.

R2 A whole series of ITU-T specs which refers to European analog and digital trunk signaling. It refers to a type of trunk found in Europe which uses compelled handshaking on every MF (multi-frequency) signaling digit.

RA Rate Area

RA Number Same as Return Material Authorization Number, or RMA. A code number provided by the seller as a prerequisite to returning product for either repair or refund. An indispensable tracking procedure, it operates like a purchase order system. If you return computer or telephone equipment without an RMA, chances are your equipment will be lost.

RAC Remote Access Concentrator. A RAS is a larger Remote Access Server. According to Mark Galvin, president of RAScom, many people are now distinguishing between RAS (Remote Acccess Server) and RAC (Remote Access Concentrator). There seems to be two different cut-offs for the transition from RAS to RAC. The first, I believe originally defined by Dataquest, is when the port count exceeds 12. The industry seems to be adopting a different cut-off at anything T-1 or bigger as a RAC. A remote access server or remote access concentrator is a piece of computer hardware which

sits on a corporate LAN and into which employees dial on the public switched telephone network to get access to their email and to software and data on the corporate LAN (e.g. status on customer orders). Remote access servers are also used by commercial service providers, such as Internet Access Providers (ISPs) to allow their customers access into their networks. Remote Access Servers are typically measured by how many simultaneous dial-in users (on analog or digital lines) they can handle and whether they can work with cheaper digital circuits, such as T-1 and E-1 connections. See also Remote Access Concentrator and Universal Edge Server.

RACE An association in the European Economic Community. RACE stands for Research and development for Advanced Communications in Europe.

Raceway Metal or plastic channel used for loosely holding electrical and telephone wires in buildings. A raceway is usually located in the floor and is encased on three or four sides by concrete. A raceway is used for interior wiring and performs the same job as a conduit but is typically larger.

Raceway Method A ceiling distribution method in which open or closed metal trays are suspended in false ceilings from the structural floor above. The raceway method is generally used in large buildings or for complex distribution systems that demand extra support. When closed metal trays are embedded in the floor, this distribution method is often called underfloor raceways. See also CEILING DISTRIBUTION SYSTEMS and UNDERFLOOR DUCT METHOD.

Rack 1. An equipment rack. In our industry, the standard equipment rack is 19 inches (48.26 cm) wide at the front. Much equipment is designed to fit into a standard rack. A rack is typically made of aluminum or steel, onto which equipment is mounted. A rack is typically attached to a building ceiling or wall. Cables are laid in and fastened to the rack. Sometimes a rack is called a tray. What a rack is to equipment, so a frame is to wiring. See also DISTRIBUTION FRAME. 2. Rack (the digits). a term which implies the storing or registering of numerical data. See Register.

Rack Unit RU. Unit of measure of vertical space in an equipment rack. One rack unit is equal to 1.75 inches (4.45 cm).

Rackmount Designed to be installed in a cabinet, usually 19" wide.

RACON RAdar transponder beaCON. Short-range navigation devices that provide target images on a ship's maritime navigation radar system. The transponder beacons transmit, either automatically or in response to a predetermined received signal, a pulsed radio signal with specific characteristics. RACONs generally operate in the 9300-9500 MHz band, and are used to identify specific locations such as hazards to navigation; think of them as replacements for lighthouses and you won't be far off. Most RACONs are operated by the U.S. Coast Guard. See also Radar.

Rad 1. The unit used to measure the absorption of ionizing radiation.
2. A British Term. Recorded Announcement Device, a device which automatically answers a line and delivers a pre-recorded message. Often used to tell a caller to a telebusiness unit

that the call is in a queue and will be dealt with soon. More sophisticated RADs gather information, take messages or work in conjunction with interactive fax machines.

3. An abbreviation for Rapid Application Development. Most relate it to a quick programming environment.

Radar RAdio Detection And Ranging. See RADAR DETECTOR.

Radar Detector Picture a trooper sitting in his car aiming his radar gun down the highway. The gun emits a beam of electrons at microwave frequency. Those beams bounce off approaching vehicles and reflect back to the trooper's radar at an altered frequency (the Doppler Effect). By measuring the change in frequency, the trooper calculates the speed of the oncoming vehicle. The trouble is the radar beam fans out like a searchlight. At a distance of 1,000 feet, the beam is about as wide as the highway itself. That makes it difficult for the trooper to know which vehicle he's tracking.

Also, his reading can be thrown off by any number of operating errors or by interference from power lines, neon lights or even the fan motor in the trooper's car. According to some estimates, Esquire Magazine reported, as many as 30% of all radar-generated speeding tickets were given in error. In 1979 a Miami TV station showed a police radar clocking a house going 28 miles per hour and a banyan tree doing 86! Radar detectors are very much like FM receivers. They can pick up radar signals more than a mile from the source. At that distance the beam is too weak to bounce all the way back to the trooper's car but strong enough to make the detector beep.

Radar Screen 1. A CRT (cathode ray tube) showing

Radial Acceleration The rate at which a track on an optical disc accelerates toward and away from the center, because it is not perfectly aligned or perfectly round.

Radiant Energy Energy as measured in joules which is transferred via electromagnetic waves. There is no associated transfer of matter. And typically the giver or energy and the receiver of energy are not touching.

Radio RF. System of communication employing electromagnetic waves propagated through space. Because of their varying characteristics, radio waves of different lengths are employed for different purposes and are usually identified by their frequency. The shortest waves are the highest frequency, or numbers of cycles per second; the longest waves have the lowest frequency, or fewest cycles per second. In honor of the German radio pioneer Heinrich Hertz, his name has been given to the cycle per second (hertz, Hz); 1 kilohertz (Khz) is 1000 cycles per second, 1 megahertz (Mhz) is 1 million cycles per second, and 1 gigahertz (Ghz) is 1 billion cycles per second. Radio waves range from a few kilohertz to several gigahertz. Waves of visible light are much shorter. In a vacuum, all electromagnetic waves (but not audio waves) travel at a uniform speed of about 300,000 km (about 186,000 miles) per second.

Radio waves are used not only in radio broadcasting but in wireless devices, telephone transmission, television, radar, navigational systems, and communication. In the atmosphere the physical characteristics of the air cause slight variations in velocity, which are sources of error in such radio-communications systems as radar. Also, storms or electrical disturbances produce anomalous phenomena in the propagation of radio waves.

Because electromagnetic waves in a uniform atmosphere travel in straight lines and because the earth's surface is spherical, long distance radio communication is made possible by the reflection of radio waves from the ionosphere. Radio waves shorter than about 10 m (about 33 ft.) in wavelength —

designated as very high, ultrahigh, and super high frequencies (VHF, UHF, and SHF) - are usually not reflected by the ionosphere; thus, in normal practice, such very short waves are received only within line-of-sight distances. Wavelengths shorter than a few centimeters are absorbed by water droplets or clouds; those shorter than 1.5 cm (0.6 in) may be absorbed selectively by the water vapor present in a clear atmosphere. A typical radio-communication system has two main components, a transmitter and a receiver. The transmitter generates electrical oscillations at a radio frequency called the carrier frequency. Either the amplitude or the frequency itself may be modulated to vary the carrier wave. An amplitude - modulated signal consists of the carrier frequency plus two sidebands resulting from modulation. Frequency modulation produces more than one pair of sidebands for each modulation frequency. These produce the complex variations that emerge as speech or other sound in radio broadcasting, and in the alterations of light and darkness in television broadcasting.

Radio Broadcast Data System RBDS. A new system designed to let radio stations broadcasters send text messages, such as emergency warnings and traffic alerts to radios equipped with special LCD screens. The system is designed ultimately to replace the Emergency Broadcast System

Radio Button 1. A call center term. A button used for selecting from a group of options that are mutually exclusive. As with a car radio, selecting a particular button de-selects the previously selected button.

2. An Internet term. Radio buttons are used in forms to indicate a list of items. Only one button can be selected at one time.

Radio Common Carrier RCC. A common carrier engaged in Public Mobile Service, which also is not the business of providing land line local exchange telephone service. These carriers were once known as Miscellaneous Common Carriers.

Radio Communication Any telecommunication by means of radio waves.

Radio Frequency That group of electromagnetic energy whose wavelengths are between the audio and the light range. Electromagnetic waves transmitted usually are between 500 KHz and 300 GHz.

Radio Frequency Filter Fit A Northern Telecom Norstar device designed to alleviate problems associated with radio frequency interference that may be experienced when a headset or external Auxiliary Ringer is used with a telephone.

Radio Frequency Flooding Radio frequency flooding turns a telephone into a room listening device by transmitting a high power radio signal down a telephone line. The high power radio frequency is able to bypass the open hookswitch in the mouthpiece circuit. Room sounds cause the carbon microphone to modulate the RF signal. Radio frequency flooding is hard to implement but can only be detected by security professionals with the right equipment.

Radio Frequency Interface shield RFI. A metal shield enclosing the printed circuit boards of the printer or computer to prevent interference with radio and TV reception.

Radio Frequency Interference The disruption of radio signal reception caused by any source which generates radio waves at the same frequency and along the same path as the desired wave.

Radio Frequency Interference Shield RFI Shield. A metal shield enclosing the printed circuit boards of the printer or computer to prevent radio and TV interference.

Radio Paging Access Provides attendant and phone user

dial access to customer-owned radio paging equipment to selectively tone-alert, or voice-page individuals carrying pocket radio receivers. The paged party can answer by dialing an answering code from a phone within the PBX.

Radio Paging Access With Answer Back Allows access to customer-provided paging systems and provides the capability in the PBX to connect the paged party when the former answers the radio page by dialing a special code from any PBX.

Radio Resource Management A management entity or subentity concerned with the operation of the radio resources management protocol. A cellular radio term.

Radio Resource Management Entity A management entity or subentity concerned with the operation of the radio resource management protocol. A cellular radio term.

Radio Wave Electromagnetic waves of frequencies between 10 KHz and 3MHz, propagated without guide in free space (air).

Radiogram A telegram sent by radio. Totally obsolete term, but cute.

Radiophone Apparatus for transmitting and/or receiving speech or music by radio. Totally obsolete term, but cute.

Radiosonde An automatic radio transmitter in the meteorological aids service usually carried on an aircraft, free balloon, kite, or parachute, and which transmits meteorological data.

Radiotelegraphy The use of a radio (instead of wire) to communicate telegraphy messages over a distance. An old term, not used much any more. See also Radiotelephony.

Radiotelephony The science, art, and act of transmitting speech by means of radio. Now called telecommunications.

RADIUS Remote Authentication Dial-In User Service. A popular security system which has become an ad hoc standard, RADIUS is a client/server-based authentication software system. The software supports remote access applications, allowing an organization to maintain user profiles in a centralized database residing on an authentication server which can be shared by multiple remote access servers. The remote access servers act as RADIUS clients which connect to the centralized authentication server. RADIUS is fully open, and easily can be adapted to work with legacy systems and protocols. RADIUS was developed by Livingston Enterprises, Inc., which subsequently was acquired by Lucent. See also Authentication and Client/Server.

Radome A plastic cover for a microwave antenna. It protects the antenna from awful weather, but has little effect on the radiation pattern of the antenna.

RADSL Rate Adaptive Digital Subscriber Line. Transmission technology that supports both asymmetric and symmetric applications on a single twisted pair telephone line and allows adaptive data rates. RADSL employs intelligent ADSL modems which can sense the performance of the copper loop and adjust transmission speed accordingly. These devices adjust dynamically as the performance of the loop varies during a session, much as does a V.34 modem. Depending on various characteristics of the subject cable plant, ADSL accommodates downstream transmission speeds of as much as seven megabits per second, plus bidirectional transmission speeds of as high as 640 Kbps over a single UTP (Unshielded Twisted Pair). Some RADSL equipment can be manually configured See ADSL.

RAID Redundant Array of Inexpensive Disks. The idea is simple: Put several disk drives into a single housing. Then write your data over the disk drives in such a way that if you lose one or more of the drives, you won't have lost any of your data. Thus the term "redundant." At its simplest, RAID mirrors data to an equal number of disk drives, e.g. two sets of two. At its most complex, RAID writes data across a bunch of drives, so that if one goes the data can be retrieved from the remaining drives. The opposite of RAID is SLED (Single Large Expensive Disk). See also REDUNDANT ARRAY OF INEXPENSIVE DISKS for a much more detailed explanation.

Rain Attenuation Signal losses due to absorption are common when radio signals encounter a heavily moisture laden atmosphere. Generally the higher the radio frequency, the more attenuation (i.e. the more losses). Since microwave signals for satellite and for land line are essentially line-of-sight, microwave radio is very susceptible to signal attenuation in heavy rain. Modern microwave paths are engineered with weather patterns in mind. In areas where heavy rainfalls occur, microwave links may be closer together or more attention is paid to diverse routing. "Rain fade" is another name for rain attenuation. See Rain Fade.

Rain Barrel Effect Signal distortion of a voice telephone line caused by the under-attenuated echoes on the return path.

Rain Fade I found this definition in an instruction manual for Direct Satellite System. Rain fade is the temporary loss of a satellite signal due to the inability of the signal to penetrate unusually heavy rain clouds or rainfall. Rain fade is usually brief, lasting only as long as the heavy rain cloud condition persists. See also Rain Attenuation for a better explanation.

Rainbow Series According to the National Security Agency (NSA), the Rainbow Series is a six-foot tall stack of books dealing with the evaluation of "Trusted Computer Systems." The term comes from the fact that each book is a different color. Colors include orange, aqua, burgundy, lavender, venice blue, pink, peach, turquoise, and violet.

Raised Floor A floor distribution method in which square, steel and wood-laminated plates resting on aluminum locking pedestals are attached to the building floor. The plates are usually covered with cork, carpet, or vinyl tiles, and each plate can be removed for easy access to the cables below. Also referred to as access floor.

Rake Receiver A wireless receiver which can support a number of tines, or fingers, which can be combined to form a stronger received signal. Each tine is an individual radio channel.

RAM Random Access Memory. The primary memory in a computer. Memory that can be overwritten with new information. The "random access" part of its name comes from the fact that the next "bit" of information in RAM can be located — no matter where it is — in an equal amount of time. This means that access to and from RAM memory is extraordinarily fast. By contrast, other storage media — like magnetic tape — have their information stored serially, one bit after another. Therefore you have to search for them. And your search time will depend on how far from the bit you're searching for you are. Floppy disks are faster than magnetic tape because their information is readily at hand, though the read/write head will have to search for it. Hard disks are even faster because there are multiple heads and because the disks spin faster and everything moves faster. RAM memory is the fastest of all. The problem with RAM memory is that it's volatile. This means when power is turned off (or power glitches occur) RAM memory is erased. RAM memory can be protected with rechargeable batteries — just remember to charge the batteries. See also DRAM, EDO RAM, SRAM, and VRAM.

RAM Base Address Random Access Memory Base

Address. Starting address for memory dedicated to a specific task.

RAM BIOS BIOS transferred to RAM so things go faster.

RAM Disk A logical device made from semiconductor (i.e. chip) memory which emulates the functioning of a disk drive as closely as possible. Since most semiconductor memory (RAM) is volatile, most RAM disks are also volatile, i.e. they lose their memory when you turn off power.

Ram Hook/Ram Horn Hardware attachment that holds ASW (Aerial Service Wire) drop clamps in aerial span applications.

Rambutan A symmetric cryptographic algorithm developed by Marconi.

RAMAC RAndoM ACcess. Built in 1956 by IBM, RAMAC was the first hard drive computer memory device ever built. Consisting of 50 fixed disk platters, each approximately two feet in diameter, RAMAC could store five million characters at a rate of about 2,000 bits per square inch and at a cost of approximately $10,000 per MB (MegaByte). Contemporary (1999) hard drives store information less than a dime ($.10) per megabyte.

Raman Scattering An optical fiber transmission term. Stimulated Raman Scattering (SRS) results from the interaction between the optical transmission signal and the silica molecules in the fiber. SRS affects broadband optical fiber transmission, and affects the overall optical spectrum involved in a DWDM (Dense Wavelength Division Multiplexing) transmission system. The SRS phenomenon manifests itself as a transfer of power from the shorter wavelengths to the longer wavelengths, resulting in a tilt of the optical spectrum. The effect increases as the power of the signal increases, and as the width (density) of the DWDM spectrum increases. See also DWDM.

RAMDAC Random Access Memory Digital-to-Analog Converter. The chip on a VGA board that translates the digital representation of a pixel into the analog information needed for display on the monitor. A RAMDAC actually consists of four different components — SRAM to store the color map and three digital-to-analog converters (DACs), one for each of the monitor's red, green, and blue electron guns.

Rampdown The process of reducing transmission power from the nominal power level to a level below a defined threshold.

RAN 1. Return Authorization Number, also called RMA, Returned Merchandise Authorization. A number you need for returning busted equipment to the factory. You call the factory, tell them what you want to return and its serial number, and the factory rep gives you an RMA, which you write on the outside of the box containing the thing you're sending back. The idea is that the factory sees the number on the box and immediately logs your busted thing into its computer system. This way, when you call, it can tell you where your thing is and when you might get it back and what it might cost you. At least that's the theory. The moral of this story: Don't send stuff back to the factory without a RAN or RMA (whatever the factory calls it.)

2. Recorded trunk ANnouncements. RAN devices are devices connected on 4-wires to older central offices (public exchanges). They are used to give recorded messages to callers, e.g. "The number you have called has been changed. Please make note of the new number..."

Random Scattered, unfocused, a non sequitur. A favorite expression of Bill Gates, chairman and founder of Microsoft, to dismiss ideas or strategies that lack logic (or he thinks lack

logic). According to Stewart Alsop writing in the February 2, 1998 issue of Fortune Magazine, "Bill Gates is the ultimate programming machine. He believes everything can be defined, examined and reduced to essentials, and rearranged into a logical sequence that will achieve a particular goal. Anything that doesn't work this way, anything illogical is 'random.' In the world of Bill Gates, being illogical is the most serious sin." See Randomness for its statistical meaning.

Random Access Usually refers to computer memory or storage. Random Access is the ability to reach any piece of data in the memory directly without having to pass by other pieces of data. In telephony, this means the ability to reach any other subscriber through the telco switching network. See SEQUENTIAL ACCESS.

Random Access Memory See RAM.

Random Noise Interference to telephone communications occurring at irregular intervals.

Random Early Detection. RED. Designed for TCP/IP, a router queuing technique which distributes traffic losses amongst transmitting devices in the event of buffer overflow. RED is accomplished by dropping packets on a random basis, which is determined statistically, when the mean queue depth exceeds a threshold over a period of time. While the queue is still relieved on the basis of FIFO (First In First Out) logic, RED is an improvement over pure FIFO, which simply drops data packets from the tail of the queue. See also FIFO.

Randomize A Microsoft made-up word. To become distracted, as in "I got heavily randomized by other stuff going on." See also Random.

Randomness The state of being random. See random. As a telephone industry assumption used in the development of blocking and delay formulas, randomness states that: All subscribers originate calls randomly, that is, without common cause such as a declaration of war, and each subscriber originates calls independently of all other subscribers.

Range The difference between the greatest and least of the items being considered. A measure of dispersion.

Range Extender There are two definitions. I don't know which one is correct. I thought the first one was correct. That is that a range extender is a device that increases the length of a local loop by boosting battery voltage being sent out from the telephone company central office. Bellcore, however, says a range extender is a device that permits a central office to serve a line that has resistance that exceeds the normal limit for signaling. A range extender does not extend transmission range, according to Bellcore. See RANGE EXTENDER WITH GAIN.

Range Extender With Gain REG. A unit that provides range extension in a loop for both signaling and transmission.

RARE Reseaux Associes pour la Recherche Europeenne. European association of research networks.

Rare Earth Doping Here is an article on rare earth doping from The Economist Magazine of July 6, 1991: Optical fiber is the darling of the telecommunications world. Because light waves can be superimposed on one another, fiber can carry thousands of laser generated messages at the same time, over longer and longer distances. The longest fibers were once those which doctors use to explore their patients innards. Now they can stretch 70 km (40 miles). But even that does not get you across a sea much bigger than the English Channel without the messages fading. So today's transatlantic and transpacific optical cables are interrupted about every 70 km so that the messages can be sorted out, passed through an electronic amplifier, and then, turned into light again. These amplifiers are costly. Soon, though, they may be replaced.

The key technique is called rare earth doping, which was developed not by crooked bookmakers but by scientists at Southampton University in England, and AT&T Bell laboratories in New Jersey. The rare earths are a group of chemical elements with particularly restless electrons in their atoms. If these electrons are stirred up by a laser, they rise to higher energy levels inside their atoms. When they fall back again, they emit light. The frequency of the light emitted depends on the element. The trick is to pick one which emits at frequency used for telecommunications. By adding the right rare earth to a stretch of fiber, you can make it amplify signals. You can also make a laser out of the fiber itself.

In the optical amplifier developed at Southampton University, a laser is used to lift electrons in the rare earth atoms in a stretch of fiber up to higher energy levels. When a light signal comes along, it may knock one of these electrons off its perch. The falling electron gives off light, which boosts the signal. The enhanced signal then knocks down more electrons, gathering strength as it goes. Rare earths in the cable can be used for other things, as the team at Bell Labs has found. Normal light waves, even those in laser beams, spread out and dissipate as they travel.

Solitons, a special kind of wave, do not. Tidal bores, the best known form of soliton, can move up rivers for miles without losing their shapes. Light that traveled in solitons could travel much farther along an optical fiber between boosts. Solitons are created either by pumping the initial signal through an optical amplifier, or by using a laser made from doped fiber. The soliton holds its shape because the passage of light through the fiber temporarily increases the speed of light in that part of the fiber, so the back of the wave is always trying to travel faster than the front. The stronger the light, the stronger the effect.

Rare earth doping, with metals called erbium and praseodymium, has resulted in fibers which can handle billions of bits of data per second, and carry them thousands of kilometers. AT&T hopes to use erbium amplifiers in its new transoceanic cables in the 1990s. Other companies — such as British Telecom and NTT — also like praseodymium, which is harder to handle, but emits light at a more commonly used frequency.

RARP Reverse Address Resolution Protocol. A low-level TCP/IP protocol used by a workstation (typically diskless) to query a node for purposes of obtaining its logical IP address.

RAS 1. Remote Access Server or Remote Access Services. See Remote Access Server.

2. Registration, Admissions and Status signaling function. See H.323.

3. Sun Raster Image File image format.

Raster A pattern of horizontal scanning lines on a TV screen. Input data causes the beam of the TV tube to illuminate the correct dots to produce the required characters. See RASTER SCANNING and RASTERING.

Raster Scanning The method of scanning in which the scanning spot moves along a network of parallel lines, either from side to side or top to bottom.

Rastering The process by which a document image is converted to a stream of bits representing either black or white, or one of sixteen levels of gray, for each element of the image. For Group 3 faxes, there are either 98 or 196 raster lines per vertical inch, with a horizontal resolution of 203 lines per inch (yielding 1.86 or 3.72 million elements per 8 1/2 by 11 inch page). Sixteen levels of grey ("halftone" setting — requiring four bits per element, rather than the one bit required for black and white)

can be specified, but are not typically used for documents containing only text and/or line drawings. The bit stream is compressed for transmission, and decompressed when received.

Rasterizing See RASTERING.

Rate The price of a particular service or piece of equipment from a telephone company. Telephone companies don't use the word "price." They use the word "rate." No one knows why, except that if they didn't cultivate their own jargon, there'd be no job for telecommunications dictionary writers. God forbid!

Rate Adaption 1. The process of converting a digital stream of data into a different format and rate. For example, rate adaption allows a 64-Kbps data channel to interoperate with a 56-Kbps channel. In this context, rate adaption also is known as flow control. As is true for much in life, the lowest common denominator rules.

2. An ISDN term for bandwidth-on-demand. Sensitive to the application and to its underlying bandwidth requirements, the Terminal Adapter (TA) in a BRI implementation, or the PBX or router in a PRI implementation, will establish some number of 64-Kbps channels. These channels are established either automatically or by conscious selection of the user. For example, a voice session requires only one channel; a videoconferencing session requires two or more channels, depending on the quality desired. On demand and as available, the proper number of channels are selected in order to support the connection; as is the case with all connection-oriented protocols, those channels remain active during the entire session, regardless of whether they are required.

Rate Area A telephone company term. A geographic area within which rate treatments are tie same.

Rate Arrangements Telephone customer prices charged by tariffs for specified telephone services.

Rate Averaging Telephone companies' method for establishing uniform pricing by distance rather than on the relative cost (to them) of the particular route. The theory is that some routes are more heavily trafficked, have huge transmission equipment and achieve great economies of scale. Some routes, on the other hand, have little traffic, small transmission equipment and achieve no economies of scale. Therefore, it costs more to provide calls on these less-trafficked routes. But the phone industry doesn't charge more to call small towns than big cities. The phone industry simply charges by distance, averaging its costs. This is called rate averaging.

Rate Base A regulated telephone company's plant and equipment which forms the dollar base upon which a specified rate of return can be earned. The total invested capital on which a regulated company is entitled to earn a reasonable rate of return.

Rate Based Flow Control A means of flow control in which devices (e.g., switched or routers) in a network control the rate of data flow from a transmitter. In an ATM network, for instance, the edge switches negotiate the rate of flow from the transmitting device in consideration of both its desired rate and the ability of the destination switch (and all intermediate switches) to handle that flow without overflowing buffers. Overflowed buffers would result in lost data and overall degradation of QoS (Quality of Service). See also FLOW CONTROL and ER. Contrast with QFC.

Rate Center Telephone company-designated geographic locations assigned vertical and horizontal coordinates between which airline mileages are determined for the charging of private lines. Or as defined by the telephone industry, rate center is that point within an Exchange Area defined by rate map coordinates used as the primary basis for the determination of toll rates. Rate Center may also be used for the determination of selected local rates. See Airlines Mileage and V & H.

Rate Chip A standard, nonvolatile memory device used to retain data base information on call pricing by Area Code and Central Office. Typically used in call accounting equipment.

Rate Design Utilities have a specific rate for every service provided. The rates must be approved by the PUC. In a major rate case, rates for many services will be changed in tandem. In a rate design hearing, different proposals as to rate levels are considered. The level of one rate can have an impact on what the level of another rate should be. The inter-relationship between rates and the impact of demand must all be considered in "designing" a rate structure.

Rate Elements The pricing structure of various telecommunications service offerings usually described in tariffs.

Rate Of Return The percentage of net profit which a telephone company is authorized (by a regulatory commission) to earn on its rate base. See RATE BASE.

Rate Period Dividing a day into various slices of time for the purpose of charging differently for long distance and local calls. There are three rate periods in force today in North America for intra-North America calls. One rate period is from 11:00 P.M. to 8:00 A.M.; one is from 8:00 A.M. to 5:00 P.M. and one is from 5:00 P.M. to 11:00 P.M. If you call outside the United States, there are different rate periods.

Rate Realignment In California's Alternative Regulatory Framework Phase III, rate realignment refers to redesigning telephone rates to reduce intraLATA toll rates and increase rates for other services to make up for the phone companies lost revenues. The Public Utility Commission (PUC) must approve all rate realignment proposals in the rate design stage of the proceeding.

Rate Stability Plan Commit yourself to keeping a Nynex service for several years and you'll pay less than if you keep it only from month to month. Other phone companies have similar schemes, typically by other names.

Rate Table A data base that contains the cost of calls referenced to the Area Code and/or number dialed plus time of day considerations. See RATE PERIOD.

Rate Zone A defined geographic division of an exchange area used as the primary basis for figuring toll rates.

Rated Temperature The maximum temperature at which an electric component can operate for extended periods without loss of its basic properties.

Rated Voltage The maximum voltage at which an electric component can operate for extended periods without undue degradation or safety hazard.

Rats Nest Terminals and connections with poor maintenance and sloppy wiring techniques.

Raw Bite Data The data channel bit rate that includes all protocol overhead and system overhead data bits.

Ray A beam of radiant energy. Ray is most energetic, responding to email requests from Harry for strange definitions at 2:00 AM. He also is an excellent teacher. He teaches courses on all aspects of networking all over the country and all over the world. Catch one of his seminars if you can. He often teaches a day-long seminar the day before a major trade show, like InterOp or Computer Telephony Expo. He's also a brilliant consultant. Ray's mother gave him his name and named his sisters Joy and Dawn. Ray is thus the only one not named after a dishwashing detergent. There is hidden significance in this. Ray writes the hard part of this dictionary. Blame him for all the mistakes. See Margaret, his wife.

Rayleigh Fading Multipath fading, arising from an ensemble of reflected signals arriving at the receiver antenna and creating standing waves.

Rayleigh Scattering Scattering due to tiny impurities in the optical fiber which are fractions of the wavelength of the infrared rays.

RB Reverse Battery.

RBDS Radio Broadcast Data System. A new system designed to let radio stations broadcasters send text messages, such as emergency warnings and traffic alerts to radios equipped with special LCD screens. The system is designed ultimately to replace the Emergency Broadcast System.

RBOC Regional Bell Operating Company, also called a Regional Holding Company or RHC. Here's the background: On January 8, 1982 AT&T signed a Consent Decree with the United States Department of Justice, stipulating that on midnight December 30, 1983, AT&T would divest itself of its 22 telephone operating companies. According to the terms of the Divestiture, those 22 operating Bell telephone companies would be formed into seven regional holding companies (RHC) of roughly equal size. The judge assigned which Bell operating companies would join which regional holding company. The seven RHCs were Ameritech, Bell Atlantic, BellSouth, NYNEX, Pacific Telesis, Southwestern Bell and US West. In early October, 1994, Southwestern Bell changed its name to SBC Communications, Inc. But its telephone companies, it said, would still operate under the Southwestern name. In April, 1996 Bell Atlantic bought NYNEX (the holding company for New York Telephone and New England Telephone) for $22.1 billion. The new company will be called Bell Atlantic. In April also, SBC Communications, Inc. (the name for the holding company owning Southwestern Bell Telephone) bought Pacific Telesis (the holding company for Pacific Bell) for $16.7 billion. As of early 1998, this left five regional Bell operating companies — Ameritech, Bell Atlantic, BellSouth, SBC Communications and US West.

Terms of the Divestiture placed business restrictions on AT&T and the BOCs. Those restrictions were threefold: The BOCs weren't allowed into long distance, equipment manufacturing, or information services. AT&T wasn't allowed into local telecommunications (i.e. to compete with the BOCs). But it was allowed to manufacture anything it wanted, including computers. The federal Judge overseeing Divestiture, Judge Harold Greene, is slowing the lifting the restrictions. He has allowed the BOCs into information services and AT&T into local service. He has stayed firm on the other two — no equipment manufacturing and no long distance for the RHCs (also called RBOCs).

RBS Robbed-Bit Signaling. See ROBBED-BIT SIGNALING.

RC Rate Center

RC-4 An encryption/decryption algorithm supported in Cellular Digital Packet Data (CDPD).

RCA 1. Regional Calling Area. The geographical area covered by a telephone company.

2. Once it stood for Radio Corporation of America.

RCA Globecom An International Telex and high-speed data communications company acquired by MCI from RCA in 1987. The acquisition of RCA Globecom gave MCI approximately 40% of the International Telex market and helped strengthen MCI's role in International data.

RCC Radio Common Carrier.

RCDD Registered Communication Distribution Designer, a title conferred on people who have acquired certain requisite education and experience by BICSI, the Building Industry

Consulting Service International. BICSI is at 813-397-1991.

RCEE Resource Control Execution Environment. A term from Bellcore Advanced Intelligent Network model.

RCL ReCaLl.

RCP The Berkeley UNIX remote copy program.

RD An ATM term. Routing Domain: A group of topologically contiguous systems which are running one instance of routing.

RDBS Routing Data Base System (database from which LERG is created).

RDC Redirect Confirm packet. Used in Cellular Digital Packet Data (CDPD) mobility packet.

RDCCH Reverse Digital Control CHannel. A digital cellular term defined by IS-136, which addresses cellular standards for networks employing TDMA (Time Division Multiple Access). The RDCCH includes all signaling and control information passed upstream from the user terminal equipment to the cell site. The RDCCH acts in conjunction with the FDCCH (Forward Digital Control CHannel), which includes all such information sent downstream from the cell site to the user terminal equipment. The FDCCH consists of the RACH (Reverse Access CHannel). See also IS-136, and TDMA.

RDF 1. Radio Direction Finding.

2. An ATM term. Rate Decrease Factor: An ABR service parameter, RDF controls the decrease in the cell transmission rate. RDF is a power of 2 from 1/32,768 to 1.

RDI Remote Defect Indication. An indication that a failure has occurred at the far end of an ATM network. Unlike FERF (Far-End Remote Failure), the RDI alarm indication does not identify the specific circuit in a failure condition. See FERF.

RDQ Redirect Query Packet. Used in Cellular Digital Packet Data (CDPD) mobility management.

RDR Redirect request packet. Used in Cellular Digital Packet Data (CDPD) mobility management.

RDS Radio Data System. A way of sending data along with a standard FM radio broadcast.

RDT 1. Recall Dial Tone.

2. Remote Digital Terminal.

RDY ReaDY.

RE-422 A high-speed electrical interface defined by the ITU-T, supporting data rates of up to 768 Kbps over up to 300 feet of cable.

Re-Engineering A term probably invented by Michael Hammer in the July-August, 1990 issue of Harvard Business Review. In that issue, he wrote "It is time to stop paving the cowpaths. Instead of embedding outdated processes in silicon and software, we should obliterate them and start over. We should 're-engineer' our business: use the power of modern information technology to radically redesign our business processes in order to achieve dramatic improvements in their performance." The term re-engineering now seems to me mean taking tasks presently running on mainframes and making them run on file servers running on LANs — Local Area Networks. The idea is to save money on hardware and make the information more freely available to more people. More intelligent companies also redesign their organization to use the now, more-freely available information. Also called VALUE DRIVEN RE-ENGINEERING.

Re-Initiation Time The time required for a device or system to restart (usually after a power outage).

Re-Installed Customer An MCI term. An MCI customer who is installed again with the same customer account number after having been previously canceled either at their, MCI's, or a third party's request.

REA Rural Electrification Administration. A federal agency within the Department of Agriculture, the REA was established in 1935 to bring electricity and, later, telephone service to rural America. The REA was one of the most successful federal government programs ever. Telephone companies loved the REA, as it offered loans to telcos at a very low rate of interest (2% or less, in many cases). Once the facilities were in place, the telcos nevertheless would have suffered huge losses, as the rates for basic telephone service would not have yielded a satisfactory rate of return on investment. However, the Universal Service Fund provided very substantial additional revenues to further subsidize service in such high-cost areas through the settlements process, which established the cross-subsidy mechanism between the LECs and the IXCs. In fact, a number of independent telephone companies, such as CONTEL, thrived specifically and only because of the combination of REA money and the settlements process. REA money largely dried up some years ago, at least for this purpose, as the definition of "high-cost" changed considerably and as the restrictions on access to and use of such funding became onerous. The REA now is known as the Rural Utilities Service (RUS). See also RUS.

Reach Through Reach through is a means of extending the data accessible to the end user beyond that which is stored in the OLAP server. A reach through is performed when the OLAP server recognizes that it needs additional data and automatically queries and retrieves the data from a data warehouse or OLTP system.

Reactance The opposition offered to the flow of an alternating current which is due to the presence of inductance or capacitance or both, in the circuit.

Read To glean information from a storage device, like a floppy disk. The opposite of READ is to WRITE. That's when you put information onto that storage device. Some storage devices can only be READ, but not written to. On a floppy disk that's called being "WRITE PROTECTED." See also WORM, which stands for Write Once, Read Many.

Read After Write Verification A means of assuring that data written to the hard disk matches the original data still in memory. If the data from the disk matches the data in memory, the data in memory is released. If the data doesn't match, the block location is recognized as "bad," and something happens. The data is transferred again. Or in Novell's NetWare, Hot Fix redirects the data to a good block location within the Hot Fix Redirection Area.

Read Before Write A feature of some videotape recorders that plays back the video or audio signal off of tape before it reaches the record heads, sends the signal to an external device for modification, and then applies the modified signal to the record heads so that it can be re-recorded onto the tape in its original position.

Read Only File A PC computer term. A read only file is a file that you can read but cannot make changes to. The read-only attribute specifies whether a file is read-only. To remove the read-only attribute, you would type the following command

ATTRIB -R FILENAME

Read Only Memory ROM.

1. A computer storage medium which allows the user to recall and use information (read) but not record or amend it (write).

2. The smaller part of a computer's memory, in which essential operating information is recorded in a form which can be recalled and used (read) but not amended or recorded (written). ROM is memory which is programmed at the factory and

whose contents thereafter cannot be altered, even by a power breakdown, or being written to, or anything else. ROM memory is also random-access, which means accessing its information is very fast. See also MICROPROCESSOR and RAM.

Read Write Cycle Time of reading and writing data onto a memory device. See READ.

Readable Frames The number of video frames received without error.

Readable Octets The number of octets (bytes) received without error.

Readdressing A cellular radio term. The process whereby the serving Mobile Data Intermediate System (MD-IS) receives the encapsulated packets, de-encapsulates them, then locates the Mobile End System (M-ES) to determine the cell and channel stream associated with the M-ES. The function is also performed by the Foreign Agent in Mobile IP. This definition come from the book "Internetwork Mobility," by Mark Taylor, William Waung and Mohsen Banan.

Reader 1. A device which converts information into a format recognized by a machine as input.
2. A device which interprets coded data in the process of transferring that data from one coded state of storage to another.

Readerboard Also called Electronic Displayboard, Electronic Wall Display or Message Display Unit (MDU). Readerboards are typically found in call centers. They are electronic displays. They are typically hooked to the call center's ACD or the PC monitoring the ACD (automatic call distributor) and they throw up information about how many people are waiting in line, how long the longest person has been in line, how well the agents are doing and, whose birthday it is. The idea is that all the agents in the call center can see the readerboards and change their behavior accordingly — speak faster if there are a lot of people in queue. Readerboards aren't TVs. They're typically large hanging electronic displays sporting red LCDs or small red lights. By lighting the correct collection of lights, you can put up a message. Some readerboards are very large with letters reaching 12 inches high.

Readyline 800 A toll-free service designed for the small business. Receive "800" dialed calls over your existing telephone lines and equipment — no new lines to install, no new equipment needed. You can still use those same lines to make and receive local and long distance calls. Choose the geographic areas you want to cover — from a single area code to an entire state or the whole country. Even decide when you want your toll-free number to be available. You pay a one-time start-up charge and a low monthly fee. Calling prices are based on the market coverage you choose. There are time-of-day and day-of-week discounts, and a volume usage discount. Calls are priced on a mileage/distance-sensitive basis.

RealAudio RealAudio (now called RealPlayer) client-server software system enables Internet and on-line users equipped with conventional multimedia personal computers and voice-grade telephone lines to browse, select, and play back audio or audio-based multimedia content on demand, in real time. Several radio stations broadcast their daily fare to anyone on the Internet who's listening. RealAudio is a real breakthrough compared to typical download times encountered with delivery of audio over conventional on-line methods, in which audio is downloaded at a rate that is five times longer than the actual program; the listener must wait 25 minutes before listening to just five minutes of audio. Download RealAudio from www.realaudio.com/products/player2.0.html. For Internet radio listings (what they call NetRadio Central) go

to www.netradio.net. See also www.audionet.com. RealAudio is produced by a company called RealNetworks, which had previously been called Progressive Networks.

Real Mode Originally there was the first IBM PC and it was powered by an Intel 8086 chip which addressed a maximum of 1MB (megabyte of RAM). Real mode is the term that later generations of Intel chips came to call their ability to run programs written for the 8086. Real mode allows 80286, 80386, or 80486 processors to emulate an 8086 processor but perform better than the 8086 because they operate at a faster clock rate. Real mode is limited to a maximum of 1MB of addressable memory because the 8086 processor uses a memory address bus of 20 bits. This is calculated thus: Since a bit can have one of two values, raising the base number of 2 to the power of 20 is equal to 1,048,576 unique memory addresses. Each memory address can store 1 byte of information (1,048,576 bytes equal 1MB). See also PROTECTED MODE.

Real Soon Now A on-line term used to describe when something will happen, maybe.

Real Time A voice telephone conversation is conducted in real time. That is, there is no perceived delay in the transmission of the voice message or in the response to it. This concept often applies to interaction between a computer and a terminal. In data processing or data communications, real time means the data is processed the moment it enters a computer, as opposed to BATCH processing where the information enters the system, is stored and is operated on a later time. See the follow definitions beginning with real time.

Real Time Adherence Adherence is a term used in telephone call centers to connote whether the people working in the center are doing what they're meant to be doing. Are they at work? Are they on break? Are they answering the phone? Are they at lunch? All these activities are scheduled by work force management software. If they're in line, the workers is "in adherence." If not, they're "out of adherence." Some automatic call distributors have a real time adherence data link which connects the ACD to an external computer which then tracks and displays current service rep activity measured against a pre-defined schedule. The idea is to give call center supervisors tools to manage the call center's work force more efficiently. Supervisors are able to define the task, the start time of each task, and the task duration. In addition, thresholds and ranges of acceptable deviations for the call center can be set for each task or service rep work state. Once the schedules have been defined and thresholds set, real-time displays inform the supervisor of discrepancies between the work schedule and actual activity. Service rep information will automatically appear should their status exceed the threshold, such as being on someone being on break for too long.

Real Time Capacity The capacity of the central computer processor of a stored program control telephone system to process the instructions coming at it. Real Time Capacity is probably the most important measure of the size of a telephone system relying on a single main processor.

Real Time Chat A program allowing live conversation between individuals by typing on a computer terminal. The most common tools are Talk and IRC (International Relay Chat).

Real Time D-Channel Status Display This maintenance enhancement allows you to assess active or failed status of ISDN D-Channels in real-time. This saves time since ports no longer need to be evaluated.

Real Time Transport Protocol RTP. Developed by the IETF (Internet Engineering Task Force) it adds a layer to the Internet protocol. It is designed to address problems caused when real-time interactive exchanges such as video are transported over LANs were designed for data. Running video on LAN means you can encounter significant end-to-end latency. RTP's approach is to give video higher priority than connectionless data. RTP resides above the IP, Datagram Protocol and ST-II protocols.

Rearrangement A fancy word for moving phone extensions around.

Reasonableness Checks Tests made on information reaching a real-time system or being transmitted from it to ensure that the data lie within a given range.

Reassembly 1. The process by which an IP datagram is "put back together" at the receiving host after having been fragmentation and MTU.

2. The process of combining a number of the Link Layer Service Data Unit (LSDU) into an SN-Data Protocol Data Unit (PDU) or SN-Unit-data PDU.

Reasserting Status An ISDN term. When the ISDN phone is being directly controlled by the application program, the set's physical status may be different from the status that has been received from the network. When direct control ends, the ISDN set reasserts the status received from the network to bring its physical condition back into conformity with the network status.

Reassignment Here is an explanation by Bill Etling, a senior planner for GTE. "Under the assigned plant concept, a pair is dedicated from the central office to the subscriber home and maintained at that address, even when idle. The likelihood of such a pair being reused, thus eliminating a field visit and extra assignment work, more than makes up for lost revenue while the pair is vacant. In areas of high cable fills, such a pair, when vacant, is often used to fill an order at a different address. Reassignment quickly snowballs, generating many installation field visits and assignment changes, increasing paperwork and the chance of errors."

Rebalancing Rebalancing is a new term. It means changing tariffs (the price of phone calling) to come closer to the actual costs of providing the service. Let me explain: Tariffs are published public documents which describe the prices and conditions of buying service from regulated telephone company. Tariffs developed over a period of many years. Tariffs may apply at a local, state, national, regional, or international level. Traditionally, tariffs were created in a complex fabric of balancing the overall costs of the service against regulatory and competitive issues. For instance, many regulatory authorities put in complex cross-subsidies. These allowed highly profitable or optional services (e.g., long distance and custom calling services) to subsidize residential service, i.e. to keep its price low. Similarly, business service rates commonly were set at high levels to cross-subsidize residential service rates — the logic, at least partially, was based on the assumed ability to pay and the legislators' obsession with "universal service," i.e. giving everyone phone service. As nations move toward deregulated, competitive telecommunications, older tariffs structure put burdens on the incumbent (read regulated) carriers and put them in a potentially bad competitive position. Hence, the concept of rebalancing, which seeks to reset tariffs at levels which are representative of the actual costs of provisioning the various services. At the extreme, rebalancing eliminates cross-subsidies. Thus each service would bear its rightful share of associated costs. As it relates to international calling costs, rebalancing would eliminate the disparity in calling costs. For example, it is much more expensive to call the U.S. from Argentina than it is to call Argentina from the U.S. See also ACCOUNTING RATE, BILLING RATE, CROSS SUBSIDIZATION, and TARIFF.

Rebiller A rebiller, also called a switchless reseller, buys long distance service in bulk from a long distance company, such as AT&T, and resells that service to smaller users. It typically gets its monthly bill on magnetic tape, then rebills the bulk service to its customers. A rebiller owns no communications facilities — switches or transmission. It has two "assets" — a computer program to rebill the tape and sales skills to sell its services to end users. The profit it makes comes from the difference between what it pays the long distance company and what it is able to sell its services at. It's not an easy business to be in, since you are selling a long distance company's services to compete against itself.

Rebooting Repeating a Boot. Turning on or resetting the telephone system or the computer. The word derives from "boot-strapping." Starting from scratch. Pulling oneself up by one's own bootstraps. Booting a telephone system or a computer means starting it from scratch, usually by turning its AC power on. Rebooting a telephone system is done by simply turning it off, counting to ten and turning it back on again. Rebooting is done to clear the volatile part of the telephone system's or computer's memory and its various processing and clock chips. You reboot typically when your PC "locks" inexplicably or when your telephone system does something you can't explain logically — like ring phones randomly or give strange error messages on the console. On a computer, "Lock" means that no matter which key or combination of keys you touch on your keyboard, you can't get your computer to do anything. In addition to "unlocking" your computer, you also reboot to clear RAM or RAM-resident programs. On an IBM or an IBM clone, rebooting is done by pressing the CONTROL, ALT and DELETE keys simultaneously. You can also reboot by pressing the reset button if your computer has one. (Not all do.)

You can reboot any computer by turning its power off, then turning it back on. This is usually not a good idea, since the surge of power that accompanies a computer being turned on and off will reduce the life of many of its electronic components. Some experts recommend leaving computers running full-time, though turning their hard disks off. They also recommend turning your screen off, or at least running a public domain program such as SCRNSAVE.COM or SCRN.COM which turn off your screen after several minutes of doing nothing (inactivity).

Rebuilding Imagine you have five hard disks in an array. Imagine that they are organized that data is being written to all five drives in such a way that if one drive fails, no data will be lost. That failed drive is now removed and replaced with a good drive. Immediately, the remaining four drives start writing data to the new, good, but empty drive. That process of rebuilding might take a few minutes, or an hour or two. It depends on how much data is in the system and how much activity is taking place. Typically, this rebuilding process happens in a system called RAID (which stands for Redundant Array of Inexpensive Disks). And typically RAID (which is not cheap) is found on servers on LANs. The process of rebuilding is also called reconstruction.

Rebundling Rebundling is the process of putting UNEs (Unbundled Network Elements) back together by a CLEC to become part of a competitive service offering by him to a customer. See UNE.

Recall The recall button on many phones provides a fresh dial tone without physically putting down and picking up the handset. Don't confuse it with REDIAL, which is a feature of a phone or phone system that allows a user to call the previously-dialed number by pressing one or a few buttons.

Recall Dial Tone A stutter or interrupted dial tone indicating to the extension user that the hookswitch flash has been properly used to gain access to system features.

Recall Key Used to get dial-tone or to transfer calls on a key system installed within a PBX. See also RECALL.

RECAPSS REmote CAble Pair Switching System is used to remotely handle cable transfers and related cable switching tasks by connecting a distribution cable pair to either an old cable pair or a new cable pair without interrupting service. The system accommodates both POTS and special services and the computer console operator can select one pair at a time or select thousands for sequential transfer.

Receipt Notification A report prepared by a recipient UA (User Agent) or Access Unit (upon request) and sent to the originating UA or Access Unit when a message is received by a recipient.

Receive Interruption The interruption of a transmission to a terminal to receive or send a higher priority message from the terminal.

Receive Only RO. Describing operation of a device, usually a page printer, that can receive transmissions but cannot transmit.

Received Line Signal Detector Modem interface signal defined in RS-232-C EIA interface which indicates to the attached data terminal equipment that it is receiving a signal from the distant modem.

Received Signal Level RSL. The strength of a radio signal received at the input to a radio receiver.

Received Signal Strength Indication The measured power of a received signal.

Receiver 1. Any device which receives a transmission signal.
2. Any portion of a telecommunications device which decodes an encoded signal into its desired form.
3. The earpiece portion of a telephone handset, which converts an alternating electric current into sound waves, usually through an electromagnet moving a diaphragm.
4. An electronic component capable of collecting radio frequency broadcasts and reproducing them in their original audio and/or video form, e.g. a TV or radio receiver.

Receiver Congestion A Token Ring error reported by any ring station that receives a frame addressed to itself, but has no room in its buffer to store the frame. The frame is then discarded, and within two seconds the station will report how many times this happened over the reporting period.

Receiver Off-Hook Tone The loud tone sent by the central office to tell the telephone user that his/her phone is off the hook.

Receiver Sensitivity The magnitude of the received signal necessary to produce objective BER or channel noise performance.

Receiving Perforator REPERFORATOR. A telegraph instrument in which the received signals cause the code of the corresponding characters or functions to be punched in a tape.

Recent Change Changes to line and trunk translations in a stored program control switching machine that have not been merged with the permanent data base.

Recipient Switch The switch to which a local number being ported is ported to. Sorry for the mouthful.

Reciprocal Agreements Also called Intercarrier Roaming Agreements. An agreement between two cellular carriers that allows the respective customers of the two carriers to use each others' systems automatically, without the necessity of registering as roamers.

Reciprocal Compensation Imagine a phone call from New York to Los Angeles. It may start with the customer of a new phone company, then proceed to a local phone company (let's say New York Telephone, now called Bell Atlantic). Then it may proceed to a long distance company before ending in Los Angeles and going through another one or two local phone companies before reaching the person dialed. Under the existing rules, all the companies carrying these phone calls have to be paid in some way for their transmission and switching services. There are programs in place such that the company doing the billing and collecting the money pays over some of those monies to the other phone companies in the chain. One such program is called "reciprocal compensation." The opposite of reciprocal compensation is called "Bill and Keep." Under this program, the company billing the call gets to keep all the money. The others in the chain (or most of the others in the chain) get nothing.

Reciprocal Link A hyperlink or link placed on one Web site to return the favor of another site putting a link on their page.

RECO A line item Profit and Loss description for a typical networking services business signifying the four major cost classifications: Resources (People), Equipment, Circuits and Other. RECO is used by countless IBMers.

Recognition Assisted Data Entry Commonly known as Forms Processing.

Recognized Private Operating Agency RPOA. The ITU-T term for a packet interexchange carrier. The status granted to a communications entity by its national government after it pledges to abide by mandatory regulations under Article 44 of the ITU (International Telecommunications Union) convention. For example, a publicly recognized VAN (Value Added Network)

Recognizer A voice recognition term. A system that attempts to classify speech (input utterances) as words from an active vocabulary.

Reconfiguration A fancy word for rearranging equipment, features and options.

Record In a database, a record is a group of related data items treated as one unit of information — for example, your name, address and phone number. Each Record is made up of several fields. A field is simply your last name.

Record Communications Any form of communication which produces a "written" record of the transmission. Teletypewriter and facsimile are examples of record communications. Companies such as RCA Globecom, ITT Worldcom, TRT and MCI, which provide international telex, are known as international record carriers. Before deregulation, that business was exceptionally profitable.

Record Head The electromagnetic device which magnetizes the surface of a magnetic recording — tape, disk, etc. — in proportion to an electrical signal.

Record Length The number of bytes in a record. See RECORD.

Record Locking Think about an airline reservation. You call up. You want to change your reservation. While the airline has your record open, your travel agent calls up to change it. You change your reservation. Your travel agent changes it. Which one ends up in the "permanent" record? Confusion reigns. Clearly it makes sense to only allow one person to access one record at once and lock everyone else out. Record locking is the most common and most sophisticated means for multi-user LAN applications to maintain data integrity. In a record locking system, users are prevented from working on the same data record at the same time. That way, users don't overwrite other users' changes and data integrity is maintained. But though it doesn't allow users into the same record at the same time, record locking does allows multiple users to work on the same file simultaneously. So multi-user access is maximized. Contrast with file locking, which only allows a single user to work on a file at a time.

Recorded Announcement Intercept Provides a recorded message to an intercepted call indicating why the call cannot be completed, as an alternative to attendant intercept or intercept one for DID and CCSA calls to restricted or unassigned numbers.

Recorded Announcement Service A special type of central office trunk which when dialed, will connect the caller to a prerecorded message.

Recorded Answering Device See RAD.

Recorded Telephone Dictation Phone users can dial into centralized telephone dictation equipment. The dictation equipment is usually handled as a trunk connection or it can be wired on an extension level.

Recorder A device many large phone users use to record conversations with their callers. Recording truck dispatches can help a company gain the upper hand in customer service. Purchasing departments may use the recorder to remind vendors of their promises. The financial department can document money transfer orders and investments. Recorders come in several sizes. There are cassette recorders with standard speed and slow extended play speed. Open or reel-to-reel recorders have features similar to cassette recorders. Cassette recorders may be voice-operated (VOX) or started by a recorder coupler. Channel capacities available today include 7, 10, 14, 20, 28, 30, 40, 56 and 60 channels, depending on the manufacturer. Some recorders can search for and recall conversations recorded with an option called "autosearch."

Recorder Warning Tone A one-half second burst of 1400 Hz applied to a telephone line every 15 seconds to indicate to the called party that the calling party is recording the conversation. This tone is required by law to be generated as an integral part of any recording device used for the purpose and is required to be not under the control of the calling party. The tone is recorded together with the conversation.

Recoverability The way a computer or telephone system resumes operation after overcoming a problem with the hardware (say a power failure) or a program error. Some phone systems recover quickly by themselves. Some recover slowly by themselves. Some loose data. Some need human intervention. What causes a system to fail and how and how fast it recovers is key to understand and verify during the test process. This definition from Steve Gladstone, author of the book "Testing Computer Telephony Systems."

Recovery The way a computer or telephone system resumes operation after overcoming a problem with the hardware (say a power failure) or a program error. Some phone systems recover quickly by themselves. Some recover slowly by themselves. Some need human intervention. These are the slowest. Check yours out. If your recovery is slow, and if you local power company is unreliable, you might consider backing your computer up with an uninterruptible power supply.

Rectifier Rectifiers are diodes designed to be placed in an alternating current circuit. When the alternating current flows in the diode's forward direction it passes with no resistance.

When the alternating current reverses direction it is blocked by the diode. Rectified current in such a circuit looks like a series of pulses which are just the positive peaks of the alternating current wave form. In short, rectifiers are used for converting Alternating Current (AC) into Direct Current (DC). AC current comes out of the commercial power supply — 120 volts, 60 Hz. DC power is what drives telephone systems and the circuits that move the transmission around. Typically that DC power ranges from 5 to 48 volts. You need rectifiers to change the AC to DC.

Recursion The ability of a programming language to be able to call functions from within themselves.

RED Random Early Detection. Designed for TCP/IP, a router queuing technique which distributes traffic losses amongst transmitting devices in the event of buffer overflow. RED is accomplished by dropping packets on a random basis, which is determined statistically, when the mean queue depth exceeds a threshold over a period of time. While the queue is still relieved on the basis of FIFO (First In First Out) logic, RED is an improvement over pure FIFO, which simply drops data packets from the tail of the queue. See also FIFO.

Red Alarm In T-1, a red alarm is generated for a locally detected failure such as when a condition like loss of synchronization exists for 2.5 seconds, causing a CGA, (Carrier Group Alarm). See T-1.

Red Black Concept The separation of electrical and electronic circuits, components, equipment, and systems that handle classified plain text (RED) information in electrical signal form from those that handle encrypted or unclassified (BLACK) information.

Red Book Another name for the CD-DA audio CD format introduced by Sony and Philips The Red Book standard defines the number of tracks on the disc that contain digital audio data and the error correction routines that prevent data loss. The format allows 74 minutes of digital sound to be transferred at a rate of 150 kilobytes per second (K/sec).

Red Books The CCITT's 1984 standards recommendations were published in books with red covers, hence the term "Red Books." The CCITT is now called the ITU, as in International Telecommunications Union. See ITU.

Red Box A device that produces tones similar to those produced by dropping coins into a pay phone to inform the operator or automatic machinery that money has been deposited. The red box is used to defraud telephone companies. It is so named because they are usually built small enough to be placed in the "crush proof box" of a packet of Marlboro cigarettes. Red boxes are illegal.

Red Light District On-line pornography.

Redirection In the context of message handling, a transmittal event in which an MTA (Message Transfer Agent) replaces a user among a message's immediate recipients with a user preselected for that message.

Redirection and Forwarding The process whereby the home Mobile Data Intermediate System (MD-IS), upon the receipt of packets encapsulates the packets with the address of the serving MD-IS and forwards them on to the serving MD-IS.

Redirector Networking software that accepts input/output requests for remote files, named pipes, or mailslots and then sends (redirects) them to a network service on another computer. Redirectors (also called network clients) are implemented as file system drivers in Windows 95. A redirector is a LAN software module loaded into every network workstation. It captures application programs requests for file- and equipment-sharing services and routes them through the network for action.

Reduce A Windows term. To minimize a window to an icon at the bottom of the desktop by using the Minimize button or the Minimize command. A minimized application continues running, and you can select the icon to make it the active application.

Redundancy 1. That part of any message which can be eliminated without losing the important information.

2. Having one or more "backup" systems available in case of failure of the main system.

Redundancy Check A technique of error detection involving the transmission of additional data related to the basic data in such a way that the receiving terminal, by comparing the two sets of data, can determine to a high degree of probability whether there has been an error in transmission.

Redundant Array Of Inexpensive Disks RAID. The idea is simple: Put several disk drives into a single housing. Then write your data over the disk drives in such a way that if you lose one or more of the drives, you won't have lost any of your data. Thus the term "redundant." At its simplest, RAID mirrors data to an equal number of disk drives, e.g. two sets of two. At its most complex, RAID writes data across a bunch of drives, so that if one goes the data can be retrieved from the remaining drives. RAID as a concept was first defined in 1987 by Patterson, Gibson and Katz of the University of California, Berkeley. As defined, RAID has three attributes:

1. It is a set of physical disk drives viewed by the user as a single logical device. 2. The user's data is distributed across the physical set of disk drives in a defined manner. 3. Redundant disk capacity is added so that the user's data can be recovered if one (but not more than one) drive fails.

The Berkeley engineers described five levels of RAID configurations called RAID-1 through RAID-5. RAID-0 and RAID-6 have since been added by industry usage. The distinguishing features among the various RAID levels are the way data is distributed and the way redundant capacity is implemented. Each RAID level represents very different trade-offs in terms of cost, availability and performance. Here's a simple explanation of the various levels of RAID:

Level 0: Disk striping across multiple disks. No error correction or redundancy provided.

Level 1: Disk mirroring or shadowing. One disk drive and an exact backup on a second disk, i.e. All data is redundantly recorded ("mirrored") on a second disk.

Level 2: Data is striped across multiple disks, and error checking and correcting (ECC) codes are written onto additional disks for use in fault recovery.

Level 3: Data is striped byte-by-byte across multiple disks and a single additional disk is dedicated to recording parity data.

Level 4: Similar to RAID-3, but stripes data in large chunks. Data is striped block-by-block across multiple disks and a single additional disk is dedicated to recording parity data.

Level 5: The most popular RAID. Data is striped block-by-block across multiple disk, and parity data is also spread out over multiple disks.

Level 6: RAID-5, plus redundant disk controllers, fans buses,etc.

A caveat: The above "levels" are overly simplistic. As Raid has appeared, most manufacturers have implemented different variations on the RAID theme. When I show them the above list, I usually get "Well, that's a beginning." And, of course, some levels are combined. The most popular RAID levels are 0/1 (Zero/One), which is an integral part of NetWare and Level 5, which is not but uses proprietary software techniques. The big difference is that level 0/1 maps the information on one drive to a second. You can always take one drive out, and read it. In Level 5, the data is spread across several drives. You can remove one drive and you won't lose any data. But you can't reconstruct your data from that removed drive. You need the others. When you replace a drive in Raid Level 5 (let's say because it is broken), the others will reconstruct the failed drive fairly quickly — often in less than an hour. Level 0/1 doesn't give you as much total storage space as Level 5.

Redundant Bits The extra bits included in a transmission for purposes of detecting and/or correcting errors. See REDUNDANCY CHECK.

Redundant Link A second connection between a repeater and some other network device like a repeater or switch. One of the connections is active while the other is disabled by the repeater. If link integrity is lost on the active link, it is disabled and the redundant link is enabled so the users are not affected. See also Route Diversity.

Reed Relay Two tiny pieces of metal encapsulated in a tiny nitrogen-filled glass tube. When a current is passed through a magnet around the nitrogen-filled glass capsule, one arm of the metal reed relay moves and makes contact with the other. In this way it acts as a "switch." Reed relay switches are reliable. Because they are metal, they can carry great amounts of data. They are rapidly becoming obsolete.

Re-Engineer To redesign a business process. Re-engineering aims to use the power of information technology to redesign business processes to improve speed, service and quality. See DOWNSIZING.

Re-Gift See regift

Reed-Solomon A means of accomplishing Forward Error Correction (FEC) in order to compensate for errors bursts in created in data transmission. Named for Messrs. Irving S. Reed and Gustave Solomon, staff members of MIT's Lincoln Laboratory, who published a paper entitled "Polynomial Codes over Certain Finite Fields" in the Journal of the Society for Industrial and Applied Mathematics (SIAM) in 1960. Reed-Solomon coding specifies a polynomial by plotting, or statistically sampling, a large number of points in a data block. The coding technique was a quantum leap in forward error correction (FEC) technology, as it allows recovery of data even if multiple errors occurred in a single block, and does so without the requirement for the embedding of redundant data within that block. The decoding process, however, also was challenging; Elwyn Berlekamp, a professor of electrical engineering at the University of California at Berkeley, invented an efficient algorithm for that purpose. Berlekamp's algorithm was used in the Voyager II spacecraft, and is the basis for decoding in CD players. Reed-Solomon is used in MPEG-II (Moving Pictures Experts Group) compression for digital television. The encoder examines the 187 bytes of the MPEG-II data packet (having removed the packet synchronization byte), samples them, and manipulates them as a block; thereby, the contents of the data block can be characterized and described in a 20-byte field appended to the data block. The receiver compares the 187-byte block to the 20-byte description in order to determine its validity. Should errors be detected, their exact location(s) can be identified, they can be corrected, and the original data packet can be reconstructed. As many as 10 byte errors per data packet can be corrected in this fashion.

Reengineer To redesign a business process. Re-engineering aims to use the power of information technology to redesign business processes to improve speed, service and quality. See DOWNSIZING.

Reference Channel Continuously keyed forward-transmission Radio Frequency (RF) channel, used for signal quality assessment.

Reference Clock A clock of high stability and accuracy that is used to govern the frequency of a network and mutually synchronize clocks of lower stability.

Reference Level The measure of a value used as a starting point for further measurements. In communications applications this term usually refers to a power level of a signal or a noise. A common reference level is 0 dBm, that is, 1 milliwatt.

Reference Line In faxing, the reference line is the first scanning line in memory. The location of each black pixel of this line is kept in memory for the next scanned line. Depending on the compression technique used, more or fewer scan lines are necessary.

Reference Number Prompting An AT&T Enhanced Fax Mail term. Reference number prompting is an option that allows you to prompt anyone sending a fax message to your mailbox for a reference number of up to 16 digits.

Reference Track A special magnetic track placed on Floptical diskettes used by the drive to calibrate the optical tracking system with respect to the magnetic recording tracks.

Reference Noise RN. A reference level of noise power.

Referential Integrity Refers to a database's ability to link data in two or more files, so that adding data to a record in one file automatically updates data in another file.

Referral Whois See RWHOIS

Reflectance The ratio of reflected light power to incident light power. Synonym for "return loss."

Reflections RF waves can reflect off of hills, buildings, moving cars, the atmosphere, and basically almost anything in the RF transmission environment. The reflections may vary in phase and strength from the original wave. Reflections are what allow radio waves to reach their targets around corners, behind buildings, under bridges, in parking garages, etc. RF transmissions bend around objects as a result of reflections.

REFLEX One of the communications protocols used between paging towers and the mobile pagers/receivers/beepers themselves. Other protocols are POCSAG, ERMES, GOLAY and FLEX. The same paging tower equipment can transmit messages one moment in GOLAY and the next moment in ERMES, or any of the other protocols.

Reflow Soldering A surface-mounting process for electronic components in which a solder paste is applied to the solder lands on the PCB and the components are properly aligned and placed on them. Upon heating the solder, it melts and forms a solder bond with the component terminals, electronically and mechanically bonding the component to the board.

Refractive Index A ratio of the velocity of light in a vacuum to the velocity of light in another medium, like glass.

Refresh Rate Also called Vertical Scan Frequency or Vertical Scan Rate. The phosphor coating on a monitor tube must be repainted or "refreshed" periodically. Typically, color displays use a low persistence phosphor that must be refreshed 60 times per second, or a rate of 60Hz to 70 Hz or more for VGA and higher resolution monitors. Generally, the faster the refresh rate, the less the flicker. Monochrome displays use a phosphor coating with longer persistence and typically are refreshed at a rate of 50 hertz; this difference accounts for the flicker sometimes seen on color monitors

operating in a monochrome mode. Above 70 Hz, color monitors are considered flicker-free.

Refurbished A term used in the secondary telecom equipment business. Refurbishing means that telephone equipment has been cleaned, polished, resurfaced and whatever else it takes to return the equipment to a "like-new" appearance. Refurbishing usually means it has been completely tested and is ready for installation. But don't take my word for it. Get a written guarantee. See USED, CERTIFIED and REMANUFACTURED.

Regenerate To restore a signal to its original shape. Signals need to be restored because they become distorted and acquire noise during transmission. Analog signals cannot be regenerated because it is very hard for telecommunications equipment to distinguish between unwanted noise and wanted noise (i.e. your voice) in an analog signal. Digital signals can be more easily regenerated since they consist of "ones" and "zeros." If digital signals are flattened or distorted, a simple logic circuit — "Is it a zero or a one?" — can restore the signal to its original clean squared shape.

Regenerated Traffic A telephone company term. Traffic caused by repeated subscriber attempts to seize blocked (busy) equipment.

Regeneration The process of receiving and reconstructing a digital signal so that the amplitudes, waveforms, and timing of its signal elements are constrained within specified limits.

Regenerative Repeater A device which regenerates incoming signals and retransmits these signals on an outgoing circuit. See REGENERATE.

Regenerator A receiver and transmitter combination used to reconstruct signals for digital transmission. In an optical regenerator, the receiver converts incoming optical pulses to electrical pulses, decides whether the pulses are "1s" or "0s," generates "cleaned up" electrical pulses, and then converts them to squared off pulses for transmission.

Regift A verb made popular by Kathleen Thomas of Veronis Suhler in New York City. Ms. Thomas is the recipient of occasional gifts from yours truly. Ms. Thomas, however, is savvy. She realizes that these gifts are, in the main, gifts I received from other people and which I was now passing onto to her, thus saving me money and making Ms. Thomas feel muchly appreciated. Thus her recent email correspondence with me: Harry: Thanks for all the hard work and the upcoming hard work Kathleen: It's been a pleasure. Congrats on a great deal. Harry: Chatkash coming. You know what Chatkash are? Kathleen: Yes. It's all the stuff you've been re-gifting to me.

Regina The queen. A deserved title. If only I were 20 years younger, more handsome and much smarter. Three traits to solve the riddley.

Regional Bell Operating Company RBOC. Also called Regional Holding Company or RHC. One of the seven (now five) Bell operating companies set up after Divestiture. each of which own two or more BOCs (Bell Operating Companies). The RBOCs were carved out of the old AT&T/Bell System by Judge Harold Greene when he signed off on the divestiture of the Bell operating companies from AT&T at the end of 1984. There is nothing magical about seven — nor the grouping of Bell Operating Companies (BOCs) into RBOCs — except the Judge wanted to keep them all roughly the same size, so he personally assigned which Bell operating company would join which Regional Bell Operating Company. The seven RBOCs are Ameritech, Bell Atlantic, BellSouth, NYNEX, Pacific Telesis, Southwestern Bell and US West. In early October, 1994, Southwestern Bell changed its name to SBC Communications, Inc. But its telephone companies, it said, would still operate under the Southwestern name. In April, 1996 Bell Atlantic bought NYNEX (the holding company for New York Telephone and New England Telephone) for $22.1 billion. The new company will be called Bell Atlantic. In April also, SBC Communications (the name for the holding company owning Southwestern Bell Telephone) bought Pacific Telesis (the holding company for Pacific Bell) for $16.7 billion. As of early 1998, this left five regional Bell operating companies — Ameritech, Bell Atlantic, BellSouth, SBC Communications and US West.

Terms of the Divestiture placed business restrictions on AT&T and the BOCs. Those restrictions were threefold: The BOCs weren't allowed into long distance, equipment manufacturing, or information services. AT&T wasn't allowed into local telecommunications (i.e. to compete with the BOCs). But it was allowed into computers. The federal Judge overseeing Divestiture, Judge Harold Greene, is slowing the lifting the restrictions. He has allowed the BOCs into information services and AT&T into local service. He has stayed firm on the other two — no equipment manufacturing and no long distance for the RHCs (also called RBOCs).

Regional Center A control center (Class 1 office) connecting sectional centers of the telephone system together.

Regional Holding Company RHC. Also called Regional Bell Operating Company. See Regional Bell Operating Company.

Register 1. See Traffic Register.

2. A temporary-memory device used to receive, hold, and transfer data (usually a computer word) to be operated upon by a processing unit. The register holds the information for manipulation by the telephone system or a computer. In an automatic telephone system, a register receives dialed pulses or pushbutton tones and then uses that information to control the switch. Computers typically contain a variety of registers. General-purpose registers perform such functions as accumulating arithmetic results. Other registers hold the instruction being executed, the address of a storage location, or data being retrieved from or sent to storage. Other words associated with "register" include buffer, fetch protection, M-sequence, read-only storage, permanent storage, random-access memory and shift register.

Register Differences The difference in traffic register reading after a specified time has elapsed. See also Traffic Register.

Registered Access In the context of message handling services, access to the service performed by subscribers who have been registered by the service provider to use the service.

Registered Jack RJ. Any of the RJ series of jacks, described in the Code of Federal Regulations, Title 47, part 68 used to provide interface to the public telephone network. See also RJ-11, RJ-45.

Registered user A user of a Web site with a recorded name and password. In a FrontPage web, you can register users with a WebBot Registration component.

Registrant See gTLD.

Registration Sequence Count An 8-bit counter maintained by the Mobile End System (M-ES) and incremented on each successful establishment of a data link connection with a serving Mobile Data Intermediate System (MD-IS). Used to prevent registration errors due to varying network transit delays between serving MD-IS and home MD-IS.

Registration Statement A statement, required by

Section 76.12 of the FCC Rules, which is used to notify the FCC that one or more broadcast stations will be carried by the cable television system in a specified Community Unit.

Registered Terminal Equipment Terminal equipment which is registered for connection to the telecommunications network in accordance with Subpart C of Part 68 of the FCC's Rules. If a terminal device has been properly registered it will have an identification number permanently affixed to it.

Registers An ISDN term. Registers are named storage areas for numbers or strings of characters that control the operation of the ISDN set.

Registration The address registration function is the mechanism by which Clients provide address information to the LAN Emulation Server.

Registration Number (FCC Part 68) Approval number given to telephone equipment to certify that a particular device passes the tests defined in Part 68 of the FCC Rules. These tests certify the phone won't cause any harm to the public network. They do not attest to the commercial value of the product, nor whether it will (or won't) sell. See REGISTRATION PROGRAM.

Registration Program The Federal Communications Commission program and associated directives intended to assure that all connected terminal equipment and protective circuitry will not harm the public switched telephone network or certain private line services. The program requires the registering of terminal equipment and protective circuitry in accordance with Subpart C of part 68, Title 47 of the Code of Federal Regulations. This includes the assignment of identification numbers to the equipment and the testing of the equipment. The registration program contains no requirement that accepted terminal equipment be compatible with, or function with, the network. In other words, a product registered under Part 68 doesn't mean that the product will actually work — i.e. make and receive phone calls (or whatever). Part 68 simply says it won't cause any harm to the network. See REGISTRATION NUMBER and PART 68.

Registration Timer Values Time values passed from Mobile Data Intermediate System (MD-IS) to a Mobile End System (M-ES) to inform the M-ES of the period of registration. The M-ES must register again prior to expiration of the registration timer.

REGNOT REGistration NOTification. A wireless term for the message sent from the VLR (Visitor Location Register) to the HLR (Home Location Register). The VLR is a SS7 database residing on the SCP (Signal Control Point) of the cellular provider in whose territory you are roaming. The REGNOT is sent to the HLR database, which resides on the SCP of your service provider of record in order to verify your legitimacy and to determine the features to which you have subscribed. Confirmation of the REGNOT is in the form of a "regnot" (lower case), sent over the SS7 network from the HLR to the VLR. See also HLR, SCP, SS7, VLR.

Regression Analysis A method of forecasting the future by plotting events in the past and assuming there'll be some similarity in the future. About as accurate as any other pseudo scientific method.

Regression Testing

Regulated 1. Controlled for uniformity. Many aspects of telecommunications are regulated — from the input voltage powering a telecom system to the output signal of a microwave system.

2. Adhering to the rules, regulations and sundry whims of a government agency. Most aspects of the telephone business

are under the control of a government agency to some degree. Their rules cover everything from certifying of expenses which may be capitalized to specifying how many seconds the subscriber can be forced to wait for dial tone (three seconds). Stripped to bare essentials, a regulatory agency can only do two things. First, it can allow the regulated entity to raise its prices to a point where nobody wants to buy anymore. Second, it can stop competitors coming into the business. The first (high price) is the reason no one (or few people, anyway) send telegrams. The second (keep out the competition) reason gets stymied because new technology — e.g. cheap local microwave — comes along to force the regulatory agency's hand. In the long run, no regulated entity survives because it has a regulated monopoly. It survives because it provides good service at a fair price.

Regulated Charger An uninterruptible power supply (UPS) definition. Without a regulated charger, batteries can be insufficiently charged or blistered with too much charge voltage. Either case can cause permanent damage.

Regulation See POWER CONDITIONING.

Regulatory Groups Refers to local, State or Federal entities that issue orders, findings, etc. that are binding upon providers and users of telecommunications and services.

Rehomes See Rehoming

Rehoming A major network change which involves moving a customer's local loop termination from one Central Office wire center to another. Rehoming generally involves the retermination of private line facilities, although it can simply involve local loop termination for purposes of access to switched services. Rehomes also can be for the purposes of the carrier, perhaps in connection with a switch upgrade or switch move/decommission.

Reinforced Concrete A type of construction in which steel reinforcement and concrete are combined, with the steel resisting tension and the concrete resisting compression.

REJ Abbreviation for REJect.

Rejection A word used in voice recognition to mean a type of recognition classification where the input utterance did not meet the criteria necessary to be classified as a word in the active vocabulary. Usually the speaker is asked to repeat the utterance.

REL RELease message. The fifth of the ISUP call set-up messages. A message sent in either direction indicating that the circuit identified in the message is being released due to the reason (cause) supplied and is ready to be put into the idle state on receipt of the Release Complete Message. See ISUP and COMMON CHANNEL SIGNALING.

Relation Synonym for table.

Relational Database A database that is organized and accessed according to relationships between data items. A relational database consists of tables, rows and columns. In its simplest conception, a relational database is actually a collection of data files that "relate" to each other through at least one common field. For example, one's employee number can be the common thread through several data files — payroll, telephone directory, etc. One's employee number might thus be a good way of relating all the files together in one gigantic data base management system (DBMS).

Relationship Routing A concept introduced by automatic call distributor manufacturer, Aspect Telecommunications, to have callers' calls routed to agents they had previously developed

Relative Transmission Level The ratio of the test tone power at one point to the test tone power at some other point

in the system chosen as a reference point.

Relative URL The Internet address of a page or other World Wide Web resource with respect to the Internet address of the current page. A relative URL gives the path from the current location of the page to the location of the destination page or resource. A relative URL can optionally include a protocol. For example, the relative URL doc/harry.htm refers to the page harry.htm in the directory doc, below the current directory.

Relay An electrically activated switch used to operate a circuit. It connects one set of wires to another. Usually, the relay is operated by low voltage electric current and is used to open or close another circuit, which is of much higher voltage. Older telephone switches used many relays to switch (i.e. complete) their calls. Relays come in many forms. There are hermetically-sealed relays, in which thin metal contacts are sealed in an airtight glass or metal enclosure. There are also mercury relays in which a small tube of mercury tilts and completes or breaks a circuit. See also REED RELAY.

Relay Rack Open iron work designed to mount and support electronic equipment. A relay rack is to electronic equipment what a distribution frame is to wire. See DISTRIBUTION FRAME.

Relayer Allows a user to open or close a solenoid via the phone system.

Relaying A function of a layer by means of which a layer entity receives data from a corresponding entity and transmits it to another corresponding entity.

Release 1. A call comes into a switchboard. The operator calls you to tell you it's for you. Then he/she "releases" the call to you. On most switchboards there's a button labelled "RLS." That's the release button. On some phones (not consoles) the release button is the "hang-up" button. Hitting this button means disconnecting the call. Be careful.
2. The ending of an inbound ACD call by hanging up.
3. The feature key on most ACD instruments labelled Release.
4. A term used in the secondary telecom equipment business. The relinquishing of a piece of equipment to a purchaser or user upon fulfillment or anticipated fulfillment of contractual obligations, whether written or oral.

Release Button The release button — found always on operator consoles and occasionally on some phones — ends a call in the same way that hanging up the receiver does.

Release Link Capability The ability for an originating switching system, on receipt of a new destination address from the current terminating switching system, to release the transmission link to that terminating switching system and continue call processing using the new destination address. Definition from Bellcore in reference to its concept of the Advanced Intelligent Network.

Release Link Trunk RLT. Telecommunications channel used with Centralized Attendant Service to connect attendant-seeking calls from a branch location to a main location.

Release With Howler If a phone stays off-hook without originating a call (or the receiver is accidentally knocked off), the system transmits a loud tone over the line and then disconnects the line and the phone. The central office effectively then ignores them (the line and the phone) until someone puts the receiver back on-hook again.

Reliable Sequenced Delivery The delivery of a set of Protocol Data Unit (PDUs) from a source to a destination with no errors in any PDU, in the order transmitted, and without gaps or duplicates.

Reliable Service Area RSA. The area specified by the field strength contour within which the reliability of communication service is 90 percent for a mobile unit.

Reliability A measure of how dependable a system is once you actually use it. Very different from MTBF (Mean Time Between Failures). And very different from availability. See MTBF.

Relief Relief refers to providing additional equipment to accommodate growth in customer demand.

Relocatable Code Machine language programs that can reside in any portion of memory.

Remailer Remailers are anonymous mail drops that computer hackers have set up on the Internet, untraceable electronic mail addresses where one can send or receive encrypted data. An article in the October, 1994 issue of High Times, a drug related magazine, offered plans for a similar security system as a remailer, adding one interesting twist. By incorporating a computer virus like Viper or Decide in the system, the computer could be programmed essentially to self-destruct as soon as it detected a security breach, thus rendering it worthless as evidence.

Remanufactured Equipment, parts and/or systems that have been repaired and upgraded to the latest higher revision level. The remanufacturing process makes the telecom equipment (used or new) into a finished product that is the latest release and ready for resale. Remanufactured is the term for the highest level of refurbishing equipment.

Remapping The practice of redefining the meaning of keys on the keyboard.

Remind Delay The period of time from when a call is put on hold to when a reminder tone is heard and a message appears on the telephone display.

Remission IBM-speak to change the mission of a product or a facility.

Remodulator In a split broadband cable system, a digital device at the headend that recovers the digital data from the inbound analog signal and then retransmits the data on the outbound frequency.

Remote Pertaining to a system or device that is accessed through a telephone line. The opposite is local. See Remote Access and RAS.

Remote Access Sending and receiving data to and from a computer or controlling a computer with terminals or PCs connected through communications (i.e. phone) links.

Remote Access Concentrator See RAC.

Remote Access Device RAD. Typically, a remote access device (also called a Remote Access Server) is a piece of computer hardware which sits on a corporate LAN and into which employees dial to get access to their files and their email. Remote access devices are also used by commercial service providers, such as Internet Access Providers (ISPs) to allow their customers access into their networks. For longer explanatiolns, see also Remote Access Server and Universal Edge Server.

Remote Access Server RAS. A remote access server (also called a Remote Access Device or in a bigger version, a Remote Access Concentrator) is a piece of computer hardware which sits on a corporate LAN and into which employees dial on the public switched telephone network to get access to their email and to software and data on the corporate LAN (e.g. status on customer orders). Remote access servers are also used by commercial service providers, such as Internet Access Providers (ISPs) to allow their customers access into their networks. Remote Access Servers are typically measured by how many simultaneous dial-in users (on analog or digital lines) they can handle and whether they can

work with cheaper digital circuits, such as T-1 and E-1 connections. See also Remote Access Concentrator and Universal Edge Server.

Remote Access Services Software that enable distant PCs and workstations to get into a Remote Access Server to get to software and data on a corporate LAN. Remote access services are provided through modems, analog telephones or digital ISDN lines. Remote access services is For a much longer explanation, see Remote Access (Ref: Hands-On Networking Essentials, M.J. Palmer, Course Technology, Cambridge, MA, 1998, p. 293)

Remote Access To PBX Services Allows a user outside the PBX to access the PBX by dialing it over a normal phone line. You dial the number. It answers. It may or may not say anything. It may just give you dial tone. You now punch in an authorization code. If your code is acceptable, the PBX gives you another dial tone. That dial tone is effectively the one all users within the PBX get. Once you have this dial tone, you can dial another extension, jump on the company's WATS network, get into the dictation unit, access its voice mail, or whatever. Suffice, you are inside the PBX. You can do whatever anyone else inside the PBX can do.

Remote Adapted Routing The adaptation of backbone routing techniques that take into account; slow-line communications links, intermittent connections, security, charity chatty routing protocols, management, and user ergonomics.

Remote Batch Processing Processing in a computer system in which batch programs and batch data are entered from a remote terminal or a remote PC (personal computer) over phone lines.

Remote Bridge A bridge between two or more similar networks on remote sites. Dial up or leased lines typically require a local bridge or gateway and a remote bridge or gateway an each end, in order to network.

Remote Call Forwarding RCF. This is a neat service. It allows a customer to have a local telephone number in a distant city. Every time someone calls that number, that call is forwarded to you in your city. Remote call forwarding is very much like call forwarding on a local residential line, except that you have no phone, no office and no physical presence in that distant city. Remote Call Forwarding exists purely in the central office. You can also think of it as measured Foreign Exchange. Companies buy Remote Call Forwarding for three reasons: 1. To encourage distant customers to call them by giving them a local number in their own city to call. (This the most obvious reason for an IN-WATS line, a FX or a RCF line); 2. They buy RCF over IN-WATS or FX lines because they don't have the volume to justify these potentially more expensive lines. 3. Companies buy RCF lines as overflow lines from IN-WATS and FX lines. They use their RCF lines when the other lines (FX and IN-WATS) get busy during peak busy periods. Remote Call Forwarding calls are typically charged at the same price as normal DDD calls (i.e. the most expensive to call). And you can't, as yet, reprogram RCF calls easily. You have to place an order with your friendly telco and wait for them to do the reprogramming. In 1994 Northern Telecom announced a service whereby remote call forwarded calls carried from its central office would carry the calling line ID number. This way when the call reached the distant number, it could be handled in an intelligent fashion.

Remote Concentrator A remote multiplexer. A device which places more than one distant user on two cable pairs. The idea of a remote concentrator is to substitute electronics for cable. It's simply cheaper to put electronics on either end of two cable pairs and drive many conversations through those wires, than running extra cables (digging streets, erecting poles, etc.)

Remote Control Remote control software allows a remote PC to connect to the network via a PC that is on the LAN. You must use such software for working from home, for sending in your work, checking on your email, etc. See REMOTE NODE.

Remote Diagnostics You own a phone system. You have a service company. There's some problem with it. Instead of sending a technician out, your service company dials your PBX from a data terminal or PC and "asks" your PBX in computerese what's wrong with it. If it isn't too broken, it will come back and give you some indication. This is called remote diagnostics. Some service companies call all their customers' phone systems every morning and run routine remote diagnostics on their switch. It's like going to the doctor for a daily physical. Sometimes this test may find a problem before the user is even aware. Sometimes the problem can be repaired on-line. If not, the service company will have to dispatch a technician. Remote diagnostics is a good idea. More phone systems should have it. To do remote diagnostics on a telephone system, you will typically need a phone line dedicated to the PBX and a modem on either end.

Remote Digital Loopback A test that checks the phone link and a remote modem's transmitter and receiver. Data entered from the keyboard is transmitted from the initiating modem, received by the remote modem's receiver, looped through its transmitter, and returned to the local screen for verification.

Remote IP A telephone company AIN term. When an SCP/Adjunct requests a local AIN switch to make a connection to an IP to which the AIN Switch does not have a direct ISDN connection, the indicated IP is referred to as a remote IP.

Remote Job Entry RJE. Remote Job Entry occurs in computer operations where work or input is sent in remotely over phone lines. That "work" might include the day's sales of a distant store.

Remote LAN Interconnection The connection of two or more LANs which are remotely located from each other so that LAN users can communicate with users and servers on any of the interconnected LANs.

Remote Line Concentrator A multiplexer. A device that "concentrates" several users' lines on a fewer number of trunks. Typically, a remote line concentrator is used because it's cheaper, easier or more flexible to substitute electronics for cable. See REMOTE CONCENTRATOR.

Remote Line Switch A line unit mounted near a cluster of users and equipped with intracalling capability.

Remote Line Unit A remote line concentrator without intracalling capability. See REMOTE CONCENTRATOR and REMOTE LINE SWITCH.

Remote Live Screening See LCS.

Remote Maintenance Facility See REMOTE DIAGNOSTICS.

Remote Monitoring A call center term. Remote Monitoring is most frequently used by service agency clients. This is the process whereby a qualified/authorized party can dial into a remote call center and monitor certain telephone calls. The process is usually administered from a specially designated room or place away from the agent's work area. The agent may or may not know that the specific call is being monitored.

Remote Node Remote Node software allows remote users to dial in to the corporate LAN and work with the applications

and data on the LAN as if they were "actually in the office." By dialing in, they become a node on the LAN. Using a PC, Mac, or UNIX workstation, a modem; and a remote access server, employees can connect from any location in the world that has an analog, a switched digital, or a wireless connection.

Remote Office Test Line ROTL. A testing device that acts in conjunction with a central controller and a responder to make two-way transmission and supervision measurements.

Remote Operations Service Element ROSE. An application layer protocol that provides the capability to perform remote operations at a remote process. Definition from Bellcore in reference to its concept of the Advanced Intelligent Network.

Remote Order Wire An order wire is a line on which maintenance and monitoring is done. A remote order wire is an order wire that has been extended to a distant point that may be more convenient.

Remote Procedures Call RPC. A message-passing facility that allows a distributed program to call services available on various computers in a network. Used during remote administration of computers, RPC provides a procedural view, rather than a transport-centered view, of networked operations.

Remote Programming Dial your phone system with your friendly personal computer, modem and a communications software package and you can change the telephone system's programming remotely. This feature is great for companies with telephone systems in many locations. They can all be run from one central point. This feature is also great if you want some changes made on your system. It's obviously a lot cheaper for your vendor to make those changes from his office rather than have to visit yours. It's also a lot faster. See REMOTE DIAGNOSTICS.

Remote Site The remote site is the person or location doing the sending in a file transfer operation. An example: Sales reps in the field typically update the central database on a periodic basis. The central database location is known as the host and the sales reps in the field are doing so from remote locations.

Remote Site Location A location for a DCE device which is not at the central or control site. A typical application would have a terminal at the remote site and the host computer at the central or control site.

Remote Station Any piece of equipment attached to a LAN by a telephone company supplied link. Technically, that includes all devices that aren't servers. Usually it refers to a workstation at a distant location, linked to the main LAN by a modem and connected through a serial port "gateway." See MODEM.

Remote Station Lamp Field For use at multi-line phones, usually manned by secretaries who answer many phone lines.

Remote Switching System An switch that is away from its host or control office. All or most of the central control equipment for the RSU is located in the host or control office. See also REMOTE CONCENTRATOR.

Remote Terminal A terminal connected to a computer over a phone line.

Remote Traffic Measurement Traffic and feature usage data can be transmitted by the system to a distant service technician.

Remote User Agent RUA. An X.400 standard user agent that interfaces with the MS (Message Store) for remote X.400 communications.

Remote Workstation A terminal or personal computer

connected to the LAN (local area network) by a modem. A remote workstation can be either a standalone computer or a workstation on another network.

Remotely-Hosted An SCSA definition. The Client and the Server are different (i.e., the application is on a different physical box than the service provider).

Removable Cartridge System A high-capacity storage system that can be removed from the PC. A removable cartridge systems consists of a drive mechanism and the cartridges used to store data. The most well-known removable cartridge system is the Bernoulli Box by Iomega Corp.

Removable Media Diskettes or cartridges that can be removed from a computer drive. For example, a Bernoulli box uses removable cartridges.

REN Ringer Equivalency Number. Part of the FCC certification number approving a telephone terminal product for direct sale to the end user as not doing harm to the network. The REN consists of a number and a letter which indicates the frequency response of that telephone's ringer. "A" = 20 Hz or 30 Hz "B" = a range from 15.3 Hz to 68 Hz. The remaining letters represent ringers that will work on very narrow ranges such as "C" = 15.3 Hz to 17.4 Hz, etc. The number indicates the quantity of ringers which may be connected to a single telephone line and still all ring. The total of all RENs of the telephones connected to the one line must not exceed the value 5 or some or all of the ringers may not operate.

Rendezvous Controls A concept introduced in TAPI 3.0. The Rendezvous Controls are a set of COM components that abstract the concept of a conference directory, providing a mechanism to advertise new multicast conferences and to discover existing ones. They provide a common schema (SDP) for conference announcement, as well as scriptable interfaces, authentication, encryption, and access control features. See TAPI 3.0.

Rent-A-Wreck In the 1980s, the advent of new technology in the telephone industry made telephone service much more reliable. Fewer people were needed to run phone companies. Telephone company managers soon discovered that they could increase their companies' profitability by firing these managers, which they dutifully did. Unfortunately, by the time the mid-1990s came around, telephone managers discovered they were lacking the experienced expertise necessary to run their phone company. Bingo, the answer: hire the old managers back as consultants. Around NYNEX, this program was known affectionately as "Rent-A-Wreck."

Reorder An announcement, or 120 interruptions per minute tone, returned to the caller when his call is blocked in the network. See REORDER TONE.

Reorder Tone The Reorder tone sounds like a busy signal but is twice as fast, i.e. a reorder tone is a tone applied 120 times per minute. The tone means that all switching paths are busy, all toll trunks are busy, there are equipment blockages, the caller dialed an unassigned code, or the digits he dialed got messed up along the way. Also called Channel Busy or Fast Busy Tone.

Reorg A shortened form of the word reorganization. It is used by people in companies which go through management reorganizations so often they don't have to figure what the latest organization means before the next one happens. And they certainly don't have the time to say the word "reorganization" in full. I first heard the word "reorg" from someone at Pacific Bell. He used the word as an excuse for not following up on something he had promised me.

Rep Repertory dialing. Speed dialing. Some cellular phones

are capable of storing 100 numbers.

REPACCS REmote cable PAir Cross-Connect System is a PC controlled, metallic, automated cross-connect system that may be applied to Automated Distribution Frames, Building Terminals, Service Area Interfaces (SAIs) or cross-connect boxes and closures/terminals. It dramatically reduces dispatches, provides 100% record accuracy, facilitates multiple line testing (MLT), and operates without local or battery power as well as keeps people out of restricted or hazardous areas.

Repair And Quick Clean RQC. A term in the industry which repairs telecom equipment. It means all equipment is repaired and fully tested with a burn-in (if required) and an operational systems test. It also includes minor cosmetic cleaning of the unit. Definition courtesy Nitsuko America. See also LIKE NEW REPAIR AND UPDATE and REPAIR, UPDATE AND REFURBISH.

Repair, Update And Refurbish RUR. A term in the industry which repairs telecom equipment. It means equipment is repaired and updated to current manufacturer's specifications. Also includes minor cosmetic cleaning of metal cabinets, a full diagnostic test with burn-in (if required) and an operational test. Definition courtesy Nitsuko America. See also LIKE NEW REPAIR AND UPDATE and QUICK CLEAN.

Repair Only A term used in the secondary telecom equipment business. Equipment is repaired to original working condition, but does not include refurbishment or recycling except where required to bring equipment to working condition. See also REFURBISHED and REMANUFACTURED.

Repair Service Answering RSA. Functions that support the initial handling and entry of subscriber reported troubles. They enable subscribers to request trouble verification tests, to initiate a trouble report and to obtain information on the status of an open trouble report. Definition from Bellcore in reference to its concept of the Advanced Intelligent Network.

Reparameterize a verb that means to change the current or default behavior of a software component or application by supplying new data (parameters) at time of execution.

Repartee Active Voice's voice processing system that combines automated attendant, voice mail, audiotext and facsimile features. Active Voice is in Seattle, WA.

Repeat The act of a station receiving a code-bit stream (frame or token) from an upstream station and placing it onto the ring to its downstream neighbor. The repeating station may examine, copy to a buffer, or modify control bits in the code-bit stream as appropriate.

Repeat Call The name of a Bell Atlantic service. Dialing a busy number over and over is as time-consuming as it is frustrating. With Repeat Call, your phone will continuously monitor a busy number every 45 seconds for up to 30 minutes, without interrupting your incoming or outgoing calls. So you and your employees can do other things until your phone alerts you with a special ring when the call got through.

Repeat Dialing A name for a Pacific Bell (and possibly other local telephone companies') service which automatically checks a busy number and when the line is free, it rings you back and completes the call.

Repeater 1. An opto-electronic device inserted at intervals along a circuit to boost, and amplify an analog signal being transmitted. A repeater is needed because the quality and strength of a signal decays over distance. You will find repeaters in cables and in microwave systems. Repeaters may also regenerate a digital signal — "squaring it" and "cleaning" it up — but not changing it. Regenerating the signal removes noise and thus reduces the likelihood of errors. You can only regenerate digital signals. You cannot regenerate analog signals. You can regenerate digital signals because a machine can tell what's a signal and what's noise in a digital signal. But no machine exists to do that with an analog signal.

2. The simplest type of LAN interconnection device. A repeater moves all received packets or frames between LAN segments. The primary function of a repeater is to extend the length of the network media, i.e. the cable.

Repeater Coil Got a long local loop carrying an analog signal? The signal gets weaker the further it goes. You have to amplify the signal. Repeater coils do that. They're usually installed every 5,000 feet.

Repeater Hop The action of a data transmission passing through a repeater in a communications circuit. IEEE 802.3 standards specify the number of repeater hops allowed for various types of repeaters. For example, Class II repeaters allow up to two repeater hops per segment.

Repeater Set A repeater unit plus its associated physical layers interfaces (MAUs or PHYs).

Repeating Coil A transformer which connects one telephone circuit with another without any DC connection between the circuits. Here's a more technical explanation: A voice-frequency transformer characterized by a closed core, a pair of identical balanced primary (line) windings, a pair of identical but not necessarily balanced secondary (drop) windings, and a low transmission loss at voice frequencies. It permits transfer of voice currents from one winding to another by magnetic induction, matches line and drop impedances, and prevents direct conduction between the line and the drop.

Reperforator In teletypewriter systems, a device used to punch a tape in accordance with arriving signals, permitting reproduction of the signals for retransmission. See also CHAD.

Reperforator/Transmitter RT. A teletypewriter unit consisting of a reperforator and a tape transmitter, each independent of the other.

Repertory Dialing Sometimes known as "memory dialing" or "speed-calling." A feature that allows you to recall from nine to 99 (or more) phone numbers from a phone's memory with the touch of just one, two or three buttons.

Replica A copy. See Replication.

Replication Also known as data replication. Replication is the process by which a file, a database or some other computer information in one location is updated to match a mirrored version on another computer in another location. Replication includes the process of duplicating and updating data in multiple computers on a network, some of which are permanently connected to the networks. Others, such as laptops, may only be connected at intermittent times. The idea is twofold: Everyone should have access to the same information in the database/s. Second, many people can make changes to the same record and somehow, all those changes will meld themselves into the database/s and thus, everyone will have access to the new, updated information. In the old days (i.e. pre-Lotus Notes), networked databases were stored in one place, e.g. an airline database of reservations. Everyone who wanted to access information in the catabase needed to be physically connected to the network through some form of phone line. That's still the case in most databases. Along came Lotus Notes whose major claim was everyone could create their own database/s and carry it with them on their laptops and everyone could put their own information in and Lotus Notes would update the central database and update everyone's database every time they logged into the network. If Lotus is confused, it sends messages out asking for clarification as to what the

right database entry was. In short, replication is a far more complex process than what the traditional English language definition of replication is, namely making copies of itself. In data replication, it's the coordination, updating and reconciling of constantly-being-changed databases. That's the hard part. As I wrote this, Lotus Notes had big lead in this process of database replication. But others, like Oracle and Microsoft, were trying to catch up. The easiest replication strategy is one-way transfer. A simple case of one-way data replication is a mobile user who needs to update the information on his laptop, but not to update any information at the corporate site. See also SYNCHRONIZATION.

Reply 1. A transmitted message which serves as a response to an original message. (What else?)

2. An SCSA definition. An event which is a service provider's response to a synchronous or asynchronous request.

Report Mining Coined by Gartner Group, Inc. for migrating legacy report data to a server so it can be accessed by desktop query tools and regenerated into a new report.

Report Program Generator A computer language for processing large data files.

Report-Only Event An event that the ASC (AIN Switch Capabilities) reports to a SLEE (Service Logic Execution Environment) but the ASC does not suspend processing events for the connection segment. Definition from Bellcore in reference to its concept of the Advanced Intelligent Network.

Repository A database of information about objects and components. Synonyms include library and encyclopedia.

Repudiable Messages that are repudiable are messages that you can deny receiving. Messages that are non-repudiable are messages that you cannot deny receiving, i.e. the system tracks that you received the message.

Request The formatted information that is sent to the switching domain as a result of a computing domain issuing a service across the service boundary.

Request For Comments A standard or a standard-defining document for the Internet. Individual RFCs define specific aspects of Internet operation. See RFC.

Request To Send RTS. One of the control signals on a standard RS- 232-C connector. It places the modem in the originate mode so it can begin to send.

Required For Service Date A telephone company term. This is the date beyond which service impairment may be expected to occur if equipment relief is not available. This date is used for the Timing Arrow. By this date all balancing, testing, rearrangements, and trunk relief must have been concluded.

Rerouting A short-term change in the routing of telephone traffic. Rerouting may be planned and recurring or a reaction to a nonrecurring situation.

RES Residential Enhanced Service.

Resale Buying local and/or long distance phone lines in quantity at wholesale rates and then selling them to someone else, hopefully at a profit.

Resale Carrier A long distance company that does not own its own transmission lines. It buys lines from other carriers and then resells them to its subscribers. Some resale carriers have their own switches. Some don't. Some have a mix of their own lines and leased lines. Most long distance carriers — including AT&T, MCI and Sprint — have a mix of their own lines and leased lines.

Resampling Reducing or increasing the number of pixels in an image to conform to a new size or resolution.

Reseller A company which purchases a block of cellular numbers from a cellular carrier for resale to its customers. Or a company which purchases a big block of long distance calling minutes or resale in smaller blocks to its customers. See AGGREGATOR.

Reserve Power A telephone system may be equipped with storage batteries to provide primary power during a commercial power failure. No loss of service will occur during transition to battery power. All this is a long way of saying your phone system is backed by batteries, typically lead acid (the same ones used in your car).

Reset To restore a device to its default or original state. To restore a counter or logic device to a known state, often a zero output. In computer lingo, to reset a computer is simply to turn its power off, wait ten seconds and then turn it on. It's also called cold booting the computer.

Reset Packet A packet that identifies error conditions on an X.25 communications circuit. The reset packet does not clear the session but rather notifies the communicating DTEs of error conditions at a known point in the data-packet transfer sequence.

Resident Command A command located in the personal computer's operating system itself, contained in the file COMMAND.COM.

Resident Program See RAM-RESIDENT PROGRAM.

Residential And Light Commercial Wiring Refers to the wiring system and all of its appurtenances required to provide convenient and useful telephone services to residences and light commercial buildings.

Residual Error Rate The ratio of the number of bits, unit elements, characters or blocks incorrectly received but undetected or uncorrected by the error-control equipment to the total number of bits, unit elements, characters or blocks sent.

Resipiscence The noun that comes from the verb to see the error of one's ways.

Resistance Resistance is the opposition to the flow of electric charge and is generally a function of the number of free electrons available to conduct the electric current. When the same amount of voltage is applied to an insulator as to a metal, less current flows through the insulator than the metal because the insulator has fewer free electrons. Resistance is a property intrinsic to a conducting material. Insulators have high resistance values while good conductors like copper have low resistance values. Other factors influence resistance as well. For example, the resistance of a conductor is directly proportional to its length: the longer the wire, the greater the resistance.

In short, any electrical conductor will resist the flow of electrical current. As it resists the flow of current, so the current becomes weaker. Resistance generates heat and occasionally light. It is technically defined as a property or a characteristic of a conductor, i.e. the metal through which the electricity flows. It is measured in Ohms.

Resistor A component made of a material (such as carbon) that has a specified resistance or opposition to the flow of electrical current. A resistor is designed to oppose but not completely obstruct the passage of electrical current.

Resort An electronic destination you return to, time and again, to escape normal business.

Resource 1. Any facility of a computing system or operating system required by a job or task, including memory, input/output devices, processing unit, data files, and control or processing programs.
2. A network component such as a file, printer, or serial device that is shared by other components of the network.
3. An SCSA term. A voice processing technology, such as

voice store and forward, fax processing, voice recognition, or text to speech. Here's the official definition: The abstraction of a standardized vendor-independent interface of a physical device used for call processing as seen by the Application. All Resources have common methods across all implementations. Examples of Resources are voice store and forward, fax send and receive, text to speech conversion, voice recognition, etc. Resources are assumed to have at a minimum one input or output of circuit switched TDM on the internal switch fabric of the system. Resources are shared among multiple applications. Once a Resource has been allocated to an application, it is locked from the use of any other application until freed. It is assumed that applications specify at some level their resource requirements to the server prior to accessing them. This may either be through explicitly attaching them to a Group, or having the server implicitly allocate them based on usage.
4. A Windows term. A resource is a program object, such as a button, menu or dialog box, that Windows treats differently than normal programs. Resources are developed either in a special resource language or using interactive tools. They can be loaded from separate files or bound directly to the executable file.

Resource Characteristics An SCSA definition. Resources have a set of characteristics that define their behavior beyond that defined by the methods on their interfaces. These may vary or be optional for this Resource Type. Resource may be selected by the application on the basis of their characteristics and a client may query for the characteristics of a resource which it has claimed. See RESOURCE.

Resource Class An SCSA definition. A set of methods (in object-oriented terms, a class) for controlling resource instances (or a resource). May be abbreviated as just "class." See RESOURCE.

Resource Group An SCSA term. A resource group is a dynamically formed group of resource units that can be made to work together as if they were a single device. See RESOURCE.

Resource Module A resource module is a card that slides into a PC and does everything from text-to-speech, to fax, to voice recognition, etc. Everything except interfacing to the network, which is done by another card or another part of the resource module card called the Network Interface Module. See NETWORK INTERFACE MODULE.

Resource Object An SCSA definition. An instance of a Resource Class.

Resource Unit An SCSA definition. An entity that communicates directly with a resource device. From the client application's point of view, it is the Resource Unit that provides services such as voice store-and-forward, fax send and receive, etc. See RESOURCE.

Resolution 1. The minimum difference between two discrete values that can be distinguished by a measuring device. High resolution does not necessarily imply high accuracy.
2. The degree of precision to which a quantity can be measured or determined.
3. A measurement of the smallest detail that can be distinguished by a sensor system under specific conditions.
4. A measure of the quality of a transmitted image. Beginning with the scan processing in the transmitter and ending with the display and/or printing process in the receiver, resolution is a basic parameter of any image transmission system. It affects the design of all its subsystems. In the scanner, the resolution is a function of the spot size which the scanner optics and

associated electronics "look" at the scene and through which the system can uniquely identify the smallest distance along the scan line. Resolution is measured terms of the density of the picture elements (pixels) and is the total number of pixels (horizontal x vertical) used to display alphanumeric characters of graphic images on the screen. High resolution images are composed of more dots per inch and appear smoother than low-resolution images. The higher the resolution, the better the display of details. See also MONITOR.

Resolver A program which connects user programs to domain name servers. see gTLD.

Resonance The condition that exists when inductive reactance equals capacitive reactance. In a series circuit it results in maximum current at the resonant frequency. In a parallel circuit it results in maximum voltage at the resonant frequency.

Resonant Cavity Closed metal container which has the characteristics of a parallel resonant circuit.

Resp Org Responsible Organization — the long distance company responsible for managing and administering the 800 subscriber's records in the 800 Service Management System (SMS/800). The SMS/800 only recognizes one RESP ORG for each 800 number. Management and record administration consists of data entry, changing records, accepting trouble reports and referring and/or clearing associated documents.

Responder A test line that can make transmission and supervision measurements through its host switch under control of a remote computer.

Response An answer to an inquiry. In IBM's SNA, the control information sent from a secondary station to the primary station under SDLC.

Response Time The time it takes a system to react to a given input. In voice recognition, response time typically refers to the amount of time required for a word (or utterance) to be recognized once the end of the word is detected. True response time is longer because silence often must occur before the end of the word can be declared. When operating a terminal connected to a computer, response time would be the time between the operator pressing the last key of a series of keys and the appearance of a response on the operator's display. In a data communications system, response time includes the transmission time, the processing time, the searching for records time and the transmission time back to the originator. Response time is very critical in applications like airline reservation systems. Here the customer is on the phone awaiting a reply. That time is critical in whether the customer perceives he's getting good or bad service. Response times of more than three seconds are not acceptable in situations where the customer is waiting on the phone to buy something. Response time is a function, inter alia, of the number of phone lines you lease or use. You can save a lot of phone line costs by cutting back on lines. But you'll extend response time. Life, as always, is a trade-off.

Responsible Organizations RespOrgs. Telecommunications providers that have responsibility for obtaining 800 Service numbers from the Service Management System and building and maintaining customer records. See EIGHTHUNDRED SERVICE.

Rest Stops Electronic malls and other online diversions provided for a small fee.

Restart 1. A central office word or Apple Macintosh word for resetting a PC without turning it off (also called "warm boot" or "soft reset"). To restart an MS-DOS or Windows machine while it is on, press Ctrl + Alt + Del once or twice or press the reset button. See also Boot.
2. In telephony, a system initiated action designed to restore overall service capacity.

Restart Packet A block of data that notifies X.25 DTEs that an irrecoverable error exists within X.25 network. Restart packets clear all existing SVCs and resynchronize all existing PVCs between X.25 DET and X.25 DCE.

Restocking Fee A fee for returning non-defective equipment. The fee is generally a percentage of the sales price, from 10 percent to 25 percent. Many buyers object to paying restocking fees. Check before you buy something — especially in the used telephone equipment business, where restocking fees are not uncommon.

Restore 1. Typically, to put a telephone system back into full operation.

Restore Button A Windows term. The small button containing both an up and down arrow at the right of the title bar. The Restore button appears only after you have enlarged a window to its maximum size. Mouse users can click the Restore button to return the window to its previous size. Keyboard users can use the Restore command on the Control menu.

Restriction Phone systems can disallow people or extensions from making certain calls. If they're not allowed to make long distance calls, this is called toll restriction. See TOLL RESTRICTION. There are other forms of restriction, like being able to only use the company's internal network.

Restriction From Outgoing Calls Phone users may be restricted from placing outgoing calls. See CLASS OF SERVICE.

Restriction Override Password A password which allows a caller to override restrictions when making an outside call. This Password must typically be entered before the call is dialed.

Restriction Services These features allow the attendant to control the restriction of phones or groups of phones. It can be very useful in hotels and motels to turn off service to room phones during the time between check out and check in of quests. Here are some examples of restriction services:
• CONTROLLED OUTWARD RESTRICTION: Phones can be restricted from making dialed outgoing calls while inward calls are completed normally.
• CONTROLLED STATION-TO-STATION RESTRICTION: Originating phone calls to other extensions in the system are blocked, however, normal incoming and outgoing calls can be completed.
• CONTROLLED TERMINATION RESTRICTION: Phones can complete outgoing calls normally, but incoming calls are directed to either the attendant or an intercept tone or recording.
• CONTROLLED TOTAL RESTRICTION: Restricted phone lines cannot make or receive any calls.

Restrictions Preselected telephones and lines may be restricted from dialing certain telephone numbers (such as long distance, directory assistance and other toll calls).

Resynchronization The process of returning Novell NetWare SFT III (Novell's System Fault Tolerance) servers to a mirrored state after a failure. SFT III checks changes to all active servers and ensures that those changes are copied to the other servers. The time it takes the servers to complete resynchronization depends on the amount of memory and the disk storage in each server. Server memory synchronization is much faster than disk mirroring because disk mirroring speed is limited by the disk channel.

Retard Coil A coil having a large inductance which retards sudden changes of the current flowing through its winding.

Retention A telemarketing term. Refers to a marketing goal to keep current customers buying. The opposite is churn.

Retractile Cord A coiled cord that springs back to its original length when you let it go. Telephone handset cords are the most common retractile cords. There are wide quality variations among retractile cords. Western Electric (oops AT&T Technologies) has set a very good standard for retractile cords. But not everyone conforms to it. If you want to quickly see the quality difference among various coil cords, take six from different manufacturers and hang them over your office door and come back in a week. You'll see a bunch touching the floor. Others will still be taut. Another way is simply to connect them to your phone, one by one, and listen to the differences. Some simply sound weaker. Cheaper ones tend to sound worse. (So what's new?)

Retraining Training is a feature of some modems which adjust to the conditions including amplitude response, delay distortions, timing recovery, and echo characteristic, of a particular telecommunications connection by a receiving modem. Retraining occurs after modems have been successfully transmitting and receiving data. Usually due to a change in line conditions.

Retransmission A method of error control in which hosts receiving messages acknowledge the receipt of correct messages and either do not acknowledge, or acknowledge in the negative, the receipt of incorrect messages. The lack of acknowledgment, or receipt of negative acknowledgment, is an indication to the sending host that it should transmit the failed message again.

Retransmissive Star In optical fiber transmission, a passive component that permits the light signal on an input fiber to be retransmitted on multiple output fibers. The signal comes in on one fiber, hits a star-type connector which splays the transmission out. A retransmissive star is formed by heating together a bundle of fibers to near their melting point. Such a device is used mainly in fiber-based local networks. It's also called star coupler. When you see one you'll be surprised how crude this device looks, despite its fancy name.

Retrial After failing to complete a call, a person tries again. This is called a "retrial." The term is used in traffic engineering. It's critical in figuring needed trunking capacity. See QUEUING THEORY, POISSON and TRAFFIC ENGINEERING.

Retrofit Kit A conversion kit which makes a standard pay phone into one which will accept credit cards.

Retry In the bisynchronous protocol, the process of resending the current block of data a prescribed number of times until it is accepted.

Return A carriage return. This key on some keyboards is also called "Enter." Touching the CR (Carriage Return) gives you two functions: a "line terminating function" and a "new line function", abbreviated "NL". Simply put, a Return at the end of a line, terminates that line and begins a new one.

Return Authorization Number See RMA.

Return Call The name of a Bell Atlantic service. Return Call automatically redials the number of the last person who called your business — whether you were able to answer the phone or not. If that number is busy, Return Call continues trying to get through for up to 30 minutes — without interrupting your incoming or outgoing calls. It signals you with a special ring when a connection is made. Besides making a return follow-up call quick and simple, this service also lets you call back anyone who hung up when you couldn't pick up the phone in time.

Return Loss A measure of the similarity of the impedance of a transmission line and the impedance at its termination. It is a ratio, expressed in decibels, of the power of outgoing signal to the power of the signal reflected back from an impedance discontinuity.

Return Material Authorization Number RMA. A code number provided by the seller as a prerequisite to returning product for either repair or refund. An indispensable tracking procedure, it operates like a purchase order system. If you return computer or telephone equipment without an RMA, chances are your equipment will be lost. RMA is also called a Return Authorization (RA) number.

Return To Zero RZ. Method of transmitting binary information such that, after each encoded bit, voltage returns to the zero level.

Return-To-Zero Code A code form having two information states called "zero" and "one" in which the signal returns to a rest state during a portion of the bit period.

Reusability The characteristic of a component that allows it to be used in more that the application for which it was created, with or without modification.

Reuse Ratio Lines of code reused per total lines of code.

Revenue Accounting Office RAO. A telephone company center using mainframe computers for billing other data processing. Functions performed include receipt and processing of AMA (Automatic Message Accounting) data and preparation of the subscriber's bill. Definition from Bellcore in reference to its concept of the Advanced Intelligent Network.

Revenue Volume Pricing Plan AT&T's Revenue Volume Pricing Plan gives discounts based on total monthly 800 and 900 billing after all other term discounts have been taken. One of two plans used by aggregators to resell 800 services, the other is Customer Specific Term Plan.

Revenue Requirement How much money a regulated phone company is allowed to earn is typically determined by its rate base (depreciated value of its assets). It is allowed to earn a percentage on its rate — just as you earn interest on your rate base (what you have deposited in the bank). So how the regulation works is (in principle) simple: Figure what the phone company's assets are. Figure what percentage you want the phone company to earn on its assets. Figure the calculation. Bingo you have the revenue requirement. Except you have to allow it to pay its expenses. So that gets added onto the revenue requirement. The formula is amount of return (rate base times rate of return — ROR) plus operations expenses. See also ROE.

Reverse Battery Signaling A type of loop signaling in which battery and ground are reversed on the tip and ring of the loop to give an "off-hook" signal when the called party answers. Some systems employ reverse battery, either for a short period or until the call is finished, to indicate that it is a toll call. In some PBXs this is used to provide toll diversion.

Reverse Battery Supervision A way of telling the originating central office that the called telephone has been answered (i.e. it has gone off-hook). The voltage of the line at the originating end is reversed. Reverse battery supervision, which puts a signal at the user's premises, is very useful for devices like call accounting systems (knowing precisely when to begin the billing cycle) and telemarketing systems (knowing precisely when to transfer the machined-dialed call over to the operator). See also ANSWER SUPERVISION.

Reverse Channel 1. A (typically) small-bandwidth channel used for supervisory or error-control signaling. Signals are transmitted in the opposite direction to the data that is sent.
2. The channel in a dial up telephone circuit from the called party to the calling party.

Reverse Interrupt In Bisync, a control character

sequence (DLE) sequence sent by a receiving station instead of ACK1 or ACK0 to request premature termination of the transmission in progress.

Reverse Matching Attaching the name and address to a phone number. A It's a job usually done by a specialized service bureau. Called "reverse" matching because the service bureaus started in business by attaching phone numbers to lists containing names and their addresses. With ANI (Automatic Number Identification), we get the phone numbers of people calling us. But we don't get their names and addresses. We need to get this information for many reasons. The obvious being that getting this information on-line and fast saves asking the caller for it and typing all the stuff in. That saves time on the phone — as much as 20 seconds. And fewer questions about boring stuff like phone number, address, city, state, zip means less typing time (also called data entry time, less clerking time) and more time to explain the specials we're selling today. In short, the fewer questions we ask, the less we type and the more stuff we can sell. Reverse matching can be done instantly on-line via a direct data hookup to a distant specialized service bureau or it can be done at the end of the month when we receive our 800 phone bill containing the phone numbers of the people who called us that month.

Reverse Operation Briefly running a shredder in reverse to clear jams.

Reverse Transfer An Inter-Tel term for a phone feature in which a call on common hold at any phone may be retrieved from any phone anywhere in the phone system.

Reverse Video A video display with all the characters reversed. Characters which are normally white on the screen appear black. And blacks appear white. Reverse video is used to emphasize or enhance things — like those characters to be printed in italics or bold.

Revertive Pulse Ground pulses sent back to the sender in the originating panel office from the various selector frames to control the selection process.

Revertive Pulsing In telephone networks, a means of controlling distant switching selections by pulsing, in which the near end receives signals from the far end.

Revisable-Form Document An electronic document with its formatting information intact, readable and modifiable.

Rewritable Optical Disks They look like CD-ROM disks but they're not. On one side you can store 284 megabytes or 335 megabytes depending on how large you make the sectors — either 512-byte sectors or 1,024-byte, respectively. All the optical disks conform to standards set up by ISO. Compared with hard drives, they are very slow.

REX Routine Exercise.

REXX REstructured eXtended eXecutor. An interpreted script language, or procedural programming language, developed by IBM originally for users of large operating systems in a mainframe environment. REXX was designed for ease of learning and ease of use for both programmers and non-programmers. REXX offers powerful character manipulation in terms of symbolic objects (e.g., words and numbers) with which people normally deal, automatic data typing and debugging capabilities. REXX can be compared with other interpreted script languages such as Microsoft's Visual Basic, Netscape's JavaScript and Larry Wall's Perl.

RF Radio Frequency. Electromagnetic waves operating between 10 kHz and 3 MHz propagated without guide (wire or cable) in free space. If you have a home computer that lets you use your home TV set as a video display device, then the computer has an rf Generator. This means that this device is generating an rf carrier to carry the video signal information. For the purposes of the FCC's regulation of cable television systems, this term includes any carrier, modulated or unmodulated, whether radiated over the air by an antenna or carried by a coaxial cable. This term dates from the early days of radio (hence, the name "radio" frequency) when the only uses for RF were AM broadcasting and ship-to-shore communications. The term is still in use today, even though it now includes video and control signals as well as audio.

RF Channel Number An identifier assigned to a Radio Frequency (RF) channel to distinguish it from other rf channels. A cellular radio term.

RF Channel Pair Two associated Radio Frequency (rf) channels, one forward and one reverse. The former is used to support forward transmissions from the Mobile Data Base Station (MDBS) to the Mobile End System (M-ES). The reverse channel carriers data information from an M-ES to an MDBS, and is a contention based communications channel.

RF Choke A coil of wire that filters our high frequencies. See RF Leakage.

RF Leakage RF Leakage is defined as the amount of energy which "leaks" from the connector and/or component. Although rf Leakage will vary with frequency, it is typically tested at only one frequency. Leakage, like Insertion Loss, is expressed in dB. Very large negative dB values indicate that the device does not radiate much energy.

RF Splitter Want to put attach two (or more) TV sets behind your satellite antenna? Easy, go down to Radio Shack and buy an RF Splitter. Screw the line into one side of the splitter and your two TVs sets into the other. Bingo, your two TVs will play the signal coming in from the splitter — which might be a satellite antenna, a CATV hookup, an outdoor antenna, etc. You can buy RF splitter than will allow you to hookup almost as many TVs as you, or want to have. See also Splitter.

RFAC Restricted Forced Authorization Code. A forced authorization code that is valid only if it is entered from a specific telephone extension.

RFC Request For Comment. The development of TCP/IP standards, procedures and specifications is done via this mechanism. RFCs are documents that progress through several development stages, under the control of IETF, until they are finalized or discarded. RFC#### documents Internet "Request For Comment" documents (i.e., RFC822, RFC1521, etc.). The contents of an RFC may range from an official standardized protocol specification to research results or proposals. A set of papers in which the Internet's standards, proposed standards and generally agreed-upon ideas are documented and published. See the RFCs below.

RFC 1144 This RFC (Request For Comment) will provide overhead compression for the TCP/IP protocol down to 5 octets. It does this by anticipating that the next packet in a file transfer sequence will have the same address as the previous and will have the same sequence number plus one. This compression technique will be useful where SDLC encapsulation, or other bridging protocol encapsulation, is being used with low-speed PVCs (Private Virtual Circuits.) In these cases, the slight increase in processing power to perform the compression is more than balanced by the increase in application performance and throughput.

RFC 1294 This Request For Comment is Inverse ARP, which allows the automatic discovery of the addresses on the router at each end of another router's DLCIs. Right now, the RFC only applies to IP, but some equipment vendors have already expanded the protocol support to include Novell, AppleTalk, Vines, and DECnet. The benefit of the RFC is to simplify network configuration.

RFC 1315. This Request For Comment is the frame relay MIB (management information database), which standardizes what management information is made available on frame relay devices and where/how that information is accessed. This simplifies the process of integrating frame relay devices into your network monitoring and management process and programs.

RFC 1490 This Request For Comment, RFC 1294, now renumbered RFC 1490, is for multiprotocol encapsulation. The bottom line benefits are to increase interoperability between frame relay devices from different vendors. This means that you can use one vendor's routers (or other equipment type) at some locations, and a different vendor's equipment at other locations. This ability to mix and match allows you to pick the best and most cost-effective tool for the job.

RFC 1577 Under control of the IETF, Request For Comments (RFCs) are documents used to develop standards, procedures and specifications for TCP/IP. RFC 1577 is the document for classical IP.

RFC 1695 Definitions of Managed Objects for ATM Management or AToM MIB.

RFD Request For Discussion. A period of time during which comments on a particular subject are solicited. An Internet term.

RFF Radio Frequency Fingerprinting. A process in which the radio signal information and characteristics produced by the transmitter are captured and analyzed by the receiver for purposes of detecting a cloned device from accessing the network. Bursts of control data are captured and analyzed using complex signaling techniques; the data is compared to the characteristics of the legitimate transmitter in order to determine whether access should be granted or denied. Primarily used in secure military applications, the technique has been evolving since WWII; it is being considered for application in cellular telephony.

RFI 1. Request For Information. General notification of an intended purchase of equipment or equipment and lines sent to potential suppliers to determine interest and solicit general descriptive product materials, but not prices or a formal request. See RFQ for a detailed explanation.
2. Radio Frequency Interference. All computer equipment generates radio frequency signals. The FCC regulates the amount of RFI a computing device can leak past its shielding. A Class A device is sufficient for office use. A Class B is a more stringent classification for home equipment use. See EMI and RADIO FREQUENCY INTERFERENCE.

RFP Request For Proposal. A detailed document prepared by a buyer defining his requirements for service and equipment sent to one or several vendors. A vendor's response to an RFP will typically be binding on the vendor, i.e. he will be obliged to deliver what he says in his RFP at the prices and following the conditions explained in that RFP. See RFQ for a detailed explanation.

RFQ Request For Quotation. A document prepared by a buyer defining his needs for service and equipment in fairly broad terms and sent to one or several vendors. The RFQ is much less detailed than the RFP. Let's start at the beginning of the buying process. We have a buyer who wants a phone system. His first step may be to issue a formal or informal RFI — Request For Information. In effect, the RFI says "Please tell me what you have. I have a vague idea of what I want but I don't know exactly what is available to suit my needs. Please send

me some information."
After a buyer gets his responses to his RFIs, he may issue an RFQ — Request For Quotation. An RFQ may include a tentative configuration of the type of phone system the user wants, plus some listing of features the buyer is interested in. In the RFQ, the buyer asks for a "ballpark" (approximation) of the possible price for such a system. Usually the "price" is within plus or minus 10% of where it will eventually be in the final configuration. In short, an RFQ's purpose is not to buy, but to find out what's out there and what it might cost. The purpose may be to allocate a budget or to put aside some money for the forthcoming purchase.
An RFP — Request For Proposal — is much more formal and definitive. Its purpose is simple. The buyer wants to buy something. The RFP contains a list of what the buyer wants, when he wants it, how it should be installed, how it should delivered, what financing may be necessary. It is now up to the vendor/s to respond with their configuration, their precise prices and their terms and conditions of sale. Whatever the vendor responds with — called a Response to an RFP — constitutes a definite offer. At this point, the buyer can negotiate the terms of the vendor's Response to his RFP. This will lead to the writing, and eventual signing of a contract. Or the buyer may simply decide to accept the Vendor's Response. Often that Response may have a line at the back of it — "I accept the terms and conditions of this response." If the buyer signs this, then the Response to the RFP becomes a valid contract.

RFS 1. Ready for Service.
2. Remote File System. The ability to mount a disk drive somewhere on a network — but it's not on your competitor.

RG-58 Coaxial cable with 50-ohm impedance used by Thinnet.

RG-59 A coaxial cable type often used in television.

RG-62 Coaxial cable with 93-ohm impedance used by ARCnet.

RG/U RG/U or RG-U is the military designation for coaxial cable. U stands for general utility.

RGB Red. Green. Blue. The three primary colors used in video processing, often referring to the three unencoded outputs of a color camera or VTR. A color model based on the mixing of red, green, and blue — the primary additive colors used by color monitor displays and TVs. The combination and intensities of these three colors can represent the whole spectrum. Color television signals are oriented as three separate pictures: red, green and blue. Typically, they are merged together as a composite signal but for maximum quality and for computer applications the signals are segregated.

RGB Cutoff An advanced color control that lets you set your monitor to maintain color balance across different gray scales.

RGB Gain An advanced color control that lets you adjust red, blue, and green levels individually.

RGC 1604 An acronym used in a BellSouth proposal to the State of GA. The salesman who wrote the proposal didn't know what it was. It relates to Frame Relay Service.

RH Request Header or Response Header.

RHC Regional Holding Company. Also called Regional Bell Operating Company. For all intents and purposes, an RHC is the same as a Regional Bell Operating Company. Here's the background: On January 8, 1982 AT&T signed a Consent Decree with the U.S. Department of Justice, stipulating that on midnight December 30, 1983, AT&T would divest itself of its 22 telephone operating companies. According to the terms of the Divestiture, those 22 operating Bell telephone companies would be formed into seven regional holding companies (RHC) of roughly equal size. The judge assigned which Bell operating companies would join which regional holding company. The seven RHCs were Ameritech, Bell Atlantic, BellSouth, NYNEX, Pacific Telesis, Southwestern Bell and US West. In early October, 1994, Southwestern Bell changed its name to SBC Communications, Inc. But its telephone companies, it said, would still operate under the Southwestern name. In April, 1996 Bell Atlantic bought NYNEX (the holding company for New York Telephone and New England Telephone) for $22.1 billion. The new company will be called Bell Atlantic. In April also, SBC Communications, Inc. (the name for the holding company owning Southwestern Bell Telephone) bought Pacific Telesis (the holding company for Pacific Bell) for $16.7 billion. As of early 1998, this left five regional Bell operating companies — Ameritech, Bell Atlantic, BellSouth, SBC Communications and US West.

Terms of the Divestiture placed business restrictions on AT&T and the BOCs. Those restrictions were threefold: The BOCs weren't allowed into long distance, equipment manufacturing, or information services. AT&T wasn't allowed into local telecommunications (i.e. to compete with the BOCs). But it was allowed into computers. The federal Judge overseeing Divestiture, Judge Harold Greene, is slowing the lifting the restrictions. He has allowed the BOCs into information services and AT&T into local service. He has stayed firm on the other two — no equipment manufacturing and no long distance for the RHCs (also called RBOCs).

Rheostat A variable resistor.

Rhetorex A manufacturer of voice processing componentry based in Campbell, CA. Inspiration for the company's name came from the word rhetoric, which is the art of effectively using speech and language.

Rhombic Antenna An antenna composed of wire radiators describing the sides of a rhombus. It is usually terminated and unidirectional; when unterminated, it is bidirectional.

RI Ring Indicator. An RS-232 control line asserted by the DCE when a call has come in for the DTE.

Ribbon Cable Multi-wire cable that is flat instead of round. In ribbon cable, the conductors are laid side by side. Ribbon cable can be more easily laid under carpeting because it is flat and thus, can extend phone and computer services to places otherwise hard to reach. There are disadvantages to ribbon cable. Because ribbon cable is flat, it's hard to twist its individual wire conductors around each other (thus humming can be a problem). It is hard to put a metal shielding around the twisted wire pairs. It is hard to put coax cable into ribbon cable. It is hard to make ribbon cable sufficiently strong to withstand thousands of high heels trampling it. It is hard to make ribbon cable which turns a corner... But there has been enormous progress in ribbon cable. And ribbon cable is finding greater use in buildings. These days it even carries commercial A.C. power.

Ribbon Fiber Cable A cable that accommodates one to 12 ribbons, each ribbon having 12 fibers for a cable size range of 12 to 144 fibers. Ribbon cables are designed for use in larger distribution systems where small cable size and high pulling strength are important.

Ribbon Of Highway Fiber optic cable.

RIF 1. Rate Increase Factor. This factor by which the cell transmission rate may increase upon receipt of an RM-cell. See also ATM and RM-Cell.
2. Routing Information Field. In the Token Ring protocol, a optional field which is used when the transmitting frame must

pass through multiple Source Routing Protocol (SRP) bridges. Within the RIF, the value of the RII, or Routing Information Indicator, (1 or 0) indicates to the bridge whether the frame should be either forwarded to another ring or confined to the local ring. See also Bridge.

RIFF Resource Interchange File Format. Platform-independent multimedia specification (published by Microsoft and others in 1990) that allows audio, image, animation, and other multimedia elements to be stored in a common format. See also Media Control Interface (MCI).

Right Hand Rule A rule for indicating the direction of magnetic effect. Grasp the wire with the right hand and with the thumb extended along the wire in the direction of current. The curved fingertips will indicate the direction of magnetic flow. Not totally relevant for including in this dictionary. But cute. Seel also Rule of Thumb.

Right To Use See RTU.

Rightsizing Another term for re-engineering. See RE-ENGINEERING.

RII Routing Information Indicator. A Token Ring term. See RIF for an explanation.

RILD Remote ISDN Line Drawer.

RIME RelayNet International Message Exchange, a multi-tier communications network which exchanges messages among member bulletin board systems.

Rimm Job A bogus academic study masquerading as legitimate science. Named after Marty Rimm, author of the dubious "cyberporn" study from Carnegie Mellon University that Time magazine gullibly took as gospel.

Ring 1. As in Tip and Ring. One of the two wires (the two are Tip and Ring) needed to set up a telephone connection.
2. Also a reference to the ringing of the telephone set.
3. The design of a Local Area Network (LAN) in which the wiring loops from one workstation to another, forming a circle (thus, the term "ring"). In a ring LAN, data is sent from workstation to workstation around the loop in the same direction. Each workstation (which is usually a PC) acts as a repeater by re-sending messages to the next PC in the ring. The more PC's, the slower the LAN. Network control is distributed in a ring network. Since the message passes through each PC, loss of one PC may disable the entire network. However, most ring LANs recover very quickly should one PC die or be turned off. If it dies, you can remove it physically from the network. If it's off, the network senses that and the token ignores that machine. In some token LANs, the LAN will close around a dead workstation and join the two workstations on either side together. If you lose the PC doing the control functions, another PC will jump

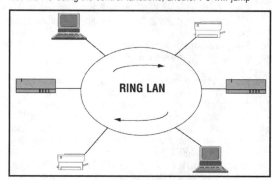

RING LAN

in and take over. This is how the IBM Token-Passing Ring works. See TOPOLOGY, BYPASS CABLE and TOKEN RING.

Ring Again The PBX remembers the last number called by a phone and will redial it when the feature is activated.

Ring Back Tone The sound you hear when you're calling someone else's phone. The tone you hear is generated by a device at your central office and may bear no relationship to the sound the phone at the other end is emitting — or not emitting. If your call didn't go through the first time, always call back at least once. See RINGING TONE.

Ring Banding A method of color coding insulated conductors by means of a small band of colored ink applied circumferentially at regular intervals along the axis of the insulated conductor.

Ring Battery Commonly unfiltered - 24 VDC source that supplies operating power to all local KSU components. Also called the B Battery.

Ring Conductor One conductor of a phone line.

Ring Cycle A ring cycle in North America is typically six seconds long, two of ringing, four of silence, then repeated.

Ring Down Box Ring down boxes, also known as CO simulators or telephone line simulators, are simple devices used for generating POTS calls — without a central office. You connect a phone to both sides of the device. When one side goes off-hook, the ring down box will "ring" the other side. When both sides are off-hook, both sides are coupled together, the line is powered and the sides can talk. Ring down boxes are available with various options and configurations. These include the ability to provide dial-tone to the caller side (required to test applications with modems, faxes, or other automated out dialing devices), caller ID, and disconnect supervision. They are generally available in one to four line sizes, although special configurations may support more. Ring-down boxes are used for giving demonstrations and testing. We use them in our test labs to test drive new computer telephony systems.

Ring Down Circuit A tie line connecting phones in which picking up one phone automatically rings the other phone. In a ringdown circuit, a ringing current (AC) is sent down the line. That current may light a lamp, set off a bell, buzz a buzzer. The idea is to alert the person at the other end to the incoming call. A ringdown circuit is often used in an elevator or other emergency situation.

Ring Down Interface A private line two-wire interface also called Loop Start Trunk.

Ring Generator A component of virtually all phone systems, ranging from large central offices to small key systems, that supplies the power to ring the bells inside phones, typically 90 volts AC at 20 Hz.

Ring Indicator Modem interface signal defined in RS-232-C which indicates to the data terminal equipment that a call is coming in.

Ring Isolator A device placed on a telephone line to disconnect the ringer when it is an idle state. It is used for noise prevention.

Ring Latency In a token-ring network, the time measured in bits at the data transmission rate, required for a signal to propagate once around the ring. Ring latency includes the signal propagation delay through the ring medium, including drop cables, plus the sum of propagation delays through each data station connected to the token-ring network. See also TOKEN-RING NETWORK.

Ring Network A network that links PBXs, computers, terminals, printers and other devices in a circular communications link. See RING.

Ring Protection Switching. RPS. Nortel's

"Introduction to SONET Networking" tutorial handbook (www.nt.com/broadband/reference/sonet_101.html) talks about "Automatic Healing of Failed or Degraded Optical Spans in a Two-Fiber BLSR." The handbook says "inthe event of failure or degradation in an optical span, automatic ring protection switching (RPS) reroutes affected traffic away from the fault within 50 milliseconds, preventing a service outage."

Ring Signal The pulse ringing voltage output of the local Interrupter KSU. Typically, this signal is 105 VAC with a duty cycle of 2 seconds on and 4 seconds off.

Ring Splash A brief "splash" of a "ring" which announces an incoming telephone call. Ring splash is a technique used by premise-based telephone switches such as PBXs and ACDs in order to reduce the size of the "glare window," which is the length of time in which "glare" can occur. Glare is a condition in a trunk simultaneously is seized by switches at both ends. For instance, a PBX or ACD user might seize a trunk for an outgoing call at precisely (or virtually so) the same time that a central office (also called public exchange) might choose to connect an incoming call over the same trunk. In order to avoid such embarrassment, the properly equipped PBX or ACD will send a ring splash in the form of a 500 ms (millisecond) ring that is splashed to the station immediately prior to the normal ring cadence. Thereby and in the context of the North American standard ringing cycle, the glare window might be reduced from a maximum of 4 seconds to a maximum of 200 ms (two hundred thousandths of a second, or one-fifth of a second). While the ring splash may make the beginning of the ring cadence sound slightly odd, the risk of embarrassment is reduced substantially. Have you ever picked up the phone, dialed in another person's ear, and then said something impressive like "Hello, Bob. Hello?" That's really embarrassing. Ring splash solves the problem. See also Glare and Ring Cycle.

Ring Tone An audible, call-progress signal connected to the calling line to indicate that the called station is being rung. the industry standard is a mixture of 440 Hz and 480 Hz, interrupted at the same rate, or ring cycle, as the ringing current being applied to the called station. See also Ring Cycle.

Ring Topology A network topology in which nodes are connected to a closed loop, no terminators are required because there are no unconnected ends.

Ring Trip The process of stopping the AC ringing signal at the central office when the telephone being rung is answered.

Ring Wiring Concentrator A site through which pass the links between repeaters, for all or a portion of a ring.

Ring Wiring Configuration The same as daisy chain wiring, except the last jack is connected to jack 1, thus completing a ring.

Ringback The tone heard by a calling device when, at the called-device's end, the telephone is ringing or the system is otherwise being alerted of the incoming call.

Ringer A bell in a telephone which indicates if a phone call is coming in. These days "ringers" are electromechanical and clunky (old-style) or small and electronic (new style). The new electronic ones are cheaper, but less interesting to listen to. Most sound like bleating sheep in heat.

Ringer Equivalence Number REN. A number required in the U.S. for registering your telephone equipment with the phone company. Add together the RENs of all the telephones on a single line. The sum of those numbers should never exceed five otherwise none of your bells will work and you won't hear an incoming call. (Your central office simply doesn't send sufficient current down the line.) The alphabetic character after the number refers to the ringing frequency of the alternating current sent down the line to ring the bell. If the letter is "A", the ringer frequency is about 20 Hertz. Most single line phones have a Ringer Equivalence of 1.0A. If the letter is "B", the ringer will respond to any current coming down the line. Any other letter, and you are probably on a party line where the ringer frequency is used for party selection. In Canada, they use the term "Load Number" instead of Ringer Equivalence. The numbers are different, but the concept is the same. See LOAD NUMBER.

Ringer Isolator A device in the phone which disconnects the ringer when ringing voltage is not present.

Ringing Alternating Current (AC) sent out from the central office along the local loop to the subscriber. It's typically 70 to 90 volts at 17 Hz to 20 Hz. You can get a mild shock if you have your hands on a telephone circuit when ringing current comes along. The rest of the time, the lines are harmless.

Ringing Generator A device in a phone system that generates the AC ringer voltage. Typically, this voltage is 90 to 115 (nominally 105) VAC at 30 Hertz.

Ringing Key A key that sends a ringing current.

Ringing Signal Any AC or DC signal transmitted over a line or trunk for the purpose of alerting someone or some thing at the distant end of an incoming call.

Ringing Tone A low tone which is one second ON and three seconds OFF. It indicates that ringing current is being sent by the central office to the person receiving the call. Ringing tone is not produced by the calling party's central office — but by the called party's central office. Thus, it is possible for you to hear ringing tone but for the person you are calling not to hear anything. As a general rule, if the person doesn't answer, call them a second time. Often, they'll say "The phone never rang." This will not be a lie, but simply a temporary glitch in their central office.

Ringing Transfer A PBX feature which allows you to choose which bells in a group of phones will ring when a call is coming in for that group.

Ringing Voltage In addition to talk battery, a Central Office provides ringing signaling. Ring Voltage is generally 70 to 90 volts at 17 Hz to 20 Hz. See also RINGING.

RIP Routing Information Protocol. RIP is based on distance-vector algorithms that measure the shortest path between two points on a network, based on the addresses of the originating and destination devices. The shortest path is determined by the number of "hops" between those points. Each router maintains a routing table, or routing database, of known addresses and routes; each router periodically broadcasts the contents of its table to neighboring routers in order that the entire network can maintain a synchronized database. See also DISTANCE VECTOR PROTOCOL.

Rip Cord A cord placed directly under the jacket of a cable in order to facilitate stripping (removal) of the jacket.

RIPE Reseaux IP Europeens, a group formed to coordinate and promote TCP/IP based networks in Europe. RIPE is responsible for management and assignment of IP (Internet Protocol) addresses in Europe, just as are ARIN and APNIC in the regions of the Americas and Asia-Pacific, respectively. RIPE holds periodic conferences to coordinate technical issues (similar to the IF) as well as running a Network Control Center (NCC) to handle operational issues such as the administration of the European domain names and routing tables. RIPE is a collaborative organization with no formal membership (over 1,000 organizations participate); all activities are performed on a voluntary basis. See also APNIC, ARIN and IP.

RIPscrip Graphics Emulation Supports RIPscrip Graphics Emulation. Popular on many bulletin board systems, RIPscrip Graphics Emulation allows users to view screens mixing text and graphics. On-screen buttons can be clicked to send commands to a remote system.

RISC Reduced Instruction Set Computing. A microprocessor architecture that favors the speed at which individual instructions execute over the robustness of the instruction set. Computers based on RISC use an unusual high speed processing technology that uses a far simpler set of operating commands. These commands greatly speed a computer's performance, especially for calculation-intensive operations such as those performed by scientists and computer-aided design (CAD) and computer-aided manufacture (CAM) engineers. RISC is a design that achieves high performance by doing the most common computer operations very quickly. In contrast, the microprocessors used in most PCs are based on a design called CISC (Complex Instruction Set Computing). CISC does not execute instructions as quickly as RISC but it has more commands and accomplishes more with each command. Programs written for RISC are typically not compatible with those written for CISC processors. RISC is the prevailing technology for workstations today. The RISC semiconductor was an IBM baby, born in its Yorktown Heights, NY lab in 1974. But internal arguments over, and even whether the chip should be used kept IBM fiddling while Sun and other companies decisively powered ahead. IBM got its first good RISC product, the RS/6000, to market in 1990. The PowerPC architecture is one example of a RISC microprocessor design. See also CISC.

RISC System/6000 An IBM family of workstations and servers designed to run applications developed for the UNIX operating system.

Riser The conduit or path between floors of a building into which telephone and other utility cables are placed to bring service from one floor to another. Your risers should be twice the size you ever think you'll need in the next 30 years. It's expensive to build risers after the building is built. Very expensive.

Riser Cable High strength cables intended for use in vertical shafts between floors in multi-story buildings. Such shafts are called riser shafts. (What else would they be called?)

Riser Closet The closet where riser cable is terminated and cross connected to either horizontal distribution cable or other riser cable. The riser closet houses cross connect facilities, and may contain auxiliary power supplies for terminal equipment located at the user work location. See also SATELLITE CLOSET.

Riser Subsystem The part of a premises distribution system that includes a main cable route and facilities for supporting the cable from an equipment room (often in the building) to the upper floors, or along the same floor, where it is terminated on a cross connect in a riser closet, at the network interface, or other distribution components of the campus subsystem. The subsystem can also extend out on a floor to connect a satellite closet or other satellite location.

Risk A potential liability, caused by a threat.

Risk Assessment The process of quantifying the potential impact on an organization from various security threats.

RISLU Remote Integrated Services Line Unit. One of the remoting arrangements that the Lucent 5ESS switch architecture permits. The RISLU terminates DSLs and connects to the switch DLTU via T-1.

RIT Rate of Information Transfer. The amount of information that can be communicated from a sender to a receiver in a given length of time.

RJ Registered Jacks. They're telephone and data plugs regis-

RJ-ICX	Single Tie Trunk, Type I or II E&M interface, 8 position.
RJ-IDC	Single-line, 4-wire, T/R, T1/R1, 6-position.
RJ-11C/W	Single-line, 2-wire, T/R, 6 position.
RJ-14C/W	Two-line, 2-wire, T/R, T(MR)/R(MR), T(OPS)/R(OPS) 6-position.
RJ-14X	Two-line, T1/R1, T2/R2, with sliding cover, 6-position.
RJ-15C	Single-line, T/R, weatherproof, 3-position.
RJ-17C	Single-line, T/R, used in hospital critical care areas, 6 positions.
RJ-18C/W	Single-line, T/R, with Make Busy leads, 6 positions.
RJ-2DX	12 lines, 4 wire, T/R, T1/R1, 50 positions.
RJ-2EX	12 Tie trunks, 2-wire, T/R, E&M Type I, 50 position.
RJ-2FX	8 Tie trunks, 2-wire, T/R, E&M SG/SB Type II 50 position.
RJ-2GX	8 Tie trunks, 4-wire, T/R, T1/R1, E&M, Type I 50 position.
RJ-2HX	6 Tie trunks, 4-wire, T/R, T1/R1, E&M, SG/SB, Type II, 50 positions.
RJ-2MB	12 lines, 2-wire, T/R, Make Busy leads, 50 position.
RJ-21X	25 lines, 2-wire, T/R, 50 position.
RJ-25C	3 lines, 2-wire, T/R, T (MR)/R(MR), T(OPS)/R(OPS), 6 position.
RJ-26X	8 lines, 2-wire, T/R, FLL, or Programmed data, 50 position.
RJ-27X	8 lines, 2-wire, T/R, Programmed Data, 50 position.
RJ-4MB	Single-line, 2-wire, T/R, MB/MB1, PR/PC, with Make Busy. 8 position, keyed and programmed.
RJ-41M	Up to 8 multiple installations of FLL or Programmed Data. 8 position, keyed.
RJ-41S	Single-line, 2-wire, T/R, FLL or Programmed Data, 8 position, keyed.
RJ-45M	Up to 8 multiple installations of Programmed Data. 8-position, keyed.
RJ-45	Single-line, 2-wire, T/R, PR/PC, programmed data, 8 position, keyed.
RJ-48C	Single-line, 4-wire, T/R, T1/R1, 1.544 Mbps, 8 position.
RJ-48H	Up to 12 lines. 4-wire, T/R, T1/R1, 1.544 Mbps, 50 position.
RJ-48M	Up to 8 lines, 4-wire, T/R, T1/R1, 1.544 Mbps, 50 position.
RJ-48S	One or two lines. T/R or T/R, T1/R1, LADC or subrate. 8-position, keyed.
RJ-48T	Up to 25 (2-wire) or 12 (4-wire), T/R OR T/R, TI/R1; LADC or Subrate,50-position.
RJ-48X	Single-line, 4-wire, T/R, T1/R1, 1.544 Mbps, 8-position with shorting bar.
RJ-61X	Up to 4 lines, T/R, 8-position.
RJ-M8	Single private line, 2/4 wire, T/R or T/R, T1/R1, Non-registered service, 8-position, keyed, w/wo loopback.

tered with the FCC. RJ-XX (where X is a number) are probably the most common plugs in the world. Here is a table of the most common registered jacks, courtesy the FCC. Following the table are descriptions of the most common RJ jacks.

RJ-11 RJ-11 is a six conductor modular jack that is typically wired for four conductors (i.e. four wires). The RJ-11 jack (also called plug) is the most common telephone jack in the world. The RJ-11 is typically used for connecting telephone instruments, modems and fax machines to a female RJ-11 jack on the wall or in the floor. That jack in turn is connected to twisted wire coming in from "the network" — which might be a PBX or the local telephone company central office. In a home installation, the red and green pair would be used for carrying the phone conversation and the black and white might be used for carrying low voltage from a plugged-in power transformer to light buttons on the phone. In many offices, the tip and ring were used for the voice conversation and the black and white were used for signaling. Increasingly, these days more and more office phone systems use only one pair, i.e. the red and green conductors. See also RJ-22 and RJ-45.

RJ-12C and RJ-12W These jacks are normally associated with one line of a key telephone system. They provide a bridged connection to the tip and ring of the telephone line and to key system A and A1 leads. The tip and ring conductors in the jack are connected ahead of the key telephone-system line circuit. The RJ-12C is surface- or flushmounted for use with desk telephone sets while the RJ-12W is for wallmounted telephone sets. Typically, these arrangements are used when registered ancillary equipment must respond to central office or PBX ringing.

RJ-13C and RJ-13W Jacks normally associated with one line of a key telephone system. They provide a bridged connection electrically behind the key-system line circuit to the tip and ring conductors and to the A and A1 leads. The RJ-13C is surface- or flushmounted for use with desk tele-phone sets while RJ-13W is for wallmounted telephone sets. These arrangements are generally used when the registered ancillary equipment does not require central office or PBX ringing to function properly.

RJ-14 A jack that looks and is exactly like the standard RJ-11 that you see on every single line telephone. Whereas the RJ-11 defines one line — with the two center, red and green, conductors being tip and ring, the RJ-14 defines two phone lines. One of the lines is the "normal" RJ-11 line — the red and green conductors in the center. The second line is the second set of conductors — black and yellow — on the outside. The RJ-14C is surface- or flushmounted for use with desk telephone sets while the RJ-14W is for walmounted telephone sets.

RJ-15C The RJ-15C is a weatherproof jack arranged to provide single-line bridged connection to tip and ring. Jack RJ-15C can be arranged for surface- or flushmounting depending upon customer needs.

RJ-16X A providing a single-line bridged tip and ring and is associated with -9 dBm (permissive) data arrangements that require mode indication for use with exclusion key telephone sets. The exclusion key telephone set requires a series jack, RJ-36X (described under 8-position jacks) as its normal means of connection.

RJ-17C A jack that provides a single-line bridged connection of tip and ring to special telephone sets or ancillary equipment (e.g., ECG machines) in hospital critical-care areas. Only registered equipment conforming to Article 517 of the 1978 National Electrical Code is permitted to connect to this jack arrangement. This jack differs from the RJ-11C in that tip and ring appear on pins 1 & 6 rather than 3 & 4.

RJ-18C A jack providing a bridged connection of single-line tip and ring with make-busy leads MB and MB1. When the registered equipment provides a contact closure between the MB and MB1 leads, a make-busy indication is transmitted to the network equipment busying out the line from further incoming calls. It's recommended that the busy indication (contact closure) be provided while the line is in the idle state to reduce the possibility of interfering with a call that is in the ringing or talking state. The RJ-18C is surface-or flush-mounted for use with desk telephone sets.

RJ-19C A jack normally associated with one line of a key telephone system. It provides a bridged connection of single-line tip and ring behind a key-system line circuit, with A and A1 lead control, and a direct connection for MB/MB1 make-busy leads. When the modem provides a contact closure between the MB and MB1 leads, a make-busy indication is transmitted to the network equipment busying out the line from further incoming calls. It's recommended that the busy indication (contact closure) be provided while the line is in the idle state in order to reduce the possibility of interfering with a call that is in the ringing or talking state. The RJ-19C is surface or flushmounted for use with desk telephone sets.

RJ-21 Same as an RJ-21X. See RJ-21X.

RJ-21X An Amphenol connector under a different name. Here's the explanation: Amphenol is a manufacturer of electrical and electronic connectors. They make many different models, many of which are compatible with products made by other companies. Their most famous connector is probably the 25-pair connector used on 1A2 key telephones and for connecting cables to many electronic key systems and PBXs. The telephone companies call the 25-pair Amphenol connector used as a demarcation point the RJ-21X. The RJ-21X connector is made by other companies including 3M, AMP and TRW. People in the phone business often call non-amphenol 25-pair connectors, amphenol connectors. The RJ-21X is often used with Traffic-Data Recording Equipment and Multiple-Lien Communications Systems. The user must specify the connection sequence for each title appearing in the jack.

RJ-22 RJ-22 is a four position modular jack that is typically used for connecting telephone handsets to telephone instruments. It is always wired with four conductors (also called wires). It is different and slightly smaller than the more common RJ-11 which is typically used for connecting telephone instruments, modems and fax machines to a female RJ-11 jack on the wall or in the floor. That jack in turn is connected to twisted wire coming in from "the network" — which might be a PBX or the local telephone company central office. See RJ-11 and RJ-45.

RJ-22X This jack is associated with a telephone company-provided key telephone system when connection to several lines is required. It provides bridged connections of up to 12 telephone lines and their associated A and A1 leads. The tip and ring conductors in the jack are wired ahead of the lien circuit in the key telephone system. This arrangement is used when the modem must respond to central office or PBX ringing.

RJ-23X This jack is normally associated with a telephone company-provided key telephone system when connection is required to several lines. It's wired to provide bridged connections of up to 12 key-system line circuits and associated A and A1 leads. It differs from and is preferred over the RJ-22X, in that tip and ring conductors in the jack are wired

behind the key-system line circuits. This arrangement is typically used when the modem doesn't require central office or PBX ringing to function properly.

RJ-24X This jack is normally associated with a telephone company-provided key telephone system. It's typically used with registered ancillary devices such as conferencing devices, music on hold, etc., and is wired to provide the same tip, ring, A, and A1 appearances as a standard five-line key telephone set.

RJ-25C RJ-25C provides for bridged connection to the tip and ring conductors of three separate telephone lines. The telephone company will wire the lines to the jack in the sequence designated by the customer. The RJ-25C is surface- or flushmounted for use with the desk telephone sets and ancillary devices.

RJ-26X An RJ-26X is a multiple-line universal data jack for up to 8 lines in a 50-position miniature ribbon connector and accommodates either fixed-loss loop (FLL) or programmed (P) types of data equipment. A switch, accessible to the customer, is provided on each line to select FLL or P type of operation. FLL equipment transmits at -4 + 1 dB with respect to one milliwatt and a pad is included in the data jack so that pad loss plus loop loss is nominally 8 dB. Programmed-type data equipment adjusts its output power in accordance with a programming resistor in the data jack. By these means, signals from either FLL or P types of registered data equipment will arrive at the local telephone company central office at a nominal -12 dB with respect to one milliwatt for optimum data transmission.

RJ-27X An RJ-27X is a multiple-line programmable data jack for up to 8 lines in a 50-position miniature ribbon connector and accommodates programmed data equipment only.

RJ-31X An RJ-31X provides a series connection to the tip and ring conductors of a telephone line. It's wired ahead of all station equipment electrically and is typically used with registered alarm-reporting devices. When there's an alarm condition, the registered device functions to cut off all station equipment wired behind it, via this jack.

RJ-32X Provides a series connection to the tip and ring conductors of a telephone line. It differs from RJ-31X in that it's wired ahead of a particular telephone set rather than ahead of all the station equipment. It's typically used with registered automatic dialers.

RJ-33X Is normally associated with a key telephone system. It provides a series connection to the tip and ring conductors of the telephone line and the key- system lien circuit A and A1 leads. The tip and ring conductors are wired ahead of the key-system line circuit. This arrangement is typically used when the modem requires central office or PBX ringing.

RJ-34X Is normally associated with a key telephone system. It's wired to provide a series connection to the key-system line circuit tip and ring conductors and it's a and A1 leads. It differs from RJ-33X in that all conductors are wired behind the key-system line circuit. This arrangement is typically used when the modem is not critical as to type of ringing signal or doesn't require central office or PBX ringing.

RJ-35X Is normally associated with a key telephone set. It's wired to provide a series connection to the tip and ring conductors of the telephone line and a bridged connection to the A and A1 leads. It differs from RJ-33X and RJ-34X in that the tip and ring leads are connected to the common wiring behind the pickup keys of the station set but ahead of the switch hook. The jack is wired to the key telephone set so that the modem functions on the line selected on the key telephone set.

RJ-36X Provides a connection for a registered telephone set equipped with an exclusion key when the telephone line is also to be used with a registered data set or registered protective circuitry. It's wired to provide a series connection to the tip and ring conductors of the telephone line and mode- indication leads MI and MIC. With this jack, the exclusion key can be used to transfer the telephone line between the modem and the telephone set. As a customer option, the exclusion key may be wired so that either the telephone set or the modem controls the line. In the former case, the exclusion key must be operated to transfer the telephone line to the modem. In the latter case, the telephone line is normally associated with the modem. Operation of the exclusion key is required to transfer the line to the telephone set. In either case, a closure on the MI and MIC leads indicates the voice mode.

RJ-37X Is used for providing two-line service with exclusion. The jack is wired to provide a bridged connection to the tip and ring conductors of two telephone lines with exclusion on line 1.

RJ-38X Provides a series connection to the tip and ring conductors of a telephone line identical to those described for RJ-31X. However, the jack also provides a continuity circuit which is used as an indication that the plug of the registered equipment is engaged with the jack. The jack is wired ahead of all station equipment electrically and is typically used with registered alarm dialers.

RJ-41M and RJ-45M Provide a multiple-mounting arrangement for mounting a number of RJ-41S or RJ-45S Single-Line Universal or Programmed data jacks. The telephone companies will terminate USOCs, RJ-41M and RJ-45M with RKM2X (which is the USOC equivalent for a mounting arrangement) and the appropriate number of RJ-41S or Rj-45S single-line data jacks as required by the user. The mounting arrangement will accommodate up to 16 single-line data jacks. In effect, this arrangement provides the features of a patch panel. The user has complete flexibility in patching the color and plug from any modem to any line. The arrangement can be mounted on a wall or on 19- or 23-inch relay racks.

RJ-41S Is a single-line universal data jack normally associated with fixed-loss loop (FLL) or programmed (P) modems. A switch, accessible to the user, is provided to select FLL to P type of operation (FLL equipment transmits at - 4 dB with respect to one milliwatt, and a pad is included in the data jack so that pad loss plus loop loss is nominally 8 dB Programmed modems adjust their output power in accordance with a programming resistor in the data jack. By these means, signals from either FLL or P types of registered modems will arrive at the local telephone company central office at a nominal -12 dB with respect to one milliwatt for optimum data transmission.) A sliding cover is provided to keep dirt and dust from entering the jack when it's not in use. The FLL/P switch selects the desired method of operation. Two matted surfaces are provided on the housing of the jack for the telephone company installer to write in the loop loss (designated LPL) and the telephone line number (designated TLN).

RJ-45 The RJ-45 is the 8-pin connector used for data transmission over standard telephone wire. That wire could be flat or twisted. And it's very important that you know what you're working with. You can easily use flat wire for serial data communications up to 19.2 Kbps. Up to that speed you're connecting with your wire to a data PBX, a modem, a printer or a printer buffer. If you wish to connect to a 10Base-T local area network, which you also do with a RJ-45, you must use twisted wire. You can typically tell the difference by looking at the

cable. If it's flat grey satin (like a typical phone wire, only bigger) than it's probably untwisted. If it's circular, then it's probably twisted and therefore good for LANs. RJ-45 connectors come into two varieties — keyed and non-keyed. Keyed means that the male RJ-45 plug has a small, square bump on its end and the female RJ-45 plug is shaped to accommodate the plug. A keyed RJ-45 plug will not fit into a female, non-keyed (i.e. normal) RJ-45. See RJ-11 and RJ-22.

RJ-45S Is a single-line data jack normally associated with programmed (P) modems. This jack is the same as the universal data jack RJ-41S described above, except that the pad for fixed loss loop (FLL) equipment and the switch to select FLL or P type of operation are omitted. Its appearance is the same as RJ-41S except that RJ-45S does not have the FLL/P switch. Both jacks provide bridged connections to the tip and ring of a telephone line and provide mode-indication leads for use with exclusion key telephone sets when required. The exclusion key telephone set requires a series jack RJ-36X as its normal means of connection.

RJ-48C An 8-position keyed plug most commonly used for connecting T-1 circuits. The RJ-48C is an 8-position plug with four-wires (two for transmit, two for receive) commonly connected. When the phone company delivers T-1 to your offices, it usually terminates its T-1 circuit on a RJ-48C. And it expects you to connect that RJ-48C to your phone system or T-1 channel bank and then to your phone system.

RJ-48S Normally associated with DDS services from the telephone company, this jack is used with DDS CSU/DSUs.

RJ-71C An RJ-71C provides a multiple series arrangement of tip and ring. It's typically used with registered series devices such as toll restrictors, etc. Jack RJ-71C can accommodate up to 12 circuits per jack (i.e., one tip and ring "in" and one tip and ring "out", 4 leads per circuit). This arrangement does not currently provide restoration upon disconnection of registered equipment. Thus, a manual bridging plug is provided in order to maintain circuit continuity upon withdrawal of a registered plug.

RJ-8 A coaxial cable with a transmission impedance of 50 ohms. It's largely used for data, not video. See also RJ-58.

RJ-A1X and RJ-A3X RJ-A1X and RJ-A3X are adapters used to adapt 4- and 12- position jacks, respectively, to a 6-position miniature bridged jack. They provide bridged connections to the tip and ring of the telephone line. If A and A1 leads are already terminated in the 4- or 12-pin jack, they will appear in positions 2 and 5 in the adapter. If A and A1 leads are not involved, positions 2 and 5 are reserved for telephone company use.

RJ-A2X An RJ-A2X is an adapter that coverts a single miniature jack to two miniature jacks. It provides a bridged connection to the tip and ring conductors of the telephone line. If A and A1 leads are already terminated in an existing miniature bridged jack, they will appear in positions 2 and 5 both miniature bridged jacks in the adapters. If A and A1 leads are not provided, positions 2 and 5 are reserved for telephone company use.

RJ-A3X See RJ-A1X.

RJE Remote Job Entry. A Remote Job Entry terminal is used for the transmission of "batch" data to a remote computer system. Processed information is then returned to the printer in the terminal. This type of processing from a remote site is a standard method of data transmission. See IBM.

RJXXX Registered Jack.

RLC ReLease Complete message. The sixth ISUP call set-up message. A message sent in either direction in response to the receipt of a Release Message, or if appropriate to, a Reset Circuit Message when the circuit concerned has been brought into the idle condition. See ISUP and COMMON CHANNEL SIGNALING.

RLCM Remote Line Concentrating Module.

RLCS Remote Live Call Screening. A Panasonic phone feature. RLCS is even better than live call screening feature. It lets you plug a single cordless phone into the External Data Port (XDP) on your Panasonic phone and monitor incoming voice mail remotely. When a call ends up in voice mail you use your cordless phone to screen your call. To take the call simply hit the same key again. This way you are always assured you will receive the important call you were waiting for — even if you're in a meeting. This feature is only available on the Panasonic Super Hybrid PBX with its integrated digital Panasonic voice mail system. But I liked the feature enough to include it in my dictionary.

RLE RLE, an acronym for Run Length Encoding, is a loseless compression algorithm, used in the BinHex version of MIME. See BinHex and MIME. Run Length Encoding is also a compression algorithm used in Group III (fax-standard) compression. A scanner reads a horizontal line of pixels — the "scan line" — and counts the black pixels. This is the "run length" or "run count." The same is then done for white pixels. A scan line might have a run of 230 white pixels, followed by a run of 23 black, followed by 45 white, etc. The result is a set of run lengths for the line. That's RLE. The different run lengths are then given codes — shorter ones for the most frequently occurring run lengths, longer ones for the unique run lengths — to represent them. Thats called "Huffman Coding." The codes are set in Group III compression; some compression schemes can assign new codes for each image.

RLL 1. Radio in the Local Loop. Another term for WLL (Wireless Local Loop). RLL/WLL is a means for CLECs to deploy local loop capability rapidly, bypassing the incumbent LEC in the process. SEE WLL AND CLEC for more detail.
2. Run Length Limited. A type of data coding used for disk drives. The term Run Length Limited derives from the fact that the techniques limit the distance (run length) between magnetic flux reversals on the disk platter. An RLL certified hard drive can use an MFM controller card but the storage capacity and the data transfer rate will be reduced.

Rlogin Rlogin is an application that provides a terminal interface between UNIX hosts using the TCP/IP network protocol. Unlike Telnet, Rlogin assumes the remote host is (or behaves like) a UNIX machine.

RLP Radio Link Protocol.

RLR Receive Loudness Rating.

RLT Release Link Trunks.

RM Resource Management. A mechanism used by the explicit-rate flow control scheme defined by the ATM Forum. Special control cells are used for explicit rate marking, with those cells taking up as much 3% of the network capacity, but serving to quickly convey information about network congestion back to the source (transmitting device). Explicit rate marking requires that the ATM switch mark the ABR RM (Available Bit Rate Resource Management) cells with the maximum transmission speed which can be supported over a VC (Virtual Circuit). Refer to RM-cell.

RM-Cell An ATM term. Resource Management Cell: Information about the state of the network-like bandwidth availability, state of congestion, and impending congestion, is conveyed to the source through special control cells called Resource Management Cells (RM-cells).

RMA Returned Merchandise Authorization. A code number provided by the seller as a prerequisite to returning product for either repair or refund. An indispensable tracking procedure, it operates like a purchase order system. If you return computer or telephone equipment without an RMA number, chances are that it will be lost. The manufacturer will deny they ever received it. And you will be out of pocket and blamed. RMA is also called a Return Authorization (RA) number.

RMAS Remote Memory Administration System.

RMATS Remote Maintenance and Test System. That equipment and programming used to run, maintain and test a telephone system remotely — usually by dialing in on a special phone line.

RMON ReMOte monitoring specification. A simple network management protocol used to manage networks remotely. It provides multi vendor interoperability between monitoring devices and management stations. RMON, is a set of SNMP-based MIBs (Management Information Bases) that define the instrumenting, monitoring, and diagnosing of local area networks at the OSI Data-Link layer. In IETF RFC 1271, the original RMON, which is sometimes referred to as RMON-1, defines nine groups of Ethernet diagnostics. A tenth group, for Token Ring, was added later in RFC 1513. RMON uses SNMP to transport data. To be RMON-compliant, a vendor need implement only one of the nine RMON groups. See MIB, RMON-2, RMON Probe, RMON Token Ring and SNMP.

RMON-2 The second Remote MONitoring MIB standard, called RMON-2, defines network monitoring above the Data-Link layer. It provides information and gathers statistics at the OSI Network layer and Application layer. Unlike the original RMON, RMON-2 can see across segments and through routers, and it maps network addresses (such as IP) onto MAC addresses. RMON-2 is currently a proposed standard under IETF RFC 2021. To be compliant with RMON-2, a vendor must implement all the monitoring functions for at least one protocol. RMON-2 does not include MAC-level monitoring, and thus it is not a replacement for the original RMON. See RMON.

RMON Groups The original IETF proposed standard for the RMON MIB, RFC 1271 defines nine Ethernet groups: Ethernet Statistics, Ethernet History, Alarms, Hosts, Host Top N ("N" indicates that it collects information on a number of devices), Traffic Matrix, Filters, Packet Capture, and Events. RFC 1513 extends this standard to support Token Ring. See MIB, RMON, RMON Token Ring. b

RMON Probe Sometimes called an RMON agent, an RMON probe is either firmware built into a specific network device like a router or switch, or a specific device built for network monitoring and inserted into a network segment. An RMON probe tracks and analyzes traffic and gathers statistics, which are then sent back to the monitoring software. Historically, an RMON probe was a separate piece of hardware, but now RMON firmware is embedded in high-end switches and routers. See RMON.

RMON Token Ring IETF proposed standard RFC 1513 is an extension to the original RMON MIB (RFC 1271), with support for Token Ring. Some sources refer to this standard as RMON TR, but it's generally considered a replacement for the older standard. In RFC 1513, the RFC 1271 Statistics and History monitoring groups have additional specifications for Token Ring, and a tenth group is added to monitor ring configuration and source routing. In 1994, the proposed standard became a draft standard under the designation RFC 1757; many vendors use the RFC 1513 and 1757 numbers interchangeably. See MIB, RMON.

RMP See Roving Monitor Port.

RMS Root Mean Square. Method of measuring amplifier power.

RMS-D1 Remote Measurement System Digital 1.

RMU Remote Mask Unit.

RNR Abbreviation for not ready to receive.

RO 1. Receive Only.

2. An ATM term. Read-Only: Attributes which are read-only can not be written by Network Management. Only the PNNI Protocol entity may change the value of a read-only attribute. Network Management entities are restricted to only reading such read-only attributes. Read-only attributes are typically for statistical information, including reporting result of actions taken by auto-configuration.

Road Kill Companies who go out of business because they couldn't figure out how to compete in the all digital, mega-channel, interactive marketplace. See also ROAD PIZZA

Road Map Software that enables easier navigation to information desired. Especially helpful for Internet users.

Road Pizza Companies who get run over by their competitors. Also a code name for a previous new product from Apple Computer.

Road Warrior A businessperson who travels...a lot. The road warrior's office is an airplane seat or a hotel room, and his weapons are a laptop with modem and communications software, a pager and a cell phone. Road warriors have a love/hate relationship with the road.

Roadblocks Slang for legislation, or lack thereof, which inhibits rather than promotes the growth of the markets using interactive multimedia.

Roadhog Any company trying to dominate the information highway through control of pieces of the infrastructure.

ROADS Robust Open Architecture Distributed Switching.

Roaming Using your cellular phone in a city besides the one in which you live. Roaming usually incurs extra charges. Roaming prices are usually at a premium to local home area prices.

Roaming Dial-back The ability of the Dialup Switch to dial-back the user at a roaming location.

Roamops The Roaming Operations working group of the IETF (Internet Engineering Task Force). Roamops is developing a standard to define an authentication method for Internet users and a billing process across ISPs. The objective is to allow travelers to use a local ISP in another city or country, much like you can roam with your cell phone for an additional charge. If you use a local ISP, you currently must either place a long distance call to your local ISP's POP (Point of Presence), where your e-mail sits in a mailbox on his mail server—much like you must place a long distance call to access your voice mail. Alternatively, you can use a national/international provider to gain access, or forward your e-mail from your local ISP to a national/international provider. All three of these alternatives are cumbersome, troublesome, and expensive, to say the least.

There also exist several roaming alliances which do the same thing, but on a proprietary basis. GRIC (Global Roaming Internet Connection) and I-Pass Alliance both act as clearing houses, much like a bank clearing house for ATM transactions, with each participating ISP setting its own rates for roaming transactions. The large telco ISPs, including both LECs and IXCs, also are establishing a proprietary roaming alliance. See also GRIC and I-PASS.

Robbed-Bit Signaling This explanation from Gary Maier of Dianatel: ISDN is the key to future sophisticated telephone

network services with its dynamic, highly configurable T-1 connection (also called PRI connection). Since T-1 is a common method of carrying 24 telephone circuits, many wonder about the uses for ISDN, especially when they learn ISDN signaling requires an entire voice channel, reducing today's T-1 from 24 voice channels to 23. But the popular signaling mechanism of "robbed bit" signaling in T-1 has serious limitations. Robbed bit signaling typically uses bits known as the A and B bits. These bits are sent by each side of a T-1 termination and are buried in the voice data of each voice channel in the T-1 circuit. Hence the term "robbed bit" as the bits are stolen from the voice data. Since the bits are stolen so infrequently, the voice quality is not compromised by much. But the available signaling combinations are limited to ringing, hang up, wink, and pulse digit dialing. In fact, the limitations are obvious when one recognizes DNIS and ANI information are sent as DTMF tones.

This introduces a problem: time. Each DTMF tone requires at least 100 milliseconds to send, which in a DNIS and ANI situation with 20 DTMFs will take at least two full seconds. There is also a margin for error in transmission or detection, resulting in DNIS or ANI failures. With the explosion of telephone related services, the telephone companies are turning to ISDN PRI to provide the more complicated and exact signaling required for new services. ISDN employs a more robust method of signaling. ISDN uses a T-1 circuit as 23 voice channels and one signaling channel. The term 23B plus D refers to 23 bearer (voice) channels and 1 Data (signaling) channel. The data channel carries the signaling information at a rate of 64 kilobits per second. This speed is many times greater than some of the most powerful modems available. Because of this high speed, telephone calls can be placed more quickly, and because of the protocol used, DNIS or ANI transmission failures are impossible.

Additionally, since no bits are "robbed" from the voice channels, the voice quality is better than that of Robbed Bit signaling on today's T-1 circuits. Also, computer modems and high speed faxes can use the voice channel for sending digital data instead of the traditional analog bit "noise." Therefore, ISDN PRI offers the end user countless new service capabilities. One channel could be used for faxing, another for modem data, several for video, another for a LAN and the remainder for voice. Suddenly, the average T-1 circuit becomes a pipeline for all communications! Increasingly long distance carriers are using ISDN PRI to provide inbound 800 calls with ANI and DNIS and re-routing skills. See BIT ROBBING.

ROBO Remote Office / Branch Office market. See also SOHO and SOMO.

Robust A term used by telecommunications switch manufacturers to describe the alleged hefty quality of their network connections — especially their switch-to-host links. The price of their link is often directly related to the number of times the manufacturer uses the word "robust" in a customer presentation.

Rock Ridge Format A set of CD-WO (write-once CDs) specifications to provide directory structures that may be updated as additional files are added. The specifications include: System Use Sharing Protocol (SUSP) and the Rock Ridge Interchange Protocol Specification (RRIPS). The specifications are extensions of the ISO 9660 format for CD-ROM. The SUSP extension to the ISO 9660 standard allows multiple file system extensions to coexist on one CD-ROM disc. The RRIP specification lets POSIX files and directories be recorded on CD-ROM without requiring modifications to files, such as shortening file and directory names.

Rodent Rubber Another term for a B connector. See B CONNECTOR.

ROE Regulatory commission authorized allowed rate of return on equity. See REVENUE REQUIREMENT.

ROFL Abbreviation for "Rolling On the Floor, Laughing;" commonly used on E-mail and BBSs (Bulletin Board Systems).

ROH Receiver Off Hook.

ROI Return On Investment.

Roll About A totally self-contained videoconferencing system consisting of the codec, video monitor, audio system, network interfaces and other components. These roll about systems can, in theory, be moved from room-to-room but in fact are not because they are electronic equipment that does not benefit from jostling. It's also heavy.

Roll Call Polling A technique in which every station is interrogated sequentially by a central computer system.

Roll Call A teleconferencing term. In a Dial-Out (operator-initiated) teleconference, the operator will announce each of the participants as he adds them to the conference bridge.

Rollover A Web design page term. Imagine you're on a web site and you move cursor over a piece of art or you click on it. Suddenly, that piece of art changes and reveals an explanation or new menu, etc. A rollover is a piece of software which changes the appearance of objects when you roll over them or click on them. See also Rollover Lines.

Rollover Lines You receive many incoming calls. You don't want to miss a call, so you ask your phone company to set your phone lines up to roll over, also called hunt, also called ISG (Incoming Service Group) in telephonese. You order five lines in hunt. The calls come into the first. If the first one is busy, the second rings. If it's busy, the third rings. If they're all busy, then the caller receives a busy. The commonest types of hunting are sequential and circular hunting. Sequential hunting starts at the number dialed, keeps trying one number after another in number order and ends at the last number in the group. It's typically descending. For example, it starts at 691-8215, goes to 691-8216, then 691-8217, etc. But it can also be ascending — from 691-8217 up. Circular hunting hunts all the lines in the hunting group, regardless of the starting point. Circular hunting, according to our understanding, circles only once (though your phone company may be able to program it circle a couple of times). The differences between sequential and circular are subtle. Circular seems to work better for large groups of numbers. You don't need consecutive phone numbers to do rollovers. Nowadays you can roll lines forwards, backwards and jump around, for example most idle, least idle. Rollovers are now done in software. This also has its downside, since software fails. For example, theoretically if a rollover strikes a dead trunk, it should bounce to the next live trunk. But sometimes it hangs on the dead trunk and many of your incoming calls never get answered. They might ring and ring. They might hit a busy. My recommendation: Test your rollovers at least twice a day. In particular, test that your callers ultimately get a busy if all your lines are busy. Nothing worse your customer should receive a ring-no-answer or a constant busy when calling your company. See also Terminal Number.

ROLM A telephone equipment manufacturer based in Santa Clara, CA, at least once upon a time. ROLM was started in 1969 by four engineers to produce computers for the military. The company introduced one of the first digital PBXs in 1975. It was a great PBX. Later, they developed a line of KTSs (Key Telephone Systems) and hybrid PBX/KTS systems. They were not so good. IBM acquired ROLM in 1984 as part of their plan to integrate the worlds of computers and communications. It didn't work...at all. And IBM lost a lot of money with Rolm. In 1989, IBM sold ROLM to Siemens, at which time it became ROLM Company. In 1994, the name was changed to Siemens Rolm Communication Inc. In 1996, the name was changed to Siemens Business Communication Systems, Inc. Siemens really doesn't use the name ROLM (or Rolm) anymore, but there are a lot of ROLM systems still in service.

Rolodex A Trademarked product which started life as paper card based device to keep names and address on. Now it has become more of a generic name to connote software to let you look up peoples' phone numbers and addresses. Software to do this is also called PIM — for Personal Information Manager.

ROM Read Only Memory. Computer memory which can only be read from. New data cannot be entered and the existing data is non-volatile. This means it stays there even when power is turned off. A ROM is a memory device which is programmed at the factory and whose contents thereafter cannot be altered. In contrast is the device called RAM, whose contents can be altered. See READ ONLY MEMORY and MICRO-PROCESSOR.

ROM Font The ROM Font is your PC's type font. It consists of a set of 256 characters which cannot be edited — unless you are running in video mode, in which case you can design your own type font.

Rom Shadowing 386 and higher CPUs provide memory access on 32 & 64 bit paths. Often they will use a 16 bit data path for system ROM BIOS info. Also some adapter cards (ie. older video, network adapters etc.) with on board BIOS may use an 8 bit path to system memory. For high end computers this is a bottleneck. Like having YIELD signs out on the lanes within a freeway. ROM is very slow, 150ns-200ns. Modern RAM is 60ns or less. Therefore when the system is waiting on this data it generates wait states. For high end computers these wait states slow the entire system down. There is a system developed to transfer the contents of all the slow 8-16 bit ROM chips through out the system into 32 bit faster main memory. "This is ROM SHADOWING". This is accomplished using the MMU, the memory management unit. The MMU takes a copy of the ROM BIOS codes and places it into RAM. To the rest of the system this RAM location looks exactly like the original ROM location. This definition courtesy Charlie Irby, chasirby@foothill.net.

Roofing Filter A low-pass filter used to reduce unwanted higher frequencies.

Room Cut-Off Hotel/motel guest telephones restricted from outgoing calls when the guest room is unoccupied.

Room Status And Selection Provides the capability to store and display the occupancy and cleaning status and the type number of each guest room. This helps housekeeping management, maid locating and room selection. Also, communications between the front desk and the housekeeper are speeded up via real-time maid activity and checkout audit printouts to indicate which rooms need cleaning next. The occupancy status is normally changed by the maid or inspector dialing from the room telephone.

Root The base of a tree. The base of a hard disk. See Root Directory.

Root Directory The top-level directory of a PC disk, hard or floppy. The root directory is created when you format the disk. From the root directory, you can create files and other directories.

Root Web The FrontPage web that is provided by the server by default. To access the root web, you supply the URL of the server without specifying a page name. FrontPage is installed with a default root web named <root web>. All FrontPage webs are contained by the root FrontPage web. <f"xb futura extrabold">ROSE Remote Operations Service Element. An application layer protocol that provides the capability to perform remote operations at a remote process. Definition from Bellcore in reference to its concept of the Advanced Intelligent Network.

Rostered Staff Factor RSF. A call center term. Alternatively called an Overlay, Shrink Factor or Shrinkage. RSF is a numerical factor that leads to the minimum staff needed on schedule over and above base staff required to achieve your service level and response time objectives. It is calculated after base staffing is determined and before schedules are organized, and accounts for things like breaks, absenteeism and ongoing training.

Rostering A call center term. The practice of rotating employees through all existing schedules in a matrix, or roster, of schedules. This "share the grief" method is prevalent in Europe and Australia, where agents work through an entire roster.

ROT13 A way to encode things that the general Internet community can't read. Each letter in a message is replaced by the letter 13 spaces away from it in the alphabet. There are online decoders to read these. For instance, Harry Newton becomes Uneel Arjgba, which sounds a lot more exotic.

ROTA A call center term. 1. An European term for a rotating shift pattern or rotating schedule, 2. Short form for roster.

Rotary Dial The circular telephone dial. As it returns to its normal position (after being turned) it opens and closes the electrical loop sent by the central office. Rotary dial telephones momentarily break the DC circuit (stop current flow) to represent the digits dialed. The circuit is broken three times for the digit 3. The CO counts these evenly-spaced breaks and determines which digit has been dialed. You can hear the "clicks". The number "seven," for example consists of seven "opens and closes," or seven clicks. You can dial on a rotary phone without using the rotary dial. Simply depress the switch hook quickly, allowing pauses in between to signify that you're about to send a new digit. It's a good party trick.

Rotary Dial Calling The telephone system will accept dialing from conventional rotary dial sets.

Rotary Hunt You buy several phone lines. Let's say 212-691-8215, 212-691-8216, 212-691-8217, 212-691-8218. Someone dials you on your main number — 212-691-8215. It's busy. (That's our number.) The central office slides the call over to 212-691-8216. If that number is busy, it slides it over to 212-691-8217, and so on. This is called rotary hunt. It hunts to the next line in the rotary group. In the old days, the phone lines you could rotary hunt to had to be in numerical sequence. But now with modern stored program control central offices, your lines in rotary hunt can be very different as long as they're all on the same exchange.

Rotary Output To Central Office Most central offices are equipped to provide tone dial service. In cases where the telephone company central office trunks are not designed to accept tone signaling, your on premise phone system (PBX, key system or single line phone) will translate the number entered by a phone in tones into rotary dial pulses which can be processed by the central office.

Rotating Cylinder (Drum) Scanner A scanning technique using a drum and a photocell scan head. The original

is attached to the drum, enabling the scan head to travel along the length of the document. Reflected light from the document is concentrated on the scanner photocell, which causes an analog signal.

Rotating Helical Aperture Scanner Original is illuminated by a lamp when fed onto the platen, via a mirror and lens system, the document's image is focused first through a fixed horizontal slot, then through a rotating spiral slit disk series, and finally onto a photocell to generate an analogous electrical current.

Rotational Mailboxes Information only mailboxes whose information is automatically changed on a time sensitive or usage sensitive basis.

ROTFL I'm "Rolling on the Floor, Laughing." Used in e-mail.

ROTL 1. Remote Office Test Line. Provides the capability to originate automatic inter office trunk transmission test calls under the automatic control of CAROT from a remote location. 2. A popular online abbreviation, shorthand for "Rolling On The Floor Laughing"; an appropriate typed response to a particularly amusing online remark. Other common Net acronyms include IMHO ("In My Humble Opinion") and IMNSHO ("In My Not-So-Humble pinion").

Rotor The rotating part of a motor or other electrical machines.

ROTS Rotary Out-Trunk Switches.

Round Cutter These are used to cut cables. The blades of the cutter are curved so that there is a space between them.

Round Robin This is a method of distributing incoming calls to a bunch of people. This method selects the next agent on the list following the agent that received the last call. See also TOP DOWN and LONGEST AVAILABLE.

Roundtrip Propagation Delay Roundtrip propagation delay from a burst modem to a burst modem will be about 470 milliseconds to 570 milliseconds (About half a second). See SATELLITE TRANSMISSION DELAY.

Routable Protocols Protocols, such as TCP/IP, DECnet, and XNS, that support Network Layer addressing. Packets constructed using these protocols contain information about how data should move through a network. This information, carried in the NLA (Network Layer Address) field of the packet, is used by internetworking devices to make routing decisions.

Route The path that a message takes. In telephone companyese, a route is the particular trunk group or interconnected trunk groups between two reference points used to establish a path for a call. This term (or the term routing) is also used as a verb to define the act of selecting a route or routes.

Route 66 A colloquial term for the Internet, with something interesting everywhere along the way. Route 66 used to be the way we drove across America before they put in concrete highways.

Route Advance This feature routes outgoing calls over alternate long distance lines when the first choice trunk group is busy. The phone user selects the first choice route by dialing the corresponding access code. The phone equipment automatically advances to alternate trunks and trunk groups, based on the user's class of service. Route advance is a more primitive form of least cost routing. See LEAST COST ROUTING.

Route Daemon A program that runs under 4.2 or 4.3BSD UNIX systems (and derived operating systems) to propagate routes among machines n a local area network. Pronounced "route-dee."

Route Discovery Process through which a brouter can learn LAN topology by passing information about its address and the LANs it connects and receiving the same information from others.

Route Indicator An address or group of characters in the heading of a message defining the final circuit or terminal to which the message is to be delivered.

Route List A sequence of trunk groups that can be searched for a particular route. This list is comprised of trunk groups and configuration attributes (e.g. Class of Service) governing the use of a particular trunk group.

Route Mile Let's say that you have two sheaths of fiber, each of which contains ten fibers and runs for one mile. That is one route mile (total distance of all fibers), two sheath miles (2 sheaths running one mile), and twenty fiber miles (20 fibers running one mile).

Route Optimization Another way of saying Least Cost Routing.

Route Server An ATM term. A physical device that runs one or more network layer routing protocols, and which uses a route query protocol in order to provide network layer routing forwarding descriptions to clients.

Route Xpander Card A board manufactured by IBM for insertion into a PC which provides the PC with a wide area interface to a frame relay network, including handling all of the necessary protocol encapsulation.

Routed Protocol A protocol that can be routed by a router. To do so a router must understand the logical internetwork as perceived by that routed protocol. Examples of routed protocols include DECnet, AppleTalk, and IP.

Router 1. As in software, router is a system level function that directs a call to an application.

2. As in hardware, routers are the central switching offices of the Internet and corporate Intranets and WANs. Routers are bought by everybody — from backbone service providers to local Internet Service Providers (ISPs), from corporations to Universities. The main provider of routers in the world is Cisco. It has built its gigantic business on selling routers — from small ones, connecting a simple corporate LAN to the Internet, to corporate enterprise wide networks, to huge ones connecting the largest of the largest backbone service providers. A router is, in the strictest terms, an interface between two networks.

Routers are highly intelligent devices which connect like and unlike LANs (Local Area Networks). They connect to MANs (Metropolitan Area Networks) and WANs (Wide Area Networks), such as X.25, Frame Relay and ATM. Routers are protocol-sensitive, typically supporting multiple protocols. Routers most commonly operate at the bottom 3 layers of the OSI model, using the Physical, Link and Network Layers to provide addressing and switching. Routers also may operate at Layer 4, the Transport Layer, in order the ensure end-to-end reliability of data transfer.

Routers are much more capable devices than are bridges, which operate primarily at Layer 1, and switches, which operate primarily at Layer 2. Routers send their traffic based on a high level of intelligence inside themselves. This intelligence allows them to consider the network as a whole. How they route (also called routing considerations) might include destination address, packet priority level, least-cost route, minimum route delay, minimum route distance, route congestion level, and community of interest. Routers are unique in their ability to consider an enterprise network as comprising multiple physical and logical subnets (subnetworks). Thereby, they are quite capable of confining data traffic within a subnet, on the basis of privilege as defined in a policy-based routing table. In a traditional router topology, each router port defines

a physical subnet, and each subnet is a broadcast domain. Within that domain, all connected devices share broadcast traffic; devices outside of that domain can neither see that traffic, nor can they respond to it. Contemporary routers have the ability to define subnets on a logical basis, based on logical address (e.g., MAC or IP address) information contained within the packet header, and acted upon through consultation with a programmed routing table. In addition to standalone routers developed specifically for that purpose, server-based routers can be implemented. Such routers are in the form of high-performance PCs with routing software. As software will perform less effectively and efficiently than firmware, such devices generally are considered to be less than desirable for large enterprise-wide application, although they do serve well in support of smaller remote offices and less-intensive applications. Routers also are self-learning, as they can communicate their existence and can learn of the existence of new routers, nodes and LAN segments. Routers constantly monitor the condition of the network, as a whole, in order to dynamically adapt to changes in network conditions. Characteristics of routers can include:

- LAN Extension
- Store & Forward
- Support for Multiple Media
- Support for Multiple LAN Segments
- Support for Disparate LAN Protocols
- Filtering
- Encapsulation
- Accommodation of Various and Large Packet Sizes
- High-Speed Internal Buses (1+ Gbps)
- Self-Learning
- Routing Based on Multiple Factors
- Route Length
- Number of Hops
- Route Congestion
- Traffic Type
- Support for a Community of Interest (VLAN)
- Redundancy
- Network Management via SNMP

Router protocols include both bridging and routing protocols, as they perform both functions. Those protocols fall into 3 categories:

1. Gateway Protocols establish router-to-router connections between like routers. The gateway protocol passes routing information and keep alive packets during periods of idleness.

2. Serial Line Protocols provide for communications over serial or dial-up links connecting unlike routers. Examples include HDLC, SLIP (Serial Line Interface Protocol) and PPP (Point-to-Point Protocol).

3. Protocol Stack Routing and Bridging Protocols advise the router as to which packets should be routed and which should be bridged.

This definition courtesy of "Communications Systems & Networks," the best-selling book by Ray Horak, my Contributing Editor. To buy the book, www.amazon.com. See also Bridges, Hubs, Internetworking and Switches.

Router-Based Firewall A router-based firewall is a packet-filtering router. Not everyone agrees that a packet-filtering router alone is a firewall. Many people insist that only a system that includes a dual-homed gateway is a firewall. However, other people argue that a packet-filtering router is a firewall because the router meets important firewall criteria: The router is a computer through which incoming and outgoing packets must pass through which only authorized packets can pass.

Router Droppings The inclusions added to e-mail messages when a server or recipient cannot be found. Cryptic and foul-looking, their meaning is usually impossible to fathom. Also called "daemon droppings."

Router Rip A Cisco term. This command enables the RIP routing process on the router for TCP/IP.

Routine A program, or a sequence of instructions called by a program, that has some general or frequent use.

Routing The process of selecting the correct circuit path for a message.

Routing Area Subdomain A cellular radio term. The combined geographic area of all Mobile Data Base Stations (MDBSs) controlled by a single Mobile Data Intermediate System (MD-IS).

Routing Code 1. Another narne for area code. See DN (as in directory number).
2. The combination of characters or digits required by the switching system to route a transmission to its desired destination.

Routing Computation The process of applying a mathematical algorithm to a topology database to compute routes. There are many types of routing computations that may be used. The Djikstra algorithm is one particular example of a possible routing computation.

Routing Constraint A generic term that refers to either a topology constraint or a path constraint.

Routing Data Base Distance table in DNA.

Routing Flexibility The ability to send information over various network paths to avoid congestion and use portions of a total network that would otherwise be idle.

Routing Information Protocol RIP is based on distance-vector algorithms that measure the shortest path between two points on a network, based on the addresses of the originating and destination devices. The shortest path is determined by the number of "hops" between those points. Each router maintains a routing table, or routing database, of known addresses and routes; each router periodically broadcasts the contents of its table to neighboring routers in order that the entire network can maintain a synchronized database. See also DISTANCE VECTOR PROTOCOL.

Routing Label The part of a signaling message identifying its destination.

Routing Metric The method by which a routing algorithm determines that one route is better than another. This information is stored in routing tables. Such tables include reliability, delay bandwidth, load, MTUs, communication costs, and hop count.

Routing Protocol A general term indicating a protocol run between routers and/or route servers in order to exchange information used to allow figuring of routes. The result of the routing computation will be one or more forwarding descriptions. In short, a protocol that accomplishes routing through the implementation of a specific routing algorithm. Examples of routing protocols include IGRP, RIP, and OSPF.

Routing switcher An electronic device that routes a user-supplied signal (audio, video, etc.) from any input to any user-selected output. Inputs are called sources. Outputs are called destinations.

Routing Table 1. Incoming Phone Calls: A routing table is a user definable list of steps which are treatment instructions for an incoming call. Ideally these steps should be addressed and the call treatment begun before the call is answered. A routing table should consist of a minimum of steps that

include agent groups, voice response devices, announcements (delay and informational) music on hold, intraflow and interflow steps, route dialing (machine based call forwarding). A significant issue in the structure of routing tables is "look-back" capability, where no one previously interrogated resource is abandoned by the system (i.e. an agent group is now ignored, even though an agent is now available, because the ACD does not consider previous steps in the routing table).
2. Outgoing Phone Calls: For a specific calling site, this table lists the long distance routing choices for each location to be dialed. There may be only one choice (route) listed for some or all destinations or there may be several choices for some destinations. (It depends how many outgoing lines and how many outgoing trunk groups you have.) If there are several choices then they will be ranked by some criteria (least cost, best quality, etc.).
3. In data communications, a routing table is a table in a router or some other internetworking device that keeps track of routes (and, in some cases, metrics associated with those routes) to particular network destinations. See ROUTING METRIC.

Routing Update A message sent from a router to indicate network and associated cost information. Routing updates are typically sent at regular intervals and after a change in network topology.

Roving Monitor Port Switch feature that lets you monitor network traffic on one or more ports via a third-party LAN packet analyzer. RMP can let you change the monitoring and monitored ports via software commands instead of via hardware changes.

Row In a table, a horizontal collection of cells.

RPC Remote Procedure Call. 1. A protocol governing the method with which an application activates processes on other nodes and retrieves the results. A popular paradigm for implementing the client-server model of distributed computing. A request is sent to a remote system to execute a designated procedure, using arguments supplied, and the result returned to the caller. There are many variations and subtleties, resulting in a variety of different RPC protocols.
2. A mechanism defined by Sun Microsystems and described in RFC-1057 that provides a standard for initiating and controlling processes on remote or distributed computer systems.

RP-125 A SMPTE parallel component digital video standard.

RPG Report Program Generator. A computer language for processing large data files.

RPM Remote Packet Module.

RPN Reverse Polish Notation. A calculator using RPN (such as a high-end Hewlett Packard business calculator) starts with the number you type in, then you hit enter, then you type in another number and the minus sign. Bingo, your screen shows the result of your calculation. In "normal" calculators you'd have to do another step, namely hit the enter button. Frankly, I prefer RPN. It's faster, easier and more logical. Most calculators don't come with RPN, sadly.

RPOA Recognized Private Operating Agency. A term used by the ITU-T to describe those companies designated as operating telephone companies — if the country's phone networks are not run by government-owned administrations, such as the PTTs in Europe. A recognized Private Operating Agency is an organization that handles internetwork communications (e.g., long distance carriers). To identify some RPOAs, you must dial a prefix before your outgoing directory number. An RPOA can also refer to one or more DNICs that will connect two X.25 endpoints. For ISDN X.25, an RPOA is usually the DNIC for the ISDN's long distance carrier. See also ITU.

RPQ Request for Price Quotation. Solicitation for pricing for a specific component, software product, service or system. See also RFQ.

RPS Ring Protection Switching. Nortel's "Introduction to SONET Networking" tutorial handbook (www.nt.com/broadband/reference/sonet_101.html) talks about "Automatic Healing of Failed or Degraded Optical Spans in a Two-Fiber BLSR." The handbook says "inthe event of failure or degradation in an optical span, automatic ring protection switching (RPS) reroutes affected traffic away from the fault within 50 milliseconds, preventing a service outage."

RQC Repair and Quick Clean. A term in the industry which repairs telecom equipment. It means all equipment is repaired and fully tested with a burn-in (if required) and an operational systems test. It also includes minor cosmetic cleaning of the unit. Definition courtesy Nitsuko America. See also LIKE NEW REPAIR AND UPDATE and REPAIR, UPDATE AND REFURBISH.

RR Abbreviation for Ready to Receive.

RRM Radio Resource Management.

RRME Radio Resource Management Entity.

RS 1. Recommended Standard, as in RS-232.

2. Record Separator, in data processing terms.

3. An ATM term. Remote single-layer (Test Method): An abstract test method in which the upper tester is within the system under test and there is a point of control and observation at the upper service boundary of the Implementation Under Test (IUT) for testing one protocol layer. Test events are specified in terms of the abstract service primitives (ASP) and/or protocol data units at the lower tester PCO.

RS-170 The EIA (Electronics Industries Association) standard for the combination of signals required to form NTSC monochrome (black and white) video.

RS-170A The EIA standard for the combination of signals required to form NTSC color video. It has the same base as RS-170, with the addition of color information.

RS-232 Also known as RS-232-C and in its latest version EIA/TIA-232-E. RS-232 is actually a set of standards specifying three types of interfaces — electrical, functional and mechanical. These are used for communicating between computers, terminals and modems. The RS-232-C standard, which was developed by the EIA (Electrical Industries Association), defines the mechanical and electrical characteristics for connecting DTE and DCE data communications devices. It defines what the interface does, circuit functions and their corresponding connector pin assignments. The standard applies to both synchronous and asynchronous binary data transmission. The most commonly used RS-232 interface is ideal for the data-transmission range of 0-20 Kbps/50ft. (15.2 m). It employs unbalanced signaling and is usually used with 25 pin D-shaped connectors (DB25) to interconnect DTEs (computers, controllers, etc.) and DCEs (modems, converters, etc.). Serial data exits through an RS-232 port via the Transmit Data (TD) lead and arrives at the destination device's RS-232-C port through the Receive data (RD) lead. RS-232-C is compatible with these standards: ITU V.24; ITU V.28, ISO IS2110. Most personal computers use the RS-232-C interface to attach modems. Some printers also use RS-232-C. You should be aware that despite the fact that RS-232-C is an EIA "standard," you cannot necessarily connect one RS-232-C equipped device to another one (like a printer to a computer) and expect them to work intelligently together. That's because different RS-232-C devices are often wired or pinned differently and may also use different wires for different functions. The "traditional" RS-232C plug has 25 pins. With the introduction of the IBM PC AT in the mid-1980s, most PCs and laptops switched to the "new" RS-232-C plug with only nine pins, called the DB-9. This smaller plug does essentially the same thing as its bigger cousin, but you need an adapter cable to connect one to another. They're widely available. See also interface and the RS-232-C diagram. See EIA/TIA-232-E and the APPENDIX for description of the pins and what they do. See also Crossover cable — the name for a specially-wired RS-232 cable which allows two DTE devices or two DCE devices to be connected through serial ports and transmit and receive information across the cable. The sending wire on one end is joined to the receiving wire on the other. In an RS-232 cable, this typically means that conductors 2 and 3 are reversed.

RS-232 Fax Server A RS-232 fax server is software which connects a network server to a fax machine via an RS-232 port attached to the fax machine. There are not many fax machines with R2-232 so you need to chose carefully. The idea of this arrangement is to let users send faxes directly from their own PC via the fax server via the attached fax machine, or directly from the fax machine. Users can also use the fax machine as a scanner.

RS-250B In telecommunications, a transmission specification for NTSC video and audio.

RS-328 October, 1966 the Electronic Industries Association issues its first fax standard: the EIA Standard RS-328, Message Facsimile Equipment for Operation on Switched Voice Facilities Using Data Communications Equipment. The Group 1 standard, as it later became known, made possible the more generalized business use of fax. Transmission was analog and it took four to six minutes to send a page.

RS-366 An EIA interface standard for auto dialing.

RS-422 Defines a balanced interface with no accompanying physical connector. Manufacturers who adhere to this standard use many different connectors, including screw terminals, DB9, DB25 with nonstandard pinning, DB25 following RS-530, and DB37 following RS-449. RS-422 is commonly used in point-to-point communications conducted with a dual-state driver. Transmissions can run long distances at high speeds. RS is a standard operating in conjunction with RS-449 that specifies electrical characteristics for balanced circuits (circuits with their own ground leads). RS-422 (now known as EIA/TIA-422) is a balanced electrical implementation of RS-449 for high-speed data transmission. RS stands for recommended standard.

RS-422-A Electrical characteristics of balanced-voltage digital interface circuits.

RS-423 A standard operating in conjunction with RS-449 that specifies electrical characteristics for unbalanced circuits (circuits using common or shared grounding techniques). Another EIA standard for DTE/DCE connection which specifies interface requirements for expanded transmission speeds (up to 2 Mbps), longer cable lengths, and 10 additional functions. RS-449 applies to binary, serial, synchronous or asynchronous communications. Half- and full-duplex modes are accommodated and transmission can be over 2- or 4-wire facilities such as point-to-point or multipoint lines. The physical connection between DTE and DCE is made through a 37-contact connector; a separate 9-connector is specified to service secondary channel interchange circuits, when used.

RS-423-A Electrical characteristics of unbalanced-voltage digital interface circuits. RS-423 (now known as EIA/TIA-423) is an unbalanced electrical implementation of RS-449 for RS-232-C compatibility.

RS-449 RS-449 (now known as EIA/TIA-449) is essentially a faster (up to 2 Mbps) version of RS-232-C capable of longer cable runs. RS-449 is another "standard" data communications connector. It uses uses 37-pins and is designed for higher speed transmission. Each signal pin has its own return line, instead of a common ground return and the signal pairs (signal, return) are balanced lines rather than a signal referenced to ground. This cable typically uses twisted pairs, while a RS-232-C cable usually doesn't.

According to Black Box Corp, RS-449 defines functional/mechanical interfaces for DTEs/DCEs that employs serial binary data interchange, and is usually used with synchronous transmissions. It identifies signals (TD, RD, etc.) that correspond with the pin numbers for a balanced interface on DB37 and DB9 connectors. RS-449 was originally intended to replace RS-232-C, in order to improve data-transmission capabilities to up to 2 Mbps/200ft. (60M), reduce electrical "crosstalk," and accommodate additional signal functions. RS-232-C and RS-449 were to become interoperable by using electrical interface standards RS-422 and RS-423. But right now RS-232 and RS-449 are incompatible in terms of mechanical and electrical specifications. RS-449 is technically compatible with these standards: RS-530, V.10, V.110, and ITUT X.21 bis.

RS-485 Resembles RS-422 except that associated drivers are tri-state, not dual-state. It may be used in multipoint applications where one central computer controls many different devices. Up to 64 devices may be interconnected with RS-485. RS-485 describes electrical characteristics of a balanced interface used as a bus for master/slave operation. Used in industry for the Process Field Bus, and in telco management networks.

RS-499-1 Addendum 1 to RS-449. (What else?)

RS-530 Supercedes RS-449 and complements RS-232. Based on a 25-pin connection, it works in conjunction with either electrical interface RS-422 (balanced electrical circuits) or RS-423 (unbalanced electrical circuits). RS-530 defines the mechanical/electrical interfaces between DTEs and DCEs that transmit serial binary data, whether synchronous or asynchronous. RS-530 provides a means for taking advantage of higher data rates with the same mechanical connector used for RS-232. However, RS-530 and RS-232 are not compatible. And RS-530 offers the benefits of RS-449 and the efficiency of a 25-pin design. It accommodates data transmission rates from 20 Kbps to 2 Mbps; maximum distance depends on which electrical interface is used. (RS-530 is compatible with these standards: ITU 10, V.11, X.26; MIL-188114; RS- 449.)

RSA 1. Rural Service (or Statistical) Area. The FCC designated 428 rural markets across the United States and licensed two service providers per RSA. See also MSA.

2.   A public key encryption algorithm invented in 1977 and named after its inventors, Rivest, Shamir and Adleman. A large number algorithm, RSA is highly secure, as each user finds two large prime numbers ("p" and "q") which are then multiplied together ("p" x "q" = "n"). The public key is "n," and the private key is "p" and "q." RSA key sizes range from 768 to 2,048 bits; as a 2,048 bit key yields 2 to the 2,048th power possible combinations, RSA is highly immune to even the most persistent brute force security attacks. RSA Data Security Inc., a wholly owned subsidiary of Security Dynamics Technologies Inc., offers a number of security tools based on the RSA core algorithm, of which over 75 million copies have been licensed. PGP (Pretty Good Privacy) incorporates the RSA core algorithm, as does SET (Secure Electronic Commerce), which is an emerging RSA standard for secure commerce over the Internet. RSA core technologies are part of existing and proposed standards of ANSI, IEEE, ISO, and ITU. See also DIGITAL CERTIFICATE, ENCRYPTION, PGP, PRIVATE KEY, PUBLIC KEY and SET.

RSC Remote Switching Center.

RSC-S Remote Switching Center-S.

RSE An ATM term. Remote Single-layer Embedded (Test Method): An abstract test method in which the upper tester is within the system under test and there is a point of control and observation at the upper service boundary of the Implementation Under Test (IUT) for testing a protocol layer or sublayer which is part of a multi-protocol IUT.

RSFG An ATM term. Route Server Functional Group: The group of functions performed to provide internetworking level functions in an MPOA System. This includes running conventional interworking Routing Protocols and providing inter-IASG destination resolution.

RSL Request and Status Links. A generic term for linking computers and PBXs. Every manufacturer of phone systems is evolving towards open architecture and their own "RSL." The term RSL, which is too passive, is being replaced with PHI (PBX Host Interface), a term coined by Probe Research. Manufacturer PHI names include:

ACL — Applications Connectivity Link — Siemens' PHI link protocol

ACT — Applied Computer Telephony — Hewlett Packard's generic application interface to PBXs

Application Bridge — Aspect Telecommunications' ACD to host computer link

ASAI — AT&T's Adjunct Switch Application Interface

CIT — Digital Equipment Corporation's Computer Integrated Telephony (works with major PBXs)

CSA — Callpath Services Architecture — IBM's Computer to PBX link

Call Frame — Harris' PBX to computer link

Callpath Host — IBM and ROLM's CICS-based integrated voice and data applications platform which links to ROLM's 9751

Callpath — IBM's announced, CICS application link to IBM's CSA, available on the AS400 in April of 1991

Callbridge — Rolm's CBX and Siemens to IBM host or non-IBM host computer link

CompuCall — Northern Telecom's DMS central office link to computer interface

CPI — Computer to PBX Interface developed by Northern Telecom and DEC

CSP — Nabnasset's Communications Services Platform

CSTA — Computer Supported Telephony Application, PHI standard from ECMA

DECags — DEC ASAI Gateway Services. Two-directional link to AT&T's Definity

DMI — AT&T's Digital Multiplexed Interface, a T-1 PBX to computer interface

HCI — Host Command Interface. Mitel's digital PBX link to DEC computer

IG — AT&T's ISDN Gateway (one direction from the switch to the host)

ITG — AT&T's Integrated Telemarketing Gateway (two directional)

ISDN/AP — NT's PHI SL1 protocol supports NT's Meridian Link PHI

Meridian Link — NT's PHI product available on the Meridian

PBX

OAI — Open Application Interface. InteCom's and NEC's PHI

ONA — Open Network Architecture (for telephone central offices)

PACT — Siemens' PBX and Computer Teaming, protocols between PBXs and computers

PDI — Telenova/Lexar's Predictive Dialing Interface

SAI — Stratus Computer Switch Application Interface

SCAI — Switch to Computer Application Interface, the name given by T1S1 to PHI

SCIL — Aristacom's Switch Computer Interface Link Transaction Link

STEP — Speech and Telephony Environment for Programmers; WANG's link

Transaction Link — Rockwell's link from its Galaxy ACD to an external computer

Solid State Applications Interface Bridge — Solid's State Systems' PHI

Teleos IRX-9000 — Teleos' Intelligent Call Distribution platform

For more information, see OAI (Open Architecture Interface.)

RSM 1. Remote Switching Module. An AT&T 5ESS switch standalone switching module that supports all line features and routes intro-RSM calls. It is either a single module or a multi module and can be situated up to 150 miles from the 5ESS switch host.

2. Radio Sub-system Management. A wireless telecommunications term. Management of radio channels including timing and frequency as well as all machines between the mobile station and the MSC.

RSRB Remote Source Route Bridging. Source-route bridging over wide-area links.

RSSI Received Signal Strength Indication.

RSU Remote Service Unit. I saw the word in a Tellabs press release in April of 1996. In that release, they talked about how Cablespan RSUs "will be used to provide integrated delivery of CATV and telephony services over a standard coaxial drop at the subscriber's home. The RSUs, which will be installed on the outside of each subscriber's home, will function as a standard network interface device for termination of the customer's CATV and telephone service. The RSU provides the RF transceiver for the CATV network, modulation/demodulation of the broadband RF signals, multiplexing/demultiplexing of the digital signals, analog-to-digital conversion (as required), signaling conversion for subscriber loop operation, and diagnostics for problem isolation. The unit provides a 'virtual twisted pair' back to a local switching interface for transparent telephony operation of CLASS services and custom calling features."

RSVP The Resource Reservation Protocol (RSVP) is an IETF standard designed to support resource (for example, bandwidth) reservations through networks of varying topologies and media. Through RSVP, a user's quality of service requests are propagated to all routers along the data path, allowing the network to reconfigure itself (at all network levels) to meet the desired level of service. The RSVP protocol engages network resources by establishing flows throughout the network. A flow is a network path associated with one or more senders, one or more receivers, and a certain quality of service. A sending host wishing to send data that requires a certain QoS will broadcast, via an RSVP-enabled Winsock Service Provider, "path" messages toward the intended recipients. These path messages, which describe the bandwidth requirements and relevant parameters of the data to be sent, are propagated to all intermediate routers along the path. A

receiving host, interested in this particular data, will confirm the flow (and the network path) by sending "reserve" messages through the network, describing the bandwidth characteristics of data it wishes to receive from the sender. As these reserve messages propagate back toward the sender, intermediate routers, based on bandwidth capacity, decide whether or not to accept the proposed reservation and commit resources. If an affirmative decision is made, the resources are committed and reserve messages are propagated to the next hop on the path from source to destination.

The idea is that for presumably a premium price, RSVP will enable certain traffic, such as videoconferences, to be delivered before e-mail. Today, all traffic on IP networks moves on a first-come-first-served basis and is charged at a flat rate. "In some ways RSVP will change what the Internet is all about, because you'll start to have different qualities of service and differential prices which are new," said Abel Weinrib, a key Internet strategist for Intel Corp. Virtually unknown among the general Internet community, RSVP has been quietly pushing ahead towards becoming acceptable and popular. It is now part of Microsoft's TAPI 3.0. It is being pushed also by Cisco Systems Inc., which makes the routers that direct most Internet traffic, and by Intel, which wants to spur demand for microprocessors by making computers and IP networks more useful for uses like phone calls and video conferencing. In an article I read, Cisco marketing manager Peter Long said RSVP technology would be included in new network software Cisco is delivering. That software controls the routers that direct Internet traffic. Cisco sells more than 80 percent of the routers used in commercial and corporate Internets. Long expects Cisco customers to start using RSVP technology to create what he calls "diamond lanes" on the Internet. "Right now, if there is congestion on the Internet, your traffic sits there, like a car stuck on an onramp," Long said. He said RSVP would act like "a big crane that picks you up and puts you over the other cars," onto these so-called diamond lanes that bypass congested parts of the Net. See TAPI 3.0.

RT 1. Reorder Tone.

2. Remote Terminal. Local loop terminates at Remote Terminal intermediate points closer to the service user to improve service reliability.

3. Remote Termination. A node at which terminates a high-capacity local distribution facility in a DLC (Digital Loop Carrier) scenario. The other end of the circuit is known as a COT (Central Office Termination). See also DLC.

RTAN Real Time ANI.

RTC RunTime Control.

RTCP Real Time Conferencing Protocol. Supports real-time conferencing for large groups on the Internet. It has source identification and support for audio and video bridges/gateways. Supports multicast-to-unicast translators.

RTF Rich Text Format. A way of encoding documents such that the messages include boldface, italics and other limited text stylings. RTF is meant to be a cross word processing platform such that if you send one RTF document (by email, for example) from one word processor to another word processor, that second word processor will be able to recreate the document's original format.

RTFM Read The Fantastic Manual. This acronym is often used when someone asks a simple or common question. The word "Fantastic" is usually replaced with one much more vulgar. Used on e-mail, newsgroups, and the Internet.

RTM 1. Ready To Manufacture.

2. Read The Manual.

RTNR A British term. Ring Tone No Reply, a telephone call which has not been answered. Typically a telebusiness system will automatically re-dial the number after a pre-determined period.

RTO-IS Ready To Order - In Service.

RTP 1. Realtime Transport Protocol. An IETF standard for streaming realtime multimedia over IP in packets. Supports transport of real-time data like interactive voice and video over packet switched networks. A thin protocol providing support for content identification, timing reconstruction, loss detection and security. The ARPA DARTnet transcontinental IP network experiments lead to RTPs popularity. Now championed by the Audio/Video Transport (AVT) Working Group. AVT is part of the IETF (Internet Engineering Task Force). RTP does not do resource reservation or quality of service control. It relies on resource reservation protocols like RSVP. See H.323.

2. Routing Table Protocol. Used in Banyan VINES routing with delay as a routing metric.

RTS Request To Send. One of the control signals on a standard RS-232-C connector. It places the modem in the originate mode so it can begin to send. See the APPENDIX.

RTSE Reliable Transfer Service Element. The OSI application service element responsible for transfer of bulk-mode objects.

RTTU Remote Trunk Test Unit.

RTU 1. Remote Termination Unit. A device installed at the service user site that connects to the local loop to provide high-speed connectivity. Also referred to as the ATU-R.

2. Right To Use. A term manufacturers have invented to stifle the used/secondary market in their equipment. Basically, the manufacturer says "Fine, you can sell your no-longer-needed product to some used equipment dealer. But if someone buys it from the dealer and wants to use it, they have to pay me a Right To Use fee." Without payment of this fee, the manufacturer won't contract to maintain the customer's equipment and certainly won't sell the customer software updates, etc. The right to use fee is exorbitant — typically considerably more than what the product actually sells on the used market for. A better approach for a manufacturer would be to innovate a little more and make the customer wants his new product more than his old price (despite the old product's lower price).

3. Remote Terminal (not terminating) Unit. RTUs are employed by utilities' SCADA (Supervisory Control And Data Acquisition) systems in electric substations or gas/water/steam pumping plants to monitor status/condition and/or metering data and to control operations at a remote site. SCADA systems are not limited to distribution systems; SCADA is also used to manage transmission facilities. Distribution is local delivery to end customers; transmission is backbone transport facilities. The analogies between the electrical, water, gas, and telecommunications infrastructures/networks go on and on. See SCADA.

RTV Real-Time Video. DVI software that implements quick-and-dirty, realtime video compression. Once called "edit-level video," it stores video as only 10 frames per second. Meant for use while developing DVI applications.

RU 1. Request Unit or Response Unit. A basic unit of data in SNA.

2. Receive Unit

3. Abbreviation of rack unit. See Rack Unit.

Rubber Bandwidth A term coined by Ascend, an inverse multiplexer manufacturer, to refer to the ability to support applications needing varying speeds. It breaks the original signal up into 56- or 64-Kbps chunks, and places these separate transmissions on the public switched digital network.

See also INVERSE MULTIPLEXER.

Rule Based System The most popular way to represent knowledge in an expert system. In general, a rule-based system's knowledge base contains both facts and IF..THEN production rules.

Rule of Thumb The phrase "rule of thumb" came from an old English law which made it illegal to beat your wife with anything wider than your thumb.

Run To start a software program.

Run/Stop On a Northern Telecom Norstar phone, this feature inserts a delay in a dialing sequence. The delay can be any length of time.

Run Length Encoding A form of data compression which is semantic-dependent in nature. Such techniques are designed to respond to specific types of local redundancy, such as image representation and processing. Run length encoding is a common technique which involves the scanning of image elements along a scan line or row. As the device scans the image, it identifies redundant data and converts it into a code corresponding to the length of the run of such redundant data. A string of identical bits is indicated by sending only one example, followed by a shorthand description of the number of times it repeats. Fax machines use run length encoding, identifying runs of black or white dots on the page, encoding the length of the run of redundant data as they scan the document, a line at a time. The data is transmitted in compressed form, using this form of data shorthand, and the process is reversed by the receiving fax machine. The advantage, of course, is that the cost of the call is reduced considerably, as the transmission time is much less. Run length encoding also is used in data processing applications to reduce the amount of processing time involved by compressing sequences of zeros or blanks in data fields. Run length encoding also is used to compress memory-intensive files, such as bitmapped graphics. In such an application, the technique is especially useful for black-and-white or cartoon-style line are; runs of the same color can be replaced with a single character. Run length encoded files are identified by the ".rle" file extension. Commonly, .pcx files are run length encoded, as are .tiff and .bmp files; even though these files retain their own file extensions and formats.

Run Time The time it takes to execute a software program. See RUNTIME.

Runt An Ethernet frame that is shorter than the valid minimum packet length, usually caused by a collision. The term is imprecise, and may indicate a collision, collision fragments, a short frame with a valid FCS checksum, or a short frame with an invalid FCS checksum. In a litter of animals, the smallest and weakest animal is typically called the runt. Because the runt is weak, he/she usually ends up sucking his/her sucking hind tit — the nipple that produces the least amount of food. A person who is small and contemptible is also called a runt. See also Runt Frame and Runt Packet.

Runt Frame A small packet received with FCS or alignment errors. Runt frames are the result of collision occurring on connected segment or among stations connected to attached repeaters. See Runt.

Runt Packet A data packet with a legal shorter than required by the IEEE 802.3 standard of 512 bytes. See also Runt.

Runtime 1. A computer term, usually used in the "heavy metal" world of mainframe computers. Runtime (or run time) is the amount of time it takes the CPU (Centralized Processing Unit) to execute a program or perform an operation.

2. A runtime environment is the software that plays back multimedia materials. The runtime material is created by the author. Examples of runtime applications are presentations are training, where the material cannot be edited but only viewed. The runtime software could be a slide show viewer, a software-only video playback application, or a hypermedia runtime document. See Runtime License.

Runtime Control RTC. An SCSA definition. The mechanism by which one Resource Object can influence the behavior of another. Typically used for things such as terminating conditions and speed/volume control.

Runtime License A one-time or royalty-based fee paid for the inclusion of runtime code in a replicated product.

RUR Repair, Update and Refurbish. A term in the industry which repairs telecom equipment. It means equipment is repaired and updated to current manufacturer's specifications. Also includes minor cosmetic cleaning of metal cabinets, a full diagnostic test with burn-in (if required) and an operational test. Definition courtesy Nitsuko America. See also LIKE NEW REPAIR AND UPDATE and QUICK CLEAN.

Rural Service Area RSA. An area not included in either an MSA or a New England Country Metropolitan Area for which a common carrier may have a license to provide cellular service.

RUS Rural Utilities Service. The successor to the REA (Rural Electrification Administration). The RUS is an agency of the U.S. Department of Agriculture (USDA). According to the USDA, "RUS is a vital source of financing and technical assistance for rural telecommunication systems." RUS also provides funding for electric and water programs through public/private partnerships designed to further rural infrastructure development. See also REA.

RVVP See REVENUE VOLUME PRICING PLAN.

RW An ATM term. Read-Write : Attributes which are read-write can not be written by the PNNI protocol entity. Only the Network Management Entity may change the value of a read-write attribute. The PNNI Protocol Entity is restricted to only reading such read-write attributes. Read-write attributes are typically used to provide the ability for Network Management to configure, control, and manage a PNNI Protocol Entity's behavior.

RWhois Referral Whois. An experimental distributed whois service intended to replace the centralized Whois model. Work began in April 1995, with an active test bed of RWhois servers established between each of the regional registries in September 1995. The RWhois Operational Development Working Group is a forum for coordinating the deployment, engineering and operation of the RWhois protocol. User authentication will be required and operational procedures will be established. See InterNIC, DNS and Whois.

RZ Return to Zero. A method of transmitting binary information where voltage returns to a zero (reference) level after each encoded bit.

RZ Code Return to zero code. A code form having two information states called "zero" and "one" in which the signal returns to a rest state during a portion of the bit period.

S Designation the sleeve or control leads in electromechanical Central Offices which are used to make busy circuits, trunks and subscriber lines, as well as to test for busy conditions. It also designates the sleeve wire on a switchboard cord.

S Band 2 GHz to 2.4 GHz. The designation S Band applied to WW2 radar in the range 1.55 - 4.2 GHz. For telecom the interesting region is 1.7 to 2.3 GigaHertz where Microwave Fixed services are giving way to PCS mobile.

S Interface For basic rate access in ISDN, the S interface is the standard four-wire, 144-Kbps (2B+D) interface between ISDN terminals or terminal adapters and the network channel termination, which is two wires. The S interface allows a variety of terminal types and subscriber networks (e.g., PBXs, LANs, and controllers) to be connected to this type of network. At the S interface, there are 4,000 frames of 48 bits each, per second, for 192 Kbps. The user's portion is 36 bits per frame, or 144 Kbps. Out of that 144 Kbps, the user gets two B channels, each of 64 Kbps, and one D channel of 16 Kbps. The local telephone company usually needs a portion of the D channel for signaling. And often it will sell you a 9.6 Kbps packet switched service carved out of the D channel. See also T INTERFACE, U INTERFACE and ISDN.

S Mail Address Snail Mail address. Your post office address.

S Port Refers to the port in an FDDI topology which connects a single attachment station or single attachment concentrator to a concentrator.

S Reference Point The reference point between ISDN user terminal equipment (i.e., TE1 or TA) and network termination equipment (NT2 or NT1).

S SEED Symmetric Self Electro-optic Effect Device. A switching device in which signals enter and exit as beams of light, not through electrical contacts. In 1990 AT&T Bell Labs built a general purpose digital optical processor/computer. The device contained 2,048 S-SEED chips which could be accessed simultaneously with separate beams of light. That means, that ultimately, such a computer could process huge amounts of information in parallel.

S Video Type of video signal used in Hi8, S-VHS and some laserdisc formats, It transmits luminance and color portions separately, using multiple wires. S-Video avoids composite video encoding, such as NTSC and the resulting loss of picture quality. Also known as Y-C Video.

S-BCCH System message-Broadcast Control CHannel. A logical channel element of the BCCH signaling and control channel used in digital cellular networks employing TDMA (Time Division Multiple Access), as defined by IS-136. See also BCCH, IS-136 and TDMA.

S-HTTP Secure Hypertext Transfer Protocol. An extension of HTTP for authentication and data encryption between a Web server and a Web browser.

S.100 A software voice processing standard established by the ECTF ((Enterprise Computer Telephony Forum). S.100, published in March, 1996 with an addendum last month, specifies a set of software interfaces that provide an effective way to develop compuyter telephony applications in an open environment, independent of underlying hardware. It defines a client-server model in which applications use a collection of services to allocate, configure and operate hardware resources. S.100 enables multiple vendors' applications to operate on any S.100-compliant platform. See ECTF, H.100, M.100 and S.100 Media Services API.

S.100 Media Services API Defines a client server model in which applications use a collection of services to allocate, configure and operate hardware resources. Independent of operating system or hardware vendor, it abstracts implementation details of call processing hardware and switch fabrics from the applications themselves. See S.100 and ECTF.

S/DMS SONET Digital Multiplex System. A family of products for Northern Telecom's DMS family of central office switches. There were three initial members of the family as we went to press:

S/DMS TransportNode — a transport vehicle which provides SONET connectivity and bandwidth management for long haul, interoffice and local applications. It includes elements for the transport of SONET over fiber or radio, including integrated bandwidth management functions.

S/DMS AccessNode — a business service access vehicle which delivers switched and special services to the customer's premises, with bandwidth ranging from narrowband to broadband.

S/DMS SuperNode — an evolution of the DMS SuperNode system, which also supports the narrowband capabilities of a SONET network, while adding future broadband switching and network management capabilities.

S/MIME Secure Multipurpose Internet Mail Extension. An emerging de facto security standard for securing all types of e-mail. Increasingly being used in lieu of PGP (Pretty Good Privacy) and PEM (Privacy Enhanced Mail) security techniques, S/MIME has received broad support from the vendor community. Developed by RSA Data Security in 1996, S/MIME was a proprietary security mechanism, which fact led the IETF to reject it for consideration as a standard. In November 1997, RSA announced that it would give up the trademark and other rights to the protocol and the underlying encryption algorithm, leading the IETF to reconsider it as a standard. S/MIME is built on the Public Key Encryption Standard. Digital signatures are used to ensure that the message has not been tampered with during network transit. Digital signatures also provide nonrepudiation, thereby denying senders the ability to deny that they sent a message. The message content is encrypted and enclosed in a digital envelope; the envelope can be opened and the message read only with the use of the recipient's public key, which is sent along with the message. See also Digital Signature, Encryption, MIME, PEM, PGP and Public Key.

S/T Reference Point An ISDN term. In the absence of the NT2, the user-network interface is usually called the S/T reference point.

S/W Software.

S10 Register Hayes, the modem people, invented their "Command Set." This command set lets you control your Hayes compatible modem. In the Command Set there are "S" registers which set how the modem responds to events like answering. Should it answer on the first, second, third, etc. ring. There are 27 registers. The most important S register is S10. This register sets the time between loss of carrier and internal modem disconnect. The factory setting is 1.3 seconds. Drop carrier for 1.3 seconds and your modem will turn itself off. This is long enough for all conditions, except the awful "call waiting" signal you get at hotels and at home. There is a solution: Get your communications software to "go local." Then type ATS10=20. That will increase your S10 register to two seconds. If you have a 300 or 1200 baud you'll have to do this every time you turn on your modem. If you have a 2400 baud modem (the only one to get), you type ATS10=20&W only once. The "&W" writes it into your 2400 baud's non-volatile memory. If you want to check to see if you did it right, type ATS10? That will reply by saying 020. That means 20 tenths of a second, or two seconds. If that still doesn't work for you, increase S10 to three seconds. Other S registers control how long your modem waits for the other end to answer, how long its dialing "pause" is, how quickly it outpulses tones for dialing, etc.

SA Source Address. The address from which the message or data originated. A six octet value uniquely identifying an end point and which is sent in an IEEE LAN frame header to indicate source of frame.

SAA 1. Systems Application Architecture. A set of specifications written by IBM describing how users should interface with applications and communications programs. The idea is to give all software "a common feel" so that training will be less burdensome. According to IBM advertising, "SAA will make it possible for everyone in an organization to access information regardless of its location. What's more, all software written to SAA specifications will provide similar screen layouts, menus and terminology." For a fuller explanation, see SYSTEMS APPLICATIONS ARCHITECTURE.
2. An AT&T Merlin term. Supplemental Alert Adapter. A device that permits 48VDC alerting equipment to be connect to an analog multiline telephone jack so that people working in noisy or remote areas of a building can be alerted to incoming calls.

SAAL Signaling ATM Adaptation Layer: This resides between the ATM layer and the Q.2931 function. The SAAL provides reliable transport of Q.2931 messages between Q.2931 entities (e.g., ATM switch and host) over the ATM layer; two sublayers: common part and service specific part.

SABME Set Asynchronous Balanced Mode Extended.

SAC Single-Attached Concentrator.

Sacrificial Host A computer server placed outside an organization's Internet firewall to provide a service that might otherwise hurt the local internal area network's security.

SADL Synchronous Auto Dial Language. Created by Racal Vadic, SADL is a public domain auto-dialing protocol which defines procedures in BSC, SDLC (SNA) and HDLC for PCs and larger computers that wish to control synchronous modems directly under program control. SADL does for synchronous dialing systems what the Hayes "AT" command set has done for the async PC dialing world.

Saddle A device for establishing the position of the raceway or raceways within the concrete relative to the screed line, and for maintaining the spacing between the raceways.

SAF-TE SCSI Accessed Fault Tolerant Enclosures specification is a non-proprietary, standardized alert detection and status reporting system for storage subsystems which can send and receive information via a standard SCSI interface. A SAF-TE compliant enclosure is designed to monitor and provide notification to the LAN administrator on the condition of disk

drives, power and cooling systems and allow for communication to server-based software agents for network notification. Under the SAF-TE specification, the enclosure is typically implemented as an assignable SCSI target using a low-cost SCSI chip and 8-bit microcontroller. The microcontroller is attached to various alarm sensors and status lights / displays on the enclosure. The enclosure target ID is periodically polled (e.g. every 10-20 seconds) by the host to detect / send changes in status. Disconnect / reconnect and asynchronous event notifications area are not used.

SAFE A store-and-forward MCI International message switching system that provides customers with control of their Telex messages by enabling them to create messages, then specify the message handling parameters in a unique customer reference file. Customers can send and receive messages from the MCI Safe computer at speeds up to 9600 bps.

Safe Area That area in the center of a video frame which is sure to be displayed on all types of receivers and monitors. Televisions and other monitors made at different times and by different companies are slightly different in size and shape, and the outer edge of the video frame (about 10 percent) of the total picture is not produced in the same way on all sets.

Safe Mode A state in which Windows 95 loads in VGA without 32-bit drivers or network support. Usually occurs when there's a hardware or driver conflict or problem, or one of the drivers doesn't work well. You can't run Windows 95 usefully in Safe Mode. You have to fix it so it can run again in what it calls "normal" mode. There are three solutions to fixing the problem. The first is to keep rebooting Windows 95 until it returns to normal mode all by itself, which it probably will eventually. The second is to contact your PC's manufacturer and find out if he has any updated drivers you can have. Or third, you can remove all the installed hardware, boot up in normal mode (it will easily) and then re-install everything. I had this problem of Windows booting up in safe mode constantly and refusing to go to normal mode. I fixed it by replacing my mouse driver with a new version, which my manufacturer provided.

SAFENET Survivable Adaptable Fiber Network. A U.S. Navy experimental fiber-based local area network designed to survive conventional and limited nuclear battle conditions.

Safety Belt A thing made of leather. It's used by outside plant workers to attach themselves to and to climb utility poles. It's also called a body belt.

Sag 1. The downward curvature of a wire or cable due to its weight.
2. The opposite of surge. When the line voltage drops far enough to affect the operation of a phone system or computer.

Sagan A large quantity. Its derivation is thought to come from Carl Sagan, the astronomer, who used the term "billions and billions" on his TV series.

SAI 1. See Serving Area Interface.
2. Stratus Computer's PBX Switch to Stratus Computer Application Interface.

SAIC Science Applications International Corporation. According to SAIC, it is the largest employee-owned research and engineering company in the U.S. On November 21, 1996, SAIC announced its agreement to acquire Bellcore. See Bellcore. SAIC also owned Network Solutions Inc. (NSI), which currently administers the traditional Internet gTLDs (generic Top Level Domains). NSI went public in early 1998. www.saic.com. The domains under NSI control include the following: .com, which designates commercial entities; .edu, which designates institutions of higher learning; .org, which is intended to designate not-for-profit organizations, but is misused and abused.

SAID SOD Speech-Activated Intelligent Dialing Stringing of Digits.

Sales Agent See AGGREGATOR.

Sales Automation See SALES FORCE AUTOMATION.

Sales Force Automation The use of computers and computer software by salespeople to boost their sales. There are two types of sales force automation — those totally self-contained on the computers of salespeople (mostly laptops) or those which communicate with headquarters computer over phone lines. There are many purposes of the phone communication — sending orders in, finding out about back orders, getting updates on "specials," dropping letters and memos in, getting new prices, new products, new technical specs, etc. Salespeople routinely show 10% to 20% sales gains armed with a laptop PC and sales automation software (also called "personal contact") software.

Salmon Day One of those days where you swim upstream all day only to get screwed in the end.

Salvo The sending of a group of commands at the same time.

Sam 1. Security Accounts Manager

2. The Newton family's excessively spoiled cat.

See SAM Technology.

SAM Technology Self Administered Maintenance. Application invented by a software engineer named Dave Tedesco. This is a technique added to large corporate web sites to allow the non-technical people to make changes to the web site without screwing up its functionality.

Sampling 1. Converting continuous signals, like voice or video, into discrete values, e.g. digital signals. See also Digital Signal Processing, PCM and Sampling Rate.

2. Examining a small percentage of the universe to determine makeup of the entire universe. A cook concludes that the entire pot of soup needs salt after sampling only one teaspoonful. the cook makes the assumption that the rest of the soup will taste the same as his sample.

Sampling Rate 1. The number of times per second that an analog signal is measured and converted to a binary number — the purpose being to convert the analog signal to a digital analog. The most common digital signal — PCM — samples voice 8,000 times a second.

2. The number of times per second that a digital audio sample is taken during recording or read during playback — expressed in kilohertz (kHz). An audio CD sampled at a rate of 44kHz has 44,100 bits of information per second.

SAN Storage Area Network. A network which links host computers to storage servers and systems. The network protocols can include FC-AL (Fibre Channel-Arbitrated Loop), SSA (Serial Systems Architecture), ATM (Asynchronous Transfer Mode) and Fast (100 Mbps or Gigabit) Ethernet-currently roughly in that order of preference. The storage technology can be JBOD (Just a Bunch Of Disks), RAID (Redundant Array of Inexpensive Disks), a bunch of servers on a network, or a more complex and expensive host storage server such as a midrange or mainframe computer.

Sandbox 1. Applications that are downloaded from a client on the Internet or an Intranet and which have the potential to damage that client. Viruses are an example of a malicious attempt to do so through damaging the hard drive, corrupting or erasing files, or perhaps damaging the operating system. Some computer programming languages or operating environments (e.g., Java) deny a distributed object access to operating system calls or to other resources. Such restricted objects are said to be "in the sandbox," according to Network magazine.

2. Many people starting high-tech companies look for outside financing from angels or venture capitalists. Some people look for the money to grow their company, by selling product and ultimately making a profit. Some people look for the money so they can continue having fun writing software, creating hardware, and doing whatever neat things amuses them. Sandboxes are an occupational hazard facing all people who are funding companies.

Sandhog An underground worker, typically those building tunnels.

Sanity Check A check to confirm the service capability of a switching system. This test has not been applied to the author of this dictionary.

SAP In local area networks, SAP is Service Access Point. The point at which the services of an OSI layer are made available to the next higher layer. The SAP is named according to the layer providing the services, e.g., Transport services are provided at a Transport SAP (TSAP) at the top of the Transport Layer. IBM defines SAP as a logical point made available by an interface card where information can be received and transmitted.

In an ATM network, a SAP is used for the following purposes:

1. When the application initiates an outgoing call to a remote ATM device, a destination_SAP specifies the ATM address of the remote device, plus further addressing that identifies the target software entity within the remote device.

2. When the application prepares to respond to incoming calls from remote ATM devices, a local_SAP specifies the ATM address of the device housing the application, plus further addressing that identifies the application within the local device. There are several groups of SAPs that are specified as valid for Native ATM Services.

SAP-Address Service Access Point Address.

SAPI Service Access Point Identifier. The SAPI identifies a logical point at which data link layer services are provided by a data link layer entity to a Layer 3 entity. ISDN jargon. See also WINDOWS TELEPHONY.

SAR Segmentation And Reassembly.

1. Generically speaking, a process of segmenting relatively large data packets into smaller packets for purposes of achieving compatibility with a network protocol relying on a smaller specific packet size. The process is often required in conjunction with ATM, SMDS and X.25 networks.

2. A sublayer of the ATM protocol stack, specifically of the ATM Adaptation Layer (AAL). The native Protocol Data Unit (PDU) associated with the transmitting device is segmented into 48-octet payload fields at this sublayer. At the target end of the data communication, the SAR serves to reassemble the native PDU by extracting and combining multiple 48-octet payloads from multiple ATM cells.

SAS 1. Simple Attachment Scheme.

2. Severly errored frame/Alarm indication Signal. A one-second period of time in which are detected multiple frame errors or an alarm indication signal over a digital circuit. See also CV, ES and SES.

SASG Special Autonomous Study Group. These ITU-T study groups are chartered to produce handbooks on basic telecommunications technical or administrative subjects for developing countries.

SASI Shugart Associates System Interface. The first SCSI interface specification defined by Shugart, a disk drive manufacturer. Later it was modified and renamed as the Small Computer System Interface (SCSI), pronounced Scuzzy. See also SCSI.

SAT Subscriber Access Termination. An SMDS term.

SATAN Security Administrator Tool for Analyzing Networks. This tool allows a network analyst to mimic a malicious hacker (or cracker) for the purpose of identifying weaknesses in system and network security. It also provides malicious hackers a nifty tool. See hacker.

Satellite 1. A microwave receiver, repeater, regenerator in orbit above the earth. Traditional communications satellites are known as GEO's, as they are in Geosynchronous Earth Orbit, which is an equatorial orbit with the satellites at high altitudes of approximately 22,300 miles. In such an orbital slot and at that altitude, they maintain their position relative to the earth's surface. More recently developed satellites are placed in Low or Middle Earth Orbits, hence the terms LEO and MEO. LEOs and MEOs vary widely in terms of orbital paths and altitudes; therefore, they are not synchronized with the earth's rotation. See GEO, MEO and LEO. See also SATELLITE TRANSMISSION. 2. Something distant to the main something. See MAIN DISTRIBUTION FRAME, SATELLITE CABINET and SATELLITE DISTRIBUTION FRAME.

Satellite Business Systems SBS. A satellite long distance carrier originally owned jointly by IBM, Aetna Insurance and Comsat, but now owned by MCI (which acquired it in 1986). SBS started out to serve the data communications transmission marketplace but found that marketplace too small to be profitable. It then started to serve the voice transmission marketplace and did somewhat better. But everyone hated satellite voice calls because of the delay and the frequent echos. Satellite Business Systems no longer exists as a separate entity. It has been merged into MCI. There are estimates on how much money SBS ost in its short history. They are substantial, ranging around $1 billion — a lot of money in those days. See SBS.

Satellite Cabinet Surface-mounted or flush-type wall cabinets for housing circuit administration hardware. Satellite cabinets, like satellite closets, supplement riser closets by providing additional facilities for connecting horizontal wiring subsystem cables from information outlets in user locations. Sometimes referred to as satellite location.

Satellite Closet A walk-in or shallow wall closet that supplements a backbone or riser closet by providing additional facilities for connecting riser subsystem cables to horizontal wiring subsystem cables from information outlets. Also referred to as satellite location. See also RISER CLOSET and BACKBONE CLOSET.

Satellite Communications The use of geostationary orbiting satellites to relay information.

Satellite Communications Control SCC. The earth station equipment that controls such communications functions as access, echo suppression, forward error correction and signaling.

Satellite Constellation The arrangement in space of a set of satellites.

Satellite Delay Compensator A device that compensates for the absolute delay in a satellite circuit communicating with data terminal equipment (DTE) with the DTE's own protocol.

Satellite Delivered Signal A television signal delivered to a cable television headend by communications satellite. This term should not be confused with the signal received from a Satellite Television Broadcast Station. Compare with Satellite Television Broadcast Station.

Satellite Digital Audio Radio Service Special satellites transmitting digitally to tiny (fewer than two inches wide) antennas in the S band. The concept is that with digital tech-

nology, satellite broadcasters can stuff dozens of channels of CD-quality, interference-resistant programming into a narrowband of frequencies.

Satellite Distribution Frame An intermediate point for connecting wires running between a group of phones and the Main Distribution Frame located elsewhere in the building. A fat multi-conductor cable comes from the main distribution frame to the satellite distribution frame, where it splits into individual cables to individual phones or workstations. The satellite distribution frame is usually located in a satellite wiring closet or cabinet. These wires are ultimately connected to the telephone system. See DISTRIBUTION FRAME.

Satellite Downlink The communications path from a satellite to its ground station. Opposite to Satellite Uplink.

Satellite Uplink The communications path from a ground station to its satellite. Opposite to Satellite Downlink.

Satellite Facility A transmission system using a satellite in a geostationary orbit above the earth and a number of earth stations.

Satellite Link Microwave link using a satellite to receive, amplify and retransmit signals. Typically that satellite is in a geosynchronous orbit.

Satellite Operation A configuration of multiple PBXs or one big PBX and several smaller PBXs. The configuration gives a company with several nearby locations a unified system of centralized trunks, centralized attendants, overall call detail recording and many of the advantages of a private network. The key advantage of satellite operation is that one big centralized telephone system can contain most of the intelligence and computer smarts for the total system. This advantage is heavily economic. A variation on satellite operation is called CENTRALIZED ATTENDANT SERVICE (CAS).

Satellite PBX A satellite PBX has no direct incoming connection from the public network. All incoming calls are routed from an associated main PBX over tie trunks. This definition places no restrictions on the handling of outgoing calls from the satellite PBX. A satellite PBX can have one-way outgoing trunks to the central office, in addition to outgoing service on trunks to the main PBX. A satellite PBX has no direct trunks to a node; however, calls to the node can be made through the main PBX.

Satellite Premises Channel This is the cable connecting arrangement between a dedicated earth station and the Customer Provided Equipment.

Satellite Processor A computer with little computing power used for operations that do not require the full processing power of the main machine.

Satellite Television Broadcast Station A United States television broadcast station which: - Operates pursuant to Part 73, Subpart E, of the FCC Rules. - Operates at full-power levels, typically thousands of watts. - Rebroadcasts all (or substantially all) of the signal of another full-power television broadcast station. Example: Stations KGMD-TV (Hilo) and KGMV (Wailuku) are satellites of station KGMB, Honolulu. Together, the three stations cover most of Hawaii. This term should not be confused with a communications satellite, or to a signal received from a communications satellite. Compare with Satellite Delivered Signal.

Satellite Transmission A form of transmission which sends signals to an orbiting satellite which receives them, amplifies them and returns those signals back to earth. Satellite transmission provides great clarity but suffers from delay. See SATELLITE TRANSMISSION DELAY.

Satellite Transmission Delay Referring to the time it

takes a signal to travel from one satellite earth station to the satellite in the sky then to the satellite earth station at the other end. Since most communications satellites orbit the earth at a distance of approximately 22,300 miles, the total distance the signal travels is 44,600 miles. Since radio waves travel at the speed of light (186,000 miles per second), simple arithmetic will show a delay of approximately one-quarter of one second thus, 44,600 divided by 186,000 = 0.239 second. If you are waiting for a reply, double this time. (You double the distance.)

SATPhone A telephone that works directly off a satellite. It comes with a small parabolic antenna which you aim at the satellite. You turn it on and talk. It's easy, though expensive — typically as much as $10 a minute. But for that you can talk from practically anywhere in the world, so long as you got lots of battery or easy close access to commercial power.

Saturation 1. The intensity of the colors in the active picture. The voltage levels of the colors. The degree by which the eye perceives a color as departing from a gray or white scale of the same brightness. A 100% saturated color does not contain any white; adding white reduces saturation. In NTSC and PAL video signals, the color saturation at any particular instant in the picture is conveyed by the corresponding instantaneous amplitude of the active video subcarrier.

2. The point on the operational curve of an amplifier at which an increase in input amplitude will no longer result in an increase in amplitude at the output.

Save A telephone feature that allows the user to put a phone number into memory for future calls, by pressing one or two buttons after dialing it the first time. See also SNR.

Save And Repeat Another way of saying "Autodial." Electronic phones may be able to save a number so you can dial it later by simply hitting one button on the phone. This feature is similar to a "Last Number Redial" button, except that button just dials the last number called. "Save and Repeat" puts a number into temporary storage for dialing at another time. Phones should have both auto-redial and save-and-repeat buttons.

SBC Communications Inc. In early October, 1994, Southwestern Bell Corporation, one of the seven original regional Bell operating companies, changed its name to SBC Communications, Inc. The company's subsidiaries continue to be Southwestern Bell Telephone Co., Southwestern Bell Mobile Systems, Southwestern Bell Yellow Pages, Southwestern Bell Telecom and SBC International. SBC continues to operate as a LEC in the states of Arkansas, Kansas, Missouri, Oklahoma and Texas. SBC merged with (read "acquired") Nevada Bell and Pacific Bell on April 1, 1997, at which point the Pacific Telesis (PacTel) holding company ceased to exist; that acquisition is complete. SBC announced in January 1998 an agreement to acquire SNET (Southern New England Telephone), which operates as a LEC in portions of Connecticut; at the time of this writing, that acquisition is pending final regulatory approvals. SBC announced its intent to merge with (read "acquire") Ameritech on May 11, 1998; at the time of this writing, that acquisition is pending final regulatory approvals. Now you know why Southwestern Bell changed its name to SBC Communications Inc. Assuming that the Ameritech acquisition (Whoops, I mean merger) is completed that will reduce the number of surviving RBOCs to four, as Bell Atlantic acquired (Whoops again, I mean merged with) NYNEX. SNET, technically speaking, was not a RBOC, as it was not wholly owned by AT&T at the time of divestiture. See also WMBTOPCITBWTNTALI.

SBM 1. DMS SuperNode Billing Manager.

2. Subnet Bandwidth Manager. The SBM provides centralized bandwidth management on shared networks. See TAPI 3.0.

SBS 1. Satellite Business Systems. A long distance satellite company that started out as a joint venture between Lockheed and MCI, was sold to IBM Aetna and Comsat and then eventually was given to MCI in exchange for shares issued to IBM. SBS never made any money. But that was irrelevant. Its job was to help IBM sell computer networks. See SATELLITE BUSINESS SYSTEMS.

2. Sick Building Syndrome. Phenomenon of employee discomfort and illness, or perceived illness, due mainly to a polluted indoor air supply. An office is diagnosed "sick" when more than 20% of its occupants exhibit typical symptoms, complaints persists for two weeks or more and disappear when sufferers are away from the building.

Sbus Sun Microsystems' resource sharing and expansion bus interface for the SPARC architecture. SBus expansion cards can communicate with each other through this interface. SBus competes with VME, PCI and EISA/ISA as industry standard I/O buses for computing platforms. Currently, there are several computer telephony vendors selling SBus compliant DSP and network interface cards for SBus. You can develop cards with an both SBus computer interface and a mezzanine for MVIP and SCSA busses. The Sbus specification has been adopted by the IEEE (Institute of Electrical and Electronic Engineers) as a new bus standard.

SC Connector A snap on fiber optic plastic connector.

SCA 1. Selective Call Acceptance.

2. Supplemental Communications Authority. The authority granted by the Federal Communications Commission to transmit on a subcarrier.

SCADA Supervisory Control And Data Acquisition. SCADA systems are used extensively by power, water, gas and other utility companies to monitor and manage distribution facilities. They also are used, although more sparingly, to monitor and control end user usage levels for purposes such as remote meter reading and load shedding. Traditionally, such systems made use of telephone lines for such purposes, although wireless technologies are now deployed widely. Some power utilities have deployed fiber optic transmission facilities (allegedly) for this purpose, although the small amount of bandwidth required for such an application clearly does not justify the cost of fiber. It is widely accepted that such fiber deployment is a preemptive strike against LECs and CATV providers who seek to place fiber on the power utility companies poles and in their conduits in a convergence scenario. In effect, the power utilities are laying information grids for resale to carriers which desire substantial bandwidth in competition for transmission of voice, data, video, image, TV and multimedia signals in a deregulated environment.

SCAI Switch to Computer Applications Interface. A protocol that defines how switches talk to outboard computers, i.e. computers which are external to the switch and contain such a database of customer buying information. Using SCAI, calls and data screens about a calling customer can be presented to the agent simultaneously. See OPEN APPLICATION INTERFACE.

Scalability Fancy way of saying size something can grow to relatively easily. See Scalable.

Scalable Something that can be made larger or smaller relatively easily and painlessly. And the cost to grow is relatively straight line, rather than stair step, as in the days of "forklift upgrades." At least that was the earlier, accepted definition. Then Microsoft started referring to Windows NT as "scalable," namely that it runs on everything from Intel to RISC processors and single- to multi-processor systems. Scalable often refers to technology applications which can be made greater or smaller without quantum leaps in cost. For instance, Virtual Private Networks (e.g., Switched 56/64, X.25, Frame Relay, SMDS and ATM networks) serve as effective replacements for dedicated, leased-line networks as their capabilities are scalable, with the costs remaining in reasonable relationship to associated functicnality.

Scalable Video Scalable video is a playback format that can determine the playback capabilities of the computer on which it is playing. Using this information, it allows video playback to take advantage of high performance computer capabilities while retaining the ability to play on a lower performance computer.

Scalable Typeface A set of letters, numbers, punctuation marks, and symbols that are a given design (i.e. of one font) but can be scaled to any size.

Scaled Point Size A point size that approximates a specified point size for use on the screen. For example, text that prints at 10 point on the printer may be represented by a slightly larger font on the screen to make up for the screen's lower resolution.

Scaling A video compression technique which involves adjustment of the transmitted image in consideration of the presentation capabilities of the receiving device. In the case of a receiving device which is less capable in terms of resolution, for instance, the codec in the transmitting device reduces the resolution of the image prior to transmission. In this fashion, the receiving device is presented with a signal which matches its display capabilities. Additionally and more importantly, transmission bandwidth is not wasted.

SCAN 1. Switched Circuit Automatic Network. 2. To examine sequentially, part by part. 3. To examine every reference or every entry in a file routinely as part of a retrieval scheme. 4. In electromagnetic or acoustic search, one complete rotation of an antenna. 5. The motion of an electronic beam through space searching for a target. Scanning is produced by the motion of the antenna or by lobe switching.

Scan Time The time between two successive polls to a workstation on a data communications network.

Scanner 1. A radio receiver which automatically skips across selected frequencies, allowing you to listen in to any of the frequencies. You can buy scanners, for example, that let you scan all police, fire, cellular frequencies and let you listen in on any conversation that is presently occurring.

2. A program on a bulletin board system which scans the message base for previously entered e-mail and pulls a copy of each message and makes them available to the BBS (Bulletin Board System) mailer program.

3. A device used to input graphic images into the computer. Scanners look at or "scan" a piece of paper and put the image's information into digital form. The information can then be recognized by the computer. Scanners come in three basic types — flat-bed, sheet-fed and as one part of a multifunction devices that prints, copies, faxes and scans. A fax machine also contains a scanner which "looks" at the original document and determines the brightness level of each pixel to be transmitted. The accuracy at which a scanner gets information from the document it is scanning and sends it to an attached computer is measured in two ways: by resolution and color information. Resolution is defined as dpi (dots per inch) or pixels, which determines the maximum size of the image. For example, a 2,400-pixel by 1,800 pixel scanned at 300 dpi (dots per inch) creates a maximum image of 8" x 6". Color

information is defined by the number of bits of information per color. Today's scanners produce images with 24, 30 and 36 bits per pixel. In a 24 bit scanner, you make your color by choosing 8 bits each of red, green and blue). The more pixels and the more bits of information per color, the larger the imaged file (often going into the millions of bytes) and the more accurate the representation will be. See also Optical Character Recognition.

Scanner Accuracy The accuracy at which a scanner inputs information is measured in two ways: by resolution and color information. Resolution is defined as dpi (dots per inch) or pixels, which determines the maximum size of the image. For example, a 2,400-pixel by 1,800 pixel scanned by a scanner at 300 dpi creates a maximum image of 8" x 6". Color information is defined by the number of bits of information per color. Most scanners produce images with 24 bits per pixel (8 bits each of red, green and blue). The more pixels and bits of information per color, the larger the imaged file and the more accurate the representation will be.

Scanning Rate In video communications, the scanning rate is the rate at which the screen is refreshed. Even numbered lines are refreshed in one scan, with odd numbered lines being refreshed in the next scan. In combination, the two scans yield a frame refreshed. The process happens so quickly that it is imperceptible to the human eye/brain. The scanning rate is related directly to the frequency of the power source. For instance, the U.S. NTSC TV standard calls for 30 fps (frames per second), related to the 60 Hz power standard. The European PAL standard, on the other hand, calls for 25 fps.

SCANNS Multiplexers which perform the vital functions of monitoring and control within System 75s and 85s Automated Building Management feature. The SCANNS continuously scan sensors and send the resulting data to local control units.

SCAPI SCSA Application Programming Interface. A high-level, object oriented hardware independent, technology independent programming model that permits the design and implementation of call processing applications.

Scattering A cause of lightwave signal loss in optical fiber transmission. The diffusion of a light beam caused by microscopic variations in the material density of the transmission medium. Scattering is a physical mechanism in fibers that attenuates light by changing its direction.

SCbus An SCSA definition. The standard bus for communication within an SCSA node. The SCbus features a hybrid bus architecture consisting of a serial Message Bus for control and signaling, and a 16-wire TDM data bus. The SCbus is a serial time division multiplexed bus for carrying information between hardware devices in a signal processing node. The SCbus can support up to 1024 b directional timeslots in a PC implementation or up to 2048 timeslots in a backplane implementation. The SCbus uses 16 synchronous data lines for carrying data and a dedicated messaging channel (SCbus Message Channel) for carrying signaling information and messages between devices. See S.100 and SCBUS MESSAGE CHANNEL.

SCbus Message Channel An SCSA definition. The SCbus message channel is a 2.048 Mbps serial line for carrying signaling information and messages between hardware devices in an SCSA compliant server. The SCbus message channel uses an HDLC (high level data link controller) protocol. The SCbus message channel is an optional element of the SCSA Hardware Model and provides faster system performance by allowing for direct communication of messaging information between devices at the firmware level and without consuming any data timeslots. See S.100.

SCC 1. Specialized Common Carrier. Another term for a long distance carrier in competition with AT&T. The word "Specialized" came about because these long distance carriers purported to provide "specialized" circuits for business customers. At one stage they were also known as OCCs, or Other Common Carriers (i.e. other than AT&T). These days, both terms have fallen into disrepute. All long distance carriers — including AT&T — are called IntereXchange Carriers (IXCs).
2. Standards Council of Canada.
3. Satellite Communications Control.
4. SuperComputing Center. There are five NSF-funded supercomputing centers (SCCs): Cornell Theory Center, National Center for Atmospheric Research, National Center for Supercomputing Applications, Pittsburgh Supercomputing Center, and San Diego Supercomputing Center.

SCCP Signaling Connection Control Part. Part of the ITU-T #7 signaling protocol. and of the SS7 protocol. It provides additional routing and management functions for transfer of messages other than call set-up between signaling points. A SS7 protocol that provides additional functions to the Message Transfer Part (MTP). It typically supports Transaction Capabilities Application Part (TCAP). See also SIGNALING SYSTEM 7 and COMMON CHANNEL SIGNALING.

SCCS Switching Center Control System.

SCDPI SCSA Device Programming Interface: A set of callable functions that allow SCSA application software to control SCSA hardware. The SCdpi consists of both common call processing services and technology specific modules for the application of particular resources to call processing tasks. See SCSA.

SCE Service Creation Environment. A Bellcore term used in the jargon of intelligent networks (INs) to allow outside developers to define and create new value-added (i.e., intelligent) services by connecting pre-existing blocks of code into a flow chart that describes the logical processes the service will use to handle calls. A critical and distinguishing feature of the AIN concept, the SCE comprises a toolkit for the creation of services which can be provided on a network basis. The carrier can develop a generic service which can be offered to multiple users; similarly, a third-party software developer or end user can develop such a service application. Once the application is developed, the application logic and supporting databases can be partitioned in order that multiple users can take advantage of it. In such a scenario, each user organization would have the ability to customize the application, which would draw on a customized, partitioned and secured database.

By way of example, routing logic for an ACD network might be centralized. Multiple organizations, each with multiple incoming call centers, could customize the generic call routing application in consideration of their specific Quality of Service parameters, cost issues, agent skill sets, and so on.

SCF Shared Channel Feedback. A digital wireless term defined by IS-136, the Interim Standard for digital cellular networks employing TDMA (Time Division Multiple Access). SCF is a logical channel which is part of the FDCCH (Forward Digital Control Channel) used to send signaling and control information from the cell site to the user terminal equipment. SCF information keeps all terminal devices advised of the level of network availability. SCF also provides each device with time slots for transmission and reception in order to avoid data collisions. See also FDCCH, IS-136 and TDMA.

Schedule A call center term. A record that specifies when an employee is supposed to be on duty to handle calls. The complete definition of a schedule is the days of week worked, start

time, break times and durations (as well as paid/unpaid status), and stop time. See the following six terms.

Schedule Exception A call center term. A specific date and period when an employee cannot handle calls or is engaged in some kind of special activity. An absence, meeting, or other work assignment creates an "exception" to the employee's daily work file schedule.

Schedule Inflexibility A call center term. A phenomenon that tends to create overstaffing in some periods when full coverage is the objective in creating a set of schedules. This is caused by the fact that it is impractical to have extremely short schedules for covering momentary peaks in call volume. To achieve a near perfect match of staff and workload at all times would require shifts of virtually every length; for example, 2-hour shifts, 45-minute shifts, even 15-minute shifts.

Schedule Preference A call center term. A description of the days and hours that an employee would like to work, used by the automatic assignment process to match the employee to a suitable schedule. In some call centers, each employee can have as many as 10 schedules preferences ordered by priority.

Schedule Test A call center term. A variation of the scheduling process that allows you to forecast the service quality that will result from using an existing set of schedules.

Schedule Trade A call center term. A situation in which two employees have agreed to work each other's schedules, or an employee has agreed to work the other's schedule, on a specific date or dates.

Scheduling Making the timetable of agent hours and shifts for your call center. Takes into account vacation days, breaks, training time, lengths of shifts and forecasting information. A call center software management package helps you do this.

Schematic A diagram of the electrical scheme of a circuit with components represented by graphic symbols.

Scheme The part of a URL that tells an HTML client, like a browser, which access method to use to retrieve the file specified in the URL. See URL.

Schlepp A Yiddish word meaning to carry around, to drag around, as in "This phone system is heavy. Schlepping it is a pain." See also Chutzpah.

Scintillation In electromagnetic wave propagation, a random fluctuation of the received field strength about its mean value, the deviations usually being relatively small. Think of scintillation as the creation of a spark, hopefully a small one. The effects of this phenomenon become more significant as the frequency of the propagating wave increases. The discovery of scintillation must have been truly scintillating.

SCM 1. Station Class Mark. A two digit number that identifies certain capabilities of your cellular phone. How the cellular network handles your call is based on these digits. The SCM tells the system if your phone transmits at standard power levels or low power levels, if it can use the full 832 channels or only the original 666 frequencies. The last attribute identified is whether your phone uses voice activated transmission (VOX).
2. Subscriber Carrier Mode.

SCM-100A Subscriber Carrier Module-100 Access; same as SMA.

Scope 1. A slang term for cathode ray oscilloscope.
2. An ATM term. A scope defines the level of advertisement for an address. The level is a level of a peer group in the PNNI routing hierarchy.

SCOTS 1. Surveillance and COntrol of Transmission Systems.
2. Switched Circuit Ordering and Tracking System. MCI's automated tracking and order processing system for Dial up products, IMTs, and the MCI switched network.

SCP 1. Service Control Point. Also called Signal Control Point. A remote database within the System Signaling 7 network. The SCP supplies the translation and routing data needed to deliver advanced network services. The SCP translates an 800-IN-WATS number to the required routing number. It is separated from the actual switch, making it easier to introduce new services on the network. See also SCP and TCAP. For a full explanation of the Advanced Intelligent Network, see AIN.
2. Northern Telecom term for a Satellite Communications Processor.

SCPC Single Channel Per Carrier. A technique used in analog satellite and certain other radio systems. SCPC supports one transmission per frequency channel. Multiple channels can be supported through Frequency Division Multiplexing (FDM). See also FDM.

SCR 1. Abbreviation for Silicon Controlled Rectifier, a semiconductor device that allows one electric circuit to control another; often replaces electromechanical relays.
2. SCSA Call Router.
3. Sustainable Cell Rate. Parameter defined by the ATM forum for ATM traffic management. The SCR is an upper bound on the conforming average rate of an ATM connection over time scales which are long relative to those for which the PCR is defined. Enforcement of this bound by the UPC could allow the network to allocate sufficient resources, but less than those based on the PCR, and still ensure that the performance objectives (e.g., for Cell Loss Ratio) can be achieved.

Scrambler A device which deliberately distorts a voice or data conversation so that only another like device can figure out the content of the message. Analog scramblers invert the frequencies of speech. Digital scramblers first convert speech to digital form and then encrypt. Both types also perform the reverse process. The sophistication (i.e. complexity) of a scrambler determines its price.

Scratchpad A part of the random access memory of a computer or telephone system which can be used to temporarily store data. In a cellular phone system, scratch pad allows storage of phone numbers in temporary memory during a call. Silent scratch pads allows number entry into scratch pad without making beep tones. See also REGISTER.

Screen Dump A reasonably exact copy of what's on your PC's screen printed out or saved as a file.

Screen Font The font that is displayed on your screen. It is, hopefully, designed to match the printer font so that documents look the same on the screen as they do when printed. You typically need a graphics interface on your company — like Windows or X-Windows — to make the font you see on your screen the same as what you see when you print it out.

Screen Pop Screen Pop presents customer data and product and service information simultaneously with the incoming telephone call. Imagine a call center. An agent's phone is ringing. As it rings, the agent's computer screen pops up with information about the caller, what he ordered last, how much he owes, etc. This is called screen pop. The technology to make it happen typically comes from caller ID or ANI — information carried on the phone call just before the voice. It can also come from an IVR (Interactive Voice Response) system which answers the phone and asks the caller to punch in his phone or account number and then passes the phone call to the agent. In Screen Pop, the phone system typically listens for incoming digits and passes them across an attached local area network. See CALLER ID.

Screen Refresh Rate The rate at which your computer screen is re-drawn every second by a horizontal beam that

scans from the top left hand corner to the bottom right hand corner. Screen refresh rates differ by the graphics standard you're running.

Screen Response Time The time it takes to refresh a terminal screen.

Screen Synch A colloquial term for sending an telephone service agent a phone call together with a screen of information about the incoming call, e.g. the customer's purchasing record or experiences with your product (if you're a help desk, for example.)

Screened Dual-Homed Gateway A screened dual-homed gateway is a dual-homed gateway that is guarded by a packet-filtering router.

Screened Subnet Also referred to as the demilitarized zone, a screened subnet is a collection of computers that are shielded from both the trusted network and the untrusted network by packet-filtering routers and gateways.

Screened Transfer You are transferring a call from your phone to your boss. You dial a code for transfer, then dial your boss. The caller you're transferring is automatically put on hold. You speak to your boss, tell her who you're putting through. She okays the transfer, then you hit another digit and the call goes right through. This is called screened transfer. An unscreened transfer occurs when you simply dial your boss's office and send the call through without announcing it. Most PBXs have the ability to do both screened and unscreened calls.

Screened Twisted Pair ScTP. A type of cabling similar to UTP but ScTP has a foil shield between the conductors and the cable jacket. It also has a drain wire (a bare conductor). ScTP is used when ordinary UTP might pick up interference that would interfere with transmission. See UTP Cable.

Screening Telephone Number The telephone number used by the phone company to bill, regardless of the number of phone lines associated with that number.

Screw Post Also called binding post. Screw posts are still used on many residential jacks. A conductor is installed on a screw post by stripping the insulation from the conductor to a half inch from the end, unscrewing the post to loosen it, wrapping the bare copper end of the conductor around the screw post between the washers and then re-tightening the screw. This doesn't make a very reliable connection, and it's easy for an installer to break the copper conductor by tightening the screw too tightly.

Script A type of computer code than can be directly executed by a program that understands the language in which the script is written. Scripts do not need to be compiled into object code to be executed.

Script Files Some communications programs had script files that automate logging onto communications services, such as MCI Mail. The files are saved on your disk and read by your communications software when connecting to a remote service. Newer communications programs will "write" their own scripts by recording what you do in response to what questions from the remote service. This typically happens using a program feature called "Learn."

Script Kiddies Aspiring hackers who use ready-made scripts, language and techniques written by more experienced hackers to break into online distant computer sites, usually via dial-up phone lines.

Script Language A software language that contains English-statements for commands. A statement might be as simple as WrapPara() for wrap paragraph. Typically a script language contains commands that are specific to the type of task it's doing. For example, VOS from Parity Software in San Francisco is a script language for voice processing using Dialogic voice processing cards. A script language is more flexible than an Applications Generator, but is more difficult to program.

Scroll Bar A bar that appears at the right and/or bottom edge of a window or list box whose contents are not completely visible. Each scroll bar contains two scroll arrows and a scroll box, which enable you to scroll through the contents of the window or list box.

Scrolling Browsing through information at a video terminal. Scrolling is the continuous movement of information either vertically or horizontally on a video screen as if the information were on a paper being rolled under it.

Scruple The scruple was a unit of weight equal to 20 grains, used by apothecaries in olden days. Apprentices were supposed to always use these weights to measure out prescriptions. However, this was very tedious and many times the apprentice would just take a pinch of whatever was supposed to go into the prescription, without weighing it. If caught by the apothecary the apprentice was often scolded, "What is the matter with you? Have you no scruples?"

SCS Structured Cabling System. What it sounds like

SCSA Signal Computing System Architecture. SCSA is a comprehensive architecture that describes how both hardware and software building blocks work together. It has now been absored by S.100, but the following words still apply: It focuses on "Signal Computing" devices, which refer to any devices that are required to transmit information over the telephone network. Information can be transmitted via data modems, fax, voice or even video. SCSA defines how all these devices work together. Signal computing systems combine three major elements for call processing. Network interfaces provide for the input and output of signals transmitted and switched in telecommunications networks. Digital signal processors and software algorithms transform the signals through low-level manipulation. Application programs provide computer control of the processed signals to bring value to the end user.

SCSA is the common set of standards that telecommunication system manufacturers and computing system manufacturers can use to create computer telephony systems. The theory is no single company today can create the total solution for all customers. SCSA represents the common ground between the two fields so that manufacturers from each area can safely develop products that will work with other manufacturers. SCSA's coverage extends from low-level bus and hardware interfaces, like the inter-board switching bus that enables boards from different suppliers to work together, to high-level application programming and software interfaces, so that software designed to work with one set of hardware products, will work with different hardware. Dialogic Corporation of Parsippany, NJ announced SCSA in the Spring of 1993. Dialogic said that SCSA was defined and created with input from a number of leading computer and switch manufacturers, call processing suppliers, and technology developers. In many cases, SCSA has drawn on existing standards, like the T.611 fax standard endorsed by the European Computer Manufacturers Association, and in other cases SCSA has extended standards to make them more useful for call processing suppliers and users.

SCSA describes all elements of the system architecture from the electrical characteristics of the SCbus and SCxbus to the high level application programming interfaces (APIs). According to TELECONNECT Magazine, this SCSA standard (and now, by extension, the S.100 standard) is remarkable for several things:

1. On the day of its announcement over 60 telecom and voice processing companies publicly endorsed SCSA. In early 1994, over 150 companies public endorsed it.

2. With SCSA — a standard for PC/LANs and VME-back-planed computers — you can build much larger telecom switches and much larger call and voice processing boxes. Previous standards, like AEB, PEB and MVIP, were basically limited to what you could do with one PC. Now PCs can be joined together. With SCSA, you can put 16 T-1 lines, or 512 voice lines in one PC and join together 16 PCs, for a total of 16 x 16 x 24 = 6,144 lines! That's a central office built out of networked PCs. A mainframe built out of a LAN. The SCSA joining is not via LAN or LAN-emulation. That would be too slow and the transmission too bursty (great for data, lousy for voice). It's via an SCbus — something that looks and works like a PBX backplane.

3. SCSA incorporates virtually every other standard in PC-based switching — including the most popular ones, Mitel's ST-Bus, MVIP, Siemens PCM Highway, AEB and PEB.

4. It's a lot faster and more reliable. All signaling is out of band. There's clock fall back and time slot bundling. It's more modular, meaning you can start with one PC and grow one at a time. That makes it more "modular" (scaleable is the new word). It's also hot pluggable. You don't have to turn off to upgrade.

5. It has applications portability. Tandem, the highly-successful fault tolerant minicomputer maker, has an SCSA application in a call center. They call it the Tandem Non-Step Call Center. It uses the Tandem 2400 VRU and the 4800 VRU.

SCSA is open, truly open. All its specs and all levels of its specs are available. To that extent, SCSA represents a remarkable gamble by its creator, Dialogic, a telecom/voice processing hardware company. It is encouraging competing manufacturers to build hardware to its specs and gambling that it won't be left in the dust, as IBM was with its PC. (Compaq, not IBM, built the first '386 PC.)

SCSA, as an idea, is revolutionary (for telecom). No one in telecom has ever promulgated an open standard everyone can adopt — hardware and software vendors. Write one application, create one applications generator, design one piece of hardware. Erector set telecom/voice processing! Build small. Build large. Just join the bits and pieces together. See also AEB, ECTF, PEB, MVIP, SIGNAL COMPUTING (for a differently-worded definition), S.100 and TAO.

SCSA Call Router An SCSA definition. A system service of SCSA which provides the basic necessities of inbound and outbound call processing and call sharing to client applications, without those applications needing to be aware of the underlying telephony interface operations. See ECTF, SCSA and TAO.

SCSA Compatible An SCSA term. Able to function in an SCSA environment in its native mode.

SCSA Hardware Model An SCSA definition. The hardware layers of the SCSA specification. The SCSA Hardware Model defines an open architectural specification for a digital intra-node communication bus (SCbus), a switching model (SCSA Switching Model), and an multimode expansion capability (multimode Network Architecture, or MNA). The SCSA Hardware Model may be implemented independently of the SCSA Telephony Application Objects Framework. See TAO.

SCSA Message Protocol The open communications protocol by which entities communicate with one another in an SCSA system. The SCSA Message Protocol (SMP) is independent of the transport layers it is built upon, computer hardware, operating system, network topology (or lack thereof), and technology vendor. All SCSA-compliant AIAs will translate the functions called by client applications (via the API) into SMP messages; these are transmitted to service providers regardless of their location. Therefore, applications written to the API will be portable from one call processing environment to another. See TAO.

SCSA Message Protocol Interface The message presentation format required by, and used by, the service provider in delivering SPM information. Contrast with Service Provider Messages. See TAO.

SCSA Server A collection of service providers (objects) which in the aggregate implement the minimum set of services required for SCSA system conformance. The assumption is that these services are at a minimum provided to remote hosted client applications via common transports such as LANs, but may also be provided to client applications which are hosted on the SCSA server itself. Note that this is a logical image which may be implemented through multiple nodes (machines). See TAO.

SCSA Telephony Application Objects Framework
The SCSA Telephony Application Objects (TAO) Framework originally defined the software layers of the SCSA open computer telephony specification. The SCSA TAO Framework defined a hardware-independent, open software architecture that simplifies design of distributed computer telephony systems. The SCSA TAO Framework includes a suite of interoperable, vendor-independent application programming interfaces (SCSA APIs), a set of System Services for handling various server management functions, and a set of messages and a standard transport for communication among various technology resources and system service providers (Service Provider Messages and SCSA Message Protocol). In 1995, TAO's development was taken over by a new organization ECTF — the Enterprise Computer Telephony Forum. The ECTF expanded the idea of TAO to make it the open software framework for the whole computer telephony world — to encompass hardware conforming to all major specifications, including SCSA and MVIP. Towards the end of January, 1996, ECTF promulgated TAO (now under a different name) as the software standard for the new computer telephony industry.

SCSI Small Computer System Interface. (Pronounced Scuzzie.) The brains of a computer is its microprocessor. That microprocessor (computer on a chip) does the computer's primary work (i.e. calculations). There must be a way for information to get into and out of the microprocessor. The history of computers could be written as a continuing race to figure new, faster and more efficient ways of getting information into and out of the microprocessor. The obsession with input/output stems from the fact that the microprocessor can work much faster than you can get information in and out and out of it. SCSI is a way for a devices such as magnetic hard disks, optical disk drives, tape drives, CD-ROM drives, printers and scanners to communicate with the computer's main processor. SCSI is a bus and an interface standard. The theory is that if you buy a SCSI device you can plug it into your computer's SCSI port and it will work — just as a parallel or serial port device will work. There are two good points about the SCSI interface — especially the newer SCSI-2 interface. First, is that it's fast. The second good point is that on one SCSI bus you can daisy chain up to seven different devices (so long as you remember to terminate the end of the chain.) (In reality, you rarely have more than four devices hooked up on one SCSI link, since protocol overhead and other factors begin to degrade system performance.) Each device will work quickly and each won't siphon excessive power from the computer's

main processor. That's because the SCSI bus typically has its own controller/microprocessor which takes care of the SCSI input/output workload. SCSI disk drives work faster than a "normal" IDE hard drive, which is why many new computers are coming with SCSI drives, not IDE drives. ANSI (American National Standards Institute) has set several guidelines for SCSI connection. There is SCSI-1 and SCSI-2. The SCSI specifications are available from www.ansi.org.

All Apple Macintosh computers come with built-in SCSI ports to which you can daisy chain one SCSI peripheral after another, until you have a total of seven. This is a fairly easy job, since Macintosh SCSI ports are standard and manufacturers of Macintosh SCSI peripherals will certify that their product works with the Macintosh SCSI standard. They wouldn't sell it if it didn't. One point: If you've removed the hard drive in your Macintosh (and replaced it with one or more SCSI-attached drives) your Macintosh may require a hard disk terminator. Some (not all) Macintoshes require a hard disk terminator (a $5 device) if their hard disk has been removed.

To add SCSI devices to a MS-DOS machine, you must first place a SCSI adapter card in your PC's bus or your MS-DOS laptop's PCMCIA slot and connect the SCSI devices to that card. Sadly, for MS-DOS machines, SCSI is not a universal plug-n-play standard. According to Keith Comer of Toshiba, when asked why Toshiba's computers didn't come with SCSI ports as they came with parallel and serial ports, said, "I an unconvinced of SCSI's universal compatibility. It's a nontrivial task to connect SCSI devices. All devices need their own drivers. And each need to be configured for the particular SCSI card you have. Further, many of the SCSI drivers are incompatible with memory managers. In short, for us as manufacturers it would be a support nightmare."

The problem is lessening slowly. Corel (Ottawa, Canada) and others are creating "standard" SCSI Interface kits (software and/or hardware). These make connecting things less of a pain. But your desired-to-connect SCSI device (e.g. CD-ROM or magnetic optical drive) must be on the Corel list of approved devices. And — this is the Catch 22 of SCSI: If your SCSI device is new, you can be sure it will not be on Corel's list and probably will not work. In short, do not even bother trying to connect your "standard" SCSI device — unless someone has assured you that they have seen it work and it is on someone's list of approved SCSI devices. Yours truly has failed to connect many new SCSI devices using "standard" SCSI software. And when I asked the manufacturers (Corel, etc.) why they didn't work, I was told that my devices were too new and they hadn't released the necessary device driver software. But it's worse. Fingerpointing prevails. Manufacturers of SCSI deny responsibility for making SCSI device drivers to make their hardware work. And manufacturers of SCSI software and SCSCI adapter cards say they haven't been able to obtain/acquire/buy one of the new devices and figure out how to connect it. Further, there's no assurances that they will ever both to figure out how to connect that particular device. As I write this entry, I have two SCSI devices on my desk which I cannot connect through several "standard" SCSI adapter cards I am testing. I am able to connect them through a SCSI cable connected to my computer's parallel port. The throughput is very slow, . An example: Using a parallel cable I was able to transfer a one meg file in 21 seconds. Using the only PCMCIA SCSI adapter card I could coax to work, I was able to transfer the same one meg file in 12.3 seconds. See Geoport, SCSI-2, SCSI Transfer Rate and USB. http://fieldnet.ne.mediaone.net/gary/scsifaq.html

SCSI-2 SCSI-2 (pronounced Scuzzie-Two) is a 16-bit implementation of the 8-bit SCSI bus. Using a superset of the SCSI commands, the SCSI-2 maintains downward compatibility with other standard SCSI devices while improving upon reliability and data throughput. SCSI-2 is capable of transferring data at rates up to 10 megabytes per second, twice as fast as SCSI-1. SCSI-2 defines more than a speed. It defines a command set and electrical characteristics. See SCSI Transfer Rate

SCSI Transfer Rate SCSI transfer rate is the speed of moving data between the SCSI adapter board and the SCSI device. Host transfer rate is the speed of moving data between the adapter board and the host PC. See SCSI and SCSI-II. Some hard disks come as SCSI. One way of distinguishing between these SCSI disks is to look at at the pinning on the SCSI hard disks. There are three basic varieties of SCSI hard disks:

• 50-Pin. Ultra SCSI, 20 megabyte per second transfer rates, standard 50-pin cable which is backwards compatible with previous SCSI connections. Maximum cable length is 4.5ft.

• 68-Pin. Ultra Wide, 40 megabyte per second transfer rate, 68-pin Wide Cable requires Ultra Wide Controller for maximum transfer rates and optimal performance.

• 80-Pin. Ultra Wide SCA, 40 megabyte per second transfer rate, Single connector Drive designed to plug into systems with 80 pin back plane. Thus no controller Card and no Cable.

SCTE Society of Cable Telecommunications Engineers, Inc. A not-for-profit professional organization organized in 1969 to promote the sharing of operational and technical knowledge in the field of cable TV and broadband communications.

ScTP 1. Screened Twisted Pair. A type of Shielded Twisted Pair (STP) cable which employs a braided screen shield to protect the signal-carrying conductors from EMI (ElectroMagnetic Interference). See also FTP and UTP.
2. Simple Computer Telephony Protocol. SCTP is an Internet protocol authored by Brian McConnell (PhoneZone.Com) and Paul Davidson (Nortel). The protocol, modeled after other Internet application protocols (such as HTTP (worldwide web), SMTP (email), etc), creates a simple, cross-platform interface for building computer telephony applications. Unlike APIs such as TAPI and TSAPI, SCTP can be implemented on any machine which is capable of talking to TCP/IP networks. APIs, on the other hand, are operating system specific. The protocol is primarily intended for use in call control and system administration software. It is not used to create interactive voice response applications. Several vendors, such as Nexpath, a PC PBX manufacturer, have used the protocol to create cross-platform Java CTI applets which will run on virtually any operating system. SCTP is public domain, meaning the specification is public, and that anybody can use the protocol freely. www.phonezone.com/sctp

ScTP RJ-45 Plug These are used to terminate four pair ScTP patch cords. They have metal areas to connect the cable's foil shield with the equipment that it is plugged into.

SCVF Single Channel Voice Frequency.

SCWID Spontaneous Call Waiting Display.

SCxbus An SCSA term. The standard SCSA bus for communication between nodes. The SCxbus features the same architecture as the SCbus. See SCxbus Adapter.

SCxbus Adapter Inter-box expansion adapter for the SCbus.

SD Starting Delimiter

SDE 1. Synchronization Distribution Expander. 2. Secure Data Exchange as defined by the IEEE 802.10 security committee.

SDF Sub Distribution Frame. Intermediate cross connect points, usually located in wiring or utility closets. A trunk cable

or LAN backbone is run from each SDF to the MDF (Main Distribution Frame).

SDH Synchronous Digital Hierarchy. A set of standard fiber-optic-based serial standards planned for use with Sonet and ATM in Europe. Some of the SDH and SONET standards are identical. Standardized by the ITU. See SONET for a much fuller explanation.

SDK Software Development Kit.

SDL Signaling Data Link.

SDLC Synchronous Data Link Control. A bit-oriented synchronous communications protocol developed by IBM, SDLC is at the core of IBM's SNA (System Network Architecture). Intended for high-speed data transfer between IBM devices of significance (read mainframes), SDLC forms data into packets known as frames, with as many as 128 frames being transmitted sequentially in a given data transfer. Each frame comprises a header, text and trailer. The header consists of Framing bits (F) indicating the beginning of the frame, Address information (A), and various Control data (C). The data payload, referred to as Text, consists of as many as 7 blocks of data, each of as many as 512 characters. The trailer comprises a Frame Check Sequence (FCS) for error detection and correction, and a set of Framing bits (F) indicating the end of the frame. SDLC is a protocol which supports device communications generally conducted over high-speed, dedicated private line, digital circuits. SDLC can operate in either point-to-point or multipoint network configurations. See also HDLC and IBM. Contrast with BINARY SYNCHRONOUS COMMUNICATIONS.

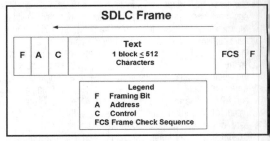

SDLC-To-Token-Ring LLC Transformation A technique to integrate SDLC link-attached SNA devices into a LAN/WAN internet. A modified remote polling process is used to make the link-attached devices appear to be LAN-attached.

SDM 1. Subrate Data Multiplexing. A European term. In North America, it's called SRDM.
2. DMS SuperNode Data Manager.

SDMA Station Detail Message Accounting. See CALL ACCOUNTING.

SDMF Single Data Message Format. See Caller ID Message Format.

SDN Software Defined Network. See SOFTWARE DEFINED NETWORK, SDN SERVING OFFICE, VIRTUAL NETWORK, VPN and the APPENDIX.

SDN Serving Office One of many AT&T-supplied switching nodes in an SDN network. See also SOFTWARE DEFINED NETWORK and the APPENDIX.

SDR Session Detail Record. Cisco's term for CDR (Call Detail Record). Refers to detail records (generated by routers) that are captured and processed by Telco Research's Enterprise Accounting system. Similar to PBX CDR that is captured and processed by a call accounting system.

SDRAM Synchronized Dynamic Random Access Memory. An emerging replacement for DRAM. SDRAM's memory access cycles are synchronized with the CPU clock in order to eliminate wait time associated with memory fetches between RAM and the CPU. See also DRAM and RAM.

SDRM Sub-rate Data Multiplexing. Refers to a service where a DSO (64 Kbps) channel may contain one 56 Kbps signal, five 9.6 Kbps signals, ten 4.8 Kbps signals or twenty 2.4 Kbps signals. Although speeds may be mixed, the highest speed determines the number of signals supported.

SDP 1. Session Description Protocol. SDP conference descriptors are stored in the ILS Dynamic Directory Conference Server, part of Active Directory in Windows NT 5.0. See TAPI 3.0 for a full description.
2. Service Delivery Point. Ed Ward of Belair, MD, writes "I am new to the formal telecommunications world. Although I have worked with circuits / modems / routers etc, up until now it was an ancillary duty. I am a government employee now working full time as a telecommunications specialist. Two of our main contractors are AT&T and SPRINT. Here is one term we use when we are ordering circuits from them. SDP - Service Delivery Point - Defined as the minimum point of presence (POP) where the commercial carrier brings the line into a structure to begin the service. (i.e. House, Basement of building, Telephone closet etc). It is further defined as the point where the "interior" wiring connection is made."

SDSAF Switched Digital Services Applications Forum, a group of manufacturers and carriers whose objective to standardize the interconnection of switched 56 kilobit and n x switched 56 channel local and long distance services. The group is based in Reston, VA. Today a switched 56 Kbps "phone" call between multiple carriers probably wouldn't get through. In short, this group is trying to bring the simplicity of the voice dial up phone system into the switched data world.

SDSL Symmetrical Digital Subscriber Line. Also known as HDSL2, although the local loop length is limited to about 10,000 feet. Speeds of up to one megabit per second are possible both ways (thus symmetrical). See also xDSL, ADSL, HDSL, IDSL, RADSL and VDSL.

SDT Structured Data Transfer: An AAL1 data transfer mode in which data is structured into blocks which are then segmented into cells for transfer.

SDTV Standard Definition TV. A term used to describe the current NTSC analog TV standard, which offers poorer resolution than HDTV (High Definition TV). In December 1996, the FCC approved standards for ATV (Advanced TV), the successor term for HDTV. ATV delivers roughly four times the number of pixels (picture elements) and offers an aspect ratio of 16:9 (horizontal:vertical). ATV is digital, while SDTV is analog. Thereby, ATV offers much better quality, while consuming much less bandwidth through the use of sophisticated compression techniques. SDTV is analog. See HDTV and ATV.

SDU-SMDS Data Unit The user payload in an SMDS L3PDU packet. The SDU can contain up to 9,188 bytes.

SDU An ATM term. Service Data Unit: A unit of interface information whose identity is preserved from one end of a layer connection to the other.

SDV Switched Digital Video.

SDVN Switched Digital Video Network.

SE 1. Systems Engineering
2. An ATM term. Switching Element: Switching Element refers to the device or network node which performs ATM switching functions based on the VPI or VPI/VCI pair.

Seagull Manager A manager who flies in, makes a lot of noise, is critical of everything, then leaves. The seagull metaphor comes from the activity that seagulls are most known for, i.e. defecating. See also Albatross Manager.

Sealed Case PC A fancy name for a PC that never needs to be opened up.

Sealing Current A designation for a powering situation that consists of a wet loop without span power.

Sealink Sealink is an error-correcting file transfer, data transmission protocol for transmitting files between PCs. It is a variant of Xmodem. It was developed to overcome the transmission delays caused by satellite relays or packet-switching networks.

SEAL Simple and Efficient Adaptation Layer: An earlier name for AAL5.

Seals A way of telling if a device has been tampered with.

Seamless The word seamless means "perfectly smooth, without awkward transitions." In software, it means that what takes place between the user and the application accessed by the user is perfectly smooth to the user and the software being used by the used will work easily with other software the user is also using. On a LAN accessing "seamless" applications, the user doesn't perceive he's on a network because his programs run as though they were on his personal computer. In fact, the word "seamless" is a very vague term, meaning now that the "seamless" software you're about to use will work easily with other software you're using. But no one knows exactly what easily is, not how easy seamless is. And no one has a technical definition of seamless. Originally, the word meant "without the stitches showing." The 15-century word got a boost from the phrase seamless stockings which filled a brief period between silk stockings — which had seams that always needed straightening — and panty hose that don't, sadly. In short, when any vendor says they offer "seamless integration" with something else, don't believe them. Check.

Search Drive A drive that is automatically searched by the operating system when a requested file is not found in the current (default) directory. A search drive allows a user working in one directory to transparently access an application or data file that is located in another directory.

Search Engine An Internet World Wide Web term. A search engine is a program that returns a list of Web Sites (URLs) that match some user-selected criteria such as "contains the words cotton and blouse." Basically, the procedure is simple. You surf to the search engine's site. You click a couple of times and type in what you're looking for. A few seconds later you get choices. You finally make your selection and you get instantly hotlinked over to that site. Search Engines are the most useful thing to come along in years. I use them daily to find everything from information on a company I'm looking for to new definitions to fill this ever-expanding dictionary. Here are the main Internet search engines and their addresses.
ALTA VISTA
http://altavista.digital.com
EXCITE
www.excite.com
LYCOS
www.lycos.com
INFOSEEK
http://guide.infoseek.com
OPENTEXT
www.opentext.com:8080
INKTOMI
www.inktomi.berkeley.edu/query.html
WWW.WORM
www.cs.colorado.edu/wwww/
WEBCRAWLER

www.webcrawler.com

YAHOO

http://akebono.stanford.edu/yahoo/

SEAS Signaling Engineering and Administration System.

Seasonality The month-to-month fluctuation in call volume that can be expected to recur each year in a call center. For example, a call center might always have peak months in the spring and early summer and slow months in the late fall and winter.

Seated Agents A call center term. See Base Staff.

SECABS Small Exchange Carrier Access Billing Specifications

SECAM Acronym for Systeme Electronique Couleur Avec Memoire or SEquential Couleur Avec Memoire. A television signal standard used in France, eastern European countries, Russia, the former Soviet Union and some African countries. SECAM uses an 819-line scan picture which provides a better resolution than PAL's 625-lines and NTSC's 525-lines (the US standard). All three systems are not compatible. You cannot view an Australian, English or French videotape or through-the-air broadcast on a US TV. See NTSC.

Second Dialtone 1. Dialtone given to the caller, on a phone system (e.g. PBX, Centrex or hybrid), after dialing an access code (e.g. 8 or 9) to make a call out of the system (e.g. a local or long distance). Sometimes referred to as outside dialtone. 2. Dialtone returned to a caller after they've dialed a local or long distance number and reached some type of switching device. That switching device might allow you to dial into a fax machine, a modem, a phone or an answering machine. It might allow you to dial into one of several cash registers or soda machines (to check if they need being refilled). It might even allow you to dial long distance through a voice mail system or through a long distance phone company. As you dial through networks you might encounter not only second, but also third and fourth dialtones.

Secondary Equipment Used telecommunications equipment. See also USED, CERTIFIED, REFURBISHED and REMANUFACTURED.

Secondary Market The market for used business telecommunications and computer equipment.

Secondary Protection Primary protection is a device that sits at your building entrance between your phone line coming in from outside and your line going into and up your building. The phone company is responsible for installing primary protection. Secondary protection sits on your floor just next to your phone system. Secondary protection is designed to protect your phone equipment from spikes, surges and high electricity that might affect your phone lines between the primary protection downstairs and the secondary protection upstairs. Secondary protection typically costs $20 to $30 a line. It's worth every penny.

Secondary Radar A radiodetermination system based on the comparison of reference signals with radio signals retransmitted from the position to be determined.

Secondary Radiation Particles (such as photons, Compton recoil electrons, delta rays, secondary cosmic rays, and secondary electrons) that are produced by the action of primary radiation on matter.

Secondary Resource An SCSA definition. Any resource that is attached to a Group after the Group has been created around a primary resource. See SCSA.

Secondary Ring One of the two rings making up an FDDI ring. The secondary ring is usually used in the event of failure of the primary ring.

Secondary Server Under NetWare, the secondary server is the SFT III NetWare server that is activated after the primary server that receives the mirrored copy of the memory and disk from the first server activated. The secondary server mirrors the disk and memory image of the primary server. Though it cannot be used to do additional work (because it uses all of its CPU cycles keeping up with the primary server), the secondary server can act as a router for the local network segments to which it is directly attached. In addition to mirroring the primary server, the secondary server provides split seeks. Either SFT III server may function as primary or secondary, depending on the state of the system. You cannot permanently designate which server is primary or secondary. System failure determines each server's role, that is, when the primary server fails, the secondary server becomes the new primary server. When the failed server is restored, it becomes the new secondary server. See also SERVER MIRRORING.

Secondary Service Area The service area of a broadcast station served by the skywave and not subject to objectionable interference and in which the signal is subject to intermittent variations in strength.

Secondary Station In a data communication network, the secondary station responsible for performing unbalanced link-level operations, as instructed by the primary station. A secondary station interprets received commands and generates responses.

Secondary Winding The minor winding on a relay having two windings. The winding on a transformer that is not connected to a AC source.

SECORD SEcure voice CORD board.

Secretarial Hunting The secretary's station number is programmed as the last number in one or more hunt groups. If all phones within a hunt group are busy the call will hunt to the secretary.

Secretarial Intercept A PBX feature. Causes calls for an executive to ring his/her secretary — even if the executive's direct extension number was dialed. The executive's phone will ring only if the secretary's phone is placed on "Do Not Disturb" or the secretary transfers the call in.

SECTEL Acronym for SECure TELephone.

Sectional Center A control center connecting primary telco switching centers. A Class 2 office. The next to the highest rank (Class

2) Toll Switching Center which homes on a Regional Center (Class 1).

Sector A pie-shaped portion of a hard disk. A disk is divided into tracks and sectors. Tracks are complete circuits and are divided into sectors. Under MS-DOS a sector is 512 bytes.

Sectoring The process of dividing a mobile cellular radio cell into sectors, or smaller patterns of coverage. Traditionally, all cell antennae were omnidirectional; that is to say that they provided coverage in a 360-degree pattern. Sectoring is applied when either the capacity of the cell site is insufficient or when interference becomes a problem. Sectoring divides the number of channels assigned to the cell into smaller groups of channels, which are assigned to a sector through the use of directional antennae. Commonly, the cell antenna is divided into three sectors, each with a 120-degree coverage pattern. You probably have noticed that many contemporary cell site antennae are very tall for better coverage and triangular in shape for purposes of sectoring.

Secure Channel A technology that provides privacy, integrity, and authentication in point-to-point communications such as a connection on the Internet between a Web browser and a Web server. You can tell if you have a secure channel with Netscape by checking out the key on the bottom left hand side of your screen. If the key is broken, your connection is insecure. If it's together in one piece, then your Internet conversation is secure, which means it's encrypted and therefore hard for someone to break into and make sense of. See Internet Security, which details the problem of security on the Internet.

Secure Electronic Transaction SET. A system designed for electronic commerce over the Internet that promises to make stealing credit card numbers much more difficult. See Digital Cash.

Secure Kernel The core of a secure operating system.

Secure Public Dial A security term. Secure Public Dial is dialup switching functiona ity that allows the service provider to offer customers the security of a private connection with the economics of public dial. Also referred to as Virtual Private Networking (VPN) allows the service provider to support multiple enterprises' dial-in requirements, securely from the same dialup switch. Each separate customer has access to their own virtual network although they may share physical ports and access paths. The service provider manages the multicustomer net with the same ease as if it were one internetwork.

Secure Sockets Layer SSL. A low-level protocol that enables secure communications between a server and FrontPage or a browser.

Secure Telephone Unit STU. A U.S. Government-approved telecommunication terminal designed to protect the transmission of sensitive or classified information — voice, data and fax.

Secure Voice Voice signals that are encoded or encrypted to prevent unauthorized listening.

Secure Voice Cord Board SECORD. A desk-mounted patch panel that provides the capability for controlling 16 wideband (50 Kbps) or narrowband (2400 bps) user lines and five narrowband trunks to AUTOVON or other DCS narrowband facilities.

Security A way of insuring data on a LAN is protected from unauthorized use. Network security measures can be software-based, where passwords restrict users' access to certain data files or directories. This kind of security is usually implemented by the network operating system. Audit trails are another software-based security measure, where an ongoing journal of what users did what with what files is maintained. Security can also be hardware-based, using the more traditional lock and key.

Security Access Manager SAM. An AT&T computer-based building access security system.

Security Accounts Manager Database Also called the Directory Services Database, stores information about user accounts, groups, and access privileges on a Microsoft Windows NT server.

Security Blanking The ability of a switch to blank out the called digits for certain extensions so no called number detail is printed. Senior executives in serious takeover negotiations find this feature useful. There have been instances of people figuring out which company another company is about to buy based on telephone calling records. If you have this information, you can buy the company's stock before the bid is announced and make a lot of money. This feature — security blanking — is designed to avoid such occurrences.

Security Cabinet A cabinet, usually on casters, used to store confidential materials under lock and key prior to shredding.

Security Code 1. A user identification code required by computer systems to protect information or information resources from unauthorized use. 2. A six-digit number used

to prevent unauthorized or accidental alteration of data programmed into cellular phones. The factory default is 000000.

Security Dots The asterisks that appear onscreen as you type in your password.

Security Equivalence A security equivalence allows one user to have the same rights as another. Use security equivalence when you need to give a user temporary access to the same information or rights as another user. By using a security equivalence, you avoid having to review the whole directory structure and determine which rights need to be assigned in which directories.

Security Management Protects a network from invalid accesses. It is one of the management categories defined by the ISO (International Standards Organization).

Security Modem A class of modem providing secure access.

Security Stud A cylindrically shaped metal finger that holds open the door to a Cash Box until the box is removed for collection.

Seedware Software designed to get demand for a product or a new market segment started. Software designed to "seed" a market. Seedware is typically a less-full featured piece of software than the software you're really trying to sell. Seedware typically costs very little. It may even be free. It's also called FEEDWARE.

Seek Time The time it takes to move a disk drive's read/write head to a given track. Seek time varies depending on where the head starts from and has to go it. Average seek time is a critical measure of the speed of a computer disk drive.

SEF 1. Source Explicit Forwarding. Security feature that allows transmissions only from specified stations to be forwarded by bridges.
2. Severely Errored Framing. A SONET defect which is the first indication of trouble in detecting valid signal framing patterns. Four consecutive errored framing patterns constitutes an SEF defect. If the SEF defect persists, an LOF defect is instituted, and if the LOF defect persists an LOF alarm is declared. See also LOF.

SEFS Severely Errored Framing Seconds. A count of the number of seconds during which at any point the SEF defect was present. See SEF.

Segment 1. 64 characters. Use as a method of data communications billing by some overseas phone companies.
2. An electronically continuous portion of a network, usually consisting of the same wire.
3. A single ATM link or group of interconnected ATM links of an ATM connection.

Segmentation and Reassembly SAR. An ATM term. See SAR.

Seize To access a circuit and use it, or make it busy so that others cannot use it.

Seizure Signal A signal used by the calling end of a trunk or line to indicate a request for service.

SEL Selector: A subfield carried in SETUP message part of ATM endpoint address Domain specific Part (DSP) defined by ISO 10589, not used for ATM network routing, used by ATM end systems only.

Select Call Forwarding A name for a Pacific Bell (and possibly other local telephone companies') service which allows you to have calls from selected numbers ring at another number.

Selection What most people call a "block" in a word processed document, Microsoft calls a "selection." If you want to print test you've blocked, tell Windows to print a selection.

Selective Call Acceptance — Permits incoming calls only from numbers you pre-select.
Call Forwarding — Forewords all calls from a pre-selected list to a specific destination.
Call Rejection — Blocks incoming calls from numbers you pre-select

Selective Calling The ability of the transmitting phone to specify which of several phones on the a line is to receive a message.

Selective Fading Fading in which the components of the received radio signal fluctuate independently.

Selective Paging To Station A phone can page to individual phone instruments.

Selective Ringing A method of ringing only the desired party on a party line.

Selective Signaling A method of inband signaling used on private networks to tell switches to switch the call.

Selectivity Ability of a tuner or receiver to get only a desired station while rejecting other adjacent stations. The higher the figure expressed in decibels (dB), the better the selectivity.

Selector The identifier (octet string) used by an OSI entity to distinguish among multiple SAPs at which it provides services to the layer above.

Self Diagnostics Your phone system tells you when something is wrong with it by sending you a "message" via the operator console or through one of the data ports on the phone system.

Self Discharge When internal chemical reactions (such as the dying out of chemicals) cause the loss of useful capacity of a cell or battery in storage.

Self Electro-Optic Effect Devices SEEDs. Switches guided by light.

Self Extinguishing The characteristic of a material whose flame is extinguished after the igniting flame is removed.

Self Healing Self healing means it fixes itself, which is really a misnomer. It really doesn't fix itself. In telecom, self-healing really means 100% connectivity. The term came into telecom use with the invention of SONET fiber rings. Here's how they work: A phone company lays down several concentric rings of fiber — typically around a city or an industrial park. You, the customer, buy service from the carrier. If one ring fails, the system knows and instantly your traffic is shunted in a different direction or to a different fiber. "Self-healing" is actually 100% fiber connectivity between business locations and telephone company serving wire centers. "Self-healing" means that SONET rings provide for automatic network backup with 100% redundancy so that if there is a point of failure on your fiber ring, your service continues.
There is another definition. Brian Livingston, author of books on Windows and InfoWorld columnist, says that self healing software upgrades itself by periodically calling a central phone number (via bulletin board) or a Web site and automatically downloading and installing any new components that may have become available since the last product upgrade.

Self Hosted An SCSA definition. The Client and the Server are on the same computer platform. A server with one or more self-hosted applications may be a standalone unit which is not connected to any other system.

Self Learning Bridge See Bridge.

Self Test The capability of a PBX to run programs at regular intervals to test its own operation and signal when failures have occurred or are about to occur without human intervention.

Selling Selling consists of three steps: Prospecting. Qualifying and Closing.

Selling Steps A telemarketing term. The steps involved in a sale: A clear call objective, identify and reach the decision maker, introduction and call justification, identify needs, present solution/benefits, answer questions or objections, close, confirm the conditions, and congratulate.

Semaphore 1. A message sent when a file is opened to prevent other users from opening the same file at that time. Its purpose is to preserve the integrity of data (i.e. stop it from being messed with) while you're using it.
2. An apparatus for visual signaling, such as by the use of flags. The term comes from the Greek "sema," meaning sign or signal, and "phoros," meaning carrying. Semaphores were in common use for message signaling prior to the invention of the telegraph. In fact, the British Admiralty Office on August 5, 1816 officially rejected the idea of the electric telegraph, which had been suggested by Mr. (later Sir) Ronald Edwards. It seems that the Admiralty preferred the semaphore, although it was useless during the night or when it was foggy.

Semi-Automatic Message Switching A network control technique whereby an operator manually switches a message to a destination according to the address information contained in the message header.

Semi-Rigid A cable containing a flexible inner core and a relatively inflexible sheathing material, such as a metallic tube but, which can be bent for coiling or spooling and placing in duct or cable run.

Semiconductor Semiconductors are materials that are midway between conductors and insulators. Semiconductors will only conduct electricity when a certain threshold voltage has been reached. Because the energy needed to "turn on" a semiconductor can be high, most semiconductors have their conductivity enhanced through a process called doping. Doping consists of adding an impurity that has either a surplus or shortage of electrons. Semiconductors with impurities that provide a surplus of electrons are called N-type semiconductors. Semiconductors with impurities that have a shortage of electrons are called P-type semiconductors. Together N-type and P-type semiconductors are the basic building blocks of nearly all solid state electronic devices. Semiconductors have a resistance to electricity somewhere between a conductor (e.g. a copper wire) and an insulator (e.g. plastic). Hence the word "semi" conductor. Silicon and germanium are the two most commonly-used semiconductor materials. The flow of current in a semiconductor can be changed by light or the presence or absence of an electric or magnetic field.

Semiconductor Laser A laser is really an oscillator, which means an amplifier plus a feedback. In a semiconductor laser, the amplification is provided by the population inversion in the active layer obtained by injecting current in it. There are two basic types of semiconductor lasers — the Fabry-Perot and DFB lasers. They use different ways to provide the feedback. In a Fabry-Perot (FP), you a have a mirror at each end (usually obtained by cleaving the semiconductor crystal), which reflects part of the light and give the feedback.
In a DFB (or Distributed FeedBack) laser, a Bragg grating is incorporated along the active layer, providing a distributed reflection of the light. Often the end facets are anti-reflection coated to avoid the Fabry-Perot effect. The grating is usually made by varying the thickness of one layer in a periodic fashion. This gives a periodic variation of the propagation constant (effective index) of the waveguide. For the right period, you get reflection of the light at a certain wavelength.
The main difference between the two types is that DFBs are often single mode (only one wavelength determined by the

Bragg grating), while FPs are usually multi-mode (4 to 20 different wavelengths at the same time and determined by the gain curve of the active material and the cavity length).

Semipermanent Connection A connection established via a service order or via network management.

Send and Pray A descriptive term for a data communications protocol that provides little or no assurance that the data gets to the destination device as intended. Send and pray first was applied to Parity Checking, also known as Vertical Redundancy Checking (VRC), which is the very poor error detection mechanism used in asynchronous protocols using the ASCII coding scheme. Parity checking not only is highly uncertain in its ability to detect an error created in transmission, but also provides no mechanism for the receiving device to advise the transmitting device of a detected error in order that it might be corrected by some means. Hence, the term "send and pray"-you just send the data and pray that it gets to the other end correctly. More recently, the term has been applied to voice and data over IP networks. While IP provides a fairly robust mechanism for error detection and correction, it was intended for true data communications applications, which have time to recover from errors created in the process of transmission and switching. Voice and video, however, have no time to recover. The data is presented to the receiving device as it exits the network. Any errored, delayed or corrupted data packets are simple discarded in this stream-oriented communication mode. If you are talking over an IP-based network, therefore, you just send your voice into the phone and pray that it all gets there.

Send and Prey A term for the method by which the misguided amongst us seek to damage our data. These poor, misguided souls embed viruses in programs which we then download. They send them to us, and then prey upon us. This term was coined by Ray Horak, my Contributing Editor.

Sender Equipment in the originating telephone system which outpulses (sends out) the routing digits and the called person's number. Senders are necessary in computerized (stored program control) switches because the switch needs to know all the digits of the numbers you are calling before it chooses and seizes a trunk.

Sendmail Sendmail is the UNIX software that handles electronic mail. It is the most common form of electronic mail software on the Internet. Sendmail provides back-end message routing and handling for Simple Mail Transfer Protocol-based electronic mail communications.

Seniority A call center term. A field in each employee record establishing that employees seniority for the purpose of automatic schedule assignment. This is typically the employee's hire date (year, month, and day) plus a three-digit "tie breaker."

Sensitivity 1. The input signal level required for a tuner, amplifier, etc., to produce a stated output. The lower the required input, the higher the sensitivity. Measured in mV (microvolts). 2. The degree to which a radio receiver responds to the wave to which it is tuned.

Sensitivity Analysis The process of rerunning a financial study to figure the degrees to which changing the assumptions changes the result of the analysis.

Sensor Applications Services involving the remote monitoring of instruments including burglar alarms, fire alarms, meter reading, energy management and load shedding.

Sensor Glove An interface device for experiencing virtual reality with the hand. Wired with sensors, it detects changes in finger, hand and arm movements and relays them to the computer, allowing users to manipulate and move things in a virtual environment. See VIRTUAL REALITY.

Sent Paid A utility industry term that describes all calls charged to the originating number or collected as coins in a pay telephone.

Separation 1. Extent to which two stereo channels are kept apart. Expressed in decibels, the larger the number, the better the separation and stereo effect.
2. The use of FDM (Frequency Division Multiplexing) to establish separation of upstream and downstream information channels. FDM is commonly used in cellular telephony and other radio technologies and applications. Channel separation using FDM also sometimes is used in high-speed data transmission systems in order to maintain separation between the primary data stream and other information streams such as POTS and ISDN. For example, it is proposed that FDM channel separation will be used in early versions of VDSL (Very-high-data-rate Digital Subscriber Line). See also FDM and VDSL.

Separations and Settlements A complex set of accounting procedures developed by the traditional telephone industry. The procedures classify telephone plant as intrastate or interstate, and return revenues from long distance phone calls to local telephone companies to compensate them for the use of local exchange facilities (e.g., switches and local loop cable plant) in the origination and termination of long distance calls. The issue is that long distance is highly profitable, while local service is not. Prior to the breakup of the Bell System, AT&T and the local telephone companies cooperated in this process. AT&T Long Lines calculated the settlement payments for the BOCs (Bell Operating Companies), which then reimbursed the independent (i.e., non-Bell) companies within their states of operation. In 1983, the FCC mandated the creation of NECA (National Exchange Carrier Association), which is charged with administering this process under Subpart G of Part 69 of the FCC's rules and regulations. The settlements pool consists of several elements. Subscribers contribute directly to the pool through the monthly Subscriber Line Charge (SLC), which is capped by the FCC at various levels for all local loops of the same type; that amount appears as a separate line item on your phone bill. (The SLC also is known variously as the Access Charge, CALC, and EUCL.) The IXCs (IntereXchange Carriers) pay into the pool a fixed amount for access trunks and termination facilities between their networks and the LEC networks; this charge is known as the CCL (Carrier Common Line Charge). The IXCs also pay per-connection and per-minute access charges into the pool. All LEC revenues from these sources are submitted to NECA, along with statements of associated costs. NECA then divides the revenues on an average-cost basis, with special consideration for "high-cost" serving areas, which are defined as having service costs in excess of 115% of the national average. In this manner, the LECs are reimbursed for the use of their networks, and "universal service" availability is assured. As deregulation and local competition have begun to alter the telecommunications landscape, the CLECs (Competitive Local Exchange Carriers) have been included in this process, as well. See also Access Charge, CLEC, NECA and Universal Service.

Separator Pertaining to wire and cable, a layer of insulating material such as textile, paper, etc. placed between a conductor and its insulation, between a cable jacket and the components it covers or between various components of a multi-conductor cable. It can be utilized to improve stripping qualities and/or flexibility, or can offer additional mechanical or electrical protection to the components it separates.

SEPP Secure Electronic Payment Protocol. An open specification for secure bank card transactions over the Internet, SEPP was developed jointly by IBM, Netscape, GTE, CyberCash and MasterCard. An embodiment of the iKP protocol, which uses public-key cryptography, SEPP is intended for HTTP (HyperText Transfer Protocol) transactions and is adapted to bank card payments in the general context of Electronic Commerce. SEPP messages are transmitted as MIME (Multipurpose Internet Mail Extensions) attachments, and are based on common syntax standards. See also Electronic Commerce, iKP, MIME and Public Key.

SEPT Signaling End Point Translator, part of Signaling System 7. See SIGNALING SYSTEM 7.

Sequence A call center term. A pattern of days on and days off as defined in either a schedule preference or shift definition.

Sequencing Sequencing is the process of dividing a data message into smaller pieces for transmission, where each piece has its own sequence number for reassembly of the complete message at the destination end. Sequencing is thus also the process of properly ordering the receipt of packet data at their destination, regardless of the time they have taken to travel the X.25 network. It's similar to packetizing. See PACKET.

Sequencing Receivers All GPS (Global Positioning System) receivers must receive information from at least four satellites to calculate accurately where they (and thus you) are. Sequencing receivers use a single channel and move it from one satellite to the next to gather this data. They usually have less circuitry so they're cheaper and they consume less power than receivers which work on four satellites simultaneously. Unfortunately the sequencing can interrupt positioning and can limit their overall accuracy.

Sequential Pertaining to events occurring in a specific time or code order.

Sequential Access The need to read data — one record after another in sequence — before getting to the information you want. Magnetic tape, for example, requires you to read the entire tape up to where your information is. This is because the computer cannot tell where on the tape your information is because records on tape files are often of variable length. Most Random Access files, usually kept on a disk drive, require records to be of a fixed length, such as 80 characters per record. Then when you seek record 23, the computer seeks character 1840 in the file (23 x 80), and takes the next 80 characters as the record you want. Using the analogy of music recorded on records and magnetic tape, a phonograph record needle has the capability of random access because the needle can be set down in the spaces between cuts on the record. With mag tape, you must fast forward past all the music you don't want before you get to the music you want to hear. Also, if the computer tape drive is fast forwarding, it cannot count characters to find the record, and must read in the data you don't want (and throw it away), before it gets to the data you need. Random access is much faster than sequential access.

Sequential Hunting See Rollover

Sequential Logic Element A device that has at least one output channel and one or more input channels, all characterized by discrete states, such that the state of each output channel is determined by the previous states of the input channels.

Sequential Packet Exchange SPX. Novell's implementation of SPP for its NetWare local area network operating system.

Serial Bus Serial Bus was the original name for Intel's standard for a type of very local, local area network that would be used for connecting peripherals to the motherboard of a PC. There'd be one plug on the back of the PC into which you'd daisy chain various peripherals, including a mouse, a keyboard, speakers, printers, a microphone and a telephone. The idea of serial bus is to clear away all the clutter on the back of the PC. In March of 1995 when the first technical specs were released, serial bus' name was changed to Universal Serial Bus (I don't know why). See USB.

Serial Call Telephone system feature set up by the attendant when an incoming calling party wishes to speak with more than one person internally. When the first party hangs up, the call automatically moves to the second person the outside party wants to speak with. When that person hangs up, then the call automatically goes to the third person, etc.

Serial Communication Networks (local and long distance) use the RS-232 serial communications standard to send information to serial printers, remote workstations, remote routers, and asynchronous communication servers. The RS-232 standard uses several parameters that must match on both systems for information to be transferred. These parameters include baud rate, character length, parity, stop bit, and XON/XOFF.

Baud rate is the signal modulation rate, or the speed at which a signal changes. Since most modems or serial printers attached to personal computers send only one bit per signaling event, baud can be thought of as bits per second. However, higher-speed modems may transfer several bits per signal change. Typical baud rates are 300, 1200, 2400, 4800, 9600 and 19,200. The higher the number, the greater the number of signal changes and, therefore, the faster the transmission.

Character length specifies the number of bits used to form a character. The standard ASCII character set (including letters, numbers, and punctuation) consists of 128 characters and requires a character length of 7 bits for transmissions. Extended character sets (containing line drawings or the foreign characters used in IBM's extended character set) contain an additional 128 characters and require a character lengths of 8 bits. Parity error checking can only be used with character lengths of 7 bits.

Parity is a method of checking for errors in transmitted data. You can set parity to odd or even, or not use parity at all. When the character length is set to 8, parity checking cannot be done because there are no "spare" bits in the byte. When the character length is 7, the eighth bit in each byte is set to 0 or 1 so that the sum of bits (Os and 1s) in the byte is odd or even (according to the parity setting). When each character is received, its parity is checked again. If it is incorrect (because a bit was changed during transmission), the communications software determines that a transmission error has occurred and can request that the data be retransmitted.

Stop bit is a special signal that indicates the end of that character. Today's modems are fast enough that the stop bit is always set to one Slower modems used to require two stop bits.

XON/XOFF is one of many methods used to prevent the sending system from transmitting data faster than the receiving system can accept the information. See also EIA/TIA-232-E, RS-232-C and serial data transmission.

Serial Data Transmission Serial data transmission is the most common method of sending data from one DTE to another. Data is sent out in a stream, one bit at a time over one channel. When a computer is instructed to send data to another DTE, the data within the computer must pass through a serial interface to exit as serial data. Then it passes through ports, cables, and connectors that link the various devices. The boundaries (physical, functional, and electrical) shared by these devices are called interfaces. See serial communications.

Serial Digital Digital information that is transmitted in serial form. Often used informally to refer to serial digital television signals.

Serial Interface The "lowest common denominator" of data communications. A mechanism for changing the parallel arrangement of data within computers to the serial (one bit after the other) form used on data transmission lines and vice versa. At least one serial interface is usually provided on all computers for the connection of a terminal, a modem or a printer. Sometimes also called a serial port. See EIA/TIA-232-E, RS-232-C, SERIAL INTERFACE CARD and the APPENDIX.

Serial Interface Card A printed circuit card which drops into one of the expansion slots of your computer and changes the parallel internal communications of your computer into the one-bit-at-time serial transmission for sending information to your modem or to a serial or nter.

Serial Memory Memory medium to which access is in a set sequence and not at random.

Serial Port An input/output port (plug) that transmits data out one bit at a time, as opposed to the parallel port which transmits data out eight bits, or one byte at a time. Most personal computers (PCs) have at least one serial and one parallel port. In a typical configuration, the serial port is used for a modem while the parallel port is used for a printer. For a diagram of a typical 25-pin RS-232-C serial port, see the APPENDIX at the back of this book.

Serial Processing Method of data processing in which only one bit is handled at a time.

Serial Transmission Sending pulses one after another rather than several at the same time (parallel). When transmitting data over a telephone line there is only one set of wires. Therefore, the only logical way to transmit it is to send the data in serial mode. It is possible to use eight different frequencies to transmit a character all at once (parallel), but these modems are ridiculously expensive. See SERIAL PORT and PARALLEL.

Serialize To change from para lel-by-byte to serial-by-bit.

Series A connection of electrical apparatus or circuits in which all of the current passes through each of the devices in succession or on after another. See also PARALLEL.

Series 11000 An AT&T private line long distance tariff created in the 1970s and designed expressly to reduce MCI's chances of selling any private lines and thus of surviving. It was thrown out by the FCC and the tariff figured in MCI's and the Federal Government's antitrust against AT&T.

Series Circuits In a series circuit, the electric current has only one path to follow. All of the electric current flows through all the components of the circuit. To calculate the resistance of a series circuit add up the resistance of each of the components in the circuit. In contrast, see parallel circuits.

Series Connection A connection of electrical apparatus or circuits in which all of the current passes through each of the devices in succession or on after another. See also PARALLEL.

Series RF Tap A bugging device. It is a radio transmitter which is installed in series with one wire of the telephone circuit. Normally a parasite (i.e. takes power from the phone line). Transmits both sides of the conversation. It transmits only when the phone is off-hook.

Server 1. Hardware definition of server. A server is a shared computer on the local area network that can be as simple as a regular PC set aside to handle print requests to a single printer. Or, more usually, it is the fastest and brawniest PC around. It may be used as a repository and distributor of oodles of data.

It may also be the gatekeeper controlling access to voice mail, electronic-mail, facsimile services. At one stage, a local area network had only one server. These days networks have multiple servers. Servers these days have multiple brains, large arrays of big disk drives (often in redundant arrays) and other powerful features. New powerful servers are called superservers. A $35,000 superserver today can match the performance of a $2 million mainframe of ten years ago. Then again, according to Peter Lewis of the New York Times, the lowliest client today has more computing power than was available to the entire Allied Army in World War II. See DOWNSIZING for some of the benefits of running servers as against mainframes. 2. Software definition of server. A server is a program which provides some service to other (client) programs. The connection between a client program and the server program is traditionally by message passing, often over a local area or wide area network, and uses some protocol to encode the client's requests and the server's responses.

Server API A SCSA term. A communications protocol that allows a call processing application running on one computer to control SCSA hardware residing in another computer.

Server Application A Windows NT application that can create objects for linking or embedding into other documents.

Server Farm Imagine a room stuffed with PCs, ranged in racks along walls, ranged in racks in lines like a library's back room. The PCs are really servers — powerful PCs containing databases and other information they are dispensing to the thousands of PCs dialing into them from afar. A server farm may be owned by one company and used by one company, or it may be owned by one company and each of the machines leased to other companies. I first heard the term when MCI described a room it had in a place called Pentagon City. There it had hundreds of servers each of which it leased to other companies who used those servers as their Web sites.

Server Message Block SMB. The protocol developed by Microsoft, Intel, and IBM that defines a series of commands used to pass information between network computers. The redirector packages SMB requests into a network control block (NBC) structure that can be sent over the network to a remote device. The network provider listens for SMB messages destined for it and removes the data portion of the SMB request so that it can be processed by a local device. In short, SMB is basically a protocol to provide access to server-based files and print queues. SMB operates above the session layer, and usually works over a network using a NetBIOS application program interface. SMB is similar in nature to a remote procedure call (RPC) that is specialized for file systems.

Server Mirroring Server mirroring means you have two servers on your networks and each exactly what the other is doing simultaneously. It's a backup method. In Novell's NetWare, server mirroring requires two similarly configured NetWare servers. They should be evenly matched in terms of CPU speed, memory, and storage capacity. The servers are not required to be identical in terms of microprocessors type (386/486), microprocessor revision level, or clock speed. However, identical servers are recommended for NetWare SFT III 3.11. If the two servers are unequal in terms of performance, then SFT III performs at the speed of the slower server. The NetWare servers must be directly connected by a mirrored server link. SFT III servers can reside on different network segments, as long as they share a dedicated mirrored server link.

Server Operating System An SCSA definition. Operating System running on the SCSA Server.

Server Push Server push is a Internet term. With server

push, the Web server sends data to display on the browser display, but leaves the connection open. At some point, the server sends additional data for the browser to display. Server push is used for displaying multimedia information on the browser.

Severely Errored Second SES. A second during which the bit error rate over a digital circuit is greater than a specific limit. During a severely errored second, transmission performance is significantly degraded. The specific definition of SES depends on the circuit involved, e.g. T-1, T-3, OC-3 and OC-48.

Service Access Code SACs are 3-digit codes in the NPA (N00) format that are used as the first three digits of a 10-digit address, and that are assigned for special network uses. Whereas NPA codes are normally used for identifying specific geographical areas, certain SACs have been allocated in the NANP (North American Numbering Plan) to identify generic services or provide access capability. Currently only four SACs have been assigned and are in use: 600, 700, 800, and 900.

Service Advertising Protocol A protocol developed by Novell so that devices attached to a network could advertise their functionality. For instance, a file server and print server advertise different functions. An SNMP agent can also advertiser itself using SAP's. All of the Compaq manageable repeaters and switches except the 50xx switches support SAP broadcasts.

Service Affecting This definition courtesy Steve Gladstone, author of the book, "Testing Computer Telephony Systems" (available from 212-691-8215). These are major bugs that significantly impact the reliability or the functionality of CT systems. Comprehensive testing must uncover all service affecting problems with the goal that no computer telephony system should be installed at a live customer installation without an acceptable workaround for the service affecting problem. The goal for transition of a product from one phase to the next is that no service affecting bugs remain, and that the bug rate for new bugs be at or approaching zero.

Service Area 1. Another term for a LATA, according to some Bell operating companies.
2. The more common usage is the geographic area served by a supplier. The area in which the supplier, theoretically, stands ready to provide his service. The service area of New York Telephone (now called Bell Atlantic) is most (not all) of New York State.

Service Boundary The boundary existing between a computing domain and a switching domain as it is established via their interconnected service boundaries over some underlying interconnection medium.

Service Bureau A data processing center that does work for others. There are many ways of bringing work to a service bureau, including mailing it and transmitting it over phone lines. If it comes over phone lines, the service is likely to be called "time sharing."

Service Charge The amount you pay each month to receive cellular service. This amount is fixed, and you pay the same fee each month regardless of how much or how little you use your cellular phone. It usually ranges from about $10 to $65 per month, depending on the carrier's tariffs and the particular plan of service you select. In addition you page air time. Service Charge doesn't usually include any air time.

Service Charge Detail A listing of all the telephone equipment installed as part of a specific telephone system. Usually provided by the vendor or maintenance organization.

Service Code A 3-digit code in general use by customers to reach telephone company service, for example, 411 (Directory

Assistance), 611 (Repair Service), 811 (Business Office) and 911 (Emergency Service).

Service Control Point SCP. The local versions of the national SMS/800 number database. SCPs contain the intelligence to screen the full ten digits of an 800 number and route calls to the appropriate, customer-designated long distance carrier. Bellcore defines SCP as the network system in the Advanced Intelligent Network Release 1 architecture that contains SLEE (Service Logic Execution Environment) functionality and communicates with AIN Release 1 Switching Systems in processing AIN Release 1 calls.

Service Creation The set of activities that must be performed to create a new service to be offered to subscribers and the associated service-specific operations capabilities to support the new service. Definition from Bellcore in reference to its concept of the Advanced Intelligent Network.

Service Creation Environment SCE. A telephone company AIN term. The surroundings, including organizational structure and computing and communications resources, in which creation of new services takes place. See Servce Creation.

Service Creation Tool What the computer telephony industry calls an applications generator, the telephone industry calls a service creation tool. It is a software tool that, in response to your input, writes code a computer can understand. In simple terms, it is software that writes software. Service creation tools have three major benefits: 1. They save time. You can write software faster. 2. They are perfect for quickly demonstrating an application. 3. They can often be used by non-programmers. See also APPLICATION GENERATOR.

Service Delivery Point See SDP.

Service Display Displays a specific service being presently in effect.

Service Entrance The point at which network communications lines (telephone company lines) enter a building.

Service Function A primary or secondary service function.

Service Identification The information uniquely identifying an NS/EP telecommunications service to the service vendor and/or service user. NS/EP (National Security and Emergency Preparedness) is federal government telecommunications services that are used to maintain a state of readiness or to respond to and manage any event or crisis (local, national, or international) that causes or could cause injury or harm to the population, damage to or loss of property, or degrade or threaten the national security or emergency preparedness of the United States. See also NS/EP Telecommunications.

Service Independent BAF The finite set of BAF (Bellcore Automatic Message Accounting Format) structures and modules needed to record usage of Advanced Intelligent Network (AIN) Release 1 services. The set is said to be robust because it will be designed to record future AIN Release 1 services that have not yet been identified. Definition from Bellcore in reference to its concept of the Advanced Intelligent Network. See also AIN and BAF.

Service Interworking A Frame Relay/ATM term. Service interworking is a means of connecting frame relay networks over an ATM backbone. The frame relay PVC (Permanent Virtual Circuit) connects to the ATM PVC through an ATM switch which accomplishes frame relay-to-ATM protocol conversion at the point of entry; at the point of exit from the ATM network, the process is reversed. This is in contrast to Network Interworking, which makes use of a tunneling protocol to provide what essentially is a cut-through path through the ATM network. Service interworking is defined in the FRF.8 specification from the Frame Relay Forum and is recognized by the ATM Forum. See NETWORK INTERWORKING.

Service Level Usually expressed as a percentage of a statistical goal. For example, if your goal is an average speed of answer of 100 seconds or less, and 80% of your calls are answered in 100 seconds or less, then your service level is 80%.

Service Level Agreement SLA. An agreement between a user and a service provider, defining the nature of the service provided and establishing a set of metrics (fancy word for measurements) to be used to measure the level of service provided measured against the agreed level of service. Such service levels might include provisioning (when the service is meant to be up and running), average availability, restoration times for outages, availability, average and maximum periods of outage, average and maximum response times, latency, delivery rates (e.g. average and minimum throughput). The SLA also typically establishes trouble-reporting procedures, escalation procedures, penalties for not meeting the level of service demanded — typically refunds to the users. An example: in May of 1998, GTE announced a new SLA that promises its "Internet Advantage dedicated Internet access customers will get a minimum packet oss guarantee from GTE. If Internet Advantage customers experience more than a 10% packet loss during any ten minute interval, they will be credited with one day of service." UUNET Technologies has a SLA that says if its' network in unavailable for one, you, the user, are credited with one full day of service.

Service Line An exchange line associated with multiple data station installations to provide monitoring and testing of both customer and Telco data equipment.

Service Logic Execution Environment SLEE. A functional group residing in an SCP (Service Control Point) or Adjunct that contains the Service Logic and Control, Information Management, AMA (Automatic Message Accounting) and Operations FEs (Functional Entity). This composite set of capabilities, which includes FC routines, provides a functionally consistent interface to SLPs (Service Logic Program) independent of the underlying operating system. Definition from Bellcore in reference to its concept of the Advanced Intelligent Network.

Service Management System SMS. An operations support system used to facilitate the provisioning and administration of service data required by the SCP. Use of this term does not imply any specific technology platform. See SMS.

Service Measurements Measurements of the actual grade of service provided to our subscribers.

Service Negotiation The functionality needed to gather subscriber Advanced Intelligent Network Release 1 business needs; provide answers, AIN Release 1 service/feature descriptions, prerequisite non AIN services, availability and costs; reserve AIN Release 1 resources (e.g. 800 number and 900 number); identify required network resources; and identify available AIN Release 1 services/features by wire center. Definition from Bellcore in reference to its concept of the Advanced Intelligent Network.

Service Node See ATM Edge Switch

Service Objectives If you're the phone company, service objectives are a statement of the quality of service that is to be provided to the customer; for example, no more than 1.5 percent of customers should have to wait more than three seconds for dial tone during the average busy hour or, the busy-hour blocking on a last choice trunk group not exceed 1 per-

cent. According to my friends in the phone industry, service objectives are the criteria used when engineering quantities of switching equipment. Certain basic principles must be kept in mind: 1. The service must be of high quality.

2. Rates for telephone service must be reasonable. 3. When costs of operating the telephone business are subtracted from revenues produced by reasonable rates, enough profit must remain to attract new capital needed to meet increasing demands for service.

If you're designing phone systems, service objectives define the functional and performance goals for how your system will work and what will be experienced by the system's users and other systems. For example, a switch might have a service objective of providing dialtone to all users in less than two seconds. Or, of answering all incoming calls within ten seconds. Or, a VRU may have a service objective of responding to all user DTMF inputs in less than one second. When we buy or design systems, we often have many service objectives, rarely just one.

Service Observation 1. A generic word used by telephone companies to check the quality of the service they're providing. Some of it is done automatically with machinery. Some of it is done by senior operators who listen in on the conversations of other operators dealing with their subscribers. In short, the senior operators "observe" the service the junior operators are providing. 2. As a feature of some telephone systems, the Service Observation (SOB) command provides the capability to automatically record data about completed calls, incomplete calls and abnormal calls for the purpose of qualitative supervision of call traffic conditions.

Service Order A Telephone Company definition. The official form on which desired customer services are recorded for processing, e.g., new connects, changes, disconnects, etc.

Service Period The time during which the telephone company furnishes a circuit.

Service Observing Period (month) A telephone company term. All business days of a month (approximately 22 days). It is recommended that the period established for measuring dial tone speed and incoming matching loss coincide as closely as possible with the Service Observing Month.

Service Points The points on the customer's premises where such channels or facilities are terminated in switching equipment used for communications with phones or customer-provided equipment located on the premises.

Service Portability A telephone company AIN term. The ability of an end user to retain the same geographic or non-geographic telephone number (NANP numbers) as he/she changes from one type of service to another. The INC Number Portability Workshop agreed that NANP numbers (e.g., 800, 500, 555, 950) should not be service portable for applications outside of their respective industry approved service definitions or guidelines, should those definitions or guidelines exist.

Service Profile Identifier See SPID.

Service Provider 1. A Windows Telephony Applications standard which lies between Windows Telephony and the network. It defines how the network — anything from POTS to T-1, from a Northern Telecom to an AT&T PBX — shall interface to Windows Telephony, which in turn talks to the Applications Programming Interface, which talks to the Windows telephony applications software. See WINDOWS TELEPHONY.

2. An organization that provides connections to a part of the Internet. If you want to connect your company's network, or even your personal computer, to the Internet, you have to talk

to a "service provider." Also called an ISP, i.e. Internet Service Provider.

3. A service provider is also a company which provides information to people who call up on a phone or on a modem.

4. An SCSA computer telephony definition. An addressable entity providing application and administrative support to the client environment by responding to client requests and maintaining the operational integrity of the server.

Service Provider Messages An SCSA definition. The message information required by, and provided by, the service provider to perform its functions in the environment in which it is installed. Contrast with SCSA Message Protocol Interface. See SERVICE PROVIDER.

Service Provider Network Identifier SPNI. An identifier for the service provider operating a particular CDPD network.

Service Provider Portability A telephone company AIN term. The ability of an end user to retain the same geographic or non-geographic telephone number (NANP numbers) as he/she change form one service provider to another.

Service Provisioning Tool What the computer industry calls a network manager, the telephone industry calls a service provisioning tool. It is a complex piece of software that allows telephone companies to contact their various switches and sundry computers dispersed over a wide geographic area, to log onto those machines and to upload, download and organize those machines so they are able to make different, new, updated software services for the telephone industry's customers. Telephone companies use various networks to get into their remote switches. Those networks might vary from dial-up to ISDN to packet switched networks to T-1. The better service provisioning tools allow one technician in one place to update and test multiple central offices and computers simultaneously.

Service Quality A call center term. A measure of how well staffing matches workload, expressed often as average delay (in answering a call).

Service Terminal The equipment needed to terminate the channel and connect to the phone apparatus or customer terminal.

Service Traffic Management STM. The platform functionality for detecting overloads associated with a specific service and for sending service-specific control messages to the appropriate entities. STM is the SLEE (Service Logic Execution Environment) functionality for detecting overloads associated with a specific service and for sending Automatic Code Gap messages to the appropriate entities. The SN&M (Service Negotiation and Management) OA (Operations Application) also provides STM (Service Traffic Management)-related capabilities.

Service Switching Point SSP. A telephone company AIN term. A switching system, including its remotes, that identifies calls associated with intelligent network services and initiates dialogues with the SCPs in which the logic for the services resides. See SSP.

Services Management System SMS. Administers 800 Data Base Service numbers on a national basis. Customer records for 800 Service are entered into the SCP through this system. See EIGHTHUNDRED SERVICE.

Services Node SN. A network system in the AIN architecture containing functions that enable flexible information interactions between an end user and the network.

Services On Demand An AT&T term for the immediate provision of almost any network service through universal ports, whenever required by a user; as opposed to provision

via an expensive, time consuming, inflexible service order process.

Serving Area Interface A serving area interface is part of a phone company's outside plant. It is a fancy name for a box on a pole, a box attached to a wall or a box in the ground that connects the phone company's feeder or subfeeder cables (those coming from the central office) to the drop wires or buried service wires that connect to the customer's premises. It's also called a cross-wire box. See also FEEDER PLANT and DROP WIRE.

Serving Closet The general term used to refer to either a riser or a satellite closet; Satellite Cabinet; Satellite Closet.

Serving Mobile Data Intermediate System A cellular radio term. The CDPD network entity that operates the Mobile Serving Function. The serving MD-IS communicates with and is the peer endpoint for the MDLP connection to the M-ES.

Serving Office An office of AT&T or its Connecting or Concurring Carriers, from which interstate communications services are furnished.

Serving Wire The term for the phone number that serves the location, referring to the phone number and terminating wire as one unit. Usually applies to a POTS number.

Serving Wire Center The wire center from which service is provided to the customer.

Servo Short for servomechanism. Devices which constantly detect a variable, and adjust a mechanism to respond to changes. A servo might monitor optical signal strength bouncing back from a disc's surface, and adjust the position of the head to compensate.

SERVORD Service Order.

SES 1. Satellite Earth Stations.

2. Severely Errored Second. A second in which a severe number of errors are detected over a digital circuit. Each error comprises a code violation (CV), such as a bipolar violation. The specific definition of SES depends on the type of circuit involved, e.g. T-1, T-3, OC-3 and OC-48. See also CV and ES.

3. Source End Station: An ATM termination point, which is the source of ATM messages of a connection, and is used as a reference point for ABR services. See DES.

Sesame Secure European System for Applications in a Multivendor Environment. Developed by the ECMA (European Computer Manufacturers Association), it is intended for very large networks of disparate origin.

Session 1. An active communications connection, measured from beginning to end, between computers or applications over a network. Often used in reference to terminal-to-mainframe connections. Also a data conversation between two devices, say, a dumb terminal and a mainframe. It is possible to have more than one session going between two devices simultaneously.

2. As defined under the Orange Book, a recorded segment of a compact disc which may contain one or more tracks of any type (data or audio). The session is a purely logical concept; when a multisession disc is mounted in a multisession CD-ROM player, what the user will see is one large session encompassing all the data on the disc.

Session Layer The fifth layer — the network processing layer — in the OSI Reference Model, which sets up the conditions whereby individual nodes on the network can communicate or send data to each other. The session layer is responsible for binding and unbinding logical links between users. It manages, maintains and controls the dialogue between the users of the service. The session layer's many functions include network gateway communications.

Session Lead-In The data area at the beginning of a recordable compact disc that is left blank for the disc's table of contents. The session lead-in uses up 6750 blocks of space. See TRACK.

Session Lead-Out The data area at the end of a session which indicates that the end of the data has been reached. When a session is closed, information about its content s written into the disc's Table of Contents, and the lead-out and the pre-gap are written to prepare the disc for a subsequent session. The lead-out and the pre-gap together take up 4650 two-kilobyte blocks (nine megabytes). See TRACK.

Set 1. Set is another name for a telephone.

2. SET. Secure Electronic Transaction. A developing, open specification for handling credit card transactions over any sort of network, with emphasis on Internet and the World Wide Web. Rather than providing the merchant with a credit card number, you send the information to them in encoded form. The merchant can't see the encrypted credit card number, but can forward the transaction request to the credit card company or clearinghouse where the information is decrypted and verified. Only the financial institution has the key to unlock the encrypted account information. The financial institution responds to the merchant request with a digital certificate which serves to verify the authenticity of the parties and the overall legitimacy of the transaction. SET ensures that no one else (e.g., hackers) can gain access to your credit card information. It also ensures that an unscrupulous ecommerce merchant can't take advantage of your credit card number. The theory is that you don't always know with whom you are dealing in Cyberspace.

Set Associative Mapping A caching technique where each block of main computer memory is assigned to a location in each cache set where the cache is divided into multiple sets.

Set Copy Set Copy allows the duplication of programming settings from one telephone to another.

Set Top Box The electronics box which sits on top of your TV, connecting it to your incoming CATV signal and your TV's incoming coaxial cable. Set-tops vary greatly in their complexity with older models merely translating the frequency received off the cable into a frequency suitable for the television receiver while newer models can be addressable with a unique identity much like a telephone. That identity can be addressed from the cable headend. This allows the CATV operator to turn individual channels on and off, such as pay channels.

SETA SouthEastern Telecommunications Association, a user group.

SETI Search for ExtraTerrestrial Intelligence. A federally funded project which uses arrays of radiotelescopes to search the heavens for signs of intelligent life as evidenced by radio transmissions. SETI is based on rationale laid out in a "Nature" article by physicists Philip Morrison and Guiseppe Cocconi and first implemented by Frank Drake, a Cornell astronomer. As radio waves propagate infinitely at the speed of light in the pure vacuum of space, the thinking is that at least traces of intelligent life can be identified, even though they will be of millennia long past since other stars and galaxies are thousands or millions of light years away. Promoted as a far less expensive technique than that of space travel, the project is interesting but in jeopardy as results have been nil over the last 25 years or so.

SF 1. Single Frequency. A method of inband signaling. Single frequency signaling typically uses the presence or absence of a single specified frequency (usually 2,600 Hz). See SIGNALING.

2. SuperFrame: A DS1 framing format in which 24 DS0 timeslots plus a coded framing bit are organized into a frame which is repeated 12 times to form the superframe.

SFBI Shared Frame Buffer Interconnect, a specification that makes it possible for hardware manufactures to produce a single-board video-graphics adapter for the PC.

SFC Switch Fabric Controller.

SFD Start Frame Delimiter. A binary pattern at the end of eight octets of timing information in an Ethernet frame that tells the receiving station that the timing information is over, and all subsequent signal represents an actual frame. The pattern is two 1s after a long string of alternating one, zero, one, zero, etc. The one octet SFD field is 10101011 in binary.

SFG Simulated Facility Group.

SFQL Structured Full-Text Query Language. A proposed standard for full-text databases. The primary focus of the proposed standard is interoperability of CD-ROMs. SFQL is based on the SQL (Structured Query Language) standard for relational databases.

SFT System Fault Tolerance. The capability to recover from or avoid a system crash. Novell uses a Transaction Tracking System (TSS), disk mirroring, and disk duplexing as its system recovery methods. System Fault Tolerance as a Novell NetWare term means data duplication on multiple storage devices. If one storage device fails, the data is available from another device. There are several levels of hardware and software system fault tolerance. Each level of redundancy (duplication) decreases the possibility of data loss.

SFTA Scalable Fault Tolerant Architecture.

SG Study Group. The ITU-T has formalized committees studying future telecommunications standards. These groups are called Study Groups.

SGCP See Simple Gateway Control Protocol

SGML Standard Generalized Markup Language. A text-based language for describing the content and structure of digital documents. HTML, which has gained fame as the language to create World-Wide Web pages on the Internet, is a descendant of SGML. SGML documents are viewed with transformers, which render SGML data the way Web browsers render HTML data. SGML was adopted by the International Standards Organization in 1986. SGML allows organizations to structure and manage information in a cross-platform, application-independent way. It tags documents as a series of data objects rather than storing them as huge files. Theoretically, SGML can reduce errors, slice costs and speed work. SCML attempts to separate the informational content of a document from the information needed to present it, either on paper or on screen.

Shadow BIOS ROM Shadow BIOS ROM is a concept I first found in Toshiba laptops which use Flash ROM to hold the machine's BIOS. When you start the machine, the BIOS copies itself from the flash ROM to the Shadow BIOS area. Accessing the BIOS from the Shadow BIOS is much faster than from flash ROM. I learned later that Compaq actually started what they called shadowing the BIOS. According to InfoWorld, Compaq did it because PC-compatible systems available at the time could have no more than 16 megabytes of RAM. Compaq decided to use the memory address at the top of the 15-megabyte physical address space for the shadow RAM.

Shadow Mask The most common type of color picture tube in which the electron beam is directed through a perforated metal mask to the desired phosphor color element.

Shadow ROM A process used in many 386 machines to map ROM BIOS activities into faster 32-bit RAM memory. Shadow memory must be loaded with BIOS routines each time the computer boots. See also SHADOW BIOS ROM.

Shannon A measurement of the quality of information in a message represented by one or the other of two equally probable, exclusive and exhaustive states. See SHANNON'S LAW.

Shannon's Law A statement defining the theoretical maximum rate at which error-free digits can be transmitted over a bandwidth-limited channel in the presence of noise.

Shannon-Fano Coding Developed by Claude Shannon of Bell Labs and R.M. Fano of MIT. A compression agorithm based on the ASCII character set, and supporting 256 values. The frequency of occurrence of each symbol in a set of symbols is tallied. The list is divided and subdivided, with the most frequent symbols being represented by the fewest number of bits. Shannon-Fano coding is used, along with other compression algorithms, in the implode (compression) algorithm of PKZIP, for instance. See also COMPRESSION.

Shaping Descriptor An ATM term. N ordered pairs of GCRA parameters (I,L) used to define the negotiated traffic shape of a connection.

Shaping Network A network inserted in a circuit for improving or modifying the wave shape of the signals.

Share To make resources, such as directories, files, printers, and ClipBook pages, available to network users.

Shared Clipboard A feature of Microsoft's NetMeeting. See NetMeeting.

Shared Lock In a database a shared lock is created by non update (read) operations. Other users can read the data concurrently, but no transaction can acquire an exclusive lock on the data until all the shared locks have been released.

Shared Logic Simultaneous use of a single computer by multiple users.

Shared Memory Portion of memory accessible to multiple processes.

Shared Modem Pools Dial-out users share resources. Any authorized user attached to the network can dial out a port on the dial-up switch, reach a modem and go for it. Benefits: Reduced costs, improved management and security; It eliminates the need for separate modem and separate modem phone lines.

Shared Resource Any device, data, file or program that is used by more than one other device, program or person. For Windows NT, Windows 95 and Windows for Workgroups, shared resources refer to any resource that is made available to network users, such as directories, files, printers, and named pipes.

Shared Screens A multimedia concept. Shared screen applications enable two or more workstations to display the same screen simultaneously. For example, two users sharing a screen can work on the same spreadsheet. Changes made by one user can be seen by the other as they are made. Shared screens can be implemented in two ways. One way enables people to view each other's screen while one person makes changes. The other way enables people to run the same application on both screens so that both users can make changes simultaneously.

Shared Services Providing PBX-based communications and processing services to the unaffiliated tenants and/or the building manager/owner of a commercial building in a stand-alone or campus environment.

Shared Tenant Services Providing centralized telecommunications services to tenants in a building or complex.

Shared Whiteboards A multimedia concept. Shared whiteboards enable you to "mark-up" a screen using a mouse or stylus input device and have the results show on other screens, often communicating over long distance telephone lines. The concept is similar to a traditional whiteboard mark-up process where everyone has a different color marking pen to circle, write, or cross out items. The background board can be a window from the workstation such as a spreadsheet, image, or blank canvas, or it can be the entire workstation screen. The shared whiteboard can be used for either real-time or store-and-forward collaboration. In the store-and-forward scenario, the mark-ups can be implemented in a time-delayed fashion so everyone can follow the entire step-by-step process.

Shareware Imagine you write software. You've just written a great program. You now want to sell it. You have two choices. You can take advertisements, sell it to retailers, get distributors to carry it, hire salespeople, etc. In other words, go the commercial route. This is expensive and requires a major marketing / sales budget. The other choice is to go the Shareware route. This involves giving away your software on various bulletin boards, on many Web sites, in "shareware" direct mail catalogs. People download the software for free and try it. If they like it, they will send you money. They will do this because you offer them an instruction manual, a new version of the software that doesn't blast "unregistered" on the splash screen when you load the software, or an upgraded version of the software, with more features, or a chit that assuages your guilt at using unpaid-for software that someone worked real hard on.

Sharon and David Good friends, still. I hope.

SHARP Self Healing Alternate Route Protection. A system typically employing redundant cables (often fiber) that carry traffic between two separate local exchange carrier offices along divergent paths.

Shear A computer imaging term. A tool for distorting a selected area vertically or horizontally.

Sheath The outer jacket (usually metal or plastic) surrounding copper and fiber cables that prevents water damage to the cables inside.

Sheath Miles Let's say that you have two sheaths of fiber, each of which contains ten fibers and runs for one mile. That is one route mile (total distance of all fibers), two sheath miles (two sheaths running one mile), and twenty fiber miles (20 fibers running one mile).

Shelf Life The useful life of components when not in use — such as being stored on a shelf as spare parts or in a warehouse awaiting shipment. Batteries tend to have the shortest shelf life of most telecommunications components. Today, the shelf life is less a problem of shelf decay and more a problem of technological obsolescence.

Shelfware Software that is bought, then placed on the shelf, but never used. See also feedware, hookemware, hyperware, meatware, seedware, shovelware, smokeware slideware and vaporware.

Shell An outer layer of a program that provides the user interface, or the user's way of commanding the computer. Instead of presenting the user with a bland C prompt, i.e. C:> the shell presents a list of programs that the user can choose from, making it easier, allegedly, to figure out which program to run. The problem with shells is that they often take up precious memory. That memory might better be used in actually running a program faster, or more efficiently.

Shell Account An Internet term. A type of interface on a dial up connection in which you log in to the host computer and use a command shell to get to the Internet. Shell accounts are typically text-based-only interfaces controlled by host servers which normally don't allow for use of graphic Web browsers.

Shield A metallic layer consisting of type, braid, wire or sheath that surrounds insulated conductors in shielded cable. The shield may be the metallic sheath of the cable or the metallic layer inside a nonmetallic sheath. Shields reduce stray electrical fields and provide for safety of personnel. See SHIELD EFFECTIVENESS.

Shield Effectiveness The relative ability of a cable shield to screen our undesirable radiation. Frequently confused with the term shield percentage, which it is not.

Shielded Pair Two insulated wires in a cable wrapped with metallic braid or foil to prevent the wires acting as antennas and picking up external interference (e.g. a local TV station).

Shielded Twisted Pair A pair of insulated wires which are twisted together in a spiral manner. In addition, the pair is wrapped with metallic foil or braid, designed to insulate (i.e. shield) the pair from electromagnetic interference.

Shielding 1. The metal-backed mylar, plastic, teflon or PVC that protects a data-communications medium such as coaxial cable from Electromagnetic Interface (EMI) and Radio Frequency Interference (RFI).

2. The process by which electrical conductors are wrapped with metallic foil or braid to insulate them from interference and thus provide high quality transmission. Many devices can cause interference to cables (i.e. multiple conductors) carrying telecommunications conversations. Such things include high voltage AC power lines, machinery with motors, machines which make rays of some type (X-Ray systems, TV sets etc.). By wrapping conductors around the cable cores, these cables are less likely to be affected by these outside forces and the noise they create on telephone lines. Shielding will also lessen the chance that the information movement along the cable will interfere with signals on other, adjacent cables. The need for shielding stems from this phenomenon: If you send an electrical signal along one pair of cables, those cables will give off a small amount of electrical energy — called magnetic radiation. That radiation will cause electromagnetic interference with a cable close by. If you "shield" the pair carrying the electrical signal, you will cut down the susceptibility of those cables to interference from other cables. LANs should always be installed with the best quality shielded cable. They will run better with shielded cable. Never skimp on the quality of the cable you're installing for LANs. Most telephones don't require shielded cable unless the cable serving them is passing through some area of high electromagnetic interference.

Shift 1. The movement of data to either the right or the left of an existing position in a data field. 2. The code control function of converting the characters from upper to lower case, or vice versa.

Shift Button This button acts exactly like a Shift button on a typewriter or computer. It gives the key you're touching a second meaning — either a capital letter or a second set of speed dial buttons, etc.

Shift Character The control character which defines the shift function.

Shift Definitions A call center term. A template from which the program can create schedules during a scheduling run. Each shift definition is a record in a Scenario giving more or less precise instructions on shift length, time of day, breaks, and how extensively such schedules can be used in meeting staffing requirements.

Shift Register A register in which a clock pulse causes the stored data to move to the right or left one bit position. See ZERO STUFFING.

Shim A shim is a piece of software. The term is associated with calls to communications protocols such as TCP or IP for communications services. The shim inserts itself into the logical space between a program asking for service and the program, such as a TCP-conforming communications program, able to provide the service. The function of a shim is to intercept calls made by higher level programs, such as applications, to translate them, and to pass them off to some other piece off software -perhaps IPX. The program requesting the service is fooled by the shim into thinking it is receiving the service from the software it addressed. This term courtesy Frank Derfler. Thank you Derf.

Ship To Shore Telephone See MARINE TELEPHONES.

Shock A sudden stimulation of the nerve and convulsive contraction of the muscles caused by a discharge of electricity through the body. The severity depends on the amount and duration of the current and whether the path of the current is through a vital organ.

Shockwave A Web browser plugin which provides for Macromedia Director movies to be viewed on World Wide Web pages. Shockwave is a component of Macromedia's solution for interactive professionals who develop digital media for the World Wide Web. If you have created an interactive movie using Macromedia Director, you will need to compress the movie through a program called "Afterburner" before you can use it as Shockwave on a Web site. www.macromedia.com/Tools/Shockwave/Info/index.html

Shoe Hickey A mark made on your neck by someone standing on your shoulders. This definition contributed by Steve Hersee, founder of Copia International, a great fax server company. Steve does circus work and trains cheer leaders as a hobby.

Shoebox A shoebox is a housing device with a power supply to support external peripherals. When the user's main computer case is filled to capacity, an external device is needed to handle the overflow. Hence a shoebox.

Short A circuit impairment that exists when two conductors of the same pair, which normally make up an operating electrical circuit, touch or are connected.

Short Bus A high-speed common channel in the AT&T ISDN packet controller over which all messages between sending and receiving devices pass.

Short Circuit A near zero resistance connection between any two wires that disrupts transmission where two pairs are involved usually called a "cross." It disrupts transmission and may cause an excessive current flow. In AC electricity, a short circuit is an unintended connection between two supply conductors (i.e.: HOT and Neutral conductors.) A short circuit will usually cause high current flow and will operate the over current protection (fuse or breakers) to interrupt the circuit.

Short Event A carrier event that occurs when the activity duration is shorter than the ShortEventMaxTime (84 bits).

Short Haul Between a few hundred yards and 20 miles. Many people would argue with this definition.

Short Haul Modem A data set designed for use in communicating data up to distances of 25 miles over a dedicated unloaded copper pair. Many people would argue with this definition.

Short Message Service Center See SMSC.

Short Reach Short Reach refers to optical sections of approximately 2 km or less in length. The sections may be interoffice or intra office sections. A SONET term.

Short Tone DTMF See Short Tones

Short Tones First, we invented touchtone, also called

DTMF, Dual Tone Multi Frequency tones. You'd punch your number with tones, instead of dialing them. Then someone thought you could control telephone response gadgets, like voice mail, interactive voice response, etc. with touch tones. For these gadgets to work, they had to "hear" the tones you sent. No one really set standards as to the minimum length tone they would hear. But it was generally conceded that they were to be 120 milliseconds. So some manufacturers of telephone equipment started to make phone equipment that, if you pushed a touchtone button, the machine would only sent a touchtone of 120 millisecond duration. That was called a short tone. It wasn't very useful because the manufacturers quickly discovered that many pieces of equipment couldn't respond that quickly. And the manufacturers got complains that their customers couldn't call their voice mail, their bank, etc. As a result, some manufacturers of equipment brought out new hardware (replacing the old) to allow you to send "long tones," which are now defined as touchtones that last for as long as you hold down the button — just as it is (and has always been) on a normal single line, non-electronic, non-digital telephone. Isn't progress wonderful? See also DTMF for a much longer explanation of tone dialing.

Shoulder Surfing You're standing at a pay phone. You punch in your credit card numbers to make your long distance call. There's a fellow standing behind you. He's carefully watching what you're doing. He is memorizing the digits you have punched in. When you are through, he will write them down and sell them to someone else, who will use them to make fraudulent long distance phone calls. Our friend is indulging in a new "occupation." It's called "shoulder surfing."

Shovelware A term used to refer to the tendency of early CD-ROM disc producers to shovel anything they could to fill up their voluminous CD-ROM discs. They did this so they could tout all the great value they were offering in their CD-ROM discs. See also HOOKEMWARE, HYPERWARE, MEAT-WARE, SLIDEWARE, VAPORWARE and BUNDLE FODDER.

SHP Signaling Handoff Point, a type of equipment which acts as a gateway between two dissimilar Signaling System 7 networks, allowing information exchange between the two networks.

Shrink Wrap Software that requires no customization. So called because it comes in a package that is plastic shrink-wrapped. Of course, so does non-shrink wrap software, or software that has to be customized for your use. But this is a dictionary that explains what words mean. It doesn't explain the lack of logic behind the naming of those words.

Shrinkage Percentages A call center term. A group of Scenario budget assumptions that define the percentage of the time employees are scheduled to work but are not available to handle calls because of absence, breaks, vacation, non-productivity, training, and other activities.

SHT Short Hold Time.

SHTTP Secure Hypertext Transfer Protocol. An extension of HTTP for authentication and data encryption between a Web server and a Web browser.

Shunt 1. A conductor joining two points in an electrical circuit so as to form a parallel or alternative path through which a portion of the current may pass. For example, a shunt might be applied to a main circuit to connect a control circuit for purposes of regulating the amount of current passing through the main circuit.

2. A means by which traffic is switched or diverted from one network to another. For instance, a number of manufacturers are developing devices intended to be positioned logically between central offices (COs) and the connected local loops.

Such devices would identify data traffic destined for the Internet or an Intranet, most obviously by means of recognizing the dialed telephone number as being associated with a remote access server supporting IP (Internet Protocol) traffic. At that point, the traffic is routed over a packet network, rather than the circuit-switched Public Switched Telephone Network (PSTN). Thereby, IP traffic's significant contribution to congestion in the PSTN are mitigated. See also Circuit Switching, IP, Packet Switching and PSTN.

Shunt Circuit An arrangement of apparatus or circuits in which the total current is subdivided. Same as Parallel Circuit.

SI Shift In.

Si-Ge Silicon Geranium, a chip technology announced by IBM in the fall of 1998. Cheaper. Faster, etc. One use for this technology will be combining a cell phone with a GPS system, so you know always where you are.

SIA Securities Industries Association.

Siamese Pair A CATV (Community Antenna TeleVision) term. A siamese pair is a connection between a coaxial cable and a twisted pair cable in order that CATV provider can connect to devices such as telephones and PCs which do not have coax interfaces and, thereby, to provide POTS (Plain Old Telephone Service) and Internet access, as well as CATV access. A number of large CATV providers have upgraded their old analog, one-way, coax-based transmission systems in order to operate as CLECs (Competitive Local Exchange Carriers), as well as entertainment TV providers. See also CATV, CLEC, ILEC, and Sidecar.

SIBB Service Independent Building Blocks, a term coined by Bellcore and the ITU-T for the Intelligent Network. Creation of SIBBs will in theory make it easier for non-software specialists to create new services by mixing and matching SIBBs. SIBBs are like software objects — capsules of reusable software that can be combined together to create wondrous and complex new computer telephony services.

SID 1. System IDentification number. A five digit number that has been assigned to identify the particular cellular carrier from whom you are obtaining service. This number identifies your "home" system.

2. Security ID.

Side By Side Monitoring A call center term. The process whereby a supervisor or other qualified party listens in on the calls of an agent by sitting at their side. The supervisor is able to listen to both sides of the conversation, usually via double jacking, which is plugging a headset into the agent's phone and listening in on the line.

Side Circuit A metallic, single pair circuit arranged to derive a phantom circuit. The phantom circuit is derived by center tapping a repeating coil in each of two side circuits.

Side Hour Any hour that is not the Busy Hour. A telephone company term: An amount of time equal to one hour that is time consistent and adjacent to the CBH (Component Busy Hour). This one hour period must have average weekly usage equal to at least 90% of the CBH during the busy season and 80% of the CBH during the nonbusy season.

Sideband The frequencies on either side of the main frequency in a telecommunications signal. In the early days these sideband frequencies were not used because they were too "noisy," unreliable and were not needed. Now, technology has improved and frequencies are in short supply, more and more transmission vendors are making use of their sidebands, and thus substantially broadening the throughput of their existing transmission paths. Sideband technology has made great strides especially in through-the-air microwave transmission.

Sidecar An add-on module for a CATV (Community Antenna TeleVision) set-top converter box. A sidecar essentially is peripheral equipment for interactive, two-way access to the CATV provider's network. The sidecar allows you to gain access to an interactive program guide, to access the Internet through a built-in cable modem, and to access the PSTN (Public Switched Telephone Network). In order to provide a full range of capabilities, the CATV provider, of course, must have enhanced the cable system to support two-way transmission, packet-switched data access to the Internet, and circuit-switched access to the PSTN. A number of CATV providers have done so in order to position themselves as CLECs (Competitive Local Exchange Carriers), in competition against the ILECs (Incumbent LECs) in voice and data. In order to improve on the quality of transmission and to provide more bandwidth, many also have upgraded their coaxial cable transmission systems to hybrid systems, incorporating fiber optic transmission facilities in the high-capacity trunk segments of their network. See also CATV, CLEC, ILEC and Siamese Pair.

Sidetone A part of the design of a telephone handset which allows you to hear your own voice while speaking. The idea is to let you know that the telephone you're speaking on is working. Too much sidetone becomes an echo and is bad. Too little sidetone makes the channel unerringly quiet and people start to think it's busted.

SIDN Security Industry Digital Network.

SIF 1. Standard Image Format. See MPEG.
2. SONET Interoperability Forum. A voluntary industry group established to define and resolve issues of SONET implementation. SIF was formed by Southwestern Bell in 1991; Bellcore and other RBOCs soon joined. SIF now is open to membership of any interested party, including vendors, service providers and end users. SIF works under the umbrella of ATIS (Alliance for Telecommunications Industry Solutions). www.atis.org/atis/sif/index.html See also ATIS.
3. Signaling Information Fields. A SS7 term. See SIGNALING INFORMATION FIELDS.

SIG 1. Special Interest Group. A SIG is an ongoing discussion group held electronically via PCs. A SIG focuses on one area of interest. Members phone in with their PCs, read messages posted, contribute their wisdom, ask questions, etc. SIGs are ways people get up-to-date accurate information on a subject. SIGs are run on most BBS (Bulletin Board Systems). SIGs are to bulletin boards what on-line services call conferences or forums.
2. Special Interest Group. In this context, a SIG is a voluntary organization which is dedicated to the advancement of a particular technology, standard or technique. Examples include the ATM Forum, CDPD Forum, Frame Relay Forum and SMDS Interest Group.
3. SMDS Interest Group. A defunct consortium of vendors and consultants who were committed to advancing worldwide SMDS as an open, interoperable solution for high-performance data connectivity. On June 16, 1997, the Board of Trustees announced that the group was disbanded, turning over all responsibilities to regional organizations. The Board declared that its mission had been fulfilled.

Sig File Signature File. A file that automatically is appended to every e-mail message you send. The sig file commonly contains your name, title, company, telephone number, fax number, and return e-mail address. See also vCARD.

Sign-On To go through the process of beginning a working session between you, your data terminal or PC and a computer system.

Sign On/Sign Off The process of identifying oneself to a machine so as to gain access. In the case of an ACD system this process allows statistics to be kept for this person individually. It also allows for the movement of the person around the system while statistics are accumulated in one logical file.

Signal 1. An electrical wave used to convey information.
2. An alert.
3. An acoustic device (e.g. a bell) or a visual device (e.g. a lamp) which calls attention. To transmit an information signal or alerting signal.
4. Bellcore defines signal as a state that is applied to operate and control the component groups of a telecommunications circuit to cause it to perform its intended function. Generally speaking, says Bellcore, there are five basic categories of "signals" commonly used in the telecommunications network. Included are supervisory signals, information signals, address signals, control signals, and alerting signals.

Signal Computing Signal computing, as it has come to be, refers to the processing of analog signals for transmission over the worldwide telephone system. According to Analog Devices, the core of the Signal Computing concept is the ability to program an open, multi function chipset with easily upgraded software. The Signal Computing model gives non-traditional audio, video, speech, and communications technology providers opportunities to embrace PC channels. The term, signal computing, picked up steam in mid-1993 when Dialogic announced a new standard called SCSA (Signal Computing System Architecture). In Dialogic's words, SCSA is a comprehensive architecture that describes how both hardware and software building blocks work together. It focuses on "Signal Computing" devices, which refer to any devices that are required to transmit information over the telephone network. Information can be transmitted via data modems, fax, voice or even video. SCSA defines how all these devices work together. Signal computing systems combine three major elements for call processing. Network interfaces provide for the input and output of signals transmitted and switched in telecommunications networks. Digital signal processors and software algorithms transform the signals through low-level manipulation. Application programs provide computer control of the processed signals to bring value to the end user.

SCSA is the common set of standards that telecommunication system manufacturers and computing system manufacturers can use to create computer telephony systems. No single company today can create the total solution for all customers. SCSA represents the common ground between the two fields so that manufacturers from each area can safely develop products that will work with other manufacturers.

SCSA's coverage extends from low-level bus and hardware interfaces, like the inter-board switching bus that enables boards from different suppliers to work together, to high-level application programming and software interfaces, so that software designed to work with one set of hardware products, will work with different hardware as well. SCSA has been defined with input from leading computer and switch manufacturers, call processing suppliers, and technology developers. In many cases, SCSA has drawn on existing standards, like the T.611 fax standard endorsed by the European Computer Manufacturers Association, and in other cases SCSA has extended standards to make them more useful for call processing suppliers and users.

Signal Computing System Architecture SCSA is an open hardware and software architecture supported by over 240 companies for developing multi-technology computer telephony systems using standard interfaces. SCSA consists of two independent layers: The SCSA Hardware Model and the SCSA Telephony Application Objects (TAO) Framework. While designed for interoperability, these two layers may be implemented independently of one another. See SIGNAL COMPUTING.

Signal Conditioning The amplification and/or modification of electrical signals to make them more appropriate for transmission over a certain medium — cable, microwave, etc.

Signal Control Point SCP. Computers that hold databases in which customer-specific information used by the "advanced intelligent network" (AIN) to route calls is stored. The AIN refers to a specific architecture — typically promulgated and created by a local or long distance phone company — that provides core capabilities in which customer-specific information held in databases within the network is used to intelligently process calls. An example of an AIN service is an 800 service that routes calls based on where the calls are coming from. The routing information is stored in the SCP, which is typically tied into the Signaling System 7 network. See SCP. For a full explanation of the Advanced Intelligent Network, see AIN.

Signal Converter The equipment which changes the data signal into a form suitable for the transmission medium, or the reverse. The converter can also work with DC/AC current. The signal converter comprises a modulator and/or demodulator.

Signal generator A test oscillator that can be adjusted to provide a test signal at some desired frequency, voltage, modulation, and waveform.

Signal Ground SGD. In RS-232-C signaling, Signal Ground establishes a common reference level for the voltages of all other signals (such as RXD/TXD), except Frame Ground.

Signal Level The strength of a signal, generally expressed in either absolute units of voltage or power, or in units relative to the strength of the signal at its source.

Signal Processing Signal processing is a combination of computer telephony call control and media processing. Call control means moving telephone calls around — answering them, hanging up on them, transferring them, conferencing them, etc. — all the stuff you do on your office phone every day. Media processing means bringing computer power to bear on the media stream, the actual voice, video or data inside the phone call. You might save it to your hard disk. You might want to have your PC software listen for key words in what you saved, for example, the caller's phone number. You might even want your software to try transcribing the conversation into text you could use in your word processor.
See CALL CONTROL, DIGITAL SIGNAL PROCESSING, MEDIA PROCESSING and NATIVE SIGNAL PROCESSING.

Signal Processing Component An SCSA definition. An atomic bundle of signal processing functionality which can be allocated to a single group. It can be capable of supporting the functionality of one or more resources or a simple coder.

Signal Processing Element An SCSA definition. That part of a Signal Processing Component which is associated with a single Resource.

Signal Processing Platform A SCSA definition. This is a software component that supports a specific hardware package. This is typically an executable program and may control one or more instances of the vendor-specific hardware package. Each SPP may support several different types of signal processing functionality clustered into SPCs.

Signal Repeaters A signal repeater does nothing more than receive a signal and retransmit it. Repeaters are used where the original transmission is very weak, or the transmis-

sion is being sent over long distances.

Signal Strength Indicator A display on a cellular radio that lets you know before you call about the relative strength of the cellular transmitter in your immediate area. On most cellular radios, the signal strength indicator has five bars, with five the strongest. It's best to call when you have four or five. Three is marginal. Below three, forget it. Go elsewhere and try again.

Signal Switching Point SSP. See SSP. For a full explanation of the Advanced Intelligent Network, see AIN.

Signal To Noise Ratio 1. The ratio of the usable signal being transmitted to the noise or undesired signal. Usually expressed in decibels. This ratio is a measure of the quality of a transmission.
2. The amount of useful information to be found in a given ratio Internet Usenet newsgroup. Often used derogatorily, for example: "the signal-to-noise ratio in this newsgroup is pretty low."

Signal Transfer Point The packet switch in the Common Channel Interoffice Signaling (CCIS) system. The CCIS is a packet switched network operating at 4800 bits per second. CCIS replaces both SF (Single Frequency) and MF (Multi-frequency) by converting dialed digits to data messages. See SCP, STP and SIGNALING SYSTEM 7. For a full explanation of the Advanced Intelligent Network, see AIN.

Signals Signal being transmitted are information. They can be spoken words such as a telephone conversation, music, or even computer data. The simplest signal we can send is a sine wave. All sound waves are combinations of simpler sine waves at different amplitudes and frequencies to produce complicated wave forms. Similarly, computer data is a series of ones and zeros which electrically look like a square wave. Square waves, like all other types of waves, can be represented by a combination of sine waves with different frequencies and amplitudes. That a complex wave is the sum of simple sine waves was discovered by the French mathematician Jean Baptiste Joseph Fourier and is called Fourier's Theorem.

Signals Intelligence SIGINT. A federal government term. A category of intelligence information comprising, either individually or in combination, all communications intelligence, electronics intelligence, and foreign instrumentation signals intelligence, however transmitted.

Signaling In any telephone system — inside an office or across the country — some form of signaling mechanism is required to set up and tear down the calls. When you call from your office desk across the country to someone else's desk, many forms of different signaling are used. There's the signaling between your office desk phone and your office phone system. There's the signaling between your office phone system and your local telephone company central office. And there's the signaling between your local central office and the central office you're trying to reach across the country. All forms of these signaling may be different. Simple examples of signalings are ringing of your phone (someone is calling), dial tone (it's OK to dial), ringing (hopefully someone will answer), etc. Originally, telephone systems such as POTS (Plain Old Telephone Service) used in-band signaling to carry signals. With in-band signaling, signals such as DTMF tones (touchtones), are carried in the same circuit as the talk path. Newer signaling (i.e. most of it today) is carried as out-of-band signaling, which uses uses a separate data network. This is much more efficient. For example, voice circuits need not be allocated for calls that do not complete. Also, this approach allows additional quantities of information to be transferred to support advanced applications such as Caller ID, roaming in wireless systems, and 800 number routing. See Signaling System 7.

Signaling Connection Control Part SCCP. Part of the SS7 protocol that provides communication between signaling nodes by adding circuit and routing information to the signaling message. The ISDN-UP (Integrated Services Digital Network User Part) and TCAP (Transaction Capabilities Application Part) use the SCCP (Signaling Connection Control Part) and the MTP (Message Transfer Part) to transport information. Definition from Bellcore in reference to its concept of the Advanced Intelligent Network.

Signaling Information Fields SIF. In SS7 (Signaling System 7) signaling messages, the SIF is a variable length field which contains all the signaling information. Also included is any routing information in the Routing Label, which the network uses to properly connect the call. Such signaling information might include the Calling Party Number (CPN) subfield, along with the Presentation Indicator (PI). The SIF contains 2-272 octets of data. See also CALLING PARTY NUMBER, PRESENTATION INDICATOR, and SS7.

Signaling Link Selection Code SLS. The part of a routing label that identifies the SS7 signaling link on which the message should be sent.

Signaling Point A node in a SS7 signaling network that either originates and receives signaling messages, or transfers signaling messages from one signaling link to another, or both. SPs are located at each switch in a Signaling System 7 network. They interface the switch with the Signal Transfer Points (STPs). See SIGNAL TRANSFER POINTS and SIGNALING SYSTEM 7.

Signaling Point Code A binary code uniquely identifying a SS7 signaling point in a signaling network. This code is used, according to its position in the label, either as destination point code or as originating point code.

Signaling Point Interface SPOI. The demarcation point on the SS7 signaling link between a LEC network and a Wireless Services Provider (WSP) network. The point established the technical interface and can designate the test point and operational division of responsibility for the signaling.

Signaling System 7 SS7. All phone systems need signaling. According to James Harry Green, author of the Dow Jones-Irwin Handbook of Telecommunications, signals have three basic functions:

1. SUPERVISING. Monitoring the status of a line or circuit to see if it is busy, idle or requesting service. Supervision is a term derived from the job telephone operators perform in manually monitoring circuits on a switchboard. On switchboards, supervisory signals are shown by a lit lamp indicating a request for service on an incoming line or an on-hook condition of a switchboard cord circuit. In the network (i.e. the automated part of the network), supervisory signals are indicated by the voltage level on signaling leads, or the on-hook/off-hook status of signaling tones or bits.

2. ALERTING. Indicates the arrival of an incoming call. Alerting signals are bells, buzzers, whcofers, tones, strobes and lights.

3. ADDRESSING. Transmitting routing and destination signals over the network. Addressing signals are in the form of dial pulses, tone pulses or data pulses over loops, trunks and signaling networks.

In the old days, signaling was mostly MF (multi-frequency) and SF (single frequency) and is inband. This means that it goes along and occupies the same circuits as those which carry voice conversations. There are two problems with this. First, about 35% of all toll calls are not completed because the phone doesn't answer or is busy, or there are equipment problems along the way. The circuit time used in signaling is sub-

stantial, expensive and wasteful. Second, inband signaling is vulnerable to fraud. So the idea of out-of-band signaling came about. It got the name of Common Channel Interoffice Signaling (CCIS) because it used a communications network totally separate from the switched voice network. In North America, CCIS started out as an AT&T packet switched network operating at 4800 bits per second. Each of the packet switches in this network (they are no longer exclusively AT&T's) are called Signal Transfer Points — STPs. CCIS has the following advantages over SF/MF signaling:

Fraud is reduced. "Talk-off" is reduced. (Talk-off occurs when your voice contains enough 2600 Hz energy to activate the tone-detecting circuits in the central office.) Signaling is faster allowing circuits and conversations to be set up and torn down (i.e. disconnected) faster. Signals can be sent in both directions simultaneously and during voice conversation if necessary. Network management information is routed over the CCIS network. For example, when trunks fail, switching systems can be told with CCIS data messages to reroute traffic around problem areas.

The older CCIS signaling has been replaced with a newer out-of-band signaling system called ITU Signaling System 7. According to an AT&T technical paper delivered at the International Switching Symposium in Spring, 1987, ITU Signaling System 7 is being required by

telecommunications administrations worldwide (i.e. all the local country-owned telephone companies) for their networks. AT&T continued with the introduction of digital switches and transmission equipment with 56 Kbps and 64 Kbps transmission rates, the International Telegraph and Telephone Consultative Committee (ITU in French) in 1980 approved the ITU 7 recommendations optimized for digital networks. This new protocol uses destination routing, octet oriented fields, variable length messages and a maximum message length allowing for 256 bytes of data. Addition of flow control, connectionless services and Integrated Services Digital Network (ISDN) capabilities were approved by ITU in 1984. A major characteristic of ITU #7 is its layered functional structure. Its transport functions are divided into four levels, three of which constitute the Message Transfer Part (MTP). The fourth consists of a common Signaling Connection Control Part (SCCP). The SS7 protocol consists of four basic sub-protocols:

• Message Transfer Part (MTP), which provides functions for basic routing of signaling messages between signaling points.

• Signaling Connection Control Part (SCCP), which provides additional routing and management functions for transfer of messages other than call setup between signaling points.

• Integrated Services Digital Network User Part (ISUP), which provides for transfer of call setup signaling information between signaling points.

• Transaction Capabilities Application Part (TCAP), which provides for transfer of non-circuit related information between signaling points.

Signaling System 7 provides two major capabilities:

1. Fast call setup, via high-speed circuit-switched connections.

2. Transaction capabilities which deal with remote data base interactions. What this means in its simplest terms and in one simple application is that Signaling System 7 information can tell the called party who's calling and, more important, tell the called party's computer. A scenario: when you call a direct mail order business, Signaling System 7 will send a signal as to which phone is calling. The agent's CRT screen will pop the caller's name and perhaps the caller's most recent buying

information. The agent may answer the phone "Good morning, Mr. Newton. Did you enjoy the three khaki pants we sent you last week?..." Signaling System 7 will be an integral part of ISDN. It will enable us to extend full PBX and Centrex-based services like call forwarding, call waiting, call screening, call transfer, etc. outside the switch to the full international network. In effect, with Signaling System 7, the entire network will acquire the "smarts" of today's smartest electronic digital PBX. See also CAPTAIN CRUNCH, ISUP, MTP, SCCP, Signaling System 7 Software Layers and TCAP.

Signaling System 7 Software Layers MTP (Message Transfer Part) Layers 1 through 3: These layers provide complete lower level functionality at the Physical, Data Link and Network Level. They serve as a signaling transfer point, and support multiple congestion priority, message discrimination, distribution and routing.

ISUP (Integrated Services Digital Network User Part): This layer provides the network side protocol for the signaling functions required to support voice, data, text and video services in an Integrated Services Digital Network (ISDN). Specifically, ISUP supports the call control function for the control of analog or digital circuit switched network connections carrying voice or data traffic.

SCCP (Signaling Control Connection Part): This layer supports higher protocol layers (such as TCAP and IS-634) with an array of data transfer services including connection-less and connection orientated services. SCCP supports global title translation (routing based on directory number or application title rather than point codes), and ensures reliable data transfer independent of the underlying hardware.

TCAP (Transaction Capabilities Application Part): This layer provides the signaling function for communication with network databases. TCAP is an SS7 (and ISDN) application protocol which provides noncircuit transaction based information exchange between network entities. For example, TCAP enables transaction based service applications such as enhanced dial-800 which must exchange information between a pair of signaling nodes in an SS7 network to access remote databases, referred to as Service Control Points (SCPs). Important applications which use TCAP include:
• 800 number routing.
• Automated credit card calling which queries the Line Information Database (LIDB) for calling card validation.
* Advanced Intelligent Network call processing (referred to as AIN call "triggers").

TUP (Telephony User Part): This layer provides the telephone signaling function for national and international telephone call control. TUP is primarily used outside of the U.S. in Europe, China, parts of Asia and Latin America, and can control all the various types of national and international connections used worldwide.

Higher Level Application Parts: These layers are highly application specific, which each designed to serve a particular application type. Examples include:
• GSM MAP (Mobile Application Part): This layer provides inter-system connectivity between wireless systems, and was specifically developed as part of the GSM standard.
• IS-41: This layer provides similar functionality to GSM MAP. It provides inter-system connectivity between wireless systems and is typically deployed in North American wireless networks. For example, it is widely used to provide interconnection between the analog AMPS (Advanced Mobile Phone System) cellular systems in the U.S.
• IS-634: This layer provides the interface for Mobile

Switching Center (MSC) to base station communications for public 800 MHz cellular networks (i.e., for AMPS).
• INAP (Intelligent Network Application Part): This layer runs on top of TCAP and provides similar functionality to MAP but for a fixed network. Note that INAP is primarily a European standard, developed by ETSI. INAP is part of the CS1, CS2 IN Capability Set, the European equivalent to the AIN specification. While the AIN and CS specifications are similar and can be deployed using SS7 for functions such as call routing, there are some differences which the standards organizations are working to converge.
• 1129/1129+/1129A: These protocols provide a direct connection between the SCP and IP and are variants of the Bellcore 1129 and AIN 0.2 standards. In some cases, SS7 may not be required to implement the direct SCP-IP connection; in other instances, SS7 is used. Using SS7 as the underlying protocol allows any SCP to communicate directly with any IP in the SS7 network. When SS7 is used, the 1129 application layer typically runs on top of TCAP.

Thanks to Natural MicroSystems for an explanation of SS7 Software Layers. See also their white paper, "SS7 and Intelligent Networking Applications." It's on their web site, www.nmss.com

Signaling System 7 Standards SS7 standards are defined in the ITU-TS documents as noted below:
• Q.700-Q709
Messaging Transfer Part (MTP)
• Q.710
PBX Application
• Q.711-Q716
Signaling Connection Control Part (SCCP)
• Q.721-Q.725
Telephone User Part (TUP)
• Q.730
ISDN Supplementary Systems
• Q.741
Data User Part (DUP)
• Q.761-Q.766
ISDN User Part (ISUP)
• Q.771-Q.775
Transaction Capabilities Application Part (TCAP)
• Q.791-Q.795
Monitoring, Operations, and Maintenance
• Q.780-Q.783
Test Specifications

Signaling Transfer Point STP. A signaling point with the function of transferring signaling messages from one signaling link to another and considered exclusively from the viewpoint of the transferrer.

Significant Hour Any hour that influences the sizing of a trunk group.

Silence Management See SILENCE SUPPRESSION.

Silence Suppression A term used in voice compression for transmission whereby silence in the voice conversation is filled with other transmissions — e.g. data, video, imaging, etc. According to AT&T, the average voice conversation is 62% quiet and 38% not quiet (i.e. actual conversation). You can figure that for yourself: One person is speaking at a time. That's 50% of the circuit silent. The person who's speaking doesn't speak continuously. He pauses, takes a breath, thinks, etc. That's another 12%. Thus 62% in silence. A company like Micom in Simi Valley, CA makes a product called Marathon which uses the 62% silence between syllables, words and sentences to transmit data, fax and video. Micom tells me that the typical talk spurt sequence is 300 milliseconds. And it can use very small time (as short as 300 milliseconds) between talk spurts to stuff data, fax or video into and transmit.

Silent Intrusion Synonymous with Silent Participation.

Silent Mode Many Japanese fax machine sport a "silent mode." People in Japan who have fax machines in their houses use it so they are not awaken in the middle of the night by incoming faxes. In silent mode, your fax machine simply answers incoming calls, grabs incoming faxes into memory, but makes no noise whatsoever. In the morning when you wake up, you see a message on the machine that says "Faxes received." You hit a button and the machine starts to print the faxes it received. According to my friend, Emiko Magoshi, the shortfall of the fax machine at her house is that its memory is limited to accepting only 10 pages.

Silent Monitoring A call center term. The process whereby a supervisor or other qualified person listens in to the calls of an agent from a specially designated room or place away from the agent's work area. The agent may or may not be aware that his call is being monitored. Often the calling user may hear an opening message "This call may be monitored for quality assurance purposes.' Typically the supervisor can hear both sides of the conversation but cannot break in on the line and say anything.

Silent Participation A feature that allows a third party, such as an ACD agent supervisor, to join the call. The joining party can hear the entire conversation, but cannot be heard by either originating party. The feature, sometimes called Silent Intrusion or Silent Monitoring, may provide a tone to one or both devices to indicate that they are being monitored.

Sili Code See CLLI Code.

Silicon A dark gray, hard, crystalline solid. Next to oxygen, the second most abundant element in the earth's surface. Transistor chips are made from silicon, and it is the basic material for most integrated circuits and semiconductor devices. Silicon, a neutral element, is found primarily in raw form as sand. See SEMICONDUCTOR.

Silicon Alley An area of Manhattan roughly corresponding to Chelsea and SOHO.

Silicon Avalanche Diode A fast-acting surge protector which features a narrow voltage clamping range.

Silicon Alley The area roughly corresponding to Chelsea, SOHO and NOHO in downtown Manhattan, New York. So called because so much of the web and Internet content is written by companies located in this area.

Silicon Bayou In and around New Orleans, there are a growing number of a high-tech companies. They say their nw high-tech area is called Silicon Bayou. See the next few definitions.

Silicon Fen The vast flat and marshy lands in and around Cambridge, England, the site of the presitigious university. According to the New York Times, Saturday January 4, 1998, "the area has become home to more than 1,000 high-technology companies that employ at least 30,000 people and produce more than $3 billion a year in revenues."

Silicon Forest The area around Microsoft in the Redmond/Seattle area of Washington State is known as Silicon Forest because of all the high-tech companies that have sprouted. See also SILICON BAYOU, SILICON MUDFLATS, SILICON VALLEY, SILICORN VALLEY and SILLYWOOD.

Silicon Gulch Austin, Texas because of the concentration of high-tech companies that have located there. See the definitions above and below.

Silicon Mudflats The part of the San Francisco area that is on the Oakland side of the bay and sits between San Francisco

and Oakland. That area is home to a bunch of high-tech companies. See also SILICON BAYOU, SILICON FOREST, SILICON VALLEY, SILICORN VALLEY and SILLYWOOD.

Silicon Valley Silicon Valley in Santa Clara County, California, south of San Francisco Bay, is known for its microelectronics innovation. It is one of the two places (the other being Dallas, Texas) where the microchip was invented and produced. There are over 3,000 microelectronic hardware and software firms within a radius of about 30 miles. Silicon Valley owes its good fortune to:

• The fact that W. Shockley, one of the three inventors of the transistor, being a native of Palo Alto, went there in 1955 after he left Bell Laboratories, to start his own industrial research center;

• The proximity of technology universities, especially Stanford University;

• A very active financial market in California and the enterprising attitude of capitalists willing to enter into "joint ventures," often very profitable;

• The West Coast orientation to Japan.
The above information courtesy of Electronics, Computers and Telephone Switching by Robert J. Chapuis and Amos E Joel, Jr. See also SILICON BAYOU, SILICON FOREST, SILICON MUDFLATS, SILICORN VALLEY and SILLYWOOD.

Silicorn Valley A region in the state of Iowa that is fostering high-tech development. See also SILICON BAYOU, SILICON FOREST, SILICON MUDFLATS, SILICON VALLEY and SILLYWOOD.

Silly code See CLLI Code.

Sillywood The convergence of Silicon Valley and Hollywood.

SILS A Standard being formulated for Interoperating LAN Security.

Silver Satin Once upon a time, phones came with cords that matched the color of the phone. This proved expensive and confusing to workers, who were color blind, who had to match the cord with the phone. So a manufacturer (we think it was AT&T) decided that the time was ripe for all phones to have a cord that matched every decor and every phone. In actuality, the color they settled upon — silver satin — matches no decor man or woman has ever created and certainly no phone has ever been produced in the silver satin color. However, the world is now stuck with every phone coming with one standard, ugly line cord, called silver satin. See also TOUCHTONE.

SIM 1. Subscriber Identity Module. A 'smart' card installed or inserted into a mobile telephone containing all subscriber-related data. This facilitates a telephone call from any valid mobile telephone since the subscriber data is used to complete the call rather than the telephone internal serial number. See GSM and SIM Card for a better description.
2. Single Interface Module (SIM). An NEC term. - The minimum equipment configuration for the NEAX2400 IMS is the Single Interface Module. The SIM, says NEC, is a fully featured, totally non-blocking digital switch capable of supporting all NEAX2400 IMS feature package application programs. Typical stations installed range from only a few to 300 lines.

SIM Card Subscriber Identity Module, also called Smart Cards. Every mobile phone that conforms to the GSM (global system for mobile communications) and including many PCS (personal communications services) handsets has something called a SIM card. GSM phones won't work without these cards. Each SIM card contains a microchip that houses a microprocessor with eight kilobytes of memory. The card stores a mathematical algorithm that encrypts voice and data transmissions and makes it nearly impossible to listen in on calls. The SIM card also identifies the caller to the mobile network as being a legitimate caller. PCS and GSM cards come in two basic varieties. Some handsets work with slot-in SIM cards that are the size of conventional credit cards. Others come with smaller cards already built into the handset. GSM is the standard mobile service for Europe and Australia and most countries outside the U.S. Some PCS operators in the U.S. have adopted the GSM standard. See GSM.

SIMD Single Instruction Multiple Data is a type of parallel processing computer, which includes dozens of processors. Each processor runs the same instructions but on different data and one chip provides central coordination. See also MIMD and MMX.

SIMM Single In line Memory Module. Used on Macs and PCs. A form of chip packaging in which leads (pins) are arranged in a single row protruding from the chip. Connectors are attached to a stiff contact strip that permits a SIMM to be inserted into a slot like an expansion adapter. On PCs, SIMM-style RAM chips have virtually replaced the dual in-line package (DIP) chips, identifiable by two rows of protruding legs, that were popular in the 1980s. The most common SIMM is the 30-pin, 9-bit wide "1 by 9", which is the standard memory upgrade for PCs. See SIMM socket.

SIMM Socket The connector inside the Macintosh that holds the SIMM and connects it to the rest of the computer electronically.

Simple Gateway Control Protocol SGCP is a protocol and an architecture that Bellcore has created to address the concept of a network that would combine voice and data on a single packet switched Internet Protocol (IP) network. SGCP largely operates at low level — level 2 in the OSI. So it will probably be combined with higher level concepts such as IPDC (IP Device Control) and MGCP (Master Gateway Control Protocol). According to Bellcore, the philosophy behind the Simple Gateway Control Protocol (SGCP) is that the network is dumb, the "endpoint" is simple, and services are provided by intelligent Call Agents, and not in the trunking gateway (TGW) or in the residential gateway (RGW). SGCP is simple to use and easy to program, according to Bellcore, but is powerful enough to support basic telephony services and enhanced telephony services like call waiting, call transfer, and conferencing. The protocol is also flexible enough to support future IP telephony services. The SGCP, according to Bellcore, is a simple UDP-based protocol, instead of TCP-based, that allows support for managing endpoints and the connections between the endpoints. The SGCP is easily scalable, has support for failovers, and processes information in real-time. There is a low CPU requirement and low memory requirement for the endpoint because the SGCP is handling a small set of simple transactions at a time. This means that the endpoint can then be mass-produced cheaply. And there is no need for expensive and resource-hungry parsers. The system is text-based, has an extensible protocol, and the connection descriptions are based on SDP. When new services are introduced by the call agent, there is no need to change or update the endpoint. SGCP controls the endpoint by hooking transactions, relying on DTMF input, and by playing tones. For full details see www.bellcore.com/SGCP/SGCPWhitePaper.rtf

Simple Mail Transfer Protocol SMTP. The TCP/IP protocol governing electronic mail transmissions and receptions. An application-level protocol which runs over TCP/IP, supporting text-oriented e-mail between devices supporting

Message Handling Service (MHS). Multipurpose Internet Mail Extension (MIME) is a SMTP extension supporting compound mail, which is integrated mail, including perhaps e-mail, image, voice and video mail.

Simple MAPI Simple MAPI is a subset of MAPI that lets developers easily create "mail-aware" applications capable of exchanging messages and data files with other network clients.

Simple Network Management Protocol SMNP. The protocol governing network management and monitoring of network devices and their functions. SNMP came out of the TCP/IP environment.

Simplegram A Versit term. A syntax for encoding the vCard in a clear-text encoding. Simplegrams are nominally, based on the ASCII, 7-bit character set.

Simplex 1. Operating a channel in one direction only with no ability to operate in the other direction. For example one side of a telephone conversation is all that could be carried by a Simplex line.
2. One-sided printing.

Simplex Loop Powering In T-1, refers to the powering of the digital signal pairs that are simplex in nature (Tip or Ring) and that may have voltage applied to maintain the required 60 mA dc current to control repeater signal regeneration, loopbacks, keep alive signals and alarms.

SIMTEL20 The White Sands Missile Range used to maintain a giant collection of free and low-cost software of all kinds, which was "mirrored" to numerous other ftp sites on the Internet. In the fall of 1993, the Air Force decided it had better things to do than maintain a free software library and shut it down.

Simulcast To broadcast simultaneously on two different channels (paths).

Simultaneous Peripheral Operations On Line SPOOL. Temporarily storing programs, data or output on magnetic tape or in RAM for later output or execution. Many PCs use a small software spooling program which accepts material to be printed very quickly, stores it in a portion of RAM, then feeds that material to the printer at a speed the printer can handle. See SPOOLING.

Simulator A program in which a mathematical model represents an external system or process. For example, an engineer can simulate the forces that act on a building during an earthquake to find out how much damage is likely to be incurred.

Simulation A technique, often involving a computer, to guess the outcome of various events in the future. Where multitudes of complex events interact, simulation may well be the only way to deal with a given problem. Simulation is often used in traffic engineering instead of or in addition to the proven formula. Many people believe that simulation should NEVER be used when standard, proven formulas (such as Poisson, Erlang B and Erlang C) are appropriate. The difficulty with simulation is actually finding out what rules to use and then programming them correctly. Once this is done, it takes time, even on a fast computer, to run thousands of simulations to get a stable statistical estimate: a single run of simulation, like a single roll of the dice, is worse than useless. Simulators were built into hardware to predict the overall behavior of AT&T's No. 5 Crossbar switch when it was developed in the late 1940s; 1ESS behavior was simulated with software in the 1960s. These were major efforts involving many people and several years, but they dealt with problems far beyond the capabilities of standard traffic equations.

Singing An undesirable whistle or howl on a transmission circuit. Singing is usually caused by feedback, excessive gain, or unbalance of hybrid coils, or by some combination. Singing is the same effect observed when you increase the volume on a public address system until the system squeals or "sings."

Singing Return Loss The loss at which a circuit oscillates or sings at the extreme low and high ends of the voice band.

Single Address Message A message which is to be transmitted to only one specific terminal, as opposed to a broadcast or group message.

Single Attached Concentrator SAC. An FDDI (or CDDI) concentrator that connects to the network by being cascaded from an M (master) port of another FDDI(or CDDI) concentrator.

Single Attachment Concentrator SAC. A concentrator that offers one S port for connection to one ring of the FDDI network and multiple M ports for attachment of devices such as workstations. A SAC provides less reliability than does a Dual Attachment Concentrator (DAC), although at lower cost, as it connects to only one, rather than, to both fiber rings of the FDDI dual counter-rotating ring architecture. See also Single Attachment Station, Dual Attachment Concentrator, Dual Attachment Station and FDDI.

Single Attachment Station SAS. A device such as a workstation which connects directly to only one of the FDDI dual counter-rotating rings, rather than attaching to both as does a Dual Attachment Station or gaining ring access through a Single or Dual Attachment Concentrator (DAC). See also Single Attachment Concentrator, Dual Attachment Concentrator, Dual Attachment Station and FDDI.

Single Channel Broadband A local network scheme in which the entire spectrum of the cable is devoted to a single transmission path; frequency-division multiplexing is not used. Also known as carrierband.

Single Day Assignment A call center term. A method of automatic assignment (of employees to Master File schedules) that operates only on schedules of exactly one workday in length. Unlike multi-day assignment (which operates on schedules of any length), this makes it possible for each assigned employee to be scheduled for different hours each day.

Single Digit Dialing Provides for single-digit dialing to reach a preselected phone or group of phones.

Single Ended Terminal Device A device which terminates only one line at a given time.

Single Fiber Cable A plastic-coated fiber surrounded by an extruded layer of polyvinyl chloride, encased in a synthetic strengthening material and enclosed in an outer polyvinyl chloride sheath.

Single Frequency Signaling SF Signaling. The use of one tone — typically 2600 Hz — to indicate if the phone line is busy or idle (supervision) and to convey dial pulse signals from one end of a trunk or line to the other, using the presence or absence of a single specified frequency. The conversion into tones, or vice versa, is done by SF signal units. See also FREQUENCY, INBAND SIGNALING, SIGNAL.

Single In Line Memory Modules Basically memory packaged so it can be slipped into a PC or laptop PC much easier than present methods of installing memory — which typically consist of pushing memory chips with legs into printed circuit boards. The problem with memory with legs is that you're likely to bend one of the legs and thus have the installation go awry.

Single Line Instrument A telephone set normally used to access only one line. However when used with advanced telephone systems, additional lines can be accessed by dialing specific codes rather than by depressing keys.

Single Master Domain A relationship in a domain or amongst domains in which trusts are set up so that management control is centralized in one domain. See Trust Relationship.

Single Mode Fiber Single-mode fiber is a fiber that allows only a single-mode of light to propagate. This eliminates the main limitation to bandwidth, modal dispersion. However, the small core of a single-mode fiber makes coupling light into the fiber more difficult, and thus lasers must be used. The main limitation to the bandwidth of a single-mode fiber is chromatic dispersion. Laser sources must also be used to attain high bandwidth, because LEDs emit a large range of frequencies, and thus chromatic dispersion becomes significant. The current bandwidth of the fastest commercial systems (as of 1995) is approximately five gigabits/second, which can be transmitted for approximately 100-200 kilometers before a repeater or fiber-optic amplifier is required.

Single Number Service 1. Allows callers to dial a single number to reach a company with multiple local stores. For instance, a pizza chain might advertise a single number. The approximate physical location of the caller can be determined by the network by virtue of the originating number. Based on that data, the location can be compared to a network database defining the location of the closest pizza parlor. The caller, therefore need not sort through the phone book for this information, which can be especially difficult without a knowledge of the area. Additionally, all calls are routed automatically and without error to the outlet in closest proximity, ensuring that the pizza arrives quickly, as well as hot and fresh.
2. An optional feature for 800 IN-WATS Services which allows a subscriber who has or wants to have both intrastate and interstate 800 service to use the same 1-800 number for both services. If you'd like to buy another copy of this dictionary, call 1-800-LIBRARY. That phone number will be answered at our office in New York City. It will work for calls from inside and from outside New York State.

Single Party Revertive Ringing A feature from a Northern Telecom central office. Single-Party Revertive Ringing provides enhanced flexibility to residential subscribers' existing telephone service by turning extension phones into intercoms. This feature creates a type of home intercom system with which subscribers can reach people in other rooms in their home and/or on a remote part of their property (i.e., barn, garage, greenhouse, etc.). To initiate a home intercom call, the subscriber dials his own number and hangs up the phone. All extension phone rings. The intended party picks up the extension and is connected.

Single Protocol Router A communications device using the same mix of protocols which is designed to make decisions about which of several paths a packet of information will take. The packets are routed according to address information contained within the packet.

Single Sideband SSB. An Amplitude Modulation (AM) technique for encoding analog or digital data using either analog or digital transmission. SSB suppresses one sideband of the carrier frequency at the source. As only one sideband is used and as the carrier signal is suppressed (i.e., carries no information), less power is used and less bandwidth (one-half) is required than is the case with DSBSC (Double SideBand Suppressed Carrier). Carrier synchronization is lost. See also Amplitude Modulation, DSBSC, DSBTC, and VSB.

Single Sideband Transmission A system of transmission which suppresses one side-band of the carrier frequency at the source. Also applied to receiving systems designed to reproduce such transmissions.

Single Slot 1. Current standard for coin phone construction that uses one slot for the deposit of all acceptable coins.
2. Cards that fit into one slot inside a Wintel (Windows/Intel) PC are called single slot cards. The significance of calling them single slot is that in the old days individual cards couldn't do very much. To build something — like an interactive voice response system — you often needed several cards. But with time, components have shrunk and cards have gotten more capabilities. Complete computer telephony systems can now be built with one single slot PC card. Thus the significance of this definition.

Single Use Batteries Batteries that aren't rechargeable — in other words, the Duracell and Eveready batteries at your local supermarket.

Single Wire Line A transmission path which uses a single conductor and a ground return to complete a circuit. Used a lot in rural areas.

Sink That part of a communications system which receives information.

SIO Serial Input/Output. The electronic methodology used in serial data transmission.

SIP 1. Single Inline Module with pins.
2. An SMDS term meaning SMDS Interface Protocol.
3. Simple Internet Protocol.
4. Session Initiation Protocol

SIP L2PDU-SMDS Standard Interface Protocol Layer 2 protocol data unit. The 53-octet unit of information processed by the second layer of the SIP.

SIP L3PDU-SMDS An SMDS term. Standard Interface Protocol layer 3 protocol data unit. A variable-length (up to 9,220 octets long) unit of information processed by the third layer of the SIP.

SIP SMDS Interface Protocol An SMDS term. The protocol defined at the interface between the SMDS network and the end user.

SIPP 1. Simple Internet Protocol Plus. One of the 3-IPng candidates.
2. SMDS Interface Protocol: Protocol where layer 2 is based on ATM, AAL and DQDB. Layer 1 is DS1 and DS3.

SIR 1. Speaker Independent Recognition. See SPEAKER INDEPENDENT RECOGNITION.
2. An SMDS term meaning Sustained Information Rate. Determined at time of subscription, the SIR defines the long term average throughout an SMDS access line can carry. SIR is enforced through a Credit Manager resident of the SMDS network switches, and the Access Class-4 Mbps, 10 Mbps, 16 Mbps, 25 Mbps or 34 Mbps- the subscriber selects.

SIT Tones 1. Standard Information Tones. These are tones sent out by a central office to a pay phone to indicate that the dialed call has been answered by the distant phone, etc. 2. Special Information Tones. These are tones for identifying network provided announcements. Here's Bellcore's explanation: Automated detection devices cannot distinguish recorded voice from live voice answer unless a machine-detectable signal is included with the recorded announcement. The ITU, which specifies signals that may be applied to international circuits, has defined Special Information Tones for identifying network provided announcement. The SIT used to precede machine-generated announcements also alerts the calling customer that a machine-generated announcement follows. Since SIT consists of a sequence of three precisely defined tones, SIT can be machine-detected, and therefore machine-generated announcements preceded by a SIT can be classified. At least four SIT encodings

have been defined: Vacant Code (VC), Intercept (IC), Reorder (RO) and No Circuit (NC). With the exception of some small stored Program Control Systems (SPCSs) and some customer negotiated announcements, Bell operating companies in North America now precede appropriate announcements with encoded SITs to detect and classify announcements.

SITA Society of International Aeronautical Telecommunications. The international data communications network used by many airlines.

Sitcom Single Income, Two Children, Oppressive Mortgage.

Site Controller An industrial grade PC located at MCI terminals and junctions. It provides the operator local visibility into alarm and performance information from the Extended Superframe Monitoring Unit and 1/0 DXC, as well as enabling the operator to interact with both devices.

Site License Companies that buy software for multiple computers typically buy one copy of the program and a license to reproduce it up to a certain number of times. This is called a site license, though it may apply to its use throughout an organization. Site licenses vary. Some require that a copy be bought for each potential user — the only purpose being to indicate the volume discount and keep tabs. Others allow for a copy to be placed on a network server but limit the number of users who can gain simultaneous access. This is called a concurrent site license. And many network administrators prefer this concurrent license, since it gives them greater control. For example, if the software is customized, it need be customized only once.

Sitekeeper See IPTC.

SiteRP SiteRP stands for Sprint Interface to External Routing Processor. It is a service which Sprint offers to enable customer premise equipment (e.g. automatic call distributors) to interact with its long distance network to give that network information on how to route incoming 800 calls. With SiteRP, a call processing query is sent to the Sprint SCP (Signal Control Point). The SCP is instructed to check a user database kept on the customer premises for instructions on how to handle the call. Once that call handling instructions have been extracted from the user database, the customer prem ses equipment sends a message back to the Sprint SCP, which, in turn, uses that information to decide which call processing instructions it will pass back to the network switch. SiteRP allows such services as real-time switching of inbound calls, to allow for load balancing among agents or for specific calls to go to specific agents, etc. There are a million reasons why inbound calls from different places and different numbers should go to different agents. To use SiteRP, users must directly interconnect their SiteRP processors, which can be a mainframe or a PC (or in between) to each of Sprint's five SCPs via 56 Kbps dedicated links. See also ACD, AIN, SCP, SSP and STP.

SIVR Speaker Independent Voice Recognition. See SPEAKER INDEPENDENT VOICE RECOGNITION.

Six Digit Translation We have a long distance number 212-691-8215. The ability of a switching system to do six digit translation means that it can "look at" 212-691 and figure how to route the phone call. The criterion of choosing which way to send the call is, most often, the least expensive way. Six digit translation is often an integral part of Least Cost Routing programs within the phone system which tell the calls to go over the lines perceived by the user to be the least cost way of getting the call from point A to point B. There are typically two types of "least cost routing" translation — that which examines the first three digits of the phone number (i.e. just the area code) and the first six digits of the phone number (i.e. the area code and the three digits of the local central office). Six digit translation is preferred because it allows you more flexibility in routing, particularly to big area codes, like 213 in LA, where there are long distance calls within the area code. See also LEAST COST ROUTING and ALTERNATE ROUTING.

Sizing A telephone company term. The Network Switching Engineering activity of determining the types and quantities of equipment needed for relief. Also see Relief.

Skew 1. The deviation from synchronization of two or more signals.

2. A computer imaging term. A tool that slants a selected area in any direction.

3. In parallel transmission, the difference in arrival time of bits transmitted at the same time.

4. For data recorded on multichannel magnetic tape, the difference in time of reading bits recorded in a single line.

5. In facsimile systems, the angular deviation for the received frame from rectangularity due to asynchronism between scanner and recorder. Skew is expressed numerically as the tangent of the angle of deviation.

See also Skew Ray and Skewed Distribution.

Skew Ray In a multimode optical fiber, any bound ray that in propagating does not intersect the fiber axis (in contrast with a meridional ray). In a straight, ideal fiber, a skew ray traverses a helical path along the fiber, not crossing the fiber axis. A skew ray is not confined to the meridian plane.

Skewed Distribution The mean and standard deviation describe many distributions quite well. However, a few distributions are markedly lopsided or skewed. When it is necessary to specify a distribution more precisely, the coefficient of skewness may be computed. It is sometimes expressed as the mean minus the mode divided by the standard deviation. That is, it is the spread between the mean and the mode as related to the spread or scatter of the total distribution.

Skills-Based Routing A call center term for routing incoming calls based on the type of service requested, assuring that calls go to agents with the skills to provide the highest quality of service to the calling customer. In other words, someone calling about a broken refrigerator should be directed to a refrigerator expert, not a vacuum cleaner expert. Skills-Based Routing takes advantage of the routing capabilities of the automatic call distributor, in consideration of the unique skills of individual agents or agent groups and the requirements of individual callers. In this manner, privileged customers with special needs can be afforded special treatment. For example, a Platinum credit card holder with a past due account balance and who prefers to conduct business in Cantonese can be directed to an agent skilled in collections involving privileged customers and who speaks fluent Cantonese. The routing process may be accomplished on the basis of a client profile stored in a database on an adjunct computer systems linked to the ACD. Prior to completing the call, the database would be queried, with the query process being initiated on the basis of the caller's touchtone entry of his account number or on the basis of Caller ID. See also SOURCE/DESTINATION ROUTING, CALENDAR ROUTING, and END-OF-SHIFT ROUTING.

Skin Surface layer in a sandwich structure.

Skin Effect The tendency of a current to pass through the outer portion rather than through the center of a conductor.

Skip Key Use "skip keys" wherever possible. Skip keys are single key-presses that take the caller to the most commonly sought places, including information message boxes. For example, a skip key is Dial 1 for Books. Dial 3 for subscriptions. Dial 4 for our fax number. Ron Acher invented this term.

Or thinks he did.

Skip Zone a ring-shaped region within the transmission range wherein signals from a transmitter are not received. It is the area between the farthest points reached by the ground wave and nearest points at which reflected sky waves come back to earth.

Skunkworks Term for usually-secret high-pressure/high-tech research group in a company or government, often populated by people who don't see much sunlight or use much soap. Hence the name, skunkworks. Original usage was from L'il Abner.

Sky Station Sky Station International Inc. (Washington, D.C.) intends to deliver wireless broadband services through solar-powered communications platforms measuring approximately 200 feet in both length and width. The platforms will be held aloft by tethered balloons floating at altitudes of about 14 miles. Each communications platform will serve an area of about 465,000 square miles, providing up to 150,000 channels of 64 Kbps; the theoretical maximum, according to Sky Station, is 650,000 channels. As many as 250 of the airships are planned, with the total cost estimated at $7.5 billion. Subject to approvals from the FCC and the ITU-R, the first test is scheduled for October 1997. Commercial service is planned for the New York City area in 1999, with coverage intended for 95% of the world's population by 2005. Company officials include former U.S. Secretary of State and NATO commander Alexander M. Haig.

SLA See SERVICE LEVEL AGREEMENT.

Slab on Grade Concrete floor placed directly on soil, without basement or crawlspace.

Slamming The practice of switching a telephone customer's long distance supplier without obtaining permission from the customer. A long distance company might do this to get itself some easy revenues. After a bitter court battle, AT&T and MCI settled on some proposed standards to allow a long distance to switch a customer's service under the following circumstances:

1. The customer initiates the switch by calling either their local telephone company, their current long distance company or the long distance company they want to switch to;

2. Someone outside the long distance company's sales department verifies the request to switch; or

3. The customer submits written authorization to the company.

Slave A device which operates under the control of a master — another device or system. Slave switching systems are common in rural areas. The master central office might be in town A. Twenty miles away there's a smaller town. It makes no economic sense to serve those subscribers each on single local dedicated loops from Town A. Best solution: place a "slave" central office in that distant town and drive its software, diagnostics and changes from the main central office in Town A. See also REMOTE CONCENTRATOR.

SLC Subscriber Line Charge. A charge on the monthly bill of a phone subscriber in the United States, which produces revenues for the local telephone company. The money collected from the subscriber line charge is used to compensate the local telephone company for a part of the cost of the installation and maintenance of the telephone wire, poles and other facilities that link your home to the telephone network. These wires, poles, and other facilities are referred to as the "local loop." The SLC also is known variously as the Access Charge, CALC, and EUCL. See also Access Charge.

SLC-96 Pronounced "Slick 96." A short haul multiplexing device which enables up to 96 analog telephone customers to be served on three pairs of wires. SLC stands for Subscriber

Loop Carrier. See SLCC.

SLCC Abbreviation for Subscriber Line (or Loop) Carrier Circuit, and pronounced "slick." It's a system that allows one pair of wires, that would normally provide one phone line, to carry multiple conversations. Various models are available, with capacity ranging from 2 to 96 lines. A SLCC is used between phone company central offices and areas where there are too many customers for the cable that is in place. It's much less expensive to install SLCCs than new cable, but the SLCC provides lower-than-normal line voltage, which may cause some phones to malfunction.

SLED Single Large Expensive Disk. The opposite of RAID (Redundant Array of Inexpensive Disks). See RAID and REDUNDANT ARRAY OF INEXPENSIVE DISKS.

SLEE Service Logic Execution Environment. An AIN term. See AIN.

Sleep Mode 1. A means of increasing the battery life of portable computers. When you are using your laptop and are running on battery power, the computer will "go to sleep," if you are not actively using it for some period of time. Effectively, the system "times out." When you move the mouse or touch a key, the system springs back into life. Some makers of PCMCIA modem cards have included something called "sleep mode" into their modems. It puts the modems into a power-saving mode when you're not communicating using your modem. In actual fact, if you really want to save battery life, the best tip is to simply remove the modem temporarily.
2. A feature of digital cellular phones, allowing the phone to remain active, but not "on," when you are not engaged in a call. The phone sleeps most of the time, awakening every few milliseconds to check for incoming calls. Sleep Mode extends battery standby time by 300% or so.

Sleep Tight In Shakespeare's time, mattresses were secured on bed frames by ropes. When you pulled on the ropes the mattress tightened, making the bed firmer to sleep on. That's where the phrase, "good night, sleep tight" came from.

Sleeves Short lengths of conduit, usually made from rigid metal pipes, used to protect cables entering a premises through a building wall or running through concrete floors between vertically aligned riser closets. Sleeves also provide for easy pulling of cable.

SLIC Subscriber's Line Interface Circuit, a device that interfaces the SLC series of Lucent's products of local subscriber pair gain or multiplexing devices. See SLC-96.

Slideware Slideware is hardware or software whose reason for existing (eventually) has been explained in 35-mm slides, foils, charts and/or PC presentation programs. Slide is best described as "virtual vaporware." Vaporware is software which has been announced, perhaps even demonstrated, but not delivered to commercial customers. Hyperware is hardware which has been announced but has not yet been delivered. Slideware is less real than vaporware or hyperware. Classic slideware is the ISDN deployment strategies of many of the telephone companies in the United States. See also HOOKEMWARE, HYPERWARE, MEATWARE, PODIUMWARE, SHOVELWARE, and VAPORWARE.

SLIP Serial Line Internet Protocol. An Internet protocol which is used to run IP over serial lines such as telephone circuits. IP is the Internet Protocol. It is the most important of the protocols on which the Internet is based. It allows a packet to traverse multiple networks on the way to its final destination. See INTERNET, SERIAL LINE INTERNET PROTOCOL and PPP.

Slip Sleeve An oversized conduit that moves easily along an inner conduit and covers a gap or missing part of the smaller conduit.

Slip/Slip Rate The loss (or rate of loss) of a data bit on a T-1 link due to a frame misalignment between the timing at a transmit node and timing at a receive node.

Slip-Controlled The occurrence at the receiving terminal of a replication or deletion of the information bits in a frame.

Slip-Uncontrolled The loss or gain of a digit position of a set of consecutive digit positions in a digital signal resulting from an aberration of the timing processes associated with the transmission or switching of the digital signal. The magnitude or the instant of the loss or gain is not controlled.

Slist Connected to a NetWare network? Type Slist in DOS on your PC and you will get a list of all the NetWare servers you are connected to.

SLM System Load Module.

Sloppy Clicker Someone who always manages to move Windows icons whenever he or she clicks on them. Definition courtesy Wired Magazine.

Slot Sizes PCMCIA cards come now in three standard physical slot and card sizes: Type I: 3.3 mm thick, mostly for memory cards. Type II: 5 mm thick, fits most of the current cards on the market, including communications and networking cards. Type III: 10.5 mm, used for cards that have rotating hard disks in them. Toshiba, in its T4600, came out with a fourth size that is 16 mm thick. There is more to compatibility than just fit. The size of the slot is independent of the version of firmware that supports it. Some cards may only work with certain operating systems, BIOS, and drivers. Apple's Newton has a Type II slot but only supports cards that have been designed specifically for it. See CARD SERVICES, PCMCIA STANDARDS, SOCKET SERVICES and SLOT SIZES.

Slots Openings, typically rectangular, in the floor of vertically aligned riser closets that enable cable to pass through from floor to floor. A slot accommodates more cables than an individual sleeve.

Slotted A Medium Access Control (MAC) protocol is slotted of attempts to transmit can only be made at times that are synchronized between contending devices. A cellular radio term.

Slotted Aloha An access control technique for multiple-access transmission media. The technique is the same as ALOHA, except that packets must be transmitted in well-defined time slots.

Slotted Ring A LAN architecture in which a constant number of fixed-length slots (packets) circulate continuously around the ring. A full/empty indicator within the slot header indicates when a workstation or PC attached to the LAN may place information into the slot. Think of a slotted ring LAN as an empty train that constantly travels in a circle, being filled and emptied at different terminals (workstations).

Slotting The process of assigning a circuit to available channel capacity across the network during the circuit design process. When a circuit is slotted, it has an assigned path from one end of the network to the other.

Slow Switching Channel A sequencing GPS (Global Positioning System) receiver channel that switches too slowly to allow the continuous recovery of the data message.

SLP Service Location Protocol. Described in the IETF's RFC 2165, SLP is a TCP/IP-based protocol which allows a computer to automatically discover and make use of network resources available over a corporate Intranet. Such resources might include printers, e-mail servers, Web servers and fax servers. It works like this. The client workstation makes use of a "user agent" to seek out the appropriate server based on its attributes, as "advertised" by "server agents." "Directory agents" serve an intermediary function between the user agents and server agents, aggregating advertisements in order to minimize the amount of time spent searching for the services across all networks and subnetworks in the enterprise Intranet.

SLR Send Loudness Rating.

SLSA Single Line Switching Apparatus.

SLT Single-Line Telephone, as opposed to a phone that has buttons to select from several lines.

SM 1. Switch module.

2. Service Module. An ADSL term for a device which performs terminal adaption functions for access to the ADSL network. Examples of terminal such devices include set-top boxes for TV sets, PC interfaces, and routers for LAN access. See ADSL.

SMA Subscriber Carrier Module-100A; same as SCM-100A.

SMA 905/906 (Subminiature type A). A former microwave connector modified by Amphenol to become the "standard" fiber optic connector. The 905 version is a straight ferrule design, where as the 906 is a stepped ferrule design and uses a plastic sleeve for alignment.

Small Computer System Interface SCSI. Pronounced scuzzy, SCSI is a bus that allows computers to communicate with any peripheral device that carries embedded intelligence. The standard is covered by the American National Standards Institute (ANSI) and has developed into from SCSI-1 into SCSI-2 and, now SCSI-3. Different types of device can be connected in a daisy chain via a 50-pin cable (68-pin for SCSI-3), both ends of which must be terminated. All signals on the cable are common to all devices. To avoid bus contention, each device (up to a maximum of seven) connected to the bus is given a unique "SCSI" address with each address being allocated a degree of priority. The SCSI-1 bus carries 8-bit data, 1-bit parity and 9-bit control lines, to provide a maximum synchronous data transfer rate of 5 Mbyte/s. The maximum bus length is dependent upon the type of bus driver/receiver used. For single-ended devices, the length is restricted to 6 meters, but you wouldn't want to go that far. A desktop is about the maximum distance for an un-amplified SCSI-1 bus. The 6 meters can be extended to 25 meters through the use of differential driver/receivers. SCSI-2 has been extended to increase the data rate to 10 Mbyte/switch via either 16 or 32-bit processors. The maximum data rate is thus nearer to 40 Mbyte/s. The protocol has also been expanded to include tagged commands. This allows the execution of queued control commands according to a prescribed sequence. For a greater explanation, see SCSI, SCSI-2 and SCSI TRANSFER RATE.

Small Vocabulary A voice recognition term. Vocabularies containing fewer than 50 words.

Smart Agent A software tool that acts like you. It is your agent and it acts intelligent on your behalf. For example, you might ask assign it a series of tasks to do on your behalf every day, or even throughout the day. For example, "Which orders that I sold today have been shipped today?" See INTELLIGENT AGENT and SMART MESSAGING.

Smart Battery A rechargeable battery equipped with a microchip that collects and communicates present, calculated, and predicted battery information to a notebook computer or cellular phone.

Smart Battery Data SBD. The information accessible across the System Management Bus between the smart battery and the device.

Smart Building Also referred to as the "intelligent" building. A centrally managed structure that offers advanced technology and perhaps shared tenant services. Central to the early ideas of a smart building was the concept of one gigantic phone system shared by all the tenants. The idea was that you needed a big switch to get lots of features and all the benefits of cheap long distance. That's no longer true. Small switches offer much the same benefits. Now buildings are often smart, but frequently they're smart, using those smarts to save energy and insure safety and less to share telephone service among the tenants, who have shown a tendency to fend for themselves.

Smart Card A credit card-sized card which contains electronics, including a microprocessor, memory and a battery. The card can be used to store the entire repair and maintenance history on the family automobile or the health history of a member of the family. Since smart cards are tamper-resistant hardware devices that store your private keys and other sensitive information, they can be used for security applications. There is no direct contact between the "smart card" and the device which reads it. This avoids the problem of wear that afflicts traditional credit cards — both their embossed numbers on the front and their magnetic strip on the back. According to the Smart Card Industry Association, a smart card is any card with a microprocessor. Non-smart chip cards are simple memory cards and hardwired logic cards. As the smart card has developed, we see new and specialized versions of it. Memory cards store information or values, such as debit or credit information. Processor cards perform calculations, complex processing and security applications. Contact cards read information when a card is inserted into a reader. Contactless cards wirelessly read information. Combicards uses both a card readeer and a wireless device.

Smart Hunt Groups Picture a call center with people answering the phone from customers. The people answering the phone have different skills, know different things about the company's products, and are thus better able to answer some questions better than others. Now picture that the phone system the people is using is "smart" and somewhere in it and in the databases attached to it, it knows who of the people (also called agents) is good at what. What skills they have. What they know about which products, etc. Then, by asking questions of incoming callers through a combination of an attached interactive voice response system, the incoming caller's phone number and the number the caller dialed, it is able to route the call to the exact correct person. If that correct person is not available, it routes the caller to another person. And to another, etc. There may be many smart hunt groups, each consisting of the same people. If you think of a call center with several hundred people you can easily see it's possible to have potentially thousands of "smart hunt groups." Just depends on how complex your product line is and how sophisticated you want to get. The term was coined by Rick Luhmann, editor of Computer Telephony Magazine, New York, NY.

Smart Jack Industry term for the device to test integrity of T-1 circuits remotely from the central office. Installed on the customer premises in the form of a semi-intelligent demarcation point (demarc), the smart jack is completely passive until activated remotely by a digital code, typically something like "FACILITY 2," sent down the T-1. This code activates a relay that breaks the T-1 circuit and closes a receive-to-transmit loop across the T-1 at the customer end, sending the signal back to the central office (CO). This allows the CO to confirm the integrity of the loop without having to dispatch a technician to the site. Once the loopback test is completed, the relay automatically resets. The advantage of a smart jack is that the carrier can accomplish the loopback without engaging the DSU/CSU, which typically is in the CPE (Customer Premise Equipment) domain-i.e., it is customer-owned and, therefore, can possess uncertain attributes. As a result, the smart jack does not short when disconnected from the DSU/CSU; therefore, it always is available to the carrier for purposes of conducting such a loop test. The smart jack commonly has the same appearance as does a RJ48 jack; indeed, a RJ48 cable can connect the smart jack and the DSU/CSU to provide a hard-wired shorting function. The smart jack also may be rack-mounted. See also CSU, Demarc, DSU, Loopback, and RJ48C.

Smart Phone 1. A phone which can send, receive, sort, prioritize and notify users of communication transmissions, combining the capabilities of a PC with the familiarity of a telephone. They differentiate themselves from traditional telephones because they feature the following components: a microprocessor, an 8 t0 32 bit CPU, memory, function keys, context-sensitive keys, and a built-in modem.

2. A phone created by AT&T Network Systems and due out some time in 1992 or 1993. The phone never appeared. But it was a neat idea. It was to have no buttons but simply a touch sensitive screen. From one moment to another the image on the screen would change, requesting information from the user — the number or person to call, the letters you'd like to transmit to an electronic mail system, the amount of money you'd like to transfer to your checking account, etc.

Smart Retries A term used in the fax blasting industry to refer to doing different retries based on the status of the failed fax. For example, you might receive a busy and then retry 3 times 10 minutes apart, then 2 times more eight hours later. Another smart retry might be to restart sending a multi page fax at the page that the failed fax stopped at. This feature might also have an alternate cover page that can say that this fax is a continuation of a previous fax that didn't make it fully.

Smart Web Browsers A Web Browser is a piece of software which allows you to search the World Wide Web for the information you're seeking for. A smart Web Browser will contain a modicum of intelligence, allowing it to find what you're looking for faster, easier and less consuming of your time.

SmartHouse The National Society of Home Builders has registered SmartHouse. BellSouth has trademarked SuperHouse. And GTE has registered SmartPark. One day, fiber optic will snake to everyone's house, bringing the potential of immense information services. Until that day comes, there'll be fantastical demonstrations at trade shows.

SMAS Switched Maintenance Access System.

SMATV Satellite Master Antenna Television. A distribution system that feeds satellite signals to hotels, apartments, etc. Often associated with pay-per-view.

SMB Server Message Block. The protocol developed by Microsoft, Intel, and IBM that defines a series of commands used to pass information between network computers. The redirector packages SMB requests into a network control block (NBC) structure that can be sent over the network to a remote device. The network provider listens for SMB messages destined for it and removes the data portion of the SMB request so that it can be processed by a local device. In short, SMB is basically a protocol to provide access to server-based files and print queues. SMB operates above the session layer, and usually works over a network using a NetBIOS application program interface. SMB is similar in nature to a remote procedure call (RPC) that is specialized for file systems.

On June 13, 1996, Microsoft said, "The SMB protocol is an open technology widely available on UNIX, VMS and other platforms. It has been an Open Group (formerly X/Open) standard for PC and UNIX interoperability since 1992 (X/Open CAE Specification C209), and it is supported in products such as AT&T Advanced Server for UNIX, Digital's PATHWORKS, HP Advanced Server 9000, IBM Warp Connect, IBM LAN Server, Novell Enterprise Toolkit, and 3Com 3+Share, among others. SMB is also the featured file and print sharing protocol of Samba, a popular freeware network file system available for LINUX and many UNIX platforms."

The above Microsoft words came in a press release announcing something called CIFS, which is essentially an enhanced version of CIFS. See CIFS.

SMBus System Management Bus. A two-wire bus for more intelligent handling of rechargeable batteries in portable stuff, like laptops.

SMC Standard Management Committee. Directs specialized working groups of the Architecture and Standards Steering Council.

SMDA Station Message Detail Accounting. Another name for telephone call accounting. See CALL ACCOUNTING SYSTEM.

SMDI Station Message Desk Interface or Simplified Message Desk Interface. The SMDI is the data link from the central office if you have ESSX, Centrex or Centron (etc.) that gives you your stutter dial tone or message waiting light. In essence, SMDI is a data line from the central office containing information and instructions to your on-premises voice mail box. With SMDI, the calling person is not required to re-enter the called phone number (or in any other way identify the called party) once the call terminates to the messaging system.

SMDR Station Message Detail Reporting. Another name for telephone call accounting. See SMDR PORT and CALL ACCOUNTING SYSTEM.

SMDR Port Modern PBXs and some larger key systems have an Station Message Detail Recording (SMDR) electrical plug, usually an RS-232-C receptacle, into which one plugs a printer or a call accounting system. The telephone system sends information on each call made from the system to the outside world through the SMDR port. That information — who made the call, where it went. what time of day, etc. — will be printed by the printer or will be "captured" by the call accounting system on a floppy or a hard magnetic disk and later processed into meaningful management reports. See CALL ACCOUNTING SYSTEM.

SME Small to medium enterprises. Commonly used by EC & EDI folks.

Smear The undesirable blurring of edges in a compressed image, often caused by the DCT which tends to eliminate the high-frequency portions of an image which represent sharp edges.

SMF-PDM The ANSI X3T9.5 PDM standard which defines the requirements for the transmission of data over single mode fiber in an FDDI topology. Also refers to the ANSI working group responsible for the development and perpetuation of the standard.

SMDS Switched Multimegabit Data Service. A connectionless high-speed data transmission service intended for application in a Metropolitan Area Network (MAN) environment. SMDS is a public network service designed primarily to for LAN-to-LAN interconnection. An offshoot of the Distributed Queue Dual Bus (DQDB) standard defined by the IEEE 802.6 standard, SMDS owes its commercial success to Bellcore, which refined and commercialized it at the request of the RBOCs as a MAN service appropriate for RBOC offering within the confines of a LATA. SMDS offers bandwidth up to T-3 (45 Mbps) and the critical advantage of excellent congestion control, which virtually ensures that data arrives intact and as transmitted. SMDS supports asynchronous, synchronous and isochronous data. Although intended as a MAN service offering, it is possible to extend its reach over the WAN. SMDS is a cell-switched service offering, based on a 53-octet cell with a 48-octet payload similar to that of ATM. SMDS involves a process of converting data into cells before presentation to the network. Each set of data in its native form can be as much as 9,188 octets in length, constituting what is known as a Protocol Data Unit (PDU). At the point of SMDS processing, that PDU becomes known as a SMDS Service Data Unit (SDU), which is then appended with a header and trailer which contain network control information. The SDU, with header and trailer, is then segmented into units of 48 octets, each of which is prepended with a header and appended with a trailer before presentation to the SMDS network in the form of a 53-octet cell. Prepended means preceded as part of the same data unit (generic "packet"), whereas appended means succeeded as part of the same data unit. In fact, SMDS was designed with ATM in mind, providing a smooth transition path to ATM by virtue of the size of both the overall cell cell and the payload, although the specifics of the cell structure and other elements of the specific protocols are quite different. SMDS garnered a good deal of interest during the early 1990s, although broad support never developed. SMDS is not widely available and is highly unlikely to ever gain a broad level of acceptance; Frame Relay and ATM, in particular, have and will continue to overshadow SMDS. See THE SMDS INTEREST GROUP.

SMDS Interest Group A defunct consortium of vendors and consultants who were committed to advancing worldwide

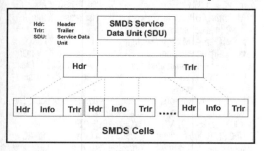

SMDS Cells

SMDS as an open, interoperable solution for high-performance data connectivity. On June 16, 1997, the Board of Trustees announced that the group was disbanded, turning over all responsibilities to regional organizations. The Board declared that its mission had been fulfilled.

SMDSU Switched Multimegabit Data Service Unit. See DSU.

SME Security Management Entity

SMF Single Mode Fiber: Fiber optic cable in which the signal or light propagates in a single mode or path. Since all light follows the same path or travels the same distance, a transmitted pulse is not dispersed and does not interfere with adjacent pulses. SMF fibers can support longer distances and are limited mainly by the amount of attenuation. Refer to MMF.

SMI Structure of Management Information. The set of rules and formats for defining, accessing and adding objects to the Internet MIB. SMI was elevated to full standard status in May 1990.

SMIL Synchronized Multimedia Integrated Language. Pronounced "SMILE." An open proposed recommendation from the World Wide Web Consortium (W3C) for a stylized layout language for the creation of Web-based multimedia pre-

sentations. SMIL, which is compliant with XML (eXtended Markup Language), defines the mechanism by which authors can compose a multimedia presentation, combining audio, video, text, graphics. SMIL allows the synchronization of these multimedia elements in terms of their presentation on screen, as well as the timing of their delivery to the client. The resulting SMIL file (.smi) offers the advantage of being in the form of a simple, markup text file that can be created with any text editor. See also W3C and XML.

SMM System Management Mode. See APM.

Smoke Test Test of new or repaired equipment by turning it on. If there's smoke, it doesn't work!

Smokeware When manufacturers announce new products, sometimes the products are not ready. Smokeware derives from the expression "smoke and mirrors."

SMOP Small Matter of Programming. A little software and it will all work. Yeah!

Smooth Call Arrival A call center term. Calls that arrive evenly across a period of time. Virtually non-existent in incoming environments.

Smooth Handoff See Mobile IP.

SMP 1. Symmetric MultiProcessing. The use of several CPUs (i.e. several Intel Pentium chips) inside a PC or (more typically) a server to achieve a performance boost. This is achieved by an operating system that lets CPUs run operations in parallel. This is not a "normal" PC operating system. Normal PC operating systems — MS-DOS, Windows, etc. — expect that their work will be done on one processor, not on more than one. See also SYMMETRIC MULTIPROCESSING.

2. SCSA Message Protocol.

3. Simple Management Protocol.

SMPI SCSA Message Protocol Interface.

SMPTE Society of Motion Picture & Television Engineers. An international society dedicated to advancing the theory and application of motion-picture technology including film, television, video, computer imaging and telecommunications. Founded in 1916 as the Society of Motion Picture Engineers, the "T" was added in 1950 in recognition of the emerging television industry. The SMPTE is an accredited ANSI Standards Developing Organization, and is recognized by both the ISO and the IEC. Current membership numbers about 8,500 in 72 countries. www.smpte.org

SMPTE Time Code A video term. Time code that conforms to SMPTE standards. It consists of an eight-digit number specifying hours:minutes:seconds:frames. Each number identifies one frame on a videotape. SMPTE time code may be of either the drop-frame or non-drop frame type. In GVG editors, the SMPTE time code mode enables the editor to read either drop-frame or non-drop frame code from tape and perform calculations for either type (also called mixed time code). SMTPE is a standardized edit time code adopted by SMPTE, the Society of Motion Picture and Television Engineers.

SMR Specialized Mobile Radio. Also known as TMR (Trunk Mobile Radio). A two-way radio telephony service making use of macrocells covering an area of up to 50 miles in diameter. The first SMR system was placed in service by the Detroit Police Department in 1921. The first commercial SMR service was offered in 1946 in St. Louis by AT&T. Subsequently, private SMR was widely used, and still is, in dispatch applications by truck and taxi fleets. Eventually, SMR took the form of IMTS (Improved Mobile Telephone Service), which used smaller frequency channels in support of many more conversations through improved efficiency. Many SMR operators have converted their SMR networks to digital so they can deliver both voice and data to a single device with improved efficiency and security, and thus compete with cellular and PCS radio providers. SMR systems have far less radio spectrum than cellular has, but the signal can reach 25 times farther, which means it's cheaper to build a national network. See also ESMR.

SMRP Simple Multicast Routing Protocol. Apple Computer's specialized network protocol for routing multimedia data streams on AppleTalk networks.

SMRT Single Message-unit Rate Timing. USA telephone company tariff under which local calls are timed in 5-minute increments — with a single message unit charge applied to each complete or partial increment.

SMS 1. Service Management System, a term coined by Bellcore for the Intelligent Network. The SMS allows provision and updating of information on subscribers and services in near-real time for billing and administrative purposes.
2. Short Message Service. A means to send or receive, short alphanumeric messages to or from mobile telephones.

SMS-R Subscriber Carrier Module-100S REmote.

SMS/800 The national database Service Management System that retains all 800 records. This database provides long distance carriers a single interface for 800 number reservations and record maintenance. Developed by Bellcore, the database has been in use by various Regional Bell Operating Companies (RBOCs) since 1988. The FCC mandated that a neutral third party administer the database after 800 portability, which occurred in May, 1993. That administration responsibility now lies with the SMS/800 Number Administration Committee (SNAC), which is part of the OBF (Ordering and Billing Forum), which operates under the auspices of the Alliance for Telecommunications Industry Solutions (ATIS). See also ATIS, OBF and SNAC.

SMSA Standard Metropolitan Statistical Area. A metropolitan area consisting of one or more cities as defined by the Office of Management and Budget and used by the FCC to allocate the cellular radio market.

SMSC Short Messaging Service Center. On a wireless network, allows short text messages to be exchanged between mobile telephones and other networks. See SMSCH.

SMSCH Short Message Service CHannel. Specified in IS-136, SMSCH carries signaling information for set up and delivery of short alphanumeric messages from the cell site to the user terminal equipment. SMSCH is a logical subchannel of SPACH (SMS (Short Message Service) point-to-point messaging, Paging, and Access response CHannel), which is a logical channel of the DCCH (Digital Control CHannel), a signaling and control channel which is employed in cellular systems based on TDMA (Time Division Multiple Access). The DCCH operates on a set of frequencies separate from those used to support cellular conversations. See also DCCH, IS-136, SMS, SPACH and TDMA.

SMT 1. Surface Mounting Technology.
2. Station Management. The part of FDDI that manages stations on a ring.

SMTA Single-line Multi-extension Telephone Apparatus.

SMTP Simple Mail Transfer Protocol. SMTP is a TCP/IP protocol for sending e-mail between servers. Virtually all e-mail systems that send mail via the Internet use SMTP to send their messages. Typically, you will send your email via SMTP to a POP3 (Post Office Protocol) server, from where your addressee will retrieve your message. Because of SMTP and POP3, you need to specify both your POP3 server and your SMTP server when you configure your e-mail client application, e.g. Microsoft Outlook, Eudora. Your email system will ask you for your SMTP server. Mine is smtp.email.msn.com while my POP3 server is pop3.email.msn.com. SMTP is actually an application-level protocol which runs over TCP/IP, supporting text-oriented e-mail between devices supporting Message Handling Service (MHS). You can send complex attachments, however, with SMTP, by simply using MIME, which stands for Multipurpose Internet Mail Extension. MIME is an SMTP extension supporting compound mail, which is integrated mail, including perhaps e-mail, a Word document, an Excel spreadsheet, an image, a voice WAV file and perhaps also a video clip in an AVI format. See POP3.

SMU 1. Subscriber Carrier Module-100 URBAN.
2. System Management Unit. The card or equipment in an xDSL unit that takes care of the management of the unit.

Smurfing A Denial of Service (DoS) attack by a hacker, usually a very young hacker ' hence, the origin of the term. Such an attack involves the sending of a stream of diagnostic "ping" messages to a list of IP servers, each of which forwards them to all LAN-attached workstations, each of which responds. The return address, however, is forged to reflect that of the target of the attack. The resulting stream of responses, which is magnified many times as the pinged servers and attached devices try over and over again to respond, effectively shuts down the targeted server. The targeted server might be a single server or it might be a complete service, i.e. a complete Web site, e.g. www.IveBeenSmurfed.com. See also Hacker and www.mcs.net/smurf.

Snow Video noise.

SN An ATM term. Sequence Number: SN is a 4 octet field in a Resource Management cell defined by the ITU-T in recommendation 1.371 to sequence such cells. It is not used for ATM Forum ABR. An ATM switch will either preserve this field or set it in accordance with 1.371.

SN Cell An ATM term. Sequence Number Cell: A cell sent periodically on each link of an AIMUX to indicate how many cells have been transmitted since the previous SN cell. These cells are used to verify the sequence of payload cells reassembled at the receiver.

SNA Systems Network Architecture. An IBM product. The most successful computer network architecture in the world. See IBM and SYSTEMS NETWORK ARCHITECTURE.

SNAC SMS/800 Number Administration Committee. SNAC is responsible for administering the Service Management System/800 (SMS/800) database for 800 and other toll-free numbers in the U.S. The SMS/800 database is the central repository for toll-free numbers, identifying the carriers to which they have been assigned, or to which they have been transferred at the request of the user. The SMS/800 database comprises all toll-free numbers, which are defined as 8NN, where NN is a set of two identical numbers (e.g., 800, 888 and 877). SNAC is a committee of the Ordering and Billing Forum (OBF), which operates under the auspices of the Alliance for Telecommunications Industry Solutions (ATIS). See also ATIS, OBF and SMS/800.

SNADS SNA Distribution Services. An IBM protocol that allows the distribution of electronic mail and attached documents through an SNA network.

SNAFU Situation Normal All Fouled Up. World War II military slang, that often describes frustrating facets of the telecom industry. Actually, the GIs used a much less genteel word than "Fouled." Actually, most of us still do.

Snail Mail A term used to reference delivery of messages by your local postal service. In short, mail that comes through a slot in your front door or a box mounted outside your house.

Snake A flexible strip of metal, typically 1/4 to 1/2" wide and 10 to 100' long, used to pull or push wire and cable through conduit, ceilings, walls or crawl spaces where it is difficult or impossible for a human to fit.

SNAP Subnetwork Access Protocol. A version of the IEEE local area network logical link control frame similar to the more traditional data link level transmission frame that lets you use nonstandard higher-level protocols. The Subnet Access Protocol is an Internet protocol that operates between a network entity in the subnet and a network entity in the end system and specifies a standard method of encapsulating IP datagrams and ARP messages on IEEE networks. The SNAP entity in the end system makes use of the services of the subnet and performs three key functions: data transfer, connection management, and quality of services selection.

Snapshot A view of something at an instant in time, rather than over time. A snapshot of a network, for instance, might be a view of the network configuration, in both physical and logical terms, as of midnight, January 1, 2000.

SNC Subnetwork Connection: In the context of ATM, an entity that passes ATM cells transparently, (i.e., without adding any overhead). A SNC may be either a stand-alone SNC, or a concatenation of SNCs and link connections.

SND Cellular language for SEND. You punch the digits for the phone number you want to dial into your phone. Check them on the screen. If they're fine, hit the SND button. Bingo, your call goes through.

SNDCF Subnetwork Dependent Convergence Function.

SNDCP Subnetwork Dependent Convergence Protocol. A Network Layer protocol that supports subnetwork convergence.

Snd Format Sound resource format. A digital audio sound resource utilized by many Macintosh applications and by the Mac OS. Double-clicking on a snd file enables playback of the sound. Also called a System 7 sound.

Sneak Currents Unwanted but steady currents which seep into a communication circuit. These low-level currents are insufficient to trigger electrical surge protectors and therefore are able to pass them undetected. They are usually too weak to cause immediate damage, but if unchecked could potentially create harmful heating effects. Sneak currents may result from contact between communications lines and AC power circuits or from power induction, and may cause equipment damage due to overheating.

Sneak Fuse A fuse operated by a low-level current and capable of preventing sneak currents on communication lines. See also SNEAK CURRENTS.

Sneakernet A "network" for moving files between computers. It is the oldest "network." It consists of copying a file to a floppy disk, walking to another machine, loading it on that machine. The term 'sneakernet' refers to the fact that the main method of moving the disks is by feet, presumably clad in sneakers.

SNET Southern New England Telephone Corporation. Originally an independent telephone company serving a substantial portion of Connecticut, SNET stock was partially held by AT&T at the time of the Modified Final Judgement (MFJ). As SNET was not a wholly-owned subsidiary of AT&T at the time, it was not affected directly by the limitations imposed on the Bell Operating Companies (BOCs) and it was able to enter the long distance business. In January 1998, SBC Communications, Inc. (Southwestern Bell) announced an agreement to acquire SNET. According to company practice, SNET is pronounced S-N-E-T, with each letter being spelled

out. See also MFJ and SBC.

SNI Subscriber Network Interface. SMDS term describing generic access to a SMDS network over a dedicated circuit which can be DS-0, DS1 or DS3.

Snickelway A narrow footpath in a medieval city where merchants sold their wares and villagers came to shop.

Sniffer Sniffer is a registered trademark owned by Network General Corporation. The Sniffer Network Analyzer is a member of the family of Network General products, that monitors traffic on a network and reports on problems on the network. The company is sensitive about the word Sniffer being used as a generic term for network monitoring. If you do use it as a generic term, their VP and General Counsel, Scott C. Neely, will write you a letter telling you about trademarks, etc.

Sniglet Any word that does not appear in the dictionary, but should. A term invented by Rich Hall of the HBO Television program "Not Necessarily The News". An example of a sniglet is the definition of "Hozone". It's obviously where socks go when they don't come back from the laundry.

Snips Heavy duty scissors, used for cutting metal.

SNMP 1. Signaling Network Management Protocol.

2. Simple Network Management Protocol. SNMP is the most common method by which network management applications can query a management agent using a supported MIB (Management Information Base). SNMP operates at the OSI Application layer. The IP-based SNMP is the basis of most network management software, to the extent that today the phrase "managed device" implies SNMP compliance. RMON and RMON-2 use SNMP as their method of accessing device MIB information. In 1988, the Department of Defense and commercial TCP/IP implementors designed a network management architecture for the needs of the average Internet (a collection of disparate networks joined together with bridges or routers). Although SNMP was designed as the TCP's stack network management protocol, it can now manage virtually any network type and has been extended to include non-TCP devices such as 802.1 Ethernet bridges. SNMP is widely deployed in TCP/IP (Transmission Control Protocol/Internet Protocol) networks, but actual transport independence means it is not limited to TCP/IP. SNMP has been implemented over Ethernet as well as OSI transports. SNMP became a TCP/IP standard protocol in May 1990. SNMP operates on top of the Internet Protocol, and is similar in concept to IBM's NetView and ISO's CMIP. In 1991, Microsoft started referring to SNMP as SubNetwork Access Protocol. In November of 1993 Cisco Systems announced that its internetwork routers will support version 2 of the Simple Network Management Protocol (SMNP) and it has licensed SNMP v2 developed by SNMP Research, Inc. of Knoxville, TN. See CMIP, MIB, RMON, RMON-2 SNMP-2.

SNMP-2 A major revision of the original SNMP, SNMP-2 is currently a proposed standard covered by RFC 1902 through RFC 1908. The SNMP-2 MIB —a superset of MIB-2 — addresses many performance, security, and manager-to-manager communication concerns about SNMP. For example, SNMP-2 supports encryption of management passwords. See MIB, MIB-2, SNMP.

Snow Random noise or interference appearing in a video picture as white specs.

Snowshoe A snowshoe shaped gadget that is used to maintain a minimum bend radius for installed fiber optic cable.

SNPA Subnetwork Point of Attachment

SNR 1. Abbreviation for saved number redial. A phone system memory feature that allows the user to store a number for as long as it is useful, as opposed to other numbers that are stored more permanently. Some phones have two buttons — one for REDIAL and one for SNR. Redial will dial the last phone number you called. Saved Number Redial will dial one you dialed earlier and chose to save because you're going to call it back. Both buttons are very useful.

2. Abbreviation of signal-to-noise ratio. The SNR relates how much stronger a signal is than the background noise. Usually expressed in decibels (dB). See signal-to-noise ratio.

Snuggling Snuggling is a method used by operators of surveillance equipment. A surveillance equipment operator, when building radio transmitters, will select a transmitter frequency close to that of a nearby high powered transmitter, usually, a commercial radio station. Most ordinary radio gear will automatically tune into the stronger of the two signals. The operator must use a specially modified receiver capable of detecting and isolating the weaker signal. "Snuggling" will be difficult to detect by low quality RF receivers when making a countermeasure sweep.

SO Serving Office. Central office where IXC (IntereXchange Carrier) has POP (Point Of Presence).

Soak A means of uncovering problems in software and hardware by running them under operating conditions while they are closely supervised by their developers.

SOC Software Optionality Control.

Social Computing A term that emerged in the summer of 1993. Defined by Peter Lewis in the New York Times of September 19, 1993, social computing is a "communications-rich brew," which is "expected to create new ways for businesses and their customers to communicate, over new types of wireless as well as wired pathways, using new types of computers called personal communicators." According to Peter Lewis, "The rise of social computing is expected to shift the emphasis of computing devices away from simple number crunching and data base management to wider-ranging forms of business communications...Where client server broke away from mainframe-based systems and distributed computing power to everyone in the organization, social computing goes the next step and extends the distribution of computing power to a company's customers."

Social Contract An arrangement between the local telephone company and its local regulatory authority whereby the telephone company's services are detariffed, but cannot be priced at less than cost. Quality of service standards apply.

Social Engineering Gaining privileged information about a computer system (such as a password) by skillful lying — usually via a phone call. Often done by impersonating an authorized user.

Socket 1. A synonym for a port.

2, A technology that serves as the endpoint when computers communicate with each other.

3. The socket in a PC which is responsible for accepting a PCMCIA Card and mapping the host's internal bus signals to the PCMCIA interface signals.

4. An operating system abstraction which provides the capability for application programs to automatically access communications protocols. Developed as part of the early work on TCP/IP.

Socket Interface The Sockets Interface, introduced in the early 1980s with the release of Berkeley UNIX, was the first consistent and well-defined application programming interface (API). It is used at the transport layer between Transmission Control Protocol (TCP) or User Datagram Protocol (UDP) and the applications on a system. Since 1980,

sockets have been implemented on virtually every platform.

Socket Number In TCP/IP, the socket number is the joining of the sender's (or receiver's) IP address and the port numbers for the service being used. These two together uniquely identifies the connection in the Internet.

Socket Services The software layer directly above the hardware that provides a standardized interface to manipulate PCMCIA Cards, sockets and adapters. Socket Services is a BIOS level software interface that provides a method for accessing the PCMCIA slots of a computer, desktop or laptop (but most typically a laptop). Ideally, socket services software should be integrated into the notebook's BIOS, but few manufacturers have done so to date. For PCMCIA cards to operate correctly you also need Card Services, which is (not are) a software management interface that allows the allocation of system resources (such as memory and interrupts) automatically once the Socket Services detects that a PC Card has been inserted. You can, however, happily operate PCMCIA cards in your laptop without using socket and card services. You simply load the correct device drivers for those cards. Such drivers always come with PCMCIA cards when you buy the cards. You will, however, have to load new drivers every time you change cards and allocate the correct memory exclusions. You will have to reboot if you disconnect your network card. Theoretically, with socket and card services loaded, you do not have to reboot every time you change cards. My experience is that this works, except with network cards, which cannot be hotswapped. See PCMCIA.

Sockets An application program interface (API) for communications between a user application program and TCP/IP. See SOCKET and SOCKET NUMBER.

SOCKS A circuit-level security technology developed by David Koblas in 1990 and since made publicly available by the IETF (Internet Engineering Task Force. SOCKSv4 was a true circuit-level proxy gateway, working to secure the flow of data at the TCP layer. SOCKSv5, recently released, also provides for security in a client/server environment, including the support of multiple means of authentication which can be negotiated between client and server. SOCKSv5 also supports the transfer of UDP data as a stream, avoiding the need to treat each packet of UDP data as an independent message. SOCKSv5 also allows protocol filtering, which offers enhanced access control on a protocol-specific basis. For example, a network administrator can add a SMTP (Simple Mail Transfer Protocol) filter command to prevent hackers from extracting from a mail message information such as a mail alias. Reference implementations exist for most UNIX platforms, as well as Windows NT. The cross-platform nature of SOCKS offers portability to Macintosh and other operating systems and browsers. See ALIAS, PROXY, PROXY SERVER SMTP and UDP.

SOF Start Of File

Soft Copy 1. A copy of a file or program which resides on magnetic medium, such as a floppy disk, or any form that is not a hard copy — which is paper.

2. Old legacy systems term reapplied to distributed computing in which reports are created on-screen from data residing within different applications.

Soft Ferrite Ferrite that is magnetized only while exposed to a magnetic field. Used to make cores for inductors, transformers, and other electronic components. See BARIUM FERRITE, FERRITE and HARD FERRITE.

Soft Font A font, usually provided by a font vendor, that must be installed on your computer and sent to the printer before text formatted in that font can be printed. Also known as downloadable font.

Soft Key There are three types of keys on a telephone: hard, programmable and soft. HARD keys are those which do one thing and one thing only, e.g. the touchtone buttons 1, 2, 3, * and # etc. PROGRAMMABLE keys are those which you can program to do produce a bunch of tones. Those tones might be "dial mother." They might be "transfer this call to my home for the evening." They might be "go into data mode, dial my distant computer, log in and put in my password." SOFT keys are the most interesting. They are unmarked buttons which sit below or above on the side of a screen. They derive their meaning from what's presently on the screen. And what's on the screen will change based on where the call is at that moment — in a conference call, about to set up a conference call, about to go into voice mail, into voice mail, programming a speed dial number, etc.

Soft Modem Which is short for software modem. It's a concept pioneered apparently by Motorola in which the modem consists of a small phoneline interface card and software that uses the PC's main CPU (e.g. a Pentium) for the main communications tasks. Allegedly such modem will cheaper than ones based on hardware, ones we're all used to.

Soft Sectored A floppy disk whose sector boundaries are marked with records instead of holes. Soft-sectored disks have typically one hole. Hard sectored disks have many holes. A soft-sectored disk won't work on disk drives which use hard-sectored drives and vice versa — even though the disks might be the same size. Soft-sectored disks are now much more common.

Soft Selectable/Soft Strappable Refers to an option that is controllable through software rather than hardware.

Softkey See SOFT KEY.

Softkey Mapping See ADSI.

Software The detailed instructions to operate a computer. The term was created to differentiate instructions (i.e. the program) from the hardware. See PROGRAM, HARDWARE and FIRMWARE.

Software Data Compression As an ISDN term, it means the ability to compress data before it arrives at the serial port of the ISDN terminal adapter. Can improve performance by as much as 400%

Software Defined Network SDN. Generically, a software defined network refers to a virtual private network. Specifically, it refers to AT&T's Software Defined Network Service, which was introduced in 1985 for AT&T's largest customers and provided only dedicated access services. In 1989, AT&T extended its SDN Network to switched access. Currently, SDN is the most commonly resold of all long distance services. The AT&T Software Defined Network Service Description of July 1986 describes SDN as a service developed for multi-location businesses which allows network managers to tailor their network to their own specific communications needs. Call processing information is stored in a database that is accessed during a call. Calls are transferred over AT&T facilities to either a location that is a dedicated part of the network (for "on-net calling") or to non-dedicated facilities that are part of the network ("off-net" calling). Any company location can become part of a network through SDN dedicated access lines to an AT&T SDN serving office. Here is further explanation of SDN by Siemens Information Systems, whose Saturn PBX has the capability to interface to and mesh neatly with an AT&T SDN network. Here is Siemens' explanation (it's good): When a company establishes an SDN, each phone on the network has a unique seven digit number. This number may or may not be the same as the Listed Directory

Number (LDN). When a call is placed at one PBX, it is sent over a dedicated access line to the long distance network. The call is received by the SDN Serving Office and digits are sent via a CCIS (Common Channel Interoffice Signaling) link to the Network Control Point (NCP) for analysis and routing. There is one NCP per SDN network. The NCP contains the unique database for the company using the SDN. The NCP analyzes the digits received against the database, determines whether it is an on-net or off-net call, and sets up the path over which the call will be rerouted on the long distance network. If it is an on-net call, the NCP translates the unique seven digit locator code to one that will be recognized by the AT&T network, sends the call over the network to another SDN Serving Office, and completes the call over dedicated lines to the PBX being called. There is a discount for any call which remains on-net throughout. If the call is off-net, the digits dialed are sent over the long distance network to a Central Office that is not part of the SDN. The call is then completed to the corporate PBX over DID (Direct Inward Dial) lines. Since the introduction of route selection features in PBXs, the caller now has the ability to dial the 10-digit LDN (Long Distance Number) and have the PBX make the decisions on the route and make the decisions on any digit translation or deletion that is necessary to route the call. The CCIS network which carries SDN call signaling is a packet switching network operating at 4,800 bits per second. It will eventually be replaced by ITU 7 Signaling, a more powerful internationally-accepted signaling system. See SIGNALING SYSTEM 7.

Software Engineering A broadly defined discipline that integrates the many aspects of programming, from writing code to meeting budgets, to produce affordable software that works.

Software Interfaces Language between programs which allows one program to call upon another for assistance in processing.

Software Metering Software that monitors the use of applications — such word processing, spreadsheets, databases, etc. All metering programs tend to take advantage of a software feature called "concurrency." Concurrency means that a company need only buy as many licenses to a program as it has people using the program at one time — concurrent users, in other words.

Software Modem Also called soft modem. Basically it's a modem which consists of a small phoneline interface card and software that uses the PC's main CPU (e.g. a Pentium) for its main communications tasks. Such modem is typically cheaper than one based on dedicated hardware (chips), i.e. ones we're all used to.

Software Only Video Playback A multimedia term. Video software playback displays a stream of video without any specialized chips or boards. The playback is done through a software application. The video is usually compressed to minimize the storage space required.

Software MPEG Playback See MPEG.

Software Supervision "Answer Supervision" is knowing when the person at the other end answers the phone. The main reason for wanting to know this is so that a phone company can start billing the call. There are two ways of doing answer supervision. You can get it from the nation's phone system, i.e. the distant office signals back across the country when the called person picks up the phone. Or you can fake it with "software supervision." Essentially this means there's electronics which "listens" to the call. If it "hears" voice or something like voice, it assumes the conversation has started and it's time to start billing the call. Software supervision is not accurate. But when you haven't got access to real answer supervision (for whatever reason) it's better than the previous alternative, which was "timeout." In timeout answer supervision, the carrier simply assumed the call had begun after a certain number of seconds — like 30 — had elapsed with the calling person hanging up. This meant, for example, if you called Grandma and she wasn't there, but you left it ringing, 'cause you knew she took time to answer the phone, then you'd be charged for the call — even though she didn't answer phone!

SoftWindows Sun Microsystems software for running PC applications on SPARC platforms.

SOH Start Of Header. A transmission control character used as the first character of the heading of an information message.
2. Section Overhead. SONET frames include 9 octets of SOH for maintenance of SONET links. SOH information includes transport status, messages and alarm indication. Without SOH data, you are SOL.

SOHO 1. Small Office Home Office. An acronym for a new telecom market opening up, which is part work at home, part commute from home. It's getting larger as companies downsize and their workers become "consultants" or small businesspeople. It's getting larger as companies close their distant sales offices ask their salespeople to work out of home. By the year 2000, some estimates show that more than half of the American workforce will work out of their home, or in some non-conventional job, but certainly not for what was typical in the 1950s through the end of the 1980s — a large, multi-office, multi-tier American corporation.
2. In New York City, there's an area called SOHO. It stands for South Of HOuston Street. See Silicon Alley.

Solaris Sun Microsystems' Unix-based 32-bit operating environment for network application servers, Internet servers, and high-end desktop systems. Solaris has multithreading, symmetric multiprocessing, integrated TCP/IP-based networking and centralized network administration tools. Solaris has an open systems architecture. It supports Sun SPARC, Intel and PowerPC processor-based platforms. Solaris supports IPX/SPX protocol stack to connect to Novell networks. Security and system administration uses NFS with compliance to the POSIX 1003.6 standard. Solaris is scalable from desktop computers to Internet servers; from single processor computers to 64-processor systems.

Solaris Teleservices See TELESERVICES.

Solder 1. An alloy of lead and tin having a low melting point.
2. To unite or join by solder.

Solenoid A coil consisting of a number of turns in cylindrical form.

Solid State Any semiconductor device that controls electrons, electric fields and magnetic fields in a solid material — and typically has no moving parts.

Solid State Applications Interface Bridge Solid's State Systems' PBX to external computer link. See OPEN APPLICATION INTERFACE.

Solid State Transfer An uninterruptible power supply (UPS) definition. Many UPSes use "mechanical relays" to switch from AC power to battery power. This technology requires 12 milliseconds switching time — slow enough to cause data loss. Solid state switching is faster and eliminates this type of problem.

Solitons Solitons are light pulses that maintain their shape over long distances. In early 1993, scientists at AT&T Bell Labs announced that they had transmitted error-free solitons at 10 gigabits per second over one channel and at 20 gigabits per second more than 13,000 kilometers using two channels. To accomplish this feat, they used sliding-frequency guiding filters. See also RARE EARTH DOPING.

On April 27, 1998, MCI said that it was using Pirelli's "Soliton" Technology to Carry Voice and Data Traffic Triple The Distance, Without Costly Regeneration Equipment. MCI announced today a successful network trial using Solitons, a technology based on a scientific theory formulated back in 1834 and adapted for modern use by Pirelli Cables and Systems, that will dramatically increase the distance voice and data traffic can be transmitted without the use of costly electrical regeneration equipment. MCI officials say the technology has the potential to reduce transmission costs by as much as 20 percent. In Soliton field trials conducted earlier this month, MCI demonstrated the ability to send a single stream of data traffic at speeds of 10 gigabits-per-second (Gbps) more than 900 kilometers over existing installed fiber without regenerators — triple the distance that 10 Gbps. traffic can be transported today. MCI also successfully transmitted Soliton data streams using dense wavelength-division multiplexing technology, carrying four data streams at 10 Gbps each, traveling more than 450 kilometers without regenerators.

Here is MCI's explanation of how the Soliton technology works: A Soliton is a type of wave or, in the case of optical fiber, a narrow pulse of light that retains its shape as it travels long distances along the fiber. The Soliton's ability to keep its shape helps to overcome the problem of lightwave dispersion, and the consequent loss of data integrity, as the data-carrying lightwave travels over long distances. Modern Soliton technology is based on a phenomenon first documented in 1834 by a Scottish engineer named John Scott Russell who, while watching a boat being drawn along a canal by a pair of horses, noticed that when the boat stopped suddenly, the wave of water created by the bow continued forward at great velocity without losing speed or shape. Russell was convinced that the Soliton was an important scientific discovery. But his theory wasn't fully borne out until the 1960s when scientists began to learn that many phenomena in such fields as physics, electronic and biology can be explained by Solitons. The key elements of Pirelli's 10 Gbps Wavelength Division Multiplexed systems are the Soliton converters, which transmit and receive traffic. The Soliton transmitter generates a pulse with the proper shape and power to allow for the transmission of data and voice traffic over very long distances without electronic regenerators. Eliminating regenerators is important as MCI continues its transition to an all-optical networking environment. Electronic regenerators are bit-rate- and protocol-dependent. Removing them creates transparency, a key tenet in the creation of the purely optical network. In the Pirelli Soliton trials, MCI also evaluated a new technology, known as Dispersion Compensation Grating (DCG). DCG overcomes the distortion of the optical signals as they are transmitted through the network. Instead of trying to compensate for large amounts of signal dispersion at the end of a network, DCG periodically removes the distortion where needed along the transmission line.

Solution IBM-speak to solve, as in "We've got to solution this problem if we're going to make the sale."

Solution Assembler Another name for a system integrator. A vendor who puts together a collection of products which purport to be the solution/s to your IS problem.

SOMO Acronym for Small Office Medium Office.

SON Service Order Number. The SON is the number issued by the local exchange carrier to confirm the order for the ISDN service. It provides a matching number for cross referencing the order to the phone company.

SONET Synchronous Optical NETwork. A family of fiber optic transmission rates from 51.84 million bits per second to 13.27 gigabits (thousand million) per second (and going higher, as we speak), created to provide the flexibility needed to transport many digital signals with different capacities, and to provide a design standard for manufacturers. SONET is an optical interface standard that allows interworking of transmission products from multiple vendors (i.e., mid-span meets). It defines a physical interface, optical line rates known as Optical Carrier (OC) signals, frame format and an OAM&P protocol (Operations, Administration, Maintenance, and Provisioning). The OC signals have their origins in electrical equivalents known as Synchronous Transport Signals (STSs). The base rate is 51.84 Mbps (OC-1/STS-1). Higher rates are direct multiples of the base rate. SONET development began at the suggestion of MCI to the Exchange Carriers Standards Association (ECSA). Bellcore then took over the project, and it ultimately came to rest at the American National Standards Institute (ANSI). Much of the development was carried out by ECSA under the auspices of ANSI. Work started on the SONET standard in the ANSI accredited T1/X1 committee in 1985, and the Phase 1 SONET standard was issued in March 1988. SONET has also been adopted by the ITU-T (International Telecommunications Union-Telecommunications Standardization Sector), nee ITU. The ITU-T version is known as SDH (Synchronous Digital Hierarchy), which varies slightly and most obviously in terms of the fact that the SDH levels begin at 155 Mbps. In SDH, the fundamental building blocks are known as STMs (Synchronous Transport Modules) and are equivalent in rate to three SONET STS-1s. SONET is intended to attain the following goals: Multi-vendor interworking, to be cost effective for existing services on an end-to-end basis, to create an infrastructure to support new broadband services and for enhanced operations, administration, maintenance and provisioning (OAM&P). SONET offers many advantages over asynchronous transport including: Opportunity for back-to-back multiplexing, digital cross-connect panels; Easy evolution to broadband transport; Compatibility with evolving operations standards; Enhanced performance monitoring and extension of OAM&P capabilities to end users. SONET/SDH offers the critical advantage of a standard to which manufacturers can build fiber optic gear in order to ensure interconnectivity and (at least some level of) interoperability. Thereby, carriers can safely acquire and deploy multi-vendor networks without being wed to a single manufacturer. This last point was, in fact, the primary impetus for SONET development. SONET transmission equipment interleaves frames of data in simple integer multiples to form a synchronous high speed signal known as a Synchronous Transport Signal (STS). This permits easy access to low speed signals (e.g. DS-0, DS-1, etc.) without multi-stage multiplexing and demultiplexing. The low speed signals are mapped into sub-STS-1 signals called Virtual Tributaries (VTs), or Virtual Containers (VCs) in SDH. SONET uses a 51.84 Mb/s STS-1 signal as the basic building block. Higher rate signals are multiples of STS-1 (e.g. the STS-12/OC-12 signal has a rate of 12 x 51.84 Mb/s or 622.080 Mb/s). The frame format consists of 90 x 9 bytes. The SONET frame format is divided into two main areas: Synchronous Payload Envelope (SPE) and Transport Overhead (TOH). The SPE contains the information being transported by the frame. The TOH supports the OAM&P functions of SONET

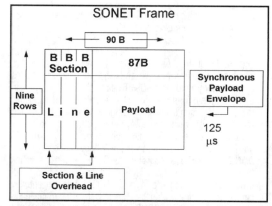

and includes a data communication channel that provides an OAM&P communication path between multiple interconnected SONET network elements. The Synchronous Payload Envelope can handle payloads in any of three ways:

1. As a continuous 50.11 Mb/s envelope for carrying asynchronous DS-3, and other payloads requiring up to 50.11 Mb/s capacity in asynchronous (byte invisible) or byte visible format;

2. In a VT (Virtual Tributary) structured envelope to accommodate DS-1, DS-1C, DS-2, European CEPT1, or future VT based services (see chart below). These signals can have either an asynchronous or byte visible format; and

3. As concatenated payloads to accommodate services requiring more than 50.11 Mb/s capacity. For example, three STS-1 SPEs may be concatenated to transport a broadband ISDN signal of 135 Mb/s. According to AT&T, the main SONET characteristics are: A family of rates at N x 51.84 Mbps; Optical interconnect allowing mid-span meet; Intra office mixed vendor interconnects; Overhead channels for OAM&P functions and Synchronous networking.

SONET rates are

OC Level	Line Rates	Capacity
OC-1	51.84 Mbps	28 DS1s or 1 DS3
OC-3	155.52 Mbps	84 DS1s or 3 DS3s
OC-9	466.56 Mbps	252 DS1s or 9 DS3s
OC-12	622.08 Mbps	336 DS1s or 12 DS3s
OC-18	933.12 Mbps	504 DS1s or 18 DS3s
OC-24	1.244 Gbps	672 DS1s or 24 DS3s
OC-36	1.866 Gbps	1008 DS1s or 36 DS3s
OC-48	2.488 Gbps	1344 DS1s or 48 DS3s
OC-96	4.976 Gbps	2688 DS1s or 96 DS3s
OC-192	9.953 Gbps	5376 DS1s or 192 DS3s
OC-255*	13.21 Gbps	

The next rate, on which development work is already underway, will be OC-768 (39.813 Gbps).

• OC-256 is a bit iffy. Bellcore standards documents state "SONET optical transmission systems support only certain values of N. Currently, these values are 1, 3, 12, 24, 48, and 192." Although OC-256 does not fit into the power of two progression above OC-3 as the rest do, it has still been considered by some as being the next rate to be implemented after OC-192, or the jump may go directly to OC-768. SONET/SDH networks typically are deployed in a ring physical topology, with multiple fibers providing redundancy. In the event that a given fiber suffers a catastrophic failure, one or more other fibers are available. The rings are of two types: Line-Switched and Path-Switched. SONET also may be deployed in a linear physical topology, in which case the system operates as a logical ring. See also ADM, Line Switched Ring, Path Switched Ring, SONET Interface Layers, SONET Ring, STM, Stratum Level and STS.

SONET Head A device on the end of a boring machine. Such machine is used to bore holes under highways, rivers and sundry obstructions. The SONET head contains sensors which can help determine what it is about to strike as it moves ahead underground. The SONET head will signal the person operating the boring machine what lies ahead and hopefully, the operator, is sufficiently intelligent to move the boring machine up or down or sideways in order to miss the potential obstruction — which might be anything from a rock to another fibre cable to a high voltage AC power line.

SONET Interface Layers The SONET standards define four interface layers. Each layer requires the services of all lower-level layers to perform its functions. While conceptually similar to layering within the Open System Interconnection (OSI) reference model, SONET itself corresponds only to the OSI Physical Layer. The SONET interface layers are:

1. Physical Layer: Handles bit transport across the physical medium; primarily responsible for converting STS (electrical) signals to and from OC (optical) signals. Once the signal has been expressed optically, this layer is sometimes referred to as the photonic layer. Electro-optical devices communicate at this layer;

2. Section Layer: Transports STS-N frames and Section Overhead (SOH) across the medium; functions include framing, scrambling, and error monitoring. Section Terminating Equipment (STE) communicate at this layer;

3. Line Layer: Responsible for the reliable transport of the Synchronous Payload Envelope (SPE) (i.e., user data) and Line Overhead (LOH) across the medium; responsibilities include synchronization and multiplexing for the Path Layer and mapping the SPE and LOH into an STS-N frame. An OC-N-to-OC-M multiplexer is an example of Line Terminating Equipment (LTE); and

4. Path Layer: Handles transport of services (e.g., DS-1, DS-3, E-1, or video) between Path Terminal Equipment (PTE); the main function is to map the services and Path Overhead (POH) information into the PTE includes SONET-capable switches with an interface to a non-SONET network, such as a T1-to-SONET multiplexer.

SONET Ring SONET transmission systems ideally are laid out in a physical ring for purposes of redundancy. In practice, the topology often is that of a linear ring, which is linear in its physical appearance, but which operates as a logical ring. See also SONET, PHYSICAL TOPOLOGY, LOGICAL TOPOLOGY, LINE-SWITCHED RING, and PATH-SWITCHED RING.

SONIA A group of technology vendors which advanced a profile for NCs (Network Computers), also known as "thin clients." SONIA is derived from the names of the members of the group: Sun Microsystems, Oracle Corp., Netscape Communications Corp., IBM, and Apple Computer Inc. See also NC and THIN CLIENT.

Sony Mini Disc A 2 1/2 inch silvery CD (compact Disc) that can record and play 74 minutes of sounds, almost as much as its five-inch forebear. To record on this disc, a laser momentarily heats a tiny spot on the disk to 400 degrees Fahrenheit, while a magnetic head writes the signal into the heated part of the magnetic layer. To play the disk an optical pickup analyzes the polarity of the light reflected from each spot.

SOP Standard Operating Procedure.

Sort To order a collection of records — for example, a telephone directory — in some specified way, say, in alphabetical

order. Computers can sort in virtually any way you ask them to. Most companies don't produce sufficient "sorts" on their telephone directories.

Sort Scheme A call center term. A list of fields that tells the program how to sort a report or a list of records. This can be simple scheme that sorts by only one field or a complex scheme consisting of sorts within sorts.

Sound The best explanation of what's happening in sound came from The Economist of August 24, 1996: Sounds are vibrations of molecules. Recording means getting something else (e.g., a microphone) to vibrate in sympathy, and turning those vibrations into electrical signals (which can then be stored as grooves on a disk, say). A loudspeaker does the reverse: electrical signals are turned into wobbles of a usually cone-shaped piece of material that batters the molecules of the air to recreate the sound. Audible sound spans a huge range of frequencies from around 20 hertz (vibrations per second) to 20kHz. Most loudspeakers need two to three cones of different sizes to cover this range. Moreover, like bells and wine glasses, cones have their own natural frequencies at which they vibrate when hit. They overrespond to electrical signals close to those frequencies. As a result, sound always loses something when being electronically regurgitated, although careful loudspeaker and amplifier design can make up for a lot.

Sound Files Files on PCs have their own extensions — the three letters which follow the name of the file. For example, a sound file of jungle noises might be called jungle.wav. Here are the typical extensions on sound files of various computers:
Microsoft Windows — .wav
Apple — .aif
NeXT — .snd
MIDI — .mid and .nni
Sound Blaster — .voc
Intel Indeo Video Movie clips — .avi

Sound Powered Telephone A telephone in which the operating power is derived from the speech input only. See SOUND.

Sound Waves The waves given off by a vibrating body, which are transmitted by an elastic material medium (such as the air) and which can be detected by the ear. See SOUND and SOUND FILES.

Source That part of a communications system which transmits information.

Source Address The part of a message which indicates who sent the message. Just like the top left-hand address on the envelope.

Source Code A set of instructions, written in a programming language, that must be translated to machine instructions before the program can be run on a computer. The program which finally runs on that computer is known as the object code.

Source Explicit Forwarding A feature that allows MAC-layer bridges on local area networks to forward packets from only source address specified by the network administrator.

Source Route A hierarchically complete source route. See SOURCE ROUTING.

Source Route Bridging SRB. Token-ring technique for establishing communications between devices on different LANs. See SOURCE ROUTING.

Source Routing A method used by a bridge for moving data between two networks. Originally developed by IBM's token ring network, it relies on information contained within the token to route the packet between the two networks. Since the information in the token is supplied by the computer that sent the data packet, that computer must know on which network the destination computer is located. IBM developed a special protocol that lets computers discover that information. For source routing to work, every computer and every bridge on all networks must support this protocol. If some computers do not use this protocol, they will not receive packets from bridges that use source routing. See BRIDGE. Compare to TRANSPARENT ROUTING.

In IBM's method of routing local area network data across bridges, IBM's bridges can be configured as either single-route broadcast or all-routes broadcast. The default is single-route broadcast. Single-route broadcasting means that only one designated single-route bridge will pass the packet and only one copy of the packet will arrive at its destination. Single-route broadcast bridges can transmit both single-route and all-routes packets. All-routes broadcasting sends the packet across every possible route in the network, resulting in as many copies of the frame at the destination as there are all-routes broadcasting bridges in the network. All-routes broadcast bridges only pass all-routes broadcast packets.

Source Routing Protocol. SRP. See Bridge.

Source Routing Transparent SRT. See Bridge.

Source Traffic Descriptor An ATM term. A set of traffic parameters belonging to the ATM Traffic Descriptor used during the connection set-up to capture the intrinsic traffic characteristics of the connection requested by the source.

Source/Destination Routing A term used in call centers for routing calls based on where they originate or terminate. See also SKILLS-BASED ROUTING, CALENDAR ROUTING and END-OF-SHIFT ROUTING.

Source/Sink Device A source/sink device is byte-synchronous with a byte orientation. Source devices originate; sink devices terminate.

Southern New England Telephone Corporation SNET. Pronounced S-N-E-T, not SNET. Originally an independent telephone company serving a substantial portion of Connecticut, SNET stock was partially held by AT&T at the time of the Modified Final Judgement (MFJ). As SNET was not a wholly-owned subsidiary of AT&T at the time, it was not affected directly by the limitations imposed on the Bell Operating Companies (BOCs) and it was able to enter the long distance business. In January 1998, SBC Corporation (Southwestern Bell) announced an agreement to acquire SNET. See also MFJ and SBC.

Southwestern Bell Corporation One of the seven Regional Holding Companies formed at Divestiture. It includes Southwestern Bell Telephone Co., Southwestern Bell Mobile Systems, Southwestern Bell Yellow Pages, Southwestern Bell Telecom and SBC International. In early October, 1994, Southwestern Bell Corporation changed its name to SBC Communications Inc., apparently feeling as though the "Bell" name no longer was of value. (The old Bell Operating Companies (BOCs) fought to retain the exclusive use of the "Bell" name at divestiture. Since that time, most have abandoned it.) In 1996, SBC announced its plan to acquire Pacific Telesis, the holding company for Pacific Bell and Nevada Bell; that merger was completed in April 1997. In January 1998, SBC announced its intent to acquire SNET (Southern New England Telephone). On May 11, 1998, SBC announced its intention to acquire (Whoops, merge with) Ameritech. Now you know why the company changed its name to SBC. And by the time you read this, it will probably have bought some more phone companies, if the U.S. government allows it.

SP 1. Support Processor. 2. Sending Program. 3. Signal Present. 4. Signal Processor. 5. Signaling Point.

SPA 1. Software Publishers Association. A not-for-profit organization formed in 1984, SPA is a principal software industry trade association which represents leading publishers, as well as start-ups. SPA supports "companies that develop and publish software applications, components, tools and digital content for use on the desktop, client-server networks and on-line." SPA's MPC (Multimedia PC) Working Group has published several versions of MPC standards over the years, with the current version being MPC3. SPA's headquarters is in Washington, DC; it also has offices in Paris, France. Membership totals over 1,200. www.spa.org. See also MPC3.
2. Shared Printer Access. An ISDN term for the sharing of a printer by multiple users. With a Terminal Adapter (TA) on the PC's serial port and another on the serial printer, a remote worker using ISDN BRI can send a print job over the 16 Kbps D channel. Once the transmission is complete, the call is terminated and the printer is available for another remote worker who is similarly equipped.

Space 1. In digital transmission, the space is equated to the zero (0) and the mark is equated to the one (1). In telecommunications, space is the absence of a signal. It is equivalent to a binary "0".
2. Space also stands for Service Creation and Customization.
3. What other people call marketplace, Microsoft calls space. You and I would say, "This product fits into the computer telephony marketplace." Microsoft would say "This product fits in the computer telephony space."

Space Brokers Companies that provide all the facilities needed to start a 976 or 900 service. Those facilities include offices, computing equipment, voice processing software, telephone lines and numbers.

Space Diversity Protection of a radio signal by providing a separate antenna located a few feet below the regular antenna on the same tower to assume the load when the regular transmission path on the same tower fades because of rain, a bird flying through it, etc.

Space Division Multiplexing Each distinct signal or message travels over a separate physical path such as its own wire or wire pair within a cable.

Space Division Switching Method for switching circuits in which each connection through the switch takes a physically separate path.

Space Hold A no traffic line condition where a steady space is transmitted.

Space Parity In data transmission, setting the parity bit so it is always zero.

Space Segment 1. The part of a satellite system that is in space.
2. This is also the imprecise term used to describe the band of frequency purchased by the satellite customer. The customer can purchase a portion of the bandwidth of a single transponder or the customer can purchase one or more entire transponder bandwidths.

Spacecraft Switched Time Division Multiple Access SSTDMA. A method of sharing the capacity of a communications satellite by on-board switching of signals aimed at earth stations.

SPACH SMS (Short Message Service) point-to-point messaging, Paging, and Access response CHannel. A digital wireless term defined by IS-136, the Interim Standard for digital cellular networks employing TDMA (Time Division Multiple Access). SPACH is a logical channel which is part of the FDCCH (Forward Digital Control CHannel) used to send signaling and control information downstream from the cell site to the user terminal equipment. SPACH is further subdivided into three logical subchannels: ARCH, SMSCH, and PCH. See also ARCH, FDCCH, IS-136, PCH, SMSCH and TDMA.

Spade Lug A metal connector attached to the end of a piece of wire, typically by soldering or by pressure. The metal spade lug is shaped like a "U." The idea is to slide the flat "U" shaped metal piece under a screw and then tighten the screw, thus making a connection. In the old days, all phones came with spade lug connectors. These days, there are other faster, more efficient ways of connecting phones — including modular jacks and punchdown tools.

SPAG 1. Europe's Standards Promotion and Application Group. 2. Standards Promotion and Application Group. A group of European OSI manufacturers which chooses option subsets and publishes these in a "Guide to the Use of Standards" (GUS).

Spaghetti Code A program written without thought, logic or structure. And whose "logic" is therefore very difficult to follow. Some would say this definition covers most software written today. That's unfair.

Spam 1. Hormel's ever popular spiced ham, consisting of leftovers from the processing of pork, plus lots of additives.
2. Junk e-mail. The term is derived from Hormel's pink, canned spiced ham that splatters messily when hurled. A milder form of spamming is called crossposting. See SPAMMING.
3. Message posted to numerous Usenet newsgroups to which it has absolutely no relevance (also a verb). See SPAMMING.

Spam Filter Software which keeps out spammed email. The filter is typically based on certain criteria, like who sent it.

Spamdexing From Wired Magazine: The practice of entering the same keyword multiple times in a Web page to force it to the top of search results in a search engine.

Spamming Random indiscriminate posting of items (often advertisements) on computer bulletin boards. The term is derived from a brand of pink, canned meat that splatters messily when hurled. A milder form of spamming is called crossposting.

Span 1. Refers to that portion of a high speed digital system than connects a C.O. (Central Office) to C.O. or terminal office to terminal office.
2. Also called a T-Span Line. A repeatered outside plant four-wire, two twisted-pair transmission line.
3. A call center term. The total duration of a schedule from start time to stop time, including all breaks.

Span Line A T-1 link.

Span Powered In T-1, refers to the application of a varying voltage (+130V to -130V) to the digital cable pairs to maintain a 60mA DC current at each repeater and at the customer premises (this power is generally used for regeneration, loop backs, keep alive signals and alarms).

Spanning Tree Algorithm STA. An algorithm, the original version of which was invented by Digital Equipment Corporation, used to prevent logic loops in a bridged network by creating a spanning tree. The algorithm is now documented in the IEEE 802.1d specifications, although the Digital algorithm and the IEEE 802.1d algorithm are not the same, nor are they compatible. When multiple paths exist, says PC Magazine's Frank Derfler, STA lets a bridge use only the most efficient one. If that path fails, STA automatically reconfigures the network to make another path become active, sustaining network operations. This algorithm is used mostly by local bridges; it is not economical for use over leased telephone circuits connecting remote bridges.

Spanning Tree Protocol STP. Inactivation of links between networks so that information packets are channeled along one route and will not search endlessly for a destination. See Bridge.

SPAP Secure Password Authentication Protocol.

SPARC Sun Microsystems' open, RISC-based (Reduced Instruction Set Computer) architecture for microprocessors. SPARC is the basis for Sun's own computer platforms and it's licensed to third parties.

SPARCengine Sun Microsystems' standard microprocessors supporting SBus expansion modules and I/O. 50Mhz and above speeds. Up to 512 MB memory on-board.

SPARC V9 Sun Microsystems' latest microprocessor architecture. It handles real time MPEG-2 decompression and other on-chip multimedia.

Spare Pairs In existing distribution systems, twisted pairs that are not being used and can be used to serve new communications devices. Spare pairs are exactly what they sound like — spare pairs of cables. Best to install as many spares as you can when you initially wire up a building or office. Remember Newton's Rule: You'll always need twice as much cabling as you ever dreamed in your wildest dreams you'd need.

Spark An arc of very short duration.

Spark Gap Terminals or electrodes designed to permit spark discharges to take place across a gap.

Spark Test A test designed to locate pin-holes in a wire's insulation by application of an electrical potential across the material for a very short period of time while the wire is drawn through an electrode field with one end of the wire grounded.

Sparse Network 1. A network concept describing an environment in which the intelligence of the End Offices (Central Offices) largely is stripped away in favor of the placement of relatively few centralized computer platforms which perform the majority of call processing. The dumb switches make calls to the centralized processors which consult associated databases, providing the switches with instructions. The concept of a Sparse Network is fundamental to that of the Advanced Intelligent Network (AIN).
2. A network concept involving many fewer End Offices than are currently deployed. Rather than a user gaining access to a local End Office, traffic would be concentrated at local points and shipped to a larger and more capable office serving a much larger geographic area. Advances in transmission technology, namely fiber optics, make this concept feasible as the cost of transmission bandwidth is dropping precipitously, while the cost of switches (particularly intelligent switches) is not. Hence the concentration of switches and switch intelligence.

SPATA SPeech And daTA. Watch for this expression to pick up steam once true integration of voice and data occurs. The expression does not come from the sentence: "Spata to integrate today than tomorrow."

Spatial Data Management A technique which allows users access to information by pointing at picture symbols on the screen.

SPC 1. Stored Program Control. All phone systems these days are SPCs. There's stored software, which is the program, which controls the computer or microprocessor which in turn controls the operation of the switch. Thus switches are stored program control.
2. Signal Processing Component.

SPC Allocation Service An SCSA definition. A service

which allocates SPCs (Signal Processing Components) to Groups.

SPCAS SPC Allocation Service.

SPCS Stored Program Controlled Switch. A digital switch that supports call control, routing, and supplementary services provision under software control. Pretty well switches made after 1970 in North America are SPCSs.

SPCL SPectrum CeLlular error-correction protocol.

SPE 1. Switch Processing Element or Signal Processing Element. 2. Synchronous Payload Envelope. A SONET term describing the envelope which carries the user data, or payload. The SPE comprises 783 octets, organized into 87 columns and 9 rows. Three different payload structures are defined to address different input requirements: 1. Direct-to-STS-1 line rate multiplexing takes 28 DS-Is, 14 DS-ICs or 7 DS-2s directly into the 51.84 Mbps rate. Each is uniquely transported within the SPE; 2. Asynchronous DS-3 Multiplexing takes a complete asynchronous DS-3 bundle (the output of an M13 for example) into the SPE; 3. Synchronous DS-3 Multiplexing maps a Syntran DS-3 signal to the SPE.

Speaker Adaptive Speech recognition which improves with use. See SPEECH RECOGNITION.

Speaker Dependent Voice Recognition Technology capable of recognizing speech from a given user or others who sound like this user after completion of an enrollment procedure. It is not voice verification although it is sometimes confused with this technology.

Speaker Identification Speaker identification is used to determine the identity of a known speaker. It is accomplished by taking spoken input and searching a database of all known system users for a match. Due to its speaker dependent recognition characteristics, you must first be enrolled as a user prior to using the system. To enroll as a user, an individual is required to speak one or more password phrases which are recorded. These phrases create a reference templates which are stored in the system user database for later use during identification sessions. When in operation, the individual using the system is prompted for a specific password or password phrase. When speaking the prompted password as input it creates a new template. This template is then compared to all reference templates in the system for that particular password. The reference template with the closest match is selected. The uniqueness of each user's voice and the finite number of users of the system makes the identification accuracy quite high. With speaker identification the speaker does not claim to be a particular individual. He or she is identified from a group of common users. For the most part, this technology is used for hands free operation of a system where messages and other information specific to that identified individual are pulled-up for use at that time.

Speaker Independent Voice Recognition SIR or SIVR. Technology capable of recognizing any user without prior training or knowledge of the user. SIR converts speech to accurate and meaningful textual information (typically ASCII). SIR is used to accept input from callers to voice processors where the callers are using rotary dial phones instead of touchtone phones. SIR can substitute for the numbers on the DTMF keypad and can add the benefit of a few basic voice commands, e.g., Yes, No, Help, etc.

Because computer processing demands are formidable with speaker independent recognition, accurate speaker independent products are created with limited vocabularies. In contrast, trainable or speaker dependent recognizers can feature larger vocabularies at lower prices. SIR has been slowly gaining acceptance in telephone applications. SIR is increasingly used in automated operator assistance applications. SIR will see increased use as system builders respond to pressures to provide voice processing functions to the enormous rotary phone installed base domestically and abroad.

Speaker Recognition Having a machine recognize human voice. This is an imprecise term.

Speakerphone A telephone which has a speaker and microphone for hands free, two-way conversation.

Special Access A dedicated line from a customer to a long distance company provided by a local phone company.

Special Billing Number A phone number assigned to certain customers for billing purposes. It cannot be called. It may be given to an operator as the calling number on an outgoing paid call, or it may be used as a "third number billed" number. It's designed as a measure of security and accounting convenience.

Special Characters Microsoft calls special characters that ones not found on your computer's keyboard. In Windows, these characters are accessible through Character Map, an application in the Accessories folder.

Special Distribution A call center term. A half-hourly or quarter-hourly call volume or average handle time distribution created for a day in which calling patterns differ significantly from those normally occurring on that day of the week.

Special Grade Access Line An AUTOVON access line specially conditioned, usually by providing amplitude and delay equalization, to give it characteristics suitable for handling special services; e.g., lower signaling rates of 600 to 2400 bits per second.

Special Information Tone SIT. A series of tones played by the telephone company at the beginning of a recorded announcement, such as indicating the telephone number dialed is no longer in service, has been changed, and so on. Automatic dialers may have the capability of recognizing the different sets of tones, allowing the user to decide whether to pass certain ones through to the agents or filter them out.

Special Night Answering Position Provides either a console or a pre-assigned single extension phone to answer all incoming night calls.

Special Routing Code A 3-digit code in the form 0XX and 1XX available for use within a network and used to modify routing or call-handling logic. End users are prevented from using system codes by the arrangement of the switching equipment to block all customer-dialed calls with a 0 or a 1 in the fourth digit of a 10-digit number, as well as 7-digit calls with a 0 or 1 in the first digit.

Special Services A variety of services that are separate from the public switched network.

Specialized Common Carrier A company providing domestic long distance telecommunications services other than AT&T. See OTHER COMMON CARRIERS.

Specific Gravity The ratio of the weight of any volume of substance to a weight of an equal volume of some substance taken as a standard, usually water for liquids and hydrogen for gases.

Specific Inductive Capacity The direct measure of the ability of a substance to store up electrical energy when used as a dielectric material in a condenser.

Speckle The bright and dark spots on the end face of a fiber caused by the interference of modes.

Spectral bandwidth In telecommunications, the spectral bandwidth for single peak devices is the difference between the wavelengths at which the radiant intensity is 50% (or 3dB) down from the maximum value.

Spectral Efficiency The efficiency of a microwave system in its use of the radio spectrum, usually expressed in bits per Hz for digital radios and KHz per voice channel in analog radios.

Spectrogram A basic research tool for the speech scientist which provides a three-dimensional visual representation of speech.

Spectrum A continuous range of frequencies, usually wide in extent within which waves have some specific common characteristics. See SPECTRUM DESIGNATION OF FREQUENCY.

Spectrum Analyzer Tunable RF instrument which displays a portion of the RF spectrum with amplitude of signals on the vertical axis and frequency on the horizontal axis on a screen. Used in TSCM to analyze transmissions for the characteristics of an illegitimate transmitter (radio bug).

Spectrum Designation Of Frequency A method of referring to a range or band of communication frequencies. In American practice the designator is a two- or three-letter abbreviation for the name. In ITU practice, the designator is numeric. These ranges or bands are:

Management is based on network management principles, the difference being that instead of network elements, it is the spectrum that is being managed. Spectrum management takes into account variables like co-channel and adjacent channel interference, RSSI (Received Signal Strength Indication) values, power levels, frequencies etc. Spectrum Management involves the gathering and using such information proactively to improve network performance. The collected data can be used to generate information that is smart, concise and meaningful. This type of information will help in quicker and smarter decisions thus reducing the delay significantly and supply the carriers with an advantage that not reduces their cost but also adds value to their networks by making them smart, more reliable and proactive.

Speech API See MIcrosoft Speech API.

Speech Concatenation A term used in voice processing for economical digitized speech playback that uses independently recorded files of phrases or file segments linked together under application program control to produce a customized response in natural sounding language. For example, order status, bank balances, bus schedules or lottery results, etc. Concatenation is done for speed and economy. It lends itself to limited and structured vocabularies that are best stored in

Spectrum Designation Of Frequency

FREQUENCY RANGE	TYPICAL AMERICAN DESIGNATOR	ITU FREQUENCY BAND DESIGNATOR
30 - 300 Hz	ELF (Extremely Low Frequency	2
300 - 3000 Hz	ULF (Ultra Low Frequency	3
3 - 30 kHz	VLF (Very Low Frequency)	4
30 - 300 kHz	LF (Low Frequency)	5
300 - 3000 kHz	MF (Medium Frequency)	6
3 - 30 MHz	HF (High Frequency)	7
30 - 300 MHz	VHF (Very High Frequency)	8
300 - 3000 MHz	UHF (Ultra High Frequency)	9
3 - 30 GHz	SHF (Super High Frequency)	10
30 - 300 GHz	EHF (Extremely High Frequency)	11

FREQUENCY RANGE	ITU FREQUENCY BAND DESIGNATOR	
3 - 30 THz	13	
30 -300 THz	14	**KEY**
300- 3000 THz	15	THz = Terahertz (10 to the 12th power hertz)
3 - 30 PHz	16	PHz = Petahertz (10 to the 15th power hertz)
30 - 300 PHz	17	EHz = Exahertz (10 to the 18th power hertz)
300 - 3000 PHz	18	
3 - 30 EHz	19	
30 - 300 EHz	20	
300 - 3000 EHz	21	

Spectrum Management Spectrum Management is a concept you find in the cell business. It is the process of managing the radio spectrum" for purposes of imparting efficiency and intelligence to the spectrum as well as monitoring the spectrum. Spectrum management not only reduces the factors that will hinder the optimal efficiency of the allocated spectrum but also improves the overall performance of each cell and consequently the overall cellular network. Spectrum

RAM (Random Access Memory) or speedily accessible from disk. Concatenation does not replace Text-To-Speech (TTS) as a method of getting the voice processor to deliver its responses. Concatenation, however, can be an excellent complement to TTS when a voice application demands broad, real time vocabulary production. See TEXT-TO-SPEECH.

Speech Digit Signaling Signaling in which digit time slots used primarily for encoded speech are periodically used for signaling (as, optionally, in ISDN). See also ISDN.

Speech Recognition Voice recognition is the ability of a machine to recognize your particular voice. This contrasts with speech recognition, which is different. It is the ability of a machine to understand human speech — yours and everyone else's. Voice recognition needs training. Speech recognition doesn't.

In the August/September, 1996 issue of a magazine called Speech Technology, David L. Basore talked about recent advantages which are making speech recognition systems much more natural. He wrote, "Speech recognition technology currently supports a wide range of viable applications, from voice controlled VCR remotes to sophisticated call center IVR applications. It can be separated conveniently into the following three categories:

1. Isolated word and phrase recognition in which a system is trained to recognize a discrete set of command words or phrases and to respond appropriately.

2. Connected word recognition n which a system is trained on a discrete set of vocabulary words (for example, digits), but is required to recognize fluent sequences of these words such as credit card numbers.

3. Continuous speech recognition in which a system is trained on a discrete set of subword vocabulary units (e.g., phonemes), but is required to recognize fluent speech. For more advanced applications, the vocabulary can be unlimited and the job of the recognizer is to understand the meaning of the spoken input.

A speech recognition system usually is made up of an input device, a voice board that provides analog-to-digital conversion of the speech signal, and a signal processing module that takes the digitized samples and converts them into a series of patterns. These patterns are then compared to a set of stored models that have been constructed from the knowledge of acoustics, language , and dictionaries. The technology may be speaker dependent (trained), speaker adaptive (improves with use), or fully speaker independent. In addition, features such as barge-in capability, which allow the user to speak at anytime, and key word spotting, which makes it possible to pick out key words from among a sentence of extraneous words, enable the development of more advanced applications.

For an explanation of when to and when not to use speech recognition, see MICROSOFT SPEECH API.

Speed Bumps When your host system processes information faster than your network can handle and forces you to slow down.

Speed Dial A feature that enables a PBX or PBX phone to store certain telephone numbers and dial them automatically when a code is entered. See SPEED DIALING.

Speed Dialing Permits fast cialing of frequently used numbers. A repertory of numbers may be stored in the instrument and/or in the telephone switch. Usually a button or one, two or three digits are dialed to activate speed dialing.

Speed Of Light In A Vacuum 299 x 106 meters per second. Used in computing index of refraction.

Speech Synthesizer A device that produces human speech sounds from input in another form.

Speech Transmission Index STI. A measure with a range from 0 to 1.0; 1.0 represents the best possible understanding of a given message. It measures how much of the message can be lost in transmission and still be understood.

Speedsync See LAPLINK.

SPF Shortest Path First.

SPI Service Provider Interface. See SERVICE PROVIDE INTERFACE and WINDOWS TELEPHONY.

SPID Service Profile IDentifier. When you order an ISDN line, your phone company will give you a SPID for every phone number you have. Each ISDN BRI line typically has two phone numbers and thus two SPIDs. The SPID is an eight to 14 digit number that identifies the services you ordered. The SPID is actually a label identifier that points to a particular location in your telephone company's central office memory where the relevant details of your ISDN service are stored. You will need your SPID number because your ISDN phone, fax or PC software asks for it and often cannot be installed without it and certainly won't run without it. Also, when you make an ISDN call, your ISDN device — phone, fax, PC software, etc. — sends that number to your central office, which uses it to validate that you can use the services and features you are requesting. When you order a single BRI service, you will usually get two phone numbers — one for each of the BRI lines. This means that you should then have one SPID number — a different one — for each of those two lines. It's easy to lose your SPID number. Your PC may blow up. Your fax may break. Please write your SPID numbers on your ISDN wall jack. Write them in a file. Tattoo them on your tushy. Make sure you can find them when you need them. You can call your phone company and beg for them. But it's not worth the aggravation. Tip: If you find that the SPIDs which your telephone company gave you don't work, add a 1 or 0 or 000 at the end of the number and try that. By the way, Northern Telecom says that SPIDs for its DMS-100 switch are normally a 12-14 digit numerical code consisting of the user's three digit code, seven digit ISDN telephone number, followed by another zero. According to my ISDN "man" at Bell Atlantic, Matt McGuire, the standard SPID format for NI-1 is now 14 digits. They now look like this

212 691 8215 0101
212 691 8216 0101

However, telephone companies follow a different standard. One company uses this format:

213 691 8215 01
213 691 8215 01

In short, no consistency, once again. Our advice remains. Choose your ISDN equipment first, then have the vendor order your ISDN line (if it will). Make sure you find out your lines' SPIDS and write them down. Do not lose them. If you do, you'll have a dickens of a time getting them out of your local phone company. When ordering ISDN phone lines, ask how much for installation? How much this month? How much next month? Some months your local telephone company gives the service away. Other times, it charges as much as $500. See AutoSpid, a new industry initiative to come up with software and hardware that automatically "negotiates" a connection to the telecom provider by automatically downloading the SPID numbers from your digital central office switch.

Spider A program that prowls the Internet, attempting to locate new, publicly accessible resources such as WWW documents, files available in public FTP archives, and Gopher documents. Also called wanderers or robots (bots), spiders contribute their discoveries to a database, which Internet users can search by using an Internet-accessible search engine.

In January, 1998, I received a solicitation for Bull's Eye Gold, which bills itself as the premier email address collection tool. "This program allows you to develop targeted lists of email addresses. Doctors, florists, MLM, biz opp...Our software uses the latest in search technology called "spidering". By simply feeding the spider program a starting website it will collect for hours. The spider will go from website to targeted website providing you with thousands upon thousands of fresh targeted email addresses. When you are done collecting, the spider removes duplicates and saves the email list in a ready to send format. No longer is it necessary to send millions of ads to get a handful of responses."

Spidering See Spider.

SPIE Society of Photometric Industry Engineers or Society of PhotoOptical Instrumentation Engineering.

Spiff A telemarketing term. An award that forms the prize for a quick motivational incentive. Spiffs can include movie tickets, pizza parties, gifts selected from a catalog and other such prizes.

Spike An in-phase impulse causing spontaneous increases in voltage. Spikes are very fast impulses, less than 100 microseconds, of high-voltage electricity ranging from 400 volts to 5,600 volts superimposed on the normal 120V AC electrical sine wave. See also METAL OXIDE VARISTOR.

Spike Markets A term developed by Apple. By "spike," the company means software and hardware combinations that allow Apple to rise (spike) through a noisy marketplace. By doing this, Apple hopes to grab attention in some reasonably horizontal niche markets and show people why a Mac is worth a few (hundred) extra bucks. The first of these spikes is a foray into home video editing, using Apple's new Performa 6400 and a piece of software called Avid Cinema.

Spike Mike Contact microphone for listening through walls.

Spikes Electrical anomalies represented as short duration, instantaneous, very high voltage fluctuations on an electrical service.

Spill-Forward Feature A service feature, in the operation of an intermediate office, that, acting on incoming trunk service treatment indications, assumes routing control of the call from the originating office. This increases the chances of completion by offering the call to more trunk groups than are available in the originating office.

SPIN Service Provider Identification Number. A number that identifies the telecommunications service provider from which schools and libraries obtain discounted service through the Schools and Libraries Universal Service Program. The program, which was created by the Telecommunications Act of 1996 and which is funded by the Universal Service Fund, provides for discounted telecommunications services, Internet access and internal connections . The Schools and Libraries Corporation approves applications for such discounted rates, which are known collectively as "E-rate."

Spin Stabilization A method of preventing a satellite from tumbling by spinning it about its axis.

SPINA Subscriber Personal Identification Number Access. A term identified in the TIA IS-53 (Interim Standard 53), addressing security in cellular telephone networks. SPINA requires the user to enter a PIN in order to gain access to the network. Access is granted for a specified period of time or until the occurrence of some event, such as the terminal's being turned off. The PIN is transmitted "in the clear," i.e. unencrypted; therefore, it is susceptible to interception. See SPINI.

SPINI Subscriber Personal Identification Number Intercept. A term identified in the TIA IS-53 (Interim Standard 53), addressing security in cellular telephone networks. Unlike SPINA, SPINI may require that the user enter a PIN before each call in order to gain access to the network. Both SPINI and SPINA typically require a 4-digit PIN. The PIN is transmitted "in the clear," i.e. unencrypted; therefore, it is susceptible to interception. See SPINA.

Spindle The rotating hub structure to which the disks in a hard disk system are attached.

Spindle Synchronization A process that coordinates all hard disks in a RAID array to use a single drive's spindle synchronization pulse.

Spinner A name given to people in the United States who regularly change their long distance carrier. Several million people each year switch their long distance carriers. The problem and cost of churn has become fairly major for the long distance industry.

Spiral Life Cycle A term used in COM development. Another term used to describe the iterative development process. Opposite of waterfall life cycle.

Spiral Wrap A term given to describe the helical wrap of a tape or thread over a core.

SPIRIT European consortium focused on standardizing telecom operators' procurement processes.

Splash Tone Distinctive sound used on some phone systems to indicate that a command has been received, or that something has to be done. Vaguely resembles water being splashed.

Splashing A "splash" happens when an Alternate Operator Service (AOS) company, located in a city different to the one you're calling from, connects your call to the long distance carrier of your choice in the city the AOS operator is in. Splashing does not imply backhauling, but it often happens. For example, let's say you're calling from Hotel Magnificent in Chicago. You ask AT&T to handle your call. The AOS, located in Atlanta, "splashes" your call over to AT&T in Atlanta. But you're calling Los Angeles. Bingo. Your AT&T call to LA is now more expensive than it would be if you had been connected directly to AT&T from Hotel Magnificent in Chicago.

Splice Verb. The joining of two or more cables together by splicing the conductors pair-to-pair.

Splice Box A box, located in a pathway run, intended to house a cable splice.

Splice Closure A device used to protect a cable or wire splice.

Splice Tray A place where you splice fiber optic cables and then leave them in the splice tray. It's an elaborate connector.

Spline A curve shape produced on a computer or video device by connecting dots or points at various intervals along the curve. In digital picture manipulators, each key frame becomes a point on a curve and the user can control how straight or curved the path of the transformed image is as it travels through the key frame points.

Splint A pejorative word for Sprint, the third largest interexchange carrier. The word was created by William G. McGowan, the driving force behind MCI for so many wonderful years.

Split A call center term. Split is an ACD routing division that allows calls arriving on specific trunks or calls of certain transaction types to be answered by specific groups of employees. Also referred to as gate or group. Same as Group. See ACD or AUTOMATIC CALL DISTRIBUTOR.

Split Access To Outgoing Trunks Two separate trunk groups provided for direct outward dialing which can be accessed by dialing the same trunk access code. Controlled on class of service basis.

Split Channel Modem A modem which divides a communications channel into separate send and receive channels. Most modems which use the dial-up phone network are split channel — meaning they can transmit and receive simultaneously over a two wire circuit. See also SPLIT STREAM MODEM, which is another term for the same thing.

Split Horizon The view a router has of a wide area network interface in a partial mesh environment where an incoming packet may need to be sent out the same interface over which it is received to reach its ultimate destination. Split horizon is normally disabled to ensure that this cannot occur and that routing loops are not created. This posed a problem with frame relay because packets should be sent back out the same physical interface over which they have been received, but not the same logical interface. Router vendors have now solved this problem, enabling routers to support partially meshed frame relay networks.

Split Link When one multiplexer uses two links to communicate to two separate multiplexers.

Split Pair Something that happens in cable splicing when one wire of a pair gets spliced to the wire of an adjacent pair. It's more accurate to call it a mistake. This error cancels the crosstalk elimination characteristics of using twisted pair wiring in which the two conductors necessary for the circuit are twisted around each other. When you have a split pair, the two conductors come from two different pairs.

Split Resplit Method Split Resplit is an inductive wire-tap. Telephone wires are twisted in bundles so as to reduce crosstalk between telephone lines. The Split-resplit method involves crossing the target line pair with an unused pair of telephone lines with the goal of increasing crosstalk with the operator's line. The operator's line can then inductively pick up conversations on the target line. Because the signal levels are low in the pick up line, audio amplifiers have to be used to clearly hear intercepted audio.

Split Seeks A process by which Novell NetWare SFT III splits multiple read requests between the two servers' disks for simultaneous processing and faster disk rads. Disk reads are split between both servers, with only one server doing a particular read and sending the data read over the MSL (if necessary) to the other server. MSL is the Mirror Server Link.

Split Stream Modem A modem which can handle multiple, independent channels over a single transmission path.

Split System A switching system which implements the functions of more than one logical node.

Split Tunneling Let's start with tunneling. Tunneling typically means to secure a secure, temporary path for your communications via the Internet. For example, a telecommuter might dial into an ISP (Internet Service Provider), which would recognize the request for a high-priority, point-to-point tunnel across the Internet to a corporate gateway. The tunnel would be set up, effectively snaking its way through other, lower-priority Internet traffic. Now imagine a a corporate network that uses both the Internet and a private Intranet or a VPN (Virtual Private Network). Split tunneling simply means the ability to send material securely (that's the critical word) over both networks. You need a device to do the switching and some software. Some devices enable simultaneous connections through the VPN and to the Internet.

Splitter 1. A network that supplies signals to a number of outputs which are individually matched and isolated from each other.

2. A coaxial cable TV device. Imagine a single coaxial cable carrying TV signals. You want to connect two subscribers or two TV sets to this single cable. You insert a splitter, a small cheap (under $10) device. You screw the incoming cable in one side of the splitter. The other side has two screw terminals into which you can screw two coaxial cables — one for each subscriber, or for each TV set. Splitters are passive devices. They don't require external electricity to work.

3. xDSL is the generic name for technology that's puts several megabits of a data transmission on a local loop — from the phone company's central office to your home. A subscriber, like you or I, would use that phone line for two purposes — first, to get onto the Internet and the Web, and second, to speak on the phone. This means that at our house we'd need a xDSL box into which we'd plug our computer and our various analog phones. The industry refers to this box as a "splitter," meaning that it splits the incoming bit stream into voice and data. This is a stupid name for it, since the device is really a multiplexer. And typically such a device would have to be installed by a phone company technician and would replace the demarcation box outside the house. It gets worse. The phone companies have created an adjective called "splitterless" to describe a xDSL box that still splits between voice and data but doesn't require a visit from a telephone company technician. In other words, you'll be able to go down to your friendly local electronics discount store, buy a splitter box, take it home, plug one side into your phone line, and into the other side you'll plug your PC and your analog phone instrument or instruments. And you won't need a visit from your friendly phone company technician. In short, the concept of "splitterless" refers to whether the phone company needs to send a technician to install the box or not. Splitter means it must send a technician. Splitterless means it doesn't have to. By the way, splitterless xDSL technology is not trivial. My friend Paul Sun talks about a splitterless xSDL box as needing to contain five million transistors and have the horsepower of a a 1,000 MHz Pentium PC. See also G.990.

Splitterless For a detailed explanation, see Splitter immediately above. See also ADSL Lite and particularly G.990.

Splitting 1. A filter which splits or separates signals on the basis of their transmission frequency. For example, a splitter can be incorporated into an ATU-R (ADSL Termination Unit-Remote) located at the subscriber premise. The splitter would serve to separate the high-frequency data transmission from the low-frequency POTS voice transmission. The data transmission would then be delivered to a TV set or PC, while the POTS transmission would be delivered to the telephones. See also FILTER.

2. A splitter is a telephone console device which permits an operator to consult privately with one party on a call without the other party's hearing. Or permits a three-party telephone conference user to consult privately with one side of the conference while the other is effectively put on hold. Jumping from one party to the other is called "Swapping."

3. See also Splitter.

SPM 1. Subscriber Private Meter.

2. An AT&T Merlin term. System Programming and Maintenance. A DOS- or UNIX-based application for programming and maintaining the Merlin communications system.

3. Service Provider Messages.

SPN Subscriber Premises Networks.

SPNI Service Provider Network Identifier

SPOI Signaling Point Of Interface. The demarcation point on the SS7 signaling link between a LEC network and a Wireless Services Provider (WSP) network. The point established the technical interface and can designate the test point and operational division of responsibility for the signaling.

Spoofing 1. In COMSEC applications, the interception, alteration, and retransmission of a cipher signal or data in such a way as to mislead the receiver.

2. In automated-information-systems applications, an attempt to gain access to an automated information system by posing

as an authorized user. In electronic mail applications, spoofing is where one person impersonates another person in order to gain access to that person's electronic mail.

3.Spoofing is a networking term. It is a method by which the client and/or router filters network traffic to keep unnecessary traffic from going over a WAN link. THe ability for a device to determine what is not "meaningful" traffic and rather than forwarding the traffic over the connection, the device responds to the source of the traffic with the response that would have been generated by the intended destination device.

SPOOL Simultaneous Peripheral Operation On Line. A program or piece of hardware that controls a buffer of data going to some output device, including a printer or a screen. A spool allows several users to send data to a device such as a printer at the same time, even when the printer is busy. The spool controls the transmission of data to the device by using a buffer and creating a temporary file in which to store the data going to the busy device. See Spooler and Spooling.

Spooler A program that controls spooling. Spooling, a term mostly associated with printers, stands for Simultaneous Peripheral Operations On Line. Spooling temporarily stores programs or program outputs on magnetic tape, RAM, or disks for output or processing.

SPOOLING Simultaneous Peripheral Operations On Line. Spooling means temporarily storing programs or program outputs on magnetic tape, RAM or disks for output or processing. The word "Spooling" is mostly associated with printers. Here's an example: Pretend that a lot of people on your Local Area Network all want to send their reports to the printer today. Instead of each person having control of the printer and relinquishing it only when they're through, each user tells the print Spooler what file they want printed. The program, called the spooler, places the print request in the print queue. When your request reaches the top of the queue, your report is printed out. Using a PC as print spooler slows it down. Best not to use it for much else.

Spot Beam Antenna A satellite antenna capable of illuminating or focusing on a narrow portion of the earth's surface.

Spot Frame See Dedicated Inside Plant.

SPOX An operating system for digital signal processors from Spectron Microsystems, Goleta, CA, now owned by Dialogic in Parsippany, NJ. Spox is a real-time, multitasking operating system that is optimized for use with fixed and floating point digital signal processors in both single- and multiprocessor systems. The SPOX environment is implemented as a library of relocatable, C-callable modules.

SPP 1. Sequenced Packet Protocol. XNS (Xerox Network Systems) protocol governing sequenced data.

2. Signal Processing Platform.

Spread Spectrum Also called frequency hopping, spread spectrum is a modulation technique used in wireless systems. The data to be transmitted are packetized, and spread over a wider range of bandwidth than demanded by the content of the original information stream. Spread spectrum takes an input signal, mixes it with FM noise and "spreads" the signal over a broad frequency range. Spread spectrum receivers recognize a spread signal, acquire and "de-spread" it and thus return it to its initial form (the original message). A large number of transmissions can be supported over a given range of frequencies, with each transmission comprising a packet stream and with each packet in a stream being distinguished by an ID contained within the packet header. The receiver is able to distinguish each packet stream from all others by virtue of that ID, even though multiple transmissions share the same frequen-

cies at the same time, with the potential for the overlapping of packets. Spread spectrum is highly secure. Would-be eavesdroppers hear only unintelligible blips. Attempts to jam the signal succeed only at knocking out a few small bits of it. So effective is the concept that it is now the principal antijamming device in the U.S. Government's Milstar defense communications satellite system. Spread spectrum technology also is used extensively in wireless LANs and in CDMA (Code Division Multiple Access), the access technique used in many PCS (Personal Communications Systems) cellular systems.

There are two versions of spread spectrum. Direct Sequence Spread Spectrum (DSSS) spreads the signal over a wide range of the 2.4 GHz frequency band. Frquency Hopping Spread Spectrum (FHSS) involves the transmission of short bursts of information over specific frequencies, with the frequency-hopping carefully coordinated between transmitter and receiver. See also CDMA, DSSS and FHSS.

Hedy Lamarr, the actress, created the concept of spread spectrum in 1940 and, two years later, received a U.S. patent for a "secret communication system." The patent was issued to her and George Antheil, a film-score composer, to whom Ms. Lamarr had turned for help in perfecting her idea. Spread spectrum was used extensively by the Allies during the World War II in the Pacific Theater, where it solved the problem of Japanese jamming of radio-controlled torpedoes. WW2 electronics were pretty primitive, and Hedy's system used a mechanical switching system, like a piano roll, to shift frequencies faster than the Nazis or the Japanese could follow them. More recently, spread-spectrum has been combined with digital technology, for spy-proof and noise-resistant battlefield communications. In 1962, Sylvania installed it on ships sent to blockade Cuba. Ms. Lamarr never received one penny for her invention. Ms. Lamarr was quite an innovator. She delighted and shocked audiences in the 1930s by dancing onsreen in the nude in the movie "Ecstacy."

Sprint The third largest IXC, behind AT&T and MCI; also a LEC of significance. Sprint began as a venture of Southern Pacific Railroad, which had the clever idea of using its right-of-way to lay a fiber optic cable network. Subsequently, Southern Pacific sold the network to GTE, at which point it became know as GTE Sprint. The company became known as US Sprint when GTE and United Telecom decided to form a (50/50) joint venture from US Telecom (United's long distance company), GTE Sprint and GTE Telnet. United Telecom bought GTE's interest, acquiring the final 19.9% in 1992. Now it's just called Sprint Corporation. From its very beginning, and under its various names, Sprint boasted a fully digital fiber optic network — the first. Through its acquisition in 1993 of Centel, Sprint currently operates as a LEC (local exchange carrier) in 19 states.

Sprite As used in computer graphics refers to a graphic image that can move over a background and other graphic objects in a non-destructive manner.

SPS 1. Signaling Protocols and Switching.

2. Standard Positioning Service. The normal civilian positioning accuracy obtained by using the single frequency C/A code in the GPS (Global Positioning System) system.

3. Solution Provider, also called Microsoft Solution Provider. See MICROSOFT SOLUTION PROVIDER.

SPTS Single Program Transport Stream: An MPEG-2 Transport Stream that consists of only one program.

Spud A special long-handled shovel used to loosen soil in a hole into which you're going to put a telephone pole.

Spudger Shaped like a pencil, it's a gadget phone technicians use to find their way through a multi-paired telephone cable on their hunt for one single pair.

Spurious A term used in voice recognition. A spurious error is said to occur when a sound that is not a valid spoken input is incorrectly accepted as an input speech utterance.

Spurious Emission Emission on a frequency or frequencies which are outside the necessary bandwidth and the level of which may be reduced without affecting the corresponding transmission of information. Spurious emissions include harmonic emissions, parasitic emissions, intermodulation products and frequency conversion products, but exclude out-of-band emissions.

Spurs 1. The sharp metal devices on the climbers used by telephone line-persons (people who climb telephone poles). Such climbing spurs make a mess of telephone poles.
2. The cowboy devices awarded by US WEST to privileged persons who have done US WEST some nice favor or are otherwise deserving of honor.

Sputnik Sputnik was the world's first artificial satellite. It was launched by the Russians on October 4, 1957. It freaked out the Americans and started the space race, which the Americans later won.

SPX Sequenced Packet eXchange. 1. An enhanced set of commands implemented on top of IPX to create a true transport layer interface. SPX provides more functions than IPX, including guaranteed packet delivery. 2. Novell's implementation of SPP for its NetWare local area network operating system.

SQE Signal Quality Error. The 802.3 specification defines this for signals from the MAU to the NIC. Also referred to as heartbeat, is a signal sent by transceivers after a frame is transmitted in order to verify the connection, and is also used by the transceiver to notify the station that a collision was detected. The SQE is primarily used in 10Base-5 environments as a test signal to reassure the station that the transceiver is still operating properly. Some older network devices will not operate properly unless SQE is enabled; almost all new devices do not require SQE. SQE should always be disabled when a transceiver is connected to a repeater (including a 10BASE-T hub), or if it is not required.

SQL Structured Query Language. Invented by IBM and first commercialized by Oracle in the early 1990s, SQL is a powerful database language used for creating, maintaining and viewing database data. It is becoming somewhat of a standard in the mainframe and minicomputer world, and it is on its way to becoming a PC standard. When it is a fully-accepted standard, different computer systems running different DBMSs will easily be able to communicate and exchange data with each other by simply trading SQL commands. SQL is commonly used with database servers, i.e. those running on a local area network. There is now an ANSI standard SQL definition for all computer systems. The largest purveyors of SQL databases are Gupta, Informix, Microsoft, Powersoft, Oracle and Sybase. See ODBC, QBE and SQL SERVER.

SQL Server Microsoft SQL Server. A Microsoft retail product that provides distributed database management. Multiple workstations manipulate data stored on a server, where the server coordinates operations and performs resource-intensive calculations.

Square Key System A "square" key system is one that has all telephone lines appearing on every telephone and each telephone has a separate button or "key" for each line. See Squared Key System for a longer explanation.

Square Operation If there are fewer than eight lines in a Merlin system, all users can access all lines. See SQUARED KEY SYSTEM.

Square Wave A term used to refer to a digital signal, which is binary in nature. This is in marked contrast to an analog "sine wave," which varies continuously in terms of its amplitude and frequency. In other words, digital signals involve only two values: "1" and "0." Computer systems speak digital. Every value (i.e., letter, number, punctuation mark, and control character) is expressed in terms of a specific and unique combination of 1s and 0s of a specific length according to a particular coding scheme. Not only do computers create and store information in such form, they also output information in that form and they expect to see information presented to them in that form. One advantage of digital communications in the form of a bit stream, or stream of 1 and 0 bits, is that computer communications is supported without the need for conversion to analog and back again for transmission across the network. The bit stream is transmitted in the form of a square wave, which consists of discrete values representing these 1s and 0s.

Within the CPE (Customer Premise Equipment) domain and in an electrically-based mode of operation, this representation generally involves a positive voltage for a "1" and a null voltage (0 voltage) for a "0" — this is an "on" and "off" approach. Alternatively, a "1" can be represented as a relatively high level of positive voltage such as +3.0 volts, and a "0" as a relatively low level of positive voltage such as +1.5 volts: in this approach, the electrical circuit continues to flow energy in waves of discrete voltage values. Further still, a "1" can be in the form of a positive voltage such as +1.5 volts, and a "0" can be a negative voltage such as -1.5 volts: again, a "wave" approach. Regardless of the approach, within your own domain you can play the square wave game anyway you (and your manufacturer) choose; after all, it's your game and you pay for the privilege of setting your own rules.

In a public network, it is quite a different matter. T-carrier systems, for instance, require that 1s be represented as alternating positive and negative voltages, while 0s are null voltages. Numerous devices in the public network depend on this electrical coding scheme to maintain synchronization. Further and as the public network serves vast numbers of users, there must be uniformity in order for the network to function at all.

In a radio system, the approach is different still, with the square wave taking the form of radio waves of relatively high and relatively low amplitude.

In an optical network, several approaches can be used. One approach calls for the square wave to take the form of light waves of different levels of intensity (i.e., bright and brighter)—which essentially differences in amplitude, or power level. The second approach calls for the laser light source to pulse on and off, with the presence of light indicating a "1" and the absence of light indicating a "0."

In any event, there are two discrete values represented in the form of "on and off," or "high" and "low," or "plus" and "minus." Digital networks, which use square waves for transmission, offer clear advantages. Most especially, they're much cleaner. Any noise they picked up on their travels across the network is disregarded as the signal is received, boosted and recreated. This is in marked contrast to an analog signal, which simply is amplified, along with any noise which might be present. Square waves are also cheaper to produce. That's good. But if you send a square ringing "wave" to a device like a high-speed modem that's expecting an analog sine wave, that high-speed modem will not respond as it doesn't speak digital at that side of the connection.

Squared Key System A "squared" key system is one that has all telephone lines appearing on every telephone and each telephone has a separate button or "key" for each line. No one quite knows where the word "squared" came from. So if our explanation bears no relation to the word "squared," sorry. But it goes like this (we think): In the old days there were 1A2 phone systems. These 1A2 phones had buttons on them. These buttons could correspond to trunks — any trunk. These were called non-squared systems. Then came electronic key systems. Each trunk had to "appear" (i.e. be) the same button on each phone. These electronic key systems were called squared systems. There are advantages and disadvantages. Squared systems are portrayed as having one advantage: You can go to any phone anywhere in the system and punch any button for any trunk and know it to be the same button for the same trunk. Thus less confusion. But this means you can only have as many trunks on your key system as you have trunk buttons on your key telephones. In a non-squared system — a 1A2, the newer hybrids or some of the newer programmable key systems — you can have more trunks than you have buttons on each phone. Some phones will have trunks that others don't have and vice versa. Thus you can have more trunks on your phone system than you have buttons on your phones. This means, for example, that four executives can have each have private lines and access to four trunks on a six button phone. (The other button is for Hold.)

Squelch A circuit function that acts to suppress the audio output of a receiver. See also squelching.

Squelching Referring to the "Rerouting of Pass-Through Traffic During Node Failures", Nortel's "Introduction to SONET Networking" tutorial handbook says, "While tributaries terminating at the failed node cannot be protected, traffic passing through that node is automatically redirected. ... In an action referred to as "squelching," nodes adjacent to the failure replace non-restorable traffic with a path layer alarm indication signal (AIS) to notify the far end of the interruption in service. The squelching feature employs automatically generated squelch maps that require no manual record keeping to maintain." See also AIS.

Squirt the Bird To transmit a signal up to a satellite. "The crew and talent are ready; when do we squirt the bird?"

SR 1. Speech Recognition. See SPEECH RECOGNITION.

2. Source Routing: A bridged method whereby the source at a data exchange determines the route that subsequent frames will use.

SRAM Static Random Access Memory. A form of RAM that retains its data without the constant refreshing that DRAM requires. SRAM is generally preferable to DRAM because it offers faster memory access times. but it also costs more to make because it has more electrical components. The most common use for SRAM is to cache data traveling between the CPU and a RAM subsystem populated with DRAM. This improves your PC's performance by reducing the number of DRAM accesses needed. See also DRAM and RAM.

SRC 1. Strategic Review Committee (ETSI).

2. Stupid Rich Customer. One who will buy anything.

SRDC SubRate Digital Cross-connect.

SRDM SubRate Data Multiplexer. The Europeans call it SDM. An SRDM typically subdivides DS-0 of 64 Kbps, into a number of circuits, each less than 64 Kbps.

SRF Specifically Routed Frame: A Source Routing Bridging Frame which uses a specific route between the source and destination.

SRM Sub-Rate Multiplexing. SRM. A technique used to combine data from a number of different digital sources into a basic rate channel, efficiently using the bandwidth on the primary rates for data circuits and/or digitized voice.

SRP 1. Soure Routing Protocol. See Bridge.

2. Suggested Retail Price.

SRS 1. Statistics Repository System.

2. Shared Registry System. A neutral, shared, and centralized repository containing the database of Internet domain name information. In conjunction with the expansion of the Domain Naming System (DNS), the Council of Registrars (CORE) has contracted with Emergent Corporation to build, maintain and operate the SRS. SRS supports up to 90 registrars, independent organizations which are authorized to assign the new TLDs (Top Level Domains), comprising .arts, .firm, .info, .nom, .rec, .shop, and .web. See also CORE, DNS, Domain, and URL for longer explanations.

3. Stimulate Raman Scattering. An optical fiber transmission term. Stimulated Raman Scattering (SRS) results from the interaction between the optical transmission signal and the silica molecules in the fiber. SRS affects broadband optical fiber transmission, and affects the overall optical spectrum involved in a DWDM (Dense Wavelength Division Multiplexing) transmission system. The SRS phenomenon manifests itself as a transfer of power from the shorter wavelengths to the longer wavelengths, resulting in a tilt of the optical spectrum. The effect increases as the power of the signal increases, and as the width (density) of the DWDM spectrum increases. See also DWDM.

SRT 1. Station Ringing Transfer.

2. Source Routing Transparent, a token ring bridging standard that is jointly sponsored by the IEEE and IBM. It combines IBM Source Routing and Transparent Bridging (IEEE 802.1) in the same unit. This provides a way for universal bridging of token ring LANs supporting IBM and all non-IBM LAN protocols. An SRT bridge examines each data packet on the ring to discover whether the packet is using a source routing or non-source routing protocol. It then applies the appropriate bridging method. See also BRIDGE, SOURCE ROUTING and TRANSPARENT ROUTING.

SRTS Synchronous Residual Time Stamp: A clock recovery technique in which difference signals between source timing and a network reference timing signal are transmitted to allow reconstruction of the source timing at the destination.

SS-CDMA Spread Spectrum Code Division Multiple Access.

SS7 Signaling System 7. See SIGNALING SYSTEM 7.

SSA In 1993, IBM, working with a committee of other major manufacturers (Conner, Western Digital, Micropolis, etc), announced an architecture, named Serial Systems Architecture (SSA). This SSA removed some of the constraints of SCSI, particularly the limitation on the attachment of storage devices. Where SCSI allows seven devices to be attached on a string, SSA attaches 127 devices on a loop. SCSI uses bulky and expensive straps where SSA employs low-cost thin cabling. Data rates are also increased from 10MB/s on SCSI to 80MB/s on SSA which is expected to increase to 160 MB/s in 1995. The use of SSA will enable faster data transfer, increase the maximum storage capacity and procedure smaller devices, all at lower costs than today's equivalents. In May of 1994, 17 companies issued a joint press release announcing their commitment to SSA. The number of companies working quietly on the development of SSA products and devices exceeds that figure. See also FIBRE CHANNEL, FIREWIRE and SCSI.

SSAP Source Service Access Point

SSB Single SideBand. See SINGLE SIDEBAND.

SSB-SC Single-S deBand Suppressed Carrier.

SSCF Service Specific Coordination Function: SSCF is a function defined in Q.2130, B- SDN Signaling ATM Adaptation Layer-Service Specific Coordination Function for Support of Signaling at the User-to-Network Interface.

SSCOP Service Specific Connection Oriented Protocol: An adaptation layer protocol defined in ITU-T Specification: Q.2110.

SSCP 1. System Services Control Point. A host based network entity in SNA that manages the network configuration, coordinates network operator and problem determination requests, maintains network address and mapping tables and provides directory support and session services.
2. Service Specific Convergence Sublayer: The portion of the convergence sublayer that is dependent upon the type of traffic that is being converted.

SSD Shared Secret Data. A secret key defined in the ANSI-41 (formerly IS-41C) standard, the SSD is used in cellular networks. In conjunction with the A-key (Authentication key), the SSD provides an authentication mechanism for cell phone security. Both keys are encrypted through the CAVE algorithm, and both are known only to the cell phone and the AC (Authentication Center). See also A-key, AuC and CAVE.

SSL Secure Sockets Layer, a transport level technology for authentication and data encryption between a Web server and a Web browser. Developed by Netscape, SSL negotiates point-to-point security between a client and a server. SSL sends data over a "socket," a secure channel at the connection layer existing in most TCP/IP applications.

SSP Service Switching Point. Also called Signal Switching Point. A PSTN switch(End Office or Tandem) that can recognize IN (Intelligent Network) calls and route and connect them under the direction of an SCP (Service ControlPoint. A computer database that holds information on IN (IntelligentNetwork) services and subscribers. The SCP is separated from the actual SCP switch, making it easier to introduce new services on the network. See SCP. For a full explanation of the Advanced Intelligent Network, see AIN.

SST Spread Spectrum Technology. See SPREAD SPECTRUM.

SSTP Switched Services Transport Protocol. See SCTP.

SSU Session Support Utility. A DEC-proprietary protocol that allows multiple sessions to run simultaneously over a single serial cable. SSU is used to allow terminals to provide two session "windows" that can display session output simultaneously.

ST 1. STart signal to indicate end of outpulsing.
2. Straight Tip. A fiber-optic connector designed by AT&T which uses the bayonet style coupling rather than screw on as the SMA uses. The ST is generally considered the eventual replacement for the SMA type connector.
3. Signaling terminal.

ST Connection An optical medium connector plug and socket.

ST Connector See STRAIGHT-TIP CONNECTOR.

ST-506/412 Interface One of several industry standard interfaces between a hard disk and hard disk controller. The "intelligence" is on the controller rather than the drive.

ST-Bus Serial Telecom Bus. Mitel Semiconductor, which makes telecom componentry, has structured its digital component product line around the ST-BUS. The ST-BUS is a high speed, synchronous serial bus for transporting information in a digital format. Whether the digital information is voice, data, or video — or a mixture of each — the ST-BUS is designed to accommodate it. The ST-BUS consists of one or several serial data streams with a framing signal and clock signals. The

framing signal always has a period of 125 us, resulting in 8,000 frames per second, with the original ST-BUS clock rate of 2.048 Mbit/s (thirty two 64 kbit/s channels). The ST-BUS standard now includes higher speed modes of 4.096 or 8.192 Mbit/s ST-BUS, resulting in 64 or 128 channels of 64kbit/s, respectively. This provides the bandwidth necessary for newer multimedia applications. According to Mitel, the advantages of using the ST-BUS are:
1. Printed circuit board area devoted to information transfer between functional modules is minimized.
2. Fewer tracks, backplane connections, and intra-shelf cables are needed compared to systems that use parallel paths.
3. The ST-BUS is designed to be divided down into individual channels of 64 kbit/s, resulting in improved efficiency and lower cost when several information paths are able to share the same ST-BUS.
4. Additional glue logic is not required when using ST-BUS compatible components.
5. From an IC perspective, the ST-BUS results in lower pin counts, improved reliability, and less power consumption.

STA Spanning Tree Algorithm. A technique based on an IEEE 802.1 standard that detects and eliminates logical loops in a bridged network. When multiple paths exist, STA lets a bridge use only the most efficient one. If that path fails, STA automatically reconfigures the network so that another path becomes active, sustaining network operations.

Stack A set of data storage locations that are accessed in a fixed sequence.

Stackable A term referring to devices/system the capacity of which can be increased through connecting (daisy-chaining) the device to additional devices. Thereby, it is not necessary to increase device/system capacity through a complete replacement, often known in the telephone equipment business bas a "forklift upgrade." LAN hubs and switches are often stackable. Such devices often can be interconnected (stacked) in a wiring closet, See also DAISY CHAIN, FORKLIFT, and SCALABLE.

Staffing Basis A call center term. The basis upon which staffing requirements are calculated. Can be either desired service quality for a given day of the week or the number of staff that will product the highest net revenue.

Staffing Requirements Forecast A call center term. A calculation of the number of employees required in each period of the day to handle the forecast call volume for that period.

Stage & Test A term used in the secondary telecom equipment business. The installation (stage) and diagnostic testing of a PBX switch, cabinet, part, or peripheral in a reconfiguration center facility — where a dealer tests the complete system as one entity before shipment.

Stagger In facsimile systems, periodic error in the position of the recorded spot along the recorded line.

Stair Stepping 1. Using a low-level account to gain ever-higher levels of unauthorized access in a network.
2. Video term. Jagged raster representation of diagonals or curves; correctively called anti-aliasing.

Stalker Site A Web site created by an obviously obsessed fan. "Have you seen that Gillian Anderson stalker site? The guy's got like 200 pictures of her!" A Wired Magazine definition.

Stakeholder Corporate stakeholders, a termed coined SAP, include employees, customers, partners and shareholders. As business competition intensifies and planning cycles speed up, corporations will need to keep all these constituents informed and involved, Kevin McKay, CEO of SAP America Inc. said in September, 1998. "We need to communicate with every stakeholder because this will open up new opportunities," McKay said.

Stand Alone Any device that can perform independently of something else.

Standard Industrial Classification SIC. The classification or segmentation of businesses that are increasingly finite based on 2, 4, 6 or more digit identifiers. Developed by the U.S. Department of Commerce in the early 1960's.

Standard Industry Practice Terminology used to indicate normal rules used within the secondary telecom equipment business. These rules have developed over time and usage, but lack formal support by industry groups or dealers. In short, there is no clear definition as to what "standard industry practice" means in the secondary business.

Standard Jack The means of connecting Customer premises equipment to a circuit as specified in the FCC Registration Program.

Standard Metropolitan Statistical Area SMSA. A metropolitan area consisting of one or more cities as defined by the Office of Management and Budget and used by the FCC to allocate the cellular radio market.

Standard Test Zone A single-frequency signal with a standardization level generally used for level alignment of single links in tandem.

Standardized Test Tone A single frequency signal at a standardized power level.

Standards Agreed principles of protocol. Standards are set by committees working under various trade and international organizations. RS standards, such as RS-232-C are set by the EIA, the Electronics Industries Association. ANSI standards for data communications are from the X committee. Standards from ANSI would look like X3.4-1967 which is the standard for the ASCII code. The ITU (now called the ITU-T) does not put out standards, but rather, publishes "recommendations", owing to the international egos involved. "V" series recommendations refer to data transmission over the telephone network, while "X" series recommendations, such as X.25 (properly pronounced "Eks dot twenty five"), refer to data transmission over public data networks. Notice that the ANSI standards have the year they were approved as part of the name of the standard, while ITU recommendations do not. The placement of the "dot" is another clue as to whose confusing standard belongs to whom.

When you're buying a phone system, at minimum it should conform to four standards:

* Emissions compliance according to the FCC Part 15.
* Telephone compliance according to the FCC Part 68.
* Safety standards set by the National Electric Code, OSHA and the Underwriters Laboratories 1459.
* Bellcore compliance (from the Network Equipment Building System publication and their Generic Physical Design Requirements for Telecommunications Products and Equipment publication. See STANDARDS BODIES.

The US Standard railroad gauge (distance between the rails) is 4 feet, 8.5 inches. That's an exceedingly odd number. Why was that gauge used? Because that's the way they built them in England, and the US railroads were built by English expatriates. Why did the English people build them like that? Because the first rail lines were built by the same people who built the pre-railroad tramways, and that's the gauge they used. Why did "they" use that gauge then? Because the people who built the tramways used the same jigs and tools that they used for building wagons, which used that wheel spacing. Okay! Why did the wagons use that odd wheel spacing? Well, if they tried to use any other spacing the wagons would break on some of the old,

long distance roads, because that's the spacing of the old wheel ruts. So who built these old rutted roads? The first long distance roads in Europe were built by Imperial Rome for the benefit of their legions. The roads have been used ever since. And the ruts? The initial ruts, which everyone else had to match for fear of destroying their wagons, were first made by Roman war chariots. Since the chariots were made for or by Imperial Rome they were all alike in the matter of wheel spacing. Thus, we have the answer to the original questions. The United State standard railroad gauge of 4 feet, 8.5 inches derives from the original specification for an Imperial Roman army war chariot. Specs and Bureaucracies live forever. So the next time you are handed a specification and wonder what (expletive deleted) came up with it, you may be exactly right. Because the Imperial Roman chariots were made to be just wide enough to accommodate the back-ends of two war horses.

Standards Bodies See the Appendix at the back of this dictionary.

Standby Processor A spare computer exists which can direct PBX operations if the primary one fails. Some standbys are just sitting there, installed but not turned on. They require someone to turn them on. Some standbys are actually running all the time, as the main one is. If the main one crashes, the standby processor is ready to take over.

Standby Time The amount of time you can leave your fully charged cellular portable or transportable phone turned on to receive incoming calls before the phone will completely discharge the batteries. See TALK TIME.

Standing Wave When you look at it on an oscilloscope the pattern of the wave is perfectly flat, i.e. horizontal. It's caused by two sine waves of the same frequency moving in opposite directions. In transmission line theory the accepted definition is simply the superposition of two waves traveling in opposite directions.

Standing Wave Ratio SWR. The ratio of the amplitude of a standing wave at an anti-node to the amplitude at a node.

Star 1. A topology in which all phones or workstations are wired directly to a central service unit or workstation that establishes, maintains and breaks connections between the workstations. Virtually all phone systems are stars configurations. ISDN BRI bus will be the first phone system to operate on a bus. In datacom language, the center of a star is called the hub. The advantage of a star is that it is easy to isolate a problem node. However, if the central node fails, the entire network fails. The star network we're all most familiar with is our local telephone exchange. At the center (the hub) rests the central office. Spanning out in a star are the lines going to the individual workstations (telephones) in peoples' houses and offices. 2. Advanced telecommunications for the industrially less advanced regions of the European Community.

Star Button The star button on the touchtone phone is often used to mean "No" in interactive voice response or computer telephony systems.

Star Coupler A device that couples multiple fibers at a central point and distributes the signal from one fiber into all others simultaneously.

Star Network A computer network with peripheral nodes all connected to one or more computers at a centrally located facility.

Star Quadded Cable Spiral-four cable. See STAR NETWORK.

Star Topology A LAN topology in which end points on a network are connected to a common central switch by point-to-point links. See STAR.

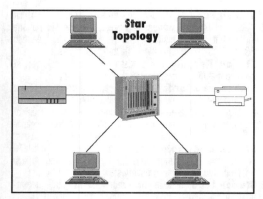

Star Topology

StarLAN An obsolete local area network developed by AT&T using twisted pair telephone wires in a star configuration.

Start Bit In asynchronous data communications, characters are sent at arbitrary intervals, i.e. when the operator hits a key. In order for the computer to make heads or tails of what's coming in, each character starts its transmission with a Start Bit. This way, if the first bit of the character to be transmitted is a 1, the fact of receiving a Start Bit (always a 0) tips off the computer that the next bit is part of a transmitted character and not just part of the inter-character gap. See STOP BIT.

Start Element 1. The start pulse of a transmission character. It is used for synchronization of the following bits in a serial transmission process. 2. One of the input or output points in a communications system. This would include a telephone set, a data terminal, a computer communications port.

Start Of Heading SOH A control character used in data communications that designates the beginning of the message header.

Start Of Heading Character SOH. A transmission control character used as the first character of a message heading.

Start Of Message SOM. A control character used in data communications that designates the beginning of the message.

Start Of Text STX. A control character used in data communications that designates the beginning of the information being transmitted.

Start Stop Transmission The technique of asynchronous data transmission wherein each character is comprised of a start element at its beginning and a stop element at its end. Start-stop elements allow the receiving device to determine where the transmitted bits for one character ends and the next begins.

Start Time Interval A call center term. A scheduling rule that governs the times at which schedules can start; for example, at 15-minute intervals as opposed to 30-minute intervals.

STAT MUX Informal for STATistical MUltipleXor.

State 1. The condition of a connection within a telephone call that reflects what the past action on that connection has been and that determines what the next set of actions may be.
2. The instantaneous properties of an object that characterize that object's current condition. See STATE MACHINE PROGRAMMING.

State Machine Programming To control multiple telephone lines in a single voice processing program, a new program structure is required. Dialogic calls this technique state machine programming. Computer Science called state machines "Deterministic Finite State Automata."

State Public Service Commission PSC. The State legislative body responsible for among other things, regulating the

operation of telephone companies and other persons involved in the furnishing of telephone service. Some states' PSCs are called Public Utilities Commissions. See the next definition.

State Tax Two Out Of Three Rule. When determining state tax jurisdiction for the purpose of figuring phone bills, there are three locations to consider: originating station, destination station, and the location that the bill is sent to. If two out of three are the same, then that state receives the tax.

State Transition The act of moving from one state to another.

State Utility Commissions Each state has a utility commission responsible for the regulation of telephone service provided wholly within that state. Regulation extends to introduction of new services, their prices, who will provide them, as well as discontinuance of existing services.

Stateful Protocols that maintain information about a user's session. FTP is a stateful protocol. Stateless is the opposite.

Stateful Inspection Firewall A stateful inspection firewall examines the contents of individual packets at all layers of the OSI model, from the network layer to the application layer. To perform this task, this firewall relies on packet-filtering algorithms to examine and compare each packet against known bit patterns of authorized packets.

Stateless Protocols that do not maintain information about a user's session. Each transmission is considered a new session. HTTP is a stateless protocol.

Statement 1. In computer programming languages, a language construct that represents a set of declarations or a step in a sequence of actions. 2. In computer programming, a meaningful expression or generalized instruction represented in a source language.

Static Interference caused by natural electric disturbances in the atmosphere, in your office, in your home. Static electricity can play havoc with telephone systems. Properly grounding your phone system to a true cold water pipe (not one that connects to a PVC plastic piper) is the most minimal protection.

Static IP Address See IP ADDRESS.

Static Object Information that has been pasted into a document. Unlike embedded or "linked" objects, static objects cannot be changed from within the document. The only way you can change a static object is to delete it from the document, change it in the application used to create it, and paste it into the document again.

Static Positioning Location determination when the GPS (Global Positioning System) receiver's antenna is presumed to be stationary in the earth. This allows the use of various averaging techniques that improve the accuracy of figuring where you are by factors over 1000.

Static RAM Static Random Access Memory chips do not require a refresh cycle like Dynamic RAM chips and thus can be accessed well over twice as quickly. Static RAM chips must have power to maintain the data they are holding. Static RAM chips cost more than Dynamic RAM chips, which are also called D-RAM.

Static Route A route that is manually entered into a routing table. Static routes take precedence over routes chosen by all dynamic routing protocols. See Static Routing.

Static Routing Static routing involves the selection of a route for data traffic on the basis of routing options preset by the network administrator. Dynamic routing, on the other hand, adjusts automatically to changes in network topology or traffic. Dynamic routing automatically accomplishes load balancing, therefore optimizing the performance of the network "on the fly." Dynamic routing is more effective, but the routers are more costly and the more complex decision-making process imposes additional delays on the subject packet traffic. See also Router.

Static Wire A grounded wire at the very top of a telephone or utility pole intended to protect lower conductors (i.e. telephone, CATV, etc.) from lightning. See Joint Pole.

Station 1. A dumb word for a telephone. Also called an instrument, or a telephone instrument. An extension station is one connected "behind" a PBX or key system. In other words, the PBX or key system is between the station and the telephone central office. We tried to remove the word "station" from this dictionary, but failed. We suspect the word comes from the very old days when the telephone industry was regulated by the Interstate Commerce Commission, (the ICC) which also regulated the railroad industry.
2. A shortened word for workstation — a name for a PC on a LAN.

Station Adapters Cables and interface assemblies for connecting Dialogic network interface and switching products to telephones or analog telephone lines.

Station Apparatus The equipment which is installed on the customer's premises, including phones, ancillary electronics and small hardware.

Station Auxiliary Power Supply This device is used to provide power to an electronic phone that is connected more than 300 meters (or 1,000 feet) or so away from the Key Service Unit.

Station Battery A separate power source which provides the necessary DC power to drive a telephone system. Individual telephones are usually powered by a central source, i.e. their PBX or central office. The batteries may also power radio and telephone equipment as well as provide emergency lighting and controls for equipment. See BATTERY.

Station Busy Lamps Lamps located on a station instrument, providing visual indication of each busy phone in the system. Busy Lamp Fields (BLFs) often come on key systems and sometimes on smaller PBXs. They're very handy.

Station Busy Override Pre selected phones have the privilege and ability to preempt busy circuits and override a private conversation.

Station Call Transfer A phone user can transfer incoming and outgoing calls to another phone without attendant assistance.

Station Camp-On Phones can camp-on to a busy extension. The camped-on phone will be notified of the camp-on by a special beep signal. The person at the other end may or may not hear the signal.

Station Clock The principal clock or alternative clock located at a particular station providing the timing reference for all major telecommunications functions at that station. A station clock may also be used to provide timing or frequency signals to other equipment.

Station Code The final four digits of a standard seven or 10 digit telephone number.

Station Conductor A wire that terminates at the equipment side of the lightning protector.

Station Direct Station Selection The phone user places a call to an extension within the PBX by pushing a single pushbutton on his phone.

Station Equipment Telephone instruments and associated equipment furnished to subscribers. We suspect that the word "station" came from early telephone industry which was regulated by the same government agency which regulated the railroad business.

Station Hunting This feature allows a calling phone which places a call to a busy phone to proceed to the next idle phone in the hunt group. This jump is done automatically. See also ROTARY HUNT, which is the same thing for trunks.

Station Keeping The process on board a satellite for keeping it at its assigned longitude and inclination.

Station Line Cards Station line cards sit inside telephone systems and drive a bunch of connected phones. These cards translate the software inside the phone system into electrical impulses which tell the phones at the other what they're capable of and let them do things, like dial, transfer, conference, etc.

Station Line Protector Circuitry that protects the telephone system from high voltage hits and lightning strikes. Such circuitry is usually on every station line card.

Station Load The total power requirements of the integrated station facilities.

Station Message Detail Recording Now refers to the RS-232-C "port" or plug found on the back of most modern PBXs and some larger key systems. See CALL ACCOUNTING SYSTEM and CALL DETAIL RECORDING.

Station Message Registers Message unit information centrally recorded on a per-station basis for each completed outgoing call.

Station Message Waiting Special light on a phone to alert hotel/motel guests of messages waiting at the front desk.

Station Monitoring Selected phones can monitor (i.e. listen in on) any other phones in the system.

Station Override Security Designated phones can be shielded against executive busy override (presumably other executives).

Station Protector A station protector protects phones and other phone-like devices ("stations") from lightning. A station protector is typically a gas discharge, carbon block or other device that short circuits harmful voltages to ground in the event of lightning strikes on the phone line. Sometimes it works. Sometimes it doesn't. If a bad thunderstorm is about to erupt over your house or office, it's a good idea to unplug your phones from AC and from phone lines.

Station Rearrangement And Change Allows a user to move phones, change the features and/or restrictions assigned to phones and administer features associated with telephones.

Station Review A study of how people in an organization use the telephones and what communications needs are not being satisfied.

Station Ringer Cutoff Allows the ringer on the telephone to be turned off. Not always a good idea, since calls may still come in for that phone, but no one may pick them up because they don't hear it ring.

Station Set Another word for a common desk telephone. Station comes from earliest days when the phone industry in the U.S. was regulated by the same agency that regulated the railroads. It made phones stations, thus easier for the government bureaucrats.

Station To Station Call A directly dialed call. No operator is used. Most calls are now directly dialed. Some long distance companies don't even have operators to help complete calls. AT&T still does.

Station Tone Ringing Electronic tone ringer that replaces the bell.

Station Transfer Security If trunk call is transferred from one phone to another, and the second phone does not answer within a predetermined time, the trunk call will automatically go to the attendant.

Station Visual Signaling Lamp on a phone which indicates flashing incoming, steady busy, and "wink" hold visual conditions associated with that phone.

Stationary Orbit An orbit, any point on which has a period equal to the average rotational period of the Earth, is called a synchronous orbit. If the orbit is also circular and equatorial, it is called a stationary or geostationary orbit.

Statistical Equilibrium A telephone company definition. A state of traffic in which, over any considerable length of time, the call arrivals and departures are essentially equal. Traffic that is in statistical equilibrium has an average value of some measure of its level (such as the number of attempts arriving in a specified time interval) that does not change with time.

Statistical Multiplexing A multiplexing technique that differs from simple multiplexing in that the share of the available transmission bandwidth allocated to a given user varies dynamically. In other words in statistical multiplexing, a channel is assigned only to devices (e.g., telephone, data terminal or fax machine) which are active and seeking to communicate. Static multiplexers, the original multiplexers, dedicated a channel to a device whether it was active or not. This was horribly wasteful, as devices commonly were inactive. As a result, statistical multiplexing is much more powerful than normal static FDM (Frequency Division Multiplexing) or static TDM (Time Division Multiplexing). In other words, Statistical Multiplexers act as contention devices, as well as multiplexers, making intelligent decisions about providing access to expensive bandwidth based on programmable parameters such as first-come-first-served, application priority (e.g., data vs. voice), and bandwidth reservations. See also Fast Packet Multiplexing, FDM and TDM.

Statistics Numbers looking for an argument. The best statistics are those for which you are the sole source or those which are repeated sufficiently often for them to become an integral part of society's conventional wisdom.

Statistics Port In network management systems, interface for reporting events and status.

STATMUX A statistical multiplexer. See STATISTICAL MULTIPLEXING.

Status Information Information about the logical state of a piece of equipment.

Status Signal Unit Signal unit of CCS used to initiate transmission on a link or to recover from loss of transmission.

Statute Mile A unit of distance equal to 1.609 km, 0.869 nmi, or 5,280 ft.

Stay or Bail Moment The precious few seconds that elapse after loading the front page of a Web site, during which one decides to either stay or leave. Defined by David Siegel in his book "Creating Killer Web Sites."

STC 1. Society of Telecommunications Consultants. A professional society for telecommunications consultants. They endeavor to set standards of behavior for the consulting community, chiefly to avoid having telecom consultants recommend to their clients equipment they receive a secret commission on from the manufacturer.
2. System Time Clock: The master clock in an MPEG-2 encoder or decoder system.

STD Subscriber Trunk Dialing. An non-North American term for direct distance dialing, i.e. dialing long distance calls directly without an operator's assistance. Pricing for long distance calls is typically done by billing a standard amount of money, e.g. a German Mark, for a length of speaking time, which shortens the further you call. For example. you might get one minute for a Mark if you're calling 50 miles. If you call

200 miles you might only get 20 seconds.

STDM Statistical Time Division Multiplexer. STDMs are TDMs (Time Division Multiplexers) with an added microprocessor that provides more intelligent data flow control and enhanced functionality, such as error control and more sophisticated user diagnostics. The major difference between TDMs and STDMs is that stat muxes dynamically allocate time slots on the link to inputting devices on an as-needed basis (rather than in round-robin fashion where all devices are polled in preordained order). Therefore, there is no idle time on the link because a device does not have information to send. Unlike TDMs, STDMs have buffers for holding data from attached devices. They can handle a combined input speed (aggregate speed) that exceeds the speed of the communications link.

STE 1. Station Terminal Equipment.
2. Section Terminating Equipment. SONET equipment that terminates a section of a link between a transmitter and repeater, repeater and repeater or repeater and receiver. This is usually implemented in wide area facilities and not implemented by SONET Lite. STE Network elements perform section functions such as facility performance monitoring. The section is the portion of a transmission facility between a lightwave terminal and a line repeater or between two line repeaters.
3. Spanning Tree Explorer: A Source Route Bridging frame which uses the Spanning Tree algorithm in determining a route through a network. Often used in ATM networks.

Steady-state Condition 1. In a communication circuit, a condition in which some specified characteristic of a condition, such as value, rate, periodicity, or amplitude, exhibits only negligible change over an arbitrarily long period of time.
2. In an electrical circuit, a condition, occurring after all initial transients or fluctuating conditions have damped out, in which currents, voltages, or fields remain essentially constant or oscillate uniformly without changes in characteristics such as amplitude, frequency, or wave shape. 3. In fiber optics, synonym for equilibrium mode power distribution.

Steerable Beam Antenna An antenna whose main beam can be directed in various directions either by an electrical or mechanical drive system.

Stentor A new name for the long distance network of nine regional Canadian phone companies. Inspiration for the name came from the Greek poet Homer. He had immortalized Stentor, a warrior in the Trojan war, "whose voice was as powerful as the voices of 50 other men." And so the adjective "stentorian" survived to be applied to someone with a powerful voice, often a politician. The nine members of Stentor are Bell Canada, British Columbia Telephone (BC Tel), AGT Limited, Manitoba Telephone Systems, SaskTel, Maritime Telephone, New Brunswick Telephone, Island Telephone and Newfound4land Telephone.

Step 1. One movement of an electromechanical switch which typically corresponds to one impulse from a rotary dial or one impulse from a touch tone phone which has been converted to a rotary dial. 2. Wang's name for its telephony link is STEP, which stands for Speech and Telephony Environment for Programmers.

Step By Step SXS. An automatic dial-telephone system in which calls go through the switching equipment by a succession of switches that move a step at a time, from stage to stage, each step being made in response to the dialing of a number. SXS is electromechanical switching. It was invented in the 1920s.

Step Call The phone user can, upon finding that the called phone is busy, call an idle nearby phone by merely dialing an additional digit.

Step Down 1. To reduce the voltage. Such reduction in voltage will increase the current.

2. A feature of fax machines that makes them drop their transmission speed when the quality of the phone lines they are transmitting over begins to deteriorate. Dropping the transmission speed is the major way of getting the faxes through on "dirty" lines. All Group III fax machines have "step-down" as a built in feature, or should have.

Step Down Transformer A transformer wound to give a lower voltage on the secondary side than that impressed on (i.e. put into) the primary. The current, however, will be stepped up. A step down (often spelled step-down or step-down) transformer has more primary than secondary turns. See also Stepdown Transformer and Joint Pole.

Step Index Fiber An optical fiber with a core having a uniform refractive index. See also Step Index Optical Fiber.

Step Index Optical Fiber A fiber that has a constant refractive index at its core but a different refractive index as the outer cladding is approached. This design minimizes loses at the core-cladding interface and is preferred for single-mode, long-distance transmission.

Step Index Profile For an optical fiber, a refractive index profile characterized by a uniform refractive index within the core and a sharp decrease in refractive index at the core-cladding interface.

Step Up Transformer A transformer wound to give a higher voltage on the secondary side than that impressed on the primary. The current, however, will be stepped down. It has fewer primary than secondary turns.

Stepdown transformer An oil-cooled transformer often mounted a telephone pole. Such transformer converts the primary voltage to the secondary voltage. Most stepdown transformers are designed for single-phase operation; if a three-phase secondary circuit is required, three physical transformers are sometimes mounted on the same pole. See Joint Pole.

Stepped Index Referring to a type of optical fiber which exhibits a uniform refractive index at the core and a sharp decrease in the refractive index at the core-cladding interface.

Stereophonic Crosstalk An undesired signal occurring in the main channel from modulation of the stereophonic channel or that occurring in the stereophonic channel from modulation of the main channel.

Stereophonic Sound Subcarrier A subcarrier within the FM broadcast baseband used for transmitting signals for stereophonic sound reception of the main broadcast program service.

Stereophonic Sound Subchannel The band of frequencies from 23 kHz to 99 kHz containing sound subcarriers and their associated sidebands.

Steve In German, he means to tailor and that he does with great attention to detail and fine craftsmanship. A more talented man you couldn't hope to meet.

STG An imaging term. Scale To Gray. STG uses gray pixels to fill in jagged edges of document images. STG improves readability. According to a study commissioner by Cornerstone and done by Dr. Jim Sheedy, the ability to read STG images was improved between 4% and 19%, depending on the resolution, and symptoms such as headaches, tired back, blurred vision were cut way down.

STFS Standard Time and Frequency Signal.

STICI Pronounced sticky. It stands for Self Teaching Interpretive Communicating Interface. It's being touted as the "next wave of user interface" (the next wave after GUI).

Apparently STICIs are found on gadgets like the Apple Newton MessagePad. See also GUI.

According to BIS Strategic Decisions, features of STICI include:

Self-Teaching: The operating system (OS) and interface use agents (special background processes) to study how the user makes use of the device. For example, agents will track exactly how a user use the OS and applications.

Interpretive: The system is able to make inferences based on the information it has collected about the user. The system does some interpretation, moving from the traditional interface approach of "Do what I say" to "Do what I mean." Unlike the static menus and dialog boxes of GUI systems, STICI systems will dynamically adapt their operation, thereby better anticipating user needs. Agents will automate common tasks, based on observed usage patterns.

Communicating: The system will be able to manage all the different communications functions offered, such as store and forward, cellular, logging on and off wireless LANs and linking to the user's desktop PC. These communications will be transparent to the user, leaving him or her free to concentrate on the task at hand.

Interface: The interface will be oriented around documents. Traditionally, interfaces have centered on applications, such as a word processor or a spreadsheet. With the STICI, the interface is centered on documents. The user creates a document, writing or drawing freely, and the various applications needed are simply tools accessed to create a chart, to write or to show numbers within that document. Users will be able to seamlessly link applications from multiple vendors, as descendent technologies evolve from OLE and Publish & Subscribe. Both Microsoft and Apple are developing document-oriented interfaces for the next generations of their respective desktop operating systems.

Sticky A sticky shift key lets you access the shifted functions (such as capital A) by pressing the shift key first and then pressing the second key. Sticky keys may stay down for a second or two. Or you may have to hit them again to unstick them — somewhat like the CapsLock key.

STID Service Termination Identifier. An ISDN Service Profile term.

Stimulated Emission Radiation emitted when the internal energy of a quantum mechanical system drops from an excited level to a lower level when induced by the presence of radiant energy at the same frequency. An example is the radiation from an injection laser diode above lasing threshold.

STL 1. Standard Telegraph Level. 2. Studio-To-Transmitter link — typically through the air microwave.

STM Synchronous Transfer Mode. A transport and switching method that depends on information occurring in regular and fixed patterns with respect to a reference such as a frame pattern. A time division multiplex-and-switching technique to be used across the user's network interface for a broadband ISDN. It gives each user up to 50 million bits per second simultaneously — regardless of the number of users. See also ATM.

STM-1 Synchronous Transport Module 1: SDH standard for transmission over OC-3 optical fiber at 155.52 Mbps.

STM-n Synchronous Transport Module "n": (where n is an integer) SDH standards for transmission over optical fiber (OC-'n x 3) by multiplexing "n" STM-1 frames, (e.g., STM-4 at 622.08 Mbps and STM-16 at 2.488 Gbps). The SONET version is known as STS (Synchronous Transport Signal), beginning at 51.84 Mbps.

STM-nc Synchronous Transport Module "n" concatenated:

(where n is an integer) SDH standards for transmission over optical fiber (OC-n x 3) by multiplexing "n" STM-1 frames, (e.g., STM-4 at 622.08 Mbps and STM-16 at 2.488 Gbps, but treating the information fields as a single concatenated payload).

STN 1. Statens Telenamd (Swedish National Telecommunicaticns Council).

2. Super Twist Nematic is a passive matrix technology now used in screens in some laptop computers. In a passive matrix color screen the current travels along transparent electrodes printed on the glass screen. These electrodes are driven by transistors placed around the edges of the display. Horizontal and vertical electrodes form a grid-like matrix, with a pixel at every intersection. A major problem with passive technology arises when current is lost in crosstalk as the electrodes criss-cross each other. This crossing over effect greatly diminishes overall display quality. See ACTIVE MATRIX, LCD and TFT.

Stop Bit The Stop Bit is an interval at the end of each Asynchronous Character that allows the receiving computer to pause before the start of the next character. The Stop Bit is always a 1. See START BIT.

Stop Element The last element of a character in asynchronous serial transmission, used to ensure recognition of the next start element.

Stop Record Signal In facsimile systems, a signal used for stopping the process of converting the electrical signal to an image on the record sheet.

Stop/Start Transmission A method of transmission in which a group of bits are preceded by a start bit and followed by a stop bit. Also called asynchronous transmission. See ASYNCHRONOUS.

Storage Unit A device in which information can be recorded and retained for later retrieval and use.

Store And Forward S/F. In communications systems, when a message is transmitted to some intermediate relay point and stored temporarily. Later the message is sent the rest of the way. Not very convenient for voice conversations, but useful for telex type, and other one-way transmission of messages. Telephone answering machines, as well as voice mailboxes are considered forms of Store and Forward message switching.

Store Locator Service See SINGLE NUMBER DIALING

Stored Procedures Compiled code on a database server that reduces the processing burden on clients.

Stored Program A telephone company definition. The instructions which are placed in the memory of common controlled switching unit and to which it refers while processing a call. Stored programs commonly use alterable magnetic marks to record the program instruction. See also Stored Program Computer and Stored Program Control.

Stored Program Computer A computer controlled by internally stored instructions, that can synthesize and store instructions, and that can subsequently execute those instructions. See also STORED PROGRAM CONTROL.

Stored Program Control SPC. The routing of a phone call through a switching matrix is handled by a program stored in a computer-like device, which may well be a special-purpose computer. Before SPC switches came along, the rotary dialing of the phone caused the elements of the switch to directly "step" through their dialing path. This was slow and cumbersome, since dialing can be slow. Also subscribers can abort half way (they made a mistake) and this can mess up the switch's efficiency. Thus the move to stored program control switches was very significant. These days virtually all switches as stored program control. Nothing happens in the switching matrix until the stored program control receives all the

dialing digits and decides what to do with them.

Stovepiping In a call center, agents typically need access to many databases. In the past they've used dumb terminals. They log into one computer, get into one database, go further into it. When they need information out of another database, they've typically had to climb out of the previous database, the previous computer, log into another and climb down into it. This is called stovepiping, because it follows the contours of a stovepipe. These days, agents have intelligent computers as terminals. They can access several databases at once, by simply having different windows open on their screen or having a front end program that populates a screen with information from several databases, most likely using a GUI interface.

STP 1. Shielded Twisted Pair. Twisted pair (TP) wiring with a metallic shield surrounding the signal-carrying conductors in order to protect them from ambient noise in the form of EMI (ElectroMagnetic Interference). The outer shield may be in the form of a thin metallic mesh in the case of ScTP (Screened Twisted Pair), or a very thin metallic foil in the case of FTP (Foil Twisted Pair). In either case, the shield effectively serves to ensure noise-free information transfer. The shield, however, acts as an antenna, converting received noise into current flowing in the shield; it must be properly electrically grounded or the shield current actually will intensify the noise problem. Any discontinuity in the shield also will result in increased noise. To function effectively, every component of a shielded cabling system must be fully shielded. See also Attenuation and STP-A.

2. Signal Transfer Point. The packet switch in the nation's emerging Common Channel Interoffice Signaling (CCIS) system. The CCIS is a packet switched network operating at 4800 bits per second. CCIS replaces both SF (Single Frequency) and MF (Multi-frequency) by converting dialed digits to data messages. It will run at 56,000 bps with the introduction of Signaling System 7. See SIGNALING SYSTEM 7. For a full explanation of the Advanced Intelligent Network, see AIN.

3. Spanning Tree Protocol. See Bridge.

STP-A Shielded Twisted Pair-A. A modification of the original STP standard, STP-A supports increased carrier frequencies and, therefore, increased transmission speeds. STP-A makes use of the same cable, although the improved connector includes a metal shield between the two conductors as a crosstalk barrier. The new connector is intermateable with the old. STP-A originally was tested up to 100 MHz, and now up to 300 MHz, in support of high-speed Token Ring. See also STP.

Straight-Through When wiring up phone and some data extensions, there are basically two ways of doing it — straight-through and crossover. Straight-through occurs when you wire both ends identically so the signals pass straight through. This is typically done with patch panels and modular EIA adapters. Crossover wiring has a reverse order of wiring. As an example, let's take a four conductor, RJ-11. In a crossover wiring (e.g. an RJ-11 phone extension cord), conductor 1 would be connected to hole 1 on one plug and one 4 on the other end. Conductor 2 would be connected to 4. And 3 would be connected to hole 2.

Straight-Tip Connector ST Connector. An optical fiber connector used to join single fibers together at interconnects or to connect them to optical cross connects.

Straightforward Outward Completion Operator can place an outgoing call for phone user. Also called "Through Supervision."

Strain Relief The connection between the cable and the termination, usually a modular plug, that bonds the cable jacket to the connector so that the individual conductors don't have to absorb tension when the cable is pulled or moved. There are two types of strain relief. The primary strain relief crimps onto the cable's outer jacket where the modular plug meets the cable, and the secondary strain relief crimps onto the rubbery insulation around each conductor inside the business end of the plug. Not all crimp dies crimp the secondary strain relief, and some crimps have a different secondary strain relief location. If the cable jacket and conductors' insulation isn't crimped, the strain of moving or pulling the cable (this is especially important at the desktop, where cables get unplugged and plugged, jostled and pulled) is all borne by your copper connection. Make sure the modular connectors (cable plugs) your technician is crimping have primary and secondary strain relief. Some dies for crimping tools don't support secondary strain relief, which anchors the insulation around the cable conductors to the plug. Strain relief is important because otherwise the fragile copper wire carrying your connection takes all of the tugging and pulling when the cable is plugged in and unplugged or moved. If the sheath of the cable is not attached to the modular plug at, (this is called primary strain relief), your connections has no strain relief at all. The cable sheath should be anchored at the end of the plug away from the connectors. Bye-bye connection.

Strand 1. A single uninsulated wire.
2. Strand (as the term applies to telephone companies) is an uninsulated and unpowered stranded steel cable, installed on telephone and utility poles and similar structures to support telephone cable. Cable is lashed to the strand; other devices are fitted with clamps which attach to the strand. On joint poles, the CATV strand is usually installed below electric power facilities and above the telephone facilities.
3. Strand (as the term applies to cable television) is an uninsulated and unpowered stranded steel cable, typically 1/4" or 3/8" diameter, installed on telephone and utility poles and similar structures to support cable television distribution devices such as hard cable, amplifiers and taps. Cable is lashed to the strand; other devices are fitted with clamps which attach to the strand. On joint poles, the CATV strand is usually installed below electric power facilities and above telephone facilities. See Hard Cable, Joint Pole, and Lashing.

Strand Lay The distance of advance of one strand of a spirally stranded conductor, in one turn, measured axially.

Stranded Conductor A conductor composed of groups of wires twisted together.

Stranded Copper A type of electrical wire conductor comprised of multiple copper wires twisted together forming a single conductor and then covered with an insulating jacket. Stranded conductors perform less well than do solid-core conductors in terms of transmission quality, and are more distance-limited. Stranded conductors have greater flex strength, however; therefore, they are commonly used in applications where the cable is flexed frequently and aggressively.

Stranded Fiber Cable A fiber optic cable in which multiple individual optical fibers contained within the same cable sheath are twisted around each other in a helix. Also twisted with the fibers are strength members, generally constructed of aramid (commonly known as Kevlar). The twisting process improves the flex strength of the cable, much as is the case with stranded copper. If the fibers were not helically stranded, each fiber essentially would stand on its own, and would be more susceptible to fatigue, which would result in the growth of surface imperfections or microcracks, and eventually fiber breakage. See also Aramid, Helix and Stranded Copper.

Strap A permanent, wired connection between two more points.

Stratum Level Stratum Level refers to the accuracy of a SONET clock (or one used for any other application for that matter). Stratum Level 1 is the Cesium beam reference located in Paris, France. Stratum 2 is slightly less accurate, but still able to keep a SONET system in sync without a reference for about 3 to 5 days. Sync is defined as plus or minus one T-1 frame (125 microseconds). Stratum 3 is less accurate than Stratum 2, etc. Stratum timing is one of the most fundamental concepts as to what makes SONET possible.

Stray Current Current through a path other than the intended path. See also SPURIOUS EMISSION.

Strawman This concept is widely used in selling. The simple idea is to set up a Buyer's Checklist and tell your prospective customer that this Checklist is objective. Any product that meets all the criteria is worth buying. Of course, there's only one product that meets all the criteria. It's yours.

Stream An SCSA term. One of 16 physical data lines making up the SCbus or SCxbus Data Bus. See S.100.

Streamer Streaming tape drive.

Streaming An Internet term. A Web page typically consists of text and graphics (still and moving) images. The text is typically fewer bytes than the graphics which are heavy on bytes. Thus, to receive the text to your PC from their Web page typically takes much less time than receiving the graphics images. So Netscape had an idea, which they first pioneered in their browser. Let's get the text up on the user's screen fast, and paint the user's screen with the images as they came in. This allowed the user to look at a new page of text on screen as the graphics came in over the phone lines. Netscape called this streaming. That was the first use of streaming. But then a company called Real Networks came along. It had an idea. Wouldn't it be nice if we could put audio recordings and video (clips, movies, etc.) on a Web site and have people click on them and start hearing or seeing them immediately — as against (in the pre-Real Networks' days) waiting to download the entire file, then playing it. See home page, Internet and streaming media.

Streaming Media After Netscape defined the concept more narrowly — see Streaming — Real Networks defined the concept more broadly to audio and video coming to you in packets over the Internet. The idea of the "stream" is that it is so fast that the audio sounds like radio and the video become full-blown 30 frames per second video, comparable in quality to commercial, over-the-air TV. We aren't there, yet. But we're getting closer. See streaming.

Streaming Tape Backup A device to back up files and programs. A streaming tape backup looks very much like a large audio cassette. It records data sequentially.

Streaming Tape Drive A magnetic tape unit especially designed to make a nonstop dump or restore magnetic disks without stopping at interblock gaps.

Streams An architecture introduced with Unix System V, Release 3.2 that provides for flexible and layered communication path between processes (programs) and device drivers. Many companies market applications and devices that can integrate through Streams protocols.

Street Price The real selling price of computers, hardware, and software. Most laptop and desktop computers sell for 25 percent below list price. Software may be discounted even more.

Street Talk The Banyan-developed protocol for discovering and maintaining resource information distributed among the servers connected to Banyan's VINES network operating system. Also known as a global naming service.

Streetsweeper A heavy duty shotgun with a revolving round magazine typically holding 18 12-gauge or 20-gauge shotgun shells. This word crept into a story the Wall Street Journal ran on cellular fraud. When the Feds rang a cellular phone store as a sting operation, one customer offered to trade his streetsweeper in on a phone. That's how dependent Detroit's drug-traffickers had become on cellular phones and beepers.

Stress Puppy A person who thrives on being stressed-out.

String A sequence of elements of the same type, such as characters, considered as a unit (a whole) by a computer. A data structure composed of a sequence of characters, usually in human-readable text.

Striping 1. A RAID (Redundant Array of Inexpensive Disks) term. The process of dividing a large logical block of data into multiple physical blocks of equal size for storage on multiple disk drives. See RAID and REDUNDANT ARRAY OF INEXPENSIVE DISKS.
2. A Fibre Channel definition. Striping is a way of achieving higher bandwidth using multiple N_ports in parallel to transmit a single information unit across multiple levels.

Strobe A signal that triggers a data reading, transfer of information or sampling. Such a sampling might be to figure if a circuit is active and, if so, what level of activity is taking place. The sampling process might allow a carrier to bill the user correctly for circuit usage.

Stroke A straight line or arc that is used as a segment of a graphic character.

Stroke Edge An imaging and OCR term. In character recognition, the line of discontinuity between a side of a stroke and the background, obtained by averaging, over the length of the stroke, the irregularities resulting from the printing and detecting processes.

Stroke Speed In facsimile systems, the number of times per minute that a fixed line perpendicular to the direction of scanning is crossed in one direction by a scanning or recording spot. In most conventional mechanical systems, this is equivalent to drum speed. In systems in which the picture signal is used while scanning in both directions, the stroke speed is twice the above figure.

Stroke Width In character recognition, the distance measured perpendicularly to the stroke centerline between the two stroke edges.

Strong text The HTML character style used for strong emphasis. Certain browsers display this style as bold.

Strowger, Armond The man who invented the telephone dial and the earliest automatic telephone switch as a method of allowing the user to complete calls without using the Operator. In Kansas City in the late 1800's, Ol' Armond was an undertaker who wasn't getting much business. That's because the girlfriend of a rival undertaker was a telephone operator, and when she got a call asking for the local undertaker, she forwarded the calls to her boyfriend. This story may or may not be apocryphal. But it's a great story.

Structured Query Language SQL. A relational database language (ANSI Standard) that consists of a set of facilities for defining, manipulating and controlling data.

Structured Programming A technique for organizing and coding (computer) programs in which a hierarchy of modules is used, each having a single entry and a single exit point, and in which control is passed downward through the structure without unconditional branches to higher levels of the structure. Three types of control flow are used: sequential, test, and iteration.

Structured Wiring As data flows have sped up in recent years and as moves, adds and changes have proliferated, so the erstwhile idea of wiring up a building with plain old analog voice telephone wire has become no longer intelligent. The idea then came up of defining wiring standards and flexible schemes so that a user could feel comfortable about choosing a complete solution for wiring phones, workstations, PCs, LANS and other communicating devices throughout the building, the campus, the network, the company and throughout his life in the place. Consistency of design, flexibile layout and logic are the keys to structured wiring systems. Typically a structured wiring system consists of two elements:
1. Manufacturer-originated standard components that link wires together in a systematic, intelligent way.
2. A set of rules for building smart wiring. The ANSI/TIA/EIA-568-A and ISO/IEC 11801 standards specify the minimum requirements for telecommunications cabling within a commercial building. The Commercial Building Telecommunications Wiring Standard is available from Global Engineering Documents, Englewood CO 314-726-0444 / 800-854-7179.

A structured cabling system will improve performance in five ways, according to Anixter, a leading supplier of structured wiring systems:
1. It eases network segmentation, the job of dividing the network into pieces to isolate and minimize traffic, and thus congestion.
2. It ensures that proper physical requirements, such as distance, capacitance, and attenuation are met.
3. It means adds, moves, and changes are easy to make without expensive and cumbersome rewiring.
4. It radically eases problem detection and isolation.
5. It allows for intelligent, easy and computerized tracking and documentation.

"Structure" brings order to what has often been an afterthought — wiring. The main pieces of a structured wiring system are:
1. Drop cable. The cable that runs from the computer to a network outlet.
2. Cable run. The cable that runs from the outlet to the wiring closet.
3. Patch panel. A board that collects all the cable runs in one place and "patches" them to different parts of the wiring concentrator. Network managers (users or their secretaries, it's that simple) change the LAN layout by plugging and unplugging "patch cables" between the patch panel and the wiring concentrators. No rewiring is necessary to move one user from one network segment to another.
4. Wiring concentrator. It makes the network connections. Some wiring concentrators are dumb, making only physical connections between network segments. Others are intelligent, making networking decisions and providing network diagnostics. A wiring concentrator can have bridges and routers that divide the network into segments. It can have the hardware necessary to change from one media, say twisted pair, to another, say fiber optic. And it can contain the hardware to change from one network type to another, say from Ethernet to Token Ring.

Here is a glossary of structured wiring words, with thanks to Anixter.

Access Method
The method of "communicating" on the wire. Examples include Ethernet, Token
Ring, AppleTalk, AS400 and 3270.
Cable Type (Media)
The type of cable used in the system. Examples are coaxial,

UTP, STP and fiber. Factors including cost, connectivity and bandwidth are important in determining cable type.

Data Speeds

Different interconnect products (cables and connectors) are capable of supporting different data rates. For instance, Level 3 cable supports data rates up to 10 Mbps. (See LEVEL).

Environment

Where the structured wiring system is found. The large majority of systems are located in office environments as opposed to factory or industrial environments.

Life Cycle

How long the cable is physically anticipated to be in place. For example, if a customer intends to be in a large office for 10 years, fiber installation may be considered.

Methodology

The physical means of getting the wiring system to the user (its distribution path). Examples include modular furniture, surface mounts, fixed wall, recessed wall, raised floor and undercarpet wiring.

Topology

The way the cable is physically laid out or configured. Examples include

star, ring, daisy chain and backbone.

Structured Wiring System See STRUCTURED WIRING.

STS Synchronous Transport Signal. The electrical equivalent of SONET OC-level. The signal begins as electrical and is converted into optical prior to presentation to the fiber optic medium. The STS frame consists of the Synchronous Payload Envelope (SPE), Section Overhead (SOH), Line Overhead (LOH), Path Overhead (POH), and Payload. SOH and LOH comprise what is known as Transport Overhead (TOH). See also STM.

STS-1 Synchronous Transport Signal level 1. An electrical signal that is converted to or from Sonet's optically based signal; equivalent to the OC-1 signal of 51.84 Mbps.

STS-3 Synchronous Transport Signal level 2. (yes, that's right. It's STS-3, though it's level 2.) ATM Physical Layer implementation supporting 155 Mbps.

STS-n Synchronous Transport Signal "n" : (where n is an integer) SONET standards for transmission over OC-n optical fiber by multiplexing "n" STS-1 frames, (e.g., STS-3 at 155.52 Mbps STS-12 at 622.08 Mbps and STS-48 at 2.488 Gbps).

STS-nc Synchronous Transport Signal "n" concatenated: (where n is an integer) SONET standards for transmission over OC-n optical fiber by multiplexing "n" STS-1 frames, (e.g., STS-3 at 155.52 Mbps STS-12 at 622.08 Mbps and STS-48 at 2.488 Gbps but treating the information fields as a single concatenated payload).

STU Secure Telephone Unit.

STU-III The third generation of secure telephone units used by the military and its suppliers.

Studio-To-Transmitter Link STL. Any communication link used for transmission of broadcast material from a studio to the transmitter. It's typically microwave radio but it may also be a conventional landline link.

Study Group 15 The ITU, a United Nations agency, coordinates the development of global communications standards. Study Group 15 of the ITU Telecommunication Standardization Sector (ITU-T) is where the work on communications specifications is carried out. It is responsible for the standards development in the area of transport networks, systems and equipment. See also G.990.

Stunt Box A device to 1. control the nonprinting functions of a teletypewriter terminal, such as a carriage return and line feed and 2. a device to recognize line control characters.

Stutter Dial Tone Stutter Dial Tone is the broken-up tone a user hears on a phone when they pick up their phone to make a call and they have a message waiting in voice mail. This is used to notify users that they have a voice mail message when the phones don't or can't have a message-waiting light. They are given stutter dial tone instead of regular dial tone.

STX Start of Text. See PACKET.

Stylus A pen-shaped instrument (usually made out of plastic) that is used to enter text, draw images, or point to choices on a computer — desktop or PDA. The pen is designed to make writing on the screen feel just like writing on paper. Yes!

SU 1. Subscriber Unit. A radio frequency modem used to acquire the airlink. A wireless term.

2. Service User. The end user at the customer premises.

Sub Substitute Character. A control character used in the place of a character that has been found to be invalid or in error.

Subaddressing A name for an ISDN service which enables many different types of terminals — phones, fax machines, PCs, etc. — to be connected to the ISDN user interface and uniquely identified during a call request. See ISDN.

Subcarrier 1. A carrier which modulates a main carrier so that two different modulating signals can be transmitted simultaneously, one on the main carrier and one on the subcarrier. See Sub Carrier Modulation.

2. In NTSC or PAL video, a continuous sine wave of extremely accurate frequency which constitutes a portion of the video signal. The subcarrier is phase modulated to carry picture hue information and amplitude modulated to carry color saturation information. The NTSC subcarrier frequency is 3.579545 MHz, and the PAL-I frequency is 4.43361875 MHz. A sample of the subcarrier, called color burst, is included in the video signal during horizontal blanking. Color burst serves as a phase reference against which the modulated subcarrier is compared in order to decode the color information.

Subcarrier Modulation Subcarrier modulation combines a signal with a single low frequency sine wave. The low frequency signal is called a sub-carrier. This combined signal is then added to a higher frequency radio signal. The resulting high frequency radio signal is very complex and the original signal is not detectable by ordinary means. To detect a signal that has been modulated by a subcarrier, it must be passed through two detector circuits, one to separate the subcarrier from the high frequency radio transmission, and a second to separate the sub-carrier from the desired information.

Subconference A teleconferencing term. During the course of a large teleconference, the moderator can hold an off-line subconference (i.e., a caucus, or closed meeting) with a number of participants. During the subconference, the other participants remain connected to the main conference. Once the subconference is completed, that group rejoins the main conference.

Sublayer A logical sub-division of a layer.

Submarining Same as CURSOR SUBMARINING. When you drag your cursor across a screen and the cursor disappears as you move it. That's called Cursor Submarining. It happens most on monochrome LCD screens because they change slowly — much slower than active matrix screens or CRTs or VDTs (glass screens).

Submission In X.400 terms, the transmission of a message or probe from a originator's UA (User Agent), MS (Message Store), o AU(Access Unit) to an MTA (Message Transfer Agent).

Submodule A small circuit board that mounts on a larger module. Also called a daughterboard.

Subnet a portion of a network, which may be a physically independent network, which shares a network address with other portions of the network and is distinguished by a subnet number. A subnet is to a network what a network is to an internet.

Subnet Mask A number used to identify a subnetwork so that an IP address can be shared on a LAN.

Subnet Number A part of the internet address which designates a subnet. It is ignored for the purposes internet routing.

Subnetwork 1. A collection of OSI end systems and intermediate systems under the control of a single administrative domain and utilizing a single network access protocol. Examples: private X.25 networks. collection of bridged LANs. 2. A token ring LAN that is used to serve the communication needs of a department. Subnetworks are normally connected to token ring backbones via token ring bridges or routers so that they can communicate with other subnetworks via the backbone or with computers directly connected to the backbone. 3. An ATM term. A collection of managed entities grouped together from a connectivity perspective, according to their ability to transport ATM cells.

Subnetwork Access Protocol SNAP. A version of the IEEE local area network logical link control frame similar to the more traditional data link level transmission frame that lets you use nonstandard higher-level protocols.

subNMS An ATM term. Subnetwork Management System: A Network Management System that is managing one or more subnetworks and that is managed by one or more Network Management Systems.

Subroutine A functionally isolated program or sequence of instructions for a specific function that is often called by a program. A piece of software that performs a useful function that will be needed often. The code for the subroutine is stored on disk (like a letter, etc.) and dropped into a larger program as needed. A nicely-written subroutine saves you "re-inventing the wheel" and allows you to re-use your code in many programs.

Subscriber A person or company who has telephone service provided by a phone company. In other industries, subscribers are called customers. Some telephone companies are beginning to call their subscribers customers. About time.

Subscriber Access Terminal SAT. A SMDS term for DTE in the context of a SMDS network. The SAT gains access to the network through either a SNI or DXI interface.

Subscriber Line The telephone line connecting the local telco central office to the subscriber's telephone instrument or telephone system.

Subscriber Line Charge SLC. A monthly charge on subscribers created by the Federal Communications Commission and paid to the local telephone company. The logic for this charge has something to do with reimbursing the local phone companies for some costs which they are allegedly not recovering elsewhere. In reality, it's just another rate increase. The SLC also is known variously as the Access Charge, CALC, and EUCL. See also Access Charge.

Subscriber Loop The circuit that connects the telephone company's central office to the demarcation point on the customer's premises. The circuit is most likely a pair of wires. But it could be three wires if some external signaling is being used. It could also be four wires if the circuit was a four-wire full duplex leased line.

Subscriber Loop Carrier See SLC-96.

Subscriber Network Interface See SNI

Subscriber Number The number that permits a user to reach a subscriber in the same local network or numbering area (same as Directory Number or DN).

Subscriber Plant Factor A planning factor used by common carriers to allocate investment in phone equipment, subscriber lines and the non traffic sensitive portion of the central office equipment.

Subscriber Premises Equipment Cable-related equipment located on the subscriber premises, whether owned by the subscriber or the cable system. This term includes:

• Subscriber-owned consumer-electronics equipment (TV sets, VCRs, FM tuners, closed-caption decoders).

• Subscriber-owned terminal devices (generic converters, digital audio tuners).

• System-owned terminal devices (generic converters, converter/descramblers, digital audio tuners, special equipment to enable simultaneous reception of multiple signals). Compare with Subscriber Terminal.

Subscriber Terminal The point at which the subscriber-owned cable television equipment is connected to the cable system; typically, a 75-ohm 'F'-connector or a 300-ohm balanced line. Compare to Subscriber Premises Equipment.

Subscriber Trunk Dialing STD. The European version of direct-distance dialing. Pricing for long distance calls is typically done by billing a standard amount of money, e.g. a German Mark, for a length of speaking time, which shortens the further you call. For example. you might get one minute for a Mark if you're calling 50 miles. If you call 200 miles you might only get 20 seconds.

Subscriber Unit SU. The Radio Frequency (RF) modem used to acquire the airlink; can be an integral part of the Mobile End System (M-ES) or a separate component.

Subscriber Drop Wire which runs from a cable terminal or distribution point to the subscriber's premises.

Subset A contraction for Subscriber Set, or telephone set.

Subsplit A method of allocating frequencies in a broadband transmission system. Transmit frequencies are in the range of 5 to 32 megahertz, and receive frequencies are in the range of 54 to 300 megahertz.

Substation An additional phone which has been established as an extension to the main phone or primary line.

Substitute Character A transmission control character used in place of a character found to be in error.

Substitution A word used in voice recognition to mean a type of error that occurs when a word within the active vocabulary is spoken correctly but classified as another word within the vocabulary. This error is usually dealt with during a verification stage in an application, i.e. " you said, 1,2,3...correct?"

Subvoice-Grade Channel A communications channel of bandwidth narrower than a standard 3Hz voice line. A subvoice-grade channel is usually used for slow data transmission such as teletype or telemetry.

Sucker Trap A feature of a security firewall. Sucker traps log access attempts, separate legitimate from illegitimate users, and maintain an audit log of the illegitimate.

Suhler A synonym for intellect heft. Industrial grade socks are necessary to support the ideas. Watch out for the padding as they creep up on you. The socks and the ideas.

Suit A pejorative term for a professional manager. The term often is used by bright, hardworking folks who work for a start-up company. Such people tend to be very creative and tend not to dress casually. When the startup becomes successful, it often goes public or is bought by a large, well-established firm. In come the "suits" — professional managers who dress in fancy, expensive suits and have sport arrogance to match. These professional managers then proceed to mess up the company because they try and install ponderous, "big company" practices - lots of budgeting, procedures and policies — on a company that succeeded because it was light of foot. The worst of the "suits" are known as "empty suits."

Suite A collection. A suite of software tools is a collection of software tools.

Summary Address An ATM term. An address prefix that tells a node how to summarize reachability information.

Summary Billing Some telecom carriers will give you one monthly consolidated phone bill — no matter how many number accounts you have in your billing area. Ask.

Summation Check A check based on the formation of the sum of the digits of a numeral. The sum of the individual digits is usually compared with a previously computed value.

Sun Microsystems Sun Microsystems is a Californian computer manufacturer. SUN stands for Stanford University Network. See Java. www.sun.com

Sun Synchronous A term describing the fact that the orbits of LEOs (Low Earth Orbiting) and MEOs (Middle Earth Orbiting) satellite systems can be adjusted such that the greatest number of satellites in the constellation are positioned over geographic areas which are in the light of day. At such times, the greatest amount of traffic originates and terminates. See LEO and MEO.

Sun Transit Outrage Satellite circuit outage caused by direct radiation of the sun's rays on an earth station receiving antenna.

Sundown Rule A rule in voice mail which says that all messages should be returned that day, before the sun goes down.

SUNOS SunOS is Sun Microsystems' implementation of UNIX.

SunPC Hardware/software solutions from Sun Microsystems for MS-DOS on SPARC platforms.

SunView SunView is Sum Microsystems' kernel-based window system.

SunXTL Server Part of Sun Microsystems' XTL Teleservices architecture. Provides multi-client and multi-device support. The server is the central point of contact for all teleservices services. Resource management and security are provided by the server. Communicates with the Sun XTL provider to place and receive telephone calls. An application may access the data associated with a call by acquiring a data stream from the API. www.sun.com

Super G3 Super G3 is a new unofficial "standard" for higher speed fax machines, which contain a 33.6 Kbps V.34 modem, V8 handshaking and the new ITU-T T.85 JBIG image compression. On most phone lines such a machine should get close to double the speed of the highest speed Group 3 fax machines, namely 14.4 Kbps. But, the JBIG image compression will speed faxing of gray scale images by as much as five to six times. In short, these machines will send faxes much faster — if they send to a Super G3 machine at the other end. Super G3 is compatible with and can communicate with older fax machines, Group 1, 2, 3 and 3 Enhanced.

Super Server A file server with more than one CPU (Central Processing Unit). At time of writing this dictionary, a high-end super server might contain four Intel Pentium chips. To take advantage of these super servers, you need an operating system capable of asymmetrical multi-processing, such as Unix and Windows NT Advanced Server.

Super Speed Calling A feature of Northern Telecom's DMS line of central offices. Subscribers can use a four-letter dialable name, preceded by an octothorpe (#), to speed-call up to 14 digits. For example, "#WORK" or "#HOME" could be used as a code for a longer number. Dialing is faster, and subscribers have the added convenience of using easier-to-remember names instead of forgettable numbers on their speed calling list. The list size can be set by the telephone operating company at initial setup on a per-system basis. Northern's default is 12 names per list.

Super-JANET The latest phase in the development of JANET, the UK educational and research network run by UKERNA. It uses SMDS and ATM to provide multi-service network facilities for many new applications including Multimedia Conferencing.

Supercomputing A term applied to a class of high-speed computers employing advanced technologies such as simplified instruction sets, wide data paths and pipelining.

Superconductors Superconductors are materials which have no resistance to the flow of electricity. They are widely believed to have great potential for dramatically faster telecommunications switches and computers. In the past, the superconducting state — zero resistance to the flow of electricity — could be achieved only by cooling certain metal alloys to temperatures of near absolute zero, or about 460 degrees below zero. Starting in 1986, researchers discovered that ceramic materials could reach superconductivity at temperatures as high as 235 degrees below zero.

Superdrive The name for Apple's 1.44 Mb floppy that can read and write MS-DOS formatted floppies and Mac formatted disks. DOS floppies require Apple File Exchange or a third party product to read the DOS format.

Superframe Format The superframe transmission structure consists of 12 DS1 frames (2316 bits). The DS1 frame comprises 193 bit positions, the first of which is the frame overhead-bit position. Frame overhead bit positions are used for the frame and signaling phase alignment only.

Supergroup Sixty voice channels. In more technical terms: the assembly of five 12-channel groups occupying adjacent bands in the spectrum for the purpose of simultaneous modulation or demodulation.

Superheterodyne A type of radio receiver operating on the heterodyne or beat principle. See HETERODYNE.

Superhouse The National Society of Home Builders has registered SmartHouse. BellSouth has trademarked SuperHouse. And GTE has registered SmartPark. One day, fiber optic will snake to everyone's house, bringing the potential of immense information services. Until that day comes they'll be lots of interesting demonstrations at distant trade shows.

Superimposed Ringing A way of stopping party line phone users from hearing each other's ring by superimposing a DC (direct current) voltage over the ringing signal and using it to alert a vacuum tube or semiconductor device in only the phone instrument that we want to ring. See also SUPERPOSED RINGING.

Superposed Circuit An additional channel obtained from one or more circuits, normally provided for other channels, in such a manner that all the channels can be used simultaneously without mutual interference.

Superposed Ringing Party-line telephone ringing in which a combination of alternating and direct currents is used, the objective being to only ring the bell in the phone of the one whose call is coming in. See also SUPERIMPOSED RINGING for a better explanation.

Superserver IBM's new name for a mainframe. The name is clearly an attempt to position the mainframe more competitively against the new extremely powerful PCs which are taking over the mainframe's role.

Supersite Toby Corey, president of USWeb Corp., defines supersite as an Internet site combining a public Internet site, an extranet for business partners and an administrative intranet. He says a supersite has a common architecture across intranet, Internet, extranet and Web sites. He says it serves multiple audiences and it implements various levels of access control.

Superstation A commercial television broadcast station which is transmitted to a cable television headend by a communications satellite and then retransmitted by the cable system to its subscribers. FCC rules provide that a cable television system may carry a superstation under the same conditions that it may carry any other television broadcast station. Examples: WGN-TV and WWOR are superstations (TBS is not a superstation, in spite of its self-proclaimed status as "Superstation TBS.") Compare Origination Cablecasting.

Supertrunk A cable that carries several video channels between facilities of a cable television company. A trunk between the master and the hub heaends in a hub CATV system.

Supervised Transfer A call transfer made by an automatic device such as voice response unit which attempts to determine the result of the transfer — answered, busy, ring no answer — by analyzing call progress tones on the time.

Supervision Supervision of a phone call is detecting when a called party has picked up his phone and when that party has hung up. Supervision is used primarily for billing purposes. Not all long distance carriers have supervision capability. It depends on how "equal accessed" they have chosen to be. See ANSWER SUPERVISION, SOFTWARE SUPERVISION and SIGNALING SYSTEM 7.

Supervisor The person responsible for day to day maintenance and operation of a phone system. Typically used in conjunction with an ACD — automatic call distributor.

Supervisory Call This service feature allows the attendant, after connecting an incoming CO line or tie line call to the wanted phone, to continuously supervise the call in progress.

Supervisory Control Characters or signals which automatically actuate equipment or indicators at a remote terminal.

Supervisory Lamp A lamp which shows the operator whether the person is speaking (off-hook) or is not speaking (on-hook). These days such lamps are called BUSY LAMP FIELDS. In some smaller key systems, all phones have them. Busy lamp fields are an operator's best friend.

Supervisory Program 1. A program, usually part of an operating system, that controls the execution of other computer programs and regulates the flow of work in a data processing system. 2. A computer program that allocates computer component space and schedules computer events by task queuing and system interrupts. Control of the system is returned to the supervisory program frequently enough to ensure that demands on the system are met.

Supervisory Relay A relay which, during a call, is controlled by the transmitter current supplied to a subscriber line to receive from the associated phone signals that control the actions of operators or switching mechanisms.

Supervisory Routine A routine that allocates computer component space and schedules computer events by task queuing and system interrupts. Control of the system is returned to the supervisory program frequently enough to ensure that demands on the system are met.

Supervisory Signal 1. Supervisory signals are the means by which a telephone user initiates a request for service; or holds or releases a connection; or flashes to recall an operator or to initiate additional features, for example, 3-way calling.

Supervisory signals are also used to initiate and terminate charging on a call. A signal also indicates whether a circuit is in use, or not in use.
2. A signal used to indicate the various operating states of circuit combinations.

Supplementary Services Telephone company talk for services above basic ability to make a phone call. Supplementary services include fast dialing, calling line ID, call waiting, call forwarding, and videoconferencing features. Here's a definitions of supplementary services as applied to ISDN service: Additional services, such as hold, conference, and call forwarding, offered to an ISDN customer. Supplementary services are always present if activated at the switch. Although the specific features and call appearances may differ between service providers, supplementary service generally provides users with the ability to connect and disconnect new calls when one (or more) call exists.

Supplementary Telephone Service The lowest level of service in Windows Telephony Services is called Basic Telephony and provides a guaranteed set of functions that corresponds to "Plain Old Telephone Service" (POTS - only make calls and receive calls). The next service level is Supplementary Telephone Service providing advanced switch features such as hold, transfer, etc. All supplementary services are optional. Finally, there is the Extended Telephony level. This API level provides numerous and well-defined API extension mechanisms that enable application developers to access service provider-specific functions not directly defined by the Telephony API. See WINDOWS TELEPHONY SERVICES.

Supply Chain Management The supply chain is are the electronic links between a company and its suppliers and distributors. Supply chain management seeks to apply software, hardware, networking and telecommunications links in order to get the right product to the right customer at the right time.

Support A verb that typically means "This hardware or software has the following feature." An example: "This software supports 32-bit file sharing."

Support Hardware The racks, clamps, cabinets, brackets, trays, and other equipment that provide the physical means to hold the transmission media and connecting hardware. An AT&T definition.

Support Strand Called a messenger. A strength element used to carry the weight of the telecommunications cable.

Suppressed Carrier Single-Sideband Emission A single-sideband emission in which the carrier is virtually suppressed and not intended to be used for demodulation.

Suppressed Carrier Transmission A transmission technique in which only the sidebands (one or both) are transmitted and the main carrier is not transmitted and thus not used.

Suppressed Voltage Ratings Several ranges are assigned by UL for grading transient suppression voltages. For instance, a 400 volt rating indicates a maximum peak voltage between 330 and 400 volts, These ratings appear between 300 volts and 6000 volts peak.

Suppression First, there were 800 toll-free numbers in North America. Then, in the fall of 1995, the FCC introduced 888 numbers — another toll-free dialing code, since the 800 code was filling up. The FCC allowed bona fide holders of 800 numbers to request '888' replicas of their 800 numbers. 1-800-FLOWERS could request 1-888-FLOWERS, etc...for all the obvious reasons. Prior to 12/1/95, then current holders of 800 numbers could submit their request that the 888 version be suppressed, i.e. not permitted to be made available for general assignment. Many 800 owners asked for suppression of their numbers. Many asked for their use. In some cases this was successful. In others, some suppressed numbers were installed by other companies later on — i.e. a screw-up. In short, you have to be ultra-careful with suppression.

Suppressors, Echo Echo is controlled in long distance circuits with devices called echo suppressors. These devices automatically insert loss in the return path of a four-wire circuit. All long distance circuits are four-wire — two wires for each of the two paths (receiving and transmitting). The echo suppressor jumps back and forth between the two transmission paths. Properly adjusted, an echo suppressor puts only sufficient loss in a circuit so a listener can interrupt the talker. With very long circuits — 22,300 miles — in satellites, a better way is needed. They're called echo cancelers.

Surcharge A charge imposed in accordance with the Commission's Access Reconsideration decision in CC Docket 78-72, Phase 1, FCC 83-356. released August 22, 1983 and updated too many times since. The monthly charge is about $2.00 and is going up to $3.50. This charge is said to compensate the local phone company for long distance commissions (called settlements and separations) lost and now replaced with per minute access charges.

Surf The word "surf" in its classic sense means to ride the crest of a wave, skimming quickly across the water underneath. The electronic world has adopted that definition of skimming quickly and included it in several definitions. For example, shoulder surfing is gazing quickly over someone's shoulder while they're making a call at a payphone and writing down the user's credit card numbers. Channel surfing is moving from one TV channel to another quickly. Surfing the Web means moving from one Web site to another, jumping around in search of knowledge or amusement, but certainly in a non-linear way. See also INTERNET and WEB BROWSER.

Surface Mount With surface mount technology, Components sit on the surface of printed circuit boards and are soldered to conductive pads. In the "thru-the-hole" process, component leads are placed through holes in the boards and are sent through wave soldering for attachment. Surface mount technology is more cost-effective, as it allows for denser packaging on the board and components can be mounted on both sides of the surface.

Surface Outlet A Communications Outlet (modular jack) that is installed on the surface of the mounting location. The premises wire serving such an outlet may or may not be concealed behind the mounting surface.

Surface Wave A wave that is guided along the interface between two different media or by a refractive index gradient. The field components of the wave diminish with distance from the interface. Optical energy is not converted from the surface wave field to another form of energy and the wave does not have a component directed normal to the interface surface. In optical fiber transmission, evanescent waves are surface waves. In radio transmission, ground waves are surface waves that propagate close to the surface of the Earth, the Earth having one refractive index and the atmosphere another, thus constituting an interface surface.

Surfing See SHOULDER SURFING and CHANNEL SURFING.

Surge An increase in line voltage that lasts longer than one cycle of the line frequency of 60Hz, the North American frequency or 50Hz in many other countries, especially those running at 240 volts.

Surge Protector A device which plugs between the phone system and the commercial AC power outlet. It is designed to protect the phone system from high voltage spikes (also called

surges) which might damage the phone system. When a surge occurs on the power line, the surge protector sends the overload to ground. How fast it sends it to ground is a subject that could fill a book. The type of surge protector that you buy will be determined mostly by the speed you need to protect your equipment.

Surges The increased flow of current through an electrical device brought about by an instantaneous change in its resistance or impedance.

Survivability A property of a system, subsystem, equipment, process, or procedure that provides a defined degree of assurance that the device or system will continue to work during and after a natural or man-made disturbance; e.g. nuclear attack. This term must be qualified by specifying the range of conditions over which the entity will survive, the minimum acceptable level or post-disturbance functionality, and the maximum acceptable outage duration.

Survivable Adaptable Fiber Network SAFENET. A U.S. Navy experimental fiber-based local area network designed to survive conventional and limited nuclear battle conditions.

Susan A nice name for a wife. Everyone should have such an incredible wife. We got married in 1976. And it's only gotten better. She is the mother of my two children, Claire and Michael, and the boss of us all. After 15 editions, she is almost used to seeing more of my monitor than me. On the other hand, if she saw more of me, it's doubtful that our marriage would have lasted so long.

Susceptiveness In telephone systems, the tendency of circuits to pick up noise and low frequency induction from power systems. It depends on telephone circuit balance, transpositions, wiring spacing, and isolation from ground.

Suspended Ceiling A ceiling that creates an area or space between the ceiling material and the structure above. See also Plenum.

Suspended Customer An MCI definition. An MCI customer who has requested service but has not yet been installed due to insufficient network capacity or some other operational/administrative constraint.

Suspended On-Demand Connection The mode of an on-demand connection where the communications line is dropped and the connection sites are actively spoofing.

Sustaining Engineering An endeavor in which a company devotes a bunch of people to maintain the quality of engineering on existing products. Such group does not focus on new product.

SUT An ATM term. System Under Test: The real open system in which the Implementation Under Test (IUT) resides.

SVC Switched Virtual Circuit. A virtual circuit connection established across a network on an as-needed basis and lasting only for the duration of the transfer. The datacom equivalent of a dialed phone call, the specific path provided in support of the SVC is determined on a call-by-call basis and in consideration of both the end points and the level of congestion in the network. SVCs are used extensively in X.25 networks, and increasingly in Frame Relay networks. SVCs are much more complex to provision than are PVCs (Permanent Virtual Circuits), but perform much better as they effectively provide automatic and dynamic network load-balancing. In other words, SVCs are set up in consideration of the load on the network, and its subnetworks, in order that the least congested path be established and, therefore, that the data transmission receive the lowest possible level of delay. See also PVC, VC and VCC.

SVCC Switched Virtual Channel Connection: A Switched VCC is one which is established and taken down dynamically through control signaling. A Virtual Channel Connection (VCC) is an ATM connection where switching is performed on the VPI/VCI fields of each cell.

SVD Simultaneous Voice Data. In the fall of 1994, the term began to apply to several techniques for putting voice conversations and data transfers on the same analog phone line. Some of these techniques involve interrupting the voice conversation while data is transferred. Others involve transferring the data and voice simultaneously on different bandwidths.

SVGA Super Video Graphics Array. An extension of the VGA video standard. SVGA enables video adapters to support resolutions of up to 1,024 by 768 pixels with up to 16.7 million simultaneous colors, which is known as true color because it's the number of colors someone once figured is in a great Kodachrome slide. See also VGA.

SVN Subscriber Verification Number. Number issued by the long-distance carrier to confirm the order for long distance service.

SVPC Switched Virtual Path Connection: A Switched Virtual Path Connection is one which is established and taken down dynamically through control signaling. A Virtual Path Connection (VPC) is an ATM connection where switching is performed on the VPI field only of each cell.

Svyazinvest A new holding company that will control much of Russia's telecommunications industry. Svyazinvest was originally conceived both as a holding company for the state's interests in 86 local telephone firms across Russia as a competitor for Rostelcom, a state-run near-monopoly in long distance and international telephony. But, later in the planning, the state's 38% shareholding in Rostelcom was dumped into Svyazinvest. According to the Economist Magazine, Svyazinvest's bosses will therefore have power over almost all of Russia's communications systems (in many regions, the local phone companies also hold the first cellular telephone licenses.)

SWACT Switch of Activity.

SWAP This came from an issue of Internet Week in the Winter of 1998. Seeking to further leverage the ability of the Web to link business partners, a group of vendors led by Netscape, Sun Microsystems and Hewlett-Packard earlier this week proposed a standard to tie together disparate workflow systems on an intranet or over the Internet. The trio, along with 20 other vendors, said they would support the Simple Workflow Access Protocol (SWAP), a proposed Internet standard that would allow disparate workflow engines to manage, monitor, initiate and control the execution of workflow processes between one another within an intranet or over the Internet.

Swap File Some operating systems and applications let you use more memory than what you have in RAM. They do this by pretending that part of your hard disk is RAM memory. They do this by creating a swap file on your hard disk and swapping memory back and forth. Some computer systems call this virtual memory. You need to be careful with swap files. Never turn your machine off when you have applications running. If you do you're likely to leave a huge swap on your hard disk, which you may not find (it's hidden) and which your system may not dispose of. To get back the space on your hard disk, you'll need to erase it separately.

SWATS Standard Wireless AT Command Set. An extension to the Hayes AT command set to support wireless modems, such as those used in standard AMPS analog cellular phones.

SWC Service Wire Center.

SWEDAC Swedish Board for Technical Accreditation. They

have established two standards, which effectively limit radiation emissions, MPR1 and MPR2. These standards specify maximum values for both alternating electric fields and magnetic fields and provide monitor manufacturers with guidelines in creating low emission monitors. There is, as yet, no definite proof of harm from normal computers monitors. But the argument goes that they weren't so sure about nicotine 30 years ago.

Sweep Acquisition A technique whereby the frequency of the local oscillator is slowly swept past the reference to assure that the pull-in range is reached.

Swell An increase from nominal voltage lasting one or more line cycles.

SWHK Abbreviation for SWITCH HOOK. Originally referred to an actual hook on older phones that held the receiver, and sprang upward to close a switch and activate the phone when the receiver was picked up. Today the term refers to any of various buttons and plungers that are pressed down and released when the handset is put down (physically "hung up" in the old days) and picked up.

SWIFT Society for Worldwide Interbank Financial Telecommunications.

SWIGS Special Working Interest Groups. see www.fiberchannel.com.

Swim Slow, graceful, undesired movements of display elements, groups, or images about their mean position on a display surface, such as that of a monitor. Swim can be followed by the human eye, whereas jitters usually appears as a blur.

Swiped Out An ATM or credit card that has been rendered useless because the magnetic strip is worn away from extensive (and expensive) use.

Swiss Army Knife The one tool to have when you can't have more than one. Be sure it has a normal screwdriver, a phillips head screwdriver and a pair of scissors. A corkscrew also is useful.

Switch A mechanical, electrical or electronic device which opens or closes circuits, completes or breaks an electrical path, or selects paths or circuits.

Switch Based Resellers Switch-based resellers lease facilities from national carriers or large private line networks. They resell services provided over those facilities under their own name and provide sales, customer service, billing and technical support. Switch-based resellers own or lease their own switching equipment and, in some cases, own their transmission facilities. they typically provide originating service on a regional basis.

Switch Busy Hour The busy hour for a single switch.

Switch Domain An SCSA definition. A single instance of a particular technology-specific connection type. See S.100.

Switch Driver Protocol Mapper Code running on a Telephony Server that translates between a particular switches proprietary switch-server protocol and one of the specified computer telephony integration (CTI) protocols. See Protocol Mapper.

Switch Fabric An SCSA definition. The facility for connecting any two (or more) transmitting or receiving Service Providers.

Switch Fabric Controller An SCSA definition. A technology-specific, replaceable ASP within the SCSA server. The SFC is designed to support both the internal connectivity within the group and the complex, multiparty call processing applications not directly addressed by the functionality of the Group.

Switch Feature A service provided by the switch that can be invoked by a computing domain or by manual telephone activity. "Do not disturb" is an example of a switch feature.

Switch Hook It is also called the Hook Switch. A switch hook or hook switch was originally an electrical "switch" connected to the "hook" on which the handset (or receiver) was placed when the telephone was not in use. The switch hook is now the little plunger at the top of most telephones which is pushed down when the handset is resting in its cradle (on-hook). When the handset is raised, the plunger pops up (the phone goes off-hook). Momentarily depressing the switch hook (under 0.8 of a second) can signal various services such as calling the attendant, conferencing or transferring calls.

In ISDN, the AT&T ISDN sets have several switch hooks; one for the handset, one for the speakerphone, a "virtual" switch hook, and if an adjunct is attached, an adjunct switch hook. If all switch hooks are "on-hook" or hung up, the ISDN set is on-hook. If any switch hook is "off-hook," then the ISDN set is off-hook. If more than one switch hook is off-hook, the ISDN set uses a complex algorithm to determine whether the handset, the speakerphone, or the adjunct has precedence (only one can be used at a time).

Switch Hook Flash A signaling technique whereby the signal is originated by momentarily depressing the switch hook. See SWITCH HOOK.

Switch Interface The Ethernet MAC controller interface. In general, a switch interface on a switch is the same as a port. However, the number of interfaces does not necessarily correspond to the number of ports. For example, a MAB port on a switch may be a 4-port repeater.

Switch Message Information that originates in a switch. A Call-Progress Event Message is one category of switch messages. Delivered is an example of a call-progress event message.

Switch Over When a failure occurs in the equipment, a switch may occur to an alternative piece of equipment.

Switch Port An SCSA definition. A resource that allows a Group to communicate with another Group. All Groups implicitly possess a Switch Port as a secondary resource, but in order to use it, the application must explicitly connect the Switch Ports of two Groups.

Switch Redirect A central office service which instantly, on command, redirects thousands of phone numbers to different phone numbers. Such a service has great use in a disaster.

Switch Room The room in which you put phone equipment. Also called the Phone Room. (What else?) The Phone Room should be large, clean and should stay at roughly seventy degrees and 50% humidity. You, the customer, are responsible for the quality and condition of your phone room. The messier it is, the hotter it is, the dirtier it is, the poorer your phone system (and its technicians) will function.

Switch Tender In the old, old days, the switch tender was the person who took care of the switch that moved trains from one track to another. That person often stayed for many hours a day in a small hut next to the track. Based on timetables and telegraph and phone communications with central dispatch, he would change the switch and thus move incoming trains to the right track. The job of being a switch tender is now obsolete as switches are now changed remotely by signals over phone lines. The expression "sleeping at the switch" came from the switch tender profession. Sleeping at the switch could train derailments and dead passengers. It was not a good idea.

Switch Train In a telecom circuit (typically a step-by-step central office), the series of switching devices which a call moves through in sequence.

Switchboard The attendant position of a PBX. Most of them don't actually have "boards" (they were big), they have consoles (they're much smaller and they fit on desks). Switchboards are desks.

Switchboard Cable A cable used within and between the central office main frames and the switchboard.

Switched 56 A switched data service which lets you dial someone else and transmit at 56 Kbps over the PSTN (Public Switched Telephone Network). Actually Switched 56/64, Switched 56 also is available at 64 Kbps, where the carrier supports non-intrusive signaling and control. It is a circuit-switched service, letting the user transmit data at 56/64 Kbps over a four-wire, digital, synchronous network. The cost of a Switched 56/64 call is calculated based on duration and time of day, with discounts for non-prime time calls; discounts also apply according to day of week, holidays, and so on.

Provided by LECs, Switched 56/64 was the first VPN (Virtual Private Network) service offered. The term "VPN" currently is applied to IXC services, with bandwidth available at 56/64 Kbps, increments of 56/64 Kbps, 384 Kbps, and 1.544 Mbps (T1). Switched 56/64 offers a high degree of redundancy, flexibility and scalability. It often is used as an alternative to private, leased-line networks — or as a backup for them.

Applications for Switched 56 include videoconferencing, high speed data transfer, digital audio broadcasting, Group IV fax and remote LAN access for telecommuters. It also is used as an access technology for IXC-provided VPNs. While it is widely deployed, Switched 56/64 faces a great deal of pressure from ISDN BRI, which offers as much as 144 Kbps, where available. Over time, Switched 56/64 will disappear for good in North America. See also VPN.

Switched Access A method of obtaining test access to telecommunications circuits by using electromechanical circuitry to switch test apparatus to the circuit.

Switched Access Line Service All residential and most businesses use this type of telephone access. It refers to the connection between your phone and the long distance companies' switch (POP) when you make a regular local or long distance telephone call over standard phone lines.

Switched Carrier In data terms, physical line specification selection indicating a half duplex line in a bisync network.

Switched Circuit Automatic Network SCAN. A service arrangement at certain Telco premises to interconnect private line telephone service channels of a switched service network provided to certain agencies of the federal Government.

Switched Circuit Ordering And Tracking System SCOTS. MCI's automated tracking and order processing system for Dial up products, IMTs, and the MCI switched network.

Switched DAL Switched Dedicated Access (Egress) Line. Dedicated trunk group (T-1, etc.) circuit(s) used to access (1+, etc.) or egress (800, etc.) through normal network switching facilities. The Switched DAL is dedicated to a particular inbound or outbound call type.

Switched Digital Services Applications Forum See SDSAF.

Switched Line A circuit which is routed through a circuit switched network, such as the telephone or telex network.

Switched Local Service You pick up the phone. You dial a local number. Bingo, you have switched local phone service. The reason this trivial definition is even in this dictionary is because many states in the U.S. now — finally — allow companies to offer local switched telephone service in competition with the established company, e.g. Nynex or Southern Bell. Previously, they had only allowed competition in leased lines. And then previous to that they had not allowed any competition in any area of local phone service.

So things are changing, albeit very very slowly.

Switched Loop In telephony, a circuit that automatically releases connection from a console or switchboard, once connection has been made, to the appropriate terminal. Loop buttons or jacks are used to answer incoming listed directory number calls, dial "0" internal calls, transfer requests , and intercepted calls. The attendant can handle only one call at a time.

Switched Loop Operation Each call requiring attendant assistance is automatically switched to one of several switched loops on an attendant position.

Switched Multibeam A type of "smart antennae" used in Wireless Local Loop (WLL) systems. Switched multibeam antennae detect signal strength in a given connection, and select a beam between an erd device and one of perhaps many WLL antennae, locking in on the strongest signal. Also in the general category of "smart antennae" systems is the phased array approach. See also PHASED ARRAY and WLL.

Switched Multimegabit Data Service SMDS. A 1.544 Mbps public data service with an IEEE 802.6 standard user interface. It can support Ethernet, Token Ring and FDDI (OC-3c) LAN-to-LAN connections. See SMDS and SMDS Interest Group.

Switched Network See PSTN.

Switched Private Line Network A network which results from combining point-to-point circuits with switches.

Switched Service Network A private line network that uses scan and/or CCSA type common control switching.

Switched Transport A name for telephone traffic between the local exchange carriers' Central Offices and an interexchange carrier's point of presence (POP). Switched transport is generally provided on a monopoly basis as part of a LEC's network.

Switched Virtual Circuit SVC. A call which is only established for the duration of a session and is then disconnected. See SVC.

Switcher Also called a production switcher. A video term. A device that allows transitions between different video pictures. May also contain special effects generators.

Switchhook A synonym for hookswitch or hook switch. Also spelled switch hook. See SWITCH HOOK.

Switching Connecting the calling party to the called party. This may involve one or many physical switches.

Switching Arrangement A circuit component which enables a Customer to establish a communications path between two phones on a network.

Switching Centers There are four levels in the North American switching hierarchy run at AT&T. They are: Class 1 — Regional Center, Class 2 — Sectional Center, Class 3 — Primary Center, Class 4c — Toll Center and Class 4P — Toll Point. In addition, the local Bell operating companies run a fifth level in the hierarchy, called the Class 5 — End Office.

Switching Equipment Premises equipment which performs the functions of establishing and releasing connections on a per call basis between two or more circuits, services or communications systems.

Switching Equipment Capacity A telephone company term. The capacity of switching equipment is expressed in network access lines. these components can be grouped into four categories. For D&F Chart purposes, the four categories are: 1. Dial Tone Equipment; 2. Talking Channels; 3. Switching Control; and 4. Trunk Terminations.

Switching Fee A one-time, per-line fee imposed by the LEC to reprogram their switching system to change your default long-distance carrier. Some resellers and IXCs will reimburse new subscribers for this fee.

Switching Hub A multiport hub that delivers the full, uncontested bandwidth between any pair of ports. An intelligent switching hub also provides bridging and multiprotocol routing capabilities.

Switching Point Same as end office and intermediate office.

Switching System 1. An assembly of equipment arranged for establishing connections between lines, lines to trunks, or trunks to trunks.

2. An ATM term. A set of one or more systems that act together and appear as a single switch for the purposes of PNNI routing.

Switchless Resellers A switchless reseller buys long distance service in bulk from a long distance company, such as AT&T, and resells that service to smaller users. It typically gets its monthly bill on magnetic tape, then rebills the bulk service to its customers. A switchless reseller owns no communications facilities — switches or transmission. It has two "assets" — a computer program to rebill the tape and some sales skills to sell its services to end users. The profit it makes comes from the difference between what it pays the long distance company and what it is able to sell its services at. Switchless resellers are also called rebillers. It's not an easy business to be in, since you are selling a long distance company's services to compete against itself. See also AGGREGATOR.

SXS Step by Step switching system. An automatic dial-telephone system in which calls go through the switching equipment by a succession of switches that move a step at a time, from stage to stage, each step being made in response to the dialing of a number.

Symbol 1. An abbreviated, predetermined representation of any relationship, association or convention.

2. In digital transmission, a recognizable electrical state which is associated with a signal element, which is an electrical signal within a defined period of time. In a binary transmission, for example, a signal element is represented as one of two possible states or symbols, i.e., 1 or 0.

An abbreviated, predetermined representation of any relationship, association or convention.

Symbolic Debugger A debugger is a wholly- or partly-memory-resident program that lets you closely monitor and control execution of an application under development. At the most basic level, a debugger lets you look at running machine code, and fiddle around with the contents of memory — great if you understand machine code (and are looking at machine code you've written from scratch). Not great if you don't know machine code, or are looking at machine code output by a high-level language compiler (e.g., C++ compiler). A basic symbolic debugger references the symbol table of an executable, providing readable variable names, function entry-points, etc., more or less as they appear in source. Easier for machine-language folks (because of the labels). Not much easier for high-level language folks, because you're still dealing with machine code. A source-level symbolic debugger references both the symbol table of an executable and various files produced during compilation; and lets you work with high-level language source directly, during target program execution. Fully-integrated debuggers like this are built into Microsoft's Visual/X products. Functions common to most debuggers include the ability to set "breakpoints" (i.e., run the program until you reach this step, then stop), "watch variables" (i.e., show me how the value of this variable changes — and possibly stop if it assumes a predetermined value), "single-step execution" (i.e., do this step and stop), change variable values in mid-execution, etc.

Symbolic Language A computer programming language used to express addresses and instructions with symbols convenient to humans rather than machines.

Symbolic Logic The discipline in which valid arguments and operations are dealt with using an artificial language designed to avoid the ambiguities and logical inadequacies of natural languages.

Symmetric Balanced in proportion. A symmetric telecom channel has the same speed in both directions. It's important to contrast symmetric with full duplex which is transmission in two directions simultaneously, or, more technically, bidirectional, simultaneous two-way communications. For example, ISDN BRI provides full duplex, symmetric bandwidth, as each of the two B channels provides 64 Kbps in each direction and the D channel operates at 16 Kbps in each direction. Symmetric also can refer to the physical topology of the network. For example, a point-to-point circuit connects one device directly to one other device. Asymmetric, on the other hand, refers to something which is not perfectly balanced. See the next several definitions. See also Asymmetric, Bps, Byte and Full Duplex.

Symmetric Connection A connection with the same bandwidth (i.e. speed) in both directions. See also Bps, Byte, Full Duplex and Symmetric.

Symmetric Multiprocessing SMP. A type of multiprocessing in which more than one processor can execute kernel-level code at the same time. The degree of symmetry can vary from limited, where there is very little concurrency of execution, to the theoretically ideal fully-symmetric system where any function can be executed on any processor at any time. Processors within the same system share all processes, including disk I/O, network I/O and memory. Compare to ASYMMETRIC MULTIPROCESSING, wherein processors in the same or different systems are dedicated to specific tasks, such as disk I/O, network I/O or memory management. They off-load these tasks from the main system CPU, which generally is responsible for running the operating system. Each processor usually has its own dedicated memory. See SMP.

Symmetrical Channel A channel in which the send and receive directions of transmission have the same data signaling rate.

Symmetrical Compression A compression system which requires equal processing capability for compression and decompression of an image. This form of compression is used in applications where both compression and decompression will be utilized frequently. Examples include: still-image databasing, still-image transmission (color fax), video production, video mail, videophones, and videoconferencing.

Symmetrical Digital Subscriber Line See SDSL.

Symmetrical Pair A balanced transmission line in a multipair cable having equal conductor resistances per unit length, equal impedances from each conductor to earth, and equal impedances to other lines.

Symocasting Sending the same audio stream out from several Web sites simultaneously. You do this because one site would not accommodate the demand for this stream.

Syn, Syn Character, Synchronous Idle In synchronous transmission. Control character in character-oriented protocols used to maintain synchronization and as a time-fill in the absence of data. The sequence of two SYN characters in succession is used to maintain synchronization following each line turnaround. Contrast with flag.

Sync 1. Synchronization character.
2. The portion of an encoded video signal that occurs during blanking and is used to synchronize the operation of cameras, monitors, and other equipment. Horizontal sync occurs within the blanking period in each horizontal scanning line, and vertical sync occurs within the vertical blanking period.

Sync Bits Synchronizing bits (more properly bytes or characters) used in synchronous transmission to maintain synchronization between transmitter and receiver.

Sync Generator A video term. A device that generates synchronizing pulses need by video source equipment to provide proper equipment or studio timing. Pulses typically produced by a sync generator include subcarrier, burst flag, sync, blanking, H & V drives, color frame identification, and color black.

Sync Pulse Timing pulses added to a video signal to keep the entire video process synchronized in time.

Synchronet Service Dedicated point to point and multipoint digital data transmission service offered by BellSouth at speeds of 2.4, 4.8, 9.6, 19.2, 56 and 64 Kbps.

Synchronization 1. A networking term which means that the entire network is controlled by one master clock and transmissions arrive and depart at precise times so that information is neither lost nor jumbled. For a bigger explanation, see NETWORK SYNCHRONIZATION and SYNCHRONOUS.
2. An uninterruptible power supply (UPS) definition. Specially designed circuitry is "synchronized" to your AC power outlet to ensure continuity of power. Without this feature, power reversal can occur on the input.
3. A multimedia term. Synchronization is very precise real-time processing, down to the millisecond. Some forms of multimedia, such as audio and video, are time critical. Time delays that might not be noticeable in text or graphics delivery, but are unacceptable for audio and video. Workstations and networks must be capable of transmitting this kind of data in a synchronized manner. Where audio and video are combined, they must be time stamped so that they can both play back at the same time.
4. Start with a database on your server. Now, take a copy of part of it on your laptop — for example, your very own sales leads. Go traveling. Come back in a week. You want to update the database with your changes. But you don't want to destroy other peoples' changes. Some people are calling this "file synchronization." Synchronization is a critical part of what is increasingly being called "Groupware." See also REPLICATION.
5. A Video term referring to the timing of the vertical and horizontal presentation of the multiple still images. Vertical synch prevents the picture from flipping, or scrolling unnaturally. Horizontal synch keeps the picture from twisting. If both vertical and horizontal are out of synch, the picture looks truly wretched.

Synchronization Bit A binary bit used to synchronize the transmission and receipt of characters in data communications.

Synchronization Bits Bits transmitted from source to destination for the purpose of synchronizing the clocks of the transmitting and receiving devices. The term "synchronization bit" is usually applied to digital data streams, whereas the term "synchronization pulse" is usually applied to analog signals.

Synchronization Code In digital systems, a sequence of digital symbols introduced into a transmission signal to achieve or maintain synchronism.

Synchronization Pulses Bits transmitted from source to destination for the purpose of synchronizing the clocks of the transmitting and receiving devices. The term "synchronization pulse" is usually applied to analog signals, whereas the term "synchronization bit" is usually applied to digital data streams.

Synchronize The word synchronize means "to cause to match exactly." When you're synchronizing, you're causing one file on one computer to precisely match another one on another computer. Why would you want to do this? Let's say you have a database of sales contacts on a file server. One of your salesman takes a copy of his sales contacts with him on his laptop. He travels and makes changes to his contacts. Now he dials into the office via modem and wants to "synchronize" his changed database with the now-changed main database, and make them both the same, i.e. into synch. This process is far more difficult than it sounds because it means allowing for the changes made at the server and by the salesman. You have to set up elaborate rules.

In operating systems, such as Windows NT, the word "synchronize" has a narrower meaning. Windows NT instruction manual defines "synchronize" as "to replicate the domain controller to one server of the domain, or to all the servers of a domain. This is usually performed automatically by the system, but can also be invoked manually by an administrator." See also REPLICATE.

Synchronizing Achieving and maintaining synchronism. In facsimile, achieving and maintaining predetermined speed relations between the scanning spot and the recording spot within each scanning line.

Synchronizing Pilot In FDM, a reference frequency used for achieving and maintaining synchronization of the oscillators of a carrier system or for comparing the frequencies or phases of the currents generated by those oscillators.

Synchronous The condition that occurs when two events happen in a specific time relationship with each other and both are under control of a master clock. Synchronous transmission means there is a constant time between successive bits, characters or events. The timing is achieved by the sharing of a single clock. Each end of the transmission synchronizes itself with the use of clocks and information sent along with the transmitted data. Synchronous is the most popular communications method to and from mainframes. In synchronous transmission, characters are spaced by time, not by start and stop bits. Because you don't have to add these bits, synchronous transmission of a message will take fewer bits (and therefore less time) than asynchronous transmission. But because precise clocks and careful timing are needed in synchronous transmission, it's usually more expensive to set up synchronous transmission. Most networks are synchronous these days. See ASYNCHRONOUS and NETWORK SYNCHRONIZATION.

Synchronous Completion A computing domain issues a service request and need not wait for it to complete. If the computing domain waits for this completion, this is known as SYNCHRONOUS, but if it is sent off to another system entity and the computing domain goes on to other activities before the function completes (and the system later sends a message to the computing domain announcing the function's completion), that completion is known as asynchronous.

Synchronous Control Character SYN. A transmission control character used in synchronous transmission to provide a signal in the absence of any other character. SYN is defined by IBM Corp's Binary Synchronous Communications (BSC) protocol.

Synchronous Data Link Control SDLC. A data communications line protocol associated with the IBM Systems Network Architecture. See SYSTEMS NETWORK ARCHITECTURE.

Synchronous Data Network A data network in which synchronism is achieved and maintained between data circuit-terminating equipment (DCE) and the data switching exchange (DSE), and between DSEs. The data signaling rates are controlled by timing equipment within the network. See NETWORK SYNCHRONIZATION.

Synchronous Data Transfer A physical transfer of data to or from a device that has a predictable time relationship with the execution of an I/O (Input/Output) request. See SYNCHRONOUS.

Synchronous Digital Hierarchy SDH. Term used by the International Telegraph and Telephone Consultative Committee to refer to Sonet.

Synchronous Idle Character A transmission control character used in synchronous transmission systems to provide a signal from which synchronism or synchronous correction may be achieved between data terminal equipment, particularly when no other character is being transmitted.

Synchronous Network A network in which all the communication links are synchronized to a common clock.

Synchronous Orbit An orbit, any point on which has a period equal to the average rotational period of the Earth. If the orbit is also circular and equatorial, it is called a stationary (geostationary) orbit.

Synchronous Payload Envelope The major portion of the SONET frame format used to carry the STS-1 signal divided into an information payload section and a transport overhead system. SPE is used to address three payload structures: direct to STS-1 line rate multiplexing; asynchronous DS-3 multiplexing; and synchronous DS-3 multiplexing.

Synchronous Request An SCSA definition. A request where the client blocks until the completion of the request. Contrast with asynchronous request .

Synchronous Satellite A satellite in a synchronous orbit. See SYNCHRONOUS ORBIT.

Synchronous TDM A multiplexing scheme in which timing is obtained from a clock that in turn controls both the multiplexer and the channel source.

Synchronous Time-Division Multiplexing STDM. A time-division multiplexing method whereby devices have access to a high-speed transmission medium at fixed time periods independent of likely load.

Synchronous Transfer Mode A proposed transport level, a time division multiplex-and-switching technique to be used across the user's network interface for a broadband ISDN. See STM.

Synchronous Transmission Transmission in which the data characters and bits are transmitted at a fixed rate with the transmitter and receiver synchronized. Synchronous transmission eliminates the need for start and stop bits. See SYNCHRONOUS and ASYNCHRONOUS.

Synchronous Transmission Mode STM. The synchronous transmission capability of a system that is capable of both synchronous and asynchronous capabilities of Broadband Integrated Services Digital Network (B-ISDN) service.

Synchronous Transport Module 1 STM-1. SDH standard for transmission over OC-3 optical fiber at 155.52 Mbps.

Synchronous Transport Signal Level 1 STS-1. The basic signaling rate for a Synchronous Optical Network (SONET) transmission medium. The STS-1 rate is 51.84 Mbps.

Synchronous Transport Signaling Level n STS-n. A definition of the transmission speed a Synchronous Optical Network (SONET) transmission medium where n is an integer between 1 and 48 and relates to the multiplier to be applied to the basic STS-1 51.8-Mbps transmission speed. STS-48 is 48 times faster than STS-1, with a speed of 2.5 gigabits per second.

Syndrome Basic element of decoding procedure. Identifies the bits in error.

Syntax The rules of grammar in any language, including computer language. Specifically, it is the set of rules for using a programming language. It is the grammar used in programming statements.

Syntax Error An error caused by incorrect programming statements according to the rules of the language being used. Sometimes the computer will throw up "SN" to indicate a syntax error.

Synthesized Voice Human speech approximated by a computer device that concatenates basic speech parts (or phonemes) together. Usually has a metallic, Germanic sound.

Syntonization The process of setting the frequency of one oscillator equal to that of another.

Syntran Synchronous Transmission A restructured DS-3 signal format for synchronous transmission at the 4.736 Mbps DS-3 level of the North American Hierarchy.

SYSGEN Acronym for SYStem GENeration.

SYSOP The SYStem OPerator of a PC-based electronic bulletin board/mail service or on-line computer service, such as CompuServe or America On Line. SYSOPs (pronounced sis-ops) typically put computers and modems on phone lines, then published the phone number, then invited people with computers to call them and leave them messages and interesting software programs which they had written. These programs then became "public domain," or freeware. And other callers were invited to download these programs for their own use. Lead Sysops are called Wizops.

SYSREQ System request; the seldom used key used to get attention from another computer.

System An organized assembly of equipment, personnel, procedures and other facilities designed to perform a specific function or set of functions.

System Administrator The person or persons responsible for the administrative and operational functions of a computer and a telecom system that are independent of any particular application. The system Administrator is likely to be a person with the best overview of all the applications. The System Administrator advises application designers about the data that already exists on the various services, makes recommendations about standardizing data definitions across applications, and so on.

System Build This is the original manufacturer system building that occurs when the order is placed by the buyer with the vendor. The basic configuration is set up to reflect the users needs at that point in time. Thereafter, if any changes occur to reflect changes in the operating environment, the manufacturer must reconfigure the system to reflect this change. There is usually a reprogramming charge and a delay associated with the change.

System Clock The clock designated as the reference for all clocking in a network of electronic devices such as a multiplexer or transmission facilities management system.

System Common Equipment The equipment on a premises that provides functions common to terminal devices such as telephones, data terminals, integrated work station terminals, and personal computers. Typically, the system common equipment is the PBX switch, data packet switch, or central host computer. Often called common equipment.

System Connect The method by which connection is physically made to the cost computer or local area network.

System Coordinator This is the title assigned to the person responsible for administration programming and the training of workers on your phone system.

System Disk A disk that has been formatted as a system disk. MS-DOS system disks have two hidden files and the COMMAND.COM file. You can start the computer using a system disk.

System Fault Tolerance SFT. The ability of computer to work fully regardless of component failures.

System Feature A telephone switch feature that is typically available all the users.

System Gain The amount of free space path loss that a radio can overcome by a combination of enhancing transmitted power and improving receiver sensitivity.

System Message Messages that are not associated with a mailbox.

System Redundancy The duplication of system components to protect against failure. For protection against failure, install redundant cabling, power supplies, disk storage, gateways, routers, network boards, printers, switches and other mission-critical network components.

System Reload A process allowing stored data to be written from a tape into the system memory. Picture: your telephone system goes dead. For whatever reason it loses all memory of its generic programming and your specific programming (whose extension gets what, etc.). You have to quickly grab the backup (hopefully you have it on tape or magnetic disk) and load it back into your telephone system's memory. This is called system reload. Sometimes it's done automatically. Sometimes you have to do it manually.

System Segment A conceptual subset of a system, usually referring to one which can be functionally replaced without damaging the capability of the system.

System Service Provider An SCSA definition. An entity that provides system wide services, such as session management and security, and the allocation and tracking of resources and groups.

System Side Defines all cabling and connectors from the host computer or local area network to the cross connect field at the distribution frame.

System Speed Dial Simplified ways of dialing. You do them by dialing several digits. System speed dial numbers can be used by everyone on the phone system — whether they are on an electronic phone or just a simply single line phone.

System Test This definition courtesy Steve Gladstone, author, "Testing Computer Telephony Systems": System test is the phase of the product life cycle that examines the entire system as a "whole" to assure it is ready to go to a true alpha or beta test. System testing is also more oriented to inter system functions as opposed to earlier phases. To pass a system test, all features and functions are expected to work correctly (function to specification) in all areas of the system — features, administration, maintenance, billing, etc. Additionally, the system must function as an "architectural whole," including all hardware and software components. Representative databases must be loaded to simulate site applications. Full load and stress testing is performed. It is in this phase that the bulk of system level testing will take place. System testing has a major focus on external load and other stimuli.

System V Interface Definition SVID. A UNIX application-to-system software interface developed and supported by AT&T. The interface is similar to POSIX.

Systems Analysis Analyzing an organization's activities to figure the best way of applying computer systems to its organization.

Systems Analyst A person who performs systems analysis and who follows through with methods, techniques and programs to meet the need.

Systems Integrator A systems integrator is a company that specializes in planning, coordinating, scheduling, testing, improving and sometimes maintaining a computing operation (sometimes companywide, sometimes just locally). In the old days, this was done almost exclusively by the International Business Machines Corporation. Somewhere along, companies discovered they could get more flexibility and computing power at a lower cost by shopping around. Today, hundreds of companies contribute various components — hardware, software, wiring, communications and so on — to a company's computer operation. But the added flexibility can bring stunning complexity. Systems integrators try to bring order to the disparate suppliers.

Systems Integration Interface SII. As used in the definition of the proposed mu tivendor integration architecture sponsored by Niopon Telegraph and Telephone (NTT) of Japan, SII specifies any set of standardized services used to connect computer based-systems.

Systems Network Architecture SNA. IBM's successful computer network architecture. At one stage the most successful computer network architecture in the world. In the days of mainframe computers, it was as successful in the computer networking world as AT&T's telephone network design was in telecommunications. The best explanation we've ever read of SNA is in James Harry Green's Dow Jones-Irwin Handbook of Telecommunications. Here is an excerpt:

"SNA is a tree-structured architecture, with a mainframe host computer acting as the network control center. The boundaries described by the host computer, front-end processors, cluster controllers and terminals are referred to as the network's domain. Unlike the switched telephone network that establishes physical paths between terminals for the duration of a session, SNA establishes a logical path between network nodes, and it routes each message with addressing information contained in the protocol. The network is therefore incompatible with any but approved protocols. SNA uses the SDLC data link protocol exclusively. Devices using asynchronous or binary synchronous can access SNA only through protocol converters...SNA works in seven layers roughly analogous to ISO's seven level OSI model. Unlike OSI, however, SNA is fully defined at each level. SNA was first announced in 1974 and is the basis for much of the OSI model, but it differs from OSI in several significant respects." For more on these differences, see page 96 in Green's Handbook.

The following is a description we received from IBM's PR department: "What is SNA?" In general, SNA is the description of the rules that enable IBM's customers to transmit and receive information through their computer networks. SNA may also be viewed as three distinct but related entities: a specification, a plan for structuring a network and a set of products. First, SNA is a specification governing the design of IBM products that are to communicate with one another in a network. It is called an architecture because it specifies the operating relationships of those products as part of system. Second, SNA provides a coherent structure that enables users to establish and manage their networks and, in response to new requirements and technologies, to change or expand them. Third, SNA may be viewed as a set of products: combinations of hardware and programming designed in accordance with the specification of SNA. In addition to a large number of computer terminals for both specific industries and general applications, IBM's SNA product line includes host processors, communication controllers, and adapters, modems and data encryption units. The SNA product line also includes a variety of programs and programming subsystems. Telecommunications access methods, network management programs, distributed applications programming and the network control program are examples.

Systems Network Interconnection SNI. A service defined by IBM that allows for the interconnection of separately defined and controlled Systems Network Architecture (SNA) networks. See Systems Network Architecture.

Systems Services Control Point SSCP. An IBM Corp. Systems Network Architecture (SNA) term for the software that manages the available connection services to be used by the Network Control Program (NCP). There is only one SSCP in an SNA network domain, and the software normally resides in the host processor, which is a member of the IBM System/370 mainframe family.

Systems Software A type of program used to enhance the operating systems and the computer systems they support.

Systemview The network management program that purports to let UNIX-based computers be managed along with other IBM systems.

T 1. Trunk. as in T-1. See T-1.

2. Tip. See Tip & Ring.

3. Tera, which is 10 raised to the 12th power, or 1,000,000,000,000.

4. An ADSL (Asymmetric Digital Subscriber Line) term for the functional interface between the Premises Distribution Network (PDN) and the Service Module(s), both of which are installed at the user premise. The PDN is the inside cable and wire system, and associated premises electronics. The Service Modules accomplish terminal (e.g., TV, telephone, PC and router) adaption functions for access to the ADSL network. When the PDN is in the form of a point-to-point, passive wiring system, the "T" may be the same as a "T-SM" interface. The "T" may be in the form of a separate physical unit, or may be embedded in a combined ATU-R/Service Module, with the ATU-R being an Asymmetric Terminating Unit-Remote. See also ADSL, ATU-R, PDN, SM and T-SM.

5. Twisted pair. As in 10BaseT, an IEEE standard for Ethernet LANs, which run at 10Mbps (million bits per second), Baseband (single channel transmission), over Twisted pair (Cat 3, 4, or 5).

T 1 See T-1

T Carrier T stands for trunk. T Carrier is a generic name for any of several digitally multiplexed carrier systems. The designators for T (Trunk) carrier in the North American digital hierarchy correspond to the designators for the digital signal (DS) level hierarchy. T carrier systems were originally designed to transmit digitized voice signals. Current applications also include digital data transmission. The table below lists the designators and rates for current T carrier systems. If an "F" precedes the "T", it's an optical fiber cable system, but the same speeds.

NORTH AMERICAN DESIGNATOR (DS LEVEL)

Designator	Rate	Channels
T1 (DS 1)	1.544 Mbps	24 voice channels
T1C	3.152 Mbps	48 voice channels
T2 (DS 2)	6.312 Mbps	96 voice channels
T3 (DS 3)	44.736 Mbps	672 voice channels
T4 (DS 4)	274.176 Mbps	4032 voice channels

JAPANESE HIERARCHY

Designator	Rate	Channels
DS 1	1.544 Mbps	24 voice channels
DS 2	6.312 Mbps	96 voice channels
DS 3	32.064 Mbps	480 voice channels
DS 4	97.728 Mbps	1440 voice channels
DS 5	400.352 Mbps	5760 voice channels

EUROPEAN HIERARCHY (CEPT)

Designator	Rate	Channels
DS 1	2.048 Mbps	30 voice channels
DS 2	8.448 Mbps	120 voice channels
DS 3	34.368 Mbps	480 voice channels
DS 4	139.268 Mbps	1920 voice channels
DS 5	565.148 Mbps	7680 voice channels

See T-1 below.

T Connection T-shaped three-way conductor for distributing an incoming signal in two outgoing ways. Same shape as a T-connection in the road.

T Connector A T-shaped device with two female connectors and one male BNC connector used with Ethernet coaxial cable and used on local area networks.

T Interface 4-wire ISDN BRI circuit. Picture this: You order an ISDN circuit from your local phone company. They deliver it on a normal phone line — one copper pair. At your offices, you plug in a small device called a network termination device. That device converts the two-wire circuit called a U interface, into a four-wire S or T interface which you'll use to plug in your ISDN terminal equipment, which might be a phone, a computer, a PBX, a videoconferencing device, etc. It may be all the above. The S or T interface is designed to allow you hook up to eight terminal devices on one ISDN line. Definitions are changing, especially in ISDN. At one stage, the T Interface (more properly the T-Reference Point) needed an NT1 rather than an NT2. These days, some people believe the T interface refers to an ISDN electrical connection to a PBX. While, the S bus refers to a connection to other devices, like phones and videoconferencing devices. There's no electrical difference between an S and a T interface. But I may be proven wrong by some phone company that changes the specs. One of ISDN's most charming features is its ability to acquire different specs and features from one provider to the next. See S INTERFACE and U INTERFACE and ISDN.

T Reference Point See T INTERFACE.

T Span A telephone term for a transmission medium through which a T-carrier system is operated. Also called a Span line. See T CARRIER and T-1.

T Span Line Also called a Span line. A repeatered outside plant four-wire, two twisted-pair transmission line. See T CARRIER and T-1.

T Tap A passive line interface used for extracting data from a circuit. Also, for extracting optical signals from a fiber cable or electrical signals from a coaxial cable.

T+T Telephone and Telegraph. I found T+T engraved on a small metal closet in a hotel in Zurich, Switzerland. When I opened the closet, I found a crude telephone line crossconnect panel. The hotel ran its phone lines up a central shaft, terminated them on a crossconnect panel, then ran the lines to each guest room.

T-1 Also spelled T1, which stands for Trunk Level 1. A digital transmission link with a total signaling speed of 1.544 Mbps (1,544,000 bits per second). T-1 is a standard for digital transmission in North America, — the United States and Canada. T-1 is part of a progression of digital transmission pipes — a hierarchy known generically as the DS (Digital Signal Level) hierarchy. (For the complete hierarchy, see the definition for T Carrier above.) In the olden days, T-1 was delivered to your business on two pairs of unshielded twisted copper wires — one pair for transmit and one pair for receive — the combination of these two simplex (unidirectional) circuits yields a full duplex (bidirectional) circuit. These days, T-1 often is delivered on fiber optic transmission systems by the CLECs (Competitive LECs) and by the ILECs (Incumbent Local Exchange Carriers), where fiber is available (or insist on it being installed as part of your order). You can lease T-1 as a channelized service (delivered as separate voice or data channels), or as an unchannelized raw bit stream (i.e., 1.536 Mbps of transmission both ways, plus .008 Mbps framing bits) and do with the bits as you wish-the framing bits are not under your control, however. North American carriers typically deliver T-1 channelized, i.e., split into 24 voice-grade channels, with each running at 56/64 Kbps (i.e., 56,000 or 64,000 bits per second), depending on the generation of the channel bank equipment involved. If you have need for a bunch of local phones, it's often cheaper to get them delivered on T-1 channels than as individual phone lines. One expensive circuit (i.e, the multi-channel T-1) is far less expensive than 24 less expensive circuits (e.g., single-channel voice circuits). While channelized T-1 was developed for and is optimized for uncompressed voice communications, it also can be used for channelized data communications. A channelized approach is required for access to the traditional PSTN, which is channelized throughout the traditional carrier networks.

On the other hand, an unchannelized approach is better for most data communications applications, and for compressed voice, video and IP telephony. The unchannelized approach provides you with 1.536 Mbps which you can split up any way you choose. If you lease a raw T-1 pipe, you could, for example, split it (i.e., multiplex it) into 12 voice grade channels to support 12 voice conversations, and use the remaining 768 Kbps for either reasonably high-speed access to the Internet or for videoconferencing with your distant office. You could also compress voice to run at speeds of perhaps 8 Kbps or less by using IP Telephony techniques and, therefore, put many more voice calls over a single T-1 pipe. Unchannelized T-1 also is commonly used for access to a frame relay or ATM network, or for Internet access. In such an application, your router or data switch or data concentrator effectively multiplexes data packets (i.e., packets, frames or cells) through the "clear" pipe. Channelization would make no sense in such an application. In addition to use in network access applications, T-1 also can be used for private, leased line networking. In a private network, you might use channelized leased T-1 PBX tie trunks to "tie" together your voice PBXs. You might use unchannelized T-1 tie trunks to directly connect your local area network routers or data switches. Note that T-1 is medium-independent. You can run it over electrical (i.e., twisted pair or coaxial cable), optical (i.e., fiber optics or infrared) or radio (i.e., microwave or satellite) transmission media.

Outside of the United States and Canada, DS-1 is called E-1, as it was developed by the CEPT (Conference of European Postal and Telecommunications Administrations) for use in Europe. E-1 runs at a total signaling rate of 2,048,000 bits per second. Only one element remains constant between it and the North American's T-1 — the DS-0, namely the 64 Kbps channel. Most often it represents a PCM voice signal sampled at 8,000 times per second, or 64,000 bits per second. However, the form of PCM encoding, also known as companding, differs between T-1 (mu-law) and E-1 (A-law). Conversion of E-1 to T-1 involves both the compression law and the signaling format. At the higher rate of 2.048 Mbps, 32 time slots are defined at the CEPT interface, but two time slots (channels) are used for non-intrusive signaling and control purposes. The remaining 30 channels are clear 64 Kbps channels for user information-voice, video, data, etc. See also CEPT, Channel Bank, Companding, Compression, ISDN PRI, PCM, TDM, and Time Division Multiplexing and the following five definitions.

T-1 Framing Digitization and coding of analog voice sig-

nals requires 8,000 samples per second (two times the highest voice frequency of 4,000 Hz) and its coding in 8-bit words yields the fundamental T-1 building block of 64 Kbps for voice. This is termed a Level 0 Signal and is represented by DS-0 (Digital Signal at Level 0). Combining 24 such voice channels into a serial bit stream using Time Division Multiplexing (TDM) is performed on a frame-by-frame basis. A frame is a sample of all 24 channels (24 x 8 = 192) plus a synchronization bit called a framing bit, which yields a block of 193 bits. Frames are transmitted at a rate of 8,000 per second (corresponding to the required sampling rate), thus creating a 1.544 Mbps (8,000 x 193 = 1,544 Mbps) transmission rate, the standard North American T-1 rate. This rate is termed DS-1. See also D-4 FRAMING and EXTENDED SUPERFRAME FORMAT. See also T CARRIER, T2, T3 and T4.

T-1 Span Line See SPAN.

T-2 The North American standard for DS-2 (Digital Signal Level 2). T-2 operates at a signaling rate of 6.312 Mbps, and is capable of handling 96 voice conversations depending on the encoding scheme chosen. T-2 is four times the capacity of T-1. It generally is used only in carrier backbone networks. See T-1.

T-3 The North American standard for DS-3 (Digital Signal Level 3). T-2 operates at a signaling rate of 44.736 Mbps, equivalent to 28 T-1s. Commonly referred to as 45 megabits per second, rounded up. Capable of handling 672 voice conversations. T-3 runs on fiber optic or microwave transmission media, as twisted pair is not capable of supporting such a high signaling rate over distances of any significance. Running on fiber, it is typically called FT-3. Both Bill Gates and George Lucas have T-3 lines coming into their houses. See T-1.

T-4 The North American standard for DS-4. T-4 supports a signaling rate of 274.176 Mbps and is capable of handling 4,032 voice conversations. T-4 has 168 times the capacity of T-1. T-4 can run on coaxial cable, microwave radio or fiber optic transmission systems. T-4 generally is used only in carrier backbone networks, and generally is not available for end-user consumption. See also T-1.

T-BERD See T-BIRD.

T-BIRD also T-BERD. Colloquial term for a T-1 carrier analyzer, used by T-1 circuit technicians. Taken from the brand name of a leading device, T Berd 90A.

T-I Channel Either of the two external ports of a TDI or RDI which provides for transmitting or receiving eight TDM channels. Do not confuse with T-1 (as in T-one).

T-SM An ADSL term for the functional interface between the ATU-R (Asymmetric Transmission Unit-Remote) and the PDN (Premises Distribution Network). The ATU-R may contain one or more T-SMs (Service Modules) for terminal adaption. The T-SM may be the same as the "T" interface where the PDN is a point-to-point, passive cable and wire system. The T-SM may be a separate physical device, or may be embedded in an integrated ATU-R/SM. See also ADSL, ATU-R, PDN, SM and T.

T.120 The most important transmission protocol standard for document conferencing (viewing, changing and moving files) over transmission media ranging from analog phone lines to the Internet. T.120 is the International Telecommunications Union (ITU-T) standards suite for document conferencing via FTP (File Transfer Protocol). Virtually all major players in the document conferencing industry have announced support for this standard. Document conferencing adds a visual dimension to voice-only conference calls by allowing groups of people to share computer documents in real-time while participating in a standard voice conference call. Whatever materials would normally be distributed in a face-to-face meeting — graphs, spreadsheets, diagrams, or documents — can be shared on-line, in real-time. Participants can easily connect to a conference anywhere in the world, with the only requirements being a Windows PC, a modem and a document conferencing software program. T.120 series standards provide a framework to enable multi-point data conferencing across LANs, WANs and the Internet. The T.120 architecture relies on a multilayered approach with defined protocols and service definitions between layers. Each layer presumes the existence of all layers below. The lower level layers (T.122, T.123, T.124 and T.125) specify an application-independent mechanism for providing multi-point data communications services to any application that can use these facilities. The upper level layers (T.126 and T.127) define protocols for specific conferencing applications, such as shared whiteboarding and binary file transfer. See also H.320, standards which extend data conferencing into video conferencing.

T.121. ITU standard for generic T.120 Application Template.

T.122 Multipoint Communications Service for Audiographics and Audiovisual Conferencing - Service Definition. (ITU approved 1993.) See T.120 and T.125.

T.123 Protocol Stacks for Audiographics and Audiovisual Teleconference Applications. (ITU approved 1993/1994.) See T.120.

T.124 Generic Conference Control for Audiovisual Services (GCC) (ITU voted approval in 3/95.) T.125: Multipoint Communication Service - Protocol Specification (MCS). (ITU approved 1994.) See T.120 and T.122.

T.125 Multipoint Communication Service - Protocol Specification (MCS). (ITU approved 1994.) See T.120.

T.126 Provides shared whiteboard and document conferencing protocols. See T.120.

T.127 Provides multipoint binary file transfer. See T.120.

T.128 ITU standard for Audio Visual Control for Multipoint Multimedia Systems.

T.134 See T.140.

T.140 In early February, 1998, the ITU-T reported that it had just completed work on three recommendations aimed at enhancing the capability of the deaf or speech-impaired to use telecommunications. The first recommendation, V.18, describes a multi-function text telephone that bridges the gap that has, according to the ITU, between several incompatible text telephones in use today. The second (T.140) adds new facilities to enable the use of different alphabets and character sets in text communications. These include Arabic, Cyrillic and Kanjii, as well as Latin-based characters. Finally, T.134 dwcribes how these facilities can be integrated in the multimedia communications systems defined by the ITU-T.

T.30 ITU-T standard. Fax handshake protocol. This standard describes the overall procedure for establishing and managing communication between two fax machines. There are five phases of operation covered: call set up, pre message procedure (selecting the communication mode), message transmission (including phasing and synchronization), post message procedure (end-of-message and confirmation) and call release (disconnection).

T.35 ITU-T recommendation proposing a procedure for the allocation of ITU-T members' country or area codes for non-standard facilities in telematic services.

T.37 New ITU standard for transferring of facsimile messages via store and forward over packet-switched IP networks — the Internet, corporate Intranets, etc.

T.38 A new ITU-T standard for sending real-time facsimile messages over packet-switched IP networks — the Internet, corporate Intranets, etc.

T.4 ITU-T standard for Group 3 fax machines, using T.30 and various V series standards. It a so describes the data compression methods MH and MR.

T.434 The concept is simple: use a fax machine equipped with a disk drive to send and receive binary files as easily as you send a fax. The system would benefit from the fax's technical sophistication and ease of use, such as calling tone and the called unit's identification of capabilities. T.434 is an evolving ITU-T recommendation which defines a format used to encode a binary file and its attributes into a set of octets. This encoded binary file can then by sent over phone lines using error-corrected T.30 fax pages. The union of these two elements (file format and T.30 ECM) is known as "Binary File Transfer." The T.434 file attribute encoding is independent of ECM's block and page segmentation. T.434 defines 27 attributes which are used to describe a file. These attributes include protocol version, filename, permitted actions, contents type, storage account, date and time of creation, date and time of last modification, date and time of last read access, identity of creator, identity of last modifier, identity of last readers, filesize, future filesize, access control, legal qualifications, private use, structure, application reference, machine, operating system, recipient, character set, compression, environment, pathname, user visible string and data file content. Fisk Communications of San Diego, CA has done extensive work in the area of T.434 and has granted an irrevocable royalty-free and compensation free license to the Telecommunications Industry Association for basically all Fisk's work on T.434. Fisk hopes to make T.434 an extensive standard so that its equipment, which sends binary file transfers, can communicate with and receive from other machines. Microsoft also has defined a binary file transfer in the fax portion of its Microsoft At Work architecture. Some observers believe the At Work binary file transfer architecture is richer and more robust than T.434, supporting password and public/private-key encryption as well as digital signature verification. As of writing, it is not clear which standard will win.

T.6 ITU-T recommendation for Group 3 fax machines using T.30 and various V series standards. It also describes compression methods (Modified Huffman and Modified READ).

T.611 Also known as Appli/Com. A messaging standard proposed by France and Germany defining a Programmable Communication Interface (PCI) for Group 3 fax, Group 4 fax, teletex and telex service.

T1.601 The ANSI specification for ISDN BRI outside wire, known as the U interface. T1.601 uses the 2B1Q line code operating at 160 Kb/s (144 Kb/s of 2B+D plus Layer 1 overhead bits). The electrical signal can tolerate a maximum loss of 42 dB at 40 KHz, which usually limits the local loop length to 18 Kft or less.

T11 A technical committee of the National Committee for Information Technology Standards , titled T11 I/O Interfaces. It is tasked with developing standards for moving data in and out of central computers.

T1 1. A committee accredited by ANSI and sponsored by the Alliance For Telecommunication Industry Solutions (ATIS). The committee's role is to establish U.S. standards for digital telephony, particularly T1. ATIS provides all the administrative and logistical support. See ALLIANCE FOR TELECOMMUNICATIONS INDUSTRY SOLUTIONS.

2. See T-1 above (as in T-ONE).

T1-606/T1-6ac/T1-gfr ANSI's frame relay service specifications.

T-120 See T.120

T1C 3.152 million bits per second. Capable of handling 48 voice conversations. T1C is further up the North American digital carrier hierarchy. See T CARRIER and T-1.

T1E1 An ANSI standards sub-committee dealing with Network Interfaces.

T1M1 An ANSI standards sub-committee dealing with T-1 Inter-Network Operations, Administration and Maintenance.

T1Q1 An ANSI standards sub-committee dealing with performance.

T1S1 T1S1 is a technical subcommittee to T-1 responsible for standards related to services,architectures, and signaling.

T1X1 T1X1 is a technical subcommittee to T-1 responsible for standards pertaining to synchronous interfaces and hierarchical structures relevant to interconnection of network transport signals.

T2 6.312 million bits per second. Capable of handling at least 96 voice conversations depending on the encoding scheme chosen. T-2 is four times the capacity of T-1. T-2 is further up the North American digital carrier hierarchy. In this dictionary we have adopted the style of writing T2 as T-2. See T-1.

T3 Twenty eight (28) T-1 lines or 44.736 million bits per second. Commonly referred to as 45 megabits per second. Capable of handling 672 voice conversations. T-3 runs on fiber optic and is typically called FT3. T-3 is further up the North American digital carrier hierarchy. In this dictionary we have adopted the style of writing T3 as T-3. See T-1.

T3POS Transaction Processing Protocol for Point of Sale. T3POS is a transaction switching and transport protocol designed to provide existing Point-Of-Sale (POS) equipment and future POS terminals with efficient and economical switching and transport service over an X.25 based packet network.

T4 274.176 million bits per second. Capable of handling 4032 voice conversations. T-4 has 168 times the capacity of T-1. T-4 can run on coaxial cable, waveguide, millimeter radio or fiber optic. T-4 is further up the North American digital carrier hierarchy. In this dictionary we have adopted the style of writing T4 as T-4. See also T-1.

TA 1. A Terminal Adapter allows existing non-ISDN terminals to operate on ISDN lines. It provides conversion between a non-ISDN terminal device and the ISDN user/network interface. 2. Technical Advisory. These documents are documents describing Bellcore's preliminary view of proposed generic requirements for products, new technologies, services, or interfaces.

3. Termination Attempt.

Table Driven Describing a logical computer process, widespread in the operation of communications devices and networks, in which a user-entered variable is matched against an array of pre defined values. Frequently used in network routing, access security and modem operation. It involves a table look up that is a reference to a collection of pre defined values.

Table Hook-Up Method An information retrieval system in which the input information and the related output information are stored as a pair. When a particular input is given, the table is accessed and the output data which coincides with the input is taken out.

Tables A collection of data in which each item is arranged in relation to the other items. Many telephony functions use "look up tables" to determine the routing of calls. These tables solve the problem, "If the call is going to this exchange in this area code, then use this trunk and this routing pattern." See TABLE DRIVEN.

TABS AT&T's Telemetry Asynchronous Block Serial protocol. A polled point-to-point or multi point "master-slave" (remote-monitored equipment) communication protocol that supports moderate data transfer rates over intra office wire pairs. The remotes send "requests" or "polls" to monitored equipment. The monitored equipment answers the request with "responses." Defines two physical interfaces for direct connection between the telemetry remote and the monitored equipment:

. RS422 Point-to-Point
. RS485 Point-to-Point or Multi-Point.

Four wire, two Tx (remote to monitored) and two to Tx (monitored to remote), 22 or 21 gauge twisted pair, max 4 kft remote-to-monitored.

TACACS Terminal Access Controller Access Control System. An IETF (RFC 1492) standard security protocol which runs between client devices on a network and against a TACACS server. TACACS is an authentication mechanism which is used to authenticate the identity of a device seeking remote access to a privileged database. Variations on the theme include TACACS+, which provides services of authentication, authorization and accounting independently. TACSAS+ supports a challenge/response system and password encryption, as well as the standard TACACS user authentication. See also Authentication.

TACS Total Access Communications System. A short-lived cellular telephone system developed for use in the U.K. TACS is a derivative of AMPS (Advanced Mobile Phone System), the analog cellular standard developed by Motorola and widely deployed in the U.S. and other parts of the world. TACS operated in the 900 MHz band, supporting 1,000 voice grade channels. TACS gave way to GSM, which is a much better digital technology. JTAC (Japanese TACS), which operated in the 800 and 900 MHz ranges, suffered a similar fate. See also AMPS and GSM.

Tactical Automatic Digital Switching System TADSS. A transportable store-and-forward, message-switching system designed for rapid deployment in support of tactical forces. A military definition.

Tactical Command And Control (C2) Systems The equipment, communication, procedures, and personnel essential to a commander for planning, directing, coordinating, and controlling tactical operations of assigned forces pursuant to the missions assigned. A military definition.

Tactical Communication A military term. A method or means of conveying information of any kind, especially orders and decisions from one command, person, or place to another within the tactical forces, normally by means of electronic equipment (including communications security equipment). Excluded from this definition are communications provided to tactical forces by DCS, to non tactical forces by DCS, to tactical forces by non tactical military commands, and to tactical forces by civil organizations.

Tactical Communication System A system configured by various types of fixed-size, self-contained assemblages, such as radio terminals and repeaters; switching, transmission, and terminal equipment; and interconnect and control facilities, that are used within or in support of tactical military forces. The system provides securable voice and data communications and among mobile users to facilitate command and control within, and in support of, tactical forces.

Tactical Data Information Link TADIL. A military term. A Joint-Chiefs-of-Staff-approved standardized communication link suitable for transmission of digital information. A TADIL is characterized by its standardized message formats and transmission characteristics.

Tactical Data Information Link — A TADIL—A. A military term. A netted link in which one unit acts as a net control station and interrogates each unit by roll call. Once interrogated, that unit transmits its data to the net. This means that each unit receives all the information transmitted. This is a direct transfer of data and no relaying is involved.

Tactical Data Information Link — B TADIL—B. A military term. A point-to-point data link between two units which provides for simultaneous transmission and reception of full duplex data.

Tactical Load A military term. That part of the operational load required by the host service consisting of weapons, detection, command control systems, and related functions.

TAD Telephone Answering Device.

TADIL Tactical Data Information Link.

TADSS Tactical Automatic Digital Switching System.

TAF Targeted Accessibility Fund. In the late Spring of 1998, the New York State Public Service Commission ordered the establishment of a Targeted Accessibility Fund (TAF), which, according to the NYPSC, "will fund the costs incurred by all local exchange carriers, including competitive carriers, for E-911, lifeline and telecommunications relay service. All telecommunications carriers will be required to contribute to the fund based upon their intra-state gross revenues net of payments made to underlying carriers...Under PSC guidelines, the TAF is to be administered by a ten person board."

TAFAS Trunk Answer From Any Station. The ability to answer an incoming phone call from any telephone attached to the system.

Tag Tags are codes used for formatting HTML documents for the World Wide Web. There are both single and compound tags. For example, the single code for a line break is

Tag Tags are codes used for formatting HTML documents for the World Wide Web. There are both single and compound tags. For example, the single code for a line break is <br>. Bold text requires compound tags. For example, if you want to bold the word help, you would mark it <b>help</b>, where <b> means to turn on bolding and </b> means to shut it off.

Tag Image File Format TIFF provides a way of storing and exchanging digital image data. Aldus Corp., Microsoft Corp., and major scanner vendors developed TIFF to help link scanned images with the popular desktop publishing applications. It is now used for many different types of software applications ranging from medical imagery to fax modem data transfers, CAD programs, and 3D graphic packages. The current TIFF specification supports three main types of image data: Black and white data, halftones or dithered data, and grayscale data. Some wags think TIFF stands for "Took It From a Fotograf." It doesn't.

Tag Switching A technique developed by Cisco Systems for high-performance packet forwarding through a router. A label or "tag" is assigned to destination networks or hosts. As a packet stream is presented to the tag edge switch, it analyzes the network-layer header prepended to each packet, selects a route for the packet from its internal routing tables, prepends the PDU with a tag from its Tag Information Base (TIB), and forwards the packet to the next-hop tag switch, which typically is a core tag switch. That core switch then forwards the packet solely on the basis of the tag, eliminating the need to re-analyze the header. As the packet reaches the tag edge router at the egress point of the network, the tag is stripped off and the packet is delivered to the target device. While this approach adds a small amount of overhead to each packet, the speed of packet processing is improved considerably, particularly in a complex network in which multiple routers or switches must act on each packet. Tag switching can be applied to IPX and other network protocols, as well as IP. Tag switching also allows Layer 2 switches to participate in Layer 3 routing, reducing the number of routing peers with which each edge router must deal, and, thereby, enhancing the scalability of the network.

According to Cisco, the first application of tag switching likely will be in the Internet, where scalability is of great importance. Additionally, tag switching's improved speed of packet processing promises to enhance overall Internet performance. Cisco has submitted the Tag Switching specification to the IETF for consideration as an Internet standard. Contrast with IP SWITCHING.

TAI inTernational Atomic tIme.

Tail An echo cancellation term. The tail, measured in milliseconds, is the amount of your conversation which returns to you in the echo, as measured in milliseconds. A tail of zero milliseconds clearly means there's no echo.

Tail Circuit 1. A feeder circuit or an access line to a network. Typically, a connection from a satellite, microwave receiver to a user's equipment location.

2. A point-to-point circuit connecting a remote terminal to a local terminal via two modems at an intermediate site. A crossover cable connects the two modems at the intermediate site.

3. A communications line from the end of a major transmission link, such as a microwave link, satellite link, or LAN, to the end-user location. A tail circuit is a part of a user-to-user connection.

Tail-End-Hop-Off TEHO. In a private network with several nodes (locations), TEHO occurs when a call placed from one location on the network to a location not on the network leaves the network at the node closest to its destination.

Tailing In facsimile systems, the excessive prolongation of the decay of the signal. Also called Hangover.

Taligent A joint venture of IBM and Apple to create a new operating system for personal computers.

Talk A service on Internet whereby you hold conversations with others by typing into your computer. And they reply by typing into their computer.

Talk Battery The DC voltage supplied by the central office to the subscriber's loop so as to allow you to have a voice conversation. Also known as A BATTERY. Typically somewhere between 5 and 25 volts. See B BATTERY.

Talk Off In voice processing, in response to questions such as "Press one for Harry," the user touchtones buttons on his phone. Those buttons generate DTMF (Dual Tone Multi-Frequency) tones. The system has to figure out what the person "said" with his touchtones. The tricky part of DTMF detection is distinguishing between tones generated from an actual "key press" and "tones" caused by speech. Mistaking a person's speech (as in leaving a message) for DTMF is called "talk-off." Mistaking a person's recorded speech (as in playing back a message) for DTMF is called "play-off."

You can imagine the havoc poor DTMF detection can cause a voice processing system. For example, if touchtoning three means "delete this message" and while playing the message, the system incorrectly detects a portion of the message playback as the touchtones for a key press three, I'll delete the message when I had intended to listen to it. On the other hand, if I'm listening to a message and want to delete it prior to finishing the message, I want the system to detect my key press three as the real thing and go ahead and delete the message.

Talk Path The tip and ring conductors of a telephone circuit.

Talk Set, Optical An instrument for talking over fibers — used when installing and testing the cable.

Talk Time 1. The amount of time agents spend on the phone, as opposed to the time between calls spent updating records, sending out literature or going to the bathroom.

2. The length of time you can talk on your portable or transportable cellular phone from a fully charged battery without standby time. The battery capacity of a cellular portable or transportable is usually expressed in terms of so many minutes of talk time or so many hours of standby time. When you are talking, the phone draws more power from the battery.

Talk-Off See TALK OFF.

Talkdown Missed signals in the presence of speech. Commonly used to describe the performance of a DTMF receiver when it fails to recognize a valid DTMF tone due to cancellation of that tone by speech.

Talker Provides interaction with a Dialogic board, allowing for Voice Mail functionality.

Talking Channel Capacity TC. The network access line capacity of talking channel equipment of an entity is the maximum number of network access lines that can be served without exceeding the percent incoming matching loss objective for that entity.

Talking Head That part of the person seen in the typical business videoconference; the head and shoulders. This type of image is fairly easy to capture with compressed video because there is very little motion in a talking head image.

Talkoff See TALK OFF.

TAM 1. Telephone Answering Machine.

2. Telecommunications Access Method.

Tandem 1. TDM. One type of Central office; it establishes trunk-to-trunk connections. functions may be combined in a single switching system.

2. The connection of networks or circuits in series. That is, the connection of the output of one circuit to the input of another. See TANDEM SWITCH.

Tandem Architecture A physical network topology where connectivity between locations is achieved by linking several locations together in a chain using private line circuits. In a tandem architecture, a packet may have to pass through several intermediate locations before reaching its final destination. A single network failure can affect connectivity between several locations, a primary weakness of the topology.

Tandem Call A call processed by two or more switches. Also used to designate this type of call at a switch where a connection is established from one trunk to another (tandem trunking). See TANDEM SWITCH.

Tandem Center In a communication system, an installation in which switching equipment connects trunks to trunks, but not any customer loops.

Tandem Data Circuit A data channel passing through more than two data circuit-terminating equipment (DCE) devices in series.

Tandem Office A major phone company switching center for the switched telephone network. It serves to connect central offices when direct interoffice trunks are not available.

Tandem PBX A main PBX is one which has a Directory Number (DN) and can connect PBX stations to the public network for both incoming and outgoing calls. A main PBX can have an associated satellite PBX, and can be part of a Tandem Tie Trunk Network (TTTN). If the main PBX provides tandem switching for tie trunks, it is called a tandem PBX.

Tandem Point An intermediate location in a tandem architecture.

Tandem Queuing A telephone company term. When for example: A shortage of receivers in a terminating office will be reflected in originating offices trunking traffic into the terminating switch. The obvious effect will be a slowing down of the originating office transmitters, i.e., Tandem Queuing.

Tandem Switch Tandem is a telephony term meaning to "connect in series." Thus a tandem switch connects one trunk to another. A tandem switch is an intermediate switch or connection between an originating telephone call location and the final destination of the call. The tandem point passes the call along. A PBX can often handle tandem calls from other/to other locations as well as process calls to, from and within its own location.

Tandem Tie Trunk Switching The PBX permits tie lines to "tandem" through the switch. This means an incoming tie line call from a distant PBX receives a dial tone instead of automatically connecting with the operator. The caller can then dial a connection with either a phone on the PBX or an outgoing line. The outgoing line can be a local trunk in which case the distant PBX has access to a form of foreign exchange service, or another tie line which links a third system. This system of tie lines is widely used to form a corporate communications system, allowing economical connections between distant offices. To provide tie line tandeming ability, the PBX must be able to detect when either tie line goes on-hook at the distant end so that it can break its tandem connection and allow the tie lines to be used for other calls.

Tandem Trunks Trunks between an end office and a tandem switching machine or between tandem switching machines. Tandem trunks can provide direct routing or alternate routing capability when direct trunks are occupied.

Tank Test A voltage dielectric test in which the wire or cable test sample is submerged in water and voltage is applied between the conductor and water as ground.

TANSTAAFL There Ain't No Such Thing As A Free Lunch.

TAO Telephony Application Object. Part of the SCSA programming framework. See S.100 and SCSA Telephony Application Objects Framework.

Tap 1. An electrical connection permitting signals to be transmitted onto or off a bus. The link between the bus and the drop cable that connects the workstation to the bus. Also a device used on CATV cables for matching impedance or connecting subscriber drops.

2. Telocator Alphanumeric Paging Protocol. A 7-bit messaging protocol which allows someone sitting at a terminal or computer to send a message to a pager (also known as a beeper). It also provides an error detecting link from the sender to the paging service provider. Invalid data, bad PINs, and other common errors are reported back to the sender. Information about TAP can be gotten from PCI in Washington DC at 202-467-4770 For a more modern 8-bit protocol, see TDP, which stands for Telocator Data Protocol.

3. A term in video compression referring to the number of pixels or lines considered in the process of averaging values through the filtering process. MPEG uses a 7-tap filter.

Tap Button A button found on single line phones behind a PBX or Centrex. The tap button gives a precisely measured Hookswitch flash. The purpose of this button is to signal the

PBX that it is about to receive a command — typically a transfer. To transfer a call on a single line phone, you typically depress the hookswitch, then punch out the extension you want to transfer the call to, announce the call when someone answers, then hang up and the PBX or Centrex transfers the call. The problem with using a hookswitch to make this transfer is that if you depress the hookswitch for too long you will cut the call off. As a result, some manufacturers put a tap button on their single line phones. This button gives the precise hookswitch signal for the precise length of time necessary — no more, no less. The Tap Button is also called a Flash button or a Tap Key.

Tap Key Also called Tap Button or Flash Key. A button on a phone that accomplishes the same function as a switch hook but is not a switch hook. See TAP BUTTON.

TAPAC Terminal Attachment Program Advisory Committee. Body which recommends telecom standards to the Canadian Federal Government.

Tape Drive The physical unit that holds, reads and writes magnetic tape.

Tape Reader A device which reads information recorded on punched paper tape or magnetic tape.

Tape Relay A method of retransmitting TTY traffic from one channel to another, in which messages arriving on an incoming channel are recorded in the form of perforated tape, this tape then being either fed directly and automatically into an outgoing channel, or manually transferred to an automatic transmitter for transmission on an outgoing channel.

Tapered Fiber An optical fiber in which the cross section, i.e., cross-sectional diameter or area, varies, i.e., increases or decreases, monotonically with length.

TAPI Telephone Application Programming Interface. Also called Microsoft/Intel Telephony API. A term that refers to the Windows Telephony API. TAPI is a changing (i.e. improving) set of functions supported by Windows that allow Windows applications (Windows 3.xx, 95 and NT) to program telephone-line-based devices such as single and multi-line phones (both digital and analog), modems and fax machines in a device-independent manner. TAPI essentially does to telephony devices what Windows printer system did to printers — make them easy to install and allow many application programs to work with many telephony devices, irrespective of who made the devices. TAPI is one of numerous high-level device interfaces that Windows offers as part of the Windows Open Services Architecture (WOSA). TAPI simplifies the process of writing a telephony application that works with a wide variety of modems and other devices supported by TAPI drivers. See also DIAL STRING, MICROSOFT FAX, TAPI 2.0, TAPI 3.0, WINDOWS 95 and WINDOWS TELEPHONY for fuller explanations.

TAPI 2.0 The following is Microsoft's explanation: The Microsoft Windows Telephony API (TAPI) 2.0 ships as part of Windows NT Server 4.0 and Windows NT Workstation 4.0. TAPI 2.0 is the latest release of the TAPI specification, introduced in 1993. TAPI helps bridge the gap between the telephone and computer. TAPI helps the PC to understand how telephone networks operate. With TAPI, programmers can exploit telephone network capabilities from within regular Windows-based applications. With TAPI, the jungle of proprietary telephony hardware is turned into standard programming interfaces. Application developers can wave goodbye to the complexity and variability of the underlying telephone network. They no longer have to hard-code their applications to a particular system's signaling or message set requirements.

Instead, applications developers write code to the Telephony Applications Programming Interface. Developers can focus on their application without worrying about the nitty-gritty programming details of connecting to a specific telephone network. TAPI supports PBXs, key telephone systems, ISDN, the analog PSTN, cellular, CENTREX and other types of telephone networks. (So long as the various providers have written an SPI — Service Provider Interface, which is code to translate their commands into code which TAPI can understand. — Harry)

TAPI 2.0 includes these enhancements:
- 32-bit architecture. All core TAPI components are now based on the Win32 architecture. Non-Intel processors running Win NT Server 4.0 or Workstation 4.0 are supported.
- 32-bit application portability. Existing Win32 apps currently running on Win 95 using TAPI 1.4 will run on NT Workstation 4.0 or Server 4.0 on Intel x86 microprocessors.
- 16-bit application portability. Existing applications currently running on Win 95 and 3.1 using TAPI 1.3 will run on NT Workstation 4.0 or NT Server 4.0 on Intel x86 microprocessors.
- Unicode support. Win32 apps can now call the existing ANSI TAPI functions or the new Unicode versions of functions. Unicode is a 16-bit, fixed-width character encoding standard. It encompasses virtually all of the characters commonly used on computers today.
- Expanded feature support for call center applications. TAPI now supports an expanded set of features to better serve call center operations with Windows. New call center features supported include ACD queues, predictive dialing, and call routing.
- Registry support. All telephony parameters are now stored in the Windows registry. All stored parameters can be updated across the LAN.
- Quality of Service (QoS) support. Applications can request, negotiate, and re-negotiate QoS performance parameters with the network. Improved QoS support reduces or eliminates latency and other negative characteristics for applications, especially voice and data apps, over various networks.
- Enhanced device sharing. Applications can restrict handling of inbound calls on a device to a single address. This supports features such as distinctive ringing.
- Additions and changes to TAPI functions. Many new TAPI functions and messages are available with TAPI 2.0. In addition, several functions and messages already supported by TAPI 1.4 were changed in some measure to make them more consistent in their operation. See TAPI 3.0

TAPI 3.0 Microsoft announced TAPI 3.0 in mid-September, 1997. At that time, Mitch Goldberg of Microsoft, said TAPI 3.0 will be available initially in NT 5 beta 1 in September, fully featured in NT 5 beta 2, and available commercially when the NT 5.0 operating system ships (around the middle of 1998). TAPI 3.0 will also be available for Windows 98 ("Memphis") in the NT 5.0 time frame. microsoft.com/communications At the same time, Microsoft released a White Paper titled "IP Telephony with TAPI 3.0." Here are key excerpts. For the full paper, www.ctexpo.com. TAPI 3.0 is an evolutionary API providing convergence of both traditional PSTN telephony and IP Telephony. IP Telephony is an emerging set of technologies which enables voice, data, and video collaboration over existing LANs, WANs and the Internet. TAPI 3.0 enables IP Telephony on the Microsoft Windows operating system platform by providing simple and generic methods for making connections between two or more machines, and accessing any media streams involved in the connection. TAPI 3.0 sup-

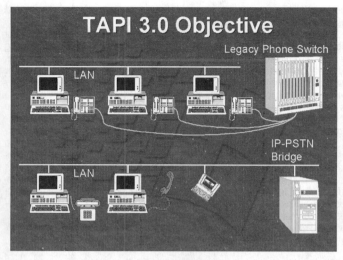

TAPI 3.0 Objective

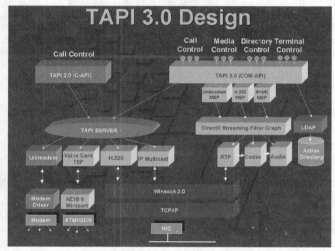

TAPI 3.0 Design

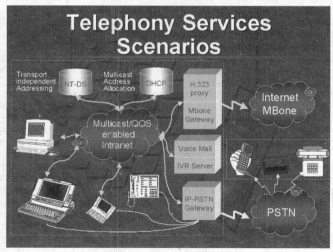

Telephony Services Scenarios

ports standards based H.323 conferencing and IP multicast conferencing. It utilizes the Windows NT 5.0 operating system's Active Directory service to simplify deployment within an organization, and includes quality of service (QoS) support to improve conference quality and network manageability.

What is IP Telephony? IP Telephony is an emerging set of technologies that enables voice, data, and video collaboration over existing IP-based LANs, WANs, and the Internet. Specifically, IP Telephony uses open IETF and ITU standards to move multimedia traffic over any network that uses IP (the Internet Protocol). This offers users both flexibility in physical media (for example, POTS lines, ADSL, ISDN, leased lines, coaxial cable, satellite, and twisted pair) and flexibility of physical location. As a result, the same ubiquitous networks that carry Web, e-mail and data traffic can be used to connect to individuals, businesses, schools and governments worldwide. TAPI 3.0 is an evolutionary API that supports convergence of both traditional PSTN telephony and telephony over IP networks.

What are the Benefits of IP Telephony? IP Telephony allows organizations and individuals to lower the costs of existing services, such as voice and broadcast video, while at the same time broadening their means of communication to include modern video conferencing, application sharing, and whiteboarding tools. In the past, organizations have deployed separate networks to handle traditional voice, data, and video traffic. Each with different transport requirements, these networks were expensive to install, maintain, and reconfigure. Furthermore, since these networks were physically distinct, integration was difficult if not impossible, limiting their potential usefulness. IP Telephony blends voice, video and data by specifying a common transport, IP, for each, effectively collapsing three networks into one. The result is increased manageability, lower support costs, a new breed of collaboration tools and increased productivity. Possible applications for IP Telephony include telecommuting, real-time document collaboration, distance learning, employee training, video conferencing, video mail and video on demand.

What is TAPI 3.0? As telephony and call control become more common in the desktop computer, a general telephony interface is needed to enable applications to access all the telephony options available on any machine. Additionally, it is imperative that the media or data on a call is available to applications in a standard manner. TAPI 3.0 is an architecture that provides simple and generic methods for making connections between two or more machines, and accessing any media streams involved in that connection. It abstracts call-control functionality to allow different, and seemingly incompatible, communication protocols to expose a common interface to applications. IP Telephony is a demand poised for explosive growth, as organizations begin an historic shift from expensive and inflexible circuit-switched public telephone networks to intelligent, flexible and inexpensive IP networks.

Inside TAPI 3.0: TAPI 3.0 integrates multimedia stream control with legacy telephony. Additionally, it is an evolution of the TAPI 2.1 API to the COM model, allowing TAPI applications to be written in any language, such as JavaT, C/C++ and the Microsoft Visual Basic programming system. Besides supporting classic telephony providers, TAPI 3.0 supports standard H.323 conferencing and IP multicast conferencing. TAPI 3.0 utilizes the Windows NT 5.0 Active Directory service to simplify deployment within an organization, and it supports quality of service (QoS) features to improve conference quality and network manageability.

There are four major components to TAPI 3.0:
1. TAPI 3.0 COM API
2. TAPI Server
3. Telephony Service Providers
4. Media Stream Providers

In contrast to TAPI 2.1, the TAPI 3.0 API is implemented as a suite of Component Object Model (COM) objects. Moving TAPI to the object-oriented COM model allows component upgrades of TAPI features. It also allows developers to write TAPI-enabled applications in any language, such as Java, Visual Basic, or C/C++. The TAPI Server process (TAPISRV.EXE) abstracts the TSPI (TAPI Service Provider Interface) from TAPI 3.0 and TAPI 2.1, allowing TAPI 2.1 Telephony Service Providers to be used with TAPI 3.0, maintaining the internal state of TAPI. Telephony Service Providers (TSPs) are responsible for resolving the protocol-independent call model of TAPI into protocol-specific call control mechanisms. TAPI 3.0 provides backward compatibility with TAPI 2.1 TSPs. Two IP Telephony service providers (and their associated MSPs) ship by default with TAPI 3.0: the H.323 TSP and the IP Multicast Conferencing TSP. TAPI 3.0 provides a uniform way to access the media streams in a call, supporting the DirectShow API as the primary media stream handler. TAPI Media Stream Providers (MSPs) implement DirectShow interfaces for a particular TSP and are required for any telephony service that makes use of DirectShow streaming. Generic streams are handled by the application.

There are five objects in the TAPI 3.0 API:
1. TAPI
2. Address
3. Terminal
4. Call
5. CallHub

The TAPI object is the application's entry point to TAPI 3.0. This object represents all telephony resources to which the local computer has access, allowing an application to enumerate all local and remote addresses. An Address object represents the origination or destination point for a call. Address capabilities, such as media and terminal support, can be retrieved from this object. An application can wait for a call on an Address object,
or can create an outgoing call object from an Address object. A Terminal object represents the sink, or renderer, at the ter-

mination or origination point of a connection. The Terminal object can map to hardware used for
human interaction, such as a telephone or microphone, but can also be a file or any other device capable of receiving input or creating output. The Call object represents an address' connection between the local address and one or more other addresses (This connection can be made directly or through a CallHub). The Call object can be imagined as a first-party view of a telephone call. All call control is done through the Call object. There is a call object for each member of a CallHub. The CallHub object represents a set of related calls. A CallHub object cannot be created directly by an application. They are created indirectly when incoming calls are received through TAPI 3.0. Using a CallHub object, a user can enumerate the other participants in a call or conference, and possibly (because of the location independent nature of COM) perform call control on the remote Call objects associated with those users, subject to sufficient permissions:

Using TAPI Objects to Place a Call:
1. Create and initialize a TAPI object.
2. Use the TAPI object to enumerate all available Address objects on a machine (for example, network cards, modems, and ISDN lines).
3. Enumerate the supported address types of each Address object (for example, a phone number, IP address, and so on).
4. Choose an Address object based on queries for support for appropriate media (audio, video, and so on) and address types.
5. Use the CreateCall() method of the Address object to create a Call object associated with a particular address.
6. Select appropriate Terminals on the Call object.
7. Call the Connect() method of the Call object to place the call.

To Answer a Call:
1. Create and initialize a TAPI object.
2. Use the TAPI object to enumerate all available Address objects on a machine (for example, network cards, modems, and ISDN lines).
3. Enumerate the supported address types of each Address object (e.g. a phone number, IP address, etc.).
4. Choose an Address object based on queries for support of appropriate media (audio, video, and so on) and address types.
5. Register an interest in specific media types with the appropriate Address object.
6. Register a call event handler (i.e. implement an ITCallNotification interface) with the Address object.
7. TAPI notifies the application of a new call through ITCallNotification and creates a Call object.
8. Select appropriate Terminals on the Call object.
9. Call the Connect() method of the Call object to place the call.
10. Call the Answer() method of the Call object to answer the call.

Media Streaming Model: The Windows operating system provides an extensible framework for efficient control and manipulation of streaming media called the DirectShow API. DirectShow, through its exposed COM interfaces, provides TAPI 3.0 with unified stream control. At the heart of the DirectShow services is a modular system of pluggable components called filters, arranged in a configuration called a filter graph. A component called the filter graph manager oversees the connection of these filters and controls the stream's data flow. Each filter's capabilities are described by a number of special COM interfaces called pins. Each pin instance can consume or produce streaming data, such as digital audio. While COM objects are usually exposed in user mode programs, the DirectShow streaming architecture includes an extension to the Windows driver model that allows the connection of media streams directly at the device driver level. These high-performance streaming extensions to the Windows driver model avoid user-to-kernel mode transitions, and allow efficient routing of data streams between different hardware components at the device driver level. Each kernel mode filter is mirrored by a corresponding user mode proxy that facilitates connection setup and can be used to control hardware-specific features. DirectShow network filters extend the streaming architecture to machines connected on an IP network. The Real-Time Transport protocol (RTP), designed to carry real-time data over connectionless networks, transports TAPI media streams and provides appropriate time stamp information. TAPI 3.0 includes a kernel mode RTP network filter. TAPI 3.0 utilizes this technology to present a unified access method for the media streams in multimedia calls. Applications can route these streams by manipulating corresponding filter graphs; they can also easily connect streams from multiple calls for bridging and conferencing capabilities.

What is H.323?: H.323 is a comprehensive International Telecommunications Union (ITU) standard for multimedia communications (voice, video, and data) over connectionless networks that do not provide a guaranteed quality of service, such as IP-based networks and the Internet. It provides for call control, multimedia management, and bandwidth management for point-to-point and multipoint conferences. H.323 mandates support for standard audio and video codecs and supports data sharing via the T.120 standard. Furthermore, the H.323 standard is network, platform, and application independent, allowing any H.323 compliant terminal to interoperate with any other.

H.323 allows multimedia streaming over current packet-switched networks. To counter the effects of LAN latency, H.323 uses as a transport the Real-time Transport Protocol (RTP), an IETF standard designed to handle the requirements of streaming real-time audio and video over the Internet. The H.323 standard specifies three command and control protocols:

1. H.245 for call control
2. Q.931 for call signaling
3. The RAS (Registration, Admissions, and Status) signaling function

The H.245 control channel is responsible for control messages governing operation of the H.323 terminal, including capability exchanges, commands, and indications. Q.931 is used to set up a connection between two terminals, while RAS governs registration, admission, and bandwidth functions between endpoints and gatekeepers (RAS is not used if a gatekeeper is not present).

H.323 defines four major components for an H.323-based communications system:

1. Terminals
2. Gateways
3. Gatekeepers
4. Multipoint Control Units (MCUs)

Terminals are the client endpoints on the network. All terminals must support voice communications; video and data support is optional.

A Gateway is an optional element in an H.323 conference. Gateways bridge H.323 conferences to other networks, communications protocols, and multimedia formats. Gateways are not required if connections to other networks or non-H.323 compliant terminals are not needed. Gatekeepers perform two important functions which help maintain the robustness of the network - address translation and bandwidth management.

Gatekeepers map LAN aliases to IP addresses and provide address lookups when needed. Gatekeepers also exercise call control functions to limit the number of H.323 connections, and the total bandwidth used by these connections, in an H.323 "zone." A Gatekeeper is not required in an H.323 system-however, if a Gatekeeper is present, terminals must make use of its services.

Multipoint Control Units (MCU) support conferences between three or more endpoints. An MCU consists of a required Multipoint Controller (MC) and zero or more Multipoint Processors (MPs). The MC performs H.245 negotiations between all terminals to determine common audio and video processing capabilities, while the Multipoint Processor (MP) routes audio, video, and data streams between terminal endpoints.

Any H.323 client is guaranteed to support the following standards: H.261 and G.711. H.261 is an ITU-standard video codec designed to transmit compressed video at a rate of 64 Kbps and at a resolution of 176x44 pixels (QCIF). G.711 is an ITU-standard audio codec designed to transmit A-law and Ê-law PCM audio at bit rates of 48, 56, and 64 Kbps. Optionally, an H.323 client may support additional codecs: H.263 and G.723. H.263 is an ITU-standard video codec based on and compatible with H.261. It offers improved compression over H.261 and transmits video at a resolution of 176 x 44 pixels (QCIF). G.723 is an ITU-standard audio codec designed to operate at very low bit rates.

The H.323 Telephony Service Provider (along with its associated Media Stream Provider) allows TAPI-enabled applications to engage in multimedia sessions with any H.323-compliant terminal on the local area network. Specifically, the H.323 Telephony Service Provider (TSP) implements the H.323 signaling stack. The TSP accepts a number of different address formats, including name, machine name, and e-mail address. The H.323 MSP is responsible for constructing the DirectShow filter graph for an H.323 connection (including the RTP, RTP payload handler, codec, sink, and renderer filters).

H.323 telephony is complicated by the reality that a user's network address (in this case, a user's IP address) is highly volatile and cannot be counted on to remain unchanged between H.323 sessions. The TAPI H.323 TSP uses the services of the Windows NT Active Directory to perform user-to-IP address resolution. Specifically, user-to-IP mapping information is stored and continually refreshed using the Internet Locator Service (ILS) Dynamic Directory, a real-time server component of the Active Directory.

The following user scenario illustrates IP address resolution in the H.323 TSP:

1. John wishes to initiate an H.323 conference with Alice, another user on the LAN. Once Alice's video conferencing application creates an Address object and puts it in listen mode, Alice's IP address is added to the Windows NT Active Directory by the H.323 TSP. This information has a finite time to live (TTL) and is refreshed at regular intervals via the Lightweight Directory Access Protocol (LDAP):

2. John's H.323 TSP then queries the ILS Dynamic Directory server for Alice's IP address. Specifically, John queries for any and all RTPerson objects in the Directory associated with Alice:

3. Armed with Alice's up-to-date IP address, John initiates an H.323 call to Alice's machine, and H.323-standard negotiations and media selection occurs between the peer TSPs on both machines. Once capability negotiations have been completed, both H.323 Media Stream Providers (MSPs) construct appropriate DirectShow filter graphs, and all media

streams are passed off to DirectShow to handle. The conference then begins.

What is IP Multicast Conferencing? IP Multicast is an extension to IP that allows for efficient group communication. IP Multicast arose out of the need for a lightweight, scalable conferencing solution that solved the problems associated with real-time traffic over a datagram, "best-effort" network. There are many advantages to using IP Multicast: scalability, fault tolerance, robustness, and ease of setup. The IP Multicast conferencing model incorporates the following key features:

1. No global coordination is needed to add and remove members from a conference.

2. To reach a multicast group, a user sends data to a single multicast IP address. No knowledge of the other users in a group is necessary.

3. To receive data, users register their interest in a particular multicast IP address with a multicast aware router. No knowledge of the other users in a group is necessary. Routers hide the multicast implementation details from the user.

Traditional connection-oriented conferencing suffers from a number of problems:

1. User complexity: Users must know the location of every user they wish to converse with, limiting scalability and fault-tolerance and rendering it difficult for users to add and remove themselves from a conference.

2. Wasted bandwidth: A user wishing to broadcast data to N users must send data through N connections.

The total bandwidth required for multiparty conferences in which all users are sending data goes up as N squared the number of parties involved, leading to huge scalability problems. IP Multicast takes advantage of the actual network topology to eliminate the transmission of redundant data down the same communications links.

IP Multicast implements a lightweight, session-based communications model, which places relatively little burden on conference users. Using IP Multicast, users send only one copy of their information to a group IP address that reaches all recipients. IP Multicast is designed to scale well as the number of participants expands. Adding one more user does not add a corresponding amount of bandwidth. Multicasting also results in a greatly reduced load on the sending server. IP Multicast routes these one-to-many data streams efficiently by constructing a spanning tree, in which there is only one path from one router to any other. Copies of the stream are made only when paths diverge.

Without multicasting, the same information must either be carried over the network multiple times, one time for each recipient, or broadcast to everyone on the network, consuming unnecessary bandwidth and processing. IP Multicast uses Class D Internet Protocol addresses to specify multicast host groups, ranging from 224.0.0.0 to 239.255.255.255. Both permanent and temporary group addresses are supported. Permanent addresses are assigned by the Internet Assigned Numbers Authority (IANA) and include 224.0.0.1, the "all-hosts group" used to address all multicast hosts on the local network, and 224.0.0.2, which addresses all routers on a LAN. The range of addresses between 224.0.0.0 and 224.0.0.255 is reserved for routing and other low-level network protocols. Other addresses and ranges have been reserved for applications, such as 224.0.13.000 to 224.0.13.255 for Net News (for more information, see RFC 1700, "Assigned Numbers" at ftp://ftp.internic.net/rfc/rfc1700.txt).

The transport protocol for IP Multicast is RTP (Real-time Transport Protocol), which provides a standard multimedia header giving timestamp, sequence numbering, and payload format information. Applications for IP Multicast include video and audio conferencing, telecommuting, database and Web-site replication, distance learning, dissemination of stock quotes, and collaborative computing. At present, the largest demonstration of the capabilities of IP Multicast is the Internet MBONE (Multicast Backbone). The MBONE is an experimental, global multicast network layered on top of the physical Internet. It has been in existence for about five years, and presently carries IETF meetings, NASA space shuttle launches, music, concerts, and many other live meetings and performances. www.mbone.com.

The IP Multicast Conferencing TSP is chiefly responsible for resolving conference names to IP multicast addresses, using the Session Description Protocol (SDP) conference descriptors stored in the ILS Dynamic Directory Conference Server. It is complemented by the Rendezvous conference controls, described later. The IP Multicast Conferencing MSP is responsible for constructing an appropriate DirectShow filter graph for an IP multicast connection (including RTP, RTP payload handler, codec, sink, and renderer filters).

TAPI 3.0 uses the IETF standard Session Description Protocol to advertise IP multicast conferences across the enterprise. SDP descriptors are stored in the Windows NT Active Directory — specifically, in the ILS Dynamic Directory Conference Server. In contrast to the Dynamic Directory servers used by the H.323 TSP, there is only one ILS Conference Server per enterprise, since conference announcements are not continually refreshed, therefore consuming little bandwidth.

TAPI 3.0's IP multicast conference mechanism is illustrated in the following scenario, in which John wishes to initiate a multicast conference:

1. John's TAPI 3.0-enabled application uses the Rendezvous Controls

to create an SDP session descriptor on the ILS Conference Server. The SDP descriptor contains, among other things, the conference name, start and end time information, the IP multicast address of the conference, and the media types used for the conference.

2. Jim queries the ILS Conference Server for SDP descriptors of conferences matching his criteria.

3. Mary and Alice perform similar queries and use the SDP information they receive to decide to participate in John's conference. Armed with the multicast IP address of the conference, they join the multicast host group.

The Rendezvous Controls are a set of COM components that abstract the concept of a conference directory, providing a mechanism to advertise new multicast conferences and to discover existing ones. They provide a common schema (SDP) for conference announcement, as well as scriptable interfaces, authentication, encryption, and access control features.

The user may add, delete, and enumerate multicast conferences stored on an ILS Conference Server via the Rendezvous Controls. These controls manipulate conference data via the Lightweight Directory Access Protocol (LDAP). The conferencing application uses the Rendezvous Controls to obtain session descriptors for the conferences that match the user's criteria. Access control lists (ACLs) protect each of the stored conference announcements, and whether or not an announcement is visible and accessible depends upon the user's credentials. Once the user has chosen a conference, the user application searches for all Address objects that support the address type "Multicast Conference Name." The application then uses the conference name from the SDP descriptor as a parameter to the CreateCall() method of the appropriate Address object, passes the appropriate Terminal objects to the returned Call object, and calls Call->Connect(). The Rendezvous Controls store the conference information on an ILS Conference Server in a format defined by the Session Description Protocol (SDP), an IETF standard for announcing multimedia conferences. The purpose of SDP is to publicize sufficient information about a conference (time, media, and location information) to allow prospective users to participate if they so choose. Originally designed to operate over the Internet MBONE (IP Multicast Backbone), SDP has been integrated by TAPI 3.0 with the Windows NT Active Directory, thereby extending its functionality to local area networks. An SDP descriptor advertises the following information about a conference.

A session description is broken into three main parts: a single Session Description, zero or more Time Descriptions, and zero or more Media Descriptions. The Session Description contains global attributes that apply to the whole conference or all media streams. Time Descriptions contain conference start, stop, and repeat time information, while Media Descriptions contain details that are specific to a particular media stream.

While traditional IP multicast conferences operating over the MBONE have advertised conferences using a push model based on the Session Announcement Protocol (SAP), TAPI 3.0 utilizes a pull-based approach using Windows NT Active Directory services. This approach offers numerous advantages, among them bandwidth conservation and ease of administration.

TAPI 3.0's conference security system addresses the following needs:
• Controlling who can create, delete, and view conference announcements. * Preventing conference eavesdropping.

TAPI 3.0 utilizes the security features of the Windows NT Active Directory and LDAP to provide for secure conferencing over insecure networks such as the Internet. Each object in the Active Directory can be associated with an Access Control List (ACL) specifying object access rights on a user or group basis. By associating ACLs with SDP conference descriptors, conference creators can specify who can enumerate and view conference announcements. User authentication is provided by the Windows NT security subsystem.

Session Descriptors are transmitted from the ILS Conference Server to the user over LDAP in encrypted form, via a Secure Sockets Layer (SSL) connection, ensuring that the SDP is safe from eavesdroppers. IP Multicast makes no provision for authenticating users. Any user may anonymously join a multicast host group. To keep conferences private, TAPI 3.0 allows an IP multicast conference to be encrypted, with the encryption key distributed from within the conference descriptor. Only users with sufficient permissions have access to a conference's SDP descriptor, and therefore the Multicast Encryption Key. Once an authenticated user fetches the encryption key, he or she can participate in the conference.

In contrast to traditional data traffic, multimedia streams, such as those used in IP Telephony or videoconferencing, may be extremely bandwidth and delay sensitive, imposing unique quality of service (QoS) demands on the underlying networks that carry them. Unfortunately, IP, with a connectionless, "best-effort" delivery model, does not guarantee delivery of packets in order, in a timely manner, or at all. In order to

deploy real-time applications over IP networks with an acceptable level of quality, certain bandwidth, latency, and jitter requirements must be guaranteed, and must be met in a fashion that allows multimedia traffic to coexist with traditional data traffic on the same network.

Bandwidth: Multimedia data, and in particular video, may require orders of more bandwidth than traditional networks have been provisioned to handle. An uncompressed NTSC video stream, for example, can require upwards of 220 megabits per second to transmit. Even compressed, a handful of multimedia streams can completely overwhelm any other traffic on the network.

Latency: The amount of time a multimedia packet takes to get from the source to the destination (latency) has a major impact on the perceived quality of the call. There are many contributors towards latency, including transmission delays, queuing delays in network equipment, and delays in host protocol stacks. Latency must be minimized in order to maintain a certain level of interactivity and to avoid unnatural pauses in conversation.

Jitter: In contrast to data traffic, real-time multimedia packets must arrive in order and on time to be of any use to the receiver. Variations in packet arrival time (jitter) must be below a certain threshold to avoid dropped packets (and therefore irritating shrieks and gaps in the call). Jitter, by determining receive buffer sizes, also affects latency.

Coexistence: In comparison with multimedia traffic, data traffic is relatively bursty, and arrives in unpredictable chunks (for instance, when someone opens a Web page, or downloads a file from an FTP site). Aggregations of such bursts can clog routers and cause gaps in multimedia conferences, leaving calls at the mercy of everyone on the network (including other IP Telephony users). Multimedia bandwidth must be protected from data traffic, and vice versa.

Public-switched telephone networks guarantee a minimum quality of service by allocating static circuits for every telephone call. Such an approach is simple to implement, but wastes bandwidth, lacks robustness, and makes voice, video, and data integration difficult. Furthermore, circuit-switched data paths are impossible to create using a connectionless network such as IP.

QoS support on IP networks offers the following benefits:
1. Support for real-time multimedia applications.
2. Assurance of timely transfers of large amounts of data.
3. The ability to share the network in a manner that avoids starving applications of bandwidth.

Quality of service in TAPI 3.0 is handled through the DirectShow RTP filter, which negotiates bandwidth capabilities with the network based on the requirements of the DirectShow codecs associated with a particular media stream. These requirements are indicated to the RTP filter by the codecs via its own QoS interface. The RTP filter then uses the COM Winsock2 GQoS interfaces to indicate, in an abstract form, its QoS requirements to the Winsock2 QoS service provider (QoS SP). The QoS SP, in turn, invokes a number of varying QoS mechanisms appropriate for the application, the underlying media, and the network, in order to guarantee appropriate end-to-end QoS. These mechanisms include:
a. The Resource Reservation Protocol (RSVP)
b. Local Traffic Control: Packet Scheduling; 802.1p; Appropriate layer 2 signaling mechanisms
c. IP Type of Service and DTR header settings
The Resource Reservation Protocol (RSVP) is an IETF standard designed to support resource (for example, bandwidth)

reservations through networks of varying topologies and media. Through RSVP, a user's quality of service requests are propagated to all routers along the data path, allowing the network to reconfigure itself (at all network levels) to meet the desired level of service. The RSVP protocol engages network resources by establishing flows throughout the network. A flow is a network path associated with one or more senders, one or more receivers, and a certain quality of service. A sending host wishing to send data that requires a certain QoS will broadcast, via an RSVP-enabled Winsock Service Provider, "path" messages toward the intended recipients. These path messages, which describe the bandwidth requirements and relevant parameters of the data to be sent, are propagated to all intermediate routers along the path. A receiving host, interested in this particular data, will confirm the flow (and the network path) by sending "reserve" messages through the network, describing the bandwidth characteristics of data it wishes to receive from the sender. As these reserve messages propagate back toward the sender, intermediate routers, based on bandwidth capacity, decide whether or not to accept the proposed reservation and commit resources. If an affirmative decision is made, the resources are committed and reserve messages are propagated to the next hop on the path from source to destination.

Packet Scheduling: This mechanism can be used in conjunction with RSVP (if the underlying network is RSVP-enabled) or without RSVP. Traffic is identified as belonging to one flow or another, and packets from each flow are scheduled in accordance with the traffic control parameters for the flow. These parameters generally include a scheduled rate (token bucket parameter) and some indication of priority. The former is used to pace the transmission of packets to the network. The latter is used to determine the order in which packets should be submitted to the network when congestion occurs.

801.2p: Traffic control can also be used to determine the 802.1 User Priority value (a MAC header field used to indicate relative packet priority) to be associated with each transmitted packet. 802.1p-enabled switches can then give preferential treatment to certain packets over others, providing additional quality of service support at the data link layer level.

Layer 2 Signaling Mechanisms: In response to Winsock 2 QoS APIs, the QoS service provider may invoke additional traffic control mechanisms depending on the specific underlying data link layer. It may signal an underlying ATM network, for instance, to set up an appropriate virtual circuit for each flow. When the underlying media is a traditional 802 shared media network, the QoS service provider may extend the standard RSVP mechanism to signal a Subnet Bandwidth Manager (SBM). The SBM provides centralized bandwidth management on shared networks.

Each IP packet contains a three-bit Precedence field, which indicates the priority of the packet. An additional field can be used to indicate a delay, throughput, or reliability preference to the network. Local traffic control can be used to set these bits in the IP headers of packets on particular flows. As a result, packets belonging to a flow will be treated appropriately later by three devices on the network. These fields are analogous to 802.1p priority settings but are interpreted by higher layer network devices. www.microsoft.com/communications

TARGA Truevision Advanced Raster Graphics Adapter.

Target A SCSI device that performs an operation requested by an initiator.

Target Token Rotation Time. TTRT. An FDDI (Fiber Distributed Data Interface)token travels along the network ring

from node to node. If a node does not need to transmit data, it picks up the token and sends it to the next node. If the node possessing the token does need to transmit, it can send as many frames as desired for a fixed amount of time.

Tariff Documents filed by a regulated telephone company with a state public utility commission or the Federal Communications Commission. The tariff, a public document, details services, equipment and pricing offered by the telephone company (a common carrier) to all potential customers. Being a "common carrier" means it (the phone company) must offer its services to everybody at the prices and at the conditions outlined in its public tariffs. Tariffs do not carry the weight of law behind them. If you or the telephone company violate them, no one will go to jail. The worst that can happen to you, as a subscriber, is that your service will be cut off, or threatened to be cut off. Regulatory authorities do not normally approve tariffs. They accept them — until they are successfully challenged before a hearing of the regulatory body or in court (usually Federal Court). Many tariffs were accepted by regulatory commissions only to be struck down in court as unlawful, discriminatory, not cost-justified, etc. Monies collected under the tariff have been refunded and unnecessary equipment removed. In these new, competitive days, many telephone companies are violating their own tariffs by charging less money than their tariffs say they should, or bundling services together at a discount. They are also providing service and equipment on terms less onerous than outlined in their tariffs. Many users now regard tariffs as starting bargaining points, rather than ending bargaining points.

Tariff 12 A user-specific long distance tariff of AT&T. Tariff 12 gives AT&T the ability to price its long distance services for one company practically any which way it feels — giving them a mix of services at stable prices over the long term with significant volume discounts. As this dictionary was going to the printer, a federal appeals court overturned the Federal Communications Commission's April 1989 decision allowing AT&T to offer custom networks and ordered the FCC to reopen its investigation into the legality of the Tariff 12 deals. There are still some users. They are "grandfathered" until we get a final say on the tariff. And, as we go to press, AT&T can offer Tariff 12 customized services to any company — but cannot include 800 services in its Tariff 12 pricing.

Tariff 15 A user-specific long distance tariff of AT&T. Tariff 15 gives AT&T the ability to price its long distance services for one company practically any way it feels. Tariff 15 is single-customer discounting. Some of AT&T competitors claim the tariff is "illegal."

Tariff Rebalancing Largely an initiative of the FCC in the US, national, regional (e.g., EU) and international (ITU) regulatory authorities and policy-making bodies are considering the rebalancing of tariffs in order that they be more closely related to the costs of providing the various telecommunications products and services. While most attention is focused on the accountings rate for long distance, both domestic and international, tariff rebalancing encompasses all tariffed products (including equipment rentals) and services (e.g., local service), and across both the business and consumer domains. Historically, tariffs for individual products and services were designed in the context of the overall tariff structure, which addressed the full range of such products and services. Within this overall structure existed a complex set of cross-subsidies which generally resulted in low tariffs for basic consumer services, such as residential local service. Relatively high tariffs existed both for optional services, such as long distance, international and custom calling features. The primary justification for this arrangement was that of the desire to gain "universal service" - a phone in everybody's home. Mixed in with this was the concept of "ability to pay." Further, individual consumers vote while companies do not. As consumer rights advocates became more vocal during the past twenty years or so, the pressure to retain these cross-subsidies increased. It generally is recognized, however, that the traditional tariff structure places incumbent carriers and service providers at a decided disadvantage in a competitive environment, as the pricing policies of the newer competitors are not constrained by tariffs, nor strange societal concepts of universal service and ability to pay. Further — and this is perhaps a major impetus — there exists a clear imbalance in the accounting rates for international long distance calls. For instance, a call from Argentina to the U.S. is much more expensive than is a call in the reverse direction, even though the costs for call origination are roughly equal to the costs of termination. This imbalance in international long distance tariffs has led to a huge imbalance of trade, to the detriment of the US. Therefore, both domestic US deregulation, and the resulting increase in competition, and the international imbalance of trade resulting from imbalance in international long distance tariffs have led to the FCC initiative. Many other nations and regions have followed the U.S. lead in terms of deregulation and competition, and are in the process of rebalancing domestic tariffs. Although the WTO (World Trade Organization) has taken the initiative in terms of the rebalancing of international long distance tariffs, many nations are resisting, citing their opinions that the WTO is unduly influenced by the U.S. Further, many developing nations rely heavily on this imbalance of trade as a major source of hard currency. See also Cross Subsidization, Separations and Settlements, Tariff and Universal Service.

TARM Telephone Answering and Recording Machines.

TAS Telephone Answering Service.

TASC Telecommunications A arm, Surveillance, and Control system. Expands the scope of maintenance from the traditional alarm monitoring and control functions to include performance monitoring and fault locating.

TASI Time Assignment Speech Interpolation. A money saving analog multiplexing procedure which keeps the connection to the circuit as long as someone is speaking and lets other conversations use the circuit during the intervals (measured in microseconds) when there's no speaking. Since a long distance circuit is usually only half used — one person speaking, one person listening — at least 50% of the circuit can be used by someone else. TASI is typically used by long distance companies on submarine cable across the Atlantic and the Pacific. Unfortunately, the flip-flopping around of circuit allocation by TASI means that the first tiny bit of a conversation is often lost. This can be disastrous for data. The key in data is to keep transmitting. The key in voice is to say something like "Ah" to seize the channel and then say what you want. TASI is very much like a very fast version of mobile dispatch radio. A modern version of TASI is called Digital Speech Interpolation (DSI). TASI is somewhat comparable to statistical multiplexing of data. Ray Horak, my Contributing Editor, recently had a gentleman from Cable & Wireless attending one of his public seminars. Apparently this gentleman was one of the engineers working on an early submarine cable between the US and the UK. The cable system used TASI, which was by no means perfect, but it certainly enhanced the voice traffic carrying capacity of that very limited 50-pair

cable system. The system worked fine most of the time, but occasionally crashed at no particular time and for no apparent reason. The engineers were baffled. Finally, they recorded the conversations until the system crashed. It seems that the TASI approach would fail when required to support a large number of conversations between women. It seems that women have a tendency to talk and listen at the same time, and to do so quite effectively. Men, it seems, either lack this skill or prefer the "I talk and you listen, then you talk and I listen" protocol. Mystery solved! (Ladies, please don't call me to complain about this definition. Ray swears that it is true. Call him.)

Task Management Allocating resources and overseeing the sequence of tasks completed by the computer.

Task Switching You have a computer and you want to have it do several tasks at once. There are two ways. One is multi tasking. The computer will keep working on several tasks at once, though you may not see them on your screen. For example, you pay start a spreadsheet recalculating. And then you may call your electronic mail system. While you're receiving your mail, your spreadsheet is still recalculating. When you're through, you can switch back to your spreadsheet and see the final results. That's called multi tasking and MS-DOS doesn't have it. MS-DOS 5.0 through its Shell has something called task switching, whose idea is that you can load several programs into your computer and switch quickly between them. But the programs you put in background won't run. They stop the moment you put them in background. When you cycle back to them, they will start running again. Sadly, MS-DOS 5.0 Shell's task switching capabilities are very weak. Your programs will lock up and you will lose your data. I do not recommend using Task Switching in MS-DOS 5.0. Windows 3.1 is alleged to have a form of multi tasking. I don't trust it either yet. Windows has locked up on me on several occasions.

Task-To-Task Communication The process whereby one computer program exchanges data with another. May also be called program-to-program communications.

Tax Identification Number A unique identifier for business organizations that is used for reporting tax payments to the government (similar to the social security number for individuals).

TAXI Transparent Asynchronous Transmitter Receiver Interface. 100-Mbps ATM physical interface specification based on the FDDI PHY.

TAT TransAtlantic Telephone cable.

TB An ATM term. Transparent Bridging: An IETF bridging standard where bridge behavior is transparent to the data traffic. To avoid ambiguous routes or loops, a Spanning Tree algorithm is used.

TB/S TeraBits per Second.

TBA To Be Announced. Pricing of a product that may exist and that may, one day be priced. You often see TBA after products which are hyperware, vaporware, mirrorware or smokeware.

TBB 1. Transnational Broadband Backbone. An international, high-speed (T-3 or faster), backbone system of transmission facilities and, perhaps, switching systems.

2. Telecommunications Bonding Backbone. An electrically conductive path for telecommunications, the metallic parts of which path are permanently joined to ensure 1) electrical continuity, 2) the capacity to conduct safely any current likely to be imposed, and 3) the ability to limit dangerous electrical potentials. This definition courtesy of BICSI. In other words, a TBB is a permanently hard-wired metallic cable system,

which is safely grounded and surge-protected in order to protect your equipment.

TBBIBC TBB Interconnecting Bonding Conductor. A conductor used specifically to interconnect metallic wires in a TBB (Telecommunications Bonding Backbone). See also TBB.

TBD To Be Determined.

TBE Transient Buffer Exposure: This is a negotiated number of cells that the network would like to limit the source to sending during startup periods, before the first RM-cell returns.

TBOS Telemetry Byte Oriented Serial protocol. TBOS is a protocol for transmitting alarm, status, and control points between NE and OS. TBOS defines one physical interface for direct connection between telemetry remote and the monitored equipment. This is a point-to-point communication, RS 422A modified four wire, two to Tx (remote to monitored) and two to Tx (monitored to remote), 26 gauge, max 4 kft remote-to-monitored. Remote sees a 100 to 180 ohms resistor at monitored terminal.

TBR Timed BReak.

TC 1. Transmission Control.

2. Transmission Convergence. The TC sublayer is a dimension of the ATM Physical Layer (PHY), working in tight formation with the Physical Medium (PM) sublayer. The TC accepts frames of data transmitted across the PM, delivering them to the ATM Layer for segmentation into cells, generates the Header Error Check (HEC), and sends idle cells when the ATM layer has none to send. On reception, the TC sublayer delineates individual cells in the received bit stream and reconstitutes the frames of data, using the HEC to detect and correct received errors.

3. Telecommunications Closet. A closet which houses telecommunications wiring and telecom wiring equipment. It contains the BHC (Backbone to Horizontal Cross-connect). It may also contain the network demarcation, or MC (Main Cross-connect). The telecommunications closet is used to connect up telecom wiring. The closet typically has a door. It's a good idea to lock the door and not put anything else in the closet, like mops, buckets and brooms.

4. Top Cat. A cartoon character from the 1960s. Top Cat was really cool.

TCA TeleCommunications Association. A not-for-profit users association of communications management professionals. Formed in 1961, TCA recently expanded its focus to position itself as "The Information Technology and Telecommunications Association." Most of its members are West-of-the-Rockies telecommunications management professionals. For many years, TCA held a great annual conference in San Diego. The TCA conference was huge, and it was lots of fun. The vendors spent fortunes on their exhibits. They spent even more on hospitality suites and lavish dinner parties. Over time, the show declined as the conference business became for-profit, as the technologies moved ever more quickly, as the number of conferences proliferated, as the conference business became more highly focused and more fragmented, and as companies downsized and could no longer afford to subsidize associations like TCA. The TCA show moved to Reno, Nevada in 1997. That was the last show. Several of the TCA chapters still hold annual events, but they are much smaller. www.tca.org

TCAM TeleCommunications Access Method. A popular telecommunications software package to run on IBM 370 computers. See IBM.

TCAP Transactional Capabilities Application Part. Provides the signaling function for network data bases. TCAP is an ISDN application protocol. In addition to PRA and ISUP, the third major ISDN protocol in the delivery of advanced network services is TCAP, a CCS7 application protocol that provides the platform to support non-circuit related, transaction-based information exchange between network entities. This capability is required by transaction-based services that must exchange information between a pair of signaling nodes in a CCS7 network. Examples of these services include enhanced dial-800 service, automated credit card calling and virtual private networking. The TCAP protocol enables these services to access remote databases called service control points (SCPs) to process part of the call. The SCP supplies the translation and routing data needed to deliver advanced network services — like translating a dial calls into the required routing number. TCAP is useful also in coordinating some enhanced call-related services. For example, network ring again requires the connection of two users when both stations become idle. In this case, TCAP is used to coordinate between the users' switches while waiting for each line to become idle. And it can do this without tying up network trunks.

One of the major advantages of TCAP is that it provides a set of protocol building blocks for use in a variety of service definitions. The TCAP building blocks are subdivided into the transaction sublayer and the component sublayer. For more on TCAP, see the 1988-3 issue of Northern Telecom's Telesis publication.

TCAS T-Carrier Administration System. Provides mechanized support for the facility maintenance and administration center to achieve centralized administration and control of the digital network.

TCF Training Check Frame. Last step in a series of signals in a fax transmission called a training sequence, designed to let the receiver adjust to telephone line conditions.

Tchotchke A New York Jewish word meaning trinkets, best exemplified by the giveaway junk we often pick up at telecommunications trade shows.

TCIC Trunk Circuit Identification Code. Only relevant to SS7. A number that uniquely identifies a trunk between an origination point code and a destination point code. An example would be between two telephone company switches.

TCIF TeleCommunications Industry Forum. A voluntary special interest group under ATIS (Alliance for Telecommunications Industry Solutions). TCIF addresses areas such as electronic commerce, including bar coding and EDI (Electronic Data Interchange). www.atis.org/atis/tcif/index.html See also ATIS.

TCM 1. Traveling Class Mark.

2. Trellis Coding Modulation.

3. Time Compression Multiplexing. A digital transmission technique that permits full duplex data transmission by sending compressed bursts of data in a "ping-pong" fashion.

4. Telecommunications Manager. The TCM is the manager of the department that plans, controls, and administers the telephony and telecommunications assets of the company. He ensures that the telephone and telecommunications systems are well-run and functioning smoothly. These assets may include the PBX and ISDN, T-1, local and long distance telephone lines, telephone sets, authorization codes, cable pairs, WANs, Fax machines, voice mail systems, automated attendants, interactive voice response systems, automatic call distribution, multiplexors, modem pools, etc. The internal data facilities such as LANs and routers may be under the administration of the TCM, or could be the responsibility of

the Management Information Systems (MIS) department. But since the TCM has responsibility for both the inside wiring and the outside Carrier facilities, close coordination would be required if the internal data facilities are controlled by the MIS department.

The following are the functions of the TCM;

• Operating, administering, monitoring, and maintaining the existing telecommunications systems.

• Dealing with the various vendors and providers, including verifying and paying the bills.

• Preparing and managing the Telecommunications budget.

• Keeping abreast of changes in technology, services, industry structure, and rates.

• Assisting company management in developing a corporate telecommunications policy that meets business objectives.

• Developing and implementing company telephone and telecommunications procedures for efficient and cost effective use, and training company employees in these procedures.

• Upgrading, procurement, selecting, contracting, or purchasing a system, new system, equipment, or services.

• Planning and analyzing for growth, new requirements, or future functionality.

The goal of the TCM is to provide good telecommunications services for an organization and its employees at the lowest possible cost. This definition courtesy, Robert J. Perillo, Perillo@dockmaster.ncsc.mil.

TCNS Thomas Conrad Networking System is a 100 million bit per second proprietary networking system (LAN) based on ARCnet that can use most standard ARCnet drivers on any network operating systems.

TCO Total Cost of Ownership. A term coined by The Gartner Group to bring attention to the actual, total cost to the enterprise of owning a PC. The most figure in 1997 was $29,353 for owning a standard, networked, Windows 95 PC for a period of three years. Gartner cites the cost of a NetPC (Thin Client) at a much lower cost. The point is clear and fairly obvious — consider not only the acquisition/implementation cost of a workstation (networked or not), but also consider the total cost, including administration, maintenance, support, software upgrades and training.

TCP 1. Transmission Control Protocol. ARPAnet-developed transport layer protocol. Corresponds to OSI layers 4 and 5, transport and session. TCP is a transport layer, connection-oriented, end-to-end protocol. It provides reliable, sequenced, and unduplicated delivery of bytes to a remote or local user. TCP provides reliable byte stream communication between pairs of processes in hosts attached to interconnected networks. It is the portion of the TCP/IP protocol suite that governs the exchange of sequential data. See TCP/IP for a much longer explanation.

2. An ATM term. Test Coordination Procedure: A set of rules to coordinate the test process between the lower tester and the upper tester. The purpose is to enable the lower tester to control the operation of the upper tester. These procedures may, or may not, be specified in an abstract test suite.

TCP/IP According to Microsoft: Transmission Control Protocol/Internet Protocol (TCP/IP) is a networking protocol that provides communication across interconnected networks, between computers with diverse hardware architectures and various operating systems. TCP (Transmission Control Protocol) and IP (Internet Protocol) are only two protocols in the family of Internet protocols. Over time, however, "TCP/IP" has been used in industry to denote the family of common Internet protocols. The Internet protocols are a result

of a Defense Advanced Research Projects Agency (DARPA) research project on network interconnection in the late 1970s. It was mandated on all United States defense long-haul networks in 1983 but was not widely accepted until the integration with 4.2 BSD (Berkeley Software Distribution) UNIX. The popularity of TCP/IP (Harry's note: it's the Internet's networking protocol) is based on:

• Robust client-server framework. TCP/IP is an excellent client-server application platform, especially in wide-area network (WAN) environments.

• Information sharing. Thousands of academic, defense, scientific, and commercial organizations share data, electronic mail and services on the connected Internet using TCP/IP.

• General availability. Implementations of TCP/IP are available on nearly every popular computer operating system. Source code is widely available for many implementations. Additionally, bridge, router and network analyzer vendors all offer support for the TCP/IP protocol family with n their products.

TCP/IP is the most complete and accepted networking protocol available. Virtually all modern operating systems offer TCP/IP support, and most large networks rely on TCP/IP for all their network traffic. Microsoft TCP/IP provides cross-platform connectivity and a client-server development framework that many software vendors and corporate developers are using to develop distributed and client-server applications in heterogeneous enterprise networks over TCP/IP.

How TCP Works: TCP is a reliable, connection-oriented protocol. Connection-oriented implies that TCP first establishes a connection between the two systems that intend to exchange data. Since most networks are built on shared media (for example, several systems sharing the same cabling), it is necessary to break chunks of data into manageable pieces so that no two communicating computers monopolize the network. These pieces are called packets. When an application sends a message to TCP for transmission, TCP breaks the message into packets, sized appropriately for the network, and sends them over the network.

Because a single message is often broken into many packets, TCP marks these packets with sequence numbers before sending them. The sequence numbers allow the receiving system to properly reassemble the packets into the original message. Being able to reassemble the original message is not enough, the accuracy of the data must also be verified. TCP does this by computing a checksum. A checksum is a simple mathematical computation applied, by the sender, to the data contained in the TCP packet. The recipient then does the same calculation on the received data and compares the result with the checksum that the sender computed. If the results match, the recipient sends an acknowledgment (ACK). If the results do not match, the recipient asks the sender to resend the packet. Finally, TCP uses port IDs to specify which application running on the system is sending or receiving the data.

The port ID, checksum, and sequence number are inserted into the TCP packet in a special section called the header. The header is at the beginning of the packet containing this and other "control" information for TCP.

How IP Works: IP is the messenger protocol of TCP/IP. The IP protocol, much simpler than TCP, basically addresses and sends packets. IP relies on three pieces of information, which you provide, to receive and deliver packets successfully: IP address, subnet mask, and default gateway.

The IP address identifies your system on the TCP/IP network. IP addresses are 32-bit addresses that are globally unique on a network. They are generally represented in dotted decimal notation, which separates the four bytes of the address with periods. An IP address looks like this: 102.54.94.97

Although an IP address is a single value, it really contains two pieces of information: (a.) Your system's network ID, and (b.) Your system's host (or system) ID.

The subnet mask, also represented in dotted decimal notation, is used to extract these two values from your IP address. The value of the subnet mask is determined by setting the network ID bits of the IP address to ones and the host ID bits to zeros. The result allows TCP/IP to determine the host and network IDs of the local workstation. Here's how to understand an IP address. For example:

When the IP address is 102.54.94.97 (specified by the user)
And the subnet mask is 255.255.0.0 (specified by the user)
The network ID is 102.54 (IP address and subnet mask)
And the host ID is 94.97 (IP address and subnet mask)

OK. the above was Microsoft's definition. Here's my definition, which covers some areas Microsoft doesn't. TCP/IP is a set of protocols developed by the Department of Defense to link dissimilar computers across many kinds of networks, including unreliable ones and ones connected to dissimilar LANs. TCP/IP is the protocol used on the Internet. It is, in essence, the glue that binds the Internet. Developed in the 1970s by the U.S. Department of Defense's Advanced Research Projects Agency (DARPA) as a military standard protocol, its assurance of multi vendor connectivity has made it popular among commercial users as well, who have adopted TCP/IP. Consequently, TCP/IP now is supported by many manufacturers of minicomputers, personal computers, mainframes, technical workstations and data communications equipment. It is also the protocol commonly used over many Ethernet LANs (as well as X.25) networks. It has been implemented on everything from PC LANs to minis and mainframes.

TCP/IP currently divides networking functionality into only four layers:

A Network Interface Layer that corresponds to the OSI Physical and Data Link Layers. This layer manages the exchange of data between a device and the network to which it is attached and routes data between devices on the same network.

An Internet Layer which corresponds to the OSI network layer. The Internet Protocol (IP) subset of the TCP/IP suite runs at this layer. IP provides the addressing needed to allow routers to forward packets across a multiple LAN inter network. In IEEE terms, it provides connectionless datagram service, which means it attempts to deliver every packet, but has no provision for retransmitting lost or damaged packets. IP leaves such error correction, if required, to higher level protocols, such as TCP.

IP addresses are 32 bits in length and have two parts: the Network Identifier (Net ID) and the Host Identifier (Host ID). Assigned by a central authority, the Net ID specifies the address, unique across the Internet, for each network or related group of networks. Assigned by the local network administrator, the Host ID specifies a particular host, station or node within a given network and need only be unique within that network.

A Transport Layer, which corresponds to the OSI Transport Layer. The Transmission Control Protocol (TCP) subset runs at this layer. TCP provides end-to-end connectivity between data source and destination with detection of, and recovery from, lost, duplicated, or corrupted packets — thus offering the error control lacking in lower level IP routing. In TCP, message blocks from applications are divided into smaller segments, each with a sequence number that indicates the order of the segment within the block. The destination device examines the message segments and, when a complete sequence of segments is received, sends an acknowledgement (ACK) to the source, containing the number of the next byte expected at the destination.

An Application Layer, which corresponds to the session, presentation and application layers of the OSI model. This layer manages the function required by the user programs and includes protocols for remote log-in (Telnet), file transfer (FTP), and electronic mail (SMTP). See OSI.

TCR 1. Transaction Confirmation Report. A report from a fax machine listing the faxes received and transmitted. It provides details about each fax, including date, time, the remote fax's number, results, total pages.

2. An ATM term. Tagged Cell Rate: An ABR service parameter, TCR limits the rate at which a source may send out-of-rate forward RM-cells. TCR is a constant fixed at 10 cells/second.

TCS 1. Transmission Convergence Sublayer: This is part of the ATM physical layer that defines how cells will be transmitted by the actual physical layer.

2. TCS is a manufacturer of excellent call center manpower scheduling software packages. It is based in Nashville, TN.

TDAS See Traffic Data Administration System.

TDC Time Division Controller.

TDD 1. Time Division Duplex. A method used in cellular and PCS networks employing TDMA (Time Division Multiple Access) to support full duplex communications. Each radio channel is divided into multiple time slots through TDMA, thereby supporting multiple conversations. TDD supports transmission in the forward direction (from the cell phone to the cell site) through one radio frequency channel and one time slot. Another radio channel and time slot supports transmission in the backward direction (from the cell site to the cell phone).

2. Telecommunications Device for the Deaf. Under the Communications Act of 1934, a TDD is defined as a machine "that employs graphic communication in the transmission of coded signals through a wire or radio." TDD devices (which typically look like simple computer terminals) use the Baudot method of communications. Most TDD devices are acoustically coupled and are slow, running at 300 baud.

There is a special TDD/TTY Operator Services number. It's 800-855-1155. Users of TDDs often abbreviate commonly used words or expressions to save time. Here are some of the most frequently used:

ANS Answer R Are
CUD Could REC Receive
GA Go Ahead SK Stop Keying
LTR Letter THRU Through
MSG Message THX Thank You
MIN Minute U You
NITE Night UR Your
PLS Please WUD Would
QQ Question XOX Hugs & Kisses

TDDRA Telephone Disclosure and Dispute Resolution Act. A US federal act passed in 1992 which required both the FCC and the FTC to prescribe regulations governing pay-per-call services. Subsequently, the FTC adopted its 900-Number Rule, which became effective November 1, 1993. Under the TDDRA, a consumer's telephone service cannot be disconnected for failure to pay charges for a 900-number call, and 900-number blocking must be made available to consumer who do not wish to have access to 900-number services. See also 900-Number Rule.

TDF 1. Trunk Distributing Frame.

2. Transborder Data Flows are movements of machine-read-

able data across international boundaries. TDF legislation began in the 1970s and has been put into effect by many countries in an attempt to protect personal privacy of citizens. This term has particular meaning as it relates to electronic commerce or EDI and is becomming more and more relevant with the use of the Internet as a means to conduct global business.

TDHS Time Domain Harmoric Scaling

TDI Transmit Division Intertie.

TDM See TIME DIVISION MULTIPLEX.

TDMA Time Division Multiple Access. One of several technologies used to separate multiple conversation transmissions over a finite frequency allocation of through-the-air bandwidth. As with FDMA (Frequency Division Multiple Access), TDMA is used to allocate a discrete amount of frequency bandwidth to each user, in order to permit many simultaneous conversations. However, each caller is assigned a specific timeslot for transmission. A digital cellular telephone system using TDMA assigns 10 timeslots for each frequency channel, and cellular telephones send bursts, or packets, of information during each timeslot. The packets of information are reassembled by the receiving telephone into the original voice components. TDMA promises to significantly increase the efficiency of cellular telephone systems, allowing a greater number of simultaneous conversations. See CDMA, FDMA.

TDMS 1. Technical Document Management Systems.
2. Time Division Multiplex System.
3. Transmission Distortion Measuring Set.

TDOA Time Difference Of Arrival. A precise method of locating a radio receiver, TDOA is being proposed to support wireless 911 services for cellular and PCS networks. Operating much like GPS (Global Fositicning Satellite Systems), although in reverse, GPOA uses three cell site antennae to lock in on the signal from the cell phone. The times of signal arrival at each cell site are compared through the use of a precise master clock. Although the differences in time of signal arrival may be only microseconds, the location of the cell phone can be determined through a process of time triangulation, allowing the exact location of the device to be plotted. GPOA offers much improved location-determination than does the old method of triangulation, which relies on signal strength. See also 911 and TRIANGULATION.

TDP Telocator Data Protocol. A new 8-bit protocol for sending messages and binary files (images, spreadsheets, word processing files, executables, etc.) to pagers (also known as beepers. The older (and more common) 7-bit messaging protocol now widely in use is called TAP, which stands for Telocator Alphanumeric Pagng Protocol. This protocol can only send simple alphanumeric messages, like "Your shares in XYZ are now $23, up 98%."

TDR Time Domain Reflectometer.

TDS MCI's name for Terrestrial Data Services, i.e. services that run through on-the-ground fiber, rather than through-the-air satellite services.

TDSAI Transit Delay Selection And Indication
1. ISDN Terminal Equipment. See the next two definitions.
2. Terminal Equipment: As an ATM term, terminal equipment represents the endpoint of ATM connection(s) and termination of the various protocols within the connection(s).

TE1 Terminal equipment type 1 that supports ISDN standards and thus can connect directly to the ISDN network. TE1 could be an ISDN telephone, a personal computer capable of working with ISDN, a videophone, etc. In short, any device that can attach to and work with ISDN.

TE2 Terminal equipment that does not support ISDN stan-

dards and thus requires a Terminal Adapter. Non-ISDN terminal equipment (e.g. analog telephone) linked at the RS-232, RS-449, or V.35 interfaces.

TEC NIS Telecommunications and Electronics Consortium in the Newly Independent States. An organization based in Moscow and administered by TIA (Telecommunications Industry Association) to assist US telecommunications and telecommunications-related electronics companies with doing business in the region.

Techinfo A common campuswide information system developed at MIT. An Internet term.

Technical Advisory TA. A Bellcore document containing a preliminary view of proposed generic requirements for a technology, equipment, service or interface. The TA document type is being replaced by the Generic Requirements (GR) document type. See Generic Requirements.

Technical Control Center A testing center for telecommunications circuits. The center provides test access and computer-assisted support functions to aid in circuit maintenance.

Technical Control Facility A federal government term . A term plant, or a designated and specially configured part thereof, containing the equipment necessary for ensuring fast, reliable, and secure exchange of information. This facility typically includes distribution frames and associated panels, jacks, and switches; and monitoring, test, conditioning, and order wire equipment.

Technical Load A military term. The portion of the operational load required for communications, tactical operations, and ancillary equipment including necessary lighting, air conditioning, or ventilation required for full continuity of communications.

Technical Means A term in the spy business which means spy satellites and electronic eavesdropping stations, typically costing tens of billions of dollars.

Technical Office Protocol TOP. A seven-layer network architecture designed for office automation that uses International Standards Organization (ISO) or ITU specifications at each level. TOP was defined by Boeing Vertol Corp. and is now controlled by the MAP/TOP (Manufacturing Automation Protocol/Technical Office Protocol) Users Group.

Technical Reference TR. A Bellcore document containing the current view or performance of a technology, equipment, service or interface. The TR document type are being replaced by the Generic Requirements (GR) document type. See Generic Requirements.

TED Trunk Encryption Device.

Teddy Bear A stuffed arimal named after President "Teddy" Roosevelt, a keen hunter who once took pity on a baby bear.

TEDIS Trade Electronic Data Interchange Systems.

Teen Service A feature of some central offices which allows two telephone numbers to be assigned to a single party phone line. Each number has a distinctive ringing pattern so that the called parties can recognize which line is ringing. The inventor of this service named t after the fact that his teenage children were always receiving phone calls. And he wanted a way for them to recognize when the calls were for them and when they were for the parents. Sadly, this phenomenon now begins earlier in life, with children as young as six receiving their own calls. We speak from experience. Teen service is now used by home businesses, roommates, boarders, college dorm suite-mates, and live-in relatives.

TEF Telecommunications Entrance Facility (also called EF or Entrance Facility).

Teflon Dupont's registered trademark for fluorinated ethylene propylene (FEP). In addition to working its wonders in the modern kitchen, Teflon is an exceptional insulating material for cable systems. Teflon is also coated on cables. See also FEP.

TEHO See TAIL END HOP-OFF (traffic engineering).

TEI Terminal Endpoint Identifier. Up to eight devices can be connected to one ISDN BRI line. The TEI defines for a given message which of the eight devices is communicating with the Central Office switch. In general, more than one of the eight may be communicating.

TEK Traffic Encryption Key.

Telabuse A term coined by John Haugh of Telecommunications Advisors in Portland, OR to include "insider" toll fraud, waste and abuse.

Telbanking Banking transactions conducted through telecommunications.

Telco 1. The local telephone company. Often a term of endearment. Americanism for telephone company.
2. In some LAN circles, a telco is known as a 25-pair polarized connector that is used to consolidate multiple voice or data lines. Also known as an amphenol connector.

Telco Farm A building housing many phone companies — typically one ILEC and many CLECs.

Telco Splice Block What some parts of the data communications industry call a 66-block, i.e. a terminating block for twisted pair voice and data cable.

Tele Tele comes from the Greek work meaning "far." Telecommunications is therefore communicating over a distance.

Tele- For any word that you think should be spelled tele-word, please check the definition below, spelled without the dash, also

Teleaction Service In ISDN applications, a telecommunications service using very short messages with very low data transmission rates between the user and the network.

Teleadmin A means in GSM (non-North America cellular digital standard) to update a Subscriber Identity Module (SIM) card via a short message sent by the network operator using Short Message Services (SMS). In GSM, the SIM card is located in the cell phone's handset and is customized to a specific subscriber's service options. TeleAdmin is also known as remote SIM card updating.

Telebusiness A British term for telemarketing. Here's a definition I found in England: Activities conducted by telephone in a planned and controlled manner. The term encompasses telesales, telemarketing, customer service and information broadcast. Telebusiness can be conducted between an organization and its customers and prospects, or conducted as in internal service.

Telecom A shortened and perfectly acceptable way of saying the word "telecommunications."

Telecoms British usage. A shortened and perfectly acceptable way of saying the word "telecommunications."

Telecommunication Architecture The governing plan showing the capabilities of functional elements and their interaction, including configuration, integration, standardization, life-cycle management, and definition of protocol specifications, among these elements.

Telecommunication Facilities The aggregate of equipment, such as telephones, teletypewriters, facsimile equipment, cables, and switches, used for various modes of transmission, such as digital data, audio signals, image and video signals.

Telecommunication Service Any service provided by a telecommunication provider. A specified set of user-information transfer capabilities provided to a group of users. The telecommunication service provider has the responsibility for the acceptance, transmission, and delivery of the message.

Telecommunicationally Challenged A politically correct term for being under-phoned, i.e. having too few phones. This definition contributed by John Warrington of Ashland University, Ashland, OH.

Telecommunications 1. The art and science of "communicating" over a distance by telephone, telegraph and radio. The transmission, reception and the switching of signals, such as electrical or optical, by wire, fiber, or electromagnetic (i.e. through-the-air) means.
2. A fancy word for "telephony," which it replaced and which many thought meant only analog voice, but didn't.

Telecommunications Act of 1981, U.K. The Telecommunications Act of the U.K. is passed. It is the first step towards liberalizing the teleconnunications market in the U.K. and has four main consequences:

• The General Post Office (the erstwhile monopoly provider of telecommunications services in the U.K.) was divided into two separate entities: The Post Office and British Telecommunications (BT), which retained the monopoly over existing telecommunications networks.

• It determined that a duopoly would be created as a first step towards the introduction of competition in telecommunications.

• The Secretary of State for Trade and Industry was empowered to license other organizations to be known as Public Telecommunications Operators (PTOs), to operate public telecommunications networks (including cellular networks) in the U.K.

• It paved the way for the gradual deregulation of equipment supply, installation and maintenance which had previously been the monopoly of the GPO.

Following the Act, Mercury Communications, majority-owned by Cable & Wireless was created to compete with British Telecommunications.

Telecommunications Act of 1984, U.K. The 1984 Telecommunications Act established British Telecommunications, now known as BT, as a public limited company which would, as such, have to apply for a PTO licence from the Secretary of State. Following the Act, BT was privatized in a series of tranches, the last held in 1993. The Act also created Optel, the office of telecommunications, to become a watchdog over all aspects of the telecommunications industry in the U.K. See also Telecommunications Act of 1981.

Telecommunications Act of 1996. U.S. A federal bill signed into law on February 8, 1996 "to promote competition and reduce regulation in order to secure lower prices and higher quality services for American telecommunications consumers and encourage rapid deployment of new telecommunications technologies." The Act required local service providers in the 100 largest metropolitan areas of the United States, the Baby Bells, to implement Local Number Portability by the end of 1998. The Act also allowed the local regional Bell operating phone companies into long distance once they had met certain conditions about allowing competition in their local monopoly areas. See Dialing Parity below. You can download a copy of this Act (all 391,861 bytes) from http://thomas.loc.gov/cgi-bin/query/1?c104:./temp/~c104iLoA:: The following are definitions contained in the Act.

Affiliate — The term `affiliate' means a person that (directly or indirectly) owns or controls, is owned or controlled by, or is under common ownership or control with, another person. For purposes of this paragraph, the term `own' means to own an equity interest (or the equivalent thereof) of more than 10 percent.

AT&T Consent Decree — The term `AT&T Consent Decree' means the order entered August 24, 1982, in the antitrust action styled United States v. Western Electric, Civil Action No. 82-0192, in the United States District Court for the District of Columbia, and includes any judgment or order with respect to such action entered on or after August 24, 1982.

Bell Operating Company — The term `Bell operating company' (A) means any of the following companies: Bell Telephone Company of Nevada, Illinois Bell Telephone Company, Indiana Bell Telephone Company, Incorporated, Michigan Bell Telephone Company, New England Telephone and Telegraph Company, New Jersey Bell Telephone Company, New York Telephone Company, U S West Communications Company, South Central Bell Telephone Company, Southern Bell Telephone and Telegraph Company, Southwestern Bell Telephone Company, The Bell Telephone Company of Pennsylvania, The Chesapeake and Potomac Telephone Company, The Chesapeake and Potomac Telephone Company of Maryland, The Chesapeake and Potomac Telephone Company of Virginia, The Chesapeake and Potomac Telephone Company of West Virginia, The Diamond State Telephone Company, The Ohio Bell Telephone Company, The Pacific Telephone and Telegraph Company, or Wisconsin Telephone Company; and

(B) includes any successor or assign of any such company that provides wireline telephone exchange service; but

(C) does not include an affiliate of any such company, other than an affiliate described in subparagraph (A) or (B).

Customer Premises Equipment — The term `customer premises equipment' means equipment employed on the premises of a person (other than a carrier) to originate, route, or terminate telecommunications.

Dialing Parity — The term `dialing parity' means that a person that is not an affiliate of a local exchange carrier is able to provide telecommunications services in such a manner that customers have the ability to route automatically, without the use of any access code, their telecommunications to the telecommunications services provider of the customer's designation from among 2 or more telecommunications services providers (including such local exchange carrier).

Exchange Access — The term `exchange access' means the offering of access to telephone exchange services or facilities for the purpose of the origination or termination of telephone toll services.

Information Service — The term `information service' means the offering of a capability for generating, acquiring, storing, transforming, processing, retrieving, utilizing, or making available information via telecommunications, and includes electronic publishing, but does not include any use of any such capability for the management, control, or operation of a telecommunications system or the management of a telecommunications service.

Telecommunications Bonding Backbone A conductor that interconnects the telecommunications main grounding busbar (TMGB) to the telecommunications grounding busbar (TGB).

Telecommunications Broker A person or an organization which buys telecommunications services at bulk rates and resells these services at below "normal" i.e. retail prices.

Telecommunications Closet A closet which houses telecommunications wiring and telecom wiring equipment. Contains the BHC (Backbone to Horizontal Cross-connect). May also contain the Network Demarcation, or MC (Main Cross-connect). The telecommunications closet is used to connect up telecom wiring. The closet typically has a door. It's a good idea to lock the door and not put anything else in the closet, like mops, buckets and brooms.

Telecommunications Information Networking Architecture Consortium See TINA-C.

Telecommunications Lines Telephone and other communications lines used to transmit messages from one location to another.

Telecommunications Network The public switched telephone exchange network.

Telecommunications Relay Service Telecommunication Relay Service (TRS), formerly called Dual Party Relay Service, is available to hearing impaired customers. Defined by the Communications Act of 1934, telecommunications relay services mean "transmission services that provide the ability for an individual who has a hearing impairment or speech impairment to engage in communication by wire or radio with a hearing incividual in a manner that is functionally equivalent to the ability of an individual who does not have a hearing impairment or speech impairment to communicate using voice communication service by wire or radio." NECA (National Exchange Carrier Association) administers the Telecommunications Relay Services Fund, collecting from approximately 3,000 companies based on interstate revenues, then disbursing it to the providers of interstate TRS. Some states, such as California, also have state-level funding mechanisms in place in the form of line-item surcharges on your telephone bill. These various funds are used to compensate the LECs (Local Exchange Carriers) for the incremental cost of providing TDDs (Telecommunications Devices for the Deaf), volume-controlled telephone handsets, and other compensating devices and features. See also NECA, SEPARATIONS AND SETTLEMENTS and TDD.

Telecommunications Resellers Association TRA. An association of approximately 500 long distance resellers in North America. TRA was formed in 1992 through the merger of the Telecommunications Marketing Association and the Interexchange Resellers Association. The association is based in Washington, D.C., acting as a collective trade group for both switchless and facilities-based resellers. TRA holds several conferences and exhibitions each year, and acts as the resale industry's lobbying group and consumer watchdog. TRA says it has a strong code of ethics, and most member resellers are viewed as having reputable business standards. In 1997, TRA absorbed NWRA (National Wireless Resellers Association. www.tra-dc.org

Telecommuting The process of commuting to the office through transferring information over a communications link, rather than transferring one's physical presence. In short, working at home on a telephone, a computer, a modem and maybe a facsimile machine, rather than going into the office. As the story goes, the concept of telecommuting was invented and the term was coined in 1973 by Jack Nilles, a spacecraft designer for The Aerospace Corp. Nilles was intrigued by the questions posed by an urban planner who wondered why we could put a man on the surface of the moon but couldn't solve the problems of vehicular traffic congestion on the surface of the Earth. Eventually, Nilles left his job to become director of interdisciplinary programs at the University of Southern California, where he studied telecommuting for the next 10 years or so. There are clear benefits to telecommuting: you can live and work somewhere charming. There are disadvantages, especially accentuated if you work with others: "When you're getting data from afar, you're not in touch with the soul of the business, anymore," according to one telecommuter interviewed by the New York Times. He went on to say, "All the electronic communications are simply backup, I just hadn't factored the importance of personal loyalty and contact into the equation. And I was very wrong." Ray Horak, my Contributing Editor, telecommutes. He works on this dictionary from his SOHO in Seattle, WA, and sends me new definitions and edits old ones electronically over the Internet. The key to this successful relationship is severalfold: What's we're both doing — writing a dictionary — is a very defined, very structured and quite simple task. We both know what our goals are — to make the dictionary the best in the world. And we know how to do it. We're not debating the design of a new automobile or selling customers. These sorts of activities are far more people-oriented and are less conducive to telecommuting. In short, the nature of the task determines its success for telecommuting.

Telecomputer Telecomputer appears to be the couch potato's ultimate toy. Peter Coy writing in Business Week of November 1, 1993 called telecomputer a "computerized television." He said "the idea is that you can watch anything in the (on-line) video library anytime. Your telecomputer lets you scroll through a menu of programs, click on your choice, and send an order up the line." James H. Clark, chairman of Silicon Graphics Inc., a manufacturer of computers with heavy video skills calls telecomputer a term for a combination computer/CATV controller that is being popularized by the new media industry. The idea is to use the telecomputer to do interactive games, choose a movie to play out of thousands of choices, buy things, send electronic mail, etc.

Teleconference A telephone conversation with three or more people. They may be distant from each other. They may all be in the same office.

Teleconferencing A term for a conference of more than two people linked by telecommunications through a conference bridge. The term is applied to voice conferencing, which also is known as audioconferencing and which can include other forms of audio, such as music. Teleconferencing, in the broader sense, also includes videoconferencing and document (data) conferencing. For years, teleconferencing has been heralded as a great coming event, and a significant replacement for travel. As corporations increasingly downsize, decentralize, and encourage telecommuting, they will continue to expect more productivity from fewer people who are geographically dispersed. Teleconferencing, clearly, is one powerful solution to this dilemma. See NetMeeting, IP Telephony, The Internet, TAPI 3.0.

TELECONNECT Magazine TELECONNECT is a monthly magazine covering developments in telecommunications equipment. It's the largest monthly telecommunications magazine. Its job is to help its readers choose, install and maintain telecom equipment. TELECONNECT Magazine contains critical reviews of new products and product comparisons. It is published by the same nice people who publish (but not write) this dictionary. Subscriptions are available from www.teleconnect.com or 615-377-3322

Telecopier A fancy word for a facsimile machine.

Telecrats High-ranking telephone company executives who speak more like government bureaucrats than businesspeople. There are many of them.

Teledensity A measure of the number of phone lines per 100 of population. Between 40 and 50 lines per 100 of population indicates pretty good density. Under 10 indicates pretty bad density. Teledensity is a measure of a country's economic development. Over 70 means your country is pretty advanced. Some towns, like Washington, D.C., are over 100. That means there is more than one phone for every person. Whether this means Washington, D.C. is more advanced than other places is an interesting question.

Teledesic Teledesic LLC is a private company which is undertaking to build a global, broadband "Internet-in-the-Sky." Teledesic will deploy a global constellation of 288 LEOs (Low-Earth-Orbiting satellites) plus spares, and will operate in the Ka-band of radio spectrum (28.6-29.1 GHz uplink and 18.8-19.3 GHz downlink). The system is proposed to support millions of simultaneous users, with each having asymmetric, two-way connectivity at rates of up to 2 Mbps on the uplink and up to 64 Mbps on the downlink. The user equipment will be in the form of small (laptop-size) antennae which will mount flat on a rooftop. Teledesic proposes to provide affordable access to a broad array of advanced information services ranging from high-quality voice channels to broadband channels supporting videoconferencing, interactive multimedia and other real-time, two-way digital data applications. Teledesic plans to begin service in the year 2001. Rather than marketing services directly to users, the company will provide an open network for the delivery of such services by others. The Teledesic network will enable local telephone companies and government authorities in host countries to cost-effectively modernize existing communications systems and bring fiber-like services to areas that would never get such capabilities through wireline means. Teledesic was founded in June of 1990 and is headquartered in Kirkland, Washington, near Seattle. The Teledesic vision was created by Craig O. McCaw, founder of McCaw Cellular Communications (now AT&T Wireless), and William H. Gates III, co-founder of Microsoft. Principal investors include McCaw, Gates, Saudi Prince Alwaleed Bin Talal ($200 million), and Boeing, which is providing the launch vehicles. Motorola and Matra Marconi Space round out the founding industrial team. Latest estimate of the required total investment-a mind-boggling $9 billion. See also LEO and Ka-band.

Telefax 1. European term for fax.

2. A high-speed, 64 kilobit per second facsimile service that uses Group 4 fax machines and one Bearer channel of an ISDN circuit, or any other 64 Kbps circuit. Group 4 fax machines take about six seconds to transmit a page. They're fast and impressive.

Telefelony Another made-up word from the people who are trying to sell you consulting services. This from Jennifer Poulsen, Consultant, High Road Communications, jpoulsen@highrd.com: "Telco fraud is a big problem, and getting bigger. In 1997, phone companies across North America lost more than $12 billion in long distance to tele-felony, a term that accurately describes one of the biggest issues facing the telephone world. The perpetrators for this crime have been dubbed tele-felons. Tele-felons hack their way into company phone systems and make lengthy and expensive long distance calls all over the world. They gain access to corporate calling card numbers for the same purpose. They even let friends make such calls from a number where they work. And the damage doesn't stop at the phone system. Once the PBX is hacked, it's merely a conduit to the computer system and a gold mine of valuable data. Here are some facts:

Fraudsters love targeting new competitive carriers first as they know (or hope) the infrastructure is not in place. An average hit by an organized fraud group costs telcos $350,000 per occurrence (two hits by an organized fraud group could wipe out a new telco's entire yearly profit). A telco's image as a quality service provider is tarnished without fraud protection. Fraud is a cause of customer churn and retaining customers is crucial to the long-term viability of CLECs."

Telegaming Using communications lines — from dial-up through local area networks through WANs — to play games interactively with someone at the other end. The upcoming expected explosion of telegaming is what's driving the growth of DSVD (Digital Simultaneous Voice Data) modems and ultimately the growth of ISDN lines.

Telegram Hard-copy information, in written, printed or pictorial form, routed to the general telegraph service for transmission and delivery to the addressee. Telegrams are dying due to the high cost of delivery.

Telegraph A system employing the interruption of, or change in, the polarity of DC current signaling to convey coded information.

Telegraph Key A type of switch for making and breaking a circuit at will for the purpose of transmitting dots and dashes.

Telegraphy Aging data transmission technique characterized by maximum data rates of 75 bits per second and signaling where the direction, or polarity, of DC current flow is reversed to indicate bit states.

Teleguerilla A term coined by Telecom Australia for "the first wave of informal and unofficially sanctioned telecommuters — those who occasionally work from home with the informal approval of their immediate boss." Says Telecom Australia, "They're the ones whose bosses say, 'I don't care where you work as long as you get the project finished.'"

Telemanagement A term for the application of computer systems to the management of the telephone and telecommunications expenses of a user organization. Telemanagement includes virtually every function which the contemporary corporate telecommunications manager performs. Ray Horak, my Contributing Editor, says that telemanagement comprises the management of costs, assets, processes and security. Cost management includes call accounting, cost allocation, bill consolidation, and bill reconciliation-costs include usage-sensitive network costs (e.g. long distance calls), nonrecurring costs (e.g., installation and repair), and recurring costs (e.g., circuits and maintenance agreements). Asset management includes the cradle-to-grave management of systems (e.g., PBXs and computer systems), terminals (e.g., phones and PCs), and inside cable and wire systems. Process management includes work order and trouble ticket management, traffic analysis, and network design and optimization. Security management includes the management of toll fraud and network abuse/misuse. Telemanagement systems generally are in the form of premise-based application software systems, which typically are modular. Such systems typically are either PC-based, or client/server (PC-LAN) in nature; very large user organizations make use of mainframe-based systems. Service bureaus also offer telemanagement services, although they tend to be limited to call accounting and cost allocation.

TeleManagement Forum TMF. Previously known as the Network Management Forum (NMF). A global not-for-profit organization of over 250 service providers, computing and network equipment suppliers, software vendors, and customers of communications services. TMF promotes the streamlining and automation of business processes through its SMART TMN (Telecommunication Management Network) program. See also TMN. www.nmf.org

Telemarketing Marketing and sales conducted via the telephone. There are two sides to telemarketing — incoming and outgoing. Incoming telemarketing is largely run through 800 toll-free IN-WATS numbers and local FX (foreign exchange) lines. Outgoing telemarketing is organized over OUT-WATS lines. An expanding range of telecom gadgetry is being developed to automate telemarketing — including automated outbound dialers, voice processing technology and automatic call distributors. The tone recognition, voice detection and transaction audiotex and transaction processing capabilities of voice processing gear can be used to enhance all telemarketing applications.

Telemarketing and Consumer Fraud and Abuse Protection Act A Congressional bill passed in August 1994 with the stated purpose of combating the growth of telemarketing fraud. The bill gave law enforcement agencies new tools, and consumers new protections and guidance to help prevent the planned, fraudulent use of the telephone. See Telemarketing Sales Rule.

Telemarketing Sales Rule This rule was adopted by the FTC December 31, 1995 pursuant to the Telemarketing and Consumer Fraud and Abuse Protection Act of 1994. Key provisions require specific disclosures, prohibit misrepresentations, set limits on the times when telemarketers may call customers, prohibit calls after a consumer requests not to be called, set payment restrictions for the use of certain goods and services, and require that specific business be kept for two years. See Telemarketing and Consumer Fraud and Abuse Protection Act.

Telematic Agent TLMA. An X.400 AU (Access Unit) serving Teletex users of other telematic services (using Teletex, Fax, etc.).

Telematics Also called mobile telematics. It involves integrating wireless communications and (usually) location tracking devices (generally GPS) into automobiles. The best known example is GM's OnStarsystem, which automatically calls for assistance if the vehicle is in an accident. These systems can also perform such functions as remote engine diagnostics, tracking stolen vehicles, provide roadside assistance, etc. www.onstar.com

Telemedicine The provision of health-care services from a distance using networks supporting audio, video, and computer data transmissions. Telemedicine traditionally uses videoconferencing to diagnose illness and provide medical treatment over a distance. Often used to view or teach surgical procedures. Used also in rural areas where health care is not readily available and to provide medical services to prisoners. At the ICA Show in 1994, Southwestern Bell demonstrated telemedicine applications including a dermatology microscope, a video scope, an electronic stethoscope and a telepathology system that allows a pathologist to exercise computer control over a remote microscope.

Telemetry A communications system for the transmission of digital or analog data which represents status information on a remote process, function or device.

Telemonkey I was given this definition by a fine gentleman, who'd just been given the job of running his company's call center. He was, as you can, quickly disgusted with the quality of the labor he had to manage. He says he didn't make this term up. It's for real. Maybe. Anyway, here goes. A perjorative word for a call center agent. Originally companies staffed their centers with highly educated, well-paid agents who were usually capable of thinking independently when dealing with a customer's inquiry, but now companies have started to replace such staff with less educated, less trained, and lower paid agents who are trained to respond to customers' inquiry by referring to a database help desk, guide book or manual. Hence the idea that monkeys could handle an agent's job.

Telenet A private, commercially available network providing both packet-switched and circuit-switched service to subscribers in North America, Europe and some parts of Asia.

Telenet Remote Login Protocol A virtual terminal service specified by the U.S. Department of Defense and implemented by most versions of UNIX.

Telenet International Quotations TIQ. A market data information subscription service operated by Telrate International Co. over a network that uses proprietary protocols to enhance security and other functions.

Teleos IRX-9000 Teleos' Intelligent Call Distribution platform. Teleos makes a programmable switch which can be controlled by an external computer. It calls the switch the IRX-9000 and the ability to write software for it, its intelligent call distribution platform.

Teleparents Parents who equip their children with pagers before allowing them to go out.

Telephone 1. The invention of the devil.
2. The most intrusive device ever invented.
3. The biggest time waster of all time, as in: "What did you do all day?" "Nothing. Just spent the day on the phone."
4. Also a truly remarkable invention. Here's a list of the eight things a telephone does, according to Understanding Telephone Electronics:
a. When you lift the handset, it signals you wish to use the worldwide phone system.
b. It indicates the phone system is ready for your wish by receiving a tone, called a dial tone.
c. It sends the number of the telephone to be called.
d. It indicates the progress of your call by receiving tones — ringing, busy, etc.
e. It alerts you to an incoming call.
f. It changes your speech into electrical signals for transmission to someone distant. It also changes the electrical signals it receives from the distant person to speech so you can understand them.
g. It automatically adjusts for changes in the power supplied to it.
h. When you hang up, it signals the phone system your call is finished.
And, most remarkably, most simple telephones cost under $50.

Telephone Account Management TAM. A telemarketing/call center term. Using the telephone channel to proactively cover an assigned group of customers with the objectives of building and retaining revenues from these customer accounts. Most often primary coverage and revenue responsibility lies with the Telephone Account Management team and individual Telephone Account Management Representatives "own" a specific set of accounts within the team. TAM coverage often requires multiple outbound and inbound contacts with assigned customers driven by information uncovered by the TAM Representatives and not by any preset campaign parameters.

Telephone Amplifier A device to amplify the sound of the receiver. Something no phone should be without. Some devices work strictly on line power. They can only increase volume by 10 dB, which is often not enough (especially if you're over 40). The best telephone amplifiers are powered by

AC/DC adapters. Newer ones are powered by nicad batteries. They will amplify to 20 dB.

Telephone Answering A feature of some voice mail systems in which incoming callers are immediately directed to the called party's voice mailbox where they hear a personalized greeting in the called party's voice and are prompted to leave a detailed message.

Telephone Channel A transmission path suitable for carrying voice signals. Defined by its ability to transmit signals in a frequency range of about 300 to 3000 Hz.

Telephone Circuit 1. All telephones are made up of just three circuits, The Ringer Circuit, The Mouthpiece Circuit, and the Earpiece Circuit. The ringer circuit and the mouthpiece circuit are connected in parallel. The hookswitch keeps the mouthpiece circuit open whenever the telephone is hung up. The earpiece circuit is coupled to the mouthpiece circuit with a transformer.
The Ringer circuit is across the incoming telephone line pair at all times. The ringer circuit consists of a ringer, some sort of bell or buzzer, and a capacitor. The capacitor serves to block DC current since the telephone company's central office determines whether a phone is on or off hook by measuring the DC resistance across it's line pair. To ring the telephone's bell, a high voltage alternating current is sent down the telephone line pair.
The mouthpiece circuit is also referred to as the primary circuit, talk circuit, and DC loop. When the handset is lifted from the cradle to the off-hook position, the hookswitch closes, and creates a closed DC circuit from the central office through the line pair and to the microphone in the handset of the telephone and the primary coil of the transformer that couples the microphone circuit to the earpiece of the phone.
The DC resistance of the mouthpiece circuit is lower than the resistance of the ringer circuit. When the telephone is off hook, the central office detects the change in resistance. When the telephone is off hook, the central office will disconnect the ringing generator so as not to send high voltage down the line to a phone that is in use.
The earpiece circuit is also referred to as the secondary circuit and listen circuit. The last of the three circuits differs from the other two in that it is never directly across the incoming lines. Instead it is coupled to the primary circuit by a transformer. The current in the primary coil of the transformer is modulated by the microphones at both ends of the telephone call. This varying current induces a current in the secondary winding of the transformer. This induced current generates sound from the loudspeaker in the earpiece of the phone.
2. Electrical connection permitting the establishment of telephone communication in both directions between two telephone exchanges.

Telephone Consumer Protection Act TCPA. Legislation passed by Congress and signed by the president in 1991. It restricts specific types of unsolicited telephone calls. Among the provisions were a prohibition on calling emergency numbers or numbers for which the recipient was charged, limiting the placement of unsolicited calls to between 8 am and 9 pm, and removing people from calling lists who request that they not be called again.

Telephone Drop-in Mouthpiece A telephone drop-in mouthpiece used for bugging looks very much like the carbon microphone in the mouthpiece of a telephone. It is installed by unscrewing the mouthpiece, removing the old microphone, and dropping in the wiretap device. The transmitter draws power from the telephone line and only operates when the telephone is off hook. Both sides of the conversation are picked up and transmitted to a remote location. The telephone line is used as an antenna. Drop-in transmitters are simple to install and hard to detect. They require access to the telephone instrument.

Telephone Equipment Order TEO. TEOs are orders placed by Central Office Engineering for telephone apparatus and equipment. These orders may be telephone company or Vendor Engineered and are usually the downstream product of a Network Design Order (NDO).

Telephone Exchange A switching center for connecting and switching phone lines. A European term for what North Americans call central office.

Telephone Frequency Any frequency within that part of the audio frequency range essential for the transmitting speech, i.e. 300 to 3000 Hz.

Telephone Line Simulator Also called ring-down box. Ring down boxes, also known as CO simulators, are simple devices used for generating calls from a POTS line to a computer telephony system (or vice versa). When one side goes offhook, the ring down box will "ring" the other side. When both sides are offhook, both sides are coupled together and the line is powered. Ring down boxes are available with various options and configurations. These include the ability to provide dialtone to the caller side (required to test applications with modems, faxes, or other automated outdialing devices), caller ID, and disconnect supervision. They are generally available in one to four line sizes, although special configurations may support more. Ring-down boxes are used for giving demonstrations and testing. We use them in our test labs to testdrive new computer telephony systems.

Telephone Management System The term originally meant a system for controlling telephone costs by:
1. Automatically selecting lower-cost long distance routes for placed calls; 2. Automatically restricting certain people's abilities to make some or all long distance calls; and 3. Automatically keeping track of telephone usage by extension, time of day, number called, trunk used and sometimes by person calling and client or account to be billed for call.

These days the terms means those three functions plus a whole lot more, typically those associated with professionally managing the corporate or government telecommunications expenses, including (but not limited to):
- Computerized inventory monitoring,
- Computerized traffic engineering and network design,
- Departmental telephone bill allocation and invoicing,
- Automated telephone directory, etc.
- Project tracking,
- Automated equipment and service ordering.

In short, all the functions of professional telecommunications management that can be automated or organized in some way on a computer. The telecommunications management system thus refers to the computer hardware and the software. For more on this subject see the latest June issue of TELECONNECT Magazine. See also CALL ACCOUNTING SYSTEM.

Telephone Manager Apple's telephony API for the Macintosh world. Here is an excerpt from Apple's Web page explaining it : "Telephony is the process of managing telephones, particularly of establishing and controlling connections between telephones on a telephone network. The Telephone Manager is the part of the Macintosh system software that you can use to develop applications and other software that provide telephony capabilities. For example, you can use the Telephone Manager to place outgoing telephone calls,

answer incoming telephone calls, place calls on hold or transfer them to other telephones, and accomplish many other similar tasks. The data transferred during a telephone call can be either voice, modem, or fax data, or indeed any kind of data that can be encoded for transmission across a telephone network.

"The Telephone Manager provides a set of simple but powerful programming interfaces that you can use to support telephony activities. The Telephone Manager operates independently of the particular telephone network or networks to which a user's computer is connected. Accordingly, your application can provide telephony services whether the Macintosh computer on which it is executing is connected to an integrated services digital network (ISDN), to a private branch exchange (PBX), or to "plain old telephone service" (POTS).

"The Telephone Manager accesses a specific telephone network using a telephone tool, a software module that manages the connection between a network and the telephony applications or other software running on a Macintosh computer. Telephone tools control the device drivers of the telephony hardware (such as an ISDN card) installed on the user's system. Each telephone tool is designed for specific hardware. For example, the Apple ISDN Telephone Tool is designed for the Apple ISDN NB Card.

For more, http://gemma.apple.com/techpubs/mac/Telephony/telephony-2.html or http://developer.apple.com/techpubs/mac/Telephony/Telephony-2.html

Telephone Pioneers of America The Telephone Pioneers of America began almost a century ago, originally consisting of the 'charter employees' of the company, or 'pioneers' in telecommunications, mainly those who served with the Bell System at its outset. As time went on, there would be fewer living or active original Pioneers, thus the TPA charter was amended to allow membership by any employee of AT&T or (as they were called) a subsidiary company who had been employed by Bell (or an independent) for at least twenty years. Membership in the Pioneers was opened to more types of telephone company people over the years (including companies that are not "Bell" or AT&T).

The Telephone Pioneers have a distinguished history of community service. Pioneers devise technical solutions to improve the lives of those with disabilities, allowing them to use telephones when this would otherwise be difficult or impossible. Pioneers also assist with general community activities such as voter registration, help those who are ill, feed those who are needy, and more. The Telephone Pioneers of America has chapters throughout the USA and Canada. At the non-Bell telcos, the same organization is known as the Independent Pioneers.

Telephone Preference Service TPS. A service offered by the Direct Marketing Association, New York, NY. The DMA keeps a list of consumers who have requested that their names be removed from telemarketing calling lists. Telemarketing companies can have the list upon request. Use of the service does not relieve companies from their obligation under the TCPA. In Europe the TPS is called a Robinson List.

Telephone Receiver Telephone earpiece. Device that converts electrical energy into sound energy, designed to be held to the ear.

Telephone Relay Service TRS. A voice/data system that enables communications with the hearing impaired.

Telephone Number Salary A salary that has seven digits, based on the fact that local North American phone numbers have seven digits.

Telephone Service Representative TSR. Another

word for agent — the person who answers the phone on an automatic call distributor. See AGENT.

Telephone Set A fancy name for a telephone.

Telephone Set Emulation The concept is simple: Emulate the proprietary electronic phone on a printed circuit card inside a PC. Let the PC do everything a human using the phone could do. Only the PC will do it more efficiently and the human will find it easier to use all his phone's features because the PC's screen is bigger and the PC's keyboard easier to use than the phone's keyboard. Attach the phone emulation card to voice and call processing cards, like voice synthesis, voice recognition, voice mail, touchtone generation and recognition, etc. And bingo, phone systems acquire all the benefits of integrated voice and call processing. It's powerful concept. As I wrote this, a handful of telephone phone emulation cards had appeared. Within a little while, there won't be a phone worldwide that you won't be able to emulate on a printed circuit card you can drop into a vacant slot inside your PC.

Telephone Set Management Imagine you have a phone attached to your computer through a telephony board inside your computer. Now imagine that you pick up the phone and dial a number. If the company knows you have dialed a number and knows which number you have dialed, that feature is called handset management. It is the ability of the computer to be aware of every button pushed on the phone. The advantage of this is obvious: You really want the PC to collect those digits, so it can, for example, add a price to each call and use them for monthly billing (lawyer, accountant, etc.). You also want to be able re-dial those numbers by simply clicking on the number one you want, hitting Enter and bingo, you're redialing that number, without having to key it in again. This term, telephone set management, used to be called handset management.

Two of the early pioneers in the field of telephone set management, David Perez and Nick Nance of COM2001 Technologies in San Diego, defined telephone set management as "the ability for seamless Integration with the phone (any 2500 set) and the modem / voice processing board and /or fax machine. The hardware must notify (send a signal or command) to the software when the phone is off hook or on hook. It must also notify the software when the user presses the numeric buttons on the phone. Ultimate integration would include additional types of button support as in: volume, hold, release, redial, conference, or any button on the telephone / fax / modem etc. The reason? Telephone integration offers true Computer Telephony integration. The ability to signal the handset and feature keys allows the user to continue to use their desktop phones but take complete advantage of the Computer Telephone software on their desktop for speed dialing, transferring, conferencing, voice mail, etc.

Telephone Signaling Device A gadget which indicates that the phone is ringing. May also be hooked up to lamps or overhead lighting to cause those lights to flash when the phone is ringing.

Telephone Tag I call you. But you're not there. I leave a message. You call me back. But I'm not there. You leave a message. And so on. We're now playing telephone tag.

Telephone Tap Telephone taps are generally defined as devices which are designed to extract audio information of intelligence from the telephone line pair. The process consists of identifying the specific telephone talk pair of interest at some accessible point, the interception of their electrical signals, and the communication of these signals to the surveillance equipment operator. Telephone companies unintentionally assist the wiretapper by installing extra telephone wires for future expansion. There are almost always extra wires that can be appropriated for use in wiretapping.

Wiretaps can be installed at the telephone company's central office if the phone company cooperates. If not, taps can be installed in splice cases or in a ready access terminal. Wiretaps should be distinguished from telephone bugs. Bugs are room audio surveillance devices that use the telephone wiring to bring the audio to the surveillance operator. Telephone bugs are used because they avoid RF interference.

Telephone Tone Audible tone generated by the network which provides call progress indications to the user. Different tones (e.g., ring back, busy) allow the human ear to interpret the progress of the call. On digital networks (such as PBX or ISDN), the network may send indication messages (e.g., billing, carrier, faxCNG, modemCNG) to the telephone to indicate the status of the call, and the telephone may generate certain tones locally, driven by those messages.

Telephony The science of transmitting voice, data, video or image signals over a distance greater than what you can transmit by shouting. The word derives from the Greek for "far sound." For the first hundred years of the telephone industry's existence, the word telephony described the business of the nation's phone companies were in. It was a generic term. In the early 1980s, the term lost fashion and many phone companies decided they were no longer in telephony, but in telecommunications — a more pompous sounding term that was meant to encompass more than just voice. The pomposity of the word may have added some value to the stock of telecommunications companies. In the early 1990s, as computer companies started entering the telecommunications industry, the word telephony was resurrected. And in a white paper on Multimedia from Sun Microsystems, the company said that telephony refers to the integration of the telephone into the workstation. For instance, making or forwarding a call will be as easy as pointing to an address book entry. Caller identification (if available from the telephone company) could be used to automatically start an application or bring up a database file. Voicemail and incoming faxes can be integrated with e-mail (electronic mail). Users can have all the features of today's telephones accessible through their workstations, plus the added benefits provided by integrating the telephone with other desktop functions. See also COMPUTER TELEPHONY.

Telephony Access Module See TAM.

Telephony Interface Control A Telephony Interface Control resource is any resource that interfaces with the telephone network (public or private). This is usually claimed as the primary member of a group.

Telephony Server A telephony server is a computer whose major function is to control, add intelligence, store, forward and manipulate the various voice, data, fax and e-mail calls flowing into and out of a computer telephony system. The traditional function of a telephony server is to move call control commands from client workstations on a LAN to an attached PBX or ACD. (This is what it does under the paradigm called "Telephony Services.") A telephony server can also be a voice response system. It can also be a fax on demand system. It can also be a conferencing device. It can also be switch. And it can be all these capabilities, which traditionally run on physically separate servers, all rolled into one machine, called generically a "telephony server." See Telephony Service Application Programming Interface And Telephony Services.

Telephony Services Application Programming

Interface. TSAPI. Described by AT&T, its inventor, as "standards-based API for call control, call/device monitoring and query, call routing, device/system maintenance capabilities, and basic directory services." For a better explanation, see Telephony Server, Telephony Server NLM and Telephony Services.

Telephony Server NLM Telephony Server NetWare Loadable Module. The main part of a software product call Telephony Services announced in early 1993 by AT&T and now marketed by Lucent (but not Novell). The Telephony Server NLM is software add-on to Novell's NetWare LAN operating software. The idea is have a NetWare server equipped with the NLM, an interface card and a cable connection to an adjoining telephone system. This would mean that anyone with a PC on the network and a PBX phone on their desk will be able to use telephone features, such as auto-dialing, conference calling and multiple call handling from their desktop PC.

A Novell White Paper in Spring of 1993 said "Telephony Services for NetWare provides benefits to three main customer segments. First, applications are being developed to provide increased productivity to everyday computer desktop users. Second, call-centers take advantage of this technology as it provides a right-sizing cost-effective solution. Finally, benefits will be available to telecommunications/IS administrators by providing the ability to reduce administrative costs through easier management of user databases.

"Computer-Telephone Integration (CTI) combines telephone and computer technology to provide access and control telephone functionality from a computer terminal. It combines the easy access and usable graphical interface of the computer desktop with the features of the telephone. CTI is not a new concept. Traditionally, however, CTI has only been available in a mini and mainframe computer environments. These solutions are expensive and can be cost-justified only in large call-center applications. Consequently, the penetration of CTI solutions has been very small.

"However, providing CTI in NetWare environment brings this technology mainstream. Not only does this solution provide a more cost-effective implementation, it also allows integration with the rich set of NetWare services. In the simplest example, a Telephony Services for NetWare application allows users to make a phone call by clicking on a name from a calling list displayed on a desktop computer and having the desktop computer dial the number. Possibilities exist for applications that will allow similar functionality with the addition of conference calling capability. Instead of clicking on a single name, the user can highlight a number of names, click on a conference-call icon and have the system place the calls to all parties. The benefits which are derived from the integration of telephony with other NetWare services is far reaching. As part of continued development efforts, applications are becoming available which allow desktop video phone calls. Callers can see each other and talk on the phone, while simultaneously viewing and editing image documents.

"Other capabilities include integrating voice-mail, fax and e-mail into a single message-management application. Possibilities also exist utilizing number recognition technology to integrate computer database records with caller-id. Administrators can manage a single user database utilized by the computer network, the PBX and the voice-mail system.

"Telephony Services for NetWare takes advantages of client/server technology to provide a broad framework for creating first-party and third-party call-control applications.

These applications answer the customer demand for integrated business tools and solutions. This technology provides a logical connection between the desktop computer and the telephone. The only physical connection is established between the PBX and a NetWare server. This architecture is cost-effective and efficient by utilizing a company or organization's existing equipment. The initial product deliverables include the following components:

- Client/Server API
- Telephony Server NLM
- PBX Driver
- PBX Link Hardware
- Passageway Application

"The Telephony Server NLM is the mechanism for passing information between the PBX and the NetWare server. As part of the NLM, an open PBX Driver Interface allows PBX manufacturers to write drivers which communicate with their respective PBX models. The client/server API provides support across multiple desktop operating systems. It also allows call control at either the client or the server. The Passageway application provides the user with basic autodialing and notes capability.

"Telephony Services for NetWare provides a key opportunity for developers. Open APIs which support multiple desktop operating systems provide a development platform for both traditional telecommunications developers and new or experienced NetWare developers."

Novell has effectively stopped marketing Telephony Services, but AT&T (now Lucent) continues to market it. See Telephony Services.

Telephony Service Provider TSP. A software encapsulation of all the services provided by a particular network interface device or line device. A line device may be a single POTS bearer channel or it may be several bearer channels; e.g., a single E-1 span with 30 network channels of 64 Kb/s bandwidth. The TSP is provided by the vendor who has developed a network interface device for SCSA. See S.100.

Telephony Services Telephony Services' real name is Telephony Services for NetWare, the local area network software from Novell, Orem, Utah. Telephony Services for NetWare basically consists of an addition to the NetWare operating system, called Telephony Server NLM. (See Telephony Server NLM.) That addition handles communications between a NetWare file server (a PC loaded with NetWare) and an attached telephone switch, e.g. a PBX or ACD (automatic call distributor). The concept is very simple. Picture your department's LAN. You're sitting in front of your PC which is on your department's LAN. You have a telephone which is an extension off your company's or department's PBX. You click on an icon that says "Phone." Bingo, a screen comes up with icons and pull down menus. You can now look up Joe, click on his name. Your PC sends a command to your NetWare file server, which in turn sends a command to your PBX which tells it to dial Joe from your phone. Once it's completed dialing, it might turn on your phone's speakerphone or your telephone headset. You'll then hear Joe say "Hello." Telephony Services for NetWare is basically the software in the NetWare file server which takes care of interpreting your PC commands into commands your switch can understand and respond to. Telephony Services requires a link to your switch. Each telecom switch manufacturer has been implementing that link in a different way technically. That's fine, because Telephony Services for NetWare insulates the user and the developer. This means that computer telephony applications written for Telephony Services for NetWare will work on any switch conforming to the Telephony Services standard. As of writing, virtually every switch manufactured in North America and many made in Japan and Europe is conforming to Novell's Telephony Services.

According to a White Paper issued by Novell in March 1994 and called NetWare Telephony Services, "The three main components of Telephony Services are call control, voice processing, and speech synthesis. Call control provides the core service for PBX-to-NetWare communication and an Application Programming Interface (API) for developing client server applications. With call control, users can enjoy features, says Novell, such as making calls, transfers, or conference calling. Voice processing functions include voice mail and interactive voice response. Speech synthesis will be a key area for integrating multiple media types. Through speech/text conversion, users can access voice mail, e-mail, and fax documents through audio or text media types. The initial products Novell is delivering include:

- Client/Server API
- Telephony Server NetWare Loadable Module (NLM)
- PBX Driver

"The Telephony Server NLM is the mechanism for passing information between the PBX and the NetWare server. As part of the NLM, an open PBX Driver Interface allows PBX manufacturers to write drivers that communicate with their respective PBXs. The client/server API provides support across multiple desktop operating systems and allows call control at either the client or the server."

Client server computer telephony, according to Novell, delivers ten benefits:

1. Synchronized data screen and phone call pop. Your phone rings. The call comes with the calling number attached (via Caller ID or ANI). Your PBX or ACD passes that number (via Telephony Services) to your server, which does a quick database look up to see if it can find a name and database entry. Bingo, it finds an entry. It passes the call and the database entry simultaneously to whoever is going to answer the phone: The attendant. The boss. The sales agent. The customer service desk. The help desk. All this saves asking a lot of questions. Makes customers happier.

2. Integrated messaging. Also called Unified Messaging. Voice, fax, electronic mail, image and video. All on the one screen. Here's the scenario. You arrive in the morning. Turn on your PC. Your PC logs onto your LAN and its various servers. In seconds, it gives you a screen listing all your messages — voice mail, electronic mail, fax mail, reports, compound documents .... Anything and everything that came in for you. Each is one line. Each line tells you whom it's from. What it is. How big it is. How urgent. Skip down. Click. Your PC loads up the application. Your LAN hunts down the message. Bingo, it's on screen. If it contains voice — maybe it's a voice mail or compound document with voice in it — it rings your phone (or your headset) and plays the voice to you. Or, if you have a sound card in your PC, it can play the voice through your own PC. If it's an image, it will hunt down (also called launch) imaging software which can open the image you have received, letting you see it. Ditto, if it's a video message. Messages are deluging us. To stop them is to stop progress. But to run your eye down the list, one line per entry. Pick the key ones. Junk the junk ones. Postpone the others. That's what integrated messaging is all about. Putting some order back into your life.

3. Database transactions. Customer look ups. There are bank account balances, ticket buys, airline reservations, catalog

requests, movie times, etc. Doing business over the phone is exploding. Today, the caller inputs his request by touchtone or by recognized speech. The system responds with speech and/or fax. Today's systems are limited in size and flexibility. The voice processing application and the database typically share the same processor, often a PC. Split them. Spread the processing and database access burden. Join them on a LAN (for the data) and on new, broader voice processing "LANs," like SCSA or MVIP. You've suddenly got a computer telephony system that knows no growth constraints. You could also get the system to front-end an operator or an agent. Once the caller has punched in all his information, then the call and the screen can be simultaneously passed to the agent.

4. Telephony work groups. Sales groups. Collections groups. Help desks. R&D. We work in groups. But traditional telephony doesn't. Telephony today is BIG. Telephony today is one giant phone system for the building, for the campus. Everyone shares the same automated attendant, the same voice mail, the same ubiquitous, universal, generic telephone features. But they shouldn't. The sellers need phones that grab the caller's phone number, do a look-up on what the customer bought last and quickly route the call to the appropriate (or available) salesperson. The one who sold the customer last time. The company's help desk needs a front end voice response system that asks for the customer's serial number, some indication of the problem and tries to solve the problem by instantly sending a fax or encouraging the caller to punch his way to one of many canned solutions. "The 10 biggest problems our customers have." When all else fails, the caller can be transferred to a live human, expert at diagnosing and solving his pressing problem. A development group might need e-mails and faxes of meeting agendas sent, meeting reminder notices phoned and scheduled video conferences set up. All automatically. The accounts receivable department needs a predictive dialer to dial all our deadbeats. The telemarketing department also needs a predictive dialer, but different programming.

5. Desktop telephony. There are two important aspects. Call control and media processing services. Call control (also called call processing) is a fancy name for using your PC to get to all your phone system's features — especially those you have difficulty getting to with the forgettable commands phone makers foist on us. *39 to transfer? Or it is *79. With attractive PC screens, you point and click to easy conferencing, transferring, listening to voice mail messages, forwarding, etc. There are enormous personal productivity benefits to running your office phone from your PC: You can dial by name, not by number you can't remember. You can set up conference calls by clicking on names and have your PC call the participants and call you only when they're all on the phone. You can transfer easily. You can work your voice mail more easily on screen, instead of having to remember "Dial 3 for rewind," "Dial 2 to save," and other obscure commands. Here's a wonderful quote from Marshall R. Goldberg, Developer Relations Group at Microsoft. He says "Voice mail systems that could benefit through integration with the personal computer largely remain isolated, difficult to use, and inflexible. Browsing, storing messages in hierarchical folders, and integration of address books — functions just about everyone could use — are either unavailable or unusable." The second benefit is media control. Media control is a fancy name for affecting the content of the call. You may wish to record the phone call you're on. You may wish to have all or part of your phone call clipped and sent to someone else —

as you often today with voice mail messages. You may wish to simply file your conversations away in appropriate folders. You may wish to be able to call your PC and get it to read you back any e-mails or faxes you received in the last day or so.

6. Applying intelligence. A PC is programmable. The typical office phone isn't. A PC can be programmed to act as your personal secretary, handling different calls differently. It can be programmed to include commands, such as "If Joe calls, break into my conversation and tell me." "If Robert calls, send him to voice mail." etc.

7. The Compound Document. The typed document lacks life. But add voice, image and video clips to it and it gets life. The LAN makes the compound document easier to achieve. The Compound Document gets attention.

8. Management of phone networks. Today, phone networks are very difficult to manage. Often the PBX is managed separately from the voice mail, which is managed separately from the call accounting, etc. It's a rare day in any corporate life when the whole system is up to date, with extensions, bills and voice mail mailboxes reflecting the reality of what's actually happening. The latest generations of LAN software — NetWare 4.1 and Windows NT — have solid enterprise-wide directories and far easier management tools. Integrate these LAN management tools with telecommunications management, and potentially all you need is to make one entry (for a new employee, a change, etc.) and the whole system — telecom and computing — could update itself automatically, including even issue change orders to the MIS and telecom departments and vendors.

9. No dedicated hardware in the PC. With only one link — from the switch to the LAN — there's no need to open the desktop PC and place specialized telephony hardware in each PC that wants to take advantage of the new LAN-based telephony features.

10. Switch elimination. The ultimate potential advantage of LAN-based telephony is to eliminate the connection to the switch (PBX or ACD) by simply populating the LAN server (now called a telephony server) with specialized computer telephony cards and run the company's or department's phones off the telephony server directly.

Novell has effectively stopped marketing Telephony Services, but AT&T (now Lucent) continues to aggressively market it. and the benefits are still as valid as Novell detailed above. See Telephony Server NLM and Telephony Services Development Tools

Telephony Workgroup A concept that says work is done in groups and those groups need special telephony features and services. This is in contrast to most telephone installations today, where one giant phone system serves the company. Everyone shares the same automated attendant, the same voice mail, the same ubiquitous, universal, generic telephone features. But the concept of a telephony is that they shouldn't. Each group has different telephony needs: The sellers need phones that grab the caller's phone number, do a look-up on what the customer last bought and quickly route the call to the appropriate (or available) salesperson. The one who sold the customer last time.

The company's help desk needs a front end voice response system that asks for the customer's equipment serial number, some indication of the problem and tries to solve the problem by instantly sending a fax or encouraging the caller to punch his way to one of many canned solutions. "The 10 biggest problems our customers have." When all else fails, the caller can be transferred to a live human, expert at diagnosing and

solving his pressing problem.

A development group might need e-mails and faxes of meeting agendas sent, meeting reminder notices phoned and scheduled video conferences set up. All automatically.

Computer telephony on a LAN can do to the workgroup what products like e-mail and Lotus Notes are doing — substantially improve productivity (or at least, the pleasure of work). Except that the telephone is still less intimidating.

There are probably as many specialized telephony workgroups features needed as there are computer workgroup features needed. And since computer workgroup features are often provided on a local area network, it makes sense to provide many telephony features for that workgroup on the same local area network.

Telepoint The British name for the new generation of cheap, digital mobile phones. They're also called CT2. Think of telepoint phones as cellular phones but using micro-cells. By having smaller cells than normal cellular cells, CT2 phones can be smaller, cheaper and lighter. The first generation of these phones didn't do well, since they weren't smaller and lighter; there weren't many micro-cells and you couldn't receive an incoming call. See CT1, CT2 and CT2+.

Teleports The definition written by Gary Stix in the August 12, 1986 issue of Computer Decisions reads, "High bandwidth telecommunications distribution systems that allow major local users to obtain local, private services and long distance services. The most notable example is the New York Teleport," which is located on Staten Island. Teleports traditionally consist of two things — a fiber optic/coaxial cable network around a city and a collection of nearby satellite antennas. The cable network collects transmissions from larger customers and takes them to the antennas for shipping to and from distant offices. Teleport companies are now more successful as local communications companies than they are as long distance gateways. Which is understandable, since the cost of local calls has gone up, while the cost of long distance calls has gone down.

Telepresence In 1985 a team of researchers at NASA invented the notion of telepresence — projecting yourself into someone else's virtual reality. In one version of telepresence, according to Discover Magazine, a computer prompts a robot to mimic your movements. As you manipulate objects in your virtual world a robot somewhere else does the same thing to real objects. Telepresence will be especially useful for hazardous jobs like repairing a nuclear reactor or satellite. Or going on a blind date? See VIRTUAL REALITY.

Teleprinter A teletypewriter. Also called a telex machine.

Teleprocessing An IBM term for data communications.

Telesales British. Sales activities conducted by telephone in a planned and controlled manner.

Telescript Trademark name of General Magic's communications language technology. An interpreted, object oriented language that creates and manages intelligent agents as they move through a network of computers and personal communicators. Telescript is implemented as a portable interpreter, called Telescript engine. Programs written in Telescript are called scripts. General Magic says that it expects Telescript to become an industry standard, licensed by major suppliers of personal intelligent communicators, computers, communication networks, and information services. For more information, ask for a brochure called "Telescript Technology; The Foundation for the Electronic Marketplace," a General Magic White Paper, Mountain View, CA. See GENERAL MAGIC.

Telescripts A call center term. Rockwell's software vectors for configuring the many options of routing a call in progress, a call in queue, a call in the system, etc.

Teleservices 1. A generic term for services offered on phone links. Includes e-mail and facsimile features.

2. A product of SunSoft, a division of Sun MicroSystems, Mountain View, CA. According to SunSoft, "Sun's vision of the impact of widespread use of teleservices suggests that the computer workstation will become the new communications center, combining many existing communication media with new ones, while creating new paradigms for the expression and sharing of ideas. Information in the form of charts and pictures, schedules and plans, and audio and video will merge through application programs that provide a collaborative vehicle for decisions in the 1990s and beyond. The desktop will become the platform for a new set of productivity tools, seamlessly integrated into the critical business activities and methodology of today's companies, and providing a competitive edge for facing the global challenges of tomorrow. Individuals will gain new freedom in where they work and how they access information. And ideas will be communicated in more expedient and creative ways." SunSoft has developed Solaris Teleservices to provide a platform for next-generation workstation applications which leverage the benefits and capabilities of the telephone network and the commonplace use of it. Teleservices applications, according to SunSoft, include:

Desktop Teleservices. Workstation based telephone and answering machine applications allow users to efficiently plant, receive and manage telephone calls.

Remote Access. Users can place calls for their workstation from any telephone and access applications and data through DTMF signaling, or perhaps through speech, using a workstation's speech recognition capabilities.

Wide Area Networking. The ubiquity of telephone networks allow for the complete connectivity of all computers. Network links can be brought up or taken down on demand merely by placing or tearing down a telephone call.

The Solaris Teleservices Platform is called XTEL, which is a multilayered software architecture based on client server computing model. XTEL consists of four key components:

A client side library. Providing a high-level, object oriented application programming interface (API) to application programmers. Using the XTEL API, an application can place and retrieve telephone calls. The API library consists of a collection of C++ objects which is linked to applications that wish to use the systems teleservices resources.

A server. Providing multi-client and multi-device support, the server is the central point of contact for all teleservices, resource management and security are provided by the server. Communication between the XTEL API and the server occurs through the XTEL Server Protocol (XTELS).

One or more providers. Manages each telecommunication device connected to the system. The Teleservices server communicates with an XTEL provider using the XTEL provider protocol (XTELP).

A data stream multiplexor (Sun's spelling). The universal multiplexor (Umux) provides a uniform means for applications to access and share data channels associated with a telephone call. Umux is a streams pseudo-device driver used to connect data channels to applications. An XTEL application is linked with the XTEL API library, through which communicates with the server using the XTELS protocol. XTELS is a synchronous, symmetric messaging protocol using the Solaris loopback transport mechanism. The XTELS protocol is essential-

ly the XTEL provider protocol with extensions to support multiple clients and multiple XTEL providers. For more information, see a document called Solaris Teleservices Architectural Overview, available from SunSoft, Mountain View, CA. See TELESERVICES API.

Teletel Terminal Emulation Teletel is a popular character-based graphics emulation for communicating with the Minitel service, found primarily in France.

Teletex An ITU-T standard for text and message transmission which is replacing Telex. Teletex operates at 2400 baud, about 50 times faster than telex. Teletex uses ASCII to encode its characters for transmission.

Teletext A data communications information service used to transmit information from

remote data banks to viewers. It was transmitted over the air in the vertical blanking interval of the TV signal of the BBS (British Broadcasting Service). Teletext was originally designed for public consumption. It gave out weather information, sports results, headlines, etc. Teletext is proving somewhat more successful among corporations for the internal dissemination of information.

Teletraffic Optimizer Program Derives data by processing actual calls instead of using an analytical model based on estimates or summaries.

Teletraining Education and training through telecommunications.

Teletype A specific type of teletypewriter.

Teletyper Input Method Teletyper Input Method (TIM) is specially designed for using pushbutton phone or mobile phone to input text, command and instruct your PC through the public telephone network. Teletyper Telephone-Input Method (TIM) and Teletyper Plus use three base keys on the phone pad with the combination of other keys to form alphabets, symbols, utility and control functions in order to command your PC remotely. Information could then be sent to fax, pager or voice output.

Teletyper Input Method (TIM)

*1 = A	#4 = L	08 = W
01 = B	*5 = M	#8 = X
#1 = C	05 = N	*9 = Y
*2 = D	#5 = O	09 = Z
02 = E	*6 = P	#9 = space
#2 = F	06 = Q	** = Enter
*3 = G	#6 = R	00 = Zero
03 = H	*7 = S	## = Backspace
#3 = I	07 = T	1 to 9 press 1 to 9
*4 = J	#7 = U	
04 = K	*8 = V	

Teletypewriter TTY. A telegraph device capable of transmitting and receiving alphanumeric information over communications channels. It may also contain a keyboard similar to that of a typewriter or computer but usually with fewer keys. See TELETYPE.

Teletypewriter Control Unit TCU. A device that serves as the control and coordination unit between teletypewriter devices and a message switching center when controlling teletypewriter operations.

Teletypewriter Exchange Service TWX. A switched teletypewriter service in which suitably arranged teletypewriter stations are provided with lines to a central office for access to other such stations. TWX and Telex are commercial teletypewriter exchange services. They are currently both owned by AT&T. These days their revenues are in decline. A computer with a modem is a lot faster than TWX or telex.

Teletypewriter Signal Distortion The shifting of signal-pulse transitions from their proper positions relative to the beginning of the start pulse. The magnitude of the distortion is expressed in percent of a perfect unit pulse length.

Television 1. TV. Translated from Greek and Latin, "far-off sight," television traditionally is thought of as broadcast TV, transmitted over the airwaves using radio broadcast frequencies. Most of us today think of TV as being provided by a CATV (Community Access TV) provider via coaxial cable. Increasingly, as many as 5 million U.S. viewers think of TV as being delivered directly via satellite. A standard analog TV channel today fits into a frequency bandwidth of 6 MHz. See TV. 2. "Chewing gum for the eyes." Frank Lloyd Wright

Telework The combination of computer and telecommunications technology which enables office workers to work at home or away from the main office on a part-time or full-time basis. See also telecommuting.

Teleworker A person who works from his home or some place distant from his company's office. A teleworker may send his completed work in and pick his new work up via a modem in his PC. A teleworker may also be on the phone at home answering calls on behalf of his company and entering the results of those calls (i.e. reservations on airlines, orders for catalogs) in on a PC connected by phone lines to his company. He may use one phone line, like an ISDN BRI line or he may simply use two analog phone lines — one for talking on and one for PC's data. Or he may simply use one analog phone line and use a protocol such as VoiceView.

Telex A worldwide switched message service. Telex service is offered in the United States by the Western Union Telegraph Company, MCI, ITT, RCA, FTCC and TRT. Telex has one gigantic advantage: Overseas it's very popular and widely used. Contacting overseas businesses by telex is often far more reliable and faster than contacting them by telephone. Telex is good for overseas time zone differences because you can send a message to an unattended telex machine. It also delivers a printed record. Telex is relatively inexpensive usually costing a little less than a phone call. Telex has one disadvantage: It's very slow and not very accurate with virtually no data communications error checking procedures. Telex is being rapidly displaced by faster, more accurate forms of data communications, including the public packet switched networks and the various electronic mail services, and most recently by massive competition from low-cost facsimile machines. See TELETEX.

Telex Access Unit TLXAU. An X.400 AU (Access Unit) serving Telex users.

Telnet A program that lets you connect to other computers on the Internet. The process by which a person using one computer can sign on to a computer in another city, state or country. Telnet is the terminal-remote host protocol developed for ARPAnet. Using Telnet, you can work from your PC as if it were a terminal attached to another machine by a hardwired line. The format of the telenet command is telnet address.domain or telnet address.domain port #. These days, most users are insulated from TELNET by GUI browsers, such as Netscape or Internet Explorer.

Telnet Port The port address on a computer which supports remote telnet access. Normally port 23 is the default telnet port.

Telpak A discontinued AT&T service that gave large customers discounts on purchases of multiple analog private lines. Telpak was to discourage users from building their own private microwave systems. Users who bought Telpak, how-

ever, were not allowed to resell any of the circuits, though they were allowed to share them. The FCC ruled Telpak as being discriminatory against competition. (It was too cheap.) It may still exist on an intrastate basis in some states. Telpak typically came in bundles of 12, 24, 60 and 240 voice lines. The bigger the bundle, the cheaper the per circuit cost.

TELRIC Total Element Long Run Incremental Cost. A method of figuring out what phone service should cost based on incremental cost of equipment and labor, not counting the embedded cost of old cost.

Telstar 1 Telstar 1 was the world's first active communications satellite. It was launched on July 10, 1962. (Sputnik was launched on October 4, 1957.) There is some argument about Telstar's claim to fame as the first. The engineers at the RCA Astro-Electronics Division, Princeton NJ (now Martin-Marietta Aerospace) claimed that they launched and successfully used a satellite to broadcast the coronation of the pope a little earlier than Telstar. But it was Telstar that got all the fame and glory, and that RCA engineers don't deny that.

Telstra Telstra is Australia's largest local and overseas phone company. The company has its origins in 1901, when the Postmaster-General's Department (PMG) was formed to manage all domestic telephone, telegraph and postal services. The Overseas Telecommunications Commission (OTC) was formed in 1946 to manage international telecommunications. The Australian Telecommunications Commission, trading as Telecom Australia, was created as a separate entity in July 1975, following the breakup of the PMG. Telecom Australia and OTC merged in February 1992 to form Telstra. On July 1, 1997, Australia's telecommunications markets were opened to full competition, with no limit on the number of carriers that own transmission infrastructure who can enter the market.

Temperature Rating The maximum temperature at which the insulating material may be used in continuous operation without loss of its basic properties.

Tempest Unclassified name referring to the investigation, study, and control of compromising emanations from electrical and electronic equipment. Devices which are tempest-secure mean they do not send emanate electromagnetic signals which can, potentially, be received by others, i.e. enemies.

Template 1. A voice processing term. A pattern of information as a function of time, which is intended to represent an entire word.

2. A Northern Telecom Norstar definition: A system wide setting assigned during System Startup. The most important effects of a template are the number of lines assigned to the telephones, and the assignment of Line Pool Access. Templates will also assign other system wide defaults, such as Prime Line and Ringing Line assignment. It is important to understand that a template is only provided as a convenience, and that any settings effected by the template can be changed.

Temporal Coding Compression that is achieved by comparing frames of video over time to eliminate redundancies between frames.

Temporary Signaling Connections On August 14, 1995, AT&T announced Temporary Signaling Connections, which it billed as the first service that lets banks, retail outlets and other data-intensive businesses link their Software Defined Network locations together on demand using virtual connections created in AT&T's national signaling networks. Businesses can use Temporary Signaling Connections to verify credit card transactions, update inventory databases, exchange data with automatic cash machines. The service uses a portion of the D channel capacity of an ISDN PRI channel and passes information to other ISDN PRI locations.

Temporary Station Disconnection Allows the attendant to completely remove selected phones from service at any time on a temporary basis.

Ten-High Day A traffic engineering term for a traffic study which considers the average of the traffic during the same clock hour on the ten busiest normally recurring days of the busy season of the year. See Ten-High Day Busy Hour and Traffic Engineering.

Ten-High Day Busy Hour A telephone company term. The hour, not necessarily a clock hour, which produces the highest average load for the ten highest business day loads in that hour. It may be a different hour from the ABS busy hour. With present data collection procedures, the ten high days are usually selected from the busy period.

Ten-High Day Data THD. Data collected during the ten-high day busy hour. See above.

Tenant Partitioning Also known a Service Bureau Capability. One computer or telephone host can provide service to many tenants in the building.

Tenant Service Some businesses acquire a telephone system too large for their needs so they sell parts of the service to smaller offices in their own building or in the surrounding community. There are two ways to make money on tenant service — renting phone equipment or re-selling long distance lines. There's more money on re-selling long distance lines.

Tensile Load Refers to the maximum load or pull force that may exerted upon a cable during installation or relocation without damage. An excessive tensile load on twisted pair cables can cause elongation or untwisting which may result in signal loss.

Tensile Strength A term denoting the greatest longitudinal tensile stress a substance can bear without tearing apart or rupturing.

TEO See Telephone Equipment Order.

Tera- A million million, or a thousand giga-. See TERABYTE.

Terabyte A unit of measurement for physical data storage on some form of storage device — hard disk, optical disk, RAM memory etc. and equal to two raised to the 40th power, i.e. 1,099,511,627,776 bytes.

MB = Megabyte (2 to the 20th power)
GB = Gigabyte (2 to the 30th power)
TB = Terabyte (2 to the 40th power)
PB = Petabyte (2 to the 50th power)
EB = Exabyte (2 to the 60th power)
ZB = Zettabyte (2 to the 70th power)
YB = Yottabyte (2 to the 80th power)

One googolbyte equals 2 to the 100th power

To put things in perspective (courtesy of Microsoft), a terabyte holds a 100-byte record for every person on earth, as well as an index of those records; or a JPEG-compressed pixel for every square meter of land on earth, which is plenty to create a high-resolution photograph; or 1 billion business letters, which would fill 150 miles of bookshelf space; or 10 million JPEG images, which would provide 10 days and nights of continuous video.

Teraflop A trillion (10 to the 12th power) floating point instructions per second. A measure of a computer's speed.

Terahertz THz. A unit denoting one trillion (10 to the 12th) hertz. See TERA-

TERENA Trans-European Research ann Education Networking Association. Formed in 1994 by a merger of RARE and EARN.

Terminal 1. The point at which a telephone line ends, or is connected to other circuits of a network.

2. An input/output device for communicating with computers. Typically has a keyboard and a CRT (TV screen) display. See TERMINALS.

Terminal Adapter The terminal adapter is a protocol converter (little black box) that adapts PCs, workstations and other equipment to the peculiar world of ISDN. Some ISDN devices include built-in terminal adapters. Some don't. In technical terms, a terminal adapter is an interfacing device employed at the "R" reference point in an ISDN environment that allows connection of a non-ISDN terminal at the physical layer to communicate with an ISDN network. Typically, this adapter will support standard RJ-11 telephone connection plugs for voice and RS-232C, V.35 and RS-449 interfaces for data. See also ISDN.

Terminal Address Where there's a terminal to punch down, there's an address. That address will have numbers on it and will enable the technician who's responsible for fixing the circuit (when it breaks) to come, find it and fix it.

Terminal Block A device used to connect one group of wires to another. Usually each wire can be connected to several other wires in a bus or common arrangement. A 66-type block is the most common type of connecting block. It was invented by Western Electric. Northern Telecom has a terminal block called a Bix block. There are others. A terminating block is also called a connecting block, a punchdown block, a quick-connect block, a cross-connect block. A connecting block will include insulation displacement connections (IDC). In other words, with a connecting block, you don't have to remove the plastic shielding from around your wire conductor before you "punch it down."

Terminal Configuration The functional interconnection of the components of a terminal. For example, a keyboard-printer may be configured to transmit keystrokes without printing them. Printing is only performed on data retrieved from the communication line. Terminals with multiple components can be configured in a variety of ways.

Terminal Emulation An application that allows an intelligent computing device such as a PC to mimic or emulate the operation of a dumb terminal for communications with a mainframe or minicomputer. It does this with special printed circuit boards inserted into its motherboard and/or special software. For example, TELECONNECT uses the communications software program called Crosstalk to emulate a DEC VT-100, a Digital Equipment Corporation VT-100 terminal. We do this because emulating a DEC VT-100 works better with certain software programs we call up remotely.

Terminal Equipment Terminal Equipment usually refers to the telephones and other equipment at the end of telephone lines. See also CPE.

Terminal Equipment Type 1 TE1. In Integrated Services Digital Network (ISDN) technology, TE1 is a type of terminal compatible with ISDN.

Terminal Equipment Type TE2. In Integrated Services Digital Network (ISDN) technology, a type of terminal that must be connected to ISDN via a specially designated point, normally an RS 232 or RS 449 interface.

Terminal Hunt Group Also called Terminated Hunt Group. It's another name for a "top down" hunt group. You have a bunch of phone lines. They are in a "hunt" group which means that, if one is busy, the switch sends the call to the next available line in the hunt group. In a terminal hunt group, the switch always starts at the top of the hunt group and goes down, searching for the first available line from the top. This contrasts with a circular hunt group, where the switch remembers the last line it connected and, starting there, hunts down to the next available line, searching basically in a circle. A terminated or "top down" hunt group puts more calls on the first lines in the group. A circular hunt group tends to distribute the calls evenly. There are reasons why you might choose one type of hunt group over another. You might choose to evenly distribute your calls over a bunch of humans answering incoming calls or you might choose to send the calls top down to a voice mail system and watch the usage statistics carefully to tell if you need more or fewer lines and more or fewer voice cards.

Terminal Impedance The impedance as measured at the unloaded output terminals of transmission equipment or a line that is otherwise in normal operating condition. The ratio of voltage to current at the output terminals of a device, including the connected load.

Terminal Node In IBM Corp.'s Systems Network Architecture (SNA), a network device that cannot be programmed by the user.

Terminal Number 1. A terminal number is one or multiple circuit numbers not identified by an individual directory telephone number (DTN) but by 001, 002, 003, etc. and is only referenced as a subset of the main trunk pilot DTN. Terminal numbers may be rented from both Incumbent Local Exchange Carriers (ILECs) or Competitive Local Exchange Carriers (CLECs), like MCI Worldcom. Terminal numbers are all organized to hunt (descending, ascending, most idle, least idle). A terminal number is another name for an auxiliary or private line that doesn't have a real number, doesn't get a listing in the phone book, but gets a monthly bill. An auxiliary number is a telephone trunk you rent from your local phone company in addition to the main number you rent. Phone systems are always set up for multiple phone lines, so that when a call comes in, it doesn't hit a "busy," but rolls over to one or more auxiliary lines. That collection of lines is called an Incoming Service Group, or ISG. For example, the publisher's main office main number is 212-691-8215. But it also has 8216, 8217, 8218 and several unmarked or coded trunks. All these numbers are auxiliary lines and don't receive their own bill or directory listing from the phone company. Costs for these lines are lumped onto the bill for the main number. See Rollover and ISG.
2. TN. A Northern Telecom definition. The physical address of a device (such as, telephone set, a truck, and attendant) on the Meridian 1 PBX. The TN is composed of the loop, shelf, card and unit IDs.

Terminal Repeater A repeater for use at the end of a trunk line.

Terminal Server A small, specialized, networked computer that connects many terminals to a LAN through one network connection. Any user on the network can then connect to various network hosts. A terminal server has a single network interface and several ports for terminal connections. The advantage of a terminal server is it allows many terminals to be connected to a host via a single existing LAN cable, rather than a variety of point-to-point cables. Digital Equipment's DECnet makes heavy use of terminal servers.

Terminal Table An ordered collection of information that identifies each line, phone, component or application program from which a message can be sent.

Terminals The screws or soldering lugs to which an external circuit can be connected.

Terminals To Long Distance Operator Commonly known as "toll terminals", they provide special trunks directly to the long distance telephone company operators. Upon completion of long distance calls, the toll operator will ring the attendant (or hotel operator) and give them "time and charges" for the phone call just ended.

Terminate 1. To connect a wire conductor to something, typically a piece of equipment.
2. To end one's telecommunications service or equipment rental.

Terminated 1. The condition of a wire or cable pair which is connected to (terminated on) binding posts or a terminal block.
2. The condition of a circuit connected to a network which has the same impedance the circuit would have if it were infinitely long.

Terminated Line A telephone circuit with a resistance at the far end equal to the characteristic impedance of the line, so no reflections or standing waves are present when a signal is entered at the near end. Compare with bridge tap.

Terminating Channel The name for the circuit in a private line channel that connects a local central office with the CBX/PBX or telephone instrument at the customer's premises.

Terminating Multiplexer TM. A type of Path Terminating Equipment (PTE) used to provide access to a SONET/SDH network. The Terminating Multiplexer is equivalent to a Time Division Multiplexer (TDM) in a T/E-Carrier context. The TM also serves to perform the signal conversion process from electrical to optical on the transmit side, reversing the process on the receiving end.

Terminating Office The switching center (i.e. the central office) of the person you're calling (the "called party").

Terminating Resistor A grounding resistor placed at the end of a bus, line, or cable to prevent signals from being reflected or echoed. Sometimes shortened to terminator.

Termination Termination involves the placement of impedance matching circuits on a bus to prevent signals from being reflected or echoed.

Termination Restriction Prevents a user from receiving any calls on the phone line. A DID call to the restricted termination routes to an attendant, an announcement or intercept tone at customer option. All other calls route to intercept tone.

Termination Of Service The end of service of a line or equipment. All pursuant to the regulations set forth in the tariff.

Terminator Some communications facilities — e.g. local area networks — are bus configurations. This means one long piece of cable with workstations connected along the way, typically with "T" connectors. For a network to work properly, you need to place resistance at the end of the cable that serves to absorb the signal on the line. A thin wire Ethernet typically requires a 50 ohm resistance at either end of the bus. You can buy these Ethernet terminators already included in a connector.

Terminus A device used to terminate an optical fiber that provides a means to locate and contain an optical fiber within a connector.

Terrestrial Long distance facilities which are entirely on land and do not use satellites. This includes microwave, coaxial cable, optical fiber, normal cable, etc. There are reasons to prefer terrestrial facilities over satellite facilities:
1. No echo or delay in voice conversation. Some people find satellite conversations disturbing because of the delay; and 2. No significant reduction in data throughput. Most data communications protocols send their data in "chunks" and require an acknowledgement from the other end when one

chunk has been received before the next chunk can be sent. When it takes a long time for an acknowledgement to be received (as in a satellite circuit), the effective throughput of data becomes very slow.

When you're trying to send one one-way signal to many locations, satellites often do a better and cheaper job.

Tesla, Nikola Nikola Tesla is regarded as one of the most mysterious and least recognized scientific pioneers in modern history. He is commonly associated with high-frequency electrical devices, radio transmission, and the invention of the multi-phase alternating current system in use today. Born on July 9, 1856 in Smiljan, Lika, Croatia, Yugoslavia, Tesla was the fourth of five children and son of a reverend of the Serbian Orthodox Church. Tesla was educated at the polytechnical school at Graz, Austria (1875), where he acquired an interest in the study of electrical engineering and mathematics. Tesla was dismayed at the inefficient design of DC motors. While working as a telegraph operator in Budapest (1881) and later as a telephone engineer for Edison in Paris (1882), he developed a multi-phase alternating current system, named the Tesla Polyphase System. Other engineers had attempted AC designs modeled after direct current systems with a single circuit, but none were ever successful. Tesla's multi-circuit system included new polyphase induction motors, dynamos, and transformers, the patents to which were purchased in 1888 by the Westinghouse Electric Company from the Tesla Electric Company.

Tesla's 60 cycle AC Polyphase System was the primary competitor to the Continental Edison Company's DC Current System, and caused Thomas Edison to personally wage a massive propaganda campaign against Tesla and the "dangers of alternating current". However, AC could travel over hundreds of miles and at much higher voltages, while DC traveled only much shorter distances and required a generator every two miles. Despite Edison's efforts, the war of the currents was won by Tesla. The Chicago World's Fair in 1893 featured the first electrically powered pavilion, designed entirely by Tesla using his polyphase system and financed by Westinghouse. This pavilion featured Tesla demonstrating many high-energy devices, including the wireless transmission of energy using a tuned circuit, a device that would come to be known as radio. Guglielmo Marconi used several of Tesla's patents in his 1901 "invention" of the wireless radio transmitter. Marconi's initial patent claims for this device were rejected based on Tesla's prior patents (645,576 and 649,621 granted in 1900) for the wireless transmission of energy. Marconi's later patent claims for a signal communications device were granted. Despite a lawsuit filed by Tesla in 1915 and other efforts, history has erroneously painted Marconi as the inventor of the radio. Some redemption to this injustice occurred when the Supreme Court ruled on June 21, 1943 in the case of Marconi Wireless Telegraph Company of America vs. the United States that Tesla's radio patents predated those owned by Marconi. In effect, this ruling declared Tesla the inventor of radio; it came five months after his death.

Tesla continued to experiment with high-frequency energy. He believed that energy in phase with the natural vibrations of thunderstorms and the Earth (7.68 Hz) could be broadcast to anyplace. Using this theory, he received backing from J.P. Morgan to build a world-wide telegraphy system. Tesla also believed that electrical power was present everywhere, in unlimited quantities, and free for the taking. Tapping into such a natural source of energy could replace all other fuels. The global wireless transmission of power, however, did not interest Morgan, who did not see the benefit in providing free elec-

tricity to humanity. Tesla created hundreds of inventions and improvements during the course of his life, including the telephone repeater, wireless communications, radio, antennas, ground connections, aerial ground circuits with inductance and capacitance, tuned circuits, emitters and receivers tuned to resonance, the electronic tube, fluorescent lighting, the electromechanical audio speaker, AND and OR logic gates, radio remote-control, robotics, radar, and diathermy.

Tesla's patented inventions includes a robotic submarine (613,809), vertical take-off and landing aircraft (1,655,114), disk (bladeless)turbine engine (1,329,559), ozone generator (568,177), electro-dynamic induction lamp (514,170), and superconduction (685,012). His more fantastic concepts include anti-gravity propulsion and a thought photography machine. The more than 700 patents Tesla was awarded during his lifetime represent only a fraction of his total number of inventions and discoveries. Tesla often didn't bother to patent many of his inventions (he received his last patent in 1928) and often failed to document his work. Tesla died on January 7, 1943 secluded in his New York City apartment at the age of 86. Tesla was intelligent, highly strung, neurotic, charismatic, germ-phobic, and always very well-dressed. Over his lifetime his financial backers were a diverse group including George Westinghouse, J.P. Morgan, and even Thomas Edison. His friends and supporters, an equally diverse group, included Albert Einstein, Samuel Clemens, and Eleanor Roosevelt. His manor an mystery as depicted by the press inspired the creation of evil comic book geniuses that did battle with Superman and Captain Marvel using a myriad of strange inventions and energy devices. Tesla's secret work on "death ray" energy devices fueled much of this, as did the U.S. government's confiscation of many of his inventions after his death.

Test And Validation Physical measurements taken to verify conclusions obtained from mathematical modeling and analysis.

Test Antenna An antenna of known performance characteristics used in determining transmission characteristics of equipment and associated propagation paths.

Test Bed A constant physical and electrical environment in which devices or programs are tested in order to measure their performance against requirements, benchmarks, or each other.

Test Board A switchboard equipped with testing apparatus.

Test Center Equipment for detecting and diagnosing faults and problems with communications lines and the equipment attached to them. If centralized, a facility where a network manager or technician can gain access to (almost) any circuit in a network for the purpose of running diagnostic testing. Also called a network control center.

Test Desk A desk equipped with equipment to test and repair subscriber lines. See also TEST CENTER.

Test Friendly Busy A test to see if a line is busy. The subscriber does not know the line is being tested. Such a test is usually performed by the operator if someone calling that number requests it. It used to be free. So did a lot of things in this world.

Test Repeatability This definition courtesy Steve Gladstone, author, "Testing Computer Telephony Systems," (available from 212-691-8215) says that a key component in any successful test program is the ability to repeat tests simply and quickly. If problems are found, tests must be rerun, both to help recreate and document the problem, as well as to verify the bug fix. Repeating tests can be very time consuming as many tests will change the state of either the test system or the computer telephony system under test. For example, checking that messages are correctly deleted means starting the test from a known state where messages are in a mailbox. When the test is completed usually there are fewer messages left than when the test started. Repeating the test may therefore necessitates reinitializing the computer telephony system to the state before the test started.

Test Set A telephone handset with extra electronics designed to test telephone circuits. Also called a butt set, since it typically hangs on the technician's tool belt near the wearer's butt... Well, that's one explanation. Another is that it's called a "butt" set because it allows the use to "butt in" to a conversation and listen to its quality, etc.

Test Shoe A device that is applied to a circuit at a distributing frame to gain test access to circuit conductors.

Test Tone A tone used to find trouble on phone lines. Also called installer's tone. A small box that runs on batteries and puts an RF tone on a pair of wires. If the technician can't find a pair of wires by color or binding post, they attach a tone at one end and use an inductive amplifier (also called a banana or probe) at the other end to find a beeping tone. A more technical explanation: A tone sent at a predetermined level and frequency through a transmission system to facilitate measurement and/or alignment of the gains and/or losses of devices in the transmission circuit.

Test, Friendly Busy A test to see if a line is busy. The subscriber does not know the line is being tested. Such a test is usually performed if someone calling that number requests it.

Tetrode A four-element vacuum tube, consisting of filament (or cathode), grid, screen grid and plate.

Text Transmitted characters which make up the body of a message.

Text Based Browser A browser that cannot handle hypermedia files.

Text Enriched The successor to MS-DOS text/richtext, is a simple text markup language for MIME that lets you mark up the document (using commands enclosed in angle brackets) without making the text unreadable to someone without the software to interpret it. See MIME.

Text File A file containing only letters, numbers, and symbols. A text file contains no formatting information (like bolding and underlining and type fonts and sizes), except possibly line feeds and carriage returns. A text file is an ASCII file. A text file can be read by every word processor and editor. A text file the lowest common denominator in the word processing world. I wrote this file with an editor called The Semware Editor, which produces only text files. I did this because this dictionary has to be sent to a Macintosh for "typesetting" and to a DEC for distribution on CD-ROM. And a text file is the form both easily recognize.

Text Telephone A machine that employs graphic communication in the transmission of coded signals through a wire or radio communication system. TT supersedes the term, "TDD" or "telecommunications device for the deaf."

Text-To-Speech Synthesis TTS. Technologies for converting textual (ASCII) information into synthetic speech output. Used in voice processing applications requiring production of broad, unrelated and unpredictable vocabularies, e.g., products in a catalog, names and addresses, etc. This technology is appropriate when system design constraints prevent the more efficient use of speech concatenation alone. See SPEECH CONCATENATION.

TFT Thin Film Transistor. A display technology which uses modern active matrix technology that operates by assigning a tiny transistor to each pixel, making it possible to control pix-

els independently of each other. TFT screens are very fast, have a high contrast ratio and a wide viewing area.

TFTP Trivial File Transfer Protocol. A simplified version of FTP that transfers files but does not provide password protection or user-directory capability. It is associated with the TCP/IP family of protocols. TFTP depends on the connectionless datagram delivery service, UDP.

TFTS Terrestrial Flight Telephone System.

TGB 1. Trunk Group Busy

2. Telecommunications Grounding Busbar

TGC Transmission Group Control in IBM's SNA.

TGW Trunk Group Warning.

TH Transmission Header, an SNA term.

THD+N Total Harmonic Distortion Plus Noise. A measure of the audio clarity of a voice system. The best measure is 0%., or close. Pretty bad is 25%. The average PC voice card is around 10%.

The Grand Alliance A consortium of seven organizations which produced a workable version of HDTV — High Definition TV. The companies are AT&T, General Instrument Corporation, The Massachusetts Institute of Technology, the David Sarnoff Research Center, Philips Consumer Electronics, Thompson Consumer Electronics and the Zenith Corporation.

The Open Group Formed in February 1996 through consolidation of the Open Software Foundation (OSF) and X/Open Company Ltd. (X/Open). The stated mission of The Open Group is to make multi-vendor open systems the preferred customer choice for the delivery of the right information to the right person at the right time. Specific goals include the following: 1) Enabling a rapid vendor response to customer requirements for open systems, 2) Innovative technology research, 3) Accelerated consensus-building around standards and technology, 4) Consensus among open systems vendors, and 5) Promoting a consistent open systems message. Examples of The Open Group's work include the X Window System specification, the Motif Toolkit API, the Distributed Computing Environment (DCE) and Network File System (NFS) specifications, and the Common Desktop Environment (CDE) specification. (www.opengroup.org)

The Phone Company Also known by the initials TPC. In the 1966 James Coburn movie "The President's Analyst", the evil worldwide conspiracy was run by TPC, which turned out to be The Phone Company. Some people believe the movie was not fiction.

Theoretical Midpoint TMP. The theoretical halfway point that divides an international private line circuit into its respective US and foreign halves. A US records carrier is responsible for the US portion of service and a foreign records carrier assumes responsibility for service to the foreign half.

Thermal Noise Noise created in an electronic circuit by movement and collision of electrons.

Thermionic Emission The emission of electrons or ions under the influence of heat, as in a vacuum tube cathode.

Thermistor A device made from mixtures of metal oxides that exhibits large negative coefficient of resistance changes as the temperature increases.

Thermocoule Two dissimilar wires joined together that generate a voltage proportional to temperature when their junction is heated.

Thermoplastic Material that will soften and distort from its formed shape when heated above a critical temperature peculiar to the material.

Thermoset A plastic material which is crosslinked by a

heating process known as curing. Once cured, thermosets cannot be reshaped.

Thermostat Thermostats are temperature-activated on/off switches that usually work on the 'bimetal' principle, in which the bimetal strip consists of two bonded layers of conductive metal with different coefficients of thermal expansion, thus causing the strip to bend in proportion to temperature and to make (or break) physical and electrical contact with a fixed switch contact at a specific temperature. In practice, the bimetal element may be in strip, coiled, or snap-action conical disco form, depending on the application, and the thermal 'trip' point may or may not be adjustable. Figures 8(b) and (c) show the symbols used to represent fixed and variable thermostats. A variety of thermostats are readily available, and can easily be used in automatic temperature control or danger-warning (fire or frost) applicatiosn. Their main disadvantage is that they suffer from hysteresis; typically, a good quality adjusted thermostat may close when the temperature rises to (say) 21 C, but not re-open again until it falls to 19.5 C.

THF Tremendously High Frequency.

THI Telephone Headset Integrator. A new form of headset manufacturer who will make headsets that do new tasks, like take a phone off hook without physically having to lift the receiver off the phone.

Thick Ethernet Cable Thick Ethernet cable is 0.4-inch diameter, 50-ohm coaxial cable. Thick Ethernet cable can be bought in pre-cut lengths, with standard N-Series male connectors installed on each end. It also is available in bulk cable without connectors. Any of the following types of connectors will work:

Belden 9880 or Belden 89889

Montrose CBL5688 or Montrose CBL5713

Malco 250-4315-0004 or Malco 250-4314-0003

Inmac 1784 or Inmac 1785

Thicknet Jargon used to describe thick Ethernet coaxial cable. See THICKWIRE.

Thickwire 0.4 inch diameter, 50-ohm, Ethernet IEEE 802.3 coaxial cable. See also THICK ETHERNET CABLE and THINWIRE.

Thin Client Clients are devices and software that request information — applications or files. A client is a fancy name for a PC or workstation which in connected to a network, such as a local area network (LAN), a company's Intranet or, the Internet. The client runs on a server which houses applications and/or files. Clients come in two varieties — Fat and Thin. A "thin client" is a relatively inexpensive workstation which is akin to a dumb terminal in a mainframe environment. The thin client, according to current definition, lacks a hard drive, modem, PCMCIA slot, CD-ROM drive, floppy drive, serial port, communications port. The thin client comprises a sealed unit with often no potential for enhancement, other than adding memory. However, it does contain RAM, a limited processing power and perhaps a burned-in chip with a program or two, perhaps its user interface. The bulk of the applications and the information it needs remain on the server. Hence, the thin client is totally dependent on the server. The advantage of a thin client is low TCO (Total Cost of Operation), including costs of acquisition, maintenance and support. The downside is that the thin client is totally reliant on the server, through the network. Should either the server or the network fail, the client effectively is rendered useless until the problem is resolved. Here's a definition of Thin Client, courtesy of Oracle Corporation, writing in early 1994: "The thin client modem stores and processes more data on the server, but keeps the user interface and application functions on the client device. Example: a television with a set-top box, Apple's Newton Personal Digital Assistant, or a low-end PC." In an Internet scenario, thin clients are known as NetPCs or Netstations. The NetPC is reliant on the server, which is provided by a service provider (e.g., America OnLine, CompuServe, or your ISP). In addition to providing some combination of content and Internet access, the service provider's server will provide your client NetPC with access to all necessary applications (e.g., word processing and spread sheet applications), will store all your personal files, will provide all significant processing power, and so on. In this Internet example, the NetPC differs from the standard thin client by virtue of the fact that it does contain a modem, a communications port and communications software, all of which are required for Internet access. See also CLIENT, CLIENT SERVER, CLIENT SERVER MODEL, FAT CLIENT, MAINFRAME SERVER and MEDIA SERVER.

Thin Computing See Thin Client

Thin Ethernet A coaxial (0.2-inch, RG58A/U 50-ohm) that uses a smaller diameter coaxial cable than standard thick Ethernet. Thin Ethernet is also called "Cheapernet" due to the lower cabling cost. Thin Ethernet systems tend to have transceivers on the network interface card, rather than in external boxes. PCs connect to the Thin Ethernet bus via a coaxial "T" connector. Thin Ethernet is now the most common Ethernet coaxial cable, though twisted pair is gaining. Thin Ethernet is also referred to as ThinNet, ThinWire or Cheapernet. See also 10BASE-T.

Thinnet Jargon used to describe thin Ethernet coaxial cable. Referred to ThinNet, ThinWire or Cheapernet.

Thinwire The 50-ohm coaxial cable listed in IEEE 802.3 specifications and used in some Ethernet local area network installations.

Third Party Call Any call charged to a number other than that of the origination or destination party. It's not a good idea to let your employees make third party calls to one or more of your phone numbers. Best to ask them to place the calls on their personal phone credit cards. This way, they will spend a modicum of time justifying their exorbitant phone calls.

Third Party Call Control A call comes into your desktop phone. You can transfer that call. When the phone call has left your desk, you can no longer control it. That is called First Party Call Control. If you were still able to control the call (and let's say, switch it elsewhere) that would be called Third Party Call Control. Some Computer Telephony links allow only first party call control. Some allow third party as well. If you control the switch — the PBX or the ACD — you will typically have Third Party Call Control. If you just control the desktop, you'll typically have only First Party Call Control. There is no such animal as Second Party Call Control.

Third Party Cookie See Cookie.

Third Wire Tap The activating of a telephone handset microphone by using a third wire, thus bypassing the hook switch.

Thread 1. A sequence of computing instructions that make up a program. A multi-threaded process can have multiple threads, each executing indepencently and each executing on separate processors. A multi-threaded program, if running on a computer with multiple processors, will run much faster than a single-threaded program running on a single processor machine. Windows NT was one of the first generally available multi-threaded PC operating systems.

2. A topic on a Usenet group; e.g., "Baby Bop's voice" as a subcategory, or topical thread, that readers may follow in the

alt.barney.die.die newsgroup.

3. A series of messages related to the same topic in a discussion group, such as an original post and related follow-ups. It makes senses to read an entire thread before contributing to it to avoid repeating something that may already have been contributed one or more times.

4. A process that is part of a larger process or program.

In its Window 95 resource manual, Microsoft defined thread as
1. An executable entity that belongs to a single process, comprising a program counter, a user-mode stack, a kernel-mode stack, and a set of register values. All threads in a process have equal access to the processor's address space, object handles, and other resources. In Windows 95, threads are implemented as objects.

2. A collection of electronic mail messages organized chronologically and hierarchically to reflect the flow of the discussion.

Threaded A method of presenting articles within a newsgroup in a way that shows which articles refer to which other ones.

Threaded Code Threaded code is also known as Threaded Pseudo Code, or threaded p-code. It was first popularized in a software language called Forth. Eventually Microsoft fixed Forth and changed it into Basic. See THREAD, THREADS and MULTITHREADING.

Threads Individual processes within a single application.

Threat Analysis Examination of all actions and events that might adversely affect a system, a network or an operation.

Three Finger Salute Ctrl Alt Delete.

Three Slot An obsolete pay phone that is identified by three separate coin slots.

Three-Tier A type of client/server architecture consisting of three well-defined and separate processes, each running on a different platform:
1. The user interface, which runs on the user's computer, also called the client.
2. The functional modules that process the data. This middle tier runs on a server. It is often called the application server.
3. A database management system (DBMS) that stores the data required by the middle tier. This tier runs on a second server called the database server.

The three-tier design has some advantages over traditional two-tier or single-tier designs. Its modularity makes it easier to change or replace one tier without affecting the other tiers. It's also better for load balancing.

Three-Way Calling A local phone company feature that allows a phone user to add another user to an existing conversation and have a three party conference call.

Three-Watt Booster Optional equipment for use with a cellular phone car-mounting kit that raises a portable phone's maximum transmission power from 0.6 watts to 3.0 watts.

Three-Way Conference Transfer A PBX feature. By depressing the switch hook, a user can dial another extension and either hang up and transfer the call, get information from the called party and then resume the first call or bridge all parties together for a three-way conference call.

Three-Way-Handshake The process whereby two protocol entities synchronize during connection establishment.

Threshold 1. The minimum value of a signal that can be detected by the system or sensor under consideration.
2. Automatic call distributors allow the definition of several different thresholds that pertain to different objectives of your organization. For instance, thresholds can be defined for the maximum length of time a customer's call should wait in queue, how long an agent should spend on each call, and how many accepting the overflow.

Threshold Of Pain 1. The present price of local telephone service.
2. Unbearable noise.

Through Dialing Allows the attendant on a phone system to select a trunk and pass dial tone to a restricted phone user so that user may directly dial an outside call.

Throughput The actual amount of useful and non-redundant information which is transmitted or processed. Throughput is the end result of a data call. It may only be a small part of what was pumped in at the other end. The relationship of what went in one end and what came out the other is a measure of the efficiency of that communications link — a function of cleanliness, speed, etc.

Thumbnail Describes the size of an image you frequently find on Web pages. Usually photo or picture archives will present a thumbnail version of its contents (makes the page load quicker) and when a user clicks on the small image a larger version will appear. Sometimes these links will be to a new page containing the larger graphic and other times right to the image directly.

Thunking A Microsoft Windows 95 term for the transformation between 16-bit and 32-bit formats, which is carried out by a separate layer in the VM. the fundamental idea is to make older 16-bit programs work better in newer 32-bit operating systems.

THZ Terahertz (10 to the 12th power hertz). See also SPECTRUM DESIGNATION OF FREQUENCY.

TIA 1. Telecommunications Industry Association. TIA represents the telecommunications industry in association with the EIA (Electronics Industry Association). TIA represents companies which provide communications materials, products, systems, distribution services and professional services in the U.S. and around the world. Activities include government relations, market support activities such as trade shows and trade missions, and standards development. TIA began as a group of equipment suppliers in the form of a committee of the USTA (United States Telephone Association), splitting off in 1979 to form the USTSA (United States Telecommunications Suppliers Association). In 1988, the USTSA merged with the EIA/ITG (Information and Telecommunications Technologies Group of the Electronic Industries Association). TIA now operates under the umbrella of the EIA as the TIA/EIA, and works in conjunction with USTA. TIA is accredited by ANSI (American National Standards Institute), and contributes voluntary standards to that body. In December, 1997, the MultiMedia Telecommunications Association (MMTA), the successor organization to NATA, announced that it had been officially combined into the Telecommunications Industry Association. www.mmta.org and www.tiaonline.org. See ACTAS and NATA. www.tiaonline.org and www.eia.org See also EIA/TIA-232-E.
2. Thanks In Advance.
3. The Internet Adapter, available from SoftAware, Inc. in Marina Del Ray, CA is software used to adapt a dial-up connection on Unix shell accounts.

TIA 568 Commercial Building Telecommunications Wiring Standard, July 91

TIA 568A Commercial Building Telecommunications Cabling Standard, October 1995

TIA 569 Commercial Building Standard for Telecommunications Pathways and Spaces, Oct 90

TIA 569A Commercial Building Standard for Telecommunications Pathways and Spaces, Aug 97

TIA 570 Residential and Light Commercial Telecommunications Wiring Standard, May 91

TIA 606 Administration Standard for the Telecommunications Infrastructure of Commercial Buildings.

TIA 607 Commercial Building Grounding and Bonding Requirements for Telecommunications.

TIA/EIA The US Telecommunications Industries Association and Electronics Industries Association, which have merged. Now just called the Telecommunications Industry Association. See TIA for a full explanation.

TIA/EIA IS-95 See TIA/EIA IS-96.

TIA/EIA IS-96 IS-96 is the Speech Service Option Standard for Wideband Spread Spectrum Digital Cellular System, a new vocoder standard. The standard supports IS-95, the North American spread spectrum digital standard based on Code Division Multiple Access (CDMA), which was published in July, 1993. The engineering effort to produce IS-96 was done in TIA Technical Subcommittee TR-45,5, Wideband Spread Spectrum Digital Technologies Standards. The specific vocoder described in IS-96 is a variable rate implementation, chosen because of its combination of high voice quality and low average transmission rate. The IS-96 vocoder provides variable vocoder rates depending on voice activity. This variability typically results in an average transmission rate of under 4 Kbps and yet provides for reasonably quality voice transmission. The IS-96 also provides a variable noise threshold which tracks and eliminates much of the background noise from the speaker's environment. You can get copies of both standards from Global Engineering Documents, 15 Inverness Way East, Englewood, CO 80112.

TIA PN-2416 Backbone Cabling Systems for Residential and Commercial Buildings

TIC 1. An AT&T term for a digital carrier facility used to transmit a DS-1 formatted digital signal at 3.152 Mbps.
2. Token-Ring Interface Coupler. An IBM device that allows a controller or processor to attach directly to a Token-Ring network. This is an optional part of several IBM terminal cluster controllers and front-end processors. See TIC CARD.
3. See TELEPHONY INTERFACE CONTROL.

TIC Card Token Ring Interface Coupler is the IBM name for a variety of token ring adapter cards used to connect IBM controllers to token ring LANs. See TIC.

Tick Tone Clicking noise heard on some PBX lines indicating that the digits dialed will shortly be repeated to the central office.

Ticker A one-way telex machine used to typically report stock or commodity prices. The machine prints on ticker tape, which is about one inch wide and perfect for throwing out windows at passing celebrities. Thus the term "Ticker Tape Parade."

Ticket A telephone industry term for a filled-out form, usually a form for billing someone for a call. There are all sorts of tickets, including ones on paper, ones on computer and ones automatically generated without human intervention.

TICL Temperature Induced Cable Loss. Pronounced "tickle." A phenomenon in which the performance of fiber optic cables is adversely affected by low temperatures. The adverse impact is a multi-dB attenuation. What's strange is that the attenuation (i.e. reduction is signal strength) is localized so that it looks like a splice on an OTDR (Optical Time Domain Reflectometer, a test and measurement device often used to check the accuracy of fusion splices and the location of fiber optic breakers.) Particularly affected are cables which have been in place for at least one summer and which operate at relatively long wavelengths. Also particularly affected are cables which do not have a strong coupling between the central members (fiber cores) and the buffer tubes (protective individual fiber sheaths). The issue is that the glass fiber (expands and) contracts with changes in temperature, and to a different degree than does the buffer tube. The result is that of changes in the geometry of the fiber. TICL is a bad thing. What's difficult is fixing it, because the attenuation loss typically goes away when temperature rises, so that diagnosis is difficult ("like chasing ghosts" to use a colorful phrase). TICL is most likely to be troublesome when upgrading an existing link from single wavelength, e.g. 1310nm traffic to DWDM 1550nm traffic.

TID Terminal Identification. Used for all National-1 ISDN services, a two digit number between 00 and 62 entered after the SPID.

TIE 1. Joining cables and/or wires together.
2. Time Interval Error.
3. Trusted Information Environment, an encryption scheme.

Tie Down Verb meaning to terminate a wire on a main, intermediate or satellite distribution frame.

Tie Line A dedicated circuit linking two points without having to dial the normal phone number. A tie line may be accessed by lifting a telephone handset or by pushing one, two or three buttons.

Tie Trunk A dedicated circuit linking two PBXs.

Tie Trunk Access Allows a phone system to handle tie lines which can be accessed either by dialing a trunk group access code or through the attendant.

Tier 1 Imagine a bunch of international Internet telephony carriers. Each one has POPs (Points of Presence) in several overseas cities. A POP consists of at least one PC containing some voice cards. When someone dials another country, their call goes across the Internet, reaches the PC at the distant POP. That PC recognizes the call as for that local city, grabs it, dials the local number and conferences the Internet call with a local dial-up call. This combination Internet/local phone call is theoretically cheaper than dialing directly across the world's telephone system. In order to provide seamless, cheap international calling over the Internet, you really need POPs in every major and minor city abroad. A number of Internet telephone companies have been banding together to create this international network. They're thinking about classifying themselves into various categories. For example a Tier 1 carrier would have over 50 POPs worldwide; have a network managed by a 7x24 NOC; have the ability to reroute and fall back to the PSTN if there is congestion or a hardware problem; have redundancy in terminating locations and have the ability to offer several levels of quality. Tier 2 might have fewer. It all hasn't been defined yet. Considering [restige and marketing issues, I bet there'll be more Tier 1 providers in coming years than Tier 2.

TIFF Tag Image File Format. TIFF provides a way of storing and exchanging digital image data. Aldus Corp., Microsoft Corp., and major scanner vendors developed TIFF to help link scanned images with the popular desktop publishing applications. It is now used for many different types of software applications ranging from medical imagery to fax modem data transfers, CAD programs, and 3D graphic packages. The current TIFF specification supports three main types of image data: Black and white data, halftones or dithered data, and grayscale data. Some wags think TIFF stands for "Took It From a FotograF." It doesn't.

TIES Time Independent Escape Sequence, a feature of

modems.

Tiger Team A group hired by an organization to defeat its own security system to learn its weaknesses.

Tight Buffer Fiber Optic Cables Tight-buffered fiber optic cables use aramid strength members inside the cable instead of gel filling, as is the case with loose-tube gel-filled fiber optic cables. One of the advantages of tight-buffered fiber optic cables having aramid strength members along every inch of the cable is that the cable can be hung vertically and the fibers are still protected for the entire length of the cable. This is not the case with loose-tube gel-filled fiber optic cables because, when they are hung vertically, all the gel filling settles to the bottom and the optical fibers are no longer protected. Tight-buffered fiber optic cables also have buffer coatings (up to 900 microns) over each optical fiber cladding for added environmental and mechanical protection, increased visibility, and ease of handling. Tight-buffered fiber optic cables can be used indoors and outdoors which allows one cable to be used instead of having to switch cable types at the building entrance. This is different from loose-tube gel-filled cables because the gel is flammable and the cable must be spliced to indoor flame-retardant cables for runs into buildings. Therefore, according to manufacturers, tight-buffered fiber optic cables reduce labor, equipment and materials cost while improving system performance and reliability. See also Aramid and Tight Jacket Buffer.

Tight Jacket Buffer A buffer construction which uses a direct extrusion of plastic over the basic fiber coating. This construction serves to protect the fiber from crushing and impact loads and to some extent from the microbending induced during cabling operations. See also LOOSE TUBE BUFFER.

Tightly Coupled Describing the interrelationship of processing units that share real storage, that are controlled by the same control program and that communicate directly with each other. Compare with loosely coupled.

Tightly Coupled CPUs Term used to describe multiple-processor computers in which several processors share the same memory and bus.

TIIAP See OTIA.

Tilde The tilde is the ~ sign, which you'll find on most keyboards. It looks like an arched eybrow. Microsoft uses it in DOS to truncate a long Windows file name. Thus c:\my documents in Windows becomes c:\mydocu~1 in DOS. Other programs use it in different ways.

Tile A surface segment of a furniture system panel, usually removable for access to cables or patch panels contained within the panel.

Tiling An unpleasant mosaic-like effect created by block-oriented video compression techniques like DCT (Discrete Cosine Transform), used in the JPEG (Joint Photographics Expert Group) standard. See DCT and JPEG.

TIM Teletyper Input Method. See TELETYPER INPUT METHOD.

Timbre The quality of tone distinctive to a particular voice.

Time-based Authoring Tool A multimedia creation tool that uses time as a metaphor for building a project. Generally, objects are set up to happen at a certain time in a project, rather than in a certain place.

Time Assignment Speech Interpolation TASI. A voice telephone technique whereby the actual presence of a speech signal activates circuit use. The result is clipping of the first bit of the speech, but more efficient use of the transmission facility. TASI is used on expensive circuits, such as long submarine cables. See TASI.

Time Congestion The time resources (outgoing trunks) are busy.

Time Divert To Attendant A system feature which automatically transfers a phone to the attendant if the phone has been left off-hook too long.

Time Diversity A method of transmission wherein a signal representing the same information is sent over the same channel at different times. Often used over systems subject to burst error conditions and with the spacing adjusted to be longer than an error burst.

Time Division Controller TDC. A device which commands functions, monitors status and connects channels of TDM cards.

Time Division Multiple Access TDMA. A technique originated in satellite communications to interweave multiple conversations into one transponder so as to appear to get simultaneous conversations. A variation on TASI. A technique now used in cellular and other wireless communications. See TDMA.

Time Division Multiplex TDM. A technique for transmitting a number of separate data, voice and/or video signals simultaneously over one communications medium by quickly interleaving a piece of each signal one after another. Here's our problem. We have to transport the freight of five manufacturers from Chicago to New York. Each manufacturer's freight will fit into 20 rail boxcars. We have three basic solutions. First, build five separate railway lines from Chicago to New York. Second, rent five engines and schlepp five complete trains to New York on one railway track. Or, third, join all the boxcars together into one train of 100 boxcars and run them on one track. The train might look like this: Engine, Boxcar from Producer A, Box Car from Producer B, Producer C, Producer D, Producer E, and then the order begins again...Boxcar from Producer A, Producer B...Moving one large train of 100 boxcars is likely to be cheaper and more efficient than moving five smaller trains each of 20 boxcars on five separate railway tracks. Time Division Multiplexing, thus, represents substantial savings over have five separate networks (five separate tracks) and sending five separate transmissions (five separate trains).

This is what Time Division Multiplexing is all about. And the analogy is perfect. Take one large train (fast communications channel) and interleave pieces (boxcars) from each conversation one after another. If you do this fast enough, you'll never notice you've broken the conversations apart, moved them separately, and then put them back together at the distant end. In TDM, you "sample" each voice conversation, interleave the samples, send them on their way, then reconstruct the several conversations at the other end. There are several ways to do the sampling. You can sample eight bits (one byte) of each conversation, or you can sample one bit. The former is called word interleaving; the latter bit interleaving. The basic goal of multiplexing — whether it be time division multiplexing, or any other form — is to save money, to cram more conversations (voice, data, video or facsimile) onto fewer phone lines. To substitute electronics for copper. See also the following three definitions.

Time Division Multiplexer TDM. A device which derives multiple channels on a single transmission facility by connecting bit streams one at a time at regular intervals. It interleaves bits or characters from each terminal or device using the time. See TIME DIVISION MULTIPLEX.

Time Division Signaling Signaling over a time division multiplex system in which all voice channels share a common signaling channel, with time division providing the separation between signaling channels. See SIGNALING SYSTEM 7.

Time Division Switching The connection of two circuits in a network by assigning them to the same time slot on a common time division switched bus.

Time Domain Reflectometer TDR. A device that measures network cable characteristics such as distance, impedance, levels of RFI/EMI, connector and terminator problems, and the presence of opens and shorts. It uses radar-like principles to determine the location of metallic circuit faults.

Time Guard Band A time interval left vacant on a channel to provide a margin of safety against interference in the time domain between sequential operations, such as detection, integration, differentiation, transmission, encoding, decoding, or switching.

Time Jitters Short-term variation or instability in the duration of a specified interval.

Time Marker A reference signal, often repeated periodically, enabling the correlation of specific events with a time scale. markers are used in some systems for establishing synchronization.

Time Multiplexed Switch The space switch of which the cross point settings are changed in each time slot.

Time Of Day Display The time and date displays on phones. Actually, it's very useful information. Sometimes it's not displayed on the operator's console. As a result, the operator may never know that every phone in the office is showing the wrong time and date.

Time Of Day Routing 1. This feature automatically changes access to certain types of lines at times when the lines change from being expensive to cheap, or vice versa. For example, it's cheaper to use WATS lines before 8:00 AM in the morning. A company has offices in New York and Los Angeles. It might be cheaper to route calls to Chicago in the morning over the tie lines to LA and then out the LA WATS lines to Chicago, than to go directly out the New York WATS lines. This is a way to allocate bandwidth for LAN traffic over corporate T-1 Networks. By programming T-1 multiplexers, customers can allocate the amount of T-1 bandwidth that can be used by voice, data, and LAN traffic on a time of day basis. For example, during the day, most of the T-1 bandwidth can be allocated for voice. At night, after employees go home, more bandwidth can be allocated to LAN and other computer data traffic so that file transfers can be done faster. This is particularly useful in IBM mainframe environments where large amounts of data needs to be transferred form remote offices/divisions to the headquarters.

Time Out In telecommunications and computer networks, an event which occurs at the end of a predetermined interval of time is called Time Out. For example if you lift the phone off the cradle and do not proceed to dial, after a certain number of seconds you will hear either a voice telling you to get on with it or a howling sound of some sort. Data networks have the same thing. Don't do anything for x minutes and the system will knock you off the air, i.e. hang up on you. In more technical terms, time out is the amount of time that hardware or software waits for an expected event before taking corrective action. In its most common form, time out is the amount of time an OCC or telephone system waits after your call goes through before it begins billing or timing the call. Also see ANSWER SUPERVISION.

Time Sharing A mode of operation that provides for the interleaving of two or more independent processes on one functional unit. Its most common use is the interleaved use of time on a computing system enabling two or more users to execute computer programs concurrently. Time sharing of

computer resources is now relatively obsolete. See also TIMESHARING below.

Time Sharing Computer System A computer system permitting usage by a number of subscribers, usually through data-communication subsystems. This is usually the case where the users have only dumb terminals that cannot process data by themselves the way a stand alone computer can. Computers are being joined together to deliver more computing power where it is most needed.

Time Slice In a multi tasking environment, each task is allotted a portion of the CPU's overall processing power. This portion is called a time-slice. And it's usually measured in milliseconds. The CPU switches between tasks, and those with higher priority receive more time-slices than lower-priority tasks. See TIME SLICING.

Time Slicing The term used to describe the dividing of a computer resource so multiple applications or tasks requesting the resource are allocated some amount of the resource's time. See TIME SLICE.

Time Slot 1. In time division multiplexing or switching, the slot belonging to a voice, data or video conversation. It can be occupied with conversation or left blank. But the slot is always present. You can tell the capacity of the switch or the transmission channel by figuring how many slots are present.
2. An SCSA term. The smallest switchable data unit on the SCbus or SCxbus Data Bus. A time slot consists of eight consecutive bits of data. One time slot is equivalent to a data path with a bandwidth of 64 Kbps. See S.100.

Time Space Time System TST. The most common form of switching matrix for small digital telephone exchanges in which a space switch is sandwiched between two time switches.

Time Switch A device incorporating a clock which arranges to switch equipment on or off at predetermined times.

Time Varying Media An SCSA definition. Time-varying media, such as audio data (as opposed to space-varying media, such as image data). See S.100.

Time Zone Calling The ability of a dialing system to start and stop calling at predetermined times to different time zones.

Timed Detection As a substitute for answer supervision, some long distance phone companies use call timing and estimate that a call is completed if the caller remains off-hook for 30 seconds or more. This is not necessarily accurate, of course. The caller might be holding, thinking the person is in the shower, out in the garden, etc. Little does the caller know he is now being charged to listen to ringing signals. A long distance phone company that is "equal accessed" doesn't have this problem. A long distance company that isn't equal accessed — one that you have to dial directly with a local call — might well have this problem. Rule: When in doubt, don't wait too long on the phone listening to endless ringing. Hang up. Count to ten. Then redial.

Timed Purge A feature of interactive voice response systems, especially fax-back systems. If the document isn't requested for x number of days or weeks or if the document ages to a certain point, the system automatically deletes the document.

Timed Recall Your PBX can be instructed to place a call at a designated time. When the time comes, your PBX rings your phone. When you answer your phone, the PBX places the call.

Timed Reminders At 20-second intervals, timed reminders will alert an attendant that a call is still waiting, a called line has not yet been answered or a call is still on hold.

Timed reminders can be made longer or shorter. They can alert attendants to all sorts of events and non-events.

Timeout Two computers are "talking." One (for any reason) to respond. The other computer will keep on trying for a certain amount of time, but will eventually "give up." This is called time-out. A time also happens in a single computer. If a device (e.g. a printer) is not performing a task or responding, timeout is the amount of time the computer should wait before detecting it as an error. That period is called time-out.

Times T A new, expanded dialing plan developed by the ITU-T. Times T increases the maximum number of dialed digits from the current 12 to 15, plus the three-digit international access code (country code).

Timesharing The use of one computer by many users at one time. Each user is typically sitting in front of a data terminal and connected to the master computer through communications lines — local or long distance. The user asks the computer to work on his task, whether it be a simple as looking up some stock prices, checking an airline reservation or doing some accounting calculations. It appears to each user as if he/she has a computer dedicated to his own task, but the computer is large and powerful, and is moving rapidly from one user's task to the next. Timesharing's advantages are twofold: 1. The user may find it cheaper to time share a computer than to buy his own. 2. The computer may have valuable and extensive information in it, which would be virtually impossible to duplicate or handle in many stand-alone computers. Timesharing was more popular when computers were more expensive.

Timeslot Management Channel TMC. A dedicated channel for sending control messages used to set up and tear down calls in a T-1 frame. In a GR-303 interface group, the primary TMC is usually in channel 24 of the first DS1, while a redundant TMC (if used) would be located in a different DS-1.

Timestamp A call center term. A background, relative time market placed on a data transaction element used for throughput and processing calculations. Can be used to determine total work time by placing one at the beginning and one at the end of a transaction.

Timing Jitter Deviation of clock recovery that can occur when a receiver attempts to recover clocking as well as data from the received signal. The clock recovery will deviate in a random fashion from the transitions of the received signal.

Timing Recovery The derivation of a timing signal from a received signal.

Timing Signal The output of a clock. A signal used to synchronize connected equipment.

Timing Slip A sudden timing delay change during high-speed digital transmission often caused by using T-1 carriers from different suppliers.

TINA Telecommunications Information Networking Architecture. A developing standard which is intended to resolve issues of integration between TMN (Telecommunications Management Network) and IN (Intelligent Network) standards and concepts. TINA focuses on the definition and validation of an open architecture for worldwide telecom services through a flexible software architecture for both end-user and network management services. See TINA-C, TMN and IN.

TINA-C Telecommunications Information Networking Architecture Consortium. An international, voluntary, not-for-profit organization of vendors and others for the purpose of promoting an open network architecture for the delivery and management of sophisticated services.

Tine Also known as a finger. An individual digital channel of a wireless rake receiver. A rake receiver can support a number of tines, which can be combined to form a stronger received signal.

Tinned Wire Copper wire coated with tin to make soldering easier.

Tinsel Wire A component of some phone line cord conductors. Tinsel wire is made by rolling copper into very thin, narrow rolls and then winding several strands of tinsel around a non-metallic core (a string) and then placing an insulating cover over the resulting conductor. A cord is then built up of two or more conductors encased in a plastic jacket. The essential reason for this type of construction is to obtain good cord flexibility and long life.

Tint Another name for hue.

Tip 1. The first wire in a pair of phone wires. The second wire is called the "ring" wire. The tip is the conductor in a telephone cable pair which is usually connected to positive side of a battery at the telephone company's central office. It is the phone industry's equivalent of Ground in a normal electrical circuit. See TIP & RING.
2. TIP. The Transaction Internet Protocol protocol ensures that multivendor transaction monitors will work with one another to complete transactions over the Internet (RFC 2371). TIP came from a joint Microsoft/Tandem effort. See Transaction Internet Protocol for a much longer explanation.

Tip & Ring An old fashioned way of saying "plus" and "minus," or ground and positive in electrical circuits. Tip and Ring are telephony terms. They derive their names from the operator's cordboard plug. The tip wire was connected to the tip of the plug, and the ring wire was connected to the slip ring around the jack. A third conductor on some jacks was called the sleeve. That's it. Nothing more sinister. Nothing more interesting. See TIP, RING & GROUND.

Tip Cable A small cable connecting terminals on a distributing frame to cable pairs in the cable vault.

Tip Conductor The first conductor of a customer line.

Tip Side That conductor of a circuit which is associated with the tip of a plug, or of a telephone circuit.

Tip, Ring, Ground The conductive paths between a central office and a phone. The tip and ring leads constitute the circuit that carries a balanced speech or a data signal. The ground path in combination with the conductor is used occasionally for signaling.

TIPHON Telephone and Internet Protocol Harmonization Over Networks. An ETSI (European Telecommunications Standards Institute) project to define the interactions between emerging Internet voice technologies will interact with PSTN, ISDN and GSM networks.

TIPI Telephone Industry Price Index.

TIQ Telrate International Quotations.

TIS Technical Information Sheets.

Titanic On December 21, 1993 Vice President, Al Gore, told the National Press Club in Washington, "There is a lot of romance surrounding the sinking of the Titanic 81 years ago. But when you strip the romance away, a tragic story emerges that tells us a lot about human beings — and telecommunications. Why did the ship that couldn't be sunk steam full speed into an ice field? For in the last few hours before the Titanic collided, other ships were sending messages like this one from the Mesaba: "Lat42N to 41.25 Long 49W to Long 50.30W. Saw much heavy pack ice and great number large icebergs also field ice." And why, when the Titanic operators sent distress signal after distress signal did so few ships respond?

The answer is that — as the investigations proved — the wireless business then was just that, a business. Operators had no obligation to remain on duty. They were to do what was profitable. When the day's work was done — often the lucrative transmissions from wealthy passengers — operators shut off their sets and went to sleep. In fact, when the last ice warnings were sent, the Titanic operators were too involved sending those private messages from wealthy passengers to take them. And when they sent the distress signals, operators on the other ships were in bed."

Titles In the language of multimedia, when an author sells what he or she has created, it is called a title. The encyclopedias, dictionaries, musical works, and games available on CD are all "titles." Someone authors the material, and sells it to users who can play it back but not change the content.

TJF Test Jack Frame.

TL Tie Line.

TL1 Transaction Language 1. A machine to machine communications language which is a subset of ITU-T's man machine language.

TLA Three Letter Acronym. A form and usage common to our acronym-happy industry.

TLB Test LoopBack. A CSU (Channel Service Unit) operating mode that loops the telco's T-1 transmission facility back towards itself at itself at the same time it loops the CPE back toward itself.

TLD Top Level Domain. See Domain.

TLDN A Temporary Local Directory Number is assigned by a visited wireless network's Mobile Switching Center to support call delivery to an idle roaming subscriber. This TLDN is used by the originating MSC (Mobile Switching Center) to establish a voice path to the serving MSC via existing interconnection protocols (i.e. SS7).

TLF Trunk Link Frame.

TLG A World Wide Web term. A Top Level Domain (TLG) is defined as the alphabetical address suffix, which identifies the nature of the organization. Currently, the central naming registry on the Internet is administered by the Internet Assigned Numbers Authority, (IANA), which includes the National Science Foundation (NSF), InterNIC, Network Solutions Inc., and the International Ad Hoc Committee (IAHC). Only a select few Top Level Domains currently can be registered, including .com (commercial), .edu (educational institution), .gov (government), .org (non-profit organization), .net (network provider) and .mil (military). In this case and for example, they will register the domain name "teleconnect.com" or "computertelephony.com." (They are different sub-domains, also known as secondary domains under the Top Level Domain.) What you put in front of your domain name is your concern. Everything that is "something.something@computertelephony.com" will come to the domain "computertelephony.com" for resolution (further routing, processing, etc.). At the time of this writing, additional TLGs are under consideration. see Domain. See URL and Web Address.

TLI Transport Layer Interface. TLI. An application program interface provided with UNIX System V Release 3.

TLP Transmission Level Point.

TLS Transparent LAN Service. You have a LAN in one office. Across town you have another office with another LAN. You go to your friendly local phone company and say "I want a telecommunications service that will let me send messages, mail and files, etc. between my two LANs." Bingo, they provide you a Transparent LAN Service, which is a high speed VPN (Virtual Private Network) service that hides the complexity associated with the wide area network. With transparent LAN service, a service provider interconnects a corporation's local area networks (LAN's) in such a way that the wide area is transparent to the end user. All of the end user's local area network appear to be on the same LAN, regardless of where those local area networks are physically located.

A loosely-defined high speed VPN (Virtual Private Network) service offering of various CAPs, LECs and IXCs, TLS provides for the interconnection of LANs over the MAN or WAN public data network (PDN). In other words, a TLS customer can establish direct Ethernet-to-Ethernet or Token Ring-to-Token Ring connectivity through the public data network without either the trouble or expense of converting to a Frame Relay, SMDS or ATM format through a router. In the case of traditional 10-Mbps Ethernet, you might be able to interconnect Ethernet sites directly through a 10-Mbps SONET link provided by the carrier. More than likely, however, your connection would be over a facility of much lesser bandwidth, such as xDSL (e.g., ADSL, ADSL Lite, SDSL or HDSL). Internally, the carrier likely will support connectivity through ATM, but that fact is transparent to you, the user. It looks like straight Ethernet to you, although generally at a slower speed. See CAP, IXN and LEC.

TM 1. Trouble Management. The responsibilities associated with receiving any network events that impact customer service whether they are generated via customer contact or from internal network elements. Trouble Management tracks all problems, groups them together (if possible), and relays trouble tickets for problem resolution. A mobile phone term.

2. Traffic Management. As an ATM term, traffic Management is the aspect of the traffic control and congestion control procedures for ATM. ATM layer traffic control refers to the set of actions taken by the network to avoid congestion conditions. ATM layer congestion control refers to the set of actions taken by the network to minimize the intensity, spread and duration of congestion. The following functions form a framework for managing and controlling traffic and congestion in ATM networks and may be used in appropriate combinations.
• Connection Admission Control
• Feedback Control
• Usage Parameter Control
• Priority Control
• Traffic Shaping
• Network Resource Management
• Frame Discard
• ABR Flow Control

TMA Telecommunication Managers Association.

TMAP A piece of software announced by Northern Telecom in the summer of 1994 and designed to map Windows Telephony commands to Novell Telephony Services (TSAPI) commands. Tmap runs on the local workstation (also called client) and translates TAPI commands into TSAPI commands that are sent through the local area network to the telephony file server and thence to the PBX, causing the PBX to dial, or conference or ring a phone, etc. The idea is that computer telephony software running on the user's desktop PC could be implemented as easily through a card in the PC or through the company PBX attached to the LAN — and the whole process would be seamless to the user.

TMC Timeslot Management Channel. A dedicated channel for sending control messages used to set up and tear down calls in a T-1 frame. In a GR-303 interface group, the primary TMC is usually in channel 24 of the first DS1, while a redundant TMC (if used) would be located in a different DS-1.

TMGB Telecommunications Main Grounding Busbar

TMN Telecommunications Management Network. A network management model defined in ITU-T recommendation M.30 and related recommendations, and intended to form a standard basis for management of advanced networks such as SDH (Synchronous Digital Hierarchy) for fiber optics in land lines and GSM (Global System for Mobile Communications) in the cellular world. TMN specifies a set of standard functions with standard interfaces, and makes use of a management network which is separate and distinct from the information transmission network. Further, standard network protocols such as the OSI CMIP (Open Systems Integration Common Management Information Protocol) are specified. Implementation of this concept involves the linking of all subject device elements to OMCs (Operation and Maintenance Centers) which, in turn, are linked together over a separate network. A centralization occurs to facilitate control, monitoring, and management of all devices in the communications network, which can include legacy systems as well as newer technologies. Operation systems functions include the full range of functions defined in the OSI model: Performance Management (PM), Fault Management (FM), Configuration Management (CM), Accounting Management (AM), and Security Management (SM).

A gentleman called James Keil who wrote his master's thesis at the University of Boulder, Interdisciplinary Telecommunications Program, on TMN compliant equipment, says that a "quick and dirty definition of TMN" would be "A network management standard which seeks to provide IT, business and network service management in a multi-domain environments (i.e. VPN, RBOC, Cellular providers)." Mr. Keil also says TMN fully implemented can retrieve resources from disparate networks like SNMP, through the use of managed objects or ANSI.1. TMN has much more functionality than SNMP. See also TINA.

TMP Test Management Protocol: As an ATM term, it is a protocol which is used in the test coordination procedures for a particular test suite.

TMR Trunk Mobile Radio. Another name for SMR (Specialized Mobile Radio). See SMR.

TMS 1. Time Multiplexed Switch. In the AT&T 5ESS switch CM, the TM provides switch paths between switching modules and passes control messages to and from the message switch, and functions as the hub for clock distribution to the switching modules.

2. TOPS Message Switch.

TMSI Temporary Mobile Station Identifier A mobile station identifier (MSID) sent over the air interface and is assigned dynamically by the network to the mobile station.

TN 1. Telephone Number.

2. Twisted Nematic. Most used display technology for calculators, watches and measuring equipment. TN uses liquid crystals sandwiched between two plates of glass with integrated transparent electrodes which can be made transparent and non-transparent by applying an electric current to them. See LCD.

TN3270 Delivery of a 3270 data stream via Telnet, provided as part of the TCP/IP protocol suite.

TNC A small connector used on coaxial cable, commonly used for cellular antennas, and some data and test equipment.

TNL Terminal Net Loss.

TNPP A protocol used to send paging messages from terminal to terminal on LANs and WANs over a wire circuit.

TNS Transit Network Selection: As an ATM term, it is a sig-

naling element that identifies a public carrier to which a connection setup should be routed.

TNSS Non-Synchronous test line provides for rapid testing of ringing, tripping and supervisory functions of toll completing trunks. This test line provides an operation test which is not as complete as the Synchronous test but which can be made more rapidly.

TOA/NPI Type Of Address/Numbering Plan Identifier.

Todd The king of stockbrokers, but the peasant of tennis players.

TOF Time Out Factor: As an ATM term, it is an ABR service parameter, TOF controls the maximum time permitted between sending forward RM-cells before a rate decrease is required. It is signaled as TOFF where TOF=TOFF+1. TOFF is a power of 2 in the range: 1/8 to 4,096.

TOFF Time Out Factor: See TOF.

Toggle 1. A flip-flop switch that changes for every input pulse.

2. Any simple two-position switch.

TOH Transport Overhead. A SONET term describing an element of signaling and control. TOH includes Section Overhead (SOH) and Line Overhead (LOH).

Token 1. In networking, a unique combination of bits used to confer transmit privileges to a computer on a local area network. It also carries important information for routing messages over the network, such as source and destination addresses, access control information, route control information, and date checking information. When a LAN-attached computer receives a token, it has been given permission to transmit. On a token ring network, the token is 24 bits long. See TOKEN PASSING and TOKEN RING.

2. Here is a Rolm definition: The floating master message which coordinates use of the CBX control packet network among the nodes connected to it.

Token Bus A local network access mechanism and topology in which all phones or workstations attached to the bus listen for a broadcast token or supervisory frame. That token confers on them the right to communicate over the share channel, the token bus. An example of a Token-Bus is IEEE 802.4. See TOKEN PASSING.

Token Latency The time it takes for a token to be passed around the local area network ring.

Token Passing A method whereby each device on a local area network receives and passes the right to use the single channel on the LAN. The key to remember is that a token passing, or token ring LAN has only one channel. It's a high-speed channel. It can move a lot of data. But it can only move one "conversation" at a time. The Token acts like a traffic cop. It confers the privilege to send a transmission. Tokens are special bit patterns or packets, usually several bits in length, which circulate from node to node when there is no message traffic. Possession of the token gives exclusive access to the network for transmission of a message. The token is generated by one device on the network. If that device is turned off or fails, another device will assume the token creation task. When the package of token and message reaches its destination, the computer copies the message. The package is then put back on the network where it continues to circulate until it returns to the source computer. The source computer then releases the token for the next computer in the sequence.

With token passing it is possible to give some computers more access to the token than others. Usually one device on the network is designated the token manager. It generates the token. If that device is turned off or fails, another device will assume management of the token. There is a complicated sequence of events that result in the generation of a token and that deal with the eventuality of token loss or destruction. The logic for this process is built into token ring cards that fit inside computers. In some manufacturers' products, the logic is slightly different and can cause incompatibilities. See TOKEN, TOKEN RING and TOKEN RING PACKET

Token Ring A ring type of local area network (LAN) in which a supervisory frame, or token, must be received by an attached terminal or workstation before that terminal or workstation can start transmitting. The workstation with the token then transmits and uses the entire bandwidth of whatever communications media the token ring network is using. A token ring is a baseband network. Token ring is the technique used by IBM, Arcnet, and others. A token ring LAN can be wired as a circle or a star, with all workstations wired to a central wiring center, or to multiple wiring centers. The most common wiring scheme is called a star-wired ring. In this configuration, each computer is wired directly to a device called a Multi-station Access Unit (MAU). These are usually grouped together in a wiring closet for convenience. The MAU is wired in such a way as to create a ring between the computers. If one of the computers is turned off or breaks or its cable to the MAU is broken, the MAU automatically recreates the ring without that computer. This gives token ring networks great flexibility, reliability, and ease of configuration and maintenance.

Despite the wiring, a token ring LAN always works logically as a circle, with the token passing around the circle from one workstation to another. The advantage of token ring LANs is that media faults (broken cable) can be fixed easily. It's easy to isolate them. Token rings are typically installed in centralized closets, with loops snaking to served workstations. Some other LANs require your going up in the ceiling or into walls and finding coax taps. All the work on a token ring can be done on one or several panels. These panels allow you to isolate workstations, and thus isolate faults.

Token Ring LANs can operate at transmission rates of either 4M bits per second or 16M bits per second. The number of computers that can be connected to a single Token Ring LAN is limited to 256. The typical installation is usually less than 100. Large installations connect multiple token ring LANS with bridges. The theoretical limit of Ethernet, measured in 64 byte packets, is 14,800 packets per second (PPS). By comparison, Token Ring is 30,000 and FDDI is 170,000. See FDDI-II and FDDI TERMS. Help on this definition courtesy Tad Witkowicz of Crosscomm, Marlboro, MA, Tim Becker, Lanquest Group, Santa Clara, CA and Elaine Jones, VP Marketing, Coral Network Corporation, Marlborough, MA. See also BRIDGE. IBM TOKEN RING, MAU, TOKEN PASSING, TOKEN RING, TOKEN RING CARD and TOKEN RING PACKET.

Token Ring Card Name given to the circuit board inserted into a computer device for connection to a token ring LAN. This board provides the physical connection to the LAN. It also participates in the collective management of the token by sending various messages to other token ring cards. Usually, one token ring card on the network is designated the token manager. It automatically generates a token as soon as it discovers one is missing, often with the help of other token ring cards. The sending of messages between token ring cards can be used to gather information about what is taking place on the network. Statistics may be collected. These may indicated that the network should be altered in some way to improve performance. This management capability is a distinct advantage of token ring LANs. One possible drawback is that vari-

ous manufacturers' token ring cards may differ slightly in how they implement token management, thereby making them incompatible in certain management features. Virtually all token ring cards will work together in basic token passing.

Token Ring Lan Service Unit The ATM TLSU provides a powerful tool for offering internetworking services over ATM networks. Emulated token rings consist of up to 64 TLSU token ring ports located anywhere in the ATM network, interconnected with PVCs. These emulated token ring networks can be completely isolated form one another to ensure security and fairness among the attached LANs. The TLSUs are designed for flexible deployment, either local to an ATM switch or at a remote site. See ATM ETHERNET LAN SERVICE UNIT.

Token Ring Packet Packets on a token ring network are made up of nine fields: starting delimiter, access control, frame control, destination address, source address, routing information, the data, frame check sequence, and ending delimiter.

Starting Delimiter (SD): This is an 8-bit binary (1s and 0s) sequence which marks the beginning of a data packet.

Access Control (AC) and Frame Control (AC): These are two 8-bit sequences that are used by the computers for maintenance purposes.

Destination Address (DA): This is a 48-bit sequence that uniquely identifies the physical name of the computer to which the data packet is being transmitted. Each computer on a ring examines this field to determine if the packet is for it.

Source Address (SA): This is a 48-bit sequence that uniquely identifies the physical name of the computer that send the data packet. This is used by the receiving computer to formulate its acknowledgement.

Routing Information (RI): This is a variable-length sequence used if the data packet is being sent to a computer located on another token ring LAN. (This information can make it impossible for some bridges to route some packets. See BRIDGE.)

Data: This is a variable-length sequence (up to 17,800 bytes) that is the actual data being sent from source to destination.

Frame Check Sequence (FCS): This 32-bit sequence is used to protect the contents of the packet from being corrupted during transmission. See FRAME CHECK SEQUENCE.

Ending Delimiter (ED): This is an 8-bit sequence that signals the end of a packet.

Token Tree LAN A type of local area network with a topology in the form of branches interconnected via active hubs. Using a token-passing scheme, the active hubs grant nodes access to the medium. See TOKEN RING.

Tokentalk The Apple Macintosh implementation of the Token Ring local area network.

Toll Call A call to any location outside the local service area. A long distance call.

Toll Center A Class Four central office.

Toll Connecting Trunk A trunk used to connect a Class 5 office (local central office) to the direct distance dialing network.

Toll Denial Permits phone user to make local calls but denies completion of toll calls or calls to the toll operator without the assistance of the attendant. See TOLL RESTRICTION.

Toll Diversion A system service feature by which users are denied the ability to place toll calls without the assistance of a human attendant. Toll diversion affects the entire switching system instead of discriminating between individual extensions.

Toll Fraud Theft of long distance service. Today's most common forms of toll fraud are DISA, voice mail and shoulder surfing. According to John Haugh of Telecommunications Advisors in Portland, OR, there are three distinct varieties of toll fraud:

"First Party" Toll Fraud, which is helped along by a member of the management or staff of a user. An example would be the telecommunications manager at the Human Resources department of New York City (an "insider") who sold his agency's internal code to the thieves, who in turn ran up unauthorized long distance charges exceeding $500,000.

"Second Party" Toll Fraud, which is facilitated by a staff member or subcontractor of a long distance carrier IXCs, vendor or local exchange telephone company selling the information to the actual thieves, or their "middlemen." An example would be a "back office clerk" working for one of these concerns who sells the codes to others.

"Third Party" Toll Fraud is facilitated by unrelated "strangers" who, though various artifices, either "hack" into a user's equipment and learn the codes and procedures, or obtain the needed information through some other source, to commit Toll Fraud.

Toll Office A central office used primarily for supervising and switching toll traffic.

Toll Quality Imagine you get a good long distance telephone voice connection. You can hear them and they can hear you — as though the two of you were in the same room. That's a "toll quality" phone call. Most "toll quality" phone calls are made circuit switched on the public switched network over fiber lines and through digital switches. When you make a phone call over the Internet or some other packet switched network that doesn't give you a dedicated phone line for the duration of your call, your conversation sounds pretty awful. It's not "toll quality." Some packet switched phone conversations do, however, sound pretty good. They're usually made over a "managed" packet network — one managed by one person (not by the free-for-all that's the Internet). Vendors of products that purport to deliver decent voice conversations over packet switched networks sometimes claim their conversations are "toll quality." That means they're pretty good. This is an untechnical definition. There really isn't a technical definition of what constitutes a good phone conversation. It's very subjective. Saying your equipment produces "toll quality" phone calls is good for sales, not much else. As usual, before you buy, speak and listen.

Toll Restriction To curb a telephone user's ability to make long distance calls. Toll restriction capability on modern PBXs and key telephone systems has been increasing in sophistication. Some PBXs now allow selective restriction based on specific extensions, users or geography. In other words, Joe Smith, the president, could call everywhere. John Doe in accounting might only be allowed to call Chicago and Houston, where our two factories are located. Mary Johnson, the seller for the western U.S., might only be allowed to call Denver and points west. There's considerable debate as to how useful toll restriction really is.

Toll Saver Feature Many answering machines — both PC based and stand alone machines — allow you, the owner, to dial in and remotely retrieve your messages. Because it makes no sense to incur toll call costs if there are no messages, many machines have "a toll saver feature." They will only answer the first message on the fourth ring. They answer each additional one on the second ring. This means if you're calling remotely, you can count the rings. If you get to three rings and the machine hasn't answered, you know that there are no new messages (i.e. ones you haven't heard) and you can safely hang up without incurring any toll costs.

Toll Station A Telco phone from which established long distance message rates are charged for all messages sent over company lines.

Toll Switching Trunk A trunk connecting one or more end offices to a toll center as the first stage of concentration for intertoll traffic. Operator assistance or participation may be an optional function. In U.S. common carrier telephony service, a toll center designated "Class 4C" is an office where assistance in completing incoming calls is provided in addition to other traffic; a toll center designated "Class 4P" is an office where operators handle only outbound calls, or where switching is performed without operator assistance.

Toll Testboard Manual test position at which toll circuits are tested and repaired.

Toll Terminal A phone only furnished with long distance service.

Toll Terminal Access Allows hotel/motel guest phones to access toll calling trunks.

Toll Ticket Ticket is the telephone company term for a bill. A toll ticket is a bill containing the calling number, called number, time of day, date and call duration. Some phone systems generate their own bills automatically. Some still need an operator. It depends on the equipment and the type of call.

Toll Trunk A communications channel between a toll office and a local central office.

Tone An audio signal consisting of one or more superimposed amplitude modulated frequencies with a distinct cadence and duration. See Tone Set and Tones.

Tone Alternator A motor-driven AC generator that produces audio-frequency tones.

Tone Dial What the Australians call tone dial, Americans call touchtone. Tone dial or touchtone dial makes a different sound (in fact, a combination of two tones) for each number pushed. The correct name for tone dial is "Dual Tone MultiFrequency" (DTMF). This is because each button generates two tones, one from a "high" group of frequencies — 1209, 1136, 1477 and 1633 Hz — and one from a "low" group of frequencies — 697, 770, 852 and 841 Hz. The frequencies and the keyboard, or tone dial, layout have been internationally standardized, but the tolerances on individual frequencies vary between countries. This makes it more difficult to take a touchtone phone overseas than a rotary phone. You can "dial" a number faster on a tone dial than on a rotary dial, but you make more mistakes on a tone dial and have to redial more often. Some people actually find rotary dials to be, on average, faster for them. The design of all tone dials is stupid. Deliberately so. They were deliberately designed to be the exact opposite (i.e. upside down) of the standard calculator pad, now incorporated into virtually all computer keyboards. The reason for the dumb phone design was to slow the user's dialing down to the speed Bell central offices of early touch tone vintage could take. Today, central offices can accept tone dialing at high speed. But sadly, no one in North America makes a phone with a sensible, calculator pad or computer keyboard dial. On some telephone/computer workstations you can dial using the calculator pad on the keyboard. This is a breakthrough. It is a lot faster to use this pad. The keys are larger, more sensibly laid out and can actually be touch-typed (like touch-typing on a keyboard.) Nobody, but nobody can "touch-type" a conventional telephone tone pad. A tone dial on a telephone can provide access to various special services and features — from ordering your groceries over the phone to inquiring into the prices of your (hopefully) rising stocks.

Tone Disabling A method of controlling the operation of communications equipment by transmitting a certain tone over the phone line.

Tone Diversity A method of Voice Frequency Telegraph (VFTG) Transmission wherein two channels of a 16-channel VFTG carry the same information. This is commonly achieved by twinning the channels of a 16-channel VFTG to provide eight channels with dual diversity.

Tone Generator A handheld device which puts a tone on a cable. The tone is picked up with an inductive amplifier at connection points or the other end of the cable. Slang for the tool is Toner. See Inductive Amplifier and Tone Probe.

Tone Probe A testing device used to detect signals from a tone generator to identify phone circuits, often the size of a fat pencil or skinny banana. Some models contain speakers; others must be used with a headset or a butt set. See also Tone Generator.

Tone Ringing Either a steady or oscillating electronic tone at the phone to tell you someone is calling.

Tone Sender 1. A printed circuit card in Rolm CBX which supplies the data bus with the digital representations of the following tones: dial, ring, busy, error, howler (off-hook time-out) and pulse (after flashing).
2. A printed circuit card which generates the following tones: dial, ring, busy, error, howler (off-hook timeout) and pulse (after flashing).

Tone Set A collection of tones which are customarily used as a set for the purposes of call setup and teardown (e.g., DTMF, R1 MF, R2 MF). In the case of DTMF, the tone set can also be used by the client application during the conversation portion of a call.

Tone Signaling The transmission of supervisory, address and alerting signals over a telephone circuit by means of tones. Typically inband. See also SIGNALING SYSTEM #7.

Tone To Dial Pulse Conversion Converts DTMF (Dual Tone Multiple Frequency) signals to dial pulse signals when trunks going to carry outgoing calls are not equipped to receive tone signals. A lot of electronic phones with touchtone dials have a sliding switch that allows you to choose whether the phone will outpulse in rotary, or whether it will touchtone out. You choose whichever your trunk line will accept.

Tone/Pulse Switchable Most phones in North America come with a pushbutton dial. Many of these phones have a switch that says "Tone/Pulse." By sliding the switch one way, the pushbutton pad will dial by sending out touchtones. By sliding the switch the other way, the pushbutton pad will dial by rotary pulses. See ROTARY DIAL.

Toner Tone generator used for identifying cable pairs.

Tones There are four basic tones which you will hear as you use the telephone. These tones are used to indicate what's going on. 1. Dial tone (also called dialing tone in Europe) is typically a continuous low frequency tone of around 33 Hz depending upon the telephone company. It indicates that the line is ready to receive dialing. 2. Busy Tone when the line or equipment is in use, engaged or occupied. This is typically 400 Hz 0.75 sec on and 0.75 sec off. 3. Ring Tone is typically 133 Hz make and break 0.4 sec On: 0.2 sec Off: Indicates called line is ringing out (17 Hz intermittent applied at called end to operate the telephone bell or buzzer). 4. Number Unobtainable continuous at 400 Hz indicates out of service or temporarily suspended. Tones vary considerably from country to country and between telephone companies.

Tonnage The unit of measurement used in air conditioning systems to describe the heating or cooling capacity of a system. One ton of heat represents the amount of heat needed to melt one ton (2000 lbs.) of ice in one hour. 12,000 Btu/hr equals one ton of heat. My office is on a 5,000 square foot floor. We use a ten ton air conditioner. It works most days. I wouldn't put more in. It would be a waste.

Tool In some computer languages, a small program executed as a shell command. In other computer languages, such as BASIC, it is called a "utility."

Toolbar A series of shortcut buttons providing quick access to commands. Usually located directly below the menu bar. Not all windows have a toolbar.

Toolkit A Dialogic word for an Applications Generator.

Toolkit Developer Program A strategic alignment by Dialogic with suppliers of voice processing applications development software to provide high-level application development tools.

Tone Dialing Same as touchtone dialing. See TOUCH-TONE.

TOP Technical Office Protocol. A version of the Open Systems Interconnection (OSI) model for use in the non-shop floor environment, i.e. in the office, developed by Boeing, the people who make the planes. TOP was designed from the outset to be compliant with the ISO OSI seven-layer model. Development has been merged with MAP, and the two functional profiles share a common integration strategy, and have a single (MAP/TOP) user group.

Top Down This is a method of distributing incoming calls to a bunch of people. It always starts at the top of a list of agents and proceeds down the list looking for an available agent. See also ROUND ROBIN and LONGEST AVAILABLE.

Top Level Domain A certain segment of a network in the Transmission Control Protocol/Internet (TCP/IP) UNIX environment. A network is segmented into a hierarchy of domains or groupings. In the Internet in the United States, there are six top-level domains: com (commercial organizations), edu (education organizations), gov (government agencies), mil (Military milnet hosts), net (networking organizations), and org (non-profit organizations). The next lower level relates to specific companies, and the level below to devices within a company.

Topology Network Topology. The configuration of a communication network. The physical topology is the way the network looks. LAN physical topologies include bus, ring and star. WAN physical topology may be meshed, with each network node directly connected to every other network node, or partially meshed. The logical topology describes the way the network works. For example, a 10Base-T LAN looks like a star, but works like a bus.

Topology Aggregation The process of summarizing and compressing topology information at a hierarchical level to be advertised at the level above.

Topology Attribute A generic term that refers to either a link attribute or a nodal attribute.

Topology Constraint An ATM term. A topology constraint is a generic term that refers to either a link constraint or a nodal constraint.

Topology Database As an ATM term, it is the database that describes the topology of the entire PNNI routing domain as seen by a node.

Topology Metric A generic term that refers to either a link metric or a nodal metric.

Topology State Parameter A generic term that refers to either a link parameter or a nodal parameter.

TOPS 1. Traffic Operator Position System. A specialized console designed for telephone company operators to help them complete toll calls.
2. A computer operating system, which originally stood for

the transcendental operating system.

3. The operating system used by Digital Equipment Corp.'s DECSYSTEM-10 and DECSYSTEM-20 computers. These computers have been discontinued, but many are still in use.

TOPS MPX Northern Telecom's Traffic Operator Position System designed on a token ring for interface between operator positions and the IBM Directory Assistance system database.

Torn Tape Relay An antiquated tape relay system in which the perforated tape is manually transferred by an operator to the appropriate outgoing transmitter. In short, it's a torn tape relay is a store and forward message switching system which uses punched paper as the storage medium.

Total Harmonic Distortion The ratio of the sum of the powers of all harmonic frequency signals (other than the fundamental) to the power of the fundamental frequency signal. This ratio is measured at the output of a device under specified conditions and is expressed in decibels.

Total Internal Reflection The reflection that occurs when light strikes an interface at an angle of incidence (with respect to the normal) greater than the critical angle.

Total Network Data System. TNDS. A telephone company term. The Total Network Data System is the overall data system for all types of switching equipment.

Total Transaction Call Processing A Rockwell term. Rockwell's philosophy. It guides their approach to call centers. It involves managing the success of a call center, not merely supplying the ACD (Automatic Call Distributor). It could include software development, CTI integration, network management, consulting services, IVR and voice processing systems. Rockwell says it will act as the prime contractor or as a single provider for a call center solution.

Tower A name for a PC in a vertical or upright case. Tower PCs (if they're correctly designed) have a big benefit. Heat rises and escapes more easily than in traditional horizontal machines. Heat and power surges are the most damaging threats to PC.

Touchtone Touchtone is not a trademark of AT&T, despite what editions one through six of Newton's Telecom Dictionary said. It is a generic term for pushbutton telephones and pushbutton telecommunications services and the term "touchtone" may be used by anyone. At one stage it was a trademark of AT&T. At divestiture in 1984, AT&T gave it to the public. And that's who owns it now. The public. If you don't believe this, call Frank L. Politano, AT&T Trademark and Copyright Counsel, in Basking Ridge, New Jersey. For a full explanation of touchtone, see DTMF, which stands for Dual Tone Multi Frequency signaling, i.e. touchtone.

Touchtone Type Ahead Also known as DTMF Cut-Through. Touchtone Typed Ahead is the ability of a voice response system to receive DTMF tones while the voice synthesizer is delivering information, i.e. during speech playback. This capability of DTMF cut-through saves the user waiting until the machine has played the whole message (which typically is a menu with options). The user can simply touchtone his response anytime during the message — when he first hears his selection number, when the message first starts, etc. When the voice processor hears the touchtoned selection (i.e. the DTMF cut-through), it stops speaking and jumps to the chosen selection. For example, the machine starts to say, "If you know the person you're calling, touchtone his extension in now." But before you hear the "If you know" you push button in 230, which you know is Joe's extension. Bingo, the message stops and Joe's extension starts ringing.

Tourists People who take training classes just to get a vacation from their jobs. "We have about three serious students in

the class. The rest are tourists."

TP Abbreviation for Transport Protocol or Twisted Pair.

TP1, TP2, TP3, TP4, TP5 The various service levels of the ISO IS 8073 Transport Protocol. TP4 is the most popular service level for information system networks and is specified in the U.S. government GOSIP architecture. TP4 stands for CSI Transport Protocol Class 4 (Error Detection and Recovery Class). This is the most powerful OSI Transport Protocol, useful on top of any type of network. TP4 is the OSI equivalent to TCP.

TP-4 Transport Protocol 4. An OSI layer-4 protocol developed by the National Bureau of Standards. See TP1.

TP-4/IP A term given to the ISO protocol suite that closely resembles TCP/IP.

TP-MIC Twisted-Pair Media Interface Connector: This refers to the connector jack at the end user or network equipment that receives the twisted pair plug.

TP-PMD Twisted-Pair Physical Media Dependent, Technology under review by the ANSI X3T9.5 working group that allow 100 Mbps transmission over twisted-pair cable. Also referred to as CDDI or TPDDI.

TPAD Terminal Packet Assembler/Disassembler linked to a cluster controller or terminal device, taking native protocol input and converting it to X.25 for transmission over a packet network.

TPC TOPS Position Controller.

TPCC Third Party Call Control: As an ATM term, it is a connection setup and management function that is executed from a third party that is not involved in the data flow.

TPDDI Twisted Pair Distributed Data Interface. Also known as ANSI X3T9.5.-TPDDI. TPDDI is a new technology that allows users to run the FDDI standard 100 Mbps transmission speed over twisted-pair wiring. Unshielded twisted-pair has been tested for distances over 50 meters (164 ft.). TPDDI is designed to help users make an earlier transition to 100 Mbps at the workstation. Also known as CDDI, Copper Distributed Data Interface.

TPDU Abbreviation for Transport Protocol Data Unit.

TPI Tracks Per Inch. A measurement of how much data can be stored on a disk.

TPON 1. Telephone Passive Optical Network. The portion of a telecommunications which uses a PON between the carrier switch and the subscriber.

2. OSI Transport Protocol Class 0 (Simple Class). This is the simplest OSI Transport Protocol, useful only on top of an X.25 network (or other network that does not lose or damage data).

TQFP Thin Quad Flat Pack, a format used in the design of PCMCIA devices. Another format is called PQFP, which stands for Plastic Quad Flat Pack.

TQM Total Quality Management. Doing what management should have been doing all along.

TR 1. Trouble Report.

2. Technical Reference.

3. Technical Requirement. These publications are the standard form of Bellcore-created technical documents representing Bellcore's view of proposed generic requirements and standards for products, new technologies, services, or interfaces. What's the difference between TRs and GRs? I asked Irvin Bingham, IBingham@carrieraccess.com. He replied: In the good old days, AT&T and Bellcore issued Technical Requirements (TRs) to dictate the way things would work in "their" phone system. After the Telecommunications Act of 1996, TRs relating to competitive products and services became General Requirements

(GRs). So in 1996, TR-303 became GR-303. Bellcore also set up GR "interest groups" to elicit industry participation in defining the specifications. Although Bellcore will listen to outside suggestions and opinions, I don't think Bellcore is obligated to act on any of them. Bellcore still has absolute control over their equipment interfaces to the outside world, but they are now required to publish those specifications and make them available to their competitors-for whatever price Bellcore wants to charge. An introduction to GR-303 is available online at http://www.bellcore.com/GR/gr303.html. Unfortunately, if you want a copy of any specification, you have to pay for it. After all, when did Bellcore ever give away anything to its competitors? Bellcore even charges exorbitant membership fees (called industry funding) for the privilege of participating in a GR interest group. See GR-303, TR-303 and also ISDN.

TR-303 A defacto standard published by Bellcore. It amounts to an industry standard high level control interface to dumb switches. It also applies to Fiber In The Loop (FITL). See GR-303 for the full explanation.

TR-444 A defacto standard published by Bellcore which spells out how the Bell regionals want long distance companies to connect to the Bell regionals' local networks. Several observers compare the TR-444 specs to simple direct dial long distance voice phone service.

TRA Telecommunications Resellers Association. TRA is a Washington, D.C. association that serves those involved in the resale of local, long distance, and wireless telecommunications services in North America. TRA was formed in 1992 through the merger of the Telecommunications Marketing Association and the Interexchange Resellers Association. The acts as a collective trade group for both switchless and facilities-based resellers. TRA holds several conferences and exhibitions each year, and acts as the resale industry's lobbying group and consumer watchdog. TRA says it has a strong code of ethics, and most member resellers are viewed as having reputable business standards. www.tra.org

TRAC Northern Telecom term for Technical Recommendations Approval Committee.

Trace Agent This is a command used in the Infoswitch product line to report all the events and transactions an agent has been involved in over a defined period of time.

Trace Block See trailer.

Trace Packet A special kind of packet in a packet-switching network which functions as a normal packet but causes a report of each stage of its progress to be sent to the network control center.

Trace Program A computer program that performs a check on another computer program by showing the sequence in which the instructions are executed and usually the results of executing the instructions.

Tracer Stripe When more than one color coding stripe is required, the first or widest stripe is the base stripe. The other, usually narrower stripes are the tracer stripes.

Traceroute Traceroute is software to help you figure out what's happening on your Internet connection. Traceroute is used to evaluate the hops taken from one end of a link to the other on a TCP/IP network, such as the Internet. Traceroute shows the full connection path between your site and another Internet address. It shows how many hops a packet requires to reach the host with the time required for the packet to get to each intermediate host or router. Traceroute is a more useful superset of PING.

Track 1. A storage channel on a disk or tape which can be

magnetically encoded.
2. On a data medium, a path associated with a single read/write head as data move past the head.
3. Every time you write to a CD, you will create at least one track, which is preceded by a pre-gap and followed by a post-gap. Any session may contain one or more tracks, and the tracks within a session may be of the same of or different types (for example, a mixed-mode disc contains data and audio tracks).

Track Access Time The time it takes to move the pickup head on a disk drive from one track to another.

Track Density The number of tracks per unit length, measured in a direction perpendicular to the tracks.

Track Speed The maximum speed which a train can travel over a section of railway tracks.

Trackball An upside-down MOUSE; a rotatable ball in a housing used to position the cursor and move images on a computer screen. A mouse needs desktop room to work, a trackball stays in one place, and can even be part of a keyboard or built into a laptop computer. It's hard to see why anyone uses a mouse instead of a trackball. This dictionary was typeset by a fine lady called Jennifer Cooper-Farrow, who used a trackball and a Macintosh computer.

Tracking 1. Figuring where a satellite is and keeping track of it. This is not an easy job, given the vastness of space.
2. The effect created in compressed video when the speed of the transmission is not great enough to keep up with the speed of the action. Tracking creates a tearing effect on the video picture.
3. A call center term. A software feature that models actual events and activities in your call center to aid you in short-term planning and evaluation of employee and call center performance. The tracking functions include employee information scheduling assignment, daily activity, and intra-day performance.

Tracon Terminal Radar Approach CONtrol. An installation in an airport or close to an airport from which approaching and departing aircraft are directed by people called controllers which sit in front of giant screens which show the movement of close aircraft. These controllers speak to the pilots in the planes instructing them where to move in order to avoid collisions and to land and takeoff safely.

Tractor Feeder A device which attaches to a computer printer and allows the printer to use continuous, sprocket-fed paper. Such paper has a row of evenly spaced holes on both sides. Those holes coincide with the pins on the tractor feeder. In all tractor-fed printers, the tractor moves the paper, not the printer's platen.

Trade Secret A trade secret can be any information, knowledge, data, or the like which is useful in business and not commonly known. A trade secret is anything from a customer list to the formula for Coke syrup. Enforcing a trade secret, for example, in order to enjoin a former employee from working for a competitor normally requires proof that the secret allegedly taken was suitably identified as such, that the employee was subject to a written contract including an obligation of confidentiality, and that physical access to the secret was suitably restricted.

Trademark A trademark can be any word, symbol, slogan, design, musical jingle, or the like capable of differentiating one party's goods or services from another's. The question is whether a member of the relevant segment of the public would be misled as to the source of the goods. Thus, a descriptive mark ("Frigidaire") is less powerful than a coined mark ("Xerox"), and the same mark can be used by different parties, if on differing goods ("Cadillac" for dog food versus "Cadillac" for automobiles.) The r symbol indicates that a mark has been registered by the Federal government, while the tm or sm symbols merely indicate that the user does not intend to waive his rights in the mark. It is not legally necessary to use the statements commonly seen that certain trademarks are the property of their owners, or to use the r symbol in text, but it does prevent any accusation of misappropriation.

Trader Turret A very large key telephone used by traders of commodities, securities, etc. Turrets typically have many line buttons. Each one corresponds to a trunk, an autodial or tie-line circuit to another trader or a financial institution. The objective of turrets is to allow the trader to be in instant communication with others who might want to buy or sell that which he is trying to sell or buy. See the June issue of TELECONNECT for an annual roundup of turrets.

Traffic Bellcore's definition: A flow of attempts, calls, and messages. My definition: The amount of activity during a given period of time over a circuit, line or group of lines, or the number of messages handled by a communications switch. There are many measures of "traffic." Typically it's so many minutes of voice conversation, or so many bits of data conversation. Note that Bellcore includes attempts in its definition of traffic. I don't. The decision is yours. But you should be aware of what you include in your calculations. See also TRAFFIC ENGINEERING and QUEUING THEORY.

Traffic Capacity The number of CCS (hundred call seconds) of conversation a switching system is designed to handle in one hour. This is the simple definition. See TRAFFIC ENGINEERING.

Traffic Carried See TRAFFIC OFFERED AND CARRIED.

Traffic Concentration The average ratio of the traffic during the busy hour to the total traffic during the day.

Traffic Data Administration System TDAS. A telephone company term. The TDAS program merges the data from various data acquisition systems and performs the following functions: a. The establishment of schedules for data collection; b. Maintenance of assignments records for all data collection devices; c. The acceptance of measurement data for any time interval. d. The reporting of measurement data to downstream processes via a set of standard interfaces. e. Performance of quality control reports specifically designed to permit effective management of the data collection effort. f. Adjustment and validation of measurement data.

Traffic Data To Customer The owner of a call accounting system can poll his PBXs daily or hourly and get traffic measurements, including peg counts, usage and overflow data. Summary reports, exception reports and complete traffic register outputs can be obtained.

Traffic Engineering The science of figuring how many trunks, how much switching equipment, how many phones, how much communications equipment you'll need to handle the telephone, voice, data, image and video traffic you're estimating. Traffic engineering suffers from several problems:
1. You are basing your future needs on past traffic.
2. Most traffic engineering is based on one or more mathematical formulas, all of which approach but never quite match the real world situation of an actual operating phone system. Computer simulation is the best method of predicting one's needs, but it's expensive in both computer and people time.
3. Many people in the telecommunications industry do not understand traffic engineering, have not worked with it sufficiently and make dumb and costly mistakes.
4. Since there are now several hundred long distance compa-

nies in the United States, and several thousand differently-priced ways of dialing between major cities, traffic engineering has become very complex.
After I wrote the above definition, Lee Goeller, disagreed with me and contributed this definition.
Traffic Engineering: The application of probability theory to estimating the number of servers required to meet the needs of an anticipated number of customers. In telephone work, the servers are often trunks, and the customers are telephone calls, assumed to arrive at random (see POISSON PROCESS). When arriving calls, upon finding all trunks busy, vanish, a "blocked call cleared" situation obtains (see ERLANG B). When a call stays in the system for a given length of time, whether it gets a trunk or not, "blocked calls held" applies (see POISSON DISTRIBUTION). If a call simply waits around until a trunk becomes available and then uses the trunk for a full holding time, the correct term is "blocked calls delayed" (See ERLANG C and QUEUING THEORY). Like any form of predicting the future on the basis of past behavior, traffic engineering has its limitations; however, when used by those who have taken the trouble to learn how it works, its track record is surprisingly good, and vastly better than most forms of simulation (see SIMULATION).

Traffic Intensity A measure of the average occupancy of a facility during a period of time, normally a busy hour, measured in traffic units (erlangs) and defined as the ratio of the time during which a facility is occupied continuously or cumulatively) to the time this facility is available. A traffic intensity of one traffic unit (one erlang) means continuous occupancy of a facility during the time period under consideration, regardless of whether or not information is transmitted. See also TRAFFIC ENGINEERING.

Traffic Load Total traffic carried by a trunk during a certain time interval.

Traffic Measurement Memory and other software in a telephone system which collect telephone traffic data such as number of attempted calls, number of completed calls and number of calls encountering a busy. The objective of traffic measurement is to enter the results into traffic engineering and so arrange one's incoming and outgoing trunks to get the best possible service. See TRAFFIC ENGINEERING.

Traffic Monitor PBX feature that provides basic statistics on the amount of traffic handled by the system.

Traffic Offered And Carried People pick up the phone and try to place their calls. This is "Traffic Offered" to the switch. The calls that get through the switch and onto lines is called "Traffic Carried." The difference between traffic offered and carried is the traffic that was lost or delayed because of congestion. There are two basic ways of measuring traffic — erlangs and CCS (or hundred call seconds).

Traffic Order TO. A telephone company term. These are requests originated by the Network Switching Engineering organization. The requests cover new systems or additions, removals and rearrangements to existing systems. The traffic order recommends types, quantities, and arrangements of local and toll equipment in accordance with the latest forecasts of trunks, network access lines and traffic studies.

Traffic Overflow Occurs when traffic flow exceeds the capacity of a particular trunk group and flows over to another trunk group.

Traffic Path A path over which individual communications pass in sequence.

Traffic Recorder A device which measures traffic activity on a transmission channel. It's a recorder, not a processor.

It's dumb.

Traffic Register A software area which records occurrences within a central office, such as peg count, overflow, all trunks busy, etc.. The types of occurrences measured vary widely according to the type of system.

Traffic Sensitive A telephone company term. Applies to equipment whose ability to provide a specific level of service varies as the calling load varies.

Traffic Shaping A generalized term for a congestion control management procedure in which data traffic is regulated in order that it conform to a specified, desirable behavior pattern. Such a behavior pattern may include reduction or elimination of excessive traffic bursts from a LAN as it is presented to a Frame Relay WAN through a router. Such bursts may exceed the CIR (Committed Information Rate) and, therefore, be marked DE (Discard Eligible). During periods of Frame Relay WAN congestion, such bursts may result in discarded frames, which require retransmission. Should the excessively long bursts be transmitted successfully across the Frame Relay WAN, surcharges may apply (such surcharges are unusual for U.S. carriers). All things considered, traffic shaping may be the best approach in such a scenario. In an ATM LAN environment, traffic shaping responsibility can be accomplished by the ATM switch, which actively would alter the traffic characteristics of a cell stream on a VCC (Virtual Channel Connection) or VPC (Virtual Path Connection). This procedure may serve to reduce the peak cell rate, limit the burst length, or minimize the cell delay variation by re-spacing the cells in time in order that traffic flow not congest the switch. This can be particularly important when dealing with long bursts of high priority traffic, as such traffic literally can bring the rest of the user traffic flow to its knees. See also ATM, Committed Information Rate, Discard Eligible, Frame Relay, VCC and CPC.

Traffic Table A computer database into which a PBX enters a count of feature activity. Certain detected operating errors are also entered in the traffic table.

Traffic Usage Recorder A device for measuring and recording the amount of telephone traffic carried by a group, or several groups, of switches or trunks.

Traffic Use Code A telephone company definition. A system standard two character alpha code designating the type of traffic offered to a trunk group. Traffic Use Codes are listed and defined in Section 795-400-100 (Common Language Circuit Identification — Message Trunks).

Trail As an ATM term, it is an entity that transfers information provided by a client layer network between access points in a server layer network. The transported information is monitored at the termination points.

Trailer 1. A nonstandard way of standard way of sending data. Trailers are used on some networks by 4BSD UNIX and some of its derivatives.

2. A block of controlling information transmitted at the end of a message to trace error impacts and missing blocks. Also referred to as a trace block.

Train The creation of word reference data by presenting words to a recognizer. A voice recognition term.

Training A feature of some modems which adjust to the conditions including amplitude response, delay distortions, timing recovery, and echo characteristic, of a particular telecommunications connection by a receiving modem. See TRAINING UP.

Training up A technique that adjusts modems to current telephone line conditions. The transmitting modem sends a special training sequence to the receiving modem, which makes necessary adjustments for line conditions.

Transaction 1. It is a completed event that can be assembled in chronological sequence for an audit trail.
2. An entry or an update in a database.

Transaction Capabilities Function that controls non-circuit-related information transfer between two or more nodes via a SS7 signaling network.

Transaction Capabilities Application Part TCAP. The application layer protocol of SS7. Transaction capabilities in the SS7 protocol are functions that control non-circuit related information transferred between two or more signaling nodes. Definition from Bellcore in reference to its concept of the Advanced Intelligent Network.

Transaction Detail The detail of a transaction record.

Transaction Internet Protocol TIP. The Transaction Internet Protocol protocol ensures that multivendor transaction monitors will work with one another to complete transactions over the Internet (RFC 2371). TIP came from a joint Microsoft/Tandem effort. I excerpted the following from a Microsoft Market Bulletin.

Two companies (Microsoft and Tandem) team have combined to publish a specification for a two-phase commit protocol to make it easier for businesses to do transaction processing across the Internet. Two-phase commit is the commonly-used application protocol used by high-end system software — including Transaction Processing (TP) Monitors and databases — to coordinate the work of multiple applications on different computers as a single unit, or transaction. Businesses want to link existing transaction processing systems together across the Internet using two-phase commit protocols, but existing implementations of two-phase commit are too complex for use on the Internet. TIP is designed to solve this problem, defining a simple protocol that existing vendors of TP Monitors and databases can easily implement into their products, solving the problem of transaction coordination across the Internet. Microsoft will implement TIP in the Distributed Transaction Coordinator (DTC), Microsoft's transaction manager that first shipped with SQL Server 6.5. DTC currently supports other open two-phase commit protocols, including OLE Transactions, the X/Open's XA protocol, and has future plans to support SNA LU 6.2 Sync Level 2. Windows NT Server 5.0 will provide native support for TIP. Tandem will support TIP in its NonStop systems. Both the reference implementation and the TIP specification can be downloaded directly from www.microsoft.com/pdc or www.tandem.com/menu_pgs/svwr_pgs/svwrnews.htm. Microsoft and Tandem have submitted the TIP specification to the Internet Engineering Task Force, who have published it at http://ds.internic.net/internet-drafts/draft-lyon-itp-nodes-00.txt.

Transaction File A collection of transaction records. A transaction data entry program allows for the creation of new transaction files used to update the data base.

Transaction Link Rockwell's link from its Galaxy ACD to an external computer. See OPEN APPLICATION INTERFACE.

Transaction Tracking Your software keeps track of each transaction as it happens. And if a component of your network fails, your transaction tracking software backs out of the incomplete transaction. This allows you to maintain your database's integrity. You may, however, lose the single transaction you were working on when your network got sick.

Transactional Integrity A term that describes how your computing/telecom system handles making sure that the transaction you just made is solid and clean and that the next time you want to get to the results of the transaction you can. "Transactional integrity" becomes critical when you're storing bits and pieces of your transactions on different media, in different places. For example, you might want to store your data on a magnetic hard drive and your associated images on a separate optical drive.

Transborder Data Flow TDF. Transborder data flows are movements of machine-readable data across international boundaries. TDF legislation began in the 1970s and has been put into effect by many countries in an attempt to protect personal privacy of citizens. This term has particular meaning as it relates to electronic commerce or EDI and is becoming more and more relevant with the use of the Internet as a means to conduct global business.

Transceiver 1. Any device that transmits and receives. In sending and receiving information, it often provides data packet collision detection as well.
2. In IEEE 802.3 networks, the attachment hardware connecting the controller interface to the transmission cable. The transceiver contains the carrier-sense logic, the transmit/receive logic, and the collision-detect logic.
3. A device to connect workstations to standard thick Ethernet-style (IEEE 802.3).

Transceiver Cable In local area networks, a cable that connects a network device such as a computer to a physical medium such as an Ethernet network. A transceiver cable is also called drop cable because it runs from a network node to a transceiver (a transmit / receiver) attached to the trunk cable. See TRANSCEIVER.

Transcoder A device that combines two 1.544 megabit per second bit streams into a single 1.544 megabit per second bit stream to enable transmission of 44 or 48 voice conversations over a DS-1 medium.

Transcoding A procedure for modifying a stream of data carried so that it may be carried via a different type of network. For example, transcoding allows H.320 video encoding, carried via circuit switched TDM systems to be converted to H.323 so that it can connect with and be transmitted across packet switched ethernet LAN.

Transcriptionist A person who listens to a tape recording and types the words he hears. The word, transcriptionist, derives from the verb to transcribe. The most common employment of transcribers is in the medical industry, where busy doctors talk into tape recorders telling good and bad news of their patients. And even busier transcriptionists type those words into the patient's medical records, or whatever.

Transducer A device which converts one form of energy into another. The diaphragm in the telephone receiver and the carbon microphone in the transmitter are transducers. They change variations in sound pressure (your voice) to variations in electricity, and vice versa. Another transducer is the interface between a computer, which produces electron-based signals, and a fiber-optic transmission medium, which handles photon-based signals.

Transfer A telephone system feature which provides the ability to move a call from one extension to another. It is probably the most commonly used and misused feature on a PBX. Before you buy a PBX, check out how easy it is to transfer a call. If you have a single line phone, you should simply hit the touch hook, hear a dial tone and then dial the chosen extension number and hang up. This sounds easy in principle, but many people find it difficult since they associate the touch hook with hanging up the phone. Some companies have got-

ten around this by putting a "hook flash" button on the phone itself. Such a button is like having an autodial button which just makes the exact short tone you make when you quickly hit the hook flash button. An even better solution is an electronic phone with a button specially marked "transfer," or a button next to a screen which lights up "transfer." Failing to efficiently transfer a call is the easiest way to give your customers the wrong impression of your firm. Think of how many times have you called a company only to be told it wasn't the fellow's job and he will transfer the call, but "If we get cut off, please call Joe back on extension 2358." There are typically four types of Transfer: Transfer using Hold, Transfer using Conference, and Transfer with and without Announcement.

Transfer Callback A phone system feature. After a specified number of rings, an unanswered transferred call will return to the telephone which originally made the transfer.

Transfer Delay A characteristic of system performance that expresses the time delay in processing information through a data transmission system.

Transfer Mode A fundamental element of a communications protocol, transfer mode refers to the functioning arrangement between transmitting and receiving devices across a network. There are two basic transfer modes: connection-oriented and connectionless. Connection-oriented network protocols require that a call be set up before the data transmission begins, and that the call subsequently be torn down. Further, all data are considered to be part of a data stream. Examples of connection-oriented protocols include analog circuit-switched voice and data, ISDN, X.25 and ATM. Connectionless protocols, on the other hand, do not depend on such a process. Rather, the transmitting device gains access to the transmission medium and begins to transmit data address to the receiver, without setting up a logical connection across the physical network. LANs (e.g. Ethernet and Token Ring) make use of connectionless protocols, as does SMDS, which actually is an extension of the LAN concept across a MAN (Metropolitan Area Network). For more detail, see CONNECTION ORIENTED and CONNECTIONLESS MODE TRANSMISSION.

Transfer Protocols Protocols are all of the packaging" that surround actual user data to tell the network devices where to send the data, who it comes from, and how to tell if it arrived. Transfer protocols are designed for the efficient moving of larger chunks of user data.

Transfer Rate The speed of data transfer — in bits, bytes or characters per second — between devices.

Transfer Time A power backup term. Transfer time can refer to either the speed to which an off-line UPS transfers from utility power to battery power, or to the speed with which an on-line UPS switches from the inverter to utility power in the event of an inverter failure. In either case the time involved must be shorter than the length of time that the computer's switching power supply has enough energy to maintain adequate output voltage. this hold-up time may range from eight to 16 milliseconds, depending on the point in the power supply's recharging cycle that the power outage occurs, and the amount of energy storage capacitance within the power supply. A transfer time of 4ms is most desirable , however, it should be noted that an oversensitive unit may make unnecessary power transfers.

Transformer Transformers are devices that change electrical current from one voltage to another. A step-up transformer increases the voltage and a step-down transformer decreases voltage. The power of an electric current must be conserved so just as voltage is increased, current is decreased. Transformers work by feeding an alternating current into a primary coil. The primary coil induces a magnetic field in a secondary coil which is connected to an energy using load. The difference between the number of coils in the primary coil versus the secondary coil determines whether the voltage will be stepped up or down. One reason for using a transformer is that commercial power is typically 120 or 240 volts while many phone systems (and other computer-type "things") work best on 48, 24 or lower voltage.

Transhybrid Loss The transmission loss between opposite ports of a hybrid network, that is between the two ports of the four-wire connection.

Transient Any high-speed, short duration increase or decrease impairment that is superimposed on a circuit. Transients can interrupt or halt data exchange on a network. See HIT.

Transient Mobile Unit A mobile unit communicating through a foreign base station.

Transit Delay 1. In ISDN, the elapsed time between the moment that the first bit of a unit of data (such as a frame) passes a given point and the moment that bit passes another given point plus the transmission time of that data unit.
2. As an ATM term, it is the time difference between the instant at which the first bit of a PDU crosses one designated boundary and the instant at which the last bit of the same PDU crosses a second designated boundary.

Transit Exchange The European equivalent of a tandem exchange.

Transit Timing A method of eliminating looping between nodes used in the network layer of some packet-switched systems. This method is used in the Internet Protocol (IP) portion of Transmission Control Protocol/Internet Protocol (TCP/IP).

Transition Probabilities Probabilities of moving from one state to another.

Transition Zone The zone between the far end of the near-field region and the near end of the far-field region. The transition is gradual.

Transistor The transistor was invented in 1947 by John Bardeen, Walter H. Brattain and William Shockley of Bell Laboratories. The first transistor comprised a paper clip, two slivers of gold, and a piece of germanium on a crystal plate. Here is an explanation of how a transistor works, taken from "Signals, The Science of Telecommunications" by John Pierce and Michael Noll:
"To understand how a transistor works, we must look at the laws of quantum mechanics. We commonly picture an atom as a positive nucleus surrounded by orbiting electrons ... Vacuum tubes rely on the ability of electrons to travel freely with any energy through a vacuum. Transistors rely on the free travel of electrons through crystalline solids called semiconductors ... Semiconductors (such as silicon or gallium arsenide) differ from pure conductors, such as metals, in how full of electrons are the energy bands that allow free travel." Depending on their design, transistors can act as amplifiers or switches. See Transistor Milestones and Transistor Radio.

Transistor	Milestones
Point-contact transistor	1948
Single-crystal Germanium	1950
Grown junction transistor	1951
Alloy junction transistor	1952
Zone melting and refining	1952
Single-crystal Silicon	1952
Diffused-base transistor	1955

Oxide masking	1957
Planar transistor	1960
MOS transistor	1960
Epitaxial transistor	1960
Integrated circuits	1961

Transistor Radio Sony unveiled the first transistor radio in 1955. See Sony for the fascinating story.

Translate To change the digits dialed on your phone into digits necessary for routing the call across the country. See TRANSLATIONS.

Translating Bridge A special bridge that interconnects different LAN types using different protocols at the physical and data link layers, such as Ethernet and Token Ring. A translating bridge supports the physical and data link protocols of both LAN types. When they forward packets from one LAN to another, they manipulate the packet envelope to conform to the physical and data link protocols of the destination LAN. For a longer explanation, see Bridge.

Translation The interpretation by a switching system of all or part of a destination code to determine the routing of a call. See TRANSLATIONS.

Translations Here is a definition from Bellcore, who works with the telephone industry: Translations is the changing of information from one form to another. Example: In common control switchintg systems employing digit storage devices and decoding devices, the dialed digits are stored in a receiver or a tone decoder. The receiver/decoder translates the dialed digits data appropriate for the completion of the call and passes to a processor. With the advent of stored program control, as exemplified in a IA ESS, 5ESS-2000, DMS-100 systems, the translation function has been greatly expanded. When a customer originates a call, for example, the system needs to know if the line is denied outgoing service, if the line is being observed, what the line class is, what special equipment features it has, etc. The line equipment number is given to the translation program as an input. The translation program performs a translation and returns the answers to these questions in a coded form suitable for use by the central processor. The important thing to remember in considering the translation function in the stored program switches is the translation function is employed many times throughout the process of a call and the interplay between the translation programs and other programs is frequent.
Here's my definition: Translations are changes made by the network to dialed telephone numbers to allow the call to progress through the network. Sometimes the translations are made automatically. Take one series of dialed numbers; convert them to another. Sometimes, translations are done with the help of "look up" tables, also called databases. Here's an example of translations done with the help of a database. TELECONNECT Magazine has a WATS line, 1-800-LIBRARY. If you dial it on the phone, you'll see it is really 1-800-542-7279. But this is not its real number. When someone in California dials 1-800-LIBRARY, MCI's long distance network recognizes the "1-800" portion of the call and sends it to a special central office somewhere out west. When the call arrives, a computer looks up the number 800-542-7279 in its database and translates that to 1-212-691-8215 and puts the call back into the network. Within seconds, that number in New York, 212-691-8215 rings.

Transliterate To convert the characters of one alphabet to the corresponding characters of another alphabet.

Translator 1. A communications device that receives signals in one form, normally in analog form at a specific frequency, and retransmits them in a different form.

2. A device that converts information from one system into equivalent information in another system.

3. In telephone equipment, it is the device that converts dialed digits into call-routing information.

4. In computers, it is a program that translates from one language into another language and in particular from one programming language into another programming language.

5. In FM and TV broadcasting, it's a repeater station that receives a primary station's signal, amplifies it, shifts it in frequency, and rebroadcasts it.

Transliterate To convert the characters of one alphabet to the corresponding characters of another alphabet.

Transmission Sending electrical signals carrying information over a line to a destination. Bellcore says that transmission has the following definitions: (a) Designates a field work, such as equipment development, system design, planning, or engineering, in which electrical communication technology is used to create systems to carry information over a distance. (b) Refers to the process of sending information from one point to another. (c) Used with a modifier to describe the quality of a telephone connection: good, fair, or poor transmission. (d) refers to the transfer characteristic of a channel or network in general or, more specifically, to the amplitude transfer characteristic. You may sometimes hear the phrase, "transmission as a function of frequency."

Transmission Block A group of bits or characters transmitted as a unit, with an encoding procedure for error control purposes.

Transmission Channel All of the transmission facilities between the input (to the channel) from an initiating node and the output (from the channel) to a terminating node. In telephony, transmission channels may be of various bandwidths: e.g. nominal 3-kHz, nominal 4-kHz, or nominal 48-kHz (group). "Transmission channel" should not be confused with the more general term "channel."

Transmission Code A code by which information is sent and received on a transmission system.

Transmission Coefficient The ratio of the transmitted field strength to the incident field strength when an electromagnetic wave is incident upon an interface surface between media with two different refractive indices. In a transmission line, the ratio of the complex amplitude of the transmitted wave to that of the incident wave at a discontinuity in the line. A number indicating the probable performance of a portion of a transmission circuit. The value of a transmission coefficient is inversely related to the quality of the link or circuit.

Transmission Control Category of control characters intended to control or help transmission of information over telecommunication networks. See TCP.

Transmission Control Characters A group of characters used to facilitate or control data transmission. Examples are NAK (Not acknowledge) and EOT (end of transmission).

Transmission Control Protocol TCP. A specification for software that bundles outgoing data into packets (and bundles incoming data), manages the transmission of packets on a network, and checks for errors. TCP is the portion of the TCP/IP protocol suite that governs the exchange of sequential data. In more technical terms, Transmission Control Protocol is ARPAnet-developed transport layer protocol. Corresponds to OSI layer 4, the transport layer. TCP is a connection-oriented, end-to-end protocol. It provides reliable, sequenced, and unduplicated delivery of bytes to a remote or local user. TCP provides reliable byte stream com-

munication between pairs of processes in hosts attached to interconnected networks. It is the portion of the TCP/IP protocol suite that governs the exchange of sequential data. See TCP/IP for a much longer explanation.

Transmission Convergence TC. Transmission Convergence Sublayer, a dimension of the ATM Physical Layer (PHY). See TC.

Transmission Electronics Any of the various devices used in conjunction with different transmission media to convert from one transmission method to another. Transmission electronics devices typically include multiplexing equipment and Asynchronous Data Units.

Transmission Facility A piece of a telecommunications system through which information is transmitted, for example, a multi pair cable, a fiber optic cable, a coaxial cable, or a microwave radio.

Transmission Frame A data structure, beginning and ending with delimiters, that consists of fields predetermined by a protocol for the transmission of user and control data.

Transmission Level The power of a transmission signal at a specific point on a transmission facility. See DECIBEL.

Transmission Level Point TLP. A designated point on a circuit where the transmission level has been specified by the designer. Referencing this point in relation to others in the network can determine the performance of the network.

Transmission Limit The wavelengths above and below which the fiber ceases to be transparent and therefore, can no longer transmit information.

Transmission Loss Total loss encountered in transmission through a system.

Transmission Media Anything, such as wire, coaxial cable, fiber optics, air or vacuum, that is used to carry an electrical signal which has information. Transmission media usually refers to the various types of wire and optical fiber cable used for transmitting voice or data signals. Typically, wire cable includes twisted pair, coaxial, and twinaxial. Optical fiber cable includes single, dual, quad, stranded, and ribbon.

Transmission Objectives A stated set of desired performance characteristics for a transmission system. Characteristics for which objectives are stated include loss, noise, echo, crosstalk, frequency shift, attenuation distortion, envelope delay distortion, etc.

Transmission Payload The interface bit rate minus the overhead bits.

Transmission Protocol 0 (TP0) OSI (Open Systems Interconnection) Transmission Protocol Class 0 (Simple Class). This is the simplest OSI Transmission Protocol, useful only on top of an X.25 network (or other network that does not lose or damage data).

Transmission Protocol 4 (tp4) OSI (Open Systems Interconnections) Transmission Protocol Class 4 (error detection and recover class). This is the most powerful OSI Transmission Protocol, useful on top any type of network. TP4 is the OSI equivalent to TCP (Transmission Control Protocol).

Transmission Security Key TSK. A key that is used in the control of transmission security processes such as frequency hopping and spread spectrum.

Transmissive The way many LCD (liquid crystal display) screens on laptops reflect light.

Transmit Bus In AT&T's Information Systems Network (ISN), the circuit on the backplane of the packet controller that transports message packets from sending device interface modules to the switch module.

Transmit Digital Intertie TDI. A 16-channel serial

converter which converts the TDM Data Bus from parallel format to serial format for transmission between nodes.

Transmittance The ratio of transmitted power to incident power. In optics, frequently expressed as optical density or percent; in communications applications, generally expressed in decibels.

Transmitter The device in the telephone handset which converts speech into electrical impulses for transmission.

Transmitter Distributor A device in a teletypewriter system which converts the information from the parallel form in which it is used in the keyboard-printer to and from the serial form which it is transmitted on the transmission line.

Transmitter Start Code A coded control character or code sequence transmitted to a remote terminal instructing that terminal to begin sending information.

Transmobile The transmobile (not to be confused with a TRANSPORTABLE) is another type of cellular phone. It is essentially a standard 3-watt mobile unit — without an external battery pack — that can be quickly and easily moved from one vehicle to another. It draws its power from the vehicle's battery via a cigarette lighter plug. See BAG PHONE.

Transmultiplexer A device that takes a bunch of voice analog phone conversations and converts them directly into a T-1 1.544 megabit per second bit stream — without the need for de-multiplexing the bunches down to individual conversations, then digitizing them, then bundling them up into a T-1 digital bit stream. A transmultiplexer does it all in one go.

Transparency 1. A data communications mode that allows equipment to send and receive bit patterns of virtually any form. The user is unaware that he is transmitting to a machine that receives faster or slower, or transmits to him faster or slower, or in a different bit pattern. All the translations are done somewhere in the network. He is unaware of the changes occurring — they are transparent. ISDN is planned to be transparent.

Transparency/Opacity An imaging term. A setting available in many image-processing functions that allows part of the underlying image to show through. 80 percent opacity is equivalent to 20 percent transparency.

Transparent When applied to telephone communications, the provision of a feature or service such as Automatic Route Selection in a such a way that the user is unaware of it and it has no affect on the way he uses the telephone. It's "transparent" to him. Translations are transparent to the telephone user. See TRANSLATIONS and TRANSPARENCY.

Transparent Bridging Transparent bridging is so called because the intelligence necessary to make relaying decisions exists in the bridge itself and is thus "transparent" to the communicating workstations. It involves frame forwarding, learning workstation addresses and ensuring no topology loops exist.

Transparent GIF Transparent GIFs are useful because they appear to blend in smoothly with the user's display, even if the user has set a background color that differs from that the developer expected. They do this by assigning one color to be transparent — if the Web browser supports transparency, that color will be replaced by the browser's background color, whatever it may be.

Transparent Image An image that has had one color, usually the background, designated as 'transparent,' so that when the image is displayed in a browser, the image's background is colored with the browser's background color. The effect is an image that does not have a visible rectangular background.

Transparent Mode 1. The operation of a digital transmission facility during which the user has complete and free use of the available bandwidth and is unaware of any intermediate processing. Generally implies out-of-brand signaling (also called Clear Channel).

2. In BSC data transmission, the suppression of recognition of control characters, to allow transmission of raw binary data without fear of misinterpretation.

3. An operational mode supported by the T3POS PAD which enables the use of existing credit authorization and data capture link level protocols. This mode requires minimal modifications to the POS (Point Of Sale) terminal, and no modification to the ISP/Credit Card Association (CCA) host system software.

Transparent Networking Transport TNT. A service for transporting of LAN data across WANs in which all responsibility for the WAN transport is assumed by the WAN and is therefore invisible to the LAN.

Transparent Routing A method used by a bridge for moving data between two networks. With this type of routing, the bridge learns which computers are operating on which network. It then uses this information to route packets between networks. It does not rely on the sending computers for its decision-making routine. A special kind of bridge combines the practice of transparent routing with source routing. It is called a source routing transparent (SRT) bridge. It examines each packet that comes by to see if it is using IBM's special source routing protocol. If so, this protocol is used to forward the packet. If not, the transparent method is used. Thus, the SRT bridge will support both IBM and non-IBM network protocols. See also BRIDGE and SRT. Compare with SOURCE ROUTING.

Transponder There are two meanings: 1. A transponder is a fancy name for radio relay equipment on board a communications satellite. Just like its domestic microwave counterpart (which you see along highways), a transponder will receive a signal, amplify it, change its frequency and then send it back to earth. Transponders typically have 36 MHz bandwidth. Full motion, full color TV video requires a 6 MHz analog channel. 2. A transponder on an airline is a slightly different kettle of fish. When a radar signal strikes a airline, it activates an electronic transmitter called a transponder. The transponder sends out a coded signal to the ground radar. The code appears next to the radar image of the plane, allowing the controller to identify each plane under his control.

Transport Driver A network device driver that implements a protocol for communicating between Lan Manager and one or more media access control drivers. The transport driver transfers Lan Manager events between computers on the local area network.

Transport Efficiency An AT&T term for the ability to carry information through a network using no more resources than necessary. Transport efficiency is achieved, for example, by statistical transport, which removes silent intervals from voice, data or other traffic and carries only the bursts of meaningful user information.

Transport Layer Layer 4 in the Open Systems Interconnection (OSI) data communications reference model that, along with the underlying network, data link and physical layers, is responsible for the end-to-end control of transmitted information and the optimized use of network resources. Layer 4 defines the protocols governing message structure and portions of the network's error-checking capabilities. Also serves the session layer. Software in the transport layer checks the integrity of and formats the data carried by the physical layer (layer 1, the network wiring and interface hardware), managed by the data link layer (layer 2) and possibly routed by the network layer (layer 1, which has the rules determining the path to be taken by data flowing through a network). See OSI.

Transport Medium The actual medium over which transmission takes place. Transport media include copper wire, fiber optics, microwave and satellites.

Transport Protocol A protocol that provides end-to-end data integrity and service quality on a network. Windows 95 Resource Kit defines transport protocol as how data should be presented to the next receiving layer in the networking model and packages the data accordingly. It passes data to the network adapter driver through the NDIS interface. See also TRANSPORT PROTOCOL CLASS FOUR.

Transport Protocol Class Four TP4. An International Standards Organization (ISO) transport layer protocol designated as ISO IS 8073 Class Four Service. TP4 has been adopted by the U.S. Department of Defense and specified in the U.S. Government OSI Profile (GOSIP).

Transportable Cellular Phone The transportable cellular phone is a standard 3-watt mobile phone that can be removed from the car and used by itself with an attached battery pack. The entire unit is generally mounted or built into a custom carrying case to make it easy to carry on your shoulder. Although technically "portable," the transportable should not be confused with the true portable one-piece cellular phone. Also known as a "bag phone" or "briefcase phone"; refers to a cellular handset that is packaged with a larger carrying case containing a full-scale power supply.

Transposed pair A wiring error in a twisted-pair cabling where a twisted pair is connected to a completely different set of pins at both ends (instead of pin 1 to pin 1, and pin 2 to pin 2, the cable is incorrectly wired pin 1 to pin 8, and pin 2 to pin 7, for example).

Transposition Interchanging the relative position of conductors at regular intervals to reduce crosstalk. In data transmission, a transmission defect in which, during one character period, one or more signal elements are changed from one significant condition to the other, and an equal number of elements are changed in the opposite sense.

Transverse Interferometry The method used to measure the index profile of an optical fiber by placing it in an interferometer and illuminating the fiber transversely to its axis. Generally, a computer is required to interpret the interference pattern.

Transverse Parity Check Type of parity error checking performed on a group of bits in a transverse direction for each frame. See PARITY CHECK.

Transverse Scattering The method for measuring the index profile of an optical fiber or preform by illuminating the fiber or preform coherently and transversely to its axis, and examining the farfield irradiance pattern. A computer is required to interpret the pattern of the scattered light.

Trap 1. See TRAP and TRACE.
2. A programming term. A programmer sets a trap for something to happen when something else happens. You might say "Wait for the mouse to come by, when it does, close the trap." A trap might be sprung when a phone rings or when someone hangs up. In network management, a trap is a mechanism permitting a device to automatically send an alarm for certain network events to a management station. Typically, network management information is gained by polling network nodes on a regular basis. This strategy can be modified when a trap is set from a network node. With traps, a node alerts the management station of a catastrophic problem. The management

station can then immediately initiate a polling sequence to the node to determine the cause of the problem. This strategy is often called trap-directed polling.
3. A video term. A circuit often called a filter, which is used to attenuate undesired signals while not affecting desired signals. Typically a signal channel trap to remove a single premium service which the subscriber is not paying for.

Trap And Trace A telephone company term. Trap and Trace is the term for equipment and procedures for determining the source of an incoming call (typically a harassing call). The phone company uses traps to trace the source of the incoming call. There are two types of traps — the Terminating Trap and the Originating Trap. A terminating trap sits on the receiving phone line. In the old days, a terminating trap was a physical piece of equipment. These days, with electronic central offices, it's basically a command to the computer running the central office to keep track of all information about the source of all incoming calls. That information might be the originating telephone number. It might be the trunk number on which the call came in on. Such trunk number might look like TGN701. Or it might be the CLLI code — which stands for the Common Language Location Identifier. The CLLI code (pronounced "silly") consists of 11 characters. A sample CLLI code is "nycmny18dso." That says the call is coming in from New York City, Manhattan from a central office called 18DSO (which I happen to know is an AT&T 5E central office located on West 18th Street). Once the terminating trap identifies the possible direction /source / incoming trunk of the offending phone calls, the phone company will work it back towards the originating line. It will attach an Originating Trap to the offending trunk, then to the offending tandem office, then to the local central office. This can be a tedious and time consuming business. With the advent of Caller ID — both local and nationwide — trapping and tracing is getting faster and easier. Now if you receive an harassing call, you simply hit *57 the moment the call is over (GTE uses *69). This "tags" the incoming call's number and other information in your central office's records. You, as subscriber, can't get access to that information. But a law enforcement agency (i.e. one investigating your annoying calls) can. See ANNOYANCE CALL BUREAU and WIRE TAP.

Trap Door Hidden software or hardware mechanism that, when triggered, allows system-protection mechanisms to be circumvented.

Trash-80 Pejorative term for the TRS-80 (Tandy Radio Shack-80), an early PC sold by the Tandy Corporation through its Radio Shack retail stores. See TRS-80.

Trashing Also referred to as dumpster diving, a term used by hackers for going through trash in an effort to get information that will facilitate breaking into computers. People often write passwords on paper, then put the paper in the trash. Be careful.

Trashware Software that is so poorly designed that it winds up in the garbage can.

TRAU Transcoder and Rate Adapter Unit. A transmission function of the BSS that converts speech from the user of a mobile station into digital representation needed for an ISDN, wireless network.

Travel Card Another name for a telephone credit calling card. Travel card calls that are placed against a travel card number issued by the service provider, typically a phone company. As each call is completed, the long distance switch increases that card's account balance by the amount of each call. During the processing of a call, if the travel card is invalid or if the caller does not respond to a system prompt, the serving switch will typically ask the caller to hold the line for a live operator, and transfers the call to an Operator Workstation. When the operator answers, the OWS screen shows call information, including card number (if already entered), destination number (if already entered), trunk identification, and a failure code.

Traveling Class Mark TCM. A code that accompanies a long distance call. When Automatic Route Selection (ARS) or Uniform Numbering/Automatic Alternate Routing (UN/AAR) selects a tie trunk to a distant tandem PBX, the traveling class mark (TCM) is sent over the tie trunk. It is then used by the distant system to determine the best available long distance line consistent with the user's calling privileges. The TCM indicates the restriction level to be used based on the phone, trunk or attendant originating the call or the authorization code, if dialed.

Tray See RACK.

Treatment A billing and collections term. The specific steps of the collection process to which an account is subject. The treatment level may begin with a "courtesy" call which may go something like "Mr. Newton, this is Mrs. Horak with your friendly telephone company. We've noticed that your account is past due. In fact, you have not paid your telephone bill for three months. When might we expect payment?" At this point, Mrs. Horak verifies employment, which is a standard step. Now the conversation takes a turn for the worse. "Mr. Newton, do I understand correctly that you no longer work for Flatiron Publishing, and that you expect me to believe that you now work for Harry Newton Enterprises? Really Mr. Newton! I must request immediate payment by cash, cashier's check or money order! Failure to comply with this demand by the end of the business day will result in the disconnection of your service. Oh, did I mention that we will require a security deposit of $5,000? That, too, will have to be paid by the end of the business day. Yes, Mr. Newton, I am fully aware that it is 4:59PM. Mr. Newton, Mr. Newton." (Aside: "Ray, those guys in the switchroom are really good! They cut Harry's service off at exactly 5:00. That'll teach him to pay his bills on time!") Note: This scenario actually is very inaccurate- the guys in the switchroom aren't nearly that good. Actually, treatment levels are highly sensitive to the size of the bill, the age of the receivable, the history of the account, and other factors. Treatment levels may begin with a courtesy call, progress through several calls of a firmer tone, a formal letter or two of successively firmer tone, suspension of service, and disconnection. Restoral of service and reconnection entail service fees and generally involve a security deposit. If you don't pay your final bill quickly, you'll be dealing with a collection agency. Pay your telephone bill on time.

Treatment Level Treatment level is a term used in some telephone companies' billing and collections processes. The phrase is used to help a telephone company identify where a particular customer is in the collections/overdue billing process and proper protocol in treating the customer. See Treatment.

Tree 1. A network topology shaped like a branching tree. (What else?) It is characterized by the existence of only one route between any two network nodes. Most CATV distribution networks are tree networks.
2. In MS-DOS, a tree describes the organization of directories, subdirectories, and files on a disk.

Tree Hugger IBM-speak for an employee who resists a

move or any other change.

Tree Mailbox A special function mailbox that provides the caller with a menu and allows selections from the menu using single digit commands.

Tree Search In a tree structure, a search in which it is possible to decide, at each step, which part of the tree may be rejected without further search.

Tree Stand Aerial Cross Box. A cross box on a pole. Used where vandals live or when there's a narrow easement.

Tree Structure Describes the organization of directories, subdirectories, and files on a disk.

Tree Topology A network cabling architecture in which nodes are connected by cables to a central, or trunk, cable with a central retransmission capability.

Treeware Slang for documentation or other printed material.

Trellis Coding A method of forward error correction used in certain high-speed modems where each signal element is assigned a coded binary value representing that element's phase and amplitude. It allows the receiving modem to determine, based on the value of the preceding signal, whether or not a given signal element is received in error. See V.32 and V.32 bis.

Trellis Coding Modulation TCM. A modem modulation technique in which sophisticated mathematics are used to predict the best fit between the incoming signal and a large set of possible combinations of amplitude and phase changes. TCM provides for transmission speeds of 14,400 bps and above on single voice grade phone lines. See V.32 and V.32 bis.

Tremendously High Frequency Frequencies from 300 GHz to 3000 GHz.

TRFR TRansFeR.

TRG Technical Review Group.

Tri-Mode Tri-Mode describes a cell phone that operates in North America on both digital bands — 800 Mhz and 1900 Mhz — along with analog AMPS in the 800 Mhz band. The reason you'd want such a phone is simple: Digital service is often cheaper better in areas you can get it. But you can't get it everywhere. If you travel you need a cell phone you can use everywhere. Thus the idea of carrying a three band cell phone and subscribing to a service that gets you access to all three. See also Tri-Mode.

TRI-CWDM On August 30, 1995, MCI Communications announced the deployment of a technology that will enable it to increase the capacity of its network by 50 percent without any additional fiber optic lines. The technology, known as Tri-Color Wave Division Multiplexing (Tri-CWDM), allows existing fiber to accommodate three light signals instead of two, by routing them at different light wavelengths, through the combined use of narrow and wide band wave division multiplexing.

With this method, lightwaves are transmitted at 1557 nanometers (nm) and 1553 nm to a wide band WDM device, where a 1310 nm signal is added. Once combined, the three signals are routed through a single fiber to the next site where they are separated and sent to the receivers. Transmitting three signals in each direction allows for three different transmit pairs on just two fibers, effectively increasing the total network capacity from 5 gigabits to 7.5 gigabits. MCI officials say the technology will be particularly valuable in major metropolitan areas, where the company is enjoying outstanding growth in voice and data traffic.

Triangulation A method of locating the source of a radio signal through the use of three receivers, each of which focuses on the direction of maximum signal strength. Through the use of three receivers, it easily is possible to plot the general location of the transmitter, even though radio signals bounce off and are absorbed by physical obstructions such as buildings, trees and cars. This process, also known as Angle of Arrival, now can be accomplished by two, or even a single, receivers employing much more sophisticated, smart-antenna technology.

Triaxial Cable A cable construction having three coincident axes, such as a conductor, first shield and second shield all insulated from one another.

Tribit Transmission A transmission technique used by some modems in which three bits are transmitted simultaneously.

Tributary The lower rate signal input to a multiplexer for combination (multiplexing) with other low rate signals to form an aggregate higher rate signal.

Tributary Circuit A circuit connecting an individual phone to a switching center.

Tributary Office A local office, located outside the exchange in which a toll center is located, that has a different rate center from its toll center.

Tributary PBX An exchange within the main PBX configuration but with its own listed number. The only difference between a satellite and a tributary PBX is that the tributary PBX has a direct incoming connection from the public network. See SATELLITE PBX.

Tributary Station In a data network, a station other than the control station. On a multi point connection or a point-to-point connection using basic mode link control, any data station other than the control station.

Tributary Unit The SDH equivalent of a Virtual Channel in SONET terminology.

Trichromatic The technical name for RGB representation of color to create all the colors in the spectrum.

Trickle The name of a BITNET mail server package which provides access to anonymous FTP archive sites via e-mail.

Trigger An application-specific process invoked by a database management system as a result of a request to add, change, delete, or retrieve a data element.

Triggering The process of detecting a word (or utterance) and capturing the speech data associated with that word (or utterance) for subsequent processing.

Triggers Uncompiled code residing on an intelligent database server.

Trinkets A low to modestly priced item. Typically these are T-shirts, hats, cups and pens used for promotional or motivational purposes. Also used to bribe and cajole software developers into working even more excessive hours. When the telex business was in full bloom, trinkets were used to motivate telex operators into preferring one supplier over another. Trinkets were necessary because the price for service and equipment was identical, since it was heavily regulated.

Triode A combination of a heated cathode, a relatively cold anode, and a third electrode for controlling the current flowing between the other two; the whole enclosed in an evacuated bulb. Variously called, audion, pliotron, radiotron, oscillion, audiotron, aerotron, electron tube, vacuum tube, etc.

Triple DES A security enhancement of single-DES encryption that employs three successive single-DES block operations. Different versions use either two or three unique DES keys. This enhancement is considered to increase resistance to known cryptographic attacks by increasing resistance to known cryptographic attacks by increasing the effective key length.

Trivial File Transfer Protocol TFTP. A UNIX-based file protocol. TFTP is a simplification of the earlier Simple File Transfer Protocol (SFTP).

TRL Transistor Resistor Logic.

TRO Temporary Restraining Order

Trojan Horse Software that appears to do something normal but which contains a trap door or attack program. A Trojan Horse program can be used to break into a network through a World Wide Web site. A Trojan Horse is dangerous software.

Troposphere The lower layers of the earth's atmosphere. You can bounce certain frequency radio signals off it and use it as an elementary transmission reflector. The troposphere is the region where clouds form, convection is active, and mixing is continuous and more or less complete. The lower layers of the Earth's atmosphere, between the surface and the stratosphere, in which about 80 percent of the total mass of air is located and in which temperature normally decreases with altitude. The thickness of the troposphere varies with season and latitude; it is usually 16 km to 18km over topical regions and 10 km or less over the poles. See TROPOSPHERIC SCATTER and TROPOSPHERIC WAVE.

Tropospheric Scatter The propagation of radio waves by scattering as a result of irregularities or discontinuities in the physical properties of the troposphere. The propagation of electromagnetic waves by scattering as a result of irregularities or discontinuities in the physical properties of the troposphere. A method of transhorizon communications using frequencies from approximately 350 MHz to approximately 8400 MHz. The propagation mechanism is still not fully understood, though it includes several distinguishable but changeable mechanisms such as propagation by means of random reflections and scattering from irregularities in the dielectric gradient density of the troposphere, smooth-Earth diffraction, and diffraction over isolated obstacles (knife-edge diffraction).

Tropospheric Wave A radio wave that is propagated by reflection from a place of abrupt change in the dielectric constant or its gradient in the troposphere. In some cases, the ground wave may be so altered that new components appear to arise from reflection in regions of rapidly changing dielectric constant. When these components are distinguishable from the other components, they are called "tropospheric waves."

Trouble Number Display The operator will know what the trouble is with the phone system by seeing a number pop up on her/his console. That number may pop up automatically or the operator may have to hit the ALM (for ALARM) or similar button.

Trouble Ticket Form used to report problems. Often incorrectly filled-in. Check.

Trouble Unit A weighting figure applied to telephone circuit or circuits to indicate expected performance in a given period.

Troubles Per Hundred Troubles per hundred is a criterion for acceptable customer service which telephone companies and public utility commissions have agreed upon. It's measured in terms of the number of complaints received per hundred telephones in one month. Six complaints per hundred is considered the maximum for acceptable service. See QUALITY OF SERVICE.

TRS-80 Tandy Radio Shack-80. One of the early PCs. It was introduced by Tandy Corporation through its Radio Shack stores. It was based on the Zilog Z80 chip and began selling in the late 1970s. Along with Apple and Commodore it proved the viability of personal computers. The TRS-80 helped ensure the success of later generations of PCs by introducing a spreadsheet called Visicalc, word processing such as

Electric Pencil, WordStar and databases such as dBASE. It also was one of Microsoft's earliest customers for their Basic language package. There were several TRS-80 models including the original Model 1, Model 2, Xenix (UNIX), Model 3 with integrated drives and Model 4. Some of the models could run CP/M or TRSDOS. Tandy later also introduced the Model 100 which was, arguably the first Laptop/Notebook, and for which Bill Gates is alleged to have written much of the software code. Detractors of the TRS-80 referred to it as the "Trash-80." The TRS-80 Model 2 actually had a cage designed for the specific purpose of accepting printed circuit cards. Sadly, Radio Shack never released the technical specs on the cage. No one (including Radio Shack itself) produced cards and the machine was quickly superseded by the IBM PC. In short, Radio Shack once had the market for PCs right at its fingertips. But blew an incredible opportunity. Sad.

TRU Inside a telephone system, the TRU (the Tone Receiver Unit) is used to retrieve and interpret touch-tone data received. Those tones might be sent by a TSU — Tone Sender Unit.

TrueSpeech TrueSpeech is a low-bandwidth method of digitizing speech, which was created by a company called DSP Group, Inc. Santa Clara, CA. TrueSpeech uses compression to drop one minute of voice down to 62 kilobytes with remarkably little degradation. It is used in many digital telephone answering devices for storing and reproducing voice. TrueSpeech's compression is not meant for high fidelity music. But it is more than acceptable for such business applications as voice mail, voice annotation, dictation, and education and training. The small file size means that it can be transferred more easily to other users by using either a corporate network. The DSP Group describe TrueSpeech an enabling technology for speech compression in personal computers and future personal communications devices. Speech compression is key technology to the effective convergence of personal computers and telephony. TrueSpeech compression is a technology based on complex mathematical algorithms which are derived from the way airflow from our lungs is shaped by the throat, mouth, and tongue when we speak. This shaping is what our ear finally hears. TrueSpeech is 5 to 15 times more efficient than other methods of digital speech compression. For example, a one minute long speech file which uses other PC audio technology would consume as much as 960 kilobytes. With TrueSpeech, the same file would be just over 60 kilobytes. TrueSpeech is used in the Microsoft Sound System, which also lets you choose the voice sampling you wish when you're recording material. Here is Microsoft Sound System's recording options:

	Sampling Rate	Technology
TrueSpeech	62K per minute	8 KHz	Proprietary
Voice	234K per minute	8 KHz	ADPCM
Radio	322K per minute	11 KHz	ADPCM
Tape	1291K per minute	22 KHz	PCM
CD	5176K per minute	44 Khz	PCM

The above is for mono recordings. For stereo, double the amount of space.

Truetype A Windows 3.1 feature. Fonts that are scalable and sometimes generated as bitmaps or soft fonts, depending on the capabilities of your printer. TrueType fonts can be sized to any height, and they print exactly as they appear on the screen. Using TrueType, you'll be able to create documents that retain their format and fonts on any Windows 3.1 machine — even if the fonts aren't installed on that computer. This makes Windows 3.1 documents portable.

Truevoice 1. In the fall of 1993, AT&T announced that it was introducing new voice quality throughout its long distance network. And that it was calling that quality "true voice." AT&T set up a demo line. Some people thought they could notice an improvement. Some thought they couldn't. I personally thought true voice sounded pretty good.
2. The trademark name of Centigram's text-to-speech product, which they acquired from SpeechPlus.

Trumpet Winsock A once-popular Windows 3.xx communications program and TCP/IP stack which allowed people to dial into the Internet and use browsers to surf the Internet. I never liked the program and had great difficulty with it. Fortunately the program has effectively been killed by dial up networking capabilities now part of every Windows 95.

Truncated Binary Exponential Back Off Another name for exponential back off used in IEEE 802.3 local area networks. In an exponential back-off process, the time delay between successive attempts to transmit a specific frame is increased exponentially.

Truncation In data processing, the deletion or omission of a leading or a trailing portion of a string in accordance with specified criteria.

Trunk A communication line between two switching systems. The term switching systems typically includes equipment in a central office (the telephone company) and PBXs. A tie trunk connects PBXs. Central office trunks connect a PBX to the switching system at the central office.

Trunk Access Number 1. The number of the trunk over which a call is to be routed.
2. The number that needs to be dialed in order to gain access to an outbound trunk. This applies to both local and local distance trunks, as the access number can be different.

Trunk Answer A phone system feature. This feature allows a ringing call to be answered from any telephone in the system. Typically the feature must be activated in phone system programming.

Trunk Answer From Any Phone A phone system feature. When a call comes in, something rings. You can now answer the incoming call from any phone. To do so, you must dial a special code or hit a special feature button on your phone. When my office phone system bells ring, all we have to do is to touch "6" on any phone and we can answer the incoming call. Typically the feature must be activated in phone system programming.

Trunk Circuit An assemblage of electronic elements located in the switching machine.

Trunk Conditioning Electrical treatment of transmission lines to improve their performance for specific uses such as data transmission. The "tuning" and/or addition of equipment to improve the transmission characteristics of a leased voice-grade line so that it meets the specifications for higher-speed data transmission. Voice-grade lines often have too much "noise" on them. By altering the equipment at both ends of the line, this noise on the line can be overcome. This allows transmission of higher-speed data, which is much more sensitive to noise than voice. See also Conditioning.

Trunk Data Module TDM. Provides the interface between the DCP signal and a modem or Digital Service Unit (DSU).

Trunk Direct Termination An option on switchboards which terminates a trunk group on one key (or button) on the console.

Trunk Encryption Device TED. A bulk encryption device used to provide secure communication over a wideband digital transmission link. It is usually located between the output of a trunk group multiplexer and a wideband radio or cable facility.

Trunk Exchange A telephone exchange dedicated primarily to interconnecting trunks.

Trunk Group A group of essentially like trunks that go between the same two geographical points. They have similar electrical characteristics. A trunk group performs the same function as a single trunk, except that on a trunk group you can carry multiple conversations. You use a trunk group when your traffic demands it. Typically, the trunks in a trunk group are accessed the same way. You dial your Band 5 WATS trunk group by dialing 62, for example. If the first trunk of that group is busy, you choose the second, then the third, etc. See TRUNK HUNTING.

Trunk Group Alternate Route The alternate route for a high-usage trunk group. A trunk group alternate route consists of all the trunk groups in tandem that lead to the distant terminal of the high-usage trunk group.

Trunk Group Multiplexer TGM. A time division multiplexer whose function is to combine individual digital trunk groups into a higher rate bit stream for transmission over wideband digital communication links.

Trunk Group Warning Alerts the attendant when a preset number of trunks in a group are busy. See TRUNK GROUP.

Trunk Holding Time The length of time a caller is connected with a voice processing system. Defined from the time when the system goes off-hock to the time the port (i.e. the trunk) is placed back on hook.

Trunk Hunting Switching incoming calls to the next consecutive number if the first called number is busy.

Trunk Make Busy A fancy name for saying that, by punching a few buttons on the console, you can make any trunks in your PBX or key system busy, effectively putting the trunk out of service. You may want to do this if your trunk is acting up. By busying it out at the console, you are effectively denying its use to anyone in the company. Thus you are protecting yourself from further complaints. Hopefully, it will be repaired promptly.

Trunk Monitoring Feature which allows individual trunk testing to verify supervision and transmission. You dial an access code and then the specific trunk number from the attendant console. You want the ability to test a specific trunk because normally you might be only accessing a trunk group when you dial an access code. Thus, each time you dial into the trunk group, you might end up on another individual trunk. Some PBXs have a variation of trunk monitoring, whereby if a user encounters a bad trunk, he can dial a specific code, then hang up. The PBX recognizes these digits and makes a trouble report on that specific trunk, possibly reporting it to the operator, keeping it in memory for later analysis or dialing a remote diagnostic center and reporting its agony.

Trunk Number Display The specific trunk number of an incoming call can be displayed on the attendant console, enabling your attendant to instantly identify the origin of certain calls. For example, if you have several tie lines to branch offices, your attendant knows immediately which office is calling. Many newer PBXs have displays on individual telephones, which show the actual trunk being used for outgoing and incoming calls. This provides an additional measure of control. You might, for example speak faster if you knew the call was coming in on your IN-WATS line. You might also answer the call differently if you know what trunk it's coming in on. For example, you might be running several, totally-separate businesses from the same console. Each business has a different number. The only way you know what to answer —

Joe's Bakery or Mary's Real Estate — is by the trunk.

Trunk Occupancy The percentage of time (normally an hour) that trunks are in use. Trunk occupancy may also be expressed as the carried CCS per trunk.

Trunk Order A document (or data system equivalent) used in an operating telephone company to request a change to a trunk group.

Trunk Queuing A feature whereby your phone system automatically stacks requests for outgoing circuits and processes those requests on, typically, a first-in/first-out basis. See QUEUING THEORY.

Trunk Reservation The attendant can hold a single trunk in a group and then extend it to a specific phone. This means, for example, that a WATS line can be held for someone special — a heavy caller, the president of the firm, etc.

Trunk Restriction Some people may not be allowed to use certain trunks at certain times. The sophistication of trunk restriction depends on the switch and the way it's programmed.

Trunk Segment The main segment of cable in an Ethernet network is called the trunk segment.

Trunk Side Connection A carrier term. Trunk side connections are within the carrier network. InterMachine Trunks (IMTs) connect carrier switches to other carrier switches. Such switches include circuit switches such as Central Offices (COs) and Tandem switches, Frame Relay switches and routers, packet switches, and ATM switches. End user organizations can lease local loops with trunk side connections, as well; such a loop would appear to the carrier network as being a part of it, and would be used for access to ANI (Automatic Number Identification) information. Compare with Line Side Connection.

Trunk Type TT. Trunks that use the same type of equipment going to the same terminating location.

Trunk Type Master File TTMF. An MCI definition. A comprehensive listing of all trunk assignments on the MCI network for shared and dedicated services, necessary for processing and billing MCI customer calls.

Trunk To Tie Trunk Connections The ability of the switching system to provide the attendant with the capability of extending an incoming trunk call to a tie trunk terminating some place else.

Trunk To Trunk By Station A PBX feature which permits the user who established a three-way conference involving himself and two trunks to drop from the call without disconnecting the trunk-to-trunk connection.

Trunk To Trunk Connections The attendant can establish connections between two outside parties on separate trunks. Call your office on your IN-WATS. Ask the operator to extend that call to the VP who happens to be at his home. The operator must place an outside call to the VP on an outside trunk and join that call to the incoming call. Sometimes it works.

Trunk To Trunk Consultations Allows a phone connected to an outside trunk circuit to gain access to a second outside trunk for "outside" consultation. No conference capability is available with this feature.

Trunk Transfer By Station Permits the user who established a three-way conference involving two lines to drop from the call without disconnecting the trunk-to-trunk connection.

Trunk Verification By Customer Provides the attendant or phone user access to individual lines in a trunk group to check their condition. See also TRUNK MONITORING.

Trunk Verification By Station Provides a warning

tone if a phone user enters a busy trunk.

Trunks In Service The number of trunks in a group in use or available to carry calls. Trunk in service equals total trunks minus the trunks broken or made busy for any reason.

Trust Relationship The trust relationship is the link between two domains (e.g. two servers on a network) that enables a user with an account in one domain to have access to resources on another domain. The trusting domain is allowing the trusted domain to return to the trusting domain a list of global groups and other information about users who are authenticated in the trusted domain. In the MIS world, a domain is "the part of a computer network in which the data processing resources are under common control." In the Internet, a domain is a place you can visit with your browser — i.e. a World Wide Web site. See domain and inter-domain trust relationships.

Truth Table An operation table for a logic operation. A table that describes a logic function by listing all possible combinations of input values and indicating, for each combination, the output value.

TRX Transmitter and Receiver (Transceiver). A wireless telecommunications term. A function of the radio channel device for receiving and transmitting signals or information on the radio channel.

TRXXX Various Bellcore standards.

TS 1. Transport Stream. As an ATM term, it is one of two types of streams produced by the MPEG-2 Systems layer. The Transport Stream consists of 188 byte packets and can contain multiple programs.
2. Traffic Shaping: Traffic shaping in an ATM network is a mechanism that alters the traffic characteristics of a stream of cells on a connection to achieve better network efficiency, while meeting the QoS (Quality of Service) objectives, or to ensure conformance at a subsequent interface. Traffic shaping must maintain cell sequence integrity on a connection. Shaping modifies traffic characteristics of a cell flow with the consequence of increasing the mean Cell Transfer Delay.
3. Time Stamp: As an ATM term, Time Stamping is used on OAM cells to compare time of entry of cell to time of exit of cell to be used to determine the cell transfer delay of the connection.

TSAC Time Slot Assigner Circuit; a circuit that determines when a CODEC will put its eight bits of data on a RCM bit stream.

TSAP Abbreviation for Transport Service Access Point in the OSI transport protocol layer.

TSAPI Telephony Server Application Programming Interface. Described by AT&T, its inventor, as "standards-based API for call control, call/device monitoring and query, call routing, device/system maintenance capabilities, and basic directory services." For a better explanation, see TELEPHONY SERVICES.

TSB Telecommunications System Bulletin. Interim changes to an interim standard. Not very interesting, but there it is. See IS.

TSB67 Part of the EIA/TIA-568-A standard. TSB67 describes the requirements for field testing an installed Category 3,4 or 5 twisted pair network cable.

TSC Two-Six Code. A trunk group reference number. The first two characters are alphabet (a-z) and the last six characters are numberic digits.

TSCM Technical Surveillance CounterMeasures. Commonly called debugging, sweeps or electronic sweeping.

TServer Telephony Server. Name of NetWare Telephony Services LAN server which is joined physically (by wire) to an adjacent PBX. See TELEPHONY SERVICES.

TSI 1. Time Slot Interchange or Interchanger. A way of temporarily storing data bytes so they can be sent in a different order than they were received. Time Slot Interchange is a way to switch calls.
2. Transmitting Subscriber Information. A frame that may be sent by the caller, with the caller's phone number, which may be used to screen calls, etc.

TSIC Time Slot Interchange Circuit; a device that switches digital highways in PCM based switching systems. In short, a digital crosspoint switch.

TSIU Time Slot Interchange Unit. Switching module hardware unit that provides the digital time switching function.

TSK Transmission Security Key.

TSM Terminal Server Manager (TSM), a program that allows terminal servers on a network to be remotely managed from another node. It is supported on VMS systems running the LAT protocol (and is incompatible with TCP/IP-only networks).

TSO Time Share Operation.

TSP 1. Telecommunication Service Priority.
2. Terminal Service Profile.
3. Telecommunications Service Provider. A new term for an Internet Service Provider.

TSPS Traffic Service Position System permits operator positions serving public phones and HOBIC operations to be located remotely from the CO which services the pay phone or the hotel, or the hospital, etc.

TSR 1. Telephone Service Representative. See also AGENT.
2. Terminate and Stay Resident. A term for loading a software program in an MS-DOS computer in which the program loads into memory and is always ready for running at the touch of a combination of keys, e.g. Alt M, or Ctrl ESC. Here's some information from Jackie Fox writing in PC Today: You can't load TSRs willy-nilly and expect them to work with each other. Some will get along with each other. Others won't. When you install a TSR, it goes to a location in RAM (Random Access Memory) called the Interrupt Vector Table.
The interrupt vector table is like a hotel lobby, and TSRs are like guests waiting for messages. The TSR watches every incoming keystroke to see if it's the special hot key combination (message), the TSR is waiting for. If it isn't, the TSR passes it back to the regular program. What if you have four or five TSRs loaded? The one you loaded last has seniority. It checks the incoming keystrokes first. If the TSR recognizes the keystroke combination as its own hot-key combination, it takes over. If not, it passes it along to the next TSR. This process is called interrupt handler chaining.
If none of the TSRs recognize that particular combination, they pass it along to DOS so it can process it as a regular keystroke combination. Not all TSRs pass instructions along the way they should. Some TSRs intercept keystrokes and never pass them on. Some TSRs never restore their original addresses. Sometimes two TSRs fight over the same hot key combination. Then you end up with a frozen keyboard. The basic problem is there are no rules for loading and running TSRs.

TSRM Telecommunication Standards Reference Manual.

TSS The Telecommunications Standards Section (TSS) is one of four organs of the ITU. Any specification with an ITU-T or ITU-TSS designation refers to the TSS organ. See ITU.

TST Time-Space-Time system.

TSTN Triple Super Twisted Nematic. A display technology often used on laptop computers which uses three layers of crystal to give better contrast and more grey scales.

TSTS Transaction Switching and Transport Services. In 1992, the Regional Bell Operating Companies (RBOCs) agreed to provide uniform transaction processing capabilities under a banner called Transaction Switching and Transport Services.

TSU Tone Sender Unit. A device inside a telephone system. The TSU passes along touch-tone digits to telephone extension cards within the phone system. See also TRU (the Tone Receiver Unit) which is used to retrieve and interpret touch-tone data received.

TSYN The SYNCHronous test line provides for testing of ringing, tripping and supervisory functions of toll completing trunks. See TNSS.

TTB Talking Total Bollocks. This is a European (specifically UK) term coined by telecom experts when noticing a particularly enthusiastic sales guy getting over excited about some new piece of technology. Chris Hall contributed this dubious definition.

TTC The Telecommunications Technology Committee, a Japanese standards committee.

TTCN Tree and Tabular Combined Notation: The internationally standardized test script notation for specifying abstract test suites. TTCN provides a notation which is independent of test methods, layers and protocol.

TTD Temporary Text Delay. The TTD control sequence (STX ENQ) is sent by a sending station in message transfer state when it wants to retain the line but is not ready to transmit.

TTI Transmit Terminal Identification. A fax machine's stupid term for its telephone number and the name of its owner. When you receive a fax from someone, the top line of the fax typically will have a phone number and a name on it. That phone number and name does NOT come from the phone or the phone company. It comes from what the person who owns the machine programmed into his machine. He typically did that by punching buttons on his fax machine. He'll do that if he can understand the instruction booklet which came with his fax (which he probably won't). The point of all this is twofold: First, don't forget to put your name and phone number into your fax machine. Second, don't assume that what you read at the top of any fax you receive is accurate.

TTIA Telecommunications Technology Investment Act of 1993.

TTL Transistor Transistor Logic.

TTR Touch Tone Receiver. A device used to decode touchtones dialed from single-line telephones or Remote Access telephones.

TTRT Target Token Rotation Time. An FDDI (Fiber Distributed Data Interface)token travels along the network ring from node to node. If a node does not need to transmit data, it picks up the token and sends it to the next node. If the node possessing the token does need to transmit, it can send as many frames as desired for a fixed amount of time.

TTS 1. Text-To-Speech. A term used in voice processing. See TEXT-TO-SPEECH.

2. Transaction Tracking System. A Novell NetWare feature that protects database applications from corruption by backing out incomplete transactions that result from a failure in a network component. When a transaction is backed out, data and index information in the database are returned to the state they were in before the transaction began.

TTTN Tandem Tie Trunk Network.

TTY A teletypewriter. Typewriter-style device for communicating alphanumeric information over telecom networks. TTY is the most widely used type of emulation for PC computer communications.

TTY/TDD A unique telecommunication device for the deaf,

using TTY principles.

TU 1. Transmit Unit. Term used in a DS-3 channel bank.
2. Tributary Unit in SDH terminology. Equivalent to a Virtual Channel.

TUA Telecommunications Users Association (UK). The TUA says it aims to support the development of UK businesses through the application of telecommunication and information technologies. It also strives to bring about a fair and competitive market within the UK. TUA holds an annual trade show in the U.K, typically around November or December. www.tua.co.uk.

TUANZ Telecommunications Users Association of New Zealand. A non-profit society of over 500 telecommunications users including major NZ corporations, small to large businesses, government departments, educational institutions and interested individuals. By the way, New Zealand is about as deregulated as you can get; hence, it is a technology testbed. All the really neat technologies that we now see being introduced in the US have been trialed in New Zealand for years. www.tuanz.gen.nz

TUBA TCP and UDP with Bigger Address. One of the three IPng candidates.

TUG Telecommunication User Group.

TUI Telephony User Interface. Like GUI, except that it means that someone is using the telephone to get to information in a computer. See GUI.

Tumbling A form of cellular fraud first appearing in late 1990. The crook alters a cellular telephone so that it "tumbles" through a series of ESNs in order to make the caller to appear to be another new customer each time a call is made. By the time the cellular phone network operator has checked with the network where the bogus telephone supposedly is registered and discovered the fraud, the crook has tumbled the telephone, or changed its electronic serial number (typically by one digit) and is ready to make more free calls. As the carriers have moved to IS-41 pre-call validation, this form of fraud has all but been eliminated. See also CLONE FRAUD.

Tunable Operating System Parameters Tuning an operating system is the same as optimizing it, in that you rewrite commands and programs so they operate faster and more efficiently. Any new operating system needs to be tuned to the specific machine on which it is running.

Tungsten A metallic element used in ceramic IC packaging to provide the traces within the package that connect the device circuitry to the external terminals pads or leads.

Tuning Adjusting the parameters and components of a circuit so that it resonates at a particular frequency or so that the current or voltage is either maximized or minimized at a specific point in the circuit. Tuning is usually accomplished by adjusting the capacitance or the inductance, or both, of elements that are connected to or in the circuit.

Tunnel Diode A tunnel diode conducts electricity very well in both directions. However, their resistance to current flowing in the forward direction is very unusual. As the voltage is increased, the current carried by the diode also increases until it reaches a peak. Increasing the voltage beyond the peak value causes the amount of current passing through the diode to decrease! The current will continue to decrease until it reaches a minimum value and then rise again with increasing voltage.

Tunneling 1. As a local area network term, tunneling means to temporarily change the destination of a packet in order to traverse one or more routers that are incapable of routing to the real destination. For example, to route through a backbone whose internal routers don't contain entries for external destinations, the entry border router must "tunnel" to the exit border router.

2. As an Internet term, tunneling means to provide a secure, temporary path over the Internet. For example, a telecommuter might dial into an ISP (Internet Service Provider), which would recognize the request for a high-priority, point-to-point tunnel across the Internet to a corporate gateway. The tunnel would be set up, effectively snaking its way through other, lower-priority Internet traffic. See Split Tunneling.

Tunneling Ray Leaky ray.

TUP Telephone User Part. An SS7 term for the predecessor to ISUP (Integrated Services User Part). TUP was employed for call control purposes within and between national networks, both wired and wireless. ISUP adds support for data, advanced ISDN, and IN (Intelligent Networks). See also ISUP.

Tuple Address In the Frame Relay (FR) network, packages are sent from switch A to switch B. Each package contains a header section. In the header section of the package, there is information on the DLCI (Data Link Connection Identifier). At switch B, the traffic information is written to a file (based on the Bay Network FR, it is called DATA) every two hours. The billing adjunct processor (AP) will TFTP or FTP the DATA files after it is generated. In each DATA file, it contains the size of the package, number of packages,
switch A's IP address (e.g., 156.52.245.1),
switch A's DLCI (e.g., 1000),
switch B's IP address (e.g., 166.100.221.25),
and switch B's DLCI (e.g., 502). The database of the billing adjunct processor contains the tuple address (switch A's IP & DLCI and switch B's IP & DLCI information) of the circuit that the customer rents. The billing AP converts the DATA files' format to a standard format file and sends it to the billing company which then bills the customer who rents that tuple address. This is very much like our phone bill we receive monthly from a telephone company based on where we called.

TUR Traffic Usage Recorder. A device which connects to a network element in order to capture and record traffic statistics. Most network elements (e.g., PBXs, ACDs, data switches and routers) have special ports to which such a device can connect, usually via a RS-232 cable. As traffic flows through the network element, various information about that traffic is output to the TUR in real time. The TUR holds that raw data in buffer memory until such time as it is polled by a centralized computer system and the data is downloaded. Subsequently, the data is processed and reports are generated by a traffic analysis application software system. This process of traffic analysis of historical data is essential to the processes of network design and optimization, which balance network performance (availability) against network costs.

Turing Machine A mathematical model of a device that changes its internal state and reads from, writes on, and moves a potentially infinite tape, all in accordance with its present state, thereby constituting a model for computer-like behavior. This is the same Alan Turing who coined the Turing test which I mention under artificial intelligence.

Turbo FAT Turbo FAT is an index NetWare v2.2 creates to group all the FAT (File Allocation Table) entries corresponding to a file larger than 262,144KB. The first entry in the turbo FAT index table consists of the first FAT number of the file. The second entry consists of the second FAT number of the file, etc. The turbo FAT enables a large file to be accessed quickly.

Turbocharging A phone fraud term. Turbocharging is the practice of increasing phone charges by adding on extra time onto each call in the hope that the customer won't notice.

TURN The Utilities Reform Network, formerly known as Toward Utility Rate Normalization. A non-profit consumer advocacy that represents the small customer in California, TURN characterizes itself as the only independent, statewide, consumer utility watchdog group. TURN represents residential and small business consumers on utility (e.g., telecommunications and electric power) before the California Public Utilities Commission (CPUC), the state legislature, and the courts. www.turn.org

Turnaround Time The actual time required to reverse the direction of transmission from sender to receiver or vice versa when using a half-duplex circuit. The turnaround time is needed for line propagation effects, modem timing and computer reaction.

Turnkey System An entire phone system with hardware and software assembled and installed by a vendor and sold as a total package. The term "turnkey" means the buyer is presented with the key to the thing he has just bought. He turns the key and the system will do everything it is supposed to do, including work. Most telephone systems are purchased Turnkey. An integral part of a contract to buy a turnkey phone system is the terms and conditions for the acceptance of the system. Someone has to define what it means for the thing to work, what you expect from it — so you, the buyer, can formally accept the system and thus incur an obligation to pay for it. Defining Acceptance Conditions is no small task on bigger phone systems.

Turnup When a circuit becomes live, it is turned up and working. Turnup is result of completing the installation of a circuit and making it available to the customer who requested it.

Turret A very large key system for financial traders, emergency teams at nuclear power stations and others who need single phone button access to hundreds of people. By simply pushing one button, the user can dial one of hundreds of people. These buttons may be connected to tie lines, foreign exchange lines. They may even be DDD lines with autodial capability. Like all good key systems, the buttons have a lamping display which shows if the particular line is idle, busy, ringing, on hold, etc.

TUV Technischer Uberwachungs-Verein. TUV. A German electrical testing and certification organization similar to Underwriters Laboratories (UL).

TVC See TRUNK VERIFICATION BY CUSTOMER.

TVRO Television Receive Only Earth Station. Earth station equipment that receives video signals from satellite or MDS-type transmissions. Such stations have only receive capability and need not be licensed by the FCC unless the owner wants protection from interference. Authority for reception and use of material transmitted must be given by the sender.

TV/PC Bill Gates' new name for a TV with PC smarts.

TVM See Time-Varying Media.

TVM Object An SCSA definition. An encapsulation of an atomic piece of time-varying media. This encapsulation may be the data itself or a reference to the data.

TVS See TRUNK VERIFICATION BY STATION.

TWAIN TWAIN is a protocol intended to make easy the communication between software applications and image-acquisition devices, such as cameras and scanners. It is, in essence, a cross-platform application interface standard for image capturing. It allows you to bring images into imaging programs (like HiJaak Pro, Hotshot Graphics, DocuWare, Documagic) from your graphics hardware — for example, desktop scanners, hand-held scanners, slide scanners, frame grabbers and digital cameras. If your hardware is TWAIN compliant and if you have installed the correct driver, you should be able to use that imaging hardware with any TWAIN complaint application. The TWAIN protocol is the most popular protocol for imaging sources and has become an industry standard. Any application that supports TWAIN can communicate with any TWAIN-compliant imaging device. TWAIN is spearheaded by Hewlett-Packard, Logitech, Eastman Kodak, Aldus, Caere and other imaging hardware and software vendors. It was previously known as CLASP and "Direct Connect" during its development stage. Apparently the term TWAIN comes from "Toolkit Without An Interesting Name." Despite its silly name, it is a very serious standard. There are several key elements to TWAIN including:

• Application Layer
This is the application that controls and uses the TWAIN resource.

• Protocol Layer
This contains the TWAIN Source Manager (the code that communicates between
the application and the Source).

• Acquisition Layer
This is software that controls the image acquisition device. The application layer is developed by the device manufacturer. It can be thought of as a hardware device driver.

• Device Layer
This is the physical device, such as a scanner.

TWAIN Working Group An industry organization dedicated to developing and advancing software standards for the imaging world. www.twain.org. See TWAIN.

Tweak Freak A computer techie obsessed with finding the root of all tech problems, regardless of the relevance. A tweak freak might spend hours trying to track down something that could instantly be fixed by reinstalling the software.

Twin Cable A cable composed of two insulated conductors laid parallel and either attached to each other by the insulation or bound together with a common covering.

Twinax Twinaxial Cable made up of two central conducting leads of coaxial cable. See TWINAXIAL CABLE.

Twinaxial Cable Two insulated conductors inside a common insulator, covered by a metallic shield, and enclosed in a cable sheath. Because it carries high frequencies, twinaxial cable is often used for data transmission and video applications, especially for cable television.

Twinning (pronounced twin-ning.) The act of paralleling systems to work together. An example is connecting a wireless system to the same CO line as a key system, so the user can use either instrument to access the trunk.

Twinplex A frequency-shift-keyed, carrier telegraphy system in which four unique tones (two pairs of tones) are transmitted over a single transmission channel (such as one twisted pair). One tone of each tone pair represents a "mark," and the other, a "space."

Twist 1. A change, as a function of temperature, in the response characteristic of a transmission line.
2. The amplitude ratio of a pair of DTMF tones. Because of transmission and equipment variations, a pair of tones that originated equal in amplitude may arrive with a considerable difference in amplitude. In short, signals at different frequencies are transmitted with differing response by the transmission system. Twist usually refers to distortion of DTMF signals.

Twisted Pair Two insulated copper wires twisted around each other to reduce induction (thus interference) from one wire to the other. The twists, or lays, are varied in length to reduce the potential for signal interference between pairs. Several sets of twisted pair wires may be enclosed in a single cable. In cables greater than 25 pairs, the twisted pairs are grouped and bound together in a common cable sheath. Twisted pair cable is the most common type of transmission media. It is the normal cabling from a central office to your home or office, or from your PBX to your office phone. Twisted pair wiring comes in various thicknesses. As a general rule, the thicker the cable is, the better the quality of the conversation and the longer cable can be and still get acceptable conversation quality. However, the thicker it is, the more it costs. Here's a historical and technical explanation from Ray Horak's best-selling book, Communications Systems & Networks:
Metallic wires were used almost exclusively in telecommunications networks for the first 80 years, certainly until the development of microwave and satellite radio communications systems. Initially, uninsulated iron telegraph wires were leased from Western Union for this purpose, although copper was soon found to be a much more appropriate medium. The early metallic electrical circuits were one-wire, supporting two-way communications with each telephone connected to ground in order to complete the circuit. In 1881, John J. Carty, a young American Bell technician and one of the original operators, suggested the use of a second wire to complete the circuit and, thereby, to avoid the emanation of electrical noise from the earth ground. This second conductor also supports common Central Office battery; as a result, your phone stills works when the lights go out. Twisted pair involves two copper conductors, which generally are solid core, although stranded wire is used occasionally in some applications. Each conductor is separately insulated by polyethylene, polyvinyl chloride, flouropolymer resin, Teflon, or some other low-smoke, fire retardant substance. The insulation separates the conductors, thereby avoiding shorting the electrical circuit which is accomplished by virtue of the two conductors, and serves to reduce electromagnetic emissions. Both conductors serve for signal transmission and reception. As each conductor carries a similar electrical signal, twisted pair is considered to be a (electrically) "balanced" medium. The twisting process involves the separately insulated conductors being twisted 90⁻ at routine, specified intervals, hence the term twisted pair. This twisting process serves to improve the performance of the medium by containing the electromagnetic field within the pair. Thereby, the radiation of electromagnetic energy is reduced and the strength of the signal within the wire is improved over a distance. Clearly, this reduction of radiated energy also serves to minimize the impact on adjacent pairs in a multi-pair cable configuration, as the other conductors absorb that radiated electromagnetic energy much as an antenna would absorb a radio signal. This is especially important in high-bandwidth applications, as higher frequency signals tend to attenuate (lose power) more rapidly over distance. Additionally, the radiated electromagnetic field tends to be greater at higher frequencies, thereby impacting adjacent pairs to a greater extent. Generally speaking, the more twists/ft., the better the performance of the wire. Several sets of twisted pair wires may be enclosed in a single cable. In cables greater than 25 pairs, the twisted pairs are grouped into "binder groups," which are contained within a common cable sheath. Twisted pair cable is the most common type of transmission media. It is the normal cabling from a central office to your home or office, or from your PBX to your office phone. Twisted pair wiring comes in various thicknesses, or gauges. As a general rule, the thicker the conductor, the bet-

ter the quality of the conversation and the longer cable can be and still get acceptable conversation quality. However, the thicker it is, the more it costs. Most twisted pair circuits are Unshielded Twisted Pair (UTP). UTP involves no special shielding-just simple insulation. Shielded Twisted Pair (STP) sometimes is used in high-noise environments, where the cable must be run in proximity to electric motors or other sources of ambient electromagnetic interference which can distort the signal. STP looks much like a coaxial cable, as the central conductors are insulated and then surrounded by an outer conductor (shield) of steel, copper alloy, or some other metal. The STP shield absorbs the ambient noise, and conducts it to ground, thereby protecting the center conductor. See also Attenuation, Cat 1-5, STP, and UTP.

Twitch Game A computer or arcade game that's all hand-eye coordination and little brain. Similar to "thumb candy."

Two Dimensional Coding A data compression scheme in facsimile transmission that uses the previous scan line as a reference when scanning a subsequent line. Because an image has a high degree of correlation vertically as well as horizontally, two-dimensional coding schemes work only with variable increments between one line and the next, permitting higher data compression. See ONE DIMENSIONAL CODING.

Two Electrode Vacuum Tube A vacuum tube having a hot cathode and a relatively cold anode, i.e., one with filament and plate only.

Two Hots In Outlets In AC electrical power, more than one HOT conductor has been incorrectly connected to the terminals in the outlet being tested. Dangers include extreme fire hazard and/or major damage to equipment plugged into the outlet.

Two Out Of Five Code A decimal code system in which each decimal digit is represented by five binary bits, two of which are ones and three are zeroes.

Two Out Of Three Rule When determining state tax jurisdiction for the purpose of figuring phone bills, there are three locations to consider: originating station, destination station, and the location that the bill is sent to. If two out of three are the same, then that state receives the tax.

Two Party Hold On Console Allows an attendant to hold a call with both a calling and a called phone (or trunk) connected. Such a feature is required for activation of Attendant Lockout, Serial Call and Trunk-to-Trunk connections features.

Two Party Station Service PBX system with two internal phones, each with selective ringing. Resembles rural two party service of old.

Two Pilot Regulation In FDM systems, the use of two pilot frequencies within a band so that the change in attenuation due to twist can be detected and compensated for by a regulator.

Two Tier Pricing A complex and now largely obsolete AT&T pricing plan which imposed two monthly "rate elements" on every hardware piece of an AT&T (now Lucent) telephone system. Tier A was a fixed rate, not subject to rate increases. It was fixed for a certain number of months, say 60. It was, allegedly, to pay for the system. At the end of the 60 months, Tier A disappeared, as though it were a full-payout lease and you now owned the equipment (which you didn't.) Tier B is the second element in this pricing scheme. It covers maintenance, and it is subject to rate increases. Neither AT&T nor Lucent don't offer two-tier pricing any longer. Many two-tier contracts are now finding their Tier A payments ceasing. Remember, the equipment still belongs to Lucent.

Two Tone Key Same as frequency shift keying.

Two Tone Keying In telegraphy systems, a system employing a transmission path composed of two channels in the same direction, one for transmitting the "space" binary modulation, the other for transmitting the "mark" of the same modulation; or that form of keying in which the modulating wave causes the carrier to be modulated with a single tone for the "marking" condition and modulated with a different single tone for the "spacing" condition.

Two Way Alternate Operation Transmission in one direction or the other but not in both simultaneously. Most often referred to as half-duplex transmission.

Two Way Simultaneous Operation Transmission and reception at the same time. More often referred to as full-duplex transmission.

Two Way Splitting PBX feature. Allows a telephone user to jump back and forth between two calls. Try this: Someone calls you. You both decide you want to speak to a third person. You call that person and conference the three of you together. Then you decide you want to consult with one of the people confidentially. So you "split" one from the other and you speak to one. Then you swap back and forth between the two, speaking to one and then the other in complete privacy. It's easier to do this sort of complicated phone transaction on a phone with a LCD screen. Fortunately, these are becoming more common these days.

Two Way Trade A call center term. A schedule trade in which both employees are working each other's schedules.

Two Way Trunk A trunk which can be seized from either end. Can be used to carry conversations into or out of a telephone system, i.e. most trunks. Some trunks are set up as one-way only. A classic one-way trunk is a IN-WATS line. It is designed to only receive calls.

Two Wire Circuit A transmission circuit composed of two wires — signal and ground — used to both send and receive information. In contrast, a four wire circuit consists of two pairs. One pair is used to send. One pair is used to receive. All trunk circuits — long distance circuits — are four wire. A four wire circuit costs more but delivers better reception. All local loop circuits — those coming from a Class 5 central office to the subscriber's phone system — are two wire, unless you ask for a four-wire circuit and pay a little more.

TWT Traveling Wave Tube.

TWX (Pronounced TWIX.) Teletype Writer eXchange. An automatic teletypewriter (i.e. telex-like) switching service where subscribers may dial any other subscriber and send and receive a message. Formerly owned by AT&T and sold to Western Union in 1972. It differed from Telex in that TWX used AT&T's normal long distance phone network, was thus more ubiquitous, was faster than Telex and was incompatible with Telex, which Western Union owned. However, Western Union, in a major accomplishment, got them to talk to each other.

TX The designation of a copper RJ-45 connection for Fast Ethernet.

Tymnet One of the first public X.25 packet switched networks. The name "Tymnet" comes from the fact that this network, like all X.25 packet networks, was established to support timeshare applications. Tymnet was founded by McDonnell Douglas Network Systems Company. BT (British Telecom) acquired the network in 1990; MCI subsequently acquired BT North America, including the North American portion of Tymnet. And MCI was later acquired by Worldcom. See also Time Sharing.

Type 1 Cable The IBM Cabling System specification for two-pair, 22 gauge, solid conductor cable protected with a braided wire shield. Tested to 16 Mbps, Type 1 is used between Token Ring MAUs and from the MAU to the wallplate.

Type 2 Cable The IBM Cabling System specification for a six-pair, 22 gauge shielded cable for voice transmission application. The six-pair version of Type 1 Cable, Type 2 also is tested to 16 Mbps. Typical application is for the two shielded pairs to be used for Token Ring or 10Base-T LANs, with the remaining four pairs being outside the shield and being used for voice transmission. Type 2 is a six-pair equivalent of Category 3 (Cat 3) cable.

Type 3 Cable IBM's term for telephone wire, Type 3 is single-pair UTP (Unshielded Twisted Pair) wire of 22 or 24 gauge, and involving a minimum of 2 twists per foot. It is used in applications such as 4 Mbps Token-Ring networks.

Type 5 Cable The IBM Cabling System specification for 62.5/125 micron multimode fiber optic cable in a two-pair configuration, which is the de facto standard for FDDI (Fiber Distributed Data Interface).

Type 6 Cable The IBM Cabling System specification for two-pair, stranded 26 gauge wire used in patch cable applications, as well for connecting LAN station adapters to wall plates. Type 6 is limited to a distance of 30 meters.

Type 8 Cable The IBM Cabling System specification for untwisted, shielded two-pair, 26 gauge wire. This flat, ribbon cable commonly is used under carpets.

Type 9 Cable The IBM Cabling System specification for two-pair, shielded, 26 gauge wire, which can be either stranded or solid core. Type 9 Cable accepts RJ-45 termination, and typically is used in connecting from the wall plate to the LAN station adapter.

Type 66 Punchdown Block A standard, solderless terminal wiring block used today. Invented by Western Electronic.

Type A Intelligent Network term describing IN (Intelligent Network) services invoked by, and affecting, a single user. Most of them can only be invoked during call setup or teardown.

Type Ahead Imagine a voice processing service. It says "punch in your zip code at the beep." If you are able to punch in your zip code before you hear the beep or before the talking stops, you have "type ahead." If you are unable to punch in your zip code before you hear the beep, you don't have "type ahead." Better interactive voice response systems have "type ahead."

Type Approval A concept in which a design is approved by an agency and all devices subsequently manufactured according to that design are automatically approved.

Type B Intelligent Network term describing IN (Intelligent Network) services invoked at any point by, and affecting directly, several users.

Type I PC Card The thinnest PCMCIA Card from factor at 3.3 mm thick. The Type I format is typically used for various memory enhancements, including RAM, Flash, OTP, SRAM, and EEPROM.

Type II PC Card A PCMCIA Card which is 5 mm thick. This card is typically used for I/O such as modem, LAN, and host communications.

Type III PC Card The thickest PCMCIA Card type at 10.5 mm thick, the Type III Format is primarily used for memory enhancements or I/O capabilities that require more space, such as rotating media and wireless communication devices.

Typebar Linear type element in a printer containing the printable symbols.

Typing Reperforator Same as receive only typing reperforator.

U Interface An ISDN term. The reference point for a BRI (Basic Rate Interface) connection between a telephone company local loop and a customer premises. The U Interface specifies a single-pair loop (physical two-wire twisted pair) over which a logical four-wire circuit is derived. The resulting digital local loop supports three full duplex channels-two B (Bearer) and one D (Data) channel, in a residential or small office application. Each of the B channels provides 64 Kbps of bandwidth. The D channel provides 16 Kbps of bandwidth, 9.6 Kbps of which can be used to support X.25 packet data traffic, and the balance of which is reserved for signaling and control purposes. The U Interface is designed to work over a maximum distance of 18,000 feet, which addresses the vast majority of US local loops. You screw the two "U" wires (local loop pair) coming in from your local ISDN CO into a black box about the size of desk printing calculator, called an NT-1. Out the side of the black box comes four wires, which are called the "S Bus." Onto these four wires you can attach, in a loop configuration (also called single bus), as many as eight ISDN terminals, telephones, fax machines, etc. See ISDN.

U Law Actually it's Mu Law, but the "Mu" symbol isn't available on conventional keyboards. Mu Law is a voice amplitude compression/expansion quasi-logarithmic curve, based on the approximation with 15 linear segments. Used for PCM encoding/decoding in North America. See Mu Law and PCM.

U Plane The user plane within the ISDN protocol architecture; these protocols are for the transfer of information between user applications, such as digitized voice, video and data; user plane information may be carried transparently by the network or may be processed or manipulated (e.g. A- to u-law conversion).

U Reference Point In the U.S., the point that defines the line of demarcation between user-owned and supplier-owned Integrated Services Digital Network (ISDN) facilities.

U-C An ADSL term for the functional interface between the "U" (standard, two-wire, twisted pair local loop) and the POTS splitter at the Centralized (i.e., Central Office) end of the network. The functional equivalent of the U-C at the premise end of the network is known as the "U-R." The asymmetric nature of this technology requires that the "C" and "R" interfaces be distinguished. See also ADSL, SPLITTER and U-R.

U-NII Unlicensed-National Information Infrastructure. A group of three frequency bands, each of 100 MHz in the 5 GHz band, set aside by the FCC in January 1997 for support of a projected family of high-speed, low-power, wireless voice and data devices. Pressuring the FCC and the Clinton administration for the allocations was the WINForum (Wireless Information Network Forum) trade association, led by Apple Computer, manufacturer of the now-defunct Newton PDA (Personal Digital Assistant). The nature of the devices and systems which will make use of the U-NII remain uncertain at the time of this writing.

U-R An ADSL term for the functional interface between the "U" (standard, two-wire, twisted pair local loop) and the POTS splitter at the Remote (i.e., customer premise) end of the network. The functional equivalent of the U-R at the premise end of the network is known as the "U-C." The asymmetric nature of this technology requires that the "C" and "R" interfaces be distinguished. See also ADSL, SPLITTER and U-C.

UA User Agent. An OSI application process that represents a human user or organization in the X.400 Message Handling System. A US creates, submits, and takes delivery of messages on the user's behalf.

UADSL Universal Asymmetric Digital Subscriber Line. A new standard from the telephone companies in order to provide faster access to the Internet for their subscribers and to compete against cable modems. See also Cable Modem and DOCSIS.

UART Universal Asynchronous Receiver/Transmitter. PCs have a serial port, which is used for bringing data into and out of the computer. The serial port is used for data movement on a channel which requires that one bit be sent (or received) after another, i.e. serially. The UART is a device, usually an integrated circuit chip that performs the parallel-to-serial conversion of digital data to be transmitted and the serial-to-parallel conversion of digital data that has been transmitted. The UART converts the incoming serial data from a modem (or whatever else is connected to the serial port) into the parallel form which your computer handles. UART also does the opposite. It converts the computer's parallel data into serial data suitable for asynchronous transmission on phone lines. UART chips control the serial port/s on personal computers. Now read the next definition. See also 16550, INTERRUPT, INTERRUPT LATENCY, INTERRUPT OVERHEAD, INTERRUPT REQUEST and UART OVERRUN.

UART Overrun UART overruns are errors received from a universal asynchronous receiver/transmitter when receiving equipment cannot match transmission speed. UART overrun occurs when the UART's receive buffer is not serviced quickly enough by the CPU, and the next incoming byte of data crashes into the previous byte. The previous byte is then lost, forcing the communications driver to report and error. Your communications software must then ask for a retransmission of the lost data. High interrupt overhead is the most common cause of a UART overrun. The easiest way of solving UART overrun is to get yourself a UART with a 16 byte buffer (like the 16550), not one of the old more typical one byte buffer UART. See 16550 and UART.

UAS UnAvailable Seconds. A count of the number of seconds that a circuit or path is unavailable.

UAWG Universal ADSL Working Group. A group aimed at accelerating the adoption and availability of high-speed digital Internet access for the mass market, the UAWG has the goal of proposing "a simplified version of ADSL which will deliver to consumers high-speed modem communications over existing phone lines based on an open, interoperable ITU standard." That standard is anticipated to be in the form of an interoperable extension of the ANSI T1.413 ADSL, and is known variously as ADSL Lite, G.lite and Splitterless ADSL. Notably, ADSL Lite will be application-specific, specifically for Internet access, which is very much unlike the original ADSL concept. ADS_ Lite will allow access to the Internet through a modem (either internal or external) operating at speeds of as much as 1.5 Mbps over existing twisted pair local loops of relatively short length and of good quality. UAWG members and supporters include approximately 300 major carriers and manufacturers of computers and other equipment. See also ADSL.

UBR Undefined Bit Rate or Unspecified Bit Rate. Traffic class defined by the ATM Forum. UBR is an ATM service category which does not specify traffic related service guarantees. Specifically, UBR does not include the notion of a per-connection negotiated bandwidth. No numerical commitments are made with respect to the cell loss ratio experienced by a UBR connection, or as to the cell transfer delay experienced by cells on the connection.

UCA Utility Communications Architecture. An architecture for networks used to monitor and control electric power distribution systems.

UCAID University Corporation for Advanced Internet Development. Here's the explanation: Internet2 is the next generation Internet, replacing the current Internet exclusively for the use of member universities, Internet2 is a UCAID project. As a result of what they saw as the deteriorating performance of the Internet, 34 U.S. universities announced in October, 1996 the formation of Internet 2. Subsequently, the central goals of the project were adopted as part of the Clinton administration's Next Generation Initiative (NGI). This second version of the Internet is a collaboration of the National Science Foundation (NSF), the U.S. Department of Energy, over 110 research universities, and a small number of private businesses. Each participating university has committed at least $500,000 to fund the project. Intended to serve as a private Internet for the exclusive use of its member organizations, it will be separate from the traditional Internet. The network eventually will operate over fiber optic transmission facilities at speeds of up to 2.4 Gbps (SONET OC-48), although current speeds of connection are at 155 Mbps (OC-3) and 622 Mbps (OC-12), but going higher to OC-48. Internet2 will connect through gigiPOPs, switches with throughput in the range of billions of packets per second, and will run the IPv6 protocol. www.internet2.edu. See also Internet.

UCC Uniform Commercial Code.

UCD Uniform Call Distributor. A device for allocating incoming calls to a bunch of people. Less full-featured than an Automatic Call Distributor. For a bigger explanation see UNIFORM CALL DISTRIBUTOR.

UCT Universal Coordinated Time. See ZULU TIME.

UDC Connector These connectors are used to terminate 2-pair STP cable. UDC connectors form a hermaphroditic connection, meaning that there is no jack (female end) or plug (male end).

UDF Universal Disk Format. See OSTA.

UDI Unrestricted Digital Information.

UDK A dumb GTE abbreviation, for Universal Dialing Keyset, a key pad that is switchable for either TONE or PULSE dialing. Outside GTE's private world, a keyset would mean a KEY TELEPHONE, not part of a phone.

UDOP The ultimate dumb, open programmable (UDOP) switch built from multi-vendor SC-based products. A term coined by Dialogic.

UDP User Datagram Protocol. A TCP/IP protocol describing how messages reach application programs within a destination computer. This protocol is normally bundled with IP-layer software. UDP is a transport layer, connectionless mode protocol, providing a (potentially unreliable, unsequenced, and/or duplicated) datagram mode of communication for delivery of packets to a remote or local user.

UDP/IP User Datagram Protocol/Internet Protocol. See UDP.

UDSL Unidirectional HDSL (High-bit-rate Digital Subscriber Line). A variation on the HDSL theme proposed by a small group of companies in Europe. See HDSL and xDSL.

UEM Universal Equipment Module. A Northern Telecom acronym. The basic unit of Meridian 1 PBX modular packaging. A UEM is a self-contained hardware cabinet housing a card cage, with a power supply, backplane, and circuit cards. If the UEM has the card cage for an AEM installed, it functions as an AEM.

UFGATE A program which enables a FIDO compatible bulletin board system to exchange UUCP mail with UUCP sites.

UG UnderGround.

UHF The Ultra High Frequency part of the radio frequency spectrum ranging between 300 Megahertz and 3 Gigahertz.

UI 1. User Interface, as in GUI, or Graphical User Interface. 2. UNIX International is a consortium of computer hardware and software vendors which is interested in the development of open software standards for the UNIX industry. Prominent members include AT&T, Sun, UNISYS and Fujitsu.

UIFN Universal International Freephone Number. In early June 1996, the ITU-T released its approval of a new standard, called E.169. This standard will allow International Freephone Service customers to be allocated a unique Universal International Freephone Number (UIFN) which will remain the same throughout the world, regardless of country or telecommunications carrier. "Freephone" is a service which permits the cost of a telephone call to be charged to the called party, rather than the calling party. In North America, 800 and 888 numbers are "freephone" numbers. According to the ITU-T, a UIFN is composed of a three digit country code for global service application, 800, and an 8-digit Global Subscriber Number (GSN), resulting in an eleven digit fixed format, which allows companies to choose the digits they wish and embed existing freephone numbers into the available number space.

UIS Universal Information Services. AT&T's vision of a single fully-integrated, user-defined digital network with a universal port of entry. Very similar to ISDN, now aggressively adopted by AT&T.

UKERNA UK Education and Research Networking Association. See JANET, Super-JANET.

UL Underwriters Laboratories, a privately-owned company that charges manufacturers a stiff fee to make sure their products meet the safety standards which UL itself develops. A UL label on a product has a very specific message. It says the product confirms to the safety standards UL has developed - nothing more. It does not affirm that the product will work. UL is now beginning to concern itself with adopting and promulgating standards (which have nothing to do with safety standards) including those relating to cabling. www.ul.com. See UL Approved, UL Cable Certification Program and UL NNNN. APPENDIX, www.ul.com

UL 1449 A method of rating and approving surge suppressors. This Underwriters Laboratories measurement is important as it tells if you're buying a true surge suppressor or just an extension cord. This listing measures how much voltage actually reaches the attached equipment after going through the surge suppressor. It's on a scale from about 330 volts to 6,000 volts. The lower the rating, the greater the protection. Decent surge suppressors tend to be rated around 400 volts for the basic units and 340 for the advanced and superior models. In short, check for UL 1449 rating on your surge arrestor before you buy it.

UL 1459 Effective 7/1/91, telephone equipment manufacturers will be required to provide protection from current overloads and power line crosses on equipment systems. Equipment systems covered under this listing requirement include single- and multi-line telephones, PBXs, key systems and central office switches. In general, the UL 1459 requirements apply to any location where wires enter a building from the public network, as well as in most IROB (In Range Out of Building) situations. See also NEC REQUIREMENTS and UNDERWRITERS LABORATORIES.

UL 1863 This requirement covers miscellaneous accessories intended to be electrically connected to the telecommunications network. The listing requirement applies to components that comprise the premises communications wiring system from the point of demarcation up to and including the final outlet providing modular plug and jack connection (or equivalent). Requirements are listed under Communication Circuit Accessories, UL 1863. Listing equipment for all other equipment will be covered under UL 1459, effective July 1, 1991. See also NEC REQUIREMENTS and UNDERWRITERS LABORATORIES.

UL 497 & 497A According to the National Electrical Code, primary and secondary protection systems that will be used on a telephone circuit must be listed for that purpose. The listing requirements are UL 497 for primary protection systems and UL 497A for secondary protection systems. See also NEC REQUIREMENTS and UNDERWRITERS LABORATORIES.

UL Approved Tested and approved by the Underwriters Laboratories. The Underwriters Laboratories, Inc. was established by the National Board of Fire Underwriters to test equipment affecting insurance risks of fire and safety. Most phone systems are tested and approved. Most of the testing focuses on the power supply feeding the phone system. The power supply is that little black box that plugs into the AC wall outlet at one end, takes 120 volt AC and converts it to low voltage DC power that the phone system typically runs on. If the power supply tests OK, then that's usually sufficient UL testing. For it is the power supply — and what happens to the commercial AC power that feeds into the power supply — that determines the potential fire hazard of your phone system. After many fire deaths in recent years, most local communities are a lot more concerned about UL Approval of installed telephone equipment. Fire departments have been known to zealously enforce these rules. In addition to the UL approval, the other major fire concern is the use of proper wire in new building construction, with especial emphasis on teflon-covered cable in plenum ceilings. See also UL, an entry which talks about UL's expanding certification business. See also UL CABLE CERTIFICATION PROGRAM.

UL Cable Certification Program United Laboratories, in conjunction with companies such as Anixter, has developed a Data-Transmission Performance-Level Marking Program that covers UL Listed communications cable or power-limited circuit cable. The UL program identifies five levels of performance. UL evaluates cable samples to all of the tests required for each level. Only Levels II through V require testing.

LEVEL I: Level I cable performance is intended for basic communications and power-limited circuit cable. There are no performance criteria for cable at this level.

LEVEL II: Level II cable performance requirements are similar to those for Type 3 cable (multi-pair communications cable) of the IBM Cabling System Technical Interface Specification (GA27-3773-1). These requirements apply to both shielded and unshielded cable constructions. Level II covers cable with two to 25 pair twisted pairs of conductors.

LEVEL III: Level III data cable complies with the transmission requirements in the Electrical Industries Association/Telecommunications Wiring Standard for Horizontal Unshielded Twisted-Pair (UTP) Cable and with the requirements for Category 3 in the proposed EIA/TIA Technical Systems Bulletin PN-2841. These requirements apply to both shielded and unshielded cables.

LEVEL IV: Level IV cable complies with the requirements in the proposed National Manufacturer Association (NEMA) Standard for Low-Loss Premises Telecommunications Cable. Level IV requirements are similar to Category 4 requirements of the proposed Electronic Industries Association / Telecommunication Industry Association (EIA/TIA) Technical Systems Bulletin PN-2841. These requirements apply to both shielded and unshielded cable constructions.

LEVEL V: Level V cable complies with the requirements in the proposed National Electrical Manufacturers Association (NEMA) Standard for Low-Loss Extended-Frequency Premises Telecommunications Cable. Level V requirements are similar to Category 5 requirements of the proposed Electronic Industries Association/Telecommunication Industry Association (EIA/TIA) Technical Systems Bulletin PN-2841. These requirements apply to both shielded and unshielded cable constructions.

UL evaluates communications and data transmission cable to one of two UL Safety Standards: UL 444, the Standard for Safety for Communications Cable; and UL 13, the Standard for Safety for Power-Limited Circuit Cable.

ULANA Unified Local Area Network Architecture. An ongoing U.S. Air Force project aimed at creating a series of interconnected local area networks using Transmission Control Protocol/Internet Protocol (TCP/IP) as the unifying transport layer.

ULP Upper Layer Protocol. In the context of the OSI Reference Model, a ULP is an application-level protocol which may reside at a higher layer than something like ATM (Layers 1 and 2) or TCP (Layer 3).

ULS User Location Service (ULS) provides a mechanism for users of Microsoft's NetMeeting to locate other people on the Internet, even if their Internet addresses change. A sample of the ULS can be found at http://uls.microsoft.com/.

ULSI Ultra Large Scale Integration, the technique of putting millions of transistors on a single integrated circuit. Compare with LSI (Large Scale Integration) and VLSI (Very Large Scale Integration).

Ultimedia IBM's product in multimedia — combining sound, motion video, photographic imagery, graphics, text and touch into a unified, natural interface representing, in IBM's words, the ultimate in multimedia solutions. Ultimedia supports both Ultimotion and Indeo video. Ultimedia was coined in the Spring of 1992.

Ultimotion IBM's video compression algorithm. Although IBM supports Indeo video technology in OS/2 and Windows systems, IBM feels several OS/2 vertical applications are adequately served by the Ultimotion algorithm. Ultimotion does not offer software scalable playback or single step compression. See ULTIMEDIA.

Ultra Hi-Res Ultra high resolution. Properly speaking, the term should be for monitors with resolutions of 1,200 x 800, 1,024 x 1024 or better, but it is sometimes used to describe monitors with 800 x 600 resolution and above.

Ultra High Frequency Frequencies from 300 MHz to 3000 MHz.

Ultrasonic Bonding The use of ultrasonic energy and pressure to join two materials.

Ultraviolet That portion of the electromagnetic spectrum in which the wavelength is just below the visible spectrum, extending from approximately 4 nanometers to approximately 400 nanometers. Some scientists place the lower limit at values between 1 and 40 nanometers, 1 nm being the upper wavelength limit of X-rays. The 400-nm limit is the lowest visible frequency, namely violet. "Light" in the ultraviolet spectrum is used for erasing EPROMS.

Ultraviolet Fiber Special fiber which extends the usable range into the UV region of the spectrum.

Ultrawide Band Radio Also known as Digital Pulse Wireless. The new technology of ultrawide band radio uses a digital transmission consisting of small on-off bursts of energy at extremely low power but over the entire radio spectrum. According to the New York Times, "by precisely timing the pulses within accuracies of up to a trillionth of a second, the designers of ultrawide band radio systems are able to create low-power communications systems that are almost impossible to jam, tend to penetrate physical obstacles easily and are almost invulnerable to eavesdropping. Police officers could use such a system "to see through" walls and doors to detect the location of people. According to the New York Times, the most promising application for ultrawide band radio might edventually be an alternative to today's wireless office network technologies that are limited in speed. "Because of its design, ultrawide band advocates," according to the Times, say the technology has the potential to deliver vastly higher amounts of data because a large number of transmitters could broadcast simultaneously in close proximity without interfering with one another. See also Bluetooth.

UM 1. Micron (10-6 meters).
2. United Messaging. See Unified Messaging.

UMA An acronym for Upper Memory Area. See UPPER MEMORY AREA.

UMB An acronym for Upper Memory Block, an area of upper memory (the area between 640KB and 1MB of RAM) in an MS-DOS PC that has been remapped with usable RAM. This allows device drivers and TSRs to be loaded high, into the UMB and out of conventional memory. See UPPER MEMORY AREA.

UME UNI Management Entity: The software residing in the ATM devices at each end of the UNI circuit that implements the management interface to the ATM network.

UMIG Universal Messaging Interoperability Group. This group is an offshoot of AMIS (Audio Messaging Interchange Specification). The UMIG is being technically facilitated (their words) by the Information Industry Association in Washington, DC. The UMIG has two top priorities: First, to foster the development of "universal messaging," which entails integrating platforms supporting different types of messaging, such as voice processing, electronic mail and facsimile messaging to allow users to move easily among the three media. The second priority entails working towards standardized addressing and directory schemes that make it easy and intuitive for users to message one another.

UMTS Universal Mobile Telecommunications System. The technology envisioned for the next generation of GSM (Global System for Mobile Communications). UMTS is intended to support data transfer rates of 144 Kbps to 2 Mbps in support of mobile access to multimedia Internet applications. ETSI (The European Telecommunications Standards Institute) currently is evaluating proposals for the UMTS standard. UMTS services are expected to be available in 2002. See also GSM.

Umux Universal Data Stream Multiplexor. Part of Sun Microsystem's XTL Teleservices architecture. Provides a uniform means for applications to access and share data channels associated with a telephone call. Umux is a streams pseudo-device driver used to connect data channels to applications and to control access to data. The server and the provider work together to deliver the stream to the application with Umux.

UN Industry jargon for UNreachable, an unsuccessful call where the agent is unable to speak to the contact or decision maker.

Unassigned Cell An ATM filler cell used to occupy available bandwidth when there are no assigned (user-generated) cells to send. Unassigned cells carry a PCI/VCI (Protocol Control Information/Virtual Channel Identifier) value of 0/0. Unassigned cells are discarded at the ATM Layer. They can be replaced by assigned cells, as required, during the process of cell multiplexing. See also ATM Layer, ATM Protocol Reference Model, PCI, VCI, and Idle Cell.

Unattended Equipment working without a human attendant or operator. There are pros and cons to operators. On the pro side, they offer a personalized service that's absolutely critical to customer goodwill. On the con side, they can be slow and cumbersome. They can be very irksome when you know you could do that task yourself, but have to wait for the operator. Some companies have only one main number. Some companies use a main number and DID — on their Centrex and their PBX. Some companies use an automated attendant and an operator. There's more flexibility with DID and a main number, or Centrex DID and a main number. Customers without knowledge dial the main number. Customers with knowledge can dial direct DID numbers. See also AUTOMATED ATTENDANTS.

Unattended Call Calls placed by a computerized dialing system in anticipation of an agent being available to answer the call. A called party is detected answering the phone and no agent is available to serve the call. The system hangs up on the party so as not to create any greater nuisance than has already occurred. The telemarketing industry does not believe that an unattended call can be queued for the next available agent.

Unbalanced Line A telephone circuit in which the voltages on the two conductors are not equal with respect to ground. Unbalanced lines give poor phone service. Lines can become unbalanced when they come from the central office or when they are in the PBX or the on site phone system. Problems can and should be repaired for decent quality results.

Unbundled 1. Services, programs, software and training sold separately from the hardware.
2. Services and products leased by local phone companies as a result of the Telecommunications Act of 1996. For a much better explanation, see Unbundled Network Element.

Unbundled Network Element UNE (pronounced you-nee). The Telecommunications Act of 1996 requires that the ILECs (Incumbent Local Exchange Carriers) unbundle their NEs (Network Elements) and make them made available to the CLECs (Competitive LECs) on the basis of incremental cost. UNEs are defined as physical and functional elements of the network, e.g., NIDs (Network Interface Devices), local loops, switch ports, and dedicated and common transport facilities. When combined into a complete set in order to provide an end-to-end circuit, the UNEs constitute a UNE-P (UNE-Platform). Unbundled Network Elements is a term used in negotiations between a CLEC (Competitive local Exchange Carrier) and the ILEC (Incumbent Local Exchange Carrier) to describe the various network components that will be used or leased by the CLEC from the ILEC. These components include such things as the actual copper wire to the customers, fiber strands, and local switching. The CLEC will lease these UNEs with pricing based on the previously-signed Interconnection Agreement between the CLEC and the ILEC. Typically, a CLEC will colocate a switch at the ILEC's wirecenter, then pay for the "unbundled" local loop to make a connection to the customer. Alternately, a CLEC might lease both an unbundled local loop and an unbundled switch, and make a connection to their network at the LEC's switch. See CLEC, ILEC, the Telecommunications Act of 1996, UNE Rate and UNE-P.

UNC Universal naming convention. See also UNC NAMES.

UNC Names Filenames or other resources names that begin with the string \\, indicating that they exist on a remote computer.

Uncontrolled Terminal A user terminal that is on line at all times and that does not contain the logic that would allow it to be polled, called, or otherwise controlled by the device to which it is connected.

Under mouse arrest Getting busted for violating an online service's rules of conduct. "Sorry I couldn't get back to you. AOL put me under mouse arrest."

Underfill A condition for launching light into a fiber in which not all the modes that the fiber can support are excited (i.e. turned on).

Underfloor Duct Method A floor distribution method using a series of metal distribution channels, often embedded in concrete, for placing cables. This method uses one or two levels depending on the complexity of the system. Sometimes referred to as underfloor raceways. See also RACEWAYS METHOD.

Underflow In computing, a condition occurring when a machine calculation produces a non-zero result that is smaller than the smallest non-zero quantity that the machine's storage unit is capable of storing or representing.

Underground Cable installed in buried conduit. Does not typically include cables buried directly in the ground.

Underlap In facsimile, a defect that occurs when the width of the scanning line is less than the scanning pitch.

Underlying Carrier A common carrier providing facilities to another common carrier which then provides services to end users.

Underrun A network error indicating that buffer checks show the buffer as empty. Underrunns shouldn't happen in a well managed network. An underrun is often a synchronization problem.

Understaffing Limit A call center term. The percentage by which you'll allow the scheduling process to fall short of the required staffing level in any period. This typically provides more economical coverage during the least-busy periods of the day.

Underwriters Laboratories, Inc. A non-profit laboratory which examines and tests devices, materials and systems for safety, not for satisfactory operation. It also has begun to establish safety standards. See UL for a longer explanation.

Undesired Signal Any signal that tends to produce degradation in the operation of equipment or systems.

Undetected Error Ratio The ratio of the number of bits, unit elements, characters, or blocks incorrectly received and undetected, to the total number of bits, unit elements, characters, or blocks sent.

Undirected Pickup A phone system feature. Undirected Pickup lets you pickup any call ringing at any extension in the pickup group in which your extension is a member. The pickup groups are pre-programmed in the switch.

Undisturbed Day A day in which the sunspot activity or ionospheric disturbance does not interfere with radio communications.

UNE (pronounced you-need). Unbundled Network Element. The Telecommunications Act of 1996 requires that the ILECs (Incumbent Local Exchange Carriers) unbundle their NEs (Network Elements), which must be made available to the CLECs (Competitive LECs) on the basis of incremental cost. This means that CLECs will pay the additional costs the ILECs incur in making these facilities available. the words "incremental cost" are meant to signal to the ILECs that they are not to inflate the price of these facilities by adding overhead costs (e.g. the salary of the ILEC's people in charge of investor relations). UNEs are defined as physical and functional elements of the network, e.g., NIDs (Network Interface Devices), local loops, switch ports, and dedicated and common transport facilities. When combined into a complete set in order to provide an end-to-end circuit, the UNEs constitute a UNE-P (UNE-Platform). Unbundled Network Elements is a term used in negotiations between a CLEC (Competitive local Exchange Carrier) and the ILEC (Incumbent Local Exchange Carrier) to describe the various network components that will be used or leased by the CLEC from the ILEC. These components include such things as the actual copper wire to the customers, fiber strands, and local switching. The CLEC will lease these UNEs with pricing based on the previously-signed Interconnection Agreement between the CLEC and the ILEC. Typically, a CLEC will colocate a switch at the ILEC's wirecenter, then pay for the "unbundled" local loop to make a connection to the customer. Alternately, a CLEC might lease both an unbundled local loop and an unbundled switch, and make a connection to their network at the LEC's switch. See CLEC, ILEC, the Telecommunications Act of 1996, UNE Rate and UNE-P.

UNE Rate The fee, set by state regulators, that an ILEC charges a CLEC to unbundle network elements as part of making the local exchange market competitive. Rebundling is the process of putting UNEs back together by a CLEC to become part of a competitive service offering by him to a customer.

UNE-P Unbundled Network Element-Platform. See UNE.

Unequal Access Refers to long distance phone companies who do not take advantage of Judge Harold Greene's Equal Access divestiture provisions. Rather than a carrier selection code, unequal access carriers require you to dial a local seven digit number and punch in an authorization code. If the carrier elected to pay for Equal Access, you would just dial directly the same 10 digits you do today, and your local telephone company would give your billing number to your long distance company.

Unerase A command for getting back files you've accidentally erased. See MS-DOS.

Ungrounded Not connected to ground. PBXs, key systems and other phone systems will not work well when not connected to a solid ground because they have no place to send high voltage spikes (static electricity, lightning strikes, etc.) Improper grounding is probably the most common cause of phone system faults. Our feeling: the better the ground, the better the phone system performance. One way of grounding is the third wire of an electrical outlet. This may be OK if you check where that wire is ultimately connected to. You can ground to the metal cold water pipe. But that may connect to a plastic PVC pipe one floor below. Best to check. A ground ultimately ending firmly routed a dozen feet below the ground is best.

UNH IOL University of New Hampshire Interoperability Lab. A testing organization affiliated with the Research Computing Center of the University of New Hampshire which tests FDDI products for vendor interoperability.

UNI User Network Interface. Specifications for the procedures and protocols between user equipment and either an ATM or Frame Relay network. The UNI is the physical, electrical and functional demarcation point between the user and the public network service provider. By way of example, the Frame Relay UNI involves both the user's FRAD (Frame Relay Access Device) and the carrier's FRND (Frame Relay Network Device) across a dedicated link. The ATM (Asynchronous Transfer Mode) UNI was developed and is promoted by the ATM Forum; the Frame Relay UNI, by the Frame Relay Forum.

UNI A User Network Interface A. A B-ISDN term for a SONET OC-3 link from the network to the premise, operating at 155 Mbps.

UNI B User Network Interface B. A B-ISDN term for a SONET OC-12 link from the network to the premise, operating at 622 Mbps.

UNI Interface See UNI.

UNIBOL A UNIX version of COBOL.

Unicast The communication from one device to another device over a network. In other words, a point-to-point communication. In ATM, for instance, Unicast describes the transmit operation of a single FDU (Protocol Data Unit) by a source interface where the PDU reaches a single destination. A PDU, by the way, is a single set of data which may be in the form of a block or frame of data comprised of a fixed number of bits, as well as control information; the specifics of the PDU vary according to the nature of the native protocol which governs the process of communications between networked devices. By way of another example, the new IPv6 (Internet Protocol, version 6) specification, supports Unicast, as well as Anycast and Multicast. Contrast with ANYCAST, BROADCAST, IP MULTICAST and MULTICAST.

Unicasting 1. Communicating from one device to another. In contrast, multicasting sends one stream of information to many.
2. As an ATM term, it is the transmit operation of a single PDU by a source interface where the PDU reaches a single destination.

Unicode Unicode is a 16 bit system for encoding letters and characters of all the world's languages. At 16 bits it can encode 65,536 characters. That's two raised to the 16th power. Work it out: Multiply 2
$x\ 2\ x\ 2\ x\ 2\ x\ 2\ x\ 2\ x\ 2\ x\ 2\ x\ 2\ x\ 2\ x\ 2\ x\ 2\ x\ 2\ x\ 2\ x\ 2$.
Sixteen-bit characters (like Unicode) are also called Wide Characters. The first 128 codes of Unicode are identical to

ASCII. Just add another zero byte to each ASCII character to convert to Unicode. Unicode contains over 20,000 Han characters, which are used to represent whole words or concepts in Chinese, Japanese and Korean.

Unidirectional The transmission of information in one direction only.

Unidirectional Bus A distribution conductor or set of conductors that can transfer information in one direction only.

Unidirectional Path Switched Ring UPSR. A SONET transport method in which working traffic is transmitted in one direction. UPSR is preferred for interconnected rings with numerous signals crossing the rings.

Unified Messaging Another name for integrated messaging. See INTEGRATED MESSAGING and TELEPHONY SERVICES.

Unified Network Management Architecture AT&T's architecture for providing end-to-end network management. There are three levels within the architecture — network elements, element management systems, and network management integration.

Uniform Access Number See UNISERV.

Uniform Call Distribution See UNIFORM CALL DISTRIBUTOR.

Uniform Call Distributor UCD. A device for distributing many incoming calls uniformly among a group of people (typically called "agents" because of the early use of these machines by the airline, hotel and car reservation industry). These days the term Uniform Call Distributor is falling into disrepute as the newer term, Automatic Call Distributor comes in. According to incoming call experts, a Uniform Call Distributor is generally less "intelligent," and therefore less costly than an ACD. A UCD will distribute calls following a predetermined logic, for example "top down" or "round robin." It will not typically pay any heed to real-time traffic load, or which agent has been busiest or idle the longest. Also, a UCD's management reports tend to be rudimentary, consisting of simple pegs counts, as opposed to an ACD, which can produce reports on the productivity of agents.

Uniform Encoding An analog-to-digital conversion process in which, except for the highest and lowest quantization steps, all of the quantization subrange values are equal.

Uniform Linear Array An antenna composed of a relatively large number of usually identical elements arranged in a single line or in a plane with uniform spacing and usually with a uniform feed system.

Uniform Numbering Plan A uniform seven-digit number assignment made to each phone in a private corporate network. Such a plan allows routing of calls to distant phones from any on-net telephone without any differences in the dialed number. Without a uniform numbering plan, you would dial your boss in New York differently if you were in the company's Chicago office and differently again if you were in your company's San Francisco office. With a uniform numbering plan, it would be the same from all locations. The nation's long distance network has, obviously, a uniform numbering plan.

Uniform Resource Locator URL. An Internet term. A standardized way of accessing various resources on the World Wide Web. See URL for a detailed explanation. I believe the correct term is Universal Resource Locator, not Uniform Resource Locator. See URL for a detailed explanation.

Uniform Service Order Code See USOC.

Uniform Spectrum Random Noise Noise distributed

over the spectrum in such a way that the power per unit bandwidth is constant. Also known as "white" noise.

Unimodem Unimodem, the "Universal Modem Driver" for Windows 95 and now Windows NT Server 4.0 and Windows NT Workstation 4.0, is both a TAPI service provider and a VCOMM device driver. It translates TAPI (Windows Telephony API) function calls into AT commands to configure, dial, and answer modems. See AT COMMAND SET and UNIMODEM V. See the following for Unimodem specifics: http://207.68.137.34/ntserver/communications/unimodem.htm

Unimodem V Unimodem stands for Universal Modem Driver. Unimodem V is Unimodem updated for voice. The V stands for voice, not five. It now replaces Unimodem. Unimodem stands for Universal Modem Driver. It is part of Windows 95 and Windows NT Server 4.0 and Windows NT Workstation 4.0. It is both a TAPI service provider and a VCOMM device driver. It translates TAPI (Windows Telephony API) function calls into AT commands to configure, dial, and answer modems. Unimodem V is the universal modem driver and telephony service provider for the Windows operating system. Included in Unimodem V are the features requested most often by users to support voice modems, including wave playback and record to/from the phone line, wave playback and record to/from the handset, and support for speakerphones, caller I.D., distinctive ringing and call forwarding. Unimodem now supports the most popular voice modems on the market. For Unimodem/V specifics: http://207.68.137.34/corpinfo/press/1995/95dec/unimdmpr.htm

Uninstalled Euphemism for being fired. Heard on the voicemail of a Vice President at a downsizing computer firm: "You have reached the number of an uninstalled Vice President. Please dial our main number and ask the operator for assistance." See also Decruitment.

Unintelligent Crosstalk Crosstalk giving rise to unintelligent signals.

Uninterruptible Power Supply UPS. A device providing a steady source of electric energy to a piece of equipment. A continuous on-line UPS is one in which the load is continually drawing power through the batteries, battery charger and invertor and not directly from the AC supply. A steady off-line UPS normally has the load connected to the AC supply. When the line is weak or down, it transfers the load without any user intervention. UPS are typically used to provide continuous power in case you lose commercial power. An UPS is typically a bank of wet cell batteries (similar to automobile batteries, but often much, much larger) engineered to power a phone system up to eight hours without any re-charging. A UPS system can also include a gasoline-powered generator. And if the generator works (make sure it has gas), you can power your phone system for much longer. According to Bell Labs, however, over 90% of all power outages last less than five minutes. The cost of Uninterruptible Power Supplies is typically a direct function of how large the battery/batteries are. The larger the batteries, the higher the cost. Many file servers on local area networks are also backed by UPSes. Many NetWare file servers, which are protected by a UPS, often are attached a printed circuit card inside the server. This card acts as an early warning system. When AC power drops, and the UPS takes over, it signals the file server through the card what has happened. The file server then will send a message to all the workstations on the network that the file server has lost AC power, is running on battery power, is running out and would

everyone kindly log off the server. This protects the network.

Unipolar Signal A two-state signal where one of the states is represented by voltage or current and the other state is represented by no voltage or no current. The current flow can be in either direction.

Unique Addressing The addressing of a node by using the software-programmable address assigned to each one upon system initialization. For example, TELECONNECT's LAN has a "unique" addressing scheme. Each workstation is known by the operator's first name.

Unisource A European provider of Virtual Network Services (VNS). Unisource was created in 1994 by three European carriers: PTT Telecom Netherlands, Swiss Telecom PTT, and Telia of Sweden. Unisource provides a wide range of voice, data, and Internet services in a number of countries through its equity partners, as well as through a group of distributors. Distributors include AT&T-Unisource Communications Services, WorldPartners, and Infonet. The three equity partners merged their international networks in June 1997 into Unisource Carrier Services (UCS), which operates a fiber optic backbone running ATM. See also VNS.

Unit Interval In a system using isochronous transmission, that interval of time such that the theoretical durations of the significant intervals of a signal are all whole multiples of this interval. The unit interval is the shortest time interval between two consecutive significant instants.

United States Telephone Association USTA. The largest trade association of telephone companies, with membership of over 1,200. USTA has its roots in the National Telephone Association, formed in 1897 to unite independent (non-Bell) telephone companies. Subsequently, the organization changed its name to the USITA (United States Independent Telephone Association). After the break-up of AT&T in 1984, the RBOCs (regional Bell Operating Companies) were admitted as members, and the name was shortened to USTA. The organization, based in Washington, lobbies the FCC, Congress and other regulatory, legislative and judicial bodies to ensure that no regulations or legislation are passed to the detriment of its members. USTA also provides a little education for its members. USTA had a sister organization, the United States Telephone Suppliers Association, which is now merged with the EIA, forming the Telecommunications Industry Association. www.usta.org See the next definition.

United States Telephone Suppliers Association USTSA. An association of suppliers — manufacturers and wholesalers — which originally was a committee of the United States Telephone Association (USTA). USTSA merged with EIA/ITG (Electronic Industries Association/Information and Telecommunications Technologies Group) in 1988 to form the TIA (Telecommunications Industry Association), which operates under the umbrella of the EIA. www.tiaonline.org and www.eia.org See also TIA and EIA.

Unity Gain Refers to the balance between signal loss on a broadband network and signal gain through amplifiers.

Universal Access Number A single number dialed from anywhere in the country which will route a customer to one or several locations for service, advice, etc. The definition varies depending on whose networking scheme you're dealing with.

Universal Addressing The addressing of a node by the use of the universal addresses which all nodes recognize.

Universal ADSL See ADSL Lite

Universal Agent A telephone agent who answers incoming calls and also makes outgoing calls. This duality feature may not seem worth of its inclusion in this dictionary. But the fact is that agents have largely been just "inbound" or just "outbound" — because managers felt that most agents were not capable of doing both. The skills were, allegedly, too different. Now the idea is to "empower" the agent with more flexibility and make them "universal," i.e. capable of being used for both inbound and outbound.

Universal Asynchronous Receiver-Transmitter UART. A device that converts outgoing parallel data from your computer to serial transmission and converts incoming serial data to parallel for reception. See UART for a bigger explanation.

Universal Circuit Card See UNIVERSAL TRUNK CARD.

Universal Device A SCSA device. A call processing device which has every conceivable resource for the handling of calls. The SCSA programming applies resources from many different physical devices to a call processing task. These then act as if they were a single universal device.

Universal Edge Server MediaGate, San Jose, CA, defines a "universal edge server" as a new breed of Remote Access Server (RAS) that offers telephony functionality combined with traditional data remote access capabilities. Such Universal Edge Server allows user to combine voice, email and pager communications into a single, secure message box, accessible via phone, fax, web browser or email client.

Universal Mailbox Allows a user of unified messaging services to have single access to all messages from internal and external electronic mail systems, fax systems and voice mail systems. A really neat idea given today's lack of standardization among electronic mail services.

Universal Name Space The set of all unique object identifiers in a domain, network, enterprise, etc. Object naming standards and methods for locating and sending messages to mobile objects are required in large-scale object-oriented distributed-computing systems.

Universal Night Answer A feature of telephone systems that permits any phone to pick up any incoming trunk call when the Attendant's console is unmanned (unpersonned?) and the phone system is set up (typically at the console) for "Night Answer."

Universal Pay Phone Description for a coin-and-credit-card phone.

Universal Ports A modern telephone system is typically an empty cabinet into which you slide printed circuit cards. Those cards have an edge connector and they slide into a connector at the rear of the cabinet. That connector connects via wires to other connectors in what is typically called the phone system's bus. In the old days, phone systems had dedicated slots — meaning you could only slide one type of printed circuit card into that particular slot. As phone systems got more advanced, they acquired "universal ports." Our definition of a universal port is that all the slots are totally flexible — namely that you can slide any trunk or phone card (either electronic or single line phone) into any slot in the phone system. The advantage of this is obviously a far more flexible phone system, able to accommodate lots of phones and few trunks or vice versa.

Universal Power Supply A power supply which you can plug into electricity ranging from 100 volts to 240 volts AC. With a universal power supply (now standard with many laptops) you can travel the world, plugging yourself into virtually any power outlet and have your device work perfectly, without the need for a transformer. What you'll need, however, is a plug that converts the plug you have into the necessary plug for that country. Such a converter plug shouldn't cost you more than $2.

Universal Resource Locator URL. An Internet term. A URL is a fancy name for an address on the World Wide Web. In more technical terms, a URL is a string expression that can represent any resource on the Internet or local TC/IP system. The standard convention for a URL is as follows:
method://host_spec {port} {path} {file} {misc}
Here's an example of a URL. http://www.harrynewton.com. Typing those letters into your browser brings you to the opening screen — or home page — of my web site. In general http:// can be safely omitted with most browsers and you'll still get to the site.

Universal Sender Allows the dialed number to be sent out by the user.

Universal Serial Bus See USB.

Universal Service Originally conceived by the first chairman of the Bell System, Theodore Vail, universal service has been the goal of the entire telephone industry, including its Federal and state regulators, since the 1920s. The idea of universal service is to have residential telephone service priced sufficiently low so anyone in the United States can afford it. Keeping residential service low has been the reason why local business service is usually priced much higher — though the two services are usually identical. Universal Service is now effectively accomplished. It has become a political rallying cry, used to justify the strange and wonderful pricing schemes the phone industry and its regulators create with little regard for the actual costs of providing those services.

The Communications Act of 1934 defined universal service policy as "To make available, so far as possible, to all the people of the United States a rapid, efficient Nationwide, and worldwide wire and radio communication service with adequate facilities at reasonable charges." The same act created the FCC (Federal Communications Commission), charging it with the responsibility to carry out this policy, as well as to regulate the telecommunications industry, in general. Prior to the breakup of the Bell System in 1983, AT&T and the BOCs (Bell Operating Companies) administered this fund through the "settlements" process, which essentially reimbursed the LECs (Local Exchange Carriers) for the use of their local networks in originating and terminating long distance calls. "High cost" (i.e., rural) LECs were compensated at very high levels, in recognition of the universal service policy. Since 1983, NECA (National Exchange Carrier Association) has been charged with this responsibility.

The Telecommunications Act of 1996 considerably expands the definition of "universal service" to include "access to advanced telecommunications and information service...in all regions of the Nation, including low-income consumers and those in rural, insular, and high cost areas...reasonably comparable to those services...and those rates...for similar services in urban areas." The Act goes on to provide for discounts to elementary and secondary schools and classrooms, health care providers, and libraries. See also FCC, HIGH COST, NECA, SEPARATIONS AND SETTLEMENTS and UNIVERSAL SERVICE FUND.

Universal Service Fund USF. Under the direction of the FCC, the National Exchange Carrier Association (NECA) administers the USF, which is a cost allocation mechanism designed to keep local exchange rates at reasonable levels, especially in "high cost" (i.e., rural) areas. NECA administers the program by collecting USF data, determining LEC eligibility, billing the IXCs (long distance phone companies), and distributing the payments. The goal of the Universal Service Fund is to provide at least once access line for basic telephone service to every household in the U.S. The fund gets money from a surcharge on phone lines. The fund's goal is to offset operating costs of small telcos. See also FCC, NECA and UNIVERSAL SERVICE.

Universal Trunk Cards Most PBXs have different circuit boards (or circuit cards - same thing) for combination trunks and for DID (direct inward dial) trunks. A Universal Circuit Board enables you to use the ports on the board for either combination trunks OR DID trunks. This capability makes the PBX more flexible. If you have spare ports on a trunk circuit board you may use them for either a combination or a DID type of trunk.

Universal Turret A very large key system for financial traders, emergency teams at nuclear power stations and others who need single phone button access to hundreds of people. By simply pushing one of the button in front of them, the user can dial one of hundreds of people. These buttons may be connected to tie lines, foreign exchange lines. They may even be DDD lines with autodial capability. Like all good key systems, the buttons have a lamping display which shows if the particular line is idle, busy, ringing, on hold, etc.

Universal Wall Jack There's really no such animal. Every manufacturer of installation gadgetry is trying to propagate the idea that their jack is universal, when it really isn't. The "universal" wall jack we installed in our new offices is actually four jacks — 1. Four pairs for two PBX voice lines (one electronic two-pair phone and one tip and ring phone) and one spare. 2. One RS-232-C 12-conductor shielded cable for connecting to centralized printers, for connecting to a dataPBX and for permanent null-modem connection of computers. 3. One for connecting to our high-speed, one megabit per second LAN, and 4. One spare twisted, shielded, stranded pair for a second LAN, or whatever comes along.

Universe A call center term. The total number of names to be attempted on an outbound call program.

UNIX An immensely powerful and complex operating system for computers for running data processing and for running telephone systems. UNIX provides multi-tasking, multi-user capabilities that allow both multiple programs to be run simultaneously and multiple users to use a single computer. On a single-user system, such as MS-DOS, only one person at a time, on an individual task basis, can use a computer's files, programs, and other resources. UNIX works on many different computers. This means you can often take applications software which runs on UNIX and move it — with little changing — to a bigger, different computer, or to a smaller, computer. This process of moving programs to other computers is known as "porting." Today, the UNIX operating system is available on a wide range of hardware, from small personal computers to the most powerful mainframes, from a multitude of hardware and software vendors. UNIX was developed in 1969 by Ken Thompson of AT&T Bell Laboratories. UNIX is also a trademark of UNIX Systems Laboratories, Inc. which used to be owned by Novell Inc. but then was sold to SCO, Santa Cruz Operations.

UNIX-To-UNIX Copy Program UUCP. A standard UNIX utility for exchanging information between two UNIX-based machines in a network. UUCP may also be referred to as the UNIX-TO-UNIX Communications Protocol and is widely used for electronic mail transfer.

Unlicensed PCS Unlicensed PCS is the name for wireless frequency in the PCS band, which in the United States is 1.920 GHz - 1.930 GHz. The advantage of unlicensed PCS is that you can install a wireless telephone system in your company in this band without having to secure licenses from the Federal Communications Commission. Such systems are often called "business wireless." Such systems are typically all digital and often hang off a PBX. And the wireless phones often will have most of the features that an electronic phone wired to the PBX would have.

Unlisted Number There are various interpretations of what constitutes an "unlisted, an "unpublished" or a "non-published" phone number in North America. Some phone companies use these words interchangeably. Some don't. In California, Pacific Bell offers unpublished phone service. Your phone number is not listed in the paper phone directories, but is listed with dial up "Directory Assistance." Pacific Bell also has a more expensive service called "Unlisted Service." Here, your phone number is not included in the paper phone directories or given out to callers to Directory Assistance. Nynex (and I presume other phone companies) has a service whereby you can leave a message for the owner of an unlisted number. "Please call me. You've won the lottery." The owner of the unlisted number then has the choice to return the call or not. He doesn't pay to receive this message. Some telephone companies confuse the definitions and some invent new ones. For example, some phone companies use the term "non-published" number. You won't find the number in a phone book or by calling Directory Assistance. Over 25% of many private phone numbers in major metropolitan areas are unlisted, unpublished or non-published — a "service" their subscribers pay extra for. To my simple brain, it's a lot easier to simply publish your name as "Apple Plumpudding." See UNPUBLISHED.

Unloaded Line A telephone line with its loading coils removed to increase the distance and speed with which data may be transmitted over the line. A fee is usually charged for removing the coils.

Unlock Code This is a three-digit number required to unlock a cellular phone when you have electronically locked it to prevent unauthorized use. You might lock it when you park your car in a hotel. The factory default is 123.

UNMA Unified Network Management Architecture. AT&T's proprietary architecture for network management.

UNMR Universal Network Management Record.

Unnumbered Command In a data transmission, a command that does not contain sequence numbers in the control field.

Unnumbered Information Frame U1 frame. A transmission frame generated by the High-Level Data Link Control (HDLC) data link protocol, where no flow control and no error control are implemented.

UnPBX An UnPBX is a server on a LAN, dedicated to communications. The UnPBX is comprised of four elements. First, it is one or several joined PCs; Second, it has boards inside the PC and software to run them; Third, it's joined to the same phone lines a PBX is joined to — analog POTS lines to digital T-1 lines; And fourth, it's joined to a local area network. Some of the cards that drop into an UnPBX are proprietary. Most are not. Most UnPBXs run on familiar operating systems — NT and Unix, etc. Most UnPBXs are open. They offer far more programming "hooks" than any other telephony device, including the "open" PBX. Most UnPBXs drop a phone on your desk via a tip and ring line. You speak on the phone. You can dial on the phone, or dial via the screen on your PC. Some UnPBXs zing your voice over ATM, Ethernet or the Internet. Some use the multimedia-equipped PC to do the dialing, speaking and phoning. You typically do everything else — from checking your messages to setting up conference calls — via software on your LAN-attached PC. Some UnPBXs have their own desktop software. Some use a browser — Netscape or Internet Explorer. Some people call the UnPBX a "total communications server." Here's what the UnPBX does:

1. Typically, the UnPBX is an office's phone system. It is a PBX in an

PC. You dial your mother from your desk through it. It will switch incoming calls to your desktop. You can call the guy in the next office on it. In short, an UnPBX has all the basic PBX functions. The UnPBX is your office's auto attendant / voice mail system. It answers incoming calls, gives out a message, listens for tones and transfers the call to you. It will alert you to the incoming call, prompt you for what it should do with the call — hold it, transfer it, conference it, dump it in voice mail, etc. If not answered, it will probably put the call in voice mail. It could also dial half a dozen outside numbers and chase you down. It can usually do solid IVR. Punch 1 to find out when we sent your order. Punch 2 to find out how much it cost. Give out and collect information from your customers at 3:00 AM. It is your office's fax and email server. It handles incoming faxes and emails. It tells you that you have mail. It lets you view your fax or email mail or listen to your voice mail. It is your one place for all your mail — fax, email, voice mail, video mail, etc. One screen showing all your mail. We call it unified messaging. The UnPBX makes you super organized. It is an ACD (automatic call distributor). It's not as sophisticated as what BIG airlines have. But it can come very close. An UnPBX's ACD can make your customers very, very happy. The UnPBX can be a predictive dialer. It can pump out telemarketing and dunning calls with the best of them.

Being open, being standards-based and being programmable, an UnPBX may be more or less than the list above. For example, it may also page you when you have a message. Or it may only page you when you have one from your largest customer.

We haven't scratched what's possible with an UnPBX. Today's UnPBX delivers many benefits. Among them:

1. It's cheaper. Don't be fooled by the sticker prices you see in Ed's Roundup. One communications system is a lot cheaper than today's multiple, cobbled-together systems.

2. It's easier to manage. It has one database. One place you assign phone extensions, email addresses, fax locations, email boxes, etc. One place to do billing for all these services. You'll pay the UnPBX off manyfold with just the administration savings.

3. It's much easier to use. You see the words, the icons, the images. Click. Drag and Drop. You now can use telephony features today's horrid telephones deny you.

4. You feel in control. You see your messages. They're all in one place. You can join voice mail messages to Excel spreadsheets and forward the multimedia message to your client. It sells.

5. You can have it your way. Write some software for your UnPBX. Get your reseller or system integrator to do it. Customize your phone system, your IVR system, your fax system, your ACD... all in one box. Gain a major competitive advantage.

Unpublished Phone Number There are various inter-

pretations of what constitutes an "unpublished" phone number in North America. Some phone companies use these words interchangeably. Some don't. In California, Pacific Bell offers unpublished phone service. Your phone number is not listed in the paper phone directories, but is listed with dial up "Directory Assistance." Pacific Bell also has a more expensive service called "Unlisted Service." Here, your phone number is not included in the paper phone directories or given out to callers to Directory Assistance. Nynex (and I presume other phone companies) has a service whereby you can leave a message for the owner of an unlisted number. "Please call me. You've won the lottery." The owner of the unlisted number then has the choice to return the call or not. He doesn't pay to receive this message. Some telephone companies confuse the definitions and some invent new ones. For example, some phone companies use the term "non-published" number. You won't find the number in a phone book or by calling Directory Assistance. Over 25% of many private phone numbers in major metropolitan areas are unlisted, unpublished or non-published — a "service" their subscribers pay extra for. To my simple brain, it's a lot easier to simply publish your name as "Apple Plumpudding." See UNPUBLISHED.

Unrestricted Digital Information An ISDN term. An information sequence of bits is transferred at its specified bit rate without alteration.

Unshielded Wiring not protected by a metal sheathing from electromagnetic and radio frequency interference, but covered with plastic and/or PVC.

Unshielded Twisted Pair UTP. A transmission medium consisting of a pair of copper conductors which are electrically balanced. Each conductor is separately insulated (typically with plastic) in order to prevent the conductors from "shorting." The conductors are twisted around each other at routine intervals in order to confine the electromagnetic field within the conductors and, thereby, to 1) maximize signal strength over a distance, and 2) minimize interference between adjacent pairs in a multi-pair cable. UTP conductors come in various gauges and various numbers of twists per foot. The thicker the conductor, the less resistance and the better the performance; the more twists per foot, the better the performance. UTP comes in various configurations; in large cable systems, multiple pairs are combined in binder groups of 25 pairs, and multiple binder groups are combined in a single insulated cable sheath. In small configurations, such as desktop applications requiring one to four pairs, UTP is relatively inexpensive to acquire and to deploy when compared to coaxial cable and fiber optic cable; hence, its increasing popularity in both voice and data applications. For longer explanations, see also UTP Cable, Category 1 through 6, and STP.

Unsolicited Event Events in switching that happen without control of a program that allegedly is controlling your phone and phone system. Such an unsolicited event might be a user picking up his phone and hanging it up or simply pushing a random button on the phone. Or it may be that the switch actually does something you or your controlling computer don't expect it to do.

Unsuccessful Call A call attempt that does not result in the establishment of a connection.

Unsupervised Transfer Someone transfers a call to someone else without telling the person who's calling. Also called Blind Transfer.

Unused A term used in the secondary telecom equipment

business. Equipment never used and still in O.E.M. original packaging with all appropriate documents and user guides. Such equipment may or may not carry the O.E.M. standard warranty. In other words, manufacturers don't want to see independent (unauthorized) dealers advertising their equipment as new, but "unused" is acceptable.

Up Sell To sell a higher value product to an existing customer. For example to lease a more sophisticated photocopier to an existing customer. In contrast, cross selling is when you buy a shirt from me. I sell you a tie. You buy a car from me. I sell you a mobile phone for your car. Up selling is when I sell you a more expensive shirt or a more expensive car.

Up-Converter A device for performing frequency translation in such a manner that the output frequencies are higher than the input frequencies.

Up-Sampling A technique used for recreating an approximation of data compressed by down-sampling. Up-sampling generally doubles the data at each iteration of the signal by inserting a value between the adjacent compressed values in the case of "down-sampling by two." This operation is fundamental in the Fast Packet Algorithm used in Wavelet Transforms, which are commonly employed in image compression. See DOWN-SAMPLING, FAST PACKET ALGORITHM and WAVELET TRANSFORM.

UPC Usage Parameter Control: As an ATM term, Usage Parameter Control is defined as the set of actions taken by the network to monitor and control traffic, in terms of traffic offered and validity of the ATM connection, at the end-system access. Its main purpose is to protect network resources from malicious as well as unintentional misbehavior, which can affect the QoS of other already established connections, by detecting violations of negotiated parameters and taking appropriate actions.

UPCS Unlicensed Personal Communications Services. The FCC designated 1890-1930 MHz to UPCS. 1920-1930 MHz is presently being assigned by UTAM, primarily for short range, wireless PBX applications. Here are the specific designations: 1890-1910 MHz isochronous, 1910-1920 MHz asynchronous, 1920-1930 MHz isochronous. Isochronous communication is good for voice and asynchronous communication is for bursty data.

Upgrade A call center term. A technique to increase the revenue of an order that is quality rated. It means getting the customer to by a better quality, more expensive version of the item sold. See also Up Sell.

Uplink 1. In satellites, it's the link from the earth station up to the satellite. The link from the satellite down to the earth station is called the downlink. The uplink and downlink operate on different frequencies so that they don't interfere with each other. The uplink is at a higher frequency. For example, C-Band satellites use the 6 GHz frequency range on the uplink and 4 GHz on the downlink. International customers often buy uplinks and downlinks from different suppliers, as each nation typically awards one or more national agencies exclusive franchise rights. See DOWNLINK.

2. In data transmission, an uplink is from a data station to the headend or mainframe.

3. As an ATM term, it represents the connectivity from a border node to an upnode.

Upload To transmit a data file from your computer to another computer. The opposite of download, which is receiving a file on your computer from another computer. Upload means the same as TRANSMIT, while DOWNLOAD means the same as receive. Before you upload or download, check at least

three times you're going the direction you want. It's very easy to erase files (weeks of work) if you make a mistake and confuse uploading and downloading. (Don't laugh. We've done it several times. Dumb!)

Upnode As an ATM term, it is the node that represents a border node's outside neighbor in the common peer group. The upnode must be a neighboring peer of one of the border node's ancestors.

UPP Universal Payment Preamble (UPP). The negotiation protocol that identifies appropriate payment methodology in the context of Joint Electronics Payments Initiative (JEPI), a specification from the World Wide Web Consortium (W3C) and CommerceNet for a universal payment platform to allow merchants and consumers to transact E-Commerce (Electronic Commerce) over the Internet. UPP works in conjunction with Protocol Extensions Protocol (PEP), an extension layer that sits on top of HTTP (HyperText Transfer Protocol). These protocols are intended to make payment negotiations automatic for end users, happening at the moment of purchase, based on browser configurations See also Electronic Commerce and JEPI.

Upper Memory Area In an IBM compatible PC, upper memory is the area between 640KB and 1MB of RAM. This area is made up of Upper Memory Blocks (UMBs) of various sizes. Access to this area is possible only with a special memory drive such as MS-DOS's EMM386.EXE.

Upper Memory Blocks See UPPER MEMORY AREA.

UPS 1. See UNINTERRUPTIBLE POWER SUPPLY.

2. United Parcel Service, a package deliverer that regularly lives up to its self-imposed, relatively speedy delivery schedule.

UPS Monitoring UPS monitoring allows a local area network file server to monitor an attached Uninterruptible Power Supply (UPS). When a power failure occurs, NetWare notifies users. After a time out specified with SERVER.CFG and ROUTER.CFG, the server logs out any remaining users, closes any open files, and shuts itself down. If you install a Novell-approved UPS, you must also install a printed circuit board in the file server to monitor the UPS. If you have a file server with a microchannel bus (as compared to the more common AT bus), the UPS is monitored through the mouse port and does not require a board.

UPSR Unidirectional Path-Switched Ring. A SONET term. Path-switched rings employ redundant fiber optic transmission facilities in a pair configuration, with one fiber transmitting in one direction and with the backup fiber transmitting in the other. If the primary ring fails, the backup takes over. See also PATH SWITCHED RING and SONET.

Upspeak According to Newsweek, upspeak is the annoying way teenagers speak.

Upstream In a communications circuit, there are two circuits — coming to you and going away from you. Upstream is another term for the name of the channel going away from you. In a broadband TV network, the definition of the upstream channel or signal is different. It is the channel from the transmitting stations to the CATV headend. See UPSTREAM CHANNEL.

Upstream Channel In a communications circuit, there are two circuits — coming to you and going away from you. Upstream is another term for the name of the channel going away from you. In a broadband TV network, the definition of the upstream channel or signal is different. It is the channel from the transmitting stations to the CATV headend. In yet another definition, in the cable TV industry, the upstream channel is a collection of frequencies on a CATV channel reserved for transmission from the terminal next to the user's TV set to (upstream to) the CATV company's computer. Such signals might be requests for pay movies. See UPSTREAM.

Upstream Operations Functions that provide a BCC (Bellcore Client Company) control of features and service configurations and subject to BCC control, some service management capabilities for subscribers. These functions include Service Negotiation and Management, Service Provisioning and Repair Service Answering/Work Force Administration. Definition from Bellcore in reference to its concept of the Advanced Intelligent Network.

UPT Universal Personal Telecommunications. According to L.M. Ericsson, Swedish telecom manufacturer, UPT is a "new service concept in the field of telecommunications which aims at making telecommunications both universal and personal. instead of calling a telephone line or a mobile terminal, you call the person you wish to get in touch with and leave it to the network to locate the line or terminal where he/she can be reached." There was an article on UTP in the 1993 No.4 issue of the Ericsson Review. An article in the June, 1996 issue of IEEE Communications Magazine described UPT as a service that enables users to access various services through personal mobility. It enables each UPT user to participate in a user-defined set of subscribed services, and to initiate and receive calls on the basis of a personal, network transparent UPT number across multiple networks on any fixed or mobile terminal, irrespective of geographical location. This service is limited only by terminal and network capabilities and restrictions imposed by the network operator. In short, UPT is still not totally defined and is under discussion by the world's major standards bodies. For more information on UPT, see ITU-T Recommendation F.850, Principles for Universal Personal Telecommunications, Geneva, 1993.

Uptime Colloquial expression for the uninterrupted amount of time that network or computer resources are working and available to a user. In short, time between failures or periods of nonavailability (as for maintenance).

Upward Compatible Any device that can be easily organized, fixed or configured to work in either a different, expanded operating environment or some enhanced mode. Software is said to be upward compatible if a computer larger than the one for which it was written can run the program.

Urban Service Any of the grades of service regularly furnished inside base or locality rate areas, or outside base or locality rate areas at base or locality rates plus zone connection charges or incremental rates. Another way of saying expanded metropolitan phone service.

URL Universal Resource Locator. An Internet term. A URL is a fancy name for an Internet address. A URL is an address that can lead you to a file on any computer connected to the Internet anywhere in the world. Thus, its name — Universal Resource Locator.

In more technical terms, a URL is a string expression that can represent any resource on the Internet or local TC/IP system. The standard convention for a URL is as follows:

method of protocol to be used://host's name/folder or directory on host/name of file or document

Here's an example of a URL: http://www.harrynewton.com/fantasy/happy.html

Let's see what it all means. The http stands for HyperText Transport Protocol. That tells your browser (e.g. Netscape or Internet Explorer) to use that protocol when searching for the

address. Http is the "default" protocol of the Internet. But it's not the only one. There are other protocols, including ftp (file transfer protocol), news (for Usenet news groups), and "mailto" (to send email to a specific address).

The www.harrynewton.com is simply the name of my computer. All Web addresses start as numbers. This one is no different. You can reach my home page by giving your browser the following command http://209.94.129.207/. The Web's own lookup tables do an instaneous translation to that number from www.harrynewton.com when you type in www.harrynewton.com. The translation mechanism is very much like the translation the phone industry does when you dial any 800 number, e.g. 1-800-LIBRARY — translate it to a real number, i.e. 212-691-8215.

The /fantasy/ means that there's a folder or subdirectory on my web site's computer disk called fanstasy and inside that folder there's a document called happy.html. And that's what we're looking for. See also Web address.

URM User request manager.

US West One of the seven Regional Holding Companies formed at divestiture. It includes Mountain Telephone, Northwestern Bell and Pacific Northwest Bell among other service entities and entrepreneurial adventures.

USACII See ASCII. The name change was a result of the name change of the standards organization. When the name changed again to ANSI, most people simply reverted to ASCII.

Usage A measurement of the load carried by a server or group of servers, usually expressed in CCS. Usage may also be expressed in erlangs.

Usage Based Usage-Based refers to a rate or price for telephone service based on usage rather than a flat, fixed monthly fee. Until a few years ago, most local phone service in the United States was charged on a flat rate basis. Increasingly, phone companies are switching their local charging over to usage-based. Flat-rate calling will probably disappear within a few years. Allegedly, usage based phone service pricing is fairer on those phone subscribers who don't use their phone much. Usage based pricing is not consistent throughout the U.S. Typically, you get charged for each call. And the charging is very much like that for long distance — by length of call, by time of day and by distance called. See also FLAT RATE.

Usage Sensitive A form of Measured Rate Service. See USAGE BASED.

USART Universal Synchronous/Asynchronous Receiver/Transmitter. An integrated circuit chip that handles the I/O (input/output) functions of a computer port. It converts data coming in parallel form from the CPU into serial form suitable for transmission, and vice versa.

USB Universal Serial Bus. In March of 1995, Compaq, Digital, IBM, Intel, Microsoft, NEC and Northern Telecom announced a new "open and freely licensed" serial bus called Universal Serial Bus — USB for short. The bus (which could also be called a special purpose local area network) is 12 megabits per seconds and supports up to 63 devices. The idea of Universal Serial Bus is to replace the PC cable clutter. USB's proponents showed a diagram of a future PC with only three ports out the back — a USB, a graphics port (for your monitor) and a LAN port. Gone were the parallel, serial, graphics, modem, sound/game and mouse ports. USB is designed to handle a broad range of devices — telephones (analog, digital and proprietary), modems, printers, mice, joysticks, scanners, keyboards, tablets. USB is designed to be "completely Plug and Play," meaning that devices will be correctly detected and configured automatically as soon they are attached. USB also has "Hot attach/detach," which allows adding and removing devices at any time, without powering down or rebooting." USB uses a connector, currently in design. The mockup I saw was the size of your pinky finger. Topology is tiered star, with up to five meters per segment. At each star is a hub or pod with connections to other devices and pods. Hubs function as repeaters, providing power for devices, routing signals in each direction and providing terminations for each line. Some devices, e.g. proprietary PBX phones, are expected to come with their own built in pods. A phone pod will allow a PBX manufacturer to pass call and media control to the desktop PC. PBX makers are likely to go for this choice, since it enables them to continue selling proprietary PBX phones, which, today, are a very profitable part of their business. USB is sophisticated in that it will handle certain "important" data streams — e.g. voice and video — with preference. Serial Bus has three basic types of data transfer:

• Isochronous or streaming real time data which occupies a prenegotiated amount of Serial Bus bandwidth with a prene-gotiated latency. (This would be for voice and video.)

• Asynchronous interactive data such as characters or coor-dinates with few human perceptible echo or feedback respon-sible characteristics (e.g. tele-gaming).

• Asynchronous block transfer data which is generated or consumed in relatively large and bursty amounts and has wide dynamic latitude in transmission constraints.

Does USB mean T-1 or E-1 To every desktop? Potentially, yes. It's certainly powerful enough. USB is 7.8 times the speed of T-1. In the meantime, ISDN may be the major ben-eficiary, according to its proponents. Timetable: 0.9 specifi-cation: call 1-800-433-3652 and for $35, they'll send you the 250-page document. Or pick it up for free through the World Wide Web. The address is www.teleport/.com/~USB.

USDLA United States Distance Learning Association. Their mission: To promote the development and application of dis-tance learning for education and training. Constituents include K-12 education, higher education, continuing educa-tion, corporate training, and military and government train-ing. www.usdla.org

Used Equipment which was previously in service (i.e. used someplace else) and may not have been tested, refurbished or remanufactured before you bought it. Used simply means it's no longer new. No more, no less. See also CERTIFIED and REFURBISHED.

Used Or Out Of Service A term used in the secondary telecom equipment business. Equipment taken from ser-vice. Can be in any condition. Generally expected to work and be complete.

USENET The USENET is an informal, rather anarchic group of computer systems that exchange "news." News is essen-tially similar to "bulletin boards" on other networks. USENET actually predates the Internet network. These days, the Internet is used to transfer much of the USENET's traffic. You can find newsgroups covering every conceivable sub-ject from nude sunbathing to molecular physics. See INTER-NET, MUDS and USENET NEWSGROUP ORGANIZATION.

USENET Newsgroup Organization USENET conver-sations are organized in hierarchical newsgroup trees. There are seven core newsgroup hierarchies or trees: comp (com-puters), misc (miscellaneous topics), news (newsgroup information), rec (recreation), sci (science), soc (society), talk (conversation). Each tree branches into different levels of newsgroup sub-topics.

User Accessible Tables There are many tables (data-bases) inside a phone system. They include the extensions with privileges and long-distance dialing selections (see LEAST COST ROUTING). In the old days, most PBX and phone system tables were not accessible to the user, on the assumptions that 1. The user would screw the tables up, and/or 2. Really didn't care about getting access. Things have changed. Users now want faster and greater control over their own destiny. So, many manufacturers are making their tables user accessible.

User Account Each user has a user account that is part of local area network security and controls the user environ-ment. Some account features are assigned to each user auto-matically, some must be assigned, some are optional.

User Agent 1. Generally refers to the windows and menus used to make interfacing to UNIX easier.

2. UA. An OSI (Open Systems Interconnection) process that represents a human user, or organization, or application in the X.400 MHS (Message Handling System). The user creates, submits, and takes delivery of messages on the user's behalf, and in some cases, can even create the message. UA is thus an X.400 electronic mail term: It is software that prepares the message for transmission to the Message Transfer Agent. The user can be an individual or a distribution list. Users are known by their originator/recipient (O/R) addresses.

User Context A user session created by an operating sys-tem in response to a logon request, and typically character-ized by privilege sets that strictly define the user's authority to access system resources and information on a LAN. Contexts restrict unauthorized access to facilities and data and protect the system itself from user and applications interference, accidental or otherwise. Contexts are a feature of most multi user operating systems, usually integrated with the security system.

User Datagram Protocol UDP. A packet format includ-ed in the Transmission Control Protocol/Internet Protocol (TCP/IP) suite and used for short user messages and control messages. The transmission of UDPs is unacknowledged.

User Data X.25 call control field used to transfer informa-tion concerning layers above X.25 between the originating and terminating DTEs.

User Event Input from the user (for example, clicking the mouse, pressing a key, etc.) that causes a multimedia project to perform a specific function (for example, play a movie, change pages, etc).

User Friendly Computer programs or systems which are designed for simple operation by non-technical users. At least that's the theory.

User ID 1. Persistent information in an ASC (AIN Switch Capabilities) that the ASC communicates to the SLEE (Service Logic Execution Environment) as a parameter in a message to the SLEE. The SLEE uses this parameter to iden-tify the set of information related to the user (e.g., customer record) that service logic needs to perform its task. If a user does not subscribe to any Advanced Intelligent Network Release 1 feature but invokes an AIN Release 1 feature, the user ID in the ASC may correlate in the SLEE to a set of default information to be used by service logic.

2. A compression of "user identification"; the unique account signature of an Internet user; that which precedes the @ (at) sign in an E-mail address.

User Loop A 2- or 4-wire circuit connecting a user to a PBX or other phone system.

User Message Part of a CPN message directing a destination node to accomplish some task.

User Plane An ATM term referring to the functions which address flow control and error control. The User Plane cuts through all 4 layers of the ATM Protocol Reference Model.

User Segment The part of a satellite system that includes the receivers on the users' premises.

User To User Messaging An ISDN service enabling voice and computer data to be transmitted simultaneously — for example, enabling one person to transmit a spreadsheet file to another so both can examine it on their individual screens, and for each to have the ability to change the spreadsheet and have the changes appear instantly on the other person's screen, and then discuss the changes.

User's Set Apparatus located on the premises of a communications user. Designed to work with other parts of his system.

UserID A compression of "user identification"; the unique account signature of an Internet user; that which precedes the "at" sign in an E-mail address.

Username The name by which you or someone else is known by on the Internet. Used when logging into an access provider or when entering a member's only area on the Web.

USITA United States Independent Telephone Association, the old name for the United States Telephone Association. See USTA.

USOC Universal Service Order Code. (Pronounced "U-Sock.") An old Bell System term identifying a particular service or equipment offered under tariff. There was nothing "Universal" or consistent about USOC codes. Many services and equipment were called different things by different Bell operating companies. Since Divestiture, there is absolutely no consistency and little relevance left in USOC codes. Each Bell operating company is busily creating its own billing codes and terms.

According to some oldtimers there was an order to USOCs designation of types of telephones. USOCs were five characters, 12345; shown 123++ meant two or more characters were to follow; 123+x meant one or more character to follow; 123xx meant this code was complete. Let's focus on the last two characters, "++." The First "+" indicated the color, "B" for black, "W" for white. "E" for beige because "B" was used for black, and so on.

The second "+" indicated the type of instrument: "K" for key (remember those clear lighted pick-up buttons and the red hold button?) "C" for combined meaning all was contained in one telephone, and sat on a desk, etc. So, a "BK" was a black key set and, a "BC" was a black desk set.

At one time, just about all Bell operating companies used the standard USOC. All but New York Telephone that is. In the late seventies/early eighties, before the big "D" (divestiture), a new order was put to USOC. What had been New York Tel USOCs, now pretty much became everyone else's USOCs. For consistency of explanation, the definitions of the "K" and the "C" remain the same (K for key and C for combined). Color never did matter to Nynex. So, CV was A Combined Voice (simple telephone) and a KV was a Key Voice (multi line telephone). Obviously the "V" was for voice.

USOC as applied to four-pair wiring can be either "Generic", the original configuration used before standardization or an RJ-61X configuration. A jack pattern gains the "RJ" prefix when it becomes registered by the Federal Communications Commission (FCC). When the generic USOC pattern was registered, pins 1 and 8 were reversed, and this became RJ-61X. The "Generic" USOC wiring scheme is, described in

more detail under Color Code.

USOP User Service Order Profile.

USOS Universal Operations Services. A software application that supports UIS by providing traditional network operations functions.

USP Usage Sensitive Pricing. A tariff for local service under which the subscriber only pays for the telephone service he uses. This is done for gas and electric service.

USPID User Service Profile Identifier. An ISDN term.

USPTO United States Patent and Trademark Office.

USRT Universal Synchronous Receiver/Transmitter. Integrated circuit that performs conversion of parallel data to serial for transmission over a synchronous data channel.

USTA See UNITED STATES TELEPHONE ASSOCIATION.

USTSA See UNITED STATES TELEPHONE SUPPLIERS ASSOCIATION.

UT 1. Universal Time.

2. Upper Tester. An ATM term. The representation in ISO/IEC 9646 of the means of providing, during test execution, control and observation of the upper service boundary of the IUT, as defined by the chosen Abstract Test Method.

3. User terminal.

UTAM Unlicensed Transition and Management for Microwave. Relocation in the 2.0 Ghz Band. UTAM Inc. is an open industry organization pledged to relocate incumbents (POFS) presently operating in the 1890 to 1930 MHz band. Designated by the FCC as the frequency coordinator for the UPCS spectrum. "Memorandum and Opinion and Order - June 1994," FCC Docket 90-314.

UTC 1. Universal Time Coordinated; also known as UCT, for Universal Coordinated Time, which is time kept by the "i" laboratory, where i is any laboratory cooperating in the determination of UTC. In the United States, the official UTC is kept by the U.S. Naval Observatory and is referred to as UTC.

2. UTC, The Telecommunications Association. UTC, formerly known as the Utilities Telecommunications Council primarily represents the interests of the utility industries (electric, gas, steam, and water) worldwide, but is open to vendors, manufacturers, and other interests. It is worthy to note that the utility industry is second only to the "telephone" industry in the procurement of telecommunications products and services. www.utc.org

UTDR Universal Trunk Data Record.

Utility In some computer languages, a small program executed as a shell command is called a "tool." In other computer languages, such as BASIC, it is called a "utility."

Utility Communications Architecture UCA. An architecture for networks used to monitor and control electric power distribution systems.

Utility Program A computer program in general support of the processes of a computer; for example, a diagnostic program.

Utility Routine A routine in general support of the operation of a computer, including input/output, diagnostic, tracing or monitoring.

Utilize An absolutely awful word created the people who believe that speaking or writing in big words is a demonstration of their superior intelligence.

UTOPIA Universal Test and Operations Interface for ATM. Refers to an electrical interface between the TC (Transmission Convergence) and PMD (Physical Medium Dependent) sublayers of the PHY (Physical) layer, which is the bottom layer of the ATM Protocol Reference Model. Utopia is the interface for devices connecting to an ATM network.

UTP see Unshielded Twisted Pair.

UTP Cable Unshielded Twisted Pair Cable. Most UTP cables have eight conductors. They are organized into four pairs. Each pair has a ring conductor and a tip conductor. The tip is colored white, usually with colored stripes. The ring is a solid color, usually with white stripes. The conductors of each pair are twisted around each other at a constant rate. However, each pair has different twist lengths. These exact lengths vary between manufacturers and types of cables. See Category of Performance. The conductors in most UTP and ScTP cables are 24 gauge copper wire. (Some manufacturers use 22 gauge. Conductor size doesn't matter if the cable is rated at the Category of Performance you've specified.) UTP cable can have solid copper or stranded conductors. Solid is less expensive than stranded, but stranded conductors are more flexible because they're made of tiny individual strands of copper. See Unshield Twisted Pair.

UTP RJ-45 Plug These are used to terminate four pair UTP patch cords. A clip on the plug holds it into the jack. See Unshield Twisted Pair.

UTR Universal Tone Receiver.

UTS Universal Telephone Service.

Utterance A word used in voice recognition to mean a vocalized sound that is typically a word.

UUCP UNIX-to-UNIX Copy Program. A standard UNIX utility for exchanging information between two UNIX-based machines in a network. UUCP may also be referred to as the UNIX-TO-UNIX Communications Protocol and is widely used for electronic mail transfer. In short, UUCP is the name of a Unix command, but it is now also used to refer to the protocols used by it to transfer files between Unix machines. There are a number of such protocols, and the two machines choose between the ones supported by each. Free implementations also exist for VMS and MS-DOS. The Internet newsgroup comp.mail.uucp has more information.

UUD Short for UUDecode, which is a software utility used to decode UUEncoded files. See UUENCODE.

UUE Short for UUEncode, which is a de facto encoding protocol used to transfer binary files across the Internet and on-line services. See UUENCODE.

UUEncode UNIX-to-UNIX encoding. Software that allows you to take a binary computer file, e.g. a Word for Windows document or a PowerPoint presentation, convert it to ASCII for sending the binary file across the Internet or some other e-mail service and then, once received, converting it back to binary. In its Uuencoded form, the file is visible as ASCII, but it makes absolutely no sense. Direct e-mail services, like America On Line, MCI Mail or CompuServe, allow you to "attach" a binary file to a message — without ever affecting the message. The person at the other end receives it as a binary file. However, once the message passes across an X.400 gateway, from one e-mail service to another (including or not including the Internet), you need Uuencoding. Also called MIME encapsulation. See MULTIPURPOSE INTERNET MAIL EXTENSION for an example of UUencoding.

UUI User-to-User Information. An ISDN term. UUI comprises information of end-to-end significance sent over the ISDN D (Data) channel in the context of UUS (User-to-User Signaling), as defined by the ITU-T and ETSI. UUI services, which remain to be defined completely, must be subscribed by the sending party; the target party side of the connection has no acceptance or rejection procedure. UUS falls into three categories. UUS1 provides for the transmission and reception of UUI during call set-up and termination, through ISUP (ISDN User Part) call-control messages. UUS2 provides for the transmission and reception of UUI subsequent to call set-up, but prior to the establishment of a connection. UUS3 provides for the transmission and reception of UUI only while the connection is established, i.e., during the active phase of the circuit-switched call. UUS1/2/3 messages comprise packets of 128 bytes, with maximum numbers of packets established for each UUS category. UUI information is sent within standard Q.931 signaling messages, as separate signaling messages which are either associated with an existing call or are sent as non-call messages. UUI information sent as an associated call message might include the caller's name or account number to be transmitted to the destination. Detailed information can be found in ITU Q.931, or related ITU documents I.257.1 and D.231. See also ISUP.

UV 1. UltraViolet.
2. A microvolt.

V Abbreviation for VOLT.

V & H V&H stands for Vertical and Horizontal. Below are vertical and horizontal coordinates of Major Continental US cities as presented in charts published by long distance carriers in North America. The monthly charge for many leased circuits provided by either an IXC (Inter-Exchange Carrier) or a LEC (Local Exchange Carrier) is billed on the basis of "airline mileage" between the two points. The two points for an IXC (long distance carrier) private line circuit are (1) IXC POP to (2) IXC POP, and the monthly charge is based on the mileage between them. While the two points for LEC (Local carrier) charges are based on the (1) customer premise SWC/CO to
(2) customer premise SWC/CO or IXC POP SWC/CO (When one end is an IXC POP, this can be referred to as a dedicated access loop), and the monthly charge is based on the mileage between them. Though it sounds as if it's the distance a crow would fly directly between the two points, in reality, it is the distance in mileage between two Rate Centers whose position is laid down according to industry standards, originally created by AT&T. The entire U.S. is divided by a vertical and horizontal grid. The coordinates — vertical and horizontal — of each rate center are defined and applied to a square root formula which yields the distance between the two points. Think back to school. There's a right-angled triangle. At the top is one Rate Center. At the side is the other Rate Center. The horizontal is the horizontal coordinate. The vertical is the vertical coordinate. The formula is simple: Square the vertical distance. Square the horizontal distance. Add the two together. Then take their square root. That will give you the distance across the hypotenuse — the side opposite the right angle in the triangle — i.e. the airline mileage. Here is the actual formula for calculating airline mileage. The airline mileage is the square root of $((V1 - V2) \times (V1 - V2) + (H1 - H2) \times (H1 - H2))$ / 10. Dividing by ten is necessary in this case to allow for the way the coordinates are presented in North American mileage charts. This formula (without the division by ten) is called the Pythagorean theorem and is known to every schoolboy. It is named for its inventor, Pythagoras, Greek philosopher, mathematician, and religious reformer, who lived 582-500 B.C. (Thank you to Andrew Funk Manager, Access Pricing Strategy, Qwest Communications International for his considerable help on this complex definition.) Here are the V & H coordinates of major American cities:

	V	H
ALABAMA		
Birmingham	7518	2304
Huntsville	7267	2535
Mobile	8167	2367
Montgomery	7692	2247
ARIZONA		
Flagstaff	8746	6760
Phoenix	9135	6748
Tucson	9345	6485
Yuma	9385	7171
ARKANSAS		
Fayetteville	7600	3872
Hot Springs	7827	3554
Pine Bluff	7803	3358
CALIFORNIA		
Anaheim	9250	7810
Bakersfield	89	8060
Fresno	8669	8239
Long Beach	9217	7856
Los Angeles	9213	7878
Oakland	8486	8695
Redwood City	8556	8682
Sacramento	8304	8580
San Bernardino	9172	7710
San Diego	9468	7629
San Francisco	8492	8719
San Jose	8583	8619
Santa Monica	9227	7920
Santa Rosa	8354	8787
Sunnyvale	8576	8643
Van Nuys	9197	7919
COLORADO		
Denver	7501	5899
Fort Collins	7331	5965
Grand Junction	7804	6438
Greeley	7345	5895
Pueblo	7787	5742
CONNECTICUT		
Bridgeport	4841	1360
Hartford	4687	1373
New Haven	4792	1342
New London	4700	1242
Stamford	4897	1388
DELAWARE		
Wilmington	5326	1485
DISTRICT OF COLUMBIA (D.C.)		
Washington	5622	1583
FLORIDA		
Clearwater	8203	1206
Daytona Beach	7791	1052
Fort Lauderdale	8282	0557
Jacksonville	7649	1276
Miami	8351	0527
Orlando	7954	1031
Tallahassee	7877	1716
Tampa	8173	1147
GEORGIA		
Atlanta	7260	2083
Augusta	7089	1674
Macon	7364	1865
Savannah	7266	1379
IDAHO		
Boise	7096	7869
Pocatello	7146	7250
ILLINOIS		
Chicago	5986	3426

Joliet	6088	3454	Helena	6336	7348
Peoria	6362	3592	Missoula	6336	7650
Rock Island	6276	3816	**NEBRASKA**		
Springfield	6539	3518	Grand Island	6901	4936
INDIANA			Omaha	6687	4595
Bloomington	6417	2984	NEVADA		
Fort Wayne	5942	2982	Carson City	8139	8306
Indianapolis	6272	2992	Las Vegas	8665	7411
Muncie	6130	2925	Reno	8064	8323
South Bend	5918	3206	**NEW HAMPSHIRE**		
Terre Haute	6428	3145	Concord	4326	1426
IOWA			Manchester	4354	1388
Burlington	6449	3829	Nashua	4394	1356
Cedar Rapids	6261	4021	**NEW JERSEY**		
Des Moines	6471	4275	Atlantic City	5284	1284
Dubuque	6088	3925	Camden	5249	1453
Iowa City	6313	3972	Hackensack	4976	1432
Sioux City	6468	4768	Morristown	5035	1478
KANSAS			Newark	5015	1430
Dodge City	7640	4958	New Brunswick	5085	1434
Topeka	7110	4369	Trenton	5164	1440
Wichita	7489	4520	**NEW MEXICO**		
KENTUCKY			Albuquerque	8549	5887
Danville	6558	2561	Las Cruces	9132	5742
Frankfort	6462	2634	Santa Fe	8389	5804
Madisonville	6845	2942	**NEW YORK**		
Paducah	6982	3088	Albany	4639	1629
Winchester	6441	2509	Binghamton	4943	1837
LOUISIANA			Buffalo	5076	2326
Baton Rouge	8476	2874	Nassau	4961	1355
New Orleans	8483	2638	New York City	4977	1406
Shreveport	8272	3495	Poughkeepsie	4821	1526
MAINE			Rochester	4913	2195
Augusta	3961	1870	Syracuse	4798	1990
Lewiston	4042	1391	Troy	4616	1633
Portland	4121	1384	Westchester	4912	1330
MARYLAND			**NORTH CAROLINA**		
Baltimore	5510	1575	Asheville	6749	2001
MASSACHUSETTS			Charlotte	6657	1698
Boston	4422	1249	Fayetteville	6501	1385
Framingham	4472	1284	Raleigh	6344	1436
Springfield	4620	1408	Winston-Salem	6440	1710
Worchester	4513	1330	**NORTH DAKOTA**		
MICHIGAN			Bismarck	5840	5736
Detroit	5536	2828	Fargo	5615	5182
Flint	5461	2993	Grand Forks	5420	5300
Grand Rapids	5628	3261	OHIO		
Kalamazoo	5749	3177	Akron	5637	2472
Lansing	5584	3081	Canton	5676	2419
MINNESOTA			Cincinnati	6263	2679
Duluth	5352	4530	Cleveland	5574	2543
Minneapolis	5777	4513	Columbus	5872	2555
St. Paul	5776	4498	Dayton	6113	2705
MISSISSIPPI			Toledo	5704	2820
Biloxi	8296	2481	**OKLAHOMA**		
Jackson	8035	2880	Lawton	8178	4451
Meridian	7899	2639	Oklahoma City	7947	4373
MISSOURI			Tulsa	7707	4173
Joplin	7421	4015	**OREGON**		
Kansas City	7027	4203	Medford	7503	8892
St. Joseph	6913	4301	Pendleton	6707	8326
Springfield	7310	3836	Portland	6799	8914
MONTANA			**PENNSYLVANIA**		
Billings	6391	6790	Allentown	5166	1585

Altoona	5460	1972
Harrisburg	5363	1733
Philadelphia	5257	1501
Pittsburgh	5621	2185
Reading	5258	1612
Scranton	5042	1715

RHODE ISLAND

Providence	4550	1219

SOUTH CAROLINA

Charleston	7021	1281
Columbia	6901	1589
Spartanburg 6811 1833		

SOUTH DAKOTA

Aberdeen	5992	5308
Huron	6201	5183
Sioux Falls	6279	4900

TENNESSEE

Chattanooga	7098	2366
Johnson City	6595	2050
Knoxville	6801	2251
Memphis	7471	3125
Nashville	7010	2710

TEXAS

Amarillo	8266	5076
Austin	9005	3996
Corpus Christi	9475	3739
Dallas	8436	4034
El Paso	9231	5655
Fort Worth	8479	4122
Houston	8938	3563
Laredo	9681	4099
Lubbock	8596	4962
San Antonio	9225	4062

UTAH

Logan	7367	7102
Ogden	7480	7100
Provo	7680	7006
Salt Lake City	7576	7065

VERMONT

Burlington	4270	1808

VIRGINIA

Blacksburg	6247	1867
Leesburg	5634	1685
Lynchburg	6093	1703
Norfolk	5918	1223
Richmond	5906	1472
Roanoke	6196	1801

WASHINGTON

Bellingham	6087	8933
Kennewick	6595	8391
North Bend	6354	8815
Seattle	6336	8896
Spokane	6247	8180
Yakima	6533	8607

WEST VIRGINIA

Clarksburg	5865	2095
Morgantown	5764	2083
Wheeling	5755	2241

WISCONSIN

Appleton	5589	3776
Eau Claire	5698	4261
Green Bay	5512	3747
La Crosse	5874	4133
Madison	5887	3796

Milwaukee	5788	3589
Racine	5837	3535

WYOMING

Casper	6918	6297
Cheyenne	7203	5958

V Band A range of radio frequencies in the 40 GHz and 50 GHz range, also known as the Q Band. The "V" is a random, arbitrarily-assigned designation with its roots in the context of military security during World War II.

V Chip Violence Chip. A type of TV filter required in the Telecommunications Act of 1996. The Act requires the FCC to prescribe regulations, in conjunction with the electronic manufacturing industry, requiring that television sets manufactured after February 1998 include "features designed to enable viewers to block display of all programs with a common rating. This is commonly referred to as the 'V Chip.'" Similar capability is available through filtering agents for screening Internet content. The V Chip and Internet filtering agents primarily are intended to allow parents to filter content which they consider offensive and inappropriate for their children. These content filtering technologies can be overridden with the proper password. See also Filtering Agent.

V Commerce See V-Commerce below.

V Fast A new higher speed over-normal-phone-line modem called V.Fast Class (V.FC) for 28,800 bits per second speed. See V.34.

V Interface The 2-wire ISDN physical interface used for single-customer termination from a remote terminal. See ISDN and U INTERFACE.

V Path Custom Network Service V PATH is a network service from Nynex for voice and data transmissions specifically designed for large, multi-location business customers within a regional calling area. V PATH is designed to provide the features of a private network using public switched lines.

V Reference Point The proposed interface point in an ISDN environment between the line termination and the exchange termination.

V Series Recommendations ITU-T standards dealing with data communications operation over the telephone network. The idea of standards is simple. If you have them and if every manufacturer conforms, then every modem can talk to every other one. That's the idea. But it's not always that simple. Sometimes you have to conform to several standards. For example, in the higher speed modems, for example those at 9,600 bps, you have to conform to speed. That's one standard. You have to conform to error control. That's another standard. And you also have to conform to data compression — if you are using data compression. ISDN terminal adapters are V series recommendations, too. ITU-T uses the term "bis" to designate the second in a family of related standards and "ter" designates the third in a family.

V-Commerce Voice Commerce. Imagine that you dial a phone number. Your bank answers. "What's your password, please?" You say, "Harry Newton." It answers, "Thank you." You have three bills to pay, but only enough money to pay two. Here are the three..." You answer "Pay the Visa bill." You get the message. Voice commerce is a fancy name for interactive voice response using speech recognition as the input mechanism, not the buttons on a touchtone phone. With touchtone buttons, you often have to keep pressing buttons to progress through zillions of menus to get what you want. Speech recognition, if it works, is faster since it gets you to where you want to go a lot faster.

V.110 Terminal rate adaptation protocols for the ISDN B channel with a V-type interface. Includes V.120.

V.120 Terminal rate adaptation protocols for the ISDN B channel with a V-type interface. Includes V.110.

V.13 ITU-T standard for simulated carrier control. Allows a full-duplex modem to be used to emulate a half-duplex modem with interchange circuits changing at appropriate times.

V.14 ITU-T standard for asynchronous-to-synchronous conversion without error control. Allows a modem that is actually synchronous to be used to carry start/stop (async) characters. If a V.42 modem connects with another modem that doesn't have error-control, it falls back to V.14 operation to work without error-control.

V.17 ITU-T standard for simplex (one-way transmission) modulation technique for use in extended Group 3 Facsimile applications only. Provides 7200, 9600, 12000, and 14400 bps trellis-coded modulation (the modulation scheme is similar to V.33), MMR (Modified Modified Read) compression and error-correction mode (ECM).

V.18 An ITU recommendation aimed at enhancing the capability of the deaf or speech-impaired to use telecommunications. The V.18 recommendation describes a multi-function text telephone that, according to the ITU, "bridges the gap that has existed between several incompatible text telephones in use today." See T.140.

V.21 ITU-T standard for 300 bit per second duplex modems for use on the switched telephone network. V.21 modulation is used in a half-duplex mode for Group 3 fax negotiation and control procedures (ITU-T T.30). Modems made in the U.S. or Canada follow the Bell 103 standard. However, the modem can be set to answer V.21 calls from overseas.

V.21 CH 2 ITU-T standard for 300 bps modem, describing the operation of modems at 300 bps, and used for critical control and handshaking functions. This low speed is highly tolerant of noise and impairments on the phone line. Fax machines use only Channel 2 of the V.21 recommendations (half duplex channel).

V.21 Fax An ITU-T standard for facsimile operations at 300 bps.

V.22 ITU-T standard for 1,200 bit per second duplex modems for use on the switched telephone network and on leased circuits. V.22 is compatible with the Bell 212A standard observed in the U.S. and Canada. See V.22 bis.

V.22 bis ITU-T standard for 2,400 bit per second duplex modems for use on the switched telephone network. "Bis: is used by the ITU-T to designate the second in a family of related standards. "ter" designates the third in a family. The standard includes an automatic link negotiation fallback to 1200 bps and compatibility with Bell 212A/V.22 modems.

The principal characteristics of the V.22bis modems are: Duplex mode of operation on the PSTN and point-to-point 2-wire leased circuits, Channel separation by frequency division, Quadrature amplitude modulation for each channel with synchronous line transmission at 600 baud, An adaptive equalizer and a compromise equalizer, Test facilities, Data signaling rates of 1200bps and 2400bps. V.22bis is compatible with the V.22 modem and includes automatic bit rate recognition. The V.22bis modem technology is used for applications that require transactions at a rate of 2400 bps or slower. Some such applications include set-top boxes, credit card transactions, fax relay systems, satellite receivers, utility meters, and network control.

V.23 V.23 is the standard for a modem with a 600 bps or 1200 bps "forward channel" and a 75 bps "reverse" channel

for use on the switched telephone network.

V.24 ITU-T definitions for interchange circuits between data terminal equipment (DTE) and data communications equipment (DCE) equipment. In data communications, V.24 is a set of standards specifying the characteristics for interfaces. Those standards include descriptions of the various functions provided by each of the pins. This standard is similar (but not identical) to the RS-232-C as established by the American TIA/EIA — Telecommunications Industry Association / Electronics Industries Association.

V.25 Automatic calling and/or answering equipment on the general switched telephone network, including disabling of echo suppressors on manually established calls. Among other things, V.25 specifies an answer tone different from the Bell answer tone. Many modems, including U.S. Robotics modems, can be set with the BO command so that they use the V.25 2100 Hz tone when answering overseas calls.

V.25 bis An ITU-T standard for synchronous communications between the mainframe or host and the modem using the HDLC or character-oriented protocol. Modulation depends on the serial port rate and setting of the transmitting clock source.

V.26 V.26 is the ITU-T standard for 2400 bps modem for use on 4-wire leased lines.

V.26 bis ITU-T standard for 1.2/2.4 Kbps modem. It is important to note that V.26 bis is a half-duplex modem (1200 or 2400 bps in only one direction at a time); it provides an optional 75 bps reverse channel as well.

V.26 ter V.26 ter is a FULL DUPLEX 2400 bps modem, like V.22 bis. The difference is that V.26 ter uses echo cancellation (like V.32) instead of frequency division (like V.22 bis), making it more expensive than V.22 bis. It was intended to serve as a fallback mode from V.32, but most manufacturers ignored it and provide V.22 bis as a fallback instead (V.26 ter is used only in a few installations in France, as far as we know).

V.27 ITU-T standard for 4,800 bits per second modem with manual equalizer for use on leased telephone-type circuits. May be full-duplex on four wire leased lines, or half-duplex on two wire lines.

V.27 bis ITU-T standard for 2,400 / 4,800 bits per second modem with automatic equalizer for use on leased telephone-type circuits. 2.4 Kbps modem for 4-wire leased circuits. Either speed (2,400 is a fallback) can be used on either 4-wire leased lines (full duplex) or 2-wire leased lines (half-duplex). It also provides an optional 75 bps reverse channel.

V.27 ter ITU-T standard for 2,400 / 4,800 bits per second modem for use on the switched telephone network. Half-Duplex only. V.27 ter is the modulation scheme used in Group 3 Facsimile for image transfer at 2400 and 4800 bps. 4800 bps is a common "fallback" speed.

V.28 V.28, entitled "Electrical Characteristics for Unbalanced Double-Current Interchange Circuits" provides the ITU-T equivalent of the electrical characteristics defined in EIA-232.

V.29 ITU-T standard for 9,600 bits per second modem for use on point-to-point leased circuits. Virtually all 9,600 bps leased line modems adhere to this standard. V.29 uses a carrier frequency of 1700 Hz which is varied in both phase and amplitude. V.29 also provides fallback rates of 4800 and 7200 bps. V.29 can be full-duplex on 4-wire leased circuits, or half-duplex on two wire and dial up circuits. V.29 is the modulation technique used in Group 3 fax for image transfer at 7200 and 9600bps.

V.3 ITU-T specification that describes communications control procedures implemented in 7-bit ASCII code.

V.32 ITU-T standard for 9,600 bit per second two wire full duplex modem operating on regular dial up lines or 2-wire leased lines. If you're buying a 9,600 bps modem for use on the normal dial up switched phone lines, make sure it conforms to V.32. If your modem also conforms to V.42 bis, you should be able to transmit and receive at up to 38,400 bps with other modems that conform to these two specifications. I personally use a number of V.32/V.42 bis modem and they work wonderfully fast. V.32 also provides fallback operation at 4,800 bps. See also V.32 bis, V.42 bis ERROR CORRECTION and V.42 bis DATA COMPRESSION and MODULATION PROTOCOLS.

V.32 bis New higher speed ITU-T standard for full-duplex transmission on two wire leased and dial up lines at 4,800, 7,200, 9,600, 12,000, and 14,400 bps. Provides backward compatibility with V.32. Modems running at V.32 bis at its highest speed of 14,400 bps are actually transmitting that many bits per seconds. They do not rely on compression to achieve that high speed. However, with data compression — such as V.42 and V.42 bis — they can achieve higher speeds. The V.32 bis standard also includes "rapid rate renegotiation" feature to allow quick and smooth rate changes when line conditions change. See MODULATION PROTOCOLS, V.42 and V.42 bis.

V.32 terbo Modulation scheme that extends the V.32 connection range: 4800, 7200, 9600, 12K and 14.4K bps. V.32 bis terbo modems fall back to the next lower speed when line quality is impaired, and fall back further as necessary. They fall forward to the next higher speed when line quality improves.

V.33 ITU-T standard for 14,400 and 12,000 bps modem for use on four wire leased lines.

V.34 V.34 is the international standard for dial up modems of up to 28,800 bits per second. Since the standard suggests speeds twice as fast as the top standard they replace, they carry the nickname "V.Fast." New V.34 modems have a feature called line probing that will allows them to identify the capacities and quality of the specific phone line and adjust themselves to allow, for each individual connection, for maximum throughput. The standard also supports a half-duplex mode of operation for fax applications. The new V.34 technology includes an optional auxiliary channel with a synchronous data signaling rate of 200 bits/second. Data conveyed on this channel consists of modem control data. V.34 modems contain multidimensional trellis coding, which is used to gain higher immunity to noise and other phone line impairments. V.34 modems are the first modems to identify themselves to telephone network equipment (handshaking). V.34 technology has been long in coming and has had to overcome many obstacles. At one point, members of the modem manufacturing industry became so impatient, that some of them began shipping their own proprietary versions of what they thought V.34/V.Fast/28,800 bps modems would be. Many of these modems are only compatible, at higher than 14,400 bps speeds, with themselves. See V.34bis.

V.34 bis Also known as V.34+. A faster version of the data communications standard, V.34, which supports up to 28,800 bps. V.34 bis adds two higher data rates to V.34. These speeds are 31,200 bits per second and 33,600 bits per second. V.34 bis is now the most common standard for PC data communications over dial-up phone lines.

V.35 ITU-T standard for trunk interface between a network access device and a packet network that defines signaling for data rates greater than 19.2 Kbps. It is an international standard termed "data transmission up to 1.544 Mbps" (i.e. T-1).

V.35 Connectors

It's typically used for DTE or DCE equipment that interface to a high-speed digital carrier. The physical interface is a 34-pin connector, which can't connect, either physically or electrically, to any other interface without a special converter. See V.36.

V.36 ITU-T recommendation for 4-wire communications at speeds greater than 48 Kbps. It is intended to replace V.35. See V.35.

V.42 Error Correction ITU-T error-correction standard specifying both MNP4 and LAP-M. The ITU-T title says "Error-correcting procedures for DCEs using Asynchronous-to-Synchronous Conversion". It also notes in the text that it applies only to full-duplex devices. The ITU-T modulation schemes with which V.42 may be used are V.22, V.22 bis, V.26 ter, and V.32, and V.32 bis. LAPM, based on HDLC, is the "primary" protocol, on which all future extensions will be based. The Alternative Protocol specified in Annex A of the Recommendation is for backward compatibility with the "installed base" of error-correcting modems. See V.42 bis.

V.42 bis Data Compression ITU-T data compression standard. It compresses files "on the fly" at an average ratio of 3.5:1 and can yield file transfer speeds of up to 9,600 bps on a 2,400 bps modem, 38,400 bits per second with a 9,600 bps modem, 57,600 bps with a 14,400 bps V.32 bis modem, or 115,600 bit/s on a 28,800 bps modem. On-the-fly data compression only has value if you use it to transfer and receive material that is not already compressed. Compressing stuff a second time yields no significant improvement in speed (assuming your compression technique worked the first time around). So the decision to buy a V.42 bis modem depends on the material you're working with and your pocketbook. V.42 bis modems are more expensive.

V.42 bis was approved by the ITU-T because of its technical merits. Existing data compression methods (MNP 5 for example) only provided up to two-to-one compression. Also, V.42 bis provides for built-in "feedback" mechanisms, so that the modem can monitor its own compression performance. If the DTE starts send pre-compressed or otherwise uncompressible data, V.42 bis can automatically suspend its operation to avoid expansion of the data. It continues to monitor performance even when sending data "in the clear," and when a performance improvement can be gained by reactivating compression, it will do so automatically.

V.42 bis was selected because it would work with a wide variety of different implementations — different amounts of memory, different processor speeds, etc. Because of this, there WILL be differences between various manufacturer's products in terms of THROUGHPUT performance (although they will all properly compress and decompress, some will do it faster than others). If maximum throughput is important, you should check published benchmark tests to find the modem that provides the best performance.

This chart, courtesy Hayes, shows the speedup that's possible. It includes information on a modem called the Hayes Optima 288, which includes a proprietary (i.e. not compatible with anyone else) Hayes enhanced implementation of V.42 bis.

V.32	9,600 + data compression	=	38,400 bit/s
V.32 bis	14,400 + data compression	=	57,600 bit/s
V.34	28,800 + data compression	= 115,600 bit/s	
Hayes Optima 288	28,800 + Hayes V.42 bis = 230,400 bit/s		

V.5 See V5

V.54 ITU-T standard for loop test devices in modems, DCEs (Data Communications Equipments) and DTEs (Data Terminal Equipment). Defines local and remote loopbacks. There are four basic tests — a local digital loopback test that is used to test the DTE's send and receive circuits; a local analog loopback test that is used to test the local modem's operation; a remote analog loopback test that is used to test the communication link to the remote modem; and a remote digital loopback test that is used to test the remote modem's operation. If a modem has V.54 capability (most V.32 and V.32 bis modems do), its manual should include documentation on performing the various tests. Version 7 of the Norton Utilities (from Symantec) also includes a local digital loopback test for your PC's COM ports, for which you will need the optional jumper plug offered with the software. Where a modem supports local digital loopback testing, it simulates the jumper plug and does not, therefore, need to be disconnected.

V.61 ITU-T V.61 is a 14.4 kbps V.32bis analog multiplexing technology standard developed by AT&T Paradyne and marketed as VoiceSpan. The data rate is reduced to 4800 bps during simultaneous voice and data. This analog Simultaneous voice and data (ASVD) standard has now been effectively obsoleted by V.70.

V.70 ITU-T standard for Digital Simultaneous Voice and Data (DSVD) modems. DSVD allows the simultaneous transmission of data and digitally-encoded voice signals over a single dial-up analog phone line. DSVD modems use for V.34 modulation (up to 33.6 kilobits per second), but may also use V.32 bis modulation (14,400 kilobits per second). The DSVD voice coder is a modified version of an existing specification and is defined as G.729 Annex A. The DSVD voice/data multiplexing scheme is an extension of the V.42 error correction protocol widely used in modems today. DSVD also specifies fallbacks that enable DSVD modems to communicated with standard data modems (i.e. V.34, V.32 bis, V.32 and V.22). See DSVD for a bigger explanation.

V.75 ITU-T recommendations which specify DSVD control procedures. See V.70.

V.76 ITU-T recommendations which define V.70 multiplexing procedures.

V.8 A way V.34 modems negotiate connection features and options.

V.8 bis New start-up sequence for multimedia modems.

V.80 V.80 is the application interface defined in the H.324 ITU video conferencing standard. A V.80 modem provides a standard method for H.324 applications to communicate over modems. A V.80 modem provides three main functions:

1. Converts synchronous H.324 streams to run on asynchronous modem connections. That is, they accept and send data in synch with a timing device ("clock"). Serial ports and modems are asynchronous, meaning they accept and receive data independent of any clocking device. V.80 converts the synchronous data stream of an H.324 application so that it can communicate through an asynchronous modem connection.

2. Allows for rate adjustments based on line conditions. Modems adjust to different line conditions throughout a call. Under bad conditions a modem will slow down. When conditions clear, a modem will resume at top speed. A V.80 modem alerts an H.324 video phone of its rate adjustments thereby allowing the application to adjust the rate at which it sends video and audio.

3. Communicates lost packets to the H.324 application. During transmission, data can be lost due to buffer overflows, phone line errors and a number of other issues. Under these conditions, a V.80 modem communicates lost data information to the H.324 application, helping it to keep real-time audio and video flowing to both sides of the call. See H.324.

V.90 V.90 is the new ITU-T standard for Pulse Code Modulation (PCM) modems running at speeds to 56 Kbps. It has been referred to informally as V.PCM until the numeric designation, V.90, was assigned in Geneva on February 6, 1998. It allows speeds of up to 56 Kbps in one direction only, from the central site equipment to the end user. The "back channel" upstream from end user to the central site remains limited to 33.6 kbps (V.34 speeds). See 56 Kbps Modem (for a longer technical explanation), V.91 and V.PCM.

V.91 The V.91, all-digital extension to V.90, allows modem signals to be transmitted through all-digital telephone connections which are configured for speech rather than data signals. Such connections, which terminate digitally at both the customer's and service provider's premises, have hitherto only been able to achieve data rates of 33,600 bit/s, however the use of V.91 modems will allow data to be transmitted on these lines at close to 64,000 bit/s. The standard is expected to be particularly useful on ISDN connections where a data bearer channel is not available or cannot be guaranteed. See V.90.

V.ASVD Analog Simultaneous Voice and Data modem.

V.AVD Alternating Voice and Data. This is the same function as provided by VoiceView products.

V.DSVD Digital Simultaneous Voice and Data.

V.Fast V.FC. An interim modem standard to support speeds to 28,800 bits per second for uncompressed data transmission rates over regular dial up, voice-grade lines. V.FAST stands for Very Fast. V.Fast was a "standard" that only a few manufacturers of modems adopted. These manufacturers adopted V.Fast because they were impatient with the ITU's slowness. Eventually, however the ITU did adopt a new standard, called V.34. See V.34 and V.34bis.

V.FC Version Fast Class. It is an interim standard that was developed for use until the ITU-T ratified V.Fast, i.e. V.34, which is the speed that a V.34 modem communicates at — namely at 28,800 bits per second. V.FC was eventually obsoleted by V.34, which the ITU-T eventually adopted. See V.34.

V.GMUX The multiplexer for V.DSVD.

V.pcm All the makers of 56 Kbps modems have been promising an interoperable 56 Kbps specification, and the International Telecommunication Union (the ITU) is overseeing its development. In February of 1998, an ITU committee voted to accept a new standard, called V.pcm (later termed V.90), that combines technologies from Lucent, 3Com, and Motorola. To reinforce that effort, Lucent and 3Com now say they will test 56 Kbps modems and modem chip sets using the new standard. As I write this in the middle of April, 1998, there are two types of 56 Kbps modems being sold worldwide — those that adhere to the X2 standard and those that adhere to the 56flex standard. These modems cannot talk to each other at 56 Kbps (or 53 Kbps, as it really is), since their standards are not compatible. They can however, talk to each other at 33.6 Kbps, which is an internationally accepted standard. See X2, 56flex and V.90.

V.Standards Standards recommended by the ITU-T. See above.

V/T Internet-speak for Virtual Time. R/T means the time it takes to download stuff. Writing in the New York Times, Charles McGrath said it was "customary in Net-speak to make a distinction between r/t, or real time — the time in which all these delays and jam-ups occur and v/t, or virtual time, which is time on the Net: a kind of external present in which it is neither day nor night and the clock never ticks. V/t is time without urgency, without priority."

V5 A standard approved by ETSI (European Telecommunications Standards Institute) in 1997 for the interface between the access network and the carrier switch for basic telephony, ISDN and semi-permanent leased lines. The V5 standard effectively provides for open access to both wired and wireless networks, thereby encouraging competition in a deregulated environment. V5 is European Telecommunications Standards Institute's (ETSI's) open standard interface between an Access Node (AN) and a Local Exchange (LE) for supporting PSTN (Public Switched Telephone Network) and ISDN (Integrated Services Digital Network). Examples of Access Nodes include Digital Loop Carrier (DLC) systems, wireless loop carrier system, and Hybrid Fiber Coax (HFC) systems. V5 also has generated interest in China and SouthEast Asia. See ETSI and www.etsi.org.

VA This is a form of power measurement called "Volt-Amps". A VA rating is the Volts rating multiplied by the Amps (current) rating. The VA rating can be used to indicate the output capacity of a UPS (Uninterruptible Power Supply) or other power source, or it can be used to indicate the input power requirement of a computer or other AC load. For loads, the VA rating multiplied by the Power Factor is equal to the Watts rating. The VA rating of a load must always be greater than or equal to the Watts rating because Power Factor cannot be greater than 1. This definition courtesy American Power Company.

VAB Value Added Business partner. A term which Hewlett-Packard uses for developers which write software for its computers. HP helps its VABs sell software. Clearly, by doing so, it helps sell more HP computers.

VAC Voice Activity Compression.

Vacant Code An unassigned area code, central office or station code.

Vacant Code Intercept Routes all calls made to an unassigned "level" (first digit dialed) to the attendant, a busy signal, a "reorder" signal or to a recorded announcement.

Vacant Number Intercept Routes all calls of unassigned numbers to the attendant, a busy signal or a prerecorded announcement.

Vacation Message See Auto Responder.

Vacation Service A service offered by local telephone companies to subscribers who will be away. A live operator or a machine intercepts the calls and delivers a message. When you come back, you get your old number. But in the meantime, while you're away, you pay less money per month than you would for normal phone service. Also known as Absent Subscriber Service.

VACC Value Added Common Carrier. A common carrier that provides some network service other than simple end-to-end data transmission. Services include least-cost routing, accounting data, and delivery clarification.

Vacuum Tubes Before there were solid state devices there were vacuum tubes. A vacuum tube is an air-evacuated glass bulb with at least two electrodes: a cathode and an anode. The cathode is heated causing the electrons to "boil off." If a voltage is placed across the cathode and the anode, the electrons will be attracted to the anode completing the electric circuit. A tube with just a cathode and anode is called a diode. If a grid is placed between the anode and cathode, a small current placed on the grid can control the much larger cathode-anode current. This type of tube is called a triode. As vacuum tubes evolved, additional grids were inserted between the cathode and anode to produce tetrodes, pentodes, etc. Today, transistors have replaced vacuum tubes in all except a few specialized applications.

The first electronic computers, Eniac and Univac, built in the wartime secrecy of the 1940s, employed vacuum tubes. They had an average life span of about 20 hours, but with thousands of hot glowing tubes in a single machine, some computers shut down every seven to twelve minutes. Vacuum tubes imposed a limit on the size and power of planned next generations of computers. The second generation of computers, never used vacuum tubes. It used transistors, which were invented in 1947.

VAD 1. Value Added Dealer. Another term for Value Added Reseller (VAR). Essentially, VARs or VADs are companies who buy equipment from computer or telephone manufacturers, add some of their own software and possibly some peripheral hardware to it, then resell the whole computer or telephone system to end users, typically corporations.
2. Voice Activated Dialing.

VADSL Very-high-speed ADSL. A variation on the theme of VDSL (Very-high-data-rate Digital Subscriber Line), which likely will support symmetric, bidirectional transmission. See also ADSL, VDSL and xDSL.

Vail, Theodore N. Theodore N. Vail began his career with the Bell System as general manager of the Bell Telephone Company in 1878. He later became the first president of the American Telephone & Telegraph Company in 1885. He left AT&T two years later. After pursuing other interests for 20 years, he returned as president of AT&T in 1907, retiring in 1919 as chairman of the board. Vail believed in "One policy, one system, universal service." He regarded telephony as a natural monopoly. He saw the necessity for regulation and welcomed it.

VAIVR Voice Activated Interactive Voice Response.

Validation 1. Generally, all long distance carriers, operator service providers and private pay phone companies will not put a call through unless they can "validate," the caller's telephone company calling card, home/business phone number or credit card. Until the advent of US West's Billing Validation Service and other similar databases in 1987, the companies who needed to validate their callers' billing requests had to turn back the caller or accept the call on faith. Validating a user's calling card is, simply, a Yes-No. If the card number is validated, it is Yes. Getting the validation involves a data call from the provider to the owner of the database. There are many ways of doing this, including a dedicated trunk and an port through an X.25 network. Here's an explanation from material put out by Harris, maker long distance switches, including the P2000V: "Validation processing starts with a check of the P2000V's own internal database of invalid 'billed to' numbers. This database contains numbers that the system administrator wishes to temporarily block. If a call's 'billed to' number does not appear in the database, the P2000V then queries the external validation service. The P2000V directly accesses external validation services via an X.25 modem connected to a leased line. The P2000V can also access Line Information Database (LIDB) through LIDB service bureau providers."

2. Tests to determine whether an implemented system fulfills its requirements. The checking of data for correctness or for compliance with applicable standards, rules, and conventions. The portion of the development of specialized security test and evaluation, procedures, tools, and equipment needed to establish acceptance for joint usage of an automated information system by one or more departments or agencies and their contractors.

3. A telephone company term. The determination of the degree of validity of a measuring device. The validation checks that can be made using output data are of six general types: 1. Compare related sets of registers; 2. Compare like groups of equipments; 3. Compare past and present data; 4. Compare usage and peg count; 5. Compare usage against grade of service.

Validity Check Any check designed to insure the quality of transmission.

Value Added Refers to a voice or data network service that uses available transmission facilities and then adds some other service or services to increase the value of the transmission.

Value Added Carrier VAC. A voice or data common carrier that adds special service features, usually computer related, to services purchased from other carriers and then sells the package of service and features.

Value Added Common Carrier VACC. A common carrier that provides some network service other than simple end-to-end data transmission. Services include least-cost routing, accounting data, and delivery clarification.

Value Added Network VAN. A data communications network in which some form of processing of a signal takes place, or information is added by the network. No one knows, however, exactly what a VAN is. The general idea is that a VAN buys "basic" transmission and sometimes switching services from local and long distance phone companies and adds something else — typically an interactive computer with a database, a computer and massive storage. In this way, the VAN adds value to basic communications services. Dial up stock market quoting services are VANs. Electronic mail providers are VANs. But VANs can also simply be basic X.25 packet switching networks which are open to the public. Such a network will use X.25 packet switching to provide error correction, redundancy, and other forms of network reliability. Private organizations (companies, universities, etc.) may set up their own value-added networks, or — as in the case of PDNs (Public Data Networks) — another fancy name for a VAN that offers its services to the public. The classic VAN is a packet-switched operation like Tymnet, GTE Telenet, MCI Mail or AT&T Mail.

A VAN can also be communication network that provides features other than transmission of information, such as translation of one type of computer signal to another type of computer signal, called protocol conversion. VAN sometimes refers to packet-switched networks with protocol conversion. The value added is referred to as dissimilar system interface capability.

Value Added Reseller See OEM and VAR.

Value Added Service A communications facility using common carrier networks for transmission and providing extra data features with separate equipment. Store and forward message switching, terminal interfacing and host interfacing features are common extras. See also VALUE ADDED NETWORK.

Value Driven Re-Engineering A fancy term for Re-Engineering, which is a term probably invented by Michael Hammer in the July-August, 1990 issue of Harvard Business Review. In that issue, he wrote "It is time to stop paving the cowpaths. Instead of embedding outdated processes in silicon and software, we should obliterate them and start over. We should 're-engineer' our business: use the power of modern information technology to radically redesign our business processes to achieve dramatic improvements in their performance." The term re-engineering now seems to me mean taking tasks presently running on mainframes and making them run on file servers running on LANs — Local Area Networks. The idea is to save money on hardware and make the information more freely available to more people. More intelligent companies also redesign their organization to use the now, more-freely available information. See RE-ENGINEERING.

Valve The original British word for an electron tube.

Vampire Tap In local area networking technology, a cable tap that penetrates through the outer shield to make connection to the inner conductor of a coax cable. The name comes from the fact that the connector pierces the insulation and outer shield by means of one or more sharp "teeth" in order to access the communications artery, much like a vampire's teeth pierce might your jugular in order to drink your blood.

VAN See Value Added Network.

Van Allen Belts Two layers of charged particles emitted from the sun that are trapped within the earth's magnetic influence. These are named after the discoverer, J. Van Allen. The inner layer exists from about 2,400 to 5,600 km altitude above the earth's surface and consists of secondary charged particles. The outer layer lies between about 13,000 and 19,000 kilometers and is thought to consist of the original particles released from the sun's surface.

Van Eck Detection Kit A receiver that monitors the electromagnetic radiation given off by a computer screen, allowing an eavesdropper to monitor the contents of a victim's screen from a distance (say in the bushes outside a company).

Vanity A telephone, often an 800 toll-free number, which spells something. By way of example, 1-800-542-7279 is advertised as 1-800-LIBRARY, the toll-free number of the publisher of this dictionary. Clearly, there is only a single set of digits which spell "LIBRARY;" therefore, such numbers can be of great value, largely due to the advantage of spontaneity of recall. The recent expansion of the North American 800 number dialing scheme to include 888 numbers has created a storm of controversy, as the publisher must now protect that vanity number from duplication via the 888 prefix.

The introduction of UIFN (Universal International Freefone Number) services further promises to infringe on the uniqueness of that number. However, it should be noted that touchtone keypads in North America always display letters as well as numbers, while that is not the case in most of the rest of the world. Yet, hotel phones, fax machines and cellular phones worldwide generally do display letters as well as numbers. See also 800 Service and UIFN.

VAPD Voice Activated Premier Dialing.

VAPN Voice Access to Private Network.

Vapor Seal A vapor seal is an essential infiltration of a critical space, such as a data processing center or other room that contains sensitive electronic instrumentation. Essentially, a vapor seal is a barrier that prevents air, moisture, and containments from migrating through tiny cracks or pores in the walls, floor, and ceiling into the critical space. Vapor barriers may be created using plastic film, vapor-retardant paint, vinyl wall coverings and vinyl floor systems, in combination with careful sealing of all openings into the room.

Vaporware A semi-affectionate slang term for software which has been announced, perhaps even demonstrated, but not delivered to commercial customers. Hyperware is hardware which has been announced but has not yet been delivered. Slideware is hardware or software whose reason for existing (eventually) has been explained in 35-mm slides, foils, charts and/or PC presentation programs. Slideware is usually less real than vaporware or hyperware, though some people would argue with this. Allegedly the term vaporware cam as a result of the many delays in releasing Windows after Bill Gates of Microsoft announced it at the Fall, 1983 Comdex show in Las Vegas. See also HOOKEMWARE, HYPERWARE, MEATWARE, SLIDEWARE and SHOVELWARE.

VAR Value Added Reseller. Typically VARs are organizations that package standard products with software solutions for a specific industry. VARs include business partners ranging in size from providers of specialty turn-key solutions to larger system integrators.

Variable Bit Rate Service VBR. A telecommunications service in which the bit rate is allowed to vary within defined limits. Instead of a fixed rate, the service bit rate is specified by statistically expressed parameters.

Variable Call Forwarding An optional feature of AT&T's 800 IN-WATS service. It allows the subscriber to route calls to certain locations based on time of day or day of week.

Variable Format Message A message in which the page format of the output is controlled by format characters embedded in the message itself. The alternative is to have the format determined by prior agreement between the origin and the destination.

Variable Length Buffer A buffer into which data may be entered at one rate and removed at another, without changing the data sequence. Most first-in, first-out (FIFO) storage devices serve this purpose in that the input rate may be variable while the output rate is constant or the output rate may be variable while the input rate is fixed. Various clocking and control systems are used to allow control of underflow or overflow conditions.

Variable Length Record A file in a database containing records not of uniform length and in which the distinctions between fields are made with commas, tabs or spaces. Records become uniform in length either because they are uniform to start with or they are "padded" with special characters.

Variable Quantizing Level VQL. A speech-encoding technique that quantizes and encodes an analog voice conversation for transmission at 32,000 bits per second.

Variable Resistor A resistance element which may be varied to afford various values.

Variable Term Pricing Plan VTPP. A rate plan developed by AT&T to replace two-tier pricing. VTPP used to provide for two, four, five or six year contracts, over which period the customer is promised stable prices for some — not all — of the equipment and/or tariffed services he uses. Generally, under VTPP, the customer does not end up owning any of the equipment. VTPP has now been replaced by more normal ways of doing commercial business — outright sale, leasing, etc.

Variable Timing Parameter Timing durations for features such as hold recall, camp-on recall, off-hook duration, and many other programmable telephone system services.

Variance The average squared deviation. To calculate the variance, determine the difference of each item from the group mean. Then square (multiply by itself) each of the differences (the deviations). Next, average the squares of the deviations to determine average square which is the variance (for samples, divide by one less than the number of items in the sample).

Variolosser A device with a variable level of attenuation which is controlled by an external signal. Often this signal is the level of the signal being attenuated, that is the higher the level of the signal the more it is attenuated.

VARTI Value Added Reseller Telephone Integrator. A term coined at Telecom Developers '92. It refers to the VARs and interconnects of the 90s that are combining telephony and personal computers to offer products that tie the telephone network to personal computer applications.

VAX A line of minicomputers made by Digital Equipment Corporation (DEC), now part of Compaq.

VBI Vertical Blanking Interval. The vertical blanking interval is the portion of the television signal which carries no visual information and appears as a horizontal black bar between the pictures when a TV set needs vertical tuning. The VBI is used for carrying close-captioned signals for the hearing impaired. Digitized data can also be inserted into the VBI for transmission at rates greater than 100,000 bps. Information services such as stock market quotations and news offerings are now available via the VBI of a CATV signal. The data embedded in the VBI signal is retrieved from a standard cable or satellite receiver wall outlet by a receiver set, which connects to a RS-232 port on a microcomputer. Software packages then allow subscribers instant access to the information, which may be displayed in a number of formats.

vBNS Very high-speed Backbone Network Service. A high-speed SONET fiber optic backbone network being developed by MCI (now MCI Worldcom) for the National Science Foundation (NSF), vBNS will serve as the backbone transport network for Internet2. Initially, vBNS will runs at a speed of 155 Mbps (OC-3); ultimately, the network will run at 2.4 Gbps (OC-48). The first deployment of vBNS connects five NSF-funded supercomputing centers (SCCs): Cornell Theory Center, National Center for Atmospheric Research, National Center for Supercomputing Applications, Pittsburgh Supercomputing Center, and San Diego Supercomputing Center. Also connected are the NSF-funded Network Access Points (NAPs) at Hayward, CA; Chicago, IL; Pennsauken, NJ; and Washington, DC. See also Internet2 and SONET.

VBR Variable Bit Rate. A voice service over a an ATM switch. Voice conversations receive only as much bandwidth as they need, the remaining bandwidth is dynamically allocated to other services that may need it more at any given moment. Northern Telecom refers to this approach as "making bandwidth elastic." VBR also refers to networking processes such as LANs which generate messages in a random, bursty manner rather than continuously.

VC 1. Virtual Channel. A SONET term. Existing with a Virtual Tributary (VT), a VC is virtually to a traditional TDM channel. Note that a TDM channel is either set aside for or prioritized for a particular transmission; in other words, it's a dedicated channel. A Virtual Channel, on the other hand is not set in such a rigid environment; rather, such channels float in a

SONET frame, available only when required in a particular transmission, and not necessarily found in the same place in time. All things considered, "virtual channel" is just a fancier term for a channel, but maintaining the "virtual" nomenclature of SONET-Virtually Aggravating at times, isn't it! Also known as a Tributary Unit in SDH (Synchronous Digital Hierarchy) terminology. See also SDH, SONET, TDM and Tributary Unit.
2. Virtual Channel. An ATM term. According to the ITU-T, a virtual channel is a unidirectional communication capability for the transport of ATM cells." A Virtual Channel Identifier (VCI) in the header of the ATM cell is assigned or removed, respectively, to either originate or terminate a Virtual Channel Link (VCL). VCLs are concatenated to form a Virtual Channel Connection (VCC), which is and end-to-end VP (Virtual Path) at the ATM layer. Once again, "virtual" is the operative word. Channels are "virtual" in the ATM world, as the extent to which they are made available depends on the priority level of the traffic, as defined in the cell header. High-priority traffic gets lot of VCs through the ATM switch, while low-priority traffic gets fewer. VCs also are defined in the UNI 3.0 specification. See also ATM, Concatenation, UNI, VCC, VCI, VCL, and Virtual.
3. Virtual Circuit. In packet switching, network facilities that give the appearance to the user of an actual end-to-end circuit. VCs define the physical path that all packets in a packet stream will follow during a session between two or more computing systems. Virtual circuits allow many users to share switches and transmission facilities, while each can enjoy the advantages of a Virtual Private Network (VPN). VCs can be provisioned as Permanent Virtual Circuits (PVCs) or Switched Virtual Circuits (SVCs). X.25 Packet Switched networks, Frame Relay and ATM all make use of VCs. See also ATM, Frame Relay, Packet Switching, PVC, SVC and X.25. EXPAND SLIGHTLY, XREF: VCC Virtual Channel Connection. As an ATM term, it is a concatenation of VCLs (Virtual Channel Links) that extend between the points where the ATM service users access the ATM layer. The points at which ATM cell payload is passed to, or received from, the users of the ATM Layer (i.e., a higher layer or ATM-entity) for processing signify the endpoints of a VCC. VCCs are unidirectional. See also Concatenation and VC.
4. Virtual Container. SDH defines a number of "containers". The container and the path overhead from a "Virtual Container" (VC) in Europe or "Virtual Tributary" (VT) in North America (ref: ITU G.709).
VCA See VOICE CONNECTING ARRANGEMENT.
vCalendar A virtual calendar specification by the IMC (Internet Mail Consortium) as an electronic exchange format for personal scheduling information. vCalendar is an open specification based on industry standards including the X/Open and XAPIA Calendaring and Scheduling API, the ISO 8601 international date and time standard, and the related MIME e-mail standards. Adoption of the standard allows software products to exchange calendaring and scheduling information in an easy, automated and consistent manner. According to the IMC, it can work like this. You are at a business meeting with representatives from various companies. Every imaginable portable computing device is present, from PDAs, to hand-held organizers, to laptops and notebook PCs. The chairperson of the meeting communicates from his device to all other devices via infrared beam. All attendees then do the same. Thus, the next meeting is scheduled electronically, with conflicts identified. At the same time, personal information is passed around via vCard technology.

vCalendar is being enhanced by a new specification called iCalendar, which is designed specifically for the Internet. See iCalendar, IMC and vCard.
vCard A totally wonderful idea. vCard is a tiny file (with the extension .vcf) that contains all the information on your business card — your street address, your phone numbers, your email address, etc. You attach this file to an email you send to someone. They receive it as an attachment. They click on it. It opens up as entry in the electronic address book or PIM (Personal Information Manager). They then click "save." And bingo, your information is added to their address book. The benefits are obvious: huge time savings, huge savings in accuracy, etc. Most browsers (Netscape was one of the first) and most email clients are now beginning to support the vCard specification. And most allow you to choose that your vCard file is attached to each and every email you send. My recommendation to everyone is simple: organize so that you attach your vCard to each of your outgoing emails.
vCard is a Versit idea. They defined it as an electronic, virtual information card that can be transferred between computers, PDAs, or other electronic devices through telephone lines, or e-mail networks, or infrared links. With vCard, according to Versit, individuals can consistently identify themselves without restating or rekeying their information. vCards include data such as name, address, phone number, e-mail user ID, with multimedia support for photographs, sound clips and company logos. vCard is the result of a collaborative industry effort between Versit and multiple vendors. The vCard SDK is available in Windows 95, OS/2, and MacOS. The SDK includes the vCard Specification, a demonstration application, vCard file examples, and free source code. A printed copy of the vCard V2.0 Specification may be obtained from the Internet Mail Consortium (IMC) (www.imc.org), which developed the specification in cooperation with leading producers of desktop software, hand-held organizers, Internet web clients, and others. The specification is open and based on industry standards, including ITU-T X.500 directory services.
VCC Virtual Channel Connection. As an ATM term, it is a concatenation of VCLs that extends between the points where the ATM service users access the ATM layer. The points at which the ATM cell payload is passed to, or received from, users of the ATM Layer (i.e., a higher layer or ATM-entity) for processing signify the endpoints of a VCC. VCCs are unidirectional.
VCEP Video Compression/Expansion Processor chip.
VCI Virtual Channel Identifier. An ATM term. The address or label of a VC (Virtual Channel). The VCI is a unique numerical tag, defined by a 16 bit field in the ATM cell header, that identifies a VC over which a stream of cells is to travel during the course of a session between devices. See also VC.
VCL Virtual Channel Link. An ATM term. A means of unidirectional transport of ATM cells between the point where a VCI (Virtual Channel Identifier) value is assigned as it is presented to the ATM network, and the point where that value is translated or removed as it exits the ATM network. VCLs are concatenated to form VCCs (Virtual Channel Connections). See also VCC and VCI.
VCN Virtual Corporate Network. Stentor's name for a service it later changed to Advantage VNet. It's similar to MCI's VNet.
VCO Voltage Controlled Oscillator: An oscillator whose clock frequency is determined by the magnitude of the voltage presented at its input. The frequency changes when the voltage changes.

VCOS Visible Caching Operating System. VCOS is a realtime multitasking DSP operating system for the AT&T DSP3210 Digital Signal Processor. Visible Caching means the programmer caches the program and the data onchip, in contrast to logic caching where state machines (implemented in silicon) perform all caching.

VCPI An acronym for the Virtual Control Program Interface, a standard developed by Quarterdeck and Phar Lap Software for running multiple programs and controlling the Virtual-86 mode of 386 microprocessors. A program that's VCPI-compatible and can run in the protected mode under DOS without conflicting with other programs in the system.

VCR VideoCassette Recorder (or Player).

VCSEL Vertical Cavity Surface Emitting Laser. A VCSEL (pronounced "VIX-els") is a tiny laser less than half the width of a human hair (10 x 10 x 2 microns) that emits light in a cylindrical beam vertically from its surface and offers several significant advantages over edge-emitting lasers. VCSELs are smaller than edge-emitting lasers (in the case of CD lasers, they're 100 times smaller); are cheaper to manufacture, as they use the same fabrication techniques as chips (so Moore's Law now applies to lasers); offer greatly reduced power consumption (requiring only 1-2 milliwatts per gigabit per second); and can be packed closely together in two-dimensional arrays, yielding an unlimited aggregate data transfer rate, depending on how many you've stacked. VCSELS and fiber-optic cabling are at the heart of Gigabit Ethernet technology. This definition courtesy Gary Clem, San Francisco.

VD Virtual Destination. See VS/VD.

vDC Volts Direct Current. For example, 12 vDC is the voltage which powers automobiles in most parts of the world, including North America.

VDE Verband Deutscher Elektrotechniker. Federation of German Electrical Engineers similar in form to the IEEE.

VDECK An 8 millimeter cassette recorder developed by Sony Corporation for use as a computer peripheral.

VDI Video Device Interface. A software driver interface that improves video quality by increasing playback frame rates and enhancing motion smoothness and picture sharpness. VDI was developed by Intel and will be broadly licensed to the industry.

VDISK Virtual DISK. Part of the computer's Random Access Memory assigned to simulate a disk. VDISK is a feature of the MS-DOS operating system.

VDM Voice Data Multiplexer.

VDO A technology that enables Internet video broadcasting and desktop video conferencing on the Internet and over regular telephone lines and private networks. VDOPhone which provides the ability to have private point to point audio/video contact is currently only available for Windows95 and requires a Pentium processor. The VDOLive player however is available for Windows and Power Macs and provides the ability as a Netscape plugin for viewing and hearing live Internet Broadcasts. To download the VDOLive Player go to www.vdo.net/download. See also REALAUDIO.

VDRV Variable Data Rate Video. In digital systems, the ability to vary the amount of data processed per frame to match image quality and transmission bandwidth requirements. DVI symmetrical and asymmetrical and asymmetrical systems can compress video at variable data rates.

VDS Vocabulary Development System.

VDSL Very-high-data-rate Digital Subscriber Line. A technology in the very early stages of definition. Initial VDSL implementation likely will be in asymmetric form, essentially being

very high speed variations on the ADSL theme. Goals are stated in terms of submultiples of the SONET and SDH principal speed of 155 Mbps. Specifically, target downstream performance is 51.84 Mbps over UTP local loops of 1,000 feet (300 meters), 25.92 Mbps at 3,000 feet (1,000 meters), and 12.95 Mbps at 4,500 feet (1,500 meters). Upstream data rates are anticipated to fall into three ranges: 1.6-2.3 Mbps, 19.2 Mbps, and a rate equal to the upstream rate.

The application for VDSL is in a hybrid local loop scenario, with FTTN (Fiber-To-The-Neighborhood) providing distribution from the CO to the neighborhood, and with VDSL over UTP (Unshielded Twisted Pair) carrying the signal the last leg to the residential premise. Clearly, the specific application is for highly bandwidth-intensive information streams such as are required for support of HDTV and Video on Demand. Early work on VDSL has begun in standards bodies including ANSI T1E1.4, ETSI, DAVIC, The ATM Forum and The ADSL Forum. See also ADSL, ADSL Forum, HDSL, IDSL and SDSL.

VDT 1. Video Display Terminal. A data terminal with a TV screen. Another name for computer monitor. VDT is the term you hear in Europe.

2. Video Dial Tone. The new concept of getting home entertainment, information and interactive services to residences over some form of new broadband network stretching into the nation's homes. Video Dial Tone is a term used by traditional telephone companies. They're the ones allegedly building this broadband network to provide "Video Dial Tone."

VDU Visual Display Unit.

Vector A quantity in the visual (video) telecommunications industry that describes the magnitude and direction of an object's movement — for example, a head moving to the right. See VECTOR IMAGES.

Vector Graphics Images defined by sets of straight lines, defined by the locations of the end points.

Vector Images Images based on lines drawn between specific coordinates. A vector image is based on the specific mathematics of lines. In contrast, a raster image is a bit-mapped (i.e. bit-drawn) image. A vector engineering image is more useful for engineering, since it can be changed easier than a bit-mapped image. A vector image can easily be converted to a raster image. But it's much more difficult to go from a raster image to a vector image. Some storage systems now store images as combination raster/vector.

Vector Processor Array Processor.

Velocity Of Light The speed of light in a vacuum is 186,280 miles per second, or 299,792 kilometers per second. The speed of light is very important because today we can measure time more accurately than length. In effect, we define the meter as the time traveled by light in 0.000000003335640952 of a second as measured by the cesium clock.

Velocity Of Propagation The speed at which a signal travels from a sender, through a transmission line and finally arrives at the receiver.

Velocity Of Sound The velocity of sound varies with the medium carrying it. In air at 0 degrees centigrade, it's 331 meters per second. In glass at 20 degrees centigrade, its 5485 meters per second.

Velveeta An Internet Usenet posting, often commercial in nature, excessively cross-posted to a large number of newsgroups. Similar to Spam, although that term is often used to describe an identical post that's been loaded onto lots of inappropriate newsgroups, one group at a time (rather than cross-posted). This definition courtesy Wired Magazine.

Vendor Code Software written by the same company that manufactured the computer system on which it is running (or not running...).

Vendor ID A Plug and Play term. Vendor ID is the 32-bit vendor ID that indicates the manufacturer, specific model, and version of a device. It is this number which helps Plug and Play configure the PC to run the device.

Vendor Independent Hardware or software that will work with hardware and software manufactured by other vendors. The opposite of proprietary.

Vendor Independent Messaging Group A group of software and software companies who are trying to create non-proprietary, standard programming interfaces to help software and corporate developers write messaging and mail-enabled applications. Ultimately, end users should be able to work together more effectively and be able to exchange information from within desktop applications in a work group environment regardless of vendor platform. Members of the group include Apple, Borland, IBM, Lotus, Novell and WordPerfect.

Vendors ISDN VIA. In June 1996, according to a story in InfoWorld, thirteen major networking vendors united behind the banner of simple, standardized ISDN access with the announcement of the Vendors ISDN Association (VIA) at the recent ISDN World trade show in Los Angeles. The association will provide forums for discussing technical issues involved in standardizing ISDN service and products across the United States. Initial members include Cisco Systems Inc., Bay Networks Inc., 3Com Corp., and Ascend Communications Inc., as well as Microsoft Corp. and Intel Corp. The association grew out of the ISDN Forum, created in January. It is affiliated with the National ISDN Users Forum and the National ISDN Council, and it will work on standards with those organizations. The group's first focus will be on automating configuration of ISDN devices. Currently, users need to configure their devices manually by entering a Service Profile Identifier. The VIA is pushing for implementation of noninitializing terminals (NITs) by manufacturers of access devices. NITs will automatically send the configuration information, said Rob Rank, an ISDN product manager at Intel and vice president of the VIA.

Vengeance Billing A term that originated with expensive restaurants in Paris and received new meaning when long distance carriers in the U.S. started billing international calls at high prices on their poor unsuspecting customers. Term contributed by Jeddy Lieber of Paris.

VENUS-P A ITU X.25 packet-switched network operated in Japan by Nippon Telephone and Telegraph (NTT) Co.

Verbose In English, verbose means too many words. The speaker spoke verbosely means he used lots of words to say the little he had to say. In computerese, verbose means you get to see on screen what's going on. If you add the switch /v to the command line of a DOS program command, the program most likely will, as it loads, give you information on screen that shows what it's doing as it loads. If you don't add that switch, it will load or try and load and you won't be any the wiser as to what went on. If the program works well, there's no reason to turn on /v. If it doesn't, turn it on and see what happens. Not all programs have the /v switch.

Verification A service of a phone company operator who dials into a busy or otherwise impossible-to-reach line and checks that line and reports on that check to the caller. Phone companies are beginning to charge for this service. As of writing, AT&T, for example, was charging 40 cents to verify

the line was busy and 70 cents additional for the operator to interrupt the conversation and say another call was coming in.

Verification Trunk A trunk to which an operator has access and which will switch through to a called line even if the line is busy.

Verified Off-Hook In telephone systems, a service provided by a unit that is inserted on each of a transmission circuit for the purpose of verifying supervisory signals on the circuit. Off-hook service is a priority telephone service for key personnel, affording a connection from caller to receiver by the simple expedient of removing the phone from its cradle or hook.

Verifier A device that checks the correctness of transcribed data, usually by comparing with a second transcription of the same data or by comparing a retranscription with the original data.

VERONICA Very Easy Rodent Oriented Netwide Index to Computerized Archives. An Internet service that allows users to search Gopher systems for documents.

Versit Versit is a loose association between Apple, AT&T, IBM and Siemens Rolm. Its "vision"? To "enable diverse communication and computing devices, applications and services from competing vendors to interoperate in all environments. Communicate and collaborate with anyone, any time, anywhere.." The products include PDAs, notebooks, phones, servers and "collaboration products." One early thrust: standardize on call control within Novell/AT&T's Telephony Services. Background: Call control among PBXs "conforming" to Telephony Services is not standard. PBXs often do the same things differently. Example: Conference a call on one PBX, the PBX may put one call automatically on hold as the other is dialed. Another PBX may expect it to be done manually. The good news: IBM has agreed to pass all its CallPath call control standards over to Versit. Versit's members (including Novell coopted for this task) are now working on making Telephony Services call control more standard. Upshot: Developers won't have to test their "standard" telephony services software on each and every PBX. What works on one will work on the others. That's the goal. In late July 1995, Versit effectively merged all its activities into another association, called ECTF. (www.versit.com) Versit@cup.portal.com. See ECTF and PBX DRIVER PROFILES.

VersitCard A vCard.

Vertex Provider of tax jurisdiction rate tables and related software.

Vertical That part of a wiring grid which connects the host computer of Main Distribution Frame (MDF) to equipment located on other fields.

Vertical And Horizontal Coordinates V & H Coordinates. For purposes of determining airline mileage between locations, vertical and horizontal coordinates have been established across the United States. These V&H coordinates are derived from geographic latitude and longitude coordinates. See V & H.

Vertical Blanking Interval The interval between television frames in which the picture is blanked to enable the trace (which "paints" the screen) to return to the upper left hand corner of the screen, from where the trace starts, once again, to paint a new screen. Several companies are eyeing the vertical blanking interval as a place to send digital data, including news and weather information. The vertical blanking interval was the basis of teletext, a 1970s technology that, with the help of a decoder, displays printed information on the TV screen. Teletext has never caught on in the U.S. in part because the amount of data that could be transmitted comfortably was small.

Vertical Cavity Surface Emitting Laser See VCSEL (pronounced "VIX-els").

Vertical Integration A firm is vertically-integrated when it owns or controls a firm in an upstream or downstream market. For example, a coal-fired power station which owns a coal mine is vertically-integrated.

Vertical Interval The portion of the video signal that occurs between the end of one field and the beginning of the next. During this time, the electron beams in the cameras and monitors are turned off (invisible) so that they can return from the bottom of the screen to the top to begin another scan.

Vertical Linearity A video term. A control that allows you to set spacing consistently across the monitor. Thus a shape intended to have a 1-inch diameter will have a 1-inch diameter wherever it appears.

Vertical Market Application An application that is industry-specific and typically very task-specific.

Vertical Redundancy Check VRC. A relatively poor method of error control used in asynchronous transmission in support of the ASCII coding scheme. A check bit, or parity bit added to each ASCII character in a message such that the number of bits in each character, including the parity bit, is odd (odd parity), or even (even parity). The term comes from the fact that the bits representing each character of data conceptually is viewed in a vertical fashion. For instance, the word "CONTEXT" consists of 7 letters, each of which consists of 7 bits, viewed as follows:

```
BIT/VALUE  C O N T E X T
1*    1 1 0 0 1 0 0
2*    1 1 1 0 0 0 0
3*    0 1 1 1 1 0 1
4*    0 1 1 0 0 1 0
5*    0 0 0 1 0 1 1
6*    0 0 0 0 0 0 0
7*    1 1 1 1 1 1 1
8**   0 0 1 0 0 0 0
* INFORMATION BIT
** PARITY BIT
```

The transmitting machine sums the bit values for each character, beginning with "nothing," which is an even value in mathematical terms. In the case of the letter "C," for instance, the next bit is a "1" bit, which creates an odd value. The next bit is a "1" bit, which creates an even value. The next four bits are "0" bits, which do not change the even value. The seventh bit is a "1" bit, which creates an odd value, once again. Assuming that the device is set for odd parity, which is the default, it will insert a "0" bit in the eighth bit position, thereby retaining the odd value. (Should the device be set for even parity, a "1" bit would have been inserted in the eighth bit position.) Should the value of the 7 information bits be an even value, the device appends a "1" bit in order to create an odd value. After the data, character-by-character, has been formatted in this fashion, each bit sequence is transmitted across the network to the target device, which also is set for odd (or even) parity. The receiving device goes through exactly the same process, examining each character for parity. If the parity does not match the expectation of the receiving device, the subject character is flagged as errored, although no remedial action is taken. As there is reasonable likelihood that two bits in a given character can be errored in the process of transmission, that the parity of the character therefore would not be affected, and that the receiving device would not detect

the fact that the character was errored, this technique is known as "send and pray." Any remedial action must be accomplished on a man-to-machine basis. LRC (Longitudinal Redundancy Checking) often is used in conjunction with VRC to improve the likelihood of detecting an error. See also LONGITUDINAL REDUNDANCY CHECKING and PARITY.

Vertical Service Options that the customer can add to his basic service such as touchtone, conference calling, speed dialing, etc. No one can explain why it's called "vertical" service.

Very High Data Rate Digital Subscriber Line See VDSL.

Very High Frequency VHF. Frequencies from 30 MHz to 300 MHz.

Very Large Scale Integration VLSI. Semiconductor chip with several thousand active elements or logic gates — the equivalent of several thousand transistors on a single chip. VLSI is the technique for making the micro chip, the so-called "computer on a chip."

Very Long Event A local area network term. A very long event is the condition that occurs when the repeater is forced to go into a jabber protection mode because of the excessive number of times a port receives a packet.

Very Low Frequency VLF. Frequencies from 3 KHz to 300 KHz.

Vertical Refresh Rate The number of times the monitor redraws its screen every second. A too-low refresh rate can result in flicker, causing eyestrain.

VESA Video Electronics Standards Association, San Jose, CA. Along with eight leading video board manufacturers, NEC Home Electronics founded VESA in the late 1980s. The association's main goal is to standardize the electrical, timing, and programming issues surrounding 800 x 600 pixel resolution video displays, commonly known as Super VGA. VESA has also issued a standard called "local bus," a new high-speed bus for the PC designed to move video between the CPU and the screen a lot faster than the conventional AT bus.

Vestigial Sideband Transmission VSB. A modified sideband transmission technique in which one sideband and the carrier are suppressed, and only a portion of the remaining sideband is transmitted. Reduced power requirements is an advantage. See also Amplitude Modulation, DSBSC, DSBTC, and SSB.

VF 1. Voice Frequency.
2. Variance Factor. An ATM term. VF is a relative measure of cell rate margin normalized by the variance of the aggregate cell rate on the link.

VFast An earlier name for the 56Kbps modem spec. See 56 Kbps Modem.

VFAT Virtual File Allocation Table. A fat file system is a file system based on a file allocation table, maintained by the operating system, to keep track of the status of various segments of disk space used for file storage. The 32-bit implementation in Windows 95 is called the Virtual File Allocation Table (VFAT). An extension of the FAT file system in DOS and Windows 3.xx, VFAT supports long filenames while retaining some compatibility with FAT volumes. See also FAT.

VFDN Voice Frequency Directory Number. A Northern Tom term.

VFG Virtual Facility Group.

VFTG Voice Frequency TeleGraph.

VG Voice Grade. A term commonly applied to a local loop. A voice grade circuit is generally analog in nature, and provides 4 KHz bandwidth. As voice is not particularly demanding in terms of either bandwidth or error performance, a voice grade circuit is not particularly capable. See also VGE.

VGA Variable Graphics Array. A graphics standard developed by IBM for the IBM PC. VGA allows the PC's screen to generate any of four levels of resolution — with one of the sharpest being 640 horizontal picture elements, known as pels or pixels, by 480 pels vertically with 16 colors. VGA is superior to earlier graphics standards, such as CGA and EGA. VGA is barely adequate for CAD-CAE. See MONITOR for all the numbers on pixels in various screens. VGA was the graphics standard introduced for IBM PS/2 line and quickly adopted by PC compatibles; supports analog monitors with a 31.5 Hz horizontal scan rate.

VGE Voice Grade Equivalent. A term commonly applied to a level of digital bandwidth sufficient to support a voice conversation using standard encoding techniques-i.e., 56/64 Kbps using PCM (Pulse Code Modulation). Voice Grade (VG) generally refers to an analog local loop circuit which provides 4 KHz bandwidth. See also VG.

VGPL Voice Grade Private Line.

VHD Very High Density. Techniques of recording 20 megabytes and more on a 3 1/2" magnetic disk.

VHF Very High Frequency. The portion of the electromagnetic spectrum with frequencies between 30 and 300 MHz.

VHI Virtual Host Interface.

VHS Video Home System using half-inch tape introduced by Matsushita/JVC in 1975 and now the most popular form of video tape. There is also a VHS at 3/4". It's often used inside ad agencies for previewing work in progress. Industrial video tape — the stuff the TV stations use — is one inch. And it shows a much better quality picture than half-inch VHS.

VI Architecture See Virtual Interface Architecture

VIA See Vendors Industry Association.

Via Net Loss VNL. A planning factor used in allocating the attenuation losses of trunks in a transmission network. A specified value for this loss is selected to obtain a satisfactory balance between loss and talker echo performance. The lowest loss in dB at which it is desirable to operate a trunk facility considering limitations of echo, crosstalk, noise and signing.

Vibratory Plow A plow that rips open the ground by vibrating a plow share.

VIDCAP Microsoft's Video For Windows program to capture video input to RAM or hard disk memory.

Video From Latin, translated as "I see," video adds the element of sight to communications. While some of us are visually-oriented, and others of us are more oriented kinesthetically-oriented (learn by doing, as in with muscles and tendons and energy and sweat), we all find a communication to be enhanced through pictures...especially motion pictures. Motion pictures are the essence of video, and video is the essence of true and full communications. Visual communications, by the way, are also extremely bandwidth-intensive. See MULTIMEDIA.

Video Capture Video Capture means converting an analog video signal into a digital format that can be saved onto a hard disk or optical storage device and manipulated with graphics software. This is accomplished with a device internal in a computer called a "frame grabber" or video capture board. Images thus captured are digitized, and can be dropped into a document or database record and may be transmitted locally on a LAN or long distance over a WAN. See VIDEO CAPTURE BOARD.

Video Capture Board To capture a single frame of motion video successfully, you need a board inside your PC that can capture the two fields comprising a single video

frame. The best source of single frame video images is a laser disk player which can pause and display a perfect frame of video without noise or jitter. Video cameras or camcorders aimed at a static, non-moving image also work well. VCR, which produces a jittery image when the tape is paused, are the poorest source. See also FRAME GRABBER.

Video Codec The device that converts an analog video signal into digital code.

Video Compression A method of transmitting analog television signals over a narrow digital channel by processing the signal digitally. You can compress an analog TV signal into one T-1 signal of 1.544 megabits per second. More advanced compression techniques will enable video signals to be compressed into fewer bits per second. One increasingly common method allows a full-color reasonably full-motion video to be compressed into two 56 Kbps channels.

Video Conference See VIDEOCONFERENCE.

Video Dial Tone Video dial tone in telco-speak means the phone company, in competition with the cable TV business, provides video to houses and offices. It does not affect the content of that video signal in any way. Thus the term video dial tone, which is like voice dial tone, whose content the phone company also does not affect or change in any way, shape or form.

Video Driver A piece of software which translate instructions from the software you are running into thousands of colored dots, or pixels, that appear on your video monitor. A video driver is also called a display Driver. Symptoms of a video driver giving trouble can range from colors that don't look right, to horizontal flashing lines to simply a black screen. In the Macintosh world, Apple rigidly defined video drivers. Windows, in contrast, is a free-for-all. Windows 3.1 defined the lowest common denominator of displays — namely 16 colors at 640 x 480 pixels. But most multimedia programs and many games won't run with only 16 colors. They require at least 256 colors.

Video Electronics Standards Association See VESA.

Video Mail Electronic mail that includes moving or still images.

Video Monitor A high quality television set (without RF circuits) that accepts video baseband inputs directly from a TV camera, videotape recorder, etc.

Video Networking I first saw the term video networking in a white paper written by L. David Passmore of Decisys, Inc. The white paper was in a press kit from Madge Networks, announcing new video hardware architecture. David wrote "When people typically think of video networking, they think of it as an application, namely video conferencing. However, video networking is really an architecture that supports a range of business applications featuring video communications. These applications can be deployed over the video network to span the LAN and WAN environments. Furthermore, if the video network is deployed properly, it can provide a consolidated (data, voice and video) WAN access solution across the enterprise."

Video On Demand VOD. Punch some buttons. Order up Gone With the Wind to start playing at your house at 8:26 P.M. on Channel 35. Bingo, you have video on demand. It's a great concept with two major problems: The equipment to provide the service is complex and expensive. Second, there's little consumer research on whether consumers are prepared to pay the high price that will be necessary. Still, it's a neat idea. In short, video on demand is a service that allows many users to request the same videos at the same time or anytime. Video on demand requires a high-end video server with hundreds of gigabytes of storage. See NVOD, VIDEO DIAL TONE and VIDEO SERVER.

Video Path The electronic path within the device that routes and processes the video signals. Video path length refers to the amount of time required for a signal to travel from input to output.

Video Processing Amplifier A device that stabilizes the composite video signal, regenerates the synchronizing signals, and allows other adjustments to the video signal parameters.

Video Server A device that could store hundreds, if not thousands of movies, ready for watching by subscribers at their individual whim. A video server could be jukebox like device that would stack several hundred movies. Or it could be a powerful, large computer with several large hard disks and/or optical disk drives. The device would be used in conjunction with the local telephone companies' service called video dial tone — providing movies over normal phone lines to their subscribers or it could be used with the CATV industry's Video On Demand service.

Video Signal Transmission of moving frames or pictures of information requiring frequencies of 1 to 6 Megahertz. A commercial quality full-color full-motion TV signal requires 6 MHz.

Video Switcher Device that accepts inputs from a variety of video sources and allows the operator to select a particular source to be sent to the switcher's output(s). May also include circuits for video mixing, wiping, keying, and other special effects.

Video Teleconferencing Also called Videoconferencing. The real-time, and usually two-way, transmission of digitized video images between two or more locations. Transmitted images may be freeze-frame (where television screen is repainted every few seconds to every 20 seconds) or full motion. Bandwidth requirements for two-way videoconferencing range from 6 MHz for analog, full-motion, full-color, commercial grade TV to two 56 Kbps lines for digitally-encoded reasonably full motion, full color, to 384 Kbps for even better video transmission to 1,544 Mbit/s for very good quality, full-color, full motion TV. See also VIDEOCONFERENCING.

Video Telephony Real time video call similar to a voice call.

Video Wall Multi-screen video system where a large number of video monitors (typically 16 monitors arrayed in a 4 x 4 matrix) or back projection modules together produce one very large image or combinations of images. Video walls come with their own software, which lets you program the video effects you want. Typically, you can feed a video wall everything from VGA computer output to moving TV (NTSC) signals. Video walls are used for exhibitions and trade shows. They're not cheap. But, when programmed properly, they ARE spectacular.

Video Windows A Bellcore invention which is basically a large, high capacity video conferencing device. Bellcore's Video Windows are connected by two optical links, each carrying 45 million bits of information per second. Though impressive, Bellcore's Video Windows is not considered "high definition" TV. For that to happen, you'd probably need 100 to 150 million bits being transmitted in both directions each second.

Videoconference Videoconference is to communicate with others using video and audio software and hardware to see and hear each other. Audio can be provided through specialized videoconferencing equipment, through the telephone, or through the computer. Videoconferencing has traditionally been done with dedicated video equipment. But, increasingly personal computers communicating over switched digital lines are being used for videoconferencing. See also VIDEO-CONFERENCING.

Videoconferencing Video and audio communication between two or more people via a videocodec (coder/decoder) at either end and linked by digital circuits. Formerly needing at least T-1 speeds (1.54 megabits per second), systems are now available offering acceptable quality for general use at 128 Kbit/s and reasonable 7 KHz audio. Factors influencing the growth of videoconferencing are improved compression technology, reduced cost through VLSI chip technology, lower-cost switched digital networks — particularly T-1, fractional T-1, and ISDN — and the emergence of standards. See VIDEOCONFERENCING STANDARDS.

Videoconferencing Standards ITU-T H.261 was the standards watershed. Announced in November 1990, it relates to the decoding process used when decompressing videoconferencing pictures, providing a uniform process for codecs to read the incoming signals. Originally defined by Compression Labs Inc. Other important standards are H.221: communications framing; H.230 control and indication signals and H.242d: call setup and disconnect. Encryption, still-frame graphics coding and data transmission standards are still being developed.

Videophone 2500 In January, 1992 AT&T introduced a product called Videophone 2500, which transmitted moving (albeit slowly-moving) color pictures over normal analog phone lines. The phone carried a price tag $1,500 a piece. It was not compatible with one MCI later introduced, made for it by GEC-Marconi of England and costing only $750 retail. Videophone 2500 relies on video compression from Compression Labs, Inc. of San Jose, CA. According to the New York Times, the phone took two years, about $10 million and 30 full-time people at AT&T to develop. The January 3, 1993 New York Times carried a quote from John F. Hanley, group VP for AT&T consumer products division, "We could make an AT&T phone talk to an MCI phone. It would be in both of our interests." The AT&T phone and the MCI phone are now effectively dead. See H.323.

Videotape Formats Videotape formats are, in general, classified by the width of magnetic tape used.
-1": Used for professional or "broadcast quality" video recording and editing. Comes in large, open reels.
-3/4": U-matic (Sony). Most industrial video uses this format, stored in inch-thick cassettes.
-1/2": Cassette based, primarily consumer format. VHS - the most popular home videotape format - is 1/2", as is Sony's Beta format. Their higher-quality counterparts (Super-VHS and Super Beta, respectively) are also in the 1/2" format.
-8mm: New consumer format that provides high-quality recording in tiny tape format. Popularly used in hand-held camera-recorders (camcorders).

Videotape recorder A device which permits audio and video signals to be recorded on magnetic tape.

Videotex Two-way interactive electronic data transmission or home information retrieval system using the telephone network. Videotex has not been successful because of its (erstwhile) need for expensive, proprietary (i.e. dedicated) equipment and lack of variety in information offered. There are various forms of videotex. The "classic" European version of interactive videotext typically works at 75 baud going out from the terminal and 1200 baud coming in from the central office. Some American versions ape the European system. Some have 1200 baud both ways. In interactive videotext, you can do everything from sending serious electronic mail to your business suppliers to holding raunchy conversations with perverts in distant cities. As long as you pay your bills, no European PTT seems to care about what you transmit or receive. In France, videotex is called Minitel. And it's a success because the French phone company funds it.

Vidicon Camera An image sensing device that uses an electron gun to scan a photosensitive target on which a scene is imaged.

View 1. In satellite communications, the ability of a satellite to "see" a satellite earthstation, aimed sufficiently above the horizon and clear of other obstructions so that it is within a free line of sight. A pair of satellite earthstations has a satellite in "mutual" view when both enjoy unobstructed line-of-sight contact with the satellite simultaneously.
2. An alternative way of looking at the data in one or more database tables. A view is usually created as a subset of columns from one or more tables.

Viewdata An information retrieval system that uses a remote database accessible through the public telephone network. Video display of the data is on a monitor or television receiver. Another name for Videotex, the original English (UK) name for it. See VIDEOTEX.

ViewFax A module of Active Voice's TeLANophy that allows users to receive and preview their faxes on their personal desktop computer. Each fax document is stored electronically until the receiver wants to see it on screen or send it to a printer. Active Voice is in Seattle, WA. See also ViewFax and ViewMail.

ViewMail A module of Active Voice's TeLANophy which allows users to see all voice, fax and e-mail message on the desktop PC, allowing users to retrieve them in practically any desired order, including LIFO (Last In, First Out) as against LILO (Last In, Last Out), which is what most voice mail systems do. Active Voice is in Seattle, WA. See also ViewCall and ViewFax.

VIM Vendor Independent Messaging. A new E-mail protocol developed by Lotus, Apple, Novell and Borland to provide a common layer where dissimilar messaging programs can share data and back-end services. A group called the Vendor Independent Messaging Group will is intent on developing an open, industry-standard interface that will allow e-mail features to be built into a variety of software products. See also MAPI, which is the E-mail protocol developed by Microsoft.

Viper A deadly computer virus.

VIPR Voice over IP Router.

VIR A video term. Abbreviation of vertical interval reference. Reference signal inserted into the vertical interval of source video. This signal is used further down the video chain to verify parameters and to automatically adjust gains and phase.

Virgil A guide to the Internet. Someone who's been there before.

Virtual In the telephone industry, "Virtual" is something that pretends to be something it isn't, but can be made to appear to be that thing. A virtual private line is effectively a dial up phone line with an auto-dialer on it. To the user, it appears to be a private line. (But the phone company can re-sell that capacity when it's not in use.) The concept of "virtual" is to give the telephone company an excuse to lower the price to the end user. See VIRTUAL NETWORK.

Virtual 8086 Mode Virtual 8086 mode allows the Intel 80386 and beyond microprocessors to emulate multiple real mode processors and still switch to and from protected mode. The processor can load and execute real mode applications (in virtual 8086 mode), then switch to protected mode and load and execute another application that requires access to the full extended memory available. The microprocessor, together with a control program like Microsoft Windows or OS/2 assumes the responsibility of protecting applications from one another. See REAL MODE and PROTECTED MODE.

Virtual Banding 1. In WATS services, virtual banding is the ability of trunks to carry traffic to all WATS bands, with billing based on the end points of the call instead of the band over which the traffic went.
2. MCI's definition: Allows customers of MCI's, PRISM, Hotel WATS, and University WATS to call nationwide while only paying for the distance to the actual area. For example, if a customer calls to a Band 1 area, Band 1 pricing is used. Similarly, if a call is placed to a Band 4 area, Band 4 pricing is used.

Virtual Bypass Virtual bypass is a way smaller users can fill the unused portion of local T-1 dedicated loops going from a user site to a local office of a long distance company, called a POP (Point of Presence).

Virtual Call Capability 1. Provides setup and clearing on a per call basis. Each call placed appears to have a dedicated connection for the duration of the call.
2. A data communications packet network service feature in which a call setup procedure and a call-clearing procedure will determine a period of communication between two DTEs. This service requires end-to-end transfer control of packets within a network. Data may be delivered to the network before the call setup has been completed but it will not be delivered to the destination address if the call setup is not successful. The user's data are delivered from the network in the same order in which they are received by the network. See also VIRTUAL CIRCUIT.

Virtual CD Image Created by dragging and dropping files into into the main window of many CD authoring programs. Can be used to write directly to CD on-the-fly, or to master a real ISO 9660 image to hard disk.

Virtual CD Player A virtual CD player which the related device driver fools the operating system into believing is a real one connected to your system. It is used to stimulate CD performance from a real ISO image residing on hard disk.

Virtual Cell A call, established over a network, that uses the capabilities of either a real or virtual circuit by sharing all or any part of the resources of the circuit for the duration of the call.

Virtual Channel Identifier VCI. A 16-bit field in the ATM cell header identifying the Virtual Circuit which the data will travel from transmitting device to target device. The Virtual Channel is contained within a Virtual Path.

Virtual Channel Switch An ATM term. A network element that connects VCLs. It terminates VPCs and translates VCI values. It is directed by Control Plane functions and relays the cells of a VC.

Virtual Circuit A communications link — voice or data — that appears to the user to be a dedicated point-to-point circuit. Virtual circuits are generally set up on a per-call basis and disconnected when the call is ended. The concept of a virtual circuit was first used in data communications with packet switching. A packetized data call may send packets over different physical paths through a network to its destination, but is considered to have a single virtual circuit. Virtual circuits have become more common in ultra-high speed applications, like frame relay or SMDS. There the connection might be permanently connected like a LAN. When the user wants to transmit he simply transmits. There's no dialing in the conventional sense, just the addition of an address field on the information being transmitted. A virtual circuit is referred to as a logical, rather than physical path for a call. A virtual voice circuit is anything from as simple as a phone with an auto dialer in it to a high-speed link in which voice calls are digitized and send on the equivalent of a ultra high-speed, wide-area equivalent of a local area network. There are two basic reasons people buy virtual circuits. They're cheaper and faster. See PERMANENT VIRTUAL CIRCUIT.

Virtual Circuit Capability A network service feature providing a user with a virtual circuit. This feature is not necessarily limited to packet mode transmission. e.g., an analog signal may be converted at its network node to a digital form, which may then be routed over the network via any available route. See VIRTUAL CIRCUIT.

Virtual Collocation There are two definitions of this evolving term. First: Imagine that you're a CLEC. Your idea is to put your switching and/or Internet equipment in the central office of an ILEC and rent some of the ILEC's circuits out to customers. You are now able to legally do this. Some of the central offices, however, are not large enough to accommodate all the equipments that the various new CLECs are trying to locate in their central office. So the ILECs have figured a new deal. It's called virtual collocation. The CLEC puts his equipment in the ILEC's central office. But the ILEC installs it, configures it, maintains it, fixes it, and does everything necessary. The CLEC can remotely monitor and remotely control his equipment as much as possible. But he can't physically go near it. Obviously, the CLEC has to train the ILEC's people and trust them to do the right thing. See also Adjacent Collocation, Collocation and Physical Collocation,

Virtual Computing A new term for software that shapes computing hardware into hardware that never was. Virtual computing uses FPGAs — Field Programmable Gate Arrays. See FPGAs.

Virtual Connection A logical connection that is made to a virtual circuit.

Virtual Container See VC and SONET.

Virtual Device A device that software can refer to but that doesn't physically exist.

Virtual Disk A portion of RAM (Random Access Memory) assigned to simulate a disk drive. Also called a ram disk. See RAM DISK.

Virtual Fax A device consisting of a personal computer and an image scanner that can duplicate the functions of a facsimile machine.

Virtual File Allocation Table VFAT. A fat file systems is a file system based on a file allocation table, maintained by the operating system, to keep track of the status of various segments of disk space used for file storage. The 32-bit implementation in Windows 95 is called the Virtual File Allocation Table (VFAT). See FAT.

Virtual Hard Drive Memory Factor The available space on a hard drive partition that Windows can address as physical memory.

Virtual Interface Architecture April 16, 1997 - Compaq Computer Corp., Intel Corp., Microsoft Corp. and other industry leaders today announced an initiative to define high-speed communication interfaces for clusters of servers

and workstations. Called the Virtual Interface (VI) Architecture specification, the initiative will enable a new class of scalable cluster products offering high performance, low total cost of ownership and broad applicability. More than 40 companies will participate in the process to complete the draft technical specification before its public release.

A cluster is a group of computers and storage devices that function as a single system. Businesses use clusters in place of individual computers for higher availability and enterprise-class scalability. It is possible to use standard local area network (LAN) and wide area network (WAN) technology to connect the machines in a cluster. However, large clusters and high-performance applications require lower latency, higher bandwidth and additional features not offered by standard LAN and WAN technology. A system area network (SAN) is a specialized network optimized for the reliability and performance requirements of clusters.

The VI Architecture specification provides standard hardware and software interfaces for cluster communications. This will spur innovation in SAN technology and make the LAN, WAN and SAN differences transparent to the applications. The VI Architecture specification will support reliable, high-performance SANs, helping clusters achieve their full potential as cost-efficient platforms for large-scale, mission-critical applications.

"Information technology industry leaders continue to lower the cost of information processing on all fronts while enabling advanced customer solutions by bringing value-added technology to the mass market," said Britt Mayo, director of information technology at Pennzoil Company. "Their efforts to drive the creation of an industry standard for the VI Architecture will make multisystem solutions widely available at new levels of price/performance."

The VI Architecture specification will be media, processor and operating system independent. The software interface will support a variety of efficient programming models to simplify development and ensure performance. The hardware interface will be compatible with standard networks such as ATM, Ethernet and Fiber Channel as well as specialized SAN products available from a variety of vendors.

Virtual ISDN This is an alternative way for a customer to get ISDN service. A customer can be serviced out of a nearby central office which has ISDN capabilities but not charged the extra mileage charges as they would with a foreign exchange. The phone company does not add on charges because the costs are recouped from the large volume of customers serviced out of the CO. A customer will usually have to change phone numbers if the CO where they receive their POTS service becomes ISDN capable.

Virtual LAN A logical grouping of users regardless of their physical locations on the network. Racal-Datacom defines a virtual LAN as "a LAN extended beyond its geographical limit and flexibly configured to add or remove locations." LANs are typically extended beyond their geographical limits (i.e. several thousand feet within a building or campus) by using telephone company facilities, like T-1, T-3, Sonet, etc.

Virtual Machine Facility VM/370. An IBM system control program, essentially an operating system that controls the concurrent execution of multiple virtual machines on a single System/370 mainframe.

Virtual Machine VM. Software that mimics the performance of a hardware device. For Intel 80386 and higher processors, a virtual machine is protected memory space that is created through the processor's hardware capabilities.

Virtual Memory 1. In computer systems, the memory as it appears to the operating programs running in the CPU. Virtual memory is typically the addition of RAM memory and swapfile memory — portion of a hard disk devoted solely to swapfile memory.

2. The term used with Apple Macintoshes to connote the ability to use disk swap files as RAM. This requires the Macintosh to be running System 7 and PMMU.

3. The space on your hard disk that various versions of Windows (including Windows for Workgroups and NT) use as if it were actually memory. Windows NT does this through the use of swap files. The benefit of using virtual memory is that you can run more applications at one time than your system's physical memory would otherwise allow. The drawbacks are the disk space required for the virtual-memory swap file and the decreased execution speed when swapping to the hard disk is required.

Virtual Memory Manager Virtual Memory Manager is a software-only approach to Expanded Memory. These work almost identically to the EMS emulators, except that they use your hard disk rather than extended memory as the storage medium for blocks of memory copied out of your program. As you can imagine, this is painfully s-l-o-w. Use this approach only as a last resort.

Virtual Network A network that is programmed, not hard-wired, to meet a customer's specifications. Created on as-needed basis. Also called Software Defined Network by AT&T. See SOFTWARE DEFINED NETWORK and VIRTUAL PRIVATE NETWORK.

Virtual Path Identifier VPI. An 8-bit field in the ATM header, identifying the Virtual Path (i.e. Virtual Circuit) over which the transmitted data will flow from the transmitting device to the target device.

Virtual Path Switch An ATM term. A network element that connects VPLs. It translates VPI (not VCI) values and is directed by Control Plane functions. It relays the cell of the VP.

Virtual Printer Memory In a PostScript printer, virtual printer memory is a part of memory that stores font information. The memory in PostScript printers is divided into banded memory and virtual memory. Banded memory contains graphics and page-layout information needed to print your documents. Virtual memory contains any font information that is sent to your printer either when you print a document or when you download fonts.

Virtual Printer Technology VPT. Virtual Printer Technology is the enterprise network printer architecture developed by Dataproducts Corporation that enables a printer to become an intelligent node in a networked computing environment and provide printing services to other network nodes through a Client/Server type relationship.

Virtual Private Network VPN. Imagine you're a company with offices all over the country or the globe. You want a cheap, secure way to talk between your offices and to send data between your offices. You can simply dial your offices over the PSTN — the Public Switched Telephone Network. You can lease dedicated, 24-hour per day, seven day a week lines between your offices. Or you can create a "Virtual Private Network." What this means is that you rent or acquire some part of someone else's network (a phone company, an Internet provider) and you make your own network out of theirs. It's calling carving out a network. To do this, you might use a combination of software, dialing codes and some equipment. You can use a virtual private network to do only voice, to do only data, to do just video or to do a combination. There are

many ways to create a VPN. You can do it using virtually any and all the circuits which phone companies and the Internet provide. The thing you will learn about VPNs, however, is that they tend to mean different things to different vendors. To you, a customer, a VPN is basically whatever you want to build. And the prices for them are negotiable between you and the various providers. Here are several definitions:

1. A public circuit-switched service offered by IXCs (long distance phone companies) and making use of the PSTN (Public Switched Telephone Network). Originally known as Switched 56, the current usage of the term "VPN" distinguishes services offered by AT&T, MCI (now MCI Worldcom) and Sprint from Switched 56/64 Kbps services offered by the LECs (local phone companies). Although the specifics vary by IXC, VPNs offer bandwidth options of 56/64 Kbps, increments of 56/64 Kbps, 384 Kbps and 1.544 Mbps (T-1). The last two options are designed with videoconferencing in mind. VPNs provide transmission characteristics similar to those of private lines, such as conditioning, error testing, priority access, and higher speed over full-duplex, four wire, dial-up circuits with a line quality adequate for reasonable speed data. As VPN services are dial-up services provided over the PSTN, they offer the advantage of the high level of PSTN redundancy, which translates into a high level of network resiliency. This network resiliency compares favorably to private, leased-line networks, which are highly susceptible to catastrophic failure. In fact, VPNs often are deployed as a backup to leased-line networks. VPNs also are extremely effective in support of enterprise data networking in organizations with large numbers of small sites. Small locations with relatively modest communications requirements often cannot be connected to long-haul, leased-line networks cost effectively. VPNs offer the advantages of flexibility and scalability, as sites can be added or deleted relatively easily, with costs maintaining a fairly reasonable relationship to enterprise network functionality. The processes of network configuration (design) and reconfiguration are greatly simplified as compared to a leased-line network. Provisioning time is also greatly reduced. The greatest disadvantage of VPNs is that all calls are priced based on an algorithm much like that of a voice call. In other words, costs are calculated by duration and time of day, with prime time calls being priced at a premium. Day-of-week and other special discounts also apply. Some carriers also consider distance in the pricing of VPN calls. Purely from a cost standpoint, leased-lines are preferred for networking large sites with intensive communications needs. Leased line networks also can support not only data and video transmission, but also voice, thereby offering the advantage of integration of all communications needs over a single network. Access to a VPN POP (Point of Presence) — the place the long distance company beings the VPN it's leasing you — can be gained directly from the IXC (Inter-eXchange Carrier), from a CAP (Competitive Access Provider), or from the LEC (Local Exchange Carrier). Appropriate access technologies include leased lines, Switched 56/64, and ISDN. See also SWITCHED 56 and PRIVATE LINE.

2. A means of augmenting a shared network on a secure basis through encryption or tunneling. Such a shared network could be an IP (Internet Protocol) network such as X.25 or the Internet, or an Intranet, or a frame relay network. Tunneling involves encapsulation of that encrypted data inside IP packets or frame relay frames. Additional security is provided through firewalls at the user sites and, perhaps, in the carrier network. VPNs also allow prioritize user data, in order to

enhance performance. Effectively VPNs are intended to be high performance extranets.

3. In a direct mail catalog, which Data Comm Warehouse (www.warehouse.com) sent out in October, 1998, they posed the question, "What a Virtual Private Network can do for you?" And they answered it thusly: A VPN is the creation of a highly secure, point-to-point connection over the Public Internet. Users make local connections to the Internet Service Provider (ISP), saving on long distance or expensive distance-based digital services, then data is transmitted over the Internet. Your data is encrypted using a key known only to the sender and the receiver, so no one can read your sensitive information. And your router can support both multiple VPN tunnels and normal open Internet access at the same time. A VPN gives you two uses for the connection to your Internet Service Provider — tapping the information potential of the Net and transmitting secure business information.

See also Encryption, Extranet, Firewall, Internet and Tunneling.

Virtual Private Office I found this definition of Virtual Private Office in the September 13, 1997 issue of the Economist, one of my favorite magazines. They had a roundup of telecommunications. Here's how they started their roundup:

In Anderson Consulting's smart new offices in Wellsley, just outside Boston, Mark Greenberg is entitled as a senior partner to three filing-cabinet drawers of storage space. In one, he keeps a bubble-wrapped package, containing the sort of personal mementoes_family photographs, shields and so on_with which businessmen like to decorate their offices, together with a diagram to show how they should be arranged. On the rare days when Mr Greenberg is not visiting a client or jetting around the world, he reserves an office. When he arrives, his treasures are neatly laid out on the desk for him to make him feel at home.

But this is, in effect, a virtual private office, his just for the day. Struck by the waste involved in maintaining expensive permanent offices for people with itinerant lives, the partners in the world's largest management consultancy have created something that feels like a cross between a hotel and a luxurious club. The Wellesley office is staffed by the cream of Boston's hotels: people who understand the business of providing services for important and self-important people. The reception desk looks like a hotel foyer; each floor has lots of little "huddle rooms" with comfortable armchairs, as well as brainstorming rooms with less comfortable ones; and there are open spaces for coffee and conversation with colleagues.

Virtual Reality VR. The publisher of Virtual Reality Report says, "Virtual reality is a way of enabling people to participate directly in real-time, 3-D environments generated by computers." Virtual reality involves the user's immersion in and interaction with a graphic screen/s. Using 3-D goggles and sensor-laden gloves, people "enter" computer-generated environments and interact with the images displayed there. Says Business Week, "Imagine the difference between viewing fish swimming in an aquarium and donning scuba gear to swim around them. That's the sensory leap between regular computer graphics and virtual reality. There are three kinds of VR (Virtual Reality) immersion. First, the toe in the water experience of beginners who stand outside the imaginary world and communicate by computer with characters inside it. next, wading up to the hips, are the "through the window" users, who use a "flying mouse" to project themselves into the virtual, or artificial, world. Then there are the hold-the-

nose plungers: "first persona interaction within the computer-generated world via the use of head-mounted stereoscopic display, gloves, bodysuits and audio systems providing binaural sound. The trick with virtual reality is not only to simulate another world but to interact with it — pouring in data affecting its plots, changing its characters and introducing real-world unpredictability into this "mirror world." Once virtual reality was called artificial reality. But artificial means "fake," while virtual means "almost." The father of virtual reality is Joran Lanier. A term close to virtual reality is telepresence. See TELEPRESENCE.

Virtual Route Virtual circuit in IBM's SNA. See SYSTEMS NETWORK ARCHITECTURE.

Virtual Route Pacing Control SNA congestion control at the path control level. See SYSTEMS NETWORK ARCHITECTURE.

Virtual Storage Storage space that may be viewed as addressable main storage to a computer user, but is actually auxiliary storage (usually peripheral mass storage) mapped into real addresses. The amount of virtual storage is limited by the addressing scheme of the computer.

Virtual Telecommunication Access Method VTAM (Pronounced "Vee-Tam.") A program component in an IBM computer which handles some of the communications processing tasks for an application program. VTAM also provides resource sharing, a technique for efficiently using a network to reduce transmission costs.

Virtual Terminal VT. A universal terminal. The ISO virtual terminal (VT) protocol is designed to describe the operation of a so-called universal terminal so any terminal can talk with any host computer.

Virtual Terminal Protocol VTP. Virtual Terminal Protocol enables computers to communicate with various types of terminals by interpreting and translating the instructions for both the computer and the terminal.

Virtual Tributary VT. A structure designed for transport and switching of sub-DS3 payloads. A unit of sub-Sonet bandwidth that can be combined, or concatenated, for transmission through the network; VT1.5 equals 1.544 Mbps; VT2 equals 2.048 Mbps; VT3 equals 3 Mbps; VT6 equals 6 Mbps. See SONET for a fuller explanation.

Virtualization The process of implementing a network based on virtual Network segments.

Virus A software program capable of replicating itself and usually capable of wreaking great harm on the system.

Viscount of Vapor During the 1980s, Bill Gates, chairman and co-founder of Microsoft, became known as the Viscount of Vapor, because so many of his announced new products failed to materialize, or when they did materialize, appeared much, much later than he said they would.

Visible Light Electromagnetic radiation visible to the human eye at wavelengths of 400-700 nm.

Visitors' Location Register VLR. A wireless telecommunications term. A local database maintained by the cellular provider in whose territory you are roaming. When you place a call in a roaming scenario, the local provider queries the HLR (Home Location Register) over the SS7 network through the use of a REGNOT (REGistration NOTification). The HLR is maintained by your cellular provider of record in order to verify your legitimacy and to secure your profile of features. The HLR responds to the REGNOT with a "regnot" (lower case), and transfers the necessary data. This information is maintained by the local provider as long as you remain an active roamer within that area of coverage. This process of query and download is accomplished via SS7 links between SCPs (Signal Control Points). SCPs typically are associated with MSCs (Mobile services Switching Centers), also known as MTSOs (Mobile Traffic Switching Offices) for registering visiting mobile station users. VLRs and HLRs are employed in a variety of cellular networks, including AMPS, GSM and PCS. See also AMPS, GSM, HLR, MSC and PCS.

Vista A videotext service offered in Canada.

Vistium A desktop videoconferencing device which AT&T introduced and then withdrew from the market towards the end of 1995.

Visual Basic A version of the programming language BASIC written by Microsoft Corporation for Windows. The new program promises to make it much easier for businesses to develop customized Windows applications. Some programmers are calling the software a major breakthrough in ease of programming. When I wrote this, Microsoft had sold over one million copies of Visual Basic.

Visual Carrier The portion of a television signal which carries the video portion of the picture.

Visual Display Unit VDU. Another term for a computer monitor. VDU is preferred in Europe.

Visual Message Waiting Indicator VMWI. You are talking to someone on the phone, or perhaps you went to lunch, or on a business trip, or got to work late. Someone else called and left a message on your voice mailbox. You get a Visual Message Waiting Indicator — a little lamp lights on your display phone says "Message Waiting." That's VMWI. Big fancy name for a very simple concept — a light or message on your phone that tells you someone called.

Visual Solutions A family of AT&T products which do videoconferencing, first announced on March 23, 1993.

Visual Voice Mail An application displaying and controlling voice messages on a desktop computer. Usually associated with unified messaging.

Visual Voice Messaging A term created by Microsoft as part of its At Work announcement in June of 1993. There'll be At Work-based visual voice messaging servers sitting on a LAN. Messages for PC users on the LAN will be able to be displayed in a list, much like electronic mail, including the caller's name or number, the time he or she called and the length of the call. This information would let the user browse all messages and select the order for listening to the messages. Administrative options, such as creating a new greeting, will be accessed with a single button. Operations that are difficult today, such as forwarding a voice message to multiple people, will be dramatically simplified, according to Microsoft. One will simply select the recipients from the phone book and broadcast the message. Using visual voice messaging, users will be able to bypass today's inconsistent, time consuming and confusing audio menus and access their voice messages with the push of a button or the click of a mouse on a Windows type icon. Messages will be able to be retrieved in any order and even delivered to a single mailbox along with other messages such as e-mail and faxes. These visual voice messaging servers will, according to Microsoft, provide applications beyond basic voice messaging, such as supporting voice annotation of PC documents or reading electronic mail over the phone to a traveler.

Visualization A combination of computerized graphics and imaging technology that provides high-resolution, video-like results on the workstation or personal computer's screen.

Visually Impaired Attendant Service Visually impaired attendant service capability is achieved by augmenting the normal visual signals provided on a standard attendant position with special tactile devices and/or audible signals which enable a visually impaired person to operate the position.

VITA VME International Trade Association. A widely supported industry trade group in Scottsdale, AZ. VITA is chartered to promote the growth and technical excellence of the VME bus and Futurebus-based microcomputer board market. VITA is chartered to submit standards for ANSI registration. See VME. www.vita.com

Visual Carrier The portion of a television signal which carries the video portion of the picture.

VITC Vertical Interval Time Code. Contains the same information as the SMPTE time code. It is superimposed onto the vertical blanking interval, so that the correct time code can be read even when a helical scanning VCR is in the pause or slow (DT) mode.

Viterbi Decoder Algorithm for decoding Trellis encoded signals.

Vitreous Silica Glass consisting of almost pure silicon dioxide.

VL-Bus A new PC bus from VESA — the Video Electronics Standards Association. The VL bus is up to 20 times as fast as an ISA bus, the most common PC bus and the one common to the original PC, the PC XT and the PC AT and clones. VL-Bus was popular on 486 PCs. Pentium-based machines now largely use the newer, PCI bus. As a result, VL-Bus is pretty obsolete. See ISA, MICROCHANNEL and PCI.

VLAN Virtual Local Area Network. A means by which LAN users on different physical LAN segments are afforded priority access privileges across the LAN backbone in order that they appear to be on the same physical segment of an enterprise-level logical LAN. VLAN solutions, which are priority in nature, are implemented in LAN switches, and VLAN membership is defined by the LAN administrator on the basis of either port address or MAC (Medium Access Control) address. Consider the following example: There are a large number of design engineers in a large office environment. They are grouped into three design groups, each of which is served by a LAN switch. Several lead engineers in each workgroup have the responsibilities for coordinating the efforts of their respective teams with those of the other teams. Those lead engineers have priority-level access through the switches and across the inter-switch transmission backbones. While they are on separate physical LAN segments, the priority level of access allows them to interwork as though they were on the same segment. They are not on the same physical LAN, but it's as though they were. It's a Virtual LAN.

VLC Variable length coding.

VLF Very Low Frequency. That portion of the electromagnetic spectrum having continuous frequencies ranging from about 3 Hz to 30 kHz.

VLM Novell Virtual Loadable Module network client architecture uses packet burst technology, so ample packets are sent without waiting for packet acknowledgement. VLM support (compared with the old IPX) improves transfer times, especially for compressible files, since there is no waiting for acknowledgement.

VLR Visitors' Location Register. A wireless telecommunications term. A local database maintained by the cellular provider in whose territory you are roaming. When you place a call in a roaming scenario, the local provider queries the HLR (Home Location Register) over the SS7 network through the use of a REGNOT (REGistration NOTification). The HLR is maintained by your cellular provider of record in order to verify your legitimacy and to secure your profile of features. The HLR responds to the REGNOT with a "regnot" (lower case), and transfers the necessary data. This information is maintained by the local provider as long as you remain an active roamer within that area of coverage. This process of query and download is accomplished via SS7 links between SCPs (Signal Control Points). SCPs typically are associated with MSCs (Mobile services Switching Centers), also known as MTSOs (Mobile Traffic Switching Offices) for registering visiting mobile station users. VLRs and HLRs are employed in a variety of cellular networks, including AMPS, GSM and PCS. See also AMPS, GSM, HLR, MSC and PCS.

VLSI Very Large Scale Integration. The art of putting hundreds of thousands of transistors onto a single quarter-inch square integrated circuit. Compare with LSI and ULSI.

VM 1. Voice Mail, Voice Messaging or Virtual memory. See VIRTUAL STORAGE and VOICE MAIL.

2. Virtual Machine. IBM's mainframe operating system.

VME Acronym for "VersaModule-Europe". A one through 21 slot, mechanical and electrical bus standard originally developed by the Munich, Germany division of Motorola in the late 70s. VME uses most of the bus structure from then current Motorola's VersaBus board standard along with the newly developed DIN 41612 standard pin-in-socket connector for enhanced reliability. After years of work, VME was finally adopted by the ANSI/IEEE in 1987 (as ANSI/IEEE-1014). VME is known in Europe as the IEC 821 bus. This makes it an open standard. The VME backplane runs at 80 Mbytes per second. It is the most common bus or big open computers (i.e. ones larger than the PC). As of writing, there were over 300 vendors offering more than 3,000 off-the-shelf VME products. The IEEE standard is soon to lapse and be replaced by an extended VME64 specification, now in ANSI ballot being conducted by VITA. See VMEBus.

VME64 An enhanced VME bus standard which includes multiplexed address and data cycles with 40 and 64 bit address modes and 64 bit data transfer modes allowing up to 80 MB/s transfer speed. This standard is under the ANSI ballot process conducted by VITA. See VME.

VME64 Extensions A VITA draft standard that provides extra functionality to VME64 including 5 row J1/P1 and J2/P2 connectors that support live insertion on both 3U and 6U VME boards. Other features: 3.3V power, more grounds, ETL (slew rate) drivers, geographic addressing (slot ID) as well as support for parity, a serial diagnostic bus, JTAG test support and lots of user I/O. Some mechanical features: locking extractors, RFI gasketing, and ESD chassis discharge strips. See VME.

VMEBus VersaModule Eurocard BUS. A 32-bit bus developed by Motorola, Signetics. Mostek and Thompson CSF. Used widely in industrial, commercial and military applications with over 300 manufacturers of VMEbus products worldwide. VME64 is an expanded version that provides 64-bit data transfer and addressing.

VMEC Voice Messaging Educational Committee. An organization formed by voice messaging manufacturers and service providers to promote a better understanding of voice mail and its business benefits, and to help business implement voice mail systems in ways that meet the needs of callers and mailbox owners alike. See VME.

VMI Voice Messaging Interface.

VMF Validation Message Fraud

VMR Violation Monitoring and Removal. The process of removing a violations which are detected, so that violations do not propagate beyond the maintenance span.

VMS Virtual Memory System

VMS OSI Transport Services VOTS. A Digital Equipment Corp. software product that modifies Digital's DECnet transport layer to conform to the International Standards Organization (ISO) Transport Protocol Class Four (TP4).

VMUF Voice Messaging User Interface Forum. A standards body formed by voice messaging end users, service providers and manufacturers to define a minimum set of common human interface specifications for voice messaging systems.

VMX Voice Message Exchange. One day in 1979, Gordon Matthews came back from lunch and noticed that he had received the usual half dozen messages that had been randomly taken down semi-correctly by a harried receptionist. Already a noted inventor, Matthews saw an opportunity to build an adjunct device to the company PBX which would these messages to be recorded by the caller without an intermediary and would allow the recipient to store these messages, forward them to others or to directly reply to them if they were generated from another internal extension. He called this device the Voice Message Exchange, and the company later became VMX, which later got bought by Octel.

VNET Virtual private NETwork. An MCI (now MCI Worldcom) term for a service it offers to customers who want to join geographically dispersed switches (typically PBXs). Instead of private lines joining the PBX, Vnet uses fast switched lines.

VNL Via Net Loss. A loss objective for trunks, the value of which has been selected to obtain a satisfactory balance between two data terminals for the duration of the call.

VNS Virtual Network Services. A VPN (Virtual Private Network) term, referring specifically to a range of international VPN services, including voice, data, Internet and multimedia. Providers of traditional VNS include AT&T, MCI and Sprint. During the recent past, groups of international carriers have formed multilateral affiliations to provide VNS on a seamless basis, providing identical feature content and functionality across national borders, regardless of the affiliated country of origination or termination. See also Concert, Global One, and WorldPartners.

VO Verification Office.

VOATM Voice over ATM

VOC In February, 1996 a newspaper called, Investor's Business Daily, ran a story entitled "Is Your Office at Home Making You Sick?" It said that more than 20 million American workers are telecommuting, with another 20 million owning home based businesses. "Experts are finding that home based offices outfitted with the latest fax machines, photocopiers, laser printers and personal computers often foster unhealthy environments. Office machinery emits air pollutants called volatile organic compounds, or VOCs, which can make your head hurt and irritate your eyes, nose and throat. VOCs are also known to cause more serious health problems, including kidney and liver damage, experts say."

Vocabulary Development Development of specific word sets to be used for speaker independent recognition applications.

Vocoder Voice coder. A device that synthesizes speech. Vocoders use a speech analyzer to convert analog waveforms into narrowband digital signals. They are used in digital cellular phones as well as in the entertainment business (e.g., the voice of Darth Vader in Star Wars). Vocoders are an early type of voice coder, consisting of a speech analyzer and a speech synthesizer. The analyzer circuitry converts analog speech waveforms into digital signals. The synthesizer converts the digital signals into artificial speech sounds. For COMSEC purposes, a vocoder may be used in conjunction with a key generator and a modulator-demodulator device to transmit digitally encrypted speech signals over normal narrowband voice communication channels. These devices are used to reduce the bandwidth requirements for transmitting digitized speech signals. There are analog vocoders that move incoming signals from one portion of the spectrum to another portion.

VOD Video On Demand. At one stage it was considered a gigantic potential money maker for the phone companies who were real interested in getting into this market. Now it doesn't seem so hot. See NVOD and VIDEO ON DEMAND.

VODAS Voice Operated Device Anti-Sing. A device used to prevent the overall voice frequency singing of a two-way telephone circuit by ensuring that transmission can occur in only one direction at any given instant.

VoFR Voice Over Frame Relay. Basically, there are two main standards regarding voice transmission over data networks: H.323 and "Voice Over Frame Relay Implementation Agreement" (FRF.11). Both specify that the following coders should be used: G.711, G.728, G.729, and G.723.1. The H.323 adds the G.722, and the VoFR (Voice over Frame Relay) adds the G.726/7 coders. The G.711 is a PCM coder that uses 64ks/s and two companding techniques: A-law and Mu-law. Recommendation G.722 describes 7 kHz audio-coding within 64 kbit/s. The G.723.1 describes a Dual rate speech coder for multimedia communications transmitting at 5.3 and 6.3 kbit/s and is based on Multi Pulse Maximum Likelihood Quantizer (MP-MLQ) (Voice frame duration of 30mSec). The G.726 describes 40, 32, 24, 16 kbit/s Adaptive Differential Pulse Code Modulation (ADPCM). The G.727 describes 5-, 4-, 3- and 2-bits sample embedded adaptive differential pulse code modulation (ADPCM). The G.728 describes Coding of speech at 16 kbit/s using Low-Delay Code Excited Linear Prediction (LD CELP). The G.729 describes Coding of speech at 8 kbit/s using Conjugate-Structure Algebraic-Code- Excited Linear-Prediction (CS-ACELP) (Voice frame duration of 10mSec).

VOGAD Voice-Operated Gain Adjusting Device. A voice-operated compressor circuit that is designed to provide a near-constant level of output signal from a range of input amplitudes. Such a circuit has a fast attack time with a relatively slow release time to avoid excess volume compression at the system output.

VOHDLC Voice over HDLC

Voice Activated Dialing A feature that permits you to dial a number by calling that number out to your cellular phone, instead of punching it in yourself. See VOICE ACTIVATED VIDEO.

Voice Activated Switching Used in multipoint video conferencing so all sites automatically see the video of the person speaking.

Voice Activated Video A microphone/camera that is activated in response to voice. Imagine you're watching a videoconference going on in four locations. You can hear what everyone is saying. What you need is to be able to see the person who is speaking the loudest, and therefore, presumably the principal speaker — the person whose attention everyone should be focused on. In voice activated video, the videoconferencing system senses who's speaking the loudest and throws that person's face up on everyone's screen.

Voice Activity Compression VAC. A method of conserving transmission capacity by not transmitting pauses in speech.

Voice Applications Program System software providing the necessary logic to carry out the functions requested by telephone system users. It is responsible for actual call processing, making the various voice connections and providing user features, such as Call Forwarding, Speed Dialing, Conference, etc.

Voice Board Also called a voice card or speech card. A Voice Board is an IBM PC- or AT-compatible expansion card which can perform voice processing functions. A voice board has several important characteristics: It has a computer bus connection. It has a telephone line interface. It typically has a voice bus connection. And it supports one of several operating systems, e.g. MS-DOS, UNIX. At a minimum, a voice board will usually include support for going on and off-hook (answering, initiating and terminating a call) notification of call termination (hang-up detection); sending flash hook; and dialing digits (touchtone and rotary). See VOICE BUS and VRU.

Voice Body Part An X.400 term. A body part sent or forwarded from an originator to a recipient which conveys voice encoded data and related information. The related information consists of parameters which are used to assist in the processing of voice data. These parameters include information detailing the duration of the voice data, the voice encoding algorithm used to encode the voice data, and supplementary information.

Voice Browser Pick up a phone, call a VoxML-enabled Web site, ask it questions using your voice and what's known as "natural voice commands" (such as "When is my plane leaving?") and hear responses to your questions. This uses an upcoming technology called Voice Markup Language (VoxML), which several manufacturers plan to develop as an open platform and submit to the World Wide Web Consortium for standards approval. The idea is to VoxML-enable your Web site so it can respond to voice recognition.

Voice Bulletin Boards These are voice mailboxes which contain pre-recorded information that can be updated as frequently as the provider of the mailboxes desires and can be accessed by the public 24 hours a day. Voice bulletin boards can be used by city or county departments which receive a large number of calls asking for routine information, e.g., summer programs for kids as listed by a parks and recreation department; jobs currently open in the city as listed by the personnel departments; etc.

Voice Bus Picture an open PC. Peer down into it. At the bottom of the PC, you'll see a printed circuit board containing chips and empty connectors. That board is called the motherboard. Fatherboards are inserted into the connectors on the motherboard. These fatherboards do things on the PC — like pump out video to your screen or material to your printer or your local area network. The motherboard controls which device does what WHEN by sending signals along the motherboard's data bus — basically a circuit that connects all the various fatherboards through their connectors. That data bus was not designed for voice. For voice you need another bus. Several voice processing manufacturers have addressed that need by creating a voice bus at the top of their PC-based voice processing cards. They have tiny pins sticking out of their cards. You attach a ribbon cable from one set of pins on one voice processing card to the next set on the adjacent card and then the next. There are several voice bus "standards." Two come from Dialogic. One is called AEB, Analog Expansion Bus. And one is called PEB, PC Expansion Bus (a digital version). One comes from a consortium of companies and is called MVIP. There are many advantages to having a voice bus. It gives you enormous flexibility to mix and match voice processing boards, like voice recognition, voice synthesis, switching, voice storage, etc. You can build really powerful voice processing systems inside today's fast '386 and '486 PCs with the great variety of voice processing now available. For more information on this exciting field, read TELECONNECT Magazine. 212-691-8215. See MVIP.

Voice Call A telephone call established for the purpose of transmitting voice, rather than data.

Voice Calling One manufacturer describes this as allowing a phone user to have calls automatically answered and connected to his phone's loudspeaker. Not a common definition. A Northern Telecom Norstar definition: This feature allows a voice announcement to be made, or a conversation to begin, through the speaker of another telephone in the system.

Voice Circuit A circuit able to carry one telephone conversation or its equivalent, i.e. the typical analog telephone channel coming into your house or office. It's the standard subunit in which telecommunication capacity is counted. It has a bandwidth between 300 Hz and 3000 Hz. The U.S. analog equivalent is 3 KHz. The digital equivalent is 56 Kbps in North American and 64 Kbps in Europe. This is not sufficient for high fidelity voice transmission. You'd probably need at least 10,000 Hz. But it's sufficient to recognize and understand the person on the other end.

Voice Coil The element in a dynamic microphone which vibrates when sound waves strike it. The coil of wire in a loudspeaker through which audio frequency current is sent to produce vibrations of the cone and reproduction of sound.

Voice Commerce See V-Commerce.

Voice Compression Refers to the process of electronically modifying a 64 Kbps PCM voice channel to obtain a channel of 32 Kbps or less for the purpose of increased efficiency in transmission.

Voice Connecting Arrangement VCA. A device that, once upon a time, was necessary for connecting your own phone system to the nation's switched telephone network. Most phones now meet FCC (and other) safety standards, so VCAs are no longer necessary. Most phone systems (as opposed to phones) do have internal protection circuitry, as shown by the "F" (for fully protected) in their FCC registration number.

Voice Coupler An interface arrangement once provided by the telephone company to permit direct electrical connection of customer-provided voice terminal equipment to the national telephone network. No longer needed because of the FCC's Registration Program.

Voice Data An SCSA definition. Encoded audio data.

Voice Dialing The ability to tell your phone to dial by talking to it. Say, "Call Police" and it will automatically dial the police. This feature has enormous benefits for handicapped people. It will have greater benefits for normal people when the technology of voice recognition improves.

Voice Digitization The conversion of an analog voice signal into binary (digital) bits for storage or transmission.

Voice Driver A Dialogic product that comes for MS-DOS, OS/2 and UNIX. In MS-DOS, it is a terminate and stay resident (TSR) program which acts as a central server for MS-DOS based applications. It provides all of the services required to support installable device drivers for each hardware component and for the application. See also Device Driver.

Voice DTMF Forms Applications This Voice DTMF (DUAL TONE MULTIPLE FREQUENCY) application allows a

use of a voice mail system to take specific information from its customers 24 hours a day. By prompting callers to respond by speaking or pressing the keys of their touchtone phones, a city department, for example, could plan service calls, building inspections or send out appropriate forms.

Voice Frame See VOICE FRAME.

Voice Frequency VF. An audio frequency in the range essential for transmission of speech. Typically from about 300 Hz to 3000 Hz. See VOICE FREQUENCIES.

Voice Frequencies VF. Those frequencies lying within that part of the audio range that is employed for the transmission of speech. In telephony, the usable voice frequency band ranges from a nominal 300 Hz to 3400 Hz. In telephony, the bandwidth allocated for a single voice frequency transmission channel is usually 4 KHz, including guard bands.

Voice Frequency Telegraph System A telegraph system permitting use of up to 20 channels on a single voice circuit by frequency division multiplexing.

Voice Grade A communications channel which can transmit and receive voice conversation in the range of 300 Hertz to 3000 Hertz.

Voice Hogging See VOICE SWITCHED.

Voice Integration Allows computer fax solutions to be store and forward hubs for both image as well as voice communication. Many of these products work on PC-based systems and offer all the capabilities of a message center.

Voice Jail A poorly designed voicemail system that has so many submenus one gets lost and has to hang up and call back. Also called Voice Mail Jail.

Voice Mail Voice Mail allows you to receive, edit and forward messages to one or more voice mailboxes in your company or in your universe of friends. With voice mail, employees can have their own private mailboxes. Here's an explanation of how it works: You call a number. A machine answers. "Sorry. I'm not in. Leave me a message and I'll call you back." It could be a $50 answering machine. Or it could be a $200,000 voice mail "system." The primary purpose is the same — to leave someone a message. After that, the differences become profound. a voice mail system lets you handle a voice message as you would a paper message. You can copy it, store it, send it to one or many people, with or without your own comments. When voice mail helps business, it has enormous benefits. When it's abused — such as when people "hide" behind it and never return their messages — it's useless. Some people hate voice mail. Some people love it. It's clearly here to stay.

In the fall of 1991, the Wall Street Journal carried a story negative on voice mail. Les Lesniak, Rolm's Senior VP Marketing disagreed. His reply published in the Journal is one of the finest explanations of voice mail's virtues:

"The writer's observations ignore the way today's voice communication technology is making communication between people easier and more convenient, and is elevating the level of service savvy companies provide their customers. Manufacturers use it to take orders after hours and on weekends. Financial services companies use it to provide account information to customers on a 24-hour basis. Colleges use it to register students. A retail executive uses it to broadcast messages to her staff. And a lawyer uses it to respond to calls when traveling.

"Voice messaging keeps calls confidential, simplifies decision making, saves time and money, eliminates inaccurate messages and "telephone tag," allows people to use their time more productively. In short, it keeps communication crisp, clear and constant. The writer's line of thinking would demand that people remain at their desk 24 hours a day. If they don't, the phone goes unanswered, a receptionist answers the phone and takes a message, or an answering machine records the message and cuts off the caller at will. None of these scenarios is ideal.

"To be successful, voice mail technology must be understood by users and supported by top management, And it must meet the needs of the customer. Training for all employees must be mandatory and the system must be administered and managed properly. 'Must answer' lines and greetings that are changed daily are only two ideas that make voice mail not just helpful, but essential to customer service and an enhanced company image.

"Contrary to the writer's view, voice mail contributes to effective business communication and is far superior to an unanswered phone call, a misplaced message or an answering machine." Here are some statistics which add weight to voice mail's logic:

• 75% of all business calls are not completed on the first attempt.

• This can easily waste $50 to $150 per employee per month in toll charges.

• Half of the calls are for one-way transfers of information.

• Two thirds of all-phone calls are less important than the work they interrupt.

• The average length of a voice mail message is 43 seconds. The average long distance call is 3.4 minutes. Voice mail is 80% faster.

Here are the standard benefits of voice mail:

1. No more "telephone tag." Voice mail improves communications. It lets people communicate in non-real time.

2. Shorter calls. When you leave messages on voice mail, your calls are invariably shorter. You get right to the point. Live communications encourage "chit chat" - wasting time and money.

3. No more time zone/business hour dilemma. No more waiting till noon (or rising at 6 A.M.) to call bi-coastally or across continents.

4. Reduce labor costs, Instead of answering phones and taking messages, employees are free to do more vital tasks.

5. Fewer callbacks. In some cases, as many as 50%.

6. Improved message content. Voice mail is much more accurate and private than pink slips. Messages are in your own voice, with all the original intonations and inflections.

7. Less paging and shorter holding times.

8. Less peakload traffic.

9. 24-hour availability.

10. Better customer service.

11. Voice mail allows work groups to stay in contact - morning, noon and night.

12. Voice mail reduces unwanted interruptions.

See Also VOICE MAIL JAIL and VOICE MAIL SYSTEM.

Voice Mail Jail What happens when you reach a voice mail message and you try and reach a human by punching "0" (zero) and you get transferred to another voice mail box and you try again by punching "0" or some other number you're told to punch...and you never reach a human. You're stuck forever inside the bowels of a voice mail machine, being instructed to go from one box to another, never reaching a real human. You're in voice mail jail.

Voice Mail System A device to record, store and retrieve voice messages. There are two types of voice mail devices — those which are "stand alone" and those which profess some integration with the user's phone system. A stand alone voice mail is not dissimilar to a collection of single person answering machines, with several added features. You can instruct the machines (voice mail boxes) to forward messages among themselves. You can organize to allocate your friends and business acquaintances their own mail boxes so they can dial, leave messages, pick up messages from you, pass messages to you, etc. You can also edit messages, add comments and deliver messages to a mailbox at a pre-arranged time. Messages can be tagged "urgent" or "non-urgent" or stored for future listening. The range of voice mail options varies among manufacturers.

An integrated voice mail system includes two additional features. First, it will tell you if you have any messages. It does this by lighting a light on your phone and/or putting a message on your phone's alphanumeric display. Second, if your phone rings for a certain number of rings (you set the number), the phone will transfer your caller automatically to your voice mail box, which will answer the phone, deliver a little "I am away" message and then receive and record the caller's message.

There are other levels of integration. You might have a phone which has "soft" buttons and an alphanumeric display. That display might label your phone's soft buttons like those on a cassette recorder — forward, reverse, slow, fast, stop, etc. so you can go through your messages any way you like. Telenova has such a phone. It's very impressive.

There are pros and cons to voice mail systems. Some employees will hide behind them, forwarding calls from their customers into voice mail boxes and never returning them. Some employees will make good use of them. They dial in for their messages, research what the customer wants and return the voice mail calls quickly. Many voice mail systems are being combined with automated attendants. Many are being combined with interactive voice processing systems, including sophisticated tie-ins to mainframe databases. Some people hate voice mail systems. Others love them. It all depends on how the system is used, managed and sold. See also VOICE MAIL, AUDIOTEX, AUTOMATED ATTENDANTS, INFORMATION CENTER MAILBOXES, ENHANCED CALL PROCESSING and VOICE PROCESSING.

Voice Markup Language See Voice Browser.

Voice Message Service A leased service typically over dial up phone lines which provides the ability for a phone user to access a voice mail system and leave a message for a particular phone user. See VOICE MAIL SYSTEM.

Voice Message Exchange See VMX.

Voice Messaging Recording, storing, playing and distributing phone messages. Essentially voice messaging takes the benefits of voice mail (such as bulk messaging) beyond the immediate office to almost any phone destination you select. Voice messaging is often done through service bureaus. Nynex has an interesting way of looking at voice messaging. Nynex sees it as four distinct areas: 1. Voice Mail, where messages can be retrieved and played back at any time from a user's "voice mailbox"; 2. Call Answering, which routes calls made to a busy/no answer extension into a voice mailbox; 3. Call Processing, which lets callers route themselves among voice mailboxes via their touchtone phones; and 4. Information Mailbox, which stores general recorded information for callers to hear.

Voice Modem A new type of modem which handles both voice and data over standard analog phone lines. A voice modem is the classic computer telephony device, since it applies intelligence to the making and receiving of normal analog phone calls. Such voice modem might be a full-duplex speakerphone and an answering machine / voice mail device. Such modem might be able to detect incoming and outgoing touchtone and other signals, such as Caller ID. Such modem might also include music on hold, pager dialing, bong and SIT tone detect, line break detect, local phone on / off detect, extension off hook detect, remote ring back detection and VoiceView. The thrust towards voice modems is coming from chip manufacturers, including Sierra Semiconductor, Rockwell and Cirrus Logic. Some standards bodies are working on voice modems. Two standards are emerging — IS-101 and PN-3131.

Voice On The Net Coalition An organization formed to stop regulatory attempts to stifle the growth of voice on the Internet. See VON Coalition. 802-878-9884 and www.von.org

Voice Operated Relay (VOX) Circuit A voice-operated relay circuit that permits the equivalent of push-to-talk operation of a transmitter by the operator.

Voice Over TIE/communications name for a totally wonderful feature on a phone system — namely that while you are speaking to someone on the phone your operator can talk to you "over" the conversation you're having. What happens is that you hear your operator in your telephone's handset receiver, but the person you're speaking with can't. You can reply to the operator (telling him/her you'll be one minute, please call back, etc.) by hitting a DND/MIC (Do Not Disturb/Microphone) button on your phone. Voice Over has major benefits. It saves on long distance calls you don't have to return. It closes deals that can't wait. And it gives customers immediate answers. In short, it improves corporate efficiency and customer satisfaction.

Voice Paging Access Gives attendants and phone users the ability to dial loudspeaker paging equipment throughout the building. An unbelievably useful feature, if your people are prone to wander.

Voice Print A voice recognition term. A voice print is a speech template used to "train" systems, in particular voice patterns. When a system is operating, the user's speech is compared to the stored voice prints. If they match, the system recognizes the word and executes the command.

Voice Processing Think of voice processing as a voice computer. Where a computer has a keyboard for entering information, a voice processing system recognizes touchtones from remote telephones. It may also recognize spoken words. Where a computer has a screen for showing results, a voice processing system uses a digitized synthesized voice to "read" the screen to the distant caller.

Whatever a computer can do, a voice processing system can too, from looking up train timetables to moving calls around a business (auto attendant) to taking messages (voice mail). The only limitation on a voice processing system is that you can't present as many alternatives on a phone as you can on a screen. The caller's brain simply can't remember more than a few. With voice processing, you have to present the menus in smaller chunks.

Voice processing is the broad term made up of two narrower terms — call processing and content processing. Call processing consists of physically moving the call around. Think of call processing as switching. Content consists of actually doing something to the call's content, like digitizing it and storing it on a hard disk, or editing it, or recognizing it (voice recognition) or some purpose (e.g. using it as input into a computer program.) See VOICE BOARD, VOICE RESPONSE UNIT and VOICE SERVER.

Voice Recognition The ability of a machine to recognize your particular voice. This contrasts with speech recognition, which is different. Speech recognition is the ability of a machine to understand human speech — yours and most everyone else's. Voice recognition needs training. Speech recognition doesn't. See SPEAKER DEPENDENT and SPEAKER INDEPENDENT VOICE RECOGNITION.

Voice Response Unit VRU. Think of a Voice Response Unit (also called Interactive Voice Response Unit) as a voice computer. Where a computer has a keyboard for entering information, an IVR uses remote touchtone telephones. Where a computer has a screen for showing the results, an IVR uses a digitized synthesized voice to "read" the screen to the distant caller. An IVR can do whatever a computer can, from looking up train timetables to moving calls around an automatic call distributor (ACD). The only limitation on an IVR is that you can't present as many alternatives on a phone as you can on a screen. The caller's brain simply won't remember more than a few. With IVR, you have to present the menus in smaller chunks. See IVR and VOICE BOARD.

Voice Ring Multiple Digital Intertie Buses connected in series to all nodes. Provides extra channels for voice data transmission when direct link (DI) channels are busy.

Voice Server A PC sitting on a LAN (Local Area Network) and containing voice files which are accessible by the PCs on the LAN. Such voice files may be transmitted on the LAN or over phone lines under the control of the PCs on the LAN. A voice server might contain voice mail. It might contain voice annotated electronic mail. Its primary function is to store voice in such a way that it's accessible easily. Voice servers are typically faster, have more disk capacity and more backup provisions than normal PCs. According to a letter I received in early May, 1993 from the lawyers for a company called Digital Sound Corporation, that company owns federal trademark registration number 1,324,258 for the mark Voiceserver, spelled as one word, not two.

Voice Store And Forward Voice mail. A PBX service that allows voice messages to be stored digitally in secondary storage and retrieved remotely by dialing access and identification codes. See VOICE MAIL SYSTEM.

Voice Switched A device which responds to voice. When the device hears a voice, it turns on and transmits it, muting the receive side. The most common voice-switched device is the desk speakerphone. With voice switching, it's easy to hog a circuit. Just keep making a noise. Watch out for voice hogging. If you're calling someone and waiting for them by listening in on your speakerphone, mute your speakerphone. This way you'll hear them when they answer.

Voice Switching Equipment used in voice and video conferences. The equipment is activated by sounds of sufficient amplitude; hopefully speech, but also loud noises. Fast switching activates microphones so that only one conference participant can speak at a time. See also VOICE ACTIVATED VIDEO.

Voice Terminal A pretentious AT&T term for a TELEPHONE.

Voice Verification The process of verifying one's claimed identity through analyzing voice patterns.

Voiceband A transmission service with a bandwidth considered suitable for transmission of audio signals. The frequency range generally is 300 or 500 hertz to 3,000 or 3,400 hertz — the frequency range the common analog home phone service is made at.

Voiceframe VoiceFrame is Harris Digital Telephone Systems' name for their open application platform for voice and call processing. According to Harris, VoiceFrame allows businesses to create a comprehensive set of applications that link computers, telephone networks and the telephone. Examples of applications that use the VoiceFrame platform include

• touchtone driven transaction processing such as telebanking

• operator services

• advanced paging systems

• intelligent call routing via host computer database inquiries

• other call center applications.

VoiceFrame serves as a communications controller whose primary responsibility is the disposition of inbound and outbound call traffic under computer control. VoiceFrame has these abilities

• It interprets call signaling information and translates it into protocols for host computer use.

• It accepts commands from the host and interprets them to switch, route and complete calls.

• It provides host computer access to private and public network services, such as DNIS, DID, SMSI, ANI, 950, 900 and 800 services.

• It allows the host computer to perform those tasks for which it is best suited — real-time call routing decision making, database look up, complex calculations and detailed billing.

VoiceSpan VoiceSpan is a new class of modem technology that combines existing modem technology with the ability to simultaneous transfer voice (without digitizing the voice). This general technology has been termed analog simultaneous voice and data (ASVD) to contrast it from digital simultaneous voice and data (DSVD) where the voice is first digitized and then multiplexed with the data. Currently V.asvd is based on technology similar to V.32bis (maximum data rate 14,400 bit/s) but the technology allows the possibility of expansion to include V.34 like capabilities (maximum data rate 33,600 bit/s). The advantages of ASVD verses DSVD are likely to include better sound quality, rapid ability to detect pauses in voice and send the highest data rate, simpler technology and reduced voice delay between speakers. The disadvantages are likely to be a lower data rate when voice is present and the possibility of less flexibility compared to the all digital approach (DSVD).

Currently two different Recommendations: V.8 and V.25 are used to define what happens when a telephone call is started in fax or data mode. These Recommendations are responsible for the initial sounds (`beeps and squawks'), you hear when your modem connects to a remote modem. However these Recommendations do not easily allow a modem to connect after a voice telephone conversation has started. Most conferencing calls would likely begin with a voice conversation and then switch to data and/or video conferencing. So there is considerable interest in defining a new way to accomplish this. Although some ITU members believe that just making some changes to V.8 would accomplish the same as V.8bis, more simply.

In the market Radish Communications Inc. has established a protocol to accomplish the start- up of data communications during a call as part of an alternating voice and data system called VoiceView(TM). Radish proposed to SG 14 the use of the VoiceView start-up sequence for V.8bis. Unfortunately, while no insurmountable technical problems were identified with the VoiceView approach, the standards committees could not agreed to make V.8bis compatible with the VoiceView start up. The proposed draft V.8bis created at this meeting of SG 14

is independent (the two start-up mechanisms will not confuse each other) of the VoiceView proposal. Therefore manufacturers can make equipment that supports both the VoiceView start-up and the proposed draft V.8bis if they wish.

The biggest remaining problem to SG 14 members is compatibility. Whenever a new sequence of `beeps and squawks' is created all the different types of fax machines, modems, voice response systems, DTMF detectors, etc. currently installed on and in the telephone networks worldwide need to be tested to make sure that the new signals do not cause any serious compatibility problems. The current draft of V.8bis just hasn't been tested yet. This makes the members of SG 14 uncomfortable about its compatibility.

As a result of this concern the US, United Kingdom and France all expressed reservations about Study Group 14 supporting the current draft of V.8bis. This will result in a delay in the approval of V.8bis until at least March 1996. Such a delay is in favor of Radish and the VoiceView partners. The delay allows them to develop an installed base of alternative voice and data VoiceView users. However the likely approval of V.8bis in March 1996 means that the simultaneous voice and data applications will use V.8bis as their start-up sequence. And over the long term markets usually move away from the proprietary solutions to the ITU standard, e.g. MNP4 to V.42, MNP5 to V.42bis, V.FC to V.34.

Two different start-up sequences, three if you count V.8, for different applications is not the best solution. But it appears that it will work. And the output of standards work is often the best that can be agreed, not the best that can be conceived.

VoiceView The family name for the concept and the product line from Radish Communications, Boulder, CO. VoiceView describes the protocol, the transaction, and the platform. Essentially, VoiceView is technology for switching between voice, data, fax and binary image transmission on the same conversation on a standard 3 Khz analog phone line — the one you have in your home. Three types of information transfer modes are available: Modem Data Mode, Fax Data Mode and VoiceView Data Mode. Imagine that you're talking to your travel agent and want to fly from Newark to Chicago. Your travel agent starts reading off the eight different flights. You begin to scribble down the information as she reads through them as quickly as they come up on her screen. As you struggle to get the information down, she says, "Do you have VoiceView?" You purchased a modem with VoiceView capability a month ago and loaded on the software just in case, so you say, "Yes!" She says, "Great! I'll just transfer the information on my screen to your PC screen and it will save both of us a lot of time and hassles — it will only take about 3 to 10 seconds (depending on the amount of information on the screen), and then we can begin talking again. Here it comes!" At your end you experience being put on hold briefly. You hear a faint chirp at the beginning of the transmit time and another a few seconds later. You can then begin talking again with the information you need right in front of you. In the same way, you can also send and receive faxes and binary files.

VoiceView is the protocol that allows all of this to happen. The VoiceView technology must be embedded in a separate box and/or modem through which the conversation and your various communications devices — PCs, fax machine, modem — etc. are connected. To use VoiceView, you must either own a ViewBridge, made by Radish, or a VoiceView certified modem. VoiceView software will be included with the hardware. Microsoft has released a dynamic link library on the Microsoft Developers Network that will help application developers more easily take advantage of VoiceView from Microsoft Windows, with which it is compatible, since VoiceView is simply implemented as a set of AT-command set extensions. The VoiceView software in Windows95 and other Windows products will allow for the transfer and receipt of the VoiceView file. It will not allow for the viewing of the file. For that, you will need extra software. Here are some typical VoiceView applications:

• Collaborative computing. Spreadsheets and documents are easily exchanged, discussed and modified.

• Call Centers can be VoiceView enabled, for example, letting agents send complex flight times, availabilities and itineraries to customers — while the agent is on the phone, or while the customer is in the queue.

• Mixed messaging systems with VoiceView combine the retrieval of voice messages, faxes and electronic mail with one standard phone call.

• Transactions, such as sales orders, processed or confirmed using both visual and audible interchange.

• Hardware and software support is simplified with VoiceView. During a single phone call, the support person can capture diagnostic files, can send diagrams showing how to fix things, etc. troubleshoot and install fixes.

• Other apps include business card exchange and remote presentations where both audio and visual materials are sent via VoiceView. See also VOICEVIEW AGENT, VOICEVIEW BRIDGE, VOICEVIEW PEER and VOICEVIEW SET.

VoFR Voice over Frame Relay.

VoIP Voice over Internet Protocol. See VoIP Forum.

VoIP Forum Voice over Internet Protocol. The Voice over IP Forum was formed in 1996 by Cisco Systems, VocalTec, Dialogic, 3Com, Netspeak and others as a working group of the International Multimedia Teleconferencing Consortium (IMTC), which promotes the implementation of the ITU-T H.323 standard. The VoIP Forum is focused on extending the ITU-T standards to provide implementation recommendations as a means of supporting Voice over IP in order that devices of disparate manufacture can support voice communications over packet networks such as the Internet. By way of example, the VoIP Forum intends to establish directory services standards in order that Internet voice users can find each other. They also plan to port touch-tone signals to the Internet to allow the use of ACDs and voice mail systems See also VON Coalition.

VolanoMark VolanoMark is a popular Java benchmark for measuring server throughput... it measures messages per second.

Volatile Storage Computer storage that is erased when power is turned off. RAM is volatile storage.

Volser An MCI term used to denote a volume of calls. Based on the words "Volume Serial." The term "Volser" can be applied to the manual collection of calls from a switch on a switch tape or through call data transmitted via NEMAS.

Volt The unit of measurement of electromotive force. Voltage is always expressed as the potential difference in available energy between two points. One volt is the force required to produce a current of one ampere through a resistance or impedance of one ohm.

Volt Meter An instrument for measuring voltages, resistance and current.

Voltage Electricity is a essentially a flow of electrons. They're pushed into a gadget — toaster, computer, phone — on one wire and they sucked out on the other wire. For this movement of electrons to occur there must be "pressure," just as there must be pressure in the flow of water. The pressure

under which a flow of electrons moves through a gadget is called the electric voltage. Voltage doesn't indicate anything about quantity, just the pressure. The amount of electricity moving through a wire is called its current and is measured in amps. You figure the power in an electron flow (i.e. in electricity) by multiplying the flow's current by the voltage under which it flows.

Voltage Drop The voltage differential across a component or conductor due to current flow through the resistance or impedance of the component or conductor.

Voltage Rating The highest voltage that may be continuously applied to a conductor in conformance with standards or specifications.

Voltage Regulator A circuit used for controlling and maintaining a voltage at a constant level.

Voltage Spike An extremely high voltage increase on an electrical circuit that lasts only a fraction of a second, but can damage sensitive electronic equipment like telephone systems or can cause it to act "funny." If your phone system starts acting "funny," one "cure" is to shut it off, count to ten, and then turn it on again. This sometimes clears the problem.

Voltage Standing Wave Ratio VSWR. The ratio of the maximum effective voltage to the minimum effective voltage measured along the length of mis-matched radio frequency transmission line.

Voltmeter A device for measuring the difference of potential in volts.

Volume 1. A volume is a partition or collection of partitions that have been formatted for use by a computer system. A Windows NT volume can be assigned a drive letter and used to organize directories and files. In NetWare a volume is a physical amount of hard disk storage space. Its size is specified during installation. NetWare v2.2 volumes, for example, are limited to 255MB and one hard disk, but one hard disk can contain several volumes. A NetWare volume is the highest level in the NetWare directory structure (on the same level as a DOS root directory). A NetWare file server supports up to 32 volumes. NetWare volumes can be subdivided into directories by network supervisors or by users who have been assigned the appropriate rights.

2. Under ISO 9660, a single CD-Rom disc.

Volume Label A name you can assign to a floppy or hard disk in MS-DOS. The name can be up to 11 characters in length. You can assign a label when you format a disk or, at a later time, using the LABEL command.

Volume Serial Number A number assigned to a disk by MS-DOS. The FORMAT command creates the serial number on a disk.

Volume Unit VU. The unit of measurement for electrical speech power in communications work. VUs are measured in decibels above 1 milliwatt. The measuring device is called a VU meter.

VOM Abbreviation for VOLT-OHM-MILLIAMETER, probably the most common form of electronic test equipment. It measures voltage, resistance and current, and may have either a digital or analog meter readout. Some VOMs have other test functions such as audible continuity signals and special tests for semiconductors.

Vomit Comet A plane used to simulate zero-G for astronaut flight training. Trainers often get motion sickness inside.

VON Voice On the Net (Internet), involving packetized voice. A recent development, initiating a VON call typically requires a multimedia PC or Mac computer with special software which matches that on the receiving device. More recently,

Internet servers have been equipped with such software, although appropriate client (workstation) software must be installed to take advantage of this approach. More recently still, VON has been demonstrated from workstation to telephone, telephone to workstation, and telephone to telephone. Additionally, new compression techniques and new DSPs have dramatically improved the quality of VON transmission, mitigating the impacts of packet delay. See Internet Telephony for a detailed explanation. See also VON COALITION and PACKETIZED VOICE.

VON Coalition The "Voice on the Net" (VON) Coalition is an Internet organization devoted to "educating consumers and the media by monitoring and supporting present and new developed telephony, video and audio technologies that are specifically designed and manufactured for the Internet community." It was formed, inter alia, to provide a forum against the ACTA (America's Carriers Telecommunications Association) petition to the FCC which sought to ban VON as a threat to the integrity of the PSTN and the concept of Universal Service. VON Coalition, www.von.org. See Universal Service Fund.

Vote ACK Also known as Mass ACK; in Usenet, the posting of the e-mail address of each person that voted for or against a newsgroup proposal.

Voting A teleconferencing term. Also known as Polling. In a large, event-style teleconference, the participants can vote on an issue via the touchtone keypad. The teleconference service provider tabulates the electronic votes and advises the conference sponsor of the results.

Voting Receivers A group of mobile base phone receivers operating on the same frequency as a control unit to pick the best signal from among them.

VOTS VMS OSI Transport Services. A Digital Equipment Corp. software product that modifies Digital's DECnet transport layer to conform to the International Standards Organization (ISO) Transport Protocol Class Four (TP4).

VOX Voice Operated eXchange. Your voice starts it. When you stop speaking, it stops. Tape recorders use it to figure when to start recording and when to stop. There are pros and cons to VOX. With VOX you often miss the beginning of the conversation. And the tape goes on for 3 or 4 seconds after you've stopped talking. Also if ambient noise is high, VOX might mistake it for speaking and turn the recorder on and keep it running. Cellular phones also use VOX to save battery. A cellular phone without VOX is continuously transmitting a carrier back to the cell cite the entire time your call is in progress. The VOX operation used in smaller phones allows the phone to transmit only when you're actually talking. This reduces battery drain and enables handheld phones to operate longer on a smaller battery.

VoxML Voice Markup Language. See Voice Browser.

Voycall An early key system manufacturer, which made a combination 1A2 handsfree intercom telephone system. It was wood grained, inlaid into black plastic. An impressive phone system. Sadly, no more.

VP 1. Virtual Path. A SONET term for an end-to-end route between 2 points. Many Virtual Paths may share a common physical path. Each Virtual Path consists of Virtual Tributaries which, in turn, consist of Virtual Channels. In the ITU-T SDH terminology, a Virtual Path is known as a Virtual Container.
2. An ATM term. Virtual Path is a unidirectional logical association or bundle of VCs, which are communications channels that provide for the sequential unidirectional transport of ATM cells.

VPC An ATM term. Virtual Path Connection: A concatenation of VPLs (Virtual Path Links) between Virtual Path Terminators (VPTs). VPCs are unidirectional.

VPDN Virtual Private Data Network. A private data communications network built on public switching and transport facilities rather than dedicated leased facilities such as T1s.

VPDS Virtual Private Data Services. MCI's equivalent of Vnet for data.

VPI Virtual Path Identifier. An ATM term. Virtual Path Identifier is an eight bit field in the ATM cell header which indicates the virtual path over which the cell should be routed. See VPI/VCI.

VPI/VCI Virtual Path Identifier/Virtual Channel Identifier. Combined, these fields identify a connection on an ATM network. See VPI.

VPIM Voice Profile for Internet Messaging, a proposed Internet messaging protocol to allow disparate voice messaging systems to automatically exchange voice mail over the Internet. VPIM also will allow a voice messaging system to communicate with other such systems outside the organization. VPIM works like this: You record a message and enter the target telephone number of the intended recipient. Your voice processing system does a directory look-up to a public electronic directory, using LDAP (Lightweight Directory Access Protocol) to find the e-mail address assigned for voice messages for that individual. Your system converts the voice message to a MIME (Multipurpose Internet Mail Extension) attachment, and routes the message through the Internet using SMTP (Simple Mail Transfer Protocol). The message is delivered to the voice messaging system supporting the target telephone number, where it is converted back into a voice message and stored in the recipient's voice mail box. The recipient can respond in the same fashion. Now let's take it a step further. As the messages are converted to MIME attachments, and as it uses SMTP over the Internet, VPIM has the potential to support compound mail consisting perhaps of voice mail, audio mail, e-mail, and video mail. See also LDAP, MIME and SMTP.

VPL An ATM term. Virtual Path Link is a means of unidirectional transport of ATM cells between the point where a VPI value is assigned and the point where that value is translated or removed.

VPN Virtual Private Network is a software-defined network offering the appearance, functionality and usefulness of a dedicated private network, at a price savings. Here's how it works: Your company buys a bunch of leased lines from your offices to the nearest local offices of your chosen long distance carrier. You're in your New York offices. You want to dial your offices in Chicago. You pick up the phone, dial perhaps seven digits. The phone rings in Chicago. What's happened is that your local PBX has recognized that call as belonging to your VPN. So it shunts the calls over the dedicated local loop to your long distance carrier. Your carrier then checks your dialed number, perhaps changing it with the aid of a database look up table, and completes the call over the carrier's own switched telephone facilities (fiber optic, microwave, copper, etc.). These are the same facilities which you and I use when we dial 1 and the long distance number (assuming we're equal accessed to that carrier).

There are several differences between a VPN and normal dial service:
1. VPN's price per minute is cheaper, often a lot cheaper.
2. You dial fewer digits with VPN. Sometimes you can get right to the distant desk, without going through the operator at the distant end.

3. You have to pay for the dedicated phone lines at the various ends of the VPN which have those dedicated phone lines. But they're often T-1, and thus not expensive on a per voice circuit basis.

4. You have to commit to use VPN for much longer than you do with normal dial up service — which is typically month-to-month.

See Virtual Private Network.

VPN 56 Sprint's Switched 56 Kbps service, supports advanced voice, data and image network communication tools including Group IV Fax, high resolution image transfer, file transfer, videoconferencing and switched data service via access to SprintNet, a large public data network.

VPOTS Very Plain Old Telephone Service. No automated switching.

VPT Virtual Path Terminator. As an ATM term, it is a system that unbundles the VCs of a VP for independent processing of each VC.

VPU 1. Virtual Physical Unit.
2. Voice Processing Unit.

VQL Variable Quantizing Level. Speech-encoding technique that quantizes and encodes an analog voice conversation for transmission, nominally at 32 Kbps.

VR 1. Voice Recognition. See VOICE RECOGNITION. 2. Virtual Reality. See VIRTUAL REALITY.

VRAM Video RAM. Memory used to buffer an image and transfer it onto the display. It is a form of DRAM specially suited for video. VRAM differs from common DRAM in that it has two data paths — a technique known as dual porting — rather than the single path of traditional RAM; thus, it can move data in and out simultaneously. Two devices can access it at once. The CRT controller, which converts bits and bytes in video memory to pixels on the screen, and the CPU, which manipulates the contents of video memory, can access VRAM simultaneously. Conventional DRAM chips allow one read or write operation at a time. Video RAM supports simultaneous read/write, read/read and write/write operations. It's most often used in graphic accelerators. In video boards fitted with the less expensive DRAM, performance suffers somewhat because the CRT controller and the CPU must takes turns getting to the video buffer held in VRAM. See also DRAM and WRAM.

VRC Vertical Redundancy Check.

VREPAIR A Novell NetWare program somewhat analogous to MS-DOS's CHKDSK program or Windows95's Scandisk. VREPAIR fixes FAT (File Allocation Table) and DIR (Directory) Tables. It's a most useful program. Highly recommended.

VRML Virtual Reality Modeling Language. A language for writing 3D HTML applications. VRML, according to PC Magazine, is an open standard for 3-D imaging on the World-Wide Web that paves the way for virtual reality on the Internet. The way VRML code describes a 3-D scene is analogous to four points describing a square, or a center point and radius describing a sphere. VRML viewers, similar to HTML Web browsers, interpret VRML data downloaded from the Web and render it on your computer. This allows the bulk of the processing to be performed locally and drastically reduces the volume of information that must be transmitted from the Web — a key consideration if rendering is to be performed in real time. See VRML Consortium.

VRPRS Virtual Route Pacing Response in SNA.

VRU See VOICE BOARD and VOICE RESPONSE UNIT.

VS 1. Virtual Scheduling. As an ATM term, it is a method to determine the conformance of an arriving cell. The virtual scheduling algorithm updates a Theoretical Arrival Time

(TAT), which is the "nominal" arrival time of the cell assuming that the active source sends equally spaced cells. If the actual arrival time of a cell is not "too" early relative to the TAT, then the cell is conforming. Otherwise the cell is non-conforming.

2. Virtual Source. Refer to VS/VD.

3. See Virtual Storage.

VS&F Voice Store and Forward. Voice is digitally encoded, sent to large storage devices and later forwarded to the recipient. See VOICE MAIL.

VS/VD Virtual Source/Virtual Destination. An ATM term, a VS/VD is an ABR connection may be divided into two or more separately controlled ABR segments. Each ABR control segment, except the first, is sourced by a virtual source. A virtual source implements the behavior of an ABR source endpoint. Backwards RM-cells received by a virtual source are removed from the connection. Each ABR control segment, except the last, is terminated by a virtual destination. A virtual destination assumes the behavior of an ABR destination endpoint. Forward RM-cells received by a virtual destination are turned around and not forwarded to the next segment of the connection.

VSAT Very Small Aperture Terminal. A relatively small satellite antenna, typically 1.5 to 3.0 meters in diameter, used for satellite-based point-to-multipoint data communications applications. While VSAT earth stations traditionally supported data rates of as much as 56 Kbps, contemporary systems can operate at rates of 1.544 Mbps. You see VSATs on top of retail stores which use them for transmitting the day's receipts and receiving instructions for sales, etc. Consider the VSAT dishes you see on the roofs of gas stations. Large numbers of gas stations share access to a single satellite which, in turn, provides connection to a centralized data processing center. At those gas stations are intelligent gas pumps equipped with credit card readers, monitors, and limited computer memory. You swipe your credit card through the card reader, with the credit card number being transmitted through the VAST dish to the satellite to the data processing center. Once the credit is verified (i.e., the card has not been reported lost or stolen, and the balance is not overdue), the transaction is authorized in return. Once the desired amount of gas has been pumped, that information is transmitted to the data processing center, with the transaction being noted in the accounts receivable system for billing purposes. Additionally, the level of inventory (i.e., gas in the tank) is noted as having been decreased. In other words, the VSAT network supports credit verification, transaction authorization, billing and inventory management.

VSB Vestigial SideBand. A form of Amplitude Modulation (AM) that compresses required bandwidth. VSB is modulation technique used to send data over a coaxial cable network. Used by Hybrid Networks for upstream digital transmissions, VSB is faster than the more commonly used QPSK, but it's also more susceptible to noise. See also 64QAM, Amplitude Modulation, DSBSC, DSBTC, SSB, Vestigial Sideband and QPSK.

VSE 1. A British Term. Voice Services Equipment, a generic term for voice response unit, interactive voice response, voice processing unit and so on.

2. Virtual Storage Extended.

VSELP Vector Sum Exited Linear Prediction. A speech coding technique used in U.S. and proposed Japanese DMR standards. Second generation European DMR will probably use some version of VSELP.

VSS Voice Server System.

VSWR Voltage Standing Wave Ratio. The ratio of the maximum effective voltage to the minimum effective voltage measured along the length of mis-matched radio frequency transmission line. Explanation: When impedance mismatches exist, some of the energy transmitted through will be reflected back to the source. Different amounts of energy will be reflected back depending on the frequency of the energy. VSWR (Voltage Standing Wave Ratio) is a unitless ratio ranging from 1 to infinity, expressing the amount of reflected energy. A value of one indicates that all of the energy will pass through, while any higher value indicates that a portion of the energy will be reflected.

VT Virtual Tributary. A structure designed for transport and switching of sub-DS3 payloads. VT1.5 equals 1.544 Mbps; VT2 equals 2.048 Mbps; VT3 equals 3 Mbps; VT6 equals 6 Mbps. These are measures of speed in Sonet. See SONET.

VT100 A terminal-emulation system. Supported by many communications program, it is the most common one in use on the Internet. VT102 is a newer version.

VT1.5 Virtual Tributary equals 1.544 megabits per second.

VT2 Virtual Tributary equals 2.048 megabits per second.

VT3 Virtual Tributary equals 3 megabits per second.

VT6 Virtual Tributary equals 6 megabits per second.

VTA Virtual Trunk Agent.

VTAC Vermont Telecommunications Applications Center. See www.vtac.org

VTAM Virtual Telecommunications Access Method. A program component in an IBM computer which handles some of the communications processing tasks for an application program. In an IBM 370 or compatible, VTAM is a method to give users at remote terminals access to applications in the main computer. VTAM resides in the host. It performs addressing and path control functions in an SNA network that allows a terminal or an application to communicate with and transfer data to another application along some sort of transmission medium. VTAM also provides resource sharing, a technique for efficiently using a network to reduce transmission costs. See SYSTEMS NETWORK ARCHITECTURE.

VTC Video TeleConference, a term invented by the U.S. Air Force.

VTNS Virtual Telecommunications Network Services.

VToA Voice Traffic over ATM.

VTP Virtual Terminal Protocol. An International Standards Organization (ISO) standard for virtual terminal service.

VTTH Video To The Home. The general ability to provide interactive multimedia services to people in their homes.

VU Meter VU is the unit of measurement for electrical speech power in communications work. VUs are measured in decibels above 1 milliwatt. The measuring device is called a VU meter, which is an abbreviation of volume-unit meter, a type of meter used to indicate average audio amplitude.

VUI First came the CLI (Command-Line Interface). Then came the GUI (Graphical User Interface). Get ready for the VUI: the Video User Interface. Actually, you don't need to get ready for it any time soon, but you might start wondering how to use it.

VW-1 A test used by Underwriters Laboratories to classify wires and cables by their resistance to burning. (Formerly designated as FR-1.)

W 1. Abbreviation for WATT.

2 The Hayes AT Command Set describes a standard language for sending commands to asynchronous modems. One of the commands is "W." If you embed a W in your dialing string, i.e. 212-691-8215-W-10045, the modem will dial 212-691-8215 and wait until it hears dial tone. When it hears dial tone, it will dial out 10045. That is the standard Hayes command set interpretation of W. There is another. When using some of the communications software products from Crosstalk (now a subsidiary of DCA) you can place a [W] in your dialing string. If you do, your modem will dial the number until it encounters a [W]. It will then wait until you hit any button on your keyboard. The purpose of W commands is to allow you to dial through private networks (your own), through public networks (MCI, Sprint, etc.), through fax/modem/telephone switches and through any other device or network.

W-CDMA Wideband Code Division Multiple Access. See CDMA.

W-DCS Wideband Digital Cross-connect System. W-DCS is an electronic digital cross-connect system capable of cross-connecting signals below the DS3 rate.

W3 An abbreviation for the Internet's World Wide Web. See WORLD WIDE WEB.

W3C World Wide Web Consortium. A consortium jointly hosted by the Massachusetts Institute of Technology (MIT), the Institut National de Recherche en Informatique et en Automatique (INRIA), and Keio University (Japan) Initially, W3C was established in collaboration with CERN, where the WWW originated, with support from DARPA and the European Commission. Tim Berners-Lee, inventor of the WWW, acts as Director. The W3C works to produce "free, interoperable specifications and sample code. Focus is on the domains of user interface, technology and society, and architecture. In some ways, its ambitions are not that different from those of the IETF (Internet Engineering Task Force), except that its members have commercial interests. The IETF's members are volunteers and tend to come from academia. W3C also focuses on the narrower Web, whereas the IETF focuses on the broader Internet. www.w3.org

Wabi Sun Microsystems software for running Microsoft Windows applications on Solaris. Runs on Intel and SPARC.

WabiServer Wabi is Sun Microsystems software for running Microsoft Windows applications on Solaris. Runs on Intel and SPARC. A WabiServer allows multiple and simultaneous users to run Wabi on Intel and SPARC.

WACK 1. Wack a T-1. Here's what it means. Imagine you're an ISP — an Internet Service Provider. Your business is answering inbound calls from customers with PCs who want to send email, surf the Internet, have fun in chat rooms, etc. You are receiving your calls on digital circuits (e.g. T-1s) from a local phone company — CLEC or ILEC. Your inbound T-1 line may have as many as 24 separate phone lines, which your customers could call on. Sometimes your phone company sends you a call over one or more of those 24 lines and it doesn't get through. Its switching equipment assumes your equipment on that particular phone line is broken. So its

switching equipment automatically takes that phone line out of service. It sends no more calls over that line. Later in the day, it may run a diagnostic program — called FISO in some instances — and that software program may try sending calls over the lines that it had earlier taken out of service. But between taking the lines out of service and running that program as much as six or seven hours may elapse. With these lines out of service, an ISP's customers will now enjoy lousy service. So the ISP's people (after bombarded with customer complaint calls) calls the phone company and says "wack my T-1, please." What happens then is that a real live technician at the phone company then goes into each line and manually tests it, often with a technician from the ISP on the other end of the phone. The advantage of wacking the T-1 is that the circuits out of service are put back into service much faster than if they were left to the machine to do it automatically later on.

2. Wait before transmitting positive ACKnowledgement. In Bisynch, this DLE sequence is sent by a receiving station to indicate it is temporarily not ready to receive.

3. Do not confuse "wack" with "whack," which has now become a colloquial expression for killing someone.

WAR Wireless Application Environment - defines the programming interface for applications and consists of WML and WML-script - a new, Java-like local scripting language

Wafer A thin disk of a purified crystalline semiconductor, typically silicon, that is cut into chips after processing. Typically, a wafer is about one fiftieth of an inch thick and four or five inches in diameter.

Wafer Fabs Wafer fabs are a slang term for ultraclean factories that fabricate chips on silicon wafers.

WAIS Wide-Area Information Servers. A very powerful system for looking up information in databases (or libraries) across the Internet. WAIS allows you to perform a keyword search. WAIS is like an index, whereas Gopher, which is sometimes used as a complement to WAIS, is like a table of contents.

Wait On Busy An English term for the American term "Camp On" or "Call Waiting." A service allowing the subscriber to make a call to a busy phone line, wait until the call is over, then be connected automatically.

Wait State A period of time when the processor does nothing; it simply waits. A wait state is used to synchronize circuitry or devices operating at different speeds. Wait states are introduced into computers to compensate for the fact that the central microprocessor might be faster than the memory chips next to it. For example, wait states used in memory access slow down the CPU so that all components seem to be running at the same speed. A wait state is a "missed beat" in the cycle of information to and from the CPU that is necessary for a memory transaction to be completed.

Wake On LAN This standard allows a PC to be powered on by a network server, so the server can perform routine tasks. See WFM.

WAL A direct dial WATS line, as compared to a WATS line connected via a T-1 line.

Walkaways People who walk away from coin phones though they owe extra money. You can tell a phone that has just been visited by a walkaway: It's typically ringing. And when you answer it, the operator will ask you to deposit some additional coins.

Walkie-Talkie Hand-held radio transmitter and receiver. Like the police carry. Probably the best named device in telecom. You walkie, you talkie.

Walk Through 1. In Computer System Development, a peer review of a system design, code, etc. The goal is to identify errors as early as possible and learn from other people's experience. Managers and people who prepare performance reviews should NOT be in the room. The concept is to invite "egoless" constructive criticism and to nurture team-oriented validation and debug responsibility.
2. In telecommunications, a formal tour and accompanying verbal description of the work that is to be done by the customer and the vendors.

Walk Time The time required to transfer permission to poll from one station to another.

Wall Outlet A phone outlet positioned at shoulder height to accept a wall telephone set. The typical installation includes a special modular jack containing two mounting bosses that insert into key-hole slots in the base of the telephone set. Electrical connection is made by a short cord or a lug element that is integral to the telephone set base.

Wall Phone A phone that is mounted on the wall. Where else would a wall phone be mounted? Some new phones — especially some key systems — come so you can use them on the desk or mount them on a wall, without extra hardware. Some desk phones cannot be mounted on a wall. This is a disadvantage when you run out of space on your desk, as you will with all the computers and workstations you'll be putting there.

Wall Thickness A term expressing the thickness of a layer of applied insulation or jacket.

Walla Walla In movie-making, a "walla walla scene" is one where extras pretend to be talking in the background-they are not, they are just repeating "walla walla" over and over again. But when they say "walla walla" it looks like they are actually holding conversations. (Sounds like Monday morning meetings at work.)

Wallpaper The area of your Windows desktop on your PC behind and around your windows and icons. The color and pattern you put on it through the desktop manager is called wallpaper.

WAN Wide Area Network. Uses common carrier-provided lines that cover an extended geographical area. Contrast with LAN. This network uses links provided by local telephone companies and usually connects disperse sites. See WIDE AREA NETWORK.

Wander Long-term random variations of the significant instants of a digital signal from their ideal position in time. Wander is a matter of synchronization errors in digital networks. If a clocking source fails or degrades, it is essential that the signals be re-clocked at the point of the next network element. Otherwise, wander will increase through cascading elements, perhaps affecting the data payload. It is especially important to control wander in very speed networks, such as SONET, where even slight synchronization failures can be catastrophic. Wander variations are usually considered to be those that occur over a period greater than 1 second. See also DIURNAL WANDER.

WAP Wireless Application Protocol. Such proposed new protocol could simplify how wireless users access electronic and voice mail, send and receive faxes, make stock trades, conduct banking transactions and view miniature Web pages on a wire-

less terminal's LCD screen. Indeed, the proposed standard is being designed in large part to make it easier for mobile users to view shrunken Web pages using Unwired Planet's Handheld Device Markup Language (HDML), said Ben Linder, vice president of marketing at Unwired Planet. HDML "lets Web sites tailor the information format to fit the screen of the phone," Linder said. "We don't try to display the graphical Web pages on such a small device," he added; rather, Webmasters could create smaller versions of sites more suitable for viewing on such units. With WAP, those modifications and optimizations would only have to be made once in order to be viewed on an Ericsson, Motorola or Nokia terminal, Linder added. Unwired Planet's technology, called UP.Link, is used by AT&T's PocketNet and GTE Wireless services. Bell Atlantic and Nynex also offer cellular digital packet data-based services combined with UP.Link. Al Haase, director of sales, GSM, at Ericsson's North American headquarters in Richardson, Texas, acknowledged that although the demand for receiving this kind of information in a wireless format is not great today, agreement on a standard may nudge wireless access deeper into the mainstream. "The systems were not designed to support data, and thus data transmission speeds are fairly low, " Hasse said. "With the new systems coming online, data access becomes more of a reality because we are more able to link the mobile user with the Net in a timely way." See Wireless Application Protocol Forum Ltd.

War Dialer A program that tries a set of sequentially changing numbers (i.e., telephone numbers or passwords) to determine which ones respond positively.

War Room Also called a "solutions room." This is an enclosed area with a large table used for decision- or strategy-making.

Warble Tone A tone changing in frequency at a slow enough rate to give the effect of warbling. A warble tone is the sound of an electronic ringer, according to many people.

WARC World Administrative Radio Conference. Sets international frequencies. Just before Telecom '87, WARC allocated important new frequencies for satellite-based land mobile (satellite to truck, etc.) and radio determination navigation services (electronic maps for your car). WARC is part of the 154-member International Telecommunication Union. ITU-T is part of the ITU. See ITU-T.

Warez Warez is commercial software that has been pirated and made available to the public via a Bulletin Board or a Web site on the Internet. Typically, the pirate has figured out a way to remove the copy-protection or registration scheme used by the software. The use and distribution of warez software is illegal.

Warm Start Restarting or resetting a computer without turning it off (also called "soft boot"); press Ctrl + Alt + Del on an IBM or IBM compatible.

Warm Swap When a system component (e.g., a computer disk drive) fails, it may be replaced without turning the system off. During this period, the system's activity is suspended, however. Also known as a Hot Plug, it is unlike a Hot Swap, during which the system remains active. See HOT SWAP and RAID.

Warranty Span of time that equipment will be repaired or replaced due to failure. Usually does not include reimbursement of engineer's fees required for replacement. May not include equipment failure due to abuse or destruction by either intentional or unintentional means. Lightning, floods and other Acts of God are not covered under warranty.

Washington D.C. Eight square miles surrounded by reality. A comment variously attributed to John F. Kennedy, J. Edgar Hoover and sundry other luminaries.

WASI Wide Area Service Identifier

WAT Date A term telephone companies use to indicate a date a day or two in advance of the promised delivery date, by which the circuit should be "up" and available for testing, so any problems can be worked out before the technician is dispatched to the customer's premises.

Watch Commands Watch Commands are found in programming. They allow you to "watch" the value of selected application variables while the application is executing (e.g., see the last-entered touchtone digits from a caller).

Watchdog Timer There is a watchdog circuit (also called special function register) in later versions of Intel Pentium chip. When things go awry with the computer the Pentium is running, it sends out a signal to an external device. That device can then take action — for example reboot the PC. According to Intel's Web site, "The Watchdog Timer is a very useful peripheral that safeguards against software failures. The Watchdog Timer (WDT) is a 16 bit ripple counter that will reset the 8XC196KC/KD if the counter overflows. When the WDT is enabled, this peripheral monitors the execution of a program. This is useful when expensive hardware is being controlled. If we operate in an electrically noisy environment, a noise spike can be potentially fatal to the operation of the microcontroller. We can maintain control in critical applications by resetting the microcontroller when a runaway software process hangs the systems. There is only one Special Function Register to concern ourselves with in this discussion. That register is called WATCHDOG."

Watching Timer A circuit used in ETHERNET transceivers to ensure that transmission frames are never longer than the specified maximum length.

Water Bore A device which bores holes underground. Pipes are then placed in the holes and cables (fiber, coax, etc.) are then pulled through. A water bore uses high-speed, pressurized water to bore through the underground.

Water Pipe Ground A water pipe to which connection is made for the ground.

Waterfall life cycle The conventional software development process, consisting of a series of steps commonly defined as analysis design, construction, testing and implementation. The underlying assumption is that each phase does not begin until the preceding phase is complete.

WATS Wide Area Telecommunications Service. Basically, a discounted toll service provided by all long distance and local phone companies. AT&T started WATS but forgot to trademark the name, so now every supplier uses it as a generic name. There are two types of WATS services — in and out WATS, i.e. those WATS lines that allow you to dial out and those on which you receive incoming calls (the typical 800 line service). You subscribe to in- and out-WATS services separately. In the old days you needed separate in and out lines to handle the in and out WATS services. But these days you can choose to have in- and out-WATS on the same line. This is not particularly brilliant traffic engineering, since you can't receive an incoming 800 call if you're making an outgoing call. But I do know someone who has an 800 line on his cellular phone!

Many users inside companies think their company's WATS lines (and thus their WATS calls) are free, so they speak longer. This can kill the idea of buying WATS lines to save money. In the old days, interstate WATS was charged at effectively a flat rate and thus, there was some reason to believe

that marginal WATS calls were 'free.' These days EVERY WATS call costs money. EVERY one! Without exception. See 800 SERVICE and PLANT TEST NUMBER.

Watt The unit of electricity consumption and representing the product of amperage and voltage. the power requirement of a device is listed in watts, you can convert to amps by dividing the wattage by the voltage (e.g., 1,200 watts divided by 120 volts, equal 10 amps). See OHM's LAW. Don't confuse WATTS (the measure of electricity) with WATS, which stands for Wide Area Telecommunications Service. See WATS.

Wave Audio Also called "waveform audio," is a digital representation of actual sound waves. Wave audio "samples" the sound waveforms at regular intervals. The three standard sampling frequencies are 11.025 KHz, 22.05 Khz, and 44.1 KHz. Higher sampling frequencies yield higher fidelity sound.

Wave Length The distance between peaks of an electromagnetic (or other) wave. The distance traveled by a wave during one complete cycle. See also WAVELENGTH.

Waveform The characteristic shape of a signal usually shown as a plot of amplitude over a period of time.

Waveform Editor A word processor for sound. You record something. Then you "play" it back on your PC's screen. Your PC screen now looks like an oscilloscope. Then you use this wave form editor to edit (i.e. change, replace, amplify, echo, fade in or out, cut out noise, cut/paste from other files, or generally muck with) the sound. A wave form editor is used in voice processing.

Waveform monitor A device used to examine the video signal and synchronizing pulses. An oscilloscope designed especially for viewing the waveform of a video signal.

Waveguide A conducting or dielectric structure able to support and propagate one or more modes. More specifically, a waveguide is a hollow, finely-engineered metallic tube used to transmit microwave radio signals from the microwave antenna to the radio and vice versa. Waveguides comes in various shapes — rectangular, elliptical or circular. They are very sensitive and should be handled very gently. Waveguides may contain a solid or gaseous dielectric material. In optical, a waveguide used as a long transmission line consists of a solid dielectric filament (optical fiber), usually circular. In integrated optical circuits an optical waveguide may consist of a thin dielectric film.

Waveguide Scattering Scattering (other than material scattering) that is attributable to variations of geometry and refractive index profile of an optical fiber.

Wavelength The length of a wave measured from any point on one wave, to the corresponding point on the next wave, such as from crest to crest. In other words, a wavelength is the distance an electromagnetic wave travels in the time it takes to oscillate through a complete cycle. There is a direct proportion between the wavelength of a radio signal and its frequency.

Wavelength Division Multiplexing WDM. A way of increasing the capacity of an optical fiber by simultaneously operating at more than one wavelength, and at as many as four wavelengths. With WDM you can multiplex signals by transmitting them at different wavelengths through the same fiber. Wavelength division multiplexing works similar to Frequency Division Multiplexing (FDM) in the analog worlds of electrical and radio transmission systems. In optical fiber communications, WDM is any technique by which two or more optical signals having different wavelengths may be simultaneously transmitted in the same direction over one strand of fiber, and then be separated by wavelength at the distant end. Each wavelength virtual channel typically effectively is a separate light pipe, which can support a given signaling rate, such as OC-48 at 2.5 Gbps or OC-192 at 9.953 Gbps. Optical EDFAs (Erbium-Doped Fiber Amplifiers), rather than optical repeaters, are spaced tens of kilometers apart, simultaneously boosting the intensity of the multiple light channels through a "pump laser." The EDFAs operate more effectively than do conventional optical repeaters, and are much less costly to acquire and operate. Beyond WDM is DWDM (Dense WDM). DWDM systems in commercial application operate over eight- and sixteen-channel multiplexers; 32-channel systems have been released. At OC-192, for instance, a 32-channel system yields an incredible 320 Gbps, rounded up. The advantage of WDM is that of increasing network capacity without deploying additional fiber; further, WDM compares favorably to upgrading SONET equipment to operate at the higher OC-n layers. The higher-speed WDM systems also are known as DWDM (Dense WDM), although the term is a loose one. The ITU-T has defined 40 standard wavelengths from which SONET equipment manufacturers can choose. See also EDFA and SONET.

Wavelet An oscillating waveform which persists for only one or a few cycles, whereas naturally occurring waveforms persist indefinitely. Wavelets are most useful for representing information with sharp discontinuities, such as images which have sharp edges in the form of variations in color and contrast. Originally developed for in image processing and compression, wavelets and wavelet analysis can be used for such delicate applications as compression and decompression of seismic data, which is critical and which, therefore is intolerant of lossy compression techniques. Wavelet analysis yields compression ratios of 100:1 or better, faithfully reproducing the image on decompression. See WAVELET PACKET.

Wavelet Packet According to the IEEE, class of time-frequency waveforms with a location (position), a scale (duration), and an oscillation (frequency). Wavelet packets are used for analyzing data with natural oscillations, such as audio signals and fingerprint images.

Wavelet Transform A compression technique which involves the representation of a discrete signal or image through wavelet functions. Wavelet transform is computed by the fast pyramid algorithm, which involves a series of linear filtering operations in combination with down-sampling by a factor of two. See DOWN-SAMPLING, FAST PYRAMID ALGORITHM and WAVELET.

Wavetable Sound systems

WAW Waiter-Actor-Webmaster. Used to describe fly-by-night graphic designers and Web consultants trying to cash in on the Web boom. "Can you believe they hired that clueless WAW for $60K a year?!" This definition courtesy Wired Magazine.

Way Operated Circuit A circuit shared by three or more phones on a party line basis. One of the phones usually operates as the control point.

Way Station One of the phones, other than the central controller, on a way operated circuit. See WAY OPERATED CIRCUIT.

WBC Wide Band Channel. An FDDI-II term. See FDDI II.

WCCP The Web Cache Control Protocol. See Cache Engine.

WCS Wireless Communications Service. Cellular, PCS and the like.

WCV See Weighted Call Value.

WBEM Web-Based Enterprise Management. A wide-ranging blueprint for unified administration of network, systems and

software resources (established my Microsoft, Intel, Compaq, Cisco, BMC Software, and others)....a schema that incorporates three new protocols and four current Internet standards to allow users to manage distributed systems and to access network resources using any Web browser.

WDM Wavelength Division Multiplexing. A means of increasing the data-carrying capacity of an optical fiber by simultaneously operating at more than one wavelength. With WDM you can multiplex signals by concurrently transmitting them at different wavelengths through the same fiber. Wavelength division multiplexing works similar to Frequency Division Multiplexing (FDM) in the analog worlds of electrical and radio transmission systems. In optical fiber communications, WDM is any technique by which two or more optical signals having different wavelengths may be simultaneously transmitted in the same direction over one strand of fiber, and then be separated by wavelength at the distant end. Each wavelength is a "virtual channel", effectively a separate "light pipe", which can support a given signaling rate, such as OC-48 at 2.49 Gbps or OC-192 at 9.95 Gbps. Although optical WDM has been a known technology since the 1980s, it was restricted to two widely separated "wideband" frequencies. (frequency=speed/wavelength, where the speed of light in glass fiber is essentially constant, so wavelength and frequency are used interchangeably in describing the multiple channels of WDM). The number of distinct wavelengths supported has increased rapidly since WDM became "narrowband" capable in the early 1990s. Initial systems operated at two or four wavelengths, and the term "WDM" is usually used to refer to these low channel-count systems. Beyond WDM is DWDM (Dense WDM), although the terms do not have absolute differentiating definitions and are used somewhat interchangeably. DWDM systems in commercial applications use at least eight-channel or sixteen-channel multiplexers, and systems as dense as 40-channels have been released and are in carrier networks carrying live traffic. The capacity is steadily increasing, both by ever-expanding channel counts and faster supported TDM rates of the individual wavelengths. Generally, existing systems trade-off channel count against maximum supported rate: the current maximum channel count of 40 (for the present) is limited to OC-48 (almost 100 Gbps net throughput). Systems which support higher signal rate (i.e., OC-192) support fewer than half as many channels, at most. At OC-192, a 40-channel system would yield an incredible 400 Gbps, rounded up. While such a system is not yet available, it may be little more than a year away. (This definition was written in December, 1998.) Within the next 2-3 years it is reasonable to expect to see systems supporting on the order of 100 wavelengths of OC-192 each, providing almost one terabit per second transport! The advantage of WDM is that of increasing network capacity without deploying additional fiber. To install enough new actual physical fiber as the equivalent virtual fiber would be enormously more expensive and often a physical impossibility. WDM also compares very favorably to exclusively upgrading SONET equipment to operate at the higher OC-n layers without WDM, especially since the rate multiplier limit of TDM is hit sooner than the channel multiplier limit of WDM. The ITU-T had initially defined 41 standard wavelengths from which WDM equipment manufacturers could choose (expressed as frequencies, the table is centered at 194.10 THz, with 100 GHz spacing, giving 41 center frequencies from 192.10 to 196.10 THz). The recently approved ITU-T recommendation G.692 implements an additional table which reduces the spacing to

50 GHz, making 40 additional frequencies available. Further, G.692 makes it plain that the table's end-points are not absolute, and future systems are fully anticipated to include frequencies beyond those limits. It is not unreasonable to expect that as the technology continues to advance, the channel spacings may also be reduced, although not as soon as the end-points are expanded. As is common, the standards' organizations often are outpaced by the technology advances. While most initial fiber optic traffic involved SONET signals, there is increasing examination of "bypassing" SONET and directly transporting other signal formats (IP, for instance) over light without wrapping it in the SONET overhead. WDM and DWDM are physical data transport terms, and not specific to SONET, or any other format standard. Long-haul amplification of WDM signals is achieved with optical EDFAs (Erbium-Doped Fiber Amplifiers), rather than optical repeaters. The EDFAs are spaced up to 100 kilometers apart, simultaneously boosting the intensity of all the multiple light channels through a "pump laser." The EDFAs operate more effectively than do conventional optical repeaters, and are much less costly to acquire and operate. See also DWDM, EDFA and SONET.

WDMA Wavelength Division Multiple Access. A technique which is used to provide access to multiple channels carried on different wavelengths on the same fiber-optic cable. Each optical input operates on a different wavelength of light. The multiple inputs are multiplexed over a common long-haul SONET fiber optic link through WDM (Wavelength Division Multiplexing) equipment. See WDM for much more detailed explanation.

WDT See WatchDog Timer.

Weather Trunk Group A trunk group used to provide customers with weather information.

Weathermaster Method A distribution method where the unused wall space inside heating and cooling units beneath windows is used for satellite location. Cables are fed from a riser or other serving closet to the location through baseboards, conduit, or underfloor system.

Web An abbreviation for the Internet's World Wide Web. See WORLD WIDE WEB and INTERNET.

Web Address Here's a typical address of an Internet Worldwide Web site
http://www.harrynewton.com/resume/bio.html
Let's look at what it means:
http is Hypertext Transfer Protocol." http lets your browser know to expect a web page (as opposed to an FTP or gopher site).
:// is what several writers have referred to as random punctuation abuse. Basically, it's text to separate the next part of the address, namely the "Sub-domain," which is www. World Wide Web servers typically use "www," but you may see other names like "web3" or "w3." The next part of the address is the "High-Level domain," which tells you either the type or location of an organization. Common high level domains: .com = commercial .edu = university .gov = government .uk = England .fi = Finland
The last part of the address is "Unique domain." What an organization or person calls its net site, namely harrynewton. Next part is /resume/ which is simply the directory or folder on the computer where the web page is stored. Finally, bio.html is simply the document you're looking for. See also URL.

Web Application Server Web application servers are special purpose, powerful PCs which sit between a Web server and a corporate database. They offload processing tasks form the repository and cache frequently requested informa-

t on. They offer much greater growability, since multiple connections can be run to the database — instead of just one, as in the case when the Web server is linked directly. In short, Web application servers speed up getting answers to requests for information out into Web site users' hands.

Web Art A definition for the artwork that you are beginning to see proliferating the Internet, especially on home pages.

Web Browser Clients software which navigates a web of interconnected documents on the World Wide Web. A Web Browser is software which allows a computer user (like you and me) to "surf" the Internet. It lets us move easily from one World Wide Web site to another. Every time we alight on a Web Page, our Web Browser moves a copy of documents on the Web to your computer. A Web Browser uses HTTP — the HyperText Transfer Protocol. Invisible to the user of a Web Browser, HTTP is the actual protocol used by the Web Server and the Client Browser to communicate over the Internet. The most famous Web Browser is currently Netscape. Many online commercial services such as AOL, CompuServe, Microsoft, Prodigy, etc. distribute Web Browsers to surf their "intranet" as well as the "internet." See INTERNET, MOSAIC, NETSCAPE and SURF.

Web Cache Control Protocol WCCP. See Cache Engine.

Web Caching See Cache Engine.

Web Casting Also spelled Webcasting. A term created by Business Week in February, 1997. Business Week described it thusly "Swamped by information on the Web? A new technology finds and delivers news for you. It also helps companies reach workers and conduct business online. Call it broadcasting, Internet-style. Call it webcasting." The basic is simple: You go to a webcasting site. You fill in a form detailing what sort of information you want to hear about. Next time you log on, information is "pushed" to you by your webcasting supplier. You might pay for this service. Or it may be advertiser-based.

Web Crawler One of the most popular search facilities on the Web. It indexes World Wide Web pages by title and URL. You can search the Internet with Webcrawler. Check out www.webcrawler.com/cgi-bin/WebQuery

Web Hosting A service performed by Internet Service Providers (ISPs) and Internet Access Providers (IAPs) who encourage outside companies to put their Web sites on computers owned by the ISPs. These computers are attached to communications links to the Internet — often high-speed links. For this Web hosting service, the ISPs typically charge their clients by equipment and transmission capacity used.

Web Master See WEBMASTER.

Web Mistress See WEBMISTRESS.

WEB Search Engines An Internet World Wide Web term. A search engine is a program that returns a list of Web Sites (URLs) that match some user-selected criteria such as "contains the words cotton and blouse." Basically, the procedure is simple. You surf to the search engine's site. You click a couple of times and type in what you're looking for. A few seconds later you get choices. You finally make your selection and you get instantly hotlinked over to that site. Here are the main Internet search engines and their addresses.
ALTA VISTA
http://altavista.digital.com
EXCITE
http://www.excite.com
LYCOS
http://www.lycos.com

INFOSEEK
http://guide.infoseek.com
OPENTEXT
http://www.opentext.com:8080
INKTOMI
http://www.inktomi.berkeley.edu/query.html
WWW.WORM
http://www.cs.colorado.edu/wwww/
WEBCRAWLER
http:///www.webcrawler.com
YAHOO
http://akebono.stanford.edu/yahoo/

Web Server A Web Server is a powerful computer which is connected to the Internet or an Intranet. It stores documents and files — audio, video, graphics or text — and can display them to people accessing the server via hypertext transfer protocol (http). A Web server derives its name because it is part of the World Wide Web. See WORLD WIDE WEB.

Web Service Provider A vendor who provides customers with Web Pages on the vendor's computer/s. Frequently, a Web Service Provider will provide additional services such as design help and usage statistics. Often they will just provide the computer space and leave the rest to you. A Web Service Provider may or may not also be an Internet Service Provider.

Web Site Any machine on the Internet that is running a Web Server to respond to requests from remote Web Browsers is a Web Site. In more common usage it refers to individual sets of Web Pages that can be visited with Web Browsers. he front page of a Web site is called its home page. It is also spelled as one word, namely WEBSITE. See also INTERNET.

Web TV See WebTV

Webcam A webcam is a digital video camera. Train the camera on a bridge, for example. Hook the camera up to a phone line and a Web site. Now people from all over the world can visit your web site via the Internet and check out the view from your webcam. Your webcam may transmit photos every second, every minute or every day. It's your choice of how much you want to spend and how exciting the action is. People are putting up Webcams to show the view on popular sites, like stadium construction sites. Many hope to earn money by selling ads on the sites. Some do.

Webcasting See Web Casting.

Weblock The Internet data traffic version of gridlock.

Webmaster An Internet term. The Webmaster is the administrator responsible for the management and often design of a company's World Wide Web site. See also WEBMISTRESS.

Webmistress An Internet term. The Webmistress is the female administrator responsible for the management and often design of a company's World Wide Web site. See also WEBMASTER.

Webphony A term created by Nick Morley, an excellent salesperson who works for Computer Telephony Magazine. Webphony is a combination of the two words Web and Telephony. It means telephony-enabling your Web site. Here's a simple example: You're checking out a Web, say L. L. Bean, the direct mail catalog company. You'd like to buy a new kayak. You need to ask a question. Click on the "Reach an Operator" button in the corner of your screen. Bingo, you're speaking live to an L.L. Bean operator via your computer's sound system, or perhaps they called you on a second phone line — one for data surfing and one for voice. Or your ISDN 2B channel just got split into one for voice and one for data.

There are a thousand variations on this theme of adding voice to the Internet and the Internet's World Wide Web segment. We're just beginning to explore them all.

Website See WEB SITE.

Webtone A slogan created by Sun Microsystems and adopted by Nortel in early 1998 in hopes of convincing the world that it (Nortel) is capable of delivering networks that combine the best elements of the World Wide Web and the global telephone network. Nortel's own definition of Webtone goes as follows: "Webtone represents rugged, secure, mission-critical networks optimized for IP (Internet Protocol) that offer customers access anywhere, at anytime; are always ready and always available; trusted with all information; scaleable and manageable; and deliver customer value to speed their success."

WebTV Well, it's finally here! Sony and Philips (Fall, 1996) have struck an alliance and licensed WebTV technology consisting of a set-top box, a wireless keyboard and a printer adapter. The set-top box costs $329 and contains a 112-MHz, 64-bit CPU. The TV set serves as the display. Internet access is provided through dial-up connection over the dial-up phone network to the WebTV Network at a monthly cost of $20 for unlimited usage. Content is provided by the WebTV Network, which also supports e-mail, help, and content lockout. Mitsubishi has taken a different approach, embedding the set-top box within its DiamondWeb TV. Sanyo and Samsung appear to be taking the same approach as Mitsubishi.

This approach to 'Net TV is a challenge to another concept of providing access through a cable modem which would be embedded in a set-top box. That box would serve as a communications controller/splitter, supporting simultaneous voice, data and TV. Regardless of the approach taken, issues abound, including lack of standards for the set-top boxes and cable modems.

Webware A term coined by Ray Horak, my Consulting Editor, for misleading content on a Web site. Great graphics, some animation, and neat audio can cause you to believe that the product or service matches the description on the Web site. Basically, it's interactive brochureware. I can cite examples — I'll bet you can, too. See BROCHUREWARE.

Webwench An employee given all the responsibility for a Web site without any of the authority (the opposite of a Webmaster).

Webzine Magazines that are published (i.e. made public) on the World Wide Web. Typically a Webzine is available for anyone to read who wants to visit the site the electronic magazine is located at. A Webzine is also called an e-zine or a Web-zine.

WECO Western Electric COmpany. The company is now called Lucent Technologies. It used to be the equipment manufacturing arm of AT&T, but in early 1996, AT&T spun it off to the public as a separate, publicly-traded company, called Lucent Technologies. Many oldtimers are sad that about the name change and new ownership. Western Electric had a wonderful reputation and is remembered with great fondness. It has an excellent reputation for high quality products and is still used as a brand name on some products.

Weight Test The test that buyers once applied to proposals, as well as purchases. According to the weight test, the weight of the proposal was directly related to the level of effort that went into it, which was directly related to the size of the company that prepared it and/or to the level of interest that the proposer had in acquiring your business (i.e. making the sale).

Weighted Average A call center term. A method of averaging several numbers in which some numbers are increased before averaging because they have more significance relative to the other numbers.

Weighted Call Value WCV. The average handling time of a call transaction. ACD vendors count this differently. Typically, a combination of the talk time and the after-call work or wrap-up time.

WESTAR Family of communications satellites owned and operated by Western Union.

Western Electric The telecommunications manufacturing subsidiary of AT&T that was divested from AT&T on September 30, 1996. Western Electric is now called Lucent Technologies. See Lucent Technologies.

Western Union International WUI. Acquired by MCI in 1982 to establish MCI in the International Telex and communications market. WUI is now part of MCI International.

WestNet One of the National Science Foundation funded regional TCP/IP networks that covers the states of Arizona, Colorado, New Mexico, Utah, and Wyoming.

Wet Circuit A circuit carrying direct current.

Wet Loop Powering Defined as local power (non-Span provided) with use of copper pairs (power is looped at the last repeater).

Wet T-1 A T-1 line with a telephony company powered interface.

Wetting Agent A chemical which reduces surface tension in a liquid, motivating the liquid to spread more evenly on a surface.

WFM Wired For Management. Intel's umbrella term for a set of management standards supported by hardware vendors. WFM communicates with network management software to help PCs send inventory data, manage power and reboot remotely. According to InfoWorld, a standard WFM-compliant PC will include a sensor to detect intruders; DMI 2.0 in the firmware; Wake on LAN 2.0 in the Ethernet chip or an Ethernet network interface card; ACPI (Advanced Configuration and Power Interface) and on the system board with software hooks to the OS, hard disk modem and monitor; and PXE (Preboot Execution Environment) on the Ethernet chip or the NIC (Network Interface Card). The idea of WFW is to give administrators more flexibility to handle clients (i.e. PCs) remotely through a central console with protocols that are vendor independent.

WFQ Weighted Fair Queuing. A variation on the CBQ (Class-Based Queuing) queuing technique used in routers. As is the case with CBQ, WFQ queues traffic in separate queues, according to traffic class definition, guaranteeing each queue some portion of the total available bandwidth. As is also the case with CBQ, WFQ recognizes when a particular queue is not fully utilizing its allocated bandwidth and portions that capacity out to the other queues on a proportionate basis. WFQ takes queuing to yet another level, portioning out available bandwidth on the basis of individual information flows according to their message parameters. See also CBQ, FIFO, RED, and ROUTER.

WFWG Windows For Workgroups. See WINDOWS.

WG An abbreviation for workstation, i.e. a computer on a desktop that isn't a server.

Whack See Wack.

Wheatstone Bridge An instrument for measuring resistances.

Whetstones How well does a computer work? Let's test it. The Whetstone benchmark program, developed in 1976, was designed to simulate arithmetic intensive programs used in scientific computing. It is applicable in CAD and other engi-

neering areas where floating-point and trigonometric calculations are heavily used. The Whetstone program is completely CPU-bound and performs no I/O or system calls. The speed at which a system performs floating point operations is measured in units of Whetstones per second or floating point operations per second (flops). Whetstone I tests 32-bit, and Whetstone II tests 64-bit operations. See also DHRYSTONES.

Whipsawing Whipsawing is an abuse of its dominant position by an incumbent telephone company operator in an unliberalised country. Here the incumbent uses its market power to insist on receiving a greater amount under the accounting rate system for receiving and terminating calls from an operator in a liberalised country than it pays to that operator for the same service in the other direction. This is harmful to consumers in the liberalised country because they pay more than they should do for calls. See also "One Way Bypass".

Whisper Technology A call comes into a call center. The voice response unit prompts the caller to the enter their account number. When the call is transferred to the agent, the VRU "whispers" the account number to the agent, who then manually types it into his computer. This technology is now obsolete, since VRUs can now transfer their account number directly into the agent's database and have the look up done automatically. And the call is transferred simultaneously.

Whistle Through A feature that allows CAMA (Centralized Automatic Message Accounting) identification from a PBX trunk to be extended through a local central Office (public exchange) connection to an outgoing trunk to a tandem office. See also CAMA.

White Board Also called a mushroom board or peg board. This is placed between termination blocks to support route crossing wire.

White Box A device which is built of standard components, often customized to the requirements of the end user. The device is packaged in a white box, rather than in a fancy box with the name and logo of a manufacturer printed all over it. White boxes (e.g., custom-built PCs from local vendors) typically cost less than do those of name-brand manufacturers, although the warranty is only as good as is the local system integrator/retailer. See also Black Box.

White Facsimile Transmission In an amplitude-modulated facsimile system, that form of transmission in which the maximum transmitted power corresponds to the minimum density of the subject copy. In a frequency-modulated system, that form of transmission in which the lowest transmitted frequency corresponds to the minimum density of the subject copy.

White Line Skip A facsimile transmission technique used to speed up the transmission time by bypassing redundant areas such as white space. (Also known as skip scan.)

White Noise A signal whose energy is uniformly distributed among all frequencies within a band of interest. Seldom occurring in nature, white noise is a useful tool for theoretical research. White noise is also used less scientifically to simply mean background noise. When the first digital PBXs came out, their intercom circuits were so "clean," they spooked users out who were used to some noise on the line. And some PBX manufacturers added a little "white noise" to their PBXs.

White Pages 1. In many countries, including the U.S., Canada and Australia, the phone company publishes two types of telephone directories. One called the "White Pages" lists all the subscribers in alphabetical order. The other, called the Yellow Pages, lists businesses by industry. On the Internet, the White Pages are the lists of Internet users that are accessible through the Internet.

White Paper Imagine an 8 1/2" x 11" stapled small booklet of 16 to 32 pages. Imagine that the paper is the quality of plain-paper photocopier paper, i.e. non-glossy and weighing in at 20lb to 24lb. Imagine that the paper is printed plainly in black ink, looks "honest" (i.e. not slick and glossy) and discusses technical issues, or contains case studies of user installations. Bingo, now you have something called a White Paper. Companies write them and distribute them to prospective customers in the hopes that the "knowledge" the White Papers contain will turn the customers more favorably to the company's products and ideas. Companies issue White Papers as part of their marketing and sales programs. But it's the "subtle" part of their programs. Done correctly, White Papers can actually be useful, explaining how how complex things work and what benefits they deliver. Done incorrectly, they look like bad marketing and turn potential customers off.

White Peak The maximum excursion of the video signal in the white direction at the time of observation.

White Signal In facsimile, the signal resulting from the scanning of a minimum-density area of the subject copy.

Whiteboard A device which lets you share images, text and data simultaneously as you speak on the phone with someone else. That someone might be in the next office. Or that someone might be 3,000 miles away. The transport mechanism might be a local area network or an analog phone line running a special modem designed for whiteboarding or it might be an ISDN digital line running special PC software and hardware. The concept of whiteboarding is new; there are no standards. As a result to do whiteboarding successfully, you typically need the same equipment (hardware and software) on either end. Whiteboarding has the potential to be one of the most successful "multimedia" applications around. Whiteboarding is a document-conferencing function that lets multiple users simultaneously view and annotate a document with pens, highlighters, and drawing tools. More advanced whiteboard programs handle multi-page documents and provide tools for delivering them as presentations.

Who-Are-You Code WRU. A control character which operates the answerback unit in a terminal (typically a telex terminal) for identification of sending and receiving stations in a network.

Whois 1. A command on some systems that reveals the user's name, based on that person's network username.
2. A tool of the InterNIC DNS (Domain Name Server). Whois allows anyone to query a database of people and other Internet entities, such as domains, network, and hosts. The data includes company/individual name, address, phone number and electronic mail address. If you have your own domain and thought your personal information was hidden from view, you are in for a shock. Don't believe me, check out www.internic.net. See also InterNIC and RWhois.

Whole Nine Yards The term "the whole 9 yards" came from World War fighter pilots in the Pacific. When arming their airplanes on the ground, the .50 caliber machine gun ammo belts measured exactly 27 feet, before being loaded into the fuselage. If the pilots fired all their ammo at a target, it got "the whole 9 yards."

Whole Person Paradigm This is one of the more fascinating telecom concepts in a while. General Magic created it as some sort of psychological basis for the product/s it is producing. Here's General Magic's definition:
A psychological or behavior model of needs that all people experience. This paradigm is the design center for General Magic's personal intelligent communication products and

services. It consists of three elements. 1. Remember - managing your internal agenda, such as things to do and people to see. 2. Communicate - maintaining relationships with your friends, family, and associates. 3. Know - getting information about the world.

WID Wireless Integration/Interface Device. Also refered to as a "Proctor" box (name of the vendor), Cell Trace Box (US West's name for it), and protocol converter. One of its functions is helping convert older cellular phone systems which support only old-style 911 service (i.e. no location transmitting) to the newer E911 service which will transmit the cellphone user's location to the correct public safety people. www.proctorinc.com

Wide Area Network WAN. An data network typically extending a LAN (local area network) outside the building, over telephone common carrier lines to link to other LANs in remote buildings in possibly remote cities. A WAN typically uses common-carrier lines. A LAN doesn't. WANs typically run over leased phone lines — from one analog phone line to T-1 (1.544 Mbps). The jump between a local area network and a WAN is made through a device called a bridge or a router. Bridges operate independently of the protocol employed. They will work, according to Jeff Weiss, of Cryptall Communications, with all present and expected future communications packages. Routers are specific to the protocol being employed. New routing software is needed for each new protocol or protocol deviation. See BRIDGE, ROUTER, CORPORATE NETWORK and DIGITAL HIERARCHY.

Wide Area Service Identifier WASI. Unique identifier for a business grouping of licensed facilities-based cellular service providers of Cellular Digital Packet Data (CDPD). It is used within CDPD for access control decisions.

Wide Area Telephone Service See WATS and 800 SERVICE.

Wide Characters 16-bit characters. See UNICODE.

Wide Frequency Tolerant Power Plant PBX power facilities are provided that will operate from AC energy sources which are not as closely regulated as commercial AC power. The wide tolerant plant will tolerate average frequency deviations of up to plus or minus 3 Hz or voltage variations of -15% to +10% as long as both of the conditions do not occur simultaneously. This feature permits operation with customer provided emergency power generating equipment.

Wide SCSI A type of SCSI that uses a 16- or 32- bit bus. It can transmit twice as much information as narrow SCSI.

Wideband The original definition for a channel wider in bandwidth than a voice-grade channel. Then it became a channel wider than 12 voice channels. Now, it means a transmission facility providing capacity greater than narrowband (T-1 at 1.544 Mbps), e.g. T-3 at 45 Mbps. many rich folks in Silicon Valley now have T-1 circuits into their home. This makes surfing the Internet and accessing the Web more pleasurable. But George Lucas, the renowned filmaker, has a T-3 in his house. He clearly is wideband. See also BANDWIDTH. Contrast with NARROWBAND and BROADBAND.

Wideband Modem A modem whose modulated output signal can have an essential frequency spectrum that is broader than that which can be wholly contained within a voice channel with a nominal 4-kHz bandwidth. A modem whose bandwidth capability is greater than that of a narrow band modem.

Wideband Packet Transport Transmission of addressed, digitized message fragments (packets) interleaved among the addressed fragments of other messages at a rate high enough to support general purpose telecommunications services.

Wideband Switch Switch capable of handling channels wider in bandwidth than voice-grade lines. Radio and TV switches are examples of wideband switches.

Wildcards Special characters you use to represent one or more characters in an MS-DCS filename. An asterisk (*) represents several characters and a question mark (?) represents a single character. For example, the command

ERASE *.BAK

would erase all the files with the suffix "BAK."

The command

ERASE *.?A?

would erase all the files with "A" as the middle letter in a three-letter suffix.

Wildfire The all-hearing, all-doing computer telephony slave from a company called Wildfire Communications, Lexington MA. The product uses very sophisticated voice recognition software so that its "master" (i.e. the user) can get Wildfire to take messages, find him, connect his calls, transfer his calls and act as a super intelligent on-line, computerized, 24-hour a day, never resting, all obedient secretary. A real breakthrough product, first introduced in the fall of 1994. And one deserving of its own definition in this illustrious dictionary.

WiLL A name Motorola uses for its Wireless Local Loop (WiLL) product, which was developed to serve the basic telephony needs of people in urban and difficult to reach rural areas. Cellular based, WiLL technology is intended to provide fixed telephony services in areas with little or no existing wireline telephone service or as a supplement to the existing wireline service. It uses very few cellular transmit/receivers — often just one at the end of the landline.

The WiLL system provides three major benefits to the telecom operator looking to expand their service area: more rapid deployment of telephone service; lower cost alternative to copper wire installation, and increased flexibility in system implementation and design. A WiLL system can be operational in weeks, compared to the huge amounts of time it would take to lay and install copper wire from an end office to each of the subscriber points in a typical local loop. Although WiLL is cellular-based, the system does not require a cellular switch. This makes the WiLL system a lower cost alternative to using "typical" cellular systems for fixed telephony applications because the total system outlay costs as well as associated backhaul and maintenance costs are reduced.

WiLL has three elements: the WiLL System Controller (WiSC), a Digital Loop Concentrator (DLC), and a Motorola cellular base station. It interfaces directly to the central office switch via 2-wire analog subscriber loops.

Willful Intercept The act of intercepting messages intended for a station experiencing a line or equipment malfunction.

WIMP Interface Stands for Windows, Icons, Mouse and Pointing Device or Pull-down menus. A derogatory reference to GUI. Some people think WIMP is on the way out. See also GRAPHICAL USER INTERFACE.

WIN 1. Wireless In-building Network. WIN is a technology from Motorola which uses microwaves to replace local area network cabling.

2. Wireless Intelligent Network, a standard, destined to become the successor to both IS-41 and GSM. IS-41 "Rev. D" is often used interchangeably with WIN.

WIN32 API A 32-bit application programming interface for both Windows 95 and Windows NT. It updates earlier versions

of the Windows API with sophisticated operating system capabilities, security, and API routines for displaying text-based applications in a window.

WIN95 See WINDOWS 95.

Winch A machine for pulling cable into conduit (in the street or in the building) or duct liner. A winch has a rotating drum that winds up the pulling line.

Winchester Disk A sealed hard disk. The Winchester magnetic storage device was pioneered by IBM for use in its 3030 disk system. It was called Winchester because "Winchester" was IBM's code name for the secret research project that led to its invention. A Winchester hard disk drive consists of several "platters" of metal stacked on top of each other. Each of the platter's surfaces is coated with magnetic material and is "read" and "written" to by "heads" which float across (but don't touch) the surface. The whole system works roughly like the old-style Wurlitzer jukebox. There are several advantages to a Winchester disk system:

1. It can store, read and write enormous quantities of information. Some Winchesters have a capacity of over 100 megabits; 2. You can access information on a Winchester faster than on most computer storage medium (RAM and ROM are obviously faster); and 3. Winchesters are reliable and relatively inexpensive. There are also disadvantages: 1. They are very sensitive to rough handling (they hate being moved); 2. They are very sensitive to the organization of their directory track (lose that and you're in big trouble); and 3. When Winchesters "crash" (i.e. the heads touch the surface of the rotating platters), you can lose an enormous amount of precious data.

Winding Coils of wire usually found in transformers and used to boost inductance.

Window 1. A band, or range, of wavelengths at which an optical fiber is sufficiently transparent for practical use in communications applications. Each window roughly corresponds to a visible color of light in the overall light spectrum. See also DWDM, Lambda, SONET and WDM.

2. A flow-control mechanism in data communications, the size of which is equal to the number of frames, packets or messages that can be sent from a transmitter to a receiver before any reverse acknowledgment is required. It's called a pacing group in IBM's SNA.

3. A box on the CRT (cathode ray tube) of your personal computer or terminal. A software program is running inside the box. It's possible with new "windows" software to run several programs simultaneously, each accessible and visible through the "window" on your CRT.

4. A technique of displaying information on a screen in which the viewer sees what appears to be several sheets of paper much as they would appear on a desktop. The viewer can shift and shuffle the sheets on the screen. Windowing can show two files simultaneously. For example, in one window you might have a letter you're writing to someone and in another window, you might have a boilerplate letter from which you can take a paragraph or two and drop it in your present letter. Being able to see the two letters on the screen makes writing the new letter easier.

5. Video containing information or allowing information entry, keyed into the video monitor output for viewing on the monitor CRT. A window dub is a copy of a videotape with time code numbers keyed into the picture.

6. A video test signal consisting of a pulse and bar. When viewed on a monitor, the window signal produces a large white square in the center of the picture.

Window Control A credit or token scheme in which a limited number of messages or calls are allowed into the system.

Window Segment Size A parameter used to control the flow of data across a connection. A wireless term.

Window Size The minimum number of data packets that can be transmitted without additional authorization from the receiver.

Window Treatment You take the world's most beautiful window and you screw it up with expensive stuff you affix around it. Paula Friesen invented the term.

Windowing A technique of running several programs simultaneously — each in running a separate window. For example, in one window you might run a word processing program. In another, you might be calculating a spreadsheet. In a third, you might be picking up your electronic mail.

WinInet API The Microsoft Win32 Internet functions. These functions provide Win32 applications with access to common Internet protocols. These functions pluck out the heart of the Internet's Gopher, FTP, and HTTP protocols and turn them into an application programming interface (API). This provides a straightforward path to making applications Internet-aware.

WinISDN WinISDN is ISDN*Tek's (San Gregorio, CA — 415-712-3000) API for talking to internal ISDN modems. It supports all of the high level functions for call setup and answering on an ISDN modem. Most of the more popular internal ISDN modems support WinISDN. One of the ways WinISDN helps increase throughput is by handling data transfers in large blocks rather than one byte at a time. The overhead on single byte transfers is much higher than handling a single block.

Windows A Microsoft operating system that hides the cryptic DOS system of typed commands behind a graphical facade (also called a Graphical User Interface, GUI). Windows let you issue commands (i.e. run programs and complete tasks within programs) by pointing (with or without a mouse) at symbols or menu items and clicking, or hitting "Enter." Most Windows programs have the same "look and feel" to them. So issuing commands becomes almost intuitive. The idea is that "use one Windows program, you can use them all." Sort of. The latest versions of Windows were 3.1 and 3.11 (also called Windows for Workgroups). These versions contained two big improvements over 3.0 — namely OLE (Object Linking and Embedding) and DLL (Dynamic Link Library). Windows 3.1 and 3.11 is now about to be obsoleted by a new version, called Windows 95. See DLL, OLE, WINDOWS 95, WINDOWS FOR WORKGROUPS, WINDOWS NT and WINDOWS TELEPHONY.

Windows 95 Windows 95 is an operating system from Microsoft which first shipped on August 24, 1995. The August, 1995 issue of our Computer Telephony Magazine said the following about Windows 95:

Win95 is the first Windows operating system designed for communications. It does for modems and phones (of all sorts — from single line analog to proprietary ISDN phones) what Windows did for printing — insulate the suffering user from the idiocies of device drivers. Win95 does wonders for fax, for sending color pictures of the kids to Grandma and for making the world one gigantic personal local area network. And, for the first time ever, an operating system is treating voice as it should be treated — another media stream no different, no more complex than printing a pretty document.

Computer Telephony (voice, fax, e-mail) is a major focus of Windows 95. It will have a revolutionary impact on computer telephony's desktop interface and CT-enabling hardware, from simple off-the-shelf SOHO apps built on inexpensive multimedia modems to full-blown unified-messaging systems humming on the LAN. One of the key improvements of Win95 over Windows 3.xx is the replacement of the latter's monolithic communications driver (COMM.DRV) with a far more flexible communications architecture that splits communication tasks into three primary areas: Win32 communications APIs and TAPI; the universal modem driver; and comm port drivers. VCOMM is the new communications device driver. It protectmodes services and lets Windows apps and drivers use ports and modems. To conserve system resources, comm drivers are loaded into memory only when in use by an app. VCOMM also uses new Plug and Play services in Windows 95 to help configure and install comm devices. The Win32 communications APIs provide an interface for using modems and comm devices in a device-independent fashion. Applications call the Win32 APIs to configure modems and perform data I/O through them. Through TAPI, meantime, apps can control modems or other telephony devices.

The universal modem driver (Unimodem) is a layer that provides services for data and fax modems and voice so that users and app developers don't have to learn or maintain difficult modem AT commands to dial, answer and configure modems. Rather, Unimodem does these tasks automatically by using mini-drivers written by modem vendors. Unimodem is both a VCOMM device driver and a TAPI service provider. Other service providers (like those supporting things such as an ISDN adapter, a proprietary PBX phone or an AT-command modem) can also be used with TAPI. Port drivers are specifically responsible for communicating with I/O ports, which are accessed through the VCOMM driver. Port drivers provide a layered approach to device communications.

For example, Win95 provides a port driver to communicate with serial communications and parallel ports, and other vendors can provide port drivers to communicate with their own hardware adapters, such as multiport voice and fax cards. With the port driver model in Win95, it's not necessary for vendors to replace the communications subsystem as they did in Windows 3.xx, whose COMM.DRV forced people to completely replace the comm driver if something new was needed by a hardware device. The Win95 driver means we no longer have to be "hard wired" to a 16550 UART. Previous versions of Windows assumed this type of port hardware. This means new ports like USB (Universal Serial Bus) can be slipped in with full apps compatibility.

Besides this strong attempt at "virtualizing" many of the communications hardware interface problems that plagued Windows developers in the past, Win95 also strengthens itself considerably by acknowledging voice as a data type, filtered into its Plug and Play world of communication device compatibility via the Windows Telephony API (TAPI). TAPI-aware apps, for example, no longer need to provide their own modem support list because interaction with a modem is now centralized by the OS. All comm services provided with Win95 use these services. (The analogy is printing under Windows 3.xx.)

TAPI provides a standard way for communications apps to control telephony functions for data, fax and voice calls. It manages all signaling between a computer and phone network, including basic functions such as dialing, answering and hanging up a call. It also includes supplementary call-handling things such as hold, transfer, conference and call park that are often found in PBXs, Centrex, ISDN and other phone systems. In general, TAPI services arbitrate requests from apps to share comm ports and devices. Win32-based apps can use TAPI to make outgoing calls while others are waiting for incoming calls. Of course, only one call can be performed at a time, but users no longer have to close apps that are using the comm port.

TAPI does not need local hardware. It can also use drivers that work on a LAN, which support multiple systems today (e.g. Genesys, Dialogic's CT Connect, Northern Telecom's Tmap, etc.) Microsoft is making the client side ubiquitous and is planning on dropping the server side in (i.e. Windows NT Server) shortly.

MODEMING: Win95 lets you install and configure a modem once to work for all communications apps, just as you do for a printer. Benefits: Centralized modem and COMM port configuration through the "modems" option in the control panel for all comm apps created for Windows 95; Support for hundreds of modems, including automating the detection of them (again, just like Windows did with printers); Modem connections and configuration using point-and-click options rather than annoying AT commands.

The Windows OS includes three tools here: 1. HyperTerminal. This lets you connect two computers through a modem and TAPI for transferring files. It also automatically detects data bits, stop bits and parity. 2. Phone Dialer. This lets you use your computer to dial numbers for voice calls. It includes a phone dialpad, user-programmable speed dials and a call log. 3. Microsoft File Transfer. This lets you send and receive files while talking on the phone. This works with VoiceView-enabled modems.

E-MAILING: Also in Win95 is its included Microsoft Exchange client capability. Exchange is a messaging app that retrieves messages into one inbox from many kinds of messaging service providers, including, Microsoft Mail, The Microsoft Network and Microsoft Fax. With Microsoft Exchange client, you can: send or receive e-mail in a Win95 workgroup; include files and objects created in other apps as part of messages; use multiple fonts, font sizes, colors and text alignments in messages (via an included OLE-compatible text editor); create a personal address book or use books from multiple service providers; and create folders for storing related messages, files and other items.

According to Microsoft, Exchange client will work with any electronic mail system or unified messaging app that has a MAPI service provider, which architects very similar to the TAPI schematic. The MAPI service provider specifies all the connections and addressing settings needed to talk with a mail server on one end and with the Exchange client on the other. MAPI is a set of API functions and OLE interface that lets messaging clients, such as Microsoft Exchange, interact with various message service providers, such as Microsoft Mail, Microsoft Exchange Server, Microsoft Fax and various computer telephony servers running under Windows NT server. Overall, MAPI helps Exchange manage stored messages and defines the purpose and content of messages — with the objective that most end users will never know or care about it.

FAXING: As part of Windows 95, Microsoft included a new technology called Microsoft Fax. With this embedded Group 3-compatible software, users with modems and running Microsoft Fax can exchange faxes and editable files (pictures, binary files, *.EXE software, etc.) as easily as printing a document or sending an electronic mail message. To use it, you must install Microsoft Exchange. Microsoft integrated the two as a messaging application programming interface (MAPI)

service provider. All faxes sent to Microsoft Fax are received in the Exchange universal inbox. You can send a fax by composing a Microsoft Exchange message or by using the Send option on the File menu of a MAPI-compatible application (such as Microsoft Excel or Microsoft Word). In addition, Microsoft Fax includes a fax printer driver so that users can "print to fax" from within any Windows-based application.

Key features: 1. Delivery By Address Type. The MAPI service provider architecture lets you mix different types of recipients in the same message. For example, it's possible to send a message simultaneously to destination addresses in Microsoft Mail, CompuServe, Internet, normal fax and Microsoft Fax as long as profiles for these destinations have been defined within Microsoft Exchange. A recipient's fax address can be selected from the Microsoft Exchange Personal Address Book or the fax can be addressed by using an address that you use just once, such as [fax:555-1212].

1. Binary File Transfer (BFT). Microsoft Fax supports Microsoft At Work BFT, which makes it possible to attach an editable document to a Microsoft Exchange mail message. These editable documents can be sent to users of Windows 95, Windows for Workgroups 3.11, and other Microsoft FAX BFT enabled platforms.

2. Security. Microsoft Fax lets you securely exchange confidential documents by using public key encryption or digital signatures. Any security specified by user is applied before the message is passed to the modem or connected fax device.

3. Network Fax Service. You can install a fax device in one computer and share it with other users within a workgroup. Individual computers can have their own fax devices installed and still use the shared fax device.

4. Microsoft Fax Viewer. The Microsoft Fax Viewer displays outgoing fax messages that have been queued to a local fax modem or to a Microsoft Fax network fax that are queued for transmission. You can also browse multipage faxes in thumbnail or full-page views.

5. Connecting to Fax Information Services. Microsoft Fax can connect to fax-on-demand systems by using a built-in, poll-retrieve feature that allows you to retrieve rendered faxes or editable documents from a fax information service.

6. "Best Available" Fax Format. When you make a fax connection in Windows 95, Microsoft Fax queries and exchanges its fax capabilities with the recipient. This exchange of capabilities determines whether the recipient is a traditional Group 3 fax machine, which can only receive rendered faxes, or if the recipient has Microsoft Fax capabilities, and can receive editable files. Windows 95, Windows for Workgroups 3.11 and Microsoft At Work fax platforms are all capable of receiving binary files and traditional faxes. Perks:

If the receiving fax device supports Microsoft Fax capabilities and an editable document is attached to a Microsoft Exchange message, then the file is transferred in its native format. If the receiving fax device is a traditional Group 3 fax machine, then Microsoft fax converts the document to the most compressed type of fax supported by the machine (MH, MR or MMR compression type) and transmits the image by using the best available communications protocol supported by the mutual connection (that is, V.17, V.29 or V.27). If Microsoft Fax sends a noneditable fax to another Microsoft Fax user, then the fax is transmitted by using the Microsoft At Work rendered fax format. This special format is much more compressed, on average, than Group 3 MMR. Therefore, the exchange of noneditable faxes between Microsoft Fax users is always faster than between Group 3 fax machines.

Overall, you can send faxes either by using the mail client or the Microsoft Fax printer driver. In each case, the message is sent to the Microsoft Fax service provider by using MAPI. If you sent the message from a mail client, it might contain text, embedded OLE formats and attachment to the mail message. MAPI allows messages to be preprocessed based on the transport protocol used to send them. The transport protocol chooses the correct modem connection, uses TAPI to create a dial string and sends the message into a fax form to be printed by a fax machine. The rendered format is attached to the original message as a message property and is deleted either when the message is sent or when the transport protocol tries to send the message but determines it cannot.

If the message does not have to be rendered, the message is converted from its original binary format to a line image (also called a linearized form), and then it is compressed. After the message is submitted, the transport protocol determines what type of recipient the message is intended for through TAPI subkey values.

Windows 98 Windows 98 is the successor to Windows 95. See WINDOWS 95.

Windows Application A term used in this document as a shorthand term to refer to an application that is designed to run with Windows and does not run without Windows. All Windows applications follow similar conventions for arrangement of menus, style of dialog boxes, and keyboard and mouse use.

Windows CE A smaller version of Microsoft's Windows operating system to be used for a range of mobile handheld communications, computing or entertainment devices. See also Microsoft At Work, an earlier attempt by Microsoft at making an operating system for devices other than PCs.

Windows Character Set The character set used in Windows and Windows applications. Most TrueType fonts have a set of about 220 characters.

Windows For Workgroups Windows for Workgroup is an obsolete local area networked version of Microsoft Windows operating system version 3.1 that offered integrated file sharing, electronic mail (Microsoft Mail) and workgroup scheduling (Schedule+), thus bringing the graphical user interface to the workgroup. Windows For Workgroups also has Network DDE, which allows users to create compound documents that share data across network. Most importantly, Windows for Workgroups 3.11 lets you do 32-bit disk access and 32-bit file access. Windows 95 superceded Windows for Workgroups.

Windows MetaFile WMF. A method of encoding files. Other methods include EPS, PCX and TIFF.

Windows NT Windows New Technology, now called Windows 2000, is a 32-bit operating system from Microsoft, designed to replace Windows95 and MS-DOS. As an operating system, Windows NT is targeted at the top 10% "power" users who need the power of a big, powerful operating system. Here are the main advantages of Windows NT, as explained by Microsoft:
• Interoperability. Windows NT delivers support for open computing benefits through its protected subsystem architecture. Windows NT was also designed to be protocol independent. As such it will interoperate with all leading network systems, regardless of the native protocol of the system.
• Portability. Windows NT was designed to be portable across a variety of hardware systems. The Hardware Abstraction Layer (HAL) limits and isolates the amount of code necessary to port Windows NT to a new platform.

Windows NT will run on processors other than those made by Intel. MS-DOS, for example, doesn't.
• Scalability. Windows NT scales to work on both single and multi processor computer systems. This scalability gives users the flexibility to implement their own solutions, today or over time, on machines that meet the performance needs of sophisticated client server solutions.
• System Management. Windows NT supports SubNetwork Access Protocol (SNMP) and NetView network management standards.
• Published Interfaces. The interfaces to the Windows NT operating system are fully documented and published. Software developers are free to add functionality to the system based on their interface definitions.
• Support of Industry Standards. These include POSIX.1, OSF DCE, TCP/IP and WOSA, which is Microsoft's Windows Open Services Architecture. WOSA is a standard set of interfaces to connect a variety of applications with a range of back-end devices and services, such as messaging, telephony, databases, etc. Windows Telephony is part of WOSA.

Windows NT File System NTFS. An advanced file system designed for use specifically with the Windows NT operating system. NTFS supports file system recovery and extremely large storage media. It also supports object-oriented applications by treating all files as object with user-defined and system-defined attributes.

Windows Open Services Architecture See WOSA.

Windows Telephony Introduced in the spring of 1993 jointly by Microsoft and Intel, Windows Telephony is a piece of software called a Windows Telephony DLL AND two standards. The first standard is the Service Provider Interface (SPI). If a hardware manufacturer's product honors that SPI, that product can happily talk to the Windows Telephony DLL. The second standard is called the Application Programming Interface and it is directed at software developers who write applications programs. If those developers' programs adhere to the API, they can take advantage of the Windows Telephony DLL to drive whatever telephony devices or services adhere to the SPI. The Windows Telephony API is affectionately called TAPI. DLL stands for Dynamic Link Library. It is a Windows feature that allows executable code modules to be loaded on demand and linked at run time.

Windows Telephony should bring about an explosion of shrink-wrapped Windows based telephone software applications — from simple personal rolodexes to power dialers, to customized phone systems for banks and for bakers. It should also bring about an explosion of new telephony hardware devices — from telephones that look more like PCs than phones, to PCs that are phones, to blackbox telephony devices that hook to laptops and transform hotel phones.

Windows Telephony effectively removes earlier overwhelming barriers to creating PC-driven telephony applications, namely the wide enormity of telephony "network" services — from the many telephone company interfaces (POTS to T-1), to the many more proprietary interfaces behind dozens of proprietary PBXs, key systems and hybrid phone systems.

The goal is to bundle the Windows Telephony DLL in the next major release of Windows, sometime in 1994. It will also be included in Windows NT and Windows NT Server. Although the Windows NT code may be different, the API and SPI interfaces will be the same, thus causing no re-write of software code or necessitating redesign of telephony hardware.

The original work on the Windows Telephony DLL was done by Herman D'Hooge, a senior software architect with Intel's

Architecture Development Lab in Hillsboro, Oregon. The final effort is a result of joint development effort with Microsoft, where the team was headed by Charles Fitzgerald. It also includes input from 40-odd companies — including virtually all major telecom switch vendors and several major telephony developers.

The goal of joint Microsoft/Intel Windows Telephony is to get rid of the bottleneck to bringing the power of the PC to telephony. Intel and Microsoft believe that the bottleneck exists because of two factors:

First, it has been incredibly difficult to interface to the variety of telecom switches in existence today. For example, no manufacturer's switch will talk to another's manufacturer's proprietary phone.

Second, it has been incredibly redundant and time consuming for software to talk to the various switches. The big analogy is word processing in the old days. In those days, each word processing software company could easily spend 99% of his R&D budget writing drivers to get his program to work with yet another new printer. That is no longer necessary under Windows. Windows takes care of interfacing the printers. All you have to do, as a developer is to make sure you conform to Windows specs.

According to Microsoft, Windows Telephony products will include ones, such as those including:

• Visual interface to telephone features.

• Personal communication management. With a graphical user interface, people can have their PCs handle incoming telephone calls, automatically controlling which calls reach them. For example, people will be able to ensure that they receive important calls by requesting that certain calls be forwarded automatically to locations at which they expect to be working. * Telephone network access. A personal information management can be used not only to look up phone numbers, but also to actually place calls to those numbers.

• Integrated messaging. People will be able to check their messages — electronic mail, fax mail and voice mail — from a single place, namely their telephony empowered PC. Also voice mail messages can be accessed randomly, which is far more efficient than the serial access provided on most telephone based voice mail-mail systems.

• Integrated meetings. Here's Microsoft's explanation: "One of the most attractive capabilities of the computer is that it can store, communicate and present information that spans the entire spectrum of media — text, data, graphics, voice and video in any combination. By itself, the telephone can communicate and present voice information only. By combining the functions of the computer and the telephone, people in geographically separate locations can participate in interactive meetings and share visual as well as audio information. That means they can hold meetings over the telephone network that are nearly as rich in information content as in-person meetings."

See AT WORK, FAX AT WORK, TELEPHONY SERVICES, WINDOWS TELEPHONY SERVICES, WINDOWS TOOLKITS and WOSA.

Windows Telephony Services Here is Microsoft's definition: "The Windows Telephony services are provided as a WOSA (Windows Open Services Architecture) component. It consists of both an application programming interface (API) used by applications and a service provider interface (SPI) implemented by service providers.

The focus of the API is to provide "personal telephony" to the Windows platform. Telephony services break down into Simple Telephony services and Full Telephony services. Simple Telephony allows telephony-enabled applications to be easily created from within these applications without these apps needing to become aware of the details of the Full Telephony services. Word processors, spreadsheets, data bases, personal information managers can easily be extended to take advantage of this.

Complete call control is only possible through the use of the Full Telephony services. Applications access the Full Telephony API services using a first-party call control model. This means that the application controls telephone calls as if it is an endpoint of the call. The application can make calls, be notified about inbound calls, answer inbound calls, invoke switch features such as hold, transfer, conference, pickup, park, etc., detect and generate DTMF for signaling with remote equipment. An app can also use the API to monitor call-related activities occurring in the system.

The fact the API presents a first-party call control model does not restrict its use to only first-party telephony environments. The Windows Telephony API can be meaningfully used for third-party call control.

The API provides an abstraction of telephony services that is independent of the underlying telephone network and the configuration used to connect the PC to the switch and phone set. The API provides independent abstractions of the PC connections to the switch or network and the phone set. The connection may be realized in a variety of arrangements including pure client based wired or wireless connections, or client/server configurations using some sort of local area network.

The Telephony API by itself is not concerned with providing access to the information exchanged over a call. Rather, the call control provided by the API is orthogonal to the information stream management. The Telephony API can work in conjunction with other Windows services such as the Windows multimedia wave audio, MCI, or fax APIs to provide access to the information on a call. This guarantees maximum interoperability with existing audio or fax applications.

The Telephony API defines three levels of service. The lowest level of service is called Basic Telephony and provides a guaranteed set of functions that corresponds to "Plain Old Telephone Service" (POTS - only make calls and receive calls). The next service level is Supplementary Telephone Service providing advanced switch features such as hold, transfer, etc. All supplementary services are optional. Finally, there is the Extended Telephony level. This API level provides numerous and well-defined API extension mechanisms that enable application developers to access service provider-specific functions not directly defined by the Telephony API.

Windows Toolkits Windows toolkits are libraries of code that implement the graphical user interface objects that every software application uses. The toolkits save time by eliminating the need for software developers to re-implement the same code repeatedly for each application. Toolkits also have the benefit of consistent user interface implementation across all applications that use the toolkit. See also WINDOWS.

Wink A signal sent between two telecommunications devices as part of a hand-shaking protocol. It is a momentary interruption in SF (Single Frequency) tone, indicating that the distant central office is ready to receive the digits that have just been dialed. In telephone switching systems, a single supervisory pulse. On a digital connection such as a T-1 circuit, a wink is signaled by a brief change in the A and B signaling bits. On an analog line, a wink is signaled by a change in polarity (electrical + and -) on the line.

Wink Operation A timed off-hook signal normally of 140 milliseconds, which indicates the availability of an incoming register for receiving digital information from the calling office. A control system for phone systems using address signaling.

Wink Pulsing Recurring pulses of a type where the off-pulse is very short with respect to the on-pulse, e.g., on key telephone instruments, the hold position (condition) of a line is often indicated by wink pulsing the associated lamp at 120 impulses per minute, 94 percent make, 6 percent break (470 ms on, 30 ms off).

Wink Release On most modern central offices when the person or device at the other end hangs up, your local central office will send you a single frequency tone. That tone is called wink release. Such a tone can be used to alert a data device that the device at the other end has hung up. (Remember it can't tell by just listening — like you and me.) When a data device hears a wink release, it usually takes it as a signal to hang up also.

Wink Signal A short interruption of current to a busy lamp causing it to flicker. Indicates there is a line on hold.

Wink Start Short duration off hook signal. See WINK OPERATION.

WINS Windows Internet Name Service. A name resolution service that resolves Windows networking computer names to IP addresses in a routed environment. A WINS server, which is a Windows NT Server computer, handles name registrations, queries, and releases.

Winsock 2 Winsock stands for Windows Sockets. Winsocks are standard APIs between Microsoft Windows (3.1, 95 and NT) application software and TCP/IP protocol software (also called a protocol stack). Winsock 2 is a network programming interface at the transport level in the ISO reference model. It is being defined by an open, industry wide workgroup, called the Winsock Forum.

Wintel A combination of the words Windows and Intel. Wintel refers to PCs that run Microsoft Windows and use Intel microprocessors. Wintel PCs are by far the biggest-selling PCs, amounting to around 80% of all PCs sold. Observers of the PC industry refer to the "Wintel standard" to refer the phenomenon of such high market dominance by one type of PC.

Wipe A transition between two video signals that takes the shape of a geometric pattern. Used also in PowerPoint.

WIPO World Intellectual Property Organization. An intergovernmental organization with headquarters in Geneva, Switzerland, WIPO is an agency of the United Nations. WIPO is responsible for the promotion of the protection of intellectual property throughout the world through cooperation of its member states, of which there were 161 as of February 1997. WIPO also is responsible for the administration of various multilateral treaties dealing with legal and administrative aspects of intellectual property. In 1998, WIPO convened an international process to develop recommendations concerning the intellectual property issues associated with Internet domain names, including dispute resolution. See also Domain Name, gTLD, Intellectual Property and TLD.

Wire Center The location where the telephone company terminates subscriber outside cable plant (i.e. their local lines) with the necessary testing facilities to maintain them. Usually the same location as a class 5 central office. One or more Switching Entities which serve the plant facilities through a single main frame (or two or more main frames joined by the cables), regardless of the number of buildings involved. A wire center might have one or several class 5 central offices, also called public exchanges or simply switches.

A customer could get telephone service from one, several or all of these switches without paying extra. They would all be his local switch.

Wire Center, Multi-Entity A North American telephone company term. A wire center which has two or more entities serving the plant facilities through a single main frame, (or two or more main frames joined by cables), regardless of the number of buildings involved. The wire center may include entities of various switching types and combinations, e.g., 3 entities, entity 1, IAESS, entity 2, 5ESS-2000, entity 3, DMS-100.

Wire Center Serving Area That area of an exchange served by a single wire center.

Wire Concentrator A conduit; a pipe within which a large number of individual wires are routed through.

Wire Pair Two separate conductors traveling the same route, serving as a communications channel.

Wire Printer A matrix printer which uses a set of wire hammers to strike the page through a carbon ribbon, generating the matrix characters.

Wire Running Tools Tools that help you run wire in and around a building. The most common form of wire running tools that help you fish wire through hollow drywalls.

Wire Speed The rate at which bits are transmitted over a cable. Ethernet's wire speed is 10 Million bits per second. Wire speed is not transmission speed. That depends on many factors, including how many devices are transmitting simultaneously on the same cable. See MULTIPLEX.

Wire Stripper A tool which takes the insulation off a wire without hurting the inside metal conductor.

Wire Tap The attaching to a phone line of a piece of equipment whose job is to record all conversations on that phone line. Wire taps are illegal. Law enforcement agencies use them, but must receive authorization from a court to apply the tap. Such authorizations are given if the law enforcement agency argues that applying the tap will prevent crime or help bring a suspected criminal to justice. Wire taps are not authorized lightly. See also TRAP and TRACE.

Wire Telephony The transmission of speech over wires.

Wire Wrap and Solder Soldering and wire wrap dominated early cable connections. Some old buildings still have large boards of wire wrapped or soldered connections. Wire wrap is still used in telephone company-related applications, but solder for cross-connects is obsolete and not seen today.

Wired For Capacity The wired-for capacity represents the upper limit of capacity for a particular configuration. To bring to a phone system to its "wired for capacity," all that's necessary is to fill the empty slots in the system's metal shelving (its cage) with the appropriate printed circuit boards. "Wired-for Capacity" is a marginally useful term, giving little indication of the type of printed circuit boards — trunk, line, special electronic line, special circuit, etc. — that can be installed. And many PBXs allow only their printed circuit boards to go into assigned slots. Your PBX cabinet might, for example, have plenty of empty space for extra printed circuit boards, but it may not have any more space for boards which service electronic phones. Thus, it is effectively maxed out.

Wired For Management WFM. See WFM.

Wired Logic A required logic function implemented in hardware, not software.

Wireless Without wires. An English and Australian word for radio and, now in the U.S., a phone system that operates locally without wires. Cellular is wireless in the strictest sense of the term. But "wireless" has come to mean wire-less systems that work within a building.

How does wireless work? I have excerpted the following from the excellent web site, www.wirelessdimension.com, which allows you to compare cell phone rates. "Wireless phones transmit conversations and data using radio waves rather than copper wires. The oldest wireless networks use analog technology, while many newer networks use the same digital technology as CD-ROMs. The following are the major U.S. consumer wireless standards:

Analog cellular operates in the 800MHz frequency range and is available across 95 percent of the United States. The operating system (called the air interface) for analog is called Advanced Mobile Phone Service (AMPS).

Digital cellular shares the 800MHz frequency band with analog and is usually available where analog service is offered. Several incompatible air interfaces are used to implement digital cellular networks, including Code Division Multiple Access (CDMA) and Time Division Multiple Access (TDMA). Personal Communications Service (PCS) is an all-digital service that operates in the 1,900MHz frequency range and is available in metropolitan areas. PCS networks are CDMA, TDMA or global system for mobile communications (GSM).

Wireless handsets generally work on just one of the three standards listed above, but some new handsets work on both analog cellular and digital cellular networks, or on both PCS and analog cellular networks. Some PCS handset makers are releasing "tri-mode" phones that work on PCS, digital cellular and analog cellular networks. When you turn on your wireless phone, it seeks out a signal from the nearest cellular antenna. The antennas are called 'cellular" because they're arranged in a honeycomb fashion across a region. As you move around with your phone, a network computer automatically hands off your call to the nearest antenna _ even if you're traveling down the freeway at 70 mph _ to maintain a connection. If you move out of range of the array of antennas or into a place where radio frequencies may be blocked (such as an underground parking garage), you'll lose your connection. Each cellular antenna is linked to a mobile telephone switching office (MTSO), which connects your wireless call to the local "wired" telephone network. MTSOs are owned by the wireless carriers. The wired telephone network charges you a small fee, often about 3 cents per minute, for "completing" your call. This charge appears on the monthly statement you get from your carrier, or is included in the per-minute charge applied to pre-paid wireless plans.

Wireless Access Controller The first component in an in-building wireless phone systems is the wireless access controller. It does many things. It provides access to the host network, be it a host PBX or the public switched telephone network (including Centrex). The access controller also manages the picocellular infrastructure of the wireless system through connections to the radio base stations. In the case of a Northern Telecom wireless business systems, base stations are connected to the controller via 144-kilobit-per-second (kbit/s) digital links that offer 2B+D interface connectivity. This digital connectivity (two 64-kbit/s channels for voice and data, and one 16-kbit/s channel for signaling information) provides the high-speed signaling capability needed by the controller to offer advanced business services and to manage mobility across several base stations. These digital links also make it possible to enhance radio system capacity by having the controller synchronize all base stations.

The controller software structures are designed so that untethered personal directory numbers and physical ports (specific interface circuits wired to a particular location) are

dynamically associated at every call and at every hand-off to another base station. This dynamic assignment makes it possible for the same personal or group directory number to be used for a variety of wireless and wireline terminals, irrespective of location. The controller also handles user registration, roaming, and hand-off.

Roaming which is the capability to redirect incoming calls to the appropriate base station, is accomplished through a combination of radio protocols, system software, and databases. The databases make it possible to locate portable terminals, through various broadcasting or polling schemes, without incurring excessive search delays.

Hand-off, on the other hand, is the capability needed in order to cope with the fact that a user will continuously move from one location, and hence one cell, to another, while communicating. As this happens, the link must be maintained in a manner transparent to the end user, always maintaining communications with the strongest base station signal in the neighborhood of the portable terminal. The controller monitors the radio signal strength of the portable and, when the signal weakens, switches it to a base station with a stronger signal. It then switches the communications link from the former base station to the new one and signals the terminal to begin radio communication on the new channel. Interference could be caused, for example, by other portable terminals in the same cell or an adjacent cell, or by external influences, such as nearby traffic or people moving partitions in an office. In such cases, the base station redirects the call rapidly to a less noisy channel in the same cell or an adjacent cell.

Wireless Application Protocol Forum In January, 1998, Ericsson, Motorola, Nokia and Unwired Planet announced the establishment of the Wireless Application Protocol Forum Ltd. This non-profit company will administer the worldwide WAP specification process and facilitate new companies contributing to WAP specification work. According to the press release announcing the establishment of the Forum, the Wireless Application Protocol (WAP) is targeted to bring Internet content and advanced services to digital cellular phones and other wireless terminals. WAP Forum aims to create a global wireless protocol specification that works across differing wireless network technology types, for adoption by appropriate industry standards bodies. Applications using WAP will be scaleable across a variety of transport options and device types. A common standard offers potential economies of scale, encouraging cellular phone and other device manufacturers to invest in developing compatible products. Cellular and other wireless network carriers and content providers will be able to develop new differentiated service offerings as a way to attract new subscribers. Consumers will benefit through more and varied choices in mobile communications applications, advanced services and Internet access. In addition to the four founding partners, new members are now welcome to join WAP Forum. Members may contribute to the current specification work, participate in driving the continuing evolution of WAP and nominate and elect additional directors to the board of WAP Forum. In order to become members of WAP Forum, interested companies need to apply to join. All the details including the application form can be found at www.wapforum.org and www.xwap.com. Ericsson, Motorola, Nokia and Unwired Planet introduced the architecture of the Wireless Application Protocol for public review and comments on 15 September 1997.

Wireless Cable An oxymoron which means that TV signals are broadcast by microwave to antennas on customers' homes. The former name for wireless cable was MMDS, short for Multichannel Multipoint Distribution Service.

Wireless Data Network A radio-based network for data transmission. Cellular Digital Packet Data (CDPD) is an example.

Wireless Digital Standards See DIGITAL WIRELESS STANDARDS.

Wireless E-911 Phase I / Phase II
Refers to the technology and services mandated by FCC Report and Order 96-264 pursuant to Notice of Proposed Rulemaking (NPRM) 94-102. The FCC requirement applies to all cellular and PCS service providers, and those Specialized Mobile Radio carriers that provide public voice service with telephone network interconnection.

Phase I defines delivery of a wireless emergency 911 call with call-back number and identification of the cell and sector from which the call originated. This allows the call to be routed to an appropriate public service answering point (PSAP) based on caller's general position. Without Phase I capabilities, wireless calls are routed to some default service agency, e.g., the state highway patrol. The required Phase I availability date was April 1998, but at this time (early 1999) many public service agencies have not upgraded their equipment to accept the Phase I information and still employ default, or non-selective, call routing.

Phase II defines delivery of a wireless 911 call with Phase I requirements, plus location of the caller within 125 meters 67% of the time. In addition, the call is routed to the appropriate PSAP based on the caller's coordinates. The required Phase II availability is October 2001. This new capability has given rise to the potential for other, non-emergency, added-value location services for wireless customers, and has instigated technological development in wireless handsets, infrastructure, network signaling, and emergency service equipment. See also Position Determination Technology

Wireless LANS The conventional local area network (LAN) uses wires or optical fiber as a common carrier medium. However, other possibilities exist. Low microwave frequencies (lower than about 10 GHz) can provide data rates as high 10 Mbit/s. Millimetric waves at around 60 GHz could support several 10 Mbit/s channels, while infra-red beams could support even greater data throughputs. The area covered by such a scheme would be restricted by the low allowable power radiation. The data rates of such systems tend to be restricted by walls, by interference and by multipath propagation problems that arise due to reflections within the building. Because of the wide bandwidth available, channeling can easily be provided by using spread spectrum methods and code division multiple access (CDMA), a technique that significantly improves the system security.

Wireless Local Loop WLL. A means of provisioning a local loop facility without wires. Employing low power, omni-directional radio systems, WLL allows carriers to provision loops up to T-1 capacity to each subscriber, with as much as 1 Gbps in aggregate bandwidth per coverage area. Such systems are being deployed widely in Asia and other developing countries where they offer the advantages of rapid deployment, and rapid configuration and reconfiguration, as well as avoidance of the costs of burying wires and cables. WLL is particularly attractive where rocky or soggy terrain make cabled systems problematic. WLL also is highly attractive to CLECs (Competitive Local Exchange Carriers), who have a compelling requirement to deploy local loop facilities, bypassing the incumbent LECs in a deregulated, competitive

environment as envisioned by the Telecommunications Act of 1996. Numerous trials also are underway in the U.S. See also ADML and LMDS.

Wireless Messaging Technology allowing the exchange of electronic messages without plugging into a wired land-based phone line. Two wireless messaging types are available: one-way, based on existing radio paging channels; and two-way, based on either radio-packet technology or cellular technology. Some people include in-room infra-red links in the term "wireless messaging." Some of the PDAs use wireless links.

Wireless Packet Switching Unlike existing cellular networks, wireless packet-switched networks are designed specifically for data communications. Packet switching breaks messages into packets and sends these packets individually over the network. Here's how a message is sent over the RAM Mobile Data Wireless Networks, one of the packet radio networks in operation today:

1) After you've written a message and turned on your modem, you enter a send command in your e-mail software.
2) The modem breaks your message into packets. A typical packet has a message space for as much as 512 bytes (about 100 words). Longer messages are divided into 512 byte sections.
3) The modem then sends each packet separately over the RAM packet radio network. Each packet includes the sender's and receiver's addresses.
4) The network routes the message to the recipient.
5) The recipient's packet radio modem reassembles the individual packets into a single message.
6) Your recipient can then read the message.

A packet radio network, typically uses a hierarchical architecture to route messages. At the lowest level, base stations exchange wireless messages with nearby mobile computers. Base stations can route messages to other users who are within its service area, or the local switch to read recipients who are in other areas, on LANs, or on public e-mail services. The local switch can either route the message to a different base station or to a regional switch. Users of these packet radio networks can typically send messages anywhere in the network — regardless of the physical distance — for the same rate per message.

Wireless Private Branch Exchange WPBX. The WPBX offers business users the ability to make and receive calls using cordless telephones anywhere on a company's premises.

Wireless Service Provider WASP. A carrier authorized to provide wireless communications exchange services (for example, cellular carriers and paging services carriers).

Wireless Switching Center WSC. A switching system used to terminate wireless stations for purposes of interconnection to each other and to trunks interfacing with the Public Switched Telephone Network (PSTN) and other networks.

WirelessReady Alliance Six companies with interests in wireless products and services came together in December, 1998 to create WirelessReady Alliance, which is aimed at creating mobile data solutions for businesses and consumers. A spokesman for the group said that it would concentrate on existing wireless technologies, including cellular packet data, GSM and CDMA for personal communications systems and cellular services. www.sierrawireless.com/alliance.

Wireline 1. Another name for a telephone company that uses cables, not radio.
2. Cellular licenses received from the FCC with initial association to telephone company. Also referred to as B-Block. See

WIRELINE CELLULAR CARRER.

Wireline Cellular Carrier Also called the Block B carrier. Under the FCC's initial cellular licensing procedures, the Block B carrier is the local telephone company. The FCC reserved one of the two systems in every market for the local telephone — or wireline company. Wireline or Block B systems operate on the frequencies 869 to 894 Megahertz. See NON-WIRELINE CELLULAR COMPANY.

Wiremap A cable test used to determine whether each pin has connectivity with the appropriate pin at the other end of the cable. A wiremap test is also required to test for split pairs.

Wiretap See Telephone Tap.

Wiretapping To listen in clandestinely to someone else's conversation. Other than scrambling, there is no known method to protect your telephone call against wiretapping, no matter what equipment you buy from companies advertising their wares nationally. Wiretapping can be accomplished without physical connection to a phone line, though technically this would be called "bugging." For all intents and purposes you should consider your telephone conversations as public and treat your conversations as such. See Telephone Tap.

Wiring Closet A central termination area for telephone and/or network cabling. Such wiring closet might serve the building or just a floor or just part of a floor. It's designed to accommodate the wiring needs of one organization or a department of a bigger company. A wiring closer can be a physical closet or a small room. It typically contains punchdown blocks and cross-connect panels.

Wiring Concentrator A wiring concentrator is an FDDI node that provides additional attachment points for stations that are not attached directly to the dual ring, or for other concentrators in a tree structure. The concentrator is the focal point of Digital's Dual Ring of Trees topology.

Wiring Density Refers to the number of wires that may be terminated on a connecting block in a given area. A high density block may terminate twice as many wires as a low density block, while a low density block may provide better wire management since fewer wires are being dressed into and out of the connecting block.

Wiring Environment A fancy term for any any building communications wiring system.

Wiring Grid The overall architecture of building wiring.

Witchcraft It is possible to trace the origin of witch hunting to an incident that occurred in 1242. Two of the Pope's inquisitors were staying in a house in Avignonet, in the south of France. They had traveled there to root heretics. In the middle of the night a dozen men with axes, who belonged to a sect known as the Cathars that believed that the Old Testament God was a demon, were admitted to the house by claiming they had information about heretics. They slaughtered the two inquisitors and their servants, hacking their bodies until they were almost unrecognizable. After the massacre, the Pope became determined to stamp out heretics at all costs. A bloody crusade followed. Cathars were dragged out of their homes and burned. In 1244, 200 of them were burned on a gigantic bon-fire at Montsegur. Those who managed to avoid capture were no longer accused or heresy, but of a strange new crime: conspiring with the devil or, as it came to be known, witchcraft.

Witting In on a secret.

WITS Wireless Interface Telephone System.

WITSA World Information Technology and Services Alliance. A global consortium of IT industry associations, WITSA positions itself as the global voice of the IT industry. Founded in

1978 as the World Computing Services Industry Association, WITSA is an active advocate in the area of public policy, working toward increasing competition through open markets and regulatory reform, protecting intellectual property, reducing tariff and non-tariff trade barriers to IT goods and services, and safeguarding the viability and continued growth of the Internet and electronic commerce. www.witsa.org.

Wizard An on-line tutor that guides you through common procedures or processes, as in hardware wizard.

WIZOP A Chief Sysop (System Operator). See SYSOP.

WLAN Wireless Local Area Network. A LAN without wires. There are major benefits, the biggest being the ability to configure and reconfigure the LAN around quickly and cheaply, as wires need not be placed and moved. Groups of people use them often in temporary situations — a team of auditors, a group of firefighters, etc. Wireless LANs are often not as fast as wired LANs. Check. See also Wireless LANs.

WLANA Wireless LAN Alliance. A consortium of manufacturers of WLANs, with the purpose of "generating awareness and creating excitement about the present and future capabilities of wireless local area networks." See WLAN. www.wlana.com

WLIF Wireless LAN Interoperability Forum. A group of more than 20 mobile computing product and service suppliers formed in 1996, the WLIF promotes the OpenAir interface specification. Through a third-party laboratory, products are tested and certified for compatibility. www.clif.com. See OpenAir.

WLL Wireless Local Loop. See WIRELESS LOCAL LOOP.

WMBTOPCITBWTNTALI We May Be The Only Phone Company In Town But We Try Not To Act Like It. An advertising slogan used by Southwestern Bell Telephone Company in the early 1970's to counter public opinion. It didn't last long — subscribers didn't believe it and SWBTC employees hated it. Or so Ray Horak, my Consulting Editor, says — he should know, he was there. Southwestern Bell, by the way, no longer is the only phone company in town.

WMF Windows MetaFile. A file that stores an image as graphical objects such as lines, circles, and polygons rather than as pixels. There are two types of metafiles, standard and enhanced. Standard metafiles usually have a .wmf file name extension. Enhanced metafiles usually have a .emf file name extension. Metafiles preserve an image more accurately than pixels when the image is resized. The reason to include this definition is that wmf files are the source of my best Microsoft PowerPoint presentation tip. I use PowerPoint a lot. So here goes. You have a PowerPoint presentation. You want to bring one or several slides in from someone else's presentation. If you paste them in, PowerPoint will make the new slide look like your own presentation's format. This will screw up all the imported slides and you'll spend eons trying to fix them. The only sure way to bring in a slide and to retain its formatting is to open the presentation you want to import and save it as a Windows Metafile (.wmf) file. It will ask you if you want to save one slide or all of them? If you say all of them, it will save the presentation in as many files as there are slides. Then you simply paste the new .wmf slides you want into your presentation. One tip: You must modify the slides before you save them as .wmf. When they're in the .wmf format, they can't be messed with. They drop into your presentation as basically uneditable bitmapped images. But they retain the original PowerPoint's format and don't adopt the format of the presentation they drop into.

WMI Windows Management Instrumentation. Microsoft's WMI is the company's implementation of WBEM in Windows

OSes. See WBEM.

WML WML is Wireless Markup Language, and is the WAP-proposed browser language. Analogous to, and based upon HDML, but new and different. www.wapforum.org

Woo Woo Tone A tone on a phone line indicating the number is unavailable. Also the words to a neat Jeffrey Osborne song, as in "Will you woo woo with me?"

Word A collection of bits the computer recognizes as a basic information unit and uses in its operation. Usually defined by the number of bits contained in it, e.g., 5, 8, 16 or 32 bits. Using DOS, the IBM PC (and compatibles) defines a word as an eight-bit byte. Such machines use the ASCII coding scheme (or a variation); such a scheme actually involves seven information bits plus a parity bit for error detection. Here's another explanation: A group of characters capable of being processed simultaneously in the processor and treated by computer circuits as an entity. See also BYTE.

Word Length The number of bits in a data character without parity, start or stop bits.

Words Per Minute WPM. The speed of printing, typing or communications. 100 WPM is 600 characters per minute (six characters per average word) or 10 characters per second. In ASCII, asynchronous transmission at this rate is also 100 or 110 bits per second, depending on the number of stop bits.

Work Location Wiring Subsystem The part of a premises distribution system that includes the equipment and extension cords from the information outlet to the terminal device connection.

Work Order A term used in the secondary telecom equipment business. Internal document used by a remarketer specifying: 1. Work to be performed; 2. Machine or item on which work is to be performed; 3. Required completion date; 4. Cost of work; 5. Customer purchase order number and/or other pertinent billing information. This document is used internally to: 1. Implement required work; 2. Monitor progress; and 3. Issue final billing. A work order is implemented once a written request has been received authorizing the work to be performed.

Word Spotting In speech recognition over the phone, word spotting means looking for a particular phrase or word in spoken text and ignoring everything else. For example, if the word to spot was "brown," then it wouldn't matter if you said "I want the brown one," or "how about something in brown?" In short, word spotting is the process whereby specific words are recognized under specific speaking conditions (i.e. natural, unconstrained speech). It can also refer to the ability to ignore extraneous sounds during continuous word recognition.

Work Station In this dictionary I spell it as one word WORKSTATION. See WORKSTATION.

Workaround A procedure or a piece of software that gets something (i.e. a computer system working). That "workaround" is typically not recommended by the manufacturer of the equipment. That manufacturer is typically surprised when your workaround actually works and often he says, "Wow, I've learned something."

Workflow The way work moves around an organization. It follows a path. That path is called workflow. Here's a more technical way of defining workflow: The automation of standard procedures (e.g. records management in personnel operations) by imposing a set of sequential rules on the procedure. Each task, when finished, automatically initiates the next logical step in the process until the entire procedure is completed.

Workflow Management The electronic management of work processes such as forms processing (e.g. for insurance policy acceptances, college admissions, etc.) or project management using a computer network and electronic messaging as the foundation. See WORKFLOW.

Workforce Management According to Jim Gordon of TCS in Nashville, Call center workforce management is the art and science of having the right number of people...agents...at the right times, in their seats, to answer an accurately forecasted volume of incoming calls at the service level you desire.

Workgroup A fancy new word for a department, except that the members of the workgroup may belong to different departments. The idea is that members of the workgroup work with themselves, so they'd be perfect candidates to buy electronic mail packages that could send messages between themselves and other software packages that would allow them to share their collective wisdoms and schedule their meeting times. Typically members of the workgroup would be on the same local area network and share the same telephone system. See WORKGROUP TELEPHONY.

Workgroup Computing An approach to the supply of computer services whereby access to computer power and information is organized on a workgroup by workgroup basis. Such systems normally consist of computers of varying capabilities connected to a local area network. See WORKGROUP.

Workgroup Manager An assistant network supervisor with rights to create and delete bindery objects (such as users, groups, or print queues) and to manage user accounts. A Workgroup Manager has supervisory privileges over a part of the bindery. When several groups share a file server, Workgroup Managers can provide autonomous control over their own users and data.

Workgroup Telephony See TELEPHONY WORKGROUPS.

Workload 1. A call center term. The total duration of all calls in a given period (half hour or quarter hour), not counting any time spent in queue. This figure is equal to the number of calls times the average handle time per call.
2. Trunk workload is call volume x average trunk hold time Verification Call A call center term. The process by which a telephone sale or other disposition is verified to ensure the details are accurate, the costs quoted are precise, the delivery terms have been explained, the customer fully understands the purchase, etc. Verification may be the responsibility of a specially trained team or may be part of the role of the supervisor.

Workstation In the telecom industry, a workstation is a computer and a telephone on a desk and both attached to a telecom outlet on the wall. The computer industry tends to refer to workstations as high-speed personal computers, such as Sun workstations, which are used for high-powered processing tasks like CAD/CAM, engineering, etc. A common PC — like the one you find on my desk — is not usually considered a workstation. The term workstation is vague.

World Numbering Zone One of eight geographic areas used to assign a unique telephone address to each telephone subscriber. See World Numbering Zones 1-9.

World Trade Organization See WTO.

World Wide Web Also called WEB or W3. The World Wide Web is the universe of accessible information available on many computers spread through the world and attached to that gigantic computer network called the Internet. The Web has a body of software, a set of protocols and a set of defined conventions for getting at the information on the Web. The Web uses hypertext and multimedia techniques to make the web easy for anyone to roam, browse and contribute to. The Web makes publishing information (i.e. making that information public) as easy as creating a "home page" and posting it on a server somewhere in the Internet. Pick up any Web access software (e.g. Netscape), connect yourself to the Internet (through one of many dial-up, for-money, Internet access providers or one of the many free terminals in Universities) and you can discover an amazing diversity of information on the Web. From weather to stock reports to information on how to build nuclear bombs to the best tennis tips, it can be posted on the Web for all to read. Invented by Tim Berners-Lee at CERN, the Web is the first true "killer app" of the Internet. See HOME PAGE, HTML and INTERNET.

World Zone 1 The area of the World Numbering Plan which is identified with the single-digit country code "1" and includes the territories of the United States and Canada (i.e. North America), and the following Caribbean countries: Antigua, Bahamas, Barbados, Bermuda, British Virgin Islands, Cayman Islands, Dominican Republic, Granada, Jamaica, Montserrat, Puerto Rico, St. Kitts, St. Lucia, St. Vincent, Virgin Islands.

World Zone 2 Africa. See World Numbering Zone.

World Zone 3 Europe. See World Numbering Zone.

World Zone 4 Europe. See World Numbering Zone.

World Zone 5 Central/South America. See World Numbering Zone.

World Zone 6 Pacific. See World Numbering Zone.

World Zone 7 Soviet Union. See World Numbering Zone.

World Zone 8 Asia. See World Numbering Zone.

World Zone 9 Middle East. See World Numbering Zone.

Worldcom The predecessor company to MCI Worldcom. Worldcom originally was LDDS (Long Distance Discount Services), a switchless reseller of long distance services started in Jackson, Mississippi by an ex-high school basketball coach. LDDS grew, acquired Wiltel Communications, grew some more, acquired MFS, grew some more, acquired Brooks Fiber and a few other companies, grew some more, and outbid BT for MCI. MCI Worldcom is the company that resulted on September 13, 1998. It's an incredible success story...so far. See MCI and MCI Worldcom.

WorldPartners An association of AT&T, KDD (Japan), and Singapore Telecom to provide international Virtual Network Services (VNS). VNS services are provided to over 32 countries which are major communications hubs. See also VNS.

WORM 1. Write Once, Read Many times. Refers to the new type of optical disks (similar to compact discs) which can be written to only once, but read many times. In other words, once the data is written, it cannot be erased. WORM disks typically hold around 600 megabytes. See also ERASABLE OPTICAL DRIVE.
2. A program that duplicates itself repeatedly, potentially worming its way through an entire network. The Internet worm was perhaps the most famous; it successfully (and accidentally) duplicated itself on many of the systems across the Internet.

WORN Write Once, Read Never (A joke).

Worst Hour Of The Year That hour of the year during which the median noise over any radio path is at a maximum. This hour is considered to coincide with the hour during which the greatest transmission loss occurs.

WOSA Windows Open Services Architecture. According to Microsoft, WOSA provides a single system level interface for connecting front-end applications with back-end services. Windows Telephony, announced in May 1993, is part of

WOSA. According to Microsoft, application developers and users needn't worry about conversing with numerous services, each with its own protocols and interfaces, because making these connections is the business of the operating system, not of individual applications. WOSA provides an extensible framework in which Windows based applications can seamlessly access information and network resources in a distributed computing environment. WOSA accomplishes this feat by making a common set of APIs available to all applications. WOSA's idea is to act like two diplomats speaking through an interpreter. A front-end application and back-end service needn't speak each other's languages to communicate as long as they both know how to talk to the WOSA interface (e.g. Windows Telephony). As a result, WOSA allows application developers, MIS managers, and vendors of back-end services to mix and match applications and services to build enterprise solutions that shield programmers and users from the underlying complexity of the system.

This is how WOSA works: WOSA defines an abstraction layer to heterogeneous computing resources through the WOSA set of APIs. Initially, this set of APIs will include support for services such as database access, messaging (MAPI), file sharing, and printing. Because this set of APIs is extensible, new services and their corresponding APIs can be added as needed. WOSA uses a Windows dynamic-link library (DLL) that allows software components to be linked at run time. In this way, applications are able to connect to services dynamically. An application needs to know only the definition of the interface, not its implementation. WOSA defines a system level DLL to provide common procedures that service providers would otherwise have to implement. In addition, the system DLL can support functions that operate across multiple service implementations. Applications call system APIs to access services that have been standardized in the system. The code that supports the system APIs routes those calls to the appropriate service provider and provides procedures and functions that are used in common by all providers.

The primary benefit of WOSA is its ability to provide users of Windows with relatively seamlessly connections to enterprise computing environments. Other WOSA benefits, according to Microsoft include:
• Easy upgrade paths.
• Protection of software investment.
• More cost-effective software solutions.
• Flexible integration of multiple-vendor components.
• Short development cycle for solutions.
• Extensibility to include future services and implementations. See also ODBC, MAPI, and TAPI.

WPM See WORDS PER MINUTE.

WPBX Wireless private branch exchange. The WPBX offers business users the ability to make and receive calls using cordless telephones anywhere on a company's premises.

WRAM Windows Random Access Memory. Similar to VRAM, but with added logic to accelerate common video functions such as bit-block transfers and pattern fills. See also VRAM.

Wrap 1. In data communications, to place your diagnostic and test equipment around parts of a network so you can monitor their use (i.e. do network diagnostics on them). You are, in essence, wrapping your products around theirs.
2 To make a connection between a flexible wire and a hard tag by tightly wrapping the cable around the tag. There are automatic wire wrapping tools available for this job.
3. Redundancy measure in IBM Token Ring LANS. Trunk

cabling used in Token Ring TCUs contains two data paths: a main and backup normally unused). If the trunk cable is faulty, the physical disconnection of the connector at the TCU causes the signal from the main path to wrap around on to the backup path, thus maintaining the loop. The term wrap is now used on FDDI networks. If a failure occurs on one of the FDDI rings, the stations on each side of the failure reconfigure. The two rings then are combined into a single ring topology that allows all functioning stations to remain interconnected.

Wrap-Up Between-call work state that an ACD agent enters after releasing a caller. It's the time necessary to complete the transaction that just occurred on the phone. In wrap-up, the agent's ACD phone is removed from the hunting sequence. After wrap-up is completed, it is returned to the hunting sequence and is ready to take the next call.

Wrap-Up Codes A call center term. Codes agents enter into the ACD to identify the types of calls they are handling. The ACD can then generate reports on call types, by handling time, time of day, etc.

Wrap-Up Data Ad hoc data gathered by an agent in the ACD system following a call.

Wrap-Up Time A call center term. The time an employee spends completing a transaction after the call has been disconnected. Sometimes it's a few seconds. Sometimes it can be minutes. Depends on what the caller wants.

Wrapping In token-ring networks, the process of bypassing cable faults without changing the logical order of the ring by using relays and additional wire circuits.

Write To record information on a storage device, usually disk or tape.

Write Head A magnetic head capable of writing only. You find write heads on everything from tape recorders to computers.

Write Protect Using various hardware and software techniques to prohibit the computer from recording (writing) on storage medium, like a floppy or hard disk. You can write protect a 5 1/4 diskette by simply covering the little notch with a small metal tag. The idea of "Write Protect" is to stop someone (including yourself) from changing your precious data or program. You can't write protect a hard disk easily. The easiest way to stop someone changing a file is to use the program ATTRIB.EXE. See ATTRIBUTES.

Write Protection A scheme for protecting a diskette from accidental erasure. 5 1/4" diskettes have a notch which must be uncovered to allow data on the diskette to be modified. 3 1/2" diskettes have small window with a plastic tab which must be slid into place to cover the window to allow data on the diskette to be modified. See WRITE PROTECT and ATTRIBUTES.

Write Protection Label A removable label, the presence or absence of which on a diskette prevents writing on the diskette.

WRT With respect to.

WSC Wireless Switching Center. A switching center designed for wireless communications services, typically fixed wireless services including data and voice.

WTAC World Telecommunications Advisory Council. WTAC is comprised of telecommunications leaders from the private and public sectors and from every region of the world. WTAC gives advice to the ITU — the International Telecommunications Union. The WTAC held its first meeting in Geneva, Switzerland, in April, 1992. In February 1993, it published a small booklet called "Telecommunications Visions of the Future."

WTN Working Telephone Number

WTNG WaiTiNG.

WTO World Trade Organization. An organization of over 130 member countries and 30 observer countries. Formed in 1995, the WTO is a formal organization with legal status, which succeeded the ad hoc GATT (Global Agreement of Tariffs and Trade); the GATT agreement is now part of WTO agreements. According to the WTO, its main functions are assisting developing and transition economies, specialized assistance in export promotion, regional trading arrangements, cooperation in global economic policy-making, and routine notification when members introduce new trade measures or alter new ones. The WTO also assists in the settlement of trade disputes between member countries. The WTO gets involved in telecommunications in a number of ways, most recently in assisting in the development of multilateral agreements for normalization of international long distance costs. Currently, there exist incredible differences in such costs, depending on where the call originates; i.e., a 10 minute call from the U.S. to Hong Kong is much less expensive than the same call from Hong Kong to the U.S. It's more a matter of national politics and economic policy than anything else, resulting in an imbalance of trade in the magnitude of billions of dollars flowing out of the U.S. to other countries. Regardless of the direction of the call, the costs to the national carriers are roughly the same, and the revenues are split 50/50 between the two carriers. As a means around this inequity, International Callback developed. International Callback is viewed as illegal by most countries outside the U.S., as they seek to protect this significant source of hard currency. www.wto.org See also INTERNATIONAL CALLBACK.

WTP The WAP equivalent of the reliability portion of HDTP. www.wapforum.org

WUI Web User Interface. Pronounced "wooey." A GUI (Graphical User Interface) for the WWW.

WWW 1. World Wide Web; a hypertext-based system for finding and accessing resources on the Internet network. For a much better explanation, see WORLD WIDE WEB and INTERNET.
2. Also referred to as World Wide Wait, which according to some frustrated observers, is the real meaning of WWW.

WWW2 See www3.

WWW3 www3 is as an alternate name to www in a bank of web server computers, i.e. a group of computers named www1.company.com, www2.company.com, www3.company.com, etc. These servers would typically be assigned to a "round robin" DNS (Domain Name Server) and share the load for the site. To get to that site, a Web surfer could type www.company.com, www1.company.com, www2.company.com or www3.company.com. He'd get the same information and hit what appeared to him as the same server.

WYPIWYF Acronym for "What You Print Is What You Fax," also "The Way You Print Is the Way You Fax." Coined by Intel to describe its one-step pop-up menu that makes sending faxes from the PC as easy as sending a document to a printer.

WYSIWYG (pronounced Whiz-i-wig) What You See Is What You Get. A word processing term meaning what you see on your computer screen is what you will see printed on paper. The exact typeface, the correct size, the right layout, etc. Some word processors do WYSIWYG. Others don't. You usually need a screen with graphics to get the full effect.

WZ1 World Zone One. The part of the earth covered by what used to be called The North American Numbering Plan. It includes the U.S., Canada, Alaska, Hawaii, and the Caribbean islands, but does not include Mexico or Cuba.

X 1. An abbreviation for the word "cross," as in crossbar (Xbar). 5XB would be the abbreviation for a No. 5 Crossbar circuit switch.
2. Generic. As in xDSL, generic Digital Subscriber Line, which is a family of access technologies including ISDN, ADSL, IDSL, HDSL, RADSL and SDSL. Also, as in 1000Base-CX, an IEEE standard for Gigabit Ethernet over some sort of Copper cable.

X Band 7 GHz and 8 GHz. Used by military satellites.

X Recommendations The ITU-T documents that describe data communication network standards. Well-known ones include: X.25 Packet Switching standard, X.400 Message Handling System, and X.500 Directory Services.

X Series Recommendations drawn up by the ITU-T to establish communications interfaces for users' Data Terminal Equipment (DTE) and Data Circuit Terminating Equipment (DCE). They govern the attachment of data terminals to public data networks (PDNs) and the Public Switched Telephone Network (PSTN). In short, a set of rules for interfacing terminals to networks.

X Terminal A networked desktop machine that displays software applications which are running on a networked server. X terminals are allegedly cheaper and easier to network than PCs or workstations. They are said to be a forerunner to thin clients, also called network computers (NCs).

X Windows X Windows is officially called X Window, or The X Window System, or X. Sometimes it is spelled with an hyphen, e.g.X-Window. It is used primarily with UNIX systems, but not exclusively. X was originally designed to be an industry-standard network windowing system, and because UNIX is widely used, X has become the windowing standard for UNIX.

X-10 Protocol A protocol found in home automation. You can use X-10 to control conforming black boxes and devices. The protocol is a command signal that rides on the AC 60 cycle sine power curve. The signal is a series of 120 kHz pulses sent on the "zero crossing" of each cycle. The signal is in a binary fashion, transmitting the letter code, unit code and command for a device. All receivers monitor the power line waiting for a command to respond. The limit of number of unique codes available is 256. This is derived by having 16 letter codes and 16 unit codes, hence 16x16=256 "addresses." More than one device can share an address. If you decide that every time you select "A1" on, you want all the front lights to come on, then you can place all the receivers to the same address. The system is easy to use and very flexible.

X-Axis Horizontal axis on a graph or chart.

X-Base A term used to describe any database application capable of generating custom programs with dBASE-compatible code.

X-Off/X-On A flow control protocol for asynchronous serial transmission. Flow control is a method of adjusting information flow. For example, in transmitting between a computer and a printer, the computer sends the information to be printed at 9600 baud. That's several times faster than the printer can print. The printer, however, has a small memory.

The computer dumps to the memory, called a buffer, at 9600 baud. When it fills up, the printer signals the computer that it is full and please stop sending. When the buffer is ready to receive again, the printer (which also has a small computer in it) sends a signal to the desktop computer (the one doing the printing) to please start sending again. X-OFF means turn the transmitter off (xmit in Ham radio terms). It is the ASCII character Control-S. X-ON means turn the transmitter on. It is the ASCII character Control-Q. You can use these characters with many microcomputer functions. For example, if you do DIR in MS-DOS and you want to stop the fast rush of files, then type Control-S.

X-Dimension Of Recorded Spot In facsimile, the center-to-center distance between two recorded spots measured in the direction of the recorded line. This term applies to facsimile equipment that responds to a constant density in the subject copy by yielding a succession of discrete recorded spots.

X-Dimension Of Scanning Spot In facsimile, the center-to-center distance between two scanning spots measured in the direction of the scanning line on the subject copy. The numerical value of this term will depend upon the type of system used.

X-Open An international consortium of computer vendors working to create an internationally supported vendor-independent Common Applications Environment based on industry standards.

X-Series Recommendations Set of data telecommunications protocols and interfaces defined by the ITU-T.

X-Windows The UNIX equivalent of Windows. What Windows is to MS-DOS, X-Windows is to Unix. A network-based windowing system that provides a program interface for graphic window displays. X-Windows permits graphics produced on one networked station to be displayed on another. Almost all UNIX graphical interfaces, including Motif and OpenLook, are based on X-Windows. X-Windows is a networked window system developed and specified by the MIT X Consortium. Members of the X Consortium include IBM, DEC, Hewlett-Packard and Sun Microsystems. Sun Microsystems has been contracted by the MIT X Consortium to implement PEX (PHIGS Extensions to X), which will be the standard networking protocol for sending PHIGS (Programmers Hierarchical Graphics System) graphics commands through X-Windows. Some people spell it X-Windows and some spell it X Windows.

X-Y A specific variety of electromechanical switch. Does the same things as a Stronger step-by-step switch but in a horizontal plane. It's so called because it's a two motion switch with horizontal and vertical movements. The first pulse sends the switch horizontally to the right place, then the next pulse sends it vertically up to the right place and so on, until it has switched the call through. One of the most reliable switches ever produced. Unfortunately, it's slow, space-consuming and unable to be programmed with many new customer pleasing features.

X.1 A ITU-T specification that defines classes of service in a packet-switched network, such as virtual-circuit, datagram, and fast-packet services.

X.110 International routing principles and routing plan for PDNs.

X.121 International Numbering Plan for public data networks. X.121 defines the numbering system used by data devices operating in the packet mode. X.121 is used by ITU-T X.25 packet-switched networks and has been proposed by several computer vendors as the future universal addressing scheme.

X.121 Address A standard O/R (Originator/Recipient) attribute that allows Telex terminals to be identified in the context of store-and forward communications.

X.130 Call setup and clear down times for international connection to synchronous PDNs.

X.132 Grade of service over international connections to PDNs.

X.150 DTE and DCE test loops in public data networks.

X.2 The X.2 series of standards which follow encompass international user services and facilities over public switched telephone networks.

X.20 Asynchronous communications interface definitions between data terminal equipment (DTE) and data circuit terminating equipment (DCE) for start-stop transmission services on public switched telephone networks.

X.20 bis Used on public data networks of data terminal equipment (DTE) that is designed for interfacing to asynchronous duplex V-series modems.

X.200 A series of ITU (International Telegraph and Telephone Consultative Committee) recommendations that defines the type of service offered by specific layers in the OSI (Open Systems Interconnection) Model, and defines the protocol to be used at those layers. ITU is now the ITU.

X.21 Interface between data terminal equipment (DTE) and data circuit-equipment (DCE) for synchronous operation on public switched telephone networks.

X.21 bis Used on public switched telephone networks of data terminal equipment (DTE) that is designed for interfacing to synchronous V-series modems.

X.21 TSS Specification for Layer 1 interface used in the X.25 packet-switching protocol and in certain types of circuit-switched data transmissions.

X.224 A ITU-T standard associated with the transport layer of the Open Systems Interconnect (OSI) architecture used in networks employing circuit-switched techniques.

X.225 A ITU-T standard associated with the session layer of the Open Systems Interconnect (OSI) architecture used in networks employing circuit-switched techniques.

X.226 A ITU-T standard associated with the Open Systems Interconnect (OSI) architecture that defines specific presentation layer services used with circuit-switched network services.

X.24 List of definitions for interchange circuits between data terminal equipment (DTE) and data circuit terminating equipment (DCE) on public switched telephone networks.

X.25 From its beginning as an international standards recommendation from ITU-T, the term X.25 has come to represent a common reference point by which mainframe computers, word processors, mini-computers, VDUs, microcomputers and a wide variety of specialized terminal equipment from many manufacturers can be made to work together over a type of data communications network called a packet switched network. On a packet switched data network (private or public), the data to be transmitted is cut up into blocks. Each block has a header with the network address of the sender and that of the destination. As the block enters the network, the number of bits in the block are put through some mathematical functions (an algorithm) to produce a check sum.

The check sum is attached as a "trailer" to the packet as it enters the network. Packets may travel different routes through the network. But, ultimately, the packets are routed by the network to the node where the destination computer or terminal is located. At the destination, the packet is disassembled. The bits are put through the same algorithm, and if the digits computed are the same as the ones attached as the trailer, there are no detected errors. An ACK, or acknowledgement, is then sent to the transmitting end. If the check sum does not match, a NAK, or Negative Acknowledgement is sent back, and the packet is retransmitted. In this manner, high speed, low error rate information can be transmitted around the country using shared telecommunications circuits on public or private data networks.

X.25 is the protocol providing devices with direct connection to a packet switched network. These devices are typically larger computers, mainframes, minicomputers, etc. Word processors, personal computers, workstations, dumb terminals, etc. do not support the X.25 packet switching protocols unless they are connected to the network via PADs — Packet Assembler/Disassemblers. A PAD converts between the protocol used by the smaller device and the X.25 protocol. This conversion is performed on both outgoing (from the network) and incoming data (to the network), so the transmission looks transparent to the terminal. (See TRANSPARENT.) There's a very good book on X.25 called X.25 Explained, Protocols for packet switching networks by R. J. Deasington of IBM. It is available from Telecom Library on 1-800-LIBRARY, or 1-212-691-8215.

X.25 Network Any network that implements the internationally accepted ITU-T standard governing the operation of packet-switching networks. The X.25 standard describes a switched communications service where call setup times are relatively fast. The standard also defines how data streams are to be assembled into packets, controlled, routed, and protected as they cross the network.

X.28 DTE/DCE interface for start-stop-mode data terminal equipment accessing the packet assembly/disassembly facility (PAD) in a public switched telephone networks situated in the same country.

X.29 Procedures for the exchange of control information (handshaking) and user data between a packet assembly/disassembly facility (PAD) and a packet mode DTE or another PAD.

X.3 ITU-T recommendation describing the operation of a Packet Assembly/Disassembly (PAD) device or facility in a public data network. X.3 defines a set of 18 parameters that regulate basic functions performed by a PAD to control an asynchronous terminal. The setting of these parameters governs such characteristics as terminal speed, terminal display, flow control, break handling and data forwarding conditions, and so on.

X.30 Support of X.20 bis, X.21 and X.21 bis DTEs by an ISDN.

X.31 Support of packet mode DTEs by an ISDN.

X.32 Interface between Data Terminal Equipment (DTE) and Data Circuit Terminating Equipment (DCE) operating in a packet mode and accessing a packet switched public data network via a public switched telephone network or a circuit switched public data network. X.32 describes the functional and procedural aspects of the DTE/DCE interface for DTEs accessing a packet switched public data network via a public switched network.

X.38 ITU-T recommendation for the access of Group 3 facsimile equipment to the Facsimile Packet Assembly/Disassembly (FPAD) facility in public data networks situated in the same country.

X.39 ITU-T recommendation for the exchange of control information and user data between a Facsimile Packet Assembly/Disassembly (FPAD) facility and a packet mode Data Terminal Equipment (DTE) or another pad, for international networking.

X.3T9.3 The ANSI committee responsible for the creation and perpetuation of Fiber Channel standards.

X.3T9.5 The ANSI committee responsible for the creation and perpetuation of FDDI Standards.

X.4 International Alphabet No.5 for character oriented data.

X.400 X.400 is an international standard which enables disparate electronic mail systems to exchange messages. Although each e-mail system may operate internally with its own, proprietary set of protocols, the X.400 protocol acts as a translating software making communication between the electronic mail systems possible. The result is that users can now reach beyond people on their same e-mail system to the universe of users of interconnected systems. One problem with e-mail sent between X.400 networks is that the sender's name is not sent. (I kid you not.) This was one element of the protocol the committees forgot! If your message crosses an X.400 network, remember to sign your name. The X.400 standard itself is an overview which is broken down under subsequent numbers:

X.402 Overall Architecture

X.403 Conformance testing

X.407 Abstract service definition conventions

X.408 Encoded information type conversion rules

X.411 Message transfer system

X.413 Message store

X.419 Protocol specifications

X.420 Interpersonal messaging system. An IPM format specification using the X.400 transfer protocol. In addition to text, it also allows CAD/CAM, graphics, Fax, and other electronic information.

X.435 An EDI (Electronic Data Interchange) format specification based on the X.400 transfer protocol. It can also allow for CAD/CAM, graphics, Fax, and other electronic information to accompany an EDI interchange.

X.440 A VM (Voice Messaging) format specification, using the X.400 transfer protocol. In addition to voice, it can also contain CAD/CAM, graphics, fax, and other electronic information.

X.445 The X.445 standard, or APS (Asynchronous Protocol Specification), lets X.400 clients and servers exchange all types of digital data over public telephone networks rather than over X.25 leased lines, which are required today. Among those backing the spec are Intel, AT&T, Microsoft Corp., Lotus and Isocor. X445 is an extension of the X.400 standard. X.400 provides an option to other messaging backbones such as System Network Architectural Distribution Services and SMTP. It supports multimedia data traffic including text, binary files, E-mail, voice images, and sound.

In its December 19, 1994 issue, PC Week said that the X.445 standard, which earlier this month gained final approval from the International Telecommunications Union, should ease messaging dramatically by allowing users to exchange X.400 data traffic over standard telephone networks.

X.50 Fundamental parameters of multiplexing scheme for the international interface between synchronous data networks.

X.500 The ITU-T international standard designation for a directory standard that permits applications such as electronic mail to access information which can either be central or distributed. The X.500 standard for directory services provides the means to consolidate e-mail directory information through central servers situated at strategic points throughout the network. These X.500 servers then exchange directory information so each server can keep all its local mail directory information current. With X.500, any e-mail user, whether on OpenVMS, Macintosh, DOS, or UNIX workstations, can be listed in a central directory that can be accessed using an X.500-compatible user agent.

X.509 A cryptography term. The part of the ITU-T X.500 Recommendation which deals with Authentication Frameworks for Directories. Within X.509 is a specification for a certificate which binds an entity's distinguished name to its public key through the use of a digital signature. Also contains the distinguished name of the certificate issuer.

X.51 Fundamental parameters of multiplexing scheme for the international interface between synchronous data networks using 10-bit envelop structure.

X.51 bis Fundamental parameters of a 48Kbit/s transmission scheme for the international interface between synchronous data networks using a 10-bit envelope structure.

X.58 Fundamental parameters of multiplexing scheme for the international interface between synchronous data networks using a 10-bit envelope structure.

X.60 Common channel signaling for circuit switched data applications.

X.61 Signaling system no. 7 — data user part.

X.70 Terminal and transit control signaling for asynchronous services on international circuits between anisochronous data networks.

X.71 Decentralized terminal and transit control signaling on international circuits between synchronous data networks.

X.75 An international standard for linking X.25 packet switched networks. X.75 defines the connection between public networks, i.e. for a gateway between X.25 networks. See X.25.

X.80 Interworking of inter-exchange signals for circuit switched data services.

X.92 Hypothetical reference connections for synchronous PDNs in a packet switched network.

X.95 ITU-T specification dealing with a number of internal packet-switched network parameters such as packet size limitations and service restrictions.

X.96 Call progress signals in PDNs.

X/Open A group of computer manufacturers that promotes the development of portable applications based on UNIX. Formed in 1984, X/Open is dedicated to the identification, agreement and wide-scale adoption of information technology standards which reduce issues of incompatibility and which help users realize the benefits of open information systems. In February 1996, X/Open and the Open Software Foundation (OSF) consolidated to form The Open Group (www.opengroup.org). See The Open Group.

X1 See Blind Dialing.

X12 ANSI (American national Standards Institute) standard that is the dominant EDI (Electronic Data Interchange) standard in the U.S. today; designed to support cross-industry exchange of business transactions. Standard specifies the vocabulary (dictionary) and format for electronic business transactions.

x2 x2 is one of two standard ways (it is NOT a standard) for running data over dial-up phone lines at up to 53,000 bits per second one way and up to 33.6 Kbps the other way. The standard was developed for use on the Internet, with the 53 Kbps channel flowing to you. The logic is that at 53 Kbps (not 56 as advertised), Web pages fill a lot faster on your screen. x2 was developed by US Robotics. The competing 56 Kbps "standard" is called 56flex. It was developed by Rockwell Semiconductor and Lucent Technologies. x2 is not compatible with 56flex. In other words, a 56flex modem cannot talk at 56Kbps to a x2 modem. The modems can, however, communicate at speeds up to 33.6 kbps, which are international standards. In February 1998, the ITU-T released the specs on a proposed 56 Kbps international standard called V.90. Neither x2 nor 56flex conform to V.90. See also 56 Kbps Modem (for a much longer explanation) and V.90.

X3.15 Bit sequencing of ASCII in serial-by-bit data transmission.

X3.16 Character structure and character parity sense for serial-by-data communications in ASCII.

X3.36 Synchronous high speed data signaling rates between data terminal equipment and data circuit terminating equipment.

X3.41 Code extension techniques for use with 7-bit coded character set of ASCII.

X3.44 Determination of the performance of data communications systems.

X3.79 Determination of performance of data communications systems that use bit oriented control procedures.

X3.92 Data encryption algorithm.

X.Windows A networked GUI developed at the Massachusetts Institute of Technology (MIT) as part of Project Athena. Based on a client/server architecture it displays information from multiple networked hosts on a single workstation.

XA Transaction management protocol.

XA-SMDS Exchange Access SMDS. An access service provided by a local exchange carrier to an interexchange carrier. It enables the delivery of a customer's data over local and long distance SMDS networks.

Xanadu An idyllic place in which "did Kubla Khan a stately pleasure-dome decree," according to Samuel T. Coleridge in his poem (1798) "Kubla Khan." Jump forward to 1941, when Orson Wells released "Citizen Kane." Charles Foster Kane (allegedly William Randolph Hearst) built Xanadu, an elaborate palace (allegedly modeled on the San Simeon mansion of William Randolph Hearst). The place (Coleridge) didn't exist, and the mansion (Wells) brought only heartache. Jump forward to 1980, when Ted Nelson starts the project and the company Xanadu, a system for the network sale of documents with automatic royalty on every byte — hypertext and copyright protection (and royalty) on every byte. Nelson is a genius, but his Xanadu never (yet) was successful (more heartache). His "hypertext" concept is what we now see on the World Wide Web. See also Hypertext; Nelson, Ted; and World Wide Web.

XAPIA X.400 Application Program Interface Association. Microsoft has a new concept. It's called an Enterprise Messaging Server (EMS). The idea is to allow users to transparently access the messaging engine from within desktop applications to route messages, share files, or retrieve reference data. According to Microsoft, corporate developers will be able to add capabilities using Visual Basic and access EMS by writing either to the X.400 Application Program Interface Association's (XAPIA's) Common Mail Calls (CMC) or to Microsoft's Messaging API (MAPI). (ftp://nemo.ncsl.nist.gov/pub/olw/dssig/xapla/) See MAPI.

Xbar Crossbar.

XC Cross connect.

XDP Ron Stadler of Panasonic dreamed this one up. It stands for eXtra Device Port. It's an analog RJ-11 equipped port on the back of a Panasonic digital telephone, which is driven by Panasonic's Digital Super Hybrid switch. The XDP is an extension line completely separate to your digital voice line. You can be speaking on the phone while receiving or sending a fax or while sending or receiving data. Or plug a cordless phone or answering machine into the XTP.

xDSL A generic — the letter x means generic — term for Digital Subscriber Line equipments and services, including ADSL, HDSL, IDSL, SDSL and VDSL. xDSL technologies provide extremely high bandwidth over the twisted-pair that runs from your phone company's central office to your office or home. Some xDSL lines are symmetrical (the same bandwidth in both directions). Some xDSL lines are (for example, ADSL) are asymmetrical — different bandwidth (and thus speed) in both directions. Many xDSL loops are already installed. xDSL technology is being installed by local phones in order to provide faster access for their subscribers to the Internet and also (and most importantly) to counter the competition from the CATV industry with its cable modem. See ADSL, G.990, G-Lite, HDSL, IDSL, RADSL, SDSL, Splitter, Splitterless and VDSL for more detailed explanations.

Xenix Microsoft name for a 16-bit microcomputer operating system derived from AT&T Bell Labs' UNIX. See also Linux.

Xerographic Recording Recording by action of a light spot on an electrically charged photoconductive insulating surface where the latent image is developed with a resinous powder.

Xerox Network Services XNS. An old multilayer protocol system developed by Xerox and adopted, at least in part, by Novell and other vendors. XNS is one of the many distributed-file-system protocols that allow network stations to use other computers files and peripherals as if they were local. XNS is used by some companies on Ethernet LANs. In local area networking technology, special communications protocol used between networks. XNS/ITP functions at the 3rd and 4th layer of the Open Systems Interconnection (OSI) model. Similar to transmission control protocol/internet protocol (TCP/IP), which, of course, is now much more popular, since it is the foundation of the Internet. See Internet.

XFN X/Open Federated Naming. Used in Sun's Enterprise Server. Provides enterprise directory name Service for simplified access to and federation among multiple naming services. Works with Distributed Computing Environment (DCE), ONC, and Internet Domain Name Service (DNS).

XFN/NIS+ Part of Sun Microsystems' ENOS Networking Solutions. It's a secure repository of network information.

XFR TransFeR.

XGA eXtended Graphics Array. A new IBM level of video graphics which has a screen resolution of 1,024 dots horizontally by 768 vertically, yielding 786,432 possible bits of information on one screen, more than two and a half times what is possible with VGA. See also MONITOR.

XGL Part of Sun Microsystems' Imaging and Graphics Solutions. Solaris' graphics library for 2-D/3-D applications development.

XID Frame A High-Level Data Link Control (HDLC) transmission frame used to transfer operational parameters between two or more stations.

XIL Part of Sun Microsystems' Imaging and Graphics Solutions, XIL is an open Imaging API used for image enhancement, scaling, compression, color conversion, and display.

XIP eXecute-In-Place. Refers to specification for directly executing code from a PCMCIA Card without first having to load it into system memory.

XIWT Cross-Industry Working Team. About 28 companies (and growing) whose work centers on the Clinton-Gore administration's document "National Information Infrastructure — Agenda for Action." The document spells out policy initiatives required to achieve the benefits of widespread, convenient and affordable access to existing and future information resources. In short, the Information SuperHighway.

XJACK A registered trademark of modem manufacturer, MegaHertz, for one of the most innovative ideas in PCMCIA modems. XJACK is the world's tiniest female RJ-11. It's about a quarter of an inch wide and the most convenient way of attaching your laptop's modem to the phone network.

XLIU X.25/X.75/X.75' Link Interface Unit.

XLL eXtensible Linking Language. A XML linking language which provides for multidirectional linking of documents written in XML. Unlike HTML-based sites, XML sites with XLL support are capable of linkage at the object level, rather than at the page level, only. See XML.

XMA eXtended Memory specificAtion. Interface that lets DOS programs cooperatively use extended memory in 80286 and higher computers. One such driver is Microsoft's HIMEM.SYS, which manages extended memory and HMA (high memory area), a 64k block just above 1Mb.

Xmit Transmit.

XML eXtensible Markup Language, a new specification being developed by the W3C. XML is a pared-down version of SGML, designed especially for Web documents. It enables designers to create their own customized tags to provide functionality not available with HTML. For example, XML supports links that point to multiple documents, as opposed to HTML links, which can reference just one destination each. Whether XML eventually supplants HTML as the standard web formatting specification depends a lot on whether it is supported by future Web browsers. The World Wide Web Consortium released the first spec on XML in February, 1998. According to Inter@ctive Week, a weekly newspaper covering the Internet, companies are seizing upon the ability of XML to allow structured exchanges of data between machines attached to the Web. That will enable one Web server to talk to another Web server, meaning manufacturers and merchants can begin to quickly swap data, such as pricing, stock-keeping numbers, transaction terms and product descriptions. See also OFX (Open Financial Exchange.)

XModem Also called "Christiansen Protocol". An error-correcting file transfer, data transmission protocol created by Ward Christiansen of Chicago for transmitting files between PCs. A file might be anything — a letter, an article, a sales call report, a Lotus 1-2-3 spreadsheet. The XModem protocol sends information in 128 bytes blocks of data. Some sums (check sums) are done on each block and the result is sent along with the block. If the result does not check out at the other end, the computer at the other end sends a request (a NAK — Negative AcKnowledgement) to re-transmit that block once again. If the block checks out, the computer sends an ACK — an Acknowledgement. In this way, relatively error-free transmission can be accomplished.

XModem was first used by computer hobbyists and then by business users of PCs. If you're buying a telecommunications

software program for your PC — IBM, Radio Shack, Compaq, Apple, etc. — it's a good idea to buy a program with XModem. It's among the most common data communications protocols. But it's not the fastest, just the most common. AT&T Mail supports XModem protocol. So does TELECONNECT Magazine's own E-mail InfoBoard system (212-989-4675). MCI Mail does not support XModem protocol. We don't know why. There are many variations of XModem including XModem 1K (which uses blocks of 1,025 bytes), Modem7, YModem, Y-Modem-G, and ZModem. Most common communications software packages only support (i.e. will handle) the original version of XModem (checksum) and the newer CRC variation. A study in Byte Magazine (March, 1989) showed ZModem to be a far more efficient file transfer protocol than XModem, YModem, or W/XModem. The author now tends to use ZModem more commonly. It is supported by most on-line services, such as CompuServe, etc. See also DATA COMPRESSION PROTOCOLS, ERROR CONTROLS PROTOCOLS, FILE TRANSFER PROTOCOL, YMODEM and ZMODEM.

XModem-1K Xmodem-1K is an error-correcting file transfer, data transmission protocol for transmitting files between PCs. It is essentially Xmodem CRC with 1K (1024 byte) packets. On some systems and bulletin boards it may also be referred to as Ymodem.

XModem-CRC Cyclic Redundancy Checking is added to XMODEM frames for increased reliability of errors detection. See XMODEM.

XMP X/Open Management Protocol; an API and software interface specified in the Open Software Foundation's Distributed Management Environment.

XMS An acronym for eXtended Memory Specification. To run this standard, your system must have 350K of extended memory. XMS creates the HMA (High Memory Area), then governs access to and the allocation of the remainder of extended memory.

XNMS Trademark for MICOM's IBM PC-based packet data network (PDN) network management system software products.

XNS Xerox Network System. The lan architecture developed at the Xerox Palo Alto Research Center (Parc). It is a five-layer architecture of protocols and was the foundation of the OSI seven-layer model. It has been adopted in part by Novell and other vendors. XNS is one of the many distributed-file-system protocols that allow network stations to use other computers files and peripherals as if they were local. XNS is used by some companies on Ethernet LANs.

XO Crystal Oscillator.

XON/XOFF XON/XOFF are standard ASCII control characters used to tell an intelligent device to stop or resume transmitting data. In most systems typing <Ctrl>-S sends the XOFF character, i.e. to stop transmitting. Some devices understand <Ctrl>-Q as XON, i.e. start transmitting again. Others interpret the pressing of any key after <Ctrl>-S as XON.

XOpen A group of computer manufacturers that promotes the development of portable applications based on UNIX. They publish a document called the X/Open Portability Guide. X/Open is the correct spelling.

XPAD An eXternal Packet Assembler/Disassembler.

XPC A set of protocols developed by British Telecom Tymnet to allow asynchronous terminals to connect to an X.25 packet-switched network.

XPM Extended Peripheral Module.

XPL A Private Line provided by a long distance carrier in the U.S.

XRB Transmit Reference Burst.

XSG X.25 Service Group.

XSL eXtensible Stylesheet Language. An extension of XML which allows the automatic transformation of XML-based data into HTML and other presentation formats. Thereby, the presentation format can vary from the underlying data structure. See XML.

XT Abbreviation for crosstalk.

XTEL API The Sun Solaris Teleservices Application Programming Interface. See TELESERVICES for a long explanation.

XTELS The Sun Solaris Teleservices protocol. The XTELS protocol is essentially the XTEL provider protocol with extensions to support multiple clients and multiple XTEL providers. See TELESERVICES for a long explanation.

XTELTool A window-based tool used to configure providers in the Sun Solaris Teleservices platform. Basically a graphical user interface. See TELESERVICES for a long explanation.

XTL Sun Microsystems' "Teleservices" architecture for Solaris. XTL is the foundation library for applications using or controlling telecom data streams. It is used for computer telephony application development. The XTL subsystem and API includes call control functions. It establishes a call or connection, and data stream access methods to control the flow of data over that connection.

XTL API Part of the Sun XTL Teleservices architecture. An object-oriented interface accessed using the C++ language. The API includes XtlProvider, XtlCall, XtlCallState, and XtlMonitor objects. These base objects define the command and callback methods of XTL teleservices. There are three methods types on these objects: synchronous requests, asynchronous commands, and asynchronous events.

XTL Call Object Part of Sun Microsystems' XTL Teleservices architecture. C++ objects created by developers with the XTL API. Each XtlCall object corresponds to a telephone call. An XtlCall object has command methods to query the current state of a call or request a change in state (to put a call on hold). An XtlCall object also has callback methods for the asynchronous notification of state changes.

XTL Provider Configuration Database Part of the Sun XTL Teleservices architecture. Means of registering third-party "providers" in an XTL-based system. The database shows the existence of each provider, how to invoke it (used by the XTL system internally) and describes the specific capabilities of that provider.

XTL Provider Interface Part of Sun Microsystems' XTL Teleservices architecture. The means for third-party developers and Independent Hardware Vendors (IHVs) to integrate provider, technology or call-type-specific features into an XTL environment. Developers use the XTL Provider Library (MPI) to do this.

XTL Provider Library Part of Sun Microsystems' XTL Teleservices architecture. Isolates the provider code from the intricacies of the XTL system services. The library supplies interfaces to the provider database, the server messaging system for commands and callbacks, and the means to make device data streams accessible to applications.

XTL System Services Part of Sun Microsystems' XTL Teleservices architecture. Acts as the intermediary between the application view of a call object and the provider's implementation of the call. The XTL server, along with the application and provider libraries, handles the interprocess message passing, object identification and creation, call ownership, security, and asynchronous event notification. The XTL subsystem also manages a database of available providers and helps manage data stream routing and access.

XTLtool A Sun Microsystems term for a GUI tool supplied to browse and edit the provider configuration database in an XTL environment.

XWindows See X-WINDOWS.

Y-Dimension Of Recorded Spot In facsimile, the center-to-center distance between two recorded spots measured perpendicular to the recorded line.

Y-Dimension Of Scanning Spot In facsimile, the center-to-center distance between two scanning spots measured perpendicular to the scanning line on the subject copy. The numerical value of this term will depend upon the type of system used.

Y/C A two channel video channel. One is for color (chrominance) and the other for black and white (luminance).

Y2K The year 2000. The millennium bug. What happens when your computer hits the year 2000. There are three happenings. First, it will move its date to 1900 and stay there. In this case programs that use date calculations — like figuring how much money you have in your insurance policy will screw up. For example, because many mainframe programs used 95, instead of 1995 in the date field. That means that when it hits 00 or 01, the computer won't be able to figure that the difference between 01 and 95 is 6. (Figure it. 2001 minus 1995 = 6). Second, it will move its date to 1900 and you'll be able to manually change it back. Third, it will move its date to 2000 flawlessly. Most PCs manufactured after 1995 will handle the Y2K problem flawlessly. Many mainframe and minicomputers manufactured before that won't. Some phone systems will also mess up. They may stop working altogether. They may work in strange ways, like producing the wrong information on calls being made or being received. For more info: www.ibm.com/IBM/year2000; www.microsoft.com/CIO/articles/YEAR2000faq.htm; www.year2000.com. See the Millennium Bug.

Yagi Antenna A type of directional antenna.

Yahoo A popular search engine and portal on the Web. Apparently Yahoo stands for "Yet Another Hierarchical Officious Oracle." See Search Engines for a list of other indexes to what's on the Web. http://akebono.stanford.edu/yahoo/

Yellow Alarm A T-1 alarm signal sent back toward the source of a failed transmit circuit in a DS-1 2-way transmission path. A yellow sends 0's (zeros) in bit two of all time slots. See also T-1. Also called Yellow Signal.

Yellow Signal In telecommunications, a signal sent back in the direction of a failure, indicating that the input of a network element has failed. The yellow signal varies with the DS framing used.

Yellow Cable A coaxial cable used in 10Base-5 networks. It is also referred to as "Thick" coax. This was the first cable used on many early LANs.

Yellow Pages A directory of telephone numbers classified by type of business. It was printed on yellow paper throughout most of the twentieth century until it was obsoleted in the late 1990s by dial up yellow page directories operated by voice processing systems and in the early 21st century by electronic directories delivered on disposable laser disks. As a concession to history, the laser disks are now painted bright yellow. Actually, yellow pages remain one of the phone companies' most lucrative sources of revenues. Advertising rates are not cheap. There is now competition.

There are many "Yellow Pages" directories, since AT&T never trademarked the term "Yellow Pages." Some "yellow page" directories are better value than others. And some are more legitimate than others. Some actually never get printed or, if they are printed, are not printed in great quantity and are not distributed as widely as their sales literature implies. Many businesses have been suckered into paying money for listings and advertisements in directories that never appeared. This fictitious directory scam also has happened with "telex" and "fax" directories. This "scam" is fraud by mail and is heavily stomped upon by the US Postal Service. As a result, many fake directories (especially the telex ones) are "published" abroad.

Yes A program which Novell created to certify products — hardware and software — that work with its products. It's a rigorous program. It takes significant work to be "Yes" certified by Novell. Novell has Yes programs for NetWare and UnixWare. It's Novell's equivalent of the Good Housekeeping Seal of Approval.

YMMV Your Mileage May Vary. A response usually given when the answer is not precise and depends on the user's own circumstances. YMMV is an acronym used in electronic mail on the Internet to save words or to be hip, or whatever.

YModem A faster transfer variation of XMODEM. In YMODEM, XMODEM's 128-byte block grew to YMODEM's 1024 bytes (1 kilobyte). YMODEM combines the 1K block and the 128-byte block modems into the same protocols. YMODEM, or 1K as it is known, became the thrifty way to send files (i.e. it saved on phone time). Enhancements were added, such as auto-fallback to 128-byte blocks if too many errors were encountered (because of bad phone lines, etc.) See XMODEM for a much larger explanation of file transfer using X, Y and ZMODEM protocols.

YModem-G Ymodem-g is a variant of Ymodem. It is designed to be used with modems that support error control. This protocol does not provide software error correction or recovery, but expects the modem to provide it. It is a streaming protocol that sends and receives 1K packets in a continuous stream until told to stop. It does not wait for positive acknowledgement after each block is sent, but rather sends blocks in rapid succession. If any block is unsuccessfully transferred, the entire transfer is canceled. See also ZMODEM, which we prefer.

Yottabyte YB. A combination of the homonymic Greek "iota," referring to the last letter of the Latin alphabet, and the English "bite," meaning "a small amount of food." A unit of measurement for physical data storage on some form of storage device-hard disk, optical disk, RAM memory etc. and equal to two raised to the 80th power, i.e.
1,208,925,819,614,600,000,000,000 bytes. Here are the others in the chain:
KB = Kilobyte (2 to the 10th power)
MB = Megabyte (2 to the 20th power)
GB = Gigabyte (2 to the 30th power)
TB = Terabyte (2 to the 40th power)
PB = Petabyte (2 to the 50th power)

EB = Exabyte (2 to the 60th power)
ZB = Zettabyte (2 to the 70th power)
YB = Yottabyte (2 to the 80th power)
One googolbyte equals 2 to the 100th power

Yuppie Food Coupons Crisp $20 bills that spew from ATM machines. Often used to split the bill after a meal. "We all owe $8, but all anybody's got is yuppie food coupons." A Wired Magazine definition.

Yurt A Mongolian circular shed. Some companies are making Yurt-like sheds for installation in the back yards of telecommuters — especially those telecommuters who don't have enough room inside for a home office.

YUV A color encoding scheme for natural pictures in which luminance and chrominance are separate. The human eye is less sensitive to color variations than to intensity variations. YUV allows the encoding of luminance (Y) information at full bandwidth and chrominance (UV) information at half bandwidth.

YUV9 The color encoding scheme used in Indeo Video Technology. The YUV9 format stores information in 4x4 pixel blocks. Sixteen bytes of luminance are stored for every one byte of chrominance. For example, a 640x480 image will have 307,200 bytes of luminance and 19,200 bytes of chrominance.

Z

Z Abbreviation for Zulu time. See Greenwich Mean Time.

Zap To eradicate all or part of a program or database, sometimes by lightning, sometimes intentionally.

ZBTSI Zero Byte Time Slot Interchange. A technique used with the T carrier extended superframe format (ESF) in which an area in the ESF frame carries information about the location of all-zero bytes (eight consecutive "0"s) within the data stream.

Zen A Japanese sect of Buddhism that stresses attaining enlightenment through intuition rather than by studying scripture.

Zen Mail Email messages that arrive with no text in the message body.

Zener Diode A particular type of semiconductor which acts as a normal rectifier until the voltage applied to it reaches a certain point. At this point — at the zener voltage or the avalanche voltage — the zener diode becomes conducting. Zener diodes can be used in circuits where it is desirable that they "turn on" only when a minimum voltage is reached. These types of circuits include computer equipment and voice activated circuits such as telephone wiretap devices. The main use of a Zener diode is to provide a reference voltage. Thus it is often called a "reference diode". The Zener diode is the device that made it possible to make digital integrated circuits. Without on-chip regulation and reference voltage, we would not be able to just "hook 'em together", as we do now.

Zero Beat Reception Also called "homodyne" reception. A method of reception using a radio frequency current of the proper magnitude and phase relation so that the voltage impressed on the detector will be of the same nature as that of the wave. An old radio term.

Zero Bit The high-order bit in a byte or a word.

Zero Byte Time Slot Interchange ZBTSI. A method of coding in which a variable address code is exchanged for any zero octet. The address information describes where, in the serial bit stream, zero octets originally occurred. It is a five-step process where data enters a buffer, zero octets are identified and removed, the nonzero bytes move to fill in the gaps, the first gap is identified and a transparent flag bit is set in front of the message to indicate that one or more bytes originally contained zeros. See ZBTSI.

Zero Code Suppression The insertion of a "one" bit to prevent the transmission of eight or more consecutive "zero" bits. Used primarily with digital T-1 and related telephone-company facilities which require a minimum "ones density" to keep the individual sub channels of a multiplexed, high-speed facility active. Several different schemes are currently employed to accomplish this. Proposals for a standard are being evaluated by the ITU-T. See also ZERO SUPPRESSION.

Zero Fill See ZEROFILL.

Zero Frequency The frequency (wavelength) at which the attenuation of the lightguide is at a minimum.

Zero Hop Routing A layer 3 switch that offers cut-through services making every end station one hop away from each other

Zero Insertion In SDLC, the process of including a bina-ry 0 in a transmitted data stream to avoid confusing data and SYN characters; the inserted 0 is removed at the receiving end.

Zero Latency Enterprise The Gartner Group defines this new management buzzword as an organization that "exploits the immediate exchange of information across geographical, technical and organizational boundaries to achieve business benefit." I guess it means the place moves quickly. Nothing like saying something in big words that can't say more easily with smaller, fewer words.

Zero Power Modem A modem that takes its power from the phone line and therefore needs no battery or external power. Such modems are often limited in their speed and capabilities.

Zero Slot LAN A Local Area Network (LAN) that uses a PC's serial port to transmit and receive data. It doesn't require a network interface card to be installed in a slot in the PC, thus the name "zero-slot" LAN. RS-232 LANs usually use standard RS-232 or phone cable to link PCs. Software does the rest of the work. Due to the slow speed of serial communications on a PC, RS-232 LANs are usually restricted to speeds of around 19.2K bits per second. What they lose in speed, however, they make up in low price.

Zero Stuffing Get a cup of coffee right now. Synchronous data transmission is done by sending what IBM and AT&T call Frames, and what everyone else calls Packets. A frame starts off by sending a bit pattern of 01111110 (notice the six 1's in a row). Synchronous transmission is for sending a bit stream, which means that the bits may (but probably do not) have any relation to the transmission of characters. This is especially true when sending digitized voice. As the bits pass to the receiver, they go through a shift register. When the flag signifying the end of a frame goes by, the last 16 bits in the shift register are the check digits.

The receiver computes the check digits based on the data bits that have gone by. As the sender sent the data, it computed the check digit, sent it after the end of the frame, and then sent the flag. If the receiver computes the same check digit that the sender sent, then one can be reasonably assured the data came through without error. But that's not what I came to talk to you about. I came to talk about Zero Stuffing. The problem is that somewhere in the bit stream, there is the possibility of there being six 1 bits in a row. To the receiving computer, six 1's means a flag. Therefore the sending computer, if it "sees" six 1 bits, will send five 1 bits, and stuff a zero in the bit stream. In fact, if it sees even five 1 bits, it will stuff a zero anyway, so there will be no ambiguity. The rule is, "If there are five ones in a row and it is NOT the end of a frame, stuff a zero into the bit stream." This way the receiver will know that this is in no way the end of the frame yet. Now if the receiver sees six 1's in a row, it knows without a doubt that it IS at the end of a frame, and should proceed with the error checking.

Zero Suppression The elimination of nonsignificant zeros from a numeral. Zero suppression is the replacement of leading zeros in a number with blanks so that when the number appears, the leading zeros are gone. The data becomes more readable. For example, the number 00023 would be displayed on the monitor or printed as 23.

Zero Test Level Point A level point used as a reference in determining loss in circuits. Analogous to using sea level when defining altitude. Written as 0 TLP.

Zero Transmission Level Point ZTLP. In telephony, a reference point for measuring the signal power gain and losses of telecommunications circuit, at which a zero dBm signal level is applied.

Zero Transmission Level Reference Point A point in a circuit to which all relative transmission levels are referenced. The transmission level at the transmitting switchboard is frequently taken as the zero transmission level reference point.

Zero Usage Customer An MCI definition. An MCI customer who has not placed a call over the network, even though he/she is an active customer. Sometimes used interchangeably, but incorrectly, with the term "no usage customer."

Zerofill 1. To fill unused storage locations with the character "0."

2. Here's definition from GammaLink, a fax board maker: A traditional fax device is mechanical. It must reset its printer and advance the pages as it prints each scan line it receives. If the receiving machine's printing capability is slower than the transmitting machine's data sending capability, the transmitting machine adds "fill bits" (also called Zero Fill) to pad out the span of send time, giving the slower machine the additional time it needs to reset prior to receiving the next scan line.

Zettabyte ZB. A unit of measurement for physical data storage on some form of storage device — hard disk, optical disk, RAM memory etc. and equal to two raised to the 70th power, i.e. 1,180,591,620,717,400,000,000 bytes.

MB = Megabyte (2 to the 20th power)
GB = Gigabyte (2 to the 30th power)
TB = Terabyte (2 to the 40th power)
PB = Petabyte (2 to the 50th power)
EB = Exabyte (2 to the 60th power)
ZB = Zettabyte (2 to the 70th power)
YB = Yottabyte (2 to the 80th power)

One googolbyte equals 2 to the 100th power

ZIF Zero Insertion Force. Intel makes a bunch of math co-processor chips which are used with their 80XXX range of microprocessors. ZIF is a special device which is typically soldered to the motherboard. You place an 80387 chip on this device, move the handle down, it grabs the chip and pulls the chip down, seating it electrically. When you want to remove the chip, you simply lift the handle and up the chip comes. The device was invented by Intel because so many people were apparently breaking the legs on their math co-processor chips each time they removed them. Apparently the problem was most prevalent in the computer rental business.

Zinc Spark Gap A spark gap having zinc as the electrode.

Zip It all started years ago with a very popular PC program called PKZIP. You could run this program on a file on your PC and bingo PKZIP would reduce the size of the file by as much as 90%. You could then transmit the zipped file over a phone line, saving 90% of the time and 90% of the cost. At the other end the recipient would run another file called PKUNZIP and bingo your file would be returned to its original size. You get the program from PKWARE, Inc. Glendale, WI 414-352-3670. www.pkware.com

Zip Code The specific codes assigned to addresses in the United States to speed up the sorting of mail and thus speed up its delivery. Zip stands for Zone Improvement Program. Zip codes were originally introduced into the United States in

1963. They were originally five digits. Then they grew to nine digits. The latest zip codes are 11 digits. Let's say you have a zip code 10036-3959-29. The 100 is the region, in this case New York City. The 36 indicates the specific post office in the region. The 3959 means the carrier route, i.e. the delivery sector in a neighborhood. The 29 is called the sequence. It indicates a specific address along a given route. Overseas, zip codes are often called postal codes. They do the same thing the American zip codes do. They allow machinery to sort the mail and thus get it to its destination faster.

Zip Cord Another name for a patch cable or jumper, often used to describe fiberoptics cable used in a zone wiring configuration.

Zip Tone Short burst of dial tone to an ACD agent headset indicating a call is being connected to the agent console.

Zipped See ZIP.

ZModem ZMODEM is an error-correcting file transfer, data transmission protocol for transmitting files between PCs. A file might be anything — a letter, an article, a sales call report, a Lotus 1-2-3 spreadsheet. Always use ZMODEM if you can. It's the best and fastest data transmission protocol to use. This is not my sole advice. Virtually every writer in data communications recommends it. Here's an explanation, beginning with XMODEM, an older, more common and less efficient protocol. Both XMODEM and YMODEM transmit, then receive, then transmit. The handshake (ACK or NAK) happens when the sender isn't sending. ZMODEM adds full duplex-transmission to the transfer protocol. ZMODEM does not depend on any ACK signals from the host computer. It keeps sending unless it receives a NAK, at which time it falls back to the failed block and starts to retransmit at that point. ZMODEM was written by Chuck Forsberg. According to PC Magazine (April 30, 1991) ZMODEM is the first choice of most bulletin boards. ZMODEM, according to PC Magazine, features relatively low overhead and significant reliability and speed. ZMODEM dynamically adjusts it packet size depending on line conditions and uses a very reliable 32-bit CRC error check. It has a unique file recovery feature. Let's say ZMODEM aborts a transfer because of a bad line (or whatever), it can start up again from the point it aborted the transfer. Other file transfer protocols have to start all over again. ZMODEM's ability to continue is a major benefit. ZMODEM in some communications program is a little more automated than other protocols. For example, ZMODEM will start itself when the other end gives a signal — thus saving a keystroke or two and speeding things up. See FILE TRANSFER PROTOCOL, XMODEM and YMODEM.

Zone 1. A telephony definition: One of a series of specified areas, beyond the base rate area of an exchange. Service is furnished in zones at rates in addition to base rates.

2. A LAN definition. Part of a local area network, typically defined by a router. A router will let you get into one part of someone else's network. They define what you are able to get access to. You might get to that router by an external telecommunications circuit — dial up, ISDN, Switched 56, T-1 etc.

Zone Bits 1. One or two leftmost bits in a commonly used system of six bits for each character.

2. Any bit in a group of bit positions that are used to indicate a specific class of items, i.e., numbers, letters, commands.

Zone Cabling According to AMP, its inventor, zone cabling is the subsystem for companies on the move. Zone Cabling subsystems allow for flexible, changeable cabling of open office areas for voice, data, video and power. The open office area is divided into zones, with feeder cables running to a distribution point within each zone and short cable runs to each outlet. Zones are wired with reusable, pre-terminated cable assemblies. Office area reconfiguration is fast and easy when using plug and play assemblies not requiring retesting. Disruption and productivity loss during moves are minimal and confined to the zones you're moving. Zone cabling can be implemented in CNA or DNA architectures.

Zone Method A ceiling distribution method in which ceiling space is divided into sections or zones. Cables are then run to the center of each zone to serve the information outlets nearby. See also CEILING DISTRIBUTION SYSTEMS.

Zone Of Silence Skip zone.

Zone Paging Ability to page a specific department or area in or out of a building. "Page John in the Accounting Department." Zone paging is useful for finding people who wander, as most of us do.

Zoomed Video ZV. A technology that allows certain streams of digital information to write directly to a laptop's screen, bypassing the CPU and its bus (ISA, PCI, EISA, etc.). Zoomed video can show full 30 frames per second movies to a laptop screen. Zoomed video used MPEG-2. To show NTSC on a laptop, you need a special ZV (Zoomed Video) CardBus PCMCIA card. It feeds NTSC video directly to the screen. Zoomed Video technology allegedly will bring full laptop screen video conferencing to laptops.

Zoomed Video Port ZV. A PCMCIA standard which adapts the PC Card slot to allow the insertion of a ZV Port Card. Inserting a ZV Port Card establishes direct communications between the PC Card controller and the audio and video controllers, allowing large amounts of multimedia data to bypass the CPU or systems bus. The ZV Port standards makes full screen, full motion video accessible to the notebook computer user.

Zulu Time Coordinated Universal Time. Another term for Greenwich Mean Time (GMT). Greenwich is a borough in SE London, England, which is located on the prime meridian from which geographic longitude is measured. Greenwich was formerly the site of the Greenwich Observatory. And for these historic reasons, Greenwich is the place from which world time starts. For example, GMT time is five hours earlier Eastern Standard Time — i.e. the time in the northern hemisphere Summer. GMT (Zulu Time) is always the same worldwide. Communication network switches are typically coordinated on GMT.

ZV See Zoomed Video.

ZZF Zentralamt fur Zulassungen im Fernmeldewessen (Approval Authority — Germany).

ZZZZZ Time The last entry in the Manhattan White Pages in the September 1997 - August 1998 issue. ZZZZZ Time is an interactive adult service. What did you expect?

Appendix

NEWTON's TELECOM DICTIONARY

15TH EDITION

STANDARDS ORGANIZATIONS
SPECIAL TELECOM INTEREST GROUPS
PUBLICATIONS,
COMPUTER AND TELECOM
INTERNATIONAL CALLING CODES
PLUGS AND CONNECTORS

STANDARDS ORGANIZATIONS

ACUTA
The Association for Telecommunications Professionals in
Higher Education
152 W. Zandale Drive, Suite 200
Lexington, KY 40503
Tel: 606-278-3338
Fax: 606-278-3268
www.acuta.org

ADSL Forum
39355 California Street, Suite 307
Fremont, CA 94538
Tel: 510-608-5905
www.adsl.com

ANSI
American National Standards Institute
11 West 42nd Street
New York, NY 10036
Tel: 212.642.4900
Fax: 212.398.0023
www.ansi.org

ARIN
American Registry for Internet Numbers
4506 Daly Drive, Suite 200
Chantilly, VA 20151
Tel: 703-227-0660
Fax: 703-227-0676
www.arin.net

ATA
American Telemarketing Association
4605 Lankershim Blvd
Suite 824
North Hollywood, CA 91602-1891
Tel: 800-441-3335
Fax: 818-766-8168

ATSC
Advanced Television Systems Committee
1750 K Street, N.W., Suite 800
Washington, D.C. 20006
Tel: 202-828-3130
Fax: 202-828-3131
www.atsc.org

ATIS
Alliance for Telecommunications Industry Solutions
1200 G St. NW, Suite 500
Washington, DC 20005
Tel: 202-628-6380
www.atis.org

ATM Forum
2570 West El Camino Real, Suite 304
Mountain View, CA 94040
Tel: 415-949-6700
Fax: 415-949-6705
www.atmforum.com

Bellcore
Bell Communications Research
8 Corporate Place
Piscataway, NJ 08854-4156
Tel: 908-699-2000
www.bellcore.com

BICSI
Building Industry Consulting Service International
8610 Hidden River Parkway
Tampa, FL 33637
Tel: 813-979-1991 or 800-242-7405
Fax: 813-971-4311
www.bicsci.org

BSI
British Standards Institution
British Standards House
389 Chiswick High Road
London W4 4AL, United Kingdom
Tel: 44(0)-181-996-9000
Fax: 44(0)-181-996-7400
www.bsi.org.uk

BTA
Business Technology Association
12411 Wornall Road
Kansas City, MO 64145
Tel: 816-941-3100
Fax: 816-941-2829 and 816-941-4838
www.btanet.org

CableLabs
Cable Television Laboratories, Inc.
400 Centennial Drive
Louisville, CO 80027-1266
Tel: 303-661-9100
Fax: 303-661-9199
www.cablelabs.com

California ISDN Users' Group
P.O. Box 27901-391
San Francisco, CA 94127
Tel: 415.241.9943
Fax: 415.753.6942
Email: info@ciug.org
Web: www.ciug.org, www.isdnworld.com

CBTA
Canadian Business Telecommunications Alliance
Canada Trust Tower
161 Bay Street, Suite 3650
P.O. Box 705
Toronto, Ontario M5J 2S1
Canada
Tel: 416-865-9993
Fax: 416-865-0859
www.cbta.ca

CCMA
Call Centre Management Association (UK)
Ranmore House
The Crescent
Leatherhead, Surrey
England KT22 8DY
Roy Bailey, Secretary
Telephone 01293 538400 Fax 01293 521313
email r.bailey@pncl.co.uk

CDG
CDMA Development Group
650 Town Center Drive
Suite 820
Costa Mesa CA 92626
714-545-5211
www.cdg.org

CDPD Forum
401 North Michigan Avenue
Chicago, IL 60611
Tel: 800-335-2373
Fax: 312-321-6869
www.cdpd.org

CEMA
The Consumer Electronics Manufacturers Association
2500 Wilson Blvd
Arlington VA 22201
Tel: 703-907-7600
Fax: 703-907-7601

CEN
European Committee for Standardization
rue de Stassart 36
B-1050 Brussels, Belgium
Tel: 32-2-519-68-11
Fax: 32-2-519-68-19

CENELEC
European Committee for Electrotechnical Standards
rue de Stassart 35
B-1050 Brussels, Belgium
Tel: 32-2-51-96-871
Fax: 32-2-51-96-919

CMA
Communications Managers Association
1201 Mt. Kemble Avenue
Morristown, NJ 07960-6628
Tel: 201-425-1700
Fax: 201-425-0777
www.cma.org

CommerceNet
4005 Miranda Avenue, Suite 175
Palo Alto, CA 94304
Tel: 650-858-1930
Fax: 650-858-1936
www.commerce.net

Competitive Telephone Carriers of New York, Inc.
One Columbia Place
Albany, NY
518-434-8112 F 518-434-3232

CompTel
Competitive Telecommunications Association
1900 M Street, N.W., Suite 800
Washington, D.C. 20036
Tel: 202-296-6650
Fax: 202-296-7585
www.comptel.org

CompTIA
The Computing Technology Industry Association
450 East 22 Street, Suite 230
Lombard, IL 60148-6158
Tel: 630-268-1818
Fax: 630-268-1384
www.comptia.org

Committee T1
www.t1.org

CSA
Canadian Standards Association
178 Rexdale Blvd.
Rexdale, Ontario M9W 1R3
Canada
Tel: 416-747-4000
www.csa.ca

CTIA
Cellular Telecommunications Industry Association
1250 Connecticut Ave., NW, Suite 800
Washington, DC 20036
202-785-0081
www.wow-com.com and www.ctia.org

DAVIC
Digital Audio Visual Council
c/o SIA Societa' Italiana Avionice
C.P. 3176-Strada Antica di Colllegno, 253
I-10146 Torino
Italy
Tel: 39-11-7720-114
Fax: 39-11-725-679
www.davic.org

ECTF
Electronic Computer Technology Forum
303 Vintage Park Drive
Foster City, CA 94404
Tel: 415-578-6852
www.ectf.org

ECMA
European Computer Manufacturers Association
114, Rue de Rhone CH-1204
Geneva, Switzerland
Tel: 41-22-846-60-00
www.ecma.ch

EFF
The Electronic Frontier Foundation
1550 Bryant Street, Suite 725
San Francisco, CA 94103-4832
Tel: 415-436-9333
Fax: 415-436-9993
www.eff.org

EIA
Electronic Industries Alliance
2500 Wilson Blvd.
Arlington, VA 22201
Tel: 703-907-7500
Fax: 703-907-7501
www.eia.org

EMA
Electronic Messaging Association
1655 North Fort Myer Drive, Suite 500
Arlington, VA 22209
Tel: 703-524-5550
Fax: 703-524-5558
www.ema.org

ETSI
European Telecommunications Standards Institute
Route des Lucioles
F-06921 Sophia Antipolis Cedex - FRANCE
Tel: +33 (0)4 92 94 42 00
Fax: +33 (0)4 93 65 47 16
www.etsi.fr

FCA
Fibre Channel Association
2570 West El Camino Real, Suite 304
Mountain View, CA 94040-1313
Tel: 650-949-6730
Fax: 650-949-6735
www.fibrechannel.com

FCC
Federal Communications Commission
1919 M Street, NW.
Washington DC
Tel: 202-418-0200
www.fcc.gov

Frame Relay Forum
303 Vintage Park Drive
Foster City, CA 94404
Tel: 415-578-6980
Fax: 415-525-0182
www.frforum.com

Gigabit Ethernet Alliance
20111 Stevens Creek Boulevard, Suite 280
Cupertino, CA 95014
Tel: 408-241-8904
Fax: 408-241-8918
www.gigabit-ethernet.org

GO-MVIP, Inc.
3220 N Street, NW, Suite 360
Washington, D.C. 20007
Tel: 508-650-1388
Fax: 508-650-1375
www.mvip.org

HomePNA
Home Phoneline Networking Alliance
Deepak Kamlani or Betsy Gillette
Interprise Ventures
Voice: (925) 277-8110
Fax: (925) 277-8111
dkamlani@inventures.com
betsy@inventures.com
www.homepna.org

ICA
International Communications Association
2735 Villa Creek Drive, Suite 200
Dallas, TX 75234
Tel: 972-620-7020
Fax: 972-488-9985
www.icanet.com

ICEA
Insulated Cable Engineers Association
P.O. Box 440
South Yarmouth
MA 02664
Tel: 508-394-4424

ICSA
International Computer Security Association
1200 Walnut Bottom Road
Carlisle, PA 17013-7635
Tel: 717-258-1816
www.ncsa.com

IEC
International Engineering Consortium
549 W. Randolph Street
Suite 600
Chicago, IL 60661
Tel: 312-559-4100
Fax: 312-559-4111
www.iec.org

IEC
International Electrotechnical Commission
3, rue de Varembe
P.O. Box 131
1211 Geneva 20
Switzerland
Tel: 41-22-919-02-11
www.iec.ch

IEEE
Institute of Electrical and Electronics Engineers, Inc.
445 Hoes Lane
P.O. Box 1331
Piscataway, NJ 08855-1331
Tel: 908-981-0060
www.ieee.org

IMC
Internet Mail Consortium
127 Segre Place
Santa Cruz, CA 95060
Tel: 408-426-9827
Fax: 408-426-7301
www.imc.org

IMTC
International Multimedia Teleconferencing Consortium,
Inc.
Tel: 510-277-1320
Fax: 510-277-8111
www.imtc.org

InterNIC
P.O. Box 1656
Herndon, VA 22070
Internet: admin@ds.internic.net
www.internic.net

IrDA
Infrared Data Association
P.O. Box 3883
Walnut Creek, CA 94598
Tel: 510-943-6546
Fax: 510-943-5600
www.irda.org

ISO
International Organization for Standardization
One Rue de Varembe CH-1211
Case Postale 56
Geneva 20, Switzerland
Tel: 41-22-749-0111
www.iso.ch

ITAA
Information Technology Association of America
1616 N. Ft. Myer Drive, Suite 1300
Arlington, VA 22209
Tel: 703-522-5055
Fax: 703-525-2279
www.itaa.org

ITIC
Information Technology Industry Council
1250 Eye Street NW   Suite 200
Washington, DC 20005
Tel: 202-737-8888
Fax: 202-638-4922
www.itic.org

ITCA
International TeleConferencing Association
100 Four Falls Corporate Center, Suite 105
West Conshohocken, PA 19428
Tel: 610-941-2020
Fax: 610-941-2015
www.itca.org

ITU
International Telecommunications Union
Place des Nations
CH-1211 Geneva 20
Switzerland
www.itu.ch

MOMA
Message Oriented Middleware Association
303 Vintage Park Drive
Foster City, CA 94404-1138
Tel: 415-378-6699
Fax: 415-525-0182
http://198.93.24.24

MMTA
MultiMedia Telecommunications Association
2500 Wilson Boulevard, Suite 300
Arlington, VA 22201-3834
Tel: 703-907-7470
Fax: 703-907-7478
www.mmta.org

NAB
National Association of Broadcasters
1771 N Street, N.W.
Washington, D.C. 20036
Tel: 202-429-5300
www.nab.org

NARTE
National Association of Radio and
  Telecommunications Engineers
P.O. Box 678
Medway, MA 02053
Tel: 508-533-8333
Fax: 508-533-3815
www.narte.org

NARUC
National Association of Regulatory Utility Commissioners
1100 Pennsylvania Avenue NW, Suite 603
Post Office Box 684
Washington, D.C. 20044-0684
Tel: 202-898-2200
Fax: 202-898-2213
www.naruc.org

NCTA
National Cable TV Association
1724 Massachusetts Avenue, N.W.
Washington, D.C. 20036
Tel: 202-775-3669
www.ncta.com

NECA
National Exchange Carrier Association
100 South Jefferson Road
Whippany, NJ 07981-8597
Tel: 973-884-8000
Fax: 973-884-8469
www.neca.org

NFPA
National Fire Protection Association
Batterymarch Park
Quincey, MA 02169
Tel: 617-770-3000
Fax: 617-770-0700
www.nfpa.org

NIST
National Institute of Standards and Technology
Gaithersburg, MD 20899
Tel: 301-975-2000
www.nist.com

NIUF
North American ISDN Users' Forum
National Institute of Standards & Technology (NIST)
Building 820, Room 445
Gaithersburg, MD 20899
Tel: 301-975-2937
Fax: 301-926-9675
www.niuf.nist.gov

NMF
Network Management Forum
1201 Mt. Kemble Avenue
Morristown, NJ 07960
Tel: 973-425-1900
Fax: 973-4251515
www.nmf.org

NANOG
North American Network Operators' Group
www.nanog.org

NATOA
National Association of Telecommunications Officers
  and Advisors
1650 Tysons Boulevard, Suite 200
McLean, VA 22102
Tel: 703-506-3275
Fax: 703-506-3266
www.natoa.org

NPA
Network Professional Association
P.O. Box 809161
Chicago, IL 60680-9161
Tel: 801-223-9444
Fax: 801-223-9486
www.npa.org

NTIA
National Telecommunications and Information
   Administration
U.S. Department of Commerce
Washington, D.C. 20230
Tel: 202-377-1880
www.ntia.doc.gov

NTIS
National Technical Information Service
Technology Administration
U.S. Department of Commerce
Springfield, VA 22161
Tel: 703-605-6000
Fax: 703-321-8547
www.ntis.gov

OBI
Open Buying on the Internet Consortium
57 Bedford Street, Suite 208
Lexington, MA 02173
Tel: 781-863-5396
Fax: 781-861-1708
www.supplyworks.com/obi

OFR
Office of the Federal Register
National Archives & Records Administration
Suite 700
800 North Capitol Street NW
Washington, D.C. 20408
Tel: 202-523-3117
Fax: 202-523-6866

OFTEL
50 Ludgate Hill
London EC4M 7JJ
Tel: +44 171 834 8700
Fax: +44 171 634 8943
www.oftel.gov.uk

OMG
Object Management Group
4902 Old Connecticut Path
Framingham, MA 01701
Tel: 508-820-4300
Fax: 508-820-4303
www.omg.org

OSTA
Optical Storage Technology Association
311 East Carrillo Street
Santa Barbara, CA 93101
Tel: 805-963-3853
Fax: 805-962-1541
www.osta.org

PCIA
Personal Communications Industry Association
500 Montgomery Street, Suite 700
Alexandria, VA 22314
Tel: 703-739-0300
Fax: 703-836-1608
www.pcia.com

PCMCIA
Personal Computer Memory Card International
   Association
2635 North First Street, Suite 209
San Jose, CA 95134
Tel: 408-433-2273
Fax: 408-433-9558
www.pcmcia.org

PICMG
PCI Industrial Computer Manufacturers Group
c/o Rogers Communications
401 Edgewater Place, Suite 500
Wakefield, MA 01880
Tel: 781-246-9318
Fax: 781-224-1239
www.picmg.org

PTC
Pacific Telecommunications Council
2454 S. Beretania Street, Suite 302
Honolulu, HI 96826
Tel: 808-941-3789
Fax: 808-944-4874
www.ptc.org

SCTE
Society of Cable Telecommunications Engineers, Inc.
140 Philips Road
Exton, PA 19341-1318
Tel: 610-363-6888
Fax: 610-363-5898
www.scte.org

SIF
SONET Interoperability Forum
c/o Alliance for Telecommunications Solutions
1200 G Street, N.W., Suite 500
Washington, DC 20005
Tel: 202-628-6380
Fax: 202-393-5453
www.atis.org/atis/sif/index.html

SMPTE
Society of Motion Picture & Television Engineers
595 W. Hartsdale Avenue
White Plains, NY 10607-1824
Tel: 914-761-1100
Fax: 914-761-3115
www.smpte.org

SPA
Software Publishers Association
1730 M Street, NW
Suite 700
Washington, DC 20036-4510
Tel: 202-452-1600
Fax: 202-223-8756
www.spa.org

TCA
Telecommunications Association
74 New Montgomery Street, Suite 230
San Francisco, CA 94105-3411
Tel: 415-777-4646
Fax: 415-777-5295
www.tca.org

TCIF
Telecommunications Industry Forum
c/o Alliance for Telecommunications Industry Solutions
1200 G Street, NW
Suite 500
Washington, DC 20005
Tel: 202-628-6380
Fax: 202-393-5453
www.atis.org/atis/tcif/index.html

Telecom Corridor Technology Business Council
411 Belle Grove
Richardson, TX 75080-5297
Fax: 972-680-9103

TIA
Telecommunications Industry Association
2500 Wilson Boulevard
Arlington, VA 22201
Tel: 703-907-7700
Fax: 703-907-7727
www.eia.org

TMF
TeleManagement Forum
1201 Mt. Kemble Avenue
Morristown, NJ 07960
Tel: 973-425-1900
Fax: 973-4251515
www.tmforum.org

TRA
Telecommunications Resellers Association
1730 K Street, N.W., Suite 1201
Washington, D.C. 20006
Tel: 202-835-9898
Fax: 202-835-9893
www.tra-dc.org

The Open Group
11 Cambridge Center
Cambridge, MA 02142
Tel: 617-621-8700
Fax: 617-621-0631
www.opengroup.org

UL
Underwriters Laboratory
333 Pfingsten Road
Northbrook, IL 60062
Tel: 847-272-8800
Fax 847-272-8129
www.ul.com

USDLA
United States Distance Learning Association
1240 Central Boulevard, Suite A
Brentwood, CA 94513
Tel: 925-513-4253
Fax: 925-513-4255
www.usdla.org

USTA
United States Telephone Association
1401 H Street, N.W., Suite 600
Washington, D.C. 20005-2164
Tel: 202-326-7300
Fax: 202-326-7333
www.usta.org

USTTI
United States Telecommunications Training Institute
1150 Connecticut Av.., N.W. Suite 702
Washington, D.C. 20036
Tel: 202-785-7373
Fax: 202-785-1930
www.telemobile.com

U.S. Department of Commerce National Technical
   Information Service
5285 Port Royal Road
Springfield, VA 22161
Tel: 703-487-4600
www.ntis.gov

UTC
The Telecommunications Association
(Originally The Utilities Telecommunications Council)
1140 Connecticut Ave., NW,
Washington, DC 20036
Tel: 202-872-0030
www.utc.org

W3C
World Wide Web Consortium
Massachusetts Institute of Technology
Laboratory for Computer Science
545 Technology Square
Cambridge, MA 02139
Tel: 617-253-2613
Fax: 617-258-5999
www.w3.org

WIPO
World Intellectual Property Organization
P.O. Box 18
CH-1211
Geneva 20
Switzerland
Tel: 41-22-338-9111
Fax: 41-22-733-54-28
www.wipo.org

WLI Forum
Wireless LAN Interoperability Forum
1111 W. El Camino Road, #109-171
Sunnyvale, CA 94087
www.wlif.com

WLANA
Wireless LAN Alliance
2723 Delaware Avenue
Redwood City, CA 94061
www.wlana.com

WSTA
Wall STreet Telecommunications Association
One West Front Street
Red Bank, NJ 07701
Tel: 908-530-8808
Fax: 908-530-0020
www.wsta.org

WTO
World Trade Organization
154, rue de Lausanne
CH-1211 Geneva 21
Switzerland
Tel: 41-22-739-5111
Fax: 41-22-739-5458
www.wto.org

Publications, Computer & Telecom

Alcatel's Electrical Communication, a quarterly technical journal from Alcatel NV, 33 rue Emeriau, 75725 Paris Cedex 15, France
America's Network, 312 West Randolph, Ste. 600, Chicago, IL 60606
Boardwatch, 8500 West Bowlds Ave, Littleton, CO 80127
Business Communications Review, 950 York Rd., Hinsdale, IL 60521
BYTE, One Phoenix Mill Lane, Peterborough, NH 03458
Cabling Business Magazine, 12035 Shiloh Road, Ste. 350, Dallas, TX 75228
CALL CENTER Magazine, 12 West 21 Street, New York, NY 10010
Communications News, 2504 N. Tamiami Trail, Nokomis, FL 34275
Communications Week (now called Internet Week) 600 Community Drive., Manhasset, NY 11030
Computer Telephony Magazine, 12 West 21 Street, New York, NY 10010
Computerworld, 500 Old Connecticut Road, Framingham, MA 01701
Data Communications, 1221 Avenue of the Americas, New York, NY 10020
Dr. Dobb's Journal c/o Miller Freeman Inc., 600 Harrison Street, San Francisco, CA 94107
Electronic Business, Cahners Bldg, 275 Washington St., Newton, MA 02158
Federal Computer Week, 3110 Fairview Park Dr, Ste. 1040, Falls Church, VA 22042
Flatiron Publishing, 12 West 21 Street, New York, NY 10010
IEEE Spectrum, 345 East 47 Street., New York, NY 10017
IMAGING Magazine, 12 West 21 Street, New York, NY 10010
InfoWorld, 155 Bovet Rd., San Mateo, CA 94402
Inform Business Systems, 10516 Summit Ave., Suite 200, Kingston, MD 20895
Information Week, 600 Community Dr., Manhasset, NY 11030
Internet Week, (used to be called Communications Week) 600 Community Drive., Manhasset, NY 11030
LAN Magazine, (now called Network Magazine) 500 Howard St., San Francisco, CA 94105
MacUser, One Park Avenue, New York, NY 10016
MacWEEK, P.O. Box 1763, Riverton, NJ 08077-9763
MacWorld, Fifth Floor, 501 Second Street, San Francisco, CA 94107
Miller Freeman, 12 West 21 Street, New York, NY 10010
MSJ Microsoft Systems Journal, 411 Borel Avenue, Suite 100, San Mateo, CA 94402
Mobile Office, 21600 Oxnard St., Suite 480, Woodland Hills, CA 91367
Network Computing, 600 Community Drive, Manhasset, NY 11030
Network Magazine, (used to be called LAN Magazine) 600 Harrison Street, San Franciso, CA 94107
Network World, 161 Worcester Road, Framingham, MA 01701-9172
Payphone Magazine, P.O. Box 42371, Houston, TX 77242
PC Magazine, One Park Avenue, New York, NY 10016
PC WEEK, One Park Avenue, New York, NY 10016
PC World, 501 Second Street, #600, San Francisco, CA 94107
PC/Computing, 950 Tower Lane, Foster City, CA 94404
Phone+, 4141 N. Scottsdale Road, Suite 316, Scottsdale, AZ 85251
Public Communications Magazine, P.O. Box 42371, Houston, TX 77242
Satellite Comm, 214 Massachusetts Ave. NE, Washington DC 20002
Speech Technology, 43 Danbury Road, Wilton, CT 06897
Telecom Gear, 15400 Knoll Trail, Dallas, TX 75248
TELECOM LIBRARY Inc, 12 West 21 Street, New York, NY 10010 212-691-8215. Free catalog.
TeleManagement, 1400 Bayly Street, Office Mall 2, Suite 3, Pickering, Ontario, Canada L1W 3R2
Telemarketing. 1 Technology Plaza, Norwalk, CT 06854
Telecommunications, 685 Canton Street, Norwood, MA, 02062
TELECONNECT Magazine, 12 West 21 Street, New York, NY 10010
TeleProfessional, 209 West Fifth Street, Suite N, Waterloo, IA 50701
UnixWorld, P.O. Box 571, Hightstown, NJ 08520-9331
VARBusiness, 1 Jericho Plaza, Jericho, NY 11753
Voice Processing Magazine, P.O. Box 42382, Houston, TX 77242
Windows Magazine, 1 Jericho Plaza, Jericho, NY 11753
Windows/DOS Developer's Journal, 1601 W. 23rd Street, Lawrence, KS 66046
Windows Sources, One Park Avenue, New York, NY 10016
Wired, 520 Third Street, Fourth Floor, San Francisco, CA 94107

International Calling Codes

Country	Code
Afghanistan	93
Albania	355
Algeria	213
American Samoa	684
Andorra	376
Angola	244
Anguilla	264
Antarctica	672
Antigua and Barbuda	1
Argentina	54
Armenia	374
Aruba	297
Australia	61
Austria	43
Azerbaijan	994
Bahamas	1
Bahrain	973
Bangladesh	880
Barbados	1
Belarus	375
Belgium	32
Belize	501
Benin	229
Bermuda	1
Bhutan	975
Bolivia	591
Bosnia and Herzegovina	387
Botswana	267
Brazil	55
Brunei Darussalam	673
Bulgaria	359
Burkina Faso	226
Burundi	257
Cambodia	855
Cameroon	237
Canada	1
Cape Verde	238
Cayman Islands	1
Central African Republic	236
Chad	235
Chile	56
China	86
Christmas Island	
Cocos (Keeling) Islands	672
Colombia	57
Comoros	269
Congo	242
Cook Islands	682
Costa Rica	506
Cote D'Ivoire (Ivory Coast)	225
Croatia (Hrvatska)	385
Cuba	53
Cyprus	357
Czech Republic	420
Czechoslovakia (former)	420
Denmark	45
Diego Garcia	246
Djibouti	253
Dominica	1
Dominican Republic	1
Ecuador	593
Egypt	20
El Salvador	503
Equatorial Guinea	240
Eritrea	291
Estonia	372
Ethiopia	251
Falkland Islands (Malvinas)	500
Faroe Islands	298
Fiji	679
Finland	358
France	33
France, Metropolitan	33
French Guiana	594
French Polynesia	689
French Antilles	596
Gabon	241
Gambia	220
Georgia	995
Germany	49
Ghana	233
Gibraltar	350
Great Britain (UK)	44
Greece	30
Greenland	299
Grenada	1
Guadeloupe	590
Guam	671
Guatemala	502
Guinea	224
Guinea-Bissau	245
Guyana	592
Haiti	509
Heard and McDonald Islands	692
Honduras	504
Hong Kong	852
Hungary	36
Iceland	354
India	91
Indonesia	62
Inmarsat	
East Atlantic Ocean	871
Indian Ocean	873
Pacific Ocean	872
West Atlantic Ocean	874
Iran	98
Iraq	964
Ireland	353

Israel	972	Norway	47
Italy	39	Oman	968
Ivory Coast	225	Pakistan	92
Jamaica	1	Palau	675
Japan	81	Panama	507
Jordan	962	Papua New Guinea	675
Kazakhstan	7	Paraguay	595
Kenya	254	Peru	51
Kiribati	686	Philippines	63
Korea North	850	Pitcairn	872
Korea South	82	Poland	48
Kuwait	965	Portugal	351
Kyrgyzstan	7	Puerto Rico	1
Laos	856	Qatar	974
Latvia	371	Reunion Island	262
Lebanon	961	Romania	40
Lesotho	266	Russian Federation	7
Liberia	231	Rwanda	250
Libya	218	Saint Kitts and Nevis	1
Liechtenstein	41	Saint Lucia	1
Lithuania	370	Saint Vincent & the Grenadines	1
Luxembourg	352	Samoa	685
Macau	853	San Marino	378
Macedonia	389	Sao Tome and Principe	239
Madagascar	261	Saudi Arabia	966
Malawi	265	Senegal	221
Malaysia	60	Seychelles	248
Maldives	960	Sierra Leone	232
Mali	223	Singapore	65
Malta	356	Slovak Republic	421
Marshall Islands	692	Slovenia	386
Martinique	596	Solomon Islands	677
Mauritania	222	Somalia	252
Mauritius	230	South Africa	27
Mayotte	356	Spain	34
Mexico	52	Sri Lanka	94
Micronesia	691	St. Helena	290
Moldova	373	St. Pierre and Miquelon	508
Monaco	377	Sudan	249
Mongolia	976	Surinam	597
Montserrat	1	Svalbard and Jan Mayen Islands	378
Morocco	212	Swaziland	268
Mozambique	258	Sweden	46
Myanmar	95	Switzerland	41
Namibia	264	Syria	963
Nauru	674	Taiwan	886
Nepal	977	Tajikistan	7
Netherlands	31	Tanzania	255
Netherlands Antilles	599	Thailand	66
New Caledonia	687	Togo	228
New Zealand	64	Tokelau	690
Nicaragua	505	Tonga	676
Niger	227	Trinidad and Tobago	1
Nigeria	234	Tunisia	216
Niue	683	Turkey	90
Norfolk Island	672	Turkmenistan	7
Northern Mariana Islands	670	Turks and Caicos Islands	649

Tuvalu	688
US Minor Outlying Islands	1
USSR (former)	7
Uganda	256
Ukraine	380
United Arab Emirates	971
United Kingdom	44
United States	1
Uruguay	598
Uzbekistan	7
Vanuatu	678
Vatican City State	396
Venezuela	58
Viet Nam	84
Virgin Islands (British)	1
Virgin Islands (U.S.)	1
Wallis and Futuna Islands	681
Western Sahara	34
Western Samoa	685
Yemen	967
Yugoslavia	381
Zaire	243
Zambia	260
Zimbabwe	263

5 PIN DIN (F).EPS

6 PIN MINI DIN (F).EPS

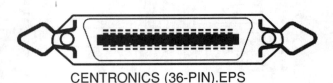

CENTRONICS (36-PIN).EPS

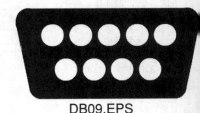

DB09.EPS

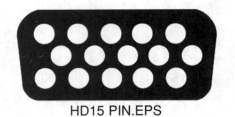

HD15 PIN.EPS

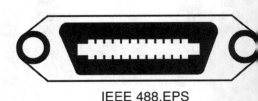

IEEE 488.EPS

RS-232 INTERFACE (MALE).EPS

RS-449.EPS

RS-530.EPS

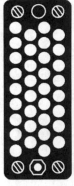

V.35 ROTATED.EPS

Black Box Corporation

The World's Source for Connectivity^{SM}

BBL SOURCE LOGO.EPS

Centronics Parallel Interface

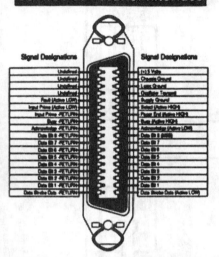

RS-449 Interface

RS-232 Interface

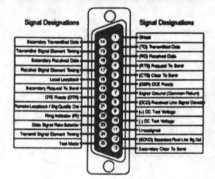

IEEE 488 Interface

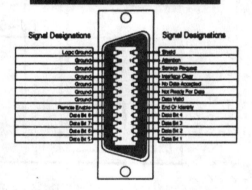

IBM PC Color Monitor Interface

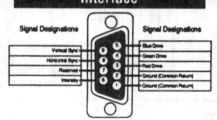

IBM PC Monochrome Monitor Interface

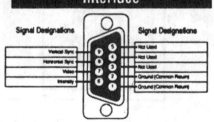

IBM PS/2 Keyboard Interface

RESERVED
+5 VOLTS
RESERVED

CLOCK
GROUND
DATA

IBM PC Keyboard Interface

KEYBOARD RESET
+5 VOLTS

KEYBOARD CLOCK
GROUND

KEYBOARD DATA

RJ-11-4

RJ-11-6

BRi-U
Pin 4&5

BRi S/t
    TE NT Polarity
3   TX   RX   +
4   RX   TX   +
5   RX   TX   −
6   TX   RX   −

RJ-45

PRi CT1/CSU

1   Receive Ring
2   "   Tip
4   Ring
5   Tip

RJ-45 KEYED

RJ-11-6 MODIFIED

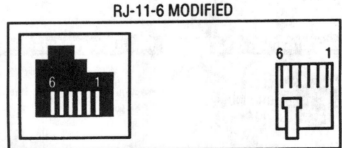

Telecom
BOOKS

TELECOM BOOKS
a division of Miller Freeman

Telecom Books publishes books on Computer Telephony, Telecommunications, Networking and Voice Processing, the Internet, Data Communications, Call Center. It also distributes the books of other publishers, making it the "central source" for all the above materials. Call or write for you FREE catalog.

337 Killer Voice Processing Applications
ATM Users' Guide
Client Server Computer Telephony
Voice Over Frame Relay
Customer Service Over the Phone
Frames, Packets and Cells in Broadband Networking
The UnPBX
Audio Teleconferencing
The Guide to T-1 Networking
UNIX Computer Telephony
Understanding Computer Telephony
Secrets of Windows Telephony
1001 Computer Telephony Tips
PC Telephony
Telecommunications Projects Made Easy
Telephony for Computer Professionals
"Which Phone System Should I Buy?"
Which Telephone Service Provider
VideoConferencing: The Whole Picture

Quantity Purchases

If you wish to purchase this book, or any others, in quantity, please contact:

Christine Kern, Manager
Telecom Books
12 West 21 Street
New York, NY 10010
212-691-8215 or 1-800-LIBRARY
facsimile orders: 212-691-1191
www.Telecombooks.com
Call 1-800-LIBRARY for your **FREE** Catalog.